RODD'S CHEMISTRY OF CARBON COMPOUNDS

ELSEVIER SCIENTIFIC PUBLISHING COMPANY
335 JAN VAN GALENSTRAAT
P.O. BOX 211, 1000 AE AMSTERDAM, THE NETHERLANDS

DISTRIBUTORS FOR THE U.S.A. AND CANADA

ELSEVIER/NORTH-HOLLAND INC.
52 VANDERBILT AVENUE, NEW YORK, N.Y. 10017

WITH 22 TABLES

ISBN: 0-444-41768-0 (VOL. IVL)
ISBN: 0-444-40664-6 (SERIES)

LIBRARY OF CONGRESS CARD CATALOG NUMBER 64-4605

PRINTED IN THE NETHERLANDS

RODD'S CHEMISTRY OF CARBON COMPOUNDS

RODD'S CHEMISTRY OF CARBON COMPOUNDS

VOLUME I

GENERAL INTRODUCTION

ALIPHATIC COMPOUNDS

*

VOLUME II

ALICYCLIC COMPOUNDS

*

VOLUME III

AROMATIC COMPOUNDS

*

VOLUME IV

HETEROCYCLIC COMPOUNDS

*

VOLUME V

MISCELLANEOUS

GENERAL INDEX

*

RODD'S CHEMISTRY OF CARBON COMPOUNDS

A modern comprehensive treatise

SECOND EDITION

Edited by
S. COFFEY
M.Sc. (London), D.Sc. (Leyden), C.Chem., F.R.I.C.
formerly of
I.C.I. Dyestuffs Division, Blackley, Manchester

VOLUME IV PART L

HETEROCYCLIC COMPOUNDS

Fused-ring heterocyclic compounds containing three or more nitrogen atoms; purines and related ring systems, nucleosides, nucleotides and nucleic acids; pteridines, alloxazines, flavins and related compounds. The biosynthesis of plant alkaloids and nitrogenous microbial metabolites

ELSEVIER SCIENTIFIC PUBLISHING COMPANY
AMSTERDAM OXFORD NEW YORK
1980

CONTRIBUTORS TO THIS VOLUME

R. B. HERBERT, B.SC., PH.D.
Department of Organic Chemistry, The University of Leeds, Leeds LS2 9JT

D. S. JONES, PH.D., C.CHEM, F.R.I.C., A.C.T.
Department of Biochemistry, The University of Liverpool, Liverpool L69 38X

K. OHTA, M.SC., PH.D.
Chemical Engineering and Construction Division, Niigata Engineering Co., Ltd., Ota-ku, Tokyo (Japan)

G. SHAW, D.SC., PH.D., A.R.C.S., D.I.C.
School of Studies in Chemistry, University of Bradford, Bradford, West Yorkshire, BD7 1DP

H. C. S. WOOD, B.SC., PH.D., F.R.I.C., F.R.S.E.
Department of Pure and Applied Chemistry, University of Strathclyde, Glasgow G1 1XL

R. WRIGGLESWORTH, B.SC., D.PHIL.
Chemical Research Laboratory, Wellcome Research Laboratories, Langley Court, Beckenham, Kent BR3 3BS

R. E. FAIRBAIRN, B.SC., PH.D., F.R.I.C.
formerly of Research Department, Dyestuffs Division, I.C.I. Ltd., Manchester 9 *(Index)*

PREFACE TO VOLUME IV L

This sub-volume (Part IV L) is the terminal volume of this revised edition of *Rodd's Chemistry of Carbon Compounds* of which, at this date, Parts IV C, IV D, IV I and IV J have still to be published. It deals systematically with two main topics intimately concerned with the chemistry of living organisms, namely (*1*) the chemistry of fused-ring heterocyclic compounds having three or more (usually four) nitrogen atoms in the fundamental heterocyclic structure, and (*2*) the biosynthesis of plant alkaloids and nitrogenous microbial metabolites.

In Chapter 57 Dr. G. SHAW describes the extensive chemistry of the purines and their derivatives, one of the most ubiquitous groups of all naturally occurring heterocyclic compounds, and compounds derived from related ring systems. This class of compounds, comprising a wide range of chemically versatile members, is extremely important in that a very large proportion occur as essential components in the more complex compounds which control the various vital processes of living organisms. Such compounds are the subject of Chapter 58, by Dr. D. S. JONES; in sections 1 and 2, respectively, are described compounds in which a purine or a pyrimidine is linked to a carbohydrate to form a *N*-glycoside or nucleoside, as such compounds are called, and their various types of phosphate esters or nucleotides, which are obtained by the chemical or enzymic hydrolysis of enzymes or more commonly nucleic acids, the two types of compounds described in sections 3 and 4 of the chapter. The most abundant nucleosides of natural origin are the D-ribosides and the 2-D-deoxyribosides. The purines adenine and guanine and the pyrimidines uracil and cytosine, occur as their ribosides adenosine, guanosine, uridine and cytidine, respectively, in ribonucleic acid (RNA). The deoxyribosides of adenine, guanine, cytosine and thymine occur as components of deoxyribonucleic acid (DNA), found principally in the nucleus of the cell, where, in association with proteins, it constitutes the chromosomes. As the genetic carrying substance of all known living cells its chemical composition determines hereditary characteristics. RNA on the other hand, is located mainly in the cytoplasm where all three types, transfer-RNA (t-RNA), ribosomal-RNA (r-RNA) and messinger-RNA (m-RNA) play a role in protein synthesis. The analysis, synthesis and methods of determining the detailed stereochemical structures of these compounds presented in this chapter provides a fascinating account of one of the outstanding recent advances made in the biomedical sciences.

The purines (imidazolopyrimidines) have a fused five- and six-membered ring heterocycle; the basic heterocyclic structure of the pteridines (pyrimido-[4,5-*b*]pyrimidines) and the related azoxazines and flavins, described by

Dr. K. Ohta, Dr. R. Wrigglesworth and Professor H. C. S. Wood in Chapter 59, consists of two fused six-membered rings, each containing two nitrogen atoms. While these compounds show a general resemblance to the purines in their methods of synthesis and in being widely distributed in nature, they show marked differences in their properties and in the role they play in biological processes. They are found as pigments in the wings of butterflies and in various other parts of insects and bacteria, they act as co-factors in biological enzymic hydroxylation processes, for example, lumazine is a precursor of the essential food constituent riboflavin (vitamin B_2), biopterin is an essential nutritional requirement of certain bacteria, while pteroylglutamic (folic) acid has a growth-promoting effect on many organisms. Rather remarkably, although pteridine derivatives analogous to the nucleosides have been prepared in the laboratory, none have been isolated from biological material. The dihydropteridines are particularly interesting as they act as mediators of cellular electron transport and many biologically active compounds are of this type.

Finally, in Chapter 60 Dr. R. B. Herbert describes how, with the advent of readily available radioactive organic compounds, organic chemists have joined forces with biochemists in the profitable exploration of the biosynthetic pathways leading to secondary metabolites, and has reviewed the results of such investigations published up to June 1976 in so far as they concern one large group of compounds, namely, the plant alkaloids and the quite often related microbial metabolites containing nitrogen.

This abbreviated contents of Part IV L of *Rodd's Chemistry of Carbon Compounds* shows it to be unique in the series in that it is concerned almost entirely with substances intimately connected with the control of processes taking place in living organisms, and the products they engender. The origins of organic chemistry are to be found in the early study of natural products. As the study progressed it came to mean the study of carbon compounds, hence the title of the original edition of this series of volumes. The present volume of this revised edition clearly emphasises the direction one of the main streams of research in organic chemistry is now taking is back to the study of natural products and the chemistry of vital processes. This volume must be of interest to a wide range of readers working in biomedical fields.

The editor is glad to place on record his appreciation of the help and cooperation he has received from each of the contributors to this volume.

December 1979 S. Coffey

CONTENTS

VOLUME IV L

Heterocyclic Compounds: Fused-ring heterocyclic compounds containing three or more nitrogen atoms; purines and related ring systems; nucleosides, nucleotides and nucleic acids; pteridines, alloxazines, flavins and related compounds. The biosynthesis of plant alkaloids and nitrogenous microbial metabolites

Chapter 58. Nucleosides, Nucleotides and Nucleic Acids
by D. S. JONES

Chapter 59. Pteridines, Alloxazines, Flavins and Related Compounds
by K. OHTA, R. WRIGGLESWORTH and H. C. S. WOOD

Chapter 60. The Biosynthesis of Plant Alkaloids and Nitrogenous Microbial Metabolites
by R. B. HERBERT

Titles of other parts of Volume IV

HETEROCYCLIC COMPOUNDS

Vol. IV A: Three-, four- and five-membered heterocyclic compounds with a single hetero-atom in the ring

Vol. IV B: Five-membered heterocyclic compounds with a single hetero-atom in the ring: alkaloids, dyes and pigments

Vol. IV C: Five-membered heterocyclic compounds with two hetero-atoms in the ring from Groups V and/or VI of the Periodic Table

Vol. IV D: Five-membered heterocyclic compounds with more than two hetero-atoms in the ring

Vol. IV E: Six-membered monoheterocyclic compounds containing oxygen, sulphur, selenium, tellurium, silicon, germanium, tin, lead or iodine as the heteroatom

Vol. IV F: Six-membered heterocyclic compounds with a single nitrogen atom in the ring: pyridine, polymethylenepyridines, quinoline, isoquinoline and their derivatives

Vol. IV G: Six-membered heterocyclic compounds with a single nitrogen atom in the ring to which are fused two or more carbocyclic ring systems, and six-membered ring compounds where the hetero-atom is phosphorus, arsenic, antimony or bismuth. Alkaloids containing a six-membered heterocyclic ring system

Vol. IV H: Six-membered heterocyclic compounds with (a) a nitrogen atom common to two or more fused rings; (b) one hetero-atom in each of two fused rings. Six-membered ring compounds with two hetero-atoms from Groups VI B, or V B and VI B of the Periodic Table, respectively. Isoquinoline, lupinane and quinolizidine alkaloids

Vol. IV I: Six-membered heterocyclic compounds with two hetero-atoms from Group V of the Periodic Table: the Pyridazine and Pyrimidine groups

Vol. IV J: Six-membered heterocyclic compounds with two hetero-atoms from Group V of the Periodic Table: the Pyrazine group. Phenoxazine, phenothiazine, phenazine and sulphur dyes. Six-membered heterocyclic compounds with three or more hetero-atoms

Vol. IV K: Six-membered heterocyclic compounds with two or more hetero-atoms one or more of which are from Groups II, III, IV, V or VII of the Periodic Table. Heterocyclic compounds with seven or more atoms in the ring

OFFICIAL PUBLICATIONS

B.P.	British (United Kingdom) Patent
F.P.	French Patent
G.P.	German Patent
Ger. Offen.	German Patent Application, open for inspection
Sw. P.	Swiss Patent
U.S.P.	United States Patent
U.S.S.R.P.	Russian Patent
B.I.O.S.	British Intelligence Objectives Sub-Committee Reports, H.M. Stationery Office, London.
C.I.O.S.	Combined Intelligence Objectives Sub-Committee Reports
F.I.A.T.	Field Information Agency, Technical Reports of U.S. Group Control Council for Germany
B.S.	British Standards Specification
A.S.T.M.	American Society for Testing and Materials
A.P.I.	American Petroleum Institute Projects
C.I.	Colour Index Number of Dyestuffs and Pigments

SCIENTIFIC JOURNALS AND PERIODICALS

With few obvious and self-explanatory modifications the abbreviations used in references to journals and periodicals comprising the extensive literature on organic chemistry, are those used in the World List of Scientific Periodicals.

LIST OF ABBREVIATED NAMES OF CHEMICAL FIRMS MENTIONED IN PATENT REFERENCES

A.G.F.A., Agfa A.G.	Aktiengesellschaft für Anilinfabrikation (Berlin)
B.A.S.F.	Badische Anilin- und Soda-Fabrik (Ludwigshafen)
Bayer	Farbenfabriken vorm. Friedrich Bayer und Co. (Leverkusen)
Cassella	Leopold Cassella und Co. (Frankfurt am Main)
C.F.M.	Compagnie française des Matières Colorantes (Paris)
CIBA	Gesellschaft für chemische Industrie (Basel)
Du Pont	E.I. Du Pont de Nemours and Co. (U.S.A.)
G.A.F.	General Anilin and Film Corporation (U.S.A.)
Geigy A.G.	J. R. Geigy S.A. (Basel)
Hoechst	Hoechst A.G. (see M.L.B.)
I.C.I.	Imperial Chemical Industries, Ltd. (London)
I.G.	(= Interessen Gemeinschaft Farbenindustrie) of the principal dyestuffs manufacturers in Germany
Kalle	Kalle und Co., A.G. (Biebrich am Rhein)
M.L.B.	Farbwerke vormals Meister, Lucius und Brüning (Hoechst)
Sandoz	Sandoz A.G. Chemische Fabrik (Basel)

LIST OF COMMON ABBREVIATIONS AND SYMBOLS USED

A	acid
Å	Ångström units
Ac	acetyl
a	axial
as, *asymm.*	asymmetrical
at.	atmosphere
B	base
Bu	butyl
b.p.	boiling point
C, mC and μC	curie, millicurie and microcurie
c, *C*	concentration
c.d.	circular dichroism
conc.	concentrated
crit.	critical
D	Debye unit, 1×10^{-18} e.s.u.
D	dissociation energy
D	dextro-rotatory; dextro configuration
DL	optically inactive (externally compensated)
d	density
dec. or decomp.	with decomposition
deriv.	derivative
E	energy; extinction; electromeric effect
*E*1,*E*2	uni- and bi-molecular elimination mechanisms
E1cB	unimolecular elimination in conjugate base
e.s.r.	electron spin resonance
Et	ethyl
e	nuclear charge; equatorial
f	oscillator strength
f.p.	freezing point
G	free energy
g.l.c.	gas liquid chromatography
g	spectroscopic splitting factor, 2.0023
H	applied magnetic field; heat content
h	Planck's constant
Hz	hertz
I	spin quantum number; intensity; inductive effect
i.r.	infrared
J	coupling constant in n.m.r. spectra
K	dissociation constant
k	Boltzmann constant; velocity constant
kcal	kilocalories
L	laevorotatory; laevo configuration
M	molecular weight; molar; mesomeric effect
Me	methyl

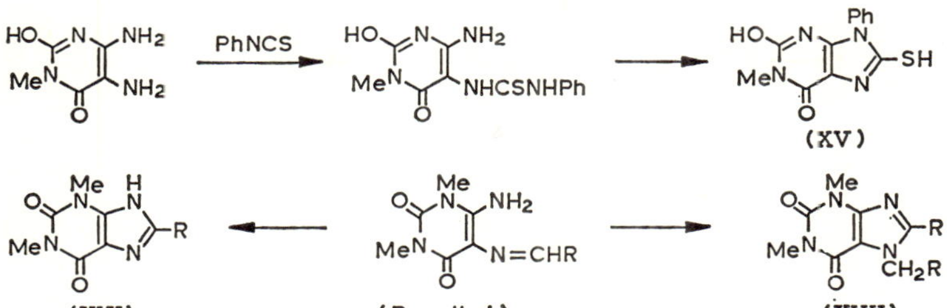

amino-6-hydroxy-2-methylthiopyrimidine gave the corresponding 8-aminopurine instead of the 8-mercapto compound (*Johns* and *Baumann*, J. biol. Chem., 1913, **14,** 387); the extension of this work by *Cook* is discussed later (pp. 16, 20). Traube condensed alkyl and aryl isocyanates with 4,5-diaminopyrimidines and obtained 2,6,8-trihydroxy-9-phenylpyrine (9-phenyluric acid) in this way. Similarly, aryl isothiocyanates give 9-aryl-8-mercaptopurines (*e.g.* XV). The Schiff base from an alkanecarbalcehyde and a 4,5-diamopyrimidine may either cyclise to an 8-alkylpurine XVI or react with a further molecule of aldehyde to give a 7,8-dialkylpurine XVII (*Traube* and *W. Nithack*, Ber., 1906, **39,** 227).

That the formation of 9-alkyl-8-mercaptopurines is not the invariable result of cyclising 5-thioureidopyrimidines was shown by *F. E.* and *T. J. King* (J. chem. Soc., 1947, 943) who obtained 6-hydroxy-8-mercapto-2-methyl-

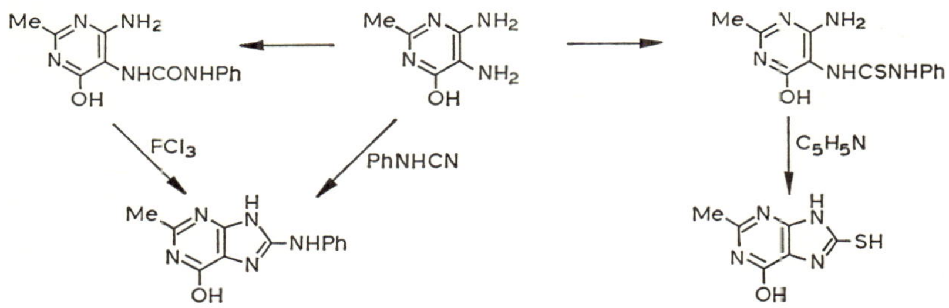

purine (XVIII) by heating 4-amino-6-hydroxy-2-methyl-5-phenylthioureidopyrimidine in boiling pyridine. Their other purine syntheses from 4,5-diamino-6-hydroxy-2-methylpyrimidine are also shown.

A considerable number of 8-arylpurines has been prepared by cyclising 5-acylamino-4-aminopyrimidines (*Elion et al.*, J. Amer. chem. Soc., 1951, **73,** 5235). Cyclisation normally occurs in hot phosphoryl chloride, and heating the amide or its sodium salt with the appropriate arylamide is sometimes effective.

m	mass; mole; molecule; *meta-*
ml	millilitre
m.p.	melting point
Ms	mesyl (methanesulphonyl)
[M]	molecular rotation
N	Avogadro number; normal
n.m.r.	nuclear magnetic resonance
N.O.E.	Nuclear Overhauser Effect
n	normal; refractive index; principal quantum number
o	*ortho-*
o.r.d.	optical rotatory dispersion
P	polarisation; probability; orbital state
Pr	propyl
Ph	phenyl
p	*para-;* orbital
p.m.r.	proton magnetic resonance
R	clockwise configuration
S	counterclockwise config.; entropy; net spin of incompleted electronic shells; orbital state
S_N1, S_N2	uni- and bi-molecular nucleophilic substitution mechanisms
S_Ni	internal nucleophilic substitution mechanisms
s	symmetrical; orbital
sec	secondary
soln.	solution
symm.	symmetrical
T	absolute temperature
Tosyl	*p*-toluenesulphonyl
Trityl	triphenylmethyl
t	time
temp.	temperature (in degrees centigrade)
tert	tertiary
U	potential energy
u.v.	ultraviolet
v	velocity
α	optical rotation (in water unless otherwise stated)
$[\alpha]$	specific optical rotation
α_A	atomic susceptibility
α_E	electronic susceptibility
ε	dielectric constant; extinction coefficient
μ	microns (10^{-4} cm); dipole moment; magnetic moment
μ_B	Bohr magneton
μg	microgram (10^{-6} g)
λ	wavelength
υ	frequency; wave number
χ, χ_d, χ_μ	magnetic diamagnetic and paramagnetic susceptibilities
$\sim$	about

(+)	dextrorotatory
(—)	laevorotatory
⊖	negative charge
⊕	positive charge

Chapter 57

Purines and Related Ring Systems

G. SHAW

1. Purines, imidazolo[4,5-*d*]pyrimidines

(*a*) *General introduction*

Purine derivatives are among the most ubiquitous of all naturally occurring heterocyclic compounds. They occur both in the free state and as nitrogenous glycosides, especially ribofuranosides or 2-deoxyribofuranosides (nucleosides) and derived nucleoside mono-phosphates (nucleotides) and cyclic phosphates such as c-AMP or polyphosphates including nucleoside di- or tri-phosphates such as ADP or ATP. The purine nucleotides together with analogous pyrimidine derivatives also form the monomer units of the two great groups of natural polymers the ribo-(RNA) and deoxyribonucleic acids (DNA). Of the simple purines the methylated xanthines are present in plants used for making the popular beverages tea and coffee. Many alkylated purines are also found in urine and other animal fluids, the hydroxypurines occur in the wings of butterflies, whilst guanine is associated with the iridescence of various fish, amphibian and reptilian skins and several substituted adenines, especially zeatin, play important parts in cell growth processes.

The chemistry of the simpler purines has been well established for many years but that of the more complex nucleotides and nucleic acids, to be described in the following chapter, dates largely from the more recent discovery of the vital roles they play both in many biochemical reactions and as units in nucleic acid production. Much current research is to a large extent increasingly directed towards the nucleic acids and nucleotides but also towards the preparation of analogous substances which may have special potential as drugs or anti-metabolites, to reaching a better understanding of the physicochemical properties of the purine-ring system, and to a study of the evolution of the system within the earth's changing environment.

TABLE 1

TRIVIAL NAMES AND STRUCTURES OF SOME NATURALLY OCCURRING PURINE DERIVATIVES

Trivial name	*Substitution at position* 2	6	7	8
(*a*) *Free bases*				
Adenine	H	NH_2	H	H
Guanine	NH_2	OH	H	H
Isoguanine	OH	NH_2	H	H
Hypoxanthine	H	OH	H	H
Kinetin	H	NHR	H	H
Triacanthine	H	NH_2	—	H
Uric acid	OH	OH	H	OH
Xanthine	OH	OH	H	H
Zeatin	H	NHR^1	H	H
(*b*) *Nucleosides and nucleoside antibiotics*				
Adenosine	H	NH_2	R^2	H
Aristeromycin	H	NH_2	R^3	H
Cordycepin	H	NH_2	R^4	H
Crotonoside	OH	NH_2	R^2	H
Decoyinine (Angustmycin A)	H	NH_2	R^5	H
Deoxyadenosine	H	NH_2	R^6	H
Deoxyguanosine	NH_2	OH	R^6	H
Guanosine	NH_2	OH	R^2	H
Inosine	H	OH	R^2	H
Nebularine	H	H	R^2	H
Nucleocidin	H	NH_2	R^7	H
Psicofuranine (Angustmycin C)	H	NH_2	R^8	H
Puromycin	H	NMe_2	R^9	H
Septacidin	H	NHR^{10}	H	H
Spongoadenosine	H	NH_2	R^{11}	H
Spongosine	OMe	NH_2	R^2	H
Xanthosine	OH	OH	R^2	H

R = furfuryl, R^1 = 4-hydroxy-3-methyl-*trans*-but-2-enyl, R^2 = β-D-ribofuranosyl, R^3 = β-(2α,3α-dihydroxy)-4β-(hydroxymethyl)cyclopentyl, R^4 = β-3-deoxy-D-ribofuranosyl, R^5 = 6-deoxy-β-D-*erythro*-hex-5-enofuran-2-ulosyl, R^6 = β-2′-deoxy-D-ribofuranosyl, R^7 = 4-fluoro-5-*O*-sulphamoyl-β-D-ribofuranosyl, R^8 = β-D-psicofuranosyl, R^9 = 3-(4-methoxy-L-phenylalanylamino)-3-deoxy-β-D-ribofuranosyl, R^{10} = 4-deoxy-*N*-isopalmitoylglycyl-4-amino-L-*glycero*-L-glucoheptosyl, R^{11} = β-D-arabinofuranosyl.

Nomenclature. Purines are bicyclic heterocycles containing a pyrimidine ring fused to an imidazole ring (I). The name and the enumeration of the system was first proposed by *Emil Fischer* (Ber., 1899, **32,** 435) but the correct ring structure was first proposed by *L. Medicus* in his formula for uric acid (Ann., 1875, **175,** 243). Because of this early origin the numbering of the structure does not conform to rule; in accordance with I.U.P.A.C. Rule B 2.11 the whole pyrimidine ring is first numbered consecutively followed by the imidazole ring.

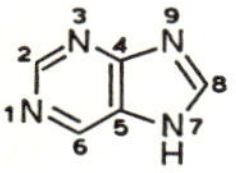

(I)

Many purine derivatives have long been known by trivial names and their systematic names are rarely used; Table 1 gives the structure of the common naturally occurring purines with their trivial names. In most of these compounds hydroxyl or amino groups occur in positions 2 or 6 and various tautomeric forms become possible; such compounds will here normally be written in the "aromatic" form since most purine derivatives patently exist in this form, the hydroxypurines which exist in the oxo form being the main exception.

The methylated purines which are found in nature are derivatives of a tautomeric form of xanthine; their trivial names and the positions in which the xanthine is substituted are given in Table 2. Purine derivatives commonly have high melting points and they often decompose on heating. They are best characterized by their ultraviolet absorption spectra (see p. 96).

TABLE 2

METHYL DERIVATIVES OF XANTHINE (2,6-DIHYDROXYPURINE)

Trivial name	*Substituents in xanthine* 1	3	7
Heteroxanthine	H	H	Me
Theobromine	H	Me	Me
Theophylline	Me	Me	H
Paraxanthine	Me	H	Me
Caffeine	Me	Me	Me

(b) *Synthetic methods*

The imidazolopyrimidine bicyclic structure of the purines immediately suggests two major routes to their synthesis, which in each case involve completion of the second ring of an appropriate monocyclic precursor pyrimidine or imidazole. Methods involving the simultaneous formation of both rings are of lesser importance and can in any case normally be placed into one of the two main groups. Of the two main routes that from pyrimidines has been most widely used perhaps because suitably substituted pyrimidines have in the past been more readily available and more stable than corresponding imidazoles. The use of imidazoles has increased, however, in recent years and this route is of special interest since it follows in many ways that of the pathway of biosynthesis *de novo* of purine nucleotides.

In addition to the main syntheses many purines are prepared by modification of preformed purines. The competition of the doubly bonded nitrogen atoms in the pyrimidine ring for the π-electrons makes the ring a π-electron-deficient-system whereas the imidazole ring with both a doubly and a singly bonded nitrogen atom is largely a π-electron-excessive-system. In the purine skeleton therefore the net electron distribution results from a combination of its two component rings forming a fairly stable bicyclic system and a result of partial localisation of the electron distribution is the production of dipoles which are normally resolved along the long and short axes of the molecule. The direction and strength of the dipole effect is modified and even reversed with the introduction of substituent groups with electron-withdrawing or electron-releasing properties (*J. H. Lister*, "Adv. in Heterocyclic Chemistry", eds. *A. R. Katritzky* and *A. J. Boulton*, Academic Press, New York, Vol. 6, 1966; *idem*, "Purines" in Heterocyclic Compounds (II), ed. *D. J. Brown*, Interscience, Wiley, New York, 1971.

The π-electron-excessive nature of the imidazole part of purine derivatives makes them a target in the anionic form (metal derivatives) for alkylation on N-7 or N-9 whereas in the unionised form alkylation occurs elsewhere and quaternisation on either ring nitrogen can occur. On the other hand, nucleophilic attack on carbon atoms 2, 6 and 8 can also occur normally in the order of reactivity of C-8 > C-6 > C-2 including halogenation at C-8 or in the anion at C-6. Purines can also be nitrated and coupled with diazonium salts. In addition to direct substitution reactions of these types, many purine syntheses involve modification of existing groups especially of OH → halogen or SH and further replacement of what are normally fairly reactive halogens by a variety of groups including OH, SR, NR_2, SO_2R. Purines may also be oxidised to *N*-oxides and a few have also been reduced to 1,6-dihydro derivatives by catalytic hydrogenation over palladium, or by electrolytic reduction at a mercury cathode (*D. L. Smith* and *P. J. Elving*, J. Amer. chem. Soc., 1962, **84**, 1412). A few 7,8- or 8,9-dihydropurines have also been prepared by desulphurisation of appropriate 8-thiopurines (*H. Biltz*, Ann., 1921, **426**, 237) or of certain quaternary derivatives (*A. V. El'tsov* and *Kh. L. Muravich-Alexander*, Tetrahedron Letters, 1968, 739).

(*i*) *Synthesis from pyrimidines*

(*1*) *From 4,5-diaminopyrimidines.* The commonest and most versatile method of purine synthesis, due to *W. Traube*, starts with a 4,5-diamino-pyrimidine. Such compounds require the insertion of only one carbon-

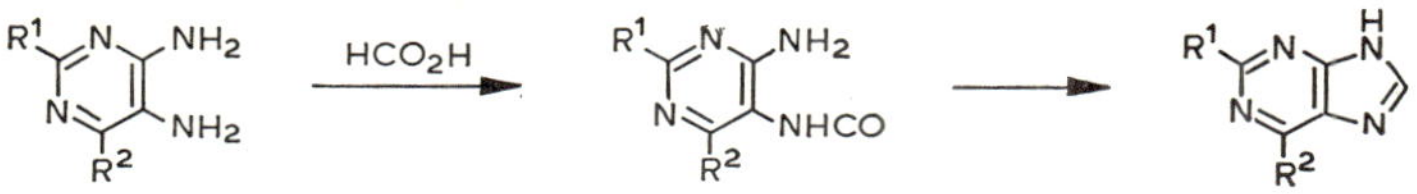

containing group between the two amino groups to complete the imidazole ring. A variety of substituent groups R^1, R^2, may be present in the 2- and 6-positions and a variety of cyclising agents can be used; pyrimidines carrying a substituent on either amino group cyclise to 7- or 9-substituted purines.

Traube used a wide range of cyclising reagents to form the imidazole ring of which the first, and until recently most important, was formic acid. When a 4,5-diaminopyrimidine is heated with formic acid, the 5-formamido derivative is formed exclusively and may often be cyclised quite readily without isolation; the isolated compound gives the purine when it or its sodium salt is heated at temperatures up to 220°. The first purine to be synthesized in this manner was guanine, which was formed directly by heating 2,4,5-triamino-6-hydroxy-pyrimidine with formic acid at 220° (Ber., 1900, **33**, 1371, 3035). The ease with which 4-amino-5-formamido-pyrimidines may be cyclised has been linked to their acidity particularly in so far as it concerns the propensity of the 5-formylamino group to function

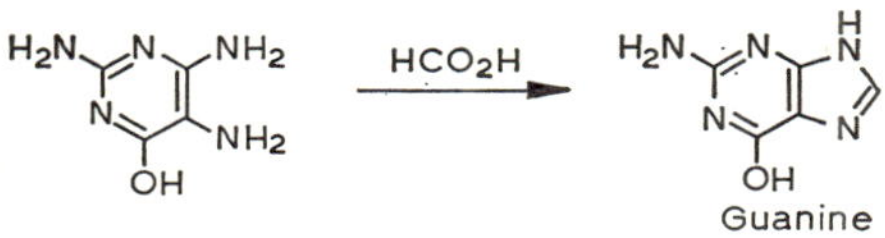

as a hydroxymethylene group and so favour the formation of the double bond in the 7,8 position, thus it is necessary to heat the sodium salt of 4-amino-5-formamido-2,6-dihydroxypyrimidine to 220° before xanthine is obtained, whereas the much weaker acid II cyclised spontaneously to give caffeine (*idem, loc. cit.*), and caffeine also results from the non-alkylated 4-amino-5-formylamino-2,6-dihydroxypyrimidine (III) on treatment with dimethyl sulphate and alkali, alkylation presumably preceding ring-closure (*B. Bobransky* and *Z. Synowiedski*, J. Amer. pharm. Assoc. Sci. Edn., 1948, **37**, 62). Theobromine is formed similarly from IV. 4-Amino-5-formylaminopyrimidine-6(*H*)-thione (V) in methanol with chlorine and hydro-

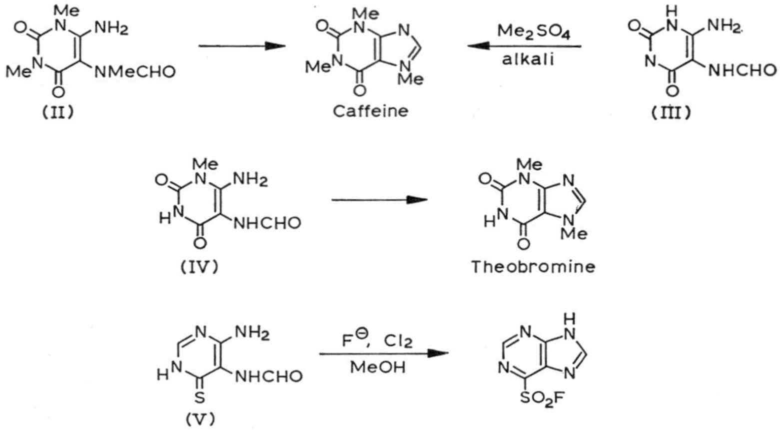

fluoric acid at 0° gives 6-fluorosulphonylpurine the ease of cyclisation resulting presumably from prior formation of a fluorosulphonylpyrimidine (*A. J. Beaman* and *R. K. Robins*, J. Amer. chem. Soc., 1961, **83**, 4038).

The 9-alkylpurines may be prepared by methods analogous to the caffeine synthesis. Heating 5-amino-2-hydroxy-4-methylamino-6-methylpyrimidine with formic acid gives 2-hydroxy-6,9-dimethylpurine (*C. O. Johns*, J. biol. Chem., 1912, **12**, 91).

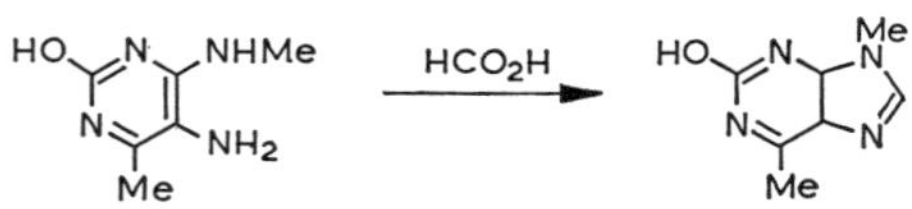

When 8-[^{14}C]adenine is prepared by this route, the use of 4-formylmorpholine as the solvent is advantageous (*V. M. Clark* and *H. M. Kalckar*, J. chem. Soc., 1950, 1029).

The use of formic acid is inappropriate if there is present in the 6-position of the pyrimidine ring either a group capable of providing an alternative method of cyclisation, 4-amino-5-formamido-6-mercaptopurine, for example, affording either the thiazolopyrimidine VI or 6-mercaptopurine according to cyclisation conditions (*G. B. Elion et al.*, J. Amer. chem. Soc., 1956, **78**, 2858), or one unstable to formic acid as when the cyclisation of 5-amino-6-chloro-4-methylaminopyrimidine afforded 9-methylhypoxanthine (*Robins*

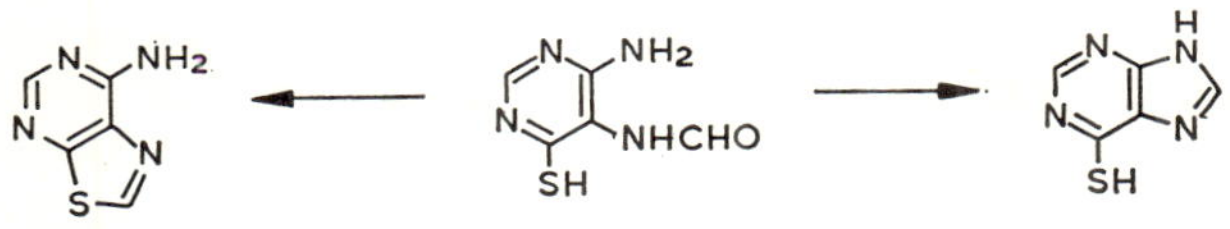
(VI)

and *H. H. Lin*, *ibid.*, 1957, **74**, 490). For these and other reasons reagents other than formic acid have been successfully sought for the cyclisation of 4,5-diaminopyrimidines to purines. These include the use of the Vilsmeier–Haack reagent $[(Me_2N^{\oplus}CHOPOCL_2)Cl^{\ominus}]$ produced from dimethylformamide and phosphoryl chloride (*A. Vilsmeier* and *A. Haack*, Ber., 1927, **60**, 119), which enables 4,5-diaminopyrimidines to be formylated and cyclised *in situ* at low temperatures (*J. Clark* and *Lister*, J. chem. Soc., 1961, 5048) and is especially useful for the production of halogenopurines from related pyrimidines when other methods frequently lead to hydrolysis or acidolysis.

Post-war unambiguous syntheses of purine nucleosides and nucleotides by *A. R. Todd* and co-workers required mild methods for completion of the imidazole ring system and this led to the use of dithioformic acid instead of formic acid (*J. Baddiley et al.*, *ibid.*, 1943, 383). Aqueous sodium dithioformate or methanolic dithioformic acid react readily with 4-alkylamino-5-amino- or 5-amino-4-glycosylamino-pyrimidines to give the corresponding

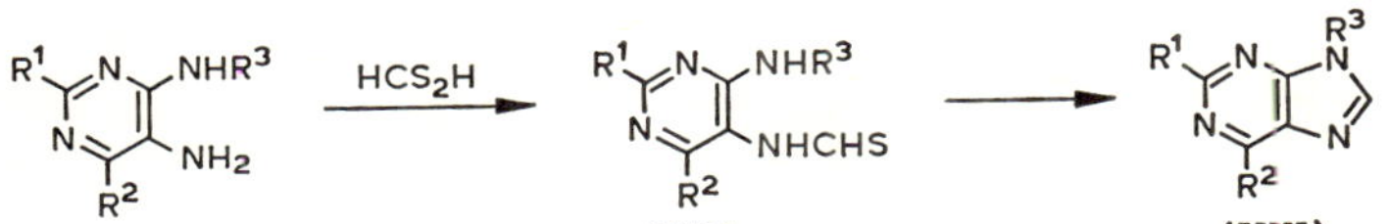
(VII) (VIII)

5-thioformamido compounds VII, which cyclise readily at their melting point, or in the presence of a base such as hot pyridine or methanolic sodium methoxide, to give 9-substituted purines VIII in excellent yield.

Of many compounds synthesized in this way may be mentioned adenine (VIII, $R^1 = R^3 = H$; $R^2 = NH_2$), 2-methylthio-9-methyladenine (VIII,

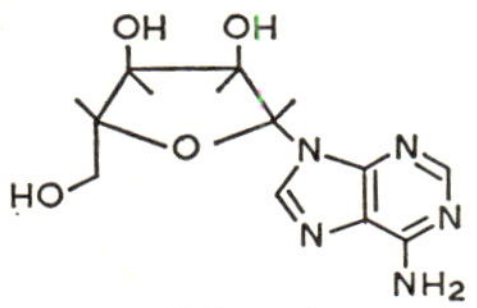
Adenosine

$R^1 = SMe$; $R^2 = NH_2$; $R^3 = Me$) (*Baddiley*, *B. Lythgoe* and *Todd*, *ibid.*, 1943, 386) and, as a culmination of the work with glycosides, adenosine itself (*G. W. Kenner*, *C. W. Taylor* and *Todd*, *ibid.*, 1949, 1620) (see p. 128).

The cyclisation of compounds of type VII is easier when R^3 = alkyl than when R^3 = H; cyclisation invariably involves the most basic nitrogen atom available and thus gives 9-substituted purines when 4-amino-5-thioformamido-6-substituted aminopyrimidines are used except when the substituent group is an acetylated glycoside. The formation of 6-acetylglycosylaminopurine in this case is attributed to bonding of the 2-acetoxy group of the sugar residue with the glycosidic amino group, causing increased basicity of the free amino group (*Kenner*, *Lythgoe* and *Todd*, *ibid.*, 1944, 654).

Although dithioformic acid is superior in many respects to formic acid, cyclisation is at times incomplete (*cf. Robins et al.*, J. Amer. chem. Soc., 1953, **75**, 263) and other modifications of the Traube method have been devised for the preparation of purines unsubstituted in the 8-position of which the most important is now probably the use of formamide introduced by *G. B. Brown* and co-workers (*A. Bendich et al.*, *ibid.*, 1948, **70**, 3109). It was then used in conjunction with formic acid, but the presence of the latter is unnecessary (*Robins et al.*, *loc. cit.*; *cf. H. Bredereck et al.*, Chem.

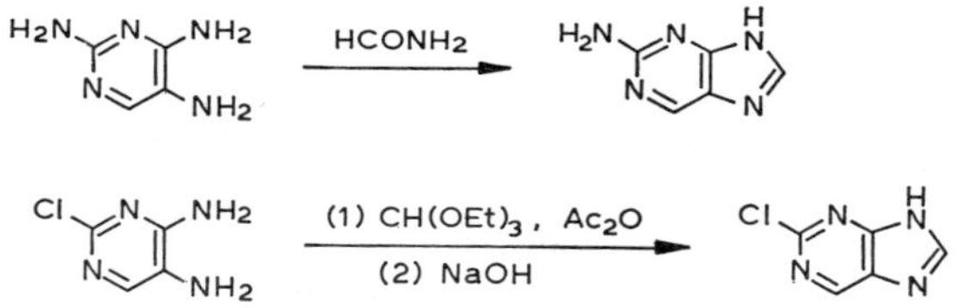

Ber., 1950, **83**, 201). Diaminopyrimidines also afford 8-unsubstituted purines when heated in acetic anhydride with triethyl orthoformate the *N*-acetyl derivative formed initially being hydrolysed by aqueous alkali (*J. A. Montgomery*, J. Amer. chem. Soc., 1956, **78**, 1928).

In this connection it is of interest that 4,5-diamino-6-furfurylaminopyrimidine gave kinetin, 6-furfurylaminopurine, *via* its acetyl derivative, with ethyl orthoformate (not the 9-*N*-furfuryl derivative to be expected with

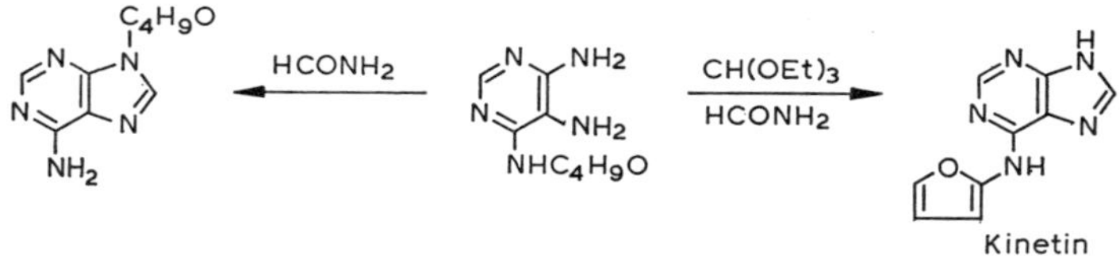

Kinetin

formic or dithioformic acids as discussed previously); but with formamide a mixture of kinetin and 6-amino-9-furfurylpurine was obtained (*R. Hull*, J. chem. Soc., 1958, 2746).

Montgomery also introduced diethoxymethyl acetate (from triethyl orthoformate and acetic anhydride) (*H. W. Post* and *E. R. Erikson*, J. org.

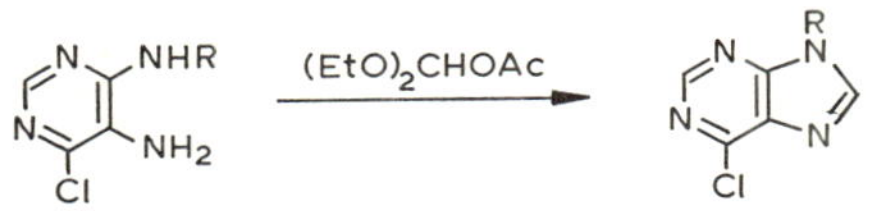

Chem., 1938, **2,** 260), as the cyclising agent, 9-substituted purines being prepared in excellent yields in this way (*Montgomery et al.*, J. Amer. chem. Soc., 1958, **80**, 404, 409).

A more unusual reagent, *symm.*-triazine, which apparently reacts only with primary amino groups, is capable of converting 4,5-diaminopyrimidines

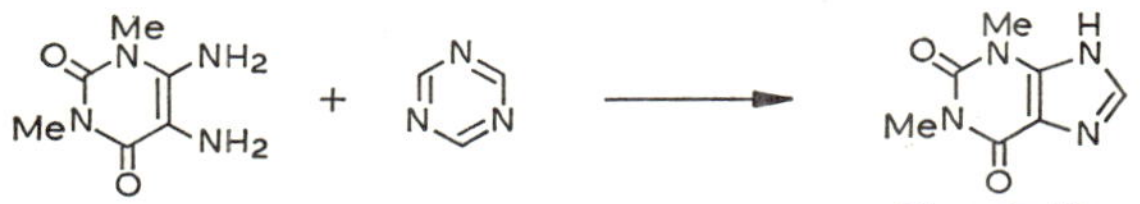

Theophylline

to purines and a 99% yield of theophylline was obtained in this way, other cyclisations investigated being rather less effective (*C. Grundemann* and *A. Kreutzberger*, *ibid.*, 1955, **77**, 6559).

Carboxylic acids other than formic acid cyclise 4,5-diaminopyrimidines to purines carrying a substituent in the 8-position. In this way 3,8-dimethylxanthine (IX) and 3-methylxanthine-8-carboxylic acid (X) are prepared by condensing the appropriate pyrimidone with acetic acid and oxalic acids,

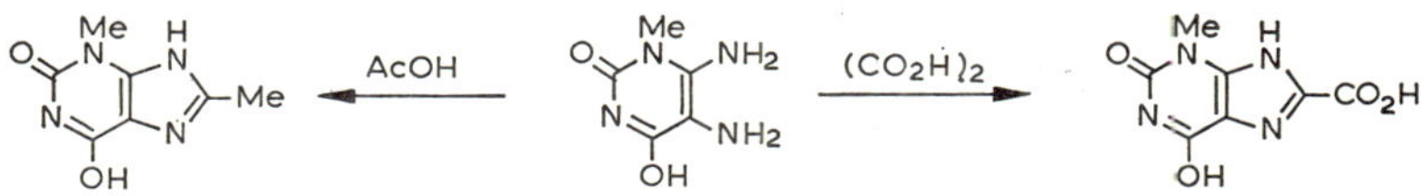

(IX) (X)

respectively (*Traube*, Ann., 1923, **432,** 266). Acid anhydrides give the same products as the acids themselves (*Johns* and *E. J. Baumann*, J. biol. Chem., 1913, **15,** 515).

Other examples are the production of 1,3,7,8-tetramethylxanthine (XII) on boiling 1,3-dimethyl-4,5-dimethylaminouracil (XI) with acetic anhydride (*Brederick et al.*, Chem. Ber., 1953, **86,** 845), the cyclisation in warm concentrated sulphuric acid of the 5-acylaminopyrimidine XIII (*W. T. Caldwell* and *C. S. Ching*, J. Amer. chem. Soc., 1955, **77,** 6631), to the purine analogue of pteroic acid XIV, and the formation of 8-trifluoromethylpurines using trifluoracetic anhydride (*A. Giner-Sorolla* and *Bendich*, *ibid.*, 1958, **80,** 5744). The sugar lactones ribonolactone and gluconolactone both react with 4,5-diamino-6-diethylamino-2-methylpyrimidine to yield the corresponding 8-polyhydroxymethylenepurine but cyclisation to the purine did not occur when 4,5,6-triamino-2-methylthiopyrimidine was used (*Hull*, J. chem. Soc.,

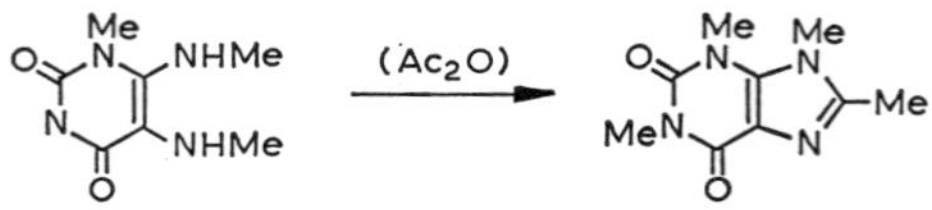

(XI) (XII)

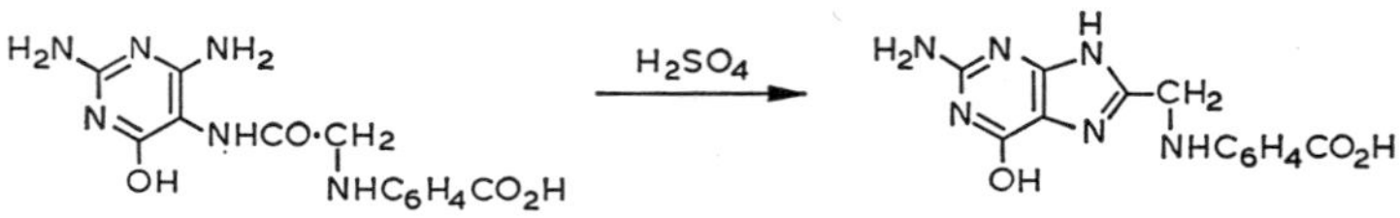

(XIII) (XIV)

1958, 4069). Carbon dioxide has been used to cyclise the fully trimethylsilylated derivative of 4,5-diamino-2,6-dioxypyrimidine to uric acid (*L. Birkofer*, *H. P. Kühlthau* and *A. Ritter*, Chem. Ber., 1960, **93**, 2810) but is normally of limited use. Among other carbonic acid derivatives ethyl chloroformate has been used to prepare 8-hydroxypurine (*W. Traube*, Ber., 1900, **33**, 3035) through the intermediate 5-ethylcarbamoyl derivative but the 8-hydroxypurines are more usually prepared by the use of other reagents. Urea was used by *S. Gabriel* and *J. Colman* (*ibid.*, 1901, **34**, 1234) for their synthesis of uric acid, potassium cyanate gives 5-ureido compounds which

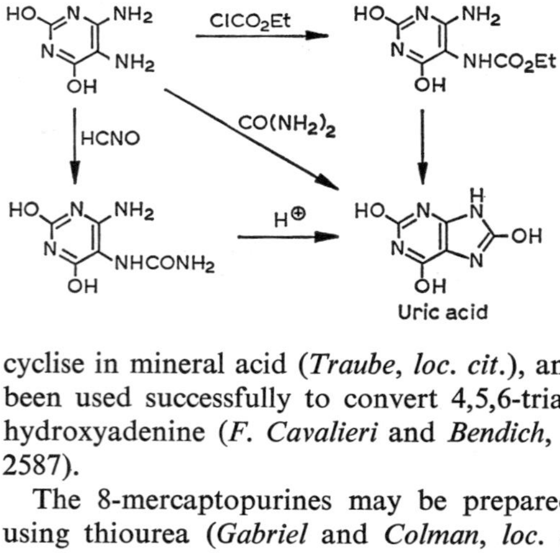

Uric acid

cyclise in mineral acid (*Traube*, *loc. cit.*), and phosgene in caustic soda has been used successfully to convert 4,5,6-triaminopyrimidine sulphate to 8-hydroxyadenine (*F. Cavalieri* and *Bendich*, J. Amer. chem. Soc., 1950, **72**, 2587).

The 8-mercaptopurines may be prepared in an analogous fashion by using thiourea (*Gabriel* and *Colman*, *loc. cit.*), thiophosgene (*Johns* and *Baumann*, *loc. cit.*), or carbon disulphide (*C. F. Boehringer and Söhne*, G.P. 142,468/1903). The last compound readily converts 5,6-diamino-4-methylamino-2-methylthiopyrimidine into 6-amino-8-mercapto-9-methyl-2-methylthiopurine (VIII, R^1 = SMe; R^2 = NH_2; R^3 = Me) (*A. H. Cook* and *E. Smith*, J. chem. Soc., 1949, 3002). However, thiourea and 4,5-di-

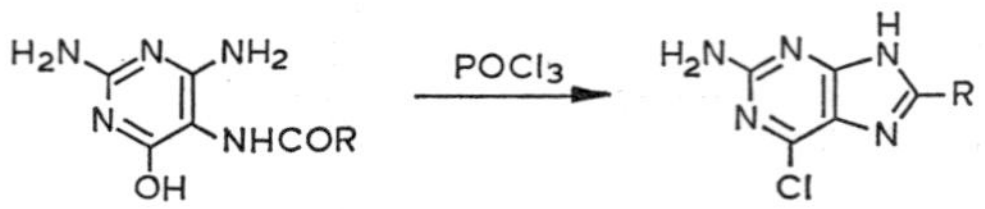

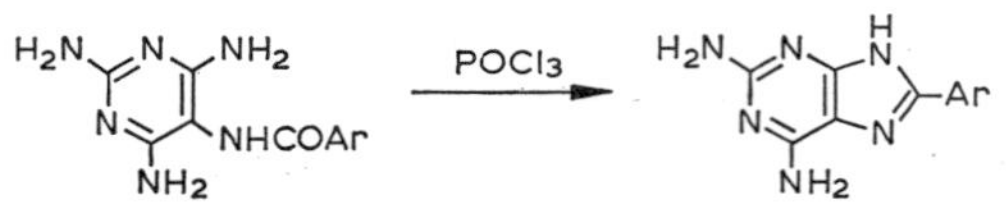

8-Arylpurines (and 8-alkyl or unsubstituted derivatives) have also been prepared by reaction of a diaminopyrimidine with an aldehyde in the presence of an oxidising agent such as ferric chloride or nitrobenzene (*D. Jerchel*, *M. Kracht* and *K. Krucker*, Ann., 1954, **590,** 232) including a synthesis of 8-phenyltheophylline (*D. Jerchel*, G.P. 955,861/1957).

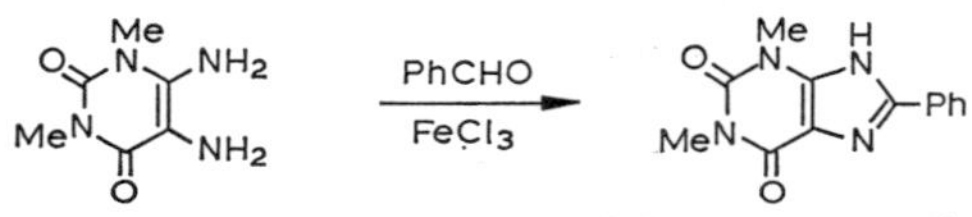

3-Phenyltheophylline

(*2*) *From 4-amino-5-nitro-*, *-nitroso-* and *-phenylazo-pyrimidines.* These methods include both reductive and (with nitroso derivatives) cyclodehydration syntheses. 4-Amino-5-nitrosopyrimidines in formic acid with Raney nickel produce the purines directly and syntheses of guanine, hypoxanthine, xanthine and 3-methylxanthine have been successful (*C. E. Liau*, *K. Yamashita* and *M. Matsui*, Agric. and biol. Chem., Japan, 1962, **26,** 624). Sodium acetate is normally required when more than one hydroxy group is present. Other reducing agents have been used including dithionite in aqueous solution with phosgene to produce uric acid (*R. Lohrmann*, *J. M. Lagowski* and *H. S. Forrest*, J. chem. Soc., 1964, 451), sulphur dioxide and sodium sulphite (*G. Scheuing* and *W. Konz*, G.P. 834,195/1952). Similar reactions using 5-nitropyrimidines have produced adenine (*Y. Fujimoto* and *N. Ono*, J. pharm. Soc., Japan, 1965, **86,** 3671) and 2-substituted purines (*G. Levin* and *M. Tamari*, J. chem. Soc., 1960, 2782), and an analogous reductive

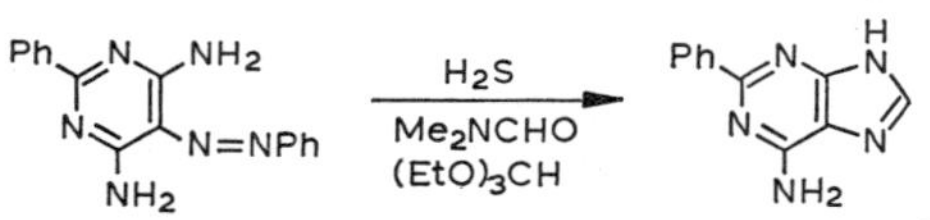

2-Phenyladenine

cyclisation of 4,6-diamino-2-phenyl-5-phenylazopyrimidine with hydrogen sulphide in the presence of dimethylformamide and triethyl orthoformate gave 2-phenyladenine (*E. C. Taylor* and *R. W. Morrison*, J. Amer. chem. Soc., 1965, **87**, 1976).

Certain 4-amino-5-nitrosopyrimidines may be cyclised to purines by dehydration. Thus theophylline or 8-phenyltheophylline are produced by heating 1,3-dimethyl-4-methyl(or benzyl)amino-5-nitrosouracil (XIX, R = H, Ph) (*H. G. Goldner*, *G. Dietz* and *E. Carstens*, Naturwiss., 1964, **51**, 137; 1966, **691**, 142).

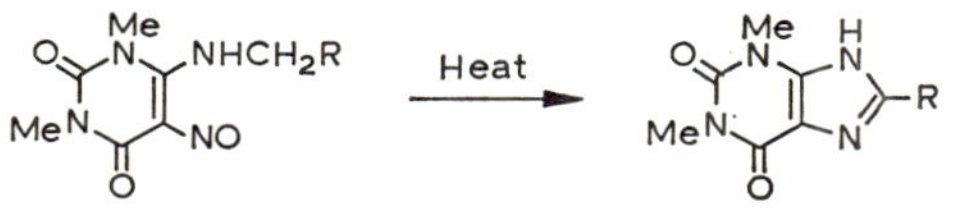

(XIX)

Prior acylation for example of 1,3-dimethyl-4-methylamino-5-nitrosouracil with acetic anhydride to produce the acyloxyimino derivative XX results in smoother cyclisation in this case to theophylline (*W. Pfleiderer* and *F. E. Kempter*, Angew. Chem. intern. Edn., 1967, **6**, 258).

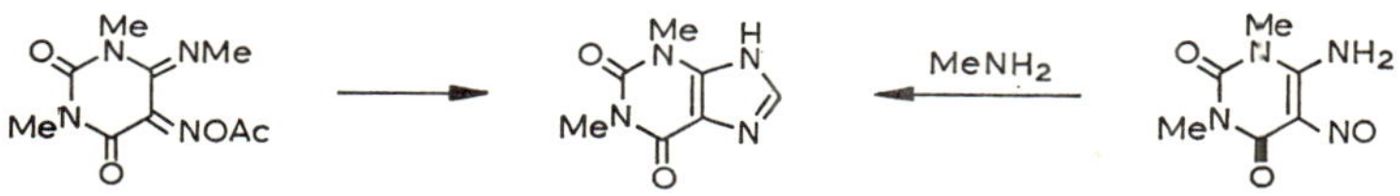

(XX) Theophylline

An extension includes reaction of the amino-nitrosopyrimidine and an amine, thus methylamine and 4-amino-5-nitroso-1,3-dimethyluracil gave theophylline (*Goldner*, *Dietz* and *Carstens*, *loc. cit.*).

In the absence of two hydrogens on the primary alkylamine alternative cyclisation may occur. Thus 4-isopropylamino-1,3-dimethyl-5-nitrosouracil gave, *via* and 8,8-dimethyl derivative, 8-methylcaffeine (*Goldner et al.*, Z. Chem., 1964, **4**, 454).

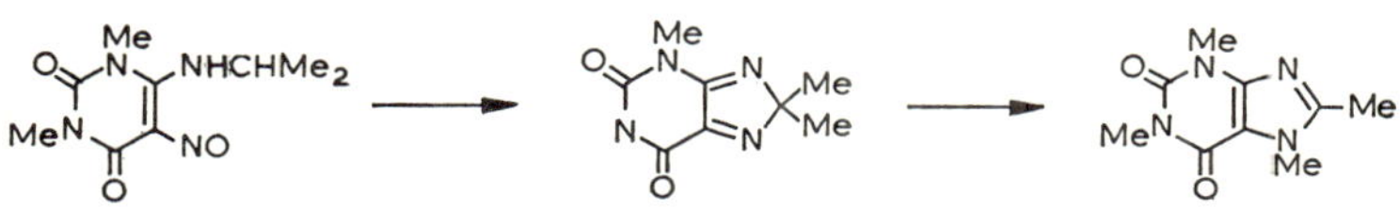

8-Methylcaffeine

A new synthesis of purines from 6-amino-5-nitrosouracil and the 1,1-dimethylhydrazones of various aldehydes, in dimethylformamide, has been reported (*F. Yaneda* and *T. Nagamatsu*, J. chem. Soc., Perkin I, 1976, 1547).

(*3*) *From 5-amino-4-hydroxypyrimidines.* The synthesis of purines from 5-amino-4-hydroxypyrimidines is due to *Emil Fischer*; the intermediate in this type of synthesis is a 5-ureido- or 5-thioureido-4-hydroxypyrimidine which is formed by treating the 5-amino-4-hydroxypyrimidine with cyanate

(*Fischer* and *L. Ach*, Ber., 1895, **28,** 2473), with an alkyl or aryl isocyanate (*Fischer*, *ibid.*, 1900, **33,** 1701; *E. F. Armstrong*, *ibid.*, 1900, **33,** 2308), when 9-substituted purines are formed, or with an alkyl or aryl isothiocyanate to give a 9-substituted-8-mercaptopurine (*H. Biltz* and *K. Strufe*, Ann., 1921, **423,** 200).

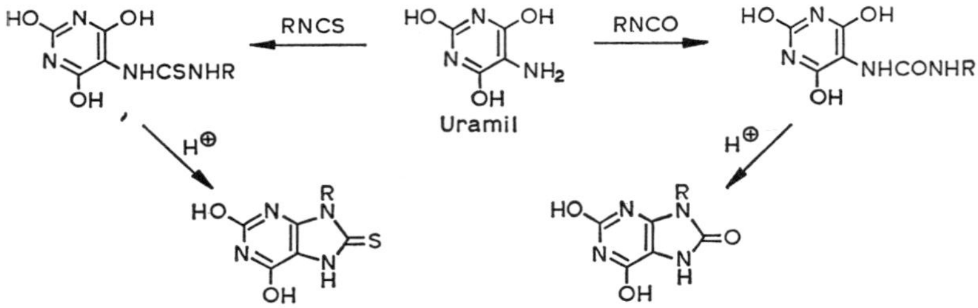

Cyclisation is best effected with mineral acids. The method appears to be limited to uramil (5-amino-2,4,6-trihydroxypyrimidine) and its derivatives. *Fischer* and *H. Tullner* (Ber., 1902, **35,** 2563) prepared 2,6-dihydroxy-8-mercaptopurine by treating 2,4,6-trihydroxy-5-cyanamidopyrimidine with hydrogen sulphide and cyclising the resultant thioureido compound. Substituted 5-aminopyrimidines also undergo this reaction (*E. Fischer*, *ibid.*, 1897, **30,** 559; *Fischer* and *H. Clemm*, *ibid.*, p. 3089). It is noteworthy that only 2,6-dihydroxypurines have so far been prepared by this route.

A method of preparing 8-substituted xanthines such as XXI developed by *F. G. Fischer et al.* (Chem. Ber., 1952, **85,** 752) may be considered as an

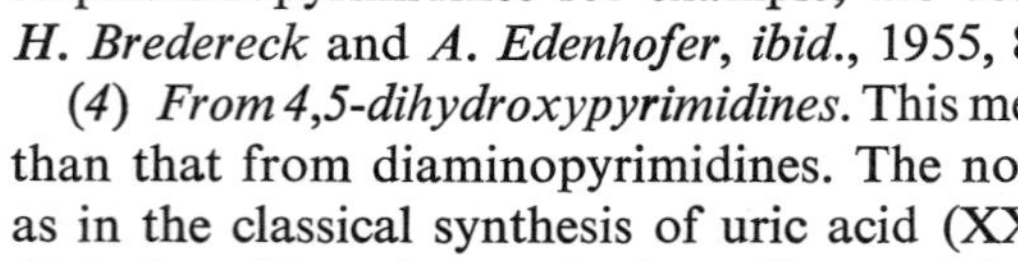

(XXI)

example of purine preparation from uramils although 2,6-dihydroxy-5-sulphamidopyrimidines for example, are used as the starting material (*cf. H. Bredereck* and *A. Edenhofer*, *ibid.*, 1955, **88,** 1306).

(*4*) *From 4,5-dihydroxypyrimidines.* This method is of much less importance than that from diaminopyrimidines. The normal cyclising agents are urea, as in the classical synthesis of uric acid (XXII, $R^1 = R^2 = H$) from isodialuric acid and urea in hot dilute sulphuric acid (*R. Behrend* and *O. Roosen*, Ann., 1888, **251,** 235). *N*-Substituted ureas may be used also. The latter give rise to 7-substituted purines; isodialuric acid and *N*-methylurea, for example, giving 7-methyluric acid (XXII, $R^1 = H$; $R^2 = Me$) (*A. Prüsse*, Ann., 1925, **441,** 203); *N,N*-dimethylurea also reacts with isodialuric acid

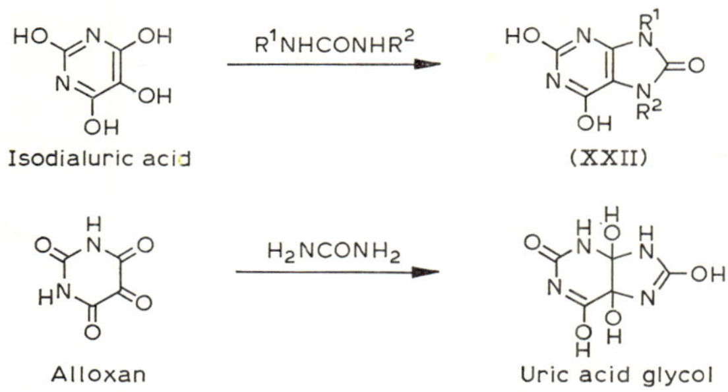

to give 7,9-dimethyluric acid (XXII, $R^1 = R^2 =$ Me). The condensation of alloxan with urea can give uric acid glycol (*H. Biltz* and *M. Heyn*, Ber., 1912, **45**, 1677; 1914, **47**, 459).

A new route to purines involves the reaction of 6-benzylamino-1,3-dimethyluracil with 4-phenyl-1,2,4-triazole-3,5-dione to produce an adduct which cyclised in hot nitrobenzene to produce 9-benzyl-1,3-dimethyl-8-phenylxanthine.

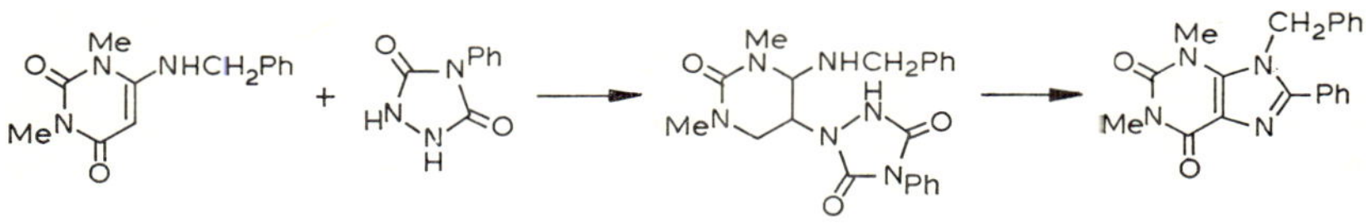

(*ii*) *From imidazoles*

The synthesis of purines from imidazole precursors was at one time of limited value due largely to the difficulty of obtaining appropriate imidazoles which in any case are much less stable than the corresponding pyrimidines and readily prone to aerial oxidation; in addition, the early workers were able only to prepare xanthine derivatives by this route. These disadvantages have been largely overcome and in recent years this type of synthesis has received increasing attention. This has arisen in part from the observation that 4-aminoimidazole-5-carboxamide (and later its nucleoside and nucleotide derivatives) accumulated in *Escherichia coli* cultures which had been inhibited by sub-toxic doses of sulphonamides (*W. Shive et al.*, J. Amer. chem. Soc., 1947, **69**, 725). It is now known that the *de novo* biosynthetic pathway to purine nucleotides proceeds in part *via* related 5-aminoimidazole nucleotides (*J. M. Buchanan* and *S. C. Hartman*, Adv. in Enzymology, 1959, **21**, 199; *idem*, Ann. Rev. Biochem., 1959, **28**, 365; *J. G. Flaks* and *L. N. Lukens*, in "Methods in Enzymology", Vol. 6, eds. *S. P. Colowick* and *N. D. Kaplan*, Academic Press, New York, 1963, p. 52). In addition, the so-called "abiogenic syntheses" of purines which proceed by reaction sequences which are purported to have been possible on a primeval earth, and consequently may have been a primary source of some of the main chemicals required for a chemical synthesis of the first living system, also appear to proceed *via* imidazole intermediates (*J. Brooks* and *G. Shaw*, "Origin and Development of Living Systems", Academic Press, New York, 1973). It will, however, be more convenient to discuss methods of this last type in a separate section (see p. 22).

The first successful synthesis of a purine from an imidazole was effected by *J. Sarasin* and *E. Wegman* (Helv., 1924, **7,** 713) who converted 4-amino-1-methylimidazole-5-carboxamide to 7-methylxanthine (XXIII) by heating it

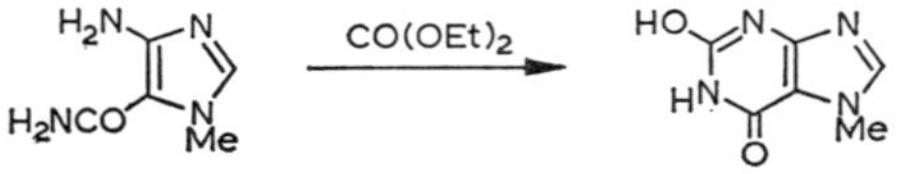

(XXIII)

at 160–170° with diethyl carbonate, and the same imidazole derivative affords 7-methylhypoxanthine with formamide (*R. N. Prasad* and *R. K. Robins*, J. Amer. chem. Soc., 1957, **79,** 6401). In a similar way *F. Montequi* and *Deaz de Plaza* prepared 7-ethyl-8-methylxanthine (Anal. Soc. fis. quim., 1926, **24,** 731). This general method has since been greatly extended by *A. H. Cook* and *E. Smith* (J. chem. Soc., 1949, 2329, 3001); thus, the substituted aminoimidazolecarboxamide XXV has been cyclised by three different reagents to give either the 6-hydroxypurine XXIV or its 2-mercapto analogue XXVI and the similar imidazoles (XXVIII, R = Me or $PhCH_2$) have also been cyclised by three different methods to give the purines XXVII, XXIX

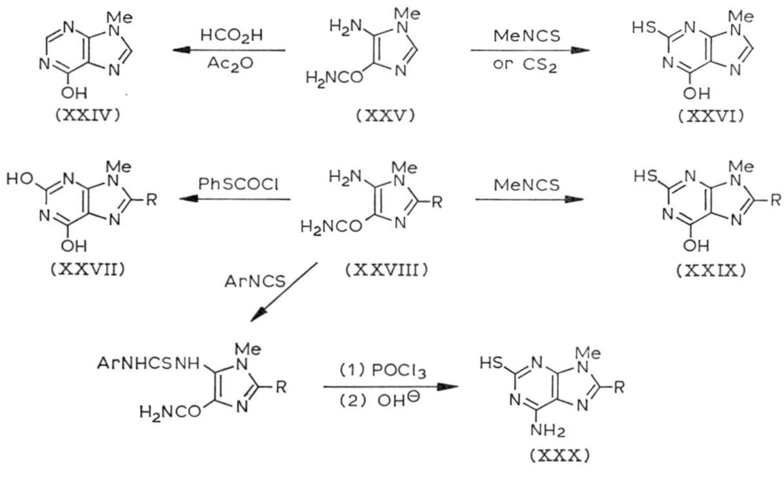

and XXX. Since the mercaptopurine (XXX, R = H) may be converted to 9-methyladenine by methylation at the mercapto group followed by desulphurisation, this synthesis affords a method for preparing the important adenine types of purine.

Among the mildest cyclisation procedures are those which involve the use of formic–acetic anhydride which produces the 5-*N*-formyl derivative XXXI, cyclisation of which occurs with aqueous sodium hydrogen carbonate

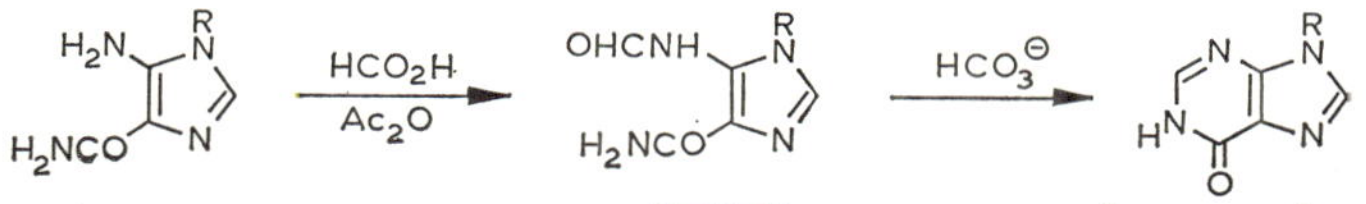

(XXXI) Hypoxanthine (R = H)

or other weak bases as in the synthesis of hypoxanthine and other purines (*E. Shaw*, J. biol. Chem., 1950, **185**, 439; *T. Ichikawa*, *T. Kato* and *T. Takenishi*, J. heterocycl. Chem., 1965, **2**, 253).

The conditions are mild enough to allow of the synthesis of purine nucleosides (*G. R. Greenberg* and *E. L. Spilman*, J. biol. Chem., 1956, **219**, 411; *J. Baddiley et al.*, J. chem. Soc., 1959, 2893; *G. Shaw* and *D. V. Wilson*, *ibid.*, 1962, 2937).

The same general type of synthesis is also effective for preparing 1-alkylpurines (*F. G. Mann* and *J. W. G. Porter*, *ibid.*, 1945, 751). The synthesis of paraxanthine from 5-amino-1-methylimidazole-4-carboxymethylamide is an example.

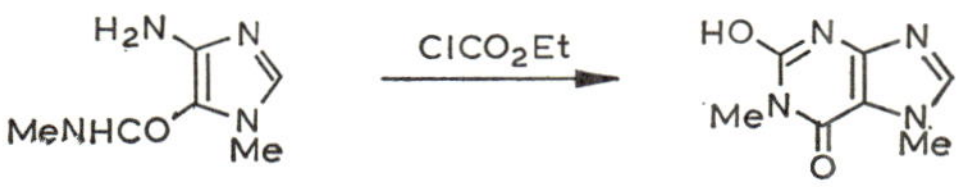

Paraxanthine

This type of synthesis has been used to prepare hypoxanthine, xanthine and, with sodium methyl xanthate in dimethylformamide, 2-thioxanthine (*A. Yamazaki*, *I. Kumashiro* and *Takenishi*, J. org. Chem., 1967, **32**, 3032), adenine and isoguanine (*E. Shaw*, *loc. cit.*). In the last two cases 5-amino-4-guanidinoimidazole (XXXII) was used as the starting material; the routes employed to these various important purines are shown in Scheme 1. The cyclisation of XXXII with phosgene has also been reported (*L. F. Cavalieri et al.*, J. Amer. chem. Soc., 1949, **71**, 3973), and xanthine has been prepared from 4-aminoimidazole-5-carboxamide, diethyl carbonate and sodium ethoxide (*Yamazaki*, *Kumashiro* and *Takenishi*, J. org. Chem., 1967, **32**, 3258). 9-Methyladenine (XXXIII) has also been prepared by reaction of 5-amino-4-cyano-2-methylimidazole with triethyl orthoformate and reaction of the resulting 5-ethoxymethyleneamino-1-methylimidazole with alcoholic

Scheme 1

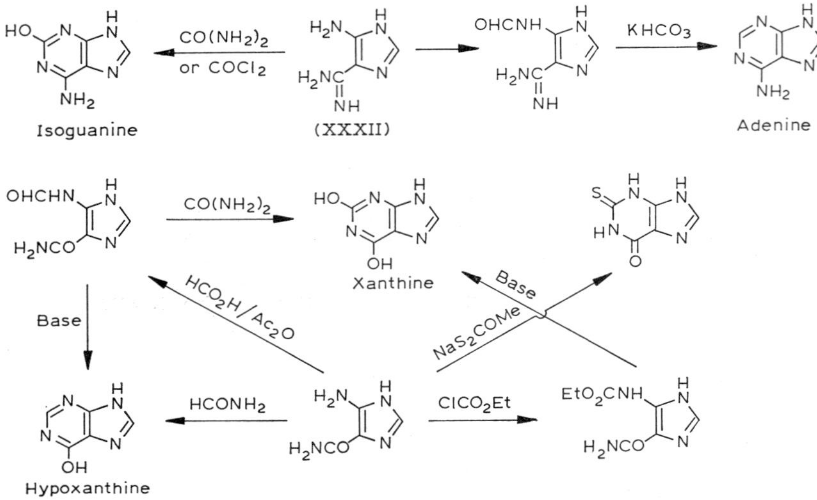

ammonia (*G. Shaw* and *D. N. Butler*, J. chem. Soc., 1959, 4040) and 7-methyladenine was similarly obtained from the isomeric 3-methylimidazole (*Taylor* and *P. K. Loeffler*, J. Amer. chem. Soc., 1960, **82,** 3147), which with methylamine in hot benzene also gave the 1,9-dimethyladenine and with urea 7-methylisoguanine (*E. Shaw*, J. org. Chem., 1962, **27,** 883). 5-Amino-4-

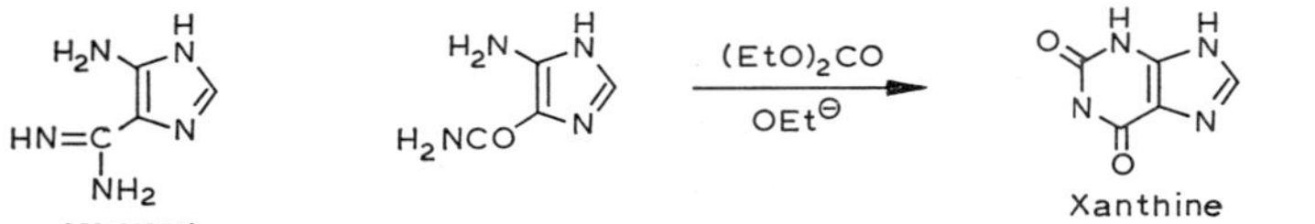

(XXXII) Xanthine

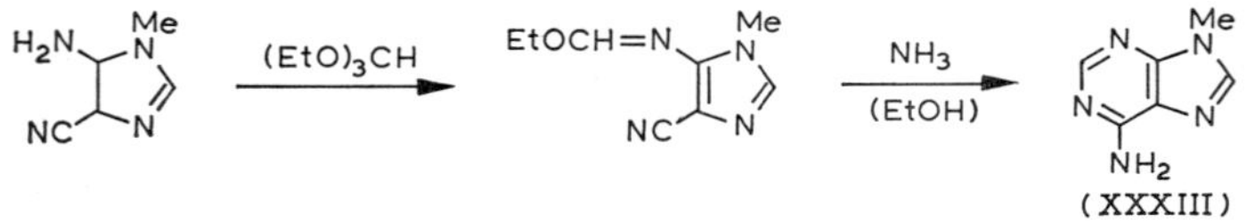

(XXXIII)

cyanoimidazole with formamide acetate in 2-methoxyethanol produced a good yield of adenine (*J. P. Ferris* and *L. E. Orgel*, J. Amer. chem. Soc., 1965, **87**, 4976; 1966, **88**, 3829).

Similarly, 9-methyl-6-thiopurine (XXXIV, R = Me) has been prepared from 5-amino-1-methyl-4-thiocarbamoylimidazole with formic–acetic anhydride and cyclisation of the intermediate *N*-formyl derivative with sodium hydrogen carbonate (*G. Shaw* and *Butler*, *loc. cit.*). The thioamide was

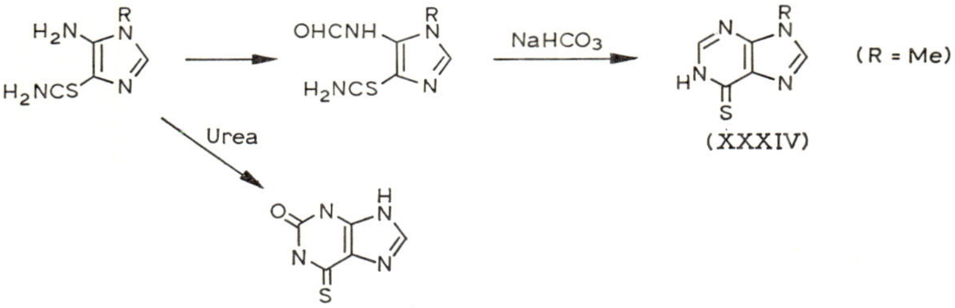

(XXXIV)

6-Thioxanthine
(R = H)

prepared by direct synthesis and others have been obtained by thiation of carboxamides with phosphorus pentasulphide (*G. H. Hitchings* and *G. B. Elion*, U.S.P., 2,756,228/1956; *M. Ikehara*, *Y. Nakazawa* and *H. Nakayama*, Chem. pharm. Bull., Japan, 1962, **10,** 660). The unsubstituted imidazolethioamide (R = H) with ethyl formate and sodium ethoxide produced 6-thiopurine and fusion with urea gave the corresponding 6-thioxanthine (*Yamazaki et al.*, *ibid.*, 1968, **16,** 2172).

Analogous reactions have been used to prepare certain purine *N*-oxides and related *N*-hydroxypurines. Thus 5-amino-4-(*N*-hydroxy)amidinoimidazole with triethyl orthoformate in dimethylformamide gave adenine 1-oxide (XXXV, R = H) and the thio analogue (XXXV, R = SH) was produced from the same imidazole and carbon disulphide in pyridine (*R. M. Cresswell* and *G. B. Brown*, J. org. Chem., 1963, **28,** 2560). Similarly, hypoxanthine 1-oxide (XXXVI) was obtained from the corresponding 5-hydroxamic acid with triethyl orthoformate (*E. C. Taylor*, *C. C. Cheng* and *O. Vogl*, *ibid.*, 1959, **24,** 2019), 7-methyladenine 3-oxide (XXXVII) from 5-cyano-4-hydroxyamino-1-methylimidazole (*Taylor* and *Loeffler*, *ibid.*, 1959, **24,** 2035) and 9-hydroxy-8-methylhypoxanthine (XXXVIII, R = H), 9-hydroxy-8-methylxanthine (XXXVIII, R = OH) and 9-hydroxy-8-methylguanine (XXXVIII, R = NH_2) by cyclisation of 5-amino-1-benzyloxy-2-methylimidazole-4-carboxamide with formate and carbonate esters or from a guanidino derivative XXXIX, respectively (*A. A. Watson* and *Brown*, J. org. Chem., 1972, **37,** 1867) and removal of the benzyl group with hydrogen bromide in acetic acid. 9-Hydroxy-hypoxanthine, -xanthine and -guanine (*Watson*, J. org. Chem., 1974, **39,** 2911; 1977, **42,** 1610) and 1-hydroxyguanine (*Watson*, *S. C. Nesnow* and *Brown*, *ibid.*, 1973, **38,** 3044) have been prepared in a similar manner using the general Shaw synthesis.

Purines and thiopurines may also be prepared from 5-aminoimidazole-4-carboxylates *via* their ureido or thioureido derivatives. This method, first used to prepare xanthine from 5-ureido-4-methoxycarbonylimidazole (*W. E. Allsebrook*, *J. M. Gulland* and *L. F. Storey*, J. chem. Soc., 1942, 232), has

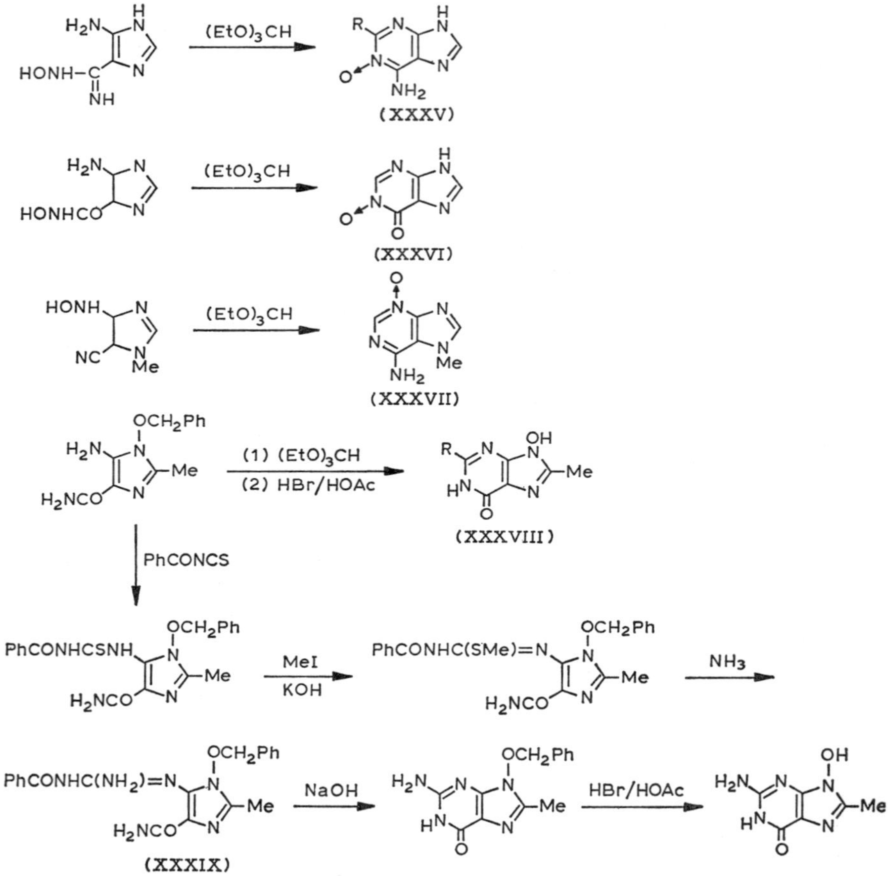

also been developed by *A. H. Cook* and his colleagues who have prepared a large number of xanthine and 2-mercaptoxanthines by these methods. The required 5-ureido- or thioureido-imidazole-4-carboxylates are readily prepared from methyl isocyanate or isothiocyanate and the corresponding 5-amino compounds, and are then cyclised in the presence of a base; a few of the many possible examples are given on p. 16; 7- or 9-alkylated purines have been prepared by similar methods (*Cook et al., ibid.*, 1949, 1069, 1071; 1950, 1884).

The preparation of guanine derivatives from 5-ethoxycarbonyl-4-thioureidoimidazoles (*Cook* and *G. H. Thomas, ibid.*, 1950, 1888) represents an interesting type of purine synthesis. These imidazoles cyclise in the presence

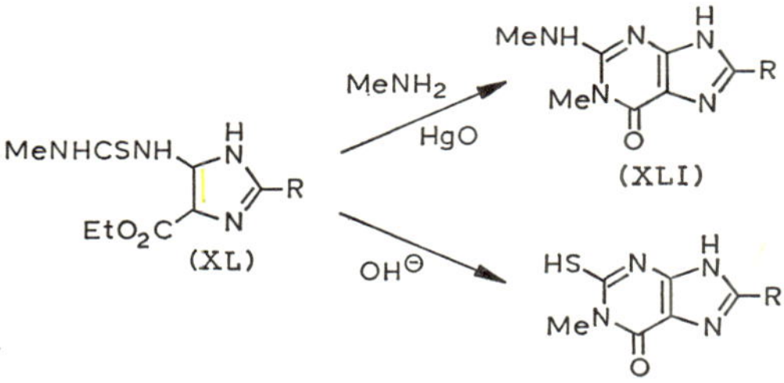

of methylamine and mercuric oxide to give 2-amino- and not 2-mercaptopurines. Thus 5-ethoxycarbonyl-4-methylthioureidoimidazoles (XL; R = H, Ph, $PhCH_2$) are cyclised under these conditions to the dimethylguanine derivative XLI. This reaction recalls the previously mentioned work of *Johns* and *Baumann* in which a 4,5-diaminopyrimidine was found to give an 8-aminopurine with thiourea (p. 11) and it was shown by *Cook* and *Thomas* that 4,5-diamino-2,6-dihydroxy- or 4,5-diamino-6-hydroxy-2-methylthiopyrimidine give, with methyl isothiocyanate, 5-thioureides which are cyclised in the presence of mercuric oxide to give 8-methylaminopurines. Guanine has also been prepared from 4-cyanamido-5-imidazolecarboxamide and alkali. The imidazole was obtained by brief treatment of 5-(*S*-methylisothiocarbamoyl)amino-4-imidazolecarboxamide with alkali (*A. Yamazaki, M. Okutsu* and *Y. Yamada*, Nucleic Acids Res., 1976, **3**, 251).

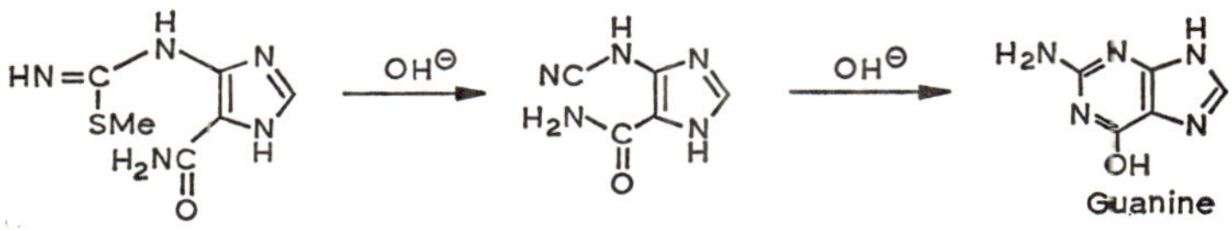

A series of 5-amino-1-glycofuranosylimidazole-4-carboxylic acid derivatives including esters and amides have been prepared by *G. Shaw* and coworkers (J. chem. Soc., 1973, 1720; Chem. Comm., 1970, 1114; 1971, 190; Experientia, 1972, **28**, 763) and these have been cyclised to produce purine

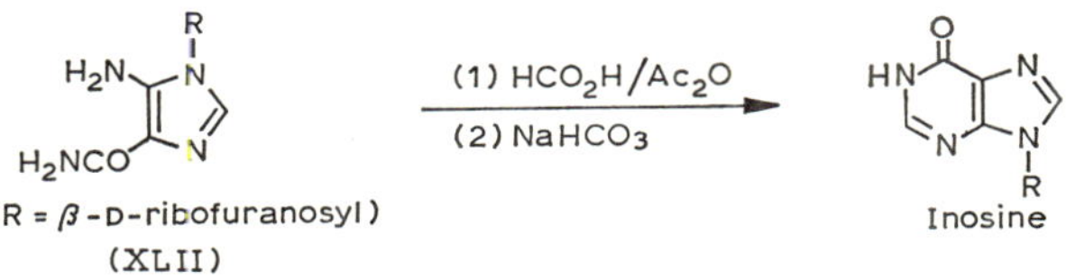

glycofuranosides including ribose derivatives. Thus inosine results from the aminoimidazole riboside XLII with formic–acetic anhydride followed by sodium hydrogen carbonate.

An earlier method which was formerly used to prepare certain 9-glycosylpurines is derived from the reaction of imidazole-4,5-dicarboxamide (R = H) with hypobromite which produces xanthine (XLIII, R = H) (*R. A. Baxter* and *F. S. Spring*, *ibid.*, 1945, 232) and the *N*-methylimidazole (R = Me) by the same reaction gives 9-methylxanthine (XLIII, R = Me), no 7-methylxanthine being formed (*idem, ibid.*, 1947, 378). This successful approach was followed by the synthesis of xanthines bearing glycosidic

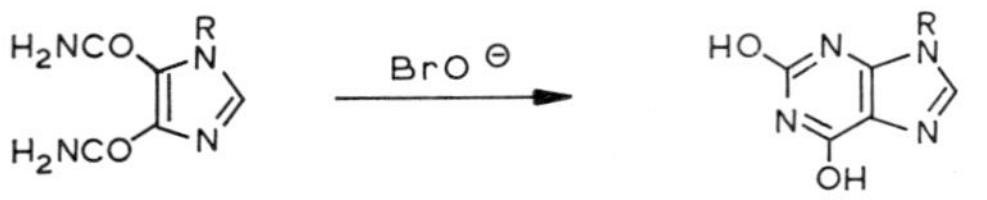

(XLIII)

groups in the 9-position culminating in the synthesis of xanthosine itself (XLIII, R = β-D-ribofuranosyl) (*G. A. Howard et al.*, *ibid.*, 1949, 232). Although, the 9-glycosylxanthines are the major product the presence of some of the 7-substituted isomer has been demonstrated in one case (*J. Baddiley et al.*, *ibid.*, 1958, 3606).

The following preparation of a dihydropurine from 4-aminomethyl-2,5-dibromoimidazole has been reported (*P. C. Miller* and *N. Chatterjee*, J. Indian chem. Soc., 1934, **11,** 867; C.A., 1935, **29,** 2953).

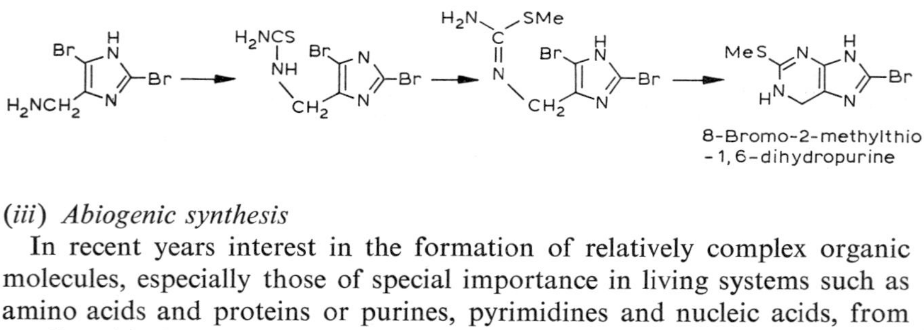

8-Bromo-2-methylthio-
-1,6-dihydropurine

(*iii*) *Abiogenic synthesis*

In recent years interest in the formation of relatively complex organic molecules, especially those of special importance in living systems such as amino acids and proteins or purines, pyrimidines and nucleic acids, from small molecules such as methane, carbon dioxide, formaldehyde, ammonia, etc., has increased markedly. Before the second world war much of this type of organic chemistry such as Robinson's classical synthesis of tropinone was largely founded on attempts to explain or understand the many diverse syntheses of naturally occurring compounds such as alkaloids, aromatic and terpenoid compounds, etc. which clearly existed in biological systems, and to attempt to duplicate them by non-enzymic methods under so-called physiological conditions. More recent work in this field, however, has stemmed from attempts to duplicate conditions on a postulated primeval earth and to provide a basis for the formation of all the important (for life production) organic chemicals from very simple molecules and reasonable energy sources. In a pioneer experiment (*S. L. Miller*, Science, 1953, **117,** 528; *idem*, J. Amer. chem. Soc., 1955, **77,** 2351) which stemmed from theories requiring a reducing (hydrogen-containing) atmosphere on the primeval earth (*H. C. Urey*, "The Planets", Yale University Press, 1952; *A. I. Oparin*, "Origin of Life", Dover Publications, New York, 1953), mixtures of hydrogen, methane, ammonia and water vapour were subjected to a continuous spark discharge for a week to produce many types of organic compounds but no purines. In 1963 however, adenine was obtained in 0.01% yield by irradiation of the Miller mixture with a 4.5 MeV electron beam (*C. Ponnamperuma et al.*, Proc. Nat. Acad. Sci., 1963, **49,** 737). Adenine had been produced earlier (*J. Oro*, Biochem. Biophys. Res. Comm., 1960, **2,** 207) in heated strong solutions of ammonium cyanide and later it was found that 0.1% yields of adenine could be obtained by heating 0.1% aqueous solutions of ammonium cyanide at 90° for 1 day or at 70° for 25 days (*Oro* and *A. P. Kimball*, Arch. Biochem. Biophys., 1961, **94,** 217) and evidence for the involvement of 5-aminoimidazole precursors in the reaction, including the 4-carbonitrile and 4-carboxamidine, came from their isolation together with formamidine from such solutions (*idem*, *ibid.*, 1962, **96,** 293); the following reaction scheme was proposed.

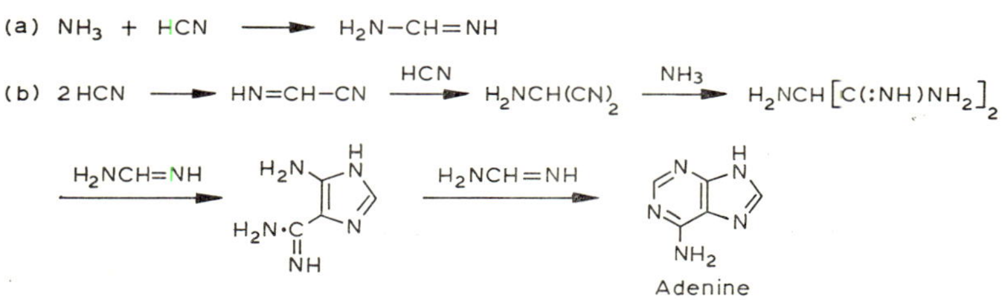

In addition to adenine the presence of hypoxanthine (but not guanine) in heated aqueous ammonium cyanide solutions together with at least fifty other ultraviolet absorbing compounds was confirmed (*C. U. Lowe*, *H. W. Rees* and *R. Markham*, Nature, 1963, **199,** 219).

Also *C. Ponnamperuma* had found (in "The Origins of Prebiological Systems and their Molecular Matrices", Academic Press, London and New York, 1965, p. 221) that both adenine and guanine could be produced in yields of 1% and 0.5%, respectively, by the ultraviolet radiation of dilute solutions of hydrogen cyanide for seven days at room temperature. A biradical scheme (*R. M. Kliss* and *C. N. Matthews*, Proc. Nat. Acad. Sci., 1962, **48,** 1300) was proposed.

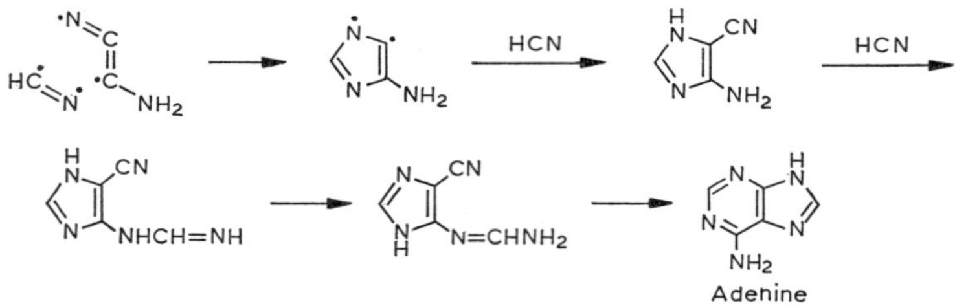

L. E. Orgel and co-workers established the formation of diaminomaleonitrile as a relatively stable tetramer of hydrogen cyanide (*R. A. Sanchez*, *J. P. Ferris* and *Orgel*, J. mol. Biol., 1967, **30,** 223) and other experiments showed that the tetramer could be produced from aminomalononitrile and hydrogen cyanide (*Ferris* and *Orgel*, J. Amer. chem. Soc., 1966, **88,** 3829). Ultraviolet photoisomerisation of the tetramer to 5-aminoimidazole-4-carbonitrile readily occurred in 80% yield (*idem, ibid.*, 1966, **88,** 1074; *Ferris, J. E. Kuder* and *A. W. Catalano*, Science, 1969, **166,** 765) *via* diaminofuranonitrile. The formation of the tetramer is enhanced in dilute solutions by freezing, and yields of 10% were produced after 11 months at —20°. The aminoimidazolecarbonitrile was found to react with ammonium cyanide solutions to produce adenine at 30–100° and extrapolation to —20° suggested that molar solutions of adenine and related heterocyclic compounds could have been formed after 10^6 years in a primeval lake which alternately thawed in summer and froze in winter.

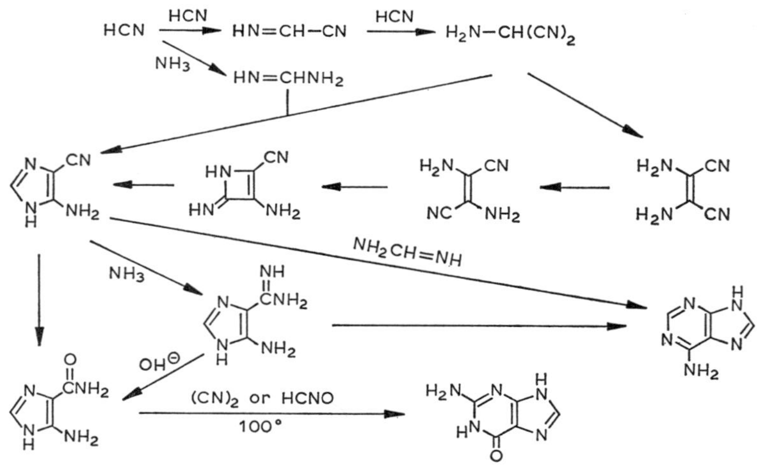

Also the carbonitrile is readily hydrolysed to the aminoimidazolecarboxamide from which a 20% yield of guanine may be obtained by reaction with cyanogen or cyanate at 100°. These reactions are summarised on p. 24.

Extension of the ammonia–hydrogen cyanide procedure to the use of acetamidine in place of hydrogen cyanide gives 2-methyl-, 8-methyl- and 2,8-dimethyl-adenines whereas liquid methylamine and hydrogen cyanide produces 7- and 9-methyl-6-dimethylaminopurines (*Y. Yamada, T. Kumashiro* and *T. Takenishi*, J. org. Chem., 1968, **33**, 642). Although the various reactions mentioned here have received much publicity it must be mentioned that similar types of reactions have frequently appeared in the earlier literature and have generally received scant attention. Thus uric acid was produced in 1882 by *J. Horbaczewski* by heating glycine (Monatsh., 1882, **3**, 796), or from trichlorolactic acid amide and urea (*idem, ibid.*, 1887, **8**, 201, 584; *R. Behrend*, Ann., 1925, **441**, 215).

(*iv*) *Other syntheses*

Apart from the so-called abiogenic syntheses outlined above, the rational synthesis of purines from acyclic precursors by more or less one-step processes has received only scattered attention. Nevertheless in recent years several interesting examples of such methods have been described and it would be fair to say that much of the mechanism and understanding of the types of abiogenic reactions mentioned under (*iii*) has arisen from the results of these generally earlier experiments. Indeed it is not possible to draw any clear line between these various types of reactions except that presumably a reaction should only be considered abiogenic if it takes place under conditions known to have existed on a primordial earth. Since there is of course much controversy about the nature of such conditions the distinctions become even less clear (*J. Brooks* and *G. Shaw*, in "Origin and Development of Living Systems", Academic Press, London and New York, 1973) and consequently popular theories of abiogenicity and chemical evolution must be treated with great caution.

Perhaps the first rational synthesis of a purine from acyclic precursors was that of hypoxanthine (*E. Shaw*, J. biol. Chem., 1950, **185**, 439) by heating formamidomalonamidinecarboxamide (XLIV) with formamide. Similar reactions involved aminomalondiamidine with ortho-esters in dimethylformamide or acetonitrile to produce various 2,8-dialkylpurines. Thus hypoxanthine was formed by 5 minutes heating with triethyl orthoformate whereas 2,8-dimethyl- and 2,8-diethyl-hypoxanthine respectively required 10 and 60 hours heating with the appropriate ortho-ester (*E. Richter, P. X. Loeffler* and *E. C. Taylor*, J. Amer. chem. Soc., 1960, **82**, 3144). Shorter reaction times using aminomalonamidine hydrochloride and ortho-esters led similarly to 2,8-dialkyladenines (*idem*, *loc. cit.*; *Taylor*, Chem. and Biol. of Purines, 20)*.

When XLV was reacted with hydroxylamine before treatment with triethyl orthoformate in acetic anhydride, hypoxanthine *N*-oxide (XLVI) was obtained together with an isomer which was either the 3- or 9-*N*-oxide.

(XLIV) Hypoxanthine

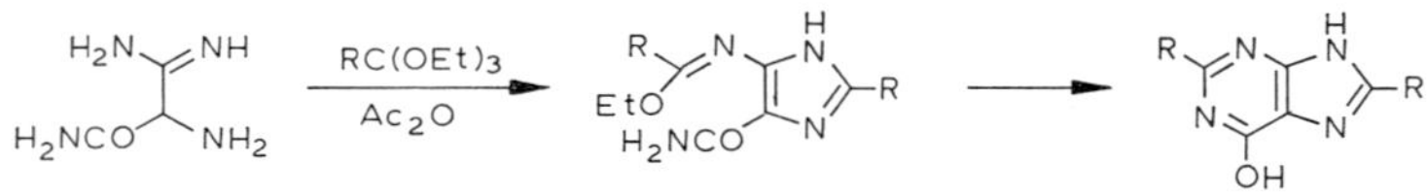

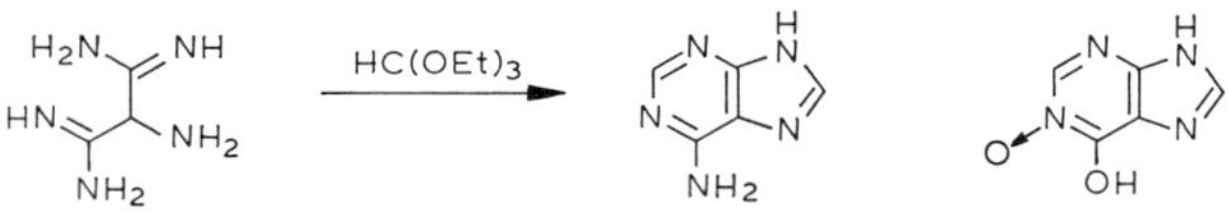

(XLV) (XLVI)

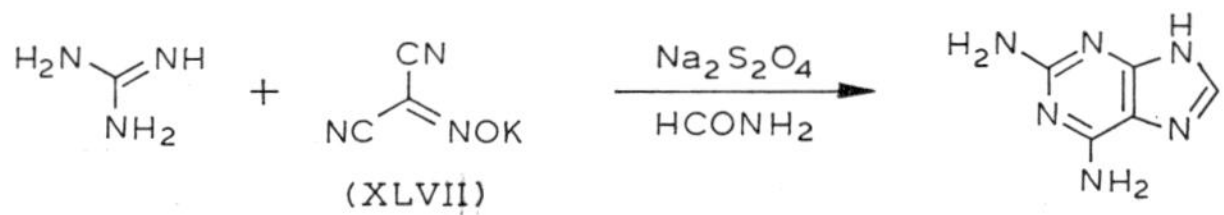

(XLVII)

The silver salt of hydroxyiminomalononitrile can be converted by means of amidines into 4,6-diamino-5-nitrosopyrimidines which, with formamide and formic acid in the presence of sodium dithionite afford 6-aminopurines. This process has been completed in one step by heating guanidine carbonate, potassium hydroxyiminomalononitrile (XLVII) and sodium dithionite in formamide to afford 2,6-diaminopurine in excellent yield (*O. Vogl* and *Taylor*, J. Amer. chem. Soc., 1957, **79**, 1518).

Purine has also been produced in 40% yield by heating *N*-cyanomethylphthalimide with trisformamidomethane or formamidine acetate in form-

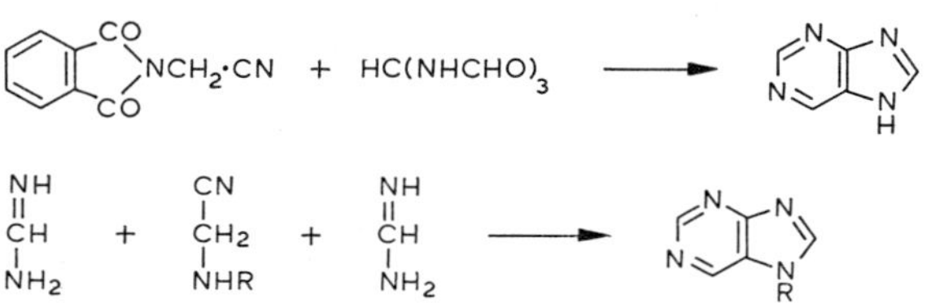

* The abbreviation "Chem. and Biol. of Purines" used in this and the following chapter refers to "The Chemistry and Biology of Purines" (C.I.B.A. Foundation Symposium) eds. *G. E. W. Wolstenholme* and *C. M. O'Connor*, Churchill, London, 1957).

amide or *n*-butanol (*H. Bredereck*, *F. Effenberger* and *G. Rainer*, Angew. Chem., 1961, **73,** 63; *idem*, G.P., 1,150,988/1963; *Bredereck et al.*, Ann., 1962, **659,** 133), and 7-substituted purines could be obtained similarly using alkyl- or aryl-aminoacetonitriles (*idem*, *ibid.*, 1964, **673,** 82).

(*c*) *Purine and alkylpurines*

(*i*) *Purine*

Purine, $C_5H_5N_4$, was first prepared in poor yield by *E. Fischer* (Ber., 1899, **32,** 493) by reducing 2,6,8-trichloropurine first with phosphonium iodide and then with zinc dust. It occurs naturally as its 9-D-ribofuranoside, nebularine (see p. 142). The first total synthesis (*O. Isay*, *ibid.*, 1906, **39,** 250, 257) used the Traube method (p. 5) which gives an excellent yield (*A. Albert* and *D. J. Brown*, J. chem. Soc., 1954, 2066). It has been obtained also by cyclising 4,5-diaminopyrimidine with 1,3,5-triazine or diethoxymethyl acetate (82% yield) (*J. A. Montgomery* and *C. Temple*, J. org. Chem., 1960, **25,** 395), by hydrogenating 6-chloro- (*A. Bendich et al.*, J. Amer. chem. Soc., 1954, **76,** 6073), 2,6-dichloro- and 2,6,8-trichloropurine (*Bredereck et al.*, Chem. Ber., 1962, **95,** 54) with a palladium catalyst, by the dethiation of purine-6-thiol- or -2,6-dithiol (*A. G. Beaman*, *ibid.*, 1954, **76,** 5633) or purine-6-thiol 3-*N*-oxide (*G. B. Brown et al.*, J. med. Chem., 1965, **8,** 190) with Raney nickel in water, by heating purine-6-carboxylic acid at 190° (*L. B. Mackay* and *G. H. Hitchings*, J. Amer. chem. Soc., 1956, **78,** 3511; *S. Cohen et al.*, J. org. Chem., 1962, **27,** 3645) or purine-8-carboxylic acid in boiling water (*Albert*, J. chem. Soc., 1960, 4705), from purine-6-sulphinic acid and formic acid (*J. J. Fox et al.*, J. org. Chem., 1961, **26,** 3401) and 6-hydrazinopurine with sodium hydroxide (*Montgomery et al.*, *ibid.*, 1965, **30,** 3601). It has also been produced from acyclic precursors (see under *iv*, p. 25 *et seq.*). Similarly 2- and 6-deuteropurine have been produced by Traube syntheses using 2- and 4-deuteropyrimidines (*S. Matsuura* and *T. Goto*, J. chem. Soc., 1965, 623) or by catalytic deuteration of 2- or 6-chloropurine (*W. C. Coburn et al.*, J. org. Chem., 1965, **30,** 1110), and 6- or 8-deuteropurine resulted from dethiation of purine-6- or -8-thiols with deuterated Raney nickel (*M. P. Schweizer et al.*, J. Amer. chem. Soc., 1864, **86,** 696).

Purine, m.p. 216°, is a colourless, crystalline, weak base which sublimes *in vacuo*. It is soluble in water (1 g in 2 ml at 20°, in 0.9 ml at 100°), alcohols and benzene but sparingly soluble in ether. It readily forms monobasic salts with acids; *nitrate*, m.p. 205° (decomp.), *hydrobromide*, m.p. 231–232°, *picrate*, m.p. 208°; its cation has pK_a 2.4. Salts are also formed with metals *e.g.* sodium and zinc; the compound is a slightly stronger acid than phenol pK_a 8.9 (*Albert* and *Brown*, *loc. cit.*). Its infrared absorption has been reported by *C. H. Willits et al.* (J. Amer. chem. Soc., 1955, **77,** 2569). Purine is stable to hot alkali and relatively stable to dilute acid (*cf.*, however, *M. P. Gordon et al.*, *ibid.*, 1957, **79,** 3245) and is oxidised slowly by cold aqueous permanganate or warm fuming nitric acid. Earlier attempts to methylate it were unsuccessful but *Bredereck* obtained 9-methylpurine with both dimethyl sulphate and diazomethane (Chem. Ber., 1956, **89,** 12) and acetylation with acetic anhydride produced a mixture of 7- and 9-acetylpurines (*Montgomery* and *Temple*, *loc. cit.*). Bromination gives an unstable adduct (*J. H. Lister*, "Purines" in "Heterocyclic Compounds", ed. *D. J. Brown*, Interscience, New York, 1971)

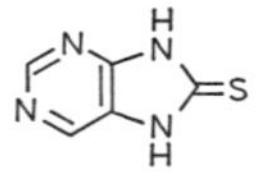

(XLVIII)

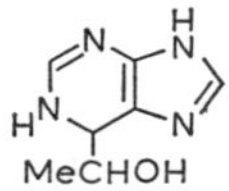

(XLIX)

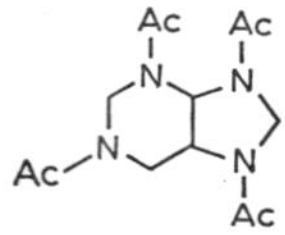

(L)

and sulphur at 245° gave 7,8-dihydro-8-thiopurine (XLVIII) in 75% yield (*A. Giner-Sorolla, E. Thom* and *A. Bendich*, J. org. Chem., 1964, **29**, 3209) whereas addition of methanol and ethanol occurs by irradiation of deoxygenated solutions to produce 1,6-dihydro-6-α-hydroxyalkylpurines such as XLIX from ethanol (*H. Linschitz* and *J. S. Connolly*, J. Amer. chem. Soc., 1968, **90**, 2979; *idem*, Photochem. and Photobiol., 1968, **7**, 791).

Oxidation with per-acids results mainly in degradation but small amounts of the 1- and 3-oxide are also formed (*M. A. Stevens et al.*, J. org. Chem., 1962, **27**, 567) whereas hydrogenation over platinic oxide in acetic anhydride gives 1,3,7,9-tetra-acetyloctahydropurine (L) (*I. Butula*, Ann., 1969, **729**, 73).

(*ii*) *C-Alkylpurines*

All the mono-*C*-methylpurines, including the previously unknown 2-*methylpurine*, m.p. 286°, all the di-*C*-methylpurines and 2,6,8-*trimethylpurine*, m.p. 220°, have been synthesized by *R. N. Prasad et al.* (J. Amer. chem. Soc., 1959, **81**, 193). Cyclisation of 4,5-diamino-6-hydroxy-2-methylpyrimidine with triethyl orthoacetate or triethyl orthoformate afforded 2,8-dimethyl- and 2-methyl-hypoxanthine, respectively which were converted into the purine homologues *via* the 6-halogeno derivative and reduction. Similar methods gave the other compounds mentioned, the melting points of the **dimethylpurines** being 2,6-, 271°; 2,8-, 217° and 6,8-, 263–265°. **8-Methylpurine,** m.p. 271–273°, has been prepared from 4,5-diaminopyrimidine and acetic anhydride (*Isay, loc. cit.*; *Albert* and *Brown*, *loc. cit.*) and from the 6-mercapto derivative (*H. C. Koppel* and *R. K. Robins*, J. org. Chem., 1958, **33**, 1457); 6-*methylpurine*, m.p. 235–236°, was first synthesized by *Gabriel* and *Colman* (Ber., 1901, **34**, 1426) and has also been prepared by condensation of 6-chloropurine with diethyl sodiomalonate and decarboxylation of the derived 6-dicarboxymethylpurine (*H. Lettre et al.*, Naturwiss., 1963, **50**, 224). The method can also be applied to 2- (*I. M. Ovcharova* and *E. S. Golovchinskaya*, Zhur. obshcheĭ Khim., 1964, **34**, 3254) and 8-chloropurines (*L. A. Nikolaeva* and *Golovchinskaya*, *ibid.*, 1964, **34**, 1137), and by alkylation of the purinyl malonic ester intermediate and subsequent decarboxylation of the dicarboxylic acid a variety of *C*-alkylpurines can be obtained.

8-Alkyl-7-hydroxyxanthines, in aqueous solution, undergo, at pH ~5, an unusual spontaneous autocatalytic disproportionation to yield the corresponding **8-alkylxanthines** as the reduction product and 6-amino-5-nitrosouracil as the oxidation product (*G. Zvilichovsky*, Tetrahedron, 1977, **33**, 2259).

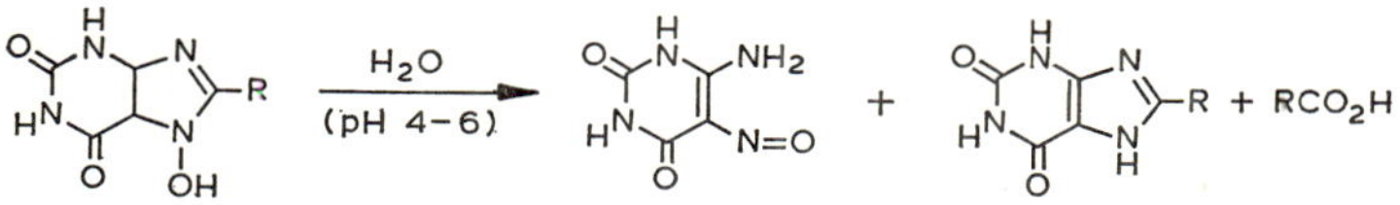

8-Methylguanine has been prepared by methylating guanine with *tert*-butyl hydroperoxide in the presence of ferrous ions; under the same conditions adenine and hypoxanthine give 2-, 8-monomethyl or the 2,8-dimethyl derivatives. Guanine and dimethylformamide under the same conditions produce 8-dimethylcarbamoylguanine (*M. Moeda, K. Nushi* and *Y. Kawazoe, ibid.*, 1974, **30,** 2677).

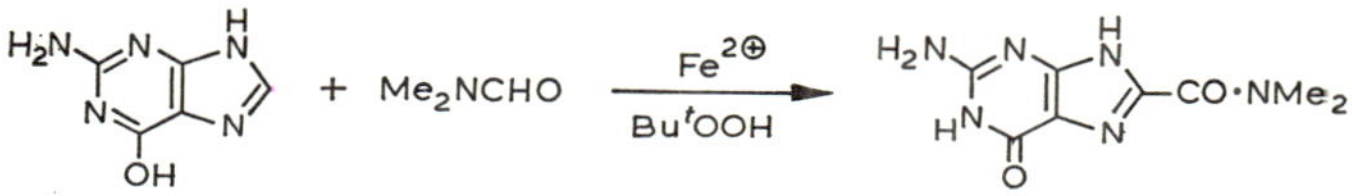

(*iii*) *N-Alkylpurines*

9-**Methylpurine,** m.p. 162–163°, is more labile than purine, owing to its inability to form an anion; alkaline hydrolysis gives a quantitative yield of 5-amino-4-methylaminopyrimidine (*E. Fischer*, Ber., 1898, **31,** 2551; *Albert* and *Brown, loc. cit.*). 9-*Ethylpurine*, m.p. 53–56°, behaves similarly (*Montgomery* and *Temple*, J. Amer. chem. Soc., 1957, **79,** 5238) and

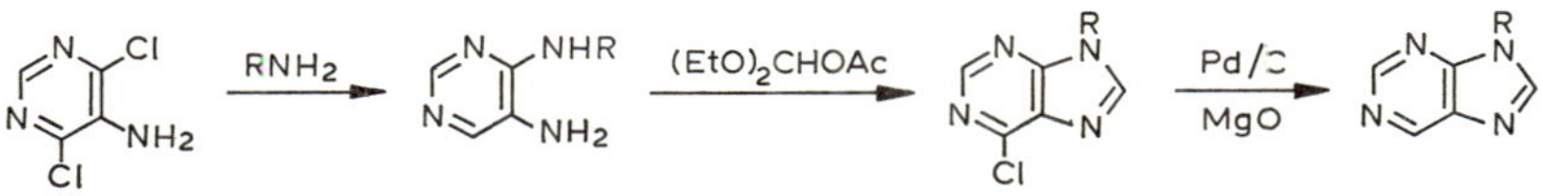

presumably also 7-**methylpurine,** m.p. 184° (*Fischer, loc. cit.*; *Bendich et al.*, J. Amer. chem. Soc., 1954, **76,** 6073). A series of 9-alkylpurines has been prepared by the following route (*Montgomery* and *Temple, loc. cit.*; J. Amer. chem. Soc., 1958, **80,** 409).

Both 7- and 9-methylpurine also result from catalytic hydrogenation of appropriate 2- or 8-chloro-9-methylpurine or 2-chloro-7-methylpurine (*A. G. Beaman* and *Robins*, J. org. Chem., 1963, **28,** 2310). 1-**Methylpurine,** m.p. 234–235°, is obtained by dethiation of 1,6-dihydro-1-methyl-6-thiopurine with Raney nickel in hot water (*L. B. Townsend* and *Robins*, J. org. Chem., 1962, **27,** 990) and 3-**methylpurine,** m.p. 184–185°, is similarly formed from 3-methyl-6-methylthiopurine with Raney nickel in hot methanol (*idem*, J. heterocycl. Chem., 1966, **3,** 241).

Various *N*-alkylated purines have been prepared by alkylation with trialkyl phosphates (*K. Yamacuchi, Hayachi* and *M. Kinoshita*, J. org. Chem., 1975, **40,** 385).

7-**Vinylpurine,** m.p. 137–138°, is obtained by reaction of purine with vinyl acetate in the presence of mercuric acetate (*J. Pitha*, J. org. Chem., 1975, **40,** 3297).

Ultraviolet absorption maxima of purine and some alkylpurines are given in Table 3. Halogenopurines are described in section (*e*) (p. 52).

TABLE 3

ULTRAVIOLET LIGHT ABSORPTION OF PURINE AND SOME ALKYLPURINES[a]

Purine derivative	λ_{max} *nm (10^{-3} ε in brackets) in aqueous solution* *Acidic*	*Neutral*	*Basic*	*Ref.*
Purine	<220 (12.3) 260 (6.17)	220 (3.02) 263 (8.91)	219 (8.32) 271 (7.76)	1
1-Methyl-	268 (6.6)		275 (6.1)	2
2-Methyl-	266 (5.4)		275 (8.0)	3
3-Methyl-	275 (9.8)		276 (7.0)	4
6-Methyl-	265 (7.59)	261 (8.32)	271 (8.51)	1
8-Methyl-	264 (8.32)	266 (10.2)	274 (8.32)	5
9-Methyl-	262 (5.5)	264 (7.7)		1
9-Ethyl-	261.5 (5.85)	263 (7.82)	263 (7.82)	6
8-Phenyl-	237 (13.5) 304 (25.7)	231 (11.5) 298 (26.3)	233 (17.4) 304 (26.3)	1
2,6-Dimethyl-	270 (6.96)		275 (9.18)	3
6,8-Dimethyl-	269 (7.7)		278 (8.9)	3
2,8-Dimethyl-	269 (7.2)		278 (9.6)	3
2,6,8-Trimethyl	273 (5.02)		278 (6.33)	3

References

1 *S. F. Mason*, J. chem. Soc., 1954, 2072; *R. T. Koda*, *J. A. Biles* and *W. Wolf*, J. pharm. Sci., 1968, **57**, 2056.
2 *K. W. Merz* and *P. H. Stähl*, Arch. Pharm., 1965, **293**, 801.
3 *R. N. Prasad et al.*, J. Amer. chem. Soc., 1959, **81**, 193.
4 *F. Cacace* and *R. Masironi*, Ann. Chem. (Italy), 1957, **47**, 366.
5 *H. C. Koppell* and *R. K. Robins*, J. org. Chem., 1958, **23**, 1457.
6 *J. A. Montgomery* and *C. Temple*, J. Amer. chem. Soc., 1957, **79**, 5238.

[a] For a more extensive list see *J. H. Lister*, "Purines" in The Chemistry of Heterocyclic Compounds, eds. *A. Weissberger*, *E. C. Taylor* and *D. J. Brown*, Wiley, New York, 1971.

(*d*) *Hydroxypurines*

The study of the chemistry of the purines began in 1776 with the discovery by *C. W. Scheele* of uric acid (2,6,8-trihydroxypurine) in gall stones and urine; uric acid has since been found to be very widely distributed in nature both as such and as its 9-β-D-ribofuranoside. Efforts to elucidate its structure extended over many decades (see Reviews by *Emil Fischer*, Ber., 1899, **32,** 435; *H. Biltz*, Ann., 1921, **423,** 319; J. pr. Chem., 1936, [ii], **145,** 65) and laid the foundation of purine chemistry. For this reason, and also because the hydroxypurines have commonly been the source of many purine derivatives, their chemistry will be considered before that of the halogenopurines. The hydroxypurines are now considered to exist in the amide form (*S. F. Mason*, Chem. and Biol. of Purines, p. 60; *F. Bergmann* and *S. Dikstein*, J. Amer. chem. Soc., 1955, **77**, 691; *J. H. Lister*, 1971, "The Purines", *loc. cit.*, Chapter VI) and this should be borne in mind throughout this section.

(*i*) *Monohydroxypurines*

2-Hydroxypurine, $C_5H_4ON_4$, prepared by cyclisation of 4,5-diamino-2-hydroxypyrimidine with formic acid (*C. O. Johns*, J. biol. Chem., 1912, **11**, 69) and by deamination of 2-aminopurine (*Albert* and *Brown*, *loc. cit.*), crystallizes as the monohydrate and forms salts with acids. The *picrate* decomposes at 245°. It is reduced to a dihydro-derivative by hydrogen over palladised charcoal in the presence of acid (*A. Bendich*, Chem. and Biol. of Purines, p. 309) and affords a pteridine with aqueous glyoxal or glyoxylic acid at low pH (*Albert*, Biochem. J., 1957, **65**, 124).

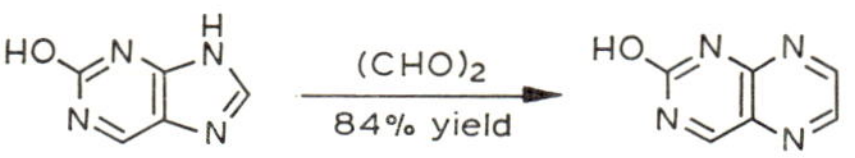

Hypoxanthine, 6-hydroxypurine, $C_5H_4ON_4$, the only monohydroxypurine of any importance, was first discovered in spleen (*W. Scherer*, Ann., 1850, **73**, 328) and has since been found in all types of living tissue, both plant and animal. It is a growth factor for some unicellular organisms (*cf. H. Hurni*, Z. Vitamin-Forsch., 1945, **16**, 69), and occurs as a 9-β-D-ribofuranoside, inosine and related nucleotide inosine 5′-phosphate which is an important intermediate in the *de novo* biosynthesis of purine nucleotides (*Hartman* and *Buchanan*, 1956, *loc. cit.*).

The structural relationship between hypoxanthine and uric acid was demonstrated by *Fischer* (Ber., 1897, **30**, 2226) who prepared hypoxanthine from trichloropurine by hydrolysing the 6-chloro group with aqueous alkali and removing the two remaining chloro groups with phosphonium iodide (see p. 27). An imidazole synthesis has been mentioned on p. 17 and the Traube method in which 4,5-diamino-2-mercaptopyrimidine is cyclised by formate and the product desulphurised with nitric acid to give hypoxanthine is worthy of note (*W. Traube*, Ann., 1904, **331**, 67).

A Traube cyclisation with the triethyl orthoformate–acetic anhydride reagent gives a 95% yield (*E. C. Taylor* and *C. C. Cheng*, J. org. Chem., 1960, **25**, 148) and a one-stage synthesis, from aminomalonamidine hydrochloride with triethyl orthoformate, and dimethylformamide a yield of 85% (*E. Richter*, *P. K. Loeffler* and *Taylor*, J. Amer. chem. Soc., 1960, **82**, 3144). Adenine is converted to hypoxanthine by nitrous acid (*A. Kossel*, Z. physiol. Chem., 1836, **10**, 258) and by adenase (*W. Jones*, *ibid.*, 1904, **42**, 35).

Hypoxanthine is appreciably soluble (70 parts) in hot water and is amphoteric; soluble salts are formed with acids and insoluble ones with heavy metals, the *picrate* has m.p. 250–254° (decomp.) (*K. Yoshimura*, Biochem. Z., 1911, **37**, 480). Heating with water at 200° hydrolyses hypoxanthine to carbon dioxide ammonia and formic acid (*Kossel*, Z. physiol. Chem., 1882, **6**, 429). Hypoxanthine is readily monobrominated (*M. Kruger*, *ibid.*, 1894, **18,** 449) and couples with alkaline diazonium solutions (*R. Burian*, *ibid.*, 1907, **51,** 432). It is converted to 6-mercaptopurine by phosphorus pentasulphide (*G. B. Elion et al.*, J. Amer. chem. Soc., 1952, **74,** 411) and to 6-chloropurine by phosphoryl chloride and dimethylaniline (*Bendich et al.*, *ibid.*, 1954, **76,** 6073) or, with phosphoryl bromide, to 6-bromopurine (*Elion* and *Hitchins*, *ibid.*, 1956, **78,** 3508). Methylation with methyl iodide in alcoholic sodium ethoxide gives the characteristic 1,7-dimethylhypoxanthine sodium iodide complex, $C_7H_8ON_4NaI,3H_2O$ (*E. Fischer*, Ber., 1897, **30,** 2231), which is formed by no other natural purine although 2-methylhypoxanthine behaves similarly in forming a sodium iodide complex of 1,2,7-trimethylhypoxanthine (*Traube et al.*, Ann., 1923, **432,** 266).

With dimethyl sulphate or methyl toluene-*p*-sulphonate in dimethylacetamide, hypoxanthine or the 7- or 9-methyl derivatives produce the 7,9-dimethylhypoxanthinium salt which gives the betaine LI (see also p. 82) with ammonia (*J. W. Jones* and *Robins*, J. Amer. chem. Soc., Soc., 1962, **84,** 1914; *Bredereck et al.*, Chem. Ber., 1966, **99,** 944).

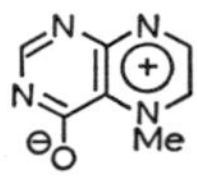

(LI)

Hypoxanthine may be determined on the micro-scale by measuring the oxygen taken up during its oxidation to uric acid by xanthine oxidase (*H. A. Krebs* and *A. Orström*, Biochem. J., 1939, **33,** 984). Like 2-hydroxypurine it affords a hydroxypteridine on reaction with ethyl glyoxylate hemiacetal (*A. Albert*, Biochem. J., 1957, **65,** 124).

8-Hydroxypurine, m.p. 305–307°, has been synthesised by *O. Isay* (Ber., 1906, **39,** 250) and by *Albert* and *D. J. Brown* (J. chem. Soc., 1954, 2070) from 4,5-diaminopyrimidine and urea (*cf.* also *Fischer*, Ber., 1899, **32,** 476) and **6-hydroxy-8-methylpurine,** m.p. >300°, is obtained by heating 4,5-diamino-6-hydroxypyrimidine with acetic anhydride (*H. C. Koppel* and *Robins*, J. org. Chem., 1958, **23,** 1437).

Table 4 gives the ultraviolet absorption data on some mono- and di-hydroxypurines.

(*ii*) *Dihydroxypurines and their derivatives*

Xanthine, $C_5H_4O_2N_4$, best known in the form of its *N*-methyl derivatives, theobromine, theophylline and caffeine, was first isolated from urinary calculi in 1817 (*Marcet*, "An Essay on the Chemical History and Medical Treatment of Calculi Disorders", London, 1817) and called "xanthic oxide" since it gave a yellow colour with nitric acid. *F. Wohler* and *J. v. Liebig*'s analytical data (Ann., 1838, **26,** 340) and finally *Emil Fischer*'s synthesis

TABLE 4

ULTRAVIOLET ABSORPTION AND IONISATION PROPERTIES OF MONOHYDROXY- AND DIHYDROXY-PURINES[1]

Purine derivative	*Basic pKa*	*Acidic pKa*	λ_{max} *nm (*10^{-3} ε *in brackets) in aqueous solution*			
			State of ionisation			*pH* = 13.0
			Cation	*Neutral*	*Anion*	
2-Hydroxy-	1.69	8.43, 11.90	264 (4.68) 322 (6.46)	238 (2.88) 315 (4.9)	271 (4.79) 313 (4.79)	219 (19.5) 265 (3.98)
6-Hydroxy-	1.98	8.94, 12.10	248 (10.47)	249 (10.47)	258 (11.22)	262 (10.96)
8-Hydroxy-	2.58	8.24, >12	280 (10.47)	235 (3.24) 277 (11.22)	285 (12.88)	284 (12.59)
2,6-Dihydroxy-		7.44, 11.12		267 (7.94)	240 (8.13) 277 (8.32)	283 (8.71)
2,8-Dihydroxy-		7.45, >12		230 (7.94) 310 (5.01)	262 (9.55) 306 (7.41)	220 (16.98) 310 (9.55)
6,8-Dihydroxy-		7.65, 9.87		257 (12.2) 280 (5.75)	265 (10.96)	271 (13.8)
2,6-Dihydroxy-8-methyl-[2]			265 (7.8)		237 (9.7) 281 (8.5)	
6-Hydroxy-8-methyl-[2]			260 (8.0)		257 (2.1)	

References

1 *S. F. Mason*, J. chem. Soc., 1954, 2071.
2 *H. C. Koppel* and *R. K. Robins*, J. org. Chem., 1958, **23**, 1457.

from 2,6,8-trichloropurine (Ber., 1897, **30,** 2226) proved the structure. It has since been found in many animal and vegetable materials both in the free state and as the riboside, xanthosine. The various effective methods of synthesis have already been considered.

Xanthine, 2,6-dihydroxypurine, decomposes on heating but may be characterised as its *perchlorate*, m.p. 262–264°, as may many of its methyl derivatives (*H. Bredereck et al.*, Chem. Ber., 1953, **86,** 850). Its ionisation is considered to be due to the loss of protons first from position 3 and then from position 7; the compound has $pK_{a_1} = 7.53$, $pK_{a_2} = 11.63$ (*L. F. Cavalieri et al.*, J. Amer. chem. Soc., 1954, **76,** 1119). In chemical properties xanthine is very similar to uric acid being soluble in, but not hydrolysed by, acid or alkali except under very vigorous conditions, and it forms salts with acids and with metals. The complex precipitated with ammoniacal copper sulphate has been used in an assay procedure (*J. N. Williams*, J. biol. Chem., 1950, **184,** 627). Xanthine is much more stable to oxidation than uric acid; with ammoniacal silver nitrate an insoluble silver salt is produced and hot dilute nitric acid does not oxidise it but probably causes nitration (*C. Wulff*, Z. physiol. Chem., 1893, **17,** 640). Xanthine is not hydrogenated even under vigorous conditions (*T. B. Johnson* and *J. C. Ambelong*, Science, 1939, **90,** 68) but it is reduced in sulphuric acid at a lead cathode to give the 3-deoxy derivative (*F. Tafel* and *L. Ach*, Ber., 1901, **34,** 1165) and by sodium amalgam or zinc and hydrochloric acid (*D. Hamer et al.*, Arch. Biochem. Biophys., 1953, **47,** 272); heteroxanthine and paraxanthine are similarly reduced electrolytically (*Tafel*, Ber., 1899, **32,** 3206). Xanthine is oxidised by potassium chlorate in hydrochloric acid to alloxan (*Fischer*, Ann., 1882, **215,** 310), by chlorine water to purpuric acid (*idem*, Ber., 1897, **30,** 2236) and to uric acid by the somewhat unspecific enzyme xanthine oxidase. The 8 position in xanthine, like that in the *N*-methylxanthines, is readily substituted by electrophilic reagents; 8-halogenoxanthines are readily obtained and xanthine couples with alkaline diazonium solutions to give 8-azo derivatives which may be reduced to 8-aminoxanthine or hydrolysed to uric acid (*R. Burian*, *ibid.*, 1904, **37,** 703; *Fischer*, Z. physiol. Chem., 1909, **60,** 73). Boiling a solution of xanthine in pyridine with phosphorus pentasulphide gives 2-hydroxy-6-mercaptopurine which is reconverted to xanthine by nitric acid (*A. G. Beaman*, J. Amer. chem. Soc., 1954, **76,** 5633). Phosphoryl chloride containing half an equivalent of water converts it into 2,6-dichloropurine.

The alkylation of xanthine (reviewed by *H. Biltz* and *A. Beck*, J. pr. Chem., 1928, [ii], **118,** 198) shows the order of alkylation at the nitrogen positions as 3, 7, 1 and substitution is easier when the first position has been alkylated. It is noteworthy, however, that methyl iodide converts xanthine and xanthosine into 3,9-dimethylxanthine hydriodide but that 3,7-dimethylxanthine is formed from xanthosine and methyl iodide in the presence of barium carbonate (*J. M. Gulland* and *T. F. Macrae*, J. chem. Soc., 1933, 662). Dimethyl sulphate readily converts xanthine or its salts to caffeine; theophylline and theobromine behave in the same manner (*P. A. W. Self* and *W. R. Rankin*, Quart. J. Pharmacol., 1931, **4,** 346). The action of diazomethane on xanthine and its derivatives is very slow as is the direct alkylation of xanthine and its salts by alkyl halides until one alkyl group has been introduced. The effect of pH and temperature on the methylation of xanthine with dimethyl sulphate has been studied and conditions found which enable excellent yields of caffeine and theobromine to be obtained; the efficient conversion of 3-methylxanthine to theobromine (3,7-dimethylxanthene) was accomplished similarly (*Bredereck et al.*, Chem. Ber., 1950, **83**, 201).

Triethylsilylated xanthine (LII) with methyl iodide gives 7-methylxanthine at room temperature or theobromine at 60° (*G. Gräfe*, Arch. Pharm., 1967, **300**, 111).

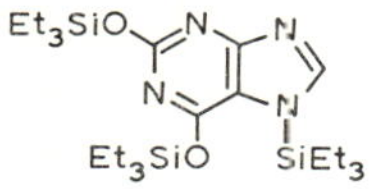

(LII)

8-**Methylxanthine,** m.p. >300° (decomp.), *perchlorate*, m.p. 292–294° (*Bredereck et al.*, Chem. Ber., 1953, **86**, 850), prepared from the diaminopyrimidine and acetic anhydride, has λ_{max} 265 nm (ε, 7800) in acidic and 237, 281 nm (ε, 9700 and 8500) in basic solution; attempts to prepare the corresponding dichloro derivative by the method suitable for xanthine itself were unsuccessful (*Koppel* and *Robins*, *loc. cit.*). These authors prepared various *N*-alkyl derivatives and *Bredereck* characterised 8-methylxanthine as its 1,3,7-*trimethyl* derivative, m.p. 206–208°, after obtaining it by heating uric acid with acetamide (Chem. Ber., 1950, **83**, 201).

A remarkable synthesis of 8-substituted xanthines involves the reaction of 3-acetoxyxanthine (LIII) with inorganic or organic nucleophiles under very mild conditions (*U. Wolcke, N. J. M. Birdsall* and *G. B. Brown*, Tetrahedron Letters, 1969, 785; *Birdsall*,

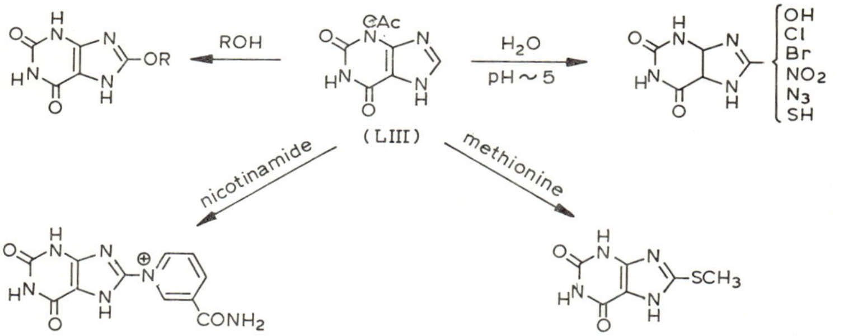

T.-C. Lee and *Wolcke*, Tetrahedron, 1971, **27**, 5961); compounds prepared include 8-halogenoxanthines and uric acid, and methionine can also be a source of an 8-methylthio group (*Brown* and *J. C. Parham*, "The Purines, Theory and Experiment", Israel Acad. Sci., Jerusalem, 1972, **4**, 550).

Vigorous methylation of 3-hydroxyxanthine with dimethyl sulphate in dimethylformamide also gave 7,9-dimethyluric acid. 9-Hydroxy-8-methylhypoxanthine (LIV) has

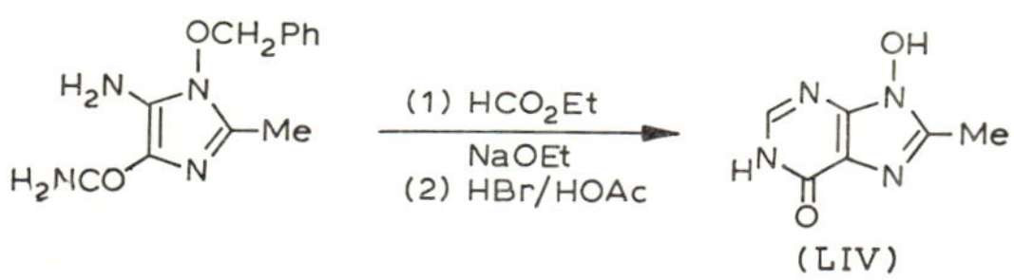

(LIV)

also been prepared from the 5-amino-1-benzyloxyimidazole with ethyl formate and sodium ethoxide and removal of the benzyl group with hydrogen in acetic acid (*A. A. Watson* and *Brown*, J. org. Chem., 1972, **37,** 1867).

2,8-**Dihydroxypurine,** $C_5H_4O_2N_4$, has been synthesized from 4,5-diamino-2-hydroxypyrimidine and urea (*Albert* and *D. J. Brown*, J. chem. Soc., 1954, 2060) as an insoluble powder having no melting point. It is soluble in both alkali and acid and gives the murexide reaction. Hydrogenation with platinic oxide in aqueous hydrochloric acid gave a dihydro derivative which cleaved to form 5-ureido-5,6-dihydrouracil (*W. L. F. Armarego* and *P. A. Reece*, Tetrahedron Letters, 1975, No. 6, 423).

6,8-**Dihydroxypurine,** $C_5H_4O_2N_4$, is the only dihydroxypurine which does not give the murexide reaction, the necessary 2-hydroxy group being absent. It was synthesized by *Fischer* and *Ach* (Ber., 1897, **30,** 2218) from 2,6-dichloro-8-hydroxypurine in the manner shown below and by heating urea with 4,5-diamino-6-hydroxypyrimidine at 170° (*Albert* and *Brown*, *loc. cit.*).

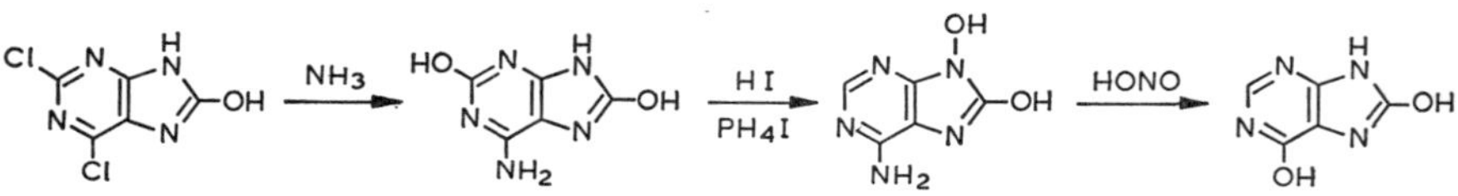

A number of 6-methyl-2,8-dihydroxypurine derivatives have been prepared by heating the appropriate pyrimidine with urea. These are all solids with either very high or no melting points (*C. O. Johns*, J. biol. Chem., 1912, **11,** 394; 1913, **15,** 120; **16,** 137; Amer. chem. J., 1909, **41,** 63): 2,8-*dihydroxy*-6-*methylpurine*, no m.p., 2,8-*dihydroxy*-1,6-*dimethylpurine*, m.p. 260–265° (decomp.); 2,8-*dihydroxy*-6,9-*dimethylpurine*, no m.p.; 9-*ethyl*-2,8-*dihydroxy*-6-*methylpurine*.

(*iii*) *Uric acid, 2,6,8-trihydroxypurine*

The correct composition of uric acid, $C_5H_4O_3N_4$, was established by *J. v. Liebig* and *E. Mitschlerich* in 1834, but many years elapsed before its structure was elucidated. The major steps in this were firstly, the oxidative degradation of uric acid to allantoin, to urea plus alloxan and to many other products by *F. Wöhler* and *J. v. Liebig* in 1838, secondly the work of *Adolf von Baeyer* in 1836–1864, which clarified the chemistry of alloxan and of barbituric acid, which he had obtained from it, and finally the syntheses of *R. Behrend* and *O. Roosen* and of *Emil Fischer* and the demonstration by the latter that there were two isomeric monomethyl uric acids, thus eliminating the old Fittig formulation and confirming that of *Medicus*. The first synthesis by *J. Horbaczewski* (Ber., 1882, **15,** 2678) preceded and did not assist, the final elucidation of structure and the first unambiguous preparation was that of *Behrend* and *Roosen* (Ann., 1888, **125,** 248, 253) who proceeded logically from ethyl acetoacetate to alloxan which affords uric acid on heating with urea; *Fischer* and *Ach* (Ber., 1895, **28,** 2473; 1897, **30,** 559) utilised barbituric acid as the starting material for a more convenient and practical synthesis shown below.

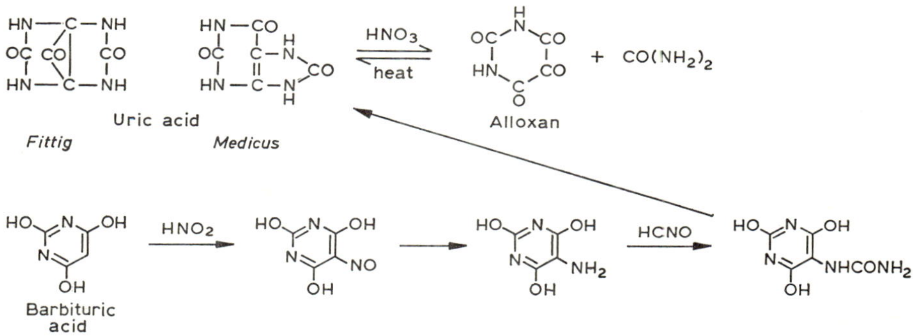

More recent syntheses, however, have used 4,5,6-triamino-2-hydroxypyrimidine as the starting material and uric acid labelled at positions 1 and 3 or at 9 with ^{15}N has been prepared in excellent yields by this method (*L. F. Cavalieri et al.*, J. Amer. chem. Soc., 1958, **70,** 1240; *cf. E. Grimeaux*, Bull. Soc. chim. Fr., 1879, **31,** [ii], 535). Uric acid has also been prepared from 5-amino-6-hydroxypyrimidine (*W. Traube*, Ber., 1900, **33,** 3045), and from 3-hydroxyxanthine (*Brown* and *Parham*, 1972, *loc. cit.*).

Uric acid (λ_{max} 285 nm in water when unionised) is essentially a dibasic acid (*Biltz* and *L. Hermann*, Ber., 1921, **54,** 1676; *F. Bergmann* and *S. Dikstein*, J. Amer. chem. Soc., 1955, **77,** 691) consider the order of ionisation to be 9 (pK_a 5.75) and then 3 (pK_a 10.3); but see p. 40. It is slightly soluble (0.08%) in boiling water but readily so in alkali to which, as to acids, it is relatively stable (*cf. Albert* and *Brown*, J. chem. Soc., 1954, 2060). Water at 200°, however, converts it in part to a mixture of pteridines and pyrimido[5,4-*g*]pteridines (*W. Pfleiderer*, Chem. and Biol. of Purines, p. 77) and hot concentrated hydrochloric acid disrupts the molecule to give three equivalents of ammonia and carbon dioxide and one of glycine, the latter presumably arising from aminomalonic acid since its carboxyl group derives equally from C-4 and C-6, the CH_2 from C-5 and the nitrogen from N-7 of uric acid (*cf. C. A. Dalgleish* and *A. Neuberger*, J. chem. Soc., 1954, 3407). Correspondingly, of the 3,9-dimethyl-, 9- and 7-methyl-uric acids only the last gave sarcosine on hydrolysis, the others producing glycine (*Fischer*, Ber., 1895, **28,** 2840).

Uric acid is stable to most reducing agents but electrolytic reduction gives "purone", "isopurone" or "tetrahydrouric acid" according to the conditions (*Tafel, ibid.*, 1901, **34,** 258, 1181). Uric acid is readily oxidised *e.g.* by hydrogen peroxide, cupric or silver salts in alkaline solution. The products of oxidation obtained vary with the conditions employed although, in general, it is the pH of the medium rather than the nature of the oxidising agent which is important.

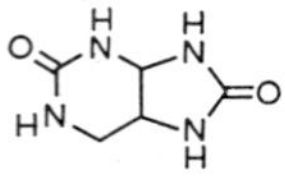

"Purone"

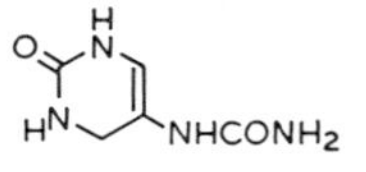

"Isopurone"

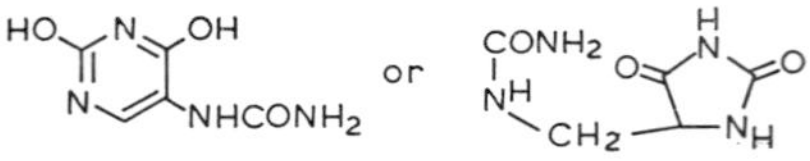

"Tetrahydrouric acid"

The oxidation of uric acid in alkaline solution can give allantoin (LV), oxonic acid, allantoxaidin or oxaluric acid according to the conditions (*Wohler* and *Liebig*, Ann., 1838, **26,** 285; *Biltz* and *R. Robl*, Ber., 1920, **53,** 1950, 1967; *Biltz* and *H. Schauder*, J. pr. Chem., 1923, **106,** 114). The reaction paths postulated by *Biltz* to explain the formation of these and other products involved hydroxyacetylenediureincarboxylic acid (LVI) as the key intermediate as in route A below (*Biltz* and *F. Max*, Ber., 1921, **54,** 2451; *W. Schuler* and

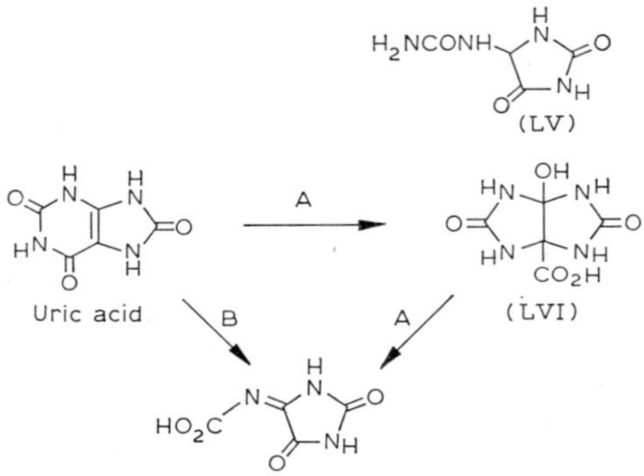

W. Reindel, Z. physiol. Chem., 1932, **208,** 248; *Biltz* and *G. Schiemann*, J. pr. Chem., 1926, **113,** 77), whereas *C. S. Venable* (J. Amer. chem. Soc., 1918, **40,** 1099) and others believed that breakdown of the pyrimidine ring of uric acid led directly to an oxonic acid of structure LVII in which the original imidazole ring survived (route B).

More recent work has rejected both of the suggested pathways and each of the structures previously assigned to oxonic acid and allantoxaidin. Samples of uric acid in which each carbon atom in turn was labelled isotopically with ^{14}C and also 7-[^{15}N]uric acid have been prepared and their oxidation studied (*H. Brandenberger*, Helv., 1954, **37,** 441; *S. C. Hartmann* and *J. Fellig*, J. Amer. chem. Soc., 1955, **77,** 1051; *Dalgleish* and *Neuberger*, *loc. cit.*). It was then apparent that the oxidation of uric acid by oxygen or hydrogen peroxide in alkaline solution caused the loss of the C-6 atom as carbon dioxide, the partial loss of the N-7 atom as ammonia and the retention of the other original atoms to give

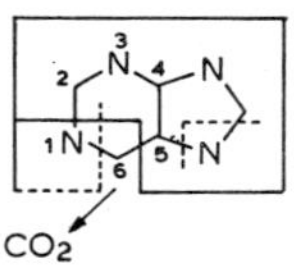

(LXVIII)

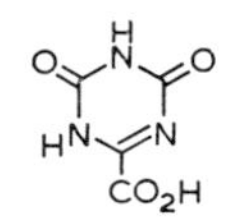

Oxonic acid
(LXIX)

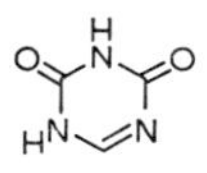

Allantoxaidin
(LXX)

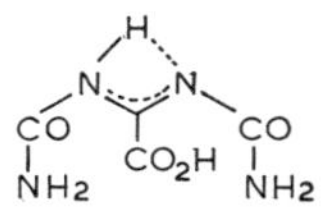

(LXXI)

oxonic acid the carboxyl group of which contained that carbon atom originally at position 5 in uric acid (*cf.* LXVIII). Allantoxaidin has now been shown to be the *s*-triazine LXX and the oxonic acid from which it derives by decarboxylation is thus presumably to be formulated as LXIX. The carbon atom in the latter to which the carboxylic acid group is attached derives from the C-4 atom in uric acid. Thus, oxidative attack of uric acid in alkali occurs in the 5-position and, formally, oxonic acid arises by the cyclisation in either way of the symmetrical intermediate LXXI with loss of ammonia. The specific enzyme uricase catalyses the oxidation of uric acid to allantoin (LV) and this oxidation involves the removal of two electrons simultaneously from the acid (*K. Bentley* and *Neuberger*, Biochem. J., 1952, **52**, 694). In both this oxidation and the chemical one with alkaline permanganate, which gives the same product, the C-6 atom of uric acid is again lost, the CH group in the hydantoin originates from the C-4 atom and the nitrogen atoms of the original uric acid are equally distributed between the ureido group and the imidazole ring

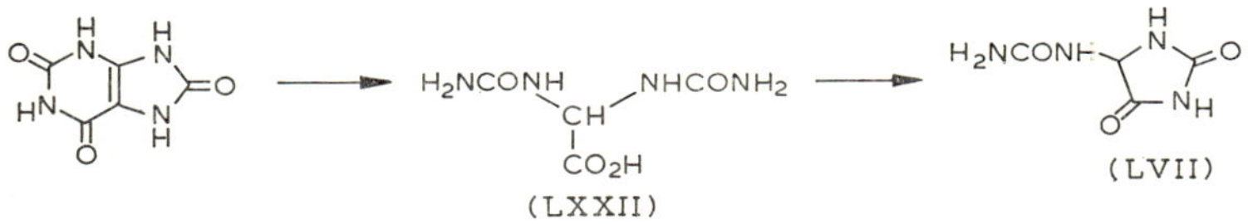

(*G. B. Brown et al.*, J. Amer. chem. Soc., 1948, **70**, 1242; *Dalgleish* and *Neuberger*, *loc. cit.*). Formally, this oxidation also proceeds through the symmetrical intermediate LXXII, the dihydro-derivative of LXXI, which may cyclise in two ways to give hydantoin.

The results of tracer studies are in complete accord with the earlier finding that both 1- and 7-methyluric acids afford 3-methylallantoin and the 3- and 9-methyl analogues 1-methylallantoin on oxidation (*Fischer* and *Ach*, Ber., 1899, **32**, 2723).

Oxidation of uric acid in acid solution gives mainly alloxan and urea (*H. Biltz*, Ann., 1909, **368**, 170) but energetic conditions give parabanic acid which is a further oxidation product of alloxan; oxidation with dilute nitric acid can give alloxantin or purpuric acid (*Wohler* and *Liebig*, Ann., 1938, **26**, 285, 319; *A. Hantzsch*, Ber., 1921, **54**, 1267). The formation of a purple colour due to the ammonium salt of purpuric acid, when ammonia is added to the solution from nitric acid oxidation, is a characteristic reaction of uric acid and related purines.

The suggested mechanism for the oxidation of uric acid with chlorine (*Biltz* and *H. Schauder*, J. pr. Chem., 1923, **106**, 124) is given below.

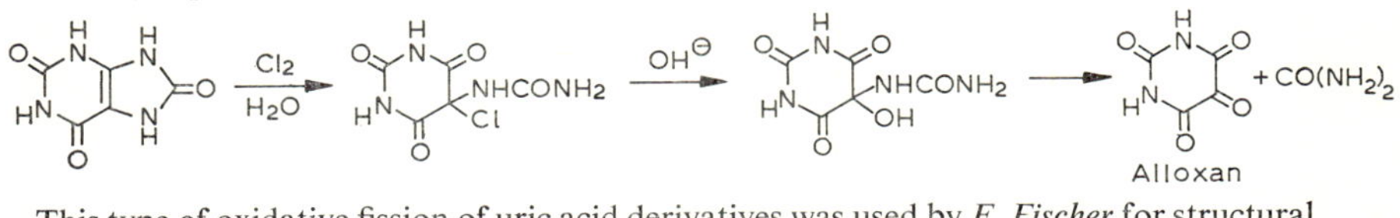

This type of oxidative fission of uric acid derivatives was used by *E. Fischer* for structural determination. The structure of caffeine, for example, was elucidated by its oxidation with chlorine to 1,3-dimethylalloxan (Ann., 1882, **215**, 257).

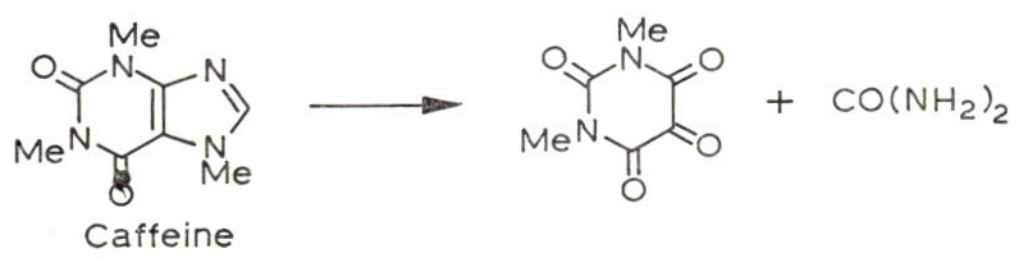

Glycols may be obtained from hydroxypurines by vigorous oxidation; their stability under these conditions was a deciding factor in excluding them finally as possible initial products of uric acid oxidations under both acid and alkaline conditions. Theobromine,

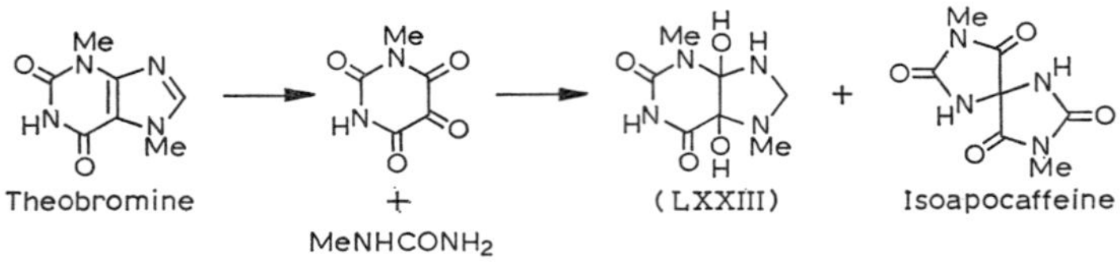

for example, is oxidised by perchloric acid in hydrochloric acid to 1-methylalloxan and methylurea which are converted by the acid present to the glycol LXXIII and isoapocaffeine (3,7-dimethylcaffolide) (*Biltz* and *L. Loewe*, Ber., 1931, **64**, 1074).

The alkylation of uric acid yields a variety of derivatives depending upon the nature of the reagent and the conditions employed (*Fischer*, Ber., 1899, **32**, 461; *Biltz* and *F. Max*, *ibid.*, 1920, **53**, 2327; *Biltz* and *L. Herrmann*, *ibid.*, 1921, **54**, 1676; *Biltz*, J. pr. Chem., 1936, **145**, 63; Ber., 1936, **69**, 2420; 1938, **71**, 1391, 2040; 1939, **72**, 807). Alkylations may be effected by treating either an alkaline solution of uric acid or a dry metal urate with an alkyl halide. The products differ according to the method adopted. *Biltz* and *Herrmann* (*loc. cit.*) showed from a study of the ionisation constants of uric acid and its methylation products, that the proton at position 3 in uric acid is the most readily removed, being followed in descending order by those at 9, 1 and 7, respectively, the latter two being removed only with great difficulty. The urates are analogous to ethyl sodioacetate and thus alkylate on the nitrogen atom adjacent to the alkoxide oxygen instead of forming oxygen ethers. Thus the direct methylation of lead urate gives the expected 3-methyluric acid and lead 7-methylurate gives 3,7-dimethyluric acid. When position 3 is blocked, position 9 is substituted. The silver salt of 1,3,7-trimethyluric acid, however, gives initially 8-alkyloxy-1,3,7-trimethylxanthine which subsequently rearranges in the normal manner to 9-alkyl-1,3,7-trimethyluric acid. The action of methyl sulphate on metal urates is normal but in alkaline solution a different course is followed as the weakly acidic, *i.e.* most basic, 1-position is the first and often the only one to be substituted. This suggests that only under these conditions is quaternisation the first step in the reaction. The introduction of an *N*-hydroxymethyl group into uric acid, which can subsequently be reduced to methyl (*Fischer* and *Ach*, *ibid.*, 1899, **32**, 250; *Biltz* and *Ach*, *ibid.*, 1921, **54**, 1093) appears to take place similarly, as does the conversion of potassium 1,3-dimethylurate into 1,3,9-trimethyluric acid with methyl toluene-*p*-sulphonate in *o*-dichlorobenzene at 150° (*Biltz* and *H. Pardon*, *ibid.*, 1930, **63**, 2876).

Alkylation of uric acid and its derivatives with diazomethane was reviewed by the latter investigators (J. pr. Chem., 1932, **134**, 310) as follows: diazomethane, which will not methylate 3,9-dialkyluric acids, converts 9-alkyluric acids into 9-alkyl-2-methoxy-1,7-dimethyl-6,8-dioxotetrahydropurines and 9-unsubstituted uric acids into 8-methoxy compounds. It is supposed that the 2- or 8-methoxy compound is first formed and that positions 9 and 3 are then substituted in that order, positions 1 and 7 being difficult to substitute. Reactions with diazoethane are similar (*Biltz* and *Max*, *loc. cit.*). The behaviour of acetyluric acids towards diazomethane was also reviewed by *Biltz* and *Pardon*. A 7-

acetyl group prevents the formation of an 8-methoxy group; a 7-acetyl-9-alkyluric acid will give a 2-methoxy derivative and an 8-acetoxy compound is alkylated on all available nitrogen atoms. In accordance with these rules 7-acetyl-3-methyluric acid gives 7-acetyl-1,3,9-trimethyluric acid (LXXIV) with diazomethane; 7-acetyl-9-methyluric acid gives 7-acetyl-2-methoxy-1,9-dimethyl-6,8-dioxotetrahydropurine which is rearranged to LXXIV

(LXXIV)

by heat. The acetylation of uric acid and its derivatives occurs in the imidazole ring (*Biltz* and *Pardon*, J. pr. Chem., 1932, **134**, 310) preferentially at position 7; if this is blocked and both position 9 and, surprisingly, also position 3 are free, an 8-acetoxyanthine is formed. Appropriate substitution (*i.e.* 3, 7 and 9 free) gives a 7,8-diacetyl compound as in the case of uric acid itself which forms 8-acetoxy-7-acetylxanthine. Boiling uric acid with acetic anhydride, however, gives unexpectedly 9-acetyl-2,6,8-trihydroxy-8-methyl-

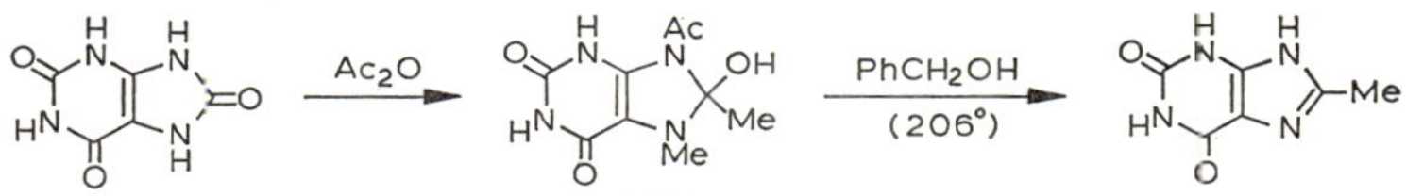

(LXXV)

7,8-dihydropurine (LXXV) which is deacetylated and dehydrated by boiling benzyl alcohol to 8-methylxanthine (*I. T. Strukow* and *L. S. Diokina*, Farmatzia, 1945, **8**, 16).

The halogenation of uric acid is considered later (see p. 54).

Heating uric acid with formamide affords xanthine in almost quantitative yield and with acetamide, 8-methylxanthine is obtained; 1,3-dimethyluric acid is similarly converted to theophylline but other *N*-methyl derivatives of uric acid fail to react with formamide (*H. Bredereck et al.*, Ber., 1950, **83**, 201). Previous attempts to convert uric acid into xanthine by reaction with chloroform, calcium formate or formic acid gave only low yields (*E. E. Sundwick*, Z. physiol. Chem., 1897, **23**, 476; 1898, **26**, 131; Chem. Ztbl., 1897, II, 776; 1898, II, 1212; Skand. Arch. f. Physiol., 1911, **25**, 256; Chem. Ztbl., 1911, I, 1411; Z. f. Physiol., 1912, **76**, 486; Chem. Ztbl., 1912, I, 1550; *Biltz*, J. pr. Chem., 1927, **118**, 166).

The *N*-alkyl derivatives of uric acid, prepared as described above, are all solids which decompose at temperatures above 300°; they normally show maximal light absorption at 283–290 nm when unionised (*F. Bergmann* and *S. Dikstein*, J. Amer. chem. Soc., 1955, **77**, 691).

Derivatives the trioxo form of uric acid, of **$\Delta^{4,9}$-isouric acid,** have been described by *Biltz*. The 5-chloro derivatives *e.g.* LXXVI are obtained by halogenation of appropriately substituted (*e.g.* 1,3,9-trimethyl-) uric acids (see p. 50) and are converted by alcohols into 5-alkoxyisouric acids, *e.g.* LXXVII, which may often be prepared also by heating 4,5-dialkoxydihydrouric acids. In this way, 5-*chloro*-1,3-*dimethyl*-, m.p. 225°, and 5-*chloro*-3,7-*dimethyl*-, m.p. 168–169°, -*isouric acids* afford 5-*methoxy*-1,3-*dimethyl*-, m.p. 207°, and 5-*methoxy*-3,7-*dimethyl*-, m.p. 205–206°, -*isouric acids*, respectively (*cf. Biltz*

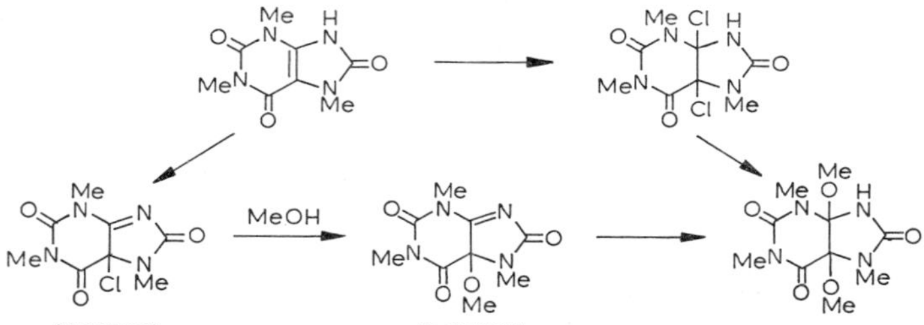

(LXXVI) (LXXVII)

et al., Ann., 1917, **413,** 117, 142, 165; 1914, **406,** 33, 39, 53; 1935, **515,** 222; Ber., 1910, **43,** 2560).

The 7,9-dimethyl-Δ^3-isouric acid prepared by *Biltz* (Ann., 1927, **457,** 109) affords 7,9-dimethyluric acid on treatment with fuming hydrochloric acid and 4,7,9-trimethyl-4,5-dihydrouric acid with boiling acetic acid.

(*iv*) *N-Alkylhydroxypurines*

Some of the *N*-alkylhydroxypurines are of great economic and social significance, on account of their occurrence in those plants which are used for preparing beverages such as tea and coffee and their responsibility for the popular physiological action of the latter. They are widely distributed in the vegetable and animal kingdoms; three of the four possible mono-methylxanthines, three of the five dimethylxanthines and one of the two trimethylxanthines have been isolated from natural sources. Caffeine is a stimulant for the central nervous system and like theobromine and theo-phylline, is an important diuretic.

The chemistry of some *N*-methylpurines has been reviewed briefly by *G. B. Elion* (Chem. and Biol. of Purines, p. 39). The alkylation of uric acid and xanthine have already been discussed and the properties of various representative compounds are collected in Table 5. They are mostly solids of high melting point which may be characterised either by their light absorption characteristics or, in some cases, as their perchlorates.

2-Hydroxy-3-methyl- (no m.p.) and **2-hydroxy-3,7-dimethyl-purine,** m.p. 256–257°, were prepared by *J. Tafel* by the unusual method of electrolytic reduction of the corresponding xanthines (Ber., 1899, **32,** 3201; 1900, **33,** 3372) and a series of mono-, di- and tri-methyl derivatives of 2-hydroxypurine (all solids decomposing at elevated temperatures) by *C. O. Johns* (J. biol. Chem., 1911, **9,** 166; 1912, **11,** 71; **12,** 93; 1913, **14,** 7; **15,** 517).

1-Methylhypoxanthine (6-hydroxy-1-methylpurine), which occurs in human urine (*B. Weissmann et al.*, J. biol. Chem., 1957, **224,** 407) has been prepared by *Elion* (*loc. cit.*) by Traube cyclisation of 4,5-diamino-1-*N*-methyl-6-oxodihydropyrimidine. The 3-, 7- and 9-methylhypoxanthines have been synthesized by *E. Fischer* (Ber., 1897, **30,** 2400; 1898, **31,** 104) and by *W. Traube* (Arch. Pharm., 1906, **244,** 11); *cf.* also *Bredereck et al.*

TABLE 5

PROPERTIES OF *N*-ALKYL-HYDROXY-PURINES

Compound	m.p. (°C)	λ_{max} nm (10^{-3} ε in brackets) in aqueous solution			pK	Ref.
		Acidic	Neutral	Basic		
2-Hydroxy-7-methylpurine	356–357 (decomp.)	249 (12.0)		261 (10.62)		1
2-Hydroxy-9-methylpurine	—	264 (4.67)	238 (288)	271 (4.79)	9.19	2
		322 (6.46)	315 (4.89)	313 (4.79)		
1-Methylhypoxanthine	—	249	251	260		3
3-Methylhypoxanthine	280 (decomp.)	253	264	265		3
7-Methylhypoxanthine	355 (decomp.)	250 (9.8)	256	261 (10.1)		3, 4
9-Methylhypoxanthine	390 (decomp.)	251 (6.3)		256 (7.35)		5, 2
9-Ethylhypoxanthine	263–265	250 (10.8)		250 (12.4)		6
1,7-Dimethylhypoxanthine	—		255 (8.71)			2
7-Methyl-8-oxodihydropurine	—	284 (7.94)	240 (3.98)	286 (11.48)	8.20	2
			278 (9.33)			
9-Methyl-8-oxodihydropurine	—	281 (11.75)	277 (10.23)	289 (10.93)	9.05	2
7,9-Dimethyl-8-oxodihydropurine	—	285 (10.72)	241 (3.63)			2
			279 (10.23)			
1-Methylxanthine	(decomp.)		266 (10.2)	241 (8.5)	7.7	7, 8
				276 (8.5)	12.05	
3-Methylxanthine	227–229[a] (decomp.)		271.5	278 (12.1)	8.8	8, 9
7-Methylxanthine	380 (decomp.)		269 (10.0)	291 (9.0)	7	8, 9
					10.7	
9-Methylxanthine	340 (decomp.)		235 (7.3)	247 (9.3)	6.3	7, 8
			264 (9.3)	278 (9.3)		
1,3-Dimethylxanthine	268 (248)[a]		272 (9.7)	278 (11.9)	8.7	8, 9
1,7-Dimethylxanthine	298–299		269 (10.1)	290 (8.8)	8.8	8, 9
3,7-Dimethylxanthine	(277–279)[a]		273 (10.3)		9.9	8, 9
3,9-Dimethylxanthine	>300	—	235 (8.1)	269 (10.2)	10.1	10
			268 (9.8)			

TABLE 5 *(continued)*

Compound	*m.p. (°C)*	λ_{max} *nm (10^{-3} ε in brackets) in aqueous solution*			*pK*	*Ref.*
		Acidic	*Neutral*	*Basic*		
7,9-Dimethylxanthine	350–370	239 (5.9)	250 (8.3)	—	3.2	
		262 (9.5)	284 (8.3)		12.1	11
8,9-Dimethylxanthine	354–355	238 (8.5)	—	245 (10.7)	6.2	10
		265 (10.5)		278 (10.0)		
1,3,7-Trimethylxanthine	236.5 (89)[a]		273 (9.7)	273 (9.7)		7, 9
1,3,9-Trimethylxanthine	285–287		239 (7.6)	240 (7.6)		7, 12
			267 (9.0)	269 (9.0)		
1,3,8-Trimethylxanthine	(105)[a]					13
1,7,8-Trimethylxanthine	344 (decomp.)					14
1,3,7,8-Tetramethylxanthine	210 (85)[a]					13, 14
1-Methyluric acid	—	260 (9.0)	260 (9.0)	260 (9.0)	5.75	9, 15
		294 (6.8)			10.6	
3-Methyluric acid	—	292 (10.3)	236 (8.1)	236 (8.1)	5.75	9, 15
			292 (10.0)	292 (10.0)	12	
7-Methyluric acid	370–372 (decomp.)	299 (9.7)	237 (8.3)	234 (7.1)		15
			290 (9.3)	283 (9.1)		
9-Methyluric acid	—	251 (6.9)	239 (6.9)	238 (6.5)		15
		300 (8.2)	291 (8.2)	287 (8.2)		
1,3-Dimethyluric acid	286				5.75	9
3,7-Dimethyluric acid	—	289			5.5	9
					12	
1,3,7-Trimethyluric acid	—	289			6.0	9
3,7,9-Trimethyluric acid	373–375	294			8.35	9
1,3,7,9-Tetramethyluric acid	228					16

[a] m.p. of *perchlorate* (ref. 13).

References (table 5)

1 *R. N. Prasad* and *R. K. Robins*, J. Amer. chem. Soc., 1957, **79,** 6401.
2 *S. F. Mason*, Chem. and Biol. of Purines, p. 64.
3 *G. B. Elion*, *ibid.*, p. 46.
4 *J. M. Gulland* and *E. R. Holiday*, J. chem. Soc., 1936, 768.
5 *Robins* and *H. H. Lin*, J. Amer. chem. Soc., 1957, **79,** 490.
6 *J. A. Montgomery* and *C. Temple*, *ibid.*, 1957, **79,** 5238.
7 *J. M. Gulland et al.*, J. chem. Soc., 1934, 1643.
8 *L. F. Cavalieri et al.*, J. Amer. chem. Soc., 1954, **76,** 1119.
9 *F. Bergman* and *S. Dikstein*, *ibid.*, 1955, **77,** 691.
10 *W. Pfleiderer* and *Nubel*, Ann., 1961, **647,** 155.
11 *Pfleiderer*, *ibid.*, 1961, **647,** 161.
12 *F. F. Blicke* and *R. L. Schaff*, J. Amer. chem. Soc., 1956, **78,** 5858.
13 *H. Bredereck et al.*, Chem. Ber., 1953, **86,** 845.
14 *H. Bader* and *J. D. Downer*, J. chem. Soc., 1952, 1641.
15 *R. Falconer* and *Gulland*, *ibid.*, 1939, 1369.
16 *R. Fischer* and *F. Ach*, Ber., 1899, **32,** 2742.

(Chem. Ber., 1947, **80,** 401; 1948, **81,** 307). A series of 9-alkyl derivatives of hypoxanthine has been synthesized by heating 4-alkylamino-5-amino-6-chloropyrimidines with formic acid (*J. A. Montgomery* and *C. Temple*, J. Amer. chem. Soc., 1958, **80,** 409).

All the mono-*N*-methyl derivatives of xanthine have been prepared (*M. Engelmann*, Ber., 1909, **42,** 177; *H. Biltz et al.*, Ann., 1921, **423,** 200; *Fischer, loc. cit.*; *Traube*, Ber., 1900, **33,** 3035; *Elion*, Chem. and Biol. of Purines, p. 39).

1-Methylxanthine, 2,6-dihydroxy-1-methylpurine, $C_6H_6O_2N_4$, is the most important purine constituent of human urine (3.1 g in 1000 l; *M. Kruger* and *G. Salamon*, Z. physiol. Chem., 1898, **24,** 364). It is soluble in acid and alkali, gives the murexide test and is generally similar in its chemical properties to xanthine itself. It is only weakly diuretic.

3-Methylxanthine, 2,6-dihydroxy-3-methylpurine, also has been found in mammalian urine after large doses of caffeine, theobromine or theophylline (*M. Albanese*, Gazz., 1895, [ii], **25,** 298). Its structure was determined by *E. Fischer* and *F. Ach* (Ber., 1898, **31,** 1981,

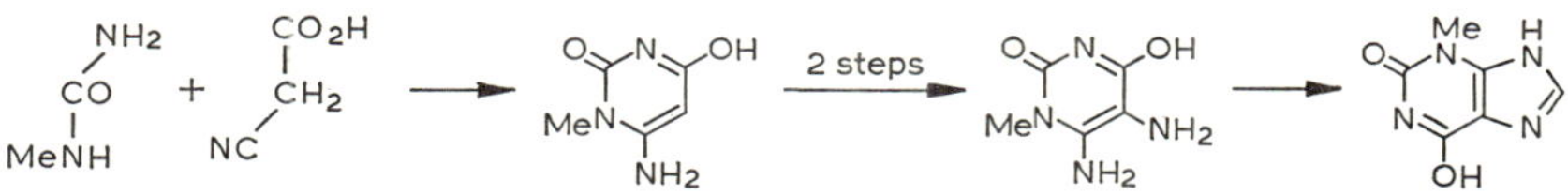

1986) who prepared it by chlorinating 3-methyluric acid and reducing the 8-chloro-3-methylxanthine obtained. The compound is an intermediate in the commercial synthesis of theobromine, the Traube method being used as shown.

The reductive cyclisation of 4,5-diamino-3-methyluracil by boiling formamide has been used to prepare 3-methylxanthine in excellent yield (*H. Bredereck et al.*, *ibid.*, 1950, **83,** 201). Methyl iodide in alkali readily converts 3-methylxanthine into theobromine and a very convenient method of methylation is described by *Bredereck et al.*

7-Methylxanthine, heteroxanthine, like 1-methylxanthine, is a constituent of human urine. It differs chemically from its isomers in forming an insoluble salt, $C_6H_5O_2N_4Na{\cdot}5H_2O$, with sodium hydroxide solution; like the other monomethylxanthines it gives a positive murexide reaction. Its structure was established by hydrolysis to sarcosine with concentrated hydrochloric acid (*Kruger* and *Salamon*, Z. physiol. Chem., 1895, **21**, 172) and its formation from theobromine (*Fischer*, Ber., 1897, **30**, 2400, 2403).

9-**Methylxanthine** does not occur naturally but has been synthesized from uramil by cyclisation with methyl isocyanate followed by desulphurisation (*Böhringer und Söhne*, G.P., 120,327/1898; *J. M. Gulland* and *F. B. Storey*, J. chem. Soc., 1938, 692).

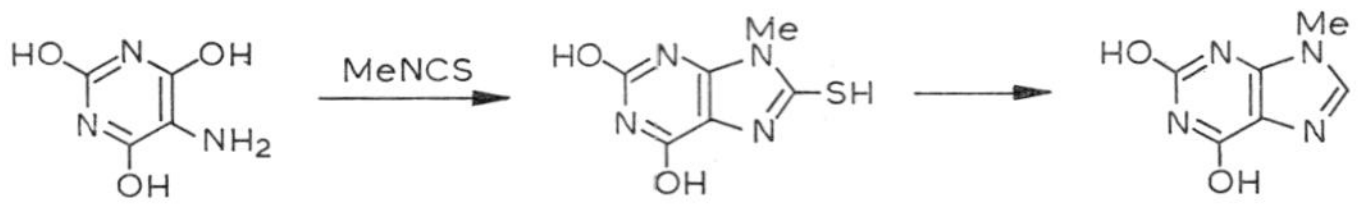

Raney nickel converts 8-mercapto-9-phenylxanthine to 9-*phenylxanthine*, m.p. > 300° (λ_{max} 260 and 275 nm, ε = 13110 and 12500 in acidic and alkaline solution) which, like xanthine, gives a 6-mercapto derivative with phosphorus pentasulphide and can be converted into the corresponding 2,6-dichloro derivative (*H. C. Koppel* and *R. K. Robins*, J. Amer. chem. Soc., 1958, **23**, 2751).

Theophylline, 1,3-dimethylxanthine, $C_7H_8O_2N_4$, m.p. (monohydrate) 272–274°, 7-*acetate*, m.p. 158°, is present, together with caffeine, in tea leaves (*cf. T. Wieland* and *H. Baur*,

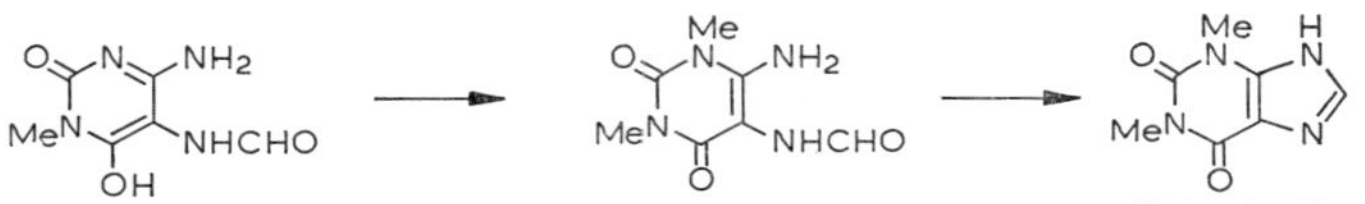

Theophylline

Angew. Chem., 1951, **63**, 511). It is conveniently obtained by the synthetic method illustrated below (*Traube*, Ber., 1900, **33**, 3041, 3052), by reducing 1,3-dimethyluric acid or cyclising 4,5-diamino-1,3-dimethyluracil with boiling formamide (*Bredereck et al.*, *loc. cit.*).

Theophylline is sparingly soluble in water (1 part in 226 at 15°, p*K* 8.7) but readily soluble in warm water or in alkali; it forms salts with acids, insoluble salts with heavy metals and soluble salts with diamines. The salt with ethylenediamine (**aminophylline**) shows the powerful diuretic action of theophylline and has been used clinically for this purpose (*Chem. Werke vorm. Dr. Heinrich Byk*, G.P. 223,695/1907; Chem. Ztbl., 1910, II, 515). The soluble complexes formed by theophylline with the sodium salts of organic acids such as acetic and benzoic acid have also found medicinal uses. The structure of theophylline is confirmed by its oxidation, with potassium chlorate in hydrochloric acid, to dimethylalloxan (*A. Kossel*, Z. physiol. Chem., 1889, **13**, 306). The murexide test is positive. Theophylline is readily oxidised; the first product often appears to be 1,3-dimethyluric acid and further oxidation is, naturally, similar to that of uric acid, discussed on p. 38. Like other xanthines theophylline couples in the 8 position with diazonium salts (*R. Burian*, Ber., 1904, **37**, 704). Methyl iodide converts theophylline into caffeine by methylation at the 7 position; other alkyl halides react similarly (*cf. J. P. Parikh* and *A. Burger*, J. Amer. chem. Soc., 1955, **77**, 2386; *D. B. Ishay*, J. chem. Soc., 1956, 3975; *J. Klosa*, Arch. Pharm., 1955, **288**, 301) although, in aqueous alkaline solution, crotyl bromide alkylates at the 8 position (*J. Donat* and *E. Carstens*, Chem. Ber., 1959, **92**, 1500). 7-Acetic acid derivatives of theophylline, and theobromine, have been prepared and used for further synthesis. Acetohalogeno-sugars react in an analogous fashion to give acetylated

7-glycosyltheophyllines (*Fischer* and *B. Helferich*, *ibid.*, 1914, **47,** 211, 217), the ease of formation of these compounds making them suitable for confirming the structure of new synthetic acetohalogeno-sugars; 7-β-D-ribofuranosyltheophylline, the first synthetic furanoid *N*-glycoside, was prepared for this purpose (*G. A. Howard, B. Lythgoe* and *A. R. Todd*, J. chem. Soc., 1947, 1052).

Paraxanthine, 1,7-dimethylxanthine, has many properties in common with theophylline; it is, for example, an efficient diuretic, soluble in hot water and forms salts with metals. It is unusual, however, in possessing antithyroid activity (*F. G. Mann* and *J. W. G. Porter*, *ibid.*, 1945, 751). The sodium salt, $C_7H_7O_2N_4Na \cdot 2H_2O$, is peculiar in being insoluble in water. Paraxanthine with phosphoryl chloride gives 2,6-dichloro-7-methylpurine, a reaction used in the determination of its structure (*Fischer*, Ber., 1899, **32,** 471). The derivation of paraxanthine from caffeine is shown on p. 48 and methods of synthesis were discussed earlier (p. 17). This was a key compound in Fischer's elucidation of the inter-relationships of the various *N*-methylpurines. It has been prepared by cyclising 4-amino-5-carboxymethylamino-1-methylimidazole with ethyl chloroformate. With alkyl or aralkyl halides in alcoholic alkali it gives 3-alkyl- or 3-aralkyl-paraxanthines (*F. F. Blicke* and *H. C. Godt*, J. Amer. chem. Soc., 1954, **76,** 3653). The 8-chloro-, and hence the 8-substituted-amino derivatives of these were prepared by these authors from 8-chloro-1,3,7-trimethyluric acid.

Theobromine, 3,7-dimethylxanthine, m.p. 277°, 1-*acetate*, m.p. 165°, was the first dimethylxanthine to be recognised. It is the main purine constituent of the cocoa bean (*cf. T. Wieland* and *Bauer*, Angew. Chem., 1951, **63,** 511) along with small amounts of theophylline (*Franzke et al.*, Naturwiss., 1968, **55,** 299) and occurs in other plants such as cascarilla (*I. G. Farben*, Rev. facultad. ciene. quim., 1947, **22,** 285; C.A., 1950, **44,** 8060). The relationship of theobromine to caffeine is shown by its methylation to the latter (*A. Strecker*, Ann., 1861, **118,** 170) and its relation to xanthine by the methylation of xanthine lead salt to theobromine with methyl iodide or dimethyl sulphate (*E. Fischer*, Ann., 1882, **215,** 311; *H. Bredereck et al.*, Chem. Ber., 1950, **83,** 201).

3-Methylxanthine can be methylated to theobromine (see p. 45) (*Bredereck*, *loc. cit.*). Although silver theobromine reacts with alkyl halides to give 1-alkyl-3,7-dimethylpurines (*A. Strecker*, Ann., 1861, **118,** 170; *H. Biltz*, *ibid.*, 1921, **423,** 318) it yields an *O*-glycoside with acetobromoglucose (*Fischer* and *B. Helferich*, Ber., 1914, **47,** 210). The oxidation of theobromine by moist chlorine to methylalloxan and methylurea (*A. Kossel*, Z. physiol. Chem., 1889, **13,** 306) indicates that the two methyl groups are in separate rings and its conversion by phosphoryl chloride to the same 2,6-dichloro-7-methylpurine obtained from paraxanthine serves to illustrate the methods by which its structure was finally elucidated (*vide supra*). The degradation of theobromine to heteroxanthine is illustrated on p. 48.

Like theophylline, theobromine is strongly diuretic, is amphoteric, readily forms salts with acids and bases and gives soluble complexes with the sodium salts of organic acids. Theobromine is hydrolysed by strong acids and alkalis to sarcosine, methylamine and ammonia (*E. Schmidt*, Ann., 1883, **295,** 299) and is readily substituted at the 8-position to

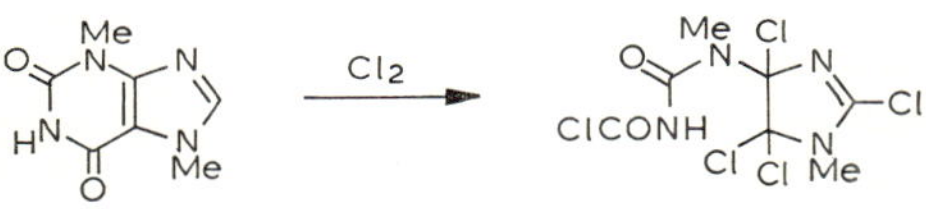

give, for example, 8-chlorotheobromine (*Biltz*, Ber., 1911, **44,** 1526). Like caffeine, theobromine is reduced by zinc and acid to the 6-deoxydihydro derivative (*T. Fichter* and *W. Kern*, Helv., 1926, **9,** 380). The chlorination of theobromine is a somewhat complex reaction but with chlorine in boiling chloroform, oxidation occurs with the production of *N*-(2,4,5,5-tetrachloro-1-methyl-Δ^2-imidazolinyl-4)-*N*-methylallophanoyl chloride (*A. R. Todd* and *N. Whitakker*, J. chem. Soc., 1946, 628).

The reactivity of the 5 position in this reaction recalls the behaviour of uric acid on oxidation in alkaline solution.

A series of 9-alkyl-1,3-dimethylxanthines has been prepared by desulphurising the corresponding 8-mercapto derivatives by nitrous acid or Raney nickel (*Blicke* and *R. L. Schaaf*, J. Amer. chem. Soc., 1956, **78,** 5857).

Chemical interrelations between the important methylxanthines are illustrated in the following reaction scheme.

Caffeine, 1,3,7-trimethylxanthine, $C_8H_{10}O_2N_4$, is the characteristic purine of the coffee bean (*cf. Wieland* and *Bauer*, *loc. cit.*). It may be synthesized by heating the 1,3-dimethyl derivative of 4-amino-5-formamide-2,6-dihydroxypyrimidine with methyl iodide in ethanolic sodium ethoxide (*W. Traube*, Ber., 1900, **33,** 3042, 3054) or by Cook's method

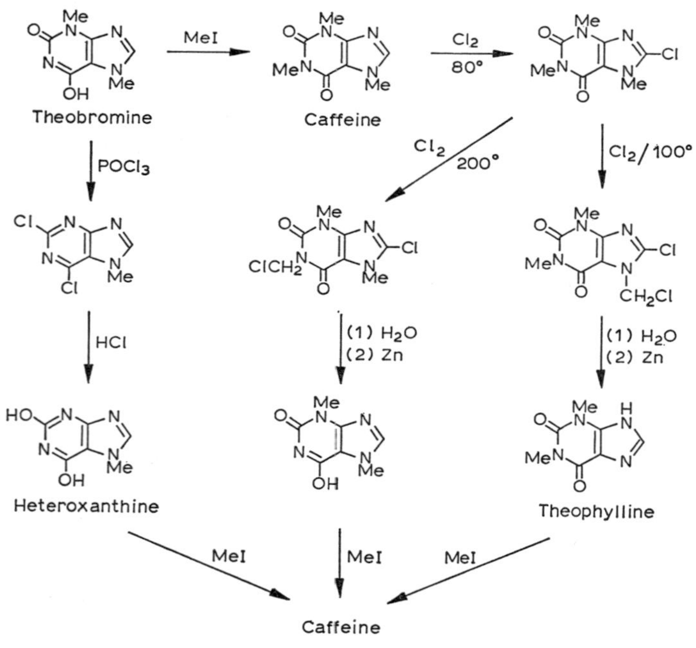

(p. 20) from 4-amino-1-methylimidazole-4-carboxamide (*H. Bader* and *J. D. Downes*, J. chem. Soc., 1952, 1641). Although normally manufactured from coffee and tea dust, it has been prepared from guanine by acid hydrolysis of this to xanthine followed by methylation with methyl acetate (*W. E. Decker*, U.S.P., 2,509,084/1946; C.A., 1950, **44,** 8366);

also by methylation of xanthine with dimethyl sulphate (p. 34). Caffeine differs from the dimethylxanthines in its ease of sublimation and much greater solubility in organic solvents and water. It is not acidic but readily forms salts with acids; it gives the murexide reaction

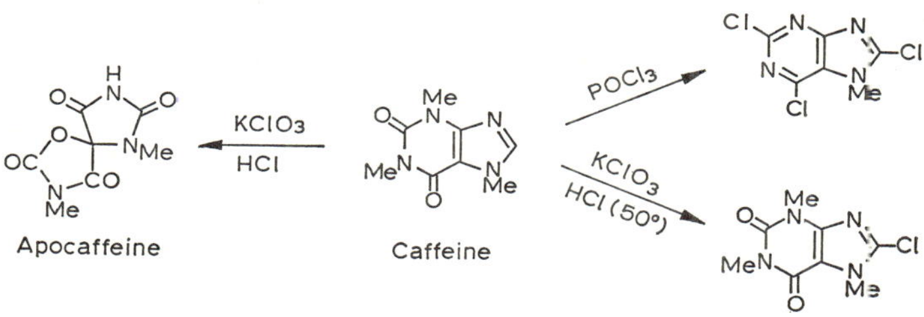

(for oxidation see p. 39). The 8-halogeno derivatives of caffeine, which were of great importance in the elucidation of its structure, are readily prepared. A hydrochloric acid solution of potassium chlorate at 50° converts caffeine into 8-chlorocaffeine (*H. Biltz*, Ber., 1910, **43,** 1622) whereas the cold reaction mixture produces **apocaffeine (1,7-dimethyl-caffolide)** (*idem, ibid.*, 1910, **43,** 1618). Reaction with phosphoryl chloride removes two methyl groups, 2,6,8-trichloro-7-methylpurine being formed (*Fischer, ibid.*, 1885, **28,** 2489).

The action of gaseous chlorine on caffeine solutions has been described by *E. S. Golovinskaya* (Zhur. priklad. Khim., 1958, **31,** 918; C.A., 1958, **52,** 20185).

These reactions, and many others, including stability to acid except under stringent conditions, show the close chemical analogy between caffeine and the other methylated xanthines; its instability to alkali is, however, in strong contrast. Cold dilute sodium hydroxide solution hydrolyses caffeine to caffeidinecarboxylic acid, $C_9H_{12}O_3N_4$ (*cf. Biltz*, Ann., 1914, **404,** 143) which is reconverted to caffeine by heating with phosphoryl chloride.

Caffeidinecarboxylic acid is hydrolysed by boiling water to carbon dioxide and caffeidine, $C_7H_{12}O_4N$, which may be oxidised (*Schmidt*, Ber., 1881, **14,** 816) to *N*-methylparabanic

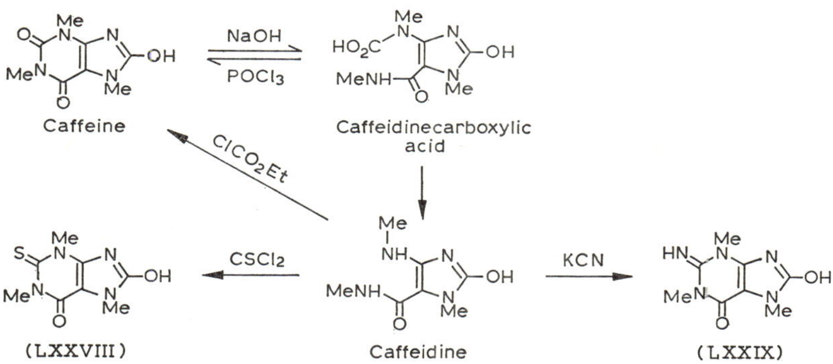

acid (*cf.* C.C.C., 2nd Edn., Vol. IV A, p. 320), reconverted into caffeine by reaction with ethyl chloroformate and to 2-thiocaffeine (LXXVIII) or the 2-imino analogue of caffeine, LXXIX by thiophosgene and potassium cyanide, respectively.

When heated with methyl iodide caffeine gives the 9-methiodide which is converted by silver oxide to the corresponding quaternary hydroxide. Treatment of caffeine with acetic anhydride at 200° gives 8-methyltheophylline (*R. C. Houston* and *W. F. Allen*, J. Amer. chem. Soc., 1917, **56,** 1793).

Caffeine and other hydroxypurine derivatives form water-soluble complexes with polycyclic aromatic compounds, some of which are crystalline and well defined (*H. W. Malherbe*, Biochem. J., 1946, **40,** 351; *J. Both et al.*, J. chem. Soc., 1954, 598).

Isocaffeine, 1,3,9-**trimethylxanthine,** m.p. 285°, was prepared by *Biltz* (Ann., 1921, **423,** 223), and synthesized by cyclising 5-amino-4-methylamino-2,6-dioxo-1,3-dimethyltetrahydroxypyrimidine with formic acid or formamide (*W. Pfleiderer* and *K. H. Schündehütte*, *ibid.*, 1958, **612,** 158).

1,3,7,9-**Tetramethyluric acid,** m.p. 228°, obtained by methylation of uric acid or its mono-, di- or tri-methyl derivatives or by heating 8-methoxycaffeine (*Biltz*, *ibid.*, 1917, **413,** 200) is converted by phosphoryl chloride at 160° into 8-chlorocaffeine and is hydrolysed to an imidazole derivative by alkali (*idem*, Ber., 1931, **64,** 1974); (see p. 60).

Epiguanine, 7-**methylguanine,** 2-**amino**-6-**hydroxy**-7-**methylpurine,** $C_6H_7O_5N$, the only naturally occurring aminohydroxy-*N*-methylpurine, is present in urine (~ 3.4 g in 10,000 l). Its structure follows from its deamination to 7-methylhypoxanthine (heteroxanthine) and its oxidation by potassium chlorate in hydrochloric acid to guanidine (*E. Fischer*, *ibid.*, 1898, **31,** 3270).

Toxoflavin, $C_6H_6O_2N_4$, one of the two toxic principles of "Bongkrek" bacteria has been assigned the structure LXXX, it gives the murexide test after reduction and shows similarities to the purines (*cf. L. Stern*, Biochem. J., 1935, **29,** 500).

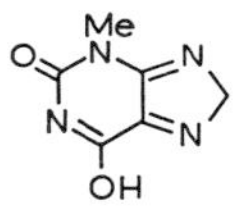

(LXXX)

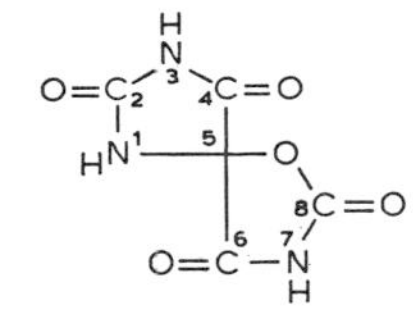

(LXXXI)

Caffolides. These compounds have been little studied for many years but deserve mention at this point since they are derived from *N*-methylhydroxypurines by various oxidative processes. The parent compound is unknown and the structure LXXXI assigned thereto is not established unequivocally but mono-, di- and tri-methyl derivatives have been described. Thus 3-**methylcaffolide,** m.p. 220° (decomp.), prepared by the hydrolysis of 9-methyl or 3,9-dimethyluric acid glycol (see pp. 15, 40) affords, on hydrolysis, 3-methyl 5-hydroxyhydantoin-5-carboxamide; 3,7-**dimethylcaffolide,** m.p. 177–178° (decomp.) (p. 40) was obtained by the hydrolysis of 4-chloro-1,9-dimethyl-$\Delta^{5,7}$-isouric acid and also by condensing methylurea with methylalloxan; 1,3,7-**trimethylcaffolide,** obtained from 1,7,9-trimethyluric acid by chlorination in aqueous solution, was also prepared by methylating 1,3-dimethyl-5-hydroxyhydantoin-5-carboxyureide with diazomethane (*Biltz et al.*, Ann., 1921, **423,** 181, 231, 290).

(*v*) *Alkoxypurines*

Alkoxypurines are normally prepared by treating a chloropurine with a sodium alkoxide and occasionally, synthetically from alkoxypyrimidines.

2-**Methoxypurine,** m.p. 205–206°, was obtained by a Traube synthesis but formic–acetic anhydride was required to affect the cyclisation (*A. Albert* and *D. J. Brown*, J. chem. Soc., 1954, 2060) and *K. L. Dille* and *B. E. Christensen* prepared 2,6-**dimethoxypurine,** m.p. >300° (decomp.) from the appropriate diaminopyrimidine and formamide (J. Amer. chem. Soc., 1954, **76,** 5087). The properties of some representative methoxypurines are shown in Table 6. These ethers often undergo the well known rearrangement of α-alkoxyamines to alkylamides (*cf. L. H. Wheeler* and *T. B. Johnson*, Amer. chem. J., 1899, **21,** 185). A typical example is the transformation of 8-methoxytheobromine (LXXXII) to 3,7,9-trimethyluric acid (LXXXIII) on heating to 290° (*Biltz* and *H. Pardon*, Ber., 1930, **63,**

(LXXXII) (LXXXIII)

2876). 8-**Ethoxypurines** have been prepared from a diaminopyrimidine and tetraethoxymethane (*D. J. Brown* and *R. K. Lynn*, J. chem. Soc., Perkin I, 1974, 349).

Acid hydrolysis of 6-methoxy-9-methylpurine produces a complex mixture of products including 5-amino-1-methylimidazole-4-carboxamide and 9-methylhypoxanthine in the ratio 7:3 (*J. L. Wang* and *D. S. Fuchs*, *ibid.*, 1974, 1284).

TABLE 6

PROPERTIES OF SOME METHOXYPURINES

Compound	*m.p. (°C)*	*λ_{max} nm (10^{-3} ε in brackets) in aqueous solution*		*Ref.*
		Acidic	*Alkaline*	
2-Methoxy-	205–206	284 (6.76)	283 (7.59)	1, 2
6-Methoxy-	194–195	254 (10.23)	261 (9.77)	2, 3
8-Methoxy-	153–154	271 (11.2)	279 (9.5)	4
6-Methoxy-7-methyl-	200	256 (11.6)	258 (9.6)	5
6-Methoxy-9-methyl-	152–153	254 (9.85)	254 (10.6)	5
8-Chloro-6-methoxy-	203–204	263 (12.0)	265 (11.4)	6
6-Amino-8-methoxy-	—	270 (13.9)	265 (11.4) 246 (4.5)	6
2-Amino-6-methoxy-	—	286 (11.2) 235 (5.0)	284 (7.9)	7
2-Amino-6-methoxy-3-methyl-	—	284 (11.8)	287 (14.0)	8

References

1 *A. Albert* and *D. J. Brown*, J. chem. Soc., 1954, 2060.
2 *S. F. Mason*, *ibid.*, 1954, 2071.

3 *G. Huber*, Chem. Ber., 1957, **80,** 698.
4 *D. J. Brown* and *Mason,* J. chem. Soc., 1957, 682.
5 *R. N. Prasad* and *R. K. Robins,* J. Amer. chem. Soc., 1957, **79,** 6401.
6 *Robins, ibid.*, 1958, **80,** 6671.
7 *R. W. Balsiger* and *J. A. Montgomery,* J. org. Chem., 1960, **25,** 1573.
8 *M. Kasha*, J. chim. Phys., 1961, **58,** 914.

(*e*) *Halogenopurines*

The relative ease of formation of halogenopurines and the reactivity of the halogens in positions 2-, 6- or 8- towards nucleophiles has made them amongst the most valuable of intermediates for the synthesis and interconversion of purines. The properties of some representatives are listed in Table 7 (p. 63).

(*i*) *Methods of preparation*

(*1*) **Chloropurines.** One of the oldest and still most widely used methods for the preparation of halogenopurines is from the corresponding hydroxypurine and phosphorus halides. Heating uric acid with phosphoryl chloride alone, for example, gives 2,6-dichloro-8-hydroxypurine which needs drastic treatment before the remaining hydroxyl group is replaced to give 2,6,8-trichloropurine (*E. Fischer* and *F. Ach*, Ber., 1897, **30,** 2208; *Fischer, ibid.*, p. 2200). The trichloropurine has been obtained in high yield (90%) under less rigorous conditions by using excess chlorinating reagent [*B. G. Boldyrev* and *R. G. Makitra*, J. appl. Chem., U.S.S.R. (Engl. Edn.), 1955, **28,** 399]. Xanthine on the other hand requires addition of a small amount of water to the phosphoryl chloride before it will produce 2,6-dichloropurine and the active agent is thought to be pyrophosphoryl chloride, $Cl_2P(O)OP(O)Cl_2$, formed by partial hydrolysis (*G. B. Elion* and *G. H. Hitchings,* J. Amer. chem. Soc., 1956, **78,** 3508); the same reagent may be formed when mixtures of phosphorus pentachloride and phosphoryl chloride are used (*H. C. Koppel* and *R. K. Robins,* J. Amer. chem. Soc., 1958, **80,** 2751). Mixtures of phosphoryl chloride and tertiary amines, usually *N,N*-dimethyl- or *N,N*-diethylaniline, have been most widely used as in the conversion of hypoxanthine to 6-chloropurine (*A. Bendich, P. J. Russell* and *J. J. Fox,* J. Amer. chem. Soc., 1954, **76,** 6073; *A. J. Beaman* and *Robins,* J. appl. Chem., 1962, **12,** 432) although *N*-ethylpiperidine is considered to produce better yields (*H. R. Schutte et al.*, Z. Chem., 1964, **4,** 430). The method has been applied to nucleosides as in the conversion of inosine to 6-chloropurineriboside (*J. F. Gerster, J. W. Jones* and *Robins,* J. org. Chem., 1963, **28,** 945). With these reagents xanthine fails to react (*Elion* and *Hitchings, loc. cit.*) but 6,8-dihydroxypurines readily give 6,8-dichloropurines (*Robins,* J. Amer. chem. Soc., 1958, **80,** 6671). Involvement of the tertiary base in these reactions is rare but may occur with aliphatic tertiary amines; thus hypoxanthine and xanthine with phosphoryl chloride and triethylamine produce 6-diethylamino- or 2,6-bisdiethylamino-purines, respectively (*Robins* and *B. E. Christensen*, *ibid.*, 1952, **74**, 3624).

Replacement of the hydroxyl groups in *N*-alkyluric acids appears to be facilitated by substitution of a neighbouring nitrogen atom but concurrent removal of *N*-alkyl groups often occurs and this may permit oxo groups originally present to tautomerise and be replaced by halogen. Thus, although 1,3-dimethyluric acid and a mixture of phosphorus trichloride and phosphoryl chloride gives 8-chlorotheophylline in the normal manner (*Fischer* and *Ach*, *ibid.*, 1895, **2**, 3135), 3,7-dimethyluric acid gives, with the same reagents at 140°, 6-chloro-2,8-dihydroxy-3,7-dimethylpurine and then, at 170°, 2,6,8-trichloro-7-methylpurine (*Fischer*, *ibid.*, 1895, **28**, 248; 1897,

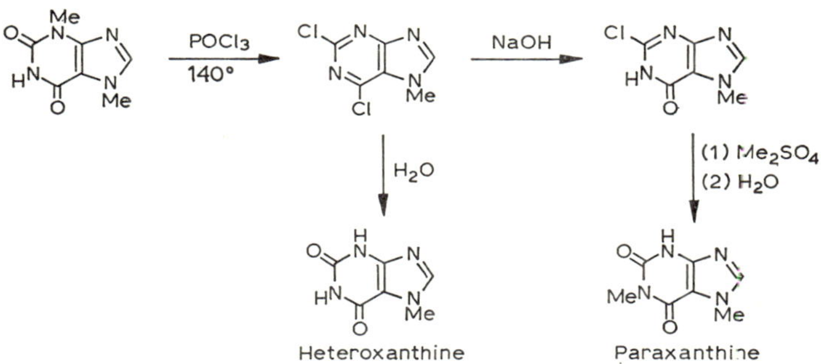

30, 1839). Theobromine is also demethylated at the 3-position in this way (*Fischer*, *ibid.*, 1897, **30**, 2400) and *Fischer* utilised this reaction to prepared heteroxanthine and paraxanthine.

This type of demethylation is effected *via* intermediate chloromethyl derivatives when phosphorus pentachloride or chlorine in phosphoryl chloride are used and caffeine may be converted into 8-chloro-(chloromethyl)dimethyl-, 8-chloro-di(chloromethyl)methyl- or 8-chloro-tri(chloromethyl)-xanthines in this way (*Fischer* and *Ach*, *ibid.*, 1906, **39**, 423).

The direct halogenation of hydroxypurines has been the subject of considerable study, chiefly by *H. Biltz*. The course of these reactions is greatly affected by the solvent used; thus, in non-polar solvents, the xanthines give the corresponding 8-halogenoxanthine (*Fischer*, Ann., 1882, **213**, 316) as in the conversion of theobromine to 8-chlorotheobromine (LXXXIV) by chlorine in chloroform (*Fischer*, Ann., 1883, **221**, 336). Theobromine in acetic acid is, however, converted by chlorine to 5-chloro-3,7-dimethylisouric acid (LXXXV) (*Biltz* and *F. Damm*, 1914, **406**, 22) which is reduced

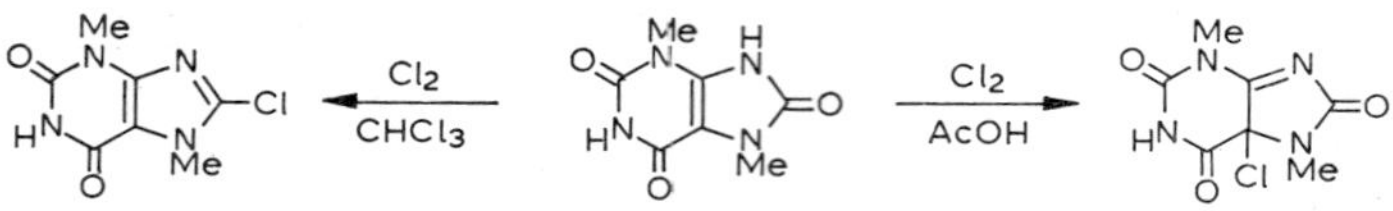

(LXXXIV) Theobromine (LXXXV)

by stannous chloride to 3,7-dimethyluric acid; in a similar way theophylline and caffeine are converted into 1,3-dimethyl- and 1,3,7-trimethyl-uric acids, respectively (*Biltz* and *K. Strufe*, *ibid.*, 1917, **413,** 157; *Biltz*, Ber., 1910, **43,** 3553). The course of the reactions between chlorine and uric acid and its methyl and acetyl derivatives is even more complex but it appears, in general, that the 4,5 double bond is first halogenated and that subsequent reactions depend upon the type and position of *N*-substitution and the solvents used (*idem*, J. pr. Chem., 1932, **134,** 335; Ann., 1935, **515,** 201).

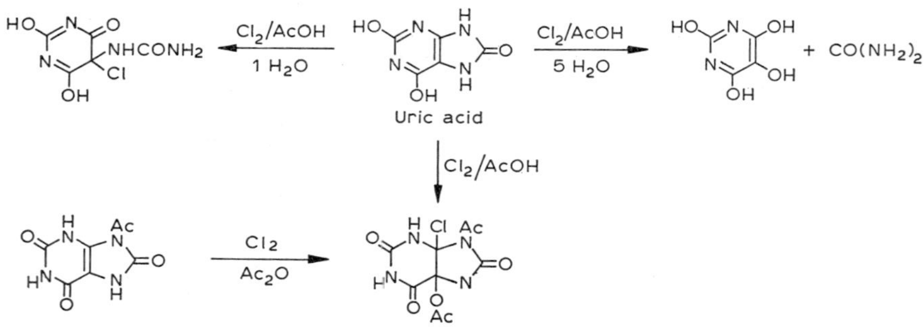

Although chloro-4,5-diaminopyrimidines fail to give chloropurines with formic acid owing to hydrolysis, the synthesis can be effected using triethyl orthoformate in acetic anhydride or, preferably, diethoxymethyl acetate (*J. A. Montgomery* and *L. B. Holum*, J. Amer. chem. Soc., 1958, **80,** 404). The former reagent affords a mixture of the compound required and its *N*-acetyl derivative, which is hydrolysed by warm dilute alkali to the chloropurine but by warm water to a 5-formamidopyrimidine.

(*2*) **Bromo-** and **iodo-purines** are all of comparatively minor importance. The former can be prepared from hydroxypurines with phosphoryl bromide (*cf. Elion* and *Hitchings*, *loc. cit.*) or by direct bromination as in the conversion of hypoxanthine (*G. Bruhns*, Ber., 1890, **23,** 225) or xanthine (*A. J. Beaman* and *R. K. Robins*, J. org. Chem., 1963, **28,** 2310) to the corresponding 8-bromo derivatives, and the latter from a chloro derivative and hydriodic acid (*Elion* and *Hitchings*, *loc. cit.*) or rarely by direct iodination as in the formation of 8-iodocaffeine from caffeine and iodine at 150° for 6 hours (*Y. Yoshitomi*, J. pharm. Soc., Japan, 1924, **508,** 460).

However, iodination of guanosine and xanthosine with *N*-iodosuccinimide in dimethyl sulphoxide proceeds to give good yields of the 8-iodo derivatives with catalysis by *n*-butyl sulphide (*D. Lipkin et al.*, J. biol. Chem., 1963, **238,** 2249).

(*3*) **Fluoropurines** are normally made by substitution of a chlorine or amino group (after diazotisation). Thus 6-fluoropurine may be obtained from 6-chloropurine by first quaternising with trimethylamine and heating the quaternary salt with potassium hydrogen fluoride in ethanol at 50° (*J. Kiburis* and *J. H. Lister*, Chem. Comm., 1969, 381). Various substituted 2-, 6- or 8-fluoropurines have also been prepared from corresponding chloro derivatives by heating with silver fluoride in toluene or xylene (*Beaman* and *Robins*, *loc. cit.*). Treatment of 2-aminopurine with sodium nitrite and 48% tetrafluoroboric acid gave 2-fluoropurine in 41% yield, whereas 2-fluoro-6-methylpurine was similarly obtained in only 6% yield; 2,6-diaminopurine similarly gave 6-amino-2-fluoropurine in poor (6%) yield (*Montgomery* and *K. Hewson*, J. Amer. chem. Soc., 1960, **82,** 463), but a better yield (22%) was obtained if the reaction was carried out in liquid hydrogen fluoride containing sodium nitrite (*C. N. Eaton* and *G. H. Denny*, J. org. Chem., 1969, **34,** 747).

Reduction of polyhalogeno derivatives has been employed for the preparation of simpler compounds, reduction of the corresponding 2,6,8-trichloropurines with zinc being used to obtain the 7-methyl- and 9-phenyl- derivatives of 2-chloropurine (*Fischer*, Ber., 1898, **31,** 2552; 1901, **33,** 2279).

(*ii*) *Properties and reactions*

The chloropurines in general derive their importance from the differing reactivities of their halogen groups and their reactions with a variety of reagents. They may be used, therefore, for the synthesis of a wide range of purine derivatives. The reactions of the chloropurines with bases have been shown to be second order and bimolecular (*G. B. Barlin* and *N. B. Chapman*, J. chem. Soc., 1965, 3017; *Barlin*, *ibid.*, 1967, B, 954).

The ease with which the halogen groups of 2,6,8-trichloropurine, for example, may be replaced by nucleophilic reagents decreases in the order 6, 2, 8 and similarly, 2-chloropurine reacts less readily with amines than 6-chloropurine (*Montgomery* and *Holum*, J. Amer. chem. Soc., 1957, **79,** 2185). With the 7- or 9- alkyl-2,6,8-trichloropurine, however, attack with strong alkali occurs initially at C-8 followed by attack at C-6 then finally at C-2 (*E. Y. Sutcliffe* and *Robins*, J. org. Chem., 1963, **28,** 1662). Results of this type have led to the suggestion that a stable anion LXXXVI is formed and this precludes attack by B⊖ at C-8 hence C-6 is first attacked. This effect is also shown by 6,8-dichloropurine which with alkali produces 8-chlorohypoxanthine (*Beaman* and *Robins*, J. appl. Chem., 1962, **12,** 432). On the other hand 2,8-dichloropurine with bases involves prior attack at C-8 rather than C-2 as might be expected from the "anion theory" (*idem, loc. cit.*; *A. F. Lewis*, *Beaman* and *Robins*, Canad. J. Chem., 1963, **41,** 1807); the lack of reactivity of the C-2

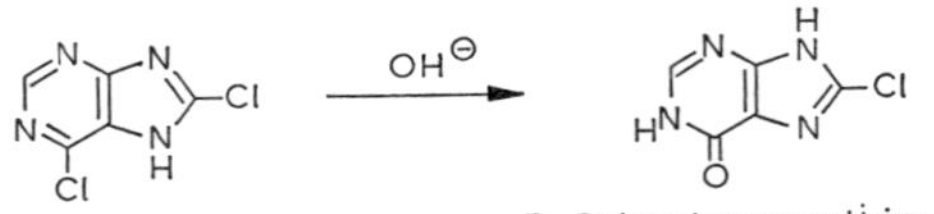

8-Chlorohypoxanthine

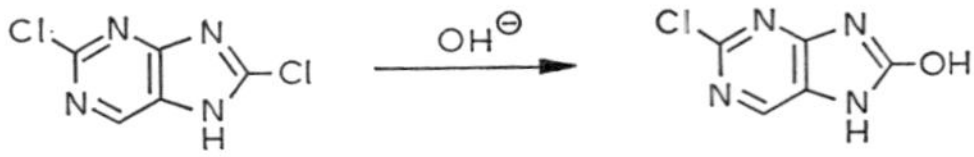

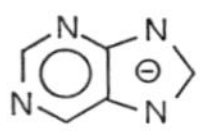

(LXXXVI)

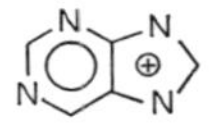

(LXXXVII)

halogen might be explained by partial fixation of the C-4–C-5 double bond (*J. H. Lister*, *op. cit.*, p. 153).

In acid conditions it is suggested that a protonated form LXXXVII may be dominant and in this case reversal of the substitution order should occur. This is generally so; thus reaction of 6,8-dichloro- (*Robins*, *loc. cit.*) and 2,6,8-trichloro-purine (*E. Fischer*, Ber., 1897, **30**, 2220) with hot acid produces 6-chloro-8-oxo- and 2,6-dichloro-8-oxo-purine, respectively.

Fischer utilised these differences to prepare purine (*ibid.*, 1898, **31**, 2550), adenine, guanine, hypoxanthine and xanthine (*ibid.*, 1897, **30**, 2226) from trichloropurine, and adenosine and guanosine have been prepared from the same compound (*J. Davoll*, *B. Lythgoe* and *A. R. Todd*, J. chem. Soc., 1948, 967, 1685).

The contrast in behaviour of an 8-halogeno group is exemplified by the rapid hydrolysis of 8-bromocaffeine by aqueous potash to which reagent 8-bromoxanthine is stable at 100° (*Fischer*, Ber., 1896, **31**, 3272). The halogenopurines react with ammonia and amines and are hydrolysed to hydroxypurines the course of the reaction varying according to pH. Thus whereas the conversion of 6-chloropurine to adenine requires reaction with ammonia in butanol under pressure (*Bendich*, *Russell* and *Fox*, *loc. cit.*, p. 6073), 6-fluoro-9-methylpurine produces the corresponding 9-methyladenine with aqueous ammonia on a water bath (*A. G. Beaman* and *Robins*, J. med. Chem., 1962, **5**, 1067). Primary and secondary aliphatic amines are more reactive than ammonia and these include amino acids such as aspartic acid which produces succinoaminopurine (see p. 77) (*C. E. Carter*, J. biol. Chem., 1956, **223**, 139) and hydroxylamine which produces hydroxylaminopurines (*Bendich* and *A. Giner-Sorolla*, J. Amer. chem. Soc., 1958, **80**, 3932; *Giner-Sorolla et al.*, Biochemistry, 1966, **5**, 3057).

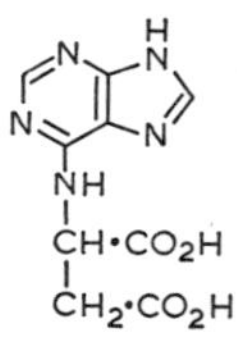

Succinoaminopurine

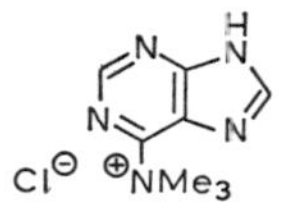

(LXXXVIII)

6-Chloropurine with trimethylamine in dimethylformamide gives a good yield of the quaternary salt LXXXVIII (*J. P. Horwitz* and *V. K. Vaitkevicius*, Experientia, 1961, **17**, 552).

In a similar manner 2,6-dichloropurine with alcoholic ammonia produces 6-amino-2-chloropurine but with aqueous ammonia it gives 2,6-diaminopurine (*G. B. Brown* and *V. S. Weliky*, J. org. Chem., 1958, **23**, 125).

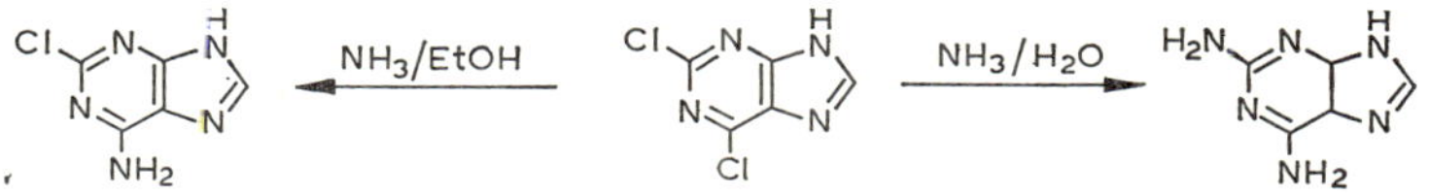

Substitution of the imidazole nitrogen results in easier replacement of the 6-chloro group; thus 2,6-dichloro-9-substituted purines readily react with alcoholic ammonia on a water bath to give 9-substituted 2-chloroadenines (*W. A. Bowles et al.*, J. med. Chem., 1963, **6**, 471). The halogen groups are similarly progressively substituted by aliphatic or alicyclic amines (*S. R. Breshears et al.*, J. Amer. chem. Soc., 1959, **81**, 3789; *K. Panagapoulos et al.*, Arzneim.-Forsch., 1965, **15**, 204). Amination of 6,8-dichloropurine also proceeds with progressive replacement of the 6- (100° for 12 hours) then the 8-chloro (135° for 6 hours) groups by amino but alkylation of the imidazole nitrogen does not in this case reverse the order of substitution; thus 9-benzyl-6,8-dichloropurine with aqueous dimethylamine gives 9-benzyl-8-chloro-6-dimethylaminopurine (*Montgomery, Hewson* and *C. Temple*, J. med. Chem., 1962, **5**, 15).

Amination of 2,6,8-trichloropurine in aqueous ammonia at 100° for 6 hours gives 6-amino-2,8-dichloropurine (*Fischer*, Ber., 1897, **30**, 2226), whereas refluxing with primary or secondary amines gives corresponding 6-alkylamino-2,8-dichloropurines and at high temperatures (150–170°) in a sealed tube it is possible to replace all the halogen atoms (*Breshears, loc. cit.*). With pyridine in ethanol, 2,6,8-trichloropurine gives the betaine LXXXIX whereas in pyridine solution the pyridinium chloride results. When the trichloropurine contains alkyl groups at N-7 or N-9, anion formation is not possible and primary attack with ammonia and amines occurs at C-8. Thus 2,6,8-trichloro-7-methylpurine (XC) with ethanolic ammonia gives 8-amino-2,6-dichloro-7-methylpurine (*Fischer*, Ber., 1897,

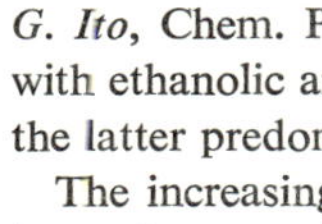

(LXXXIX)

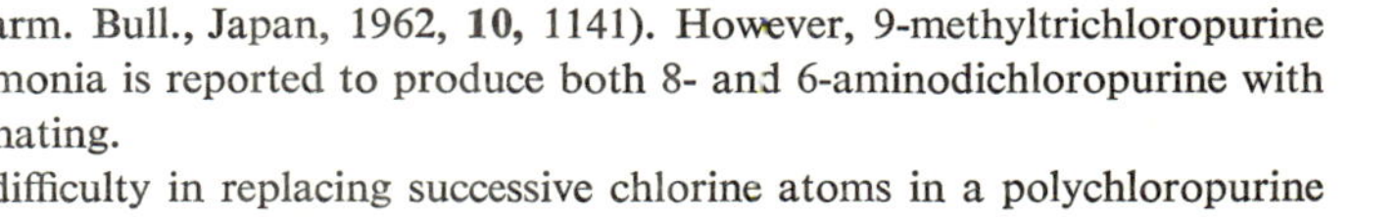

(XC)

30, 1845) or with aqueous ammonia 6,8-diamino-2-chloro-7-methylpurine (*T. Itai* and *G. Ito*, Chem. Pharm. Bull., Japan, 1962, **10**, 1141). However, 9-methyltrichloropurine with ethanolic ammonia is reported to produce both 8- and 6-aminodichloropurine with the latter predominating.

The increasing difficulty in replacing successive chlorine atoms in a polychloropurine by amino or substituted amino groups is echoed in the halogeno-oxopurines. Thus 2-chlorohypoxanthine is very resistant to attack by ammonia, and aliphatic amines require forcing conditions. Methylamine in ethanol requires heating to 130° for several hours (*J. F. Gerster* and *R. K. Robins*, J. Amer. chem. Soc., 1965, **87,** 3752) although histamine reacts in boiling butanol (*H. Ballweg*, Ann., 1961, **649,** 114). On the other hand, 2-chloro-1,9-dimethylhypoxanthine is aminated by aqueous ammonia at 80° (*I. M. Ovcharova et al.*, Zhur. obshcheĭ Khim., 1962, **32**, 2010). Similarly, difficulties in aminating 8-halogeno-caffeines resulted in use of the Gabrielle phthalimide procedure to produce 8-phthalimido-caffeine, hydrolysis of which gave 8-aminocaffeine (*H. Zimmer* and *J. B. Mettalia*, J. org. Chem., 1959, **24,** 1813) but an improved yield (43%) was obtained by use of hydrazine and heating the derived 8-hydrazinocaffeine in dimethylformamide.

Hydrazine also reacts with other chloropurines. Thus with 6-chloropurine 6-hydrazino-purine is produced at room temperature whereas 2-chloropurine only gives 2-hydrazino-purine after heating with hydrazine at 80° for 12 hours, but 2-chlorohypoxanthine (2-chloro-6-oxo-1,6-dihydropurine) gives 2-hydrazinohypoxanthine at waterbath temperature (*Montgomery and Holum, loc. cit.*).

8-Hydrazinopurines tend to degrade to 8-amino derivatives (see above) but derivatives of theobromine and 1,7-dimethylxanthine (*H. Priewe* and *A. Poljack*, Chem. Ber., 1955, **88,** 1932) and caffeine (*F. Cramer*, Ber., 1894, **27,** 3089) have been obtained. 2,6-Dichloro-purine with hydrazine at room temperature gives 2-chloro-6-hydrazinopurine and on heating with hydrazine 2,6-dihydrazinopurine is obtained (*Montgomery* and *Holum, loc. cit.*). All chlorine atoms in 2,6,8-trichloropurine may be replaced by heating with hydrazine hydrate (*Breshears et al., loc. cit.*).

In a similar manner 6-hydroxyaminopurine and its 9-β-D-riboside may be produced from 6-chloropurine (or the 9-β-D-riboside) and ethanolic hydroxylamine (*C. C. Chang*, J. med. Chem., 1965, **8,** 884; *A. Giner-Sorolla, L. Medrek* and *A. Bendich, ibid.*, 1966, **9,** 143).

2-Fluorohypoxanthine with hydroxylamine gives 2-hydroxyaminohypoxanthine (*Gerster* and *Robins*, J. org. Chem., 1966, **31,** 3258) and 2,6-dihydroxyaminopurine is produced from 6-chloro-2-fluoropurine and hot ethanolic hydroxylamine (*Giner-Sorolla et al.*, J. med. Chem., 1968, **11,** 521).

With metal alkoxides the corresponding ethers are obtained as in the conversion of 8-chlorocaffeine to the 8-ethoxy derivatives (*E. Fischer*, Ann., 1882, **215,** 266), or of 6-chloropurine to 6-methoxypurine and related compounds (*G. Huber*, Chem. Ber., 1957, **90,** 698) by heating in alkaline solution under reflux with the sodium alkoxide. On the other hand, substitution at N-3 results in much enhanced activity of the chlorine; thus 6-chloro-3-methylpurine hydrochloride when warmed in methanol produced 6-methoxy-3-methylpurine directly (*F. Bergmann, Z. Neiman* and *M. Kleiner*, J. chem. Soc., C, 1966, 10). On the other hand, 2,6-dichloropurine with sodium methoxide requires prolonged reaction to produce 2-chloro-6-methoxypurine (*W. C. Coburn et al.*, J. org. Chem., 1965, **30,** 1110). If ionisation of the purine is prevented by using the readily removed substituent tetra-hydropyran-2-yl groups at N-9 then the corresponding 2,6-dichloropurine gives ultimately 2,6-diethoxypurine under reflux conditions (*F. Cassidy, R. K. Olsen* and *Robins*, J. hetero-cycl. Chem., 1968, **5,** 461). In contrast 2,8-dichloropurine with methanolic sodium methoxide under reflux produces 2,8-dimethoxypurine (*A. F. Lewis, Beaman* and *Robins*, Canad. J. Chem., 1963, **41,** 1807) whereas, under the same conditions, 6,8-dichloropurine gives only 8-chloro-6-methoxypurine (*Robins*, J. Amer. chem. Soc., 1958, **80,** 6671). 2,6,8-Trichloropurine with sodium ethoxide in ethanol at room temperature gives 2,8-dichloro-6-ethoxypurine and at 100° 8-chloro-2,6-diethoxypurine (*Fischer*, Ber., 1897, **30,** 2226; *D. W. MacCorquodale*, J. Amer. chem. Soc., 1929, **51,** 2245). Alkylation at N-7 or N-9 results in enhanced activity. Thus 7-methyl-2,6,8-trichloropurine with ethanolic potassium hydroxide at 3° gave 2,6-dichloro-8-ethoxypurine and at 40° 2-chloro-6,8-diethoxypurine (*Fischer*, *loc. cit.*).

Halogeno groups are converted into mercapto groups by the use of thiourea (*Temple* and *Montgomery*, J. org. Chem., 1966, 31, 1417) or thioacetamide (*Giner-Sorolla* and *Bendich*, J. med. Chem., 1965, **8,** 667), or aqueous potassium hydrosulphide. The ease of replacement varies and whereas 2,6,8-trichloro- and 2,6-dichloro-7-methylpurine react completely in the cold with the latter reagent it is necessary to heat 2,6-dichloro-8-hydroxypurine at 140° for 3 h to prepare the dimercapto compound (*P. C. Ray et al.*, J. chem. Soc., 1923, **123,** 1957). Stepwise reaction is illustrated by the behaviour of 2,6-dichloro-7,9-dimethyl-8-oxodihydropurine, which rapidly gives the 6-mercapto derivative with cold ethanolic hydrogen sulphide and the 2,6-dimercapto compound only after 3 h at 130° with aqueous potassium hydrosulphide. Thiols react to give purine thioethers and an extensive series has been prepared from 8-chlorocaffeine (*L. M. Long*, J. Amer. chem. Soc., 1947, **69,** 2939) and 6-chloropurine (*H. C. Koppel*, *D. E. O'Brien* and *R. K. Robins*, J. org. Chem., 1959, **24,** 259). 6-**Selenopurine** is similarly obtained from 6-chloropurine with hydrogen selenide or better selenourea in sodium ethoxide under reflux (*H. G. Mautner*, J. Amer. chem. Soc., 1956, **78,** 5292).

Halogenopurines can react normally with malonic ester and caffeine-8-acetic acid has

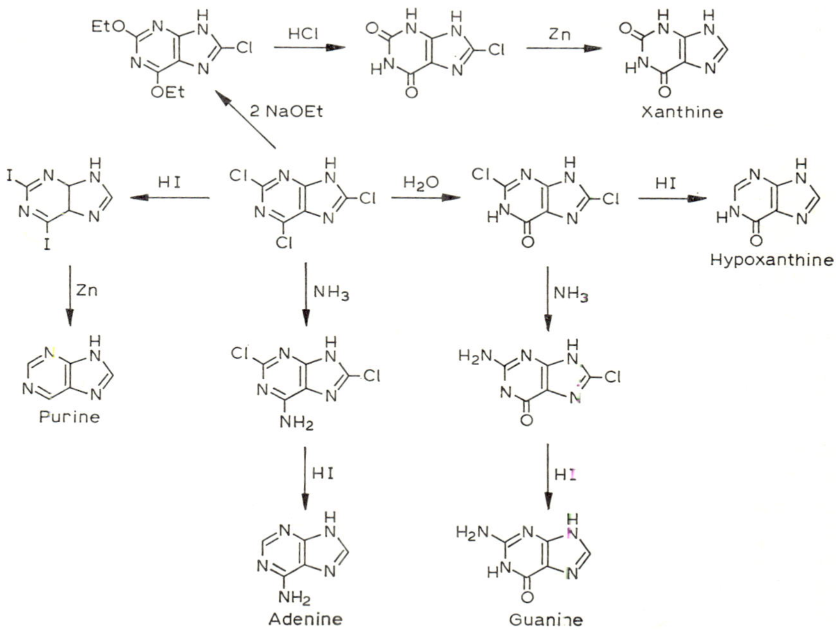

been prepared in this way from 8-bromocaffeine (*G. Bargioni*, Boll. Chem. Farm., 1935, **64,** 869). Incidentally it is of interest that the preparation of the latter compound from 1,3,7,9-tetramethyluric acid was the first occasion on which the uric acid and xanthine series of purines were inter-related (*Fischer*, Ber., 1897, **30,** 3010).

The reduction of halogenopurines has received considerable attention since it was used so successfully by *Fischer* for the preparation of many naturally occurring purines from 2,6,8-trichloropurine (Ber., 1897, **30,** 3226; 1898, **31,** 2250). The ways in which he used reductions with hydrogen iodide and zinc to prepare adenine, guanine, hypoxanthine, xanthine and purine are illustrated on p. 59.

Electrochemical reduction in acid solution converts 8-chlorotheophylline into theophylline (*Y. Yoshitoni*, J. pharm. Soc., Japan, 1924, **510,** 649), and caffeine chlorinated at 8 and on the N-7 methyl into deoxycaffeine when mercury electrodes are used and into caffeine and its 8-chloro derivative using lead electrodes (*idem, ibid.*, 1924, **512,** 839). The same dichlorocaffeine is also converted into 8-chlorocaffeine by hydrogenation in the presence of colloidal palladium in alkaline solution (*idem, ibid.*, 1924, **512,** 839). 8-Chlorocaffeine is reduced by hydrogen in the presence of palladium to caffeine (*K. W. Rosenmund* and *Z. Zetzsche*, Ber., 1918, **51,** 582) and 8-chloro-theophylline or -paraxanthine are reduced to the parent compounds in good yield by iodine in tetralin (*M. Göttler*, G.P. 576,604/1933; C.A., 1933, **27,** 5757). The reduction of halogenopurineglycosides by Mozingo nickel has been used for preparing purine glycosides including the natural ones (*J. Davoll, B. Lythgoe* and *A. R. Todd*, J. chem. Soc., 1946, 833; 1948, 967, 1685).

(*iii*) *Individual halogenopurines*

2-**Chloropurine,** m.p. 231°, was synthesized from 4,5-diamino-2-chloropyrimidine and triethyl orthoformate (*J. A. Montgomery*, J. Amer. chem. Soc., 1956, **78,** 1928) and its *7-methyl-*, m.p. 200–201°; *9-methyl-*, m.p. 135° and *9-phenyl* derivatives by reducing the corresponding trichloropurines with zinc (*E. Fischer*, Ber., 1898, **31,** 2552; 1900, **33,** 2279). *2-Iodo-7-methyl-*, m.p. 229°, and *2-iodo-9-methylpurine*, m.p. 171–172°, were prepared from the corresponding chloro analogues (*idem, ibid.*, 1898, **31,** 2551).

6-**Chloropurine,** m.p. 175–177°, obtained by the halogenation of hypoxanthine (*A. Bendich et al.*, J. Amer. chem. Soc., 1954, **76,** 6073) or synthetically from the appropriate diaminopyrimidine (*Montgomery, loc. cit.*) is converted by hydrogen iodide to 6-**iodopurine,** m.p. 167° (decomp.) (*G. B. Elion* and *G. H. Hitchings, ibid.*, 1956, **78,** 3508); it is stable in boiling water but is hydrolysed rapidly to hypoxanthine in the presence of either acid or alkali. It is stable to aqueous ammonia at 100°, but adenine is formed when the compound is heated to 150°, with ammonia in butanol. With thiourea in boiling ethanol 6-mercaptopurine is formed readily. 6-Chloro-2-phenylpurine (no m.p.) is obtained from 2-phenylhypoxanthine (*W. Traube*, Ber., 1904, **37,** 2268). *6-Chloro-9-methylpurine*, m.p. 140–141°, which has antitumour activity is formed from 5-amino-6-chloro-4-methylaminopyrimidine and triethyl orthoformate (*R. K. Robins* and *H. H. Lin*, J. Amer. chem. Soc., 1957, **79,** 490); *6-chloro-9-ethylpurine*, m.p. 81–84°, by a corresponding cyclisation using diethoxymethyl acetate (*Montgomery* and *C. Temple, ibid.*, 1957, **79,** 5238); *6-chloro-7-methylpurine*, m.p. 195–197°, is obtained by halogenating 7-methylhypoxanthine (*R. N. Prasad* and *Robins, ibid.*, p. 6401).

6-**Bromopurine,** m.p. 200° (decomp.), arises from the action of phosphoryl bromide on hypoxanthine (*Elion* and *Hitchings, loc. cit.*).

6-**Fluoropurine,** m.p. 125–126°, is obtained from 6-chloropurine with trimethylamine followed by heating the resulting quaternary salt with potassium hydrogen fluoride in ethanol at 50° (*J. Kiburis* and *J. H. Lister*, Chem. Comm., 1969, 381).

2-*Fluoro-9-methylpurine,* m.p. 151°; 8-*fluoro-9-methylpurine,* m.p. 111–112°, 6-*fluoro-9-methylpurine,* m.p. 135–136° and 2,6-*difluoro-7-methylpurine,* m.p. 154–161°, are prepared from the corresponding chloropurines and silver fluoride in hot toluene or xylene (*A. J. Beaman* and *Robins*, J. org. Chem., 1963, **28,** 2310).

Copper cyanide in boiling pyridine converts 6-iodopurine into 6-**cyanopurine,** m.p. 177–178°, λ_{max} 289 nm (ε = 7510) at pH 1 and 292 nm (ε = 6530) at pH 11 in water, from which were obtained the corresponding amide, thioamide, and alkyl imidate by reaction with hot aqueous sodium hydroxide, hydrogen sulphide in alcoholic ammonia and alcoholic hydrogen chloride, respectively; hot dilute sulphuric acid hydrolysed the nitrile to hypoxanthine (*L. B. MacKay* and *Hitchings, ibid.*, 1956, **78,** 3511).

The conversion of xanthine to 2,6-**dichloropurine,** m.p. 181° (decomp.) and 2,6-**dibromopurine** has been mentioned previously (p. 34). 2,6-**Diiodopurine,** m.p. 224°, is obtained from 2,6,8-trichloropurine (*vide infra*). Hydrazine reacts in a stepwise manner with 2,6-dichloropurine giving first 2-chloro-6-hydrazinopurine, whereas hydrolysis of the dichloro compound with acid gives xanthine, partial hydrolysis occurs under basic conditions to give 2-chlorohypoxanthine as would be expected (*Montgomery* and *L. B. Hollum*, J. Amer. chem. Soc., 1957, **79,** 2185).

2,6-*Dichloro-7-methylpurine,* m.p. 199–200°, is obtained from theobromine (see p. 47) (*cf. R. R. Adams* and *F. C. Whitmore*, J. Amer. chem. Soc., 1945, **67,** 1274; *Prasad* and *Robins, loc. cit.*) and the 9-*methyl* analogue, m.p. 153°, by chlorination of 9-methylxanthine (*J. M. Gulland* and *F. B. Storey*, J. chem. Soc., 1938, 692). The halogen atoms in these two compounds react in the same order (6 before 2) as in their parent; the first with aqueous ammonia gives 6-amino-2-chloro-7-methylpurine (*E. Fischer*, Ber., 1897, **30,** 2214; 1898, **31,** 117) and the second by consecutive treatments with aqueous alkali and aqueous ammonia is converted into 9-methylguanine (*Gulland* and *Sorey, loc. cit.*). *Adams* and *Whitmore* (*loc. cit.*) prepared in this way dialkylaminoalkyl derivatives from 7-methyl-purine.

6,8-**Dichloropurine,** m.p. 178° (decomp.), is obtained by heating 6,8-dihydroxypurine with phosphoryl chloride in diethylaniline (*Robins*, J. Amer. chem. Soc., 1958, **80,** 671).

2,6,8-**Trichloropurine,** m.p. 187–189° (decomp.), has received more attention than any other halogenopurine. It is acidic, giving aqueous solutions which turn litmus red and liberate carbon dioxide from sodium carbonate. This acidity is associated with its differing behaviour towards acid and alkali, already described (p. 55). Ethanolic potash gives 2,8-dichloro-6-ethoxypurine at room temperature and 8-chloro-2,6-diethoxypurine at 100°. With hydrogen iodide and phosphonium iodide at 0°, 2,6-diiodopurine is obtained (*vide supra*) (*E. Fischer*, Ber., 1898, **31,** 2561). Methylation of the compound with methyl iodide under basic conditions gives rise to a mixture of 2,6,8-*trichloro-7-methyl-*, m.p. 159–161°, and 2,6,8-*trichloro-9-methyl-purine,* m.p. 176° (*idem, ibid.*, 1897, **30,** 2224) and these are converted to the corresponding 8-ethoxy derivatives with ethanolic potash in contrast to the behaviour to trichloropurine itself. 2,6,8-*Trichloro-9-phenylpurine,* m.p. 210–211°, was prepared from 2,6-dichloro-8-hydroxy-9-phenylpurine (*idem, ibid.*, 1900, **33,** 2278).

(*iv*) *Chlorohydroxypurines and their N-substituted derivatives*

Boiling N-sodium hydroxide hydrolyses 2,6-dichloropurine to **2-chloro-6-hydroxypurine,** m.p. > 300° (*Montgomery* and *Holum, loc. cit.*) and hydrolysis of 2,6,8-trichloro-9-methylpurine affords **2-chloro-6,8-dihydroxy-9-methylpurine,** m.p. 320° (decomp.), which is converted by methyl iodide in alkali to a mixture of 7,9-*dimethyl-*, m.p. 312° (decomp.) and 1,7,9-*trimethyl-*, m.p. 258–259°, derivatives of 2-chloro-6,8-dihydroxypurine but to the 1,9-*dimethyl* analogue, m.p. 291°, by formaldehyde (*Fischer*, Ber., 1899, **32,** 251).

The 3-*methyl-*, m.p. > 300° (decomp.) and 3,7-*dimethyl-*, m.p. 290° (decomp.), derivatives of 6-*chloro-*2,8-*dihydroxypurine* were prepared by the action of phosphoryl chloride on 3-methyl- and 3,7-dimethyl-uric acid, respectively (*Fischer*, Ber., 1897, **30,** 1851; 1899, **32,** 2733) and similarly the 3-*methyl-*, m.p. 340–345° (decomp.), 7-*methyl-*, m.p. 340° (decomp.), 1,3-*dimethyl-*8-*chlorotheophylline*, m.p. 300° (decomp.), 1,7-*dimethyl-*, m.p.295° and 3,7-*dimethyl-*, m.p. 304° (decomp.), derivatives of 8-*chloro-*2,6-*dihydroxypurine* (*idem, ibid.*, 1898, **31,** 1988, 2622; 1899, **32,** 2741; 1906, **39,** 430; Chem. Ztbl., 1898, **II,** 1192). **8-Chloro-2,6-dihydroxypurine** (no m.p.), is formed by the acid hydrolysis of 8-chloro-2,6-diethoxypurine and affords 8-*chlorocaffeine*, m.p. 188°, on methylation (*Fischer*, Ber., 1897, **30,** 2236).

2,8-Dichloro-6-hydroxypurine, m.p. > 350°, is formed by hydrolysing trichloropurine with potash at 100°, whereas acid hydrolysis gives **2,6-dichloro-8-hydroxypurine,** m.p. 250°, also obtained by the action of phosphoryl chloride on potassium urate (*idem, ibid.*, 1897, **30,** 2223). Acid hydrolysis of 2,6,8-trichloro-7-methylpurine similarly affords 2,6-*dichloro-*8-*hydroxy-*7-*methylpurine*, m.p. 278°, which is formed also by halogenation of the corresponding methyluric acid and the 9-*methyl* analogue, m.p. 280–281°, is obtained by both these routes.

Methylation of 2,6-dichloro-8-hydroxypurine gives its 7,9-*dimethyl* derivative, m.p. 187–188°, and the 9-*phenyl* derivative, m.p. 323°, is obtained by acid hydrolysis of 9-phenyltrichloropurine (*idem, ibid.*, 1895, **28,** 2490; 1897, **30,** 2211, 2224; 1899, **32,** 271; 1900, **33,** 2279).

Properties, including light absorption data, of some halogenopurines are listed in Table 7.

(*f*) *Mercaptopurines*

(*i*) *Synthesis*

Mercaptopurines (purinethiols) are readily prepared by the action of metallic hydrosulphides in solution on halogenopurines (*E. Fischer*, Ber., 1898, **31,** 431). In some cases reaction occurs in the cold, trichloropurine giving 2,6,8-trimercaptopurine in this way but higher temperatures are usually required. Thus 2,6-dichloro-7,9-dimethyl-8-oxodihydropurine gives the 6-thiol with cold ethanolic hydrogen sulphide but to obtain the dithiol it is necessary to heat with aqueous potassium hydrosulphide at 130–140° for 3 hours (*P. C. Ray et al.*, J. chem. Soc., 1923, 1957). Halogeno groups are also often conveniently replaced by employing thiourea or thioacetamide (see p. 60). Aminopurines have been converted directly into mercapto-

TABLE 7

PROPERTIES OF HALOGENOPURINES

Substance	*m.p. (°C)*	*λ_{max} (nm) in aqueous solution (10^{-3} ε in brackets)*			*Ref.*
		Acidic	*Neutral*	*Basic*	
2-Chloro-	231	270.5 (8.07)	272 (7.99)	278.5 (8.15)	1
6-Chloro-	175–177	244 (8.95)	265.5 (9.53)	274 (8.72)	1, 2, 3
6-Bromo-	200 (decomp.)	266 (9.7)		275 (8.55)	3
6-Iodo-	167 (decomp.)	276 (10.4)		280 (9.3)	3
2,6-Dichloro-	181 (decomp.)	275 (8.9)	277 (7.86)	280 (8.5)	1, 3
6,8-Dichloro-	178 (decomp.)	271 (13.0)		278 (12.4)	4
2,6-Dibromo-		247 (3.9) 279 (8.0)		283 (7.75)	3
6-Chloro-9-methyl-	143–144	265 (8.1)		268 (21.1)	5
6-Chloro-9-ethyl-	81–84	265 (9.46)	266 (9.40)	266 (9.40)	6
6-Chloro-7-methyl-	198–199	268 (8.42)		271 (8.46)	7
2-Fluoro-		264 (8.3)	268 (8.4)	273 (8.8)	8
6-Chloro-2-fluoro-	216	270 (9.7)	272 (0.5)	275 (9.0)	8
2-Fluoro-6-methyl-	194	262 (9.9)	266 (9.8)	272 (10.4)	8
2-Fluoro-9-methyl-	151	261 (9.7)		207 (8.6)	9
8-Fluoro-9-methyl-purine	111–112	261 (6.4)		280 (9.3)	9

References

1 *J. A. Montgomery*, J. Amer. chem. Soc., 1956, **78,** 1929.
2 *S. F. Mason*, J. chem. Soc., 1954, 2073.
3 *G. B. Elion* and *G. H. Hitchings*, J. Amer. chem. Soc., 1956, **78,** 3508.
4 *R. K. Robins*, *ibid.*, 1958, **80,** 6671.
5 *Robins* and *H. H. Lin*, *ibid.*, 1957, **79,** 490.
6 *Montgomery* and *C. Temple*, *ibid.*, p. 5238.
7 *R. N. Prasad* and *Robins*, *ibid.*, p. 6401.
8 *Montgomery* and *K. Hewson*, J. Amer. chem. Soc., 1960, **82,** 463.
9 *A. J. Beaman* and *Robins*, J. org. Chem., 1963, **28,** 1310.

purines by reaction with liquid hydrogen sulphide (*T. Ueda, M. Imazawa* and *K. Miura*, Tetrahedron Letters, 1971, No. 27, 2507).

The direct conversion of hydroxypurines to mercaptopurines may be accomplished in many cases by the use of phosphorus pentasulphide, preferably in boiling pyridine, as for example with hypoxanthine (*G. B. Elion et al.*, J. Amer. chem. Soc., 1952, **74,** 411; *cf.* "Synthetic Procedures in Nucleic Acid Chemistry", Vol. I, eds. *W. W. Zorbach* and *R. S. Tipson*, Interscience, New York, 1968). Phosphorus trisulphide has also been used in this way to prepare 6-mercapto- and 6,8-dimercapto-caffeine (*A. M. Khaletskii* and *M. S. Eshman*, J. gen. Chem., U.S.S.R., 1948, **18,** 2116). In some early German work, 8-mercapto derivatives were prepared by heating theophylline and 3-methyluric acid with carbon disulphide and alkali at 150° (*C. F. Boehringer und Söhne*, G.P., 133,300/1901; Chem. Ztbl., 1902, II, 314). The 8-mercaptopurines are, however, more usually prepared by cyclising a 4,5-diaminopyrimidine with thiourea or an isothiocyanate; these and other general synthetic methods for mercaptopurines have been discussed previously (pp. 6, 10, 11, 16, 19, 21).

Miscellaneous syntheses include the direct introduction of sulphur; thus purine with sulphur at 245° gave a 75% yield of 8-mercaptopurine (*A. Giner-Sorolla, E. Thom* and *A. Bendich*, J. org. Chem., 1964, **29,** 3209).

8-Mercapto-1,9-dimethyl-, -7,9-dibenzyl- and -7,9-dimethyl-purines were produced similarly (*Z. Nieman*, Chem. Comm., 1968, 200) and 1,3,7-trimethylhypoxanthinium nitrate when heated with sulphur in pyridine gave the 2-mercapto derivative (*R. Walentowski* and *H. W. Wanzlick*, Chem. Ber., 1969, **102,** 3000).

(ii) Properties and reactions

The mercapto group on a purine ring may be replaced by hydrogen either by reaction with nitrous acid (*Fischer*, Ber., 1899, **32,** 486) or Raney nickel (*D. J. Brown*, J. Soc. chem. Ind., 1950, **69,** 353). Iodine in sodium bicarbonate solution, however, converts 8-mercapto-

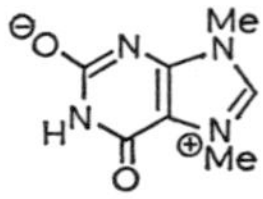

(XCI)

purines into 8-iodopurines (*H. Biltz* and *A. Beck*, J. pr. Chem., 1928, **118,** 149), whereas an 8-thione group is reduced to give the corresponding deoxypurine; 7,9-dimethyl-8-thiouric acid thus affords "7,9-dimethyldeoxyuric acid", $C_7H_8O_2N_4$ (XCI) (*H. Bredereck et al.*, Ber., 1959, **92,** 566; *Biltz* and *R. Bulow*, Ann., 1922, **426,** 250).

S. F. Mason has suggested that mercaptopurines exist in the thione form (Chem. and Biol. of Purines, p. 70). Alkaline hydrogen peroxide converts a mercaptopurine into the

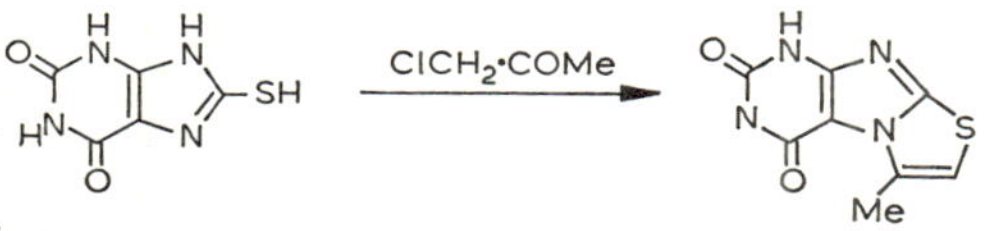

corresponding hydroxypurine and paraxanthine has been obtained from the corresponding 2-mercapto compound in this way (*A. H. Cook* and *G. H. Thomas*, J. chem. Soc., 1950, 1885). Since the mercaptopurines are normally solids of high melting point it is common to characterise them by methylation to the methylthio derivatives. This reaction proceeds readily with dimethyl sulphate (*Biltz* and *Bulow*, Ann., 1922, **426**, 306) or methyl iodide in basic solution. The methylthio derivatives are reduced easily to the purines by Raney nickel (*A. R. Todd et al.*, J. chem. Soc., 1945, 556). Alkylation of 8-mercaptopurines with an α-chloroketone gives a thiazolopurine (*Todd* and *F. Bergel*, *ibid.*, 1936, 1539).

Direct replacement of a thiol group by ammonia or amines is normally difficult but ammonia converts 3-methyl-6-mercaptopurine to 3-methyladenine and 3-methyl-2,6-dimercaptopurine gives 3-methyl-2-mercaptoadenine (*Elion*, J. org. Chem., 1962, **27**, 2478). Reaction of 6-methylsulphonylpurine with ammoniacal hydrogen sulphide gives 6-mercaptopurine (*F. Bergmann*, *M. Rahat* and *I. Tamur*, J. chem. Soc., Perkin I, 1974, 470).

In contrast, alkylthiopurines react somewhat more readily with ammonia or amines but even so forcing conditions are normally required. Thus whereas 2-methylthiopurine is unreactive with an aqueous alkylamine at 140°, 2-methylthiohypoxanthine gave the 2-alkylamino derivative (*Elion*, *W. Lange* and *G. H. Hitchings*, J. Amer. chem. Soc., 1956, **78**, 217) and 3-methyladenine was produced from 3-methyl-6-methylthiopurine (*J. W. Jones* and *R. K. Robins*, *ibid.*, 1962, **84**, 1914) with methanolic ammonia at 100°. However, 8-methylthiopurine requires prolonged heating with ammonia solution at 160° in the presence of copper acetate and copper bronze to produce 8-aminopurine (*A. Albert* and *Brown*, J. chem. Soc., 1954, 2060).

(*iii*) *Individual purinethiols*

2-Mercaptopurine, m.p. ~250° (decomp.), prepared by Traube's method, gives on methylation with methyl iodide in aqueous sodium hydroxide, *2-methylthiopurine*, m.p. 250–255° (decomp.), which is unusually stable to aqueous methylamine (and sodium hydroxide) at 100° but is hydrolysed by this and dimethylamine at 150° to 4,5-diamino-2-methylthiopyrimidine (*Albert* and *Brown*, *loc. cit.*). Traube's original preparation of 6-**hydroxy-2-mercaptopurine**, m.p. 325–340° (decomp.), has been improved by *Elion et al.* (J. Amer. chem. Soc., 1956, **78**, 217; *cf. Robins et al.*, *ibid.*, 1953, 263) who used its *2-methylthio* derivative (m.p. >300°) to prepare various 2-substituted aminohypoxanthines. *6-Hydroxy-2-mercapto-9-methylpurine*, m.p. 315°, is formed when 5-amino-1-methylimidazole-4-carboxamide is heated with carbon disulphide in alkali (*Cook* and *E. Smith*, J. chem. Soc., 1949, 2333), and heating 4-amino-1-methylimidazole-5-carboxamide with thiourea at 150° gives *6-hydroxy-2-mercapto-7-methylpurine*, m.p. >360°, also prepared from the corresponding 2-chloropurine and thiourea in boiling ethanol (*R. N. Prasad* and *Robins*, J. Amer. chem. Soc., 1957, **79**, 6401).

6-**Amino-2-mercaptopurine** (**2-mercaptoadenine,** no m.p.) was made by *Traube* (Ann., 1904, **331,** 77; *cf. Bendich et al.*, J. Amer. chem. Soc., 1948, **60,** 3109); a formamide cyclisation was used by *Robins et al.* to prepare both this and 2-*methylthioadenine*, m.p. 294–295° (*cf. J. Baddiley et al.*, J. chem. Soc., 1943, 308); the latter is oxidised by aqueous chlorine to the *sulphoxide*, m.p. >350° (*K. J. M. Andrews et al.*, *ibid.*, 1949, 2495). 2-**Mercaptocaffeine,** m.p. 205°, which has a bitter taste and is a diuretic, is formed by the cyclisation of caffeidine with thiophosgene (p. 49) (*Biltz*, Ber., 1928, **61,** 1418).

6-**Mercaptopurine,** m.p. 313–314° (decomp.), is formed by heating hypoxanthine with phosphorus pentasulphide in boiling pyridine and material labelled at position 8 with ^{14}C and with ^{25}S has been made by synthesis from 4,5-diamino-6-mercaptopyrimidine (*Elion et al.*, J. Amer. chem. Soc., 1952, **74,** 411; *ibid.*, 1956, **76,** 4027); 6-*methylthio-*, m.p. 218–220° (decomp.), and 6-*benzylthiopurine*, m.p. 188–189°, were derived by alkylation and the latter compound also by synthesis.

6-Mercaptopurine has found considerable clinical application (synonyms **Puri-Nethol** and **Leukerin**) in the treatment of leukaemia, especially in children, and a side-effect which involves inhibition of the normal immune response has been of value in human organ transplant procedures including kidneys. The drug is thought to interfere with purine nucleotide *de novo* biosynthesis (for a general review see *C. C. Stock*, "Experimental Chemotherapy", eds. *Schnitzer* and *Hawking*, Academic Press, New York, 1966, Vol. 4).

A large number of 6-mercaptopurine ethers have also been prepared and tested for their kinetin-like activity (*C. G. Skinner et al.*, Arch. Biochem. Biophys., 1956, **65,** 569; *B. S. Gorton et al.*, *ibid.*, 1957, **66,** 493). Acid hydrolysis of the 2-ethoxy derivative affords 6-*mercapto-7-methylpurine*, m.p. 308–310° (*Fischer*, Ber., 1898, **31,** 431) which is formed also by thiation of 7-methylhypoxanthine and of 6-chloro-7-methylpurine and gives 6-*ethylthio-7-methylpurine*, m.p. 158.5°, on alkylation (*Prasad* and *Robins*, *loc. cit.*). 6-*Mercapto-9-methylpurine* has m.p. >300°, *methylthio* derivative, m.p. 171–172° (*Robins* and *H. H. Lin*, J. Amer. chem. Soc., 1957, **79,** 490). 9-*Ethyl-6-mercaptopurine* has m.p. 333–337° (decomp.), *methylthio* derivative, m.p. 116–118° (*J. A. Montgomery* and *C. Temple*, *ibid.*, 1957, **79,** 5238).

Reaction of 6,8-dichloropurine with thiourea, potassium hydrosulphide or methanethiol gives 8-chloro-6-mercapto- (no m.p.) or 8-*chloro-6-methylthio-purine*, m.p. 194–196° (decomp.) (*Robins*, *ibid.*, 1958, **80,** 6671). Phosphorus pentasulphide and xanthine afford **2-hydroxy-6-mercaptopurine,** m.p. >325° (*A. G. Beaman*, *ibid.*, 1954, **76,** 5633); the 7-*methyl* derivative, m.p. 343–344°, was prepared similarly (*Prasad* and *Robins*, *loc. cit.*). This same reaction converts 6,8-dihydroxypurine to 8-**hydroxy-6-mercaptopurine** (no m.p.), also synthesized by cyclisation with urea of the diaminopyrimidine (*Robins*, *loc. cit.*).

2-**Amino-6-mercaptopurine (thioguanine),** m.p. >360°, *methylthio* derivative, m.p. 239.5–240°, is an inhibitor of bacterial growth and for this reason several substituted-amino derivatives and its 7-*methyl* derivative, m.p. >300°, have been prepared (*Elion et al.*, J. Amer. chem. Soc., 1955, **77,** 1676; 1956, **78,** 217; 1959, **81,** 1898; Arch. Biochem. Biophys., 1957, **71,** 358; *Montgomery* and *L. B. Hobson*, J. Amer. chem. Soc., 1957, **79,** 2185; *Prasad* and *Robins*, *loc. cit.*). By the only recorded use of free guanidine as a cyclising agent in this field, *Robins* (*loc. cit.*) prepared 8-**amino-6-mercaptopurine** (no m.p.) from 4,5-diamino-6-mercaptopyrimidine. 6-**Thiouric acid** (2,8-**dihydroxy-6-mercaptopurine**), m.p. >300°, a metabolic product of 6-mercaptopurine is obtained by reaction of uric acid with phosphorus pentasulphide and by synthesis (*L. T. Loo et al.*, J. Amer. chem. Soc., 1959, **81,** 3039; *Elion et al.*, *ibid.*, 1959, **81,** 3042).

Sodium hydroselenide or selenourea convert 6-chloropurine into 6-**selenopurine** (*H. G. Mautner*, *ibid.*, 1956, **78,** 5292), m.p. 281–281.5°, which gives a *methylseleno* derivative, m.p. 193–194°. It is of interest that the pK_a and wavelength of maximal ultraviolet light absorption increase along the series 6-hydroxy-, 6-mercapto-, 6-seleno-purine (*cf.* Table 8). 8-**Mercaptopurine,** m.p. 300–310° (decomp.) (*methylthio* derivative, m.p. 257–259°) was prepared from thiourea and 4,5-diaminopyrimidine (*O. Isay*, Ber., 1906, **39,** 250; *A. Albert* and *D. J. Brown*, J. chem. Soc., 1954, 2070), its 6-*methyl* derivative, m.p. ~340° (decomp.), similarly (*S. Gabriel* and *J. Colman*, Ber., 1901, **34,** 1248) and the 7-*methyl* derivative, m.p. 248–249°, by reacting trichloro-7-methylpurine with aqueous potassium hydrosulphide at 0° and reducing the product with phosphonium iodide (*Fischer*, Ber., 1898, **31,** 432). Reaction of the requisite diaminopyrimidine with thiourea afforded the yellow 6-*ethyl*-8-*mercaptopurine*, m.p. >300° (*Todd et al.*, J. chem. Soc., 1936, 1559; 6-**hydroxy-8-mercaptopurine** (no m.p.), *methylthio* derivative, m.p. 220–222° (decomp.); 6-**amino-8-mercaptopurine** (no m.p.), *methylthio* derivative, m.p. 288–290° (decomp.) (*Robins*, *loc. cit.*); 6-*hydroxy*-8-*mercapto*-2-*methylpurine*, m.p. >300°, *methylthio* derivative, m.p. >310°, 6-*chloro*-8-*methylthiopurine* m.p. 231° (*F. E. King* and *T. J. King*, J. chem. Soc., 1947, 943) and 2-**hydroxy-8-mercaptopurine** (no m.p.) (*C. O. Johns*, J. biol. Chem., 1916, **21,** 321). Carbon disulphide in aqueous potassium hydroxide at 150° converted 3-methyluric acid and 1,3-dimethyluric acid into their 8-*mercapto* analogues, m.p. ~340° (decomp.), and ~320° (decomp.), respectively (*C. F. Boehringer und Söhne*, *loc. cit.*) and treatment of the corresponding 8-bromo derivatives with potassium hydrosulphide afforded 8-**mercaptoxanthine** (no m.p.) (*C. O. Johns*, Ber., 1898, **31,** 445) and 8-**mercaptocaffeine,** m.p. 316° (decomp.) (*Fischer*, *ibid.*, 1899, **32,** 485).

Refluxing 6-hydroxy-2-mercaptopurine with phosphorus pentasulphide (*Beaman*, J. Amer. chem. Soc., 1945, **76,** 5633) and a Traube synthesis (*K. L. Dille* and *B. E. Christensen*, *ibid.*, p. 5088) have both been used to prepare 2,6-**dimercaptopurine,** m.p. >330°; the *di*(*methylthio*) derivative, m.p. 253–254°, was obtained by cyclising 4,5-diamino-2,6-di-(methylthio)pyrimidine sulphate with formamide (*Dille* and *Christensen*, *loc. cit.*).

6,8-**Dimercaptopurine** (no m.p.), *di*(*methylthio*) derivative, m.p. 257–258°, resulted from the reaction between 6,8-dichloropurine and 2 moles of thiourea; its 6- and 8-monomethylthio analogues (no m.p.) have been prepared by a series of reactions from 2,8-dichloropurine (*Robins*, *ibid.*, 1958, **80,** 6671).

Potassium hydrosulphide at 100° or thiourea in boiling ethanol converts 2,6-dichloro-7-methylpurine into 2,6-*dimercapto*-7-*methylpurine*, m.p. >360° (decomp.), the *di*(*methylthio*) derivative has m.p. 176–178° (*Fischer*, Ber., 1898, **31,** 439; *Prasad* and *Robins*, *loc. cit.*). *Johns* (*loc. cit.*) cyclised the appropriate diaminopyrimidine with thiourea to prepare **6-hydroxy-2,8-dimercaptopurine** (no m.p.) and various 6-amino and 6-alkylamino derivatives of 2,8-dimercaptopurines have been synthesised by *Cook* and *Smith* (J. chem. Soc., 1949, 2329, 3001) and by *B. R. Baker et al.* (J. Amer. chem. Soc., 1954, **76,** 631).

Trimercaptopurine and its 7-methyl derivative (neither has a m.p.) were prepared from the corresponding trichloropurine by *Fischer* (Ber., 1898, **31,** 443).

Absorption properties of mercaptopurines and their alkylthio analogues are recorded in Table 8.

TABLE 8

ULTRAVIOLET ABSORPTION OF MERCAPTOPURINES

Purine derivative	λ_{max} *(nm) in aqueous solution (10^{-3} ε in brackets)*			*Ref.*
	Acid	*Neutral*	*Basic*	
2-Mercapto-	227–232 (7.41) 287 (18.62) 382 (1.82)	241 (12.59) 285 (17.78) 346 (1.51)	235 (13.18) 263 (15.49) 328 (3.09)	1
6-Mercapto-	225 (6.61) 324 (14.79)	225 (7.41) 325 (18.62)	228 (9.55) 312 (14.45)	1
8-Mercapto-	238 (14.79) 280[a] (4.07) 331 (18.20)	231 (10.23) 310 (28.84)	228 (13.49) 313 (23.44)	1
2-Methylthio-	241 (13.49) 250[a] (12.30) 314 (4.37)	232 (16.60) 250[a] (8.51) 305 (6.03)	240 (19.05) 301 (6.17)	1
6-Methylthio-	222 (12.02) 313 (12.88)	255[a] (4.57) 290 (22.39)	222 (18.62) 290 (20.42)	1
8-Methylthio-	232 (10.96) 305 (20.89)	246 (3.89) 290 (19.95)	220 (16.98) 296 (18.62)	1
6,8-Dimercapto-	270 (11.9) 235 (27.8)		238 (10.5) 259 (11.2) 334 (23.0)	2
2,6-Dimercapto-			346 (13.49)	3
6,8-Di(methylthio)-	252 (12.1) 335 (25.4)		240 (18.0) 312 (25.7)	2
6-Mercapto-8-methylthio-	259 (18.0) 342 (33.3)		251 (19.6) 328 (27.7)	2
8-Mercapto-6-methylthio-	253 (18.2) 326 (36.2)		248 (15.4) 326 (28.7)	2
6-Mercapto-9-methyl-	324 (19.6)		310 (16.8)	4
9-Methyl-6-methylthio-	222 (12.0) 295 (19.5)		287 (20.8)	4
8-Hydroxy 6-mercapto-	238 (13.3) 332 (20.6)		235 (24.0) 311 (24.2)	2
6-Hydroxy-8-mercapto-	234 (8.7) 290 (27.2)		234 (17.6) 290 (18.8)	2

TABLE 8 *(continued)*

Purine derivative	λ_{max} *(nm) in aqueous solution* *(10^{-3} ε in brackets)* Acid		*Neutral*	*Basic*	*Ref.*
2-Hydroxy-6-mercapto-				340 (21.38)	3
6-Hydroxy-2-mercapto-	285 265	(21.6) (16.4)		278 (16.80)	5
6-Amino-2-mercapto-			230 (9.64) 285 (12.60)		6
2-Amino-6-mercapto-	258 347	(8.1) (20.9)		242 (8.7) 270 (7.2) 322 (16.0)	5
6-Amino-8-mercapto-	242 310	(12.0) (24.7)		229 (18.4) 301 (23.4)	2
8-Amino-6-mercapto-	238 332	(17.9) (25.4)		312 (26.9) 240 (21.0)	2
6-Seleno-			361[a] (14.55)		7
6-Methylseleno-purine			226[a] (9.44) 300[a] (16.9)		7

[a] In ethanol.

References

1 *S. F. Mason*, J. chem. Soc., 1954, 2071.
2 *R. K. Robins*, J. Amer. chem. Soc., 1958, **80**, 6671.
3 *A. G. Beaman*, *ibid.*, 1954, **76**, 5633.
4 *J. A. Montgomery* and *C. Temple*, *ibid.*, 1957, **79**, 5238.
5 *G. B. Elion et al.*, *ibid.*, 1952, **74**, 411; 1955, **77**, 1676; 1956, **78**, 217.
6 *A. Bendich et al.*, *ibid.*, 1948, **70**, 3109.
7 *H. G. Mautner*, *ibid.*, 1956, **78**, 5292.

(*g*) *Aminopurines*

The commonest naturally occurring purine derivatives are adenine (6-aminopurine) and guanine (2-amino-6-hydroxypurine) which are found in all living matter as components of the nucleic acids.

They also occur widely as adenosine and guanosine mono-, di- and tri-phosphates and as adenosine cyclic (3-, 5-)phosphates, which play vital roles in many metabolic processes, and adenine is a constituent of several

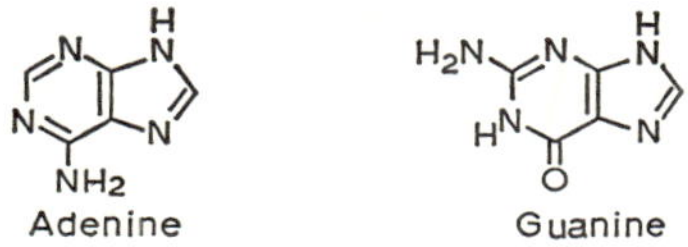

nucleoside antibiotics including cordycepin and nucleocidin (see Table 1, p. 2) (*cf. R. J. Suhadolnik*, Nucleoside Antibiotics, Wiley-Interscience, London, New York, 1970).

Puromycin and yeast nucleic acid contain 6-dimethylaminopurine and in this nucleic acid are found also 6-methylaminopurine and the 2-*N*-methyl and 1-methyl derivatives of guanine. Some deoxyribonucleic acids contain aminopurines other than these and the 5,6-dimethylbenzimidazole residue of Vitamin B_{12} is replaced by adenine or another aminopurine in some micro-organisms. Furthermore, 6-succinaminopurine can be a constituent of bacterial cell walls, and 6-*N*-isopentenyl- and 4-hydroxyisopentenyl-adenine (zeatin) are not only constituents of certain t-RNAs but also occur naturally in plants where they function as cell growth-promoting agents (cytokinins).

Adenine and guanine are important and even essential growth factors for some micro-organisms and synthetic 2,6-diaminopurine is active against types of leukaemia and sarcomas. The ubiquity of aminopurines in nature and the wide variety of biological action they exhibit has stimulated greatly their chemical study.

(*i*) *Methods of preparation*

Aminopurines are readily prepared either by direct synthesis using methods already described or by replacement of other substituents. The amination of a halogenopurine is commonly employed and an alternative is the replacement of a methylthio group. Other methods, such as the preparation of 8-aminopurines by coupling the appropriate xanthine with a diazonium compound and reducing the resultant azo compound, are less frequently used.

Illustrative examples are afforded by the methods which have been used for preparing guanine and adenine. The relationship of guanine to uric acid was demonstrated (*E. Fischer*, Ber., 1897, **30**, 2251) by converting the latter into 2,8-dichloro-6-hydroxypurine which was aminated in the 2 position and the product reduced by hydrogen iodide to guanine. The first direct synthesis of guanine was from 2,4,5-triamino-6-hydroxypyrimidine and formic acid (*W. Traube, ibid.*, 1900, **33**, 1371) and this method has been used to prepare isotopically labelled guanine (*F. Weygand* and *O. F. Gresskinsky, ibid.*, 1951, **84**, 839). Adenine was prepared by amination of trichloropurine to 6-amino-2,8-dichloropurine followed by reduction with hydriodic acid (*Fischer, ibid.*, 1897, **30**, 2226, 2238) and *Traube*, Ann., 1904, **331**, 69, 86) effected his synthesis by desulphurisation with peroxide of the 2-mercaptoadenine formed by cyclisation of 4,5,6-triamino-2-thiopyrimidine with formic acid. In a modification of the latter method, 4,5,6-triaminopyrimidine-2-sulphinic acid is cyclised directly to adenine with formic acid (*M. Hoffer*, C.A., 1947, **41**, 4108). The convenient synthssis of adenine by heating 5-thioformamido-4,6-diaminopyrimidine in water (*J. Baddiley, B. Lythgoe* and *A. R. Todd*, J. chem. Soc., 1943, 387) has received attention as a route to isotopically labelled adenine and useful modifications of the method have been found (*e.g. L. F. Cavalieri et al.*, J. Amer. chem. Soc., 1949, **71**, 533). The preparation of 8-aminopurines is exemplified by that of 8-aminocaffeine from 8-chlorocaffeine (*Fischer*, Ann., 1882, **215**, 265) and by reducing 8-benzeneazocaffeine. The aminolysis of 2,6-dichloro- and 2,6,8-trichloro-purine by various bases, which indicated stepwise reaction with preference for replacement at the 6 position, has been described (*B. E. Christensen et al.*, J. Amer. chem. Soc., 1959, **81**, 3789).

Chloro or methylthio groups in the 2-position of the purine ring also react with amines and with hydrazine (*J. A. Montgomery* and *L. B. Holum*, J. Amer. chem. Soc., 1957, **79**, 2185). The formation of 2-amino-6-imino-1-methyldihydropurine by heating XCII with ammonia shows that the 6-

(XCII)

thione group is replaceable by amines (*G. B. Elion*, Chem. and Biol. of Purines, p. 43), although the 6-mercapto group is said not to be and the 6-methylthio group with some difficulty (*Montgomery*, J. Amer. chem. Soc., 1956, **78**, 1928).

8-Aminopurines have usually been obtained from the 8-chloropurines but 8-aminopurine and the corresponding methylamino and dimethylamino derivatives have been obtained from the 8-methylthiopurine (*A. Albert* and *D. J. Brown*, J. chem. Soc., 1954, 2060). The fusion of guanidine with 4,5-diamino-6-hydroxypyrimidine at 200° gives 8-amino-6-hydroxypurine in good yield (*R. K. Robins*, J. Amer. chem. Soc., 1958, **80**, 6672). The direct conversion of oxopurines to dialkylaminopurines also has been achieved by heating the oxo derivative, *e.g.* hypoxanthine, with an *N,N′,N″*-trialkylphosphoramide (*Aroutyanyan et al.*, Izvest. Akad. Nauk S.S.S.R., Ser. Khim., 1969, 655).

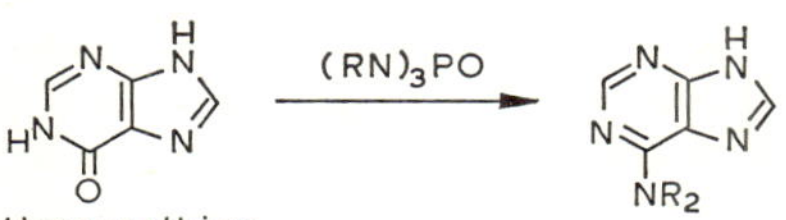

Hypoxanthine

1-, 3-, 7- and 9-*N*-Aminopurines have also been prepared. Thus hydrazine with 5-cyano-4-ethoxymethyleneamino-1-methylimidazole gave 1-amino-1,6-dihydro-6-imino-7-methylpurine (*E. C. Taylor* and *P. K. Loeffler*, J. Amer. chem. Soc., 1960, **82,** 3147) and substituted 3-aminopurines result

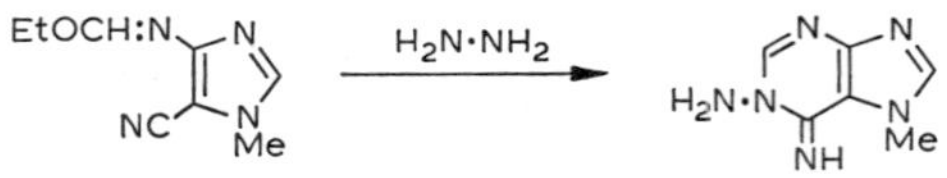

from cyclisation of 3-substituted amino-4,5-diaminopyrimidines with triethyl orthoformate (*H. G. Kazmirowski, G. Dietz* and *E. Carstens*, J. pr. Chem., 1963, **19,** 162).

9-Aminopurine may be prepared from 5-amino-4-hydrazinopyrimidine and formic acid (*Montgomery* and *C. Temple*, J. Amer. chem. Soc., 1960, **82,** 4592) whereas cyclisation of 1,5-diamino-4-carbamoyl-2-methylimidazole with triethyl orthoformate–acetic anhydride gave 9-amino-8-methylhypoxanthine (*R. Naylor et al.*, J. chem. Soc., 1961, 4845). An interesting reaction involves the conversion of 6-chloropurine 3-oxide to 9-cyanoaminopurine by heating at 85° for 30 min with cyanamide (*A. Giner-Sorolla et al.*, J. org. Chem., 1969, **34,** 2157) and equally interesting is the formation of 6,7-diamino-7,8-dihydro-8-oxopurine-9-β-D-riboside by amination of 6-amino-7,8-dihydro-8-oxopurine riboside with hydroxylamine-*O*-sulphonic acid in sodium hydroxide solution. 1-Aminopurines are similarly formed from inosine and guanosine (*A. D. Broom* and *Robins*, *ibid.*, 1969, **34,** 1025). Amination of adenine failed with hydroxylamine-*O*-sulphonic acid but with mesitylenesulphonylhydroxylamine produced 1-aminoadenine (*D. F. Wiener* and *N. J. Leonard*, J. org. Chem., 1974, **39,** 3438).

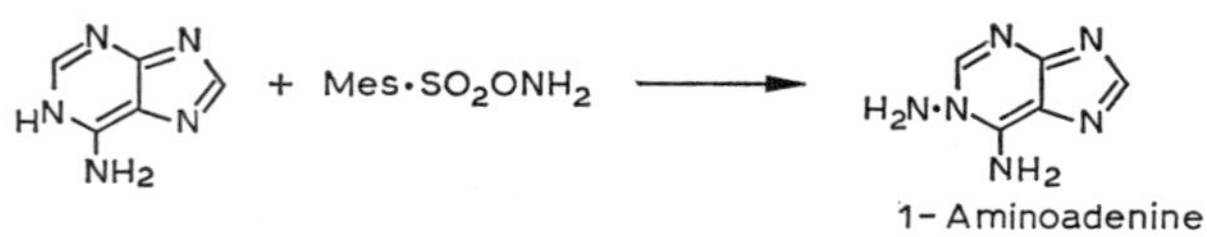

The "abiogenic" formation of aminopurines, especially adenine from simple acyclic precursors under so-called abiogenic conditions has been discussed earlier (p. 22).

(*ii*) *Properties*

The aminopurines are mostly solids of high melting points, sparingly soluble in water. They are considered to exist in the amino rather than the imino form (*S. F. Mason*, Chem. and Biol. of Purines, p. 71; *J. H. Lister*, Adv. in heterocycl. Chem., 1966, **6,** 1). However, 1-methyladenine appears to exist largely in the imino form XCIII, whereas 3-methyladenine exists largely in the amino form XCIV (*B. Pullman, H. Berthod* and *M. Dreyfus*, Theor. Chim. Acta, 1969, **15,** 265).

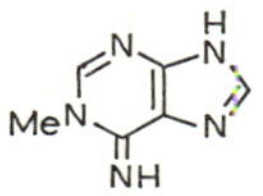
(XCIII)

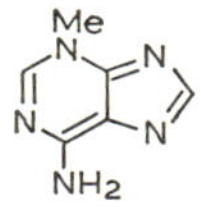
(XCIV)

The amino groups have normal reactions; although treatment with nitrous acid usually affords hydroxypurines, 8-aminocaffeine was shown by *M. Gomberg* to form a diazonium salt (Amer. chem. J., 1900, **23,** 51).

(*iii*) *Individual aminopurines*

2-Aminopurine (isoadenine), no m.p., has been obtained by the oxidation of deoxyguanine (p. 83) (*B. Tafel* and *L. Ach*, Ber., 1901, **34,** 1175) and was synthesized by *O. Isay* (*ibid.*, 1906, **39,** 250) from 2,4,5-triaminopyrimidine and formic acid. *Albert* and *Brown* (J. chem. Soc., 1954, 2068) showed that this condensation is improved by being conducted in boiling 4-formylmorpholine. 2-*Dimethylaminopurine*, m.p. 222–223°, is prepared by a similar route. 2-Aminopurine is more soluble in water than is usual in this series (1 part in 3.5 at 100°) (*idem, loc. cit.*); it dissolves also in alcohol, acids and alkali, reduces permanganate and gives 2-hydroxypurine with nitrous acid. 2-*Hydrazinopurine*, m.p. >300°, from 2-chloropurine and hydrazine (*Montgomery* and *Holum*, *loc. cit.*).

6-Aminopurine, adenine, m.p. 360–365° (decomp.), may be obtained as its picrate from acid hydrolysates of nucleic acid (*H. Bredereck* and *G. Richter*, Ber., 1938, **71,** 718) but the previously described synthetic methods are simple and convenient. From water (1 part dissolves in 1100 parts at 20° and 40 at 100°); it crystallises as the trihydrate; it forms readily crystallisable salts with mineral acids, picric and chloroauric acids. It is readily soluble in alkali, forms a mono-silver salt with ammoniacal silver nitrate and gives a red

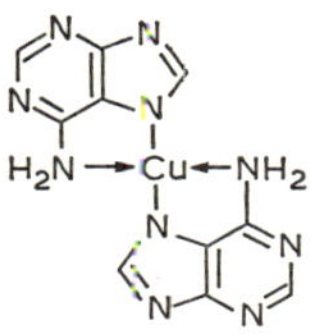
(XCV)

colour with ferric chloride solution. It forms metal complexes, *e.g.* the copper complex XCV (*T. R. Harkins* and *H. Freiser*, J. Amer. chem. Soc., 1958, **80,** 1132), and a crystalline chloromercuri complex with covalent attachment to N-9 (*H. J. Thomas* and *Montgomery*, J. org. Chem., 1966, **31,** 1413).

Adenine gives no murexide reactions. Although it is comparatively stable to alkali and acids, 6-*N*-hydrochloric acid at 150° converts it into 4-aminoimidazole-5-carboxamide and concentrated hydrochloric acid at 180° for 18 hours causes glycine to be formed from the 4, 5 and 7 positions of the purine molecule (*Cavalieri*, *J. F. Tinker* and *G. B. Brown*, J. Amer. chem. Soc., 1949, **71,** 3973). Oxidation with acidic permanganate gives urea, carbon dioxide and ammonia; formic acid is produced when adenine is hydrolysed by hot sulphuric acid (*H. Biltz* and *K. Sedlatschek*, Ber., 1924, **57,** 175), and bromination of adenine produces 8-bromoadenine (*G. Bruhns*, *ibid.*, 1890, **23,** 225; *Bruhns* and *Kossel*, Z. physiol. Chem., 1892, **16,** 1; *M. Kruger*, *ibid.*, p. 329).

Adenine is readily reduced by zinc and acid to give a purple solution (*E. Fischer*, Ber., 1897, **30,** 2241; *D. Hamer et al.*, Arch. Biochem. Biophys., 1953, **47,** 272) and the coupling of this reduction product has been used for its determination (*D. L. Woodhouse*, Arch. Biochem., 1950, **25,** 347); adenine itself also couples with diazobenzenesulphonic acid in alkali (*R. Burian*, Z. physiol. Chem., 1907, **51,** 425). Nitrous acid (*Kruger*, *ibid.*, 1894, **18,** 445) and adenase (*cf. C. Schmidt*, *ibid.*, 1933, **219,** 191) convert adenine into hypoxanthine and the rat or xanthine oxidase will convert it into 2,8-dihydroxyadenine (*A. Bendich et al.*, J. biol. Chem., 1950, **183,** 267; *H. Klenkow*, Biochem. J., 1952, **50,** 404). Treatment with boiling acetic anhydride gives a diacetyl derivative presumably containing one acetyl group attached to a ring nitrogen since boiling with water gives 6-acetamidopurine (*L. Birkofer*, Ber., 1943, **76,** 769). The last compound is reduced by lithium tetrahydridoaluminate to 6-*ethylaminopurine*, m.p. 235–236°; other *N*-acyl derivatives are similarly reduced (*H. Lettre* and *H. Ballweg*, Chem. Ber., 1958, **91,** 345). 3- and 9-Substituted adenines have been reduced with sodium tetrahydridoborate in acid solution to produce dihydroadenines. Thus 9-substituted *N*-benzoyladenine gave the 7,8-dihydro derivative whereas the 3-benzyl-*N*-benzoyladenine produced the 1,2-dihydroadenine (*Y. Maki*, *M. Suzuki* and *K. Ozeki*, Tetrahedron Letters, 1976, No. 15, 1199). Hydrogenation of 2,8-diaminopurine with platinic oxide resulted in ring-opening to produce imidazole derivatives (*M. M. Wegner* and *H. Rapoport*, J. org. Chem., 1977, **42,** 3065).

Alkylation of adenine depends largely on the presence or absence of base and also on the solvent. In neutral aqueous solution or in dimethylformamide or dimethylacetamide the major product is the 3-alkyladenine together with smaller amounts of the 1- and 9-isomers (*B. C. Pal*, Biochemistry, 1962, **1,** 558). Using alkyl bromides in dimethylacetamide *N. J. Leonard* and *T. Fujii* (J. Amer. chem. Soc., 1963, **85,** 3719) obtained alkyladenines in yields of the order 3-alkyl- (55–66%), 9-alkyl- (10–14%) and 1-alkyladenines (7–13%).

In aqueous alkaline solution or in aprotic media with contained base, alkylation of adenine produces the 9-alkyl derivative as the major product; thus both 9-benzyl- and 9-allyl-adenine resulted from treating the alkyl (aralkyl) halide with adenine in dimethylacetamide containing potassium carbonate (*Montgomery* and *Thomas*, J. heterocycl. Chem., 1964, **1,** 115; J. org. Chem., 1965, **30,** 3235). Further alkylation of an already substituted adenine can also be achieved usually by slightly more forcing conditions or use of a base. Thus 3-methyladenine with methyl iodide in methanolic potassium hydroxide in the cold gave 3,7-dimethyladenine and the same compound was also produced from 7-methyladenine by methylation in dimethylformamide at 100° (*Broom et al.*, Biochemistry, 1964, **3,** 494) whereas benzylation of 3-benzyladenine in the presence of sodium carbonate

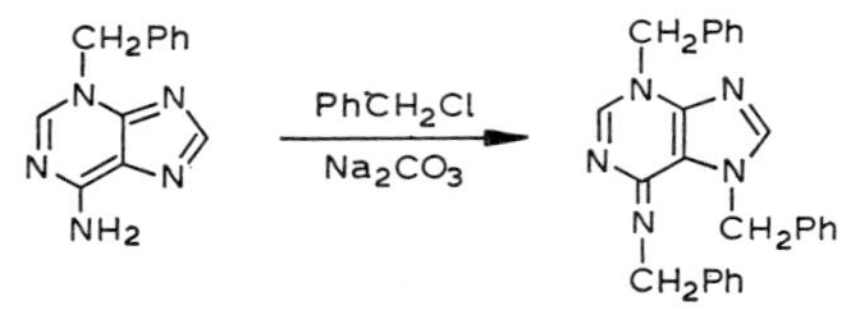

gave the tribenzyl derivative (*Montgomery* and *Thomas*, *loc. cit.*) formed presumably *via* a Dimroth rearrangement of an intermediate quaternary salt.

Further alkylation of a 9- or 1-alkyladenine produces the 1,9-dialkyl derivative (*idem*, *loc. cit.*). *N*-1- or *N*-7-Alkyl derivatives which may be difficult to prepare pure directly can frequently be obtained *via* an appropriate dialkyl derivative in which one of the alkyl groups may be readily removed. Thus benzyl groups can be removed by hydrogenation from 1-alkyl-9-benzyl- or 7-alkyl-3-benzyl-adenines to produce 1- or 7-alkyladenines (*Leonard* and *Fujii*, *loc. cit.*). Alternatively, a cyanoethyl group can be used and removed

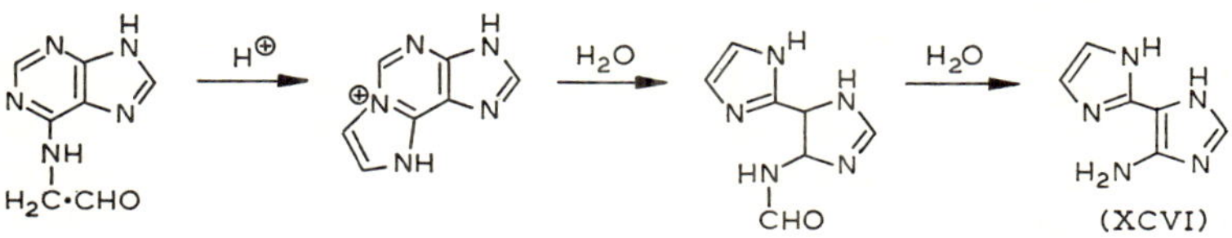

(XCVI)

by heating with methanolic potassium *tert*-butoxide (*E. P. Lira*, J. heterocycl. Chem., 1968, **5**, 863). Intramolecular alkylation can also occur. Thus 6-*N*-formylmethyladenine with dilute acid rapidly produced the bis-imidazole XCVI presumably *via* initial alkylation at $N_{(1)}$ (*G. Shaw* and *B. M. Smallwood*, J. chem. Soc., C, 1970, 2206).

An internally cyclised guanine derivative XCVII occurs as the "Y" base in t-RNA(Phe) of yeast (*K. Nakanishi et al.*, J. Amer. chem. Soc., 1970, **92**, 7617; *R. Thiebe et al.*, Biochim.

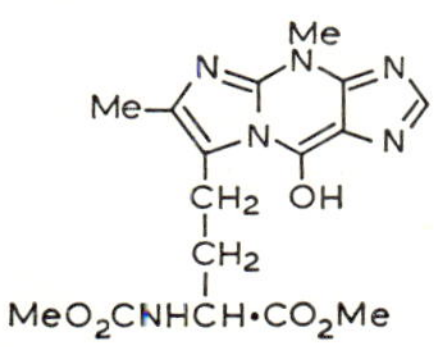

(XCVII)

Biophys. Acta, 1971, **240**, 163; *L. M. Fink et al.*, Biochemistry, 1971, **10**, 1873; *D. Yoshikami* and *E. B. Keller*, *ibid.*, 1971, **10**, 2969).

8-**Allylguanine** has been formed by a double (3,3)sigmatropic shift *via* C-5 from O^6-allylguanine (*N. J. Leonard* and *C. R. Frihart*, J. Amer. chem. Soc., 1974, **96**, 5894). Guanine and substituted guanines react with malondialdehydes to produce compounds of the type XCVIII (*R. Moschel* and *N. J. Leonard*, J. org. Chem., 1976, **41**, 294). 9-(2-**Hydroxyethoxymethyl) guanine (acycloguanine)** has been found to have anti-viral activity against herpes viruses (*H. J. Schaeffer et al.*, Nature, 1978, **272**, 583) and a related 9-(2,3-dihydroxypropyl)adenine has also broad spectrum anti-viral activity (*E. de Clercq et al.*, Science, 1978, **200**, 563; *A. Holy*, Nucleic Acids Res., Spec. Pub. No. 4, 1978, 13).

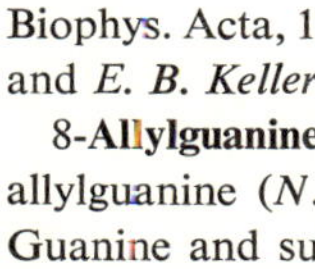

(XCVIII)

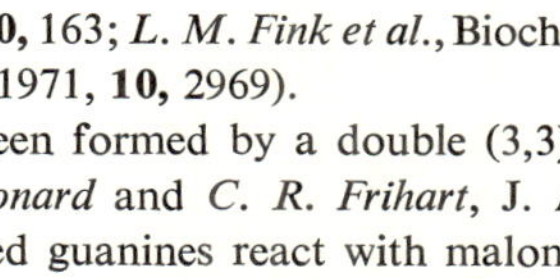

(Acycloguanine)

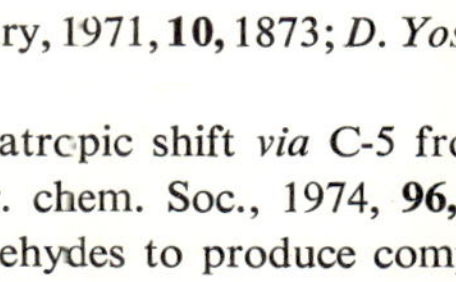

9-(2,3-Dihydroxypropyladenine)

The silver salt of adenine and 1-chloro-5-iodopentane yield 1,5-bis-9′-adenylpentane; incidentally other bispurine derivatives have been derived in this way from the reaction of 6-chloropurine with diamines (*J. R. Parikh* and *A. Burger*, J. Amer. chem. Soc., 1955, **77,** 2386; *R. N. Prasad* and *R. K. Robins*, *ibid.*, 1957, **79,** 6401; *Lettre* and *Ballweg*, Naturwiss., 1958, **45,** 364).

The general alkylation methods have also been widely adapted to the synthesis of glycosyladenines ("Synthetic Procedures in Nucleic Acid Chemistry", eds. *W. W. Zorbach* and *R. S. Tipson*, Interscience, New York, Vol. 1, 1968) and are reviewed in the next chapter.

The structure of adenine follows from its conversion by nitrous acid to hypoxanthine and is confirmed by the various syntheses already mentioned. The preparation of adenine containing radioactive or isotopic atoms has received considerable attention as the labelled product is of great value for studying the chemistry and metabolism of the compound. The methods used are all of the well known types but are worthy of mention because of the attention paid to yields in this type of synthesis. Adenine has thus been prepared from labelled thiourea (*A. Bendich et al.*, J. Amer. chem. Soc., 1948, **70,** 3109) and from labelled *N*-formyl compounds for preparing adenine labelled in the 8 position (*R. Abrams* and *L. Clark*, *ibid.*, 1951, **73,** 4609; *V. M. Clark* and *H. M. Kalckar*, J. chem. Soc., 1950, 1029; *M. Gordon*, *ibid.*, 1954, 757). Adenine labelled in the 8-position with ^{14}C is, however, best obtained by cyclising 4,5,6-triaminopyrimidine itself with labelled formic acid using 4-formylmorpholine as the solvent (*M. E. Balis* and *C. H. Winestock*, Biochemical Preparations, Wiley, New York, 1957, **5,** 62). The reduction of 6-hydroxylaminopurine also affords adenine (*Giner-Sorolla* and *Bendich*, J. Amer. chem. Soc., 1958, **80,** 3932). Crystallographic study of the hydrochlorides of adenine and guanine has provided a precise picture of the structures of both these compounds and provides support for the existence of NH_2 groups as such in aminopurines but for the formulation of hydroxypurines in the tautomeric

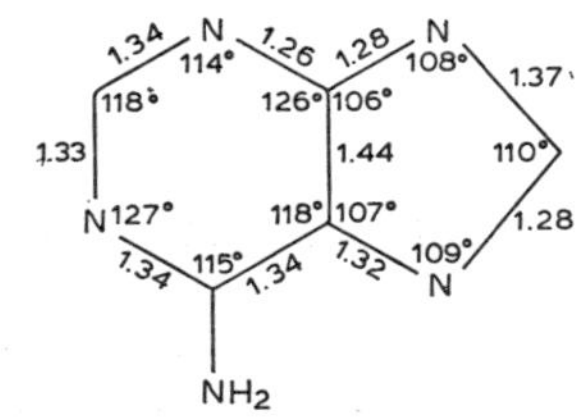

Adenine

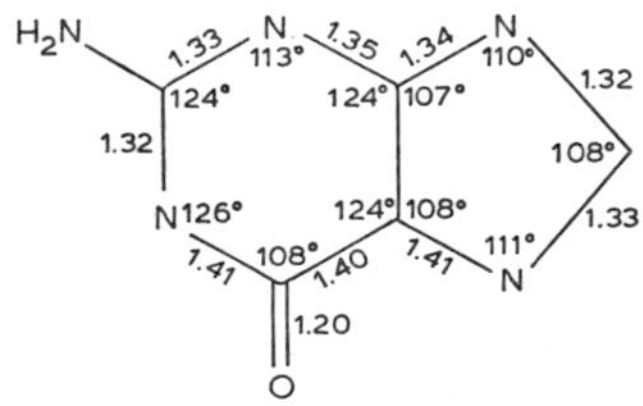

Guanine

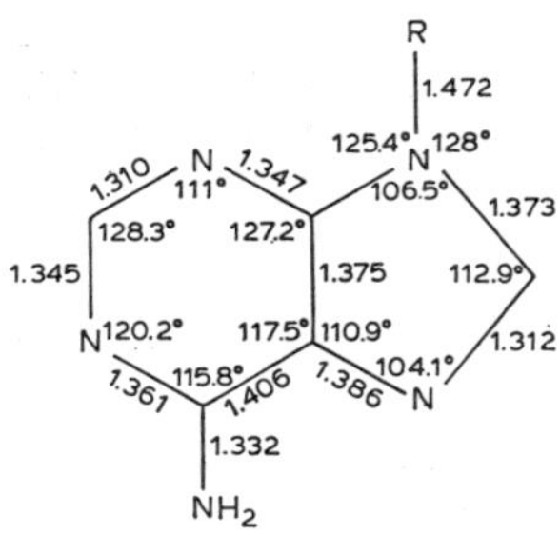

9-substituted adenine

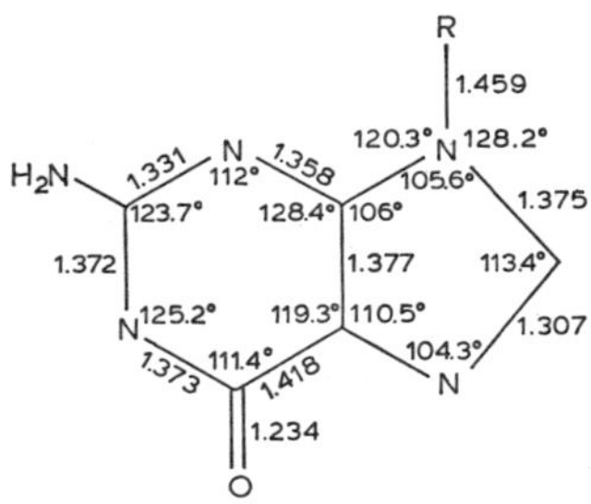

9-substituted guanine

purinone form (*J. M. Broomhead*, Acta Cryst., 1948, **1**, 324; 1951, **4**, 92; *W. Cochran, ibid.*, 1951, **4**, 81). The structural dimensions found (*H. G. Ringertz*, "The Purines", Israel Academy of Sciences, 1972, p. 61) are given below.

6-**Methylaminopurine,** m.p. 312–314° (decomp.), which is found, together with 1-methylguanine and 2-methylamino-6-hydroxypurine in yeast ribonucleic acid (*M. Adler et al.*, J. biol. Chem., 1958, **230,** 717), is obtained by heating 6-methylthiopurine and aqueous methylamine at 130° (*G. B. Elion et al.*, J. Amer. chem. Soc., 1952, **74,** 411) and is also formed when 1-methylpurine-6-thione is heated at 160° with alcoholic ammonia. With aqueous ammonia at 140°, the pyrimidine ring is opened and 5-aminoimidazole-4-carboxyamide is formed.

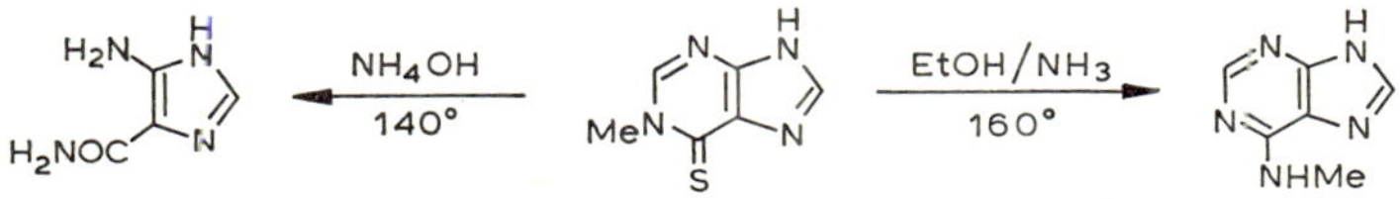

6-**Dimethylaminopurine,** m.p. 257°, picrate, m.p. 244–245° (decomp.) occurs naturally in the antibiotic puromycin (6-dimethylamino-9-[3-*p*-methoxy-1-phenylalanylamino)-*β*-D-ribofuranosyl]purine (*cf. P. M. Fryth et al.*, J. Amer. chem. Soc., 1958, **80,** 2736) and in the deoxyribonucleic acid from yeast and certain bacteria in which 2-methyladenine and 6-methylaminopurine have also been found (*J. M. Littlefield* and *D. B. Dunn*, Nature, 1958, **181,** 254). It is formed by heating aqueous dimethylamine at 140° for 24 hours with 6-methylthiopurine or methanolic dimethylamine at 100° for 1 hour with 6-chloropurine (*Elion et al., loc. cit.*; *A. Albert* and *D. J. Brown*, J. chem. Soc., 1954, 2069), or by the action of Raney nickel on the 2-methylthio derivative (*B. R. Baker et al.*, J. org. Chem., 1954, **19,**

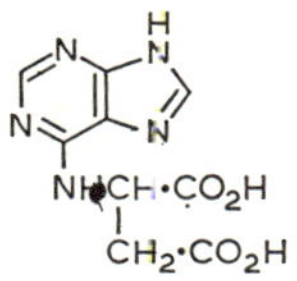

631). 6-*Succinoaminopurine* (see p. 56), chars 190°, evolves gas at 225°, is a constituent of the cell walls of certain bacteria and moulds and of mammalian liver (*A. Battio* and *G. S. Crescenzi*, Nature, 1957, **179,** 154). Hydrolysis with acid affords 5-aminoimidazole-4-carboxamide. The synthesis proceeds from trichloropurine along classical lines (*J. Baddiley et al.*, J. chem. Soc., 1956, 4659).

Zeatin, trans-6-(4-*hydroxy*-3-*methylbut*-2-*enyl*)*aminopurine*, m.p. 212–213°, is a powerful promoter of cell division in plants (*cf. F. C. Steward* and *A. D. Krikorian*, "Plants, Chemicals and Growth", Academic Press, New York, 1971) and was the first of the so-called cytokinins to be isolated from a natural source. It was first produced from extracts of *Zea mays* (*D. S. Letham*, Life Sciences, 1963, **2,** 569) by initial precipitation of the contained purines as silver salts and, after liberation from the salt, purified by preparative paper chromatography. From 60 kg of corn 0.7 mg of zeatin was obtained. It's structure was determined (*Letham, J. S. Shannon* and *I. R. C. McDonald*, Proc. chem. Soc., 1964, 230; *idem*, Tetrahedron, 1967, 23, 479) and confirmed by synthesis (*Shaw* and *D. V. Wilson*, Proc. chem. Soc., 1964, 231; *Shaw, Smallwood* and *Wilson*, J. chem. Soc., C, 1966, 921) from 6-chloropurine and *trans*-4-hydroxy-3-methylbut-2-enylamine (XCIX). The enylamine may be prepared in several ways including: routes [1]–[4], while the *cis*-isomer was obtained *via* route [5]:

[1]

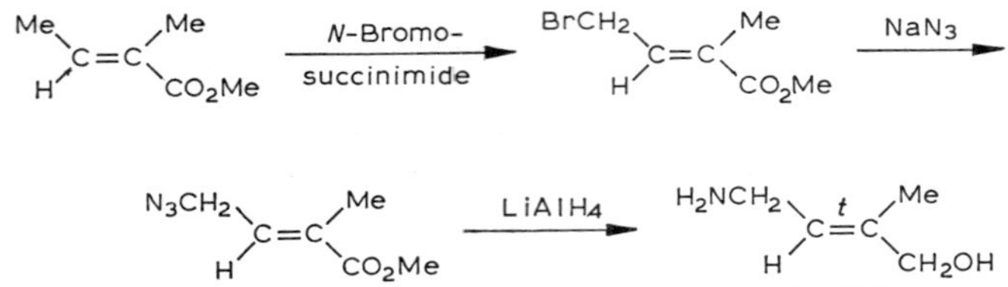

(XCIX)

(*Shaw* and *Wilson*, *loc. cit.*);

[2]

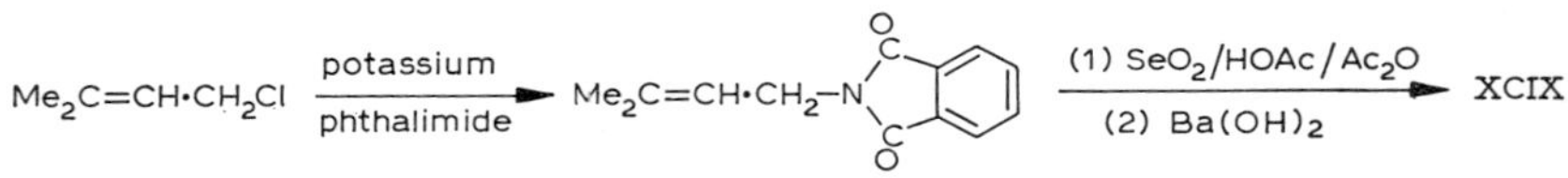

(*M. Olomucki et al.*, Compt. rend., 1965, **260,** 4519);

[3]

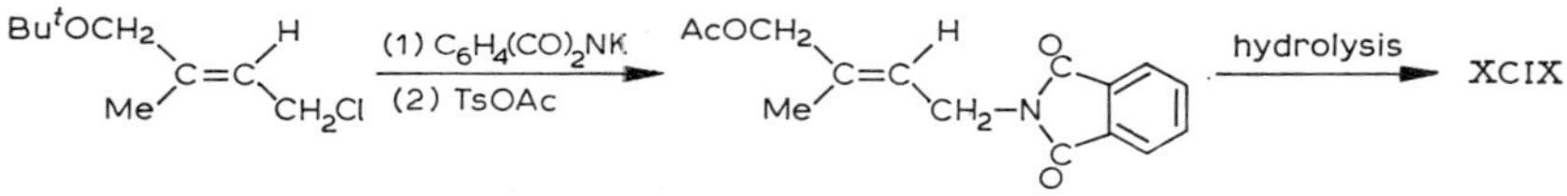

[4]

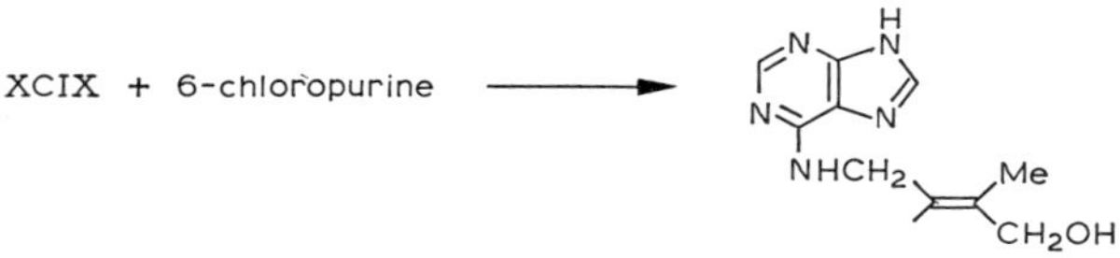

trans-Zeatin

[5]

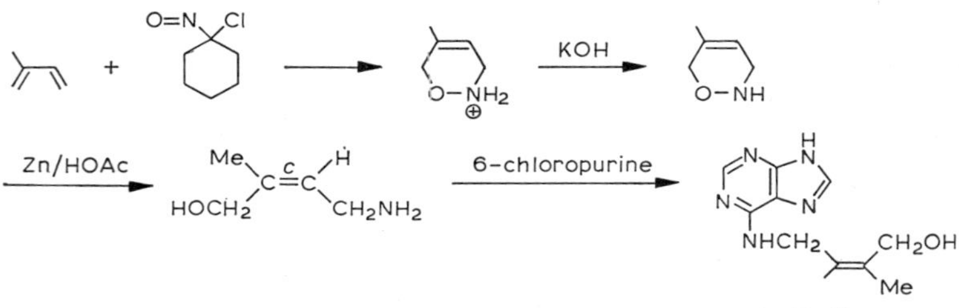

cis-Zeatin

(*J. Corse* and *J. Kuhnle*, Biochim. Biophys. res. Comm., 1972, 618).

Zeatin readily gives an O-*acetate*, m.p. 168–169° (*Shaw*, *Smallwood* and *Wilson*, *loc. cit.*) and an O-*toluene*-p-*sulphonyl* derivative, m.p. 193–194° (*idem*, J. chem. Soc., C, 1968, 2999) but with methanesulphonyl chloride in dimethylformamide it produced the *allylic chloride* C, m.p. 194° and in pyridine the *quaternary chloride* CI, m.p. 188–190°, which gave a *hydrochloride*, m.p. 244–245° and a *dipicrate*, m.p. 223–225°.

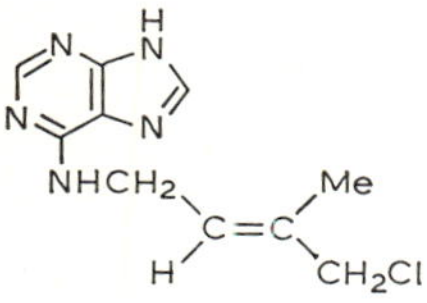

(C)

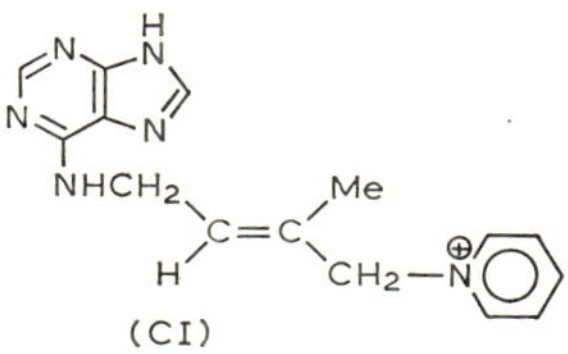

(CI)

The quaternary salt CI was also obtained from zeatin and phosphorylating agents in pyridine (*Shaw et al.*, J. chem. Soc., C, 1968, 2999). 3-*Methylzeatin*, m.p. 220–231°, 7-*methylzeatin*, m.p. 228–229° and 9-*methylzeatin, monohydrate*, m.p. 92–94°, have also been prepared by standard procedures as have 2-*aminozeatin*, m.p. 202–205°, 9-*methoxymethylzeatin*, m.p. 119–120°, 9-*ethylzeatin*, m.p. 109–111° (*Shaw*, *Smallwood* and *Wilson*, Experientia, 1967, **23**, 515; *Smallwood*, Ph.D. Thesis, University of Bradford, 1968).

Zeatin also occurs naturally as the 9-β-D-ribonucleoside (*C. O. Miller*, Proc. nat. Acad. Sci., U.S.A., 1965, **54**, 1052; *D. S. Letham*, Life Sciences, 1966, **5**, 551) synthesised by *Shaw et al.* (J. chem. Soc., C, 1966, 921) and the corresponding ribonucleotide (*Letham*, Life Sciences, 1966, **5**, 1999) which was shown to be identical with synthetic material (*Shaw*, *Smallwood* and *Wilson*, J. chem. Soc., 1968, C, 1516).

In addition (—)**dihydrozeatin,** m.p. 165–166°, was isolated from immature seeds of *Lupinus luteus* (*K. Koshimizu*, *T. Kusaki* and *T. Mitsui*, Tetrahedron Letters, 1967, No. 14, 1317), and a *cis*-**zeatin,** m.p. 206–208°, isolated as a 9-β-D-ribonucleoside from peas, spinach and corn (*R. H. Hall et al.*, Science, 1967, **156**, 69; *Hall* and *B. I. S. Srivastava*, Life Sciences, 1968, **7**, 7; *D. F. Babcock* and *R. O. Norris*, Biochemistry, 1970, **9**, 3701) was synthesised (*cf.* eqn. [5], p. 78) in a stereospecific manner (*N. J. Leonard et al.*, J. Amer. chem. Soc., 1971, **93**, 3056). In a standard tobacco callus bioassay the *trans*-zeatin was found to be at least 50 times more active than *cis*-zeatin (*Leonard et al., loc. cit.*). Other compounds related to zeatin have been isolated from wheat germ transfer ribonucleic acid including ribonucleosides of zeatin, 6-(3-methyl-2-butenyl)aminopurine (CIIa), 6-(3-methyl-2-butenylamino)-2-methylthiopurine (CIIb) and 2-methylthiozeatin (CIIc) (*M. J. Burrows et al.*, Biochemistry, 1970, **9**, 1867).

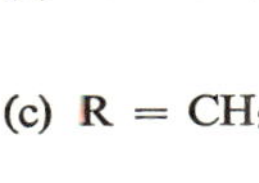

(CII)

(a) $R = CH_2 \cdot CH{=}CMe_2$; $R^1 = H$

(b) $R = CH_2 \cdot CH{=}CMe_2$; $R^1 = SMe$

(c) $R = CH_2 \cdot CH{=}C\begin{smallmatrix} \diagup Me \\ \diagdown CH_2OH \end{smallmatrix}$; $R^1 = SMe$

(d) $R = CH_2 \cdot CH(OH) \cdot CMe(OH)(CH_2OH)$

(e) $R = CH(CO_2H) \cdot CH(OH)Me$

The isopentenyl derivative CIIa has also been reported to occur as a "nucleoside" in the *t*-RNA of yeast, calf liver (*Hall et al.*, J. Amer. chem. Soc., 1966, **88,** 2614; *H. G. Zachau, D. Dutting* and *H. Feldman*, Angew. Chem., 1966, **78,** 392), spinach and garden peas (*Hall et al.*, Science, 1967, **156,** 69) and as the free base in *Corynebacterium fasciens* (*D. Dlambt, G. Thies* and *F. Skoog*, Proc. nat. Acad. Sci., U.S.A., 1966, **56,** 52; *J. P. Helleson* and *Leonard, ibid.*, p. 60). Recently (*D. S. Letham*, Phytochemistry, 1973, **12,** 2445) the presence of various other adenines related to zeatin have been reported from *Zea mays*. They include the hydroxylated derivatives 6-(2,3,4-*trihydroxy*-3-*methylbutyl-amino*)*purine* (CIId), m.p. 224–227 °, prepared from zeatin and ammoniacal potassium permanganate, and the carboxy derivative CIIe.

Kinetin, 6-*furfurylaminopurine*, m.p. 266–267 °, sublimes at 220 °, also strongly promotes cell division and was first isolated from autoclaved solutions of DNA, and has subsequently been found to be an artefact. It is amphoteric, forming a silver salt, is almost insoluble in water and is hydrolysed by acid to adenine and laevulic acid (*Miller et al.*, J. Amer. chem. Soc., 1956, **78,** 1375). Although it cannot be acetylated directly the synthesis described on p. 8 gives intially an *acetyl* derivative, m.p. 151 °, which is readily hydrolysed by alkali. 6-*Amino-9-furfurylpurine*, m.p. 191–192 °, is a by-product of this synthesis.

Kinetin and related compounds has been prepared by the reaction of 6-methylthio-purine with an amine as shown (*C. G. Skinner et al., ibid.*, 1956, **78,** 5097; 1957, **79,** 2843);

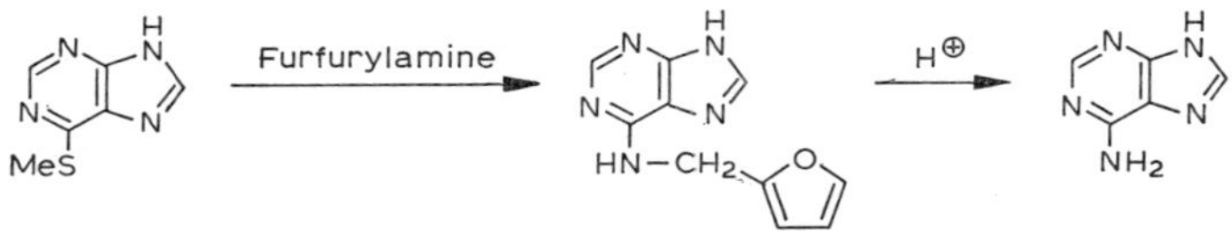

Kinetin

6-chloro-7-methylpurine has been used for preparing *N*-methyl analogues (*R. N. Prasad* and *R. K. Robins, ibid.*, 1957, **79,** 640).

Many 6-substituted-aminopurines have been prepared and tested for activity like that of kinetin and zeatin, by heating 6-chloropurine with amines in various solvents (*M. W. Bullock et al., ibid.*, 1956, **78,** 3693; *M. Sutherland* and *B. E. Christensen, ibid.*, 1957, **79,** 2251; *J. N. Daly* and *Christensen*, J. org. Chem., 1956, **21,** 177) and also from 6-methylthio-purine (*G. B. Elion et al.*, J. Amer. chem. Soc., 1952, **74,** 411; *Skinner et al., ibid.*, 1955, **77,** 6692; 1956, **78,** 2843; *L. Amirante*, Ann. Chim. Ital., 1959, **49,** 333). Structure–activity studies include those of *Steward* and *Krikorian* (*loc. cit.*), *Steward, P. L. Mott* and *Shaw* (Phytochemistry, 1973, **12,** 2335), *Shaw, Smallwood* and *Steward* (*ibid.*, 1971, **10,** 1; Experientia, 1968, 24, 1089), *H. J. Burrows et al.* (Biochemistry, 1970, **9,** 1867), *S. M. Hecht et al.* (Phytochemistry, 1970, **9,** 1173), *F. Skoog* and *D. J. Armstrong* (Ann. Rev. Plant Physiol., 1970, **21,** 359), *Skoog et al.* (Phytochemistry, 1967, **6,** 1169).

A naturally occurring tetrahydrodiaminopurine, **saxitoxin,** has the structure shown (*E. J. Schantz et al.*, J. Amer. chem. Soc., 1975, **97,** 1238; *J. Bordner et al., ibid.*, 1975, **99,** 6008).

Discadenine is a spore germination inhibitor isolated from a cellular slime mould and its structure has been shown to be 3-(3-amino-3-carboxypropyl)-6-(3-methyl-2-butenyl-amino)purine (*H. Abe et al.*, Tetrahedron Letters, 1976, 3807).

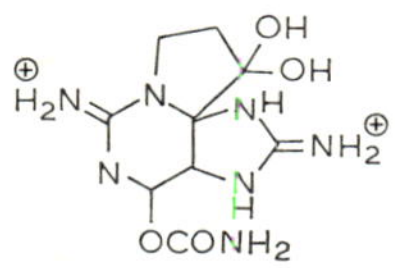

Saxitoxin hydrate cation

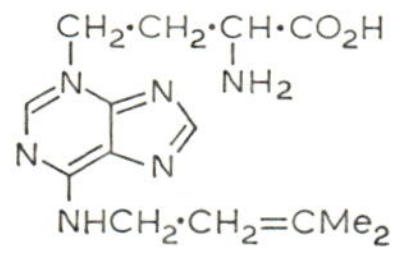

Discadenine

Nuclear N-alkyl derivatives of aminopurines. Aminopurines alkylated on nitrogen atoms 7 or 9 are, like the corresponding hydroxypurines, more labile to acid than the unsubstituted compound. For obvious reasons, they have rarely been obtained by alkylation of the aminopurines themselves. **7-Methyl-**, m.p. 283° and **9-methyl-2-aminopurine,** m.p. 247° (sublimes), were obtained by heating the corresponding 2-chloropurine with ethanolic ammonia at 150° (*E. Fischer*, Ber., 1898, **31,** 2570) but *6-amino-7-methylpurine* (*7-methyladenine*), m.p. 351° (decomp., sublimes) and the *9-methyl* isomer, m.p. 308–310°, were prepared by reducing the appropriate 2-chloro or 2,8-dichloro derivative with phosphonium iodide (*idem, ibid.*, 1897, **30,** 2249; 1898, **31,** 109; 1899, **32,** 268). Methylation of adenine with methyl iodide in alcoholic potash was shown by *Kruger* to give 9-methyladenine (Z. physiol. Chem., 1896, **18,** 434). *Prasad* and *Robins* synthesized 7-methyladenine and other 6-substituted-amino-7-methylpurines from 6-chloro-7-methylpurines (J. Amer. chem. Soc., 1957, **79,** 6401) and *Elion*, although unable to obtain 1-methyladenine, has converted both 3-methyl-6-thio- and 6-imino-2-mercapto-3,6-dihydropurine into 3-methyladenine (Chem. and Biol. of Purines, p. 39). The 7- and 9-methyl derivatives of 2,6,8-trichloropurine were also used by *Fischer* to prepare 8-amino- (no m.p.), 8-amino-2,6-dichloro- (no m.p.), 6-amino-2,8-dichloro- (no m.p.), *6-amino-2-chloro-*, m.p. 284° (see also *Prasad* and *Robins*, J. Amer. chem. Soc., 1957, **79,** 6403), *2-chloro-6-methylamino-*, m.p. 269° (decomp.),

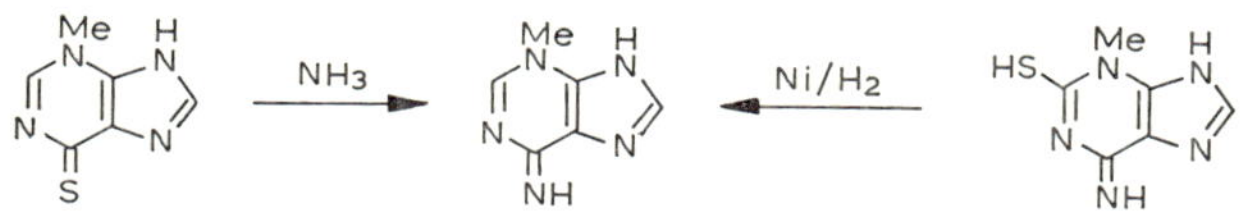

3-Methyladenine

6-amino-2,8-dihydroxy- (decomp. above 320°) and 8-amino-2,6-dihydroxy-7-methylpurine (no m.p.) and *6-amino-2,8-dichloro-9-methylpurine*, m.p. 270° (Ber., 1897, **30,** 1846; 1898, **31,** 111; 1899, **32,** 267, 488). 9-Methyladenine has also been prepared from a 5-aminoimidazole-4-carbonitrile (*R. Naylor et al., loc. cit.*).

In addition to those mentioned, many other methods have been used to prepare 9-methyladenine; *cf.* refs. cited by *R. K. Robins* and *N. H. Lin*, J. Amer. chem. Soc., 1957, **79,** 490, who used 6-chloro-9-methylpurine for the synthesis of many 6-substituted-amino-9-methylpurines. A similar series of 9-ethyl analogues has been described (*J. A. Montgomery* and *C. Temple*, *ibid.*, 1957, **79,** 5238) and 9-alkyl derivatives of adenine, together with their 8-azapurine analogues were investigated by *C. L. Leese* and *G. M. Timmis* (*ibid.*, 1958, 4107).

A naturally occurring 9-substituted adenine is **eritadenine** isolated from a Japanese mushroom, *Lentinus edodes* (*S. Takeyama, K. Okumura et al.*, J. med. Chem., 1974, **17,** 846) has high hypolipidemic activity and has been synthesised (*M. Kawazu et al.*, J. org.

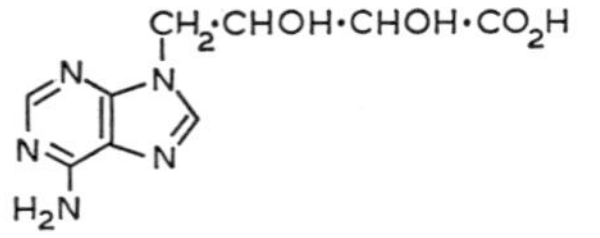

Eritadenine

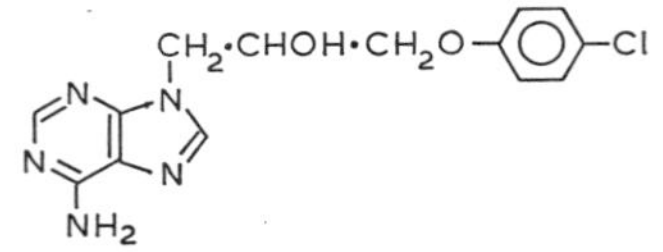

9-(4-Chlorophenoxy-2-hydroxypropyl)-adenine

Chem., 1973, **38,** 2887; *N. Takamura et al., ibid.*, 1973, **38,** 2891). A large series of eritadenine analogues have been prepared of which the most active hypolipidemic reagent was 9-(4-*chlorophenoxy-2-hydroxypropyl)adenine*, m.p. 160–161°, prepared from adenine with *O*-(4-chlorophenyl)glycidol, potassium carbonate in dimethyl sulphoxide (*W. S. D. Menna, C. Piamtadoni* and *R. G. Lamb*, J. med. Chem., 1978, **21,** 1073).

7,9-**Dimethyladenine** has been prepared as a *perchlorate* salt, m.p. 276–277° (decomp.), by methylation of 6-methoxyamino-9-methylpurine with methyl iodide, conversion of the

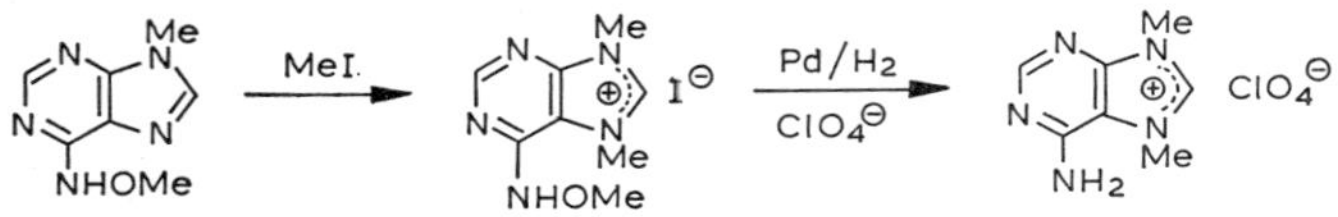

quaternary *iodide*, m.p. 250–251° (decomp.), to a perchlorate and hydrogenation of this with a palladium catalyst (*T. Fujii et al.*, Tetrahedron Letters, 1973, No. 49, 4873) (see also p. 32).

The 8-amino derivatives of theophylline, paraxanthine, theobromine and caffeine are all solids which decompose at high temperatures and are normally obtained by amination of the 8-chloro compounds, which are prepared with comparative ease. Coupling of theophylline with a diazo compound followed by reduction, and reduction of 8-nitrotheobromine with sodium amalgam have however been used as alternative methods of preparation (*C. F. Boehringer und Söhne*, G.P. 133,300/1901; 156,900/1909; Chem. Ztbl., 1902, II, 1475; 1905, I, 59; *H. Fischer*, Z. physiol. Chem., 1909, **60,** 72; *C. Brunner*, Ber., 1897, **30,** 2586; *A. Gomberg*, Amer. chem. J., 1900, **23,** 58). Coupling followed by reduction has also been used successfully to introduce an 8-amino group into xanthine, guanine and isoguanine (*Cavalieri* and *Bendich*, J. Amer. chem. Soc., 1950, **72,** 2593; *J. P. Spies* and *T. H. Harris, ibid.*, 1939, **61,** 351).

2,6-**Diaminopurine,** $C_5H_5N_6$, is of interest for its antileukaemic activity (*H. E. Skipper et al.*, Cancer, 1950, **10,** 166) although it is probably too toxic for clinical use. It was first prepared by *W. Traube* (Ber., 1904, **37,** 4547) by cyclisation of 2,4,5,6-tetraaminopyrimidine. By the use of 2,6-diaminopurine containing ^{15}N in the 1 and 3 positions and in the two amino groups (prepared by the Traube method) it has been shown that the compound is a precursor of nucleic acid and guanine in the rat (*Bendich et al.*, J. biol. Chem., 1950, **185,** 423, 435).

2,6-**Diamino-8-hydroxypurine** (no m.p.) was prepared by aminating the dichlorohydroxypurine (*E. Fischer* and *L. Ach*, Ber., 1897, **30,** 2217), 2,8-**diamino-6-hydroxypurine** (no m.p.), by coupling guanine with 2,5-dichlorobenzenediazonium chloride and reducing the product with dithionite (*H. Fischer*, Z. physiol. Chem., 1909, **60,** 76). A series of 2,6-di(alkylamino)purines was prepared from 2,6-dichloropurine by *J. A. Montgomery* and *L. B. Holum* (J. Amer. chem. Soc., 1957, **79,** 2185) and of 6-alkylamino-8-substituted purines from 2,8-dichloropurine by *R. K. Robins* (*ibid.*, 1958, **80,** 6671).

(*h*) *Aminohydroxypurines*

Guanine, 2-amino-6-hydroxypurine, $C_5H_5ON_5$, was first discovered as a constituent of guano and given its correct composition in 1846 (*B. Unger*, Ann., **58,** 18; **59,** 58); it was later shown to be the main purine constituent of faeces (*M. Kruger* and *A. Schittenhelm*, Z. physiol. Chem., 1902, **35,** 153). It is readily obtained from nucleic acid hydrolysates (*H. Bredereck* and *G. Richter*, Ber., 1938, **71,** 718) or by the synthetic methods previously described (pp. 5, 56, 59). Guanine labelled at the 3 and 1 positions with ^{15}N has been prepared from guanidine (*A. A. Plentl* and *R. Schoenheimer*, J. biol. Chem., 1944, **153,** 203).

Unlike adenine, guanine is almost insoluble in water but, like the former, it is readily soluble in acids and alkali although it is insoluble in dilute ammonium hydroxide solution. The latter property is important for isolation purposes.

The hydrolysis of guanine by acid has been studied in detail by *G. Hunter* (Biochem. J., 1936, **30,** 1183). Heating at 160° for 90 min with 3.4 *N* hydrochloric acid converts guanine into xanthine (52%), guanidinoimidazole (21.5%), glycine (14%) and guanidine (2%). Isotope studies show that the glycine formed in this hydrolysis arises from positions 4, 5 and 7 of the purine ring and the same is true of the glycine formed by hydrolysing xanthine and uric acid (*L. F. Cavalieri, J. E. Tinker* and *G. B. Brown*, J. Amer. chem. Soc., 1949, **71,** 3973). The conversion of guanine into xanthine may be more readily accomplished by nitrous acid. Guanine is much more readily oxidised than adenine and is converted by potassium chlorate in hydrochloric acid to guanidine, which was first discovered in this way, and parabanic acid (*A. Strecker*, Ann., 1861, **118,** 155, 161). This oxidation, the conversion to xanthine and its behaviour on hydrolysis establish the structure of guanine. Electrolytic reduction of guanine gives *deoxyguanine* (*2-amino-5,6-dihydropurine*), m.p. 204°, which may be oxidised to 2-aminopurine (see p. 73).

Like adenine, guanine couples with diazonium salts, is converted enzymatically to xanthine (*H. Fischer, loc. cit.*; *R. W. Martin*, Z. physiol. Chem., 1934, **226,** 53) and may be identified by paper chromatography (*e.g. E. Vischer* and *E. Chargaff*, J. biol. Chem., 1947, **168,** 781).

2-Substituted-amino-6-hydroxypurines are formed by the reaction of 6-hydroxy-2-methylthiopurine with amines (*G. B. Elion et al.*, J. Amer. chem. Soc., 1956, **78,** 217; *Montgomery* and *Holum, loc. cit.*).

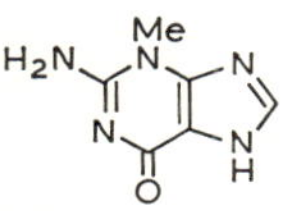

3-Methylguanine

3-Methylguanine has been shown by X-ray crystal structure analysis to exist in the crystalline state as the N-7 tautomer (*J. E. Abola, D. J. Abraham* and *L. B. Townsend*, Tetrahedron Letters, 1976, No. 39, 3483).

Herbipoline, isolated from the giant silica sponge (*D. Ackermann*, Angew. Chem., 1958, **70,** 80; *H. Bredereck et al.*, Ber., 1959, **92,** 566) has been assigned the following structure:

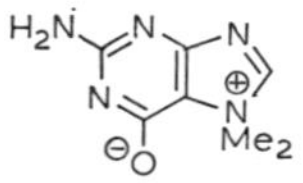

Herbipoline

Isoguanine, 6-amino-2-hydroxypurine (no m.p.), which occurs naturally as the glycoside cordycepin (see p. 2), was obtained by reducing 6-amino-8-chloro-2-ethoxypurine with phosphonium iodide and a similar reduction of 6-amino-2-chloro-8-hydroxypurine gave **6-amino-8-hydroxypurine** (no m.p.) (*E. Fischer*, *ibid.*, 1897, **30,** 2215, 2245). The synthesis of isoguanine containing ^{15}N at the 1 and 3 positions (Traube method) was accomplished by *A. Bendich et al.* (J. Amer. chem. Soc., 1948, **70,** 1240) but 6-amino-8-hydroxypurine is better prepared by fusing 4,5,6-triaminopyrimidine with urea (*R. K. Robins*, *ibid.*, 1958, **80,** 6672). Acid hydrolysis of 6-amino-2,8-dichloro- and 2-amino-6-chloro-8-hydroxy-purine gave, respectively, **2-amino-6,8-dihydroxy-**, m.p. >380° (decomp.) and **6-amino-2,8-dihydroxy-purine** (no m.p.), the latter being a metabolic oxidation product of adenine in the rat (*Fischer*, Ber., 1897, **30,** 2243; 1898, **31,** 2631).

Absorption spectra and m.p's of amino- and aminohydroxypurines are given in Table 9.

(*i*) *Nitropurines and related compounds*

Nitration of purines by conventional methods tends to produce only the 8-nitro derivatives. Thus 8-nitrocaffeine (CIIIa) (*O. Schultzen*, Z. Chem., 1867, **10,** 614) or 8-nitrotheobromine (CIIIb) (*H. Brunner* and *H. Leins*, Ber., 1897, **30,** 2584) are produced by nitration without a solvent. 8-Nitrotheophylline (CIIIc) has been prepared by nitration with fuming nitric acid in aqueous ethanol (*G. Serchi*, *L. Sancio* and *G. Bichi*, Farmaco Ed. Sci., 1955, **10,** 733). Xanthine does not react readily with nitric acid but 9-methylxanthine gives the 8-nitro

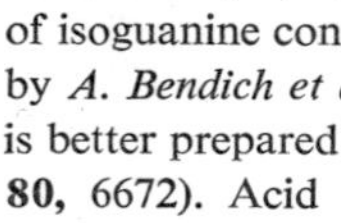

(CIII)

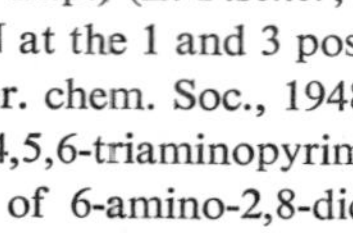

(CIV)

(a) R = R^1 = Me
(b) R = H, R^1 = Me
(c) R = Me, R^1 = H

derivative (CIV) with 50% nitric acid and the same compound results from 8-mercapto-9-methylxanthine and 50% nitric acid (*J. W. Jones* and *R. K. Robins*, J. Amer. chem. Soc., 1960, **82,** 3773).

A recent synthesis of 8-nitroxanthine involves reaction of 3-acetoxyxanthine with nitrite; this and similar reactions are discussed on p. 35.

8-Nitropurines have also been prepared by diazotisation of the 8-amino derivative and coupling with sodium nitrite. In this way 8-nitrohypoxanthine may be prepared from 8-aminohypoxanthine (*idem*, *loc. cit.*) and similar reactions produce 8-nitro-xanthine, -guanine and -theophylline. Other nitropurines are rare. However, 2-nitroinosine has been prepared in low (5%) yield by the action of nitrite on guanosine (*R. Shapiro*, J. Amer. chem. Soc., 1964, **86,** 2948).

TABLE 9

PROPERTIES OF SOME AMINO- AND AMINOHYDROXY-PURINES

Purine derivative	m.p. (°C)	λ_{max} nm (10^{-3} ε in brackets) in aqueous solution: Acidic	Neutral	Basic	Ref.
2-Amino-		235[a] (6.46) 235 (4.17)	236[a] (5.01) 304 (6.03)	267 (12.02)	1, 2
6-Amino- (adenine)	360–365 (decomp.)	262 (13.1)	261 (12.6)	267 (12.02)	2, 3
8-Amino-	>360	288 (15.85)	241 (3.24) 283 (14.79)	230[a] (9.55) 290 (11.75)	1, 2
2-Dimethylamino-	222–223	228 (31.62) 248[a] (9.33) 340 (3.02)	223 (15.7) 248 (10.47) 332 (5.13)	232 (24.0) 327 (4.68)	1, 2
2-Methylamino-	276	223 (35.5) 244 (inf.) (6.5) 326 (3.7)	219 (26.3) 240 (7.6) 319 (5.5)	226 (23.4) 272 (5.2) 316	4
6-Methylamino	312–314	267 (15.3)	266 (16.22)	273 (15.85)	5
6-Dimethylamino	249–252	276 (15.49)	275 (17.78)	281 (17.78)	5, 6
8-Methylamino	332–334 (decomp.)	230[a] (7.94) 296 (17.38)	245 (3.16) 290 (16.60)	230[a] (10.0) 298 (14.45)	1, 2
8-Dimethylamino	292 (decomp.)	230[a] (10.96) 305 (19.50)	250 (2.95) 296 (18.62)	230[a] (8.13) 306 (16.60)	1, 2
6-Amino-2-methyl-		266 (12.9)		271 (10.7)	16, 5
6-Amino-3-methyl-		274		267	6
6-Amino-7-methyl-	344–346 (decomp.)	272 (15.05)		270 (10.55)	8

TABLE 9 *(continued)*

Purine derivative	*m.p. (°C)*	λ_{max} *nm (10^{-3} ε in brackets) in aqueous solution*			*Ref.*
		Acidic	*Neutral*	*Basic*	
6-Amino-8-methyl-	> 300	269 (13.4)		266 (14.9)	9
6-Amino-9-methyl-	310	261 (14.6)		262 (11.9)	10
6-Amino-9-ethyl-	194–195	264[a] (13.1)			
		259 (13.8)	262 (14.1)	262 (14.1)	11
6-Amino 1-*N*-oxide	297–307 (decomp.)	258.5(11.5)	231 (41.5)	233 (46.2)	3
			262.5(8.1)	275 (7.4)	
6-Furfurylamino- (kinetin)	266–267	274 (17.0)	267 (18.8)	273.5(17.6)	12
6-Furfurylamino-7-methyl-	214–215	183 (18.5)		274 (16.2)	8
6-Furfurylamino-9-methyl-	175–177	267 (20.0)		269 (18.7)	10
7-Methyl-6-methylamino-	300	280 (16.0)		275 (14.3)	8
9-Methyl-6-methylamino-	190–191	269 (14.4)		277 (16.8)	10
6-Dimethylamino-9-methyl-	119–120	269 (14.4)		277 (16.8)	10
6-Dimethylamino-9-ethyl-	82–84	270 (17.4)	275 (20.6)		11
2-Hydrazino-	> 300	297 (5.0)	309 (4.7)		13
6-Hydrazino-	246–247.5	267 (13.7)	262 (9.4)		13
6-Hydrazino-7-methyl-	243	272 (12.3)		263 (7.7)	8
6-Hydrazino-9-methyl-	210–211	263 (16.1)			10
9-Ethyl-6-hydrazino-	160–162	263 (15.8)	265 (14.3)		11
6-Succinamino-	190 (chars)	275 (17.38)		274 (17.38)	14
6-*N*-Hydroxylamino-	254 (decomp.)	271 (13.3)	268 (11.8)		15
6-Azido-	185				15
6,8-Diamino-		280 (12.0)		224 (17.5)	17
				280 (14.8)	
2,6-Diamino-	300–302	241 (9.1)	246 (7.1)	243 (4.7)	1
		282 (10.5)	279 (8.9)	284 (9.3)	
2-Amino-6-hydroxy- (guanine)		247 (10.47)		245 (8.73)	18
				275 (9.48)	

6-Amino-8-hydroxy-		280 (11.5)		279 (16.0)	17
				223 (17.7)	
8-Amino-6-hydroxy-		254 (14.8)		222 (15.3)	17
				270 (13.4)	
6-Amino-2-hydroxy-		282 (13.34)		282 (12.45)	18
6-Amino-2-hydroxy-9-methyl-		268 (9.8)		238 (4.1)	19
				274 (6.5)	
6-Amino-3-(3-methylbut-2-enyl)- (triacanthine)	228–231	274 (17.5)	273 (13.8)		20
6-(4-Hydroxy-3-methylbut-*trans*- -2-enyl)amino- (zeatin)	210–211	274 (15.5)	269 (16.15)	222 (12.15)	21, 22
		205 (15.5)	210 (18.4)	275 (15.1)	
2-Aminozeatin	202–205	217 (14.3)	224 (15.8)	228 (13.9)	22, 23
		248 (12.2)	248 (sh.)		
		282 (10.1)	286 (8.2)	290 (9.0)	
3-Methylzeatin	229–231	211 (7.4)	219 (14.3)	224 (13.2)	22, 23
		286 (20.4)	290 (15.9)	290 (15.9)	
7-Methylzeatin	228–229	203–222 (11.1)	214 (12.0)	224 (6.7)	22, 23
		284 (20.1)	278 (15.1)	278 (16.6)	
9-Methylzeatin	92–94 (monohydrate)	216 (14.9)	217 (16.6)	223 (6.6)	22, 23
		267 (15.9)	270 (16.5)	270 (16.1)	
6-(4-Prop-2-ynyl)amino-	285–287	209 (12.0)	211 (17.0)	225 (12.7)	22, 24
		275 (16.6)	266 (16.45)	273 (15.4)	
6-(3,3-Dimethylallyl)amino-	213–214	207 (13.25)	213 (10.7)	226 (8.2)	22, 24
		276 (16.2)	269 (15.4)	277 (16.0)	
6-Crotylamino-	236–239	216 (16.6)	218 (13.45)	229 (14.6)	
		275 (17.2)	270 (16.4)	275 (16.75)	22, 24
6-(But-3-enyl)amino-	225–227	210 (10.1)	215 (7.15)	227 (7.7)	
		275 (15.65)	269 (14.2)	276 (15.1)	22, 24
6-(Cinnamylamino)-	240–242	209 (22.3)	210 (27.15)	224 (11.2)	22, 24
		254 (15.75)	254 (sh.)	253 (15.7)	
		276 (19.55)	270 (24.0)	276 (18.05)	

[a] Ethanol.

References (table 9)

1 *A. Albert* and *D. J. Brown*, J. chem. Soc., 1954, 2060.
2 *S. F. Mason*, *ibid.*, 1954, 2071.
3 *M. A. Stevens et al.*, J. Amer. chem. Soc., 1958, **80,** 2755.
4 *H. G. Goldner*, *G. Dietz* and *E. Carstens*, Tetrahedron Letters, 1965, 2701.
5 *J. W. Littlefield* and *D. B. Dunn*, Nature, 1958, **181,** 254.
6 *G. B. Elion*, Chem. and Biol. of Purines, p. 39.
7 *B. R. Baker et al.*, J. org. Chem., 1954, **19,** 631.
8 *R. N. Prasad* and *R. K. Robins*, J. Amer. chem. Soc., 1957, **79,** 6401.
9 *H. C. Koppel* and *Robins*, J. org. Chem., 1958, **23,** 1457.
10 *Robins* and *H. H. Lin*, J. Amer. chem. Soc., 1957, **79,** 490.
11 *J. A. Montgomery* and *C. Temple*, *ibid.*, 1957, **79,** 5238.
12 *C. O. Miller et al.*, *ibid.*, 1956, **78,** 1375.
13 *Montgomery* and *L. B. Holum*, *ibid.*, 1957, **79,** 2185.
14 *J. Baddiley et al.*, J. chem. Soc., 1956, 4659.
15 *A. Bendich et al.*, Chem. and Biol. of Purines, p. 3.
16 *Baddiley et al.*, J. chem. Soc., 1944, 318; *O. Vogl* and *E. C. Taylor*, J. Amer. chem. Soc., 1957, **79,** 1518.
17 *Robins*, *ibid.*, 1958, **80,** 6671.
18 *L. F. Cavalieri et al.*, *ibid.*, 1948, **70,** 3878.
19 *R. Falconer et al.*, J. chem. Soc., 1939, 1784.
20 *R. Denayer*, Bull. Soc. chim., 1962, 1358.
21 *G. Shaw*, *B. M. Smallwood* and *D. V. Wilson*, J. chem. Soc., C, 1966, 921.
22 *Smallwood*, Ph. D. Thesis, University of Bradford, 1968.
23 *Shaw*, *Smallwood* and *F. C. Steward*, Experientia, 1968, **24,** 1089.
24 *Idem*, Phytochemistry, 1971, **10,** 2329.

The nitropurines are readily reduced with various reducing agents and the nitro group can also be replaced by nucleophiles; thus 8-nitrotheophylline with hydrochloric acid gives 8-chlorotheophylline.

C-Nitrosopurines are not common. However, 6-**nitrosopurine** has been prepared from 6-hydroxyaminopurine by oxidation with activated manganese dioxide in water (*A. Giner-Sorolla*, J. heterocycl. Chem., 1970, **7,** 75; *idem, et al.*, "The Purines", Israel Acad. Sci. Humanities, 1972, p. 478).

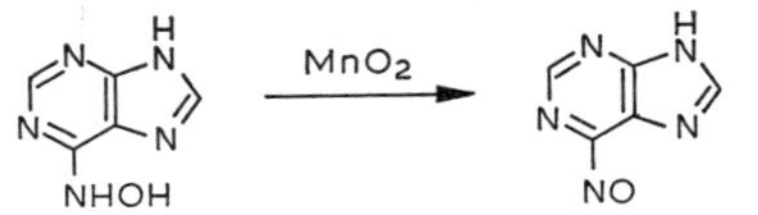

(*j*) *Purine N-oxides*

(*i*) *Purine 1-oxides*

Purine produces **purine 1-oxide** when treated with perbenzoic acid over two weeks (*M. A. Stevens et al.*, J. org. Chem., 1962, **27,** 567) but peroxide–acetic acid mixtures were used to convert adenine and related purines to their 1-oxides (*idem*, J. Amer. chem. Soc., 1958, **80,** 2755; *H. W. Smith et al.*, *ibid.*, 1960, **82,** 1148; *T. Fujii et al.*, Chem. and Ind., London, 1966, 1598). On the other hand, **hypoxanthine 1-oxide** is better prepared by deamination of adenine 1-oxide rather than by direct oxidation (*J. C. Parham*, *J. Fissekis* and *G. B. Brown*, J. org. Chem., 1966, **31**, 966). Hypoxanthine 1-oxide has also been prepared unambiguously by cyclisation of 5-amino-4-hydroxycarbamoylimidazole with triethyl orthoformate (*E. C. Taylor*, *C. C. Cheng* and *O. Vogl*, *ibid.*, 1959, **24**, 2019).

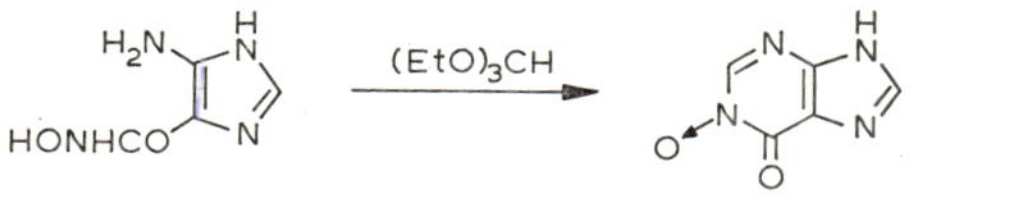

Hypoxanthine 1-oxide

A similar route to **xanthine 1-oxide** (*L. Bauer* and *D. Dhawan*, J. heterocycl. Chem., 1965, **2**, 220) and related 7-alkyl compounds including the thioxanthine analogues involved a Lossen rearrangement of 4,5-di(*N*-hydroxycarbamoyl)imidazoles with benzenesulphonyl chloride (*Bauer et al.*, *ibid.*, 1964, **1**, 275). **Adenine 1-oxide** may also be formed by cyclisation of 5-amino-4-hydroxyformamidinoimidazole with triethyl orthoformate in dimethyl-

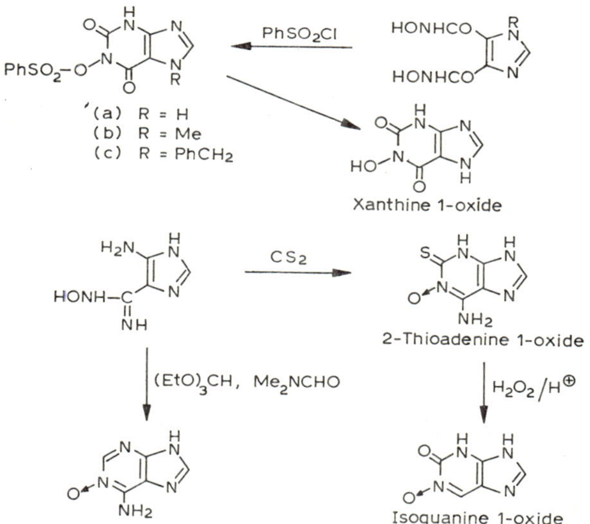

formamide (*R. M. Cresswell* and *G. B. Brown*, J. org. Chem., 1963, **28**, 2560) whereas with carbon disulphide it produces **2-thioadenine 1-oxide**, which may be oxidised to **isoguanine 1-oxide** by alkaline peroxide followed by acid (*Parham*, *Fissekis* and *Brown*, *ibid.*, 1967, **32**, 1151).

1-Alkoxy-9-alkyladenines undergo a Dimroth rearrangement to produce 6-alkoxyamino-9-methyladenines and the rates have been determined (*T. Itaya*, *F. Tanaka* and *T. Fujii*, Tetrahedron, 1972, **28**, 535).

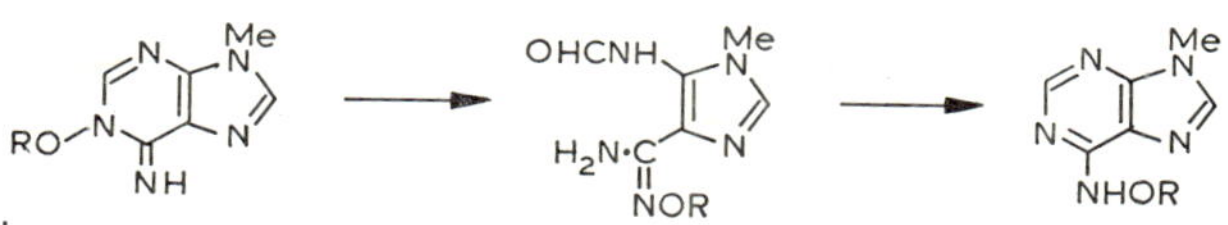

(ii) Purine 3-oxides

Oxidation of **guanine** with hydrogen peroxide in trifluoroacetic acid gives the **3-oxide** and this when boiled with 6 *N* hydrochloric acid produces **xanthine 3-oxide** (*Brown*, Prog. Nucleic Acid Res. Mol. Biol., 1968, **8,** 209; *U. Wolcke* and *Brown*, J. org. Chem., 1969, **34,** 978). 6-Chloro-, 6-methoxy- and 6-ethoxy-purine 3-oxides were similarly produced from the respective purines by treatment with perphthalic or peracetic acid in ethanol (*H. Kawashima* and *I. Kumashiro*, Bull. chem. Soc., Japan, 1969, **42,** 750; *I. Scheinfeld*

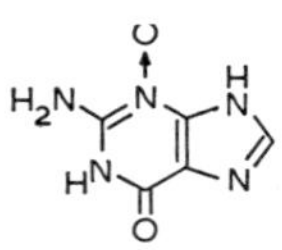

Guanine 3-oxide

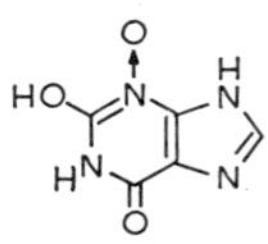

Xanthine 3-oxide

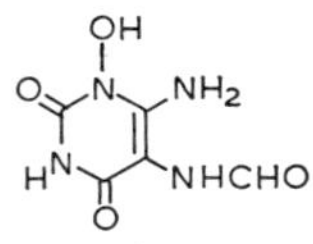

(CV)

et al., J. org. Chem., 1969, **34,** 2153). Purine and 6-methylpurine with peracids produced mixtures of the 1- and 3-oxides whereas 6-cyanopurine produced the 3-oxide which was hydrolysed to purine-6-carboxylic acid 3-oxide and this in turn could be decarboxylated to purine 3-oxide (*A. Giner-Sorolla et al.*, *ibid.*, 1971, **36,** 1228).

Xanthine 3-oxide may also be obtained by heating solutions of 6-amino-5-formamido-1-hydroxyuracil (CV) (*Wolcke* and *Brown*, *loc. cit.*) whereas reaction of 4-cyano-5-hydroxy-amino-1-methylimidazole with ethanolic formamide acetate gave 7-methyladenine 3-oxide.

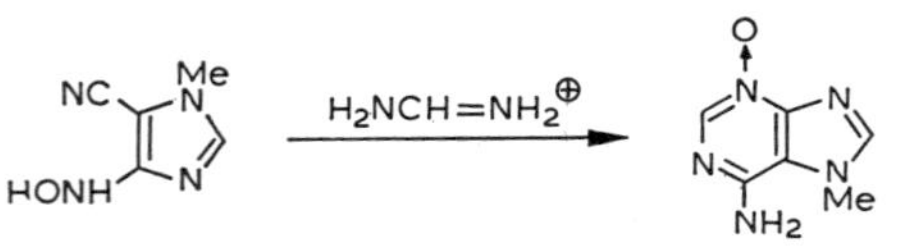

3-Hydroxy-2,3-dihydro-2-oxopurine has been prepared from 5-aminocytosine 1-oxide and triethyl orthoformate (*T. Chuyh*, *F. L. Lam* and *G. B. Brown*, J. org. Chem., 1975, **40,** 1547).

6-Chloropurine 3-oxide can be converted into hypoxanthine 3-oxide by hydrolysis with sodium hydroxide, 6-methoxy-, 6-bromo-, 6-iodo- and 6-sulpho-purine 3-oxides by reaction with sodium methoxides, bromine–hydrobromic acid, hydriodic acid and aqueous sodium sulphite, respectively (*Kawashima* and *Kumashiro*, *loc. cit.*; *Giner-Sorolla et al.*, J. org. Chem., 1969, **34,** 2157), and **6-selenopurine 3-oxide** by reaction with selenourea (*idem*, J. heterocycl. Chem., 1971, **8,** 651). 6-**Mercaptopurine** 3-**oxide** may be obtained by

(CVI)

rearrangement of the thiazolopyrimidine CVI with alkali (*G. Levin* and *Brown*, J. med. Chem., 1963, **6,** 825) or from the 6-chloro- derivative and ammonium dithiocarbonate (*Giner-Sorolla*, *loc. cit.*) and is readily converted to the 6-methylthio derivative or oxidised to the disulphide or, with chlorine, converted to the 6-chloro derivative.

A useful route to **6-hydroxyaminopurine 3-oxide** is from the 6-sulpho derivative and ethanolic hydroxylamine (*Giner-Sorolla*, J. med. Chem., 1969, **12,** 717), whereas **adenine 3-oxide** similarly results when ammonia is used under forcing conditions (*Scheinfeld et al., loc. cit.*).

Oxidation of 1-, 7-, 8- and 9-mono-*N*-methylguanines and 1,7-di-*N*-methylguanine with pertrifluoroacetic acid produced the corresponding 3-oxides and the corresponding methylated 3-hydroxyxanthines were produced by acid hydrolysis of the guanine derivatives

(*N. J. Birdsall et al.*, J. org. Chem., 1971, **36,** 2635) whereas 3-hydroxyuric acid has been synthesised by cyclisation of 6-amino-5-ethoxycarbonylamino-1-hydroxyuracil (*Brown et al.*, Biochemistry, 1971, **10,** 4463).

Many of the purine 3-oxides including those of guanine and xanthine are potent oncogens (cancer producing) in the rat (*cf. A. Myles* and *Brown*, J. biol. Chem., 1969, **244,** 4072; *G. Stohrer* and *Brown*, Science, 1970, **167,** 1622) whereas 3-hydroxyuric acid is only a weak oncogen, attributed to its *in vivo* partial reduction to 3-hydroxyxanthine (*Brown et al., loc. cit.*).

Tautomeric structures in various 1- and 3-*N*-oxides have been studied by ultraviolet absorption spectral techniques (*Parham, T. G. Winn* and *Brown*, J. org. Chem., 1971, **36,** 2639 and refs. therein); the *N*-hydroxy tautomer predominates in the neutral species of 3-hydroxyxanthine whereas guanine 3-oxide is a mixture of mostly the 3-hydroxy tautomer.

The conversion of 3-acyloxypurines into 8-substituted purines, which involves attack at C-8 by inorganic or organic nucleophiles, has been discussed on p. 35.

(*iii*) *Purine 7- and 9-oxides*; *7-hydroxypurines*

Since direct oxidative techniques do not appear to lead to purine 7- and 9-oxides, most syntheses have involved cyclisation of appropriate pyrimidines, or imidazole derivatives. Until recently few syntheses had been reported. They include the preparation of 7-hydroxytheophylline (*H. G. Goldner, G. Dietz* and *E. Carstens*, Ann., 1965, **691,** 142; 1966, **693,** 233; **698,** 145; **699,** 145) or 8-aryl or alkyl derivatives (*E. C. Taylor* and *E. E. Garcia*, J. Amer. chem. Soc., 1964, **86,** 4721) by ring-closures of 6-amino- or 6-alkylamino-1,3-dimethyl-5-nitrosouracils, and the synthesis of 2,6-diamino-7-hydroxy-8-phenylpurine from phenyliminobenzene with 2,4,6-triamino-5-nitropyrimidine (*G. M. Timmis et al.*, Chem. and Biol. of Purines, p. 134), and of 7-hydroxyxanthine from 6-amino-5-nitrosouracil and formaldehyde (*G. Zvilichovsky* and *Brown*, J. org. Chem., 1972, **37,** 1871). In contrast, attempts to produce 9-hydroxypurines by Traube type synthesis have not been successful but 9-hydroxy-8-methylhypoxanthine (CVIIa) and xanthine (CVIIb) have recently been synthesised by cyclisation of 5-amino-1-benzyloxy-2-methylimidazole-4-carboxamide, with formate or carbonate esters and debenzylation of the intermediate benzyloxy derivatives with hydrogen bromide in acetic acid. The related 9-hydroxy-8-

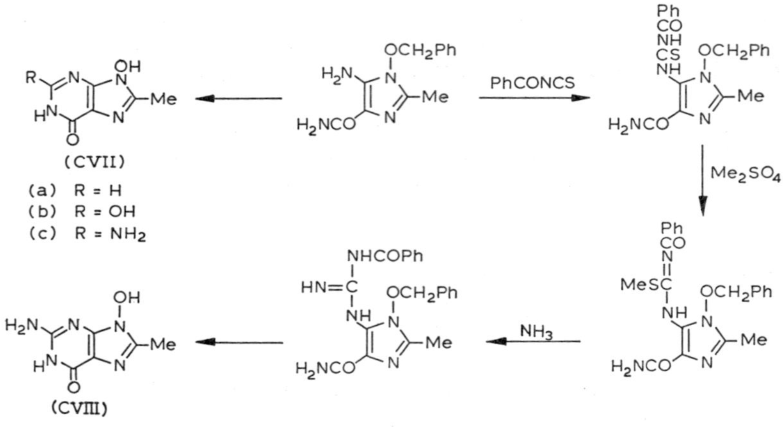

(CVII)
(a) R = H
(b) R = OH
(c) R = NH_2

(CVIII)

methylguanine derivative CVIII was similarly produced *via* a guanidinoimidazole intermediate (*A. A. Watson* and *Brown*, *ibid.*, 1972, **37,** 1867).

The *N*-oxides are normally readily reduced to the parent purines. Thus adenine is obtained by catalytic hydrogenation of adenine 1-oxide in dilute ammonia over Raney nickel (*Brown et al.*, J. Amer. chem. Soc., 1958, **80,** 2755) or from the same oxide with phosphorus pentasulphide or phosphorus trichloride (*idem*, J. org. Chem., 1965, **30,** 408) or by reduction of 6-hydroxyaminopurine 3-oxide with Raney nickel (*Giner-Sorolla*, J. med. Chem., 1969, **12,** 717).

Guanine oxides are reduced in a similar fashion using Raney nickel to produce guanine (*idem*, Cancer Res., 1965, **25,** 986). On the other hand, hypoxanthine 1-oxide is not reduced by Raney nickel but is slowly reduced in the presence of platinic oxide (*Taylor et al.*, J. org. Chem., 1959, **24,** 2019) whereas the 3-oxide reduces slowly (22 hours) with Raney nickel in aqueous ammonia (*Kawashima et al.*, Bull. chem. Soc., Japan, 1969, **42,** 750), 1- and 3-oxides of alkylated xanthines are readily reduced (*Brown et al.*, J. org. Chem., 1969, **34,** 978; *H. G. Goldner et al.*, Ann., 1965, **691,** 142) but not apparently xanthine 1-oxide (*Brown et al.*, J. org. Chem., 1967, **32,** 1151). Purine oxides can also lose their oxygen on warming and this may occur even during a reaction. Thus 8-phenyltheophylline 7-oxide when heated in dimethylformamide produces 8-phenyltheophylline (*Taylor et al.*, J. Amer. chem. Soc., 1964, **86,** 4721), amination of 6-methylthiopurine 3-oxide with morpholine gives 6-morpholinopurine (*Brown et al.*, J. med. Chem., 1965, **8,** 190) and similar reactions of 6-chloropurine 3-oxide with ammonia, hydrazine or hydroxylamine give the parent aminopurine derivative (*Giner-Sorolla et al.*, J. org. Chem., 1968, **34,** 2157). Conditions are however of special importance in these types of reductions. Thus whereas 6-mercaptopurine 3-oxide is reduced to purine with Raney nickel in aqueous ammonia (*Brown et al.*, J. org. Chem., 1967, **32,** 1151), adenine 1-oxide is produced by reduction of 2-mercaptoadenine 1-oxide with Raney nickel in aqueous sodium hydroxide (*Brown et al.*, *ibid.*, 1963, **28,** 2560). Purine *N*-oxides have also been reduced by xanthine oxidase in the presence of an electron donor (*G. Stohrer* and *Brown*, J. biol. Chem., 1969, **244,** 2498).

Alkylation of adenine 1-oxide or its 9-methyl, ethyl or benzyl derivatives with methyl iodide in dimethylacetamide produces the corresponding 1-methoxy derivatives (*T. Fujii, T. Itaya* and *S. Yamada*, Chem. and Pharm. Bull., Japan, 1965, **13,** 1017; Chem. and Ind., London, 1966, 1598). Similarly theophylline 7-oxides are methylated by dimethyl sulphate or diazomethane to produce 7-methoxytheophyllines (*Goldner et al.*, Ann., 1966, **693,** 233;

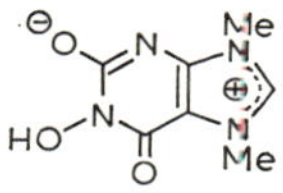

(CIX)

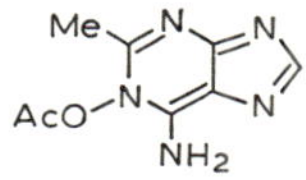

(CX)

A. D. McNaught and *Brown*, J. org. Chem., 1967, **32,** 3689). In contrast, alkylation of xanthine 1-oxide with dimethyl sulphate in dimethylformamide gives the 1-hydroxy-7,9-dimethylbetaine CIX (*U. Wolcke* and *Brown*, J. org. Chem., 1969, **34,** 978); the 3-oxide behaved similarly.

Acylation of *N*-oxides generally produces a reactive *O*-acyl derivative. Thus adenine 1-oxide gives an unstable diacetate whereas the 2-methyl analogue produces an *O*-acetate CX (*M. A. Stevens, H. W. Smith* and *Brown*, J. Amer. chem. Soc., 1960, **82,** 1148).

Both *N*-Oalkyl- and *N*-Oacyl-purines readily tend to undergo rearrangements when heated or treated with nucleophiles. The conversion in this way of purine 3-oxides to 8-substituted derivatives has been dealt with on p. 35 (*Brown* and *J. C. Parham*, The Purines, Jerusalem Symp. on Quantum Chem. and Biochem., Israel Academy of Sciences and Humanities, 1972, p. 550).

Rearrangements also occur on ultraviolet radiation. Thus adenine 1-oxide gave as major products adenine and isoguanine (*Brown, G. Levin* and *S. Murphy*, Biochemistry, 1964, **3,** 880) whereas 6-mercaptopurine 3-oxide gave 6-mercaptopurine with a little 6-thioxanthine (*Brown et al.*, J. med. Chem., 1965, **8,** 190).

1-Alkoxy-9-alkyl-6-iminopurines (CXI) undergo Dimroth type rearrangements in aqueous solution at 100° to produce 6-alkoxyamino-9-alkylpurines (CXII) (*Fujii et al.*,

(CXI) (CXII)

(CXIII) (CXIV)

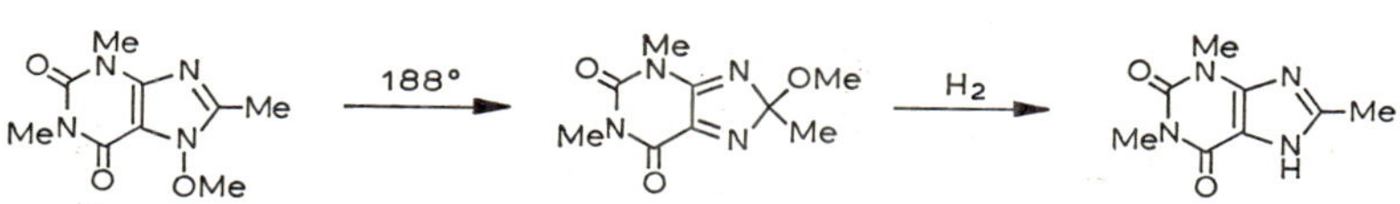

(CXV) (CXVI)

TABLE 10

PROPERTIES OF SOME PURINE *N*-OXIDES

Compound	*m.p. (°C)*	λ_{max} *nm (10^{-3} ε in brackets) in aqueous solution*			*Ref.*
		Acidic	*Neutral*	*Basic*	
Xanthine 3-*N*-oxide	> 350	230 (16.2) 284 (10.3)		216 (22.0) 295 (12.0)	1
Adenine 1-*N*-oxide	297–301	258 (11.5)	231 (41.5) 263 (8.1)	233 (46.2) 275 (7.4)	2
2,6-Diaminopurine 1-*N*-oxide		248 (8.6) 290 (8.7)	230 (31.3) 290 (7.0)	228 (26.8) 295 (7.9)	2
Guanine 3-oxide		213 (12.5) 245 (7.8) 267 (9.5)	217 (2.3) pH 4.8 270 (8.8)	226 (31.0) 283 (9.7)	3
7,9-Dimethyl-3-hydroxyxanthine		245 (5.4) 270 (6.8)	218 (18.3) 276 (6.6) 313 (5.3)	227 (19.8) 285 (4.4) 308 (3.8)	3
7,9-Dimethyl-1-hydroxyxanthine		262 (15.6)	247 (3.9) pH 4 283 (4.8)	204 (10.9) 238 (9.1) 291 (4.1)	3
1-Hydroxyisoguanine		283 (9.9)	243.4(5.3) 288 (7.6)	225 (21.0) 260 (4.8) 300 (8.1)	4 4
1-Hydroxyxanthine	350	267 (9.5)	242 (6.8) 276 (7.5)	225 (20.0) 283 (7.1)	4

TABLE 10 *(continued)*

6-Methylaminopurine 3-oxide	240 (decomp.)	291 (29.8)	228 (26.2) 300 (29.2)	233 (17.4) 297 (13.0)	5
6-*N*-Methylhydroxylaminopurine 3-oxide	195	230 (10.0) 300 (15.8)	233 (13.9) 310 (16.2)	234 (10.5) 335 (13.3)	5
6-Ethoxypurine 3-oxide	250 (decomp.)	264 (13.9)	225 (24.7) 282 (10.9)	276 (9.1)	5
Purine-6-sulphonamide 3-oxide	360 (decomp.)	232 (10.8) 320 (12.9)		231 (23.8) 318 (9.2)	5
Xanthine 7-*N*-oxide		272 (10.2)	272 (6.0)	224 (24)	6
Guanine 7-*N*-oxide		245 (7.8) 267 (9.45)		227 (31) 283 (9.7)	6
6-Mercaptopurine 3-*N*-oxide	230 (decomp.)	250 (4.5) 268 (4.7) 335 (11.9)	256 (6.3) 343 (21.5)	248 (11.8) 334 (21.9)	7
6-Methylmercaptopurine 3-*N*-oxide	246–249	250 (6.7) 318 (20.5)	245 (15) 312 (17.2)	245 (16.3) 312 (17.0)	7
6-Cyanopurine 3-*N*-oxide	305–310		231 (14.3) 300 (6.2) 326 (9.0)	231 (15.6) 307 (5.3) 335 (6.1)	8
Purine-6-carboxylic acid 3-*N*-oxide	285–287 (decomp.)	234 (11.5) 303 (6.7) 327 (9.1)	228 (14.0) 298 (7.2) 316 (9.0)	224 (18.7) 300–315 (6.2)	8
Purine 3-*N*-oxide	288–290 (decomp.)	224 (17.6) 295 (8.4)		225 (25.4) 300 (7.1)	8

References (table 10)

1 *R. M. Cresswell et al.*, J. org. Chem., 1965, **30,** 408.
2 *M. A. Stevens et al.*, J. Amer. chem. Soc., 1958, **80,** 2755.
3 *O. Wolcke* and *G. B. Brown*, J. org. Chem., 1969, **34,** 978.
4 *J. C. Parham, J. Fissekis* and *Brown, ibid.*, 1967, **32,** 1151.
5 *A. Giner-Sorolla*, J. heterocycl. Chem., 1971, **8,** 651.
6 *T. J. Delia* and *Brown*, J. org. Chem., 1966, **31,** 178.
7 *Brown et al.*, J. med. Chem., 1965, **8,** 190.
8 *Giner-Sorolla et al.*, J. org. Chem., 1971, **36,** 1228.

Chem. and Ind., London, 1966, 1967). The action of heat on 7-alkoxy-8-methyltheophyllines (CXIII) gave 8-alkoxy derivatives CXIV of caffeine (*Goldner et al., loc. cit.*). On the other hand, when 7-methoxy-8-methyltheophylline (CXV) is melted (188°) the product may be hydrogenated to 8-methyltheophylline (CXVI).

Adenine 1-*N*-**oxide,** m.p. 297–307°, is obtained in excellent yield by oxidising adenine with cold 30% hydrogen peroxide in acetic acid. Its ultraviolet light absorption closely resembles that of adenine and adenosine *N*-oxide and this, together with its degradation to 3-aminoimidazole-4-carboxamide and 5-amino-4-formamidinoimidazole establishes its structure (*Stevens* and *Brown*, J. Amer. chem. Soc., 1958, **80,** 2755, 2759). N-*Oxides* of

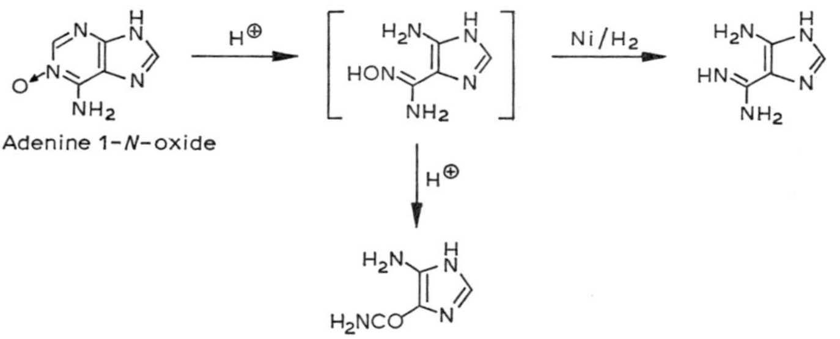

adenosine, m.p. 156° (decomp.), and 2,6-*diaminopurine*, m.p. > 280° (decomp.), are obtained similarly as is 8-*hydroxyadenine* N-*oxide*, m.p. 325° (decomp.), which is also formed by the enzymic oxidation of adenine 1-*N*-oxide in the presence of xanthine oxidase (*Brown et al.*, J. biol. Chem., 1958, **233,** 1515). Absorption spectral data and m.p.s of some purine *N*-oxides are given in Table 10.

(*k*) *Spectra of purines*

(*i*) *Ultraviolet*

Purines show characteristic absorption spectra which are invaluable for identification in the absence of characteristic melting points, and are useful for the determination of structure. This property has been used extensively for the location of purines on paper or thin-layer chromatograms (see below), in eluates from chromatography columns and also in living cells (*cf. T. Caspersson*, Symposium Soc. Exptl. Biol., I, 1947, 127). The usually relatively simple spectra of purines often comprise two main bands, *e.g.* purine itself absorbs just below 220 nm and at 263 nm in neutral solution. Since the imidazole ring has no characteristic ultraviolet light absorption the spectra of analogous purines and pyrimidines show resemblance, *e.g.* the 2,6-disubstituted purines resemble the 2,6-disubstituted 4,5-diaminopyrimidines (*L. F. Cavalieri et al.*, J. Amer. chem. Soc., 1948, **70,** 3875).

The position and intensity of the peaks is often affected by pH since many derivatives are amphoteric. The position of substitution also usually affects the spectrum, the order of increasing bathochromic shift usually being $6 < 8 < 2$ and of increasing hyperchromic effect $2 > 6 > 8$ for the longer wavelength band. In polysubstituted compounds, the effects of the various groups are not additive. Of the various substituents commonly encountered, the effect of the chloro group is negligible and that of mercapto groups is the most marked but otherwise the order of ranking of substituents by their bathochromic and hyperchromic effects alters according to the position; the order of increasing bathochromic shift in the 2 position is MeS < NH_2 < HO < NMe_2 < HS and HO < NH_2, NHMe = MeS < NMe_2 < HS in the 8 position for unionised molecules (*S. F. Mason*, J. chem. Soc., 1954, 2073; Chem. and Biol. of Purines, p. 60). Evidence that the hydroxypurines are in the oxo and the mercaptopurines in the thio form is provided by the fact that, in the unionised state, 8-hydroxypurine and its 7- and 9-methyl and 7,9-dimethyl derivatives all show similar ultraviolet absorption and that the spectra of 8-mercaptopurine and its 9-methyl derivative closely resemble one another and differ markedly from that of 8-methylthiopurine (*Mason, loc. cit.*).

J. M. Gulland and his collaborators were the first to use ultraviolet light absorption to elucidate the structure of purine derivatives. A comparison of the spectra of natural and synthetic derivatives of adenine, xanthine, isoguanine and uric acid showed that the natural glycosides carried the carbohydrate moiety in the 9 position and similar studies in the theophylline series indicated that the synthetic theophylline glycosides were substituted at position 7 (*Gulland et al.*, J. chem. Soc., 1934, 1539; 1936, 765; 1937, 1912; 1938, 692; 1939, 1369, 1784).

Since the spectra are often markedly affected by pH, the ionisation constants (*cf. A. Albert* and *E. P. Serjeant*, "Ionization constants of acids and bases", Methuen, London, 1962), and the order in which ionisation occurs in polysubstituted compounds have been studied by spectroscopic methods, *e.g.*, by *Cavalieri et al.* (J. Amer. chem. Soc., 1954, **76,** 1119), *F. Bergmann* and *S. Dikstein* (*ibid.*, 1955, **77,** 691) and by *Mason* (*loc. cit.*).

(*ii*) *Infra-red*

The polar, water-soluble nature of many purine derivatives particularly the naturally occurring examples, has limited the use of this technique as a routine although the use of pressed alkali halide discs largely removes the difficulty. The most useful region of the spectrum is probably from 1700 to 1500 wave numbers which includes NH, C=C and C=N deformation frequencies but indications of hydrogen bonding (2500–3500 cm^{-1}) are also useful as for example when the absence of hydroxyl groups in hypoxanthine and its analogues confirmed the views on structure reached on the basis of ultraviolet light absorption. Useful studies have been made in this field by, for example, *E. R. Blout* and *M. Fields* (J. Amer. chem. Soc., 1950, **72,** 479), *J. A. Montgomery et al.* (*ibid.*, 1956, **78,** 1928; 1957, **79,** 2185, 5238; 1958, **80,** 407), *C. H. Willits et al.* (*ibid.*, 1955, **77,** 2569) and *Mason* (Chem. and Biol. of Purines, p. 60).

For more detailed reviews see articles by *P. D. Lawley* (u.v.) and *R. Lumley Jones* (i.r.) in *J. H. Lister*, "Purines", ed. *D. J. Brown*, Wiley-Interscience, New York, 1971; *idem*, Adv. in heterocycl. Chem., eds. *A. R. Katritzky* and *A. J. Boulton*, Vol. 6, 1966, Academic Press, New York, and papers by *Y. Kyogoku et al.* (p. 223), *A. Novak et al.* (p. 258), *F. Bergmann et al.* (p. 264) in "The Purines", Jerusalem Symp. on Quantum Chem. and Biochem., Israel Acad. Sci. and Humanities, Jerusalem, 1972; *R. J. Badger*, *D. J. Brown* and *J. H. Lister*, J. chem. Soc., Perkin I, 1974, 152; *Brown* and *R. K. Lynn*, *ibid.*, 1974, 349.

(*iii*) *Proton magnetic resonance spectra*

Simple purines at least are, in general, rather insoluble in non-polar organic solvents and hence p.m.r. spectra are frequently measured in solvents such as D_2O or d_6DMSO using as standards, compounds such as tetramethylsilane, sodium trimethylsilylpropane-1-sulphonate, deuterochloroform or deuteroacetonitrile.

In neutral aqueous solutions purine shows three peaks corresponding to H-8 at highest field followed downfield by H-2 and H-6 and the same order is maintained although with changes in chemical shift to lower field in non-aqueous solutions (*W. C. Coburn et al.*, J. org. Chem., 1965, **30,** 1110; *J. Bullock* and *O. Jardetzky*, *ibid.*, 1964, **29,** 1988; *M. P. Schweizer et al.*, J. Amer. chem. Soc., 1964, **86,** 696; *S. Matsuura* and *T. Goto*, J. chem. Soc., 1965, 623; *F. E. Hruska et al.*, Biochemistry, 1968, **7,** 3721). With substituted purines chemical shifts can vary according to the type of substituent. Thus D_2O with 2- and 6-methylpurine H-8 occurs at higher field than the other proton with the methyl group resonance band at very high field. In the case of 8-methylpurine however, H-6 has the higher field position (*Matsuura* and

Goto, loc. cit.). With 6-substituted purines the relative field odrer of H-8 and H-2 are normally unchanged but with strong electron-donating groups such as amino- (adenine) or iodo- the field order is reversed (*Coburn et al., loc. cit.*).

In basic solution purine (anion) shows three peaks at higher field but in the same order as for the neutral molecule whereas in acid solutions (cation) the peaks, which occur in the same order, are moved downfield, and the lowest fields are attained in the strongest acid solution with H-6 showing maximum pH effect and H-8 the minimum. These results suggest that protonation may occur equally between N-1, N-3 and N-7 (*S. I. Chan et al.*, J. Amer. chem. Soc., 1964, **86,** 4182). With substituted purines much variation is possible. With adenine no evidence for protonation of the amino group can be found (*Coburn et al., loc. cit.*) whereas protonation of the amino group in 6-dimethylamino-3-methylpurine occurs (*Z. Neiman* and *F. Bergmann*, Chem. Comm., 1968, 1002). See also *Chan* and *C.C.Y. Lee* in "Purines", Jerusalem Symp. Quantum Chem. and Biochem., 1972, 277. ^{13}C chemical shifts of 6-substituted purines have been reported (*E. Breitmaier* and *W. Voelter*, Tetrahedron, 1974, **30,** 3941). The kinetics of tritium–hydrogen exchange at C-8 between xanthine, theophylline, caffeine and water have been examined (*M. Jelinska* and *J. Sobkowski, ibid.*, 1977, **33,** 803).

(*iv*) *Mass spectra*

The relative stability of the purine ring system ensures, at least in the simple derivatives, that the highest intensity peak is due to the molecular ion. Subsequent fragmentation patterns depend very much on the nature of the substituents. Thus purine fragments to hydrogen cyanide derived from $C_{(2)}$–$N_{(3)}$ and $C_{(6)}$–$N_{(1)}$, whereas 2-, 6- and 8-methylpurines produce hydrogen cyanide and methyl cyanide (*A. Tatematsu, Goto* and *Matsuura*, Nippon Kagaku Zasshi, 1966, **87,** 71). Adenine similarly produces three molecules of hydrogen cyanide (*J. M. Rice* and *G. D. Dudek*, J. Amer. chem. Soc., 1967, **89,** 2719) with only minor initial involvement of the amino group, whereas the substituted adenine zeatin, [*trans*-6-(4-hydroxy-3-methylbut-2-enyl)aminopurine], shows initial fragmentation at the amino group (*D. S. Letham, J. S. Shannon* and *I. R. C. McDonald*, Chem. Comm., 1964, 230; *Shannon* and *Letham*, New Zealand Journal of Science, 1966, **9,** 833). Fragmentation of the oxo-purines, hypoxanthine, xanthine and uric acid are similar and appear to involve loss of hydrogen cyanide or (xanthine uric acid) hydrocyanate followed by decarbonylation (*Rice* and *Dudek, loc. cit.*; *G. Spiteller* and *M. Spiteller-Friedmann*, Monatsh., 1962, **93,** 632) whereas guanine fragments to cyanamide, derived from N-1–C-2 (*Rice* and

Dudek, loc. cit.). See also *J. Deutsch, Z. Neiman* and *F. Bergmann* ("The Purines", Jerusalem Symp. Quantum Chem. and Biochem., 1972, p. 402) for a mass spectral study of *N*- and *S*-methylpurines.

(*l*) *Paper- and thin-layer chromatography of purines*

Large numbers of solvent systems have been used for the paper- and thin-layer chromatography of purines. They include *n*-butanol alone or in conjunction with ammonia or formic acid, isopropanol with either ammonia or hydrochloric acid and mixtures of isoamyl alcohol and disodium hydrogen phosphate solutions (*cf. E. Chargaff et al.*, J. Amer. chem. Soc., 1949, **71,** 1513; 1952, **74,** 4274; J. biol. Chem., 1947, **168,** 781; *T. Wieland* and *L. Bauer*, Angew. Chem., 1951, **63,** 511; *R. Markham* and *J. D. Smith*, Biochem. J., 1949, **45,** 294; 1950, **46,** 509; *A. Bendich et al.*, J. Amer. chem. Soc., 1954, **76,** 6073).

The detection of purines on paper- and thin-layer chromatograms has also been effected in many ways, *e.g.* by ultraviolet light absorption (*Markham* and *Smith, loc. cit.*), *via* silver, cobalt and copper salts (*E. Vischer* and *Chargaff*, J. biol. Chem., 1947, **168,** 781; *H. Jatkewitz*, Z. physiol. Chem., 1953, **292,** 93; *V. Blazsek*, Naturwiss., 1958, **45,** 42), the use of indicators (*K. Michl*, Naturwiss., 1953, **40,** 390; *Wieland* and *L. Wirth*, Angew. Chem., 1951, **63,** 141), by reagents which react *inter alia* with phenols (*H. Boser*, Z. physiol. Chem., 1954, **298,** 145; *G. Harris* and *R. Parsons*, Chem. and Ind., London, 1956, 1312; *E. Hjilt et al.*, Analyst, 1955, **80,** 706) and by permanganate and chlorine (*E. Gerlach* and *H. J. Döring*, Naturwiss., 1955, **42,** 344).

2. Thiazolopyrimidines

Both major types are known namely thiazolo[5,4-*d*]pyrimidine (I) and thiazolo[4,5-*d*]pyrimidines (II).

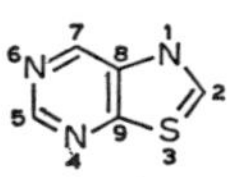

(I)

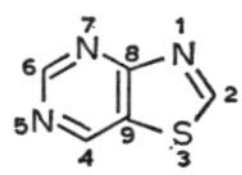

(II)

(*a*) *Thiazolo*[*5,4*-d]*pyrimidines*

(*i*) *Synthesis*

(*1*) *From 5-formamido-4-mercaptopyrimidines.* Heating 5-amino-4-mercaptopyrimidines with formic acid affords the corresponding thiazolopyrimidine (*S. J. Childress* and *R. L. McKee*, J. Amer. chem. Soc., 1951, **73,** 3862; *F. L. Rose*, J. chem. Soc., 1952, 3448; *G. P. Hager* and *C. Kaiser*, J. Amer. pharm. Assoc., 1954, **44,** 193; *G. B. Elion et al.*, J. Amer. chem. Soc., 1956, **78,** 2858). The last authors noted that the strength of formic acid used and the temperature of reaction are of considerable importance since incomplete cyclisation is possible.

(*2*) A 5-amino-4-chloropyrimidine has been converted into a 2-aminothiazolopyrimidine by heating with potassium thiocyanate in hydrochloric acid (*Rose, loc. cit.*).

In a similar manner *F. F. Blicke* and *R. L. Schaaf* (J. Amer. chem. Soc., 1956, **78,** 5857) prepared a series of 2-substituted amino-4,6-dimethyltetra-

hydrothiazolopyrimidine-5,7-diones by the reaction of alkyl and aralkyl thiocyanates with 4,5-diamino-1,3-dimethyltetrahydropyrimidine-2,6-dione, some of the corresponding isoxanthine being formed simultaneously.

In a similar manner, reaction of 5-amino-4,6-dichloropyrimidine with sodium hydrogen sulphide produced 5-amino-4-chloro-6-mercaptopyrimidine, which with triethyl orthoformate gave 7-*chlorothiazolo*[5,4-*d*]*pyrimidine* (III) (*R. Marchal* and *R. Promel*, Bull. Soc. chim. Belges, 1957, **66,** 406), m.p. 152–154° (decomp.). Further reaction of III with amines, methanol, etc. produced corresponding 7-*furfurylamino*-, m.p. 93.5–95°, 7-*benzyl*-

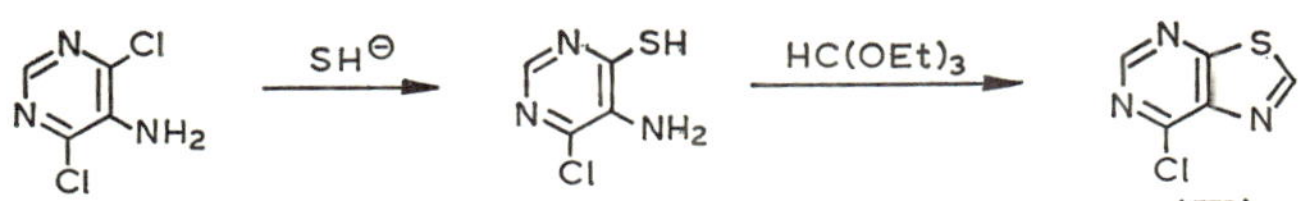

(III)

amino-, m.p. 129–130°, 7-*hydrazino*-, m.p. 200–260° (decomp.), 7-*methoxy*-, m.p. 164–166°, 7-*hydroxy*-, m.p. 300° and 7-*mercapto*-, m.p. 330°, -*thiazolo*[5,4-d]*pyrimidines* (*L. Marchal et al., ibid.*, 1960, **69,** 177; *J. A. Montgomery et al.*, J. org. Chem., 1969, **34,** 3161).

(*3*) Proof of structure of the products prepared by method (*2*) was obtained by *Blicke* and *Schaaf* by synthesis from thiazoles in the following manner:

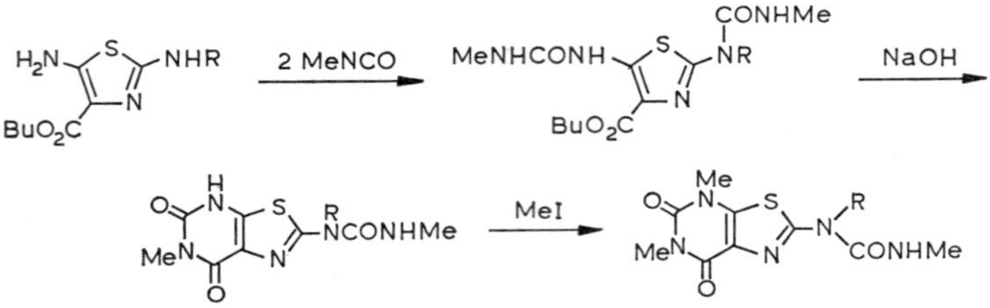

Similarly 5-amino-2-mercaptothiazole-4-carboxamide (IV) was obtained from aminocyanoacetamide and carbon disulphide (*A. H. Cook*, *I. M. Heilbron* and *P. Smith*, J. chem. Soc., 1949, 1440), methylated to V, which was cyclised to various thiazolopyrimidines including VI (*Cook et al.*, *ibid.*, 1949, 1064, 1069, 1071). The corresponding aminothiazolethiocarboxamide VII, similarly prepared from aminocyanothioacetamide and carbon

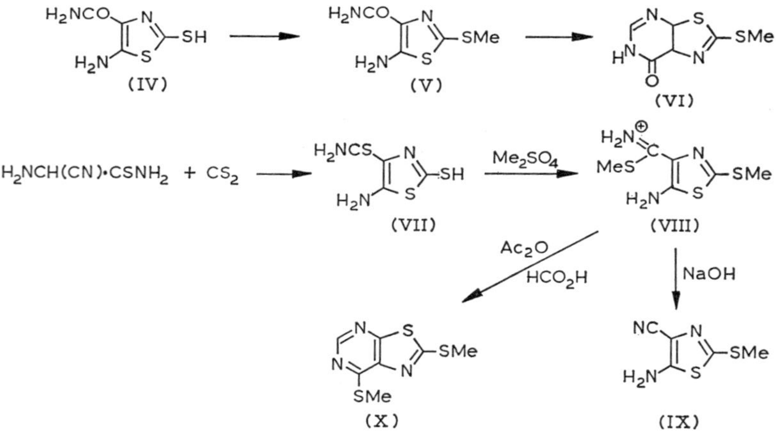

disulphide, when methylated produced the carboxythioimidate VIII which with sodium hydroxide gave the aminocarbonitrile IX and with formic–acetic anhydride 2,7-*dimethylthiothiazolo*[5,4-d]*pyrimidine* (X), m.p. 120° (*G. Shaw* and *D. N. Butler*, *ibid.*, 1959, 4040).

(*b*) *Thiazolo*[*4,5*-d]*pyrimidines*

Only a few derivatives of this system have been prepared. Thus 2,4-diamino-6-hydroxy-5-thiocyanatopyrimidine cyclised with acetic anhydride followed

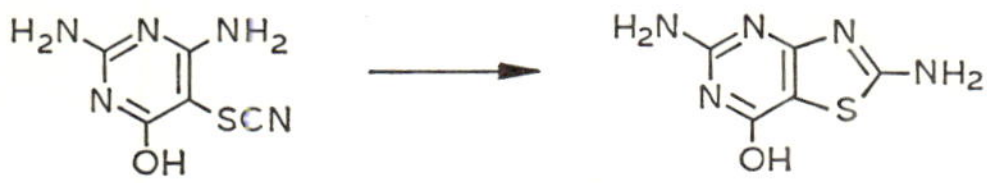

(XI)

by sodium carbamate to 2,6-*diamino-4-hydroxythiazolo*[4,5-d]*pyrimidine* (XI), m.p. 350° (*J. A. Baker* and *P. V. Chatfield*, J. chem. Soc., C, 1970, 2478; *ibid.*, p. 603).

Another route involves the reaction of thioureas with bromocyanoacetamide to produce aminothiazoles of type XII, which may be cyclised by acetic anhydride to thiazolo[4,5]pyrimidines of type XIII (*A. Singh, H. Singh* and *K. S. Narang*, Indian J. Chem., 1971, **9,** 651).

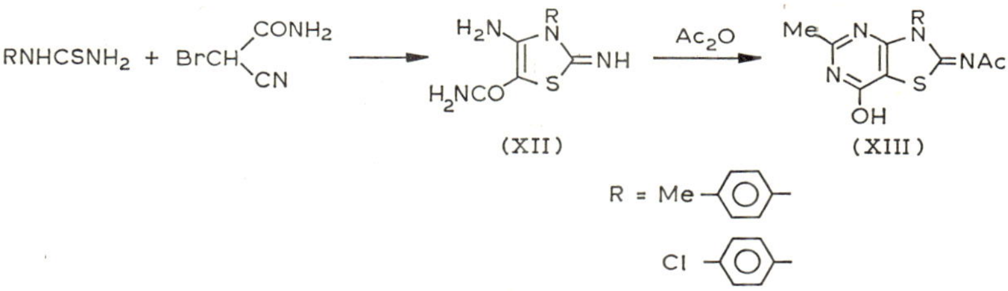

3. Pyrazolopyrimidines

A number of derivatives of pyrazolo[4,3-*d*]- and -[3,4-*d*]pyrimidines have been prepared.

(*a*) *Pyrazolo*[*3,4-*d]*pyrimidines or 1,2,5,7-tetra-azaindenes* (*I*)

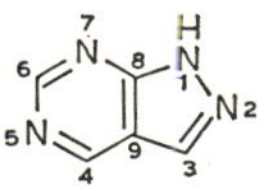

(I)

(*i*) *Synthesis*

(*1*) *I. R. Justoni* and *R. Fusco* (Gazz., 1938, **68,** 66) prepared the first pyrazolo[3,4-*d*]pyrimidine by heating 5-acetamido-1,3-diphenylpyrazole-4-carboxamide. Subsequently, derivatives have usually been synthesized from substituted pyrazoles using cyclisation methods developed originally for the preparation of purines from imidazoles as illustrated by the following examples starting from 5-amino-4-cyanopyrazole (II) (*R. K. Robins*, J.

Amer. chem. Soc., 1956, **78,** 784; J. org. Chem., 1958, **23,** 191; *cf. P. Schmidt et al.*, Helv., 1958, **41,** 1052).

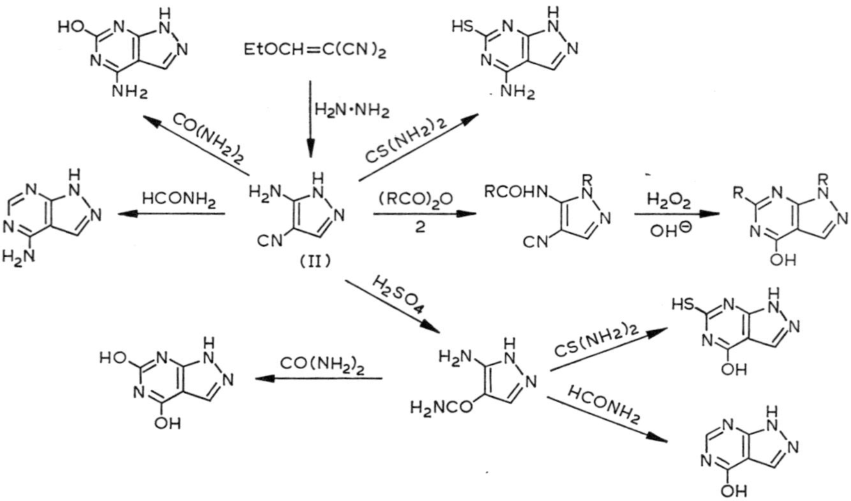

It is not without interest that the oxidation of pyrazole-4,5-dicarboxamide with hypochlorite afforded 4,6-dihydroxypyrazolo[3,4-*d*]pyrimidine and not the [4,3-*d*]-isomer (*E. A. Falco* and *G. H. Hitchings*, J. Amer. chem. Soc., 1956, **78,** 3143).

(*2*) Pyrazolo[3,4-*d*]pyrimidines including the antimetabolites allopurinol and oxyallopurinol and their derivatives have also been prepared by the reaction of various α-cyano-β-ethoxyacrylamides with hydrazine or substi-

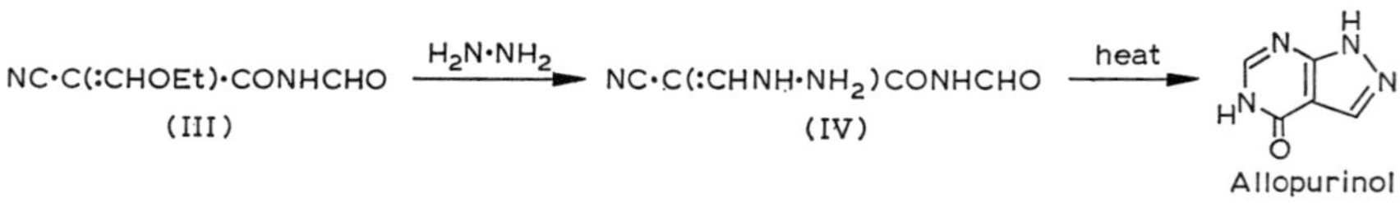

tuted hydrazines. Thus α-cyano-β-ethoxy-*N*-formylacrylamide (III) and hydrazine gave the hydrazinoacrylamide (IV), which when heated gave an almost quantitative yield of **allopurinol** (*B. G. Hildick* and *G. Shaw*, J. chem. Soc., C, 1971, 1610).

Oxyallopurinol was similarly prepared from ethyl α-cyano-β-ethoxyacryloylcarbamate and hydrazine followed by ring closure of the intermediate hydrazinomethylene derivative whereas the use of methyl- or phenyl-hydrazines with acryloyl-carbamates or -ureas produced corre-

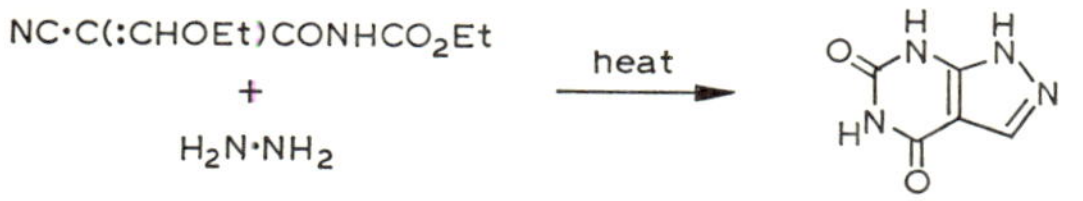

Oxyallopurinol

spondingly substituted pyrazolopyrimidines (*Hildick* and *Shaw*, *loc. cit.*; *P. Lees* and *Shaw*, J. chem. Soc., C, 1968, 1519).

In addition, reaction of phenylhydrazine with the cyanothiazine V (*M. R. Atkinson et al.*, J. chem. Soc., 1956, 3847) gave, *via* an acyclic dithio-

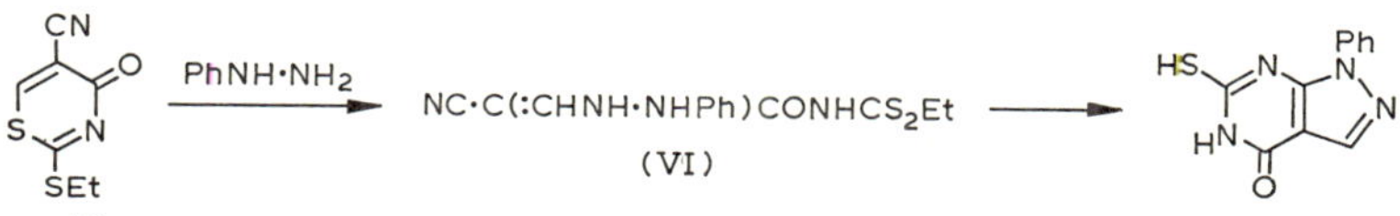

(V) (VII)

carbamate VI, 6-mercapto-1-phenylpyrazolopyrimidone (VII) (*Hildick* and *Shaw*, *loc. cit.*).

Mercapto compounds analogous to VII including VIIIa and VIIIb have also been prepared by heating a chloropyrazolopyrimidine, *e.g.* IX, with thiourea (*C. C. Cheng* and *Robins*, J. org. Chem., 1958, **23,** 852).

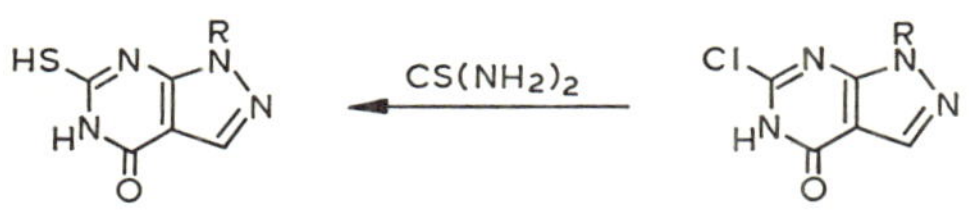

(VIII) (IX)

(a) R = H
(b) R = Me

(*3*) *W. Pfleiderer* and *K. H. Schundehutte* (Ann., 1958, **615,** 42) have prepared a series of 5,7-dimethyl-4,6-dioxotetrahydropyrazolo[3,4-*d*]-pyrimidines by the cyclisation of 6-hydrazino-1,3-dimethyl-2,4-dioxotetrahydropyrimidine with formamide, acetic anhydride and formic-acetic anhydride.

(*ii*) *Properties*

Some physical properties are listed in Tables 11 and 12.

As in the purine series, hydroxyl groups are convertible to chloro with phosphoryl chloride and to mercapto with phosphorus pentasulphide (only the 4-hydroxyl group of 4,6-dihydroxypyrazolopyrimidine so reacting) whilst the 4-chloro derivatives react with amines, sodium ethoxide, methyl iodide and with hydrogen over palladised charcoal in the normal manner (*Robins*, *loc. cit.*; *Falco* and *Hitchings*, *loc. cit.*; *Schmidt et al.*, *loc. cit*). The ultraviolet absorption properties and ionisation constant of pyrazolo[3,4-*d*]- and -[4,3-*d*]-pyrimidines have been compared with those of corresponding purines (*B. M.*

TABLE 11

PROPERTIES OF PYRAZOLO[3,4-*d*]PYRIMIDINES

Substituent	*m.p. (°C)*	λ_{max} *nm in aqueous solution (ε × 10⁻³ in brackets)*		*Ref.*
		Acidic	*Basic*	
4-Hydroxy-		250	261	1
4-Hydroxy-6-methyl-	336–338	252 (8.55)	259 (8.85)	2
4-Hydroxy-1,6-dimethyl-	277–278	253 (10.3)	268 (10.8)	2
4,6-Dihydroxy-	> 320	240 (4.75)	255 (5.6)	1, 3
4-Chloro-	130–135 (decomp.)	261		1
4-Chloro-6-methyl-	140 (decomp.)	256 (5.7)	265 (4.7)	2
4-Chloro-1,6-dimethyl-	74	267 (8.65)	268 (9.15)	2
4-Amino-	> 300	261 (8.1)	265 (7.4)	1, 3
4-Amino-6-methyl-	> 300	259 (8.65)	265 (8.8)	2
4-Mercapto-	> 300	321	315	1
4-Methoxy-	193–194	240	282	1
Parent	213–214	261	268	1

References

1 *R. K. Robins*, J. Amer. chem. Soc., 1956, **78,** 784.
2 *C. C. Cheng* and *Robins*, J. org. Chem., 1958, **23,** 191.
3 *E. A. Falco* and *G. H. Hitchings*, J. Amer. chem. Soc., 1956, **78,** 3143.

TABLE 12

PROPERTIES OF 4,6-DIOXO-4,5,6,7-TETRAHYDROPYRAZOLO[3,4-*d*]-PYRIMIDINES

Substituent	*m.p. (°C)*	λ_{max} *nm in aqueous solution (ε × 10⁻³ in brackets)*		*Ref.*
		Neutral	*Basic*	
1,5,7-Trimethyl-	227–228	239 (6.6)* 254 (6.4)*		1
2,5,7-Trimethyl	205	243 (5.9) 263 (7.9)		2
3,5,7-Trimethyl-	288	246 (8.5)	258 (9.55)	2
2,3,5,7-Tetramethyl-	230	254 (6.8)		2

* In ethanol. Inflexion.

References

1 *P. Schmidt et al.*, Helv., 1958, **41,** 1052.
2 *W. Pfleiderer* and *K. H. Schundehutte*, Ann., 1958, **615,** 42.

Lynch et al., J. chem. Soc., 1958, 2973), the pyrazolopyrimidines being slightly stronger bases and slightly weaker acids than the purines, [4,3-*d*] compounds being stronger bases than the [3,4-*d*] isomers. This is consistent with a nitrogen atom in the pyrimidine ring being the most basic centre in all these systems (*cf. Cochran*, Acta Cryst., 1951, **4**, 81).

The pyrazolo[3,4-*d*]pyrimidines show remarkably similar ultraviolet light absorption to the corresponding purines though they are somewhat less soluble in water. The chlorine in 4-chloropyrazolo[3,4-*d*]pyrimidine is, however, rather more reactive than that in 6-chloropurine as shown by its hydrolysis with boiling water and ready reaction with hydroxylamine. The methylation of 4,6-dihydroxypyrazolo[3,4-*d*]pyrimidine with dimethyl sulphate and alkali affords the 2,5,7-trimethyl derivative, the 1,3,7-analogue being prepared from a 1-*N*-substituted pyrazole (*Schmidt et al., loc. cit.*).

Several of the pyrazolo[3,4-*d*]pyrimidines have useful biological properties. Of special interest is the hypoxanthine analogue, allopurinol which inhibits xanthine oxidase *in vivo*; this property makes the compound valuable for the treatment of gout and similar diseases where diminished oxidation of purines is required (*G. B. Elion et al.*, Biochem. Pharmacol., 1963, **12**, 85; *R. W. Rundles et al.*, Trans. Assoc. Amer. Physicians, 1963, **76**, 126; *Elion et al.*, Abstracts of 6th Internat. Congress of Biochemists, New York, 1964, **4**, 305; *J. O. Alexander et al.*, Biochem. Pharmacol., 1966, **15**, 881). Many compounds of this type have also shown anti-tumor and anti-leukaemic activity (*H. E. Skipper et al.*, Cancer Res., 1957, **17**, 579; Proc. Soc. expt. Biol. Med., 1955, **89**, 594).

(*b*) *Pyrazolo[4,3-*d*]pyrimidines or 1,2,4,6-tetra-azaindenes (X)*

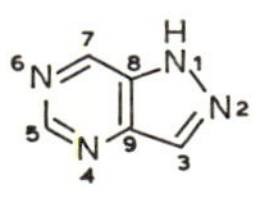

(X)

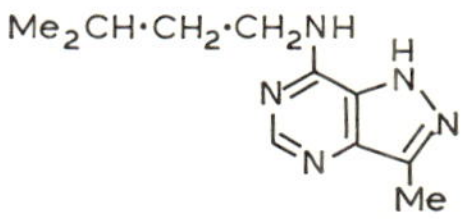

(XI)

3-Methyl-7-(3-methylbutylamino)pyrazolo[4,3-*d*]pyrimidine (XI) has been shown to be an inhibitor of cytokinins (*cf.* p. 77) when used in 100–200 fold excess (*S. M. Hecht et al.*, Proc. Nat. Acad. Sci. U.S.A., 1971, **68**, 2608).

(*i*) *Synthesis*

(*1*) *R. Behrend* (Ann., 1888, **245**, 213) prepared 5,7-dihydroxypyrazolo-[4,3-*d*]pyrimidine by the diazotisation and cyclisation of 5-amino-6-methyluracil and this approach has been extensively employed by *F. L. Rose* (J. chem. Soc., 1952, 3448; 1954, 4116) for the synthesis of substituted 5-amino-, 5-chloro and 5-methylthio-pyrazolo[4,3-*d*]pyrimidines.

Rose noted that when the procedure was applied to 2-amino-6-ethyl-4-methylpyrimidine, roughly equal amounts of 5-amino-3,7-dimethyl- and 5-amino-7-ethyl-pyrazolo[4,3-*d*]pyrimidine were formed.

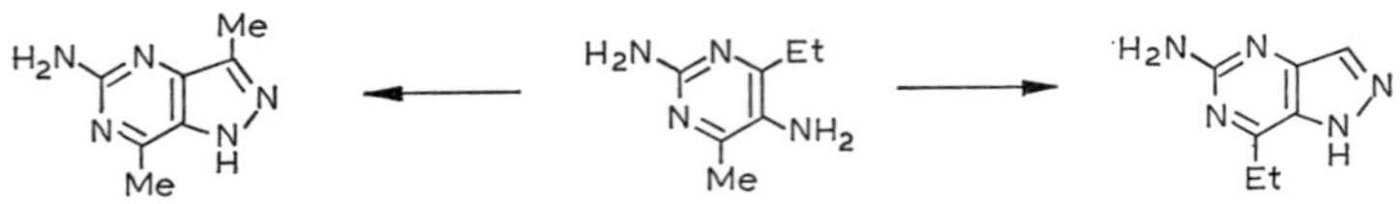

A modification of this method (*Rose, loc. cit.*) leads from 4-chloro-5-nitropyrimidines to 3-ethoxycarbonyl derivatives in the following way:

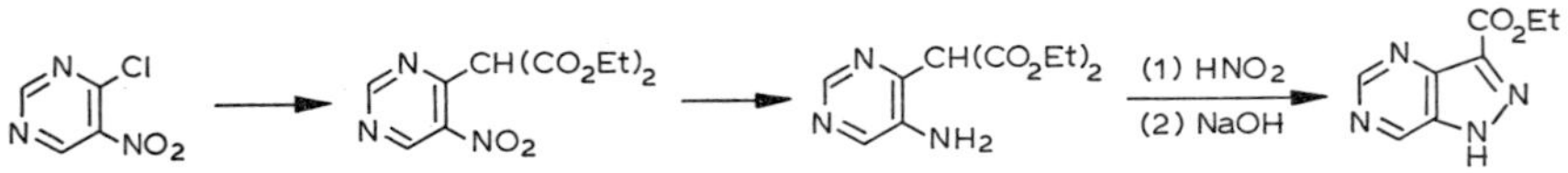

(*2*) The synthesis of a series of pyrazolo[4,3-*d*]pyrimidines from 5-nitropyrazole-5-carboxylic acid has been described by *Robins et al.* (J. Amer. chem. Soc., 1956, **78**, 2418).

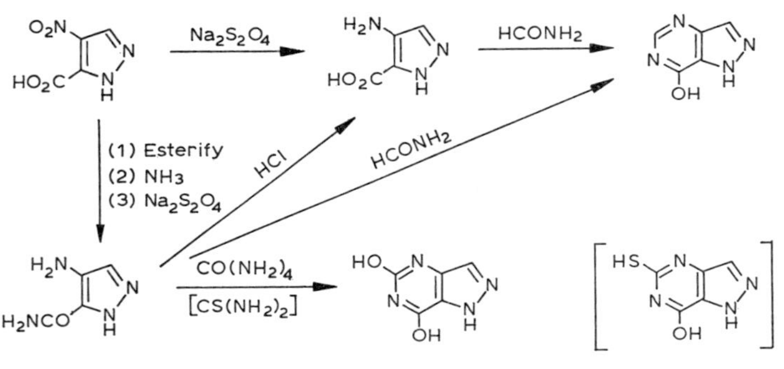

(*ii*) *Properties*

Some of the properties of the pyrazolo[4,3-*d*]pyrimidines were discussed in the section dealing with the [3,4-*d*]isomer (p. 103). They are in general slightly less soluble in water than the corresponding purines and differ from the latter also in their ultraviolet light absorption in having a general bathochromic shift of from 10 to 30 nm accompanied by a hypochromic shift. The conversion of hydroxyl to mercapto groups with phosphorus pentasulphide, the hydrolysis of a 5-chloro- to the 5-hydroxy-derivative under acidic conditions and the replacement of chloro or methylthio groups by amines proceeds as in the purine series. Chlorination or bromination of 5,7-dihydroxypyrazolo[4,3-*d*]pyrimidine affords the 3-halogeno derivative as does also the halogenation of its 1,4,6-trimethyl derivative (*Robins et al., loc. cit.*). Antitubercular and antitumour activity has been demonstrated in compounds of this group (*Lynch et al., ioc. cit.*; *Rose, loc. cit.*).

4. *vic.*Triazolo[4,5-*d*]pyrimidines (8-azapurines) (I)

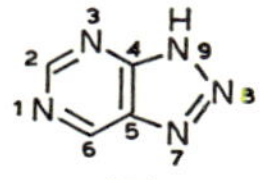

(I)

Synthesis

(*1*) A compound of this type was first prepared by treating 4,5-diamino-6-methylpyrimidine with nitrous acid (*S. Gabriel* and *J. Colman*, Ber., 1901,

34, 1249) and this method has since been used extensively (*e.g. J. Davoll*, J. chem. Soc., 1958, 1593; *C. L. Leese* and *G. M. Timmis*, *ibid.*, 1958, 4107).

The cyclisation of 4,5-diamino-6-β-D-glucopyranosylamino-2-methylthio-pyrimidine with nitrous acid afforded the 6-amino-9-β-D-glucopyranosyl-

(II)

R = β-D-glucosyl

triazolo[4,5-*d*]pyrimidine (II) and not the corresponding 6-glucosylamino derivative (*Davoll, loc. cit.*; *cf. J. Baddiley et al.*, J. chem. Soc., 1958, 3606).

Also (*E. Richter* and *E. C. Taylor*, J. Amer. chem. Soc., 1956, **78,** 5848) a 4-amino-5-benzeneazopyrimidine was oxidatively cyclised by cupric sulphate to give a 2-*N*-phenylpyrazolopyrimidine as follows:

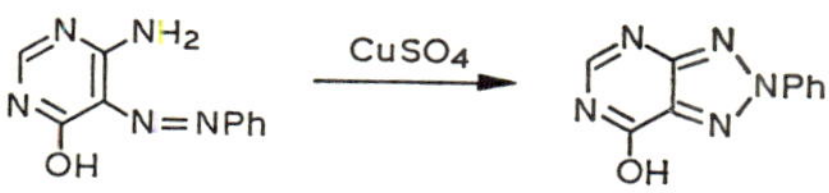

The 8-azapurine, **pathocidin,** 2-amino-6-hydroxy-8-azapurine or 8-azaguanine occurs naturally in *S. Aalbus* (*K. Anzai* and *S. Suzuki*, J. Antibiotics, Japan, 1961, A, **14,** 253; *Anzai, J. Nagatsu* and *Suzuki*, *ibid.*, p. 340;

Pathocidin

idem, ibid., 1962, A, **15,** 103; *R. J. Suhadolnik*, "Nucleoside Antibiotics", Wiley-Interscience, New York, 1970).

It was synthesised by *R. O. Roblin et al.* (J. Amer. chem. Soc., 1945, **67,** 290) by the reaction of 2,4,5-triamino-6-hydroxypyrimidine with nitrous acid.

(*2*) The formation of the triazolo[4,5-*d*]pyrimidine ring-system from substituted triazoles has been affected in ways analogous to those developed for purine synthesis. Thus, aminotriazole-carboxamides and -carboxythioamides have been cyclised by acetic anhydride or orthoformates (*Richter*

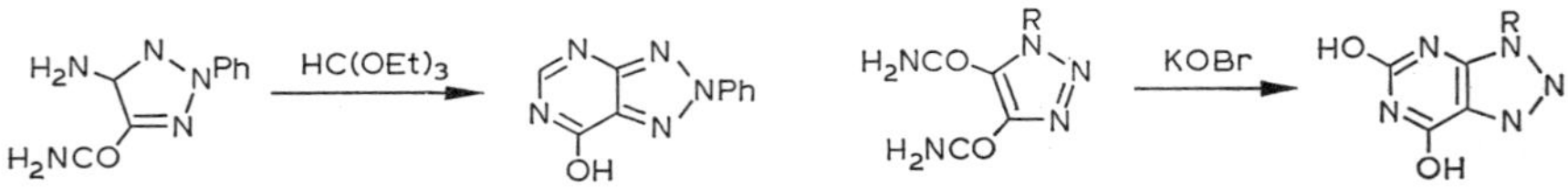

and *Taylor, loc. cit.*) and by cyanoacetamide, and triazoledicarboxamides have been cyclised with hypobromite to afford 7- and 9-glycosido-2,6-dihydroxytriazolo[4,5-*d*]pyrimidines (*Baddiley et al., loc. cit.*).

Methylation of 4-amino-3-benzyltriazole-5-carboxamide (III) produced from benzyl azide and cyanoacetamide, with methyl toluene-*p*-sulphonate, produced a quaternary derivative IV, hydrogenation of which gave the 3-methyltriazole V, which was readily cyclised by formamide to 6-hydroxy-7-methyl 8-azapurine (VI). This in turn with phosphoryl chloride and dimethyl-

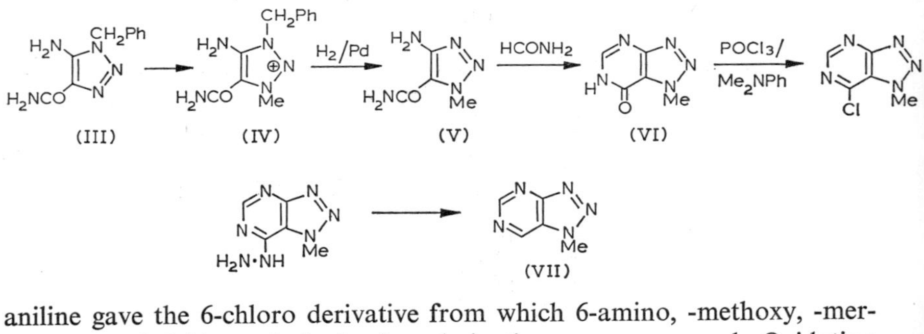

aniline gave the 6-chloro derivative from which 6-amino, -methoxy, -mercapto, -methylthio and -hydrazino derivatives were prepared. Oxidation of the 6-hydrazino derivative with silver oxide gave 7-methyl-8-azapurine (VII) (*A. Albert* and *K. Tratt, ibid.*, 1968, C, 344).

2-Methyl-8-azapurines have been prepared by an unusual route. Formylation of 4-amino-triazole-5-carboxamide (VIII) gave a formylated dimer which produced 5-formamido-2-methyl-1,2,3-triazole-4-carboxamide (IX) with dimethyl sulphate; cyclisation of IX with formamide gave 6-hydroxy-8-methyl-8-azapurine (X).

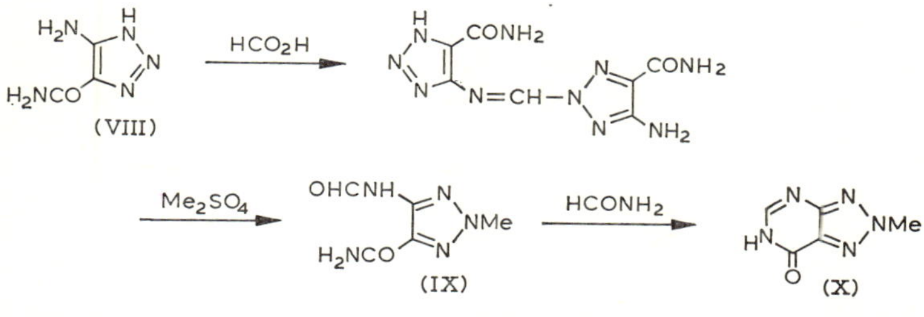

(VIII) (IX) (X)

The latter compound with phosphoryl chloride gave the 6-chloroazapurine which with hydrazine gave the 6-hydrazino derivative oxidation of which with silver oxide produced 8-methyl-8-azapurine (XI) (*Albert*, J. chem. Soc., 1968, C, 2076).

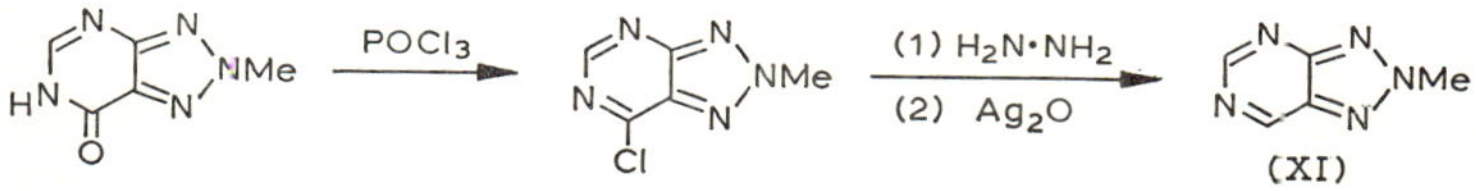

(XI)

Attempts to cyclise 5-amino-1-methyl-1,2,3-triazole-4-carboxamide (XII) with guanidine were unsuccessful but with thiourea 9-methyl-6-hydroxy-2-mercapto-8-azapurine (XIII) was produced (*Albert* and *H. Taguchi*, J. chem. Soc., Perkin I, 1972, 449).

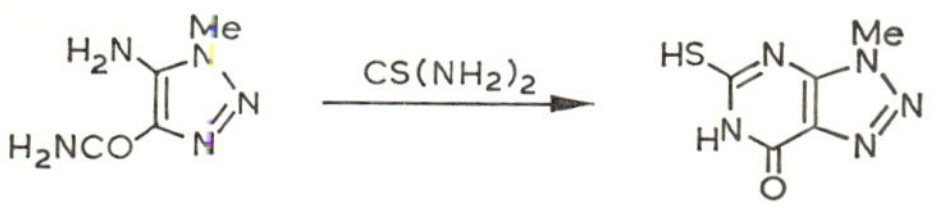

(XII) (XIII)

In a similar manner 5-amino-3-methyl-1,2,3-triazole-4-carboxamide with phosphoryl chloride and dimethylformamide produced the formamidine XIV which cyclised when heated to give 6-hydroxy-7-methyl-8-azapurine (XV) (*Albert*, *ibid.*, p. 461).

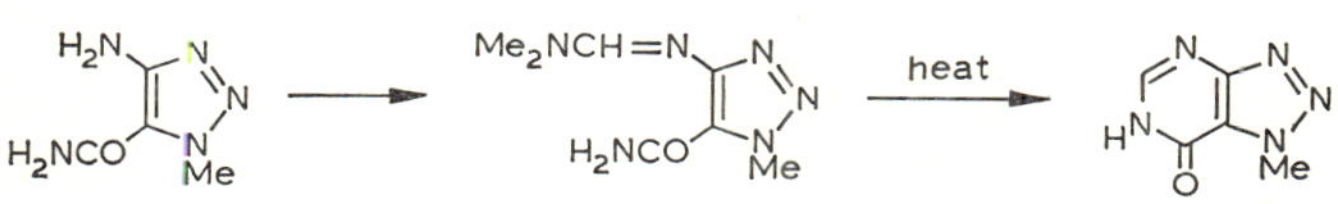

(XIV) (XV)

8-Benzyl-6-hydroxy-8-azapurine (XVI) was obtained by benzylation and subsequent cyclisation of 5-formamido-1,2,3-triazole-4-carboxamide (*Albert* and *D. Thacker*, *ibid.*, p. 468).

(XVI)

Properties

The *vic*-triazolo[4,5-*d*]pyrimidines have generally similar properties to the equivalent purines, including high melting points and solubility in water. The ultraviolet light absorptions of corresponding purines and triazolopyrimidine are very similar and are similarly affected by pH. The 1- and 3-*N*-glycosides of these compounds may, for example, be distinguished by their ultraviolet light absorption properties as were the corresponding 7- and 9-glycosidopurines (*Baddiley et al., loc. cit.*). Successive treatments of 5-amino-7-hydroxytriazolo[4,5-*d*]pyrimidine with phosphorus pentasulphide in pyridine afforded, firstly the 5-amino-7-mercapto and then the 5,7-dimercapto derivative (*C. T. Bahner et al.*,

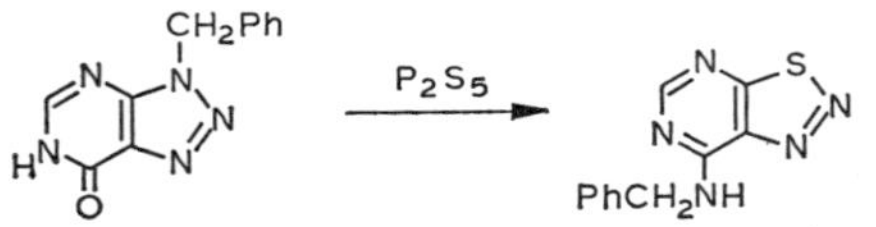

(XVII)

J. Amer. chem. Soc., 1954, **76,** 1370). On the other hand, reaction of 9-benzyl-6-hydroxy-8-azapurine with phosphorus pentasulphide gave, in addition to the 6-mercapto-8-azapurine, 7-benzylamino-1,2,3-thiadiazolo[5,4-*d*]pyrimidine (XVII) (*Albert*, J. chem. Soc., C, 1969, 152) as the major product.

(XVIII)

Hydrolysis of 6-methylthio-8-azapurine with *N* hydrochloric acid gave a high (90%) yield of methyl 5-amino-1,2,3-triazole-4-thiocarboxylate (XVIII) (*idem, ibid.*, 1969, 2379).

Methylation of 8-azapurine with methyl sulphate gives a mixture of the 7-, 8- and 9-methyl derivatives (*Albert, W. Pfleiderer* and *Thacker, ibid.*, C, 1969, 1084).

Several glycofuranosyl (including ribo-, xylo-, arabino- and deoxy-) derivatives of 8-azapurines and related 2-aza- or 2,8-diaza-purines have been prepared and their biological properties summarised (For a review see *J. A. Montgomery* and *H. J. Thomas*, "The Purines", Jerusalem Symp. Quantum Chem. Biochem., 1972, **4**, 431).

Analogues of adenosine, guanosine and xanthosine in this series were prepared by *Davoll* (*loc. cit.*) from the *N*-chloromercuri derivative of the heterocyclic compound with an acetylglycosyl halide. Thus 6-acetamido-triazolo[4,5-*d*]pyrimidine affords 6-amino-9-*β*-D-ribofuranosyltriazolo[4,5-*d*]pyrimidine (8-aza-adenosine) which was deaminated to the inosine analogue. Similarly 2,6-diacetamino-*vic*-triazolopyrimidine (XIX) gave the corresponding 9-*β*-D-triacetylribofuranosyl derivative from which deacetylation and deamination afforded the guanosine analogue, 2-amino-6-hydroxy-9-*β*-D-ribofuranosyl-*vic*-triazolo-pyrimidine (XX, R = *β*-D-ribofuranosyl).

TABLE 13

PROPERTIES OF SOME TRIAZOLO[4,5-*d*]PYRIMIDINE GLYCOSIDES

Triazolo[4,5-d]*pyrimidine*	*m.p. (°C)*	$[\alpha]_D^0$ *in* H_2O	λ_{max} *(nm) in aqueous solution* $(\varepsilon \times 10^{-3}$ *in brackets)*			*Ref.*
			Acidic	*Neutral*	*Basic*	
6-Amino-7-β-D-glucopyranoside			244 (5.3) 286 (11.6)	246 (5.5) 299 (10.3)	245 (5.3) 299 (9.8)	1
6-Amino-9-β-D-glucopyranoside	241 (decomp.)	− 22	263 (12.5)	280 (11.7)	280 (11.2)	1
6-Amino-7-β-D-ribofuranoside		− 92	278 (8.9)	278 (5.0)	310 (5.1)	2
6-Amino-9-β-D-ribofuranoside	218–219	− 79	255 (8.9) 240 (6.7)	252 (8.6) 277 (7.9)	250 (5.6) 280 (8.4)	1, 2
6-Hydroxy-9-β-D-ribofuranoside	199–200	− 54	255 (9.4)	256 (8.8)	277 (0.5)	1
2,6-Dihydroxy-9-β-D-ribofuranoside	198–199 (decomp.)	−103[a]	240 (5.9) 256 (9.5)	252 (9.7) 277 (18.8)	251 (7.1) 280 (9.5)	1
2-Amino-6-hydroxy-7-β-D-ribofuranoside	200 (decomp.)	− 75	285 (6.3)	283 (8.4) 297 (7.1)	254 (5.6) 256 (5.6) 304 (8.8)	1
2-Amino-6-hydroxy-8-β-D-ribofuranoside	sinters 230–235	− 79	238 (7.5) 296 (4.7)	237 (8.4) 301 (5.6)	303 (6.3)	1
2-Amino-6-hydroxy-9-β-D-ribofuranoside	250–252 (decomp.)	− 97[a]	255 (13.6) 269 (10.3)	256 (13.7) 275 (9.5)	222 (24.6) 279 (11.9)	1

[a] In 0.1 *N* sodium hydroxide

References

1 *J. Davoll*, J. chem. Soc., 1958, 1593.
2 *J. Baddiley et al.*, *ibid.*, p. 3606.

TABLE 14

PROPERTIES OF SOME TRIAZOLO[4,5-*d*]PYRIMIDINE (8-AZAPURINE) DERIVATIVES

Substituent	*m.p. (°C)*	λ_{max} *(nm) in aqueous solution (ε × 10⁻³ in brackets)*			*Ref.*
		Acidic	*Neutral*	*Basic*	
6-Hydroxy-	> 308 (explodes)	253.5	258.5	267.5	1
6-Amino-	> 310	270	275	275	1
9-Methyl-	320 (decomp.)	255	278	245	3
				278	
2-Amino-6-methyl-	284–285				7
2-Amino-6-hydroxy-	> 300	247.5	277	240–250	6
2-Amino-6-mercapto-	270 (decomp.)		231 (12.5)	224 (13.55)	4
			341 (8.41)	325 (8.91)	
2,6-Dihydroxy-	270	265	265	235	1
				285	
2,6-Dimercapto-			283 (14.22)	283 (11.91)	4
			343 (10.05)	343 (6.32)	
2,6-Diamino-3-methyl-	294–295	254 (9.1)	258 (5.1)	259 (5.1)	6
		284 (6.7)	287 (8.8)	287 (8.8)	
6-Hydroxy-8-phenyl-	281		240[a] (8.51)		5
			300[a] (14.45)		
			307[a] (14.45)		
6-Amino-8-phenyl-	340		286[a] (129)		5
			320[a] (132)		
6-Mercapto-8-phenyl-	323		255[a] (107.2)		5
			268[a] (107.2)		
			366[a] (129)		
			374[a] (135)		
8-Benzyl-	117	256 (4.08)	273 (4.11)		8
2,6-Diamino-7-methyl-		218 (4.13)	< 230 (> 3.5)		
		258 (4.01)	249 (3.78)		9
		282 (3.90)	309 (3.79)		
2,6-Diamino-8-methyl-		259 (4.11)		257 (3.77)	9
		259 (4.11)		257 (3.77)	9
		284 (4.08)		307 (3.91)	
9-Benzyl-6-hydroxy-		255 (4.03)		275 (4.06)	10
8-Methyl-		254 (4.0)	270 (4.06)		11
7-Methyl-		243 (3.81)	276 (3.73)		12
			285 (3.97)		

[a] In ethanol.

References (table 14)

1 *R. O. Roblin et al.*, J. Amer. chem. Soc., 1945, **67,** 290.
2 *S. Gabriel* and *J. Colman*, Ber., 1901, **34,** 1249.
3 *J. Baddiley et al.*, J. chem. Soc., 1958, 1651.
4 *C. T. Bahner et al.*, J. Amer. chem. Soc., 1954, **76,** 1370.
5 *E. Richter* and *E. C. Taylor*, *ibid.*, 1956, **78,** 5848.
6 *J. Davoll*, J. chem. Soc., 1958, 1593.
7 *F. L. Rose*, *ibid.*, 1952, 3448.
8 *A. Albert* and *D. Thacker*, *ibid.*, Perkin I, 1972, 468.
9 *Albert* and *H. Taguchi*, *ibid.*, p. 449.
10 *Albert*, J. chem. Soc. C, 1969, 152.
11 *Idem*, *ibid.*, 1968, 2076.
12 *Albert* and *K. Tratt*, *ibid.*, 1968, 344.

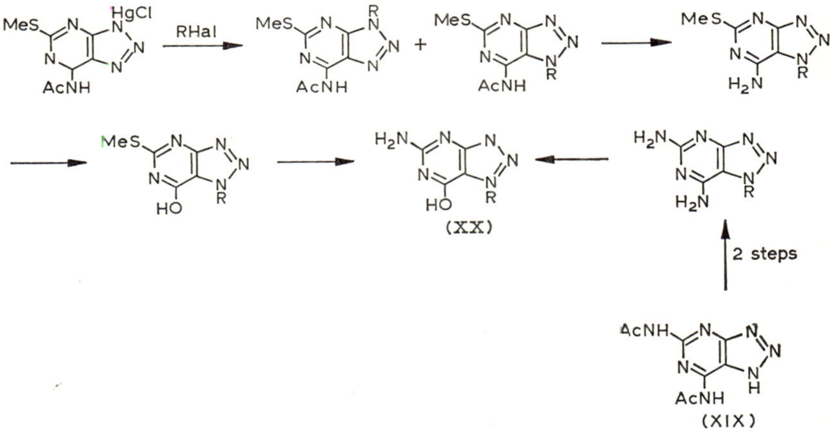

With other compounds a mixture of 8- and 9-*N*-glycosides, which differ in their absorption of ultraviolet light, were produced, *e.g.* by substitution of 6-acetamido-2-methylthio-*vic*-triazolopyrimidine (XXI). The diacetyl derivative of the 2-amino-6-hydroxy compound XX, R = H gave, however, the 8- and 9-glycoside derivatives and thus the 7-, 8- and 9-glycosides of the 2(?)-amino-6-hydroxy compound have all been prepared.

As mentioned above, treatment with hypobromite of 1-glycosido-1,2,3-triazole-4,5-dicarboxamides affords a mixture of 7- and 9-glycosido-2,6-dihydroxytriazolo[4,5-*d*]-pyramidines; these, separated by ion-exchange chromatography on Dowex-2, differ not only in their ultraviolet light absorption but also by the fact that the 9-*N*-glycoside shows a blue fluorescence in ultraviolet light under acidic conditions whilst the 7-substituted analogue shows this phenomenon under basic conditions (*Baddiley et al., loc. cit.*).

Properties of some triazolo[4,5-*d*]pyrimidines are given in Table 14 and of glycosides in Table 13.

Chapter 58

Nucleosides, Nucleotides and Nucleic Acids

D. S. JONES

1. Nucleosides

Compounds in which a purine or pyrimidine is linked to a carbohydrate to form *N*-glycosides are called *nucleosides*. The term may be applied in a general sense to all *N*-glycosides of heterocyclic bases and in exceptional cases to a few *C*-glycosides. Phosphoric esters of nucleosides are called *nucleotides*. The most abundant nucleosides of natural origin are the D-ribosides and the 2-D-deoxyribosides. The purines adenine and guanine, and the pyrimidines uracil and cytosine, occur as their ribosides adenosine, guanosine, uridine and cytidine, respectively, in ribonucleic acid (RNA). The deoxyribosides of adenine, guanine, cytosine and thymine occur as components of deoxyribonucleic acid (DNA). Smaller amounts of other purines and pyrimidines are found in some nucleic acids.

Isolation. Nucleosides are usually isolated by hydrolysis of the appropriate nucleic acid. Methods of isolating ribonucleosides include the treatment of RNA with concentrated ammonia at 175–180° (*P. A. Levene* and *W. A. Jacobs*, Ber., 1910, **43**, 3150); boiling in 50% aqueous pyridine (*H. Bredereck et al.*, *ibid.*, 1941, **74**, 694) or treatment with 8 *M* ammonium formate solution pH 4, at 140° (*D. H. Hayes*, J. chem. Soc., 1960, 1184). These reactions, however, are accompanied by considerable deamination of cytidine [*R. M. Bock*, "Methods in Enzymology", Vol. 12A, *L. Grossman* and *K. Moldave* (eds.), Academic Press, New York, 1967, p. 224]. RNA can be hydrolysed to nucleosides under milder conditions by heating with anhydrous ethylene glycol (*Bock*, *loc. cit.*). Uridine is obtained by acid hydrolysis of RNA but under the conditions of hydrolysis (1 *N* hydrochloric acid at 100°) purine nucleosides are hydrolysed to the bases, and cytidine is deaminated (*H. S. Loring* and *M. McT. Ploeser*, J. biol. Chem., 1949, **178**, 439).

Chemical hydrolysis of DNA is of little use for the preparation of deoxyribonucleosides since under the conditions required for hydrolysis, in acid purine deoxynucleosides are hydrolysed, and in alkali deoxycytidine and purine deoxynucleosides are deaminated (*R. O. Hurst*, Canad. J. Biochem.

Physiol., 1956, **34**, 265). Deoxyribonucleosides may be obtained from DNA by enzymic hydrolysis (*W. Anderson, C. A. Dekker* and *A. R. Todd*, J. chem. Soc., 1952, 2721). Degradation to deoxynucleotides can be achieved by treatment of DNA with micrococcal nuclease and spleen phosphodiesterase (*J. Josse, A. D. Kaiser* and *A. Kornberg*, J. biol. Chem., 1961, **236**, 864). The deoxynucleotides are further degraded to the deoxynucleosides with phosphomonoesterase.

Ribonucleosides and deoxyribonucleosides can be separated by a number of chromatographic techniques. The major nucleosides have been separated in preparative amounts by ion-exchange resins (*W. E. Cohn*, Science, 1949, **109**, 377) or by ion-exchange combined with crystallisation (*Anderson, Dekker* and *Todd, loc. cit.*). Smaller amounts of the four major ribo- and deoxyribo-nucleosides are separated by electrophoresis at pH 3.5 (*J. D. Smith*, "Methods in Enzymology", Vol. 12A, eds. *Grossman* and *Moldave*, Academic Press, New York, 1967, p. 350). Methods, which include partition chromatography (*R. H. Hall, ibid.*, p. 305), thin-layer chromatography (*K. Randerath* and *E. Randerath, ibid.*, p. 323) and more recently high-pressure liquid chromatography (*P. R. Brown, S. Bobick* and *F. L. Hanley*, J. Chromatog., 1974, **99**, 587) have been developed for the separation of nucleosides.

(*a*) *Structure of nucleosides*

The purine nucleosides adenosine and guanosine and the pyrimidine nucleosides uridine and cytidine are the main components of RNA. They have been assigned the structures shown below.

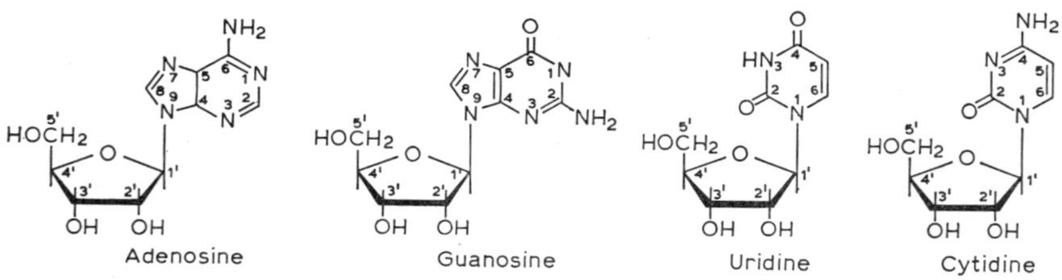

The numbering of the pyrimidine ring has caused some confusion since the official I.U.P.A.C. system came into use (as shown). Some texts still use the old (Fischer) system where the glycosidic nitrogen is numbered as 3. Besides the nucleosides shown above, many other nucleosides have been identified as minor components of RNA. Transfer RNA (tRNA) in particular contains many minor nucleosides. One minor nucleoside which is found in most, if not all, tRNA molecules is 5-β-D-ribofuranosyluracil, commonly

TABLE 1

MINOR NUCLEOSIDES FOUND IN RNA

1-Methyladenosine	1-Methylguanosine
2-Methyladenosine	*N*^{2}-Methylguanosine
N^6-Methyladenosine	7-Methylguanosine
2^1-*O*-Methyladenosine	2^1-*O*-Methylguanosine
N^6,*N*^{6}-Dimethyladenosine	*N*^{2},*N*^{2}-Dimethylguanosine
N^6-(Δ^2-Isopentenyl)adenosine	Inosine
N^6-(*cis*-4-Hydroxy-3-methylbut- -2-enyl)adenosine	1-Methylinosine
N^6-(Δ^2-Isopentenyl)-2-methyl- thioadenosine	3-Methyluridine
N-[9-(β-D-Ribofuranosyl-9*H*-purin- -6-yl)-carbamoyl]-L-threonine	5-Methyluridine
2^1(3^1)-Ribosyladenosine	2^1-*O*-Methyluridine
3-Methylcytidine	5-(β-D-Ribofuranosyl)
5-Methylcytidine	5-(2^1-*O*-Methylribosyl)
2^1-*O*-Methylcytidine	2-Thio-5-carboxymethyluridine methyl ester
N^4,*O*^{2}-Dimethylcytidine	4-Thiouridine
N^4-Acetylcytidine	5-Hydroxyuridine
2-Thiocytidine	2-Thio-5-(*N*-methylaminomethyl)- -uridine
	5-Carboxymethyluridine
	5,6-Dihydrouridine
	5,6-Dihydro-5-methyluridine

called pseudouridine, (Ψ), in which the nucleoside has a *C*-glycoside linkage. Table 1 lists some of the minor nucleosides found in RNA. Their properties are comprehensively covered elsewhere (*Hall*, "The Modified Nucleosides in Nucleic Acids", Columbia University Press, New York, 1971). Perhaps the most eccentric base of the minor nucleosides found in RNA is base "Y" which occurs in RNA^{Phe} from yeast, wheat germ and rat liver. It is a tricyclic base and has been assigned the following structure (*K. Nakanishi et al.*, J. Amer. chem. Soc., 1970, **92**, 7617) and nomenclature (*J. E. Ladner et al.*, Nucleic Acids Res., 1975, **2**, 1629).

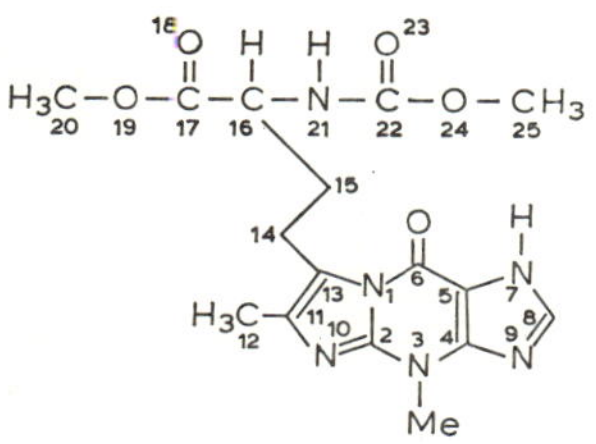

Base Y

The major components of DNA are the deoxyribosides of adenine, guanine, cytosine and thymine. These are assigned the following formulae.

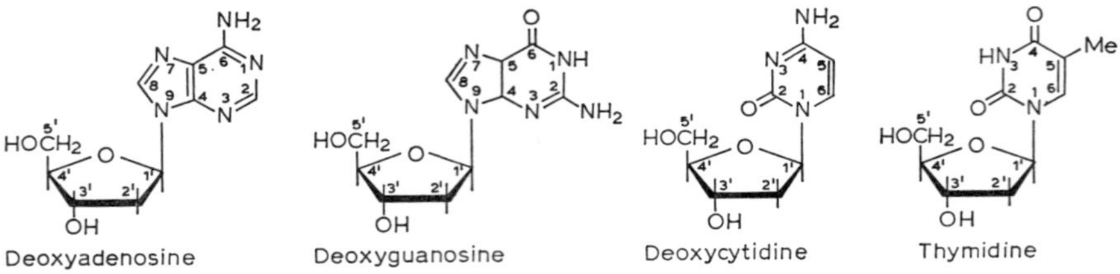

Some minor nucleosides are also found in DNA. 5-Hydroxymethyldeoxycytidine completely replaces cytidine in the DNA of the T even bacteriophages (*G. R. Wyatt* and *S. S. Cohen*, Biochem. J., 1953, **55**, 774), while in the DNA of some phages of *B. subtilis* deoxyuridine or 5-hydroxymethyldeoxyuridine totally replaces thymidine (*R. G. Kallen, M. Simon* and *J. Marmur*, J. mol. Biol., 1962, **5**, 248). 5-Methyldeoxycytidine is found in plant and animal DNA (*Wyatt*, Biochem. J., 1951, **48**, 581, 584).

Several other naturally occurring *C*-nucleosides (*e.g.* formamycin, showdomycin and pyrazomycin) have been identified some of which are antibiotics. Their chemistry is reviewed in detail elsewhere (*G. D. Davis* and *C. C. Cheng*, Prog. med. Chem., 1976, **13**, 303).

(*i*) *The nature of the sugar*

Acid hydrolysis of either RNA or the purine nucleosides obtained therefrom liberates a pentose, first crystallised and identified by *Levene* and *Jacobs* (Ber., 1908, **41**, 2703; 1909, **42**, 1198; 1911, **44**, 746). Although it differed from pentoses known at that time, its close relation to D-arabinose was soon recognised, and its identity as D-ribose was fully established by synthesis (*W. A. van Ekenstein* and *J. J. Blanksma*, Chem. Weekblad, 1909, **6**, 373; 1913, **10**, 664). Identification of the carbohydrate component of the pyrimidine nucleosides is more difficult. The glycosidic linkage in these compounds is very stable to hydrolysis and vigorous acidic conditions decompose the sugar to furfural. However, reduction of the 5,6-double bond in the pyrimidine ring labilises the glycosidic linkage and D-ribose can subsequently be isolated by gentle acidic hydrolysis (*Levene* and *F. B. la Forge*, Ber., 1912, **45**, 608). Methods used for reduction of the 5,6-double bond include catalytic hydrogenation with a rhodium catalyst (*W. E. Cohen* and *D. G. Doherty*, J. Amer. chem. Soc., 1956, **78**, 2863); reduction with sodium and ethanol in liquid ammonia (*D. C. Burke*, Chem. and Ind., 1954, 1393) or reduction with sodium amalgam and water (*L. Haavaldsen et al.*, Biochim. Biophys. Acta, 1959, **33**, 201). Initially, D-ribose was the only sugar identified in RNA nucleosides but since then, small amounts of 2^1-*O*-methyl ribonucleosides have been detected in various types of RNA (*Hall*, Biochem.

Biophys. Res. Comm., 1963, **12**, 429; *M. Honjo et al.*, Biochim. Biophys. Acta, 1964, **87**, 698).

The sugar component of DNA is less stable than D-ribose and therefore proved more difficult to isolate and study. It was isolated in crystalline form by mild acid hydrolysis of purine deoxyribonucleosides (*Levene* and *T. Mori*, J. biol. Chem., 1929, **83**, 803), and shown to be 2-deoxy-D-ribose since it was identical in all respects except sign of optical rotation with synthetic 2-deoxy-L-ribose (*Levene*, *L. A. Mikeska* and *Mori*, *ibid.*, 1930, **85**, 785). Conditions required to hydrolyse pyrimidine deoxyribonucleosides cause deoxyribose to rearrange to laevulic acid. However, deoxyribose can be obtained if the pyrimidine deoxyribosides are reduced to the dihydro compounds before hydrolysis (*Burke*, J. org. Chem., 1955, **20**, 643). When DNA is hydrolysed in the presence of phenylmethanethiol, the crystalline dibenzyldithioacetal of 2-deoxy-D-ribose can be isolated directly (*P. W. Kent*, Nature, 1950, **166**, 442).

(ii) The size of the sugar ring

Methylation of the ribonucleosides followed by acid hydrolysis gives 2,3,5-trimethylribose. From this it follows that the furanose ring structure is present in these nucleosides (*Levene* and *R. S. Tipson*, J. biol. Chem., 1932, **94**, 809; 1932, **97**, 491; 1933, **101**, 529). Tritylation studies support this conclusion and show that the 5′-OH group is the only primary hydroxyl group (*idem, ibid.*, 1934, **105**, 419; *H. Bredereck et al.*, Ber., 1932, **65**, 1830; 1933, **66**, 198; 1940, **73**, 269, 1124). Further evidence for the furanose ring structure in nucleosides comes from its behaviour towards periodate (*B. Lythgoe* and *Todd*, J. chem. Soc., 1944, 592). Each natural ribonucleoside consumes one mol of periodate giving a dialdehyde without the production of formic acid and so must be a furanoside.

Both pyrimidine and purine deoxyribonucleosides are stable to periodate and so cannot possess a *cis*-1,2-diol system. This is in agreement with a furanose structure (*D. M. Brown* and *Lythgoe*, *ibid.*, 1950, 1990).

Support for the five-membered ring structure of the sugar component of the nucleosides as well as information concerning the conformation of the ring has been obtained by X-ray structural analysis (*S. Furberg*, Acta Chem. Scand., 1950, **4**, 751; Acta Cryst., 1965, **18**, 313; *D. G. Watson*, *D. J. Sutor* and *P. Tollin*, *ibid.*, 1965, **19**, 111), and nuclear magnetic resonance spectroscopy (*C. D. Jardetzky*, J. Amer. chem. Soc., 1960, **82**, 229; 1961, **83**, 2919).

(iii) Position and configuration of the glycosidic linkage

Since purine nucleosides are readily hydrolysed by acid, a C–N linkage between sugar and base is suggested. A C–C linkage would be stable to acid.

The primary exocyclic amino groups of adenine and guanine are not involved since they are deaminated by nitrous acid without affecting the glycosidic bond (*Levene* and *Jacobs*, Ber., 1910, **43**, 3150; *J. M. Gulland* and *T. F. Macrae*, J. chem. Soc., 1933, 662; *Levene et al.*, J. biol. Chem., 1923, **55**, 437; 1935, **111**, 313). That the u.v. spectra of the natural purine nucleosides resemble those of the corresponding 9-alkylpurines so very closely and differ so markedly from those of the 7-alkylpurines suggests that the nucleosides are 9-glycosides (*Gulland et al.*, J. chem. Soc., 1936, 765; 1934, 1639; 1938, 259, 692). This has been confirmed by oxidation with periodate which also establishes the configuration of the glycosidic linkage. The dicarbaldehyde obtained in this way from adenosine is identical with that from synthetic 9-β-D-mannopyranosyladenine available by an unambiguous synthetic route. Oxidation of a 9-D-glucosyladenine with periodate gives the same dicarbaldehyde. This glucoside, synthesised from acetobromoglucose is almost certainly a β-glucoside since a Walden inversion is assumed during its synthesis.

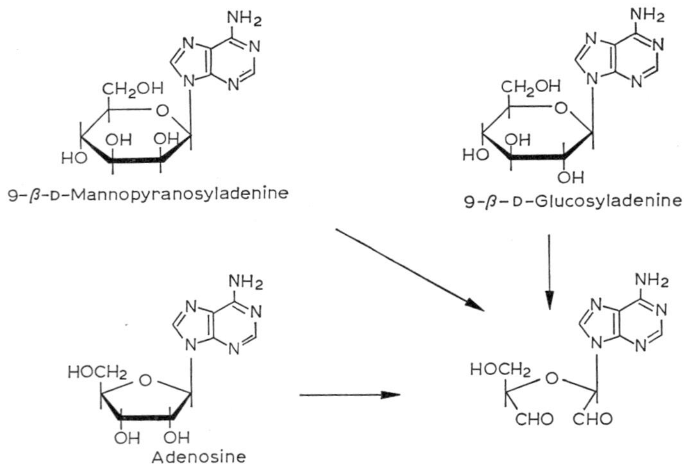

The configuration of the glycosidic linkage in the purine nucleosides is therefore β (*J. Daroll*, *Lythgoe* and *Todd*, *ibid.*, 1946, 833; *Lythgoe*, *H. Smith* and *Todd*, *ibid.*, 1947, 355).

Independent proof of the β-configuration of the natural nucleosides is provided by the ready formation from 2′,3′-isopropylidene-5′-toluene-*p*-sulphonyladenosine of a cyclonucleoside salt (*V. M. Clark*, *Todd* and *J. Zussman*, *ibid.*, 1951, 2952). Steric requirements for this transformation are only satisfied by β-glycosides.

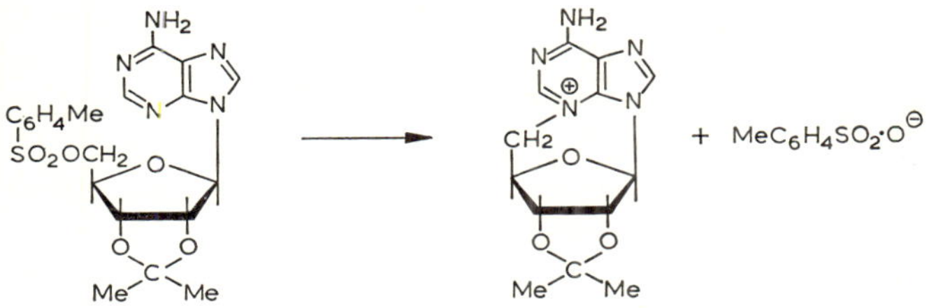

The position of the glycosidic linkage in pyrimidine nucleosides has been shown by methylation. Methylation of uridine gives a mono-*N*-methyluridine, whereas methylation of uracil gives the dimethyl derivative 1,3-dimethyluracil. Hydrolysis of the methyluridine gives 3-methyluracil, demonstrating that the ribose is attached to $N_{(1)}$ (*Levene* and *Tipson*, J. biol. Chem., 1934, **104**, 385).

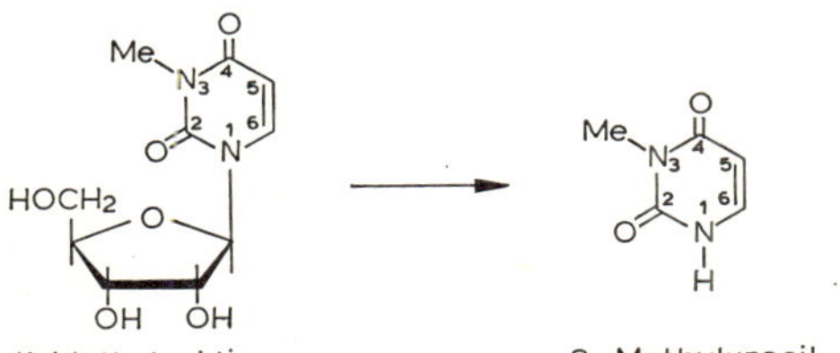

N-Methyluridine 3-Methyluracil

Since cytidine is converted into uridine by deamination, it follows that cytidine is also an N^1-glycoside. Synthetic cytosine- and uracil-1-β-D-glucopyranosides are oxidised to dicarbaldehydes identical with those from cytidine and uridine, respectively. The glycosidic bonds in the two pyrimidine nucleosides must therefore be in the $N_{(1)}$ position and have a β-configuration. Confirmation of the β-linkage is provided by the ready conversion of 2′,3′-isopropylidene-5′-toluene-*p*-sulphonylcytidine to a cyclonucleoside (*A. M. Michelson* and *Todd*, J. chem. Soc., 1955, 816).

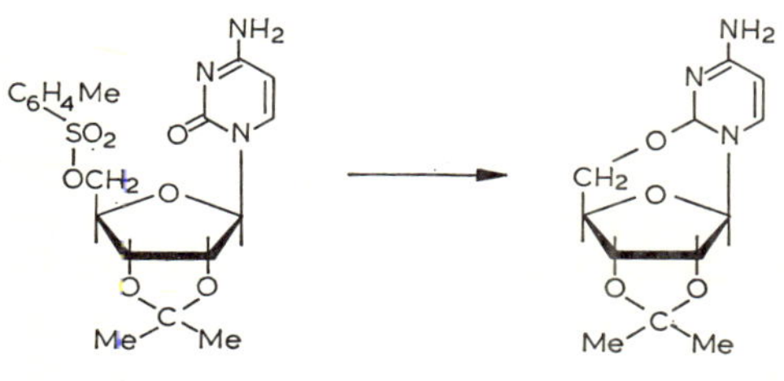

For steric reasons, this reaction is only possible with a glycoside bond in the β-configuration.

A nucleoside with an unusual structure has been isolated in small amounts from transfer RNA. The nucleoside is isomeric with uridine and has been called pseudouridine. The u.v. spectra at pH 7 and pH 12 are similar to those of 5-hydroxymethyluracil and indicate that the ribose cannot be attached to either nitrogen or oxygen. The n.m.r. spectrum and degradation to 5-hydroxymethyluracil by the sequence of reactions shown below demonstrated that the ribose was attached to $C_{(5)}$ of uracil (*W. E. Cohn*, J. biol. Chem., 1960, **235**, 1488).

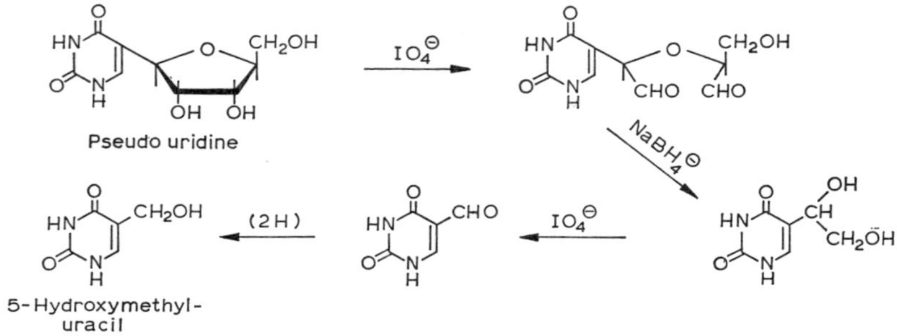

The configuration of the glycosidic link was shown to be β by the ready conversion of 2′,3′-isopropylidene-5′-toluene-*p*-sulphonylpseudouridine to O^4,5′-cyclopseudouridine (*Cohn* and *Michelson*, Biochemistry, 1962, **1**, 490). The structure of pseudouridine has been confirmed by synthesis (*R. Shapiro* and *R. W. Chambers*, J. Amer. chem. Soc., 1961, **83**, 3920; *T. L. V. Ulbricht*, Angew. Chem. internat. Edn., 1962, **1**, 476).

The configuration of the pyrimidine and purine deoxyribonucleosides follows from similar experiments to those described for the ribonucleosides and from synthesis. It has been shown that thymidine is an N^1-glycoside, since methylation and hydrolysis of DNA from thymus gives 3-methylthymine (*Bredereck et al.*, Ber., 1940, **73**, 1058). Toluene-*p*-sulphonyl derivatives of deoxyadenosine, deoxycytidine and thymidine have been converted into cyclic nucleosides (*W. Anderson et al.*, J. chem. Soc., 1954, 1882; *Michelson* and *Todd*, *ibid.*, 1955, 816), thus confirming the β-configuration of these nucleosides.

Analysis of nucleosides by physical methods has confirmed the structures described above and provided more information concerning their physical properties and conformation. Early X-ray crystal analysis showed that the glycosidic linkage is in the same plane as the flat purine or pyrimidine ring whereas the sugar ring is approximately perpendicular (*Furberg*, Acta Chem. Scand., 1950, **4**, 751; Acta Cryst., 1950, **3**, 329; *M. Huber*, *ibid.*, 1957, **10** 128).

Detailed information about bond angles, bond lengths and tautomeric forms of nucleosides has been obtained (*W. Saenger*, Angew. Chem. internat. Edn., 1973, **12**, 591). The five atoms of the ribose or deoxyribose ring are arranged in an envelope or half-chair conformation. X-Ray crystal analysis (*S. Arnott*, Prog. Biophys. mol. Biol., 1970, **21**, 265; *M. Sundaralingam*, Biopolymers, 1969, **7**, 821) and spin–spin coupling analysis of high resolution nuclear magnetic resonance spectra of nucleosides in solution (*Jardetzky*, J. Amer. chem. Soc., 1960, **82**, 222, 229; 1961, **83**, 2919; *F. E. Hruska, A. A. Grey* and *I. C. P. Smith*, *ibid.*, 1970, **92**, 4088) shows that four of the sugar ring atoms form a plane to within $\pm$ 0.04 Å and the fifth atom is displaced by about 0.5 Å. The four planar atoms are $C_{(1')}$, $C_{(4')}$, $O_{(1')}$ and either $C_{(2')}$ or $C_{(3')}$. The atom out of plane is designated *endo* if it is on the same side of the plane as the $C_{(5')}$ atom and *exo* if it is on the other side. Nucleosides with $C_{(2')}$ *endo* and $C_{(3')}$ *endo* are found most commonly. $C_{(3')}$ *exo* has only been observed in a few deoxyribonucleosides. Crystallographic and spectroscopic evidence shows that the preferred conformation for the $C_{(5')}$–$O_{(5')}$ bond is *gauche gauche* (*A. E. V. Haschemeyer* and *A. Rich*, J. mol. Biol., 1967, **27**, 369) and gives data for calculation of the dihedral angle which defines the rotation of the base around the glycosidic bond. The nucleoside can exist in either the *syn* or *anti* conformation depending upon whether the $O_{(2)}$ atom in pyrimidine or $N_{(3)}$ atom in purine nucleosides lies over the plane of the sugar (*syn*) or points away from the sugar (*anti*).

^{13}C-Magnetic resonance analysis of nucleosides has provided information concerning the electron distribution in the bases and the positions of substituents (*A. J. Jones et al.*, J. Amer. chem. Soc., 1970, **92**, 4079; Tetrahedron Letters, 1969, 5197). Mass spectrometry has been used also for identification of nucleosides (*K. Bieman* and *J. A. McCloskey*, J. Amer. chem. Soc., 1962, **84**, 2005; *S. J. Shaw et al., ibid.*, 1970, **92**, 250, 2510; *S. M. Hecht, A. S. Gupta* and *N. J. Leonard*, Anal. Biochem., 1969, **30**, 247).

(*b*) *Synthesis of nucleosides*

(*i*) *Methods of synthesis*

Three different approaches were developed originally for the synthesis of purine nucleosides.

(1) *E. Fischer* and *B. Helferich* (Ber., 1914, **47**, 210) performed the first synthesis of nucleosides by showing that the silver salts of certain purines react with acetylglycosyl halides giving glycosides. While the glycosides of theobromine thus obtained are *O*-glycosides, those of theophylline are known to be $N_{(7)}$-glycosides (*J. M. Gulland et al.*, J. chem. Soc., 1934, 1639). However, the silver salt of 2,8-dichloroadenine reacts with acetylglycosyl

halides giving acetylated $N_{(9)}$-glycosides. Repetition of the above reaction using acetyl-D-ribofuranosyl chloride, as the glycosyl derivative followed by conversion of the 2,8-dichloroadenine riboside obtained to adenine or guanine ribosides (as shown) resulted in the first total synthesis of adenosine and guanosine (*Daroll, Lythgoe* and *Todd, ibid.*, 1948, 967, 1685).

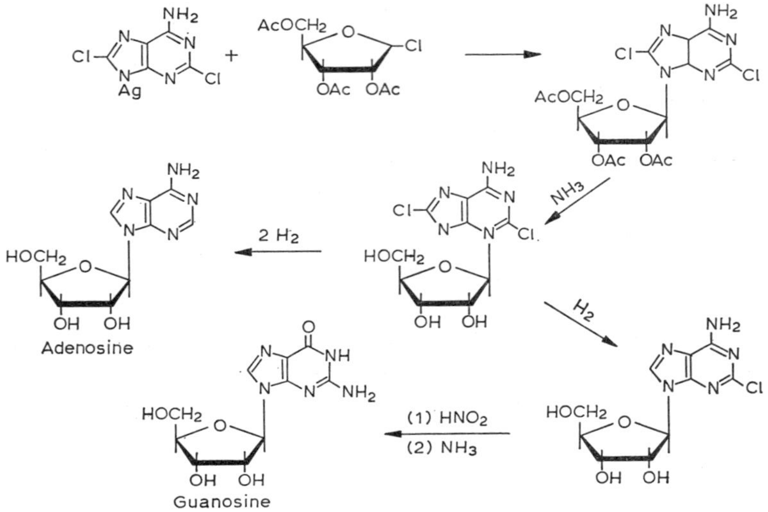

These are $N_{(9)}$-glycosides as is shown by their absorption spectra and by oxidation with periodate. 2,8-Dichloroadenine is used in the above procedure because silver adenine itself is sufficiently basic to dehydrohalogenate the sugar derivative. The basicity problem may also be overcome by acetylation or benzoylation of the amino group. Chloromercury salts of purines give better yields than silver salts. Thus adenosine, guanosine and crotonoside may be synthesised from a chloromercury salt of 2,6-diacetamidopurine (*Davoll* and *B. A. Lowry*, J. Amer. chem. Soc., 1951, **73**, 1650; *Davoll, ibid.*, 1951, **73**, 3174). Direct condensation of purine derivatives with acylglycosyl halides, which avoids the necessity of preparing the heavy metal salt of the purine, can be achieved by heating the reactants in nitromethane in the presence of a hydrogen halide acceptor *e.g.* mercurous cyanide (*N. Yamacka, K. Aso* and *K. Matsuda*, J. org. Chem., 1965, **30**, 149). Other methods of direct condensation between the purine and protected sugar halide have been developed.

Fully *O*-acetylated glycosides when heated with purines in the presence of a catalyst give varying yields of nucleosides. Effective catalysts include zinc chloride/sulphuric acid (*Y. Ishido* and *T. Sato*, Bull. chem. Soc., Japan, 1961,

34, 1347); sulphamic acid (*Ishido et al., ibid.*, 1964, **37**, 1389); chloroacetic acid (*M. J. Robins, W. A. Bowles* and *R. K. Robins*, J. Amer. chem. Soc., 1964, **86**, 1251) and phosphate esters (*T. Hashizume* and *H. Iwamura*, Tetrahedron Letters, 1965, 3095). Fusion of trimethylsilyl derivatives of purines with protected glycosyl halides gives good yields of the protected nucleosides. The trimethylsilyl groups are readily removed after synthesis with aqueous ethanol (*T. Nishimura et al.*, Chem. Pharm. Bull., 1964, **12**, 252; Agric. biol. Chem., 1964, **28**, 224). Thus adenosine can be prepared by fusion of N^6-benzoyl N^6,9-bis(trimethylsilyl)adenine and 2′,3′,5′-tri-*O*-benzylribosyl chloride at 150–160° followed by removal of the protecting groups (*Nishimura, B. Shimizu* and *I. Iwai, ibid.*, 1964, **12**, 1471). The general methods of direct condensation between derivatives of the purines and sugars are the most commonly used and convenient routes to nucleosides.

(2) A synthesis of purine nucleosides, which, although less convenient in operation than the direct reaction between purine and sugar derivatives, has the advantage of being unambiguous with respect to the location of the glycoside linkage, follows the lines of the 9-alkylpurine synthesis (see p. 7). The synthesis, which is capable of wide application, starts from aminopyrimidines which are converted directly into *N*-glycosides by reaction with a sugar (*J. Baddiley, Lythgoe* and *Todd*, J. chem. Soc., 1943, 571). An amino group is introduced at position 5 *via* the azo compound, and the imidazole ring is formed by thioformylation and cyclisation in pyridine.

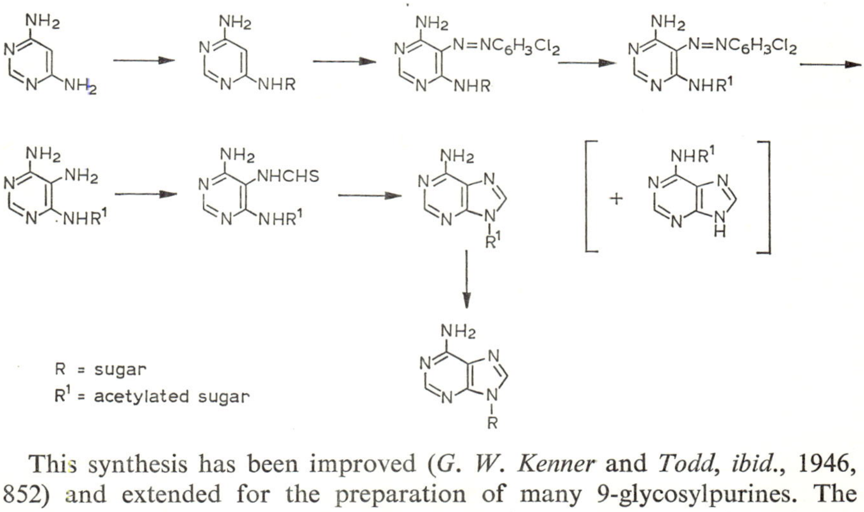

This synthesis has been improved (*G. W. Kenner* and *Todd, ibid.*, 1946, 852) and extended for the preparation of many 9-glycosylpurines. The glycosides synthesised in this way are all β-pyranosides and in order to ensure a furanose ring structure, it is necessary to start with a suitably protected aldehydo-sugar derivative. This method has been employed for the synthesis of adenosine as follows.

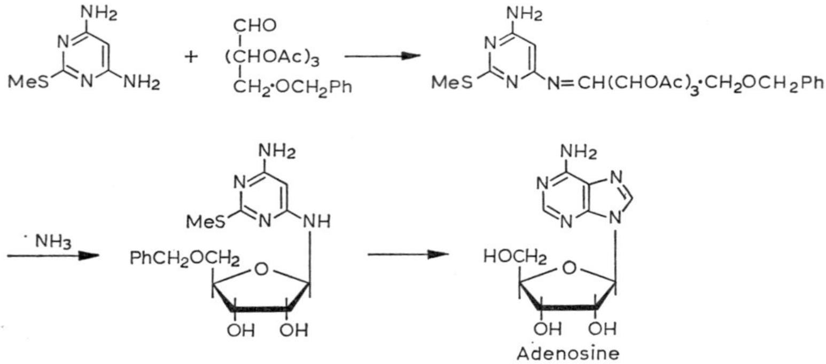

A methylthio group at $C_{(2)}$ is necessary to increase the reactivity of the pyrimidine. The aldehydo-sugar in this case is 2,3,4-triacetyl-5-benzyl-D-ribose. A Schiff's base is first formed which on deacetylation, isomerises to the normal glycoside. Introduction of an amino group at position 5, cyclisation, etc. are as before. The methylthio group is removed eventually by hydrogenolysis (*Kenner, C. W. Taylor* and *Todd, ibid.*, 1949, 1620).

(3) A third type of purine nucleoside synthesis starts from iminazoles. Although not very flexible, this route can be employed for the preparation of xanthine glycosides (*R. A. Baxter* and *F. S. Spring, ibid.*, 1947, 378; *Baxter, A. C. McLean* and *Spring, ibid.*, 1948, 523; *cf. J. Baddiley, J. G. Buchanan* and *G. O. Osborn, ibid.*, 1958, 3606). Starting from acetylribofuranosyl chloride, xanthosine itself is produced (*G. A. Howard et al., ibid.*, 1949, 232).

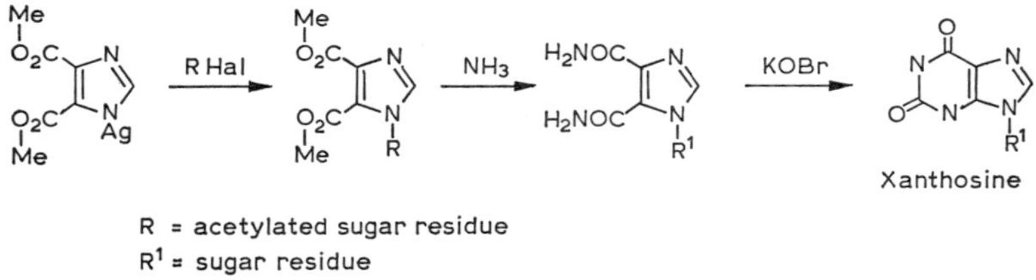

The pyrimidine nucleosides, unlike the purine nucleosides, were not prepared by treating silver salts of appropriate pyrimidines with acetylglycosyl halides, since *O*-glycosides were obtained.

G. Hilbert and *T. B. Johnson* noticed however, that 2,4-di-alkoxypyrimidines (I) react with alkyl halides to give 4-alkoxy-1-alkylpyrimidines (II) which are hydrolysed readily by acid to 1-alkyluracils and also react with ammonia to give 1-alkylcytosines (III) (J. Amer. chem. Soc., 1930, **52**, 2001). Acetylglycosyl halides behave like alkyl halides in this reaction and the pyranose analogue of uridine (IV, R^1 = sugar residue) was prepared in this way by *Hilbert* and *C. E. Rist* (J. biol. Chem., 1937, **117**, 371).

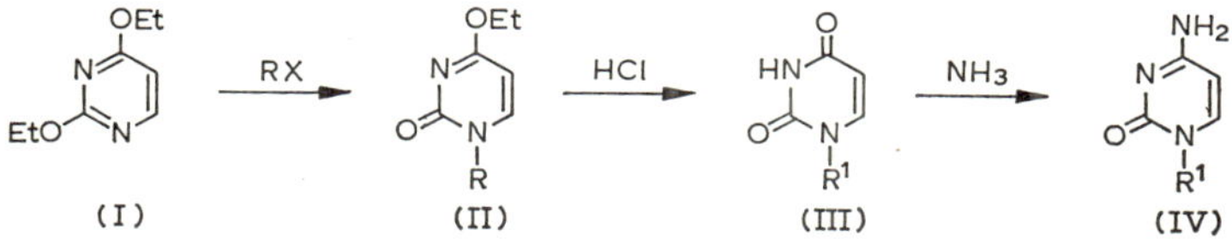

R = alkyl or acetylglycosyl
R^1 = glycosyl

The first synthesis of a natural nucleoside was accomplished by this method using acetylribofuranosyl bromide and 2,4-diethoxypyrimidine, the product of condensation being aminated and deacylated to give cytidine (*Howard, Lythgoe* and *Todd*, J. chem. Soc., 1947, 1052). The reactivity of the dialkoxypyrimidine toward the protected ribosyl halide is influenced by the nature of the alkoxy groups. Reactivity decreases in the order 2,4-dibenzyloxy-> 2,4-dimethoxy-> 2,4-diethoxy-> 2,4-diisopropoxy-pyrimidine (*M. Prystas* and *F. Šorm*, Coll. Czech. chem. Comm., 1966, **31**, 1035).

A modification of the Hilbert–Johnson reaction uses the bis(trimethylsilyl) derivatives. The bis(trimethylsilyl) derivatives of uracil (V), thymine (VI) and acetylcytosine (VII) are readily accessible (*Nishimura* and *Iwai*, Chem. Pharm. Bull., Japan, 1964, **12**, 352).

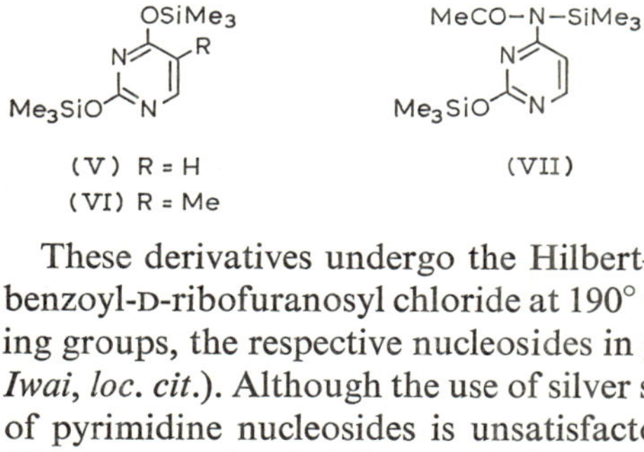

(V) R = H
(VI) R = Me

(VII)

These derivatives undergo the Hilbert–Johnson reaction with 2,3,5-tri-*O*-benzoyl-D-ribofuranosyl chloride at 190° to give, after removal of the protecting groups, the respective nucleosides in good yield (*Nishimura, Shimizu* and *Iwai, loc. cit.*). Although the use of silver salts of pyrimidines for the synthesis of pyrimidine nucleosides is unsatisfactory, other metal salts can be used. Chloromercuripyrimidine, pyrimidylmercury and dipyrimidylmercury salts have been used. For example, thymine riboside has been prepared *via* the reaction between dithyminylercury and 2,3,5-triacetylribofuranosyl chloride (*J. J. Fox et al.*, J. Amer. chem. Soc., 1956, **78**, 2117), and similarly, cytidine is formed from *N*-acetylcytosinylmercury (*idem, ibid.*, 1957, **79**, 5060). It is not possible, however, to synthesise uridine by the mercuri process (*Fox* and *I. Wempen*, Advan. Carbohydrate Chem., 1959, **14**, 283). Excellent yields of nucleosides are obtained by using a Friedel–Crafts catalysed Hilbert–Johnson reaction (*U. Niedballa* and *H. Vorbrüggen*, J. org. Chem., 1974, **39**, 3654, 3660, 3672; 1976, **41**, 2084). When 1-acetoxy-2,3,5-tri-*O*-benzylribose (VIII) is reacted with a silylated base (*e.g.* VII) in the presence of stannic chloride in 1,2-dichloroethane or acetonitrile the *β*-anomer of the protected nucleoside is obtained.

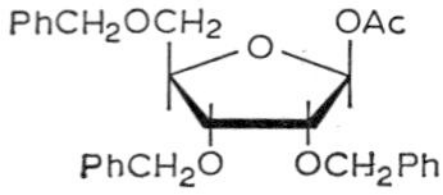

(VIII)

The synthesis of pseudouridine is a special case since it requires the formation of a *C*-glycoside linkage. It was first synthesised in poor yield from 2,3,5-tribenzoyl-D-ribofuranosyl chloride and 5-lithio-2,4-dimethoxypyrimidine (*R. W. Chambers* and *R. Shapiro*, J. Amer. chem. Soc., 1961, **83**, 3920). Better yields are obtained by coupling 5-lithio-2,4-di-*tert*-butoxypyrimidine (IX) with 2,4 : 3,5-di-*O*-benzylidene-D-ribose (X) to give XI, which is converted to pseudouridine under acidic conditions. Pseudouridine is obtained in 18% yield and its α-anomer in 8% yield (*D. M. Brown, M. G. Burdon* and *R. P. Slatcher*, J. chem. Soc., C, 1968, 1051; *U. Lerch, Burdon* and *J. G. Moffat*, J. org. Chem., 1971, **36**, 1507).

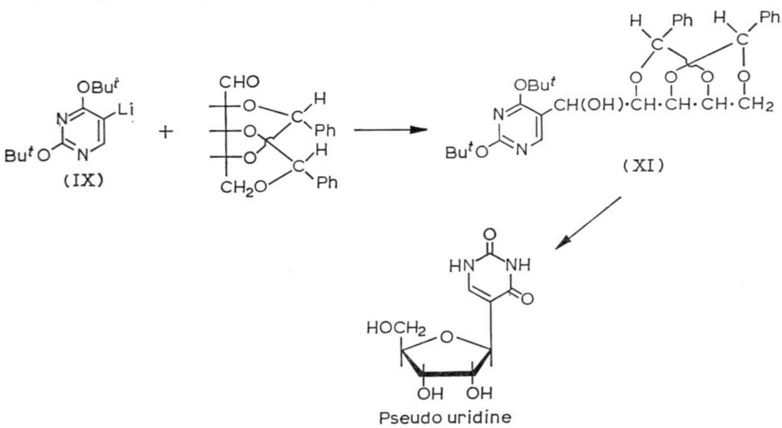

Initial attempts to prepare 2′-deoxynucleosides by direct reaction between a 2′-deoxyglycosyl halide and the metal salt of the purine or pyrimidine were unsatisfactory (*J. Davoll* and *B. Lythgoe*, J. chem. Soc., C, 1949, 2527). Synthesis of deoxyuridine and thymidine was achieved by conversion of suitable derivatives of the ribonucleosides to the 2′-deoxyribonucleosides.

The cyclonucleoside intermediate 5′-*O*-acetyl-2′-*O*-toluene-*p*-sulphonyluridine (XII) and sodium iodide give an iodo compound which may be reduced and deacylated to yield deoxyuridine (XIII). Similar operations on thymidine riboside produce thymine (*A. R. Todd et al., ibid.*, 1958, 3035, 4242).

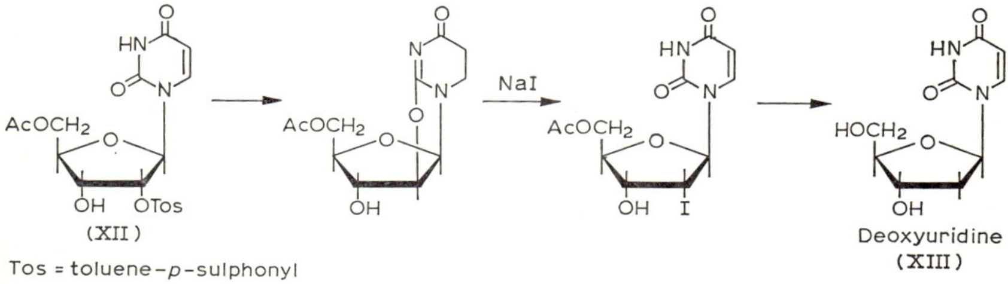

An alternative method for the synthesis of thymidine involves the trityl derivative of 5-methyl-2-thiouridine which reacts with methanesulphonyl chloride to form a thiocyclonucleoside. After removal of the trityl group and treatment with Raney nickel in alkali, a low yield of thymidine is obtained (*G. Shaw* and *R. N. Warrener*, *ibid.*, 1959, 50).

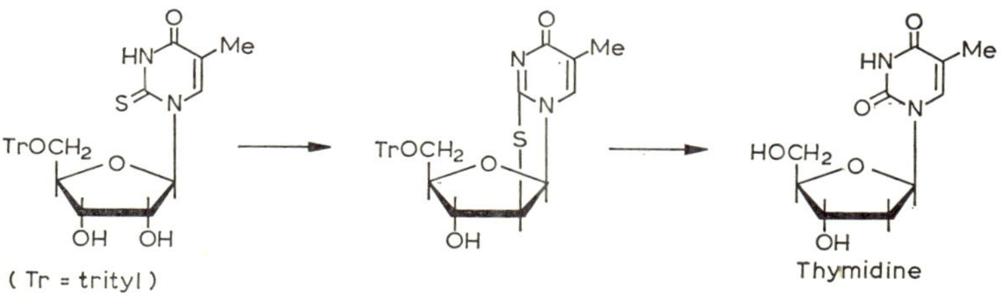

3′-*p*-Nitrobenzenesulphonyladenosine reacts with sodium iioide to give the 3′-deoxyiodo compound which on reduction is quantitatively reduced to 3′-deoxyadenosine (*Todd* and *T. L. V. Ulbricht, ibid.*, 1960, 3275). Conversion of the 3′-deoxy- to the 2′-deoxy-adenosine can be accomplished by migration of an ethylthio group from the 3′ to the 2′ position involving a 2′,3′-episulphonium ion intermediate (*C. D. Anderson, L. Goodman* and *B. R. Baker*, J. Amer. chem. Soc., 1959, **81**, 3967). The synthesis of stable derivatives of 2′-deoxyhalogeno-sugars led to the preparation of 2′-deoxynucleosides by direct methods. Reaction of chloromercuri-N^6-benzoyladenine with 3,5-di-*p*-nitrobenzoyl-2-deoxyribosyl chloride (*R. K. Ness* and *H. G. Fletcher, ibid.*, 1959, **81**, 4752; 1960, **82**, 3434) or of silver 2,8-dichloroadenine with 3,5-di-*O*-acetyl-2-deoxyribosyl chloride (*H. Venner*, Chem. Ber., 1960, **93**, 140), gives, after removal of protecting groups, 2′-deoxyadenosine. The latter reaction gives the natural *β*-anomer whereas the former reaction gives 10% of the β-anomer and 19% of the α-anomer. Purine deoxynucleosides may be prepared also by fusion of 1,3,5-tri-*O*-acetyl-2-deoxyribose with suitably blocked purines (*M. J. Robins* and *R. K. Robins*, J. Amer. chem. Soc., 1965, **87**, 4934). The method yields both anomers of the nucleoside. Both anomers of 2′-deoxyguanosine (*idem*, J. org. Chem., 1969, **34**, 2160) and 2′-deoxyadenosine (*M. J. Robins, T. A. Kwaja* and *R. K. Robins, ibid.*, 1970, **35**, 636) have been obtained.

Pyrimidine 2′-deoxynucleosides are obtained by reaction between the appropriate pyrimidylmercury compound and 3,5-di-*O*-*p*-chlorobenzoyl (or *p*-toluoyl)-2-deoxyribosyl chloride (*M. Hoffer et al.*, J. Amer. chem. Soc., 1959, **81**, 4112; 1961, **83**, 4066; 1960, **93**, 2777). Pyrimidine 2′-deoxyribonucleosides may also be prepared by using the Hilbert–Johnson reaction. Reaction of 3,5-di-*O*-*p*-toluoyl-2-deoxy-D-ribosyl chloride (XV) with various 5-substituted (*e.g.* H, Me, I, Cl, F)-2,4-dimethylpyrimidines (XIV) gives an anomeric mixture of the respective protected 4-methoxy-5-substituted pyrimidine 2′-deoxyribonucleosides (XVI) from which the respective 5-substituted pyrimidine 2′-deoxyribonucleosides (XVII) are obtained. Ammonolysis of XVI (R = H) followed by removal of the protecting groups gives 2′-deoxycytidine (XVIII) and its α-anomer (*Prystas, J. Farkas* and

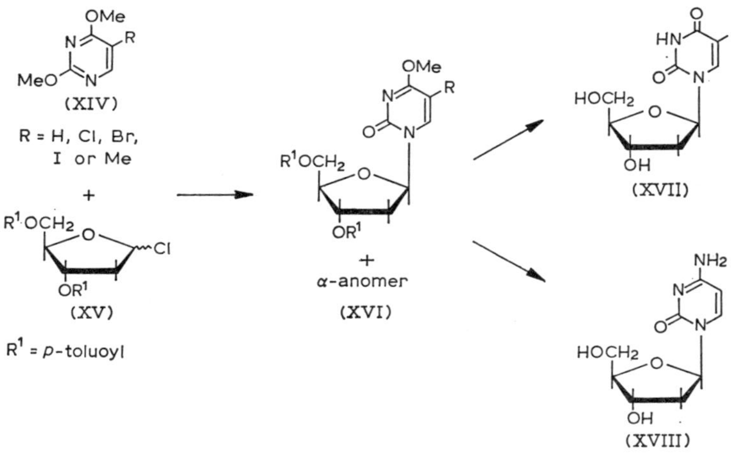

F. Šorm, Coll. Czech. chem. Comm., 1963, **28**, 3140; *Prystas* and *Šorm, ibid.*, 1964, **29**, 121, 131; 1965, **30**, 1900). 2,4-Bis-*O*-trimethylsilyl derivatives of pyrimidines react with 3,5-protected 2-deoxyribosyl chloride to give a mixture of 2′-deoxynucleoside anomers (*T. J. Bardos, M. P. Kofick* and *C. Szankay*, Tetrahedron Letters, 1966, 1759). The nucleoside 5-hydroxymethyl-2′-deoxycytidine which occurs in the DNA of some bacteriophages in place of deoxycytidine can be prepared starting from 3,5-di-*O*-*p*-toluoyl-2-deoxyribosyl chloride and either the mercury salt of N^4-benzyloxymethyl-pyrimidine (*R. Brossmer* and *E. Röhm*, Angew. Chem. intern. Edn., 1963, **2**, 742), or 5-benzyloxymethyl-2,4-dimethoxypyrimidine (*Prystas* and *Šorm*, Coll. Czech. chem. Comm., 1966, **31**, 1053).

The unusual nucleosides spongouridine and spongothymidine which occur in certain sponges are arabofuranosides of uracil and thymine (*W. Bergmann* and *R. J. Feeney*, J. org. Chem., 1951, **16**, 981; *Bergmann* and

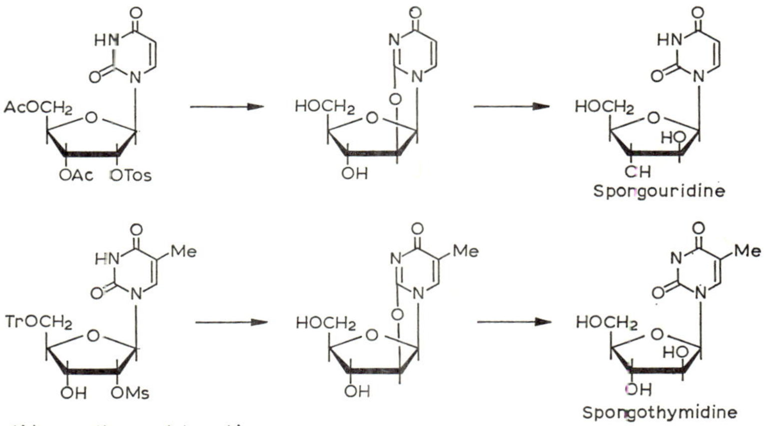

(Ms = methanesulphonyl)

D. F. Burke, *ibid.*, 1955, **20**, 1510). Their synthesis from uridine and ribothymidine establishes their structure (*Brown*, *Todd* and *S. Varadarajan*, J. chem. Soc., 1956, 2388; *Fox et al.*, *loc. cit.*).

(*ii*) *Stereochemical and mechanistic considerations*

Ribonucleosides prepared from reaction between the heavy metal salt of a purine or pyrimidine and a ribosyl halide derivative have a β-configuration irrespective of whether the ribosyl halide is α or β. To account for this *Baker*, in: "The Ciba Foundation Symposium on the Chemistry and Biology of the Purines", *G. E. W. Wolstenholme* and *C. M. O'Connor*, eds. (Churchill, London, 1957, p. 120), proposed that the product of such reactions will have a $C_{(1)}$–$C_{(2)}$ *trans*-configuration in the sugar, regardless of the con-

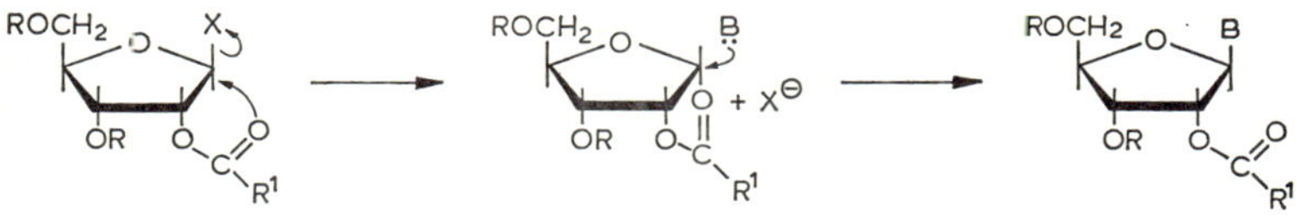

(B = purine or pyrimidine)

figuration at $C_{(1)}$–$C_{(2)}$ in the ribosyl halide. Where the ribosyl halide has the $C_{(1)}$–$C_{(2)}$ *cis*-configuration (α), a β-ribonucleoside will result from a direct S_N2 reaction with inversion at $C_{(1)}$. However, where the ribosyl halide has the $C_{(1)}$–$C_{(2)}$ *trans*-configuration (β) it is necessary to propose two S_N2 reactions, the first of which is intramolecular, involving participation of the acyloxy groups at $C_{(2)}$.

As a consequence of this phenomenon, arabinosyl halides should give α-arabinonucleosides and 2′-deoxyribosyl halides would be expected to give a mixture of α- and β-2′-deoxyribonucleosides. In general this is found to be the case. The ratio of the α- and β-2′-deoxyribonucleoside anomers found in the products of reaction is influenced by the reaction conditions. For example, in syntheses using silyl derivatives of pyrimidines, if the reaction is performed in boiling benzene or by fusion, the β-anomer is the predominant product whereas the α-anomer predominates if the reaction is carried out in benzene at 37° (*Bardos, Kofich* and *Szantay, loc. cit.*). If the Hilbert–Johnson reaction is used for the synthesis of pyrimidine 2′-deoxynucleosides, the α-anomer always predominates, which can be explained by the steric effect and transannular participation effect of the substituent at position 5 of the sugar. However, the α/β anomeric ratio is affected by the nature of the solvent. Reactions performed in the more ionic solvent acetonitrile give a higher α/β ratio of products than reactions performed in benzene (*Prystas* and *Šorm*, Coll. Czech. chem. Comm., 1965, **30**, 2960).

Mercury derivatives of purines used in the synthesis of nucleosides are covalent. $N_{(7)}$-Mercuripurines give $N_{(7)}$-nucleosides, and $N_{(9)}$-mercuripurines give $N_{(9)}$-nucleosides which indicates that synthesis involves a direct replacement mechanism (*J. A. Montgomery* and *H. J. Thomas*, J. org. Chem., 1966, **31**, 1411).

As observed earlier, the silver and mercury derivatives of pyrimidines yield different products with the same glycosyl halides. To account for this it has been suggested that chloromercuripyrimidines, dipyrimidylmercury and pyrimidine silver salts are *O*-derivatives. When these compounds react with glycosyl halides, *O*-glycosides are initially formed, which may, under suitable reaction conditions, rearrange to *N*-glycosides (*Ulbricht*, Angew. Chem. intern. Edn., 1962, **1**, 476). Mercuric bromide facilitates this *O*- → *N*-glycosyl rearrangement (*Ulbricht* and *G. T. Rogers*, J. chem. Soc., 1965, 6125, 6130; *T. Ukita, H. Hayatsu* and *Tomita*, Chem. and pharm. Bull., Japan, 1963, **11**, 1068). *N*-Deoxyribonucleosides cannot be obtained by an *O* → *N* rearrangement since *O*-deoxyribosides are too unstable. They are prepared using a different type of mercury derivative, pyrimidylmercury, which is an *N*-mercuripyrimidine and hence yields an *N*-glycoside directly without rearrangement (*Ulbricht, loc. cit.*).

(*c*) *Properties of nucleosides*

The chemical properties of the nucleosides can be grouped into two classes: those involving reactions of the base and those involving reactions of the sugar moiety.

(*i*) *Reactions of the bases*

Halogenated derivatives of the nucleoside bases are useful intermediates in the preparation of a wide variety of nucleoside analogues.

Direct halogenation of purine nucleoside bases leads to substitution at position 8. Reaction of 2′,3′,5′-tri-*O*-acetyladenosine or 3′,5′-di-*O*-acetyl-2′-*O*-deoxyadenosine with *N*-bromoacetamide, followed by deacylation, yields 8-bromoadenosine or 8-bromo-2′-deoxyadenosine, respectively (*R. E. Holmes* and *R. K. Robins*, J. Amer. chem. Soc., 1964, **86**, 1242). Alternatively, treatment of 2′,3′,5′-tri-*O*-acetylguanosine with bromine in acetic acid, followed by deacylation, gives 8-bromoguanosine (*idem, loc. cit.*). Direct halogenation of pyrimidine nucleosides gives substitution at position 5. Reaction of uridine with bromine water yields 5-bromouridine (*T. K. Fukuhara* and *D. W. Visser*, J. biol. Chem., 1951, **190**, 95), and treatment of either cytidine or 2′-deoxycytidine with chlorine in a mixture of pyridine and glacial acetic acid yields 5-chlorocytidine (*idem*, J. Amer. chem. Soc., 1955, **77**, 2393) or 5′-chloro-2′-deoxycytidine (*D. M. Frisch* and *Visser*, *ibid.*, 1959, **81**, 1756), respectively. The chemotherapeutic reagent 5-fluorouridine is obtained after treatment of uridine with trifluoromethyl hypofluorite (*M. J. Robins* and *S. R. Naik*, *ibid.*, 1971, **93**, 5277).

Replacement reactions. Halogens can be introduced into other positions of bases by replacement of other functional groups. For example, hydroxyl groups can be converted into chloro groups using the Vilsmeier–Haack reagent (thionyl chloride/dimethylformamide); thus 2′,3′,5′-tri-*O*-benzoyl-inosine is converted to the 6-chloropurine derivative (*M. Ikehara, H. Uno* and *F. Ishikawa*, Chem. Pharm. Bull., Japan, 1964, **40**, 639), and 2′,3′,5′-tri-*O*-benzoyluridine is converted into the 4-chlorouridine derivative (*J. Zemlicka* and *F. Šorm*, Coll. Czech. chem. Comm., 1965, **30**, 2052).

Amino groups in nucleosides are converted to hydroxyl groups with nitrous acid. Most commonly known are the conversions of adenosine to inosine (*A. Kossel*, Ber., 1885, **18**, 1928), and cytidine to uridine (*P. A. Levene* and *W. A. Jacobs*, *ibid.*, 1910, **43**, 3150).

Sulphur derivatives of nucleosides can be prepared by thiolation of an oxo group. On treatment of the 2′,3′,5′-tri-*O*-benzoyl derivative of guanosine and uridine with phosphorus pentasulphide in pyridine the 6-thioguanosine derivative (*J. J. Fox et al.*, J. Amer. chem. Soc., 1958, **80**, 1669) and the 4-

thiouridine derivative (*idem*, *ibid.*, 1959, **81**, 178), respectively, are obtained.

Reduction. The 5,6-double bond in pyrimidine nucleosides can be reduced by several methods. Uridine is converted into 5,6-dihydrouridine by reduction with hydrogen over palladium (*M. Green* and *S. S. Cohen*, J. Biochem., 1957, **225**, 397) or rhodium (*A. R. Hanze*, J. Amer. chem. Soc., 1967, **89**, 6720) catalysts, or by photocatalyzed reduction with sodium tetrahydridoborate (*P. Cerutti et al.*, *ibid.*, 1968, **90**, 771).

(*ii*) *Reactions of the sugar moiety*

Substitution and elimination reactions at the 2′- and 3′-positions of nucleosides are achieved only with difficulty. Treatment of 3′-*O*-methylsulphonyl-2′,5′-*O*-ditrityluridine (XIX) with sodium ethanethiolate, followed by acid treatment and desulphurization, gives only a low yield of 3′-deoxyuridine (*G. Kowollik* and *P. Langen*, Chem. Ber., 1968, **101**, 235).

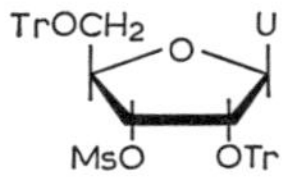

(XIX)

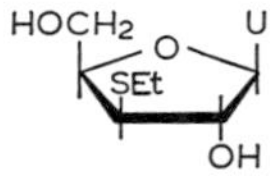

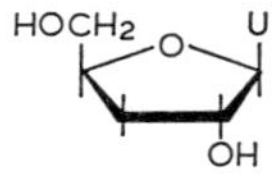

3′-Deoxyuridine

A 2′,3′-double bond can be introduced into nucleosides by several methods. 2′,3′-Didehydro-2′,3′-dideoxyuridine (XXa) is obtained on treatment of 3′-methylsulphonyl-4′-*O*-trityl-2′-deoxyuridine (XXI) with potassium butoxide followed by removal of the protecting group with acid (*J. P. Horwitz et al.*,

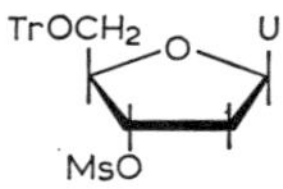

(XXI)

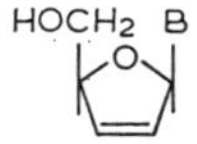

(XXa, B = uracil-1-yl)
(XXb, B = adenin-9-yl)

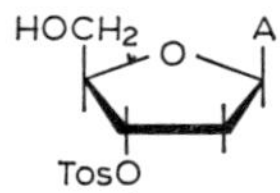

(XXII)

J. Amer. chem. Soc., 1964, **86**, 1896). In a similar way 2′,3′-didehydro-2′,3′-dideoxyadenosine (XXb) is obtained from 3′-*O*-(*p*-tolylsulphonyl)-2′-deoxyadenosine (XXII) (*J. R. McCarthy et al.*, *ibid.*, 1966, **88**, 1549; *Horwitz*, *J. Chua* and *M. Noel*, Tetrahedron Letters, 1966, 1343).

Substitution occurs more readily at the 5′-position. Halogen derivatives of the nucleosides can be prepared by direct substitution. Thus 2′,3′-*O*-isopropylideneuridine can be converted to the 5′-chloro-5′-deoxyuridine derivative by treatment with thionyl chloride in dimethylformamide (*R. F. Dods* and *J. S. Roth*, *ibid.*, 1969, 165). Similarly 2′,3′-*O*-isopropylidineadenosine gives 5′-chloro-5′-deoxy-2′,3′-*O*-isopropylideneadenosine on treatment with thionyl chloride (*K. Kikugawa* and *M. Ichino*, *ibid.*, 1971, 87). These halogen derivatives are useful intermediates for the synthesis of other

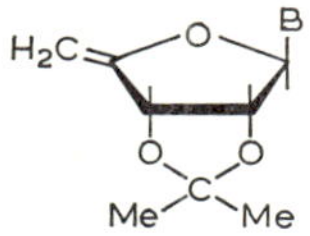

(XXIII, B = uracil-1-yl)
(XXIV, B = adenin-9-yl)

5′-substituted nucleosides. 4′,5′-Unsaturated nucleosides are obtained on base elimination of a good leaving group from the 5′-position. Treatment of either 2′,3′-*O*-isopropylidene-5′-*O*-(*p*-tolysulphonyl)uridine (*M. J. Robins, McCarthy* and *R. K. Robins*, J. heterocycl. Chem., 1967, **4**, 313) or -adenosine (*J. G. Moffatt, I. D. Jenkins* and *J. P. H. Verheyden*, J. Amer. chem. Soc., 1971, **93**, 4323) with potassium *tert*-butoxide gives either XXIII or XXIV, respectively.

Oxidation. Treatment of ribonucleosides with oxygen in the presence of a platinum catalyst oxidises the 5′-hydroxyl group to a carboxylic acid group to yield uronic acid derivatives (*A. R. Todd et al.*, J. chem. Soc., 1963, 1149; *K. Iwai* and *M. Honjo*, Chem. Pharm. Bull., Japan, 1965, **13**, 7). Uronic acid derivatives of the deoxynucleosides can be obtained by treatment of the deoxynucleoside with chromic anhydride (*A. S. Jones, A. R. Williamson* and *M. Winkley*, Carbohydrate Res., 1964, **1**, 187). 5′-Carbaldehyde derivatives are prepared by treatment of 2′,3′-isopropylidenenucleosides with dimethyl sulphoxide (*K. E. Pfitzner* and *J. G. Moffatt*, J. Amer. chem. Soc., 1963, **85**, 3027). Dimethyl sulphoxide mixed with a dehydrating reagent is also used to prepare the highly unstable 2′- and 3′-oxonucleosides (*A. F. Cook* and *Moffatt*, *ibid.*, 1967, **89**, 2696).

Cyclonucleosides are formed by intramolecular reaction of an atom of the heterocyclic base with a carbon atom of the sugar. As has already been mentioned these compounds have helped to confirm the conformation of nucleosides. $N_{(3)}$-5′-Cyclonucleosides (*e.g.* XXV) are readily formed on heating 2′,3′-*O*-isopropylidene-5′-*O*-(*p*-tolylsulphonyl) derivatives of the purine nucleosides (*V. M. Clark, Todd* and *J. Zussman*, J. chem. Soc., 1961, 2952).

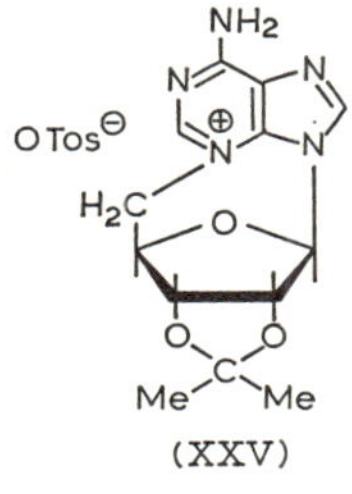

(XXV)

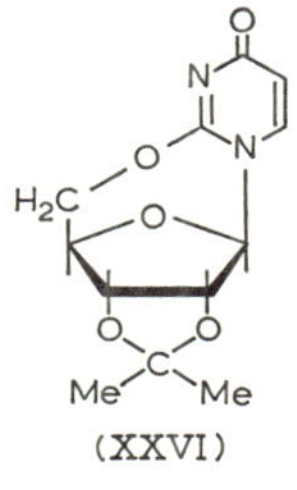

(XXVI)

$O_{(2)}$,5′-Cyclonucleosides (*e.g.* XXVI) are obtained in a similar reaction with pyrimidine nucleosides (*idem, loc. cit.*).

$O_{(2)}$,3′-Cyclonucleoside derivatives of uridine and cytidine are obtained, albeit with more difficulty, by heating either 3’-*O*-(*p*-tolylsulphonyl)uridine (*R. Letters* and *A. M. Michelson,* J. chem. Soc., 1961, 1410) or 3′-*O*-(methylsulphonyl)cytidine (*Y. Mizuno* and *T. Sasaki,* Tetrahedron Letters, 1965, 4579) with sodium *tert*-butoxide.

Pyrimidine $O_{(2)}$,2′-cyclonucleosides are important since they are intermediates in the conversion of nucleosides into 1-(*β*-D-arabinofuranosyl)-nucleosides. Treatment of uridine with thiocarbonylbis(imidazole) gives $O_{(2)}$,2′-cyclouridine (*J. J. Fox* and *I. Wempen, ibid.*, 1965, 643), which on acid hydrolysis is converted into arabinosyluracil (*idem*, Advan. Carbohyd. Chem., 1959, **14**, 349). Arabinosylcytosine is obtained by treatment of cytidine with polyphosphoric acid followed by enzymic dephosphorylation (*W. K. Roberts* and *C. A. Dekker,* J. org. Chem., 1967, **32**, 816). $O_{(2)}$,2′-Cyclocytidine has been identified as an intermediate in the reaction (*E. R. Walwich, Roberts* and *Dekker,* Proc. chem. Soc., 1959, 84).

Hydrolysis. The mechanism by which nucleosides are hydrolysed by acid has remained a puzzle for some time, but it has been proposed that pre-equilibrium protonation of the heterocyclic base followed by rate-limiting cleavage of the glycosyl bond is involved. Deoxyribonucleosides are more susceptible than ribonucleosides to hydrolysis because of differences in ΔS values of the intermediates (*R. K. Robins et al.*, J. Amer. chem. Soc., 1972, **94**, 4708; *L. Hevesi et al., ibid.*, 1972, **94**, 4715).

Selective acylation of functional groups in the nucleosides is of great importance in the synthesis of oligonucleotides, and methods have been found for selective *N*-acylation of the bases and *O*-acylation of the primary or secondary hydroxyl groups of the sugars. Selective acylation of the amino group in cytidine or 2′-deoxycytidine may be achieved by gradual addition of an acid anhydride to a boiling alcoholic solution of the nucleoside (*B. A. Otter* and *Fox,* "Synthetic Procedures in Nucleic Acid Chemistry", *W. W. Zorbach* and *R. S. Tipson* (eds.), Interscience, New York, 1968, Vol. I, p. 285). *O*-Acylation of nucleosides can be accomplished by reaction with an anhydride or acid chloride in anhydrous pyridine. By this method the completely *O*-acylated derivatives of adenosine (*H. Bredereck* and *A. Martini,* Chem. Ber., 1947, **80**, 401; *D. H. Hayes, A. M. Michelson* and *Todd,* J. chem. Soc., 1955, 808), guanosine (*Bredereck* and *Martini, loc. cit.*; *Hayes, Michelson* and *Todd, loc. cit.*; *F. Weyand* and *W. Sigmund,* Chem. Ber., 1953, **86**, 160; *Fox et al.*, J. Amer. chem. Soc., 1958, **80**, 1669; *H. G. Khorana et al., ibid.*, 1963, **85**, 3821), uridine (*P. A. Levene* and *Tipson,* J. biol. Chem.,

1933, **101**, 529; *Fox et al.*, J. Amer. chem. Soc., 1959, **81**, 178), and thymidine (*Fox, loc. cit.*; *Michelson* and *Todd*, J. chem. Soc., 1955, 816) can be obtained. Because of the ease of amino-acylation in cytidine, selective *O*-acylation is more difficult. However, this may be achieved by carrying out the acylation in the presence of an acid catalyst (*Sasaki* and *Mizuno*, Chem. Pharm. Bull., Japan, 1967, **15**, 894).

The primary 5′-hydroxyl group of the sugar of a nucleoside is more reactive than the secondary 2′- or 4′-hydroxyl groups, but this difference in reactivity is, in general, not sufficiently great to allow a selective acetylation or benzoylation of the 5′-hydroxyl group. The selectivity may be increased, however, by using an acylating reagent possessing a bulky group. For example, reaction of nucleosides with one equivalent of trityl (triphenylmethyl) chloride or other substituted trityl chlorides gives mono-5′-trityl nucleosides (*W. Anderson et al.*, J. chem. Soc., 1954, 1882). The trityl group may be removed under mild acid conditions. Substitution of methoxy groups in phenyl residues of the trityl group increases the facility of removal. With every additional 4-methoxy group the rate of acid hydrolysis of trityl group is increased approximately ten-fold (*Khorana et al.*, J. Amer. chem. Soc., 1962, **84**, 430). Treatment of 5′-trityl-nucleosides with acetic anhydride, followed by mild acid hydrolysis gives the 2′,3′-diacetyl-nucleoside (*Bredereck*, *E. Berger* and *J. Ehrenberg*, Ber., 1940, **73**, 269). Selective reaction with the 2′- and 3′-hydroxyl groups of ribonucleosides can be performed by formation of 2′,3′-isopropylidene derivatives. These were initially prepared by reaction of the nucleoside with acetone in the presence of an acid catalyst (*Levene* and *Tipson*, J. biol. Chem., 1937, **121**, 131), but the reaction is best carried out using 2,2-dimethoxypropane in the presence of di-(4-nitrophenyl)phosphate as catalyst (*A. Hampton*, J. Amer. chem. Soc., 1961, **83**, 3640). These acetals are readily hydrolysed under mild acid conditions.

Monoacetylation of 5′-*O*-substituted nucleosides gives a mixture of 2′- and 3′-*O*-acetates in which the 2′-*O*-acetate initially predominates (*G. A. R. Johnson*, Tetrahedron, 1968, **24**, 6987). These isomers are readily converted from one to the other, and after boiling in pyridine for 1 h, the 3′-*O*-isomer predominates (*C. B. Reese* and *D. R. Trentham et al.*, Tetrahedron Letters, 1965, 2467; Biochemistry, 1966, **5**, 3638). It appears, therefore, that acylation takes place at the 2′-position and this is followed by isomerisation to give the more stable 3′-isomer.

(*d*) *Individual nncleosides*

Adenosine, 9-β-D-ribofuranosyladenine, $C_{10}H_{13}O_4N_5$ (formula 1, p. 118), needles, m.p. 229° (anhydrous), is usually obtained by hydrolysis of ribonucleic acid. It occurs also as a component of many nucleotide coenzymes. It forms a *picrate* (plates, m.p. 180–185° decomp.). With acetone and zinc chloride it gives 2′,3′-*O*-isopropylideneadenosine (*P. A. Levene* and *R. S. Tipson*, J. biol. Chem., 1937, **121**, 131; *J. Baddiley*, J. chem. Soc., 1951, 1348). 5′-Trityladenosine is obtained by the action of trityl chloride on adenosine in pyridine. From it, by reaction with toluene-4-sulphonyl chloride, $N_{(6)}$,2′,3′-tritosyl-5′-trityladenosine is obtained and subsequent hydrolysis of this, in 80% acetic acid, gives $N_{(6)}$,2′,3′-tritosyladenosine (*Levene* and *Tipson*, *loc. cit.*). 5′-Trityladenosine forms a 2′,3′-diacetyl derivative which, on hydrolysis, gives 2′,3′-diacetyladenosine (*H. Bredereck*, *E. Berger* and *J. Ehrenberg*, Ber., 1940, **73**, 269). $N_{(6)}$,5′-Ditrityladenosine also forms a diacetyl derivative which, on partial hydrolysis, gives 2′,3′-diacetyladenosine.

Guanosine, 9-β-D-ribofuranosylguanine, $C_{10}H_{13}O_5N_5$ (formula 2, p. 118), m.p. 237–240°, forms needles containing 1 H_2O. Sparingly soluble in cold water, it dissolves readily in acids and alkalis and forms a *triacetyl* derivative, m.p. 226° (*H. Steudel* and *R. Freise*, Z. physiol. Chem., 1922, **120**, 126). 2′,3′-*O*-Isopropylideneguanosine and 2′,3′-*O*-benzylideneguanosine are prepared in the usual manner (*A. M. Michelson* and *A. R. Todd*, J. chem. Soc., 1949, 2476; *J. M. Gulland* and *W. G. Overend*, *ibid.*, 1948, 1380; *D. M. Brown*, *L. J. Hayes* and *Todd*, *ibid.*, 1950, 408, 2299). Methylation of acetylated guanosine gives trimethyl-*N*-methylguanosine (*Levene* and *Tipson*, J. biol. Chem., 1932, **97**, 491).

Inosine, 9-β-D-ribofuranosylhypoxanthine, $C_{10}H_{12}O_5N_4$, m.p. 218°, is best obtained by deaminating adenosine. It has been isolated as a minor component of several species of tRNA (*D. B. Dunn*, Biochem. J., 1963, **86**, 14P; *R. H. Hall*, Biochem. Biophys. Res. Comm., 1963, **13**, 394). With acetic anhydride it gives *triacetylinosine*, m.p. 236°, soluble in chloroform (*F. Haiser* and *F. Wenzel*, Monatsh., 1908, **29**, 157) and like adenosine, it forms an isopropylidene derivative (*Levene* and *Tipson*, J. biol. Chem., 1935, **111**, 313) which with toluene-4-sulphonyl chloride gives 5′-tosyl-2′,3′-*O*-isopropylideneinosine, convertible to the 5′-iodo compound with sodium iodide in acetone.

Xanthosine, 9-β-D-ribofuranosylxanthine, $C_{10}H_{12}O_6N_4$ (formula p. 128) is sparingly soluble in cold water. It is prepared by the action of nitrous acid on guanosine (*Levene* and *W. A. Jacobs*, Ber., 1910, **43**, 3163); methylation gives 1,3-dimethylxanthosine (*Levene*, J. biol. Chem., 1923, **55**, 437).

5′-Methylthioadenosine (*5′-deoxy-5′-methylthioadenosine*, *adenine thiomethylpentoside*), $C_{11}H_{15}O_3N_5S$, m.p. 212–213°, fine needles from water, is found in yeast (*J. A. Mandel* and *E. K. Dunham*, *ibid.*, 1912, **11**, 85; *R. L. Smith et al.*, Arch. Biochem. Biophys., 1953, **42**, 72). On hydrolysis it gives adenine and 5-deoxy-5-methylthioribose (*Levene* and *H. Sabotka*, J. biol. Chem., 1925, **65**, 551; *G. Wendt*, Z. physiol. Chem., 1942, **275**, 152; *K. Satoh* and *K. Makino*, Nature, 1950, **165**, 769; *F. Weygand*, *O. Tranth* and *R. Löwenfeld*,

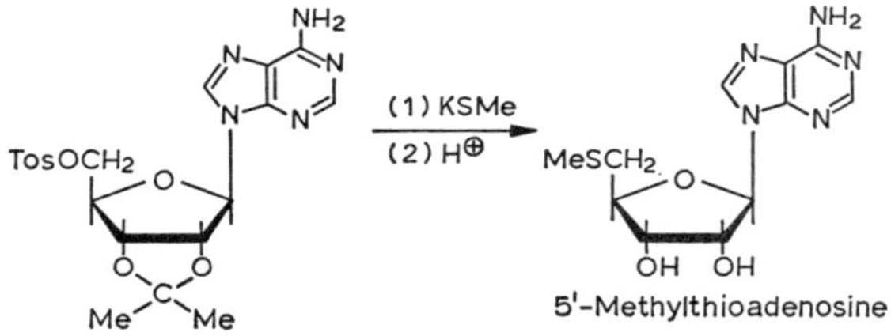

Chem. Ber., 1950, **83**, 563). Spectroacopic evidence favours an *N*-glycoside structure (*R. Falconer* and *Gulland*, J. chem. Soc., 1937, 1912) and the formula is established by its synthesis and a synthesis of its deaminization product (*Baddiley, loc. cit.*; *Weygand* and *Trauth*, Chem. Ber., 1951, **84**, 633; *Satoh* and *Makino*, Nature, 1951, **167**, 238). The synthesis from 2′,3′-*O*-isopropylideneadenosine involves tosylation at the 5′-position, substitution of the tosyl group by SMe and then hydrolysis of the isopropylidene group.

5′-Methylthioadenosine arises in nature through chemical or enzymic decomposition of *S*-adenosylmethionine (active methionine) (*Baddiley, G. L. Cantoni* and *G. A. Jamieson*, J. chem. Soc., 1953, 2662; *S. H. Mudd*, J. biol. Chem., 1959, **234**, 87).

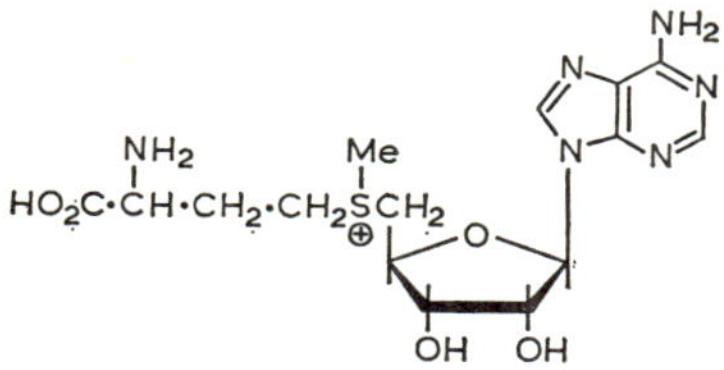

S-Adenosylmethionine

S-Adenosylmethionine is a biological methyl donor formed from methionine and adenosine triphosphate (*Cantoni, ibid.*, 1951, **189**, 745; 1953, **204**, 403). Its structure is confirmed by synthesis from 5′-methylthioadenosine and α-aminobromobutyric acid hydrobromide (*Baddiley* and *Jamieson*, J. chem. Soc., 1954, 4280; 1953, 1085). A convenient preparation from adenosine triphosphate and methionine involves an enzyme from rabbit liver (*Cantoni*, Biochemical Preparations, Vol. **5**, p. 58).

Crotonoside, isoguanine-9-β-D-riboside, $C_{10}H_{13}O_5N_5$, m.p. 237–252° (decomp.) $[\alpha]_D$ −71° in NaOH, is isomeric with guanosine. It has been isolated from the croton bean (*Croton tiglium*); on hydrolysis it gives isoguanonine and D-ribose (*E. Cherbuliez* and *K. Bernhard*, Helv., 1932, **15**, 464) and with nitrous acid, xanthosine. The sugar is situated at $N_{(9)}$ as shown by its absorption spectrum (*Falconer*, *Gulland* and *L. F. Story*, J. chem. Soc., 1939, 1784) and by the previously mentioned synthesis (p. 125 *et seq.*).

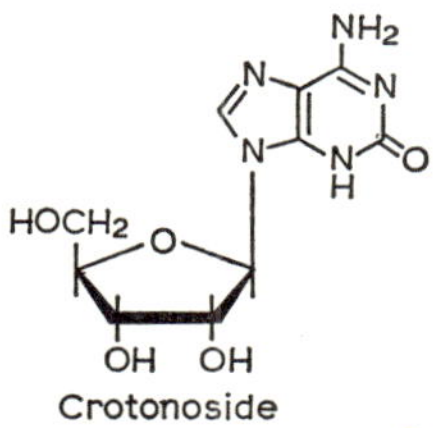

Crotonoside

Formycin, **7-amino-3-β-D-ribofuranosylpyrazole**[4,3-α]**pyrimidine**, $C_{10}H_{13}O_4N_5$ and **formycin B,** 3-**β-D-ribofuranosylpyrazole**[4,3α]6(*H*)-**pyrimidin**-7-**one**, $C_{10}H_{12}O_5N_4$, are antibiotics found in the actinomycete, *Nocardia interforma* (*H. Umezawa et al.*, J. Antibiot., Japan, 1964, **17A**, 96; 1965, **18A**, 175). Treatment of formycin with nitrous acid produces formycin B (*idem, ibid.*, 1965, **18A**, 175, 178). The structures of formycin B have been assigned on the basis of chemical synthesis and physical studies (*idem*, Tetrahedron Letters, 1966, 597; *R. K. Robins et al.*, J. heterocycl. Chem., 1966, **3**, 110). They are *C*-glycosides.

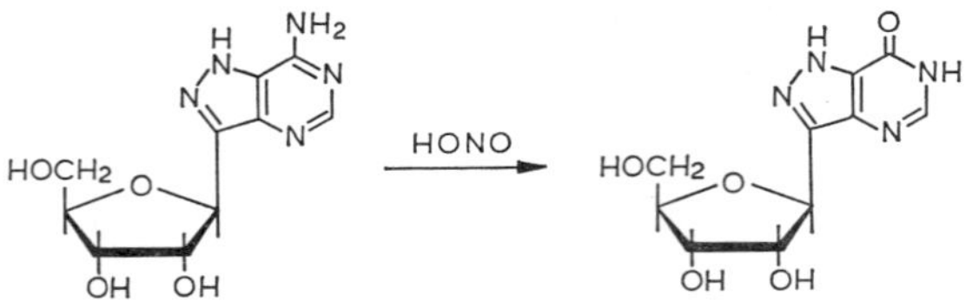

Formycin Formycin B

Purine riboside, nebularine, 9-β-D-ribofuranosylpurine, $C_{10}H_{12}O_4N_4$, is present in a mushroom (*N. Löfgren et al.*, Svensk kem. Tidskr., 1946, **58**, 269; Acta Chem. Scand., 1953, **7**, 15) and gives purine and D-ribose on hydrolysis. Its structure follows from a synthesis from purine (*G. B. Brown* and *V. S. Weliky*, J. biol. Chem., 1953, **204**, 1019). It is of interest in view of its antibacterial and antimitotic activity (*cf.* "The Chemistry and Biology of Purines", Ciba Foundation Symposium, *G. E. W. Wolstenholme* and *C. M. O'Connor*, Churchill, London, 1957).

Nucleocidin, 9-(4-fluoro-5-O-sulphamoylpentofuranosyl)adenine, is found in *Streptomycin calvus*.

The structure was elucidated from chemical studies (*C. W. Waller et al.*, J. Amer. chem. Soc., 1957, **79**, 1011), but this initial structure was corrected when it was found to contain fluorine (*W. E. Meyer et al.*, *ibid.*, 1969, **91**, 1535).

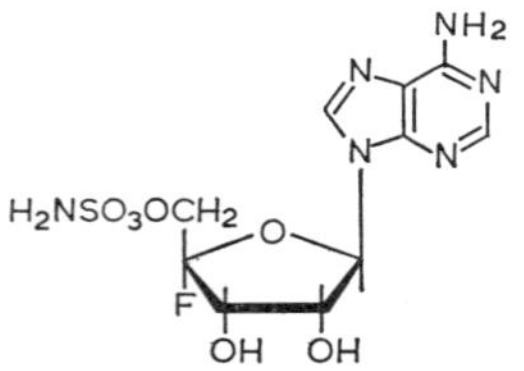

Nucleocidin

Puromycin, 6-dimethylamino-9-(3′-p-methoxy-L-phenylalanylamino-3′-deoxy-β-D-ribofuranosyl)purine, is an antibiotic produced by *Streptomyces alboniger* (*Waller et al.*, J. Amer. chem. Soc., 1953, **75**, 2025). On alcoholysis it yields 6-dimethylaminopurine, O-

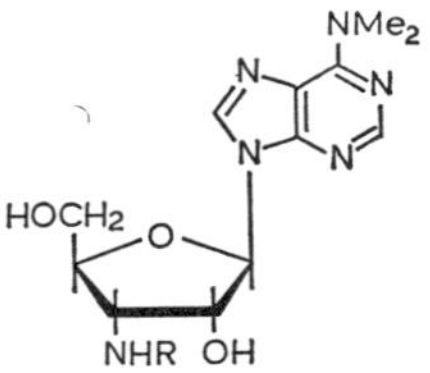

Puromycin

R = p-methoxy-L-phenylalanyl

methyl-L-tyrosine and 3-amino-3-deoxyribose. It is stable towards periodate and has a spectrum similar to that of a 9-substituted purine. The structure shown has been confirmed by synthesis.

The synthesis starts from methyl D-xylofuranoside which, through a series of conversions involving the formation and opening of 2,3-anhydro derivatives, gives 3-acetamido-

1-*O*-acetyl-2,5-di-*O*-benzoyl-3-D-ribofuranose. With titanium chloride this gives a complex of the 1-chloro-sugar which, in turn, is condensed with the chloromercuri derivative of 6-dimethylamino-2-methylthiopurine. The product was desulphurised and debenzoylated and the *p*-methoxyphenylalanyl group was introduced by standard peptide synthetic methods (*B. R. Baker*, *R. E. Schaub* and *J. H. Williams*, J. Amer. chem. Soc., 1955, **77**, 7; *Baker*, *J. P. Joseph* and *Williams*, *ibid.*, 1955, **77**, 1).

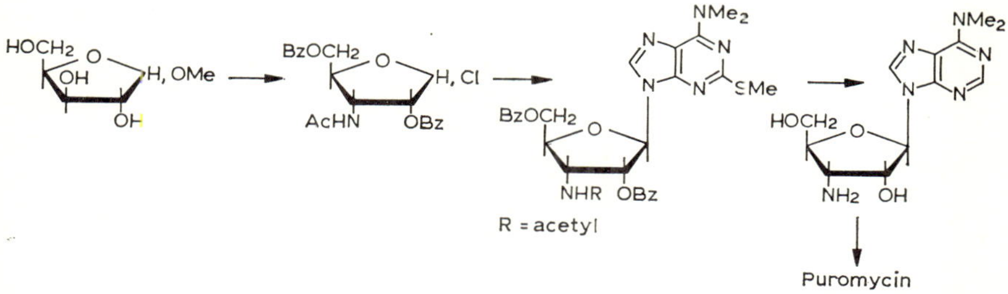

A number of analogues have been synthesised in connection with biological studies (*Baker et al.*, *ibid.*, 1955, **77**, 12, 15, 2396, 5900, 5905, 5911). These are comprehensively reviewed elsewhere (*J. A. Montgomery* and *H. J. Thomas*, Advan. Carbohydrate Chem., 1962, **17**, 301). **3′-Amino-3′-deoxyadenosine** and the nucleoside antibiotics, **homocitrullylaminoadenosine** and **lysylaminoadenosine,** which are structurally related to puromycin, are found in the mould *Cordyceps militoris* (*N. M. Kredich* and *A. J. Guarins*, J. biol. Chem., 1961, **236**, 3300; Biochim. Biophys. Acta, 1963, **68**, 317).

Tubercidine, 7-deaza-adenosine, $C_{11}H_{14}O_4N_4$ (*K. Anzai*, *G. Nakamura* and *S. Suzuki*, J. Antibiot., Japan, 1957, **10A**, 201; *Suzuki* and *S. Marumo*, *ibid.*, 1961, **14A**, 34). **Toyocamycin, 7-cyano-7-deaza-adenosine,** $C_{12}H_{13}O_4N_5$ (*H. Nishimura*, *ibid.*, 1956, **9A**, 60; *K. Ohkuma*, *ibid.*, 1961, **14A**, 343) and **sangivamycin, 7-carboxamido-7-deaza-adenosine,** $C_{12}H_{15}O_5N_5$ (*K. V. Ras* and *D. W. Renn*, "Antimicrobial Agents and Chemotherapy", 1963, p. 77), are three structurally related 7-deazapurine nucleoside antibiotics, found in

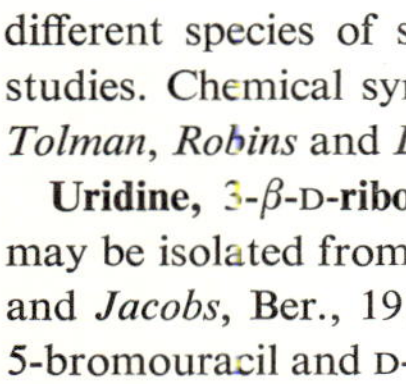

Tubercidin R = H
Toyokamycin R = CN
Sangivamycin R = CONH2

different species of streptomyces, whose structures have been determined by chemical studies. Chemical synthesis of all three nucleosides has confirmed their structure (*R. L. Tolman*, *Robins* and *L. B. Townsend*, J. Amer. chem. Soc., 1968, **90**, 5337).

Uridine, 3-β-D-ribofuranosyluracil, $C_9H_{12}O_6N_2$ (formula p. 118), needles, m.p. 165°, may be isolated from ribonucleic acid hydrolysates or by deamination of cytidine (*Levene* and *Jacobs*, Ber., 1910, **43**, 3159). Oxidation with bromine and hydrobromic acid gives 5-bromouracil and D-ribonic acid (*Levene* and *F. B. La Forge*, *ibid.*, 1912, **45**, 608) whereas, with bromine water, 5-bromouridine is formed, converted by lead oxide to 5-hydroxyuridine. Similarly, nitric acid gives 5-nitrouridine and phenylhydrazine, 4,5-di(phenylhydrazino)uridine. With hydrogen and palladium, uridine is converted into 4,5-dihydrouridine which is labile to acids giving D-ribose, and which forms a triacetyl derivative. 5′-Trityluridine may be converted into 2′,3′-dimethyluridine and 2′,3′-ditosyl-5′-trityluridine (*H. Bredereck*, Ber., 1932, **65**, 1830; *Levene* and *Tipson*, J. biol. Chem., 1934, **104**, 385; 1934, **105**, 419). Uridine forms a 2′,3′-isopropylidene derivative which can be converted into the corresponding 5′-tosyl compound and crystalline 5′-iodo-2′,3′-isopropylideneuridine (*idem, ibid.*, 1934, **106**, 113).

Cytidine, 3-β-D-ribofuranosylcytosine, $C_9H_{13}O_5N_3$ (formula p. 118), needles, m.p. 230°, very soluble in water, is converted into uridine by nitrous acid (*Levene* and *Jacobs*, Ber., 1910, **43**, 3150). Cytidine forms a *picrate*, m.p. 185–187°, a sulphate, hydrochloride and nitrate. With trityl chloride it gives 5′-tritylcytidine (*Bredereck et al., ibid.*, 1940, **73**, 269).

5-Methyluridine, 1-β-D-ribofuranosylthymine, $C_{10}H_{14}N_2O_6$, m.p. 183–185°, has been found in nearly all species of tRNA whose sequences have been determined. Exceptions are the eukaryotic initiator tRNAs.

Pseudouridine, 5-β-D-ribofuranosyluracil, $C_9H_{12}N_2O_6$ (formula p. 130), m.p. 223-224°, is the only nucleoside component of nucleic acids which has a *C*-glycoside linkage (*W. E. Cohn*, J. biol. Chem., 1960, **235**, 1488; *A. M. Michelson* and *Cohn*, Biochemistry, 1962, **1**, 490). It is found in nearly all species of tRNA.

Orotidine is a riboside of orotic acid. It has been isolated as its cyclohexylamine salt from a mutant of *Neurospora crassa* and on hydrolysis in acid gives orotic acid and ribose (*Michelson, W. Drell* and *H. K. Mitchell*, Proc. nat. Acad. Sci., 1951, **37**, 396). Orotidine-

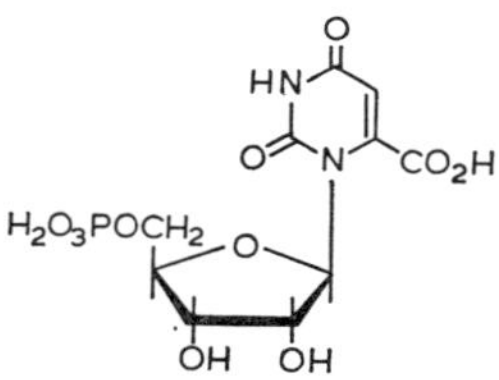

Orotidine 5′-phosphate

5′-phosphate is an important intermediate in pyrimidine nucleotide biosynthesis, where it is formed by enzymic condensation of orotic acid with ribose 5-phosphate 1-pyrophosphate (*A. Kornberg, I. Lieberman* and *E. S. Simms*, J. biol. Chem., 1955, **215**, 389, 403).

Amicetin, $C_{29}H_{42}O_9N_6$, is found in *Streptomyces* cultures and was the first pyrimidine nucleoside antibiotic to be reported (*C. DeBoer, E. L. Caron* and *J. W. Hinman*, J. Amer. chem. Soc., 1953, **75**, 499, 5864). The complete structure has been determined by chemical studies and synthesis by many workers, whose work has been reviewed (*J. J. Fox, K. A. Watanebe* and *A. Block*, "Propgress in Nucleic Acid Research and Molecular Biology", *J. W. Davison* and *Cohn*, eds., Academic Press, New York, 1966, Vol. 5, p. 272).

Vicine, $C_{10}H_{16}O_7N_4$, was first isolated from legumes of the species *Vicia* (*H. Ritthausen*, J. pr. Chem., 1873, [ii], **7**, 374; 1899, **59**, 480, 482; Ber., 1876, **9**, 301; 1896, **29**, 2108) but occurs also in peas and beet. On hydrolysis in acid it yields D-glucose and a pyrimidine, divicine (*E. Fischer, ibid.*, 1914, **47**, 2611). Although it was once thought to be a 3-glucoside of 2,5-diamino-4,6-dihydroxypyrimidine (*T. B. Johnson*, J. Amer. chem. Soc., 1914, **36**,

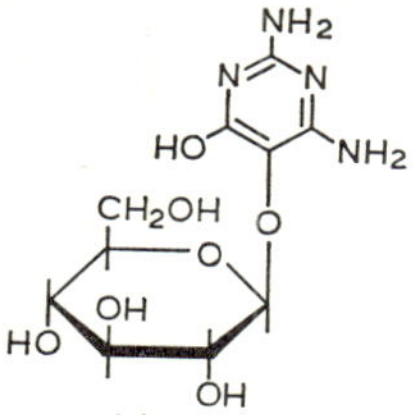

Vicine

337; *Johnson* and *C. O. Johns*, *ibid.*, 1914, **36**, 543; *P. A. Levene*, J. biol. Chem., 1914, **18**, 305; *Levene* and *J. K. Senior*, *ibid.*, 1916, **25**, 607), it in fact has the structure given above (*A. Bendich* and *G. C. Clements*, Biochim. Biophys. Acta, 1953, **12**, 462; *J. Y. Lin* and *K. H. Ling*, C. A., 1966, **65**, 4143g).

2′-Deoxyadenosine, 9-*β*-D-2′-**deoxyribofuranosyladenine,** $C_{10}H_{13}O_3N_5$ (formula p. 120), is isolated from enzyme digests of deoxyribonucleic acid (*W. Klein*, Z. physiol. Chem., 1934, **224**, 244; 1938, **255**, 82; *T. G. Brady*, Biochem. J., 1941, **35**, 855). It is very labile to acids giving adenine and D-2-deoxyribose (or laevulic acid).

2′-Deoxyguanosine, 9-*β*-D-2′-**deoxyribofuranosylguanine,** $C_{10}H_{13}O_4N_5$ (formula p. 120), is also isolated from enzyme hydrolysates of deoxyribonucleic acid (*O. Schindler*, Helv., 1949, **32**, 979). It crystallises as needles and plates.

Cordycepin, 3′-**deoxyadenosine,** $C_{10}H_{13}O_3N_5$, m.p. 225–226°, was the first known nucleoside antibiotic. It was isolated from the mould *Cordyceps militaris* (Linn.) (*F. S. Spring et al.*, J. chem. Soc., 1951, 2299). On hydrolysis it yields adenine and a deoxypentose which originally was wrongly identified and led to an incorrect structure being assigned to the nucleoside (*Spring et al.*, *ibid.*, p. 2301). It was not until 1964 when synthetic 3′-deoxy-

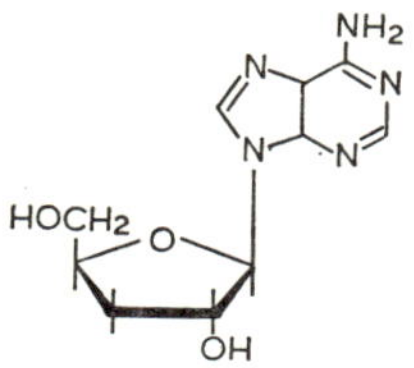

Cordycepin (3′-deoxyadenosine)

adenosine (*W. W. Lee et al.*, J. Amer. chem. Soc., 1961, **83**, 1906) was shown to be identical with cordycepin (*E. A. Kaczka et al.*, Biochem. Biophys. Res. Comm., 1964, **14**, 456) that the true structure was determined.

Thymidine, 1-*β*-D-2′-**deoxyribofuranosyl-5-methyluracil,** $C_{10}H_{14}C_5N_2$ (formula p. 120), m.p. 186°, is isolated along with cytosine deoxyriboside from enzyme hydrolysates of deoxyribonucleic acid. It gives thymine on hydrolysis. 5′-Tritylthymidine and its tosyl esters are known (*Levene* and *Tipson*, Science, 1935, **81**, 98).

Deoxycytidine, 1-*β*-D-2′-**deoxyribofuranosylcytosine,** $C_9H_{13}O_4N_3$ (formula p. 120), prisms, m.p. 210°.

Deoxyuridine, 1-*β*-D-2′-**deoxyribofuranosyluracil,** $C_9H_{12}O_5N_2$, (formula p. 131) needles, m.p. 163°, is present in small amounts in hydrolysates of deoxyribonucleic acid (*C. A. Dekker* and *A. R. Todd*, Nature, 1950, **166**, 557). However, in the transducing bacteriophage of *Bacillus subtilus*, deoxyuridine completely replaces thymidine in DNA (*I. Takahashi* and *J. Marmur*, *ibid.*, 1963, **197**, 794).

5-Methyldeoxycytidine, **1-(β-D-2′-deoxyribofuranosyl-5-methylcytosine,** $C_{10}H_{15}O_4N_3$, m.p. 211–212°, is present in plant DNA and to a lesser extent in mammalian DNA. In the Bacteriophage XP 12 in *Xanthomonas oryzae* it completely replaces deoxycytidine (*T. T. Kuo*, *T. C. Huang* and *M. H. Ting*, J. mol. Biol., 1968, **34**, 373). It is converted to thymidine by nitrous acid (*Dekker* and *D. T. Elmore*, J. chem. Soc., 1951, 2864).

5-Hydroxymethyldeoxycytidine, **1-(β-D-2′-deoxyribofuranosyl-5-hydroxymethylcytosine,** $C_{10}H_{15}O_5N_3$, m.p. 203° (decomp.). In the DNA of the T even bacteriophages, T_2, T_4 and T_6, 5-hydroxymethyldeoxycytidine completely replaces cytidine (*G. R. Wyatt* and *S. S. Cohen*, Biochem. J., 1953, **55**, 774), and many of these residues contain glucosyl or diglucosyl residues (*I. R. Lehman* and *E. A. Pratt*, J. biol. Chem., 1960, **235**, 3254).

Details of the synthesis and properties of many of the modified nucleosides found as minor components of DNA and RNA can be found elsewhere (*R. H. Hall*, "The Modified Nucleosides in Nucleic Acids", Columbia University Press, New York, 1971).

2. Nucleotides

Chemical or enzymic hydrolysis of nucleic acids gives, depending on the experimental conditions, either nucleosides, their degradation products (purines, pyrimidines, etc.) or their phosphoric esters, known as *nucleotides*. Nucleotides are also produced by hydrolysis of certain coenzymes but, from the practical viewpoint, nucleic acids are the usual source for these substances. Since nucleotides can be hydrolysed further to the corresponding nucleosides, the main structural considerations described in this section concern the location of the phosphate group.

Isolation. The first nucleotide to be isolated from natural sources was inosinic acid (*J. von Liebig*, Ann., 1847, **62**, 257). Although this is a minor component of tRNA, the bulk of it arises as a product of enzymic deamination of "muscle adenylic acid" which occurs in the free state in many extracts from natural sources (*G. Embden* and *M. Zimmerman*, Z. physiol. Chem., 1927, **167**, 137). The ribonucleotides can be obtained from nucleic acids either by mild alkaline hydrolysis (*H. Steudel* and *P. Peiser*, *ibid.*, 1922, **120**, 292) or by the action of ribonucleases (*C. B. Anfinsen* and *F. H. White*, "The Enzymes" [2nd Edn.], eds. *P. D. Boyer*, *H. Lardy* and *K. Myrbäck*, Academic Press, New York, 1961, Vol. **5**, p. 65). Nucleotides isolated in this way are not homogenous and differ from those nucleotides isolated directly from natural sources. Adenylic acid isolated by alkaline hydrolysis of yeast RNA is a mixture of two isomers which are the nucleoside 2′- or 3′-phosphates. The nucleotides guanylic, uridylic and cytidylic acids are also mixtures of isomers, which are separated by ion-exchange chromatography

(*W. E. Cohn*, Science, 1949, **109**, 377; J. Amer. chem. Soc., 1949, **71**, 2275; 1950, **72**, 2811; *Cohn* and *C. E. Carter*, *ibid.*, 1950, **72**, 2606; *Carter*, *ibid.*, 1950, **72**, 1466). Isomeric nucleotides may also be separated by fractionation of their brucine or cyclohexylamine salts (*H. S. Loring* and *N. G. Luthy*, *ibid.*, 1951, **73**, 4215; *P. Reichard, Y. Takinaka* and *Loring*, J. biol. Chem., 1952, **198**, 599). Nucleoside 5′-phosphates are obtained from RNA by digestion with phosphodiesterase from snake venom (*Cohn* and *E. Volkin*, *ibid.*, 1953, **203**, 319).

Purine deoxyribonucleotides are labile to acid and alkali and therefore their isolation by chemical hydrolysis of DNA is impractical. As in the preparation of the corresponding nucleosides, enzymic methods must be used (*W. Klein* and *S. J. Thannhauser*, Z. physiol. Chem., 1933, **218**, 173; 1934, **224**, 252; 1935, **231**, 96). 5′-Deoxyribonucleoside phosphates are obtained when DNA is digested with pancreatic deoxyribonuclease (*Volkin, J. X. Khym* and *Cohn*, J. Amer. chem. Soc., 1951, **73**, 1533). In contrast, digestion with calf spleen deoxyribonuclease (*M. E. Maver* and *A. E. Greco*, J. Nat. Cancer Inst., 1956, **17**, 503; *J. F. Koerner* and *R. L. Sinsheimer*, J. biol. Chem., 1957, **228**, 1049), or micrococcal nuclease (*L. Cunningham, B. W. Catlin* and *M. Privat de Carihle*, J. Amer. chem. Soc., 1956, **78**, 4624; *Cunningham*, Ann. N.Y. Acad. Sci., 1959, **81**, 788) gives 3′-deoxynucleoside phosphates. The pyrimidine deoxyribonucleosides, being more stable, can be prepared by chemical hydrolysis of DNA (*P. A. Levene* and *H. Mandel*, Ber., 1908, **41**, 1905; *Thannhauser* and *B. Ottenstein*, Z. physiol. Chem., 1921, **114**, 39; *C. A. Dekker, A. M. Michelson* and *A. R. Todd*, J. chem. Soc., 1953, 947).

Ion-exchange methods for the separation of mononucleotides and their preparation from both types of nucleic acid have been improved and simplified (*Cohn*, "Methods in Enzymology", eds. *S. P. Colowick* and *N. O. Kaplan*, 1957, Vol. III, p. 724). Separation of nucleotides can also be achieved by using paper chromatography (*R. Markham*, *ibid.*, 1955, Vol. III, p. 743), thin-layer chromatography (*K. Randerath* and *E. Randerath*, "Methods in Enzymology", eds. *L. Grossman* and *K. Moldave*, Academic Press, New York, 1967, Vol. XIIA, p. 323) and paper electrophoresis (*J. D. Smith*, *ibid.*, p. 350). Borate buffers can be used in chromatographic or electrophoretic procedures to separate *cis*-1,2-glycol containing nucleotides since these alone form a borate complex (*Khym*, *ibid.*, p. 93). An extension of this procedure uses a dihydroxyboryl-substituted polymer (*H. Schott*, Angew. Chem. internat. Edn., 1972, **11**, 824).

Adenosine 3′,5′-cyclic phosphate (cyclic AMP, cAMP) has been found in biological tissue (*E. W. Sutherland* and *T. W. Rall*, J. biol. Chem., 1958, **232**, 1065, 1077), and is also obtained on degradation of adenosine 5′-triphosphate by aqueous barium hydroxide (*D. Lipkin, Markham* and *W. H. Cook*, J. Amer. chem. Soc., 1959, **81**, 6075). This nucleotide has remarkable biological properties and has been designated the role of a *second messenger*, a hypothesis which states that the cyclic nucleotide mediates the effects of a variety of hormones and other biologically active agents (*G. A. Robinson, R. W. Butcher* and *Sutherland*, Ann. Review Biochem., 1968, **37**, 149). Guanosine 3′,5′-cyclic phosphate (cGMP) (*D. F. Ashman et al.*, Biochem. Biophys. Res. Comm., 1963, **11**, 330), cytidine 3′,5′-cyclic phosphate (cCMP) (*A. Block, ibid.*, 1974, **58**, 652) and uridine 3′,5′-cyclic phosphate (cUMP)

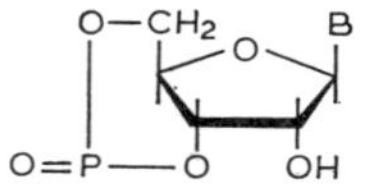

cAMP, B = adenin-9-yl
cGMP, B = guanin-9-yl
cCMP, B = cytosin-1-yl
cUMP, B = uracil-1-yl

(*Block, ibid.*, 1975, **64**, 210) have been isolated from biological material also.

Two unusual nucleotides which were originally named "magic spot" I (MSI) and "magic spot" II (MSII) have been found in cells where RNA synthesis is limited (*M. Cashel* and *J. Gallant*, Nature, 1969, **221**, 838; *Cashel*, J. biol. Chem., 1969, **244**, 3133). Chemical and enzymic studies show MSI to be guanosine 5′-diphosphate-3′-diphosphate (guanosine tetraphosphate) (*Cashel* and *B. Kalbacher, ibid.*, 1970, **245**, 2309). This structure has been confirmed by ^{13}C-nuclear magnetic resonance spectroscopy (*G. R.*

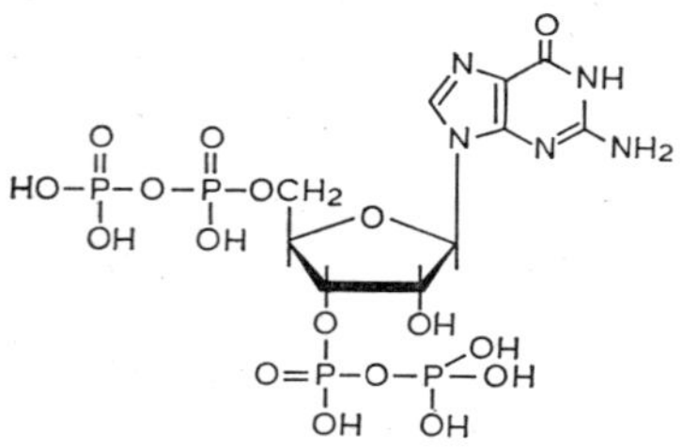

Guanosine tetraphosphate (MSI)

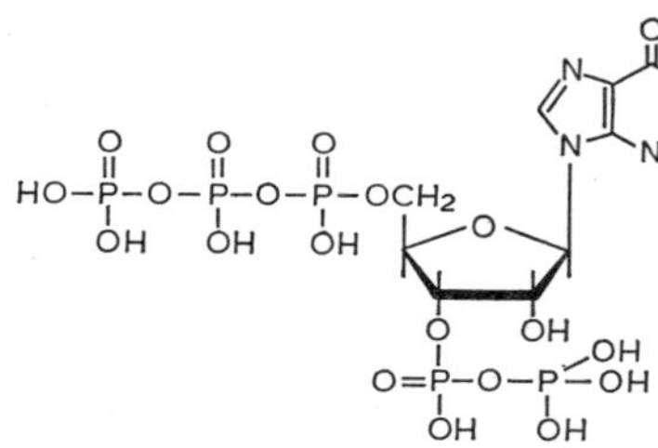

Guanosine pentaphosphate (MSII)

Gray et al., Proc. Nat. Acad. Sci. Wash., 1973, **70**, 2563). MSII is closely related to MSI and is thought to be guanosine 5′-triphosphate-3′-diphosphate (guanosine pentaphosphate) (*Gallant* and *G. Margason*, J. biol. Chem., 1972, **247**, 2289).

(*a*) *Structure of nucleotides*

Nucleotides are phosphoric esters of the nucleosides. In the case of ribonucleotides there are three possible isomers, the 2′-, 3′- or 5′-ribonucleoside phosphates and, in the case of deoxyribonucleotides just two isomers, the

3′- or 5′-deoxynucleoside phosphates. Adenylic acid isolated from muscle on deamination with nitrous acid gives inosinic acid, consequently the relationship between these compounds is equivalent to that between adenosine and inosine (*G. Schmidt*, Z. physiol. Chem., 1928, **179**, 243). Acid hydrolysis of inosinic acid liberates D-ribose 5-phosphate, which has been obtained synthetically (*P. A. Levene* and *W. A. Jacobs*, Ber., 1909, **42**, 335; *Levene* and *E. T. Stiller*, J. biol. Chem., 1934, **104**, 299; *Michelson* and *Todd*, J. chem. Soc., 1949, 2476). From these considerations and from periodate oxidation studies it follows that muscle adenylic acid (AMP) and inosinic acid (IMP) are the 5′-phosphates of adenosine and inosine respectively. This is confirmed by chemical synthesis (see later, p. 152 *et seq.*).

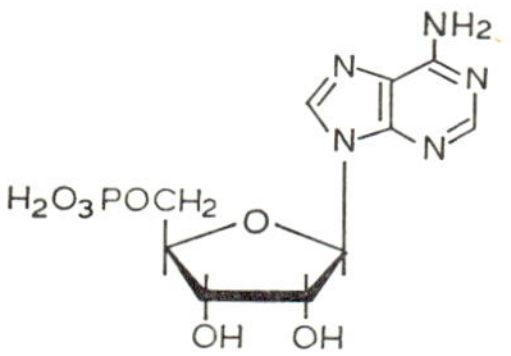

Adenosine 5′-phosphate (AMP)
(Adenylic acid)

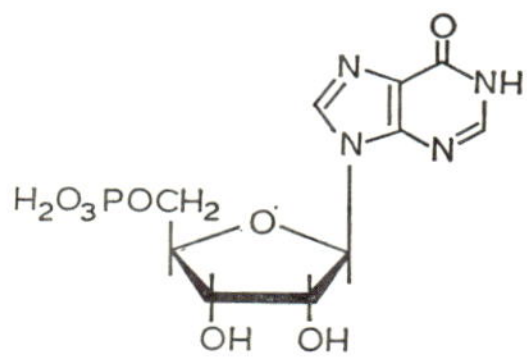

Inosine 5′-phosphate (IMP)
(Inosinic acid)

The purine nucleotides obtained by hydrolysis of yeast RNA (yeast adenylic and guanylic acid) are deaminated with nitrous acid to an inosinic acid and xanthanylic acid respectively (*Levene* and *S. A. Harris*, J. biol. Chem., 1933, **101**, 419; *Levene* and *A. Dmochowski*, *ibid.*, 1931, **93**, 563), both of which on mild hydrolysis give a D-ribose phosphate. Similarly, the reduced pyrimidine nucleotides dihydrouridylic and dihydrocytidylic acids are readily hydrolysed to D-ribose phosphates (*Levene* and *E. Jorpes*, *ibid.*, 1929, **81**, 575). The ribose phosphate obtained as a hydrolysis product of RNA differs from that obtained from muscle AMP or IMP. Therefore the nucleotides from RNA hydrolysis must be 2′- or 3′-phosphates.

Initially it was thought that the nucleotides were 3′-phosphates (*Levene* and *Harris*, *ibid.*, 1932, **98**, 9), but later *C. E. Carter* and *W. E. Cohn* (J. Amer. chem. Soc., 1950, **72**, 2606) showed that the hydrolysis products were mixtures of two isomers which are interconvertible in dilute acid (*D. M. Brown*, *L. J. Haynes* and *Todd*, J. chem. Soc., 1950, 2299), suggesting that they are 2′- and 3′-phosphates. The precise determination of the position of the phosphate group in these two nucleotides is particularly difficult because of the ease of isomerisation.

By hydrolysis of the adenylic acid isomers with the acid form of a sulphonic acid resin followed by determination of the structure of either the ribose 2-phosphate or ribose 3-phosphate liberated, adenosine 2′-phosphate

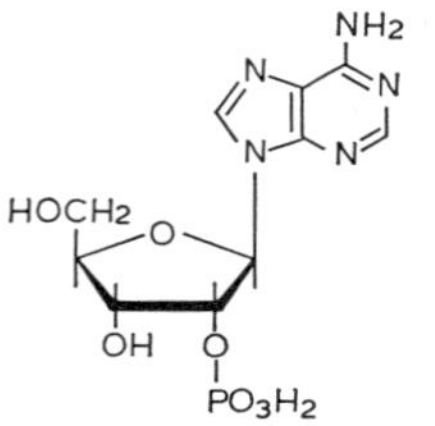

Adenosine 2'-phosphate
(2'-AMP)

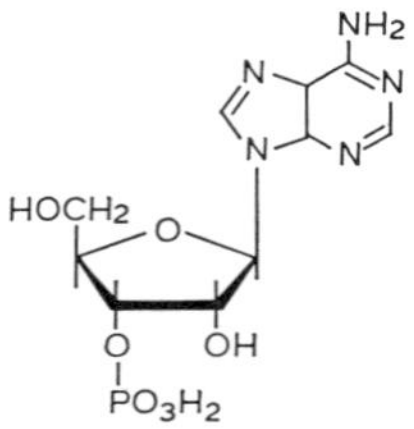

Adenosine 3'-phosphate
(3'-AMP)

and adenosine 3′-phosphate have been identified (*J. X. Khym et al.*, J. Amer. chem. Soc., 1953, **75**, 1262). An unambiguous synthesis of adenosine 2′-phosphate (*Todd et al.*, J. chem. Soc., 1954, 1448) and X-ray crystallographic studies on adenosine 3′-phosphate (*idem*, Nature, 1953, **172**, 1184) have confirmed their identities.

Two guanylic cids and two interconvertible forms of cytidylic acid have been isolated from RNA hydrolysates (*Cohn*, J. Amer. chem. Soc., 1950, **72**, 1471). Measurements of the p*K* values and densities of the cytidylic acid isomers show that one isomer has a greater density and lower p*K* value of the amino group, which suggests that it has a greater charge separation and

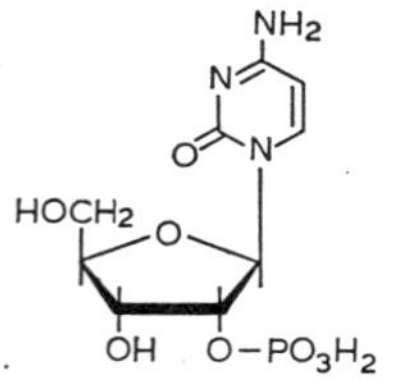

Cytidine 2'-phosphate
(2'-CMP)

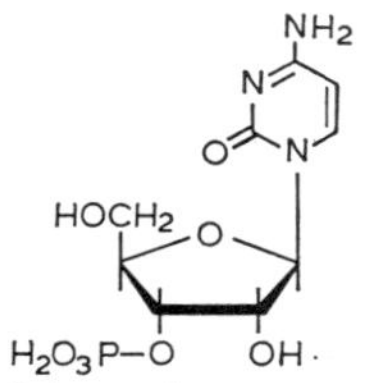

Cytidine 3'-phosphate
(3'-CMP)

hence the 3′-phosphate (*L. F. Cavalieri*, *ibid.*, 1952, **74**, 5804). Alkaline deamination of cytidine 3′-phosphate gives an isomer of uridylic acid which is assumed to be uridine 3′-phosphate (*Brown*, *Dekker* and *Todd*, J. chem. Soc., 1952, 2715).

The structures of the isomers have been confirmed by degradation of the pyrimidine ring with hot hydrazine. Under these conditions no phosphate migration occurs and the structure of the isomer is determined by identification of the product of degradation which is either ribose 2-phosphate or ribose 3-phosphate (*F. Baron* and *Brown*, *ibid.*, 1955, 2855). Unambiguous synthesis of uridine 2′-phosphate (*Brown*, *Todd* and *S. Varadarjan*, *ibid.*, 1956, 2388) and X-ray crystallographic studies on cytidine 3′-phosphate (*E. Alver* and *S. Furberg*, Acta Chem. Scand., 1959, **13**, 910) have confirmed the structures assigned to these compounds.

The monophosphates of adenine-, guanine-, cytosine- and thymine-deoxyribosides obtained from DNA by enzymic hydrolysis (*W. Klein* and *S. J. Thannhauser*, Z. physiol. Chem., 1933, **218**, 173; 1934, **224**, 252; 1935, **231**, 96) are dephosphorylated by a specific 5′-nucleotidase which suggests that they are 5′-phosphates (*Carter*, J. Amer. chem. Soc., 1951, **73**, 1537). This has been confirmed by comparison of the thymidine nucleotide with thymidine 5′-phosphate synthesised from the 3′-acetyl compound (*Michelson* and *Todd*, J. chem. Soc., 1953, 951). The same deoxyribonucleoside 5′-phosphates are obtained on digestion of DNA with pancreatic deoxyribonuclease (*E. Volkin*, *Khym* and *Cohn*, J. Amer. chem. Soc., 1951, **73**, 1533), or phosphodiesterase from snake venom (*R. O. Hurst*, *J. A. Little* and *G. C. Butler*, J. biol. Chem., 1951, **188**, 705; *R. L. Sinsheimer* and *J. F. Koerner*, *ibid.*, 1952, **198**, 293). The deoxyribonucleotides obtained by digestion of DNA with calf spleen deoxyribonuclease (*M. E. Maver* and *A. E. Greco*, J. Nat. Cancer Inst., 1956, **17**, 503; *Koerner* and *Sinsheimer*, J. biol. Chem., 1957, **228**, 1049), micrococcal nuclease (*L. Cunningham*, J. Amer. chem. Soc., 1958, **80**, 2546) or phosphodiesterase from spleen (*Koerner* and *Sinsheimer*, *loc. cit.*) are unaffected by 5′-nucleotidase, and by comparison with synthetic samples are shown to be deoxyribonucleoside 3′-phosphates.

Under certain conditions of acid hydrolysis, DNA gives rise to diphosphates of thymidine and deoxycytidine (*Levene*, *ibid.*, 1938, **126**, 63; 1921,

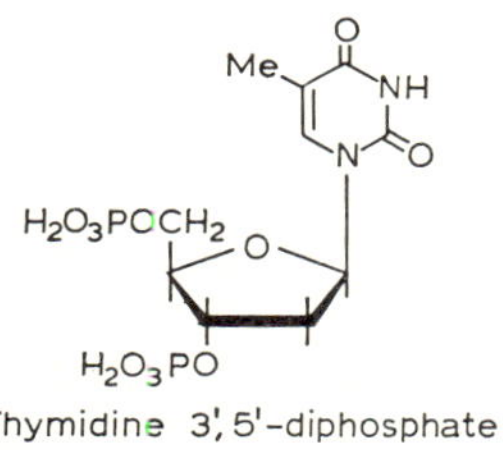

Thymidine 3′,5′-diphosphate

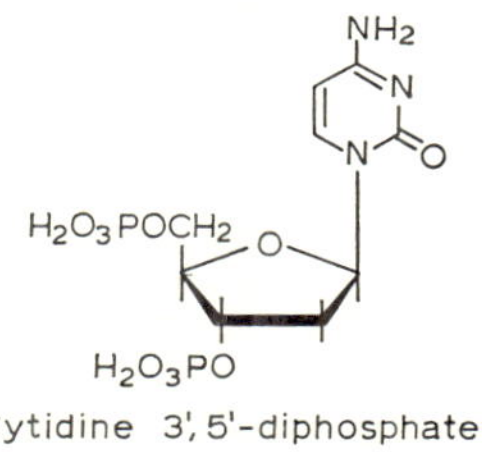

Cytidine 3′,5′-diphosphate

48, 119). Catalytic hydrogenation of the thymidine nucleotide followed by hydrolysis liberates a deoxyribose diphosphate and so the nucleotides may be represented as shown.

At pH 7 the phosphate group of the nucleotide is completely ionised so that replacing a hydrogen atom of a nucleoside by a phosphate does not introduce any new rotatable bonds nor bring about any major change in conformation. Nucleotide conformations in crystal have been deduced from X-ray studies (*S. Arnott*, *S. D. Dover* and *A. J. Wonacott*, Acta Cryst., 1969, **B25**, 2192; *Arnott* and *D. W. L. Hukins*, Nature, 1969, **224**, 886; *M. Sundaralingam*, Biopolymers, 1969, **7**, 821) and in solution by n.m.r. studies (*S. S. Danyluk et al.*, Biochemistry, 1967, **8**, 1038; 1974, **13**, 4417; 1975, **14**, 543;

P. O. P. Ts'O et al., *ibid.*, 1969, **8**, 997; Ann. N.Y. Acad. Sci., 1969, **158**, 256; J. Amer. chem. Soc., 1968, **90**, 1042; *J. H. Prestegard* and *S. I. Chan*, *ibid.*, 1969, **91**, 2843; *P. A. Hart* and *J. D. Davis*, *ibid.*, 1969, **91**, 512; *F. E. Hruska*, *A. A. Grey* and *I. C. P. Smith*, *ibid.*, 1970, **92**, 214; *D. E. Dorman* and *J. D. Roberts*, Proc. Nat. Acad. Sci. U.S.A., 1970, **65**, 19). Nucleotide conformations have also been determined by studying n.m.r. spectra of molecules in which perturbations have been introduced by specifically bound lanthanide cations (*R. J. P. Williams et al.*, Biochem. Biophys. Res. Comm., 1972, **47**, 166; J. mol. Biol., 1974, **84**, 471, 491). These and other studies (*Sundaralingam*, The Jerusalem Symposium on Quantitative Chemistry and Biochemistry, 1973, **5**, 417, in "Structure and Conformations of Nucleic Acids and Protein-Nucleic Acid Interactions", eds. *Sundaralingam* and *S. T. Rao*, University Park Press, Baltimore, U.S.A., 1975, p. 487) have led to the suggestion that nucleotides are more rigid molecules than nucleosides due to the influence of the phosphate ("The phosphate effect"). More recently these conclusions have been put in doubt since determinations of bond angles of nucleosides and nucleotides by n.m.r. studies show them to have the same time average conformations (*F. E. Evans* and *R. H. Sarma*, Nature, 1976, **263**, 567). Nucleotide structures can be determined by mass spectrometry of their trimethylsilyl derivatives (*A. M. Lawson et al.*, J. Amer. chem. Soc., 1971, **93**, 1014).

(*b*) *Synthesis of nucleotides*

Nucleotides are synthesised either by combining a purine or pyrimidine with a phosphorylated sugar or more commonly by phosphorylation of a nucleoside. A detailed account of the properties of various reagents used for the phosphorylation of nucleosides can be found elsewhere (*H. G. Khorana*, "Some Recent Developments in the Chemistry of Phosphate Esters of Biological Interest", Wiley, New York, 1961).

(*i*) *Ribonucleoside 5′-phosphates*

These can be prepared by phosphorylation of a 2′,3′-protected nucleoside. Early syntheses used phosphoryl chlorides as the phosphorylating reagents. Low yields of inosinic acid (*P. A. Levene* and *R. S. Tipson*, J. biol. Chem., 1935, **111**, 313) or adenosine 5′-phosphate (*idem*, *ibid.*, 1937, **121**, 131) were obtained by treatment of 2′,3′-*O*-isopropylideneinosine or 2′,3′-*O*-isopropylideneadenosine with phosphoryl chloride followed by cautious acid hydrolysis to remove the isopropylidene group. A low yield of adenosine 5′-phosphate was also obtained by treatment of 2′,3′-diacetyladenosine with diphenylphosphochloridate followed by hydrogenolysis and acid hydrolysis

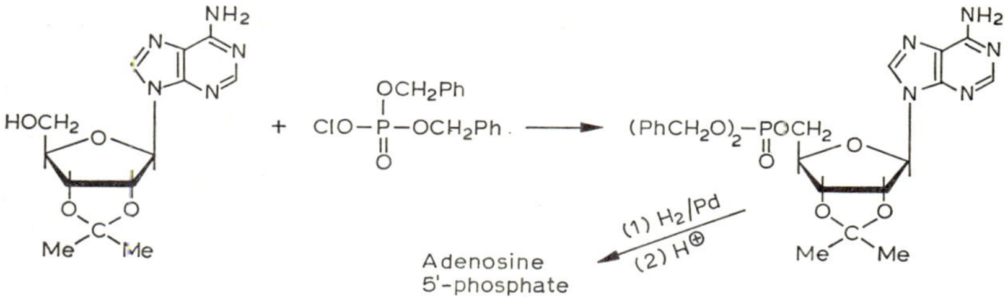

(*H. Bredereck*, *E. Berger* and *J. Ehrenberg*, Ber., 1940, **73**, 269). The first satisfactory nucleotide synthesis involved the phosphorylation of 2′,3′-*O*-isopropylideneadenosine with dibenzyl phosphorochloridate which gave the crystalline neutral ester in high yield, from which the benzyl group may be removed by hydrogenolysis and the isopropylidene group by cautious hydrolysis (*J. Baddiley* and *A. R. Todd*, J. chem. Soc., 1947, 648).

Uridine 5′-phosphate and cytidine 5′-phosphate may also be obtained in high yield from their isopropylidenenucleosides by applying the method above using dibenzyl phosphochloridate as the phosphorylating agent (*Michelson* and *Todd*, *ibid.*, 1949, 2476). A good yield of uridine 5′-phosphate is obtained by the action of a mixture of phosphoric acid and phosphorus pentoxide on 2′,3′-*O*-isopropylideneuridine, followed by acid treatment (*R. H. Hall* and *Khorana*, J. Amer. chem. Soc., 1955, **77**, 1871). Cytidine 5′-phosphate may be prepared from the benzylidene derivative in a somewhat similar manner (*Baddiley*, *J. G. Buchanan* and *A. R. Sanderson*, J. chem. Soc., 1958, 3107).

Guanosine 5′-phosphate is not readily prepared by the above methods but is obtained in good yield by treatment of 2′,3′-isopropylideneguanosine with tetra-*p*-nitrophenyl pyrophosphate, followed by removal of protecting

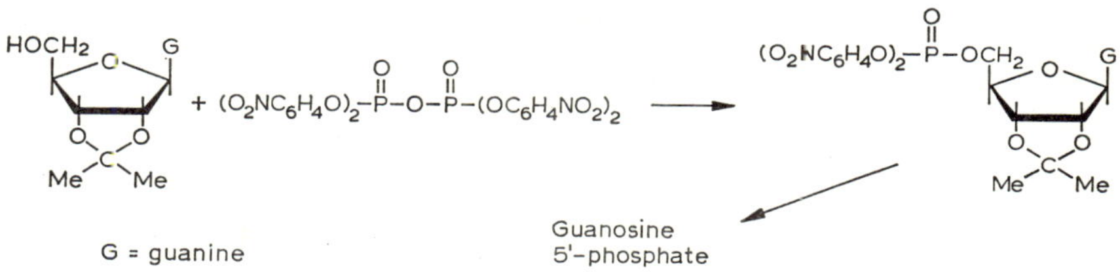

G = guanine

groups from the neutral di-*p*-nitrophenyl ester first formed, with dilute alkali, snake venom phosphodiesterase and mild acid (*R. W. Chambers*, *J. G. Moffatt* and *Khorana*, J. Amer. chem. Soc., 1957, **79**, 3747).

5′-Ribonucleotides are obtained in good yields by phosphorylation of isopropylidenenucleosides with phosphorus trichloride in acetone. The acetone and air are necessary for the reaction which proceeds *via* an intermediate dichlorophosphite (*M. Hongs et al.*, Tetrahedron Letters, 1966, 3851).

An alternative and improved approach to the synthesis of nucleotides uses a mixture of an orthophosphate ester and dicyclohexylcarbodiimide when the phosphorylating reagent is probably a cyclic trimetaphosphate (*G. Weimann* and *Khorana*, *ibid.*, 1962, **84**, 4329; *G. M. Blackburn, M. J. Brown* and *M. R. Harris*, Chem. Comm., 1966, 611). β-Cyanoethyl phosphate has been used extensively as the phosphate moiety since, after phosphorylation of a nucleoside, the cyanoethyl group is removed easily by mild alkaline conditions (*P. T. Gilham* and *G. M. Tener*, Chem. and Ind., London, 1959, 542). Many nucleotides, including guanosine 5′-phosphate, have been obtained in excellent yield by this procedure (*Tener*, J. Amer. chem. Soc., 1961, **83**, 159).

5′-Nucleoside phosphates are obtained in good yield by treatment of 2′,3′-isopropylidenenucleosides with orthophosphoric acid in the presence of mercuric chloride and *N*-methylimidazole, followed by mild acid treatment to remove the protecting groups (*H. Takaku, Y. Shimada* and *H. Oka*, Chem. and Pharm. Bull., Japan, 1973, **21**, 1844). By increasing the amount of mercuric chloride and imidazole in the reaction nucleoside 5′-diphosphates and nucleoside 5′-triphosphates are obtained.

Phosphorylation of unprotected nucleosides with phosphoryl chloride leads to 80–90% yields of 5′-nucleotides if the reaction is carried out in the presence of a wet trialkyl phosphate (*M. Yoshikawa, T. Kato* and *T. Takenishi*, Tetrahedron Letters, 1967, 5065) or aqueous pyridine (*T. Sowa* and *S. Ouchi*, Bull. chem. Soc., Japan, 1975, **48**, 2084). In the latter case the phosphorylating agent is thought to be a complex of tetrachloropyrophosphate and pyridinium chloride. The 5′-phosphates of adenosine, guanosine, uridine, cytidine and inosine have been prepared in this way. Only very small amounts of the 2′- or 3′-nucleotides are formed.

Nucleoside 5′-phosphates have been obtained in low over-all yield by reaction between a suitably protected phosphorylated sugar and a purine or pyrimidine. Reaction between 2,3-di-*O*-benzoyl-5-diphenylphosphoryl-D-ribofuranosyl bromide (I) and dithyminylmercury followed by hydrogenolysis of the reaction product and then treatment with sodium methoxide gives 1-β-D-ribofuranosylthymine 5′-phosphate (*T. Ukita* and *H. Hayatsu*, *ibid.*, 1962, **84**, 1879). A similar reaction using chloromercuri-6-benzamidopurine leads to adenosine 5′-phosphate. In the latter synthesis, however, the vigorous hydrogenolysis required for removal of the phenyl residues cannot be tolerated. The *O*- and *N*-benzoyl groups are removed under alkaline conditions which also remove one of the phenyl groups from the 5′-phosphate. The remaining phenyl group is removed by treatment with snake venom phosphodiesterase (*idem, loc. cit.*)

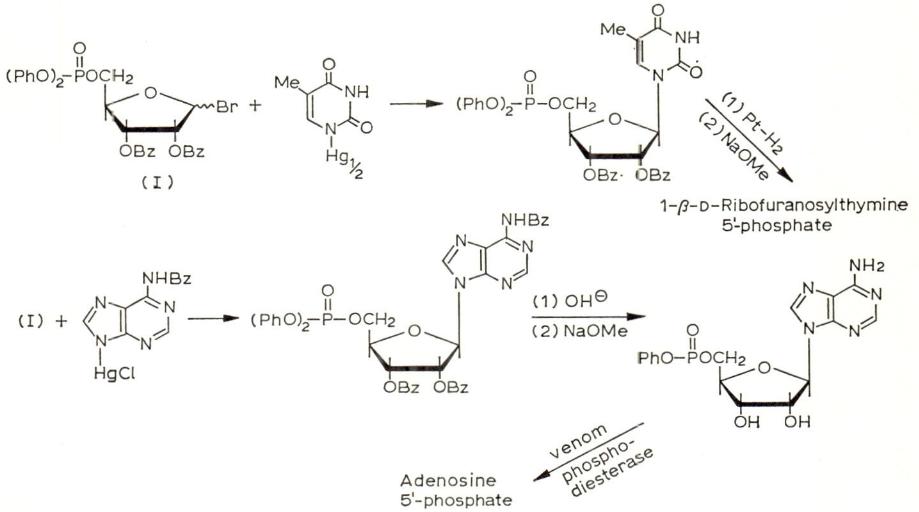

Trimethylsilyl derivatives of purines or pyrimidines can be used as an alternative to the mercuri compounds for condensation with the glycosyl halide I. The trimethylsilyl group is removed by treatment with aqueous ethanol prior to removal of the phenyl and benzoyl groups (*B. Shimizu, M. Asai* and *T. Nishimura*, Chem. and Pharm. Bull., Japan, 1965, **13**, 230).

(*ii*) *Ribonucleoside 2'- or 3'-phosphates*

The synthesis of a mixture of ribonucleoside 2'- and 3'-phosphates can be achieved by phosphorylation of 5'-*O*-tritylribonucleosides followed by removal of the trityl group under acid conditions. Adenosine 2'- and 3'-phosphates were synthesised in this manner using dibenzyl phosphorochloridate as the phosphorylating reagent (*D. M. Brown* and *A. R. Todd*, J. chem. Soc., 1953, 2708). Selective phosphorylation at the 2'- and 3'-hydroxyl groups of an unprotected nucleoside can be achieved by the reaction of ethyl trichloromethanephosphonate in the presence of triethylamine on the *cis*-diol grouping to give 2'- and 3'-ribonucleotide esters, which can then be hydrolysed with alkali to give a mixture of the free 2'- and 3'-ribonucleotides (*A. Holy*, Tetrahedron Letters, 1972, 157). The 2'- and 3'-phosphate isomers can be separated by ion-exchange chromatography (*W. E. Cohn*, "The Nucleic Acids", eds. *E. Chargaff* and *J. N. Davidson*, Academic Press, New York, 1955, Vol. 1, p. 228).

Because of the ease of migration of the phosphate group between the 2'- and 3'-hydroxyl group of a ribonucleotide, unambiguous synthesis of either a 2'- or 3'-phosphate requires that nucleosides suitably protected at the 3'-

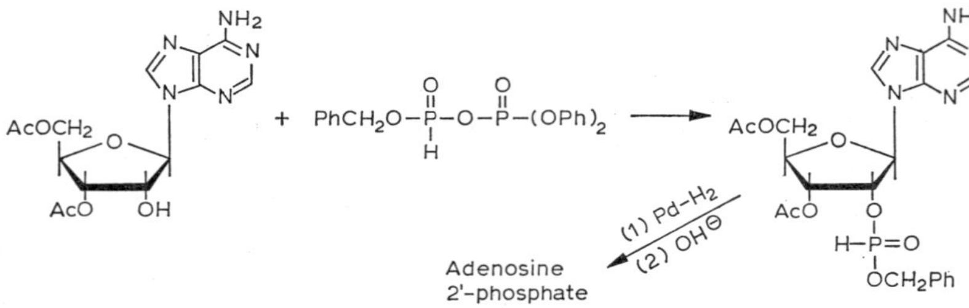

and 5′-positions or 2′- and 5′-positions must be used in the phosphorylation reactions. An unambiguous synthesis of adenosine 2′-phosphate starts from a di-*O*-acetyladenosine which was shown to be the 3′,5′-isomer. This is phosphorylated with *O*-benzylphosphonic *O,O*-diphenylphosphoric anhydride to give the nucleoside 2′-benzyl phosphite as the initial product, from which the protecting groups are removed by hydrogenolysis and treatment with alkali (*Todd et al.*, J. chem. Soc., 1954, 1448).

The synthesis of uridine 2′-phosphate from 3′,5′-*O*-diacetyluridine has been accomplished in a similar manner but the phosphorylating reagent used was dibenzyl phosphorochloridate (*D. M. Brown*, *Todd* and *S. Varadarajan*, *ibid.*, 1956, 2388). It is important in these syntheses that in order to avoid isomerisation of the phosphate group, the phosphorylated products must be converted to the monoesters before alkaline treatment to remove the acetyl groups.

(*iii*) *Ribonucleoside 2′,5′- and 3′,5′-diphosphates*

These compounds are of interest as components of certain coenzymes. The 2′,5′- and 3′,5′-diphosphates of adenosine are formed, together with some 2′,3′,5′-triphosphates by reaction of excess dibenzyl phosphorochloridate on adenosine followed by removal of the benzyl groups (*Todd et al.*, *ibid.*, 1957, 3297). The two diphosphates are separated by anion exchange chromatography (*J. Baddiley*, *J. G. Buchanan* and *R. Letters*, *ibid.*, 1958, 1000). The pyrimidine 2′,5′- and 3′,5′-diphosphates can be obtained by treating uridine or cytidine with a mixture of phosphoric acid and phosphorus pentoxide (*R. H. Hall* and *H. G. Khorana*, J. Amer. chem. Soc., 1955, **77**, 1871).

(*iv*) *Deoxyribonucleoside 5′-phosphates*

Deoxynucleotides are synthesised by somewhat similar procedures to those adopted for ribonucleotides. Thymidine 5′-phosphate is conveniently prepared by phosphorylation of 3′-*O*-acetylthymidine with dibenzyl phosphorochloridate (*A. M. Michelson* and *Todd*, J. chem. Soc., 1953, 951), or a mixture of β-cyanoethyl phosphate and dicyclohexylcarbodiimide (*G. M. Tener*, J. Amer. chem. Soc., 1961, **83**, 159), followed by removal of the protecting groups. Deoxycytidine 5′-phosphate can be synthesised by an analogous reaction with N^6,3′-*O*-diacetyldeoxycytidine (*Michelson* and *Todd*, J. chem. Soc., 1954, 34; *Tener*, *loc. cit.*). Although phosphorylation of 3′-*O*-acetyldeoxyadenosine can be accomplished using dibenzyl phosphorochloridate, the 5′-hydroxyl group of 3′-*O*-acetyldeoxyguanosine is inert to this reagent but can be phosphorylated with *O*-benzylphosphonic *O,O*-diphenylphosphoric anhydride (*D. H. Hayes, Michelson* and *Todd*, J. chem. Soc., 1955, 808). Deoxyadenosine and deoxyguanosine 5′-phosphates can be prepared by phosphorylation of the 3′-*O*-acetyldeoxynucleosides with a mixture of β-cyanoethyl phosphate and dicyclohexylcarbodiimide (*Tener*, *loc. cit.*).

(*v*) *Deoxyribonucleoside 3′-phosphates*

Dibenzyl phosphorochloridate is not a very satisfactory reagent for the phosphorylation of the secondary hydroxyl group at the 3′-position of a deoxyribonucleoside, although thymidine 3′-phosphate was first synthesised using this reagent (*Michelson* and *Todd*, J. chem. Soc., 1953, 951). Good

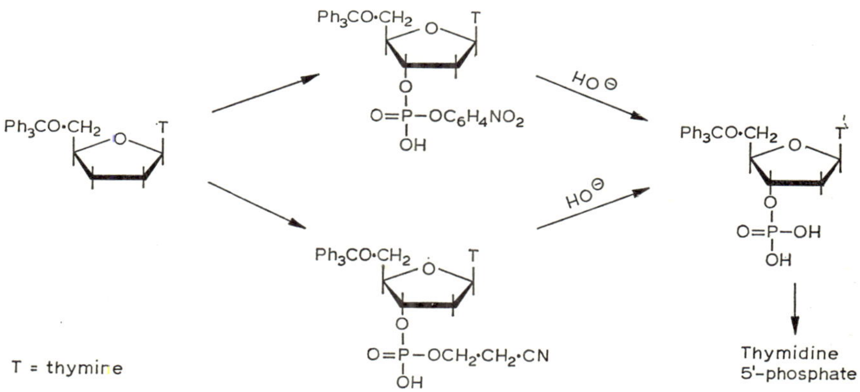

yields of thymidine 3′-phosphate can be obtained by phosphorylation of 5′-tritylthymidine with either mono-*p*-nitrophenyl phosphorochloridate (*A. F. Turner* and *Khorana*, J. Amer. chem. Soc., 1959, **81**, 4651) or the Tener reagent, followed by removal of the protecting groups.

Deoxycytidine 3′-phosphate is obtained by an analogous procedure from 5′-*O*-trityldeoxycytidine (*Tener*, *loc. cit.*) or, in improved yield, using *N*-anisoyl-5′-*O*-di-*p*-methoxytrityldeoxycytidine as the starting protected nucleoside (*Khorana et al.*, J. Amer. chem. Soc., 1963, **85**, 3821).

Because the glycosidic linkage in purine deoxyribonucleosides is sensitive to the acid conditions used for the removal of the 5′-trityl group, other means of protecting the 5′-hydroxyl group during the phosphorylation of purine deoxyribonucleosides are required. 5′-*O*-Acetyldeoxy-adenosine and -guanosine which are obtained by partial hydrolysis of the diacetates (*Todd et al.*, J. chem. Soc., 1954, 1882) can be used for the phosphorylation reaction. More conveniently, the 5′-hydroxyl group can be protected by a 5′-di- or tri-*p*-methoxytrityl group, since these substituted trityl groups are much more acid-labile than the unsubstituted trityl group (*M. Smith et al.*, J. Amer. chem. Soc., 1962, **84**, 430). The 3′-hydroxyl group of the purine deoxyribonucleosides is relatively unreactive and therefore only the more powerful phosphorylating reagents produce good yields of the protected deoxyribonucleotides. Deoxyadenosine 3′-phosphate is prepared by treatment of 5′-*O*-di-*p*-methoxytrityldeoxyadenosine with a mixture of mono-(*β*-cyanoethyl) phosphate and dicyclohexylcarbodiimide, followed by removal of the cyanoethyl group with alkali and the di-*p*-methoxytrityl group with 80% acetic acid. Deoxyguanosine 3′-phosphate is prepared by the same procedure starting from *N*,5′-*O*-bis(di-*p*-methoxytrityl)deoxyguanosine (*Khorana et al.*, *ibid.*, 1963, **85**, 3821).

(vi) Cyclic phosphates

Nucleoside 2′,3′-cyclic phosphates may be prepared from either 2′- or 3′-phosphates by treatment with trifluoroacetic anhydride, followed by removal of trifluoroacetyl groups with ammonia (*D. M. Brown*, *D. I. Magrath* and *Todd*, J. chem. Soc., 1952, 2798). Cyclisation is very readily brought about

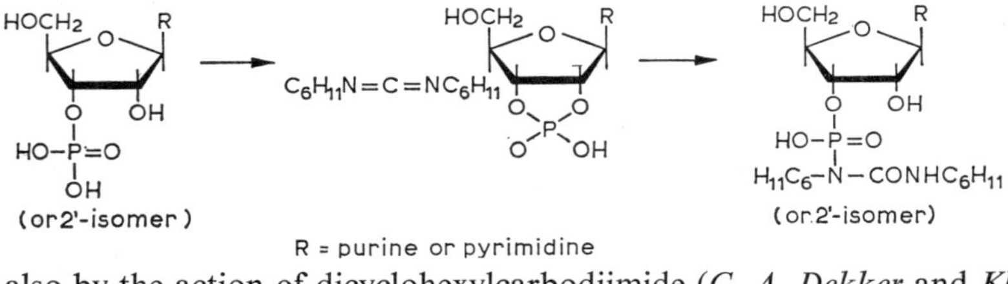

also by the action of dicyclohexylcarbodiimide (*C. A. Dekker* and *Khorana*, J. Amer. chem. Soc., 1954, **76**, 3522; *M. Smith*, *J. G. Moffatt* and *Khorana*, *ibid.*, 1958, **80**, 6204). Further reaction of these five-membered cyclic phosphates gives *N*-phosphorylureas.

Nucleoside 2′,3′-cyclic phosphates are formed during the alkaline hydrolysis of RNA, and are then hydrolysed further to both nucleoside 2′- and 3′-phosphates (*Brown* and *Todd*, J. chem. Soc., 1952, 52). Ribonucleases also hydrolyse RNA *via* 2′,3′-cyclic phosphate intermediates, which are then cleaved specifically to ribonucleoside 3′-phosphates (*Khorana*, "The Enzymes", eds. *P. D. Boyer*, *H. Lardy* and *K. Myrbäck*, Academic Press, New York, 1961, Vol. 5, p. 79).

Ribonucleoside 3′,5′-cyclic phosphates can be synthesised by cyclisation of nucleoside phospho-mono-, -di- or -tri-esters. Thus good yields of adenosine, guanosine, cytidine and uridine 3′,5′-cyclic phosphates are obtained either by treatment of ribonucleoside 5′-phosphates with dicyclohexylcarbodiimide under dilute conditions (*Smith*, *G. I. Drummond* and *Khorana*, J. Amer. chem. Soc., 1961, **83**, 698); by treatment of the *p*-nitrophenyl ester of ribonucleoside 5′-phosphates with potassium *tert*-butoxide (*P. K. Borden* and *Smith*, J. org. Chem., 1966, **31**, 3247); or by treatment of a nucleoside 2′-*O*-methoxytetrahydropyranyl 3′-diphosphate with potassium *tert*-butoxide followed by mild acid treatment to remove the protecting group (*J. H. van Boom et al.*, Chem. Comm., 1974, 618).

Purine–glycosyl bonds are rendered remarkably stable to acid on formation of the 3′,5′-cyclic phosphate whereas, in contrast, the pyrimidine–glycosyl bond of nucleoside 3′,5′-cyclic phosphates is much more labile than that of other pyrimidine nucleotides (*Smith*, *Drummond* and *Khorana*, *loc. cit.*).

The considerable interest in adenosine 3′,5′-cyclic phosphate because of its role as an intermediary in the action of various harmones has led to the synthesis and study of many derivatives (*J. G. Falbriard*, *T. Posternak* and *E. W. Sutherland*, Biochim. Biophys. Acta, 1967, **148**, 99).

Deoxyribonucleoside 3′,5′-cyclic phosphates can be synthesised either by treatment of deoxyribonucleoside 5′-phosphates with dicyclohexylcarbodiimide (*Drummond et al.*, J. Amer. chem. Soc., 1964, **86**, 1626), or by treatment of a *p*-nitrophenyl deoxyribonucleoside 5′- or 3′-phosphate with potassium *tert*-butoxide (*Smith*, *ibid.*, 1964, **86**, 3586). The purine–glycosyl bond of deoxynucleotides becomes considerably stabilised to acid on formation of the 3′,5′-cyclic phosphate (*Drummond et al.*, *loc. cit.*).

(c) *Identification of nucleosides and nucleotides*

Nucleosides and nucleotides, like many of the parent pyrimidines and purines are not readily identified by melting points or crystalline appearance. The most satisfactory methods involve a combination of physical properties, notably, ultraviolet absorption spectra and chromatographic behaviour. Table 2 records the spectral properties of the more common nucleosides and nucleotides and Table 3 includes some specific rotations and pK_a values. Data for the less common nucleosides can be found elsewhere (*R. H. Hall*, "The Modified Nucleosides in Nucleic Acids", Columbia University Press, New York, 1971; "Handbook of Biochemistry", ed. *H. A. Sober*, The Chemical Rubber Co., 1970).

TABLE 2

ULTRAVIOLET ABSORPTION PROPERTIES OF NUCLEOSIDES AND NUCLEOTIDES

	pH	λ_{max} (nm)	λ_{min} (nm)	$\varepsilon_{max} \cdot 10^{-3}$
Adenosine	1	257	230	14.6
	6	260	227	14.9
	11	259	227	15.4
Adenosine 3′-phosphate	2	257	229	14.4
	7	259	227	15.4
	12	259	227	15.4
Adenosine 5′-phosphate	2	257	230	15.0
	7	259	227	15.4
	12	259	227	15.4
Guanosine	1	256	228	12.3
	6	253	223	13.6
	11	257	230	11.3
Guanosine 3′-phosphate	1	257	228	12.2
	7	252	227	13.4
	10.8	257	230	11.25
Guanosine 5′-phosphate	1	256	228	12.2
	7	252	224	13.7
	11	258	230	11.6
Inosine	3	248	223	12.2
	6	248.5	223	12.3
	11	253	224	13.1
Cytidine	1	280	242	13.4
	7	229	226	8.3
		271	250	9.1
	13	272.5	251	9.15
Cytidine 3′-phosphate	2	279	240	13.0
	7	270	250	9.0
	12	272	250	8.9
Cytidine 5′-phosphate	2	280	241	13.2
	7	271	249	9.1
	11	271	249	9.1
Uridine	1	262	230	10.1
	7	262	230	10.1
	12	262	243	7.45
Uridine 3′-phosphate	1	262	230	10.0
	7	262	230	10.0
	13	261	241	7.8
Uridine 5′-phosphate	2	262	230	10.0
	7	262	230	10.0
	11	261	241	7.8

TABLE 2 (continued)

ULTRAVIOLET ABSORPTION PROPERTIES OF NUCLEOSIDES AND NUCLEOTIDES

	pH	λ_{max} (nm)	λ_{min} (nm)	$\varepsilon_{max} \cdot 10^{-3}$
Pseudouridine	2	262	233	7.9
	7	263	233	8.1
	12	286	245	7.7
Pseudouridine 3′(2′)-phosphate	2	263	233	8.4
	7	263	233	—
	12	286	246	8.4
Deoxyadenosine	2	258	228	14.1
	7	259	225	15.0
	13	260	—	14.9
Deoxyadenosine 5′-phosphate	2	258	230	14.3
	7	—	—	15.3
Deoxyguanosine	1	255	—	12.1
	7	254	223	13.0
	12	260	230	9.2
Deoxyguanosine 5′-phosphate	1	255	228	11.8
Deoxycytidine	1	280	241	13.2
	7	271	250	9.0
	13	271.5	250	9.1
Deoxycytidine 5′-phosphate	2	280	239	13.5
	7	271	249	9.3
Thymidine	1	267	235	9.65
	7	267	235	9.65
	13	267	246	7.4
Thymidine 3′-phosphate	7	267	—	9.5
Thymidine 5′-phosphate	2	267	—	10.2
	7	267	—	10.2
5-Methyldeoxycytidine	1	287	245	12.4
	7	277	255	8.5
	14	279	255	8.8
5-Methyldeoxycytidine 5′-phosphate	2	287	244	—
	7	278	254	—
	12	278	—	—
5-Hydroxymethyldeoxycytidine	1	283	243	12.6
	7	272	247	—
	13	274	252	—
5-Hydroxymethyldeoxycytidine 5′-phosphate	1	284	245	12.5
	7	275	254	7.7
	12	275	254	7.7

TABLE 3

SPECIFIC ROTATIONS AND *pK* VALUES OF NUCLEOSIDES AND NUCLEOTIDES

	Specific rotation		
	$[\alpha]_D^\circ$	*c* (g/dl)	pK_as
Adenosine	−61.7	0.706, H_2O	3.5, 12.5
Adenosine 3′-phosphate	−45.4	0.50, 0.5 *M* Na_2HPO_4	3.65
Adenosine 5′-phosphate	−26	1.0, 10% HCl	3.8, 6.2
Guanosine	−72	1.4, 0.1 *N* NaOH	1.6, 9.2, 12.4
Guanosine 3′-phosphate	−57	1.0, 0.5 *N* NaOH	0.7, 2.3, 5.9, 9.7
Guanosine 5′-phosphate	—	—	2.4, 9.4
Inosine	−58.8	2.5 H_2O	1.2, 8.8, 12.3
Cytidine	+34.2	2.0, H_2O	4.15, 12.5
Cytidine 3′-phosphate	+49.4	1.0, H_2O	0.8, 4.3, 6.0
Cytidine 5′-phosphate	+27.1	0.54, H_2O	4.5, 6.3
Uridine	+ 9.6	2.0, H_2O	9.2, 12.5
Uridine 3′-phosphate	+22.3	2.0, H_2O	1.02, 5.9, 9.4
Uridine 5′-phosphate	+ 3.4	1.02, 10% HCl	6.4, 9.5
Pseudouridine	− 3.0	1.0, H_2O	8.9, > 13
Pseudouridine 3′-phosphate	—	—	9.6
Deoxyadenosine	−26	1.0, H_2O	3.8
Deoxyadenosine 5′-phosphate	−38	0.23, H_2O	4.4, 6.4
Deoxyguanosine	−30.2	0.2, H_2O	2.5
Deoxyguanosine 3′-phosphate	−57	1.0, 0.5 *N* NaOH	0.7, 2.3, 5.92
Deoxyguanosine 5′-phosphate	−31	0.43, H_2O	2.9, 9.7
Deoxycytidine	+82.4	1.31, 1 *N* NaOH	4.3, > 13
Deoxycytidine 3′-phosphate	+57.0	1.35, H_2O	—
Deoxycytidine 5′-phosphate	+35	0.2, H_2O	4.6, 6.6
Thymidine	+32.8	1.04, 1 *N* NaOH	9.8, > 13
Thymidine 3′-phosphate	+ 7.3	1.5, H_2O	
Thymidine 5′-phosphate	− 4.4	0.4, H_2O	1.6, 6.5, 10.0
5-Methyldeoxycytidine	+43	1.4, H_2O	4.4, > 13
5-Methyldeoxycytidine 5′-phosphate	—	—	4.4
5-Hydroxymethyldeoxycytidine	+51	H_2O	3.5

3. Nucleotide coenzymes

The coenzymes are a small group of relatively low molecular weight compounds which are required in catalytic amounts in certain enzyme reactions. These coenzymes possess no enzymic properties themselves but combine with enzymes to form complexes which catalyse the appropriate reaction. Many coenzymes are catalytic only in multi-enzyme systems where processes are available to regenerate the coenzyme continually. In the absence of coenzyme regeneration, the coenzymes act as substrates and are required in stoicheiometric amounts. Many coenzymes contain a nucleotide residue and in some cases the nucleotide is linked to a vitamin. The nucleotide coenzymes catalyse a wide range of biochemical reactions which include the synthesis of certain macromolecules and certain oxidation-reduction processes.

(*i*) *Nucleoside di- and tri-phosphates*

Adenosine triphosphate (ATP) was first isolated from muscle by *K. Lohmann* (Naturwiss., 1929, **17**, 624) and by *C. H. Fiske* and *Y. Subbarow* (Science, 1929, **70**, 381). It plays an important role in transphosphorylation where it donates a phosphate group to a substrate. At the same time the ATP is converted to ADP. This reaction is fundamental to energy transfer during cellular metabolism. The structure of ATP has been deduced from several pieces of evidence. On hydrolysis with alkali it gives adenosine 5′-phosphate and inorganic pyrophosphate and, with acid, adenine, ribose 5-phosphate and 2 mols. of phosphoric acid (*Lohmann*, Biochem. Z., 1931, **233**, 460). ATP consumes one mole of periodate (*B. Lythgoe* and *A. R. Todd*, Nature, 1945, **155**, 695) and can be deaminated to inosine triphosphate. Enzymic dephosphorylation of ATP, which has three primary and one secondary phosphate dissociations, liberates only one phosphate group giving **adenosine**

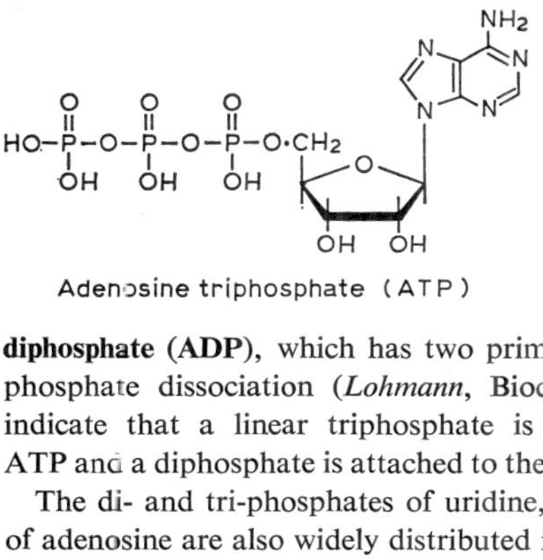

Adenosine triphosphate (ATP)

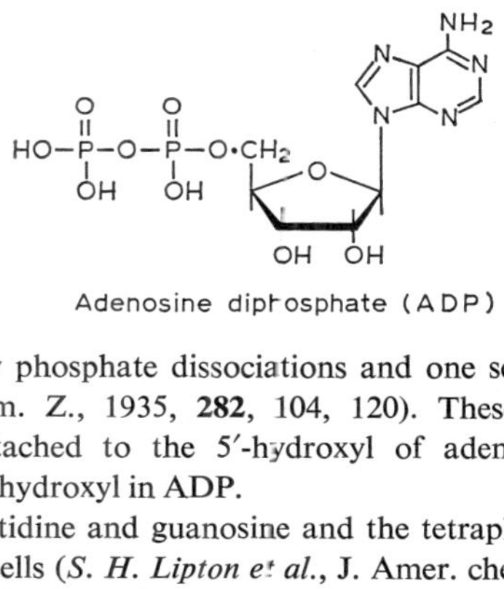

Adenosine diphosphate (ADP)

diphosphate (ADP), which has two primary phosphate dissociations and one secondary phosphate dissociation (*Lohmann*, Biochem. Z., 1935, **282**, 104, 120). These results indicate that a linear triphosphate is attached to the 5′-hydroxyl of adenosine in ATP and a diphosphate is attached to the 5′-hydroxyl in ADP.

The di- and tri-phosphates of uridine, cytidine and guanosine and the tetraphosphate of adenosine are also widely distributed in cells (*S. H. Lipton et al.*, J. Amer. chem. Soc., 1953, **75**, 5449; *R. Bergkvist* and *A. Deutsch*, Acta Chem. Scand., 1953, **7**, 1307; *D. H. Marrian*, Biochim. Biophys. Acta, 1953, **12**, 492; *H. Schmitz et al.*, Cancer Research, 1954, **14**, 66). They are concerned with the metabolism of nucleotide coenzymes and RNA biosynthesis.

Guanosine 5′-triphosphate in particular plays an essential role in the processes of initiation and elongation in protein biosynthesis (*R. Haselkorn* and *L. B. Rothman-Denes*, Ann. Review Biochem., 1973, **42**, 397). Their structures follow from enzymic and chemical hydrolysis. Deoxynucleoside triphosphates have been prepared enzymically by the incubation of the corresponding monophosphate with an enzyme from *Escherichia coli* (*A. Kornberg et al.*, J. biol. Chem., 1958, **233**, 163). These are the precursors of DNA synthesis (*idem, ibid.*, p. 171; Proc. Nat. Acad. Sci. Wash., 1958, **44**, 633, 641). Diphosphates of the deoxynucleosides are obtained on hydrolysis of deoxynucleoside diphosphate sugars (*Y. Sugino*, J. Amer. chem. Soc., 1957, **79**, 5074; *R. L. Potter* and *V. Buettner-Janusch*, J. biol. Chem., 1958, **233**, 462).

The total synthesis of ADP and ATP confirmed these proposed formulae. They were first synthesised by the phosphorochloridate method. Treatment of the silver salt of adenosine 5′-benzyl phosphate (I) with dibenzyl phosphorochloridate gives the tribenzyl pyrophosphate of adenosine (II), from which the benzyl groups are removed by hydrogenolysis to yield ADP (*J. Baddiley* and *Todd*, J. chem. Soc., 1947, 648). Partial debenzylation

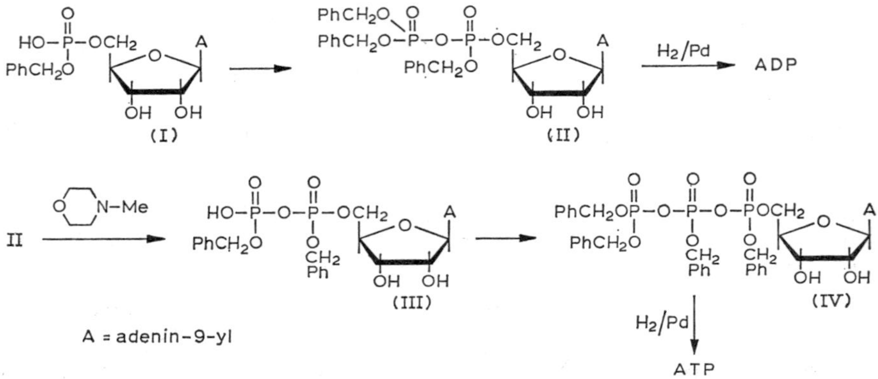

of II with *N*-methyl morpholine gives a dibenzyl ester III which, as its silver salt, is phosphorylated with a further mol. of dibenzyl phosphorochloridate to the tetrabenzyltriphosphate IV from which the benzyl groups are removed by hydrogenolysis to give ATP (*Baddiley*, *A. M. Michelson* and *Todd*, *ibid.*, 1949, 582).

The over-all yields, using the above method, have been extremely low because the fully protected intermediates are unstable. Improvement can be achieved by reacting the disilver salt of adenosine 5′-phosphate with 2 mols. of dibenzyl phosphorochloridate (*Michelson* and *Todd*, *ibid.*, 1949, 2487) or by replacement of the silver salt with a tertiary ammonium salt, when the reaction can be performed in an inert solvent, *e.g.* benzene (*Michelson*, *ibid.*, 1958, 1957).

An alternative method for the synthesis of nucleoside di- and tri-phosphates involves the use of dehydrating agents such as dicyclohexylcarbodiimide for the condensation between a nucleoside 5′-phosphate and orthophosphoric acid. A mixture of products is obtained which includes symmetrical and unsymmetrical pyrophosphates. These can be separated by ion-exchange chromatography. The formation of the desired product can be promoted by using an excess of one of the components. The di- and tri-phosphates of adenosine (*H. G. Khorana*, J. Amer. chem. Soc., 1954, **76**, 3517), uridine (*R. H. Hall* and *Khorana*, *ibid.*, 1954, **76**, 5056) and guanosine (*R. W. Chambers* and *Khorana*, *ibid.*, 1957, **79**, 3752) have been synthesised using the carbodiimide method. This procedure has been improved by using the more soluble trialkyl ammonium salts of phosphomonoesters and orthophosphoric acid which has permitted the development of completely general procedures for the synthesis of all the ribo- as well as the deoxyribo-nucleoside- 5′-triphosphates (*M. Smith* and *Khorana*, *ibid.*, 1958, **80**, 1141; *Smith*, "Biochemical Preparations", Wiley, New York, 1961, Vol. 8, p. 2). Substituted cyanamides (N≡C-NMe_2) or cyanamide itself reacts with nucleotides and phosphate to give pyrophosphates. Cytidine diphosphate and dinucleoside pyrophosphates may be prepared in this way. The reagent is closely analogous in its action to the carbodiimides (*G. W. Kenner*, *C. B. Reese* and *Todd*, J. chem. Soc., 1958, 546).

The main disadvantage of the previous methods *i.e.* formation of symmetrical pyrophosphates is overcome by the use of phosphoramidates. In one procedure, a salt of the nucleotide, *e.g.* adenosine 5′-phosphate, is treated with an ester of phosphoramidic acid, *e.g.*

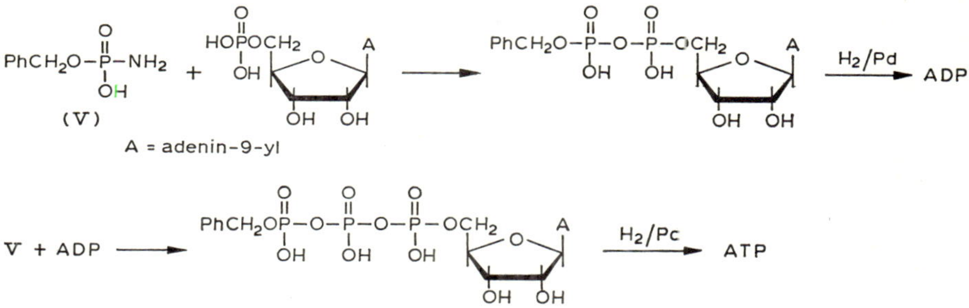

benzyl phosphoramidate (V) in dimethylformamide containing some water (*V. M. Clark*, *G. W. Kirby* and *Todd*, *ibid.*, 1957, 1497). Hydrogenation of the benzyl group gives ADP in good yield. Similarly, ADP is converted into ATP in high yield.

An alternative approach, which is particularly useful, involves nucleoside phosphoramidates. Adenosine 5′-phosphoramidate (VI) was first prepared from 2′,3′-di-*O*-acetyladenosine by reaction with phenyl phosphorodichloridate, treatment of the product with ammonia, and removal of the protecting groups with alkali (*Chambers* and *Khorana*, Chem. and Ind., London, 1956, 1022; J. Amer. chem. Soc., 1958, **80**, 3749). A more con-

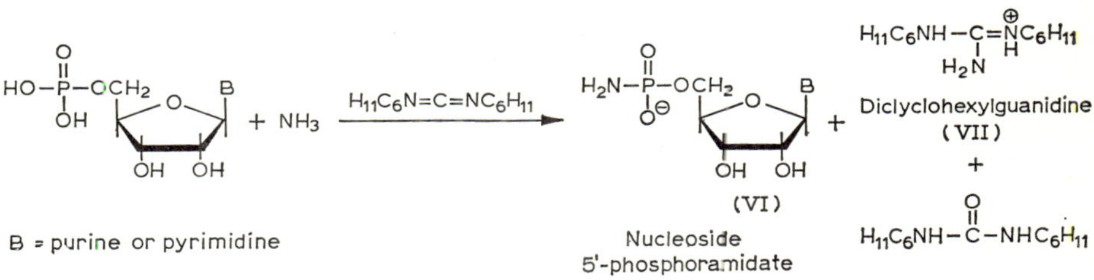

venient synthesis involves direct reaction between the nucleoside 5′-phosphate and ammonia in the presence of dicyclohexylcarbodiimide (*idem*, *ibid.*, 1958, **80**, 3752). A side reaction leading to the formation of the dicyclohexylguanidine (VII) also occurs and, as a result, the product is isolated as the guanidinium salt.

Nucleoside 5′-phosphoramidates which, because of their reactivity and solubility properties, have proved very useful in nucleotide coenzyme synthesis are the nucleoside 5′-phosphoromorpholidates VIII (*J. G. Moffatt* and *Khorana*, *ibid.*, 1961, **83**, 649). Originally pyridine was the preferred reaction medium, but it has been found that the products of reaction are more stable in dimethyl sulphoxide than pyridine. A general procedure for the synthesis of ribo- and deoxyribo-nucleoside 5′-diphosphates and triphosphates in high yield has been developed. Reaction between ribo- or deoxyribo-nucleoside 5′-phosphoro-

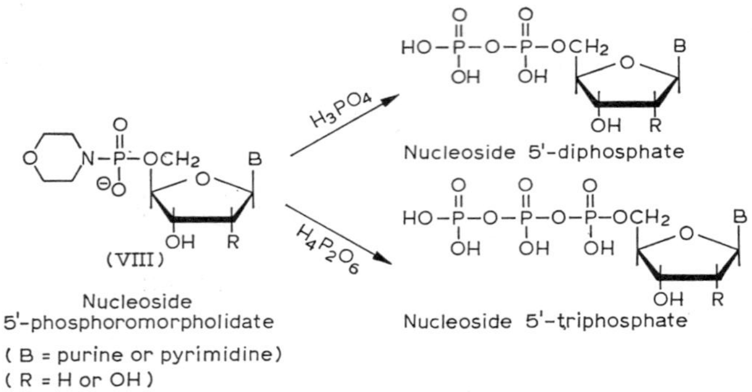

morpholidates (VIII) and the tributylammonium salt of either orthophosphoric acid or pyrophosphoric acid in dimethyl sulphoxide gives the nucleoside 5′-diphosphate or 5′-triphosphate, respectively (*Moffatt*, Canad. J. Chem., 1964, **42**, 599).

Another good method of synthesis of nucleoside 5′-triphosphate and also diphosphates and other nucleotide anhydrides involves anion exchange. In pyridine the more acidic diphenyl phosphate is displaced from P^1-nucleoside 5′-(P^2P^2-diphenyl) pyrophosphates by

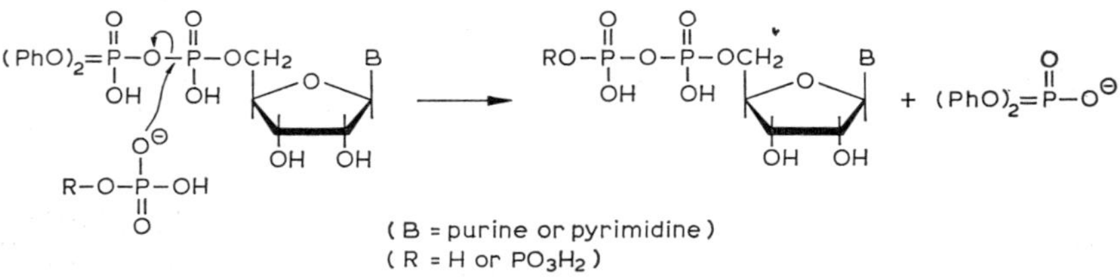

another anion. Displacement with orthophosphate gives nucleoside 5′-diphosphate and with pyrophosphate gives nucleoside 5′-triphosphates in high yield (*A. M. Michelson*, Biochim. Biophys. Acta, 1964, **91**, 1).

(*ii*) *The pyridine coenzymes*

Nicotinamide-adenine dinucleotide, NAD$^{\oplus}$ (Coenzyme I, sometimes called diphosphopyridine nucleotide, DPN) was detected in 1906 by *A. Harden* and *W. J. Young* (Proc. roy. Soc., B., 1906, **77**, 405) and was the first coenzyme to be discovered, but thirty years passed before it was isolated (*H. von Euler*, *H. Albers* and *F. Schlenk*, Z. physiol. Chem., 1936, **240**, 113; *O. Warburg* and *W. Christian*, Biochem. Z., 1936, **237**, 291).

Nicotinamide-adenine dinucleotide phosphate, NADP$^{\oplus}$ (Coenzyme II, sometimes called triphosphopyridine nucleotide, TPN), a closely related coenzyme, was isolated 25 years after the detection of NAD$^{\oplus}$ (*idem*, Biochem. Z., 1932, **254**, 238). These components play an essential part in hydrogen transport in living systems. Both of these coenzymes are easily and reversibly reduced to their dihydro derivatives, NADH and NADPH (see p. 168).

Hydrolysis of NAD$^{\oplus}$ with acid gives 1 mol. of adenine, 1 mol. of nicotinamide, 2 mols. of pentose and 2 mols. of phosphoric acid (*von Euler* and *K. Myrbäck*, J. physiol. Chem., 1931, **198**, 236; 1931, **203**, 143; 1932, **212**, 7; 1935, **233**, 95; *von Euler*, *Albers* and *Schlenk*, *ibid.*, 1937, **237**, 1; 1936, **240**, 113). NADP$^{\oplus}$ gives similar hydrolysis products with an additional mol. of phosphoric acid (*Warburg*, *Christian* and *A. Griese*, Biochem. Z., 1935, **282**, 157). In both coenzymes the amino group of adenine is unsubstituted (deamination with nitrous acid). The dihydro derivatives, unlike the parent coenzymes are very unstable to acid but relatively stable to alkali. This behaviour, and the accompanying change in ultraviolet absorption spectra, are typical of quaternary pyrimidine compounds. Furthermore, synthetic nicotinamide *N*-glycosides are similar in chemical and physical properties

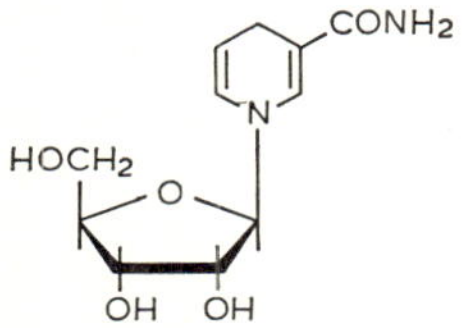

(IX)

to the two coenzymes (*P. Karrer et al.*, Helv., 1936, **19**, 811; 1937, **20**, 55). Enzymic hydrolysis of NAD$^{\oplus}$ gives *N*-ribosyl nicotinamide (*Schlenk*, Svensk Vet. Akad. Arkrv. Kemi, 1936, **12B**, 17; 1941, **14A**, 13; Naturwiss., 1940, **28**, 46; Arch. Biochem., 1943, **3**, 93). The dihydro derivative IX has been synthesised (*L. J. Haynes* and *A. R. Todd*, J. chem. Soc., 1950, 303) and is identical with that isolated from the dihydrocoenzyme.

Two equivalents of ribose 5-phosphate can be obtained on controlled hydrolysis of NAD$^{\oplus}$ (*von Euler*, *Karrer* and *B. Becker*, Helv., 1936, **19**, 1060) and the presence of a pyrophosphate linkage can be shown by titration experiments (*R. Vestin*, *Schlenk* and *von Euler*, Ber., 1937, **70**, 1369). These results led to the proposal that NAD$^{\oplus}$ has the following structure (*von Euler* and *Schlenk*, Z. physiol. Chem., 1937, **246**, 64).

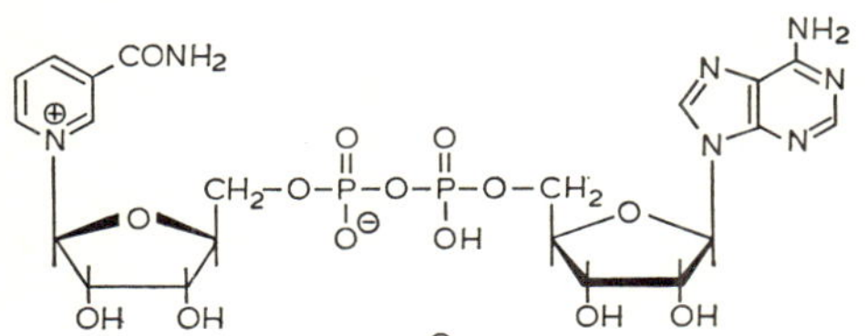

NAD$^{\oplus}$

The structure of NAD$^{\oplus}$ has been confirmed and that of NADP$^{\oplus}$ clarified by enzyme studies. NADP$^{\oplus}$ is obtained from NAD$^{\oplus}$ by enzymic phosphorylation (*A. Kornberg*, J. biol. Chem., 1950, **182**, 779). Degradation of NAD$^{\oplus}$ with pyrophosphorylase gives nicotinamide mononucleotide and adenosine 5′-phosphate (*Kornberg* and *W. E. Pricer*, *ibid.*, 1950, **182**, 763) whereas pyrophosphate degradation of NADP$^{\oplus}$ gives nicotinamide mononucleotide and adenosine-2′,5′-diphosphate (*idem*, *ibid.*, 1950, **186**, 557; *T. P. Wang*,

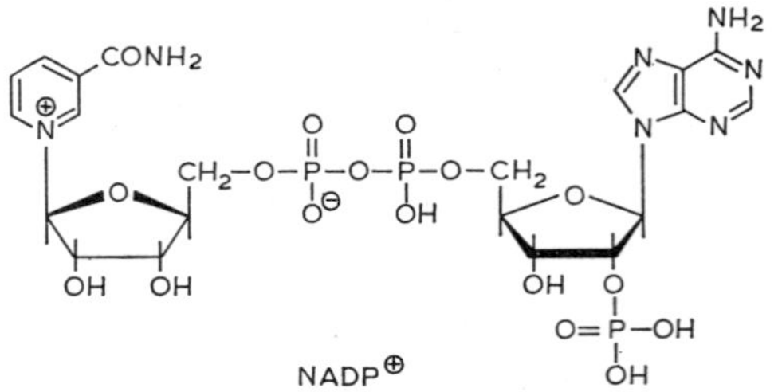

NADP$^{\oplus}$

L. Shuster and *N. O. Kaplan*, *ibid.*, 1954, **206**, 299). The structure of NADP$^{\oplus}$ is therefore as shown.

NAD$^{\oplus}$ and NADP$^{\oplus}$ are reduced to their dihydro derivatives **NADH** and **NADPH**, respectively, when they participate in biological oxidation–reduction processes. It has been shown both by deuterium labelling (*M. E. Pullman*, *A. San Pietro* and *S. P. Colowick*, J. biol. Chem., 1954, **206**, 129) and by n.m.r. spectroscopy (*W. E. Dubb*, *M. Saunders* and

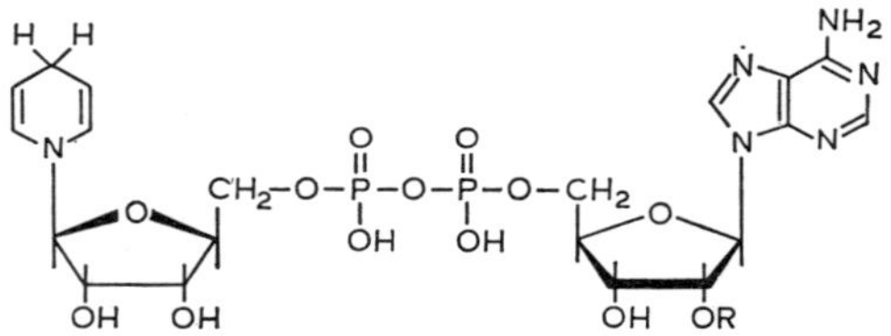

NADH, R = H
NADPH, R = PO_3H_2

J. H. Wang, J. Amer. chem. Soc., 1958, **80**, 1767; *R. F. Hutton* and *F. H. Westheimer*, Tetrahedron, 1958, **3**, 73) that reduction is at the 4 position of the nicotinamide ring. The structures of NADH and NADPH therefore become those shown.

The configuration of the ribosyl-nicotinamide glycosidic link in NAD$^{\oplus}$ has been shown to be β from n.m.r. spectroscopy (*R. V. Lemieux* and *J. W. Lown*, Canad. J. Biochem., 1963, **41**, 889). Very high frequency n.m.r. spectroscopy shows NAD$^{\oplus}$ and NADH to have two folded conformations in aqueous solution. The adenine and nicotinamide rings are stacked in right- and left-handed helical conformations (called the M and H forms), that are in rapid equilibrium (*R. H. Sarma* and *N. O. Kaplan*, Biochemistry, 1970, **9**, 539, 557). Also proton spin-lattice relaxation times show the nicotinamide interconverts rapidly between the *syn* and *anti* conformers with respect to the adjacent ribose ring (*A. P. Zens et al.*, J. Amer. chem. Soc., 1975, **97**, 2850).

The ultraviolet absorption spectra of NAD$^{\oplus}$ and NADH can be used in studying reactions in which the coenzyme participates. NAD$^{\oplus}$ has a strong absorption at 259 nm due to the nicotinamide and adenine components. In comparison NADH has a weaker absorption at 259 nm solely due to the adenine. It also has an absorption at 338 nm due to the dihydronicotinamide component.

Polarographic analysis has been used to study the oxidation–reduction behaviour of the coenzyme (*B. Janik* and *P. J. Elving*, Chem. Reviews, 1968, **68**, 259).

Direct phosphorylation of nicotinamide nucleoside with phosphoryl chloride gives the nucleotide (*A. R. Todd et al.*, J. chem. Soc., 1957, 3727). This nucleotide and adenosine 5′-phosphate react in the presence of dicyclohexylcarbodiimide giving a mixture from which NAD⊕ can be isolated (*N. A. Hughes*, *Kenner* and *Todd*, *ibid.*, 1957, 3733). A similar synthesis using a mixture of adenosine 2′,5′- and 3′,5′-diphosphate and nicotinamide nucleotide gives NADP⊕ together with other products. An impurity in synthetic nicotinamide nucleoside is its α-anomer. This gives rise to some α-anomer of NAD⊕ in the above synthesis, and it has been shown that this is indistinguishable from a compound of this presumed structure present in some natural samples of NAD⊕ (*Kaplan et al.*, J. Amer. chem. Soc., 1955, **77**, 815).

NAD⊕ can be purified by ion-exchange chromatography on DEAE cellulose which is able to separate the α and β anomers (*K. Dalziel* and *F. M. Dickinson*, "Biochemical Preparations", Wiley, New York, 1966, Vol. 11, p. 84).

Many analogues of NAD⊕ have been prepared and the effects of these structural changes on the biological activity of the coenzyme have been studied. Changes have been made in many different parts of the molecule. The 3-carboxamide group of the pyridine ring has been replaced by a large number of substituents, which include acetyl, aldehyde, ethyl, carboxylate, carboxylic acid and amino groups (*Kaplan et al.*, J. biol. Chem., 1954, **209**, 467; 1958, **231**, 685; 1963, **238**, 2823; "Methods in Enzymology", eds. *Colowick* and *Kaplan*, Academic Press, New York, 1957, Vol. 3, p. 899). Other groups which have been substituted in the nicotinamide moiety include 4-methyl, 5-methyl and 4-amino residues (*P. Walter* and *Kaplan*, J. biol. Chem., 1963, **238**, 2823). The pyridine ring has been replaced by, for example, imidazole, benzimidazole, hypoxanthine (*S. G. A. Alivisatos* and *L. LaMantia*, Biochem. Biophys. Res. Comm., 1960, **2**, 164), and adenine (*Kaplan et al.*, J. biol. Chem., 1955, **215**, 195). The adenine ring has been replaced by nicotinamide (*idem*, *ibid.*), hypoxanthine (*Kaplan et al.*, *ibid.*, 1952, **174**, 579) and uracil (*C. P. Fawcett* and *Kaplan*, *ibid.*, 1962, **237**, 1709). The ribose moiety has been replaced by deoxyribose (*idem*, *ibid.*) and a third phosphate group has been inserted (*idem*, *ibid.*). A fuller list of analogues and their effect on the biological activity of the coenzyme can be found elsewhere (*S. P. Colowick*, *J. van Eys* and *J. H. Park*, "Comprehensive Biochemistry", eds. *M. Florkin* and *E. H. Stotz*, Elsevier, Amsterdam, 1966, Vol. 14, p. 1).

NAD⊕ reacts with a number of ions to form addition or substitution compounds, *e.g.* cyanide, bisulphite, dithionite, hydroxylamine, thiols, nitromethane. In general the spectral properties of these products resemble those of NADH (*Kaplan et al.*, J. biol. Chem., 1951, **191**, 447; 1954, **211**, 447; 1957, **228**, 305).

NADH is unstable in dilute acid and rearranges to a modified pyrimidine coenzyme referred to as the primary acid product (*O. Warburg*, *W. Christian* and *A. Griese*, Biochem.

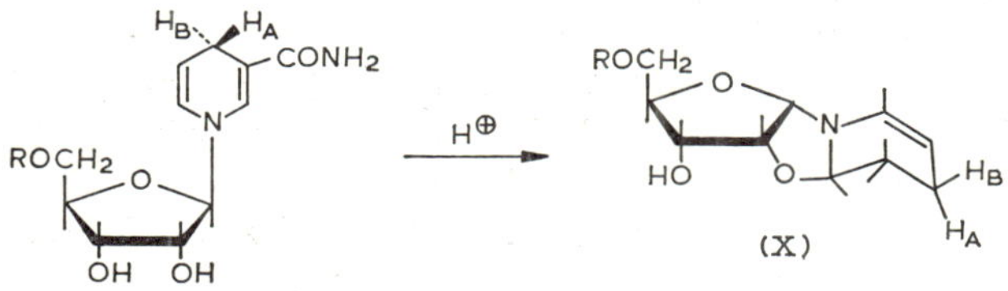

NADH (X)

R = Adenosine 5′-pyrophosphatidyl

Z., 1935, **282**, 157; *E. Haas*, *ibid.*, 1936, **288**, 123). By using proton magnetic resonance spectroscopy, circular dichroism, chemical modifications and specific deuterium labelling it has been shown that the rearrangement is stereospecific and the product, an $O^{2'}$-6B-cyclotetrahydronicotinamide nucleotide (X) has been identified (*N. J. Oppenheimer* and *Kaplan*, Biochemistry, 1974, **13**, 4765).

The use of affinity chromatography (*P. Cuatrecasas*, *M. Wilchek* and *C. B. Anfinsen*, Proc. Nat. Acad. Sci., Wash., 1968, **61**, 636; *Cuatrecasas* and *Anfinsen*, "Methods in Enzymology", ed. *W. B. Jakoby*, Academic Press, New York, 1971, Vol. 22, p. 345) for the purification and study of enzymes is becoming increasingly important. Recently nucleotides including adenosine 5′-phosphate and NAD$^{\oplus}$ have been immobilised on solid matrices in order to purify enzymes for which they are cofactors. Two methods of coupling of the nucleotides to the solid matrix have been used. In one, the mode of attachment of the nucleotide is undefined (*K. Mosbach et al.*, Biochem. J., 1972, **127**, 625; *C. R. Lowe* and *P. D. G. Dean*, FEBS Letters, 1971, **14**, 313; *Dean et al.*, Biochem. J., 1973, **133**, 507, 515). In the other method, the linkage between the matrix and the nucleotide is defined and

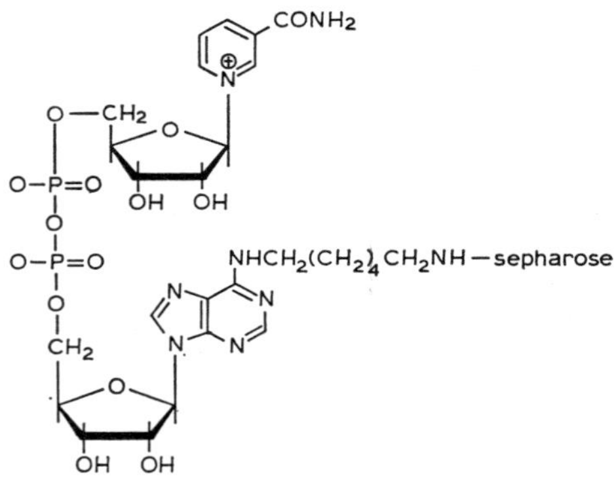

N^6-(6-Aminohexyl)-NAD$^{\oplus}$—sepharose

increases the specificity of the support. A defined immobilised NAD$^{\oplus}$ derivative has been obtained by reaction between *N*^{6}-(6-aminohexyl)-NAD$^{\oplus}$ and cyanogen bromide activated sepharose to give *N*^{6}-(6-aminohexyl)-NAD$^{\oplus}$–sepharose (*D. B. Craven*, *M. J. Harvey* and *Dean*, FEBS Letters, 1974, **38**, 320).

(*iii*) *Flavin–adenine dinucleotide* (*FAD*)

This coenzyme also participates in hydrogen transport. It was isolated by *Warburg* and *Christian* (Biochem. Z., 1938, **298**, 150) from liver, kidney and yeast. Like NAD$^{\oplus}$ it is a molecule composed of a nucleoside pyrophosphate linked to a vitamin, which in this case is vitamin B_2 or riboflavin.

Hydrolysis gives 1 mol. of adenine, the amino group of which is substituted since the coenzyme can be deaminated with nitrous acid. Careful alkaline hydrolysis liberates adenosine 5′-phosphate and acid hydrolysis

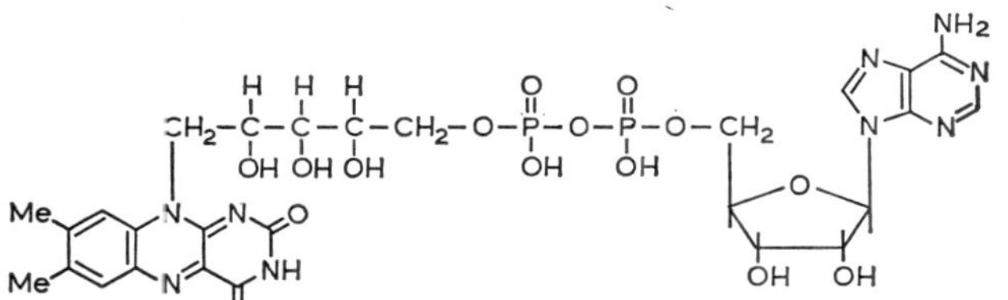

FAD
(XI)

gives riboflavin 5′-phosphate (*E. P. Abraham*, Biochem. J., 1939, **33**, 543), which is given the trivial name of flavin mononucleotide (FMN) although strictly speaking it is not a nucleotide. No formaldehyde is produced on oxidation of FAD with periodate (*P. Karrer* and *M. Frank*, Helv., 1940, **23**, 948). The coenzyme has the structure XI which has been confirmed by synthesis.

The coenzyme has been synthesised by several methods. In the first method the monosilver salt of FMN was condensed with isopropylideneadenine 5′-benzylphosphorochloridate and the isopropylidene residue removed from the product by careful hydrolysis. Debenzylation occurred during the reaction (*S. M. H. Christie*, *G. W. Kenner* and *A. R. Todd*, J. chem. Soc., 1954, 46). The yield in this synthesis is low. The carbodiimide method also gives only small amounts of FAD (*F. M. Huennekens* and *G. L. Kilgour*, J. Amer. chem. Soc., 1955, **77**, 6716), and a condensation of adenosine 5′-phosphate and FMN in the presence of trifluoroacetic anhydride results in the formation of only very small amounts of FAD (*C. DeLuca* and *N. O. Kaplan*, J. biol. Chem., 1956, **223**, 569). Better yields of FAD have been obtained by reaction of FMN with adenosine 5′-phosphoramidate (*J. G. Moffatt* and *H. G. Khorana*, J. Amer. chem. Soc., 1958, **80**, 3756) or adenosine 5′-phosphoroimidazole (*F. Cramer et al.*, Angew. Chem. intern. Edn., 1962, **1**, 331). FAD is obtained in good yield by the anion-exchange method, in which FMN displaces diphenyl phosphate from diphenyl adenosine 5′-pyrophosphate (*A. M. Michelson*, Biochim. Biophys. Acta, 1964, **91**, 1).

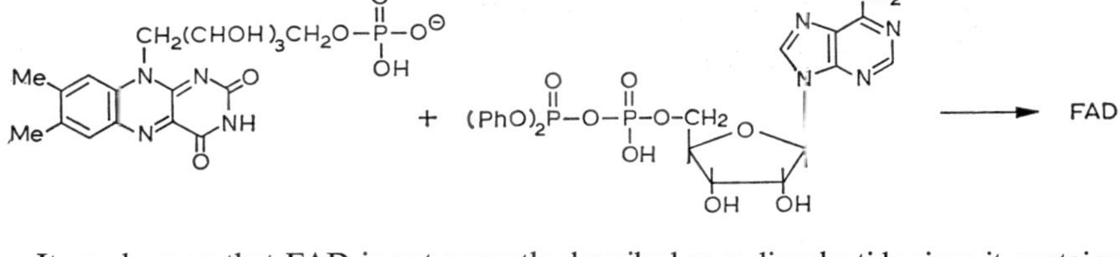

It can be seen that FAD is not correctly described as a dinucleotide since it contains only one nucleotide residue. Riboflavin 5′-phosphate (FMN), which itself shows coenzyme activity, has been synthesised by phosphorylation of 2′,3′,4′-triacetylriboflavin with phosphoryl chloride (*R. Kuhn*, *H. Rudy* and *F. Weygand*, Ber., 1936, **69**, 1543) or by a novel method which consists of the acid-catalysed reaction of riboflavin with catechol cyclic phosphate (*T. Ukita* and *K. Nagasawa*, Chem. and Pharm. Bull., Japan, 1959, **7**, 465).

FAD, in some instances, is attached to the apoenzyme by a covalent bond (*E. P. Kearney* and *T. P. Singer*, Biochim. Biophys. Acta, 1955, **17**, 596; *Kearney*, J. biol. Chem., 1960, **235**, 865). The flavoproteins have ultraviolet absorption maxima at about 280, 350–380 and 450 nm. Complete reduction of the flavin results in loss of the band at 450 nm. Studies using electron-spin resonance spectroscopy and absorption fluorescence show that in the flavoproteins of succinate dehydrogenase (SD–flavin) and amino acid oxidase

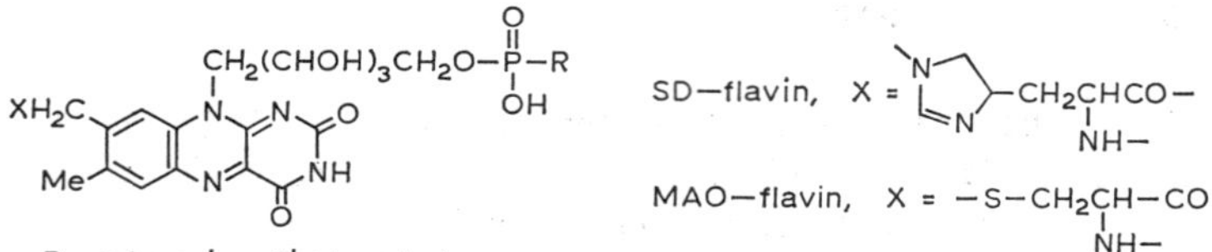

R = adenosine 5'-phosphate

(MAO–flavin) the protein is attached to the FAD by an 8α linkage. In the case of SD–flavin the linkage is to the imidazole of a histidine residue of the protein, and in MAO–flavin the linkage is to the thiol of a cysteine residue (*S. Ghisla et al.*, Angew. Chem. intern. Edn., 1971, **10**, 833).

(*iv*) *Coenzyme A*

The coenzyme involved in acetylation, coenzyme A (CoA), was discovered by *F. Lipmann* and his collaborators (J. biol. Chem., 1945, **160**, 173; 1946, **162**, 743; 1947, **167**, 869). Chemical and enzymic hydrolysis show that the coenzyme contains 1 mol. of pantothenic acid (a member of the B group of vitamins), 1 mole of adenosine, 1 mole of 2-amino-ethanethiol and 3 mols. of phosphate (*idem*, J. Amer. chem. Soc., 1950, **72**, 4838; *G. D. Novelli et al.*, Fed. Proc., 1951, **10**, 229; *J. Baddiley* and *E. M. Thain*, J. chem. Soc., 1951, 2253). A growth factor for *Lactobacillus bulgaricus* can be converted in part to CoA by liver enzymes thus indicating a chemical relationship between these substances. This factor is pantothenic acid 2-mercapto-ethylamide and is called pantetheine (*E. Snell et al.*, J. Amer. chem. Soc., 1950, **72**, 5349).

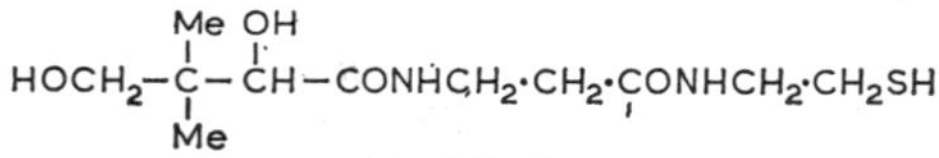

Pantetheine

The presence of a pyrophosphate linkage in the coenzyme is indicated by the action of pyrophosphatase which liberates a phosphorylated pantetheine and adenosine 3′,5′-diphosphate (*T. P. Wang*, *L. Shuster* and *N. O. Kaplan*, J. biol. Chem., 1954, **206**, 229). The position of this linkage follows from the formation of adenosine 5′-phosphate and pantothenic acid 4′-phosphate after gentle hydrolysis of CoA (*Baddiley* and *Thain*, *loc. cit.*, pp. 2253, 3421; 1952, 3783). Chemically synthesised pantetheine 4′-phosphate (*idem*, *ibid.*, 1953, 1680) is readily converted into CoA by a liver-enzyme preparation (*Baddiley et al.*, Nature, 1953, **171**, 76). From the above evidence a structure of CoA was deduced, which has been confirmed by total chemical synthesis.

The enzyme-catalysed reaction of pantetheine 4′-phosphate with one mole of ATP gives

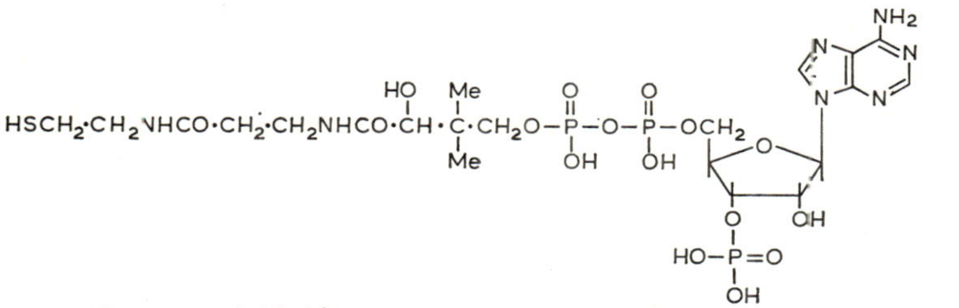

Coenzyme A (CoA)

"dephospho-CoA" which is phosphorylated by a further mole of ATP to give CoA (*M. B. Hoagland* and *G. D. Novelli*, J. biol. Chem., 1954, **207**, 767; *L. Levington* and *Novelli*, *ibid.*, 1954, **207**, 761).

Chemical synthesis of CoA involves the satisfactory preparation of the pantetheine- and adenosine-phosphate moieties and their combination in a pyrophosphate linkage. D-Pantetheine 4′-phosphate is prepared by direct phosphorylation, with dibenzyl phosphorochloridate, of D-pantethine (the oxidised form of D-pantetheine) followed by removal of the benzyl groups. These are preferably removed stepwise, the first with 50% acetic acid and the second with sodium in liquid ammonia during which step the pantethine is reduced to pantetheine. Reaction between pantetheine 4′-phosphate and adenosine 5′-phosphoromorpholidate gives "dephospho-CoA". For the synthesis of CoA it is necessary to use adenosine 2′,3′-cyclic phosphate 5′-phosphoromorpholidate as the adenosine moiety.

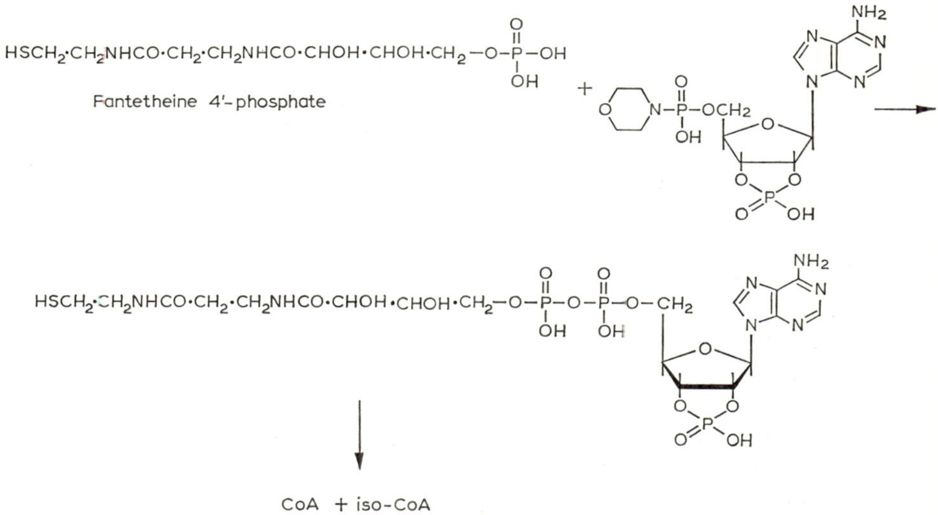

When this is reacted with pantetheine 4′-phosphate followed by treatment of the product with acid to open the 2′,3′-cyclic phosphate a mixture of CoA and iso-CoA (the adenosine 2′-phosphate derivative) is obtained, from which CoA can be separated by ion-exchange chromatography (*J. G. Moffatt* and *H. G. Khorana*, J. Amer. chem. Soc., 1961, **83**, 663).

An improved yield of CoA is obtained by using the anion-exchange method for formation of the pyrophosphate bond. Reaction of P'-adenosine-(2',3'-cyclic phosphate)-5'-$P^2.P^2$-diphenyl pyrophosphate and pantethine 4,4'-diphosphate gives the 2',3'-cyclic phosphate of oxidised CoA, from which CoA is obtained exclusively by cleavage of the cyclic phosphate with ribonuclease, and reduction to the thiol (*A. M. Michelson*, Biochim. Biophys. Acta, 1961, **50**, 607; 1964, **93**, 71). The pyrophosphate bond in CoA can also be formed using pyrophosphoryl chloride (*W. Gruber* and *F. Lynen*, Ann., 1962, **659**, 139).

The "trans-acetylation" function of coenzyme A involves the terminal thiol group which in its thioacetate form is capable of transferring an acetyl group to the substrate (*Lynen*, *E. Reichert* and *L. Rueff*, *ibid.*, 1951, **574**, 1).

CoA is involved in a large number of enzymic acylations and condensation reactions, which involve the formation of *S*-acyl derivatives of the coenzyme, It is involved in both the biosynthesis and catabolism of fatty acids. *S*-Malonyl CoA is formed by the biotin-catalysed reaction of *S*-acetyl CoA and bicarbonate (*J. Knappe*, *E. Ringelmann* and *Lynen*, Biochem. Z., 1961, **335**, 168) and is an important intermediate in the biosynthesis of fatty acids (*R. O. Brady*, *R. M. Bradley* and *E. G. Trams*, J. biol. Chem., 1960, **235**, 3093; *R. Bressler* and *S. J. Wakil*, *ibid.*, 1961, **236**, 1643). The acyl compounds are normally formed enzymically, but chemical syntheses are available from appropriate anhydrides and CoA in the presence of weak bases or from thiol acids or esters. Acetyl, propionyl, butyryl, benzoyl, crotonyl, β-hydroxybutyryl, acetoacetyl and succinyl are some of the derivatives readily pepared in this way (*I. B. Wilson et al.*, Biochim. Biophys. Acta, 1954, **12**, 315; *Lynen* and *S. Ochoa*, *ibid.*, p. 299; *D. Schachter* and *J. V. Taggart*, J. biol. Chem., 1954, **208**, 263; *S. T. Wahil* and *H. R. Mahler*, *ibid.*, 1954, **207**, 125; *Wilson*, J. Amer. chem. Soc., 1952, **74**, 3205; *T. Wieland* and *Rueff*, Angew. Chem., 1953, **65**, 186; *E. T. Simon* and *D. Shemin*, J. Amer. chem. Soc., 1953, **75**, 2520; Biochemical Preparations, New York, Vol. 5, p. 30).

In order to relate the biological activity of CoA to its molecular structure several analogues of CoA have been synthesised and the effects of these structural alterations on the biological properties have been studied. The adenine moiety has been replaced with guanine (*M. Shimizu et al.*, Chem. Pharm. Bull., Japan, 1966, **14**, 683) and a number of analogues have been prepared in which the cysteamine moiety has been modified. These include the replacement of the sulphur atom with selenium (*W. H. H. Günther* and *H. G. Mautner*, J. Amer. chem. Soc., 1965, **87**, 2708) and oxygen (*C. T. Stewart* and *T. L. Miller*, Biochim. Biophys. Res. Comm., 1965, **20**, 433; *Miller*, *G. L. Rowley* and *Stewart*, J. Amer. chem. Soc., 1966, **88**, 2299), and complete removal of the sulphur to give the desulpho-CoA (*J. F. A. Chase*, *B. Middleton* and *P. K. Tubbs*, Biochem. Biophys. Res. Comm., 1966, **23**,

208). Methyl groups have been added to the cysteamine moiety at the α- and β-positions (*Shimizu et al.*, Chem. Pharm. Bull., Japan, 1966, **14**, 681), and also within the carbon atom chain to give the homocysteamine derivative (*idem*, Tetrahedron, 1968, **24**, 5241).

(v) *Nucleoside diphosphate sugars*

Many nucleoside diphosphate sugars are known. They play an important role in the metabolism of sugars and the biosynthesis of oligosaccharides. Uridine and guanosine are the most common nucleoside components found in these compounds, but some diphosphate sugars of cytidine, adenosine, thymidine, deoxyuridine and pseudouridine are known. Many of these are listed elsewhere (*E. F. Neufield* and *W. Z. Hassid*, Advan. Carbohyd. Chem., 1963, **18**, 309; *H. Nickaido* and *Hassid*, *ibid.*, 1971, **26**, 351) and will not be described here.

Uridine diphosphate-D-glucose (UDPG) was discovered by *L. F. Leloir* and his collaborators as a factor required for the conversion of galactose 1-phosphate into glucose 1-phosphate (J. biol. Chem., 1950, **184**, 333; Biochem. J., 1952, **51**, 426). Acid hydrolysis liberates uridine 5′-phosphate, inorganic orthophosphate and D-glucose; very gentle acid

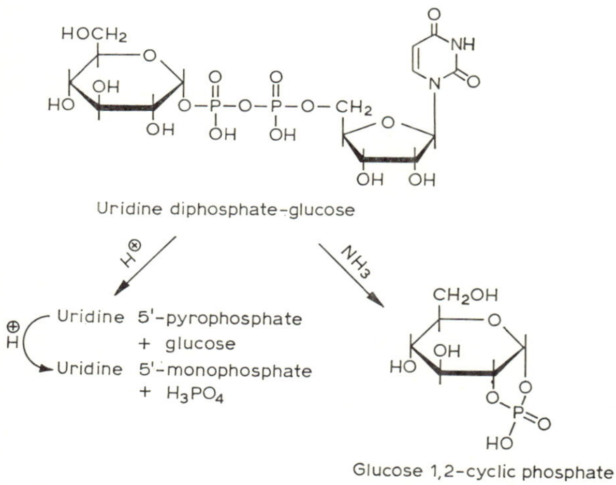

hydrolysis gives uridine 5′-pyrophosphate (UDP) and glucose. It follows from this lability that UDPG is probably a derivative of glucose 1-phosphate and this is substantiated by the isolation of the cyclic glucose 1,2-phosphate on treatment of UDPG with ammonia.

That the configuration of the glycosidic glucose 1-phosphate linkage is α is shown by chemical (see below) and enzymic synthesis (*Hassid et al.*, Arch. Biochem. Biophys., 1957, **69**, 602; Proc. Nat. Acad. Sci., U.S.A., 1959, **45**, 905), which also confirms the total structure.

UDPG was first synthesised from uridine 5′-phosphate and α-D-glucose 1-phosphate by the action of dicyclohexylcarbodiimide (*G. W. Kenner*, *A. R. Todd* and *R. F. Webb*, J. chem. Soc., 1954, 2843). It is synthesised much more satisfactorily by the reaction of α-D-glucose

1-phosphate with uridine 5′-phosphoromorpholidate (*H. G. Khorana et al.*, J. Amer. chem. Soc., 1961, **83**, 659) or by the anion-exchange method whereby *P′*-uridine 5′-(P^2P^2-diphenyl)pyrophosphate is treated with glucose 1-phosphate (*A. M. Michelson*, Biochim. Biophys. Acta, 1964, **91**, 1). Both of the latter procedures give UDPG in yields of greater than 60%.

UDPG participates in a number of enzymic reactions. The epimerisation of glucose and galactose 1-phosphates occurs while these sugars are joined in the nucleotide (*Leloir*, Arch. Biochem. Biophys., 1951, **33**, 186):

UDPG $\rightleftharpoons$ UDP-galactose

The enzyme UDPG epimerase catalyses the reaction in which a Walden inversion of the hydroxyl group at $C_{(4)}$ occurs. NAD$^{\oplus}$ is a cofactor for the enzyme and the reaction probably involves a 4-oxo intermediate (*H. M. Kalchar* and *E. S. Maxwell*, Biochim. Biophys. Acta, 1955, **18**, 444; *Maxwell* and *H. de Robichon-Szulmajster*, J. biol. Chem., 1960, **235**, 308; *D. B. Wilson* and *D. S. Hogness*, *ibid.*, 1964, **239**, 2469; *G. Nelsestuen* and *S. Kirkwood*, Fed. Proc., 1970, **29**, 337). A further important function of UDPG is as a glucose donor in the biosynthesis of oligo- and poly-saccharides. With fructose 6-phosphate it gives sucrose phosphate (*Leloir* and *C. E. Cardini*, J. Amer. chem. Soc., 1953, **75**, 6084; *Leloir et al.*, J. biol. Chem., 1955, **214**, 149, 157). Similar reactions in which the fructose is replaced by D-glucose, D-xylulose or D-sorbose give the corresponding disaccharides. UDP-Galactose and D-glucose 1-phosphate give lactose phosphate (*J. E. Gandar et al.*, Arch. Biochem. Biophys., 1956, **60**, 259; 1957, **69**, 85). UDPG reacts with oligosaccharides in the biosynthesis of cellulose (*L. Glaser*, Biochim. Biophys. Acta, 1957, **25**, 436), glycogen (*Leloir* and *Cardini*, J. Amer. chem. Soc., 1957, **79**, 6340) and starch (*Leloir et al.*, J. biol. Chem., 1961, **236**, 636).

Many other uridine diphosphate sugars have now been found which appear to be involved in the biosynthesis of oligo- and poly-saccharides.

Uridine diphosphate–glucuronic acid is involved in the biosynthesis of glucuronides (*G. J. Dutton* and *I. D. E. Storey*, Biochem. J., 1954, **57**, 275; 1955, **59**, 279; *Dutton*, *ibid.*, 1956, **64**, 693). It is obtained enzymically by oxidation of UDPG with NAD$^{\oplus}$ (*J. L. Strominger et al.*, *ibid.*, 1957, **66**, 567; J. biol. Chem., 1957, **224**, 79).

UDI–glucuronic acid can be synthesised by reaction between uridine 5′-phosphoramidate and glucuronic acid 1-phosphate (*T. Tanaka et al.*, Chem. and Pharm. Bull., Japan, 1962, **10**, 750) or by the anion-exchange method using *P′*-uridine 5′-(P^2P^2-diphenyl)-pyrophosphate and glucuronic acid 1-phosphate (*Michelson* and *F. Wold*, Biochemistry, 1962, **1**, 1171).

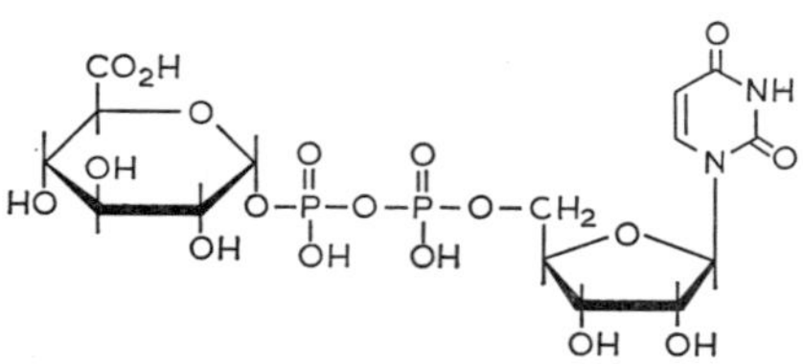

UDP-glucuronic acid

Uridine diphosphate–*N*-acetyl-D-muramic acid is involved in the synthesis of peptidoglycan which is a component of cell walls. Its biosynthesis and structures have been elucidated mainly through the work of *J. T. Park* and *J. L. Strominger*. (For a review of this work see, for example, *J. M. Ghuysen*, Bacteriol. Review, 1968, **32**, 425.) The antibiotics penicillin and D-cycloserine inhibit peptidoglycan synthesis and hence cells such as *Staphylococcus aureus* grown in the presence of these compounds accumulate the nucleotide precursors

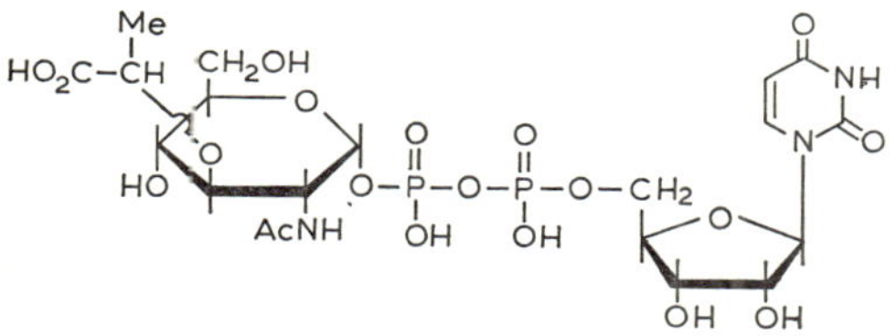

UDP–*N*-acetylmuramic acid

(*Park*, J. biol. Chem., 1952, **194**, 877, 885; *U. Roze* and *Strominger*, Mol. Pharmacol., 1966, **2**, 92). The structure of UDP–*N*-acetylmuramic acid follows from its hydrolysis in dilute acid to UDP and *N*-acetylmuramic acid (*Park* and *Strominger*, Science, 1957, **125**, 99).

Guanosine diphosphate–D-manose (GDPM) occurs in micro-organisms and in animal tissues (*E. Cabib* and *Leloir*, J. biol. Chem., 1954, **206**, 779). Its behaviour towards acids is similar to that of UDPG, first mannose and GDP, then GMP and inorganic phosphate being formed. It is synthesised enzymically from GTP and α-D-mannose 1-phosphate (*A. Munch-Petersen*, Arch. Biochem. Biophys., 1955, **55**, 592).

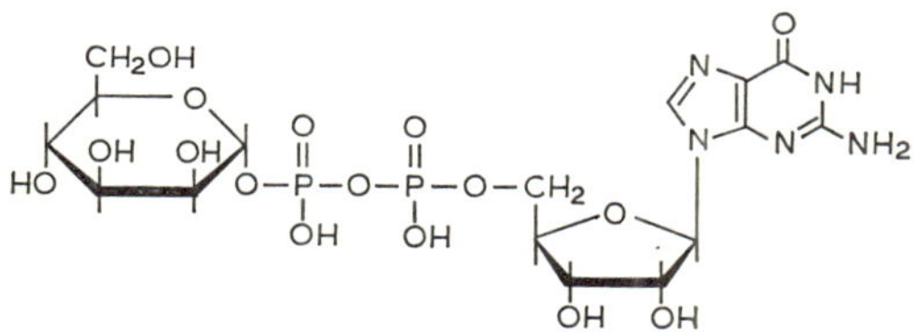

Guanosine diphosphate-D-mannose

GDP mannose can be synthesised by reaction between guanosine 5′-phosphoromorpholidate and D-mannose 1-phosphate (*H. G. Khorana et al.*, J. Amer. chem. Soc., 1961, **83**, 659) or by the anion-exchange method (*Michelson*, Biochim. Biophys. Acta, 1964, **91**, 1).

GDPM is involved in the synthesis of mannan ("yeast germ") which is found in yeast (*Cabib et al.*, Biochem. Biophys. Res. Comm., 1963, **12**, 504; J. biol. Chem., 1968, **243**, 502).

Several other guanosine diphosphate sugars have been isolated from various natural sources and include GDG–D-glucose, GDP–D-fructose (*H. G. Pontis, A. L. James* and *J. Baddiley*, Biochem. J., 1960, **75**, 428), GDP–L-fucose (*R. Denamur et al.*, Compt. rend. 1958, **246**, 2820), GDP–D-galactose and GDP–D-xylose (*R. R. Selvendran* and *F. A. Isherwood*, Biochem. J., 1967, **105**, 723). GDPM is a precursor of GDP–D-fucose which is formed *via* GDP–4-oxo-6-deoxy-D-mannose (*V. Ginsburg*, J. biol. Chem., 1961, **236**, 2389).

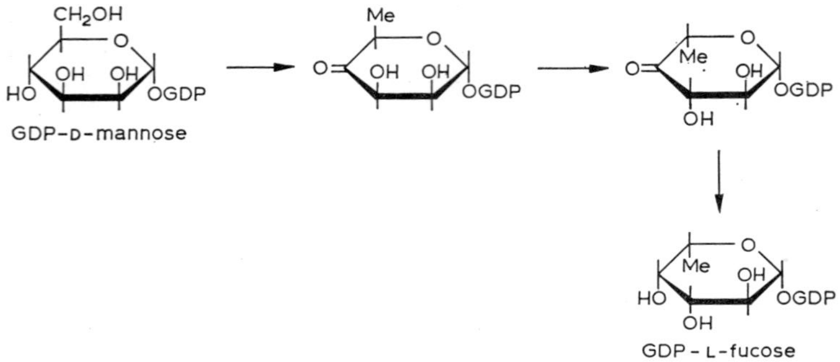

Cytidine diphosphate–D-glycerol and cytidine-diphosphate–L-ribitol occur in *Lactobacillus arabinosus* and other bacteria (*Baddiley* and *A. P. Mathias*, J. chem. Soc., 1954, 2723; *Baddiley et al.*, Biochem. J., 1956, **64**, 599). Both yield cytidine 5′-phosphate and isomeric polyol phosphates on acid hydrolysis, and with rattlesnake venom, which contains both a pyrophosphatase and 5′-nucleotidase, they are hydrolysed to cytidine, inorganic phosphate and either L-α glycerophosphate or D-ribitol 5-phosphate (*idem*, J. chem. Soc., 1956, 4186, 4583; 1957, 1869, 4058). With ammonia the products are cytidine 5′-phosphate and a cyclic phosphate of glycerol or ribitol. The stereochemistry of the ribitol phosphate follows from degradation of its cyclic phosphate to a mixture of D-2- and 3-phosphoglyceric acids.

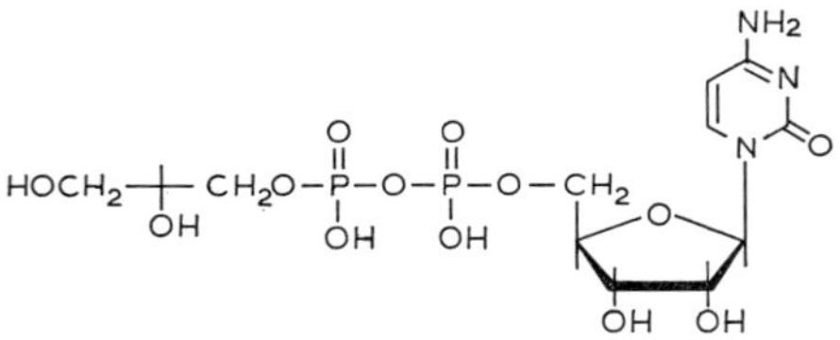

CDP–D-glycerol

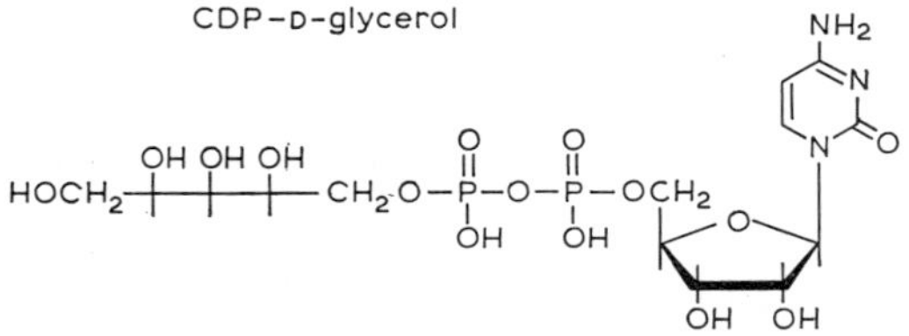

CDP–L-ribitol

CDP–glycerol was first synthesised from cytidine 5′-phosphate and isopropylidene-α-glycerophosphate by the carbodiimide method, and CDP–ribitol through CDP–ribose, followed by reduction with sodium tetrahydridoborate (*Baddiley et al.*, *ibid.*, 1958, 3107; 1959, 2192). They may be synthesised more conveniently by the phosphoromorpholidate (*Khorana et al.*, J. Amer. chem. Soc., 1961, **83**, 659) or the anion-exchange methods (*Michelson*, Biochim. Biophys. Acta, 1964, **91**, 1).

CDP–glycerol and CDP–ribitol are synthesised enzymically from CTP and the appropriate alcohol phosphate (*D. R. D. Shaw*, Biochem. J., 1962, **82**, 297). These nucleotides are involved in the formation of teichoic acids, which are found mainly in the cell walls of a number of gram-positive bacteria (*Baddiley et al.*, J. chem. Soc., 1958, 4344. For review, see *idem*, "Essays in Biochemistry", 1972, **8**, 35). In teichoic acids, glycerol and ribitol residues are joined through phosphate diester linkages to form a chain, to which sugar and D-alanine ester residues are attached through the hydroxyl groups of the alditols. The structures vary between different micro-organisms and different strains of the same micro-organism (*idem*, Advan. carbohydrate Chem., 1966, **21**, 323; Advan. Enzymology, 1968, **30**, 223).

The transfer of the sugar from a nucleoside diphosphate sugar into polysaccharide involves a lipid intermediate. This was first demonstrated for the synthesis of peptidoglycan from UDP–*N*-acetylmuramic acid and UDP–*N*-acetylglucosamine (*J. L. Strominger et al.*, Proc. Nat. Acad. Sci. Wash., 1965, **53**, 881), where it has been shown by mass spectrometry that the lipid involved is the phospho-monoester of a C_{55} polyisoprenoid alcohol (*Y. Higashi*, *Strominger* and *C. C. Sweeley*, *ibid.*, 1967, **57**, 1878). The involvement of polyisoprenols in the synthesis of other polysaccharide derivatives such as mannan, capsular polysaccharide, and teichoic acids of micro-organisms, and also in plant and mammalian glycoprotein synthesis is now known (*W. J. Lennarz* and *M. G. Scher*, Biochim. Biophys. Acta, 1972, **265**, 417; *C. J. Waechter* and *Lennarz*, Ann. Review Biochem., 1976, **45**, 95).

Two other cytidine diphosphate derivatives, which are well characterised, but which are not sugars, are **cytidine diphosphate-choline** which is an intermediate in the biosynthesis of lecithin (*E. P. Kennedy* and *S. B. Weiss*, J. biol. Chem., 1956, **222**, 193; *Kennedy*, *ibid.*, p. 185) and **cytidine diphosphate-ethanolamine,** which is involved similarly in the formation of phosphatidylethanolamine.

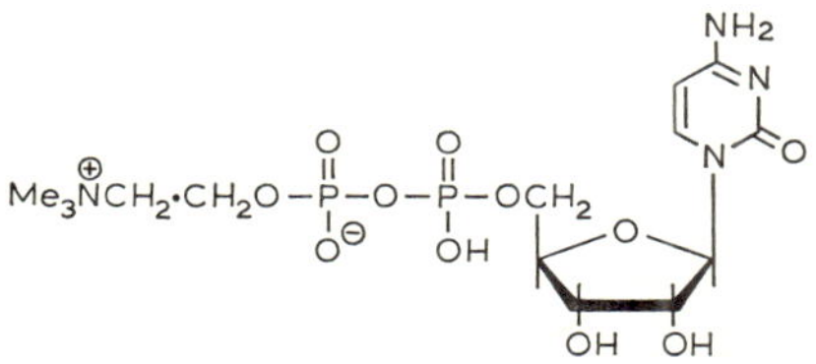

CDP-choline

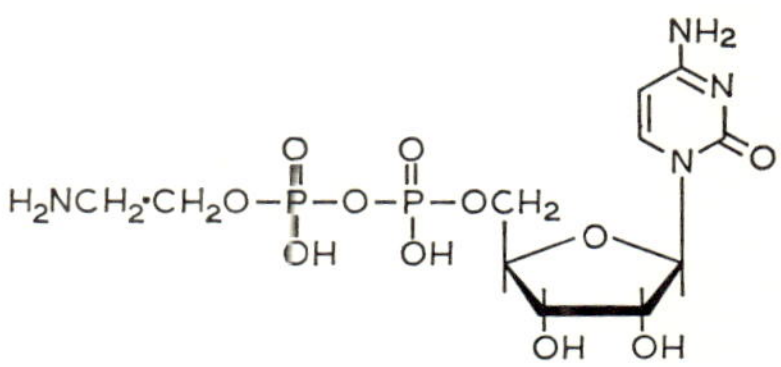

CDP-ethanolamine

They are formed enzymically from CTP and the phosphate of either choline or ethanolamine.

CDP–choline or –ethanolamine is prepared by reaction of choline or ethanolamine phosphate with cytidine 5′-phosphoramidate (*Y. Sanno* and *K. Tanaka*, Chem. and Pharm. Pharm. Bull., Japan, 1962, **10**, 231), cytidine 5′-phosphoromorpholidate (*T. Tanaka et al.*, Yakugaku Zasshi, Japan, 1965, **85**, 863; C.A., 1966, **64**, 3660f), or P^1-cytidine 5′-(P^2P^2-diphenyl)pyrophosphate (*R. Letters* and *Michelson*, Bull. Soc. chim. Biol., 1963, **45**, 1353).

These nucleotides yield phosphatides by reaction with $\alpha\beta$-diglycerides (*Kennedy et al.*, Nature, 1956, **178**, 594).

Both deoxynucleosides **deoxycytidine diphosphate–choline** and **deoxycytidine diphosphate–ethanolamine** are known (*Y. Sugino*, J. Amer. chem. Soc. 1957, **79**, 5074). *d*-CDP–choline is as active as CDP–choline in the biosynthesis of lecithin, but *d*-CDP–

ethanolamine is used only poorly in the biosynthesis of phosphatidylethanolamine (*Kennedy, L. F. Borkenhagen* and *S. W. Smith*, J. biol. Chem., 1959, **234**, 1998).

Several thymidine and adenosine diphosphate-sugars have been found in micro-organisms and plants, these include **thymidine diphosphate–D-glucose** (*S. Kornfield* and *L. Glaser*, *ibid.*, 1961, **236**, 1791), **thymidine diphosphate–D-mannose** and **thymidine diphosphate–D-rhamnose** (*N. L. Blumson* and *Baddiley*, Biochem. J., 1961, **81**, 114). The latter is involved in streptococcal cell-wall polysaccharide biosynthesis (*L. D. Zeleznick*, Science, 1963, **140**, 400).

Adenosine diphosphate–D-glucose has been isolated from plants (*E. Recondo, M. Dankert* and *L. F. Leloir*, Biochem. Biophys. Res. Comm., 1963, **12**, 204) and transfers glucose into starch (*Recondo* and *Leloir*, *ibid.*, **6**, 85). Other adenosine diphosphate sugars include **ADP–D-fructose, ADP–D-mannose** and **ADP–D-galactose** (*Leloir et al.*, *ibid.*, 1964, **14**, 358; *D. F. Cumming*, Biochem. J., 1970, **116**, 189).

An intermediate in the biosynthesis of sulphuric esters has been identified as **adenosine 3′-phosphate 5′-sulphatophosphate** (active sulphate) (*H. Hiltz* and *F. Lipmann*, Proc. Nat. Acad. Sci. Wash., 1955, **41**, 880). It is dephosphorylated by rye-grass 3′-nucleotidase to

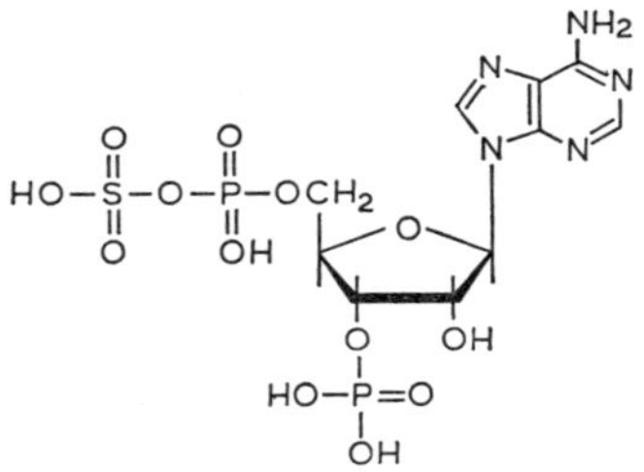

Adenosine 3'-phosphate
5'-sulphatophosphate

give adenosine 5′-sulphatophosphate and with dilute acid it gives adenosine 3′,5′-diphosphate and inorganic sulphate (*P. W. Robbins* and *Lipmann*, J. biol. Chem., 1957, **229**, 837).

Both adenosine 5′-sulphatophosphate and active sulphate are synthesised by the action of (1-pyridinio)sulphate on adenosins 5′-phosphate and 3′,5′-diphosphate, respectively (*J. Baddiley et al.*, J. chem. Soc., 1957, 1067; 1959, 1731). The carbodiimide method may be used for adenosine 5′-sulphatophosphate (*P. Reichard* and *N. R. Ringertz*, J. Amer. chem. Soc., 1957, **79**, 2025), but better yields are obtained using the anion-exchange method (*A. M. Michelson* and *F. Wold*, Biochemistry, 1962, **1**, 1171).

The nucleotide is synthesised enzymically from ATP and inorganic sulphate. The initial product, adenosine 5′-sulphatophosphate, is phosphorylated at the 3′-position with a second mole of ATP (*Robbins* and *Lipmann*, J. biol. Chem., 1958, **233**, 681, 686; *L. G. Wilson* and *R. S. Bandurski*, Arch. Biochem. Biophys., 1956, **62**, 503; *idem*, J. biol. Chem., 1958, **233**, 975). It can donate its sulphate group to a variety of compounds, including phenols and their derivatives.

4. Nucleic acids

Oligonucleotides, polynucleotides and nucleic acids are polymers of ribonucleotides or deoxyribonucleotides. Nucleic acids which are essential components of all living cells were first discovered by *Friedrich Miescher* ("Die Histochemischen und Physiologischen Arbeiten", Leipzig, 1897) during the latter part of the nineteenth century. There are two types of nucleic acid: deoxyribonucleic acids (DNA) are polymers of monodeoxyribonucleotides and ribonucleic acids (RNA) of monoribonucleotides. Originally it was thought that cells from animal tissue contained DNA and cells from plant tissues contained RNA, but it is now known that all cells contain both DNA and RNA. The majority of DNA is found in the nucleus of the cell where, in association with proteins, it constitutes the chromosome. Smaller amounts are found in components of the cytoplasm such as the chloroplasts of photosynthetic organisms and the mitochondria. DNA is the genetic carrying substance (*O. T. Avery*, *C. M. McLeod* and *M. McCarthy*, J. expl Med., 1944, **79**, 137) of all known living cells and it is therefore the chemical composition of DNA which determines hereditary characteristics. This obviously means that DNA must be a large and complex molecule if it is to store the vast amount of information required to characterise a cell. Genes are, in fact, discrete pieces of DNA which contain a unique sequence of deoxyribonucleotides.

Although some RNA is also found in the nucleus, where its biosynthesis takes place, it is mainly located in the cytoplasm. Here three types of RNA are found: transfer RNA (tRNA), ribosomal RNA (rRNA) and messenger RNA (mRNA), all of which play a role in protein synthesis. Transfer RNA, which is the smallest species of RNA, is aminoacylated and carries the amino acid to the site of protein synthesis. Protein synthesis takes place on ribosomes, which are large nucleoproteins composed of 60% ribosomal RNA of which there are several species. The sequence of amino acids in the protein synthesised on the ribosome is determined by the genetic information in DNA. This information is carried from the DNA to the ribosome by messenger RNA molecules. The translation of information from the four-letter nucleotide alphabet of the nucleic acids into the twenty-letter amino acid alphabet of protein is carried out according to the genetic code whereby a specific trinucleotide sequence in the nucleic acid directs the insertion of a unique amino acid into protein. tRNA is intimately involved in this translation process. There are many different species of tRNA in each cell. Each one recognises a specific trinucleotide sequence and can be amino-acylated by only one amino acid. There is therefore a unique relationship between that particular trinucleotide sequence (the codon) and the amino acid.

The processes of DNA, RNA and protein biosynthesis in higher organisms (eukaryote cells) and lower organisms (prokaryote cells) over all are very similar, although in prokaryotes there is no organised nucleus, and the major species of ribosomal RNA are somewhat smaller than those of eukaryotes. The tRNA molecules of both types of organism, however, are of the same size. A detailed account of the biological functions and properties of the nucleic acids is outside the scope of this book; however such texts as *Davidson's* "The Biochemistry of the Nucleic Acids", 8th Edn., revised by *R. L. P. Adams, R. H. Burdon, A. M. Campbell* and *R. M. S. Smellie*, Chapman and Hall, London 1976, cover the subject adequately.

(*a*) *Isolation of nucleic acids*

Great care is required in the isolation of DNA and RNA since these molecules are sensitive to extremes of pH and temperature, and in the case of DNA, degradation can be brought about by mechanical shear forces during mixing or the drawing of a solution of the DNA into a pipette. The stresses which arise in these long, linear molecules are sufficient to break the covalent bonds (*P. F. Davison*, Proc. Nat. Acad. Sci. Wash., 1959, **45**, 1560).

After lysis of the cell the DNA is usually isolated as a deoxyribonucleoprotein complex (*A. E. Mirsky, ibid.*, 1971, **68**, 2945). This nucleoprotein can be extracted into 1 *M* sodium chloride and when this solution is shaken with chloroform containing isoamyl alcohol the DNA remains in the aqueous phase while the protein is precipitated as a gel at the interphase between the chloroform and water (*M. G. Sevag, D. B. Lackmann* and *J. Smolens*, J. biol. Chem., 1938, **124**, 425). A general method for the isolation of DNA from micro-organisms uses this method of deproteinisation (*J. Marmar*, J. mol. Biol., 1961, **3**, 208). Detergents such as sodium dodecyl sulphate can be used as an alternative for the separation of nucleic acid from protein (*E. R. M. Kay, N. S. Simmons* and *A. L. Dounce*, J. Amer. chem. Soc., 1952, **74**, 1724). Today the most common procedure for the isolation of DNA uses phenol, which, for most organisms, both lyses the cell and separates nucleic acid from protein (*K. S. Kirby*, Biochem. J., 1957, **66**, 495; 1958, **70**, 206).

Methods for the isolation of RNA are similar to those used for DNA, however, before extraction of the RNA, it is sometimes advantageous to separate the nucleus from cytoplasmic material. Extraction of the cells or cellular fractions with phenol is the usual method of obtaining RNA (*idem*, Biochim. Biophys. Acta, 1962, **55**, 545). RNA molecules are usually much smaller than those of DNA and therefore degradation by mechanical shearing is much less. However, extreme care needs to be taken to prevent the degradation of RNA by ribonucleases which are released from the cells, or are found on the operator's fingers (*R. W. Holley*, *J. Apgar* and *S. H. Merrill*, J. biol. Chem., 1961, **236**, PC 42).

Separation of DNA from RNA can be achieved either by adjusting the extraction conditions so that the DNA remains in the interphase between the aqueous phase and the phenol, while the RNA is found in the aqueous phase or by fractional precipitation of the isolated material (*Kirby*, "Progress in Nucleic Acid Research", eds. *J. N. Davidson* and *W. E. Cohn*, Academic Press, New York, 1964, Vol. 3, p. 1). Alternatively, purified deoxyribonuclease can be used to degrade the contaminating DNA. Conversely, purified ribonuclease can be used to remove contaminating RNA from DNA preparations. Detailed experimental methods for the isolation of DNA and RNA from various tissues and organisms and sub-cellular organelles can be found elsewhere ("Methods in Enzymology", eds.-in-chief *S. P. Colowick* and *N. O. Kaplan*, Academic Press, 1957, Vol. 3, pp. 684, 687, 692, 696; 1963, Vol. 6, p. 726; 1967, Vol. 12A, pp. 533, 538, 541, 543, 545, 550, 554, 557, 581, 588, 596, 607, 609, 613; 1968, Vol. 12B, pp. 87, 99, 100, 112, 115, 120, 129, 156; 1971, Vol. 20, p. 3).

DNA isolated from cells is usually partially degraded although if care is taken the molecular weight exceeds 10^6. Intact DNA, however, can be isolated from some micro-organisms, bacteriophages and viruses. That from *Escherichia coli* has been shown to be circular (*J. Cairns*, J. mol. Biol., 1963, **6**, 208), and has a molecular weight of approximately 2.2×10^9. The whole DNAs isolated from bacteriophages and viruses have lower molecular weights and may be linear or circular structures. Most DNA molecules exist as double-stranded structures (see p. 192) but some of the bacteriophages (*e.g.* φ1 74, fd, M 13) contain a single-stranded infective DNA of low molecular weight (1.6×10^6), which during replication in the host cell is converted to a double-stranded form with twice the molecular weight (*R. L. Sinsheimer*, J. mol. Biol., 1959, **1**, 43; *A. K. Kleinschmidt*, *A. Burton* and *Sinsheimer*, Science, 1963, **142**, 961).

DNA molecules may be fractionated in a number of ways. They way be separated according to their size and shape by zone centrifugation through sucrose gradients. A number of empirical equations relate the sedimentation rate to the molecular weight (*D. M. Crothers* and *B. H. Zimm*, J. mol. Biol., 1965, **12**, 525; *B. Hudson*, *D. A. Clayton* and *T. Vinograd*, Cold Spring Harbor Symp. Quant. Biol., 1968, **33**, 435). Alternatively, DNA molecules may be separated according to their buoyant densities, a property which is directed by a number of factors *e.g.* whether the DNA is single-stranded or double-stranded, and the amount of guanine and cytosine in the DNA. Separation according to buoyant density is accomplished by equilibrium centrifugation in gradients of either caesium chloride or caesium sulphate

(*W. Szybalski*, "Methods in Enzymology", eds. *L. Grossman* and *K. Moldave*, Academic Press, New York, 1968, Vol. 12B, p. 330). Chromatographic methods may also be used for fractionation of DNA. Columns of methylated albumin–kieselguhr (MAK) fractionate DNA according to its size (*L. S. Lerman*, Biochim. Biophys. Acta, 1955, **18**, 132), content of hydrogen bonding (*J. D. Mandell* and *A. D. Hershey*, Anal. Biochem., 1960, **1**, 66; *N. Sueoka* and *T. Y. Cheng*, J. mol. Biol., 1962, **4**, 161) or base composition (*Cheng* and *Sueoka*, Science, 1963, **141**, 1194). MAK columns also separate RNA from DNA (*Mandell* and *Hershey*, *loc. cit.*). Other chromatography media used include calcium phosphate (hydroxyl apatite) (*R. K. Main, M. J. Wilkins* and *L. J. Cole*, J. Amer. chem. Soc., 1959, **81**, 6490; *G. Bernardi*, Biochem. Biophys. Res. Comm., 1961, **6**, 54), ion-exchange celluloses (*S. Kit*, Arch. Biochem. Biophys., 1960, **87**, 318) and polylysine kieselguhr (*S. R. Ayad* and *J. Blamire*, J. Chromatogr., 1969, **42**, 248).

RNA isolated from whole cells contains mainly ribosomal RNA and transfer RNA with possibly minute quantities of messenger RNA. In the case of eukaryote organisms, if care has been taken in the extraction procedure, small quantities of high molecular weight heterogeneous nuclear RNA may also be found.

Separation of the RNA species can be accomplished by several methods, one of the most commonly used being centrifugation, through a sucrose density gradient, which separates the molecules according to size. The two major species of ribosomal RNA are separated from each other, and from messenger RNA (if present) and transfer RNA (*E. H. McConkey*, "Methods in Enzymology", eds. *Grossman* and *Moldave*, 1967, Vol. 12A, p. 620; *F. Gros et al.*, Nature, 1961, **190**, 581; *H. Chantrenne*, A. *Burny* and *G. Marbaix*, "Progress in Nucleic Acid Research and Molecular Biology", eds. *Davidson* and *Cohn*, Academic Press, New York, 1967, Vol. 7, p. 173). Column chromatography may also be used. The 23S and 16S ribosomal RNA species from *Escherichia coli* are separated from each other and lower molecular weight RNA species by reversed-phase chromatography (*B. Z. Egan, J. E. Caton* and *A. D. Kelmers*, Biochemistry, 1971, **10**, 1890) or on lysine–agarose (*D. S. Jones, H. K. Lundgren* and *F. T. Jay*, Nucleic Acids Res., 1976, **3**, 1569). In the latter case 5S ribosomal RNA is also separated from tRNA. Ion-exchange chromatography is particularly useful in separating short-chain oligoribonucleotides. If elution is carried out with concentration gradients of urea (*R. V. Tomlinson* and *G. M. Tener*, J. Amer. chem. Soc., 1962, **84**, 2644) or salt (*G. W. Rushizky, E. M. Bartos* and *H. A. Sober*, Biochemistry, 1964, **3**, 626), oligonucleotides containing up to eight nucleotides are separated according to their chain length. Separation of RNA species with high resolution can be achieved by electrophoresis on gels of polyacrylamide (*V. E. Loening*, Biochem. J., 1967, **102**, 251; *A. C. Peacock* and *C. W. Dingman*, Biochemistry, 1967, **6**, 1818; 1968, **7**, 668; *R. DeWachter* and *W. Friers*, "Methods in Enzymology", eds. *Grossman* and *Moldave*, 1971, Vol. 21, p. 167) or agarose (*W. McIndoe* and *H. Munro*, Biochim. Biophys. Acta, 1967, **134**, 458). Originally these methods were used as an analytical tool only, but they have since been developed as preparative procedures (*W. G. Lanyon, J. Paul* and *R. Williamson*, FEBS Letters, 1968, **1**, 279; *P. A. Weil* and *A. Hampel*, Biochemistry, 1973, **12**, 4361). Specific messenger RNA molecules have recently been obtained from mammalian and plant cells. These are most conveniently prepared by isolation of polysomes, which are ribosomes strung together on a messenger RNA molecule. When polysomes are dissociated with sodium dodecyl sulphate (*J. D. Lingrel et al.*, Series Haematol., 1971, **4**, 37), EDTA (*Williamson, R. Clayton* and *D. E. S. Truman*, Biochem. Biophys. Res. Comm., 1972, **46**, 1936) or phenol (*G. Brawerman, J. Mendicki* and *S. Y. Lee*, Biochemistry, 1972, **11**, 637) messenger RNAs are released. A unique feature of messenger RNA molecules is the polyadenylic acid sequence which is found at their 3′-end (*R. Sheldon et al., ibid.*, 1972, **11**, 3829). The polyadenylic acid sequence which is 150–200 nucleotides long is found in nearly all messenger RNAs except the messenger RNA for histones (*J. E. Darnell et al.*, J. mol. Biol., 1972, **71**, 21). It is also found in heterogeneous nuclear RNA which is the messenger RNA precursor (*idem*, Science, 1971, **174**, 507; *R. N. Bastos* and *H. Aviv*, Cell, 1977, **11**, 641). Use has been made of the polyadenylic acid sequence in the purification of messenger RNA by using chromatographic columns of either cellulose–oligothymidylic acid (*Aviv* and *P. Leder*, Proc. Nat. Acad. Sci., Wash., 1972, **69**, 1408) or sepharose–polyduridylic acid (*V. Lindberg* and *T. Persson*, Eur. J. Biochem., 1972, **31**, 246) which retard the messenger RNA molecules.

Three classes of ribosomal RNA are found in the ribosomes of bacteria, or in the chloroplasts and mitochondria of higher organisms. These are designated as the 23S, 16S and 5S RNAs according to their sedimentation coefficients and have molecular weights of 1.1×10^6, 0.55×10^6 and 4×10^4, respectively (*C. G. Kurland*, J. mol. Biol., 1960, **2**, 83; *R. Rosset, R. Monier* and *J. Julien*, Bull. Soc. chim. Biol., 1964, **46**, 87). Ribosomal RNAs isolated from the cytoplasmic ribosomes of higher organisms are somewhat larger, the major classes having sedimentation coefficients of 28S, –32S, 18S and 5S corresponding to molecular weights of approximately 1.8×10^6, 0.7×10^6 and 4×10^4, respectively. Another small species of ribosomal RNA (5.8S) (originally termed 7S RNA) has also been found in higher organisms (*T. T. Pene, E. Knight* and *Darnell*, J. mol. Biol., 1968, **33**, 609; *G. M. Rubin*, Eur. J. Biochem., 1974, **41**, 197).

Transfer RNA may be fractionated into its individual species by a number of methods. The first pure species of tRNA to be isolated, the tRNA from yeast, which accepts alanine ($tRNA^{Ala}$), was obtained by counter distribution (*Apger*, *Holley* and *Merrill*, J. biol. Chem., 1962, **237**, 796) and the method is still used when large quantities of pure material are required. However, it is practically more convenient to use methods of column chromatography. Various ion-exchangers which include DEAE–Sephadex or cellulose (*R. M. Bock* and *J. Cherayil*, "Methods in Enzymology", eds. *Grossman* and *Moldave*, 1967, Vol. 12A, p. 638; *S. Nishimura et al.*, Biochim. Biophys. Acta, 1967, **142**, 133), benzoylated DEAE cellulose (*I. Gillam et al.*, Biochemistry, 1967, **6**, 3043), which gives the column an increased affinity for aromatic groups, and the dipolar ion-exchanger arginine–sepharose (*Jay* and *Jones*, Prep. Biochem., 1973, **3**, 517; *Jay*, *C. Coultas* and *Jones*, Nucleic Acids Res., 1976, **3**, 177) have been used. Reverse-phase chromatography (RPC) is also widely used (*A. D. Kelmes et al.*, "Methods in Enzymology", eds. *Moldave* and *Grossman*, 1971, Vol. 20, p. 9; Biochim. Biophys. Acta, 1971, **228**, 770). For the isolation of specific amino acid isoaccepting tRNA it is possible to modify chemically one isoaccepting tRNA species in order to facilitate its isolation (*M. L. Stephenson* and *P. C. Zamecnik*, "Methods in Enzymology", eds. *Grossman* and *Moldave*, Academic Press, New York, 1967, Vol. 12A, p. 670; *I. Gillam et al.*, Biochemistry, 1968, **7**, 2459; *T. F. McCutchan*, *P. T. Gillam* and *D. Soll*, Nucleic Acids Res., 1975, **2**, 853). The choice of method is largely dependent upon which species and the amount of that species of tRNA required.

(*b*) *Structure of nucleic acids*

Hydrolysis of nucleic acids yields only nucleosides, nucleotides or their component parts. D-2-Deoxyribose is found in DNA and D-ribose in RNA. The original proposal that nucleic acids were polymers of tetranucleotide units, which implied that there would be equimolar amounts of each of the four constituent bases was shown to be false as soon as accurate base composition analysis established that the molar properties of the bases varied widely according to the source of the material. Nucleic acids are in fact polymers of deoxyribonucleotides (DNA), or ribonucleotides (RNA).

(*i*) *Primary structure of DNA*

The bases found in the major deoxyribonucleotide units of DNA are adenine, guanine, cytosine and thymine. A small amount of 5-methylcytosine is also found in the DNA from some sources (*G. R. Wyatt*, Biochem. J., 1951, **48**, 581, 584), and in some phages cytosine is replaced by 5-hydroxy-

TABLE 4

MOLAR PROPORTIONS OF BASES IN DNA FROM DIFFERENT SOURCES

Source	*Adenine* (*A*)	*Guanine* (*G*)	*Cytosine* (*C*)	*Thymine* (*T*)	5-*Methyl-cytosine*
Bovine thymus	1.0	0.76	0.75	0.99	0.05
Bovine spleen	1.0	0.81	0.75	0.98	0.05
Bovine sperm	1.0	0.77	0.72	0.95	0.05
Yeast	1.0	0.60	0.55	1.05	—
Wheat germ	1.0	0.83	0.62	0.99	0.22
E. coli	1.0	0.96	0.97	0.92	—
φX174	1.0	1.01	0.75	1.29	—

methylcytosine or deoxyuridine (*R. G. Kallen, M. Simon* and *J. Marmur*, J. mol. Biol., 1962, **5**, 248). The molar proportions of these bases vary according to the source of the DNA. Table 4 shows the molar proportions of the bases in some DNA samples.

DNAs have been classified as A–T rich which are more commonly found (see Table 4), and the much rarer G–C rich species. It will be seen from Table 4 that adenine and thymine are present in equimolar amounts and guanine and cytosine are also present in equimolar amounts. These common properties of DNAs from different species were first noted by *E. Chargaff* (Experientia, 1950, **6**, 201), but it was only later that the significance of the latter observation in relation to the secondary structure of DNA became apparent (see p. 194). The base composition has a marked effect on some of the properties of DNA. The buoyant density of DNA is found to have a linear relationship to the content of guanine plus cytosine in the DNA (*N. Sueoka, Marmur* and *P. Doty*, Nature, 1959, **193**, 1429).

Of prime importance in studying the structure of nucleic acids is an understanding of the nature of the internucleotide linkage. Potentiometric titrations of mononucleotides, showed there to be an equal number of primary and secondary phosphoryl dissociations, but in DNA and RNA the primary to secondary ratio was shown to be much greater than unity (*P. A. Levene* and *H. S. Simms*, J. biol. Chem., 1925, **65**, 519; 1926, **70**, 327). Enzymic hydrolysis products of DNA are deoxyribonucleoside 3′-phosphates or deoxyribonucleoside 5′-phosphates (see p. 147) and when DNA is hydrolysed with acid, among the products deoxycytidine and thymidine 3′,5′-diphosphates are found (*Levene, ibid.*, 1921, **48**, 119; 1938, **126**, 63; *C. A. Dekker, A. M. Michelson* and *A. R. Todd*, J. chem. Soc., 1953, 947). These results led to the concept of DNA being a polynucleotide in which deoxyribonucleosides are joined through phosphodiester linkages. The only positions available in the deoxyribonucleoside for the phosphodiester bonds are at the 3′- and 5′-

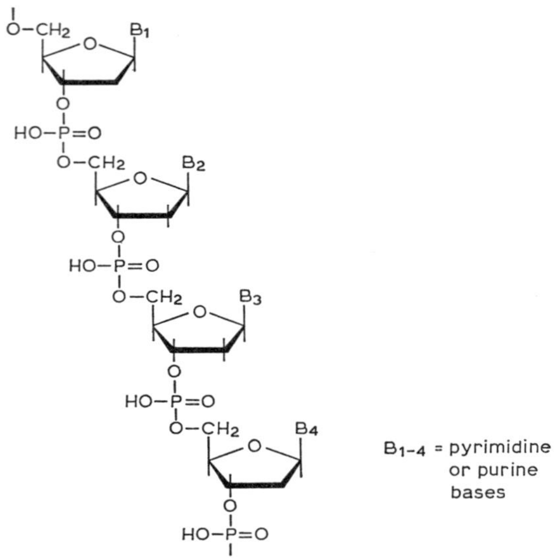

Deoxyribonucleic acid (DNA)

hydroxyl positions. The structure for DNA, which was first suggested by *D. M. Brown* and *Todd* (*ibid.*, 1952, 52), therefore becomes as shown.

This structure has been confirmed by synthesis. Deoxyribo-oligonucleotides specifically synthesised with 3′→5′-phosphodiester linkage are found to be biologically active (*H. G. Khorana*, Biochem. J., 1968, **109**, 709).

It will be seen that the drawing of oligonucleotide structures becomes tedious and abbreviated formulae are used. The DNA structure shown becomes:

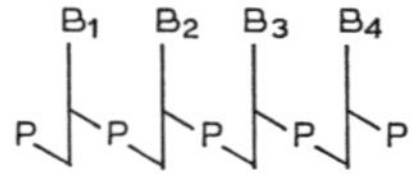

The left-hand side is the 5′-end of the chain and the right-hand side is the 3′-end.

The nucleosides are linked by 3′→5′-phosphodiester bridges. The common bases, guanine, cytosine and thymine are represented by their first letter, so, for example, the oligonucleotide 5′-*O*-phosphoryldeoxyadenylyl-(3′→5′)-deoxycytidylyl-(3′→5′)-thymidylyl-(3′→5′)-deoxyguanosine becomes:

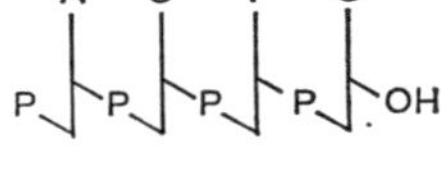

or may be further abbreviated to d-pApCpTpG-OH, or simply d-pA-C-T-G-OH. The prefix "d-" indicates that the chain is a *deoxy*ribo-oligonucleotide.

In order to understand more of the primary structure of DNA a knowledge of the sequence of bases is required. The technique of nearest-neighbour frequency analysis gives the frequency of dinucleotide sequences in newly synthesised ^{32}P-labelled DNA (*J. Josse, A. D. Kaiser* and *A. Kornberg*, J. biol. Chem., 1961, **236**, 864). The principle of the method is the incorporation of the ^{32}P-label into the DNA from an α-^{32}P-labelled deoxyribonucleoside 5′-triphosphate, followed by degradation of the DNA to give deoxyribonucleoside 3′-phosphate products. This leads to the transfer of the ^{32}P-label from the original nucleotide to its ajacent nucleotide on its 5′-(left-hand)side, and hence the identity of the bases adjacent to the originally labelled deoxyribonucleoside triphosphate is deduced.

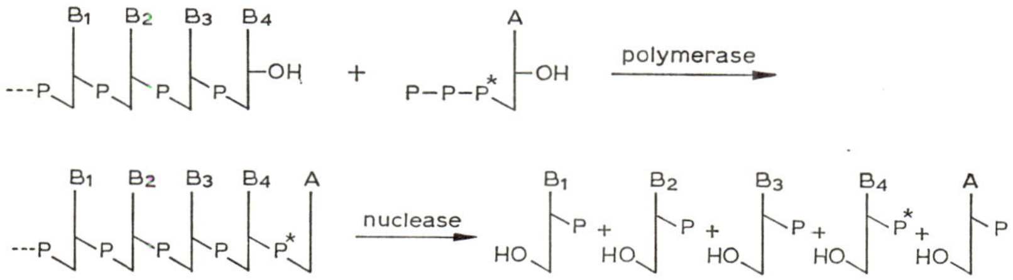

P indicates the ^{32}P-labelled phosphate group.
Illustration of the method of nearest-neighbour frequency analysis.

Initially the sequential analysis of DNA molecules lagged behind that of RNA molecules for two reasons. Firstly, the smallest RNA molecules, tRNAs, contain approximately 75 nucleotide residues, whereas the smallest DNA molecules (some of the bacteriophage DNAs) contain more than 5000 nucleotide residues. Secondly, whereas there are ribonucleases available which cleave at specific ribonucleotide residues, the common deoxyribonucleases have little base specificity. Short sequences of the DNAs from bacteriophages fl and φ174 were originally deduced by a variety of techniques (*E. B. Ziff, J. W. Sedat* and *F. Galibert*, Nature, New Biology, 1973, **241**, 106; *F. Sanger et al.*, Proc. Nat. Acad. Sci., U.S.A., 1973, **70**, 1209). One method relied on the partial specificity of a deoxyribonuclease isolated from bacteriophage T4 infected *Escherichia coli* whereby oligonucleotides with pC--- at the 5′- end can be obtained. In another approach fl DNA was used as a template for the synthesis of short complementary ^{32}P-labelled DNA fragments. The use of a synthetic deoxyribo-oligonucleotide primer caused initiation at a single site. In the latter approach the technique of nearest-neighbour frequency analysis was used to obtain dinucleotide sequences and in both methods depurination was used to obtain short pyrimidine sequences (*K. Burton* and *G. B. Peterson*, Biochem. J., 1960, **75**, 17). The mechanism for this cleavage is described later (see p. 230). The separation and sequential analysis of the short oligonucleotide fragments is achieved using two-dimensional ionophoretic methods developed for RNA (see p. 203).

Considerable advances in methods for the sequential analysis of DNA have been made recently. These methods allow the sequences of long DNA fragments to be determined relatively quickly. Most notably the sequence of nucleotides in the DNAs of the entire genomes of the bacteriophages φ174 (*Sanger et al.*, Nature, 1977, **265**, 678) and G4 (*G. N. Godson et al.*, *ibid.*, 1978, **276**, 236) and the virus SV40 (*W. Friers et al.*, *ibid.*, 1978, **273**, 113; *V. B. Reddy et al.*, Science, 1978, **200**, 494) have been determined. (All have DNA genomes containing more than 5000 nucleotides.) The advances which have made the sequential analysis of long DNA molecules feasible have been the isolation and availability of the restriction endonucleases and methods for the sequential analysis of large DNA fragments.

The restriction endonucleases cleave DNA at specific sites dictated by short specific sequences of deoxyribo-oligonucleotides. Restriction enzymes isolated from different organisms cleave at different sequence specific sites. (For review of the properties of restriction enzymes, see for example *D. Nathans* and *H. O. Smith*, Ann. Reviews, Biochem., 1975, **44**, 273). By use of these enzymes a large DNA molecule can be cleaved into several series of distinct fragments of a suitable size for sequential analysis. Also since the different restriction enzymes cleave at different sites on the DNA a series of overlapping fragments can be obtained which can be assembled in a unique order.

Two methods, which have common features, have been developed independently for sequence determination of large DNA fragments. The "plus and minus" method (*Sanger* and *A. R. Coulson*, J. mol. Biol., 1975, **94**, 441) is illustrated in Fig. 1. A "primer" is extended by copying the complementary DNA strand using a DNA polymerase and all four deoxyribonucleotide triphosphates (one of which is ^{32}P-labelled) under conditions such that successive nucleotides are added, according to the base-pairing rules, to produce extension products of every possible chain length. (See upper part of Fig. 1). These products are then sub-divided. One portion of the extended molecule is separated according to size by electrophoresis on a polyacrylamide gel (Column O). Other portions are subjected to the "minus system". Here, extension is continued in a system similar to that described above but in four separated experiments where one of the triphosphates is omitted in turn. This leads to a series of products in which the extended chain is terminated at the position immediately preceding that of the omitted nucleotide. (The

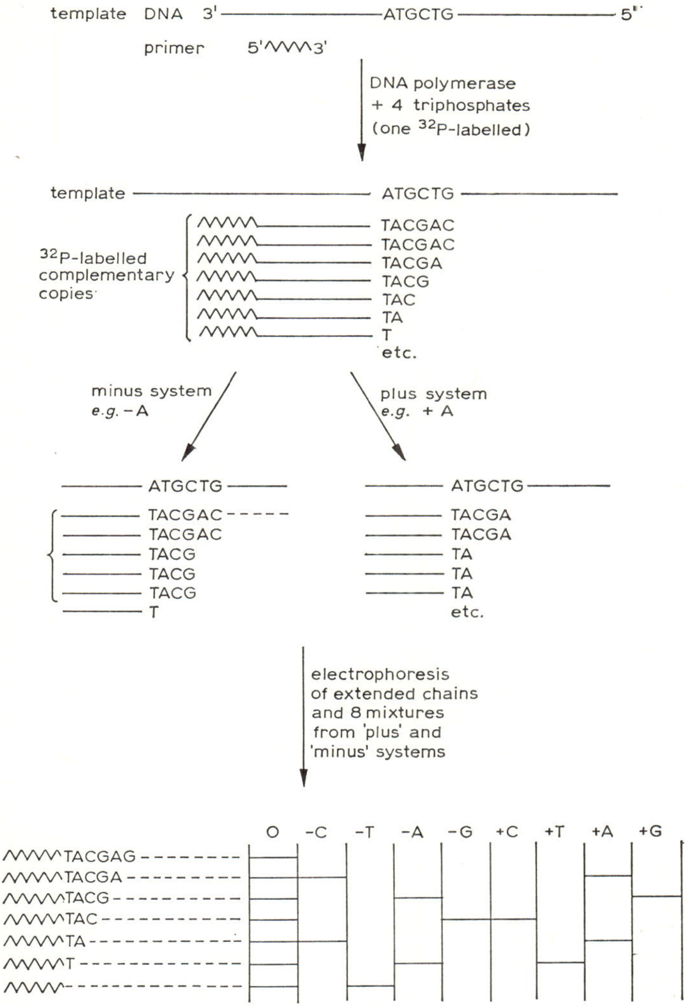

Fig. 1. Taken from *Sanger* and *Coulson, loc. cit.*

"minus A system" is illustrated in Fig. 1). These products are also separated by gel electrophoresis (Columns –C, –T, –A, –G). Further portions are subjected to the "plus system". Here the original extended chains are subjected to an exonuclease activity which degrades the chains sequentially from their 3′-ends in the absence of triphosphates. A series of experiments of this type is carried out. In each case one of the trinucleotides is added and the extended chains are degraded until this nucleotide is reached. (The "plus A system" is illustrated in Fig. 1). These products are separated by gel electrophoresis (columns +C, +T, +A, +G). From the patterns obtained by autoradiography of the gels, it is possible to read off the sequence of the DNA fragment. The sequence of fragments of up to 200 nucleotides long can be deduced in this way.

The second method relies on a chemical approach (*A. M. Maxam* and *W. Gilbert*, Proc. Nat. Acad. Sci., U.S.A., 1977, **74**, 506). The fragment of DNA whose sequence is required is radioactively labelled at one end. Samples of the labelled DNA are then degraded by different chemical methods so that in each case a series of labelled products of different chain length are obtained. Since the methods of degradation involve the removal of a specific base from the DNA, it can be deduced that for each product of the series the next residue of the chain will be the one removed by the specific chemical cleavage. The products of different chain length are separated and identified by gel electrophoresis in a manner similar to that of the alternative method described above. Specific cleavage at guanine and adenine residues is brought about by treatment of the DNA with dimethyl sulphate which methylates guanines at the $N_{(7)}$ position and adenines at the $N_{(3)}$ position (*P. D. Lawley* and *P. Brookes*, Biochem. J., 1963, **89**, 127). The glycosidic bonds of the methylated purines are unstable and break on heating at neutral pH, leaving the free sugar. Treatment with base cleaves the sugar from the neighbouring phosphate groups. Since the guanines methylate five times faster than the adenines, bands on the gels which arise from cleavage at guanines are more intense. Specific cleavage at thymine and cytosine residues is achieved with hydrazine, which cleaves the base leaving ribosyl area (*A. Temperli et al.*, Biochim. Biophys. Acta, 1964, **91**, 462; *D. H. Hayes* and *F. Hayes-Baron*, J. chem. Soc., 1967, 1528; *A. R. Cashmore* and *G. B. Peterson*, Biochim. Biophys. Acta, 1969, **174**, 591). Treatment with piperidine displaces the hydrazine products from the sugars and catalyses the β-elimination of the phosphates. Bands which arise from cleavage at thymine or cytosine are of similar intensity. The presence of 2 *M* NaCl preferentially depresses the reaction of hydrazine at thymidine residues. Therefore when cleavage is carried out under these conditions the bands which occur are due to cleavage at cytosine only.

(*ii*) *Secondary structure of DNA*

The DNA molecule consists of two polydeoxynucleotide chains wound round each other in the form of a double helix. Both helices are right-handed and are held together by hydrogen bonds. They are wound round a common axis running centrally along the length of the molecule. This double helical structure was proposed by *J. D. Watson* and *F. H. C. Crick* (Nature, 1953, **171**, 737) following the building of models based on information gained from X-ray diffraction studies (*M. H. F. Wilkins, A. R. Stokes* and *H. R. Wilson, ibid.*, 1953, **171**, 737; *R. Franklin* and *R. G. Gosling, ibid.*, 1953, **171**, 740; **172**, 156). A diagrammatic representation of the Watson and Crick model is shown in Fig. 2. The outer sides of the two helices represent the phosphate-deoxyribose chains of the polynucleotides with the bases represented by the horizontal rods holding the two chains together by hydrogen bonding. *Watson* and *Crick* showed from their model building that the bases will fit the arrangement if they are paired purines with pyrimidines in each case.

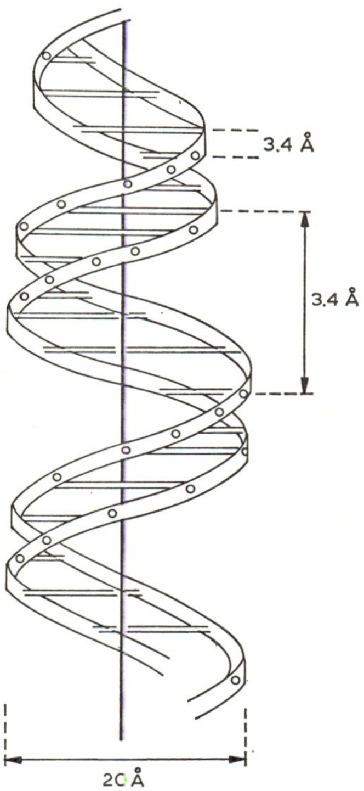

Fig. 2. The DNA molecule as represented by *Watson* and *Crick* (Nature, 1953, **171**, 737).

As mentioned previously, *E. Chargaff* (Experientia, 1950, **6**, 201) had pointed out the equivalence of adenine with thymine, and guanine with cytosine in DNA (see Table 4). Therefore the most obvious base-pairs are adenine and thymine, and guanine and cytosine, with the hydrogen bonding as shown.

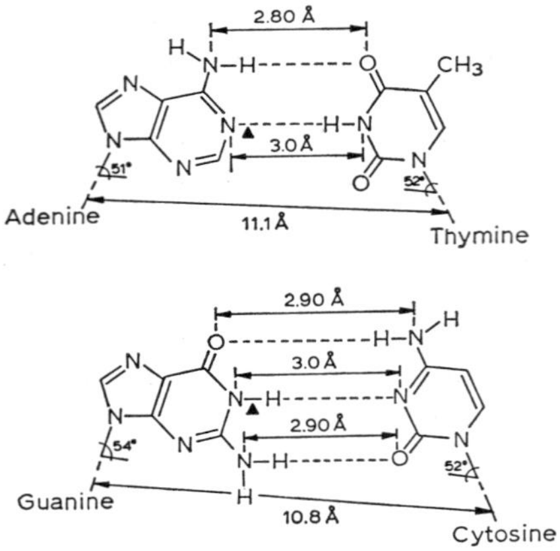

The hydrogen bonding of the base pairs in DNA. The dotted lines represent the hydrogen bond and ▲ shows the position of the longitudinal axis in the double helix.

It should be noted that the molar ratios of the bases in the bacteriophage φ174 (see Table 4) show that the amount of adenine and thymine, and guanine and cytosine are not equivalent in this DNA. This is because φ174 DNA is single-stranded and therefore pairing between the bases of different strands cannot occur.

In their original proposal for the base-pairing, *Watson* and *Crick* suggested that the bases in each of the base pairs were joined by two hydrogen bonds but later the guanine–cytosine base pair was modified to have three hydrogen bonds between the bases (*L. Pauling* and *R. B. Carey*, Arch. Biochem. Biophys., 1956, **65**, 164). Several physical methods, notably i.r. and n.m.r. spectroscopy, and X-ray structural analysis, have been used to provide experimental evidence which verifies the structure of the base-pairs. The results from i.r. spectroscopy (*R. M. Hamlin, R. C. Lord* and *A. Rich*, Science, 1965, **148**, 1734; *J. Pitha, R. Jones* and *P. Pithova*, Canad. J. Chem., 1966, **44**, 1045) and n.m.r. spectroscopy (*R. R. Sharp, H. T. Miles* and *E. D. Becker*, Biochem. Biophys. Res. Comm., 1966, **23**, 194; *L. Katz* and *S. Penman*, J. mol. Biol., 1966, **15**, 220; *K. H. Scheit*, Angew. Chem., 1967, **79**, 90) obtained by examination of the spectra of derivatives of adenine and thymine or uracil mixtures, and guanine and cytosine mixtures in solution are able to confirm unequivocally only the structure of the guanine–cytosine base pair. X-Ray structural analysis of mixtures of derivatives of guanine and cytosine again confirms the Watson–Crick base pairing, but analyses of derivatives of adenine and thymine or uracil (*K. Hoogstein*, Acta Cryst., 1959, **12**, 822; 1963, **16**, 907; *A. E. V. Hashemeyer* and *H. M. Sobell*, *ibid.*,

1965, **18**, 525) show different types of base pairing. However, *Rich* and his collaborators (Proc. Nat. Acad. Sci. Wash., 1973, **70**, 349; Nature, 1973, **243**, 150) have shown by X-ray structural analysis that crystals of adenylyl-(3'→5')-uridine (ApU) when paired with itself and also guanylyl-(3'→5')-cytidine (GpC) paired with itself show the Watson–Crick base-pairing.

The base-pairs lie in planes perpendicular to the axis of the double helix and are stacked 3.4 Å apart. The two chains are of opposite polarity *i.e.* one strand is 3'→5' and the other is 5'→3', but both are right-handed helices and form a shallow and a wide groove. One complete turn of the double helix comprises 10 base-pairs and is 34 Å long. The diameter of the double helix is approximately 20 Å. Interactions between the stacks of bases, sometimes called "base stacking forces", play an important role in the stability and

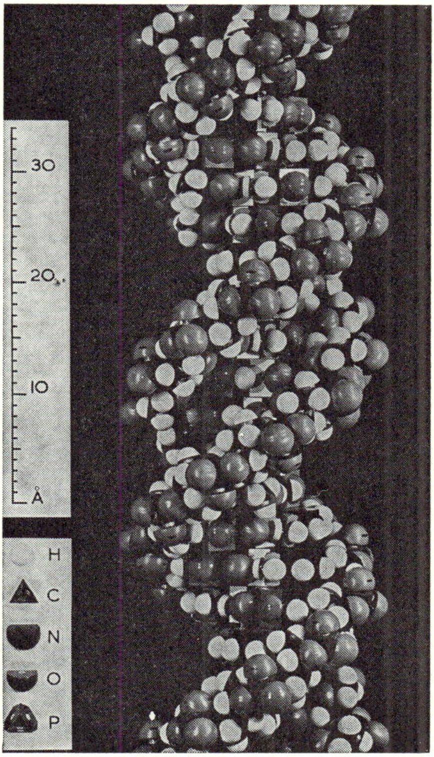

Fig. 3. A photograph of a model of part of the DNA molecule in the B-form. The shallow and deep grooves can be seen clearly. (By courtesy of Professor *M. H. F. Wilkins*, Biophysics Department, King's College, London.)

configuration of the DNA molecule (see "Organic Chemistry of Nucleic Acids", eds. *N. K. Kocketkov* and *E. I. Budovskii*, Plenum, New York, 1971, p. 183).

X-Ray analysis shows that there are three main crystalline forms of DNA dependent upon the relative humidity, and type and ionic strength of the cation. The B-form, which is the one described already, is found in fibres of the sodium salt at 92% humidity. This is the form which exists *in vivo* (*M. H. F. Wilkins* and *J. T. Randall*, Biochim. Biophys. Acta, 1953, **10**, 192) and c.d. studies show that it is the form adopted by DNA in solution (*M. J. B. Tunis-Schneider* and *M. F. Maestre*, *ibid.*, 1970, **52**, 521). A photographic model of a model of part of the DNA molecule in the B-form is shown in Fig. 3. The A-form of DNA is found in fibres of the sodium, potassium or caesium salts of DNA at 75% humidity. In this form there are eleven base pairs in each turn of the double helix which is right-handed. The pitch is 28 Å and the base pairs are at an angle of 20° to a line perpendicular to the axis of the helix (*Wilkins et al.*, *ibid.*, 1965, **12**, 60). The C-form is found in fibres of the lithium salt at 66% relative humidity. Here there are 9.3 base-pairs in each turn of the helix which has a pitch of 31 Å. The base-pairs in this form are inclined at an angle of 6° to a line drawn perpendicular to the axis of the helix (*Wilkins et al.*, *ibid.*, 1961, **3**, 547).

(iii) Tertiary structure of DNA

DNA isolated in degraded form is linear and usually double-stranded. However, DNA which has been isolated intact from several sources, *e.g.* bacteria, viruses, chloroplasts and mitochondria, is found to be circular. The circular structures can be either single- or double-stranded and the double-stranded structures may have both strands totally covalently bonded in two completely closed circles, or only one strand may be totally covalently bonded while the other strand has a single break and is an "open" circle. As another alternative neither strand may be totally covalently bonded as in the bacteriophage λ. In the host cell it is circular but when released from the cell it has a linear form (*V. C. Bode* and *A. D. J. Kaiser*, J. mol. Biol., 1965, **14**, 399; *E. J. Young* and *R. L. J. Sinsheimer*, *ibid.*, 1964, **10**, 562). This DNA has what is known as "cohesive ends". These are single-stranded regions at each end of the molecule which are complementary to each other, *i.e.* can form base pairs (*A. D. Hershey et al.*, Proc. Nat. Acad. Sci., Wash., 1963, **49**, 748; 1965, **53**, 325), and when these two single-stranded regions overlap a circular duplex is formed. Some low molecular weight bacteriophage DNAs, *e.g.* φ174 and f2, exist in several forms. In the phage particle it exists as a totally covalently linked, circular, single-stranded molecule containing about 5500 nucleotides (*W. Friers* and *Sinsheimer*, J. mol. Biol., 1962, **5**,

408, 424). After infection of the host bacteria the DNA goes through two double-stranded intermediates (replicative forms) during synthesis. RF II is a double-stranded circular form in which one strand has a single break and is in the "open" form (*Sinsheimer, R. Knippers* and *T. Komano*, Cold Spring Harbor Symp. Quant. Biol., 1968, **33**, 443), whereas in the RF I form there are no breaks in either strand and the DNA has a completely "closed" double-stranded structure (*A. Burton* and *Sinsheimer*, Science, 1963, **142**, 962).

Double-stranded circular DNAs which have no breaks in either strand have an interesting tertiary structure. Any modification to the DNA which affects the number of base pairs in one turn of the helix will put a strain on the molecule since the strands are not free to unwind. An alteration of this kind caused the imposition of a superhelical structure on the initial double helix in order to compensate for the strain. Evidence for the superhelical structure in this type of DNA is shown in electron micrographs (*J. B. David* and *D. R. Wolstenholme*, J. mol. Biol., 1967, **28**, 233) and by its greater density (*J. Vinograde* and *J. Lebowitz*, J. gen. Physiol., 1966, **49**, 103). The strands of linear double-stranded DNA separate when the DNA is treated with mild alkali or subjected to high temperature, but the topology of covalently closed double-stranded DNA prohibits separation of the strands. If, however, the base-pairs are broken by raising the temperature higher, the supercoiled structure collapses into two interwoven random coils which form a heavy, compact molecule (*Vinograd et al.*, Proc. Nat. Acad. Sci. Wash., 1965, **53**, 1104).

As mentioned above, high temperature causes the strands of DNA to separate. If a solution of double-stranded DNA (or double-stranded RNA or a hybrid of one DNA strand and one RNA strand) is gradually heated, over a small temperature range which is specific for a given DNA, the physical properties of the solution change. The optical rotation and viscosity decrease and the optical density increases, reflecting the change taking place in the structure of the DNA in which the ordered double-stranded helical structure is disrupted and the strands separate as two random coils. This phenomenon, usually called the melting or denaturation of the DNA, is followed by measuring the optical density of the solution over a given temperature range, when a pattern similar to that shown (Fig. 4) is obtained. If it is assumed that the optical density at low temperature corresponds to DNA in its completely double-stranded form and the optical density at high temperature corresponds to DNA in which this secondary structure has been completely disrupted, then the optical density midway between these values represents the temperature at which the DNA is half in the double helical form and half in the random coil form. This temperature is known as the

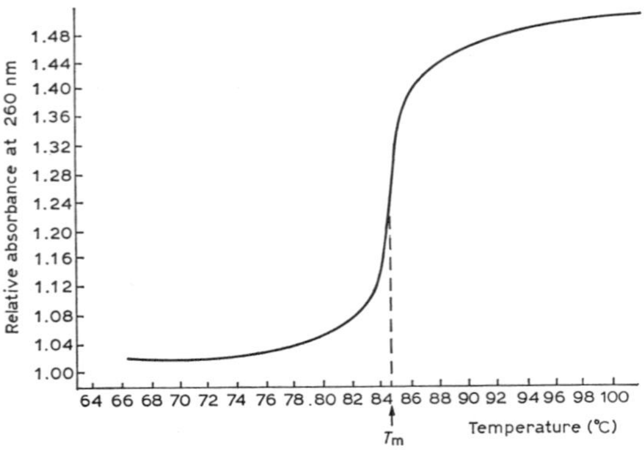

Fig. 4.

melting temperature, T_m, and is constant under given conditions of pH and salt concentration for a given DNA sample. The shape of the melting curve and the T_m reflect the properties of the DNA sample. A homogeneous DNA sample will melt over a small temperature range whereas heterogeneous DNA will have a broader range. The T_m of a polynucleotide increases with molecular weight (*M. N. Lipsett, L. A. Heppel* and *D. F. Bradley*, Biochim. Biophys. Acta, 1960, **41**, 175; J. biol. Chem., 1961, **236**, 857; *G. Felsenfeld* and *H. T. Miles*, Ann. Reviews Biochem., 1967, **36**, 407), and shows a linear relationship with increases in the content of guanine–cytosine base pairs (*J. Marmur* and *P. Doty*, J. mol. Biol., 1962, **5**, 109). The T_m is also affected by the ionic strength of the solution (*Doty et al.*, Proc. Nat. Acad. Sci. Wash., 1959, **45**, 482), and the nature of salts and solvents present.

The process of denaturation of DNA by heating can be reversed (renaturation) to give a product similar in properties to the original (native) double-stranded DNA (*Marmur* and *Doty*, J. Mol. Biol., 1961, **3**, 585). If, after melting DNA, conditions are carefully controlled and the solution is cooled very slowly the optical-density pattern obtained with decrease in temperature follows closely the melting curve, and the product obtained resembles the original native DNA. However, if after melting, the solution is cooled rapidly, the strands remain separate, which can be seen both from the optical density, which decreases very little with decrease in temperature, and by structural examination of the product.

Some of the cyclic DNA molecules obtained from viruses (*M. G. Rush et al.*, Proc. Nat. Acad. Sci. Wash., 1967, **58**, 1676; *Rush, R. Eason* and *Vinograd*, Biochim. Biophys. Acta, 1971, **228**, 585) and from mitochondria (*B. Hudson* and *Vinograd*, Nature, 1969, **221**, 332) are found as dimers or trimers. The cyclic molecules are interlocked in the form of catenanes.

(*iv*) *Primary structure of RNA*

The major ribonucleotide units obtained on hydrolysis of RNA contain the bases adenine, guanine, cytosine and uracil. The molar properties of these bases found in total RNA isolated from different sources (see Table 5) vary considerably and it can be seen that in contrast to the molar proportions of bases found in DNA there is no equivalence of adenine and uracil (which replaces thymine in RNA), and guanine and cytosine. It was pointed out by *D. Elson* and *E. Chargaff*, Biochim. Biophys. Acta, 1955, **17**, 367), however,

TABLE 5

MOLAR PROPORTIONS OF BASES IN TOTAL RNA FROM DIFFERENT SOURCES

Source	*Adenine(A)*	*Guanine(G)*	*Cytosine(C)*	*Uracil(U)*
Ox liver	1.00	1.60	1.92	1.27
Ox kidney	1.00	1.32	1.69	1.04
Rat kidney	1.00	1.52	1.58	1.05
Sea urchin embryo	1.00	1.30	1.36	0.92
Yeast	1.00	0.97	0.89	1.08
E. coli	1.00	1.14	0.98	0.83

The data is taken from the results of *Elson* and *Chargaff*, *loc. cit.*

that the number of bases carrying an amino group in the "6" position (*i.e.* adenine and cytosine) is approximately equal to the number of bases carrying an oxo group in that position (*i.e.* guanine and uracil). The molar proportions of the bases in different types of RNA from the same source also differs, which is shown in Tables 6 and 7, where the molar proportions of the major bases found in transfer RNA, ribosomal RNA and messenger RNA from *E. coli*, and these RNAs plus their nuclear precursor RNAs from Hela cells are

TABLE 6

MOLAR PROPORTIONS OF BASES IN DIFFERENT TYPES OF RNA FROM *E. coli*

Type of RNA	*Adenine*	*Guanine*	*Cytosine*	*Uracil*	*Ref.*
Transfer RNA	1.0	1.58	1.43	1.09	1
Ribosomal RNA	1.0	1.25	0.86	0.86	2
Messenger RNA	1.0	1.08	0.96	0.95	3

References
1 *G. Zubay*, J. mol. Biol., 1962, **4**, 347.
2 *P. F. Spahr* and *A. Tissières*, *ibid.*, 1959, **1**, 237.
3 *F. Gros et al.*, Cold Spring Harbor Symp. Quant. Biol., 1961, **26**, 111.

TABLE 7

MOLAR PROPORTIONS OF BASES IN DIFFERENT TYPES OF RNA FROM HeLa CELLS

Type of RNA	*Adenine*	*Guanine*	*Cytosine*	*Uracil*
	Cytoplasmic RNA			
28s Ribosomal RNA	1.0	2.25	2.0	1.0
18s Ribosomal RNA	1.0	1.43	1.29	1.05
5.8s Ribosomal RNA	1.0	1.33	1.33	1.10
5s Ribosomal RNA	1.0	1.89	1.45	1.22
Transfer RNA	1.0	1.23	1.23	1.10
Messenger RNA	1.0	0.81	0.92	1.08
	Nuclear RNA			
45s Ribosomal RNA Precursor	1.0	2.84	2.54	1.31
32s Ribosomal RNA Precursor	1.0	2.64	2.36	1.15
DNA-like RNA Messenger RNA Precursor	1.0	0.81	0.85	1.19

Data is taken from *J. E. Darnell* (Bact. Review, 1968, **32**, 262).

shown. Transfer RNA and, to a lesser extent, ribosomal RNA and messenger RNA also contain a number of minor nucleosides (see Table 1, p. 119) but these have not been included in Table 6.

As with DNA, potentiometric titrations of RNA show that it is a polymer containing phosphodiester linkages. In DNA, as we have already seen, the only positions available for the formation of phosphate ester bonds are the 3- and 5-OH groups and DNA has the structure of a polymer in which deoxyribonucleosides are linked through 3′→5′-phosphodiester linkages. In RNA there is a third hydroxyl group to be considered at position 2 in the ribose moiety. Involvement of the 5′-hydroxyl group of the ribose in the formation of the phosphodiester linkage is demonstrated by the quantitative production of ribonucleoside 5′-phosphates when RNA is hydrolysed with snake venom phosphodiesterase (*W. E. Cohn* and *E. Volkin*, J. biol. Chem., 1953, **203**, 319). Alkaline hydrolysis of RNA yields a mixture of ribonucleoside 2′- and 3′-phosphates which may lead one to the conclusion that either the 2′- or 3′-OH groups are involved in the phosphodiester linkage. It has been shown, however, that the 2′-OH group participates in the hydrolysis of the 3′,5′-phosphodiester linkage (*D. M. Brown* and *A. R. Todd*, J. chem. Soc., 1952, 52). RNA, unlike DNA, is rather unstable to alkali, but both RNA and DNA are hydrolysed readily by acid. These differences can be interpreted by the involvement of the 2′-OH group in the alkaline hydrolysis of RNA. The ready interconversion of 2′- and 3′-ribonucleotides under acid conditions proceeds through a 2′,3′-cyclic phosphate intermediate. Furthermore, it is known that mono- and di-alkyl esters of phosphoric acids are stable to alkali, whereas tri-alkyl esters are readily hydrolysed. If in RNA it is possible for the dialkyl phosphate to form a trialkyl ester by cyclisation with an adjacent hydroxyl group, then it would become labile to alkali. The alkaline degradation of RNA would occur therefore as shown.

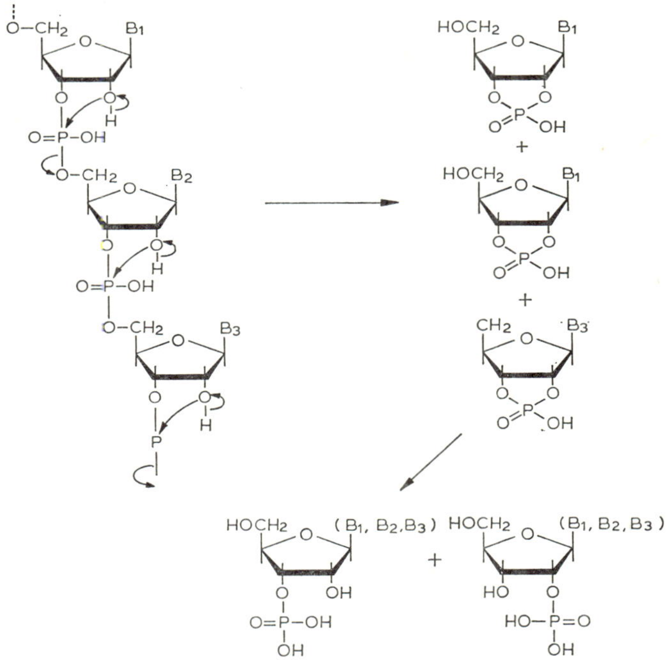

Scheme for the alkaline hydrolysis of RNA *via* cyclic 2,3′-phosphate intermediates.

The mechanism is supported by isotope studies (*D. Lipkin*, *P. T. Talbert* and *M. Cohn*, J. Amer. chem. Soc., 1954, **76**, 2871), and by the resistance to alkali of the internucleotide bond adjacent to a ribose residue methylated at the 2-position. 2′-*O*-Methylnucleosides are found in ribosomal RNA, transfer RNA (*R. H. Hall*, Biochemistry, 1964, **3**, 876) and messenger RNA (*R. P. Perry* and *D. E. Kelly*, Cell, 1974, **1**, 37; *M. J. Adams* and *S. Cory*, Nature, 1975, **255**, 28). Proof that it is the 3′-OH group of the ribose which is involved in phosphodiester internucleoside linkage comes from the results of enzymic hydrolysis. Digestion of RNA with ribonuclease leads to pyrimidine nucleoside 3′-phosphate and oligonucleotides terminating in 3′-phosphates. With synthetic substrates it is found that this enzyme will only hydrolyse alkyl esters of nucleoside 3′-phosphates whereas the esters of nucleoside 2′-phosphates are not attacked (*Brown* and *Todd*, J. chem. Soc., 1953, 2040). Support for the 3′→5′-phosphodiester linkage in RNA comes also from the action of the exonucleoase spleen phosphodiesterase. The sole products of degradation of RNA are nucleoside 3′-phosphates. In this reaction there is no cyclic phosphate formation and so migration of the phosphate group cannot take place (*L. A. Heppel* and *P. R. Whitfield*, Biochem. J., 1955, **60**, 1; *Whitfield*, *Heppel* and *R. Markham*, *ibid.*, 1955, **60**, 15). Oligoribonucleotides which are synthesised in an unambiguous manner to give 3′→5′-phosphodiester linkages (*H. G. Khorana et al.*, J. Amer. chem. Soc., 1966, **88**, 819) are found to have the same biological activity as the same oligonucleotides prepared either by degradation of RNA or by enzymic

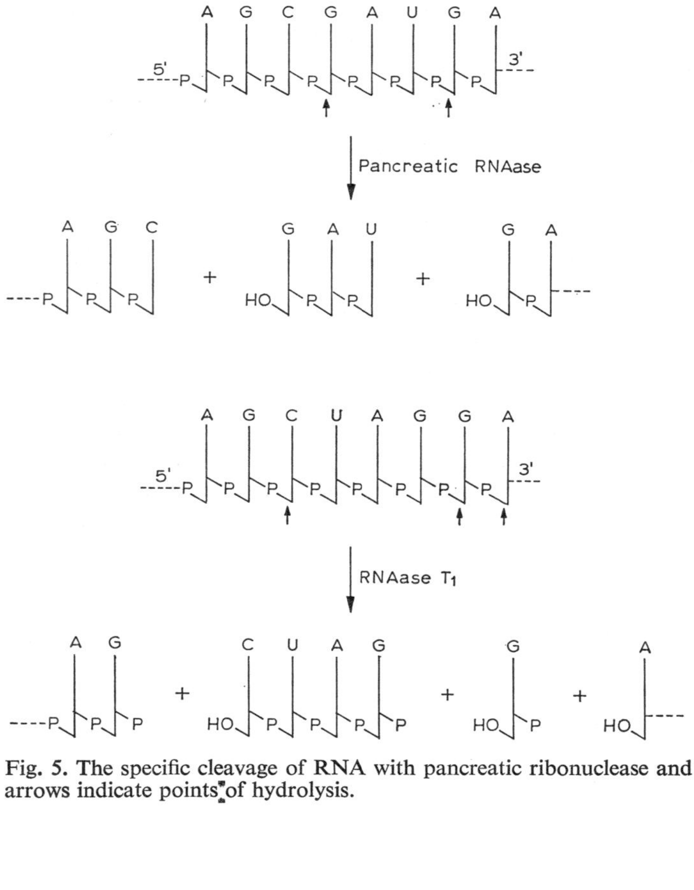

Fig. 5. The specific cleavage of RNA with pancreatic ribonuclease and ribonuclease T1; arrows indicate points of hydrolysis.

synthesis (*idem*, Proc. Nat. Acad. Sci., Wash., 1965, **54**, 1378; *M. Nirenberg et al.*, *ibid.*, 1965, **53**, 1161). This provides confirmation of the structure of the internucleotide linkage.

The primary sequence of bases in many RNA molecules is now known, and a comprehensive record of those known up to 1974 ("Handbook of Nucleic Acid Sequences", ed. *B. G. Barrell* and *B. F. C. Clark*, Joynson and Bruvvers, Oxford, 1974) and of tRNAs, S5 and S8.5 RNAs up to 1977 (*V. A. Erdmann*, Nucleic Acids Res., 1978, **5**, r1) has been compiled.

The first RNA molecule whose sequence was deduced was the transfer RNA from yeast which may be esterified with alanine, $tRNA^{Ala}$*. The structural analysis was carried out by *R. W. Holley* and his coworkers (Science, 1965, **147**, 1462) who used analogous techniques to those used for sequential analysis of polypeptides. The tRNA molecule was degraded with either pancreatic ribonuclease, which cleaves specifically the internucleotide bonds on the 3′-side of pyrimidine containing nucleotides or ribonuclease T_1 (takadiastase) obtained from *Aspergillus oryzae*, which cleaves specifically on the 3′-side of guanine-containing nucleotides (Fig. 5).

The fragments obtained were separated by ion-exchange chromatography and paper chromatography or electrophoresis and the sequence of bases deduced by using other enzymes such as phosphodiesterase from snake venom. These techniques gave the total sequence of two series of oligonucleotides. Overlaps of sequence of oligonucleotides from the two series gave the sequence of several large oligonucleotides and many small fragments, and allowed the tRNA molecule to be described totally in a series of oligonucleotides of defined sequence. The ordering of these nucleotides in an unambiguous sequence was accomplished by splitting the whole molecule either into several large fragments or into two halves by controlled digestion with ribonuclease T_1 followed by analysis of these large fragments. This work including the isolation and purification of the tRNA species took many years and required relatively large amounts of material. The identification of the fragments was greatly aided by the presence of modified bases which were used as labels (in a non-radioactive sense). The sequence of $tRNA^{Ala}$ from yeast is shown later (see p. 207). The sequences of several other tRNA molecules have been deduced using the general method developed by *Holley*. However, *F. Sanger*, who previously had laid the foundation for the sequential analysis of proteins, has introduced modifications into this

*tRNA molecules are identified by the specific amino acid with which they may be enzymically esterified. $tRNA^{Ala}$ refers to that tRNA which is esterified with alanine but it does not imply that the amino-acyl bond has been formed. Ala-$tRNA^{Ala}$ describes the amino-acyl-tRNA which has alanine attached to it through an ester linkage.

general method which have greatly speeded up the analysis of nucleic acid and reduced considerably the amount of material required. Radiolabelled RNA is used and the nucleotide fragments from enzymic digestion are separated by a "fingerprinting" technique. This involves electrophoresis of the oligoribonucleotide fragments at pH 3.5 on cellulose acetate in one direction followed by either electrophoresis at pH 1.9 on diethylaminoethyl-cellulose paper or chromatography on diethylaminoethyl-cellulose paper using non-radioactive oligonucleotides as the eluting ions ("homochromatography") at right angles to the first direction (*Sanger, G. G. Brownlee* and *B. G. Barrell*, J. mol. Biol., 1965, **13**, 373; *Brownlee, Sanger* and *Barrell, ibid.*, 1968, **34**, 379; *Brownlee* and *Sanger*, Europ. J. Biochem., 1969, **11**, 395). To use this method it is necessary to be able to label the RNA radioactively. For RNA from micro-organisms and viruses it is possible to do this by growing the organism in media containing ^{32}P-labelled phosphate. For RNAs which cannot be labelled in this manner, several methods for the *in vitro* end-group labelling of oligonucleotides or tRNAs, which make possible sequential analysis of oligonucleotides on a small scale, are available. After digestion of the RNA the oligonucleotides obtained can be labelled enzymically with ^{32}P at the 5′-end (*M. Szekely* and *Sanger*, J. mol. Biol., 1969, **43**, 609; *M. Simsek et al.*, Proc. Nat. Acad. Sci., U.S.A., 1973, **70**, 1041) or the 3′-end (*K. S. Szeto* and *D. Söll*, Nucleic Acids Res., 1974, **1**, 171). Alternatively, the oligonucleotides can be labelled with ^{3}H at the 3′-end by a chemical method (*U. L. RajBhandary*, J. biol. Chem., 1968, **243**, 556). After digestion of the RNA with enzymes, the 2′,3′-diol end groups of the fragments are oxidised with periodate and the 2′,3′-dialdehydes obtained are reduced with sodium [^{3}H]tetrahydridoborate to yield 3′-[^{3}H]-labelled dialcohol derivatives. This enables small amounts of the fragments to be detected throughout the sequential analysis (*K. Randerath* and *E. Randerath*, Methods Cancer Res., 1971, **9**, 3; *K. Randerath et al.*, Nucleic Acids Res., 1974, **1**, 1121).

The method of sequential analysis developed by Sanger has been modified further in order to apply it to the analysis of other types of RNA which do not contain so many modified bases as are found in tRNA and from which therefore fragments are more difficult to identify. Chemical modification of bases in ribosomal 5s RNA aided its structural elucidation (*Brownlee, Sanger* and *Barrell, loc. cit.*). Treatment of the RNA with a water soluble carbodiimide derivative, *N*-cyclohexyl-N^1-[morpholinyl-(4)-ethyl]-carbodiimide methyl-*p*-toluenesulphonate causes modification of uridine residues although cytidine does not react (*P. T. Gilham*, J. Amer. chem. Soc., 1962, **84**, 687). After treatment uridine residues are not susceptible to cleavage by pancreatic ribonuclease so that specific cleavage at cytidine residues only can

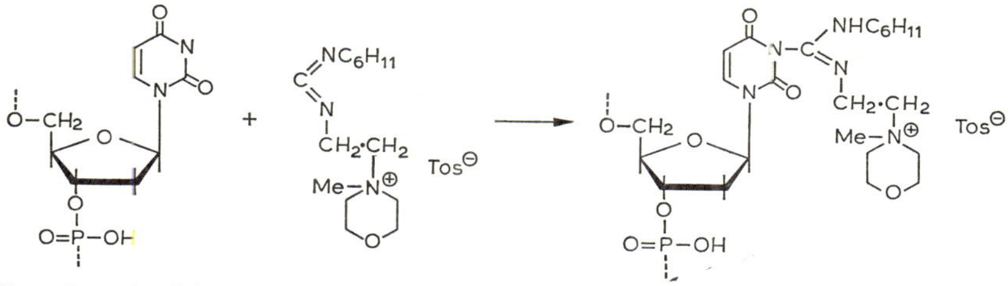

Reaction of uridine with *N*-cyclohexyl-N^1-[morpholinyl-(4)-ethyl]carbodiimide methyl-*p*-toluenesulphonate.

be achieved (*J. C. Lee, N. W. Y. Ho* and *Gilham*, Biochim. Biophys. Acta, 1965, **95**, 503).

Methylation of RNA with dimethyl sulphate preferentially methylates guanosine residues to give 7-methylguanosine which is resistant to ribonuclease T_1 (*P. D. Lawley* and *P. Brookes*, Biochem. J., 1963, **89**, 127). If the RNA is only partially methylated so that not all the guanosine residues are modified, cleavage with ribonuclease T_1 gives rise to larger fragments than are obtained from the untreated RNA.

Rapid methods exactly analogous to those devised for the sequential analysis of large DNA fragments (see pp. 190 *et seq.*), have been devised for RNA sequencing. The "plus and minus" method involves the synthesis of a ^{32}P-labelled DNA molecule complementary to the RNA molecule using the enzyme reverse transcriptase, followed by "plus and minus" reactions with appropriate enzymes (*G. G. Brownlee* and *E. M. Cartwright*, J. mol. Biol., 1977, **114**, 93). Alternatively, a method analogous to the DNA chemical cleavage method of *Maxam* and *Gilbert* has been devised. For RNA, the series of fragments of different chain length terminating in a specific base are obtained by enzymic rather than chemical means (*H. Denis-Keller, Maxam* and *Gilbert*, Nucleic Acids Res., 1977, **4**, 2527; *A. Simonestis et al.*, Nature, 1977, **269**, 833).

Practical details of some of the methods for the sequential analysis of nucleic acids can be found elsewhere (*Brownlee*, "Determination of Sequences in RNA", Laboratory Techniques in Biochemistry and Molecular Biology, eds. *T. S. Work* and *E. Work*, North-Holland, Amsterdam, 1972, Vol. 3).

To date the most notable achievements in this field have been the elucidation of the complete nucleotide sequence of the genome RNA of MS2 virus, 3569 nucleotide residues long (*W. Friers et al.*, Nature, 1972, **237**, 82; 1975; **256**, 273; 1976, **260**, 500) and *Escherichia coli* 16S ribosomal RNA, 1554 residues long (*J. P. Ebel et al.*, Nucleic Acid Res., 1975, **2**, 265; FEBS Letters, 1978, **94**, 152; *J. Brosius et al.*, Proc. Nat. Acad. Sci. Washington, 1978, **75**, 4801).

(v) Secondary structure of RNA

Although most RNA molecules are single-stranded, a number, in particular those isolated from viruses *e.g.* reovirus, are found to be double-stranded. These RNA molecules are distinguished from single-stranded RNA in a number of ways. They obey Chargaff's rules in that the molar proportion of adenine is equal to uracil, and guanine equal to cytosine; they have a low molar extinction; on heating they are converted from a helical form into a random coil; they do not react with formaldehyde which is a characteristic property of double-stranded hydrogen-bonded helices, and they have greater resistance to hydrolysis by ribonuclease. The structure of these double-stranded RNA molecules has been examined by X-ray structural analysis (*M. H. F. Wilkins et al.*, Nature, 1966, **211**, 227; J. mol. Biol., 1967, **27**, 507, 525, 535, 549). Conclusions about the structure of the double-stranded RNA are not so definite as for DNA but it is thought most likely that its double helical structure in solution, and also that of DNA–RNA hybrids, resembles the A-form of DNA (*M. J. B. Tunis* and *J. E. Hearst*, Biopolymers, 1968, **6**, 1218).

Before the primary sequence of the small RNA molecules had been deduced, it was known that single-stranded RNAs, particularly tRNAs, contained a certain amount of secondary structure. This knowledge came mainly from observations of increase in optical density with increase in temperature (*A. Tissiers*, J. mol. Biol., 1960, **1**, 365), a process which is reversible on cooling even when performed rapidly (*P. Doty*, J. Polymer Sci., 1961, **55**, 1). Also treatment of tRNA with formaldehyde indicated that about two-thirds of the bases participate in the formation of base-pairs by hydrogen bonding (*J. T. Penniston* and *Doty*, Biopolymers, 1963, **1**, 145; *G. Zubay* and *R. Marciello*, Biochem. Biophys. Res. Commun., 1964, **14**, 272). When *Holley* deduced the primary structure of $tRNA^{Ala}$, he suggested three possible secondary structures which gave the maximum possible base-pairing (*R. W. Holley et al.*, Science, 1965, **147**, 1462) (Fig. 6).

Of the three possible suggestions the "clover-leaf" is the preferred structure for several reasons, not least of which is that all the primary tRNA sequences which have been deduced can be formed into a clover-leaf arrangement. A few of these structures are shown (Figs. 7–10) and a generalised clover-leaf structure for tRNA has been established (*M. Levitt*, Nature, 1969, **224**, 759) (Fig. 11). This generalised structure has five base-paired regions, a–e, and four loops I–IV. The base-paired region d and the loop III form an "extra arm" which is of variable length. Bases which are inserted in the generalised structure are ones which are located in the same relative positions in most tRNA molecules. The sequence G–T–ψ–C in loop IV is replaced by G–A–U–C in $tRNA^{Met}$ species which act as initiators of protein

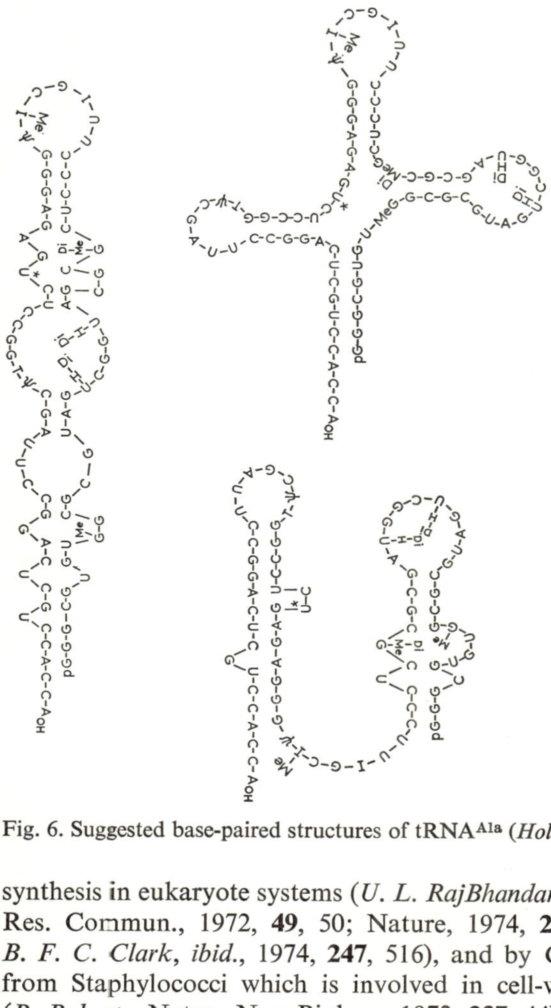

Fig. 6. Suggested base-paired structures of tRNA^{Ala} (*Holley*).

synthesis in eukaryote systems (*U. L. RajBhandary et al.*, Biochem. Biophys. Res. Commun., 1972, **49**, 50; Nature, 1974, **247**, 518; *P. W. Piper* and *B. F. C. Clark*, *ibid.*, 1974, **247**, 516), and by G–U–G–C in the tRNA^{Gly} from Staphylococci which is involved in cell-wall glycoprotein synthesis (*R. Roberts*, Nature New Biology, 1972, **237**, 44). The anticodon trinucleotide, which is found in loop II, is the portion of the tRNA which base-pairs, in an antiparallel manner, with a unique trinucleotide (the codon) in the messenger RNA. The molecule terminates at the 3′-end in a free 2′,3′-*cis*-diol, and it is to this portion of the tRNA that the amino acid is covalently joined by an ester linkage. Originally the amino acid was thought to be attached to the 3′-OH, but it is now known that some amino acids are esterified to the 2′-OH position (*T. H. Frazer* and *A. Rich*, Proc. Nat. Acad.

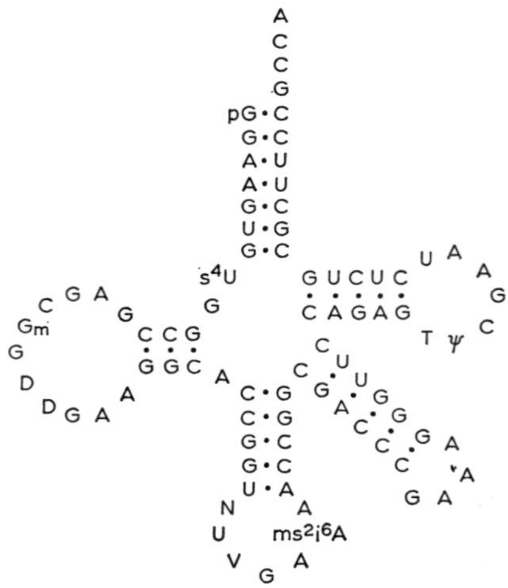

Fig. 7. The nucleotide N is unknown. $tRNA_1^{Ser}$ from *E. coli* (*H. Ishikura*, *Y. Yamada* and *S. Nishimura*, FEBS Letters, 1971, **16**, 68).

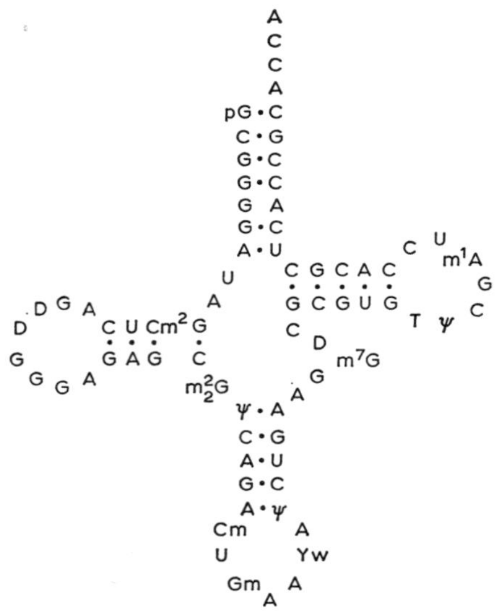

Fig. 8. Yw is a derivative of Y base (see p. 119), $tRNA^{Phe}$ from wheat germ (*R. W. Holley et al.*, Proc. Nat. Acad. Sci., Wash., 1969, **62**, 941).

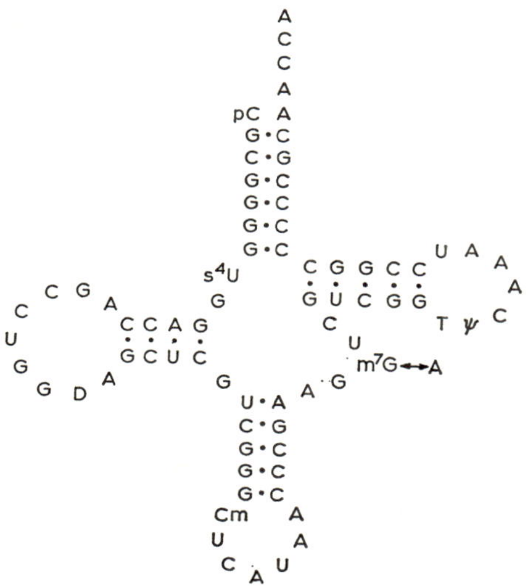

Fig. 9. tRNA$_f^{Met}$ from *E. coli* (*S. K. Dube et al.*, Nature, 1968, **218**, 232). The subscript "f" in tRNA^{111} indicates that this species of tRNA can be formylated enzymically and it is the protein-synthesis initiator tRNA.

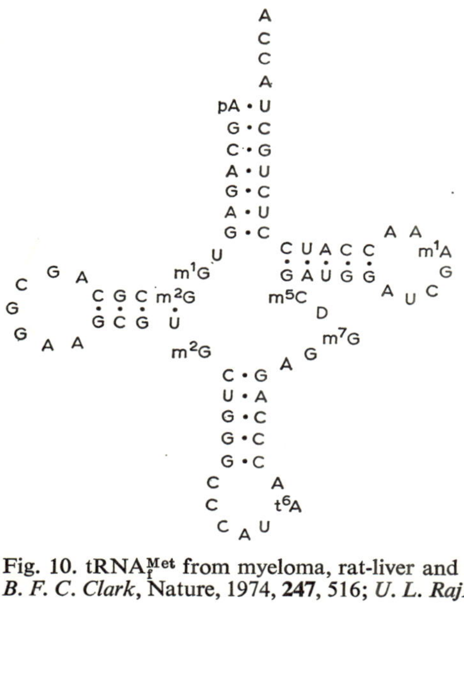

Fig. 10. tRNA$_f^{Met}$ from myeloma, rat-liver and sheep-mammary gland (*P. W. Piper* and *B. F. C. Clark*, Nature, 1974, **247**, 516; *U. L. RajBhandary et al.*, *ibid.*, 1974, **247**, 518).

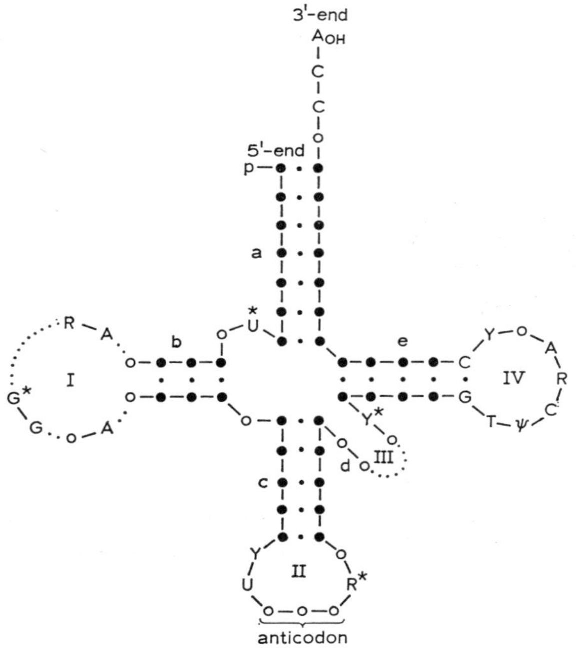

Fig. 11. Generalised tRNA clover-leaf structure.

Sci., Wash., 1975, **72**, 3044; *M. Sprinzl* and *F. Cramer*, *ibid.*, p. 3049).

Support for the clover-leaf structure is provided by the reaction of tRNA with specific chemical reagents such as monoperphthalic acid which reacts specifically with A residues (*Cramer et al.*, *ibid.*, 1968, **61**, 1384), water-soluble carbodiimide (G and U residues) (*N. W. Y. Ho* and *P. T. Gilham*, Biochemistry, 1967, **6**, 3249), kethoxal (G residues) (*M. Litt*, *ibid.*, 1969, **8**, 3249), methoxyamine (C residues) (*A. R. Cashmore*, *D. M. Brown* and *J. D. Smith*, J. mol. Biol., 1971, **59**, 395) and acrylonitrile (pseudouridine and thiouridine) (*M. Yoshida* and *T. Ukita*, Biochim. Biophys. Acta, 1966, **123**, 214). These reagents react preferentially with the bases in the loops which are not base-paired. However the common T–ψ–C sequence is unreactive.

As the primary structures of other types of RNA molecules have been elucidated these too have been designated secondary structures which permit extensive base-pairing. Two structures, one proposed for 16S ribosomal RNA (Fig. 12) and the other for part of the sequence of MS2 RNA which codes for the coat protein (Fig. 13), are shown. It will be seen that in addi-

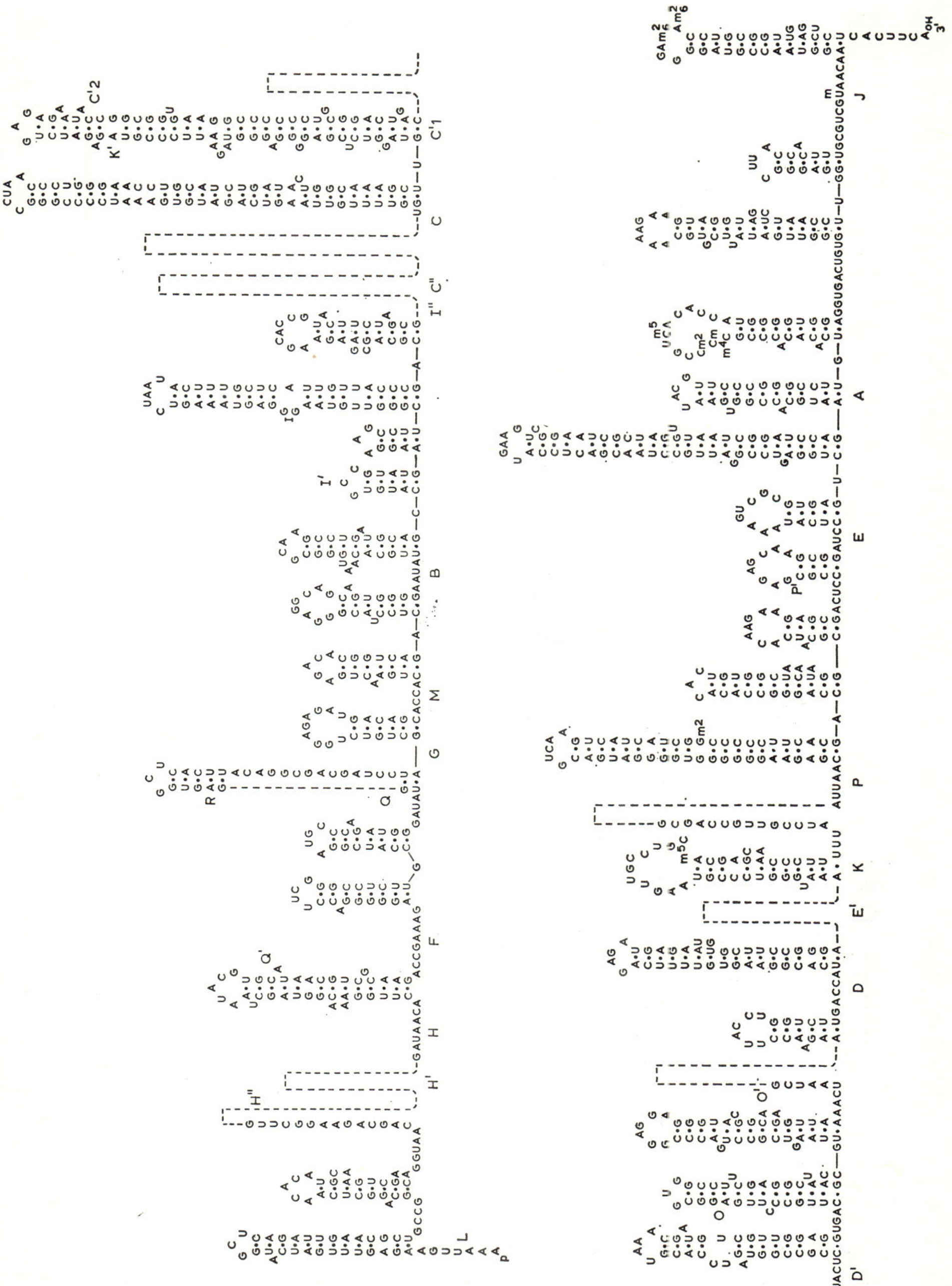

Fig. 12. The sequence of bases in 16S ribosomal RNA from *E. coli* showing the suggested secondary structure. The dashed lines show postulated loops for ambiguous areas (*J. P. Ebel et al.*, Nucleic Acids Res., 1975, **2**, 256).

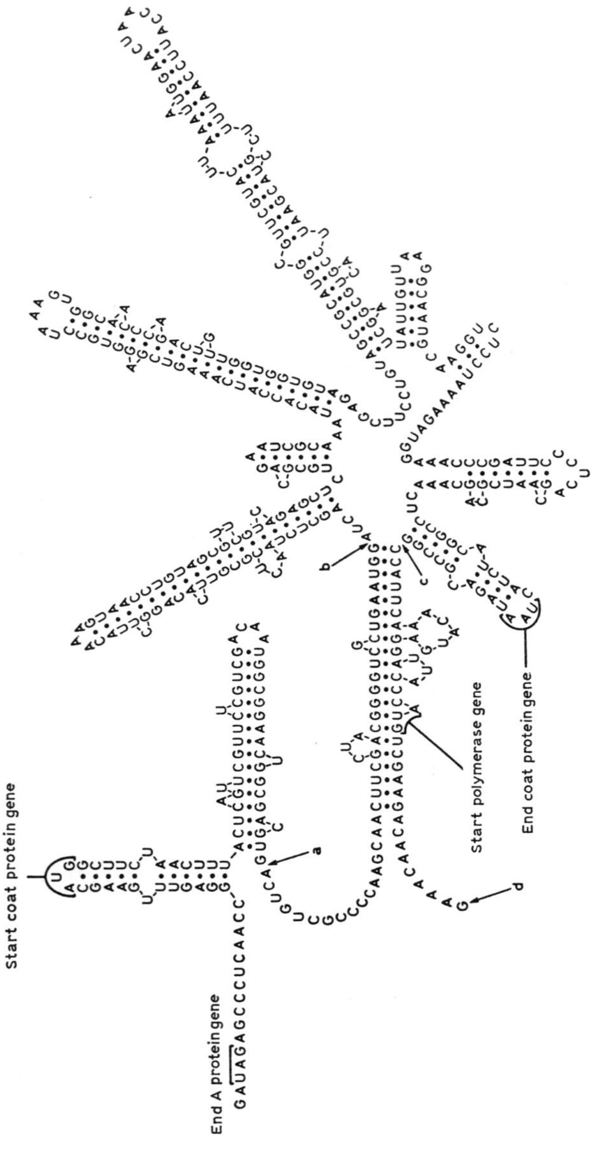

Fig. 13. The primary sequence of nucleosides in the part of MS2 RNA which codes for coat protein showing the proposed secondary structure (*W. Friers et al.*, Nature, 1972, **237**, 82).

tion to the usual "Watson–Crick" base-pairing, *i.e.* adenine with uracil and guanine with cytosine, a base-pair between guanine and uracil is suggested in each of the types of RNA mentioned above.

(vi) The tertiary structure of RNA

The RNA molecule for which most progress has been made in our understanding of the tertiary structure is transfer RNA. Information has been provided from both chemical and physical studies. Examination of data obtained shortly after the clover-leaf secondary structure was proposed suggested that the transfer RNA has a folded tertiary structure. As mentioned previously the common sequence G–T–ψ–C although assigned to a loop in the generalised clover-leaf structure is not available for chemical reaction which suggests that it is shielded in the tertiary structure. Also when tRNA is thermally denatured it is found that changes in the viscosity and sedimentation properties, which are determined by the shape and volume of the molecule, occur before significant changes in the optical density, which is determined by the interaction between the base-pairs in double-stranded regions (*J. R. Fresco et al.*, Proc. Nat. Acad. Sci., U.S.A., 1966, **55**, 191; Cold Spring Harbor Symp. quant. Biol., 1966, **31**, 527). These results sug-

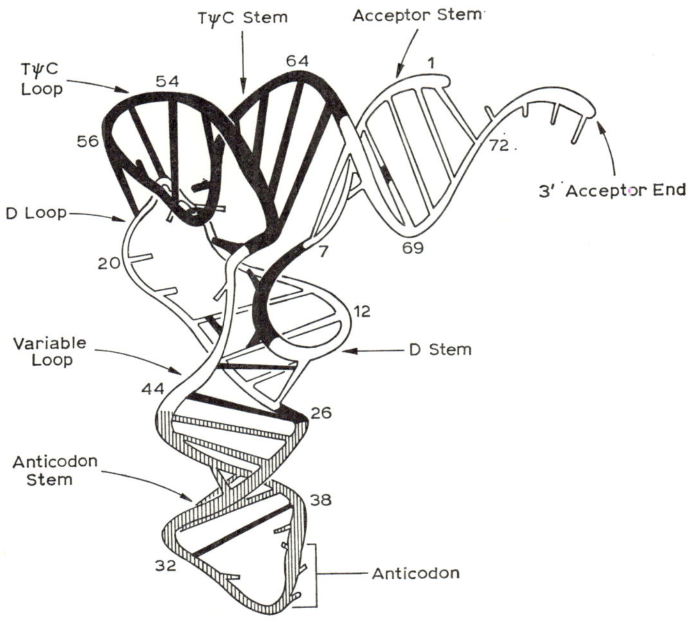

Fig. 14. Taken from *Rich et al.* (Nucleic Acids Res., 1975, **2**, 2329).

gest that there is a change in over-all shape before breakdown of the helical portions of the molecule.

X-Ray diffraction analyses at 2.5 Å resolution of two different crystal forms of yeast $tRNA^{Phe}$ has made available the atomic coordinates, from which the three-dimensional structures of the tRNA in the orthorhombic crystal (Fig. 14) (*A. Rich et al.*, Nucleic Acids Res., 1975, **2**, 2329; Proc. Nat. Acad. Sci., U.S.A., 1975, **72**, 4866) and monoclinic crystal (Fig. 15) (*A. Klug et al.*, Nucleic Acids Res., 1975, **2**, 1629; Proc. Nat. Acad. Sci., U.S.A., 1975, **72**, 4414; J. mol. Biol., 1976, **108**, 619) have been determined.

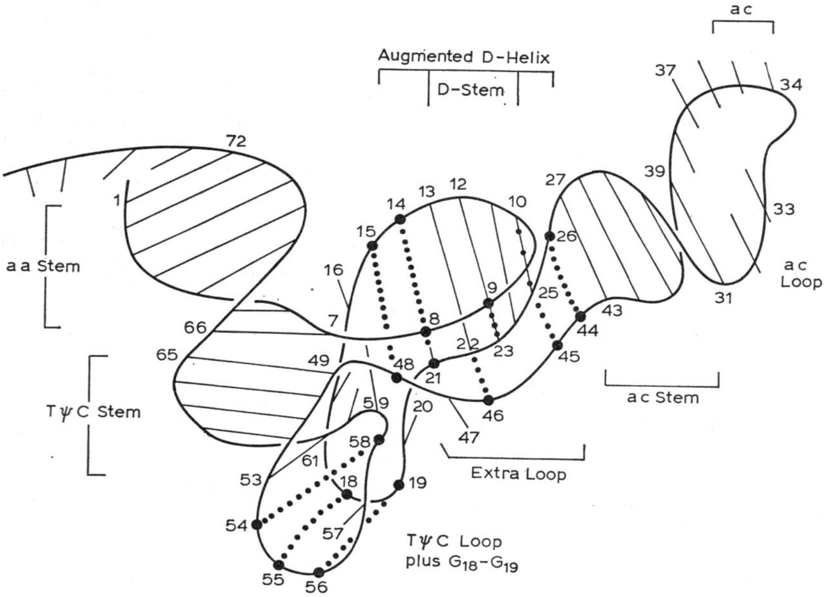

Fig. 15. Taken from *Klug et al.* (Proc. Nat. Acad. Sci., U.S.A., 1975, **72**, 4414).

The three-dimensional structures thus determined are refinements of structures determined earlier using analyses from X-ray diffraction at lower resolution. (References to this earlier work are cited in the references given above.) The two structures are almost identical, the minor differences being in the conformation of the 3′-terminal residues $C_{(75)}$ and $A_{(76)}$. The structure takes the shape of an L, the two arms being formed by helices. One of these is composed in the main, of the a and e base-paired regions (see Fig. 11) which takes the form of an RNA A-type helix. The other is composed in the main of the b and c base-paired regions. All but four of the bases stack in these two helical regions. Loop IV folds over loop I at the corner of the

L-shape. The molecule is stabilised by many tertiary interactions, for example, hydrogen bonding between the invarient Gs in loop I and ψ and C in loop IV. Since in loop IV the T and the A are also hydrogen-bonded this explains why the loop is resistent to chemical reaction.

Evidence from chemical and other physical methods indicates that the three-dimensional structure of the molecule in solution is consistent with that determined in the crystal. An examination of the bases which are able to react with specific chemical reagents (see p. 210) shows that these are all in exposed positions (*J. D. Robertus et al.*, Nucleic Acids Res., 1974, **1**, 927).

Transfer RNA has been extensively analysed by n.m.r. spectroscopy. Hydrogen-bonded ring N–H hydrogen atoms can be detected (*D. R. Kearns, D. Patel* and *R. G. Schulman*, Nature, 1971, **229**, 338; J. mol. Biol., 1971, **61**, 265), and this enables the number of hydrogen bonds in tRNA to be determined. A value of 26 $\pm$ 1 was obtained for yeast $tRNA^{Phe}$ (*B. R. Reid* and *G. T. Robillard*, Nature, 1975, **257**, 287). Since the number of base pairs in the clover-leaf secondary structure is 21, it suggests that there are at least six tertiary base pairs, which is in general agreement with the crystallographic three-dimensional structure. A review of the use of n.m.r. spectroscopy in tRNA structural analysis can be found elsewhere (*Kearns*, Prog. Nucleic Acid Res., mol. Biol., 1976, **18**, 91).

Although the three-dimensional structure of only one tRNA molecule has been determined, both chemical and physical studies on other tRNAs suggest that they all have similar structures. The results of chemical modification work on other tRNAs are similar to that of yeast $tRNA^{Phe}$. In *Escherichia coli* $tRNA^{Val}$ a photo-induced covalent cross link occurs between $C_{(13)}$ and S^4U_8 (*M. Yaniv*, *A. Favre* and *B. G. Barrell*, Nature, 1969, **223**, 1331). In the yeast $tRNA^{Phe}$ structure bases in these positions are suitably sited for chemical interaction. Approximately the same number of N–H hydrogen bonds are detected in *E. coli* $tRNA^{Met}$ (*W. E. Daniel* and *M. Cohen*, Proc. Nat. Acad. Sci., U.S.A., 1975, **72**, 2582) as are found in yeast $tRNA^{Phe}$ as determined by n.m.r. spectroscopy. Mixtures of tRNA molecules form high quality crystals which suggest that they all have a similar tertiary structure (*Fresco, R. D. Blake* and *R. Longridge*, Nature, 1968, **22**, 1285; 1970, **225**, 32). It is interesting to note that many of the bases which are located in the same relative positions in most tRNA molecules are ones which are involved in tertiary base-pairing interactions.

(c) *Synthesis of polynucleotides*

The general approach used for the synthesis of polynucleotides is similar to that for other polymeric compounds. Consideration has to be given to the protection of functional groups in the mononucleotides and nucleosides which are not to take part in the formation of phosphodiester linkages to suitable condensing reagents and conditions and to selective removal of protecting groups at different stages of the synthesis. Much greater progress has been made in the synthesis of deoxyribonucleotide polymers than in the synthesis of ribonucleotide polymers for two reasons. Firstly in the ribo-series, the protection of the 2′-hydroxyl group in addition to the other functional groups common to the deoxyribonucleotides has presented a more complex problem and, secondly, care is required to prevent cyclisation at the 2′,3′-positions of the ribonucleotide.

The preparation of thymidylyl–thymidine (TpT) and thymidylyl–thymidine 5-phosphate (pTpT) by condensation of appropriately protected thymidine- or thymidine 5′-phosphate-3′-benzylphosphorochloridates with 3′-*O*-acetyl-thymidine (*A. M. Michelson* and *A. R. Todd*, J. chem. Soc., 1955, 2632) represented the first chemical synthesis of a dinucleotide containing a specific 3′→5′-internucleotide linkage. Since that time synthetic methods have been developed, to the extent that polydeoxyribonucleotides representing the double-stranded DNA sequence for $tRNA^{Ala}$ (*H. G. Khorana et al.*, Nature, 1970, **227**, 27) and for $tRNA^{Tyr}$ precursor molecule (*idem*, J. biol. Chem., 1976, **251**, 565) have been synthesised.

The most common condensation reagents used for the formation of the phosphodiester bond between the nucleotides are dicyclohexylcarbodiimide (DCC) (I) and aromatic sulphonyl chlorides. The mechanism of activation of the phosphate is complex and in the case of dicyclohexylcarbodiimide the initial phosphorylating species is a trimetaphosphate (II) (*G. Weimann* and *Khorana*, J. Amer. chem. Soc., 1962, **84**, 4329). Phosphodiester bond formation then takes place by nucleophilic attack on the trimetaphosphate by the hydroxyl group of the second component.

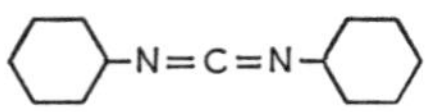

Dicyclohexylcarbodiimide
(DCC)
(I)

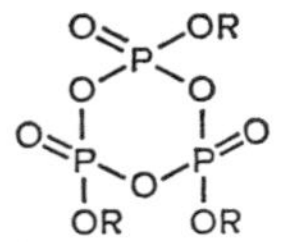

Trimetaphosphate
(II)

The most common aromatic sulphonyl chloride type condensation reagents used are mesitylenesulphonyl chloride (MS) (III) (*T. M. Jacob* and *Khorana*, *ibid.*, 1964, **86**, 1630) and 2,4,6-tri-isopropylbenzenesulphonyl chloride (TPS) (IV) (*R. Lohrmann* and *Khorana*, *ibid.*, 1966, **88**, 829). The latter is preferable since sulphonation of nucleoside residues is prevented because of the steric hindrance of the bulky groups attached to the benzene

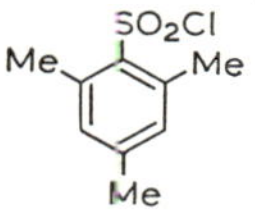

Mesitylenesulphonyl chloride (MS)

(III)

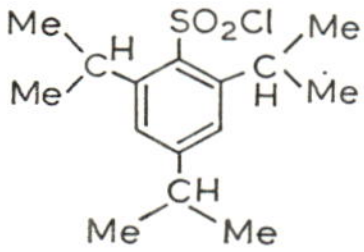

2,4,6-Triisopropylbenzene-sulphonyl chloride (TPS)

(IV)

ring. Trimetaphosphates are not intermediates when the phosphate group is activated by these sulphonyl chlorides (*Jacob* and *Khorana*, *loc. cit.*). The condensation reactions are carried out in anhydrous conditions, pyridine being the usual solvent.

Other reagents which have been used for the synthesis of oligonucleotides include trichloroacetonitrile (*F. Cramer*, *W. Rittersdorf* and *W. Böhm*, Ann., 1962, **654**, 180), Woodward's reagent (V) (*Cramer et al.*, Angew. Chem. intern. edn., 1962, **1**, 331), picryl chloride (*idem*, *ibid.*, 1963, **2**, 43), a mixture of triphenylphosphine and 2,2′-dipyridyl disulphide (*T. Mukaiyama* and

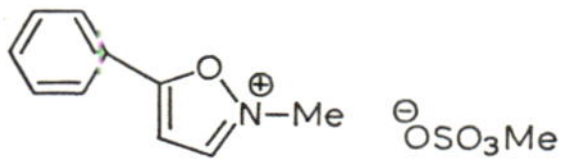

Woodward's reagent
N-methyl-5-phenylisoxazolium methosulphate

(V)

M. Hashimoto, J. Amer. chem. Soc., 1972, **94**, 8528) and, more recently, mesitylene- and 4-nitrobenzene-sulphonyltriazoles (*N. Katagiri*, *K. Itakura* and *S. A. Narang*, *ibid.*, 1975, **97**, 7332).

Two approaches may be taken for the formation of the internucleotide bond. Either a nucleoside 3′-phosphate can be polymerised or reacted with the 5′-hydroxyl group of the second (nucleoside) component, or a nucleoside 5′-phosphate can be used for polymerisation or reacted with the 3′-hydroxyl group of the second component of the reaction. For the synthesis of oligodeoxyribonucleotides, the latter approach is more common. The synthesis of the simplest dideoxyribonucleoside phosphate, thymidylyl–thymidine (TpT) is illustrated on p. 218 (*P. T. Gilham* and *Khorana*, *ibid.*, 1958, **80**, 6212).

It will be seen that the 5′-hydroxyl group of the nucleoside components is blocked by the bulky trityl group which, when desired is removed under acid conditions, and the 3′-hydroxyl group of the 5′-mononucleotide components is blocked by an acetyl group. The acetyl group is removed under alkaline conditions when required. When nucleosides and nucleotides other than thymidine and thymidylic acid are used it is necessary to block their amino groups, otherwise they will react with the activated phosphates. These pro-

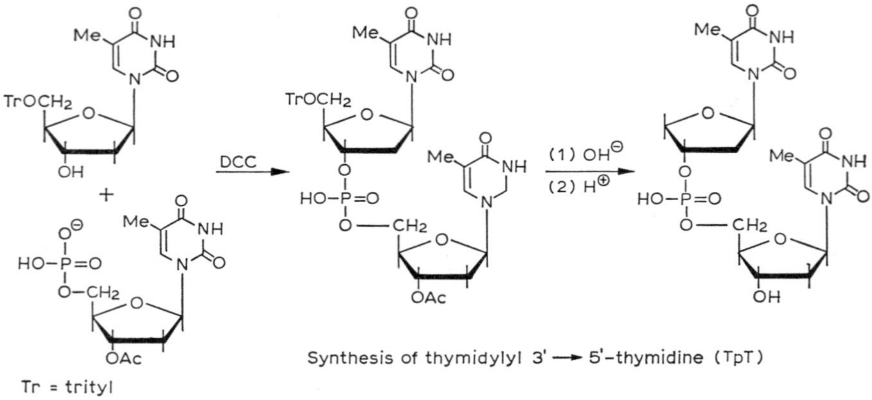

Synthesis of thymidylyl 3' ⟶ 5'-thymidine (TpT)

tecting groups must be stable to conditions required for the removal of the 3'- and 5'-hydroxyl blocking groups and be able to survive all the reactions involved in the synthesis of the oligonucleotide. Also conditions for their removal must not affect the internucleotide linkage or the glycosyl bonds. Amino-protecting groups commonly used are anisoyl for cytosine (*Khorana, A. F. Truner* and *J. P. Vizsolyi*, J. Amer. chem. Soc., 1961, **83**, 686), benzoyl for adenine (*R. K. Ralph* and *Khorana*, ibid., 1961, **83**, 2926), and acetyl, isobutyryl and 2-methylbutyryl (*Khorana et al., ibid.*, 1963, **85**, 1983; *H. Büchi* and *Khorana*, J. mol. Biol., 1972, **72**, 251) for guanine. The protecting groups are removed satisfactorily by treatment with concentrated ammonium hydroxide. While the trityl group is a satisfactory blocking group for the 5'-hydroxyl group of the nucleoside component when oligonucleotides containing pyrimidine bases only are being prepared, the acidic conditions required for its removal cause some cleavage of glycosyl bonds in purine nucleotides. Introduction of *p*-methoxy groups into the trityl moiety increases its lability to acid (see p. 158) and both monomethoxy- and dimethoxy-trityl groups have proved satisfactory blocking groups for protection of the 5'-hydroxyl group (*Khorana et al.*, J. Amer. chem. Soc., 1963, **85**, 3821). The most common 3'-hydroxyl blocking group is acetyl which is removed selectively with alkali under carefully controlled conditions. This selective hydrolysis is possible because ionisation of the amide group in the aromatic purine or pyrimidine rings at alkaline pH extends the resonating system and stabilises the acyl group on the amino function (*idem*, Angew. Chem. intern. Edn., 1972, **11**, 451).

Using the protecting groups described above, and a stepwise procedure whereby suitably protected deoxyribonucleoside 3'-phosphate derivatives are added to a growing oligonucleotide "nucleoside component", thus extending the chain from the 5′-end of the molecule, many oligonucleotides have been synthesised (*Khorana*, Pure and Applied Chem., London, 1968, **17**, 349). During the synthesis, chromatographic methods are used to purify the products after each coupling reaction. The scheme below for the synthesis of

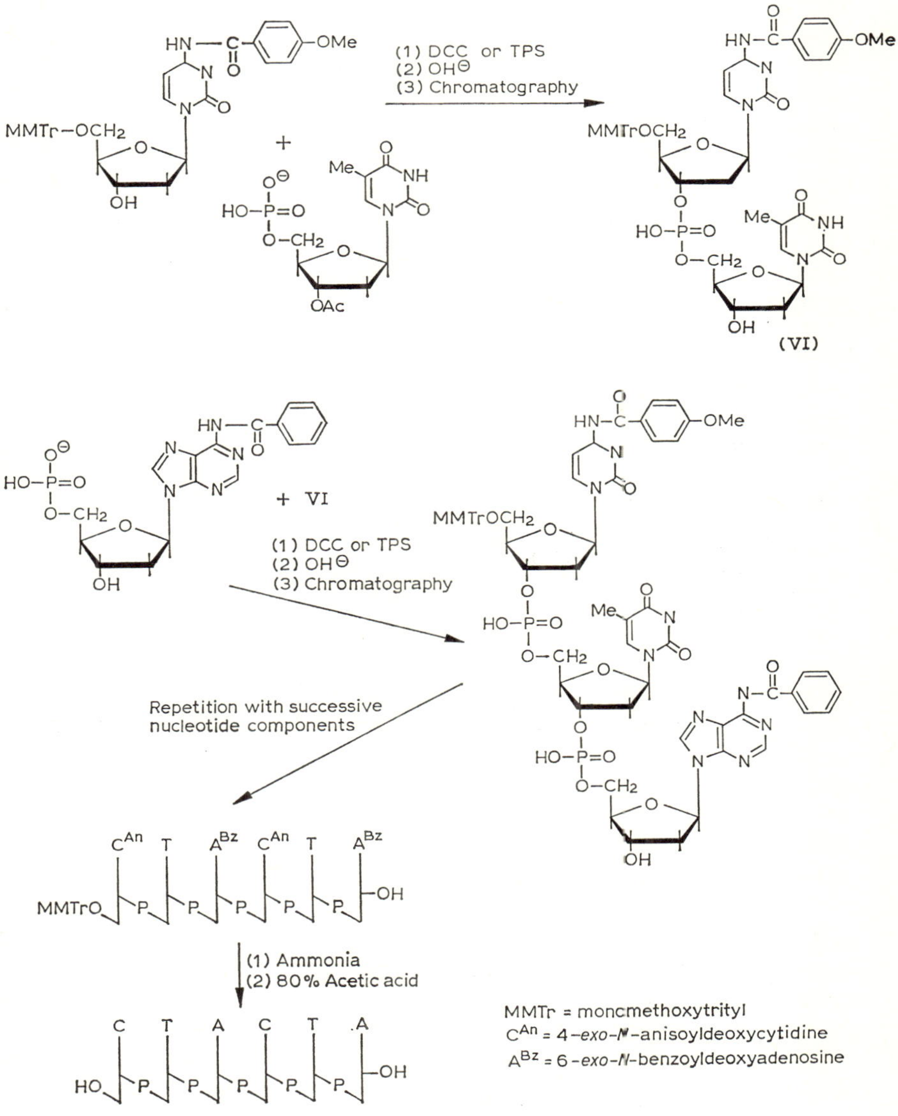

a hexanucleoside pentaphosphate containing the deoxyribonucleoside residues cytosine, thymidine and adenosine illustrates the use of protecting groups in the stepwise synthesis of an oligonucleotide of defined sequence (*T. M. Jacob* and *Khorana*, J. Amer. chem. Soc., 1965, **87**, 2971; *Khorana*, Angew. Chem. intern. Edn., 1972, **11**, 451).

One difficulty which sometimes arises during the synthesis of oligonucleotides by methods outlined above, is the activation of internal phosphodiester bonds which leads to the formation of side-products such as pyrophosphates (*Khorana*, "Some Recent Developments in the Chemistry of Phosphate Esters of Biological Interest", Wiley, New York, London, 1961, p. 100). One approach which overcomes this difficulty is to use the "phosphotriester" method of synthesis. Here an activated phosphodiester is used as the nucleotide component instead of a monoesterified component which, after reaction with the nucleoside component, yields a product having the nucleotide linkage in the form of a triester. The "nucleotide" component

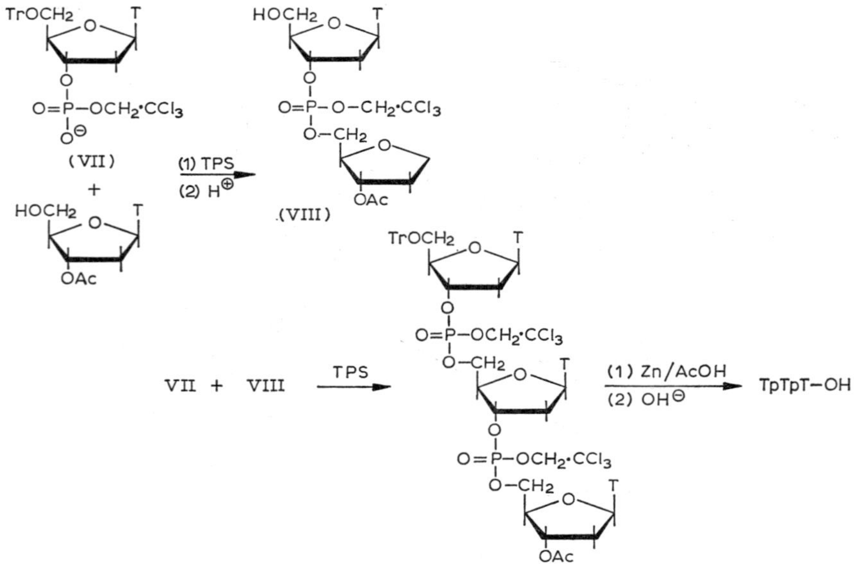

in this method is a nucleoside 3′-phosphate in which a second acidic group of the phosphate is esterified. The synthesis of thymidylyl-(3′→5′)-thymidylyl-(3′→5′)-thymidine (TpTpT) has been synthesised by this method as shown (*F. Echstein* and *I. Rizk*, Angew. Chem. intern. Edn., 1967, **6**, 695).

In this method 5′-*O*-tritylthymidine 3′-(*β*,*β*,*β*-trichloroethyl) phosphate (VII) is prepared by treatment of 5′-tritylthymidine 3′-phosphate with *β*,*β*,*β*-

trichloroethylphosphodichloride (*Echstein* and *Rizk*, *ibid.*, 1967, **6**, 362). The trichloroethyl group can be selectively removed from an oligonucleotide by treatment with zinc or zinc/copper in acetic acid or dimethylformamide. Other groups which have been used to protect the internucleotide bond in the phosphotriester synthetic method are cyanoethyl (*R. L. Letsinger* and *K. K. Ogilvie*, J. Amer. chem. Soc., 1967, **89**, 4801), phenyl (*C. B. Reese* and *R. Saffill*, Chem. Comm., 1968, 767) and phenylthio-groups (*T. Hata* and *M. Sekine*, J. Amer. chem. Soc., 1974, **96**, 7363), which can be removed under mild alkaline conditions. Besides protecting the phosphate group of the internucleotide bond, the triester method offers the advantage that the products are soluble in organic solvents. This means that ion-exchange chromatography can be replaced by chromatography on silica gel which has a higher capacity. For analytical purposes the products can be separated on t.l.c. or by high performance liquid chromatography (HPLC) when speed of running and resolution of compounds are improved.

Homo-oligonucleotides can be prepared by condensation of a suitably protected mononucleotide 5′-phosphate with dicyclohexylcarbodiimide or picryl chloride. Using the latter reagent polythymidylic acid containing 30 residues can be obtained (*F. N. Hayes* and *E. Hansbery*, J. Amer. chem. Soc., 1964, **86**, 4172). Using DCC, shorter oligomes of deoxyadenylic acid (*R. F. Ralf* and *H. G. Khorana*, *ibid.*, 1961, **83**, 2926), deoxycytidylic acid (*Khorana*, *A. F. Turner* and *J. P. Vizsolyi*, *ibid.*, 1961, **83**, 686) and deoxyguanylic acid (*Khorana et al.*, *ibid.*, 1963, **85**, 1983) can be obtained from the protected mononucleotides. After condensation, the protecting groups are removed by treatment of the protected oligomer with concentrated ammonia.

Oligonucleotides which contain repeating di- (*E. Ohtsuka*, *M. W. Moon* and *Khorana*, *ibid.*, 1965, **87**, 2956), tri- (*S. A. Narang*, *T. M. Jacob* and *Khorana*, *ibid.*, 1967, **89**, 2167; *F. Cramer*, *W. Frolke* and *H. Matzwa*, Angew. Chem., 1967, **79**, 580) and tetra-nucleotides (*Jacob*, *Narang* and *Khorana*, J. Amer. chem. Soc., 1967, **89**, 2177; *H. Kossel*, *H. Büchi* and *Khorana*, *ibid.*, 1967, **89**, 2185) can be prepared by polymerisation of suitably protected di-, tri- and tetra-nucleotides having a 5′-phosphate, followed by removal of the protecting groups. An example is the oligonucleotide containing thymidine and guanosine in alternating sequence which is prepared by treating 5′-phosphothymidylyl-(3′→5′)-2-*exo*-*N*-acetyldeoxyguanosine with dicyclohexylcarbodiimide.

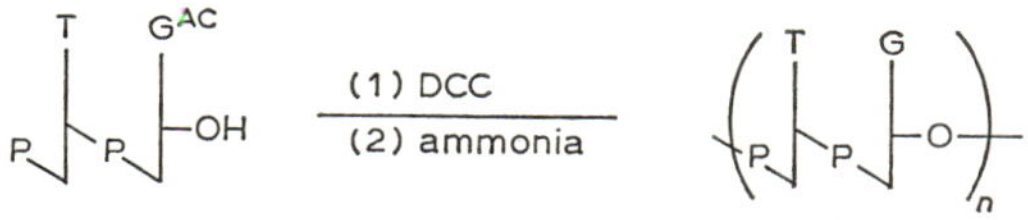

It will be seen that the oligonucleotides used in these polymerisation experiments have a 5′-phosphate group. For the synthesis of oligonucleotides of this type it is necessary to use a 5′-mononucleotide in which the phosphate group is protected in place of the "nucleoside component" used in the stepwise procedure described previously. This can be done by condensation of the amino-protected nucleotide with cyanoethanol using DCC. The cyanoethylated protected 5′-mononucleotide is then condensed with a suitable second nucleotide using DCC or an aromatic sulphonyl chloride. The fully protected dinucleotide is treated with mild alkali which selectively removes the cyanoethyl group and acetyl group from the 3′-hydroxyl posi-

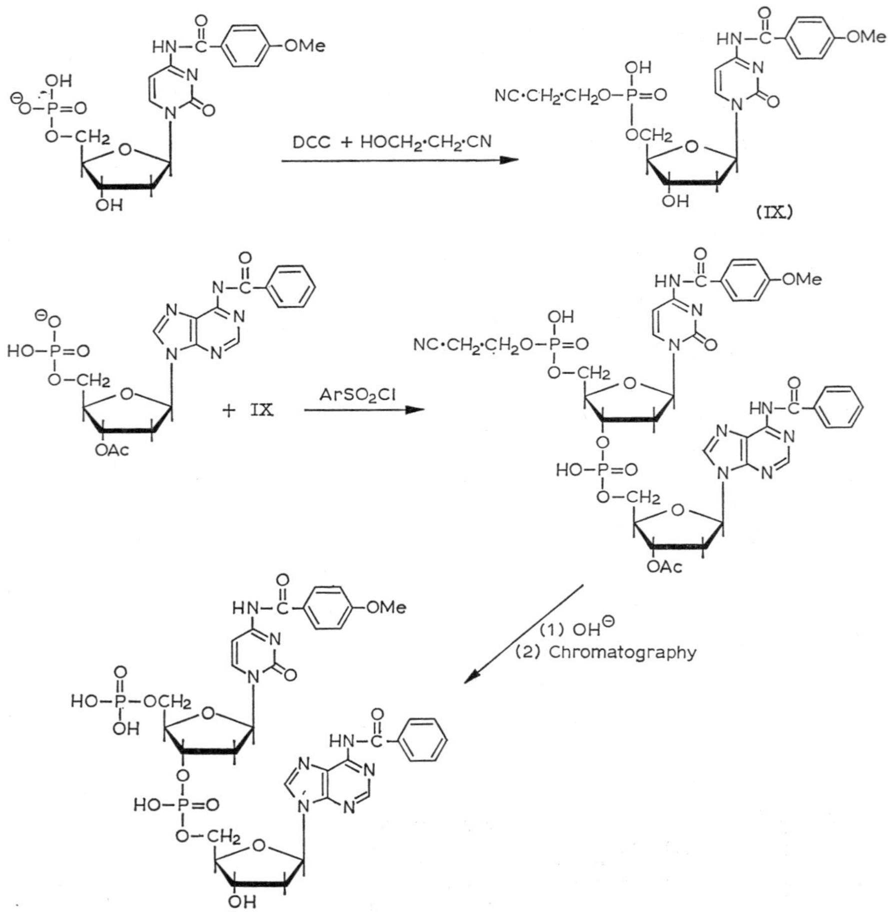

tion. The synthesis shown illustrates the procedure (*Ohtsuka, Moon* and *Khorana, loc. cit.*).

Other phosphate protecting groups are the trichloroethyl group (*R. B. Woodward et al.*, J. Amer. chem. Soc., 1966, **88**, 852; *Eckstein*, Angew. Chem. intern. Edn., 1966, **5**, 671) or aromatic phosphoramidates (*Ohtsuka et al.*, J. Amer. chem. Soc., 1970, **92**, 3441). The amidate can be removed at pH 7 under mild conditions with isoamyl nitrite. An extension of the latter method

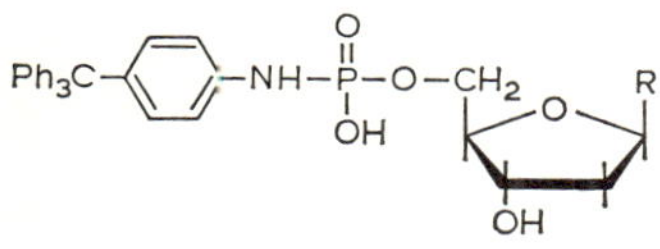

(X)

R = thymine, 4-*exo*-*N*-anisoylcytosine, 6-*exo*-*N*-benzoyladenine or 2-*exo*-*N*-isobutyrylguanine

of protection uses the highly lipophilic amine *p*-(triphenylmethyl)aniline which when condensed with a suitably protected nucleoside 5′-phosphate derivative gives compounds of the type X.

When this type of phosphoramidate derivative is used, the mononucleotide and dinucleotide products obtained during synthesis can be purified by extraction into organic solvents, which avoids the time-consuming chromatographic procedures (*K. L. Agarwal, A. Yamazaki* and *Khorana, ibid.*, 1971, **93**, 2754).

Another protecting group which has gained popularity is the 2-arylthioethyl group (XI) (*Narang et al., ibid.*, 1972, **94**, 6183). In its reduced form the group shows considerable stability to a variety of reaction conditions, however, when oxidised to the sulphoxide XII with peroxide or more preferably to the sulphone XIII with *N*-chlorosuccinimide (NCS) (*Khorana et al., ibid.*, 1973, **95**, 2020) the oxidised form of the protecting group is removed under

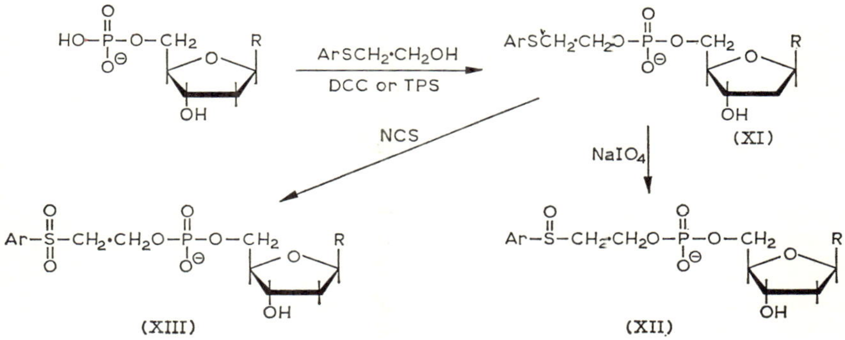

mild basic conditions. If 2-(4-tritylphenyl)thioethanol, 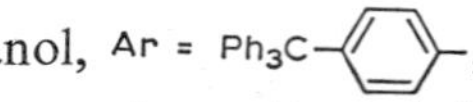, is used as the protecting group again the products may be extracted into organic solvents (*Khorana et al., ibid.*, 1976, **98**, 1065).

Besides being used in polymerisation experiments, the short oligonucleotide derivatives which have 5′-phosphate groups are used as preformed blocks in the synthesis of polynucleotides of defined seqence (*Kössel, Moon* and *Khorana, ibid.*, 1967, **89**, 2148). This method is faster than the stepwise approach but the yields at the condensation steps are generally lower. It is the preferred method provided that the preformed, protected oligonucleotides are readily prepared in quantity. This method was largely used in the chemical synthesis of protected polynucleotide derivatives which lay the foundation for the synthesis of the genes for $tRNA^{Ala}$ (*Khorana et al.*, Nature, 1970, **227**, 27; J. mol. Biol., 1972, **72**, 209, 219, 251, 289, 309, 329, 351, 375, 407, 427, 445, 457, 475) and pre-$tRNA^{Tyr}$ (*idem*, J. biol. Chem., 1976, **251**, 565, 571, 587, 599, 609, 624, 634, 642, 651, 658, 667, 676). The genes are double-stranded DNA sequences, in which one strand has the same base sequence and the other the complementary sequence to the tRNAs except that thymine in DNA replaces uracil in RNA and the bases are not modified. Khorana's approach was to chemically synthesise large oligonucleotides corresponding to fragments in each of the DNA strands in such a way that the fragments in each strand of the DNA overlapped by 4–6 nucleotides. These oligonucleo-

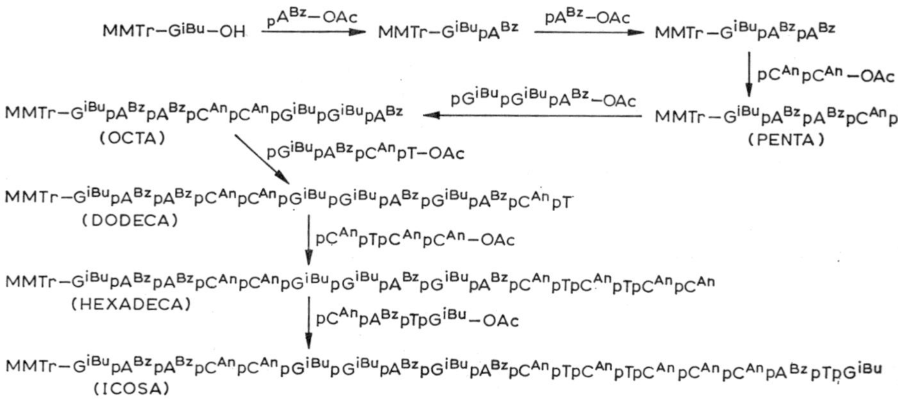

Steps in the blockwise chemical synthesis of an icosadeoxyribonucleotide. MMTr indicates a monomethoxytrityl group at the 5′-hydroxyl end of the terminal nucleoside. The protecting groups on the heterocyclic rings of different deoxyribonucleosides are shown by superscripts on the nucleoside initial: An = anisoyl; Bz = benzoyl; iBu = isobutyl; OAc at the right-hand end of oligonucleotides stands for 3′-*O*-acetyl. (Taken from *Khorana et al.*, Nature, 1970, **227**, 27.)

tides were then deprotected and phosphorylated enzymically at the 5′-end using polynucleotide kinase. After being positioned in turn by base-pairing the fragments were covalently joined enzymically using the bacteriophage T4 DNA ligase. The method is illustrated on p. 224, where the synthesis of the icosanucleotide complementary to residues 21–40 from the 3′-end of tRNA^{Ala} (XIV), and below where the joining of the fragments of the tRNA^{Ala} gene (XV), are outlined.

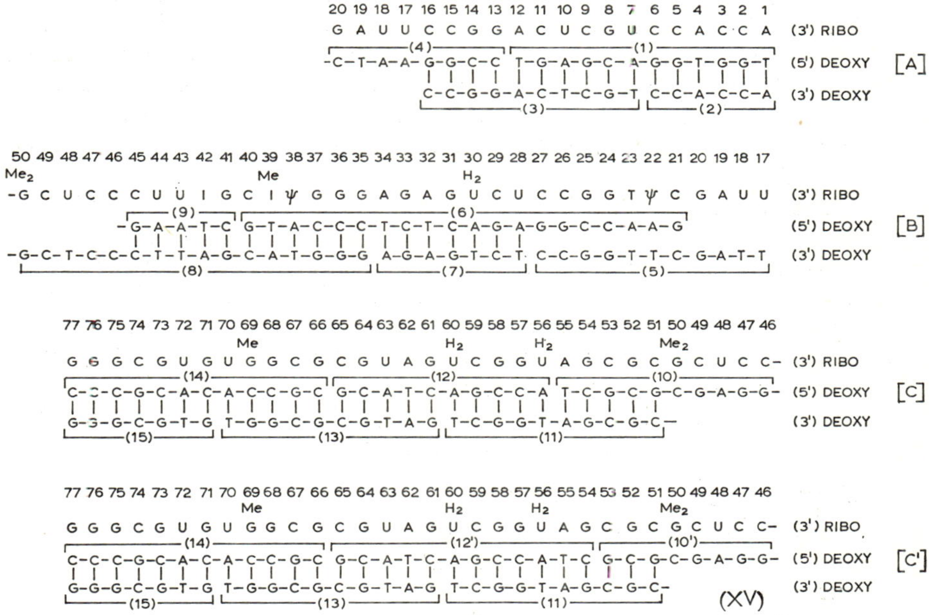

(XV)

Total plan for the synthesis of a yeast alanine tRNA gene. (Taken from *Khorana et al., loc. cit.*)

Following the success in the synthesis of peptides on polymer supports (*R. B. Merrifield*, Advan. Enzymol., 1969, **32**, 221) this approach has been investigated for the synthesis of oligonucleotides. Most of the polymers used in the earlier work were of one of the polystyrene types. These are either insoluble with a low degree of cross-linking so that they swell in the reaction solvent (*R. L. Letsinger* and *V. Mahadevan*, J. Amer. chem. Soc., 1965, **87**, 3526; 1966, **88**, 5319), insoluble with a high degree of cross-linking, which do not swell (*F. Cramer* and *H. Koster*, Angew. Chem. intern. Edn., 1968, **7**, 473) or soluble (*Cramer et al.*, *ibid.*, 1966, **5**, 601; *H. Hayatsu* and *Khorana*, J. Amer. chem. Soc., 1967, **89**, 3880). The latter soluble types are soluble in pyridine, but not in water, alcohol or ether so that the product attached to the polymer can be precipitated at the end of the reaction by pouring the reaction mixture into one of these solvents. The use of a polymer support has been applied to both the diester (*idem, loc. cit.*) and the triester method of synthesis (*Letsinger* and *Mahadevan, loc. cit.*). One difficulty with the solid-phase method of synthesis is that yields in the formation of into-nucleotide bonds are lower than those obtained in solution, which results in complex mixtures being obtained even after only a few coupling steps. This is partly due to the incompatibility of the non-polar polymer and the polar reaction media. In order to overcome this several more polar polymers have been introduced. One of these is a polydimethylacrylamide resin, which has been used to synthesise successfully two heptadeoxyribonucleotides using the phosphodiester approach (*M. J. Gait* and *R. C. Sheppard*, Nucleic Acids Res., 1977, **4**, 1135).

The synthesis of oligoribonucleotides is not as advanced as that of oligodeoxyribonucleosides, however, all the possible 16 diribonucleoside monophosphates and 64 triribonucleoside diphosphates have been prepared. Two approaches can be made to the synthesis of the specific (3′→5′) internucleotide bond. Either a fully protected nucleoside 3′-phosphate is condensed with another protected ribonucleoside carrying a free 5′-hydroxyl group such as XVI, or a protected nucleoside 5′-phosphate is condensed with a ribonucleoside or nucleotide having only the 3′-hydroxyl group free, and all other functional groups protected *e.g.* XVII.

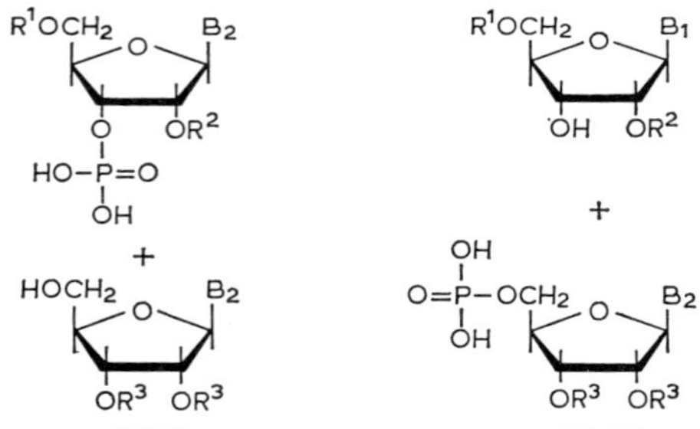

(XVI) (XVII)

R^1, R^2, R^3 = protecting groups
B_1, B_2 = suitably protected bases

The availability of ribonucleoside 3′-phosphates and the finding of conditions which allowed the 2′-OH to be acetylated quantitatively with avoidance of 2′,3′-cyclic phosphate formation (*Y. Lapidot* and *Khorana*, J. Amer. chem. Soc., 1963, **85**, 3852), prompted Khorana and his colleagues to choose the former approach (XVI). Using a 5′-*O*-methoxytrityl-2′-*O*-acetyl-3′-nucleo-

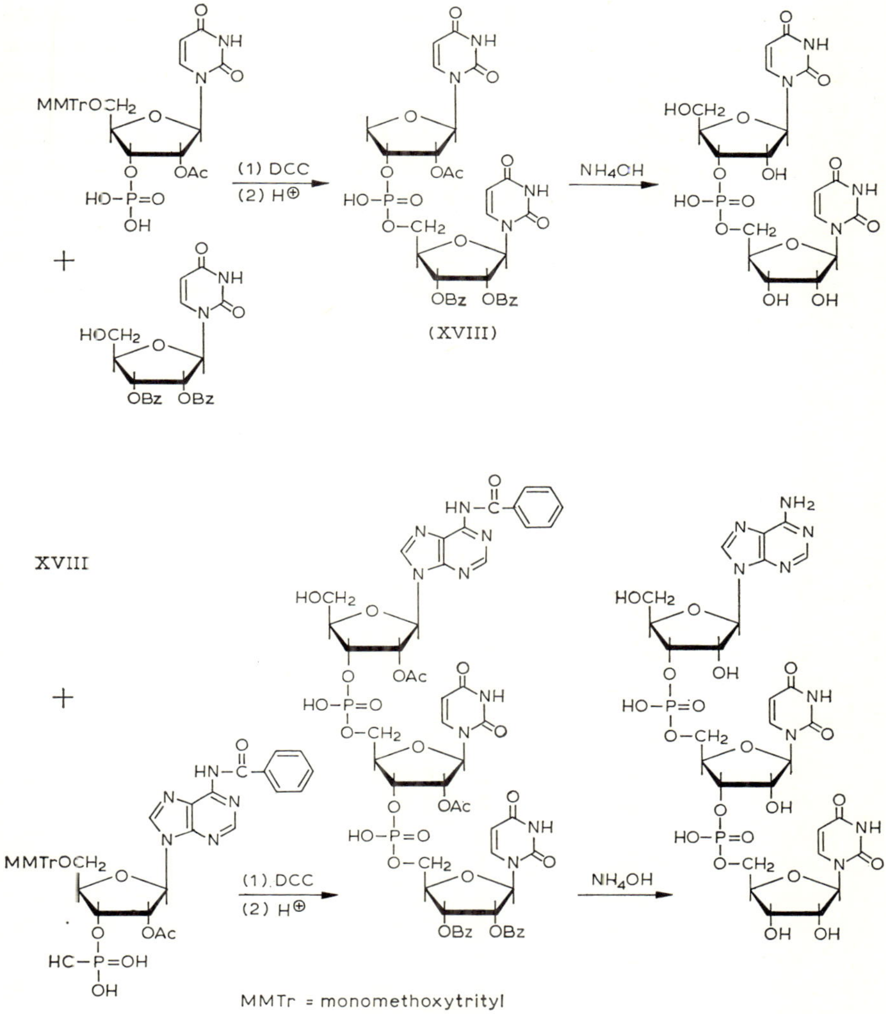

tide as the fully protected ribonucleoside 3′-phosphate moiety and a 2′,3′-di-*O*-benzoylnucleoside as the protected ribonucleoside carrying a free 5′-hydroxyl group the scheme for the synthesis of di- and tri-nucleotides is shown above (*idem*, *loc. cit.*).

Using this general approach with some minor modifications, the synthesis of all of the 64 ribotrinucleotides derivable from the four common mononucleotides has been achieved (*Khorana et al.*, J. Amer. chem. Soc., 1966,

88, 819). In the synthesis of oligoribonucleotides in addition to the considerations given to the synthesis of oligodeoxyribonucleotides there is the problem of protection of the 2′-OH group. In the above synthesis the 2′-OH of the nucleoside 3′-phosphate was protected by an acetyl group which can be removed under alkaline conditions. A major problem in using base-labile acyl groups for protection of the 2′-OH group is the ease of migration to an unprotected 3′-OH group. This means that these protecting groups are unsuitable when the approach *via* nucleoside 5′-phosphates (XXI) is used. Here protecting groups removed under mild acid or neutral conditions are used, as is illustrated below in the synthesis of a dinucleoside monophosphate.

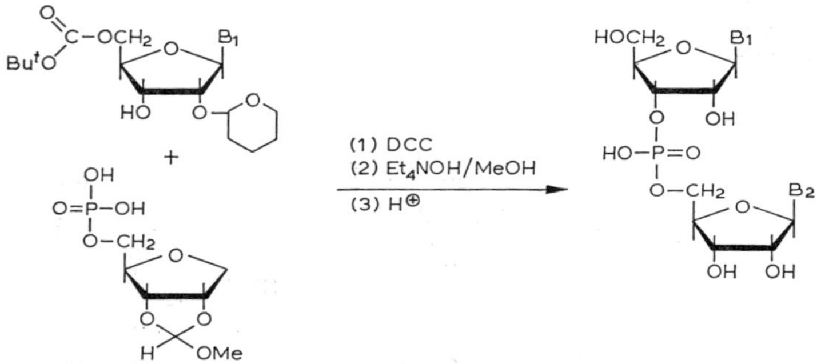

B$_1$, B$_2$ = suitably protected purine or pyrimidine bases

Synthesis of a dinucleoside monophosphate.

The tetrahydropyran-2-yl group which can be removed under mild acid conditions can be introduced readily into the 2′-position of a nucleotide or nucleoside by reaction of either a 3′,5′-cyclic nucleotide (*M. Smith* and *Khorana, ibid.*, 1959, **81**, 2911; *Khorana et al., ibid.*, 1962, **84**, 430) or a 3′,5′-suitably protected nucleoside (*B. E. Griffin, M. Jarman* and *C. B. Reese*, Tetrahedron, 1968, **24**, 639), respectively, with 2,3-dihydro-4*H*-pyran. The synthesis of the anticodon loop of *Escherichia coli* tRNA^{Met} has been achieved successfully using the tetrahydropyranyl group to protect the 2′-OH groups (*T. Neilson* and *E. S. Werstuik*, J. Amer. chem. Soc., 1974, **96**, 2295). One disadvantage of the tetrahydropyran-2-yl group is that with an asymmetric nucleoside it gives rise to a pair of diastereoisomers. To overcome this the symmetrical 4-methoxytetrahydropyran-4-yl group (XIX) has

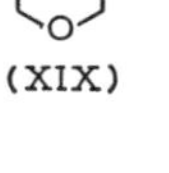

(XIX)

been introduced as an alternative to the tetrahydropyran-2-yl group (*Reese et al.*, *ibid.*, 1967, **89**, 3366; J. chem. Soc., C, 1971, 3230).

Two 2′-OH protecting groups which can be removed under neutral conditions are the benzyl group (*Khorana et al.*, J. Amer. chem. Soc., 1962, **84**, 430), which can be removed by catalytic hydrogenolysis and the 2-nitrobenzyl group which can be removed by photolysis at 350 nm. 2′-*O*-(-2-Nitrobenzyl)-nucleosides can be prepared successfully either by treatment of an unprotected nucleoside with (2-nitrophenyl)diazomethane followed by separation of the 2′- and 3′-isomers (*D. G. Bartholomew* and *A. D. Broom*, Chem. Comm., 1975, 38) or by treatment of the 2′,3′-*O*-(dibutylstannylene) derivative of a

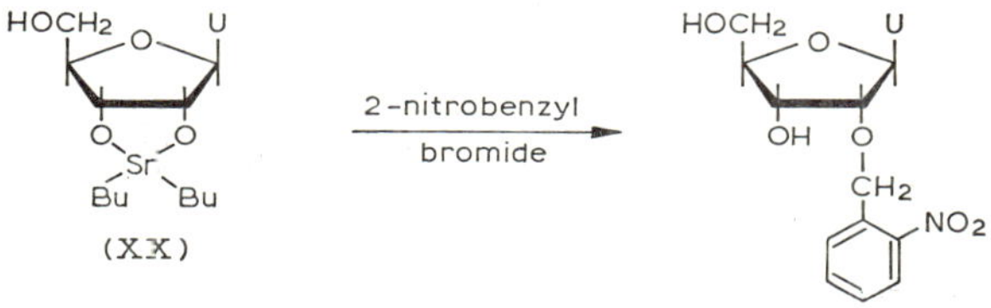

(XX)

nucleoside (XX), with 2-nitrobenzyl bromide when only the 2′-isomer is isolated (*E. Ohtsuka*, *S. Tanaka* and *M. Ikehara*, Nucleic Acids Res., 1974, **1**, 1351; *D. Wagner*, *J. P. H. Verheyden* and *J. G. Moffatt*, J. org. Chem., 1974, **39**, 24).

The 2-nitrobenzyl 2′-protecting group has been used successfully in the synthesis of oligoribonucleotides corresponding to fragments of *Escherichia coli* $tRNA^{Met}$ (*Ohtsuka*, *Tanaka* and *Ikehara*, Chem. Pharm. Bull., Japan, 1977, **25**, 949).

The phosphotriester method has been applied to the synthesis of oligoribonucleotides. Using the 2-chlorophenyl group, which can be removed under very mild alkaline conditions, to protect the phosphate group, UpUpU is obtained in high yield after purification of the products on silica chromatography (*Reese et al.*, Chem. Comm., 1971, 869). The β-cyanoethyl and β,β,β-trichloroethyl groups which are used in oligodeoxyribonucleotide synthesis have also been used in oligoribonucleotide synthesis. It is found that although the trichloroethyl phosphate group gives higher yields of product. its removal from the completed oligonucleotide is difficult so that the cyanoethyl group is preferred (*J. Smart*, Coll. Czech. chem. Comm., 1972, **37**, 1870; 1973, **38**, 3189; 1975, **40**, 1043).

(*d*) *Properties of nucleic acids*

Although many of the chemical and physical properties of the nucleic acids and polynucleotides are simply related to the properties of the constituent components, *e.g.* the purine and pyrimidine bases, the sugar moiety and the phosphate, some properties which are specific to the polymers need to be considered. Reference has already been made to the greater susceptibility of DNA than RNA to acid hydrolysis (p. 147). In the hydrolysis of DNA the first step involves the cleavage of the labile purine glycosyl linkages. The mechanism involves protonation of the nucleoside to give firstly an *N*-protonated species which, on isomerisation to the *O*-protonated species, is followed by a rate-limiting cleavage of the *N*-glycosyl bond (*R. K. Robins et al.*, J. Amer. chem. Soc., 1972, **94**, 4708).

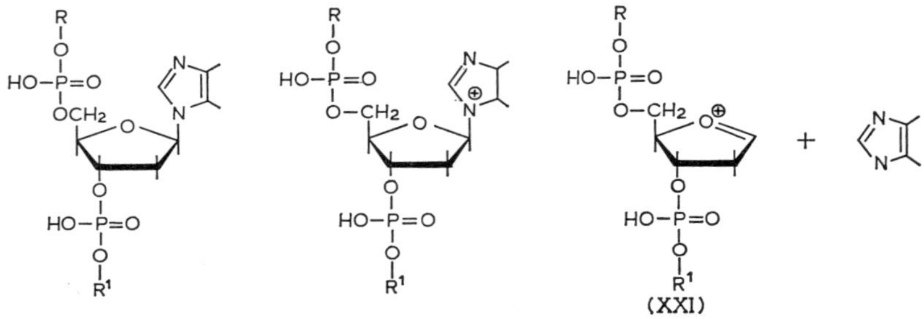

(XXI)

R, R[1] represent the continuation of the polynucleotide chain.

The previously suggested mechanism through a Schiff's base (*A. M. Michelson*, "The Chemistry of Nucleosides and Nucleotides", Academic Press, New York, 1963, p. 26) appears to be incorrect. It is not clear why the *N*-glycosyl bond on purine nucleosides is so much more easily hydrolysed than in the pyrimidine nucleosides, but it may be differences in ΔS value which contribute to the deoxyribonucleosides being more susceptible to acid hydrolysis than the ribonucleosides (*O. B. Nagy et al.*, J. Amer. chem. Soc., 1972, **94**, 4715). The initial product obtained on treatment of DNA with dilute acid under mild conditions is apurinic acid, which exhibits the aldehydic properties of 2-deoxyribose and therefore contains sugar residues in which the $C_{(1)}$ position is unsubstituted (*C. Tamm, M. E. Hodes* and *E. Chargaff*, J. biol. Chem., 1952, **195**, 49). More vigorous hydrolysis conditions produce oligomers of pyrimidine nucleotides which possess a phosphomonoester group at each end of the chain. These oligomers arise by cleavage of the apurinic acid by a β-elimination mechanism. This occurs when an activating group is in a position β to a phosphate, which is a good leaving group. The activating group in this case is a carbonyl at position $C_{(1)}$ formed by hydration and rearrangement of XXI to give a compound of the type XXII. Alternative conditions for degradation of this type are 66.7% formic acid in the presence of diphenylamine (*K. Burton* and *G. B. Peterson*,

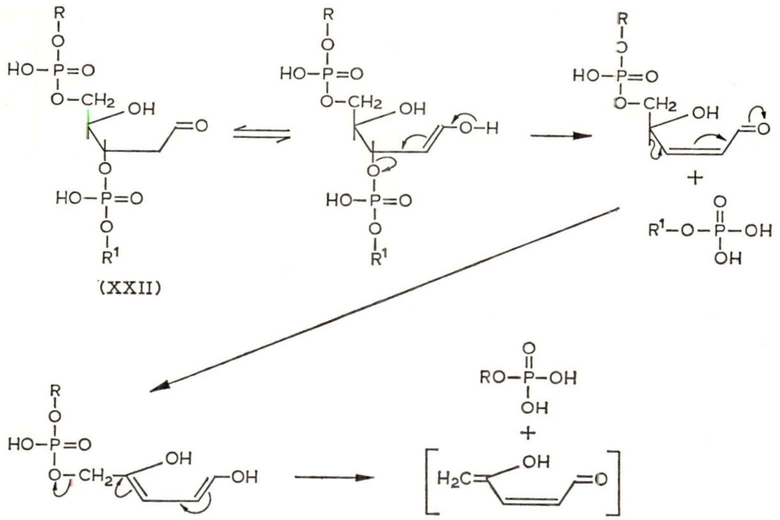

Mechanism of cleavage of the phosphodiester bond during acid hydrolysis of DNA.

Biochim. Biophys. Acta, 1957, **26**, 608). These conditions are more efficient in the cleavage of the phosphodiester linkage. Formic acid alone produces apurinization of DNA, but the diphenylamine is essential for rupture of the phosphodiester bond (*Peterson* and *Burton*, Biochem. J., 1964, **92**, 666). It evidently forms an enamine XXIII with the carbonyl group of the deoxyribose moiety, thereby facilitating the β-elimination of the 3′-phosphate group.

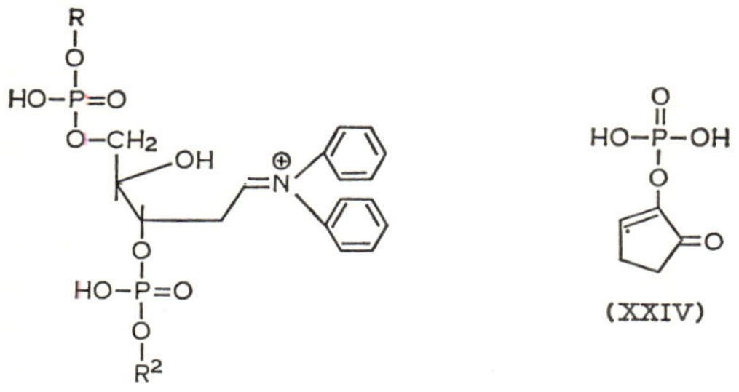

(XXIII) (XXIV)

As might be expected, treatment of apurinic acid with alkali also brings about the rupture of the phosphodiester linkages by β-elimination (*A. S. Jones et al.*, Biochim. Biophys. Acta, 1968, **166**, 732; J. chem. Soc., C, 1968, 2042), a mechanism supported by the isolation of 2-oxocyclopentenyl 1-phosphate (XXIV) as a reaction product.

It will be clear from the above that other reactions on DNA which involve the removal of any of the bases are likely to lead to the degradation of the polynucleotide chain. Treatment of DNA with anhydrous hydrazine causes degradation of the thymine and cytosine bases to substituted pyrazole derivatives and urea and the formation of apyrimidinic polymers (*S. Takamura*, Biochim. Biophys. Acta, 1958, **29**, 447; Bull. chem. Soc., Japan, 1959, **32**, 920). The apyrimidinic acid produced is degraded to oligonucleotide fragments with dilute alkali (*V. Habermann*, Coll. Czech. chem. Comm., 1961, **26**, 3147; 1963, **28**, 510). Oxidation of DNA (*C. R. Bayley* and *Jones*, Trans. Faraday Soc., 1959, **55**, 492; *Jones* and *R. T. Walker*, Nature, 1964, **202**, 24, 1108; *Jones et al.*, J. chem. Soc., 1964, 373; Nature, 1967, **216**, 793) and RNA (*J. J. Holbrook*, *Jones* and *M. J. Welch*, J. chem. Soc., 1965, 3998), with potassium permanganate causes the quantitative degradation of guanine, cytosine and thymine and leads to a polyphosphate–sugar chain containing only adenine. Under alkaline conditions adenine oligonucleotides are obtained. Oxidation with osmium tetroxide is more specific since it reacts only with the pyrimidine bases (*Burton* and *W. T. Riley*, Biochem. J., 1966, **98**, 70) however, since it reacts with thymidine 45 times as rapidly as it does with cytosine, it can be used to remove thymine selectively from DNA. Again the oxidised residues can be cleaved from the DNA by treatment with alkali or by treatment with diphenylamine and formic acid. At neutral and acid pH values, hydroxylamine reacts mainly with cytosine but at higher pH reacts with uracil which means that the uridine residues in RNA can be modified selectively (*D. W. Verwoerd*, *H. Kohlhage* and *W. Zillig*, Nature, 1961, **192**, 1038; Z. physiol. Chem., 1963, **332**, 184). Some degradation of the polynucleotide chains containing uridine takes place owing to the existence of the open form of the ribose oxime from which the 3′-phosphate group is eliminated by β-elimination (*E. I. Budowsky*, *N. A. Simukova* and *L. I. Guskova*, Biochim. Biophys. Acta, 1968, **166**, 755). The hydroxylamine group can be removed to leave free ribose units in the uracil-free RNA by mild acid hydrolysis (*Budowsky et al.*, Biochem. Biophys. Res. Comm., 1965, **152**, 1005). Specific cleavage at these free ribose units is then carried out in the presence of aromatic amines which act as β-elimination catalysts. Under these conditions the bases in DNA are unreactive towards hydroxylamine, however, by prior treatment of DNA with nitrous acid whereby the cytosine residues are converted to uracil, specific cleavage of DNA at these positions can be obtained (*H. S. Shapiro* and *E. Chargaff*, Biochemistry, 1966, **5**, 3012).

A number of compounds interact specifically with polynucleotides by forming a complex with the polynucleotide structure prior to covalent bond formation. Of particular interest are compounds which intercalate between the base pairs of the DNA double-stranded helix and then interact with the

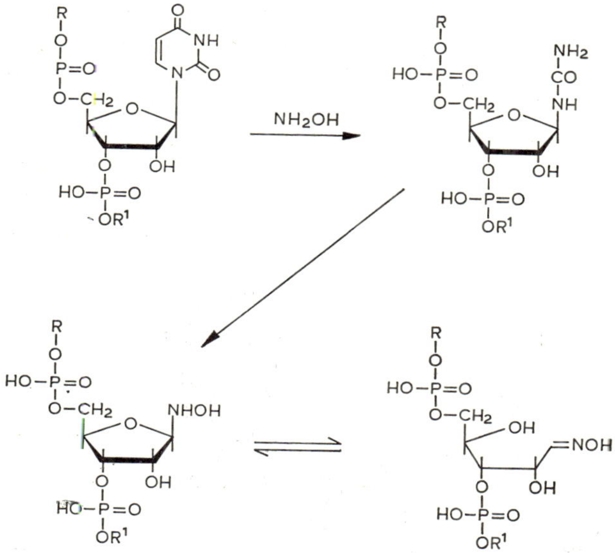

Modification of the uridine residue in RNA by hydroxylamine.

bases. Many carcinogenic polycyclic hydrocarbons react in this way. They form intercalation complexes with DNA and on activation with light or by oxidation covalent bonds are formed (*J. A. Lesko et al.*, Prog. Mol. Subcell. Biol., 1971, **7**, 434). The acridine dyes cause frame-shift mutations in DNA by intercalation (*L. S. Lerman*, J. mol. Biol., 1961, **3**, 18; *H. J. Li* and *D. M. Crothers*, *ibid.*, 1969, **39**, 461). When these dyes carry a functional side-chain of the nitrogen mustard type they alkylate a neighbouring base in the DNA (*B. N. Ames* and *H. J. Whitfield Jr.*, Cold Spring Harbor Symp. Quant. Biol., 1966, **31**, 221; *Ames*, *P. Sims* and *P. L. Grover*, Science, 1972, **176**, 47). The antibiotic mitomycin C (XXV) inhibits DNA synthesis in the cell by cross-linking the complementary DNA strands following intercalation. It is firstly reduced in the cell to an active bifunctional alkylating product which interacts with guanine residues (*W. Szybalski* and *V. N. Iyer* in "Antibiotics, Vol. 1, Mechanisms of Action", eds. *D. Gottlieb* and *P. D. Shaw*, Springer, Berlin, 1967, p. 211).

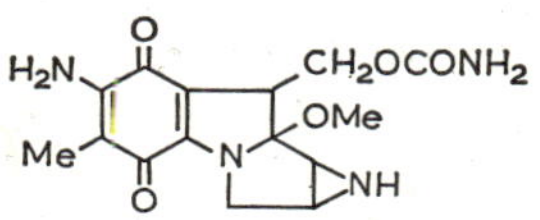

(XXV)

The pyrimidines of the nucleic acid undergo photochemical conversions which are, in the main, addition reactions, the most common of which being the photohydration of uracil (*A. M. Moore*, Canad. J. Chem., 1957, **35**, 163) and the photodimerisation of uracil and thymine (*R. Beukers* and *W. Berends*, Biochim. Biophys. Acta, 1960, **38**, 573; **41**, 550; 1961, **49**, 181). Irradiation of DNA gives dimers similar to those obtained on irradiation of thymine (*A. Wacker, H. Dellweg* and *D. Weinblum*, Naturwiss., 1960, **47**, 477) and irradiation of RNA gives a product similar to the uracil dimer (*H. G. Zachan*,

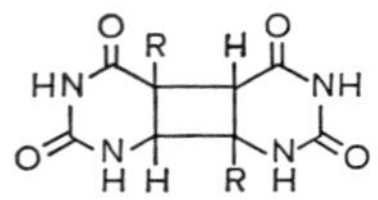

(XXVI)

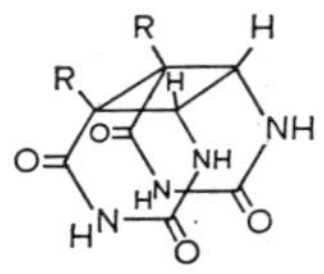

(XXVI a)

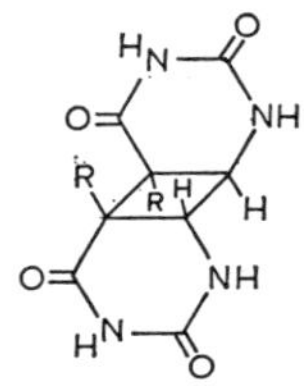

(XXVIb).

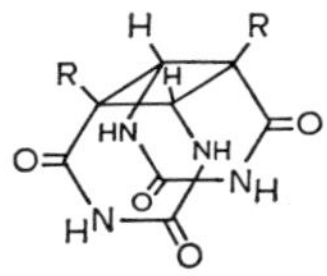

(XXVIc)

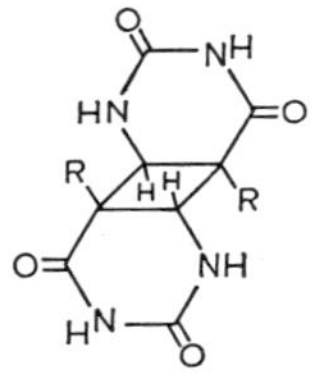

(XXVId)

R = H, uracil dimer
R = Me, thymine dimer

Hoppe-Seyler's Z. physiol. Chem., 1964, **336**, 176). These dimers have a cyclobutane structure (XXVI) and there are four possible stereoisomers for each dimer (XXVIa–XXVId).

All four isomers have been found in dimers of thymine and their structures have been confirmed by infrared and nuclear magnetic resonance spectroscopy, X-ray analysis and chemical methods (*Weinblum* and *Johns*, Biochim. Biophys. Acta, 1966, **114**, 450; *N. Camerman, S. C. Nyburg* and *Weinblum*, Tetrahedron Letters, 1967, 4127; J. Amer. chem. Soc., 1969, **91**, 982; *Camerman* and *Nyburg*, Acta Cryst., 1970, **B25**, 388; *G. M. Blackburn* and *R. J. H. Davies*, J. chem. Soc., 1966, 2239). The thymine dimer which is formed by photochemical reaction between adjacent thymidine residues in the same ordered strand of the DNA (*R. A. Deering* and *R. B. Setlow*, Biochim. Biophys. Acta, 1963, **68**, 526) has the structure XXVIa (R = Me) (*Blackburn* and *Davies*, Biochem. Biophys. Res. Comm., 1966, **22**, 704; J. Amer. chem. Soc., 1967, **89**, 5941; *A. J. Varghese* and *S. J. Wang*, Nature, 1967, **213**, 909; *Weinblum*, Biochem. Biophys. Res. Comm., 1967, **27**, 384). The formation of thymine dimers in DNA is reduced as the temperature of the solution in which the photochemical reaction is taking place approaches the melting temperature of the DNA. This suggests that the secondary structure of the DNA favours dimer formation, presumably because of the stacked conformation of the bases (*J. L. Hazzyn* and *R. O. Rahn*, *ibid.*, 1967, **29**, 327). The presence of thymine dimers blocks the action of DNA polymerase and so prevents replication in the cell (*R. B. Setlow*, "Regulation of Nucleic Acid and Protein Biosynthesis", eds. *V. V. Koningsberger* and *L. Bosch*, 1967, Vol. 10, p. 51). There are two ways in which the cell corrects this fault. Either, in the presence of visible light, an enzyme cleaves the bond between the pyrimidine rings and restores the bases to their original form (*V. Nüsslein et al.*, Nature, New Biology, 1971, **234**, 386), or by a series of enzymic reactions the DNA is "repaired". This involves the excision of the thymine along with a number of adjacent nucleotide residues followed by reinsertion of the correct residues in the gap by copying of the opposite strand of the DNA and covalent rejoining of the new fragment to the remainder of the DNA strand from which the thymine dimer was removed (*A. Kornberg et al.*, Nature, 1969, **224**, 495).

As mentioned on p. 234, irradiation of $tRNA^{Val}$ causes a photochemical reaction between the 4-thiouridine residue at position 8 and the cytosine residue at position 13, which leads to covalent formation between the bases of these residues, with the elimination of sulphur. After enzymic cleavage of the tRNA molecule the product of reaction of the two nucleosides can be

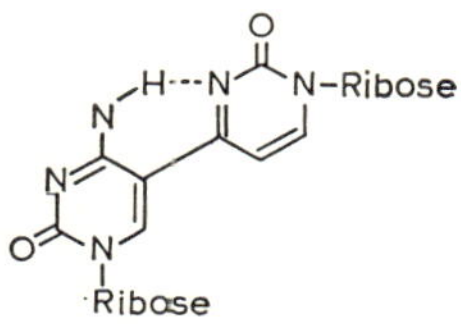

(XXVII)

isolated. It has been identified as the diriboside of 5-(4′-pyrimidin-2′-one)-cytosine (XXVII) (*A. Favre*, *B. Roques* and *J. L. Fourrey*, FEBS Letters, 1972, **24**, 209).

Many physicochemical methods have been used to study the properties and some of the finer points of the structure of oligonucleotides, polynucleotides and nucleic acids. Brief mention of these studies has been made in parts of this chapter but detailed coverage can be found elsewhere, *e.g.* *V. A. Bloomfield*, *D. M. Crothers* and *I. Tinoco Jr.*, "Physical Chemistry of Nucleic Acids", Harper and Row, New York, 1974; "Basic Principles in Nucleic Acid Chemistry", ed. *P.O.P. Ts'O*, Academic Press, New York, 1974, Vols. I and II.

Chapter 59

Pteridines, Alloxazines, Flavins and Related Compounds

K. OHTA, R. WRIGGLESWORTH AND H. C. S. WOOD

1. Pteridines

The story of the pteridines goes back to the last century when *F. Gowland Hopkins* (Nature, 1889, **40**, 335) isolated the pigments from the wings of the English Brimstone and Cabbage White butterflies. Thirty years later, unsuccessful attempts to characterise the pigments were made by German chemists (*H. Wieland* and *C. Schöpf*, Ber., 1925, **58**, 2178; 1926, **59**, 2067) who named the yellow and white pigments xanthopterin and leucopterin, respectively. Fifteen years after this, *R. Purrmann* showed (Ann., 1940, **544**, 182; **546**, 98; 1941, **548**, 284) xanthopterin to be 2-aminopteridine-4,6-dione (I) and leucopterin to be 2-aminopteridine-4,6,7-trione (II). Another pigment isoxanthopterin was assigned the structure 2-aminopteridine-4,7-dione (III).

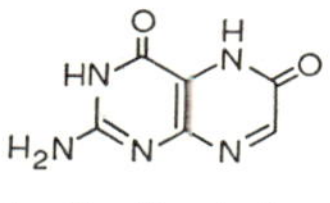

Xanthopterin
(I)

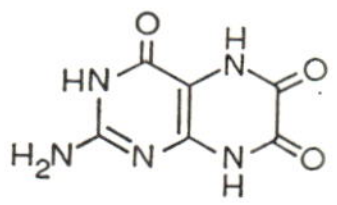

Leucopterin
(II)

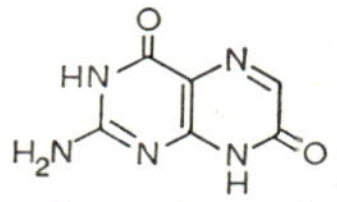

Isoxanthopterin
(III)

In 1894 the first pteridine was synthesised by oxidation of tolualloxazine (*O. Kühling*, Ber., 1894, **27**, 2116). The synthesis of the pteridines was little explored, however, until 1940–1950, when many important naturally occurring substances (*e.g.* folic acid) were shown to be pteridine derivatives.

The systematic name for pteridine (IV) is pyrimido[4,5-*b*]pyrazine and the I.U.P.A.C. approved method of numbering 1 to 8 is usually employed and

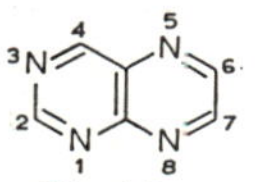

Pteridine
(IV)

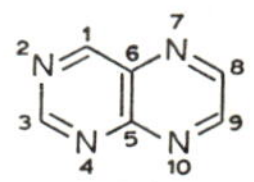

(V)

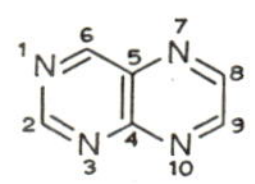

(Va)

is adopted in this chapter. Many old publications used an alternative numbering system 1 to 10 (V) similar to that used for the purines, while *G. R.*

Ramage and *T. S. Stevens* in *Rodds*' C.C.C., 1st Edn., Vol. IV C, p. 1760, numbered the formula as shown in Va. The early name 'pterin' is now used only for 2-aminopteridin-4-one and similarly, the trivial name 'lumazine' is given to pteridine-2,4-dione.

(*a*) *Synthesis and preparative methods*

There are two basic types of synthesis of pteridines, starting either from a suitably substituted pyrimidine or from a pyrazine derivative. The former method has been used more extensively as the appropriate pyrimidines are readily available.

(*i*) *From pyrimidines*

(*1*) Condensation of 4,5-diaminopyrimidines with α,β-dicarbonyl compounds is known as the Isay reaction.

When the dicarbonyl compound is unsymmetrical, *e.g.* methylglyoxal, this method can, and often does, give a mixture of two isomers which leads to difficulty in separation and identification. Attempts have therefore been made to direct the substituent by the use of sodium hydrogen sulphite or hydrazines (*H. S. Forrest* and *J. Walker*, J. chem. Soc., 1949, 79). A successful attempt was made using mercaptoethanol for the directed condensation of 2,5,6-triaminopyrimidin-4-one with methylglyoxal which gave the 6-methylpteridine as the major product (*C. B. Storm*, *R. Shiman* and *S. Kaufman*, J. org. Chem., 1971, **36**, 3925):

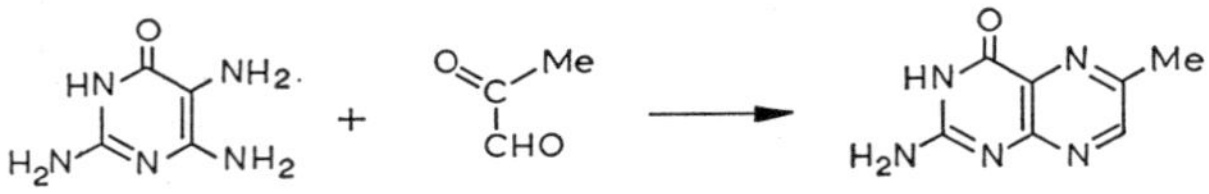

In contrast, phenylglyoxal with 5,6-diaminopyrimidine-2,4-dione or methylglyoxal with 4,5-diaminopyrimidine gave only the 7-isomers in the presence or absence of sodium hydrogen sulphite or hydrazine (*G. P. G. Dick*, *H. C. S. Wood* and *W. R. Logan*, J. chem. Soc., 1956, 2131; *A. Albert*, *D. J. Brown* and *Wood*, *ibid.*, 1954, 3832). 2,5,6-Triaminopyrimidin-4-one when condensed with phenylglyoxal gave the 7-isomer as the major product, but with dichloroacetophenone or nitroacetophenone the 6-isomer was obtained (*R. B. Angier*, J. org. Chem., 1963, **28**, 1398).

The direction of substitution was also affected by pH. Thus in strongly acid solution, α-oxocarboxylic acids when condensed with 4,5-diaminopyrimidine gave pteridin-6-ones whereas in weakly acid, or neutral solution, the main product was the 7-isomer (*G. B. Elion, G. H. Hitchings* and *P. B. Russell*, J. Amer. chem. Soc., 1950, **72**, 78; *Albert, Brown* and *G. Cheeseman*, J. chem. Soc., 1952, 1620).

A rationalisation of this phenomenon has been presented by *W. Pfleiderer* (Chem. Ber., 1959, **92**, 3190) who suggested that the 5-amino group of the pyrimidine reacted with the more reactive carbonyl group to form the corresponding Schiff base which subsequently cyclised to the pteridin-7-one in neutral or weakly acid solution. In strongly acid solution the Schiff base rearranged to give the 6-isomer.

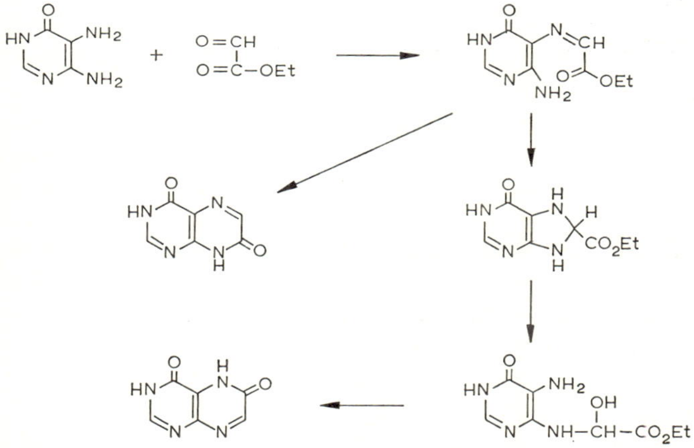

Dichloroacetic acid or chloral have been used as the two-carbon fragment to give 2-aminopteridine-4,6-dione (*Purrmann*, Ann., 1940, **546**, 98; 1941, **548**, 284; *W. E. Fidler* and *Wood*, J. chem. Soc., 1956, 3311; *F. Bergmann, M. Tamari* and *H. Ungar-Waron, ibid.*, 1964, 565):

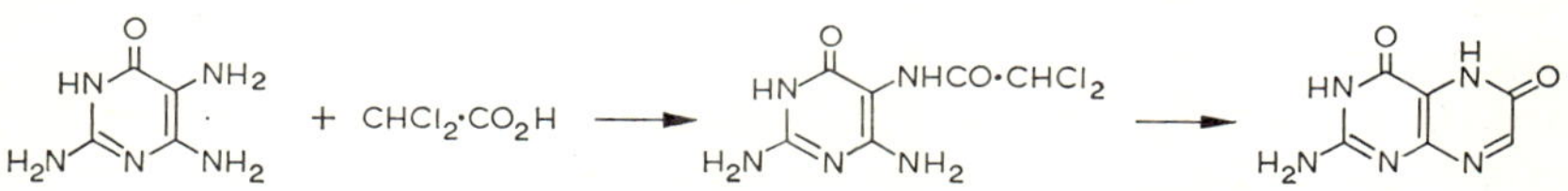

α-Hydroxycarbonyl compounds such as dihydroxyacetone or sugars, also have been condensed with diaminopyrimidines (*H. J. Backer* and *A. C. Houtman*, Rec. Trav. chim., 1948, **67**, 260; *Forrest* and *Walker*, *loc. cit.*):

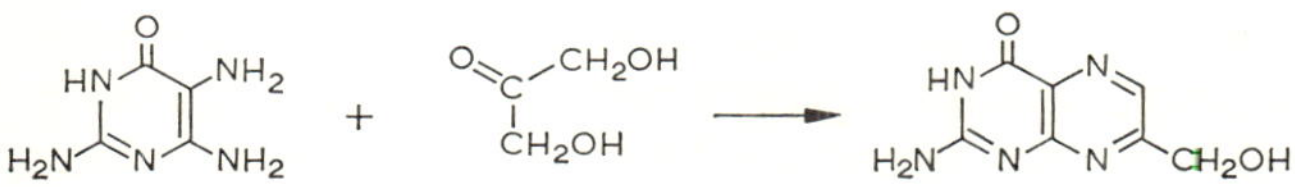

A further modification of this basic scheme has been the condensation of 4,5-diaminopyrimidines with aldehydes and hydrogen cyanide which yields the corresponding 7-aminopteridines (*F. F. Blicke* and *H. C. Godt*, J. Amer. chem. Soc., 1954, **76**, 2798; *M. Polonovski* in "Ciba Foundation Symposium on Chemistry and Biology of Pteridines", Churchill, London, 1954, p. 171; *I. J. Pachter*, J. org. Chem., 1963, **28**, 1191).

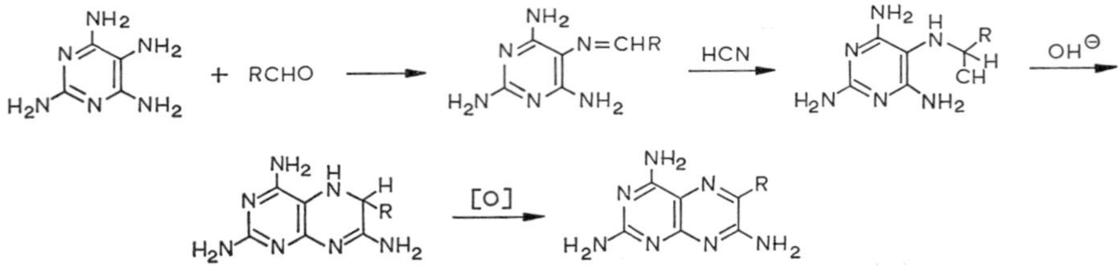

An alternative cyclisation of the intermediate 6-amino-5-arylideneamino-pyrimidine using triethyl orthoformate in dimethylformamide and subsequent cyclisation of the 6-ethoxymethyleneamino product has been described for the preparation of 6-arylpteridines (*F. Yoneda* and *M. Higuchi*, J. chem. Soc., Perkin I, 1977, 1336).

(*2*) An alternative and unambiguous approach to pteridine synthesis is that of *G. M. Timmis* (Nature, 1949, **164**, 139), which involves condensation of a 4-amino-5-nitrosopyrimidine with an α-oxomethylene compound.

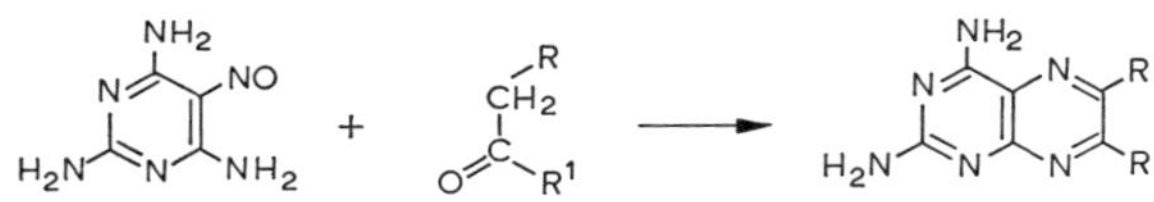

The scope of the reaction has been extended to include α-substituted acetonitriles and acetyl chlorides which give the 7-amino- and 7-hydroxy-pteridines, respectively (*R. G. W. Spickett* and *Timmis*, J. chem. Soc., 1954, 2887). Further extensions of this reaction have used cyanoacetic acid, cyano-acetamide, diethyl malonate (*T. S. Osdene* and *Timmis*, *ibid.*, 1955, 2036, 2038), phosphonomethylene compounds (*E. C. Taylor* and *B. E. Evans*, Chem. Comm., 1971, 189) and phenacylidenetriphenylphosphoranes (*K. Senga*, *H. Kanazawa* and *S. Nishigaki*, *ibid.*, 1976, 588).

Further application of this method to the reaction of compounds having methylene groups activated by adjacent acyl moieties with 4,6-diamino-5-nitroso-2-phenylpyrimidine in the presence of sodium cyanide results in the incorporation of cyanide, deacylation and cyclisation to give the 6-substituted 7-aminopteridines (*Pachter* and *P. E. Nemeth*, J. org. Chem., 1963, **28**, 1187). Other examples have been described using phenacylpyridinium and aceton-ylpyridinium salts (*F. Kröhnke*, Angew. Chem., 1953, **65**, 605; *Pachter*,

Nemeth and *A. J. Villani*, J. org. Chem., 1963, **28**, 1197) and enamines (*J. Weinstock* and *V. D. Wiebelhaus*, in "Pteridine Chemistry" Pergamon Press, Oxford, 1964, p. 37).

The corresponding cyclisation of substituted nitropyrimidines has produced pteridinone *N*-oxides; thus 4-(2-cyanoacetamido)-5-nitropyrimidines cyclise in aqueous sodium hydroxide at 40° to give 6-cyanopteridin-7(8*H*)-one 5-*N*-oxides, whereas under reflux the pteridin-6(5*H*),7(8*H*)-diones or their 5-*N*-hydroxy derivatives are formed (*G. Tennant* and *C. W. Yacomeni*, Chem. Comm., 1975, 819).

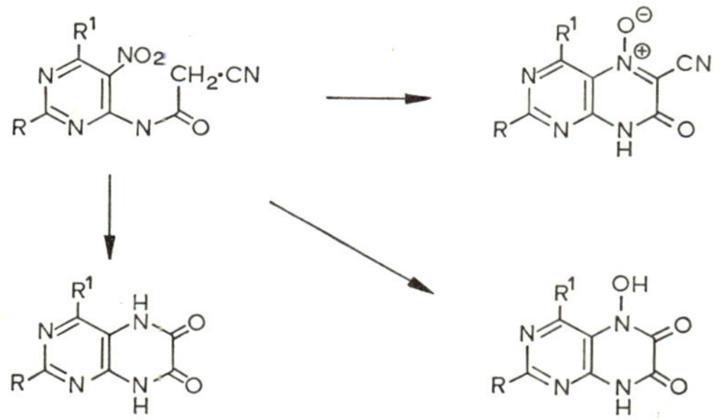

(*ii*) *From pyrazines*

A smaller number of pteridines have been prepared from pyrazine intermediates. The first example was the conversion of pyrazine-2,3-dicarboxamide by Hofmann reaction into lumazine (pteridine-2,4-dione) (*S. Gabriel* and *A. Sonn*, Ber., 1907, **40**, 4850).

Lumazine

(*1*) Cyclisation of 3-aminopyrazine-2-carboxamide or its thio- analogue with triethyl orthoformate in acetic anhydride gave pteridin-4-one or -4-thione, respectively (*Albert*, *Brown* and *Cheeseman*, J. chem. Soc., 1951, 474).

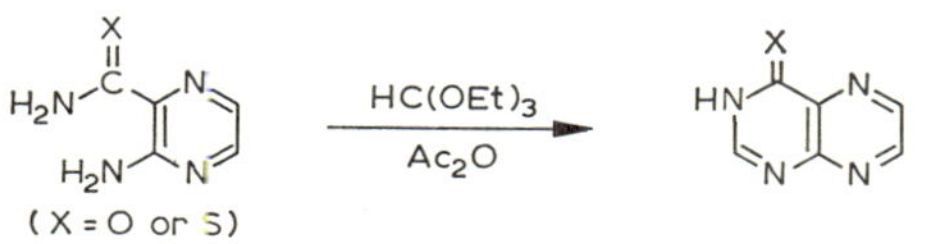

Extensions of this reaction have involved replacement of triethyl orthoformate with formic acid, formamide or acyl chlorides.

(*2*) 2,4-Disubstituted pteridines have been synthesised from 3-chloro-5,6-

diphenylpyrazine-2-carbonitrile (*Taylor* and *W. W. Paudler*, Chem. and Ind., 1955, 1061), or from methyl 3-chloropyrazine-2-carboxylate (*Dick* and *Wood*, J. chem. Soc., 1955, 1379) by condensation with urea, thiourea or guanidine.

Cyclisation of pyrazine *N*-oxides to the corresponding pteridine *N*-oxides has formed the basis of a general and versatile synthetic route to pteridines. Condensation of α-aminocarbonitriles with α-hydroxyiminocarbonyl compounds produces the intermediate pyrazine 1-oxides which are subsequently converted to the pteridines by a variety of reagents (*Taylor* and *K. Lenard*, J. Amer. chem. Soc., 1968, **90**, 2424; Ann., 1969, **726**, 100; *Taylor et al.*, J. Amer. chem. Soc., 1973, **95**, 6407, 6413; *Taylor* and *T. Kobayashi*, J. org. Chem., 1973, **38**, 2817).

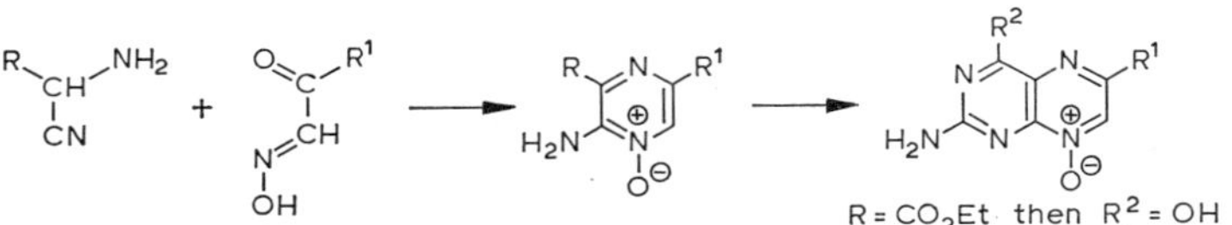

R = CO_2Et then R^2 = OH
R = CN then R^2 = NH_2

A review of the important applications of this synthesis has appeared recently (*Taylor*, in "Chemistry and Biology of Pteridines", ed. *W. Pfleiderer*, De Gruyter, Berlin, 1975, p. 543).

The synthesis has been extended to pteridine hydroxamic acids by the condensation of α-aminomalononitrile with α-oxohydroxamic acids and subsequent cyclisation of the *N*-hydroxypyrazines (*Taylor* and *P. A. Jacobi*, J. org. Chem., 1975, **40**, 2332).

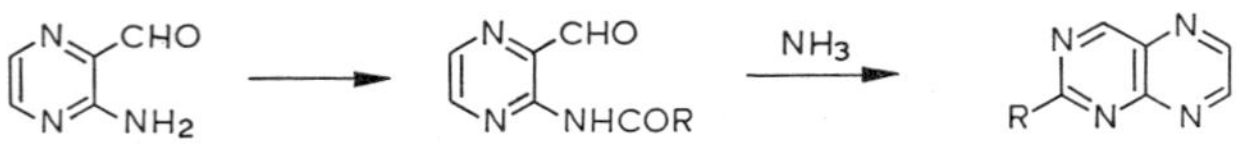

4-Unsubstituted pteridines have been prepared by a modification of this route using pyrazinecarbaldehydes as starting materials (*Albert* and *K. Ohta*, J. chem. Soc., C, 1971, 2357).

(*b*) *General properties*

(*i*) *Physical properties*

A comprehensive list of solubilities, u.v. spectra and ionisation constants has been drawn up by *Albert*, *Brown* and *Cheeseman* (*loc. cit.*; J. chem. Soc., 1952, 1620, 4219; Quart. Reviews, 1952, 197), and this has been supplemented by data from *Albert* and *Pfleiderer* ("Physical Methods in Heterocyclic Chemistry", *A. R. Katritzky*, ed., Academic Press, New York, 1963, Vol. 1, pp. 1 and 177).

Pteridine and its simple alkyl derivatives are soluble in polar and non-polar solvents and have sharp melting points. However, pteridines with hydrophilic substituents, such as amino or hydroxy groups, are insoluble in common organic solvents, sparingly soluble in water and decompose without melting. The latter tendency increases with the increasing degree of such substitution. This phenomenon has been attributed to intermolecular hydrogen bonding resulting in abnormally strong crystal-lattice forces (*Albert, Brown* and *Cheeseman*, J. chem. Soc., 1952, 4219).

[1]*H-n.m.r. spectra* of 2-aminopteridin-4-ones have been measured in trifluoroacetic acid which gives the monocations protonated at N-1 and in fluorosulphonic acid which gives the dications protonated at N-1 and N-8 (*A. Dieffenbacher* and *W. von Philipsborn*, Helv., 1969, **52**, 743). Similar results have been obtained by study of ^{13}C-n.m.r. spectra (*G. Müller* and *von Philipsborn*, *ibid.*, 1973, **56**, 2680).

Many other ^{13}C-n.m.r. studies have been described recently (*U. Ewers, H. Guenther* and *L. Jaenicke*, Chem. Ber., 1974, **107**, 876, 3275; 1973, **106**, 3951; *W. Frick, R. Weber* and *M. Viscontini*, Helv., 1974, **57**, 2658; *J. A. Lyon, B. R. Dunlap* and *P. D. Ellis*, J. magn. Reson., 1975, **18**, 293; *J. P. Geerts, A. Nagel* and *H. C. van der Plas*, *ibid.*, 1976, **8**, 607).

Mass spectra of simple substituted pteridines have been reported (*T. Goto, A. Tatematsu* und *S. Matsuura*, J. org. Chem., 1965, **30**, 1844; *I. Uemara* and *Goto*, Nippon Kagaku Zasshi, 1968, **89**, 520). Measurement of the mass spectra of the naturally occurring 2-aminopteridin-4-ones (*Y. Iwanami* and *M. Akino*, Tetrahedron Letters, 1972, 3219) and related compounds, which are often involatile, has been facilitated by trimethylsilylation or acetylation (*J. A. Blair* and *C. D. Foxall*, Org. Mass. Spect., 1969, **2**, 293; *P. Haug*, Anal. Biochem., 1970, **37**, 285; *K. Kobayashi* and *Goto*, in "Chemistry and Biology of Pteridines", International Academic Printing Co., Tokyo, 1970, p. 57; *T. Lloyd, S. Markey* and *N. Weiner*, Anal. Biochem., 1971, **42**, 108; *W. J. A. Vandenheuvel et al.*, J. heterocycl. Chem., 1972, **9**, 451; *Weber, Frick* and *Viscontini*, Helv., 1973, **56**, 2919).

I.r. spectra of simple substituted pteridines have been described by *S. F. Mason* (J. chem. Soc., 1955, 2336).

U.v. spectra have proved to be useful in the characterisation of naturally occurring pteridines and in studies of tautomeric structure. A comprehensive list of references and discussion of typical spectra are given by *W. L. F. Armarego* ("Physical Methods in Heterocyclic Chemistry", *Katritzky*, ed., Academic Press, New York, 1971, Vol. III, p. 67).

(ii) General chemical properties

Pteridine has a high ratio of nitrogen to carbon in the nucleus which reduces the aromatic character and shows the characteristics of a π-deficient *N*-heteroaromatic derivative. Pteridines seldom undergo electrophilic substitution but are very susceptible to nucleophilic substitution and addition reactions.

(*1*) *Covalent hydration* (*A. Albert* and *Armarego*, "Advances in Heterocyclic Chemistry", 1965, **4**, 1; *D. D. Perrin*, *ibid.*, p. 43). Pteridine in neutral aqueous solution is predominantly anhydrous, the ratio of the anhydrous to the hydrated species (3,4-dihydro-4-hydroxypteridine) being 3.5 (*idem*, J. chem. Soc., 1962, 645). However, in dilute acid solution pteridine undergoes rapid covalent hydration across the 3,4-double bond to give the cation of the 3,4-monohydrate, which slowly equilibrates with water to give a mixture of the 3,4-monohydrate and the 5,6:7,8-dihydrate (5,6,7,8-tetrahydro-6,7-dihydroxypteridine) cation. A detailed kinetic analysis of the hydration of pteridine over a wide range of pH has been described (*Y. Pocker, D. Bjorkquist* and *W. Schaffer*, J. Amer. chem. Soc., 1975, **97**, 5540). Methyl-group substitution discourages the hydration (*Albert, T. J. Batterham* and *J. J. McCormack*, J. chem. Soc., B, 1966, 1105) but the presence of electron-withdrawing substituents causes the hydration to go to completion even in neutral solution (*J. Clark, ibid.*, C, 1967, 1543; 1968, 313; *Clark* and *W. Pendergast, ibid.*, 1968, 1124; 1969, 1751).

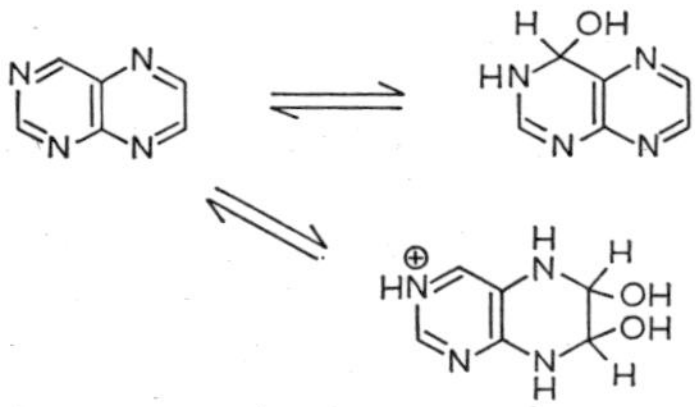

(*2*) *Nucleophilic addition.* The addition of "Michael reagents" to substituted pteridines has been extensively studied (*Albert* and *F. Reich, ibid.*, 1961, 127; *Albert* and *C. F. Howell, ibid.*, 1962, 1591; *Albert* and *McCormack, ibid.*, 1965, 6930; *ibid.*, C, 1966, 1117; 1968, 63).

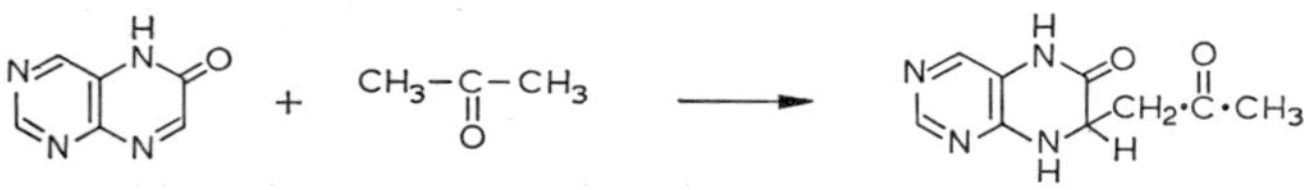

Addition also occurs to the dihydropteridines as exemplified by the reaction of diethyl malonate with 2-amino-7,8-dihydropteridin-4-one which gives the corresponding 5,6,7,8-tetrahydro derivative (*A. Stuart, H. C. S. Wood* and *D. Duncan, ibid.*, 1966, 285).

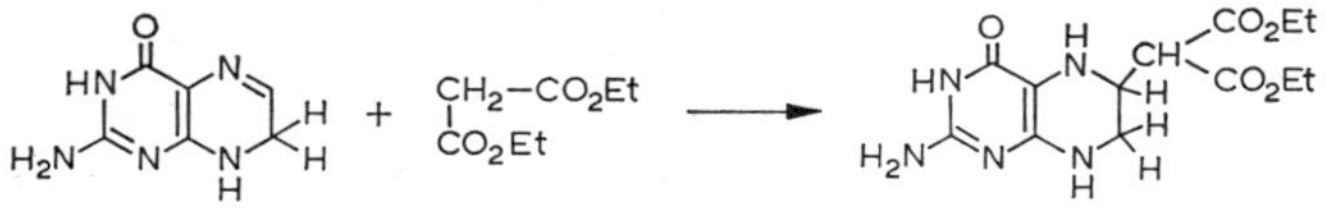

(*3*) *Nucleophilic substitution.* Many 2,4-disubstituted pteridines undergo nucleophilic substitution with alkylamines giving the corresponding alkylaminopteridines *via* the ring-cleaved product shown below which subsequently recyclises to the new product (*E. C. Taylor* and *C. K. Cain*, J. Amer. chem. Soc., 1951, **73**, 4384; 1952, **74**, 1644; *Taylor, ibid.*, p. 1648; *W. R. Boon*, J. chem. Soc., 1957, 2146).

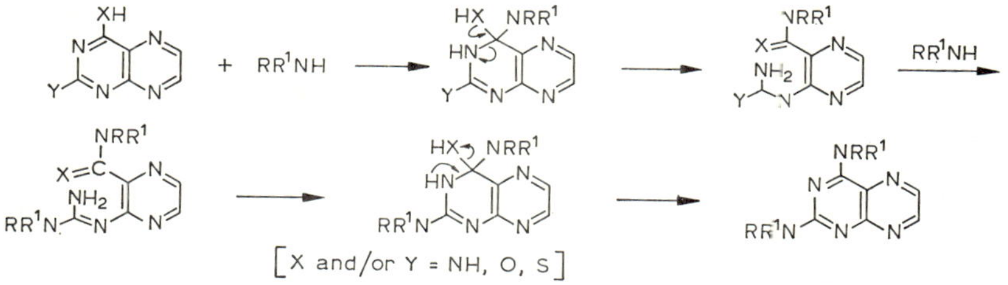

2- and 4-Alkylthiopteridines readily undergo substitution with a variety of nucleophiles (*W. V. Curran* and *R. B. Angier*, J. org. Chem., 1963, **28**, 2672; *McCormack* and *H. G. Mautner*, *ibid.*, 1964, **29**, 3370).

(*4*) *Ring cleavage reactions.* Pteridine and its simply substituted derivatives are easily hydrolysed in acid or alkaline solution. However, more highly substituted pteridines, such as pteridine-2,4,6,7-tetraone are resistant to hydrolysis. The presence of electron-donating groups in the nucleus reduces the π-deficient character and thus increases the aromatic stability (*Albert*, Quart. Reviews, 1952, 197).

(*5*) *Tautomerism.* The tautomerism of amino-, hydroxy- and amino-hydroxy-pteridines has been investigated by comparison of their u.v. spectra and ionisation constants with those of the corresponding methylated derivatives. These experiments have shown that in the hydroxypteridines (*Albert*, *loc. cit.*; *W. Pfleiderer*, Chem. Ber., 1957, **90**, 2582, 2588, 2604, 2631) and in the aminohydroxypteridines, which include many of the naturally occurring pteridines (*Pfleiderer et al., ibid.*, 1960, **93**, 2015; *Pfleiderer* and *M. Rukwied, ibid.*, 1961, **94**, 1; 1962, **95**, 755; *Pfleiderer* and *H. Fink, ibid.*,

1963, **96**, 2950, 2964, 2977), the hydroxyl group exists predominantly as the oxo tautomer and the amino group remains as such (*D. J. Brown* and *N. W. Jacobsen*, J. chem. Soc., 1961, 4413).

(*c*) *Pteridine and its derivatives*

(*i*) *Pteridine*

Pteridine, pale yellow plates, m.p. 140°, is prepared by the condensation of 4,5-diaminopyrimidine with glyoxal bisulphite in aqueous solution

(*W. G. M. Jones*, Nature, 1948, **162**, 524) or better with syrupy glyoxal or polyglyoxal in alcohol (*A. Albert, D. J. Brown* and *G. Cheeseman*, J. chem. Soc., 1951, 474; *Albert* and *H. Yamamoto, ibid.*, C, 1968, 2289). It has also been prepared by oxidation of 3,4-dihydropteridine (*Albert* and *K. Ohta, ibid.*, 1970, 1540). It readily sublimes at 125–130°/20 mm and is crystallised from benzene/light petroleum (b.p. 60–80°). The partition coefficient between water and chloroform is 1.98, between water and benzene 0.21 and between water and ether 0.05.

The crystal structure of pteridine showed the molecule to be planar to within 0.06 Å, with angles and bond lengths as shown in the diagram (*T. A. Hamer* and *J. M. Robertson*, J. chem. Soc., 1956, 3586).

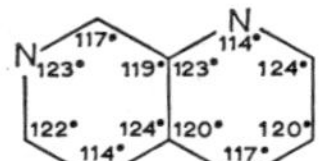

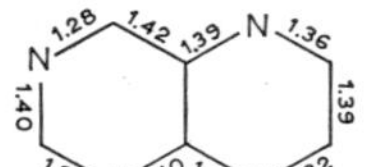

The electron distribution around the pteridine nucleus has been calculated by *O. Chalver* and *C. Sandorfy* (Compt. rend., 1949, **228**, 566) and by *T. H. Goodwin* and *A. L. Porte* (J. chem. Soc., 1956, 3595).

The ^{1}H-n.m.r. spectrum of pteridine has been measured in deuterochloroform and the chemical-shift assignments are as shown (*Matsuura* and *Goto, ibid.*, 1963, 1773; 1965, 623).

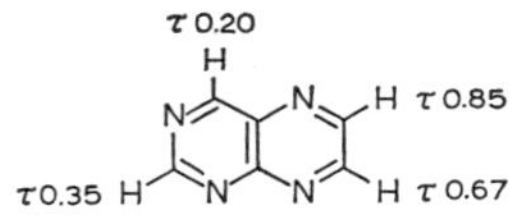

Pteridine is a very weak base and the estimated pK_a of the anhydrous species is about −1.1 (*J. W. Bunting* and *D. D. Perrin*, J. chem. Soc., B, 1966, 436), however, the measured value is 4.12 (*Albert, Brown* and *Cheeseman, loc. cit.*) because of the covalent hydration phenomenon described above. Besides adding the elements of water, pteridine also covalently adds amines (*Albert* and *Ohta, loc. cit.*; *B. E. Evans*, J. chem. Soc., Perkin I, 1974, 357) and alcohols (*Albert* and *H. Mizuno*, J. chem. Soc., B, 1971, 2423).

In acid solution pteridine is readily hydrolysed to 3-aminopyrazine-2-carbaldehyde (*Albert, Brown* and *H. C. S. Wood, ibid.*, 1956, 2066; *Albert* and *Yamamoto, loc. cit.*). On reduction of pteridine with lithium tetrahydridoaluminate 5,6,7,8-tetrahydropteridine is formed (*E. C. Taylor* and *W. R. Sherman*, J. Amer. chem. Soc., 1959, **81**, 2464).

(*ii*) *Pteridine homologues*

The methylpteridines are prepared by the Isay reaction with the exception of the 6-methyl derivative which is prepared *via* 3,4-dihydro-6-methylpteridine from 2-amino-3-aminomethyl-5-methylpyrazine.

2-Methyl-, m.p. 141–142° (*Albert, Brown* and *Wood*, J. chem. Soc., 1954, 3832); 4-**methyl-**, m.p. 152–153° (*idem, ibid.*; *J. H. Lister, G. M. Ramage* and *E. Coates, ibid.*, 1954,

4109); 6-**methyl-**, m.p. 130° (*Albert* and *Ohta*, *loc. cit.*); 7-**methyl-**, m.p. 133–134°; 6,7-**dimethyl-**, m.p. 148–149° (*Albert*, *Brown* and *Wood*, *loc. cit.*); 4,6,7-**trimethyl-**, m.p. 139–140° (*Albert* and *Yamamoto*, J. chem. Soc., C, 1968, 2292); 2,4,6,7-**tetramethyl-**, m.p. 124–125° (*idem, ibid.*); 2,6,7-**trimethyl-pteridine**, m.p. 134–135° (*Albert*, *Brown* and *Wood*, *loc. cit.*).

(*iii*) *Halogenopteridines*

The chloropteridines have been most extensively studied as they have proved useful as intermediates for preparing amino- or mercapto-pteridines. Very few bromo-pteridines have been prepared (*J. Roch*, U.S.P., 2,940,972/1960; C.A., 1960, **54**, 24824f; *J. H. Jones* and *E. J. Cragoe*, J. med. Chem. 1968, **11**, 322).

Methods of preparation. Chloropteridines may be prepared from a suitably substituted pyrimidine and a dicarbonyl compound but more commonly they are obtained by chlorination of the corresponding pteridinone by the use of phosphorus pentachloride in phosphoryl chloride, phosphorus trichloride, pentachloroethane or benzoyl chloride.

Properties

The chloropteridines readily undergo nucleophilic substitution although the reactivity of the chlorine atom varies with the position of substitution. This latter fact makes these derivatives especially useful in synthesis. Thus 4,7-dichloropteridine reacts with amines at room temperature to give the 7-alkylamino-4-chloropteridine which can be further aminated with a different amine under more forcing conditions to give the mixed di(alkylamino)pteridine (*Brown* and *N. W. Jacobsen*, J. chem. Soc., 1965, 1175). Similarly 6,7-dichloropteridine reacts with nucleophiles at the 7-position to give 6-chloro-7-substituted pteridines which may undergo further nucleophilic substitution under different conditions (*Albert* and *J. Clark*, *ibid.*, p. 27). Further examples are the reaction of 2,4-dichloro-6,7-dimethylpteridine with ammonia at room temperature to give 4-amino-2-chloro-6,7-dimethylpteridine (*J. W. Daly* and *B. E. Christensen*, J. Amer. chem. Soc., 1956, **78**, 225) and the reactions of 2,4,6,7-tetrachloropteridine which reacts with dilute sodium hydroxide to give 2,4-dichloropteridine-6,7-dione and with ammonia at —70° to give 6,7-diamino-2,4-dichloropteridine. Further amination of the latter at 150° for 8 hours gives 2,4,6,7-tetra-aminopteridine (*Taylor* and *Sherman*, *loc. cit.*).

A few examples of reductive elimination of the chlorine atom have been reported; thus reduction of 2,4,6,7-tetrachloropteridine with lithium tetrahydridoaluminate gave 2,4-dichloro-5,6,7,8-tetrahydropteridine (*idem, ibid.*).

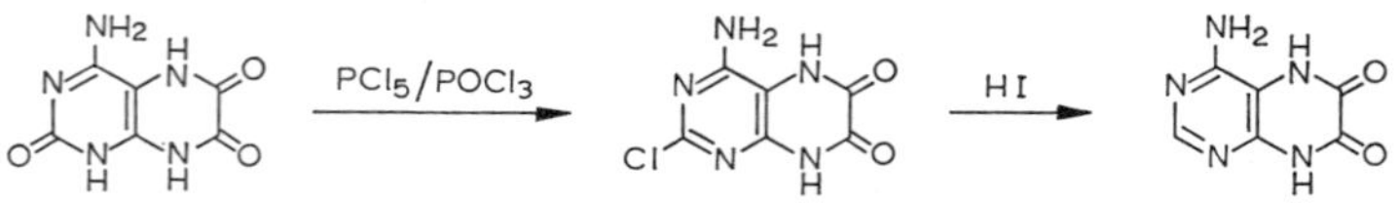

This reaction has proved useful for the elimination of the hydroxyl moiety. Chlorination of 4-aminopteridine-2,6,7-trione and subsequent reductive elimination of the chlorine atom gave 4-aminopteridine-6,7-dione (*H. Wieland* and *R. Liebig*, Ann., 1944, **555**, 146).

Similarly, the 7-hydroxyl group of 2-amino-4,7-dihydroxypteridine-6-carboxylic acid was removed (*J. H. Mowat et al.*, J. Amer. chem. Soc., 1948, **70**, 14).

2-**Chloropteridine,** m.p. 106–107° (decomp.), is obtained by condensing 4,5-diamino-2-chloropyrimidine with dry polyglyoxal in methanol. It is readily hydrolysed to pteridin-2-one by boiling water or by dilute acid solution. Strong acid causes a violet colouration (*Albert, Brown* and *Cheeseman, loc. cit.*). 2-*Chloro*-6,7-*dimethylpteridine*, m.p. 135–137°, is easily hydrolysed, thionated and aminated (*Daly* and *Christensen, loc. cit.*). 2-*Chloro*-4,6,7-*trimethylpteridine*, m.p. 129° (*Brown, B. T. England* and *J. M. Lyall*, J. chem. Soc., C, 1966, 226).

4-**Chloropteridine** is obtained by cyclisation of 4,5-diamino-6-chloropyrimidine with glyoxal (*Albert, Brown* and *Cheeseman*, J. chem. Soc., 1952, 4219). It decomposes at about 140° when placed in a preheated bath, it darkens on exposure to light and gives a red colour on prolonged boiling in petroleum ether. 4-*Chloro*-6,7-*dimethylpteridine*, m.p. 149–151° (*Daly* and *Christensen, loc. cit.*).

6-**Chloropteridine,** m.p. 149–150°, is prepared from pteridin-6-one by treatment with phosphorus pentachloride and phosphoryl chloride (*Albert, Brown* and *Cheeseman, loc. cit.*; *Albert* and *Clark*, J. chem. Soc., 1964, 1666). It is easily hydrolysed in aqueous sodium bicarbonate solution at room temperature. On treatment with ammonia in benzene at 20° it gives 6-aminopteridine, with sodium methoxide in methanol at 20° it gives 6-methoxypteridine and with 2 *M* sodium hydroxide saturated with hydrogen sulphide it gives pteridine-6-thione.

7-**Chloropteridine,** m.p. 100–101°, is prepared in a similar manner to the 6-isomer (*Albert, Brown* and *Wood, ibid.*, 1954, 3832; *Albert* and *Clark, loc. cit.*).

2,4-**Dichloropteridine,** m.p. 150° (decomp.) (*Taylor* and *Sherman, loc. cit.*); 4,6-dichloropteridine has been reported but not characterised (*Albert, J. H. Lister* and *C. Pedersen*, J. chem. Soc., 1956, 4621); 4,7-*dichloropteridine*, m.p. 127–128° (*Brown* and *Jacobsen, loc. cit.*); 6,7-*dichloropteridine*, m.p. 141–142° (*Albert* and *Clark, loc. cit.*); 2,4,7-**trichloropteridine,** m.p. 133° (*Albert, Lister* and *Hedersen, loc. cit.*); 4,6,7-*trichloropteridine*, m.p. 179–181° (*idem, ibid.*); 2,4,6,7-**tetrachloropteridine,** m.p. 161–162° (*C. Schöpf, R. Reichert* and *K. Riefstaht*, Ann., 1951, **548**, 82; *Taylor* and *Sherman, loc. cit.*).

(*iv*) *Aminopteridines*

Many aminopteridines, especially 2,4-diamino- and 2,4,7-triamino-pteridines have been prepared for use in drugs, particularly in the fields of antibacterial agents and diuretics.

(*1*) *Methods of preparation.* Generally, aminopteridines are prepared from suitably substituted 4,5-diaminopyrimidines and dicarbonyl compounds or from 3-aminopyrazine-2-carbonitriles, or the corresponding *N*-oxides (*Taylor* and *Kobayashi*, J. org. Chem., 1976, **41**, 1299), but amination of the corresponding chloro- or mercapto-pteridines has also been used. A direct conversion of 4-hydroxypteridines to their 4-amino analogues has been reported using phenyl phosphorodiamidate (PPDA) (*G. R. Gapski* and *J. M. Whitely*, in "Chemistry and Biology of Pteridines", ed. *Pfleiderer*, De Gruyter, Berlin, 1975, p. 627).

Many 7-aminopteridines, especially those which are highly substituted in the 2-, 4- or 6-positions, have been prepared from 4-amino-5-nitroso-pyrimidines and acetonitriles or from 4,5-diaminopyrimidines and alkane-carbaldehydes and sodium cyanide.

(*2*) *Properties.* Amino substituents at positions 2 and 4 of the pteridine nucleus undergo nucleophilic substitution reactions with alkylamines (*E. C. Taylor* and *C. K. Cain*, J. Amer. chem. Soc., 1951, **73**, 4384; *W. R. Boon*, J. chem. Soc., 1957, 2146).

2-Aminopteridine, yellow crystals, m.p. 275° (decomp.), is prepared from 2,4,5-triaminopyrimidine and glyoxal bisulphite; the precipitated bisulphite adduct gives the desired product on shaking with sodium hydroxide solution (*Albert, Brown* and *Cheeseman, ibid.*, 1951, 474). It is virtually insoluble in the common organic solvents and it decomposes in cold 10 *M* hydrochloric acid or boiling 1 *M* sodium hydroxide to give coloured products. It is easily acetylated by acetic anhydride to give 2-*acetamidopteridine*, m.p. 229–231°. 2-*Amino*-4,6,7-*trimethylpteridine*, m.p. 312–313° (*F. L. Rose, ibid.*, 1952, 3448).

4-Aminopteridine, prepared from 4,5,6-triaminopyrimidine and glyoxal, crystallises from water in plates, m.p. 309–311° (decomp.) (*Albert, Brown* and *Cheeseman, loc. cit.*; *R. M. Evans et al.*, J. chem. Soc., 1956, 4106). 4-*Acetamidopteridine*, m.p. 191–192°. 4-Aminopteridine on boiling with 0.1 *N* sodium hydroxide gives pteridin-4-one; with aqueous ammonia at 170° it gives ammonium 3-aminopyrazine-2-carboxylate, and with hydrazine it gives 2-*aminomethyleneaminopyrazine*-3-*carboxamidrazone* (VI), m.p. 173–174° (*Evans et al., loc. cit.*). 4-*Dimethylaminopteridine*, m.p. 164–165° (*Albert, Brown* and *Cheeseman*, J. chem. Soc., 1952, 4219).

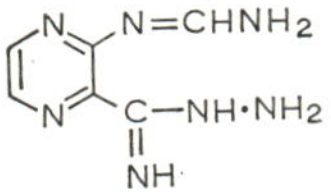

(VI)

6-**Aminopteridine,** m.p. > 300° (decomp.), is formed as long yellow needles by passing dry ammonia into 6-chloropteridine in benzene (*idem, ibid.*, 1952, 1620). It is decomposed on refluxing for 30 min with acetic anhydride. Hydrolysis to pteridin-6-one occurs at room temperature at pH 2. 6-*Dimethylaminopteridine*, m.p. 212°.

7-**Aminopteridine,** m.p. > 320° (decomp.), is prepared from the 7-methylthio- or 7-chloropteridine and ammonia (*Albert, Brown* and *H. C. S. Wood*, J. chem. Soc., 1954,

3832). The 7-amino compound is more stable to hydrolysis than the 6-isomer and readily forms an *acetyl* derivative, m.p. > 250° (decomp.). 7-*Dimethylaminopteridine*, m.p. 319–320°.

2,4-**Diaminopteridine,** which decomposes on heating, has been prepared from 2,4,5,6-tetra-aminopyrimidine (*M. F. Mallette, Taylor* and *Cain*, J. Amer. chem. Soc., 1947, **69**, 1814; *Albert, Brown* and *Cheeseman*, J. chem. Soc., 1952, 4219; *G. Konrad* and *W. Pfleiderer*, Chem. Ber., 1970, **103**, 722) and from 2-amino-4-benzylthiopteridine (*J. J. McCormack* and *H. G. Mautner*, J. org. Chem., 1964, **29**, 3370). Alternatively it is produced by oxidation of 2,4-diamino-5,6,7,8-tetrahydropteridine with manganese dioxide (*Taylor* and *W. R. Sherman*, J. Amer. chem. Soc., 1959, **81**, 2464). It is readily hydrolysed by 6 *N* hydrochloric acid to 2-aminopteridin-4-one (*Taylor* and *Cain*, *ibid.*, 1949, **71**, 2538). 2,4-Diamino-6,7-dialkyl- and -6,7-diphenyl-pteridines have shown antibacterial activity (*L. J. Daniel et al.*, J. biol. Chem., 1947, **169**, 689; *N. R. Campbell, J. H. Dunsmuir* and *M. E. H. Fitzgerald*, J. chem. Soc., 1950, 2743; *Cain, Taylor* and *Daniel*, J. Amer. chem. Soc., 1949, **71**, 892). 2-, 4- and 2,4-bis-alkylaminopteridine derivatives have also been examined in this context (*G. B. Elion* and *G. H. Hitchings*, J. biol. Chem., 1951, **188**, 611; *B. Roth, J. M. Smith* and *M. E. Hultquist*, J. Amer. chem. Soc., 1951, **73**, 2864; *Taylor* and *Cain*, *ibid.*, 1952, **74**, 1644; *Taylor*, *ibid.*, p. 1648; *M. D. Potter* and *T. Henshall*, J. chem. Soc., 1956, 2000; *J. I. Degraw et al.*, J. med. Chem., 1971, **14**, 866; *J. P. Jonak et al.*, *ibid.*, 1972, **15**, 1331).

Other diaminopteridines include 4,5-*diamino*- (*Albert, J. H. Lister* and *C. Pedersen*, J. chem. Soc., 1956, 4621), 4,7-*diamino*-, m.p. > 300° (*T. S. Osdene* and *G. M. Timmis*, *ibid.*, 1955, 2036) and 6,7-*diamino-pteridine*, m.p. > 250° (*Albert* and *J. Clark*, *ibid.*, 1965, 27).

2,4,7-**Triaminopteridine,** m.p. > 300°, was first prepared from 2,4,6-triamino-5-nitro-pyrimidine and cyanoacetic acid which initially gave 2,4,7-triaminopteridine-6-carboxylic acid, which on decarboxylation gave the above compound (*Osdene* and *Timmis*, *loc. cit.*).

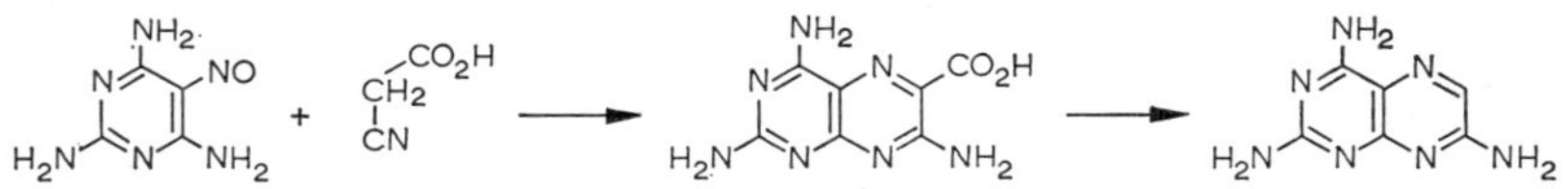

Another route has been described by *Albert, Lister* and *Pedersen* (*loc. cit.*), from 2,4,7-trichloropteridine and ammonia. Other 2,4,7-triaminopteridines have been prepared (*H. J. Jones* and *E. J. Cragoe*, J. med. Chem., 1968, **11**, 322).

Other aminopteridines which have been reported include 4,6,7-**triamino**- (*Albert, Lister* and *Pedersen*, *loc. cit.*) and 2,4,6,7-**tetraamino-pteridine,** isolated as the *monohydrochloride*, m.p. > 360°, which is slightly soluble in boiling water and loses one amino group to give a triaminopteridone by refluxing in 0.3 *N* sodium hydroxide.

(v) *Oxo(hydroxy)pteridines*

The hydroxypteridines in general exist as the cyclic amide (lactam) tautomers with the hydrogen present on the adjacent nitrogen (*D. J. Brown* and *S. F. Mason*, J. chem. Soc., 1956, 3443; *W. Pfleiderer*, Chem. Ber., 1957, **90**, 2582, 2588, 2604, 2617, 2624, 2631). Insertion of hydroxyl groups into the nucleus markedly increases the acidic properties, increases the stability towards hydrolysis and decreases the solubility.

Hydroxypteridines are commonly prepared from the corresponding 4,5-diaminopyrimidine and a dicarbonyl compound.

Pteridin-2-one monohydrate is prepared by condensation of 4,5-diaminopyrimidine-2-one and glyoxal or by hydrolysis of 2-chloropteridine (*Albert*, *Brown* and *Cheeseman*, J. chem. Soc., 1951, 474). Its formation from 2-hydroxypurine and glyoxal in water at 37° has been claimed to be of possible biological significance (*Albert*, Biochem. J., 1957, **65**, 124). It is insoluble in the common organic solvents and crystallises from water as needles of the 3,4-*monohydrate* (m.p. > 240°, decomp.), which loses water only on heating above 180°.

Adducts with methanol and ethanol are formed on prolonged heating of pteridin-2-one with these solvents. Michael-type reagents add in a similar manner (*Albert* and *C. F. Howell*, J. chem. Soc., 1962, 1591). It is easily reduced to the 3,4-dihydropteridine (*Albert* and *S. Matsuura*, *ibid.*, 1961, 5131) and it dimerises in alkaline solution to give [1,1′,4,4′-tetrahydro-4,4′-bipteridinylidene]-1,2′-diol (VII) (*Albert* and *F. Reich*, J. chem. Soc., 1960, 1370). Methylation gives 1-methylpterid-2-one (*Albert*, *Brown* and *H. C. S. Wood*, *ibid.*, 1956, 2066).

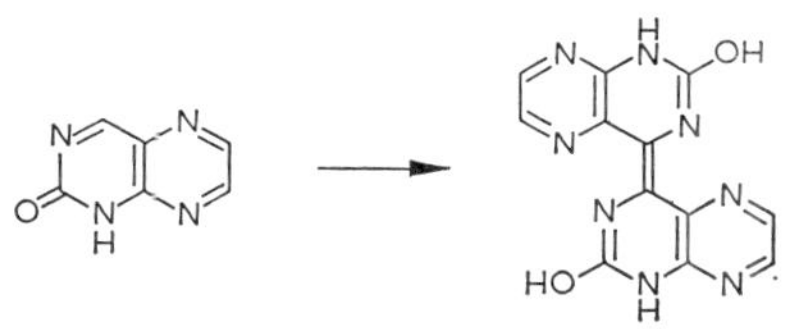

(VII)

Pteridin-4-one, m.p. > 350°, is obtained from 5,6-diaminopyrimidin-4-one and glyoxal or alternatively by cyclisation of 2-aminopyrazine-3-carboxamide with acetic anhydride and either ethyl orthoformate or formic acid (*Albert*, *Brown* and *Cheeseman*, *loc. cit.*). Oxidation of pteridine and hydrolysis of the corresponding amino and methoxy derivatives also give pteridin-4-one.

Pteridin-4-one many be crystallised from water, formamide or diethylformamide and forms stable chelate complexes across the hydroxyl group and $N_{(5)}$. It is degraded by hot dilute acid to 2-aminopyrazine-3-carboxylic acid, its amide and 2-aminopyrazine. Similarly, alkaline cleavage gives 2-aminopyrazine-3-carboxylate and 5,6-diaminopyrimidine-4-one (*Albert*, J. chem. Soc., 1955, 2690). Methylation with diazomethane gives a mixture of 4-methoxypteridine and 3-methylpteridin-4-one, with dimethyl sulphate gives a mixture of 1- and 3-methylpteridin-4-ones (*Albert*, *Brown* and *Wood*, *loc. cit.*) and with methyl iodide gives 1,3-dimethyl-4-oxopteridinium iodide (*Z. Neiman*, *F. Bergmann* and *A. K. Meyer*, J. chem. Soc., C, 1969, 2415).

Pteridin-6-one monohydrate, m.p. > 240° (decomp.), is formed by condensation of 4,5-diaminopyrimidine and ethyl glyoxylate hemiacetal in 2 *N* sulphuric acid. It is distinguishable from other hydroxypteridines by titration which produces a hysteresis loop caused by hydration of the 7,8-double bond.

Pteridin-6-one is reduced by sodium amalgam to the 7,8-dihydro derivative, which can be reoxidised to pteridin-6-one by potassium permanganate; further oxidation by hydrogen peroxide gives pteridine-6,7-dione. It is stable in strong acid solution but with 1 *N* sodium hydroxide at 20°, complete decomposition occurs in five days to give pteridine-6,7-dione and 7,8-dihydropteridin-6-one and a dimer (*Albert*, J. chem. Soc., 1955, 2690).

Pteridin-7-one, prepared from 4,5-diaminopyrimidine and ethyl glyoxylate hemiacetal at pH 6, decomposes above 230°. Reduction with dithionite gives 4-*amino-5-carboxymethylaminopyrimidine*, m.p. 310° (decomp.).

Pteridine-2,4-dione (lumazine) was first prepared by oxidation of tolualloxazine (*O. Kühling*, Ber., 1894, **27**, 2116). Many other methods of preparation have been described. Thus the action of potassium hypobromite on pyrazine-2,3-dicarboxamide (*S. Gabriel* and *A. Sonn*, *ibid.*, 1907, **40**, 4850) or the interaction of 4,5-diaminouracil and glyoxal (*R. Kuhn* and *A. H. Cook*, *ibid.*, 1937, **70**, 761) both gave lumazine. It melted at 348–349° (*J. Weijlard*, *M. Tishler* and *A. E. Erickson*, J. Amer. chem. Soc., 1945, **67**, 802). Lumazine is also the product of acid or alkaline hydrolysis or diazotisation of 4-aminopteridin-2-one. Alternatively, oxidation of pteridin-2-one with potassium permanganate gives lumazine (*Brown* and *Mason*, *loc. cit.*). Recently it has been isolated from the ant *Formica polyctena*, along with several other lumazine derivatives (*G. H. Schmidt* and *M. Viscontini*, Helv., 1967, **50**, 34). It is hydrolysed by concentrated sulphuric acid to 2-aminopyrazine and by 20% sodium hydroxide or alkylamines to 3-aminopyrazine-2-carboxylic acid or its amides, respectively (*E. C. Taylor et al.*, J. Amer. chem. Soc., 1953, **75**, 1904). Methylation with diazomethane gives 1,2-dimethyllumazine (*Albert*, *Brown* and *Wood*, *loc. cit.*).

6,7-**Dimethyl**-8-D-**ribityllumazine,** m.p. 273–274°, $[\alpha]_D^{20}$ −164°, was first isolated from *Eremothecium ashybii* (*T. Masuda*, Chem. Pharm. Bull., Japan, 1956, **4**, 71) and the structure was determined by degradation (*idem*, *ibid.*, 1956, **4**, 375; 1957, **5**, 28). The lumazine was also isolated from *Ashbya gossypii* and the structure confirmed by synthesis (*G. F. Maley* and *G. W. E. Plaut*, Fed. Proc., 1958, **17**, 168; J. biol. Chem., 1959, **234**, 641). Several other syntheses have been described (*Pfleiderer* and *G. Nübel*, Chem. Ber., 1960, **93**, 1406; *Masuda et al.*, Chem. Pharm. Bull., Japan, 1959, **7**, 361; *J. Davoll* and *D. D. Evans*, J. chem. Soc., 1960, 5041; *C. H. Winestock* and *Plaut*, J. org. Chem., 1961, **76**, 4456). *Plaut* (J. biol. Chem., 1960, **235**, PC 41; 1963, **238**, 2225) has established this compound as the biosynthetic precursor of riboflavin (see riboflavin section below).

6-**Methyl**-7-**oxo**-8-D-**ribityllumazine,** m.p. 267° (decomp.), was isolated from *E. ashybii* (*Masuda*, Chem. Pharm. Bull., Japan, 1956, **4**, 72) and from *A. gossypii* (*Plaut* and *Maley*, J. biol. Chem., 1959, **234**, 3010). The structure was determined by degradation (*Masuda*, *T. Kishi* and *M. Asai*, Chem. Pharm. Bull., Japan, 1958, **6**, 29) and confirmed by synthesis (*Plaut* and *Maley*, Arch. Biochem. Biophys., 1959, **80**, 219; *W. S. McNutt*, J. Amer. chem. Soc., 1960, **82**, 217). The compound has also been obtained by oxygenation of hot alkaline solutions of 6,7-dimethyl-8-D-ribityllumazine, a process which proceeds *via* intermediate hydration at the 7 position of the dialkyllumazine (*T. Rowan* and *H. C. S. Wood*, J. chem. Soc., C, 1968, 452). There is evidence for this transformation occurring biologically also (*S. Kuwada et al.*, Chem. Pharm. Bull., Japan, 1958, **6**, 619). A similar oxidation of 6,7,8-trimethyllumazine using potassium permanganate has been described (*K. Oyama* and *R. Stewart*, Canad. J. Chem., 1974, **52**, 3879).

8-Alkyl-lumazines add one molecule of water in alkaline solution to give 7,8-dihydro-7-

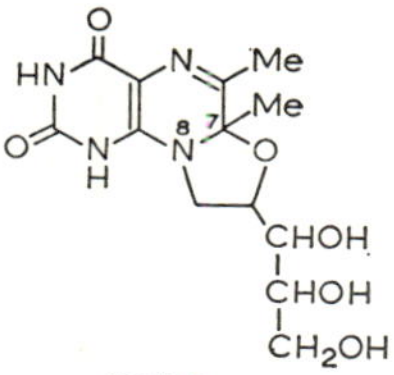

(VIII)

hydroxylumazines and in the case of 8-D-ribityllumazines subsequent cyclisation occurs to give oxazine derivatives of the type VIII (*Pfleiderer et al.*, Chem. Ber., 1966, **99**, 3503).

T. Paterson and *Wood* (Chem. Comm., 1969, 290) showed that the 7-methyl protons of 6,7-dimethyl-8-D-ribityllumazine and related compounds were exchanged for deuterium by refluxing for 15 min in deuterium oxide or by standing at room temperature with sodium deuteroxide (pH 13). Similarly the 7-methyl protons of 6,7,8-trimethyllumazine were readily exchanged with deuterium, whereas, under acid catalysis the 6-methyl protons of 7,8-dihydro-6,7,8-trimethyllumazine were exchanged (*J. M. McAndless* and *R. Stewart*, Canad. J. Chem., 1970, **48**, 263).

The methyl group in 7-methyllumazine is ionised by aqueous base to give the carbanion α- to the aromatic pyrazine ring. This carbanion condenses with aromatic aldehydes to give the 7-aralkylidene derivatives. This reaction has been described also for the 2,4-diaminopteridine series (*E. M. Kaiser* and *S. L. Hartzell*, J. org. Chem., 1977, **42**, 2951).

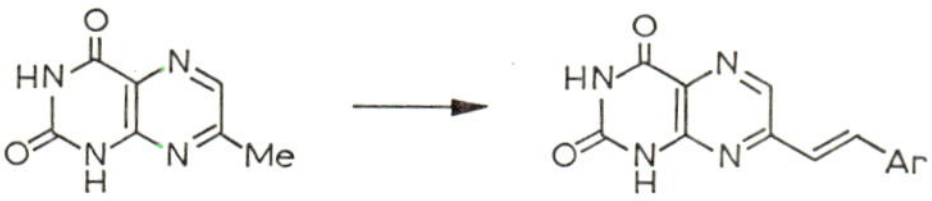

Pteridine-2,6-dione monohydrate, m.p. > 250° (decomp.), is prepared from 4,5-diaminopyrimidine-2-one and ethyl glyoxylate hemiacetal in 1 *N* hydrochloric acid or alternatively by oxidation of the 7,8-dihydro compound. The yellow aqueous solution of its sodium or potassium salt has an intense green fluorescence which fades in daylight (*A. Albert*, *J. H. Lister* and *C. Pedersen*, J. chem. Soc., 1956, 4621).

Pteridine-2,7-dione monohydrate is prepared using the above condensation reaction in water, and not hydrochloric acid, as the solvent. The product is purified *via* the sodium salt; the contaminating product is the 2,6-isomer which is formed in equal amounts.

Pteridine-4,6-dione is prepared from pteridine-4,6,7-trione by reduction with sodium amalgam followed by oxidation of the resulting 7,8-dihydro compound. This method is specific for the removal of a 7-hydroxyl group in the presence of the 6-hydroxyl. An alternative preparation is from ethyl glyoxylate and 5,6-diaminopyrimidin-4-one (*Albert* and *Brown*, J. chem. Soc., 1953, 74; *Pfleiderer*, Chem. Ber., 1959, **92**, 3190).

Pteridine-4,7-dione, m.p. > 350°, is obtained from the same pyrimidine and ethyl oxaloacetate followed by hydrolysis and decarboxylation (*Albert* and *Brown*, *loc. cit.*). Other methods using ethyl glyoxylate (*Pfleiderer*, *loc. cit.*) and chloral (*Albert*, *Lister* and *Pedersen*, *loc. cit.*) have been described.

Pteridine-6,7-dione may be purified *via* the potassium salt as large plates, m.p. > 350°. It is prepared from 4,5-diaminopyrimidine and excess of crystalline oxalic acid by slowly heating to 160° and maintaining this temperature for 2 hours (*Albert*, *Brown* and *Cheeseman*, J. Chem. Soc. 1952, 1620). Oxidation of either pteridin-6-one or pteridin-7-one gives the 6,7-dione.

Other polyhydroxypteridines which have been reported are **pteridine-2,4,6-trione** (*S. Nawa et al.*, J. Amer. chem. Soc., 1953, **75**, 4450; *Albert*, *Lister* and *Pedersen*, *loc. cit.*); **pteridine-2,4,7-trione** (*R. Tchesche* and *F. Korte*, Chem. Ber., 1951, **84**, 801); **pteridine-4,6,7-trione** (*Albert*, *Lister* and *Pedersen*, *loc. cit.*); **pteridine-2,4,6,7-tetraone** (*A. Bertho* and *M. Bentler*, Ann., 1950, **570**, 127).

(*vi*) *Methoxypteridines*

The methoxypteridines are prepared either from the pyrimidines or from the corresponding chloropteridines with sodium methoxide.

2-**Methoxypteridine,** m.p. 149–151° (*Albert*, *Brown* and *Cheeseman*, J. chem. Soc., 1952, 4219) is prepared from the methoxypyrimidine. 4-**Methoxypteridine,** m.p. 195°, is prepared similarly. It is soluble in 40 parts of boiling or 100 parts of cold benzene. On treatment with 1 *N* sodium hydroxide for 2 hours pteridin-4-one is formed. 6-**Methoxy-**, m.p. 124–125° (*idem*, *ibid.*, 1952, 1620) and 7-**methoxy-pteridine,** m.p. 130° (*Albert*, *Brown* and *Wood*, *ibid.*, 1954, 3832) are prepared from the corresponding chloro compounds. The 7-isomer is instantly converted to the insoluble sodium salt of pteridin-7-one by cold 1 *N* sodium hydroxide. 2,4-**Dimethoxypteridine,** m.p. 195–197°, is prepared from the 2,4-dimethoxy-5,6,7,8-tetrahydropteridine by oxidation with manganese dioxide or alternatively from 2,4-dichloropteridine and sodium methoxide (*E. C. Taylor* and *W. Sherman*, J. Amer. chem. Soc., 1959, **87**, 2464).

(*vii*) *Mercapto- and alkylthio-pteridines*

The mercapto- and alkylthio-pteridines are useful intermediates for use in synthesis because the sulphur-containing groups are easily replaced by other groups such as alkylamino or hydroxyl. They are prepared from mercapto- or alkylthio-4,5-diaminopyrimidines by the Isay reaction, or from the corresponding hydroxy- or chloro-derivative by treatment with phosphorus pentasulphide or sodium hydrogen sulphide, respectively. The mercapto compounds are readily *S*-methylated by methyl iodide in aqueous alkali. The synthesis of several thiolumazines has been described using the Isay reaction; the correct tautomeric structures of the products were established on the basis of pK_a values and u.v. spectra (*H. J. Schneider* and *Pfleiderer*, Chem. Ber., 1974, **107**, 3377).

Pteridine-2-thione, m.p. 200–205° (decomp.), is prepared from 4,5-diaminopyrimidine-2-thione and glyoxal (*G. B. Elion* and *G. H. Hitchings*, J. Amer. chem. Soc., 1947, **69**, 2553). This pteridine is easily reduced to the 3,4-dihydro derivative by sodium tetrahydridoborate and also reacts readily with sodium hydrogen sulphite and "Michael reagents" (*Albert* and *J. J. McCormack*, J. chem. Soc., C, 1968, 63). 2-**Methylthiopteridine,**

m.p. 133–136°, may be prepared either from pteridine-2-thione and methyl iodide or from the corresponding pyrimidine.

Pteridine-4-thione, m.p. 290° (decomp.), is obtained as yellow crystals from 2-aminopyrazine-3-thioamide with ethyl formate and acetic anhydride or better from 5,6-diaminopyrimidine-4-thione and glyoxal (*Albert, Brown* and *Cheeseman*, J. chem. Soc., 1951, 474). On methylation 4-**methylthiopteridine**, m.p. 191°, is formed (*Albert, Brown* and *Wood, loc. cit.*). Neither of these compounds is desulphurised by Raney nickel. Pteridine-4-thione is converted by sodium carbonate/thioacetic acid to 2-aminopyrazine-3-carbonitrile. In dilute acid, pteridin-4-one is obtained (*Taylor et al.*, J. Amer. chem. Soc., 1960, **82**, 6058).

Pteridine-6-thione, m.p. > 160°, is prepared from 6-chloropteridine and sodium hydrogen sulphide (*Albert* and *J. Clark*, J. chem. Soc., 1965, 27). This is methylated by methyl iodide to give 6-*methylthiopteridine*, m.p. 163°, which is easily hydrolysed to pteridin-6-one.

Pteridine-7-thione is prepared from the corresponding hydroxy compound by treatment with phosphorus pentasulphide (*Albert, Brown* and *Wood, loc. cit.*).

7-**Methylmercaptopteridine** has m.p. 142–143°.

Pteridine-6,7-dithione, m.p. > 250°, is prepared from 6,7-dichloropteridine and sodium hydrogen sulphide. It is unstable in boiling water, slightly unstable at pH 0 and pH 6.8 but more stable at pH 13 (*Albert* and *Clark*, J. chem. Soc., 1965, 27).

The thionation of 2-chloro-, 4-chloro- and 2,4-dichloro-6,7-dimethylpteridines, with sodium hydrogen sulphide yields the corresponding *pteridine-2-thione*, -4-*thione* and -2,4-*dithione*, which have m.p.'s 220° (decomp.), 300° and 250–254° (decomp.), respectively (*J. W. Daly* and *B. E. Christensen*, J. Amer. chem. Soc., 1956, **78**, 225).

(*viii*) *Aminopteridinones*

The majority of the naturally occurring pteridines are derivatives of 2-aminopteridin-4-one. They are widely distributed but the role of the simple compounds has not been clarified. 2-Aminopteridine-4,6-dione (xanthopterin), 2-aminopteridine-4,7-dione (isoxanthopterin) and 2-aminopteridine-4,6,7-trione (leucopterin) are found as pigments in certain insects. Biopterin is a nutritional requirement for *Crithidia fasciculata*, a trypanosomid parasite of mosquitoes (*H. A. Nathan* and *J. Cowperthwaite*, J. Protozool., 1955, **2**, 37). Pteridines are also known to be cofactors in the enzymic hydroxylation of phenylalanine to tyrosine, tyrosine to "dopa" (*S. Kaufman*, Trans. N.Y. Acad. Sci., 1964, **26**, 977), cinnamic acid to *p*-coumaric acid (*P. M. Nair* and *L. C. Vining*, Phytochem., 1965, **4**, 161) and in the oxidation of long-chain alkyl esters of glycerol to fatty acids (*A. Tietz, M. Lindberg* and *E. P. Kennedy*, J. biol. Chem., 1964, **239**, 4081).

Pterin, 2-*aminopteridin*-4-*one*, m.p. > 360°, is commonly synthesised from 2,5,6-triaminopyrimidin-4-one and glyoxal (*C. K. Cain, M. F. Mallette* and *E. C. Taylor*, J. Amer. chem. Soc., 1946, **68**, 1996; *W. Pfleiderer et al.*, Chem. Ber., 1960, **93**, 2015) but it has also been made from the pyrazine intermediate, methyl 3-chloropyrazine-2-carboxylate and guanidine. The product is best purified *via* the sodium salt (*G. P. G. Dick* and *H. C. S. Wood*, J. chem. Soc., 1955, 1379). Methylation of pterin gives the unexpected $N_{(8)}$-methylated product shown (*D. J. Brown* and *N. W. Jacobsen*, *ibid.*, 1961, 4413).

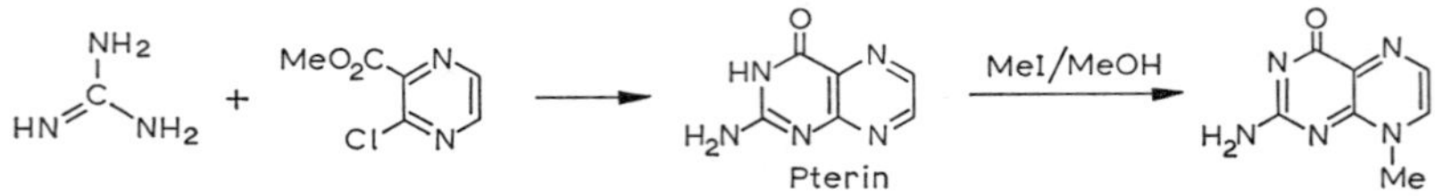

Sepiapterin, a yellow pigment obtained from the wing of the butterfly and from the eye of the fruit fly, has the structure 2-amino-6-(*S*)(2-hydroxypropionyl)-7,8-dihydropteridin-4-one. The structure was established by oxidative degradation (*S. Nawa*, Bull. chem. Soc., Japan, 1960, **33**, 1555) and by n.m.r. spectroscopy (*H. S. Forrest* and *Nawa*, Nature, 1962, **196**, 372).

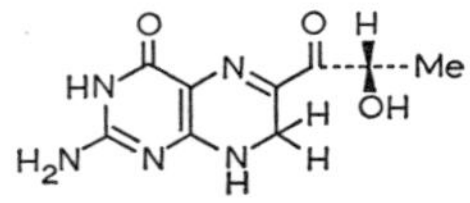
Sepiapterin

A synthesis of sepiapterin, in 20% isolated yield, by autoxidation of 5,6,7,8-tetrahydrobiopterin, at pH 4, has been described. This interconversion provides evidence for the *S*-configuration at the asymmetric centre; this evidence was supported also by c.d. studies (*Pfleiderer*, in "Chemistry and Biology of Pteridines", De Gruyter, Berlin, 1975, p. 941). The absolute stereochemistry has now been confirmed by degradation of sepiapterin to L-lactic cid (*S. Matsuura*, *T. Sugimoto* and *M. Tsusue*, Bull. chem. Soc., Japan, 1977, **50**, 2168).

Xanthopterin, 2-*aminopteridine*-4,6-*dione*, m.p. > 350°, is a yellow pigment which was first isolated from butterfly wings and later found in other insects, in the skin of amphibia and in human urine. The first synthesis was from 2,5,6-triaminopyrimidin-4-one by condensation with dichloroacetic acid, followed by cyclisation of the resulting amide with silver carbonate (*R. Purrmann*, Ann., 1950, **546**, 98; 1941, **548**, 284). Instead of dichloroacetic acid, glyoxylates or chloral have been used (*W. Koschara*, Z. physiol. Chem., 1943, **277**, 159; *F. Korte* and *E. G. Fuchs*, Chem. Ber., 1953, **86**, 114; *W. E. Fidler* and *Wood*, J. chem. Soc., 1956, 3311).

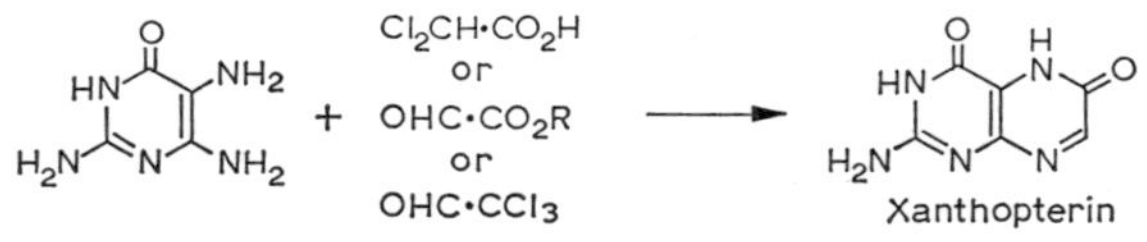

As an alternative route, reduction of leucopterin with sodium amalgam giving 7,8-dihydroxanthopterin followed by oxidation provided xanthopterin (*J. R. Totter*, J. biol. Chem., 1944, **154**, 105; *G. B. Elion et al.*, J. Amer. chem. Soc., 1949, **71**, 741; *Albert* and *Wood*, J. appl. Chem., 1952, **2**, 591). *A. Stuart* and *Wood* (J. chem. Soc., 1963, 4186) reported the formation of xanthopterin from 1-(2-amino-6-hydroxy-5-nitropyrimidin-4-ylamino)-1-deoxy-D-*erythro*pentulose by cyclisation and reduction to give 7,8-dihydroxanthopterin followed by oxidation and they also suggested a possible mode of biosynthesis. Regiospecific syntheses of xanthopterin and isoxanthopterin have been reported using the Taylor method from the appropriately substituted pyrazines, however, the method of choice for the synthesis of xanthopterin involves a quantitative isomerisation of pterin 8-*N*-oxide with trifluoroacetic acid/trifluoroacetic anhydride (*E. C. Taylor et al.*, J. org. Chem., 1975, **40**, 2341).

Isoxanthopterin, m.p. > 350°, is a colourless substance first isolated from the wings of the cabbage white butterfly. It was first synthesised in 1941 by *Purrmann* (Ann., 1941, **548**, 284) who condensed 2,5,6-triaminopyrimidin-4-one with ethyl oxomalonate to give isoxanthopterin-6-carboxylic acid, decarboxylation of which at 260° gave isoxanthopterin. An alternative method using ethyl glyoxylate hemiacetal has been described (*Albert* and *Wood*, J. appl. Chem., 1953, **3**, 521). The structure was determined as 2-aminopteridine-4,6-dione, from u.v. spectra and pK_a data (*Pfleiderer* and *M. Rukwied*, Chem. Ber., 1961, **94**, 1).

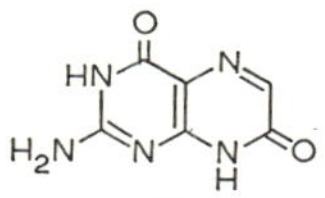

Isoxanthopterin

Leucopterin, m.p. > 350°, is prepared from 2,5,6-triaminopyrimidin-4-one and oxalic acid by fusion (*Purrmann*, Ann., 1940, **544**, 182; *Totter, loc. cit.*; *Albert* and *Wood*, J. appl. Chem., 1952, **2**, 591). Catalytic oxidation of xanthopterin produces leucopterin (*Purrmann, loc. cit.*).

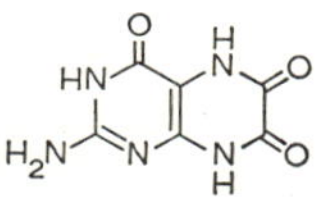

Leucopterin

With diazomethane in the presence of methanol leucopterin gives two isomeric trimethyl derivatives (*H. Wieland* and *P. Decker*, Ann., 1941, **547**, 180; *Pfleiderer* and *Rukwied*, Chem. Ber., 1961, **94**, 118). Phosphorus pentachloride and phosphoryl chloride convert leucopterin to the 4-chloro derivative and hydriodic acid gives 2-aminopteridine-6,7-dione, also available by the condensation of 2,5,6-triaminopyrimidin-4-one with oxalic acid (*Purrmann, loc. cit.*).

Asperopterin B is the aglycone of **asperopterin A** both of which have been isolated from *Aspergillus oryzae* (*Y. Keneko* and *M. Sanda*, Hakko Kogaku Zasshi, 1969, **47**, 8). A

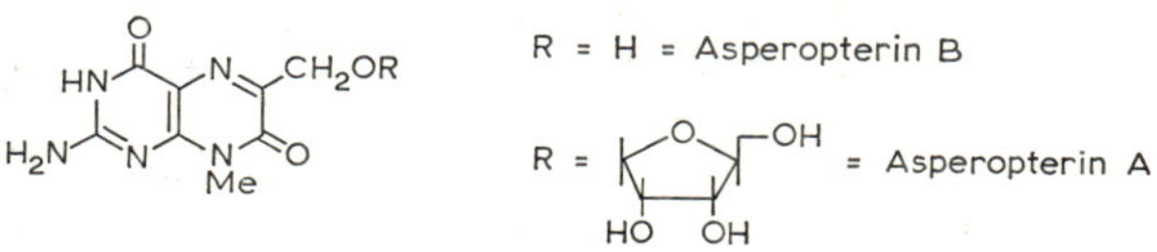

synthesis of asperopterin B has been described using the pyrazine *N*-oxide approach (*Taylor*, *R. F. Abdulla* and *P. A. Jacobi*, J. org. Chem., 1975, **40**, 2336).

L-*erythro*-Biopterin, isolated from human urine and also from *Drosophila melanogaster* (*Forrest* and *H. K. Mitchell*, J. Amer. chem. Soc., 1955, **77**, 4865), has the structure 2-amino-6-(L-*erythro*-1,2-dihydroxypropyl)pteridin-4-one (*E. L. Patterson*, *R. Milstrey* and *E. L. R. Stokstad*, *ibid.*, 1956, **78**, 5866; *Patterson*, *M. H. V. Saltra* and *Stokstad*, *ibid.*, p. 5871). The first synthesis of biopterin involved condensation of 2,5,6-triaminopyrimidin-4-one with 5-deoxy-L-arabinose. This preparation produced a difficultly separable mixture and a more convenient synthesis was developed later from 1-(2-amino-6-hydroxy-5-nitro-pyrimidin-4-ylamino)-1,5-dideoxy-L-*erythro*-pentulose by reduction of the nitro group, followed by spontaneous cyclisation and oxidation which gave biopterin (*K. J. M. Andrews*, *W. E. Barber* and *B. P. Tong*, J. chem. Soc., C, 1969, 928). An unequivocal total synthesis of biopterin using the pyrazine *N*-oxide route has been described by *Taylor* and *Jacobi* (J. Amer. chem. Soc., 1974, **96**, 6781). The key step involves the condensation of the α-oxoaldoxime (generated from 5-deoxy-L-arabinose *in situ*) with benzyl α-amino-cyanoacetate methanesulphonate to give the pyrazine *N*-oxide. This is then elaborated in the standard manner in two steps to give L-*erythro*biopterin.

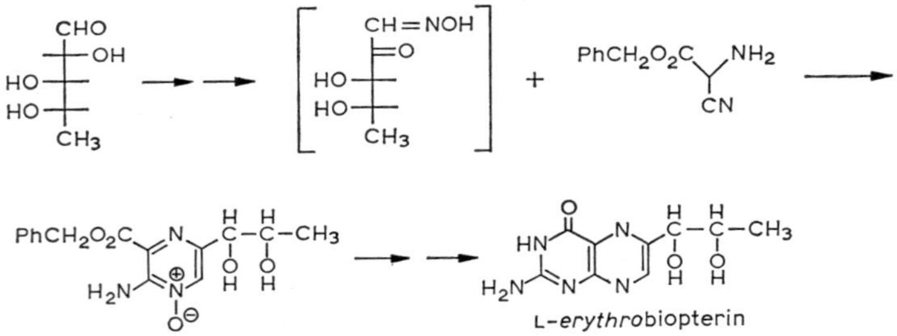

Two other syntheses based on condensations with 2,5,6-triaminopyrimidin-4-one have been reported (*T. Sugimoto* and *S. Matsuura*, Bull. chem. Soc., Japan, 1975, **48**, 3767; *B. Schircks*, *J, H. Bieri* and *M. Viscontini*, Helv., 1977, **60**, 211).

Kaufman reported in 1959 that certain synthetic pteridines could replace the essential naturally-occurring co-factor for the enzymatic conversion of phenylalanine to tyrosine (J. biol. Chem., 1959, **234**, 2677). Later the naturally occurring pteridine was shown to be **dihydrobiopterin** which, in the process of hydroxylation, is first reduced to tetrahydrobio-pterin by the enzyme dihydrofolate reductase. The tetrahydro derivative is oxidised by molecular oxygen to the quinonoid dihydro derivative shown, during the coupled hydroxy-lation of phenylalanine to tyrosine by phenylalanine hydroxylase. The dihydropteridine is reduced back to the tetrahydro derivative by the NADPH-dependent enzyme dihydro-pteridine reductase (*idem*, Ann. Review Biochem., 1967, **36**, 171).

Ichthyopterin, isolated from carp (*R. Hüttel* and *G. Sprengling*, Ann., 1943, **554**, 69) was identified as 6-(1,2-dihydroxypropyl)isoxanthopterin by synthesis from 6-acetoxyiso-xanthopterin (*R. Tschesche* and *A. Glaser*, Chem. Ber., 1958, **91**, 2081). The identity of ichthyopterin and 7-hydroxybiopterin has not been established unequivocally; this will

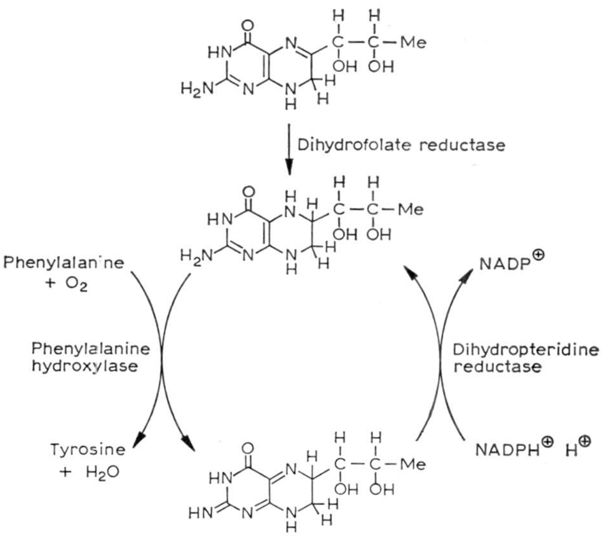

await confirmation of the hypothetical L-*erythro* configuration of the side-chain in the former.

Neopterin was first isolated from the larvae of the honey bee and from "royal jelly" (*H. Rembold* and *L. Buschman*, Ann., 1963, **662**, 72; Chem. Ber., 1963, **96**, 1406). The structure was shown to be 2-amino-6-(D-*erythro*-1,2,3-trihydroxypropyl)pteridin-4-one. The configuration was established by comparison with the four synthetic diastereoisomers which were made by condensation of 2,5,6-triaminopyrimidin-4-one with the appropriate pentoses.

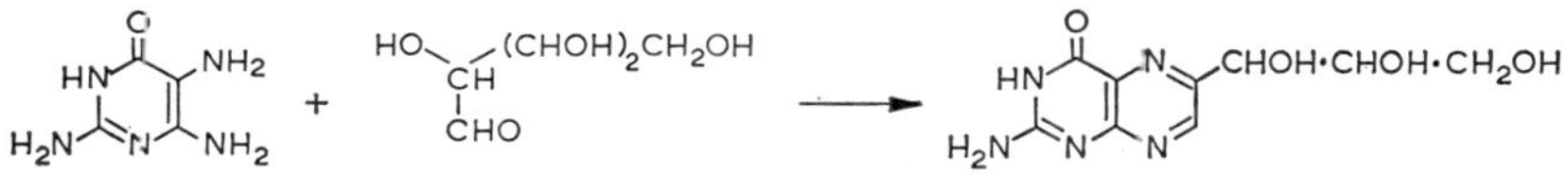

Neopterin

This procedure suffered from the problem of separation of the product from the corresponding 7-isomer which was also formed and it was subsequently improved by *J. M. Lagowski*, *Forrest* and *Wood* (Proc. chem. Soc., 1963, 343) who cyclised 1-(2-amino-6-hydroxy-5-nitropyrimidin-4-ylamino)-1-deoxy-L-*erythro*pentulose to L-dihydroneopterin

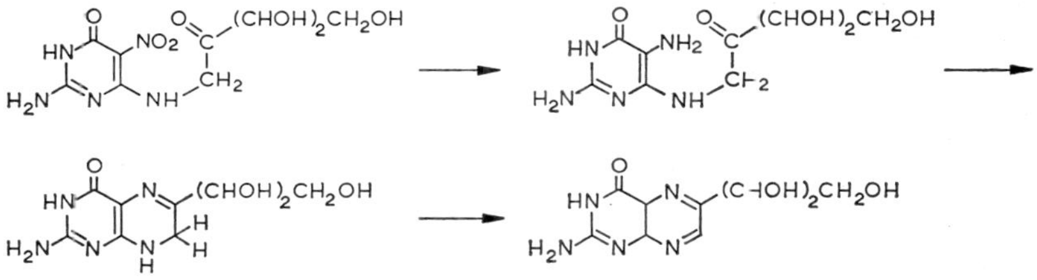

which was oxidised to L-neopterin in acid solution. A similar synthesis has provided L- and D-neopterins (*Andrews*, *Barber* and *Tong*, *loc. cit.*).

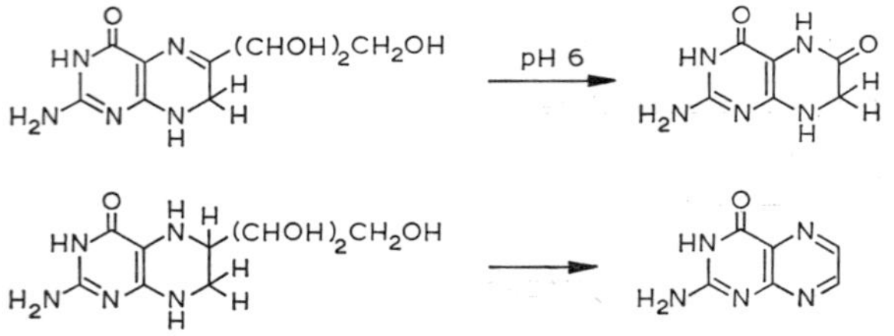

Oxidation of 7,8-dihydroneopterin at pH 6 afforded dihydroxanthopterin (*Stuart* and *Wood*, *loc. cit.*) and aerobic oxidation of tetrahydroneopterin formed pterin (*Rembold et al.*, Biochim. Biophys. Acta, 1969, **184**, 386).

Erythropterin was degraded to 7-methylxanthopterin and oxalic acid by *Pfleiderer* (Chem. Ber., 1962, **95**, 2195). This degradation and n.m.r. studies (*W. von Philipsborn*, *H. Stierlin* and *W. Traber*, in *Pfleiderer* and *Taylor*, eds., "Pteridine Chemistry", Pergamon Press, London, 1964, p. 169) established the structure shown.

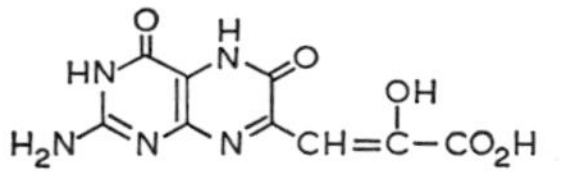

Erythropterin

Erythropterin has been synthesised from xanthopterin and oxaloacetic acid (*C. Schöpf* and *K. H. Gänshirt*, Angew. Chem. intern. Edn., 1962, **1**, 115) and from xanthopterin and pyruvic acid (*M. Viscontini* and *Stierlin*, Helv., 1961, **44**, 1783; 1963, **46**, 51).

Pterorhodin is a red pigment found in the wings of the butterfly *Appias nero* (*Pfleiderer*, Z. Naturforsch., 1963, **18B**, 420). It has been synthesised from xanthopterin and also from 7-methylxanthopterin (*Purrmann* and *M. Maas*, Ann., 1944, **556**, 186; *P. B. Russell et al.*, J. Amer. chem. Soc., 1949, **71**, 3412).

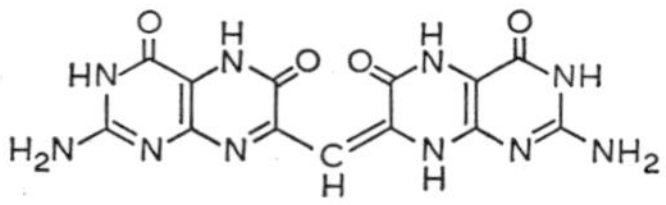

Pterorhodin

Drosopterins were first isolated from the eyes of the fruitfly *Drosophila melanogaster* and consist of three kinds of red-pigments which are separable on cellulose and are named **drosopterin, neodrosopterin** and **isodrosopterin** (*Viscontini et al.*, Helv., 1955, **38**, 397). The

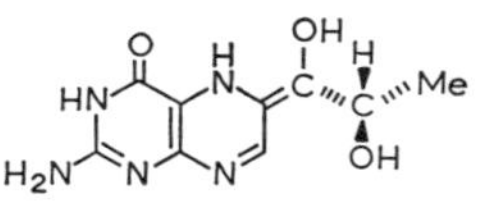

Drosopterin

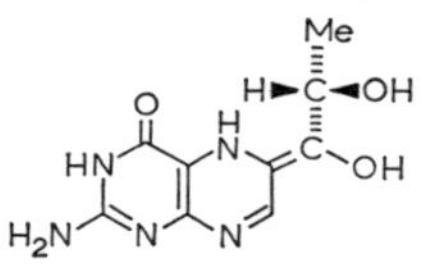

Isodrosopterin

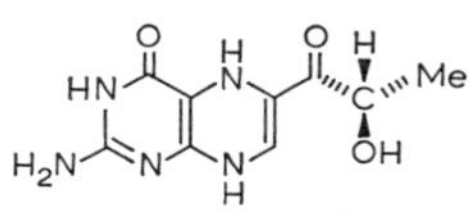

Neodrosopterin

isolation and structural determination of the red pigments were pursued by *Viscontini* and coworkers (*ibid.*, 1957, **40**, 579, 968; 1958, **41**, 922, 1299; 1959, **42**, 1679) who proposed the biopterin-like structures shown. These, however, have not been fully established.

An attempted synthesis of **sepiapterin** from reduced 2-aminopteridin-4-one and 3-hydroxy-2-oxobutanoic acid (*Forrest* and *Nawa*, ih "Pteridine Chemistry", *Pfleiderer* and *Taylor*, eds., Pergamon Press, Oxford, 1964, p. 281) unexpectedly produced drosopterins. Later the structures of the drosopterins were investigated by Japanese workers (*K. Sugiura*, *M. Goto* and *Nawa*, Tetrahedron Letters, 1969, 2963; *Sugiura* and *Goto*, *ibid.*, 1970, 4059) and by German workers. With the aid of synthetic materials, the latter group proposed the atropisomeric dimeric pteridine structure for drosopterin and isodrosopterin (*H. Schlobach* and *Pfleiderer*, Angew. Chem. intern. Edn., 1971, **10**, 414; Helv., 1972, **55**, 2518, 2525, 2533, 2541; *K. Rokos*, in *Pfleiderer*, *op. cit.*, pl. 931).

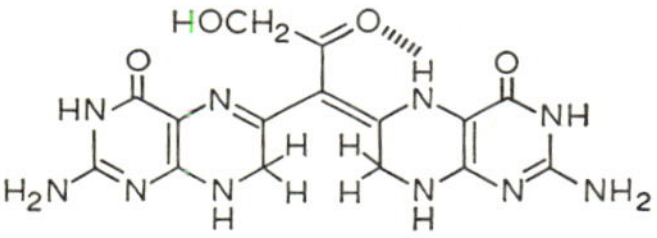

Recently, however, *Sugiura* and *Goto* have claimed that the terminal methyl group of 3-hydroxy-2-oxobutanoate is not incorporated into the α-carbon atom of the side-chain of the drosopterins as would be required by the above structure (Tetrahedron Letters, 1973, 1187). Because of this observation and also as a consequence of further chemical and spectroscopic investigations, a new structural assignment has been made of the enantiomeric droso- and isodroso-pterins (*N. Theobald* and *Pfleiderer*, *ibid.*, 1977, 841).

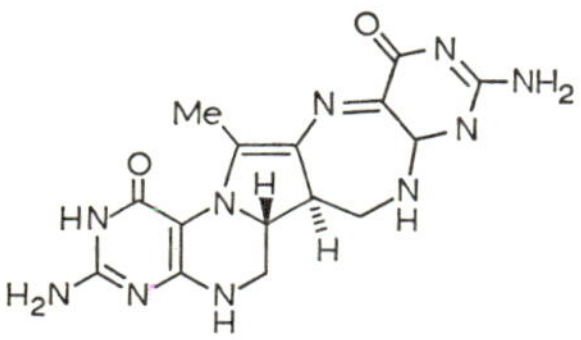

Urothione, isolated from human urine (*W. Koschara*, Z. physiol. Chem., 1940, **263**, 78), is a sulphur-containing pteridine, the structure of which was determined by *R. Tschesche*, *F. Korte* and *G. Heuschkel* (Chem. Ber., 1955, **88**, 1251) as

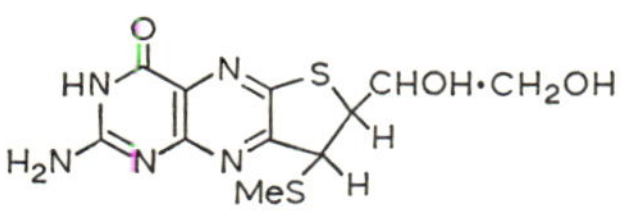

Many other simply substituted aminopteridinones have been reported.

2-Aminopteridin-7-one (*Pfleiderer* and *Rukwied*, Chem. Ber., 1961, **94**, 1); **2-amino-pteridine-6,7-dione** (*Wieland*, *A. Tartter* and *Purrmann*, Ann., 1940, **545**, 209; *Albert*, *Lister* and *Pedersen*, J. chem. Soc., 1956, 4621); **4-aminopteridin-2-one** (*Taylor* and *C. K. C Cain*, J. Amer. chem. Soc., 1949, **71**, 2538; *Albert*, *Brown* and *G. Cheeseman*, J. chem. Soc., 1952, 4219; *Pfleiderer* and *H. Fink*, Chem. Ber., 1963, **96**, 2950); **4-aminopteridin-6-one**

(*D. Soll* and *Pfleiderer*, *ibid.*, 1963, **96**, 2977); **4-aminopteridin-7-one** (*idem. ibid.*); **4-amino-pteridine-2,7-dione** (*Albert, Lister* and *Pedersen, loc. cit.*); **4-aminopteridine-6,7-dione** (*Wieland* and *R. Liebig*, Ann., 1944, **555**, 146); **4-aminopteridine-2,6,7-trione** (*idem, ibid.*); **6-aminopteridin-7-one** (*Albert* and *J. Clark*, J. chem. Soc., 1965, 27); **7-aminopteridin-4-one** (*Brown* and *Jacobsen, ibid.*, 1965, 1175); **7-aminopteridin-6-one** (*Albert* and *Clark, ibid.*, 1965, 27); **2,4-diaminopteridin-6-one** and **2,4-diaminopteridin-7-one** (*G. Konrad* and *Pfleiderer*, Chem. Ber., 1970, **103**, 735).

(*ix*) *Pteridinecarboxylic acids*

Several methods of preparation of these derivatives, which are important for the characterisation of pteridines, have been described as follows:

(*1*) Oxidation of alkylpteridines with alkaline potassium permanganate solution (*J. H. Mowat et al.*, J. Amer. chem. Soc., 1948, **70**, 14).

(*2*) From 4,5-diaminopyrimidines, *e.g.* 5,6-diaminopyrimidin-4-one and ethyl sodio-oxaloacetate give 6-methoxycarbonylpteridine-4,7-dione (*A. Albert* and *D. J. Brown*, J. chem. Soc., 19653, 74).

(3) Condensation of 4-amino-5-nitrosopyrimidines with ethyl malonate or cyanoacetic acid (*T. S. Osdene* and *G. M. Timmis, ibid.*, 1955, 2036).

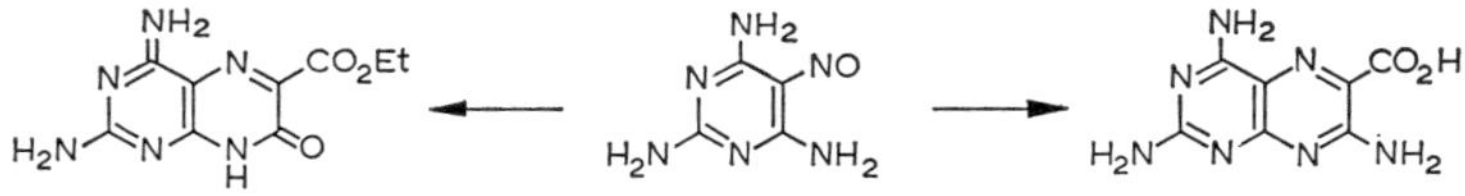

Ethyl pteridine-2-carboxylate, m.p. 128.5°, is prepared from 3-ethoxalylaminopyrazine-2-carbonitrile by reduction and simultanelous cyclisation to give ethyl 3,4-dihydro-pteridine-2-carboxylate which is oxidised to the desired product (*Albert* and *K. Ohta*, J. chem. Soc., C, 1970, 1540).

Ethyl pteridine-4-carboxylate, m.p. 76°, is prepared from the corresponding pyrimidine with glyoxal (*J. Clark, ibid.*, 1967, 1543). The pteridine covalently adds water (and ethanol) to give the 6,7-dihydroxy-5,6,7,8-tetrahydropteridine.

Aminopteridinecarboxylic acids. **Ethyl 2-aminopteridine-4-carboxylate,** m.p. 189°, is prepared from the corresponding pyrimidine with polyglyoxal. This pteridine also adds water (*Clark* and *W. Pendergast, ibid.*, 1968, 1124).

2,4-Diaminopteridine-6-carboxylic acid is prepared by the oxidation of "aminofolic acid" (*R. L. Dion* and *T. L. Loo*, J. org. Chem., 1961, **26**, 1857) or by oxidation of the corresponding 6-methylpteridine (*J. Weinstock et al.*, J. med. Chem., 1968, **11**, 618). **2,4-Diaminopteridine-7-carboxylic acid,** the **-6,7-dicarboxylic acid, 4,7-diaminopteridine-6-carboxylic acid** and **2,4,7-triaminopteridine-6-carboxylic acid** have all been prepared by oxidation of the corresponding methyl derivatives (*C. K. Cain, M. F. Mallette* and *E. C. Taylor*, J. Amer. chem. Soc., 1948, **70**, 3026; *Osdene* and *Timmis*, J. chem. Soc., 1955, 2038).

Pteridinonecarboxylic acids. **Pteridine-2,4-dione-6-carboxylic acid** has been prepared by oxidation of 2-hydroxy-2-deaminofolic acid (*R. B. Angier et al.*, J. Amer. chem. Soc., 1952, **74**, 408; *T. Masuda, T. Kishi* and *M. Asai*, Chem. Pharm. Bull., Japan, 1958, **6**, 291) or by hydrolysis of 2-aminopteridine-4-one-6-carboxylic acid in 6 *N* hydrochloric acid (*Dion* and *Loo, loc. cit.*).

Pteridine-2,4-dione-7-carboxylic acid and the -6,7-**dicarboxylic acid** are prepared by oxidation of the corresponding methyl derivatives (*Cain, Mallette* and *Taylor, loc. cit.*; *Masuda, Kishi* and *Asai, loc. cit.*; *R. Mengel* and *W. Pfleiderer*, in "Chemistry and Biology of Pteridines", De Gruyter, Berlin, 1975, p. 617).

Dibromination of 6-methylpteridine-4,7-dione followed by hydrolysis gives pteridine-4,7-dione-6-carbaldehyde which on oxidation with permanganate gives **pteridine-4,7-dione-6-carboxylic acid** (isolated as the monohydrate) which is decarboxylated at 260° (*Albert* and *Brown, loc. cit.*).

Pteridine-4,6-dione-7-carboxylic acid is prepared from 5,6-diaminopyrimidin-4-one with sodium mesoxalate in dilute acetic cid (*Pfleiderer*, Chem. Ber., 1969, **92**, 3190).

Pteridine-2,4,6-trione-7-carboxylic acid is similarly prepared from the corresponding pyrimidine (*R. Tschesche* and *F. Korte, ibid.*, 1951, **84**, 801) or alternatively by condensation of 5,6-diaminouracil with 1,3-dimethylalloxan in water, which gives pteridine-2,4,6-trione-7-carboxymethylamide, and subsequently hydrolysed to the 7-carboxylic acid (*Pfleiderer, ibid.*, 1957, **90**, 2624).

Pteridine-2,4,7-trione-6-carboxylic acid is prepared from 5,6-diaminouracil with sodium mesoxalate in acid solution (*Tschesche* and *Korte, loc. cit.*) or with diethyl mesoxalate (*Pfleiderer*, Chem. Ber., 1957, **90**, 2617). *Taylor* and *H. M. Loux* reported that condensation of 5,6-diaminouracil with alloxan in alkaline solution gave the 6-carboxylic acid also (J. Amer. chem. Soc., 1959, **81**, 2474).

Aminopteridinonecarboxylic acids. 2-**Aminopteridin-4-one-6-carboxylic acid** is obtained from the degradation of pteroylglutamic acid. It may also be prepared from isoxanthopterin-6-carboxylic acid by treatment with phosphorus pentachloride and phosphoryl chloride and subsequent dehalogenation of the resulting 7-chloro compound by hydriodic acid. Another method of preparation involves condensation of 2,5,6-triaminopyrimidin-4-one with 1-bromo-2,2-diethoxypropionate or alternatively with 3,3-dimethoxyacetoacetate and subsequent oxidation of the intermediate 6-pteridineacetic acid (*Mowat et al., loc. cit.*). Oxidation of the 6-methyl derivative also yields the 6-carboxylic acid (*H. S. Forrest* and *J. Walker*, J. chem. Soc., 1959, 2077; *Pfleiderer, H. Zondler* and *R. Mengel*, Ann., 1970, **741**, 64).

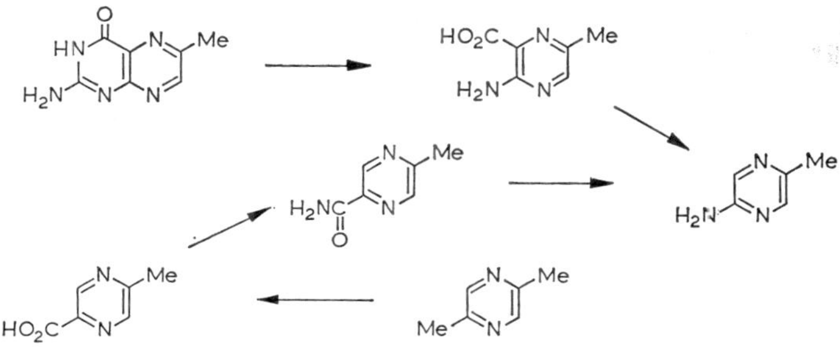

2-**Aminopteridin-4-one-7-caroxylic acid** is prepared by oxidation of the corresponding 7-methylpteridine with alkaline permanganate (*Mowat et al., loc. cit.*; *Cain, Mallette* and *Taylor, loc. cit.*; *Pfleiderer, Zondler* and *Mengel, loc. cit.*). The 6- and 7-carboxylic acids may be identified by their characteristic ultraviolet absorption spectra (*Mowat et al.,*

loc. cit.) and may be distinguished by degradation of the precursor of the former carboxylic acid, namely the 6-methylpteridine, to 2-amino-5-methylpyrazine which has been prepared unambiguously from 2,5-dimethylpyrazine.

2,5,6-Triaminopyrimidin-4-one when condensed with ethyl mesoxalate in dilute acetic acid with subsequent hydrolysis gives an 85% yield of **isoxanthopterincarboxylic acid** (2-aminopteridine-4,7-dione-6-carboxylic acid). However, the same reaction carried out in 2 *N* sulphuric cid gives **xanthopterincarboxylic acid** (2-aminopteridine-4,6-dione-7-carboxylic acid). The former acid decarboxylates at 260° but the latter is stable until reduced by hydriodic acid to the dihydro derivative which then decarboxylates at 240° to dihydroxanthopterin (*R. Purrmann*, Ann., 1941, **548**, 284). An alternative method of preparation of isoxanthopterincarboxylic acid has been described by *Taylor* and *Loux* (*loc. cit.*) who condensed 2,5,6-triaminopyrimidin-4-one with alloxan.

(*x*) *Pteridine glycosides*

A number of pteridine glycosides have been synthesised and are of interest because of their close structural similarity to the purine nucleosides. No naturally occurring compounds of this type have been reported, however.

These compounds can be prepared by (*1*) reaction of a pyrimidine aminoglycoside with a suitable two-carbon fragment (*Pfleiderer* and *D. Söll*, J. heterocycl. Chem., 1964, **1**, 23; *Pfleiderer* and *E. Bühler*, Chem. Ber., 1966, **99**, 3022; *H. Rokos* and *Pfleiderer*, *ibid.*, 1971, **104**, 770; *D. Autenrieth et al.*,

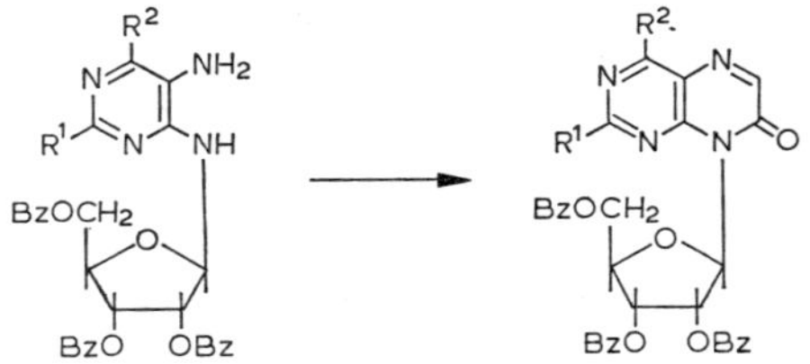

Angew. Chem. intern. Edn., 1971, **10**, 927), and (*2*) by reaction of a suitably protected pteridine, preferably as its trimethylsilyl derivative with a fully acylated sugar (*Pfleiderer et al.*, Chem. Ber., 1973, **106**, 317, 1401, 1952; *K. Harzer* and *Pfleiderer*, Helv., 1973, **56**, 1225; *Pfleiderer* and *F. Reisser*, Chem. Ber., 1966, **99**, 536; 1962, **95**, 1621; *M. Ott* and *Pfleiderer*, *ibid.*, 1974, **107**, 339; *I. Southon* and *Pfleiderer*, in "Chemistry and Biology of

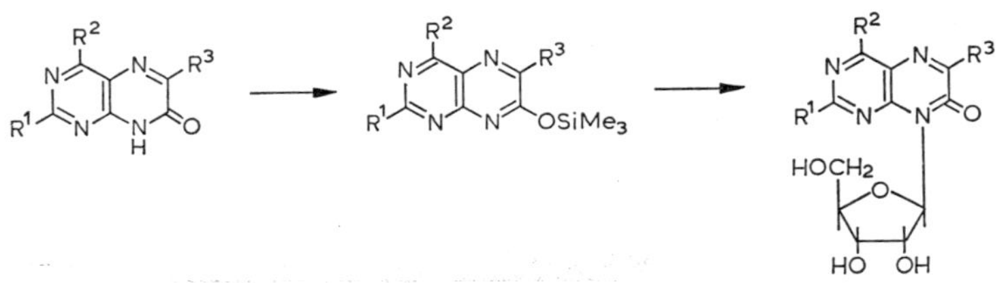

Pteridines), De Gruyter, Berlin, 1975, p. 783; *T. Itoh* and *Pfleiderer*, Chem. Ber., 1976, **109**, 3228; *K. Kobayashi* and *Pfleiderer*, *ibid.*, pp. 3159, 3175, 3184, 3194; *K. Eistetter* and *Pfleiderer*, *ibid.*, p. 3208; *W. Hutzenlaub* and *Kobayashi*, *ibid.*, p. 3217).

Chemical and enzymic syntheses of the nucleotide analogues have been described (*H. Rokos* and *G. Harzer*, in "Chemistry and Biology of Pteridines), p. 795).

(xi) Pteroylglutamic acid and pteroic acid and related compounds

Pteroylglutamic acid (also known as folic acid) is a naturally occurring pteridinecarboxylic acid which has a growth-promoting effect on many organisms. It has been isolated from yeast (*E. E. Snell* and *W. H. Peterson*, J. Bact., 1940, **39**, 273), from spinach (*H. K. Mitchell*, *Snell* and *R. J. William*, J. Amer. chem. Soc., 1941, **63**, 2284; 1944, **66**, 267), and from liver (*J. J. Pfiffner et al.*, Science, 1943, **97**, 404; *E. L. R. Stockstad*, J. biol. Chem., 1943, **149**, 573). The structure was determined by degradation and synthesis to be *N*-[4-(2-amino-6-pteridin-4-onylmethylamino)benzoyl]glutamic acid ($n=1$), by the American Cyanamid group (J. Amer. chem. Soc., 1948, **90**, 5, 14, 19, 23, 25, 27).

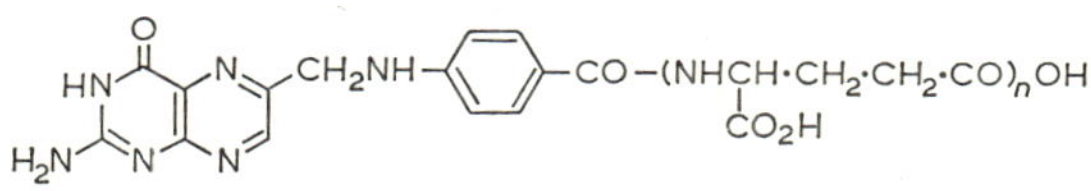

At the same time, pteroylglutamic acids having two additional glutamate residues ($n = 3$) (*Stockstad et al.*, *ibid.*, 1948, **70**, 5, 10), and six additional glutamate residues ($n = 7$) (*Pfiffner et al.*, Science, 1945, **102**, 228; J. Amer. chem. Soc., 1946, **68**, 1392) were isolated. The latter yields the monoglutamyl derivative on enzymic hydrolysis.

Other naturally occurring folic acid derivatives include the following: **$N_{(10)}$-formylpteroylglutamic acid,** which has little or no *Citrovorum* factor growth activity but is as active as folic acid in the growth of *Streptococcus faecalis* was isolated from horse liver along with **$N_{(5)}$-formyltetrahydrofolic acid** and **$N_{(5)}$-methyltetrahydrofolic acid** (*M. Silverman, J. C.*

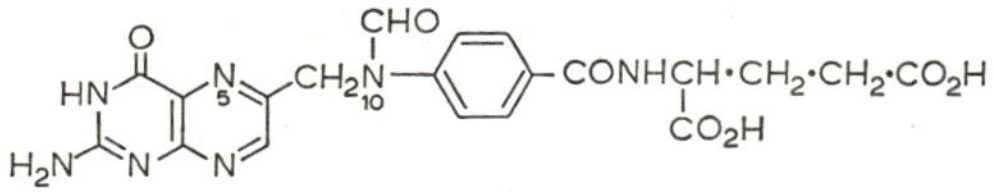

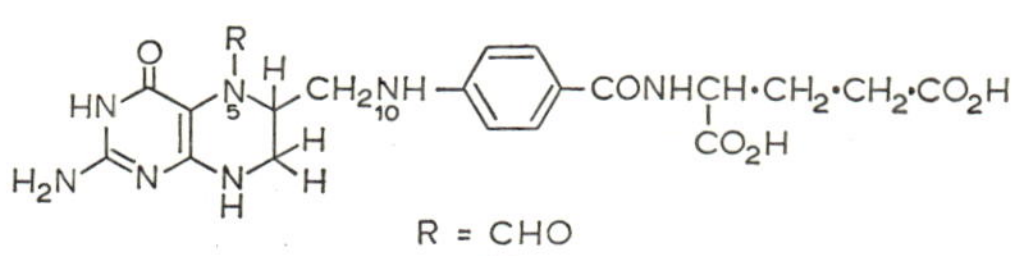

Keresztesy and *G. J. Koval*, J. biol. Chem., 1954, **211**, 53; *H. E. Sauberlich* and *C. A. Baumann*, *ibid.*, 1948, **176**, 165; *Keresztesy* and *Silverman*, J. Amer. chem. Soc., 1951, **73**, 5510; *K. O. Donaldson*, *Keresztesy* and *Silverman*, J. biol. Chem., 1959, **234**, 3235; Biochem. Biophys. Res. Comm., 1961, **5**, 286, 289). The latter compound was shown to be involved in the biosynthesis of methionine (*W. Sakami* and *I. Ukstins*, J. biol. Chem., 1961, **236**, RC50).

Pteroylglutamic acid (folic acid) is an orange-yellow crystalline solid obtained as a dihydrate which loses water at 75° *in vacuo*. It is almost insoluble in water, alcohol and other organic solvents but the sodium salt is water-soluble. The acid is decomposed by sunlight, especially in the presence of riboflavin, to give 2-aminopteridin-4-one-6-carbaldehyde. This undergoes oxidation to the carboxylic acid which in turn is photolytically decarboxylated (*S. Scheindlin*, *A. Lee* and *I. Griffith*, J. Amer. pharm. Assoc., 1952, **41**, 420).

Folic acid is stable to alkali in the absence of air but under aerobic conditions the side-chain is cleaved to give pterin-6-carboxylic acid and 4-aminobenzoylglutamic acid. Acidic hydrolysis gives 6-methylpterin (*Stockstad et al., loc. cit.*), whereas sulphurous acid causes cleavage to pterin-6-carbaldehyde and 4-aminobenzoylglutamic acid (*C. W. Waller et al.*, J. Amer. chem. Soc., 1950, **72**, 4630).

Asymmetric catalytic reduction of the 5–6 carbon – nitrogen bond in folic acid using an optically active rhodium catalyst has been reported (*P. H. Boyle* and *M. T. Keating*, Chem. Comm., 1974, 375).

Folic acid is formylated by formic acid to yield $N_{(10)}$-formylfolic acid (*M. Gordon et al.*, J. Amer. chem. Soc., 1948, **70**, 878). On reduction, the latter yields $N_{(10)}$-formyltetrahydro-

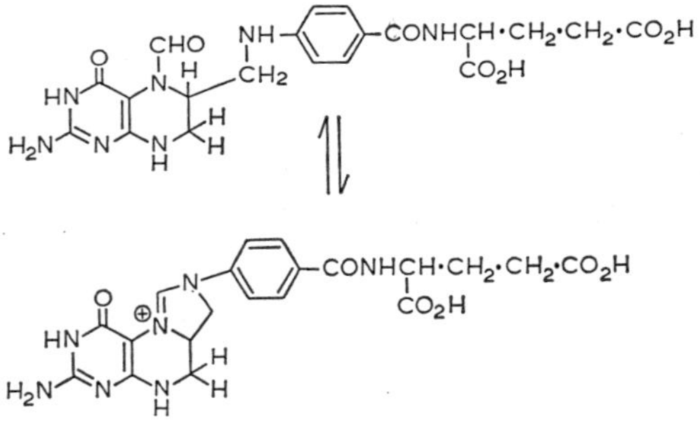

folic acid, which on heating at pH 12 is partially rearranged to $N_{(5)}$-formyltetrahydrofolic acid (*M. May et al., ibid.*, 1951, **73**, 3067). This is fairly stable under alkaline conditions but at pH 2 rearranges to 5,10-methenyltetrahydrofolic acid, a change which is reversed at pH 13 (*idem, loc. cit.*; *D. B. Cosulich et al.*, J. Amer. chem. Soc., 1952, **74**, 3252).

Two types of synthesis of folic acid have been described in the literature:

(*1*) *From the pyrimidine.* 2,5,6-Triaminopyrimidin-4-one has been condensed with several aliphatic reagents in the presence of 4-aminobenzoylglutamic acid. In the first synthesis, using 2,3-dibromopropanal, the intermediate dihydro compound spontaneously oxidised to give a precipitated product which contained 20.8% of pteroylglutamic acid on the basis of microbiological activity. A crystalline product was obtained from this material by repeated fractional precipitation at a controlled pH (*Waller et al.*, J. Amer. chem. Soc., 1948, **70**, 19).

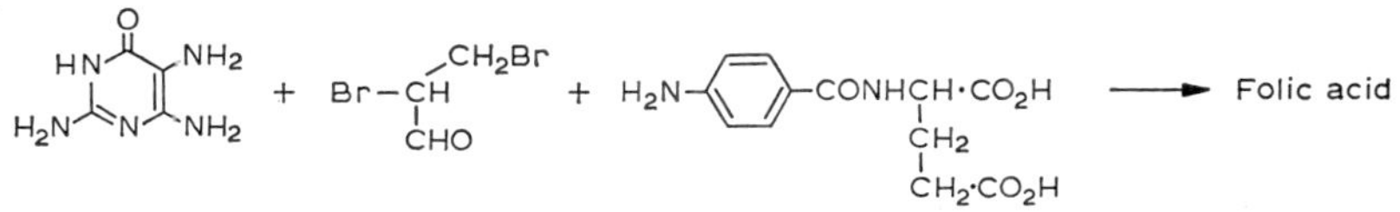

Similar syntheses have used 2,2,3-tribromopropanal (*F. Weygand* and *V. Schmied-Kowarzik*, Chem. Ber., 1949, **82**, 333), and 1-bromo-3,3-dichloroacetone (*F. E. King* and *P. C. Spensley*, J. chem. Soc., 1952, 2144) as the component for formation of the pyrazine ring.

Condensation of the pyrimidine with ethyl *N*-tosyl-*N*-(3-acetoxy-2-oxopropyl)-4-aminobenzoylglutamate also provides folic acid (*D. I. Weisblat et al.*, J. Amer. chem. Soc., 1953, **75**, 3625).

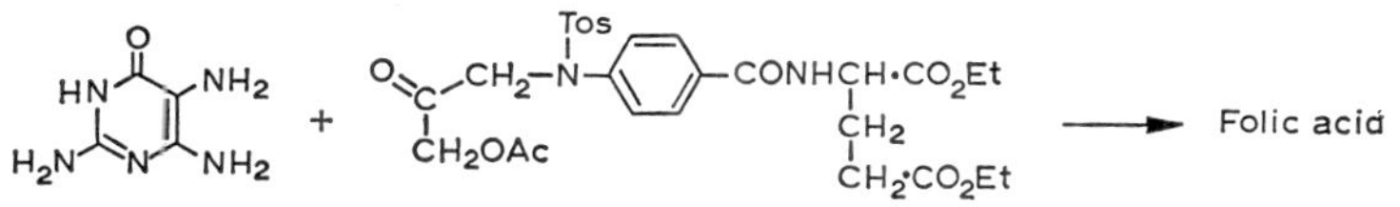

Reductone (2,3-dihydroxyacrylaldehyde) when condensed with diethyl *N*-(4-aminobenzoyl)glutamate in dilute hydrochloric acid gave diethyl *N*-[4-(2,3-dihydroxy-2-ene-propylideneamino)benzoyl]glutamate which on further reaction with the above pyrimidine produced pteroylglutamic acid (*Angier et al.*, J. Amer. chem. Soc., 1948, **70**, 25).

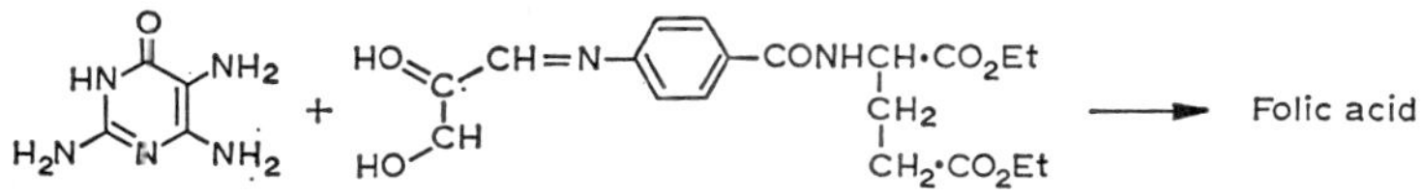

(*2*) *From pteridines*. Several pteridine precursors have been condensed with 4-aminobenzoylglutamate to give folic acid. Thus reaction of *N*-(2-aminopteridin-4-onyl)methylpyridinium iodide, prepared from 2,5,6-triaminopyrimidin-4-one by condensation with the product obtained from pyridine and 2,3-dibromopropanal in the presence of potassium iodide with 4-aminobenzoylglutamate produced folic acid (*M. H. Hultquist et al.*, *ibid.*, 1948, **70**, 23). Alternative pteridine precursors are 2-amino-6-bromomethylpteridin-4-one

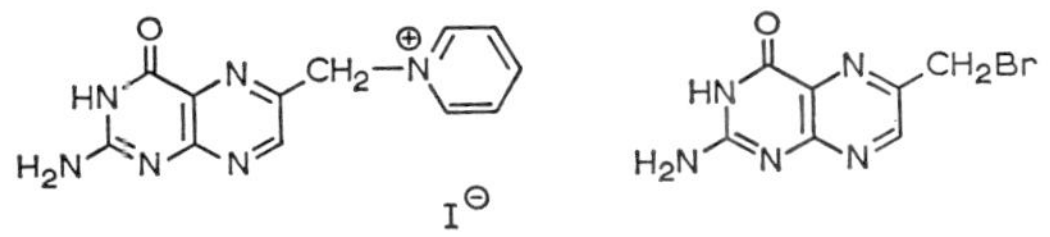

(*J. H. Boothe et al.*, *ibid.*, 1948, **70**, 27) and 2-aminopteridin-4-one-6-carbaldehyde. The intermediate Schiff base formed using the latter reagent is reduced by sodium tetrahydridoborate to folic acid (*E. C. Roberts* and *Y. F. Shealy*, J. med. Chem., 1973, **16**, 697; *J. H. Bieri* and *M. Viscontini*, Helv., 1973, **56**, 2905; *E. Khalifa*, *Bieri* and *Viscontini*, *ibid.*, p. 2911; *Khalifa*, *P. K. Sengupta* and *Bieri*, *ibid.*, 1976, **59**, 242).

A much improved method based on the nucleophilic substitution reactions of 2-amino-6-bromomethylpteridin-4-one has been described for the synthesis of folic acid, and the method has been extended also to the synthesis of analogues by the use of 6-bromomethyl-

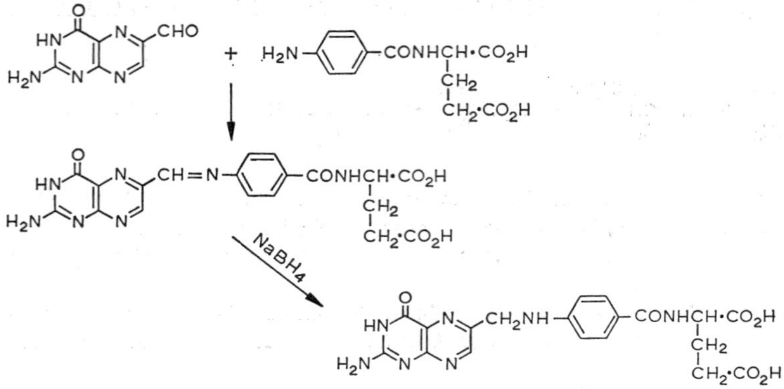

pteridine-2,4-diamine hydrobromide which is readily made in reasonable yield (*J. R. Piper* and *J. Montgomery*, J. heterocycl. Chem., 1974, **11**, 279; *Montgomery et al.*, in "Chemistry and Biology of Pteridines", De Gruyter, Berlin, 1975, p. 485; *Piper* and *Montgomery*, J. org. Chem., 1977, **42**, 208).

A synthesis of pteroic acid (cf. p. 9) using the pyrazine *N*-oxide route has been described by *E. C. Taylor et al.* (J. org. Chem., 1975, **40**, 2347). This procedure should be applicable also to the synthesis of folic acid.

Folic acid analogues. Soon after the structure determination and synthesis of folic acid, the 2,4-diamino analogue of folate — **aminopterin** — was synthesised (*D. R. Seeger, J. M. Smith Jr.* and *M. E. Hultquist*, J. Amer. chem. Soc., 1947, **96**, 2567). This substance proved to be active against leukaemia in children (*S. Farber et al.*, New England J. Med., 1948, **238**, 787). The $N_{(10)}$-methyl derivative of aminopterin — **methotrexate (amethoperin)** — synthesised by a direct extension of the folate synthesis described above (*Seeger et al.*, J. Amer. chem. Soc., 1949. **71**, 1753) proved to be an even more potent anti-cancer agent and is still in clinical use. *Piper* and *Montgomery* (*loc. cit.*) have described an improved synthesis of methotrexate by a modification of their synthesis of folic acid (*vide supra*), which is outlined below.

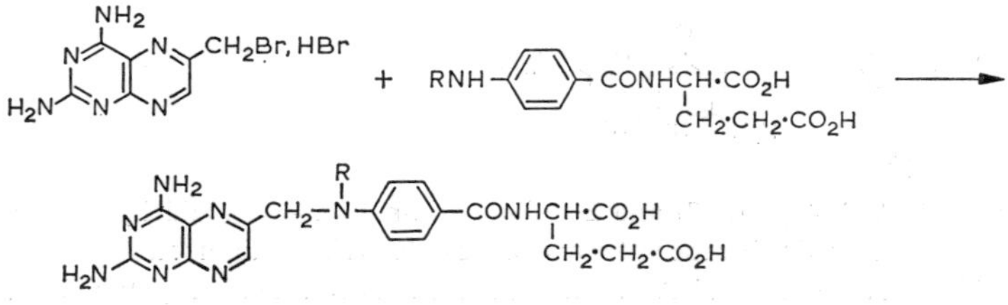

R = H, Aminopterin
R = Me, Methtrexate

7,8-Dihydro-8-methylmethotrexate has been synthesised from 7,8-dihydromethotrexate by a novel $N_{(8)}$-alkylation procedure using *n*-butyl-lithium in dimethyl sulphoxide followed by treatment with methyl iodide. The method appears to be of general application to the

$N_{(8)}$-alkylation of 2,4-diamino-7,8-dihydropteridines (*M. Chaykovsky*, J. org. Chem., 1975, **40**, 145).

As a consequence of the biological activity of these folate analogues a large amount of synthetic effort has gone into this area. A list of the recent synthetic compounds is given below. The references contain citations to earlier work.

10-Dezaa-aminopterin (*J. I. Degraw*, *R. L. Kisliuk* and *Y. Gaumont*, J. med. Chem., 1974, **17**, 552).

1-Deaza-$N_{(10)}$-methylfolic acid (*R. D. Elliott* and *Temple Jr.*, *ibid.*, p. 553).

10-Thiofolic acid (*M. G. Nair*, *P. T. Campbell* and *C. M. Baugh*, J. org. Chem., 1975, **40**, 1745; *Y. H. Kim*, *Gaumont* and *Kisliuk*, J. med. Chem., 1975, **18**, 776; *H. G. Mautner*, *Kim* and *Gaumont*, in "Chemistry and Biology of Pteridines", ed. *W. Pfleiderer*, p. 515).

$N_{(10)}$-Methyl-4-thiofolic acid (*Elliott* and *Temple Jr.*, J. med. Chem., 1975, **18**, 492).

Isoaminopterin (*Nair*, *L. P. Mercer* and *Baugh*, *ibid.*, 1974, **17**, 1268).

10-Thioaminopterin (*Nair*, *Campbell* and *E. Braverman*, Tetrahedron Letters, 1975, 2745).

7-Aza-aminopterin (*Temple jr.* and *L. Kussner*, J. org. Chem., 1975, **40**, 2205).

5-Methyldihydrofolate (*T. L. Deits et al.*, in "Chemistry and Biology of Pteridines", p. 525).

10-Oxafolic acid and **10-oxa-aminopterin** (*Nair* and *Campbell*, J. med. Chem., 1976, **19**, 825).

(*d*) *Hydropteridines*

Various types of reduced pteridines are known, the stability of these derivatives depending upon the position and nature of the functional groups attached to the nucleus. The role of reduced pteridines as mediators of cellular electron transport has been reviewed (*H. Rembold*, in "Chemistry and Biology of Pteridines", ed. *Pfleiderer*, De Gruyter, Berlin, 1975, p. 359).

(*i*) *Dihydropteridines*

3,4-, 5,6- and 7,8-Dihydropteridines are known. They are all readily oxidised to the corresponding pteridines and on reduction 5,6,7,8-tetrahydropteridines are obtained from the 5,6- and 7,8-dihydro derivatives, the 3,4-dihydro derivative being resistant to reduction. The 7,8-dihydropteridines are of special importance as many naturally occurring biologically active compounds are of this type.

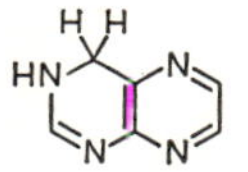

3,4-

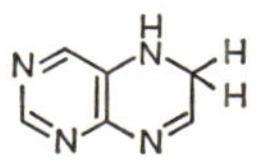

5,6-

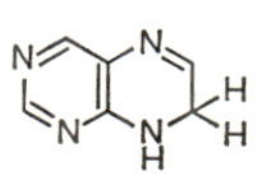

7,8-

Methods of preparation

(*1*) *Direct synthesis. W. R. Boon* and his co-workers developed a method of unambiguous synthesis of 7,8-dihydropteridines which involved condensation of 4-chloro-5-nitropyrimidines or 4-chloro-5-phenylazopyrimidines with α-amino-esters of α-aminoketones (*Boon, W. G. M. Jones* and *G. R. Ramage*, J. chem. Soc., 1951, 96; *Boon* and *Jones, ivid.*, p. 591; *Boon* and *T. Leigh, ibid.*, p. 1497; *Boon, ibid.*, 1957, 2146).

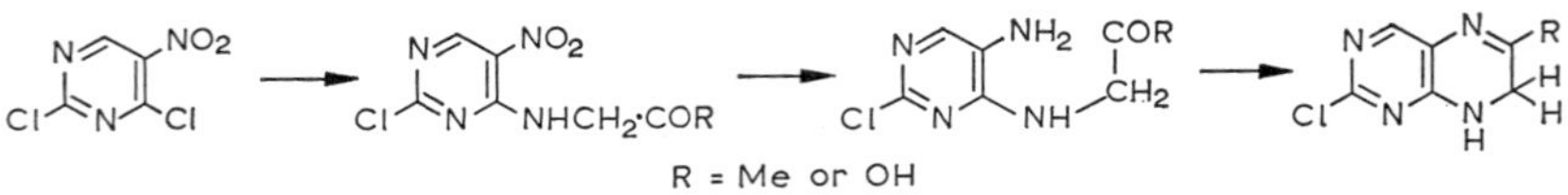

(*2*) *Reduction of aromatic pteridines.* Various types of reducing agents such as Raney nickel, sodium dithionite, sodium amalgam and sodium tetrahydridoborate have been used for this reaction.

3,4-*Dihydropteridines.* 3,4-**Dihydropteridine,** m.p. 181° (decomp.), the parent compound of this series, has been prepared from 2-amino-3-aminomethylpyrazine and triethyl orthoformate. This compound resisted catalytic hydrogenation to 1,2,3,4-tetrahydropteridine but was readily oxidised to pteridine with manganese dioxide (*A. Albert* and *K. Ohta*, J. chem. Soc., C, 1970, 1540).

Pteridines substituted at the 2-position with amino-, hydroxy-, mercapto- or methylthio-groups may be reduced to the corresponding 3,4-dihydro derivatives with tetrahydridoborate (*Albert* and *S. Matsuura*, J. chem. Soc., 1961, 5131; *Albert* and *J. J. McCormack, ibid.*, C, 1966, 1117; *idem, ibid.*, 1968, 63).

5,6-*Dihydropteridines.* 5,6-**Dihydropteridin-7-one** has been prepared by cyclisation of 4-amino-5-carboxymethylaminopyrimidine in hydrochloric acid (*Albert, D. J. Brown* and *G. Cheeseman*, J. chem. Soc., 1952, 1620) and also by reduction of the corresponding pteridine with potassium tetrahydridoborate or potassium amalgam.

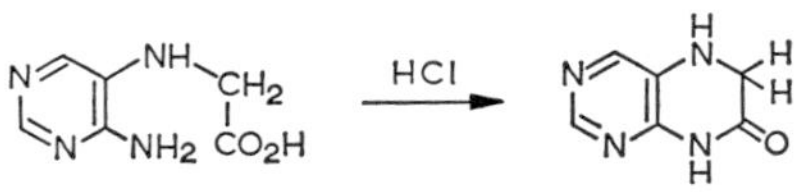

5,6-**Dihydropteridin-4-one** has been prepared from 5,6-diaminopyrimidin-4-one by reductive condensation with glyoxal monoacetal (*Albert* and *Matsuura, ibid.*, 1962, 2162).

5,6-**Dihydropteridine-4,7-dione** has been prepared by reduction of the corresponding pteridine with potassium tetrahydridoborate, potassium amalgam or palladium and hydrogen, and alternatively by cyclisation of the appropriate pyrimidine (*Albert* and *Matsuura, loc. cit.*).

7,8-*Dihydropteridines.* These dihydro derivatives may be prepared by reduction of the corresponding pteridine or by the Boon synthesis described above.

7,8-**Dihydro-4,6-dimethylpteridine,** m.p. 135°, has been prepared from the corresponding 2-chloro compound by catalytic hydrogenation (*idem, loc. cit.*).

7,8-**Dihydropteridin-2-one,** m.p. 190° (decomp.), has been prepared by the reductive cyclisation of 4-methylformamido-5-nitropyrimidin-2-one (*idem*, J. chem. Soc., 1961, 5131).

Other reported 7,8-dihydropteridinones are:

7,8-**Dihydropteridin**-4-**one,** m.p. 263–265° (decomp.) (*Albert* and *Matsuura, loc. cit.*); 7,8-**dihydropteridin**-6-**one,** m.p. > 400°; 7,8-**dihydropteridine**-2,6-**dione,** m.p. > 200°; 7,8-**dihydropteridine**-4,6-**dione,** m.p. > 300° (*Boon, Jones* and *Ramage, loc. cit.*); 7,8-**dihydro-pteridine**-2,4,6-**trione** (*Boon* and *T. Leigh,* B.P. 677,342/1952).

The removal of the oxo-function of pteridin-7-ones by reduction with sodium or potassium amalgam results in 7,8-dihydropteridines (*Albert, Brown* and *Cheeseman, loc. cit.*; *Albert* and *Brown,* J. chem. Soc., 1953, 74; *Albert, J. Lister* and *C. Pedersen, ibid.,* 1956, 4621).

2-**Amino**-7,8-**dihydropteridin**-4-**one** was prepared by the Boon method. It was isolated as the bisulphite salt which was subsequently converted to the sodium salt by refrigeration of the former in aqueous sodium hydroxide solution (*A. Stuart, H. C. S. Wood* and *D. Duncan, ibid.,* C, 1966, 285).

2-**Amino**-6-**methyl**-7,8-**dihydropteridin**-4-**one** was first prepared by *J. H. Boothe et al.,* from (2-amino-6-pteridin-4-onylmethyl)pyridinium iodide by reduction with zinc dust (J. Amer. chem. Soc., 1948, **70,** 27).

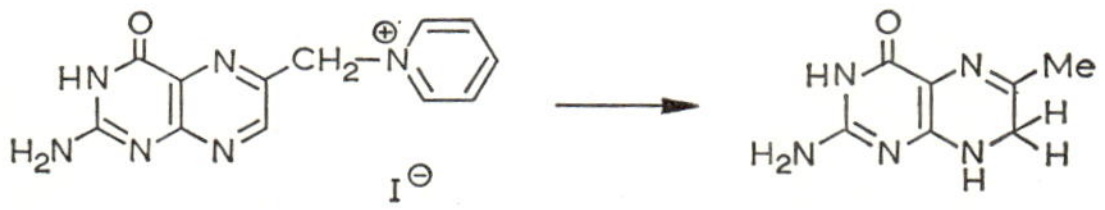

Later *Boon* and *Leigh* synthesised this pteridine from the phenylazopyrimidine shown by reductive cyclisation (J. chem. Soc., 1951, 1497).

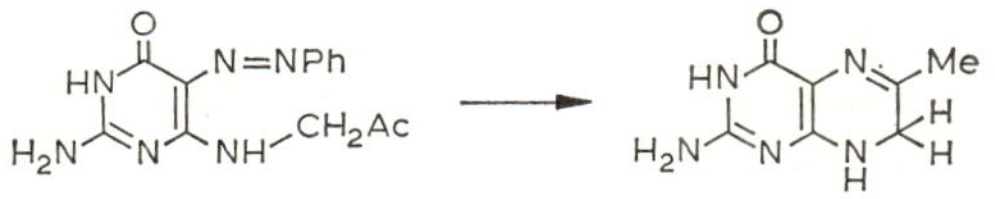

Replacement of the phenylazo group by nitro allows the reaction to proceed in the same manner (*T. Neilson* and *Wood, ibid.,* 1962, 44). The location of the hydrogen atoms was confirmed by the unambiguous synthesis of 2-amino-7,8-dihydro-6,7,7-trimethylpteridin-4-one, the u.v. spectrum of which was very similar to that of the 7,8-dihydropteridine produced by reduction of 2-amino-6-methylpteridin-4-one (*Pfleiderer* and *H. Zondler,* Chem. Ber., 1966, **99,** 3008). The reduction of 2-amino-6-methylpteridin-4-one 8-*N*-oxide also gives this dihydropteridine (*E. C. Taylor* and *K. Lenard,* J. Amer. chem. Soc., 1968, **90,** 2424).

Dihydroxanthopterin (2-amino-7,8-dihydropteridine-4,6-dione) is available by reduction with sodium amalgam of xanthopterin (*W. Koschara,* Z. physiol. Chem., 1937, **250,** 161; *B. L. O'Dell et al.,* J. Amer. chem. Soc., 1947, **69,** 250), leucopterin (*J. R. Totter,* J. biol. Chem., 1944, **154,** 105; *G. B. Elion et al.,* J. Amer. chem. Soc., 1949, **71,** 741) or xanthopterincarboxylic acid since the dihydro-acid decarboxylates readily (*R. Purrmann,* Ann., 1941, **548,** 284).

The unambiguous synthesis of dihydroxanthopterin by *Boon* and *Leigh* (*loc. cit.*), who reductively cyclised 2-amino-6-ethoxycarbonylmethylamino-5-phenylazopyrimidin-4-one, confirmed the structure of the former compound. A similar synthesis has more recently made use of the corresponding 5-nitropyrimidine (*Pfleiderer*, Chem. Ber., 1974, **107**, 785). The structural assignment was also supported by the similarity in properties of the 7,7-dimethyl analogue (*Stuart, D. W. West* and *Wood*, J. chem. Soc., 1964, 4769).

2-Amino-7,8-dihydropteridin-4-ones having substituents at the 6-position may be prepared by addition of the substituent to the 2-amino-7,8-dihydropteridin-4-one which results in the corresponding 5,6,7,8-tetrahydro derivative which is then readily oxidised back

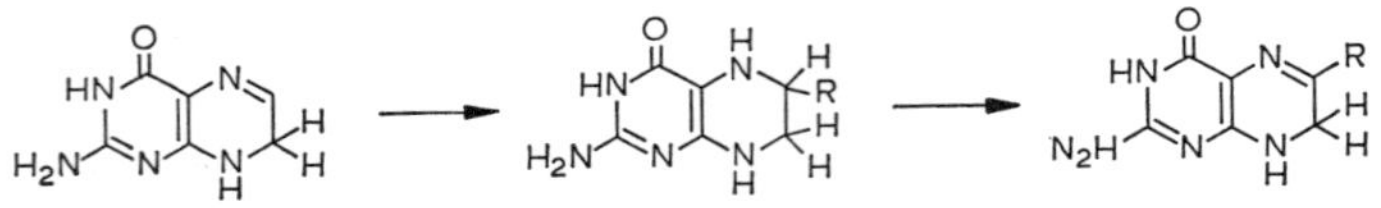

to the dihydro oxidation level (*Stuart, Wood* and *Duncan, loc. cit.*; *C. van Baalen* and *H. S. Forrest*, J. Amer. chem. Soc., 1959, **81**, 1770; *M. Viscontini* and *H. R. Weilenmann*, Helv., 1959, **42**, 1854; *Forrest et al., ibid.*, 1960, **43**, 1005).

Various blocked dihydropteridines, which thus represent stabilised analogues of the parent dihydropteridines, have been described. The synthesis of $N_{(8)}$-alkyl-7,8-dihydro-2,4-diaminopteridines by alkylation of the dihydropteridines (*Chaykovsky, loc. cit*). (see p. 268) represents a significant advance in synthetic methodology.

An alternative approach to the synthesis of 8-methyl-7,8-dihydropterins involves the reduction and subsequent cyclisation of [2-amino-6-hydroxy-5-phenylazopyrimidin-4-yl-(methyl)amino]acetaldehyde (*R. Brown et al.*, J. chem. Soc., Perkin I, 1977, 1003; *Pfleiderer* and *Zondler, loc. cit.*).

(*ii*) *Tetrahydropteridines*

Reduction of aromatic pteridines affects only the pyrazine ring system resulting in 5,6,7,8-tetrahydropteridines. Other tetrahydropteridines have been prepared by direct cyclisation.

1,2,3,4-Tetrahydropteridine, m.p. 146°, has been prepared from 2-amino-3-aminomethylpyrazine by condensation with formaldehyde (*Albert* and *Ohta, loc. cit.*).

5,6,7,8-Tetrahydropteridine, m.p. 146–147°, was first prepared from pteridine by reduction with lithium tetrahydridoaluminate (*Taylor* and *W. R. Sherman*, J. Amer. chem. Soc., 1959, **81**, 2464). Alternatively it has been prepared from 5-amino-2-chloro-4-(*N*-2-hydroxyethylbenzylamino)pyrimidine by cyclisation with phosphoryl chloride to give the 8-benzyl-

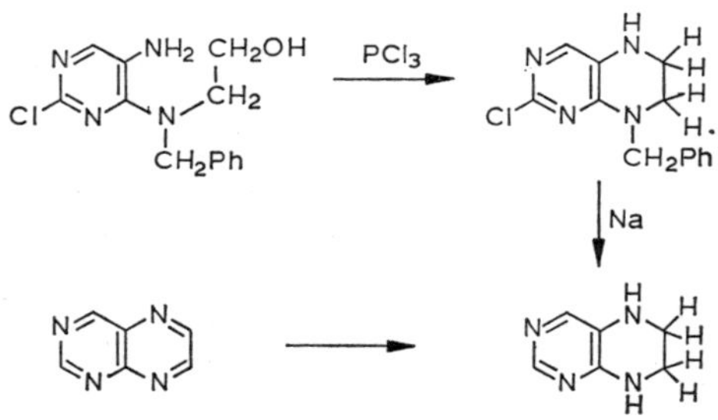

2-chloro-5,6,7,8-tetrahydropteridine which can then be dehalogenated and debenzylated by sodium metal and liquid ammonia (*P. R. Brook* and *Ramage*, J. chem. Soc., 1957, 1).

In contrast to pteridine which is rapidly destroyed by acid, base or light, 5,6,7,8-tetrahydropteridine is stable to boiling 1 *M* hydrochloric acid or 1 *M* sodium hydroxide for at least one hour and appears to be completely stable to light. An attempted oxidation to pteridine under various conditions was unsuccessful (*Taylor* and *Sherman*, *loc. cit.*).

5,6,7,8-Tetrahydropteridin-2-one, m.p. 220° (decomp.), is readily prepared from 7,8-dihydropteridin-2-one by catalytic reduction using Adams catalyst.

5,6,7,8-Tetrahydropteridin-4-one, m.p. 234° (decomp.), can be prepared either by reduction of the parent pteridine or by reduction and subsequent cyclisation with phosphorus trichloride of 4-(*N*-benzyl-2-hydroxyethylamino)-6-ethoxy-5-nitropyrimidine to 8-benzyl-5,6,7,8-tetrahydropteridin-4-one, which is then debenzylated to the product by sodium in liquid ammonia (*Albert* and *Matsuura*, J. chem. Soc., 1961, 5131).

2,4-Dichloro-5,6,7,8-tetrahydropteridine, m.p. 210°, is obtained from 2,4-dichloro- or 2,4,6,7-tetrachloro-pteridine by lithium tetrahydridoaluminate reduction. The tetrahydropteridine yields 2,4-diamino-5,6,7,8-tetrahydropteridine on high pressure amination but on exposure to air, the latter is readily oxidised to the dihydropteridine (*Taylor* and *Sherman*, J. Amer. chem. Soc., 1959, **81**, 2464).

2-Amino-5,6,7,8-tetrahydropteridin-4-one can be prepared from 2-aminopteridin-4-one by catalytic hydrogenation (*Viscontini* and *Weilenmann*, Helv., 1958, **41**, 2170) or by reduction of the corresponding 7,8-dihydropteridone with sodium tetrahydridoborate (*S. Kaufman*, J. biol. Chem., 1964, **239**, 332). The tetrahydropteridine is enzymically oxidised to the 7,8-dihydropteridine in air by a process which is reported to involve the quinonoid dihydropteridine as an intermediate which is subsequently (and non-enzymically) rearranged to the 7,8-dihydropteridine (*Kaufman*, *loc. cit.*).

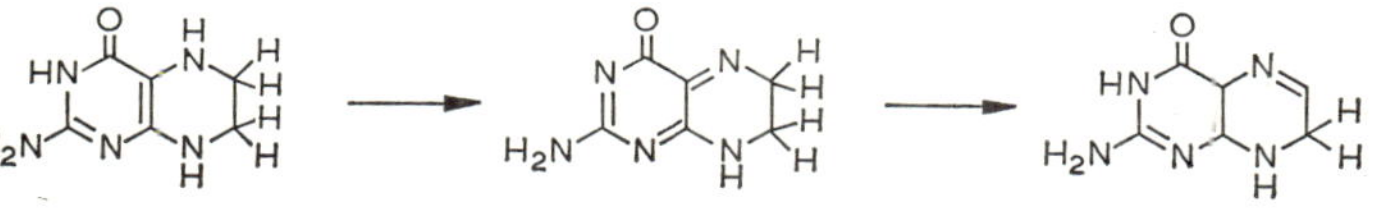

(*e*) *Biosynthesis of dihydropteridines*

It has been shown from *in vitro* studies that 7,8-dihydroneopterin-3′-triphosphate is formed from guanosine-5′-triphosphate. This process involves first the elimination of one carbon atom ($C_{(8)}$) from the purine ring followed by an Amadori rearrangement of the ribose moiety to produce a deoxypentulose derivative which then recyclises to give 7,8-dihydroneopterin-3′-triphosphate (*A. W. Burg* and *G. M. Brown*, J. biol. Chem., 1968, **243**, 2349; *Brown et al.*, in "Chemistry and Biology of Pteridines", ed. *W. Pfleiderer*, De Gruyter, Berlin, 1975, p. 219). This series of reactions is catalysed by the enzyme GTP cyclohydrolase I. After elimination of the triphosphate residue, two carbon atoms of the side-chain are cleaved by a reverse-aldol reaction to give 2-amino-6-hydroxymethyl-7,8-dihydropteridin-4-one (*J. B. Mathis*

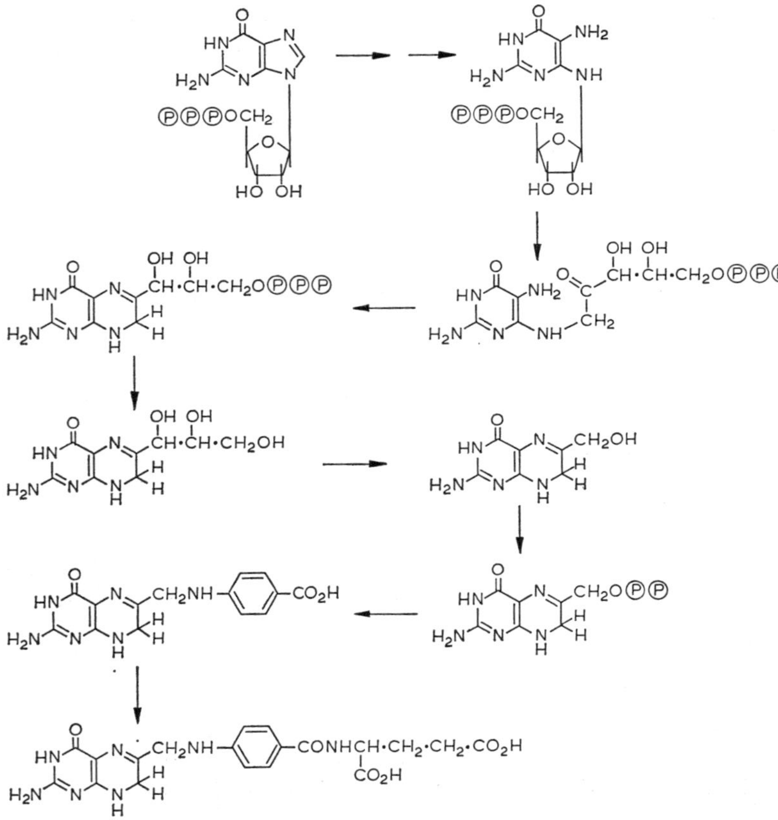

and *Brown*, *ibid.*, 1970, **245**, 3015), the diphosphate of which can sequentially add 4-aminobenzoic acid and glutamate to give dihydrofolic acid (*T. Shiota*, Arch. Biochem. Biophys., 1959, **80**, 155; *D. P. Richey* and *Brown*, J. biol. Chem., 1969, **244**, 1582). Dihydrofolate is reduced to tetrahydrofolate, one of the active co-factor forms of folate, by the enzyme dihydrofolate reductase.

Progress in the clarification of the biosynthetic pathway leading to biopterin is being made. Radioactive incorporation experiments have established that guanosine or one of its phosphorylated derivatives (*e.g.* GTP) is a precursor of biopterin and have suggested that sepiapterin is an intermediate product along this pathway. This then leads to the hypothesis that there is a branch point in mainstream folate biosynthesis at dihydroneopterin (triphosphate?) which in consequence acts as a precursor to both biopterin and the folates (*I. Fukushima* and *T. Shiota*, J. biol. Chem., 1974, 249, 4445;

C. L. Fan, *G. G. Krivi* and *Brown*, Biochem. Biophys. Res. Comm., 1975, **67**, 1047; *I. Eto*, *Fukushima* and *Shiota*, J. biol. Chem., 1976, **251**, 6505):

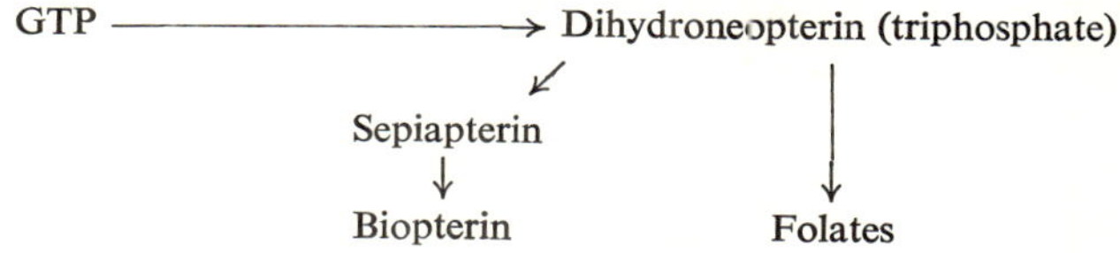

The conversion of pterins into lumazines is a well documented enzymic reaction (*B. Levenberg* and *O. Hayaishi*, *ibid.*, 1959, **234**, 955; *C. C. Levy* and *W. S. McNutt*, Biochemistry, 1962, **1**, 1161). Pterin, lumazine and xanthopterin are all hydroxylated at the 7-position by the enzyme xanthine oxidase to give isoxanthopterin, pteridine-2,4,7-trione and leucopterin, respectively (*F. Bergmann* and *H. Kwietny*, Biochim. Biophys. Acta, 1958, **28**, 613; 1959, **33**, 29; *Y. Hayashi*, *ibid.*, 1962, **58**, 351).

2. Alloxazines and flavins

The alloxazine ring system (I) may be considered as a benzoderivative of the pteridine structure (benzo[*g*]pteridine-2,4-dione) or as a pyrimidoquinoxaline derivative (2,4-dioxo-1,2,3,4-tetrahydropyrimido[4,5-*b*]quinoxaline).

The original numbering system (used in *Rodd*, C.C.C., 1st Edn.) is shown

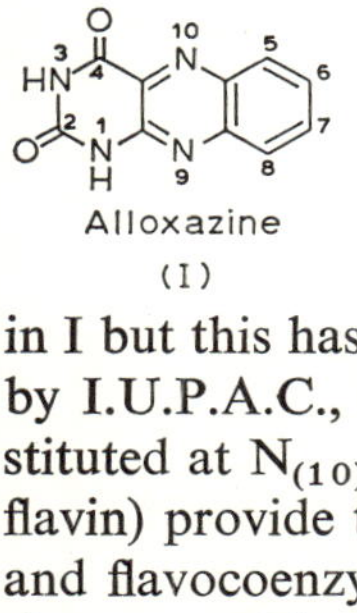

Alloxazine

(I)

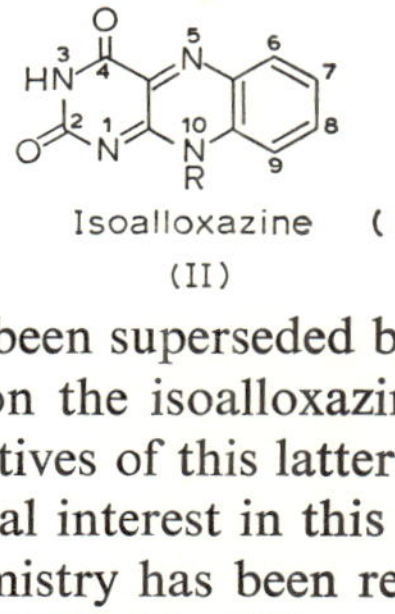

Isoalloxazine (R = H)

(II)

in I but this has largely been superseded by the alternative system approved by I.U.P.A.C., shown on the isoalloxazine ring system (II) which is substituted at $N_{(10)}$. Derivatives of this latter system such as Vitamin B_2 (riboflavin) provide the central interest in this series. The present state of flavin and flavocoenzyme chemistry has been reviewed (*P. Hemmerich*, Fortschr. chem. org. Naturstoffe, 1976, **33**, 451).

(*a*) *Synthesis*

The first synthesis of alloxazine was achieved by condensation of alloxan with *o*-phenylenediamine by *O. Kühling* (Ber., 1891, **24**, 2363, 3029).

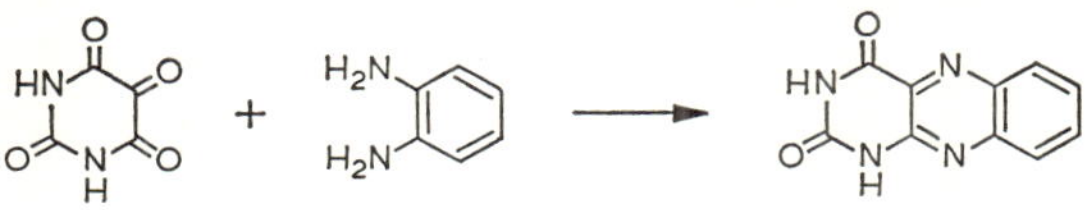

Since that time several other synthetic routes to the alloxazine ring system have been developed:

(*1*) Instead of alloxan, 5,5-dichlorobarbituric acid may be condensed with diamines (*M. Tishler, J. W. Wellman* and *K. Ladenburg*, J. Amer. chem. Soc., 1945, **67**, 2165).

(*2*) Condensation of 5,6-diaminouracil with 3,4-dimethyl-*o*-benzoquinone leads to 7,8-dimethylalloxazine (lumichrome) (*T. J. Bardos, D. B. Olsen* and *T. Enkoji, ibid.*, 1957, **79**, 4704).

(*3*) Condensation of an *ortho*-arylazoaniline with barbituric acid produces alloxazine (*Hemmerich, S. Fallab* and *H. Erlenmeyer*, Helv., 1956, **39**, 1242).

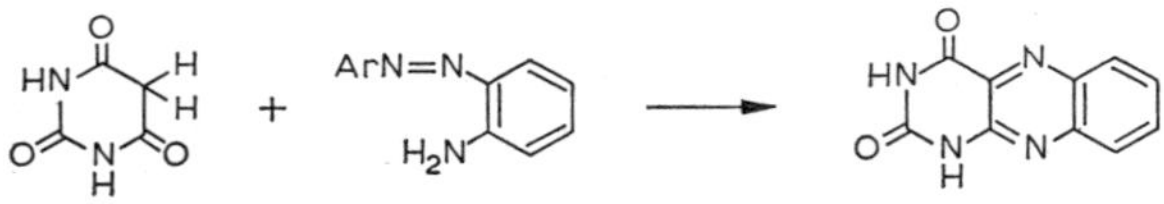

(*4*) In contrast, condensation of 5-nitrosobarbituric acid with *N*-methylaniline produces isoalloxazine (*Hemmerich, B. Prijs* and *Erlenmeyer, ibid.*, 1959, **42**, 1604).

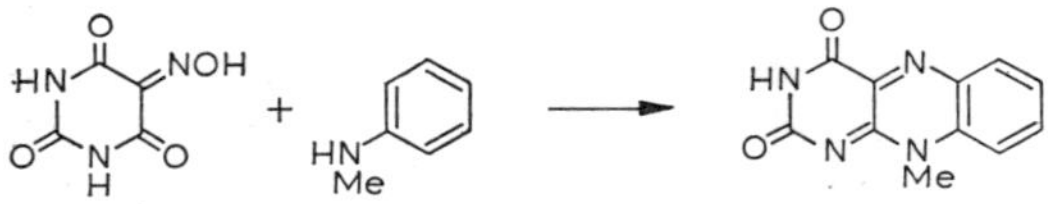

(*5*) 4-Phenylaminouracil, on nitrosation and subsequent cyclisation, gives a mixture of alloxazine and the corresponding 5-*N*-oxide. The latter may be converted to alloxazine by reduction with the aid of a Raney nickel catalyst (*H. Goldner, G. Dietz* and *E. Carstens*, Ann., 1966, **694**, 142).

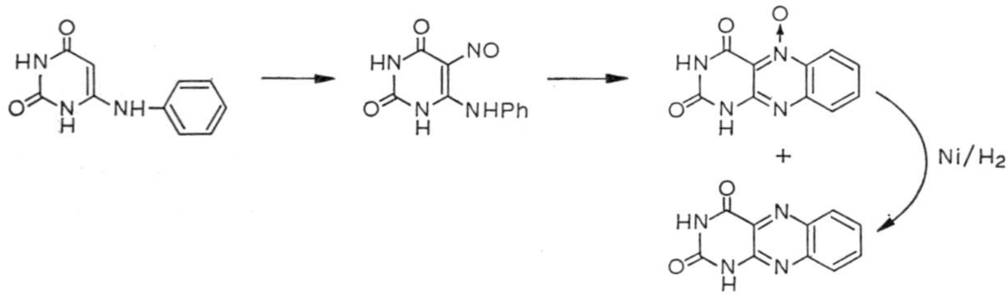

The nitrosative cyclisation reaction has also been carried out using a mixture of *N*-nitrosodimethylamine and phosphoryl chloride (*F. Yoneda, K. Shinozuka* and *Y. Sakuma*, Heterocycles, 1977, **6**, 1179).

Instead of the nitroso derivative the 5-diethyl hydrazinodicarboxylate (*Yoneda* and *Fukuzawa*, Chem. Comm., 1972, 503) and the 5-nitro compounds (*Yoneda, Sakuma* and *T. Nagamatsu, ibid.*, 1975, 977) can be used as the source of the alloxazine 5-nitrogen atom.

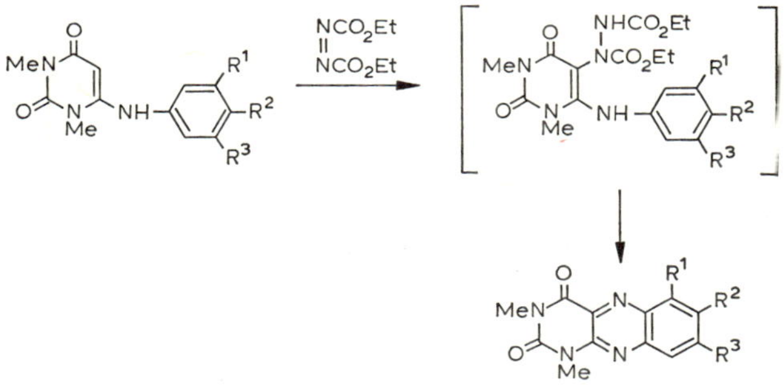

(*6*) 5-Nitro-4-phenylaminouracil and 5-amino-4-anilinouracil can be cyclised by triethyl phosphite and lead tetra-acetate, respectively, to alloxazine (*E. C. Taylor et al.*, J. Amer. chem. Soc., 1967, **89**, 3369).

(*7*) Condensation of 4-amino-1,3-dimethyluracil with nitrosobenzene gives 1,3-dimethylalloxazine (*idem, loc. cit.*; *Yoneda et al., ibid.*, 1976, **98**, 830).

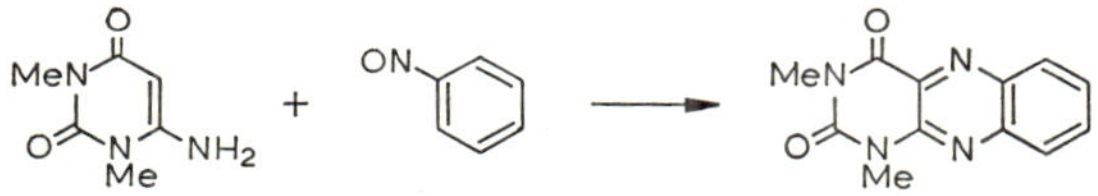

(*8*) 5,6-Dimethylbenzoxadiazole *N*-oxide on condensation with malononitrile gives 2-amino-3-cyanoquinoxaline 1,4-dioxide which may be converted in several steps into 7,8-dimethylalloxazine as shown (*F. Seng* and *K. Ley*, Angew. Chem. intern. Edn., 1972, **11**, 1010).

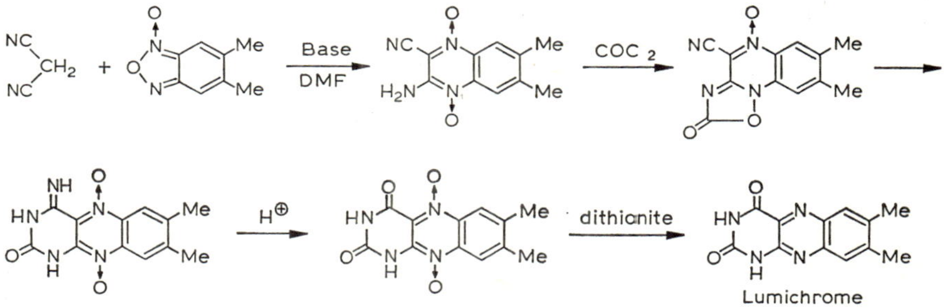

(*9*) Condensation of 5,6-diaminouracil with diacetyl dimer (as its hemiacetal) gives lumichrome (*A. J. Birch* and *C. J. Moye*, J. chem. Soc., 1957, 412; 1958, 2622).

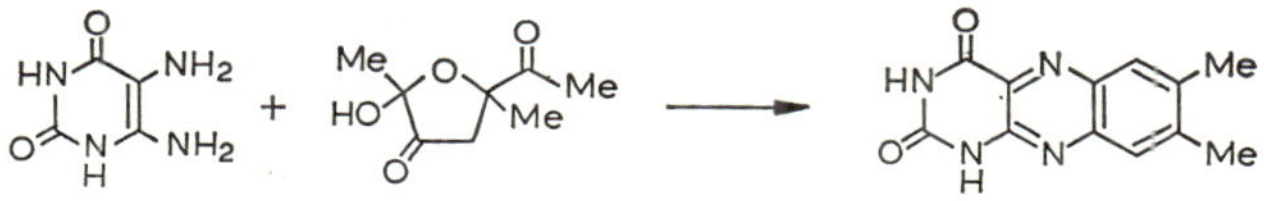

(*10*) Condensation of pyrimidineoxadiazole *N*-oxides with aromatic amines gives alloxazines (*Yoneda, Sakuma* and *S. Matsumoto*, Heterocycles, 1975, **3**, 113).

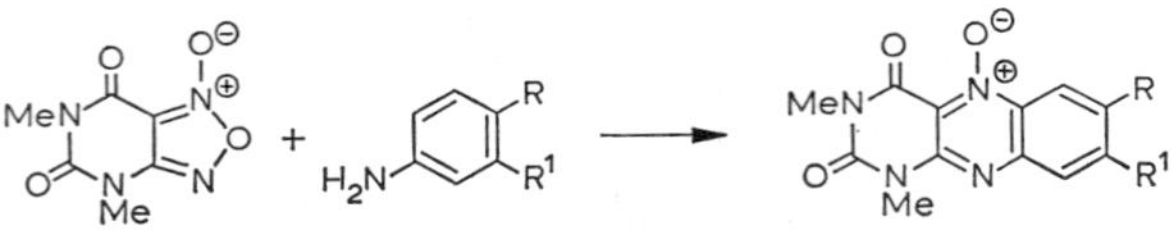

(*11*) A novel route to isoalloxazines has been described using a tetrahydroquinoxaline derivative as an intermediate. Thus reaction of 1-morpholinocyclohexane with the oxime III produced the azadiene IV. Subsequent reaction of this with amines gave the key intermediate 1-alkyl-2-amino-3-methoxycarbonylpyrazine, which, on reaction with phenyl isocyanate and subsequent cyclisation with trimethylamine produced the tetrahydroisoalloxazine. Dehydrogenation of the latter to the isoalloxazine was achieved using palladium on charcoal in boiling decalin (*D. Clerin, A. Lacroix* and *J. P. Fleury*, Tetrahedron Letters, 1976, 2899).

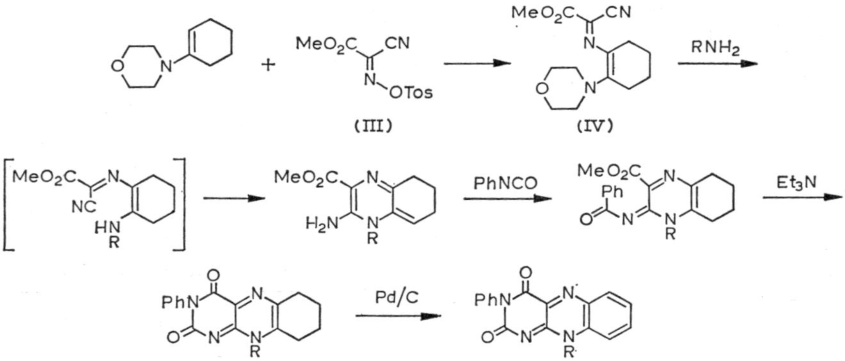

(*b*) *Chemical properties*

Alloxazines are yellow, weakly acidic solids which melt at a high temperature with decomposition. They undergo direct methylation on the nitrogen atoms of the pyrimidine ring. This ring is cleaved by alkaline hydrolysis, the resulting product being the corresponding quinoxalonecarboxylic acid. Reduction of alloxazine in acid solution gives first a deep red radical ion followed by the

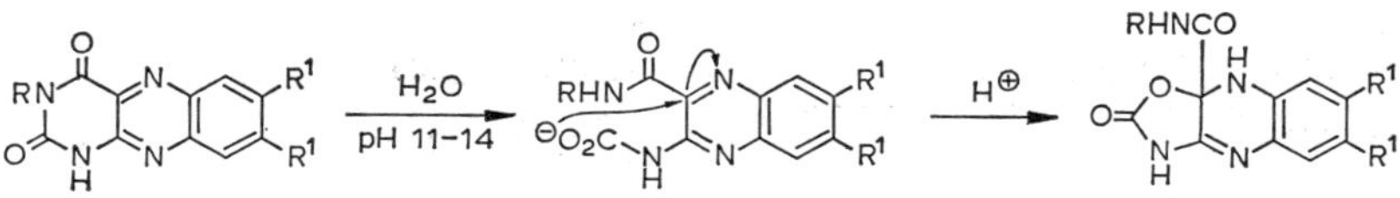

dihydroalloxazine (*R. Kuhn* and *F. Bär*, Ber., 1934, **67**, 898) whereas oxidation gives pteridine-2,4-dione-6,7-dicarboxylic acid (*Kühling*, *ibid.*, 1894, **27**, 2116). Covalent hydration of alloxazines occurs *via* $C_{(2)}$–$N_{(2)}$ cleavage at pH 11–14 to generate the carbamate anion which on acidification recyclises to give the stable alloxazine hydrate (*J. Koziol* and *P. Hemmerich*, Ann., 1976, 1276).

Isoalloxazines are yellow to orange-brown, feebly amphoteric compounds which are reduced catalytically or with dithionite to colourless 5,10-dihydro derivatives (leuco compounds). The latter are readily oxidised back to the isoalloxazines. The isoalloxazines are converted to the thermodynamically more stable alloxazines under a variety of alkylating conditions (*L. Hevesi*, Tetrahedron Letters, 1976, 1389).

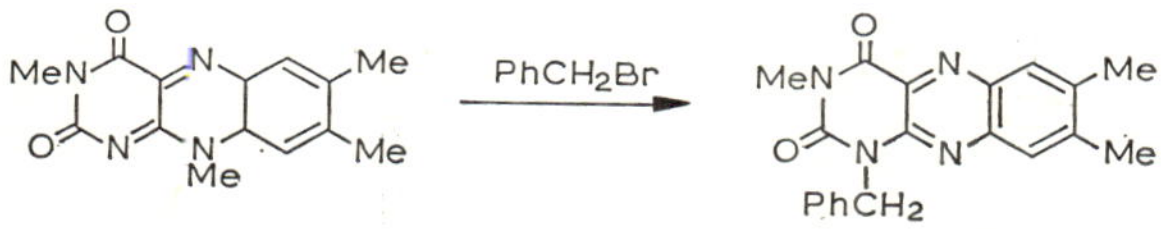

Alkaline hydrolysis of isoalloxazines gives a urea and a quinoxalonecarboxylic acid by initial attack at the 4-position giving the intermediate ureidoquinoxalinecarboxylic acid (*T. C. Bruice et al.*, J. Amer. chem. Soc., 1971, **93**, 7327).

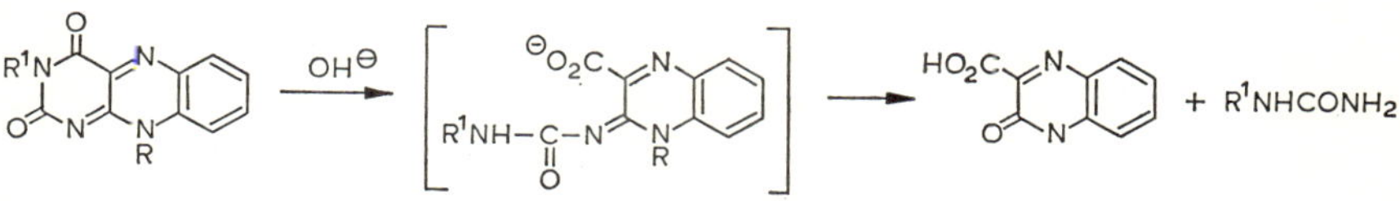

In a photochemical reaction, 3-benzyllumiflavin (V) adds cyclopentadiene in acetonitrile resulting in the 4a-(1-cyclopentadienyl)-4a,5-dihydrolumiflavin whereas cyclohexadiene under similar conditions reduces the flavin to

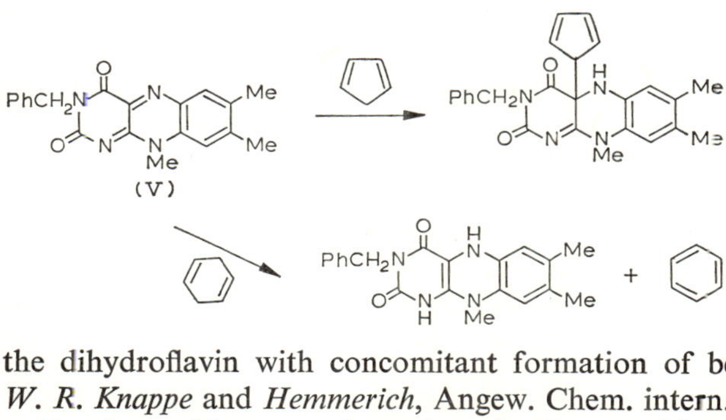

the dihydroflavin with concomitant formation of benzene (*M. Brustlein*, *W. R. Knappe* and *Hemmerich*, Angew. Chem. intern. Edn., 1971, **10**, 804).

Covalent adducts are formed at the $C_{(4a)}$ position in the photochemical oxidative decarboxylation of phenyl acetate in the presence of isoalloxazines (*W. H. Walker*, *Hemmerich* and *V. Massey*, Eur. J. Biochem., 1970, **13**, 258) and in the inactivation of the enzyme, L-lactate oxidase, by the suicide substrate 2-hydroxy-3-butynoic cid (*A. Schonbrunn et al.*, Biochemistry, 1976, **15**, 1798). In contrast the irreversible inhibition of the flavo-enzyme, monoamine-oxidase, with 3-dimethylamino-1-propyne, occurs by $N_{(5)}$-adduct formation (*A. L. Maycock et al.*, *ibid.*, p. 114). The $C_{(4a)}$ position has also been implicated as the reactive electrophilic centre in non-photochemical reactions by *L. E. Brown* and *G. A. Hamilton* (J. Amer. chem. Soc., 1970, **92**, 7225) as shown below. However, the proposal of this mechanism as a general one for flavin-catalysed "biological" redox reactions has been criticised by *S. Shinkai* and *Bruice*, *ibid.*, 1973, **95**, 7526).

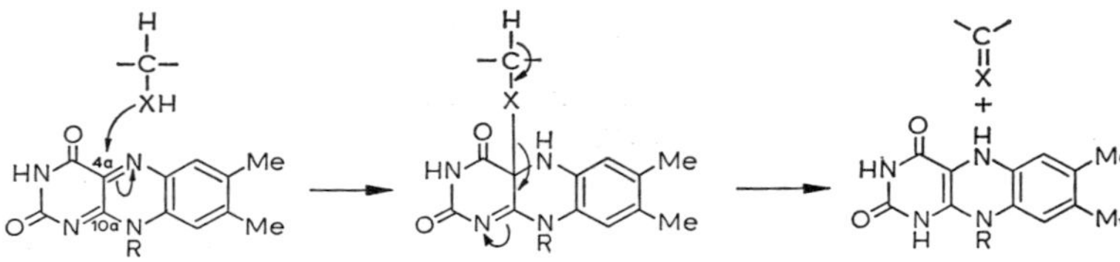

Addition to the alternative $C_{(10a)}$ position in reactions of the above type and also in the oxygen-requiring oxidation of reduced flavins (*Hamilton*, in "Progress in Bio-organic Chemistry", eds. *E. T. Kaiser* and *F. J. Kézdy*, Vol. I, Interscience, New York, 1971, Ch. 2) has been discounted by steric blocking of this position (*L. Main*, *G. J. Kasperek* and *Bruice*, Chem. Comm., 1972, 847), although the $C_{(10a)}$ position has been implicated in hydroperoxide formation in a model system for bacterial bioluminescence (*F. McCapra* and *P. Leeson*, Chem. Comm., 1976, 1037; *C. Kemal* and *Bruice*, J. Amer. chem. Soc., 1977, **99**, 7065). A critical appraisal of the controversial area of the mechanism of flavin-linked reactions in biological systems and in models has been presented by *Bruice* (Progr. bio-org. Chem., 1976, **4**, 1).

A quantitative evaluation of the electrophilicity of the $C_{(8)}$ position in the isoalloxazines has recently been reported (*S. B. Smith*, *Brustlein* and *Bruice*, J. Amer. chem. Soc., 1974, **96**, 3696).

(*c*) *Alloxazines and homologues*

Alloxazine exists as yellow needles which decompose above 300° without melting. It is relatively insoluble in organic solvents and is precipitated by carbon dioxide from its solutions in alkali.

7,8-**Dimethylalloxazine (lumichrome)** is obtained from the photodealkylation of riboflavin in dilute acid solution (*P. Karrer et al.*, Helv., 1934, **17**, 1010). The ultraviolet absorption spectrum in 0.1 *M* aqueous sodium hydroxide has max (log ε) of 263 (4.64) and 340 (3.88) nm (*Bardos*, *Olsen* and *Enkoji*, J. Amer. chem. Soc., 1957, **79**, 4704).

Alloxazine is methylated by dimethyl sulphate to give **1,3-dimethylalloxazine,** m.p. 247–247.5° (*H. Bredereck* and *W. Pfleiderer*, Chem. Ber., 1954, **87**, 1119). The latter can also be prepared from dimethylalloxan and *o*-phenylenediamine. Other derivatives include the 1-*methyl*-, m.p. 356–358° and the 3-*methyl*-, m.p. 343–344°, compounds which are prepared from the corresponding quinoxalinecarboxamide (*Pfleiderer*, Chem. Ber., 1956, **89**, 1148) and also **1,3,7,8-tetramethylalloxazine,** m.p. 252° (*Kuhn* and *H. Rudy*, Ber., 1934, **67**, 1826).

10-Methylisoalloxazine has m.p. 392° (decomp.) (*Kuhn* and *F. Weygand*, *ibid.*, p. 1409).

7,8,10-Trimethylisoalloxazine (lumiflavin) is produced by photochemical degradation of riboflavin in alkaline solution. It forms orange needles, m.p. 328°, having max (log ε) 270 (4.87), 340 (4.21), 360 (4.25) and 445 (4.41) nm.

Alkylation takes place on the pyrimidine nitrogen atom giving the 3-*methyl* derivative, m.p. 326° (*Kuhn* and *Rudy*, Ber., 1934, **67**, 1125). Lumichrome and lumiflavin or their corresponding lower homologues are also produced by the photolysis of 10-(*β-hydroxyethyl*)*isoalloxazine*, m.p. 310° (decomp.), 10-(*β,γ-dihydroxypropyl*)*isoalloxazine*, m.p. 300° and the 10-(*β-dihydroxypropyl*)-7,8-*dimethylisoalloxazine*, m.p. 294° (decomp.), under various conditions (*Karrer et al.*, Helv., 1934, **17**, 1165, 1516). Although photolysis does not cleave the 10-methyl group, this is readily removed by heating with hydroxylamine sulphate in acetic cid (*Hemmerich*, *Prijs* and *Erlenmeyer*, *ibid.*, 1960, **48**, 372).

(*d*) *Riboflavin and related compounds*

Riboflavin (Vitamin B_2) was characterised as the thermostable moiety of the "water-soluble Vitamin B" complex necessary for growth in young rats; it was first isolated from albumen and whey and later from a wide variety of animal organs and vegetable sources (*Kuhn et al.*, Ber., 1933, **66**, 576, 1034, 1950; *P. Ellinger* and *W. Koschara*, *ibid.*, pp. 315, 808, 1411; *Karrer* and *K. Schöpp*, Helv., 1934, **17**, 735, 771, 1013). The identity of the flavins from different sources was accepted after some hesitation, the term lactoflavin being given to the structure until this was established as 7,8-dimethyl-10-D-ribitylisoalloxine (*Karrer*, *Schöpp* and *F. Benz*, *ibid.*, 1935, **18**, 426; *Kuhn et al.*, Naturwiss., 1935, **23**, 260).

Riboflavin forms orange-brown needles, m.p. 292° (decomp.); $[\alpha]_D$ + 110° in sodium hydroxide, + 360° in the presence of borax; max (log. ε), 220 (4.85), 265 (4.89), 365 (4.33) and 445 (4.38) nm. For purification purposes, use may be made of its adsorption on to lead sulphide or its conversion into the thallous salt. It is feebly amphoteric, pH–fluorescence curves indicating pK_a values of 1.7 and 10.2 (*Kuhn* and *G. Meruzzi*, Ber., 1934, **67**, 888).

Riboflavin forms a chloroform-soluble *tetra*-O-*acetyl* derivative, m.p. 242°, and a *diisopropylidene* derivative, m.p. 293° (*Kuhn et al.*, *ibid.*, 1935, **68**, 625). It yields one molecule of formaldehyde when treated with lead tetra-acetate. All these observations support the presence of the carbohydrate-derived side-chain — $(CHOH)_3CH_2OH$.

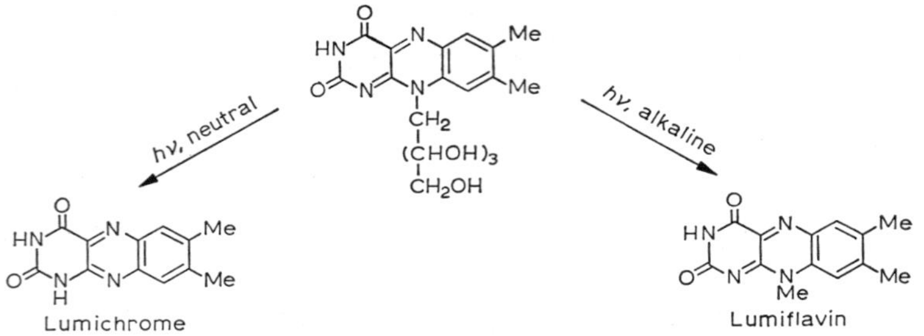

Photochemical decomposition of the vitamin in neutral methanolic solution gives lumichrome, the constitution of which was established by synthesis (*Karrer et al.*, Helv., 1934, **17**, 1010; *Kuhn* and *Rudy*, Ber., 1934, **67**, 1936).

When the irradiation is conducted in alkaline solution the product is lumiflavin, the ultraviolet and visible absorption spectrum of which closely resembles that of riboflavin. Further degradation of lumiflavin took the following course:

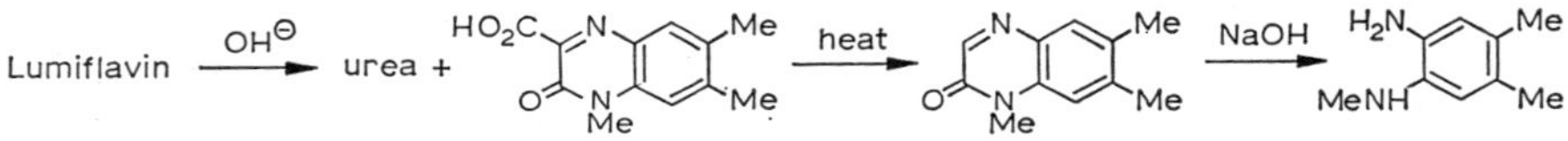

On the probable assumption that riboflavin differed from lumiflavin only in containing a 2,3,4,5-tetrahydroxy-*n*-pentyl group in place of the methyl group at position 10, various 7,8-dimethyl-10-pentitylisoalloxazines were synthesised. The easily accessible L-arabityl compound, L-araboflavin, showed biological activity and the D-ribityl isomer proved to be identical with the natural vitamin. The total synthesis of riboflavin was effected by *Karrer et al.* (Helv., 1935, **18**, 426, 1130, 1435; 1936, **19**, 264) as shown below.

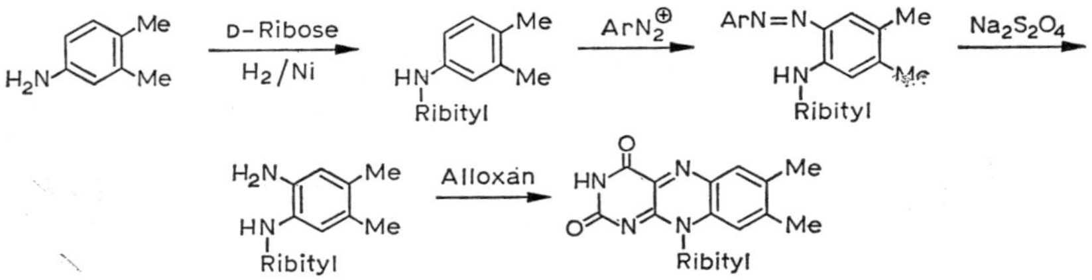

An alternative synthesis was published concurrently by *Kuhn et al.* (Ber., 1935, **68**, 1765).

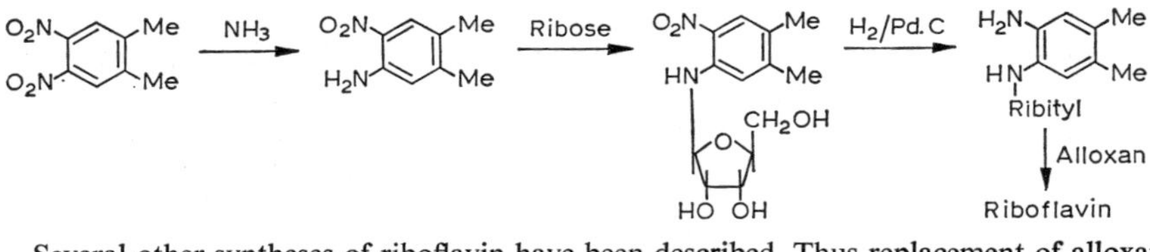

Several other syntheses of riboflavin have been described. Thus replacement of alloxan by 5,5-dichlorobarbituric acid in the above syntheses results in riboflavin (*Tishler, Wellman*

and *Ladenburg*, *loc. cit.*). Condensation of barbituric acid with *N*-ribityl-3,4-dimethyl-6-(4-nitrophenylazo)aniline also produced riboflavin (*Tishler et al.*, J. Amer. chem. Soc., 1947, **69**, 1487).

An alternative approach has involved condensation of 5-amino-4-D-ribitylaminouracil with 3,4-dimethyl-1,2-benzoquinone (*J. Davoll* and *D. D. Evans*, J. chem. Soc., 1960, 5041; *R. M. Cresswell, T. Neilson* and *H. C. S. Wood, ibid.*, 1961, 476).

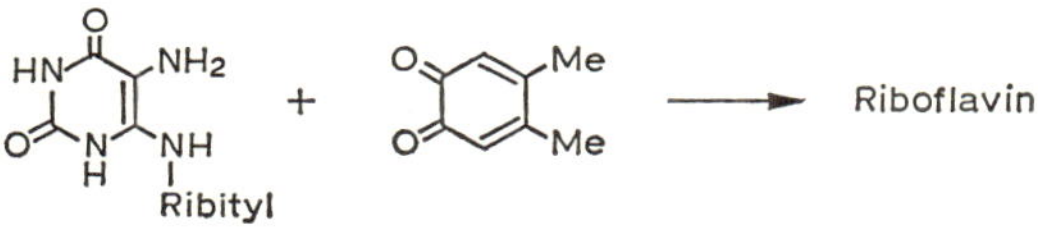

Riboflavin is also obtained from condensation of the above uracil derivative with the hemi-acetal of diacetyl dimer (*Cresswell* and *Wood*, Proc. chem. Soc., 1959, 387).

Numerous riboflavin analogues have been synthesised in order to investigate the structural requirements for the biological activity which is manifested by promoting growth in rats and in the micro-organisms which cause lactic acid fermentation. Activity is destroyed by *N*-methylation at position 3 or by acetylation of the alcoholic hydroxyl groups; the D-configuration of the ribityl moiety is also obligatory for activity. Alkyl substitution at $C_{(7)}$ and/or $C_{(8)}$ in the benzene ring is also a requirement although alkylation in any other position causes complete loss of activity. 5-Deazariboflavin was synthesised in 1970 by *D. O'Brien*, *L. Weinstock* and *C. Cheng* (J. heterocycl. Chem., 1970, **7**, 99) and subsequently has been the subject of much study because of its conversion to deaza-FMN and deaza-FAD which are able to replace FMN and FAD for normal redox function in several enzyme systems (*R. Spencer*, *J. Fisher* and *C. Walsh*, Biochemistry, 1976, **15**, 1043; *L. B. Hersh* and *M. Schuman-Jorns*, in "Flavins and Flavoproteins", ed. *T. P. Singer*, Elsevier, Amsterdam, 1976, p. 362). 5-Deazariboflavin also has potent anticoccicidal activity, *in vivo*, as a riboflavin antagonist (*D. W. Graham et al.*, Experientia, 1977, **33**, 1274). Several later syntheses have been described. These are all based on the cyclisation of 6-D-ribityl-3,4-xylidinouracil derivatives (*M. Janda* and *P. Hemmerick*, Angew. Chem., 1976, **88**, 475; *F. Yoneda* and *Y. Sakuma*, Chem. Comm., 1976, 203).

The synthesis and biological properties of other deaza analogues of riboflavin have been described recently (*W. I. Ashton et al.*, Tetrahedron Letters, 1977, 2551; *R. Spencer*, *J. Fisher* and *C. Walsh*, Biochemistry, 1977, **16**, 3586, 3594).

Alloxazine-10-glycosides are much less stable than the 10-pentityl and 10-hexityl-isoalloxazines and they are biologically inactive (*Kuhh* and *Rudy*, Ber., 1936, **69**, 2557; *E. E. Snell* and *F. M. Strong*, Enzymologia, 1939, **6**, 186).

A list of riboflavin analogues is given in Table 1.

TABLE 1

RIBOFLAVIN ANALOGUES

Substituents in positions					
6	7	8	9	10	*Refs.*
—	—	—	—	D-ribityl	1
—	Me	—	—	D-ribityl	2
—	—	Me	—	D-ribityl	1
—	—	Et	—	D-ribityl	3
—	Et	Me	—	D-ribityl	3
—	Me	—	Me	D-ribityl	4
Me	—	Me	—	D-ribityl	4
—	Et	Br	—	D-ribityl	12
—	Br	Et	—	D-ribityl	12
—	Me	Me	—	L-ribityl	5
—	Me	Me	—	D-arabityl	6
—	Me	Me	—	L-arabityl	7,8
—	$CH_2{\cdot}CH_2{\cdot}CH_2$		—	L-arabityl	8
—	$-(CH_2)_4-$		—	L-arabityl	9
—	Me	Me	—	D-yxlityl	10,11
—	Me	Me	—	D-lyxityl	5
—	Me	Me	—	D-deoxyribityl	1
—	Me	Me	—	L-rhamnityl	11
—	—	Me	—	D-sorbityl	7 11
—	—	Me	—	D-mannityl	11
—	—	Me	—	D-dulcityl	11

References
1 *P. Karrer et al.*, Helv., 1935, **18**, 1143.
2 *Karrer* and *F. M. Strong*, *ibid.*, p. 1343.
3 *Karrer* and *T. H. Quibell*, *ibid.*, 1936, **19**, 1034.
4 *R. Kuhn et al.*, Ber., 1937, **70**, 1293.
5 *Karrer et al.*, Helv., 1935, **18**, 908.
6 *Kuhn* and *F. Weygand*, Ber., 1935, **68**, 1282.
7 *Karrer et al.*, Helv., 1935, **18, 69.**
8 *Kuhn* and *Weygand*, Ber., 1934, **67**, 2084.
9 *Kuhn et al.*, *ibid.*, 1937, **70**, 1302.
10 *Karrer et al.*, Helv., 1935, **18**, 426.
11 *Idem*, *ibid.*, p. 522.
12 *J. P. Lambooy*, J. med. Chem., 1974, **117**, 227.

Structural identification of the schizoflavins isolated from *Schizophyllum commune* has been reported (*S. Tachibana*, *T. Murakami* and *T. Ninomiya*, J. nutr. Sci. Vitaminol., 1975, **21**, 347). They are oxidation products of riboflavin where the $C_{(5)}$ position of the ribityl group has been modified, thus schizoflavin-1 is 10-(4-carboxy-2,3,4-trihydroxybutyl)-7,8-dimethylisoalloxazine and schizoflavin-2 is 10-(4-formyl-2,3,4-trihydroxybutyl)-7,8-dimethylisoalloxazine.

Roseoflavin isolated from *Streptomyces* culture broths has been identified as 8-dimethylamino-7-methyl-10-D-ribitylisoalloxazine by chemical and spectroscopic methods (*S. Otani* in "Flavins and Flavoproteins", ed. *T. P. Singer*, Elsevier, Amsterdam, 1976, p. 323).

3-Methylriboflavin has been isolated from the toadstool *Panellus serotinus* (*W. Steglich* and *L. Zechlin*, Z. Naturforsch., C. 1977, **32C**, 520).

(*i*) *Biosynthesis*

6,7-Dimethyl-8-D-ribityllumazine ("G" compound) isolated from *Eremothecium ashbyii* and *Ashbya gossypii*, was shown to be the biosynthetic precursor of riboflavin (*G. F. Maley* and *G. W. E. Plaut*, J. biol. Chem., 1959, **234**, 641; J. Amer. chem. Soc., 1959, **87**, 2025). Later, *Plaut* discovered that the additional four-carbon unit required to form the *o*-xylene ring moiety of riboflavin was supplied from a second molecule of the lumazine precursor (*ibid.*, 1960, **235**, PC41; 1963, **238**, 2225).

In turn, recent evidence of incorporation of radioactivity from ^{14}C-labelled guanosine into "G" compound (*B. Mailander* and *A. Bacher*, J. biol. Chem., 1976, **251**, 3623) and also the isolation of an enzyme (GTP cyclohydrolase II) from *E. coli* which, it is suggested, effects this transformation (or some part of it) (*F. Foor* and *G. M. Brown*, *ibid.*, 1975, **250**, 3545), supports the hypothesis that the biosynthesis of riboflavin occurs from the GTP and thus follows the same general pathway as the pteridines.

T. Rowan and *Wood* subsequently obtained riboflavin non-enzymically from the 6,7-dimethyl-8-D-ribityllumazine in 56% yield by refluxing a buffered aqueous solution at pH 7.3 (Proc. chem. Soc., 1963, 21). The precise mechanism of formation of riboflavin from two molecules of the lumazine has been investigated using deuterium labelling (*T. Paterson* and *Wood*, Chem. Comm., 1969, 290) and a mechanism to explain the unusual labelling pattern has been postulated.

The same pattern of labelling was subsequently observed in the corresponding enzymic conversion (*Plaut*, *R. L. Beach* and *T. Aogaichi*, Biochem., 1970, **9**, 771).

7-Oxo-8-D-ribityllumazines are potent competitive inhibitors of the enzyme (riboflavin synthetase) which catalyses the conversion of two molecules of "G" compound to riboflavin and this property has been utilised to effect a purification of the enzyme using affinity chromatography with an immobilised 7-oxo-8-D-ribityllumazine ligand (*H. C. S. Wood et al.*, Chem. Comm., 1975, 464; *R. Wrigglesworth et al.*, in "Chemistry and Biology of Pteridines", ed. *W. Pfleiderer*, De Gruyter, Berlin, 1975, p. 93).

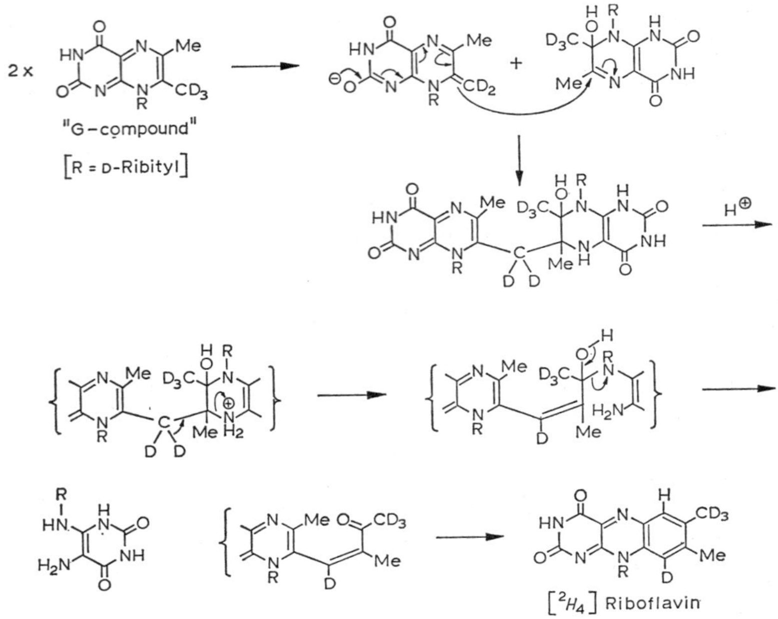

Riboflavin is further metabolised to riboflavin 5′-phosphate (FMN) and to flavin–adenine-dinucleotide (FAD) which are found as co-factors and prosthetic groups in enzymic reactions. Recently, riboflavin has been implicated from labelling studies as the biosynthetic precursor of the 5,6-dimethylbenzimidazole moiety of vitamin B_{12} (*P. Renz*, FEBS Letters, 1970, **6**, 187; *Renz* and *R. Weyhenmeyer*, *ibid.*, 1972, **22**, 124; *S.-H. Lu, M. F. Winkler* and *W. L. Alworth*, Chem. Comm., 1971, 191; *Alworth, M. F. Dove* and *N. H. Baker*, Biochemistry, 1977, **16**, 526; *Renz, R. Wurm* and *J. Hoerig*, Z. Naturforsch., C, 1977, **32C**, 7; *Hoerig* and *Renz*, FEBS Letters, 1977, **80**, 337).

(*ii*) *Chemical properties*

(*1*) *Photodealkylation of riboflavin.* As discussed above, riboflavin is photolytically cleaved in alkaline solution to give lumiflavin, and in neutral or acid solution, to lumichrome. This reaction is not specific to riboflavin but to any isoalloxazine the $C_{(10)}$ alkyl group of which has a hydroxy substituent in the β-position.

The course of this photolysis has been clarified as shown below (*W. M. Moore et al.*, J. Amer. chem. Soc., 1963, **85**, 3367; *E. C. Smith* and *D. Metzler*, *ibid.*, p. 3285; *M. Gladys* and *W. R. Knappe*, Chem. Ber., 1974, **107**, 3658).

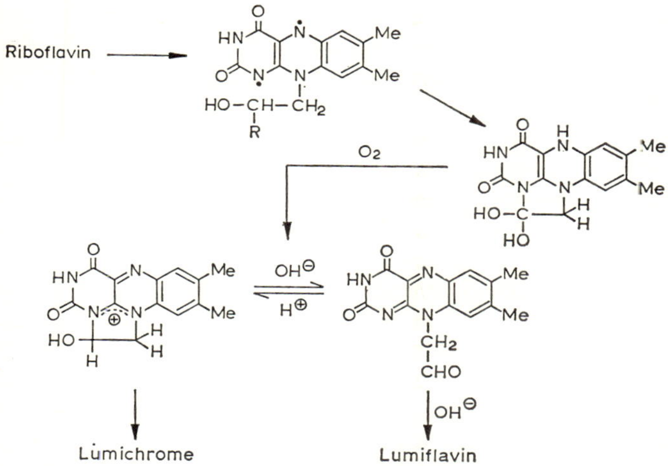

(*2*) *Reduction* of riboflavin occurs *via* two consecutive one-electron transfers involving an intermediate radical. Each of the three states (oxidised form, radical and reduced form) of the flavin is amphoteric resulting in a total of nine distinct monomolecular species existing at different pH's (*Hemmerich*, *C. Veeger* and *Wood*, Angew. Chem. intern. Edn., 1965, **4**, 671). This review gives a list of ultraviolet and visible spectral absorptions of these various flavin species.

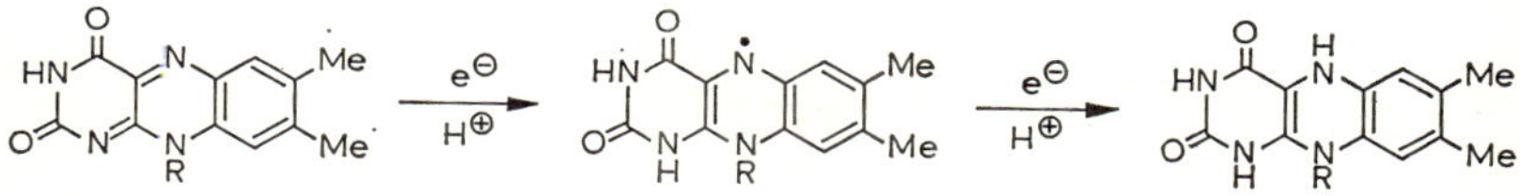

(*3*) *Reactivity of $C_{(8)}$-methyl group*. The acidity of the protons of the $C_{(8)}$-methyl group of riboflavin and lumiflavin is shown by the base-catalysed dimerisation of these compounds (*Hemmerich*, *Prijs* and *Erlenmeyer*, Helv., 1959, **42**, 2164) and also by their ability to condense with aromatic aldehydes

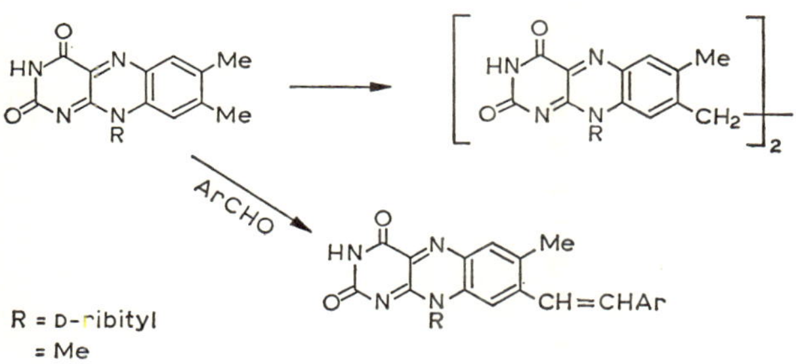

to produce 8α-substituted derivatives as shown below (*Hemmerich, ibid.*, 1960, **43**, 1942).

Similarly, tetra-acetylriboflavin is brominated at the $C_{(8)}$ methyl group giving the dibromomethyl derivative (*D. B. McCormick*, J. heterocycl. Chem., 1970, **7**, 447). The reactivity of the 8-methyl group is biologically important in that riboflavin attaches covalently to specific "flavoproteins" such as succinate dehydrogenase at this position (*Hemmerich et al.*, FEBS Letters, 1969, **3**, 37; *T. P. Singer* and *D. E. Edmondson*, *ibid.*, 1974, **42**, 1).

(*e*) *Flavin phosphoric acid derivatives*

The essential biological function of Vitamin B_2 appears to be in its involvement as the prosthetic group of the "yellow enzymes" which are concerned with electron transport processes. These systems can be decomposed into a protein apoenzyme and a riboflavinphosphoric acid and may be reconstituted from these components (*H. Theorell*, Biochem. Z., 1934, **272**, 155; 1935, **278**, 263; *R. Kuhn* and *H. Rudy*, Berl, 1936, **69**, 1974). The phosphate group has been located on the terminal hydroxyl (5′) of the ribityl side-chain of riboflavin since the acid gives no formaldehyde on oxidation with periodate (*P. Karrer et al.*, Helv., 1937, **20**, 79).

The flavinphosphoric acid has been synthesised from riboflavin by the following steps (*Kuhn et al.*, 1936, **69**, 1543):

$$F-CH_2(CHOH)_3CH_2OH \longrightarrow F-CH_2(CHOH)_3CH_2OCPh_3$$

$$\downarrow$$

$$F-CH_2(CHOAc)_3CH_2OH \longleftarrow F-CH_2(CHOAc)_3CH_2OCPh_3$$

$$\downarrow$$

$$F-CH_2(CHOAc)_3CH_2OP(O)(OH)_2 \longrightarrow F-CH_2(CHOH)_3CH_2OP(O)(OH)_2$$

It may be prepared more effectively by direct phosphorylation of riboflavin (*H. S. Forrest* and *A. R. Todd*, J. chem. Soc., 1950, 3295).

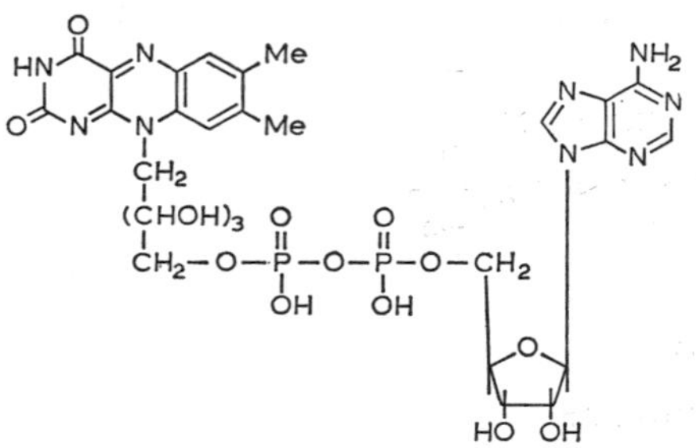

Flavine adenine dinucleotide

Enzymes isolated which contain riboflavinphosphoric acid (flavin mononucleotide, FMN) as the prosthetic group often show low activity and may represent degradation products of more active material containing "flavin adenine dinucleotide" (FAD), which has the structure shown (*O. Warburg* and *W. Christian*, Biochem. Z., 1938, **298**, 150).

An enzyme found in yeast synthesises FAD from riboflavin monophosphate (FMN) and adenosine triphosphate (ATP) according to the following reaction (*A. W. Schrecker* and *A. Kornberg*, J. biol. Chem., 1950, **182**, 795):

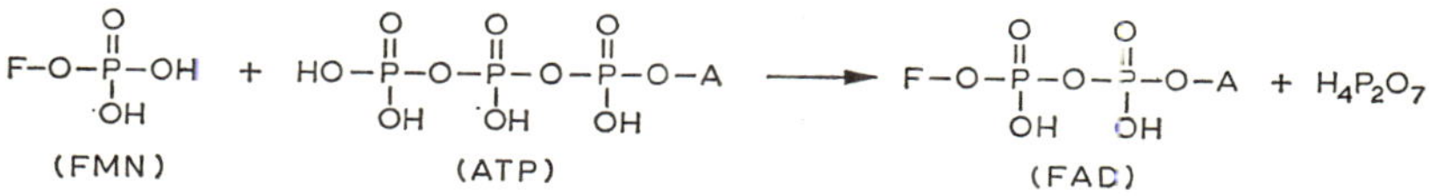

This coenzyme has been synthesised by condensation of the sodium salt of riboflavin phosphoric acid with a protected adenosine derivative and subsequent deprotection (*Todd et al.*, J. chem. Soc., 1954, 46; *F. M. Huennekens* and *G. L. Kilgour*, J. Amer. chem. Soc., 1955, **77**, 6716).

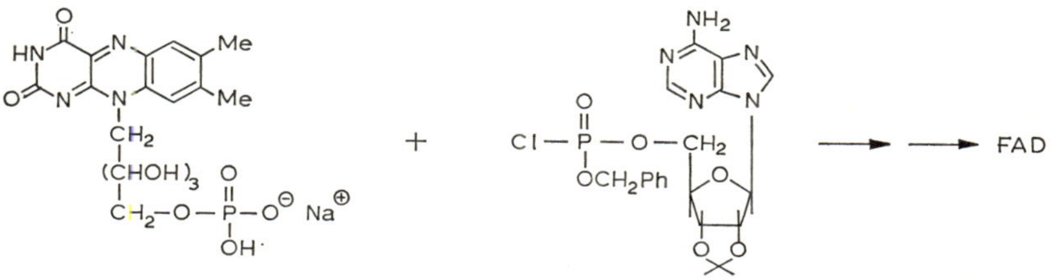

Chapter 60

The Biosynthesis of Plant Alkaloids and Nitrogenous Microbial Metabolites

R. B. HERBERT

1. Introduction

"Organic Chemists have often been tempted to leave the security of their own proper pastures and to graze, albeit speculatively, in the attractive fields of biochemistry" (*R. Robinson*, "The Structural Relations of Natural Products", Oxford University Press, London, 1954). With the advent of readily available radioactive organic compounds, organic chemists have been able to go further than speculation and have joined biochemists in the profitable exploration of the biosynthetic pathways leading to secondary metabolites, of which plant alkaloids are one group. This review is concerned with a discussion of the results of these investigations into the biosynthesis of alkaloids as well as those of, the quite often related, microbial metabolites containing nitrogen. The literature is surveyed up to the end of June 1976. (For other reviews see: "Biosynthese der Alkaloide", ed. *K. Mothes* and *H. R. Schütte*, VEB Deutscher Verlag der Wissenschaften, Berlin, 1969; *I. D. Spenser*, in "Comprehensive Biochemistry", eds. *M. Florkin* and *E. H. Stotz*, Elsevier, Amsterdam, 1968, Vol. 20, p. 231; *E. Leete*, in "Biogenesis of Natural Compounds", ed. *P. Bernfeld*, 2nd edn., Pergamon, Oxford, 1967, p. 953. For more recent material see: "The Alkaloids" and "Biosynthesis" which are Specialist Periodical Reports published by The Chemical Society, London — material post-dating this review will be found in volume 8 and succeeding volumes of "The Alkaloids").

2. Pyrrolidine alkaloids

(*a*) *Tropane and related alkaloids*

Although the synthesis of tropinone (V) from succindialdehyde, methylamine, and acetonedicarboxylic acid (*R. Robinson*, J. chem. Soc., 1917, **111**, 762; *C. Schöpf* and *G. Lehmann*, Ann., 1935, **518**, 1) provides one of the most spectacular examples of organic synthesis under "physiological" conditions, the biosynthesis of hyoscyamine (VII) and related alkaloids like hygrine (III) follows another course (see Scheme 1), but one which was broadly anticipated over 50 years ago (*Robinson*, J. chem. Soc., 1917, **111**, 876).

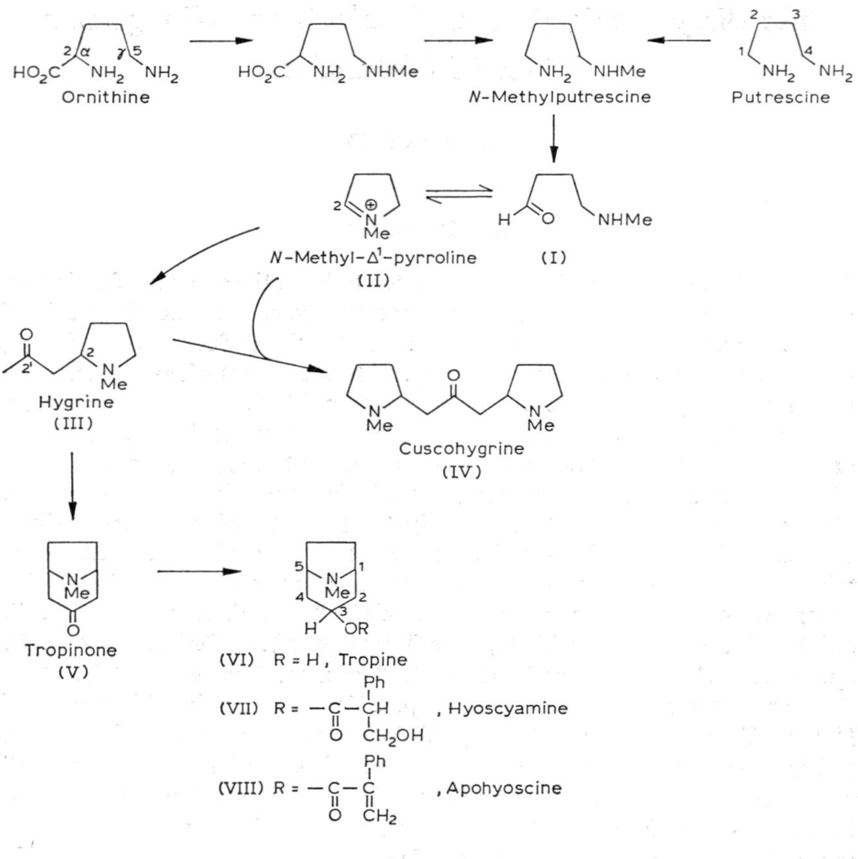

Scheme 1

Tracer experiments with [2-^{14}C]labelled ornithine showed that hyoscyamine (VII), scopolamine (XI), and meteloidine (XIII) were specifically derived from this amino-acid (*E. Leete*, *L. Marion* and *I. D. Spenser*, Nature, 1954, **174**, 650; Canad. J. Chem., 1954, **32**, 1116; *Leete*, J. Amer. chem. Soc., 1962, **84**, 55; *H. W. Liebisch et al.*, Z. Naturforsch., 1965, **20B**, 1183; *F. A. Turner* and *J. E. Gearien*, J. pharm Sci., 1964, **53**, 1309; *Leete* and *S. J. Nelson*, Phytochemistry, 1969, **8**, 413). Similar results were obtained for hyoscyamine and scopolamine with [1,4-$^{14}C_2$]putrescine and the label was found to be confined to the bridgehead carbon atoms (*J. Kaczkowski* and *Marion*, Canad. J. Chem., 1963, **41**, 2651; *Liebisch, H. R. Schütte* and *K. Mothes*, Ann., 1963, **668**, 139; *Liebisch et al., loc. cit.*). Whilst equal labelling of the bridgehead carbons by the putrescine is expected, the single label in the ornithine might turn up in one or both of these carbon atoms. A most elegant degradative sequence was consequently developed to establish the ornithine labelling pattern. Hyoscyamine was converted to the racemate IX (see Scheme 2), which was resolved.

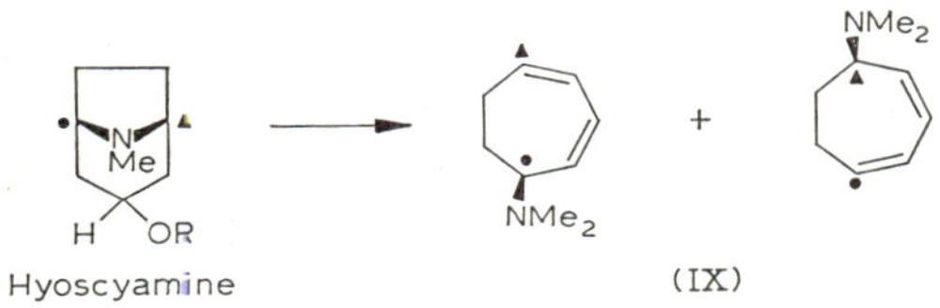

Scheme 2

Further degradation allowed isolation of the carbon atom to which the amino group, in each enantiomer, was attached; these carbons correspond to the two bridgehead atoms in hyoscyamine and a distinction between them was thus achieved. It was found that [2-^{14}C]ornithine labelled only one of the bridgehead positions (*A. A. Bothner-By et al.*, J. Amer. chem. Soc., 1962, **84**, 52; *Leete, loc. cit.*; see also other refs. above for specific incorporation of [2-^{14}C]-ornithine). It follows from this stereospecific and unsymmetrical incorporation that putrescine cannot be an intermediate between ornithine and hyoscyamine. The results are in contrast with those for nicotine where a symmetrical incorporation of ornithine was observed (see section 2b, p. 299).

So far the biosynthesis of these pyrrolidine alkaloids might appear to resemble that of the piperidine bases (section 3, p. 309). Further evidence for the derivation of the five-membered heterocycles indicates a quite divergent evolution, however. The results of [2-^{14}C, δ-^{15}N]- and [2-^{14}C, α-^{15}N]-ornithine incorporation into hyoscyamine (VII), scopolamine (XI) and the related base, cuscohygrine (IV), indicate that the δ-amino group is retained and the α-amino group lost (*Liebisch, A. S. Radwan* and *Schütte*, Ann., 1969, **721**, 163; *Liebisch* and *Schütte*, Z. Pflanzenphysiol, 1967, **57**, 434). ^{15}N-Incorporations can be criticized on the grounds that the results may be distorted by facile transamination reactions, so the next set of evidence is more convincing. *N*-Methylputrescine is more efficiently converted than ornithine and putrescine into cuscohygrine (IV) and hyoscyamine (VII), and no loss of the methyl group occurs (*Liebisch, W. Maier* and *Schütte*, Tetrahedron Letters,

1966, 4079; *Liebisch, Radwan* and *Schütte, loc. cit.*). Further, whilst labelled δ-*N*-methylornithine serves as a good precursor for these alkaloids and is utilized intact without loss of the *N*-methyl group, labelled α-*N*-methylornithine gave a negative result (*F. E. Baralle* and *G. G. Gross*, Chem. Comm., 1969, 721; *A. Ahmed* and *Leete*, Phytochemistry, 1970, **9**, 2345).

In summary the results show that the tropane bases arise from ornithine *via* *N*-methylornithine which on decarboxylation yields *N*-methylputrescine. In this way the individuality of C-2 and C-5 of ornithine is maintained. In spite of the unsymmetrical incorporation of ornithine it appears that putrescine may still be a primary source of the pyrrolidine ring since inactive putrescine, it has been shown, dilutes radioactivity from the alkaloids being synthesized from [2-^{14}C]ornithine. An analogous result was observed with radioactive putrescine and inactive ornithine (*Liebisch et al., loc. cit.*). Putrescine presumably enters the pathway *via* its *N*-methyl derivative.

Model condensations, *e.g.* of *N*-methyl-Δ^1-pyrroline (II) with acetoacetic acid which yields hygrine (III) (*E. F. L. J. Anet, G. K. Hughes* and *E. Ritchie*, Austral. J. Scientific Res., 1949, **Ser. 2A,** 616), provide an apparently reliable guide to the biosynthetic steps which follow *N*-methylputrescine. They begin most plausibly with deamination to (I) $\leftrightharpoons$ (II). [Biosynthesis of the pyrrolidine ring of nicotine (see Section 2b, p. 299) is broadly similar to that of the tropane alkaloids and it was shown there that the pyrroline (II) is a nicotine precursor].

As expected, acetate is a specific precursor for hygrine (III) and also cuscohygrine (IV) and hyoscyamine (VII); hyoscyamine was labelled at C-3 by [1-^{14}C]acetate, and at C-2 and C-4 by [2-^{14}C]acetate (*D. G. O'Donovan* and *M. F. Keogh*, J. chem. Soc., C, 1969, 223; *Liebisch, Radwan* and *Schütte, loc. cit.*; *Baralle* and *Gros*, Phytochemistry, 1969, **8**, 849, 853; *Kaczkowski, Schütte* and *K. Mothes*, Naturwissenschaften, 1960, **47**, 304; Biochim. Biophys. Acta, 1961, **46**, 588). The labelling data so far adduced link hygrine and hyoscyamine biogenetically and strongly suggest that cyclisation of hygrine may occur *in vivo* to give hyoscyamine. Proof of this relationship came with the specific and efficient *in vivo* transformation of [N-*methyl*-^{14}C,2′-^{14}C]hygrine (as III) into hyoscyamine without change in the ratio of the two labels (*O'Donovan* and *Keogh, loc. cit.*). Tropinone (V) and tropine (VI) are implicated as intermediates in this transformation and accordingly doubly labelled tropine (bearing a tritium label at C-3) was incorporated into hyoscyamine (VII) and also XI, XIII and XIV; no loss of tritium was observed so interconversion with tropinone does not occur (*Leete*, Phytochemistry, 1972, **11**, 1713).

Further, hyoscyamine (VII) is a precursor of scopolamine (XI) (*A. Romeike* and *G. Fodor*, Tetrahedron Letters, 1960, No. 22, 1). The intriguing

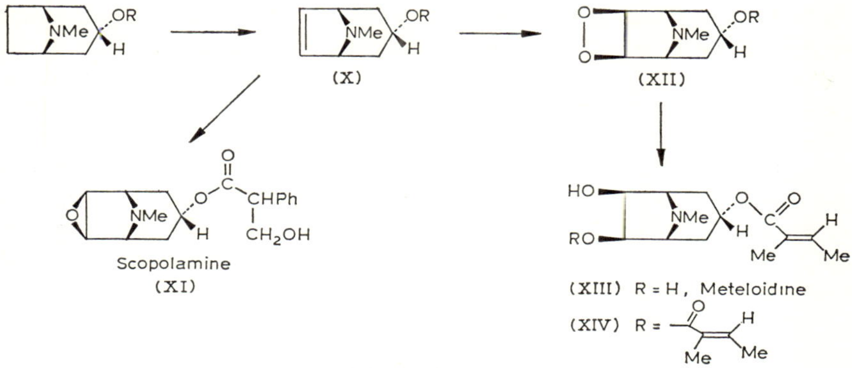

Scheme 3

problem of the mechanism of hydroxylation of the tropine nucleus to give meteloidine and scopolamine remains. The stereochemistry of the hydroxy-groups in meteloidine precludes an epoxide intermediate such as scopolamine and suggests the involvement of the dioxetane XII. This intermediate and scopolamine could arise by different oxygenation mechanisms from X.

[N-*methyl*-^{14}C,2′-^{14}C]Hygrine was efficiently and specifically incorporated into cuscohygrine (IV). The *specific* activity in the *N*-methyl groups of the cuscohygrine was one half of that required to maintain the *N*-methyl: C-2′ ratio of radioactivity of the hygrine. Therefore the second pyrrolidine ring of cuscohygrine arises not by degradation of hygrine but by condensation of hygrine (III) with a further molecule of *N*-methyl-Δ^1-pyrroline (II) (*O'Donovan* and *Keogh, loc. cit.*).

The results obtained for the biosynthesis of the tropane and related alkaloids are summarized in Schemes 1 and 3.

It appears that the **tropic acid moieties** of hyoscyamine (VII) and scopolamine (XI) can arise by three different pathways. Evidence for derivation of tropic acid from phenylalanine is particularly clear cut although the mechanism of the transformation is obscure. The labelling pattern deduced is shown in Scheme 4, from which it is clear that rearrangement involves migration of the carboxy-group of phenylalanine (*Leete*, J. Amer. chem. Soc., 1960, **82**, 612; *M. L. Louden* and *Leete*, *ibid.*, 1962, **84**, 4507; *E. W. Underhill* and *H. W. Youngken Jr.*, J. pharm. Sci., 1962, **51**, 121; *D. Gross*

Scheme 4

and *Schütte*, Arch. Pharm., 1963, **296**, 1); results with [U-^{14}C]phenylalanine indicate that the amino acid is incorporated as an intact C_6-C_3 unit.

Further support for the idea that phenylalanine affords tropic acid by rearrangement, rather than by fragmentation and recombination, comes from the incorporation of [1,3-^{14}C]phenylalanine where it was found that the ratio of activity at C-1/C-2 was the same as the ratio at C-1/C-3 in the administered phenylalanine (*C. A. Gibson* and *Youngken Jr.*, J. pharm. Sci., 1967, **56**, 854). Proof that the rearrangement is an intramolecular one was obtained when it was observed that [1,3-$^{13}C_2$]phenylalanine (81% doubly labelled molecules) gave hyoscyamine (VII) and scopolamine (XI), the n.m.r. spectra of which showed ^{13}C–^{13}C coupling for signals corresponding to C-1 and C-2 of tropic acid. This coupling demands the presence of two contiguous ^{13}C atoms and can only have arisen as the result of intramolecular rearrangement. (Intermolecular rearrangement would have led to singly labelled products due to dilution of label with unlabelled materials in the plant) (*Leete, N. Kowanko* and *R. A. Newmark*, J. Amer. chem. Soc., 1975, **97**, 6826). Tritium at C-2 of phenylalanine is lost on incorporation into tropic acid but tritium at C-3 is apparently retained. In line with loss of tritium from C-2, which suggests that phenylpyruvic acid is involved in tropic acid biosynthesis, labelled phenylpyruvic acid was found to be an efficient precursor (*Liebisch et al.*, Abstracts of the Seventh International Symposium on the Chemistry of Natural Products, Riga, 1970, p. 557; in "Biochemie und Physiologie der Alkaloide", Fourth International Symposium, 1969, eds. *Mothes, K. Schreiber* and *Schütte*, Akademie-Verlag, Berlin, 1972, p. 233; *Schütte* and *Liebisch*, Z. Pflanzenphysiol., 1967, **57**, 440). Phenyllactic acid is a more efficient precursor than phenylalanine for tropic acid and atropic acid (apohyoscine: VIII). Moreover, label is specifically incorporated and in the same manner as label from phenylalanine (*W. C. Evans* and *J. G. Woolley*, Phytochemistry, 1976, **15**, 287). Rearrangement does not appear to involve an ester of tropine since neither the phenyllactate of tropine (littorine: XV) nor cinnamoyltropine were incorporated intact into tropane alkaloids (*Leete* and *E. P. Kirven*, *ibid.*, 1974, **13**, 1501).

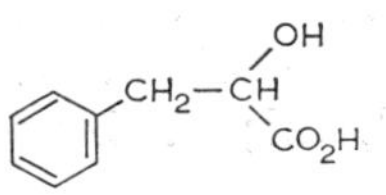

Phenyllactic acid

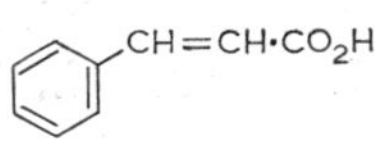

Cinnamic acid

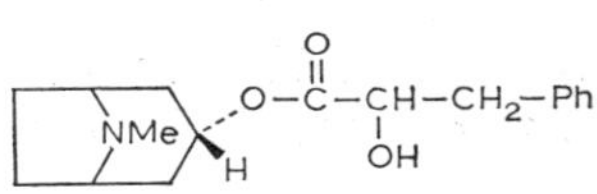

Littorine
(XV)

Although cinnamic acid, an important metabolite of phenylalanine, has been linked on chemical grounds with tropic acid biosynthesis *via* its epoxide, neither of these compounds had until recently been found to act as a tropic acid precursor (*Liebisch et al., loc. cit.*; *Evans, J. G. Woolley* and *V. A. Woolley*, in "Biochemie und Physiologie der Alkaloide", Fourth International Symposium, 1969, p. 227). The recent evidence is that cinnamic acid is a similarly efficient precursor for atropine (racemic hyoscyamine: VII) as is phenylalanine and the way of labelling the alkaloid is the same too (*B. V. Prabhu, Gibson* and *L. C. Schramm*, Lloydia, 1976, **39**, 79). This result has important consequences for the mechanism of the rearrangement which affords tropic acid and stands now in need of verification (see p. 335 for a related rearrangement in the biosynthesis of the microbial metabolite, tenellin).

Phenylacetic acid can also serve as a tropic acid precursor (*Underhill* and *Youngken, loc. cit*; *N. W. Hamon* and *J. L. Eyolfson*, J. pharm. Sci., 1972, **61**, 2006) with manifest inclusion of a C_1 unit (as the carboxyl group of tropic acid). Likely precursors of this single atom were incorporated with randomization of activity, however, although more satisfactory results were obtained with formate and serine in *Datura* root-tissue culture (*Hamon* and *Eyolfson, loc. cit*). The third source of tropic acid appears to be tryptophan (*A. M. Goodeve* and *E. Ramstad*, Experientia, 1961, **17**, 124; *Hamon* and *Eyolfson, loc. cit.*). The validity of this pathway has, however, been questioned [*Leete* in "Biosynthesis", ed. *T. A. Geissman* (Specialist Periodical Reports), The Chemical Society, London, 1973, Vol. 2, p. 106], and negative results with phenylacetic acid have been obtained recently (*Leete, Kowanko* and *Newmark, loc. cit.*).

The **phenyllactic acid residue** of littorine (XV) has been shown to be derived from phenylalanine (*Evans* and *V. A. Woolley*, Phytochemistry, 1969, **8**, 2183).

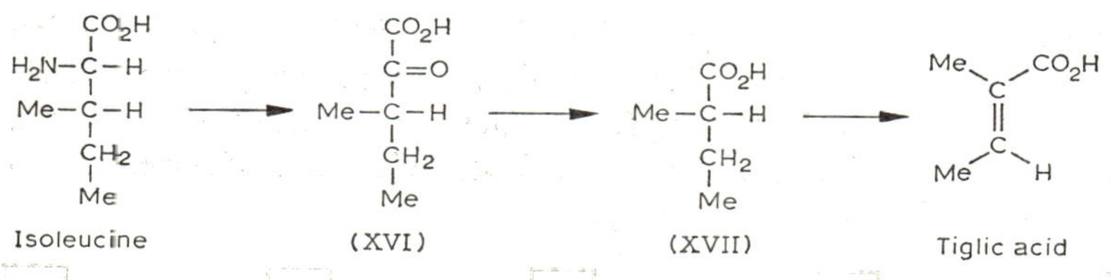

Scheme 5

The origin of the **tigloyl moieties** of meteloidine (XIII) and related plant alkaloids has been found to be similar to that of tiglic acid in animals (*M. J. Coon et al.*, J. biol. Chem., 1956, **218**, 391; and refs. cited). Thus isoleucine serves as a specific precursor for the tigloyl moieties of meteloidine (XIII) and XIV with loss of its carboxy-function (Scheme 5). (*Evans* and *J. G. Woolley*, J. Pharm. Pharmacol., 1965, **17**, 37S; *Leete* and *J. B. Murrill*, Tetrahedron Letters, 1967, 1727; *P. J. Beresford* and *Woolley*, Phytochemistry, 1974, **13**, 2143). In further accord with tiglic acid biosynthesis in animals, 2-methylbutyric acid (XVII) was found to be a specific precursor for the tigloyl moieties of XIII and XIV; and a further likely intermediate is XVI (*K. Basey* and *Woolley*, Phytochemistry, 1973, **12**, 2197; *Leete*, *ibid.* p. 2203). Indication that introduction of the double bond was by dehydrogenation rather than dehydration followed from the failure of 2-hydroxy-2-methylbutyric acid to serve as a precursor (*Basey* and *Woolley*, *loc. cit.*). Further corroboration of the route shown in Scheme 5 was obtained by showing that L-isoleucine is a precursor of the 2-methylbutanoyl moiety of XVIII which notably is of the same configuration as L-isoleucine (*Beresford* and *Woolley*, Phytochemistry, 1974, **13**, 2511).

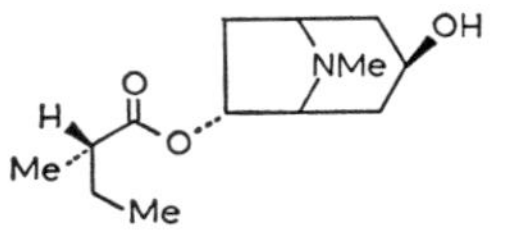

(XVIII)

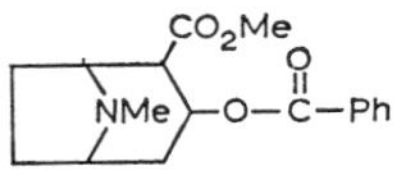

Cocaine

Cocaine is an example of a tropane-like base with a vestigial carboxyl-group, which co-occurs with hygrine and tropine esters (*H. G. Boit*, "Ergebnisse der Alkaloide bis 1960", Akademie-Verlag, Berlin, 1961). The biosynthesis of cocaine is obscure for, so far, only minute incorporations of acetate, ornithine, putrescine and methionine have been observed (*P. Tamprateep*, *E. H. Taylor* and *E. Ramstad*, Lloydia, 1963, **26**, 203). The benzoyl moiety has been found to derive specifically from phenylalanine (incorporation of label from [3-^{14}C]phenylalanine into the carboxyl group) (*Gross* and *Schütte*, *loc. cit.*).

The betaine **stachydrine** is biosynthesized from ornithine by a route which differs from that to the pyrrolidine alkaloids. The pathway proceeds *via* proline and hygric acid (Scheme 6); (*A. V. Robertson* and *L. Marion*, Canad. J. Chem., 1959, **37**, 1197; *J. M. Essery*, *D. J. McCaldin* and *Marion*, Phytochemistry, 1962, **1**, 209, and refs. cited).

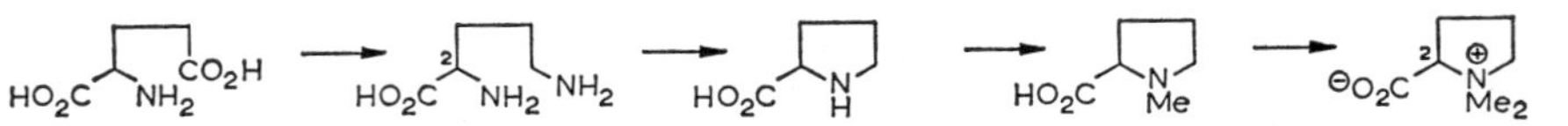

Glutamic acid — Ornithine — Proline — Hygric acid — Stachydrine

Scheme 6

Unusual in the structure for **cochlearine** is the benzoic acid residue with a *meta*-hydroxy group. Good incorporations of phenylalanine, *m*-tyrosine and putrescine have been recorded; acetate and shikimate were poor precursors (*Liebisch*, *H. Bernash* and *Schütte*, Z. Chem., 1973, **13**, 372).

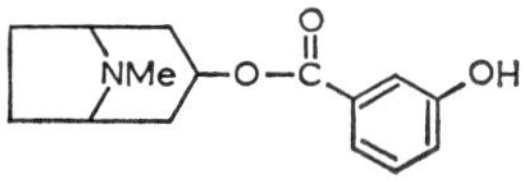

Cochlearine

(*b*) *Nicotine*

(*i*) *The pyrrolidine ring*

The biosynthesis of the pyrrolidine ring of nicotine (XX) is similar to that of the tropane alkaloids (section 2a), the only apparent difference lying in the manner in which ornithine is incorporated. Thus for nicotine this amino-acid is utilized by way of a symmetrical intermediate, whilst on incorporation into the tropane alkaloids the identities of C-2 and C-5 of ornithine are retained.

Various C_1 donors have been found to provide the *N*-methyl group of nicotine. Of interest is the incorporation of [*Me*-^{14}C,$^{2}H_3$]methionine with essentially no loss of deuterium, proving that the methyl group originates directly from methionine and not, for example, from formaldehyde (*L. J. Dewey, R. U. Byerrum* and *C. D. Ball*, J. Amer. chem. Soc., 1954, **76**, 3997).

It is well established that [2-^{14}C]ornithine is incorporated into nicotine (XX) with in each case equal distribution of label over C-2′ and C-5′ (*E. Leete, ibid.*, 1958, **80**, 2162; Chem. and Ind., London, 1955, 537; *Dewey, Byerrum* and *Ball*, Biochim. Biophys. Acta, 1955, **18**, 141; *Leete* and *K. J. Siegfried*, J. Amer. chem. Soc., 1957, **79**, 4529; *B. L. Lamberts, Dewey* and *Byerrum*, Biochim. Biophys. Acta, 1959, **33**, 22; *A. A. Liebman, B. P. Mundy* and *H. Rapoport*, J. Amer. chem. Soc., 1967, **89**, 664). Ornithine is thus a specific precursor, and a symmetrical intermediate is clearly implicated; the diamine, putrescine, is an obvious candidate. In agreement a specific incorporation of [1,4-$^{14}C_2$]putrescine was recorded with similar distribution of label (*Leete, ibid.*, 1958, **80**, 2162; *H.-B. Schröter*, Abh. Deut. Akad. Wiss. Berlin, Klasse Chem. Geol. Biol., 1966, No. 3, p. 157; *D. Yoshida* and *T. Mitake*, Plant Cell Physiol., 1966, **7**, 301). Results similar to those obtained with ornithine were recorded for [2-^{14}C]glutamic acid (XXIII) (*Leete, loc. cit. Lamberts* and *Byerrum*, J. biol. Chem., 1958, 233, 939).

The incorporation of intermediates of the tricarboxylic acid cycle and close relatives was found, in most cases, to be consistent with a pathway *via* glutamic acid (XXIII) and a later symmetrical intermediate. In some experiments, however, results inconsistent with simple theory were obtained (*Byerrum et al.*, Biochemistry, 1965, **4**, 1628; J. biol. Chem., 1962, **237**, 887; *D. R. Christman* and *R. F. Dawson*, Biochemistry, 1963, **3**, 182).

Important further results come from studies with [2-^{14}C,α-^{15}N]- and

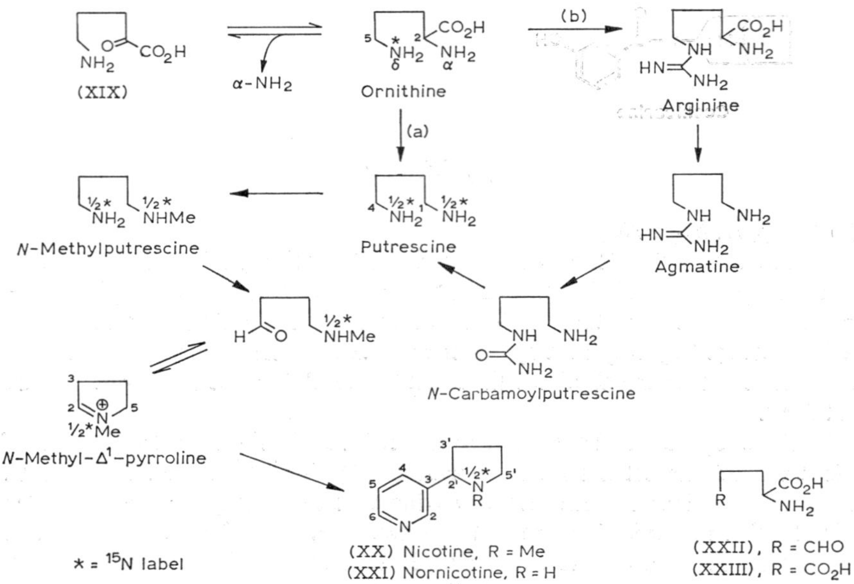

Scheme 7

[2-^{14}C,δ-^{15}N]-ornithine. ^{15}N was incorporated into the pyrrolidine ring of nicotine from the latter but not the former precursor and the $^{14}C/^{15}N$ ratio of the product was found to be twice that of the compound fed, and may be explained in terms of the pathway shown in Scheme 7 (*Leete*, *E. G. Gros* and *T. J. Gilbertson*, Tetrahedron Letters, 1964, 587). These results eliminate as intermediates glutamic semialdehyde (XXII), which would result in loss of the δ-amino-group; and free putrescine arising by direct decarboxylation of ornithine as this would result in the incorporation of label from both the α- and the δ-amino group. The results will allow putrescine to be an intermediate, however, if prior to the generation of putrescine there is rapid equilibration between ornithine and δ-amino-α-oxovaleric acid (XIX) resulting in loss of label from the α-amino group (Scheme 7).

Putrescine can arise formally either by decarboxylation of ornithine (path a, Scheme 7) or by path b. The validity of this latter pathway for nicotine biosynthesis was supported when it was shown that all the compounds along this pathway were reasonably efficiently incorporated into nicotine (*Yoshida* and *Mitake*, *loc. cit.*).

The incorporation of [2-^{14}C,δ-^{15}N]ornithine into nicotine with loss of half the ^{15}N-label can be rationalised by the pathway from putrescine indicated in the rest of Scheme 7; it is fundamental to this scheme that methylation of putrescine to give N-methylputrescine occurs non-specifically, *i.e.* no distinction is made between the two amino groups (*Leete*, *Gros* and *Gilbertson*, *loc. cit.*; *I. D. Spencer*, in "Comprehensive Biochemistry", eds. *M. Florkin* and *E. H. Stotz*, Elsevier, Amsterdam, 1968, Vol. 20, p. 242).

N-Methyl-Δ^1-pyrroline is implicated as a late intermediate in this scheme. Strong support for its involvement in nicotine biosynthesis comes from an efficient incorporation of [2-^{14}C]-N-methyl-Δ^1-pyrroline into nicotine with localization of the label at C-2′. (This result shows further that no double bond shift to C-5 occurs in N-methyl-Δ^1-pyrroline, for otherwise equilibration of label between C-2 and C-5 would have occurred) (*Leete*, J. Amer. chem. Soc., 1967, **89**, 7081). Further support for a role for N-methyl-Δ^1-pyrroline in nicotine biosynthesis comes from the isolation of this heterocycle in radioactive form from *Nicotiana* after feeding radioactive ornithine, putrescine and methionine (*S. Mizusaki et al.*, Arch. Biochem. Biophys., 1966, **117**, 677; Plant Physiol., 1968, **43**, 93).

[*Me*-^{15}N,^{14}C]-N-Methylputrescine has been found to give nicotine without change in isotope ratio. Similar results were obtained with [1-^{14}C,*Me*-^{15}N]N-methylputrescine and, as expected, the ^{14}C-label was confined to C-5′ of nicotine. Label from [1-^{3}H]N-methylputrescine also appeared, as expected, at C-5′ (*H. R. Schütte*, *W. Maier* and *K. Mothes*, Acta Biochem. Polon., 1966, **13**, 401; *Schütte*, *Maier* and *U. Stephan*, Z. Naturforsch., 1968, **23b**, 1426). This evidence clearly indicates that N-methylputrescine is incorporated intact into nicotine, and it may thus be included as a likely intermediate in nicotine biosynthesis.

Powerful support for any biosynthetic scheme can be obtained by isolating the enzymes which carry out the various reactions of the pathway proposed. In connection with the biosynthesis of nicotine two enzymes have been isolated from tobacco roots, one of which, putrescine N-methyltransferase, specifically catalysed the formation of N-methylputrescine from putrescine. The other enzyme isolated was N-methylputrescine oxidase which catalysed the conversion of N-methylputrescine into N-methyl-Δ^1-pyrroline. A further important point was that this enzyme was much more specific for N-methylputrescine than for putrescine (*Mizusaki et al.*, Phytochemistry, 1972, **11**, 2757; Plant Cell Physiol, 1971, **12**, 633).

The foregoing evidence provides convincing support for the biosynthesis of nicotine following the pathway indicated in Scheme 7. Evidence which is incompatible with this pathway was obtained with the isomeric N-methylornithines. [2,*Me*-^{14}C$_2$]-δ-N-Methylornithine was found to be incorporated intact into nicotine (it was a better precursor than [2,*Me*-^{14}C$_2$]-α-N-methylornithine which suffered demethylation prior to incorporation). However, the necessary unsymmetrical incorporation which was observed, excludes δ-*N*-methylornithine as a normal intermediate *between* ornithine and nicotine (*Gilbertson* and *Leete*, J. Amer. chem. Soc., 1967, **89**, 7085). [Results from a less rigorous study indicated that α-*N*-methylornithine rather than the δ-*N*-methyl isomer was a precursor for nicotine (*Schröter* and *D. Neumann*, Tetrahedron Letters, 1966, 1279).]

The relatively simple picture of nicotine biosynthesis so far outlined is complicated by the results of $^{14}CO_2$ feeding experiments. One set of data indicates that formation of the pyrrolidine ring occurs by way of at least one symmetrical intermediate (*H. R. Zielke et al.*, J. biol. Chem., 1968, **243**, 4757) whilst the other is consistent with a route involving only unsymmetrical intermediates (*Rapoport et al.*, J. Amer. chem. Soc., 1964, **86**, 3375; 1965, **87**, 4399; 1967, **89**, 664). It is interesting to note that under conditions when $^{14}CO_2$ gave unsymmetrically labelled nicotine, ornithine was incorporated symmetrically.

More rigorous experimentation, including administration of the $^{14}CO_2$ under steady-rate conditions have been applied in an attempt to resolve this dichotomy. (*M. L. Rueppel*, *Mundy* and *Rapoport*, Phytochemistry, 1974, **13**, 141). In a series of experiments in *N. glutinosa* different results were obtained. Some of the labelling patterns were symmetrical whereas others were unsymmetrical. This suggests that elaboration of nicotine occurs by two different pathways. An alternative way of explaining the results, however, is that the biosynthetic pathway to the pyrrolidine ring of nicotine resembles (in part) the route to the piperidine alkaloids (see section 3), *i.e.* variable/incomplete desorption and readsorption of enzyme-bound putrescine (arising by decarboxylation of ornithine) leading to nicotine formed in part unsymmetrically (from enzyme-bound putrescine) and in part symmetrically (from putrescine which had been desorbed from, and readsorbed to, the enzyme surface prior to further reaction).

Either of these explanations is consistent with the observation that nornicotine (XXI) isolated from root cultures of *N. rustica* after feeding [2-^{14}C]-ornithine, contained two-thirds of the activity at C-2′, with one-third at C-5′ (by difference) (*Mizusaki*, *T. Kisaki* and *E. Tamaki*, Agric. biol. Chem., Japan, 1965, **29**, 714). In view of the intermediacy of *N*-methyl derivatives in nicotine biosynthesis it is unlikely that nornicotine is also an intermediate. Indeed, it has been shown, by kinetic studies with $^{14}CO_2$, that nornicotine arises by demethylation of nicotine (*W. L. Alworth* and *Rapoport*, Arch. Biochem. Biophys., 1965, **112**, 45) but may also arise independently of nicotine (*idem, loc. cit.*; *T. C. Tso* and *R. N. Jeffrey*, Plant Physiol, 1957, **32**, 86). The course of nicotine metabolism in *Nicotiana* has been examined and has been found to proceed *via* nornicotine (XXI), myosmine (XXIV) and nicotine acid (*Leete* and *M. R. Chedekel*, Phytochemistry, 1974, **13**, 1853).

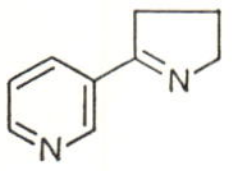

(XXIV)

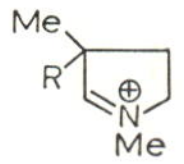

(XXV), R = H
(XXVI), R = Me

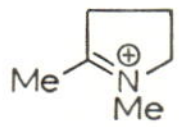

(XXVII)

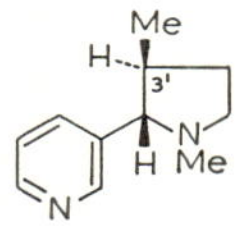

(XXVIII)

The ultimate step in nicotine biosynthesis involves the coupling of nicotinic acid with *N*-methyl-Δ^1-pyrroline (see Scheme 7). Experiments with *N*-methyl-Δ^1-pyrroline derivatives, bearing methyl groups at C-2 or C-3, have allowed some delineation of the stereochemistry and specificity of this coupling reaction. Marked differences in the efficiency of conversion of XXV, XXVI, and XXVII into the corresponding nicotine derivatives was observed (360:20:1, respectively), which was in accord with the amount of steric hindrance which each would be expected to exert in the coupling step. It is of interest that of the possible optical isomers of 3′-methylnicotine, only XXVIII was isolated (*Rueppel* and *Rapoport*, J. Amer. chem. Soc., 1971, **93**, 7021).

(ii) The pyridine ring

The labelling pattern found in nicotine (XX) after feeding [2,3,7-$^{14}C_3$]-nicotinic acid (as XXIX) demonstrates quite clearly that (a) the pyridine ring of the alkaloid is derived specifically from nicotinic acid, (b) no symmetrical intermediate is involved, and (c) the pyrrolidine ring is attached to the position from which the carboxyl group is lost (*G. R. Waller et al., ibid.*, 1965, **87**, 4184; J. biol. Chem., 1967, **242**, 887; *T. A. Scott* and *J. P. Glynn*, Phytochemistry, 1967, **6**, 505). [2,3,7,8-^{14}C]Quinolinic acid (as XXX) was incorporated similarly, thus establishing the biosynthetic sequence XXX → XXIX → nicotine (*Waller et al., loc. cit.*).

The question of the mechanism of the coupling reaction between nicotinic acid and *N*-methyl-Δ^1-pyrroline is an intriguing one, one which has been almost fully answered by the results of feeding experiments in *N. tabacum* root cultures with tritiated and deuteriated samples of nicotinic acid. [2-^{3}H]-, [4-^{2}H]-, and [5-^{3}H]-Nicotinic acids afforded nicotine with at least ten times more activity than did [6-^{3}H]nicotinic acid. This suggested that hydrogen was lost from C-6 on formation of nicotine, it being shown that such loss did not happen in the culture medium prior to incorporation into nicotine. The failure of [^{15}N]-6-hydroxynicotinic acid to act as a precursor for nicotine excluded hydroxylation at C-6 as the cause of tritium loss (*Dawson et al.*, J. Amer. chem. Soc., 1960, **82**, 2628; *Leete* and *Y.-Y. Liu*, Phytochemistry, 1973, **12**, 593).

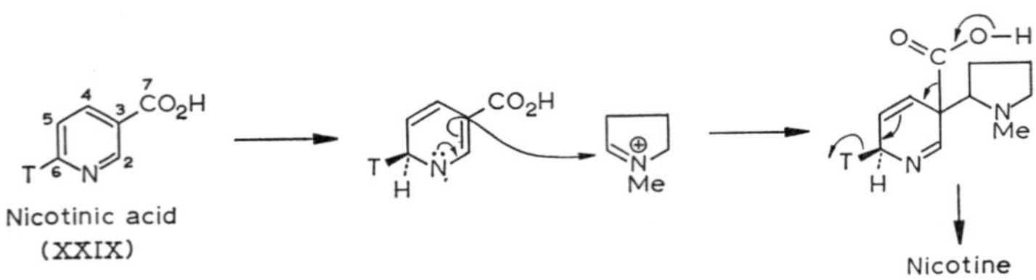

Scheme 8

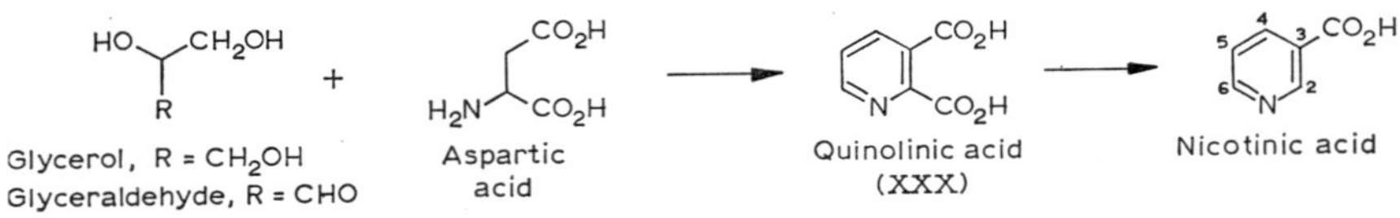

Scheme 9

The foregoing results may be accommodated most pausibly by a mechanism (shown in Scheme 8 for [6-^{3}H]nicotinic acid) in which a dihydronicotinic acid is an intermediate. It is necessary to postulate that addition and removal of hydrogen is stereospecific otherwise tritium would be preferentially retained at C-6 by a primary isotope effect. This is reasonable as such reactions would probably be enzyme mediated. This mechanism requires complete loss of tritium from C-6, but the experimental results are that some label is retained. Degradation showed, however, that only half the residual tritium was at C-6 (*idem, loc. cit.*). Further experiments are needed to resolve this problem but it is possible that hydrogen shifts occur which are not connected with formation of the nicotine skeleton.

The biosynthesis of nicotinic acid in animals and some micro-organisms is well established to be by degradation of tryptophan. In plants on the other hand, nicotinic acid does not originate from tryptophan. Instead it has its origins in aspartic acid and glycerol (Scheme 9) (*Spenser*, in "Comprehensive Biochemistry", 1968, Vol. 20, p. 270).

Nicotine isolated after feeding labelled tryptophan (and other intermediates implicated on the route to nicotinic acid *via* tryptophan) was inactive (*Leete*, Chem. and Ind., London, 1957, 1270; *J. Grimshaw* and *L. Marion*, Nature, 1958, **181**, 112; *L. M. Henderson et al.*, J. biol. Chem., 1959, **234**, 93; *K. Bowden*, Nature, 1953, **172**, 768). On the other hand, labelled aspartic acid and glycerol were found to give radioactive nicotine (XX), glycerol providing, it was shown, carbons 4, 5 and 6 (*R. U. Byerrum et al.*, J. Amer. chem. Soc., 1965, **87**, 4184; J. biol. Chem., 1965, **240**, 4099). [3-^{14}C]Aspartic acid gave nicotine with unequal distribution of label between C-2 and C-3 with a greater proportion of the label at C-2, and a similar result was obtained with [3-^{14}C]malic acid. From the specifity of the labelling and the low dilution observed, it follows that these acids are closely related to the immediate precursors of the pyridine ring of nicotine (*T. M. Jackanicz* and *Byerrum*, *ibid.*, 1966, **241**, 1296). The unequal distribution of label between C-2 and C-3 excludes as later intermediates symmetrical compounds such as succinic acid, which is also incorporated (*T. Griffith* and *Byerrum*, Biochem. Biophys. Res. Comm., 1963, **10**, 293). Results with [3-^{14}C]-glyceraldehyde indicate that it may be a more immediate precursor than glycerol for nicotine, as it is more efficiently incorporated, C-4 of nicotine being the site of heaviest labelling (*J. Fleeker* and *Byerrum*, J. biol. Chem., 1967, **242**, 3042). The plausible biosynthetic pathway suggested by these results is summarized in Scheme 9.

The conclusion that the pyridine ring of nicotine is formed from a C_2 unit derived from aspartic acid and a C_3 unit derived from glyceraldehyde is supported by the results of experiments with $^{14}CO_2$. The greatest incorporation of $^{14}CO_2$ occurred at positions 4, 5 and 6, which were labelled to a similar extent; the incorporation at C-2 was much lower and that at C-3 was similar (*H. R. Zielke, C. M. Reinke* and *Byerrum*, *ibid.*, 1969, **244**, 95). Studies of the biosynthesis of ricinine (p. 333), anabasine (p. 312) and nicotinic acid itself provide further corroborative evidence.

(*c*) *Pyrrolizidine alkaloids*

Retronecine (XXXI) is an example of the basic fragment found in the ester alkaloids, *e.g.* XXXII, of *Senecio* and *Crotalaria* species. It is derived perhaps not surprisingly from ornithine and putrescine. The manner, however, in which these precursors are utilized is as yet poorly understood, the available evidence being in conflict. On the one hand [2-^{14}C]- and [5-^{14}C]-ornithine and [1,4-$^{14}C_2$]putrescine have been found to give retronecine in which a quarter of the label in each case was located at C-1′ (*W. Bottomley* and *T. A. Geissman*, Phytochemistry, 1964, **3**, 357). From this it may be concluded that two molecules of ornithine/putrescine are utilized as two C_4 units (see dotted line in XXXI and that at least for the unit containing the proven site of labelling, C-1′, ornithine is incorporated through a symmetrical intermediate (same labelling by both ornithine precursors) which may be putrescine. As far as the results go they suggest a pathway related to that which affords lupinine (see p. 326).

On the other hand [2-^{14}C]ornithine has been found to give retronecine with labelling as follows: 71% at C-7 + C-8, 26% at C-1′ and 0% at both C-3 and C-5 (*C. A. Hughes, R. Letcher* and *F. L. Warren*, J. chem. Soc., 1964, 4974). This is consistent only with incorporation of two different ornithine-derived units along a pathway which does not involve symmetrical intermediates for either. (Although it is the conclusion of both studies that a quarter of the activity from [2-C]^{14}ornithine resides at C-1′, it is the similar incorporation of [5-^{14}C]ornithine in the one study and the absence of label at C-3 from the [2-^{14}C]ornithine fed in the other, which brings them into disagreement. As elsewhere, the incorporation of putrescine does not conflict with unsymmetrical incorporation of ornithine.)

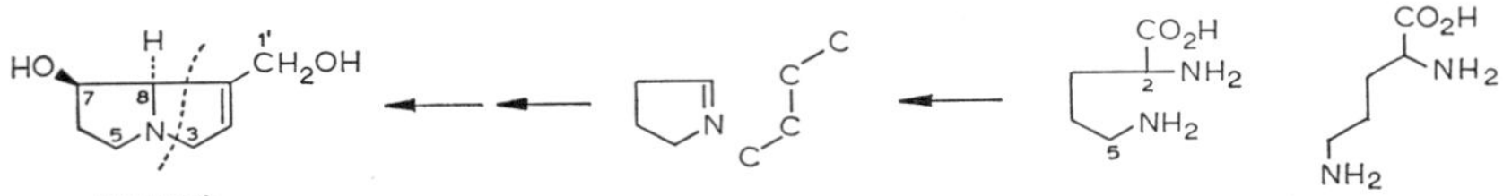

(XXXI)

Ornithine

Scheme 10

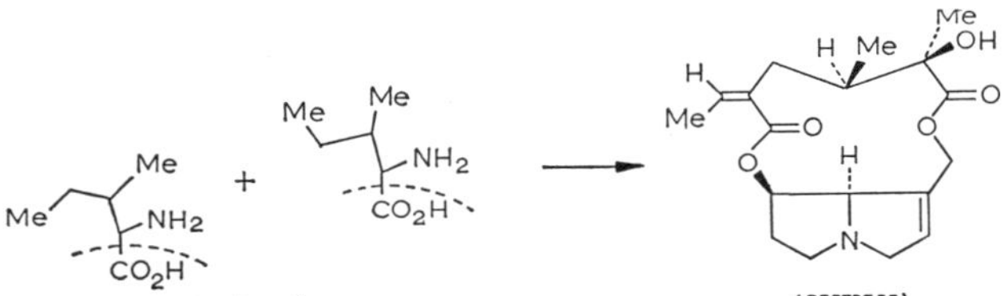

Isoleucine

(XXXII)

Scheme 11

Incorporation of [2-^{14}C]acetate into retronecine (*idem, loc. cit.*) is in accord with the utilization of two different ornithine-derived units for the formation of the pyrrolizidine nucleus but the labelling observed does not fit for normal metabolism to this amino-acid through the tricarboxylic or glyoxylic acid cycles (*Spenser*, in "Comprehensive Biochemistry", 1968, Vol. 20, p. 250). Ornithine and putrescine are normal metabolites of arginine. Accordingly labelled arginine has been found to afford radioactive retronecine but it is slightly less efficiently incorporated than is ornithine (*N. M. Bale* and *D. H. G. Crout*, Phytochemistry, 1975, **14**, 2617). Bearing in mind the mechanism of biosynthesis of other pyrrolidine alkaloids the above results can probably best be accommodated by the pathway shown in Scheme 10; one can speculate that the different ways in which ornithine was observed to be utilized are reflections of a pathway in which unbound putrescine equilibrates variably with enzyme-bound putrescine derived from ornithine (which would be unsymmetrical) (*cf.* the discussion on the biosynthesis of piperidine alkaloids and of nicotine).

Available evidence on the biosynthesis of the necic acids associated with retronecine and its congeners points to derivation of these acids from branched-chain amino acid precursors. The C_{10} necic acids of the senecic acid type, which appear as diesters in alkaloids like senecionine (XXXII) are formed from two units of L-isoleucine (Scheme 11); threonine is also a precursor, the labelling pattern observed being consistent with metabolism *via* isoleucine (*Crout et al.*, J. chem. Soc., Perkin I, 1972, 671; 1974, 2079). The monocrotalic acid component of monocrotaline also derives from L-isoleucine (and L-threonine), this amino acid accounting for C-1, -2, -3, -6, and 7- (*D. J. Robins*, *Bale*, and *Crout*, *ibid.*, 1974, 2082). The remaining atoms (C-4, -5, and -8) plausibly derive from propionic cid (*E. K. Nowacki* and *R. U. Byerrum*, Life Sci., 1962, **1**, 157; *Robins*, *Bale*, and *Crout*, *loc. cit.*).

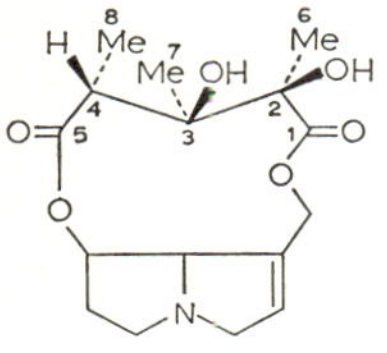

Monocrotaline

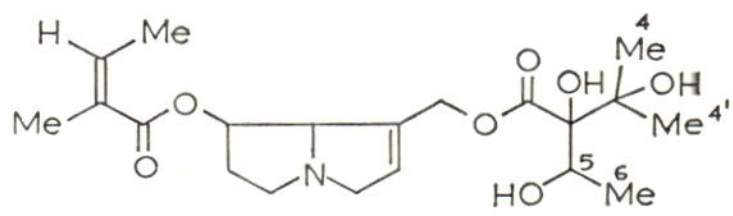

Angelic acid

Heliosupine

Echimidinic acid

The angelic acid moiety found in heliosupine has similar origins to its geometrical isomer, tiglic acid, found in the tropane alkaloids (see p. 297). This origin is in isoleucine (*Crout*, J. chem. Soc. C, 1967, 1233). The echimidinic acid fragment of heliosupine on the other hand derives, partially, from valine: label from [4-^{14}C]valine appears at C-4/C-4′ of the echimidinic acid and it may be adduced therefore that valine provides five of the carbon atoms of this acid, a further two carbon atoms (C-5 and -6) being added at some stage (*idem*, *ibid.*, 1966, 1968).

(*d*) *Phenanthroindolizidine and related alkaloids*

The phenanthroindolizidine alkaloid tylophorine (XXXIII) has been shown to derive specifically from phenylalanine and tyrosine as shown in Scheme 12.

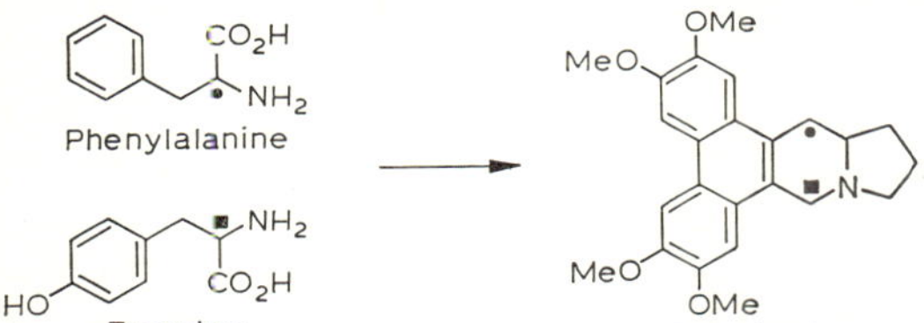

Tylophorine
(XXXIII)

Scheme 12

The remaining atoms are most probably accounted for by ornithine (*N. B. Mulchandani*, *S. S. Iyer* and *L. P. Badheka*, Pyhtochemistry, 1969, **8**, 1931: 1971, **10**, 1047). Important delineation of the pathway to tylophorine and its congeners was gained when it was shown that the 2-phenacylpyrrolidines XXXVI, XXXVII, and XXXVIII were intact precursors for tylophorinine (XXXIX). The oxo-acids XXXIV and XXXV, but not XL, were also incorporated. Taken together these results define the part-sequence XXXIV → XXXVI → XXXVII → XXXVIII for tylophorinine biosynthesis; the incorporation of XXXV indicates that hydroxylation can also occur before formation of the phenacylpyrrolidine skeleton (*R. B. Herbert*, *F. B. Jackson* and *I. T. Nicolson*, Chem. Comm., 1976, 865).

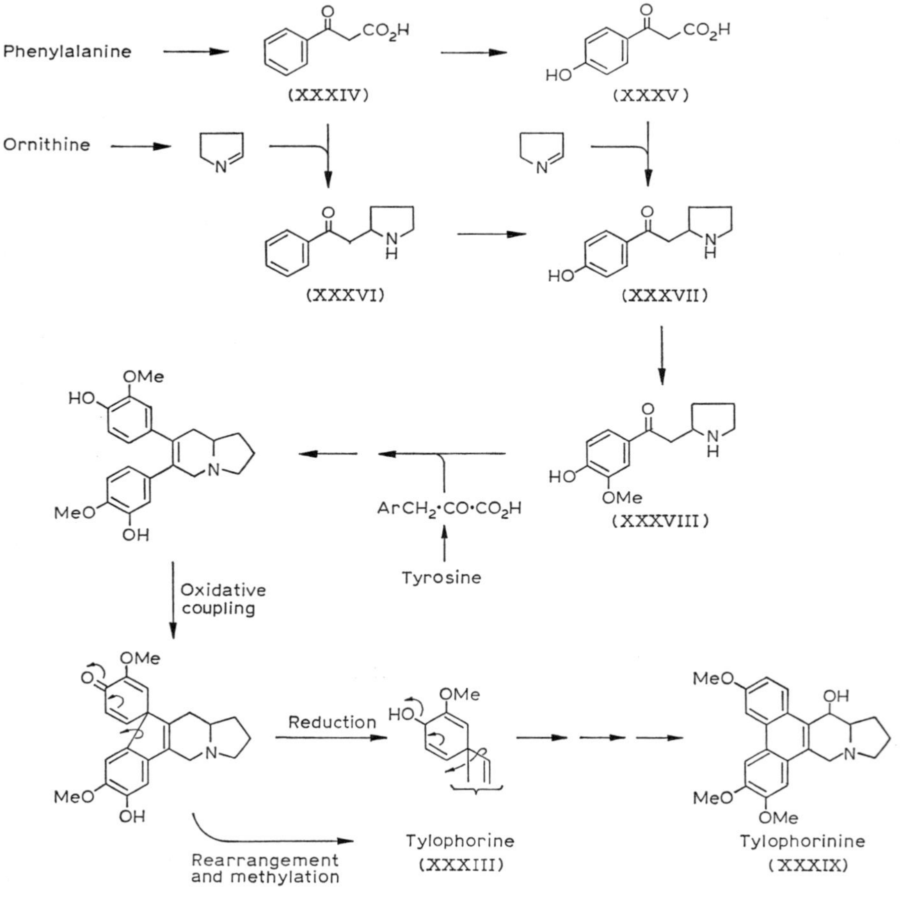

Scheme 13

Later stages in biosynthesis are suggested by the natural occurrence of septicine (XLI) and the variation in structure represented by tylophorine (XXXIII) and tylophorinine (XXXIX) in the light of knowledge on probable steps involving phenol oxidative coupling (for a discussion of phenol oxidative coupling see p. 386). The delineated and speculative steps are summarized in Scheme 13.

The biosynthesis of the phenanthroquinolizidine alkaloids, *e.g.* cryptopleurine (XLII), has not been studied but it is likely to be similar to that outlined for phenanthroindolizidine alkaloids (with substitution of lysine for ornithine as one of the precursor amino-acids) because of similar structures and the natural occurrence of analogues of XXXVI and XLI (*N. K. Hart, S. R. Johns* and *J. A. Lamberton*, Austral. J. Chem., 1968, **21**, 2579, 1397).

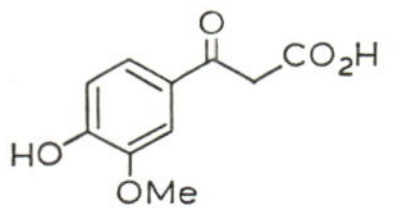

(XL)

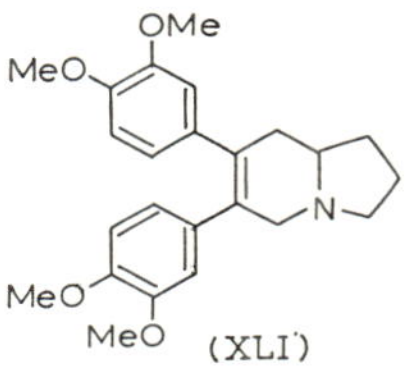

(XLI)

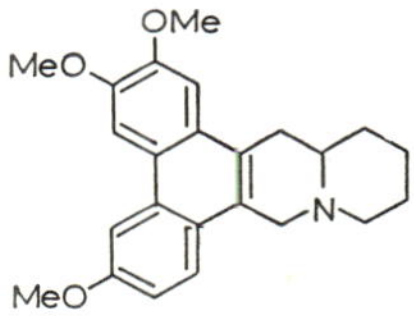

(XLII)

3. Piperidine alkaloids

A large group of alkaloids, whose structures contain a piperidine ring, are formed as variations of a common structural element I. They are the subject of review in this section. The biological origin of this system is in the majority of cases lysine plus appropriate β-ketocarboxylic acid and, only rarely, an acetate-derived polyketide plus ammonia.

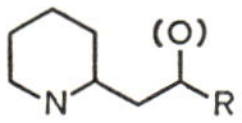

(I)

(*a*) *Alkaloids derived exclusively from acetate*

Although lysine and Δ^1-piperidine have been reported as precursors for hemlock alkaloids, *e.g.* **coniine** (VI), subsequent experiments have failed to support these findings. Instead a specific incorporation of [1-^{14}C]-acetate has been observed. Since the label appeared on the alternate carbons, C-2′, -2, -4, and -6, it must be concluded that coniine is formed directly from an acetate-derived polyketide or equivalent, and not at all from lysine metabolism (*E. Leete*, J. Amer. chem. Soc., 1964, **86**, 2509; 1970, **92**, 3835).

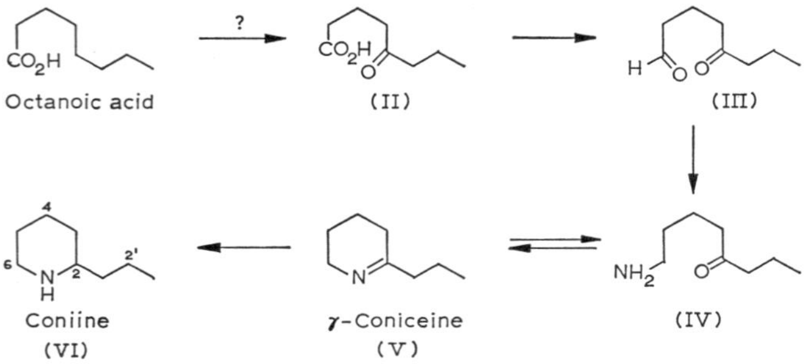

Scheme 14

A large number of secondary metabolites derived from acetate can be constructed formally by alternative cyclisation and reduction reactions of poly-β-keto-acids/polyketides. An intermediate between acetic acid and coniine was likely to be 5-oxo-octanoic acid (II). Before feeding this substrate, however, readily available [1-^{14}C]octanoic acid was administered to hemlock plants for no profound reason, but with serendipitous results. The acid was an excellent and specific precursor of coniine. A similar if less definitive result was obtained with [8-^{14}C]octanoic acid. In both experiments randomization of label increased with the interval between feeding and isolation and it could be shown that this was in accord with breakdown of the octanoic acid by consecutive β-oxidations to precursor acetate. Labelling of coniine by acetate from this source and from administered [1-^{14}C]- and [2-^{14}C]-labelled material was further consistent with partial incorporation *via* the Krebs cycle which results in loss of label from C-1 whilst label from C-2 ultimately is spread over C-1 and C-2 (*cf. I. D. Spenser*, in "Comprehensive Biochemistry", eds. *M. Florkin* and *E. H. Stotz*, Elsevier, Amsterdam, 1968, Vol. 20, p. 364). A rare chain elongation was observed in the specific incorporation of [1-^{14}C]hexanoic acid into coniine.

Both 5-oxo-octanoic acid (II) and the corresponding aldehyde (III) when tested were found to be efficiently and specifically incorporated into coniine (*Leete* and *J. Olsen*, J. Amer. chem. Soc., 1972, **94**, 5472). Moreover, radioactive II was isolated after feeding [1-^{14}C]octanoate and thus II is a normal intermediate in coniine biosynthesis. Whether octanoate or the corresponding polyketide is a normal intermediate is arguable — certainly there are no other secondary metabolites known to be formed from fatty acids rather than polyketides (*cf. B. E. Cross* and *P. Hendley*, Chem. Comm., 1975, 124; *J. C. Vaderas et al.*, Helv., 1975, **58**, 1886. See also the biosynthesis of pinidine below). The next biosynthetic intermediate after III is *a priori* likely to be the transamination product IV $\leftrightharpoons$ γ-coniceine (V). This is strongly supported, as is the intermediacy of III, by the isolation of an enzyme from hemlock leaves which catalyses the conversion of III into γ-coniceine (V) in the presence of alanine (*M. F. Roberts*, Phytochemistry, 1971, **10**, 3057). The transamination reaction is irreversible *in vivo* (*Leete* and *Olsen, loc. cit.*). On the other hand, rapid equilibration between coniine and γ-coniceine occurs in hemlock (*J. W. Fairburn* and *P. N. Suwal*, Phytochemistry, 1961, **1**, 38) and it has been observed that natural (+)coniine is more efficiently converted into its biological precursor γ-coniceine than is the (—)isomer (*Leete* and *Olsen, loc. cit.*). An enzyme which catalyses the conversion of γ-coniceine into coniine has been isolated as has one which *N*-methylates coniine (*Roberts*, Phytochemistry, 1974, **13**, 1847, 1975, **14**, 2393).

Experiments with $^{14}CO^{2}$ confirm that the sequence of alkaloid formation in hemlock is γ-coniceine (V) → coniine (VI) → *N*-methylconiine (*S. M. C. Dietrich* and *R. O. Martin*, Biochemistry, 1969, **8**, 4163).

The pathway to coniine so far deduced is summarized in Scheme 14.

The labelling of **pinidine** (VII) by [1-^{14}C]acetate implies for this alkaloid, as for coniine, a simple linear combination of acetate units (labelling of C-2, -4, -6 and -9) (*Leete* and *K. N. Juneau*, J. Amer. chem. Soc., 1969, **91**, 5614). Analogy with coniine biosynthesis suggests VIII or IX, and perhaps decanoic acid, as likely intermediates in pinidine biosynthesis, but [10-^{14}C]-5,9-dioxodecanoic acid (as VIII), [1-^{14}C]nonadione, the decarboxylation product of IX, and [9-^{14}C]- and [10-^{14}C]-decanoic acids labelled pinidine only to an insignificant extent. A derivative of VIII but not IX may still be implicated, however, since inactive sodium acetate fed at the same time as diethyl [1-^{14}C]malonate diluted activity from C-2 which must thus be the "starter" acetate unit (*Leete* and *R. A. Carver*, J. org. Chem., 1975, **40**, 2151; *Leete, J. C. Lechleiter* and *Carver*, Tetrahedron Letters, 1975, 3779).

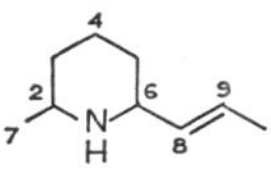

Pinidine
(VII)

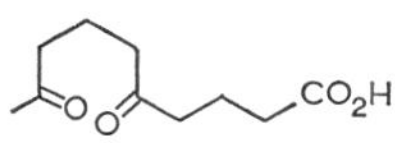

(VIII)

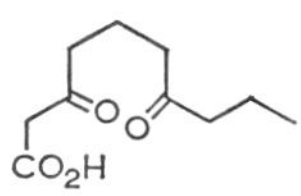

(IX)

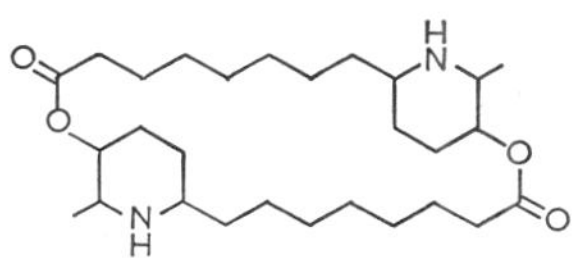

(X)

Acetic acid has been found to be a more efficient precursor than lysine for **carpaine** (X) consistent with its formation from a pair of C_{14} poly-β-oxo acids (*C. W. L. Bevan* and *A.U. Ogan*, Phytochemistry, 1964, **3**, 591; *M. Spiteller-Friedman* and *G. Spiteller*, Monatsh., 1964, **95**, 1234).

The biosynthesis of the microbial metabolite **nigrifactin** (XI) seems to follow a pathway which resembles that of coniine. A linear combination of acetate units is indicated by the results of experiments with [^{14}C]- and [^{13}C]-acetic acid. Moreover specific incorporations of XII and XIII, as well as XIV and XV, were recorded. Particularly the latter pair showed significantly lower incorporations than acetate, perhaps for solubility reasons. Nonetheless the relatively low incorporations cast some doubt on their roles as normal intermediates in nigrifactin biosynthesis (*T. Terashima et al.*, Chem. Comm., 1973, 75).

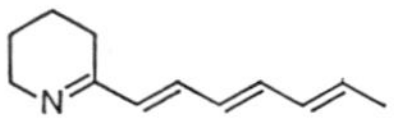
Nigrifactin
(XI)

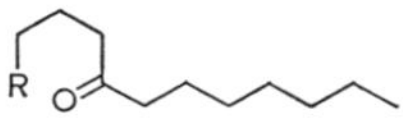
(XII) R = CO_2H
(XIII) R = CHO

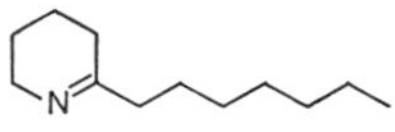
(XIV)

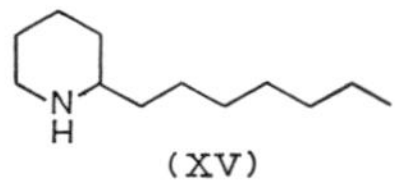
(XV)

(*b*) *Alkaloids derived from lysine*

The majority of alkaloids which contain the piperidyl unit (I) have their genesis in lysine in accord with early prediction (*R. Robinson*, J. chem. Soc., 1917, **111**, 876). As will be noted in the sequel this amino acid (XVI) is transformed into alkaloidal piperidine rings by pathways in which one or more symmetrical intermediates are involved or in which the identity of C-2, and of C-6, of lysine is retained. It appears that the unsymmetrical nature of any intermediates is not associated with the sort of early *N*-methylation deduced for pyrrolidine alkaloids from ornithine (see above). Instead, the first step from lysine is reasonably accommodated within a hypothesis of enzymatic deaminative decarboxylation of the amino acid which is bound to pyridoxal as enzyme-linked co-factor (see p. 315). Cadaverine is generally found to be specifically incorporated into piperidine alkaloids and it has been suggested that cadaverine may be implicated *via* the cadaverine–enzyme complex formed on decarboxylation of lysine (*E. Leistner* and *Spenser*, J. Amer. chem. Soc., 1973, **95**, 4715).

(*i*) *Simple piperidine bases*

Because of interlocking data on the biosynthesis of **sedamine** (XVII), ***N*-methylpelletierine** (XVIII) and **anabasine** (XIX) these alkaloids are conveniently discussed together.

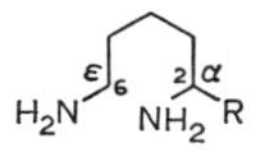
Cadaverine, R = H
(XVI) Lysine, R = CO_2H

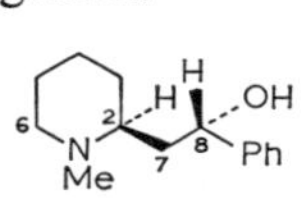
Sedamine
(XVII)

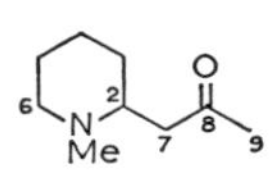
N-Methylpelletierine
(XVIII)

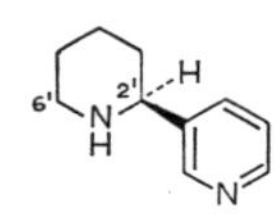
Anabasine
(XIX)

The way in which lysine is transformed into sedamine (XVII) is defined by the results of experiments with [^{14}C)- and [^{3}H]-labelled samples. These results are that C-2 and C-6 of the amino acid become, respectively, C-2 and C-6 of the alkaloid, and that tritium is lost neither from C-2 nor C-6 (nor from C-4 and C-5) (*R. N. Gupta* and *Spenser*, Phytochemistry, 1970, **9**, 2329; Canad. J. Chem., 1967, **45**, 1275; *Spenser*, unpublished results). Similar results were obtained for the biosynthesis of *N*-methylpelletierine (XVIII) (*Gupta* and *Spenser*, Phytochemistry, 1969, **8**, 1937; *M. F. Keogh* and *D. G. O'Donovan*, J. chem. Soc., (C), 1970, 1792). In analogous manner [2-^{14}C]-lysine was found to label C-2′ of anabasine (XIX) (*Leete*, J. Amer. chem. Soc., 1956, **78**, 3520; *M. L. Solt*, *R. F. Dawson* and *D. R. Christman*, Plant Physiol., 1960, **35**, 887). It is clear then that lysine is incorporated into these alkaloids unsymmetrically, the retention of tritium at C-6 indicating that the amino group at this position is retained; loss of the amino group from C-2 must be associated with a deaminative decarboxylation (see below) which does not result in loss of hydrogen from C-2 either.

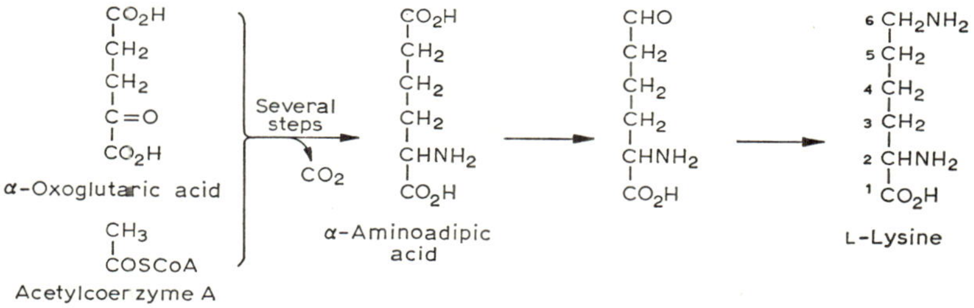

Scheme 15

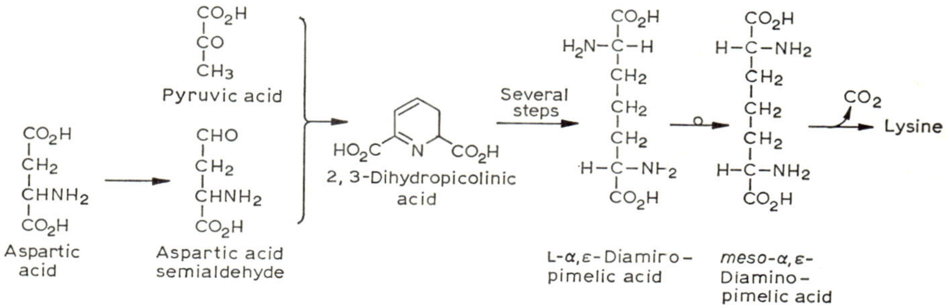

Scheme 16

Lysine itself may be formed by either of two pathways, *i.e.* either *via* α-aminoadipic acid (Scheme 15) or α,ε-diminopimelic acid (Scheme 16). The way in which [4-^{14}C]aspartate, [2,3-$^{14}C_2$]succinate, [2-^{14}C]acetate, and

[2-^{14}C]alanine label the piperidine ring of anabasine has led to the conclusion that anabasine is formed in *Nicotiana glauca* from lysine which is derived by the diaminopimelic route (*T. Griffith* and *G. D. Griffith*, Phytochemistry, 1966, **5**, 1175).

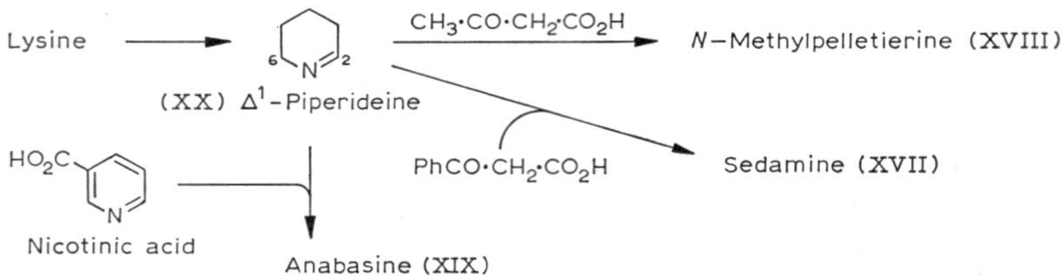

The formation of XVII, XVIII and XIX, by fusion of a lysine-derived unit and another appropriate fragment can reasonably be thought of as involving nucleophilic attack of this fragment on to Δ^1-piperideine (XX) which has its genesis in lysine (*cf.* pyrrolidine alkaloid biosynthesis above). In support [6-^{14}C]-Δ^1-piperideine (as XX) was found to be a better precursor than lysine for anabasine (XIX). Moreover, the incorporation of label was specifically into C-6′, the expected site; notably C-2′ was devoid of label (*Leete*, J. Amer. chem. Soc., 1969, **91**, 1697). In accord with this ^{15}N label from C-6 rather than C-2 of lysine was transferred to the piperidine nitrogen of anabasine (*Leete*, *E. G. Gros* and *T. J. Gilbertson*, *ibid.*, 1964, **86**, 3907).

The biosynthetic origins of the skeletal fragments of XVII, XVIII and XIX not accounted for by lysine are as follows. A specific incorporation of [2-^{14}C]acetate into C-9 (and C-7) and of [1-^{14}C]acetate into C-8 of *N*-methylpelletierine (XVIII) established that the C_3 side chain of the alkaloid is derived specifically from acetate, probably by way of acetoacetate (*Keogh* and *O'Donovan*, *loc. cit.*). Incorporation of label from C-2 and C-3 of the side chain of phenylalanine into C-7 and C-8, respectively, of sedamine (XVII) established this amino acid as an intact precursor. Its incorporation probably occurs by way of cinnamic and benzoylacetic acids (*cf.* lobeline below) [*Gupta* and *Spenser*, *loc. cit.* (1967)]. The pyridine ring of anabasine (XIX) arises like that of nicotine (p. 299) from nicotinic acid, which quite naturally is formed from the same precursors as those for nicotine (*Solt*, *Dawson* and *Christman*, *loc. cit.*; *Leete* and *A. R. Friedman*, J. Amer. chem. Soc., 1964, **86**, 1224; 1963, **85**, 2141).

Cadaverine has been shown, like lysine, to be a precursor for the piperidine rings of anabasine, sedamine, *N*-methylpelletierine and pseudopelletierine (XXI) (*Leete*, *ibid.* 1958, **80**, 4393; *H. W. Liebisch*, *N. Marekov* and *H. R. Schütte*, Z. Naturforsch., 1968, **23b**, 1116; *Leistner* and *Spenser*, *loc. cit.*). Such a symmetrical molecule cannot be an *intermediate* in the conversion of lysine into these piperidine alkaloids for otherwise the identities of C-2 and C-6 of the amino acid would be lost, in contradiction of the tracer evidence. The suggestion that these identities could be retained by early *N*-methylation *i.e.* with *N*-methylcadaverine and ε-*N*-methyllysine as intermediates (*cf.* pyrrolidine alkaloid biosynthesis, p. 291) was not supported by the results of subsequent experiments.

Although ε-*N*-methyllysine was shown to be present in *Sedum acre*, a plant which also produces sedamine, and was labelled by [6-^{3}H]lysine and [*Me*-^{14}C]methionine the isotope ratio of the ε-*N*-methyllysine differed from that of the sedamine isolated in the same experiment. Moreover, [2-^{14}C,ε-*N*-*Me*-^{3}H]-ε-*N*-methyllysine was a poor precursor for sedamine and the transformation resulted in alteration of the isotope ratio. In addition attempts to demonstrate the presence of *N*-methylcadaverine in *Sedum* species and in

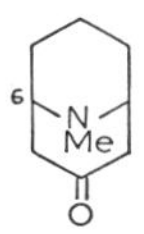

(XXI)

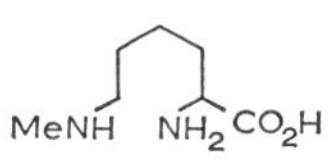

ε-*N*-Methyl-lysine

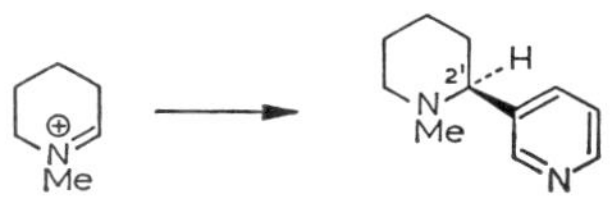

N-Methyl-Δ^{1}-piperideine

(XXII)

Nicotiana glauca (a plant which produces anabasine) were unsuccessful (*Leistner* and *Spenser*, *loc. cit.*; *P. Korzan* and *Gilbertson*. Phytochemistry, 1974, **13**, 435). Early *N*-methylation implies that *N*-methyl-Δ^{1}-piperideine is a precursor for, *e.g.*, anabasine and [2-^{14}C]-labelled material was efficiently and specifically incorporated into anabasine (XIX) and *N*-methylanabasine (XXII). However, the specific activity of the *N*-methylanabasine, unlike that of the anabasine, was the same as the specific activity of the precursor, *i.e.* no dilution by inactive alkaloid had occurred in the plant. Consequently, *N*-methylanabasine is not a normal constituent of the *Nicotiana* plants used in the experiment, a conclusion which was confirmed by careful examination of plant extracts. Thus conversion of *N*-methyl-Δ^{1}-piperideine into alkaloids is an aberrant transformation, a reflection of the plant's ability to deal with foreign substrates. Interestingly, the *N*-methylanabasine was optically active, having the same configuration at C-2′ as anabasine (*Leete* and *M. R. Chedekel*, *ibid.*, 1972, **11**, 2751). [Other foreign substrates have also been transformed into unnatural alkaloids by *Nicotiana* species (*M. L. Rueppel* and *H. Rapoport*, J. Amer. chem. Soc., 1971, **93**, 7021; *Leete*, *G. B. Bodem*, and *M. F. Manuel*, Phytochemistry, 1971, **10**, 2687)]. The results of these two studies clearly refute the hypothesis of early *N*-methylation beginning with *N*-methyllysine.

A brilliantly simple solution to the observations on lysine and cadaverine incorporation has been proposed which fits all the evidence (*Leistner* and

Spenser, *loc. cit.*; see also *Herbert*, in "The Alkaloids" (Specialist Periodical Reports), ed. *J. E. Saxton*, The Chemical Society, London, 1974, Vol. 4, p. 6). Enzymatic decarboxylation of lysine is envisaged as proceeding *via* an orthodox pyridoxal-linked intermediate as shown in Scheme 17: the proposal is more than mechanistically attractive for L-lysine decarboxylase (EC 4.1.1.18 L-lysine carboxy-lyase) and diamine oxidase [EC 1.4.3.6, amine:oxygen oxidoreductase (deaminating)] the two enzymes whose participation in the conversion of lysine into Δ^1-piperideine is likely, both require pyridoxal phosphate as a cofactor (*M. Dixon* and *E. C. Webb*, "The Enzymes", 2nd edn., Longmans, London, 1967, pp. 688, 762). Intermediate XXIII can be seen to be derivable not only from lysine but also from cadaverine which explains the incorporation of this base into XVII, XVIII and XIX. However, whilst the cadaverine formed from lysine remains enzyme-bound it is unsymmetrical so for these alkaloids no equilibration with unbound cadaverine can occur. But, as will be discussed later, some piperidine alkaloids are derived from lysine in such a way that C-2 and C-6 of the amino acid become equivalent. This is simply explained on the above model by allowing extensive equilibration of XXIII with unbound cadaverine. (One wonders at this stage whether the pyrrolidine alkaloids on closer examination might not turn out to have a similar genesis from ornithine as that outlined for piperidine alkaloids in Scheme 17.)

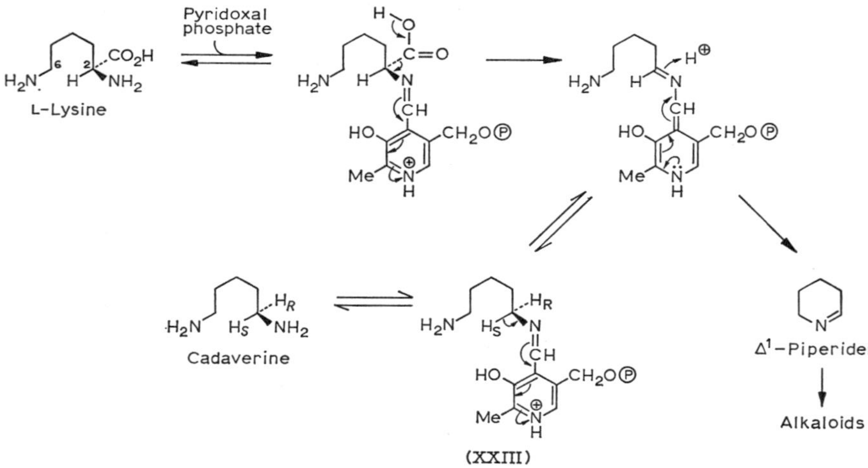

Scheme 17

On the basis of the above hypothesis cadaverine assumes importance as a precursor for piperidine alkaloids and it has been shown, using chirally tritiated samples of cadaverine, that incorporation of the diamine into *N*-

methylpelletierine (XVIII) results in retention of the *pro-R* proton at C-1 of XVIII and loss of the 1-*pro-S* hydrogen atom (tritium appears at C-6 also, but there is no loss of label). This is consistent with a stereospecific oxidative deamination of cadaverine to Δ^1-piperideine (*Leistner* and *Spenser*, *loc. cit.*; *idem*, Chem. Comm., 1975, 378). The loss of 25% of the tritium label from non-chirally labelled [1-^{3}H]cadaverine on incorporation into *N*-methylpelletierine, puzzling at the time, is now explained in terms of the above model, the tritium loss being exactly as predicted. It seems that subsequent elaboration of *N*-methylpelletierine to pseudopelletierine (XXI) is accompanied by preferential retention of tritium at C-6 by a primary isotope effect (*Liebisch*, *Marekov* and *Schütte*, *loc. cit.*; *Leistner* and *Spenser*, *loc. cit.*).

Although amino acids have been administered to plants on occasions legion in number, rarely has attention been paid to the question of whether there is any selectivity for the D- or L-amino acid in alkaloid biosynthesis. Exceptions are found in the biosynthesis of *Amaryllidaceae* alkaloids where it was shown that D- and L-tyrosine were equally well utilized in lycorine and norpluviine formation (*I. T. Bruce* and *G. W. Kirby*, Chem. Comm., 1968, 207), generally in the study of microbial metabolites, and also in the study of piperidine alkaloids. L-lysine was found to be a much better precursor than the D-isomer in the biosynthesis of sedamine (XVII), anabasine (XIX), *N*-methylpelletierine (XVIII), *N*-methylallosedridine (XXIV) (notably of opposite configuration at C-2 to sedamine) and lycopodine (XXXVI, p. 323) (*Gilbertson*, Phytochemistry, 1972, **11,** 1737; *Leistner*, *Gupta* and *Spenser*, J. Amer. chem. Soc., 1973, **95,** 4040; *W. D. Marshall et al.*, Canad. J. Chem., 1975, **53,** 41). In contrast pipecolic acid (L, p. 329) which is elaborated in a different way from lysine (see below) is preferentially derived from D-lysine in the same plants.

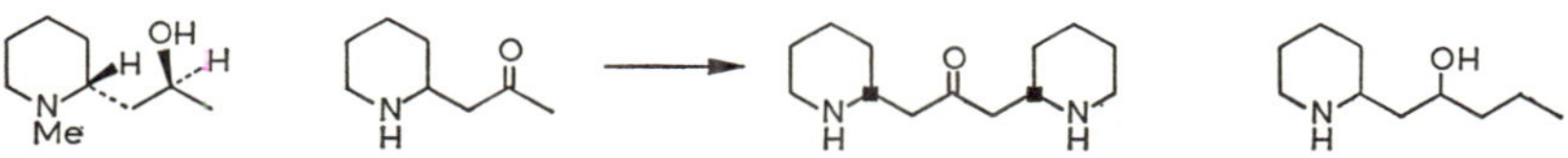

(XXIV) Pelletierine Anaferine (XXV)

By analogy with pyrrolidine alkaloid biosynthesis, **anaferine** and **pseudopelletierine** (XXI) might be expected to have their genesis through an intermediate of the *N*-methylpelletierine type. In accord *N*-methylpelletierine (XVIII) is a precursor for pseudopelletierine (XXI); and pelletierine is incorporated into anaferine. Moreover, the appropriate specific labelling by lysine (unsymmetrical intermediates: labelling of ■ positions in anaferine) and acetate have been recorded (*Keogh* and *O'Donovan*, J. chem. Soc., C 1970, 1792). The base XXV, related to pelletierine, also derives from lysine in an unsymmetrical way. The C_5 side-chain should derive from acetate but

[2-^{14}C]-labelled material has been found not to label the alkaloid (*O'Donovan* and *P. D. Creedon*, Tetrahedron Letters, 1971, 1341).

8-Phenyllobelol-I, which differs from sedamine (XVII) only in the configuration at C-8, occurs together with **lobeline** (XXVIII) in *Lobelia inflata.* As discussed above sedamine is derived from lysine by way of unsymmetrical intermediates so a similar pathway for 8-phenyllobelol-I and hence lobeline might be expected. Incorporation of [2-^{14}C]lysine, however, gave lobeline with equal labelling of C-2 and C-6, indicating at least one symmetrical intermediate is involved (*Keogh* and *O'Donovan*, J. chem. Soc., C 1970, 2470). Inspection of the structure of lobeline (XXVIII) though, immediately suggests that symmetrization of the label could occur at a late stage of biosynthesis and that lobelanine (XXVII) might be an intermediate. Accordingly [*N-Me*-^{14}C]lobelanine was found to be incorporated with specificity and high efficiency establishing this base as an immediate precursor (*O'Donovan* and *T. Forde*, *ibid.*, 1971, 2889). This does not exclude cadaverine as an earlier symmetrical precursor, but it was found that this base was much less efficiently incorporated than lysine and Δ^1-piperideine which argues strongly that it is not an obligatory intermediate in lobeline biosynthesis and thus symmetrization of the lysine label occurs *via* lobelanine (XXVII) (*O'Donovan et al.*, J. chem. Soc., Perkin I, 1975, 415).

It is a reasonable assumption that the two C_6–C_2 moieties of lobeline would arise by successive additions of benzoylacetic acid to Δ^1-piperideine (XX). This is supported by specific incorporations of labelled samples of phenylalanine, cinnamic acid, and 3-hydroxy-3-phenylpropionic acid (XXVI) (*O'Donovan et al., loc. cit*; *Keogh* and *O'Donovan, loc. cit.*). The probability of the last compound being an intermediate is increased by its isolation from *L. inflata* (*H. Wieland et al.*, Ann., 1939, **540**, 103). The biosynthetic pathway deduced from the above results is summarized in Scheme 18.

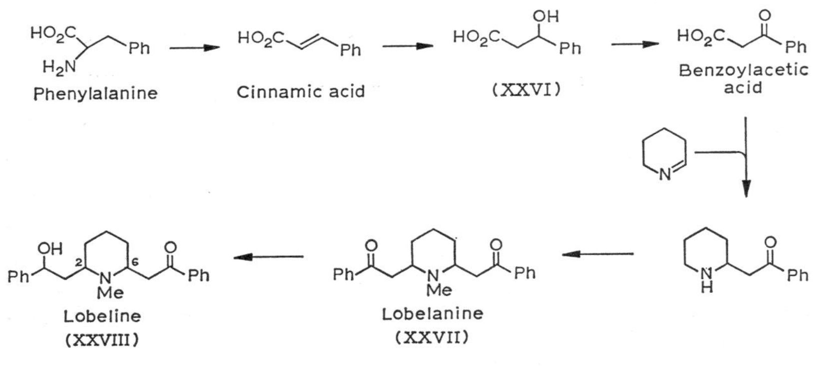

Scheme 18

The biosynthesis of *Lobelia* alkaloids with C_4 side chains can be accounted for in terms of a pathway related to lobeline with truncation of an intermediate XXIX leading to, *e.g.*, XXX, on the grounds of the labelling patterns observed after feeding lysine and acetate (*O'Donovan et al., loc. cit.*).

(XXIX) (XXX)

(XXXI) Lobinaline

A further alkaloid found in *L. inflata* is **lobinaline.** The structure of this alkaloid may be broken visually along the dotted line shown into two units of the type seen in sedamine (XVII). In accord with a pathway involving dimerization of two sedamine-like intermediates, phenylalanine (and cinnamic acid) and lysine, were found to be specific precursors for both "halves" of the alkaloid and label appeared at the expected sites (unsymmetrical incorporation of lysine) with apparently equal distribution of radioactivity between the two "halves". Further substantiation came with the efficient incorporation of [4-^{14}C]-2-phenacylpiperidine (XXXI) into lobinaline with equal labelling of C-2 and C-6′ (*R. N. Gupta* and *I. D. Spenser*, Canad. J. Chem., 1971, **49**, 384).

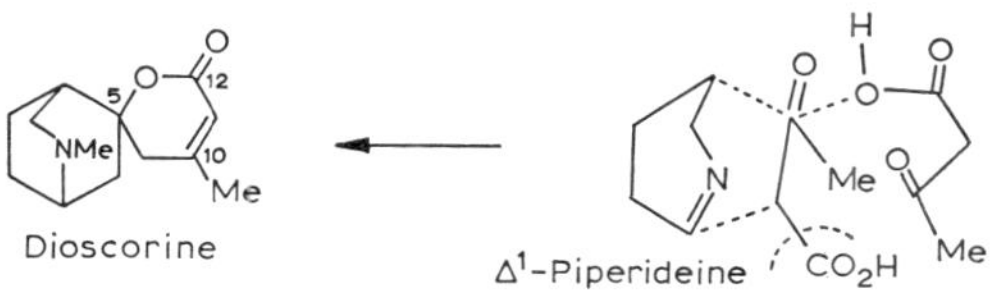

Dioscorine Δ^1-Piperideine

The incorporation of [1-^{14}C)acetate into **dioscorine** with equal labelling of C-5, -10, and -12, is consistent with a genesis as shown, the Δ^1-piperideine moiety arising in the usual way from lysine. Unfortunately results from feeding radioactive lysine and Δ^1-piperideine are negative (*Leete* and *A. R. Pinder*, Phytochemistry, 1972, **11**, 3219).

A unique tetracyclic structure is a feature of the alkaloids, *e.g.* **securinine,** of the *Securinega* genus. Appropriately its genesis, at least as far as rings C and D are concerned, is also unique. Carbon atoms 6–13 of securinine derive from tyrosine. Transformation of this amino acid into securinine involves

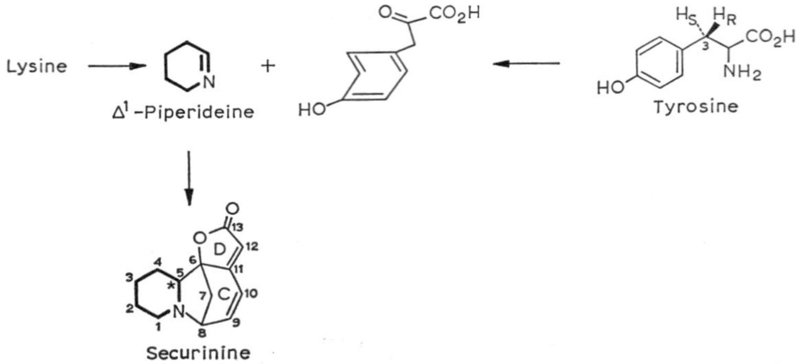

loss of one of the C-3 hydrogen atoms (C-3 become C-12 in securinine) which has been shown to be stereospecific and involving loss of the *pro-S* atom (*R. J. Parry*, Tetrahedron Letters, 1974, 307; Chem. Comm., 1975, 144; *U. Sankawa, K. Yamasaki* and *Y. Ebizuka*, Tetrahedron Letters, 1974, 1867). The remaining C_5N unit (heavy bonding) of securinine can be formed from cadaverine, [1,5-$^{14}C_2$]-labelled material affording alkaloid with equal labelling of the expected positions, C-1 and C-5 (*idem, loc. cit.*). Lysine and Δ^1-piperideine are also precursors, [2-^{14}C]-labelled samples of each yielding securinine with label confined to the asterisked carbon atom. Since [*RS*-6-^{3}H,6-^{14}C]-DL-lysine gave securinine without loss of tritium, C-6 cannot undergo oxidation in the course of biotransformation to alkaloid. Therefore it is the ε-amino group rather than the α-amino function which is retained. These observations are all consistent with the pathway outlined above (Scheme 17) for other piperidine alkaloids. It is interesting to note that securinine is the first among those alkaloids with a nitrogen atom common to two rings which is known to avoid a symmetrical intermediate (*W. M. Golebiewski, P. Horsewood* and *Spenser*, Chem. Comm., 1976, 217).

(*ii*) *Tetrahydroanabasine alkaloids*

The tetrahydroanabasine skeleton is found in the *Adenocarpus* alkaloids **adenocarpine** (XXXIV) and **santiaguine** (XXXII).3′-Piperidyl-Δ^2-piperideine (XXXIII) is efficiently incorporated into adenocarpine (*H. R. Schütte et al.*, Phytochemistry, 1964, **3**, 249).

Cadaverine is also a precursor for adenocarpine (*Schütte et al., loc. cit.*) and the incorporation of [2-^{14}C]- and [6-^{14}C]-lysine into santiaguine is specific and unsymmetrical (Scheme 19). Labelled Δ^1-piperideine was incorporated in like manner and is thus, most probably, an intermediate in the formation of each piperidine ring of santiaguine *via* XXXIII (*O'Donovan* and *Creedon*, J. chem. Soc., Perkin I, 1974, 2524). The final link in the biosynthetic chain

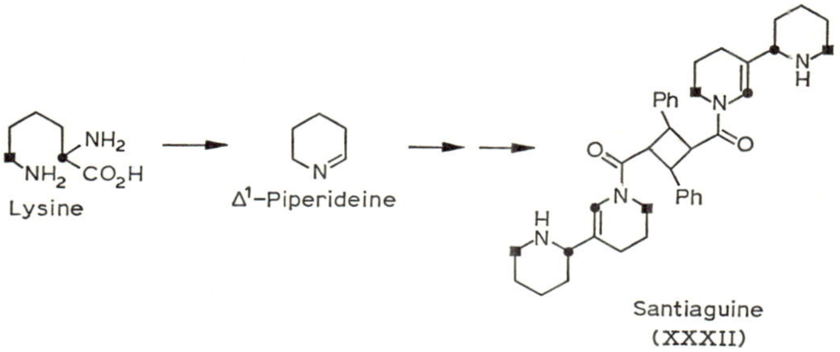

Scheme 19

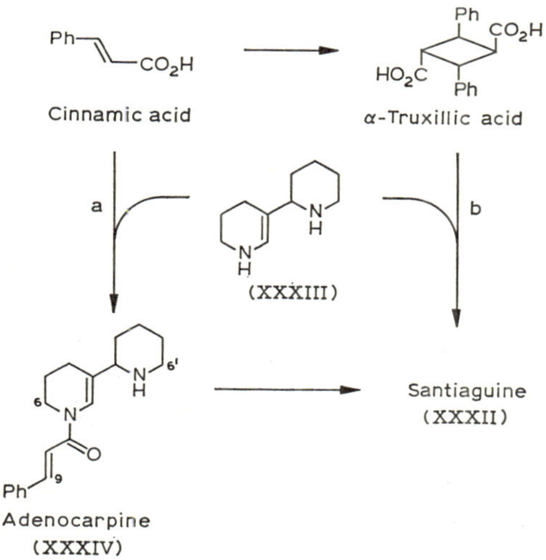

Scheme 20

comes with the efficient incorporation of [9-^{14}C,6,6′-$^{3}H_2$]adenocapine into santiaguine without change in isotope ratio, which demonstrates the intact incorporation of this precursor (*idem*, J. chem. Soc., C, 1971, 1604). These results lead to the pathway a to adenocarpine and santiaguine shown in Scheme 20, which involves dimerization of two adenocarpine units in the formation of santiaguine. The alternative pathway, b, in which cinnamic acid first dimerizes affording α-truxillic acid is supported by results which show that α-truxillic acid exists free in *Adenocarpus foliosus*, is specifically labelled by radioactive cinnamic acid (also, together with phenylalanine, a specific precursor for santiaguine), and is itself incorporated into santiaguine. However, on the basis of relative incorporation efficiencies, it appears that the route *via* adenocarpine is, at least, the major one (*idem*, J. chem. Soc., Perkin I, 1974, 2524).

Although α-truxillic acid can be formed photochemically *in vitro* from cinnamic acid, the observation that the incorporation of labelled adenocarpine into santiaguine was unaffected by growing the plants in darkness or light, argues against such a pathway *in vivo* (*idem*, *loc. cit.*).

A final point of interest in the biosynthesis of these tetrahydroanabasine alkaloids is the contrast between the biosynthetic pathway which affords these bipiperidyl bases and those which lead to the related systems in anabasine (p. 312) and anatabine (p. 331).

(iii) Lythraceae alkaloids

A C_6–C_3 unit hypothetically derivable from 4-hydroxycinnamic acid is manifest in the Lythraceae alkaloids **cryogenine** and **decodine.** Although this unit can, with little doubt, be linked by oxidative coupling (see p. 386) to a second oxygenated aromatic unit, it is not clear by inspection how many carbon atoms of the side-chain of this latter unit are used in the biosynthesis of the quinolizidine ring of these alkaloids. Attendant on this is the question of the biosynthesis of the other atoms of the ring system. Clarification of these points has been achieved by feeding experiments with phenylalanine and lysine.

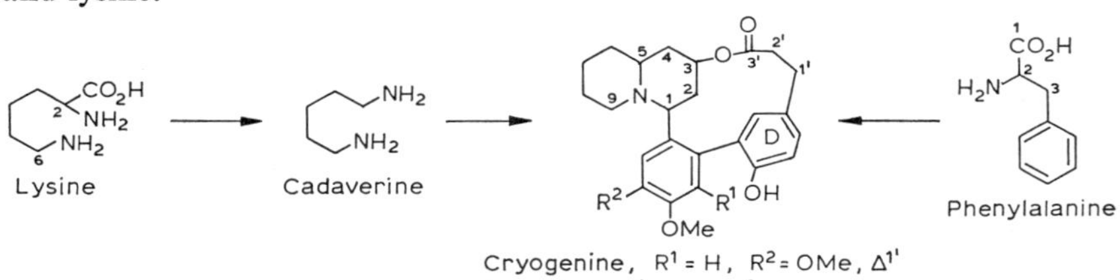

Cryogenine, R^1 = H, R^2 = OMe, $\Delta^{1'}$
Decodine, R^1 = OH, R^2 = H
Decinine, R^1 = H, R^2 = OMe

Label from [3-^{14}C]phenylalanine was incorporated into cryogenine in *Heimia salicifolia*, and appeared at C-1′ and C-1. It is interesting to note that significantly more of the label appeared at C-1′ than at the other position intimating that the two phenylalanine units implicated are different (*A. Rother* and *A. E. Schwarting*, Phytochemistry, 1972, **11**, 2475). More definitive evidence comes from a feeding experiment with [1,3-$^{14}C_2$]phenylalanine in *Decodon verticillatus*. Decodine was found to be labelled at C-1′, -3′, and -1; the ratio of activity at C-1′ and -3′ was the same as in the precursor. Thus an intact C_6–C_3 unit is established as the precursor for ring D plus C-1′, -2′, and -3′. As C-1 but not C-3 was labelled a C_6–C_3 unit cannot be involved here and a C_6–C_1 or C_6–C_2 unit must be implicated. The former alternative allowing generation of the remaining carbons from pelletierine (XXXV) is *a priori* the more attractive (*S. H. Koo, F. Comer* and *I. D. Spenser*, Chem. Comm., 1970, 897). Both [2-^{14}C]- and [6-^{14}C]-lysine serve as progenitors for decodine and decinine. Label in each case was deduced to be equally over C-5 and -9, which establishes that lysine is a specific precursor, providing a C_5 unit (C-1 of lysine is not incorporated) and its utilization must be *via* a symmetrical intermediate, probably cadaverine (*Koo et al., ibid.*, 1970, 396). This observation stands in contrast with those on other piperidine alkaloids (see above) but agrees with those on *Lycopodium* alkaloids (see below). However, as discussed above these differing observations can be accommodated within a simple hypothesis (Scheme 17).

(iv) Lycopodium alkaloids

The *Lycopodium* alkaloids exemplified by **lycopodine** (XXXVI) and **cernuine** (XXXVII) have been regarded as modified dimers of pelletierine (XXXV). Supporting evidence comes from the specific incorporations of acetate, lysine, and Δ^1-piperideine into lycopodine in *Lycopodium tristachyum* and cernuine in *L. cernuum*; the results are summarized in Scheme 21, as is the hypothetical derivation through two molecules of pelletierine. It is notable that these alkaloids, like the Lythraceae alkaloids above, are generated from lysine by way of a symmetrical intermediate. That this intermediate could be cadaverine was supported by incorporation of this diamine into both alkaloids (*Spenser et al.*, J. Amer. chem. Soc., 1968, **90**, 1360;

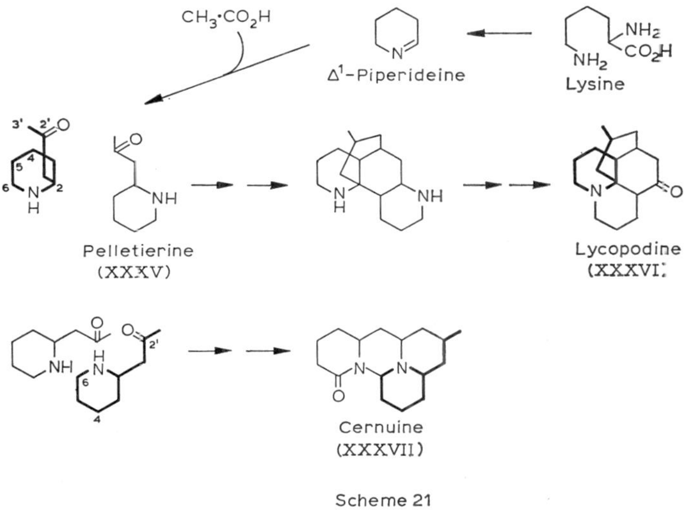

Scheme 21

Canad. J. Chem., 1972, **50**, 2591; 1970, **48**, 1893; 1970, **48**, 2911; 1971, **49**, 3352).

The most striking feature of these results was the equal incorporation of lysine, cadaverine and Δ^1-piperideine into both putative pelletierine (C_8N) units of lycopodine and cernuine. Strong evidence is provided, therefore, for the dimer hypothesis but the crucial experiment was to test the incorporation of pelletierine itself. When this was done the results were quite unexpected. Although [4-^{3}H,2′,6-$^{14}C_2$]-pelletierine gave radioactive cernuine without change in isotope ratio, thus indicating intact incorporation, degradation showed that only one of the C_8N units was derived from pelletierine (shown by heavy bonding in XXXVII). Similarly, incorporation of [4,5-$^3H_2$,2-^{14}C] pelletierine (without change in isotope ratio), [6-^{14}C)pelletierine and [2,3′-$^{14}C_2$]pelletierine into lycopodine was observed and once again label was incorporated into only one half (heavy bonding in XXXVI) of lycopodine (*idem, loc. cit.*, 1970, p. 2911).

The equal labelling of both C_8N units by radioactive lysine, cadaverine, and Δ^1-piperideine requires that either the two pelletierine-like units be

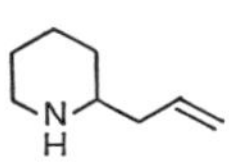

(XXXVIII)

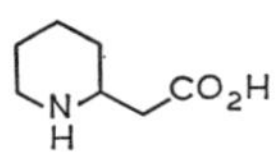

(XXXIX)

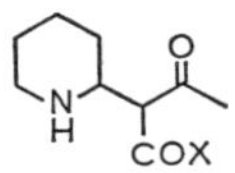

(XL)

identical or, if different, they must be derived from Δ^1-piperideine with the same overall dilution by unlabelled pools. The former possibility excludes pelletierine as a normal intermediate and accommodates exogenous material as a substitute for one of the C_8N intermediates but not for the other. The two C_8N intermediates were suggested as being derived from 2-allylpiperidine (XXXVIII) but the failure of [2-^{14}C]-labelled material to act as a specific precursor for lycopodine provides evidence against this hypothesis. Additional evidence was more telling. Radioactive pelletierine was isolated after feeding radioactive cadaverine and Δ^1-piperideine to *L. tristachyum* and thereby the presence of pelletierine in the plant was demonstrated, as well as its synthesis from lycopodine precursors. Proof that pelletierine is a normal intermediate in lycopodine biosynthesis was obtained when it was shown that pelletierine very effectively diluted activity from one C_8N unit (heavy bonding in XXXVI) of lycopodine on feeding radioactive cadaverine or Δ^1-piperideine (*idem, ibid.*, 1972, **50**, 2591). To account for the various results a hypothesis was advanced central to which was the intermediacy of the CoA ester of piperidine-2-acetic acid. Subsequently, rigorous experimentation has shown that piperidine-2-acetic acid (XXXIX) does not play a part in lycopodine biosynthesis and the hypothesis thus needs rethinking (*idem*, Canad. J. Chem., 1975, **53**, 41). In summary it is clear that pelletierine is a normal intermediate in the biosynthesis of lycopodine and cernuine and in the course of biosynthesis it becomes linked to another C_8N unit which closely resembles it and which is not separated from it by any intermediates which exist in the plant in significant concentrations (equal labelling of both C_8N by lysine and Δ^1-piperideine). The only easily imagined intermediates not so far examined foı the C_8N unit, which do not arise from pelletierine, is the product XL of acetoacetate condensation with Δ^1-piperideine and perhaps Δ^1-piperideine itself, to which an acetoacetate unit is added after condensation with pelletierine.

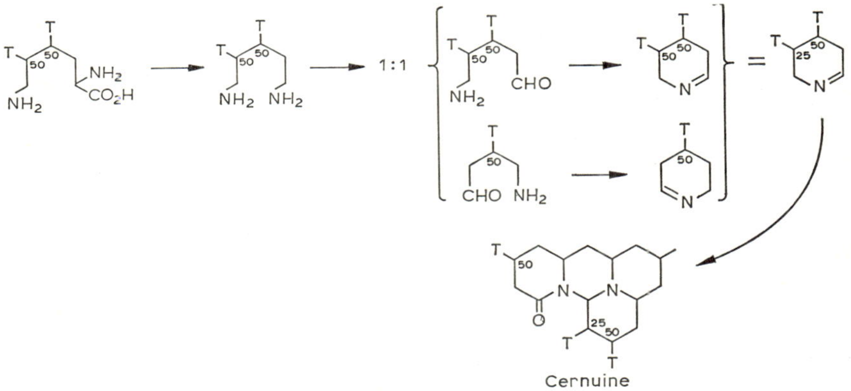

Scheme 22

The fate of tritium atoms on transformation of tritiated ornithine into *Lycopodium* alkaloids is interesting. [2-^{14}C,4,5-$^3H_2$] Lysine was incorporated into cernuine with loss of 39% of the tritium label compared with 37.5% predicted according to Scheme 22, where it was assumed that tritium on carbon atoms adjacent to carbonyl groups would be completely lost by exchange. The close correspondence of the observed and theoretical values further supports the outlined pathway for cernuine biosynthesis (*idem, ibid.*, 1971, **49**, 3352). Incorporation of [6-^{14}C,4,5-$^3H_2$]lysine has been found to give lycopodine with 22% loss of tritium. This was rationalized in terms of a route, similar to that which leads to cernuine, in which half the tritium at C-5 is lost and confirmed when it was shown that [6-^{14}C,4-^{3}H]lysine was incorporated without tritium loss (*idem, ibid.*, 1970, **48**, 1893; 1975, **53**, 41). The relative incorporations of isotope from doubly labelled samples of lysine (DL-[4,5-$^3H_2$]-/D-[6-^{14}C]lysine and L-[4,5-$^3H_2$]-/DL-[6-^{14}C]lysine) have shown that lycopodine is derived from L-lysine in *L. tristachyum* whereas pipecolic acid is formed from D-lysine, an observation in accord with others (see p. 317) (*idem, loc. cit.*).

(*v*) *Quinolizidine alkaloids*

The incorporation of [1,5-$^{14}C_2$]cadaverine into **lupinine** (XLI), structurally the simplest lupin alkaloid, with 25% of the activity deduced as being localized at each of C-2, -6, -10, and -11, provides good evidence in support of the hypothesis that lupinine is derived from 2 molecules of cadaverine (Scheme 23) (*cf.* p. 306 for a possibly related pathway in the biosynthesis of pyrrolizidine alkaloids) (*H. R. Schütte*, Arch. Pharm., 1960, **293**, 1006; *M. Souček* and *Schütte*, Angew. Chem. internat. edn., 1962, **1**, 597; *Schütte* and *H. Hindorf*, Z. Naturforsch., 1964, **19b**, 855).

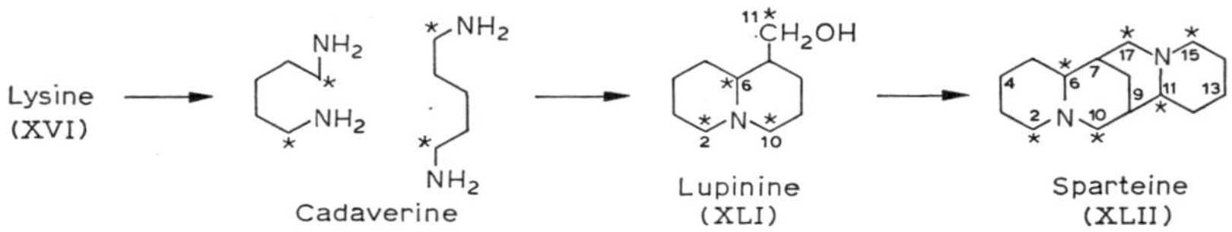

Scheme 23

The tetracyclic base **sparteine** (XLII), like lupinine, is a major alkaloid of the bitter form of the yellow lupin, *Lupinus luteus*. This suggests a related genesis from cadaverine, which requires labelling by [1,5-$^{14}C_2$]cadaverine of C-2, -6, -10, -11, -15 and -17 (see Scheme 23). Convincing support for this hypothesis was adduced by showing that a sixth of the label in sparteine derived from this precursor was located at each of C-2, -15 and -17 (carbons 3, 4, 5, 12, 13 and 14 were devoid of radioactivity) (*Schütte, F. Bohlmann* and *W. Reusche*, Arch. Pharm., 1961, **294**, 610; *K. Mothes et al.*, Ann., 1964, **680**, 93). These results were further supported when it was shown that lupinine was a precursor for sparteine (*Schütte*, Atompraxis, 1961, **7**, 91; *E. Nowacki, D. Nowacka* and *R. U. Byerrum*, Bull. Acad. Polon Sci., Ser. Sci. Biol., 1966, **14**, 25).

Degradation of lupinine and sparteine obtained after feeding [2-^{14}C]-lysine allowed the sites of labelling to be determined as the same as from [1,5-^{14}C]cadaverine. Thus lysine is incorporated *via* a symmetrical intermediate, most probably cadaverine (see discussion on p. 315) (refs. as for cadaverine above). Consistent with this view, [2-^{14}C,ε-^{15}N]- and [2-^{14}C,α-^{15}N]-lysine both gave sparteine with three times as much ^{14}C as ^{15}N, which requires decarboxylation of lysine before deamination, *i.e.* passage *via* cadaverine (in considering the ratio of ^{15}N:^{14}C expected account is taken of the necessary incorporation of three lysine carbon chains but only two nitrogen atoms) (*Mothes et al., loc. cit.*; *Schütte* and *G. Seelig*, Ann., 1968, **711**, 221).

Further support for the intermediacy of cadaverine comes from the isolation of radioactive cadaverine from *L. lupineus* following administration of radioactive lysine (*Schütte* and *D. Knöfel*, Z. Pflanzenphysiol., 1968, **59**, 80).

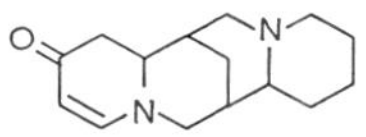

Multiflorine

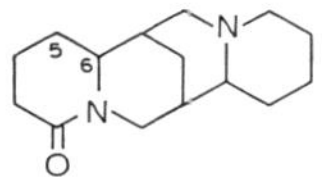

Lupanine
(XLIII)

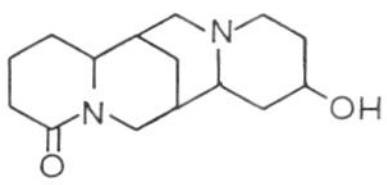

Hydroxylupanine

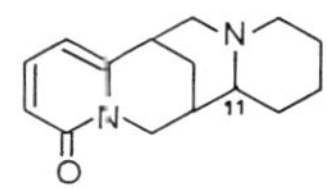

Thermopsine

The alkaloids **lupanine** (XLIII), **hydroxylupanine** and **multiflorine** have been shown to derive from three molecules of lysine *via* cadaverine (*Schütte et al.*, Naturwiss., 1961, **48**, 669; Ann., 1965, **685**, 187; 1968, **711**, 221; Arch. Pharm., 1962, **295**, 20; Z. Pflanzenphysiol., 1967, **57**, 188; *Nowacki et al.*, Bull. Acad. Polon. Sci., Ser. Sci. Biol., 1964, **12**, 489; 1966, **14**, 97).

In young *L. nanus* plants added cadaverine but not putrescine or hexamethylenediamine lowered the incorporation of lysine into thermopsine, thus providing further evidence for the intermediacy of cadaverine in the biosynthesis of alkaloids of this type from lysine (*E. K. Nowacki* and *G. R. Waller*, Phytochemistry, 1975, **14**, 155).

Results of experiments with labelled quinolizidine alkaloids have suggested that in *L. angustifolius* and *L. albus* the sequence of biosynthesis is sparteine → lupanine → hydroxylupanine which is also irreversible (*E. Nowacki, Nowacka* and *Byerrum, loc. cit.*; *Schütte et al.*, Arch. Pharm., 1963, **296**, 438). In accord, in *L. arboreus* plants, $^{14}CO_2$ was taken up into sparteine but not into lupanine, which indicates that lupanine is not a precursor for sparteine. On the other hand, in *L. polyphyllus*, lupanine was labelled by $^{14}CO_2$ but no trace of (radioactive) sparteine could be detected, which tends to suggest that sparteine is not a precursor here for lupanine unless the pool size is indetectably small (*Y. D. Cho, R. O. Martin* and *J. N. Anderson*, J. Amer. chem. Soc., 1971, **93**, 2087).

Both lysine and cadaverine are specific precursors for the C_{14} base **angustifoline**. Incorporation of labelled samples of the tetracyclic bases, sparteine (XLII) and lupanine (XLIII), into angustifoline indicated that this

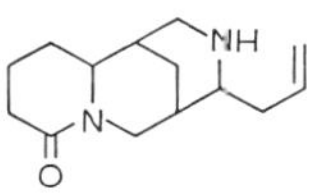

Angustifoline

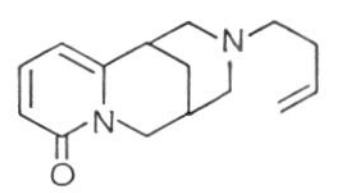

Rhombifoline

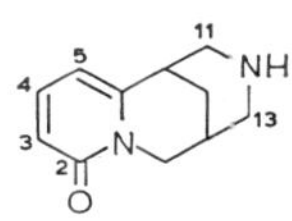

Cytisine

alkaloid is formed by degradation of alkaloids with a tetracyclic skeleton (*Schütte* and *Seelig*, unpublished results; quoted by *Schütte*, in "Biosynthese der Alkaloide", ed. *Mothes* and *Schütte*, VEB Deutscher Verlag der Wissenschaften, Berlin, 1969, p. 334). That the tricyclic skeleton is formed from the tetracyclic one rather than *vice versa*, is supported strongly by several pieces of other evidence. Firstly, conversion of say angustifoline to hydroxylupanine requires addition of a C-1 unit. Failure of formate or formaldehyde to act as the appropriate precursor argues against such a conversion (*E. Nowacki* and *Byerrum*, Bull. Acad. Polon. Sci., Ser. Sci. Biol., 1964, **12**, 483). Secondly, experiments with $^{14}CO_2$ in *Thermopsis rhombifolia* and *T. caroliniana* showed that the tetracyclic alkaloid lupanine (XLIII) was the first alkaloid to be labelled, followed by 5,6-dehydrolupanine and anagyrine (the C-11 epimer of thermopsine). Then the first of the tricyclic bases, rhombifoline, was labelled. Rhombifoline appeared to play a primary role in the subsequent formation of **cytisine** and *N*-methylcytisine. The sequence of labelling strongly suggests that tricyclic bases are derived from tetracyclic ones, and a role for 5,6-dehydrolupanine in this connection was suggested. Thermopsine appeared after rhombifoline. As with similar experiments on the biosynthesis of the alkaloids of *L. angustifolius* no evidence could be found for the formation of sparteine (XLII) (*Cho* and *Martin*, Canad. J. Biochem., 1971, **49**, 971). A similar study using lysine as a source of radioactivity for five Leguminosae species has indicated not only that lysine is a precursor for all the alkaloids examined, but also further relationships between alkaloids. In particular it was concluded that saturated alkaloids are precursors for those with a pyridone ring, this being supported by the incorporation of sparteine (XLII) and lupanine (XLIII) into more highly oxidized alkaloids (*E. K. Nowacki* and *Waller*, Phytochemistry, 1975, **14**, 155, 161, 165). Finally, rhombifoline has been found not to be a precursor for lupanine and hydroxylupanine (*E. Nowacki*, *Nowacka* and *Byerrum*, *loc. cit.*).

In accord with the preceding discussion cytisine has been found to be derived specifically from lysine and cadaverine, a fifth of the activity from [2-^{14}C]lysine and [1,5-$^{14}C_2$]cadaverine being found at each of carbons 2, 11 and 13 (C-3, -4, and -5 were inactive). In addition the incorporation of lupinine and sparteine into cytisine has been observed (*Schütte* and *J. Lehfeldt*, Z. Naturforsch., 1964, **19b**, 1085; J. prakt. Chem., 1964, **24**, 143).

An alternative tetracyclic skeleton to the one found in sparteine is to be seen in **matrine**, and a similar derivation from three molecules of lysine *via* cadaverine has been demonstrated (*Schütte et al.*, Arch. Pharm., 1962, **295**, 34; Ann., 1965, **685**, 194; *S. Shibata* and *U. Sankawa*, Chem. and Ind., London, 1963, 1161). Cadaverine is also a precursor for sophocarpine and related alkaloids [*B. A. Abdusalamov et al.*, Biochemistry, (U.S.S.R.), 1971,

36, 239]. Matrine, sophocarpine and their *N*-oxides appear to be interconvertible. Neither matrine nor sophocarpine is, however, converted into sophoridine, indicating that the configurational differences between sophoridine, indicating that the configurational differences between sophoridine and the matrine-type are fixed at an early stage of biosynthesis (*J. K. Kuschmuradov*, *D. Gross* and *Schütte*, Phytochemistry, 1972, **11**, 3441).

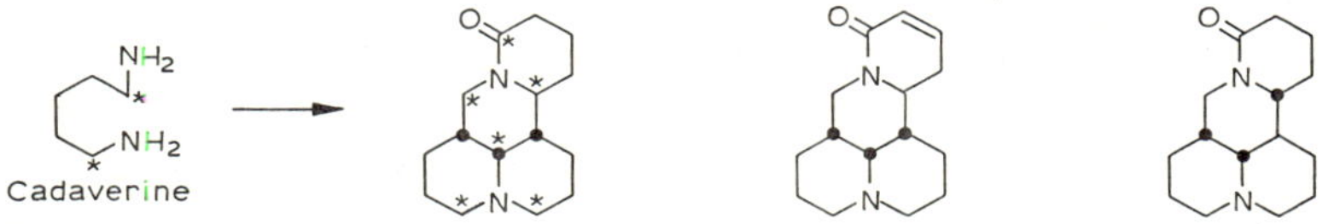

Cadaverine Matrine Sophocarpine Sophoridine

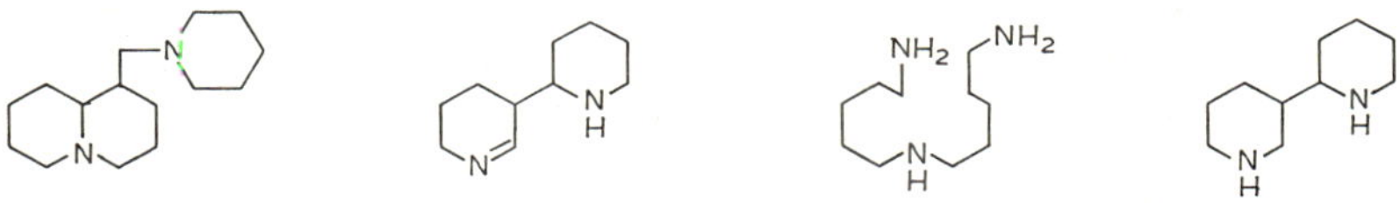

(XLIV) (XLV) (XLVI) (XLVII)

The foregoing allows one to build up a reasonable picture of the biosynthesis of lupin and matrine alkaloids which includes interrelationships. An outstanding problem of some magnitude is the mechanism by which lysine/cadaverine is transformed into the various alkaloids and, associated with it, the various intermediates so far undetected. In this regard it is to be noted that piperidyllupinine (XLIV), tetrahydroanabasine (XLV), bis(5-aminopentyl)amine (XLVI), $\alpha\beta$-bipiperidine (XLVII), and 1,5-dipiperidinopentane are poor alkaloid precursors (*Schütte et al.*, Ann., 1965, **685**, 187; Arch. Pharm., 1964, **297**, 118; Z. Pflanzenphysiol., 1968, **59**, 80; *E. Nowacki et al.*, Bull. Akad. Polon. Sci., Ser. Sci. Biol., 1966, **14**, 97). On the other hand, a specific incorporation of [2-^{14}C]piperideine into matrine has been recorded, although activity appeared to be very unequally distributed between the labelling sites (*Shibata* and *Sankawa*, *loc. cit.*).

(*vi*) *Pipecolic acid and slaframine*

Pipecolic acid (L) is found widely in plants, animals and micro-organisms. In the representative species of each examined it has been found that DL-[6-^{3}H,6-^{14}C]lysine is incorporated into pipecolic acid without tritium loss, whereas DL-[2-^{3}H,6-^{14}C]lysine gives pipecolic acid devoid of tritium (*R. N.*

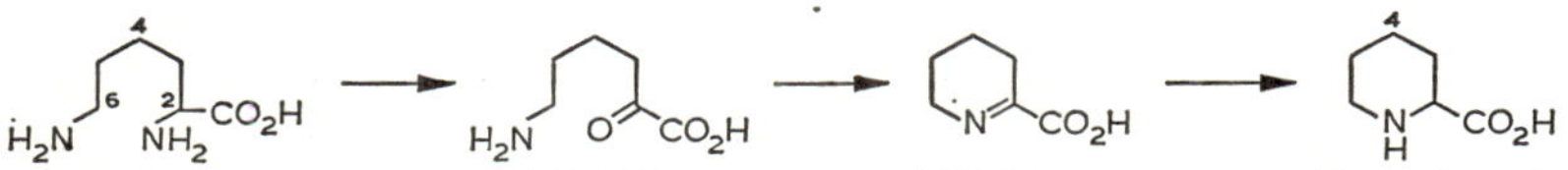

Lysine (XLVIII) (XLIX) Pipecolic acid (L)

Gupta and *I. D. Spenser*, Phytochemistry, 1970, **9**, 2329; J. biol. Chem., 1969, **244**, 88). This is consistent with a pathway in which pipecolic acid is formed in plants *via* ε-amino-α-oxocaproic acid (XLVIII) and Δ^1-piperideine-2-carboxylic acid (XLIX). This means that the biosynthetic pathway to pipecolic acid is quite different from that which leads to the piperidine alkaloids (see p. 309). A further difference lies in the preferential incorporation of D- rather than L-lysine into pipecolic acid in rats, *Pseudomonas putida*, and various plants, whereas L-lysine is the preferred progenitor of piperidine alkaloids (in the same plants) (*J. A. Grove et al.*, Biochim. Biophys. Acta, 1969, **184**, 329; *R. H. Aldag* and *J. L. Young*, Planta, 1970, **95**, 187; *T. J. Gilbertson*, Phytochemistry, 1972, **11**, 1737; *Spenser et al.*, J. Amer. chem. Soc., 1973, **95**, 4040; Canad. J. Chem., 1975, **53**, 41; *Y.-F. Chang* and *E. Adam*, Biochem. Biophys. Res. Comm., 1971, **45**, 570; *D. L. Miller* and *V. W. Rodwell*, J. biol. Chem., 1971, **246**, 2758). Pipecolic acid is a precursor for 4-hydroxy- and 4-amino-pipecolic acid in *Strophanthus scandens* (*Schütte et al.*, Flora, 1962, **152**, 590; 1963, **153**, 426).

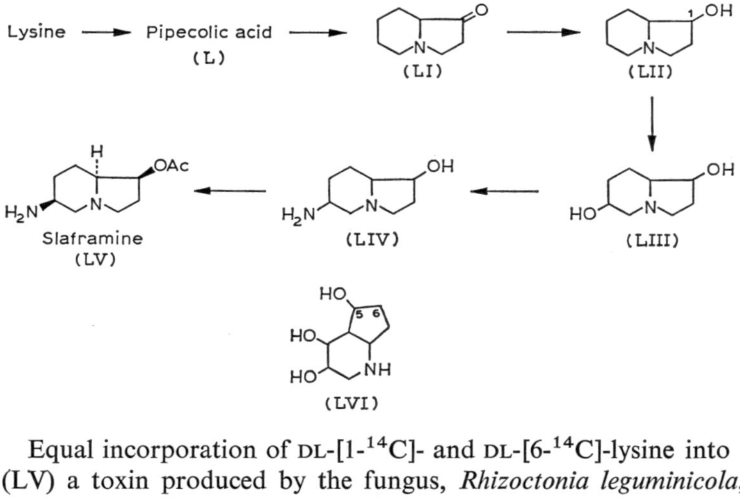

Equal incorporation of DL-[1-^{14}C]- and DL-[6-^{14}C]-lysine into **slaframine** (LV) a toxin produced by the fungus, *Rhizoctonia leguminicola*, indicates that the amino acid is incorporated as a C_6 moiety. Addition of inactive pipecolic acid to the cultures diluted the lysine label in the derived slaframine and pipecolic acid was also labelled by the radioactive lysine. Further, carboxyl- and ring-labelled pipecolic acids were both well incorporated into the metabolite but 2-hydroxymethylpiperidine was not. These results define the pathway as lysine → pipecolic acid → slaframine. It is further defined by the incorporation with increasing efficiency of labelled samples of LII, LI and LIII. Tritium at C-1 of LII is retained on formation of slaframine and so

LII must follow LI on the pathway, which is further supported by the discovery that a cell-free extract of *R. leguminicola* would catalyse the NADPH-dependent transformation of LI into LII as well as catalysing the conversion of LIV into slaframine in the presence of acetyl-CoA. A curious feature of the investigation is the observation that pipecolic acid is derived from L-lysine rather than D-lysine in *R. leguminicola* (*cf.* above) (*F. P. Guengerich et al.*, Biochemistry, 1973, **12**, 4264, 4270). A similar derivation from lysine *via* pipecolic acid is apparent for LVI, which is also elaborated by *R. leguminicola*. The origin of the two carbon atoms not accounted for by pipecolic acid (C-5 and C-6) is unknown: DL-[1-^{14}C]- and DL-[3-^{14}C]-serine were not incorporated (*Guengerich, S. J. DiMari* and *H. P. Broquist*, J. Amer. chem. Soc., 1973, **95**, 2055).

4. Pyridine alkaloids

This section covers the biosynthesis of pyridine alkaloids apart from nicotine and anabasine which were discussed in sections 2 and 3, respectively.

The known derivation of anabasine (XIX, p. 312) from nicotinic acid and lysine leads one to expect a similar origin for **anatabine**, which after all is found in the same plants as anabasine.

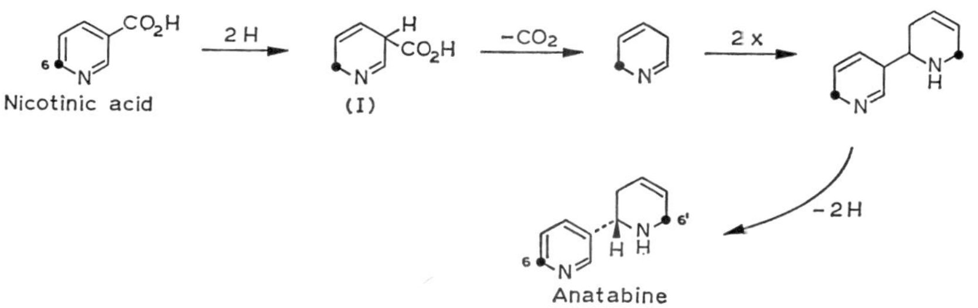

Scheme 24

However, [2-^{14}C]lysine and [2-^{14}C]-4-hydroxylysine were found not to label anatabine and results with $^{14}CO_2$ indicate that anabasine is not a precursor for anatabine (*T. Kisaki, S. Mizusaki* and *E. Tamaki*, Phytochemistry, 1968, **7**, 323; *W. L. Alworth* and *H. Rapoport*, Arch. Biochem. Biophys., 1965, **112**, 45; *E. Leete*, Chem. Comm., 1975, 9). On the other hand [6-^{14}C]-nicotinic acid gave radioactive anatabine, but surprisingly the activity was found to be equally divided between C-6 and C-6′. It follows that nicotinic acid is the source of both rings of the alkaloid and the equal distribution of label indicates that the two units from which the anatabine is formed are closely related, if not identical. The pathway suggested (Scheme 24) plausibly includes I, a likely intermediate from nicotinic acid in anabasine biosynthesis (*idem, loc. cit.*).

If the piperidine ring of anatabine has an unorthodox genesis from nicotinic acid then equally unorthodox is the genesis of the pyridine rings of mimosine, desmosine, and dipicolinic acid from lysine. Radioactive **mimosine** was isolated after administration of DL-[2-^{14}C]-, DL-[6-^{14}C]-and L-[6-^{14}C]-lysine to *Leucaena glauca* and *Mimosa pudica*. In each case radioactivity was confined to the pyridone nucleus of the alkaloid.

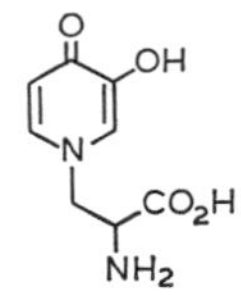

Mimosine

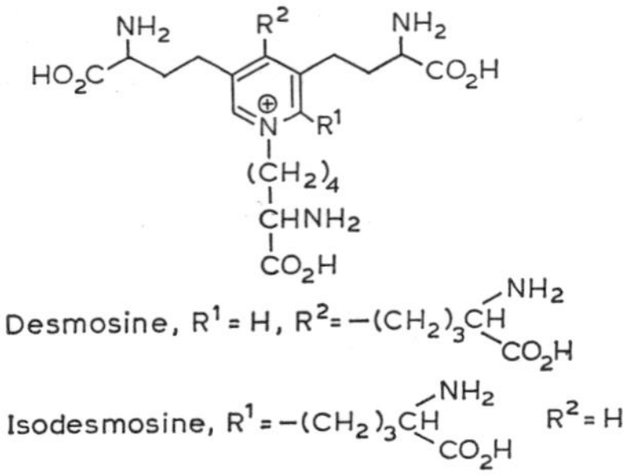

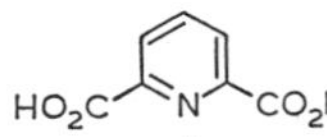

Dipicolinic acid

Radioactivity from C-3 but not from C-1 and C-4 of aspartic acid was incorporated specifically into the pyridone ring of mimosine in *M. pudica* as was [6-^{14}C]α-aminoadipic acid (*J. Hylin*, Phytochemistry, 1964, **3**, 161; *H. P. Tiwari* and *I. D. Spenser*, Canad. J. Biochem., 1965, **43**, 1687; *A. D. Notation* and *Spenser*, *ibid.*, 1964, **42**, 1803; *Tiwari, W. R. Penrose*, and *Spenser*, Phytochemistry, 1967, **6**, 1245). This evidence is consistent with formation of mimosine from lysine by way of α-aminoadipic acid (Scheme 15, p. 313), but needs confirmation in more detailed degradations to determine the precise sites of labelling because elaboration of a secondary metabolite with a pyridine ring by this route is otherwise unknown. The alanyl side-chain not unexpectedly arises from serine (*idem, loc. cit.*).

Demosine and **isodesmosine** serve as cross-linking centres in the animal protein elastin. Their formation is accounted for in terms of four lysine units. Both [6-^{14}C]- and [1-^{14}C]-lysine were incorporated and with the latter precursor the activity was located in the four carboxyl groups. As in the case of the plant base mimosine it seems that the α-aminoadipic acid pathway (Scheme 15, p. 313) may be involved (*R. A. Anwar* and *G. Oda*, J. biol. Chem., 1966, **241**, 4638; Biochim. Biophys. Acta, 1967, **133**, 151; *E. J. Miller et al.*, Biochem. Biophys. Res. Comm., 1967, **26**, 132).

Dipicolinic acid is produced by aerobic spore-forming bacteria, *e.g. Bacillus megaterium*, during sporulation. The pattern of incorporation of alanine, glutamic acid, and aspartic acid suggested a biosynthetic pathway involving combination of C_3 and C_4 units *via* 2,3-dihydropicolinic acid (see Scheme 16, p. 313) (*H. H. Martin* and *J. W. Foster*, J. Bact., 1958, **76**, 167). This compound is an intermediate *en route* to lysine *via* diaminopimelic acid [for the elaboration of lysine it is the alternative pathway (Scheme 16) to the one *via* α-aminoadipic acid (Scheme 15)]. This suggests that dipicolinic acid is a byproduct of lysine biosynthesis, a suggestion which is supported by other evidence. Firstly, extracts of *Bacillus* species were found to give dipicolinic acid on incubation with dioxopimelic acid in the presence of ammonia, and with aspartic acid semialdehyde and pyruvate (*J. Powell* and *R. E. Strange*, Nature, 1959, **184**, 878; *M. L. Bach* and *C. Gilvarg*, J. biol. Chem., 1966, **241**, 4563). A second piece of evidence comes from the conclusion of a study with lysine auxotrophs of *B. cereus* (*A. I. Aronson*, *E. Henderson* and *A. Tincher*, Biochem. Biophys. Res. Comm., 1967, **26**, 454). Diaminopimelic acid itself does not appear to be an intermediate for dipicolinic acid (*Martin* and *Foster*, *loc. cit.*; *A. J. Finlayson* and *F. J. Simpson*, Canad. J. Biochem. Physiol., 1961, **39**, 1551). In the unrelated organism, *Penicillium citreo-viride*, on the other hand, *meso*- and DL-[1,7-$^{14}C_2$]diaminopimelic acids serve as specific precursors for the metabolite, activity appearing only in the carboxyl groups (*S. W. Tanenbaum* and *K. Kaneko*, Biochemistry, 1964, **3**, 1314). Incorporation of other precursors was found to be consistent with elaboration of dipicolinic acid from an intermediate of the diaminopimelate pathway to lysine. It is to be noted, however, that in this mould lysine is produced by the α-aminoadipic acid pathway, and α-aminoadipic acid is also a precursor for dipicolinic acid. Further, diaminopimelic acid could not be identified in the microbial cultures (*idem.*, *loc. cit.*; *Foster et al.*, J. Bact., 1966, **91**, 562, 570). It seems that α-oxoadipic acid may stand at a branch-point for the normal biosynthesis in this organism of lysine *via* α-aminoadipic acid (Scheme 15, p. 313) and for the formation of dioxopimelic acid by chain-elongation. The dioxopimelic acid is then thought of as being transformed into dipicolinic acid, and it is at this stage of biosynthesis that exogenous diaminopimelate can be implicated.

A return to orthodoxy is found in the biosynthesis of **ricinine**. Like the pyridine rings of nicotine and anabasine, that of ricinine has its genesis in aspartic acid and glycerol in *Ricinis communis* (*L. Marion et al.*, Canad. J. Chem., 1963, **41**, 1142; 1966, **44**, 23; *K. S. Yang* and *G. R. Waller*, Phytochemistry, 1965, **4**, 881; *U. Schiedt* and *H. Boeckh-Behrens*, Z. physiol. Chem., 1962, **330**, 58); the *N*-methyl group arises from methionine (*M. Dubeck* and *S. Kirkwood*, J. biol. Chem., 1952, **199**, 307). Moreover quinolinic acid (XXX, p. 303), nicotinic acid, nicotinamide, *N*-methylnicotinamide, trigonellin, nicotinic acid mononucleotide, and the adenine dinucleotide derivatives of nicotinic acid and nicotinamide are all utilized for the biosynthesis of ricinine; it is clear that the nitrile function arises by dehydration of an amide group because of the observed incorporation of [$CO^{15}NH_2$]-nicotinamide (*Yang* and *Waller*, *loc. cit.*; *Waller et al.*, Plant Physiol., 1964, **39**, 244; 1963, **38**, 318; J. biol. Chem., 1961, **236**, 1186; 1966, **241**, 4411; Phytochemistry, 1967, **6**, 1637; *Leete* and *F. H. B. Leitz*, Chem. and Ind., London, 1957, 1572; *D. Gross et al.*, Z. Pflanzenphysiol., 1968, **58**, 410).

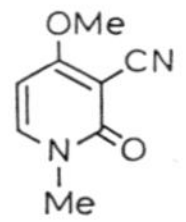

Ricinine

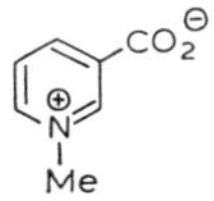

Trigonellin

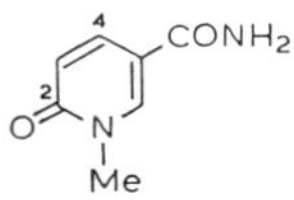

(II)

These results, particularly those with the nucleotide derivatives, may be taken as indicating that the pyridine–nucleotide cycle is associated with ricinine biosynthesis. Affirmation of this has come from studies using azaserine and azaleucine as inhibitors for NAD^+ synthetase: Both inhibitors produced a significant decrease in incorporation of [6-^{14}C]quinolinic acid into ricinine and into nucleotide-cycle intermediates (*R. D. Johnson* and *Waller*, Phytochemistry, 1974, **13**, 1493).

Trigonellin, which is found in a number of higher plants, has itself been shown not to arise from tryptophan (*R. U. Byerrum et al.*, J. biol. Chem., 1959, **234**, 93; *Leete*, *Marion* and *Spenser*, Canad. J. Chem., 1955, **33**, 405). Nicotinic acid is, not unexpectedly, a precursor; the quaternary *N*-methyl group arises by transfer from *S*-adenosylmethionine (*J. G. Joshi* and *P. Handler*, J. biol. Chem., 1960, **235**, 2981).

Nicotinic acid and nicotinamide have been found to be progenitors of *N*-methyl-2-pyridone-5-carboxamide (II), an alkaloid found in young greenhouse-grown *Trewia nudiflora* (*S. D. Sastry* and *Waller*, Phytochemistry, 1972, **11**, 2241). A similar derivation for this metabolite and also 1-methyl-4-pyridone-3-carboxamide from nicotinic acid and *N*-methylnicotinamide has been demonstrated in mamalian systems (*K. V. Rajagopalan*, *I. Fridovich* and *Handler*, J. biol. Chem., 1962, **237**, 922; *S. Chaykin*, Biochim. Biophys. Acta, 1964, **82**, 633; *G. P. Quinn* and *P. Greengard*, Arch. Biochem. Biophys., 1966, **115**, 146). In the hydroxylation of *N*-methylnicotinamide which affords 1-methyl-4-pyridone-3-carboxamide, and the one in which nicotinic acid is

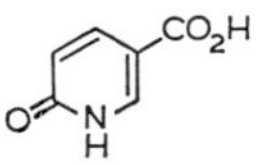

(III)

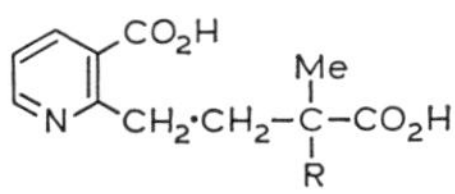

Wilfordic acid, R = H
Hydroxywilfordic acid, R = OH

transformed into III in the bacterium *Pseudomonas fluorescens*, the oxygen originates from water and not molecular oxygen (*idem, loc. cit.*; *A. L. Hunt, D. E. Hughes* and *J. M. Lowenstein*, Biochem. J., 1957, **66**, 2P). Consequently these reactions are not mediated by an oxidase and they represent so far unique examples of such hydroxylation in aromatic systems. Whether this applies to the biosynthesis of related plant alkaloids remains to be seen.

Hydrolysis of the complex ester alkaloids of *Tripterygium wilfordii* yields either **wilfordic acid** or **hydroxywilfordic acid**, for which nicotinic acid and nicotinamide serve as efficient precursors (*H. J. Lee* and *Waller*, Phytochemistry, 1972, **11**, 2233).

The origins of the fungal pyridine derivative, **fusaric acid**, are acetate for the side-chain plus C-5 and -6, and aspartic acid for part of the ring (C-2, -3, and -4, and carboxyl group), thus bearing some resemblance to the non-tryptophan origins of nicotinic acid (*L. C. Vining et al.*, Canad. J. Biochem., 1967, **45**, 809; *R. D. Hill, A. M. Unrau* and *D. T. Canvin*, Canad. J. Chem., 1966, **44**, 2077). The evidence rests on the specific incorporations of [1-^{14}C]- and [2-^{14}C]-aspartic acid, the clear incorporation of [1-^{14}C]- and [2-^{14}C]-acetate into the remaining atoms, and an interpretation of acetate labelling of the "aspartate unit" as being the result of normal passage through the tricarboxylic acid cycle (see *I. D. Spenser*, in "Comprehensive Biochemistry", eds. *M. Florkin* and *E. H. Stotz*, Elsevier, Amsterdam, 1968, Vol. 20, p. 364).

The picolinic acid moiety of **proferrorosamine A**, a metabolite of *Pseudomonas roseus fluorescens*, derives efficiently from picolinic acid, lysine, and glutamic acid (*M. Pouteau-Thouvenot et al.*, Helv., 1972, **55**, 2295; 1973, **56**, 1067).

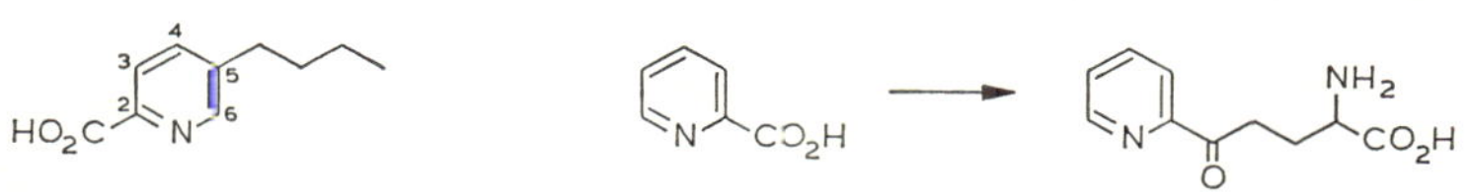

Fusaric acid Picolinic acid Proferrorosamine A

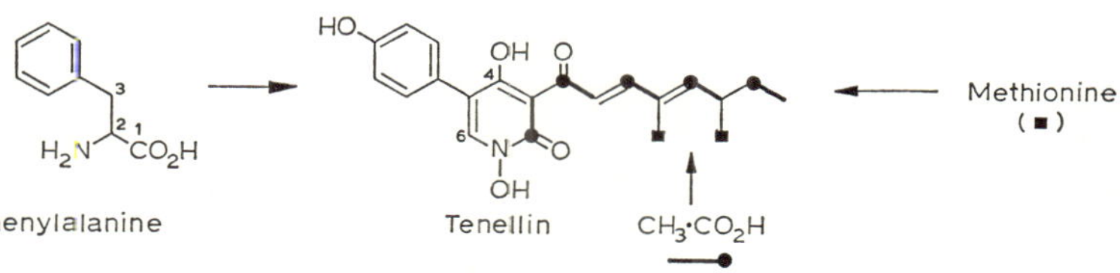

Phenylalanine Tenellin

The fungal metabolite **tenellin** is, the results of ^{13}C-labelled experiments establish, derived from phenylalanine, methionine and acetate; feeding [1,2-$^{13}C_2$]acetate gave tenellin with the expected satellite resonances due to ^{13}C–^{13}C spin–spin coupling, from which the intact C_2 acetate derived units could be discerned (*A. G. McInnes et al.*, Chem. Comm., 1974, 282). Moreover, C-1 of phenylalanine appears at C-4 of tenellin and C-2 at C-6. The implied rearrangement of the phenylalanine skeleton finds analogy in the biosynthesis of tropic acid from this amino acid in plants (see p. 296). DL-[1,3-$^{13}C_2$]Phenylalanine gave tenellin which showed ^{13}C–^{13}C coupling between C-4 and C-5 which is consistent only with an intramolecular rearrangement of the skeleton of the amino acid. (An intermolecular rearrangement would have led to separation of the labels.) Such an intramolecular rearrangement accords with observations on tropic acid biosynthesis (*Leete et al.*, Tetrahedron Letters, 1975, 4103).

5. Bases derived from the tryptophan pathway

The aromatic amino acid, tryptophan, (I), serves as the starting point for the elaboration by plants and micro-organisms of secondary metabolites whose number is legion. The amino acid is generally to be recognised as an indole unit but may appear in extensively modified form as in actinomycin and in the *Cinchona* alkaloids, *e.g.* quinine.

Anthranilic acid is an intermediate on the biosynthetic pathway to tryptophan and it serves as a starting point for the elaboration of a number of metabolites of plant and microbial origin. One group of microbial metabolites, based on the phenazine ring system, rather than being derived from anthranilic acid as the structures suggest, is formed directly from shikimic acid the precursor for aromatic amino acids. This acid is also the precursor for the plant base shihunine.

(*a*) *Gramine*

One of the simplest modifications of tryptophan is found in **gramine** (V), the indole alkaloid of barley. Study of the biosynthesis of this alkaloid is a milestone in such studies because demonstration that label from [3-^{14}C]-tryptophan appeared specifically at the side-chain methylene of gramine was the first such experiment to be performed on alkaloids outside the study of methylation (*K. Bowden* and *L. Marion*, Canad. J. Chem., 1951, **29**, 1037). It was rigorously established in subsequent experiments that the indole nucleus and C-3 together with its two hydrogen atoms were incorporated as an intact unit into gramine (*E. Leete* and *Marion*, *ibid.*, 1953, **31**, 1195; *D. O'Donovan* and *Leete*, J. Amer. chem. Soc., 1963, **85**, 461; *D. Gross*, *H. Lehmann* and *H. R. Schütte*, Tetrahedron Letters, 1971, 4047). Various indole derivatives, of which tryptamine is the most notable, were found to be less

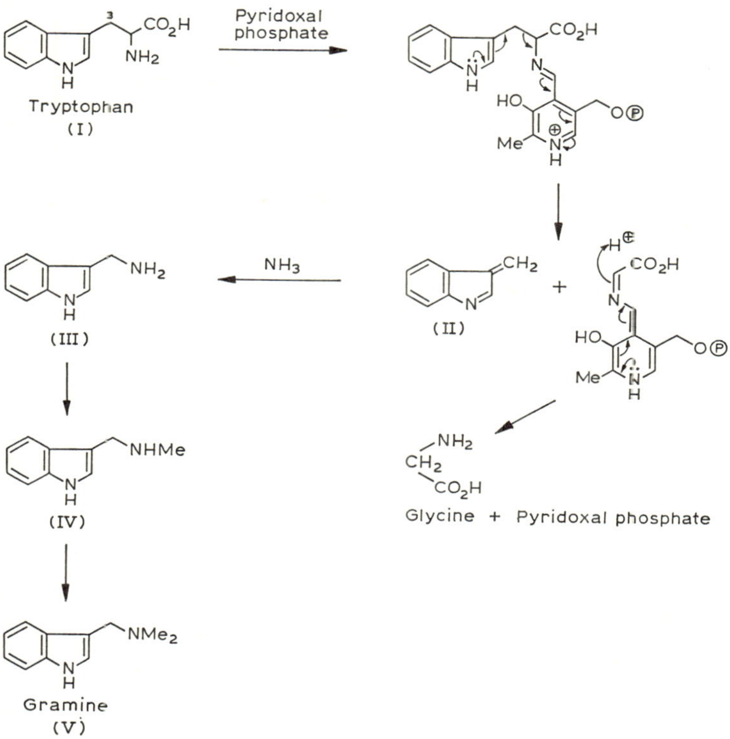

Scheme 25

effective precursors for gramine than tryptophan (*A. Breccia* and *Marion*, Canad. J. Chem., 1959, **37**, 1066; *F. Wightman, M. D. Chisholm* and *A. C. Neish*, Phytochemistry, 1961, **1**, 30). The *N*-methyl groups of gramine arise by transmethylation from *S*-adenosylmethionine (*Leete* and *Marion*, Canad. J. Chem., 1954, **32**, 646; *S. H. Mudd*, Biochim. Biophys. Acta, 1960, **37**, 164; Nature, 1961, **189**, 489). 3-Aminomethylindole (III) and 3-methylamino-methylindole (IV), putative intermediates in a methylation sequence to gramine, are both constituents of barley and generate gramine *in vivo* (*B. G. Gower* and *Leete*, J. Amer. chem. Soc., 1963, **85**, 3683), and when incubated with *S*-adenosylmethionine in the presence of barley shoot extract (*Mudd, loc. cit.*, 1961). An enzyme preparation from barley shoots also catalyses the conversion of tryptophan into gramine (*Breccia, A. M. Crespi*, and *M. A. Rampi*, Z. Naturforsch., 1966, **21b**, 1243).

These observations are consistent with a plausible mechanism for the biosynthesis of gramine which involves a pyridoxal linked adduct (see

Scheme 25) (*E. Wenkert*, J. Amer. chem. Soc., 1962, **84**, 98). The observation that the amino group of tryptophan can serve as the source of the amino group of gramine (*Gross*, *A. Nemeckova* and *Schütte*, Z. Pflanzenphysiol,, 1967, **57**, 60) is not immediately reconcileable with the above hypothesis, but experiments with ^{15}N-labelled aspartic acid, glutamic acid and glutamine have demonstrated that the nitrogen atom of gramine does not originate significantly from the general nitrogen pool. It is possible then, if the pyridoxal-mediated fragmentation of tryptophan operates, that the reaction takes place in a multi-enzyme complex within which the amino group is transferred from tryptophan *via* glycine to II.

Recently, a specific incorporation of [3-^{14}C]tryptophan into gramine has been observed in a plant [*Lupinus hartwegii* (Leguminosae)] from a different family to barley. Similar results in such different plants argues for a biosynthesis of gramine which is species-independent. In the experiments with *L. hartwegii* radioactive indole-3-carbaldehyde was isolated, and it seems that it might be a metabolite of gramine (*Leete*, Phytochemistry, 1975, **14**, 471).

(*b*) *Dimethyltryptamine derivatives*

Comparison of incorporation efficiencies of various possible intermediates between tryptophan and **psilocybin**, the mushroom hallucinogen, has suggested tentatively the following biosynthetic sequence: Tryptophan → tryptamine → *N*-methyltryptamine → *N*,*N*-dimethyltryptamine → psilocin (VI) → psilocybin (VII) (*S. Agurell*, *S. Blomkvist* and *P. Catalfomo*, Acta Pharm. Suecica, 1966, **3**, 37; *Agurell* and *J. L. G. Nilsson*, Acta Chem. Scand., 1968, **22**, 1210).

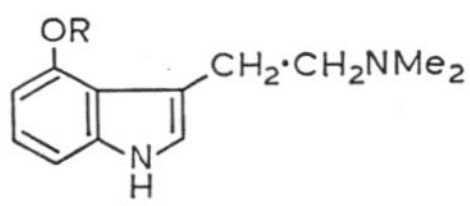

(VI) R = H
(VII) R = PO_3H_2

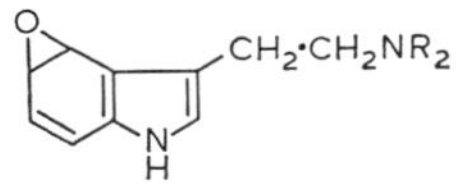

(VIII)

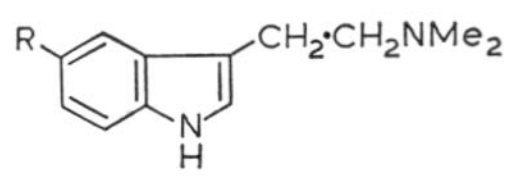

(IX) R = OMe
(X) R = H

Hydroxylation in the course of the biosynthesis of psilocybin is likely to proceed *via* an arene oxide of the type VIII. Alternative ring-opening of this oxide would give 5-hydroxytryptamine derivatives of which serotonin (5-hydroxytryptamine) is an example widely distributed in the animal kingdom. (The biosynthesis of serotonin has been studied: *A. Meister*, "Biochemistry of the Amino Acids", 2nd edn., Academic Press, New York, 1965, Vol. 2, p. 870). A further example is 5-methoxy-*N'N*-dimethyltryptamine (IX) which is a major alkaloid of *Phalaris tuberosa*, the other major alkaloid being *N'N*-dimethyltryptamine.

It might seem that the pathway to these two alkaloids would be a simple one but a detailed investigation has shown that almost all of the possible pathways from tryptophan (made up from alternative sequences of decarboxylation, hydroxylation, and *O*- and *N*-methylation) appear to operate (*C. Baxter* and *M. Slaytor*, Phytochemistry, 1972, **11**, 2767). Such results are a caution against the simplistic belief in a single biosynthetic pathway to secondary metabolites.

(*c*) *Indolmycin*

A point of particular interest in the biosynthesis of **indolmycin**, a metabolite of *Streptomyces griseus*, is the introduction of the *C*-methyl group. The skeleton itself is simply derived from tryptophan (and its precursors, anthranilic acid and indole), arginine and methionine (Scheme 26) (*H. G. Floss et al.*, J. Amer. chem. Soc., 1971, **93**, 3028).

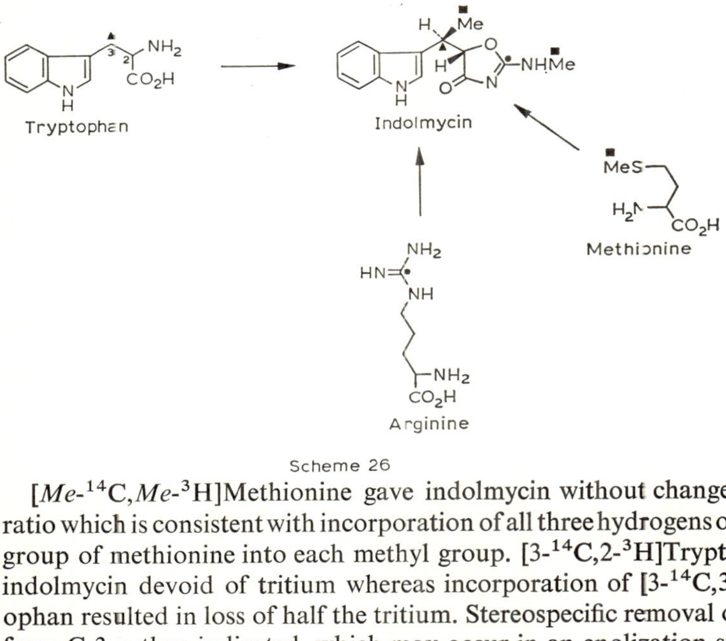

Scheme 26

[*Me*-^{14}C,*Me*-^{3}H]Methionine gave indolmycin without change in isotope ratio which is consistent with incorporation of all three hydrogens of the methyl group of methionine into each methyl group. [3-^{14}C,2-^{3}H]Tryptophan gave indolmycin devoid of tritium whereas incorporation of [3-^{14}C,3-$^{3}H_2$]tryptophan resulted in loss of half the tritium. Stereospecific removal of hydrogen from C-3 is thus indicated, which may occur in an enolization step prior to methylation of the likely substrate, indolepyruvic acid (XI), the formation of which from tryptophan would result in loss of tritium from C-2.

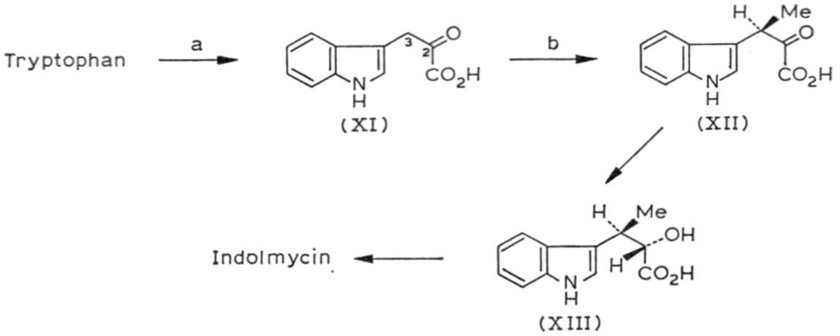

Scheme 27

Indolmycenic acid (XIII) is a co-metabolite of indolmycin and has been shown to be a precursor for it. This result and the above observations are accommodated in the biosynthetic pathway to indolmycin shown in Scheme 27. This pathway is substantiated in the first instance by the observation that the steps a and b can be carried out with cell-free extracts, and tryptophan cannot substitute for indolepyruvic acid (XI) in step b (*idem*, Biochem. Biophys. Res. Comm., 1970, **39**, 594; and unpublished results); the *in vitro* reaction b yields (*R*)-3-methylindolepyruvic acid (XII) which could be transformed into indolmycin *in vivo*; the (*S*)-isomer was not utilized (*idem*, *loc. cit.*, 1971). Furthermore, fractionation of the extracts has given an enzyme capable of effecting step a in the presence of pyridoxal phosphate, which was found to be specific for the L-isomer of tryptophan. Also obtained was a *C*-methyltransferase which catalyses the transfer of a methyl group from *S*-adenosylmethionine to the 3-position of indolepyruvic acid; no cofactors were required. Crude enzyme did not methylate L-tryptophan so although 3-methyltryptophan is an efficient precursor for indolmycin it is unlikely to be a natural intermediate. This is supported by the failure to identify this compound in *S. griseus* cultures. Since 3-methyltryptophan undergoes transamination in the presence of the first enzyme mentioned, this suggests that exogenous material enters the pathway to indolmycin at this point (*idem*, J. biol. Chem., 1975, **250**, 7819).

(*d*) *Pyrrolnitrin*

Elaboration of **pyrrolnitrin**, a metabolite of *Pseudomonas aureofaciens*, from tryptophan manifestly involves disruption of the indole nucleus of the precursor. A detailed study of the fates of the individual atoms in tryptophan reveals that no rearrangement occurs, and only the carboxyl group is lost. Experiments with ^{14}C-,^{15}N-labelled tryptophan have shown that the indole nitrogen (and C-2 of the indole ring) is incorporated intact but appreciable

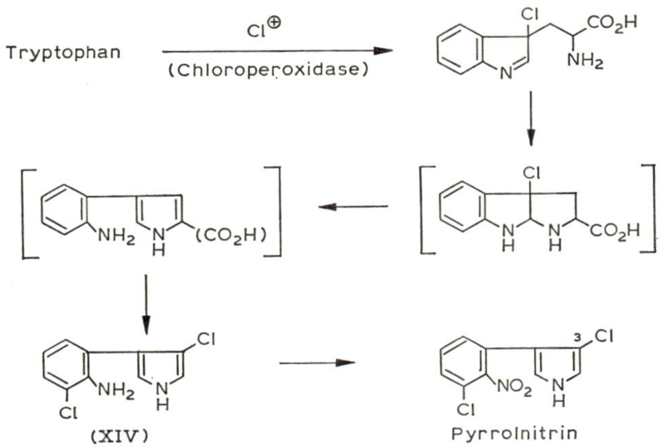

Scheme 28

loss occurs of label in the side-chain nitrogen atom (by transamination); residual label was, however, confined to the pyrrolic nitrogen of pyrrolnitrin (*Floss et al.*, Biochem. Biophys. Res. Comm., 1971, **45**, 781). D-Tryptophan rather than its enantiomer is probably the primary source for the biosynthesis of the metabolite (*D. H. Lively et al.*, Antimicrobial Agents Chemotherapy, 1966, 462; *R. Hamill et al.*, *ibid.*, 1967, 388). The ^{13}C n.m.r. spectrum of pyrrolnitrin derived from DL-[3-^{13}C]tryptophan showed an enhanced signal for C-3, to be expected if tryptophan was utilized without rearrangement (*Floss et al.*, J. Amer. chem. Soc., 1972, **94**, 8942).

The amino compound XIV has been isolated from *Ps. aureofaciens* and is efficiently converted into pyrrolnitrin (*Hamill et al.*, *loc. cit.*). This provides valuable additional information and together with the above evidence leads to the pathway illustrated in Scheme 28.

(*e*) *Sporidesmin*

It has been concluded that **sporidesmin A** originates from tryptophan, alanine and methionine in the fungus *Pithomyces chartarum* (*N. R. Towers* and *D. E. Wright*, New Zealand J. agric. Res., 1969, **12**, 275). Incorporation of tryptophan has been confirmed (*G. W. Kirby* and *M. J. Varley*, Chem. Comm., 1974, 833). Samples of (3*R*)- and (3*S*)-[3-^{3}H]-DL-tryptophan were fed in admixture with tryptophan labelled with ^{14}C in the side-chain (C-1 or C-3) or the nucleus (C-2′). The level of tritium loss was essentially independent of the position of the ^{14}C label, thus establishing an intact incorporation for the amino acid. (This is supported by the incorporation of DL-

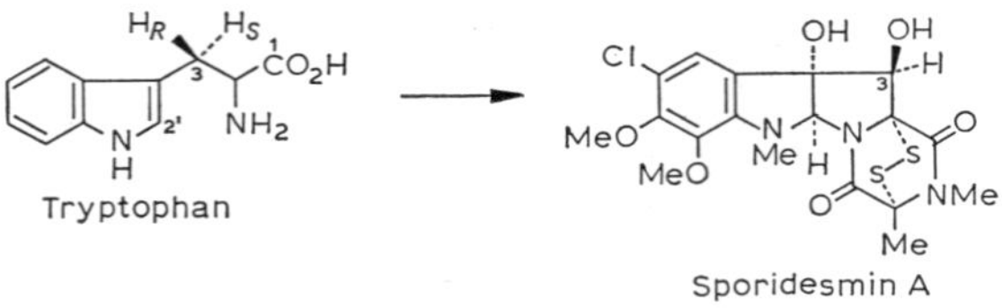

[2′-^{3}H,3-^{14}C]tryptophan into sporidesmin without change in isotope ratio.) The results, moreover, indicate that hydroxylation at the carbon atom which corresponds to C-3 of tryptophan, occurs with loss of the *pro-R* proton, *i.e.* hydroxylation with normal retention of configuration.

(*f*) *β-Carboline alkaloids*

There is an apparent structural resemblance between the *β*-carboline alkaloids, *e.g.* **eleagnine**, and the simple isoquinoline alkaloids, *e.g.* anhalonidine (see Section 6*d*, p. 400 *et seq.*). Anhalonidine (XLVI) is known to be derived from the amino acid XLVIII, formed from the corresponding phenethylamine, and pyruvic acid (see p. 402); the imine L, is implicated as the intermediate after XLVIII. It is likely that this pathway, involving intermediates of the type XLVIII, is a common one for the biosynthesis of isoquinoline alkaloids, and a similar pathway to *β*-carboline bases could thus be expected. In support harmalan, the analogue of L, was efficiently and specifically incorporated into harman in the plant, *Passiflora edulis* (*M. Slaytor* and *I. J. McFarlane*, Phytochemistry, 1968, **7**, 605). But, so was eleagnine, which may argue rather for a sequence of dehydrogenation steps, eleagnine → harmalan → harman.

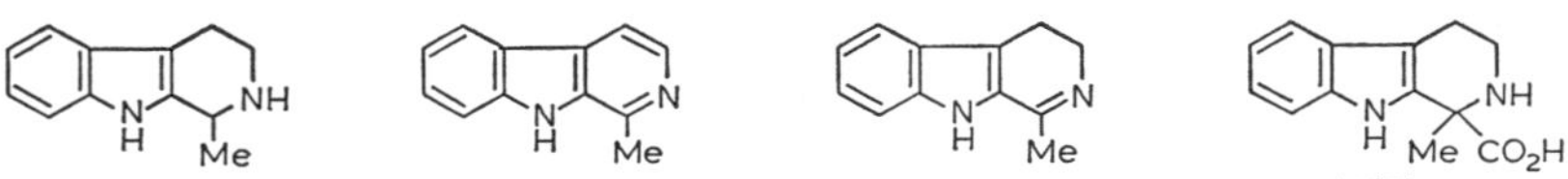

N-Acetylphenethylamine derivatives have been found not to be precursors for simple isoquinoline alkaloids (see p. 400; also *McFarlane* and *Slaytor*, Phytochemistry, 1972, **11**, 229), but [2-*acetyl*-^{14}C]-*N*-acetyltryptamine was efficiently and specifically incorporated into harman in *P. edulis*. However, *N*-acetyltryptamine was not incorporated into eleagnine in *Eleagnus angustifolia*. It has been suggested therefore that the two alkaloids may have a different genesis in the two plants, supported by the successful isolation by radioisotope dilution of *N*-acetyltryptamine from *P. edulis* and failure under similar conditions to isolate it from *E. angustifolia* (*idem, loc. cit.*). Further experimentation to delineate more clearly the biosynthesis of *β*-carboline alkaloids is required and in this connection the amino acid XV is an obvious candidate for examination as a precursor.

As to be expected, the β-carboline alkaloids, derive from tryptophan and tryptamine (*idem, loc. cit.*; *McFarlane* and *Slaytor, loc. cit.*).

(g) *Calycanthine and folicanthine*

Calycanthine is found in plants accompanied by dimeric indole alkaloids like **folicanthine**. Both of these alkaloids incorporate label from [2-^{14}C]-tryptophan, and in the case of folicanthine the label has been shown to be confined to the starred sites in accord with a simple hypothesis involving dimerization of methyltryptamine. It follows that calycanthine probably arises along a similar pathway (*D. G. O'Donovan* and *M. F. Keogh*, J. chem. Soc., 1966, 1570; *H. R. Schütte* and *B. Maier*, Arch. Pharm., 1965, **298**, 459).

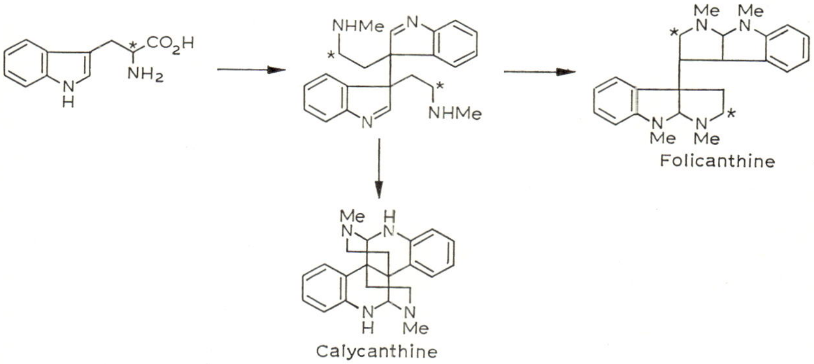

(h) *Isoprenoid indole bases*

(i) *Echinulin and brevianamide*

Echinulin, a metabolite of *Aspergillus amstelodami*, is derived from tryptophan (*J. C. MacDonald* and *G. P. Slater*, Canad. J. Microbiol., 1966, **12**, 455; *A. J. Birch* and *K. R. Farrar*, J. chem. Soc., 1963, 4277), alanine and mevalonic acid (*Birch et al., ibid.*, 1961, 3128); and the dipeptidic tryptophan derivative *cyclo*-L-alanyl-L-tryptophyl (XVI) is a more immediate precursor for echinulin than is tryptophan (*Slater*, *MacDonald* and *R. Nakashima*, Biochemistry, 1970, **9**, 2886). By nice fusion of biochemical and tracer experiments clear evidence has been obtained that XVII lies between XVI and echinulin. Firstly, a partially purified enzyme preparation from *A. amstelodami* was found to catalyse the transfer of an isoprene unit from dimethylallyl pyrophosphate to *cyclo*-L-alanyl-L-tryptophyl (XVI) to yield XVII. Secondly, doubly labelled XVII was incorporated with high efficiency into echinulin without change in isotope ratio (*C. M. Allen, Jr.*, Biochemistry, 1972, **11**, 2154; J. Amer. chem. Soc., 1973, **95**, 2386). This dipeptide is also a precursor for several other relatives of echinulin (*R. Marchelli, A. Dossena* and *G. Casnati*, Chem. Comm., 1975, 779).

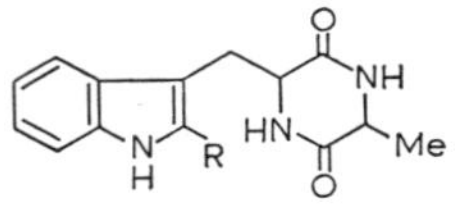

(XVI) R = H

(XVII) R = 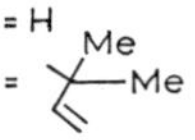

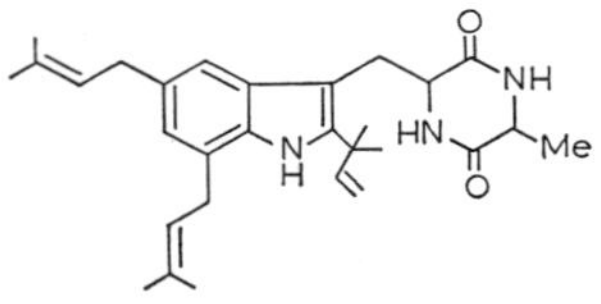

Echinulin

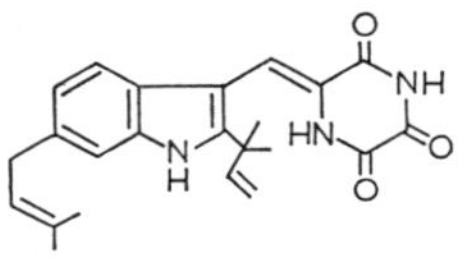

Neoechinulin

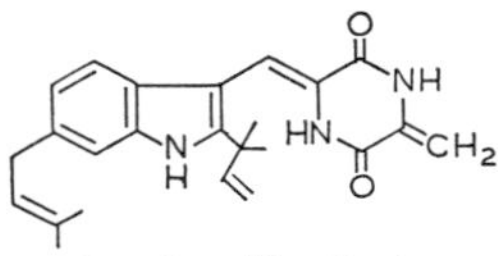

Cryptoechinulin A

The pattern of incorporation of samples of tryptophan, variously tritiated on the benzene ring, into echinulin and neoechinulin establish that prenylation of the benzene nucleus occurs with loss of hydrogen from the site of substitution but does not involve loss of hydrogen from any of the other positions (*Casnati et al.*, J. chem. Soc., Perkin I, 1974, 2397). Transformation of tryptophan into the dehydroechinulin relative, **cryptoechinulin A**, has been shown to involve stereospecific removal of the 3-*pro-S* hydrogen atom of the amino acid. From the stereochemistry of the double bond in the metabolite it follows that hydrogen loss occurs with formal *cis* stereochemistry (*C. Fuganti et al.*, Chem. Comm., 1975, 778).

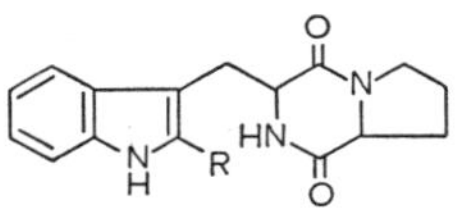

(XVIII) R = H

(XIX) R = 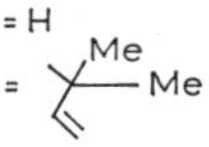

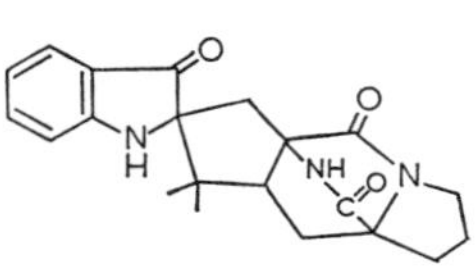

Brevianamide A

Brevianamide A, isolated from *Penicillium brevicompactum*, is structurally similar to echinulin and accordingly appears to be derived in a similar way. Radioactive mevalonate, proline and tryptophan gave labelled brevianamide A. Incorporation of *cyclo*-L-[*methylene*-^{14}C]tryptophyl-L-[5-^{3}H]-proline (as XVIII) (*cf.* XVI in echinulin biosynthesis) without change in isotope ratio indicated that this precursor was utilized intact for brevianamide A production (*J. Baldas, Birch* and *R. A. Russell*, J. chem. Soc., Perkin I, 1974, 50). As further evidence for its role in brevianamide A biosynthesis, XVIII has been isolated from *P. brevicompactum* cultures (*Birch* and *Russell*, Tetrahedron, 1972, **28**, 2999). By analogy with echinulin biosynthesis, (XIX) could lie between XVIII and brevianamide A. Its isolation (from *A. ustus*) provides support for this suggestion (*P. S. Steyn, ibid.*, 1973, **29**, 107).

(*ii*) *Cyclopiazonic acid*

α-Cyclopiazonic acid is formed in *Penicillium cyclopium* from tryptophan, mevalonate, and an acetate-derived C_4 unit [heavy bonding in XXI] (*C. W. Holzapfel* and *D. C. Wilkins*, Phytochemistry, 1971, **10**, 351) by way of β-cyclopiazonic acid (*J. C. Schabort*, International Symposium on the Control of the Human Environment, Abstract, p. 62, Johannesburg 1969; quoted in Chem. Comm., 1973, 812). The involvement of XX, isolated from a cell-free extract of *P. cyclopium*, is also likely (*R. M. McGrath, Steyn* and *N. P. Ferreira*, Chem. Comm., 1973, 812). This compound is transformed together

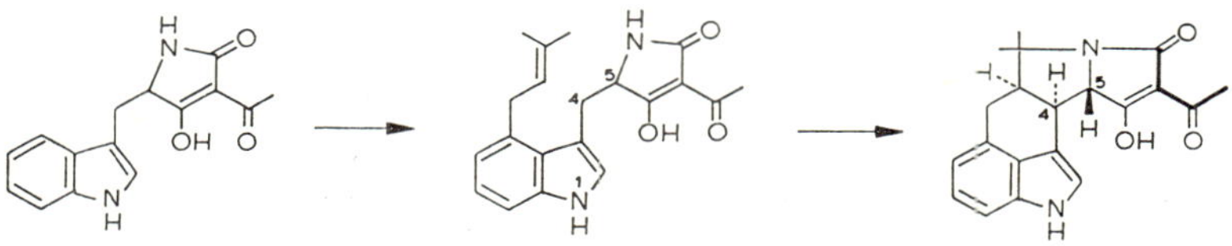

(XX) β-Cyclopiazonic acid α-Cyclopiazonic acid (XXI)

with dimethylallylpyrophosphate into β-cyclopiazonic acid. Moreover, γ,γ-dimethylallyltryptophan is not a precursor for α-cyclopiazonic acid. Dimethylallylpyrophosphate was not only a better precursor for the cyclopiazonic acids than either the parent alcohol or the monophosphate, but was also found to stimulate their production (*McGrath et al.*, Bio-org. Chem., 1976, **5**, 11).

Experiments with (3*RS*)-, (3*R*)-, and (3*S*)-[3-^{3}H]tryptophan have shown that tritium is retained as far as β-cyclopiazonic acid (at C-4) but that subsequent conversion into α-cyclopiazonic acid involves stereospecific removal of what was the 3-*pro-S* hydrogen atom in tryptophan. It follows from this, and the stereochemistry of α-cyclopiazonic acid, that C–C bond formation takes place from the opposite side of the molecule to proton removal. Further information relating to the cyclization reaction is that β-cyclopiazonic acid formed biosynthetically from [2-^{3}H]tryptophan (label at C-5 in the acid) is converted into α-cyclopiazonic acid without tritium loss and so a 1,4- but not a 4,5-dehydro-intermediate is possible (*Steyn et al.*, Chem. Comm., 1975, 465).

*(iii) Ergot alkaloids**

The ergot fungus (*Claviceps purpurea*), a parasite of rye and other grasses, is a major source of a group of indole alkaloids with a common ergoline skeleton seen, for example, in **chanoclavine-I** (XXIII) and **elymoclavine** (XXV). Not surprisingly, tryptophan serves as a specific precursor for these alkaloids (*D. Gröger et al.*, Z. Naturforsch., 1959, **14b**, 355; 1960, **15b**, 266; *W. A. Taber* and *L. C. Vining*, Chem. and Ind., London, 1959, 1218; *H. Plieninger et al.*, Ann., 1961, **642**, 214; Angew. Chem., 1959, **71**, 383; *R. M. Baxter, S. I. Kandel* and *A. Okany*, Chem. and Ind., London, 1960, 266), experiments with doubly labelled precursors allowing conclusions to be reached about the fates of atoms from the tryptophan side-chain (*H. G. Floss, U. Mothes* and *H. Günther*, Z. Naturforsch., 1964, **19b**, 784). Thus both C-2 and C-3 of the side-chain are retained as are the hydrogen atoms at C-3, but the C-2 proton which appears at C-5 of elymoclavine is only partially retained (approximately 50% with DL-precursor) or completely lost (D-tryptophan); loss of ^{15}N from labelled tryptophan paralleled those observed

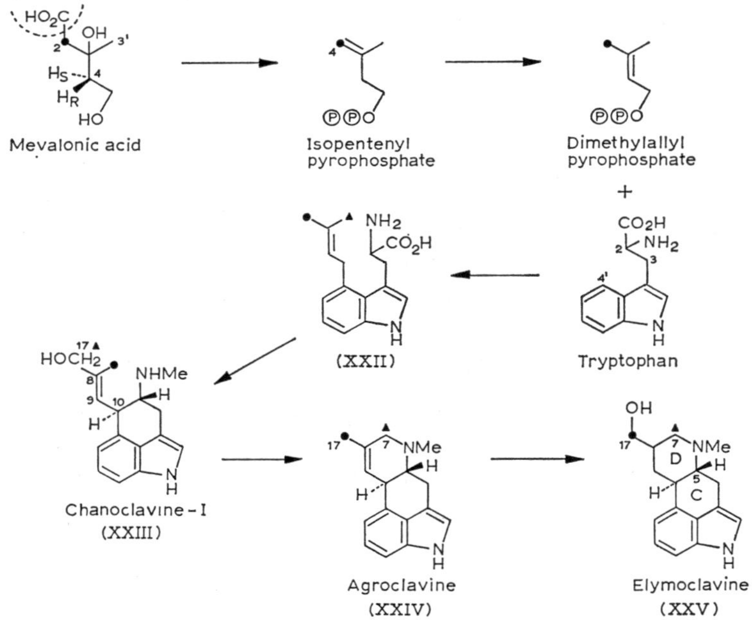

Scheme 29

* A recent comprehensive review has appeared (*H. G. Floss*, Tetrahedron, 1976, **32**, 873).

for C-2 tritium. These results indicate that it is from the L-isomer of tryptophan that the ergoline skeleton is elaborated, D-tryptophan entering the pathway presumably *via* indolepyruvic acid. Since the ergot alkaloids have the *R*-configuration at C-5 (corresponding to D-amino acid) construction of the ergoline skeleton involves an inversion of configuration at the site corresponding to C-2 of L-tryptophan. This inversion must, moreover, occur without loss of hydrogen or nitrogen from C-2.

Label from [*carboxyl*-^{14}C]tryptophan is not incorporated (*Gröger et al., loc. cit.*, 1960) but loss of the carboxyl group from tryptophan is not the next step in biosynthesis because radioactive tryptamine (and *N*-methyltryptamine) is not incorporated either (*Baxter et al.*, Chem. and Ind., London, 1961, 1453; *Floss* and *Gröger*, Z. Naturforsch., 1963, **18b**, 519). Although a significant incorporation of *N*-methyltryptophan was recorded, this was shown to be the result of demethylation to tryptophan (*idem, ibid.*, 1964, **19b**, 393). These results point to attachment of the remaining five carbons of the ergoline skeleton on to tryptophan as the next step. These carbons have been proved to originate from mevalonic acid, C-2 of this precursor becoming C-17 of the tetracyclic ergoline skeleton; C-1 of mevalonate is not incorporated (*Gröger et al., ibid.*, 1960, **15b**, 141; *Birch et al.*, Tetrahedron Letters, 1960, No. 7, p. 1; J. chem. Soc., 1962, 421; *E. H. Taylor* and *E. Ramstad*, Nature, 1960, **188**, 494; J. pharm. Sci., 1961, **50**, 681; *Baxter, Kandel*, and *Okany*, J. Amer. chem. Soc., 1962, **84**, 2997). As for example in the biosynthesis of squalene it is the 3*R*-enantiomer of mevalonic acid which serves as the precursor for Ergot alkaloids (*Floss et al.*, J. Amer. chem. Soc., 1968, **90**, 6500; Lloydia, 1972, **35**, 272; *M. Seiler, W. Acklin* and *D. Arigoni*, Chem. Comm., 1970, 1394). As in the biosynthesis of terpenes mevalonate was deduced to be incorporated *via* dimethylallylpyrophosphate and isopentenyl pyrophosphate: Both of these compounds when added to cultures diluted the incorporation of [2-^{14}C]mevalonate and labelled samples of both were incorporated into the ergoline skeleton (*H. Plieninger et al.*, Ann., 1961, **642**, 214; 1967, **706**, 223; *Baxter, Kandel* and *Okany*, Tetrahedron Letters, 1961, 596). Moreover, γ,γ-dimethylallyl pyrophosphate accumulates when alkaloid synthesis is inhibited (*J. E. Robbers* and *Floss*, Arch. Biochem. Biophys., 1968, **126**, 967; *S. Agurell* and *J.-E. Lindgren*, Tetrahedron Letters, 1968, 5127).

The foregoing is consistent with the biosynthesis of Ergot alkaloids proceeding from tryptophan by way of isoprenylation at C-4′ to give γ,γ-dimethylallyltryptophan (XXII). One difficulty with such a route is that C-4′ is not the most reactive site towards electrophilic substitution, but alternative hypotheses have not been supported by subsequent chemical and biochemical examination (see *Floss*, Tetrahedron, 1976, **32**, 886). On the other hand convincing evidence has been obtained which points to γ,γ-dimethylallyltryptophan as an intact precursor for the ergoline skeleton. This evidence includes results of experiments with double-labelled precursors, the isolation of an enzyme from *Claviceps* which catalyses the transformation of L-tryptophan and dimethylallylpyrophosphate into dimethylallyltryptophan, and the observation that this compound accumulates in cultures in which oxygen exclusion or addition of ethionine leads to inhibition of alkaloid synthesis (*Plieninger et al.*, Ann., 1964, **672**, 223; *Floss et al.*, Z. Naturforsch., 1964, **19b**, 202; Biochem. Biophys. Res. Comm., 1971, **44**, 1244; Arch. Biochem. Biophys., 1968, **126**, 967; *Agurell*, Acta Pharm. Suecica, 1966, **3**, 11; *Agurell* and *Lindgren*, *loc. cit.*).

Early experiments on later stages of biosynthesis allowed some conclusions to be reached on interconversions of Ergot alkaloids (*Agurell*, Acta Pharm. Suecica, 1966, **3**, 71; *Agurell* and *Ramstad*, Arch. Biochem. Biophys., 1962, **98**, 457; *Baxter et al.*, J. Amer. chem. Soc., 1962, **84**, 4350). The most important outcome was the establishment of the oxidative route agroclavine → elymoclavine → lysergic acid (XXVI) as a main biosynthetic sequence. Conflicting evidence was obtained, however, on the role of the tricyclic base chanoclavine, evidence which probably results from working unknowingly with different stereoisomers of the compound. Only one, chanoclavine-I, as later showed, is in fact related biosynthetically to other alkaloids, being an important precursor for the tetracyclic bases, *e.g.* elymoclavine and lysergic acid derivatives (*Floss et al.*, *ibid.*, 1968, **90**, 6500; Z. Naturforsch., 1966, **21b**, 827; *T. Fehr*, *Acklin* and *Arigoni*, Chem. Comm., 1966, 801; *R. Voigt et al.*, Pharmazie, 1967, **22**, 326). The incorporation of chanoclavine-I is in accord with the sequence: chanoclavine-I → agroclavine → elymoclavine.

Label from [2-^{14}C]mevalonate (see Scheme 29) was found to be located in the *C*-methyl group of the chanoclavines, irrespective of stereochemistry (*Fehr*, *Acklin* and *Arigoni*, *loc. cit.*). Curiously the mevalonate label appears at C-17 of elymoclavine and agroclavine. This suggests that ring-closure of chanoclavine-I involves *cis–trans* isomerization of the 8,9-double bond. Proof that this did occur was obtained when it was shown that a ^{14}C-label on the methyl group of a chanoclavine-I precursor appeared at C-17 of agroclavine and elymoclavine whereas precursor with label on the hydroxymethyl group afforded alkaloids with label at C-7 (*idem*, *loc. cit.*; *Floss et al.*, *loc. cit.*, 1968).

C-2 of mevalonate is known to give rise to C-4 of isopentenyl pyrophosphate and the normal steric course of the subsequent isomerization results in the formation of dimethylallyl pyrophosphate with C-2 of mevalonate as the *E*-methyl group. This labelling pattern is not observed in chanoclavine-I which means that either isomerization of isopentenyl pyrophosphate takes a different steric course in ergot or another *cis–trans* double-bond isomerization occurs between dimethylallyltryptophan and chanoclavine-I. Evidence supporting the latter explanation (illustrated in Scheme 29) was obtained by tracing the fates of the two diastereotopic protons at C-4 of mevalonic acid. It is known (*cf. G. Popjak* and *J. W. Cornforth*, Biochem. J., 1966, **101**, 553) that in both the isopentenyl pyrophosphate isomerase and prenyl transferase reactions formation of a *trans* double bond correlates with loss of the 4-*pro-S* proton and retention of the 4-*pro-R* hydrogen atom, the reverse being true for formation of a *cis* double bond. The observation that the 4-*pro-S* hydrogen atom of mevalonate, rather than the 4-*pro-R* one, is lost on formation of chanoclavine-I with "*cis*" double-bond stereochemistry is in accord with normal formation of dimethylallyl pyrophosphate and subsequent double-bond isomerization of dimethylallyltryptophan (*Floss et al., loc. cit.*, 1959). Definitive evidence comes from feeding dimethylallyltryptophan labelled as shown in (XXII; ▲ = ^{14}C). The chanoclavine-I, agroclavine and elymoclavine isolated were shown to be labelled (with ▲) as indicated in Scheme 29 (*Arigoni et al.*, Chimia, 1975, **29**, 526). Conversion of dimethylallyltryptophan through chanoclavine-I into elymoclavine and agroclavine involves then two double-bond isomerizations which it is logical to associate with the actual ring-closure reactions.

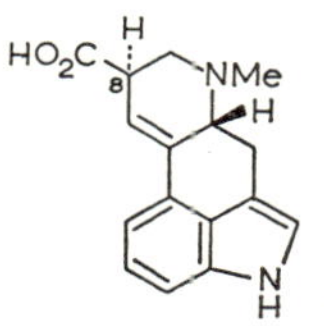

Lysergic acid
(XXVI)

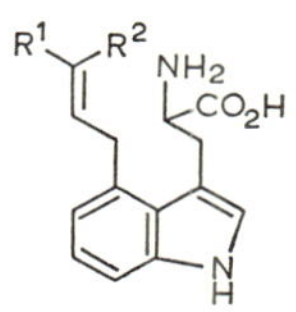

(XXVII) $R^1 = Me$, $R^2 = CH_2OH$
(XXVIII) $R^1 = CH_2OH$, $R^2 = Me$

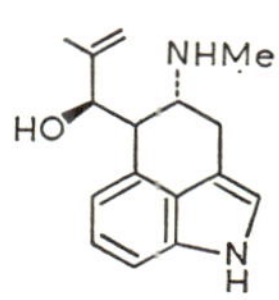

(XXIX)

In connection with these reactions the possibility that allylic alcohols might be intermediates in biosynthesis has been examined with labelled samples of XXVII, XXVIII, and the naturally occurring alcohol, paliclavine (XXIX). The results indicate that none of them is normally involved in biosynthesis (*Arigoni et al., loc. cit.*; *Ackлin, Fehr* and *P. A. Stadler*, Helv., 1975, **58**, 2492). On another tack, clavicipitic acid (XXX), which can under some conditions be a major product of dimethylallyltryptophan metabolism, was examined as a precursor for elymoclavine. The results indicate that in all probability it is not (*R. S. Bajwa et al.*, Phytochemistry, 1975, **14**, 735). The functionalization of C-1 of the isoprenoid residue apparent in both paliclavine and clavicipitic acid, however, is in accord with such a reaction being associated with an early stage of biosynthesis, these metabolites being byproducts. Entry of the hydroxyl group present on the side-chain of chanoclavine-I seems to occur before formation of this base since labelled desoxychanoclavine-I, its *N*-nor derivative, and its *N*-methyl derivative were not incorporated into agroclavine, elymoclavine or paspalic acid (XXXI) (*Arigoni*, paper presented at Symposium on Organic Chemical Approaches to Biosynthesis, London, 1965; *Fehr*, Dissertation No. 3967, ETH Zürich, 1967) but other evidence on allylic alcohols discussed above seems to be in conflict.

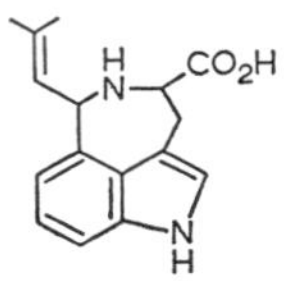

(XXX)

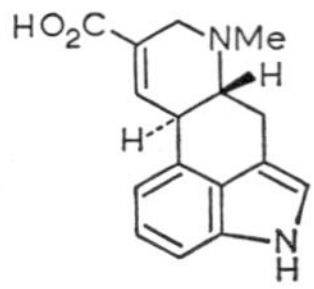

(XXXI)

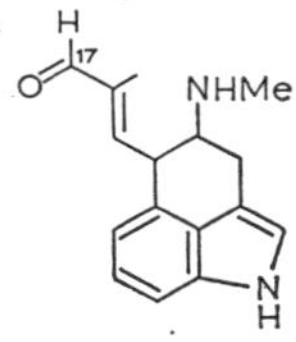

(XXXII)

The mechanism of formation of the ergoline ring C (first ring-closure and double-bond isomerization) remains unknown. Formation of ring D, however, can be accounted for satisfactorily. In the conversion of chanoclavine-I (XXIII) into elymoclavine (XXV) one of the C-17 hydrogen atoms is lost; the other appears as expected at C-7 (*Floss et al.*, J. Amer. chem. Soc., 1974, **96**, 1898). This suggested that the aldehyde XXXII might be an intermediate. Strong support for this came with the efficient incorporation of labelled chanoclavine-I-aldehyde (XXXII) into elymoclavine; tritium label from C-17 in the precursor appeared at C-7 in the product. Moreover, two deuterium atoms at C-3′ of a mevalonate sample used as a precursor were retained in the derived chanoclavine-I (C-3′ of mevalonate becomes C-17 of chanoclavine-I) and only one of these was retained in the formation of elymoclavine. Changes in oxidation level which occur at this centre in the course of biosynthesis appear to be the result of normal physiological processes since proton addition is stereospecific — the deuterium in the elymoclavine was largely equatorial. An enzyme system isolated from *Claviceps* which catalyses the transformation of chanoclavine-I (but not isochanoclavine-I) into agroclavine, was also able to transform chanoclavine-I-aldehyde (*D. Erge*, *W. Maier* and *D. Gröger*, Biochem. Physiol. Pflanzen, 1973, **164**, 234).

In considering the transformation of chanoclavine-I into elymoclavine and the attendant isomerization of the allyl double bond knowledge of the fates of the hydrogen atoms at C-9 and C-10 of XXIII is important. That at C-10 has been found to be retained completely whereas only incomplete ($\simeq$ 70%) retention of the C-9 proton was observed (*S. Agurell*, Acta Pharm. Suecica, 1966, **3**, 71; *Floss et al.*, J. Amer. Chem. Soc., 1968, **90**, 6500). A 1:1 mixture of [2-^{13}C]- and [4-$^{2}H_2$]-mevalonate afforded chanoclavine-I which contained only singly labelled species (^{13}C or ^{2}H) whereas elymoclavine showed an appreciable percentage of double labelled molecules *i.e.* molecules which contained both ^{13}C *and* deuterium. Analogous results were obtained with [3′, 4-$^{2}H_5$]mevalonate (*Floss et al., ibid.*, 1974, **96**, 1898; *Floss*, Lloydia, 1972, **35**, 399). These observations are nicely accommodated in a mechanism of ring closure in which *inter*molecular hydrogen transfer occurs at C-9 on enzyme-bound chanoclavine-I-aldehyde (see Scheme 30). After addition of some X–H group across the double bond, rotation about the 8,9-bond allows transfer of the original hydrogen at C-9 to the enzyme in the subsequent

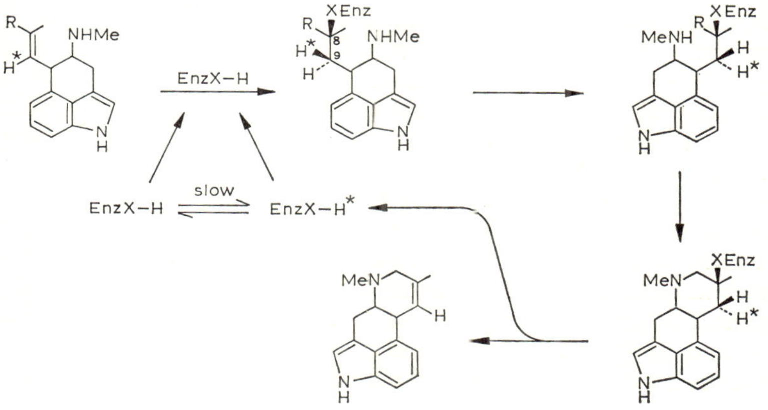

Scheme 30

elimination step. Assuming slow exchange of the hydrogen attached to the enzyme, most of it would be transferred to another molecule of chanoclavine-I-aldehyde in the next cycle. (This accounts not only for the formation of double-labelled molecules above but also the incomplete retention of the C-9 proton). In order to avoid anti-Michael addition to the $\alpha\beta$-unsaturated aldehyde it is necessary that the aldehyde function is masked in some way.

Incorporation of [2-^{14}C,4-^{3}H]mevalonate into chanoclavine-I, in contrast to incorporation into other alkaloids, *e.g.* elymoclavine, results in the isolation of material which is appreciably enriched in tritium (*Floss et al., loc. cit.*; *Arigoni et al.*, unpublished results). This observation can be explained in terms of an unexceptional primary isotope effect which favours the consumption of chanoclavine-I with a hydrogen atom at C-9 (this material will include ^{14}C-labelled species) and can be explained in terms of the mechanism in Scheme 30 provided the steps which lead up to the elimination are reversible.

The origin of the *N*-methyl group present in ergot bases is unexceptionally methionine; formate was also incorporated (*R. M. Baxter et al.*, Canad. J. Chem., 1964, **42**, 2936). However, the point at which methylation occurs is as yet unknown, although it is apparent that the reaction occurs after isoprenylation of tryptophan (see above). Neither tritiated *N*-demethylchanoclavine-I or -II were incorporated into tetracyclic ergolines (*J. M. Cassady, C. I. Abou-Chaar* and *Floss*, Lloydia, 1973, **36**, 390). Nor was *N*-noragroclavine methylated *in vivo*. It may be that *N*-methylation is associated with the first ring-closure which forms chanoclavine-I.

The **rugulovasines**, *e.g.* rugulovasine, are elaborated by *Penicillium concavorugulovasum*. As expected from their close structural resemblance to the ergolines, rugulovasine has its genesis in tryptophan and mevalonic acid (in the biosynthetic experiments chanoclavines-I and -II were also labelled) (*M. Abe*, "Features and Perspectives of Production of Indole Alkaloids by Fungi other than *Claviceps* Species", paper presented at the first International Symposium on Genetics of Industrial Micro-organisms, Prague, 1970). Moreover, tritiated dimethylallyltryptophan (XXII) also afforded radioactive samples of all three alkaloids, but chanoclavine-I (XXIII) was not converted into either rugulovasine or chanoclavine-II. On the other hand, rugulovasine was converted into dihydrorugulovasine and dihydrorugulovasine lactam, and dihydrorugulovasine was transformed into the lactam. These observations suggest the pathway outlined in Scheme 31.

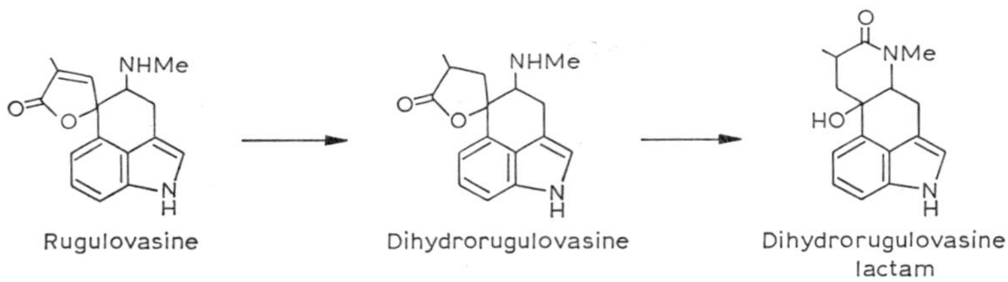

Scheme 31

A most interesting discovery was made when it was found that ergoline alkaloids were produced not only by fungi, but also by higher plants in the Convolvulaceae family (*A. Hofmann* and *H. Tscherter*, Experientia, 1960, **16**, 414). Of further interest is the observation that tryptophan and mevalonic acid are alkaloid precursors here too (*Gröger et al.*, Z. Naturforsch., 1963, **18b**, 1123).

Elymoclavine has been found to be the precursor of an important group of Ergot bases, the derivatives of **lysergic acid** (XXVI) (*K. Mothes et al.*, Tetrahedron Letters, 1962, 933; *Gröger, H. R. Schütte* and *K. Stolle*, Z. Naturforsch., 1963, **18b**, 850; *E. Eich*, Planta Med., 1973, **23**, 330; *Floss et al.*,

J. org. Chem., 1973, **38**, 2249; and refs. cited). The details of the formation of lysergic acid have not been fully elucidated but it has been shown that penniclavine (XXXIII) and lysergol (XXXIV) are not involved. The observation with the latter compound excludes migration of the double bond prior to oxidation of the hydroxymethyl group (*Agurell* and *M. Johansson*, Acta Chem. Scand., 1964, **18**, 2285; *Floss et al.*, Z. Naturforsch., 1966, **21b**, 128).

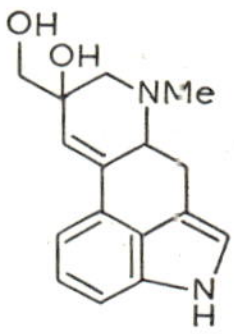

(XXXIII)

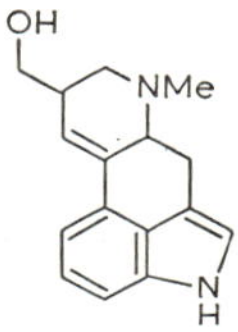

(XXXIV)

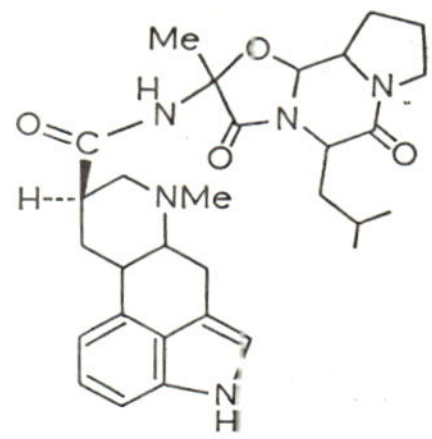

(XXXV)

Instead, since paspalic acid (XXXI), the $\Delta^{3,9}$-isomer of lysergic acid, is very efficiently converted into lysergic acid amides, it is most probable that the shift of the double bond occurs at the oxidation stage of the acid, although it may occur at the aldehyde oxidation level (*T. Ohashi, S. Aoki* and *M. Abe*, Agric. Biol. Chem., Japan, 1970, **34**, A37; *Floss et al., loc. cit.*, 1973; *Agurell*, Acta Pharm. Suecica, 1966, **3**, 65). Conversion of lysergic acid into its amide derivatives involves no further shift of the double bond since [8-^{3}H]lysergic acid is incorporated without tritium loss (*idem, loc. cit.*). In *Sphacelia sorghi*, by contrast, it appears that the dihydrolysergyl derivative XXXV is formed from dihydroclavine alkaloids, saturation of the double bond occurring before dihydroelymoclavine, possibly at the chanoclavine stage (*K. D. Barrow et al.*, Tetrahedron Letters, 1974, 1557).

The simplest amide derivatives of lysergic acid are exemplified by lysergic acid α-hydroxyethylamide (XXXVI) and ergonovine (= ergometrine) (XXXVII). In spite of this simplicity the biosynthesis of these compounds is as yet incompletely understood. The nitrogen atom and all the carbon atoms of alanine except C-1 are used for the elaboration of the α-hydroxyethylamide moiety of the former compound [lysergic acid amide, isolysergic acid amide (C-8 epimer), ethylamine and acetamide are not utilized for biosynthesis] (*Agurell*, Acta Pharm. Suecica, 1966, **3**, 33, 71; *E. Kleinerová* and *J. Kybal*, Folia Microbiol., 1973, **18**, 390; *Gröger, D. Erge* and *Floss*, Z. Naturforsch., 1968, **23b**, 177; *A. Minghetti* and *F. Arcamone*, Experientia, 1969, **25**, 926; *N. Castagnoli, Jr. et al.*, Biochem. J., 1970, **117**, 451). Alanine is also the progenitor of the side-chain of ergonovine (α-methylserine is not incorporated) (*Minghetti* and *Arcamone, loc. cit.*; *U. Nelson* and *Agurell*, Acta Chem. Scand., 1969, **23**, 3393). Alaninol is a potential intermediate here but

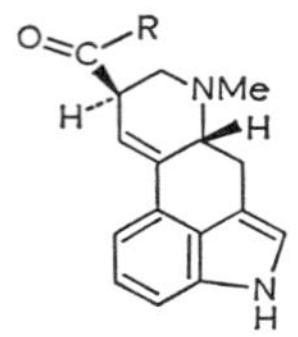

(XXXVI) R = —NH—CH(Me)—OH

Ergonovine, (XXXVII) R = —NH—C(Me)(H)—CH_2OH

(XXXVIII) R = —NH—C(Me)(H)—CO_2H

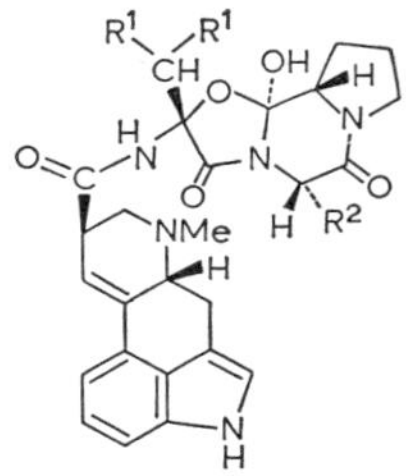

Ergotamine, (XXXIX) R^1 = H, R^2 = CH_2Ph

Ergocornine, (XL) R^1 = Me, R^2 = $CHMe_2$

Ergocryptine, (XLI) R^1 = Me, R^2 = $CH_2{\cdot}CHMe_2$

whilst one group of workers recorded a higher incorporation of L-alaninol compared with L-alanine, two others failed to observe an incorporation (*Minghetti* and *Arcamone, loc. cit.*; *Nelson* and *Agurell, loc. cit.*; *J. Majer, Kybal* and *J. Komersová*, Folia Microbiol., 1967, **12**, 489). Lysergylalanine (XXXVIII) is an attractive possibility (*Agurell*, Acta Pharm. Suecica, 1966, **3**, 71) as an intermediate in the biosynthesis of ergonovine as well as of some peptidic ergot bases. However, only an unsatisfactorily low incorporation into ergonovine, and also lysergic acid α-hydroxyethylamide (XXXVI), was recorded (*Floss et al.*, Chem. Comm., 1969, 418; Lloydia, 1971, **34**, 449), and it has been found to be a non-intact precursor for the peptidic alkaloid ergotamine (see below).

As expected the peptide moieties of peptidic ergot alkaloids are formed from lysergic acid (*Minghetti* and *Arcamone, loc. cit.*; *W. Maier, Erge* and *Gröger*, Biochem. Physiol. Pflanzen, 1971, **162**, 559) and the appropriate amino acids: phenylalanine (*L. C. Vining* and *W. A. Taber*, Canad. J. Microbiol., 1963, **9**, 291; *R. A. Bassett, E. B. Chain* and *K. Corbett*, Biochem. J., 1973, **134**, 1), proline (*idem, loc. cit.*; *Gröger* and *Erge*, Z. Naturforsch., 1970, **25b**, 196), leucine (*Maier, Erge* and *Gröger, loc. cit.*). and valine (*idem, loc. cit.*; *Floss et al.*, Lloydia, 1971, **34**, 446).

The α-hydroxy-amino residues of ergotamine (XXXIX), ergocornine (XL), and ergocryptine (XLI) likewise have their genesis in the appropriate α-amino acids: valine (*Maier, Erge* and *Gröger, loc. cit.*; *Floss et al., loc. cit.*; *G. P. Basmadjian* and *Floss*, J. pharm. Sci., 1971, **60**, 949) and alanine (*Bassett, Chain* and *Corbett, loc. cit.*; *Floss et al.*, Lloydia, 1971, **34**, 442; *Minghetti* and *Arcamone, loc. cit.*; *Majer, Kybal* and *Komersová, loc. cit.*). As in the case of the simple lysergyl amides the ways in which the various units are assembled is not clear. Thus lysergylalanine (XXXVIII) and ergonovine (XXXVII) are not effective precursors for ergotamine (*Floss et al., loc. cit.*; *G. P. Vicario et al.*, J. labelled Compounds, 1967, **3**, 492) and although more satisfactory incorporations of lysergylvaline into ergotoxine (a 3:1 mixture of ergocornine and ergocryptine) were recorded, it appears that incorporation occurs only after degradation to the constituent fragments (*Maier*, *Erge* and *Gröger*, Biochem. Physiol. Pflanzen, 1974, **165**, 479; *Floss et al.*, Lloydia, 1971, **34**, 446).

An alternative to the above (which suggests that amino acids are added stepwise to lysergic acid) is that cyclic di- or tri-peptides combine with lysergic acid to give the peptidic alkaloids. Certainly the importance of cyclic peptides in the biosynthesis of many microbial metabolites is increasingly

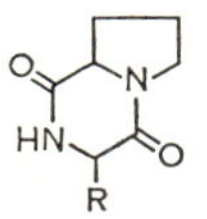

(XLII) R = $CHMe_2$

(XLIII) R = CH_2Ph

being realised. In support such peptides have been isolated from ergot cultures, in particular L-leucyl-D-proline lactam (XLII) and L-phenylalanyl-D- and -L-proline-lactam (XLIII). Their appearance in relation to alkaloid production suggested that they had a role in the biosynthesis of peptide ergot alkaloids. Moreover, tritiated samples of L-leucyl-D- and -L-proline-lactam (as XLII) were satisfactorily incorporated into ergocryptine/ergocryptinine (C-8 epimers) and of L-phenylalanyl-D- and -L-proline-lactam (as XLIII) into ergotamine/ergotaminine (C-8 epimers) (*M. Abe* et al., Agric. and Biol. Chem., Japan, 1971, **35**, A1, A6). However, the almost equal incorporations of D- and L-proline lactams into the alkaloids (which have the L-configuration) suggests that the precursors may have been hydrolysed prior to incorporation. Such a conclusion is supported by other work with di- and tri-peptides which has shown that incorporation was only with prior fragmentation into constituent amino acids (*Gröger* and *S. Johne*, Experientia, 1972, **28**, 241; *Floss et al., ibid.*, 1974, **30**, 1369). With the evidence pointing away from the involvement of these peptidic fragments it has been suggested that assembly of the constituents occurs on a multi-enzyme complex without the formation of free intermediates (*Floss et al., loc. cit.*).

Significant progress is being made with the isolation and study of the **enzymes** responsible for the various steps in ergot biosynthesis. Such information is invaluable for a proper understanding of the processes involved, but can only be alluded to briefly here. Cell-free systems have been obtained which are capable of the *de novo* synthesis of ergolines from simple precursors (*F. L. Cavender* and *J. A. Anderson*, Biochim. Biophys. Acta, 1970, **208**, 345; *T. Ohashi*, *N. Shibuya* and *Abe*, Agric. and biol. Chem., Japan, 1972, **36**, A16) and such systems have been used for the study of alkaloid interconversions (*Abe et al.*, *ibid.*, 1970, **34**, A36, A37; 1971, **35**, A6; 1972, **36**, A39).

The enzyme responsible for the first step in ergot biosynthesis, dimethylallyltryptophan synthetase, has been isolated (*Floss et al.*, Biochem. Biophys. Res. Comm., 1971, **44**, 1244) as have enzyme systems which catalyse the cyclisation of chanoclavine-I to give tetracyclic ergolines. Attempts to purify the latter enzyme or enzyme systems has only met with limited success (*E. Ramstad et al.*, Chem. Comm., 1970, 775; *Erge*, *Maier* and *Gröger*, Biochem. Physiol. Pflanzen, 1973, **164**, 234; *Gröger* and *P. Sajdl*, Pharmazie, 1972, **27**, 188). An ammonium sulphate fraction of a cell-free *Claviceps* preparation has been obtained which catalyses the hydroxylation of agroclavine to elymoclavine, the enzyme responsible behaving as a mixed-function oxidase (*J. C. Hsu* and *Anderson*, Chem. Comm., 1970, 1318; Biochim. Biophys. Acta, 1971, **230**, 518).

(*iv*) *Terpenoid indole alkaloids**

Corynanthe-Strychnos, Aspidosperma and Iboga alkaloids. The rich structural variation within the family of terpenoid indole alkaloids, *e.g.*

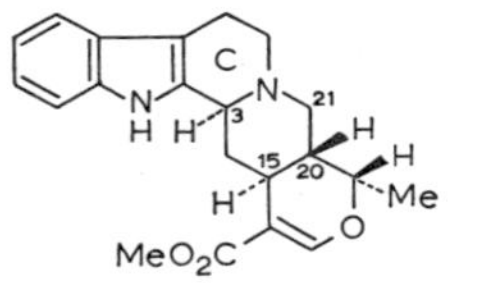

Ajmalicine
(XLIV)

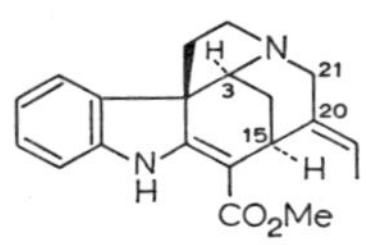

Akuammicine
(XLV)

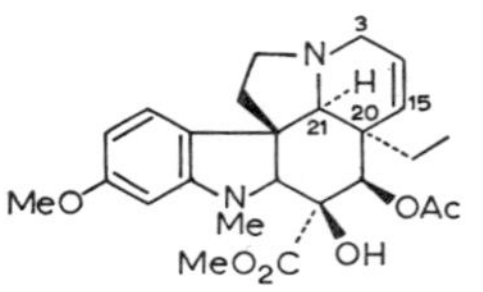

Vindoline
(XLVI)

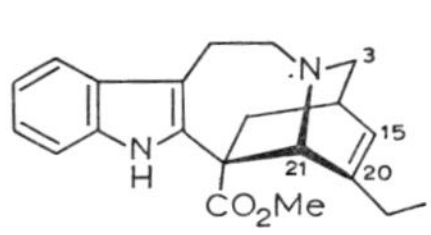

Catharanthine
(XLVII)

* A detailed and authoritative review of work carried out up to June 1970 has been written by Professor *A. R. Battersby*. It appears in "The Alkaloids" (Specialist Periodical Reports) ed. *J. E. Saxton*, The Chemical Society, London, Vol. 1, 1971. Details of more recent work will be found in subsequent volumes of the same series. For a recent comprehensive review see *G. A. Cordell*, Lloydia, 1974, **37**, 219.

XLIV–XLVII brings the total number to over 1000. The tryptamine moiety, with additional C_9–C_{10} unit, serves to identify the group and where tested this moiety has been found to be derived from tryptophan and tryptamine (*E. Leete*, J. Amer. chem. Soc., 1960, **82**, 6338; Tetrahedron, 1961, **14**, 35; *Leete et al.*, J. Amer. chem. Soc., 1965, **87**, 4168; Tetrahedron Letters, 1964, 1499; *D. Gröger et al.*, Arch. Pharm. 1967, **300**, 393; Tetrahedron Letters, 1964, 2579; *A. R. Battersby, A. R. Burnett*, and *P. G. Parsons*, J. chem. Soc., C, 1969, 1193; *J. P. Kutney et al.*, J. Amer. chem. Soc., 1968, **90**, 3566). The different arrangements of the remaining nine or ten skeletal carbon atoms accounts for the large number of alkaloids in the group. Although at first sight the structural variation might seem bewildering, closer inspection reveals three main groups of bases: (a) *Corynanthé–Strychnos* type, *e.g.* ajmalicine and akuammicine, with C_9–C_{10} unit, L,; (b) *Aspidosperma* type, *e.g.* vindoline, with C_9–C_{10} unit, LI; (c) *Iboga* type, *e.g.* catharanthine, with a third C_9–C_{10} variation, XLIX; where only nine carbon atoms are present in the non-tryptamine residue it is invariably the carbon indicated by the dotted line which is lost.

Early speculation on the genesis of the ubiquitous C_9–C_{10} units which implicated, on the one hand the products of the shikimic acid pathway up to and including aromatic amino-acids, and acetate* on the other have not stood the experimental test (for reviews see *I. D. Spenser*, in "Comprehensive Biochemistry", eds. *M. Florkin* and *E. H. Stotz*, Elsevier, Amsterdam, 1968, Vol. 20, p. 340; *D. Gröger*, in "Biosynthese der Alkaloide", eds. *K. Mothes* and *H. R. Schütte*, VEB Deutscher Verlag der Wissenschaften, Berlin, 1969, p. 459; *A. R. Battersby*, Pure and applied Chem., London, 1967, **14**, 117). The conception that these units are monoterpenoid in origin (*R. Thomas*, Tetrahedron Letters, 1961, 544; *E. Wenkert*, J. Amer. chem. Soc., 1962, **84**, 98) has however survived to become established fact. It was correctly envisaged that each C_9–C_{10} unit could be derived by scission of a cyclopentane ring of a monoterpene with further modification as indicated (Scheme 32). The crucial experiment was the examination of mevalonate as a specific precursor.

---

* Acetate can be incorporated into many terpenoid compounds *via* mevalonate (*D. V. Banthorpe, B. V. Charlwood* and *M. J. O. Francis*, Chem. Reviews, 1972, **72**, 115; *J. W. Cornforth*, Chem. in Britain, 1968, **4**, 102) but it is not a specific precursor for the indole alkaloids in spite of their partial mevalonoid origins (*E. Leete*, Adv. Enzymol., 1969, **32**, 373). A search for a source for mevalonate other than acetate has led to the accumulation of some controversial evidence in favour of glycine (*A. K. Garg* and *J. R. Gear*, Phytochemistry, 1972, **11**, 689 and refs. cited; *D. Gröger, W. Maier* and *P. Simchen*, Experientia, 1970, **26**, 820; *J. P. Kutney et al.*, J. Amer. chem. Soc., 1970, **92**, 2174; *Maier* and *Gröger*, Arch. Pharm., 1971, **304**, 351).

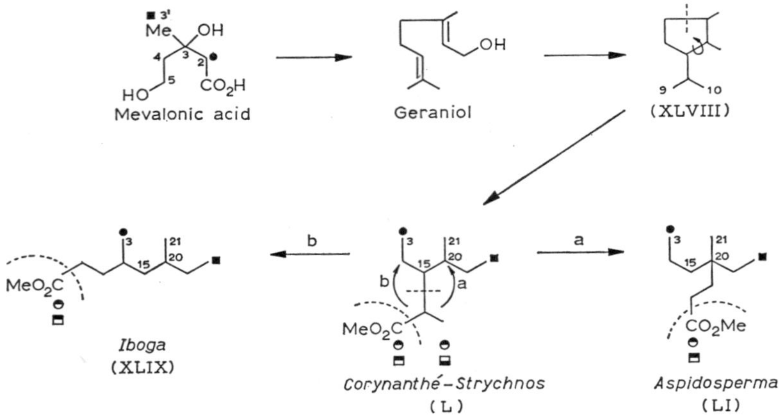

Scheme 32

In the first experiments no incorporation of mevalonate could be observed. Satisfactory incorporations were recorded later, however, with different plants and under different conditions which led to better use of exogenous mevalonate for alkaloid synthesis. An impressive number of experiments with various ^{14}C-labelled mevalonates was carried out and the results quite clearly demonstrate that the C_9–C_{10} unit of representatives of the *Corynanthé–Strychnos*, *Aspidosperma*, and *Iboga* types were each derived along a pathway involving the head-to-tail linkage of two mevalonate units passing through a cyclic monoterpenoid of type XLVIII (*Battersby et al.*, Chem. Comm., 1966, 46, 888; *A. I. Scott et al.*, J. Amer. chem. Soc., 1968, **90**, 4144; *H. Goeggel* and *D. Arigoni*, Chem. Comm., 1965, 538; *Gröger et al.*, Arch. Pharm., 1967, **300**, 393). Moreover, it is (3-*R*)-mevalonate not the (3-*S*)-isomer which is involved, consistent with observations in other systems, *e.g.* Ergot alkaloids (see above) (*Battersby et al.*, Chem. Comm., 1968, 951). In the incorporation of mevalonate loss of identity of the two terminal positions (C-9 and -10) of the stylized intermediate XLVIII was apparent: [2-^{14}C]-mevalonate yielded reserpinine, vindoline, ajmalicine, and catharanthine with approximately 25% of the activity in the respective ester groups, rather than 0 or 50% (refs. above). This was confirmed by obtaining a similar result with [3′-^{14}C]mevalonate incorporation into the latter three alkaloids (*cf.* Scheme 32) (*S. Escher*, *P. Loew* and *Arigoni*, *ibid.*, 1970, 823). A similar result had been observed for the biosynthesis of the cyclopentane monoterpene plummieride (*D. A. Yeowell* and *H. Schmid*, Experientia, 1964, **20**, 250). Appropriate to the monoterpene origins of the C_9–C_{10} unit, geraniol (and also its stereoisomer nerol) was found to be a specific precursor for all three types of alkaloid (*Battersby et al.*, Chem. Comm., 1966, 346 888; *Scott et al., ibid.*, 1966, 348; *Leete* and *S. Ueda*, Tetrahedron Letters, 1966, 4915; *Loew, H. Goeggel* and *Arigoni*, Chem. Comm., 1966, 347).

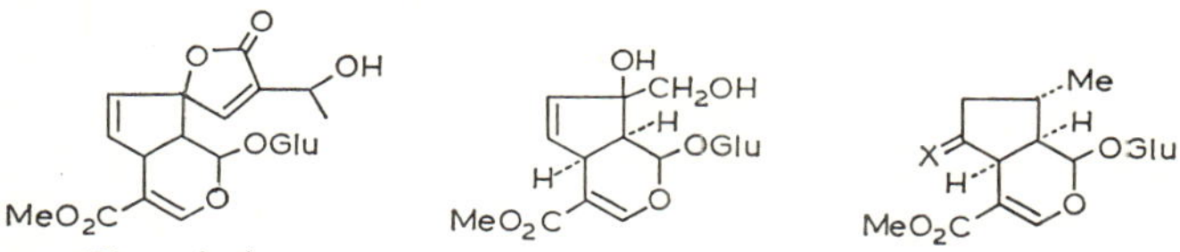

Plummieride (LII) (LIII) X = O (LIV) X = H, OH

Loganin (LV) which occurs alongside indole alkaloids in *Strychnos* species (*A. J. Birch* and *J. Grimshaw*, J. chem. Soc., 1961, 1407; *K. Sheth, E. Ramstad* and *J. Wolinsky*, Tetrahedron Letters, 1961, 394 and refs. cited) has been shown to be the crucial cyclopentane monoterpene intermediate XLVIII in the biosynthesis of all three groups of terpenoid indole alkaloids. Labelled loganin was specifically incorporated (see Scheme 33) whereas the monoterpenes LII, LIII and LIV were not precursors (*Battersby et al.*, Chem. Comm., 1966, 890; 1968, 133; *Loew* and *Arigoni, ibid.*, 1968, 137). Moreover, loganin is a natural constituent of *Vinca rosea*, the plant used for most of the above experiments (*Battersby et al.*, Chem. Comm., 1966, 890; *Loew* and *Arigoni, loc. cit.*) and is itself derived specifically from geraniol (see Scheme

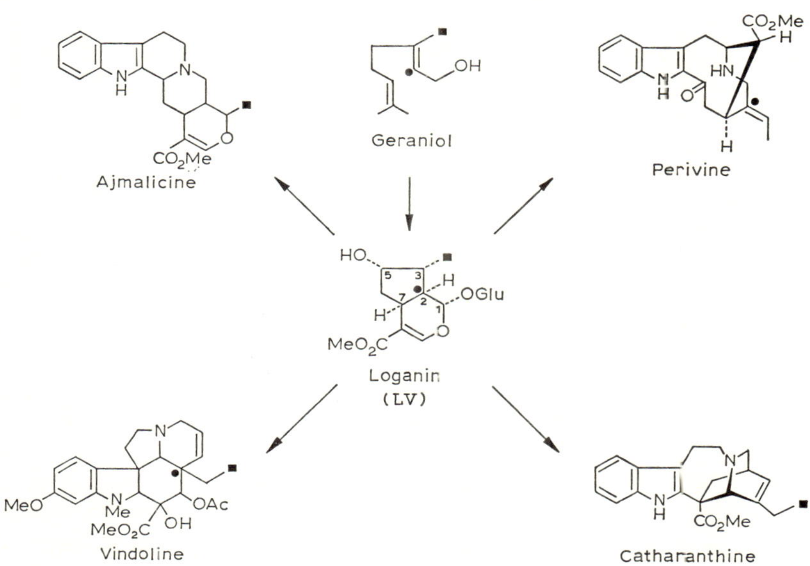

Scheme 33

33) in *V. rosea* (*idem, loc. cit.*; *Battersby*, *E. S. Hall* and *R. Southgate*, J. chem. Soc., C, 1969, 721). The likelihood that deoxyloganin (LIX) is an intermediate between geraniol and loganin was confirmed when it was shown that [4-^{3}H]deoxyloganic acid (as LVIII) was incorporated specifically into loganin in *Lonicera japonica*, and that [O-*methyl*-^{3}H]deoxyloganin (as LIX) was a specific precursor in *V. rosea* for the indole alkaloids included in Scheme 33, and serpentine (XLIV, ring C aromatized) as well as loganin; deoxyloganin is moreover a natural constituent of *V. rosea* (*Battersby*, *A. R. Burnett* and *P. G. Parsons*, Chem. Comm., 1970, 826).

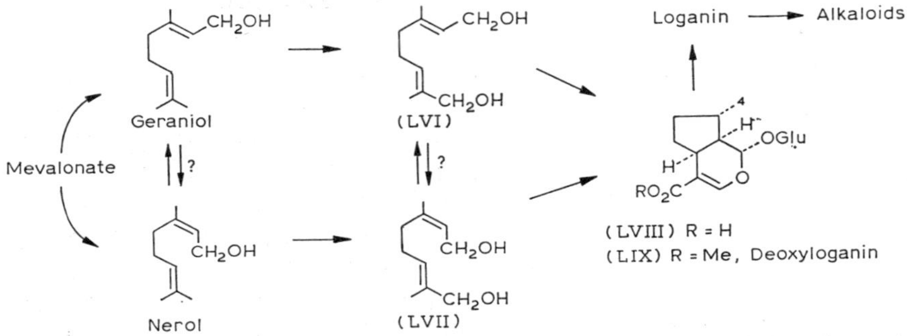

The C-10 hydroxylated derivatives LVI and LVII of geraniol and nerol, respectively, can be included on the biosynthetic pathway to the indole alkaloids, for labelled samples of each were found to be specific precursors for representatives of each alkaloid group and loganin, with the *cis*-compound LVII being generally more effective than its *trans*-isomer; label at C-9 of LVI and LVII became randomized over the alkaloid equivalents of C-9/C-10 in the course of biosynthesis (*Escher*, *Loew* and *Arigoni*, *ibid.*, *loc. cit.*; *Battersby*, *S. H. Brown* and *T. G. Payne*, *ibid.*, p. 827). The failure of various other derivatives of geraniol and nerol to act as precursors restricts the range of possible intermediates beyond these terpenes and leads to a plausible mechanism for cyclisation shown in Scheme 34 (*Escher*, *Loew* and *Arigoni*, *loc. cit.*; *Yeowell* and *Schmid*, *loc. cit.*) in which the observed equivalence of C-9 and C-10 label is accommodated.

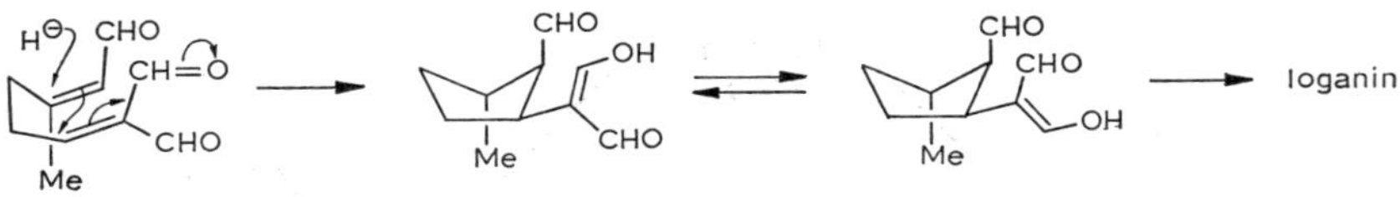

Scheme 34

Conversion of loganin (LV) into the indole alkaloids must involve cleavage of the cyclopentane ring (*cf.* Scheme 32) which may be rationalized in terms of the mechanism shown in Scheme 35, to give secologanin (LX) which is

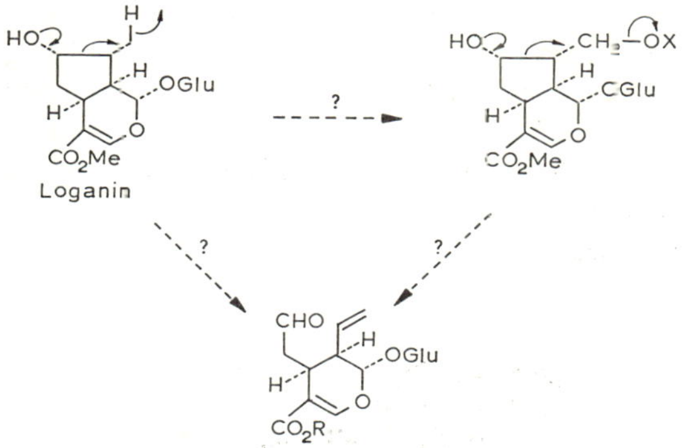

(LX) R = Me, Secologanin
(LXI) R = H

Scheme 35

found in masked form in two natural products, foliamenthin and menthiafolin (*Arigoni et al.*, Chem. Comm., 1968, 1276; *Battersby et al.*, *ibid.*, p. 1277). The latter compound could be simply transformed into secologanin and this compound with a tritium label on the ester function was shown to be a specific precursor for representative examples of all three families of indole bases. It is like loganin, moreover, a natural constituent of *V. rosea* (*Battersby*, *Burnett* and *Parsons*, J. chem. Soc., C, 1969, 1187). Secologanic acid (LXI) and secologanoside (characterized as its methylation acetylation product LXII) have also been isolated from *V. rosea* (*R. Guarnaccia* and *J. Coscia*, J. Amer. chem. Soc., 1971, **93**, 6320). Sweroside (LXIII) is a precursor for vindoline (*H. Inouye, S. Ueda* and *Y. Takeda*, Chem. Pharm. Bull., Japan, 1971, **19**, 587) probably being incorporated *via* secologanin with which it is interconvertible *in vivo* (*Battersby et al.*, unpublished work).

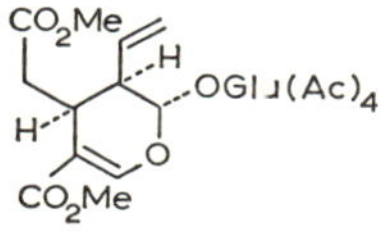

(LXII)

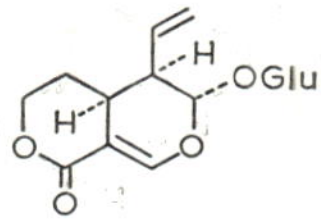

(LXIII)

Condensation of tryptamine with secologanin gave two glucosides, vincoside (LXIV) and its C-3 epimer isovincoside (LXV) which were then shown to be present in *V. rosea* (*Battersby, Burnett* and *Parsons*, J. chem. Soc., C, 1969, 1193). Isovincoside was also found to be identical with strictosidine isolated from *Rhazia stricta* (*R. T. Brown et al.*, Tetrahedron Letters, 1968, 4349; Chem. Comm., 1971, 910; *K. T. D. DeSilva, G. N. Smith* and *K. E. H. Warren, ibid.*, p. 905).

[O-*methyl*-H][3]Loganin was specifically incorporated into strictosidine/isovincoside (LXV) (*Brown et al.*, Tetrahedron Letters, 1968, 4349). This base, very recent experiments show, is the parent terpenoid indole alkaloid, the base from which the large *Corynanthé–Strychnos*, *Aspidosperma*, and *Iboga* families are derived (*M. Rueffer*, *N. Nagakura* and *M. H. Zenk*, Tetrahedron Letters, 1978, 1593 and refs. cited therein).

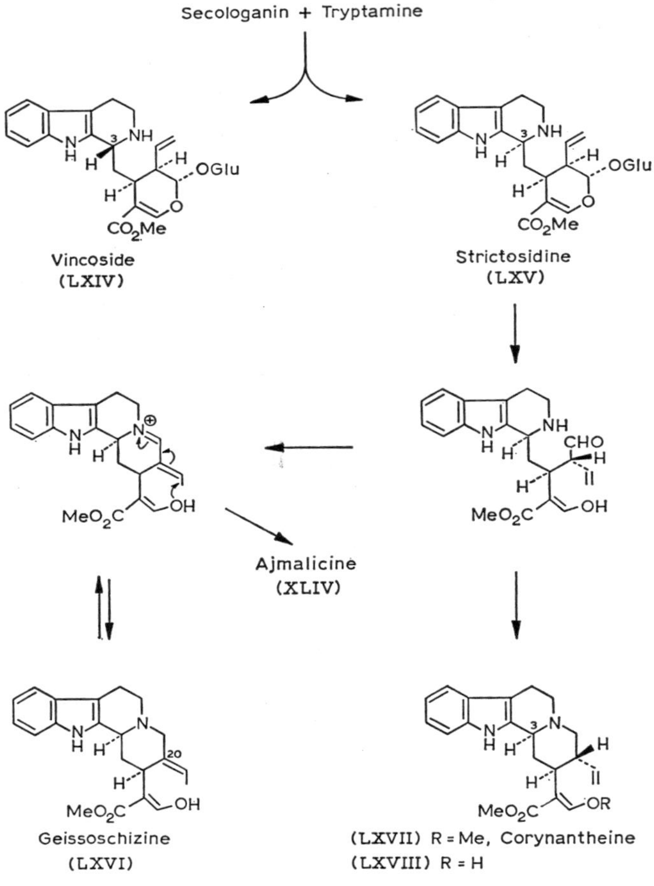

Scheme 36

The formation of the *Corynanthé*-type bases, *e.g.* corynantheine and ajmalicine, from strictosidine (LXV) requires no skeletal rearrangement, and is regarded as involving enzymatic cleavage of glucose as a first step with plausible following transformations (Scheme 36) (*Battersby*, *Burnett* and

Parsons, *loc. cit.*). It is reasonable to consider this skeletal type as the precursor for alkaloids with rearranged skeleton, and this is supported by the observation that alkaloids with *Corynanthé*-type structures appear in *V. rosea* seedlings before *Aspidosperma* and *Iboga* bases (*A. A. Qureshi* and *A. I. Scott*, Chem. Comm., 1968, 948). Feeding experiments showed that whereas corynantheine aldehyde (LXVIII) was not incorporated into other alkaloids in *V. rosea* (*Battersby et al.*, *ibid.*, 1968, 951), its double-bond isomer, geissoschizine (LXVI), which was shown to be present in *V. rosea*, was a precursor for *Iboga* and *Aspidosperma* alkaloids: [O-*methyl*-^{3}H,*Ar*-^{3}H]geissoschizine was incorporated intact into ajmalicine (XLIV), serpentine (XLIV; ring C aromatized), akuammicine (XLV), vindoline (XLVI) and catharanthine (XLVII) (*Battersby* and *E. S. Hall*, *ibid.*, 1969, 793); [*Ar*-^{2}H$_4$]geissoschizine was incorporated into akuammicine (XLV) and coronaridine (XLVII $\Delta^{15,20}$ reduced) (*Scott*, *P. C. Cherry* and *Qureshi*, J. Amer. chem. Soc., 1969, **91**, 4932). These results establish the biological rearrangement of the *Corynanthé* skeleton (geissoschizine) into the *Aspidosperma* (vindoline) and *Iboga* (catharanthine) skeletons and also the less drastic modification which affords the *Strychnos* skeleton (akuammicine).

Geissoschizine (LXVI) is also a precursor for strychnine in *Strychnos nux vomica* as is the Wieland–Gumlich aldehyde (LXXI) (*S. I. Heimberger* and

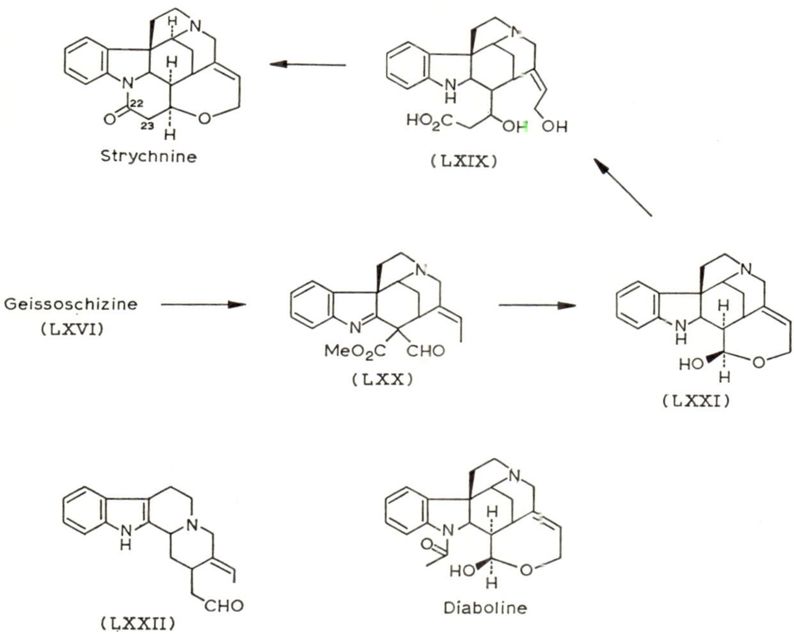

Scott, Chem. Comm., 1973, 217). Like other terpenoid indole alkaloids strychnine is formed from mevalonate *via* geraniol and tryptophan, the 2-carbon bridge (C-22 and -23) arising from acetate (*D. Gröger et al.*, Helv., 1969, **52**, 776). Although geissoschizal (LXXII) could be isolated from *S. nux vomica* it is not a strychnine precursor, suggesting that the necessary loss of a C_1 unit from geissoschizine occurs after rearrangement to the *Strychnos* skeleton, and that LXX may be an intermediate (*Heimberger* and *Scott, loc. cit.*).

In contrast to the incorporation of LXXI which confirms pioneering ideas on strychnine biosynthesis (*R. B. Woodward*, Nature, 1948, **162**, 155; *R. Robinson*, "The Structural Relations of Natural Products", Oxford University Press, 1954, p. 112) it is clear that diaboline is not involved in strychnine biosynthesis. This suggests that LXIX may be an intermediate instead and some experimental evidence was obtained that this compound was present and biosynthesized in *S. nux vomica* and was a precursor for strychnine (*Heimberger* and *Scott, loc. cit.*).

If geissoschizine is to be a normal intermediate in the biosynthesis of more complex alkaloids then tritium label at C-2 of loganin (LV, p. 359) must be lost since the equivalent site, C-20, in geissoschizine (LXVI) has no hydrogen atom. In accord, the appropriate loss was observed in the generation of the three alkaloid types (*Battersby et al.*, Chem. Comm., 1968, 951).

Stemmadenine (LXXIV), tabersonine (LXXVII) and preakuammicine (LXXIII), isolated from *V. rosea*, have proved to be further important biosynthetic intermediates. Thus stemmadenine is a precursor for tabersonine, vindoline (XLVI), and catharanthine (XLVII), whilst labelled tabersonine was incorporated into catharanthine (irreversibly) and into vindoline. These results led to the conclusion that catharanthine was derived from stemma-

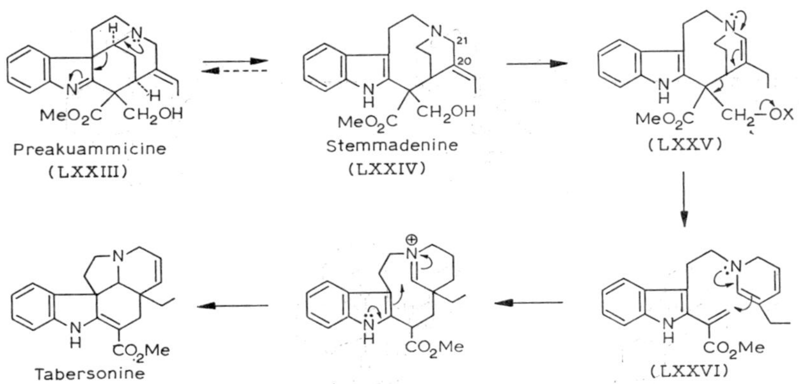

Scheme 37

denine *via* tabersonine (*Qureshi* and *Scott, loc. cit.*; *J. P. Kutney et al.*, J. Amer. chem. Soc., 1968, **90**, 3566). The results of feeding labelled tabersonine to *V. rosea* seedlings and observing the formation of radioactive alkaloids in relation to time, strengthened the conclusion that tabersonine is an intermediate in the biosynthesis of *Iboga* alkaloids. Similar experiments with [2-^{14}C]tryptophan supported roles for geissoschizine (LXVI), preakuammicine (LXXIII), and tabersonine (LXXVII) as important ("dynamic") intermediates whereas ajmalicine (XLIV), catharanthine (XLVII), coronaridine (XLVII; $\Delta^{15,20}$ reduced) and vindoline (XLVI) appeared as the end-products of biosynthesis. This study also revealed the early formation, with high turnover, of an unknown base which could be the intermediate (LXXVIII) lying between strictosidine (LXV) and geissoschizine (LXVI) (*Scott et al.*, Bio-org. Chem., 1971, **1**, 157). In these experiments with tryptophan, stemmadenine did not show the characteristics of a dynamic intermediate in conflict with the tracer evidence, and also mechanistic considerations which place it centrally on the pathway to *Aspidosperma* and *Iboga* bases.

An attractive rationalization for the transformation of stemmadenine (or close relative) into *Aspidosperma* and *Iboga* alkaloids is one which involves reactions of enamines. Central to these reactions, illustrated for tabersonine in Scheme 37, is the enamine LXXVI arrived at by fragmentation of stemmadenine. The necessary rearrangement of the monoterpene unit can be allowed for neatly in terms of this compound, alternative cyclisation paths leading to *Aspidosperma* and *Iboga* alkaloids (*Qureshi* and *Scott*, Chem. Comm., 1968, 945, 948). The foregoing evidence, however, suggests that the *Aspidosperma* type (tabersonine) is a precursor for the *Iboga* type (catharanthine) so these ideas need modification.

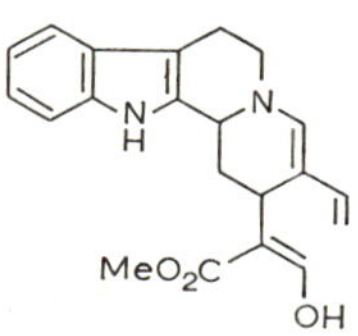

(LXXVIII)

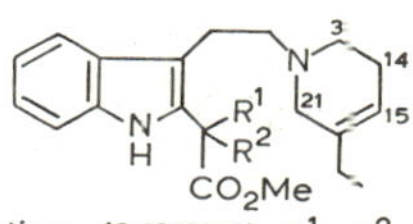

Secodine, (LXXIX) $R^1 = R^2 = CH_2$
(LXXX) $R^1 = CH_2OH$, $R^2 = H$

The validity of LXXVI as an intermediate is supported by the isolation of simple derivatives of secodine (LXXIX) from *Rhazia* species and by the specific incorporation of labelled secodine into vindoline (XLVI) [low incorporations were recorded for *inter alia*, catharanthine and ajmalicine; and for all alkaloids using 16,17-dihydrosecodin-17-ol (LXXX)]. [*Carboxy*-^{14}C;3,14,15,21-$^3H_4$]secodine was converted into vindoline with loss of 60% of the tritium label, consistent with the involvement of secodine in biosynthesis *via* a more highly oxidised intermediate, (like LXXVI) (*Kutney et al.*, J. Amer. chem. Soc., 1971, **93**, 7322; Bio-org. Chem., 1971, **1**, 194).

As already seen in part observation of tritium retention and loss from particular sites during the course of biosynthesis provides a test of conclusions reached. [In the following the numbers in brackets refer to the site of labelling in loganin (LV, p. 359)]. Experiments were carried out with geraniol (C-1 and -2), nerol (C-2), mevalonate (C-2 and -7) and loganin (C-5) and the results are that tritium from C-1, C-5 and C-7 is retained in each of the alkaloid types whereas that from C-2 is lost. This is completely in accord with the pathway so far delineated. In addition to the points made earlier it can be noted that the double-bond migration in going from stemmadenine LXXIV to LXXV is stereospecific and leaves unaffected the hydrogen at C-21, which is derived from C-1 of loganin (*Battersby et al.*, Chem. Comm., 1968, 951, 1582; *Battersby*, in "The Alkaloids" (Specialist Periodical Reports), ed. *J. E. Saxton*, The Chemical Society, London, 1971, Vol. 1, p. 31).

Vincamine. Support for the proposal (*E. Wenkert* and *B. Wickberg*, J. Amer. chem. Soc., 1965, **87**, 1580) that vincamine is derived from the *Aspidosperma* skeleton has come in a provisional way with the reported incorporation of tritiated tryptophan, stemmadenine (LXXIV), and tabersonine (LXXVII) into the alkaloid (*Kutney et al.*, *ibid.*, 1971, **93**, 255). No further results have been reported so far.

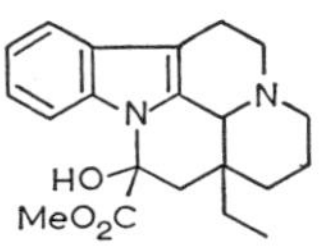

Vincamine

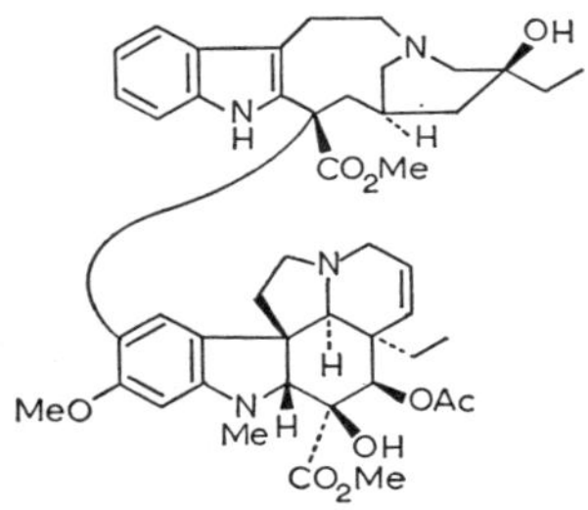

Vincaleucoblastine

Vincaleucoblastine. It may be presumed that dimeric indole alkaloids like vincaleucoblastine are formed by joining together bases like catharanthine (XLVII) and vindoline (XLVI). So far it has been shown only that tritiated loganin (LV) is incorporated into vincaleucoblastine with retention of tritium at a level which is the average of the values obtained for XLVI and XLVII (*P. E. Daddona* and *C. R. Hutchinson*, *ibid.* 1974, **96**, 6806) but catharanthine has been observed not to be a precursor for the alkaloid (*Scott*, *J. D. Michael* and *C. L. Yeh*, unpublished work quoted by *Daddona* and *Hutchinson*).

Apparicine and **uleine.** Modification of the tryptophan side-chain during biosynthesis is apparent in the structures of apparicine and uleine. There is no definite evidence on the course of biosynthesis of the latter (*Kutney*, Heterocycles, 1976, **4**, 429) but a little more is known about apparicine biosynthesis.

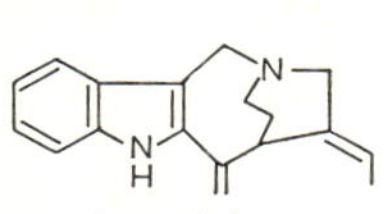

Apparicine Uleine

The evident loss of C-2 of tryptophan in the genesis of the alkaloid has been affirmed by comparative feeding experiments with [2-^{14}C,*Ar*-^{3}H]- and [3-^{14}C, *Ar*-^{3}H]-tryptophan. [*Ar*-^{3}H]Stemmadenine (as LXXIV) was incorporated so it appears that loss of this carbon atom occurs late in biosynthesis (*Kutney, V. R. Nelson* and *D. C. Wigfield*, J. Amer. chem. Soc., 1969, **91**, 4279). Secodine (LXXIX) is a precursor for apparicine, the carboxyl group apparently being retained but loss of approximately half the tritium located at C-3, -14, -15, and -21 was observed consistent with incorporation *via* a more highly oxidized intermediate (*Kutney et al.*, Bio-org. Chem., 1971, **1**, 194). Similar results were obtained for vindoline (see above).

Cinchona alkaloids. The early stages in the evidently complex biosynthesis of the *Cinchona* alkaloids, *e.g.* quinine, are similar to those delineated for

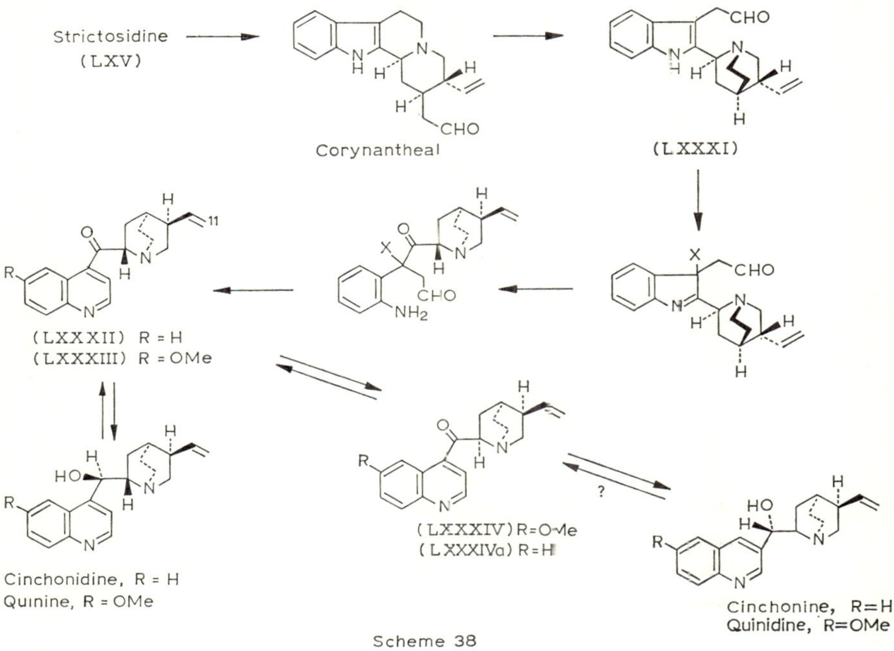

Scheme 38

Coryananthé and related alkaloids (see above). Thus quinine arises from tryptophan and from geraniol by way of loganin (LV) (*E. Leete et al.*, J. Amer. Chem. Soc., 1962, **84**, 4919; 1969, **91**, 2698; *Battersby et al.*, Chem. Comm., 1966, 888; 1970, 194); and strictosidine (LXV) was incorporated into the *Cinchona* alkaloids (see *Rueffer*, *Nagakura* and *Zenk*, Tetrahedron Letters, 1978, 1593). At least one further intermediate of the *Corynanthé* type is implicated, namely corynantheal, but it is at this stage that the biosynthesis of *Cinchona* alkaloids diverges from that of *Corynanthé* and related bases; corynantheal was an efficient precursor for the three *Cinchona* alkaloids but corynantheine aldehyde (LXVIII) was not incorporated (*Battersby* and *R. J. Parry*, Chem. Comm., 1971, 30) (*cf.* Scheme 36, p. 362).

Circumstantial evidence indicates that the aldehyde (LXXXI) may be an intermediate after corynantheal: the alcohol corresponding to LXXXI was not incorporated and although the carboxylic acid was not tested its intermediacy was ruled out by the observation that [1-^{14}C,1-$^{3}H_2$]tryptamine was incorporated with retention of half the tritium (*idem, loc. cit.* p. 31).

Genesis of the *Cinchona* skeleton from LXXXI requires a drastic skeletal rearrangement, and one particularly involving those atoms associated with the indole nucleus. So far there is no evidence available on the intriguing reactions involved, but the pathway may proceed plausibly as shown in Scheme 38.

The very late stages of biosynthesis involve minor modification of the *Cinchona* skeleton and are quite well understood (*idem, loc. cit.*). The hypothetical route outlined above indicates that cinchonidinone (LXXXII) may be the first base with the *Cinchona* skeleton. Its presence in *Cinchona ledgeriana* shoots was tested for by adding inactive material as carrier during isolation after allowing the shoots to metabolize [1-^{14}C]tryptamine. Isolation of radioactive alkaloid proved that cinchonidinone is a naturally occurring alkaloid. Moreover, [11-^{3}H]-labelled material was incorporated efficiently into cinchonidine and cinchonine; in the latter case the incorporation was proved to be specific. A low incorporation into quinine was observed. If significant it indicates that the methoxy group of this base is introduced after formation of the cinchonidinone skeleton. The conversion of cinchonidinone into cinchonidine is apparently reversible since feeding a radioactive sample of the latter led to isolation of a radioactive sample of the former. A further detail is that LXXXIV has been identified as a natural product. The foregoing provides then good evidence for the later biosynthetic steps illustrated in Scheme 38; whether all of these steps are reversible remains to be seen.

Camptothecin. The suggested relationship between camptothecin and the terpenoid indole alkaloids (*E. Wenkert et al.*, J. Amer. chem. Soc., 1967, **89**,

6741; *E. Winterfeldt*, Ann., 1971, **745**, 23) implies a reorganization of the skeleton as extensive as that in *Cinchona* biosynthesis (see above). That such a relationship does exist has been shown by obtaining incorporations of radioactive tryptophan, tryptamine, mevalonate, geraniol/nerol, secologanin, and vincoside/isovincoside into camptothecin. [2-^{14}C,*Ar*-$^{3}H_4$]Tryptophan was incorporated without change in isotope ratio so no hydrogen (tritium) is lost from the aromatic ring during biosynthesis and the tryptophan skeleton (apart from C-1) is incorporated intact (*C. R. Hutchinson et al.*, J. Amer. chem. Soc., 1974, **96**, 5609; *G. M. Sheriha* and *H. Rapoport*, Phytochemistry, 1976, **15**, 505).

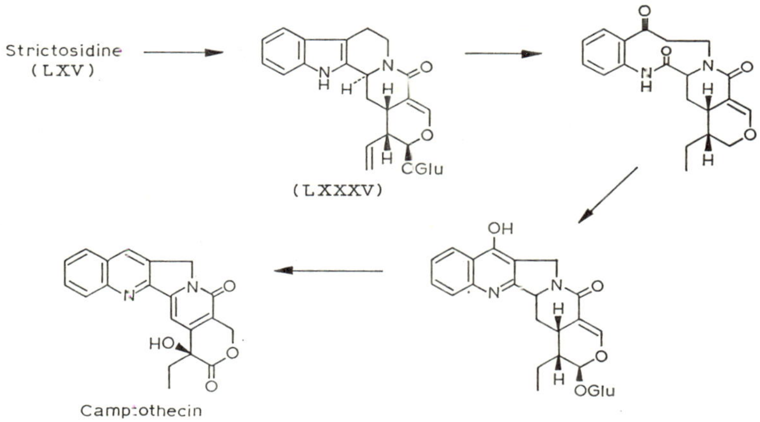

However, the low level at which geissoschizine (LXVI), the next intermediate after vincoside in terpenoid indole alkaloid biosynthesis, was incorporated indicates that the branchpoint for camptothecin biosynthesis is at vincoside/isovincoside, and it was shown that the lactam LXXXV derived from strictosidine (LXV) is involved in camptothecin biosynthesis (in these experiments of *Hutchinson et al.*, use was made of ^{13}C-labelled LXXXV, one of the first such applications of ^{13}C-labelling in plant alkaloid biosynthesis).

(*i*) *Phenoxazinones*

Although formally the phenoxazinone skeleton, *e.g.* **actinocinin**, is more closely related to anthranilic acid than tryptophan, the available evidence indicates the reverse. Thus the insect pigments, the ommochromes, *e.g.* **xanthommatin**, derive from products of tryptophan catabolism namely kynurenine and 3-hydroxykynurenine (*A. Butenandt* and *R. Beckmann*, Z. Physiol. Chem., 1955, **301**, 115; *Butenandt* and *E. Neubert*, *ibid.*, p. 109).

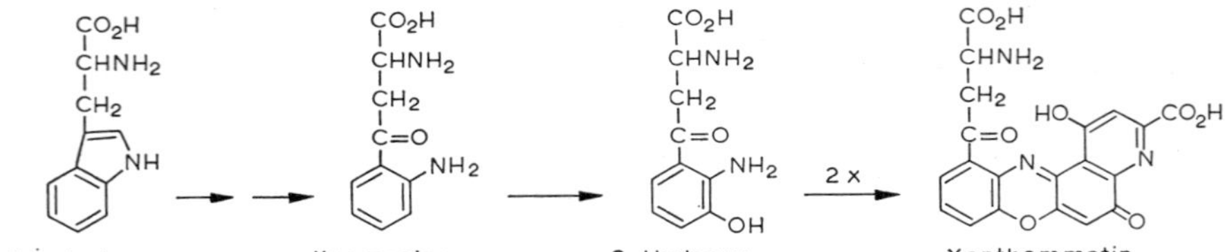

Tryptophan Kynurenine 3-Hydroxy-kynurenine Xanthommatin

The **actinomycins**, a group of peptidic antibiotics produced by *Streptomyces antibioticus* are similarly derived from tryptophan and the products of its catabolism, kynurenine and 3-hydroxykynurenine (*E. Katz et al.*, Biochim. Biophys. Acta, 1962, **57**, 283; 1962, **62**, 80; *R. B. Herbert*, Tetrahedron Letters, 1974, 4525). Depression of the incorporation of radioactive tryptophan into actinomycin by 3-hydroxy-4-methylkynurenine (LXXXVI) (*D. Perlman et al.*, J. Antibiotics, 1973, **26**, 289) and 3-hydroxy-4-methylanthranilic acid (LXXXVII) (*Katz et al.*, J. biol. Chem., 1965, **240**, 4377; 1967, **242**, 657) indicate that these too are biosynthetic intermediates. Indeed oxidative dimerization of 3-hydroxy-4-methylanthranilic acid (LXXXVII) to yield actinocinin has been achieved using a crude enzyme preparation of *S. antibioticus*, a process which must reflect the normal biosynthetic sequence (*Katz* and *H. Weissbach*, *ibid.*, 1962, **237**, 882; 1961, **236**, PC16).

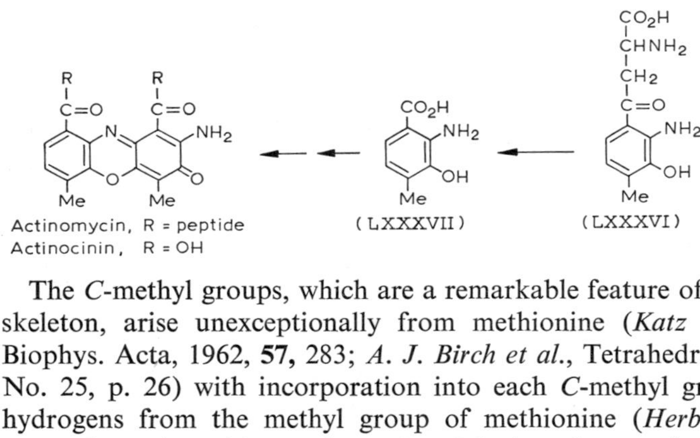

Actinomycin, R = peptide
Actinocinin, R = OH
(LXXXVII) (LXXXVI)

The *C*-methyl groups, which are a remarkable feature of the actinomycin skeleton, arise unexceptionally from methionine (*Katz et al.*, Biochim. Biophys. Acta, 1962, **57**, 283; *A. J. Birch et al.*, Tetrahedron Letters, 1960, No. 25, p. 26) with incorporation into each *C*-methyl group of all three hydrogens from the methyl group of methionine (*Herbert*, *loc. cit.*). It seems from the evidence above that 3-hydroxykynurenine is the substrate on which the interesting *C*-methylation occurs.

(*j*) *Damascenine*

The biosynthesis of the plant alkaloid **damascenine**, in contrast to that of the phenoxazinones, proceeds directly from anthranilic acid and not through tryptophan.

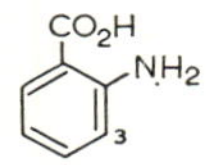

Anthranilic acid (LXXXVIII)

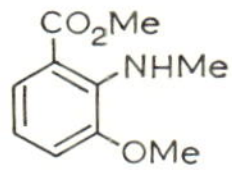

Damascenine

Thus shikimic acid, [*carboxyl*-^{14}C]-labelled anthranilic acid (as LXXXVIII), 3-hydroxyanthranilic acid and 3-methoxyanthranilic acid were incorporated into the alkaloid. The result with anthranilic acid precludes the intermediacy of tryptophan since the transformation of anthranilic acid into tryptophan occurs with loss of the carboxyl group, and in this case the label. In support tryptophan was found to be a less effective precursor for damascenine than anthranilic acid or shikimic acid (*D. Munsche* and *K. Mothes*, Phytochemistry, 1965, **4**, 705).

(*k*) *Quinoline alkaloids*

(*i*) *Echinorine*

The simplest plant quinoline is **echinorine**. Its biosynthesis is equally simple: the mode of anthranilic acid and acetate incorporation in *Echinops ritro* is illustrated.

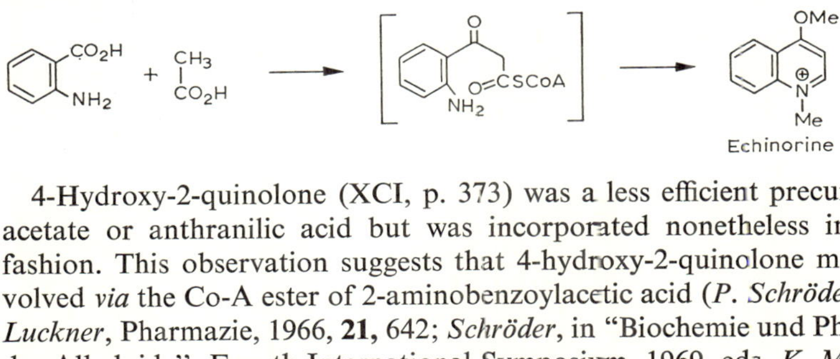

Echinorine

4-Hydroxy-2-quinolone (XCI, p. 373) was a less efficient precursor than acetate or anthranilic acid but was incorporated nonetheless in specific fashion. This observation suggests that 4-hydroxy-2-quinolone may be involved *via* the Co-A ester of 2-aminobenzoylacetic acid (*P. Schröder* and *M. Luckner*, Pharmazie, 1966, **21**, 642; *Schröder*, in "Biochemie und Physiologie der Alkaloide", Fourth International Symposium, 1969, eds. *K. Mothes*, *K. Schreiber* and *H. R. Schütte*, Akademie-Verlag, Berlin, p. 519).

(*ii*) *Pseudans*

The pseudans are a group of simple quinoline derivatives with aliphatic side-chains of varying length (C_7–C_{12}), *e.g.* **pseudan-VII**, produced by the bacterium *Pseudomonas aeruginosa*. The molecular skeleton is accounted for by anthranilic acid, acetate and malonate as indicated, labelling by acetate and malonate being consonant with that expected for the biosynthesis of a fatty acid unit.

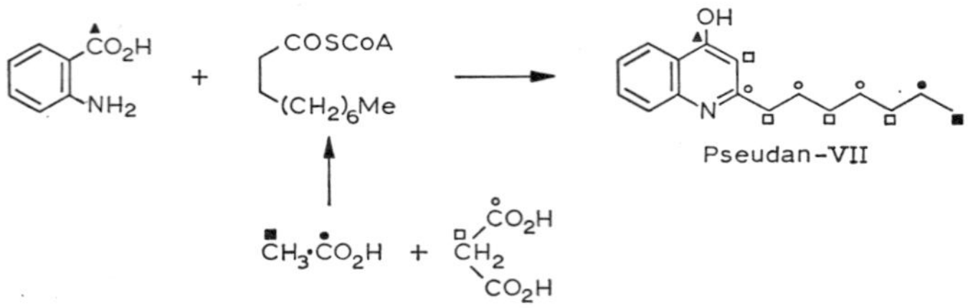

(An incidental but remarkable feature of the results was the high level of incorporation recorded: 62% for anthranilic acid) (*C. Ritter* and *Luckner*, European J. Biochem., 1971, **18**, 391). 4-Hydroxy-2-quinolone (XCI) was not incorporated. This indicates that the pseudans are formed by condensation between anthranilic acid and the complete fatty acid, which its incorporation establishes as *β*-oxodecanoic acid for pseudan-VII (*idem, loc. cit.*).

(*iii*) *Furoquinoline and related alkaloids*

Furoquinoline alkaloids, which are found in plants of the Rutaceae family, may be exemplified by **dictamnine** and **skimmianine**. The origins of the quinoline ring in these alkaloids are the same as those found for the simpler base echinorine. Thus labelled anthranilic acid (but not tryptophan), acetate and methionine were well incorporated. Degradation was used to establish the specificity of the incorporations (*M. Matsuo* and *Y. Kasida*, Chem. Pharm. Bull., Japan, 1966, **14**, 1108; *Matsuo, M. Yamasaki* and *Kasida*, Biochem. Biophys. Res. Comm., 1966, **23**, 679; *I. Monkovic, I. D. Spenser* and *A. O. Plunkett*, Canad. J. Chem., 1967, **45**, 1935; *Spenser*, unpublished work).

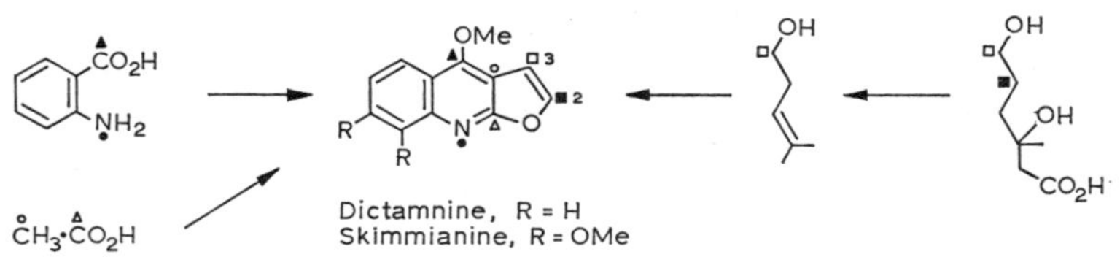

Scheme 39

Carbon atoms 2 and 3 of the alkaloid skeleton were not derived from the precursors fed. The clue to their origin comes from the observation that alkaloids of the type LXXXIX and XCV are also found in the Rutaceae. This suggests an origin for these atoms from mevalonic acid. Although this could not be demonstrated in *Skimmia japonica* and *Dictamnus albus*, satisfactory incorporations were recorded after feeding [4-^{14}C]- and [5-^{14}C]-

mevalonic acid and [1-^{14}C]dimethylallyl alcohol to *Fagara coco*. The activity was confined essentially to C-2, C-3 (mevalonic acid), and C-3 (dimethylallyl alcohol) of the skimmianine isolated (*A. O. Colonna* and *E. G. Gros*, Phytochemistry, 1971, **10**, 1515). The results for this and the other precursors are illustrated in Scheme 39.

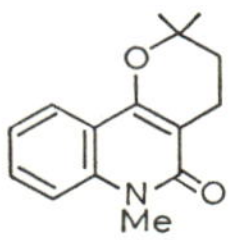

(LXXXIX)

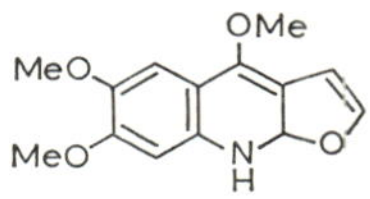

(XC)

4-Hydroxy-2-quinolone which is in pH regulated equilibrium with 2-aminobenzoylacetic acid and derivable from anthranilic and acetic acids is the next established intermediate after anthranilic acid in the biosynthesis of furoquinoline alkaloids. This follows from incorporation into kokusaginine (XC), dictamnine, platydesmine metho-salt (XCVI), and skimmianine (XCVIII) (*M. Cobet* and *Luckner*, European J. Biochem., 1968, **4**, 76; Phytochemistry, 1971, **10**, 1031; *J. F. Collins*, *W. J. Donnelly* and *M. F. Grundon*, J. chem. Soc., Perkin I, 1974, 2177). Prenylation of the quinoline skeleton may occur as well with a methoxyl group at C-4 as a hydroxyl group, since XCII is as efficient a precursor for the alkaloids as XCI. Methy-

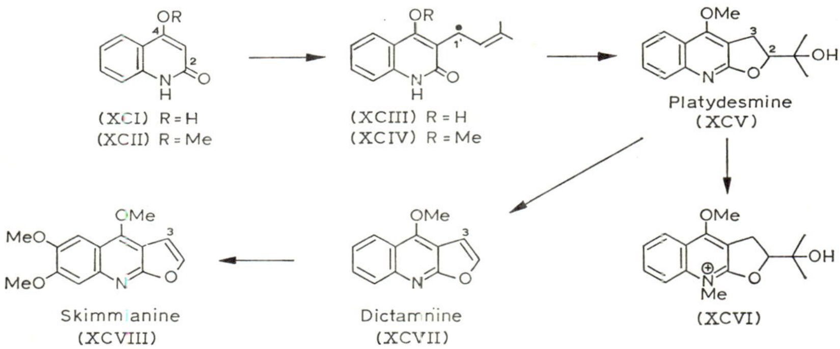

Scheme 40

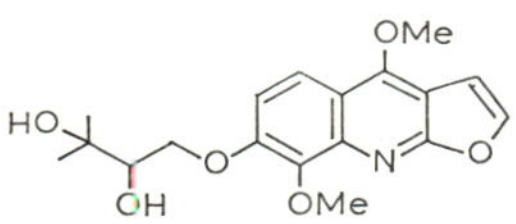

Evoxine

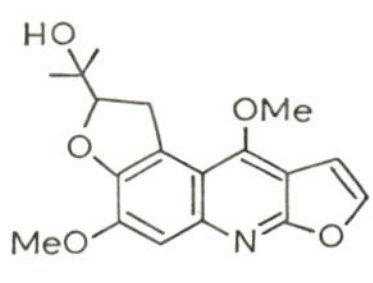

Choisyine

lation of the oxygen function at C-2 as well, however, blocks further reaction, as seen in the isolation of inactive alkaloid after feeding [2,4-$^{14}C_2$]-2,4-dimethoxyquinoline. The prenylquinolines XCIII and XCIV are equally efficient precursors for dictamnine and platydesmine metho-salt, and are thus presumably successive intermediates in furoquinoline biosynthesis (*idem, loc. cit.*) (Scheme 40). That the *O*-methyl group of XCIV was not labile was proved when [1′-^{14}C,O-*methyl*-^{3}H]-2-hydroxy-4-methoxyprenylquinoline was incorporated into dictamnine without change in isotope ratio (*Grundon, D. M. Harrison* and *C. G. Spyropoulos,* J. chem. Soc., Perkin I, 1975, 302).

The next likely intermediate in the formation of dictamnine was platydesmine XCV and accordingly efficient and specific conversions of labelled material were recorded into dictamnine and platydesmine metho-salt (XCVI) and, at a lower level, skimmianine in *S. japonica.* Further, strengthening its position as an intermediate, platydesmine was found to be a natural contituent of this plant. Platydesmine metho-salt (XCVI) was but poorly incorporated into dictamnine and is thus probably not an intermediate in its formation (*Collins, Donnelly* and *Grundon, loc. cit.*).

The conversion of platydesmine (XCV) into dictamnine (XCVII) involves the loss of the isopropyl side-chain from C-2. Loss of half the tritium on incorporation of [1′-^{14}C,1′-^{3}H]-4-methoxy-3-prenyl-2-quinolone (as XCIV) into skimmianine, evoxine and choisyine provides support for a mechanism for this fragmentation reaction, in which stereospecific allylic hydroxylation occurs at C-3 of platydesmine followed by loss of the isopropyl group as shown. The results are also consistent with initiation of the fragmentation reaction on platydesmine by hydride extraction from C-3. They do not permit the formation of an intermediate with a carbonyl group at C-1′ (*Grundon, Harrison* and *Spyropoulos, loc. cit.*).

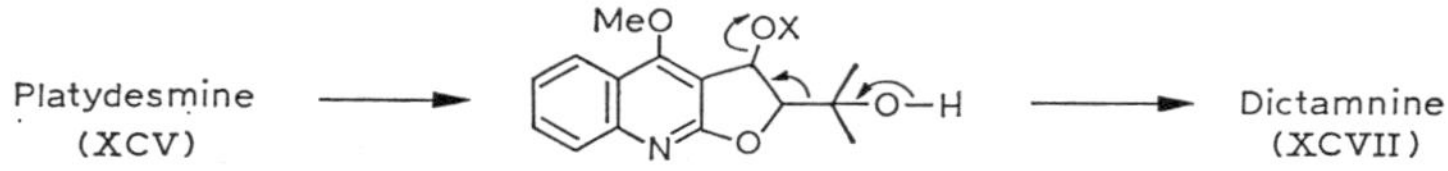

The biosynthesis of the furocoumarins involves a similar fragmentation reaction and there is a further similarity in that hydroxylation of the benzene ring occurs at a late stage of biosynthesis (*D. J. Austin* and *S. A. Brown,* Phytochemistry, 1973, **12**, 1657, and refs. cited). Incorporation of tracers into the oxygenated bases skimmianine and kokusaginine have tended to be much lower in general than into dictamnine, suggesting that hydroxylation of the benzene ring is a late step. This was confirmed when it was shown that dictamnine was efficiently converted into skimmianine (XCVIII) in *Choisya ternata,* which in contrast to *S. japonica* contains skimmianine as a major alkaloid and dictamnine as a minor one. Dictamnine was also found to be a precursor for evoxine and choisyine (*Grundon, Harrison* and *Spyropoulos*, J. chem. Soc. Perkin I, 1974, 2181).

Supporting evidence for the biosynthetic pathway to the furoquinoline alkaloids gleaned in whole plants comes from an investigation using cell-suspension cultures of *Ruta graveolens*, which produced furoquinoline bases and edulinine (*D. Boulanger, B. K. Bailey* and *W. Steck*, Phytochemistry,

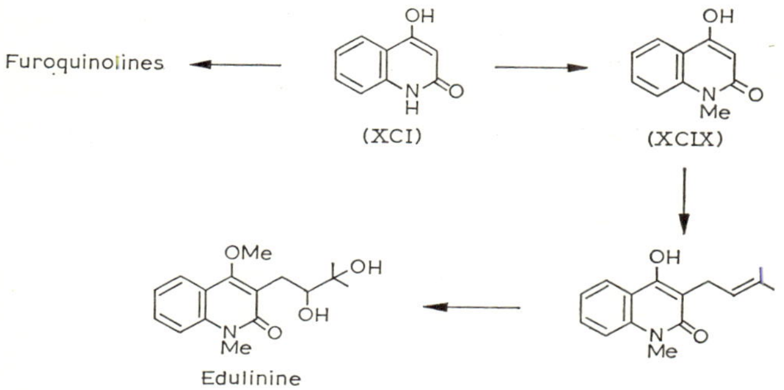

1973, **12**, 2399). Production of both dictamnine and edulinine was stimulated by added 4-hydroxy-2-quinolone (XCI). Subsequent experiments indicated that this compound was at a branch point for biosynthesis of the two alkaloid types, with competitive methylation, which affords XCIX, causing diversion from furoquinoline to edulinine biosynthesis, a conclusion which may be extendable to whole plants. The pathway deduced for the culture is illustrated.

The alkaloids of *Ravenia spectabilis* and *Flindersia ifflaiana* exhibit interesting structural variation within the quinoline group, and there is some evidence for their genesis by biological Claisen rearrangements. **Ifflaiamine** and XCIXa can be isolated from *F. ifflaiana* while **ravenoline** and **ravenine**

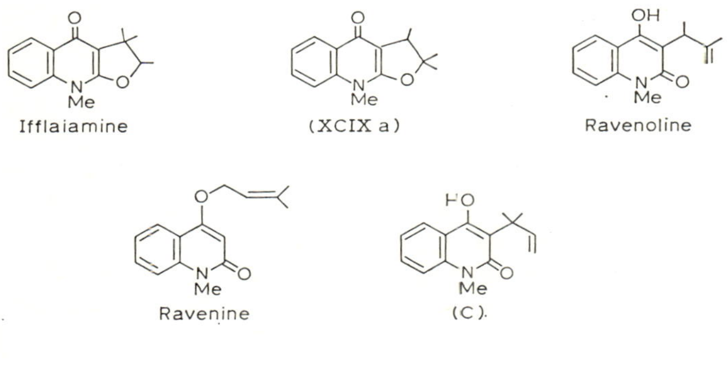

occur in *R. spectabilis*. Ravenine has been shown to be a precursor for ravenoline. An intermediate in this transformation could be C, ring closure of which would give ifflaiamine. Similarly ravenoline could afford XCIXa (*T. R. Chamberlain, J. F. Collins* and *Grundon*, Chem. Comm., 1969, 1269).

(*l*) *Acridone alkaloids*

Acridone alkaloids are found, like the furoquinoline alkaloids, in the Rutaceae. It is therefore no surprise that anthranilic acid serves as a source for part of the skeleton of these alkaloids too. The experimental evidence is that tritiated anthranilic acid is a specific precursor for ring A of **melicopicine, melicopine** and **melicopidine** in *Acronychia baueri* and *Evodia*

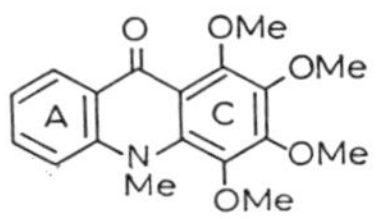

Melicopicine

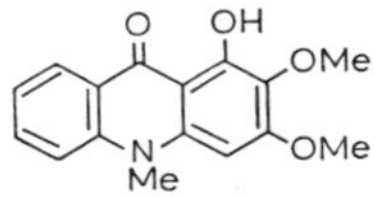

Arborinine

xanthoxyloides, and arborinine in *Glycosmis arborea* (*R. H. Prager* and *H. M. Thredgold*, Austral. J. Chem., 1969, **22**, 2627; *D. Gröger* and *S. Johne*, Z. Naturforsch., 1968, **23b**, 1072). 4-Hydroxy-2-quinolone (XCI) (a furoquinoline precursor) and its *N*-methyl derivative were incorporated into melicopicine. Further evidence for their normal intermediacy in acridone biosynthesis came from their isolation by radioisotope dilution from *A. baueri*. The incorporation of *N*-methylanthranilic acid indicates that early methylation may be important in the biosynthesis of acridone alkaloids (*Prager* and *Thredgold, loc. cit.*).

Some support for the likely derivation of ring C from acetate (*R. Robinson*, "The Structural Relations of Natural Products", Clarendon Press, Oxford, 1954, p. 93) comes from the recording of incorporation of acetate into melicopicine; mevalonic acid gave inactive alkaloid (*Prager* and *Thredgold, loc. cit.*).

(*m*) *Graveoline*

Graveoline, the major alkaloid of *Ruta angustifolia*, arises from two aromatic amino acids, anthranilic acid and phenylalanine. The anthranilic acid is incorporated as an intact unit whilst the phenylalanine suffers loss of its carboxyl group, being incorporated as a C_8 unit. Further experiments indicate that cinnamic and benzoylacetic acids and oxygenated derivatives may well be intermediates (*M. Blaschke-Cobet* and *M. Luckner*, Phytochemistry, 1973, **12**, 2393).

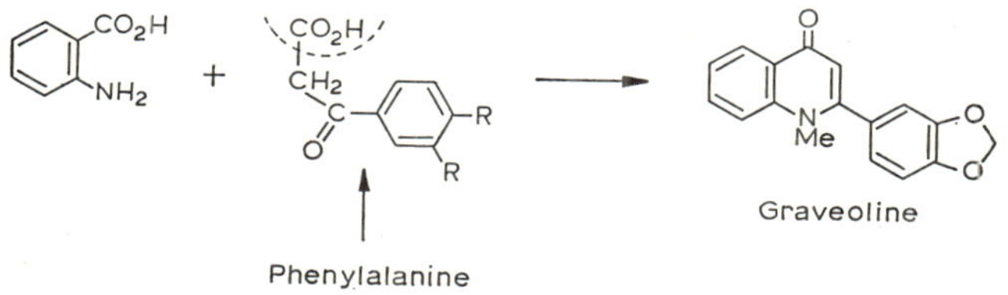

(*n*) *Quinazoline alkaloids*

Like graveoline (previous section), **arborine** is derived from anthranilic acid and phenylalanine. Again anthranilic acid is incorporated intact (by way of *N*-methylanthranilic acid) and phenylalanine is incorporated as a C_8 unit. The nature of the intermediates which lie between phenylalanine and graveoline is not absolutely clear, however. Low but specific incorporations of [1-^{14}C]-phenethylamine and [1-^{14}C]phenylacetic acid have been recorded. Doubly

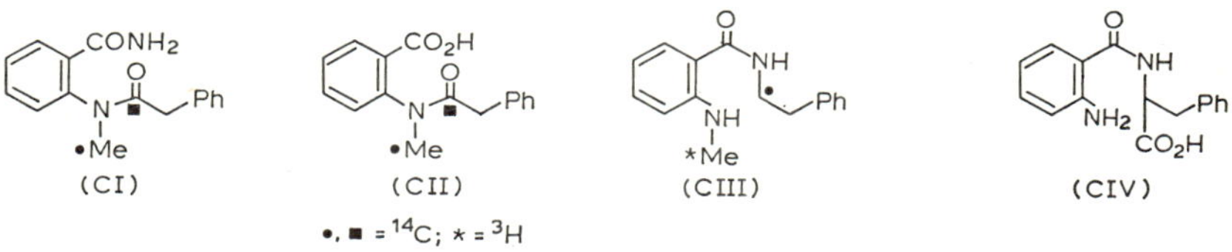

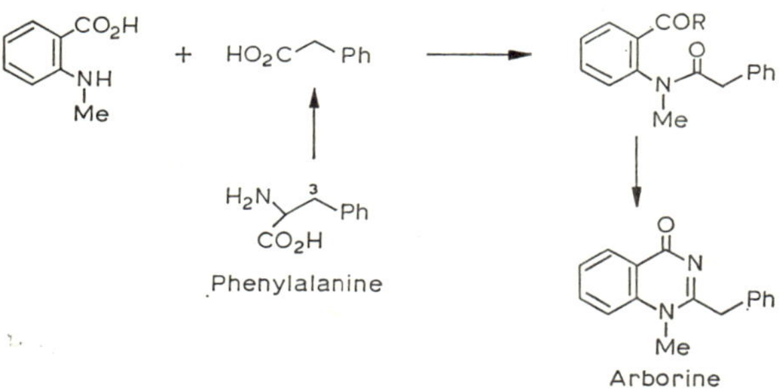

labelled samples of CI and CII (labels as shown) were well incorporated into arborine without change in isotope ratio whilst CIII and CIV were not incorporated. These results are consistent with the biosynthetic pathway to arborine illustrated. One piece of evidence is in conflict with this pathway, however. [3-^{14}C,^{15}N]Phenylalanine has been found to give arborine without change in isotope ratio (*D. G. O'Donovan* and *H. Horan*, J. chem. Soc., C, 1970, 2466; *D. Gröger et al.*, European J. Biochem., 1970, **15**, 415; Z. Naturforsch., 1968, **23b**, 1072).

The biosynthesis of **peganine** (= vasicine) has been studied in two plants from different families, and two different pathways have been deduced (*D. R. Liljegren*, Phytochemistry, 1968, **7**, 1299; 1971, **10**, 2661; *Gröger et al., ibid.*, 1968, **7**, 429; 1972, **11**, 2263; Z. Pflanzenphysiol., 1969, **61**, 353). Only anthranilic acid is common to both pathways (*Gröger* and *K. Mothes*, Arch. Pharm., 1960, **293**, 1049; *Gröger*, *Johne* and *Mothes*, Experientia, 1965, **21**, 13; *Johne*, *Gröger* and *G. Richter*, Arch. Pharm., 1968, **301**, 721).

In *Peganum harmala* (Zygophyllaceae) specific incorporation into peganine of both [2- ^{14}C]-and [5-^{14}C]-ornithine with equal labelling of C-1 and C-10 has been found. A symmetrical intermediate lying between ornithine and vasicine was thus indicated, which could be putrescine. Curiously, however, although putrescine was incorporated, label was unequally distributed between C-1 and C-10. More random labelling was obtained with DL-[5-^{14}C]-glutamic acid and DL-[5-^{14}C]proline although C-1 and C-10 were marginally more heavily labelled. The pathway deduced for peganine in *P. harmala* may be summarized as path *a* (*Liljegren, loc. cit.*).

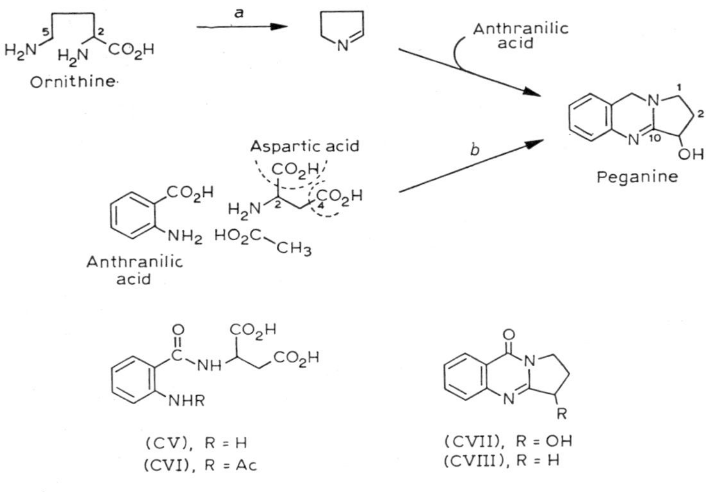

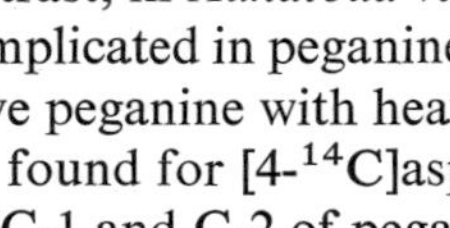

(CV), R = H
(CVI), R = Ac

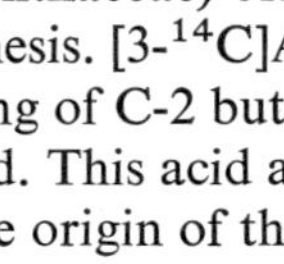

(CVII), R = OH
(CVIII), R = H

In contrast, in *Adhatoda vasica* (Acanthaceae) ornithine and its congeners are not implicated in peganine biosynthesis. [3-^{14}C]Aspartic acid on the other hand gave peganine with heavy labelling of C-2 but only random incorporation was found for [4-^{14}C]aspartic acid. This acid appears then to provide a C_2 unit (C-1 and C-2 of peganine). The origin of the remaining two carbons has been tested for with acetate, pyruvate, and glycine. Scrambling of label was observed but [3-^{14}C]pyruvate was found to label the expected site, C-3, heavily and a similar, if less specific, result was obtained with [2-^{14}C]acetate. Proof that these two carbons were acetate-derived came with the specific incorporation, without change in isotope ratio, of [*acetyl*-2-^{14}C,^{15}N]-*N*-acetylanthranilic acid (*Gröger et al., loc. cit.*). Anthranoylaspartic acid (CV) is also a fairly specific precursor for peganine (*Johne, K. Waiblinger* and *Gröger*, Pharmazie, 1973, **28**, 403). In view of the intact incorporation of *N*-acetylanthranilic acid the most important precursor for peganine may be CVI. The pathway deduced for peganine in *A. vasica* is shown as path *b*.

Vasicinone (CVII) and deoxyvasicinone (CVIII) co-occur with peganine. Attempts to observe either interconversion or the order of biosynthesis of these three alkaloids have been without success (*Liljegren, loc. cit.*; *Johne, Waiblinger* and *Gröger, loc. cit.*).

(*o*) *Rutaecarpine and evodiamine*

Inspection of the structures of these two alkaloids of *Evodia rutaecarpa* suggests an origin from one unit each of anthranilic acid and tryptophan. The available evidence is in support of this. [3-^{14}C]Tryptophan was specifically incorporated into the *β*-carboline moiety of the alkaloids. Tritiated

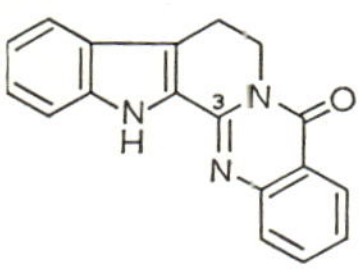

Rutaecarpine

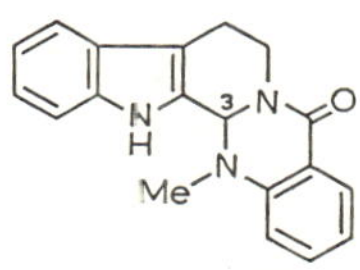

Evodiamine

anthranilic acid was also incorporated. The origin of C-3 in these two bases was found to be in methionine, this precursor also supplying the *N*-methyl group of evodiamine (label from radioactive formate was similarly incorporated, but some scrambling was apparent) (*M. Yamazaki et al.*, Tetrahedron Letters, 1967, 3317; 1966, 3221).

(*p*) *Benzodiazepine bases*

Cyclopenin, cyclopenol, viridicatin and **viridicatol** are metabolites of *Penicillium cyclopium*. A highly specific enzyme cyclopenase isolated from *P. cyclopium* converts cyclopenin and cyclopenol into viridicatin and viridicatol, respectively. These reactions are also catalysed by acid (or alkali) and in both the acid- and enzyme-catalysed reactions C-5 of the substrate is lost and label is incorporated from neither $H_2{}^{18}O$ nor ^2H_2O. From this it seems that the mechanism of enzyme and acid catalysis is similar (*M. Luckner*, European J. Biochem., 1967, **2**, 74; *Luckner*, *K. Winter* and *J. Reisch*, *ibid.*, 1969, **7**, 380; *Luckner* and *L. Nover*, "Biochemie und Physiologie der Alkaloide", Fourth International Symposium, 1969, eds. *K. Mothes*, *K. Schreiber* and *H. R. Schütte*, Akademie-Verlag, Berlin, p. 525).

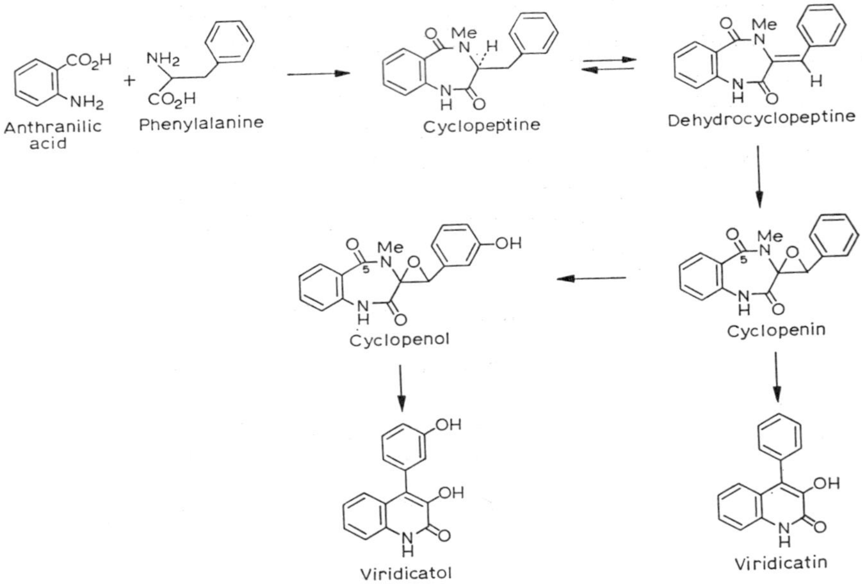

Cyclopenin and cyclopenol are derived from intact molecules of anthranilic acid and phenylalanine, the *N*-methyl groups arising from methionine (*Nover* and *Luckner*, European J. Biochem., 1969, **10**, 268). The epoxide and aromatic oxygens arise from molecular oxygen (*idem*, F.E.B.S. Letters, 1969, **3**, 292) and cyclopenin is a precursor of cyclopenol (*idem*, "Biochemie und Physiologie der Alkaloide", *loc. cit.*, p. 535). The phenylalanine moiety of cyclopenin suffers aromatic hydroxylation in the unusual *meta* position with partial "NIH" shift (*idem*, F.E.B.S. Letters, 1969, **3**, 292); *m*-tyrosine and tyrosine are only unspecific precursors (*Nover*, Pharmazie, 1969, **24**, 365). An enzyme capable of catalyzing the hydroxylation has been isolated from *P. cyclopium* cultures. It shows typical properties of a mixed-function oxidase, requiring both an oxygen and a hydrogen donor as cofactors (*I. Richter* and *Luckner*, Phytochemistry, 1976, **15**, 67).

Cyclopeptine and dehydrocyclopeptine are intermediates in cyclopenin and cyclopenol biosynthesis: they are incorporated into these bases, are present in *P. cyclopium* cultures, and are themselves formed from phenylalanine and anthranilic acid. In addition they are reversibly transformed into each other *in vivo*. Negative results with compounds likely to be implicated earlier on the pathway suggest that the formation of cyclopeptine from precursor amino acids takes place with enzyme-bound intermediates (*J. Framm et al.*, European J. Biochem., 1973, **37**, 78). An enzyme which catalyses the interconversion of cyclopeptine and its dehydro-derivative has been isolated. It shows a high degree of substrate specificity and since cyclopeptine, *m*-hydroxylated in ring B, was not dehydrogenated it follows that introduction of the hydroxy-group characteristic of cyclopenol, must occur after dehydrogenation, in accord with the observation (see above) that cyclopenin is a precursor for cyclopenol (*E. A. Aboutabl* and *Luckner*, Phytochemistry, 1975, **14**, 2573).

(*q*) *Benzoxazinones*

1,4-Benzoxazinones, *e.g.* DIMBOA (CIX), have been isolated from maize, wheat and rye. DIMBOA has been deduced to be built up from an aromatic unit derived from shikimic acid, and a C_2 unit derived from ribose, carbons 2 and 3 of the oxazine ring arising respectively from carbons 2 and 1 of ribose (*J. E. Reimann* and *R. U. Byerrum*, Biochemistry, 1964, **3**, 847).

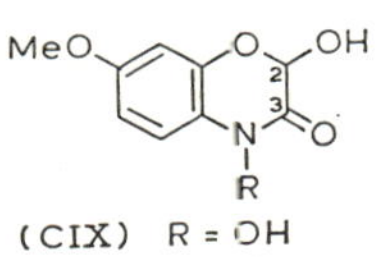

(CIX) R = OH
(CX) R = H

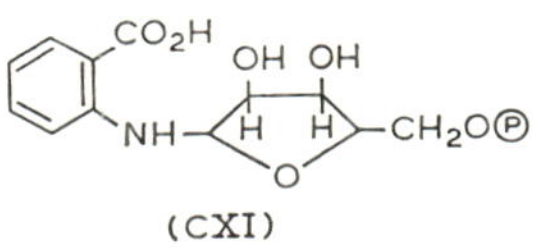

(CXI)

Anthranilic acid is also a precursor, but [^{3}H]-3-hydroxyanthranilic acid and [^{14}C]-2-aminophenol were not incorporated. Thus anthranilic acid appears to be an immediate precursor and the next intermediate may be *N*-(5′-phosphoribosyl)anthranilate (CXI), also an intermediate in tryptophan biosynthesis (*C. L. Tipton et al.*, Phytochemistry, 1973, **12**, 347). A study of the relationships of the benzoxazinones (as CX) and *N*-hydroxybenzoxazinones (as CIX) has shown that they are interconvertible and the reactions probably proceed *via* the glucosides (*Tipton et al., loc. cit.*).

(*r*) *Phenazines*

The phenazine ring system is common to approximately thirty microbial metabolites, some of which bear carbon substituents at C-1 and C-6, *e.g.* phenazine-1,6-dicarboxylic acid (CXII) (*N. N. Gerber*, J. heterocyclic Chem., 1969, **6**, 297) and **griseolutein A** (CXIII) [*S. Nakamura et al.*, J. Antibiotics

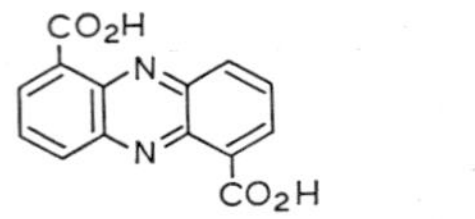

Phenazine-1,6-dicarboxylic acid
(CXII)

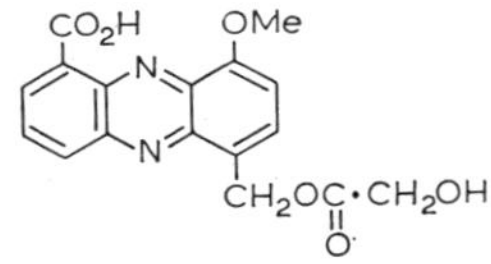

Griseolutein A
(CXIII)

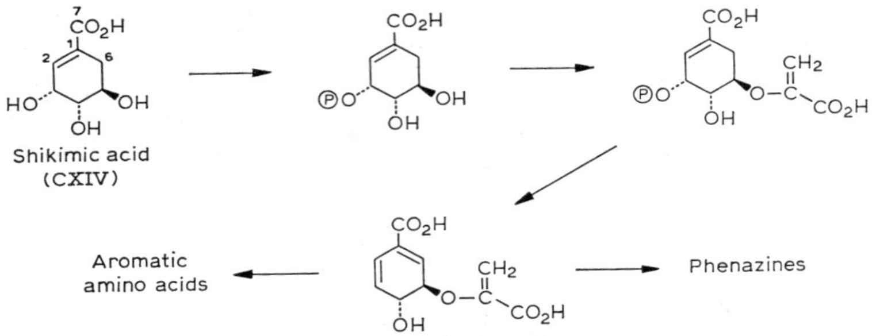

Shikimic acid
(CXIV)

Chorismic acid
(CXV)

(A), 1959, **12**, 55]. This skeletal arrangement suggests a derivation for phenazines from two molecules of anthranilic acid and this idea is made more plausible by the proven involvement of this amino acid in the biosynthesis of a wide variety of secondary metabolites in plants and microorganisms. In particular, the pseudans produced, like many naturally occurring phenazines, by a *Pseudomonas* species, have been shown to incorporate labelled anthranilic acid with high efficiency (see above). There is, however, a strong body of evidence which points firmly away from anthranilic acid as a precursor for the phenazine ring system. In particular, two independent studies with mutants of *P. aeruginosa* have shown that the biosynthetic pathway to pyocyanin (CXXV, p. 384) diverges from that which leads to anthranilic acid and other aromatic amino-acids between shikimic acid (CXIV) and chorismic acid (CXV) inclusive, and most probably it is at chorismic acid that diversion occurs (*D. H. Calhoun, M. Carson* and *R. A. Jensen,* J. gen. Microbiol., 1972, **72,** 581; *R. P. Longley et al.*, Canad. J. Microbiol., 1972, **18,** 1357). [It is notable that in the biosynthesis of naphthoquinones derived from shikimic acid, the branch-point is also apparently at chorismic acid (*P. Dansette* and *R. Azerad*, Biochem. Biophys. Res. Comm., 1970, **40**, 1090)]. In a related study, evidence was obtained that all the carbon atoms of the phenazine nucleus arise from shikimic acid *i.e.* two molecules are involved (*W. M. Ingledew* and *J. J. R. Campbell,* Canad. J. Microbiol., 1969, **15**, 535). Proof that two molecules of shikimic acid are utilized for the construction of the phenazine nucleus was obtained when it was shown that the incorporation of [2-^{2}H]shikimic acid into iodinin (CXXVIII), gave material some molecules of which were dideuteriated (*R. B. Herbert, F. G. Holliman* and *J. B. Sheridan*, Tetrahedron Letters, 1976, 639).

Experiments with radioactive shikimic acid have been further directed to determine the manner in which the precursor molecules are built into the phenazine ring system (chorismic acid cannot be used as exogenous material is not transported to the site of phenazine synthesis). Shikimic acid labelled with ^{14}C (1,6-$^{14}C_2$,1,6,7-$^{14}C_3$,1-^{14}C, and 6-^{14}C) has been incorporated into pyocyanin (CXXV), phenazine-1-carboxylic acid (CXVIII) and iodinin (CXXVIII). The results, of which those obtained with singly labelled materials are particularly definitive, are consistent with arrangement of two shikimic acid units as CXVI or CXVII. The former is intrinsically more appealing for this is the arrangement expected for phenazine-1,6-dicar-

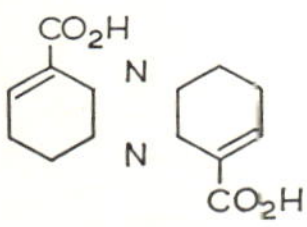
(CXVI)

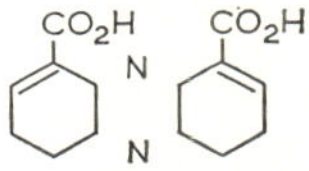
(CXVII)

boxylic acid (CXII) and griseolutein A (XCIII) and allows for entry of both amino groups at C-6 (of shikimic acid) as in the biosynthesis of anthranilic acid (*U. Hollstein* and *L. G. Marshall*, J. org. Chem., 1972, **37**, 3510; *Hollstein* and *D. A. McCamey*, *ibid.*, 1973, **38**, 3415; *Herbert, F. G. Holliman* and *J. B. Sheridan*, Tetrahedron Letters, 1974, 4201). Proof that CXVI was the way in which the shikimic acid units were linked together, for iodinin at least, was obtained when it was shown that on incorporation of [2-^{2}H]-shikimic acid into iodinin (CXXVIII), all the deuterium (and some molecules were dideuteriated) was in positions *ortho* to the hydroxyl groups (*idem*, *ibid.*, 1976, 639).

As the phenazine metabolites pyocyanin (CXXV) and the aeruginosins A and B are produced by different strains of the same organism, *P. aeruginosa*, it was to be expected that they would arise by related pathways. This is indeed so. The betaine CXXII is a precursor for both aeruginosin A (CXXIII) (*G. S. Hansford, Holliman* and *Herbert*, J. chem. Soc., Perkin I, 1972, 103) and pyocyanin (CXXV) (*M. E. Flood, Herbert* and *Holliman*, *ibid.*, 1972, 622). In experiments with pyocyanin the deuteriated precursors, 1-carboxy-5-methyl-[6,7,8,9-$^2H_4$]phenazinium betaine (as CXXII) and similarly labelled phenazine-1-carboxylic acid (CXVIII) were used as precursors. Both were efficiently incorporated and without loss of deuterium. It follows from this that hydroxylation of CXXII occurs as a specific decarboxylative reaction, and the arene oxide CXXIV seems a likely intermediate. As 1-hydroxy-

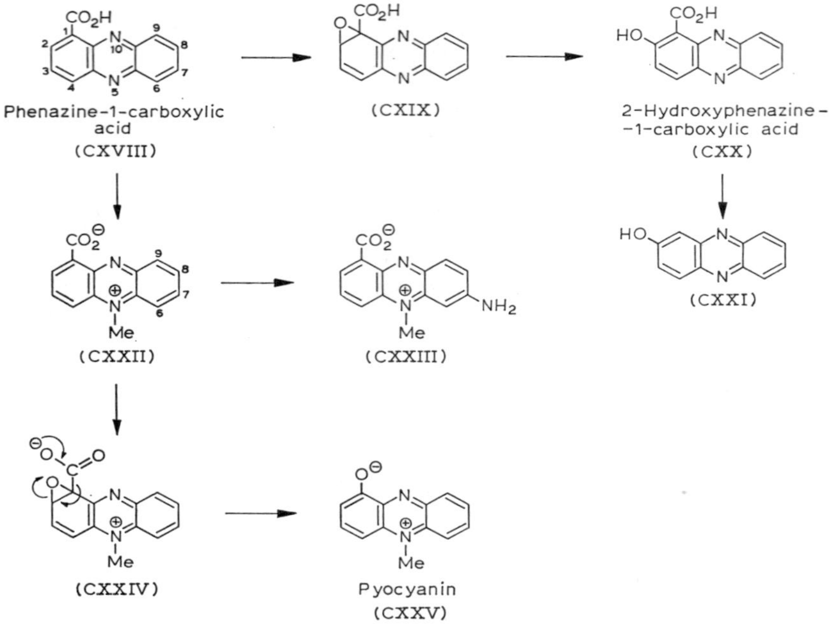

Scheme 41

phenazine was not incorporated the biosynthesis of pyocyanin follows the sequence shown in Scheme 41.

Alternative opening of the epoxide ring of CXXIV would give the *N*-methyl derivative of another phenazine metabolite, 2-hydroxyphenazine-1-carboxylic acid (CXX). (This acid, together with phenazine-1-carboxylic acid and 2-hydroxyphenazine are produced by *P. aureofaciens*.) This particular epoxide was excluded as an intermediate when it was found that the betaine CXXII was not incorporated. On the other hand, [6,7,8,9-$^2H_4$]phenazine-1-carboxylic acid (as CXVIII) was incorporated, tetradeuterio-samples of 2-hydroxyphenazine-1-carboxylic acid (CXX) and 2-hydroxyphenazine (CXXI) being isolated. The respective levels of incorporation were consonant with the expected sequence shown in Scheme 41. Mechanistically an arene oxide intermediate is still attractive and would accordingly be CXIX (*Flood*, *Herbert* and *Holliman, loc. cit.*).

Coupling of two shikimic acid derivatives could lead through phenazine-1,6-dicarboxylic acid (CXII) to all other phenazines. Attempts to demonstrate this have been unsuccessful, however. Labelled phenazine-1,6-dicarboxylic acid was not incorporated into pyocyanin (*idem, loc. cit.*) or iodinin (*Herbert, Holliman* and *P. N. Ibberson*, Chem. Comm., 1972, 355).

Iodinin showed no incorporation, in *Brevibacterium iodinum*, of deuteriated CXVIII, CXXII or 1-hydroxyphenazine. On the other hand, the *N*-oxide CXXVI showed efficient incorporation of [6,7,8,9-$^2H_4$]-phenazine-1-carboxylic acid and deuteriated 1-hydroxyphenazine, showing that the negative results were not due to transport difficulties. Since the carboxyphenazine precursor was incorporated without deuterium loss its conversion into CXXVI is analogous to its conversion into pyocyanin (*Herbert, Holliman* and *Ibberson, loc. cit.*).

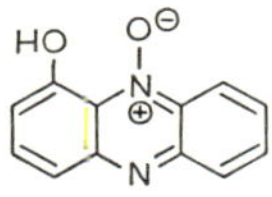

(CXXVI)

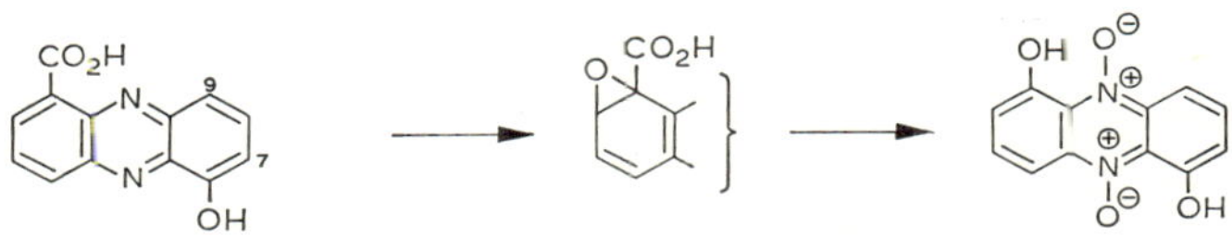

6-Hydroxyphenazine-1-carboxylic acid (CXXVII)

Iodinin (CXXVIII)

On the other hand, highly efficient incorporation of 6-hydroxy-[7,9-$^2H_2$]-phenazine-1-carboxylic acid (as CXXVII) into iodinin was observed (*idem, loc. cit.*). It thus seems probable that *B. iodinum* converts phenazine-1-carboxylic acid and 6-hydroxyphenazine-1-carboxylic acid by the same or similar enzyme(s) into CXXVI and iodinin, respectively, *N*-oxide formation occurring after hydroxylative decarboxylation since 1,6-dihydroxyphenazine (*M. Podojil* and *N. N. Gerber*, Biochemistry, 1967, **6**, 2701; *Herbert, Holliman* and *Ibberson, loc. cit.*) and its *N*-oxide (*Podojil* and *Gerber, loc. cit.*) are also iodinin precursors.

(*s*) *Shihunine*

The diacid CXXIX is an important intermediate in the biosynthesis of several 1,4-naphthoquinones, *viz.* the plant quinones lawsone, juglone and the bacterial menaquinones [*R. Bentley*, in "Biosynthesis", ed. *T. A. Geissman* (Specialist Periodical Reports), The Chemical Society, London, 1975, Vol. 3, p. 181]. It has also been found to be an efficient and specific precursor for the alkaloid shihunine in the orchid, *Dendrobium pierardii* (*E. Leete* and *G. B. Bodem*, Chem. Comm., 1973, 522). The suggested pathway is illustrated.

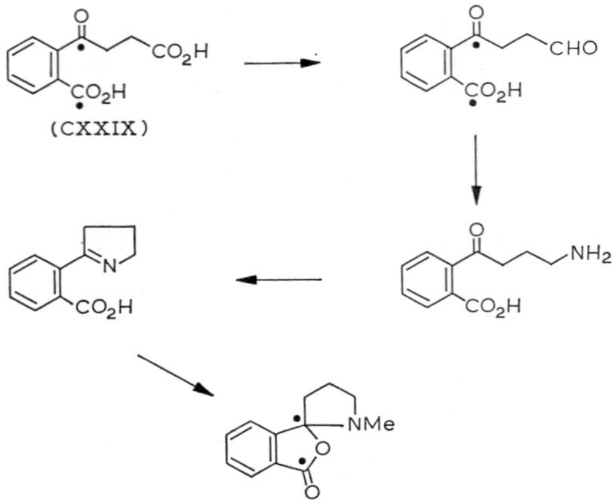

Shihunine

The naphthoquinones are products of the shikimic acid pathway to aromatic amino acids but the path which leads to these naphthoquinones branches from the main pathway before the formation of aromatic compounds probably no later than chorismic acid (*Bentley, loc. cit.*). It will be of great interest to see whether the biosynthesis of shihunine also follows this route since all the other bases of plant origin which arise from products of the shikimic acid pathway derive from aromatic precursors.

6. Bases derived from phenylalanine and tyrosine

(*a*) *Introduction*

Many of the complex alkaloids which have their origins in the aromatic amino acids phenylalanine and tyrosine arise by a combination of two units derived from these precursors. This is followed by skeletal modification in which phenol-oxidative coupling appears to play a key rôle. This reaction in which coupling of phenol radicals occurs was long recognised as of probable importance in the biogenesis of many natural products but it was *D. H. R. Barton* and *T. Cohen* (Festschrift A. Stoll, Birkhäuser Basel, 1957, p. 117; see also *H. Erdtman* and *C. A. Wachtmeister*, *ibid.*, p. 144) who applied these ideas in a unified way to alkaloid biogenesis, and their hypothesis has had a seminal influence on subsequent studies on the biosynthesis of *inter alia* benzylisoquinoline and Amaryllidaceae alkaloids. The hypothesis is that various structural types may be simply derived from radicals formed by one electron oxidation of a biphenol (I).

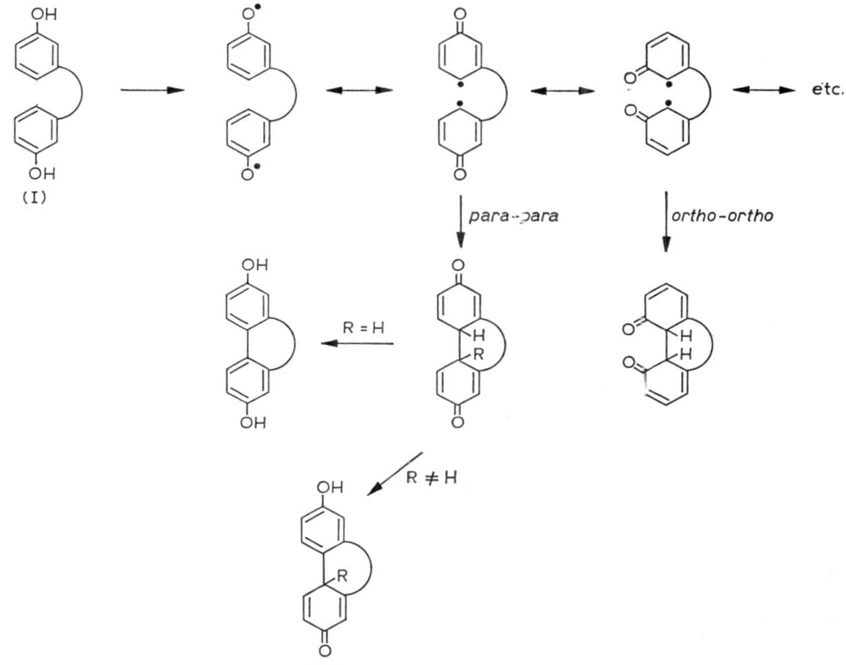

Scheme 42

Since the unpaired electrons are delocalized, coupling of the radicals may occur in one of several ways: by joining the oxygen atom of one mesomeric radical to one of sites on the aromatic ring of the other, or by joining two aromatic carbon atoms together, and this may be achieved by an *ortho–ortho*, *ortho–para* or *para–para* linkage (as illustrated in part in Scheme 42). After coupling, simple prototopic shift regenerates the aromatic ring unless there is a substituent on one (or both) of the coupling sites when the product of coupling will be the dienone (as II). An important chemical illustration is provided by the oxidation of *p*-cresol to Pummerer's ketone.

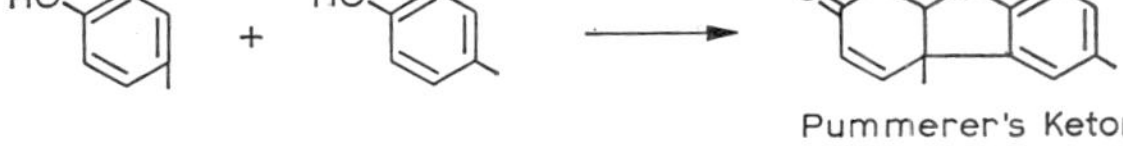

Pummerer's Ketone

Studies in the biosynthesis of alkaloids described below have provided examples of all the various coupling modes, and dienones or compounds derived by further modification of them have been isolated, of which morphine (p. 414) is the most eminent example (for extensive treatment of this topic see "Oxidative Coupling of Phenols", eds. *W. I. Taylor* and *A. R. Battersby*, Arnold, London, 1967).

(*b*) *Amaryllidaceae alkaloids*

At first sight the Amaryllidaceae alkaloids represented by **lycorine** (VII), **haemanthamine** (X), and **galanthamine** (XIII) appear to be of unrelated structure. However, *Barton* and *Cohen* (*loc. cit.*; *Barton*, Hugo Muller

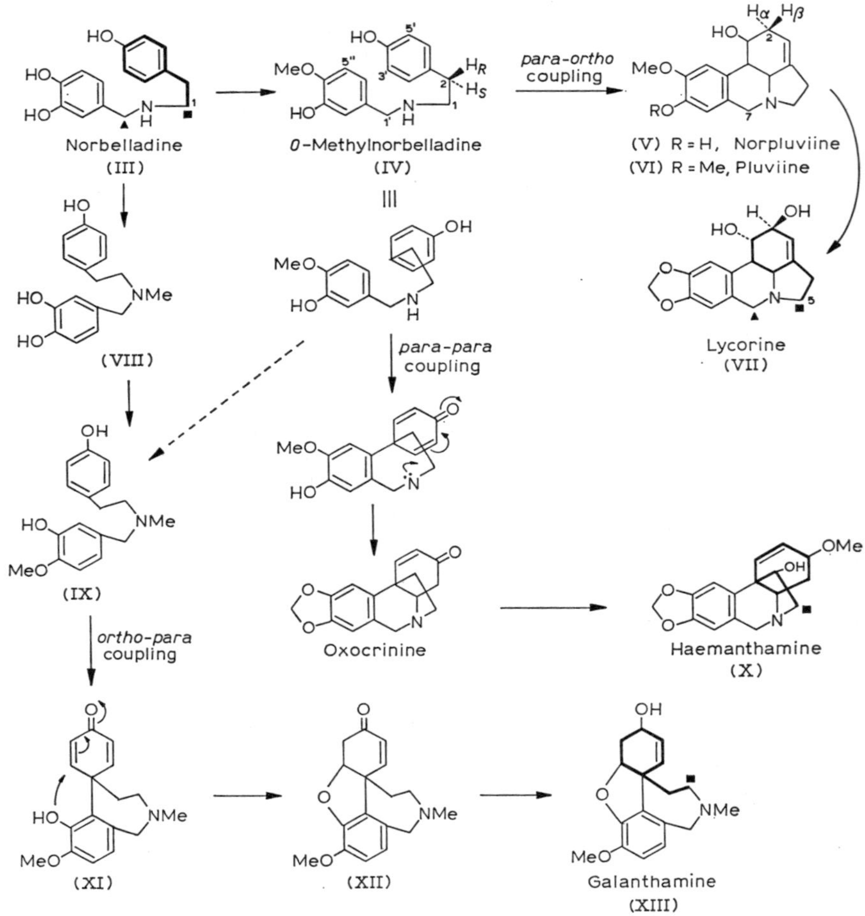

Scheme 43

Lecture, Proc. chem. Soc., 1963, 293), in a beautiful illustration of their hypothesis based on the oxidative coupling of biphenols (see p. 386) recognized that these alkaloids could arise from a single compound, norbelladine (Scheme 43). Subsequently the correctness of these ideas has been firmly established and further expanded in a series of important tracer experiments.

Examination of the structure of norbelladine (III) indicates an origin from a C_6–C_1 and a C_6–C_2 unit. These units as seen in VII, X and XIII have been shown to have their origins in tyrosine (C_6–C_2; heavy bonding in Scheme 43) and phenylalanine (C_6–C_1). Thus [2-^{14}C]tyrosine gave lycorine with specific labelling of C-5 (in accord with hypothesis: see Scheme 43); both D- and L-tyrosine were equally well utilized. As expected for a C_6–C_2 unit, label from [1-^{14}C]tyrosine was not incorporated. As further definition of this unit labelled tyramine was found to be a good precursor for lycorine and norpluviine (V) and was incorporated in the same way as tyrosine (*Battersby et al.*, J. chem. Soc., 1964, 1595; *R. J. Suhadolnik*, *A. G. Fischer* and *J. Zulalian*, J. Amer. chem. Soc., 1962, **84**, 4348; *I. T. Bruce* and *G. W. Kirby*, Chem. Comm., 1968, 207). [2-^{14}C]Tyrosine was specifically incorporated into galanthamine (XIII) and haemanthamine (X), again in a manner which accords with prediction (Scheme 43) (*Barton et al.*, J. chem. Soc., 1963, 4545; *Battersby*, *H. M. Fales* and *W. C. Wildman*, J. Amer. chem. Soc., 1961, **83**, 4098). Further definition of the origins of the C_6–C_2 unit comes from the demonstration that [3-^{14}C]tyrosine is a specific precursor for this fragment in lycorine (VII), haemanthamine (X), haemanthidine (XV), tazettine (XVI)

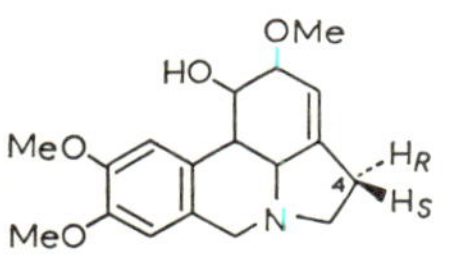

Galanthine
(XIV)

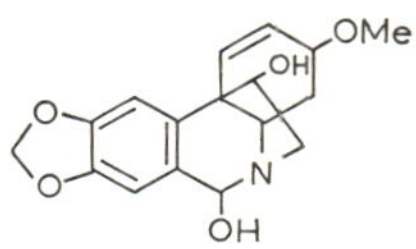

Haemanthidine
(XV)

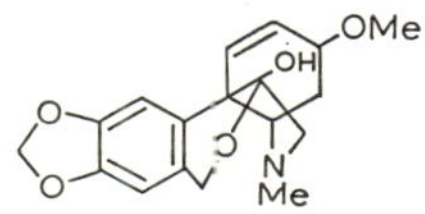

Tazettine
(XVI)

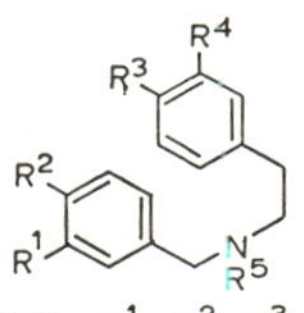

(XVII) $R^1 = R^2 = R^3 = OMe$, $R^4 = H$, $R^5 = Me$

(XVIII) $R^2 = R^3 = OH$, $R^1 = OMe$, $R^4 = R^5 = H$

(XIX) $R^1 = R^2 = R^4 = R^5 = H$, $R^3 = OH$

(XX) $R^1 = R^2 = R^3 = R^4 = OH$, $R^5 = H$

(XXI) $R^1 = R^3 = OH$, $R^2 = R^4 = OMe$, $R^5 = H$

and belladine (XVII) (*Suhadolnik*, *Fischer* and *Zulalian*, *loc. cit.*; *Wildman et al.*, J. Amer. chem. Soc., 1962, **84**, 681; J. chem. Soc., 1964, 1595; *P. W. Jeffs*, Proc. chem. Soc., 1962, 80). It is notable that label from [3-^{14}C]-tyrosine did not appear in the C_6–C_1 unit of any of these alkaloids. A different origin for this fragment is thus indicated and this has been found to be phenylalanine manifestly by way of cinnamic acid.

Label from [3-^{14}C]phenylalanine and [3-^{14}C]cinnamic acid enters the C_6–C_1 unit of representative alkaloids, belladine (XVII), lycorine (VII), and haemanthamine (X) (*Suhadolnik et al.*, J. Amer. chem. Soc., 1962, **84**, 4348; Proc. chem. Soc., 1963, 132, 216; *Wildman et al., loc. cit.*, 1964). It is clear from the results that phenylalanine does not serve as a source for the C_6–C_2 unit, so the C_6–C_1 and C_6–C_2 units have separate origins, and as found elsewhere in biosynthesis in higher plants phenylalanine is not converted into tyrosine.

Incorporations of 4-hydroxycinnamic acid (into haemanthamine), 3,4-dihydroxycinnamic acid (into lycorine and haemanthamine) and protocatechualdehyde (XXIII) (into lycorine and haemanthamine) delineate the steps after cinnamic acid as those shown in Scheme 44 (4-hydroxybenzaldehyde and protocatechuic acid were not precursors) (*Suhadolnik* and *Zulalian*, Proc. chem. Soc., 1963, 216; 1964, 422; *Suhadolnik*, *Fischer* and *Zulalian*, Biochem. Biophys. Res. Comm., 1963, **11**, 208; Proc. chem. Soc., 1963, 132; *A. R. Battersby et al.*, J. chem. Soc., 1964, 1595).

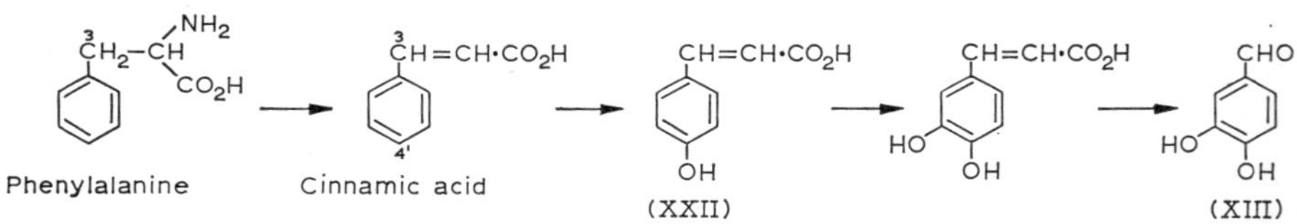

Scheme 44

Incorporation of [4′-^{3}H,3-^{14}C]cinnamic acid into norpluviine (V) was found to occur with a 50% tritium retention and with essentially all of this label confined to the position *ortho* to the methoxyl group (*W. R. Bowman*, *Bruce* and *Kirby*, Chem. Comm., 1969, 1075). This agrees well with the predicted value resulting from *para*-hydroxylation of cinnamic acid with 1,2-tritium migration and retention by primary isotope effect (the so-called "NIH" shift). Subsequent hydroxylation then results in loss of half the residual tritium. Retention of 28% of the tritium from [3′-^{3}H]cinnamic acid suggests that loss of hydrogen/tritium is stereospecific but the correctness of this observation has been questioned by other workers (*B. J. Auret et al.*, *ibid.*, 1971, 1585).

[*Dimethyl*-$^{14}C_2$]-3-hydroxy-4-methoxy-*N*-methylbenzylamine gave radioactive haemanthamine and galanthamine but the *N*-methyl group was lost, from which it follows that incorporation *via* 3-hydroxy-4-methoxybenzaldehyde is likely; negative results, however, have been obtained with this compound (*Barton et al.*, J. chem. Soc., 1963, 4545; *Battersby et al., ibid.*, 1964, 1595). If the amine is really a normal alkaloid precursor, then the oxidation which leads to the aldehyde is an unusual non-stereospecific one since tritium at the benzylic carbon of *N*-methyl-3-hydroxy-4-methoxybenzylamine is retained by primary isotope effect on incorporation into alkaloids (*C. Fuganti, D. Ghiringhelli* and *P. Grasselli*, Tetrahedron Letters, 1974, 2261).

The conversion of phenylalanine into cinnamic acid involves elimination of ammonia and is catalysed by phenylalanine ammonia lyase. The presence of the necessary enzyme in Amaryllidaceae plants has been demonstrated. Elimination of ammonia catalysed by an enzyme isolated from potatoes occurs stereospecifically in an anti-periplanar fashion with loss of the 3-*pro-S* hydrogen atom on phenylalanine. It might be expected that in Amaryllidaceae plants phenylalanine would afford, say, haemanthamine with retention of the 3-*pro-R* proton. However, it has been found that tritium is lost from C-3 of phenylalanine on conversion into haemanthamine (X) and oduline *in vivo* whatever the configuration. Clearly fragmentation of the cinnamic acid skeleton involves oxidation of C-3 to the ketone or acid level (*Battersby et al.*, J. chem. Soc., Perkin I, 1972, 2355; *R. Ife* and *E. Haslam*, J. chem. Soc., C, 1971, 2818; *Suhadolnik*, *Fischer* and *Zulalian*, Biochem. Biophys. Res. Comm., 1963, **11**, 208).

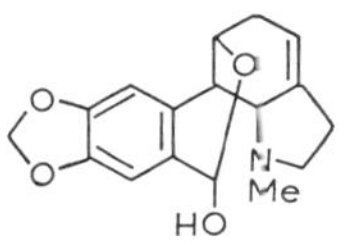

Oduline

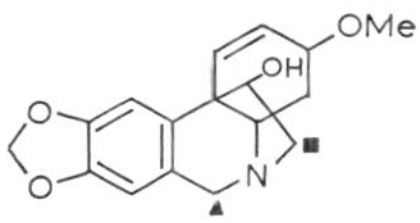

Crinamine

The foregoing results point directly to norbelladine (III) as an intermediate in the biosynthesis of Amaryllidaceae alkaloids. Direct evidence of its role as an intermediate came when it was shown that this base doubly labelled (▲, ■) as shown in III was incorporated intact into lycorine (VII), and that [1-^{14}C]norbelladine was also incorporated into lycorine (VII), haemanthamine (X), and galanthamine (XIII) with labelling confined to the expected site in each alkaloid (*cf.* Scheme 43) (*Battersby et al.*, J. chem. Soc., 1964, 1595; J. Amer. chem. Soc., 1961, **83**, 4098; *Barton et al.*, J. chem. Soc., 1963, 4545). Moreover, *O*-methylnorbelladine (IV) and *NO*-dimethylnorbelladine (IX) were shown to be natural constituents of daffodils (Amaryllidaceae) by dilution analysis after feeding radioactive tyrosine (*Battersby et al., loc. cit.*, 1964, 1595; *Barton et al., loc. cit.*). *N,O*-Dimethylnorbelladine

(IX) has also been shown to be an intact precursor for galanthamine (XIII). Not only was the skeleton incorporated intact but the *N*- and *O*-methyl groups also (*Barton et al., loc. cit.*). [1,O-*Me*-^{14}C]-*O*-methylnorbelladine (as IV) was incorporated specifically into haemanthamine. The observation that the *O*-methyl group of the precursor provided the methylenedioxy group of the alkaloid established for the first time that a methylenedioxy group originates by oxidation of an *O*-methoxyphenol; a plausible mechanism is illustrated (*Barton et al., loc. cit.*; Proc. chem. Soc., 1962, 340). It is interesting to note in regard to the pattern of biosynthesis surrounding norbelladine derivatives that an enzyme has been isolated from *Nerine bowdenii* (a plant used in some of the above experiments) which, when incubated with norbelladine and adenosylmethionine, yields *O*-methylnorbelladine (IV) almost

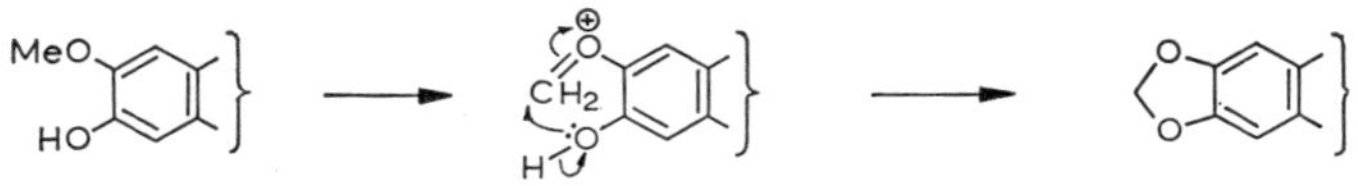

entirely. A very small amount of the isomeric methyl ether XVIII was produced. Significantly this compound is not a precursor for Amaryllidaceae alkaloids (*H. M. Fales, J. Mann* and *S. H. Mudd*, J. Amer. chem. Soc., 1963, **85**, 2025; *G. W. Kirby* and *H. P. Tiwari*, J. chem. Soc., C, 1966, 676). Other methylation patterns appear to be important in the biosynthesis of individual alkaloids. Thus although norbelladine is a precursor for galanthamine and haemanthamine, *O*-methylnorbelladine was incorporated only into haemanthamine and *N*-methylnorbelladine (VIII) and *N,O*-dimethylnorbelladine (IX) only into galanthamine (*Barton et al.*, J. chem. Soc., 1963, 4545). However, *O*-methylnorbelladine (IV) examined as a precursor for galanthamine in a different plant was found to be a comparably efficient precursor as it was for lycorine (*Fuganti*, Chim. Ind., 1969, **51**, 1254). This indicates for galanthamine that although *N,O*-dimethylnorbelladine is likely to be the norbelladine derivative on which coupling occurs there is not necessarily a single sequence of methylation leading to it. It is clear further, as the structures of the alkaloids suggest, that *N*-methylation of the norbelladine skeleton gives a compound which can be used for the biosynthesis of the galanthamine type but not those of the lycorine and haemanthamine kind.

The oxygenation pattern required of a norbelladine derivative for it to be a precursor is exclusive. None of the precursors XVIII, XIX, XX, and XXI were alkaloid precursors (*Battersby et al.*, J. chem. Soc., 1964, 1595; Proc. chem. Soc., 1963, 168; *Kirby* and *Tiwari, loc. cit.*). All the foregoing evidence firmly locates norbelladine (III) and some of its methylated derivatives as the key intermediates in the biosynthesis of various Amaryllidaceae alkaloids and in so doing gives substance to the hypothesis that these alkaloids could be derived by alternative cyclisation reactions of bisphenolic radicals (*cf.* Scheme 43; see also p. 386).

Incorporation into **norpluviine** of *O*-methylnorbelladine (labelled with tritium as shown in XIV) occurred with retention of two of the four tritium atoms, one of which was located at C-2.

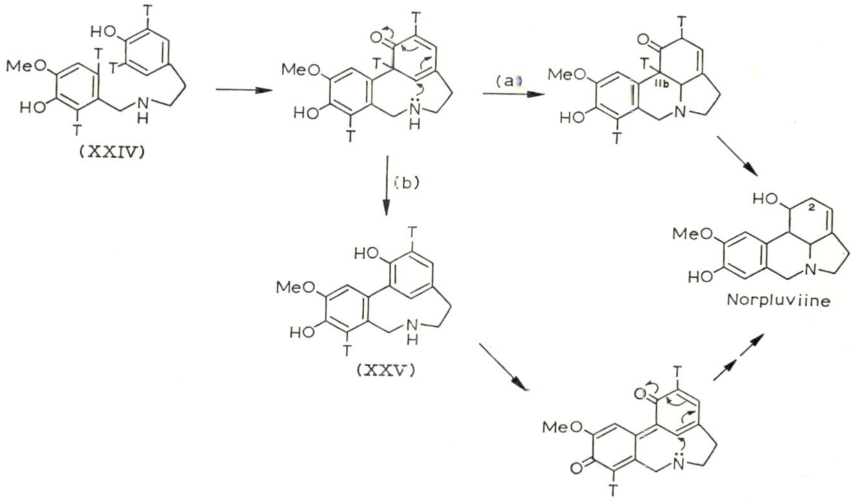

This is inconsistent with an otherwise reasonable pathway (*a*) for the conversion of *O*-methylnorbelladine into norpluviine which requires retention of three tritium atoms unless tritium is lost from C-11b as a result of, say, the intermediacy of the bisphenol XXV. A further oxidative step (and later reductive step) would then be required for the generation of the norpluviine skeleton (pathway *b*) (*Kirby* and *Tiwari*, *loc. cit.*).

Norpluviine serves as an efficient and specific precursor of **lycorine** (VII) and the reaction is not reversible. This *in vivo* conversion establishes that the extra hydroxyl group in lycorine arises by allylic oxidation of norpluviine and in support the tetrahydroxynorbelladine (XX) was not significantly utilized in lycorine biosynthesis (*Battersby et al.*, J. chem. Soc., 1964, 1595). The biological conversion of norpluviine into galanthine (XIV), pluviine (VI), and lycorine indicates that it is an intermediate in the elaboration of more highly oxidized alkaloids of the same skeletal type (*Battersby et al.*, *loc. cit.*; *Kirby* and *Tiwari*, *loc. cit.*).

In the conversion of *O*-methylnorbelladine into lycorine (VII) in daffodils tritium from C-5′ of the precursor appears at C-2 of norpluviine (V), which is formed as an intermediate, and the configuration of this tritium is apparently β (*Kirby* and *Tiwari*, *loc. cit.*). This tritium is retained in the lycorine (VII) subsequently formed, indicating that hydroxylation takes the unusual course in a biological system of occurring with inversion of configuration. This suggests a novel mechanism and the epoxide XXVII is a potential intermediate. Ring-opening followed by allylic rearrangement of the resulting alcohol would give lycorine (VII) (*I. T. Bruce* and *Kirby*, Chem. Comm., 1968, 207; Chimia, 1968, **22**, 314). Supporting evidence comes from the reported incorporation of [2β-^{3}H]caranine (as XXVIII) into lycorine in *Zephyranthes candida*; no incorporation would have been observed unless hydroxylation occurred with at least partial inversion of configuration (*W. C. Wildman* and *N. E. Heimer*, J. Amer. chem. Soc., 1967, **89**, 5265).

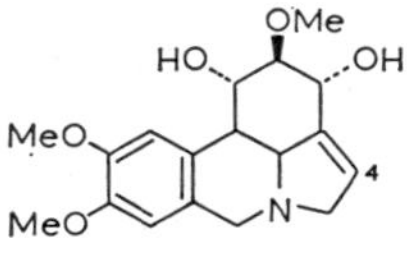

Narcissidine
(XXVI)

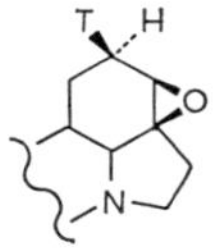

(XXVII)

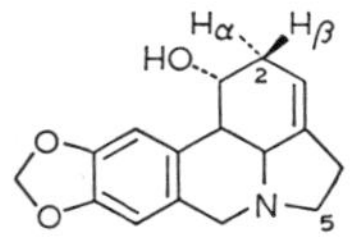

(XXVIII)

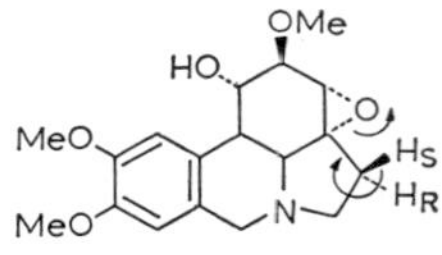

(XXIX)

In *Clivia miniata*, on the other hand, completely opposite results have been obtained. [2α-^{3}H,5-^{14}C]Caranine (as XXVIII) afforded lycorine with high tritium retention and [1-^{14}C]-*O*-methylnorbelladine (as IV) with equal tritium labelling of positions 3′,5′ and 5″ was incorporated into the alkaloid with retention of the C-5″ tritium label only (*cf.* opposite results for the C-5′ proton above). It follows that, in this plant, hydroxylation is stereospecific (essentially complete tritium loss/retention) and occurs with normal retention of stereochemistry. (The caranine result further indicates that a C-2 ketone is not an intermediate) (*C. Fuganti* and *M. Mazza*, Chem. Comm., 1972, 936).

In considering the biosynthesis of **narcissidine** (XXVI) it seemed possible that it might be formed *via* the epoxide XXIX which is analogous to XXVII discussed above. The naturally occurring base galanthine (XIV) was further considered a likely substrate for epoxidation. Results of experiments with *O*-methylnorbelladine (IV) samples chirally and randomly labelled at C-2, however, established that in the formation of narcissidine the 2-*pro-R* hydrogen of the *O*-methylnorbelladine (corresponding to the 4-*pro-S* proton of XXIX) was retained at C-4 of the alkaloid and the 2-*pro-S* hydrogen was lost. The stereochemical consequences of this result which is the opposite of that expected argue against an intermediate of type XXIX in narcissidine biosynthesis. The efficient conversion of galanthine (formed from the *O*-methylnorbelladine without tritium loss) into narcissidine does, however, establish it as a precursor for narcissidine (*C. Fuganti*, *D. Ghiringhelli* and *P. Grasselli*, Chem. Comm., 1974, 350).

The biological conversion of protocatechualdehyde into **lycorenine** involves the formal addition of hydrogen to the aldehyde carbon atom and subsequent removal on hydroxylation of the pluviine intermediate. [1-^{14}C,1′-^{3}H]-*O*-Methylnorbelladine was incorporated into pluviine and lycorenine, the former alkaloid being formed without tritium loss. Lycorenine was, however, formed with loss of half the tritium, and this accords with a stereospecific hydroxylation at C-7 (in a non-stereospecific reaction tritium would largely be retained as a result of a primary isotope effect).

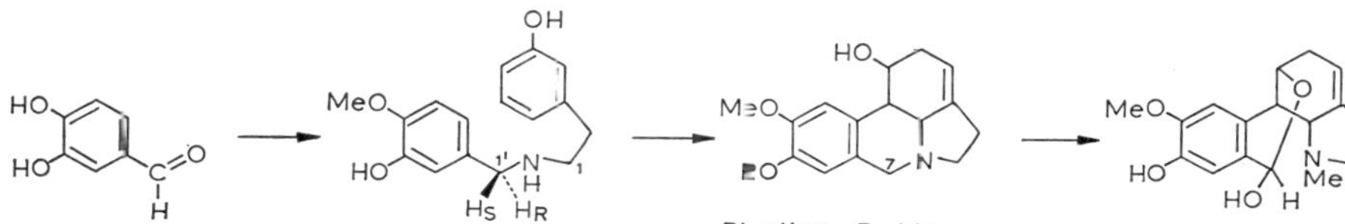

Protocatechualdehyde — *O*-Methylnorbelladine — Pluviine, R = Me; Norpluviine, R = H — Lycorenine

[*Formyl*-^{3}H]protocatechualdehyde gave [7-^{3}H]norpluviine, which was then transformed by the plant into lycorenine with almost complete tritium retention at C-7. Both hydrogen addition and removal are stereospecific therefore. Moreover, it is the hydrogen initially introduced which is later removed (an exactly similar conclusion was reached for the formation of haemanthidine - see below) (*Fuganti* and *Mazza*, Chem. Comm., 1971, 1196). (1′-*R*)-[1′-^{3}H,1-^{14}C]-*O*-Methylnorbelladine gave norpluviine without loss of tritium. This material was converted, *in vivo*, into lycorenine with loss of the tritium label (*idem*, J. chem. Soc., Perkin I, 1973, 954). It follows from the combined results that in the sequence protocatechualdehyde → *O*-methylnorbelladine → norpluviine → lycorenine, hydrogen addition and removal takes place on the *re*-face (see *K. R. Hanson*, J. Amer. chem. Soc., 1966, **88**, 2731) of the molecules concerned.

Alkaloids, which can arise by *para–para* coupling (see Scheme 43) of a norbelladine derivative as an alternative to the *ortho–para* coupling associated with the biosynthesis of norpluviine and its congeners, may be exemplified by **haemanthamine** (X), **haemanthidine** (XV) and **tazettine** (XVI). They are all produced by the plant *Sprekelia formosissima*, and incorporation and rate studies establish the sequence of biosynthesis as haemanthamine → haemanthidine → tazettine. (*Wildman* and *Fales*, J. Amer. chem. Soc., 1964, **86**, 294). Further experiments have established that in the conversion of protocatechualdehyde through *O*-methylnorbelladine and haemanthamine (X) to haemanthidine (XV), hydrogen addition and abstraction from what is formally the carbonyl carbon of the aldehyde, occurs in stereospecific fashion. Moreover it is the hydrogen initially introduced which is later removed (see above for a more detailed discussion of the exactly similar evidence adduced for an analogous pathway to lycorenine) (*Fuganti* and *Mazza*, Chem. Comm., 1971, 1196).

The stereochemistry of the hydroxylation reaction in haemanthamine biosynthesis is of interest particularly in view of the exceptional and differing results obtained for the step of hydroxylation which converts norpluviine into lycorine (see above). Complementary results obtained with *O*-methylnorbelladine (IV) stereospecifically labelled with tritium at C-2 and tyrosine similarly labelled at C-3, establish nicely that hydroxylation occurs with retention of configuration (*Battersby et al.*, J. chem. Soc., Perkin I, 1973, 1609; *Kirby* and *J. Michael*, Chem. Comm., 1971, 415; see also *Fuganti*, *Ghiringhelli* and *Grasselli*, *ibid.*, 1974, 350). The results accord with a substantial body of evidence associated with hydroxylation at saturated carbon (*R. Bentley*, "Molecular Asymmetry in Biology", Academic Press, New York, 1970, Vol. 2; see also refs. cited by authors above).

Narciclasine is found in *Narcissus* plants together with haemanthamine, lycorine and norpluviine. As we have seen haemanthamine derives from

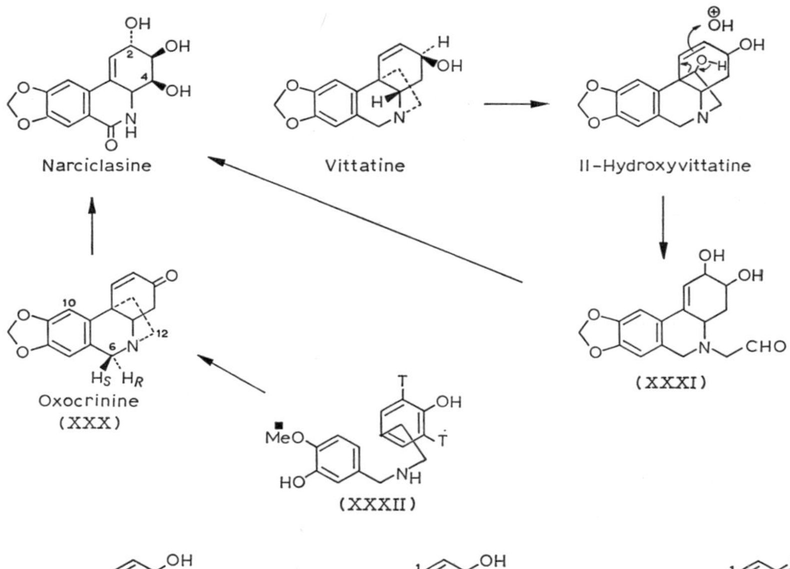

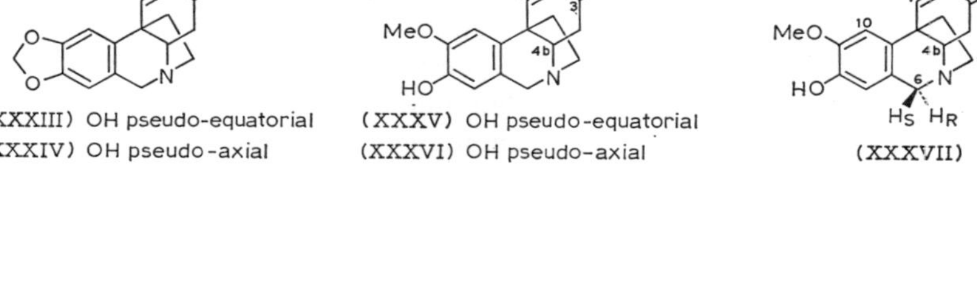

norbelladine in a manner which differs from that which leads to norpluviine and lycorine. The structure of narciclasine is such that it could *ab ovo* have a genesis along either of these pathways, and it thus presents an interesting biosynthetic problem.

O-Methylnorbelladine labelled as shown in XXXII (p. 396) was specifically incorporated into narciclasine with a tritium retention (75%) significantly higher than for norpluviine (V) and lycorine (VII) (50%, as expected) although lower than for haemanthamine (X) (100%, as expected); the tritium in narciclasine was confined to C-2 and C-4. Consideration then of the two possible routes to narciclasine leads to the conclusion that it must be formed by a route involving *para–para* coupling of *O*-methylnorbelladine as for haemanthamine with loss of the two carbon bridge seen in the latter. This is corroborated by the finding that [^{3}H]oxocrinine was a specific precursor for narciclasine (*C. Fuganti, J. Staunton* and *A. R. Battersby*, Chem. Comm., 1971, 1154). Similar results were obtained with (±)-[3-^{3}H]crinine (XXXIV), (±)-[1,3,4b-$^{3}H_3$]normaritidine (XXXVI), and (±)-[1,4b-$^{3}H_2$]noroxomaritidine (XXXVII) but labelled epicrinine (XXXIII) and epinormaritidine (XXXV) were not significantly incorporated. Further it is vittatine and not its enantiomer crinine (as XXXIV) which is the correct intermediate in narciclasine biosynthesis. The results lead to the conclusion that narciclasine biosynthesis passes through intermediates of the crinine (XXXIV) type with a pseudo-axial hydroxy group at C-3 and that oxidation of a hydroxyl group once formed at this centre does not occur (retention of tritium at C-3). Further, formation of the methylenedioxy group can be independent of the particular oxidation level at C-3. Analogous incorporations were recorded for haemanthamine in these experiments (*Fuganti* and *Mazza*, Chem. Comm., 1971, 1388; 1972, 239). Loss of the two-carbon bridge from vittatine could plausibly proceed by way of a retro-Prins reaction on 11-hydroxyvittatine to give XXXI, analogy for hydroxylation of vittatine coming from the biosynthesis of the related alkaloid haemanthamine. Strong support for this pathway to narciclasine was obtained when it was shown that tritiated 11-hydroxyvittatine was an efficient precursor for this alkaloid (*Fuganti*, Gazz., 1973, **103**, 1255). It is interesting to note that collapse of a similar *β*-hydroxyamine is proposed for a similar step in the biosynthesis of colchicine (see p. 426).

The alkaloid **ismine,** like narciclasine, is derived from oxocrinine (XXX) and noroxomaritidine (XXXVII). In particular [6,10-$^{3}H_2$,12-^{14}C]noroxomaritidine gave ismine with loss of the ^{14}C label and half the tritium from C-6 which indicates, respectively, that C-12 was not retained as the *N*-methyl group of ismine and that hydrogen abstraction from C-6 was likely to be stereospecific. The latter conclusion was confirmed with noroxomaritidine

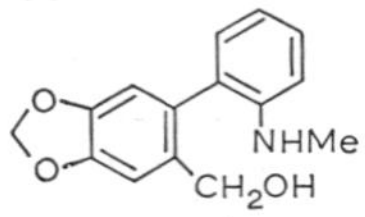

Ismine

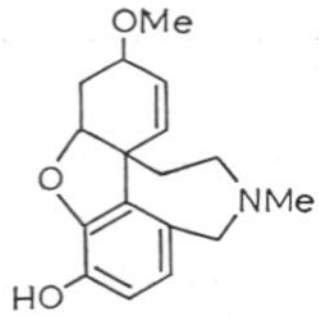

Chlidanthine

stereospecifically tritiated at C-6 and it was shown that, as in the conversion of haemanthamine into haemanthidine and norpluviine into lycorenine, the *pro-R* benzylic hydrogen was lost (*Fuganti* and *Mazza*, Chem. Comm., 1970, 1466; *Fuganti*, Tetrahedron Letters, 1973, 1785).

N,O-Dimethylnorbelladine (IX) is probably the last of the norbelladine derivatives in the biosynthesis of the third group of Amaryllidaceae alkaloids, exemplified by **galanthamine** (XIII) (*Barton et al.*, J. chem. Soc., 1963, 4545). Appropriate *ortho–para* oxidative coupling of the dimethylnorbelladine leads to the dienone intermediate XI (p. 388). The intermediacy of this compound is substantiated by the fact that the simple ether derivative narwedine (XII) is a naturally occurring alkaloid. Moreover, narwedine is an efficient and specific precursor for galanthamine (Scheme 43). In turn this alkaloid is a precursor for chlidanthine (*J. G. Bhandarkar* and *G. W. Kirby*, J. chem. Soc., C, 1970, 1224).

(*c*) *Mesembrine alkaloids*

There is an apparent structural kinship between **mesembrine** (XXXVIII) and Amaryllidaceae alkaloids of the haemanthamine (X) kind, but careful investigation has shown that the only aspect of biosynthesis common to these two groups of alkaloids is their origin from the same amino acids, phenylalanine and tyrosine. Thus in feeding experiments with a variety of radioactive norbelladine derivatives (key intermediates in Amaryllidaceae alkaloid biosynthesis) incorporation into mesembrine alkaloids was obtained only after fragmentation of the precursors. (In these experiments powerful use was made of doubly labelled precursors; singly labelled ones would have led to the wrong conclusions) (*P. W. Jeffs et al.*, Phytochemistry, 1974, **13**, 933). Tracer experiments with phenylalanine and tyrosine on the other hand clearly establish skeletal origins for mesembrine bases which are similar to those for the Amaryllidaceae alkaloids. Tyrosine provides the C_6–C_2 unit: both [2-^{14}C]- and [3-^{14}C]-tyrosine were incorporated with label from the latter confined to C-2 and/or C-3 of mesembrine. The former precursor was proved to label C-2 of mesembrenol. Tyrosine is thus the precursor of the octahydroindole moiety of these alkaloids (*idem*, J. Amer. chem. Soc., 1971, **93**, 3752). Later experiments showed that tyramine and *N*-methyltyramine were also precursors, which defines biosynthesis from tyrosine as through tyramine and then *N*-methyltyramine (*idem.*, Chem. Comm., 1976, 82). The unusual, if not unique, mesembrine C_6 unit, derives from the aromatic nucleus of phenylalanine as seen with the incorporation of [*Ar*-U-^{14}C]phenylalanine; phenylalanine labelled in the side-chain gave inactive alkaloid. Radioactive methionine labelled the methylenedioxy and *N*-methyl groups of mesembrine (*idem, loc. cit.*, 1971).

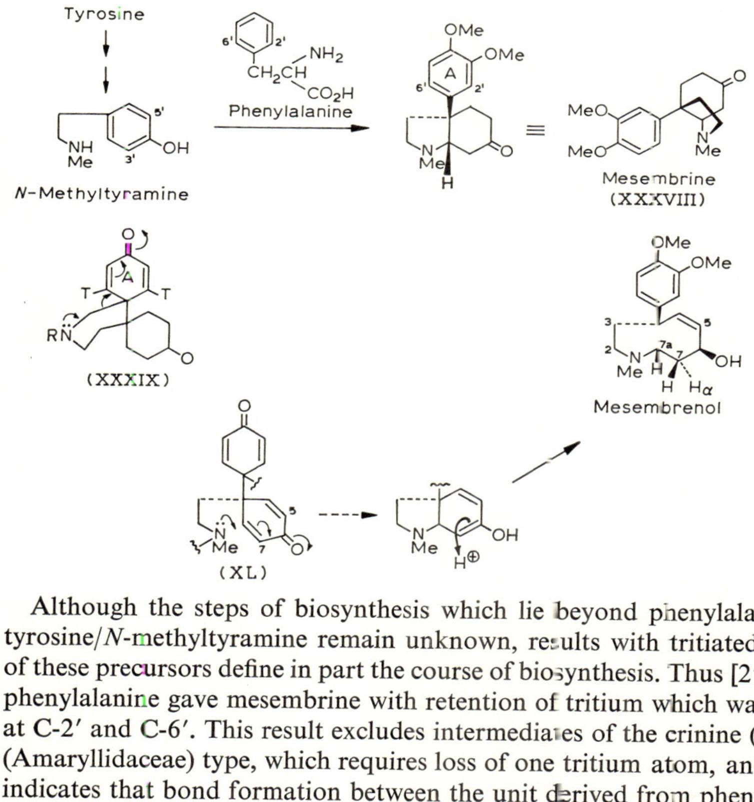

Although the steps of biosynthesis which lie beyond phenylalanine and tyrosine/*N*-methyltyramine remain unknown, results with tritiated samples of these precursors define in part the course of biosynthesis. Thus [2′,6′-$^3H_2$]-phenylalanine gave mesembrine with retention of tritium which was located at C-2′ and C-6′. This result excludes intermediates of the crinine (XXXIV) (Amaryllidaceae) type, which requires loss of one tritium atom, and further indicates that bond formation between the unit derived from phenylalanine and the one arising from tyrosine must occur at the carbon atom corresponding to C-1′ of phenylalanine. This leads logically to a ring A spirodienone (XXXIX) which, in order to retain both tritiums, must undergo aromatization by a process of the type illustrated (a notable analogy exists in the biosynthesis of *Erythrina* alkaloids: see p. 422) (*Jeffs et al.*, Phytochemistry, 1974, **13**, 933). This dienone formulation can be reconsidered as a bisdienone of the type (XL) (*idem*, J. org. Chem., 1974, **39**, 2703) to accommodate the cyclisation of nitrogen on to C-7a and some results bearing on this have been obtained.

Tyrosine, tyramine, and *N*-methyltyramine with tritium at C-3′ and C-5′ afforded mesembrenol with loss of 50% of the tritium, residual label being equally distributed between H-5 and H-7α. The results are consistent with the pathway shown involving internal conjugate addition of nitrogen followed by stereospecific β-protonation of the resulting enolate at C-7. Since tritium is equally distributed between C-5 and C-7α in mesembrenol tritium loss probably occurs while the ring is still aromatic (on dienone formation these positions become non-equivalent) and it may be that substitution by a group X occurs with later replacement by hydrogen (*idem*, Chem. Comm., 1976, 82).

(*d*) *Isoquinoline alkaloids*

(*i*) *Phenethylamines and simple isoquinolines*

N-**Methyltyramine** and **hordenine,** found in the roots of the barley plant, are the simplest of the phenethylamine alkaloids. Their genesis is from tyrosine *via* tyramine with methionine serving as the source of the *N*-methyl groups (*L. Marion et al.*, Canad. J. Chem., 1952, **30**, 749; 1953, **31**, 126; 1954, **32**, 646); it appears that demethylation of hordenine to give *N*-methyltyramine can occur (*A. W. Frank* and *Marion*, *ibid.*, 1956, **34**, 164). [In a different plant (*Echinocereus merkei*) no incorporation of dopamine into hordenine was observed nor into salsoline (XLII) (similar result with tyrosine) but tyrosine and dopamine were incorporated into 3,4-dimethoxyphenethylamine. On the other hand 3-hydroxy-4-methoxyphenethylamine served as a precursor for both salsoline and 3,4-dimethoxyphenethylamine (*I. J. McFarlane* and *M. Slaytor*, Phytochemistry, 1972, **11**, 235)]

Mescaline, which is one of the earliest known hallucinogenic substances, has the phenethylamine structure XLIII. It is elaborated along with simple isoquinoline alkaloids, *e.g.* anhalamine (XLV), by the peyote cactus, *Lophophora williamsii*. Results of extensive experiments with various hydroxylated/methylated precursors, particularly phenethylamines, have provided a detailed description of the pathways to both groups of alkaloids. Further substantiation has been obtained by identifying the precursors as naturally occurring bases (for a review see *J. Lundström*, Acta Pharm. Suecica, 1971, **8**, 275). Tyrosine is a precursor for mescaline in this plant and in *Trichocereus pachanoi* as is dopamine, with tyramine as an intermediate (and alternatively, dopa) (*E. Leete*, Chem. and Ind., London, 1959, 604; *H. Rosenberg*, *J. L. McLaughlin* and *A. G. Paul*, Lloydia, 1967, **30**, 100; *Lundström et al.*, Tetrahedron Letters, 1967, 2433; 1969, 3371; Acta Pharm.

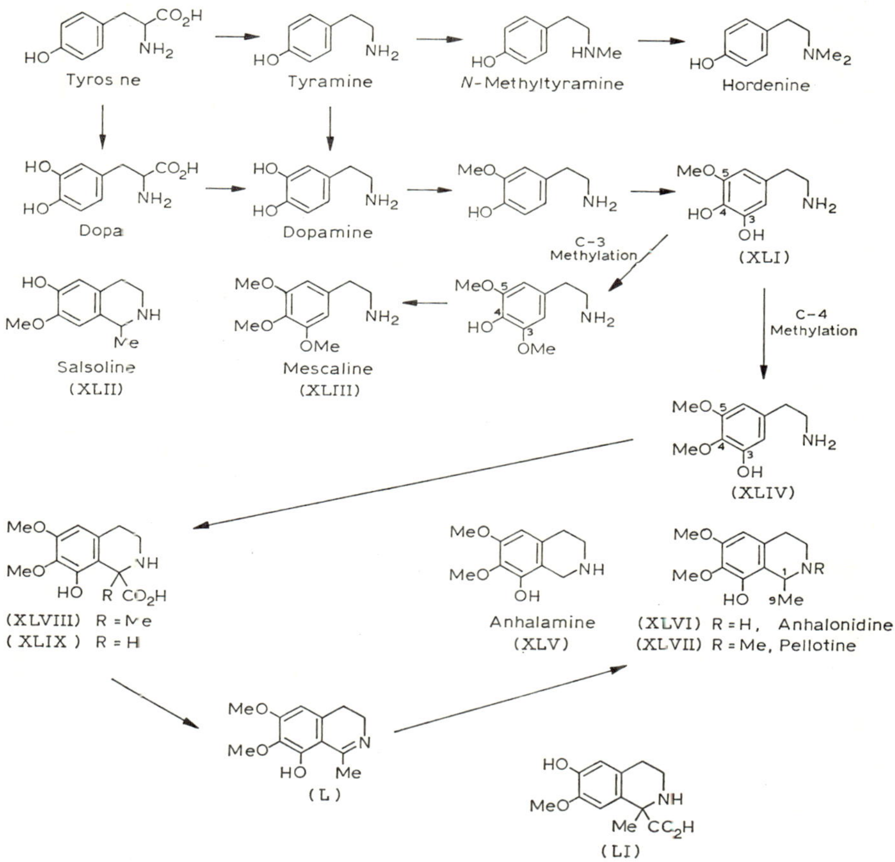

Suecica, 1970, **7**, 651; Acta Chem. Scand., 1971, **25**, 3489). Methylation of dopamine is the next step in mescaline biosynthesis which is followed by hydroxylation to give 3,4-dihydroxy-5-methoxyphenethylamine (XLI) (an alternative minor sequence has 3,4,5-trihydroxyphenethylamine as an intermediate). Finally, *O*-methylation of the C-3 hydroxyl group of this compound gives 4-hydroxy-3,5-dimethoxyphenethylamine, the immediate precursor for mescaline (*Rosenberg et al.*, Lloydia, 1969, **32**, 36, 334; *Lundström* and *S. Augurell*, Tetrahedron Letters, 1969, 3371; *Lundström*, Acta Chem. Scand., 1971, **25**, 3489).

3,4-Dihydroxy-5-methoxyphenethylamine (XLI) stands at the junction of mescaline and isoquinoline biosynthesis for methylation of the C-4 hydroxyl group leads to **anhalonidine** and **anhalamine** but not mescaline

(*Lundström* and *Agurell, loc. cit.*; Acta Pharm. Suecica, 1971, **8**, 261; *A. G. Paul et al.*, Chem. Comm., 1969, 838). Less satisfying results were obtained for **pellotine** (XLVII) but 3-hydroxy-4,5-dimethoxy-*N*-methyl-phenethylamine is an intact and efficient precursor for this alkaloid (*Lundström*, Acta Pharm. Suecica, 1971, **8,** 485). These observations on phenylethylamine and isoquinoline biosynthesis have been supported by the results of a study on the course of methylation on various precursors using an *O*-methyltransferase isolated from *L. williamsii* (*G. P. Basmadjian* and *Paul*, Lloydia, 1971, **34,** 91).

Delineation of the biosynthetic pathways to the isoquinoline bases as far as the phenethylamine (XLIV) may be considered largely predictable. However, the next step, that of including a C_1 unit (anhalamine) or a C_2 unit (anhalonidine and pellotine) in the construction of the isoquinoline skeletons, was for a long time quite elusive. Although activity from [1-^{14}C]- and [2-^{14}C]-acetate was incorporated into C-1 and -9 of pellotine it was scrambled over these atoms (*A. R. Battersby, R. Binks* and *R. Huxtable*, Tetrahedron Letters, 1967, 563). [3-^{14}C]Pyruvate gave anhalonidine again with scatter of label over the C_2 unit but with a major portion confined to C-9, as expected if pyruvate was the source of this C_2 unit (*Leete* and *J. D. Braunstein, ibid.*, 1969, 451; *Lundström, loc. cit.*). Following on from these results, *N*-acetyl-3-hydroxy-4,5-dimethoxyphenethylamine was examined as a precursor for anhalonidine but with negative results (a similar result was obtained for the incorporation of the *N*-formyl compound into anhalamine) (*G. J. Kapadia et al.*, J. Amer. chem. Soc., 1970, **92**, 6943; *Lundström, loc. cit.*).

The solution to the problem was found to be hidden in the old literature. It had been suggested that for, *e.g.*, anhalonidine, the sequence of biosynthesis involved condensation of a phenethylamine with pyruvic acid leading to the amino-acid XLVIII, which was decarboxylated to give the alkaloid (*G. Hahn et al.*, Ann., 1935, **520**, 107; Ber., 1936, **69**, 2627; 1938, **71**, 2141). Examination of XLVIII as the potential key intermediate in anhalonidine biosynthesis has proved conclusively the essential correctness of this hypothesis. It is specifically incorporated into anhalonidine, is a natural constituent of peyote cactus, and is decarboxylated to the imine L on incubation with fresh peyote slices. Anhalamine is formed in a similar way from the amino acid XLIX derived from glyoxylic rather than pyruvic acid (*Kapadia et al., loc. cit.*). The amino acid LI, which is found in peyote was also shown to be a biosynthetic intermediate, this time in the formation of **salsoline** (XLII) (*I. J. McFarlane* and *M. Slaytor*, Phytochemistry, 1972, **11**,

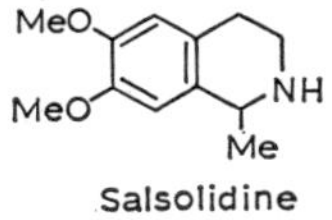

Salsolidine

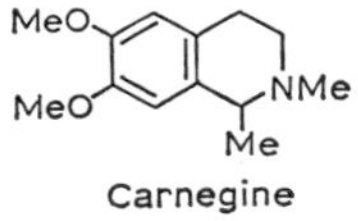

Carnegine

235). A similar (same) amino acid is a likely intermediate in the biosynthesis of salsolidine and carnegine whose early stages of biosynthesis have been delineated along the expected lines (*J. Bruhn, U. Svensson* and *S. Agurell*, Acta Chem. Scand., 1970, **24**, 3775). It is interesting that following the establishment of the significance of these amino acids in simple isoquinoline biosynthesis, similar results have been obtained for benzylisoquinolines (Section 6d, iii).

Lophocerine is unusual among the cactus isoquinoline alkaloids in having an isobutyl group at C-1. This group together with C-1 forms a C_5 unit which is specifically labelled by mevalonic acid and leucine, with the former precursor being slightly more efficiently utilized (tyrosine serves as the source

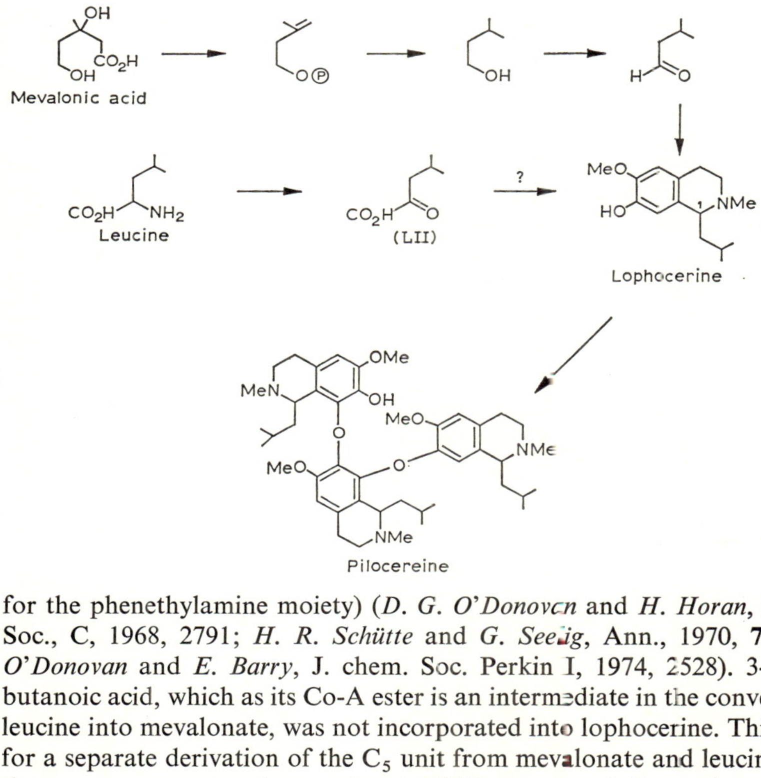

for the phenethylamine moiety) (*D. G. O'Donovan* and *H. Horan*, J. chem. Soc., C, 1968, 2791; *H. R. Schütte* and *G. Seelig*, Ann., 1970, **730**, 186; *O'Donovan* and *E. Barry*, J. chem. Soc. Perkin I, 1974, 2528). 3-Methylbutanoic acid, which as its Co-A ester is an intermediate in the conversion of leucine into mevalonate, was not incorporated into lophocerine. This argues for a separate derivation of the C_5 unit from mevalonate and leucine rather than a common one *via* mevalonate (*O'Donovan* and *Barry*, *loc. cit.*); the route from mevalonate suggested is illustrated. It is interesting to note that this perforce includes an aldehyde in contrast to the α-oxo-carboxylic acids utilized in the biosynthesis of other isoquinolines discussed above. On the other hand leucine is simply implicated *via* the oxo acid LII, its transamination product.

Pilocereine occurs naturally with lophocerine and is transparently the product of phenol oxidative coupling (see Section 6a) of three lophocerine units. In support [N-*methyl*-^{14}C]lophocerine was specifically incorporated into pilocereine (*O'Donovan* and *Horan*, J. chem. Soc., C, 1969, 1737).

The naturally occurring **β-hydroxyphenethylamines** appear to have a genesis from tyrosine similar to that of the phenethylamines. Synephrine arises in *Citrus* from tyrosine *via* tyramine and *N*-methyltyramine (*T. A.*

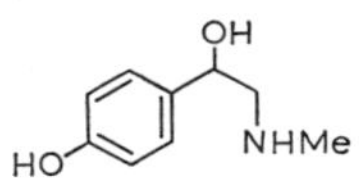

Synephrine

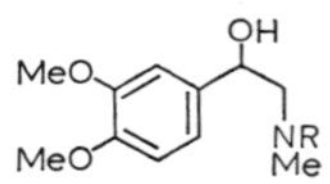

Normacromerine, R = H
Macromerine, R = Me

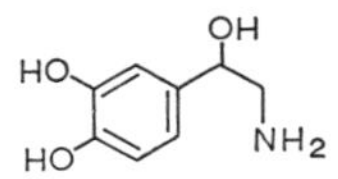

Norepinephrine

Wheaton and *I. Stewart*, Phytochemistry, 1969, **8**, 85). Normacromerine and macromerine, found in the cactus *Coryphantha macromeris* var. *runyonii*, incorporate label specifically from radioactive tyrosine, tyramine, dopa, and dopamine. Norepinephrine was a reasonably efficient precursor for normacromerine which indicates that the distinguishing β-hydroxylation can be an early step whilst the occurrence of synephrine in this cactus indicates that *N*-methylation can likewise occur early. Very low incorporations of the precursors into macromerine suggested that *N*-methylation of normacromerine is slow (*W. J. Keller*, *L. A. Spitznagel* and *L. R. Brady*, Lloydia, 1973, **36**, 397).

The plant base **ephedrine** is unusual among the phenethylamines in having an extra *C*-methyl, as well as a hydroxyl group, on the side-chain. In spite of its structural kinship with the other phenethylamines its genesis is dif-

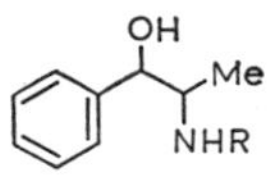

Ephedrine, R = Me
Norpseudoephedrine, R = H

ferent. It is formed from phenylalanine (*via* cinnamate) but with incorporation only as a C_6–C_1 unit; specific incorporations at a higher level of [*carboxyl*-^{14}C]benzoate and [*carbonyl*-^{14}C]benzaldehyde were also obtained. From this it is possible that ephedrine like chloramphenicol derives directly from shikimic acid (*via* benzoic acid in this case) rather than through phenylalanine. Methionine serves as the source of the *N*-methyl group of ephedrine but the origin of the remaining C_2N group is obscure. On the one hand results with C_2 and C_3 amino-acids were not definitive and on the other, relatively high incorporations of radioactive glucose and aspartate into this fragment indicates that aspartate or close relative may be its source (*S. Shibata et al.*, Phytochemistry, 1973, **12**, 2877).

The apparently related base, norpseudoephedrine, has been shown, like ephedrine, to incorporate activity from [3-^{14}C]phenylalanine (*E. Leete*, Chem. and Ind., London, 1958, 1088).

Neither phenylalanine nor tyrosine serves as a direct precursor for **chloramphenicol** in *Streptomyces venezuelae*. Instead it derives from the aromatic amino acid precursor, shikimic acid (*D. Gottlieb et al.*, J. Bact., 1962, **84**, 888; *L. C. Vining* and *D. W. S. Westlake*, Canad. J. Microbiol., 1964, **10**,

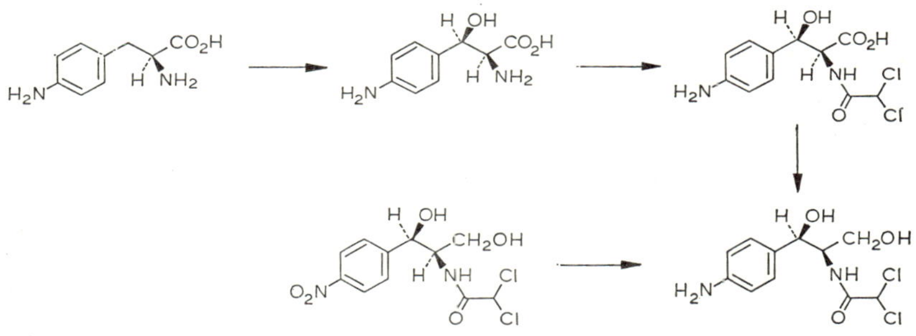

705; *Vining et al.*, Canad. J. Microbiol., 1974, **20**, 347); chorismic and prephenic acids, potential intermediates after shikimic acid, were unable to penetrate the cell walls (*idem, loc. cit.*). Further experiments establish the later stages of biosynthesis as those shown (*Vining*, *V. S. Malik*, and *Westlake*, Lloydia, 1968, **31**, 355).

(*ii*) *Phenylisoquinoline alkaloids*

A unique 1-phenyl-substituent is found in the cryptostyline alkaloids, *e.g.* **cryptostyline-I**, which makes them a group of alkaloids of particular biosynthetic interest. The biosynthesis of the skeleton, apart from the phenyl group and C-1, is accounted for in terms of an evidently orthodox pathway from tyrosine and dopa through dopamine and the phenethylamine LIII (with the hydroxyl group *para* to the position at which ring closure eventually occurs) but not LIV (*Agurell et al.*, Acta Chem. Scand., 1974, **B28**, 239; 1175).

Both dopamine and vanillin (but not isovanillin) can serve as precursors for the remaining C_6-C_1 residue in cryptostyline-I (*idem, loc. cit.*). However, the low level of vanillin incorporation observed suggests that the C_6–C_1 precursor for the Amaryllidaceae alkaloids, namely protocatechualdehyde

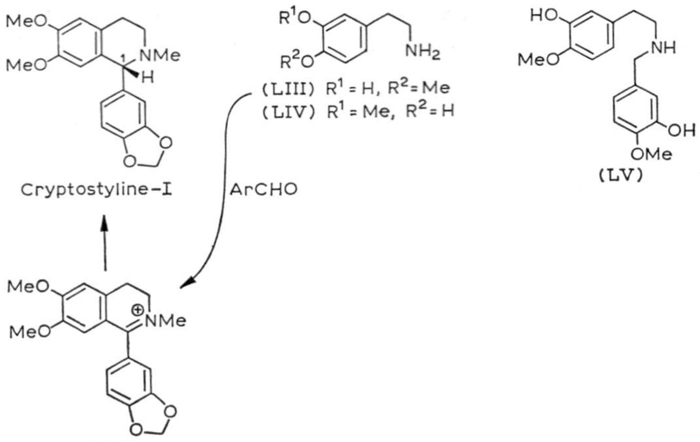

(Section 6b), may be the true precursor for the cryptostyline C_6–C_1 unit. (It has not yet been tested). It is interesting to note that cryptostyline-I unlike many other isoquinoline alkaloids evidently arises by combination of a phenethylamine with an aldehyde rather than an oxo acid (*cf.* lophocerine though, p. 403).

The immonium salt LVI is a highly efficient precursor for cryptostyline-I and a role for it as an intermediate in biosynthesis is thereby suggested. The amine LV of lower oxidation level than cryptostyline-I is an inefficient precursor (*idem, loc. cit.*).

(iii) Benzylisoquinoline alkaloids

Reticuline and **papaverine.** Of all the variations in oxygenation and methylation which can occur on the 1-benzyltetrahydroisoquinoline skeleton none is more important for the formation of alkaloids based on this skeleton than that found in reticuline LVII. As a result of studies with alkaloids derived *via*

Reticuline, (LVII) R = Me
Nor-reticuline, (LVIII) R = H
(LIX) R = CO_2H
(LX) R = H

reticuline much is known about its biosynthesis too (see below). Confirmation of its mode of genesis from two molecules of tyrosine has been obtained (*S. Tewari, D. S. Bhakuni* and *R. S. Kapil*, Chem. Comm., 1975, 554). Dopa is incorporated essentially only *via* dopamine which is the precursor for ring A plus attached C_2 unit (*cf.* Scheme 44).

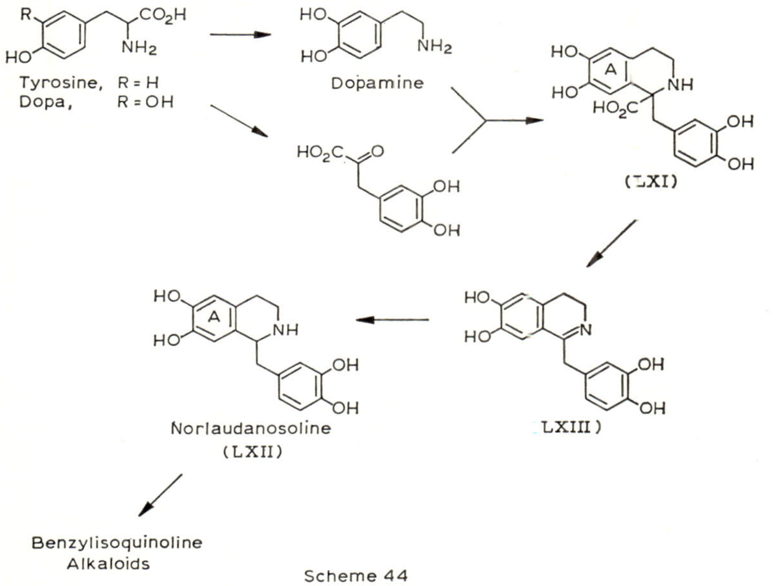

Scheme 44

The tetrahydroxy-amino acid (LXI) is an intermediate in benzylisoquinoline biosynthesis but probably not the trihydroxy compound LIX; also involved is LXIII (*A. R. Battersby, R. C. F. Jones* and *R. Kazlauskas*, Tetrahedron Letters, 1975, 1873; *M. L. Wilson* and *C. J. Coscia*, J. Amer. chem. Soc., 1975, **97**, 431). In support, neither LIX nor LX is a precursor for reticuline. Examination of methylated derivatives of norlaudanosoline (LXII) as precursors suggests that appropriate *O*-methylation precedes *N*-methylation in reticuline biosynthesis (*Tewari, Bhakuni* and *Kapil, loc. cit.*).

An early hypothesis for papaverine biosynthesis (*E. Winterstein* and *G. Trier*, "Die Alkaloide", Bornträger, Berlin, 1910, p. 307) was that the alkaloid was derived by way of norlaudanosoline (LXII) from two molecules of dopa, in turn formed from tyrosine. The specific incorporation of two molecules of [2-^{14}C]tyrosine and of [1-^{14}C]norlaudanosoline provides con-

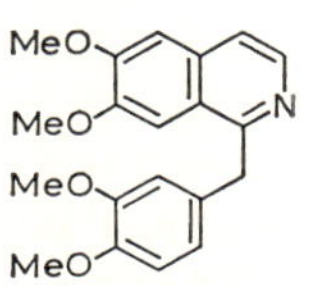

Papaverine

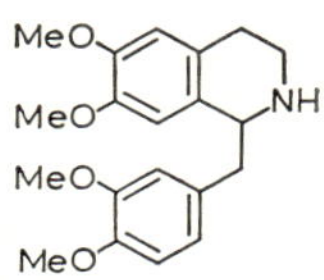

Tetrahydropapaverine

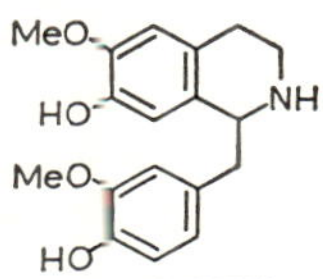

(LXIV)

vincing support for the essential correctness of the hypothesis; as elsewhere dopa may be involved essentially in the formation of one half only (*Battersby* and *B. J. T. Harper*, J. chem. Soc., 1962, 3526; *Battersby et al., ibid.*, 1964, 3600).

Detailed examination of various *O*- and *N*-methylated derivatives of norlaudanosoline points to tetrahydropapaverine as an important intermediate probably formed from nor-reticuline (LVIII) or (LXIV). It is interesting to note that in this case the methylation pattern is apparently important for deciding pathways when elsewhere in the biosynthesis of alkaloids a particular methylation pattern is generally essential only for phenol oxidative coupling. More conventionally the stereochemistry of the precursors is important — here the preferred stereoisomers are (—)norreticuline and (—)tetrahydropapaverine (*E. Brochmann-Hanssen et al.*, J. chem. Soc., Perkin I, 1975, 1531; *H. Uprety, Bhakuni* and *Kapil*, Phytochemistry, 1975, **14**, 1535).

Aporphine alkaloids. A large number of alkaloids with the aporphine skeleton, *e.g.* bulbocapnine, are found in plants. The kinship with the benzylisoquinoline skeleton of say norlaudanosoline is close, a single new bond needing to be forged in the conversion of benzylisoquinoline into aporphine type. This can reasonably be envisaged as occurring by phenol oxidative coupling (Section 6a), but consideration of plausible reaction sequences leads to several possibilities for the formation of this simple common bond. Interestingly, examples of most of these possibilities have been deduced for the biosynthesis of one or more aporphine alkaloids.

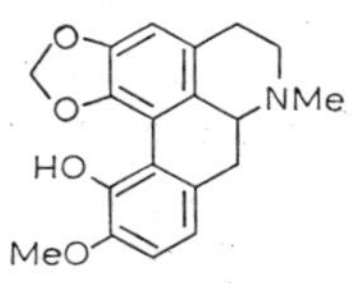

Bulbocapnine

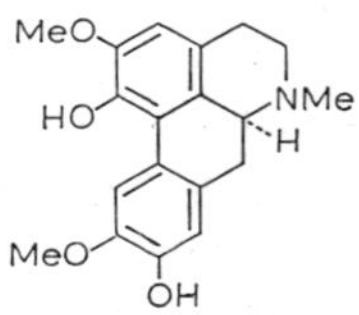

Isoboldine
(LXV)

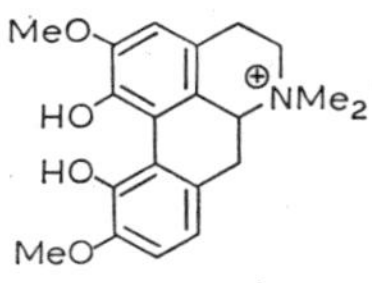

Magnoflorine

In general studies of aporphine alkaloid biosynthesis begin with examination of possible benzylisoquinoline precursors, on the no doubt correct assumption that the derivation of the benzylisoquinoline skeleton is the same as that deduced elsewhere (see other alkaloids in Section 6d, iii). The simplest conversion of a benzylisoquinoline skeleton into the aporphine type is seen in the conversion of reticuline into **bulbocapnine** manifestly by *ortho–ortho* phenol-oxidative coupling (a dienone intermediate is not, however, excluded). This was confirmed by showing that tritium from C-6′ (but not of course the equivalent C-2′) of reticuline in *Corydalis cava* was retained in the derived bulbocapnine. Moreover, reticuline is a natural constituent of *C. cava* (*G. Blaschke*, Arch. Pharm., 1968, **301**, 432; 1970, **303**, 358; *Blaschke et al.*, *ibid.*, 1974, **307**, 122).

Reticuline is also a precursor for **isoboldine** (LXV), a minor alkaloid of *Papaver somniferum* (reticuline is also the precursor of the hydrophenanthrene bases in this plant; see p. 414). Orientaline (LXVII) with an alternative methylation pattern is not incorporated, thus indicating a unique

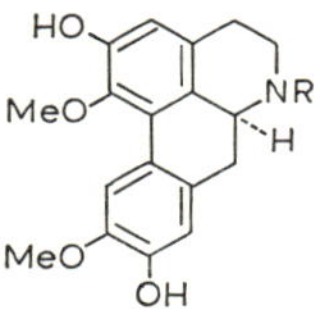

Boldine, R = Me (LXVI)
Norboldine, R = H

pathway *via* reticuline, a pathway which involves simple *ortho–para* coupling (*Brochmann-Hanssen et al.*, J. pharm. Sci., 1971, **60**, 1880; 1973, **62**, 1291). Similar results have been obtained for **magnoflorine**: reticuline but not norprotosinomenine was incorporated (*idem*, Chem. Comm., 1972, 1269).

Rigorous distinction between the possible benzylisoquinoline precursors has been made for the biosynthesis of **boldine** in *Litsea glutinosa*, and indeed the methylation pattern in the alkaloid is misleading — norprotosinomenine LXXII was not incorporated (*cf. Dicentra* bases below). Of six bases examined only reticuline (LVII) and norreticuline (LVIII) acted as boldine precursors. Further it is (+)reticuline (with the same stereochemistry as boldine) and not its enantiomer which is involved. It follows that isoboldine (LXV) is the aporphine intermediate formed by coupling and it (and norboldine) was incorporated into boldine; it is also a natural constituent of the plant. This conversion of LXV into boldine largely involves demethylation and remethylation rather than methyl transfer since boldine derived from[6-$O^{14}CH_3$,1-^3H]-reticuline showed a 64% loss of ^{14}C label (similar results have been obtained in the related case of crotonosine; see below) (*S. Tewari, Bhakuni* and *Kapil*, Chem. Comm., 1974, 940).

Thebaine and **isothebaine** (LXX) are both found in *Papaver orientale* and, although they are both elaborations of the norlaudanosoline (LXII) skeleton, thebaine is derived from reticuline (see below) whereas isothebaine is formed from orientaline (LXVII) — an excellent example of methylation pattern coding the course of phenol coupling. Further substantiation of a role for orientaline (LXVII) in isothebaine biosynthesis was obtained by showing that orientaline was incorporated without loss of the *O*-methyl group from ring D and by observing that (+)orientaline with the same absolute stereochemistry as isothebaine was a much better precursor for the alkaloid than

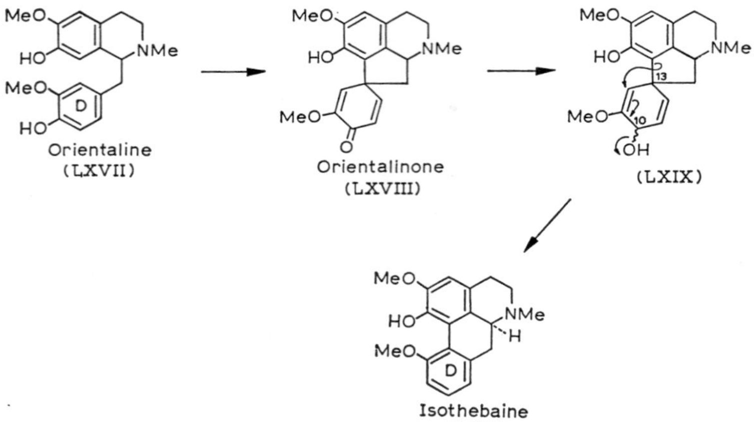

Orientaline (LXVII) → Orientalinone (LXVIII) → (LXIX) → Isothebaine (LXX)

the (—)isomer (*Battersby et al.*, Chem. Comm., 1965, 230). It is likely from this that the pathway to isothebaine involves coupling to give a dienone, orientalinone, subsequent reduction to orientalinol and then rearrangement to give the alkaloid.

Proof of the correctness of this pathway was obtained as follows: The putative key intermediate, orientalinone, was found to be a specific precursor for isothebaine and a natural constituent of *P. orientale*, as the (—)-isomer (expected stereochemistry) (*Battersby* and *T. H. Brown*, *ibid.*, 1966, 170). Reduction of orientalinone gave orientalinols-I and -II (LXIX) epimeric at C-10. Orientalinol-I, but not -II, was an efficient precursor for isothebaine, low incorporation of the latter being accompanied by loss of tritium label from C-10 in contrast to retention for orientalinol-I. (These observations are consistent with a redox conversion of isomer-II into -I *via* orientalinone). It is clear from this that rearrangement of orientalinol to isothebaine is stereospecific and may therefore like the preceding steps be attributed to enzyme catalysis (*Battersby*, *T. J. Brocksom* and *R. Ramage*, *ibid.*, 1969, 464).

Results so far discussed for aporphine biosynthesis are essentially those which might have been anticipated at the outset of the work. One set of aporphine alkaloids, however, did not conform to expectation, namely the *Dicentra eximia* alkaloids, **corydine, glaucine** and **dicentrine.** The methylation pattern of corydine suggested a pathway proceding through the oxidative coupling of reticuline (LVII) or orientaline (LXVII), but neither of these bases was incorporated, although norlaudanosoline (LXII) was. On the other hand, *N*-methylnorlaudanosoline was not, suggesting that *O*-methylation rather than *N*-methylation of norlaudanosoline was the next step in

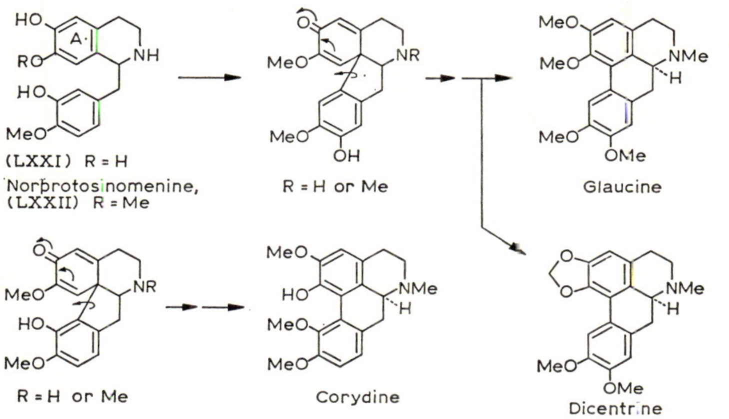

Scheme 45

biosynthesis. Of the four possible mono-methyl ethers only (LXXI) was incorporated (without loss of its *O*-methyl group). *O*-Methylation in ring A was a probable next step and of the two isomers here only norprotosinomenine (LXXII) was a precursor. It has not been established when *N*-methylation occurs but the *O*-methylation pattern of LXXII leads to the pathway shown in Scheme 45. (It will be seen below that norprotosinomenine is also a key intermediate in the biosynthesis of *Erythrina* alkaloids.) Boldine (LXVI) was now a logical intermediate in the biosynthesis of glaucine and dicentrine and it was shown to be efficiently incorporated into both bases (*Battersby et al.*, Chem. Comm., 1971, 985). It is interesting to note that a

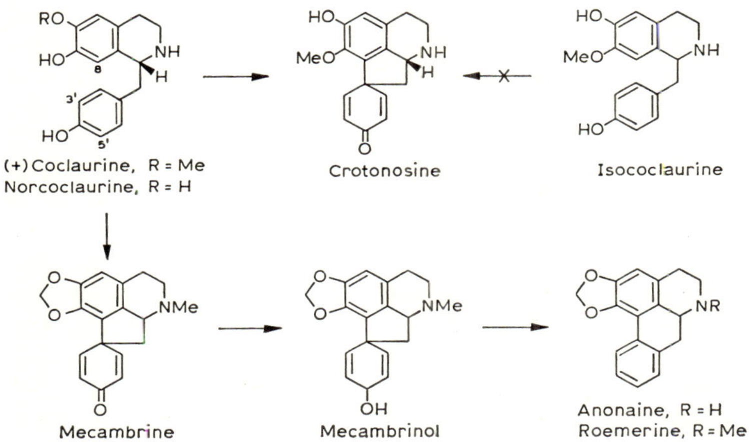

different pathway to boldine (LXVI) operates in *D. eximia* to that in *L. glutinosa* (see above).

The foregoing ascribes an important role to dienones in aporphine biosynthesis, an importance which is emphasized by their discovery as natural plant constituents. One such is **crotonosine** which is formed from (+)-coclaurine (same stereochemistry) rather than the (—)isomer, and not from isococlaurine. Demethylation and remethylation is again apparent here, an observation confirmed as occurring after coupling by the less efficient incorporation of norcoclaurine compared to coclaurine (*D. H. R. Barton et al., ibid.*, 1965,141).

An extension of the coclaurine/crotonosine pathway is found in the biosynthesis of **roemerine, mecambrine** and **anonaine** (*idem*, J. chem. Soc., C, 1967, 2134). Examination of the incorporation into roemerime of coclaurine, isococlaurine, norcoclaurine, (+)- and (—)-*N*-methylcoclaurine, and armepavine gave the following results: (*a*) all except isococlaurine, with the "wrong" methylation pattern for oxidative coupling, and the "over-methylated" armepavine were incorporated; (*b*) (+)-*N*-methylcoclaurine but not the (—)isomer was utilized, thus demonstrating the biological stereospecificity of the incorporation of this base; (*c*) some loss of the *O*-methyl group was apparent in the conversion of *N*-methylcoclaurine into roemerine

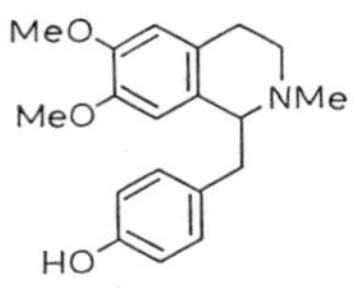

Armepavine

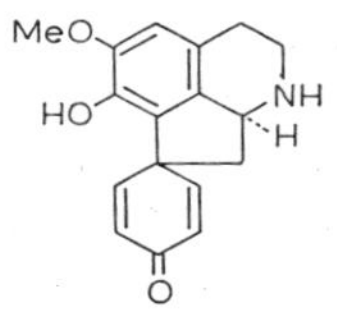

Crotsparine

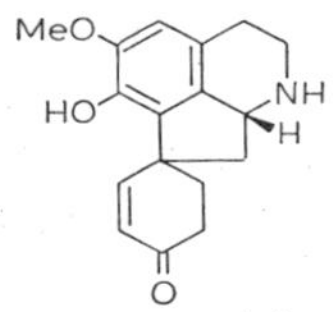

Crotsparinine

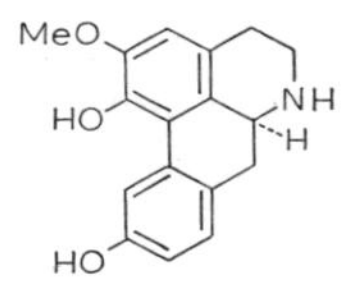

Sparsiflorine

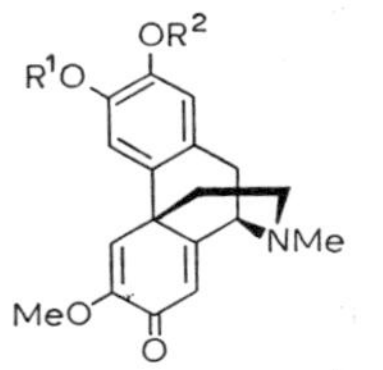

Flavinantine, $R^1 = H$, $R^2 = Me$
(LXXIII), $R^1 = Me$, $R^2 = H$

(and also mecambrine), which indicates partial demethylation and remethylation prior to formation of the methylenedioxy functions; (*d*) the different levels of incorporation suggested that *N*-methylation preceded *O*-methylation and that *N*-methylcoclaurine is the intermediate which undergoes coupling affording a dienone of the mecambrine type. A good incorporation of mecambrine into roemerine was obtained thus validating its intermediacy and establishing that the methylenedioxy group is present at the dienone stage of biosynthesis. The conversion of mecambrine into roemerine inevitably involves a dienol, mecambrinol (*cf.* isothebaine above).

A similar pathway to anonaine was indicated by the efficient incorporation of coclaurine and norcoclaurine (*idem, loc. cit.*). The alkaloids of *Croton sparsiflorus*, **crotsparine, crotsparinine** and **sparsiflorine** have been shown to have their origins also in coclaurine (isococlaurine was not incorporated). The biosynthesis of the aporphine base thus proceeds *via* a dienone, but manifestly not *via* a dienol. All three bases were found to be further transformed into their respective *N*-methyl derivatives (*D. S. Bhakuni et al.*, Phytochemistry, 1974, **13**, 2767).

Reticuline (LVII) has been found to be a much better precursor than orientaline (LXVII) for **flavinantine,** consistent with a derivation *via* LXXIII (*K. L. Stuart* and *L. Graham, ibid.*, 1973, **12**, 1967).

Bisbenzylisoquinolines. Intermolecular oxidative coupling between two *N*-methylcoclaurine molecules plausibly gives rise to the bisbenzylisoquinoline alkaloids. These may be exemplified by aztequine and dauricine formed by simple C–O and C–C links. Additional C–O linkages are to be found in, *e.g.*, epistephanine, the only base in this family so far to have been examined with tracers. [2-^{14}C]Tyrosine was incorporated equally into both halves of epistephanine (*Barton, G. W. Kirby* and *A. Wiechers*, J. chem. Soc., C, 1966, 2313). Both [N-*methyl*-^{14}C]-*N*-methylcoclaurine and [^{3}H]coclaurine were incorporated but the former was more efficiently utilized; the label from the

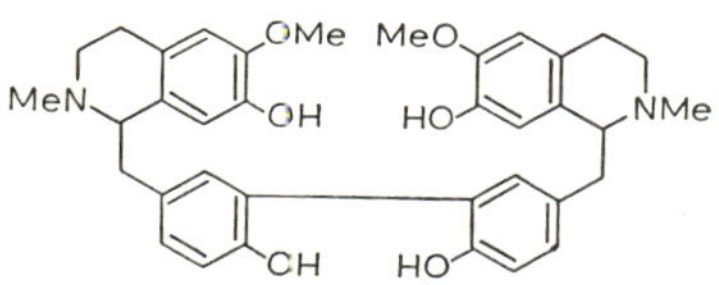

Aztequine

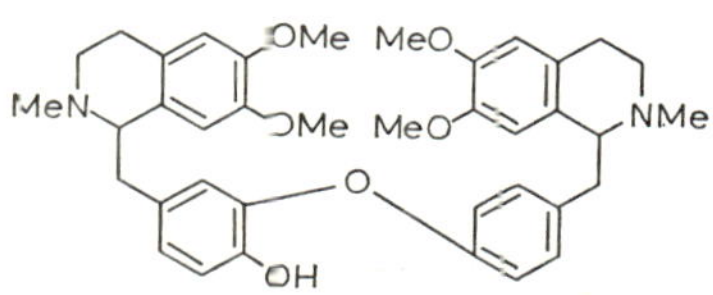

Dauricine

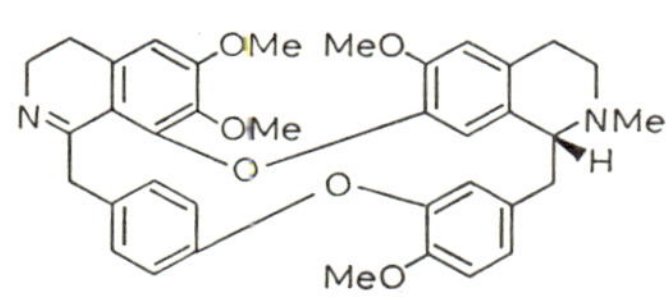

Epistephanine

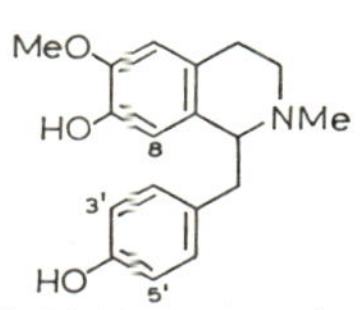

N-Methylcoclaurine

N-methylcoclaurine was located in the epistephanine methyl group consistent with specific incorporation. [8,3′,5′-$^{3}H_3$](—)-*N*-Methylcoclaurine (required stereochemistry) was much more efficiently incorporated than its antipode and only into the *N*-methylated half of epistephanine. Thus neither ready epimerization of *N*-methylcoclaurine occurs during incorporation nor demethylation to coclaurine (which is probably the precursor of the imine half of the molecule).

Morphine and related alkaloids. Morphine, thebaine, and codeine, the hydrophenanthrene alkaloids of *Papaver somniferum*, have been the subject of one of the key studies in alkaloid biosynthesis. Their implied structural relationship with other benzylisoquinoline alkaloids helped originally with the assignment of the correct structure to morphine; twisting of the benzylisoquinoline skeleton, LVII, into that shown for reticuline in Scheme 46 illustrates this relationship and also indicates a possible biogenesis by way of this skeleton.

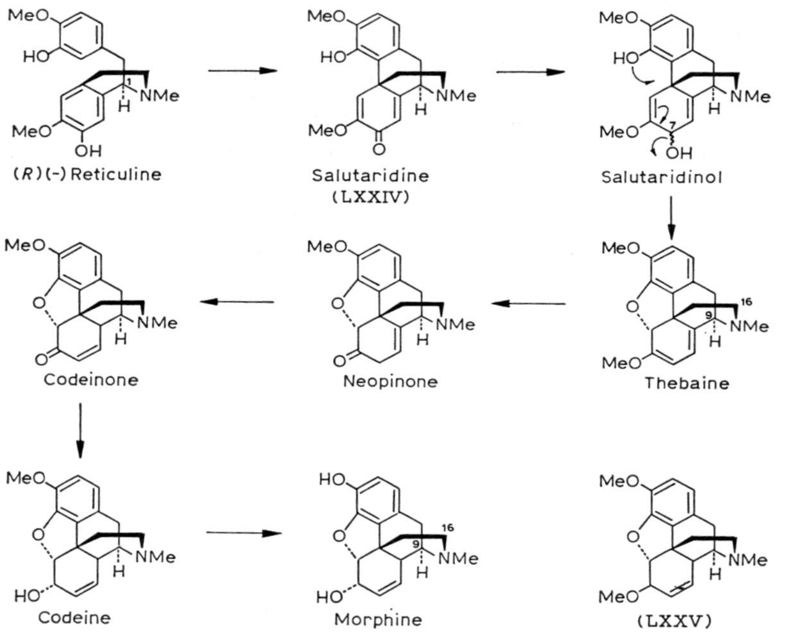

Scheme 46

Incorporation of tyrosine and dopamine provided preliminary support for this possibility. [2-^{14}C]Tyrosine gave morphine and thebaine with radioactivity located as required at C-9 and C-16 (*A. R. Battersby*, *R. Binks* and *B. J. T. Harper*, J. chem. Soc., 1962, 3534; *E. Leete*, J. Amer. chem. Soc., 1959, **81**, 3948; *D. Neubauer*, Arch. Pharm., 1965, **298**, 738). [1-^{14}C]-Dopamine on the other hand afforded morphine and codeine labelled only at C-16 (*Battersby* and *R. J. Francis*, J. chem. Soc., 1964, 4078; *Leete* and *J. B. Murrill*, Tetrahedron Letters, 1964, 147).

The crucial test of the benzylisoquinoline hypothesis lay with examination of such a molecule as a morphine precursor. In the first experiments ever to be carried out with complex plant-alkaloid precursors, $(\pm)[1\text{-}^{14}C]$- and $(\pm)[3\text{-}C^{14}]$-norlaudanosoline (as LXII, p. 407) was fed to *P. somniferum*. The results were specific incorporations into morphine, thebaine and codeine, thus establishing a genesis for these alkaloids by modification of the benzylisoquinoline skeleton (*Battersby* and *Binks*, Proc. chem. Soc., 1960, 360; *Battersby et al.*, J. chem. Soc., 1964, 3600).

Examination of the sequence in which thebaine, codeine and morphine are formed has been carried out by examining the fate of labelled samples of each and by observing the sequence of alkaloid labelling by $^{14}CO_2$ and $[2\text{-}^{14}C]$tyrosine (*G. Kleinschmidt*, Pharmazie, 1960, **15**, 663; *Neubauer*, Planta Med., 1964, **12**, 43; *H. Rapoport et al.*, J. Amer. chem. Soc., 1961, **83**, 4045; 1960, **82**, 2765; *Battersby* and *Harper*, Tetrahedron Letters, 1960, No. 27, 21). The sequence is one of demethylation: thebaine → codeine → morphine; and it is not reversible. Further experiments with $^{14}CO_2$ have shown that codeinone and neopinone are to be included as intermediates in this sequence (Scheme 46); the observed conversion of codeine methyl ether (LXXV) into morphine is probably not a normal biosynthetic process (*H. I. Parker*, *G. Blaschke* and *Rapoport*, J. Amer. chem. Soc., 1972, **94**, 1276).

Consideration must now be turned to how the benzylisoquinoline skeleton is transformed into that of thebaine the first of the hydrophenanthrene alkaloids to be formed. *Ortho–para* coupling of the diphenol, reticuline, itself a simple methylation product of norlaudanosoline, could be conceived of as giving the dienone (LXXIV) which in chemically unexceptional steps could afford thebaine. In accord reticuline served as an excellent and specific precursor for the hydrophenanthrene alkaloids. The specificity of the methylation pattern for the coupling reaction was clearly shown by the failure of the methyl isomers of reticuline, including orientaline (LXVII), to be incorporated, as well as the incorporation of reticuline itself without loss of *O*- and *N*-methyl groups. An excellent example is thereby provided in support of the oxidative phenol-coupling ideas of *Barton* and *Cohen* (see Section 6a). Moreover, the more fully methylated compounds LXXVI and

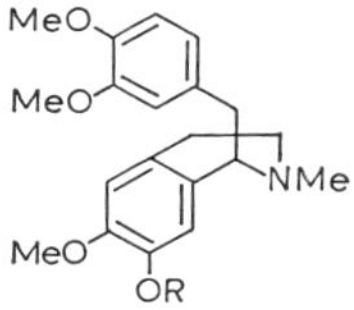

(LXXVI) R = H
(LXXVII) R = Me

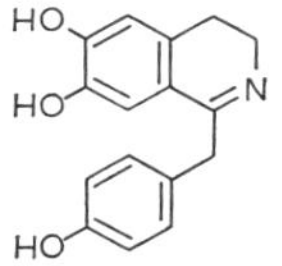

(LXXVIII)

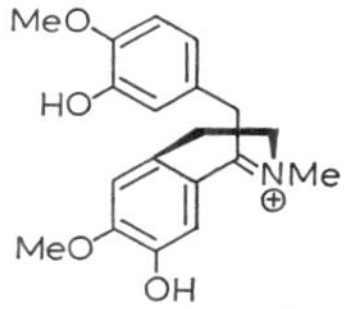

1,2-Dehydroreticuline
(LXXIX)

LXXVII were not incorporated. On the other hand [3-^{14}C]norreticuline [as LVIII] was found to be incorporated at a level intermediate between norlaudanosoline and reticuline (*Battersby et al.*, J. chem. Soc., 1964, 3600; 1965, 3323, 2423; unpublished work). *P. orientale* plants produce both isothebaine (see above) and thebaine. In confirmation of the above results with reticuline and orientaline a labelled sample of the former but not of the latter gave radioactive thebaine whereas the reverse was observed for isothebaine (*idem*, Chem. Comm., 1965, 230). Further evidence on the intermediacy of reticuline in morphine alkaloid biosynthesis is its isolation from *P. somniferum* and its labelling by $^{14}CO_2$ relative to thebaine consistent with a role as a thebaine precursor (*R. O. Martin, M. E. Warren* Jr. and *Rapoport*, J. Amer. chem. Soc., 1964, **86**, 4726; *E. Brochmann-Hanssen et al.*, J. pharm. Sci., 1964, **53**, 575; Tetrahedron Letters, 1965, 1271; *Battersby et al., ibid.*, 1965, 1275).

Experiments recently in the study of isoquinoline alkaloid biosynthesis have focussed on the first isoquinoline to be formed (*cf.* p. 402) and in the case of benzylisoquinolines this is the intermediate formed by condensation of two tyrosine derived residues. In *P. orientale*, radioactive tyramine and dopa afforded radioactive samples of norlaudanosoline and the previously unknown amino-acid LXI which was also a specific precursor for norlaudanosoline and morphine. The amino acid LIX was a very poor morphine precursor, which indicates that both aromatic building blocks must be dihydroxylated before they are joined together. Decarboxylation of LXI affords LXIII which was also found to be a morphine precursor (incorporation was observed at a lower level for LXXVIII) and was itself labelled by dopa. The biosynthetic pathway which follows from these results is shown in Scheme 44; it seems that little, if any, dopamine arises from dopa (*M. L. Wilson* and *C. J. Coscia*, J. Amer. chem. Soc., 1975, **97**, 431; *Battersby, R. C. F. Jones* and *R. Kazlauskas*, Tetrahedron Letters, 1975, 1873).

Returning to the steps in morphine biosynthesis which lie beyond reticuline (Scheme 46), the dienone LXXIV is a potential key intermediate. Shortly after its synthesis was completed a new alkaloid, salutaridine, from *Croton salutaris* was reported, which proved to be identical with LXXIV. Salutaridine was also shown to be present in *P. somniferum* by feeding [2-^{14}C]tyrosine and [3-^{14}C]norlaudanosoline in separate experiments and adding inactive salutaridine at the work-up stage. Isolation and purification gave radioactive salutaridine from which it follows that tyrosine and norlaudanosoline are precursors for this dienone as for thebaine *etc.* (*Barton et al.*, J. chem. Soc., 1965, 2423). Labelled salutaridine was fed to *P. somniferum* and was found to be a highly efficient and specific precursor for the hydrophenanthrene alkaloids (up to 15% incorporation — high for plants).

Its siting therefore as an intermediate in the biosynthesis of these alkaloids is proven (*idem, loc. cit.*).

The transformation of salutaridine into thebaine requires a further ring-closure which occurs chemically under mild conditions when the two epimeric dienols, salutaridinol-I and -II, are treated with acid. Both dienols, each with a tritium label at C-7 and an internal ^{14}C reference, were fed to *P. somniferum*. In contrast to the purely chemical reaction, only salutaridinol-I was efficiently converted (20–30 times better than isomer-II) into thebaine *in vivo*; such stereoselectivity indicates that the reaction is enzyme mediated. Moreover, the tritium label was completely retained indicating that no reconversion to salutaridine occurred (*idem, loc. cit.*).

On the basis of general observations on the modification of the benzylisoquinoline skeleton, further stereoselectivity might be expected in the conversion of reticuline into salutaridine. However, both (*R*)- and (*S*)-reticuline were similarly well incorporated indicating an additional step in the biosynthetic sequence, that of epimerization logically *via* LXXIX. Tritium loss from C-1 of both isomers indicates that both equilibrate through LXXIX. In the case of (*R*)-(—)reticuline this was less complete consistent with its having the same absolute configuration as morphine and thus being the correct isomer for further transformation. Incorporation of the (+)-isomer occurs then only by way of 1,2-dehydroreticuline (LXXIX) and (—)reticuline with inevitable loss of tritium. 1,2-Dehydroreticuline is itself an efficient morphine precursor (*Battersby, D. M. Foulkes* and *Binks*, J. chem. Soc., 1965, 3323).

Conversion of morphine into normorphine ends a sequence of demethylation reactions which begins at thebaine. The presence of normorphine in *P. somniferum* has been established as has the irreversible conversion of morphine into it, and it seems as if the major, if not the only, breakdown pathway for morphine involves normorphine (*R. J. Miller, C. Jolles* and *Rapoport*, Phytochemistry, 1973, **12**, 597). The enzyme responsible for codeine demethylation has been found to show wide substrate tolerance provided the codeine skeleton is present (*G. W. Kirby, S. R. Massey* and *P. Steinreich*, J. chem. Soc., Perkin I, 1972, 1642).

Berberine and related alkaloids. Berberine is an example from a group of alkaloids based on the benzylisoquinoline skeleton, the singular feature of which is an additional carbon atom (C-8): the so-called "berberine bridge". Less oxidized than berberine, the protoberberine alkaloids, *e.g.* stylopine (LXXXV), serve as starting points for the elaboration of various other alkaloids, *e.g.* chelidonine and narcotine.

Feeding experiments showed that berberine was specifically derived from two molecules of tyrosine, one of which was incorporated *via* dopamine

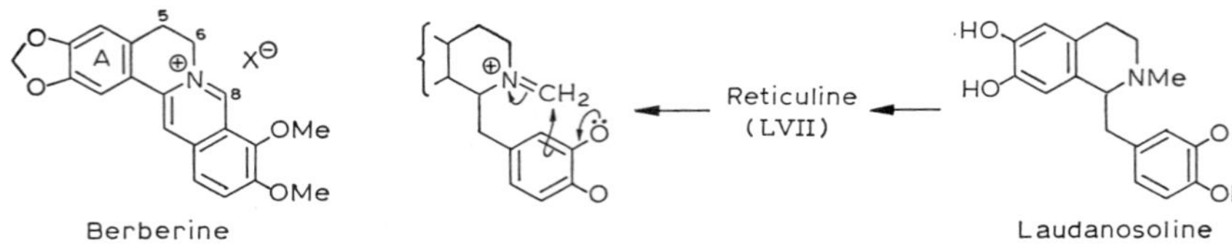

Scheme 47

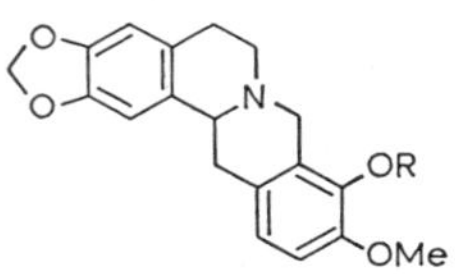

Canadine, R = Me
(LXXX), R = H

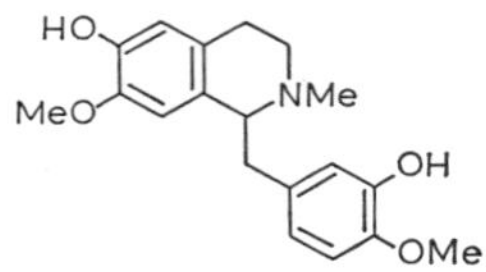

Protosinomenine
(LXXXI)

(ring A plus C-5 and -6) (*I. D. Spenser et al.*, Canad. J. Chem., 1963, **41**, 783; 1965, **43**, 2017). Radioactive methionine (and formate) labelled C-8 and the methylenedioxy and methoxy groups (*R. N. Gupta* and *Spenser*, *ibid.*, 1965, **43**, 133).

The important suggestion was made that perhaps the berberine-bridge arose by oxidative cyclisation of an *N*-methyl group in a manner (Scheme 47) analogous to the formation of a methylenedioxy-function (see p. 392) (*D. H. R. Barton*, Proc. chem. Soc., 1963, 293; *A. R. Battersby*, *ibid.*, 1963, 189; *Barton et al.*, J. chem. Soc., 1965, 6379). Validation of this came with the demonstration that *N*-methyl-labelled laudanosoline and reticuline (LVII, p. 406) were incorporated into berberine, all the radioactivity of the precursors appearing, without loss, at C-8 of the alkaloid. It was further shown that (+)(*S*)-reticuline was preferred over the (—)(*R*)-isomer as a substrate for berberine biosynthesis (*Barton*, *R. H. Hesse* and *G. W. Kirby*, J. chem. Soc., 1965, 6379; *Battersby et al.*, Proc. chem. Soc., 1963, 268; J. chem. Soc., Perkin I, 1975, 1140). The protoberberine canadine served as a good precursor for berberine but protosinomenine (LXXXI) and LXXX were poor precursors (*Barton et al.*, *loc. cit.*).

Robinson suspected that alkaloids of the **chelidonine**-type (LXXXIX) were to be included in the group of modified protoberberine bases (*R. Robinson*, "The Structural Relations of Natural Products", Oxford University Press, London, 1954, p. 89). The co-occurrence of **stylopine** with chelidonine in *Chelidonium majus* supported this idea and indicated that stylopine might be an intermediate in chelidonine biosynthesis. Conversion of the one skeleton into the other can be envisaged as involving cleavage of the C-6–N bond of stylopine followed by reclosure at C-13 (*cf.* Scheme 48, LXXXV–LXXXVIII). The way in which [2-^{14}C]tyrosine was incorporated into chelidonine is con-

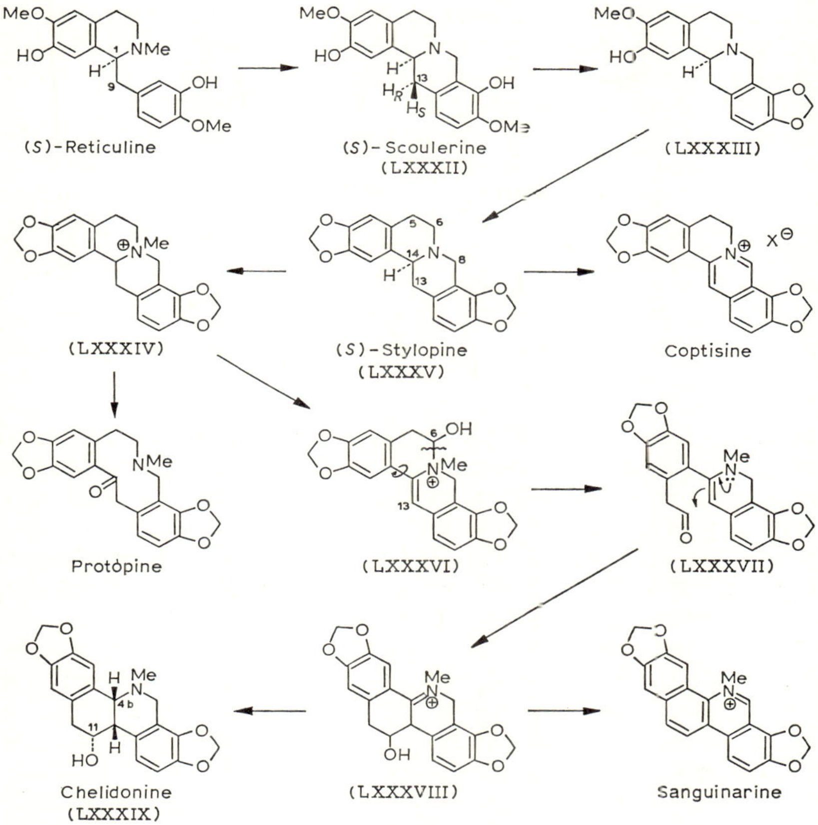

Scheme 48

sistent with this (labelling of C-11 and C-4b) — [1-^{14}C]dopamine gave chelidonine with activity confined to C-11 (*E. Leete*, J. Amer. chem. Soc., 1963, **85**, 473; *Leete* and *J. B. Murrill*, Tetrahedron Letters, 1964, 147).

Tracer experiments have established that not only is stylopine involved in chelidonine biosynthesis but **scoulerine** and reticuline also. The sequence which this indicates, reticuline → scoulerine → stylopine → chelidonine, is also one which involves the (*S*)-isomer in each case. The methylenedioxy group in ring D of stylopine was shown to arise, as in other cases, from the *ortho*-hydroxy-methoxy-system of reticuline.

Tritium labelling results established that C-1 and C-9 of reticuline were unaffected by transformation through to scoulerine and stylopine. The conversion of the *N*-methyl group of reticuline into the "berberine-bridge" atom

(C-8) of stylopine, which necessarily involves loss of one hydrogen/tritium atom, was found to occur with very high tritium retention consonant with the expected tritium-isotope effect. Subsequent reaction to give chelidonine occurs with retention of tritium at C-8 and C-5 of stylopine but loss of tritium at C-14. This reaction is, in addition, accompanied by loss of the 13-*pro-S* hydrogen atom and retention of the 13-*pro-R* proton, consistent with which is enzyme-mediated proton removal from C-13 at a stage after stylopine, and taken with the observed loss of hydrogen from C-14 gives credance to the possibility of an enamine intermediate (see LXXXVI and LXXXVII) between stylopine and chelidonine.

(*S*)-Scoulerine has also been shown to be a precursor for sanguinarine, chlereryanthine, and **protopine**, (*S*)-stylopine a precursor for coptisine and protopine, and stylopine methochloride (LXXXIV) a precursor for chelidonine and protopine; (*S*)-(+)reticuline is also a protopine precursor. On the other hand, radioactive nandinine (LXXX) failed to label chelidonine, stylopine or protopine (or berberine, see above), suggesting that the alternative, LXXXIII, may be an intermediate instead. Combination of these results leads to the pathways shown in Scheme 48 (*A. R. Battersby et al.*, J. chem. Soc., Perkin I, 1975, 1140, 1147, 1162; *C. Tani* and *K. Tagahara*, Chem. Pharm. Bull., Japan, 1974, **22**, 2457; *Barton*, *Hesse* and *Kirby*, *loc. cit.*, *Leete* and *Murrill*, Phytochemistry, 1967, **6**, 231).

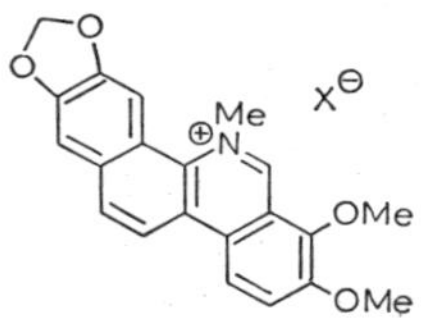

Cheleryanthine

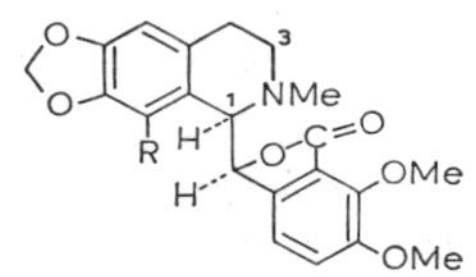

Hydrastine, R = H
Narcotine, R = OMe
Narcotoline, R = OH

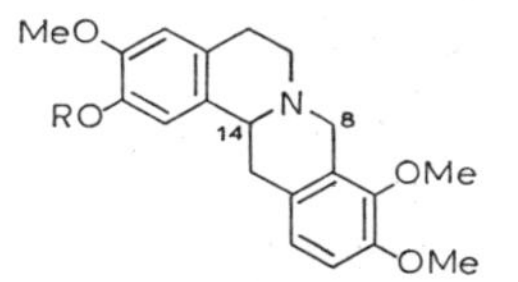

Isocorypalmine, R = H
Tetrahydropalmatine, R = Me

Tyrosine serves as a good precursor for the phthalideisoquinoline alkaloids **narcotoline, hydrastine** and **narcotine,** [2-^{14}C]-labelled material affording the last two bases with radioactivity confined to C-1 and C-3. These results accord with others obtained elsewhere for the biosynthesis of the benzylisoquinoline skeleton as does the labelling of C-3 only by [1-^{14}C]dopamine (*Battersby et al.*, J. chem. Soc., C, 1968, 2163; *J. R. Gear* and *I. D. Spenser*, Canad. J. Chem., 1963, **41**, 783; *G. Kleinschmidt* and *K. Mothes*, Z. Naturforsch., 1959, **14b**, 52). Radioactive methionine (and formate) was incorporated into the lactone carbonyl of hydrastine and narcotine (*Battersby et al., loc. cit.*; *Gupta* and *Spenser*, Canad. J. Chem., 1965, **43**, 133).

Reticuline — and the (*S*)-(+)isomer was preferred over the (*R*)-isomer — was shown to be a narcotine precursor [narcotine has the (*S*)-stereochemistry]; norlaudanosoline (LXII) was also a precursor. Of particular importance was the demonstration that the *N*-methyl group of reticuline became the lactone carbonyl of the alkaloid, thus making narcotine (and its structural relatives) a modified protoberberine. Substantiation of this conclusion and further delineation of the pathway to narcotine as being through protoberberines comes from positive results of feeding experiments with labelled samples of scoulerine (LXXXII) [(*S*)-(—)isomer preferred], isocorypalmine and canadine. As in the transformation of scoulerine into chelidonine (see above), its conversion into narcotine involves removal of the 13-*pro-S*-hydrogen atom. Introduction of the oxygen atom at C-13 thus proceeds with orthodox retention of configuration (*Battersby et al., loc. cit.*; J. chem. Soc., Perkin I, 1975, 1162).

Alkaloids of the **alpinigenine** type are plausibly modified protoberberines. In support [8-^{14}C]tetrahydropalmatine and its *N*-methyl derivative (XC), labelled as shown, were specifically incorporated into alpinigenine. A role

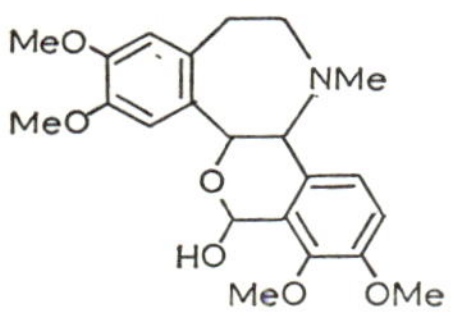

Alpinigenine

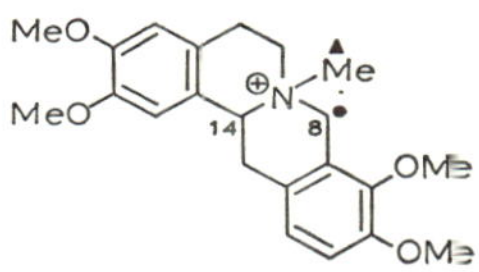

(XC)

is thus indicated for the quaternary *N*-methyl function in aiding the ring-opening reaction (cleavage of N–C-14 or N–C-8 bond) which is a necessary step in the conversion of protoberberine into alpinigenine (*H. Rönsch*, Europ. J. Biochem., 1972, **28**, 123).

Corydaline and **ochotensimine** can be seen as yet further variants on the protoberberine theme. The more obvious correlation with corydaline has been substantiated by the incorporation of reticuline (LVII, p. 406) (*G. Blaschke*, Arch. Pharm., 1968, **301**, 438; 1970, **303**, 358). Further evidence has been obtained from the labelling pattern produced by [3-^{14}C]tyrosine and [*Me*-^{14}C]methionine. The results are consistent with both alkaloids being modified protoberberine (as XCI) alkaloids (Scheme 49) — methionine labelling of C-8 (corydaline) and C-9 (ochotensimine) is that expected for the ostensible "berberine-bridge" carbons. The point at which *C*-methylation occurs in a corydaline precursor is speculative as is the skeletal rearrangement necessary for formation of ochotensimine, although chemical analogy for this does exist (*Spenser et al.*, Canad. J. Chem. 1974, **52**, 2818).

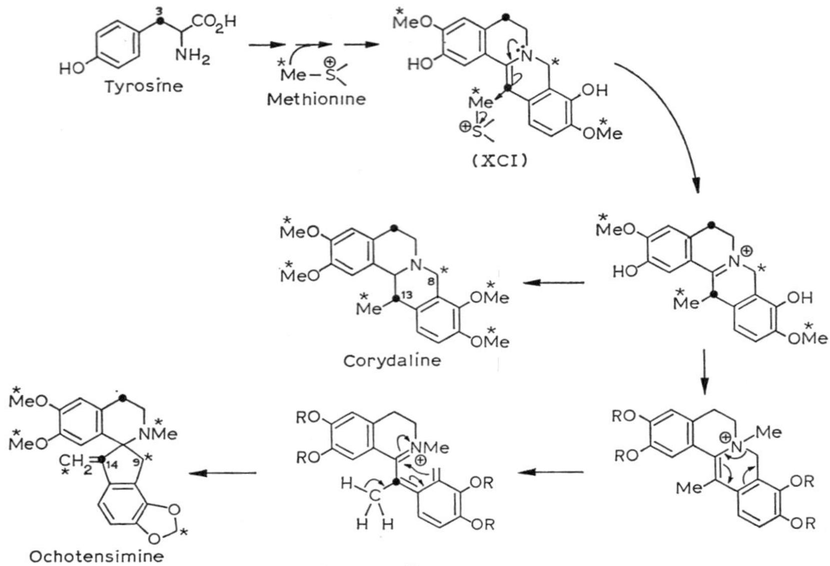

Scheme 49

Erythrina alkaloids. The structures of these alkaloids, *e.g.* erythraline, suggest an unusual genesis. This has certainly been shown to be the case but only in regard to the late stages of biosynthesis, for these bases have turned out to be but modifications of the ubiquitous benzylisoquinoline skeleton.

Equal labelling of C-8 and C-10 of β-erythroidine by [2-^{14}C]tyrosine indicated that this alkaloid is derived from two molecules of tyrosine plausibly *via* a symmetrical intermediate; phenylalanine was not incorporated (*E. Leete* and *A. Ahmad*, J. Amer. chem. Soc., 1966, **88**, 4722). Although the symmetrical intermediate could plausibly have been XCIII this base was but poorly incorporated into erythraline and erythratine. In contrast the benzylisoquinoline base *N*-norprotosinomenine was better utilized and it could be shown that it was the (+)(*S*)-isomer which served as the precursor of these *Erythrina* alkaloids. [Neither (+)-*N*-nororientaline, as LXVII, nor (±)-*N*-nor-reticuline (LVIII), p. 410 was significantly incorporated into erythraline or the erythroidines]. In the pathway which could be advanced to account for these results (Scheme 50) the symmetrical intermediate XCII was implicated as was erysodienone. In accord both were incorporated without fragmentation. Moreover, only (—)erysodienone, which has the (5*S*)-chirality of the natural alkaloids, is involved in biosynthesis. The ^{14}C labels on the methoxyl groups of XCII were retained in the erythraline generated, being equally divided between the methylenedioxy

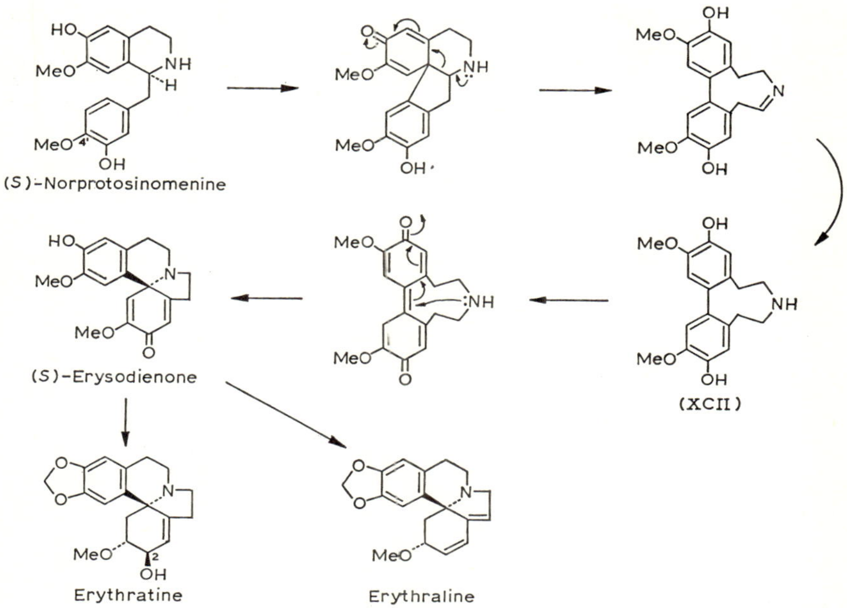

Scheme 50

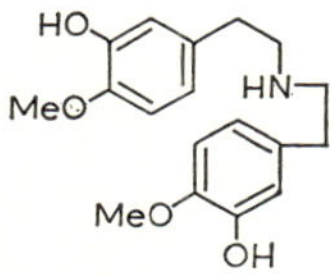

(XCIII)

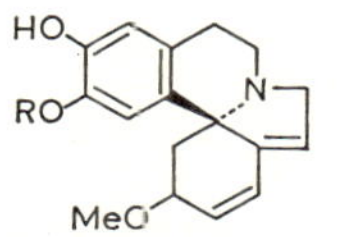

Erysopine, R = H
Erysodine, R = Me

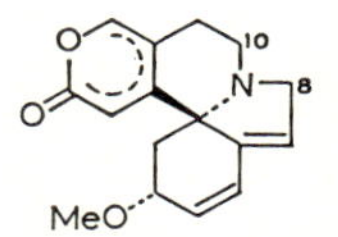

α- and β-Erythroidine

and methoxy groups (*D. H. R. Barton et al.*, J. chem. Soc., C, 1968, 1529; 1970, 1213; J. chem. Soc., Perkin I, 1974, 346, 2278). [4'-*Methoxy*-^{14}C]-*N*-norprotosinomenine gave erythraline with an equal distribution of label between methoxyl and methylenedioxy groups and so clearly the pathway involves a symmetrical intermediate (as XCII) and the asymmetry of (*S*)-norprotosinomenine is not transferred to (*S*)-erysodienone.

In contrast to the unique methylation pattern required for oxidative phenol coupling, experiments with variously *O*-methylated *Erythrina* bases have demonstrated that for these alkaloids the methylation pattern during the late stages of biosynthesis is not important. A further point is that it is not erythratine but its C-2 epimer which is a precursor for erythraline (*Barton et al., loc. cit.*, 1968/1970). Not unexpectedly the erythroidines are derived like erythraline from erysodienine (and also erysopine and erysodine) as experiments with labelled compounds showed (*Barton et al.*, J. chem. Soc., Perkin I, 1974, 2278).

Ipecac alkaloids. The alkaloids cephaeline and emetine are structurally unique among the isoquinoline alkaloids in containing, apart from two manifest C_6–C_2–N units, a C_9 fragment (heavy bonding) which familiarity with terpenoid indole alkaloid biosynthesis points to as terpenoid in origin. This is indeed so and the pathway is clearly similar*. Specific incorporations of radioactive mevalonate (into cephaeline), geraniol (into cephaeline and the even more obviously terpenoid-like ipecoside), loganin (into cephaeline

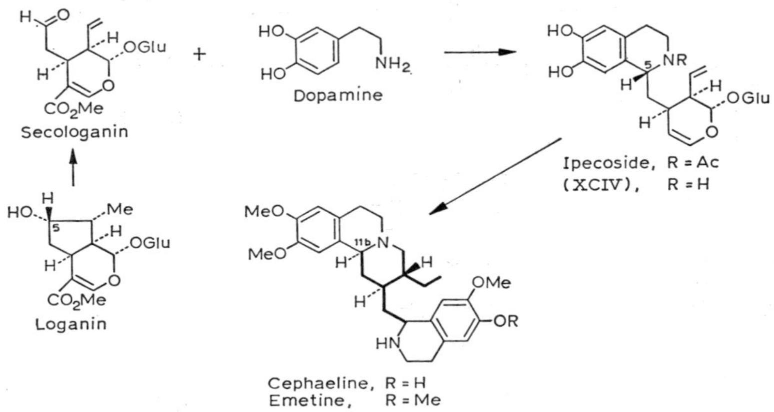

and ipecoside), and secologanin (into cephaeline, emetine, and ipecoside) have been recorded. Other possible origins for the C_9 fragment have also been excluded experimentally (*A. R. Battersby et al.*, J. chem. Soc., 1965, 7459; *Battersby* and *B. Gregory*, Chem. Comm., 1968, 134; *Battersby* and *R. J. Parry*, *ibid.*, 1971, 901).

Examination of the C-5 epimeric isoquinolines (as XCIV) as precursors revealed that, only the epimer XCIV corresponding to ipecoside in stereochemistry was incorporated (specifically and efficiently) into cephaeline, emetine (with opposite configuration at the equivalent position — C 11b), and ipecoside. Moreover, retention of tritium from C-5 of [5-^{3}H]loganin [corresponds to C-5 in XCIV] in the derived alkaloids establishes that the epimerization which necessarily occurs in the conversion of XCIV into cephaeline and emetine takes places without proton (tritium) loss from this position (*idem, loc. cit.*). [Recent results on the biosynthesis of terpenoid

* The reader should refer to the full discussion in Section 5h,iv on the biosynthesis of terpenoid C_9–C_{10} units.

indole alkaloids Section 5h,iv) show that a similar inversion does not occur in contradiction of earlier observations. Thus the validity of the observation here is to be questioned.]

As might be expected, tyrosine (but not phenylalanine) serves as the precursor for the phenethylamine fragments of these alkaloids as exemplified by cephaeline (*Battersby et al., loc. cit.*). The evidence so far adduced leads to the pathway to the *Ipecac* alkaloids illustrated, in which the phenethylamine residues are assumed to be dopamine.

Hasubanonine and protostephanine. Hasubanonine and protostephanine, both produced by *Stephania japonica*, might seem to be variations of the benzylisoquinoline skeleton but attempts to establish this or any other pathway have been frustrated. Putative precursors XCV–IC and the *O*-methyl

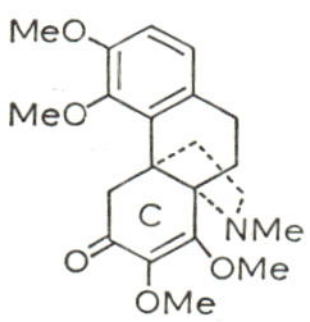

Hasubanonine

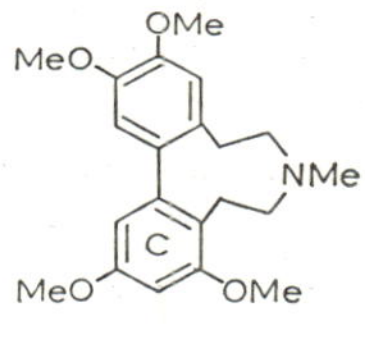

Protostephanine

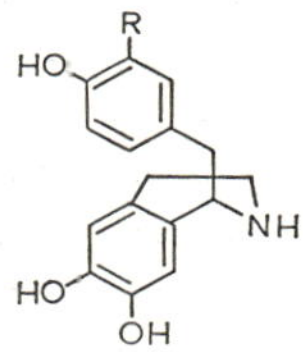

(XCV) R = H
(XCVI) R = OH

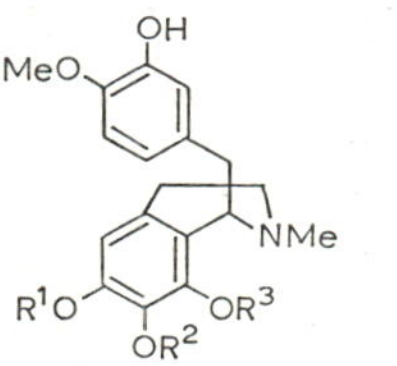

(XCVII) R^1 = H, R^2 = R^3 = Me
(XCVIII) R^2 = H, R^1 = R^3 = Me
(IC) R^3 = H, R^1 = R^2 = Me

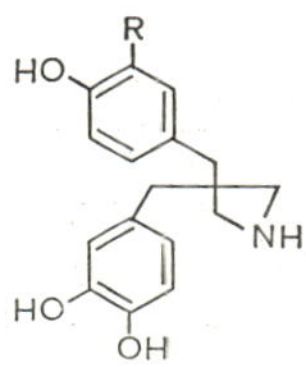

(C) R = OH
(CI) R = H

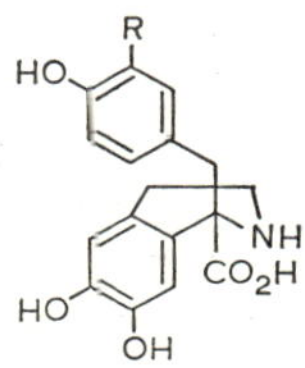

(CII) R = OH
(CIII) R = H

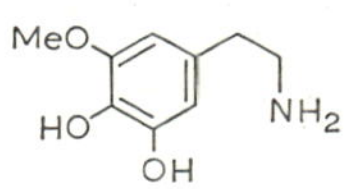

(CIV)

and *N*-methyl derivatives of XCV and XCVI, together with the bisphenethylamines C and CI were incorporated to an insignificant extent. Similar results were obtained with CII and CIII and the 1,2-dehydro-derivatives of XCV and XCVI.

On the other hand [2-^{14}C]-labelled tyrosine, dopa, tyramine and dopamine were specifically incorporated into both alkaloids. From the results it is clear that (*a*) both alkaloids are constructed from two C_6–C_2 units, for which tyrosine is the source, and (*b*) that dopa, dopamine and tyramine only serve as the source of ring C with attached ethanamine residue. This unit is manifestly a phenethylamine and the results of feeding various phenethylamines indicates that this unit is a trioxygenated but monomethylated derivative, probably CIV. Combination of this with the other C_6–C_2 unit gives a set of benzylisoquinoline and bisphenethylamine precursors whose roles in protostephanine and hasubanonine biosynthesis have yet to be tested (*Battersby et al.*, Chem. Comm., 1974, 773).

(*iv*) *Phenethylisoquinolines*

Colchicine. The tropolone ring of the non-basic alkaloid colchicine is the most notable feature in an unusual structure. The biosynthetic pathway to colchicine, at first thought likely to be unusual too, has turned out paradoxically to be quite simple (*Battersby*, Pure and appl. Chem., London, 1967, **14**, 117). The studies were carried out in *Colchicum autumnale* and *C. byzantinum.*

The first experiments established rigorously the origin of the carbon skeleton. Phenylalanine, by way of cinnamic acid, serves as the source of a C_6–C_3 unit which accounts for ring A of colchicine plus carbons 5, 6 and 7 (*Battersby et al.*, J. chem. Soc., 1964, 4257; *E. Leete* and *P. E. Nemeth*, J. Amer. chem. Soc., 1960, **82**, 6055; *Leete*, *ibid.*, 1963, **85**, 3666). Phenylalanine is not utilized for the genesis of the tropolone ring which is derived independently as a C_6–C_1 unit from tyrosine: Labels from [4′-^{14}C]- and [3-^{14}C]-tyrosine were incorporated into colchicine and were located at C-9 and C-12, respectively; [1-^{14}C]- and [2-^{14}C]-tyrosines were not specifically incorporated (*Battersby et al.*, J. chem. Soc., Perkin I, 1972, 1730; J. chem. Soc., 1964, 4257; *Leete*, Tetrahedron Letters, 1965, 333). Formation of the tropolone ring of colchicine is thus by expansion of the aromatic ring of a tyrosine derivative with inclusion of the benzylic carbon atom. It is interesting to note that the fungal tropolones, *e.g.* stipitatonic acid, have different origins being derived from acetate, but their formation resembles that of colchicine in that the tropolone ring is formed apparently by expansion of an aromatic six-membered ring (*A. I. Scott et al.*, J. Amer. chem. Soc., 1971, **93**, 3534; Chem. Comm., 1972, 1075).

Further progress in unravelling the biosynthesis of colchicine depended on an unsuspected structural relationship. The plant *Androcymbium melanthioides* is a close botanical relative of *Colchicum autumnale*, a rich source of colchicine. A dienone alkaloid, androcymbine, was isolated from this plant and its structure was established as CVIII; significantly the absolute stereochemistry was the same as that of colchicine (*Battersby et al.*, J. chem. Soc., Perkin I, 1972, 1736). Manifestly this dienone appeared to be produced through oxidative phenol-coupling of the phenethylisoquinoline CVII. Of singular value was the idea that colchicine could be derived from the

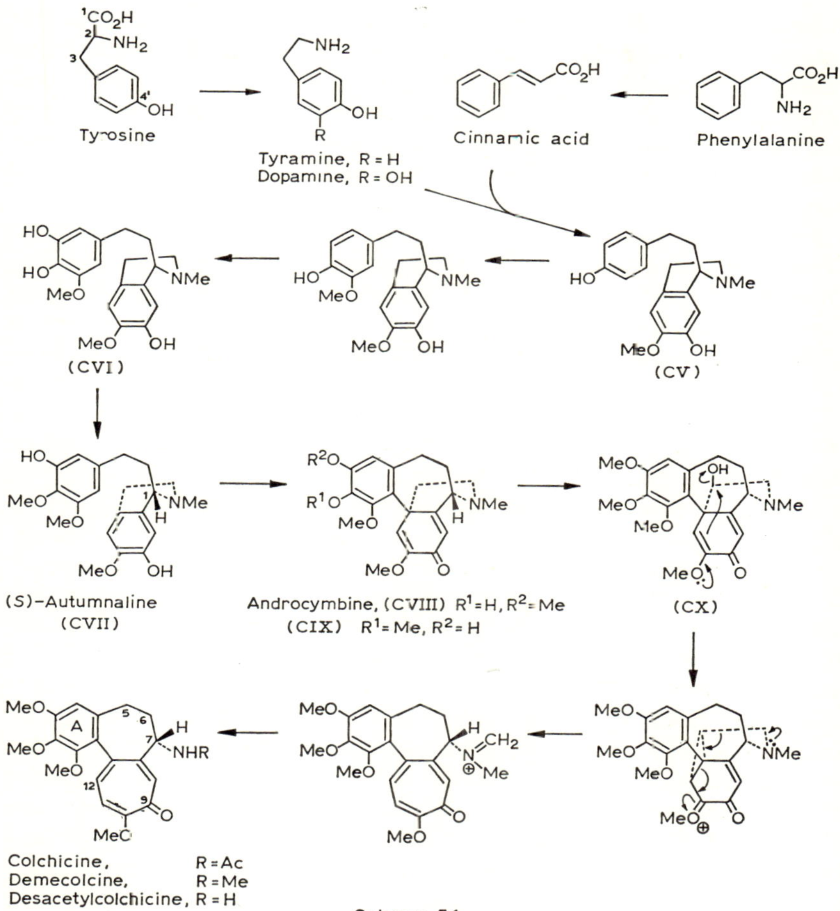

Scheme 51

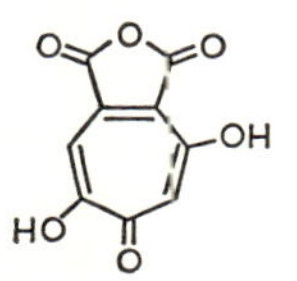

Stipitatonic acid

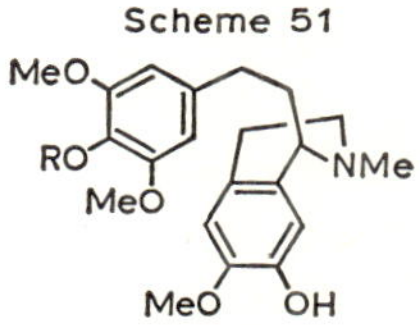

(CXI) R = H
(CXII) R = Me

androcymbine skeleton and was therefore a modified phenethylisoquinoline. Plausibly the transformation of the androcymbine skeleton to that of colchicine could occur by hydroxylation to give CX. Homoallylic ring expansion (*cf. O. Chapman* and *P. Fitton*, J. Amer. chem. Soc., 1963, **85**, 41) and subsequent minor modification would yield colchicine (see Scheme 51).

Validation of these ideas came with a remarkably high and specific incorporation of tritiated *O*-methylandrocymbine into colchicine (*Battersby et al.*, J. chem. Soc., Perkin I, 1972, 1741). The efficient and specific incorporation of CVII, called autumnaline, further established the correctness of these ideas and proved colchicine to be a modified isoquinoline. This unusual alkaloid is brought then into the large family of alkaloids derived from 1-substituted isoquinolines (*idem, loc. cit.*). Affirmation of this conclusion was obtained by showing that (*RS*)-[1-^{13}C]autumnaline afforded cochicine with enhancement of the n.m.r. signal for C-7 only. (This experiment is one of the first successful ones using ^{13}C-labelling in plant alkaloid biosynthesis; *Battersby, P. W. Sheldrake*, and *J. A. Milner*, Tetrahedron Letters, 1974, 3315.)

The poor utilization of the isoquinolines CXI and CXII in colchicine formation further strengthened the status of autumnaline as an intermediate (*Battersby et al., loc. cit.*). Moreover, only the isomer CVII of autumnaline, which has the same absolute configuration as colchicine, is implicated in the pathway. Tritium-labelling results showed that formation of the androcymbine skeleton was by *para–para*-coupling of CVII rather than *ortho–para* coupling; and hydroxylation to generate CX in turn was stereospecific (*idem*, Chem. Comm., 1967, 390). Results of experiments with various substituted phenethylisoquinolines allowed definition of the stages before autumnaline as CV–CVI (*idem*, J. chem. Soc., Perkin I, 1972, 1741).

The steps of biosynthesis which precede formation of the simplest phenethylisoquinoline have been explored (*idem, loc. cit.*). The results allow further delineation of the pathway to colchicine. Both tyramine and dopamine acted as precursors and although cinnamic acid was incorporated into colchicine none of several hydroxy-methoxycinnamic acids were, indicating that some reaction other than hydroxylation (reduction?) occurs as a first step. The conversion of phenylalanine into cinnamic acid, mediated by phenylalanine ammonia lyase, normally involves loss of the 3-*pro-S*-hydrogen. A similar result might be expected for conversion of phenylalanine through cinnamic acid into colchicine, and this was shown to be true by appropriate tracer experiments (*idem*, J. chem. Soc., Perkin I, 1972, 2355).

Examination of the final stages of biosynthesis showed that colchicine could be effectively derived from demecolcine and desacetylcolchicine. (Indeed many of the tracer results for colchicine have been backed up by similar ones for demecolcine; *idem*, Chem. Comm., 1967, 390.) All the results for colchicine biosynthesis are illustrated in Scheme 51. [Androcymbine itself is not on the pathway.]

Homoaporphines. It might be anticipated that the biosynthesis of homoaporphine alkaloids would be analogous to that of the aporphine alkaloids (see p. 408). Accordingly, the homoaporphine bases of *Kreysigia multiflora*, floramultine, multifloramine and kreysigine, have been found to be derived from a phenethylisoquinoline precursor. The efficient and specific incorporation of the colchicine precursor, autumnaline, and by contrast the poor

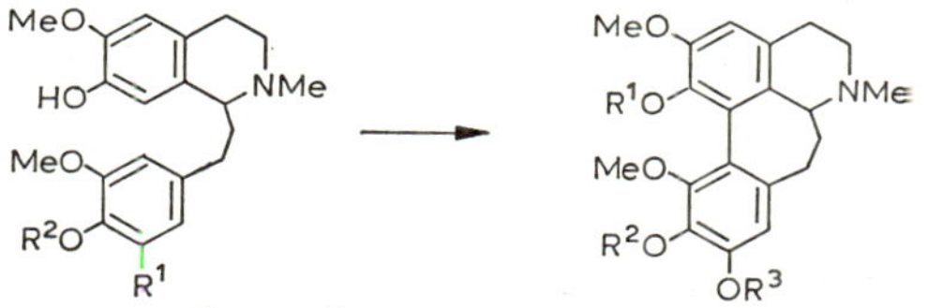

Autumnaline, R^1=OH, R^2=Me
(CXIII) R^1=R^2=H

Floramultine, R^1=R^3=H, R^2=Me
Multifloramine, R^1=R^2=H, R^3=Me
Kreysigine, R^1=H, R^2=R^3=Me

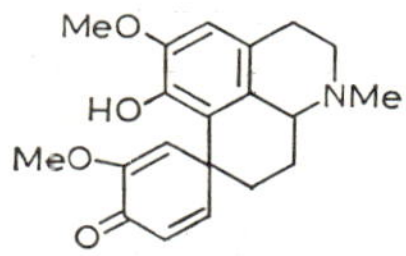

Kreysiginone

utilization of CXIV showed that these bases arise by direct coupling, with floramultine presumably the first alkaloid to be formed (*A. R. Battersby et al.*, J. chem. Soc., Perkin I, 1974, 1399). Dienone intermediates are not involved but it is interesting to note that the dienone kreysiginone and its dihydro derivative have been isolated from *K. multiflora* (*idem, ibid.*, p. 1394). Kreysiginone is manifestly derived from the base CXIII which also served as a precursor for the homoaporphines.

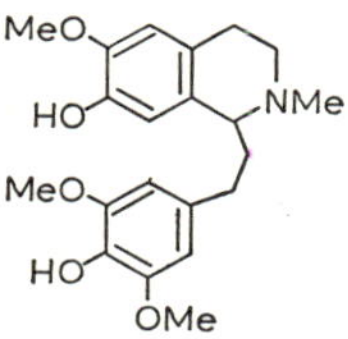

(CXIV)

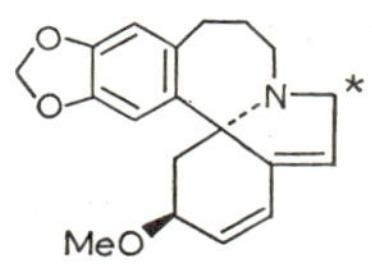

Schelhammeridine

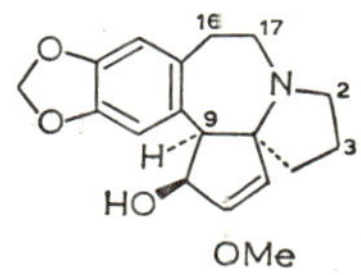

Cephalotaxine

Schelhammeridine and cephalotaxine. Schelhammeridine is structurally related to the *Erythrina* alkaloids, differing essentially in the possession of an extra skeletal carbon atom. The derivation of *Erythrina* alkaloids *via* the 1-benzylisoquinoline skeleton (see p. 422) indicates a similar derivation for schelhammeridine through a 1-phenethylisoquinoline. Preliminary results support this view. DL-[2-^{14}C]Tyrosine specifically labelled the asterisked site in the alkaloid, and radioactive phenylalanine, dopamine, and cinnamic acid were also incorporated — the results parallelling the observations on the biosynthesis of the modified phenethylisoquinoline, colchicine (see above) (*Battersby et al.*, Tetrahedron Letters, 1975, 3419).

The co-occurrence of **cephalotaxine** with alkaloids of the schelhammeridine type and an apparent structural kinship suggests a similar origin for cephalotaxine from tyrosine and phenylalanine through a 1-phenethylisoquinoline progenitor. Tyrosine was indeed found to be a cephalotaxine precursor but the labelling pattern is inconsistent with a 1-phenethylisoquinoline pathway: [3-^{14}C]tyrosine gave cephalotaxine with 68% of the activity at C-16 and 32% at C-9 whilst [2-^{14}C]tyrosine labelled C-17 to the extent of 37% with no label at C-2 or C-3; [1-^{14}C]tyrosine was only incorporated at a very low level. The novel pathway proposed which takes account of these observations has yet to be substantiated (*R. J. Parry* and *J. M. Schwab*, J. Amer. chem. Soc., 1975, **97**, 2555).

(*v*) *Miscellaneous*

Annuloline. The unique oxazole alkaloid, annuloline, produced by the annual rye grass, *Lolium multiflorum*, is elaborated from two molecules of phenylalanine. In this plant phenylalanine is converted into tyrosine which is incorporated *via* tyramine into one half of annuloline. The other half of the molecule is formed *via* cinnamic acid and its hydroxy derivative (*D. G. O'Donovan* and *H. Horan*, J. chem. Soc., C, 1971, 331).

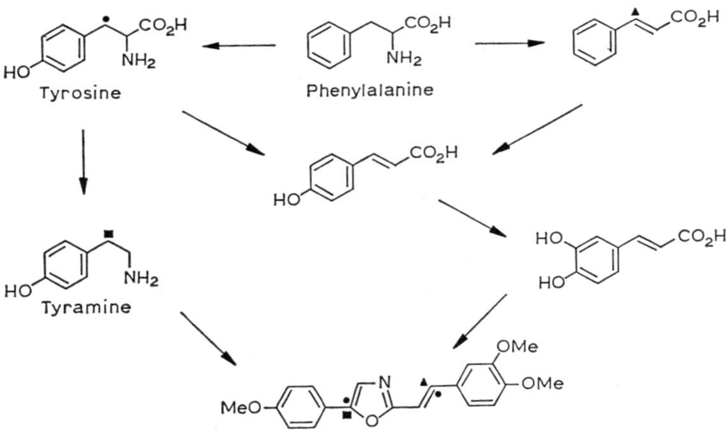

Withasomine. Low incorporations of DL-[3-^{14}C]phenylalanine and DL-[2-^{14}C]ornithine into the pyrazole alkaloid, withasomnine, have been recorded. The specificity of the labelling has not been established but the re-

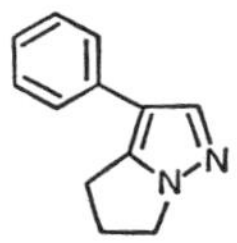

Withasomnine

sults are consistent so far with an expected origin in these two amino-acids (*O'Donovan* and *T. J. Forde*, Tetrahedron Letters, 1970, 3637).

Mycelianamide. The conversion of tyrosine into the mould product mycelianamide involves the formal desaturation of this precursor molecule. Information on the steric course of this desaturation follows from experiments with tyrosine variously labelled with tritium at C-3. Retention of almost half the tritium in the conversion of (3-*RS*)-[2-^{14}C, 3-^{3}H]tyrosine into mycelianamide in *Penicillium griseofulvum* indicated that the reaction

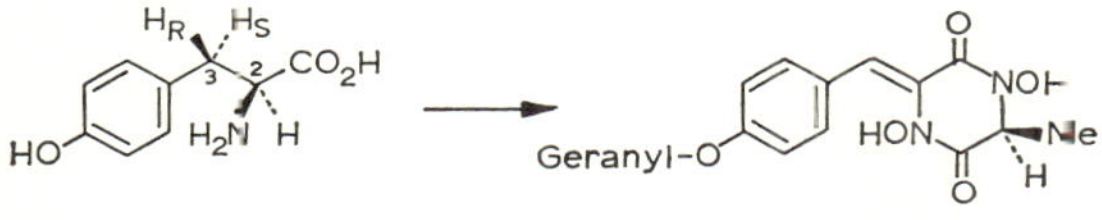

Tyrosine Mycelianamide

was stereospecific, and from the complete loss of activity from (3-*S*)-[2-^{14}C, 3-^{3}H]tyrosine in contrast to a high retention of tritium from the (3-*R*)-isomer, the reaction could be regarded formally as requiring the *cis* removal of hydrogen from the L-form of tyrosine; it was clear incidentally from the ^{14}C/^{3}H values obtained that tyrosine was built into mycelianamide in its entirety (*G. W. Kirby* and *S. Nayaranaswami*, Chem. Comm., 1973, 322).

Mycelianamide, like echinulin (p. 343), gliotoxin (see below), and cyclopenin (p. 379), is at least formally a modified dipeptide. Unlike echinulin and gliotoxin labelled samples of the appropriate dipeptides, *cyclo*-L-alanyl-L-tyrosyl and *cyclo*-L-alanyl-D-tyrosyl were not incorporated into mycelianamide. On the other hand, similar negative results have been obtained for cyclopenin. In both cases the negative results may be attributed to non-equilibration of unbound (radioactive) material with material formed on the enzyme and only desorbed after further reaction (*J. C. MacDonald* and *G. P. Slater*, Canad. J. Biochem., 1975, **53**, 475).

Gliotoxin. Phenylalanine is a precursor for the mould metabolite, gliotoxin, whilst *m*-tyrosine, *o*-tyrosine or 2,3-dihydroxyphenylalanine are not

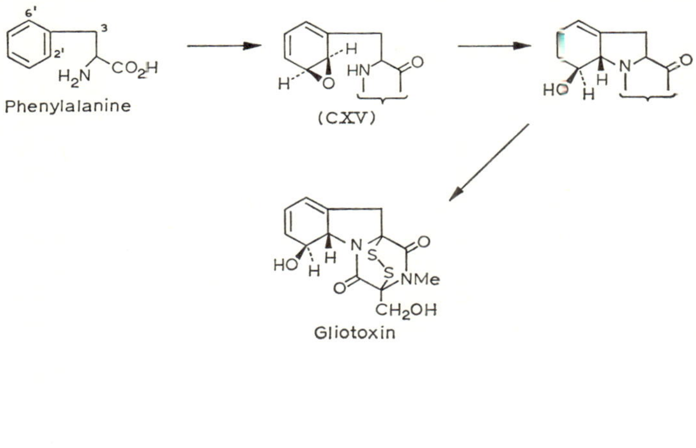

Phenylalanine (CXV) Gliotoxin

incorporated. The incorporation of phenylalanine was sufficiently high to allow the use of deuterium-labelled precursors and it could then be shown that all five aromatic hydrogens are retained in gliotoxin formation (*J. D. Bu'Lock* and *A. P. Ryles*, Chem. Comm., 1970, 1404; *D. R. Brannon et al.*, Biochem. Biophys. Res. Comm., 1971, **43**, 588). The phenol CXVIII has been reported to be a precursor for gliotoxin (*M. S. Ali, R. Rahman* and *A. Taylor*, 5th International Symposium on the Chemistry of Natural Products, IUPAC, London, 1968, Abstracts, p. 94) but in view of the foregoing evidence it cannot be an obligatory intermediate.

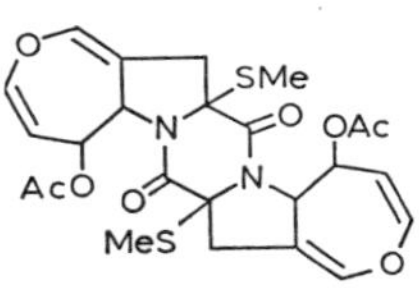

(CXVI)

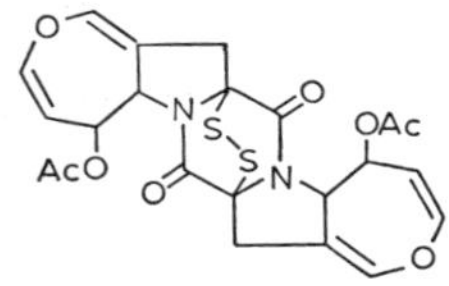

(CXVII)

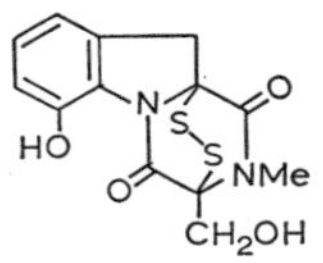

(CXVIII)

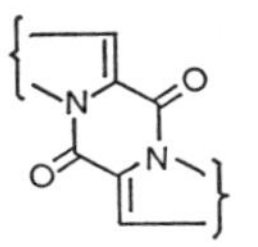

(CXIX)

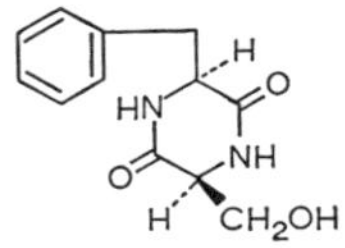

(CXX)

The results obtained for bisdethio-di(methylthio)acetylaranotin (CXVI) are similar to those obtained for gliotoxin. Thus phenylalanine but not *m*-tyrosine was found to be a precursor. Further DL-[3-^{14}C, 2′,6′-$^{3}H_2$]-phenylalanine was incorporated with retention of 80% of the tritium label and L-[$^{2}H_8$]phenylalanine was incorporated into acetylaranotin (CXVII) with retention of the aromatic and C-3 deuterium atoms. Retention of the latter excludes from acetylaranotin biosynthesis intermediates of partial structure CXIX (*Brannon et al., loc. cit.*).

The incorporation of phenylalanine into these metabolites most reasonably involves the intermediate CXV arising by 2,3-epoxidation of the aromatic ring, as illustrated for gliotoxin; the appropriate arene oxides can serve also as plausible intermediates on the way to the aranotins (*N. Neuss et al.*, Antimicrobial Agents Chemotherapy, 1968, 213; Tetrahedron Letters, 1968, 4467).

Incorporation of phenylalanine into gliotoxin proceeds with retention of both hydrogens at C-3, but in a reaction independent of gliotoxin biosynthesis exchange of hydrogen occurs at C-3 with loss of the 3-*pro-R* hydrogen atom (*G. W. Kirby et al.*, J. chem. Soc., Perkin I, 1975, 383).

Inspection of the structure of gliotoxin suggests that this compound is one of a number of metabolites based on a cyclic dipeptide. In accord, *cyclo*-phenylalanyl-seryl (CXX), with labels in both constituent amino acids, was incorporated into gliotoxin without change in isotope ratio (*Bu'Lock* and *C. Leigh*, Chem. Comm., 1975, 628).

Lunarine. In view of the importance the concept of oxidative phenol coupling has had on biosynthetic theory (see section 6a) it is of considerable interest to find a close analogue of Pummerer's ketone occurring naturally. This analogue is to be seen in the *Lunaria biennis* alkaloid, lunarine, for

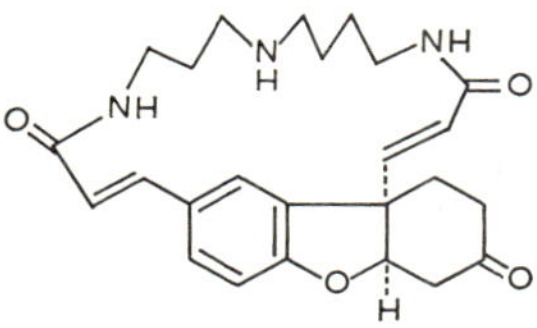

Lunarine

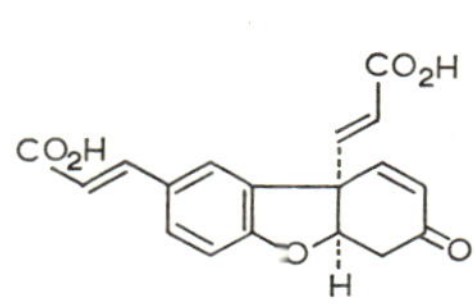

(CXXI)

$H_2N(CH_2)_3NH(CH_2)_4NH_2$

Spermidine

which CXXI is a probable precursor, being formed plausibly by the oxidative coupling of two 4-hydroxycinnamic acid molecules. In accord, phenylalanine (but not tyrosine) is incorporated into lunarine. The remaining atoms come from spermidine (*C. Poupat* and *G. Kunesch*, Compt. rend., 1971, **273C**, 433).

Betalaines. The betalaines are a group of red-violet and yellow pigments found abundantly and uniquely in plants of the Centrospermae order. A red-violet example, betanine, has been found to be derived in a unique way from two complete molecules of tyrosine through dopa, incorporation of one of these units involving aromatic ring scission. DL-[1-^{14}C]Dopa was incorporated into betanine with almost all the activity localized in the carboxyl groups at C-2 and C-15. Activity was unequally divided between these two sites and as much as 90% of it was located in the dihydropyridine moiety (*A. S. Dreiding et al.*, Helv., 1968, **51**, 1470; 1972, **55**, 642). Betalamic acid, which is most plausibly an intermediate in betalaine biosynthesis, is labelled as expected by [1-^{14}C]- and [2-^{14}C]-dopa, label from C-1 being confined to the carboxyl groups of betalamic acid (presumably only C-19 is labelled, see below) (*C. Chang*, *L. Kimler* and *T. J. Mabry*, Phytochemistry, 1974, **13**, 2771).

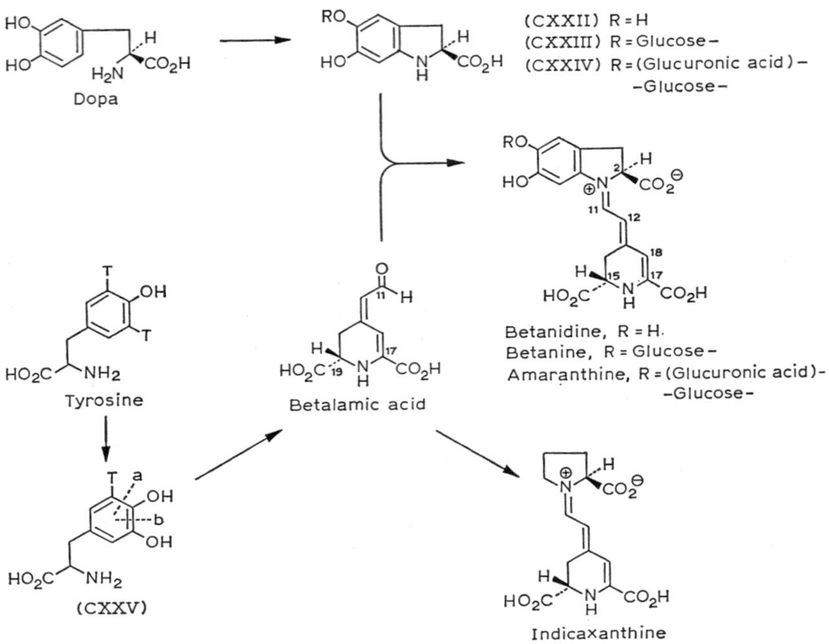

Formation of the dihydropyridine (betalamic acid) residue in betanine could result in principle from ring cleavage of dopa in one of two ways: a or b in CXXV, there being enzymes which catalyse either of these cleavages in several phenols (*O. Hayaishi* and *M. Nozaki*, Science, 1969, **164**, 389; *Hayaishi* in "Biological Oxidations", ed. *T. Singer*. Interscience, New York, 1968, p. 581). The mode of cleavage has been examined for the biosynthesis of betanine and indicaxanthine with [1-^{14}C,3′,5′-$^{3}H_2$]tyrosine. Conversion of the tyrosine into dopa was expected to result in loss of half the tritium, no "NIH" shift (*G. Guroff et al.*, Science, 1967, **157**, 1524) of tritium being expected. Otherwise C-12 (path *b*) or C-18 (path *a*) (of betanine) would have been labelled with tritium. That these positions were devoid of tritium was proved when acid catalysed exchange of C-12 and C-18 in betanine caused no loss of tritium. The dopa was thus labelled as shown in CXXV.

Cleavage along dotted line *b* would result in loss of tritium (no hydrogen atom at C-17) whilst cleavage along dotted line a would give betalamic acid with retention of the tritium label (at C-11). The results for betanine and indicaxanthine are that half the tritium present in the tyrosine fed is retained and therefore ring cleavage occurs along dotted line a (*N. Fischer* and

Dreiding, Helv., 1972, **55**, 649; *G. Impellizzeri* and *M. Piattelli*, Phytochemistry, 1972, **11**, 2499).

Betanidine is converted into betanine indicating glycosylation is a late step in biosynthesis. But the conversion of betanine into amaranthine was not at all efficient. On the other hand (*S*)-*cyclo*-dopa (CXXII) and (*S*)-*cyclo*-dopa-5-*O*-*β*-D-glucoside (CXXIII) were efficiently converted into amaranthine. The major pathway which these results indicate then is *cyclo*-dopa (CXXII) → CXXIII → CXXIV → amaranthine (*S. Sciuto et al., ibid.*, 1972, **11**, 2259; 1974, **13**, 947).

Anthramycin and related antibiotics. Tyrosine suffers minor modification on incorporation into securinine (p. 320) whilst its utilization in betalaine biosynthesis involves cleavage of the aromatic ring (see above). Similar degradation of tyrosine occurs in the biosynthesis of anthramycin and lincomycin.

Both L-[1-^{14}C]tyrosine and L-[1-^{14}C]dopa were efficiently incorporated into anthramycin and the activity was confined to the portion of the molecule shown with heavy bonding. A 64% tritium retention on incorporation of [U-^{14}C,3′,5′-$^{3}H_2$]tyrosine was consistent with the transfer of seven of the nine carbons of tyrosine. Half the tritium was lost from [1-^{14}C,3′,5′-$^{3}H_2$]-tyrosine on conversion into anthramycin, in accord with the cleavage mechanism a (Scheme 52) (the other pathways result in complete loss of tritium).

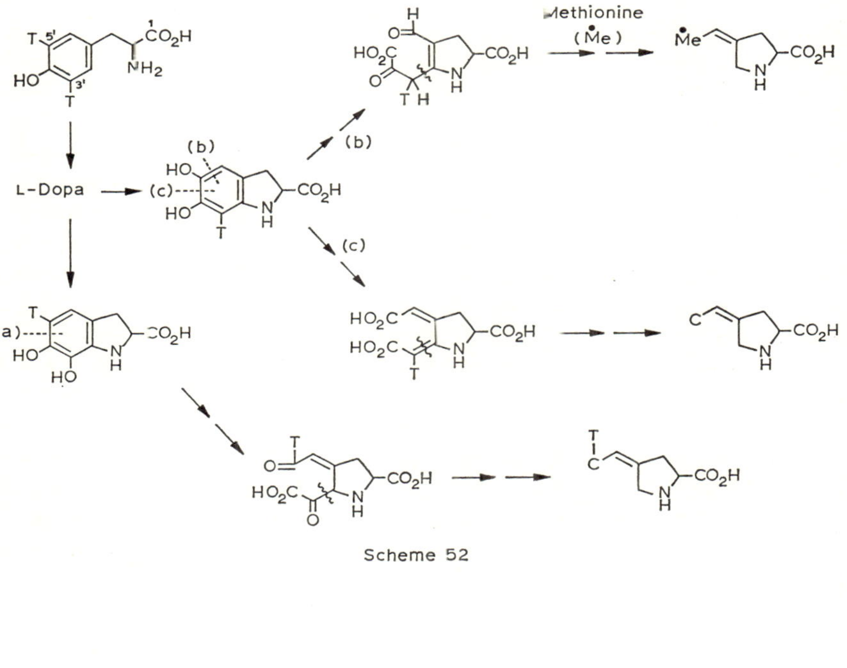

Scheme 52

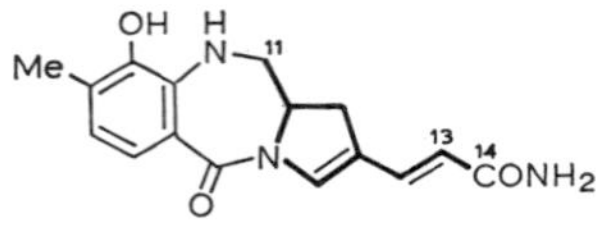

Anthramycin

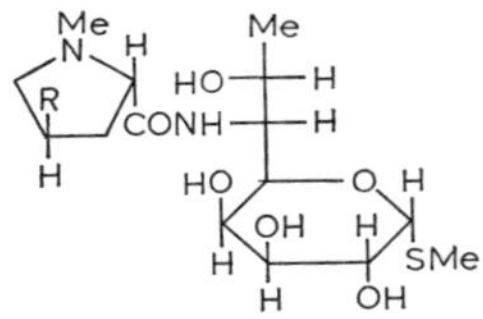

Lincomycin, R = Et and Pr[n]

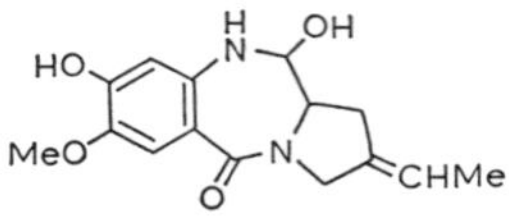

Demethyltomaymycin

Confirmation of this came with the demonstration that the incorporation of L-[3′,5′-$^2H_2$]tyrosine was with retention of one of the deuterium atoms, which was located at C-13 as required. Further important substantiation of the proposed pathway was obtained when it was shown that assimilation of the labels of [1-^{13}C]tyrosine and of [*Me*-^{13}C]methionine were respectively, into C-11 and C-14; the aromatic methyl group was also labelled by the methionine.

The remaining atoms in anthramycin can be identified with 4-methyl-3-hydroxyanthranilic acid derivable from tryptophan (*cf.* actinomycin biosynthesis, p. 370). Accordingly, [7a-^{14}C]tryptophan was well utilized as a precursor (*L. H. Hurley, M. Zmijewski* and *C.-J. Chang*, J. Amer. chem. Soc., 1975, **97**, 4372).

Structures analogous to the tyrosine-derived portion of anthramycin are to be seen in the propyl- and ethyl-proline moieties of **lincomycin.** Their origin is likewise in tyrosine: label from [1-^{14}C]- and [^{15}N]-tyrosine appears at the carboxamide group and proline nitrogen atom, respectively, of lincomycin; the amino acid appears to be incorporated as a C_7 unit. Methionine provides the *N*-methyl and *C*-methyl groups but not the methylene carbons in either formulation. This indicates that the lincomycin with the ethyl group is not the precursor of the propyl substituted antibiotic, but rather that they derive similarly from intermediates differing by one carbon atom (*D. F. Witz, E. J. Hessler* and *T. L. Miller*, Biochemistry, 1971, **10**, 1128).

There is a close resemblance between the structures of anthramycin and (11-demethyl)**tomaymycin,** and an appropriately similar biogenesis has been deduced for this metabolite. L-Tryptophan, L-methionine, L-tyrosine, and L-dopa were all efficiently utilized for demethyltomaymycin production. Incorporation of tyrosine involves ring cleavage in a similar way to that observed for anthramycin: [3′,5′-$^3H_2$,1-^{14}C]Tyrosine gave tomaymycin with tritium label confined to the *C*-methyl group which was also shown not to be derived from methionine (*cf.* Scheme 52) (transformation of the tyrosine through dopa into tomaymycin involves inevitable loss of half the tritium label; see discussion on betalaines above) (*Hurley, C. Gairola* and *Zmijewski* Jr., Chem. Comm., 1975, 120).

Cytochalasins. The biologically interesting cytochalasins have been found to have unexceptional origins in phenylalanine (heavy bonding in cytochalasin D) and acetate (*via* a C_{16}/C_{18} polyketide) with methionine providing additional C_1 units. It is further apparent that L-phenylalanine rather than the D-isomer is the preferred, or primary, precursor of cytochalasin D. Incorporation of either isomer is with complete exchange of hydrogen and

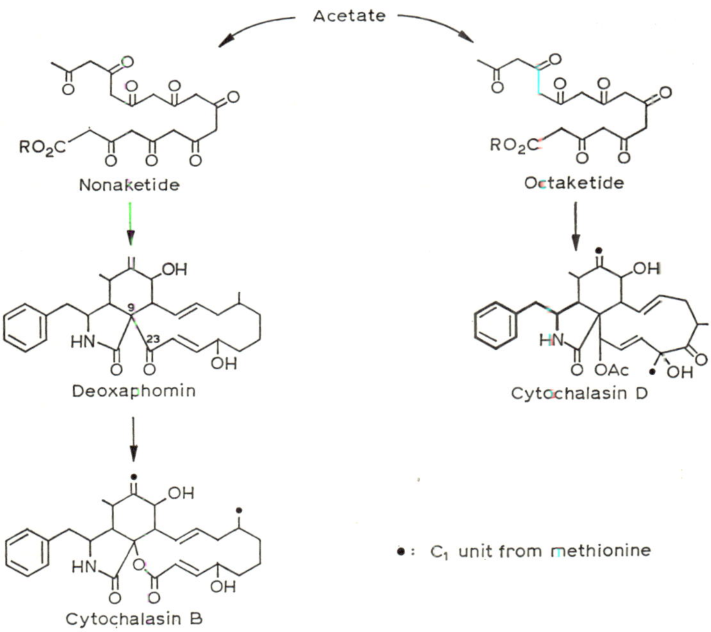

nitrogen from C-2 and extensive exchange of hydrogen from C-3 (*Ch. Tamm et al.*, Helv., 1970, **53**, 1797; 1974, **57**, 1785, 1801; 1975, **58**, 1886; 1976, **59**, 558). Deoxaphomin is a potentially important intermediate in the biosynthesis of cytochalasin B. It was shown to be derived, like other cytochalasins, from phenylalanine but more importantly it was found to be a highly efficient precursor for cytochalasin B. It is manifest that this conversion occurs by insertion of an oxygen atom between C-9 and C-23, and thus the original continuity of the fragmented polyketide chain in cytochalasin B is clear (*cf.*

rifamycin biosynthesis, p. 447) (*J.-L. Robert* and *Tamm*, *ibid.*, 1975, **58**, 2501).

Capsaicin. The active principle of red peppers, capsaicin, has been found to derive in part from phenylalanine, with tyrosine being incorporated at a much lower level (*via p*-coumaric acid?). The pathway further includes cinnamic acid and its hydroxy derivatives, *p*-coumaric and caffeic acids. Also implicated is vanillylamine which probably arises from caffeic acid *via*

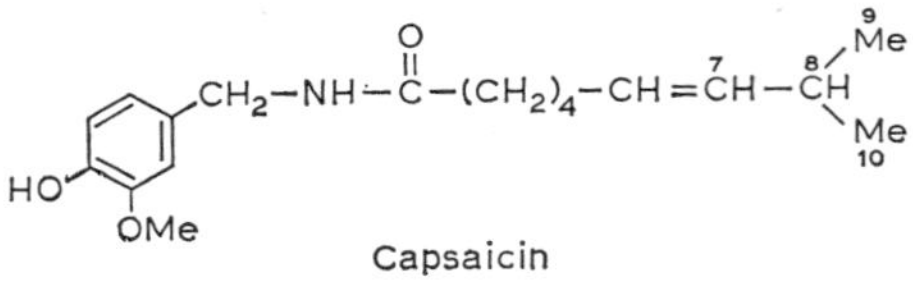

Capsaicin

protocatechualdehyde an important precursor for the Amaryllidaceae alkaloids, see p. 396 (*D. J. Bennett* and *Kirby*, J. chem. Soc., C, 1968, 442; *E. Leete* and *M. C. L. Louden*, J. Amer. chem. Soc., 1968, **90**, 6837). As observed for one Amaryllidaceae alkaloid, aromatic hydroxylation of cinnamic acid results in unusual stereospecific proton loss in the course of the attendant "NIH" shift (*W. R. Bowman*, *I. T. Bruce* and *Kirby*, Chem. Comm., 1969, 1075).

Valine (and not leucine) is a precursor for part (C-7, -8, -9 and -10) of the remaining atoms of capsaicin. It is probably utilized by way of isobutyryl-CoA to which three acetate units are added to give the acid moiety of the alkaloid (*Leete* and *Louden*, *loc. cit.*).

7. Miscellaneous Types

(*a*) *Dolichotheline*

The histamine residue in dolichotheline makes it an unusual alkaloid. Not unexpectedly though, this plant base arises in specific manner (in the cactus *Dolichothele sphaerica*) from histidine and histamine. The isovaleroyl unit was specifically labelled in turn by leucine and isovalerate. Further, L-valine may be utilized for the isovaleryl moiety, being incorporated *via* leucine and

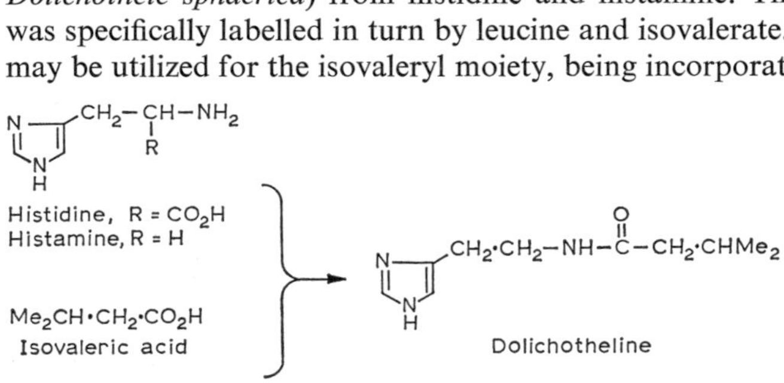

not by direct conversion to isovalerate (*H. Horan* and *D. G. O'Donovan*, J. chem. Soc., C, 1971, 2083; *H. Rosenberg* and *A. G. Paul*, Lloydia, 1971, **34**, 372).

The potential of *D. sphaerica* to synthesize unnatural alkaloids following administration of analogues of histamine and isovaleric acid has been examined in subsequent experiments and the results demonstrate the versatility of enzymes involved in the biosynthesis of secondary metabolites. Of several compounds tested, two, isocaproic acid and 4(5)-aminomethylimidazole, were transformed into unnatural alkaloids, I and II, respectively. When isocaproic acid and 4(5)-aminomethylimidazole were fed together to

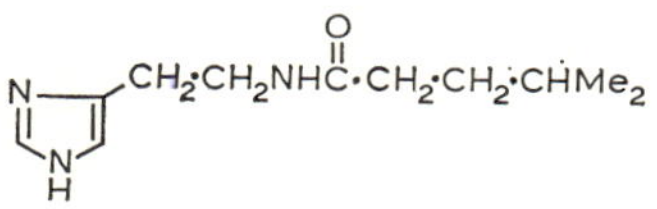

(I)

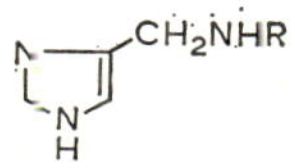

(II) R = $Me_2CH{\cdot}CH_2{\cdot}CO-$
(III) R = $Me_2CH{\cdot}CH_2{\cdot}CH_2{\cdot}CO-$

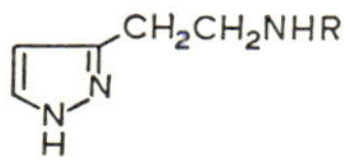

(IV) R = H
(V) R = $Me_2CH{\cdot}CH_2{\cdot}CO-$

the plant, not only was the expected unnatural alkaloid III produced but interestingly the efficiency of the conversion was improved ten-fold (14% conversion) compared with when the substrates were fed separately. It seems as if both substrates must be in sufficient concentration to trigger or promote appropriate enzyme action. In a similar experiment, isovaleric acid and the pyrazole isomer IV of histamine afforded the unnatural alkaloid V (*idem*, J. pharm. Sci., 1973, **62**, 403; *Rosenberg*, *S. J. Stohs* and *Paul*, Phytochemistry, 1974, **13**, 823). The yield of this base relative to dolichotheline was improved by simultaneous feeding of histidine decarboxylase inhibitors, an observation of some significance if plants are to be used for the synthesis of useful unnatural products (*Rosenberg* and *Stohs*, *ibid.*, 1976, **15**, 501).

In another variation of this theme cinnamic acid, but not benzoic acid, was found to act as a suitable substrate for alkaloid production in this cactus (*idem*, Lloydia, 1974, **37**, 313; *Rosenberg* and *Paul*, *loc. cit.*). *N*-Cinnamoylhistamine which is found naturally in species of the Australian Leguminosae was produced in 0.22% yield.

(*b*) *Tenuazonic acid*

Tenuazonic acid is generated in the fungus *Alternaria tenuis* from L-isoleucine and acetate. Butyric acid is also a specific precursor, being implicated presumably *via* acetoacetate. The likely mode of linkage between acetoacetate and L-isoleucine is indicated by the isolation of VIII from the cultures (*C. E. Stickings* and *R. J. Townsend*, Biochem. J., 1961, **78**, 412; *S. Gatenbeck* and *J. Sierankiewicz*, Acta Chem. Scand., 1973, **27**, 1825).

In experiments analogous to those discussed above for dolichotheline, *A. tenuis* was found to be able to substitute added L-valine and L-leucine for L-isoleucine and elaborate the tenuazonic acid analogues VII and VI, respectively (*idem*, Antimicrob. Agents Chemother., 1973, **3**, 308).

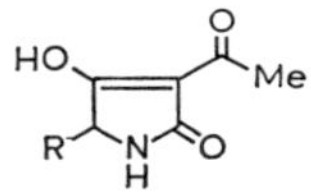

Tenuazonic acid, R =

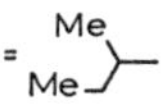

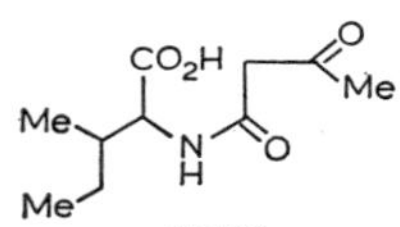

(VIII)

(VI) R =

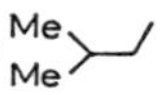

(VII) R =

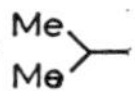

(*c*) *Azetidine-2-carboxylic acid*

The unusual heterocyclic acid, azetidine-2-carboxylic acid, has been found to derive from the entire carbon skeleton of methionine. Radioactive homoserine (XI) was also incorporated, and when a relatively large amount of inactive homoserine was fed together with labelled methionine incorporation of the latter decreased. Thus homoserine is a likely intermediate.

Azetidine-2-carboxylic acid

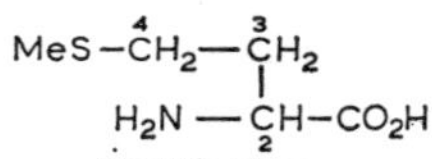

Methionine

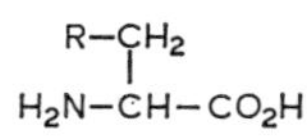

(IX) R = CHO
(X) R = CO_2H
(XI) R = CH_2OH
(XII) R = CH_2NH_2

(XIII)

Incorporation of [1-^{14}C,4-^{3}H]methionine into azetidine-2-carboxylic acid without change in isotope ratio showed that the oxidation level at C-4 was unaffected by the biotransformation of methionine into azetidine-2-carboxylic acid thus excluding as intermediates aspartic-β-semialdehyde (IX) and aspartic acid (X). On the other hand, [1-^{14}C,2-^{3}H]methionine was incorporated with almost complete loss of tritium and so methionine is probably not the ultimate precursor for azetidine-2-carboxylic acid. Further, 2,4-

diaminobutyric acid (XII) has been found to be a better precursor than methionine.

The results can be accommodated within a scheme where 4-amino-2-oxobutyric acid is an intermediate derivable from 2,4-diaminobutyric acid (with loss of tritium from C-2). Ring-closure of the aminoketone would give azetine-2-carboxylic acid (XIII) which in turn would afford azetidine-2-carboxylic acid on reduction (*E. Leete*, J. Amer. chem. Soc., 1964, **86**, 3162; Phytochemistry, 1975, **14**, 1983; *E. F.-W. Su* and *B. Levenberg*, Acta Chem. Scand., 1967, **21**, 493; *Leete et al.*, Phytochemistry, 1974, **13**, 427; *M.-L. Sung* and *L. Fowden*, *ibid.*, 1971, **10**, 1523).

(*d*) *Prodigiosin*

By use mainly of ^{13}C-labelling clear indication has been obtained that the tripyrrolic metabolite, prodigiosin, is constituted in the unprecedented way

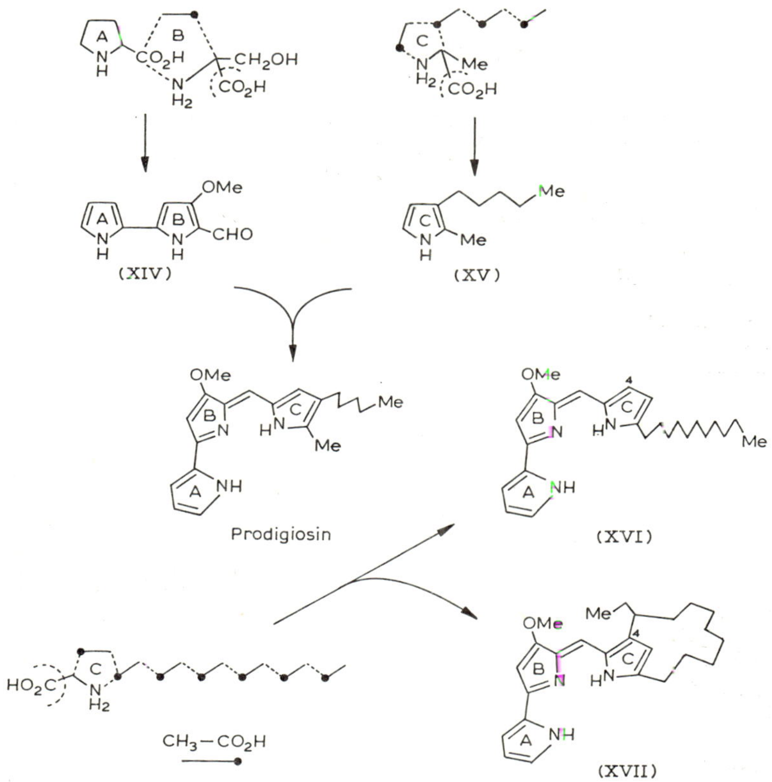

illustrated. (The biosynthetic pattern revealed by the ^{13}C-labelling was of such clarity that secondary as well as primary routes could be discerned) (*H. H. Wasserman et al.*, J. Amer. chem. Soc., 1971, **93**, 6284; 1973, **95**, 6874). The prodigiosin skeleton is also found in two other metabolites, XVI and XVII. Results of similar experiments indicate a common origin for rings A and B in all three metabolites but for ring C, where the structural differences lie, appropriately different origins, as illustrated. It is probable that the side chain seen in XVI undergoes cyclisation on to C-4 at some point in biosynthesis to give XVII (*Wasserman, C. K. Shaw* and *R. J. Sykes*, Tetrahedron Letters, 1974, 2787).

The final step in the biosynthesis of prodigiosin is condensation of XIV with XV (*Wasserman, J. E. McKeon* and *U. V. Santer*, Biochem. Biophys. Res. Comm., 1960, **3**, 146). It seems probable, considering the common origins of rings A and B of the three metabolites, that the same aldehyde precursor XIV condenses *in vivo* with the appropriate pyrroles to give XVI and XVII.

(*e*) *Phomazarin*

Phomazarin is an orange pigment produced by *Phoma terrestris* cultures. Experiments with [1-^{14}C]- and [2-^{14}C]-acetate show that it is probably derived from a polyketide made up of eight or nine acetate units (*A. J. Birch*, Proc. chem. Soc., 1962, 3; *Birch et al.*, Nature, 1961, **190**, 441).

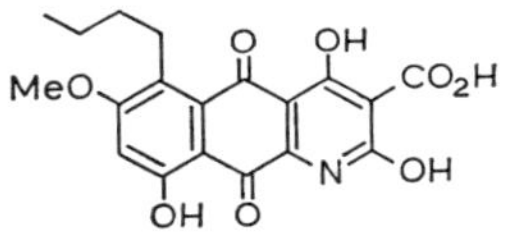

Phomazarin

(*f*) *Nybomycin*

The biosynthesis of nybomycin, a *Streptomyces* metabolite, has been studied using both ^{14}C- and ^{13}C-labelled precursors, the latter providing information that was not obtainable by degradation in the classical manner with ^{14}C. [2-^{14}C]Acetate could be shown to label only C-6′ and C-8′, and [1-^{14}C]-

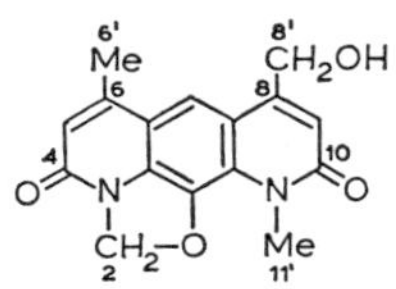

Nybomycin

acetate C-6 and C-8, whereas the ^{13}C-n.m.r. spectrum of nybomycin, after feeding [1-^{13}C]acetate, revealed labelling of C-4, C-6, C-8, and C-10 and each to a similar extent. Thus the exterior carbon atoms of both pyridone rings are each derived equally from identical C_4 acetate-derived units. It was further apparent from the ^{13}C-n.m.r. spectrum that the central ring is not acetate-derived.

Methionine is the source of C-2 and C-11′ (*W. M. J. Knöll, R. J. Huxtable* and *K. L. Rinehart Jr.*, J. Amer. chem. Soc., 1973, **95**, 2703).

(*g*) *β-Lactam antibiotics*

The highly important *β*-lactam antibiotics, the penicillins and cephalosporins, contain an almost unique *β*-lactam ring. The biosynthesis of these compounds is imperfectly understood but it is known to involve the amino acids valine, cysteine and, probably, α-aminoadipic acid, and a tripeptide derivative of them is a likely intermediate (for a comprehensive review on these antibiotics, one which includes an extensive discussion of biosynthesis, see "Cephalosporins and Penicillins, Chemistry and Biology", ed. *E. H. Flynn*, Academic Press, New York, 1972).

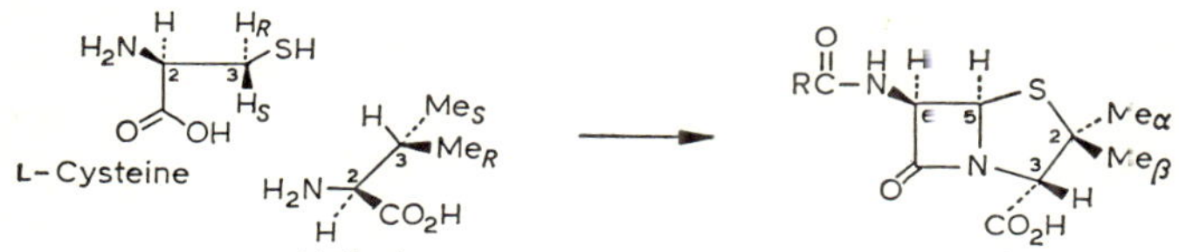

L-Cysteine L-Valine Penicillin

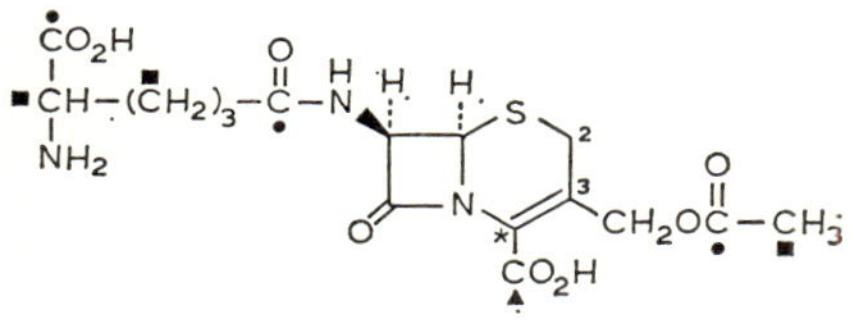

Cephalosporin C

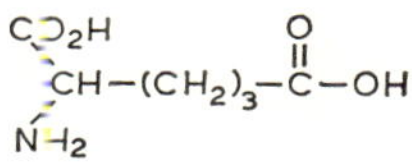

α-Aminoadipic acid

•: $Me^{13}CO_2H$
▪: $^{13}MeCO_2H$
▴: $Me_2CH{\cdot}CH(NH_2){\cdot}{}^{13}CO_2H$
*: $Me_2CH{\cdot}{}^{13}CH(NH_2){\cdot}CO_2H$

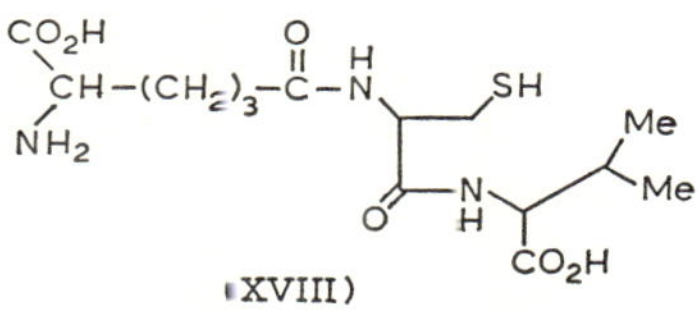

(XVIII)

DL-Valine is incorporated intact into penicillin (*H. R. V. Arnstein et al.*, Biochem. J., 1954, **57**, 353; 1957, **65**, 618). L-[^{15}N]valine, containing also a ^{14}C marker, is incorporated into penicillin without nitrogen loss, whereas tritium is lost from C-2 of the amino acid. The conversion of L-valine into penicillin involves an inversion of configuration at C-2 and from the above clearly does not involve loss of the C-2 nitrogen atom. D-Valine is of the same configuration as the associated centre (C-3) in the penicillins and it now seems, after a period of uncertainty, that it may be an equally effective precursor with L-valine for the antibiotics (*B. W. Bycroft et al.*, Chem. Comm., 1975, 923; 1976, 110; *C. M. Stevens* and *C. W. DeLong*, J. biol. Chem., 1958, **230**, 991).

Cysteine provides the remaining skeletal carbons of the penicillins (seen as an ostensible cysteine unit). L-Cysteine was a more effective precursor than the corresponding D-isomer. Incorporation of DL-[3-^{14}C,^{15}N,^{35}S]cystine occurred to give penicillin with the same isotope ratio as cystine isolated from the mycelium. An intact incorporation is thus established (*Arnstein* and *P. T. Grant*, Biochem. J., 1954, **57**, 353, 360). Deuterium or tritium present at C-2 of cysteine is incorporated into penicillin with little loss, being located exclusively at C-6. This result means that hypothetical 2,3-dehydrocysteine intermediates are not involved in biosynthesis. Carbon atom 5 of the penicillins arises from C-3 of cysteine and in the course of the transformation of amino-acid to antibiotic 3-*pro-S* hydrogen atom of cysteine is lost. The transformation thus occurs with overall retention of stereochemistry at C-3, and it is interesting to note that if oxidation of the C-3 thio-group to, say, a thiocarbaldehyde occurs in the course of biosynthesis, then the stereochemistry is of the opposite sense to that normally observed in the enzymatic conversion of primary alcohols into aldehydes (*Bycroft et al.*, Chem. Comm., 1975, 123; *D. J. Morecombe* and *D. W. Young*, *ibid.*, 1975, 198; *Arnstein* and *J. C. Crawhall*, Biochem. J., 1957, **67**, 180).

The cephalosporin skeleton is obviously related to that of the penicillins, and an attractive possibility is that it is formed by ring expansion of the penicillin skeleton. Appropriately, genesis of cephalosporins closely parallels that of the penicillins. Cystine and valine are both precursors. Recent work with ^{13}C precursors, summarized in the formula for cephalosporin C, confirmed and extended earlier ^{14}C results. The α-aminoadipoyl side-chain it was suggested, from a consideration of [2-^{13}C]acetate labelling levels, arose by way of α-oxoglutarate and acetylcoenzyme A (*N. Neuss et al.*, J. Amer. chem. Soc., 1971, **93**, 2337; Proc. roy. Soc., 1971, **B179**, 335; *P. W. Trown et al.*, Biochem. J., 1963, **86**, 280, 284; 1962, **84**, 157). It appears that *O*-acetylation is the terminal step in the biosynthesis of cephalosporin C (*Y. Fujisawa et al.*, Agric. biol. Chem., 1975, **39**, 2043, 1295, 1303).

Further work on the incorporation of valine into both penicillins and cephalosporin C has more extensively defined the way this amino acid is used. Results on the incorporation of valine stereospecifically labelled with ^{13}C on the methyl groups nicely establish that it is the 3-*pro-R* methyl group which gives C-2 of cephalosporin C and the β-methyl group of penicillin (*Neuss et al.*, J. Amer. chem. Soc., 1973, **95**, 3797; *H. Kluender et al., ibid.*, p. 6149). Valine completely labelled on both or either methyl groups with deuterium gave penicillin without loss of label. Thus no oxidation of either methyl group occurs during penicillin biosynthesis. On the other hand oxidation of one of the methyl groups must occur during the formation of the cephalosporin skeleton. In accord (2*S*,3*R*)-[4,4,4-$^2H_3$]valine was incorporated into cephalosporin C with retention (at C-2) of two deuterium atoms (such a result excludes a $\Delta^{2,3}$-cepham intermediate) (*idem*, J. Amer. chem. Soc., 1974, **96**, 4054; *D. J. Aberhart et al.*, Chem. Comm., 1974, 564).

When considering how the β-lactam antibiotics might be formed from cysteine and valine it seems reasonable that a di- or tri-peptide might be an intermediate. The isolation from *Penicillium* species of the tripeptide XVIII and also penicillin N and isopenicillin N, which have α-aminoadipoyl side chains (they are epimeric at C-5′ in this residue) as in the cephalosporins, provides corroborative evidence for such a tripeptide intermediate leading through penicillin N/isopenicillin N to other penicillins and the cephalosporins (*Arnstein* and *D. Morris*, Biochem. J., 1960, **76**, 357; *E. H. Flynn et al.*, J. Amer. chem. Soc., 1962, **84**, 4594; *E. P. Abraham et al.*, Biochem. J., 1954, **58**, 94; see footnote, *ibid.*, 1962, **84**, 157). So far attempts to establish such a route have been unsuccessful. If an intermediate tripeptide is accepted, the mechanism by which it might cyclise is unknown, even though the ingenious labelling studies discussed above closely define the fates of individual atoms in the starting amino acids. However, there are several attractive mechanisms which have been proposed. One suggestion is that cyclisation occurs by a variation on the ene reaction XIX (*J. E. Baldwin et al.*, Chem. Comm., 1973, 790; see also *A. I. Scott et al., ibid.*, 1974, 47). This is an alternative to the simpler idea of nucleophilic attack of amide nitrogen on the sp^2 carbon of a

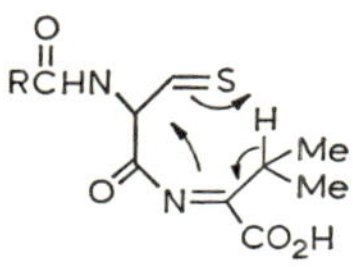

(XIX)

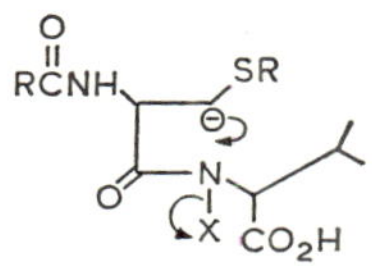

(XX)

thiocarbaldehyde or its sp^3 receptor equivalent (*Arnstein* and *Crawhall, loc. cit.*; *Arnstein*, Ann. Reports, 1957, 339). A quite different possibility is that cyclisation is associated with oxidation of the amide nitrogen; nucleophilic displacement at this centre by an anion generated at the β-carbon of the cysteine fragment follows, see XX (*A. I. Scott et al.*, Tetrahedron Letters, 1976, 1137).

(*h*) *Mitomycins, geldanomycin, rifamycins and streptovaricins*

The **mitomycins,** *e.g.*, mitomycin A, are a group of antitumour antibiotics produced by several *Streptomyces* species. Their biosynthetic origins are only partially clear. Thus methionine is the source of the *O*-methyl groups but not the *C*-methyl group. Both [*ureido*-^{14}C]citrulline and [*guanidino*-^{14}C]arginine label the antibiotic carbamoyl function, although the former is more readily

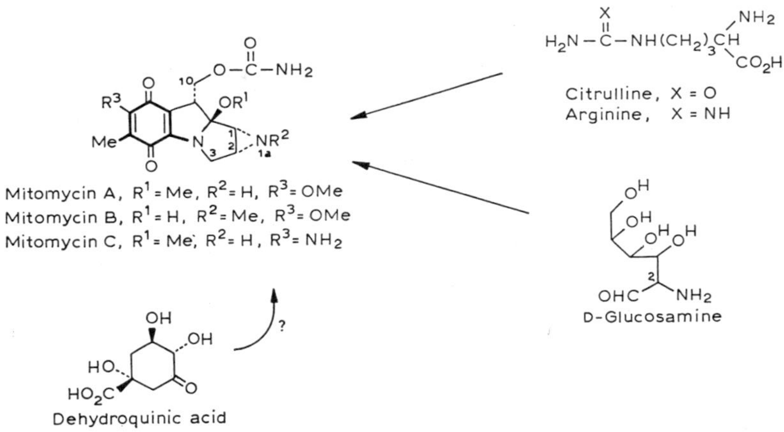

assimilated into the mitomycins. Neither tryptophan, phenylalanine and shikimic acid, were incorporated, nor mevalonate, acetate, propionate, D-arabinose, D-xylose and various C_1 compounds. On the other hand, D-[1-^{14}C]glucosamine was efficiently incorporated into mitomycin A and at a higher level than the related hexosamines, D-galactosamine and D-mannosamine; L-glucosamine was incorporated to an insignificant extent. D-[6-^{14}C]-Glucosamine labelled C-10 of mytomycin C. Incorporation of D-[1-^{14}C, ^{15}N]-glucosamine into mitomycin B occurred with little loss of ^{15}N-label and most of it was located within the aziridine nitrogen. D-[1-^{13}C,1-^{14}C,^{15}N]Glucosamine gave mitomycin B the mass spectrum of which showed a fragment ion (which arises from C-1, C-2, C-3, and N-1a and attached methyl group) corresponding to $C_3{}^{13}CH_8{}^{15}N$ with low-intensity ions corresponding to $C_3{}^{13}CH_8N$ and $C_2{}^{13}C_2H_8N$. An intact incorporation is indicated and it follows that, in the conversion of glucosamine into mitomycin B which necessarily involves inversion at C-2, loss of nitrogen from C-1 does not occur. These results define the origins of the mitomycin skeleton apart from a C_7–N unit (shown with heavy bonding). Aside from the negative results already noted [U-^{14}C]ribose and [U-^{14}C]glucose do in part label this unit, the remaining activity being incorporated *via* glucosamine. More definitive results on the incorporation of labelled glucose (in the presence of an abundance of inactive glucosamine) together with the labelling observed on incorporation of pyruvic acid indicated a biosynthetic pathway to the C_7–N unit in the mitomycins which involves part of the pathway to aromatic amino acids (the shikimic acid pathway). Negative results with shikimic acid indicates that the intermediate at which branching to the mitomycins occurs lies earlier, *e.g.* dehydroquinic acid (*E. J. Kirsch* and *J. D. Korshalla*, J. Bact., 1964, **87**, 247; *U. Hornemann et al.*, Chem. Comm., 1971, 301; 1973, 88; 1974, 1045; J. Amer. chem. Soc., 1974, **96**, 320; Developments Ind. Microbiol., 1974, **15**, 82; *G. S. Bezanson* and *L. C. Vining*, Canad. J. Biochem., 1971, **49**, 911).

The results of biosynthetic studies with **geldanomycin, streptovaricin D** and **rifamycin S** reveal the presence of a C_7–N unit similar to the one found in the mitomycins. Thus, the results of a study of geldanomycin biosynthesis, mainly with ^{13}C-labelled precursors, indicate a sequence of acetate and propionate units from C-14 through C-1, see XXI; the C_7–N unit was not labelled by either propionate or methionine (*R. D. Johnson, A. Haber* and *K. L. Rinehart Jr.*, J. Amer. chem. Soc., 1974, **96**, 3316).

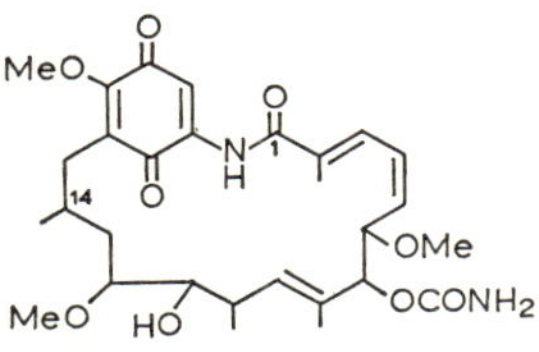

Geldanomycin

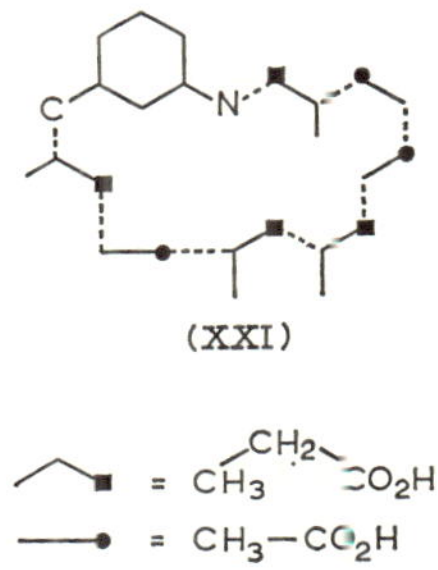

(XXI)

Similar mixed acetate–propionate pathways were deduced similarly for streptovaricin D and rifamycin S. Again negative results for the origin of the C_7–N unit (heavy bonding) were obtained. Neither the usual precursors of aromatic rings, such as shikimate or acetate, served as precursors for this unit in rifamycin S nor was label from propionate incorporated into the C_7–N unit of streptovaricin D. On the other hand, the C-25 methyl group of streptovaricin D was shown to derive from methionine (*R. J. White et al.*, Nature, 1973, **243**, 273; Tetrahedron Letters, 1974, 1367; Europ. J. Biochem., 1974, **47**, 251; *Rinehart Jr., et al.*, J. Amer. chem. Soc., 1973, **95**, 5793).

Of several radioactive compounds tested as precursors of the C_7N unit of rifamycin S [3,4-$^{14}C_2$]glucose and [1-^{14}C]glycerate were found to be sub-

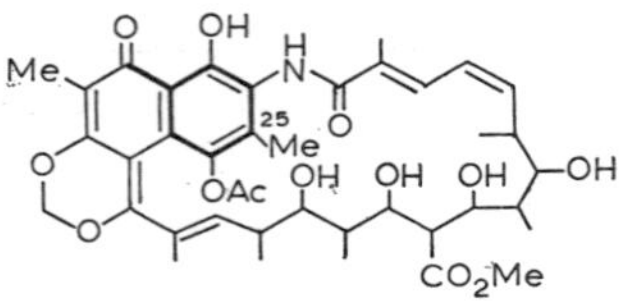

Streptovaricin D

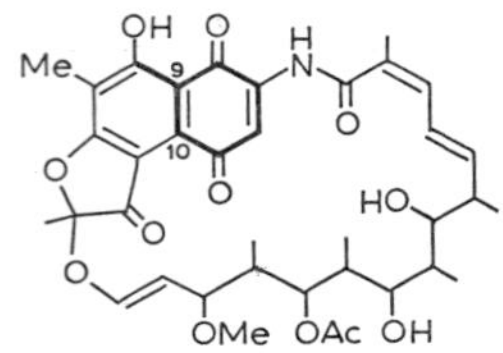

Rifamycin S

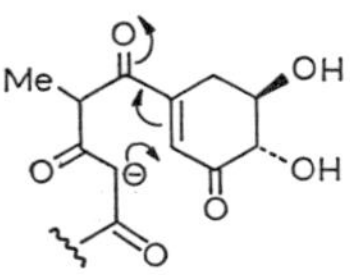

(XXII)

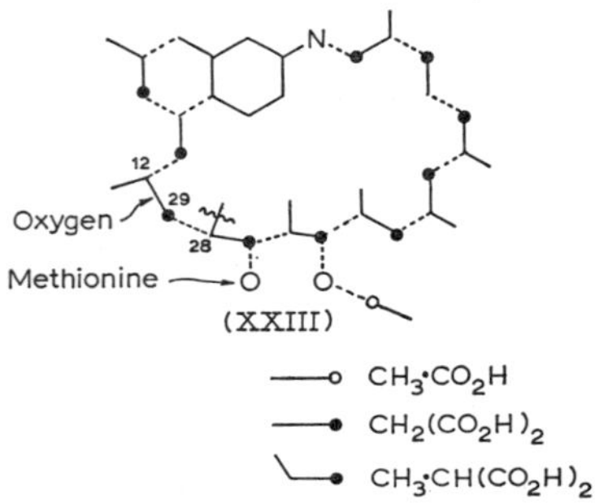

(XXIII)

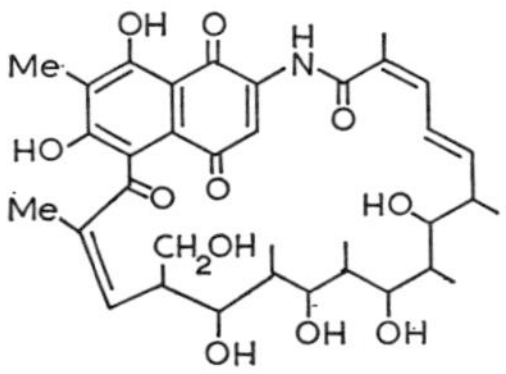

Rifamycin W

stantially the best. Further definitive results were obtained by observing the pattern of [1-^{13}C]glucose and [1-^{13}C]glycerate incorporation. They point, as in the case of mitomycin biosynthesis, to a pathway which branches from the shikimic acid pathway before the formation of shikimic acid itself. C-9 and C-10 of rifamycin S were deduced to be the location of a dehydroshikimate double bond. Michael addition to this double bond, as in XXII, allows completion of the naphthoquinone moiety of rifamycin S in an analogous fashion to the formation of the menaquinones (*White et al.*, Europ. J. Biochem., 1974, **47**, 251; F.E.B.S. Letters, 1974, **49**, 233).

The acetate–propionate labelling pattern in rifamycin S (XXIII) has two unusual features: loss of the propionate-derived methyl group from C-28 and insertion of an oxygen atom between C-12 and C-29 of the precursor. Isolation of rifamycin W with unmodified skeleton, its biosynthesis according to pattern XXIII, and its ready biotransformation into rifamycin B (as rifamycin S with oxygen-substituted hydroquinone ring B) confirms the labelling pattern deduced for rifamycin S and the presumed subsequent skeletal modification (*White, E. Martinelli* and *G. Lancini*, Proc. Nat. Acad. Sci. U.S.A., 1974, **71**, 3260).

(*i*) *Steroidal alkaloids*

Steroidal alkaloids appear as but minor modifications of the steroid skeleton. Entirely, as expected then, mevalonate and acetate have been found to be incorporated into the *Solanum* alkaloids such as tomatine and solasodine

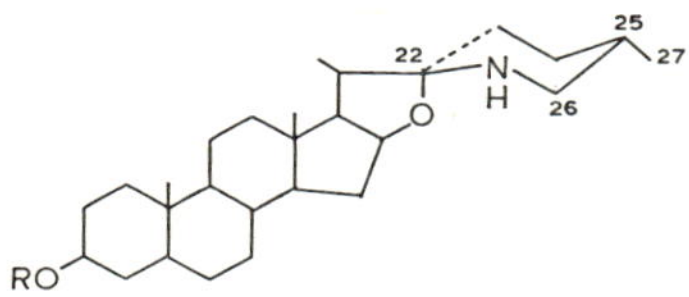

Tomatidine, R = H
Tomatine, R = glycoside

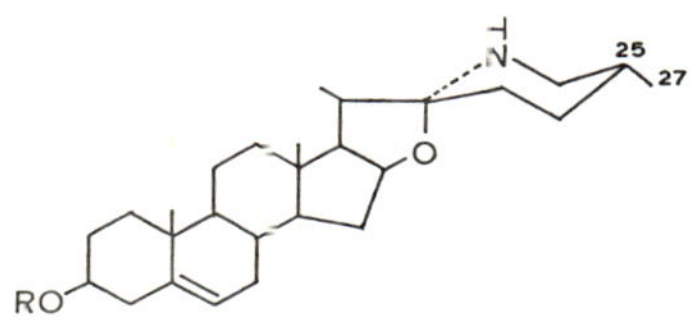

Solasodine, R = H
Solasonine, R = D-galactose < D-glucose, L-rhamnose
Solamargine, R = D-glucose < L-rhamnose, L-rhamnose

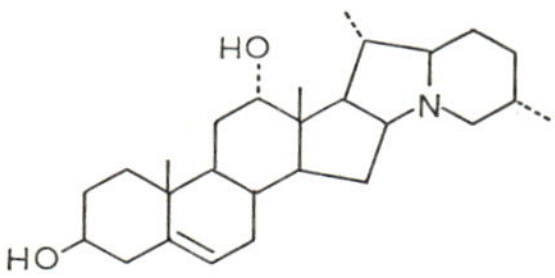

Rubijervine

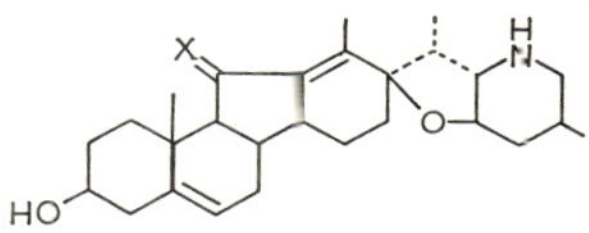

Jervine, X = O
Deoxyjervine, X = H_2

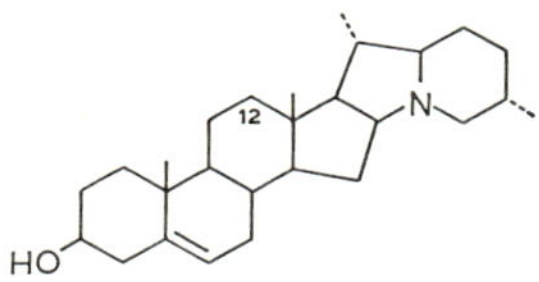

Solanidine

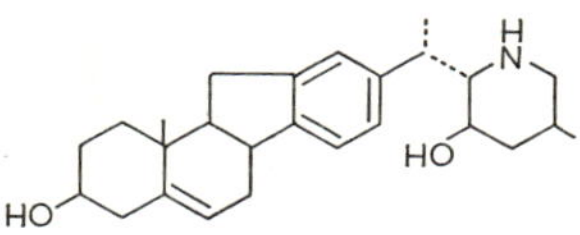

Veratramine

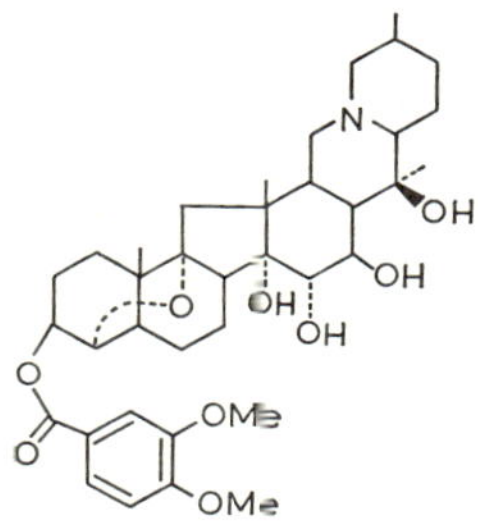

Veratroylzygadenine

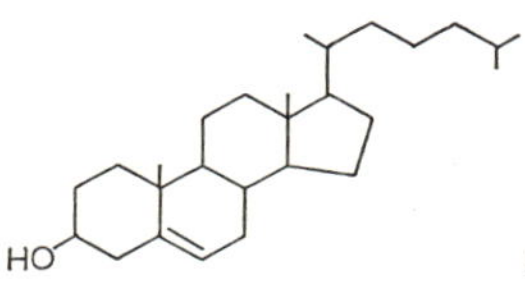

Cholesterol

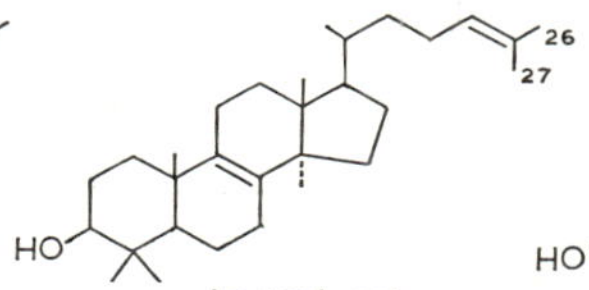

Lanosterol

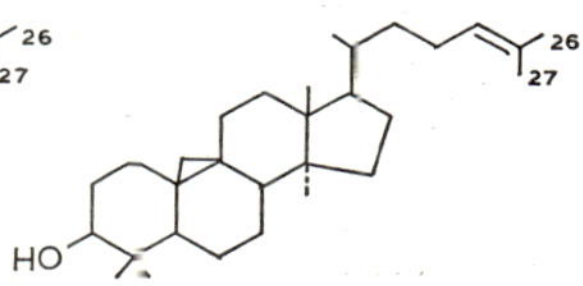

Cycloartenol

(*H. Sander* and *H. Grisebach*, Z. Naturforsch., 1958, **13b**, 755; *A. R. Guseva* and *V. A. Paseshnichenko*, Biochemistry, U.S.S.R., 1962, **27**, 721). Moreover, cholesterol is a precursor for tomatidine, solanidine, and the *Veratrum* bases rubijervine, jervine, veratramine and veratroylzygadenine (*E. Heftmann, E. R. Lieber* and *R. D. Bennett*, Phytochemistry, 1967, **6**, 225; *R. Tschesche* and *H. Hulpke*, Z. Naturforsch., 1966, **21b**, 893; 1967, **22b**, 791; *K. Kaneko et al.*, Phytochemistry, 1970, **9**, 2489). Cholesterol and pregnenolone, but not progesterone, are precursors for the 3-aminopregnenone alkaloids holaphylline and holaphyllamine (*Bennett* and *Heftmann*, Arch. Biochem. Biophys., 1965, **112**, 616; Phytochemistry, 1965, **4**, 873). [26,27-$^{14}C_2$]Cycloartenol and [26,27-$^{14}C_2$]lanosterol were found to be precursors, albeit poor ones, for solanidine, solanocapsine, and tomatidine. In the case of the last mentioned alkaloid the incorporation was shown to be specific (*H. Ripperger, W. Moritz* and *K. Schreiber*, Phytochemistry, 1971, **10**, 2699).

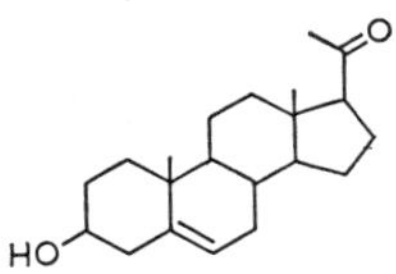

Pregnenolone

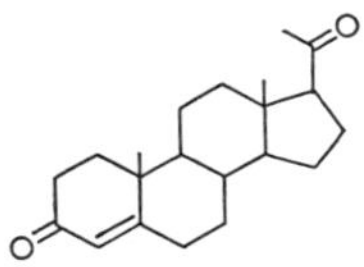

Progesterone

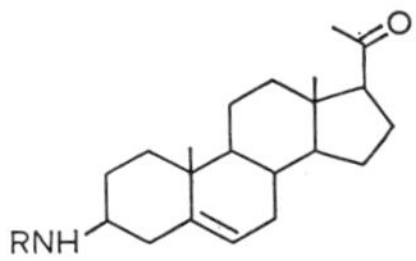

Holaphylline, R = Me
Holaphyllamine, R = H

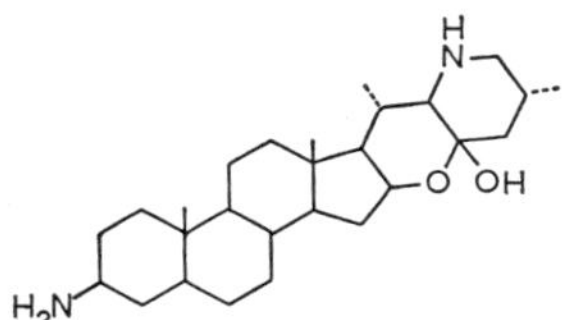

Solanocapsine

Modification of the steroidal side chain which leads to steroidal alkaloids like tomatidine and solasodine may be thought of as occurring by way of the sequence shown. As the C-27 methyl group of solasodine, a (25-*R*)-alkaloid derives from C-2 of mevalonate (*Guseva* and *Paseshnichenko*, *loc. cit.*) it follows that saturation of a Δ^{24}-intermediate as shown, occurs by addition of hydrogen to the 24-*si*, 25-*si* face (path *a*). Carbon atom-26 of tomatidine

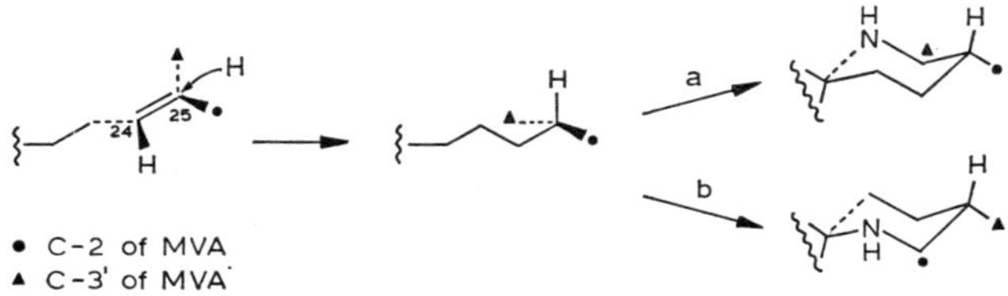

derives from C-2 of mevalonic acid, and so saturation must occur in the same sense (path *b*) to generate this (25-*S*)-alkaloid (*F. Ronchetti* and *G. Russo*, Chem. Comm., 1974, 785). Further detail on tomatidine biosynthesis is that (25-*S*)-5α-cholestan-3β,26-diol (XXIV) but not the corresponding furostan derivative XXV is a precursor. Taken with the observation (*R. Tschesche*, "Euchem. Conference on Chemistry and Biosynthesis of Steroids and Terpenoids", La Laguna, Canary Isles, Spain, 1972) that the 26-amino

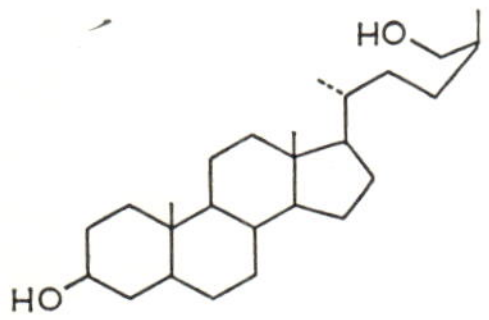

(XXIV)

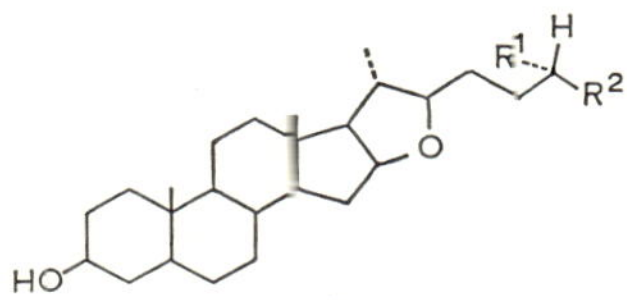

(XXV) R^1 = Me, R^2 = CH_2OH
(XXVI) Δ^5, R^1 = CH_2NH_2, R^2 = Me

compound XXVI is a precursor for solasodine, and if it is accepted that the biosynthesis of (25-*R*)- and (25-*S*)-spirosolanes will prove to be similar, it follows that introduction of the amino group occurs before formation of the tetrahydrofuran ring (*Ronchetti et al.*, Phytochemistry, 1975, **14**, 2423).

Solasodine is a precursor for the glycoalkaloids solasonine and solamargine, and an enzyme has been isolated from *Solanum nigrum* which will carry out a stepwise removal of the sugar residues from these glycoalkaloids. It may be concluded from these observations that glycoside formation is a late biosynthetic step and that a stepwise addition of sugar units occurs (*D. R. Liljegren, ibid.*, 1971, **10**, 3061; and unpublished results).

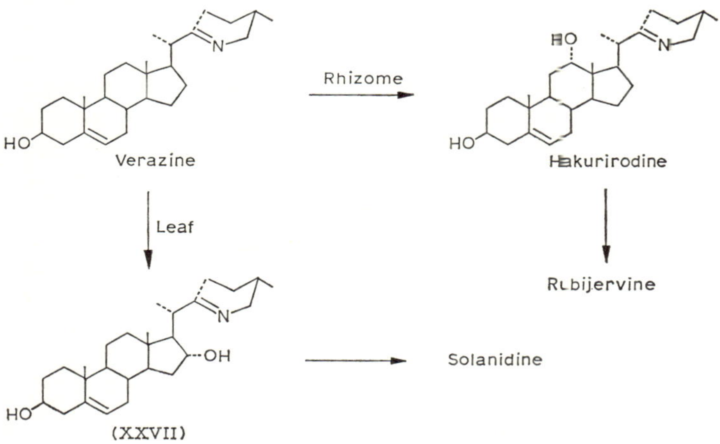

(XXVII)

A ring system different from that just discussed is seen in alkaloids like solanidine, although as indicated earlier its origins are similar. The aglycone, etioline, of one of two glycosides isolated from budding *Veratrum grandiflorum*, was shown to be XXVII. The two glycosides appear to be of relevance to solanidine biosynthesis since their concentration decreased as that of solanidine increased. Further delineation of solanidine-type biosynthesis was obtained by observing the incorporation of labelled etioline and verazine into alkaloids in dormant rhizome slices of *V. grandiflorum*, and also during budding. It was concluded that different pathways to solanidine and rubijervine operate (*Kaneko et al., ibid.*, 1975, **14**, 1295; Tetrahedron Letters, 1971, 4251).

As indicated earlier the interesting C-nor-D-homosteroidal alkaloids, *e.g.* jervine, originate from a normal steroid skeleton. It is an attractive possibility that the distinctive ring expansion–contraction sequence could be initiated on a solanidine derivative with an equatorial hydroxy-group at C-12. The appearance of radioactivity from [1-^{14}C]acetate in solanidine glycoside in *V. grandiflorum* grown in the dark, and subsequent transfer of radioactivity to jervine and veratramine upon illumination gives substance to the idea of solanidine as an intermediate (*idem*, Phytochemistry, 1972, **11**, 3199). 11-Deoxyjervine appears further to be an intermediate in jervine biosynthesis (*idem, ibid.*, 1970, **9**, 2497).

(*j*) *Terpenoid alkaloids*

The monoterpenoid alkaloids, *e.g.* skytanthine, almost by definition, originate from mevalonate and these compounds with terpenoid-like structures have been shown to be labelled by radioactive mevalonate and acetate and so far as is known arise by the normal terpenoid pathway. In accord with observations on plumieride, C-3 and C-9 of skytanthine formed from [2-^{14}C]mevalonate in 1.3 year old *Skytanthus acutus*, had become equivalent; no equivalence was, however, observed in three-year-old plants. Actinidine was labelled by [1-^{14}C]geranyl pyrophosphate and although loganin was a poor precursor for some *Tecoma stans* bases, it is likely that this skeleton is still involved in the form of deoxyloganic acid (*H. Auda et al.*, J. Amer. chem. Soc., 1967, **89**, 2476; J. biol. Chem., 1967, **242**, 4157; *C. G. Casinovi et al.*, Gazz., 1964, **94**, 1356; *D. Gross, W. Berg* and *H. R. Schütte*, Biochem. Physiol. Pflanzen, 1972, **163**, 576).

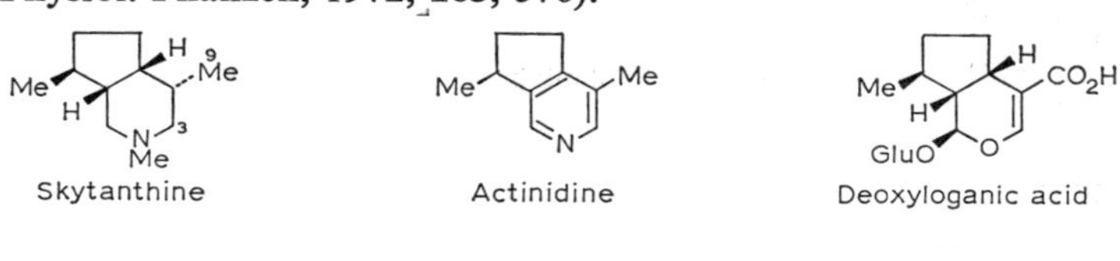

Skytanthine Actinidine Deoxyloganic acid

In contrast to the monoterpenoid alkaloids the pathway to the sesquiterpenoid alkaloid dendrobine is quite well delineated. It is labelled by [2-^{14}C]-mevalonate in the expected way. Both of the tritium atoms from [5-$^{3}H_2$]-mevalonate, which would appear at an intermediate stage at C-1 of farnesol,

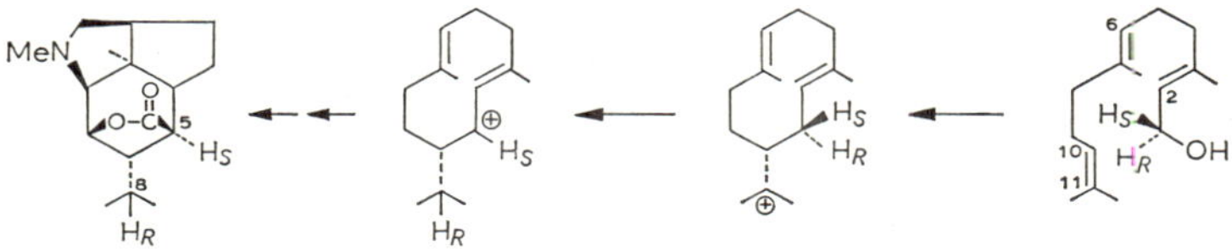

Dendrobine (XXVIII) Farresol

are retained in the formation of the alkaloid. One of the tritium atoms appears unexceptionally at C-5 whereas the other undergoes a 1,3-shift in an intermediate which may be represented as XXVIII, and thus appears at C-8. Incorporation of (1-*RS*)-[1-$^{3}H_2$]- and (1-*S*)-[1-^{3}H]-2-*trans*,6-*trans*farnesol but not the corresponding 2-*cis* isomer (similarly labelled) points to the 10-membered-ring intermediate XXVIII being as shown, rather than having the 2-*cis* geometry. The results further show that the hydrogen migration is stereospecific and involves the 1-*pro-R* hydrogen atom (*A. Corbella et al.*, Chem. Comm., 1973, 729; 1975, 288; *M. Yamazaki*, *M. Matsuo* and *K. Arai*, Chem. Pharm. Bull., Japan, 1966, **14**, 1058). Interestingly the biosynthesis of the related sesquiterpene culmorin involves migration from C-1 to C-10 rather than to C-11 (farnesol numbering) (*J. R. Hanson* and *R. Nyfeler*, Chem. Comm., 1975, 171, 824).

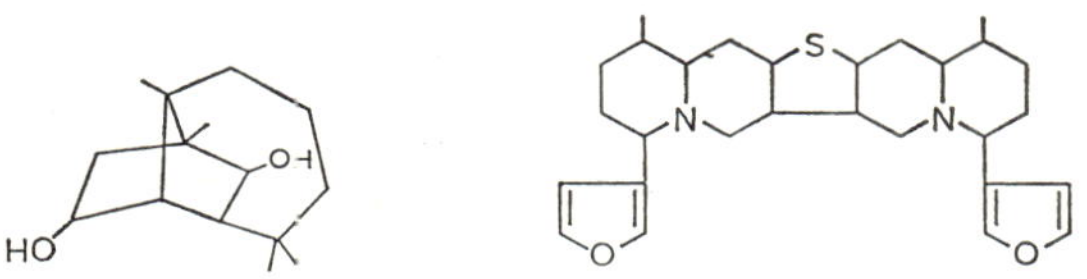

Culmorin Thiobinupharidine

[3,4-$^{14}C_2$]Mevalonate is a specific precursor for thiobinupharidine, in a pattern which accords with terpenoid hypothesis; [1,5-$^{14}C_2$]cadaverine is an unspecific precursor (*Schütte* and *J. Lehfeldt*, Arch. Pharm., 1965, **298**, 461).

The complexity and sparse functionality of the *Daphniphyllum* alkaloids, *e.g.* daphniphylline, makes them appear forbidding from an experimental viewpoint. However, by using the α-acetoxyketone function in a simple degradation of the daphniphylline isolated after administration of [2-^{14}C]- and [5-^{14}C]-mevalonate useful information has been gained. The molecule was cleaved into two fragments (see dotted line) and from the ratio of activities in these fragments (4:2 and 5:1, respectively) it was clear that six

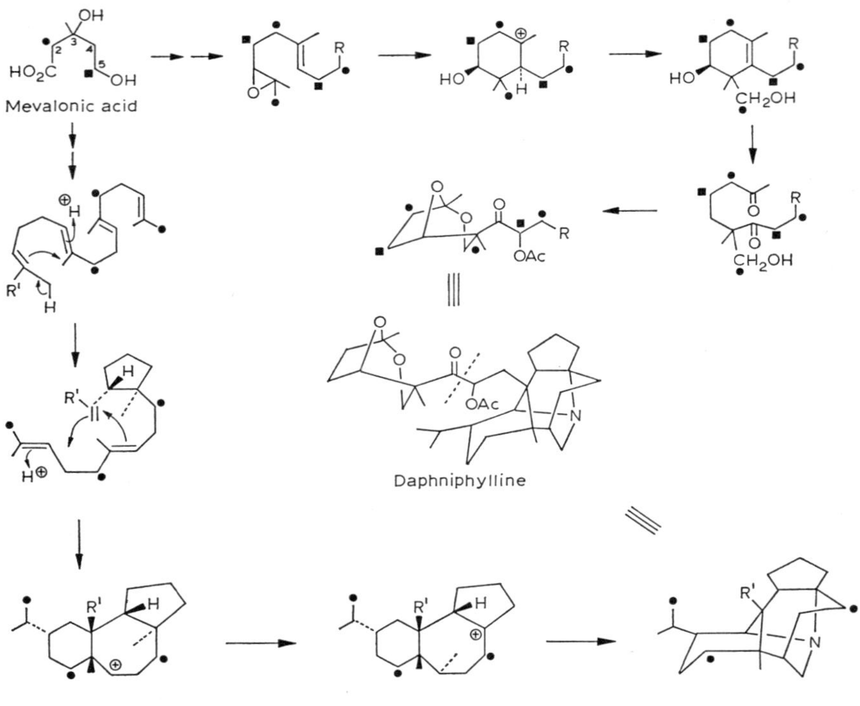

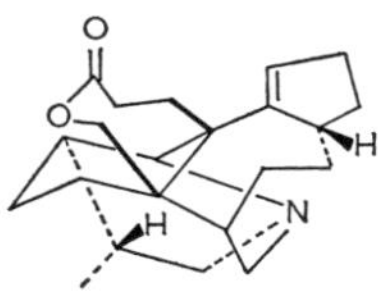

Daphnilactone B

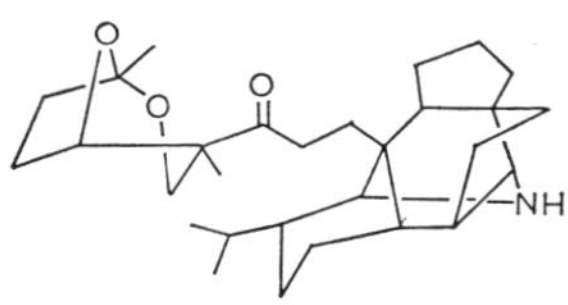

Secodaphniphylline

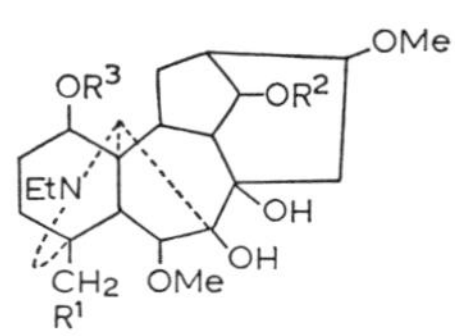

Brownine, R^1 = OMe, R^2 = H, R^3 = Me
Lycoctonine, R^1 = OH, R^2 = R^3 = Me
Delcosine, R^1 = OMe, R^2 = R^3 = H

mevalonate units are involved and a possible mode of biosynthesis was suggested (*K. T. Suzuki et al.*, Tetrahedron Letters, 1973, 799).

The biosynthesis of C_{22} triterpenoid alkaloids like daphnilactone B proceeds plausibly *via* C_{30} intermediates such as secodaphniphylline and this has been supported in preliminary experiments with [2-^{14}C]mevalonate (*H. Niwa et al.*, *ibid.*, 1973, 2129).

Acetate and mevalonate have been shown to be precursors for brownine and lycoctonine (*M. H. Benn* and *J. May*, Experientia, 1964, **20**, 252) and mevalonate and glycine were incorporated into delcosine (*G. M. Frost et al.*, Chem. and Ind., London, 1967, 320).

Guide to the Index

This index is constructed in a similar manner to the volume indexes of the first edition of the Chemistry of Carbon Compounds. However, to make the index easier to use, more descriptive entries have been made for the commonly occurring individual, and groups of chemicals.

The indexes cover primarily the chemical compounds mentioned in the text, and also include reactions and techniques, where named, and some sources of chemical compounds such as plant and animal species, oils, etc.

Chemical compounds have been indexed alphabetically under the names used by authors, editing being restricted to ensuring uniformity of entries under the same heading. In view of the alternative nomenclature that can often be used, a limited amount of cross-referencing has been done where it is considered to be helpful, but attention is particularly drawn to Convention 2 below.

For this and the succeeding volumes, the indexing conventions listed below have been adopted.

1. *Alphabetisation*

(a) The following prefixes have not been counted for alphabetising:

n-	*o-*	*as-*	*meso-*	D-	*C-*
sec-	*m-*	*sym-*	*cis-*	DL-	*O-*
tert-	*p-*	*gem-*	*trans-*	L-	*N-*
	vic-		*endo-*		*S-*
		lin-	*exo-*		*Bz-*
					Py-

Some prefixes and numbering have been omitted in the index, where they do not usefully contribute to the reference.

(b) The following prefixes have been alphabetised:

Allo	Epi	Neo
Anti	Hetero	Nor
Cyclo	Homo	Pseudo
	Iso	

(c) A letter by letter alphabetical sequence is followed for entries, firstly for the main entry, followed by the descriptive entry. The only exception to this sequence is the placing of plural entries in front of the corresponding individual entries to prevent these being overlooked by a strict alphabetical sequence which could lead to a considerable separation of plural from individual entries. Thus "butanes" will come before *n*-butane, "butenes" before 1-butene, and 2-butene, etc.

2. *Cross references*

In view of the many alternative trivial and systematic names for chemical compounds, the indexes should be searched under any alternative names which may be indicated in the main body of the text. Only a limited amount of cross-referencing has been carried out, where it is considered that it would be helpful to the user.

3. *Esters*

In the case of lower alcohols esters are indexed only under the acid, *e.g.* propionic methyl ester, not methyl propionate. Ethyl is normally omitted *e.g.* acetic ester.

4. *Derivatives*

Simple derivatives are not normally indexed if they follow in the same short section of the text.

5. *Collective and plural entries*

In place of "— derivatives" or "— compounds" the plural entry has normally been used. Plural entries have occasionally been used where compounds of the same name but differing numbering appear in the same section of the text.

6. *Main entries*

The main entry of the more common individual compounds is indicated by heavy type. Where entries relate to sections of three pages or more, the page number is followed by "ff".

Index

RODD'S CHEMISTRY OF CARBON COMPOUNDS

ELSEVIER SCIENTIFIC PUBLISHING COMPANY
335 JAN VAN GALENSTRAAT
P.O. BOX 211, 1000 AE AMSTERDAM, THE NETHERLANDS

DISTRIBUTORS FOR THE U.S.A. AND CANADA

ELSEVIER/NORTH-HOLLAND INC.
52 VANDERBILT AVENUE, NEW YORK, N.Y. 10017

WITH 43 TABLES

ISBN: 0-444-41647-1 (VOL. IVK)
ISBN: 0-444-40664-6 (SERIES)

LIBRARY OF CONGRESS CARD CATALOG NUMBER 64-4605

PRINTED IN THE NETHERLANDS

RODD'S CHEMISTRY OF CARBON COMPOUNDS

ADVISORS

The late Professor Sir ROBERT ROBINSON, O.M., M.A. (Oxon.), D.SC. (Manc.), HON.D.SC. (Lond., Liv., Wales, Dunelm, Sheff., Belfast, Bris., Oxon., Nott., Strath., Delhi, Sydney, Zagreb), HON. SC.D. (Cantab.), HON. LL.D. (Manc., Edin., Birm., St. Andrews, Glas., Liv.), HON. D. PHARM. (Madrid and Paris), HON. F.R.S.E., F.R.S., *London*

Professor A. R. BATTERSBY, M.SC. (Manc.). PH.D. (St. Andrews), D.SC. (Bris.), M.A., SC.D. (Cantab.) F.R.S., *Cambridge*

Professor R. N. HASZELDINE, M.A., PH.D., SC.D. (Cantab.), PH.D., D.SC. (Birm.), C.CHEM., F.R.I.C., F.R.S., *Manchester*

Professor R. D. HAWORTH, D.SC., PH.D. (Manc.), B.SC. (Oxon.), C.CHEM., F.R.I.C., F.R.S., *Sheffield*

The late Professor Sir EDMUND HIRST, M.A., PH.D. (St. Andrews), D.SC. (Birm.), HON.LL.D. (St. Andrews, Aberdeen, Birm., Strath.), HON. D.SC. (Dublin), F.R.I.C., F.R.S., *Edinburgh*

Professor Lord TODD, O.M., M.A. (Cantab.), D.SC. (Glas.), D.PHIL. (Oxon.), DR. PHIL. NAT. (Frankfurt), HON.LL.D. (Glas., Edin., Melb., Calif.), HON. DR. RER. NAT. (Kiel), HON. D. MET. (Sheff.), HON. D.SC. (Oxon., Dunelm, Lond., Exe., Leic., Liv., Adel., Alig., Madrid, Stras., Wales, Strath.), C.CHEM., F.R.I.C., F.R.S., *Cambridge*

RODD'S CHEMISTRY OF CARBON COMPOUNDS

VOLUME I

GENERAL INTRODUCTION

ALIPHATIC COMPOUNDS

*

VOLUME II

ALICYCLIC COMPOUNDS

*

VOLUME III

AROMATIC COMPOUNDS

*

VOLUME IV

HETEROCYCLIC COMPOUNDS

*

VOLUME V

MISCELLANEOUS

GENERAL INDEX

*

RODD'S CHEMISTRY OF CARBON COMPOUNDS

A modern comprehensive treatise

SECOND EDITION

Edited by
S. COFFEY
M.Sc. (London), D.Sc. (Leyden), C.Chem., F.R.I.C.
formerly of
I.C.I. Dyestuffs Division, Blackley, Manchester

VOLUME IV PART K

HETEROCYCLIC COMPOUNDS

Six-membered heterocyclic compounds with two or more hetero-atoms one or more of which are from Groups II, III, IV, V or VII of the Periodic Table. Heterocyclic compounds with seven or more atoms in the ring

ELSEVIER SCIENTIFIC PUBLISHING COMPANY
AMSTERDAM OXFORD NEW YORK
1979

CONTRIBUTORS TO THIS VOLUME

R. S. EDMUNDSON, B.SC., PH.D.
School of Chemistry, University of Bradford, Bradford BD7 1DP

J. M. F. GAGAN, M.A., D. PHIL.
The Open University, North West Region, Manchester M21 1PQ

R. E. FAIRBAIRN, B.SC., PH.D., F.R.I.C.
formerly of Research Department, Dyestuffs Division, I.C.I. Ltd., Manchester 9 *(Index)*

PREFACE TO VOLUME IV K

The first four sub-volumes (Parts) of Volume IV of this revised edition of *Rodd's Chemistry of Carbon Compounds* are concerned with the chemistry of organic compounds possessing in their structure an isolated three-, four- or five-membered ring in which a single hetero-atom is present. Parts E–J inclusive deal with the chemistry of the wider and more varied types of compound based on six-membered heterocyclic systems containing one or more atoms of nitrogen and/or elements of Group VI of the Periodic Table. The first two chapters of the present volume, IV K, complete this theme on six-membered heterocycles, while the rest of the subject matter, which comprises the greater part of the volume, is concerned with compounds with still larger isolated heterocyclic rings with seven or more constituent atoms.

The whole nine chapters are the work of two authors, Dr. R. S. EDMUNDSON and Dr. J. M. F. GAGAN, who are to be congratulated on their presentation of the chemistry of a very wide range of classes of compounds, many of them of topical interest for various reasons, on which the major studies have been published since the first edition of *Rodd* was compiled.

Dr. EDMUNDSON has contributed Chapters 48, 49, 53 and 54 all of which are concerned with compounds containing isolated heterocyclic rings of six or more atoms, one or more of which are "unusual" elements from Groups II, III, IV, V and VII of the Periodic Table. Chapters 48 and 49 deal, respectively, with six-membered heterocycles, with heterocyclic compounds with (a) two or more unusual atoms in the ring and (b) with one or two such atoms as well as nitrogen, while Chapters 53 and 54 describe similar types of compounds with still larger heterocyclic rings, in the latter case, along with nitrogen or an atom from Group VI.

Dr. GAGAN describes compounds with larger heterocyclic rings consisting of seven or more atoms including nitrogen and/or elements of Group IV, namely oxygen, sulphur, selenium and tellurium, as follows: Chapter 50, seven-membered ring compounds containing an element from Group VI; Chapter 51, similar seven-membered compounds containing one or more nitrogen atoms; Chapter 52, similar compounds with two or more different hetero-atoms in the ring; Chapter 55, eight-membered heterocyclic compounds; and Chapter 56, compounds with rings of nine or more atoms. The subject matter of this volume is of particular interest in that it deals with a very wide range of novel and important types of compounds only relatively few examples of which had been isolated and identified when the first edition of *The Chemistry of Carbon Compounds* was compiled. As a perusal of the cited literature references shows, this holds for most sections of the text but it is especially true for the chapters dealing with the chemistry of the larger heterocycles with seven or more ring-atoms, which occupies about four times as many pages of text as the same topics did originally.

As Dr. GAGAN indicates in his general introduction to the subject (see p. 139 *et seq.*), the recent study of the parent ring compounds and their simply substituted derivatives has had far-reaching consequences on various theoretical, synthetic, mechanistic and stereochemical aspects of organic chemistry, as well as a strong commercial interest, as witness the facts that one of the most important monomers, caprolactam, used for making "Nylon 6", and one of the most heavily prescribed drugs, "Librium", are seven-membered heterocyclic compounds. The volume should be of particular interest to both organic and inorganic chemists, biochemists, pharmacologists and readers interested in any of the biomedical sciences.

The perusal of the extensive, recent literature on the subjects dealt with in this volume, and the subsequent preparation of the material for publication, set the authors a formidable task; it is a pleasure to congratulate them on its completion and to thank them for their cooperation with the editor at all times as the final scripts were prepared for the press.

APRIL 1979 S. COFFEY

CONTENTS

VOLUME IV K

Heterocyclic Compounds: Six-membered heterocyclic compounds with two or more hetero-atoms one or more of which are from Groups II, III, IV, V or VII of the Periodic Table. Heterocyclic compounds with seven or more atoms in the ring

Chapter 50. Compounds with Larger Heterocyclic Rings: Seven-Membered Ring Compounds with an Element from Group VI in the Ring
by J. M. F. GAGAN

Chapter 51. Compounds with Larger Heterocyclic Rings: Seven-Membered Ring Compounds with an Atom from Group V of the Periodic Table in the Ring. Nitrogen
by J. M. F. GAGAN

Chapter 52. Compounds with Larger Heterocyclic Rings: Seven-Membered Ring Compounds with Two or More Different Elements in the Ring
by J. M. F. Gagan

Titles of other parts of Volume IV

HETEROCYCLIC COMPOUNDS

Vol. IV A:	Three-, four- and five-membered heterocyclic compounds with a single hetero-atom in the ring
Vol. IV B:	Five-membered heterocyclic compounds with a single hetero-atom in the ring: alkaloids, dyes and pigments
Vol. IV C:	Five-membered heterocyclic compounds with two hetero-atoms in the ring from Groups V and/or VI of the Periodic Table
Vol. IV D:	Five-membered heterocyclic compounds with more than two hetero-atoms in the ring
Vol. IV E:	Six-membered monoheterocyclic compounds containing oxygen, sulphur, selenium, tellurium, silicon, germanium, tin, lead or iodine as the heteroatom
Vol. IV F:	Six-membered heterocyclic compounds with a single nitrogen atom in the ring: pyridine, polymethylenepyridines, quinoline, isoquinoline and their derivatives
Vol. IV G:	Six-membered heterocyclic compounds with a single nitrogen atom in the ring to which are fused two or more carbocyclic ring systems, and six-membered ring compounds where the hetero-atom is phosphorus, arsenic, antimony or bismuth. Alkaloids containing a six-membered heterocyclic ring system
Vol. IV H:	Six-membered heterocyclic compounds with (a) a nitrogen atom common to two or more fused rings; (b) one hetero-atom in each of two fused rings. Six-membered ring compounds with two hetero-atoms from Groups VI B, or V B and VI B of the Periodic Table, respectively. Isoquinoline, lupinane and quinolizidine alkaloids
Vol. IV I:	Six-membered heterocyclic compounds with two hetero-atoms from Group V of the Periodic Table: the Pyridazine and Pyrimidine groups
Vol. IV J:	Six-membered heterocyclic compounds with two hetero-atoms from Group V of the Periodic Table: the Pyrazine group. Phenoxazine, phenothiazine, phenazine and sulphur dyes. Six-membered heterocyclic compounds with three or more hetero-atoms

OFFICIAL PUBLICATIONS

B.P.	British (United Kingdom) Patent
F.P.	French Patent
G.P.	German Patent
Ger. Offen.	German Patent Application, open for inspection
Sw. P.	Swiss Patent
U.S.P.	United States Patent
U.S.S.R.P.	Russian Patent
B.I.O.S.	British Intelligence Objectives Sub-Committee Reports, H.M. Stationery Office, London.
C.I.O.S.	Combined Intelligence Objectives Sub-Committee Reports
F.I.A.T.	Field Information Agency, Technical Reports of U.S. Group Control Council for Germany
B.S.	British Standards Specification
A.S.T.M.	American Society for Testing and Materials
A.P.I.	American Petroleum Institute Projects
C.I.	Colour Index Number of Dyestuffs and Pigments

SCIENTIFIC JOURNALS AND PERIODICALS

With few obvious and self-explanatory modifications the abbreviations used in references to journals and periodicals comprising the extensive literature on organic chemistry, are those used in the World List of Scientific Periodicals.

LIST OF ABBREVIATED NAMES OF CHEMICAL FIRMS MENTIONED IN PATENT REFERENCES

A.G.F.A., Agfa A.G.	Aktiengesellschaft für Anilinfabrikation (Berlin)
B.A.S.F.	Badische Anilin- und Soda-Fabrik (Ludwigshafen)
Bayer	Farbenfabriken vorm. Friedrich Bayer und Co. (Leverkusen)
Cassella	Leopold Cassella und Co. (Frankfurt am Main)
C.F.M.	Compagnie française des Matières Colorantes (Paris)
CIBA	Gesellschaft für chemische Industrie (Basel)
Du Pont	E.I. Du Pont de Nemours and Co. (U.S.A.)
G.A.F.	General Anilin and Film Corporation (U.S.A.)
Geigy A.G.	J. R. Geigy S.A. (Basel)
Hoechst	Hoechst A.G. (see M.L.B.)
I.C.I.	Imperial Chemical Industries, Ltd. (London)
I.G.	(= Interessen Gemeinschaft Farbenindustrie) of the principal dyestuffs manufacturers in Germany
Kalle	Kalle und Co., A.G. (Biebrich am Rhein)
M.L.B.	Farbwerke vormals Meister, Lucius und Brüning (Hoechst)
Sandoz	Sandoz A.G. Chemische Fabrik (Basel)

LIST OF COMMON ABBREVIATIONS AND SYMBOLS USED

A	acid
Å	Ångström units
Ac	acetyl
a	axial
as, *asymm.*	asymmetrical
at.	atmosphere
B	base
Bu	butyl
b.p.	boiling point
C, mC and μC	curie, millicurie and microcurie
c, *C*	concentration
c.d.	circular dichroism
conc.	concentrated
crit.	critical
D	Debye unit, 1×10^{-18} e.s.u.
D	dissociation energy
D	dextro-rotatory; dextro configuration
DL	optically inactive (externally compensated)
d	density
dec. or decomp.	with decomposition
deriv.	derivative
E	energy; extinction; electromeric effect
E1,E2	uni- and bi-molecular elimination mechanisms
E1cB	unimolecular elimination in conjugate base
e.s.r.	electron spin resonance
Et	ethyl
e	nuclear charge; equatorial
f	oscillator strength
f.p.	freezing point
G	free energy
g.l.c.	gas liquid chromatography
g	spectroscopic splitting factor, 2.0023
H	applied magnetic field; heat content
h	Planck's constant
Hz	hertz
I	spin quantum number; intensity; inductive effect
i.r.	infrared
J	coupling constant in n.m.r. spectra
K	dissociation constant
k	Boltzmann constant; velocity constant
kcal.	kilocalories
L	laevorotatory; laevo configuration
M	molecular weight; molar; mesomeric effect
Me	methyl

m	mass; mole; molecule; *meta-*
ml	millilitre
m.p.	melting point
Ms	mesyl (methanesulphonyl)
[M]	molecular rotation
N	Avogadro number; normal
n.m.r.	nuclear magnetic resonance
N.O.E.	Nuclear Overhauser Effect
n	normal; refractive index; principal quantum number
o	*ortho-*
o.r.d.	optical rotatory dispersion
P	polarisation; probability; orbital state
Pr	propyl
Ph	phenyl
p	*para-*; orbital
p.m.r.	proton magnetic resonance
R	clockwise configuration
S	counterclockwise config.; entropy; net spin of incompleted electronic shells; orbital state
S_N1, S_N2	uni- and bi-molecular nucleophilic substitution mechanisms
S_Ni	internal nucleophilic substitution mechanisms
s	symmetrical; orbital
sec	secondary
soln.	solution
symm.	symmetrical
T	absolute temperature
Tosyl	*p*-toluenesulphonyl
Trityl	triphenylmethyl
t	time
temp.	temperature (in degrees centigrade)
tert	tertiary
U	potential energy
u.v.	ultraviolet
v	velocity
α	optical rotation (in water unless otherwise stated)
[α]	specific optical rotation
α_A	atomic susceptibility
α_E	electronic susceptibility
ε	dielectric constant; extinction coefficient
μ	microns (10^{-4} cm); dipole moment; magnetic moment
μ_B	Bohr magneton
μg	microgram (10^{-6} g)
λ	wavelength
v	frequency; wave number
χ, χ_d, χ_μ	magnetic, diamagnetic and paramagnetic susceptibilities
$\sim$	about

(+)	dextrorotatory
(−)	laevorotatory
⊖	negative charge
⊕	positive charge

Chapter 48

Compounds with Six-Membered Heterocyclic Rings possessing Two or More Unusual Hetero Atoms

R. S. EDMUNDSON

1. Mercury compounds

The action of sodium amalgam on 1,2-dibromobenzene dissolved in petroleum ether–ethyl acetate at 120° gives a substance formulated as **dibenzo-1,4-dimercuracyclohexa**-2,5-**diene (phenomercurin, dibenzo**[1,4]-**dimercurin)**.

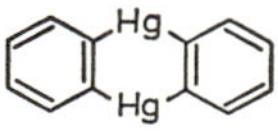

The compound is converted into 1,2-di(acetoxymercuri)benzene when treated with acetic acid, and to 1,2-diiodobenzene when acted upon by potassium iodide–iodine solution (*L. Vecchiotti*, Nature, 1930, **126**, 313; Ber., 1930, **63B**, 2275). The compound has been employed in the synthesis of a dibenzo[1,4]diborin (Section 2). Other compounds of the same series are obtained when sodium thiosulphate reacts with derivatives of 1,2-di(acetoxymercuri)benzene (*idem*, Gazz., 1928, **58**, 712; 1930, **60**, 904; CA., 1929, **23**, 3447; 1931, **25**, 1831.

2. Boron compounds

No 1,2- or 1,3-diborins appear to have been reported; 1,4-diborins are known only in partially reduced form. Of the possible triborins, only the 1,3,5-triborin ring system has been recorded (in fully reduced form).

(*a*) 1,4-*Dihydro*-1,4-*diborins*

Interaction of acetylene and boron monofluoride at −190° yields a yellow polymer together with volatile compounds including $C_4H_4B_2F_2$ and $C_4H_4B_3F_5$ (the combined yield of these two compounds being ~8%); at room temperature the latter loses boron trifluoride to give $C_4H_4B_2F_2$. The

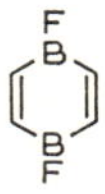

(I)

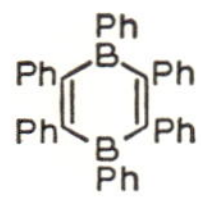

(II)

compound is rapidly destroyed by air and water, and mass spectrometry together with ^{1}H and ^{19}F n.m.r. spectroscopy lend support to the proposed 1,4-difluoro-1,4-dihydro-1,4-diborin structure I. A mixture of methyl derivatives of I is formed in the same way from propyne (*P. C. Timms*, J. Amer. chem. Soc., 1968, **90**, 4585). Condensation of diphenylacetylene with phenylboron dibromide in the presence of potassium leads to a product, m.p. 146–149°, thought to be **hexaphenyl**-1,4-**dihydro**-1,4-**diborin** (II), although the unusual high stability of the compound towards oxygen and the lack of i.r. olefinic absorption do not preclude a carborane type of structure (*F. Joy, M. F. Lappert* and *B. Prokai*, J. organometal. Chem., 1966, **5**, 506).

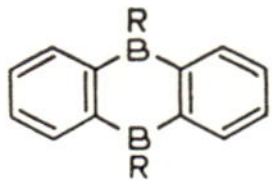

(III)

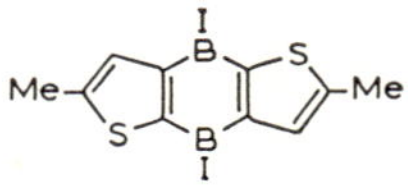

(IV)

5,10-**Diiodo**-5,10-**dihydroboranthrene(-dihydrodibenzo**[1,4]-**diborin**) (III, R = I), m.p. 208–212°, has been obtained as a by-product in the reaction leading to diphenyliodoborane from triphenylborane and phenyldiiodoborane, from the latter by reaction with sodium tetraphenylborate (*W. Siebert, M. Schmidt* and *E. Gast, ibid.*, 1969, **20**, 29), and in other attempts to prepare 1,2-bis-(diiodoboryl)benzene (*W. Siebert, K. J. Schafer* and *A. Asgarouladi*, Z. Naturforsch., 1974, B **29**, 642), and also by interaction of boron triiodide and 1,2-diiodobenzene at 180° (*Asgarouladi et al.*, Chem. Ber., 1974, **107**, 34). The corresponding 5,10-dichloro compound has been converted to 5,10-bis-(*dimethylamino*)-5,10-*dihydroboranthrene*, b.p. 148–150°/13 mm (*R. Clement*, Compt. rend., Ser. D, 1975, **261**, 4436). The *diiododimethyl*-1,4-*diborino*-[2,3-b:5,6-b′]*dithiophene* (IV), subliming at 230–240°/0.1 mm, is obtained, together with other products, when boron triiodide reacts with 2-iodo-5-methylthiophene (*Siebert*, Chem. Ber., 1970, **103**, 2308).

(*b*) *Perhydro*-1,3,5-*triborins* (1,3,5-*triborinanes*)

1,3,5-**Triethyl**-2,4,6-**trimethyl**-1,3,5-**triborinane** (V), b.p. 70–71 °/0.01 mm, is produced, together with a polycyclic compound $C_{14}H_{31}B_5$, when chlorodiethylborane is treated with lithium in tetrahydrofuran, or when ethylene is incorporated into such a reaction (*R. Köster* and *G. Benedikt*, Angew. Chem.

internat. Edn., 1964, **3**, 515; *Köster*, *Benedikt* and *M. A. Grassberger*, Ann., 1969, **719**, 187). The use of halogenoborolanes in this reaction leads to polycyclic systems, *e.g.* the trisborolotriborinane VI:

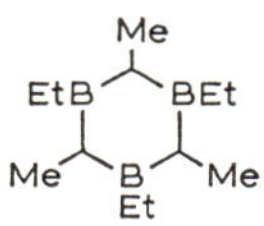

(V)

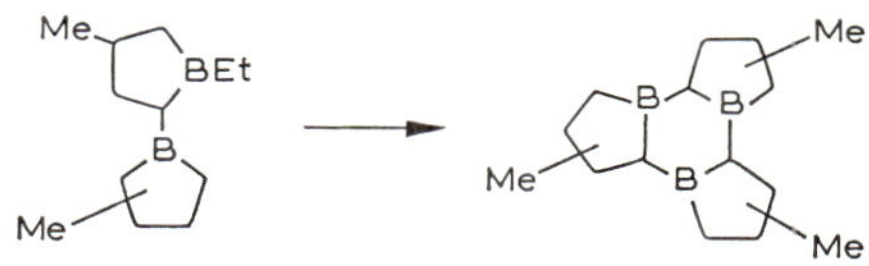

(VI)

In the same way, dilithiomethane and chlorobis(dimethylamino)borane yield 1,3,5-tris(dimethylamino)-1,3,5-triborinane, which with boron trichloride furnishes 1,3,5-trichloro-1,3,5-triborinane, hydrolyzed in aqueous acetone to 1,3,5-*trihydroxy*-1,3,5-*triborinane*, m.p. 142° (*P. Krohmer* and *J. Goubeau*, Chem. Ber., 1971, **104**, 1347).

Pyrolysis of trimethylborane at 450° in a sealed tube affords 25% of 2,4,6,8,9,10-*hexamethyl*-2,4,6,8,9,10-*hexabora-adamantane* (VII, R = Me), m.p. 85°, a compound which is air-sensitive but relatively stable to water (*M. P. Brown*, *A. K. Holliday* and *G. M. Way*, J. chem. Soc., Dalton Trans., 1975, 148). By contrast, the *hexachloro* compound (VII, R = Cl), subliming at 66–67°/0.1 mm, is very sensitive to water but has high thermal stability and is formed by thermolysis of 1,2-bis(dichloroboryl)ethane (*M. S. Reason et al.*, J. organometal. Chem., 1974, **77**, C9):

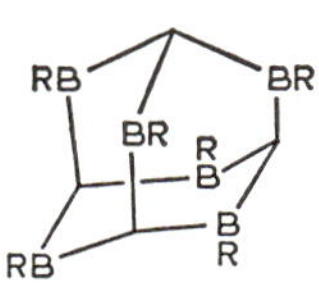

(VII)

3. Compounds of silicon, germanium and tin

(*a*) *Compounds with two hetero atoms*

Much interest has centred around the constitution of substances formed by the interaction of acetylenes with halides of silicon, germanium and, to some extent, of tin. Early work suggested a metallirene structure I for these products (see *R. Livingstone*, in C.C.C., 2nd Edn. Vol. IV A, p. 71), but more recently, physical evidence, particularly X-ray data, has proved that for some of the compounds, at least, the planar six-membered ring II is more appropriate. Nevertheless, in some cases, doubt still remains since electron diffraction studies of compounds in the germanium series have indicated some dependence of ring size on the nature of the substituent X:

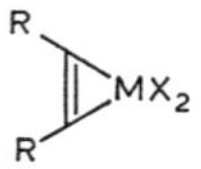

(I)

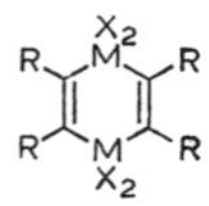

(II)

The original hopes that p_π–d_π electron interactions would be large enough to allow extensive delocalisation around the (three-membered) ring in spite of high angle strain and bond compression, and so impart some degree of aromatic character, have not been fulfilled.

As a consequence of the tetrahedral nature of bonding around a silicon atom, the molecular geometry of ring systems containing silicon (and other atoms of Group IV) should parallel that of the carbon analogues, with the potential existence of geometrical isomers; such isomers have been obtained for both saturated and unsaturated ring systems.

(*i*) 1,2,3,6-*Tetrahydro*-1,2-*disilins* (1,2-*disilacyclohex*-4-*enes*), 1,2-*dihydro*-1,2-*disilins* (1,2-*disilacyclohexadienes*), *their benzologues and germanium analogues*

Silicon difluoride condenses with unsaturated hydrocarbons at very low temperatures. The reaction with trifluoromethylacetylene at −196° provides a product which, with methanolic potassium hydroxide, is converted into III and which was therefore formulated as 1,1,2,2-**tetrafluoro**-4,5-**bis(trifluoro-**

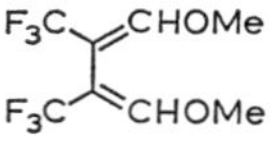

(III)

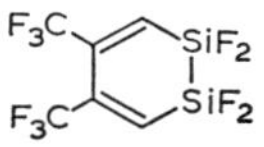

(IV)

methyl)-1,2-**disilacyclohexa**-3,5-**diene** (IV) (*C. S. Liu* and *J. C. Thompson*, Inorg. chem., 1971, 10, 1100). 1,1,2,2-**Tetramethyl**-3,6,**diphenyl**-1,2-**disilacyclohexa**-3,5-**diene** (V, R = H) undergoes isomerization on radiolysis (Hg arc). The initial product, 2,2,6,6-tetramethyl-1,3,4,5-tetraphenyl-2,6-disilabicyclo-[3.1.0]hex-3-ene (VI) is very sensitive to air and moisture, and when heated is converted to 5,5,6,6-tetramethyl-1,2,3,4-tetraphenyl-5,6-disilabicyclo-[2.1.0]hex-2-ene (VII). The reversion of VII ot VI is also brought about by irradiation. Noteworthy is the fact that for the diphenyl series (VI, VII, R = H), the means of interconversion should be reversed (*Y. Nakadaira*, *S. Kanouchi* and *H. Sakurai*, J. Amer. chem. Soc., 1974, **96**, 5623):

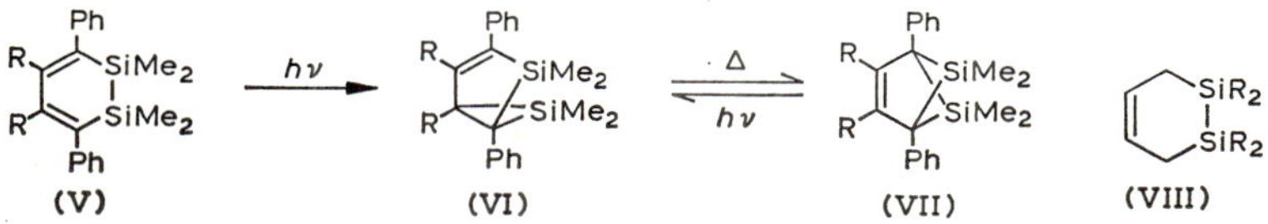

For ring expansion reactions of this system, see p. 333.

"Disilane" (from the pyrolysis of trisilane) is reported to react with butadiene at 385° to give a mixture of 1-silylsila-3-cyclopentene and **1,2-disila-4-cyclohexene** (VIII, R = H); digermane reacts similarly at the rather lower temperature of 240° (*R. L. Jenkins et al.*, J. organometal. Chem., 1975, **86**, 347). With silicon difluoride, butadiene yields 1,1,2,2-**tetrafluoro**-1,2-**disila-cyclohex-4-ene** (VIII, R = F) (*Thompson* and *J. L. Margrave*, Inorg. Chem., 1972, **11**, 912).

1,1,2,2-*Tetramethyl*-3,4,5,6-*tetraphenyl*-1,2-*disilacyclohexa*-3,5-*diene* (V, R = Ph), m.p. 174–175°, results together with 1,1-dimethyl-2,3,4,5-tetraphenylsilole, when 2,3-dichloro-2,3-dimethyl-2,3-disilabutane reacts with 1,4-dilithiotetraphenylbutadiene; it is readily oxidized, undergoing oxygen incorporation and ring expansion when treated with 3-chloroperbenzoic acid or *tert*-butyl hydroperoxide to yield a 1,2,7-oxadisilacycloheptadiene (p. 364) (*Nakadaira* and *Sakurai*, J. organometal. Chem., 1973, **47**, 61).

1,1,2,2-**Tetramethyl-1,2-disilacyclohex-4-ene,** from 1,4-dibromobutene-2 and 2,3-dichloro-2,3-dimethyl-2,3-disilabutane in the presence of magnesium, reacts with tetraphenylcyclopentadienone to yield eventually its *tetraphenyl* benzologue (IX), mp. 247°, (*L. Birkofer* and *W. Weniger*, Chem. Ber., 1973, **106**, 3595).

P. Mazerolles et al. (J. organometal. Chem., 1973, **60**, C3) have prepared examples of the 1,2-digermacyclohex-4-ene system in ways similar to those used for the silicon analogues.

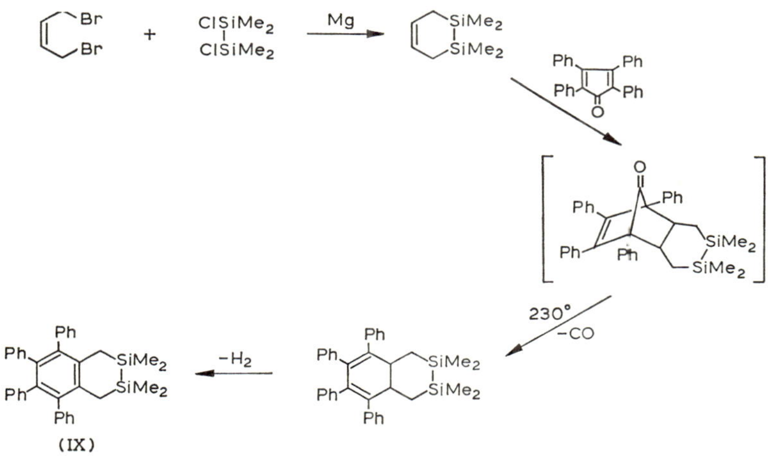

The same dichlorodisilabutane condenses with the diradicals from biphenyl, naphthalene and anthracene, so leading to 1-*phenyl*-7,7,8,8-*tetramethyl*-7,8-*disilabicyclo* [2.2.2]-*octa*-2,5-*diene* (X), m.p. 118–121°, 7,7,8,8-*tetramethyl*-2,3-*benzo*-7,8-*disilabicyclo* [2.2.2]-*octa*-2,5-*diene* (XI), m.p. 64–66°, and the 2,3,5,6-*dibenzo* compound XII, R = Me, m.p. 162–164°. Compound XII, R = Me is stable at temperatures below 450° but at 500° decomposes into anthracene together with a polymer, a similar course of decomposition being followed by the tetraphenyl compound (XII, R = Ph), obtained from diphenyldichlorosilane and anthracene in tetrahydrofuran in the presence of magnesium and magnesium bromide. When pyrolyzed in the presence of naphthalene, X yields XI in 96% yield, and when heated with 1,4-diphenylbutadiene at 260°, 1,1,2,2-*tetramethyl*-3,6-*diphenyl*-1,2-*silacyclohex*-4-*ene*, m.p. 154.5–156°, in 29% yield (*D. N. Roark* and *C. J. D. Peddle*, J. Amer. chem. Soc., 1972, **94**, 5837; *Peddle et al., ibid.*, 1969, **91**, 2807; *C. L. Smith* and *J. Pounds*, Chem. Comm., 1975, 910).

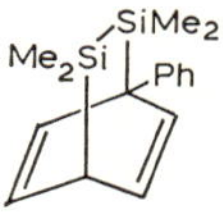

(X)

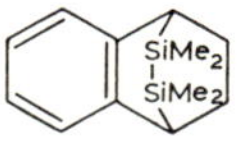

(XI)

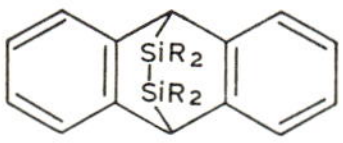

(XII)

Although *Roark* and *Peddle* suggested that the formation of, and decomposition of, adducts such as XI and XII might proceed through sp^2 hybridized species having Si=Si bonds, rather more conclusive evidence for the transitory existence of species possessing such bonds has come from a study of the copyrolysis of these compounds with benzaldehyde at 500° in a nitrogen atomosphere, when the products are cyclopolysiloxanols, *trans*-stilbene, and biphenyl (from XI) or naphthalene (from XII), materials consistent with the intermediary formation of $Me_2Si{=}SiMe_2$ (or the tetraphenyl analogue) (*T. J. Barton* and *J. A. Kilgour*, J. Amer. chem. Soc., 1974, **96**, 2278; *Smith and Pounds*, *loc.cit.*).

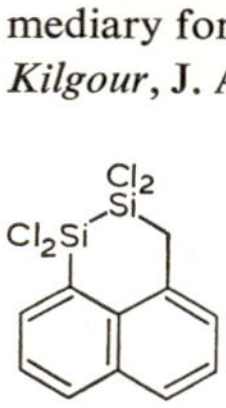

(XIII)

1,1,2,2-Tetrachloro-1,2-disiladihydrophenalene (XIII) is obtained in about 8% yield, together with other products, when a mixture of trichlorosilane and 1-naphthylmethylsilyl dichloride is heated at 650° (*E. A. Chernyshev et al.*, Zhur. obshcheĭ Khim., 1971, **41**, 122).

(*ii*) *Perhydro*-1,2-*disilins and their germanium analogues* (1,2-*disilacyclohexanes*)

1,1,2,2-Tetrafluoro-1,2-disilacyclohexane is formed, together with a substituted 1,2-disilacyclobutane, when silicon difluoride reacts with ethylene (*Thompson*, *Margrave* and *P. L. Timms*, Chem. Comm., 1966, 566) while the corresponding tetramethyl compound has been prepared by reduction of the disilacyclohexene (*Birkofer* and *Weniger*, *loc.cit.*). Japanese workers have investigated two, more practicable routes, (*a*) the reaction between butane-1,4-di(magnesium bromide) and disilanes of the type ClRR′Si·SiRR′Cl, and (*b*) treatment of 1,4-di(chlorosilyl)butanes with sodium–potassium alloy. The products of these reactions consist of mixtures of *cis* and *trans* isomers which may be separated by fractional crystallization, distillation or preparative g.l.c.

The following 1,2-**disilacyclohexanes** are prepared in a similar way: 1,1,2,2-*tetramethyl-*, b.p. 174°, n_D^{20} 1.4722; 1,1,2,2-*tetraphenyl-*, m.p. 110–111°; 1,2-*dichloro*-1,2-*diphenyl-*, m.p. 68–69.5°, b.p. 148–151°/0.03 mm; cis-1,2-*dimethyl*-1,2-*diphenyl-*, b.p. 158°/3 mm, n_D^{20} 1.5847 (trans *isomer*, b.p. 158°/3 mm, n_D^{20} 1.5888); cis-1,2-di(4-*chlorophenyl*)-1,2-*dimethyl-*, m.p. 43–43.5°, (trans *isomer*, b.p. 202–206°/2 mm); 1,2-*bis*(*tetramethylene*)-5,6-*disilabispiro*-[4.0.4.4]*tetradecane* (XIV), b.p. 103°/0.2 mm (*H. Gilman*, *W. H. Atwell* and *F. K. Cartledge*, Adv. organometal. Chem., 1966, **4**, 1; *K. Tamao*, *M. Kumada* and *M. Ishikawa*, J. organometal. Chem., 1971, **31**, 17; *Tamao*, *Kumada* and *A. Noro*, *ibid.*, 1971, **31**, 169; *Mazerolles et al.*, *ibid.*, 1975, **87**, 269).

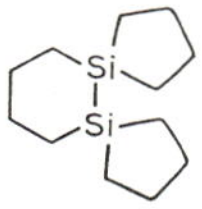

(XIV)

When 1,2-dimethyl-1,2-diphenyl-1,2-disilacyclohexane is treated with sulphuric acid followed by $NH_4F{\cdot}HF$, there results a mixture, b.p. 90–95°/100 mm, of *cis* and *trans*-1,2-difluoro-1,2-dimethyl-1,2-disilacyclohexane, separable into its components by g.l.c.; the ratio of products, 1:1.9, is not dependent upon the geometry of the starting material. The analogous chlorodephenylation process involving the use of ammonium chloride or aluminium chloride/hydrogen chloride gives an inseparable mixture of the dichloro compounds, b.p. 90/18 mm, the ratio of components being ~1:3. These results, together with other experimental observations, for example the rapid equilibration of mixtures of 1,2-difluoro compounds in benzene–sulphuric acid (although the process is much slower for initially pure stereoisomers of the same compounds), have been interpreted in terms of the concept of stereomutation of bonds (pseudorotation) in a trigonal–bipyramidal arrangement around a central silicon atom (*Tamao*, *Kumada* and *Ishikawa*, *loc.cit.*).

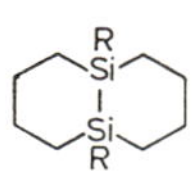

(XV)

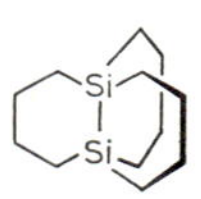

(XVI)

The reaction between butane-1,4-di(magnesium bromide) and *cis*-1,2-difluoro-1,2-dimethyl-1,2-disilacyclohexane in tetrahydrofuran under nitrogen gives *cis*-4a,8a-**dimethyl**-4a,8a-**disiladecalin** (XV, R = Me), b.p. 109°/17 mm, n_D^{20} 1.5123, shown to be conformationally labile by ^1H n.m.r. spectroscopy. The *trans* disilacyclohexane gives a mixture of cis- and trans-*disiladecalins* in the ratio 1: 0.6, separable by g.l.c.: the trans-*dimethyldisiladecalin* has m.p. 93.5–94.5°, shown to be conformationally rigid. Cis-4a,8a-*diphenyl*-4a,8a-*disiladecalin*, m.p. 104–105°, b.p. ~ 190°/4 mm, is prepared in a similar way. When treated with aluminium/hydrogen chloride, the diphenyl disiladecalin yields the *dichloro* compound (XV, R = Cl), m.p. 29.5–31°, b.p. 90°/4 mm; reaction of this with butane-1,4-di-(magnesium bromide) yields **1,6-disilatricyclo**[4.4.4.$0^{1,6}$]**tetradecane** (**1,6-disila**[4.4.4]-**propellane**) (XVI), m.p. 114.5–115.5°, admixed with other products (*Tamao, Kumada* and *Noro, loc.cit.*).

Stereospecific oxidative cleavage of the Si–Si bond in 1,2-disilacyclohexanes is achieved on treatment with perbenzoic acid and similar reagents, so yielding seven-membered ring systems (p. 364). The stability of the Si–Si bond to sulphuric acid depends on the nature of the attached groups; that in 1,2-dimethyl-1,2-diphenyl-1,2-disilacyclohexane is stable, whereas that in the tetramethyldisilacyclohexane is broken (*L. H. Sommer et al.*, J. Amer. chem. Soc., 1965, **87**, 1619; 1967, **89**, 1756; 1969, **91**, 7067).

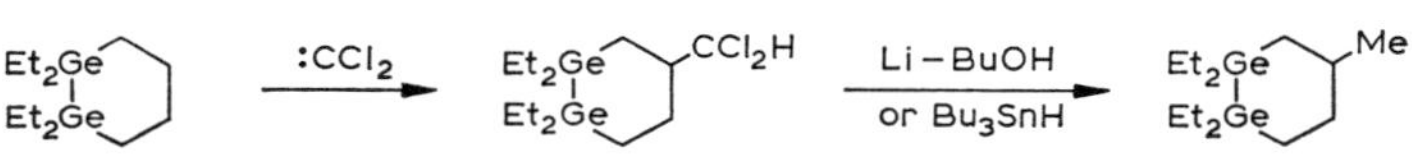

(XVII) (XVIII)

1,1,2,2-Tetramethyl-1,2-digermacyclohexane (XVII), b.p. 112°/25 mm, n_D^{20} **1.5159**, undergoes cleavage of the Ge–Ge bond when treated with bromine, the product being 3,8-dibromo-3,8-diethyl-3,8-digermadecane; the reaction with dichlorocarbene provides an unusual route to its 4-methyl derivative (XVIII) (*Mazerolles, M. Lesbre* and *M. Joanny*, J. organometal Chem., 1969, **16**, 227; *D. Seyforth et al., ibid.*, 1971, **29**, 371). Reduced derivatives of 2,3,1-digermaphosphorin are claimed to be formed when 2-germaphospholanes are treated with GeX_2 or RGeX (X = halogen) (*C. Couret et al., ibid.*, 1975, **84**, 191).

(*iii*) *Tetrahydro*-1,3-*disilins* (1,3-*disilacyclohexenes*) *and related compounds*

Pyrolysis of tetramethylsilane at 700° produces a complex mixture of products (including 39 having a boiling point below 200°) from which 1,1,3,3-**tetramethyl-1,3-disilacyclohex-4-ene** (XIX), b.p. 165°, has been isolated (*G. Fritz* and *N. Gotz*, Z. anorg. allgem. Chem., 1970, **375**, 171). 1,1,3,3-*Tetrachloro*-1,3-*disila*-1,2-*dihydrophenalene* (XX, R = Cl), obtained (*a*) by

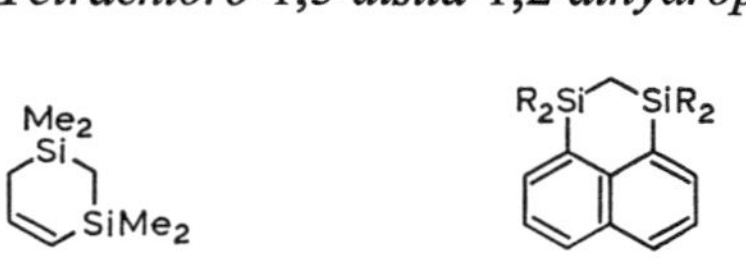

(XIX) (XX)

the action of benzyldichlorosilane on 1-methylnaphthalene at 675°, and (b) from trichlorosilane and 1,1-dichloro-1-sila-acenaphthene at the same temperature probably by insertion of :$SiCl_2$, is reduced by lithium tetrahydridoaluminate to 1,3-**disila**-1,2-**dihydrophenalene** (XX, R = H), m.p. 26°, b.p. 112–114°/2 mm, and with methylmagnesium chloride is converted into the *tetramethyl* compound (XX, R = Me), m.p. 80°, b.p. 107–108°/1.5 mm (*E. A. Chernyshev et al.*, Zhur. obshcheî Khim., 1971, **41**, 122; *I. G. Makarov et al.*, Zhur. strukt. Khim., 1969, **10**, 595; C.A., 1969, **71**, 118240).

(*iv*) *Perhydro*-1,3-*disilins* (1,3-*disilacyclohexanes*) *and related compounds*

1,1-**Dichloro**-3,3-**dimethyl**-1,3-**disilacyclohexane** (XXI, M = Si, R = Cl), b.p. 97°/20 mm, n_D^{20} 1.4792, is obtained by reaction between silicon tetrachloride and the diGrignard reagent from 1,5-dichloro-2,2-dimethyl-2-silapentane (*J. Fessenden* and *M. D. Coon*, J. org. Chem., 1964, **29**, 1607), and also by allowing 1,1,6-trichloro-5,5-dimethyl-1,5-disilapentane to react with magnesium under ether (*E. D. Babich et al.*, Izvest. Akad. Nauk S.S.S.R.,

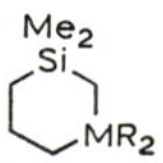

(XXI)

M = Si, Ge

Ser. Khim., 1968, 1389). Other known derivatives include the 1,1,3,3-*tetramethyl*-, b.p. 168–170°; 1-*chloro*-1,3-*dimethyl*-3-*phenyl*-, b.p. 140–141°/5 mm, and the 1,1-*dimethyl*-3,3-*trimethylene*-, b.p. 70–71°/2–3 mm, -1,3-*disilacyclohexanes*. 1,1,3,3-**Tetramethyl**-1,3-**silagermacyclohexane** (XXI, M = Ge, R = Me), b.p. 57°/10 mm, n_D^{20} 1.4714, is obtained using the latter procedure starting from 1,6-dichloro-1,1,5,5-tetramethyl-1,5-silagermahexane (*V. F. Mironov*, *T. K. Gar* and *S. A. Mikhailyants*, Doklady Akad. Nauk S.S.S.R., 1969, **188**, 120).

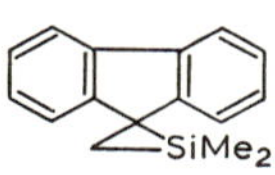

(XXII)

(XXIII)

In an attempt to obtain the spirocyclic siliran (XXII), *Peddle* (J. organometal. Chem., 1967, **9**, 171) treated 9-lithiofluorene with chloromethyldimethylsilyl chloride, but obtained instead the *tetramethyldispiro*(*fluorene*-9,2′-[1,3]*disilacyclohexane*-5′,9″-*fluorene*) (XXIII), m.p. 228–230°, together with its 1,4-disilacyclohexane *isomer*, m.p. 322°.

(*v*) 1,4-*Dihydro*-1,4-*disilins* (1,4-*disilacyclohexa*-2,5-*dienes*). *Germanium and tin analogues*

M. E. Vol'pin et al. observed that a mixture of germanium diiodide and diphenylacetylene heated at temperatures above 200° produces a *substance*, m.p. 300–303° which, originally, they formulated as 1,1-diiodo-2,3-diphenylgermirene (XXIV, M = Ge, R = Ph, X = X^1 = I). The iodine was shown to be labile to alkali resulting in the formation of a dihydroxy compound

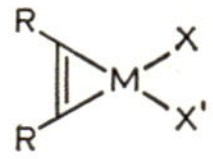

(XXIV)

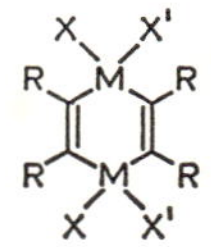

(XXV)

M = Si, Ge, Sn

which could be reconverted into the dihalide by HX; the dihalogen compound reacted normally with Grignard reagents. One other interesting feature demonstrated at the time was the apparent dimeric nature of the substances as evidenced by ebullioscopical measurements for benzene solutions (*Vol'pin* and *D. N. Kursanov*, Izvest. Akad. Nauk S.S.S.R., Ser. Khim., 1960, 1903; *Vol'pin et al.*, Tetrahedron, 1962, **18**, 107). At about the same time, these investigators observed the formation of a *substance* m.p. 324–329°, when dichlorodimethylsilane was heated in xylene solution with diphenylacetylene in the presence of sodium. In the light of the marked chemical similarity of the two substances, the second was also formulated as a silirene (XXIV, M = Si, R = Ph, X = Me) (*Vol'pin*, *Yu. D. Koreshkov* and *Kursanov*, Izvest. Akad. Nauk S.S.S.R., Ser. Khim., 1961, 1355). Thereafter, work on the silicon series paralleled that on the germanium compounds.

The first suggestion that the two substances discussed above would be best represented as derivatives of the 1,4-disila(germa)cyclohexa-2,5-diene system (XXV, M = Si, Ge; R = Ph) seems to have been made by *F. Johnson* and *R. S. Gohlke* (Tetrahedron Letters, 1962, 1291) and was based upon mass spectral measurements. X-Ray and electron diffraction measurements have also played an important part in the elucidation of the structures of these silicon and germanium compounds, and a number of studies have produced evidence for the planar, six-membered ring structure XXV (*N. G. Bokii* and *Yu. T. Struchkov*, Zhur. strukt. Khim., 1965, **6**, 751; 1966, **7**, 133; C.A., 1966, **64**, 4372; **65**, 122; *Vol'pin et al.*, Izvest. Akad. Nauk S.S.S.R., Ser. Khim., 1963, 2067; and *Bokii*, *G. N. Zakharova* and *Struchokv*, Zhur. strukt. Khim., 1965, **6**, 476; C.A., 1965, **63**, 17248). Further molecular weight measurements are in agreement (*R. West* and *R. E. Bailey*, J. Amer. chem. Soc., 1963, **85**, 2871). Very recently, however, electron diffraction experiments have suggested that, for the series of compounds obtained from acetylene and germanium halides, the size of ring might be governed by the nature of the substituent X. Thus, for $(C_2H_2GeX_2)_n$, n was found to be 1 for X = Me but 2 for X = Cl; when X = I, two sets of data were obtained dependent on temperature of vaporization of the sample (*L. V. Vilkov et al.*, Zhur. strukt. Khim., 1970, 11, **3**; C.A., 1970, **72**, 126065).

Acetylene is more reactive than is diphenylacetylene towards germanium diiodide, the reaction proceeding at 140°. Dialkoxysilanes may, in some cases, be used instead of silyl halides although higher reaction temperatures may then be required (*W. H. Atwell* and *D. R. Weyenberg*, J. organometal. Chem., 1966, **5**, 594; B.P. 1,201,517/1970; C.A. 1970, **73**, 88014).

P. G. Harrison (Inorg. Nucl. Chem. Letters, 1972, **8**, 555) has obtained similar adducts from stannous halides and acetylenedicarboxylic esters.

An alternative, and novel, method for the formation of the 1,4-disilacyclohexa-2,5-diene ring involves the insertion of acetylenes into 1,2-disilacyclobutenes, a process catalyzed at room temperature by nickel tetracarbonyl; a mixture of products is formed from a terminal alkyne (*C.-S. Liu* and *C.-W. Cheng*, J. Amer. chem. Soc., 1975, **97**, 6746):

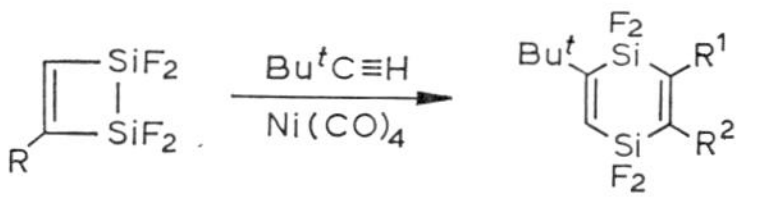

(a) R^1 = Bu^t, R^2 = H
(b) R^1 = H, R^2 = Bu^t

A further illustration of the versatility of acetylenes in the formation of metalloid-containing cyclohexa-2,5-dienes is the addition of tin hydrides to metalloid diacetylenes; 4-diethylamino-1,1,2,6-**tetramethyl-4-bora-1-stanna-2,5-cyclohexadiene** (XXVI), so formed, has b.p. 91–95°/0.1 mm (*B. Wrackmeyer* and *H. Nöth*, Z. Naturforsch., 1974, **B29**, 564):

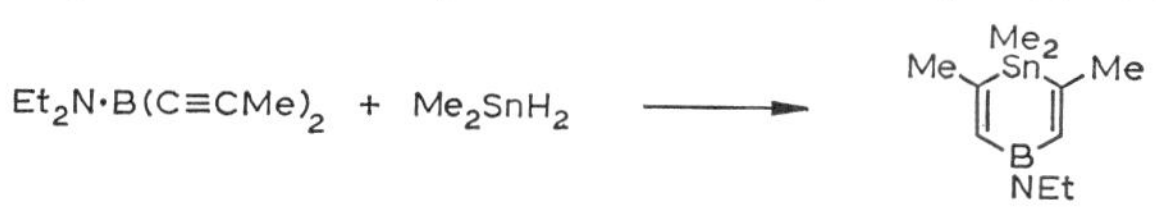

(XXVI)

The early work in both the silicon and germanium series suggested a considerable lack of reactivity of the compounds towards bromine, hydrogen and to oxidising agents (*Vol'pin et al.*, Tetrahedron, 1962, 18, 107) which led to considerations of possible aromatic character of the heterocyclic ring. However, *Johnson* and *Gohlke* (*loc.cit.*) showed that the germanium compounds do indeed react with bromine, albeit slowly. Later Russian work confirmed this and showed, moreover, that the ring could be fully reduced. Bromine and

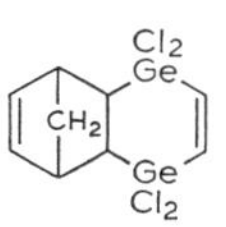

(XXVII)

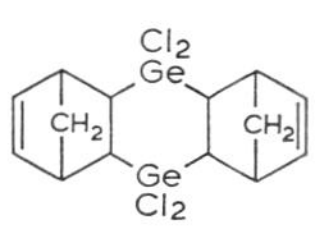

(XXVIII)

1,1,4,4-tetramethyl-1,4-digermacyclohexa-2,5-diene react to give 2,3-*dibromo*-1,1,4,4-*tetramethyl*-1,4-*digermacyclohex*-5-*ene* m.p. 72–74°, followed by 2,3,5,6-*tetrabromo*-

1,1,4,4-*tetramethyl*-1,4-*digermacyclohexane*, m.p. 145–149° (*Vol'pin et al.*, J. organometal. Chem., 1967, **8**, 87). The tetrabromo compound is rather prone to decomposition, but the intermediate stages leading to bromovinyldimethylgermanium bromide, $BrGeMe_2CH=CHBr$, have not been detected (*Johnson, Gohlke* and *W. A. Nasutavicus, ibid.*, 1965, **3**, 233). The dienophilic character of the germanium-containing ring is demonstrated by the reaction of the tetrachloride with cyclopentadiene to give the 1:1 *adduct* XXVII m.p. 175–177°, and the 2:1 *adduct* XXVIII, m.p. 264–265° (*Vol'pin et al., ibid.*, 1967, 8, 87).

In an attempt to prepare a silene, $:SiX_2$ (in this case, X = Ph), *H. Gilman et al.* examined the behaviour of 1,4,5,6,7,7-hexaphenyl-2,3-benzo-7-silabicyclo[2.2.1]heptane (XXIX) in a retro-Diels–Alder pyrolysis reaction. When diphenylacetylene was introduced during the pyrolysis, octaphenyl-1,4-disilacyclohexa-2,5-diene was isolated; its formation could be explained assuming the initial formation of diphenylsilene followed by addition to the acetylenic bond followed by dimerization and valence tautomerism (*Gilman, S. Cottis* and *Atwell*, J. Amer. chem. Soc., 1964, **86**, 1596, 5584; *R. Marucca*, J. org. Chem., 1971, **36**, 1626):

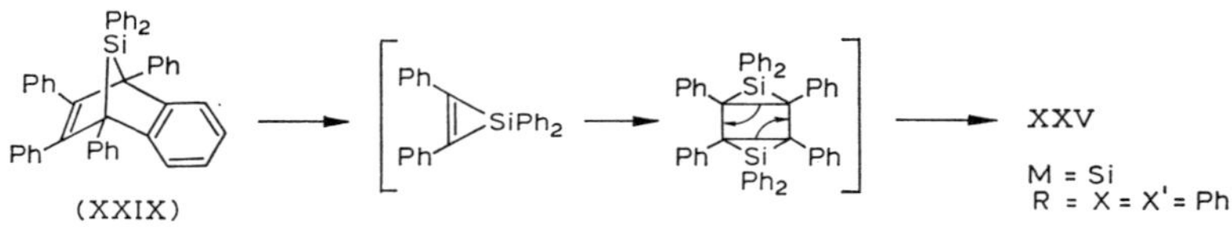

Such a mechanism appears to be inconsistent with the composition of mixtures of products formed during the pyrolysis of dialkoxydisilanes [other than $(MeO)_2Si{\cdot}Si(OMe)_3$] in the presence of mixtures of acetylenes (see Scheme 1). A diradical mechanism was rejected since siloles, the five-membered ring product expected from the interaction of a diradical with an acetylene, were not detected, and it was concluded that the formation of disilacyclohexadienes probably proceeds through a highly specific dimerization of silirene intermediates (*Atwell* and *Weyenberg*, J. Amer. chem. Soc., 1968, **90**, 3438).

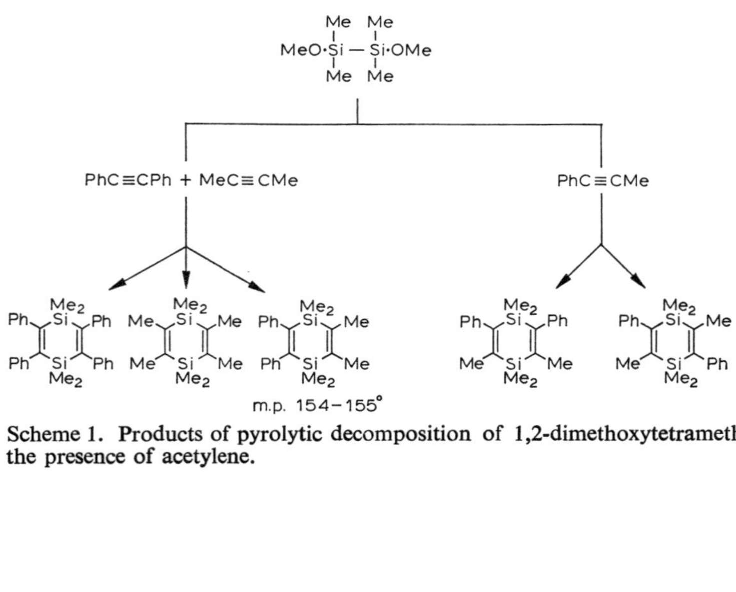

Scheme 1. Products of pyrolytic decomposition of 1,2-dimethoxytetramethyldisilane in the presence of acetylene.

Some 1,4-disila- and 1,4-digerma-cyclohexa-2,5-dienes are listed in Table 1.

TABLE 1

1,4-DISILA(GERMA)CYCLOHEXA-2,5-DIENES (XXV)

R	*X*	*X'*	*M.p.(°C) (b.p.(°C)/mm)*	
			M = Si	*M = Ge*
H	Me	MeO	(100–101/40) (1)	
H	Me	Me	(158–159) (1)	(67/14) (2)
H	Ph	Ph		149–150 (3)
H	Cl	Cl		145–146 (4)
H	I	I		225–226 (2)
Me	H	H		62–63 (5)
Me	H	Me		48–50[a] (5)
Me	H	Cl		94–97[a] (5)
Me	Me	Me	71–72.5 (6)	
Me	Cl	Cl		134–137 (5)
Me	I	I		273–275 (5)
Ph	Me	Me	323–325 (2)	305 (7)
Ph	Et	Et		239–240 (2)
Ph	Ph	Ph	293–296 (8)	
Ph	Me	MeO	211–214 (6)	
Ph	Cl	Cl		290–292 (2)
Ph	Br	Br		318 (2)
Ph	I	I		302–303 (2)

[a] *Trans* configuration.

References:
1 *E. G. Janzen, J. B. Pickett* and *W. H. Atwell*, J. Amer. chem. Soc., 1968, **90,** 2719.
2 *M. E. Volpin et al.*, Tetrahedron, 1962, **18,** 107.
3 *Idem*, J. organometal. Chem., 1967, **8,** 87.
4 *E. A. Chernysheva, N. G. Komalenkova* and *S. A. Bashkirova*, Zhur. obshcheĭ Khim., 1971, **41,** 1175.
5 *J. V. Scibelli* and *M. D. Curtis*, J. organometal. Chem., 1972, **40,** 317.
6 *Atwell* and *D. R. Weyenberg*, J. Amer. chem. Soc., 1968, **90,** 3438.
7 *F. Johnson et al.*, J. organometal. Chem., 1965, **3,** 233.
8 *H. Gilman et al.*, J. Amer. chem. Soc., 1964, **86,** 5584.

5,5,10,10-Tetrasubstituted-5,10-**dihydrosilanthrenes,** 5,10-**dihydrodibenzo** [*be*] **disilins** (XXX), *tetraphenyl-*, m.p. 234–236°, *tetrabenzyl-*, m.p. 174–176°, *tetramethyl-octafluoro-*, m.p. 229–230°, are formed in rather low yield when the appropriate 1,2-dilithiobenzene and dichlorosilane are allowed to react together (*Gilman* and *E. A. Zuech*, J. Amer. chem. Soc., 1960, **82**, 3605; *I. Haiduc* and *Gilman*, J. organometal. Chem., 1968, **11**, 55), and when

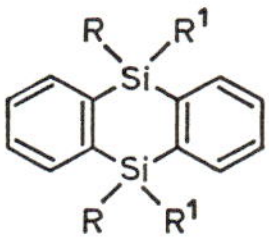

(XXX)

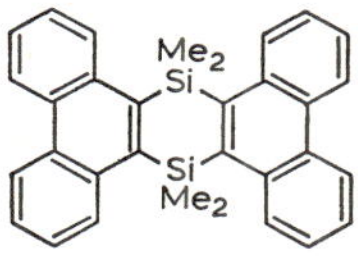

(XXXI)

2-chlorophenylsilanes are heated with sodium under xylene (*Gilman, Zuech* and *W. Steudel*, J. org. Chem., 1962, **27**, 1836). Pyrolysis of phenyldichlorosilane in a quartz tube at 680° gives *inter alia* ~1% of the 5,5,10,10-tetrachloro compound, also formed in 15% yield by heating phenyl(2-chlorophenyl)dichlorosilane with trichlorosilane at 680° for 35 sec. Reduction of the tetrachloride with lithium tetrahydridoaluminate yields 5,10-dihydrosilanthrene itself; other compounds are obtained from the chloride and Grignard reagents (*Chernysheva et al.*, Zhur. obshcheĭ Khim., 1971, **41**, 843). The 5,5,10,10-tetramethyl compound gives a red, comparatively stable radical anion, whereas the tetraphenyl compound forms an unstable radical anion showing chemiluminescence in air (*E. G. Janzen, W. B. Harrison* and *J. B. Pickett*, J. organometal. Chem., 1969, **20**, P13).

The molecular geometry of the dihydrosilanthrene system lends itself to the potential existence of geometrical isomers when R and R^1 are different. Chromatography of reaction products on alumina has allowed the separation of forms described as *cis* and *trans* isomers which differ little in m.p. (*Gilman, Zuech* and *Steudel, loc.cit.*).

Treatment of phenanthrene with lithium in dimethoxyethane followed by addition of dichlorodimethylsilane yields the tetrabenzo-5,10-dihydrosilanthrene XXXI (*A. I. Nogairelu, N. I. Tabashidze* and *R. Sh. Tkesholashvilli*, Zhur. obshcheĭ Khim., 1972, **42**, 1764).

In a correction to a previous communication, *R. Calas* and *J. Dunogues* have reported the interaction of diphenylacetylene and 2,5-dichloro-2,5-dimethyl-2,5-disilahexane in hexamethylphosphoramide with magnesium at 100° to give 1,1,4,4-*tetramethyl*-2,3-*diphenyl*-1,4-*disilacyclohex*-2-*ene* (XXXII), m.p. 122° (Compt. rend., Ser C, 1971, **272**, 554).

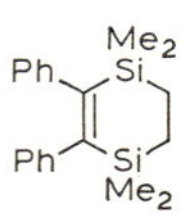

(XXXII)

(vi) Perhydro-1,4-disilins (1,4-disilacyclohexanes). Germanium and tin analogues

1,1,4,4-Tetrachloro-1,4-disilacyclohexane, m.p. 89–90°, b.p. 220°, is formed (*a*) during the polymerization of vinylsilanes at 100–150° in the presence of chloroplatinic acid–palladium/charcoal (*K. I. Kobrakov et al.*, Doklady Akad. Nauk S.S.S.R., 1970, **191**, 607); (*b*) together with other compounds in the synthesis of ethyldichlorosilane, $EtSiHCl_2$, from calcium silicide (*E. K. Chaplygina et al.*, Khim. Prom., Moscow, 1971, **47**, 821; C.A., 1972, **76**, 99749); and (*c*) by passing chloroethane over a silicon–copper composite (*J. Joklík* and *V. Bažant*, Coll. Czech. chem., 1965, **30**, 2928).

1,1,4,4-Tetramethyl-1,4-disilacyclohexane, b.p. 172–173°, n_D^{20} 1,4530, is obtained (*a*) by the action of methylmagnesium bromide on the tetrachloro compound (*idem, loc.cit.*); (*b*) in low yield when 2,2,5,5-tetramethyl-2,5-disilahexane is heated with aluminium bromide at 90° (*Tu. P. Egorov et al.*, Izvest. Akad. Nauk S.S.S.R., Ser. Khim., 1963, 822); and (*c*) together with 1,1,-dimethylsilacyclopentane when dichlorodimethylsilane is treated with ethylene in the presence of lithium in tetrahydrofuran at 0–10° (*O. M. Nefedoc* and *M. N. Manakov*, Zhur. obshcheĭ Khim., 1964, **34**, 2465). Tetra-alkyl-1,4-disilacyclohexanes are also formed together with 1,3-disilacyclopentanes during the dimerization of dialkyl-(vinyl)silanes (*J. W. Curry*, J. org. Chem., 1961, **26**, 1308). 2-*Phenyl-*, 2,2-*diphenyl-* and 2,3-*diphenyl*-1,1,4,4-*tetramethyl*-1,4-*disilacyclohexanes*, m.p.s 44°, 78° and 155°, respectively, are the compounds obtained when 2,5-dichloro-2,5-dimethyl-2,5-disilahexane in hexamethylphosphoramide reacts with styrene, 1,1-diphenylethylene and *trans*-stilbene in the presence of magnesium (*Calas* and *Dunogues*, *loc.cit.*). Carboranes containing the 1,4-disilacyclohexane ring system have been prepared (*S. Papetti et al.*, Inorg. Chem., 1964, **3**, 1444).

Substituted **1,4-digermacyclohexanes,** 1,1,4,4-*tetramethyl-*, b.p. 70–72°/12 mm, 1,1-*dibutyl*-4,4-*diethyl-*, b.p. 120–121°/0.7 mm, are obtained by reduction of the corresponding 1,4-digermacyclohexa-2,5-dienes in the presence of Raney nickel (*Vol'pin et al.*, J. organometal. Chem., 1967, **8**, 87), and by the dibenzoyl peroxide-catalyzed addition of germanium dihydrides to divinylgermanes (*Mazerolles*, Bull. Soc. chim. Fr., 1962, 1907). This type of reaction, in addition to those described earlier for 1,4-disilacyclohexanes, has been employed for the preparation of **1,1,4,4-tetraphenyl-1,4-distannacyclohexane,** m.p. 134–135° (*M. C. Henry* and *J. G. Noltes*, J. Amer. chem. Soc., 1960, **82**, 558); **1,1,4,4-tetraphenyl-4-stanna-1-silacyclohexane,** m.p. 134°, **1,1,4,4-tetraphenyl-4-stanna-1-germacyclohexane,** m.p. 124° [*Noltes* and *G. J. M. van der Kerk*, Chimia (Switz.) 1962, 16; C.A. 1963, **58,** 4658]; 4-*tert*-**butyl-1,1-dimethyl-4-bora-1-silacyclohexane,** b.p. 42°/2 mm (*M. F. Hawthorne*, J. Amer. chem. Soc., 1960, **82**, 748).

(*b*) *Perhydro*-1,3,5-*trisilins* (1,3,5-*trisilacyclohexanes*)

G. Fritz and *N. Gotz* (Z. anorg. allgem. Chem., 1970, **375**, 171) pyrolyzed tetramethylsilane at 700° and were able to detect 39 compounds having boiling points lower than 200°. By employing preparative g.l.c. the components of the pyrolysate were shown to be largely 1,3,5-**trisilacyclohexanes,** including 1,1,3,3-*tetramethyl-* admixed with 1,1,3,5-*tetramethyl-*, b.p. 193°; 1,1,3,3,5-*pentamethyl-*, b.p. 198°; 1,1,3,3,5,5-*hexamethyl*-1,3,5-*trisilacyclohexane*, together with 1,3,5,7-tetrasila-adamantanes.

Compounds possessing the 1,3,5-trisilacyclohexane skeleton are obtained by (a) a "trimerization" process involving units containing a single silicon atom, as exemplified by the formation of 1,3,5-triethoxy-1,3,5-trimethyl-1,3,5-trisilacyclohexane (XXXIII, R = Me, R^1 = OEt) (together with 1,3-*diethoxy*-1,3-*dimethyl*-1,3-*disilacyclobutane*), b.p. 68–69°/0.2 mm, n_D^{25} 1.4478 (*W. A. Kriner*, U.S.P. 3,178/1965; C.A., 1965, **63**, 1817):

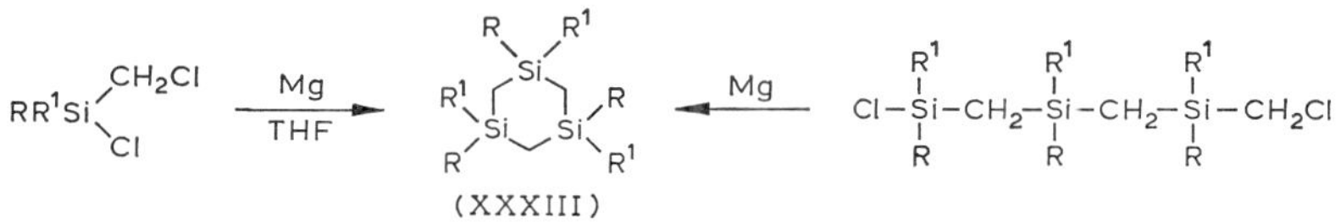

(XXXIII)

(b) By cyclization of 1,6-dihalogeno-1,3,5-trisilahexanes with magnesium or lithium under ether, as exemplified by the formation of 1-bromo-1,3,3,5-tetramethyl-5-phenyl-1,3,5-trisilacyclohexane (*Fritz* and *W. Kammerling*, Z. anorg. allgem. Chem., 1963, **322**, 34) and also of 2-methylene-1,1,3,3,5,5-hexamethyl-1,3,5-trisilacyclohexane (XXXIV) (*Fritz* and *M. Hähnke*, *ibid.*, 1972, **390**, 104).

(c) By cyclization of trisilaheptynes. Treatment of 3,3,5,5,7,7-hexamethyl-3,5,7-trisilaheptyne with chlorplatinic acid in isopropanol also yields XXXIV together with hexamethyl-1,3,5-trisilacyclohept-6-ene (*Fritz* and *Hähnke*, *loc.cit.*):

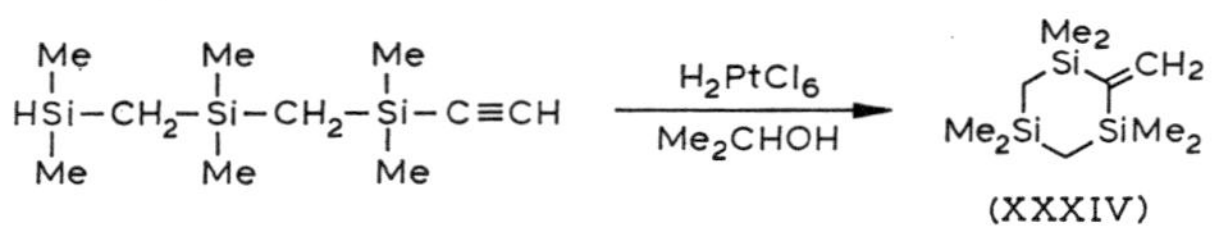

(XXXIV)

(d) They are obtained also from 1,3-disilacyclobutanes by the action of aluminium chloride (*E. D. Babich et al.*, Izvest. Akad. Nauk, S.S.S.R., Ser. khim., 1975, 953) and

(e) by the thermal isomerization of trisilacyclopentanes (*Fritz* and *B. Grunert*, Z. anorg. allgem. Chem., 1976, **419**, 249), e.g.

When 1,3,5-triethoxy-1,3,5-trimethyl-1,3,5-trisilacyclohexane is acted upon by phosphorus trichloride, there results 1,3,5-**trichloro**-1,3,5-**trimethyl**-1,3,5-**trisilacyclohexane** (XXXIII, R = Cl, R^1 = Me), b.p. 93–94°/2 mm, m.p. 58° (*Kriner*, *loc.cit.*). This and other chloro compounds in the 1,3,5-trisilacyclohexane series are converted to the hydrides on reaction with lithium tetrahydridoaluminate (*Fritz*, *R. Haase* and *D. Kummner*, Z. anorg. allgem. Chem., 1969, **365**, 1), *e.g.* 1,3,5-*trisilacyclohexane*, b.p. 159°/766 mm, 1,1,3,3,5-*pentamethyl*-1,3,5-*trisilacyclohexane*, b.p. 201°/767 mm and are hydrolyzed to the hydroxy compounds, *e.g.* 1,1-*dihydroxy*-3,3,5,5-*tetramethyl*-1,3,5-*trisilacyclohexane*, m.p. 170–172°, with aqueous sodium carbonate (*W. S. Lindsay*, B.P. 927, 129/1963; C.A., 1964, **60**, 3010). Zinc fluoride may be used to replace chlorine by fluorine (*Fritz*, *M. Berndt* and *R. Huber*, Z. anorg. allgem. Chem., 1972, **391**, 219) and complete, radiation-catalyzed halogenation is feasible. Photobromination of the 1,1,3,3,5,5-hexachloro compound yields its 2,2-dibromo derivative, while similar bromination of 1,3,5-trisilacyclohexane itself yields the 1,3-dibromo- and 1,3,5-tribromo- compounds, each as a mixture of *cis* and *trans* isomers (*Fritz* and *I. Arnasan*, *ibid.*, 1976, **419**, 213). *A. V. Topchiev* and *N. S. Nametkin* (C.A., 1963, **58**, 6852) record that reaction between such chloro compounds and Grignard reagents proceeds normally in a stepwise fashion, thus 1,1,3,3,5,5-hexachloro-1,3,5-trisilacyclohexane and methylmagnesium bromide give initially the 1,1-dichloro-3,3,5,5-tetramethyl compound and then the hexamethyl derivative; the same dichloride and phenylmagnesium bromide yield 1,1,3,3-*tetramethyl*-5,5-*diphenyl*-1,3,5-*trisilacyclohexane*, b.p. 200°/11 mm. However, Fritz and his co-workers note that the hexa- and dodeca-chloro compounds both behave unusually towards methylmagnesium chloride and methyl-lithium giving rise to linear products together with disilapentene compounds; the formation of the methylene compound XXXIV from 1,1,3,3,5,5,6,6-octachloro-1,3,5-trisilacyclohexane by this route has been claimed (*Fritz* and *N. Szczepanski*, Angew. Chem. internat. Edn., 1967, **6**, 1077; Z. anorg. allgem. Chem., 1969, **367**, 44; *Fritz*, *P. Böttinger* and *N. Braunagel*, Z. Naturforsch. B, 1971, **26**, 478; *Fritz* and *Braunagel*, Z. anorg. allgem. Chem., 1973, **399**, 280). Compounds with free Si–H bonds undergo methylation at the less sterically hindered silicon atoms, or they undergo cleavage, or fail to react completely.

Passage of dichloromethane over a copper–germanium composite at 370–390° gives a mixture of linear **germanium compounds** together with 19% of 1,1,3,3,5,5-*hexachloro*-1,3,5-*trigermacyclohexane*, b.p. 150–152°/5 mm, m.p. 91–92°, converted to the *hexamethyl* compound, b.p. 71°/4 mm by the action of methylmagnesium chloride (*V. F. Mironov* and *T. K. Gar*, Izvest. Akad. Nauk S.S.S.R., Ser. Khim., 1964, 1887).

1,1,3,3,5,5-Hexamethyl-1,3,5-tristannacyclohexane is preparable from $Me_2Sn(CH_2I)_2$ and magnesium (*D. Seyferth* and *S. C. Vick*, Synth. React. Inorg. met. org. Chem., 1974, **4**, 515).

Syntheses of 1,1,3,3,4a,6,6,8,8-nonamethyl-1,3,4a,6,8-pentasiladecalin (XXXV), as a mixture of *cis* and *trans* isomers, and 2,2,3a,5,5,6a,8,8,9a-nonamethyl-2,3a,5,6a,8,9a-hexasilaperhydrophenalene (XXXVI) are illustrated (*Fritz* and *Hähnke*, Z. anorg. allgem. Chem., 1972, **390**, 137, 157):

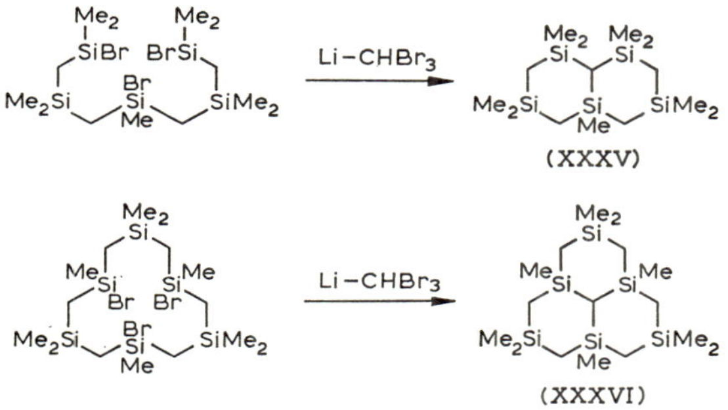

(XXXV)

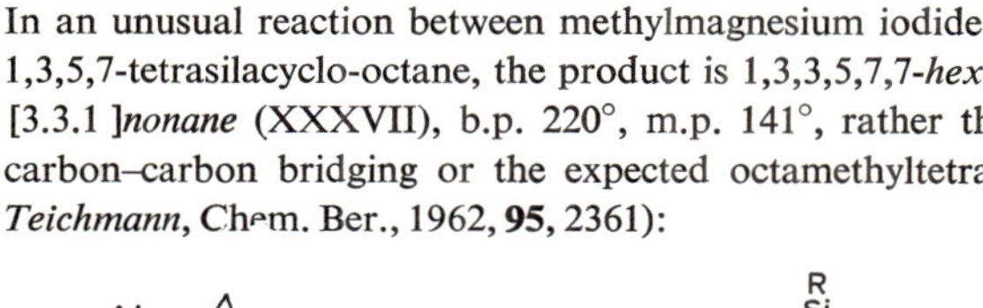

(XXXVI)

In an unusual reaction between methylmagnesium iodide and 1,1,3,3,5,5,7,7-octachloro-1,3,5,7-tetrasilacyclo-octane, the product is 1,3,3,5,7,7-*hexamethyl*-1,3,5,7-*tetrasilabicyclo*-[3.3.1]*nonane* (XXXVII), b.p. 220°, m.p. 141°, rather than an isomeric structure with carbon–carbon bridging or the expected octamethyltetrasilacyclo-octane (*Fritz* and *G. Teichmann*, Chem. Ber., 1962, **95**, 2361):

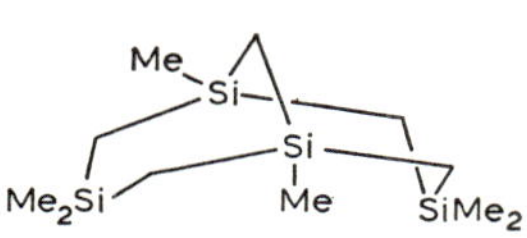

(XXXVII)

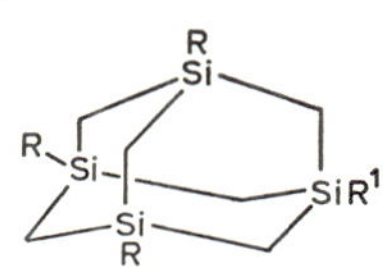

(XXXVIII)

Compounds of the tetrasila-adamantane series, XXXVIII, show greatly reduced reactivity reminiscent of those of the carbon analogue. 1,3,5,7-**Tetrachloro**-1,3,5,7-**tetrasila-adamantane** (XXXVIII, R = R^1 = Cl) is produced in only ~0.1% yield when silicon tetrachloride and chlorotrimethylsilane are heated with aluminium chloride at 500° (*A. L. Smith* and *H. A. Clark*, J. Amer. chem. Soc., 1961, **83**, 3345) and by the pyrolysis of the latter at 700° (*Fritz*, *W. König* and *H. Scheer*, Z. anorg. allgem. Chem., 1970, **377**, 240).

By using adsorption chromatography and gel chromatography, *Fritz et al.* (*loc.cit*). were able to purify the parent compound XXXVIII (R = R^1 = H), obtained by reduction of the tetrachloride with lithium tetrahydridoaluminate. Although the 1,3,5,7-tetramethyl compound may be obtained by the pyrolysis of tetramethylsilane (*Fritz* and *J. Grobe*, Z. anorg. allgem. Chem., 1962, **315**, 157) a better procedure involves treatment of 1,1,3,3,5,5-hexamethyl-1,3,5-trisilacyclohexane with aluminium bromide (*C. L. Frye*, *J. M. Klosowski* and *D. R. Weyenberg*, J. Amer. chem. Soc., 1970, **92**, 6379), a method also used for the preparation of the corresponding **tetramethyltetragerma-adamantane** (*T. K. Gar*, *A. A. Buyakov* and *V. F. Mironov*, Zhur. obshcheĭ Khim., 1972, **42**, 1521).

Typical of the lack of reactivity of compounds in this series to nucleophilic attack, is the resistance to acid of the Si–N bond in amino compounds formed by reaction of the chlorides with lithio-amines (*Frye* and *Klosowski*, J. Amer. chem. Soc., 1972, **94**, 7186), but a much higher degree of reactivity is observed in the reduction of XXXVIII (R = Me, R^1 = OMe) by diisobutylaluminium hydride, and in the conversion of the chloride XXXVIII (R = Me, R^1 = Cl) into the methoxy compound XXXVIII (R = Me, R^1 = OMe) (86% yield in 15 min using sodium methoxide in methanol) and into the tetramethyl compound using methyl-lithium in the presence of tetramethylethylenediamine (*G. D. Horner* and *L. H. Sommer*, J. organometal. Chem., 1973, **49**, C13; 1975, **84**, 297.

1,3,3,5,7,9,9,11-**Octachloro**-1,3,5,7,9,11-**hexasilapentacyclo** $[6.4.0.0^{2,7}0^{4,11}0^{5,10}]$**dodecane** (XXXIX) is formed during the pyrolysis of methyltrichlorosilane in a nitrogen atmosphere; it sublimes at 210° at atmospheric pressure under nitrogen (*Fritz*, *Haase* and *Scheer*, Angew. Chem., 1968, **80**, 397), and its structure has been confirmed by

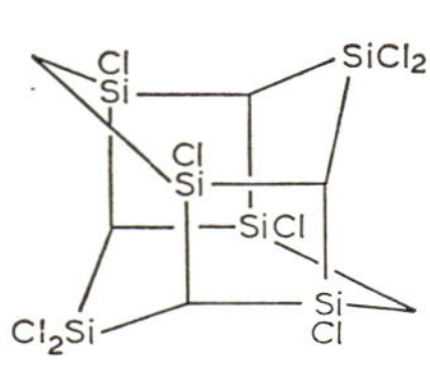

(XXXIX)

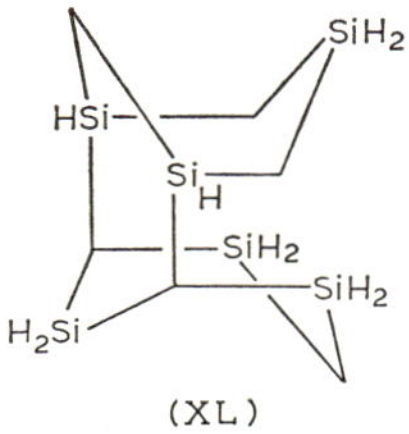

(XL)

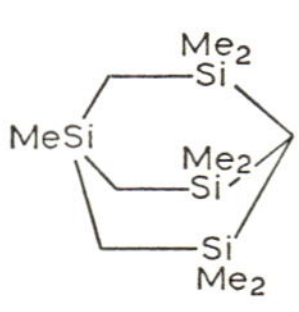

(XLI)

X-ray methods (*G. Sawitzki* and *H. G. von Schnering*, Z. anorg. allgem. Chem., 1973, **399**, 257). Reduction of the octachloro compound with lithium tetrahydridoaluminate leads through $C_6H_{12}Cl_4Si_6$ to $C_6H_{16}Si_6$ and ultimately to a ring-opened compound thought to be XL. When treated with methylmagnesium chloride, chlorine in XXXIX is replaced by Me (*Fritz*, *H. J. Dannappel* and *E. Matern*, *ibid.*, 1973, **399**, 263).

1,3,3,5,5,8,8-**Heptamethyl**-1,3,5,8-**tetrasilabicyclo**[2.2.2]**octane** (XLI), m.p. 55–56°, has been isolated from the reaction of bromoform with $MeSi(CH_2SiMe_2Br)_3$ in the presence of lithium (*Fritz* and *Hähnke*, *ibid.*. 1972, **390**, 191).

Development of the chemistry of these and other complex polycyclic polysilanes has been somewhat restricted because of the very limited availability of many of the compounds. Determination of structure has relied heavily on chromatographic purification techniques, particularly g.l.c., coupled with proton and silicon nuclear resonance spectroscopy and high resolution mass spectrometry.

(c) *Perhydrotetrasilins (tetrasilacyclohexanes)*

A **decamethyl-1,2,3,4-tetrasilacyclohexane,** m.p. 44–45°, is claimed as a product from the reaction between isobutene and dichlorodimethylsilane in the presence of lithium at 0–10° in tetrahydrofuran (*O. M. Nefedov* and *M. N. Manekov*, Angew. Chem., 1964, **76**, 270).

1,1,2,2,4,4,5,5-**Octamethyl-**1,2,4,5-**tetrasilacyclohexane** is known as its 3,6-*di(trimethylphosphine) ylid* (XLII), m.p. 126–128°, prepared from $Me_4Si_2Cl_2$ and trimethylphosphinemethylene (*M. Schmidbaur* and *W. Vornberger*, Angew. Chem. internat. Edn., 1970, **9**, 737).

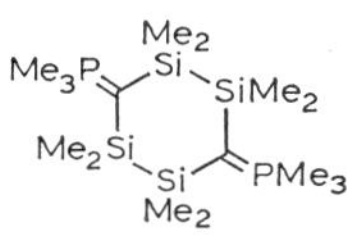

(XLII)

4. Compounds containing arsenic and phosphorus

Comparatively little information is available on the chemistry of six-membered rings possessing arsenic or phosphorus atoms in the 1,2 and 1,3 positions, and indeed, 1,2-diarsenin and 1,2-diphosphorin (I, M = As, P) appear not to have been recorded, and even partially reduced systems have received scant attention. On the other hand, the chemistry of partially reduced 1,4-diphosphorin and its arsenic analogue and of the fully reduced system, 1,2-diphosphorinane, is quite extensive.

(a) *Compounds with two hetero atoms*

(i) *Tetrahydro- and hexahydro(perhydro)-*1,2-*diphosphorins and -arsenins* (1,2-*diphosphorinanes and* 1,2-*diarsenanes*)

Buta-1,3-dienes are reported to react with polyphosphines at 145–160° to yield mixtures of phosphol-3-enes and 1,2,3,6-tetrahydro-1,2-diphosphorins (II, M = P); 1,2-*dimethyl-*, b.p. 82–83°/11 mm, (*disulphide*, m.p. 176–177°); 1,2-*diethyl-*, b.p. 114°/9 mm, (*disulphide*, m.p. 94°); 1,2,4,5-*tetramethyl-*, b.p. 100°/8 mm, (*disulphide*, m.p. 193°); and 1,2-*dimethyl-*4,5-*diphenyl-*1,2,3,6-*tetrahydro-*1,2-*diphosphorin*, b.p. 150°/0.005 mm, (*disulphide*, m.p. 178–179°) (*U. Schmidt et al.*, Chem. Ber., 1968, **101**, 1381; *A. Ecker, I. Boie* and *Schmidt*, Angew. Chem. internat. Edn., 1971, **10**, 191).

1,2,4,5-**Tetraphenyl-**1,2,3,6-**tetrahydro-**1,2-**diphosphorin** 1,2-**disulphide,** m.p. 233–234°, may be obtained from 2,3-diphenylbutadiene and phenylperthiophosphonic anhydride, but a similar reaction with phenylphosphonic dichloride does not take place (*S. Nakayama et al.*, Bull. chem. Soc. Japan, 1975, 48, 546). **1,2,4,5-Tetramethyl-1,2,3,6-tetrahydro-1,2-diarsenin** (II, M = As), prepared in a similar fashion, has b.p. 70–72°/0.25 mm.

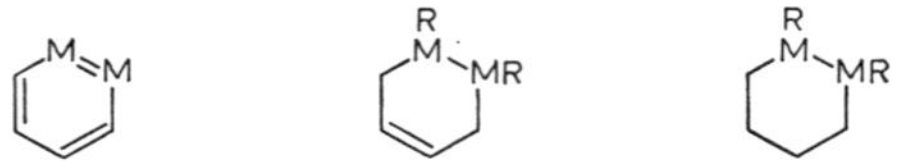

(I) (II) (III)

M = P, As

1,2-Diethyl-1,2-diphosphorinane, (III, M = P, R = Et), b.p. 70–71 °/2 mm, (*disulphide*, m.p. 115°) and the 1,2-*diphenyl* analogue, m.p. 90°, (*disulphide*, m.p. 171°) are formed when 1,4-dichlorobutane reacts with the dipotassium salt of the diethyldiphosphine or diphenyldiphosphine (*K. Issleib* and *K. Krech*, Chem. Ber., 1965, **98**, 2545; 1966, **99**, 1310).

1,4-Dichlorobutane is also used in conjunction with potassium phenylarsine in benzene–dioxane, or *via* initial conversion to butane-1,4-di(phenylarsenic acid) or to butane-1,4-di(phenyllithioarsine), the last named being heated with 1,2-dibromoethane, to prepare **1,2-diphenyl-1,2-diarsenane** (III, M = As, R = Ph), b.p. 196–198°/0.4 mm, 142–144°/0.1 mm, m.p. 58–59°; when treated with methyl iodide, ring cleavage occurs with the formation of $PhAsI(CH_2)_4AsMe_2Ph$ I (*K. Sommer*, Z. anorg. allgem. Chem., 1970, **377**, 278; *A. Tzschach* and *G. Pacholke*, Chem. Ber., 1964, **97**, 419; Z. anorg. allgem. Chem., 1965, **336**, 270).

(ii) *Di- and tetra-hydro-*1,3-*diphosphorins and benzologues. Hexahydro-*1,3-*diphosphorins*(1,3-*diphosphorinanes*). *Arsenic analogues*

The interrelationships between the known examples of derivatives of 1,3-diphosphorin are shown in the accompanying scheme.

Reaction of methylenebis(diphenylphosphine) with 1,3-dibromopropane in tetrahydrofuran yields **1,1,3,3-tetraphenyl-1,3-diphosphorinanium dibromide** (IV) (*diperchlorate*, m.p. 345–347°) and a similar reaction with 1,3-dibromoacetone leads to the *cyclic ketone* V, m.p. 230–250°. The latter is reduced by sodium tetrahydridoborate to the alcohol VI (*diperchlorate*, m.p. 287–288°) which, when heated with phosphoric acid at 180–200°, is dehydrated to a product, isolated as 1,1,3,3-*tetraphenyl*-1,2,3,6-*tetrahydro*-1,3-*diphosphorinium diperchlorate* (VII), m.p. 320–321°. Sodium carbonate converts this to the *mono-ylid*, VIII, m.p. 236–237°, and subsequent conversion to the di-ylid, IX, requires sodium hydroxide, also required to produce the mono-ylid X as the *perchlorate*, m.p. 255–257°, from IV (*G. Märkl*, Z. Naturforsch. B, 1964, **18**, 1136).

An infrared spectroscopic study of halide salts of type V has revealed that, in methanol, enolic forms predominate whereas for solutions in trifluoroacetic acid, a keto form is most important (*T. A. Mastryukova et al.*, Zhur. obshcheĭ Khim., 1973, **43**, 1195, 2613). Further, it might be expected that the *diphosphadienone* XI, m.p. 188.5–192°, would possess phenolic properties, but a ^{1}H-n.m.r. spectroscopic investigation of the system has found such properties lacking (*idem*, *ibid.*, 1974, **44**, 950).

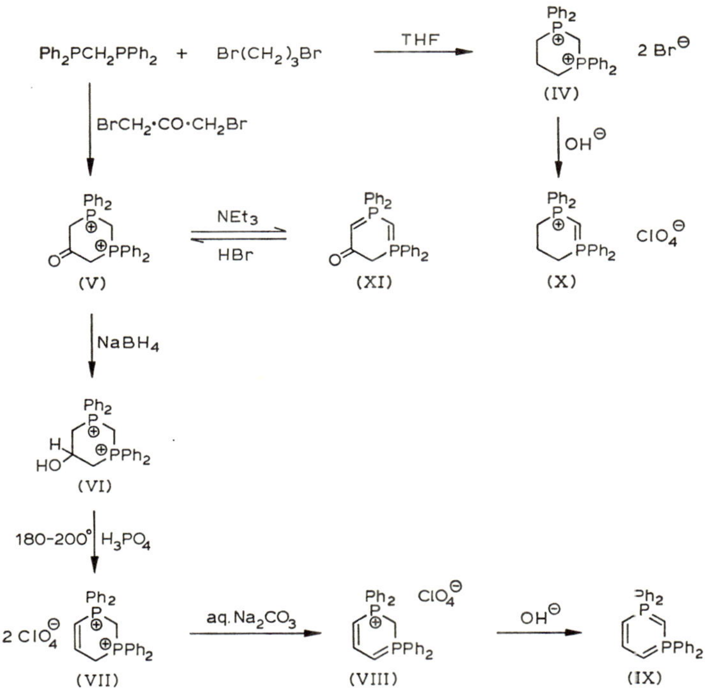

The tetrafluoroborate of V undergoes ring-opening upon treatment with water or alkali, although the products of these reactions differ *(Mastryukova, Kh. A. Suerbaer* and *M. I. Kabachnik, ibid.*, 1974, 44, 950):

$$\underset{BF_4^{\ominus}}{Ph_2MeP^{\oplus}CH_2{\cdot}CO{\cdot}CH_2P(O)Ph_2} \xleftarrow{H_2O} V{\cdot}2BF_4 \xrightarrow{OH^{\ominus}} Ph_2MeP{=}CH{\cdot}CO{\cdot}CH_2P(O)Ph_2$$

1,3-Diphenyl-1,3-diphosphorinane, m.p. 75–79°, b.p. 150°/0.05 mm, is obtained from dichloromethane and 1,3-bis(phenylphosphino)propane (*Issleib* and *W. Böttcher*, Z. anorg. allgem. Chem., 1974, **406**, 178).

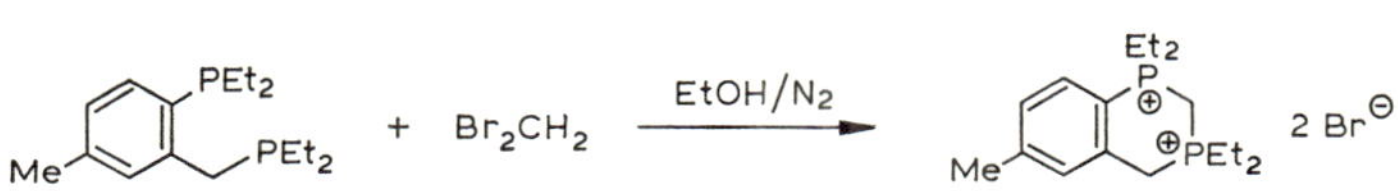

(XII)

1,1,3,3-*Tetraethyl*-6-*methyl*-1,2,3,4-*tetrahydro*-1,3-*benzodiphosphorinium dibromide* (XII), m.p. 308–309°, appears to be the only example recorded of an arene-annelated 1,3-diphosphorin (*M. Davis* and *F. G. Mann*, J. chem. Soc., 1964, 3786).

A series of 1,3-**diphenyl-perhydro**-1,3-**diphosphorinanes** (XIII) (M = *SiMe*$_2$, m.p. 122–125°, b.p. 222°/2 mm; *SnEt*$_2$, b.p. 120°/10^{-5} mm; *AlPh*, m.p. 135°; *Zn*, m.p. 298–303°), has

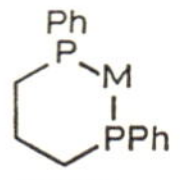

(XIII)

been prepared by interaction of a suitable metalloid or metal dichloride and a metal derivative of a 1,3-diphosphine (*Issleib* and *Böttcher*, *loc.cit.*).

2,3,3,4,8,9,9,10-**Octaphenyl**-2,4,8,10-**tetraphospha**-3,9-**disilaspiro**[5.5]-**undecane** (XIV), m.p. 126°, is obtained from tetrakis(phenylphosphinomethyl)methane *via* its sodium or potassium salt as indicated (*J. Ellerman* and *F. Poersch*, Angew. Chem. intern. Edn., 1967, **6**, 355):

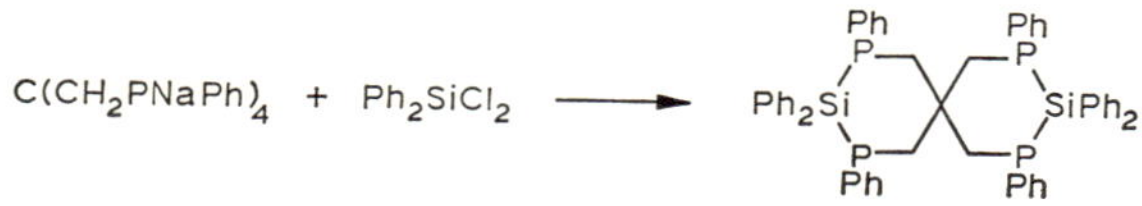

(XIV)

(*iii*) 1,4-*Diphosphorins*, 1,4-*diarsenins and their di- and tetra-hydro derivatives. Perhydro*-1,4-*diphosphorins* (1,4-*diphosphorinanes*) *and* -1,4-*diarsenins* (1,4-*diarsenanes*)

Early attempts by *A. M. Aguair* and *H. Aguair* (J. Amer. chem. Soc., 1966, **88**, 4090) to prepare a 1,4-diphosphorin XVI (R = H) met with no success. The action of a base upon the salt XV resulted in the elimination of acetylene and the formation of 1,2-diphenylphosphinoethane monoxide. However, reaction between dimethyl acetylenedicarboxylate and *cis*-1,2-diphenylphosphino-ethene yields an adduct which, on the basis of its mass spectrum and other spectroscopic data, and its hydrolysis to a linear phosphine oxide with liberation of dimethyl succinate, has been described as 2,3-di(methoxycarbonyl)-1,1,4,4-tetraphenyl-1,4-diphosph(V)orin (XVI, R = CO_2Me) (*M. A. Shaw et al.*, J. chem. Soc., C, 1970, 504):

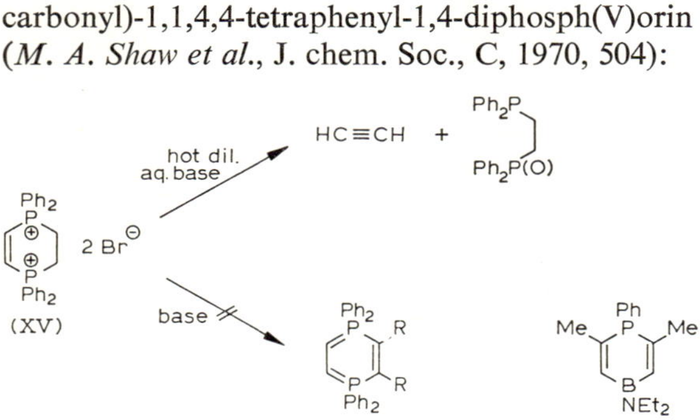

(XV) (XVI) (XVII)

Azaisobutyronitrile catalyses the radical addition of phenylphosphine and phenylarsine to a variety of di(alkynyl)metalloids.

Addition of phenylphosphine to *N*-diethylaminodipropynylborane yields **4-diethylamino-3,5-dimethyl-4-bora-1-phosphacyclohexa-2,5-diene** (XVII), b.p. 104–106°/10^{-3} mm (*H. O. Berger* and *H. Noeth*, Z. Naturforsch., B. Anorg. Chem., Org. Chem., 1975, **30**, 641).

Reaction between phenylphosphine or phenylarsine and dialkynyl *tert*-butylphosphines may be radical-initiated or may be brought about by lithium amide/liquid ammonia; the manner of ring-closure governs the ratio of *cis* and *trans* isomers of the resulting products. *trans*-1-*tert*-**Butyl-4-phenyl-1,4-dihydro-1,4-diphosphorin** (XVIII, M = P, R = H) has m.p. 56–57° (*dioxide*, m.p. 259–261°). The corresponding 1,4-*phospharsenin* has m.p. 50–51° (P-*oxide*, m.p. 128–130°). Both *cis* and *trans* forms of the dimethyl derivatives of both systems (XVIII, M = P, As, R = Me) are isolable (*Markl et al.*, Tetrahedron Letters, 1975, 3171).

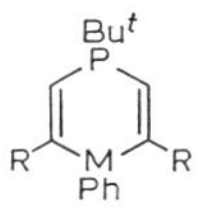

(XVIII)

Other 1,4-dihydro-1,4-diphosphorins are reported to be formed when 2,3-dichloromaleic anhydride reacts with phenyl bistrimethylsilylphosphine, $PhP(SiMe_3)_2$, (*D. Fenske et al.*, Chem. Ber., 1976, **109**, 359).

1,4-Dihydro-1,4-diphosphorinium bromides (XIX) (1,4-diphosphoniacyclohexa-2,5-dienes) are the cyclic products obtained when alkynylphosphines are allowed to react with hydrogen bromide in acetic acid at or below room temperature, *e.g.* 1,1,4,4-**tetraphenyl-1,4-dihydro-1,4-diphosphorinium dibromide**, m.p. 306–308°, and the 1,1,2,4,4,5-*hexaphenyl*-, m.p. 286–290°; 2,5-*diethyl*-1,1,4,4-*tetramethyl*-, m.p. 370–373°; 1,1,4,4-*tetraethyl*-2,5-*dimethyl*-, m.p. 318–320°, analogues. Linear addition products, XX, may be formed also but

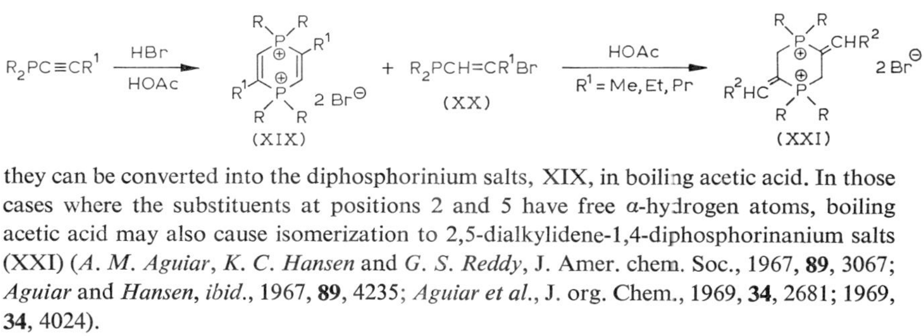

(XIX) (XX) (XXI)

they can be converted into the diphosphorinium salts, XIX, in boiling acetic acid. In those cases where the substituents at positions 2 and 5 have free α-hydrogen atoms, boiling acetic acid may also cause isomerization to 2,5-dialkylidene-1,4-diphosphorinanium salts (XXI) (*A. M. Aguiar*, *K. C. Hansen* and *G. S. Reddy*, J. Amer. chem. Soc., 1967, **89**, 3067; *Aguiar* and *Hansen*, *ibid.*, 1967, **89**, 4235; *Aguiar et al.*, J. org. Chem., 1969, **34**, 2681; 1969, **34**, 4024).

Interestingly, 1,1,4,4-*tetraphenyl*-1,4-*dihydro*-1,4-*diphosphorinium dichloride monohydrate* (XXII), m.p. 251–253°, is obtained when *cis*-1,2-bis(diphenylphosphino)ethene is quaternized with di(chloromethyl) ether (*Aguiar* and *Hansen*, *loc.cit.*):

(XXII)

The molecular structure of the 1,1,4,4-tetraethyl-2,5-dimethyl compound XIX (R = Et, R^1 = Me) appears to be that of a localized, non-conjugated diene with a planar ring (*R. L. R. Towns et al.*, J. heterocycl. Chem., 1970, **7**, 835) but this view seems to conflict with that held by *Aguiar et al.*, who, on the basis of ^{1}H-n.m.r. data for the hexaphenyl compound in trifluoroacetic acid, suggest delocalization of charge (*Aguiar*, *Hansen* and *Reddy*, *loc.cit.*).

When hexafluorobut-2-yne is heated with red phosphorus and a trace of iodine at 200° there is formed 2,3,5,6,7,8-*hexakis*(*trifluoromethyl*)-1,4-*diphosphabicyclo* [2.2.2]*octa*-2,5,7-*triene* (XXIII, M = P), m.p. 119–120°, completely unreactive to oxygen, methyl iodide at room temperature, bromine in carbon tetrachloride, and benzyl chloride at 100°. The *arsenic analogue* (XXIII, M = As), m.p. 139–140°, is obtained in a similar fashion (*C. G. Krespan*, *B. C. McKusick* and *T. L. Cairns*, J. Amer. chem. Soc., 1960, **82**, 1515; *Krespan*, *ibid.*, 1961, **83**, 3432).

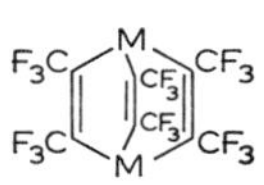

(XXIII)

M = P, As

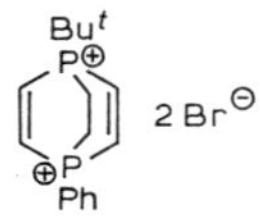

(XXIV)

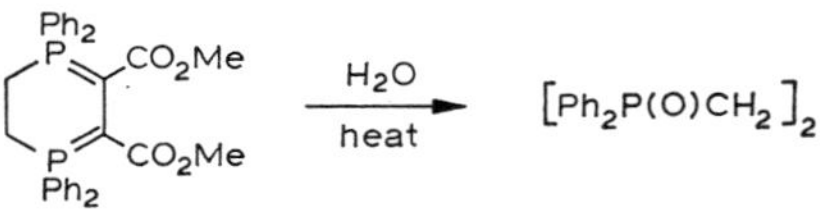

(XXV)

The 1,4-*diphosphonia salt* (XXIV), prepared conventionally, has m.p. 211–213° (*G. Markl, et al.*, Tetrahedron Letters, 1975, 3171).

Acetylenedicarboxylic esters react readily with 1,2-bis(diphenylphosphino)ethane in benzene at room temperature to give yellow 2,3-di(methoxycarbonyl)-1,1,4,4-tetraphenyl-1,4-diphosphacyclohexa-1,3-diene (XXV), unstable at temperatures higher than 150°, and readily acted upon by water (*A. N. Hughes* and *S. W. S. Jafry*, J. heterocycl. Chem., 1969, **6**, 991).

Derivatives, *e.g.* the salt XV, of 1,2,3,4-**tetrahydro**-1,4-**diphosphorin,** are obtained by the quaternization reaction indicated (*Aguiar* and *Aguiar*, *loc.cit.*) and by partial hydrogenolysis of 1,4-dihydro-1,4-diphosphorinium salts (*Aguiar*, *Hansen* and *Reddy*. *loc.cit.*); in undergoing ring cleavage with alkali, the course of the reaction depends on the relative proportions of salt and alkali (*J. J. Brophy* and *M. J. Gallagher*, Austral. J. Chem., 1969, **22**, 1385):

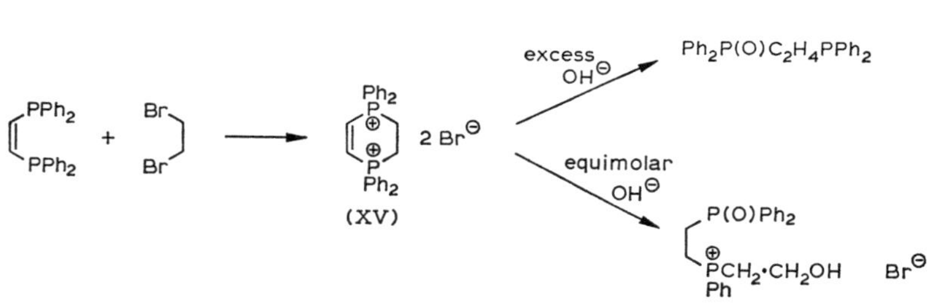

Quaternization of 1,2-bis(diethylphosphino)benzene with dibromoethane yields **1,1,4,4-tetraethyl-1,2,3,4-tetrahydro-1,4-benzodiphosphorinium dibromide** (XXVI, M = P, R = Et) (*F. A. Hart* and *Mann*, J. chem. Soc., 1967, 3939). The pyrolysis of **1,1,4,4-tetramethyl-1,2,3,4-tetrahydro-1,4-benzodiarseninium** *dibromide* (XXVI, M = As, R = Me), m.p. 255°, under nitrogen gives **1,4-dimethyl-1,2,3,4-tetrahydro-1,4-benzodiarsenin** (XXVII), b.p. 94–97°/0.015 mm, (*monomethiodide*, m.p. 250° decomp., *dimethiodide*, m.p. 245–246°) along with some of the *bromo* compound XXVIII, b.p. 120°/0.3 mm. The diarsine XXVII quarternizes with ω,ω-dibromoxylene giving **5,12-dimethyl-5,12-ethano-6,11-dihydro-dibenzo**[*b,f*] [1,4]**diarsinium dibromide** (XXIX) as the *monohydrate*, m.p. 214–216°, and

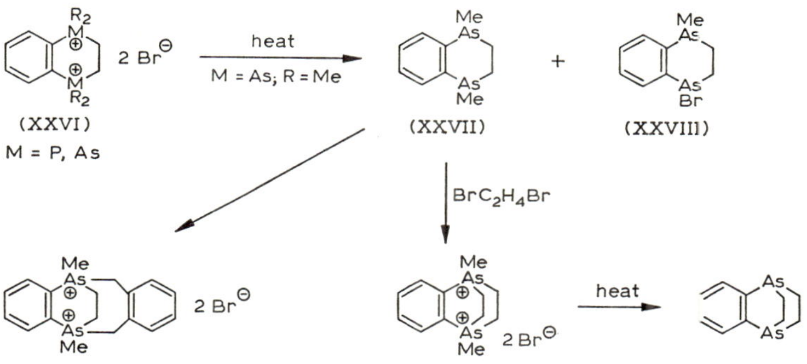

(XXVI) (XXVII) (XXVIII) (XXIX) (XXX) (XXXI)

with dibromoethane giving **1,4-dimethyl-1,4-ethano-1,2,3,4-tetrahydro-1,4-benzodiarsen-inium dibromide** (XXX), m.p. 240°, which on pyrolysis at 270° loses methyl bromide and leaves **1,4-ethano-1,2,3,4-tetrahydro-1,4-benzodiarsenin** (XXXI) b.p. 95–96°/0.01 mm (*monomethiodide*, m.p. 240° decomp., *dimethiodide*, m.p. 260° decomp.); this is readily oxidized by hydrogen peroxide to the dioxide (*R. H. Glauert* and *Mann*, J. chem. Soc., 1950, 682; *Mann* and *F. C. Baker*, *ibid.*, 1952, 4142).

The rupture of the As–C bond and formation of the As–Br bond in the pyrolysis of the quaternary salt (XXVI, M = As, R = Me) is one example of the way in which quaternary salts of phosphorus and arsenic may differ in their reactivity. A further example of such differences is illustrated by the selective dealkylation at arsenic when **1,1,4,4-tetraethyl-1,2,3,4-tetrahydro-1,4-benzophospharseninium dibromide**, m.p. 240°, is pyrolyzed at 270/15 mm leaving the mono-quaternary salt XXXII, m.p. 167–169° (*E. R. H. Jones* and *Mann*, *ibid.*, 1955, 4472):

(XXXII)

5,10-Diphenyl-5,10-dihydrophosphanthrene (dihydrodibenzo[1,4]-diphosphorin) (XXXIV, R = Ph), m.p. 184–187°, has been obtained by reaction between 2-bromo-1-iodobenzene and dilithiophenylphosphine in tetrahydrofuran (*Mann* and *M. J. Pragnall*, *ibid.*, C, 1966, 916) and also from interaction of 2-chlorophenylmagnesium bromide and phenyldichlorophosphine *via* the dilithium compound XXXIII (X = Li) (*Davis* and *Mann*, *ibid.*, 1964, 3770):

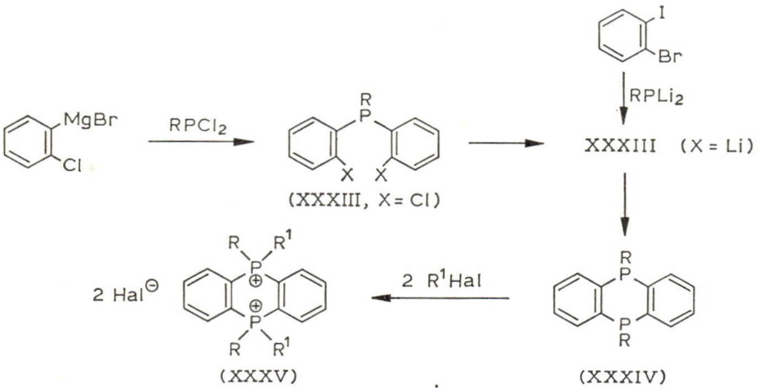

The 5,10-diphenyl compound, and the analogous 5,10-diethyl compound, form geometrically isomeric dioxides and diquaternary salts, and in certain cases of the latter, conversion to the parent diphosphine is relatively easy. The geometry of such compounds is best described in terms of butterfly conformations, the two configurations XXXVI A and B being flexible about the P–P axis to an extent which depends on the size of substituents at phosphorus. In spite of flexing, the two configurations can be distinguished by dipole moment measurements, although in practice these have been limited by the availability of the isomeric forms; the *trans* forms should possess zero dipole moment, whereas the *cis* forms will have a finite moment.

(XXXVI)

(A) *cis* form (B) *trans* form

(X = lone electron pair, O or S)

5,10-Diethyl-5,10-dihydrophosphanthrene (XXXIV, R = Et) forms separable stereoisomeric dibenzyl diquaternary phosphonium iodides when treated with benzyl iodide. The *salt* XXXV (R = Et, R^1 = $PhCH_2$, Hal = I), m.p. 326°, with alkali yields a *dioxide* having m.p. 235° and a finite dipole moment, and therefore has *cis* geometry, presumably XXXVI A, and the salt is also reduced by lithium tetrahydridoaluminate to a *diethyldiphosphine* (XXXIV, R = Et) of m.p. 52–53°, presumably of the same geometry, and which is oxidized by hydrogen peroxide to the same dioxide. The geometry of this form of the phosphine is also suggested by its reaction with dibromoethane to give 5,10-*ethano*-5,10-*diethyl*-5,10-*dihydrophosphanthrenium dibromide* (XXXV, R = Et, R^1 = –CH·CH_2–, Hal = Br), m.p. 325–326°, a similar salt also being formed with ω,ω-dibromoxylene (*Davis* and *Mann*, *Mann* and *Pragnall*, *loc.cit*). In the same way, the *dibenzyl* diquaternary *diiodide*, m.p. 320–321°, yields a *dioxide*, m.p. 257°, and on reduction, a *diphosphine*, m.p. 96–97°, presumably both of *trans* geometry (XXXVI B). The geometry of derivatives of 5,10-diphenyl-5,10-dihydrophosphanthrene has also been investigated as have also that of other fused tricyclic 1,4-dihydro-1,4-diphosphorins, which have been isolated in *cis*- and *trans*-forms (*J.-P. Lampin* and *F. Mathey*, J. organometal. Chem., 1974, **71**, 239).

By contrast to the recent discovery of, and investigations on phosphanthrenes, analogous diarsines have been known since 1921 when *L. Kalb* (Ann., 1921, **423**, 39) prepared 5,10-**dichloro**-5,10-**dihydroarsanthrene** (XXVII), m.p. 182–183°, by the route shown; he claimed that this compound could be reduced to arsanthrene itself (XXXVIII) through a series of highly coloured intermediates. However, reduction of the dichloride with zinc and ethanolic hydrogen chloride has provided a dimeric form of arsanthrene consisting of parallel ring systems with the four arsenic atoms arranged in the form of a square (*H. Veermer* and *F. Bickelhaupt*, Tetrahedron Letters, 1970, 1323, 3255).

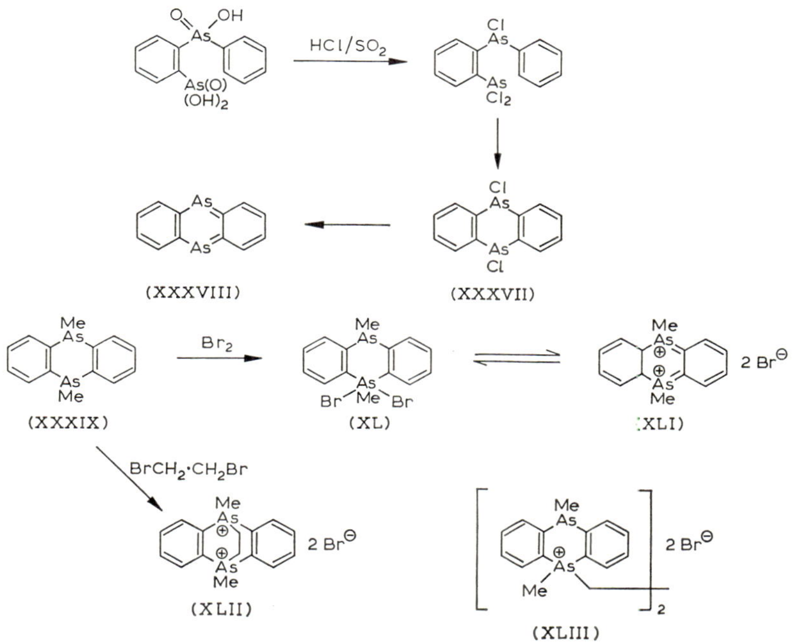

5,10-**Dimethyl-5,10-dihydroarsanthrene** (XXXIX), m.p. 191–192.5°, (*monomethiodide*, m.p. 222° decomp., *dimethiodide*, m.p. 263° decomp.) (*Jones* and *Mann*, J. chem. Soc., 1955, 411) reacts with bromine in alcoholic solution to give the *dibromide* XL, which is in equilibrium with the true salt XLI; the geometry around the arsenic atoms has been clearly demonstrated by X-ray techniques (*D. J. Sutor* and *F. R. Harper*, Acta Cryst., 1959, **12**, 585). On heating the dibromide at 0.1 mm it is converted to 5-bromo-10-methyl-5,10-dihydroarsanthrene. When 5,10-dimethyl-5,10-dihydroarsanthrene is treated with dibromoethane and dibromoxylene, quaternization takes place leading to either the bridged *salt*, *e.g.* XLII, m.p. 272° decomp., or the *bis-arsanthrenium salt*, *e.g.* XLIII, m.p. 195°, the actual product depending on experimental conditions.

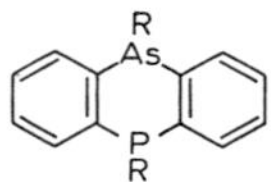

(XLIV)

(XLV) (XLVI)

Phenyldi(2-chlorophenyl)phosphine reacts with phenyldichlorarsine *via* the dilithium compound XXXIII (X = Li) to give 5,10-**diphenyl-5,10-dihydrobenzo** [1,4]**phospharsenin** (XLIV, R = Ph), m.p. 189–190°, together with other compounds. When treated with excess methyl iodide, quaternization takes place selectively at phosphorus, and the *mono-methiodide*, m.p. 300–302°, then reacts with alkali giving the phosphine *oxide*, m.p. 214–216°, oxidized to the *dioxide*, m.p. 360°, by hydrogen peroxide (*Davis* and *Mann*, *loc.cit.*).

Phosphorus-, arsenic- and antimony-containing analogues of trypticene have been reported; that from arsenic XLVI, is obtained when the dichlorodiphenyldiarsine (XLV) is heated *in vacuo*. 5,10-*o*-**Benzeno-5,10-dihydroarsanthrene**, m.p. 295–296°, forms an orange-red crystalline *tetrabromide*, m.p. 255–256°, which is readily hydrolyzed to a colourless *oxide*, m.p. > 360°, but it fails to form a methiodide, and cannot be nitrated (*N. P. McCleland* and *J. B. Whitworth*, J. chem. Soc., 1927, 2753). The *phosphorus analogue*, m.p. 323–325°, is formed when 1,2-dichlorobenzene is heated with white phosphorus together with ferric chloride as catalyst in a sealed tube at 280°; with excess benzyl bromide it yields only the monobenzyl quaternary bromide (*K. G. Weinberg* and *E. B. Whipple*, J. Amer. chem. Soc., 1971, **93**, 1801). Perfluoro derivatives of both **diarsatrypticene**, m.p. 205–206°, and of **distibatrypticene**, m.p. 245–246°, are preparable from the element itself upon heating with 1,2-diiodobenzene, although the yields are ~ 12 % only (*C. M. Woodard*, *C. Hughes* and *A. G. Massey*, J. organometal. Chem., 1976, 112, 9).

(*iv*) *Perhydro*-1,4-*diphosphorins* (1,4-*diphosphorinanes*) *and their arsenic analogues*

A number of methods have been employed to obtain derivatives of 1,4-diphosphorinane and 1,4-diarsenane as illustrated in the following scheme:

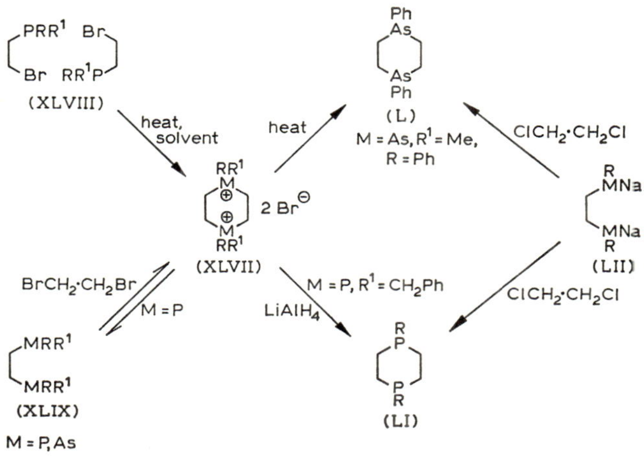

The 1,4-diphosphorinanium salts XLVII (M = P) are obtained by the self-quaternization of the 2-bromoethylphosphines XLVIII when these are heated in a suitable solvent, and also when the phosphine XLIX (M = P) is heated with dibromoethane (*C. H. S. Hitchcock* and *Mann*, J. chem. Soc., 1958, 2081). The latter reaction is also employed to prepare 1,4-diarsenanium salts (XLVII, M = As) (*Jones* and *Mann*, *ibid.*, 1955, 401), but the reaction only succeeds when performed in methanol solution. The phosphorus and arsenic salts differ in their behaviour towards heat; the phosphonium salts undergo loss of a two-carbon fragment and revert to the starting phosphine XLIX as the hydrobromide, whereas the arsonium salts lose alkyl monohalide and yield 1,4-diarsenanes (L); nevertheless, 1,4-diphosphorinanes (LI) can be prepared by reduction of benzyl quaternary phosphonium salts XLVII ($R^1 = PhCH_2$) with lithium tetrahydridoaluminate.

K. Issleib and *H. R. Roloff* (J. pr. Chem., 1970, **312**, 578) prepared six- (and eight)-membered rings by the action of phosphorus pentahalides on 2-hydroxyalkylphosphines, the bromoalkylphosphine being prepared in *situ*:

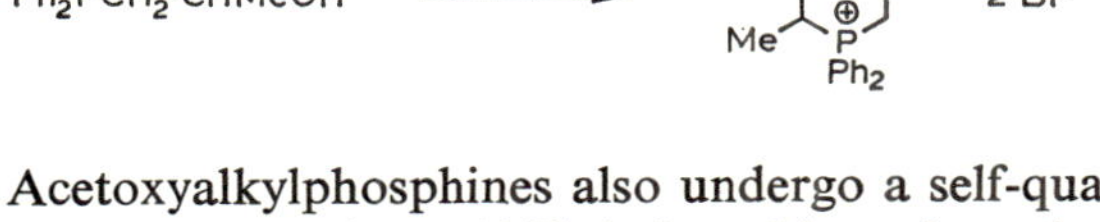

Acetoxyalkylphosphines also undergo a self-quaternization when heated at temperatures above 120°; below this, polymerization becomes more important (*M. M. Rauhut*, *G. Borowitz* and *H. C. Gillham*, J. org. Chem., 1963, **28**, 2565).

In a second type of synthesis, the disodium or dilithium derivatives of ethane-1,2-di(phosphines) or -arsines (LII) in dioxane or liquid ammonia react with 1,2-dichloroethane (*Issleib* and *K. Standtke*, Chem. Ber., 1963, **96**, 279; *Issleib* and *G. Döll*, *ibid.*, p. 1544; *Issleib* and *H. Weichmann*, *ibid.*, 1968, **101**, 2197).

Reduction of the quaternary salts XLVII (R = Ph, R^1 = $PhCH_2$), m.p. 375°, and (R = R^1 = $PhCH_2$), m.p. 287–292°, with lithium tetrahydridoaluminate yields **1,4-diphenyl-1,4-diphosphorinane** (LI, R = Ph), m.p. 92–95°, (*dioxide*, m.p. 250°), and the 1,4-*dibenzyl* compound (LI, R = $PhCH_2$), m.p. 128–130°, (*monoxide*, m.p. 302–305°). Such dibenzyl diquaternary salts undergo a non-specific alkaline hydrolytic removal of the benzyl groups to produce a *cis–trans* mixture of phosphine dioxides (*G. E. Driver* and *M. J. Gallagher*, Chem. Comm. 1970, 150). The first description of geometric isomers of the phosphonium salts seems to have been given by *L. Horner* and *J. P. Bercz* (Tetrahedron Letters, 1966, 5783) who showed that the *meso* form of the phosphine XLIX, (R = Me, R^1 = Ph) gives a *cis* di-quaternary salt XLVII, while the racemic phosphine yields the *trans* salt XLVII.

The formation of 1,4-diphosphorinanium salts from acetylenic phosphines either directly, or indirectly by isomerization of 1,4-dihydro-1,4-diphosphorinium salts has already been discussed (p. 22) but further examples of the use of acetylenic phosphines in the synthesis of 2,5-dialkylidene-1,4-diphosphorinanium compounds have been recorded by *M. S. Chatta* and *A. M. Aguiar* (J. org. Chem., 1973, **38**, 1611); the use of vinyl phosphines in the same reaction leading directly to 1,4-diphosphorinanium salts has also been mentioned briefly (*G. Märkl*, Angew. Chem. internat. Edn., 1965, **4**, 1023).

When treated with alkali, 1,4-diphosphorinanium salts undergo ring opening, the nature of the product depending on the relative amounts of substrate and alkali (*J. J. Brophy* and *Gallagher*, Austral. J. Chem., 1969, **22**, 1385):

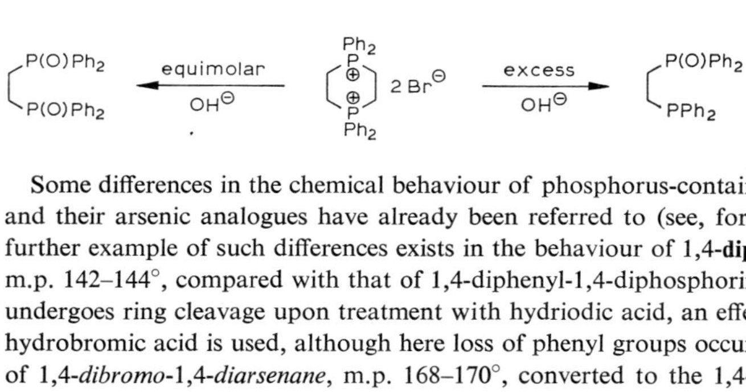

Some differences in the chemical behaviour of phosphorus-containing ring compounds and their arsenic analogues have already been referred to (see, for example, p. 25). A further example of such differences exists in the behaviour of **1,4-diphenyl-1,4-diarsenane,** m.p. 142–144°, compared with that of 1,4-diphenyl-1,4-diphosphorinane. The diarsenane undergoes ring cleavage upon treatment with hydriodic acid, an effect not observed with hydrobromic acid is used, although here loss of phenyl groups occurs with the formation of 1,4-*dibromo*-1,4-*diarsenane*, m.p. 168–170°, converted to the 1,4-*dimethyl* compound,

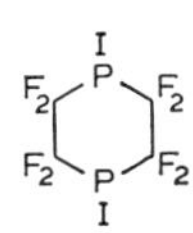

(LIII)

b.p. 113–114°/24 mm by methylmagnesium iodide, and to the parent **1,4-diarsenane,** b.p. 87–89°/14 mm, by lithium tetrahydridoaluminate. Arsenic trichloride converts the diphenyl compound into the 1,4-*dichloro*-1,4-*diarsenane*, b.p. 112–115°/0.14 mm (*K. Sommer*, Z. anorg. allgem. Chem., 1971, **383**, 136).

2,2,3,3,5,5,6,6-**Octafluoro**-1,4-**diiodo**-1,4-**diphosphorinane** (LIII), m.p. 73–75.5°, is formed when tetrafluoroethylene is heated with iodine and red phosphorus at 220°: n.m.r. spectroscopy has shown that the ring has a puckered conformation (*C. G. Krespan* and *C. M. Langkammerer*, J. org. Chem., 1962, **27**, 3584).

Quaternization reactions provide a route to 1,4-diphosphabicyclo[2.2.2]octane (LIV) from 1,4-dibenzyl-1,4-diphosphorinane; the *ethanodiphosphonium* salt, m.p. 340–344°, is reduced with lithium tetrahydridoaluminate in tetrahydrofuran to give the *diphosphine*, m.p. 252° (under nitrogen), *dimethiodide*, m.p. 375–380°, (*R. C. Hinton* and *Mann*, J. chem., Soc., 1959, 2835):

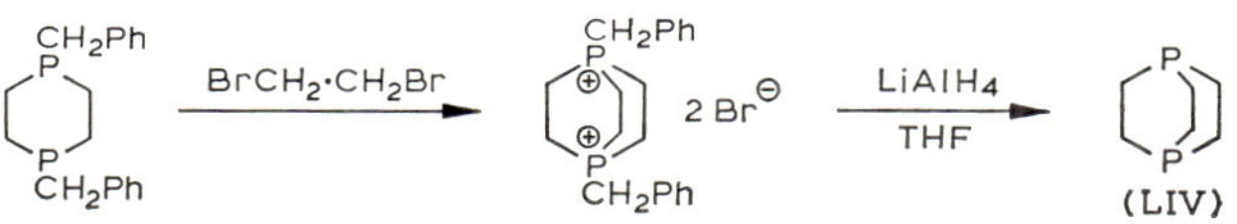

(*b*) *Compounds containing four hetero atoms*

(*i*) 1,2,4,5-*Tetra-arsenanes*

Brief descriptions have appeared of 1,2,4,5-**tetramethyl**-1,2,4,5-**tetra-arsenane,** b.p. 114–116°/0.2 mm, *tetraphenyl* compound, m.p. 156–158°. These have been obtained by the routes indicated (*Sommer*, Z. anorg. allgem. Chem., 1970, 377, 278; 1970, 377, 120).

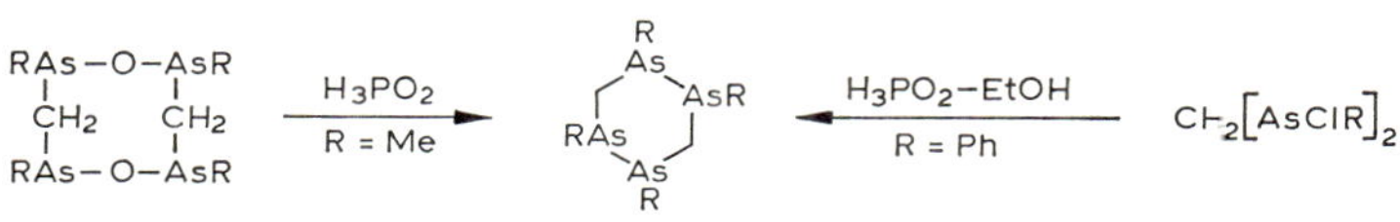

The analogous **tetraphenyl**-1,2,4,5-**tetraphosphorinane** is formed in low yield when tetraphenyltetraphospholane is treated with potassium under tetrahydrofuran followed by dichloromethane, and has been characterized by physical methods only (*M. Baudler*, *J. Vesper* and *H. Sandmann*, Z. Naturforsch., 1973, B **28**, 224).

Chapter 49

Compounds with Six-Membered Heterocyclic Rings containing One or More Unusual Hetero Atoms together with Nitrogen or Elements of Group VI

R. S. EDMUNDSON

1. Aluminium and boron compounds

Claims have been made for the successful preparation of several compounds thought to possess ring systems containing aluminium together with other hetero atoms, particularly oxygen and nitrogen, but few physical data are available. On the other hand, many compounds containing oxygen, nitrogen or sulphur as well as boron in a ring have been described in detail, and their extensive chemistry is derived largely from an interest in their actual, or potential, aromatic character; in the use of the borate ester grouping as a protecting entity, particularly in carbohydrate chemistry; and in the hydroboration process, as well as in certain technological areas.

The replacement of two adjacent carbon atoms in an aromatic hydrocarbon ring system by a unit consisting of adjacent boron and nitrogen atoms provides a heterocyclic system which is isoconjugate with the parent hydrocarbon, and the expectations that such heterocyclic compounds should then display a chemical character similar to that of the parent hydrocarbon have been realised to a considerable extent. The results of these studies are widely described, particularly in an extensive series of papers by *M. J. S. Dewar* and his collaborators from about 1958 onwards.

Examples of compounds prepared by Dewar's group are 10,9-borazarophenanthrene (I) and 9,10-borazaronaphthalene (II), these designations being those made by Dewar. These compounds, and others, for example, 4,3-borazaroisoquinoline (III) absorb infrared radiation within a broad region (1500–1350 cm^{-1}) associated with the boron–nitrogen linkage, and this, taken in conjunction with ^{31}B-n.m.r. data, strongly suggest multiple bond character between boron and nitrogen. Some degree of electron delocalisation therefore appears likely in the heterocyclic ring, be it complete as in benzene, or restricted, to some extent, as in phenanthrene. Available

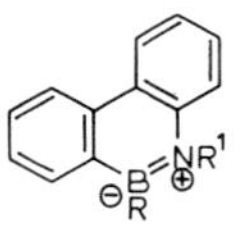
(I)

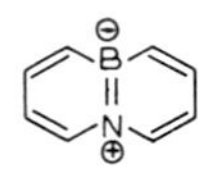
(II)

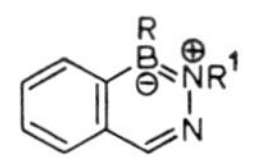
(III)

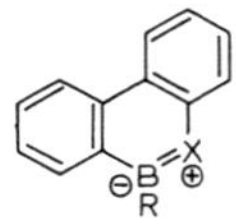
(IV)

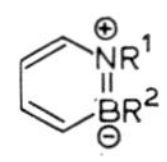
(V)

structural data derived from X-ray and similar studies do indicate planarity or near-planarity for many of the systems discussed.

These arguments apply also to the oxygen and sulphur analogues, 10,9-boroxarophenanthrene (IV, X = O) and 10,9-borathiarophenanthrene (IV, X = S) but because of the greater electronegativity of the Group VI atom, the participation of this in the electron delocalisation process is reduced, and hence aromatic character lessened.

The consequences of conjugation are particularly apparent in the ultraviolet absorption spectra; the electronic absorption curves for the boron-containing compounds resemble very closely those of the isoconjugate hydrocarbon (see *B. G. Ramsey*, "Electronic Transitions in Organometalloids", Academic Press, New York, 1969).

The conjugated boron–nitrogen compounds show remarkably high chemical stability, the majority being resistant to acids, bases and oxidizing agents. They undergo electrophilic substitution reactions, and in some cases, these proceed more easily, and further, than for the parent hydrocarbon. An exception to this rule appears to be V (R^1 = Me, R^2 = H) which, according to *H. Wille* and *J. Garbeau* (Chem. Ber., 1974, **107,** 110), is totally unlike benzene but is more like a conjugated butadiene, being highly reactive and readily polymerizing. This contrasts with V (R^1 = H, R^2 = Ph) which is relatively inert to oxidation and solvolysis (*D. G. White*, J. Amer. chem. Soc., 1963, **86,** 3634).

Further evidence for aromatic character lies in the acidity of compounds possessing the B–OH group as demonstrated by ^{31}B-n.m.r. spectroscopy; hydroxyborazarophenanthrene shows a downfield displacement of the boron resonance by 3–7 p.p.m. together with a broadening of signal characteristic of a protic acid, cf. phenolic, behaviour. Such behaviour is quite different from that shown by many other boron-containing compounds, both cyclic and acyclic, which, in neutral or alkaline-ethanolic solution, display an upfield displacement sometimes greater than 13 p.p.m. with no broadening of signal (*Dewar* and *R. Jones*, J. Amer. chem. Soc., 1967, **89,**

2408). More recently, however, *B*-hydroxyboroxarophenanthrene has been shown to display a different behaviour, consistent with it being a Lewis acid rather than a protic acid; this behaviour is possibly the result of ring strain which is reduced on coordination to give a tetrahedral arrangement around the boron atom (*F. A. Davis* and *Dewar*, J. org. Chem., 1968, **33**, 3324).

The aromatic character of the heterocyclic ring may be destroyed (*a*) by addition of, for example, water across the boron–nitrogen bond:

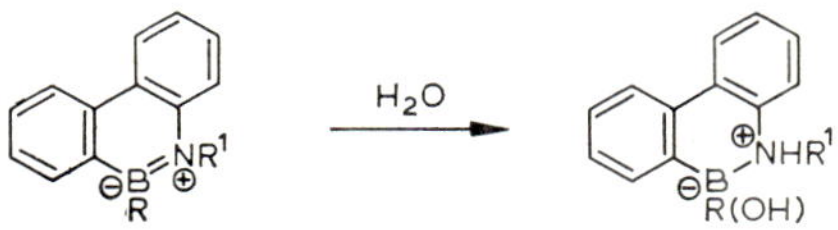

(I)

(*b*) by reactions leading to a tetrahedral boron atom. Thus, *B*-hydroxy-boroxaroisoquinoline (VI) may be said to be aromatic by virtue of the canonical form VI B. Esterification of the compound with 2-aminoethanol yields VII, in which the planarity of the heterocyclic ring has been destroyed by the resultant tetrahedral geometry at boron, and the ester VII therefore is non-aromatic.

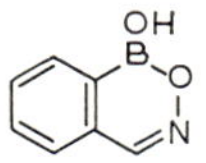

(A)

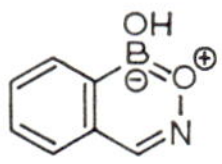

(B)

(VI)

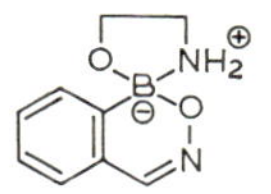

(VII)

(*c*) by reduction of the carbon–carbon double bond conjugated with the boron–nitrogen linkage:

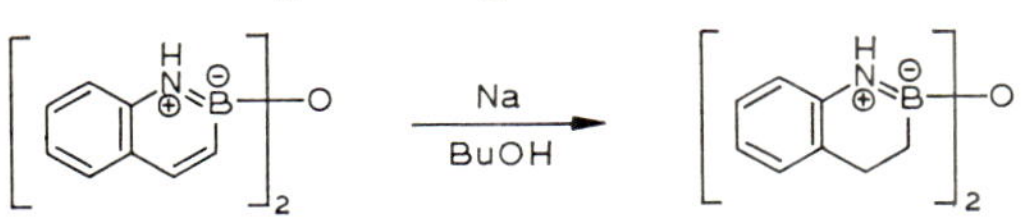

(VIII) (IX)

The anhydride (the term ether is used by *Dewar*) VIII is unreactive towards acetic anhydride; reduction with sodium and butanol leads to the "dihydro" compound IX which is acetylatable under the same conditions;

(*d*) by alteration in electron charge density at nitrogen or boron. *B*-Methyl- and *B*-phenyl-borazarophenanthrene (I, R^1 = H, R = Me, Ph) are unaffected by air or by alkali, and are relatively stable to acids. By contrast, the electron-withdrawing ability of the ethoxycarbonyl group in the compounds (I, R^1 = EtOCO, R = Me, Ph) is sufficient to render the heterocyclic ring labile to atmospheric oxidation, and with alkali in ethanol to loss of ethoxycarbonyl group and formation of (I, R^1 = H, R = Me, Ph).

A second, different cause of stability is exemplified by the borazaro-pyrimidines (boratriazines) (X). These possess the typical B=N grouping and are capable of addition of water to give XI (R = Ph, R^1 = OH) which then undergoes reversible ring opening under aqueous conditions. The stability of the system XI (R = R^1 = Ph) to hydrolysis cannot be ascribed

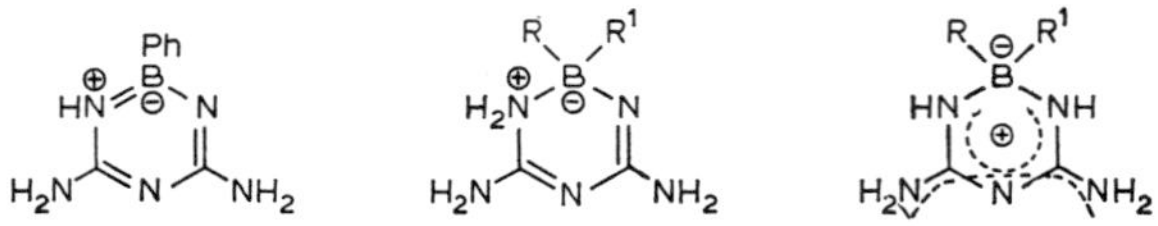

(X) (XI) (XII)

to complete electron delocalization in the ring since this is affectively blocked by the tetrahedral boron atom, and the stability is probably the result of the partial delocalization as formulated in XII (*J. E. Milks*, *G. W. Kennerly* and *J. H. Polevy*, J. Amer. chem. Soc., 1962, **84,** 2529).

A similar situation exists for the oxaoxoniaborata salts (XIII) formed from *β*-dicarbonyl compounds, and the compounds XIV obtained from *β*-aminoketones. In the former case, replacement of a ring hydrogen atom by bromine takes place by an addition–elimination mechanism, and the relative reactivity of the compound to nucleophilic reagents but stability to acidic and other conditions is consistent with electron delocalization which must be restricted to the -ene chain:

(XIII) (XIV)

(*a*) *Oxa- and thia-borins*

No derivatives of the 1,3-oxaborin and 1,3-thiaborin systems have been recorded.

(*i*) *Benz- and dibenz-*[1,2]*oxaborins; dibenzo*[1,2]*thiaborin*

The isomerization of tolan-2,2′-diboronic acid in alkaline solution, discovered by *R. L. Letsinger* and *J. R. Nazy* (J. Amer. chem. Soc., 1959, **81,** 3013), has been commented upon by *Dewar* and *R. Dietz* (J. chem. Soc., 1960, 1344) who suggested that the final product is 2-(1-**hydroxy**-2*H*-2,1-**benz**[*c*]**oxaborin-3-yl)benzeneboronic acid** (XV):

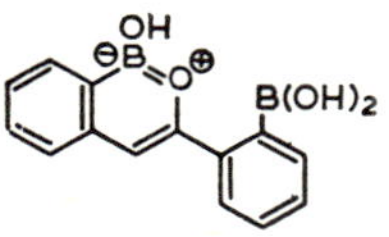

(XV)

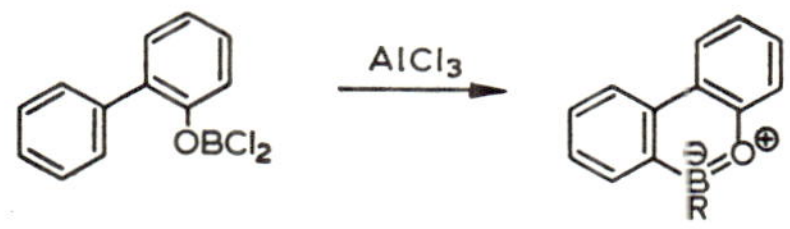

(XVI)

2-Hydroxybiphenyl reacts with an excess of boron trichloride under nitrogen at −70° to give biphenylyl dichloroboronite; when a solution of this in benzene or petroleum at 50° is treated with aluminium chloride, cyclization to the rather unstable chloro compound XVI (R = Cl) occurs; this, after hydrolysis, yields **6-hydroxy-6*H*-dibenz**[*c*,*e*][1,2]**oxaborin** (10-**hydroxy**-10,9-**boroxarophenanthrene**) (XVI, R = OH), m.p. 205–206.5° (*idem, loc. cit.; R. F. Bridger, L. J. McCabe* and *A. L. Williams*, U.S.P. 3,320,167/1967; C.A., 1967, **67,** 64512).

The hydroxy compound, which undergoes reversible dihydration to 6,6′-oxybis(6*H*-dibenz[*c*,*e*][1,2]oxaborin) when heated, is stable to boiling 6*N* hydrochloric acid and to boiling 10% aqueous potassium hydroxide, although the compound loses boron when warmed with concentrated sulphuric acid. Fusion of the hydroxy compound with potassium hydroxide yields 2-hydroxybiphenyl; degradation to 2,2′-dihydroxybiphenyl occurs upon treatment with hydrogen peroxide. The hydroxy and chloro compounds both yield the 6-*methoxy* compound, m.p. 63.5–64.5°, when treated with methanol. Reaction between the hydroxy compound and bromine in acetic acid furnishes initially its 2,4-*dibromo* derivative, m.p. 274–277°, followed by the 2,4,8-*tribromo*-6-*hydroxy* compound, m.p. 310° (the corresponding *tribromo*-6-*ethoxide* has m.p. 141–142°), the structures of which are confirmed by alternative synthesis from the corresponding brominated hydroxybiphenyl.

Another interesting synthesis of the 6-hydroxy compound (XVI, R = OH) involves reaction between the analogous azaborine and nitrous acid in acetic acid (*Dewar* and *P. Maitlis*, Chem. and Ind., 1960, 1620).

The formation of **6-phenyl-6*H*-dibenz**[*c*,*e*][1,2]**oxaborin** (XVI, R = Ph), m.p. 80–82°, by photolysis of diphenyl 2-iodophenylborinite, exemplifies a potentially important synthesis of the system because of the relative ease of formation of the required intermediates (*P. J. Grisdale* and *J. L. R. Williams*, J. org. Chem., 1969, **34,** 1675).

Derivatives of **benzo**[1,2-*c*:4,5-*c*′]**bis**[2,1]**benzoxaborin** (XVII, R = H) have been recorded (*R. D. Offenhauer* and *P. G. Rodewald*, J. Polymer Sci., B, 1968, **6,** 572) and the

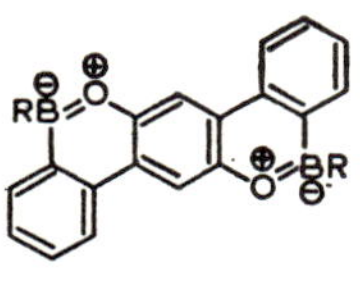

(XVII)

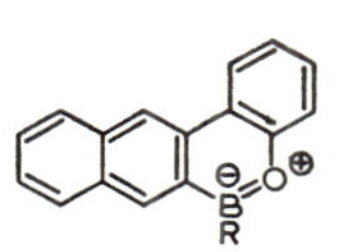

(XVIII)

yellow 6-**hydroxy**-6*H*-**benzo**[*e*]**naphth**[2,3-*c*][1,2]**oxaborin** (XVIII, R = OH), m.p. 279.5–281°, has been obtained from the corresponding *B*-methoxy-azaborine by the action of nitrous acid (*Dewar* and *W. H. Poesche*, J. Amer. chem. Soc., 1963, **85,** 2253).

Hydrolysis, with wet ether, of the chloro compound XIX (R = Cl) prepared from biphenyl-2-thiol, boron trichloride and aluminium chloride as described for the analogous

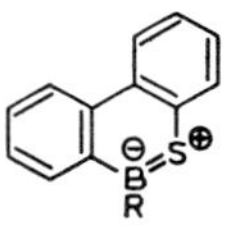

(XIX)

dibenzoxaborin, yields 6-*hydroxy*-6H-*dibenzo*[c,e][1,2]*thiaborin* (XIX, R = OH) isolated as the *anhydride*, m.p. 170–171° (*Davis* and *Dewar*, *ibid.*, 1968, **90,** 3511). The 6-*phenyl* compound XIX (R = Ph), m.p. 129–130°, is formed photosynthetically as for, and in higher yields than, the dibenzoxaborin, although only when 2,6-lutidine is present in the reaction mixture (*Grisdale* and *Williams*, *loc. cit.*).

(*ii*) *3,4-Dihydro-1,2-benzoxaborins*

1-**Hydroxy**-3,4-**dihydro**-1*H*-2,1-**benzoxaborin** (XXI, R = OH), m.p. 60–61°, is formed from the borolactam XX by the route illustrated (*J. C. Catlin* and *H. R. Snyder*, J. org. Chem., 1969, **34,** 1660):

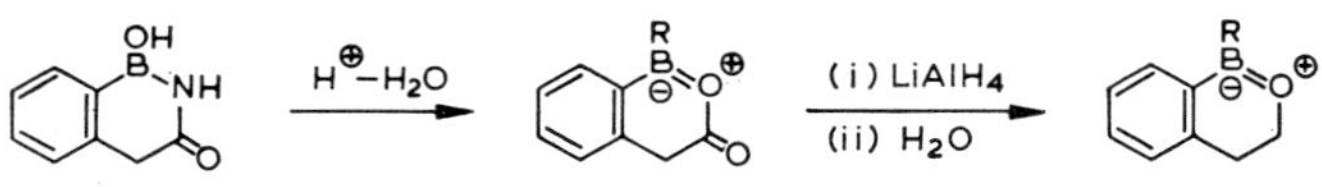

(XX) (XXI)

(*iii*) *1,2-Oxaluminanes; 1,2-oxaborinanes*

The product from the reaction between triisobutylaluminium and but-3-en-1-ol is reported to be distillable, and has been formulated as 2-**isobutyl**-1,2-**oxaluminane** (XXII) (*L. I. Zakharkin* and *L. A. Savina*, Izvest. Akad. Nauk S.S.S.R., Ser. Khim., 1962, 824):

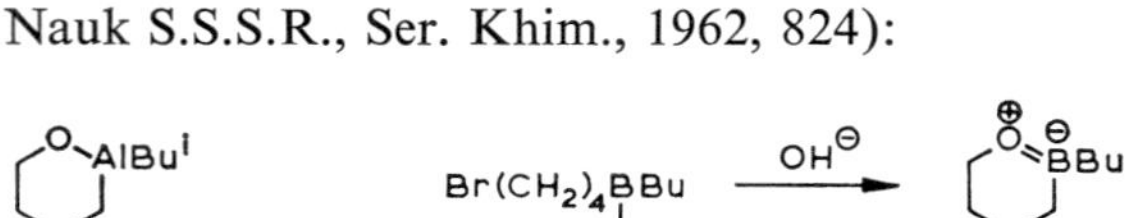

(XXII) (XXIII)

Cyclization of methyl 4-bromobutaneboronates with weak alkali leads to 2-**butyl**-1,2-**oxaborinane** (XIII), b.p. 50–51°/10 mm, n_D^{20} 1.4338; the ring is cleaved by methanolic potassium hydroxide, and also by phosphorus pentachloride yielding butyl (ω-chlorobutyl)boron chloride, $Cl(CH_2)_4$-BBuCl (*B. M. Mikhailov*, *L. S. Vasil'ev* and *V. P. Dmitrikov*, *ibid.*, 1970, 198). 2-**Hexyl**-3,6-**dimethyl**-1,2-**oxaborinane** is obtained in very low yield, together with 2-hexyl-7-methyl-1,2-oxaborepane, in reactions between allylacetone and dihexyldiborane or trihexylborane (*Vasil'ev et al.*, Zhur. obshcheĭ Khim., 1972, **42,** 1540).

An interesting but relatively unexplored reaction sequence for the synthesis of 1,2-oxaborinanes commences with the treatment of 2-lithiofuran with a trialkylborane. Hydroboration of the resultant product and subsequent acidification yields 2,3,3-trisubstituted-1,2-oxaborinanes (XXIV), as evidenced by the nature of the products of their oxidation by alkaline hydrogen peroxide (*A. Suzuki*, *N. Miyaura* and *M. Itch*, Tetrahedron, 1971, **27,** 2775):

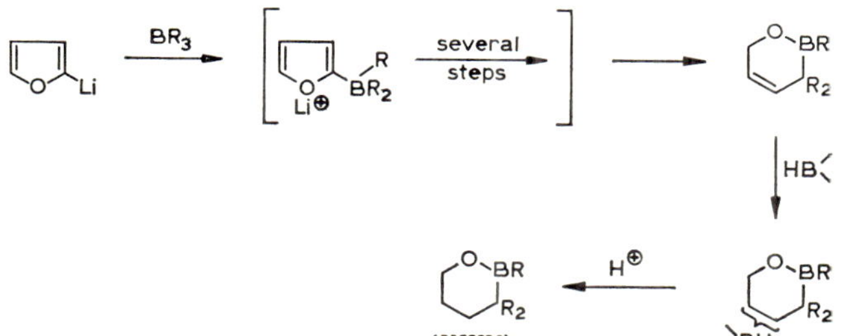

(*iv*) *Dibenz[1,4]oxaborins*

10-**Hydroxy**-10*H*-**dibenz**[*b,e*][1,4]**oxaborin** (10-**hydroxyphenoxaborin**) (XXV, R = OH), m.p. 285°, is obtained when 2,2′-dilithiodiphenyl ether is treated with butyl metaborate (*J. M. Davidson* and *C. M. French*, J. chem. Soc., 1960, 191) and also, in very low yield, when a mixture of diphenyl ether and boron tribromide is photolysed and the resulting bromo

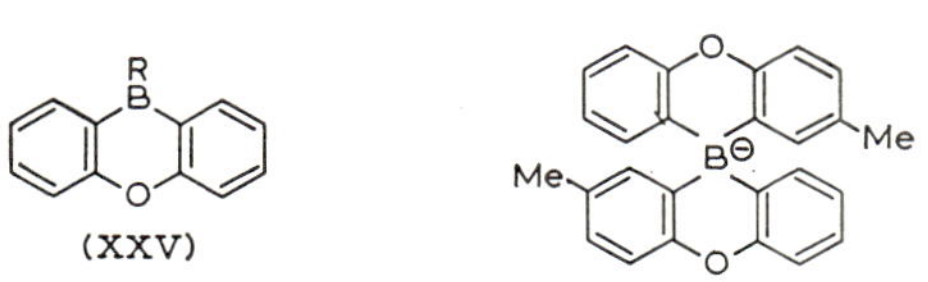

(XXV) (XXVI)

compound XXV (R = Br) hydrolyzed (*R. A. Bowie* and *A. C. Musgrave*, *ibid.*, C, 1970, 485). The *chloride* XXV (R = Cl) b.p. 108°/0.14 mm, and the *anhydride*, m.p. 288–289°, of the acid are amongst other known derivatives of this ring system (*Davidson* and *French loc. cit.*; *H. J. Roth* and *B. Miller*, Arch. Pharm., 1964, **297,** 602, **617,** 744).

The use of Grignard reagents in a similar reaction scheme is exemplified by that from 2,2′-dibromo-4-methyldiphenyl ether, which with boron trifluoride etherate gives the anion XXVI (*K. Torssell*, Acta. Chem. Scand., 1962, **16,** 87).

(*v*) *1,4-Oxa- and 1,4-thia-borinanes*

Hydroboration of divinyl ether with RBH_2,NR_3 gives 1,4-**oxaborinanes** (XXVII, X = O), 4-*tert-butyl-*, b.p. 90°/56 mm (*M. F. Hawthorne*, J. Amer. chem. Soc., 1960, **82,** 748; 1961, **83,** 2541). Derivatives of the **thia analogue**

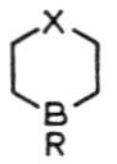
(XXVII)

(XXVII, X = S), *e.g.* 4-*butoxy-*, b.p. 45–46°/0.1 mm, are formed by free radical addition of hydrogen sulphide to divinyl borinates. The sulphur-containing ring system is degraded to ethylene when acted upon by acidified butanol or by potassium hydroxide at 120°.

(*b*) *Dioxaborins*

(*i*) *Benzo-* and *dibenzo-*[1,3,2]*dioxaluminins* and -[1,3,2]*dioxaborins*

Few examples of such ring systems have been reported.

2-**Phenyl-**1*H*,3*H*-**naphtho**[1,8-*d,e*][1,3,2]**dioxaborin** (XXVIII, R = Ph) b.p. 140°/10 mm, m.p. 154°, is formed from phenylboron dichloride and 1,8-dihydroxy-naphthalene (*M. Pailer* and *W. Fenzy*, Monatsh., 1961, **92,** 1294).

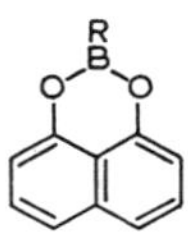
(XXVIII)

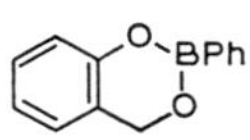
(XXIX)

2-**Phenyl-**4*H*-1,3,2-**benzodioxaborin** (XXIX), b.p. 116–119°/0.25 mm, m.p. 42°, is formed when benzeneboronic anhydride is heated with 2-hydroxybenzyl alcohol; the latter and tris(diethylamino)borane together yield the 2-*diethylamino* compound, b.p. 100–102°/0.8 mm, useful for the preparation of other derivatives (*Bowie* and *Musgrave*, J. chem. Soc., 1963, 3945; *R. H. Cragg* and *M. Nazery*, *ibid.*, Dalton Trans., 1974, 162).

A conventional reaction between dichloroboranes and salicyclic acid leads to **4-oxo-**4*H*-1,3,2-**benzodioxaborins,** 2-*methyl-*, m.p. 150° (*M. Wieber* and *W. Künzel*, Z. anorg. allgem. Chem., 1974, **403,** 107).

(*ii*) *1,3,2-Dioxa- and 1,3,2-dithia-borinanes; 1,3,2-oxaoxoniaborata salts*

Saturated, cyclic, organic derivatives of orthoboric acid, (XXX), the 1,3,2-dioxaborinanes, are formed (*a*) when mixtures of propane-1,3-diols and trialkyl borates are heated; the latter may be replaced by 1,3,2-dioxaborolanes in an equilibrium which lies in favour of free 1,2-diol and 1,3,2-

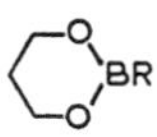
(XXX)

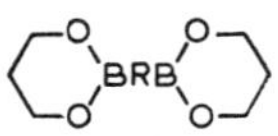
(XXXI)

dioxaborinane, and the process therefore provides a useful way of removing borate-type protecting groups from carbohydrates; it is interesting to note that competitive esterification of a mixture of ethylene and trimethylene glycols with boric acid yields largely 2-(**2-hydroxyethyl**)-1,3,2-**dioxaborinane** (XXX, R = $OCH_2 \cdot CH_2OH$) (*A. J. Hubert*, *B. Hargitay* and *J. Dale*, J. chem. Soc., 1961, 931): (*b*) when a 1,3-diol is heated with boric acid in the presence of a monohydric alcohol. The free acids (XXX, R = OH) can be obtained from the 1,3-diol (other than trimethylene glycol itself) and boric acid or trimeric boroxide with azeotropic removal of water; the reaction is catalyzed by traces of sodium alkoxide or toluene-4-sulphonic acid (*C. A. Lane*, U.S.P. 2,994,713/1961; C.A., 1962, **56,** 329; *J. L. Delarue* and *R. L. Fallard*, Belg. P. 640,031/1964; C.A., 1965, **63,** 631). In both reaction systems (*a*) and (*b*), the diborates XXXI, where R is $O(CH_2)_3O$, may also be formed.

The cyclic acids XXX (R = OH) are themselves readily esterified with alcohols (*U.S. Borax* and *Chem. Co.*, B.P. 960,809/1964; C.A., 1964, **61,** 9514) or diols or triols (*Inst. de Serotherapie hemopoietique S.A.*, B.P. 885,766/1960; C.A., 1962, **56,** 15546: *Pailer* and *M. Huemer*, Monatsh., 1964, **95,** 373), and they may also be dehydrated to the anhydrides (XXXI, R = O) [oxybis(1,3,2-dioxaborinanes)] (*Standard Oil Co.*, B.P. 917,654/1963; C.A., 1963, **59,** 644). Such acids have an acidity comparable with that of boric acid itself; many may be crystallized from water.

The chemistry of the **phenyl-1,3,2-dioxaborinanes** (XXX, R = Ph) and their butyl analogues has been investigated quite extensively in view of the value of benzene- and butane-boronic acids for protection purposes in carbohydrate chemistry. In the former case, their formation often takes place in aqueous medium or when a benzeneboronic anhydride–diol mixture is shaken, while butaneboronic acid reacts with diols in dimethylformamide even at room temperature. (For a short review of the preparation and analytical examination of boronates, see *C. J. W. Brooks* and *I. Maclean*, J. Chromatographic Sci., 1971, **9,** 18.) According to *J. C. Lockhart* (J. chem. Soc., A, 1968, 869) no great thermodyanamic differences exist between phenylboronates from 1,2- and 1,3-diols with regard to their formation in the liquid phase from the diol and anhydride; the effect of ring substitution on the reaction is as great as that of ring size. Where stereochemistry is fixed, cyclic boronates are formed only from *cis*-1,3-diols and this affords a means of separating *cis*- and *trans*-1,3-diols, the latter leading to other ring systems (*H. C. Brown* and *G. Zweifel*, J. org. Chem., 1962, **27,** 4708; *J. M. Sugihara* and *C. M. Bouman*, J. Amer. chem. Soc., 1958, **80,** 2443; *W. V. Dahlhoff* and *B. Köster*, Ann., 1975, 1625).

2-Ethyl-1,3,2-**dioxaborinanes** are produced from ethaneboronic acid and a 1,3-diol, or by thermolysis of esters of the type $Et_2BO(CH_2)_3OBEt_2$ at 190–250° (*idem, ibid.*, 1975, 1914).

Other, 2-tertiary-alkyl-1,3,2-dioxaborinanes are obtained by allowing 1,3-diols to react with triorganoboranes in the presence of carbon monoxide at 650–950 atmospheres pressure and 150° (*M. E. D. Hillman*, J. Amer. chem. Soc., 1963, **85,** 982; U.S.P. 3,206,446/1965; C.A. 1965, **63,** 16381) and from triorganoboranes and esters [XXXI, R = $O(CH_2)_3O$] at 120° in the presence of traces of borane in tetrahydrofuran (*Brown* and *S. K. Gupta*, J. Amer. chem. Soc., 1970, **92,** 6983).

1,3,2-**Dioxaborinane** (XXX, R = H) is unstable, liberating diborane on being heated. The colourless liquid 4,4,6-trimethyl-1,3,2-dioxaborinane is more stable, remaining unchanged at ambient temperature and decomposing only slightly at elevated temperatures; it is obtained in 85% yield when 2,2′-oxybis(4,4,6-trimethyl-1,3,2-dioxaborinane) is treated with lithium tetrahydridoaluminate in "tetraglyme" (*R. H. Fish* and *H. C. Newsom*, Fr. P. 1,536,699/1968; C.A., 1969, **71,** 91544) and in 50% yield when 2-chloro-4,4,6-trimethyl-1,3,2-dioxaborinane is reduced with sodium tetrahydridoborate in "diglyme" (*W. G. Woods* and *P. L. Strong*, J. Amer. chem. Soc., 1966, **88,** 4667), and undergoes *cis* addition to alkenes, dialkenes and acetylenes at 100–120° during 24–48 h (*Fish, ibid.*, 1968, **90,** 4435; J. org. Chem., 1973, **38,** 158).

2-Chloro-1,3,2-dioxaborinane (XXX, R = Cl) is obtained when the triglycol diborate (XXXI, R = $O(CH_2)_3O$) is treated with boron trichloride, the 5,5-dimethyl and 4,4,6-trimethyl derivatives being obtained in a similar fashion (*G. H. Birum* and *J. L. Dever*, U.S.P., 3,064,032/1962; C.A., 1963, **59,** 2840), and from boron trichloride and 2-(3-chloropropoxy)-1,3,2-dioxaborinane, or when trimethylene glycol is treated with boron trichloride in dichloromethane at —80° when some triglycol diborate may be formed simultaneously (*A. Finch, J. C. Lockhart* and *J. Pearn*, J. org. Chem., 1961, **26,** 3250). In dilute solutions the chloride is monomeric, unlike the 2-chloro-1,3,2-dioxaborolane (*Finch* and *P. J. Gardner*, J. inorg. nucl. Chem., 1963, **25,** 927), and undergoes conventional displacements with amines, $NCS^{\ominus}$, and organometallic reagents (*Finch et al.*, J. chem. Soc., 1962, 1428; *W. G. Woods* and *P. L. Strong*, J. organometal. Chem., 1967, **7,** 371; *Fish*, J. org. Chem., 1969, **34,** 1127; *W. G. Woods, I. S. Bengelsdorf* and *D. L. Hunter*, *ibid.*, 1966, **31,** 2766; *Woods* and *I. S. Bengelsdorf, ibid.*, 1966, **31,** 2769). It reacts normally with lower alcohols, but the simple esters then undergo further reaction to give the triglycol diborate; the 2-chloroethoxy compound does not undergo this further reaction, possibly because of internal coordination.

Physical data for a selection of 1,3,2-dioxaborinanes are listed in Table 1.

Tetrakis(1,3,2-**dioxaborin)yl-methane** (XXXII), m.p. 217–218°, formed by reaction between 4 mol. of propane-1,3-diol and terakis-(dimethoxyboryl)methane, is superior to the latter for the preparation of tetrametallomethanes (*D. S. Matteson* and *R. J. Wilasek*, J. organmetal. Chem., 1973, **57,** 231). With butyllithium it forms an anion, both pyrophoric and sensitive to moisture. The anion reacts with aldehydes, and further reaction with an alkyllithium, alkaline iodine, and alkaline peroxide, successively, yields ketones (*Matteson* and *L. A. Hagelee*, *ibid.*, 1975, **93,** 21):

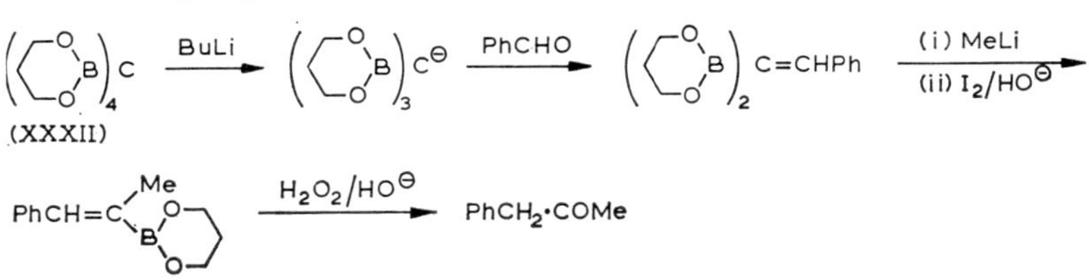

(XXXII)

TABLE 1

1,3,2-DIOXABORINANES (XXX)

R	*B.p., °C/mm, (m.p., °C) with the ring substituents*			
	None	*4-Me*	*5,5-diMe*	*4,4,6-triMe*
H				52–52.5/46 (1) 55–56/61 (2)
Cl	30/0.15 (3)	42–44/0.8 (3)	50/1.5 (3)	39–40/0.1 (3)
CH	147/151/3 (4)	(32–33) (5)	137–142/1 (6)	(78) (7)
Anhydride	115–117/1 (8)	181–184/17 (9)	134–140/0.05 (10)	120/1.5 (7)
OEt				70/9 (7)
OPh				102/0.8 (7)
Et	135/760 (14)	127–135/760 (14)	158–159/760 (14)	57.5–58.5/22–21 (11)
Ph	80–82/0.2 (12) (26–28)	73–75/0.5 (12) (19.5–20.5)	65–65.5 (12)	74–76/0.2 (12) (20–21)
NMe_2	24/0.1 (13)			
Triglycol diborate	125/0.5 (8)	207–213/17 (9)	(123–124) (6)	150/2 (7)

References

1 *R. H. Fish*, J. Amer. chem. Soc., 1968, **90,** 4435.
2 *W. G. Woods* and *P. L. Strong*, *ibid.*, 1966, **88,** 4667.
3 *G. H. Birum* and *J. L. Dever*, U.S.P. 3,064,032/1962; C.A., 1963, **59,** 2840.
4 *R. E. Rippere* and *V. K. LaMer*, J. phys. Chem., 1943, **47,** 204.
5 *Standard Oil Co.* (Ohio), B.P. 917,653/1963; C.A., 1963, **59,** 644.
6 *R. M. E. Washburn et al.*, Adv. in Chem., Ser 23, Amer. chem. Soc., 1959, 129.
7 *R. C. Mehrotra* and *G. Strivastava*, J. chem. Soc., 1962, 1032.
8 *A. Finch et al.*, J. org. Chem., 1961, **26,** 3250.
9 *H. Steinberg et al.*, unpublished work.
10 *A. C. McCloskey*, unpublished work.
11 *W. G. Woods et al.*, J. org. Chem., 1966, **31,** 2766.
12 *R. A. Bowie* and *O. C. Musgrave*, J. chem. Soc., 1963, 3945.
13 *A. Finch et al.*, *ibid.*, 1962, 1428.
14 *W. V. Duhlhoff* and *R. Köster*, Ann., 1975, 1625.

2-Chloro-1,3,2-**dithiaborinane** (XXXIII, R = Cl), b.p. 52°/0.1 mm, n_D^{25} 1.5862, and the 2-*phenyl* compound (XXXIII, R = Ph), b.p. 110–111°/0.05 mm, n_D^{25} 1.6375, are both monomeric in dilute benzene solutions, and are obtained from the dithiol and the dichloroborane in dichloromethane at −80° (*A. Finch*, and *P. J. Gardner*, Tetrahedron, 1964, **20,** 173; *Wieber* and *Künzel*, Z. anorg. allgem. Chem., 1974, **403,** 107) and in the latter case, from the dithiol and bis(dimethylamino)phenylborane, $PhB(NMe_2)_2$ (*K. Niedenzu et al.*, Chem. Ber., 1967, **100,** 1898). These compounds are hydrolytically stable but thermally unstable. Similarly prepared are 2-*phenyl*-1,3,2-*oxathiaborinane*, b.p. 89–90°/0.5 mm, and the 2-*diethylamino*-, b.p. 52–54°/0.1 mm, and 2-*butoxy*-, b.p. 55–60°/0.01 mm, compounds (*Cragg* and *Nazery*, J. chem. Soc., Dalton Trans., 1974, 1438).

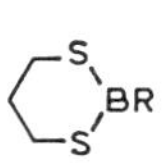

(XXXIII)

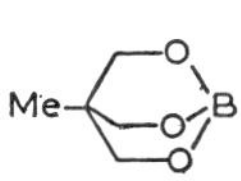

(XXXIV)

7-Ethyl-6,8-dioxa-7-borabicyclo[3.3.1]**nonane** (XXXV), b.p. 47°/0.001 mm, prepared as indicated (*Dahloff* and *Köster, loc. cit.*) may be methanolised to pure *cis*-cyclohexane-1,3-diol:

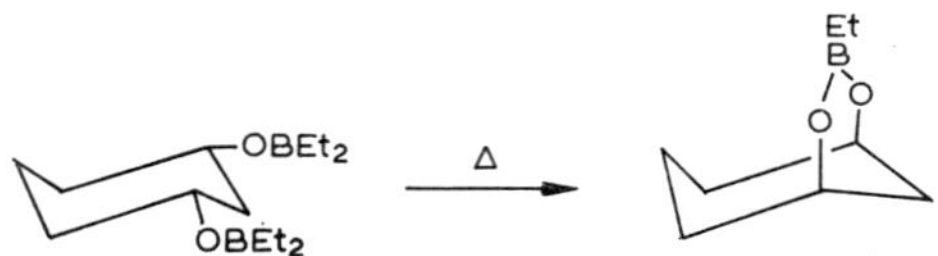

(XXXV)

The sterically improbable 4-*methyl*-2,6,7-*trioxa*-1-*borabicyclo*-[2.2.2.]*octane* (XXXIV) has m.p. 215–217° (*R. J. McManime*, U.S.P. 2,909,560/1959; C.A., 1960, **54,** 2371).

Crystallization of a mixture of triol, boric acid and metal hydroxide yields complexes of type XXVI (*A. A. Terauda, E. Shvarts* and *A. F. Tevin'sh*, Zhur. obscheĭ Khim., 1973, **43,** 2277):

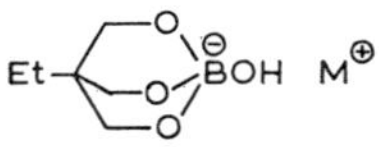

(XXXVI)

It is interesting that, although evidently 2,8,9-trioxa-1-boraadamantane is not reported, benzeneboronic anhydride and pentane-1,3,5-triol react together to give a complex XXXVII which, because of internal complexing, has a molecular geometry very similar to that expected for the adamantane type (*E. J. Bourne, I. R. McKinley* and *H. Weigel*, Carbohydrate Res., 1974, **35,** 141; *Cragg* and *Nazery*, J. organometal. Chem., 1975, **99,** 41).

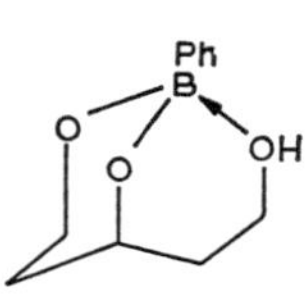

(XXXVII)

In addition to the 1,3,2-dioxaborinanes discussed, a variety of simple boron compounds react with β-dicarbonyl compounds possessing a free enolisable proton to produce internal ionic complexes known as **oxaoxoniaborata salts** (XXXVIII), and also a series of external ionic complexes known as "borinium" salts, or according to the Dewar terminology, boroxaropyrylium salts (XXXIX):

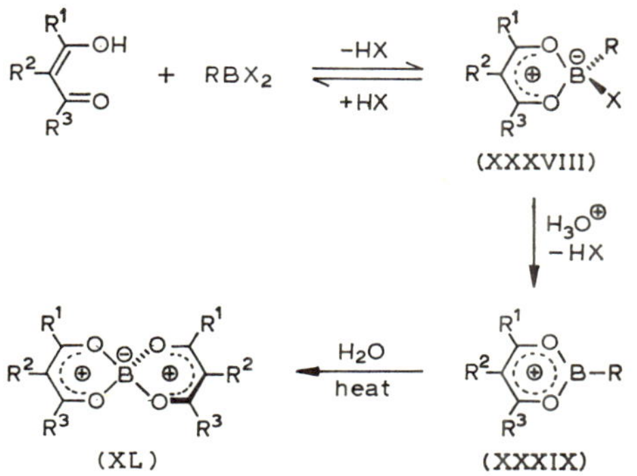

(XXXVIII) (XL) (XXXIX)

The initial condensation to form XXXVIII is possible with trialkylboranes, acyclic borate esters, cyclic borate esters, 2-aminoethylborinic esters, boronic esters and boron hydrides and fluorides. (See *A. Arsene et al.*, Spectrochim. Acta, A, 1967, **23,** 1373, for a survey of the earlier literature.) When X is OH or O-alkyl, the borata salts XXVIII cannot be isolated, but when treated with perchloric acid in dichloromethane they readily afford the crystalline 1,3,2-dioxaborinium perchlorate (XXXIX), some of which explode when heated.

2,2-Dibutyl-4,6-dimethyl-1,3,2-oxaoxoniaboratacyclohexadiene (XXXVIII, R = X = Bu, R^2 = H, R^1 = R^3 = Me), b.p. 86°/0.1 mm, prepared from tributylborane and acetylacetone, reacts readily with bromine, but even at room temperature the product loses hydrogen bromide spontaneously leaving a monobromo compound which possesses no vinylic hydrogen atoms and has equivalent methyl groups (*Hawthorne* and *M. Reintjes*, J. Amer. chem. Soc., 1964, **86,** 5016: *R. Köster* and *G. W. Rotermund*, Ann., 1965, **689,** 40).

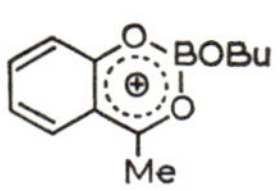

(XLI)

Solid state thermolysis of 2,4,6-triphenyl-1,3,2-boroxaropyrilium perchlorate (XXXIX, R^2 = H, R = R^1 = R^3 = Ph), from dibenzoylmethane and benzeneboronic acid, yields **tetraphenyl-5,7-dioxa-1,11-dioxonia-6-borataspiro[5.5]undecatetraene** (XL, R^2 = H, R^1 = R^3 = Ph), m.p. 334° decomp., described originally in 1906 by *W. Dilthey*; this substance is stable in boiling neutral or acidic media, but with hot alkali decomposes to dibenzoylmethane. 2-Hydroxyacetophenone and tributyl borate give the cation XLI as the yellow, hygroscopic, *perchlorate*, m.p. 223° (*A. T. Balaban*, *E. Barabas* and *Arsene*, Tetrahedron Letters, 1964, 2721).

(*c*) *Dibenz*[2.1.3]*oxadiborins*

Naphthalene-1,8-diboronic acid, obtained by hydrolysis of the diboronic tetrabutyl ester in turn formed by treatment of 1,8-dilithionaphthalene with tributyl borate, is dehydrated when heated to give what is effectively an inner anhydride:

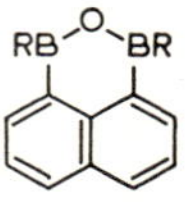

(XLII)

2,6-**Dihydroxy**-1*H*,3*H*-**naphtho**[1,8-*c,d*][1,2,6]**oxadiborin** (XLII, R = OH), m.p. 280–285°, forms an *ammonia adduct*, m.p. 293–296°, yields 1,8-diiodonaphthalene when treated with iodine and alkali, and with mercuric chloride forms 1,8-di(chloromercuri)naphthalene (*R. L. Letsinger et al.*, J. org. Chem., 1965, **30,** 807).

(*d*) *Dioxadiborins*

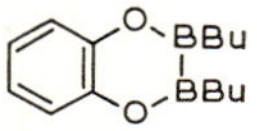

(XLIII)

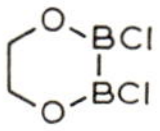

(XLIV)

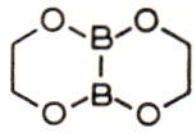

(XLV)

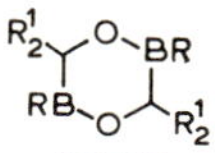

(XLVI)

(*i*) *1,4,2,3-Dioxadiborins*

2,3-**Dibutyl**-1,4,2,3-**benzodioxadiborin** (XLIII) is formed, as the bis-dimethylamine adduct, when catechol is heated with 2,3-dibutyl-1,1,4,4-tetramethyltetrazene, $(Me_2N \cdot NBu)_2$; it yields 2-butylbenzoboroxole when heated in benzene (*H. Nöth* and *P. Fritz*, Z. anorg. allgem. Chem., 1963, **324,** 129).

Diboron tetrachloride reacts with ethylene glycol to form 2,3-**dichloro**-1,4,2,3-**dioxadiborinane** (XLIV) m.p. 175–178° decomp., monomeric in chloroform. Its further reaction with the same glycol yields [1,4,2,3]**dioxadiborino**[2,3,*b*][1,4,2,3]**dioxadiborin** (XLV), m.p. 168–170°, *tetrathia*- analogue, m.p. 170–175° decomp. Unlike propane-1,3-dithiol, which reacts with diboron tetrachloride to give only polymeric material, ethanedithiol yields the monomeric *dithia analogue* of XLIV, m.p. 175–180° (*C. N. Welch* and *S. G. Shore*, Inorg. Chem., 1968, **7,** 225).

(*ii*) *1,4,2,5-Dioxaborinanes*

2,5-**Dihydroxy**-1,4,2,5-**dioxadiborinane** (XLVI, R = OH, R^1 = H), m.p. 147–148°, is also an acid anhydride, *viz.* that of hydroxymethylboronic acid (*D. S. Matteson* and *T. Chun Chang*, J. org. chem., 1966, **31,** 726). Other examples of this ring system (XLVI, R = R^1) *hexaethyl-*, b.p. 114°/10 mm, *hexabutyl-dioxadiborinane*, b.p. 154°/0.7 mm, have been obtained in high yield from trialkylboranes, carbon monoxide and water at ~50° and ~700 atmospheres (*M. E. D. Hillman*, J. Amer. chem., Soc., 1962, **84,** 4715).

(e) *Trioxadiborinanes*

Benzeneboronic acid and its anhydride condense with carbonyl compounds containing electron-withdrawing groups *e.g.* perfluorocyclobutanone and chloral hydrate; the latter yields 2,4-**diphenyl**-6-**trichloromethyl**-1,3,5,2,4-**trioxadiborinane** (XLVII), m.p. 116–117° (*Letsinger* and *S. B. Hamilton*, J. org. Chem., 1960, **25**, 592; *G. W. Parshall*, Inorg. Chem., 1965, **4**, 52).

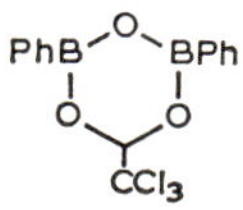

(XLVII)

(f) *Azalumines and azaborines*

Derivatives and benzologues of 1,2-azaborine and 1,4-azaborine are well characterized; those of 1,3-azaborine are not known.

(*i*) *1,2-Dihydro-1,2-azaborines; perhydro-1,2-azaborines; 1,2-dihydrobenz*[*c*]-*azaborines; 1,2-azaborino*[1,2-a][1,2]*azaborine;* 1,2,3,4-*tetrahydro*-2,1-*benz*[c]*azaborine*

The first recorded synthesis of a monocyclic 1,2-dihydro-1,2-azaborine is evidently that of 6-**methoxycarbonylethyl**-2,3-**diphenyl**-1,2-**dihydro**-1,2-**azaborine** (XLVIII), m.p. 110–110.5° (the corresponding *carboxylic acid* has m.p. 205°). The indicated manner of preparation was successful only when an aged sample of Raney nickel was employed (*M. J. S. Dewar* and *P. A. Marr*, J. Amer. chem. Soc., 1962, **84**, 3782):

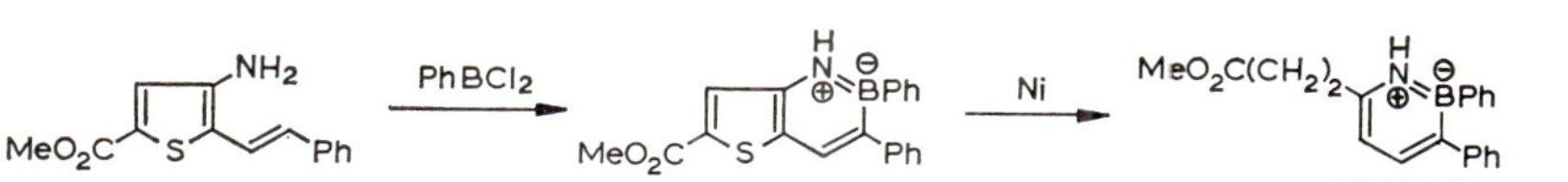

(XLVIII)

Other 1,2-dihydro-1,2-azaborines, *e.g.* the 2-*phenyl* compound, (L, R^1 = H, R = Ph), m.p. 117–118.5°, have been obtained by dehydrogenation of the corresponding perhydro compounds XLIX in the presence of palladized charcoal:

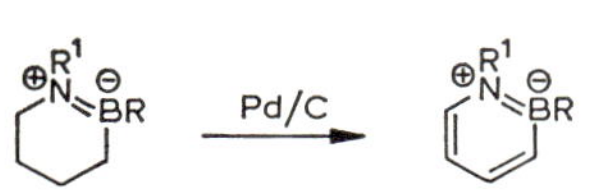

(XLIX) (L)

Two methods are used for the synthesis of **perhydro**-1,2-**azaborines** (see Table 2): they are (*a*) the heating of aminoborinic esters $Br(CH_2)_4BR(NHR^1)$

TABLE 2

PERHYDRO-1,2-AZABORINANES (XLIX)

R	*R*[1]	*B.p., °C/mm (m.p., °C)*
H	Me	106.5
Ph	H	64–68/0.4
Ph	$PhCH_2$	136/0.4
OMe	$PhCH_2$	99/0.5
OH	$PhCH_2$	(107.5–109)

with primary amines at 100–150°, a method recently extended to include other similar esters $X(CH_2)_4BR(OBu)$ where X is Br, NMe_2, and R is Oalk and Salkyl (*B. M. Mikhailov, L. S. Vasil'ev* and *V. P. Dmitrikov*, Izvest. Akad. Nauk S.S.S.R., Ser. Khim., 1968, 2164; *Vasil'ev et al.*, Zhur. obshcheĭ Khim., 1972, **42,** 1318), and (*b*) the hydroboration of 4-aminobutenes with trimethylamine-borane complex in "diglyme" at 120°, although considerable further reaction may occur with the formation of dodecahydrotris[1,2]azaborino[1,2,*a*:1′,2′,*c*:1″,2″,*e*]triazatriborine (*D. G. White*, J. Amer. chem. Soc., 1963, **85,** 3634; *K. M. Davies, Dewar* and *P. Rona*, *ibid.*, 1967, **89,** 6294; *H. Wille* and *J. Goubeau*, Chem. Ber., 1972, **105,** 2156; 1974, **107,** 110).

Perhydro-1,2-azaborines (XLIX, R = H) are readily oxidized in hexane solution, and they react with methanol, liberating hydrogen and yielding the 2-methoxy compounds (*Wille* and *Goubeau, loc. cit.*).

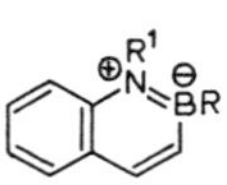

(LI)

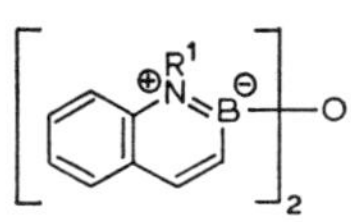

(LII)

A low yield of 2-*chloro*-1,2-*dihydro*-1,2-*benz*[c]*azaborine* (LI, R[1] = H, R = Cl), m.p. 72–74°, results when 2-aminostyrene and boron trichloride are heated with aluminium chloride in benzene (*Dewar* and *R. Dietz*, J. chem. Soc., 1959, 2728). Reduction with lithium tetrahydridoaluminate of the chloride or simple ethers (LI, R[1] = H, R = Oalkyl), or of 2,2′-oxybis(1,2-dihydro-1,2-benz[*c*]azaborine) (LII) gives **1,2-dihydro-1,2-benz[*c*]-azaborine** (LI, R = R[1] = H) m.p. 100–101° (*Dewar et al.*, J. Amer. chem. Soc., 1961, **83,** 1754).

Compound LII, (R[1] = H), m.p. 198–200°, also described as **bis(1,2-dihydro-1,2-benz[*c*]azaborin-2-yl) ether,** is prepared by hydrolysis of the aforementioned chloro compound: if crystallized from lower alcohols, the 2-alkoxy compounds LI (R[1] = H), 2-*methoxy*-, m.p. 57–58°, 2-*ethoxy*-, m.p. 54–55°, b.p. 168–170°/0.5 mm, result. Treatment of the same "ether" with a large amount of sodium and butanol causes partial reduction of the heterocyclic ring giving 2,2′-*oxybis*(1,2,3,4-*tetrahydro*-1,2-*benz*[c]*azaborine*), m.p.

87.5–89°, which, unlike LII, forms a *diacetate*, m.p. 280–282°, readily hydrolyzable to 2-(2-acetamidophenyl)ethaneboronic acid (*Dewar* and *Dietz*, Tetrahedron 1961, **15**, 26).

2-Chloro-1,2-dihydro-1,2-benz[*c*]azaborine reacts with appropriate Grignard reagents to give the 2-*methyl*-, m.p. 73–74°, and 2-*phenyl*-, m.p. 137.5–139°, compounds, the latter being formed in better yield from phenylboron dichloride and 2-aminostyrene (*idem, loc. cit.*); similar displacements take place with the 2-alkoxy compounds. 2-Methyl-1,2-dihydro-1,2-benz[*c*]azaborine takes up chlorine and bromine readily from acetic acid or carbon tetrachloride solutions to give, *via* π-complexes involving the 3,4-bond, the 3-*chloro*-, m.p. 119–120.5°, and the 3-*bromo*-, m.p. 128–129.5°, derivatives, the former also being obtainable from 2-amino-ω-chlorostyrene, or from 2,2-oxybis(3-*chloro*-1,2-*dihydro*-1,2-*benz*[c]*azaborine*), m.p. 236–239°, by reaction with methylmagnesium iodide. Extensive deboronation takes place during the course of the halogenation reactions, yielding 2-amino-ω-halogenostyrenes (*idem*, J. org. Chem., 1961, **26**, 3253).

The dipolar addition of acetylenes to borimines presents another route, not yet thoroughly investigated, for the preparation of derivatives of 1,2-dihydro-1,2-benz[*c*]azaborine. When 6-methoxy-2-pentafluorophenyl-4-*phenyl*-1,2-*dihydro*-1,2-*benz*[c]*azaborine*, (LIII) m.p. 161–162°, prepared as indicated, is treated with potassium hydroxide, the pentafluorophenyl group is lost; conversion of the resultant *B*-hydroxy compound to the 2-methoxy "ether" and subsequent dissolution in acetic acid yields 2-acetamido-5-methoxy-α-phenylstyrene (*P. T. Paetzold et al.*, Chem. Ber., 1968, **101**, 2881):

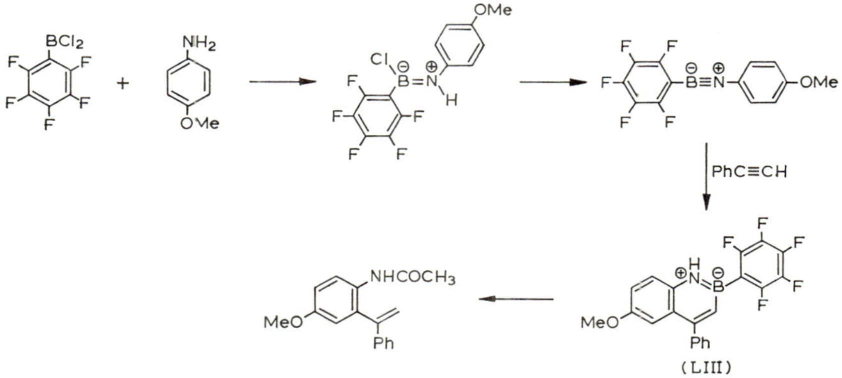

(LIII)

Hydroboration of di(3-butenyl)amine with trimethylamineborane in hot toluene gives **octahydro-1,2-azaborino[1,2,c][1,2]azaborine** (LIV) (named *octahydro*-10,9-*borazaronaphthalene* or 9,10-*azaboradecalin* by *Dewar*), b.p. 73–75°/20 mm, n_D^{24} 1.4837, readily oxidizable in air (*Dewar, G. Y. Gleicher* and *B. P. Robinson*, J. Amer. chem. Soc., 1964, **86**, 5698); addition of water yields the 10-*hydroxydecahydro* compound LV, m.p. 119–121°.

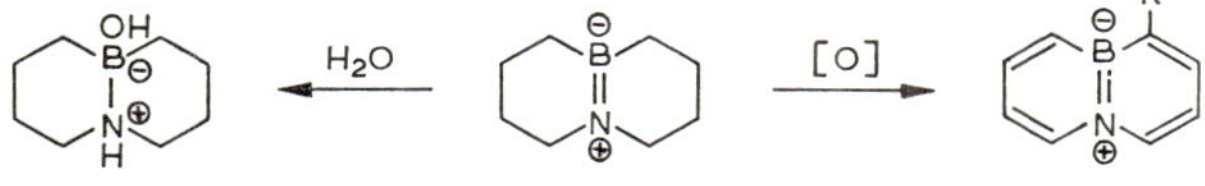

(LV) (LIV) (LVI)

Dehydrogenation of the octahydro compound with palladized charcoal at 300° in the presence of hexene or octene as hydrogen acceptor, gives 25–30% of 1,2-**azaborino**[1,2,*a*][1,2]**azaborine** (LVI, R = H) ("10,9-**borazaronaphthalene**"), m.p. 48–49°, a colourless solid possessing a naphthalenic odour and appearance, resistant to hydrolysis and to

(A) (B)

(LVII)

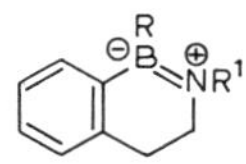

(LVIII)

aerial oxidation even in the presence of acid or alkali. It is isomorphous with naphthalene and forms a π-complex with tetracyanoethylene. It undergoes deuterium exchange with D_2SO_4 in dioxane at the 1 and 8 positions; with bromine in acetic acid, the *bromo* derivative (LVI, R = Br), m.p. 62.5–64°, is formed (*Dewar* and *R. Jones*, J. Amer. chem. Soc., 1968, **90,** 2137).

1-**Hydroxy**-1,2,3,4-**tetrahydro**-2,1-**benz**[*c*]**azaborin**-3-**one,** (LVIIA), m.p. 202–204°, is formed when 2-cyanomethylbenzeneboronic acid is warmed with 5% aqueous potassium hydroxide. Bearing in mind the nature of the boron–nitrogen bond, the formulation LVIIB for the borolactam then seems attractive, but no spectral evidence seems to have been presented for the presence, or otherwise, of the carbonyl group. Lithium tetrahydridoaluminate reduces the borolactam to 1,2,3,4-**tetrahyidro**-2,1-**benz**[*c*]**azaborine** (LVIII, R = R^1 = H) (*J. G. Catlin* and *H. R. Snyder*, J. org. Chem., 1969, **34,** 1660). 2-Methyl-1,2,3,4-tetrahydro-2,1-benz[*c*]azaborine (LVIII, R = H, R^1 = Me) is formed, with other products, when $Ph(CH_2)_2NMeBH_2$ is heated at 220–230° (*R. Köster et al.*, Ann., 1969, **720,** 23).

(*ii*) *1,2-Dihydrodibenz*[c,e][1,2]*azaluminine;* 1,2-*dihydrodibenz*[c,e][1,2]-*azaborine*

6-**Methyl**-5-**phenyl**-5,6-**dihydrodibenz**[*c,e*][1,2]**azalumine** (LIX) appears to be one of the few reasonably well characterized aluminium-containing heterocyclic compounds. It is formed when triphenylaluminium is heated with 2-methylaminobiphenyl; during the course of the reaction, benzene is

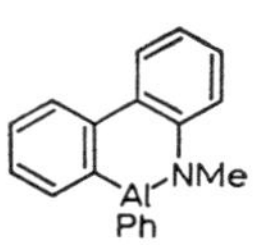

(LIX)

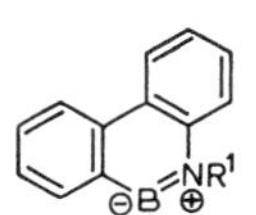

(LX)

liberated at 160°, and further heating to 240° causes cyclization of the biphenylyl intermediate to LIX, which decomposes at temperatures higher than 350°. Reaction of LIX with iodine yields 2-methylamino-2′-iodobiphenyl (*J. J. Eisch* and *M. E. Healy*, J. Amer. chem. Soc., 1964, **86,** 4221).

The chemistry of 5,6-dihydrodibenz[*c,e*][1,2]azaborine has been developed more extensively than that of either the oxygen or sulphur analogues. Studies on its reaction with electrophilic reagents have proved particularly interesting; many structures of the compounds so formed, have been confirmed by alternative synthesis starting from the appropriately substituted 2-aminobiphenyl, and also by acidolysis with concentrated sulphuric acid, when the corresponding 2-aminobiphenyls are produced.

6-**Chloro**-5,6-**dihydrodibenz**[*c,e*][1,2]**azaborine** (LX, R = Cl, R^1 = H) ("10-**chloro**-10,9-**borazarophenanthrene**") is obtained when 2-aminobiphenyl is treated with boron trichloride and the intermediate dichloroaminoborane reacted upon by aluminium chloride, cyclization to the chloride then occurring; the crude material may be purified by sublimation at 160–170°/0.05 mm or used directly. Reduction of the chloro compound with lithium tetrahydridoaluminate affords 5,6-**dihydrodibenz**[*c,e*][1,2]**azaborine** (LX, R = R^1 = H), also formed by similar reduction of the simple ethers (LX, R^1 = H, R = Oalkyl) (*Dewar, V. P. Kubba* and *R. Pettit*, J. chem. Soc., 1958, 3073; *Dewar et al.*, J. Amer. chem. Soc., 1961, **83**, 1754). Reactions of the chloride with Grignard reagents proceed normally, but the 6-phenyl compound is obtained in better yield from 2-aminobiphenyl and phenyl dichloroborane–aluminium trichloride, $PhBCl_2$–$AlCl_3$ (*Dewar, Kubba* and *Pettit, loc. cit.*) and by photolysis of $Ph_2B^{\ominus}{=}N^{\oplus}HC_6H_4I[2]$ (*P. J. Grisdale* and *J. L. R. Williams*, J. org. Chem., 1969, **34**, 1675). Hydrolysis of the chloride affords the 6-hydroxy compound, which may be reversibly dehydrated to 6,6′-oxybis(5,6-dihydrodibenz[*c,e*][1,2]azaborin) ["bis(5,6-dihydrodibenz[*c,e*][1,2]azaborin-6-yl) ether"].

Chlorination of 6-hydroxy-5,6-dihydrodibenz[*c,e*][1,2]azaborine takes place very readily at the 2 and subsequently the 4 positions (the mono-substituted compound is not isolated). Bromination takes place in a similar fashion and the high reactivity exhibited here may be contrasted with that shown by derivatives of phenazaborin (p. 56). Further halogenation proceeds through to the 2,4,8-trisubstituted pattern (*Dewar* and *Kubba*, J. org. Chem., 1960, **25**, 1722). The increased reactivity of position 4 over position 2 is also found for 6-methyl-5,6-dihydrodibenz[*c,e*][1,2]-azaborine which, in acetic acid, is chlorinated to a mixture of 4-chloro-6-methyl- and 4-chloro-6-hydroxy- compounds in a novel demethylation process.

Nitration of 6-hydroxy- and 6-methyl-5,6-dihydrodibenz[*c,e*][1,2]-azaborine gives, in each case, a mixture of the 2- and 4-mono-nitro derivatives in the ratio of 1:2 (*idem*, Tetrahedron, 1959, **7**, 213). The nitro compounds are reduced with hydrazine–Pd/C to the corresponding amino derivatives, but these fail to participate in Sandmeyer-type reactions. Interestingly (in view of the results of acetylation experiments described in the next paragraph), aminonitrobiphenyls react normally with boron trichloride, but the benzene ring is evidently too deactivated to allow cyclization to the nitro-heterocycle.

Acetylation of 6-hydroxy- and 6-methyl-5,6-dihydrodibenz[*c,e*][1,2]azaborines with one molar proportion of acetyl chloride in carbon disulphide in the presence of aluminium chloride gives, in each case, a mixture of 2-acetyl- and 2,4-diacetyl- derivatives in the ratio of 2.5:1, the introduction of the second acetyl group into an already deactivated ring being a noteworthy feature of this example of Friedel–Crafts acylation (*idem*, J. Amer. chem. Soc., 1961, **83**, 1757).

6-Methyl-5,6-dihydrodibenz[*c,e*][1,2]azaborine does not give a sodium or potassium salt in the usual way, nor can it be methylated at nitrogen with either dimethyl sulphate and alkali, or with diazomethane, consistent with the NH group being less acidic than that in pyrrole and indole. Addition of one mol. of methyl- or phenyl-lithium results in the formation of the lithio derivatives (LX, R^1 = Li, R = Me, Ph) which may then be methylated normally to give the 5,6-*dimethyl-*, m.p. 117–118 °, and the 5-*methyl-6-phenyl-*, m.p. 122.5–123.5 °, compounds (*Dewar* and *P. M. Maitlis*, J. Amer. chem. Soc., 1961, **83,** 187; *Dewar, J. Hashmall* and *Kubba*, J. org. Chem., 1964, **29,** 1755). The lithio derivatives may be acylated with ethyl chloroformate, but not with benzoyl chloride, and the *N*-ethoxycarbonyl compounds, unlike the *N*-alkyl series, are readily oxidized with loss of group from boron and concomitant boron–nitrogen bond fission (*Dewar* and *Maitlis*, Tetrahedron, 1961, **15,** 35):

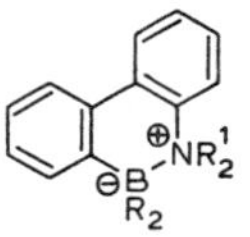

(LXI)

TABLE 3

5,6-DIHYDRODIBENZ[*c,e*][1,2]AZABORINES (LX, R^1 = H)

R	*Ring substituent*	*M.p., °C (m.p.)*	*Ref.*	*R*	*Ring substituent*	*M.p., °C (m.p.)*	*Ref.*
H	H	69.5–70	1	OH	(2)Cl	205–206*	6
Cl	H	93–94	2		(4)Cl(?)	182–183	6
Me	H	103–104	2		(8)Cl	192–193*	6
	(2)F	54–55	3		(2,4)Cl_2	263–264*	6
	(3)F	89–90	3		(2,4,8)Cl_3	283–284**	6
	(8)F	80–81	3		(2,4)Br_2	295*	6
	(9)F	52–55	3		(2)NO_2	360 dec.	4
	(2)Cl	86–87	4		(4)NO_2	266–267	4
	(4)Cl	170–175/0.5 mm	4		(2)NH_2	218–220	4
	(8)Cl	98–99	4		(4)NH_2	195–196*	4
	(2)NO_2	209–210	4		(2)MeCO	233–235	5
	(4)NO_2	165–166	4		(2,4) $(MeCO)_2$	>350	5
	(2)NH_2	125–126	4	OMe	(2)NO_2	237–238	4
	(4)NH_2	137–138	4		(4)NO_2)	155–156	4
	(2)MeCO	164–165	5	OEt	H	54–55	1
	(2,4) $(MeCO)_2$	205–207	5		(2)NO_2	172–174	4
Et	H	77–78	2		(4)NO_2	125–126	4
Ph	H	110–111.5	2		(2)MeCO	210–211	5
					(2,4) $(MeCO)_2$	258–260	5

* *Anhydride* ("ether").
** *Anhydride*, m.p. 285–286 °.

References

1 *M. J. S. Dewar et al.*, J. Amer. chem. Soc., 1961, **83,** 1754.
2 *Dewar, V. P. Kubba* and *R. Pettit*, J. chem. Soc., 1958, 3073.
3 *Dewar* and *P. J. Grisdale*, J. org. Chem., 1963, **28,** 1759.
4 *Dewar* and *Kubba*, Tetrahedron, 1959, **7,** 213.
5 *idem*, J. Amer. chem. Soc., 1961, **83,** 1757.
6 *idem*, J. org. Chem., 1960, **25,** 1722.

When 6-methyl- and 6-phenyl-5,6-dihydrodibenz[*c*,*e*][1,2]azaborines are treated with two molar proportions of methyl- or phenyl-lithium, strongly coloured dilithio compounds (LXI, R^1 = Li) are formed which may also be methylated to give, for example, *compound* LXI (R = R^1 = Me), m.p. 108–109°. Of interest, and some importance, is the displacement of methyl by phenyl when 6-methyl-5,6-dihydrodibenz[*c*,*e*][1,2]azaborine is treated with phenyllithium, a similar sequence being employed in one type of synthesis used for the preparation of polycyclic boron–nitrogen compounds (*idem*, J. Amer. chem. Soc., 1961, **83**, 187).

Physical data for a number of derivatives of 5,6-dihydrodibenz[*c*,*e*][1,2]-azaborine are given in Table 3.

(*iii*) *Polybenzazaborines*

Dewar et al. have prepared several polycyclic compounds containing the 1,2-azaborine ring system (Table 4) by general methods which include (*a*) the formation and cyclization of Grignard reagents from 4-bromobutyl derivatives of simpler ring systems, followed by dehydrogenation with sulphur or palladium on charcoal, and (*b*) the reaction of boron trichloride with a suitable amino-arene precursor, followed by aluminium chloride induced cyclization to *B*-chloro compounds. In general, the properties of the polycyclic compounds resemble those of 5,6-dihydrodibenz[*c*,*e*][1,2]-azaborine, but the following additional points may be noted. Compound LXIII, in addition to being stable to acids, alkalis and moist air, does not react with maleic anhydride. The anhydride from LXIV is degraded by the action of nitrous acid in acetic acid to a structurally analogous dibenzocinnoline while the same reagents convert the *B*-methoxy derivative of LX into the corresponding "boroxaro" compound. Whereas 2,6-diaminobiphenyl reacts normally and sequentially with phenylboron dichloride and aluminium chloride to give LXVI (R = Ph), 2,2′-diaminobiphenyl fails to react in this fashion.

The interaction of 1,2-azaborino[1,2,*a*][1,2]azaborine and 1,3-propanedial bis(diethylacetal) in the presence of trifluoroacetic acid gives the violet

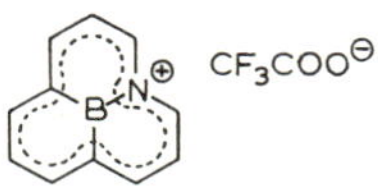

(LXVII)

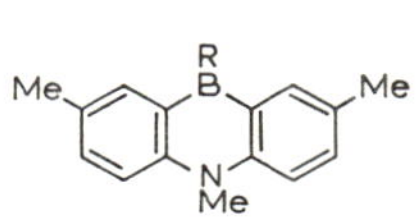

(LXVIII)

3a-**azonia-9b-boraphenalene** ("12,11-**borazarophenalenium**") **trifluoroacetate** (LXVII), stable indefinitely at −78°, but decomposing at room temperature (*Dewar* and *R. Jones*, Tetrahedron Letters, 1968, 2707).

TABLE 4

POLYCYCLIC 1,2-AZABORINES

System		*Derivatives*	*Ref.*
Structure	*Name*		
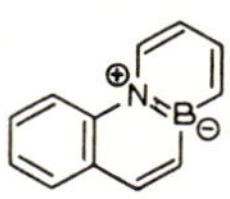 (LXII) m.p. 75–77°	[1,2]Azaborino[1,2,*a*]-[1,2]benzazaborine ("12,11-borazaro phenanthrene")	1,2,3,4-Tetrahydro-, b.p. 150–155 °/0.5 mm, m.p. 26–28 °	1
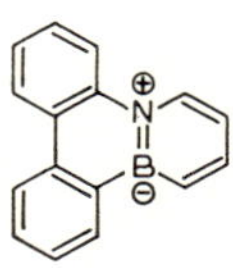 (LXIII) m.p. 185–186°	[1,2]Azaborino[1,2,*a*]-dibenz[*c,e*][1,2]azaborine ("14,13-borazaro triphenylene")	1,2,3,4-Tetrahydro-, m.p. 117 °	2
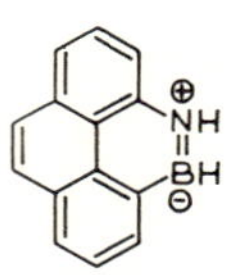 (LXIV) m.p. 129–130°	4,5-Dihydrophenanthro-[4,5,*c,d,e*][1,2]azaborine ("5,4-borazaropyrene")	5-Methoxy-, m.p. 143–45 ° 5-Methyl-, m.p. 155–156.5 ° 5-Hydroxy (as "ether") m.p. 246–247 ° 9,10-Dihydro-5-methyl-, m.p. 99.5–101.5 ° 9,10-Dihydro-5-methoxy-, m.p. 135–136° 9,10-Dihydro-5-hydroxy-(as "ether") m.p. 223 °	3
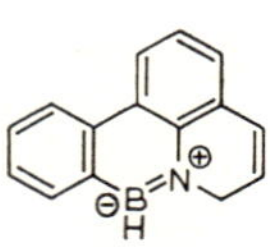	4*H*,8*H*-Quino[1,8,*b,c*]-[2,1]benzazaborine	4,5,6-Trihydro-8-hydroxyquino[1,8,*b,c*]-[2,1]benzazaborine (as "ether"), m.p. 237–240 °	1
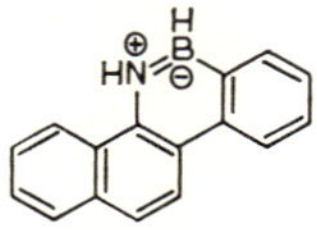 m.p. 193–195°	5,6-Dihydrobenzo[*d*]-naphth[2,1,*e*][1,2]-azaborine ("6,5-borazarochrysene")	6-Methoxy, m.p. 108–110 ° 6-methyl, m.p. 150 ° dec. 6-hydroxy (as "ether"), m.p. 288–290.5 °	3
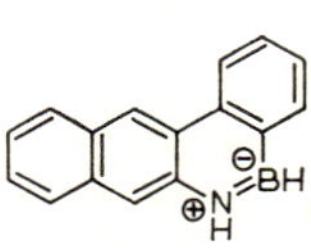 m.p. 139°	5,6-Dihydrobenzo[*c*]-naphth[2,3,*e*][1,2]-azaborine ("5,6-borazarobenz[*a*]-anthracene")	5-Methoxy, m.p. 159 ° 5-Methyl, m.p. 160–161 ° 5-hydroxy (as "ether") m.p. 322 °	4

Table 4 (continued)

System		*Derivatives*	*Ref.*
Structure	*Name*		
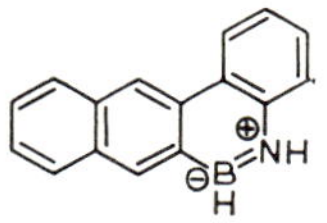 (LXV) m.p. 136.5–137°	5,6-Dihydrobenz[*e*]naphth-[2,3,*c*]azaborine ("6,5-borazarobenz[*a*]-anthracene")	6-Methoxy, m.p. 183–184° 6-methyl, m.p. 141–143° 6-hydroxy (as "ether"), m.p. 275°	4
	8*H*-Isoquino[2,1,*b*][2,1]-benzazaborine ("7,17-borazaro-5,6-dihydrobenz[*a*]anthracene")	8-Phenyl, m.p. 178–180° 8-hydroxy (as "ether"), m.p. 196–198°	1
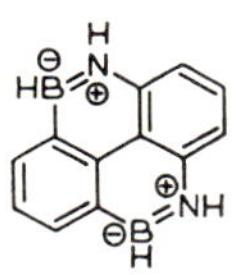 (LXVI)	4,5,9,10-Tetrahydro-4,10-diaza-5,9-diborapyrene	5,9-Diphenyl, m.p. 200–202°	5
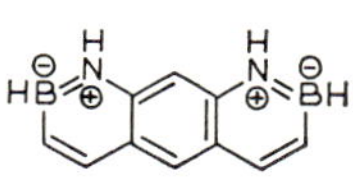	1,2,8,9-Tetrahydro-benzo[1,2,*e*:5,4,*e*′]bis-[1,2]azaborine	2,3,7,8-Tetraphenyl, m.p. 270°	5
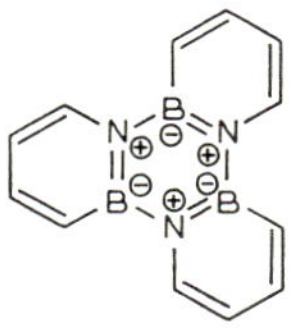 m.p. 193–195°	Tris(1,2)azaborino[1,2,*a*:1′,2′,*c*:1″,2″,*e*]triaza-triborine ("14,16,18-tribora-13,15,17-triazarotriphenylene")	Dodecahydro, m.p. 184–186° tris(dibenz[3,4:5,6,1,2]), m.p. 400–403°	2,6 7

References

1 *M. J. S. Dewar, C. Kaneko* and *M. K. Bhattacharjee*, J. Amer. chem. Soc., 1962, **84,** 4884.
2 *G. C. Culling, Dewar* and *P. A. Marr, ibid.*, 1964, **86,** 1125.
3 *Dewar* and *W. H. Poesche*, J. org. Chem., 1964, **29,** 1757.
4 *Idem*, J. Amer. chem. Soc., 1963, **85,** 2253.
5 *S. S. Chissick, Dewar* and *P. M. Maitlis*, Tetrahedron Letters, 1960, **8**.
6 *K. M. Davies, Dewar* and *P. Rona*, J. Amer. chem. Soc., 1967, **89,** 6294.
7 *R. Köster, et al.*, Ann., 1969, **720,** 23.

(iv) 5,10-Dihydrodibenz[1,4]azaborine

Hydrolysis of the initial reaction product from 2,2′-dilithio-4,4,*N*-trimethyldiphenylamine and tributyl borate, or better, butyl metaborate, yields 10-**hydroxy**-2,5,8-**trimethyl**-5,10-**dihydrodibenz**[*b*,*e*][1,4]**azaborine** (the ring system is also referred to as 10,9-borazaroanthracene and phenazaborin) (LXVIII, R = OH), isolated as the *anhydride* $[(MeC_6H_3)(NMe)B]_2O$, m.p. 244–245°. The latter yields the 10-ethoxy compound on crystallization from ethanol, but suffers degradation to 2,2′-dibromo-4,4′,*N*-trimethyldiphenylamine on attempted bromination. Reaction of the anhydride from LXVIII (R = OH) with phenylmagnesium bromide proceeds normally so giving the 10-*phenyl* compound, m.p. 114–115°, but the same anhydride fails to react with alkyl Grignard reagents (*Maitlis*, J. chem. Soc., 1961, 425).

(g) *Diazaluminines and diazaborines*

(i) 1,2,3-Diazaborines and 2,3,1-benzodiazaborines

2,3-**Dihydro**-1,2,3-**diazaborines** ("3,2-**borazaropyridines**") (LXIX) are weakly basic compounds, stable to dilute acid solutions and *N* sodium hydroxide solution, and they are spectroscopically similar to pyridine. They are ob-

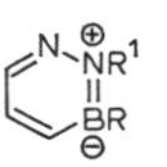

(LXIX)

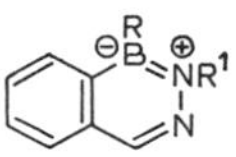

(LXX)

tained by desulphurization of fused thienodiazaborines (p. 57); 4-*ethyl*-3-*hydroxy*-, m.p. 109–112°, 5-*ethyl*-3-*hydroxy*-, m.p. 106–110°, 3-*hydroxy*-2,4,5-*trimethyl*-, m.p. 64–70°, 5-*ethyl*-2,3-*dimethyl*-, b.p. 177–180°, 4-*ethyl*-2,3-*dimethyl*- derivatives, b.p. 72–73.5°/15 mm.

By contrast with the analogous pyridine derivatives, those of 1,2,3-diazaborine are not attacked by *N,N*-dibromoisocyanuric acid in fuming sulphuric acid, nor by potassium nitrate–sulphuric acid mixture, although here some demethylation at boron is evident. The ring thus appears to be more strongly deactivated than the pyridine ring under similar acid conditions: no deuterium exchange takes place in 98% D_2SO_4 at 80° during 50 h. Nevertheless, substitution by electrophilic reagents is possible; 4-ethyl-2,3-dimethyl-2,3-dihydro-1,2-diazaborine brominates (bromine–pyridine in carbon tetrachloride) in position 6, while the analogous 5-ethyl compound brominates in positions 4 and 6. 5-*Ethyl*-3-*hydroxy*-2-*methyl*-2,3-*dihydro*-1,2,3-*diazaborine* yields the 6-*nitro* derivative, m.p. 126–127°, with fuming nitric acid in acetic anhydride. Under such strongly acid conditions compounds containing methyl groups attached to boron are liable to be demethylated. In these cases, *N*-nitropyridinium tetrafluoroborate may be used for successful nitration (*J. Namtvedt* and *S. Gronowitz*, Acta Chem, Scand., 1968, **22,** 1373;

Gronowitz and *A. Maltesson*, *ibid.*, 1971, **25,** 2435; Chem. Scr., 1972, **2,** 79; C.A.; 1972, 77, 126732; Acta Chem. Scand., 1975, **B29,** 457).

Nuclear substituted halogen may be exchanged for lithium treated with butyllithium at $-70°$ (*idem*, *ibid.*, p. 1036).

The benzo derivatives and their hetero analogues are more conveniently obtained by condensations between formyl-areneboronic acids and hydrazines. Hydrazine condenses with 2-formylbenzeneboronic acid to give **1-hydroxy-1,2-dihydro-2,3,1-benzo[*d*]diazaborine ("4-hydroxy-4,3-borazaroisoquinoline")** (LXX, R = OH, R^1 = H) isolated as the *anhydride*, m.p. 243–243.5°; substituted hydrazines behave similarly. The anhydride is unaffected by hot strong acids and alkalis, and behaves as a protic acid rather than a Lewis acid. With methylmagnesium iodide, it gives the 1-*methyl* compound (LXX, R = Me, R^1 = H), m.p. 94–96°, which undergoes reaction with dimethyl sulphate and alkali to give the 1,2-dimethyl compound (LXX, R = R^1 = Me) m.p. 18–20°; although the dimethyl compound is stable in the absence of air, it undergoes autoxidation giving 1-hydroxy-2-methyl-1,2-dihydro-2,3,1-benzo[*d*]diazaborine. Interestingly, while the 2-methoxy compound is very readily reduced by lithium tetrahydridoaluminate, 1,1′-oxybis(1,2-dihydro-2,3,1-benzo[*d*]diazaborine) requires more drastic conditions to bring about reduction of the heterocyclic ring (*M. J. S. Dewar* and *R. C. Dougherty*, J. Amer. chem. Soc., 1964, **86,** 433: *P. Tschampel* and *H. R. Snyder*, J. org. Chem., 1964, **29,** 2168; *Chemie Greuenenthal GmbH.*, Fr.P. 1,522,647/1968; C.A., 1969, **71**, 39019):

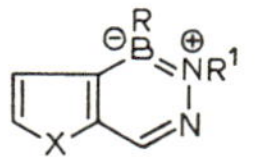
(LXXI)

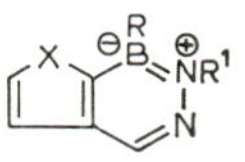
(LXXII)

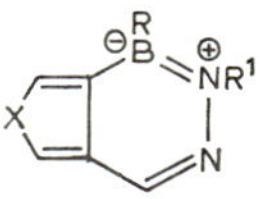
(LXXIII)

X = O, S

The fused bicyclic diazaborines LXXI–LXXIII are prepared from appropriate formylfuranboronic acids (*Gronowitz* and *M. Uri*, Ark. Kemi, 1970, **32,** 283; C.A., 1970, **73,** 109824; *B. Roques*, *D. Florentin* and *J.-P. Juhasz*, Compt. rend., C, 1971, **270,** 1898; *Florentin et al.*, Bull. chem. Soc. Fr., 1974, 2620) and formylthiopheneboronic acids (*Gronowitz*, *Namtvedt* and *Bugge*, Acta Chem. Scand., 1965, **19,** 1271; 1967, **21,** 2151; *Gronowitz* and *Maltesson ibid.*, 1971, **25,** 2435), the thiophene compounds being used in conjunction with Raney nickel as a source of 2,3,1-diazaborines. Nitration (*Gronowitz* and *Namtvedt*, Tetrahedron Letters, 1966, 2967; *Gronowitz* and *C. Roos*, Acta Chem. Scand., 1975, **B29,** 990; *Namtvedt*, *ibid.*, 1968, **22,** 1611), bromination (*Gronowitz* and *Roos*, *loc. cit.*). and metal–halogen interchange (*Gronowitz* and *Maltesson*, Acta Chem. Scand., 1975, **B29,** 1036) in these series, have been examined.

Selenium-containing analogues of LXXI–LXXIII have been recorded by *D. Semard*, *et al.*, Internat. J. Sulphur Chem., 1972, **A2,** 257).

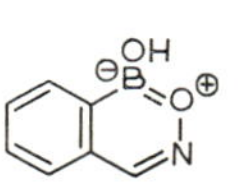
(LXXIV)

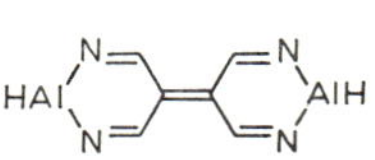
(LXXV)

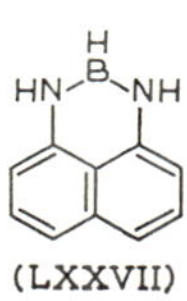
(LXXVII)

1-*Hydroxy*-1H-2,3,1-*benzo*[d]*oxazaborine* (LXXXIV), m.p. 264–265°, from 2-formylbenzeneboronic acid and hydroxylamine, also appears to be a protic rather than a Lewis acid (*Dewar* and *Dougherty*, *loc. cit.*; *K. Torssell*, Ark. Kemi, 1957, **10,** 507).

(*ii*) *1,3,2-Diazaluminines; 1,3,2-diazaborines, perhydro derivatives and benzologues*

A compound produced when aluminium hydride reacts with tetracyanoethylene has been formulated as *Δ*-5,5′-(2*H*,2′*H*)**di**(1,3,2-**diazaluminine**) (LXXV) (*D. L. Schmidt*, U.S.P. 3,408,378/1968; C.A., 1969, **70,** 20093), but a compound of a more definite constitution, *viz.* 2-*ethyl*-1,3,4-*triphenyl*-1,2,3,4-*tetrahydro*-1,3,2-*benzo*[d]*diazaluminine* (LXXVI), m.p. 233°, has been isolated from the reaction shown; it is decomposed by water and ethanol, liberating ethane (*H. Hoberg* and *A. Milchereit*, Ann., 1972, **766,** 146):

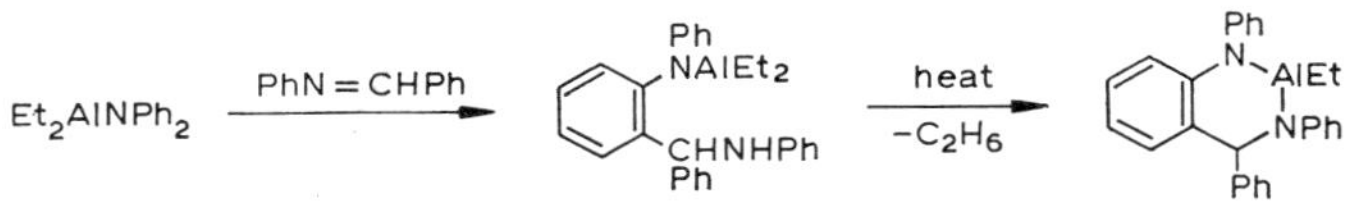

(LXXVI)

The reaction between naphthalene-1,8-diamine and sodium tetrahydridoborate in tetrahydrofuran at room temperature produces 2,3-**dihydro-1***H***-naphtho-**[1,8-*d,e*][1,3,2]**diazaborine** (LXXVII), m.p. 100–100.5° (*J. Goubeau* and *H. Schneider*, *ibid.*, 1964, **675,** 1), and it is possible that a tricyclic diazaborine is formed when the same reducing agent reacts with di-2,2′-pyridylmethane difluoroborate at 200° (*J. E. Douglas, P. M. Barelski* and *R. M. Blankenship*, J. heterocycl. Chem., 1973, **10,** 253).

The condensation between 2-aminobenzenecarboxamides and boronic acids (*J. Fried et al.*, U.S.P. 3,293,252/1966; C.A., 1967, **66,** 115783) or orthoboric esters (*H. L. Yale et al.*, J. Amer. chem. Soc., 1962, **84,** 688) in non-aqueous media, affords 1,2,3,4-**tetrahydro**-1,3,2-**benzodiazaborin**-4-**ones** (LXXVIII); other condensations between similar boron compounds and 2-aminopyridine- and 2-aminothiophenecarboxamides (*A. Dornow* and *D. Wille*, Chem. Ber., 1965, **98,** 1505; *V. P. Arya*, Indian. J. Chem., 1971, **9,** 1167) are reported. Generally, these compounds undergo solvolysis in ethanol, form stable anions in alkali, and the *compound* LXXVIII (R^1 = CH_2Ph, R = Ph, R^2 = H), m.p. 190–192°, is unique in being recrystallizable from water, others being labile under neutral aqueous conditions. Treatment of LXXVIII (R = Ph, R^1 = R^2 = H) m.p. 214–215°, with phosphoryl chloride gives a complex which is degraded to 2-aminobenzonitrile by direct interaction with concentrated ammonia solution, but which under alternative conditions is converted to the 4-*ethoxy*-, m.p. 118–119°, and 4-*amino*-, m.p. 260–263°, compounds (LXXIX, R = OEt, NH_2) (*Yale et al.*, *loc. cit.*; *Yale*, J. heterocycl. Chem., 1971, **8,** 193, 205):

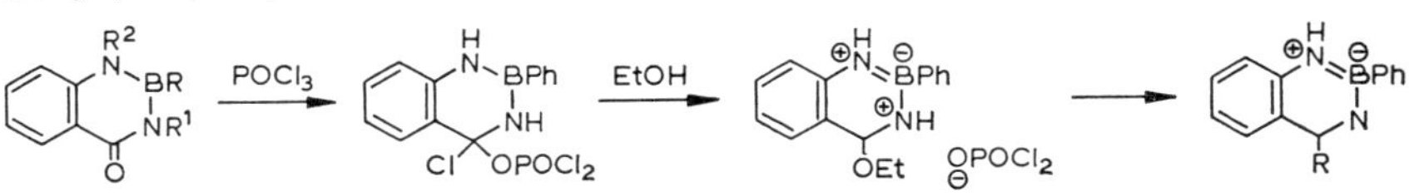

(LXXVIII) (LXXIX)

The linear products from 2-aminobenzonitriles and dihalogenoboranes cyclize when heated *in vacuo* to give nuclear halogen-substituted products (*A. Meller, W. Maringgele* and *G. Maresch*, Monatsh., 1974, **105**, 637):

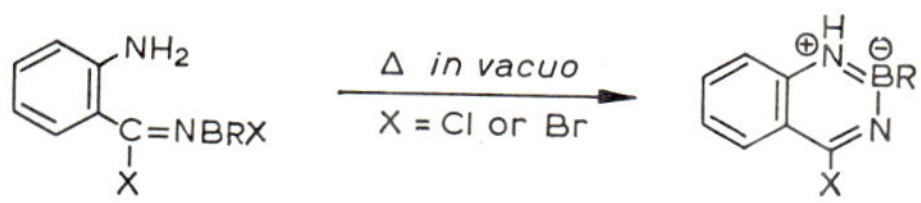

Examples of other mixed systems so far recorded are LXXX, m.p. 133° (*K. Lang, K. Neutxel* and *F. Schubert*, G.P. 1,130,445/1962; C.A., 1963, **58**, 1488), and LXXX (*Yale et al., loc. cit.*).

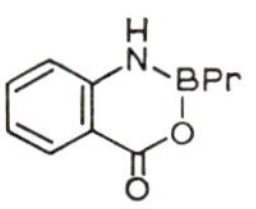

(LXXX)

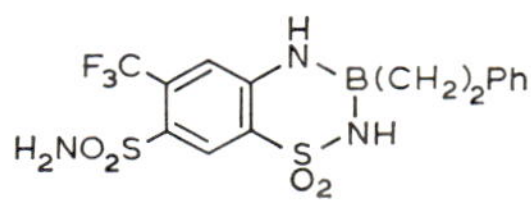

(LXXXI)

Perhydro-1,3,2-**diazaborines** (LXXXII, R = H) are obtained by treatment of *N,N'*-dialkylpropanediamines with sodium tetrahydridoborate and iodine in boiling tetrahydrofuran, 2-*methyl*-, b.p. 116–118° (*E. F. Rothgery, P. J. Busse* and *K. Niedenzu*, Inorg. Chem., 1971, **10**, 2343) but 2-substituted compounds (LXXXII, R^1 = H), 2-*ethyl*-, b.p. 44–45°/15 mm, are generally formed by reactions between diamines and aminoboranes, $RB(NMe_2)_2$ or $B(NMe_2)_3$ (*W. Weber, J. W. Dawson* and *Niedenzu, ibid.*, 1966,

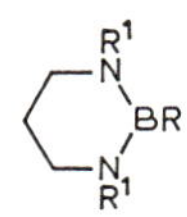

(LXXXII)

5, 726; *Niedenzu, P. Fritz* and *Dawson, ibid.*, 1964, **3**, 1077; *Niedenzu* and *Fritz*, Z. anorg. allgem. Chem., 1965, **340**, 329; *W. R. Bamford* and *S. Fordham*, Scc. Chem. Ind., Monograph No. 13, 1961, 320; *M. P. Brown, A. E. Dann* and *D. W. Hunt*, J. chem. Soc., 1962, 4648).

The results of an electron diffraction study of 2-chloro-1,3-dimethylperhydro-1,3,2-diazaborine have been interpreted as indicating extensive π-bonding between boron and nitrogen (*R. Seip* and *H. M. Seip*, J. Molec. Structure, 1975, **28**, 441). In this compound, replacement of Cl by SMe is achieved when it is treated with $Pb(SMe)_2$; the resultant methylthio compound acts as a source of the anhydride LXXXIII (*W. R. Conway, P. F. Litz* and *Niedenzu*, Synth. React. Inorg. met. org. Chem., 1975, **5**, 37).

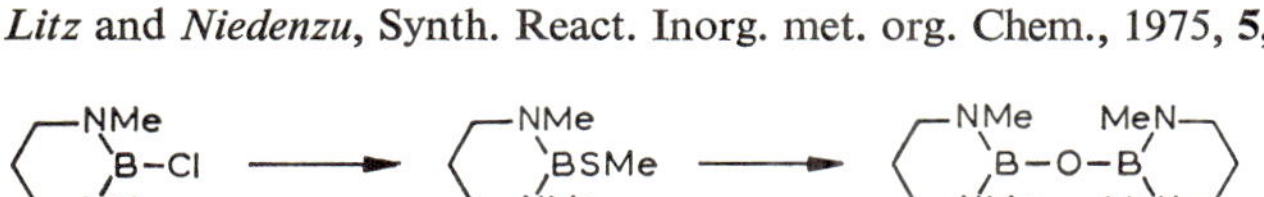

(LXXXIII)

This versatile transamination reaction is used to prepare *octahydro*-1,3,2-*diazaborino*-[1,2-a][1,3,2]*diazaborine* (1,4a,8,8a-*triazaboradecalin*) (LXXXIV), b.p. 62°/1 mm, m.p.

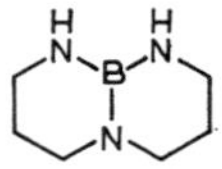

(LXXXIV)

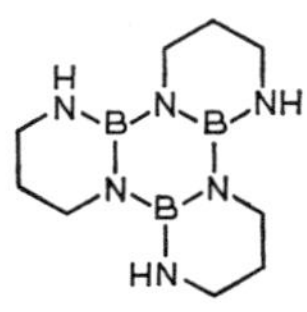

(LXXXV)

38–41°, from di(3-aminopropyl)amine (*Niedenzu* and *W. Weber*, Z. Naturforsch., B, 1966, **21**, 811; *Niedenzu, Fritz* and *Dawson, loc. cit.*), and the related LXXXV, b.p. 190–195°/1 mm, m.p. 154–155°, from 1,3-diaminopropane (*Niedenzu* and *Fritz, loc. cit.*).

Some similar systems also containing oxygen, are illustrated in the *compounds* LXXXVI, b.p. 80–90°/0.4 mm (*M. Pailer* and *H. Huemer*, Monatsh., 1964, **95**, 373) and LXXXVII (*L. S. Povarov* and *O. D. Ryazanova*, Zhur. obshcheĭ Khim., 1972, **42**, 1010):

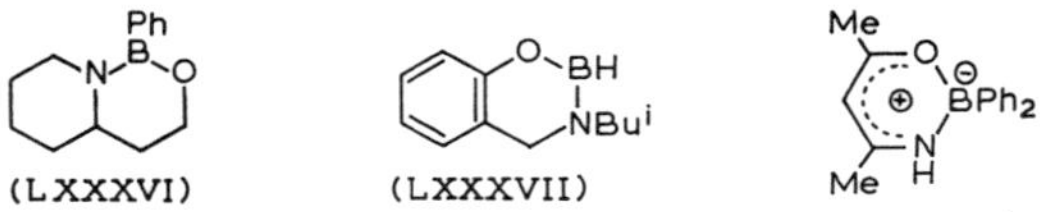

(LXXXVI) (LXXXVII) (LXXXVIII)

The azonia-borata *compound* LXXXVIII has m.p. 101–101.5° (*Dewar* and *R. C. Dougherty*, Tetrahedron Letters, 1964, 907; *R. Köster* and *W. Fenzl*, Angew. Chem. intern. Edn., 1968, **7**, 735; *I. Bally et al.*, Tetrahedron, 1973, **29**, 3185).

Other derivatives of this system have been obtained in an unusual rearrangement reaction examplified by the formation of compound LXXXIX, m.p. 159–161° (*R. E. Lyle* and *D. A. Walsh*, J. organometal. Chem., 1974, **67**, 363):

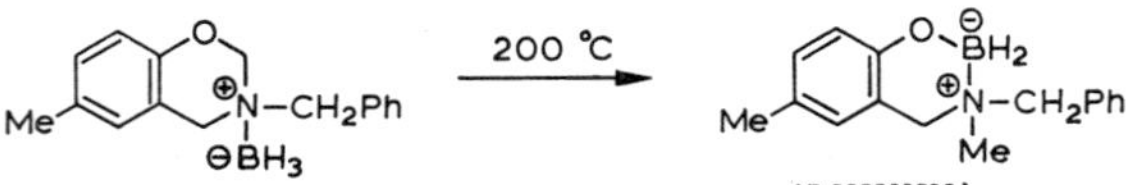

(LXXXIX)

(*iii*) *1,3,4-Diazaborines*

Examples of the monocyclic system are, as yet, unknown. Heating the product from tripropylborane and MeC(=NPh)NHPh at 230° results in the formation of 2-methyl-3-phenyl-4-propyl-3,4-dihydrobenzo[*e*][1,3,4]-diazaborine (XC) (*B. M. Mikhailov* and *V. A. Dorkhov*, Izvest. Akad. Nauk S.S.S.R., Ser. Khim., 1973, 2649).

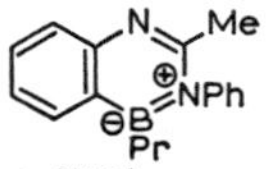

(XC)

(*h*) *Triazaborines*

3,5-Diisobutyl-2-methyl-4,6-dioxo-perhydro-1,3,5,2-triazaborine (XCI) is formed from isobutyl isocyanate and tri-*B*-methyl triorganoborazole (*Y. Proux* and *R. Clement*, Compt. rend. C, 1967, **264,** 2123; *J. L. Boone*, U.S.P. 3,060,234/1962; C.A., 1963, **58,** 5704) and the *diimine* XCII, m.p. 187°, is the product from the interaction of di-*p*-tolylcarbodiimide and $(Pr_2^iN)_2BNHBu^t$ at 220° *in vacuo* (*R. Jefferson et al.*, J. chem. Soc., A, 1966, 1584). Condensation of trialkylboranes or orthoboric esters and guanidines or

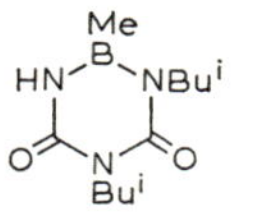

(XCI)

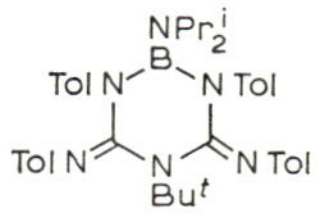

(XCII)

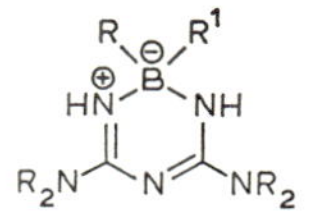

(XCIII)

biguanides affords the high melting **diamino-3-azonia-2-borata-1,5-diazacyclohexadienes** (XCIII) (*A. Joos, H. Prietzel* and *L. Strassburger*, G.P. 1,233,875/1967; C.A., 1967, **67,** 64451; *Joos* and *Strassburger*, G.P. 1,236,516/1967; C.A., 1968, **68**, 13161; G.P. 1,243,202/1967; C.A., 1968, **68,** 69053; *J. E. Milks, G. W. Kennerly* and *J. H. Polevy*, J. Amer. chem. Soc., 1962, **84,** 2529; *B. M. Mikhailov, V. A. Dorokhov* and *V. I. Seredenko*, Doklady Akad. Nauk S.S.S.R., 1971, **199,** 1328). The thermolysis of carboxamidine–borane complexes also appears to give the same azonia-borata salts (*Dorokhov* and *Mikhailov*, Zhur. obshcheĭ Khim., 1971, **41,** 2702; *Mikhailov, Dorokhov,* and *V. I. Sededenko, ibid.*, 1973, **43,** 862).

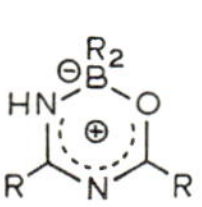

(XCIV)

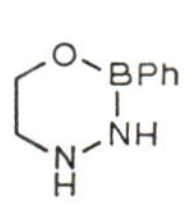

(XCV)

Similar mixed ring systems XCIV (*Mikhailov, Dorokhov* and *L. I. Lavrinovitch, ibid.*, 1972, **42,** 1553; *Dorokhov et al., ibid.*, 1973, **43,** 1115) and XCV, b.p. 90–100°/0.1 mm (*M. Pailer* and *H. Huemer*, Monatsh., 1964, **95,** 373) *W. Kleigel*, Ann., 1969, **721,** 231) are known.

(*i*) *Azadiborines*

Perhydro-1,2,6-azadiborines (XCVI), R = H, m.p. 63–66°, R = Me, b.p. 90–91°/7 mm, are formed when $(CH_2)[CH_2B(SEt)_2]_2$ is heated with ammonia or with primary amines in chloroform (*Mikhailov* and *Poznev*, Doklady Akad. Nauk S.S.S.R., 1963, **151,** 346).

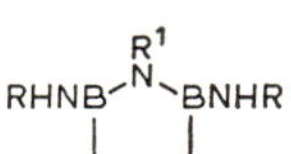

(XCVI)

(*j*) *Diazadialuminines and diazadiborines, and their benzologues*

(*i*) *1,4,2,3-Diazadiborines and 1,2,3,6-diazadiborines*

Transaminations involving $B_2(NMe_2)_2$ or di(perhydro-1,3,2-diazaborin-2-yls) take place in two stages, but with the appropriate diamine, 1,4,6,9-**tetramethylperhydro**[1,4,2,3]**di-azadiborino**[2,3,*b*][1,4,2,3]-**diazadiborin** (XCVII), m.p. 45–47°, the *dibenzo* compound

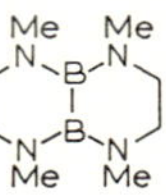
(XCVII)

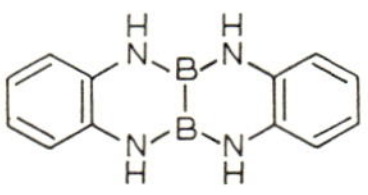
(XCVIII)

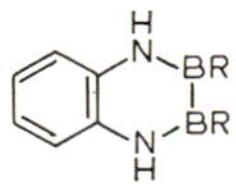
(XCIX)

XCVIII, m.p. >280°, and the 1,2,3,4-**tetrahydro**-1,4,2,3-**benzodiazadiborines** (XCIX), 2,3-*dimethyl*, m.p. 135°, 2,3-*diethyl*, m.p. 50–54°, 2,3-*diphenyl*, m.p. 100–101°, have been obtained (*C. N. Welch* and *S. G. Shore*, Inorg. Chem., 1968, **7,** 225; *H. Nöth*, and *P. Fritz*, Z. anorg. allgem. Chem., 1963, **324,** 129).

An example of the isomeric system, 1,4-**dimethyl**-1,2,3,4-**tetrahydrobenzo**-[*d*][1,2,3,6]**diazadiborine** (C), m.p. 56–57°, has been obtained by heating 1,3-dimethyl-1,3-dihydrobenzo[*c*][1,2,5]thiadiborole with hydrazine. Examples are also known in which the diazadiborine ring is fused to the thiophene ring. (*B. Asgarauladi et al.*, chem. Ber., 1974, **107,** 34).

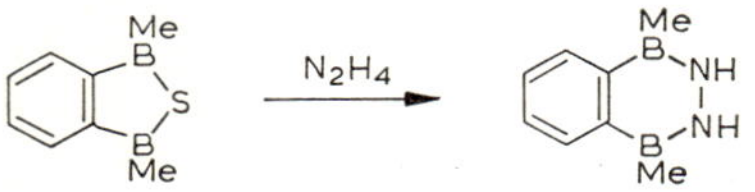

(C)

(*ii*) *1,2,3,4-Tetrahydro*[*1,3,2,4*]*benzodiazadiborine*

When *o*-toluidine is added to boron trichloride in toluene, followed by triethylamine, and the solution heated, there results a mixture containing ***B,B,B*-trichloro-*N,N,N*-tris-*o*-tolyltriazatriborine,** and from which 2,4-**dichloro**-8-**methyl**-3-*o*-**tolyl**-1,2,3,4-**tetrahydro**-1,3,2,4-**benzodiazadiborine** (CI, R = Cl) m.p. 120–122°, may be separated. This compound

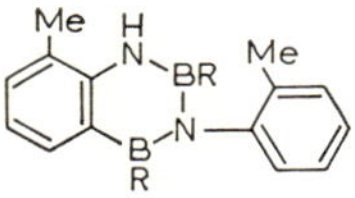
(CI)

is rapidly decomposed by water liberating *o*-toluidine, boric acid and an aminobenzene-boronic acid (*R. K. Bartlett et al.*, J. chem. Soc., A, 1966, 479) but reacts conventionally with methylmagnesium iodide (*B. Frange*, Bull. Soc. chim. Fr., 1973, 1216, 2160). A related compound has been obtained from 1-naphthylamine (*J. Cueilleron* and *Frange*, *ibid.*, 1972, 584).

(*iii*) *Perhydro-1,4,2,5-diazadialuminines and -1,4,2,5-diazadiborines*

The formation of **perhydro-1,4,2,5-diazadialuminines,** *e.g.* CII, m.p. 187–193°, is evidently accomplished only when the starting material is di-*ortho*-substituted. The product liberates ethane when treated with ethanol, and when acted upon by deuterium oxide yields $C_6H_3Me_2(CH_2NDPh)$ (*H. Hoberg*, Ann., 1971, **746,** 86).

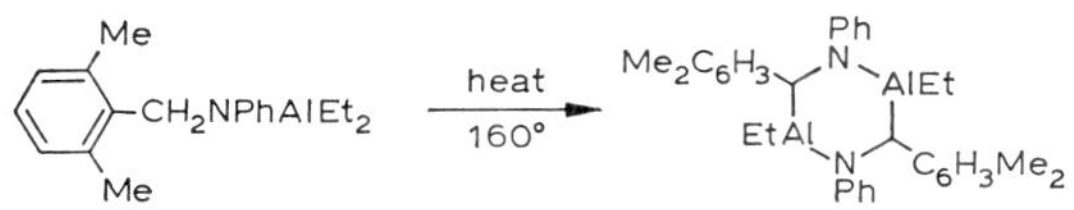

(CII)

The thermodynamically unstable 1:1 adducts from trialkylboranes and isonitriles (hydrogen cyanide also reacts) dimerize in ether at room temperature affording derivatives of 1,4,2,5-diazadiborine. The example, 2,3,3,5,6,6-*hexamethyl*-1,4-*diphenyl*-1,4,2,5-*diazadiborine* (CIII), m.p. 199°, from phenyl isocyanide and trimethylborane, is typical; when such compounds are heated at 210°, a rearrangement to 1,4-*diphenyl*-2,3,3,5,6,6-*hexamethylperhydro*-1,4,2,5-*diazadiborine* (CIV), m.p. 171°, takes place. Such products are readily oxidized by chromic oxide to ketones, and at higher temperatures, a ring

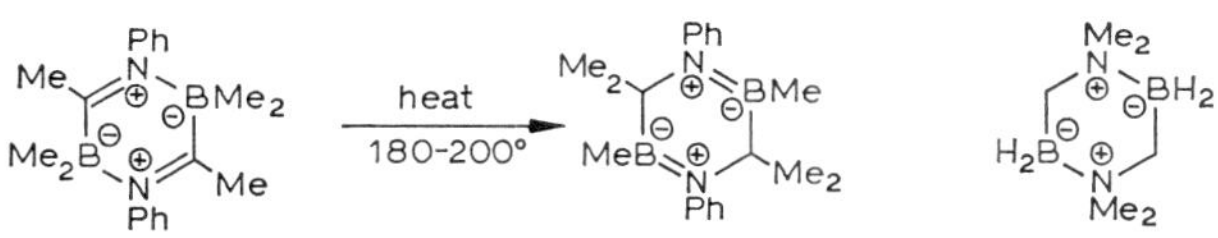

(CIII) (CIV) (CV)

contraction process may take place (*G. Hesse* and *H. Witte*, Ann., 1965, **687,** 1; *Hesse*, *Witte* and *G. Bittner*, *ibid.*, 1965, **687,** 9; *J. Casanova et al.*, Tetrahedron Letters, 1965, 703; *J. Casanova* and *H. R. Kiefer*, J. org. Chem., 1969, **34,** 2579). Compound CV is reported as being formed from sodamide and $(Me_2N)_2B^{\oplus}H_2\ I^{\ominus}$ in dimethoxyethane (*B. R. Cragg* and *G. E. Ryschkewitsch*, Inorg. Chem., 1976, **15,** 1209).

(*k*) *Triazadiborines*

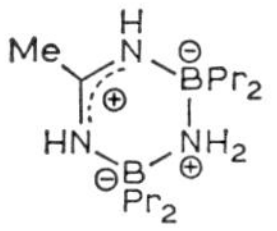

(CVI)

Structure CVI has been proposed for one of the products obtained when a mixture of $Pr_3BNH{=}CMeNH_2$ and Pr_2BNH_2 is heated at 80° (*V. A. Dorokhov* and *B. M. Mikhailov*, Zhur. obshcheĭ Khim., 1971, **41,** 2702).

2. Silicon, germanium, tin and lead compounds

(a) *Oxa-silins, -germanins and -stannins, and their thia analogues*

(i) *1,2-Oxasilins, benz[1,2]oxasilins and dibenz[1,2]oxasilins*

2,2-Dimethyl-3-phenyl-1,2-oxasilacyclohex-5-ene (I), m.p. 40–41, is formed *inter alia* to the extent of about 16% during the co-pyrolysis of 1,1-dimethyl-2-phenyl-1-silacyclobutane with propenal (*P. B. Valkovich* and *W. P. Weber*, J. org. Chem., 1975, **40,** 229).

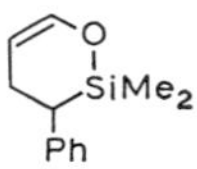
(I)

2,2,4-Trimethyl-1*H*-3,4-dihydrobenz[1,2]silin (II), b.p. 84–85.5°/5 mm, n_D^{20} 1.5212, is obtained in 72% yield when 2-propenylphenyloxydimethylsilane is treated with chloroplatinic acid (*V. F. Mironov*, *V. L. Kozlikov* and *N. S. Fedotov*, Zhur. obshcheĭ Khim., 1969, **39,** 966). 1,1-*Dichloro*-1,2,3,4- *tetrahydrobenz*[c][2,1]*oxasilin* is reported to have b.p. 81°/1 mm, n_D^{20} 1,5695 (*E. A. Chernyshev*, *T. L. Krasnova* and *E. F. Shchipanova*, ibid., 1974, **44,** 1652):

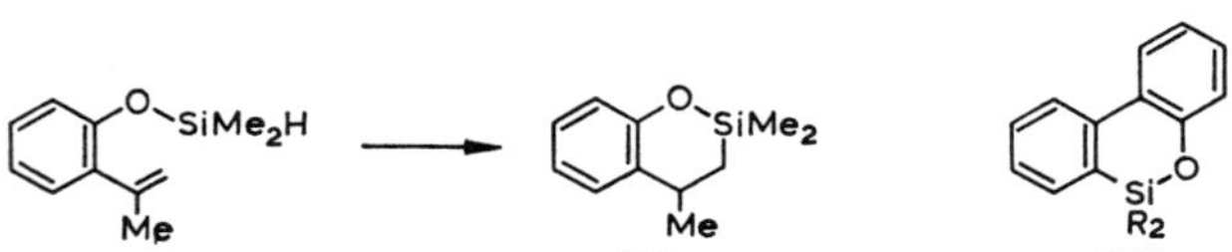
(II) (III)

6*H*-Dibenz[*c,e*][1,2]oxasilin (III, R = H) is formed during the rapid pyrolysis of chlorohydrosilyloxybiphenyl at 600–700° (*Chernyshev et al.*, U.S.S.R. Pat., 214,535/1968; C.A., 1968, **69,** 52282; U.S.S.R. Pat., 365,359/1973; C.A. 1973, **79,** 5451; *Chernyshev*, *Krasnova* and *Shehipanova*, *loc. cit.*). The *dichloride* (III, R = Cl), b.p. 150°/1 mm, m.p. 75°, is formed similarly; it can be converted into the corresponding difluoride with hydrogen fluoride and reduction with lithium tetrahydridoaluminate yields (III, R = H). The chloride is also reported to react conventionally with Grignard reagents (*idem*, Zhur. obshcheĭ Khim., 1970, **40,** 1735).

(ii) *Perhydro-1,2-oxasilins* (1,2-*oxasilacyclohexanes*) *and their germanium analogues*

2,2-Dimethyl-1,2-oxasilacyclohexane (IV), b.p. 125°/760 mm, 30°/20 mm, n_D^{20} 1.4275, is formed (*a*) when the siloxane $[EtO_2C(CH_2)_3SiMe_2]_2O$ in tetrahydrofuran is reduced with sodium tetrahydridoborate (*H. Niederpreuem*, *E. M. Harss* and *W. Simmler*, Belg. Pat. 663,783/1965; C.A., 1966, **64,** 19676); (*b*) when dichlorodimethylsilane is allowed to react with magnesium under the same solvent (*R. P. Anderson*, *U.S. Pat.*, 3,083,219/1963;

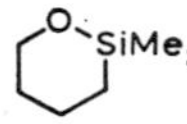
(IV)

C.A., 1963, **59,** 8789; *G. Schott* and *H. G. Bünger*, Z. anorg. allgem. Chem., 1968, **362, 71**). This reaction appears to depend on the cleavage of the tetrahydrofuran ring by the action of magnesium halide with the formation of a halogenobutoxysilane which then ring closes after initial reaction with metallic magnesium (*C. L. Smith* and *R. Gooden*, J. organometal. Chem., 1974, **81,** 33); and (*c*) when $(EtO)Me_2Si(CH_2)_4OSiMe_3$ is treated with either hydrogen chloride or sodium ethoxide or merely allowed to stand (*J. Pola, V. Bazant* and *V. Chvalovsky*, Coll. Czech. chem. Comm., 1973, **38,** 1528). Other reactions leading to this ring system include the chloroplatinic acid-catalyzed cyclization of butenyloxysilanes (*Mironov, Kozlikov* and *Fedotov, loc. cit.*) and the use of methylmethoxy(3-chloropropoxy)chlorosilane and 3-hydroxypropyldiethoxymethylsilane as starting materials in the presence of sodium or lithium (*G. Rossmy* and *G. Koerner*, Makromol, Chem., 1964, **73,** 85). Other known compounds include the 2,2,4-*trimethyl-*, b.p. 137–138°, n_D^{20} 1,4280; 2,2,6-*trimethyl-*, b.p. 122–124°, n_D^{20} 1.4490; 2,2,4,6-*tetramethyl-*, b.p. 123–125°, n_D^{20} 1.4270; 2,2-*diphenyl*-1,2-*oxasilacyclohexane*, b.p. 215–216°/34 mm, n_D^{20} 1.5722.

The 1,2-oxasilacyclohexane ring is cleaved by the action of Grignard reagents, lithium tetrahydridoaluminate and acyl fluorides so providing a valuable route to linear compounds.

2,2-**Diethyl-1,2-oxagermacyclohexane,** (V) b.p. 83°/16 mm, n_D^{20} 1.4731, has been obtained in 75% yield using two conventional reactions (*P. Massol, J. Barrau* and *J. Satgé*, J. heterocycl. Chem., 1970, **7,** 783):

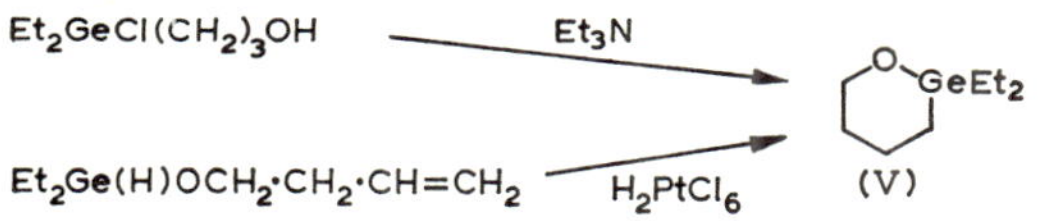

(V)

The chloroplatinic acid-catalyzed cyclization of but-3-enoic dimethylsilyl and dimethylgermyl esters leads to sila- and germa-lactones (VI).

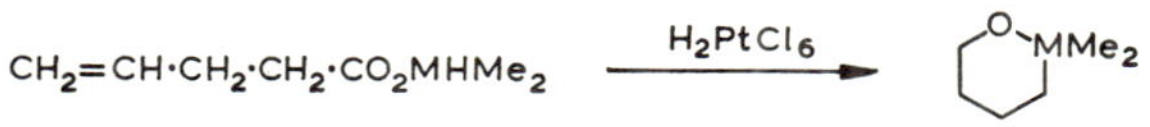

(VI)

2,2-**Dimethyl-1,2-oxasilacyclohexan-6-one** (VI, M = Si), has b.p. 69–71°/1.5 mm, n_D^{20} 1.4640 (*Mironov* and *Fedotov*, Khim. geterotsikl. Soedin., 1967, 179; C.A., 1967, **67,** 43862); it is also obtained by acidolysis of the disiloxane $[Me_3SiO_2C(CH_2)_3SiMe_2]_2O$ (*Mironov, Fedotov* and *I. G. Rybalka, ibid.*, 1969, 440; C.A., 1970, **72,** 12809). Other known derivatives of this system include the 2,2-*dichloro-*, b.p. 99–100°/1.5 mm, n_D^{20} 1.4859; 2-*chloro-2-phenyl-*, b.p. 158–159°/2.5 mm, n_D^{20} 1.5460, and 2-*chloro-2-methyl-* derivatives, b.p. 96–97°/2 mm, n_D^{20} 1.4753. 2,2-**Dimethyl-1,2-oxagermacyclohexan-6-one** (*Mironov et al.*, Zhur. obshcheĭ Khim., 1971, **41,** 2245; C.A., 1972, **76,** 113324).

(*iii*) *Perhydro-1,3-oxasilins and perhydro-1,3-thiasilins* (*1,3-oxasila- and 1,3-thiasila-cyclohexanes*)

The slow distillation of 3-hydroxypropyl(chloromethyl)silanes leads to derivatives of 1,3-**oxasilacyclohexane** (VII), 3,3-*dimethyl-*, b.p. 135–138°, n_D^{22} 1.4444; 3,3,6-*trimethyl-*, b.p. 147–148°; 6-*ethyl*-3,3-*dimethyl*, b.p. 163–169°; 3,3-*dimethyl-6-propyl-*, b.p. 190–191°.

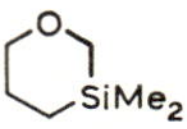
(VII)

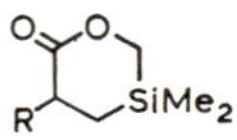
(VIII)

1,5-Dibromo-2,2-dimethyl-2-silapentane, conveniently obtained by treating 3,3-dimethyl-1,3-oxasilacyclohexane with sulphuric acid and 48% aqueous hydrobromic acid, reacts with sodium sulphide in ethanol to give 3,3-**dimethyl**-1,3-**thiasilacyclohexane,** b.p. 181°, n_D^{26} 1.4978 (*R. J. Fessenden* and *M. D. Coon*, J. org. Chem., 1964, **29,** 1607).

After preliminary addition of chloromethyldimethylsilane to trimethylsilyl acrylate, subsequent chloroplatinic acid-catalyzed cyclization yields 3,3-**dimethyl**-1,3-**oxasilacyclohexan-6-one** (VIII, R = H), b.p. 66–68°/3 mm, n_D^{20} 1.4555, which readily forms a *dimer*, m.p. 76–77.5° (*Mironov*, *Fedotov* and *Kozlikov*, Khim. geterotsikl. Soedin., 1969, 221; C.A., 1969, **71,** 50042). The 3,3,5-*trimethyl* compound VIII, (R = Me) b.p. 93–94°/3 mm, n_D^{20} 1.4655, is obtained by treatment of the appropriate 2-(chloromethyldimethylsilyl)-butyric acid with anhydrous sodium carbonate at 100–120° (*Mironov* and *Fedotov*, Khim. geterotsikl. Soedin., Akad. Nauk Latv., 1966, 453; C.A., 1967, **66,** 76077).

(*iv*) *Dibenz*[1,4]*oxa-silins, -germanins, and -stannins, and their thia analogues*

Dichlorosilanes, RR^1SiCl_2, react readily with 2,2′-dilithiodiphenyl ether in ether, or, better, tetrahydrofuran–ether mixture, under nitrogen (*K. Oita* and *H. Gilman*, J. Amer. chem. Soc., 1957, **79,** 339; *C. H. S. Hitchcock*, *F. G. Mann* and *A. Vanterpool*, J. chem.

M = Si, Sn
(IX)

Soc., 1957, 4537; *Gilman* and *W. J. Trepka*, J. org. Chem., 1961, **26,** 5202) or with 2,2′-dibromodiphenyl ether in the presence of sodium (*Tung-Yin Yu*, *Lin-Yuen Hsu* and *Shih-Huei Wu*, *Hua Hsueh Hsueh Pao*, 1958, **24,** 170; C.A., 1959, **53,** 6233) to yield 5,5-di-substituted-10*H*-**dibenz**[*b*,*e*][1,4]**oxazilins** (**phenoxasilins,** formerly named phenoxasilanes, and silaxanthenes) (IX, M = Si). In addition the 10,10-diphenyl compound has been obtained in about 2% yield by heating diphenylsilane with phenoxathiin for 6 days (*Gilman* and *D. Wittenberg*, J. Amer. chem. Soc., 1957, **79,** 6339; *Wittenberg*, *H. A. McNinch* and *Gilman ibid.*, 1958, **80,** 5418), and in 20% yield when the dilithiodiphenyl ether is acted upon in tetrahydrofuran by diphenylsilane (*Gilman* and *Trepka loc. cit.*). Some 10,10-disubstituted-phenoxasilins are listed in Table 5.

Reactions of the phenoxasilin system have been investigated by *Gilman* and *Trepka* (J. org. Chem., 1962, **27,** 1418). The action of alkali on the 10-phenyl compound produces

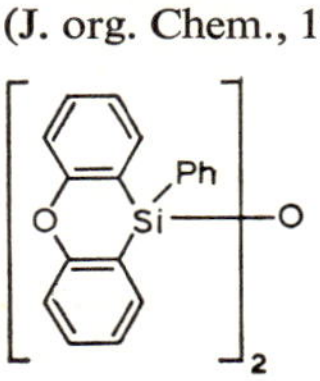
(X)

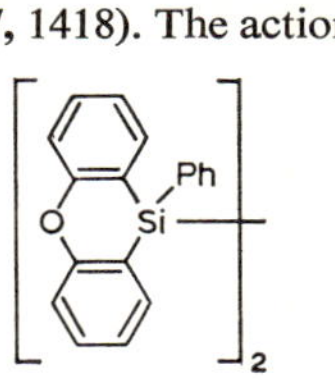
(XI)

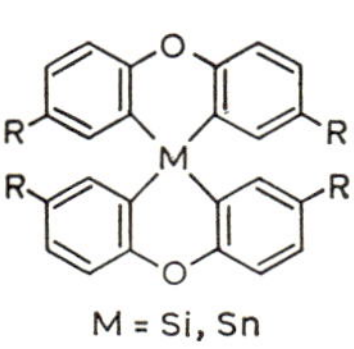
M = Si, Sn
(XII)

TABLE 5

PHENOXASILINS (IX, M = Si)

Substituents at M		*Phenoxasilins m.p. (°C)*	*Ref.*
Me	Me	74.5–77 (1) (b.p. 159–161/9 mm)	1
Ph	H	81–83	2
Ph	Br	96–99	2
Ph	OH	124–125	2
Ph	Me	56.5–58	2
Ph	Ph	176–178	2,3
Ph	*o*-Tolyl	116–117.5	2
Ph	*p*-Tolyl	155–157	2
Ph	CH_2Ph	89–90	2

References
1 *K. Oita* and *H. Gilman*, J. Amer. chem. Soc., 1957, **79**, 339.
2 *D. Wittenberg et al., ibid.*, 1958, **80**, 5418.
3 2,8-*Dimethyl*- derivative, m.p. 174° (*Tung-Yin Yu et al.*, Hua Hsueh Hsueh Pao, *loc. cit.*); 2-*methyl*-, m.p. 153–155; 3-*methyl*-, m.p. 161–162 *(Hitchcock et al., loc. cit.)*.

the 10-hydroxy-10-phenyl derivative, which may be dehydrated to the *disiloxane* X, m.p. 189.5–191°, when treated with formic acid. The 10-phenyl compound also reacts with *N*-bromosuccinimide in carbon tetrachloride yielding the 10-bromo-10-phenyl compound; this, when acted upon by sodium, gives 10,10′-**bi**(10-**phenylphenoxasilin**) (XI), m.p. 231–233°.

Reaction of 2,2′-dilithiodiphenyl ether with silicon tetrachloride yields 10,10′-**spirobi-(phenoxasilin)** (XII, M = Si, R = H), m.p. 284–285° *(Hitchcock et al., loc. cit.)*. The heterocyclic ring is cleaved by the action of lithium in dioxane, otherwise, the compound appears to be very stable. Direct electrophilic substitution appears to be impracticable, for the compound is recovered unchanged after treatment with fuming nitric acid in hot acetic anhydride, with bromine in boiling chloroform, and with *N*-bromosuccinimide in boiling carbon tetrachloride.

Because of the tetrahedral nature of the bonding around the silicon atom and probable folding of the phenoxasilin molecule about the silicon–oxygen axis, an unsymmetrically-substituted derivative should exist in enantiomeric forms. In attempts to obtain evidence for this molecular geometry, *Hitchcock et al.* converted the 2-methylphenoxasilin into the 2-bromomethyl compound by the action of *N*-bromosuccinimide, with the intention

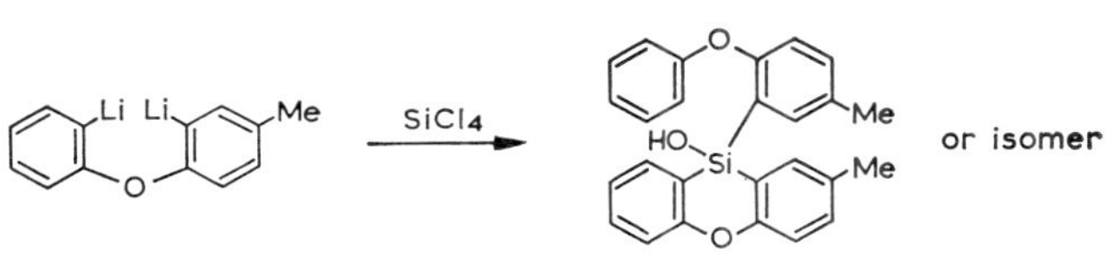

(XIII)

of converting the bromomethyl derivative into suitable model compounds for resolution purposes. However, the bromomethyl compound failed to react with ammonia and with potassium phthalimide, and would give no decisive product with either potassium cyanide or diethyl sodiomalonate. Interestingly, the action of silicon tetrachloride on 2,2′-dilithio-4-methyldiphenyl ether is such as to give a 10-*aryl*-10-*hydroxyphenoxasilin*, (XIII), m.p. 198–200°, rather than the expected spiran (XII, M = Si, R = Me).

The reactions used in the preparation of the phenoxastannins (IX, M = Sn) and the spirans (XII; M = Sn) are similar to those used for the silicon analogues. Recently reported work by *J. K. Terloww et al.* (J. organometal. Chem., 1974, **64,** 205) and *H. A. Meinema* and *J. G. Noltes* (*ibid.*, 1973, **63,** 243) would appear to show, however, that the supposedly monomeric nature of compounds described earlier by *E. J. Kupchik*, *J. A. Ursino* and *P. R. Boudjouk* (*ibid.*, 1967, **10,** 269) are, in point of fact, dimeric. Such results thus demonstrate, once again, the tendency of tin to provide polymeric products rather than the expected monomeric ring compounds.

The reaction between 2,2′-dilithiodiphenyl ether and dialkyltin dichlorides in a solvent mixture provides, at best, only moderate yields of the monomeric **phenoxastannins,** 10,10-*dimethyl*-, m.p. 49° (*Kupchik et al.* recorded m.p. 259–260°), 10,10-*diethyl*-, b.p. 116°/10^{-4} mm (*Kupchik et al.* reported m.p. 161–162°), together with smaller amounts of substances having the physical properties reported by *Kupchik* and now identified, particularly by mass-spectral measurements, as 12-membered ring compounds (p. 368) (*Meinema* and *Noltes, loc. cit.*; *Terloww et al., loc. cit.*). Although the latter work was limited to a study of the dimethyl and diethyl compounds, it must necessarily cast some doubt on the nature of the other compounds reported by *Kupchik.*

A nuclear magnetic resonance study of the 10,10-dimethylphenoxastannin, reported by the latter authors, seems to indicate that the molecule exists as rapidly inverting butterfly conformations.

A notable chemical feature is the formation of tetraphenyltin when diphenyltin dichloride is employed in the above reactions, a result ascribed by *Kupchik* to decomposition of the heterocyclic system.

When the same dilithiodiphenyl ether is treated with stannic chloride, there results a small yield of 10,10′-**spirobi(phenoxastannin)** (XIII, M = Sn, R = H), m.p. 220–221°, together with, once again, a dimeric *bispiro compound*, in this case having a bispiran structure and of m.p. 350°. 2,2-Dichlorophenoxastannin is formed only when 10,10′-spirobi(phenoxastannin) is heated strongly with more stannic chloride; it reacts normally with Grignard and lithium reagents (*Kupchik et al., loc. cit.*).

For the synthesis of 10,10-**diphenylphenothiasilin** (XIV, M = Si, R = Ph), m.p. 157–158°, *Gilman* and *Wittenberg (loc. cit.)* employed a novel exchange reaction consisting of heating together thianthrene and diphenylsilane at 250° for a long period; the yield

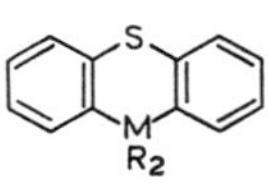

M = Si, Ge, Sn
(XIV)

is only 5%. The compound is oxidized by hydrogen peroxide in acetic acid to the *sulphone*, m.p. 208–209°, identical with that obtained by interaction of dichlorodiphenylsilane with 2,2′-dilithiodiphenylsulphone (*Oita* and *Gilman*, J. org. Chem., 1957, **22,** 336).

10,10-**Diphenylphenothiagermanin** (XIV, M = Ge R = Ph) and 10,10′-**spirobi(pheno-**

thiagermanin) are known only as their *octa-* and *perfluoro* derivatives, m.p. 140–143° and 229.5–231.5°, respectively, obtained from 2,2′-dilithioperfluorodiphenyl sulphide (*S. C. Cohen* and *A. G. Massey*. J. organometal. Chem., 1968, **12,** 341). The spiro-germanin is stable towards sulphur up to 360°, unlike **octafluoro-10,10-diphenylphenothiagermanin,** which at 230° loses germanium as diphenylgermanium sulphide, and thus yielding octafluorothianthrene.

It is worth recording that 2,2′-dilithiodiphenyl sulphone and sulphide, unlike the ether, react normally with diphenyl- and dimethyltin dichloride; the former yields 10,10-*dimethylphenothiastannin* 5,5-*dioxide*, m.p. 164–165° [also by oxidation of 10,10-*dimethylphenothiastannin*, m.p. 95–96°, with 3-chloroperoxybenzoic acid (*Meinema* and *Noltes*, *loc. cit.*)], and with stannic chloride to give the expected 10,10′-*spirobi*(*phenothiastannin* 5,5-*dioxide*), m.p. 315–320° (*Kupchik et al., loc. cit.*).

(v) *Perhydro-1,4-oxasilins (1,4-oxasilacyclohexanes) and germanium analogues*

Remarkably little is known about the chemistry of 1,4-oxasilacyclohexane and its analogues.

4,4-**Diphenyl**-1,4-**oxasilacyclohexane** (XV), m.p. 178–179°, has been obtained as a by-product, yield 2%, in reactions between diphenylsilane and sulphur-containing heterocycles (*Gilman* and *Wittenberg, loc. cit.*). 4,4-*Dimethyl*-1,4-*oxagermacyclohexan*-2-*one* (XVI) has b.p. 102°/2 mm, n_D^{20} 1.4927 (*Mironov, S. A. Mikhailyants* and *T. K. Gar*, Zhur. obshcheĭ Khim., 1969, **39,** 2281).

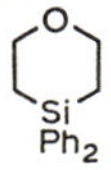

(XV)

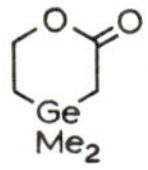

(XVI)

(b) *Dioxasilins, dioxagermanins, dioxastannins and their thia analogues*

(i) *Benzo- and dibenzo-[1,3,2]dioxasilins and analogues*

Dichlorodimethylsilane and 2-hydroxybenzyl alcohol afford 2,2-**dimethyl**-4*H*-1,3,2-**benzodioxasilin** (XVII), b.p. 52–54°/0.05 mm, (*diphenyl*, b.p. 166–168°/0.05 mm). Dichlorodibutylgermane yields **2,2-dibutyl**-4*H*-1,3,2-**benzodioxagermanin,** b.p. 122–126°/0.1 mm (*diphenyl*, b.p. 182–186°/0.1 mm) (*R. H. Cragg* and *M. Nazery*, J. organometal. Chem., 1974, **71,** 225).

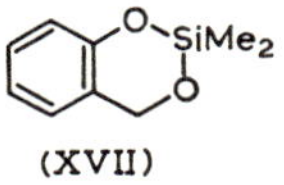

(XVII)

1,8-Dihydroxynaphthalene condenses with dichlorodimethylsilane to yield monomeric **2,2-dimethylnaphtho**[1,8,*d,e*][1,3,2]**dioxasilin** (XVIII) m.p. 53.5–55°, in 7% yield; the *diphenyl* compound has m.p. 110–111. **2,2′-Spirobi(naphtho**[1,8,*d,e*][1,3,2]**dioxasilin)**

(XIX, M = Si), m.p. 329–331°, is obtained similarly, using silicon tetrachloride (*C. M. Silcox* and *J. J. Zuckerman*, *ibid.*, 1966, **5**, 483):

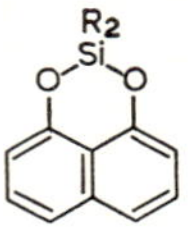

(XVIII)

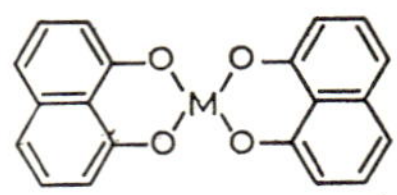

M = Si, Ge
(XIX)

The *germanium analogue* (XIX, M = Ge), m.p. 335–337°, is very reactive towards water (*idem.*, Inorg. Chem., 1967, **6**, 163). The tin analogue (XIX, M = Sn) is formed from tetraphenyltin and 1,8-dihydroxynaphthalene at 150°, with elimination of benzene (*K. Andrä* and *H. R. Hoppe*, Z. anorg. allgem. Chem., 1975, **413**, 97).

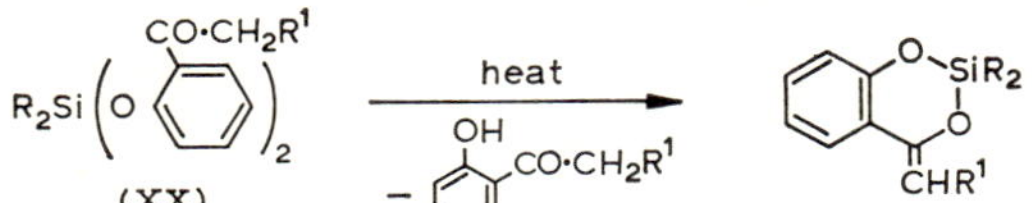

(XX) (XXI)

When alkyl 2-hydroxyaryl ketones react with dichlorosilanes in the presence of pyridine, the products are the di(aryloxy)silanes of type XX, or the hydrolytically unstable 4*H*-1,3,2-benzodioxasilins XXI (Table 6) depending upon the ratio of reactants; moreover, the "linear" silanes XX themselves undergo conversion to the benzosilins XXI when heated (*R. M. Ismail*, J. organometal. Chem., 1968, **11**, 49). Salicylanilides give similar cyclic products but only when the ratio of reactants is 1:1 (*Ismail* and *E. Bessler*, *ibid.*, 1968, **13**, 253).

TABLE 6

4-ALKYLIDENE-4*H*-1,3,2-BENZODIOXASILINS (XXI)

R	R^1	*Boiling point (°C/mm)*	n_D^{20}
Me	Me	80/0.2	1.5371
Ph	H	—	1.6118
Me	Me	80/0.2	1.5358
Ph	Me	—	1.6026
Me	Et	81/0.5	1.5296
Ph	Et	—	1.5747

The ready interaction of salicyclic acid and dichlorodimethylsilane, or tetramethoxysilane in the presence of toluene-4-sulphonic acid, to yield *2,2-dimethyl-4H-1,3,2-benzodioxasilin-4-one* (XXII, M = Si), b.p. 119°/4 mm, and the 2,2-dimethoxy analogue, is of particular interest since no reaction takes place between tetraethoxysilane and phenol or benzoic acid, nor between diethoxydimethylsilane and either catechol or phthalic acid (*M. Wieber* and *M. Schmidt*, Chem. Ber., 1963, **96**, 1561; *R. C. Mehrotra* and *R. P. Narain*, Indian J. Chem., 1968, **6**, 110; *Mehrotra* and *B. C. Pant*, J. Indian chem. Soc., 1963, **40**, 623; *J. Lukasiak*, *A. Radeoki* and *J. Halkiewicz*, Rocz. Chem., 1974, **48**, 1099).

2,2-Dimethyl-4*H*-1,3,2-benzodioxagermanin-4-one (XXII, M = Ge) has b.p. 160°/3 mm,

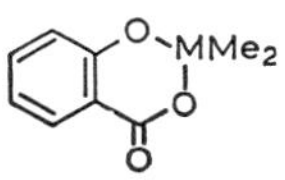

(XXII)

m.p. 156–162°. The 2,2-**diethoxytitanium** compound, m.p. 166–168°, is described as an orange, hexameric compound (*G. H. Dahl* and *B. P. Block*, Inorg. Chem., 1966, **5**, 1394); **zirconium** compounds have also been reported as have some of **tin**(II) but not of tin(IV) (*D. E. Fenton et al.*, Inorg. Chim. Acta, 1970, **4**, 235).

2,2-Dimethyl-4*H*-3,1,2-benzoxathiasilin-4-one, b.p. 158°/10 mm, m.p. 100°, and its *germanium* analogue, b.p. 160°/3 mm, m.p. 166–169°, are obtained from 2-mercapto-benzoic acid (*Wieber* and *Schmidt*, *loc. cit.* and Z. Naturforsch., B, 1963, **18**, 848).

(*ii*) *Perhydro*-1,3,2-*dioxa-silins* (1,3,2-*dioxasila-cyclohexanes*), *germanins, -stannins, -plumbins, -titanins, and -zirconins and their thia analogues*

1,3-Diols react (*a*) with dichlorosilanes in the presence of suitable organic bases, and (*b*) by transesterification with dialkoxysilanes, generally in the presence of sodium alkoxide or toluene-4-sulphonic acid, to yield 1,3,2-

TABLE 7

1,3,2-DIOXASILACYCLOHEXANES

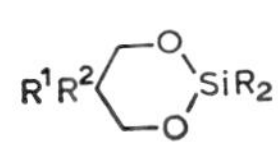

R	*R*¹	*R*²	*B.p.* (°C/mm)	n_D^{20}	*Ref.*
Me	H	H	122/763	1.4130	1
	H	OH	147/18	1.4554	1
	H	Me	29/0.8	—	2
	H	Bu*t*	69–71/9.5	—	2
	Me	Me	24–25/5	1.1417	3
	Me	NO_2	109–110/12		1
	Me	NH_2	67–59/9		1
	Me	Pr	37–39/2	1.429	3
	Et	Et	45/3	1.434	3
	Et	Bu	72–73.5/4	1.438	3
Et	Me	Me	42–43/5	1.4333	1
	Et	Et	72–74/4	1.444	3
	Me	Pr	64–66/4	1.440	3
Bu	H	H	87/2	1.4398	4
Ph	Me	Me	149–150.5/2.5 m.p. 63–64		3

References

1 *M. G. Voronkov* and *Yu. P. Romadane*, Khim. geterotsikl. Soedin., 1966, 879; C.A., 1967, **67**, 32775.
2 *P. Allriktsen* and *S. Heggelund*, Acta Chem. Scand., 1974, **B28**, 573.
3 *N. Spassky*, Compt. rend., 1960, **251**, 2371.
4 *J. C. Pommier*, *R. Calas* and *J. Valade*, Bull. Soc. chim. Fr., 1968, 1475.

dioxasilacyclohexanes (XXIII, M = Si) (Table 7), (*N. Spassky*, Compt. rend., 1960, **251,** 2371; *M. G. Voronkov* and *Yu. P. Romadane*, Khim. geterotsikl. Soedin., 1966, 879; C.A., 1967, **67,** 32775; *R. H. Kriebel* and *C. A. Burkhard*, J. Amer. chem. Soc., 1947, **69,** 2689). Similar reactions have

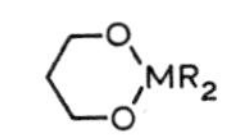

M = Si, Ge, Sn
(XXIII)

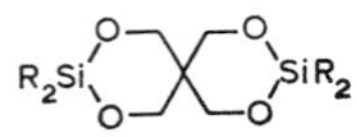

(XXIV)

been employed for the germanium (XXIII, M = Ge) and tin (XXIII, M = Sn) series (Table 8), but in addition, the use of diaminometal compounds $R_2M(NEt_2)_2$ (M = Ge, Sn) should also be noted (*S. Mathur et al.*, J. organometal. Chem., 1965, **4,** 371; *R. C. Mehrotra and Mathur, ibid.*, 1966, **6,** 425; *G. Korb et al., ibid.*, 1970, **23,** 437; *Mehrotra* and *V. P. Gupta, ibid.*, 1965, **4,** 145). Titanium and zirconium compounds are similarly prepared using the less reactive tetra-alkoxy-titanium and -zirconium (*Dahl* and *Block, loc. cit.*). A method applied to the preparation of tin compounds consists in the azeotropic removal of water from mixtures of diol and organotin oxide in hot toluene or xylene (*Voronkov* and *I. Romadane*, Khim. geterotsikl. Soedin., 1966, 892; C.A., 1967, **67,** 32751).

TABLE 8

1,3,2-DIOXAGERMACYCLOHEXANES AND
1,3,2-DIOXASTANNACYCLOHEXANES (XXIII, M = Ge, Sn)

M	*R*	*Ring substituents*	*B.p. (°C/mm)*	*Ref.*
Ge	Et	None	114–118/7	1
		4-Me	82/16	2
	Bu	None	112–115/8	3
		4-Me	101–105/3	4
		4,4,6-Me_3	104–107/2	4
		4,4,6,6-Me_4	105–108/2.5	4
	Ph	None	150/0.6	5
		4-Me	135/0.8	5
Sn	Et	None	m.p. 169–173	6
	Bu	None	184–185/0.3	6
		4-Me	165/0.25	6

References

1 *R. C. Mehrotra* and *G. Chandra*, J. chem. Soc., 1963, 2804.
2 *J. Satgé* and *G. Doussie*, Helv., 1972, **55,** 2406.
3 *G. Korb et al.*, J. organometal. Chem., 1970, **23,** 437.
4 *S. Mathur et al., ibid.*, 1965, **4,** 371.
5 *Mehrotra* and *S. Mathur, ibid.*, 1966, **6,** 425.
6 *V. G. Voronkov* and *I, Romadane*, Khim. geterotsikl. Soedin, 1966, 892; C.A., 1967, **67,** 32751.

trans-Cyclohexane-1,3-diol does not form a cyclic dimethylsilyl ester of the present type, and the ready hydrolysis of the monomeric *cis* ester, 7,7-**dimethyl**-6,8-**dioxa**-7-**silabicyclo**[3.3.1]**nonane,** b.p. 46–48/1.3 mm, n_D^{20} 1.4560, to the free diol, thus permits a ready separation of *cis*- and *trans*-cyclohexane-1,3-diols (*J. C. Pommier, R. Calas* and *J. Valade*, Bull. Soc. chim. Fr., 1968, 1475). The formation of 3,3,9,9-**tetramethyl**-2,4,8,10-**tetraoxa**-3,9-**disilaspiro**[5.5]**undecane** (XXIV), m.p. 136, b.p. 243–244/749 mm, from pentaerythritol and diethoxydimethylsilane or, better, diacetoxydimethylsilane (*L. W. Breed, W. J. Haggerty* and *J. Harvey*, J. org. Chem., 1960, **25,** 1804; *V. P. Davydova* and *Voronkov*, Zhur. obshcheĭ Khim., 1958, **28,** 1879) has been questioned (*L. Maijes* and *B. Pavare*, Latv. P.S.R., Zinat Akad. Vestis, Khim. Ser., 1968, 95; C.A., 1968, **69,** 59311). Dipole moment and ^{1}H-nuclear resonance measurements for 1,3,2-dioxasilacyclohexanes are interpreted in terms of predominant chair conformations (*L. K. Yuldasheva et al.*, Izvest. Akad. Nauk S.S.S.R., Ser. Khim., 1974, 323).

By and large, the silicon and germanium compounds (XXIII, M = Si, Ge) appear to be monomeric; dialkyltin compounds are evidently dimeric, although the nature of this association is by no means certain (the esters from 1,2-diols are also "dimeric" but are thought to possess 10-membered rings and are considered briefly elsewhere). 2,2-*Diethoxy*-4,4,6-*trimethyl*-1,3,2-*dioxatitanacyclohexane*, b.p. 136–144°/0.06 mm, n_D^{25} 1.5100, is trimeric in boiling benzene.

All the esters (XXIII and XXIV) appear to be labile under aqueous conditions, undergoing ring-opening. The tin esters participate in an unusual reaction (although in common with esters from 1,2- and 1,4-diols) when treated with carbon disulphide at 100–110°; repetition of a process thought to consist in ring-opening and insertion of carbon disulphide, followed by elimination of dialkyltin sulphide leads finally to 1,5,7,11-**tetraoxaspiro**[5.5]**undecane** in 92% yield (*S. Sakai, Y. Kobayashi* and *Y. Ishii*, J. org. Chem., 1971, **36,** 1176).

An unusual instance of the transesterification process is that which takes place between 1,3-diols and tetra-acetoxysilane in 1:1 ratio, when the products are not the expected 2,2-diacetoxy-1,3,2-dioxasilacyclohexanes, but actually the 1,5,7,11-**tetraoxa**-6-**silaspiro**-[5.5]**undecanes** (XXV, M = Si), half of the tetra-acetoxysilane being recoverable:

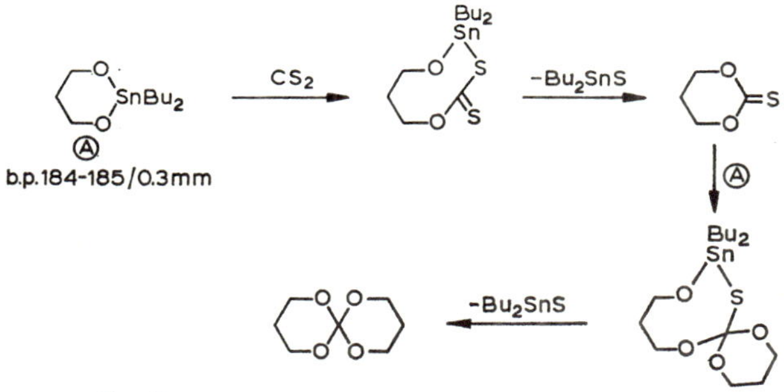

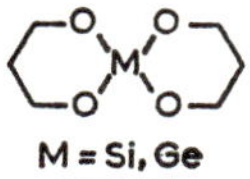

M = Si, Ge
(XXV)

Such esters (Table 9) are also formed by the depolymerization of polymers obtained from tetraethoxysilane and 1,3-diols (*Mehrotra* and *Pant*, J. Indian chem. Soc., 1964, **41,** 563). The ease of the depolymerization process depends on the number and positions of the substituents, as does the stability of the ring system (*Voronkov*, *I. Romadane* and *V. A. Pestunovich*, Khim. geterotsikl. Soedin, 1968, 972; C.A., 1969, **70,** 68820; *C. L. Frye*, J. org. Chem., 1969, **34,** 2496). Another notable feature of the chemistry of this class of compound is the ease with which a mixture of symmetrically substituted esters equilibrates to a single unsymmetrically substituted compound.

TABLE 9

1,5,7,11-TETRAOXA-6-SILA(GERMA)SPIRO[5,5]UNDECANES(XXV)

Ring substituents	*B.p. °C/mm (m.p., °C)*	
	Si(1)	*Ge (2)*
2,8-Me_2	67/2	93/0.4
3,3,9,9-Me_4	—	90–95/0.8–0.9
2,2,4,8,8,10-Me_6	(73–77)	—
3,3,9,9-Et_4	130–133/0.8	—
3,9-Et_2-3,9-Bu_2	150–155/0.5	—

References

1 *R. C. Mehrotra* and *B. C. Pant*, J. Indian. chem. Soc., 1964, **41,** 563.
2 *Mehrotra* and *G. Chandra*, J. chem. Soc., 1963, 2804.

Germanium analogues are produced in a similar way (*Mehrotra* and *G. Chandra*, J. chem. Soc., 1963, 2804).

5-Substituted-2,2-**diethyl**-1,3,2-**dioxasilacyclohexan**-4-**ones** (XXVI) are obtained by reaction between diethylsilane and the appropriate β-hydroxycarboxylic acid in the presence of colloidal nickel or palladium at 50–60° (*I. I. Lapkin*, *T. N. Povarnitsyna* and *T. Yu. Denisova*, Zhur. obshcheĭ Khim., 1972, **42,** 2032).

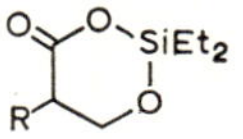

(XXVI)

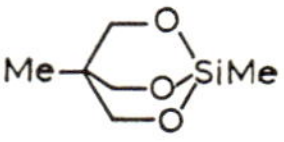

(XXVII)

Condensation between methyltrimethoxysilane and trimethylolethane is catalyzed by methoxide and provides a low yield of 1,4-*dimethyl*-2,6,7-*trioxa*-1-*silabicyclo*[2.2.2]*octane* (XXVII), m.p. 106–108° (*J. W. Ratkke*, *J. W. Guyer* and *J. G. Verkade*, J. org. Chem., 1970, **35,** 2310).

A similar reaction with *cis*-phloroglucitol, yields 1-*methyl*-2,8,9-*trioxa*-1-*sila-adamantane* (XXVIII), m.p. 54–55° (1-*phenyl*, m.p. 172.5–173°; 1-*vinyl*-, m.p. 66–67°) as hygroscopic powders (*A. I. Kuznetsov et al.*, Zhur. org. Khim. 1976, **12,** 227).

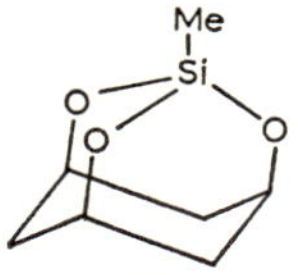

(XXVIII)

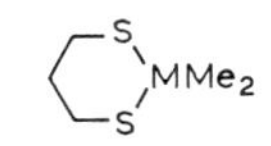

M = Si, Ge, Sn
(XXIX)

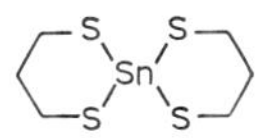

(XXX)

2,2-Dimethyl-1,3,2-dithiasilacyclohexane (XXIX, M = Si), b.p. 93°/12 mm, m.p. 6–8°, (*Wieber* and *Schmidt*, J. organometal. Chem., 1964, **1**, 336), from dichlorodimethylsilane and propane-1,3-dithiol, readily hydrolyses to that dithiol, and like the corresponding five-membered ring *compound*, undergoes a remarkable ring expansion reaction when treated with perfluoroketones; perfluoroacetone yields the monocyclic *compound* $Me_2Si{\cdot}OC(CF_3)_2{\cdot}SMe_2{\cdot}SC(CF_3)_2O$, b.p. 120°/10 mm, (*E. W. Abel, D. J. Walker* and *J. N. Wingfield*, J. chem. Soc., A, 1968, 2642). **2,2-Dimethyl-1,3,2-dithiagermacyclohexane** (XXIX, M = Ge), b.p. 103–105°/10 mm, m.p. 24–27°, is also obtained conventionally (*Wieber* and *Schmidt*, *loc. cit.*). The same authors also report **2,2-dimethyl-1,3,2-dithiastannacyclohexane** (XXIX, M = Sn), b.p. 113–117°/3 mm, but other compounds, 2,2-*dibutyl-*, m.p. 63–64°, and 2,2-*diphenyl-*, m.p. 103–104°, colourless materials soluble in organic solvents, were obtained in a study of the red colours produced during the reduction of sulphonyl chlorides by tin and hydrochloric acid (*R. C. Poller* and *J. A. Spillman*, J. chem. Soc., A, 1966, 958).

2-*Methyl-*2-*phenyl-*1,3,2-*dithiasila-*, b.p. 180–181°/0.7 mm, 2,2-*diphenyl-*1,3,2-*dithiagerma-*, m.p. 65–66°, and 2-2-*diphenyl-*1,3,2-*dithiastanna-*, m.p. 90–92°, *-cyclohexanes*, have also been prepared from the metalloid dichloride and the synthetically useful lead salt of the dithiol (*R. H. Cragg* and *A. Taylor*, J. organometal. Chem., 1975, **99**, 391).

Treatment of benzylidenedithiols with lead(II) acetate gives **1,3,2-dithiaplumba(II) cyclohexanes,** cleaved by hydrogen sulphide to the parent dithiol (*V. A. Portnyorgina et al.*, Khim. geterotsikl. Soedin., 1970, 605; C.A., 1970, **73**, 87881).

Stannic chloride and propane-1,3-dithiol react together to give **1,5,7,11-tetrathia-6-stannaspiro[5.5]undecane** (XXX) m.p. 156–157°, insoluble in organic solvents, but which forms a polymeric *pyridine complex*, m.p. 100–101°, which loses pyridine when dissolved in chloroform (*Poller* and *Spillman*, *loc. cit.*).

A few 1,3,2-oxathia- compounds have been reported (*M. I. T. Chemicals Inc.*, Neth. Appl. 6,516,434/1966; C.A. 1966, **65**, 13763).

(iii) 1,4,2-Benzodioxasilins; 1,4,2-benzodioxagermanins

When a solution of catechol and bromomethyl(ethoxy)dimethylsilane in xylene is heated, ethanol is liberated in a transesterification step; subsequent addition of triethylamine results in ring-formation yielding **2,2-dimethyl-1*H*,4*H*-2,3-dihydro-1,4,2-benzo-**

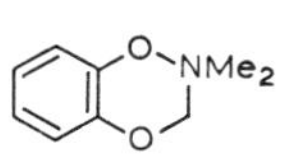

M = Si, Ge
(XXXI)

dioxasilin (XXXI, M = Si), b.p. 76–79°/0.7 mm, 102–106°/10 mm, n_D^{20} 1.5277, also produced from halogenomethyldimethylhalogenosilanes and catechol in the presence of triethylamine (*W. Simmler*, Chem., Ber., 1963, **96**, 349; *V. P. Kuznetsov* and *R. M. So-*

kolovskaya, Zhur. obshcheĭ Khim., 1969, **39,** 1997; *M. Wieber* and *M. Schmidt*, J. organometal. Chem., 1963, **1,** 22; *K. Friederich, A. de Monigny* and *W. Wimmler*, Kolloid. Zeit. Z. Polymer, 1967, **218,** 7; C.A., 1967, **67,** 91292). The monomeric **germanium** analogue has b.p. 95–96°/1 mm (*Wieber* and *C. D. Frohning*, J. organometal. Chem., 1967, **8,** 459).

Homocatechol and homodithiocatechol afford mixtures of isomeric products when treated with dichloro-silanes and -germanes (*Wieber* and *Schmidt*, *ibid.*, 1964, **2,** 129; *Wieber* and *Frohning*, *loc. cit.*).

(*iv*) *Perhydro-1,4,2-dioxasilins* (1,4,2-*dioxasilacyclohexanes*), 1,4,2-*dioxagermacyclohexanes and their thia analogues*

2,2-**Dimethyl**-1,4,2-**dioxasilacyclohexane** (XXXII), b.p. 135°, n_D^{20} 1.4288, is prepared by interaction of ethylene glycol containing dissolved sodium and chloromethyldimethylsilyl chloride, and also from the same glycol and bromomethyldimethylethoxysilane by sequential elimination of ethanol and hydrogen bromide (*K. A. Andrianov* and *L. I.*

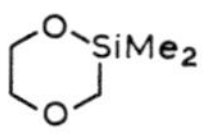

(XXXII)

Makarova, Doklady Akad. Nauk S.S.S.R., 1959, **127,** 1213; *Simmler*, Chem. Ber., 1963, **96,** 349). In the presence of sodium in dimethyl sulphoxide at 190°, the compound reacts with triethylsilanol to give linear siloxanes (*Andrianov*, *Makarova* and *V. A. Dmitriev*, Izvest. Akad. Nauk S.S.S.R., Ser. Khim., 1971, 1991). Compounds from cycloalkane-1,2-diols have been prepared more recently (*Z. Lasocki*, Syn. Inorg. Metal-org. Chem., 1973, **3,** 29; C.A., 1973, **78,** 97756).

2,2-**Dimethyl**-1,4,2-**oxathiasilacyclohexane** has b.p. 41°/2 mm, n_D^{20} 1.4800, and 2,2-**dimethyl**-1,4,2-**oxathiasilacyclohexan-6-one,** b.p. 95–97°/2 mm hydrolyses to $HO_2C\cdot CH_2SCH_2SiMe_2)_2O$ (*W. Simmler* and *H. Niederpieum*, Belg. P. 634,120/1963; C.A., 1964, **61,** 1893; *Wieber* and *Schmidt*, Chem. Ber., 1963, **96,** 2822).

2,2-**Dimethyl**-1,4,2-**dithiasilacyclohexane,** b.p. 88–91°/12 mm, n_D^{20} 1.5471, from ethanedithiol in the usual way, undergoes ring-opening when treated with dry hydrogen chloride to give 2-(chlorodimethylsilylmethylthio)ethanethiol (*Wieber and Schmidt, loc. cit.*).

(*v*) *1,3,4-Dioxagermacyclohexanes*

4,4-**Diethyl**-2-**methyl**-1,3,4-**dioxagermacyclohexane** (XXXIII, R = H, R^1 = Me) b.p. 75°/18 mm, and 2,2-*bis(trifluoromethyl)*-4,4-*diethyl*-1,3,4-*dioxagermacyclohexane* (XXXIII, R = R^1 = CF_3), b.p. 62°/0.7 mm, are prepared by insertion of acetaldehyde and hexafluoroacetone, respectively, into 2,2-diethyl-1,2-oxagermacyclobutane (*J. Barrau et al.*, Rec. Trav. chim., 1973, **92,** 321).

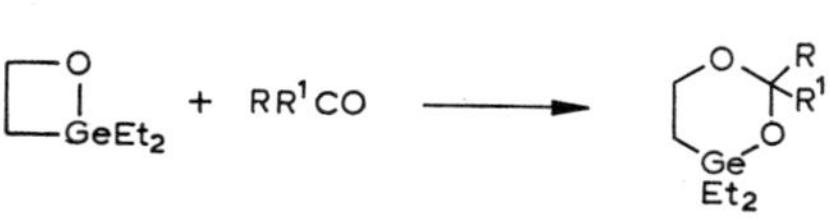

(XXXIII)

(*vi*) 1,2,6-*Oxathiasila- and* 1,2,6-*oxathiagerma-cyclohexane*

An interesting aspect of the chemistry of sila- and germa-cyclobutanes is the way in which, when treated with sulphur trioxide (free or as the dioxane complex) or with sulphur dioxide, they suffer ring-opening followed by non-stereospecific insertion of the sulphur

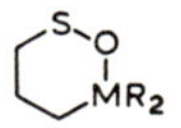

M = Si, Ge
(XXXIV)

oxide to give 1,2,6-oxathiasila- and 1,2,6-oxathiagerma-cyclohexane 6-monoxides and 6,6-dioxides (Table 10); when one or both of the substituents on the metalloid atom is chlorine or phenyl, the normal ring expansion reaction is accompanied by some exocyclic substitution (*J. Dubac et al.*, Compt. rend. C, 1968, **267,** 411; Bull. Soc. chim. Fr., 1969, 3608; J. organometal. Chem., 1970, **22,** C7; *ibid.*, 1970, **25,** 367, C20).

TABLE 10

1,2,6-OXATHIASILACYCLOHEXANES AND 1,2,6-OXATHIAGERMACYCLOHEXANES (XXXIV)

Substituents	*M = Silicon*		*M = Germanium*	
	6-Monoxide	*6,6-Dioxide*	*6-Monoxide*	*6,6-Dioxide*
		m.p. (b.p./mm)		
6,6-Me_2	(62/0.01)	73–75 (100/0.05)	—	—
6,6-Et_2	—	(142/0.4)	(103/0.5)	51–53 (170/0.2)
6,6-Bu_2	—	(130/0.01)	(111/0.07)	(160/0.02)
5,6,6-Me_3	—	56 (106/0.1)	—	—
4,6,6-Me_3	(60/0.05)	87 (115/0.2)	—	—
4-Me-6,6-Et_2	—	—	(104/0.2)	—
6,6-$(CH_2)_4$	—	24 (131/0.05)	—	—
6,6-$(CH_2)_5$	—	36 (145/0.1)	—	—
6,6-Cl_2	—	(110/0.04)	—	—

(*c*) *Trioxasilacyclohexanes; trithiastannacyclohexanes*

1,3,5-Trioxasilacyclohexanes (XXXV) have been briefly reported in connection with polymerization studies, being obtained from methylformal and methyltrichlorosilane in the presence of pyridine (*H. H. Ender*, J. appl. Polymer Sci., 1963, **7,** 1133). The 2,2-**dibutyl-1,3,5-trithiastannacyclohexane** (XXXVI), m.p. 120°, is formed by the azeotropic removal of water from a mixture of dibutyltin oxide and the appropriate dimercapto-thiadiazolidine (*C. C. Greco*, U.S.P. 3,201,408/1965; C.A., 1965, **63,** 13315).

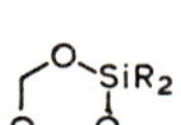

(XXXV)

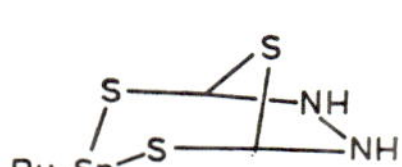

(XXXVI)

(*d*) *Oxadisilins, oxadigermanins and thia analogues*

(*i*) *2,1,3-Benzoxadisilins and related compounds*

Monocyclic 2,1,3-oxadisilins have not yet been prepared, but 1,1,3,3-tetramethyl-1*H*-2,1,3-benzoxadisilin (XXXVII), b.p. 120°/30 mm, n_D^{20} 1.4961, is obtained from 2-chlorobenzyl chloride by the route shown (*W. Piccoli*, U.S.P. 3,387,015–1968; C.A., 1968, **69,** 44015).

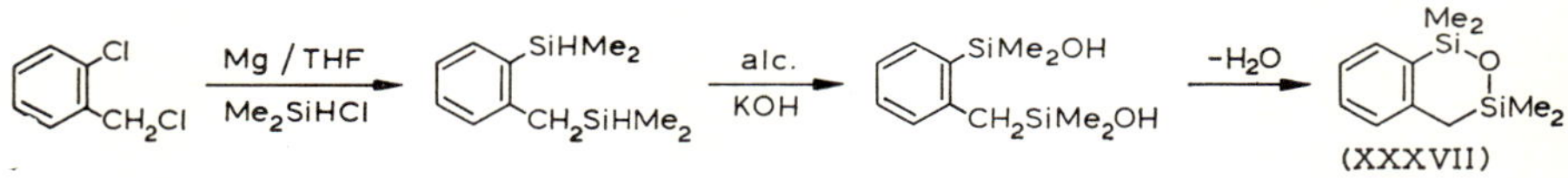

(XXXVII)

Pyrolysis of suitable naphthyldisiloxanes (XXXVIII) yields **2-oxa-2,3-dihydro-1,3-disilaphenalenes** (XXXIX), 1,3-dichloro-1,3-*dimethyl-*, b.p. 150–152°/4 mm, m.p. 127°, 1,1,3,3-*tetrachloro-*, b.p. 143–145°/1.5 mm, m.p. 133°; the former product can also be obtained from 1-chloronaphthalene and $(MeHClSi)_2O$. When R = H, a major product of the reaction is XL (*Chermyshev et al.*, Zhur. obshcheĭ Khim., 1975, **45,** 2229):

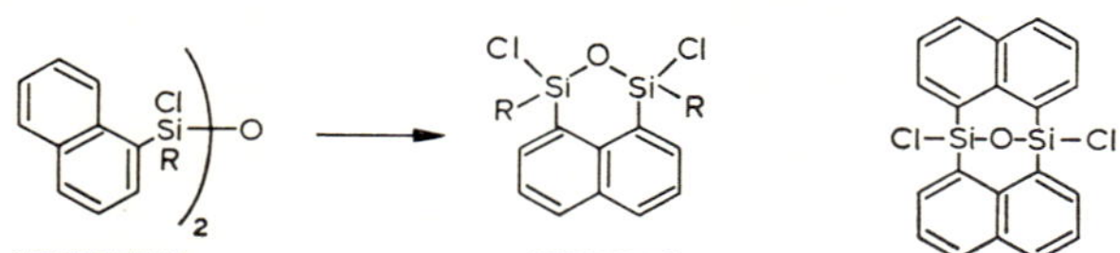

(XXXVIII) (XXXIX) (XL)

(*ii*) *1,2,6-Oxadisila-, 1,2,6-oxadigerma-cyclohexanes and thia analogues*

Several highly practicable syntheses of the 1,2,6-oxadisilacyclohexane ring system, its germanium counterpart, and their thia analogues, have been developed.

(*a*) The condensation between 1,3-di(halogenomethyl)disiloxanes and active methylene carbanions leads to 4,4-disubstituted-1,2,6-oxadisilacyclohexanes (*L. H. Somer*, U.S.P. 3,111,534/1963; C.A., 1964, **60,** 3011; *H. Niederpreuem*, *W. Simmler* and *E. M. Horn*, Belg. P. 667,783/1965; C.A., 1966, **64,** 19676; *Simmler*, *Niederpreuem and H. Walz*, Chem. Ber., 1964, **97,** 1047; *Niederpreuem* and *Simmler*, Z. anorg. allgem. Chem., 1966, **345,** 53), representative examples of which are listed in Table 11.

(*b*) The same compounds are also produced by intramolecular hydrolysis of di(ethoxysilyl)-ated active methylene compounds:

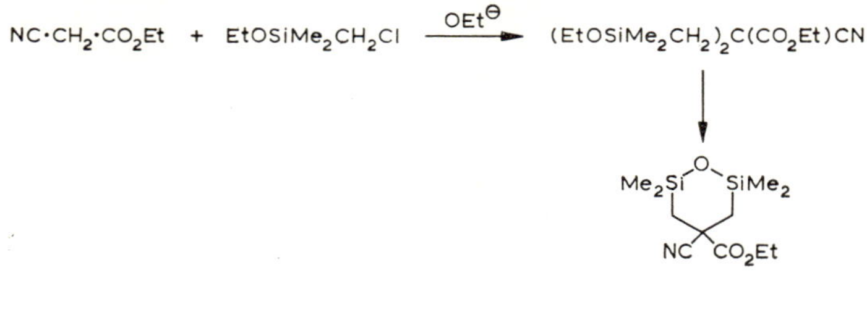

TABLE 11

1,2,6-OXADISILACYCLOHEXANES

R = R³ = Me					*R¹ = R² = H*			
R¹	*R²*	*M.p.(°C)*	*B.p. (°C/mm)*	*Ref.*	*R*	*R³*	*B.p. (°C/mm)*	*Ref.*
H	CN	55–56	105–108/1	1	Me	Cl	50–52/7	4
H	CO_2H	142–143	—	2	Me	OMe	49–50/10	4
H	CO_2Et	—	134/7	3				
H	$CONH_2$	139–140	170–180/12	2	Me	OPh	115–116/7	4
H	CH_2NH_2	—	87–88/12	2	Me	OAc	76–80/14	4
H	CH_2OH	—	68/0.4	1	Me	Et_2N	96–100/7	4
CN	CO_2Me	29–30	115–117/3	2	Me	OCH_2Ph	130–132/10	4
CN	CO_2Et	31–32	90/0.3	2	Cl	Cl	64–67/7	4
CN	$CONH_2$	166–167.5	—	2	OMe	OMe	59–61/7	5
CN	CO_2H	131–131.5	—	2	OPh	OPh	138–141/1	5
CN	Me	89–91	98–102/13	1	$OSiEt_3$	$OSiEt_3$	137–139/2	5
CN	CN	138–138.5	—	1	OMe	Ph	92–95/1	5
CN	CH_2OH	99–101	—	1	Ph	Ph	135–143/1	5
CO_2H	CO_2H	144–145	—	1				
CO_2Et	Co_2Et	—	95–97/1.5	1				
CH_2OH	CH_2OH	90–92	—	1				

References

1 *H. Niederpreuem* and *W. Simmler*, Z. anorg. allgem. Chem., 1966, **345,** 53.
2 *Simmler*, *Niederpreuem* and *H. Walz*, Chem. Ber., 1964, **97,** 1047.
3 *L. H. Somer*, U.S.P. 3,111,534/1963; C.A., 1964, **60,** 3011.
4 *K. A. Andrianov et al.*, Khim. geterotsikl. Soedin., 1968, 222; C.A. 1969, **71,** 3425.
5 *idem*, *ibid.*, 1967, 438; C.A., 1968, **68,** 22003.

Similar hydrolytic procedures are used to obtain more complex examples based on the same ring system (*K. A. Andrianov*, *L. M. Volkova* and *N. V. Delazari*, Khim. geterotsikl. Soedin., 1968, 222; C.A., 1969, **71,** 3425; *Andrianov et al.*, *ibid.*, 1967, 438; C.A., 1968, **68,** 22003; Doklady Akad. Nauk S.S.S.R., 1965, **160,** 1307; C.A., 1965, **62,** 14717):

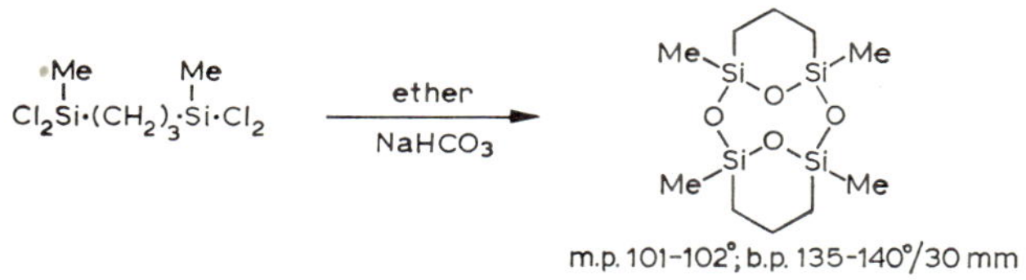

m.p. 101–102°; b.p. 135–140°/30 mm

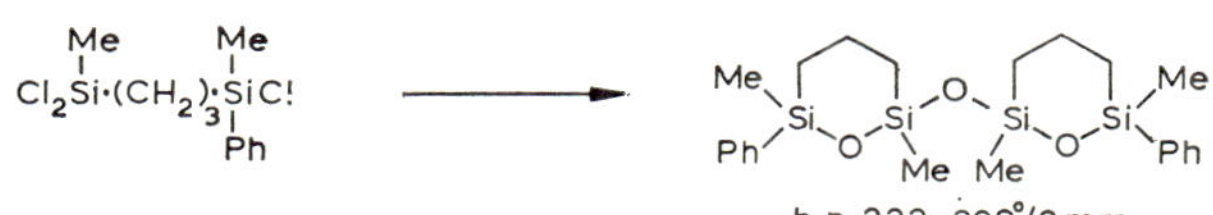

b.p. 222–228°/2 mm

Reaction between 2,6-dichloro-2-methyl-6-phenyl-2,6-disilaheptane and hydrogen sulphide affords 2,2,6-**trimethyl-6-phenyl**-1,2,6-**thiadisilacyclohexane,** b.p. 89°/1.5 mm, n_D^{20} 1.5340, (*Chi-Tao Wang, Hsiu-Chung Chou* and *Man-Shui Hung*, Hua Hsueh Hsueh Pao, 1964, **30,** 91; C.A., 1964, **61,** 1888).

The alkaline hydrolysis of 3,7-dibromo-3,7-diethyl-3,7-digermanonane yields 2,2,6,6-**tetraethyl**-1,2,6-**oxadigermacyclohexane** (XLI, X = O), b.p. 77°/0.6 mm, n_D^{20} 1.4820.

(*c*) The last named compound, its **thia**- analogue, b.p. 98°/0.5 mm, n_D^{20} 1.5316 (XLI, X = S), and its **selena**- analogue (XLI, X = Se), b.p. 112°/0.5 mm, n_D^{20} 1.5545, are also prepared by insertion of the Group VI element into 1,2-digermacyclopentanes, a procedure which, insofar as it leads to compounds with six-membered rings, has not been applied to those containing silicon (*P. Mazerolles, M. Lesbre* and *M. Joanny*, J. organometal. Chem., 1969, **16,** 227):

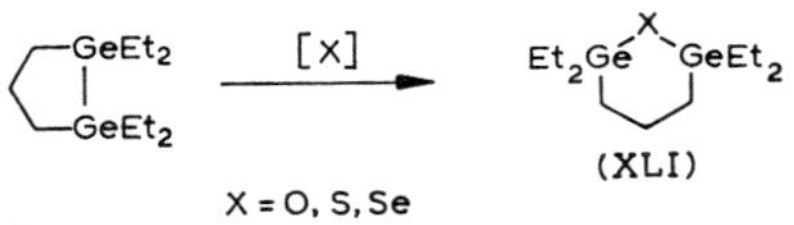

(XLI)

The 1,2,6-oxadisilacyclohexane ring system withstands the action of potassium hydroxide and hydrochloric acid in acetic acid during hydrolysis of substituents, but this ring, and presumably the germanium-containing ring also, is cleaved by the action of thionyl chloride–hydrogen chloride giving propane-1,3-di(silyl chloride) (*Niederpreuem* and *Simmler*, Fr. P. 1,416,915/1965; C.A., 1966, **64,** 19678).

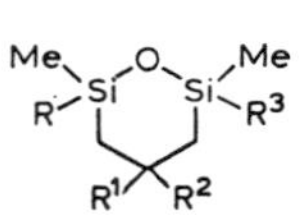

(*e*) *Dioxadisilins, dioxagermasilins and dioxadigermalins*

(*i*) 1,4,2,5- *and* 1,4,2,6-*Dioxadisilacyclohexanes*

The isomeric dioxadisilacyclohexanes XLII and XLIII are formed by the dehydration, of the disiloxane, $(HOCH_2SiMe_2)_2O$, with or without molecular rearrangement, the route followed being dependent upon the nature of the reagents employed:

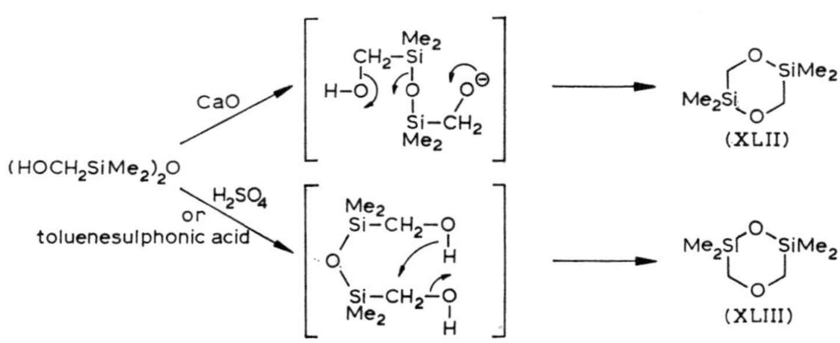

An additional method for the preparation of 2,2,5,5-**tetramethyl-1**,4,2,5-**dioxadisilacyclohexane** (XLII), b.p. 67–70°/21 mm, n_D^{20} 1.4302, consists in heating $MeCO_2CH_2SiMe_2OEt$ with methanol in the presence of toluene-4-sulphonic acid. 2,2,5,5-**Tetramethyl-1**,4,2,5-**dithiadisilacyclohexane,** m.p. 81–83°, is formed from di(chloromethyldimethyl)disilazane, or chloro(chloromethyl)dimethylsilane, and hydrogen sulphide in the presence of triethylamine; the ring system is opened to mercapto(mercaptomethyl)dimethylsilane, $HSCH_2SiMe_2SH$, by aqueous ethanol, while dry hydrogen chloride yields chloro(mercaptomethyl)dimethylsilane (*M. Schmidt* and *M. Wieber*, Inorg. Chem., 1962, **1,** 909).

2,2,6,6-**Tetramethyl**-1,4,2,6-**dioxadisilacyclohexane** (XLIII) is obtained in quantitative yield by the action of water on di(fluorodimethylsilylmethyl) ether, $(FSiMe_2CH_2)_2O$, in benzene, and of calcium oxide on di(bromomethyldimethyl)siloxane, $(BrCH_2SiMe_2)_2O$, in dimethylformamide. Its **dithia** analogue is produced by reaction between $(ClCH_2SiMe_2)_2S$ and sodium sulphide in ethanol; indeed, this type of reaction provides a practicable route to a variety of mixed oxygen–sulphur–selenium/silicon–germanium ring systems XLIV (Table 12).

The dioxa- systems tend to polymerize rapidly (*J. L. Speier, M. P. David* and *B. A. Enyon*, J. org. Chem., 1960, **25,** 1637; *Enyon,* U.S.P. 2,989,346/1959; C.A., 1960, **54,** 1576). Sulphur- and selenium-containing compounds form monomethiodides listed in Table 12.

TABLE 12

1,4,2,6-DIOXADISILACYCLOHEXANES AND ANALOGUES

(XLIV)

M	*M*[1]	*X*	*Y*	*B.p. (°C/mm)*	*Methiodide m.p.(°C)*	*Ref.*
Si	Si	O	O	30–35/2.5 57/24 153/753		1
		O	Se	193/720	125	2
		S	S	62–64/0.1	137–139	3
		S	Se	37–38/0.1	127–129	2
Ge	Ge	O	S	80/0.1		4
		O	Se	64–66/0.1	162–163	2
		S	S	72–73/0.1 (m.p. 76)		4
		S	Se	80–82/0.1	141–143	2
Si	Ge	O	S	49–50/0.1	141–143	3
		O	Se	48–51/0.1	130	2

References

1 *W. Simmler, H. Niederprüm* and *H. Sattlegger*, Chem. Ber., 1966, **99,** 1368.
2 *M. Wieber* and *G. Schwarzmann*, Monatsh., 1969, **100,** 74.
3 *idem, ibid.*, 1969, **100**, 68.
4 *idem, ibid.*, 1968, **99,** 255.

(ii) *1,2,4,6-Oxathiadisilacyclohexane*

Treatment of 1,1,3,3-tetramethyl-1,3-disilacyclobutane with sulphur trioxide in chloroform yields a *product*, b.p. 83–84°/0.2 mm, m.p. 35–37°, hydrolysis of which gives the sulphonic acid, $HOSiMe_2CH_2$-$SiMe_2$-$CH_2SO_2H_3$ suggesting that the original product is

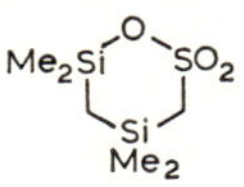

(XLV)

4,4,6,6-**tetramethyl**-1,2,4,6-**oxathiadisilacyclohexane** 2,2-**dioxide** (XLV) rather than one of several isomeric structures (*H. Schmidbaur*, *L. Sechser* and *Schmidt*, Chem. Ber., 1969, **102,** 376). More recently, the 2-**monoxide** has been obtained using sulphur dioxide in an analogous fashion (*J. Dubec et al.*, J. heterocyclic Chem., 1970, **25,** C20).

(iii) *1,4,2,3-Dioxadigermacyclohexenes*

Initial reaction of *symm*-tetraphenyldigermane with dibutylmercury, followed by biacetyl, has provided a substance formulated as 5,6-dimethyl-2,2,3,3-tetraphenyl-1,4,2,3-dioxadigermacyclohexene (XLVI), m.p. 231° (*P. Riviere*, *J. Satgé* and *D. Soula*, Compts. rend., 1973, C, **277,** 895).

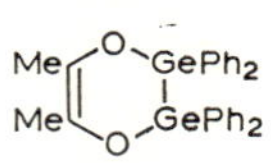

(XLVI)

(*f*) *Oxa- and dioxa-trisilacyclohexanes*

A few derivatives of oxa- and dioxa-trisilacyclohexanes have been prepared, mainly by Russian workers.

4,4-**Dichloro**-2,2,6,6-**tetramethyl**-1,2,4,6-**oxatrisilacyclohexane** (XLVII, R = Cl) is formed when di(chloromethyldimethyl)disiloxane is passed over a silicon–copper composite at 370–390°, or when it is treated with stannic iodide or tin and triethylamine (*V. F. Mironov*, *E. M. Stepina* and *V. I. Shiryaev*, Zhur. obshcheĭ Khim., 1972, **42,** 631).

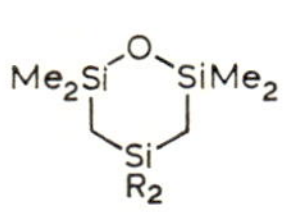

(XLVII)

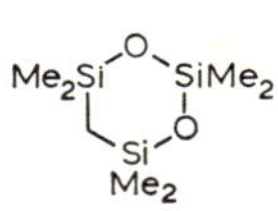

(XLVIII)

The first of these procedures is difficult to carry out since the starting material and product are difficult to separate by distillation, and it is necessary to use freezing techniques. The pure 4,4-*dichloro* compound has m.p. 34–34.5° (*A. A. Buyakov et al.*, *ibid.*, 1975, **45,** 2672).

Derivatives of 1,2,6,4-oxadisilagermacyclohexane have been described (*T. K. Gar et al.*, *ibid.*, 1976, **46,** 837).

The *hexamethyl* compound XLVII, R = Me, b.p. 90–92°/45 mm, n_D^{20} 1.4385, is obtained from the dichloride by the action of methylmagnesium chloride, by the hydrolysis of 2,6-dichloro-2,4,4,6-tetramethyl-2,4,6-trisilaheptane (*Mironov*, *Gar* and *Buyakov*, *ibid.*, 1972, **42,** 1361), and also by condensation of dichlorodimethylsilane and di(chloromethyldimethyl)disiloxane by sodium in boiling toluene (*Andrianov* and *S. E. Yakushina*, Izvest. Akad. Nauk S.S.S.R., Otd. khim. Nauk, 1962, 1396; *E. I. Fedin et al.*, Doklady Akad. Nauk S.S.S.R., 1969, **186,** 843).

Hexamethyl-1,2,4,6-**oxatrisilacyclohexane,** b.p. 179–180°, and **hexamethyl**-1,3,2,4,6-**dioxatrisilacyclohexane** (XLVIII), b.p. 156–157°, are the minor (15%) and major (45%) products of thermolysis of 2,2,4,4,6,6,8,8-octamethyl-1,5-dioxa-2,4,6,8-tetrasilacyclooctane at 700°. Hexamethyl-1,3,2,4,6-dioxatrisilacyclohexane is also formed when dihydroxydimethylsilane condenses with 2,4-dichloro-2,4-dimethyl-2,4-disilapentane in the presence of potassium hydroxide (*N. S. Nametkin et al.*, Izvest. Akad. Nauk S.S.S.R., Otd. khim. Nauk, 1968, 1329; *C. R. Sporch*, U.S.P. 3,452,072/1969; C.A., 1969, **71,** 91633) and when a mixture of dichlorodimethylsilane and bis(chloro-methylsylyl)methane, $(MeSiHCl)_2CH_2$, in ether is treated with water (*D. J. Cooke* and *W. J. Owen*, B.P. 1,220,445/1971; C.A., 1971, **74,** 88458).

In the presence of ferric chloride, both XLVII and XLVIII are converted by thionyl chloride into $Cl_2SiMeCH_2SiMeCl_2$ (*S. D. Lyukas*, *N. P. Smetankina* and *V. P. Kuznetsova*, Zhur. obshcheĭ Khim., 1966, **36,** 2003).

Hydrolysis of 1,5-dibromo-1,3,3,5,7,7-**hexamethyl**-1,3,5,7-**tetrasilacyclo-octane** yields 1,3,3,5,7,7-**hexamethyl-9-oxa**-1,3,5,7-**tetrasilabicyclo**[3.3.1]**nonane** (XLIX), m.p. 36–39°, (*G. Fritz* and *M. Hähnke*, Z. anorg. allgem. Chem., 1972, **390,** 185).

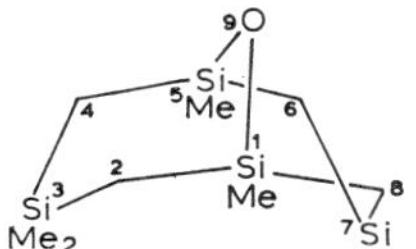

(XLIX)

(g) *Azasilines*

(i) *Dibenz[1,2]azasilines*

6,6-**Dimethyl**-5,6-**dihydrodibenz**[*c,e*][1,2]**azasiline** (L), m.p. 143°, is formed in 35% yield when 2-biphenylyldimethylsilyl azide is photolyzed with a mercury arc (*J. M. Gaidis* and *R. West*, J. Amer. chem. Soc., 1964, **86,** 5699).

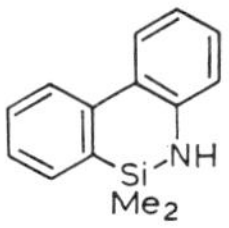

(L)

(ii) *Perhydro-1,2-azasiline (1,2-azasilacyclohexane)*

Reaction between 2,6-dichloro-2-methyl-2-silahexane and methylamine at 100° affords a moderate yield of 1,2,2-**trimethyl-perhydro**-1,2-**azasiline** (LI), b.p. 64°/55 mm, n_D^{25} 1.4451, very reactive to water yielding 3-methylaminopropyldimethylsilanol (*J. L. Speier, C. A. Roth* and *J. W. Ryan*, J. org. Chem., 1971, **36,** 3120; *Speier*, F.P. 1,335,705/1963; C.A., 1964, **60,** 1794). A similar reaction with acid chlorides of the type $ClSiMe_2CH_2 \cdot CHRCH_2 \cdot COCl$, furnishes a series of silalactams LII readily hydrolyzed at the Si–N bond giving $[RNHCO(CH_2)_3SiMe_2]_2O$ (*V. F. Mironov, N. S. Fedotov* and *I. G. Rybalka*, Khim. geterotsikl. Soedin., 1968, 1124; C.A., 1964, **70,** 115217).

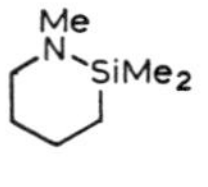

(LI)

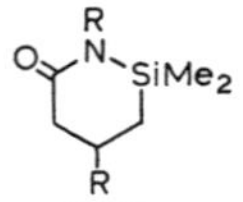

(LII)

(*iii*) *Benz*[d][*1,3*]*azasilines*

1,1-Dimethyl-1,2,3,4-**tetrahydrobenz[*d*][1,3]azasilines** (LIV), 3-*methyl*, b.p. 94–95°/6 mm, 3-*phenyl*, b.p. 121–126°/0.2 mm, 3-*benzyl*, b.p. 119–123°/0.1 mm, are prepared from primary amines and suitable 2-chloromethylsilylbenzyl bromides:

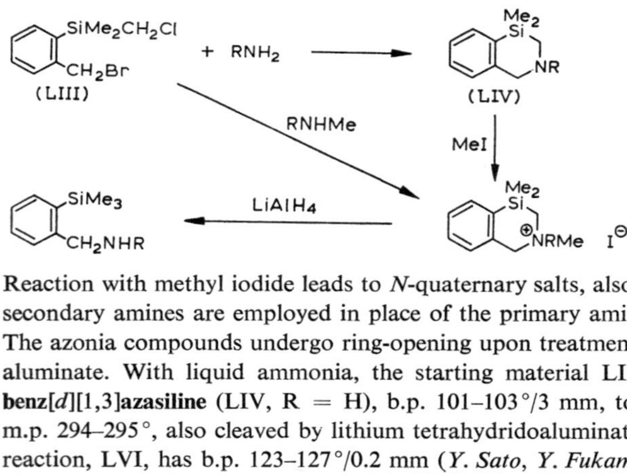

Reaction with methyl iodide leads to *N*-quaternary salts, also obtainable when *N*-methyl secondary amines are employed in place of the primary amines in the ring-closure step. The azonia compounds undergo ring-opening upon treatment with lithium tetrahydridoaluminate. With liquid ammonia, the starting material LIII yields 1,2,3,4-**tetrahydrobenz[*d*][1,3]azasiline** (LIV, R = H), b.p. 101–103°/3 mm, together with the *spiran* LV, m.p. 294–295°, also cleaved by lithium tetrahydridoaluminate; the product of this latter reaction, LVI, has b.p. 123–127°/0.2 mm (*Y. Sato, Y. Fukami* and *H. Shirai*, J. organometallic Chem., 1974, **78,** 75):

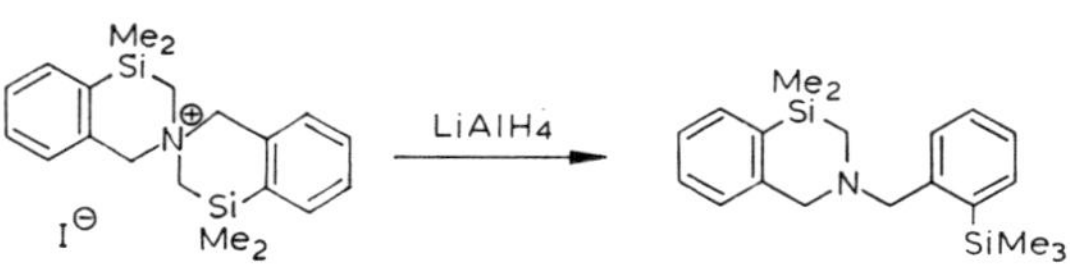

(LV) (LVI)

(*iv*) *Perhydro-1,3-azasilines (1,3-azasilacyclohexanes)*

Butylamine and 1,5-dibromo-2,2-dimethyl-2-silapentane give 1-**butyl**-3,3-**dimethylperhydro**-1,3-**azasiline** (LVII) as its *hydrobromide*, m.p. 210–211° (*R. J. Fessenden* and *M. D. Coon*, J. org. Chem., 1964, **29,** 1607). 3,3,5-*Trimethylperhydro*-1,3-*azasilin*-6-*one*,

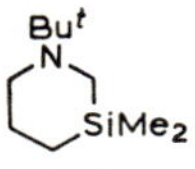

(LVII)

m.p. 55–56°, is obtained from 5-chloro-2,4,4-trimethyl-4-silapentanoamide by treatment with sodium methoxide (*Mironov* and *Fedotov*, Khim. geterotsikl. Soedin, A. Nauk Latv., 1966, 453; C.A., 1967, **66**, 76077).

(*v*) *Dibenz*[1,4]*aza-silines, -germanines, -stannines and -plumbines*

The reaction between a 2,2′-dilithiodiphenylamine and the appropriate metalloid chloride leads either to the 5,10-dihydrophenazametalline (LVIII) or to the spirobi(5,10-dihydrophenazametalline) (LIX):

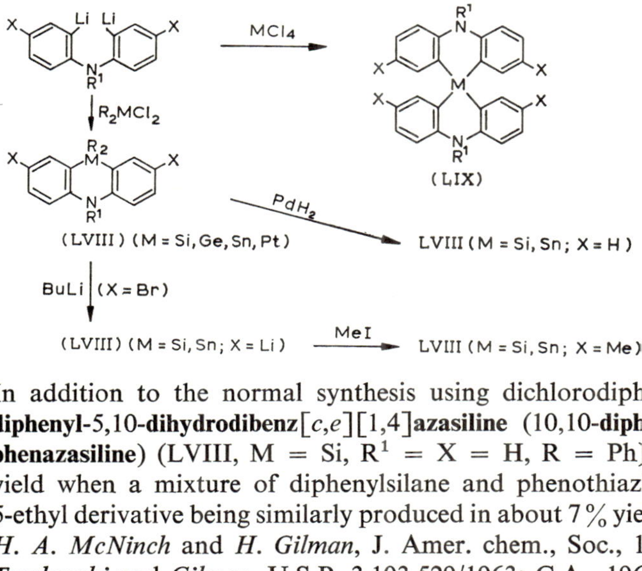

In addition to the normal synthesis using dichlorodiphenylsilane, **10,10-diphenyl-5,10-dihydrodibenz**[*c,e*][1,4]**azasiline** (**10,10-diphenyl-5,10-dihydrophenazasiline**) (LVIII, M = Si, R^1 = X = H, R = Ph) is formed in 1% yield when a mixture of diphenylsilane and phenothiazine is heated, the 5-ethyl derivative being similarly produced in about 7% yield (*D. Wittenberg, H. A. McNinch* and *H. Gilman*, J. Amer. chem., Soc., 1958, **80**, 5418; *C. Tamborski* and *Gilman*, U.S.P. 3,103,529/1963; C.A., 1964, **60**, 552); other synthetic procedures include the use of chlorotriphenylsilane, when tetraphenylsilane is a by-product, and of tetraphenyldisilane (*Gilman* and *E. A. Zuech*, J. org. Chem., 1962, **27**, 2897).

Bromination of 5-ethyl-10,10-diphenyl-5,10-dihydrophenazasiline in carbon disulphide gives the 2,8-dibromo derivative (LVIII, M = Si, X = Br, R = Ph, R^1 = Et) in 20–25% yield, together with traces of silylated diphenylamines and silicon-free diphenylamines; when the bromination is carried out in acetic acid-containing media, the yield of the heterocyclic dibromo compound falls markedly, and the brominated diphenylamine becomes the principle product (*idem, ibid.*, 1961, **26,** 3481). The 2,8-dibromide so formed, may be lithiated using butyllithium, and the lithio compounds LVIII (X = Li) methylated to the 2,8-dimethyl derivatives; they may also be hydrogenolyzed over a palladium catalyst to give the bromine-free compounds (LVIII, X = H). Indeed, this latter process gives better yields of the 10,10-disubstituted-5,10-dihydrophenazasilines, and of the corresponding spirans (LIX), together with compounds in the tin series (LVIII, LIX, M = Sn), than the standard procedure described earlier (*D. Wasserman et al.*, J. org. Chem., 1965, **30,** 3248; F.P. 1,315,605/1963; C.A., 1963, **58,** 11400; Belg. P. 613,915/1962; C.A., 1963, **58,** 1490).

The general synthetic method, outlined above, has been applied to the preparation of compounds of germanium, tin and lead (references under Table 14). Whereas silicon and germanium tetrachlorides afford the 10,10′-spirobi(5,10-dihydrophenazametalline) with some ease, it is possible to stop the reaction between 2,2′-dilithiodiphenylamines and stannic chloride at the 10,10-dichloro-5,10-dihydrophenazastannine stage, a situation reminiscent of the low yield of 10,10′-spirobiphenoxastannine from stannic chloride and dilithiodiphenyl ether. The melting points of representative 5,10-dihydrophenazametallines are 10,10′-spirobis derivatives are listed in Tables 13 and 14 (refs. in parenthesis).

TABLE 13

5,10-DIHYDROPHENAZAMETALLINES (LVIII)

			M (m.p., °C)			
R^1	R_2	*X*	*Si*	*Ge*	*Sn*	*Pb*
H	Ph_2	H	197–198 (1)			
Me	Me_2	H			133–134 (4)	
		Me	117–119 (1)		119–120 (4)	
		Br	148–149 (2)		122–123 (4)	
	Ph_2	H	186–188 (2)		148.5–150 (4)	
		Me	163–165 (1)		146–148 (4)	
		Br	200–202 (2)		191–193 (4)	
	$(CH_2Ph)_2$	Me	124–125.5 (1)			
	Cl_2	Br			164.5–166.5 (4)	
Et	H,Ph	H	75–78			
Et	Me_2	H	66–67 (5)			
		Br	132.5–134 (2)			
	Ph_2	H	122–123 (2,3)	120–121((3)	117–119 (3)	121–123 (3)
		Me	162.5–163 (2)			
		Br	206.5–207.5 (2)		177–178 (2)	
	Ph, Me	H	62–64 (3)			

References follow Table 14.

TABLE 14

10,10′-SPIROBIS(5,10-DIHYDROPHENAZAMETALLINES) (LIX)

		M (m.p., °C)		
R^1	X	Si	Ge	Sn
Me	H			164–166 (4)
	Me	230–233 (1)		209–211 (4)
	Br			259–260 (4)
Et	H	215.5–217 (2)	212–214 (3)	210–211 (3)
	Br	275–276.5 (2)		

References (Tables 13 and 14)
1 *H. Gilman* and *E. A. Zuech*, J. org. Chem., 1962, **27**, 2879.
2 *D. Wasserman et al., ibid.*, 1965, **30**, 3248.
3 *C. Tamborski* and *Gilman*, U.S.P. 3,079,414/1963; C.A., 1963, **59**, 5796.
4 *E. J. Kupchik* and *V. A. Perciaccante*, J. organometal. Chem., 1967, **10**, 181.
5 *V. O. Reikhsfel'd* and *Yu. S. Finogemov*, Zhur. obshcheĭ Khim., 1974, **44**, 289.

(*h*) *Diazasilines; diazastannines*

(*i*) *Perhydro*-1,3,2-*diazasilines* (1,3,2-*diazasilacyclohexanes*) *and their tin analogues and benzologues*

Perhydro-1,3,2-diazasilines (LX, M = Si) are prepared from the appropriate 1,3-diaminopropane by reaction with dichlorosilanes in the presence of triethylamine, or by using silylhydrazines (*G. S. Gol'din*, *L. S. Baturina* and

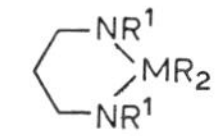

M = Si, Sn
(LX)

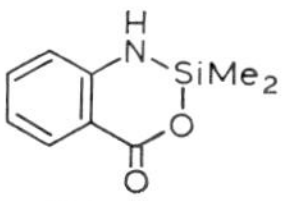

(LXI)

C. I. Trubnikova, Zhur. obshcheĭ Khim., 1972, **42**, 1043). Those compounds with silicon-containing substituents attached to nitrogen may be aminated further (*L. W. Breed, R. L. Elliott* and *A. F. Ferris*, J. Polymer Sci., **A2**, 1964, 45).

1,3,2,2-**Tetramethylperhydro**-1,3,2-**diazastannine**, b.p. 88–90°/18 mm, and 2,2-**dimethyl**-1,3-**bis(trimethylsilyl)perhydro**-1,3,2-**diazastannine**, b.p. 136–138°/16 mm, result when a diaminoalkane is heated with a diaminomethyl-tin (*O. J. Scherer et al.*, Z. Naturforsch., B, 1965, **20**, 183; *Gol'din* and *Baturina*, Zhur. obshcheĭ Khim., 1973, **45**, 2245).

2,2-Diethyl-perhydro-1,3,2-oxazastannine may be obtained from 1,3-propanolamine and di-*tert*-butoxy-diethylstannane as an undistillable oil (*D. P. Gaur*, *G. Srivastava* and *R. C. Mehrotra*, J. organometal. Chem., 1973, **63**, 213; 1974, **65**, 195).

2,2-*Dimethyl*-1,2,3,4-*tetrahydro*-3,1,2-*benz*[d]*oxazasibine*-4-*one* (LXI) from anthranilic acid has m.p. 120° (*M. Wieber* and *M. Schmidt*, Chem., Ber., 1963, **96**, 1561).

(ii) *1,4,2-Benzodiazasilines; perhydro-1,4,2-diazasilines*

2,2-**Dimethyl**-1,2,3,4-**tetrahydro**-1,4,2-**benzodiazasiline** (LXII) b.p. 142–144 °/4 mm, m.p. 50–52 °, is prepared from *o*-phenylenediamine and chloro(chloromethyl)dimethylsilane $ClCH_2SiMe_2Cl$. 2,2-**Dimethyl**-1,2,3,4-**tetrahydro**-1,4,2-**benzoxazasiline** (LXIII, M = Si), b.p. 125–131 °/9 mm, m.p. 21 ° (*Wieber* and *Schmidt*, J. organometal Chem., 1963, **1**, 22), and LXIII (M = Ge), b.p. 122 °/1 mm, (*Wieber* and *C. D. Frohning*, *ibid.*, 1967, **8**, 459), are obtained in an analogous fashion.

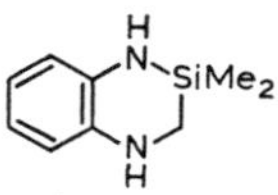
(LXII)

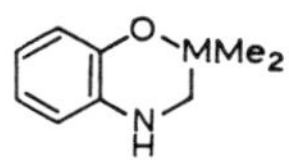
M = Si, Ge

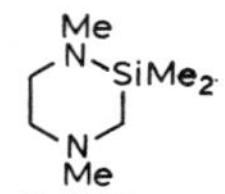
(LXIV)

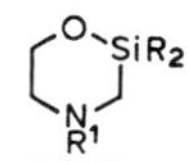
(LXV)

1,2,2,4-**Tetramethylperhydro**-1,4,2-**diazasiline** (LXIV) has b.p. 64 °/53 mm, n_D^{20} 1.4508 (*W. Simmler*, Chem. Ber., 1963, **96**, 349); a few "silamorpholines" are recorded, *e.g.* 2,2,4-*trimethyl*-, (LXV, R = R¹ = Me), b.p. 140 °, 63 °/47 mm (*idem. loc. cit.*; *W. Naegele*, *A. DeMontigny* and *Simmler*, Z. Naturforsch., B, 1970, **25**, 11), 2,2-*dimethyl*-4-*phenyl*-, deriv. (LXV, R = Me, R¹ = Ph) b.p. 148 °/3 mm.

(*i*) *Triazasilines*

W. Fink investigated the synthesis and stereochemistry of a series of perhydro-triazasilins (Chem. Ber., 1964, **97**, 1424; J. chem. Soc., 1964, 4958). When hexamethyltrisilazane or octamethyltetrasilazane is heated with phenyl isocyanate in 1:4 or 1:6 ratio, respectively, at 60–90° in toluene, it is possible to isolate a triazasila *compound*, m.p. 188–190°; since this yielded dichlorodimethylsilane and $(PhNHCO)_2NH$ when treated with dry hydrogen chloride in boiling toluene, the 2,2-**dimethyl**-1,3-**diphenyl**-**perhydro**-1,3,5,2-**triazasilin**-4,6-**dione** structure LXVI was assigned to it:

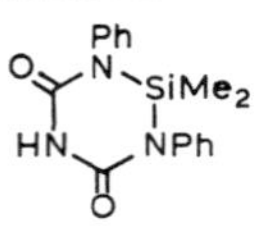
(LXVI)

Transamination reactions lead to unsymmetrical triazasilines; 1,3,3,4-**tetramethyl**-**perhydro**-1,2,4,3-**triazasiline** (LXVII), b.p. 64–65 °/16 mm (*Gol'din et al.*, Zhur. obshcheĭ Khim., 1974, **44**, 566):

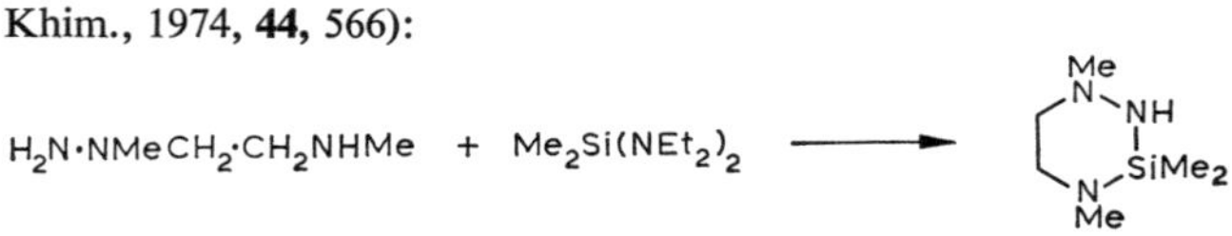
(LXVII)

(*j*) *Azadisilines*

2,2,6-**Trimethyl**-6-**phenylperhydro**-1,2,6-**azadisiline** (LXVIII), b.p. 100 °/3 mm, n_D^{20} 1.5194, is formed by ammonolysis of 2,6-dichloro-2-methyl-6-phenyl-2,6-disilaheptane (*Chi-Tao Wang*, *Hsiu-Chung Chou* and *Man-Shui Hung*, Hua Hsueh Hsueh Pao, 1964,

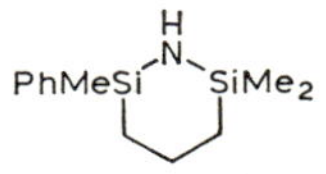

(LXVIII)

30, 91; C.A., 1964, **61,** 1888; *K. A. Andrianov et al.*, Doklady Akad. Nauk S.S.S.R., 1974, **216,** 1041).

(*k*) *Diazadisilines and analogues*

(*i*) *Perhydro*-1,2,3,6-*diazadisilines*

Hydrazinolysis of 2,5-dichloro-2,5-dimethyl-2,5-disilahexane gives a mixture of two products in approx. 1 : 1 ratio. Mixtures of different compositions are obtained by other methods and it would appear that the two products LXIX and LXX are in equilibrium at room temperature (*J. W. Connolly et al.*, Inorg. Chem., 1970, **9,** 93; *C. G. Pitt* and *K. R. Skiller, ibid.*, 1967, **6,** 865).

(LXIX) (LXX)

(*ii*) *Perhydro*-1,4,2,5-*diazadisilines and* -1,4,2,6-*diazadisilines*

The ammonolysis and aminolysis of bromomethylchlorodimethylsilane yields one or both of the isomeric, perhydro-1,4,2,5-diazadisiline and -1,4,2,6-diazadisiline systems.

Ammonolysis of bromomethylchlorodimethylsilane (LXXI) in dry benzene gives a low yield of 2,2,6,6-**tetramethylperhydro**-1,4,2,6-**diazadisiline** (LXXII, $R^1 = R^2 = H$) isolated as the *hydrobromide*, m.p. 270° dec., together with **di(bromomethyldimethyl)disilazane** (LXXIII, R = H). The latter product itself reacts with amines in 1 :3 ratio to give 4-substituted-2,2,6,6-*tetramethylperhydro*-1,4,2,6-*diazadisilines* (LXXII, R^1 = H), 4-*methyl*-,

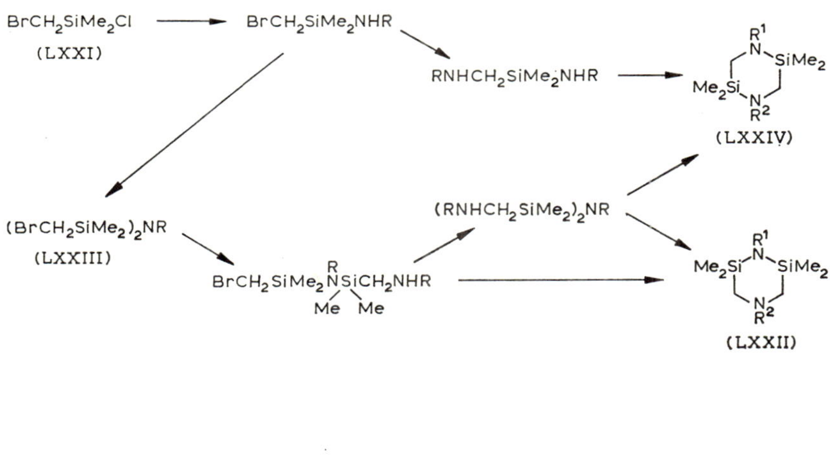

b.p. 66 °/15 mm, 4-*propyl-*, b.p. 81 °/11 mm, 4-*allyl-*, b.p. 48 °/0.4 mm; when the ratio of amine to LXXIII (R = H) is 4:1, the 1,4-disubstituted compounds are formed, *e.g.* 1,4-*dimethyl-*, b.p. 74 °/18 mm.

On the other hand, aminolysis of LXXI with methylamine gives 1,2,2,4,5,5-**hexamethyl-perhydro**-1,4,2,5-**diazadisiline** (LXXIV, $R^1 = R^2 =$ Me) as the *hydrobromide* m.p. 240° decomp., with or without 15% of the isomeric compound LXXII ($R^1 = R^2 =$ Me); similarly for the 1,4-*dipropyl-*, b.p. 125 °/17 mm, and 1,4-*diallyl-*, b.p. 83–85 °/3 mm, compounds.

To account for the concomitant formation of LXXII and LXXIV, *H. Niederprüm* and *W. Simmler* (Chem. Ber., 1963, **96,** 965) proposed the above scheme in which the conversion of LXXI via LXXIII to LXXIV takes place with rearrangement of $(RNHCH_2SiMe_2)_2NR$ in a manner akin to that of disiloxanes in the formation of isomeric dioxadisilins.

The reaction between $ClSiMe_2CH_2NHCO_2Me$ and triethylamine in 2:1 ratio also provides a route to the 1,4,2,5-diazadisilacyclohexane skeleton (*V. D. Sheludyakov et al.*, Zhur. obshcheĭ Khim., 1971, **41,** 2340; 1972, **42,** 1650; 1972, **42,** 1651; 1974, **44,** 1496). A number of mixed systems and analogues containing germanium are listed in Table 15.

TABLE 15

PERHYDRO-OXAZA- AND -THIAZA-2,6-DIMETALLINES

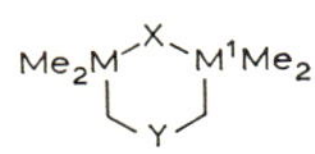

M	*M*1	*X*	*Y*	*B.p.°/mm (m.p.)*	*Methiodide m.p.(°C)*
Si	Si	O	NMe	40–41/10	
		O	$NCH_2 \cdot CH{:}CH_2$	87–88/30	
		NH	S	60–61/0.1	139–141
		NH	Se	59–61/0.1	122–123
Ge	Ge	O	NEt	92/20	
		NH	S	172	
		NH	Se	61–63/0.1	139

References

Data from *V. F. Mironov, S. A. Mikhailyants* and *T. S. Gar*, Zhur. obshcheĭ Khim., 1969, **39,** 2281; *M. Wieber* and *G. Schwarzmann*, Monatsh., 1968, **99,** 255; *ibid.*, 1969, **100,** 68, 74.

Reported reactions of perhydro-1,4,2,5-diazadisilines include ring-opening by the action of phosgene, the products being of the types $ClSiMe_2CH_2NRCOCl$ and $(ClSiMe_2CH_2NR)_2CO$ (*V. F. Mironov, Sheludyakov* and *E. S. Rodionov*, *ibid.*, 1974, **44,** 1502); ethanolysis to give $EtOSiMe_2CH_2NHR$, and hydrolysis of solutions in THF (the compounds are evidently stable to atmospheric moisture) to give $(RNHCH_2SiMe_2)_2O$ (*Sheludyakov et al.*, *ibid.*, 1974, **44,** 1506).

(*l*) *Triazasilines and diazatrisilines*

1,2,2,3,4,4,6,6-**Octamethylperhydro**-1,3-**diaza**-2,4,6-**trisiline** (LXXV) b.p. 46°/3 mm, n_D^{20} 1.4573, is obtained by the action of $Me_2Si(NMeLi)_2$ on $(BrSiMe_2)_2CH_2$; the yield is about 30% (*L. Lienhard* and *E. G. Rochow*, Angew. Chem., 1963, **75,** 638; Z. anorg. allgem. Chem., 1964, **331,** 307).

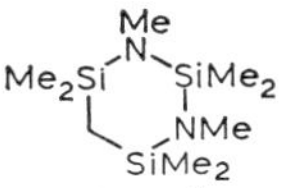

(LXXV)

Two interesting equilibria have been observed for disiladioxazine and disiloxadiazine type systems:

(LXXVI) (LXXVII)

The purely inorganic ring system, **tetramethylcyclodisiloxazane,** as its N-*acetyl* derivative (LXXVI) b.p. 28–29°/6 mm, is in equilibrium with 2,2,4,4,6-**pentamethyl**-1,2,3,4-**tetrahydro**-1,3,5-**dioxaza**-2,4-**disiline** (LXXVII), b.p. 34–35°/6 mm; they may be separated by rapid distillation (*B. Dejak* and *Z. Lasocki*, J. organometal. Chem., 1972, **44,** C39).

By reaction between carboxamides and dichlorosilanes in the presence of triethylamine, *J. F. Klebe* (J. Amer. chem. Soc., 1968, **90,** 5246) obtained a series of 2,2,4,4-tetrasubstituted-3-acyl-1,2,3,4-tetrahydro-1,3,5-oxadiaza-2,4-disilines (LXXVIII); in certain cases it was found possible to separate *cis* and *trans* isomers, sometimes by fractional crystallization.

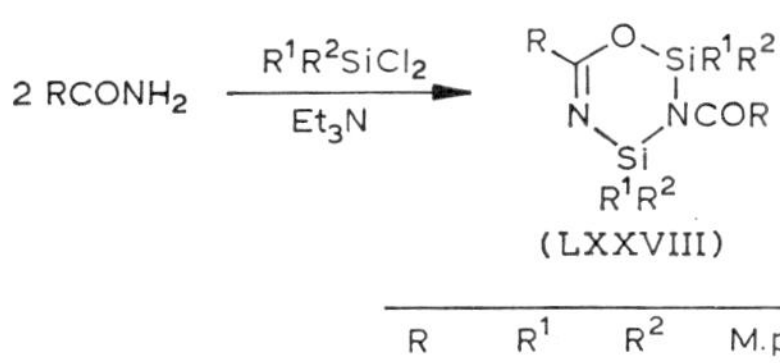

(LXXVIII)

R	R^1	R^2	M.p. (°C)
Me	Me	Me	44– 46
Me	Ph	Ph	140–145 dec.
Ph	Me	Ph	116–122
Me	Ph	Me.	108–113
Ph	Me	Me	127–128
Me	Me	OMe	45– 48

The compounds LXXVIII undergo reversible intramolecular rearrangement at two temperature levels. At 20–50°, the amide and imino ether grouping interconvert at a rate causing coalescence of ^{1}H-n.m.r. signals. With a further rise in temperature, a second coalescence is observed and which corresponds to equivalence of substitution at silicon. *Klebe* pictured these rearrangements as being based on a planar tetragonal pyramid with the ring Si–O bond forming the axial bond in the pyramid and the imino nitrogen atom being situated at the base, but an X-ray analysis of 2,2,4,4-tetramethyl-3-benzoyl-6-phenyl-1,3,5-oxadiaza-2,4-disiline suggests that pentaco-ordination at one of the silicon atoms is achieved *via* the exocyclic benzoyl carbonyl group rather than by transannular Si–N interactions; the ring oxygen atom and carbonyl oxygen atom occupy approximately axial positions in an almost planar ring structure (*F. P. Boer* and *F. P. van Remoortere*, J. Amer. chem. Soc., 1969, **91,** 2377; 1970, **92,** 801).

3. Phosphorus, arsenic and antimony compounds

(*a*) *Oxaphosphorins, oxarsenins and their thia-analogues*

(*i*) *Dibenz*[1,2]*oxaphosphorins*

The formation of 6-chloro-6*H*-dibenz[*c*,*e*][1,2]oxaphosphorin (II, R = Cl) from the phosphorodichloridite I takes place in an inert solvent in the presence of zinc chloride or aluminium chloride (*T. Saito*, G.P. 2,034,887/1972; C.A., 1972, **76,** 99823; *V. V. Pomomarev et al.*, U.S.S.R., 319,606/1971; C.A., 1972, **76,** 46302; *E. A. Chernyshev et al.*, Zhur. obshcheĭ Khim., 1972, **42,** 93):

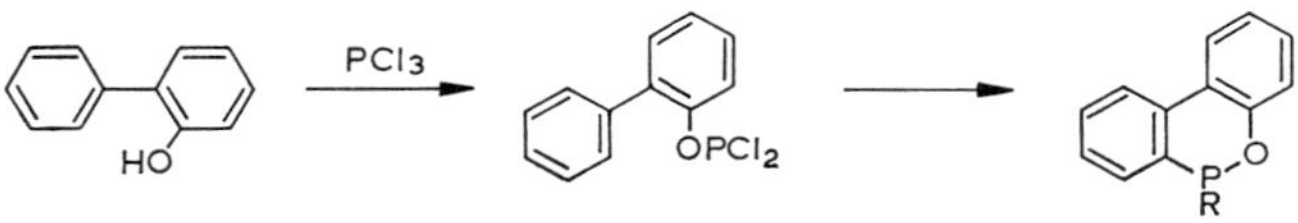

(I) (II)

It reacts in the usual way with alcohols, and the resultant esters undergo the Arbuzov reaction with alkyl halides with no concomitant ring-opening.

The *P*-oxide of II can be prepared by passing biphenylyl phosphorodichloridate, or better, di(*o*-biphenylyl) phosphorochloridate, through a hot quartz tube (*Chernyshev*, *V. I. Aksenov* and *E. F. Bugarenko*, *ibid.*, 1971, **41,** 2189).

Rapid pyrolysis of *o*-biphenylyl phosphorodichloridothionate at 550–600° yields only 15% of the P-*sulphide* of II, b.p. 203–208°/7 mm, m.p. 140–141°, the main product being dibenzothiophene (75%) (*idem*, *ibid.*, 1970, **40,** 1423).

A much more convenient procedure appears to be that which involves the cyclization of the phosphorodichloridothionate in the presence of aluminium chloride at 200° (*M. S. Bhatin* and *P. Jit*, Chem. and Ind., London, 1975, 1058).

(*ii*) *Partially reduced 1,2-oxaphosphorins and 1,2-thiaphosphorins*

The few known compounds in this series have been obtained as individual products in widely differing types of reactions; standard routes have so far not been developed for their synthesis.

Heating dimethyl 4-bromobut-2-enylphosphonate results in loss of methyl bromide and formation of 2-methoxy-2,3-dihydro-6*H*-1,2-oxaphosphorin 2-oxide (III), which when pyrolyzed at 600° for 0.02 sec liberates butadiene together with a species thought to be methyl metaphosphate $MeOP(O)_2$ (*C. H. Clapp* and *F. H. Westheimer*, J. Amer. chem., Soc., 1974, **96**, 6710):

(III)

Removal of the carbonyl protecting group from the phosphonium salt IV and treatment, subsequently, with *tert*-butoxide, yields 2,2,2-**tributyl-6-methyl-1,2,3,6-tetrahydro-6*H*-1,2-oxaphosphorin** (V), together with the 6-methylene-1,3-*dioxa* compound VI (*L. P. Davydova et al.*, Zhur. obshcheĭ Khim., 1968, **38**, 2091):

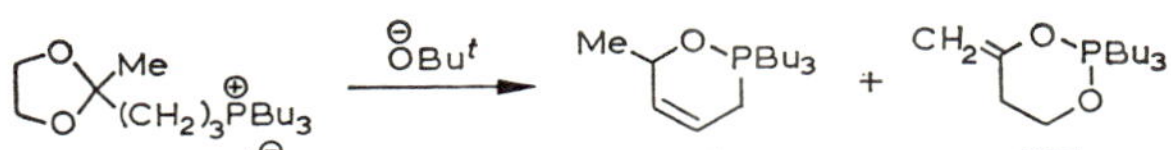

(IV) (V) (VI)

Photolysis of the diazophosphine oxide VII and reaction of the product *in situ* with α,β-unsaturated ketones gives 1,2-**oxaphosphetanes** VIII (not necessarily isolated) and reaction of these with more of the ketone results in ring expansion, the product consisting of a mixture of stereoisomers of the **pentaphenyl-2,3-dihydro-4*H*-1,2-oxaphosphorin 2-oxide** (IX), (*H. Eckes* and *M. Regitz*, Tetrahedron Letters, 1975, 447):

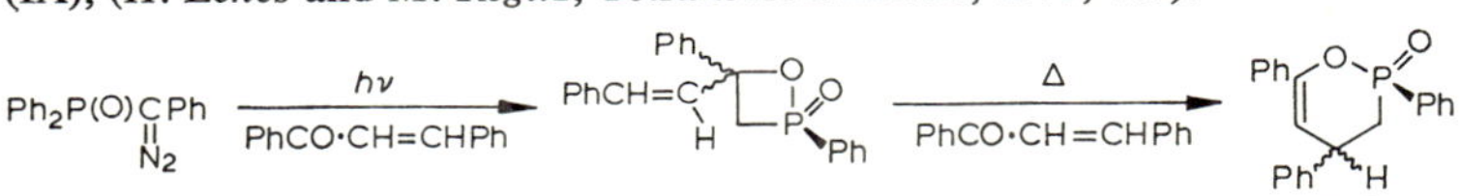

(VII) (VIII) (IX)

Various polycyclic **benz[*e*][1,2]oxaphosphorins,** *e.g.* X, b.p. 150–160°/0.025 mm, have been prepared from appropriately substituted 2-(diethylaminomethyl)phenols and tervalent phosphorus amides (*B. E. Ivanov, A. B. Ageeva* and *R. R. Shagidullin*, Izvest. Akad. Nauk S.S.S.R., Ser. Khim., 1967, 1994; *Ivanov* and *S. V. Samurina*, *ibid.*, 1974, 2079):

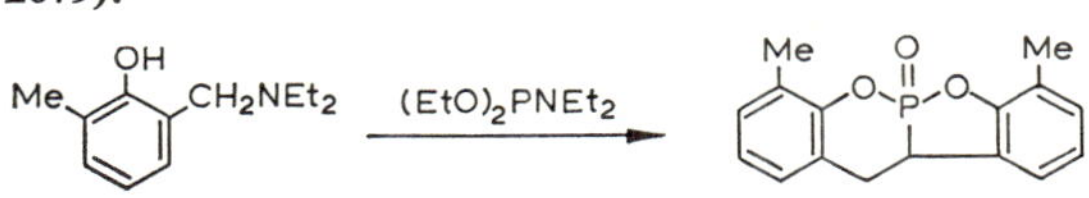

(X)

Reaction between phosphinodithioic anhydrides and 1,3-dienes yields 1,2,3,6-**tetrahydro-6*H*-1,2-thiaphosphorin 2-sulphides** (XI), 2-*phenyl-*, m.p. 70°, 3-*methyl-2-phenyl-*, m.p. 33°, 2-*phenyl-*3,4-*dimethyl-*, m.p. 62° (*A. Ecker, I. Boie* and *U. Schmidt*, Angew. Chem. intern. Edn., 1971, **10**, 191; Monatsh., 1973, **104**, 503):

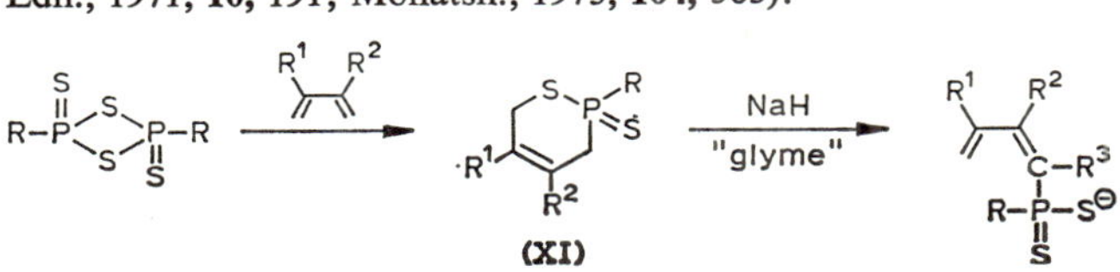

(XI)

Alternatively, phenylthiophosphoric dichloride, $PhP(S)Cl_2$, in the presence of magnesium may be employed (*S. Nakayama et al.*, Bull. chem. Soc., Japan, 1975, **48,** 546).

The ring is cleaved when these compounds are acted upon by sodium hydride under dimethoxyethane.

(*iii*) *Perhydro*-1,2-*oxaphosphorins* (*1,2-oxaphosphorinanes*) *and 1,2-oxarsenanes and their analogues*

The 1,2-oxaphosphorinane ring is known only in the form of its quinquivalent phosphorus derivatives which for the most part are prepared by thermal or base-induced cyclization of 4-substituted butanephosphonic acid derivatives.

1,2-**Oxaphosphorinane**-2-**oxide** (XII, R = H) is the product when 4-hydroxybutyl hydrogen phosphonates are heated at 100–150° (*C. W. Smith*, U.S.P. 2,648,695/1953; C.A., 1954, **48,** 8252); the *cyclic phosphonic acid* (XII, R = OH), m.p. 104.5–106°, is obtained by the action of lithium hydroxide on 4-bromobutanephosphonic acid, and by the action of heat or dicyclohexylcarbodiimide on the lithium or barium salts of 4-hydroxybutanephosphonic acid (*A. Eberhard* and *F. Westheimer*, J. Amer. chem. Soc., 1965, **87,** 253)

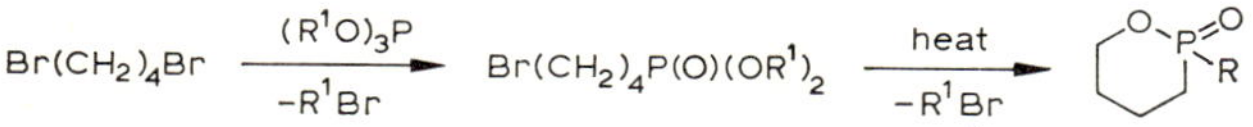

(XII)

2-**Hydroxy**-2-**oxo**-4,6-**diphenyl**-1,2-**oxaphosphorinane** is thought to be the product when styrene interacts with phosphorus in the presence of air in benzene at room temperature (*R. W. Cummings*, Chem. and Ind., 1967, 918). 2-*Ethyl*-2-*oxo*-1,2-*oxaphosphorinane* (XII, R = Et), b.p. 94–95°/1 mm, n_D^{20} 1.4710, is obtained *via* an Arbuzov reaction on bis(4-chlorobutyl) ethylphosphonite, $EtP(OCH_2CH_2CH_2CH_2Cl)_2$, at 200° (*A. N. Pudovik, E. M. Faizallin* and *S. V. Yakovleva*, Zhur. obshcheĭ Khim., 1967, **37,** 460).

More conveniently, derivatives of 1,2-oxaphosphorinane 2-oxide are prepared by *Garner's* method (*A. Y. Garner*, U.S.P. 2,916,510/1959; C.A., 1960, **54,** 5571) in which 1,4-dibromobutanes are heated with trialkyl phosphites, when the 2-alkoxy compounds are obtained; *J. Songstad* (Acta Chem. Scand., 1967, **21,** 1681) showed that the same compounds are formed when diethyl sodiophosphites are heated with 4-chlorobutanols. *K. Bergesen et al.* have shown that the products from such reactions may very often be separated into sterochemically isomeric components, designated *cis* and *trans*, by preparative g.l.c. The 2-*ethoxy*-5-*methyl* compound, b.p. 130°/2 mm, (stereochemical mixture) was separated into two fractions having n_D^{20} 1.4597 and 1.4541; the 2-*ethoxy*-6-*methyl* compound, b.p. 95°/0.5 mm, (stereochemical mixture) was separated into components having n_D^{20} 1.4492 and 1.4525; the first of each of these pairs is thought to be the *cis* isomer, having, in each case the lower i.r. $\nu_{P=O}$ (*Bergesen*, Acta Chem. Scand., 1967, **21,** 578; 1970, **24,** 2019; *Bergesen* and *A. Berge*, *ibid.*, 1970, **24,** 1844; 1972, **26,** 2975; *Bergesen* and *T. Vikane*, *ibid.*, 1971, **25,** 1147; 1972, **26,** 1794, 2153).

2-**Chloro**-1,2-**oxarsenane** has b.p. 72–74°/15 mm (*B. A. Arbuzov et al.*, Doklady Akad. Nauk S.S.S.R., 1971, **200,** 97; *K. Sommer*, Z. Anorg. allgem. Chem., 1970, **376,** 101);

2-chloro-1,2-**thiaarsenane** (X = S), b.p. 89–91 °/0.2 mm, is obtained from 4-chlorobutyl-arsenic dichloride and thiourea *(Sommer, loc. cit.)*.

(*iv*) *Perhydro-1,3-oxaphosphorins (1,3-oxaphosphorinanes) and their analogues*

As a structural component of the 2,4,8-trioxa-6-phospha-adamantane ring system, the 1,3-oxaphosphorinane ring system has been known for some time. Spectroscopic and chemical investigations of the *product*, $C_{10}H_{17}O_3P$, m.p. 88–90°, from phosphine and pentane-2,4-dione, led *M. Epstein* and *S. A. Buckler* (J. Amer. chem. Soc., 1961, **83**, 3279) to propose structure XIII for the substance; primary phosphines afforded other compounds in this series:

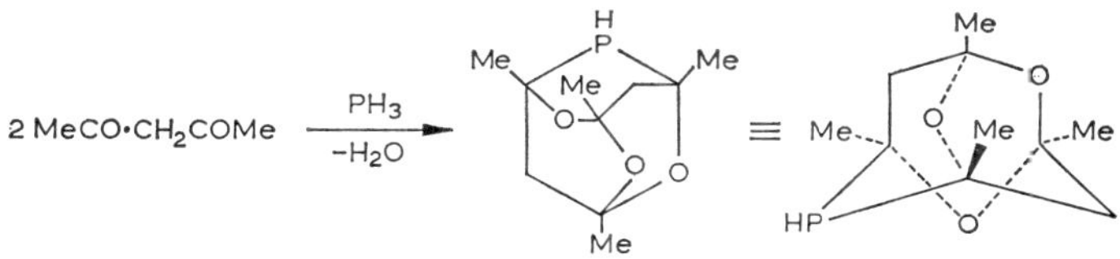

(XIII)

Monocyclic 1,3-**oxaphosphorinanes** (XIV), 5-*methyl*-2-*phenyl*-, b.p. 115–118 °/3 mm, 4,6-*dimethyl*-2-*phenyl*-, b.p. 117–119 °/3 mm, 4,6-*dimethyl*-2,4-*diphenyl*-, b.p. 190–193 °/3 mm, have been obtained by the acid-catalyzed condensation of aldehydes or ketones with 3-hydroxypropylphosphines (*K. Issleib*, *H. Oehme* and *M. Scheibe*, Syn. inorg. metal.-org. Chem., 1972, **2**, 223; *Oehme* and *H. Leissring*, Z. Chem., 1973, **13**, 291). N.m.r. data have been reported for compounds in the series (*A. Zschunke*, *A. Hauser* and *Oehme*, *ibid.*, p. 310):

(XIV)

1,3-**Thiaphosphorinanes,** *e.g.* 2,3-*diphenyl*-1,3-*thiaphosphorinane*, (XV), m.p. 78–83°, readily forming a methiodide are prepared similarly (*Issleib* and *H. J. Hannig*, Z. anorg. allgem. Chem., 1973, **402**, 189):

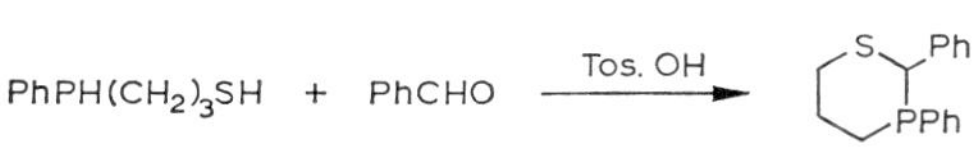

(XV)

(*v*) *1,4-Dihydro-1,4-oxaphosphorins,* 1,4-*dihydro*-1,4-*thia(selena, tellura)-phosphorins and their arsenic analogues*

4-**Phenyl**-1,4-**dihydro**-1,4-**oxaphosphorin** 4-**oxide** (XVI), m.p. 143°, is one of the products, isolated by chromatography over basic alumina, from the interaction of phenyl diethynylphosphine oxide and methylamine in diethyl ether (*M. Maumy*, Bull. Soc. chim. Fr., 1972, 1600):

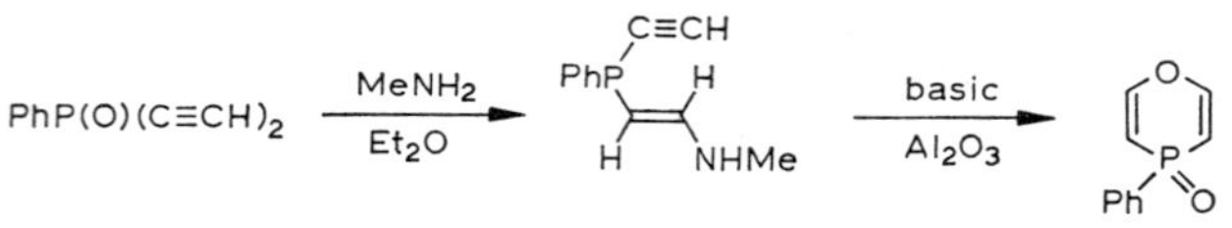

(XVI)

In an essentially similar process (presumably this time through the enol phosphine oxide, as opposed to the eneamine), the 2,4,6-*triphenyl compound*, m.p. 145° is obtained from phenacyl phenyl(phenylethynyl)phosphine oxide, PhP(O)(C≡CPh)(CH_2COPh).

Modifications to these routes include the direct hydration of phenyldialkynylphosphine oxides; the 2,6-*dimethyl*-, m.p. 185–186°, and 2,6-*di*-tert-*butyl*-, m.p. 131.5–132°, derivatives of XVI are so preparable (*A. Neaktgeboren et al.*, Rec. Trav. chim., 1975, **94**, 92). In addition, dehydration of diphenacylphosphinic acid with phosphorus pentoxide yields 4-*hydroxy*-2,6-*diphenyl*-1,4-*dihydro*-1,4-*oxaphosphorin* 4-*oxide* (XVII), m.p. 238–239°, (*L. S. Moskalevskaya* and *G. K. Fedorova*, Zhur. obshcheĭ Khim., 1975, **45,** 950):

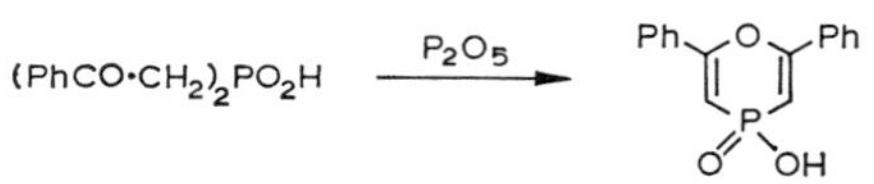

(XVII)

Under acid conditions, a novel rearrangement of 1,4-dihydro-1,4-oxaphosphorins to 1,2,5,6-tetrahydrophosphorin-5-ones (XVIII) occurs (*G. Markl* and *D. Metthes*, Tetrahedron Letters, 1974, 4385, 4388):

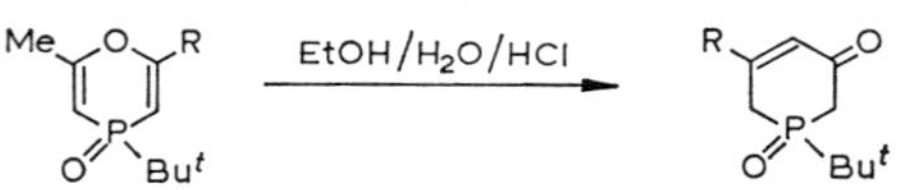

(XVIII)

The reaction between alkynylphosphines and halogenomethyl ketones affords cyclic phosphonium salts XIX *via* "linear" quaternary salts; the yields of "phosphonia-pyrans" depend on the composition of the solvent (ether, benzene, "glyme", hexamethylphosphoramide, in various proportions), and the reaction in benzene is catalyzed by triethylamine (*M. Simalty* and *M. H. Mebazaa*, Bull. Soc. chim. Fr., 1972, 3532; Tetrahedron, 1972, **28,** 3343):

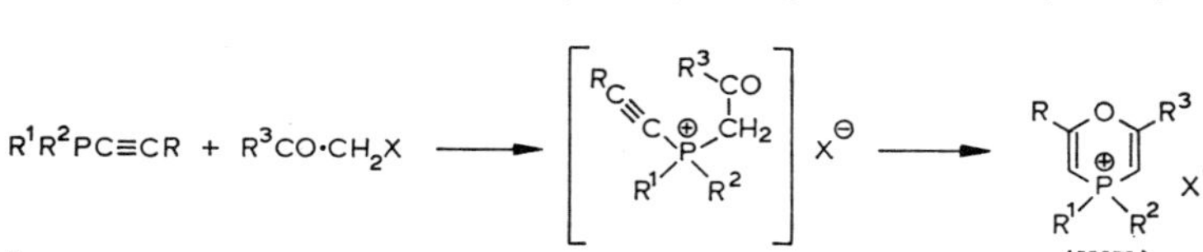

(XIX)

Providing that these phosphonium salts do not possess a phenyl group at position 3, hydrogen or methyl being "permissible", they are converted under alkaline conditions into the oxaphosphorine 4-oxides XX. If the cyclic phosphonia compound has a phenyl group at position 3, ring-opening takes place:

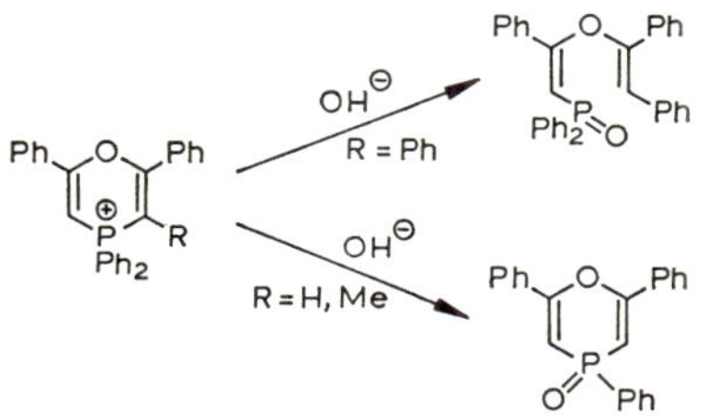

(XX)

The exothermic reaction between dialkylaminoalkynes and phosphinodithioic anhydrides yields 1,4-**dihydro**-1,4-**thiaphosphorin** 4-**sulphides,** *e.g.* 2,6-*bis(diethylamino)*-3,4,5-*tri-methyl*-1,4-*dihydro*-1,4-*thiaphosphorin* 4-*sulphide*, m.p. 65–67° (*N. Schindler* and *W. Plöger*, Synthesis, 1972, 421).

Other methods for the preparation of 1,4-dihydro-1,4-thiaphosphorins parallel those used for the oxygen analogues, and include the addition of hydrogen sulphide, selenide and telluride (as their sodium salts) to di-alkynylphosphine oxides and dialkynylphosphinic esters (*Neaktgeboren, loc. cit.*), and the action of diphosphorus pentasulphide on diphenacyl-phosphine oxides and related compounds (*M. S. Chattha et al.*, Phosphorus, 1973, **3**, 65).

So prepared are **4-phenyl-1,4-dihydro-1,4-thiaphosphorin** 4-**oxide,** its 2,5-*dimethyl* derivative, m.p. 115.5–116.5°, and its 2,6-*diphenyl* derivative; **4-phenyl-1,4-dihydro-1,4-selenaphos-phorin 4-oxide,** m.p. 155–157°; and **4-phenyl-1,4-dihydro-1,4-telluraphosphorin 4-oxide,** m.p. 193–194°:

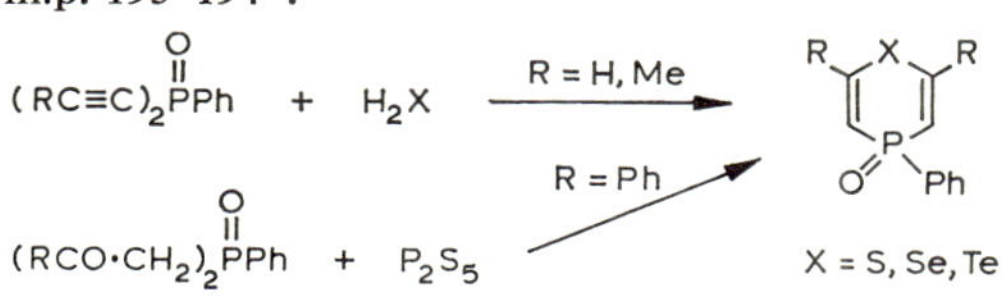

Isomeric nuclear substituted compounds are formed by an alternative addition process involving phenylphosphine or phenylarsine. Thus **3,5-dimethyl-4-phenyl-1,4-dihydro-1,4-thiaphosphorin** (XXI, M = P), is isolated as its *oxide*, m.p. 129–130°, *sulphide* m.p. 118°, and *selenide*, m.p. 132–133°; **3,5-dimethyl-4-phenyl-1,4-dihydro-1,4-thiarsine** (XXI, M = As) is isolated as *oxide*, m.p. 169–171°, and *sulphide*, m.p. 115–117° (*M. Schoufs et al.*, Rec. Trav. chim., 1974, **93**, 241):

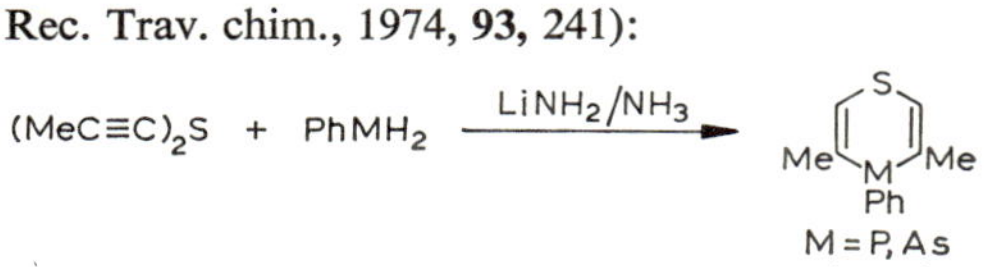

(XXI)

(*vi*) *Dibenz*[1,4]-*oxaphosphorins, -oxarsenin, -oxantimonin and their thia analogues*

Two principle methods are employed for the preparation of 10*H*-**dibenz**[*b,e*][1,4]**oxaphosphorin (phenoxaphosphine)** (XXII, M = P, X = O), the corresponding arsenic **(phenoxarsine)** and antimony **(phenoxantimonin)** compounds, and their thia analogues **(phenothiaphosphine, phenothiarsine** and **phenothiantimonin).** These are (*a*) the treatment of a 2,2′-dilithiodiphenyl ether or sulphide with a metalloid dihalide $RMCl_2$, and (*b*) the cyclization

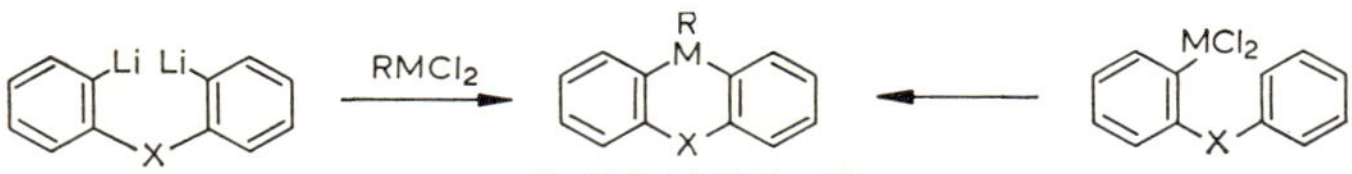

X = O, S; M = P, As, Sb
(XXII)

of 2-phenoxy- or 2-phenylthio-phenylmetalloid dichloride with aluminium chloride or similar reagent. As a variant on the latter process, direct interaction of the ether, MCl_3 and aluminium chloride is sometimes feasible. Unfortunately, 10-chlorophenoxantimonin is not produced from antimony trichloride and diphenyl ether, unlike the corresponding cyclic chloroarsine. Other preparative methods developed for the arsenic series also prove unsatisfactory in the antimony series, largely because the Bart–Schmidt reaction and its various modifications give negligible yields when applied to 2-aminodiphenyl ethers; the arsenic dichlorides are conveniently obtained from the arsonic acids in turn synthesized using the Bart reaction. Satisfactory syntheses of the antimony ring system involve the use of the diGrignard reagent from 2,2′-dibromodiphenyl ether, and by exchange reactions using analogous tin compounds:

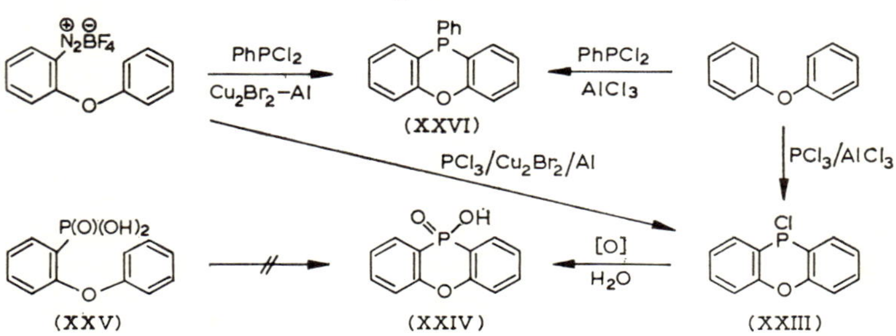

10-Chlorophenoxaphosphine (XXIII), m.p. 62–64°, is prepared (*a*) by treatment of 2-phenoxybenzenediazonium tetrafluoroborate with phosphorus trichloride and cuprous bromide followed by aluminium powder (*G. O. Doak* and *L. D. Freedman*, J. Amer. chem. Soc., 1951, **73**, 5658; *Freedman* and *Doak*, *ibid.*, 1952, **74**, 2884; *J. B. Levy*, *Doak* and *Freedman*, J. org. Chem., 1965, **30**, 660), and (*b*) by direct interaction of diphenyl ether, phosphorus trichloride and aluminium chloride, when the 10-chloro compound need not be isolated, but may be oxidatively hydrolyzed directly to the cyclic phosphinic acid XXIV; m.p. 231–234°, which could not be prepared by direct cyclization of 2-phenoxyphenylphosphonic acid (XXV) (*Freedman, Doak* and *J. R. Edmiston, ibid.*, 1961, **26,** 284).

10-**Phenylphenoxaphosphine** (XXVI), b.p. 191–193°/0.5 mm, m.p. 94.5–95° (*oxide*, m.p. 176–179°; *sulphide*, m.p. 169–170°) is obtained from 2,2′-dilithiodiphenyl ether and phenyldichlorophosphine in organic solvents under nitrogen (*Freedman* and *Doak, loc. cit.*; *I. Granoth, J. B. Levy* and *G. Symmes*, J. chem. Soc., Perkin II, 1972, 697); it forms a *methiodide*, m.p. 236–237°, conventionally, and when treated with bromobenzene and aluminium chloride under vigorous conditions, conversion to 10,10-*diphenylphenoxaphosphonium iodide*, m.p. 195–197° *(hydrate)*, takes place.

Other nuclear substituted derivatives have been listed, particularly by *Freedman*, and *Doak* and co-workers, but also by others (*S. J. Stryker*, U.S.P. 3,576,863/1971; C.A., 1971, **75,** 20618; *Levy, Doak* and *Freedman, loc. cit.*; *I. G. M. Campbell*, J. chem. Soc., C, 1968, 3026; *D. W. Allen* and *I. T. Miller, ibid.*, 1969, 252; *Granoth et al.*, Tetrahedron, 1970, **26,** 813; *Granoth, Y. Segall* and *A. Kalir*, J. chem. Soc. Perkin I, 1973, 1972; *Granoth, Israel*, J. Chem., 1975, **12,** 779).

The heterocyclic ring system is exceptionally stable, the compounds being unaffected on heating with concentrated halogen acids, differing in this respect from the arsenic and antimony compounds; indeed the acid XXIV has been nitrated.

10-**Hydroxy**-2,8-**dimethylphenothiaphosphine,** as the *oxide* XXVII, m.p. > 300°, is isolated from the reaction between di-4-tolyl sulphide and phosphorus trichloride and aluminium chloride; it forms a *S,S-dioxide*, m.p. > 300°, (*L. D. Freedman* and *G. O. Doak*, J. org. Chem., 1964, **29,** 1983; *R. D. Cook* and *H. K. Norrish*, Tetrahedron Letters, 1973,

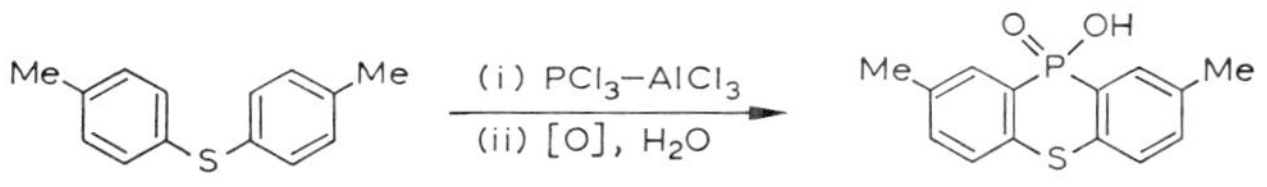

(XXVII)

521); other compounds are formed by the interaction of 2,2′-dilithiodiphenyl sulphide and phosphonic esters (*R. A. Wiley* and *J. H. Collins*, J. med. Chem., 1969, **12,** 146). Nuclear halogen-substituted compounds have been listed (*Granoth et al.*, Tetrahedron, 1969, **25,** 3919).

10-**Chlorophenoxarsine** (XXVIII, R = H), m.p. 125°, was first described by *Sir William Pope* and *E. E. Turner* in 1918 in connection with chemical warfare; the method of synthesis, involving heating a mixture of diphenyl ether, arsenic trichloride and aluminium chloride to 200°, was later used by *W. L. Lewis et al.* (J. Amer. chem. Soc., 1921, **43,** 891). The structure is established through its formation by ring closure of 2-phenoxyphenyldichloroarsine (XXIX, R^1 = H) during distillation under reduced pressure (*Turner* and *A. B. Sheppard*, J. chem. Soc., 1925, **127,** 544). 10-*Bromophenoxarsine*, m.p. 128°, and the corresponding *iodo* compound, m.p. 144°, and *thiocyanato* compound, m.p. 129°, are obtained from the chloride by metathesis with sodium or potassium salts in methanol. The chloride is stable in boiling water or hot dilute alkali, but it reacts with alcoholic alkali to give 10,10′-**oxybi(diphenoxarsine),** m.p. 182° (*C.-S. Wang* and *T. W. McGee*,

U.S.P. 3,544,610/1970; C.A., 1971, **74,** 64308; U.S.P. 3,363,024/1972; C.A., 1972, **76,** 127155) and with ammonia to give **tri(phenoxarsinyl)amine** (XXX) (*F. M. Khusnutdinova, V. I. Gavrilov* and *B. D. Chernokal'skii*, Izv. Vyssh. Ucheb. Zaved. Khim., Khim. Tekhnol, 1973, **16,** 1522; C.A., 1974, **80,** 48114).

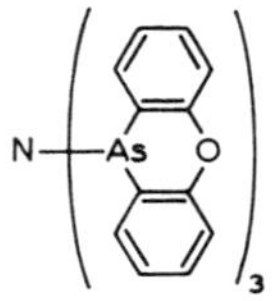

(XXX)

Interaction of the chloro compound with Grignard reagents or lithium reagents, or with lead alkyls, has afforded 10-alkyl and 10-aryl compounds, for example, the 10-*methyl-*, b.p. 185°/20 mm, 10-*ethyl-*, b.p. 194°/20 mm, 10-*phenyl-*, m.p. 107°, and 10-(4-*biphenyl*)-*phenarsenine*, m.p. 152–154°, (*J. A. Aeschlimann*, J. chem. Soc., 1925, **127,** 811; 1927, 413; *F. F. Blicke* and *E. L. Cataline*, J. Amer. chem. Soc., 1938, **60,** 419; *V. I. Gavrilov et al.*, Zhur. obshcheĭ Khim., 1974, **44,** 2506).

Further reaction of such compounds (XXXI, R^1 = H) with phenylmagnesium bromide yields 10-**phenylphenoxarsonium salts** (XXXII, R^1 = H), 10-*phenyl-*, *(bromide)* m.p. 229–230°, 10-*methyl-*, *(iodide)* m.p. 170–171°. Interestingly, 10-phenylphenoxarsine

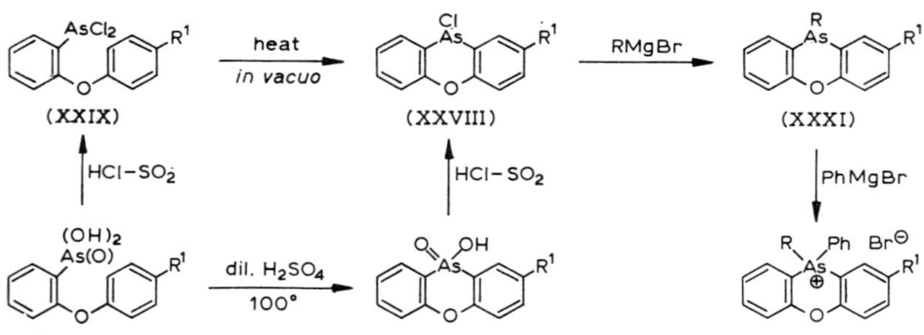

(XXXI, R^1 = H, R = Ph) cannot be obtained by reaction of phenyldichloroarsine with diphenyl ether and aluminium chloride presumably because the catalyst promotes the equilibrium

$$2\ PhAsCl_2 \rightleftharpoons Ph_2AsCl + AsCl_3$$

and the latter reacts rapidly with the ether to give 10-chlorophenoxarsine.

10-Methylphenoxarsine undergoes acyclation by acetyl chloride in the presence of aluminium chloride to give the 2-*acetyl-*, m.p. 86°, and 2,8-*diacetyl-*, m.p. 131°, derivatives (*G. Saint-Ruf, M. Clement* and *J. P. Coie*, Bull. Soc. chim. Fr., 1975, 1361). 20% Hydrogen peroxide oxidises 10-alkyl- and 10-arylphenoxarsines to give hydrates of the corresponding oxides, from which water may be removed azeotropically with benzene (*Gavrilov, G. R. Gavrilova* and *B. D. Chernokalskii*, Zhur. obshcheĭ Khim., 1975, **46,** 72).

10-**Chlorophenothiarsine,** m.p. 129–130°, is prepared similarly by ring closure of 2-phenylthiophenyl dichloroarsine, a process considerably more difficult than for the phenoxarsine series; the compound cannot be made by reaction of arsenic trichloride with diphenyl sulphide (*E. Roberts* and *Turner*, J. chem. Soc., 1926, 1207).

Little is known about **phenoxantimonin** and its sulphur analogue. Reaction of phenylantimony diiodide and the diGrignard reagent from 2,2′-dibromo-4-methyldiphenyl ether gives 2-*methyl*-10-*phenylphenoxantimonin*, b.p. 62–63°, which forms an adduct, m.p. 140–142°, with chlorine (*I. G. M. Campbell*, *ibid.*, 1947, 4).

An interesting and potentially useful synthesis consists in the instantaneous interaction of antimony trichloride and dihydrophenoxastannin or dihydrophenothiastannin compounds.

10-*Chlorophenoxantimonin*, m.p. 150–151°, 10-*chlorophenothiantimonin*, m.p. 130°, and its 5,5-*dioxide*, m.p. 199–200°, all monomeric (*H. A. Meinema*, *C. J. R. Crispim Romao* and *J. G. Noltes*, J. organometal. Chem., 1973, **55**, 139) are obtained in this way:

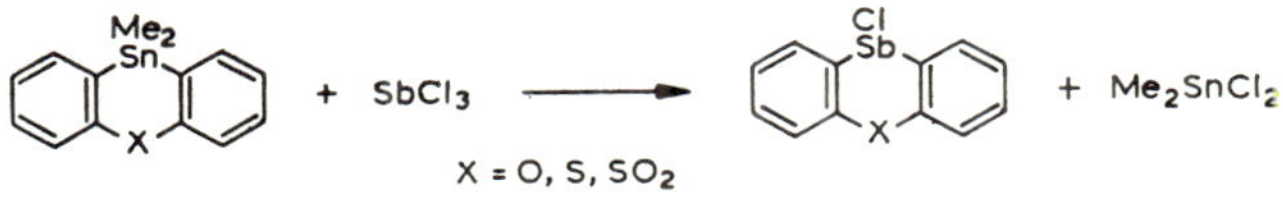

The phenoxantimonin ring system is less stable, particularly towards oxidizing agents than is the corresponding phenoxarsine ring, *e.g.* 2-methyl-10-phenylphenoxarsine may be oxidized to the 2-carboxylic acid by hot aqueous potassium permanganate, but in the antimony series, the same conditions bring about disrupture of the ring. Further, although the phenoxantimonins are stable to alkali, they are sensitive to halogen acids and elimination of antimony frequently occurs. Cleavage of the phenoxantimonin ring by phenyllithium, followed by water, has been noted for the example, XXXV → XXXVI, m.p. 180–183° (*D. Hellwinkel* and *M. Birch*, *ibid.*, 1971, **28**, 349):

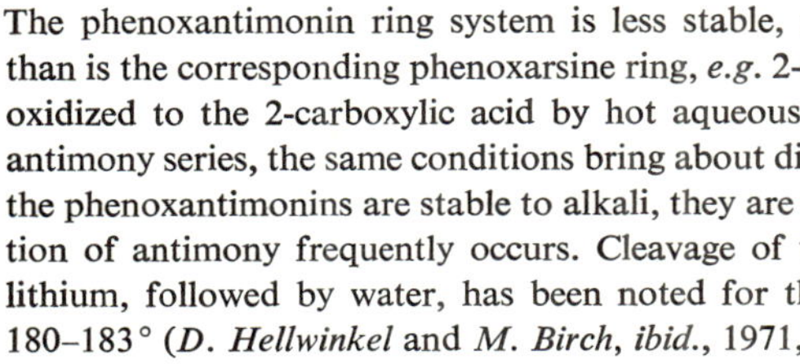

(XXXV) (XXXVI)

On the assumption that the valency angles in the centre ring in phenoxaphosphine and phenoxarsine are similar to those encountered in other phosphines and arsines, *i.e.* CPC and CAsC about 98°, and COC about 112° as in dioxane, then the tricyclic ring system should be folded about the O–P or O–As axis when the element, P or As, is in the tervalent state. As a consequence of such geometry, suitably substituted phenoxarsines and phenoxaphosphines (and also their sulphur analogues) should be resolvable into enantiomeric forms. The first attempts to obtain evidence on the molecular geometry of phenoxaphosphine and phenoxarsine were by *E. E. Turner* and his colleagues. Using intermediates prepared as described above (XXVIII–XXXIV) (*Roberts* and *Turner*, J. chem. Soc., 1925, **127,** 2004; *J. D. C. Mole* and *Turner*, *ibid.*, 1939, 1720; *M. S. Lesslie* and *Turner*, *ibid.*, 1934, 1170; 1935, 1268; 1936, 730), *Lesslie* and *Turner* prepared model compounds, the syntheses of two of which are as indicated below; the acids XXXVIII and XXXIX were resolved satisfactorily as their phenylethylamine, strychnine or cinchonine salts:

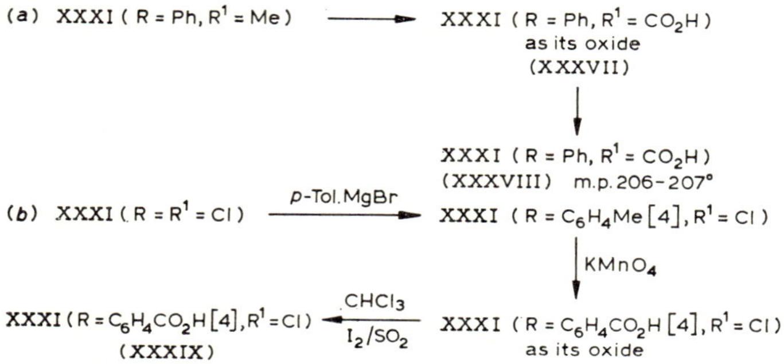

The cyclic arsines show considerable optical stability after resolution; thus XXXVIII in ethanol at 100° (sealed tube) is stable for at least 4 h; XXXIX in boiling ethanol is stable for at least 1 h, and in boiling 0.5 *N* sodium hydroxide solution is stable for at least 3 h.

To explain these results, Lesslie and Turner suggested that, either (*a*) the tricyclic phenoxarsine system is coplanar, in which case dextro- and laevo-rotatory forms would have the substituent on the arsenic atom on opposite sides of the plane, and racemization should be a relatively easy process, or alternatively, and more likely in view of the marked optical stability just commented on, (*b*) the molecule is folded about the As–O axis in what are now referred to as butterfly conformations; flexing motions (considered later by *K. Mislow*, *A. Zimmerman* and *J. T. Melillo*, J. Amer. chem. Soc., 1963, **85,** 594) would not lead to any interconversions between enantiomers and no loss of optical activity would ensue. Loss of optical activity is observed, however, when a suitable phenoxarsine is dissolved in methyl iodide. In the phenoxarsonium salt so formed, *e.g.* 2,10,10-trimethylphenoxarsonium iodide, the valency angle at arsenic would be expected to rise to about 109° and the heterocyclic ring should become planar, thus leading to loss of activity; the arsenic-methyl groups should become equivalent. The equivalence of methyl groups attached to phosphorus in 10,10-dimethylphenoxaphosphonium iodide has been demonstrated using ^{1}H-n.m.r. spectroscopy (*D. W. Allen* and *I. T. Millar*, J. chem. Soc., C, 1969, 252). Such evidence would also seem to preclude a rigidly folded structure.

The phenoxarsine oxide XXXVII has been resolved using morphine, but optical activity is readily lost, presumably by hydration at the As=O bond.

Attempts have also been made to resolve phenoxaphosphines, but as yet with little success; a typical synthesis of a model compound is indicated below (*Campbell*, *ibid.*, 1968, 3026):

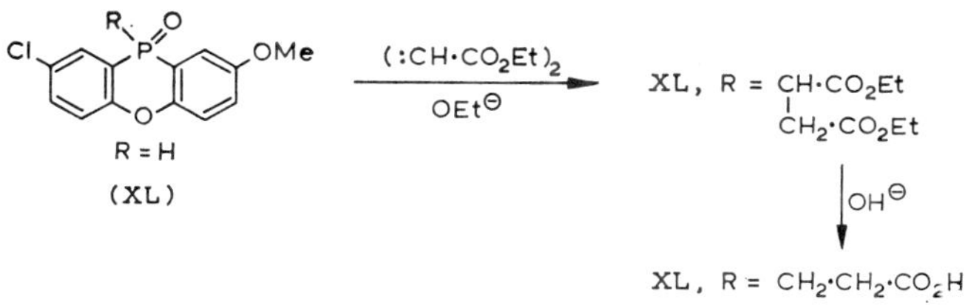

(XL)

(*vii*) *Tetrahydro*-1,4-*oxaphosphorins.* 1,4-*Oxaphosphorinanes and* 1,4-*thiaphosphorinanes and their arsenic analogues*

2,4,6-**Triphenyl**-1,2,3,4-**tetrahydro**-1,4-**oxaphosphorin** 4-**oxide** (XLI), m.p. 145°, is said to be formed by the action of sodium methoxide on phenacylphenyl(phenylethynyl)phosphine oxide, PhP(O) (C≡CPh) (CH_2COPh) (*M. Maumy*, Bull. Soc. chim. Fr., 1972, 1600):

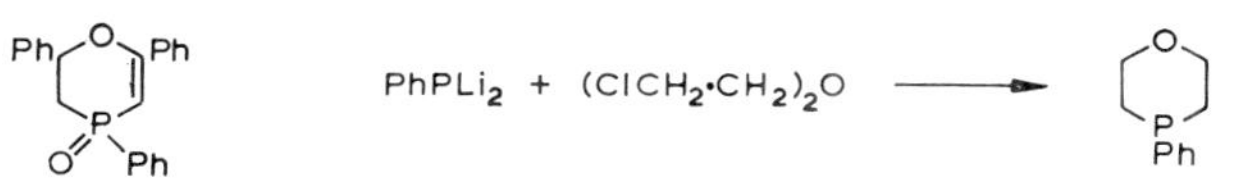

(XLI) (XLII)

4-**Phenyl**-1,4-**oxaphosphorinane** (XLII), b.p. 116–118°/4 mm, (*methiodide*, m.p. 149–150°), is prepared by the action of di(2-chloroethyl) ether on phenyldilithiophosphine. Its 2,6-*dimethyl* derivative, b.p. 154–156°/2 mm, is obtained when bis(2-hydroxypropyl)phenylphosphine, $PhP(CH_2CHMeOH)_2$, is passed over alumina at 280–300° (*K. Issleib* and *H. R. Roloff*, J. pr. Chem., 1970, **312**, 578).

Treatment of diallyldiphenylphosphonium bromide with alumina yields diphenyldipropenylphosphonium bromide which, with alkali, then forms 2,6-**dimethyl**-4,4-**diphenyl**-1,4-**oxaphosphorinanium bromide,** (XLIII), m.p. 241°; the analogous 1,4-**thiaphosphorinanium bromide,** prepared using sodium hydrogen sulphide has m.p. 218–220°. In the same way, allyldiphenylvinylphosphonium bromide yields 2-*methyl*-4,4-*diphenyl*-1,4-*thiaphosphorinanium bromide*, m.p. 212–215°, and tetraallylphosphonium bromide affords the *spiran salts* XLIV (X = O) of m.p. 242–243°, and (X = S), m.p. 234–238° decomp.

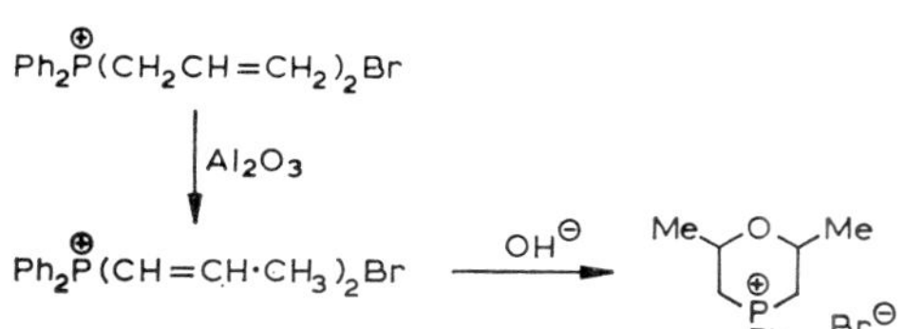

(XLIII)

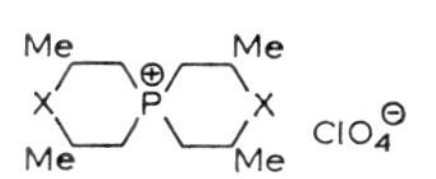

(XLIV)

The isomerization of diallyldiphenylarsonium bromide to diphenyldipropenylarsonium bromide and the ultimate formation of 2,6-**dimethyl**-4,4-**diphenyl**-1,4-**oxarsenanium bromide,** m.p. 208°, is brought about directly by strong alkali, in contrast to the process for the phosphonium salts which tend to decompose under such forcing conditions. Further differences between the phosphorus and arsenic series are found in the action of alkali on tetraallylarsonium bromide when, in addition to the spiran XLVI, *perchlorate*, m.p. 207–208°, the *arsenanium salt* XLV, m.p. 153–154°, can be isolated, and in the action of alkali on phenyltriallylarsonium bromide, when the products are 2,6-*dimethyl-4-phenyl*-1,4-*oxarsenane* (XLVII), b.p. 69–71/0.05 mm, and allyl alcohol (*L. Horner* and *S. Samaan*, Phosphorus, 1974, **4,** 1; *Samaan*, Tetrahedron Letters, 1974, 3927):

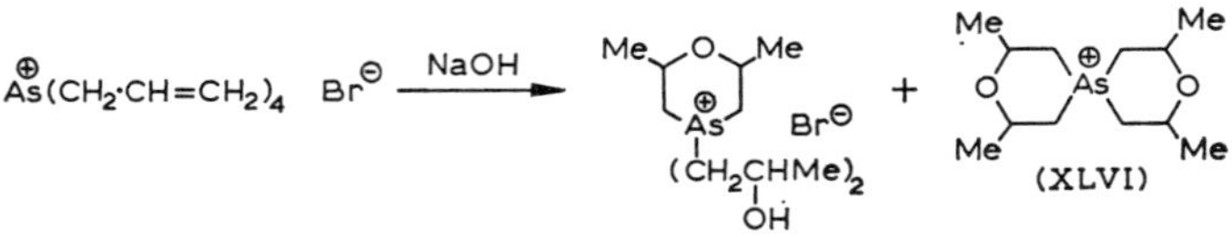

(XLV) (XLVI)

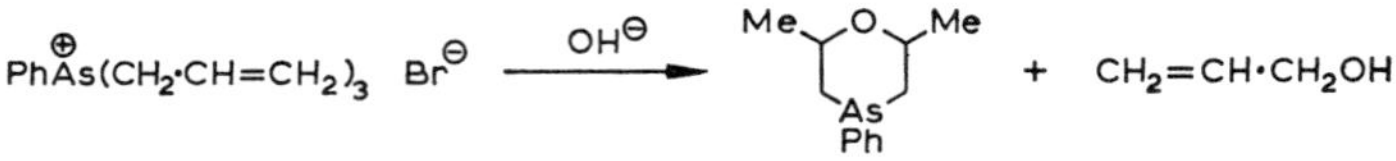

(XLVII)

The use of di(2-chloroethyl) ether or sulphide with phenylarsinedi(magnesium bromide) allowed *R. Reich et al.*, (Bull. Soc. chim. France, 1924, **35,** 1414) to prepare **4-phenyl-1,4-oxarsenane,** b.p. 149–151°/18 mm, (*methiodide*, m.p. 162–162.5°, *sulphide*, m.p. 101.5–102°) and **4-phenyl-1,4-thiarsenane,** m.p. 38°, b.p. 134°/4 mm, which reacts with methyl iodide at the arsenic atom to give a quaternary *methiodide* m.p. 226°.

(*b*) *Dioxaphosphorins, dioxarsenins, dioxantimonins and their thia analogues*

(*i*) 4H-1,3,2-*Dioxaphosphorins*, 4H-1,3,2-*benzodioxaphosphorins*, 4H-1,3,2-*benzarsenins*, *dibenzo*[1,3,2]*dioxaphosphorins*

Ring-closure of the enolic phosphoric acid XLVIII is brought about by the action of dicyclohexylcarbodiimide, when the product is **2-ethoxy-6-phenyl-4*H*-1,3,2-dioxaphosphorin-4-one 2-oxide** (XLIX), m.p. 57.5–61°:

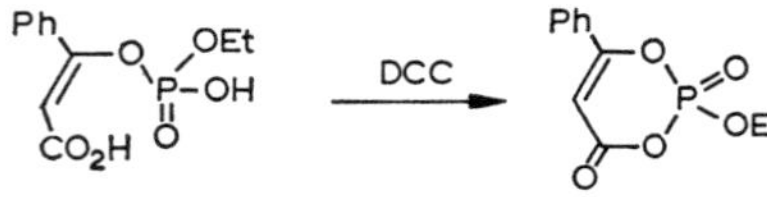

(XLVIII) (XLIX)

This compound is really a cyclic mixed carboxylic-phosphoric anhydride, and as such reacts with toluidine to yield a linear carboxylic toluidide (*J. F. Maracek* and *D. L. Griffith*, J. Amer. chem. Soc., 1970, **92,** 917).

No 4*H*-1,3,2-**benzodioxaphosphor(III)ins** (L) appear to be recorded; transesterification between 2-hydroxybenzyl alcohol and trialkyl phosphites is not feasible since the reaction leads to five-membered ring compounds. There is considerable interest, however, in

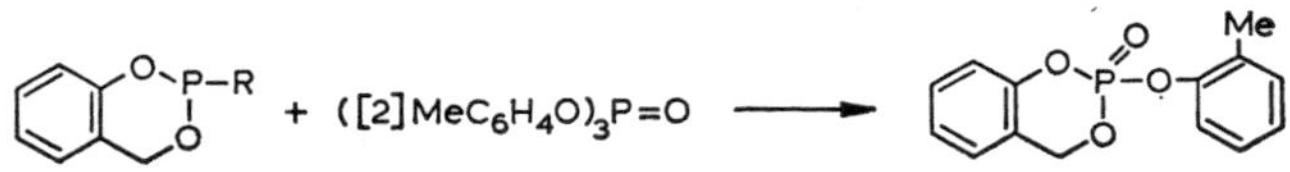

(L) (LI)

the system, stemming as it does from the observation (*M. Eto, J. E. Casida* and *T. Eto*, Biochem. Pharm., 1962, **11,** 337) that tri-*o*-cresyl phosphate is metabolized in rats with the formation *inter alia* of 2-(*o*-cresyl)-4*H*-1,3,2-benzodioxaphosphorin 2-oxide (LI), Since 1962, a large number of *P-oxides* and *P-sulphides* have been prepared and studied extensively with regard to their biological properties. These compounds may be prepared by the standard procedures [see Sect. (ii), p. 106], but the condensations between hydroxybenzyl alcohol and quinquevalent phosphorus dichlorides are best brought about under Schotten–Baumann conditions or by the use of potassium carbonate–copper powder rather than by the conventional use of organic tertiary bases, nevertheless the heterocyclic ring is opened at the aryloxyphosphorus bond by the action of alkali. The work is described in an extensive series of papers by *M. Eto* and his collaborators (Agric. biol. Chem. Tokyo, 1962, **26,** 452, 630, 834; 1963, **27,** 723, 789, 870; 1966, **30,** 181; 1968, **32,** 1056; J. Agric. chem. Soc. Japan, 1965, **39,** 311; 1966, **40,** 315; Botyu-Kagaku, 1968, **33,** 73; C.A., 1969, **70,** 28895; *ibid.*, 1969, **34,** 66; C.A., 1970, **71,** 112896). 2-Methylthio-4*H*-1,3,2-benzodioxaphosphorin 2-oxide is a phosphorylating agent for alcohols, and undergoes ring-opening in the process (*Eto et al.*, Tetrahedron Letters, 1971, 4263; Agric. biol. Chem. Tokyo, 1973, **37,** 115; *M. Sasaki, H. Ohkawa* and *Eto*, J. Fac. Agric., Kyushu Univ., 1973, **17,** 173; *M. Iio, et al., ibid.*, 1975, **19,** 73):

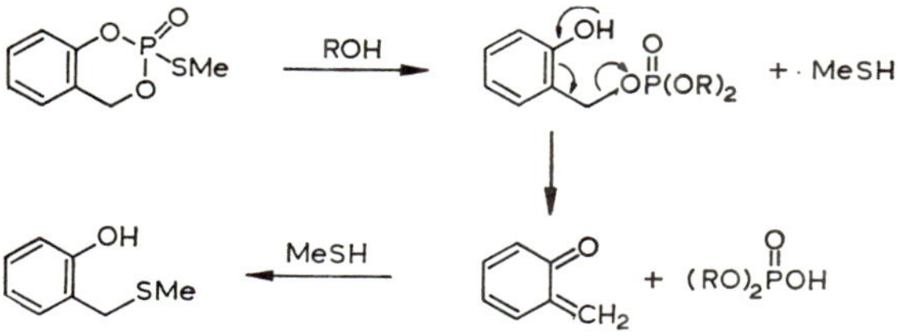

Similar compounds have been prepared from 2-hydroxymethyl-3-hydroxypyridines (*D. Pillon, J. Dueret* and *D. Emozay*, G.P., 2,251,035/1973; C.A., 1973, **79,** 18769; *K. Ruefenacht*, G.P. 2,259,953/1973; C.A., 1973, **79,** 73382).

Salicyclic acid reacts with phosphorus trichloride, with or without added organic tertiary base, to give **2-chloro-4*H*-1,3,2-benzodioxaphosphorin-4-one** (LII, M = P, R = Cl), b.p. 127–128 °/11 mm, m.p. 36–37 °, (*bromide*, b.p. 143 °/9 mm, *butoxy ester*, b.p. 99–100 °/ 0.03 mm) readily oxidized to the 2-*oxide*, m.p. 95 °, also from salicyclic acid and phosphoryl

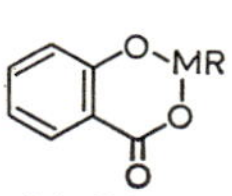

M = P,As
(LII)

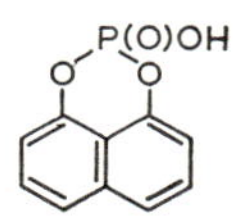

(LIII)

chloride (*J. A. Cade* and *W. Gerrard*, J. chem. Soc., 1960, 1249; *Farbwerke Hoechst A.-G.*, Belg. P. 667,036/1966; C.A., 1966, **65,** 13741). The use of phosphorus pentachloride leads to an acyclic phosphorus chloride ("Couper's compound"). Esters (LII, M = P, R = Oalkyl) react with alcohols to give trialkyl phosphites (*L. V. Nesterov* and *R. Sabirova*, Zhur. obshcheĭ Khim., 1961, **31,** 897; 1964, **35,** 1976) (Table 16).

TABLE 16

4H-1,3,2-BENZODIOXAPHOSPHORIN 2-OXIDES AND 2-SULPHIDES

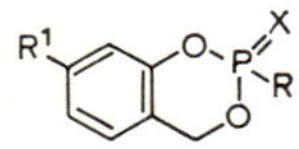

R[1]	*R*	*B.p., °C/mm (m.p., °C)*	
		X = O	*X = S*
H	Me	140/0.5 (1)	130/0.6 (1)
	Ph	(146–148) (2)	(37) (2)
	MeO	110–112/0.05 (2)	(51–53) (2)
	PhO	(77–79) (2)	(36) (2)
	SMe	144–145/0.1 (3)	(69–70) (4)
	NHMe	(87) (5)	120–123/0.2 (5)
	NMe_2	(121) (5)	118–122/0.2 (5)
Me	Ph	(93–95) (2)	—
	MeO	109/0.05 (2)	—

References
1 *M. Eto et al.*, Agr. biol. Chem. Tokyo, 1966, **30**, 181
2 *Y. Oshima* and *M. Eto*, Jap. P., 17,395/65; C.A., 1965, **63**, 17971.
3 *K. Kobayashi et al.*, Boitiu Kagaku, 1969, **34**, 165; C.A., 1970, **72**, 100196.
4 *Sumimoto* Chem. Co., Neth. Appl. 6,615,246/1967; C.A., 1967, **67**, 116710.
5 *idem.*, F. P. 1,408,408/1965; C.A., 1966, **64**, 4995.

2-Chloro-4*H*-1,3,2-benzodioxarsenin-4-ones (LII, M = As, R = Cl) m.p. 120–122°, is formed in a similar way (*H. Funk* and *H. Köhler*, J. pr. Chem., 1961, **13**, 322).

2-Hydroxy-1*H*,3*H*-naphtho[1,8,*c*,*d*][1,2,6]**dioxaphosphorin 2-oxide** (LIII) (*naphthalene-1,8-diol cyclic phosphate*), m.p. 280–284°, is described by *M. L. Bender* and *J. M. Lawlor* (J. Amer. chem. Soc., 1963, **85**, 3010).

(*ii*) *1,3,2-Dioxa-phosphorinanes, -arsenanes, and -antimonanes, and their thia analogues*

While general methods for the preparation of 4*H*-1,3,2-dioxaphosphorins are not yet available, a large number of general methods have been developed for the synthesis of 1,3,2-dioxaphosphorinanes (LIV) and their arsenic and antimony analogues. Such methods include the utilization of 1,3-diols and related compounds as starting materials in the following reactions: (*a*) trans-esterification with trialkyl phosphites giving cyclic P(III) esters (*D. C. Ayres* and *H. N. Rydon*, J. chem. Soc., 1957, 1109; *G. Kamai* and *E. T. Mukmenev*, Zhur. obshcheĭ Khim., 1963, **33**, 3197), dialkyl hydrogen phosphonates giving cyclic hydrogen phosphonates (oxides of LIV, R = H) (*A. A. Ostwald*, Canad. J. Chem., 1959, **37**, 1498; U.S.P., 3,152,164/1964; C.A., 1964, **61**, 16082; *A. Zwierzak*, Canad. J. chem., 1967, **45**, 2501) and dialkyl dithio-

TABLE 17

1,3,2-DIOXAPHOSPHORINANES

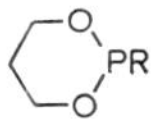

(LIV)

R		*B.P., °C/mm (m.p., °C) with ring substituents:*		
		None	*4-Me*	*5,5-diMe*
H			63/50 (3)	45–50/25 (5)
	Oxide	97–98/2.5 (1)	82–88/0.25(50–53) (4)	114–115/0.9(49–51) (6)
	Sulphide	60/0.02(34) (2)	84–86/0.1 (3)	(81–82) (7)
Cl		67/15 (8)	65/12 (11)	66/13 (12)
	Oxide	106–109/0.3(39–42) (9)	101–102/0.1 (4)	(104.5–106) (6)
	Sulphide	(39–40) (10)	110–112/2(95–97) (10)	(90.5–91.5) (6)
OH		—*	—*	—*
	Oxide	(91–92) (13)	(48–50) (11)	(174–176) (6)
	Sulphide			
SH		—*	—*	—*
	Oxide			
	Sulphide		130–131/0.4 (14)	(81–82) (6)
OEt		77/25 (15)	63–64/8 (11)	76–77/16 (16)
	Oxide	110/0.2 (17)		94–95/0.3 (16)
	Sulphide			(62–63) (16)
SEt				
	Oxide			(42–44) (16)
	Sulphide		111–112/0.6 (14)	(69.5–70.5) (16)
OPh		81–82/0.3(18)(46.5) (8)	125–130/8 (20)	90–95/1 (20)
	Oxide	(76–76.5) (19)		(133–135) (21)
	Sulphide			(98–99) (22)
Me			85.5–86.5/41 (23)	
	Oxide	91/0.01(98–99) (24)	71/0.25(40–41) (25, 26)	(119–121) (16)
	Sulphide	(81) (27)		
Ph		72–74/0.15 (28)	subl. 50/0.05 (23)	100/1.5(82–83) (30)
	Oxide	170–174/0.1(33) (29)		(103–105) (21)
	Sulphide	137/0.45 (30)		
NH_2				
	Oxide			(125–127) (31)
	Sulphide			
NMe_2			71/13.5 (32)	64–66/6 (33)
	Oxide			(104.5–106) (16)
	Sulphide			(73–73.5) (33)

* These compounds do not exist as such, but as the oxides and sulphides of the 2-proto compounds.

References

1 *A. A. Ostwald*, Canad. J. Chem., 1959, **37,** 1498.
2 *C. Z. Krawiecki* and *J. Michalski*, J. chem. Soc., 1960, 881.
3 *W. J. Stec*, *B. Uznanski* and *Michalski*, Phosphorus, 1973, **2,** 237.
4 *C. L. Bodkin* and *P. Simpson*, J. chem. Soc., Perkin II, 1973, 676.
5 *E. E. Nifant'ev* and *L. M. Matreeva*, Zhur. obshcheĭ Khim., 1969, **39,** 1555.
6 *R. S. Edmundson*, Tetrahedron, 1965, **21,** 2379.

7 *Idem.*, Chem. and Ind., 1965, 1220.
8 *K. Bergesen* and *P. Albriktson*, Acta Chem. Scand. 1971, **25**, 2257.
9 *Edmundson*, Chem. and Ind., 1962, 1828.
10 *N. I. Zemlyanskii* and *V. P. Kalashnikov*, Zhur. obshcheĭ Khim., 1967, **37**, 1141.
11 *A. E. Arbuzov* and *V. M. Zoroastrova*, Izvest. Akad. Nauk S.S.S.R., 1952, 779.
12 *Edmundson*, Chem. and Ind., 1965, 1220.
13 *E. Cherbuliez et al.*, Helv., 1959, **42**, 1397.
14 *A. N. Pudovik et al.*, Zhur. obshcheĭ Khim., 1968, **38**, 307.
15 *H. J. Lucas*, *F. W. Mitchell* and *C. N. Scully*, J. Amer. chem. Soc., 1950, **72**, 5491.
16 *Edmundson*, Tetrahedron, 1964, **20**, 2781.
17 *J.-P. Vives et al.*, Bull. Soc. chim. Fr., 1965, 2544.
18 *L. Keay* and *E. M. Crook*, J. chem. Soc., 1961, 710.
19 *H. G. Khorana et al.*, J. Amer. chem. Soc., 1957, **79**, 430.
20 *I. Kleckenbleikner*, *W. Pauze* and *F. C. Lanoue*, U.S.P. 2,834,798/1958; C.A., 1958, **52**, 17297.
21 *R. L. McConnell* and *H. W. Coover*, J. org. Chem., 1959, **24**, 630.
22 *J. P. Majoral* and *J. Navech*, Bull. Soc. chim. Fr., 1971, 95.
23 *D. W. White*, Phosphorus, 1971, **1**, 33.
24 *A. F. McKay et al.*, J. Amer. chem. Soc., 1952, **74**, 5540.
25 *Idem.*, J. Amer. chem. Soc., 1954, **76**, 3546.
26 *Arbuzov*, *Zoroastrova* and *M. K. Saikina*, Iavext. Akad. Nauk S.S.S.R., Ser. Khim., 1959, 1579.
27 *M. Wieber* and *H. U. Werther*, Monatsh., 1968, **99**, 1153.
28 *T. Mukaiyama et al.*, J. org. Chem., 1964, **29**, 2572.
29 *Edmundson* and *J. O. L. Wrigley*, Tetrahedron, 1967, **23**, 283.
30 *H. J. Harwood*, U.S.P. 3,270,092/1966; C.A., 1966, **65**, 15425.
31 *A. Zwierzak* and *A. Koziana*, Tetrahedron, 1970, **26**, 3521.
32 *Arbuzov* and *Zoroastrova*, Izvest. Akad. Nauk S.S.S.R., Ser. Khim., 1952, 789.
33 *K. D. Bartle*, *Edmundson* and *D. W. Jones*, Tetrahedron, 1967, **23**, 1701.

phosphoric acids giving (LIV, R = SH, as sulphides) (*R. Rätz* and *A. D. Bliss*, J. heterocycl. Chem., 1966, **3**, 14); (*b*) cyclization with phosphorus trichloride, phosphoryl chloride or thiochloride with, or without the addition of hydrogen chloride scavenger, giving cyclic chlorides, their oxides and sulphides (*H. J. Lucas*, *F. W. Mitchell* and *C. N. Scully*, J. Amer. chem. Soc., 1950, **72**, 5491; *A. F. McKay et al.*, *ibid.*, 1952, **74**, 5540; 1954, **76**, 3546; *Rätz* and *O. J. Sweeting*, J. org. Chem., 1963, **28**, 1608; 1965, **30**, 438; *N. I. Zemlyanskii* and *V. P. Kalashnikov*, Zhur. obshcheĭ Khim., 1967, **37**, 1141); (*c*) the use of related dichlorides $RPCl_2$, $RP(O)Cl_2$ and $RP(S)Cl_2$ (*J.-P. Vives et al.*, Bull. Soc. chim. Fr., 1965, 2544; 1966, 2350, 2355; *M. Wieber* and *H. U. Werther*, Monatsh., 1968, **99**, 1153; (*d*) the use of phosphorus pentoxide and polyphosphoric acid yielding cyclic hydrogen phosphates (*E. Cherbuliez*, *G. Cordahi* and *J. Rabinowitz*, Helv., 1959, **42**, 595; *Cherbuliez*, *H. Probst* and *Rabinowitz*, *ibid.*, 1960, **43**, 464; *A. M. Meston*, J. chem. Soc., 1963, 6059); (*e*) cyclization of hydroxyalkylphosphoric acids by dicyclohexylcarbodiimide giving cyclic hydrogen phosphates and their anhydrides (*H. G. Khorana et al.*, J. Amer. chem. Soc., 1957, **79**, 430; *Cherbuliez*, *Probst* and *Rabinowitz*, Helv., 1959, **42**, 1377) and (*f*) the use of diphosphorus pentasulphide to give cyclic dithio acids (*A. N. Pudovik*,

R. A. Cherkasov and *R. M. Kondrat'eva*, Zhur. obshcheĭ Khim., 1968, **38,** 307; *N. I. Zemlyanskii, V. P. Kalashnikov* and *A. P. Gupalo, ibid.*, 1968, **38,** 855).

A selection of 1,3,2-dioxaphosphorinanes prepared by these methods is listed in Table 17.

Following the first systematic study of the ^{1}H-n.m.r. spectra of 1,3,2-dioxaphosphorinanes (*K. D. Bartle, R. S. Edmundson* and *D. W. Jones*, Tetrahedron, 1967, **23,** 1701), and the realization that it was possible to isolate geometric isomers of derivatives of the system (*D. Z. Denny* and *D. B. Denny*, J. Amer. chem. Soc., 1966, **88,** 1830; *G. Aksnes, R. Eriksen* and *K. Mellingen*, Acta Chem. Scand., 1967, **21,** 1028), this class of compound sprang into prominence resulting in a large number of conformational studies, particularly those by *Edmundson et al.* (J. chem. Soc., Perkin I, 1972, 1660, and earlier papers), *J. P. Majoral et al.*, (Phosphorus, 1972, **2,** 111; Bull. Soc. chim. Fr., 1973, 3149 and earlier papers), *W. G. Bentrude et al.* (J. Amer. chem. Soc., 1973, **95,** 4666 and earlier papers), *J. G. Verkade et al.* (*ibid.*, 1973, **95,** 4659 and earlier papers), *D. Gagnaire et al.* (Bull. Soc. chim. Fr., 1969, 3496 and earlier papers), *M. Kainosho et al.* (Bull. chem. Soc. Japan, 1969, **42,** 1713 and other papers), *B. A. Arbuzov et al.*, (Zhur. obshcheĭ Khim., 1975, **45,** 1432), *E. E. Nifant'ev et al.*, (*ibid.*, 1973, **43,** 71 and other papers); *W. J. Stec et al.*, (Z. Naturforsch., B, 1976, **31,** 354 and earlier papers), and many other authors.

The data so acquired have allowed the use of many 1,3,2-dioxaphosphorinanes as model compounds in the study of mechanistic aspects of phosphorus chemistry, including (*a*) the stereospecificity of oxidation reactions (*Stec, A. Okruszek* and *J. Michalski*, J. org. Chem., 1976, **41,** 233), (*b*) exocyclic chlorination (*A. Skowronska, J. Mikolajczak* and *Michalski*, Chem. Comm., 1975, 986), (*c*) desulphurization (*Edmundson*, J. chem. Soc., C, 1967, 1635), (*d*) the Arbuzov reaction (*P. Simpson* and *C. L. Bodkin*, J. chem. Soc., B, 1971, 1136) and related reactions (*K. Lesiak, B. Uznanski* and *W. J. Stec*, Phosphorus, 1975, **6,** 65), and (*e*) the mechanism of displacement reactions by nucleophilic reagents (*W. S. Wadsworth* and *R. L. Wilde*, J. org. Chem., 1976, **41,** 1264; Chem. Comm., 1976, 93) and the systems used include the 4-methyl-, 5-*tert*-butyl-, 4,4,6-trimethyl- and 5-halogenomethyl-5-methyl-1,3,2-dioxaphosphorinanes. Latterly, however, very elegant studies have been carried out using bicyclic 1,3,2-dioxaphosphorinanes based upon carbohydrates particularly D-glucose and D-galactose and the potential of such compounds in organophosphorus synthetic work has been realized (*T. D. Inch et al.*, J. chem. Soc., Perkin I, 1974, 1043, 1049, 1053, 1058).

In many respects, the chemistry of the 1,3,2-dioxaphosphorinane ring resembles that of the analogous five-membered ring system. The ring is cleaved by the action of thiourea (*L. I. Mizrakh et al.*, Zhur. obshcheĭ Khim., 1972, **42,** 1700), and by Grignard reagents to give 3-hydroxypropylphosphinic esters (*Edmundson* and *J. O. L. Wrigley*, Tetrahedron, 1967, **23,** 283). Tervalent phosphorus chlorides yield open chain compounds on treatment with acryloyl chlorides (*K. I. Novitskii, N. A. Razumova* and *A. A. Petrov*, Zhur. obshcheĭ Khim., 1967, **37,** 512), with trichloracetaldehyde (*N. P. Grechkin, R. R. Shagidullin* and *L. N. Grishina*, Izvest. Akad. Nauk S.S.S.R., Ser. Khim., 1968, 854) and with 1,3-dienes (*N. A. Razumova, A. A. Petrov* and *Zh. L. Evtikhov*, Zhur. obshcheĭ Khim., 1968, **38,** 2542), many of these reactions probably taking place *via* pentacoordinate intermediates. Compounds of tervalent phosphorus having lightly or completely unsubstituted rings undergo ring opening on treatment with chlorine or bromine. The 1,3,2-dioxaphosph(III)orinane ring is sometimes opened in reactions with alkyl, aralkyl halides or *N*-chloroamines, depending on the substituents on phosphorus, ring substituents on carbon, and on the nature of the attacking halide (*A. E. Arbuzov* and *V. M. Zoroastrova*, Izvest. Akad. Nauk S.S.S.R., Ser. Khim., 1951, 536; 1952, 779; *B. A. Arbuzov, Zoroastrova* and *M. K. Saikina, ibid.*, 1959, 1579) but the ring is retained in reactions with azides (*Stec* and *A. Lopusinski*, Tetrahedron, 1973, **29,** 547) and with sulphenyl chlorides (*D. B. Denny* and *M. Moskal*, Phosphorus, 1974, **4,** 77).

In a number of studies the reactivity of the dioxaphosphorinane ring has been compared with that of analogous five-membered ring compounds (1,3,2-dioxaphospholanes). Exocyclic P–N bonds are cleaved by hydrochloric acid; in *N*-HCl some ring opening may take place. Generally, the greater the degree of ring substitution, the greater the stability of the ring towards acid and alkali, sulphides being more stable than oxides (*T. R. Fukuto* and *R. L. Metcalf*, J. med. Chem., 1965, **8,** 759; *Edmundson* and *A. J. Lambie*, J. chem. Soc., B, 1967, 577). The results of conformational differences have been observed; thus, 2-*tert*-butoxy-1,3,2-dioxaphosph(III)orinanes with axial butoxy group suffer thermolysis with elimination of isobutene whereas the stereoisomer with equatorial butoxy group is stable to heat (*A. A. Borisenko* and *E. E. Nifant'ev*, Zhur. obshcheĭ Khim., 1970, **40,** 2765). The 1,3,2-dioxaphosph(V)orinane ring is cleaved by action of trialkylamines yielding analogues of choline (*N. H. Phuong, N. T. Thuong* and *P. Chabrier*, Bull. Soc. chim. Fr., 1975, 2326; *Thuong* and *Chabrier, ibid.*, 1974, 667).

In common with five-membered ring compounds containing tervalent phosphorus, 1,3,2-dioxaphosph(III)orinanes form 1:1 and 1:2 adducts with 1,3-dienes, *vic.*-dicarbonyl compounds and unsaturated carbonyl compounds, *e.g.* LV (*B. A. Arbuzov et al.*, Doklady Akad. Nauk S.S.S.R., 1972, **205,** 843, 1370; *A. Kh. Voznesenskaya* and *N. A. Razumova*, Zhur. obshcheĭ Khim., 1975, **45** 1235):

(LV)

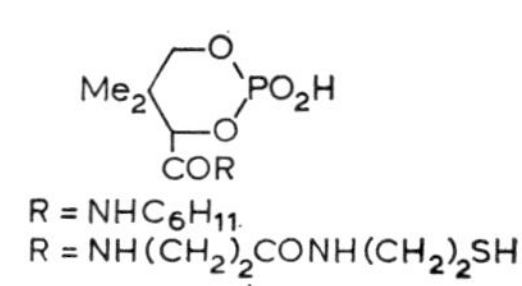

(LVI)

Some cyclic esters of actual or potential biological interest have been prepared. *T. G. Halsall* and *A. R. Hands* (J. chem. Soc., 1961, 3251) considered the cyclic phosphate from 2-methylbutane-2,4-diol as a possible intermediate in the sequence of events leading to the biochemical formation of isopentenyl pyrophosphate and thence to terpenes; the cyclic phosphate from mevalonic acid has also been prepared (*H. Machleidt, E. Cohnen* and *R. Tschesche*, Ann., 1962, **655,** 70). *J. Baddiley* and *A. R. Thain* (J. chem. Soc., 1953, 903) obtained the compounds LVI during the course of their studies on coenzyme A.

Also obtained by conventional means from propane-1,3-dithiol are 1,3,2-**dithiaphosphorinanes,** 2-*chloro-* (*oxide*, m.p. 57–59 °), 2-*methyl-*, b.p. 102 °/5 mm, (*oxide*, m.p. 114 °; *sulphide*, m.p. 121 °); 2-*chloro*-4-*methyl-*, b.p. 75–

76°/1 mm, (*oxide*, b.p. 120–125°/10^{-4} mm) and other cyclic tetrathiophosphates (*Wieber*, *J. Otto* and *M. Schmidt*, Angew. Chem., 1964, **76,** 648; *Wieber* and *Werther*, *loc. cit.; N. I. Zemlyanskii et al.*, Zhur. obshcheĭ Khim., 1972, **42,** 1647; *V. S. Blagoveshchenskii et al.*, Vestn. Mosk. Univ., Khim., 1975, **16,** 227; C.A., 1975, **83,** 179212).

Also preparable conventionally, 1,3,2-**oxathiaphosphorinane** 2-**oxides** may, in addition, be obtained by treatment of 1,3,2-dioxaphosphorinane 2-sulphides with triphenylphosphine (*N. H. Phuong et al.*, Compt. rend., C, 1970, **271,** 1465; 1971, **272,** 1145):

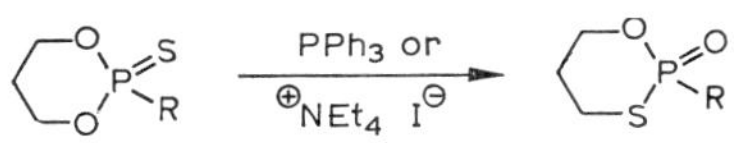

Reaction between hydroxythiophene carboxylic acids and esters and diphosphorus pentasulphide yields **thieno**-1,3,2-**oxathiaphosphorinane** 2-

TABLE 18

1,3,2-DIOXARSENANES AND 1,3,2-DIOXANTIMONANES

Ring substituents	*As*		*Sb*	
	B.p., °C/mm (m.p., °C)	*Ref.*	*m.p. (°C)*	*Ref.*
2-OH (anhydride)	93–95/0.1	1		
2-Cl	66–67/10	2	198–200 decomp.	5
2-Me	62–64/14	1		
2-Et	74–75/10	2		
2-Ph	135/12	2		
2-OMe			103–104	5
2-Cl-4-Me	34–38/5.5(20–21)	3	202–204 decomp.	5
2-OMe-4-Me	43–44/5.5	3	94–95	5
2-Cl-4,6-Me_2*cis*	60/5.5	3		
trans	66/5.5	3		
2-OMe-4,6-Me_2*cis*	40–50/5.5	3		
trans	52–53/5.5	3		
2-Cl-4,4,6-Me_3	45–47/5.5	3		
2-OMe-4,4,6-Me_3	39–40/5.5	3		
5-Me-5-CH_2OH	107/0.4	4		
2-Cl-5,5-Me_2			200–204 decomp.	5
2-OMe-5,5-Me_2			194–195	5

References
1 *K. Sommer*, Z. anorg. allgem. Chem., 1969, **370,** 227.
2 *G. Kamai* and *N. A. Chadaeva*, Doklady Akad. Nauk S.S.S.R., 1951, **81,** 837; Zhur. obshcheĭ Khim., 1956, **26,** 2468.
3 *D. W. Aksnes* and *S. Tøgessen*, Acta Chem. Scand., 1975, **29A,** 376.
4 *W. J. Ruehl* and *F. Kober*, Z. Naturforsch. anorg. Chem. org. Chem., 1976, **318,** 307.
5 *B. A. Arbuzov, Yu. Yu. Samitov* and *Yu. M. Mareev*, Doklady Akad. Nauk S.S.S.R., 1974, 219, 863.

sulphides (*J. Brelivet, P. Appriou* and *J. Teste*, Bull. Soc. chim. Fr., 1971, 1344).

1,3,2-**Dioxarsenanes** (Table 18) are prepared from the appropriate propane-1,3-diol and (*a*) arsenic trichloride (*b*) an arsonic acid or oxide $(RAsO)_n$ with azeotropic removal of water (*G. Kamai* and *N. A. Chadaeva*, Doklady Akad. Nauk S.S.S.R., 1951, **81,** 837; Zhur. obshcheĭ Khim., 1956, **26,** 2468; *V. K. Kuzmin, B. D. Chernokal'skii* and *R. R. Shagidullin*, *ibid.*, 1973, **43,** 1937), or (*c*) bis(dimethylamino) arsines (*W. J. Ruehl* and *F. Kober*, Z. Naturforsch., anorg. Chem. org. Chem., 1976, **31B**, 307) and (*d*) from the oxide $(RAsO)_n$ by reaction with trimethylene carbonate (*K. Sommer*, Z. anorg. allgem. Chem., 1969, **370,** 227).

1,3,2-Dioxantimonanes are obtained by similar methods.

1,3,2-**Dithiarsenanes** and 1,3,2-**dithiantimonanes** are recorded (*L. A. Stockten*, J. chem. Soc., 1947, 592; *E. H. Friedheim*, B.P. 655,435/1961; C.A., 1953, **47,** 144; U.S.P. 2,659,723/1953; C.A., 1953, **47,** 144).

(iii) 1,3,2-Dioxaphosphorinanes with cage structures; arsenic and antimony analogues.

Phosphorus, arsenic and antimony form cage-like polycyclic structures LVII–LIX formally based on the 1,3,2-dioxaphosphorinane ring system.

Although not the first such compound to be prepared, 2,8,9-**trioxa**-1-**phospha-adamantane** (LVII, M = P), m.p. 207°, (*oxide*, m.p. 267–268°, *sulphide*, m.p. 250–251°) (*H. Stetter* and *K. H. Steinacker*, Chem. Ber., 1952, **85,** 451) obtained from *symm*-trihydroxycyclohexane and phosphorus

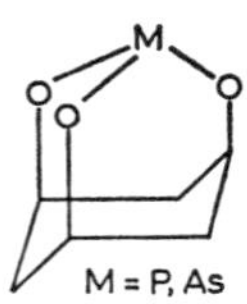

M = P, As
(LVII)

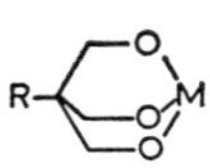

M = P, As, Sb
(LVIII)

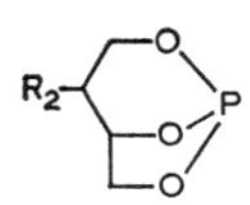

(LIX)

trichloride, received further attention in its use to demonstrate the stereospecific ring-opening in the course of reaction with benzylic-type halides (*K. D. Berlin et al.*, Tetrahedron, 1964, **20,** 323). It forms an ozonide at −78° which is a source of singlet oxygen at temperatures higher than 0° (*A. P. Schapp, K. Kees* and *A. L. Thayer*, J. org. Chem., 1975, **40,** 1185).

2,8,9-**Trioxa**-1-**arsa-adamantane** (LVII, M = As), m.p. 215–216°, is obtained similarly, and also from the triol and arsenious oxide in hot toluene (*R. D. Bertrand, R. D. Compton* and *J. G. Verkade*, J. Amer. chem. Soc., 1970, **92,** 2702; *J. F. Deffner*, U.S.P. 3,706,836/1972; C.A., 1973, **78,** 72228).

2,6,7-Trioxa-1-phosphabicyclo[2.2.2]**octane** has been known since 1926 when *F. Zetsche* and *E. Zurbrugg* described the 4-nitro compound LVIII, (M = P, R = NO_2) as its *P*-oxide, obtained from phosphoryl chloride and 1,1,1,-tris(hydroxymethyl)methane. Since then, many other derivatives isolated as their oxides, sulphides, or sometimes selenides, have been described; their structural rigidity has proved useful in allowing meaningful ^1H-^{31}P n.m.r. coupling constants to be determined. The 4-methyl compound LVIII (M = P, R = Me) was first described by *Verkade* and *L. T. Reynolds* (J. org. Chem., 1960, **25,** 663) who made it from 1,1,1-tris(hydroxymethyl)ethane and phosphorus trichloride in the presence of pyridine in tetrahydrofuran, and later from the same triol and trialkyl phosphites, but more recently it has been obtained from the same triol and tris(diethylamino)phosphine, $P(NEt_2)_3$ (*E. E. Nifant'ev* and *I. M. Petrova*, Zhur. obshcheĭ Khim., 1970, **40,** 2196), by heating 5-methyl-5-benzyloxymethyl-2-proto-1,3,2-dioxaphosphorinane 2-oxide at 250° thus eliminating benzyl alcohol in the cyclization (*idem, ibid.*, 1968, **38,** 2341), and by chromatography of 5-methyl-5-hydroxymethyl-2-(2-hydroxyphenoxy)-1,3,2-dioxaphosphorinane on alumina (*I. V. Fusenko, G. T. Bakhvalov* and *Nifant'ev*, Zhur. obshcheĭ Khim., 1968, **38,** 2528). It forms a stable ozonide below −78° (*M. E. Brennan*, Chem. Comm., 1970, 956), a convenient preparative source of singlet oxygen, and a bromobromonium salt (*G. K. McEwen* and *Verkade*, *ibid.*, 1971, 668), and undergoes stereospecific ring-opening when treated with alkyl halides and toluene-4-sulphonates.

Data for 2,6,7-trioxa-1-phosphabicyclo[2.2.2]octanes are collected in Table 19.

TABLE 19

4-SUBSTITUTED-2,6,7-TRIOXA-1-PHOSPHABICYCLO[2.2.2]OCTANE (LVIII)

R	*M.p. (°C)*	*Oxide m.p. (°C)*	*Sulphide m.p. (°C)*
H	126–127 (1)	245–247 (1)	
Me	97–98 (2)	249–250 (2)	224–225 (2)
Et	55–56 (3)	202 (3)	178–181 (4)
Ph		247–248 (5)	207–209 (5)
NO_2	81–82 (6)	243 (7)	219–220 (8)
$HOCH_2$	61 (3)		161 (9)

References

1 *E. J. Boros et al.*, J. Amer. chem. Soc., 1966, **88,** 1140.
2 *J. G. Verkade* and *L. T. Reynolds*, J. org. Chem. 1960, **25,** 663.
3 *W. S. Wadsworth* and *W. D. Emmons*, J. Amer. chem. Soc., 1962, **84,** 610.
4 *M. A. Gradsten*, U.S.P. 3,164,662/1964; C.A., 1967, **61,** 2973.
5 *W.-Hsuan Chang*, J. org. Chem., 1964, **29,** 3711.
6 *D. G. Hendricker*, J. heterocyclic Chem., 1967, **4,** 385.
7 *F. Zetsche* and *E. Zurbrugg*, Helv., 1926, **9,** 297.
8 *R. S. Edmundson* and *E. W. Mitchell*, unpublished work.
9 *R. Rätz* and *O. J. Sweeting*, J. org. Chem., 1965, **30,** 438.

Interestingly, 4-methyl-2,6,7-trioxa-1-arsabicyclo[2.2.2]octane does not form an oxide or sulphide; its antimony analogue is prepared from $SbEt_3$. The *tri-thia* analogues are prepared from the trithiol and $M(NMe_2)_3$(M = P, As). Details of these compounds are collected in Table 20.

In the 2,7,8-**trioxabicyclo**[3.2.1]**octane** series, only the 1-*phospha* compounds LIX have been described, R = *H*, b.p. 83°/35 mm, R = Me, b.p. 76–78°/0.8 mm. When treated with reactive halogen compounds, the five-membered ring opens stereospecifically to give 4-substituted-1,3,2-dioxaphosph(V)orinanes (*Edmundson* and *Mitchell*, J. chem. Soc., C, 1971, 3179; *Edmundson*, *ibid.*, 1972, 1660).

TABLE 20

ANALOGUES OF LVIII

	R	*M.p. (°C)*		
		M = P	*M = As*	*M = Sb*
2,6,7-Trioxa	Me	97–98 (1)	41–42 (1)	
	$HOCH_2$		b.p. 120/0.1 mm (5)	(b.p. 278–280/1 mm) (4)
2,6,7-Trithia	Me	187–189 (2,3)	118–120 (2)	

References

1 *J. G. Verkade* and *L. T. Reynolds*, J. org. Chem., 1960, **25,** 663.
2 *A. C. Vandenbroucke*, *E. J. Boros* and *Verkade*, Inorg. Chem., 1968, **7,** 1469.
3 *Oxide*, m.p. 236–239°; *sulphide*, m.p. 286–289°.
4 *G. Kamai* and *E. T. Mukmenev*, Tr. Kazansk Khim. Teckhnol, 1962, **30,** 296; C.A., 1964, **60,** 3993.
5 *T. E. Stevens*, U.S.P., 3,787,465/1974; C.A., 1974, **80,** 82115; *W. J. Ruehl* and *F. Kober*, Z. Naturforsch. anorg. Chem. org. Chem., 1976, **31B,** 307.

(*iv*) *1,4,2-Dioxaphosphorins and their benzo derivatives*

Unlike 2-amino-1,3,2-dioxaphospholanes, their benzologues undergo ring enlargement when acted upon by aldehydes and ketones. The products are 2-**amino**-1,2,3,4-**tetrahydro**-1,4,2-**benzodioxaphosphorin** 2-**oxides** (LX) *e.g.* 2-*diethylamino*-3-*ethyl*-, m.p. 167–168°, 2-*diethylamino*-3-*phenyl*-, m.p. 165–166°, (*V. S. Abramov* and *N. A. Il'ina*, Doklady Akad. Nauk S.S.S.R., 1959, **125,** 1027; Neketorye Vopr. organ. Khim., 1964, 256; C.A., 1966, **65,** 3902):

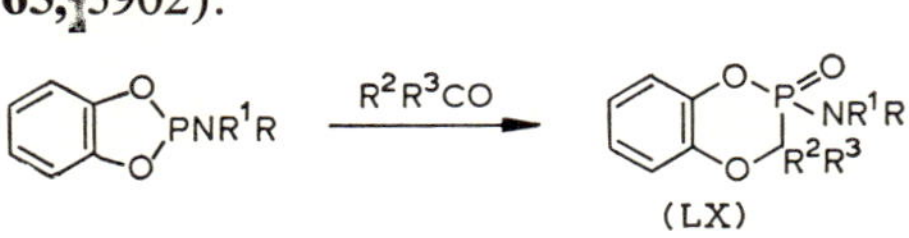

(LX)

Other compounds are obtained by reaction between catechols and (chloromethyl)methylphosphinic chloride, e.g. the 2-*methyl* compound, b.p. 114°/0.001 mm, m.p. 52–57°. Substituted benzene-1,2-dithiols(dithio-catechols) yield mixtures of 1,4,2-**benzodithiaphosphorins** (*M. Wieber* and *B. Eichhorn*, Monatsh., 1968, **99,** 261).

The reaction between 4-bromobenzaldehyde and the betaine from ethyl diphenylphosphinite and 1,2-dibenzoylethene in tetrahydrofuran under nitrogen at room temperature, proceeds with liberation of ethanol and formation of 3(4-**bromophenyl**)1,1,5,8-**tetraphenyl**-1-**phospha**-2,4,9-**trioxabicyclo**[4.3.0]**nona**-5,7-**diene** (LXI), m.p. 170–172°, (*D. D. Swanke et al.*, J. Amer. chem. Soc., 1971, **93,** 5236):

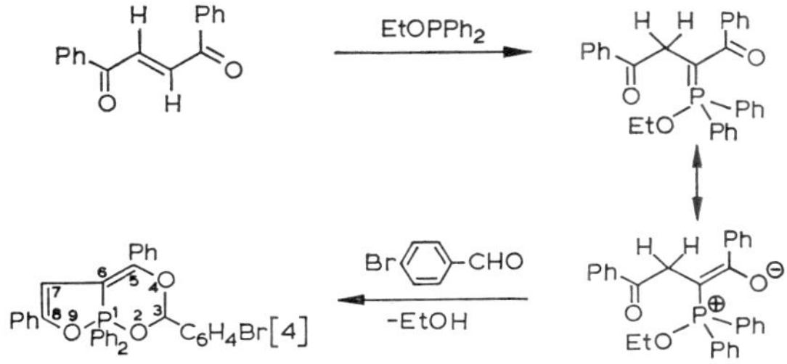

(LXI)

(*v*) *1,3,5-* and *1,4,2-Dioxaphosphorinanes;* 3,5,8-*trioxa-1-phosphabicyclo-[2.2.2]octanes*

The condensation between aldehydes and phosphine in tetrahydrofuran containing hydrogen chloride leads to derivatives of 4*H*-1,3,5-**dioxaphosphorinane,** the formation of one such compound, LXII, b.p. 110°/14 mm, being illustrated.

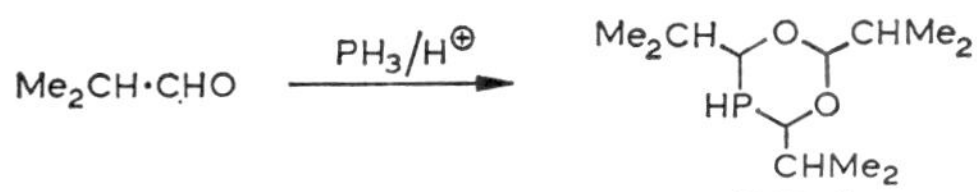

(LXII)

This compound is relatively stable to atmospheric oxygen and to acids, but when treated with oxygen, an isopropanol solution of the compound yields 5-*hydroxy*-2,4,6-*triisopropyl*-1,3,5-*dioxaphosphorinane 5-oxide*, m.p. 159–160°; this is degraded by dilute acid to the starting aldehyde. 2,4,5,6-*Tetraphenyl*-1,3,5-*dioxaphosphorinane*, m.p. 195–198° (*oxide* m.p. 275–278°), is obtained from three equivalents of benzaldehyde with phenylphosphine in acetonitrile (*S. A. Buckler* and *V. P. Wystrach*, J. Amer. chem. Soc., 1958, **80,** 6454; 1961, **83,** 168; *M. Epstein* and *Buckler*, Tetrahedron, 1962, **18,** 1231). Reaction between $PhCH_2P(O)(CH(OH)Ph)_2$ and benzaldehyde and treatment of the reaction product with toluenesulphonic acid for 25 h results in an equilibrium mixture of the DL and *meso* stereoisomers, m.p. 199–201° and 220–222°, respectively, of 5-*benzyl*-2,4,6-*triphenyl*-1,3,5-*dioxaphosphorinane 5-oxide* (LXIII), separable by utilising their differences in solubility in ethyl acetate (*A. B. Pepperman* and *T. H. Siddall*, J. org. Chem., 1973, **38,** 160; *Pepperman, G. J. Boudreaux* and *Siddall*, *ibid.*, 1975, **40,** 2056):

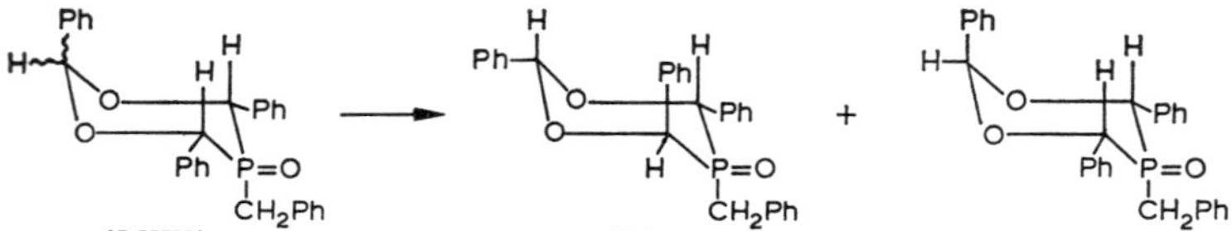

(LXIII) DL *meso*

Under acid conditions, aldehydes and ketones react with di(mercaptomethyl)phosphinic acids, and related compounds, to give 1,3,5-**dithiaphosphorinane** 5-**oxides** (LXIV) (*N. V. Ivasyuk* and *I. M. Shermergorn*, Zhur. obshcheĭ Khim., 1971, **41,** 1247):

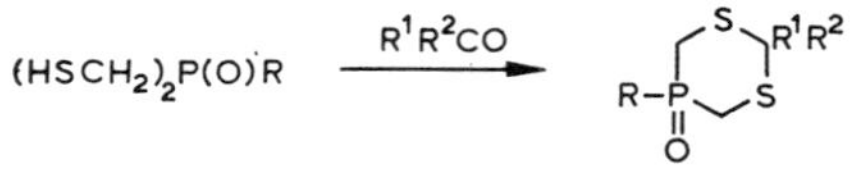

(LXIV)

Heating a mixture of tris(hydroxymethyl)phosphine and trimethyl orthoformate at 90° in the presence of triethylamine results in the formation of 3,5,8-**trioxa-1-phosphabicyclo-[2.2.2]octane** (LXV), m.p. 88–89°, *oxide* m.p. 151–155° (*E. J. Boros et al.*, J. Amer. chem. Soc., 1966, **88,** 1140).

An example of a 1,4,2-dioxaphosphorinane (LXVI, b.p. 115–116°/1 mm) is synthesized as shown (*Arbuzov* and *D. Kh. Yarmukhametova*, Izvest. Akad. Nauk. S.S.S.R., Ser. Khim., 1960, 1767):

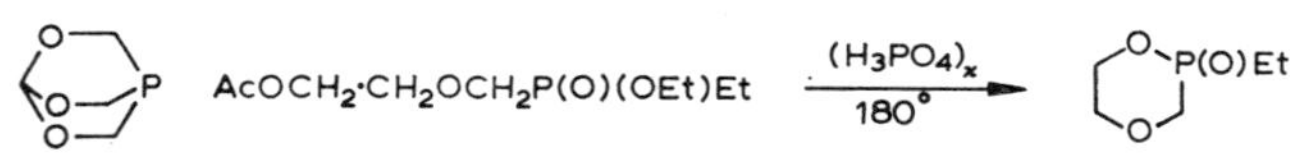

(LXV) (LXVI)

(*c*) *Perhydro-oxadiphosphorins (oxadiphosphorinanes)*

The *diphosphonia salt* (LXVII), m.p. 235–241°, derived from 3,5-diphenyl-1,3,5-oxadiphosphorinane, is made by the two processes shown (*A. M. Aguiar*, *K. C. Hansen* and *J. T. Magu*, J. org. Chem., 1967, **32,** 2383):

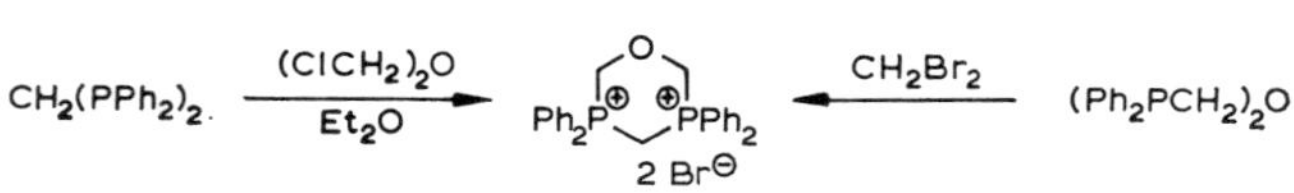

(LXVII)

(*d*) *Dioxadiphosphorinanes and arsenic analogues*

(*i*) *1,4,2,5-Dioxadiphosphorinanes and 1,4,2,5-dioxadiarsenanes*

Simple compounds of the 1,4,2,5-dioxadiphosphorinane series (LXVIII) are said to be the products formed when dialkyl hydrogen phosphonates are heated at 170–200° with paraformaldehyde or other aldehydes, together

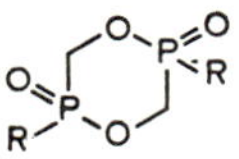

(LXVIII)

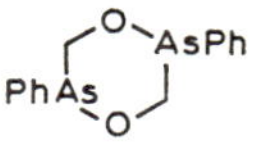

(LXIX)

with trialkyl phosphites, and also when dialkyl phosphorochloridites react with alkyl pyruvates at room temperature (*B. E. Ivanov et al.*, Izvest. Akad. Nauk S.S.S.R., 1971, 1497; *Yu. A. Levin, I. P. Goznan* and *E. E. Siderova, ibid.*, 1970, 173; *A. N. Pudovik et al.*, Zhur. obshcheĭ Khim., 1970, **40,** 2374).

A tetraphosphonic acid formed from phosphorous acid and acetic anhydride or acetyl chloride, has been shown by X-ray techniques to have the structure LXX and not LXXI which could not be differentiated by means of ^{1}H and ^{31}P-n.m.r. spectroscopy; the compound has found application in dentistry (*J. B. Prentice et al.*, J. Amer. chem. Soc., 1972, **94,** 6119; *G. Brun* and *C. Blanchard,* Rev. Chim. miner., 1972, **9,** 435, C.A., 1973, **78,** 70896; *B. Blaser et al.*, Z. anorg. allgem. Chem., 1971, **381,** 247; *E. Philippot et al.*, Cryst. Struct. Comm., 1971, **1,** 103):

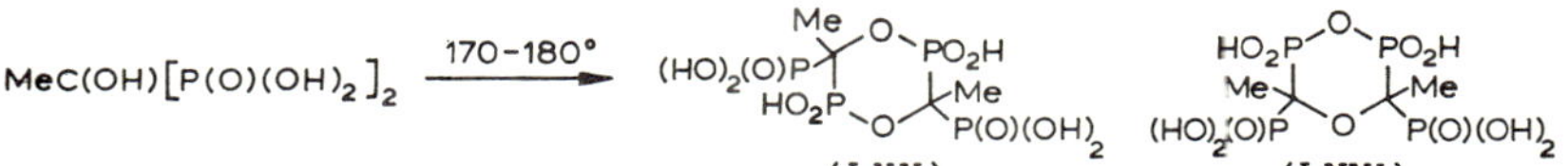

(LXX) (LXXI)

Dehydration of hydroxyethylidenediphosphonic acid also yields the acid LXX, as demonstrated by an X-ray analysis of its tetra-ammonium salt (*A. J. Collins et al.*, J. chem. Soc., Dalton Trans., 1974, 960).

Arsenic analogues, *e.g.* 2,5-**diphenyl**-1,4,2,5-**dioxadiarsenane** (LXIX), b.p. 215–216°/9 mm, are similarly obtained from primary arsines and aldehydes in the presence of hydrogen chloride (*C. S. Palmer* and *R. Adams,* J. Amer. chem. Soc., 1922, **44,** 1356).

(*ii*) *2,6,7-Trioxa-1,4-diphosphabicyclo*[2.2.2]*octane and analogues*

In the transesterification between trimethyl phosphite and tris(hydroxymethyl)phosphine in hot tetrahydrofuran to give 2,6,7-trioxa-1,4-**diphospha-**

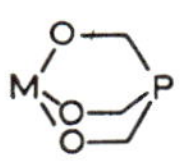

M = P, As or SiMe
(LXXII)

bicyclo[2.2.2]**octane** (LXXII, M = P), m.p. 75–76° (*dioxide*, m.p. 210–213°), care must be exercised since explosions have occurred during the reaction. With sulphur, only the 1-*sulphide*, m.p. 235–237°, is formed.

Formed in the same way are 3,5,8-**trioxa**-4-**arsa**-1-**phosphabicyclo**[2.2.2]**octane** (XXII, M = As), m.p. 91–92° (*P-oxide*, m.p. 175° dec.), and 4-**methyl**-3,5,8-**trioxa**-4-**sila**-1-**phosphabicyclo**[2.2.2]-**octane** (LXII, M = SiMe), m.p. 61–62°, (*K. J. Coskran* and *J. G. Verkade*, Inorg. Chem., 1965, **4,** 1655; *J. Rathke, J. W. Guyer* and *Verkade*, J. org. Chem., 1970, **35,** 2310).

(*e*) *Azaphosphorines and their arsenic analogues*

(*i*) *Dibenz[1,2]azaphosphorines*

Simple derivatives of the monocyclic 1,2-azaphosphorine are not known, and even the more readily available tricyclic system LXXIV has been investigated relatively little.

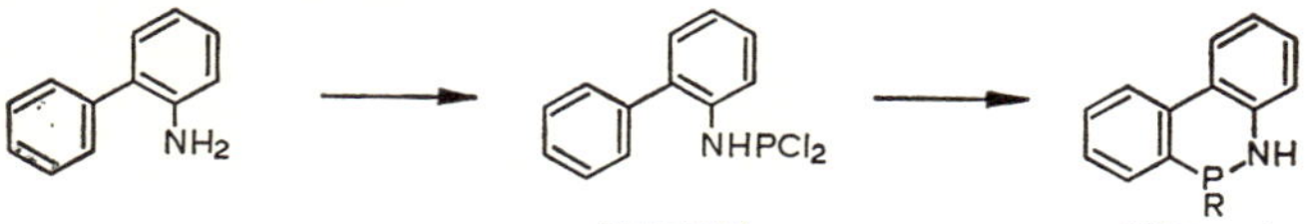

(LXXIII) (LXXIV)

6-Chloro-5,6-dihydrodibenz[*c*,*e*][1,2] azaphosphorine (LXXIV, R = Cl), m.p. 132–134°, may be sublimed from the crude product obtained when the aminodichlorophosphine LXXIII is acted upon by aluminium chloride at 210–280°. The chloride reacts normally with aryl Grignard reagents and lithium compounds to give the 6-aryl derivatives, *e.g.* 6-*phenyl*, m.p. 181° (*methiodide*, m.p. 214°), readily reacted upon by ethanolic hydrogen peroxide to give the 6-*oxide*, m.p. 289–290°. The *oxide*, m.p. 258–260°, of 6-(4-*dimethylaminophenyl*)-5,6-*dihydrodibenz*[c,e][1,2]*azaphosphorine*, m.p. 148–153°, has been resolved using camphor-10-sulphonic acid (*M. J. S. Dewar* and *V. P. Kubba*, J. Amer. chem. Soc., 1960, **82,** 5685; *I. G. M. Campbell* and *K. J. K. Way*, J. chem. Soc., 1960, 5034).

A further example of a polycyclic 1,2-azaphosphorine, *viz.* 8-*ethoxy*-2,3-*dimethoxy*-5,6-*dihydro*-9H-*indolo*[2,3-c]*isoquino*[1,2-f][1,2]*azaphosphorine* 8-*oxide* (LXXVI), m.p. 285.5–286.5°, is formed during the Bischler–Napieralski cyclization of the amide LXXV (*G. C. Morrison, R. O. Waite* and *J. Shavel*, J. heterocycl. Chem., 1966, **3,** 540):

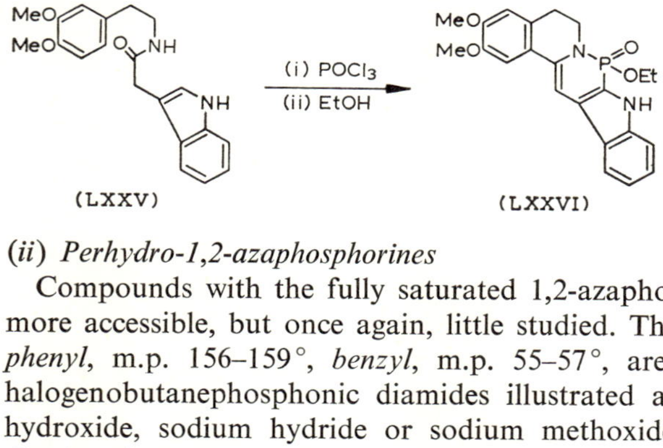

(LXXV) (LXXVI)

(*ii*) *Perhydro-1,2-azaphosphorines*

Compounds with the fully saturated 1,2-azaphosphorine ring are much more accessible, but once again, little studied. The amides LXXVII R = *phenyl*, m.p. 156–159°, *benzyl*, m.p. 55–57°, are obtained when the 4-halogenobutanephosphonic diamides illustrated are treated with sodium hydroxide, sodium hydride or sodium methoxide (*B. Helferich* and *U. Curtius*, Ann., 1962, **655,** 59; *Helferich* and *L. Schröder*, *ibid.*, 1963, **670,** 48; *D. J. Hewitt* and *G. L. Newland*, Austral. J. Chem., 1977, **30,** 579):

(LXXVII)

(iii) Tetrahydrobenz[e][1,3]azaphosphorines

1,2,3,4-**Tetrahydrobenz**[e][1,3]**azaphosphorines** (LXXVIII) are said to be the products when aldehydes and ketones react with 2-aminobenzylphosphine (*K. Issleib, H. Winkelmann* and *H. P. Abicht*, Synth. React. inorg. metal.-org. Chem., 1974, **4,** 191):

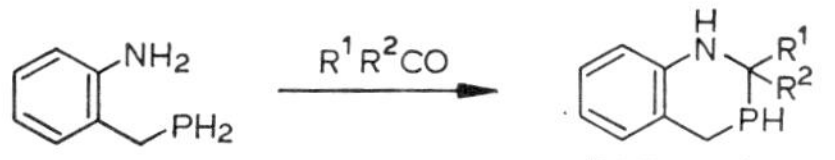

(LXXVIII)

(iv) Perhydro-1,3-azaphosphorines

3-**Phenylperhydro**-1,2,3-**azaphosphorine** (LXXIX, $R^1 = R^2 = R^3 = H$), b.p. 127–130 °/1.5 mm, (*sulphide*, m.p. 51 °), is formed when phenyl-(3-aminopropyl)phosphine reacts with formaldehyde. Other known compounds include the 1,3-*diphenyl-*, m.p. 68–70 °, b.p. 148–153 °/0.1 mm, (*hydrochloride*, m.p. 259–262 °; *sulphide*, m.p. 147–149 °) (*Issleib, H. Oehme* and *E. Leissring*, Chem. Ber., 1968, **101,** 4032; *Oehme* and *R. Thamm*, J. pr. Chem., 1973, **315,** 526):

(LXXIX)

In the same way are prepared the analogous **perhydro**-1,3-**azarsenines,** 3-*phenyl,* b.p. 98–99 °/10^{-4} mm, (*hydrochloride* m.p. 154–155 °), 1,3-*diphenyl-*, m.p. 113–115 °. When treated with methyl iodide, these compounds quaternize at arsenic (*A. Tzschack* and *P. Franke*, J. organometal. Chem., 1974, **81,** 187).

(v) 1,4-Azaphosphorines and 1,4-dihydro-1,4-azaphosphorines

Derivatives of 1,4-**dihydro**-1,4-**azaphosphorine** are formed from di(2-oxoalkyl)phosphine oxides, and from di(2-oxoalkyl)phosphonium salts (which may themselves be prepared from 1,4-dihydro-1,4-oxaphosphorins by the action of acid) by fusion of the phosphorus compound with ammonium carbonate (*J. C. Williams et al.*, Tetrahedron Letters, 1971, 4749) or treatment with ammonium acetate in acetic acid (*R. Fugnitto, M. H. Mebazaa* and *M. Simalty*, Compt. rend., C, 1972, **274,** 2206). The addition of primary amines to dialkynylphosphine oxides has also been proposed as a means of preparing 1-alkyl-1,4-dihydro-1,4-azaphosphorine-4-oxides (*A. Naaktgeboren et al.*, Rec. Trav. chim., 1975, **94,** 92).

4-tert-*Butyl*-2,6-*diphenyl*-1,4-*dihydro*-1,4-*azaphosphorine* 4-*oxide*, m.p. 360–362 °, is reduced by phenylsilane in boiling toluene to the corresponding P(III) compound, (LXXX, m.p. 258–261 ° decomp.) (*methiodide*, m.p. 229–230 °); when heated at 260–280 °, LXXX

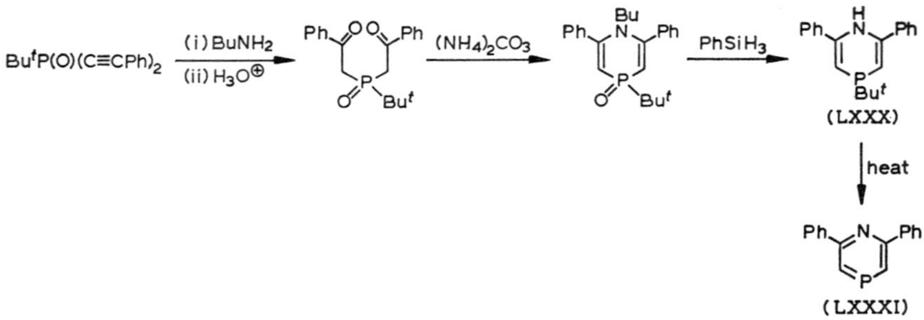

(LXXX) (LXXXI)

yields 2,6-diphenyl-1,4-azaphosphorine (LXXXI), an oil. Unlike phosphabenzene, the azaphosphorine is markedly active towards nucleophilic reagents; ethanol and phenol yield, respectively, the 4-*ethoxy*-, m.p. 117–120°, and 4-*phenoxy*-, m.p. 197–199°, 2,6-*diphenyl*-1,4-*dihydro*-1,4-*azaphosphorines*. Reaction with phenyllithium followed by methyl iodide gives 1,4-*dimethyl*-2,6-*diphenyl*-1,4-*dihydro*-1,4-*azaphosphorine* 4-*oxide*, m.p. 194–196° (*G. Markl* and *D. Matthes*, Angew. Chem. intern. Edn., 1972, **11,** 1019).

An X-ray analysis of LXX 4-oxide has demonstrated that the ring is planar with carbon–carbon bond length 1.41 Å.

1-*Ethyl*-4-*phenyl*-1,4-*dihydro*-1,4-*azaphosphorine* 4-*oxide* (LXXXII), m.p. 139–141°, is obtained from phenyldiethynylphosphine oxide and ethylamine (*Naakgeboren et al., loc. cit.*).

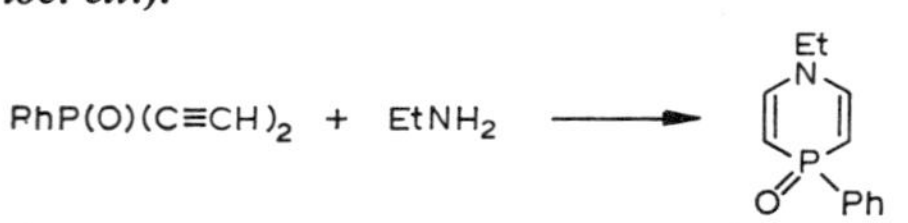

(LXXXII)

When treated with sodium hydride in dimethyl sulphoxide, 2,4,4,6-*tetraphenyl*-1,4-*dihydro*-1,4-*azaphosphorinium perchlorate* (LXXXIII), m.p. 216.5°, yields its conjugate base, the anion LXXXIV, conveniently formulated as the charge-free hybrid structure LXXV, m.p. 189.5°, (*M. H. Mebazaa* and *M. Simalty*, Tetrahedron Letters, 1972, 4363):

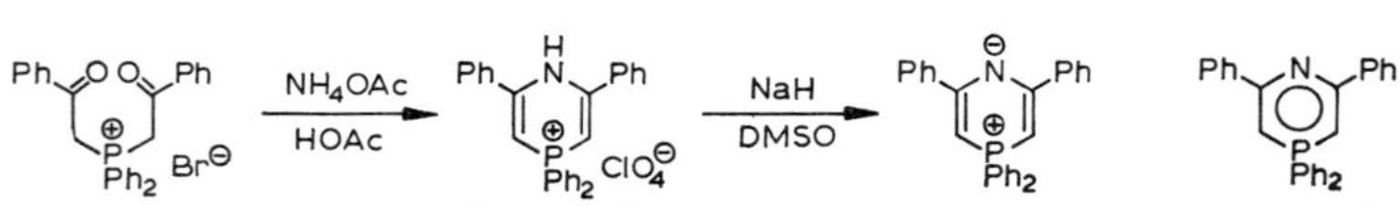

(LXXXIII) (LXXXIV) (LXXXV)

(*vi*) *Dibenz[1,4]azaphosphorine and dibenz[1,4]azarsenine*

Initial investigations, by *Michaelis* and *Schenk* (Ann., 1890, **260,** 1), on the reaction between diphenylamine and phosphorus trichloride, showed that, in the presence of zinc chloride at 250° (sealed tube), a compound $C_{12}H_{10}NOP$ was formed and to which they ascribed the structure (LXXXVI, R = OH, R^1 = H), now known to be LXXXVI (R = R^1 = H, as the oxide). A further investigation of the same reaction shows that the 10,10′-

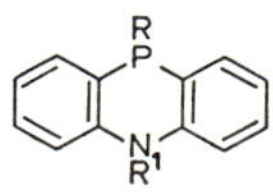

(LXXXVI)

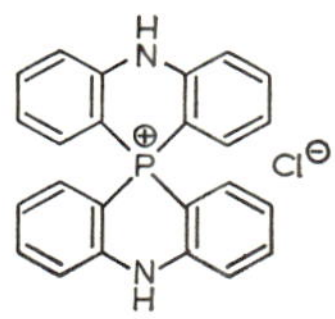

(LXXXVII)

(5*H*,5*H*′)**spirophosphazinium chloride** (LXXXVII) is also formed (*R. N. Jenkins, L. D. Freedman,* and *J. Bordner,* Chem. Comm., 1971, 1213; *Jenkins* and *Freedman,* J. org. Chem., 1975, **40,** 766). Interestingly, the latter investigators have also shown that diphenylmethylamine will not form a phenophosphazine in this way, whereas methyldi-4-tolylamine will do so.

The preparation, by essentially the same process, and properties of 5,10-**dihydrodibenz**[*b*,*e*][1,4]**azaphosphorine** (5,10-**dihydrophenophosphazine**) 10-**oxide,** (LXXXVI, R = R^1 = H, oxide), m.p. 214.5–216.5°, have been studied further, particularly by *M. Häring* (Helv., 1960, **43,** 1826).

The oxide is readily oxidized to 10-*hydroxy*-5,10-*dihydrophenophosphazine* 10-*oxide,* also known as *phenophosphazinic acid,* having p*K* 3 and m.p. 277°, acid *chloride,* m.p. 184–185.5°, *methyl ester,* m.p. 223–224°. 5,10-Dihydrophenophosphazine 10-oxide adds sulphur to give 10-*hydroxy*-5,10-*dihydrophenophosphazine* 10-*sulphide,* m.p. 213°, forming a *methyl ester,* m.p. 184–185.5°, and an acid *chloride,* m.p. 229.5°, also produced when diphenylamine and thiophosphoryl chloride are heated together under reduced pressure (*G. V. McHattie,* B.P. 860,629/1961; C.A., 1962, **57,** 2256).

5,10-**Dialkyl(aryl)**-5,10-**dihydrophenophosphazines,** 5-*methyl*-10-*phenyl*-, m.p. 159–160°, 5-*ethyl*-10-*phenyl*-, m.p. 96.5–98°, 2,5,8-*trimethyl*-10-*phenyl*-, m.p. 144–146°, are formed when 2,2′-dilithiodiphenylamines or the analogous di-Grignard reagents react with dichloro(phenyl)phosphine (*idem, loc. cit.*; *C. Baum, K. A. Lloyd* and *C. Tamborski,* J. org. Chem., 1964, **29,** 3410), and when heterocyclic analogues containing tin undergo exchange reactions with dichloro(phenyl)phosphine (*E. J. Kupchik* and *V. A. Perciaccante,* J. organometal. Chem., 1967, **10,** 181).

The first description of 10-**chloro**-5,10-**dihydrodibenz**[*b*,*e*][1,4]-**azarsine** (10-**chloro**-5,10-**dihydrophenarsazine**) (LXXXVIII, R = Cl, R^1 = H) appeared in 1915 (*Farbenfabriken vorm. F. Bayer,* G.P. 281,049/1915). The compound was obtained independently by *R. Adams* in 1918 and details of its preparation by heating diphenylamine with arsenic trichloride are given by *H. Wieland* and *W. Rheinheimer* (Ann., 1921, **423,** 1). The ease

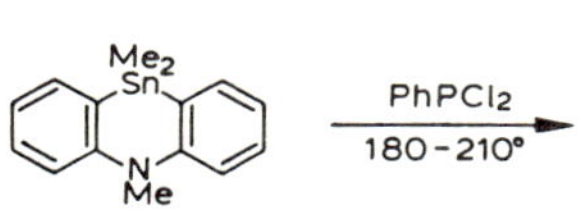

(LXXXVI, R = Ph, R^1 = Me)

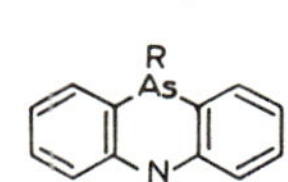

(LXXXVIII)

of preparation and the physiological properties of the compound gave it importance as a potential war gas, referred to as DM or Adamsite (*Sartori*, "The War Gases", J. and A. Churchill Ltd., London 1939, 318). The Italian method of manufacture (*Sartori, loc. cit.)* depends on the dissociation of diphenylamine hydrochloride into the base and hydrogen chloride at 100°, and the conversion by the latter of arsenic oxide into the trichloride for reaction with the diphenylamine at 130–200°; the yield of the cyclic arsenic chloride is 95%. It is noteworthy that no spiroarsonium salts are formed under conditions under which phosphorus trichloride yields spirophenophosphazinium salts (*Freedman* and *Y. L. Styles*, J. org. Chem., 1975, **40,** 2684).

10-*Chloro*-5,10-*dihydrophenarsazine* exists as a canary-yellow, stable, orthorhombic form, m.p. 195°, as a metastable monoclinic form, m.p. 186°, and as a triclinic form, m.p. 182°.

The structure XXXVIII (R = Cl, R^1 = H) received support from the work of *H. Burton* and *C. S. Gibson* (J. chem. Soc., 1926, **128,** 450) who prepared the chloro compound by ring closure of 2-amino-2′-bromodiphenylarsinic acid in *o*-dichlorobenzene, and reduction of the resultant 10-hydroxy-5,10-dihydrophenarsazine 10-oxide (phenarsazinic acid) (LXXXIX), the latter being described independently by *J. H. Schmidt* (J. Amer. chem. Soc., 1921, **43,** 2449):

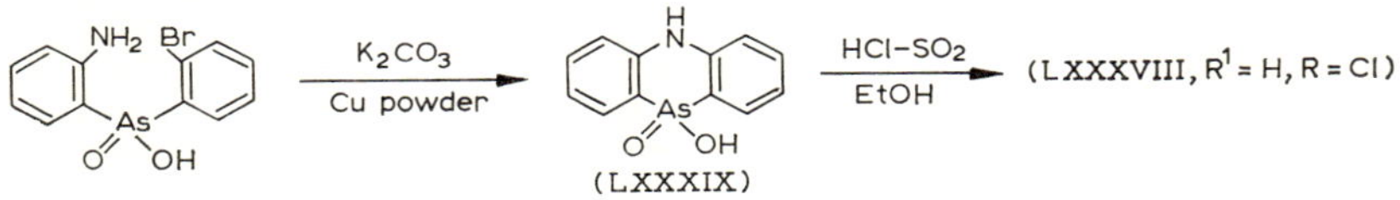

(LXXXIX)

The structure of the chloro compound has nevertheless been the subject of some controversy in view of its unusual physical properties, *e.g.* high m.p. and sparing solubility. However, an X-ray analysis has shown the proposed structure to be correct (*H. Vermeer, R. Lourens* and *F. Bickelhaupt*, Tetrahedron, 1975, **31,** 2529). 10-Chloro-5,10-dihydrophenarsazine is slowly hydrolyzed by water and the chlorine atom may also be replaced by CN, SCN, giving a crystalline *cyanide*, m.p. 223–224°, and *thiocyanate*, m.p. 229–230°, respectively. With aqueous alkalis, di(5,10-*dihydrophenarsazine*)*oxide* $(NH{:}(C_6H_4)_4{:}As)_2O$, m.p. > 350°, is formed; when the chloride is acted upon by hydrogen peroxide or chloramine-T, **phenarsazinic acid** (LXXXIX) is formed, and with hypophosphorous acid, coupling to give 10,10′-(5*H*,5′*H*-**biphenarsazine**), m.p. 304–305°, occurs (*Burton* and *Gibson*, J. chem. Soc., 1924, **125,** 2257; *Vermeer, Lourens* and *Bickelhaupt, loc. cit.*). The 10-chloro compound in acetic acid may be nitrated with fuming nitric acid, and amongst the various nitration products are the scarlet 4-*nitro*-, m.p. 156°, the pale-yellow 2,8-*dinitro*-, m.p. > 300°, and the greenish-yellow 2-*nitro*-, derivatives, all readily oxidizable to the nitrophenarsazinic acids, which may then be reduced to the corresponding aminophenarsazinic acids. Treatment of the 10-chloro compound with four equivalents of bromine in hot acetic acid results in the cleavage of the heterocyclic ring with loss of arsenic to give di(2,4-dibromophenyl)amine (*L. A. Elson, Gibson* and *J. D. A. Johnson*, J. chem. Soc., 1929, 1080).

The secondary amino group in 10-chloro-5,10-dihydrophenarsazine cannot be nitrosated, but that in phenarsazinic acid may be acetylated (N-*acetyl*, m.p. 244°) and benzoylated (N-*benzoyl*, m.p. 250°), but not methylated. It is doubtful, in fact, if *N*-alkylated phenarsazines can be obtained in this way, for, although *Wieland* and *Rheinheimer (loc. cit.)* claimed that interaction of arsenic trichloride with diphenylmethylamine gave the *N*-methyl compound (LXXXVIII, R = Cl, R^1 = Me), *Burton* and *Gibson* found that demethylation occurred with production of 10-chloro-5,10-dihydrophenarsazine, a result confirmed by *Freedman* and *Styles (loc. cit.)*, and further, that ring closure of 2-bromo-2′-methylaminodiphenylarsinic acid could not be effected.

Introduction of a phenyl group into the 10-position cannot be achieved by heating diphenylamine with dichloro(phenyl) arsine, the products then being the chloro compound LXXXVIII, (R = Cl, R^1 = H) (*Burton* and *Gibson*, J. chem. Soc., 1926, **128,** 464), but the halogen atom may be replaced by an alkyl or aryl group by reaction with the appropriate Grignard reagent. In this way, 10-*methyl*-, m.p. 105°, 10-*ethyl*-, m.p. 75°, 10-*phenyl*-, m.p. 148–149°, and 10-(1-*naphthyl*)-5,10-*dihydrophenarsazine*, m.p. 154–155°, (and others) are obtained (*J. A. Aeschlimann*, *ibid.*, 1927, 413; *O. Seide* and *J. Gorski*, Ber., 1929, **62,** 2186; *G. Kamai*, *V. I. Gavrilov* and *B. D. Chernokal'skii*, Zhur. obshcheĭ Khim., 1972, **42,** 1530; *Kamai*, *N. A. Chadaeva* and *K. A. Mamakov*, Izvest. Akad. Nauk S.S.S.R., 1970, 1092). Under carefully controlled conditions, the 10-alkyl and 10-aryl derivatives, unlike the halogeno compounds, may be nitrosated. Contact with alkyl halides brings about quaternization at arsenic, leaving the NH group unaffected. The action of concentrated mineral acids, of trichloracetic acid, and of pentachlorophenol, yields adducts (*Gavrilov et al.*, Zhur. obshcheĭ Khim., 1971, **41,** 564). 10-Alkyl-5,10-dihydrophenarsazines are converted into the 10-*iodo* compound, m.p. 219–221°, by the action of hydroiodic acid, while the 10-phenyl compound undergoes degradation to diphenylamine; in addition, 10-ethyl-5,10-dihydrophenarsazine with $EtAsBr_2$ at 220° yields the 10-*bromo* compound, m.p. 210–212°, (*Gavrilov*, *Chernokal'skii* and *Kamai*, *ibid.*, 1971, **41,** 560).

Attempts to prepare **dibenz[*b*,*e*][1,4]azarsine (phenarsazine)** (XC) from 5-chloro-5,10-dihydrophenarsazine by elimination of hydrogen chloride with pyridine or quinoline leads to the *chlorotriphenarsazine*

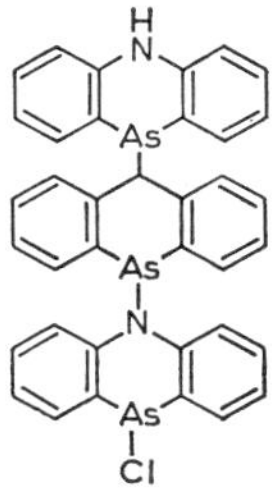

m.p. 260–263°, but on heating 10-*methoxy*-5,10-*dihydrophenarsazine*, m.p. 194°, in molten diphenylmethylamine in a current of carbon dioxide, a yellow solid, m.p. > 310°, analysing for **phenarsazine** (XC), is obtained:

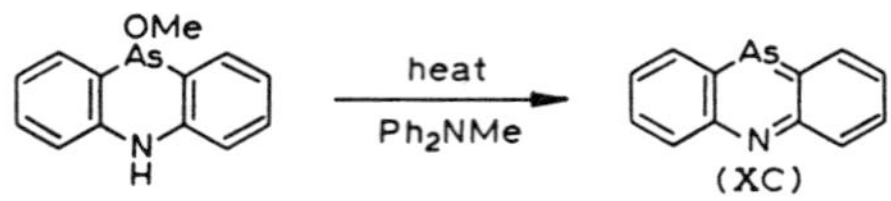

(XC)

The compound is very reactive, giving an oxide in air and regenerating 10-chloro-5,10-dihydrophenarsazine with dilute hydrochloric acid.

A detailed spectroscopic study (*Vermeer*, *Lourens* and *Bickelhaupt*, *loc. cit.*) has clearly demonstrated that Wieland and Rheinheimer's "phenarsazine" and the product from the reduction of 10-chloro-5,10-dihydrophenarsazine with hypophosphorous acid, obtained first by Burton and Gibson in 1926, are the same compound, viz. **10,10′(5*H*,5′*H*-biphenarsazine)** (XCI):

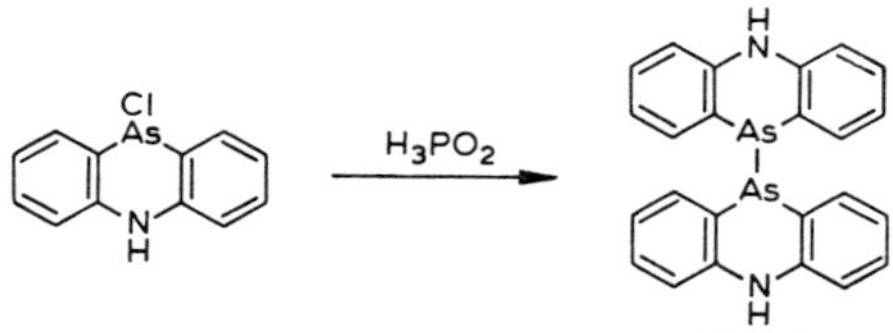

(XCI)

The chemistry of many substituted and more complex, polycyclic phenarsazines has been described in a long series of papers by *Gibson* and his collaborators (J. chem. Soc., 1927, 247, 2499; 1928, 2204; 1929, 767, 1080, 1229, 1473, 1621, 2743; 1930, 1124, 1622; 1931, 294, 2381, 2518, 3270). Other compounds, including **pyrido-** and **aza-phenarsazines,** and **thienophenarsazines** have been obtained by *W. L. Lewis* and *C. S. Hamilton* (J. Amer. chem. Soc., 1921, **43,** 2218), *R. H. Slater* (J. chem. Soc., 1931, **107,** 1939), and by *N. P. Bui-Hoi*, *M. Dufour* and *P. Jacquignon*, (*ibid.*, C, 1971, 2428) in the search for chemotherapeutics active against trypanosomes.

Both 5,10-*o*-**benzenophenophosphazine (azaphosphatrypticene)** (XCII) and its arsenic analogue are known. The reaction between tri(2-lithiophenyl)amine and triphenyl phosphite in an ethereal solvent yields the *phosphorus compound*, m.p. 255.5–258.5°, (*oxide*, m.p.

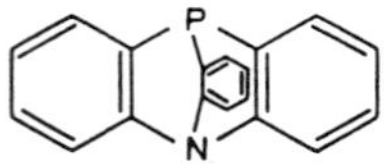

(XCII)

ca. 300°, *phosphonium methiodide*, m.p. 279–285° decomp.) in 30% yield (*D. Hellwinkel* and *W. Schenk*, Angew. Chem. int. Edn., 1969, **8,** 987). The **arsenic analogue,** m.p. 233°, is formed when 10-(2-*chlorophenyl*)-5,10-*dihydrophenarsazine*, m.p. 110°, is treated with lithium diethylamide under ether; it does not form quaternary salts, and is degraded to triphenylamine with Raney nickel in ethanol.

(*vii*) *Perhydro-1,4-azaphosphorines and perhydro-1,4-azarsenines*

The reaction between di(2-bromoethyl)aniline and phenylphosphinedi(magnesium bromide) yields 1,4-**diphenylperhydro**-1,4-**azaphosphorine** (XCIII, M = P, R = R^1 = Ph), m.p. 89–90° (*F. G. Mann* and *I. T. Millar*, J. chem. Soc., 1952, 3039).

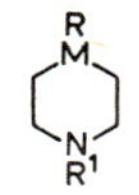

(XCIII)

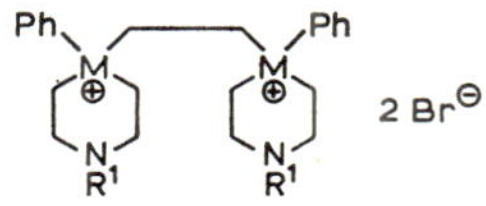

(XCIV)

The compound shows little activity towards air, but is oxidized to a *monoxide*, m.p. 143–144° by hydrogen peroxide, and by the same reagent in acetic acid to a *dioxide*, m.p. 149°, which is solvated with hydrogen peroxide. Methyl iodide gives firstly the *phosphonium methiodide*, m.p. 155–156°, and then the *ammonium-phosphonium dimethiodide*, m.p. 118°, an unstable compound; quaternization with dibromoethane consequently yields the diquaternary *bromide* (XCIV M = P, R = R^1 = Ph), m.p. 273–274°.

K. Issleib and *H. Oehme* (Chem. Ber., 1967, **100**, 2685) prepared 4-*phenylperhydro*-1,4-*azaphosphorine* (XCIII, M = P, R^1 = H, R = Ph), b.p. 145–147°/0.3 mm, (*sulphide*, m.p. 114–115°) by heating phenyldi(2-aminoethyl)phosphine; the same compound is made by reduction of the lactam XCV with lithium tetrahydridoaluminate (*H. Oehme* and *E. Leissring*, Z. Chem., 1970, **10**, 305):

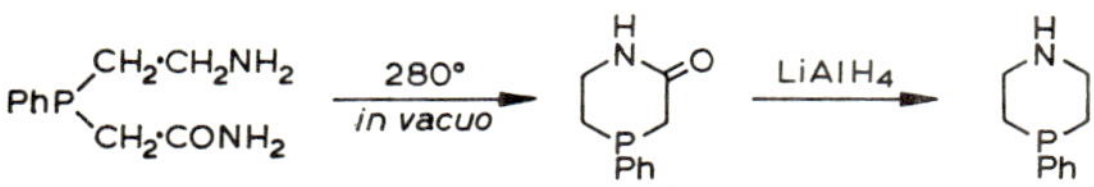

(XCV)

A more general route to compounds in this series consists in the addition of primary amines (but not ammonia) to dialkenylphosphine oxides and sulphides, and to divinylphosphinic acid esters (*S. Samaan*, Tetrahedron Letters, 1974, 3927; *Y. A. Levin, R. I. Pyrkin* and *I. V. Galyametdinova*, U.S.S.R. Pat. 414,264/1974; C.A., 1974, **80**, 133628; *D. J. Collins, R. E. Rowley* and *J. M. Swan*, Austral. J. Chem., 1974, **27**, 841), as well as to dialkenylphosphonium salts *(Samaan loc. cit.)*.

Thus prepared are 1,2,6-*trimethyl*-4-*phenylperhydro*-1,4-*azaphosphorine* 4-*sulphide* (XCVI), m.p. 145°, 1-*methyl*-4-*phenylperhydro*-1,4-*azaphosphorine* 4-*oxide*, m.p. 115–116°, (*sulphide* m.p. 103–104°):

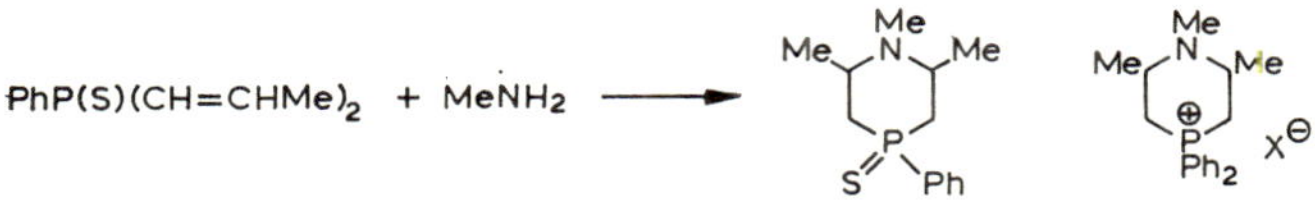

(XCVI) (XCVII)

Reduction of the cyclic phosphine oxide to the corresponding phosphine may be achieved by reaction with trichlorosilane *(Collins et al., loc. cit.)*; 1-*methyl*-4-*phenylperhydro*-1,4-*azaphosphorine* has b.p. 101–102°/0.4 mm.

Electrochemical reduction of the phosphonium salt (XCVII) also affords a phosphine, 1,2,6-*trimethyl*-4-*phenylperhydro*-1,4-*azaphosphorine*, b.p. 84–89°/0.4 mm *(Samaan loc. cit.)*.

4,8-Methano-1-methyl-1,2,3,4,7,8-hexahydrobenz[*e*][1,4]azaphosphocine 4-oxide (XCVIII), m.p. 123–124°, b.p. 160°/0.005 mm, prepared in the sequence shown, is reduced by

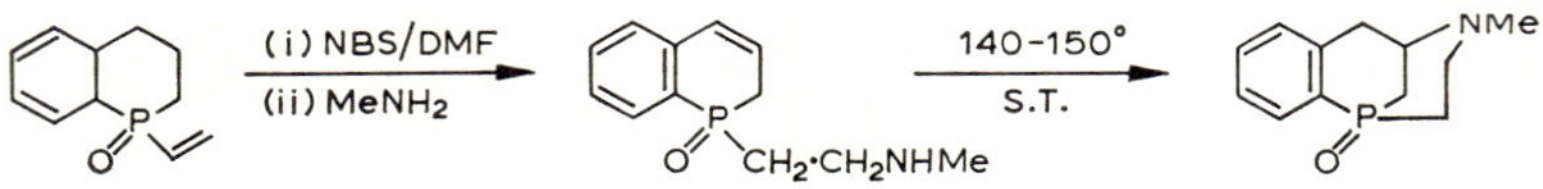

(XCVIII)

trichlorosilane to the corresponding *phosphine*, b.p. 76°/0.005 mm, which adds sulphur to give the P-*sulphide*, m.p. 137–138°, (*Collins*, *Rowley* and *Swan*, Austral. J. Chem., 1974, **27**, 815).

1,4-**Diphenylperhydro-**1,4-**azarsenine** (XCIII, M = As, R = R^1 = Ph) is made in the same way as the phosphorus analogue; it has m.p. 96–97.5°, and forms a *methiodide*, m.p. 181–182°, and also differs from the phosphorus compound in that with hydroiodic acid it yields the hydrogen iodide salt of 4-iodo-1-phenylperhydro-1,4-azarsenine (*M. H. Beeby* and *Mann*, J. chem. Soc., 1951, 886). The *salt* XCIV (M = As, R = R^1 = Ph) has m.p. 226–227°.

(*f*) *Diazaphosphorines*

(*i*) *1,3,2-Diazaphosphorines*

Extensive investigations are reported, particularly in the Russian literature, on the action of phosphorus chlorides on nitriles. The extension of this reaction to dinitriles yields diazaphosphorines:

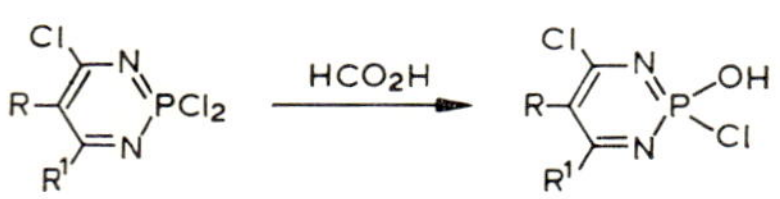

(XCIX) (C)

Thus when treated with one mole of phosphorus pentachloride in benzene at room temperature, malononitrile itself yields 2,2,4,6-**tetrachloro-**1,3,2-**diazaphosphorine** (XCIX, R = H, R^1 = Cl), b.p. 108–109°/13 mm, m.p. 40–45°. This, and other derivatives (XCIX, R^1 = Cl), R is *methyl*, m.p. 116–117°, *ethyl*, m.p. 59–65°, are also formed when dry hydrogen chloride acts upon RC(CN)=C(Cl)N=PCl_3 (*V. I. Shevchenko* and *P. P. Kornuta*, Zhur. obshcheĭ Khim., 1966, **36**, 1254). When two moles of phosphorus pentachloride are employed in the system, 2,2,4,5 6-**pentachloro-**1,3,2-**diazaphosphorine** (XCIX, R = R^1= Cl), m.p. 150–152°, is obtained (*A. V. Kirsanov*, *M. D. Bodnarchuk* and *V. I. Shevchenko*, Dopovidi Akad. Nauk Ukrain R.S.R., 1963, **221**, C.A., 1963, **59**, 12666).

1,3,2-Diazaphosphorines (XCIX, R = Ph, CN, R^1 = CCl_3, CF_3) are obtained, together with linear products, when phosphorimidic trichlorides react with suitable nitriles (*V. P. Kuchar* and *A. P. Baiko*, Zhur. obshcheĭ Khim., 1974, **44**, 1247):

$$R^1CCl_2N{=}PCl_3 + RCH_2\cdot CN \rightarrow NC\cdot CR{=}CR^1N{=}PCl_3 + \text{XCIX}$$

2,2-Dichloro-1,3,2-diazaphosphorines undergo reaction at phosphorus when treated with aniline; with alcohols or phenols in the presence of triethylamine; and with chlorine. Reaction with formic acid leads to partial hydrolytic removal of chlorine; the product is

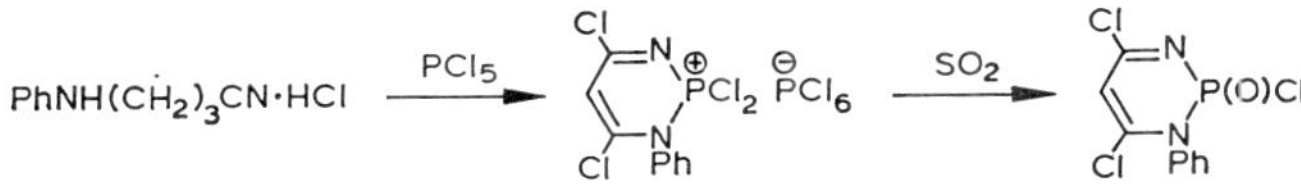

(CI) (CII)

formulated as C in a way suggestive of electron delocalization around the ring in spite of the possibility of tautomeric rearrangement.

The reaction between 3-anilinobutyronitrile hydrochloride and phosphorus pentachloride in dichloroethane yields the *salt* CI, m.p. 137–138°, converted by sulphur dioxide into **2,4,6-trichloro-1-phenyl-1,2-dihydro-1,3,2-diazaphosphorine 2-oxide** (CII), m.p. 103.5–105°, (*A. M. Pinchuk* and *I. M. Kosinskaya*, *ibid.*, 1970, **40,** 546). Similar reactions are reported for aminoalkylidenemalononitriles (*Kornuta et al. ibid.*, 1971, **41,** 988; *N. D. Bodnarchuk* and *B. B. Gavrilenko*, *ibid.*, 1969, **39,** 1961; *Kornuta*, *A. I. Kalenskaya* and *V. I. Shevchenko*, *ibid.*, 1971, **41,** 2390), while compounds of the type ArCH=(CN)CONH·P(O)Cl_2 undergo cyclization when heated in chlorobenzene to give similar type products (*Shevchenko*, *M. El Dik* and *Pinchuk*, *ibid.*, 1970, **40,** 1949).

(*ii*) *Benzo*[1,3,2]*diazaphosphorines and related compounds*

In a similar process, 2-aminobenzonitriles afford the **benzodiazaphosphorines** (CIII, R = H, m.p. 233–235°; R = Me, m.p. 118–121°); the products are referred to also as **2-phosphaquinazolines** (*A. Schmidpeter* and *N. Schindler*, Chem. Ber., 1969, **102,** 2201):

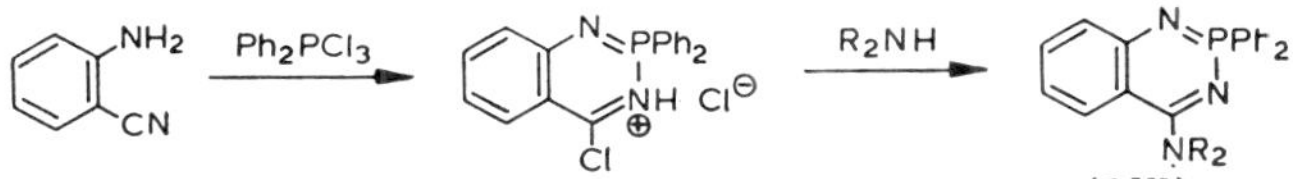

(CIII)

2-*Phenoxy*-2,3-*dihydro*-1H-*naphtho*[1,8,d-e][1,3,2]-*diazaphosphorine* (CIV) is a pink compound, m.p. 126–127°, obtained in about 13% yield when 1,8-diaminonaphthalene is warmed with triphenyl phosphite at 180° under reduced pressure (*K. Pilgram* and *F. Korte*, Tetrahedron, 1963, **19,** 137).

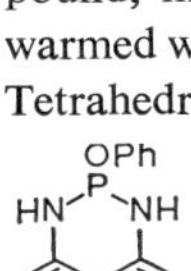

(CIV)

(*iii*) *Tetrahydro-1,3,2-diazaphosphorines and oxaza analogues and their benzologues*

Phosphorus-containing analogues of uracil (CV) are prepared by reaction between diamides of phosphoric acids and diketene (*M. N. Preobrazhenskaya et al.*, Khim. geterotsikl. Soedin., 1974, 1433; C.A., 1975, **82,** 43369):

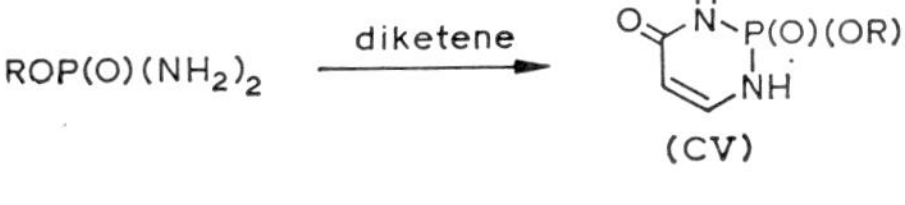

(CV)

4-Ethoxy-2-methyl-1,2,5,6-**tetrahydro-**1,3,2-**oxazaphosphorine 2-oxide** (CVI), b.p. 105–107 °/1 mm, is obtained by a standard procedure (*V. E. Shishkin et al.*, Zhur. obshcheĭ Khim., 1975, **45,** 2554):

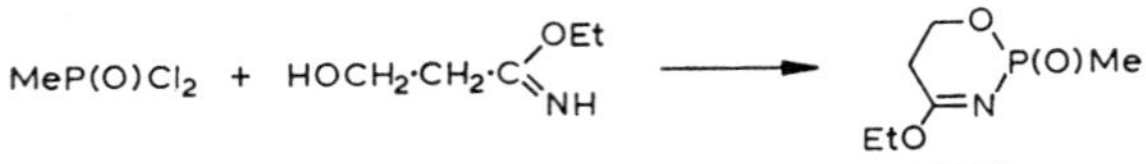

(CVI)

Some oxygen and/or sulphur-containing, bicyclic, tetrahydroazaphosphorine analogues of the diazaphosphorines discussed above, are listed in Table 21.

TABLE 21

1,2,3,4-TETRAHYDRO-1,3,2-BENZOXAZA- AND THIAZAPHOSPHORINES

Structure	*Starting materials*	*Ref.*
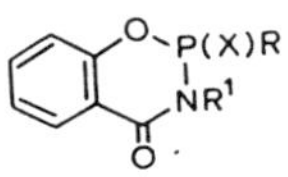	Salicylamides and $RP(X)Cl_2$	1,2
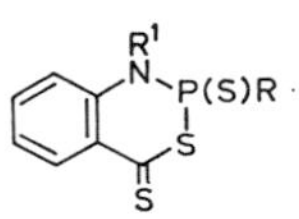	2-Aminobenzoate esters and P_2O_5	3
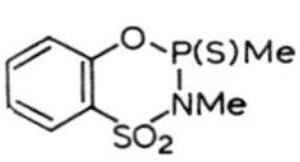 m.p. 80°	2-Hydroxy-N-methylbenzene-sulphonamide and $MeP(S)Cl_2$	4
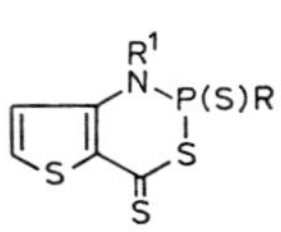	3-Aminothiophene 2-carboxylic esters and P_2S_5	5

References
1 *Y. Oshima* and *M. Eto*, Japan, P 17,395/65; C.A., 1965, **63,** 17971.
2 *K. Kobayashi* and *T. Wakamori*, Japan. P. 18,755/71; C.A., 1971, **75,** 48720.
3 *L. Legrand* and *N. Lozac'h*, Bull. Soc. chim. Fr., 1969, 1173.
4 *M. Nakagawa et al.*, Japan. P, 1428/67; C.A., 1967, **66,** 105052; *W. V. Farrar*, Chem and Ind., London 1974, 876.
5 *J. Brelivet* and *J. Teste*, Bull. Soc. chim. Fr., 1972, 2289.

(iv) *Perhydro-1,3,2-diazaphosphorines and their 1,3,2-oxaza and 1,3,2-thiaza analogues*

The reaction between phosphorus dichlorides and 1,3-diaminopropanes yields **perhydro-**1,3,2-**diazaphosphorines** (CVII): this reaction takes place cleanly when chlorides of quinquevalent phosphorus are used, when the reaction may frequently be carried out in an aqueous medium. If a dichloride of tervalent phosphorous is employed, then a *N,N'*-disubstituted propanediamine is necessary.

Typical examples of phosphorus(III) compounds (CVII, X = lone pair) are 2-*chloro*-1,3-*dimethyl*-, b.p. 45–47°/0.2 mm, 2-*chloro*-1,3,5,5-*tetramethyl*-, b.p. 49–50°/0.45 mm, 2-*methoxy*-1,3-*dimethyl*-, b.p. 66–68°/11 mm, 1,2,3-*trimethyl*-, b.p. 64°/13 mm, 1,3-*dimethyl-2-phenyl*-, b.p. 119–123°/0.3 mm, and 2-*phenyl*-1,3,5,5-*tetramethyl*-, b.p. 82–84°/0.02 mm; *perhydro*-1,3,2-*diazaphosphorines*. Many examples of phosphorus(V) compounds have been recorded (*R. S. Edmundson*, J. chem. Soc., C, 1969, 2730; *R. T. Dickerson*, U.S.P. 3,463,813/1969; C.A., 1969, **71**, 81439; *J. H. Billman, J. L. Meisenheimer* and *R. A. Awl*, J. med. Chem., 1964, **7**, 366; *R. Kraemer* and *J. Navech*, Bull. Soc. chim. Fr., 1971, 3580):

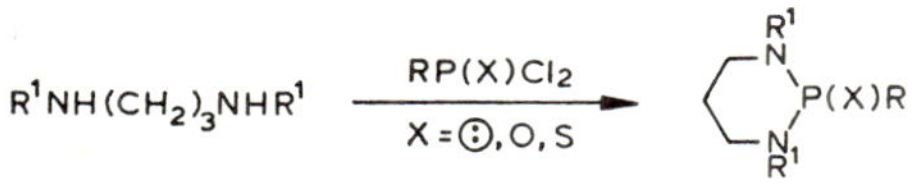

(CVII)

In the search for potential carcinostatic agents, *Billman et al.* (J. med. Chem., 1965, **8**, 264; 1966, **9**, 772; 1967, **10**, 486) prepared many *N,N'*-diphenyl- and *N,N'*-dibenzyl compounds with little success.

The interaction of phosphorus diisocyanates and methylene bases or ketene acetals yields the **perhydrodiazaphosphorine 2-oxides** (XVIII). Exposure of CVIII (Z = $C(OEt)_2$, R = $ClCH_2$) to the open air gives 2-*chloromethyl-5-ethoxycarbonyl*-4,6-*dioxoperhydro*-1,3,2-*diazaphosphorine* 2-*oxide* (*Zh. M. Ivanova et al.*, Zhur. obshcheĭ Khim., 1969, **39**, 1504).

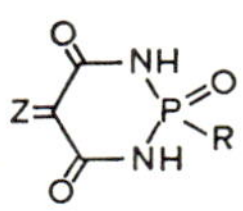

(CVIII)

Cyclic compounds possessing the N–P–O arrangement were obtained by *Petitcolas* and *R. Roe* in 1954 (Bull. chim., Fr., 1954, 378) from 3-hydroxy-2-naphthanilide and phosphoryl dichlorides, but the real impetus to the further study of the oxazaphosphorine ring system came with the discovery of the marked cytostatic activity of **2-di(2-chloroethyl)-aminoperhydro-1,3,2-oxazaphosphorine 2-oxide** (CIX), known as "**Endoxan**" and "**Cyclophosphamide**" (*H. Arnold* and *F. Bourseaux*, Angew. Chem., 1958, **70**, 539; *Arnold, Bourseaux* and *N. Brock*, Naturwiss., 1958, **45**, 64; "Cyclophosphamide", Ed. *G. H. Fairley* and *J. M. Simister*, J. Wright and Sons, 1964). The compound m.p. 41–45° *(monohydrate)* is produced by reaction between *N,N*-di(2-chloroethyl)phosphoramidic dichloride, $(ClC_2H_4)_2NP(O)Cl_2$, and propanolamine in the presence of a tertiary base in inert solvents; it is reported to be in soluble water to the extent of 50 g l^{-1}, but in aqueous media, hydrolysis to open chain compounds occurs. Using the same synthesis, Endoxan labelled with ^{32}P, and ^{14}C in various positions, has been prepared (*K.-H. Schweher*,

Radiochim. Acta, 1963, **2,** 46; Wiss. Tech. Tagung. Deut. Atomforums, Munich, 1963, 146; C.A., 1967, **66,** 2543). Many analogues and derivatives have been prepared in the search for compounds with greater biological activity (*Z. P. Papanastassiou et al.*, J. med. Chem., 1966, **9,** 357; *C. Williamson et al., ibid.*, 1967, **10,** 511; *O. M. Friedman et al., ibid.*, 1963, **6,** 50; *E. Sandrin* and *S. Guttmann*, Swiss P. 492,738/1970; C.A., 1971, **74,** 22898; *K. Marita* and *H. Miyaji*, Japan P. 17,164/1970; C.A., 1970, **73,** 66728; *Asta Werke A.-G.*, B.P. 1,188,159/1970; C.A., 1970, **73,** 44892; B.P. 822,119/1959; C.A., 1961, **55,** 3431; *A. Takamizawa* and *T. Iwata*, Japan. P. 76,081/75; C.A., 1976, **84,** 31142; *L. D. Griggs* and *M. Jarman*, J. med. Chem., 1975, **18,** 1102; *P. B. Farmer* and *P. J. Cox, ibid.*, **18,** 1106), and their stereochemistry determined (*A. Camerman, H. W. Smith* and *N. Camerman*, Biochem. Biophys. Res. Comm., 1975, **65,** 828; *W. Egan* and *G. Zon*, Tetrahedron Letters, 1976, 813; *Zon, ibid.*, 1975, 3139; *R. Kinas, K. Pankiewicz* and *W. J. Stec*, Bull. Acad. Polon. Sci., Ser. Sci. Chim., 1975, **23,** 981).

The reaction between propanolamine and phosphoryl chloride gives 2-*chloroperhydro-*1,3,2-*oxazaphosphorine* 2-*oxide*, b.p. 85–90°/0.015 mm, m.p. 80–83°, in extremely low yield (*R. H. Iwamoto et al.*, J. org. Chem., 1961, **26,** 4743); the P-*sulphide*, formed similarly, has m.p. 84–86° (*O. V. Kil'disheva, A. Puzerauskas* and *E. Ya. Pervova* Izvest. Akad. Nauk S.S.S.R., Ser. Khim., 1963, 398).

Other perhydro-1,3,2-oxazaphosphorines are obtained by heating solutions of the anions obtained from 3-bromopropyl phosphoramidates, *e.g.* CX b.p. 130°/0.01 mm (*P. Savignac, N. Thanh-Thuong* and *Chabrier*, Compt. rend., C, 1968, **266,** 1791; **267,** 183;

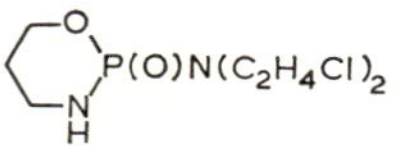

(CIX)

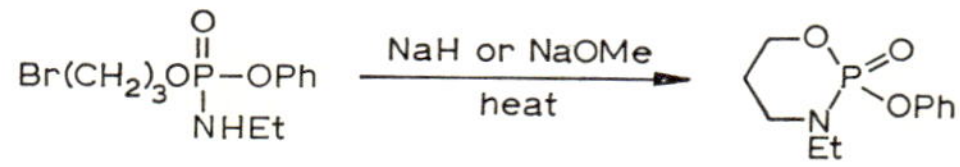

(CX)

Savignac, M. Dreux and *J. Chenault, ibid.*, 1971, **272,** 2189). The method has also been employed for the preparation of perhydro-1,3,2-thiazaphosphorines (*Chabrier* and *Savignac, ibid.*, 1968, **267,** 1166).

The *compounds* CXI (X = NH, R = NEt_2), b.p. 141–145°/0.04 mm, (X = O, R = Cl), m.p. 63–65°, (X = O, R = NEt_2), b.p. 130–132°/0.04 mm, are prepared from the appropriate 2-(2-aminoethyl)- or 2-(2-hydroxyethyl)-piperidine (*K. Winterfield* and *W. Mueller*, Pharm. Helv. Acta, 1963, **38,** 731; C.A., 1964, **60,** 8031).

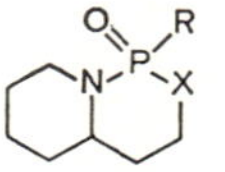

(CXI)

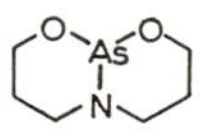

(CXIII)

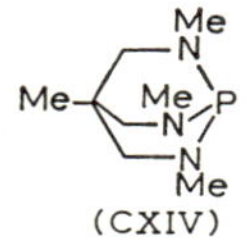

(CXIV)

Recent papers by *T. D. Inch et al.* describe bicyclic perhydro-1,3,2-oxazaphosphorines based on methylated D-glucose and D-galactose in which the ring system is conformationally rigid so providing useful models for n.m.r. studies and for studying mechanisms of reactions at phosphorus (Chem. Comm., 1975, 720; Carbohydrate Resch., 1975, **45,** 65).

2-**Benzoyl-2-phenyltetrahydro**[1,3,2]**oxazaphospholidino**[2,3,*b*][1,3,2]**oxazaphosphorin-3-**(2*H*)-**one** 9-**oxide** (CXII), m.p. 118°, is formed in an adaptation of a reaction studied extensively for linear isocyanates (*I. V. Konovalova, L. A. Burnaeva* and *A. N. Pudovik*, Zhur. obshcheĭ Khim., 1974, **44**, 261; *Pudovik et al., ibid.*, 1973, **43**, 553):

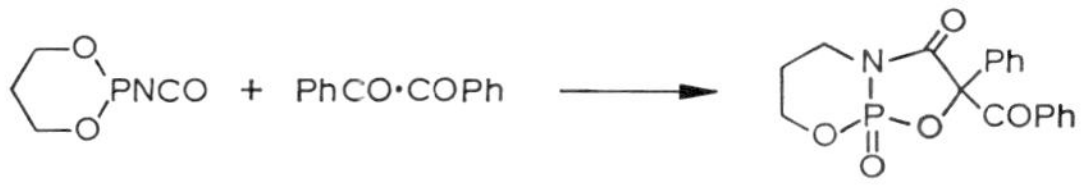

(CXII)

Diethanolamine and arsenic trichloride react together in the presence of triethylamine to give 1,8,9,10-**dioxazarsadecalin** (CXIII), b.p. 54–56°/0.2 mm (*K. Sommer, W. Lauer* and *M.-B. Goehring*, Zeit. anorg. allgem. Chem., 1970, **379**, 48).

Heating a mixture of tris(dimethylamino)phosphine and tris(methylamino)ethane causes liberation of methylamine and leads to 2,4,6,7-**tetramethyl**-2,6,7-**triaza-1-phosphabicyclo**[2.2.2]**octane** (CXIV), b.p. 82–86°/10 mm, *sulphide*, m.p. 92.5–94.5°, *oxide*, m.p. 97–99°; the **arsenic** analogue, b.p. 91–94°/10 mm, is formed similarly (*B. L. Laube et al.*, Inorg. Chem., 1967, **6**, 173).

Other examples of perhydro-1,3,2-diazaphosphorines include 9-*phenyl*-1,5-*diaza*-9-*phosphabicyclo*[3.3.1]*nonane* (CXV), b.p. 98–101°/0.01 mm; 9-*phenyl*-1,3,2-*diazaphospha-*

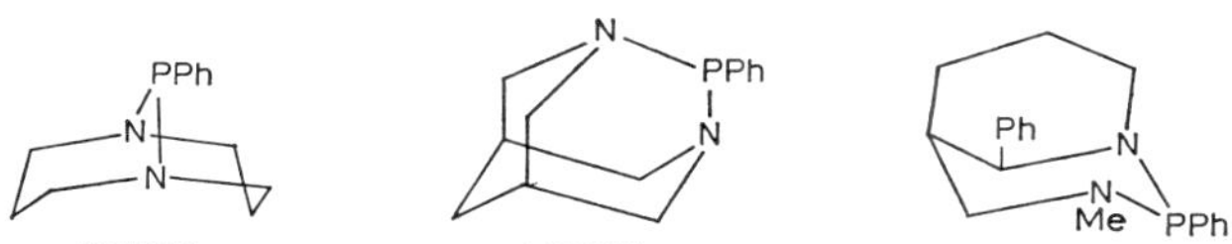

(CXV) (CXVI) (CXVII)

adamantane (CXVI), sublimes; and 3-*methyl*-9-*phenyl*-1,3-*diaza*-2-*phosphabicyclo*[2.3.1]-*nonane* (CXVII), b.p. 120°/0.01 mm (*R. O. Hutchins et al.*, J. Amer. chem. Soc., 1972, **94**, 9151).

(v) *1,2,4- and 1,3,4-Diazaphosphorines: perhydro-1,2,4-diazaphosphorines*

G. Märkl (Angew. Chem. intern. Edn., 1965, **4**, 1023) observed the dipolar addition of nitrilimines to vinylphosphines to give 1,4,5,6-**tetrahydro**-1,2,4-**diazaphosphorinium salts** (CXVIII).

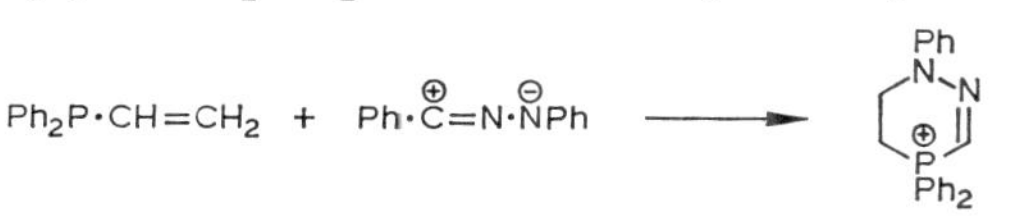

(CXVIII)

This reaction has been extended by Russian workers to include the use of ethynylphosphonites and similar acetylenic compounds (*A. A. Petrov et al.*, Zhur. obshcheĭ Khim., 1970, **40**, 2570; 1971, **41**, 2643, 2649; 1972, **42**, 1864; 1972, **42**, 1926; *L. Tanin et al., ibid.*, 1974, **44**, 1468):

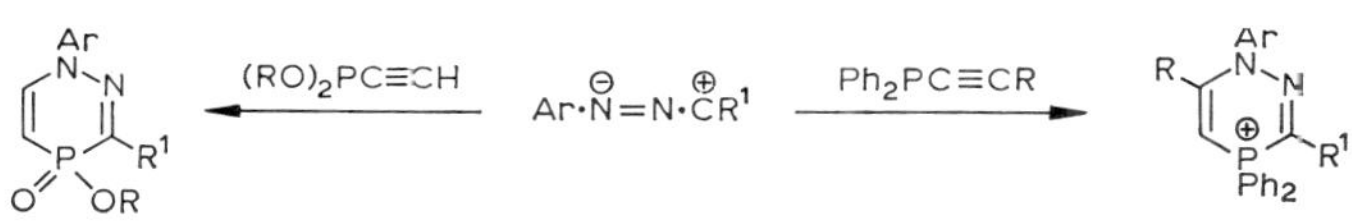

4,4-Dichloro-1,3-dimethyl-2,6-dioxo-1,2,3,6-tetrahydro-1,3,4-diazaphosphorine (CXIX) is obtained in the manner indicated, as a yellow powder or brownish-yellow crystals decomposing at 98–106° (*N. Singh* and *H. P. Latsche*, Angew. Chem. intern. Edn., 1972, **11,** 528):

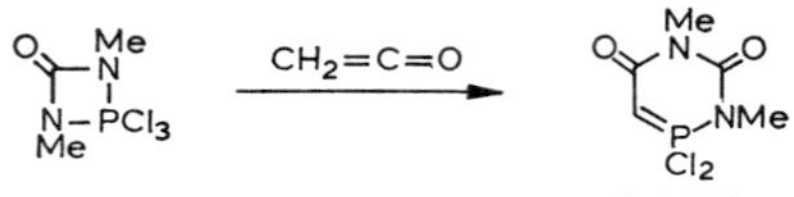

(CXIX)

2-Methyl-1,2,3,4-tetrahydro-1,4,2-benzoxazaphosphorine (CXX, X = O) as its *oxide*, m.p. 68–74°, and *sulphide*, m.p. 96–100°, and the analogous 1,4,2-benzothiazaphosphorine (CXX, X = S), as its *oxide*, m.p. 192–196°, and *sulphide* m.p. 102–109°, are made from 2-aminophenol and 2-aminobenzenethiol with $ClCH_2P(O)MeCl$ (*M. Wieber* and *N. B. Eichhorn*, Monatsh., 1968, **99,** 261).

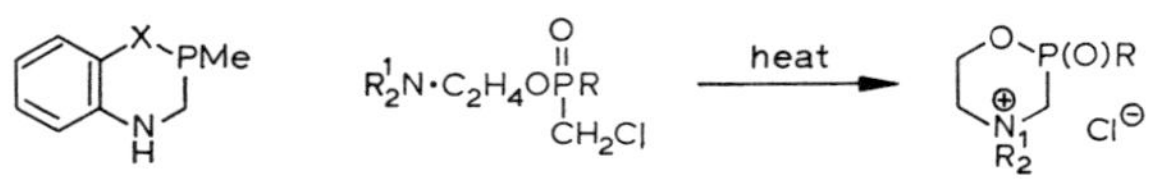

Compounds in the **perhydro-1,4,2-oxazaphosphorine** series (CXXI) are prepared by heating 2-dialkylaminoethyl chloromethylphosphonates in inert solvents (*L. I. Dulalaeva et al.*, U.S.S.R. 341,803/1972; C.A., 1973, **78,** 16241) and when ethanolamine reacts with phosphoric acid in the presence of formaldehyde (*K.-H. Worms* and *K. Woolmann*, Z. anorg. allgem. Chem., 1971, **381,** 260).

A more novel route consists in the reaction between aldehydes or ketones and 1,3,2-oxazaphosph(III)olidines; thus benzaldehyde and 2-isopropoxy-3-methyl-1,3,2-oxazaphospholidine afford a mixture of stereoisomers of 2-isopropoxy-4-methyl-3-phenylperhydro-1,4,2-oxazaphosphorine 2-oxide (CXXII) (*Yu. Yu. Samitov et al.*, Zhur. obshcheĭ Khim., 1975, **45,** 2134; *M. A. Pudovik, L. K. Kibardina* and *A. N. Pudovik*, *ibid.*, 1975, **45,** 470):

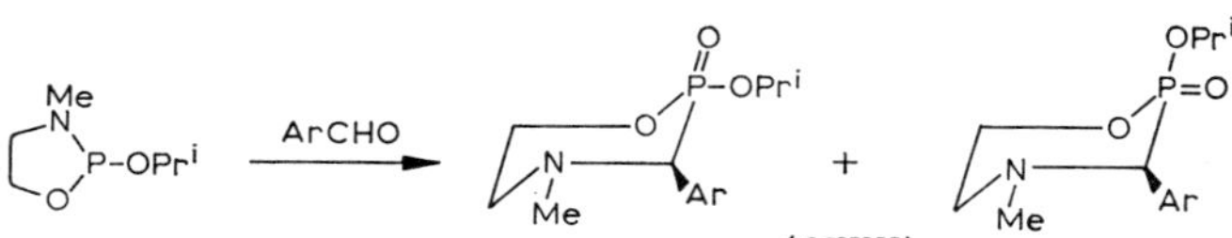

(CXXII)

Although details are not readily available, it has been claimed that α-aminomethylphosphonic monoesters react with bromoacetyl halides to give perhydro-1,4,2-oxazaphosphorine-5-ones (CXXIII) (Z. N. *Ivanova et al.*, U.S.S.R. Pat., 480,713/1975; C.A., 1975, **83,** 193, 511; Zhur. obshcheĭ Khim., 1976, **46,** 1697):

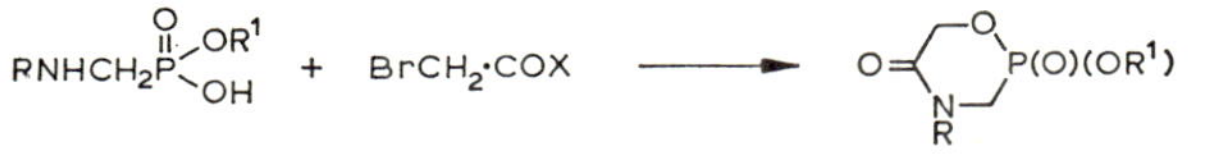

(CXXIII)

(*vi*) *Perhydro*-1,3,5-*diazaphosphorines*

Reaction between hexamethylenetetramine and a 4:1 mixture of formaldehyde and tris(hydroxymethyl)phosphine yields 1,3,5-**triaza**-7-**phospha-adamantane** (CXXIV) decomposing at 360°, and forming a *methiodide*, m.p. 203–204°, by direct reaction. It is oxidized to the phosphoryl compound which also forms a *methiodide*, m.p. 213–214°, (*D. J. Daigle, A. B. Pepperman Jr.* and *S. L. Vail*, J. heterocycl. Chem. 1974, **11,** 407).

(CXXIV) (CXXV)

In a related reaction, the use of sulphamide affords 2,1,3,5-**thiatriaza**-7-**phospha-adamantane** 2,2-**dioxide** (CXXV), m.p. 274–275°, forming a P-*oxide*, m.p. 245–246°, and a P-*methiodide*, m.p. 202–203°, (*Daigle, Pepperman Jr.* and *G. Boudreaux, ibid.*, 1974, **11,** 1085).

(*g*) *Triazaphosphorines*

(*i*) *1,3,5,2-Triazaphosphorines*

The interaction of diphenyl phosphorocyanidite and *N*-chlorobenzamidine yields a linear intermediate which when acted upon by triethylamine leads to 2,2-**diphenoxy**-4-**oxo**-6-**phenyl**-3,4-**dihydro**-1,3,5-**triaza**-2-**phosphorine,** (CXXVI), m.p. 193–195°; this, when treated with phosphorous pentachloride followed by pyridine, is converted into 4-*chloro*-2,2-*diphenoxy*-6-*phenyl*-1,3,5-*triaza*-2-*phosphorine* (CXXVII), m.p. 147–149°, (*M. V. Kolotilo, A. G. Matyusha* and *G. I. Derkach*, Zhur. obshcheĭ Khim., 1969, **39,** 188):

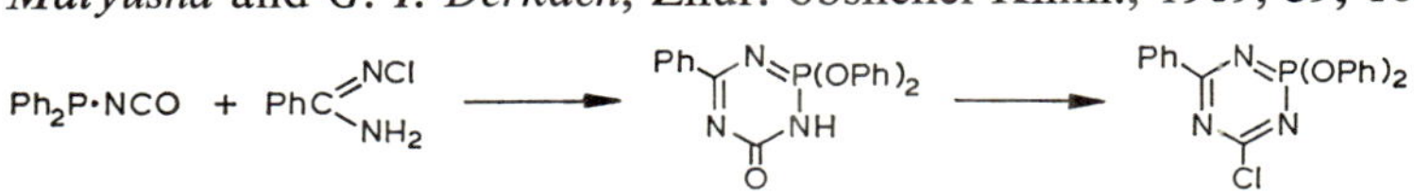

(CXXVI) (CXXVII)

4,6-**Diamino**-2,5-**dihydro**-1,3,5,2-**triazaphosphorines** are formed by treating phosphorylated guanidines with amines, and also by reaction between phosphorochloridates and biguanides:

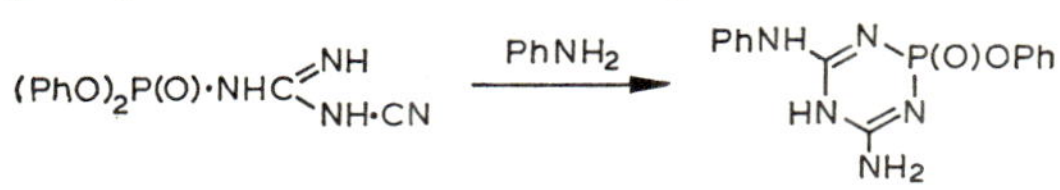

(CXXVIII)

4-*Amino-6-phenylamino-2-phenoxy-2,5-dihydro-1,3,5,2-triazaphosphorine2-oxide*(CXXVIII) has m.p. 272°; the corresponding *sulphide*, m.p. 221°, (*H. Beyer, T. Pyl* and *H. Lemke*, J. pr. Chem., 1962, **16,** 137). The products from biguanides and diphenoxy phosphorus esters condense with aldehydes and ketones suggesting that they are capable of undergoing tautomeric transformation (*J. Ebeling et al.*, Z. Naturforsch., 1971, **268** 650):

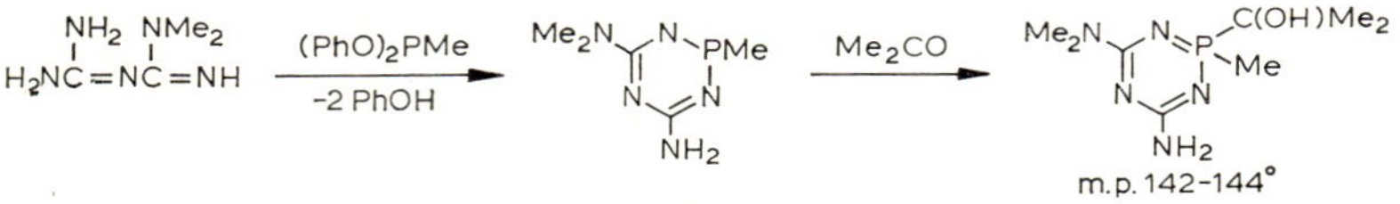

2-Chloro-1,3,5-trimethyl-4,6-dioxoperhydro-1,3,5,2-triazaphosphorine 2-oxide (CXXIX, X= O), m.p. 77–79°, subliming at 80–100°/0.01 mm, is obtained (*a*) admixed with 2,2,2-trichloro-1,3-dimethyl-1,3,2-diazaphosphetidin-4-one, when *N,N'*-dimethylurea is heated with phosphorus pentachloride in carbon tetrachloride, (*b*) when the same diazaphos-

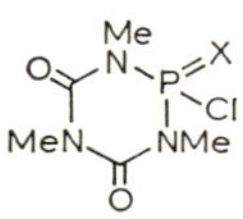

(CXXIX)

phetidinone is heated *in vacuo* and (*c*) when methyl isocyanate reacts with phosphorus pentachloride (*H. P. Latscha*, Z. anorg. allgem. Chem., 1966, **346,** 166; 1968, **355,** 73). The corresponding P-*sulphide*, m.p. 86°, is made in the same way using methyl isothiocyanate (*P. B. Hormuth* and *Latscha, ibid.*, 1969, **369,** 59).

In a related synthesis, *N*-chloroamidines are used in association with tervalent phosphorus isocyanates, although the success of this process seems to depend upon the nature of R, working well with R = Et, less so for R = Ph (*M. V. Kolotilo, A. G. Matyuscha* and *C. I. Derkach*, Zhur. obshcheĭ Khim., 1969, **39,** 188; *Derkach et al., ibid.*, 1966, **36,** 2215):

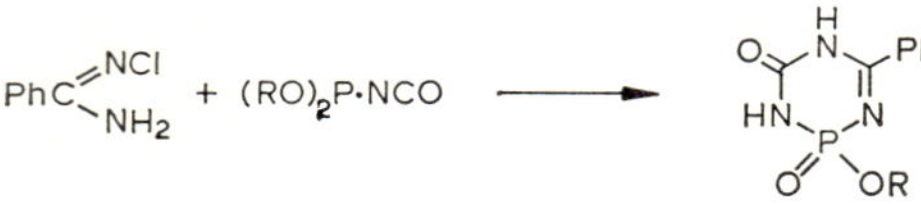

Other P-substituted derivatives of the same ring system are formed when compounds of the type $ArP(O)(NHCONHR)_2$ are heated, and when phosphonic dicyanidates are treated with primary amines (*G. Tomaschewski, C. Berseck* and *G. Hilgetag*, Chem. Ber., 1968, **101,** 2037).

(*ii*) *1,2,4,5-Triazaphosphorines*

A dihydro derivative of the 1,2,4,5-triazaphosphorine (CXXX) undergoes ring cleavage when treated with aqueous ethanol or with ethanolic sulphuric acid (*V. A. Galishev, V. N. Chistokletov* and *A. A. Petrov*, Zhur. obshcheĭ Khim., 1972, **42,** 1876; *ibid.*, 1973, **43,** 1473):

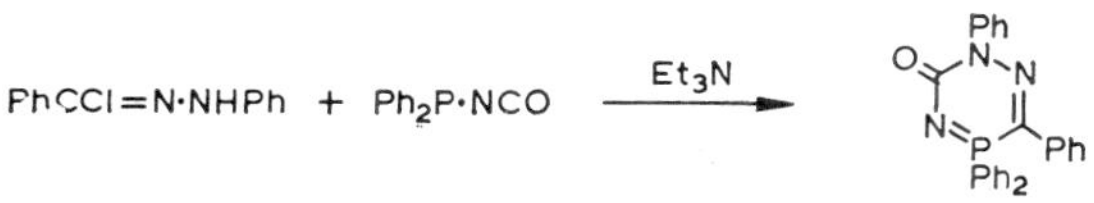

(CXXX)

(*iii*) *Perhydro-1,2,4,3-triazaphosphorines and -1,3,4,2-oxadiazaphosphorines*

1,4-**Dimethyl**-3-**phenylperhydro**-1,2,4,3-**triazaphosphorine** (CXXXI), b.p. 119–123°/1 mm, m.p. 58° (*G. S. Gol'din et al., ibid.*, 1974, **44**, 2668), and 4-*methyl*-2-*phenoxyperhydro*-1,3,4,2-*oxadiazaphosphorine* 2-*oxide* (CXXXI), m.p. 118–119° (*L. Cates*, J. heterocycl. Chem., 1973, **10**, 111) are prepared by the standard methods illustrated:

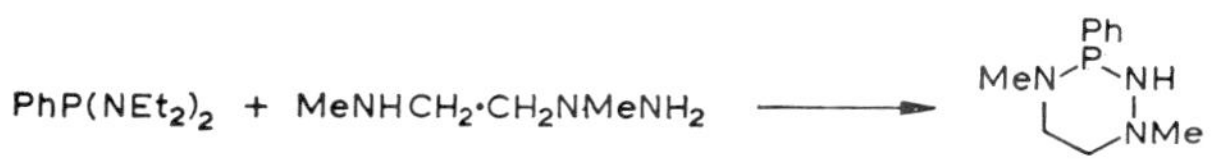

(CXXXI)

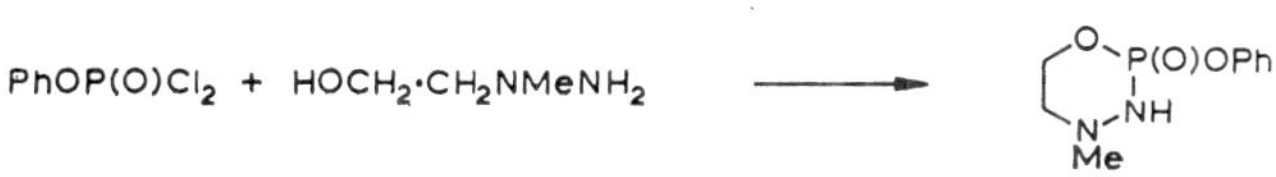

(CXXXII)

(*h*) *Perhydrotetra-aza-phosphorines and -arsenines*

Perhydro-1,2,4,5-**tetra-aza**-3-**phosphorines** (CXXXIII) are prepared from "phosphylated" dihydrazides, $RP(X)\ (NMeNH_2)_2$ (X = O or S; R = Ph, PhO or Me_2N) and aldehydes or ketones.

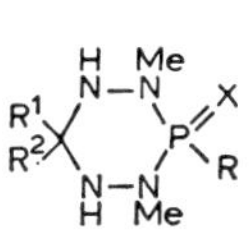

(CXXXIII)

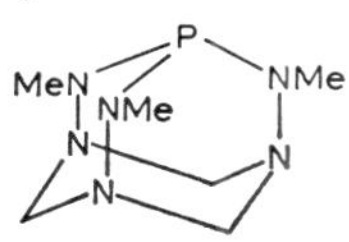

(CXXXIV)

Phosphoryl tris(hydrazides), $(X)P(NMeNH_2)_3$ (X = O or S), similarly react with formaldehyde to give the oxide and sulphide of 2,8,9-**trimethyl**-2,3,5,7,8,9-**hexa-aza**-1-**phospha-adamantane** (CXXXIV) (*J. P. Majoral et al.*, Tetrahedron 1976, **32**, 2633; J. Chem. Soc., Perkin I, 1976, 2093).

(*i*) *Diaza- and triaza-diphosphorines; diazatriphosphorines; diazatriarsenines*

2,2,4,4-**Tetraphenyl**-1,3,5-**triaza**-2,4-**diphosphorine** (CXXXV, R = H) (2,6-**diphospha**-*symm***triazine**), m.p. 235°, and its 6-*phenyl*-, m.p. 232–233°, 6-*ethoxy*-, m.p. 116–117°, 6-*dimethylamino*-, m.p. 190°, derivatives result when carboxamidines react with bis(diphenylphosphino)amine in carbon tetrachloride (*R. Appel* and *G. Saleh*, Ann., 1972, **766**, 98; *A. Schmidpeter* and *Ebeling*, Angew. Chem. intern. Edn., 1967, **6**, 87, 565; Chem.

Ber., 1968, **101**, 2602; *Schmidpeter* and *N. Schindler*, Z. anorg. allgem. Chem., 1970, **372**, 214):

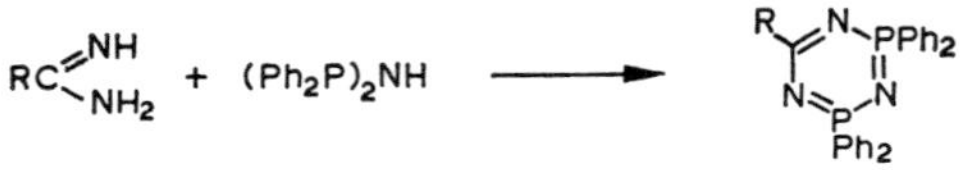

(CXXXV)

The expected reaction between bis(diphenylphosphino)methane and diazidodimethylsilane, $Me_2Si(N_3)_2$, at 150° yields 6,6-**dimethyl**-2,2,4,4-**tetraphenyl**-1,5,2,4,6-**diazadiphospha(V)silin** (CXXXVI), m.p. 121°. This, when treated with phosphorus pentafluoride affords 2,2-*difluoro*-4,4,6,6-*tetraphenyl*-1,3,2,4,6-*diazatriphosphor(V)inium hexafluorophosphate* (CXXXVII), m.p. 203° (*Appel* and *I. Ruppert*, *ibid.*, 1974, **406**, 131).

(CXXXVI) (CXXXVII)

The same bisphosphine is useful in the preparation of compounds possessing other complex ring systems (*Appel*, *R. D. Hanssgen* and *B. Ross*, Z. Naturforsch., 1967, **22B**, 1354):

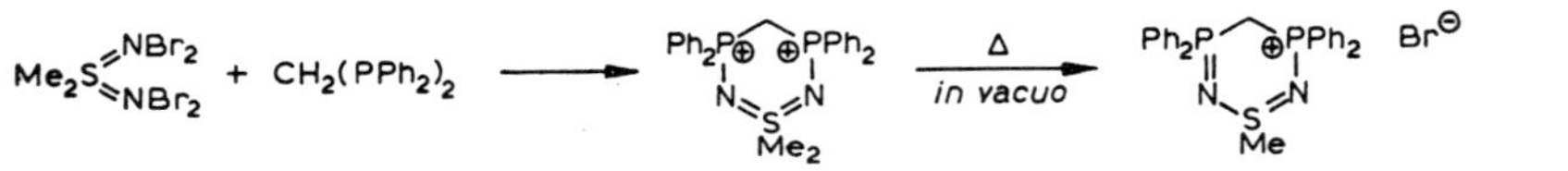

2,4,6,8-**Tetramethyl**-2,4,6,8-**tetra-aza**-1,3,5,7-**tetra-arsa-adamantane** (CXXXVIII) has b.p. 130°/0.1 mm (*F. Kober*, J. organometal. Chem., 1975, **94**, 393):

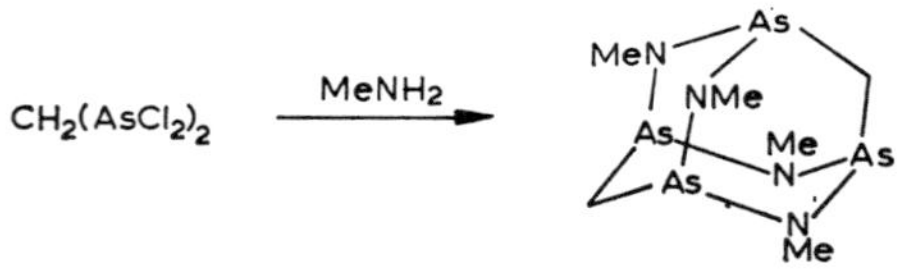

(CXXXVIII)

4. Halogen compounds

Cyclic **halogenium salts** (I) are conveniently prepared by heating aqueous, acidic solutions of 2-[2-halogenoaryloxy(thio)]benzenediazonium salts. Systematically, such salts are known as dibenz[*b*,*e*][1,4]oxachlorinium (I, Y = O, hal = Cl), dibenz[*b*,*e*][1,4]-thiabrominium (I, Y = S, hal = Br), etc. salts, but are also referred to as phenochloraoxin-5-ium, phenobromathiin-5-ium salts. The methods applied to the thiinium compounds are also applicable to their 5-oxides and 5,5-dioxides. The iodine compounds

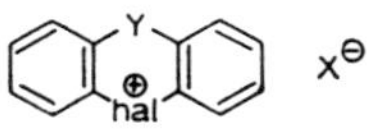

(I)

TABLE 22

DIBENZ[*b,e*][1,4]OXA(THIA)HALOGENIUM SALTS(I)
M.p. (°C) (with decomp.)

X	*Y = O*		*Y = S*		
	hal = I	*hal = Cl*	*hal = Br*	*hal = I*	*hal = Cl*
Cl	249–250 (1)			240 (3)	
Br					
I		102–103 (2)	131–132 (2)		
HSO_4					
$PtCl_6$		175–176 (2)	216–217 (2)		155–157 (2)
HgI_3					119–120 (2)
picrate					

	Y = SO		*Y = SO_2*		
	hal = Br	*hal = I*	*hal = Cl*	*hal = Br*	*hal = I*
Cl					
Br	148–149 (2)			203–204 (2)	191–192 (3)
I	154–154 (2)	154–155 (3)			144–145 (3)
HSO_4					
$PtCl_6$			160–162 (2)	179–180 (2)	
HgI_3			110–113 (2)		
picrate					193 (3)

References

1 *Eli Lilly & Co*, Belg. P. 652,071/1965; C.A. 1966, **64,** 8210.
2 *A. N. Nesmeyanov et al.*, Doklady Akad. Nauk S.S.S.R., 1968, **178,** 860.
3 *Nesmeyanov, T. P. Tolstaya* and *I. N. Lisichkina, ibid.*, 1968, **179,** 604.

are obtained by the action of phosphoric acid on aryloxyiodosobenzenes. A selection of salts is given in Table 22 with their m.p.s. mostly with decomposition.

When heated at 140–180°, cyclic halogeninium salts give high yields of di(2-halogenodiphenyl) ethers, sulphides and their oxides:

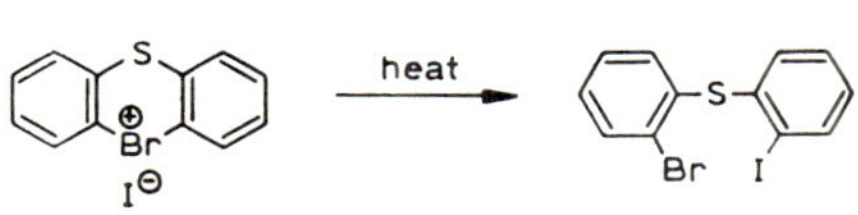

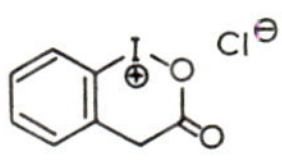

(II)

In 1894, *W. Raum* (Ber., 1894, **27,** 3232) described the chlorination of 2-iodophenylacetic acid, and hydrolysis of the chlorine-containing product; the structure II was proposed for the final material; further chemical and spectroscopic evidence has since been presented to support this early structural assignment (*G. P. Baker et al.*, J. chem. Soc., 1965, 3721; *W. Wolf, J. C. J. Chen* and *L. L. J. Hsu*, J. pharm. Sci., 1966, **55,** 68). **1-Chloro-4*H*-1,2-benziodoxin-3-one** (II) has m.p. 118° with decomposition.

Oxidation of 2-iodophenylene-1,3-diacetic acid with peroxyacetic acid in acetic anhydride–acetic acid yields [1,2]**iodoxino**[5,6,1-*i,j*][1,2]**benziodoxin**-2,8-**dione** (III), m.p. 198–199°, a molecule apparently possessing iodine in the interesting valency state of 3. The molecule is planar, possibly the result of the time-averaged equilibrium indicated (*W. C. Agosta*, Tetrahedron Letters, 1965, 2681):

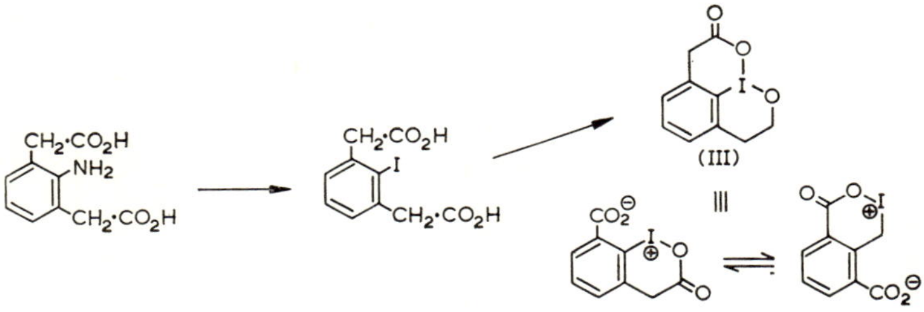

Chapter 50

Compounds with Larger Heterocyclic Rings: Seven-membered Ring Compounds with an Element from Group VI in the Ring

J. M. F. GAGAN

1. General introduction to heterocyclic compounds with rings comprising seven or more atoms

The whole character of the chemistry discussed in this and subsequent chapters has undergone a profound change since the writing of the first edition. Before 1950, heterocyclic compounds having rings with more than six members were frequently the product of accidental discovery. Since then, some of these heterocyclic systems have received intensive systematic study and some correlation of scattered results has also been possible. The expansion of the subject is reflected in the literature surveyed – approximately 375 papers in the first edition compared with nearly 3050 papers in Chapters 50–52 and 55, which cover published work to the end of 1972.

P. M. Maitlis has described the origin of interest in medium and large rings containing heteroatoms (C.C.C. lst Edn., Vol. IV, p. 1581) which developed from work on large alicyclic systems after the introduction of the high dilution technique (see also *K. Ziegler*, Houben-Weyl, "Methoden der Organischen Chemie", Vol. 4, Part 2, p. 731, G. Thieme, Stuttgart, 1955; *R. Huisgen*, Angew. Chem., 1957, **69**, 341).

Nitrogen heterocycles with their generally easier syntheses and greater stability are still more widely investigated than those containing oxygen or sulphur, but the earlier emphasis on compounds which were saturated, or relied for their unsaturation on annellated benzene rings, has shifted towards unsaturated monocyclic compounds. Much of the chemistry described for medium and large heterocycles is concerned with their synthesis and, although a large amount of spectroscopic and other physical data has also been obtained, their chemical reactions have been much less thoroughly investigated.

The present interest in these ring systems has a variety of sources. The parent ring systems or their simply substituted derivatives have provided a means of checking theoretical predictions concerning "aromaticity", stability, physical properties and reactivity, and this has been an additional encouragement towards the synthesis of fundamental or novel heterocyclic systems as an end in itself. As the range of these heterocyclic compounds increased, so did the scope for studies specific to cyclic systems, including reactions to increase or decrease ring size, ring-bridging reactions and ring–chain or transannular tautomerism. The application of orbital symmetry ideas in this field has expanded interest both in synthesis by cycloaddition, and in the photolysis and thermolysis of the heterocyclic products. New stereochemical problems relating to conformation, ring-inversion and functional group interactions are also to be found in this area. Even inorganic chemists are interested to look for new complexing agents among the larger polyheteroatom rings.

There is also a strong commercial interest and much of the information about medium sized heterocyclic compounds is to be found in the patent literature. One of the most important monomers, caprolactam, and one of the most heavily prescribed drugs, "Librium", each contains a seven-membered heterocyclic ring, and an enormous number of medium sized heterocycles has been synthesised for evaluation as polymer intermediates, dyestuffs, pharmaceuticals or agricultural chemicals.

With such a large area to cover, certain topics have necessarily been omitted or have received only cursory treatment. The *chemistry* of carbon compounds is our main concern and so theoretical, physico-chemical, stereochemical and spectroscopic work is rather neglected, as are speculative and mechanistic aspects of studies in this field, in favour of established preparative routes and chemical behaviour. Reference to the original articles will readily make good these deficiencies.

Two major omissions are the bridged ring systems, which, being mainly saturated, have properties more characteristic of aliphatic or alicyclic systems, and compounds with annellated heterocyclic rings, which are frequently prepared by analogous routes to those used for benzoannellated systems and often show similar chemical behaviour. Neither are the heterocyclophanes discussed in these chapters.

Heterocycles with more than one type of heteroatom (see p. 283) are only considered in the seven-membered ring series, even though the chemistry of the larger ring systems is extensive if rather fragmentary. Little is said either about natural products containing medium or large heterocyclic rings. The corrin ring system, the seven-membered ring sugars and the cyclic peptides are not included, but brief mention is made of a number of alkaloids and microbial metabolites. Some of these topics are covered more comprehensively in other volumes of this work.

(*a*) *Nomenclature*

The long and irregular development of the chemistry of medium and large heterocyclic rings has given rise to a literature abounding in trivial names. Where these remain in current use or are used in key references they are indicated in parentheses in the text. The systematic nomenclature adopted in this chapter is outlined below.

Monocyclic compounds containing the maximum number of non-cumulative double bonds are given a stem according to the number of atoms in the ring: for seven-membered rings -epin(e), eight-membered -ocin(e), nine-membered -onin(e) and ten-membered -ecin(e). The heteroatoms are indicated by prefixes: for oxygen ox-, for sulphur thi-, for selenium selen- and for nitrogen az- *in this order of precedence* both in the name and the numbering. Rings containing nitrogen usually retain the final (e) (see C.C.C., Vol. IV A, p. 551); oxygen, sulphur and selenium heterocycles omit it.

The degree of unsaturation is shown by a prefix dihydro-, tetrahydro- etc. A locant for one hydrogen atom is required for rings with an odd number of members and an odd number of nitrogen atoms, or an even number of members and an odd number of oxygen, sulphur or selenium atoms. The prefix perhydro- may be used to indicate complete saturation, as may the modification of the stem to -epan(e), -ocan(e), -onan(e), -ecan(e).

A heterocyclic ring fused to one benzene ring is designated l-benz(o) if a heteroatom is attached directly to the benzene ring; a corresponding number is given to heteroatoms in other positions, to give the smallest sum. The order of precedence of heteroatoms applies only to the name not to the numbering, *e.g.* compound I is 1*H*-4,1-benzothiazocine not 6*H*-3,6-benzothiazocine. The numbering of the monocyclic system is not necessarily retained, *e.g.* compound II is named 3*H*-1,5-benzodiazepine, not as a derivative of 1,4-diazepine. Other bicyclic systems are named using a lower case italic letter in square brackets to indicate the site of annellation, or, like bridged systems, by inserting the heteroatom prefix, numbered accordingly, into the systematic name of the corresponding homocycle. This latter nomenclature is still occasionally used for saturated monocyclic systems.

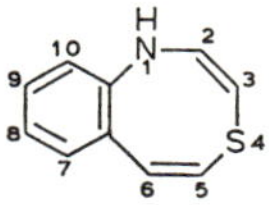
(I)

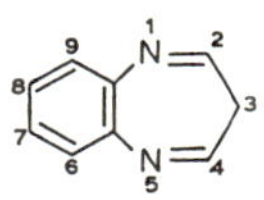
(II)

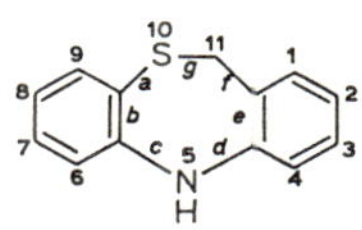
(III)

If more than one benzene ring is annellated to the heterocyclic ring, italic lower case letters show where the rings are attached to the monocyclic system, whose heteroatoms are numbered according to their precedence. The overall structure is then numbered using the Ring Index numbering system which may not give the principal heteroatom the lowest possible number, *e.g.* compound III is 5,11-dihydrodibenzo[*b*,*e*]-1,4-thiazepine not the 6,11-dihydro compound.

2. Seven membered heterocyclic compounds with oxygen in the ring

(*a*) *Monocyclic compounds with one oxygen atom*

Interest in oxepin (I) has stemmed from two suggestions made by *J. Meinwald* and *H. Nozaki* (J. Amer. chem. Soc., 1958, **80**, 3132) that it might bear a similar relationship to cyclo-octatetraene that furan does to benzene, and that it could be in equilibrium with its valence tautomer, the epoxide II.

(I) (II)

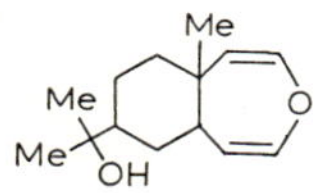

(III)

The oxepin ring has a very limited occurrence in nature. The strychnine and cularine alkaloids show this structural feature as do the limonoids, as ε-lactones, and some simpler terpenes *e.g.* occidenol (III).

(*i*) *Synthesis*

No example of a fully unsaturated monocyclic oxepin was known when the first edition of this book was published, but several oxepins including **oxepin** itself (I), b.p. 27°±14 mm, 124°±atm., n_D^{20} 1.5163, have now been prepared. Outstanding for synthetic elegance is the route from "Dewar benzene" (*E. E. van Tamelen* and *D. Carty*, *ibid.*, 1967, **89**, 3922) suggested by *E. Vogel* and *H. Günther* (Angew. chem. intern. Edn., 1976, **6**, 385), but the initial approach of *Meinwald* and *Nozaki* (*loc.cit.*) using epoxides IV (R = Br, OAc) failed only because the conditions employed were too vigorous.

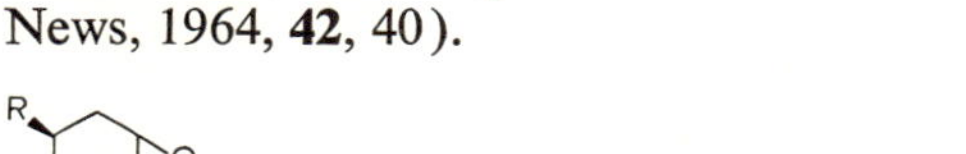

Use of 1,5-diazabicyclo[3.4.0]non-5-ene or sodium methoxide in ether allows of elimination in IV (R = Br) without promoting the thermal rearrangement to phenol which occurs above 70° (*Vogel et al.*, Tetrahedron letters, 1965, 609; Angew. Chem. intern. Edn., 1964, **3**, 510; Chem. and Eng. News, 1964, **42**, 40).

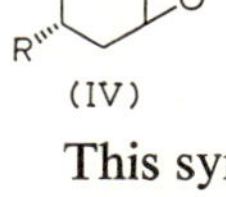

(IV)

This synthesis has also proved effective for substituted oxepins. The ready preparation of 2-*methyloxepin*, b.p. 34°/11 mm; 2,7-*dimethyloxepin*, b.p.

51°/14 mm, n_D^{20} 1.5045; and 2-*acetyloxepin*, b.p. 51°/0.4 mm, indicates that the stability of the monocyclic triene I is associated with a substituent in position 2, which may exert a steric or electronic effect (*V. R. Gaertner*, Tetrahedron, 1967, **23**, 2123). Only the last of these shows no evidence for the presence of the epoxy tautomer II (*Vogel* and *Günther*, Z. naturforsch., 1967, **22b**, 25; Angew. Chem. intern. Edn., 1965, **4**, 975; 1967, **6**, 385). 4-Methyl- and 4,5-dimethyloxepins are also prepared in this way (*B. Witkop et al.*, J. Amer. chem. Soc., 1968, **90**, 6523). The ring expansion of the heavily substituted cyclohexadiene V is a similar route (*A. Rieker*, Angew. Chem. intern. Edn., 1971, 10, 425):

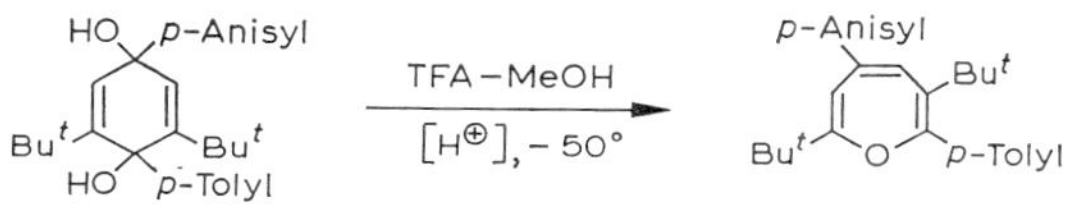

(V)

Addition of dichlorocarbene to 3-acetoxy-2,3-dihydropyran (VI; R = OAc) followed by hydrolysis, oxidation and thermal elimination of hydrogen chloride gave the oxepin tautomer VII (*S. Masamune* and *N. T. Catellucci*, Chem. and Ind., London, 1965, 184). Analogous reactions with dihydropyran VI (R = H) give 3-chloro-6,7-dihydrooxepin (*C. B. Reese et al.*, Tetrahedron, 1964, **20**, 2091; *E. E. Schweizer* and *W. E. Paraham*, J. Amer. chem. Soc., 1960, **82**, 4085):

(VI) (VII)

4,5-Di(methoxycarbonyl)oxepin has been prepared by successive photochemical and thermal rearrangements on the 1:1 adduct of furan and dimethyl acetylenedicarboxylate as well as by Vogel's epoxide route (*H. Prinzbach et al.*, Angew. Chem. intern. Edn., 1966, **5**, 1039; 1967, **6**, 1070; Chimia 1967, **21**, 469; Helv., 1969, **52**, 396).

(*ii*) *Structure and reactions*

The existence of the tautomeric equilibrium I ⇌ II is shown by the temperature dependence of the nuclear magnetic resonance spectrum [as the temperature drops more benzene oxide (II) is found] and the variation of the ultraviolet absorption spectrum with solvent (*e.g.* 30% of II in iso-octane; 90% of II in aqueous methanol). Benzene oxide (II) and oxepin (I) are of about equal stability; oxepin (I) has therefore no appreciable aromatic stabilisation and exists in a "boat" form [*Günther et al.*, Angew. Chem. intern. Edn., 1965, **4**, 975; 1087; Tetrahedron Letters, 1965, 609, 4085; *Prinzbach et al.*, *loc.cit.*].

The chemistry of oxepin is dominated by its ready conversion to phenol. Apart from the π-complexes formed with iron pentacarbonyl (*E. O. Fischer et al.*, Chem. Ber., 1967, **100**, 1905) only in catalytic reduction to perhydro-oxepins does the seven-membered ring survive, and then cyclohexanols are found as by-products. The reaction of oxepin (I) with lithium tetrahydridoaluminate or methyl-lithium yields the cyclohexadienols VIII (R = H or Me) and sodium azide and sulphide give respectively the azide (VIII; R = N_3) and the bisdisulphide (*R. M. DeMarinis* and *G. A. Berchtold*, J. Amer. chem. Soc., 1969, **91**, 6525).

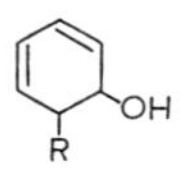

(VIII)

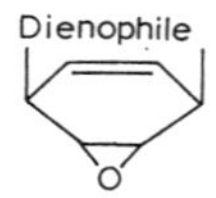

(IX)

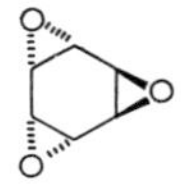
(X)

Triphenylphosphine at 120–150° removes the oxygen, leaving benzene. Ring-opening does not always give phenols; oct-4-en-2-one and octa-4,6-dien-2-one result on reduction of 2,7-dimethyloxepin with an alkali metal (*L. A. Paquette* and *T. McCreadie*, J. org. Chem., 1971, 36, 1402).

Diels–Alder additions to oxepins take place *via* the epoxy tautomer to give adducts IX (*Vogel, Günther, loc.cit.*) and singlet oxygen reacts with oxepin (I) to give *trans*-**benzene trioxide** (X) (*Vogel* and *H. J. Altenbach*, Angew. Chem. intern. Edn., 1972, **10**, 937; *C. H. Foster* and *G. A. Berchtold*, J. Amer. chem. Soc., 1972, **94**, 7939).

Irradiation of oxepins in ether causes transannular cyclization to *cis*-2-oxabicyclo-[3.2.0]hepta-3,6-dienes but in acetone, phenols are the principal photoproducts; benzene and phenol are also produced in ethereal solutions at low temperatures (*Paquette et al.*, Trans, N.Y. Acad. Sci., 1966, **28**, 397; J. Amer. chem. Soc., 1965, **87**, 3417; 1966, **88**, 1718; *J. M. Holovka* and *P. D. Gardner*, *ibid.*, 1967, **89**, 6390; *Rieker, loc.cit. O. L. Chapman*, Adv. Photochem., 1963, **1**, 323):

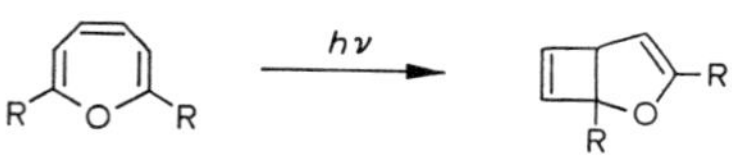

All isomeric dihydro-oxepins except 2,7-dihydro-oxepin (XI) are now known.

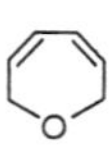
(XI)

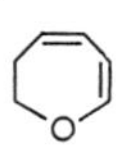
(XII)

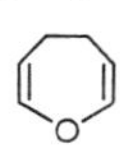
(XIII)

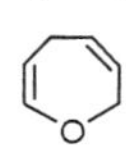
(XIV)

2,3-Dihydro-oxepin (XII), b.p. 118–120°, 61°/100 mm, n_D^{25} 1.4965, has been prepared by an addition–dielimination process from 2,3,6,7-tetrahydro-oxepin (*Meinwald et al.*, J. Amer. chem. Soc., 1960, **82**, 4087) and by heating the *exo/endo*-2-oxa-7-chloronorcarane mixture formed when chlorocarbene reacts with 2,3-dihydropyran:

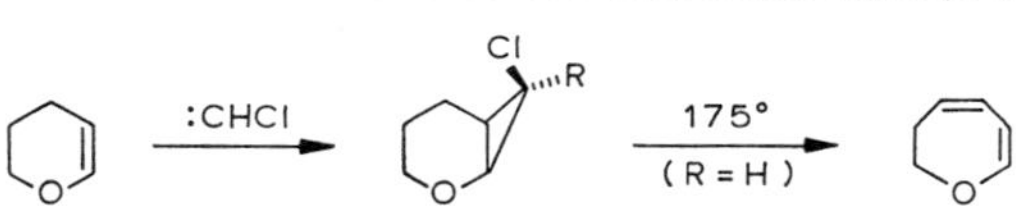

(XV) (XII)

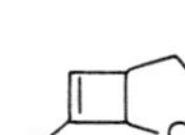

(XVI)

In quinoline, the *endo* compound is more easily pyrolysed (*E. E. Schweizer* and *W. E. Parham*, *ibid.*, 1960, **82**, 4085; *C. H. De Puy et al.*, *ibid.*, 1966, **88**, 3343; *T. Ando et al.*,

Tetrahedron Letters, 1967, 2587). Dihalogenocarbenes gave 6-halogeno analogues, *e.g.* 6-*chloro*-2,3-*dihydro-oxepin*, b.p. 75°/25 mm, n_D^{25} 1.5167–1.5178, and 3-acetoxy-2,3-dihydro-4*H*-pyran could be similarly converted into 2,3-dihydro-6-chloro-oxepin-3-one (*Masamune* and *Castelluci*, *loc.cit.*). Analogous reactions with chromene failed to give benzoxepins (*Parham* and *L. D. Heustis*, J. Amer. chem. Soc., 1962, **84**, 813).

Compound XII behaves as a typical enol ether, undergoing facile ring-opening with acids, and as a typical diene towards tetracyanoethylene (*Meinwald et al.*, *loc.cit.*). On catalytic reduction, the parent XII gives hexamethylene oxide, but its 6-chloro analogue suffers ring-opening to hexanol (*Schweizer* and *Parham*, *loc.cit.*); reduction by potassium in liquid ammonia gives the fully delocalised anion $[CH_2{\cdot}CH{\cdot}CH{\cdot}CH{\cdot}CH{\cdot}CH{\cdot}O]^{\ominus}$ (*H. Kloosterziel* and *J. A. A. Van Drunen*, Rec. Trav. chim., 1970, **89**, 667). On irradiation, a transannular isomerisation parallel to, but more facile than that shown by oxepins yields XVI (*Paquette et al*, J. Amer. chem. Soc., 1965, **87**, 3417). The 2,3-dihydro-oxepin-2-one XVII (R = Me, R[1] = H) is a product when singlet oxygen reacts with 6,6-dimethylfulvene (*S. Skorianetz et al.*, Helv, 1971, **54**, 1913). This seems to be a general reaction (*N. Harada et al.*, J. Amer. chem. Soc., 1972, **94**, 1777). Acid chlorides react with acetylene in the presence of nickel carbonyl to give 7-*phenyl*-2,3-*dihydro-oxepin*-2-*one* (XVII; R[1] = Ph, R = H), b.p. 84–85²/0.03 mm, and analogues (*M. Foa et al.*, Tetrahedron Letters, 1968, 1357).

Routes to 4,5-**dihydro-oxepin** (XIII), b.p. 113°, n_D^{25} 1.4632, are available from the thermal isomerisation of *cis*-hexatriene 3,4-oxide (XVIII) by a Cope rearrangement (*R. A. Braun*, J. org. Chem., 1963, **28**, 1383; *E. L. Stogryn et al.*, *ibid.*, 1964, **29**, 1275; *J. C. Pommelet et al.*, Compt. rend., 1970, **270C**, 1894; and the pyrolysis of 2-acetoxy-2,3,4,5-tetrahydro-oxepin (XIX; R = OAc) (*G. Pohl*, from Houben-Weyl 4th Edn., 1966, **6/4**, 666).

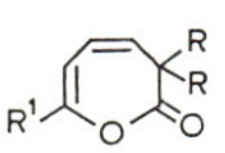

(XVII)

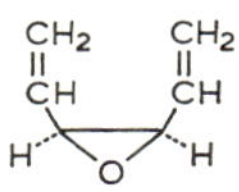

(XVIII)

Warm aqueous acetic acid converts XIII to cyclopentene-1-carbaldehyde, and catalytic reduction over Adam's catalyst or Raney nickel gives a high yield of hexamethylene oxide (*Pohl*; *Braun*, *loc. cit*). Irradiation of XIII gives cyclopent-2-enecarbaldehyde, hexa-2,5-dienal and hexa-5-ynal (*R. D. Cockroft et al.*, Tetrahedron Letters, 1970, 3539).

The unstable 2,5-**dihydro-oxepin** (XIV) is prepared by heating *cis*-2-vinylcyclopropane-carbaldehyde, and it readily reverts to this compound at room temperature (*S. J. Rhoads* and *Cockroft*, J. Amer. chem. Soc., 1969, 91, 2815). A hexamethyl derivative is obtained as shown (*P. Scribe et al.*, Tetrahedron Letters, 1970, 4375):

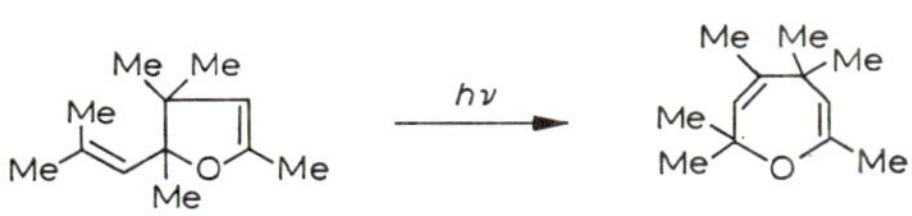

All three parent tetrahydro-oxepins (XIX; R = H, XX; R, R[1] = H; XXI) have been characterised, and substituted derivatives of XIX and XX are known.

2,3,4,5-**Tetrahydro-oxepin** (XIX; R = H), b.p. 113–115°, n_D^{25} 1.5445, is obtained as

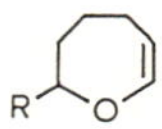
(XIX)

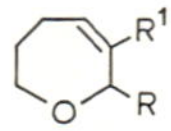
(XX)

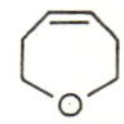
(XXI)

the principal product in the catalytic dehydrodehydrogenation of 1,6-hexanediol or 2-hydroxymethyltetrahydropyran over copper–chromium at 250° (*D. R. Larkin*, J. org. Chem., 1965, **30**, 335; *A. Misono et al.*, Bull. chem. Soc. Japan, 1969, **41**, 2447) and in 70% yield when the oxepane XXII (R = $OCH_2.DH_2Cl$) is heated in quinoline (*F. Nerdel et al.*, Tetrahedron Letters, 1966, 5385). It is an enol ether undergoing ring-opening in acids.

7-Isopropyl-4-methyl-2,3,4,5-tetrahydro-oxepin, b.p. 63°/10 mm, n_D^{15} 1.4529, was obtained by heating 8-hydroxy-2,6-dimethyloctan-3-one in phosphoric acid (*P. Teisseire* and *B. Corbier*, Recherches, Paris, 1963, **13**, 78; C.A., 1964, **61**, 7049). Ozonolysis of the product yields a mixture of acids including 3-methylglutaric acid.

2-Acyloxy-2,3,4,5-tetrahydro-oxepins (XIX; R = OCOR) are formed when cyclohexene is oxidised by air in the presence of an acyl anhydride and a radical initiator (*R. H. Snyder et al.*, J. Amer. chem. Soc., 1958, **80**, 3064; 1959, **81**, 4299; *H. J. Shine* and *Snyder*, B.P., 796,906/1958; C.A., 1959, **53**, 2188). They also show facile ring-opening reactions giving glutaric acid with neutral potassium permanganate, derivatives of adipaldehyde and cyclopentene-1-carbaldehyde with hydroxylamine and 2,4-dinitrophenylhydrazine, and hexan-1,6-diol on reduction by lithium tetrahydridoaluminate or hydrogen over platinum. When palladium was used, the ring survived and the product was 2-acetoxyoxepane (XXII; R = OAc).

A mixture of 2,3,4,7-**tetrahydro-oxepin** (XX; R,R[1] = H) b.p. 98–100°, n_D^{20} 1.4562, and 2,3,6,7-**tetrahydro-oxepin** (XXI), b.p. 118–119°, $n_D^{25.5}$ 1.4548, is produced by treating

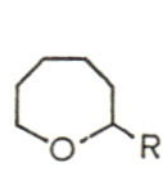
(XXII)

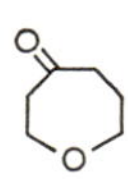
(XXIII)

tetrahydropyran-4-one with diazomethane, and reducing and dehydrating the ring-expanded ketone, oxepane-4-one (XXIII) (*S. B. Maerov*, Diss. Abs., 1954, **14**, 765; *Meinwald et al., loc.cit.*; *S. Olsen* and *R. Bredoch*, Chem. Ber., 1958, **91**, 1589). 7-*tert*-Butyl-6-chloro-2,3,4,7-tetrahydro-oxepins (XX, R = Bu*t*O, R[1] = Cl) are produced in the ring-expansion of dihydropyrans, with dichlorocarbene generated from chloroform and potassium *tert*-butoxide; the cyclopropane ring of the intermediate norcaradiene (XV; R = Cl) opens with cine-substitution rather than elimination (*J. d'Angelo* and *V. M. Thuy*, Bull. Soc. chim. Fr., 1968, 2823; *L. Skatteboll*, J. org. Chem., 1970, **35**, 3200). Norcaradiene (XV; R = Cl) also reacts with ethylene oxide to give the essential intermediate in the preparation of XXII (R = $OCH_2.CH_2Cl$) and XIX (R = H). With dilute acid, XX (R = ButO, R[1] = Cl) contracts to 3,4-dihydro-2*H*-pyran-5-carbaldehyde.

2,3,6,7-Tetrahydro-oxepin (XXI) was prepared separately by ring closure of *cis*-6-iodohex-3-en-1-ol with silver oxide in ether (*Meinwald* and *Nozaki*, *loc.cit.*). It shows little tendency towards ring-opening, being reduced to XXII (R = H) over platinum, and giving an epoxide, a *cis*-dihydroxy derivative and a *trans*-dibromo derivative, respectively, with perbenzoic acid, osmium tetroxide and bromine in carbon tetrachloride.

Oxepane *(hexahydro-oxepin, oxacycloheptane, hexamethylene oxide)* (XXII; R = H), b.p. 119–120°, n_D^{25} 1.4365, has long been known as a product in the dehydration of 1,6-hexanediol with concentrated sulphuric acid (*A. Franke et al.*, Monatsh., 1914, **35**, 1431; 1936, **69**, 167). Anhydrous hydrogen bromide and a variety of high temperature vapour phase techniques have also been used for this dehydration (*A. Mueller* and *W. Vanc*, Ber., 1944, 77, 669; *L. Kh. Friedlin et al.*, C.A., 1965, **63**, 16303; Izvest. Akad. Nauk S.S.S.R., 1961, 373; *N. I. Shnikin et al.*, *ibid.*, 1962, 1868; *C. Sehuster* and *H. Lattermann*, G.P., 840,844/1952; C.A., 1958, **52**, 16388) but the best method (up to 70% yield) is undoubtedly the use of hot dimethyl sulphoxide as dehydrating agent (*V. J. Traynelis et al.*, J. org. Chem., 1964, **29**, 123). Equally high yielding is the ferric chloride catalysed cyclisation of 1-bromo-6-methoxyhexane (*A. Kirrmann* and *N. Hamaida*, Bull. Soc. chim. Fr., 1957, 789; *Hamaida*, F.P. 1,165,673/1958; C.A., 1960, **54**, 19731).

Intramolecular cyclisation routes to oxepanes other than the parent are rare, but 2,2,7,7-tetramethyloxepane is made this way (*L. Michiels*, Bull. Soc. chim. Belges, 1913, **27**, 25; C.A. 1913, **7**, 3602). Condensation reactions between a bifunctional compound already containing an ether linkage and a second component have been used to prepare 3,3,6,6- and 4,4,5,5-tetraethoxycarbonyloxepanes (*I. Ali-Zade* and *B. A. Arbuzov*, Zhur. obshcheĭ Khim., 1943, **13**, 113; *I. W. Matschinskaya et al.*, *ibid.*, 1960, **30**, 2362):

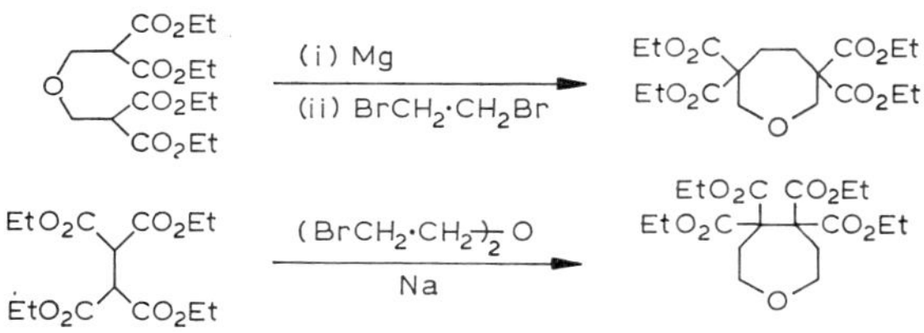

2,6-Dimethyltetrahydropyran-4-one undergoes ring expansion when treated with diazomethane to 2,7-dimethyloxepan-4-one (*Pohl. loc.cit.*). Both 3- and 4-oxo derivatives of 2,2,7,7-tetramethyloxepane are found as by-products in a similar ring expansion of 2,2,5,5-tetramethyltetrahydrofuran-3-one (*I. K. Korobitsyna* and *K. K. Pivnitskii*, Zhur. obshcheĭ Khim., 1960, **30**, 4008).

Reactions of oxepane giving ring-opening include normal ether cleavage with concentrated hydrobromic acid (*Mueller* and *Vanc*; *Olsen* and *Bredoch*, *loci cit.*) and a sealed tube reaction with methyl iodide (*A. Mueller et al.*, Monatsh., 1953, **84**, 1206). In these reactions, the product is a 1,6-dihalogenohexane, but with *n*-butyllithium the product is hex-5-en-1-ol (*R. B. Bates et al.*, J. org. Chem., 1972, **37**, 560). Oxepane also undergoes polymerisation, alone or as a copolymer (*K. Hamann et al.*, Makromol. Chem., 1967, **101**, 246; J. Polymer Sci. Part C, 1967, **16**, 2131; *B.A.S.F.*, Neth. Appl., 6,514,413/1966; C.A., 1966, **65**, 10687). The saturated seven-membered ring, however, has the stability to survive in the presence of bases and reducing agents.

Most other monocyclic oxepanes are cyclic acetals, lactones or anhydrides. The cyclic hemiacetal 2-*hydroxyoxepane* (XXII; R = OH), b.p. 115°/15 mm, n_D^{25} 1.4460, is one of the products of deamination of 2-aminomethyltetrahydropyran with nitrous acid (*J. Colonge* and *P. Corvet*, Compt. rend., 1958, **247**, 2144; Bull. Soc. chim. Fr., 1960, 287). At equilibrium, only about 15% is in the cyclic form, the remainder exists as the open chain 6-hydroxyhexanal (*C. D. Hurd* and *W. H. Saunders*, J. Amer. chem. Soc., 1952, **74**, 5324). This aldehyde has been made by cleavage of 2-hydroxycyclohexanone with lead tetra-acetate in methanol, reduction with lithium tetrahydridoaluminate and hydrolysis of the acetal, and more recently from 1-benzyloxy-6,6-diethoxyethane *via* 2-ethoxyoxepane (*T. D. Perrine*, J. org. Chem., 1953, **18**, 1356).

2-Methoxy-5-methyl- and 2-methoxy-6-methyl-oxepane are made by the action of methanolic hydrogen chloride on the corresponding ε-hydroxyaldehyde, and their 2-methyl analogue is prepared similarly from the ε-hydroxyketone, 1-hydroxyheptan-6-one. Acid hydrolysis yields a mixture of oxepane and hydroxyaldehyde or ketone (*R. Criegee* and *W. Schnorrenberg*, Ann., 1948, **560**, 141; *B. Helferich* and *G. Sparmberg*, Ber., 1931, **64**, 104; *H. E. Eschinazi*, J. Org. Chem., 1961, **26**, 3072). 2-Acetoxy-2-methyloxepane, the product of ring-expansion of 1-hydroperoxy-1-methylcyclohexane, is hydrolysed similarly (*N. J. Leonard*, U.S.P., 2,594,322/1952; C.A., 1953, **47**, 1196). Other cyclohexane peroxides also show ring-expansion (*H. Kropf* and *C. R. Bernert*, Ann., 1971, **751**, 121).

Monoperphthalic acid oxidises 1,2-benzoquinones to **2,7-dihydro-oxepin-2,7-diones** (muconic anhydrides) XXIV, which are hydrolysed to *cis, trans*-muconic acids and form a transannular bond on irradiation, yielding bicyclic anhydride XXV (*P. Karrer et al.*, Helv., 1948, **31**, 1210; *R. P. Linstead et al.*, J. chem. Soc., 1951, 3398; *G. J. Fonken*, Chem. and Ind., 1961, 1575).

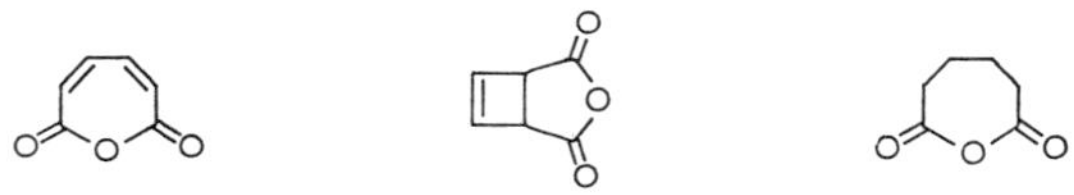

(XXIV) (XXV) (XXVI)

Oxepane-2,7-dione *(adipic anhydride)* (XXVI), m.p. 22°, b.p. 98–100°/0.1 mm, is formed by catalytic depolymerisation of the linear polymeric anhydride. The monomer repolymerises with a trace of water (*J. W. Hill*, J. Amer. chem. Soc., 1930, **52**, 4110).

A. Baeyer and *V. Villiger* (Ber., 1899, **32**, 3625; 1900, **33**, 858) converted six-membered cyclic ketones into seven-membered ring lactones by oxidation with persulphuric acid:

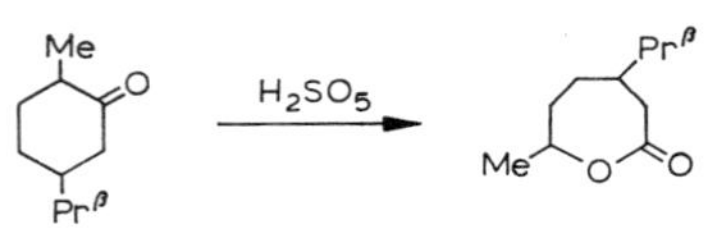

Other peracids have been used subsequently (*R. Huisgen* and *H. Ott*, Angew. Chem., 1958, 70, 312). This method has been used to synthesise steroidal oxepins and some ε-lactones of known absolute configuration (*C. G. Overberger* and *H. Kaye*, J. Amer. chem. Soc., 1967, **89**, 5640). It has been proposed that naturally occurring ε-lactones (limonoids, depsidones) are formed in a biological process analogous to this Baeyer–Villiger reaction.

Oxepan-2-one (*ε-caprolactone*) (XXVII; X = O), b.p. 98–100°/9 mm, has not been made by this method but by dehydration of ε-hydroxycaproic acid using benzenesulphonic acid in benzene (*M. Stoll* and *A. Rouve*, Helv., 1935, **18**, 1087). Seven-membered ring lactams have also been converted into ε-caprolactones by sodium nitrite in a mixture of acetic acid and acetic anhydride (*P. Bladon* and *W. McMeekin*, J. chem. Soc., 1961, 3504; *D. E. V. Ekong et al., ibid.*, Perkin I, 1972, 1943).

(XXVII) (XXVIII)

Cyclisation of 5-cyanopentanol with hydrogen chloride gives the iminolactone (XXVII; X = NR) hydrochloride (*H. Nohira et al.*, Bull. chem. Soc. Japan, 1965, **38**, 897). The pyrolysis products of the sodium salt of the tosylhydrazone (XXVII; X = N·NHTosyl), which include 2,3,4,5-tetrahydro-oxepin (XIX; R = H) suggest that carbene XXVIII is formed as an intermediate (*A. M. Foster* and *W. C. Agosta*, J. Amer. chem. Soc., 1972, **94**, 5777).

(*b*) *Bicyclic compounds including an oxepin ring*

Compounds are known in which a seven-membered ring containing one oxygen atom is fused to three-, four-, five- or six-membered rings and there are also spiro and bridged ring systems containing oxepin or reduced oxepin rings. The parent systems of the fully unsaturated 1-benzoxepin (*vide infra*) and 3-benzoxepin (XXX) series have been prepared but for the 2-benzoxepin system (XXXI), which no longer includes the benzene nucleus, only partially reduced compounds are known.

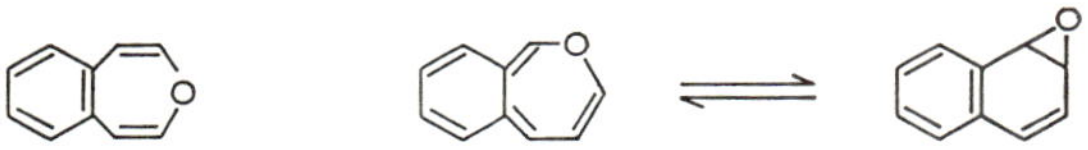

(XXX) (XXXI) (XXXIa)

A molecular orbital calculation suggests that 1- and 3-benzoxepin have approximately the same resonance energy and should be equally stable (*M. J. S. Dewar* and *N. Trinajstic*, Tetrahedron, 1970, **26**, 4269). However, investigation of the potentially tautomeric system XXXI ⇌ XXXIa) showed the equilibrium to be totally on the naphthalene oxide side and a rapid rearrangement to 1-naphthol occurred (*E. Vogel* and *F.-G. Klärner*, Angew. Chem. intern. Edn., 1968, **7**, 374).

1-Benzoxepin (XXIX), b.p. 101–102°/14 mm, 50°/0.5 mm, is obtained from 9,10-oxido-1,4,5,8,9,10-hexahydronaphthalene as shown below; 1,6-oxido[10]annulene (XXXII) (see p. 150) is formed in the same reaction (*F. Sondheimer* and *A. Shani*, J. Amer. chem. Soc., 1964, **86**, 3168; *Vogel et al.*, Angew. Chem. intern. Edn., 1964, **3**, 642).

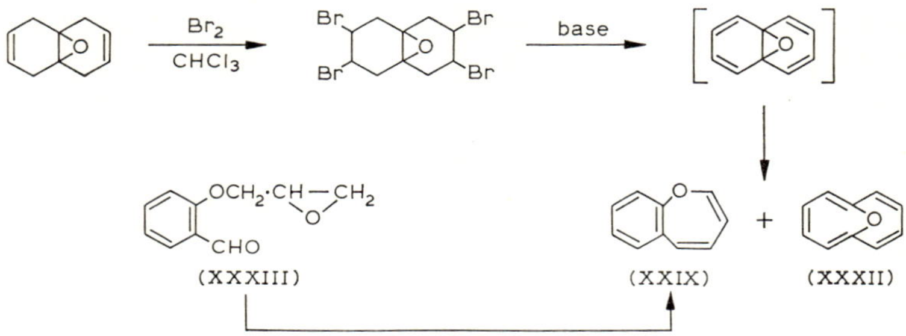

(XXXIII) (XXIX) (XXXII)

An alternative route involves an intramolecular Wittig reaction on the epoxyaldehyde XXXIII; only a 5% yield is obtained (*E. E. Schweizer et al.*, J. org. Chem., 1968, **33**, 2591).

Treatment of the enamine XXXIV with dimethyl acetylenedicarboxylate gave a tricyclic adduct which isomerised on heating to 100° to the 1-benzoxepin (XXXV; R = *N*-pyrrolidinyl). The amine is cleaved with acid leaving dimethyl 5-hydroxy-l-benzoxepin-3,4-dicarboxylate (XXXV, R = OH) (*D. N. Reinhoudt* and *C. G. Kouwenhoven*, Tetrahedron Letters, 1972, 5203).

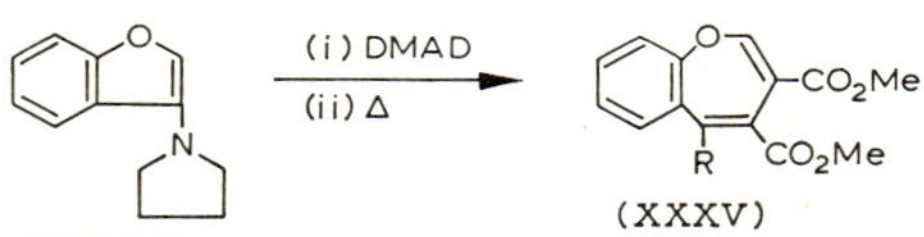

(XXXIV) (XXXV)

4-Phenyl-1-benzoxepin-3,5-dione (XXXVII), formed by treating epoxide XXXVI with acid, exists in the diketo not the dienol form, but acetylation gives 3,5-diacetoxy-4-phenyl-1-benzoxepin (XXXVIII). More vigorous treatment with acetic anhydride results in ring-contraction to 3-phenyl-1,2,4-triacetoxynaphthalene (*H. Hofman* and *H. Westernacher*, Angew. Chem. intern. Edn., 1965, **4**, 872; Chem. Ber., 1969, **102**, 205):

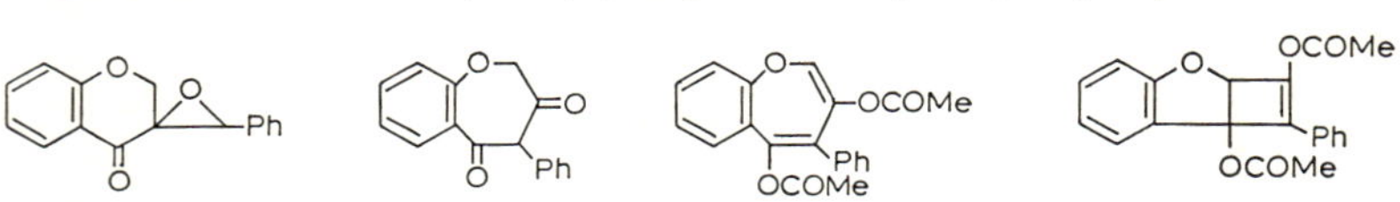

(XXXVI) (XXXVII) (XXXVIII) (XXXIX)

Catalytic reduction of 1-benzoxepins gives 2,3,4,5-tetrahydro-1-benzoxepins, and irradiation of XXXVIII gives a transannular bond formation to tricyclic XXXIX which is not reversed by heating (*H.* and *P. Hofmann*, Tetrahedron Letters, 1971, 4055).

2,3-Dihydro-1-benzoxepins XLII are prepared by dehydrating 5-hydroxytetrahydro-1-benzoxepins XLI made from the corresponding tetra-

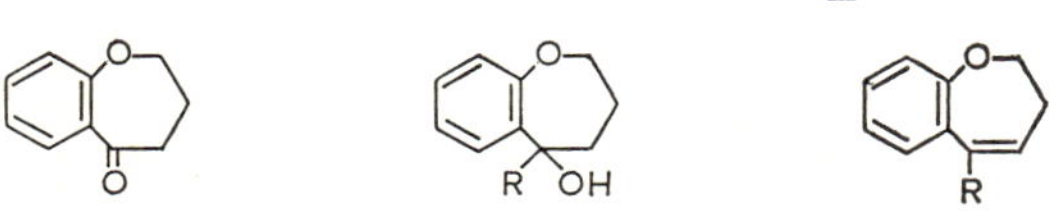

(XL) (XLI) (XLII)

hydro-1-benzoxepin-5-ones XL (*O. Dann* and *W. D. Arndt*, Ann., 1954, **587**, 38; *V. J. Traynelis et al.*, J. org. Chem., 1962, **27**, 2377; *G. Fontaine* and *P. Maitte*, Compt. rend., 1964, **258**, 4583; Ann. chim., Paris, 1968, **3**, 179). Less effective routes to XLII using a Wittig reaction for ring closure are also reported (*Schweizer et al*, Tetrahedron Letters, 1963, 979; J. org. Chem., 1966, **31**, 2907; 1968, **33**, 336; 1969, **34**, 207) and a high yielding route to dihydrobenzoxepin XLIII is as shown (*Fontaine et al.*, Compt. rend, 1968, **267**, 846):

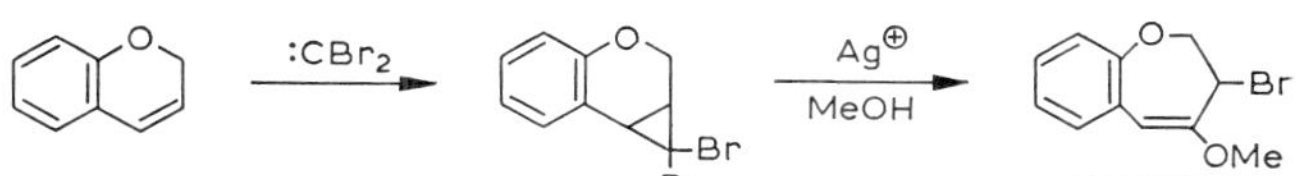

(XLIII)

2,3-Dihydro-1-benzoxepin (XLII; R = H), b.p. 113–114°, n_D^{23} 1.5877 (1.5926), gives 2,3,4,5-tetrahydro-1-benzoxepin on catalytic reduction, the 4,5-dihydroxy compound on oxidation with permanganate and tetrahydro-1-benzoxepin-4-one (XLIV) with copper sulphate (*ibid.*, Ann. chim., Paris, 1968, **3**, 469). Bromine also adds to the 4,5-double bond. In refluxing alcoholic ethoxide, compound XLII (R = H) is converted into 2*H*-2-methylchromene (*Schweizer*, Tetrahedron Letters., 1963, 979).

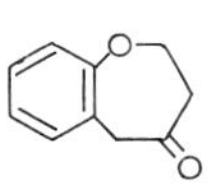

(XLIV)

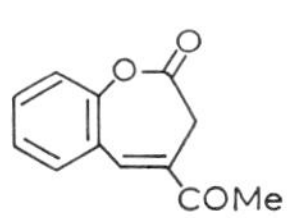

(XLV)

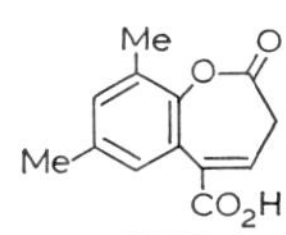

(XLVI)

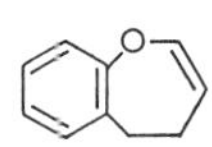

(XLVII)

2,3-Dihydro-1-benzoxepin-2-ones XLV and XLVI have been made respectively by condensing salicylaldehyde with laevulic acid (*R. N. Sen* and *B. C. Roy*, J. Indian chem. Soc., 1930, **7**, 401), and 2,4-dimethylphenol with diethyl 2-oxoglutarate (*R. V. Smith* and *M. D. Bealor*, J. org. Chem., 1962, **27**, 3092). Catalytic hydrogenation reduces the double bonds in the oxepin rings.

Irradiation of isopentenylquinones gives 2,3-dimethylchromenes, but also 2,5-**dihydro-1-benzoxepins** (*D. Creed et al.*, J. Amer. chem. Soc., 1971, **93**, 502):

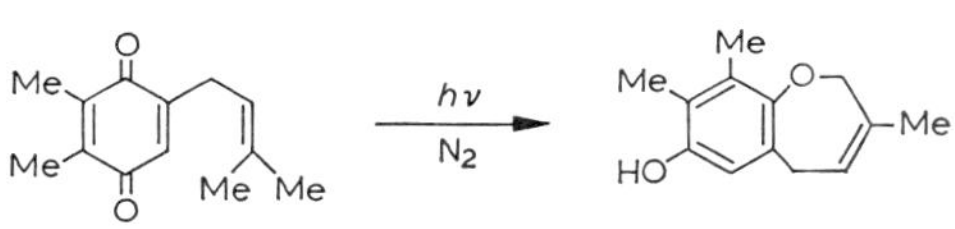

4,5-Dihydro-1-benzoxepin (XLVII) is formed by heating 2-acetoxy-2,3,4,5-tetrahydro-1-benzoxepin, which is the product of photo-oxidation of tetralin in acetic anhydride (*Fontaine*, Ann. chim. Paris, 1968, **3**, 469).

2,3,4,5-**Tetrahydro-1-benzoxepin** *(homochroman)* (XLVIII; R, R^1 = H), m.p. 28–30°; b.p. 86–88°/7 mm, 102°/18 mm; n_D^{21} 1.5398, and its 7-*methyl* homologue (XLVIII; R = Me, R^1 = H), b.p. 109°/11 mm, $n_D^{14.5}$ 1.5379, have been known since 1949. They are obtained by cyclisation of 4(2′-methoxyaryl)butan-1-ol with hydrobromic acid (*P. Cagniant*, Compt. rend., 1949, **229**, 889) or a 1-bromo-4-(2′-hydroxyaryl)butane with base (*M. S. Newman* and *A. B. Meker*, J. org. Chem., 1961, **26**, 336). High dilution technique improves the yield (*G. Baddeley et al.*, J. chem. Soc. 1956, 2455).

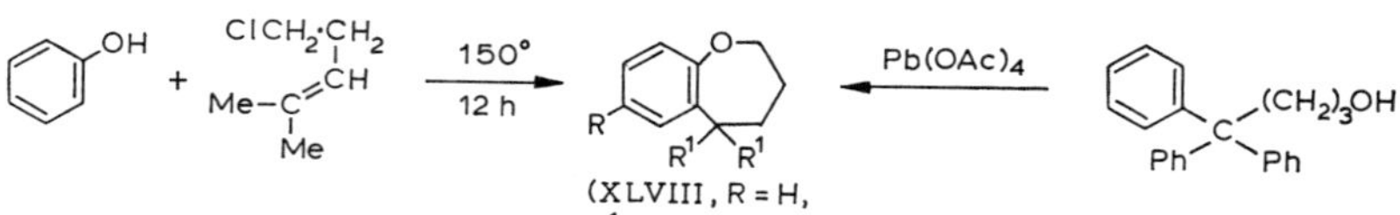

(XLVIII, R = H,
R^1 = Me, Ph)

Homochroman undergoes Friedel–Crafts acetylation at position 7 and the acetyl group may be oxidised with hypochlorite, esterified, reduced with lithium tetrahydridoaluminate and treated with gaseous hydrogen chloride without disruption of the oxepin ring. It does not react with either bromine or permanganate. Direct routes to the substituted homochromans (XLVIII; R = H, R^1 = Me) (*H. Hart et al.*, J. Amer. chem. Soc., 1962, **84**, 1740; 1963, **85**, 3269) and XLVIII (R = H, R^1 = Ph) (*W. H. Starnes*, J. org. Chem., 1968, **33**, 2767) are shown. In XLVIII (R = H, R^1 = Me), the delocalisation of the oxygen lone pairs is prohibited by the twisting of the ring (*E. M. Aarnett* and *C. Y. Wu*, Chem. and Ind., London, 1959, 1488).

An interesting five-membered to seven-membered ring expansion using sodium alkoxides has been used to convert lactone XLIX to 5-phenyl-2,3,4,5-tetrahydro-1-benzoxepin-5-carboxylic acid (XLVIII, R = H, R^1 = Ph, CO_2H) and its esters. Further transformations converted the acid group to amine and aminomethyl groups (*H. E. Zaugg et al.*, J. org. Chem., 1961, **26**, 4821, 4828; Belg. P., 618,072; 618,528 and 618,529/1962; C.A., 1963, **59**, 3897, 6372, 7500; (U.S.P., 3,122,551; 3,156,688/1962; C.A., 1964, **61**, 1844; 1965, **62**, 2764).

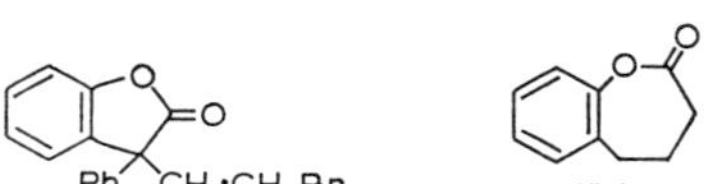

(XLIX) (L)

2,3,4,5-**Tetrahydro-1-benzoxepin-2-one** (L), m.p. 130°, may be made from 1-tetralone by oxidation with persulphate and thermal cyclisation of the intermediate, 4(2′-hydroxyphenyl)butyric acid. It is chlorinated at position 3 (*G. Schroeter*, G.P., 562,827/1928; C.A., 1933, **27**, 1224). A Baeyer–Villiger oxidative rearrangement using perbenzoic acid in chloroform, has also been used (*O. Jeger et al.*, Helv., 1954, 37, 2295).

2,3,4,5-**Tetrahydro-1-benzoxepin-3-one** (LI), b.p. 100–105°/0.4–0.6 mm, is made by a Dieckmann cyclisation at high dilution. It yields the 3-hydroxy compound on catalytic reduction and the 3-amino compound on reductive amination (*I. M. Lockhart et al.*, J. chem. Soc. Perkin I, 1972, 2425):

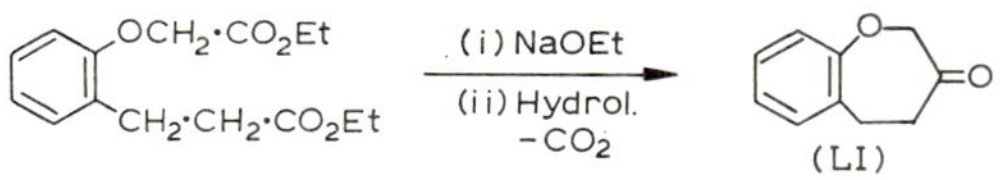

(LI)

The cyclisation of 4-phenoxybutyric acids to 2,3,4,5-tetrahydro-1-benzoxepin-5-ones may be effected by phosphorus pentoxide (*S. G. Powell* and *L. Anderson*, J. Amer. chem. Soc., 1931, **53**, 811), polyphosphoric acid or liquid hydrogen fluoride (*Dann* and *Arndt*, *loc.cit.*) and stannic chloride (*Fontaine* and *Maitte*, *loc.cit.*); *para*-linked polymers may be formed also. Many compounds with methyl-, methoxy- and chloro- substituents in the benzene ring have been prepared this way. The parent compound, 2,3,4,5-**tetrahydro-1-benzoxepin-5-one** (XL), has b.p. 106°±2 mm, and n_D^{20} 1.564. It forms an *oxime*, m.p. 99°, and a *semicarbazone*, m.p. 228–229°.

The 5-oxo group behaves as a typical ketone. It is reduced to methylene by zinc and hydrochloric acid, and to a secondary alcohol with aluminium isopropoxide. It reacts with Grignard reagents and gives a 5,5-dichloro derivative with phosphorus pentachloride. It also takes part in reductive amination, acetal formation, the Reformatsky and Wittig reactions and gives the chlorovinylaldehyde LII with the Vilsmeier reagent, phosphoryl chloride and dimethylformamide (*M. Weissenfels et al.*, Z. Chem. 1966, **6**, 471).

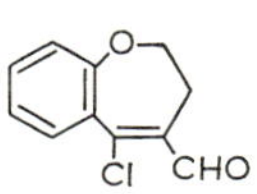

(LII)

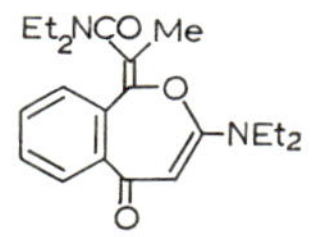

(LIII)

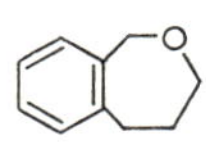

(LIV)

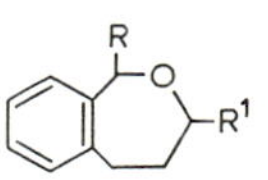

(LV)

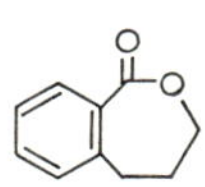

(LVI)

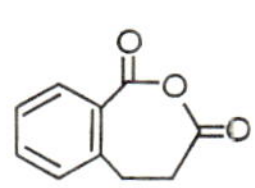

(LVII)

Mono- and di-bromination and Mannich condensation take place at the 4-position, and the 4-bromo atom may be substituted by hydroxy or methoxy groups. Benzaldehyde, selenium dioxide and amyl nitrite also attack the 4-position to give benzal, oxo and oximino derivatives, respectively.

The most highly unsaturated 2-benzoxepin known is the 1,5-**dihydro**-2-**benzoxedin**-5-**one** (LIII) obtained from the reaction of phthalic anhydride with 2 moles of the ynamine, 1-diethylaminopropyne. On heating, LIII isomerises to an indanone, and on acid hydrolysis an indenone is formed (*G. Höfle* and *W. Steglich*, Chem. Ber., 1972, **105**, 1368).

Other members of the 2-benzoxepin series are tetrahydro derivatives. 1,3,4,5-**Tetrahydro-2-benzoxepin** (LIV), m.p. 54–56°, b.p. 95–99°/0.01 mm, is synthesised by an intramolecular Friedel–Crafts reaction on chloromethyl 3-phenylpropyl ether (*A. Rieche* and *H. Gross*, *ibid.*, 1962, **95**, 91). 4-Phenyl and 4-methyl derivatives are also known (*H. W. Bersch et al.*, Arch. Pharm., 1958, **291**, 82).

The ether ring is opened by hydrogen bromide, catalytic hydrogenation and the oxidising agents, alkaline permanganate, nitric acid, chromium trioxide and selenium dioxide. The alternative cyclic hemiacetals (LV; R^1 = H, R = OH) and (LV; R^1 = OH, R = H) appear to be the respective intermediates with the last two reagents (*K. Dimroth* and *H. Freyschlag*, Chem. Ber., 1957, **90**, 1623; *C. Normant*, Compt. rend, 1967, **264**, 1665).

Photo oxidation gives the 1-lactone LVI and *N*-bromosuccinimide also brominates the 1-position. This bromo-group is replaced by diethylamine, sodium ethoxide and ethylmagnesium bromide. The anhydride, **1,3,4,5-tetrahydro-2-benzoxepin-1,3-dione** (LVII), m.p. 152°, is a product of the oxidation of 1,2-naphthoquinone with peracids (*J. Boeseken* and *G. Slooff*, Rec. Trav. chim., 1930, **49**, 100; *P. Karrer* and *L. Schneier*, Helv., 1947, **30**, 859).

3-**Benzoxepin** (XXX), m.p. 83–84°, *picrate*, m.p. 125–127°, was first prepared from *o*-phthalaldehyde by a double Wittig reaction (*Dimroth et al.*, Angew. Chem., 1961, **73**, 436; Chem. Ber., 1966, **99**, 634) and a small yield was obtained by irradiating 1,4-dihydronaphthalene 1,4-oxide (*H. Prinzbach et al.*, Helv., 1968, **51**, 911; *G. R. Ziegler* and *G. S. Hammond*, J. Amer. chem. Soc., 1968, **90**, 513; 1969, **91**, 446):

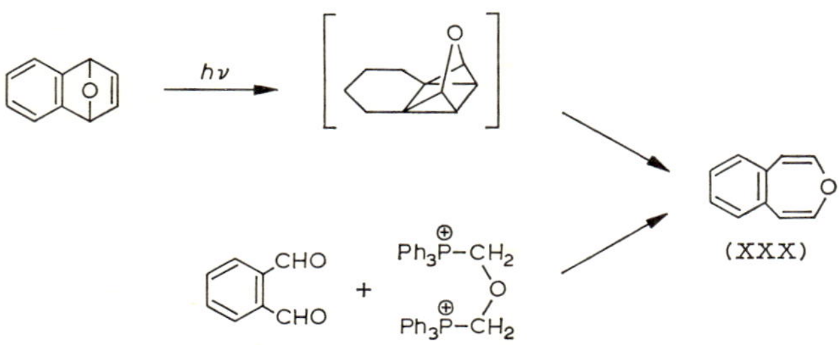

Pyrolysis of 1,2-dihydronaphthalene-2-hydroperoxide also yields XXX (*A. M. Jeffrey* and *D. M. Jerina*, ibid., 1972, **94**, 4048).

3-Benzoxepin is unaffected by dilute acid but in concentrated hydrochloric acid it undergoes ring contraction to indene-1-carbaldehyde. Catalytic reduction hydrogenates both double bonds of the seven-membered ring, yielding **1,2,4,5-tetrahydro-3-benzoxepin,** b.p. 103°/12 mm, n_D 1.5450.

3-*Benzoxepin*-2,4-*dicarboxylic acid* (LVIII; X = CO_2H), m.p. 258–260°, *dimethyl ester*, m.p. 108–111°, is made by the condensation of *o*-phthalaldehyde with dimethyl diglycollate, $O(CH_2{\cdot}CO_2Me)_2$, in the presence of base, followed by basic hydrolysis (*Dimroth* and

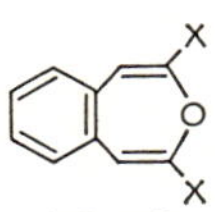

(LVIII)

Freyschlag, Chem. Ber., 1957, **90**, 1623; Angew. Chem., 1957, **69**, 95; *R. Huisgen et al.*, Ann., 1960, **630**, 128; *M. J. Jorgenson*, J. org. Chem., 1962, **27**, 3224). Diazomethane regenerates the methyl ester, and the ethyl ester is obtained from the dicarbonyl chloride.

Both acid and esters have remarkable stability towards strong acids, not even showing deuterium exchange in the rings, and the 1,2-dibromide was made only under forcing conditions. Catalytic reduction of the ester gives a 1,2-dihydro product but the acid is further reduced to the 1,2,4,5-tetrahydro compound. Lithium tetrahydridoaluminate reduces LVIII (X = CO_2Me) to the dialcohol LVIII (X = CH_2OH) which decomposes readily in ethanolic hydrogen chloride, and is oxidised to the monoaldehyde LVIII (X = CH_2OH, CHO) with manganese dioxide.

Heating the dicarboxylic acid (LVIII; X = CO_2H) results not in decarboxylation but in ring contraction, probably *via* a naphthalene oxide tautomer. At 300°, 2-hydroxynaphthoic acid is obtained, but on distillation from copper powder, the product is naphthalene-2,3-dicarboxylic acid (*Huisgen et al., loc.cit.*; Chem. Ber., 1958, **91**, 18).

The dicarbaldehyde (LVIII; X = CHO) prepared by oxidising LVIII (X = CH_2OH) with manganese dioxide reacts with dimethyl ether α,α'-bis(triphenylphosphonium bromide), and its sulphur analogue to give the oxygen bridged benzoxa- and benzothia-[11]-annulenes (LIX; X = O or S); neither shows a paramagnetic ring current (*H. Ogawa et al.*, Tetrahedron Letters, 1971, 1793; 1972, 4129).

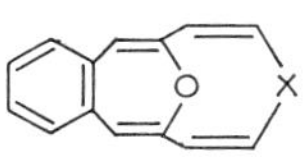

(LIX)

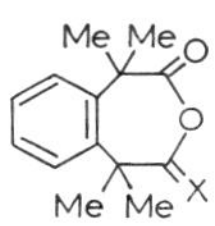

(LX)

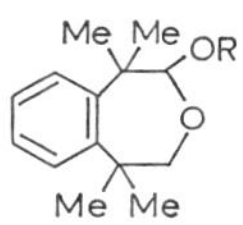

(LXI)

A number of derivatives of 1,1,5,5-tetramethyl-1,2,4,5-tetrahydro-3-benzoxepin have been prepared from *o*-phenylenediisobutyraldehyde (*A. W. Burgstahler* and *H. O. Abdel-Rahman*, J. Amer. chem. Soc., 1963, **85**, 173). Potassium hydroxide brings about cyclisation to the lactone LX (X = H_2), and lithium tetrahydridoaluminate at –40° yields the 2-hydroxy derivative LXI (R = H); the latter is converted into the methyl ether LXI (R = Me) in methanolic acid. The anhydride LX (X = O) results when *o*-phenylenediisobutyric acid is refluxed with acetic anhydride, or oxidised with periodic acid or lead tetra-acetate (*L. R. C. Barclay et al.*, Canad. J. chem., 1962, **40**, 1664); it is reduced to lactone LX (X = H_2) with sodium amalgam.

Interesting examples of two-ring oxepin systems in which the additional ring is not benzene are known. Oxepane *cis*-fused to a cyclopropane ring across the 3,4-positions (LXII) is prepared as shown:

(LXII)

and its isomer, LXIII, is obtained by reducing the dichlorocarbene adduct of 2,3,6,7-tetrahydro-oxepin with sodium in liquid ammonia. Formic acid opens the oxepane ring of LXII, but with LXIII it is the cyclopropane ring that opens to give formates LXIV and LXV (*L. A. Paquette* and *M. K. Scott*, J. Amer. chem. Soc., 1972, **94**, 6751):

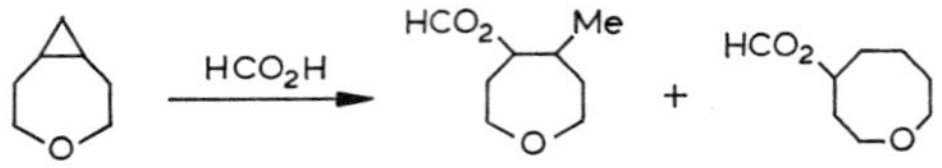

(LXIII) (LXIV) (LXV)

The mould metabolite **carolic acid** (LXVI) is a furo[3,4-*b*]oxepin (*H. Raistrick et al.*, Biochem. J., 1935, **29**, 300; 1947, **41**, 569; *J. R. Plimmer*, J. org. Chem., 1964, **29**, 511); it has recently been synthesised (*R. Sudo et al.*, *ibid.*, 1967, **32**, 1844).

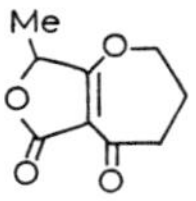

(LXVI)

11-Oxabicyclo[4.4.1]undecanes having an oxygen function at position 1 (LXVII) have been made by the rearrangement of decalin-9-peroxide alone (gives LXVII; R = H) and in the presence of acetic anhydride (R = *Ac*; m.p. 104–105°) or aroyl chlorides (R = ArCO) (*R. Criegee et al.*, Ber., 1944, **77**, 22; 722; Ann., 1948, **560**, 127; 135; 141; Fortschr. chem. Forsch., 1950, **1**, 508). This rearrangement has been extensively studied (*P. D. Bartlett et al.*, Rec. Chem. Prog., 1950, **11**, 47; J. Amer. chem. Soc., 1953, **75**, 5591; *H. L. Goering* and *A. C. Olson*, *ibid.*, 5853; *D. B. Denny et al.*, *ibid.*, 1955, **77**, 1706; 1957, **79**, 4806; 1961 **83**, 1726).

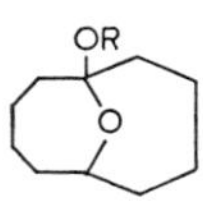

(LXVII)

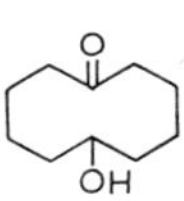

(LXVIII)

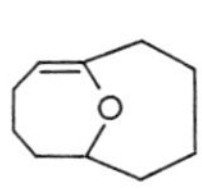

(LXIX)

With warm alkali, bicyclic compounds LXVII give 6-hydroxycyclodecanone (LXVIII) which appears to be in equilibrium with LXVII (R = H) (*W. J. Mijs et al.*, Rec. Trav. chim., 1968, **87**, 580) and which reverts to compound LXVII (R = Me) in methanolic hydrochloric acid.

11-**Oxabicyclo**[4.4.1]**undecane,** b.p. 86°/8 mm, m.p. 22–28°, can be obtained by hydrogenation of 6-hydroxycyclodecanone (*C. W. Theobald*, U.S.P., 2,544,737/1950; C.A., 1951, **45**, 8553) or the enol ether LXIX, which is an alternative product of rearrangement of decalin-9-peroxide (*Bartlett et al.*, J. Amer. chem. Soc., 1953, **75**, 5591; C.A., 1964, **61**, 9414). The preparation of 1,6-oxido[10]annulene (XXXII), the fully unsaturated parent of the 11-oxabicyclo[4.4.1]undecanes, has been described earlier (p. 149). It is relatively inert towards heat, light, oxygen and bases but with acids it rearranges to a mixture of 1-benzoxepin, 1- and 2-naphthols (*F. Sondheimer* and *A. Shani*, J. Amer. chem. Soc., 1964, **86**, 3168; 1967, **89**, 6310; *Vogel et al.*, Angew. Chem. intern. Edn., 1964, **3**, 642; 1966, **5**, 734; 1967, **6**, 385). Catalytic reduction or treatment with lithium tetrahydridoaluminate or triphenylphosphine yields naphthalene. The "aromaticity" of XXXIII is shown chemically by the formation of 1- and 2-nitro derivatives with cupric nitrate in acetic anhydride; Friedel–Crafts and bromination reactions both give naphthalene derivatives.

(c) *Oxepins with two or more annellated ring systems*

Two benzene rings may be fused to the oxepin nucleus in five ways, and examples from each class are known.

Dibenz[*b*,*d*]oxepins LXX are rare (*A. Sieglitz* and *H. Koch*, Ber., 1925, **58**, 78).

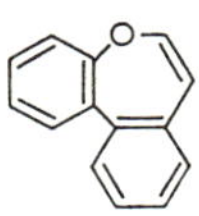
(LXX)

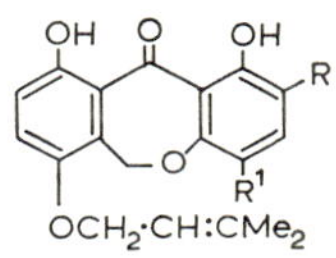

(LXXI)

Natural products in arugosin are found to have 6,11-dihydrodibenz[*b*,*e*]oxepin-11-one structures (LXXI; R, R^1 = H, CH_2·CH·CMe_2) (*J. A. Ballantine et al.*, J. chem. Soc, C, 1970, 1175). In the laboratory, 6,11-*dihydrodibenz*[b,e]*oxepin*-11-*one* (LXXII; X = O, Y = H_2), m.p. 71–72°, b.p. 142–145°/0.2 mm (*K. Stach et al.*, Angew. Chem. intern. Edn., 1962, **1**, 50; 752; Monatsh., 1962, **93**, 883, 896; *Stach* and *F. Beckelhaupt*, C.A., 1965, **62**, 7726; *S. O. Winthrop et al.*, J. med. pharm. Chem., 1962, **5**, 1207; *Böchringer* und *Söhne*, *Belg.* P., 23,259/1961; C.A., 1964, **60**, 10659; *B. M. Bloom* and *J. B. Tretter*, Belg. P., 641,498/1962; C.A., 1966, **64**, 719), -6-*one* (LXXII; X = H_2, Y = 0), m.p. 98–99°, (*W. Baker et al.*, J. chem. Soc., 1952, 1452; *M. Hubacher*, J. org. Chem., 1958, **23**, 1400) and -6,11-*dione* (LXXII; X, Y = O), m.p. 84–85° or 179–180°, (*Baker et al.*, *loc.cit.*; *M. Lamchen*, J. chem. Soc., 1962, 4695; *W. R. Orndorff* and *E. Kline*, J. Amer. chem. Soc., 1924, **46**, 2276) and many substituted derivatives are prepared by cyclodehydration of *o*-(phenoxymethyl)-, *o*-(2-hydroxybenzyl)- and *o*-(2-hydroxybenzoyl)-benzoic acids, respectively. With the last of these materials, dimeric condensation products (lactides) with a fourteen-membered ring containing two oxygen atoms are also formed. Cyclising reagents include anhydrides, phosphorus reagents, thionyl chloride and liquid hydrogen fluoride while for compound LXXII (X = O, Y = H_2), aluminium chloride has been used with the acid chloride. Compound LXXII (X, Y = O) has also been made by the ring expansion of 3-phenylphthalan (LXXIII) (*M. G. Sarngadharan* and *T. R. Seshadri*, Indian J. Chem.,

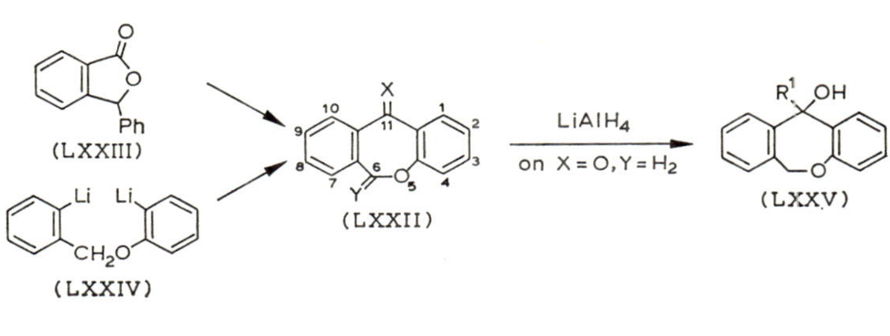

1966, **4**, 503) and an ingenious route to dibenz[*b,e*]oxepins (LXXII; X = CHR, Y = H_2) with an exocyclic double bond rather than an oxo group at position 11 is the condensation of an ester with the dilithium compound LXXIV (*C. F. Böhringer und Söhne*, B.P., 1,004,683/1963; C.A., 1966, **64**, 8158).

The oxepin rings of LXXII (X = H_2, Y = O) open very easily under acid or basic conditions to hydroxy acids or their derivatives (*Baker et al., loc.cit.*). The 11-oxo group of LXXII (X = O, Y = H_2) is reduced with lithium tetrahydridoaluminate or sodium tetrahydridoborate giving the secondary alcohol LXXV (R^1 = H) (*M. A. Davis* and *S. O. Winthrop*, U.S.P., 3,234,235/1962; C.A., 1966, **64**, 12654), and Grignard reagents give the corresponding tertiary alcohols, which may be dehydrated to a mixture of *cis*- and *trans*-11-alkylidene derivatives (LXXII; X = CHR, Y = H_2) by hot hydrochloric acid (*Stach et al., Winthrop et al., Bloom* and *Tretter*, loci cit.). If three equivalents of Grignard reagent are used on LXXII (X = O, Y = H_2), ring contraction to a 9,10-dihydroanthracene-9,10-diol occurs (*Winthrop*, Tetrahedron Letters, 1963, 1113), and ring-contraction to indene derivatives occurs when compounds LXXII (X = CHR, Y = H_2) are treated with hydrobromic acid in acetic acid (*Stach et al.*, Monatsh., 1964, **95**, 485). The hydroxy group of LXXV may be replaced by chloro on treatment with hydrogen chloride, and this chloro group subsequently replaced by amines (*M. Protiva, I. Jirkovsky* and *I. Melys*, Czech. P., 123,846/1966; C.A., 1968, 69, 43934; *P. Yonan*, U.S.P., 3,401,165/1965; C.A., 1969, **70**, 4144).

Discovery that the alkaloid cularine (LXXVI) contains the dibenz[*b,f*]oxepin system revealed by Hofmann degradation (*R. H. F. Manske*, J. Amer. chem. Soc., 1950, **72**, 55) prompted further investigations.

The most general route to **dibenz[*b,f*]oxepins** is *via* 10,11-dihydrodibenz[*b,f*]oxepin-10-ones LXXVII. A great many of these ketones are synthesised by the cyclisation of *o*-phenoxyphenylacetic acids using liquid hydrogen fluoride (*M. Kulka* and *Manske, ibid.*, 1953, **75**, 1322), polyphosphoric acid (*J. D. Loudon* and *L. A. Summers*, J. chem. Soc., 1957, 3809; *T. Kametani et al.*, Yakugaku Zasshi, 1962, **82**, 1307; 1964, **84**, 642; C.A., 1963, **58**, 13913; 1964, **61**, 13278; *J. R. Geigy, A. G.*, Neth. Appl. 6,508,284/1965; 1966, **64**, 19574; J. chem. Soc. 1968, 2877) or *via* the acid chlorides

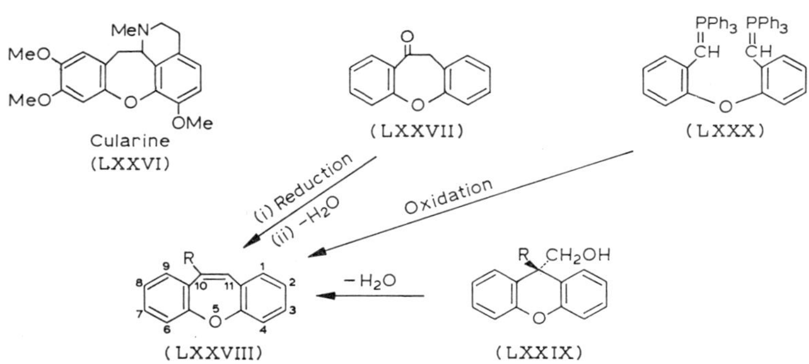

(*Manske* and *A. E. Ledingham*, J. Amer. chem. Soc., 1950, **72**, 4797; *S. Kimoto et al.*, J. pharm. Soc. Japan, 1954, **74**, 1041; C.A., 1955, **49**, 5373).

The corresponding phenoxyphenylpyruvic acids yield the unsaturated acid LXXVIII (R = CO_2H). This route is not available for nitrogen analogues as formation of a five-membered nitrogen ring is favoured. Steric factors seem to determine the direction of ring closure. The parent ketone, 10,11-*dihydrodibenz[b,f]oxepin*-10-*one*, (LXXVII), m.p. 56°, b.p. 162–165°/3 mm, *oxime*, m.p. 135–137°, and many derivatives have been converted to *dibenz[b,f]oxepin* (LXXVIII; R = H), m.p. 111°, and its analogues by reduction to 10,11-dihydrodibenz[*b,f*]oxepin-10-ol followed by dehydration (*Manske* and *Ledingham*, *loc.cit.*; *S. N. Alam* and *D. B. Mclean*, Canad. J. Chem., 1965, **43**, 3433).

An alternative route to LXXVIII is the ring expansion of xanthene derivatives (*C. L. Zirkle*, U.S.P., 3,100,207/1963; C.A., 1964, **60**, 1719). The alcohol LXXIX (R = H) yields LXXVIII (R = H) on treatment with phosphorus pentoxide (*F. A. L. Anet* and *P. M. G. Bavin*, Canad. J. chem., 1957, **35**, 1084; *V. Boekelheide et al.*, J. Amer. chem. Soc., 1967, **89**, 2746) and the compounds LXXVIII (R = *Me*), m.p. 57–58° and LXXVIII (R = *Ph*), m.p. 103°, have been made similarly (*E. D. Bergmann* and *M. Rabinowitz*, Israel J. Chem., 1963, **1**, 125; C.A., 1964, **60**, 7997). Dibenz[*b,f*]oxepin also results when xanthylium perchlorate reacts with diazomethane; if ethyl diazoacetate is used, the *ester* LXXVIII (R = CO_2Et), m.p. 72–73°, is obtained (*H. W. Whitlock et al.*, Tetrahedron Letters, 1961, 593; J. org. Chem., 1972, **37**, 2161). An exocyclic methylene carbonium ion is a probable intermediate in all these reactions.

Another direct route to LXXVIII (R = H) is the oxidation of the bis-triphenylphosphorylidene compound LXXX (*H. J. Bestmann et al.*, Angew. Chem. intern. Edn., 1964, **3**, 226) and condensation of salicylaldehyde trinitrotoluene gives 1,3-dinitrodibenz[*b,f*]oxepin, in which one nitro group has been diplaced, possibly *via* a spiro intermediate (*H. Hoyer* and *M. Vogel*, Monatsh., 1962, **93**, 766).

The oxepin ring of dibenz[*b,f*]oxepin undergoes both addition and substitution reactions. Nitration of LXXVIII (R = H) yielded 10-*nitrodibenz*[b,f]*oxepin* (LXXVIII; R = NO_2), m.p. 102–103°, which was reduced by hydrazine and palladium to the oxime of ketone LXXVII. Osmium tetroxide converted LXXVIII (R = H) to 10,11-*dihydroxy*-10,11-*dihydrodibenz*[b,f]*oxepin* (*Loudon* and *Summers*, *loc.cit.*) and bromination gave the 10,11-dibromide which could be dehydrobrominated to 10-*bromodibenz*[b,f]*oxepin* (LXXVIII; R = Br), b.p. 205–210°/10 mm, (*D. W. Jones et al.*, J. heterocycl. Chem., 1965, **5**, 327). This monobromide undergoes substitution by cyanide, giving the *nitrile* LXXVIII (R = CN), m.p. 159–160°, and eliminates a further molecule of hydrogen bromide to give a benzyne-like intermediate (*W. Tochtermann et al.*, Ann., 1967, **701**, 117).

Catalytic hydrogenation of dibenz[*b,f*]oxepin gives 10,11-**dihydrodibenz[*b,f*]oxepin** (LXXXI; R = H), m.p. 154°, b.p. 100–103°/0.2 mm (*Boekelheide, loc.cit.*) which is prepared directly by the action of phenyl-lithium on bis(2-bromomethylphenyl) ether (*Bergmann et al.*, Tetrahedron Letters, 1969, 2007).

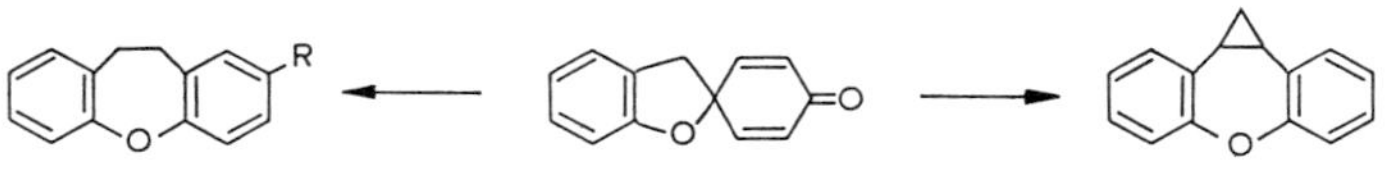

(LXXXI) (LXXXII) (LXXXIII)

The ring expansion of spiro compound LXXXII, which gives LXXXI (R = OAc), shows exclusive aralkyl migration despite the possibility of aryloxy migration (*K. Schofield et al.*, J. chem. Soc., C, 1970, 2543). Lithium tetrahydridoaluminate in tetrahydrofuran reduced both the 10,11-double bond and the ester group of LXXXVIII (R = CO_2Et) but in refluxing dioxane the product was the tetracyclic compound LXXIII, also obtained from LXXVIII (R = H) and iodomethylzinc iodide (*H. S. Kasmai* and *Whitlock*, J. org. Chem., 1972, **37**, 2161). The oxepin ring of 10,11-dihydrodibenz[*b,f*]oxepin is opened by sodium in liquid ammonia (*Manske, loc.cit.*).

Oxidation of LXXVIII with permanganate gives diphenyl ether-2,2′-dicarboxylic acid (*Manske* and *Ledingham, loc.cit.*), but oxidation of ketone LXXVII with selenium dioxide yields 10,11-*dihydrodibenzo* [b,f]*oxepin*-10,11-*dione*, m.p. 118–119°, (*F. Mathys et al.*, Helv., 1956, **39**, 1095; *A. H. Rees*, Chem. and Ind., 1957, 76). This dione undergoes a benzilic acid rearrangement to a xanthene with base or sodium sulphite.

The 10-oxo group of LXXVII shows normal ketonic behaviour with primary amino compounds (*Kametani* and *K. Fukumoto*, J. chem. Soc., 1963, 4289; *K. Kawai*, Nippon Yakurigaku Zasshi, 1960, **56**, 971; C.A., 1961, **55**, 21374) and Grignard reagents (*J. R. Geigy A.G.*, Neth. Appl., 6,574,187/1966; C.A., 1966, **65**, 12183). Position 11 may be alkylated (*idem*, Neth. Appl., 6,508,284/1965; 6,515,702/1966; C.A., 1966, **64**, 19574; 1966, **65**, 13672), brominated (*idem*, Neth. Appl., 6,605,741/1966; C.A., 1967, **66**, 65507) and converted into the hydroxyimino derivative (*Kawai, loc.cit.*).

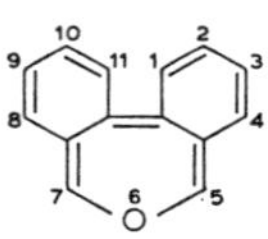

(LXXXIV)

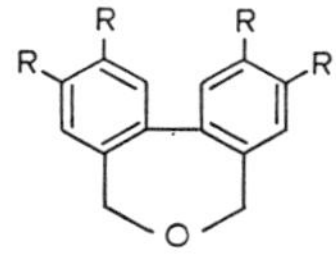

(LXXXV)

Fully unsaturated dibenz[*c,e*]oxepins (LXXXIV) in which both annelated rings would assume the quinonoid form are not known, but many of their 5,7-dihydro derivatives (LXXXV), the diphenanes, have been synthesised. A 2,3-dioxole derivative is found in the degradation of the alkaloid tazzetine but is not present in the natural product (*T. Ikeda et al.*, J. chem. Soc., 1956, 4749).

The most general route is the dehydration of 2,2′-bis(hydroxymethyl)-biphenyl or its substituted derivatives, using acetic acid (*G. Wittig et al.*,

Ber., 1931, **64,** 2395; Ann., 1933, **505**, 17), sulphuric acid (*E. E. Turner et al.*, J. chem. Soc., 1951, 3072; 1952, 854; 1954, 131; 1956, 3475; *D. M. Hall* and *F. Minhaj, ibid.*, 1957, 4584; *G. M. Bennett* and *R. L. Wain, ibid.*, 1936, 114; *M. Oki et al.*, Bull. chem. Soc. Japan, 1963, **36**, 1542; 1964, **37**, 1865; *E. Matarasso-Tchiroukhine*, Compt. rend. 1957, **244**, 476; Ann. chim. Paris, 1958, **3**, 405), hydrobromic acid (*A. J. Weinheimer et al.*, J. org. Chem., 1953, **18**, 801) and toluenesulphonic acid (*Wittig et al.*, Chem. Ber., 1951, **84**, 627). *p*-Toluenesulphonyl chloride in pyridine is also effective if acid conditions are to be avoided (*N. K. Kochetkov et al.*, Tetrahedron Letters, 1961, 730; Izvest. Akad. Nauk S.S.S.R., Otdel. Khim. Nauk, 1962, 800; C.A., 1962, **57**, 13704; *B. Pecherer* and *A. Brossi*, Helv., 1966, **49**, 2261). 2,2′-Bis(halogenomethyl)biphenyls have also been used as precursors of the 5,7-dihydrodibenz[*c,e*]oxepin system (*K. Mislow et al.*, J. Amer. chem. Soc., 1961, **83**, 2780; 1964, **86**, 1710; *R. I. T. Cromartie et al.*, J. chem. Soc., 1958, 1982).

The planar representation of LXXXV belies its molecular asymmetry and suitably substituted members of this system have been resolved into optical isomers. 5,7-*Dihydrodibenz* [c,e]*oxepin* (LXXXV; R = H), m.p. 72–73°, inverts too rapidly even at −90° for detection of discrete isomers (*Oki et al., loc. cit.: R. J. Kurland et al.*, J. chem. Phys., 1964, **40**, 2426; *I. O. Sutherland* and *M. V. J. Ramasy*, Tetrahedron, 1965, **21**, 3401; *P. A. Brown* and *Hall*, J. chem. Soc. Perkin I, 1972, 2717).

The dihydro-oxepin ring of diphenane (LXXXV; R = H) shows ring-contraction to 9-hydroxy-9,10-dihydrophenanthrene on treatment with phenyl-lithium (*Wittig et al.*, Chem. Ber., 1951, **84**, 627) or potassium in liquid ammonia (*Weinheimer et al., loc. cit.*), and the 2,3,9,10-tetramethoxy derivative (LXXXV; R = OMe) gives 2,3,8,9-tetramethoxyanthraquinone with sodium dichromate in acetic acid.

Oxidative ring cleavage of LXXXV (R = OMe) gives the corresponding 2,2′-diformylbiphenyl (*Tchiroukhine et al., loc. cit.*) and ring cleavage has also been observed with lithium tetrahydridoaluminate (*Sutherland* and *Ramsay, loc. cit.*) and acetic anhydride (*W. V. Farrar*, J. appl. Chem., London, 1964; **14**, 389).

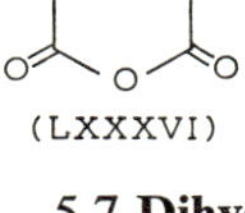

(LXXXVI)

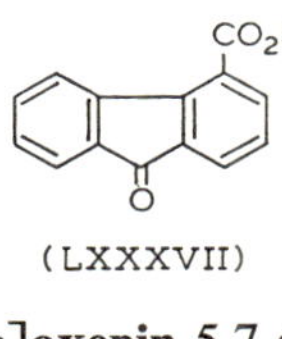

(LXXXVII)

5,7-**Dihydrodibenz**[*c,e*]**oxepin**-5,7-**dione** *(diphenic anhydride)* (LXXXVI), m.p. 224–246°, has been known since 1877, when 2,2′-dicarboxybiphenyl (diphenic acid) was cyclised with acetic anhydride (*R. Auschütz*, Ber., 1877, **10**, 1884). This has remained the reagent of choice for the parent compound (*C. Craebe* and *C. Aubin*, Ann., 1888, **247**, 260; *E. B. Hunn*, J. Amer. chem. Soc., 1923, **45**, 1024; *H. W. Underwood* and *E. L. Kochmann*, *ibid.*, 3071;

H. Shioda and *S. Kato*, C.A., 1959, **53**, 8058), and also for halogeno (*E. H. Huntress et al.*, J. Amer. chem. Soc., 1933, **55**, 2559, 4262; *Ch. Courtot* and *J. Kronstein*, Chim. Ind., Paris, 1941, **45**, 66; *Underwood* and *L. A. Clough*, J. Amer. chem. Soc., 1929, **51**, 583; *H. Brauking et al.*, Chem. Ber., 1967, **100**, 880; *C. D. Weis*, Helv., 1968, **51**, 1582), nitro (F. *Bell* and *P. H. Robinson* J. chem. Soc., 1927, 1695; *F. Pufahl*, Ber., 1929, **62**, 2817; *F. J. Moore* and *Huntress*, J. Amer. chem. Soc., 1927, **49**, 1324), amide (*H. Adkins et al.*, *ibid.*, 1924, **46**, 1917) and methoxy derivatives (*R. A. Barnes* and *R. W. Fässinger*, J. org. Chem., 1961, **26**, 4544; *B. M. Bogoslovskii* and *V. S. Krasnova*, J. gen. Chem., U.S.S.R., 1937, **7**, 1543; C.A., 1937, **31**, 8527). Other cyclodehydrating agents, phosphoryl chloride or phosphorus pentachloride and sulphuric acid also give quantities of 9-fluorenone-4-carboxylic acid (LXXXVII).

Diphenic anhydride may be prepared also from phenanthraquinone (*O. Diels* and *W. E. Thiele*, J. pr. Chem., 1940, **156**, 186; *F. Ramirez et al.*, J. Amer. chem. Soc., 1961, **83**, 492), and by the vapour phase oxidation of phenanthrene (*J. D. Brooks*, J. appl. Chem., London, 1955, **5**, 250).

Treatment of LXXXVI with Lewis acids gives LXXXVII (*F. Bell* and *F. Briggs*, J. chem. Soc., 1938, 1561; 1941, 282) and an analogous ring contraction with decarboxylation occurs on heating (*Graebe* and *C. Mensching*, Ber., 1880, **13**, 1302; *Huntress et al.*, *loc. cit.*). Ring-opening of the oxepindione ring occurs readily with alkali (*H. P. Fogelberg*, Ann. acad. sci Fennicae, 1927, **29A**, No. 4; C.A., 1928, **22**, 1153; *Hunn; Moore* and *Huntress*; *Underwood* and *Clough*, *loci cit.*), alcohols (*D. E. Adelson et al.*, J. Amer. chem. Soc., 1936, **58**, 871; *Courtot* and *Kronstein*; *Graebe* and *Aubin*; *Pufahl*; *Shioda* and *Kato*, *loci cit*; *H. R. Patel et al.*, J. Amer. pharm. Assoc., 1957, **46**, 51), ammonia and amino compounds (*Bell* and *Briggs*, *loc.cit.*; *A. Bistrzycki* and *K. Fassler*, Helv., 1923, **6**, 519; *M. Dominikiewicz*, Arch. Chem. Farm., 1937, **3**, 141; C.A., 1938, **32**, 2916; *A. Shimomura*, Mem. Coll. Sci. Kyoto, Imp.-Univ., 1925, **A8**, 19; C.A., 1925, **19**, 2196; *R. A. Selimidt* and *W. Wenner*, U.S.P., 2,693,465/1954; C.A., 1955, **49**, 12546), to give diphenic acids or their derivatives.

Diphenic anhydrides react with phenols in the presence of zinc or stannic chloride to give phenolphthalein-like indicators (*S. Dutt*, J. chem. Soc., 1923, 225; *Underwood et al.*, J. Amer. chem. Soc., 1923, **45**, 3071; 1930, **52**, 4028; 1936, **58**, 642; 643; *H. Adkins et al.*, *ibid.*, 1924, **46**, 1917); a range of other compounds is also found (*Bell* and *Briggs*, J. chem. Soc., 1938, 1561). Aromatic hydrocarbons are acylated with ring-opening giving biphenyl mono- or di-aryl ketones (*idem*, *ibid.*, 1941, 282; *D. V. Nightingale et al.*, J. Amer. chem. Soc., 1953, **75**, 4701).

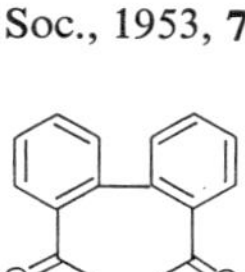

(LXXXVIII)

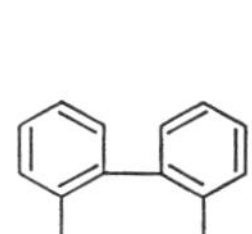

(LXXXIX)

Compounds with an active methylene group may replace the oxepin oxygen with carbon in a ring-opening and recyclisation sequence. 2-Picolines (*A. Taurins*, J. pr. Chem., 1939, **153**, 189) or diethyl malonate (*idem.*, Canad. J. Chem., 1952, **30**, 208) with zinc chloride, and phenylacetic acid with a basic catalyst give derivatives of LXXXVIII.

Reduction of diphenic anhydride with sodium tetrahydridoborate (*B. Belleau* and *R. Chevalier*, J. Amer. chem. Soc., 1968, **90**, 6865) gives **dibenz**[*c,e*]**oxepin**-5(7*H*)-**one** (LXXXIX), m.p. 136°. Other reducing agents also lead to ring-opening. Lactone LXXXIX is also made directly from 2,2′-bis(formyl)biphenyl *via* an internal Cannizzaro reaction and spontaneous lactonisation (*R. G. R. Bacon* and *R. Bankhead*, J. chem. Soc., 1963, 839; *J. Kenner* and *E. G. Turner*, *ibid.*, 1911, 2101; *E. F. M. Stephenson*, *ibid.*, 1954, 2354; *M. G. Sturrock et al.*, J. org. Chem., 1963, **28**, 2340). Substituted derivatives have been made similarly (*B. Pecherer* and *A. Brossi*, Helv., 1966, **49**, 2261). 7-Substituted derivatives are prepared by the action of organometallic reagents on diphenic anhydride or its precursors (*Nightingale*, *loc.cit.*; *A. Corbellini* and *C. Vigano*, Gazz., 1935, **65**, 735; *P. M. Everitt* and *E. E. Turner*, J. chem. Soc., 1957, 3477; *P. G. Sergeev*, J. russ. phys. chem. Soc., 1929, **61**, 1421; C.A. 1930, **24**, 1365).

Tribenz[*b,d,f*]**oxepin** (XC), m.p. 115–116°, is prepared by the pyrolysis of 2′-phenylbiphenyl-2-sulphonyl chloride in octachloronaphthalene (*E. B. McCall* and *T. J. Rawlings*, B.P., 1,016,373/1963; C.A., 1966, **64**, 9697; Tetrahedron, 1965, **21**, 1299) and diazotisation of 2-amino-2′-phenoxybiphenyl in acetic acid also gives a small yield of tribenzoxepin (*D. Hellwinkel* and *H. Seifert*, Chem. Ber., 1972, **105**, 880).

The route to the 1,2,3,4-tetraphenyl derivatives *via* reaction of the benzyne-like derivative XCI with tetraphenylbutadiene could be general (*Tochter-*

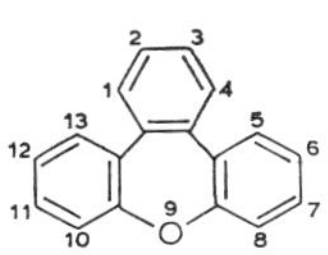

(XC)

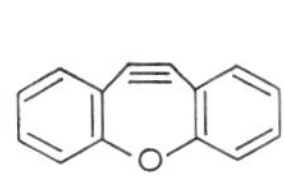

(XCI)

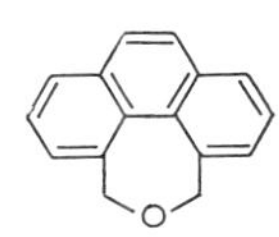

(XCII)

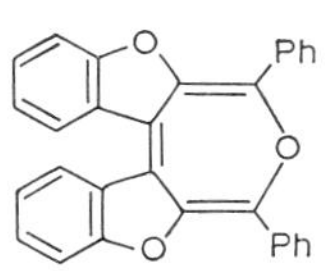

(XCIII)

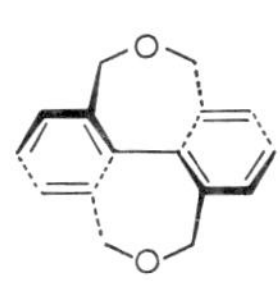

(XCIV)

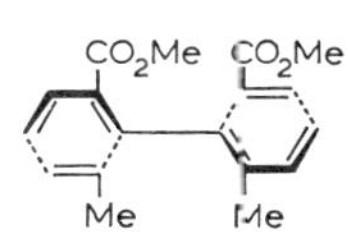

(XCV)

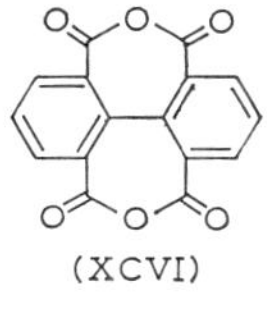

(XCVI)

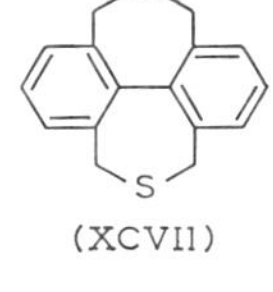

(XCVII)

mann et al., Ann., 1967, **701**, 117). *Tochtermann* and *C. Franke* (Angew. Chem. intern. Edn., 1969, **8**, 68) have obtained 1-*methyltribenz*[b,d,f]*oxepin-3-carboxylic acid*, m.p. 220°, as an optically active brucine salt.

The other oxepin known with three annellated benzene rings is 4,6-**dihydrophenanthro**[4,5-*c,d,e*]**oxepin** (XCII) (*G. M. Badger et al.*, J. chem. Soc., 1950, 2326; *M. S. Newman* and *H. S. Whitehouse*, J. Amer. chem. Soc., 1949, **71**, 3664). Routes to XCII and its 4- and 4,6-dioxo derivatives parallel those to benz[*c,e*]oxepins (*R. H. Martin et al.*, Bull. Soc. chim. Belges, 1963, **72**, 289; *H. Medenwald*, Chem. Ber., 1953, **86**, 287).

The 1*H*-benz[6.7]oxepino[2,3,4-*i,j*]isoquinoline nucleus found in **cularine** (LXXVI) is also present in other members of this group of alkaloids (*Manske*, J. Amer. chem. Soc., 1950, 72, 55) to which *Kametani* and co-workers have successfully investigated synthetic routes (J. chem. Soc., 1963, 4289; Chem. Comm., 1967, 546; Yakugaku Zasshi, 1967, 87, 196, 197; C.A., 1967, **67**, 54313, and earlier papers). The natural pigment **Flavanol Red** has structure XCIII (*T. Kubota et al.*, Tetrahedron Letters, 1966, 5385); the oxepin ring is fully unsaturated.

4*H*,6*H*-[2]**Benzoxepino**[6,5,4-*d,e,f*]**benzoxepin** (XCIV) exists as two enantiomeric forms due to the "twist" along the biphenyl bond. The *R*-form is shown. Racemic XCIV is prepared from 2,6,2′,6′-tetrakis(bromomethyl)biphenyl with moist silver oxide, but optically active material is obtained directly if the resolved biphenyl XCV is treated successively with *N*-bromosuccinimide, silver oxide, lithium tetrahydridoaluminate and concentrated sulphuric acid at −20°. It has an inversion half-life of about 10 min at room temperature, and decomposes in air near its melting temperature (*Mislow et al.*, J. Amer. chem. Soc., 1961, **83**, 2780; 1964, **86**, 1710; *Oki et al.*, Tetrahedron, 1968, **24**, 2377; Bull. chem. Soc. Japan, 1963, **36**, 1542; 1964, **37**, 1865). A tetra-oxo derivative XCVI results when pyrene is oxidised at 450° over vanadium and silver oxides (*M. M. Azanovskaya et al.*, U.S.S.R. Pat., 165,690/1963; C.A., 1965, **62**, 6447).

The mono-sulphur analogue XCVII is prepared similarly (*Mislow*, *loc.cit.*). Optically active XCVII racemises much less readily than XCIV and is stable at its melting point.

(*d*) *Seven-membered heterocyclic ring compounds containing two or more oxygen atoms in the ring*

(*i*) 1,2-*Dioxepins*

No monocyclic 1,2-dioxepins are known, and even when stabilized by annellated benzene rings cyclic peroxides with seven-membered rings are extremely rare. The non-explosive, waxy solid formed by the action of ozone

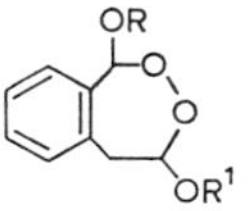

(XCVIII)

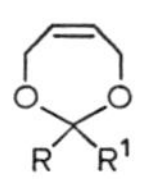

(XCIX)

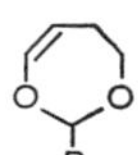

(C)

on an ethanolic solution of indene is probably a mixture of compounds XCVIII (R = H, R^1 = Et) and XCVIII (R = Et, R^1 = H) as it is converted into 1,4-*diethoxy*-4,5-*dihydro*-1H-2,3-*benzodioxepin* (XCVIII; R, R^1 = Et), m.p. 80–81°, in dilute ethanolic sulphuric acid.

The dioxepin ring is opened to homophthalic acid, homophthalaldehyde and the corresponding alcohol, β-(2-carboxyphenyl)ethanol, respectively, by the action of hydrogen peroxide, zinc and acetic acid and lithium tetrahydridoaluminate and to 2-carboxyphenylacetaldehyde with alkali (*J. L. Warnell* and *R. L. Shriner*, J. Amer. chem. Soc., 1957, **79**, 3165).

(*ii*) 1,3-*Dioxepins*

By contrast, many 1,3-dioxepins have been prepared. *cis*-2-Butene-1,4-diol reacts with aldehydes or ketones in the presence of acid to give the cyclic acetals XCIX (*K. C. Brannock* and *G. R. Lappin*, J. org. Chem., 1956, **21**, 1366; *D. B. Pattison*, *ibid.*, 1957, **22**, 662); with paraformaldehyde, 4,7-**dihydro**-2*H*-1,3-**dioxepin** (XCIX; R, R^1 = H), b.p. 124–128°, n_D^{20} 1.4570, is formed. A large excess of alkyl ketone must be used to avoid the formation of 2,5-dihydrofuran from the butenediol, but aryl ketones give low yields of 1,3-dioxepins without this precaution (*W. Kimel* and *W. Leimgruber*, F.P., 1,384,099/1962; C.A., 1965, **63**, 4263). Improved yields are possible if acetals from lower alcohols are used in exchange reactions (*C. E. Pawloski* U.S.P., 3,116,298/1963; C.A., 1964, **60**, 6856).

2-*Ethyl* (XCIX; R = H, R^1 = Et), b.p. 95°/95 mm, n_D^{25} 1.4489, 2,2-*dimethyl*- (XCIX; R, R^1 = Me), b.p. 41°/6 mm, n_D^{25} 1.4458, and 2-*phenyl*-4,7-*dihydro*-2*H*-1,3-*dioxepin* (XCIX; R = H, R^1 = Ph), b.p. 114°/3.5 mm, n_D^{25} 1.5387, are prepared in this way. Cyclic ketones or acetals give spiro derivatives useful as pesticides (*Pawloski* and *G. B. Sterling*, U.S.P., 3,117,134/1962; C.A., 1964, **60**, 8050) and trialkyl orthoformates give dioxepins XCIX (R = H, R^1 = alkoxy) *e.g.* 2-*methoxy*-4,7-*dihydro*-2H-1,3-*dioxepin* (XCIX; R = H, R^1 = OMe), b.p. 48°/4.2 mm, n_D^{25} 1.4530, under these conditions. The reaction of methyl vinyl ether with *cis*-2-butene-1,4-diol also gives a high yield of 2-*methyl*-4,7-*dihydro*-2H-1,3-*dioxepin* (XCIX; R = H, R^1 = Me), b.p. 137–138°, n_D^{25} 1.4478, (*A. Seib*, G.P., 855,864/1952, C.A., 1958, **52**, 9197).

2-*H*-1,3-Dioxepins of type XCIX are fairly stable in alkali and on heating (*C. S. Rondestvedt* and *C. J. Martell*, J. Amer. chem. Soc., 1962, 84, 3307) but with strong bases compounds XCIX (R = H, R^1 = alkyl) give the isomeric 6,7-*dihydro*-2H-1,3-*dioxepins* (C), *e.g.* 2-*methyl* b.p. 76°/135 mm, n_D^{22} 1.4431, and 2-*phenyl*, b.p. 82–84°/0.1 mm, n_D^{25} 1.540, while other 2-aryl analogues undergo ring contraction to 2-aryl-3-formyltetrahydrofurans (*W. M. Thuy* and *H. Normant*, Compt. rend., 1971, **273**, 1655); pyrolysis eventually leads to non-cyclic products (*F. Mutterer et al.*, Tetrahedron 1970, **26**, 477). Several 1,3-dioxepins of type XCIX copolymerise with trioxane (*Farbwerke Hoechst*, Czech. Pat., 108,520/1962; Belg. P., 631,682/1962, C.A., 1964, **60**, 14683; **61**, 4511; *B. E. Jennings* and *J. B. Rose*,

U.S.P., 3,385,832/1962; C.A., 1968, **69**, 19996) or dienes (*G. B. Stirling*, U.S.P., 3,219,629/1961; C.A., 1966, **64**, 5262 and earlier patents).

4,7-**Dihydro**-2*H*-1,3-**dioxepin** (XCIX; R, R^1 = H) forms Diels–Alder adducts with cyclopentadienes, one of which, CI, is a potent insecticide (*A. Hausweiler et al.*, G.P., 1,075,373/1960; 1.088,066/1958; C.A., 1961, **55**, 18003; 1962, **56**, 3498; *M. S. Mulla* and *I. W. Tzack*, *J. Econ.* Etomol, 1963, **56**, 184; C.A., 1963, **59**, 1041); with isoxazoles oxygen is eliminated from the first formed adducts, CII, leaving 1,5-dihydro-3*H*-1,3-dioxepino-[5,6-*c*]pyridines (*Kimel* and *Leimgruber*, *loc.cit.*: C.A., 1965, **62**, 7733, 11819). 4,7-Dihydro-2*H*-dioxepin adds dihalogenocarbenes to give the bicyclic componud CIII, which gives a mixture of compounds including CIV with methyl-lithium (*K. G. Taylor* and *J. Chaney*, J. Amer. chem. Soc., 1972, **94**, 8924).

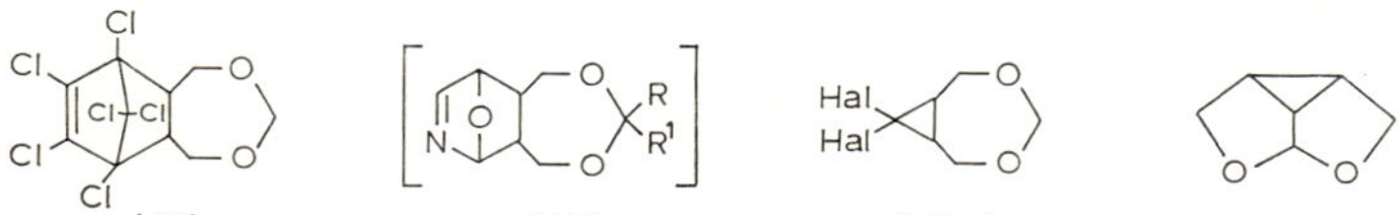

(CI) (CII) (CIII) (CIV)

Chlorine and bromine add readily to the double bond of 4,7-dihydro-2*H*-1,3-dioxepin. 5,6-*Dibromo*-1,3-*dioxepane*, m.p. 36–37°, on dehydrobromination with sodium methoxide gives not 2*H*-1,3-dioxepin, but a mixture of 5-*bromo*-4,7-*dihydro*-2H-1,3-*dioxepin*, b.p. 61–62°/7 mm, and 6-*methoxy*-6,7-*dihydro*-2H-1,3-*dioxepin*, b.p. 48–50°/7 mm.

Catalytic hydrogenation of dihydrodioxepin (CXIX; R, R^1 = H) gives 1,3-**dioxepane** (CV), b.p. 118°, n_D^{22} 1.4303, which is hydrolysed to tetrahydrofuran and formaldehyde by dilute sulphuric acid (*Brannock* and *Lappin*, *loc.cit.*). The ease of hydrolysis of cyclic

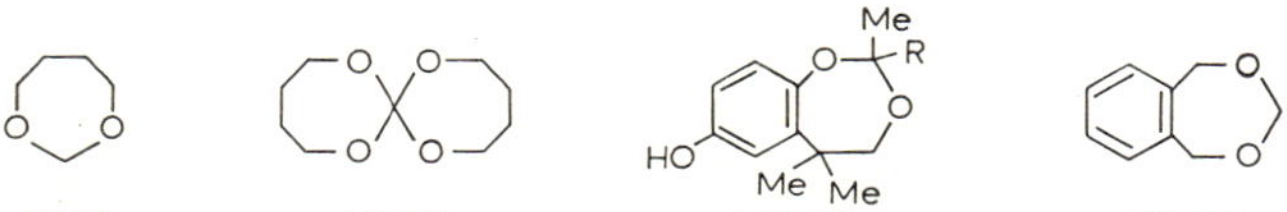

(CV) (CVI) (CVII) (CVIII)

acetals decreases from the five- to the seven-membered ring and 1,3-dioxepane (CV) hydrolyses less readily than its 2-methyl homologue (*R. Leutner*, Monatsh., 1932, **60**, 317). 1,3-Dioxepane (CV) polymerises readily at 150° (*J. W. Hill* and *W. H. Carothers*, J. Amer. chem. Soc., 1935, **57**, 925). 2,2′-*Spiro bis*-1,3-*dioxepane* (CVI), m.p. 109–111°, b.p. 110°/0.1 mm, is obtained in the reaction of $(Bu_3SnOCH_2{\cdot}CH_2{-})_2$ or its cyclic analogue with carbon disulphide (*S. Sakai et al.*, J. org. Chem., 1970, **35**, 2347; 1971, **36**, 1176; Chem. Comm., 1970, 235).

Irradiation of an acetone solution of *tert*-butylhydroquinone gives the 1,3-benzodioxepin (X, R = Me) (*S. Farid*, Chem. Comm., 1970, 303); in dry acetaldehyde, the analogue, 7-*hydroxy*-2,5,5-*trimethyl*-4,5-*dihydro*-2H-1,3-*benzo-dioxepin* (CVII; R = H), m.p. 150°, is obtained (*J. M. Bruce* and *A. Chaudry*, J. chem. Soc. Perkin I, 1972, 372). The acetal ring opens with both Brady's reagent and methanolic hydrogen chloride.

1,5-**Dihydro**-3*H*-2,4-**benzodioxepin** (XVIII), m.p. 37–38°, b.p. 97–107°/0.7 mm, is prepared in the acid catalysed condensation of 1,2-di(hydroxymethyl)benzene with para-formaldehyde (*P. A. Small* (I.C.I. Ltd.), Belg. P., 626,151/1961; C.A., 1964, **61**, 4511; *F. Dallacker et al.*, Ann., 1961, **643**, 91); it exhibits copolymerisation with trioxane. Acetals also condense with 1,2-di(hydroxymethyl)benzene to give 3-substituted derivatives of CVIII (*R. Grewe* and *A. Struve*, Chem. Ber., 1963, **96**, 2819). Several synthetic methods effective for 1,5-benzodioxepins may also be used for making 1,5-dihydro-3*H*-2,4-benzodioxepins of type XI.

Many 6-substituted derivatives of 6*H*-**dibenzo**[*d,f*][1.3]**dioxepin** (CIX) including the 6-*methyl* (CIX; R = Me, R^1 = H), m.p. 48–49°, the 6,6-*dimethyl* (CIX; R, R^1 = Me), m.p. 78–79°, and the 6,6-*diphenyl* (CIX; R, R^1 = Ph), m.p. 168–171°, homologues are prepared by condensing aldehydes, ketones, acetals, or *gem*-dichlorides with 2,2′-dihy-

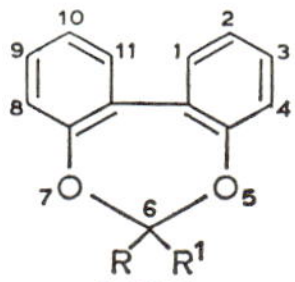
(CIX)

droxybiphenyl (*F. R. Hewgill* and *D. G. Hewitt*, J. chem. Soc., 1965, 1536; *C. M. S. Yoder* and *J. J. Zuckermann*, J. heterocycl. Chem., 1967, **4**, 166; *R. Breslow* and *E. Mohacsi*, J. Amer. chem. Soc., 1963, **85**, 431). The readily polymerised *dibenzo*[*d,f*]-[1.3]*dioxepin*-6-*one*, m.p. 101–102°, is prepared from phosgene and 2,2′-dihydroxybiphenyl (*R. J. Prochaska*, Belg. P., 626,345/1961; C.A., 1964, **60**, 13186). Thiophosgene gives the analogous thione, which on heating gives 6,6′-*spirobi*(*dibenzo* [d,f] [1.3]*dioxepin*) (*Yoder* and *Zuckermann*, *loc.cit.*).

(*iii*) 1,4-*Dioxepins*

Although many benzo derivatives are known, monocyclic 1,4-dioxepins are uncommon and the 6,7-**dihydro**-5*H*-1,4-**dioxepin** [CX; Ar = (10-phenyl-ethynyl)-9-anthryl] is one of the very few unsaturated members of this series (*G. Rio*, Compt. rend., 1954, **239**, 982). 1,4-**Dioxepanes** of type CXI are formed by ring-expansion when vinyl ethers add to the cyclic acetals from ethylene glycol in the presence of boron trifluoride (*B. M. Mikhailov* and

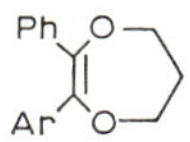
(CX)

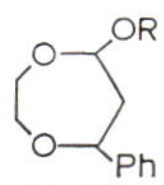
(CXI)

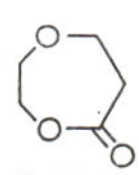
(CXII)

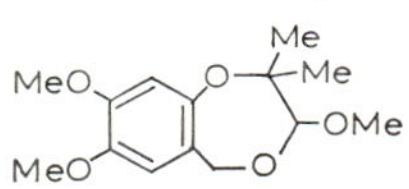
Ageratochromene

L. S. Pavarov, Izvest. Akad. Nauk S.S.S.R., Otdel, Khim, Nauk, 1960, 1903; C.A., 1961, **55**, 13409). Another ring expansion yields **1,4-dioxepan-5-one** (CXII) when bromoform is treated with silver nitrate in aqueous dioxane (*F. Baden* and *C. D. Nenitzescu*, Angew. Chem., 1960, **72**, 415). This ketone is cleaved to 1,2-dibromoethane and 3-bromopropionic acid in hydrobromic acid.

The natural product **ageratochromene** has a 2,3-dihydro-5*H*-1,4-benzodioxepin structure (*A. R. Alertsen*, Acta Polytech. Scand., 1961, Ser. 13, No. 10; C.A., 1962, **57**, 16619).

8,9-*Dimethoxy*-2,3-*dihydro*-5H-1,4-*benzodioxepin*-3,5-*dione* (CXIII), m.p. 195°, is prepared by heating 2-carboxy-5,6-dimethoxyphenoxyacetic acid with acetic anhydride (*W. H. Perkin* and *J. Yates*, J. chem. Soc., 1902, **81**, 242), and 2,3-*dihydro*-5H-1,4-*benzodioxepin*-2,5-*dione* (CXIV), m.p. 115°, results when iodoacetylsalicylic acid is cyclised using a

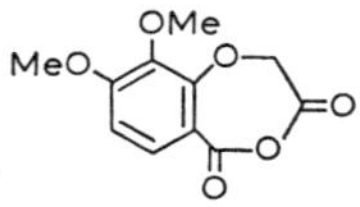

(CXIII)

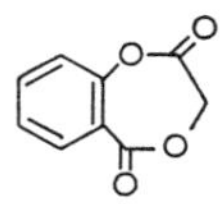

(CXIV)

tertiary amine (*F. Kagan* and *R. D. Birkenmeyer*, J. Amer. chem. Soc., 1959, **81**, 1986; U.S.P., 2,956,064/1960; C.A., 1961, **55**, 7450). The dioxepin ring is opened by water or primary and secondary amines to give, respectively, salicylglycolic acid or its amides.

The 3,4-**dihydro-2*H*-1,5-benzodioxepins** are the best represented members of this group. The parent compound (*catechol trimethylene ether*) (CXV; R, R^1 = H), b.p. 100–105°/11 mm, is formed when 1,3-dibromopropane condenses with catechol in the presence of a base at high dilution. Catechol tetra- to deca-methylene ethers are also prepared in this way (*K. Ziegler et al.*, Ann., 1937, **528**, 162; C.A., 1939, **33**, 6532; *R. B. Thompson*, U.S.P., 2,688,329/1954; C.A., 1956, **50**, 1911; *R. C. Cass et al.*, J. chem. Soc., 1958, 2595) as shown

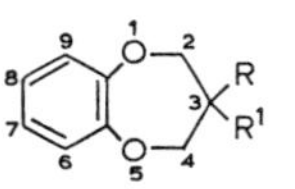

(CXV)

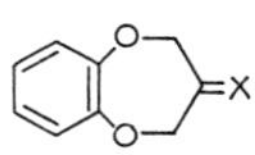

(CXVI)

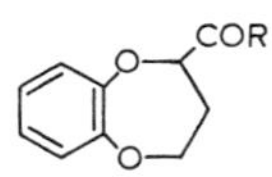

(CXVII)

in Table 1. Substituted catechols also undergo this reaction (*M. Tomita* and *T. Takahoshi*, Yakugaku Zasshi, 1957, **77**, 1041; C.A., 1958, **52**, 3816). Catechol reacts similarly with 1,3-dichloro-2-methylenepropane to give CXVI (X = CH_2), which is reduced to CXV (R = H, R^1 = Me), b.p. 111–112°/12 mm, n_D^{20} 1.5310, and with the amide and ester of 2,4-dibromobutyric acid to give the 2-substituted dioxepins CXVII (R = NH_2 and OEt), respectively (*F. Leonard* and *J. Koo*, Belg. P., 613,132/1962; C.A., 1962, **57**, 16640).

TABLE 1

POLYMETHYLENE CATECHOL ETHERS

Ring size	7	8	9	10	11	12	13	14
CH_2 groups	3	4	5	6	7	8	9	10
% yield	50	40	69	90	71	70	42	51
b.p./10 mm	103°	112°	122°	140°	156°	171°	185°	197°

Nitration of CXV (R, R^1 = H) in acetic acid gives 7-*nitro*-3, 4-*dihydro*-2H-1,5-*benzodioxepin*, m.p. 106–107°, (*Thompson, loc.cit.*).

2,4-*Dihydro*-3H-1,5-*benzodioxepin*-3-*one* (CXVI; X = O), m.p. 40–42°, *semicarbazone*, m.p. 230–233° decomp., is formed on ozonolysis of CXVI (X = CH_2) (*R. Hartmut et al.*, Z. Chem., 1968, **8**, 220) and as a by-product in the oxidation of 2-hydroxymethyl-1,4-benzodioxane. It reacts with methylmagnesium iodide to give the alcohol CXV (R = OH, R^1 = Me) (*V. Rosnati* and *F. De Marchi*, Tetrahedron, 1962, **18**, 289).

11*H*-**Dibenzo**[*b*,*e*][1,4]-**dioxepin** *(Depsidan)* (CXVIII), b.p. 112–114°/0.2 mm, is the parent ring system of a family of about 20 lichen metabolites, found as the lactones CXIX (depsidones). Depsidones are unusual among natural products in that many of them contain chlorine (up to 4 atoms). Their classical chemistry has been reviewed (*F. M. Dean*, "Naturally occur-

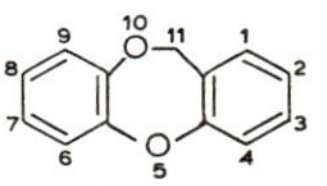

(CXVIII)

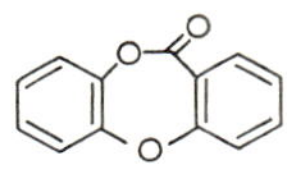
(CXIX)

ring oxygen ring compounds", Butterworth, London 1963, 564; *A. Asahina* and *S. Shibata*, "The Chemistry of Lichen substances", Jap. Soc. for the promotion of Sci., Tokyo, 1954). Depsidan is prepared by treating 2-hydroxy-2′-(hydroxymethyl)diphenyl ether with hydrogen bromide followed by base (*Y. Inubushi*, J. pharm. Soc. Japan, 1952, **72**, 1223; *M. Tomita*, *T. Ujii* and *S. Tanaka*, Yakugaku Zasshi, 1960, **80**, 358; C.A., 1960, **54**, 18432).

Depsidone (CXIX), m.p. 65–66°, and many of its derivatives are prepared by the lactonisation of 2-(2-hydroxyphenoxy)benzoic acid; thionyl chloride in pyridine is the reagent of choice (*Tomita et al.*, J. pharm. Soc. Japan, 1944, **64**, 173; *D. S. Noyce* and *J. W. Weldon*, J. Amer. chem. Soc., 1952, **74**, 401; *H. E. Ungnade* and *L. Tubin*, J. org. Chem., 1951, **16**, 1311). The possible biosynthetic route is *via* intramolecular oxidative phenol coupling of depside (phenyl salicylide) (*D. H. R. Barton* and *T. Cohen*, Festschrift A. Stoll, Birkhäuser, Basel 1957, 117; *H. Erdtman* and *C. A. Wachtmeiter*, *ibid.*, 1957, 144). **Diploicin** (CXX), m.p. 232–233°, the simplest natural depsidone was also first synthesised this way (*W. D. Ollis et al.*, Proc. chem. Soc., 1960, 393), but the synthesis shown (*J. B. Hendrickson et al.*, J. Amer. chem. Soc., 1972, **94**, 6834) from a 2,2′-dihydroxybenzophenone may be more general:

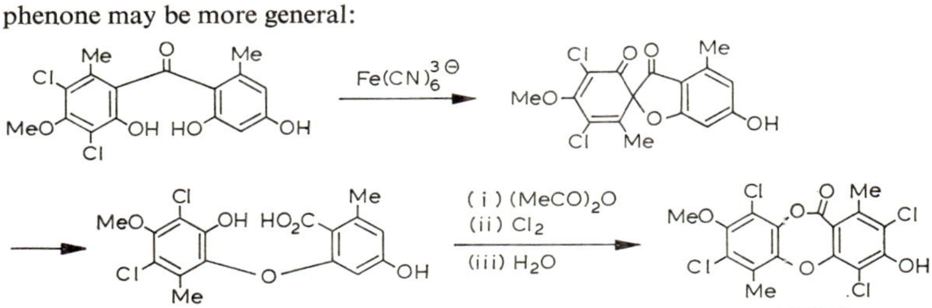

(CXX)

The lactone ring opens readily with acids or alkalis to give derivatives of 2-(2-hydroxyphenoxy)benzoic acid (*Noyce* and *Weldon*, J. Amer. chem. Soc., 1952, **74**, 5144; *V. C. Barry* and *D. Twomey*, Proc. roy. Irish Acad., 1951, **53B**, 55; C.A., 1951, **45**, 9500). Sodium in liquid ammonia opens the ring, converting the carboxylic acid group to a methyl group.

The 11*H*-dibenzo[*b*,*e*] [1.4]dioxepin system CXVIII is present in the alkaloids **insularine** and **insulanoline** (*Tomita* and *T. Kikuchi*, Yakugaku Zasshi, 1957, **77**, 1041; C.A., 1958, **52**, 3832; *Kikuchi* and *K. Bessho*, *ibid.*, 1958, **78**, 1413; C.A., 1959, **53**, 7220).

(iv) Trioxepins

Several 1,2,4-trioxepanes CXXI have been prepared by the toxic acid catalysed cyclic condensation of γ-hydroperoxy alcohols of the type $HOCH_2.CH_2.CRR^1O{\cdot}OH$, with acetone; 3,3,7,7-*tetramethyl*-1,2,4-*trioxepane* (CXXI; R, R^1 = Me) has b.p. 83–85°±4 mm (*W. Adam* and *N. Duran*, Chem. Comm., 1972, 798).

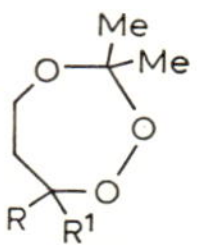

(CXXI)

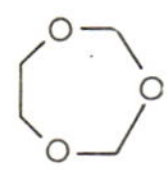

(CXXII)

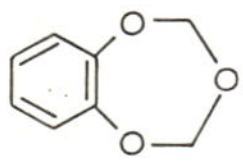

(CXXIII)

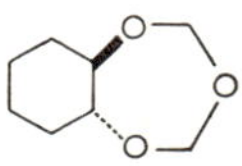

(CXXIV)

1,3,5-**Trioxepane** (CXXII), b.p. 129–131°, is obtained from 1,3-dioxolane and formaldehyde in the presence of sulphuric acid (*W. F. Gresham* and *C. D. Bell*, U.S.P., 2,475,610/1949; 2,625,569/1953; C.A., 1950, **44**, 175; 1953, **47**, 4132). Various methyl and trichloromethyl-1,3,5-trioxepanes have also been described (*A. N. Meldrum* and *G. M. Vad*, J. Indian Chem. Soc., 1937, **13**, 118; *A. C. Neish* and *F. J. Macdonald*, Canad. J. Res., 1947, **25B**, 70; *E. F. Schroeder*, U.S.P., 2,532,340/1950; 2,595,304/1952; C.A., 1951, **45**, 3867; 1953, **47**, 4377).

2,4-*Dimethyl*-2H,4H-1,3,5-*benzotrioxepin* (CXXIII), m.p. 34°, b.p. 118°/20 mm, is obtained when catechol is condensed with acetaldehyde at −5° in the presence of phosphorus pentoxide (*G. Sloof*, Rec. Trav. chim., 1935, **54**, 995).

cis-Cyclohexane-1,2-diol with *N*-bromosuccinimide in dry dimethyl sulphoxide gives methylenedioxycyclohexane, but the *trans* isomer gives *trans*-cyclohexano-1,3,5-trioxepane (CXXIV) (*S. Hanession et al.*, J. Amer. chem. Soc., 1972, **94**, 8930). *trans*-Cyclohexane-1,2-diol also reacts with formaldehyde (*F. S. H. Head*, J. chem. Soc., 1960, 1778) or dibromomethane in the presence of sodium hydride (*J. S. Brimacombe et al.*, *ibid.*, C, 1967, 2404) to give CXXIV.

3. Heterocyclic seven-membered ring compounds containing sulphur

(a) Compounds with a single sulphur atom in the ring

(i) Thiepin derivatives

It is only recently that monocyclic compounds with a fully unsaturated seven-membered ring containing divalent sulphur have been prepared, overcoming predictions of instability based on molecular orbital calculations. **Thiepin** (I) is expected to be antiaromatic if planar (*M.J. S. Dewar* and *N. Trinajstic*, J. Amer. chem. Soc., 1970, **92**, 1453; *R. Zahradnik*, Adv. heterocycl. Chem., 1965, **5**, 21). The supposed 2-methylthiepin-4,6-dithiol(II) (*A. A. Bothner-By* and *G. Traverso*, Chem. Ber., 1957, **90**, 453) probably has the thiophene structure III (*F. Arndt* and *W. Walter*, *ibid.*, 1961, **94**, 1757),

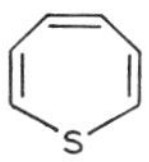
(I)

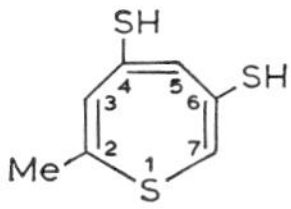
(II)

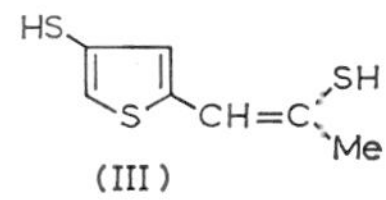
(III)

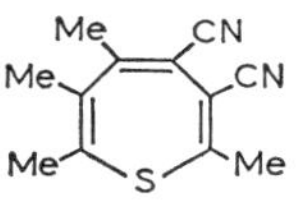
(IV)

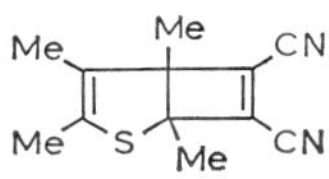
(V)

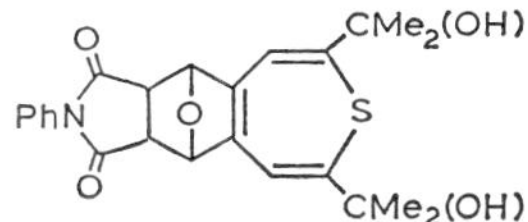
(VI)

and the supposed dinitrile IV (*H. Wynberg* and *R. Helder*, Tetrahedron Letters, 1972, 3647) has been identified as the bicyclic thiophene V (*Wynberg et al.*, *ibid.*, 1972, 5269).

A near approach to thiepin (I) without the stabilisation of a fused aromatic ring is VI (*J. M. Hofmann* and *R. H. Schlessinger*, J. Amer. chem. Soc., 1970, **92**, 5263) but true monocyclic thiepins (VII; R = alkyl) prepared as shown have recently been characterised at $-30°$, before extruding sulphur at the same temperature (*D. N. Reinhoudt* and *C. G. Kouwenhoven*, Chem. Comm., 1972, 1233):

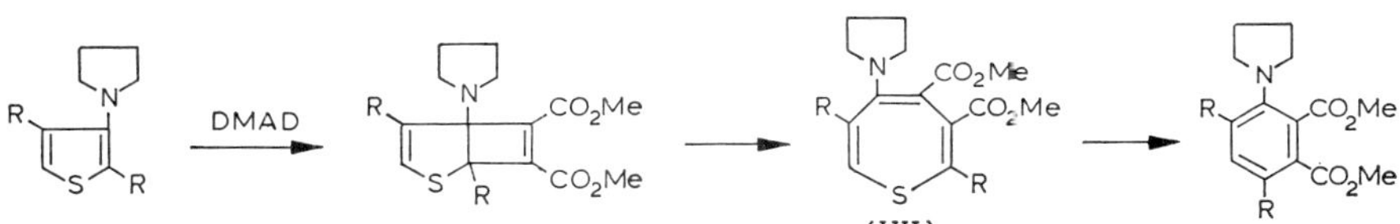

(VII)

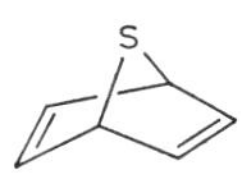
(VIII)

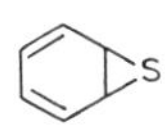
(IX)

Sulphur extrusion from thiepins to give benzene derivatives, which is a general reaction (*U. Eisner* and *T. Krishnamurthy*, Int. J. sulphur Chem., 1971, **B6**, 267), presumably thwarted attempts to rearrange 7-thianorbornadiene (VIII) and its isomer IX to thiepin (I) (*T. J. Barton et al.*, J. org. Chem., 1972, **37**, 552).

By contrast, *thiepin* 1,1-*dioxide* (XI), m.p. 117–118°, has been synthesised by the sequence shown, *via* 2,7-*dihydrothiepin* 1,1-*dioxide* (X), m.p. 107–108° (*W. L. Mock*, J. Amer. chem. Soc., 1967, **89**, 1281):

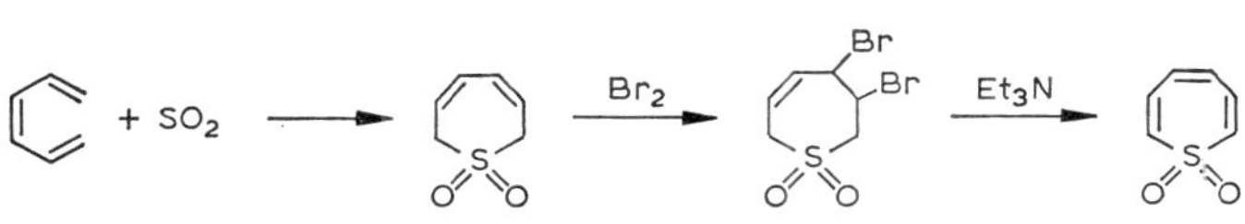

(X) (XI)

The dioxide XI decomposes to benzene and sulphur dioxide when heated above its m.p. and is reduced catalytically to *thiepane* 1,1-*dioxide*, m.p. 70–71°. Sulphone XI exists in a flattened boat conformation with shorter bonds and larger angles than expected (*Mock et al.*, *ibid.*, 1968, **90**, 4501).

2,3- and 2,7-**dihydrothiepins** have been prepared by an analogous route to that shown for VII from 4,5- and 2,5-dihydrothiophene enamines. A temperature of 100° is required to open the bicyclic adducts. The amino group of the thiepin is replaced by hydroxyl on acid hydrolysis, and these enol, or keto, derivatives are oxidised to the corresponding sulphones with peracids (*Reinhoudt* and *C. G. Liliveld*, Tetrahedron Letters, 1972, 3119).

4,5-**Dihydrothiepin** (XII) is obtained in the Cope rearrangement of *cis*-1,2-divinylthiirane at 100° (*E. L. Stogryn* and *S. J. Brois*, J. Amer. chem. Soc., 1967, **89**, 605) but the temperature required is higher than in the equivalent isomerisations of the oxygen and nitrogen analogues.

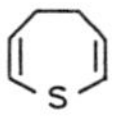

(XII)

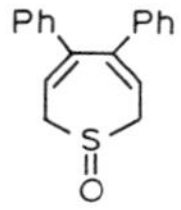

(XIII)

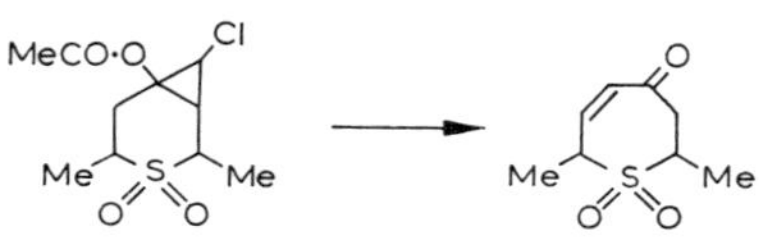

(XIV) (XV)

Reaction between 3-diazopropene and sulphur dioxide gives 4,5-*dihydrothiepin* 1,1-*dioxide*, m.p. 111–112°, which isomerises to 2,7-dihydrothiepin 1,1-dioxide (X) in refluxing benzene-triethylamine, and yields thiepin 1,1-dioxide (XI) on treatment with *N*-bromosuccinimide followed by dehydrobromination. A Cope rearrangement is proposed for this reaction pathway too (*Mock*, Chem. Comm., 1970, 1254; *L. A. Paquette* and *S. Maiorama*, *ibid.*, 1971, 313). This 1,1-dioxide of XII has unusually short double bonds (*H. L. Ammoh et al.*, Acta Cryst., 1972, **28B**, 246).

2,7-Dihydrothiepin 1,1-dioxide (X), the intermediate in the preparation of XI, is formed in ether at room temperature; the reverse reaction occurs at 150–160°. The *cis*- and *trans*-2,7-dimethyl analogues are unusual in opening in a conrotatory manner (*Mock*, J. Amer. chem. Soc., 1969, **91**, 5682). Catalytic hydrogenation gives thiepane sulphone, and bromination the 3,4-dibromide. A similar reaction takes place between 3,4-diphenyl-1,3,5-hexatriene and sulphur monoxide yielding 4,5-*diphenyl*-2,7-*dihydrothiepin* 1-*oxide* (XIII), m.p. 162°, which is oxidised by 3-chloroperbenzoic acid to the *dioxide*, m.p. 190°, and reduced by lithium tetrahydridoaluminate to 4,5-*diphenyl*-2,7-*dihydrothiepin*, m.p. 142–143°, (*R. M. Dodson* and *J. P. Nelson*, Chem. Comm., 1969, 1159).

A mixture of 2,5,6,7- and 2,3,6,7-*tetrahydrothiepin* 1,1-*dioxides* is obtained in the pyrolysis of 4-acetoxythiepane 1,1-dioxide obtained from 4-hydroxythiepan (*C. G. Overberger* and *A. Katchman*, J. Amer. chem. Soc., 1956, **78**, 1965). Tetrahydrothiepin dioxides with 2,3-double bonds are less stable than those with 3,4-double bonds (*S. B. Maerov*, Diss. Abs., 1954, **14**, 765). Ring expansion of the bicyclic sulphone XIV to the tetrahydrothiepin sulphone XV occurs on reduction with lithium tetrahydridoaluminate and treatment with triethylamine (*Mock*, *loc.cit.*); reduction with tetrahydridoborate and dehydration gives 2,7-dimethyl-2,7-dihydrothiepin 1,1-dioxide.

The first known compound with an acetylenic bond in a seven-membered ring is the thiepin XVI formed by oxidising the bishydrazone of 3,3,6,6-tetramethylthiepane-4,5-dione (XVII). The acetylenic compound XVI acts as a dienophile, and regenerates XVII with oxygen. Addition of 4-nitrophenyl isocyanide or dichlorocarbene to the triple bond, followed by hydrolysis yields the cyclopropenone XVIII (*H. Kimling* and *A. Krebs*, Tetrahedron Letters., 1970, 761; Angew. Chem. intern. Edn., 1971, **10**, 409; 509; 1972, **11**, 932; *J. Haase* and *Krebs*, Z. Naturforsch., 1972, **27**, 624):

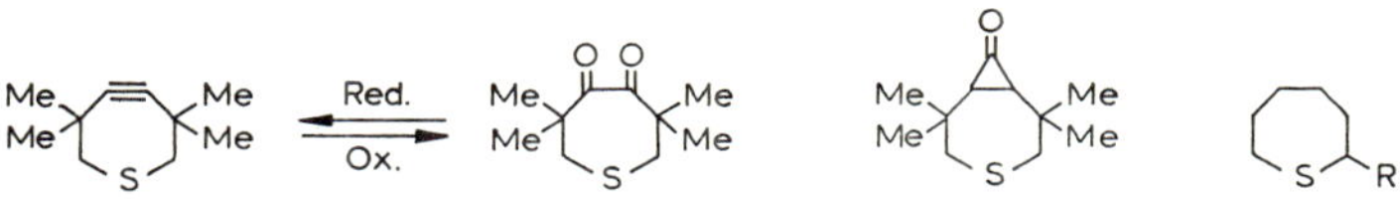

(XVI) (XVII) (XVIII) (XIX)

Introduction of a sulphur atom greatly increases the stability of the seven-membered ring acetylene in comparison with cycloheptyne, which has a half-life of less than a minute at −25°. Compound XVI does not dimerise even at 140°, but on photolysis yields thietanone and thiapyranone derivatives (*P. Y. Johnson*, Tetrahedron Letters, 1972, 1991).

Thiepane *(thiacycloheptane, hexamethylene sulphide, hexahydrothiepin)* (XIX; R = H), b.p. 174°, n_D^{20} 1.5138, is usually prepared from 1,6-dibromohexane and sodium sulphide (*J. M. Cox* and *L. N. Owen*, J. chem. Soc. C, 1967, 1130; *E. Grishkevich-Trokhimovskii*, J. Russ. phys. Chem. Soc., 1916, **48**, 944; C.A., 1917, **11**, 786; *A. Muller et al.*, Monatsh., 1953, **84**, 1206; *R. D. Obolentsev et al.*, C.A., 1960, **54**, 249; 1962, **56**, 5933). The reaction of 1,5-hexadiene with sulphur dichloride gives impure 3,6-dichlorothiepin (*Dunlop Rubber Co. Ltd.*, F.P., 1,427,429/1966; C.A., 1966, **65**, 12180), and ethyl 6-chlorohexyl sulphide cyclised to ethyl thiepanesulphonium chloride on heating (*G. M. Bennett* and *E. G. Turner*, J. chem. Soc., 1938, 813).

Thiepane is oxidised by hydrogen peroxide (1 mol.) to *thiepane* 1-*oxide*, b.p. 144–145°/14 mm, (*A. Cerniani et al.*, Gazz., 1960, **90**, 382) and by permanganate or peracids to *thiepane* 1,1-*dioxide*, m.p. 70–71°, (*Mock*, J. Amer. chem. Soc., 1967, **89**, 1281; *Grishkevich-Trokhimovskii, loc.cit.*). With hydrazoic acid the sulphoxide forms a sulphoximine, which is *N*-methylated by trimethyloxonium tetrafluoroborate (*C. R. Johnson* and *L. J. Pepoy*, J. org. Chem., 1972, **37**, 671). Thiepane gives a *methiodide*, m.p. 141–142°, (*N. J. Leonard* and *J. Figueras*, J. Amer. chem. Soc., 1952, **74**, 917) but also shows ring-opening with methyl iodide (*Muller et al., loc.cit.*); the ring is also opened by alkyl hypochlorites (*Johnson* and *M. P. Jones*, J. org. Chem., 1967, **32**, 2014). Conversion to the *mercuric chloride complex*, m.p. 152–153°, and subsequent regeneration of thiepane may be used as a purification procedure (*Obolentsev et al.*, C.A., 1962, **57**, 7218).

On treatment with *tert*-butyl acetate and cuprous chloride thiepane is converted into 2-acetoxythiepane (XIX; R = OAc), and thence by hydrolysis to 2-hydroxythiepane (XIX; R = OH). This cyclic hemithioacetal exhibits ring–chain tautomerism (*Cox* and *Owen, loc.cit.*). With *N*-chlorosuccinimide, thiopane gives 2-chlorothiepane (XIX; R = Cl), characterised as the sulphone, which reacts with appropriate Grignard reagents to form 2-*methylthiepane* (XIX; R = Me), b.p. 94–96°/52 mm, and 2-*phenylthiepane* (XIX;

R = Ph), b.p. 100–101°/0.35 mm, (*D. L. Tuleen* and *R. H. Bennett*, J. heterocycl. Chem., 1969, **6**, 115).

All three possible thiepanones are known. 2-**Thiepanone** (ε-thiocaprolactone) (XX), b.p. 74–75°/1 mm, n_D^{26} 1.5257, and the 4- and 5-methyl homologues distil when the corresponding 6-mercaptohexanoic acids are heated with phosphorus pentoxide (*Overberger* and *J. K. Weise*, J. Polymer Sci. Part B, 1964, **2**, 329; J. Amer. chem. Soc., 1968, **90**, 3525, 3533, 3538). With strong bases polythioesters are formed.

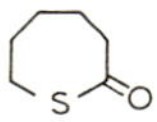
(XX)

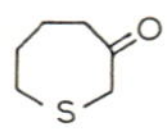
(XXI)

A. Dieckmann cyclisation of 4-ethoxycarbonylbutyl ethoxycarbonylmethyl sulphide, followed by hydrolysis and decarboxylation gives 3-*thiepanone* (XXI), m.p. 20–21°, n_D^{20} 1.5285, (oxime, m.p. 96–97°) (*Leonard* and *Figueras*, *loc.cit.*).

Diazoalkanes add to thiacyclohexan-3-one to give mixtures of 3- and 4-thiepanones separable by distillation (*R. Borsdorf et al.*, Angew. Chem. intern. Edn., 1967, **6**, 872) and 2,2-diphenylthiepan-3-one 1,1-dioxide (XXII) is obtained in the ring-expansion shown (*T. Durst* and *K. C. Tin*, Tetrahedron Letters, 1970, 2369):

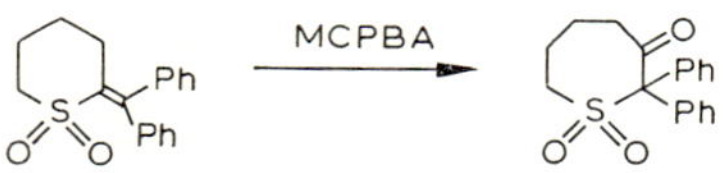

(XXII)

Reduction of 3-thiepanones with lithium tetrahydridoaluminate gives 3-thiepanols, but on Clemmensen reduction, ring-contraction to 2-methylthiacyclohexane occurs.

4-**Thiepanone** (XXIII), b.p. 72–75°/1.5 mm, n_D^{23} 1.5299, (*phenylhydrazone*, m.p. 101–103°), together with a spiro-epoxide is formed by the ring-expansion of tetrahydro-1,4-thiapyrone with diazomethane (*Overberger and Katchmann*, *loc.cit.*). Irradiation of thiacyclohexan-3-one in Freon 113 also gave XXIII together with tetrahydrothiophen-2-one (*K. K. Maheshwari* and *G. A. Berchtold*, Chem. Comm., 1969, 13). 4-Thiepanone may be oxidised to the *sulphoxide*, m.p. 75–76°, (*Leonard* and *W. C. Rippie*, J. org. Chem., 1963, **28**, 1957) and the *sulphone* (XXIV), m.p. 121–123°, which is also formed when the acetate

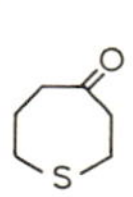
(XXIII)

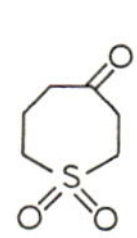
(XXIV)

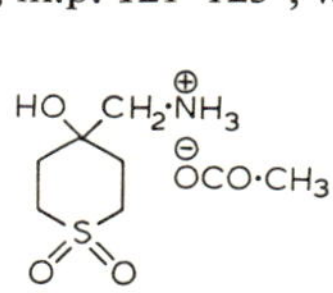
(XXV)

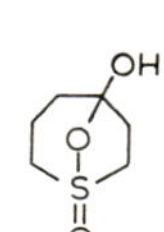
(XXVI)

XXV is treated with nitrous acid at low temperature. The 4-oxo group reacts to give normal carbonyl derivatives and Grignard reagents give 4-alkyl-4-hydroxy thiepanes (*V. A. Zagorevskii et al.*, Zhur. org. Khim., 1965, **1**, 366, 2215). Reduction of XXIII with lithium tetrahydridoaluminate gives the alcohol 4-*hydroxythiepane*, b.p. 75–77°/1 mm, n_D^{24} 1.5379, (*P. Y. Johnson* and *Berchtold*, J. org. Chem., 1970, **35**, 584). When XXIV is treated with perchloric acid transannular reaction takes place giving XXVI (*Leonard* and *Rippie*, *loc.cit.*).

3,3,6,6-**Tetramethylthiepane-4,5-dione** (XVII), b.p. 68°/0.35 mm, n_D^{20} 1.4954, is formed by oxidising the acyloin condensation product from $S(CH_2{\cdot}CMe_2{\cdot}CO_2Et)_2$ (*A. de Groot* and *Wijnberg*, Chem. Com., 1965, 171; J. org. Chem., 1966, **31**, 3954). The acyloin forms a cyclic sulphite with thionyl chloride (*Wijnnberg et al.*, Chem. Comm., 1968, 347).

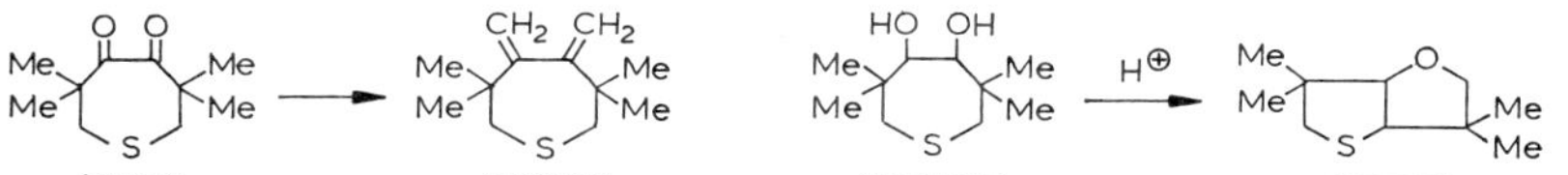

(XVII) (XXVII) (XXVIII) (XXIX)

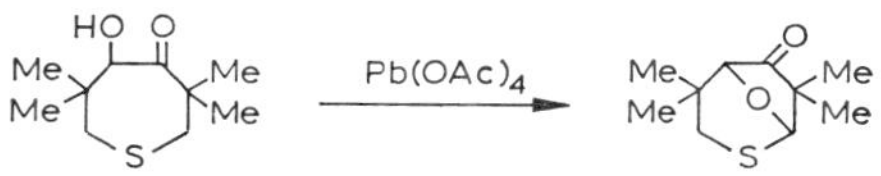

(XXX) (XXXI)

Wolff–Kishner reduction of 3,3,6,6-tetramethylthiepane-3,6-dione (XVII) monohydrazone yields 3,3,6,6-tetramethylthiepane-4-one, and reduction cf the acyloin XXX with lithium tetrahydridoaluminate gives the diol XXVIII.

Dione XVII undergoes a double Wittig reaction with excess methylene triphenylphosphorane to give the 1,3-diene XXVII, which does not form Diels–Alder adducts because of the buckling of the ring. Most other nucleophilic reagents attack only one of the oxo groups of XVII (*idem. ibid.*, 1968, **33**, 2214, 3337). Dione XVII uhdergoes a normal benzilic acid rearrangement with ring contraction (*idem, ibid.*, p. 4025) but its reduced derivative XXVIII does not behave like pinacol with acid but gives the bicyclic compound XXIX. A thiophene derivative is also obtained when hydroxyketone XXX is treated with triphenylphosphine bromide (*idem*, Tetrahedron Letters, 1968, 2365; Rec. Trav. chim., 1969, **88**, 994). The oxidation products of XXX with lead tetra-acetate are XVII and the oxygen-bridged compound XXXI (*Krebs* and *Kimling*, Ann., 1970, **740**, 126).

(*b*) *Thiepins with annellated rings*

The nearest approach to a **thiazulene** is the carboxylic acid XXXII, prepared by hydrolysis and decarboxylation of condensation products from 6-dimethylaminofulvene-2,3-dicarbaldehyde and diethyl thiodiglycollate or thiodiacetonitrile. It could not be converted into a fully conjugated cation (*C. Konig*, Ph. D. Thesis, Univ. of Marburg, 1961).

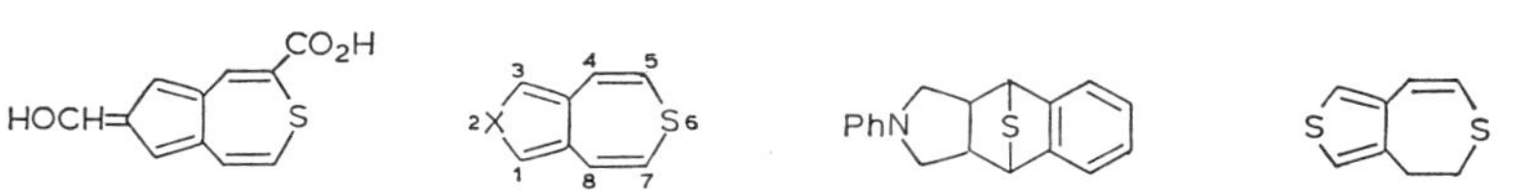

(XXXII) (XXXIII) (XXXIV) (XXXV)

Thieno[3,4-*d*]**thiepin** (2,6-*dithiazulene*) (XXXIII; X = S), m.p. 149–151°, and its 2-oxa analogue (XXXIII; X = O) have, however, been prepared by the dehydration of the sulphoxide of the 4,5-dihydro derivative XXXV with acetic anhydride. Like azulene,

XXXIII (X = S) is planar and appears to have a delocalised aromatic system, but it reacts with *N*-phenylmaleimide to give adduct XXXIV in which sulphur has been extruded from the thiepin ring. The 2-*oxa* analogue (XXXIII; X = O), m.p. 116°, the 6-*monoxide*, m.p. 75–95° and 6,6-*dioxide*, m.p. 183°, of XXXIII (X = S), prepared by oxidising dithiazulene with 3-chloroperbenzoic acid all show greater reactivity in the Diels–Alder reaction. Catalytic hydrogenation saturates the seven-membered ring but leaves the thiophene ring intact (*Schlessinger et al.*, J. Amer. chem. Soc., 1967, **89**, 7138; 1969, **91**, 3995; Tetrahedron Letters, 1968, 3017; 1969, 4361).

Treatment of 1,6-dithiacyclodeca-3,8-diyne with potassium *tert*-butoxide results in a transannular ring contraction to 4,5-*dihydrothieno*[3,4-d]*thiepin* (XXXV), m.p. 54-56°, (*G. Eglington et al.*, J. chem. Soc., 1964, 1154).

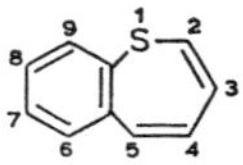

(XXXVI)

Attempts have been made to prepare 1-**benzothiepins** by ring-expansion of five- or six-membered sulphur-containing rings, and by treating functional groups in a saturated seven-membered ring to introduce double bonds into it, but the parent compound XXXVI, is still unknown.

Several attempted syntheses have given naphthalene derivatives due to the ready extrusion of sulphur from the 1-benzothiepin system, which is presumed to be a transient intermediate (*Wijnberg et al.*, Tetrahedron Letters, 1969, 2913; *F. G. Bordwell* and *B. M. Pitt*, J. Amer. chem. Soc., 1955, **77**, 572; *W. E. Truce et al., ibid.*, 1952, **74**, 3594; *Paquette et al.*, J. org. Chem., 1968, **33**, 1080; *V. J. Traynelis et al., ibid.*, 1962, **27**, 2377; 1964, **29**, 1092).

The first stable 1-**benzothiepin** was prepared by the route shown:

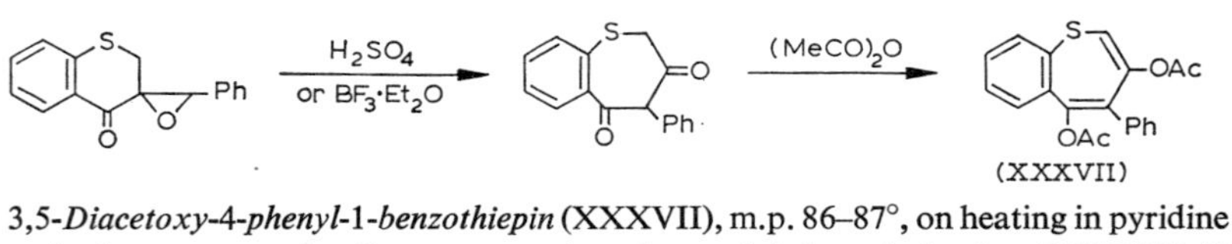

(XXXVII)

3,5-*Diacetoxy-4-phenyl-1-benzothiepin* (XXXVII), m.p. 86–87°, on heating in pyridine alone or in the presence of sodium acetate gives the naphthalene derivatives XXXVIII (R = H) and (R = SAc), respectively (*H. Hofmann* and *H. Westernacher*, Angew. Chem. intern. Edn., 1966, **5**, 958; 1967, **6**, 255; Chem. Ber., 1969, **102**, 205). 5-*Acetoxy-4-phenyl-1-benzothiepin*, m.p. 99–101°, prepared similarly, was converted into the 5-*methoxy* analogue, m.p. 77–79°, by treatment with dimethyl sulphate and potassium *tert*-butoxide (*Hofmann et al.*, Angew. Chem. intern. Edn., 1972, **11**, 423).

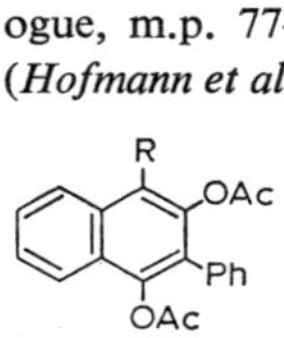

(XXXVIII)

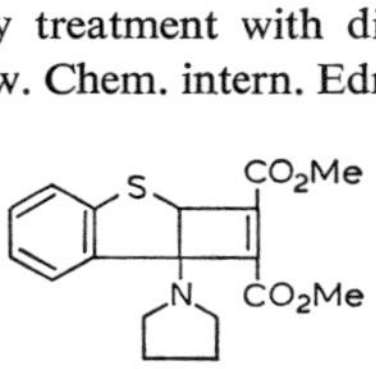

(XXXIX)

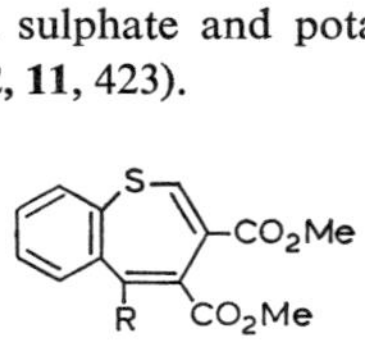

(XL)

Ring-expansion of the adduct XXXIX from 3(*N*-pyrrolidinyl)benzo[*b*]thiophene and dimethyl acetylenedicarboxylate gives the 1-benzothiepin XL (R = *N*-pyrrolindinyl). Acid hydrolysis gives the hydrogen-bond stabilised enol XL (R = OH), which extrudes sulphur at 100° and with 3-chloroperbenzoic acid the 1,1-dioxide is formed (*Reinhoudt* and *Kouwenhoven*, *loc.cit.*). Irradiation of 1-benzothiepins reverses this reaction and yields benzo[*b*]-4,5-dihydrothiophene[*d*-3,4]cyclobutenes (*Hofmann* and *B. Meyer*, Tetrahedron Letters, 1972, 4597).

Attempts to expand 4*H*- and 2*H*-1-benzothiapyran using dichlorocarbene and a strong base failed to give 1-benzothiepins, but gave instead 2-chloronaphthalene, probably after extrusion of sulphur. Tetrahydro-1-benzothiepinones could be obtained from 4-ethoxy-2-methyl-2*H*-1-benzothiapyran (*Parham et al.*, *loc.cit.*; J. org. Chem., 1964, **29**, 1575; 1969, **34**, 56).

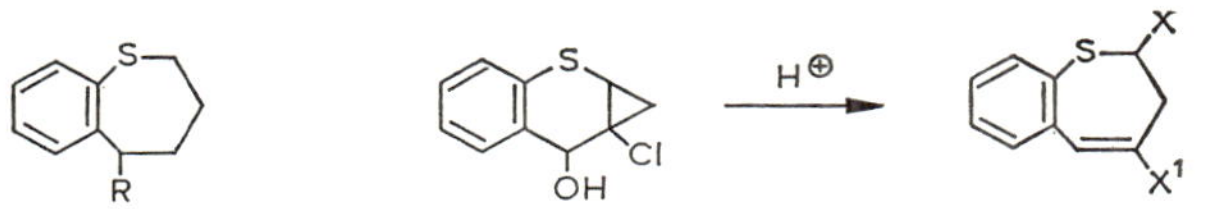

(XLI) (XLII) (XLIII)

2-Chloro- and 4-chloro-1-benzothiepins are unstable oils which extrude sulphur to give chloronaphthalenes at room temperature but the latter may be oxidised to its 1,1-dioxide. They are prepared by dehydrohalogenation of dichlorobenzothiepins (XLV; X = Cl) and XLIII (X, X^1 = Cl), respectively (*Traynelis* and *J. C. Sih*, unpublished).

1-Benzothiepin-1,1-dioxides are much more stable than 1-benzothiepins and may be prepared by *S*-oxidising a precursor before ring expansion (*e.g.* as shown above) or forming a substituted tetrahydro-1-benzothiepin 1,1-dioxide before introducing the two double bonds. 1-*Benzothiepin* 1,1-*dioxide*, m.p. 140–141°, is prepared from the 1,1-dioxide of 5-hydroxytetrahydro-1-benzothiepin (XLI; R = OH) by dehydration, allylic bromination and dehydrobromination; it distils with slight decomposition at about 250°. Catalytic reduction gives 2,3,4,5-*tetrahydro*-1-*benzothiepin* 1,1-*dioxide*, m.p. 77–78°, but addition of bromine takes place only on irradiation, first to the 2,3- and then to the 4,5-double bonds. Vigorous nitration gives the 6- and 8-mononitro derivatives (*Traynelis* and *R. F. Love*, Chem. and Ind., 1958, 439; J. org. Chem., 1961, **26**, 2728; 1964, **29**, 366). This sulphone may be prepared also directly by Friedel–Crafts cyclisation of 4-phenylbutanesulphonyl chloride (*Truce* and *J. P. Miloinis*, J. Amer. chem. Soc., 1952, **74**, 974).

2,3-**Dihydro-1-benzothiepin** (XLIII; X, X^1 = H), b.p. 95–96°, n_D^{20} 1.6470, is prepared by dehydration of alcohol XLI (R = OH) in dimethyl sulphoxide (*Traynelis et al.*, *loc.cit.*) or dehydrochlorination of the corresponding chloride (XLI; R = Cl) with collidine (*K. Sindelar* and *M. Protiva*, Coll. Czech. chem. Comm., 1968, **33**, 4315). The dihalogenocarbene route, effective in the oxepin series is reported to fail with 2,3- and 3,4-thiochromenes; only tricyclic adducts are obtained (*Parham et al.*, *loc.cit.*; J. Amer. chem. Soc., 1962, **84**, 813). Hydrogen and bromine add readily to the double bond (*Love*, Ph. D. Dissertation, University of Notre Dame, 1960) and fuming nitric acid gives 4-*nitro*-2,3-*dihydro*-1-*benzothiepin* (XLIII; X = H, X^1 = NO_2), m.p. 116–117°, *sulphoxide*, m.p. 197–198°. Compound XLIII (X, X^1 = H) forms alkylsulphonium tetrafluoroborates, a 1-oxide with sodium metaperiodate and a 1,1-*dioxide*, m.p. 106–107°, with hydrogen peroxide.

Better yields of 1,1-dioxide are obtained by dehydrating the 1,1-dioxide of alcohol XLI (R = OH), but the 1-oxide cannot be made from the sulphoxide of XLI (R = OH) (*J. R. Livingston, ibid.*, 1962).

3-*Methyl*-2,5-*dihydro*-1-*benzothiepin* (XLIV) is obtained as a by-product of the Pummerer reaction thus (*R. B. Morin et al.*, Tetrahedron Letters, 1969, 849):

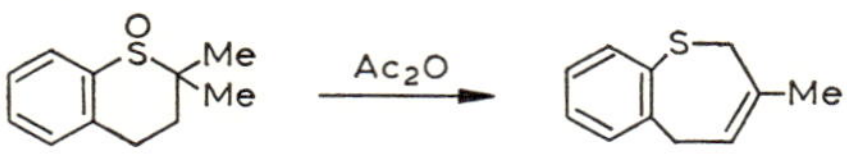

(XLIV)

2,4-*Dichloro*-2,3-*dihydro*-1-*benzothiepin* (XLIII); X, X^1 = Cl), m.p. 55–57°, is obtained in the ring expansion of intermediate XLII with acid, and 2,5-*dichloro*-4,5-*dihydro*-1-*benzothiepin* (XLV; X = Cl), is prepared by the action of thionyl chloride on 2-*chloro*-5-*hydroxy*-4,5-*dihydro*-1-*benzothiepin* (XLV; X = OH), m.p. 98–99°, which is made by treating the 4-nitrobenzoate of alcohol XLI (R = OH) with sulphuryl chloride and heating and hydrolysing the 2,2-dichloro derivative so formed (*Traynelis* and *Sih*, *loc.cit.*;

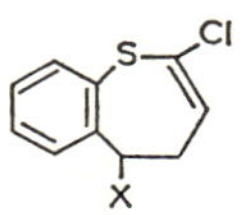

(XLV)

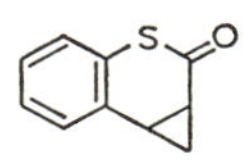

(XLVI)

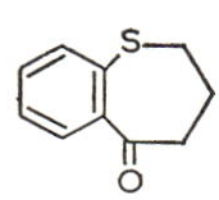

(XLVII)

Traynelis and *Livingston*, J. org. Chem., 1964, **29**, 1092). Treatment of XLV (X = OH) with strong acid yields not 2-chloro-1-benzothiepin but the product XLVI of transannular bonding. Formation of the 1,1-dioxide of XLV (X = OH) allows of easy dehydration to 2-chloro-1-benzothiepin 1,1-dioxide, but the 1-oxide of XLV (X = OH) yields 1-chloronaphthalene on dehydration (*Traynelis* and *D. M. Borgnaes*, *ibid.*, 1972, **37**, 3824).

The key intermediate in the preparation of reduced 1-benzothiepins is 2,3,4,5-**tetrahydro**-1-**benzothiepin**-5-**one** *(homothiochromanone)* (XLVII), b.p. 120°/1.5 mm, n_D^{20} 1.6230, *semicarbazone*, m.p. 222°, prepared by the cyclodehydration of 4-phenylthiobutyric acid with polyphosphoric acid (*Traynelis* and *Love*, *loc.cit.*) or the Friedel–Crafts cyclisation of the corresponding acid chloride (*P. Cagniant* and *A. Deluzarche*, Compt. rend., 1947, **223**, 667). Some substituted ketones have been made similarly (*P.* and *D. Cagniant*, Bull. Soc. chim. Fr., 1966, 3674; *H. Morren*, B.P. 1,112,681/1965; C.A., 1968, **69**, 35979).

Clemmensen reduction of XLVII gives 2,3,4,5-**tetrahydro**-1-**benzothiepin** *(homothiachroman)* (XLI; R = H), b.p. 140–141°/21 mm, n_D^{20} 1.6052, together with some 2,3-dihydro-1-benzothiepin (*Truce* and *Miloinis*, *loc.cit.*); the Wolff–Kishner reduction also gives XLI (R = H) (*Love*; *Traynelis* and *Borgnaes*, *loc.cit.*). Oxidation of homothiachroman gives the 1,1-*dioxide*, m.p. 77–78°, and Friedel–Crafts acetylation the 7-*acetyl* derivative, b.p. 217°/19 mm, (*Cagniant*, *loc.cit*; *M. J. Y. Foley* and *N. H. P. Smith*, J. chem. Soc., 1963, 1899).

Sodium tetrahydridoborate reduces XLVII to 5-*hydroxy*-2,3,4,5-*tetrahydro*-1-*benzothiepin* (XLI; R = OH), m.p. 70–71°, (*Traynelis* and *Love*, *loc.cit.*) and Grignard or Reformatsky reagents, and aryl-lithiums add to the carbonyl bond to give 5-substituted analogues of XLI (R = OH). Dehydration of these 5-alkyl-5-hydroxy compounds gives tetrahydro-1-benzothiepins with exocyclic double bonds at 5, some of which are obtained from Wittig reactions on XLVII (*F. Bickelhaupt et al.*, Chem. Ber., 1965, **98**, 685; *R. J. Mohrbacher*, U.S.P., 3,287,367/1965; C.A., 1967, **67**, 11439; *Morin*; *Sindelar* and *Protiva*, *loc.cit.*). Hydrochloric acid converts XLI (R = OH) into 5-*chloro*-2,3,4,5-*tetrahydro*-1-*benzothiepin*, b.p. 105–110°/0.4 mm, n_D^{20} 1.6159, and the hydroxy group may be alkylated and acylated also (*Love, loc.cit.*; *Morbacher* and *V. Paragamian*, U.S.P. 3,287,369/1965; C.A., 1967, **66**, 28691).

Oxidation of XLVII also gives the *sulphoxide*, m.p. 72–73°, (*Livingstone, loc.cit.*) and the 1,1-*dioxide*, m.p. 155–156°. Bromination gives 4-bromo substitution (*Love, loc.cit.*), while *N*-chlorosuccinimide or sulphuryl chloride gives the tricyclic ketone corresponding to XLII (*Traynelis* and *Sih, loc.cit.*). Position 4 is reactive in the Mannich reaction (*Mohrbacher*, U.S.P., 3,287,370/1965; C.A., 1967, **66**, 28754) and the Vilsmeier reaction (*M. Weissenfels et al.*, Z. Chem., 1966, **6**, 471), the latter yielding the chlorovinyl aldehyde

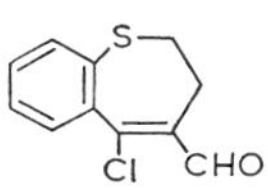

(XLVIII)

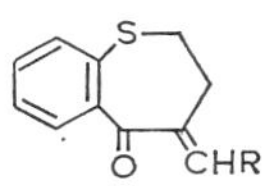

(XLIX)

XLVIII. Although benzaldehyde gives a 4-benzylidene derivative XLIX (R = Ph) and ethyl formate and morpholine an enamine XLIX (R = *N*-morpholinyl) with ketone XLVII under basic conditions, these exocyclic double bonds could not be isomerised into the thiepin ring (*Borgnaes*, Ph. D. Dissertation, University of Notre Dame, 1968).

2,3,4,5-*Tetrahydro*-1-*benzothiepin*-5-*one oxime*, m.p. 98–100°, is reduced by lithium tetrahydridoaluminate to 5-*amino*-2,3,4,5-*tetrahydro*-1-*benzothiepin* (XLI; R = NH_2), b.p. 115–117°/0.1 mm, n_D^{20} 1.6135, and the oxime tosylate rearranges with base to 4-amino-2,3,4,5-tetrahydro-1-benzothiepin-5-one, (*N. V. Dudykina* and *V. A. Zagorevskii*, C.A., 1966, **65**, 683; Zhur. org. Khim., 1966, **2**, 2222; *Mohrbacher* and *Paragamian*, *loc. cit.*; *Mohrbacher* and *E. L. Casson*, U.S.P., 3,243,439/1965; C.A., 1966, **64**, 19623). 5-Alkylamino derivatives (XLI; R = NHR^1) are formed by reducing *N*-acyl derivatives XLI (R = $NHCOR^2$), substitution of chlorine in XLI (R = Cl) by amino groups or from Leuckart reactions on ketone XLVII (*Sindelar* and *Protiva, loc.cit.*).

2-**Benzothiepin** (L) requires the benzene ring to adopt a quinonoid structure, and it is not surprising that neither the parent nor substituted examples of this system are known.

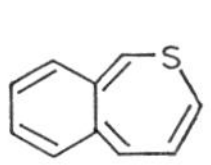

(L)

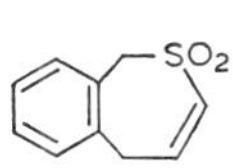

(LI)

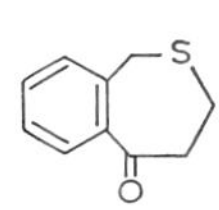

(LII)

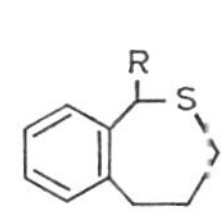

(LIII)

Treatment of benzylidene sulphone (phenylsulphene, PhCH = SO_2) with vinyl diazomethane gives 1,5-*dihydro*-2-*benzothiepin* 2,2,-*dioxide* (LI), m.p. 157–158°, probably *via* the episulphone (*Paquette* and *Maiorana, loc.cit.*).

2-**Benzothiepan**-5-**one** (LII), b.p. 183°/14 mm, n_D^{20} 1.6037, is obtained in 51% yield by the ring expansion of the product from 2-thia-1-indanone and vinyl-lithium on acid hydrolysis (*W. C. Lumma et al.*, J. org. Chem., 1970, **35**, 3442). A yield of less than 5% results on Friedel–Crafts cyclisation of 3-benzylthiopropionyl chloride (*P.* and *D. Cagniant*, Bull. Soc. chim. Fr., 1959, 1998). Wolff–Kishner reduction gives 2-benzothiepane.

2-Benzothiepane *(homoisothiachroman)* (LIII; R = H), b.p. 143°/15 mm, n_D^{24} 1.5810, is prepared also by the action of sodium sulphide on 2-(3-bromopropyl)benzyl bromide (*J. v. Braun*, Ber., 1925, **58**, 2165) or by the cyclisation of chloromethyl 3-phenylpropyl sulphide with aluminium chloride (*H. Bohme* and *B. Haack*, Chem. Ber., 1968, **101**, 2971). Two m.p.s (48–49° and 95–96°) are reported.

The latter method has also been used to prepare 1-*methyl*- (LIII; R = Me), b.p. 137°/12 mm, 1-*ethyl*- (LIII; R = Et), b.p. 95–97°/0.2 mm, and 1-*cyano*- (LIII; R = CN, m.p. 79–81°) -2-*benzothiepanes*. 1-Methyl-2-benzothiepane is also obtained in the anomalous Clemmensen reduction of 3-benzothiocin-1-one, and after loss of carbon monoxide during the cyclisation of $Ph(CH_2)_3SCH(Me)COCl$ (*P.* and *D. Cagniant*, *loc.cit.*). A little 3-*phenyl-2-benzothiepane*, m.p. 143–144°, is obtained when *cis*-2,4-diphenylthietane 1-oxide reacts with methylmagnesium iodide (*R. M. Dodson* and *P. D. Hammen*, Chem. Comm., 1968, 1294).

Chlorination of LIII (R = H) under ultraviolet radiation gives-1-*chloro-2-benzothiepane* (LIII; R = Cl); this chloro group could be replaced by the alkyl or aryl groups of Grignard reagents, or by alkoxy or cyanide ions. Most 2-benzothiepanes form 2,2-dioxides readily; they are alkylated at position 1, as are compounds LIII (R = CN), by the action of sodamide followed by an alkyl halide (*Böhme* and *Haack*, Arch., Pharm., 1969, **302**, 72).

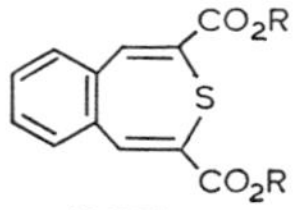

(LIV)

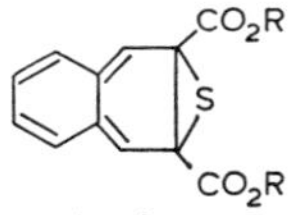

(LV)

Base-catalysed condensation of *o*-phthalaldehyde with diethyl thiodiacetate ($EtO_2C{\cdot}CH_2SCH_2{\cdot}CO_2Et$) gives 3-**benzothiepin-2,4-dicarboxylic acid** (LIV; R = H) which readily extrudes sulphur on heating to give naphthalene-2,3-dicarboxylic acid (*G. P. Scott*, J. Amer. chem. Soc., 1953, **75**, 6332; *J. D. Loudon* and *A. D. B. Sloan*, J. chem. Soc., 1962, 3262; *F. Dallacker et al.*, Ann., 1961, **643**, 82).

Esters LIV (R = Me), m.p. 95–97°, and LIV (R = Et), m.p. 75–77°, are more stable than the free acid, surviving solution in concentrated sulphuric acid and forming a *dibromo compound*, possibly the S-*bromo bromide*, m.p. 130–131°. Hydrolysis and loss of sulphur occurs with alkali, and diazomethane adds to one or both double bonds forming pyrazolines (*K. Dimroth* and *G. Lenke*, Angew. Chem., 1956, **68**, 519; Chem. Ber., 1956, **89**, 2608). The alternative structure LV (*A. Schönberg* and *M. B. E. Fayez*, J. org. Chem., 1958, **23**, 104) is shown to be incorrect by n.m.r. spectroscopy (*M. J. Jorgenson ibid.*, 1962, **27**,

3224); it may be a decomposition intermediate when sulphur is extruded (*R. Huisgen* and *M. Appl*, Chem. Ber., 1958, **91**, 18).

3-*Benzothiepin* 3,3-*dioxide* (LVIII) has been synthesised by the route shown (*Truce* and *F. J. Lotspeich*, J. Amer. chem. Soc., 1956, **78**, 848):

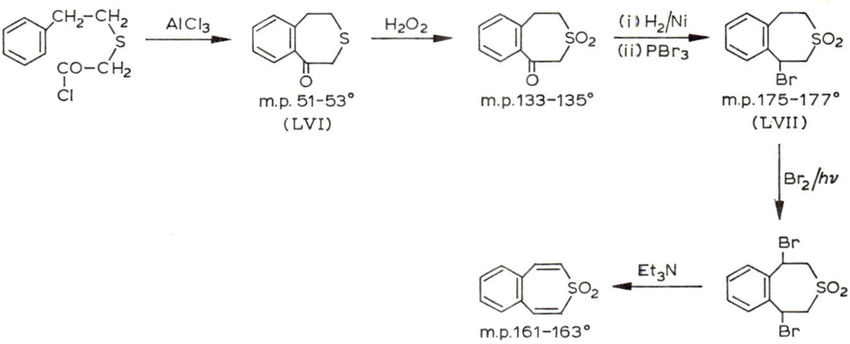

The initial ring closure has also been used to prepare substituted 1,2,4,5-**tetrahydro-3-benzothiepin-1-ones** (*v. Braun* and *K. Weissbach*, Ber., 1929, **62**, 2416; *P.* and *D. Cagniant*, *loc.cit.*; Bull. Soc. chim. Fr., 1960, 1798). In contrast to 1-benzothiepin 1,1-dioxide, sulphone LVIII extrudes sulphur dioxide at 250° to give naphthalene. The double bonds are readily oxidised by permanganate, brominated or reduced, and fuming nitric acid introduces a nitro group into the benzene ring, suggesting a puckered ring with little resonance stabilisation.

The action of triethylamine on bromosulphone LVII gives 1,2-*dihydro*-3-*benzothiepin* 3,3-*dioxide*, m.p. 135–137°, yielding the 1,2-dibromo derivative with bromine in chloroform (*Lotspeich*, J. org. Chem., 1963, **30**, 2068). It could not be dehydrogenated to LVIII.

Wolff–Kishner reduction of ketone LVI gives 1,2,4,5-**tetrahydro-3-benzothiepin,** b.p. 136–137°/15 mm, n_D^{20} 1.5907, (*P.* and *D. Cagniant*, *loc.cit.*), which is also prepared by the action of sodium sulphide on 1,2-bis(2-bromoethyl)benzene (*Truce* and *Lotspeich*, *loc.cit.*). Friedel–Crafts acylation gives 7-acyl derivatives (*Cagniant et al.*, Bull. Soc. chim. Fr., 1961, 1931).

6,7-*Dihydrodibenzo*[b,d]*thiepin*-7-*one* 5,5-*dioxide* (LIX), m.p. 167–168°, prepared by Dieckmann cyclisation of ethyl 2′-(methylsulphonyl)biphenyl-2-carboxylate is unlikely to exist in the enol form to an appreciable extent (*D. D. Emrick* and *Truce*, J. org. Chem., 1960, 25, 1103).

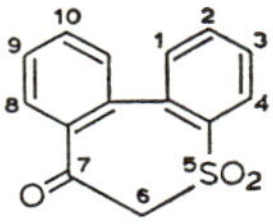

(LIX)

The physiological activity of "**Prothiaden**" [LXII; R = $(CH_2)_2NMe_2$] prompted investigation of the **dibenzo**[*b*,*e*]-**thiepin** system and a very large number of compounds of types LXII and LXI with substituted aminoalkyl side chains is known. A simple synthetic route which begins with Friedel–Crafts cyclisation or cyclodehydration of 2-carboxybenzyl phenyl sulphide to 6,11-*dihydrodibenzo*[b,e]*thiepin*-11-*one* (LX), m.p. 86–88°, is shown below (*Protiva et al.*, Experientia, 1962, **18**, 326; Czech. P., 105,590/1961; C.A., 1963, **59**, 10010; Coll. Czech. chem. Comm., 1964, **29**, 2161; *K. Stach et al.*, Monatsh., 1962, **93**, 889; Angew. chem. intern. Edn., 1962, **1**, 50; *S. O. Winthrop et al.*, J. med. pharm. Chem., 1962, **5**, 1207; *F. Gadient et al.*, Helv., 1962, **45**, 1860; *E. Jucker*, *A. J. Lindemannn* and *Gadient*, Swiss, P., 385,866/1965; C.A, 1965, **63**, 2962).

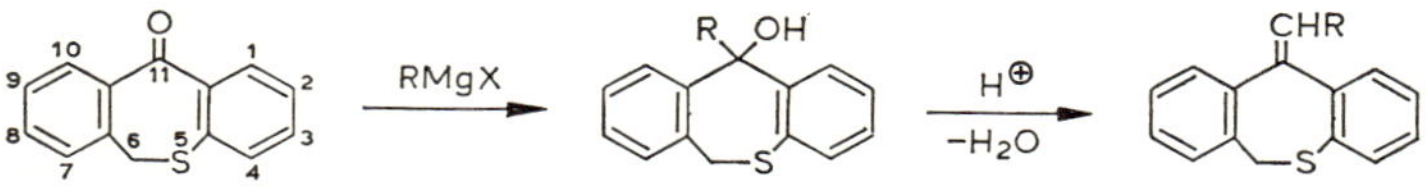

(LX) (LXI) (LXII)

Prothiaden forms a sulphoxide with hydrogen peroxide which undergoes the Pummerer reaction to the 6-acetoxy derivative of LXII [R = $(CH_2)_2NMe_2$] (*K. Sindelar* and *Protiva*, Coll. Czech. chem. Comm., 1970, **35**, 3328) and both LX and LXII form sulphones (*M. Rajsner* and *Protiva*, Czech. Pharm., 1962, **11**, 404; C.A., 1963, **59**, 2772). Base-catalysed condensations of LX with *N*,*N*-dimethylacetamide or acetonitrile at position 11 gives 11-hydroxy compounds LXI, which also dehydrate to compounds with structure LXII (*C. F. Böhringer und Söhne*, Belg. P., 23,259/1961; C.A., 1964, **60**, 10659; 1969, **70**, 19951, 37664). The exocyclic double bond is hydrogenated catalytically or chemically (*Protiva et al.*, Monatsh., 1965, **96**, 650).

Ketone LX forms an *oxime*, m.p. 223–225°, is reduced by sodium tetrahydridoborate or lithium tetrahydridoaluminate to 6,11-*dihydrodibenzo* [b,e]*thiepin*-11-*ol* (LXI; R = H), m.p. 107–108°, and also forms a bis-ether. When ketone LX or its oxime is reduced with zinc and acetic acid, 6,11-**dihydrobenzo** [*b*,*e*]**thiepin**, m.p. 103–105°, is the product (*Protiva et al.*, *ibid.*, 1965, **96**, 650). Alcohol LXI (R = H) may be converted into ethers, and with hydrogen chloride it gives 11-*chloro*-6,11-*dihydrobenzo* [b,e]*thiepin*, m.p. 81–82°. This chlorine atom is replaced in reaction with silver cyanide or amines (*Protiva et al.*, C.A., 1966, **64**, 17560; 1968, **68**, 21859; *P. Yonan*, U.S.P., 3,401,165/1965; C.A., 1969, **70**, 4144; *Sandox Ltd.* Neth. Appl., 6,608,716/1965; C.A., 1968, **68**, 49472).

(LXIII) (LXIV)

Dibenzo[*b*,*f*]**thiepin** (LXIV; R = H), m.p. 87–88 °, has been prepared by ring expansion of 10-hydroxymethylthioxanthene (LXIII; R = H) or its tosyl derivative; several substituted derivatives have been prepared similarly (*e.g.* LXIV; R = Ph), m.p. 150°, (*E. D. Bergmann* and *M. Rabinowitz*, J. org. Chem., 1960, **25**, 828; Israel. J. Chem., 1963, **1**, 125; *M. M. Urberg* and *E. T. Kaiser*, J. Amer. chem. Soc., 1967, **89**, 5931; *L. L. Zirkle*, U.S.P., 3,100,207/1963; C.A., 1964, **60**, 1719): A similar route to dibenzo[*b,f*]thiepin and some 10,11-dihydro derivatives is through the reaction of diazomethane with thioxanthene perchlorate (*H. W. Whitlock* Tetrahedron Letters, 1961, 593; *H. Seidl* and *K. Biemann*, J. heterocycl. Chem., 1967, **4**, 209).

10-*Chlorodibenzo* [b,f]*thiepin* (LXIV; R = Cl), m.p. 91–93°, is made by cyclising 2-phenylthiophenylacetic acid with phosphoryl chloride and zinc chloride (*Protiva et al.*, Coll. Czech. chem. Comm., 1967, **32**, 3186; 1968, **33**, 1852) and 2-*nitrodibenzo* [b,f]*thiepin*, m.p. 111–112°, is prepared by decarboxylation of the 2-nitro derivative of LXIV (R = CO_2H) made by cyclising 4-nitro-2-phenylthiophenylpyruvic acid (*J. D. Loudon et al.*, J. chem. Soc., 1957, 3814). Acid-catalysed eliminations of water, alcohols or amines (*Protiva et al.*, Monatsh., 1965, **96**, 182; *J. C. L. Fouche* F.P., 1,516,783/1960; *R. Gaumont* and *C. A. G. Gueremy*; C.A., 1969, **70**, 106407) and base-catalysed elimination of hydrogen halides (*J. R. Geigy A.G.*, Neth. Appl. 6,407,588/1965; 6,514,187/1966; C.A., 1965, **63**, 1777; 1966, **65**, 12183) from 10-substituted-10,11-dihydrodibenzo [*b,f*]thiepins are other general routes to LXIV.

In contrast to benzothiepins, LXIV extrudes sulphur only at high temperatures in the presence of copper bronze; phenanthrene derivatives are produced (*Loudon et al., loc.cit.*). Catalytic reduction, or sodium in alcohol hydrogenates the 10,11 double bond of LXIV but a reducing mixture of hydrogen iodide, phosphorus and calcium hypophosphite causes ring-contraction to 10-methylthioxanthene (*Fouche, loc.cit.*; *Protiva et al.*, Coll. Czech. chem. Comm., 1967, **32**, 3186).

Dibenzo [*b,f*]thiepin is converted to its 5-oxide by iodosobenzene diacetate (*A. Trifunac* and *E. T. Kaiser*, J. phys. Chem., 1970, **74**, 2236) and to its 5,5-*dioxide* (LXV; R = H), m.p. 171–172°, by peracids. Oxidation with permanganate gives bis(2-carboxyphenyl) sulphide or brings about ring-contraction to thioxanthone (*Bergmann* and *Rabinowitz*, *loc.cit.*; *Protiva*, Coll. Czech. chem. Comm. 1968, **33**, 1831).

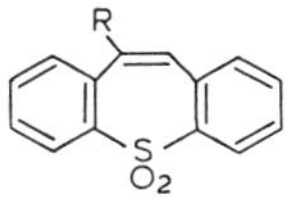

(LXV)

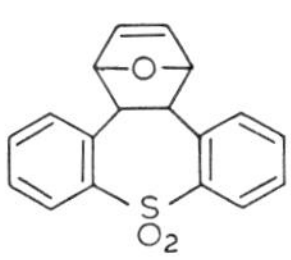

(LXVI)

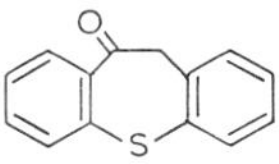

(LXVII)

The dioxide adds bromine to give 10,11-dibromo-10,11-dihydrodibenzo [*b,f*]thiepin 5,5-dioxide which, on treatment with triethylamine, yields 10-*bromodibenzo* [b,f]*thiepin* 5 5-*dioxide* (LXV; R = Br), m.p. 179–180°. With strong bases, in the presence of furan, LXV (R = Br) gives the Diels-Alder adduct LXVI, suggesting that a cyclic acetylene is an intermediate. Hydrogenation and dehydration of LXVI gives **tribenzo** [*b,d,f*]**thiepin** 9,9-**dioxide** (*W. Tochtermann et al.*, Angew. Chem. intern. Edn., 1967, 6, 370; Ann., 1967, 701, 117).

10,11-**Dihydrodibenzo**[*b,f*]**thiepin**-10-**one** (LXVII), m.p. 72–74°, and many of its derivatives have been prepared by the cyclisation of 2-(phenylthio)-phenylacetic acid. These compounds are the key intermediates of dibenzo-[*b,f*]thiepin chemistry (*Protiva et al.*, Coll. Czech. chem. Comm., 1968, **33**, 1831, 1852, 2666; 1969, **34**, 3936; 1970, **35**, 276; Monatsh., 1965, **96**, 182; *Loudon et al., loc.cit.*; *S. Kimoto et al.*, Yakugaku Zasshi, 1968, **88**, 1323; C. A., 1969, **70**, 47273). The ease of conversion of LXVII into 10-chloro-dibenzo[*b,f*]thiepin with phosphoryl chloride and its reluctance to react with complex Grignard reagents suggests that it is extensively enolised.

Its *oxime*, m.p. 143°, is reduced by sodium and alcohol to 10-*amino*-10,11-*dihydrodibenzo*-[*b,f*]*thiepin* (LXVIII; R = NH_2], m.p. 56–57°, but with lithium tetrahydridoaluminate the reduction product is the aziridine LXIX. Nitrous acid on LXIX gives dibenzo[*b,f*]-thiepin, but mineral acid causes ring-contraction to 10-formylthioxanthene (*Fouche*, Bull. Soc. chim. Fr., 1970, 1376).

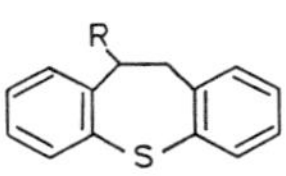
(LXVIII)

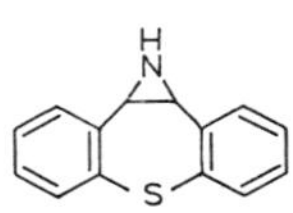
(LXIX)

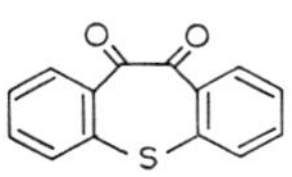
(LXX)

Wolff–Kishner reduction of LXVII gives 10,11-**dihydro-dibenzo**[*b,f*]-**thiepin** (LXVIII; R = H), m.p. 143–144°, b.p. 125–130°/0.1 mm, and reduction with sodium tetrahydridoborate or lithium tetrahydridoaluminate gives 10-*hydroxy*-10,11-*dihydrodibenzo*[b,f]*thiepin* (LXVIII; R = OH), m.p. 99–100°, (*J. R. Geigy A.G., loc.cit.*; *Protiva et al.*, Monatsh., 1965, **96**, 182).

The hydroxy group of LXVIII (R = OH) is replaced by chloro on treatment with hydrogen chloride and calcium chloride, and 10-*chloro*-10,11-*dihydrodibenzo*[b,f]*thiepin* (LXVIII; R = Cl), m.p. 84–85°, reacts with amines to give physiologically active 10-amino derivatives of LXVIII, including **"Perathiepin"** (LXVIII; R = 4-methylpiperazino), m.p. 134–136°, (*Protiva et al.*, Coll. Czech. chem. Comm., 1968, **33**, 1831, 1852).

The 10-oxo group of LXVII activates position 11 to attack by bromine, nitrous acid and alkyl halides with sodamide (*W. Schindler* and *H. Blattner*, Swiss. P., 443,344/1964; C.A., 1968, **69**, 52035); LXVII also undergoes cyanoethylation and the Mannich reaction. Oxidation of LXVII gave the 5,5-dioxide of 10,11-**dihydrodibenzo**[*b,f*]**thiepin-10,11-dione** (LXX), m.p. 135–136°, 5,5-*dioxide*, m.p. 278–280°. Both LXX and its dioxide undergo benzilic acid rearrangements to 10-thiaxanthone, and its 5,5-dioxide (*Protiva et al.*, Coll. Czech. chem. Comm., 1967, **32**, 3186) and LXX and its mono-oxime undergo a similar ring contraction in acidified aqueous ethanol to ethyl thioxanthene-9-carboxylate (*A. Lüttringhaus* and *G. Creutzburg*, Angew. Chem. intern. Edn., 1968, **7**, 128). The 5,5-dioxide does not undergo this ring contraction.

Dibenzo[*c,e*]**thiepin** (LXXI) has not yet been prepared, but 5,7-*dihydro-dibenzo*[c,e]*thiepin* (LXXII; R = H), m.p. 89–90°, is found in the reaction between 2,2′-bis(bromomethyl)biphenyl and sodium sulphide. Several substituted derivatives are made similarly (*Truce* and *Emrick*, J. Amer. chem. Soc., 1956, **78**, 6130; *K. Mislow et al., ibid.*, 1964, **86**, 1710; *L. A. Paquette, ibid.*, 1964, **86**, 4085; *R. L. Taber et al.*, J. heterocycl. Chem., 1965, **2**, 181). Selective desulphurisation of dibenzo[*d,f*]1,2-dithiocin also gives LXXII (R = H) (*D. N. Harpp* and *J. G. Gleason*, J. Amer. chem. Soc., 1971, **93**, 2437).

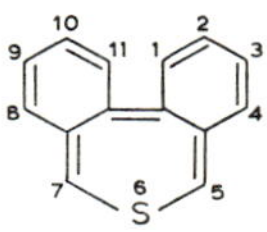

(LXXI)

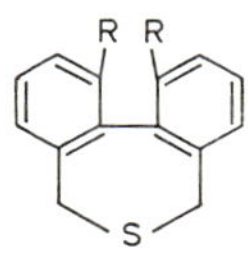

(LXXII)

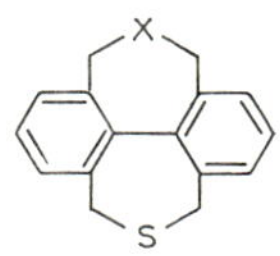

(LXXIII)

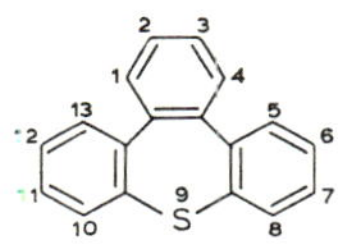

(LXIV)

Oxidation of LXXII (R = H) with peracids gives the 6,6-*dioxide*, m.p. 209–210°, and derivatives are also oxidised this way (*Paquette*, J. Amer. chem. Soc., 1964, **86**, 4089; *R. R. Fraser* and *F. J. Schuber*, Canad. J. Chem., 1970, **48**, 633).

The two phenyl rings of LXXII and its sulphone are forced out of coplanarity by the dissymetrically twisted sulphur-containing ring and resolution into optical isomers has been possible for derivatives of both series. Introduction of substituents (*R*) at positions 1 and 11 increases both the degree of twisting and the optical stability of the enantiomers (*I. O. Sutherland* and *M. V. J. Ramsay*, Tetrahedron, 1965, 21, 3401; *R. J. Kurland et al.*, J. chem. Phys., 1964, **40**, 2426).

Nitration of LXXII gives the 3- and 3,9-nitro derivatives and with methyl iodide 6-methylthiepinium iodides are formed. Raney nickel desulphurisation yielded 2,2′-biphenyl and some anthracene (*G. M. Badger et al.*, J. chem. Soc., 1962, 3241) and with sulphuryl chloride 5-, 5,5- and even 5,7-chloro substitutents are introduced. 5-*Chloro*-5,7-*dihydrodibenzo* [c,e]*thiepin* 6,6-*dioxide*, m.p. 204–205°, extruded sulphur dioxide when treated with base to give anthracene presumably *via* an intermediate with a 5,7-bond. This route is sterically unfavourable for 1,11-substituted derivatives and reaction does not occur (*Paquette, loc.cit.*). If compound LXXII (R = CH_2OH) reacts with *p*-toluenesulphonic acid, or LXXII (R = CH_2Br) reacts with sodium sulphide the respective products are the tetracyclic compounds LXXIII (X = O) and LXXIII (X = S) (*Mislow et al., loc.cit.*; J. Amer. chem. Soc., 1961, **83**, 2780).

Tribenzo[*b,d,f*]**thiepin** (LXXIV), m.p. 117–118 °, is made by the thermal decomposition of 2-chlorosulphonyl-2′-phenyldiphenyl sulphide at 250–260°.

It is stable to 400°, but extrudes sulphur both in the presence of Raney nickel, giving *o*-terphenyl, and copper bronze, when triphenylene is the product (*E. B. McCall* and *T. J. Rawlings*, Tetrahedron, 1965, **21**, 1299; C.A., 1966, **64**, 9697). With peracids, LXXIV gives the 9,9-*dioxide*, m.p. 200–201°, also prepared by *Tochtermann's* route (*vide supra*, p. 183).

(c) *Seven-membered rings containing two or more sulphur atoms*

(*i*) 1,2-*Dithiepins*

Only fully reduced 1,2-dithiepins are known, even though the anion of 3*H*(5*H*)-1,2-dithiepin(LXXV) is considered the most accessible of the sulphur-containing cyclononatetraenide anion analogues (*R. Zahradnik* and *C. Parkanyi*, Coll. Czech. chem. Comm., 1965, **30**, 3016). 3,7-Dimethyl-1,2-dithiepin-4-thione (LXXVI), suggested as a product from diacetylacetone and phosphorus pentasulphide (*F. Arndt et al.*, Chem. Ber., 1956, **89**, 730; *G. Traverso et al.*, *ibid.*, 1958, **91**, 1224 and earlier papers) is now considered to be the thiothiophthene LXXVII (*S. Benzi et al.*, Nature, 1958, **182**, 247; Gazz., 1958, **88**, 1226).

(LXXV)

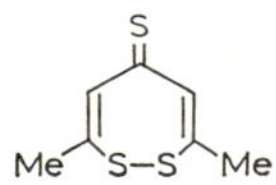
(LXXVI)

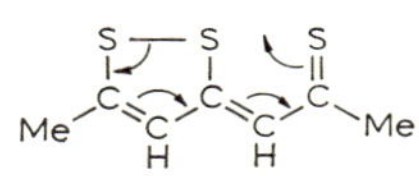
(LXXVII)

Cyclic disulphides, including 1,2-**dithiepane** (LXXVIII) b.p. 82°/14 mm, n_D^{25} 1.570, have been made by the reaction of "Bunte salts", $NaO_3S{\cdot}S(CH_2)_nS{\cdot}SO_3Na$ (from sodium thiosulphate and the appropriate α,ω-dibromoalkane) with hot aqueous cupric chloride (*J. E. Affleck* and *G. Dougherty*, J. org. Chem., 1950, **15**, 865) or distilling the "Bunte salts" or α,ω-dithiocyanates from aqueous sodium sulphide (*N. A. Rosenthal*, U.S.P., 5,284,466/1963; C.A., 1968, **68**, 13015). Much better yields are obtained when alkane-α,ω-dithiols are oxidised by ferric chloride using high dilution techniques, (*A. Schöberl* and *H. Gräfje*, Angew. Chem., 1957, **69**, 713; Ann., 1958, **614**, 66).

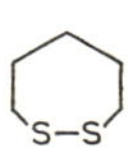
(LXXVIII)

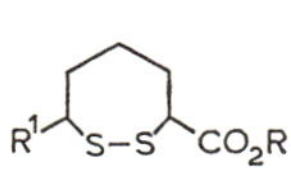
(LXXIX)

Other oxidising agents, hydrogen peroxide, iodine, oxygen, give lower yields (*H. F. Herbrandson* and *R. H. Wood*, J. med. Chem., 1969, **12**, 617, 620; *L. Schotte*, Arkiv Kemi, 1956, **9**, 413; *L. Field* and *R. B. Barbee*, J. org. Chem., 1969, **34**, 36).

1,2-*Dithiepane*-3,7-*dicarboxylic acid* (LXXIX; R = H, $R^1 = CO_2H$), racemic, m.p. 171–173°, meso, m.p. 193–194°, is prepared by treating 2,6-dibromopimelic acid with sodium disulphide (*Schotte*, Arkiv Kemi, 1956, **9**, 441), and ethyl 2,6-dibromohexanoate cyclises similarly with potassium hydrogen sulphide to LXXIX (R^1 = H, R = Et) (*C.-E. Yuan* and *M. N. Shchukina*, Zhur. obshcheĭ Khim., 1957, **27**, 1103).

Attempts have been made to assess the relative reactivity of the cyclic 1,2-disulphides based on ease of polymerisation, with or without catalysts, ease of cleavage by cyanide ion to dithiocyanates and estimated ring strain. Reactivity increases steadily from six- to ten-membered rings, after which it decreases (*Affleck* and *Dougherty*; *Schöberl* and *Gräfje*, *loci cit.*; *F. S. Dainton et al.*, Trans. Faraday Soc., 1960, **56**, 1784).

1,2-Dithiepane (LXXVIII) is oxidised to the 1,1-*dioxide*, m.p. 25°, by hydrogen peroxide, and to 1,5-pentanedisulphonic anhydride by ozone. The 1,1-dioxide is converted to 1,2-*dithiepane*-1,1,2,2-*tetroxide*, m.p. 159–160°, by hydrogen peroxide in the presence of tungstic acid (*Field* and *Barbee*, *loc.cit.*). Ease of hydrolysis of the sulphur–sulphur bond increases on going from the cyclic disulphide LXXVIII to its 1,1-dioxide, and again from 1,1-dioxide to 1,1,2,2-tetroxide (*idem*, J. org. Chem., 1969, **34**, 1792). Irradiation of LXXVIII gives pentane-1,5-dithiol in neutral, and thiane in alkaline solution together with much polymer (*V. Ramakrishnan et al.*, Photochem. Photobiol, 1965, **4**, 907) but 1,2-dithiepanes are insensitive to photo-oxidation (*R. W. Murray* and *S. L. Jindal*, J. org. Chem., 1972, **37**, 3516).

(*ii*) 1,3-*Dithiepins*

4,7-*Dihydro*-2H-1,3-*dithiepin* (LXXX; R = H) and its 2,2-*dimethyl* homologue (LXXX; R = Me), b.p. 58–61°/0.4 mm, have been reported (*A. Luttringhaus et al.*, Tetrahedron Letters, 1964, 1929). Photolysis of LXXX gives complex ring-contracted products (*R. E. Kohrmann* and *G. A. Berchtold*, J. org. Chem., 1971, **36**, 3971). Oily samples of 1,3-**dithiepane** (LXXXI; R, R^1 = H) and its 2-ethyl derivative (LXXXI; R = H, R^1 = Et), and 2,2-*dimethyl*-1,3-*dithiepane* (LXXXI; R, R^1 = Me), b.p. 94°/7 mm, have been prepared in Lewis acid-catalysed condensations between 1,4-butanedithiol and formaldehyde acetal, propionaldehyde acetal and acetone, respectively (*S. Oae et al.*, Tetrahedron, 1964, **20**, 427, 437).

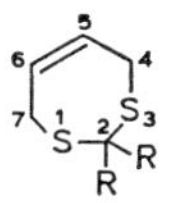

(LXXX)

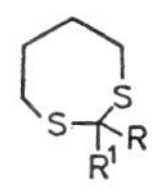

(LXXXI)

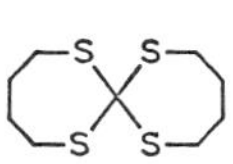

(LXXXII)

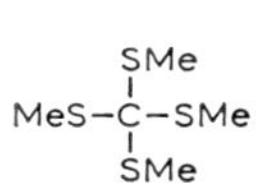

(LXXXIII)

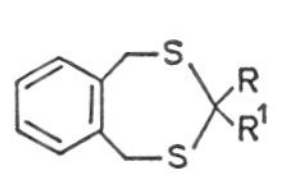

(LXXXIV)

Compound LXXXI (R = H, R^1 = CO_2Et) is made similarly; hydrolysis gives the 2-*carboxylic acid* (LXXXI) (R = H, R^1 = CO_2H), m.p. 85–86°, which may be decarboxylated (*idem*, *ibid.*, 1968, **24**, 5283, 5293). *Spiro-bis*-1,3-*dithiepane* (LXXXII), m.p. 165°, is obtained in the acid-catalysed condensation of tetra(methylthio)methane (LXXXIII) with butane-1,4-dithiol (*D. L. Coffen*, J. heterocycl. Chem., 1970, **7**, 201).

Only brief accounts are found in the literature of a few members of the benzodithiepin series. Several 1,5-**dihydro**-3*H*-2,4-**benzodithiepins** of type LXXXIV have been obtained by condensing 1,2-di(mercaptomethyl)benzene with carbonyl compounds in the presence of dry hydrogen chloride; aldehydes condense much more readily than ketones (*W. Autenrieth* and *R. Hennings*, Ber., 1901, **34**, 1772; 1902, **35**, 1388; *I. Shahak* and *E. D. Bergmann*, J. chem. Soc, C, 1966, 1005).

1,5-*Dihydro*-3H-2,4-*benzodithiepin* (LXXXIV; R, R^1 = H), m.p. 152–153°, results when the dithiol reacts with formaldehyde or diiodomethane (*A. Kotz*, Ber., 1900, **33**, 729). Its thioacetal ring is stable to acids, bases, organolithium or magnesium reagents, sodium tetrahydridoborate and lithium tetrahydridoaluminate, but is opened by mercuric salts and desulphurised with Raney nickel. Compound LXXXIV (R, R^1 = Me) has m.p. 136–137° (*Kohrmann* and *Berchtold, loc.cit.*).

Routes to 6H-**dibenzo**[*b*,*d*]-1,3-**dithiepins** LXXXV, resemble those used to prepare the thioacetals LXXXIV; aldehydes, ketones and *gem*-dihalides condense with biphenyl-2,2-dithiol to give the seven-membered ring system (*H. J. Barber* and *S. Smiles*, J. chem. Soc., 1928, 1141; *R. Breslow* and *E. Mohacsi*, J. Amer. chem. Soc., 1963, **85**, 431).

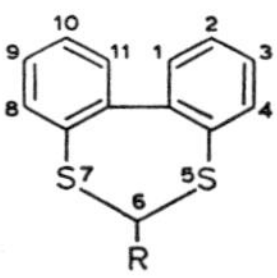

(LXXXV)

Compounds LXXXV (R = CO_2R^1) are acidic and lose the 6-proton to give a 1,3-dithiepinyl anion even with weak bases. This contrasts with unsuccessful attempts to form the corresponding cation (*D. Dürr*, Ph. D. Dissertation, University of Basel, 1961).

(*iii*) 1,4-*Dithiepins*

A little 2-*methyl*-6,7-*dihydro*-5H-1,4-*dithiepin* (LXXXVI; R = Me), b.p. 73–75°/0.5 mm, n_D^{26} 1.5881, is formed by cyclising 2-chloroallyl 3-mercaptopropyl sulphide, $CH_2{=}CCl{\cdot}CH_2S(CH_2)_3SH$, with potassium *tert*-butoxide (*A. T. Bottini* and *E. F. Böttner*, J. org. Chem., 1966, **31**, 586), and an aryl analogue (LXXXVI; R = Ar) is formed in the phosphorus pentoxide-induced ring-expansion of the thioacetal LXXXVII (*J. E. Hodgekins et al., ibid.*, 1970, **35**, 823).

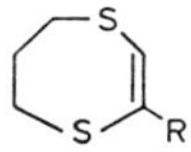

(LXXXVI)

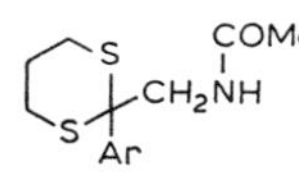

(LXXXVII)

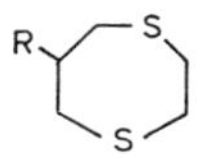

(LXXXVIII)

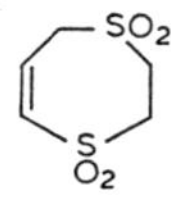

(LXXXIX)

Low yields of 1,4-**dithiepane** (LXXXVIII; R = H), m.p. 47°, b.p. 221–222°, have been prepared by the reactions both of ethylene bromide with 1,3-propanedithiol and 1,2-ethanedithiol with 1,3-dibromopropane. Oxidation with permanganate yields the 1,1,4,4-*tetroxide*, m.p. 287–288°, (*W. Autenrieth* and *K. Wolff*, Ber., 1899, **32**, 1375; *E. E. Reid et al.*, J. Amer. chem. Soc., 1933, **55**, 775; 1934, **56**, 2177; *R. C. Fuson* and *A. J. S. Speziale*, *ibid.*, 1949, **71**, 823). If either 1,3-dibromo-2-propanol or 2,3-dibromo-1-propanol reacts with 1,2-ethanedithiol in the presence of sodium ethoxide, 6-*hydroxy*-1,4-*dithiepane* (LXXXVIII; R = OH), m.p. 64–65°, is the product. Oxidation and dehydration of this alcohol gives 2,3-*dihydro*-5H-1,4-*dithiepin* 1,1,4,4-*tetroxide* (LXXXIX), m.p. 280°, while treatment with thionyl chloride gives ring-contraction to 2-chloromethyldithiane possibly *via* a thiiranium intermediate (*idem, ibid.*, p. 1582).

Dieckmann cyclisation of the diester ($(EtO_2C{\cdot}CH_2SCH_2\text{–})_2$) followed by hydrolysis and decarboxylation (*A. Lüttringhaus* and *H. Prinzbach*, Ann., 1959, **624**, 79) or condensation of 1,2-ethanedithiol with 1,3-dichloroacetone (*E. G. Howard Jr.* and *R. V. Lindsey*, J. Amer. chem. Soc., 1960, **82**, 158; *Howard*, U.S.P. 2,965,650/1960; C.A., 1961, **55**, 14492) gives good yields of 1,4-*dithiepan*-6-*one* (XC), m.p. 14°, b.p. 75°/0.1 mm, n_D^{25} 1.5925.

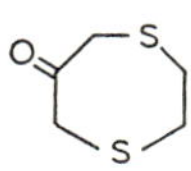
(XC)

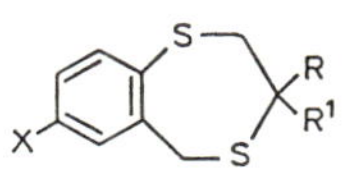
(XCI)

Oxidation with hydrogen peroxide gives first the 1,4-dioxide then the 1,1,4,4-*tetroxide*, m.p. 252–253° (decomp.), also prepared from LXXXVIII (R = OH). The carbonyl group gives a semicarbazone and an acetal and activates the adjacent methylene groups to attack by benzaldehyde.

The reaction of 2-mercaptomethylbenzenethiol with chloroacetone yields 3-*hydroxy*-3-*methyl*-2,3-*dihydro*-5H-1,4-*benzodithiepin* (XCI; R = Me, R^1 = OH, X = H), m.p. 95–98°, with phosphoryl chloride this alcohol is dehydrated to 3-*methyl*-5H-1,4-*benzodithiepin*, m.p. 100–102°. Catalytic reduction gives XCI (R = Me, R^1, X = H) (*Kohrmann* and *Berchtold, loc.cit.*).

The action of base on 2-(2-chloroethylthio)-5-chlorobenzylisothiouronium chloride gave 7-*chloro*-2,3-*dihydro*-5H-1,4-*benzodithiepin* (XCI; R, R^1 = H, X = Cl), m.p. 91–92°, b.p. 180–182°/12 mm (*M. Kulka*, Canad. J. Chem., 1958, **36**, 750) and 4,5-dimethyl-1,2-benzenedithiol condenses with 1-chloro-2-chloromethyl-2-propene to give 7,8-*dimethyl*-2-*methylene*-1,5-*benzodithiepin* (XCII) (*H. Richter et al.*, Z. Chem., 1968, **8**, 220):

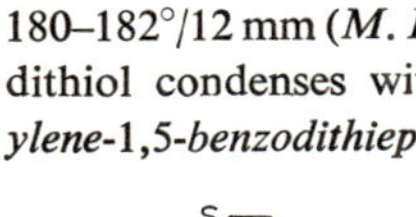
(XCII)

(iv) *Trithiepins*

4,7-**Dihydro**-1,2,3-**trithiepin** (XCIII) and 1,2,3-**trithiepane** have been prepared in low yields by treating the bis-"Bunte salts" with sodium sulphide (*Lüttringhaus et al.*, Z. Naturforsch., 1966, **21B**, 320). This route is also used to prepare 1,5-**dihydro**-2,3,4-**benzotrithiepin** (XCIV), m.p. 101–102°, (*idem*, Tetrahedron Letters, 1966, 719; *B. Milligan* and *J. M. Swan*, J. chem. Soc., 1965, 2901).

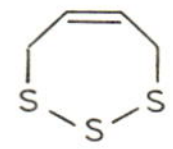

(XCIII)

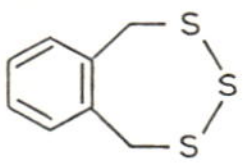

(XCIV)

Monoperphthalic acid gives the 2-*oxide*, m.p. 133°. The insertion of a third sulphur atom appears to increase the conformational stability of the heterocyclic ring (*Lüttringhaus et al.*, Angew. Chem., intern. Edn., 1964, **3**, 650).

The cleavage of 5-amino-4-chloro-1,2-dithiacyclopenten-3-ones (XCV) by alcoholic potassium hydroxide is found to give 1,2,5-trithiepins of the type XCVI (R = Aryl, R^1, R^2 = Me or Et). The trithiepin ring is sensitive to alkali at room temperature, but opens only at the boiling point of aqueous acids (*F. Boberg*, *ibid.*, 1964, **3**, 645; Ann., 1965, 683, 132).

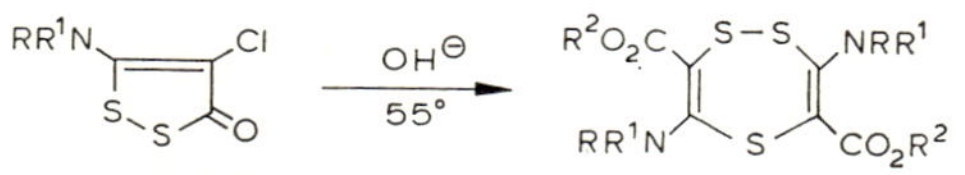

(XCV) (XCVI)

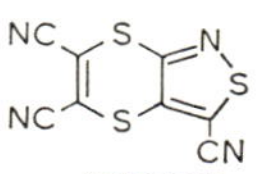

(XCVII)

Desulphurisation with Raney nickel gave mixtures of *N*-substituted 3-amino-acrylic and -propionic esters. A compound reported as 3,4,6,7-tetracyano-1,2,5-trithiepin (*A. v. Schorr et al.*, U.S.P., 3,000,780/1959; G.P., 1,095,582/1959; C.A., 1962, **56**, 5162, 12038) has been identified as the isothiazole XCVII (*T. L. Cairns*, *et al.*, J. Amer. chem. Soc., 1962, **84**, 4746).

Authentic 1,2,5-**trithiepane** (CXVIII), b.p. 61–63°/0.2 mm, n_D^{25} 1.6424, has recently been prepared by oxidation of di(2-mercaptoethyl) sulphide with ferric chloride (*Field* and *C. H. Foster*, J. org. Chem., 1970, **35**, 749).

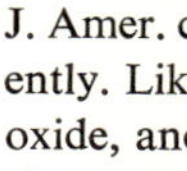

(XCVIII)

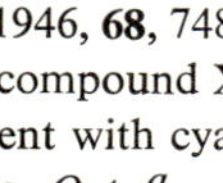

(XCIX)

Compounds previously given this structure must now be regarded as polymers (*E. Fromm* and *H. Jörg*, Ber., 1925, **58**, 304; *P. C. Ray*, J. chem. soc., 1924, 1141; *H. E. Westlake et al.*, J. Amer. chem. Soc., 1946, **68**, 748). The sulphide and disulphide functions react independently. Like thiepane, compound XCVIII forms a 5-methylsulphinium iodide and 5-sulphoxide, and on treatment with cyanide ion it resembles 1,2-dithiepane in its reluctance to cleave and polymerise. *Octafluoro*-1,2,5-*trithiepane*, b.p. 40–42°/20 mm, is formed when sulphur reacts with tetrafluoroethylene (*C. G. Krespan* and *W. R. Brasen*, J. org. Chem., 1962, **27**, 3995). On strongly heating, this trithiepane gives octafluoro-1,4-dithiane, and with base catalysts it gives low molecular weight polymers. Chlorine cleaves the disulphide link giving the sulphenyl chloride $ClS(CF_2)_2S(CF_2)_2SCl$ (*idem*, U.S.P., 3,099,688/1963; C.A., 1964, **60**, 1597).

Disodium dimercaptomaleonitrile reacts with bis(chloromethyl) sulphide to give 6,7-*dicyano*-1,3,5-*trithiepin* (XCIX), m.p. 112–114°.

A number of seven-membered rings containing more than three sulphur atoms is known. 1,2,4,6-**Tetrathiepane** (C), m.p. 83–84°, is obtained when the precipitate in the acid-catalysed reaction of formaldehyde and hydrogen sulphide is cyclised with iodine (*E. Baumann*, Ber., 1890, **23**, 1869), or when dichloromethane reacts with $Na_2S_{2.2}$ at pH 12 (*K. Morita* and *S. Kobayashi*, Chem. pharm. Bull. Japan, 1967, **15**, 988).

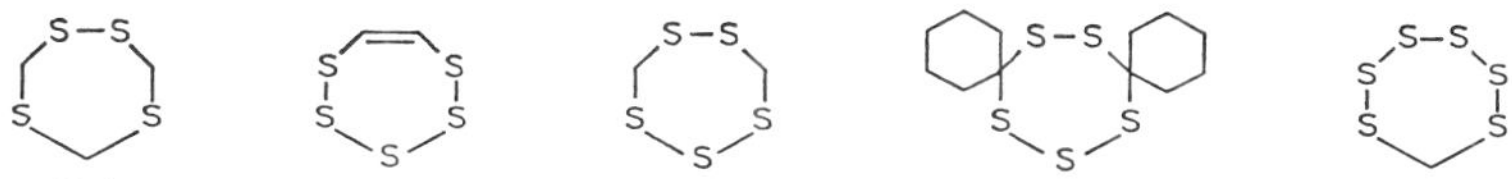

(C) (CI) (CII) (CIII) (CIV)

The unsaturated 1,2,3,4,5-**pentathiepin** (CI) is prepared in 80% yield from *cis*-mercaptoethylene and dichlorotrisulphane (S_3Cl_2), and though unstable, forms a Diels–Alder adduct with cyclopentadiene (*F. Feher* and *M. Langer*, Tetrahedron Letters, 1971, 2125).

Lenthionine (CII), m.p. 60–61°, the naturally occurring 1,2,3,5,6-**pentathiepane** is the odoriferous principle of the edible mushroom *Lentinus edodes* (Shiitake). Low yield total syntheses have been described from sodium polysulphide with formaldehyde or methylene chloride (*Morita et al.*, *ibid.*, 1966, 573; J. food Sci., 1967, **32**, 559). Other 1,2,3,5,6-pentathiepanes with 4,4,7,7 substituents have been isolated from the reaction of ketones with ammonium disulphide (*A. Fredga*, Acta Chem. Scand., 1958, **12**, 891; *B. Magnusson*, *ibid.*, 1959, **13**, 1031, 1715; *F. Asinger et al.*, Ann., 1959, **627**, 195). Reduction of the bis-(pentamethylene)pentathiepine CIII with zinc and acid gives cyclohexane-1,1-dithiol, and with lithium tetrahydridoaluminate gives cyclohexanethiol (*F. O.* and *P. A. Bobbio*, Chem. Ber., 1965, **98**, 998).

Like C and CII, **hexathiepane** (CIV), sinters at 79°, is found in a species of mushroom, and may be isolated from dichloromethane–$Na_2S_{2.5}$ reaction mixtures (*Morita et al.*, *loc.cit.*).

4. Heterocyclic seven-membered ring compounds containing selenium

The chemistry of selenepin and its derivatives has attracted little attention from organic synthetic chemists. The parent system is unknown, but the perhydro derivative **selenepane** (I), b.p. 105–106°/68 mm, n_D^{18} 1.5470, is obtained by treating 1,6-dibromohexane with sodium selenide. 1,8-Diselenacyclotetradecane and low molecular weight poly-(selenohexamethylene) are found as by-products.

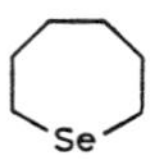
(I)

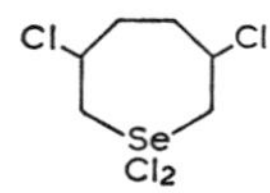
(II)

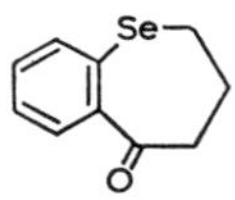
(III)

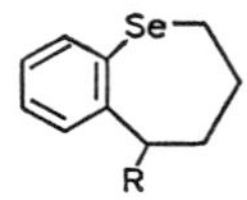
(IV)

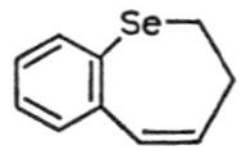
(V)

Selenepane forms a mercuric chloride complex and 1,1-dihalides with halogens (*G. T. Morgan* and *F. H. Burstall*, J. chem. Soc., 1931, 173). The 1,1-dihalide (II) is also formed in the reaction between 1,5-hexadiene and selenium dioxide in the presence of concentrated hydrochloric acid (*H. Funk* and *W. Papenroth*, J. pr. Chem., 1959, **8**, 256). Treatment of the dihalides with alkali yields selenapanes.

Cyclisation of 4(phenylseleno)butyric acids with polyphosphoric acid gave 2,3,4,5-*tetrahydro*-1-*benzoselenepin*-5-*one* (III), b.p. 112–115°/0.1 mm, and some substituted analogues. Reductive cyclisation was also effective with some acids. Reduction of III with tetrahydridoborate afforded 2,3,4,5-*tetrahydro*-1-*benzoselenepin*-5-*ol* (IV; R = OH), m.p. 71–72°, which was dehydrated with polyphosphoric acid to 2,3-**dihydro-1-benzoselenepin** (V), b.p. 174–175°/18 mm, n_D^{20} 1.5913. Under Wolff–Kishner conditions, 2,3,4,5-*tetrahydro*-1-*benzoselenepin* (IV; R = H), b.p. 146°/17 mm, n_D^{20} 1.615, was produced (*M. Renson et al.*, Bull. Soc. chim. Belges, 1964, **73**, 483; Compt. rend. 1969, **269**, 532; Bull. Soc. chim., 1971, 2689; *M. Protiva et al.*, Coll. Czech. chem. Comm., 1969, **34**, 3801).

2-(Phenylselenomethyl)benzoic acid is cyclised by polyphosphoric acid to 6,11-**dihydrobenzo**[*b,e*]**selenepin-11-one** (VI), m.p. 92–93°. Grignard reagents, which attack the oxo group, provide a means for introducing substituents at position 11 (*idem, ibid.*, 1969, **34**, 2122; Czech. P., 144,402/1968; C.A., 1972, **77**, 164542).

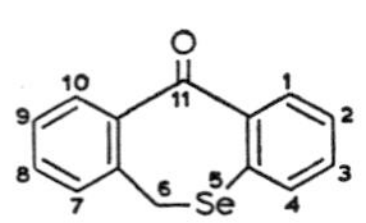
(VI)

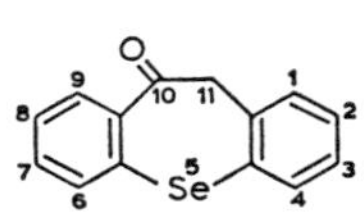
(VII)

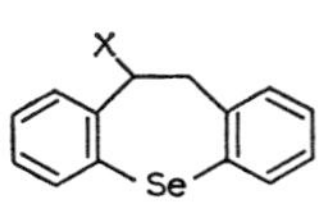
(VIII)

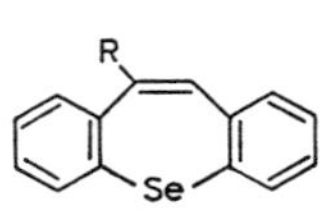
(IX)

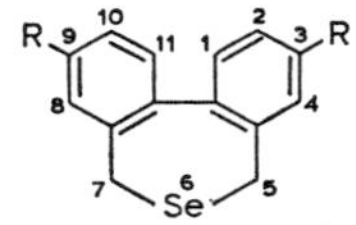
(X)

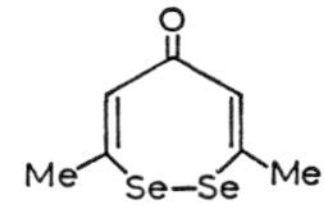
(XI)

10,11-*Dihydrodibenzo*[b,f]*selenepin*-10-*one* (VII), m.p. 88–89°, is obtained when 2(phenylseleno)phenylacetic acid undergoes cyclodehydration with polyphosphoric acid. Reduction to the corresponding *alcohol* (VIII; X = OH), m.p. 112–113°, and treatment with dry hydrogen *chloride* gave the chloride (VIII; X = Cl), m.p. 77–78°. This chloride undergoes both substitution and elimination with 1-methylpiperazine, the product of the latter reaction being **dibenzo**[*b,f*]**selenepin** (IX; R = H), m.p. 120–121° (*Protiva et al., loc.cit.*; Czech. P., 142,870/1969; C.A., 1972, **77**, 126687). Ketone VII forms the enamine IX (R = 4-methylpiperazinyl) with 1-methylpiperazine (*idem*, Naturwiss., 1969, **56**, 374; Coll. Czech. chem. Comm., 1970, **35**, 3721).

5,7-*Dihydrodibenzo*[c,e]*selenepin* (X; R = H), m.p. 65–66°, and its 3,9-*diphenyl* derivative (X; R = Ph) are prepared by treating the appropriate 2,2′-di(bromomethyl)-biphenyl with sodium selenide (*W. E. Truce* and *D. D. Emrick*, J. Amer. chem. Soc., 1956, **78**, 6130).

The "1,2-diselenepane", said to be formed when 1,5-pentane diselenocyanate is treated with alkali, is probably a low molecular weight polymer (*Morgan* and *Burstall, loc.cit.*; J. chem. Soc., 1929, 2197) but 3,7-*dimethyl*-5H-1,2-*diselenepin*-5-*one* (XI), m.p. 83°, may be prepared by the oxidative cyclisation of di(selenoacetyl)acetone with air (*G. Traverso*, Ann. chim., 1957, **47**, 3).

Chapter 51

*Compounds with Larger Heterocyclic Rings: Seven membered Ring Compounds with an Atom from Group V of the Periodic Table in the Ring. Nitrogen**

J. M. F. GAGAN

1. Seven-membered rings containing a nitrogen atom

(*a*) 1H-*Azepines*

1*H*-**Azepine** (I; R = H), a thermally unstable material spectroscopically resembling pyrrole, has been obtained in solution after alkaline hydrolysis of *N*-ethoxycarbonylazepine (I; R = CO_2Et). This 1*H*-tautomer is probably the least stable form in direct contrast to pyrrole of which it is the vinylogue,

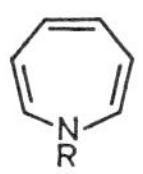

(I)

(II)

and it rapidly rearranges to 3*H*-azepine (*K. Hafner*, Angew. Chem. intern. Edn., 1964, **3**, 165). However, electronegative substituents, especially at positions 1 and 3, stabilise the azepine ring and have allowed the synthesis and investigation of several members of this series.

(*i*) *Methods of synthesis*

(*1*) The most frequently used synthetic method is the generation of a stabilised singlet nitrene in a benzenoid hydrocarbon. Insertion into the benzene ring by a concerted cycloaddition mechanism (*W. Lwowski* and *R. L. Johnson*, Tetrahedron Letters, 1967, 891) gives a 1-azanorcaradiene (II), which is unstable with respect to the azepine. Photolysis or thermolysis

* For a general introduction to heterocyclic compounds with rings comprising seven or more atoms and for an explanation of the nomenclature used, see Chapter 50, pp. 139 *et seq.*

of azidoformates (*Hafner* and *C. König*, Angew. Chem. intern. Edn., 1963, **2,** 96; *Lwowski* and *T. J. Maricich*, J. Amer. chem. Soc., 1965, **87,** 3630; *R. J. Cotter* and *W. F. Beach*, J. org. Chem., 1964, **29,** 751), carbonyl azide (*L. E. Chapman* and *R. F. Robbins*, Chem. and Ind., 1966, 1266), cyanogen azide (*F. D. Marsh* and *H. E. Simmons*, J. Amer. chem. Soc., 1965, **87,** 3529) and mesyl azide (*R. A. Abramovitch et al.*, Canad. J. Chem., 1965, **43,** 3407; Chem. Comm., 1968, 797), deoxygenation of nitrosobenzene with triethyl phosphite in the presence of trifluoroethanol (*R. J. Sundberg* and *R. H. Smith* Tetrahedron Letters, 1971, 267; *Abramovitch et al.*, J. Amer. chem. Soc., 1972, **94,** 1374) and the base-catalysed α-elimination of ethyl *N-p*-nitrobenzenesulphonyloxycarbamate (*Lwowski et al.*, *loc. cit.* and J. Amer. chem. Soc., 1964, **85,** 1200) provide the nitrene. This method used with substituted benzenes gives approximately equal amounts of the possible isomers (*Hafner et al.*, Tetrahedron Letters, 1964, 1733; *J. E. Baldwin* and *L. A. Smith*, J. Amer. chem. Soc., 1965, **87,** 4819; 1967, **89,** 1886). Separation is difficult, but has been achieved (*J. M. Photis*, J. heterocycl. Chem., 1970, **7,** 1249; 1971, **8,** 167).

(*2*) 1*H*-Azepines have been prepared from 1,4-dihydrobenzenes by the route shown:

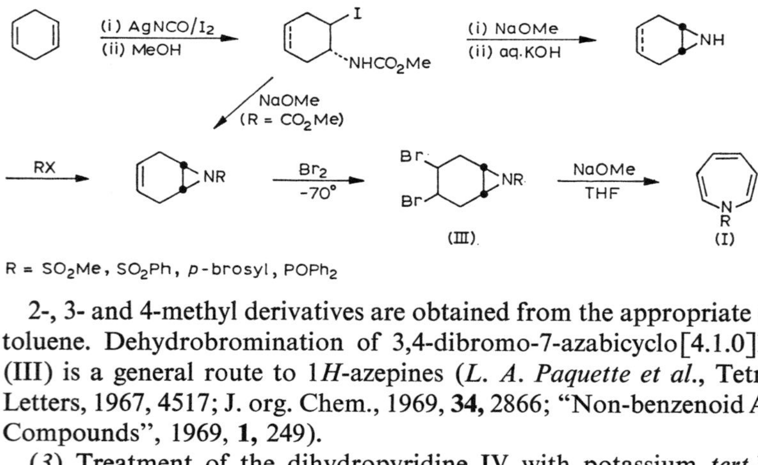

(III) (I)

R = SO_2Me, SO_2Ph, *p*-brosyl, $POPh_2$

2-, 3- and 4-methyl derivatives are obtained from the appropriate dihydrotoluene. Dehydrobromination of 3,4-dibromo-7-azabicyclo[4.1.0]heptanes (III) is a general route to 1*H*-azepines (*L. A. Paquette et al.*, Tetrahedron Letters, 1967, 4517; J. org. Chem., 1969, **34,** 2866; "Non-benzenoid Aromatic Compounds", 1969, **1,** 249).

(*3*) Treatment of the dihydropyridine IV with potassium *tert*-butoxide gives *dimethyl* 1,2,7-*trimethyl*-1H-*azepine*-3,6-*dicarboxylate* (V), m.p. 105–107°, previously believed to be VII. The bicyclic VI is obtained as a by-product but readily isomerises to V (*R. F. Childs* and *A. W. Johnson*, Chem. Comm., 1965, 95; J. chem. Soc. C, 1966, 1950). Analogues with alkyl substituents at positions 4- and 5-, and cyano groups in place of methoxycarbonyl groups are obtained similarly. Dihydroazepines with methylene

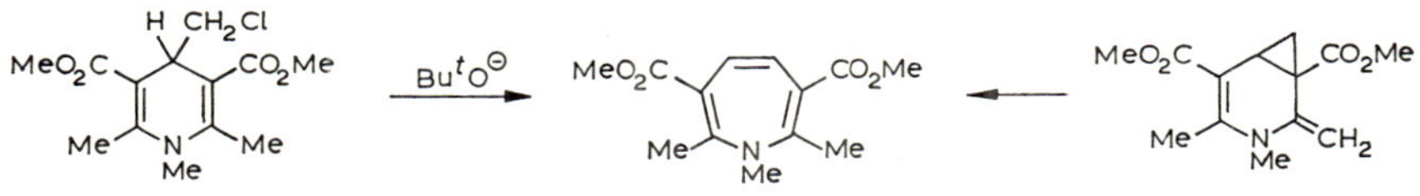

(IV) (V) (VI)

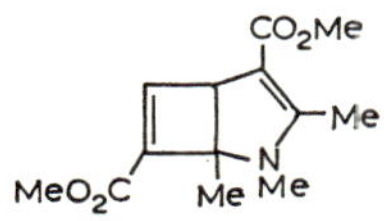

(VII)

substituents are found as intermediates, but they isomerise to 1*H*-azepines with acid (*Johnson et al., ibid.*, 1968, 1675).

(*4*) *Dimethyl* N-*acetyl*-1H-*azepine*-4,5-*dicarboxylate* (VIII; R = Ac), b.p. 125–130°/10^{-3} mm, n_D^{25} 1.5112, and the N-*tosyl* analogue (VIII, R = Tosyl), m.p. 111°, are obtained as shown from pyrroles and dimethyl acetylenedicarboxylate (*H. Prinzbach et al.*, Angew. Chem. intern. Edn., 1968, **7,** 67; 727; *A. G. McInnes et al.*, Canad. J. Chem., 1969, **47,** 2391):

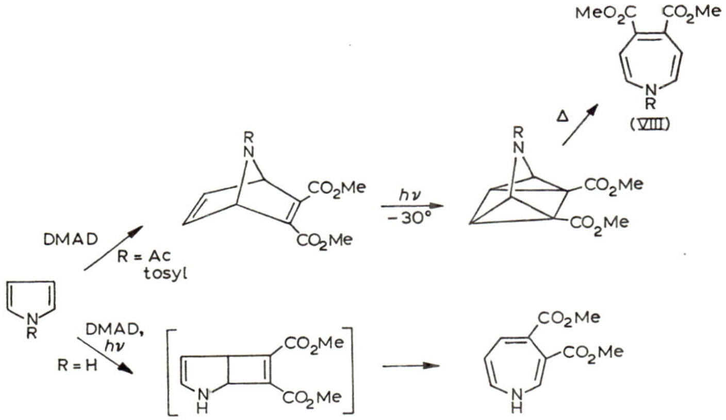

(VIII) (IX)

If large amounts of aluminium chloride are present higher yields of azepine are obtained. A similar route is said to yield *dimethyl* 1H-*azepine*-3,4-*dicarboxylate* (IX), m.p. 104°, which has no substituent on nitrogen (*R. P. Gandhi* and *V. K. Chadha*, Indian J. Chem., 1971, **9,** 305).

For the physical properties of some 1*H*-azepines see Table 1.

TABLE 1

1-*H*-AZEPINES(I)

R	*B.p. (°C/mm) (m.p.)*	n_D^{25}	*Fe(CO)₃ complex m.p. (°C)*
Me	$20/10^{-3}$		
CO_2Me	59–61/0.2	1.5379	115
CO_2Et	55–56/0.1	1.5222	
Ph	(49)		
CN	48/0.2	1.5520	(hexafluoro, m.p. 51–52)
CO_2Ph	(67–68)		108–109
CONH	(118–119)		
SO_2Me	(192–193)		95–96
SO_2Ph	(132–133)		
$POPh_2$	(152–153)		

(ii) Structure

Molecular orbital calculations show a system of strongly localised double bonds with no resonance stabilisation in the seven-membered ring of 1*H*-azepines and their annelated derivatives (*M. S. Dewar* and *N. Trinajstic*, Tetrahedron, 1970, **26,** 4269; *R. W. Schmid*, Helv., 1962, **45,** 1982). X-ray analysis confirms that the 1*H*-azepines, both free and complexed across the 2,5-positions with iron tricarbonyl, exists in a monocyclic "boat" conformation with alternating single and double bonds, which may invert quite slowly (*Paquette et al.*, J. Amer. chem. Soc., 1968, **90,** 5023; *S. M. Johnson* and *I. C. Paul*, J. chem. Soc. B, 1970, 1783; *M. G. Waite* and *G. A. Sim*, ibid., A, 1971, 1009; *H. J. Lindner* and *B. von Gross*, Chem. Ber., 1972, **105,** 434; *U. Svanholm*, Acta chem. Scand., 1971, **25,** 640). In contrast to oxepin there is indication neither of tautomerism to azanorcaradienes II, nor of homoaromatic character in the n.m.r. spectrum between −110° and 130°C (*H. Günther* and *H. H. Hinrichs*, Tetrahedron Letters, 1966, 787).

(iii) Reactions

(*1*) *Reduction.* Catalytic reduction of 1*H*-azepines usually gives perhydroazepine (azepane) derivatives, but in methanol, azepine V loses the nitrogen atom and gives dimethyl 2,3-dimethylterephthalate. Reduction of I (R = CO_2Et) with lithium tetrahydridoaluminate at low temperatures gives I (R = CH_2OH) but in refluxing ether I (R = Me) is obtained (*Hafner* and *J. Mondt*, Angew. Chem. intern. Edn., 1966, **5,** 839). Reduction of I (R = CON_3) with hydrogen sulphide yields I (R = $CONH_2$) (*Chapman* and *Robbins loc. cit.*).

(*2*) *Hydrolysis.* On treatment with acids, 1*H*-azepines readily undergo ring contraction to benzene derivatives *via* the protonation of the *N*-substituent and an irreversible valence isomerisation to an azanorcaradiene. Azepine (I, R = CO_2Et) yields ethyl *N*-phenylcarbamate, which is also the product on heating to 100°; I (R = CN) rearranges to *N*-phenylcyanamide which is hydrolysed to phenylurea, but the hexafluoro analogue is stable in acid. Decarboxylation accompanies the hydrolysis of I (R = CO_2Et) in alkali, but I (R = CN) is hydrolysed to *N*-carbamoylazepine (I, R = $CONH_2$) in alkaline hydrogen peroxide (*Hafner*, *loc. cit.*; *Marsh* and *Simmons*, *loc. cit.*).

(*3*) *Cycloaddition*. Unlike oxepin, 1*H*-azepines undergo cycloaddition and dimerisation reactions in the monocyclic triene form I, but the non-planar ring so restricts conjugation that the azepine behaves predominantly as a diene. *N*-Ethoxycarbonylazepine (I; R = CO_2Et) reacts with tetracyanoethylene, 4-phenyl-1,2,4-triazoline-3,5-dione, 1,4-phthalazinedione and *N*-phenylmaleimide, but not with dimethyl acetylenedicarboxylate or maleic anhydride, in a $(2 + 4)\pi$ cycloaddition across positions 2 and 5 (not 2 and 7 as originally proposed) (*Hafner*, *loc. cit.*; *Paquette et al.*, J. org. Chem., 1969, **34**, 2888; *A. S. Kende et al.*, J. Amer. chem. Soc., 1965, **87**, 5044, Chem. Comm., 1965, 384; *W. S. Murphy* and *J. P. McCarthy*, *ibid.*, 1970, 1129; *T. Sasaki et al.*, *ibid.*, 1970, 82; Bull. chem. Soc. Japan, 1970, **43**, 2893; *Photis*, *loc. cit.*).

Tetracyanoethylene has been used to trap transient azepines as adducts (X) the structures of which have been determined by X-ray analysis (*Baldwin* and *Smith*, *loc. cit.*; *I. C. Paul et al.*, J. Amer. chem. Soc., 1966, **88**, 3653; J. chem. Soc. B, 1967, 112). Diethyl *trans*-azodicarboxylate and nitrosobenzene both add across positions 2 and 7 (*e.g.* as in adduct XI). As concerted $(2 + 6)\pi$ cycloadditions are thermally forbidden, a two step route to these adducts is proposed (*Murphy* and *McCarthy*, *loc. cit.* and Chem. Comm., 1968, 1155).

(X)

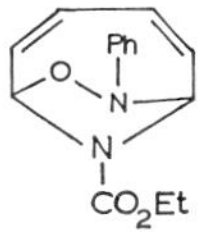

(XI)

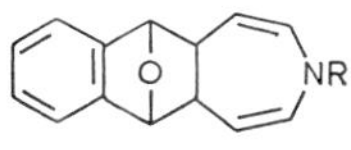

(XII)

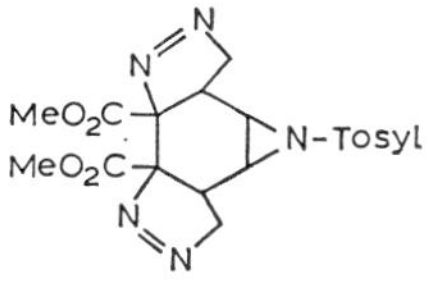

(XIII)

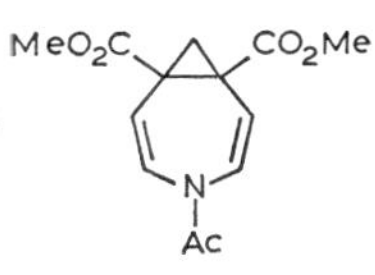

(XIV)

Azepines I (R = CO_2Et; SO_2Me) also behave as dienophiles in reaction with substituted cyclopentadienes or 1,3-diphenylisobenzofuran to give adducts across positions 4 and 5, *e.g.* XII (R = CO_2Et; SO_2Me) and as both diene and dienophile with 2,5-dimethyl-3,4-diphenylcyclopentadiene (*J. R. Wiseman* and *B. P. Chong*, Tetrahedron Letters, 1969,

1619; *Paquette, loc. cit.*). 1,3-Dipolar addition of diazomethane to azepine VIII (R = Tosyl; Ac) gives adducts XIII and XIV, respectively. The isolation of XIII provides the only evidence for the existence of azanorcaradiene tautomers (*Prinzbach et al.*, Angew. Chem. intern. Edn., 1970, **9,** 377). The formally similar iron tricarbonyl complexes are formed when 1*H*-azepines are heated with iron enneacarbonyl or irradiated in the presence of iron pentacarbonyl. These complexes greatly increase the stability of the azepines (*E. O. Fischer* and *H. Rühle*, Z. anorg. allg. Chem., 1965, **341,** 137; *Paquette et al.*, J. org. Chem., 1969, **34,** 2866).

(*4*) *Dimerisation.* The two series of dimers formed on heating 1*H*-azepines appear to have symmetrical (XV) and unsymmetrical structures (XVI). The latter is the product of a thermally allowed (6 + 4)π cycloaddition, and as it is formed when lower temperatures are employed, it is probably the kinetically favoured precursor of the thermodynamically

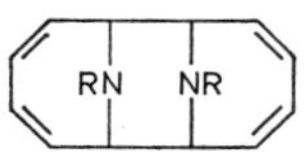

(XV)

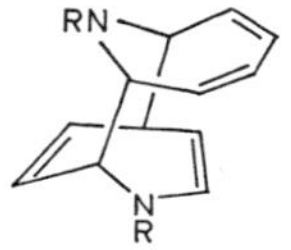

(XVI)

stable XV. Photolysis of dimers XV regenerates the azepine. Azepines with 2-substituents do not form dimers (*idem*, J. Amer. chem. Soc., 1966, **88,** 2590; 1969, **91,** 3616; J. org. Chem., 1969, **34,** 2879; *Hafner* and *Mondt*, *loc. cit.*; *G. Habermehl* and *S. Göttlicher*, Angew. Chem. intern. Edn. 1967, **6,** 805; Chem. Ber., 1971, **104,** 524; *A. L. Johnson* and *H. E. Simmons*, J. Amer. chem. Soc., 1966, **88,** 2591; 1967, **89,** 3191) (see Table 2).

TABLE 2

AZEPINE DIMERS

	R								
	H	*Me*	*Et*	*CO_2Me*	*CO_2Et*	*CN*	*$CONH_2$*	*SO_2Me*	*Ac*
XV m.p. (°C)	147–148 decomp.	171	93–95	297–298	196–197	>400	>300	Wax	317–319 decomp.
XVI m.p.(°C)		66		137–138	78	220–221*			

*The lower melting dimer of I (R = CN) may not have structure XVI.

(*5*) *Photolysis and thermolysis.* Irradiation of 1*H*-azepines I (R = CO_2Et; CO_2Me) yields the bicyclic compounds XVII (R = Et; Me); the 2-, 3- and 4-methyl homologues give a mixture of the two possible isomers. Azepin V probably gives an analogue VII, which subsequently eliminates methyl propiolate leaving pyrrole XVIII (*Paquette et al.*, J. Amer. chem. Soc.,

1966, **88,** 1718; J. org. Chem., 1969, **34,** 2885; *Johnson* and *Simmons, loc. cit.*; *Childs* and *Johnson, loc. cit.*; J. chem. Soc. C, 1966, 1950).

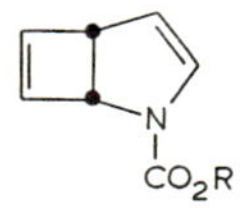

(XVII)

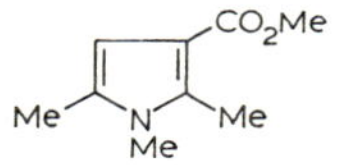

(XVIII)

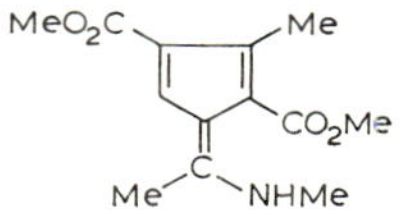

(XIX)

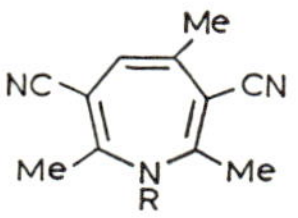

(XX)

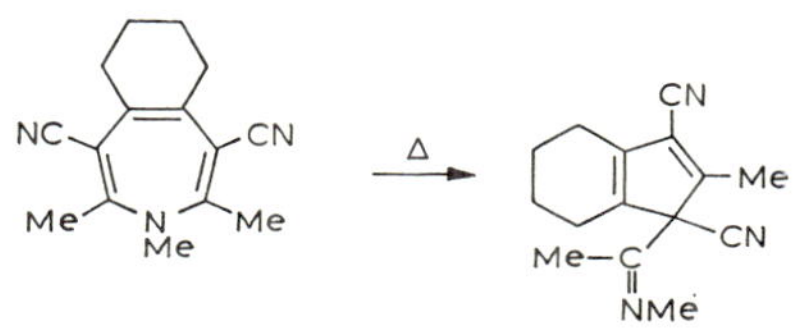

(XXI) (XXII)

Refluxing a benzene solution of azepine V gives the fulvene XIX; the similar azepines XX (R = Me; Ph) give analogous fulvenes under more vigorous conditions, but the fully substituted 4,5-tetramethyleneazepine XXI gives the cyclopentadiene XXII (*A. W. Johnson et al.*, J. chem. Soc. C, 1967, 201; 1971, 1237; Chem. Comm., 1965, 95; 1966, 442; 1970, 10).

(*6*) *Addition and substitution.* Iodine azide reacts with I (R = CO_2Et) to give dihydro-2,3- and -4,5-bisazides (*T. Sasaki et al.*, J. org. Chem., 1972, **37,** 890). The iron tricarbonyl adduct of I (R = CO_2Et) is acylated and formylated at position 3, and the 3-acetyl complex is decarboxylated on alkaline hydrolysis, allowing *N*-methylation by methyl iodide and sodium hydride (*Johnson et al.*, Chem. Comm., 1969, 631).

(*b*) 2H-*Azepines*

Fully unsaturated 2*H*-azepines have not yet been obtained, but a few 2*H*-azepin-2-ones (2-azatropones), *e.g.* XXIII, are known. Photolysis of *ortho*-substituted phenyl azides in the presence of oxygen (*Sundberg et al.*, J. Amer. chem. Soc., 1972, **94,** 513) and *O*-alkylation of 2,5-dihydro-1*H*-azepin-2,5-diones with trialkyloxonium tetrafluoroborate are the synthetic routes employed.

2-Azatropones show no evidence of 6π delocalisation. Catalytic reduction of *compound* XXIII (R = Me), m.p. 52°, gave the perhydro derivative, but XXIII (R = Et), m.p. 71–72°, yielded the dihydrodione XXIV (*D. Misiti et al.*, Tetrahedron, 1966, **22,** 1201; *E. J. Moriconi* and *I. A. Maniscalco*, J. org. Chem., 1972, **37,** 208):

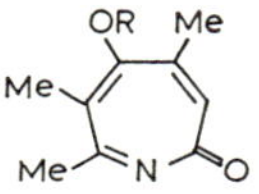

(XXIII)

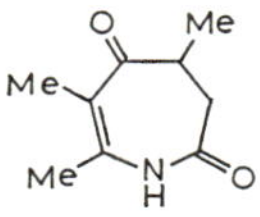

(XXIV)

(*c*) 3H-*Azepines*

Many members of the 3*H*-azepine series are known, one of which, "dibenzamil" (XXV; R = H, R^1 = Ph) was the first monocyclic unsaturated azepine to be described (*L. Wolff*, Ann., 1912, **394,** 59).

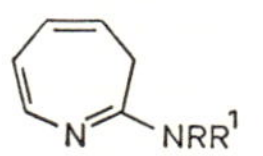

(XXV)

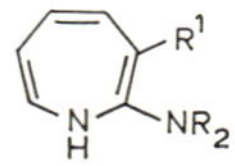

(XXVI)

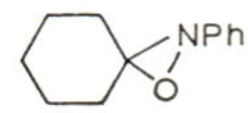

(XXVII)

3*H*-Azepine is probably the most stable tautomeric form as 4*H*-azepines are converted to their 3*H*-tautomers on heating alone or with bases (*J. Sauer et al.*, Tetrahedron Letters, 1969, 5171; *M. Anderson* and *A. W. Johnson*, J. chem. Soc., 1965, 2411).

(*i*) *Methods of synthesis*

(*1*) A frequently used route to 3*H*-azepines is the generation of aryl nitrenes in the presence of primary aromatic or secondary aliphatic amines. Pyrolysis (*Huisgen et al.*, Chem. Ber., 1958, **91,** 1, 12; 1959, **92,** 2961; *E. Vogel et al.*, Ann., 1965, **682,** 1) or photolysis (*W. von E. Doering* and *R. A. Odum*, Tetrahedron, 1966, **22,** 81; *Sundberg et al., loc. cit.*) of aryl azides, and deoxygenation of nitrobenzenes with triethyl phosphite (*J. I. G. Cadogan et al.*, Chem. Comm., 1968, 1107; 1967, 178) or nitrobenzenes (*Odum* and *E. Brenner*, J. Amer. chem. Soc., 1966, **88,** 2074) are sources of the aryl nitrene. In the photolysis of *ortho*-substituted phenyl azides ring expansion takes place on the unsubstituted side, and 1*H*-azepines XXVI are shown to be the products first formed. They revert to 3*H*-azepines XXV if oxygen is absent, but in the presence of oxygen, or oxygen and cupric ions, respectively, 2- and 4-azepinones are obtained. A ^{14}C-label at position 1 of phenyl azide is found at position 2 of the azepine after pyrolysis. The photolysis of oxaziridine XXVII in diethylamine similarly yields XXV (R, R^1 = Et) (*E. Meyer* and *G. W. Griffin*, Angew. Chem. intern. Edn. 1967, **6,** 634).

In the absence of amines, the use of phosphorus reagents may lead to the isolation of phosphorus-substituted 3*H*-azepines, *e.g.* XXVIII; (*Cadogan et al.*, J. chem. Soc. C, 1969, 2813; Chem. Comm., 1968, 736). *Ortho*- and *para*-substituents in the aromatic reagent are found at positions 2 and 4 of the azepine.

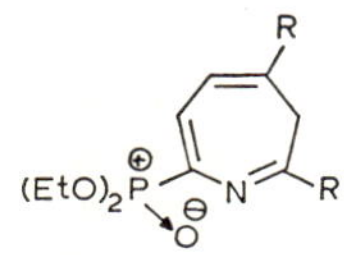

(XXVIII)

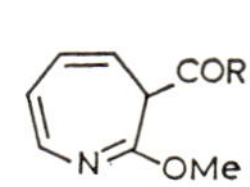

(XXIX)

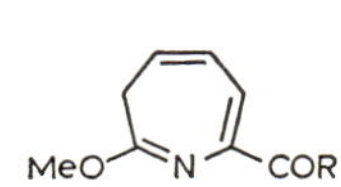

(XXX)

Azepines are also postulated as intermediates in the gas phase pyrolysis of phenyl azides which leads to cyanocyclopentadiene and pyridines (*W. D. Crow et al.*, Tetrahedron Letters, 1972, 2239, and papers cited earlier).

The photolysis of 2-azidoacetophenone in methanol gives 2-methoxy-3*H*-azepines (XXIX; XXX; R = Me) (*M. A. Berwick*, J. Amer. chem. Soc., 1971, **93,** 5780) but anthranil (3-phenylbenzo[*c*]isoxazole) gives only XXIX (R = Ph) under these conditions (*M. Ogata et al.*, Chem. Comm., 1968, 397).

(*2*) A series of 3*H*-azepines of type XXXIII is obtained by reacting 2-phenylaziridines with tetrasubstituted cyclopentadienones; tricyclic cycloaddition products XXXI and 2*H*-azepines XXXII are postulated as intermediates (*V. Nair*, J. org. Chem., 1972, **37,** 802; *A. Hassner* and *D. J. Anderson*, J. Amer. chem. Soc., 1972, **84,** 8255; 1971, **93,** 4339):

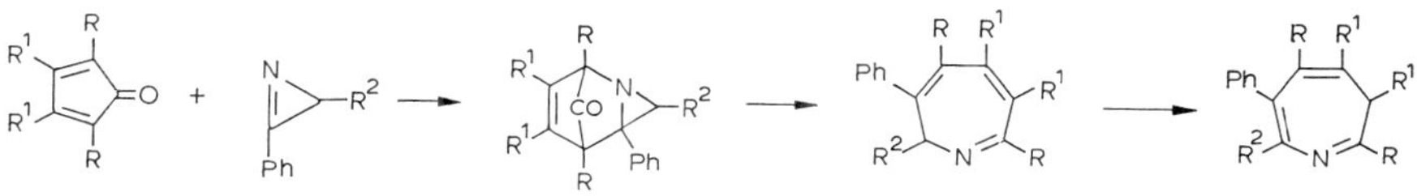

(XXXI) (XXXII) (XXXIII)

(*3*) Although there seems to be no tendency for dihydroazepinones to exist in the enol form, 2-oxo-2,3-dihydro-1*H*-azepines and their thio analogues may be *O*- and *S*-alkylated with trialkyloxonium tetrafluoroborates giving the 3*H*-azepines XXXIV (X = O, S) (*Vogel, loc. cit.*; *Paquette*, J. Amer. chem. Soc., 1964, **86,** 4096).

The thermal ring expansion of 1,2-dihydropyridine XXXV yields *dimethyl* 2,7-*dimethyl*-3H-*azepine*-4,6-*dicarboxylate* (XXXVI) (*T. J. van Bergen* and *R. M. Kellogg*, J. org. Chem., 1971, **36,** 978). *Dimethyl* 2,7-*dimethyl*-3H-*azepine*-3,6-*dicarboxylate* (b.p. 130–132°/1.5 mm, n_D^{24} 1.5268) and the *diethyl ester* (b.p. 122–123°/0.2 mm, n_D^{24} 1.5094) are obtained by heating the corresponding 4*H*-azepines (*Anderson* and *Johnson, loc. cit.*).

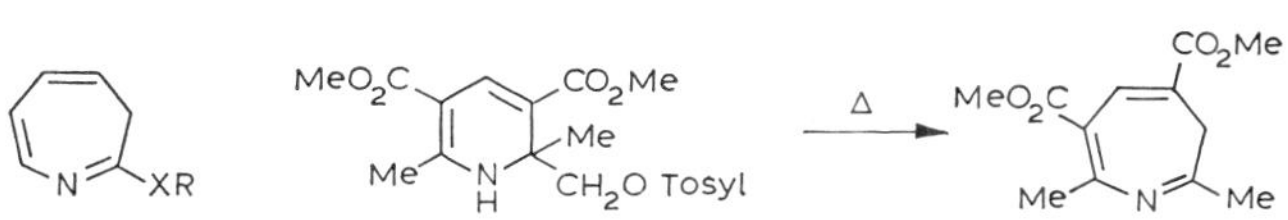

(XXXIV) (XXXV) (XXXVI)

(*ii*) *Reactions*

The chemical behaviour of the 3*H*-azepines has not been investigated as extensively as that of the 1*H*-azepines.

2-*Anilino*(3H)-*azepin* (XXV; R = H, R¹ = Ph), m.p. 151–152°, acts as a monoacidic base, is stable to acid and alkali, but undergoes ring contraction with 4-nitrobenzoyl chloride to 1-phenyl-2(4-nitrophenyl)benzimidazole. Stabilisation due to the amidine resonance

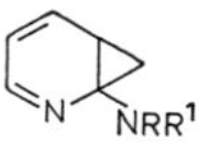

(XXXVII)

appears to favour the cross conjugated structure XXV as there is no tendency to form the bicyclic structure XXXVII and catalytic hydrogenation ceases at 2-*anilino*-4,5,6,7-*tetrahydro*-3H-*azepine*, m.p. 105°. Methylation takes place not on the ring nitrogen atom but on the nitrogen atom of the exocyclic anilino group.

The ether XXXVIII has low basicity and has not been *N*-alkylated but in acid undergoes ring contraction to tetrahydrofuranone XXXIX *(Paquette, loc. cit.)*:

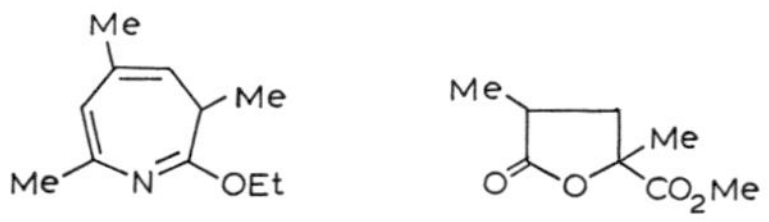

(XXXVIII) (XXXIX)

Ring contraction to pyridines occurs when azepine XXIX or its 2-amino analogues are pyrolysed or irradiated (*Ogata et al.*, Tetrahedron, 1969, **25**, 5217). The halogen atoms at all positions of 2-anilinohalogenoazepines, obtained from halogen-substituted phenyl azides (method 1) may readily be replaced by anilino groups (*R. K. Smalley* and *H. Suschitzky*, J. chem. Soc., 1964, 5922). The 2-methyl group in XXXIII (R = Me) condenses with benzaldehyde, and its hydrogen atoms exchange in deuterium oxide.

(*d*) 4H-*Azepines*

4*H*-**Azepines** XL (R, R^1 = Ph, m.p. 75–79°; R, R^1 = CO_2Me, m.p. 94–95°; R = Ph, R^1 = H, m.p. 61–62°) are prepared by the reaction of cyclopropene with 1,2,4-triazines *via* the tricyclic intermediate XLI (*J. Sauer et al.*, J. Amer. chem. Soc., 1972, **94,** 2770; Tetrahedron Letters, 1969, 5171). Both XL and its 3*H*-isomer XLII possess non-planar "boat" structures (*E. Corstensen-Oeser*, Chem. Ber., 1972, **105,** 982).

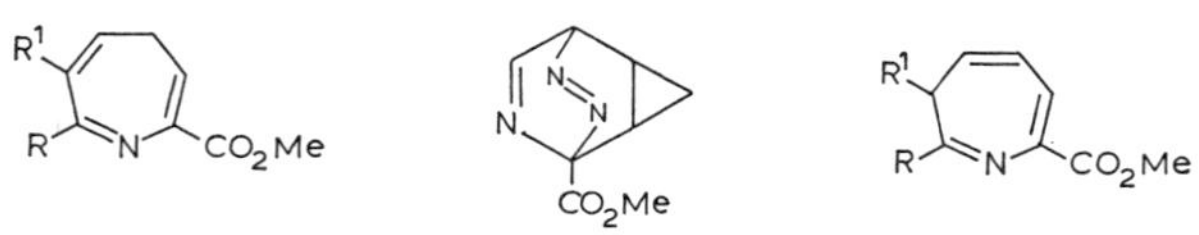

(XL) (XLI) (XLII)

Diethyl 2,7-*dimethyl*-4H-*azepine*-3,6-*dicarboxylate* (XLIII), b.p. 142–145°/0.1 mm; *dimethyl ester*, m.p. 76–77° is obtained by heating the dihydroazepines XLV (R = H, X = OEt, CN). It is in equilibrium with a bicyclic tautomer XLIV which is intermediate in the photolytic or alkali-induced ring contraction to pyrrole XLVI. A ring contraction to 4-halogenomethyl-1,4-dihydropyridines occurs with halogen acids or bromine, reversing the original synthesis of XLV. Azepine XLIII adds hydrogen cyanide to give

XLV (R = H, X = CN) (A. W. *Johnson et al.*, Proc. chem. Soc., 1964, 263; Chem. Comm., 1965, 95; J. chem. Soc. C, 1965, 2411; 1966, 1950).

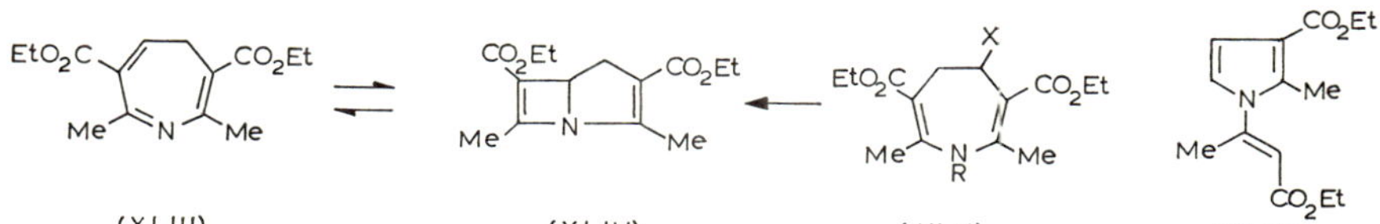

(XLIII) (XLIV) (XLV) (XLVI)

2-Mesityl-5-methylphenyl azide is unusual in giving 4*H*-azepin XLVII as major product on photolysis in diethylamine. 3*H*-Azepines, and migration to the unsubstituted *ortho* position are more usual (*R. J. Sundberg et al.*, J. Amer. chem. Soc., 1972, **94**, 513).

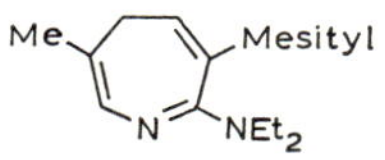

(XLVII)

4-Azatropones have been prepared from reduced dioxoazepines and by irradiating *ortho*-substituted arylazides in the presence of oxygen and cupric ion (*Misiti, loc. cit.*; *Moriconi* and *Maniscalco, loc. cit.*; *Sundberg, loc. cit.*). The conversion of XLVIII to XLIX using sulphur in a high boiling amine is one of the few successful dehydrogenations in the azepine series (*W. Treibs* and *A. Lange*, J. pr. Chem., 1961, **14**, 208). 4-Azatropones, like 2-azatropones lack 6π delocalisation, and like tropone fail to form hydrazones.

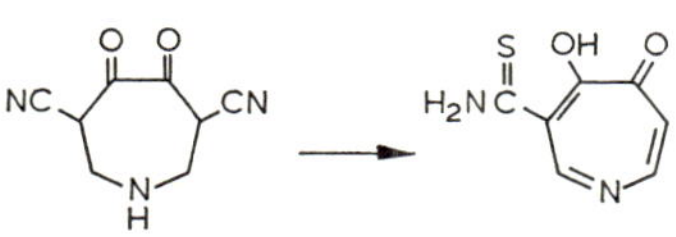

(XLVIII) (XLIX)

(*e*) *Dihydroazepines*

Several members of the 4,5-dihydro-1*H*-azepine and the 2,3-dihydro-1*H*-azepin-2-one series are known but other dihydrodiazepines are rare. Dihydroazepinones show no evidence for the existence of an enol form, supporting the conclusion that no resonance stabilisation is associated with the azepine structure (*P. Rosenmund* and *W. H. Haase*, Chem. Ber., 1966, **99**, 2504) and no tendency towards conjugation between oxo group and the diene system is observed either.

The ring expansion of 4-halogenomethyl-1,4-dihydropyridines L with nucleophiles under mild conditions is a useful general route to 4,5-dihydro-1*H*-azepines XLV. With compounds L (R = H), pyrroles, *e.g.* LI, are obtained under vigorous conditions, and 4*H*-azepines are postulated as the common precursor (*E. Benary*, *ibid.*, 1920, **53**, 2218; *H. Williams et al.*, Proc. chem. Soc., 1962, 122; J. chem. Soc., 1963, 4819; 1965, 4226; *Johnson*

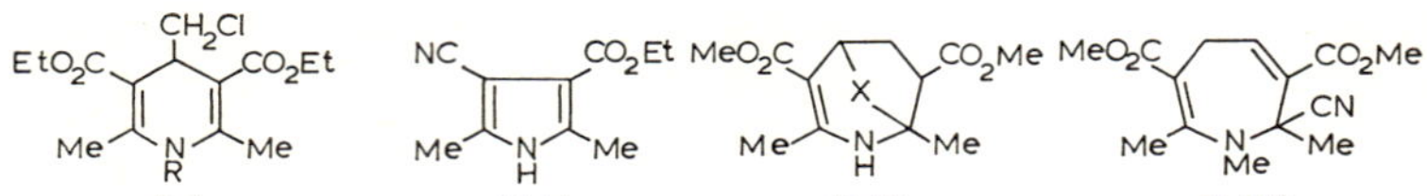

et al., ibid., 1965, 2411). The dimethyl ester and the 3,5-diacetyl analogue of L (R = H) also give 4,5-dihydro-1*H*-azepines with cyanide anion, but if water is present the latter yields a furan (*R. C. Allgrove* and *U. Eisner*, Tetrahedron Letters, 1967, 499). With primary amines, hydrogen sulphide or hydrogen selenide, XLV (R = H) reacts further to give the bridged azepines LII (X = NR, S, Se) (*Eisner et al.*, J. chem. Soc. C, 1967, 1706; 1968, 2311; *ibid.*, Perkin I, 1972, 357). In reaction with cyanide anion, L (dimethyl ester; R = Me) gives both a 4,5-dihydro-1*H*-azepine (XLV; R = Me, X = CN) and the alternative 2,5-dihydro-1*H*-azepine LIII (*Johnson et al.*, J. chem. Soc. C, 1966, 1950).

A novel route to *dimethyl* 1-*methyl*-2,7-*diphenyl*-4,5-*dihydro*-1H-*azepine*-3,6-*dicarboxylate*, m.p. 120°, is shown in the following reaction scheme (*H. D. Martin* and *M. Hekman*, Angew. Chem. intern. Edn., 1972, **11**, 926):

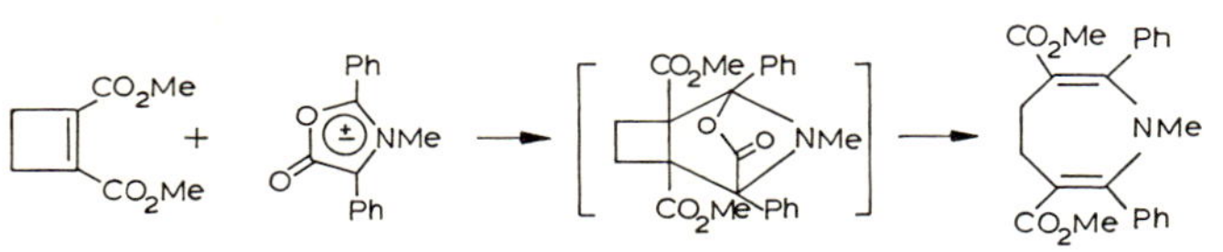

As expected XLV (R = H, X = CN) undergoes ring contraction to LI when treated with base (*Johnson et al.*, Proc. chem. Soc., 1962, 122; J. chem. Soc., 1963, 4819), but in nitrous acid protonation at position 6 is followed by ring opening and recyclisation to furo- or pyrrolo-pyridines (LIV; X = O or NH) (*idem*, J. Amer. chem. Soc., 1962, **84**, 2260; J. chem. Soc., 1964, 1632):

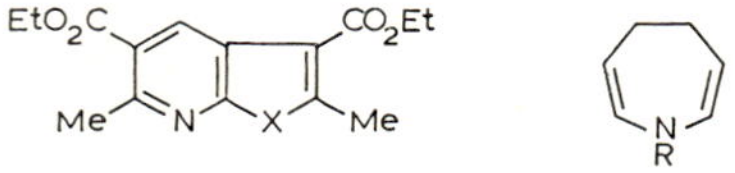

The simplest known member of this series is 1-*ethyl*-4,5-*dihydro*-1H-*azepine* (LV; R = Et), b.p. 69–70°/20 mm, obtained in the Cope rearrangement of 1-ethyl-*cis*-2,3-divinylaziridine. It collapses on heating with a trace of moisture to the ethylimine of cyclopentenaldehyde. The thermal ring expansion of the *N*-substituted vinylaziridine LVI gives 2,3-*di(trifluoromethyl)*5,6-*dihydro*-2H-*azepine* (LVII), b.p. 75°/43 mm (*E. L. Stogryn* and *S. J. Brois*, J. org. Chem., 1965, **30**, 88; J. Amer. chem. Soc., 1967, **89**, 605):

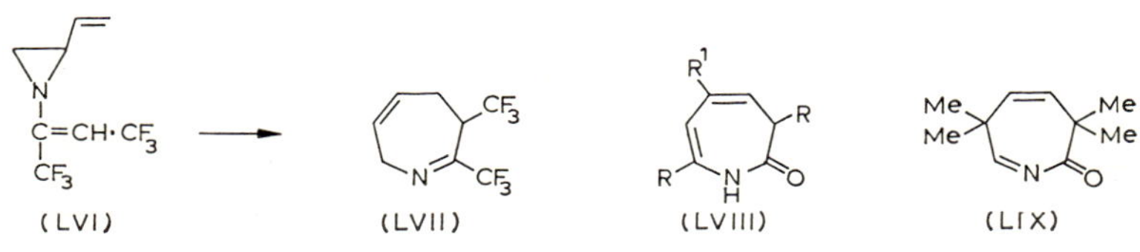

Similar rearrangement of appropriate *cis*-2-vinylcyclopropyl isocyanates gives **2,3-dihydro-1*H*-azepin-2-one** (LVIII; R, R^1 = H), m.p. 48–50° (*E. Vogel et al.*, Angew. Chem. intern. Edn., 1962, **1**, 53; Ann., 1965, **682,** 1; *W. von E. Doering* and *M. I. Goldstein*, Tetrahedron, 1959, **5,** 53) and 3,3,6,6-*tetramethyl*-2,3-*dihydro*-6H-*azepin*-2-*one* (LIX), b.p. 79°/2 mm (*T. Sasaki et al.*, J. Amer. chem. Soc., 1970, **92,** 3192; J. org. Chem., 1972, **37,** 466); irradiation reverses these reactions.

Generation of pentafluorophenyl nitrene in anisole yields 1-*pentafluorophenyl*-2,3-*dihydro*-1H-*azepin*-2-*one*, b.p. 128–130°, after chromatography (*R. A. Abramovitch et al.*, J. Amer. chem. Soc., 1972, **94,** 1374). A more general alternative route to 2,3-dihydro-1*H*-azepin-2-ones LVIII is the action of chloramine on the sodium salts of 2,6-disubstituted phenols. Prepared in this way are *azepinones* LVIII (R = *Me*, R^1 = *H*), m.p. 121–122°; (R = *Et*, R^1 = *H*), m.p. 89°; (R, R^1 = *Me*), m.p. 132° and others. If the 2- and 6-substituents differ, nitrogen attack and insertion occurs preferentially at the less hindered *ortho*-position of the ambident phenoxide ion (*W. Theilacker et al.*, Angew. Chem., 1960, **72,** 131, 498; Angew. Chem. intern. Edn., 1963 **2,** 154; *L. A. Paquette et al.*, J. Amer. chem. Soc., 1962, **84,** 4987; 1963, **85,** 3288; 1964, **86,** 4092; 4096; 1967, **89,** 3595; J. org. Chem., 1963, **28,** 3590; 1964, **29,** 3447; 1967, **32,** 2725; J. med. Chem., 1963, **6,** 771).

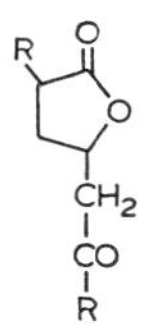

(LX)

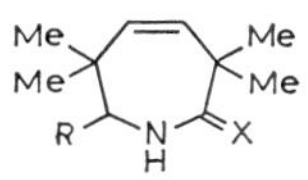

(LXI)

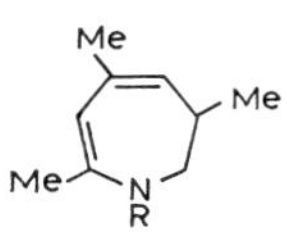

(LXII)

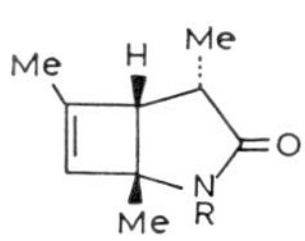

(LXIII)

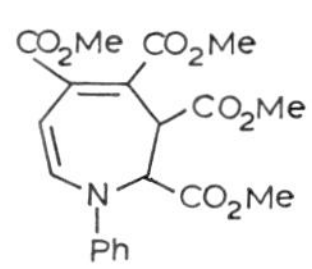

(LXIV)

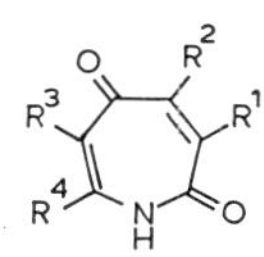

(LXV)

Acid hydrolysis of LVIII yields the butyrolactone LX. Monochlorination of LVIII occurs with several reagents, position 6 being attacked if R^1 = H, R = alkyl, but position 4 where R, R^1 = alkyl. Catalytic hydrogenation of azepinones LVIII gives caprolactams, but only the 1,7-double bond of LIX is reduced yielding *compound* LXI (R = H, X = O), m.p. 113–114°. Nucleophilic reagents also add to this double bond, to give tetrahydro-azepines LXI (X = O, R = OH, OMe, SPh, m.p. 144–145°). Lithium tetrahydrido-aluminate reduces LVIII (R, R^1 = Me) to the unstable solid LXII (R = H) but LIX is reduced further to LXI (R = H, X = H_2). With phosphorus pentasulphide in pyridine LVIII (R, R^1 = Me) is converted into 3,5,7-*trimethyl*-2,3-*dihydro*-1H-*azepine*-2-*thione*, m.p. 137°.

N-Methylation of LVIII (R, R^1 = Me) with methyl iodide and sodium hydride yields 1,3,5,7-*tetramethyl*-2,3-*dihydro*-1H-*azepin*-2-*one*, b.p. 121–122°/12 mm, n_D^{25} 1.5262, also obtained from 2,4,6-trimethylphenol and *N*-methylchloramine. Lithium tetrahydridoaluminate reduces this compound to 1,3,5,7-*tetramethyl*-2,3-*dihydro*-1H-*azepine* (LXII; R = Me), b.p. 50–54°/1 mm.

Base-catalysed deuterium exchange occurs in LVIII at position 3, and in the methyl group at position 7. Azepinones LVIII and their *N*-alkyl derivatives readily form Diels–Alder adducts with tetracyanoethylene across positions 4 and 7. On irradiation a trans-annular bond links these positions and bicyclic compounds LXIII are obtained (*idem*, Tetrahedron Letters, 1963, 2027; J. Amer. chem. Soc., 1964, **86,** 500; *O. L. Chapman* and *E. D. Hoganson*, *ibid.*, p. 498).

Tetramethyl 1-*phenyl*-2,3-*dihydro*-1H-*azepine*-2,3,4,5-*tetracarboxylate* (LXIV), m.p. 127°, is obtained by the desulphurisation of the adduct between 2-methylbenzthiazole and dimethyl acetylenedicarboxylate with Raney nickel (*R. M. Acheson et al.*, J. chem. Soc., 1965, 3200). Cyclic dienamines LXII (R = alkyl) are protonated at position 4 and react exothermically with *N*-phenylmaleimide and dimethyl acetylenedicarboxylate to give Diels–Alder adducts. Photolysis of LXII yields non-oxo analogues of LXIII.

2,5-**Dihydro-1*H*-azepin-2,5-diones** LXV, *e.g.* 4,7-*dimethyl*-, m.p. 216–217° and 3,4,6,7-*tetramethyl*-, m.p. 214–215°, are obtained when hydrazoic acid reacts with substituted 1,4-benzoquinones at 0°. The Schmidt reagent attacks the less hindered carbonyl group, but it is the substituted not the unsubstituted carbon atom that migrates to nitrogen (R^1 = H; R^4 = alkyl *not* R^4 = H; R^1 = alkyl). The structure follows from the products of alkaline hydrolysis, which disrupts the ring (*G. Caronna et al.*, Gazz. 1945, **75,** 91; 1953, **83,** 315; *D. Misiti et al.*, Tetrahedron, 1966, **22,** 1201; Tetrahedron Letters, 1965, 1071; *R. W. Rickards* and *R. M. Smith ibid.*, 1966, 2361; *G. Jones et al.*, *ibid.*, p. 2367). The product from the unsubstituted quinone is not LXV (R^1, R^2, R^3, R^4 = H) but 4,4′-dihydroxyazoxybenzene (*R. A. Raphael* and *Vogel*, J. chem. Soc., 1952, 1958; *E. Beckmann* and *O. Liesche*, Ber., 1926, **56,** 1).

(*f*) *Tetrahydroazepines*

The tetrahydroazepine structure is rather unstable and alternative products are often favoured in attempted synthetic routes. Cyclic enamines LXVI are formed by distilling 7-(alkylamino)-2-heptanones (*R. Lukeš et al.*, Coll. Czech. chem. Comm., 1961, **26,** 1105) and 7-amino-2-heptanone is

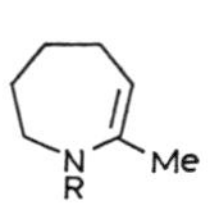

(LXVI)

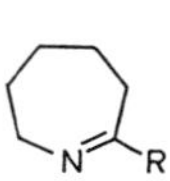

(LXVII)

said to exist as the anhydrous base LXVII (R = Me) (*S. Gabriel*, Ber., 1909, **42,** 1249, 1259). Both compounds readily revert to acyclic amino-ketones with a trace of acid. The 2-*ethyl* derivative LXVII (R = Et) has b.p. 63–65°/7 mm.

This sensitivity limits the synthetic usefulness of ring expansions of tertiary cyclohexyl azides to LXVII in strong acid (*J. H. Boyer et al.*, J.

Amer. chem. Soc., 1955, **77,** 3287; 1956, **78,** 325). Photolysis of the azides (*A. Pancrazi et al.*, Tetrahedron Letters, 1972, 5015) and the related ring-expansions found when 1-alkyl-1-nitrocyclohexanes are reduced with lithium tetrahydridoaluminate (*H. J. Barber* and *E. Lunt*, J. chem. Soc., 1960, 1187) and when 1-alkyl-1-nitrosocyclohexanes are treated with triethyl phosphite (*F. Sklarz* and *M. K. Sultan*, Tetrahedron Letters, 1972, 1319), are more satisfactory. A low yield of LXVII (R = Bu) results when butyl-lithium or butylmagnesium bromide reacts with caprolactam (*K. Heyns* and *W. Pyrus*, Chem. Ber., 1955, **88,** 678). Grignard reagents react better with caprolactim ethers to give 2-alkyltetrahydroazepines but *N*-methylaminoketones are also obtained (*O. Cervinka*, Coll. Czech. chem. Comm., 1959, **24,** 1146).

(*i*) *Tetrahydroazepinones*

Beckmann rearrangement of cyclohexanone oximes generally gives low yields of tetrahydroazepinones due to aromatisation of the starting materials (*W. Semmler*, Ber., 1892, **25,** 3352; *L. Wolff*, Ann., 1902, **322,** 351; *F. M. Beringer* and *I. Ugelow*, J. Amer. chem. Soc., 1953, **75,** 2635).

The *syn* oximes LXIX react much more readily than their *anti*- isomers, only 2,5,6,7-*tetrahydro*-1H-*azepin*-2-*one* (LXX; R = H), b.p. 60–65 °/0.5 mm, n_D^{25} 1.5238, and its 4,6-*dimethyl* analogue (LXX; R = Me), m.p. 71–73 °, being obtained from LXIX (R, R^1 = H) and (R, R = Me), respectively (*F. J. Donat* and *A. L. Nelson*, J. org. Chem.,

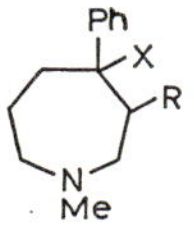

(LXVIII)

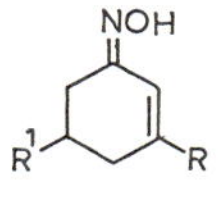

(LXIX)

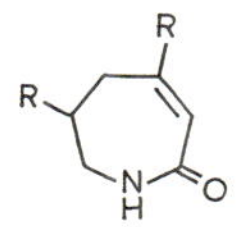

(LXX)

1957, **22,** 1107; *E. C. Horning et al.*, J. Amer. chem. Soc., 1952, **74,** 5153), but both products of the stereospecific rearrangement of LXIX (R = Me; R^1 = Ph) have been obtained: 4-*methyl*-6-*phenyl*-2,5,6,7-*tetrahydro*-1H-*azepin*-2-*one*, m.p. 118–119 °, (from *syn*) and 3-*methyl*-5-*phenyl*-4,5,6,7-*tetrahydro*-1H-*azepin*-7-*one*, m.p. 146–147 °, (from *anti*) (*R. S. Montgomery* and *G. Daugherty*, J. org. Chem., 1952, **17,** 823; *R. H. Mazur*, *ibid.*, 1961, **26,** 1289; Tetrahedron, 1971, **27,** 5381). An alternative route to tetrahydroazepinones is shown (*T. Durst et al.*, J. Amer. chem. Soc., 1972, **94,** 9261):

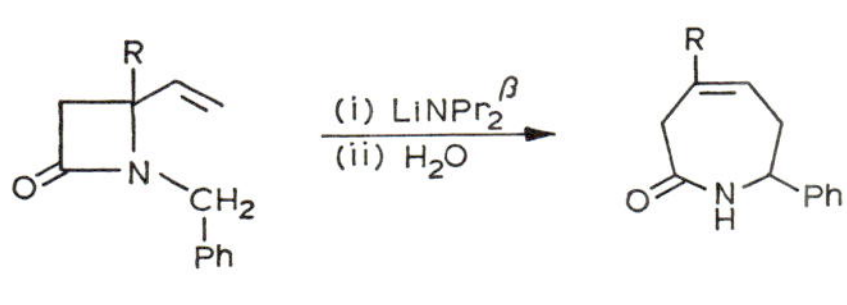

A double bond has been introduced into the 4,5-position of perhydroazepin LXVIII (R = Me; X = OCOEt) and the methyl quaternary salt of LXVIII (R = H; X = SO_2Et) by thermolysis. If LXVIII (R = H; X = COPh) is treated with phenyl-lithium a 3,4-

double bond is introduced (*J. Diamond et al.*, J. org. Chem., 1961, **26,** 2058; 1965, **30,** 1840).

(*ii*) *Caprolactim ethers and related compounds*

Caprolactim ethers are the most studied tetrahydroazepines. The O-*methyl* ether (LXXI; R = Me), b.p. 65–67°/24 mm, n_D^{25} 1.4610, and O-*ethyl* ether (LXXI; R = Et) are prepared by treating caprolactam (LXXII; R = H) with alkyl sulphates, but *N*-alkylcaprolactam results on heating or with excess reagent (*R. E. Benson* and *T. L. Cairns*, J. Amer. chem. Soc., 1948, **70,** 2115; Org. Synth., 1951, **31,** 72; *K. F. Schmidt* and *P. Zutavem.* B.P. 345,397/1931; C.A., 1932, **26,** 257; *H. Bredereck et al.*, Chem. Ber., 1959, **92,** 329; 1963, **96,** 1350). Tertiary oxonium salts (*e.g.* trialkyloxonium tetrafluoroborate) give better yields, even in the presence of other functional groups (*E. Profft* and *F. J. Becker*, J. pr., Chem., 1965, **30,** 18; *H. Meerwein et al., ibid.*, 1937, **147,** 257; 1939, **154,** 83; Ann., 1960, **632,** 38; *V. G. Granik* znd *R. G. Glushkov*, Zhur. org. Khim., 1968, 724).

Alternatively caprolactam is converted by phosphorus halides into the imidoyl chloride LXVII (R = Cl), which is treated with alcohols to give caprolactim ethers. This imidoyl chloride, which is easily hydrolysed and so not isolated, is a probable intermediate when phosgene, benzenesulphonyl chloride or ethyl chloroformate is used in this synthesis, but other intermediates have also been identified (*P. Oxley et al.*, J. chem. Soc., 1948, 1618; *W. Hechelhammer*, G.P. 940,588/1956; 948,973/1956; C.A., 1957, **51,** 12967; 1959, **53,** 6088; *M. T. Tetenbaum*, J. org. Chem., 1966, **31,** 4298). Treatment of LXVII (R = Cl) with amines gives cyclic amidines LXXIII. Beckmann or Schmidt rearrangements of cyclohexanone derivatives in the presence of alcohols or amines also yield LXXI and LXXIII, respectively (*Schmidt* and *Zutavem*, G.P. 488,447/1926; 532,969/1929; Friedl., 1933, **18,** 3048, 3050; *Oxley* and *W. F. Short*, J. chem. Soc., 1948, 1514; *G. Schroeter et al.*, Ber., 1930, **63,** 1308; *W. Z. Heldt*, J. Amer. Chem., Soc., 1958, **80,** 5880; *K. K. Kelly* and *J. S. Matthews*, J. org. Chem., 1971, **36,** 2159; *R. E. Lyle* and *St. S. Palosi*, Chem. Eng. News, 1964, **42,** 45).

(LXXI) (LXXII)

Caprolactim ethers show two general types of reaction; alkyl-oxygen fission giving caprolactam derivatives, and nucleophilic substitution of the alkoxy group. The *ether* LXXI (R = Me) rearranges to LXXII (R = Me) on heating, but the *O*-ethyl analogue (LXXI; R = Et) loses ethylene to give caprolactam (*S. Petersen* and *F. Tietze*, Chem. Ber., 1957, **90,** 909; J. med. Chem., 1963, **7,** 262; *J. W. Ralls* and *C. A. Eliger*, Chem. and Ind., 1961, 20). Markedly acidic substances are methylated by LXXI (R = Me) and caprolactim ethers have been used to esterify acids (*W. Konz*, U.S.P., 2,767,182/1956; C.A., 1957, **51,** 8150; *R. J. Leary*, *ibid.*, 14763). Treatment of caprolactim ethers with acyl chlorides yields not *O*-acyl lactims, but *N*-acyl lactams (LXXII; R = COR) (*R. Sehring* and *Konz*, G.P. 949,056/1956; C.A., 1959, **53,** 6275; *B. Stoll* and *W. Griehl*, Helv., 1965, **48,** 1805). With peracids the *hydroxamic acid* LXXII (R = OH), m.p. 80–81°, is formed (*A. M. Wade et al.*, Tetrahedron Letters, 1971, 4519) previously obtained by the reaction of *N*-(benzenesulphonyl)hydroxylamine with cyclohexanone under strongly alkaline conditions (*L. Panizzi et al.*, Ricerca Sci., 1961, **1,** 312). This system is present in the growth factors mycobactin P and T (*C. A. Snow et al.*, Biochem. J., 1965, **97,** 166; **94,** 160; 1961, **81,** 48; 1953, **55,** 596: J. chem. Soc., 1954, 2588, 4080).

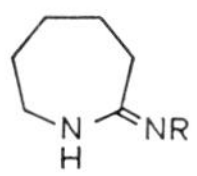

(LXXIII)

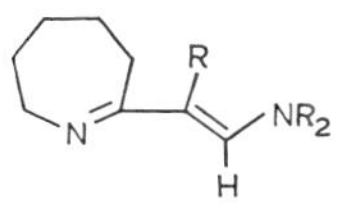

(LXXIV)

The alkoxy group of LXXI is readily replaced by amines to give amidines LXXIII; with ammonium sulphate LXXIII (R = H) is obtained (*Petersen*, Angew. Chem., 1952, **64,** 602; *E. Proft* and *F. J. Becker*, J. pr. Chem., 1965, **30,** 18; *E. Etienne* and *J. Correia*, Compt. rend., 1964, **259C,** 2660; *G. Heller*, Ber., 1907, **40,** 1296). Hydroxylamine (*idem, ibid.*, 1916, **49,** 2773; *H. Behringer* and *H. Meir*, Ann., 1957, **607,** 80) and hydrazine (*Glushkov* and *E. S. Golovchinskaya*, Zhur. priklad. Khim., 1959, **32,** 920; *R. Stolle*, Ber., 1930, **63,** 1032) give caprolactam oxime and hydrazone (LXXIII; R = OH, NH_2), respectively. The subsequent reactions of these amidines, or the use of bifunctional amino reagents, results in the annellation of heterocyclic rings across positions 1, 2 of the azepine (*Stolle et al.*, J. pr. chem., 1934, **140,** 159; *J. Korosi*, *ibid.*, 1964, **23,** 212; *Heller* and *W. Benade*, Ber., 1922, **55,** 1011; *D. Borman*, Chem. Ber., 1970, **103,** 1797; *Glushkov et al.*, Zhur. obshcheĭ Khim., 1961, **31,** 189; 1173; 1960, **30,** 649; *Petersen et al.*, Ann., 1959, **623,** 166; Chem. Ber., 1957, **90,** 909; *Petersen*, *E. Tietze* and *W. Wirth*, U.S.P. 2,852,525/1958; C.A., 1959, **53,** 6256; *E. Kagan* and *R. D. Berkenmeyer*, U.S.P. 2,956,064/1960; C.A., 1961, **55,** 7450); a substituent at position 3 of the azepine allows 2,3-annellation (*Glushkov et al.*, Khim. geterotsikl. Soedin., 1965, 240; Zhur. org. Khim., 1968, 724). Anions of active methylene compounds also replace the alkoxy group of LXXI forming a carbon–carbon bond to position 2 (*idem*, Zhur. obshcheĭ Khim., 1960, **30,** 1855) and enamines react with iminosulphonic esters (LXXI; R = SO_2alkyl) to give conjugated products LXXIV (*S. Hünig et al.*, Angew. Chem. intern. Edn. 1963, **2,** 213).

Catalytic hydrogenation of caprolactim ethers or reduction with lithium tetrahydridoaluminate yields perhydroazepine (*O. Cervinka*, Chem. Listy, 1958, **52,** 1145). *O*-Allyl and *O*-cinnamyl ethers undergo Claisen rearrangement on heating to the 3-substituted capro-

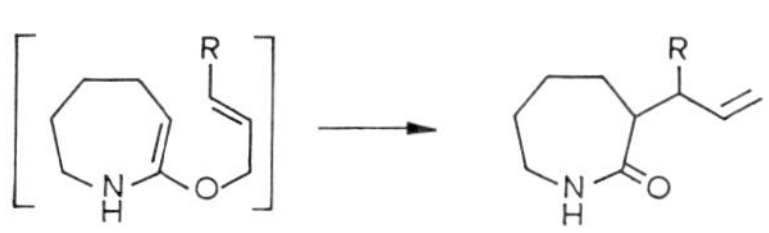

(LXXV) (LXXVI)

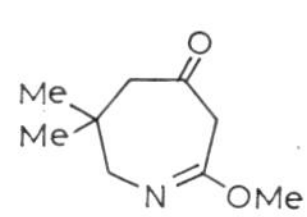

(LXXVII)

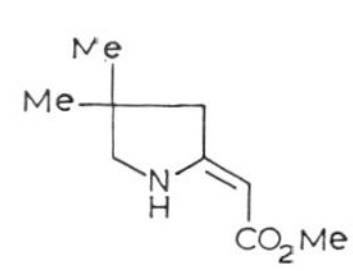

(LXXVIII)

lactams LXXVI, presumably *via* the enamine tautomer LXXV (*D. St. C. Black* and *A. M. Wade*, Chem. Comm., 1970, 871). Transannular interaction is observed when lactim ether LXXVII undergoes ring-contraction in acid to pyrrolidine LXXVIII (*Y. Tamura et al.*, Chem. and Pharm. Bull. Japan, 1971, **19,** 1068).

(g) *Perhydroazepine (azepane, hexamethyleneimine, azacycloheptane) and its derivatives*

(*i*) *Methods of synthesis*

The reactions of the azepane ring are essentially those of an aliphatic secondary amine, but much interesting chemistry occurs in the varied and ingenious methods employed for its synthesis.

(*1*) *Reduction of azepines.* Many unsaturated azepines and azepinones especially the readily available caprolactam derivatives, may conveniently be reduced to perhydroazepines. Catalytic hydrogenation saturates double bonds in the seven-membered ring and chemical methods are used to reduce carbonyl groups. Caprolactam (LXXII; R = H) and its derivatives are best reduced by lithium tetrahydridoaluminate (*L. Ruzicka et al.*, Helv., 1949, **32,** 544; *L. Lukeš* and *J. Malek*, Chem. Listy, 1951, **45,** 72; *M. Borovicka* and *M. Protiva, ibid.*, 1957, **51,** 1344; *F. F. Blicke* and *N. J. Doorenbos*, J. Amer. chem. Soc., 1954, **76,** 2317), but sodium in alcohol (*O. Wallach*, Ann., 1900, **312,** 187; 1902, **324**; 293; 1905, **343,** 45; 1906, **346,** 249; *A. Müller et al.*, Monatsh., 1928, **50,** 399; 1932, **61,** 212; *J. v. Braun* and *H. Reich*, Ann., 1925, **445,** 225), sodium tetrahydridoborate and aluminium chloride (*R. P. Mull et al.*, J. Amer. chem. Soc., 1958, **80,** 3769) and electrolytic reduction (*Ruzicka et al.*, Helv., 1935, **18,** 659) have also been used. The thiolactams are also readily reduced (*Y. Iwakura et al.*, Chem. High Polymers, Japan, 1951, **8,** 193; C.A., 1954, **48,** 130).

(*2*) *Reductive cyclisation.* Catalytic reduction of some aliphatic 1,4-dinitriles leads to azacycloheptanes, probably *via* the imino compounds LXXIII (R = H); the major product from adiponitrile itself is however 1,6-diaminohexane (*P. Ruggli et al.*, Helv., 1935, **18,** 1388; *W. A. Lazier*, U.S.P., 2,200,282/1940; C.A., 1940, **34,** 5860; *J. O. Halford* and *B. Weissmann*, J. org. Chem., 1952, **17,** 1276; *E. N. Zilberman* and *Z. D. Skorikova*, C.A.,

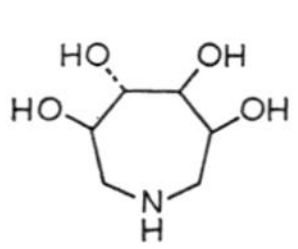

(LXXIX)

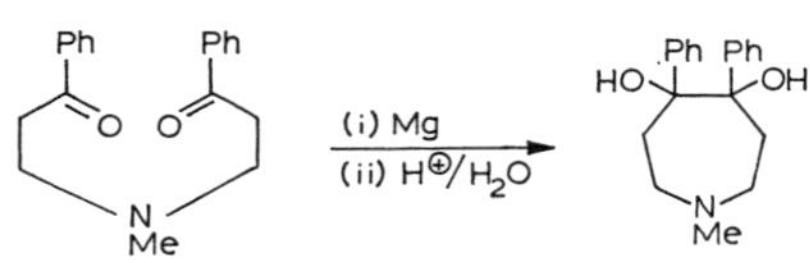

1958, **52,** 16185). Reductive cyclisation of a keto cyanide to give an aza-cycloheptane ring has been successful in the steroid series (*L. Knof*, Ann., 1961, **642,** 194), and the catalytic reduction of 6-aminohexose sugars gives tetrahydroxyazepanes, *e.g.* 6-aminoglucose gives LXXIX (*H. Paulsen* and *K. Todt*, Chem. Ber., 1967, **100,** 512). Pinacol reduction of iminodiketones gives azepanediols (*C. Mannich* and *G. Heilner*, Ber., 1922, **55,** 356).

(*3*) *Ring-closure at nitrogen.* The cyclisation of 6-bromohexylamines in the presence of base is a more valuable synthetic route for highly substituted or polycyclic azepanes than for simple compounds (*J. v. Braun et al.*, Ber., 1905, **38,** 3083; 3091; 1912, **45,** 3381; 1923, **56,** 690; 1925, **58,** 2164; *A. Müller et al.*, Monatsh., 1927, **48,** 731; 1928, **50,** 399; 1932, **61,** 219, 226; *R. Takamoto*, J. pharm. Soc., Japan, 1928, **48,** 366; C.A., 1928, **22,** 3162). Polymerisation and *N*-alkylation both compete with cyclisation, and yields are low unless high dilution techniques are employed (*G. Salomon* and *Ruzicka*, Helv., 1933, **16,** 1361; 1934, **17,** 851; 882). *N*-Alkylation is avoided by using tosylamino bromides and removing the protecting group after cyclisation by hydrolysis or reductive cleavage (*Müller* and *A. Sauerwald*, Monatsh., 1927, **48,** 727; *K. Ziegler* and *P. Orth*, Ber., 1933, **66,** 1867), or by cyclising the dialkylamino bromide to a quaternary salt and obtaining the azepane by dealkylation (*E. R. Litman* and *C. S. Marvel*, J. Amer. chem. Soc., 1930, **52,** 287). The analgesic 4-ethoxycarbonyl-1-methyl-4-phenylazepane is prepared this way from 4-dimethylamino-2-phenylbutyronitrile and 1-bromo-3-chloropropane (*F. F. Blicke* and *E-P. Tsao*, *ibid.*, 1953, **75,** 3999; *J. Diamond et al.*, J. org. Chem., 1957, **22,** 399).

1,6-Dihalides cyclise to *N*-substituted azepanes with primary amines (*Müller* and *E. Rölz*, Ber., 1928, **61,** 570; *V. Prelog et al.*, Helv., 1943, **26,** 1180; *R. G. Rice et al.*, J. org. Chem., 1958, **23,** 1352). Yields are generally better using a secondary amine and dealkylating the quaternary product (*Litman* and *Marvel*, *loc. cit.*). 1,6-Hexamethyleneglycol cyclises with primary amines in the presence of Raney nickel (*H. Adkins et al.*, J. Amer. chem. Soc., 1936, **58,** 2487; 1938, **60,** 1033; *Rice et al.*, *loc. cit.*).

1,6-Hexamethylenediamines cyclise to hexamethyleneimines when passed over alumina at 380°; 1,6-diamino-2,5-dimethylhexane gives a mixture of *cis* and *trans* isomers of 4,6-dimethyl azepane (*H. Raab*, G.P. 738,448/1943; C.A., 1944, **38,** 4273; *E. Niemers et al.*, Tetrahedron Letters, 1970, 3791).

(*4*) *Ring closure by carbon–carbon bond formation.* Acyloin and Dieckmann cyclisations have been used for the syntheses of *N*-substituted perhydroazepinones shown in the following reaction scheme where $m + n = 4$ and 5, respectively:

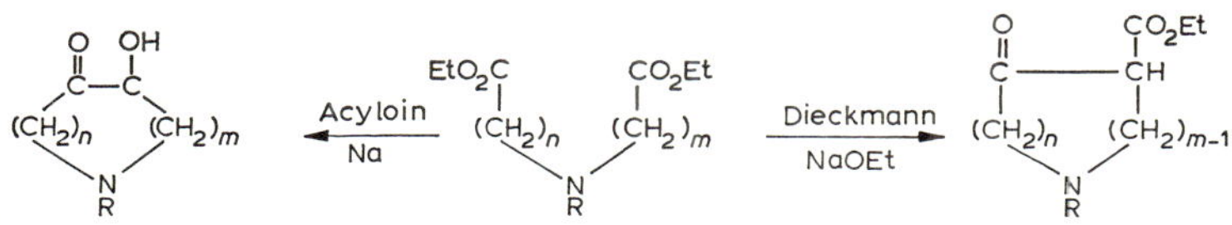

The Dieckmann method gives yields comparable with those found for carbocyclic analogues — up to 60% if high dilution techniques are used; but the acyloin synthesis, though effective for eight-membered rings and larger, gives very poor yields of azepanones (*E. A. Prill* and *S. M. McElvain*, J. Amer. chem. Soc., 1933, **55,** 1233; *N. J. Leonard et al., ibid.*, 1949, **71,** 3098; 1950, **72,** 3632; 1951, **73,** 5210; 1952, **74,** 6251; 1953, **75,** 6249; 1954, **76,** 5708; *A. Yokoo* and *S. Morosawa*, Bull chem. Soc. Japan, 1956, **29,** 631; *G. R. Proctor* and *R. H. Thompson*, J. chem. Soc., 1957, 2312). Dinitriles of the type $NC(CH_2)_nNR(CH_2)_mCN$ (R = Me, Ph; $m + n = 5$), undergo similar ring closure to give cyanoimines LXXX, which like the Dieckmann products give azepanones after hydrolysis and decarboxylation (*H. J. Nitschke* and *G. Faerber*, Chem. Ber., 1954, **87,** 1635; *R. H. Martin et al.*, Bull. soc. chim. Belges, 1958, **67,** 256; *J. Diamond et al.*, J. org. Chem., 1965, **30,** 1840).

(*5*) *Ring-expansion reactions.* The piperidine LXXXI undergoes a pinacol rearrangement in cold sulphuric acid to give the azepanone LXXXII (*R. E.* and *G. G. Lyle*, J. Amer. chem. Soc., 1954, **76,** 3536). The aminomethylpiperidine LXXXIII undergoes a similar carbonium ion ring expansion

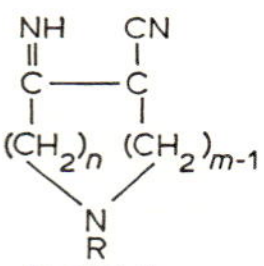

(LXXX)

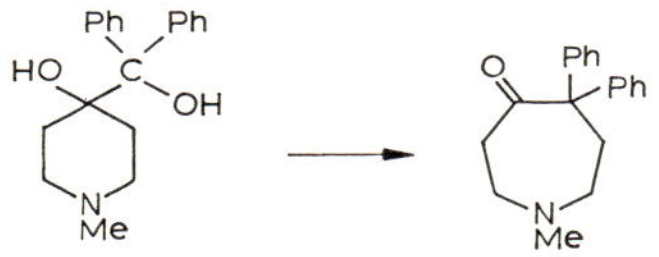

(LXXXI) (LXXXII)

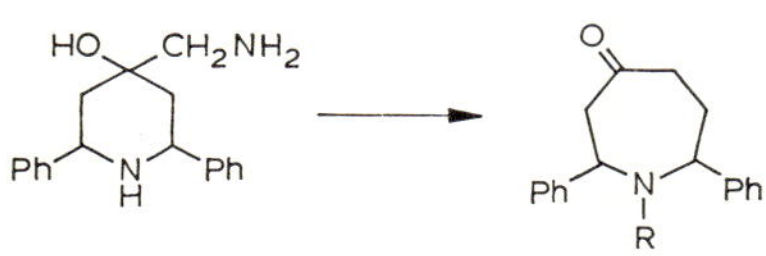

(LXXXIII) (LXXXIV)

when treated with nitrous acid at low temperature giving LXXXIV (R = NO), and 1-amino-3-methylaminopiperidine gives a mixture of tetrahydro- and perhydro-azepines under these conditions (*C. G. Overberger et al.*, J. org. Chem., 1970, **35,** 138; *L. B. Dmitriev et al.*, C.A., 1968, **69,** 43770). In the solvolysis of 4-quinuclidinemethanol tosylate the ring-expanded carbonium ion undergoes a fragmentation reaction leading to azepane LXXXV (*C. A. Grob et al.*, Helv., 1962, **45,** 1119, 1672, 1823):

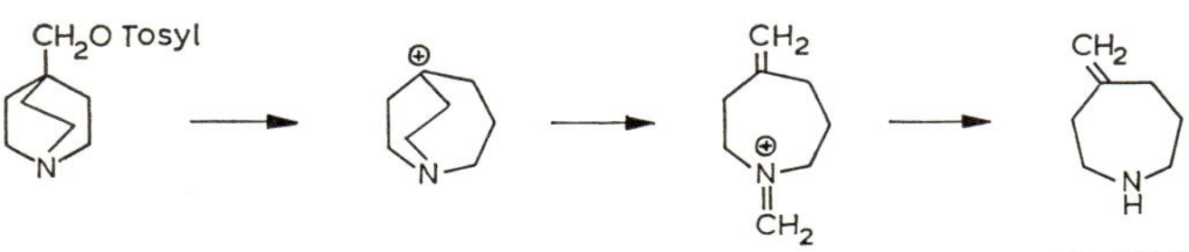

(LXXXV)

Treatment of 1-benzyl-4-piperidone with diazomethane gives 1-*benzyl-4-azepanone, hydrochloride*, m.p. 186–187° (decomp.) (*S. Morosawa*, Bull. chem. Soc. Japan, 1958, **31,** 418); this is a better route than Dieckman cyclisation. The benzyl group is removed by catalytic hydrogenolysis to leave 4-*azepanone, hydrochloride*, m.p. 171° (decomp.). Diazomethane also inserts into the double bond of tetrahydropyridinium salts (*Leonard et al.*, J. org. Chem., 1963, **28,** 1499):

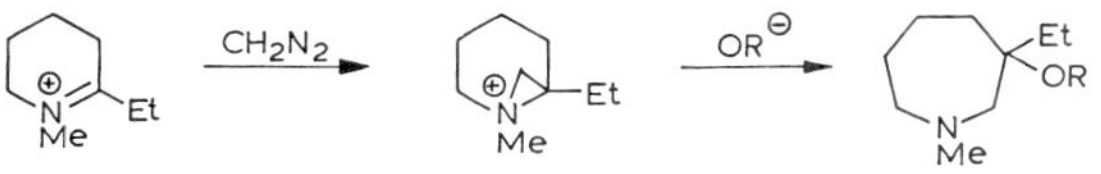

The nucleophile-induced ring expansion of 2-(2-chloroethyl)-1-methylpyrrolidine to 4-substituted 1-methylazepanes (LXXXVII probably goes through the intermediate azetidinium salt LXXXVI (*A. Ebnöther* and *E. Jucker*, Helv., 1964, **47,** 745). Reduction of 1-nitro-1-methylcyclohexane with lithium tetrahydridoaluminate gives 2-*methylperhydroazepine, picrate*, m.p. 130–131° (*H. J. Barber* and *E. Lunt*, J. chem. Soc., 1960, 1187):

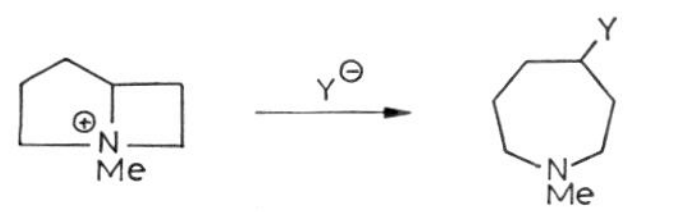

(LXXXVI) (LXXXVII)

Azepane *(hexamethyleneimine, azacycloheptane)*, b.p. 138°, n_D^{23} 1.4654; *hydrochloride*, m.p. 236°, *picrate*, m.p. 146–147°, is most conveniently prepared by reduction of ε-caprolactam with lithium tetrahydridoaluminate. It has a similar stability, and smell, to piperidine and reacts as a typical secondary amine in acylation, alkylation and enamine formation. Oxidation of the *N*-benzoyl derivative with permanganate gives 6-benzamidocaproic acid, and treatment with cyanogen bromide also opens the ring (*Ruzicka et al.*, Helv., 1949, **32,** 544; *Müller et al.*, Monatsh., 1927, **48,** 731; Ber., 1928, **61,** 568; *W. Siefken*, Ann., 1949, **562,** 102). Catalytic carbonylation at high temperature gives *N*-formylazepane (*W. Reppe et al., ibid.*, 1953, **582,** 71).

N-*Methylazepane*, b.p. 140°, n_D^{20} 1.4511, *picrate*, m.p. 217°, is made by direct methylation, or the action of formaldehyde and formic acid on azepane (*Müller et al.*, Monatsh., 1952, **83,** 386; *R. Lukeš* and *J. Malek*, Chem. Listy, 1951, **45,** 72). Other *N*-alkylazepanes have been described (*T. L. Cairns* and *B. C. McKusick*, J. org. Chem., 1950, **15,** 790; *D. Jerchel et al.*, Chem. Ber., 1952, **85,** 1130; Ann., 1952, **575,** 162; *Blicke et al.*, J. Amer. chem. Soc., 1954, **76,** 2203, 2317, 2422, 2498, 4615; J. Amer. pharm. Assoc., 1956, **45,** 440). N-*Phenylazepane*, b.p. 146–148°/12 mm, *hydrochloride*, m.p. 194°, results when azepane is treated with sodamide and diphenylsulphone (*H.-J. Nitzchke* and *H. Budka*, Chem. Ber., 1955, **88,** 264) or *N*-phenylazepine is hydrogenated (*R. J. Sundberg* and *R. H. Smith*, Tetrahedron Letters, 1971, 267).

Quaternary salts LXXXVIII (R = Me) are stable at high temperatures, and like piperidinium salts undergo demethylation not ring-opening in ethanolamine (*Müller et al.*, Monatsh., 1953, **84,** 1206; *Hünig* and *W. Baron*, Chem. Ber., 1957, **90,** 403). Ring-opening occurs with silver oxide, potassamide or phenyllithium *via* the ylid LXXXVIII (R = $C^{\ominus}H_2$) (*G. Wittig* and *T. F. Burger*, Ann., 1960, **632,** 85) and the thermal ring-opening of

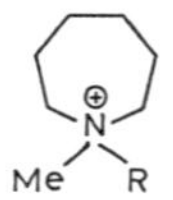

(LXXXVIII)

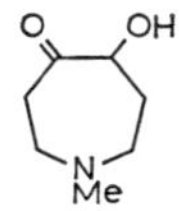

(LXXXIX)

N-oxides LXXXVIII (R = O$^{\ominus}$) is analogous (*W. Carruthers* and *R. A. W. Johnstone*, J. chem. Soc., 1965, 1653; *A. C. Cope* and *N. A. LeBel*, J. Amer. chem. Soc., 1960, **82,** 4656).

No evidence for transannular interaction was found in the acyloin LXXXIX (*Leonard et al., ibid.*, 1954, **76,** 630), however, attempted Clemmensen reduction of azepanone XC, and Wolff–Kishner reduction of azepanone LXXXIV (R = H) brings about ring contraction to the piperidine XCI and pyrrole XCII, respectively (*idem, ibid.*, 1949, **71,** 3098; 1950, **72,** 3632; *C. G. Overberger et al.*, J. org. Chem., 1970, **35,** 138):

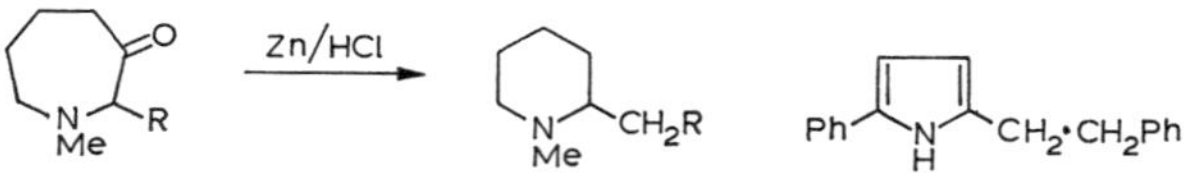

(XC) (XCI) (XCII)

(*ii*) *Syntheses of perhydroazepin-2-ones (ε-caprolactams)* (XCIII)

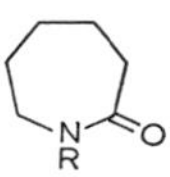

(XCIII)

(*1*) *Beckmann rearrangement.* A large number of cyclohexanone oximes undergo facile rearrangement to the corresponding seven-membered lactams in high yield (*C. S. Marvel* and *J. C. Eck*, Org. Synth., Coll. Vol. II, 1943, p. 371; *Ruzicka*, Helv., 1921, **4,** 477; *O. Wallach*, Ann., 1900, **312,** 171; *R. Huisgen et al., ibid.*, 1954, **586,** 30). Polyphosphoric acid is the reagent of choice (*E. C. Horning* and *V. L. Stromberg*, J. Amer. chem. Soc., 1952, **74,** 2680) but other strong acids and Lewis acids have been used.

This reaction has been intensively investigated due to the industrial importance of caprolactams for the preparation of polymers (Nylon 6, Perlon). This is a general ring-expansion reaction used in the preparation of the lactam series up to the twenty-membered ring (*L. G. Donarume* and *W. Z. Heldt*, Org. Reactions, 1960, **11,** 1; *Ruzicka et al.*, Helv., 1949, **32,** 544).

Azepane-2,5-dione, 2,4-*dinitrophenylhydrazone*, m.p. 234°, is prepared by Beckmann rearrangement of the oxime of 1,4-cyclohexandione trimethylene acetal (*A. H. Rees*, J. chem. Soc., 1962, 3097).

Several alkyl derivatives have also been prepared in this way (*Wallach*, Ann., 1900, **312,** 187; 1905, **343,** 45; 1906, **346,** 249). 2-Alkylcyclohexanones give predominantly *anti*-oximes, rearranging to 7-alkyl-2-azepanones (*J. v. Braun* and *A. Heymons*, Ber., 1930, **63,** 502; *J. G. Hildebrand* and *M. T. Bogert*, J. Amer. chem. Soc., 1936, **58,** 650) but 3-substituted cyclohexanones give a mixture of products (*H. E. Ungnade* and *A. D. McLaren*, *ibid.*, 1944, **66,** 118; J. org. Chem., 1945, **10,** 29). A mixture also results from 2-methylcyclohexanone in oleum, probably due to *anti–syn* isomerisation before rearrangement (*A. Schäffler* and *W. Ziegenbein*, Chem. Ber., 1955, **88,** 1374). Alkaline hydrolysis of cyclohexanone oxime arylenesulphonates also gives caprolactams.

Photo-induced Beckmann rearrangements, probably involving a spiro-oxazirane (XCIV; R = H) are especially suitable for highly hindered oximes (*e.g.* XCV) which resist chemical ring-expansion (*M. Cunningham et al.*, Canad. J. chem., 1971, **49,** 2891; Tetrahedron Letters, 1972, 1151):

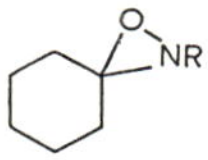

(XCIV)

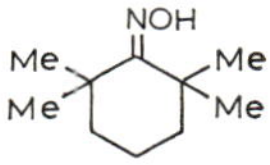

(XCV)

Oxaziranes XCIV independently prepared by treating cyclohexanone imines with peracid, readily give caprolactams on thermolysis (*H. Krimm*, Chem. Ber., 1958, **91,** 1057). Aldonitrones rearrange to amides in the opposite sense to oximes providing a useful alternative to the Beckmann rearrangement (*D. H. R. Barton et al.*, Chem. Comm., 1971, 945):

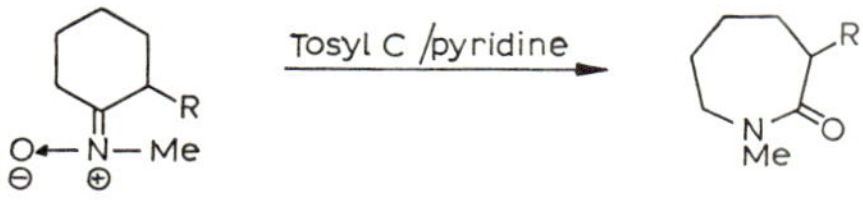

(*2*) *Schmidt reaction.* Treatment of solutions of cyclohexanones with hydrogen azide in strong acids, preferably polyphosphoric acid, yields 2-azepanones in a single step (*K. F. Schmidt.*, Ber., 1924, **57,** 704; *H. Wolff*, Org. reactions, 1946, **3,** 307; *P. A. S. Smith*, J. Amer. chem. Soc., 1948, **70,** 320; *Blicke* and *N. J. Doorenbos*, *ibid.*, 1954, **76,** 2317; *Schmidt*, G.P. 427,858/1923; Friedl., 1930, **15,** 1634; *I. Noishiki* and *M. Watanabe*, Jap. P. 5160/1951; C.A., 1953, **47,** 9998; J. chem. Soc. Japan (Ind. Chem. Sect.) 1953, **56,** 100; *R. T. Conley*, J. org. Chem., 1958, **23,** 1330). This is a general reaction which has been used to form azahomosteroids (*H. Singh*, J. chem. Soc. Perkin I. 1972, 993). Cyclohexanones with 2-substituents give only 7-substituted-2-azepanones, but mixtures of the two possible lactams are found with 3- or 4-substituted cyclohexanones (*H. Shechter* and *J. C. Kirk*, J. Amer. chem. Soc., 1951, **73,** 3087). Treatment of *aci*-nitrocyclohexane with hydrogen azide also gives caprolactam (*L. G. Donaruma* and *M. L. Huber*, J. org. Chem., 1956, **21,** 965).

If excess hydrogen azide is used in the Schmidt reaction with cyclohexanones *tetrazolo-*[1,5-a]*azepane* (Metrazole) (XCVI), m.p. 58°, is also formed.

Boron trifluoride catalyses this reaction (*L. Ruzicka et al.*, Helv., 1933, **16,** 1335; *W. B. Turner et al.*, J. chem. Soc. Perkin I, 1972, 2136; *H. Singh et al.*, *ibid.*, 990). XCVI is also obtained when lactim ethers, *O*-acyl-lactims or lactam imidoyl chlorides react with

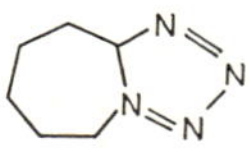

(XCVI)

hydrogen azide (*Schmidt*, G.P. 521,870/1928, 545,850/1927; Friedl., 1932, **17,** 2603, 2605) or caprolactam hydrazone is treated with nitrous acid (*R. C. Glushkov* and *E. S. Golovchinskaya*, Zhur. priklad. Khim., 1959, **32,** 920; *R. Stolle*, Ber., 1930, **63,** 1032). The chemistry of metrazole has been reviewed (*J. Marcus*, Revista de Chemie, Romania, 1957, **8,** 29; see also C.C.C., Vol. IVC, Chap. 17).

(*3*) Caprolactams can be made by cyclisation of the appropriate ε-amino acids or esters, frequently just by heating to above the melting point (*S. Gabriel* and *T. A. Maaz*, Ber., 1899, **32,** 1266; *J. v. Braun*, *ibid.*, 1907, **40,** 1834; *G. Schroeter et al.*, *ibid.*, 1930, **63,** 1308; *O. M. Friedmann et al.*, J. Amer. chem. Soc., 1960, **82,** 5202). Yields are poor, and often lactams from alternative routes are used as a source of ε-aminocaproic acids. Cyclohexanecarboxylic acid is converted into caprolactam in a mixture of fuming nitric and sulphuric acids in liquid sulphur dioxide. Nitrosylsulphuric acid, used directly in a similar synthesis (SNIA method) is shown to be present (*N. Tokura et al.*, Bull. chem. Soc., Japan, 1965, **38,** 849).

Azepanedione (XCVII) (R = OH) is obtained by irradiating 2-nitrocycloheptanone (*S. T. Reid* and *J. N. Tucker*, Chem. Comm., **1971,** 1609); the imide XCVII (R = H) results from the action of metal *tert*-butylperoxides on ε-caprolactam (*A. R. Doumaux* and *D. J. Trecker*, J. org. Chem., 1970, **35,** 2121):

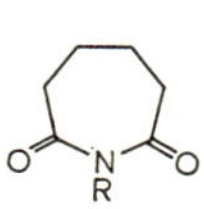

(XCVII)

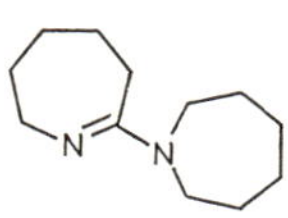

(XCVIII)

2-Azepanone (*ε-caprolactam, 2-oxohexamethyleneimine*), m.p. 69°, b.p. 139°/12 mm (*Müller* and *P. Bleier*, Monatsh., 1928, **50,** 400; *Wallach*, Ann., 1900, **312,** 187) is produced on an enormous scale for the manufacture of polymers (Nylon 6) and is possibly the most important industrial heterocyclic compound.

Reduction of caprolactam is more difficult than the reduction of caprolactim ethers. Hydrogenation over Raney nickel, ruthenium oxide and barium–copper chromite gives hexamethyleneimine as does lithium tetrahydridoaluminate. Catalytic reduction over platinum gives the amidine XCVIII (*R. E. Benson* and *T. L. Cairns*, J. Amer. chem. Soc.,

1948, **70,** 2115; *A. Etienne* and *J. Correia*, Compt. rend., 1964, **259C,** 2660; *O. Cervinka*, Coll. Czech. chem. Comm., 1959, **24,** 1146).

With nitrogen oxides in acetic acid N-*nitrosocaprolactam* (XCIII; R = NO), m.p. 13–14°, is obtained, which decomposes with loss of nitrogen and ring-opening (*R. Huisgen* and *J. Reinertshofer*, Ann., 1952, **575,** 174; *G. Nischk* and *E. Muller*, Ann., 1952, **576,** 232; *K. Heyns* and *O. F. Woyrsch*, Chem. Ber., 1953, **86,** 76). N-*Methylcaprolactam* (XCIII; R = Me), b.p. 120°/25 mm, is prepared by treating caprolactam with sodium, followed by dimethyl sulphate (*Ruzicka et al.*, Helv., 1921, **4,** 478); other *N*-alkyl derivatives are obtained with alkyl halides in the presence of sodium (*A. P. Swain et al.*, J. org. Chem., 1953, **18,** 1087; *Marvel* and *W. W. Moyer*, *ibid.*, 1957, **22,** 1065). Iodobenzene and copper powder yield N-*phenylcaprolactam* (XCIII; R = Ph), m.p. 75–77° (*H. K. Hall et al.*, J. Amer. chem. Soc., 1958, **80,** 6404). Acetic anhydride gives *N*-acetylcaprolactam (XCIII; R = Ac) and formaldehyde and vinyl cyanide the *N*-hydroxymethyl- (XCIII; R = CH_2OH) and *N*-cyanoethyl- (XCIII; R = $CH_2 \cdot CH_2 \cdot CN$) derivatives, respectively; reduction and dehydration of XCIII (R = $CH_2 \cdot CH_2 \cdot CN$) gives the useful dehalogenohydrogena-

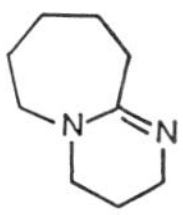

(XCIX)

tion agent 1,5-diazabicyclo[5.4.0]undecan-5-ene (XCIX) (*H. Oediger* and *Fr. Moeller*, Angew. Chem. intern. Edn., 1967, **6,** 76). Other olefins add to 2-azepanone in the presence of base affording an alternative route to *N*-alkylcaprolactams (*Benson* and *Cairns*, *loc. cit.*); similarly acetylene will add in the presence of potassium to give N-*vinylcaprolactam*, m.p. 34–35°, b.p. 95°/4 mm (*M. F. Shostakovskii et al.*, Izvest. Akad. Nauk S.S.S.R., Otdel. Khim. Nauk, 1952, 690; C.A., 1953, **47,** 9917; 1954, **48,** 584).

Azepane-2-thione, b.p. 123–128°/0.1 mm, is formed by the action of phosphorus pentasulphide on caprolactam (*Ruzicka*, Helv., 1935, **18,** 659). Grignard reagents also attack the carbonyl group giving 2,2-dialkylazepanes but ring-opened products are also found (*Heyns* and *W. Pyrus*, Chem. Ber., 1955, **88,** 678; *Lukes et al.*, Coll. Czech. chem. Comm., 1950, **15,** 303; 1961, **26,** 1105).

Caprolactam and its *N*-methyl derivative are easily hydrolysed but the difficulty of hydrolysis increases with the size of the *N*-alkyl substituent (*Marvel* and *Moyer*, *loc. cit.*). Most monoalkyl ε-caprolactams, other than those carrying *N*-substituents polymerise readily in the melt, catalysed by water, sodium hydride and many other materials. Ease of polymerisation does not correspond to the rate of hydrolysis but varies inversely with the ease of cyclisation of the parent amino acid. Highly substituted caprolactams where equilibrium favours the cyclic form therefore polymerise less readily *(Hall, loc. cit.)*.

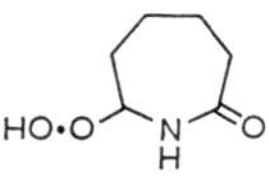

(C)

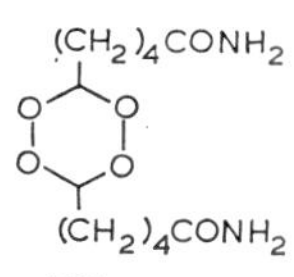

(CI)

Polymerisation of lactams is a general reaction (*K. Dachs* and *E. Schwartz*, Angew. Chem. Intern. Edn., 1962, **1,** 430).

On heating in oxygen, caprolactam is oxidised to *caprolactam hydroperoxide* (C), m.p. 139° (decomp.), and further to the cyclic imide XCVII (R = H) (*A. Reiche* and *W. Schön*, Z. Chem., 1963, **3,** 451). Reduction of C gives the ring–chain tautomer 5-formylpentanamide, and treatment with acid yields the tetroxane CI (*E. Höft et al.*, Ann., 1969, **725,** 241).

A list of methyl homologues of caprolactam is given in Table 3.

TABLE 3

METHYL HOMOLOGUES OF ε-CAPROLACTAM

	M.p. (°C)	*Ref.*
α-Methylcaprolactam	97–98	1
β-Methylcaprolactam	105–106	2
γ-Methylcaprolactam	41–42	3
δ-Methylcaprolactam	68–69	2
ε-Methylcaprolactam	90–91	1
	b.p. 142–145/15 mm	4
β,δ-Dimethylcaprolactam	122–123	3
β,ε-Dimethylcaprolactam	125–126	3
γ,ε-Dimethylcaprolactam	133.5–134.5	3
α,γ,ε-Trimethylcaprolactam	73.5–75	3
β,δ,ε-Trimethylcaprolactam	137–138	3

References

1 *A. Schäffler* and *W. Ziegenbein*, Chem. Ber., 1955, **88,** 1374.
2 *O. Wallach*, Ann., 1899, **309,** 2; 1900, **312,** 191; 1906, **346,** 253.
3 *H. Ungnade* and *A. D. McLaren*, J. org. Chem., 1945, **10,** 29.
4 *A. Müller* and *P. Krauss*, Monatsh., 1932, **61,** 206.

2. Bicyclic compounds including an azepine ring

(*a*) *Azepines fused to three-, four- or five-membered rings*

(*i*) *Homoazepines*

The 2,3- and 4,5-**cyclopropeno-1*H*-azepines** II (R = Et), m.p. 46°, b.p. 59–61°/0.1 mm, and IV (R = Et), b.p. 69–71°/0.1 mm, are obtained by the thermolysis of azidoformates or dimethyl azodicarboxylate in the presence of cycloheptatriene; unstable aziridines I and III are probable intermediates. No valence isomerisation is observed between II and IV, nor does either show cycloaddition. Irradiation leaves IV unaffected but II undergoes ring-contraction (*L. A. Paquette* and *R. J. Haluska*, Chem. Comm., 1968, 1370; J. org. Chem., 1970, **35,** 132; *W. H. Okamura* and *W. H. Snider*, Tetrahedron

Letters, 1968, 3367). Pyrolysis of II and IV leads to the different ring-contraction products shown (*Okamura, ibid.*, 1969, 4717):

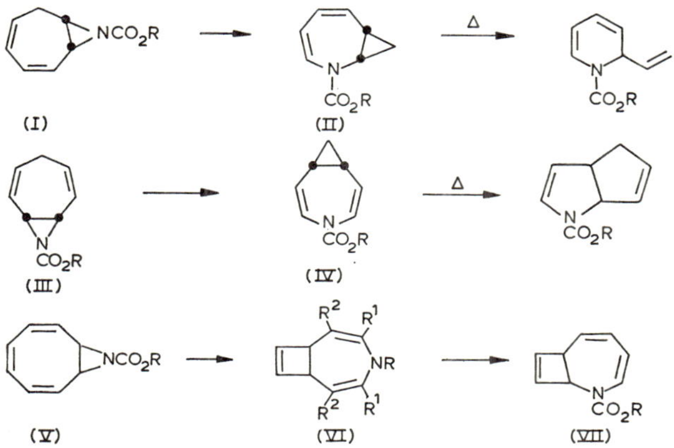

Cyclobuteno[3,4-*d*]-4,5-**dihydroazepines** VI (R = CO_2Et; R^1, R^2 = H, b.p. 65–67°/0.07 mm) and (R = CO_2Me; R^1,R^2 = H) are obtained in the thermal rearrangement of the cyclo-octatetraene nitrogen insertion product V; on further heating VI rearranges to dihydroazepine VII and indole derivatives (*S. Masamune et al.*, Accounts chem. Research, 1972, **5**, 272; Angew. Chem. internat. Edn., 1964, **3**, 582; *A. G. Anastassiou et al.*, Chem. Comm., 1972, 531; Tetrahedron Letters, 1972, 4569). Decarboxylative ring-opening of the adduct between dimethoxycarbonyl "Dewar benzene" and the mesoionic compound shown in the reaction scheme on p. 206 gives compound VI (R = Me; R^1 = Ph, Ph, R^2 = CO_2Me) (*H. D. Martin* and *M. Hekman*, Angew. Chem. intern. Edn., 1972, **11**, 926).

(*ii*) *Aza-azulenes*

A single carbon atom of the azulene nucleus VIII can be replaced to give six isomeric aza-azulene systems — 1,2,3a,4,5 and 6; representatives of all four systems with the nitrogen atom in the seven-membered ring are known.

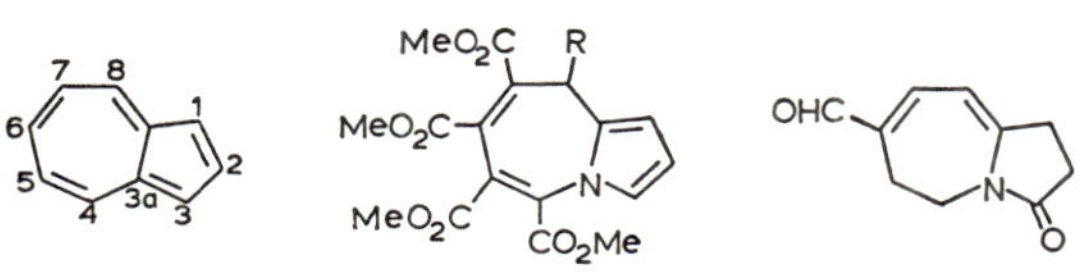

3a-Aza-azulenes IX are obtained by irradiating 1,2,3,4-tetramethoxycarbonyl-9a*H*-quinolizines, presumably by successive ring-opening and recyclisation (*R. M. Acheson* and *J. K. Stubbs*, J. chem. Soc. C, 1969, 2316; 1971, 3285). Similarly, irradiation of *N*-

chloroacetyl-*p*-*O*-methyltyramine gives the reduced derivative X (*B. Witkop et al.*, J. Amer. chem. Soc., 1967, **89,** 1039; 1968, **90,** 776). *Perhydro*-3a-*aza-azulene* (1-*azabicyclo*-[5.3.0]*decane*), b.p. 79–80°/17 mm, n_D^{20} 1.4822, *picrate*, m.p. 215°, is obtained by cyclisation of 1,4,9-tribromononane with methanolic ammonia (*R. Seiwerth et al.*, Ber., 1939, **72,** 1638; Arkiv Kemi, 1952, **24,** 21; Monatsh., 1953, **84,** 777) reductive cyclisation of diethyl δ-hydroxyiminoazelate with copper chromite (*N. J. Leonard* and *W. E. Goode*, J. Amer. chem. Soc., 1950, **72,** 5404) or in the Clemmensen or electrolytic reduction of 1-quinolizidinone (*G. R. Clemo et al.*, J. chem. Soc., 1931, 437; 1936, 1429). Similar rearrangements are found in many saturated bicyclic systems containing a nitrogen atom; the ring containing the halogen always undergoes ring contraction (*Leonard et al.*, J. Amer. chem. Soc., 1949, **71,** 3089; 3100; 1950, **72,** 4931; 1951, **73,** 5210; 1952, **74,** 1700, 6251). Perhydro-3a-aza-azulene is also obtained by reduction of *compound* XI, m.p. 109–110°, prepared as shown with lithium tetrahydridoaluminate (*L. A. Cohen* and *Witkop*, J. Amer. chem. Soc., 1955, **77,** 6595):

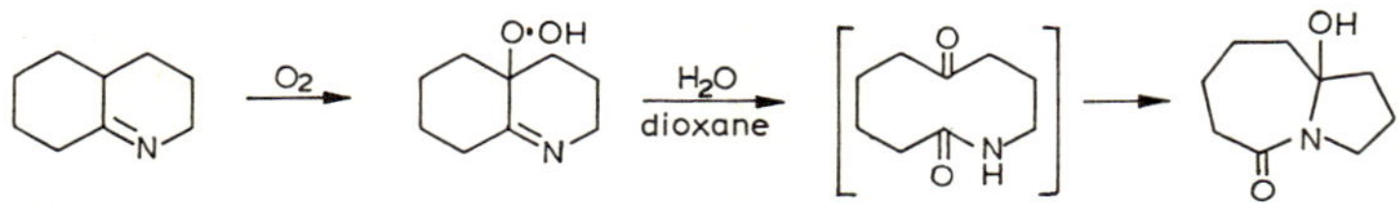

(XI)

An azepine ring may be built on to a cyclopentane derivative to give 1,2,3-**triphenyl-4-aza-azulene** (XII), m.p. 224–226° (*M. K. Connor* and *E. LeGoff*, Tetrahedron Letters, 1970, 2687):

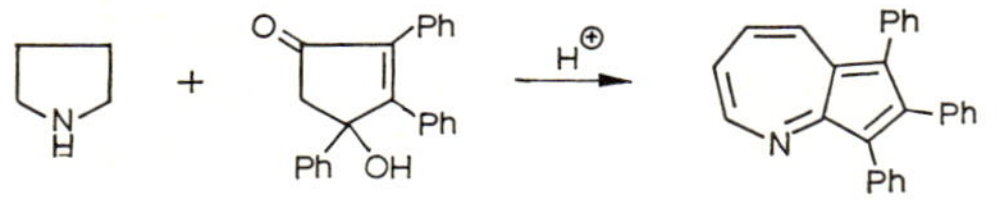

(XII)

Heating the bridged azepine XIII also gives a 4-aza-azulene XIV (*Paquette et al.*, J. org. Chem., 1969, **34,** 2879):

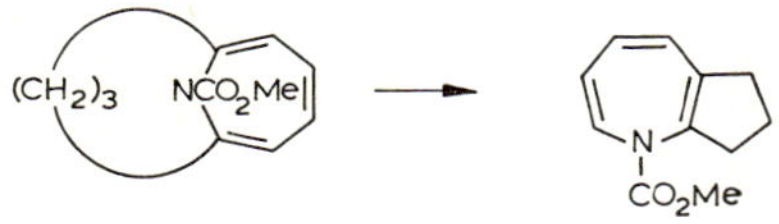

(XIII) (XIV)

The parent system, 5-**aza-azulene** (XV; R = H), m.p. 34–35°, *picrate*, m.p. 203–204°, is synthesised by condensing 6-dialkylaminofulvene and 4-dibutylaminocrotonaldehyde with oxalyl chloride, and treating the product with sodium perchlorate and base followed by ammonia (*K. Hafner et al.*, Angew. Chem., 1961, **73,** 657; intern. Edn., 1963, **2,** 123):

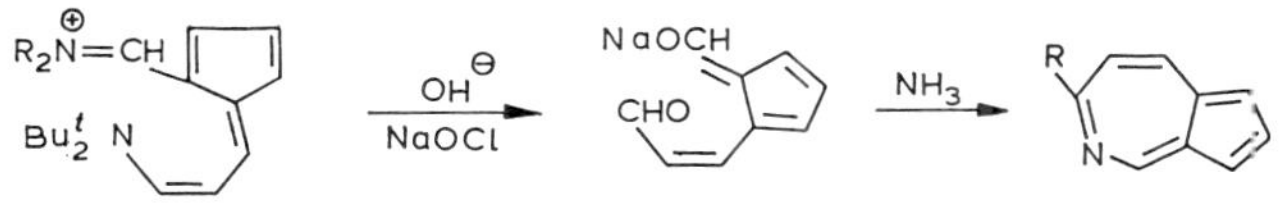

(XV)

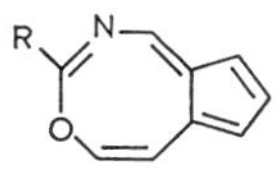

(XVI)

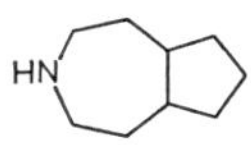

(XVII)

6-Phenyl-5-aza-azulene (XV; R = Ph), m.p. 158–159°, prepared similarly adds acetylenes, in contrast to azulene itself, both across positions 4, 7 and 3a, 6 (*idem, ibid.*, 1970, **9**, 159). Oxidation of XV with hydrogen peroxide gives bicyclic oxazocins XVI probably *via* an oxaziridine intermediate (*G. Schaden*, Chem. Ber., 1972, **105**, 3128).

Perhydro-6-aza-azulene (4-*azabicyclo*[5.3.0]*decane*) (XVII), cis-*form* b.p. 93–95°/20 mm, *picrate*, 169–171°, trans-*form picrate*, m.p. 150–151°, is prepared by the hydrogenation of 1,2-bis(cyanomethyl)cyclopentane. Unlike carbocyclic azulenes, neither perhydro-6- nor perhydro-5-aza-azulene is dehydrogenated by palladium (*J. O. Halford* and *B. Weissman*, J. org. chem. 1952, **17**, 1276; *M. Pailer* and *I. Allmer*, Monatsh., 1955, **86**, 819; *A. Bertho*, Chem. Ber., 1957, **90**, 29).

(*b*) *1-Benzazepines*

Although 1-benzoxepine is formed as a by-product in the preparation of 1,6-oxido[10]annulene, no 1*H*-1-benzazepine is found in the analogous preparation of 1,6-imino[10]annulene (*E. Vogel et al.*, Angew. Chem. intern. Edn., 1964, **3**, 642; Tetrahedron Letters, 1965, 3613). The 1-benzazepine XVIII (R = H) is however a product of the ring-expansion of 4-ethoxy-1-tosyl-1,2-dihydroquinoline with dibromocarbene:

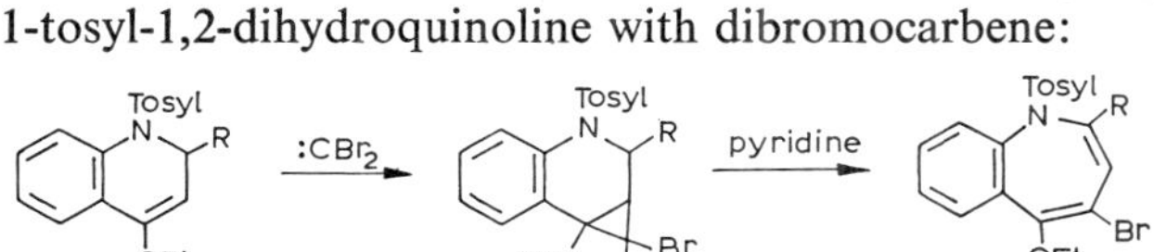

(XVIII)

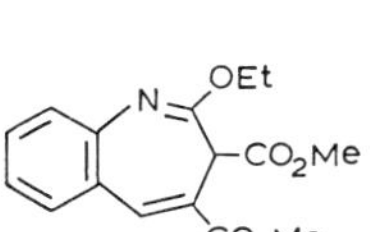

(XIX)

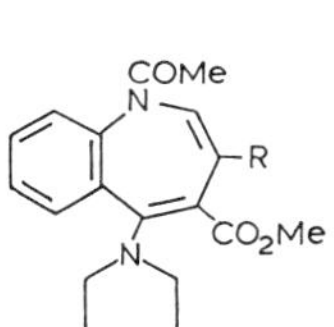

(XIX a)

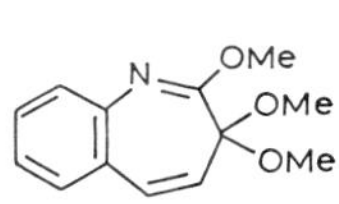

(XX)

With the analogue (R = Me) the tosyl group causes steric hindrance and the *N*-mesyl derivative must be used (*A. Cromarty* and *G. R. Proctor*, Chem. Comm., 1968, 842; J. chem. Soc. C, 1971, 3536).

Dimethyl 2-ethoxy-3*H*-1-benzazepine-3,4-dicarboxylate (XIX), m.p. 109–110°, results from the spontaneous ring-opening of the cycloaddition product between 2-ethoxyindole and dimethyl acetylenedicarboxylate (*H. Pleininger* and *D. Wild*, Chem. Ber., 1966, **99,** 3070) and an analogous reaction is shown by 1-acetyl-3(*N*)piperidinoindole (*M-S. Lin* and *V. Snieckus*, J. org. Chem., 1971, **36,** 645) to give XIXa (R = CO_2Me), or XIXa (R = H) if methyl propiolate is used. 1-Methylindole yields *dimethyl* 1-*methyl*-(1H)-1-*benzazepine*-3,4-*dicarboxylate*, m.p. 105–107°. A further molecule of diester will add across the diene system of the azepine ring (*R. M. Acheson et al.*, J. chem. Soc. Perkin I, 1972, 968; Chem. Comm.; 1971, 1225). The acetal 2,3,3-*trimethoxy*-3H-1-*benzazepine* (XX) obtained in the reduction of methyl *o*-nitrobenzylidenepyruvate is not hydrolysed to the azabenzotropolone (*M. Look*, Diss. Abs., 1957, **17,** 36), and attempts to prepare 1-benzazepin-3-one as shown below led to isolation of the *dimer* XXI, m.p. 158–159°, only (*Proctor et al.*, J. chem. Soc. C, 1968, 1280):

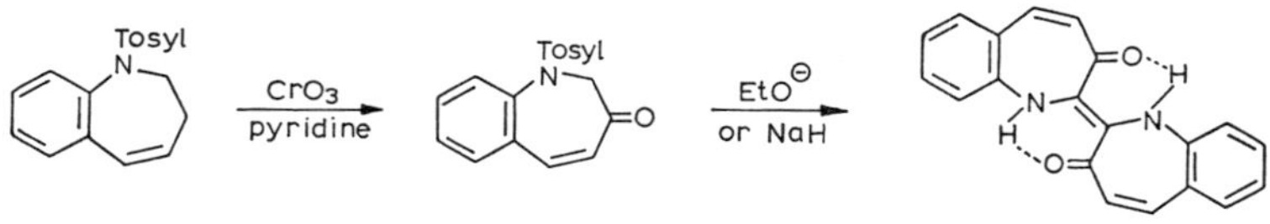

(XXI)

The 5*H*-1-**benzazepine** XXII is also obtained by a cycloaddition–ring-opening sequence using 3,3-dimethyl-3*H*-indole and 1-dimethylamino-2-phenylacetylene in the presence of boron trifluoride etherate (*H. G. Viehe et al.*, Angew. Chem. intern. Edn. 1966, **5,** 585). Several 1-benzazepine-5-ones of type XXIII have been obtained from other 1-benzazepines, principally 2,5-**dihydro-1*H*-1-benzazepine-2,5-dione** (XXIV; R = H), m.p. 222–223°, and its derivatives. Reaction of a 7,8-dimethyl derivative of XXIV with phosphoryl chloride afforded a 2-chloro-7,8-dimethyl-1-benzazepin-5-one (*A. H. Rees et al.*, J. chem. Soc., 1959, 3111; 1962, 3097; Canad. J. chem. 1969, **47,** 1227), while with triethyloxonium

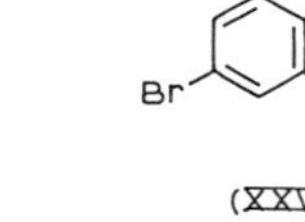

(XXII)

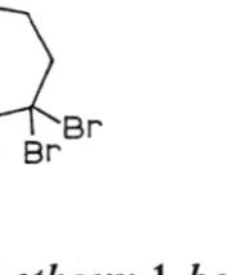

(XXIII)

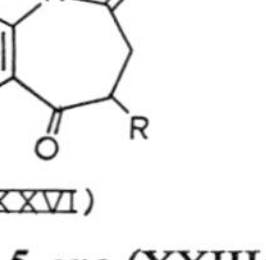

(XXIV)

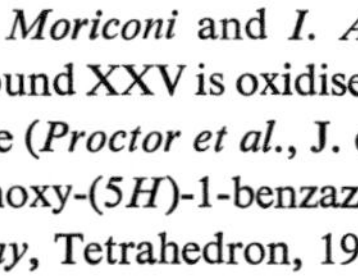

(XXV)

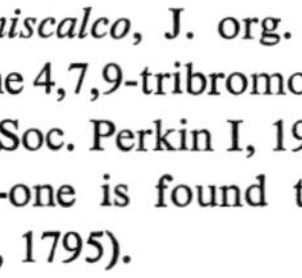

(XXVI)

tetrafluoroborate 2-*ethoxy*-1-*benzazepin*-5-*one* (XXIII; R = OEt), m.p. 46°, is obtained (*E. J. Moriconi* and *I. A. Maniscalco*, J. org. Chem., 1972, **37,** 208). The tetrabromo compound XXV is oxidised to the 4,7,9-tribromo-5-one and -2,5-dione by active manganese dioxide (*Proctor et al.*, J. chem. Soc. Perkin I, 1972, 2012). The azepine ring of 4-hydroxy-2-methoxy-(5*H*)-1-benzazepin-5-one is found to be planar (*W. A. Denne* and *M. F. Mackay*, Tetrahedron, 1972, **28,** 1795).

The ketolactam XXIV (R = H) is prepared from 1,4-naphthoquinone in a Schmidt reaction (hydrogen azide) (*G. Jones et al.*, J. chem. Soc., C, 1967, 1808; Tetrahedron Letters, 1966, 2367; *R. W. Rickards* and *R. M. Smith, ibid.*, 1966, 2361; *J. W. Moore et al., ibid.*, 1965, 1071; Tetrahedron, 1966, **22,** 1201); a similar reaction yields XXIV (R = OH), m.p. 260° decomp., from 4-azido-1,2-naphthoquinone (*idem, ibid.*, 1969, 1243). An alternative route to the 7,8-dimethyl derivative of XXIV (R = OH) (the enol form of the 2,4,5-trione) is the acid hydrolysis of the condensation product of the tetrahydrodione XXVI (R = H, R¹ = Me) with 4-nitrosodimethylaniline (*Rees, loc. cit.*; *Moriconi* and *Maniscalco, loc. cit.*).

Dione XXVI (R = H, R¹ = Me), m.p. 199°, is prepared by the thermal cyclisation of the 2-succinoylarylamine. Although the unsaturated amino acid does not cyclise, a 3,4-double bond is readily introduced in a bromination–dehydrobromination sequence. The *acid* XXVI (R = CO_2H, R¹ = H), m.p. 322–323°, and its esters are prepared by Dieckmann cyclisation (*T. A. Geissmann* and *A. K. Cho*, J. org. Chem., 1959, **24,** 41; *H. B. Macphillamy et al.*, J. Amer. chem. Soc., 1958, **80,** 2172; *R. G. Cooke* and *H. F. Haynes*, Austral. J. chem., 1958, **11,** 225). Derivatives of both XXIII and XXIV undergo ring-contraction to quinoline-4-carboxylic acids on hydrolysis *(Rees, loc. cit.)* and after reduction with lithium tetrahydridoaluminate followed by catalytic hydrogenation, give tetrahydro-1-benzazepines *(Proctor, loc. cit.)*. The oxo groups at 2 and 5 may also be reduced to methylene independently (*J. Witte* and *V. Bockelheide*, J. org. Chem., 1972, **37,** 2849).

Reduction of 2,3-dihydro-1*H*-1-benzazepin-2-ones XXVII gives **2,3-dihydrobenzazepines** XXVIII. Low yields of compounds XXVII result from the condensation of aniline with oxocarboxylic acids in the presence of aniline hydrochloride (*A. Bertho*, Chem. Ber., 1957, **90,** 29). Dehydration of 5-hydroxy-2,3,4,5-tetrahydro-1*H*-1-benzazepines also gives compounds XXVIII (*Rees, loc. cit*; *B. D. Astill* and *Boekelheide*, J. Amer. chem. Soc., 1955, **77,** 4079).

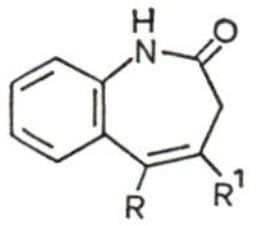

(XXVII)

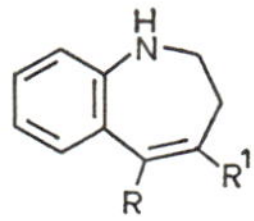

(XXVIII)

(XXIX)

2,3-*Dihydro*-1-*benzazepine* itself (XXVIII; R, R¹ = H), m.p. 48°, b.p. 95–100°/0.4 mm, is most conveniently obtained by the sodium in liquid ammonia detosylation, tetrahydroborate reduction and acid-catalysed dehydration of XXXVI (R = tosyl) (p. 227) (*Proctor et al.*, J. chem. Soc. C, 1968, 1280; 1964, 4926). Compounds of type XXVIII (R = Cl, R¹ = H) may be obtained from or converted into tetrahydro-1-benzazepin-5-one. A 2,5-dihydro-1*H*-1-benzazepine is produced in the following thermolysis (*P. Scheiner*, J. org. Chem., 1967, **32,** 2628):

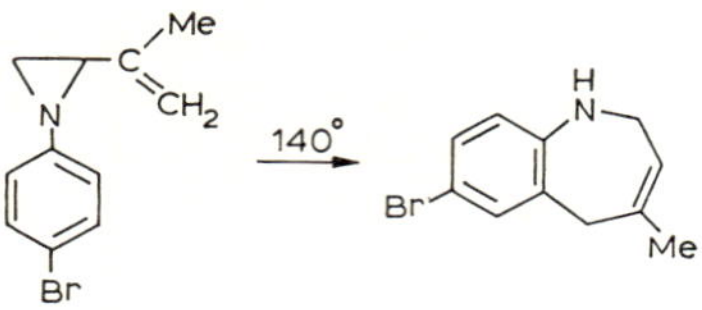

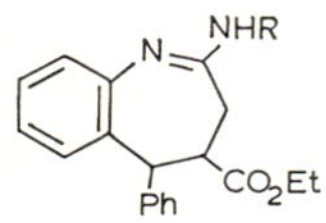

(XXX)

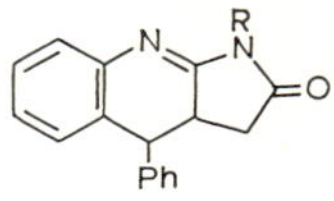

(XXXI)

and the 2,5-dihydro-(1*H*)-1-benzazepin-5-one XXIX by treating the dibromocarbene adduct shown in the reaction scheme on p. 223 with ethanolic silver nitrate.

4,5-**Dihydro-**3*H*-1-**benzazepines** XXX, formed by the action of phosphorus pentachloride and a primary amine on the corresponding lactam, undergo rearrangement to XXXI on hydrolysis (*A. Vogel et al.*, Helv., 1969, **52,** 1929).

2,3,4,5-**Tetrahydro-**(1*H*)-1-**benzazepine** (2,3-*benzazacycloheptane, homotetrahydroquinoline*) (XXXII; R = H), b.p. 253–255 °, m.p. 32 °, is prepared by the reduction of its oxidised or unsaturated derivatives or more conveniently by the action of aqueous alkali on 2-(4-chlorobutyl)aniline hydrochloride; the analogous route to the monocyclic azepane is much less effective. It is stable, strongly basic and readily *N*-alkylated or acylated. The *benzoyl* derivative (XXXII; R = COPh), m.p. 96 °, is hydrolysed to 2-(4-chlorobutyl)-

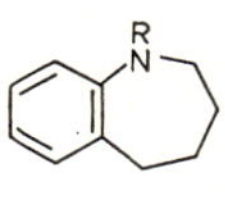

(XXXII)

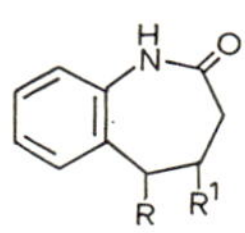

(XXXIII)

aniline, but quaternary salts and XXXII (R = Me) undergo demethylation rather than ring-opening. Reactions at nitrogen occur less readily with XXXI than with analogous tetrahydroquinolines (*J. v. Braun et al.*, Ber., 1912, **45,** 3376; 1922, **55,** 3811; *A. Müller, ibid.*, 1928, **61,** 568; *W. G. Brown et al.*, J. Amer. chem. Soc., 1939, **61,** 2597; 1943, **65,** 1841; *Proctor et al.*, J. chem. Soc., 1964, 4926).

2,3,4,5-**Tetrahydro-**(1*H*)-1-**benzazepin-2-one** *(dihydrohomocarbostyril)* (XXXIII; R, R^1 = H), m.p. 140–141 °, is readily obtained by heating *o*-aminophenylbutyric acid. The presence of a second ring both reduces the amount of polymeric by-product and increases the stability of the lactam ring (*v. Braun*, Ber., 1907, **40,** 1834; *G. Schroeter et al., ibid.*, 1930, **63,** 1308). Dihydrohomocarbostyril is also prepared by Beckmann rearrangement of α-tetralone oxime using polyphosphoric acid (*E. C. Horning et al.*, J. Amer. chem. Soc., 1952, **74,** 5153, 1954, **76,** 3651; 1955, **77,** 5932; *R. A. Barnes* and *M. T. Beachem, ibid.*, 1955, **77,** 5388; *R. Huisgen et al.*, Ann., 1954, **586,** 30; 1951, **574,** 171) and by the Schmidt reaction on α-tetralone (*L. H. Briggs* and *G. C. DeAth*, J. chem. Soc., 1937, 456; *W. L. Berry* and *P. A. S. Smith*, J. org. Chem., 1961, **26,** 27; *T. Shingu et al.*, Chem. and Ind., 1962, 1191; *R. T. Conley*, J. org. Chem., 1958, **23,** 1330). A Schmidt reaction on 4-phenylbutyric acid surprisingly gives a high yield of XXXIII (R, R^1 = H) (*S. K. Datta et al.*, J. chem. Soc., C, 1970, 2058). Compounds XXXIII (R = CO_2Me; R^1 = H) and (R = H; R^1 = CO_2Me) both undergo ring-contraction by ester–amide exchange under acid conditions yielding oxindole and quinolone derivatives, respectively *(Horning, loc. cit.)*. Lithium tetrahydridoaluminate reduces XXXIII (R, R^1 = H) to XXXII.

1-*Tosyl*-2,3,4,5-*tetrahydro*-(1H)-1-*benzazepin*-3-*one* (XXXV; R = H), m.p. 117°, results from the acid cyclisation of alcohol XXXIV (R = H). The ketone is reduced to the corresponding *alcohol*, b.p. 220°/0.5 mm, by lithium tetrahydridoaluminate, and its acetal is detosylated by sodium in liquid ammonia (*G. R. Proctor et al.*, J. chem. Soc. C, 1970, 2191):

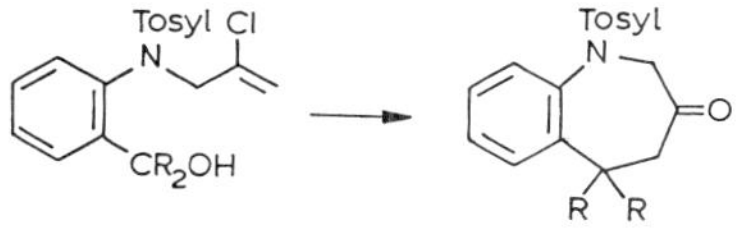

(XXXIV) (XXXV)

In contrast to the synthesis of tetrahydroquinolones, routes to 2,3,4,5-**tetrahydro-(1*H*)-1-benzazepin-5-ones** (XXXVI; R = H) *via* the formation of an aryl–carbon bond are generally unsuccessful. Dieckmann cyclisation of the *N*-substituted anthranilic esters XXXVII, followed by hydrolysis and decarboxylation is a more reliable route (*B. D. Astill* and *V. Boekelheide*, J. Amer. chem. Soc., 1955, **77,** 4079; *J. T. Braunholtz* and *F. G. Mann*, Chem. and Ind., 1957, 266; J. chem. Soc., 1957, 4174; 1958, 3377; *Proctor* and *R. H. Thompson*, *ibid.*, 1957, 2302, 2312).

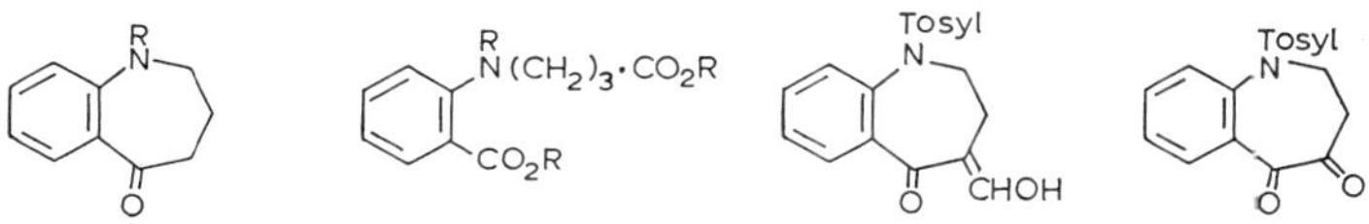

(XXXVI) (XXXVII) (XXXVIII) (XXXIX)

The *parent compound* (XXXVI; R = H), m.p. 69°, b.p. 135–140°/0.1 mm, is obtained in the acid hydrolysis of the N-*acetyl* compound (XXXVI; R = Ac), m.p. 122°, or the N-*tosyl* derivative (XXXVI; R = tosyl), m.p. 126°, both prepared in this way. The *N*-acetyl derivative is also obtained from XIXa (R = H) by hydrogenation and hydrolysis (*Lin* and *Snieckus, loc. cit.*). N-*Methyl*-2,3,4,5-*tetrahydro*-(1H)-1-*benzazepin*-5-*one* (XXXVI; R = Me), b.p. 112°/0.1–0.2 mm, n_D^{29} 1.5992, which is reduced by the Wolff–Kishner procedure to N-*methyl*-2,3,4,5-*tetrahydro*-(1H)-*benzazepine* (XXXII; R = Me), b.p. 110–115°/15 mm, is also prepared from the diester XXXVII (R = Me). Earlier reports that these cyclic aminoketones fail to form normal carbonyl derivatives appear to be false, and these compounds also show enolisation, and condensation at position 4; for example XXXVI (R = tosyl) condenses with ethyl formate at position 4 to give hydroxymethylene compound XXXVIII. The presence of a methyl group at position 2 makes both cyclisation and subsequent removal of the *N*-substituent more difficult (*Proctor et al.*, J. chem. Soc. Perkin I, 1972, 2012).

Some *N*-methyl and *N*-phenyl derivatives of 4-anilinobutyric acid may however by cyclised under Friedel–Crafts conditions and N-*phenyl*-2,3,4,5-*tetrahydro*-(1H)-1-*benzazepin*-5-*one* (XXXVI; R = Ph), m.p. 65–66°, is obtained by this route, (*idem, ibid.*, 1803).

Bromination of XXXVI (R = tosyl) with a large excess of bromine gives the detosylated tetrabromo compound XXV, which on hydrolysis yields a mixture of 7,9-dibromo derivatives. With zinc dust or sodium iodide both the bromine atoms at position 4 are replaced by hydrogen. If less bromine is used 4-bromo- and then 4,4-dibromo-1-tosyltetrahydro-(1*H*)-1-benzazepin-5-ones are formed. With XXXVI (R = H) the 4,7,9-tribromo- and then the 4,4,7,9-tetrabromotetrahydrobenzazepinones are obtained, and both tri- and tetrabromides are found to undergo ring-contraction to quinoline-4-carbaldehydes (*idem*, *ibid.*, p. 2012; J. chem. Soc. C, 1968, 1280; J. chem. Soc. 1964, 4926; 1961, 3989).

N-*Tosyl*-2,3,4,5-*tetrahydro*(1H)-1-*benzazepine*-4,5-*dione* (XXXIX), m.p. 148 °, is obtained in the oxidation of XXXVI (R = tosyl) with selenium dioxide, the ozonolysis of XXXVIII, or the hydrolysis of the 4,4-dibromo derivative of XXXVI (R = tosyl). This last reaction is reversed when XXXIX is treated with bromine (*idem*, J. chem. Soc., 1961, 3989).

(*c*) *2-Benzazepines*

The 2-benzazepine system has been studied less intensively than either the 1- or 3-benzazepine systems and few fully unsaturated derivatives are known. The 3*H*-2-**benzazepines** of the type XL are obtained by the addition of 2-phenyl-1-azirines to 1,3-diphenylinden-2-one (*A. Hassner* and *D. J. Anderson*, J. Amer. chem. Soc., 1972, **94,** 8255) and the 1H-2-*benzazepin*-1-*one* XLII (X = Cl), m.p. 117 °, is the product of the ring-opening–reclosure sequence shown (*G. Simchen*, Angew. Chem. intern. Edn., 1968, **7,** 464):

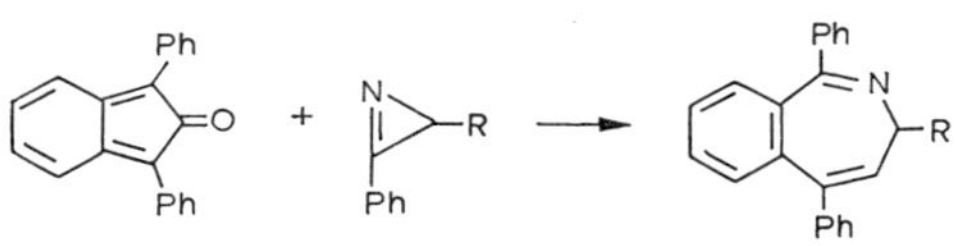

(XL)

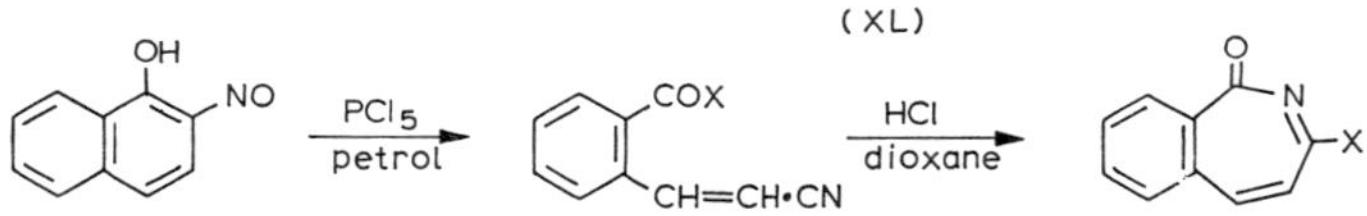

(XLI) (XLII)

If XLI (X = OH) is cyclised with polyphosphoric acid 1,3-*dihydro*-(2H)-2-*benzazepine*-1,3-*dione* (oxo form of XLII, X = OH), m.p. 177–179 °, results (*G. N. Walker et al.*, J. org. Chem., 1972, **37,** 3955; 1971, **36,** 461). Several 4-substituted analogues have been prepared similarly. *N*-alkylation allows reduction of both oxo groups to methylene by lithium tetrahydridoaluminate.

4,5-*Dihydro*-(3H)-2-*benzazepine* (XLIII; R = H), *picrate*, m.p. 159–160 °, from the cyclisation of *N*-formyl-3-phenylpropylamine with polyphosphoric acid–phosphorus pentoxide, or the oxidation of the tetrahydro compound XLIV (R = H) with manganese dioxide dimerises to XLV with a trace of acid; analogues (R = Me, Ph) are known. Reduction with tetrahydridoborates saturates the 1,2-imino bond (*I. M. Goldman et al.*, J. Amer. chem. Soc., 1969, **91,** 4941; *Y. Kanaoka et al.*, Tetrahedron Letters, 1964, 2419).

1,3,4,5-*Tetrahydro*-(2H)-2-*benzazepine* (XLIV; R = H), b.p. 120 °/17 mm, *hydrochloride*, m.p. 223–225 °, is also obtained by treating 2-(3-chloropropyl)benzylamine with dilute alkali (*v. Braun et al.*, Ber., 1923, **56**, 690; 1925, **58**, 2164, 2766), and by the action of lithium tetrahydridoaluminate on the 1,2-tetrazolo-2-benzazepine obtained in a Schmidt reaction on α-tetralone (*N. S. Hjelte* and *T. Agback*, Acta chem. Scand., 1964, **18**, 191). Thermal cyclisation of 2-(3-aminopropyl)benzoic acids gives the 1,3,4,5-*tetrahydro*-2H-2-*benzazepin*-1-*one* XLVI (R = H), m.p. 100°, and its derivatives (*A. Brossi et al.*, Helv., 1971, **54**, 743; *v. Braun* and *H. Reich*, Ann., 1925, **445**, 225). Routes to tetrahydro-2-benzazepin-5-ones XLVII *via* Friedel–Crafts cyclisation of 3-benzylaminopropionic acids

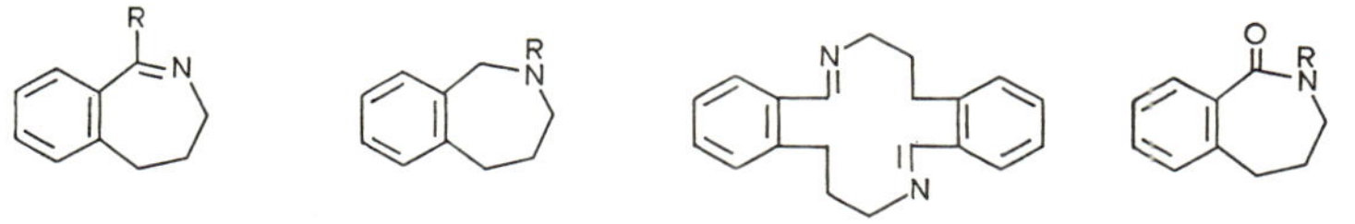

(XLIII) (XLIV) (XLV) (XLVI)

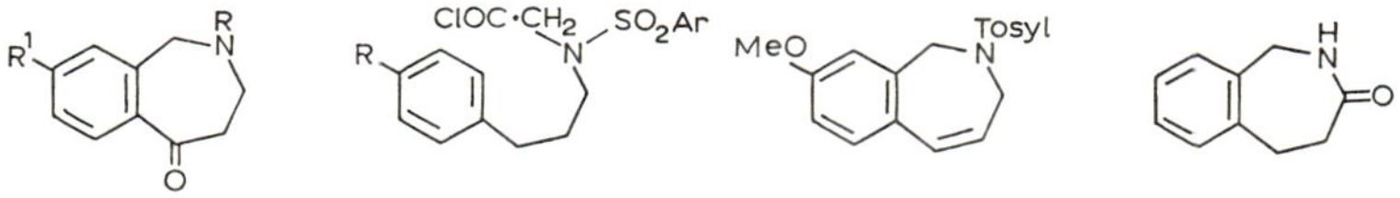

(XLVII) (XLVIII) (XLIX) (L)

are only effective if a *meta*-activating substituent (*e.g.* methoxyl) is present, as cleavage of benzylamine and loss of carbon monoxide compete (*G. R. Proctor et al.*, J. chem. Soc. C, 1970, 1461; J. chem. Soc., 1957, 2302). Loss of carbon monoxide is exploited, however, in the preparation of XLIV (R = *tosyl*), m.p. 135°, and 8-*methyl*-1,3,4,5-*tetrahydro*-(2H)-2-*benzazepine*, m.p. 27°, b.p. 137–139°/19 mm, from XLVIII (R = H or Me) (*v. Braun et al.*, Ber., 1924, **57**, 908; 1927, **60**, 102). Tetrahydridoborate reduces XLVII (R = tosyl; R = OMe) to the 5-hydroxy compound which is readily dehydrated to XLIX. XLVII (R = tosyl; R[1] = OMe) is brominated at position 4, and dehydrobromination is effected with triethylamine.

1,3,4,5-*Tetrahydro*-(2H)-2-*benzazepin*-3-*one* (L), m.p. 109–110°, is the major product of a Schmidt reaction on β-tetralone, in contrast to the 3-benzazepin-2-one produced in the Beckmann reaction of β-tetralone oxime (*J. L. Knunyants* and *B. P. Fabrichnyi*, Doklady Akad. Nauk S.S.S.R., 1949, **68**, 523; C.A., 1950, **44**, 1469).

1,3,4,5-Tetrahydro-(2*H*)-2-benzazepine-1,3-dione (LI), m.p. 119–121°, and several derivatives are made by reduction and cyclisation of cyano acid XLI (X = OH, p. 228), or catalytic reduction of the dihydrodiones XLII (X = OH) (*Walker, loc. cit.*). Photo-

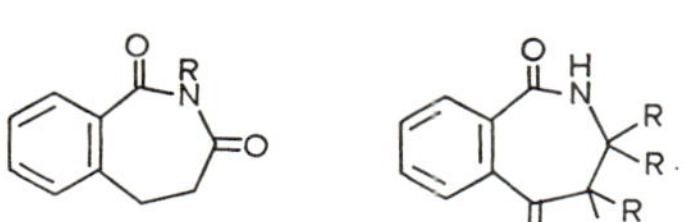

(LI) (LII)

addition of olefins to 1-ethoxyisoindol-3-one yields, after acid hydrolysis of the tricyclic intermediate, 1,3,4,5-tetrahydro-(2*H*)-2-benzazepin-1,5-dione derivatives (LII) (*T. H. Koch* and *K. H. Howard*, Tetrahedron Letters, 1972, 4035).

(*d*) *3-Benzazepines*

A dihydro-3-benzazepinone is found free in the alkaloid **stepinonine** (*T. Ibuka et al., ibid.*, 1972, 4001), and the 3-benzazepine system is present in several alkaloids from *Papaver* species, and salamander skin venom.

The base-catalysed condensation of *o*-phthalaldehyde with *N*-substituted dialkyliminodiacetates is a general route to (3*H*)-3-benzazepines LIII, *e.g.* 2,4-*dimethoxycarbonyl*-3-*methyl*-3H-3-*benzazepine* (LIII; R = CO_2Me; R^1 = Me), m.p. 146°.

The dicarboxylic acid is unstable, and on acid hydrolysis ring contraction to LIV takes place. On catalytic hydrogenation a dihydro and then a tetrahydro derivative are formed, the latter giving 1,2,4,5-*tetrahydro*-(3H)-3-*benzazepine*-2,4-*dicarboxylic acid*, m.p. 81°, on hydrolysis. The *dicarboxylic* acid LIII (R = CO_2H; R^1 = Ph), m.p. 205–207°, is more stable, forming a dimeric anhydride with thionyl chloride, and undergoing electrophilic substitution in the phenyl ring without disruption of the azepine nucleus (*K. Dimroth* and

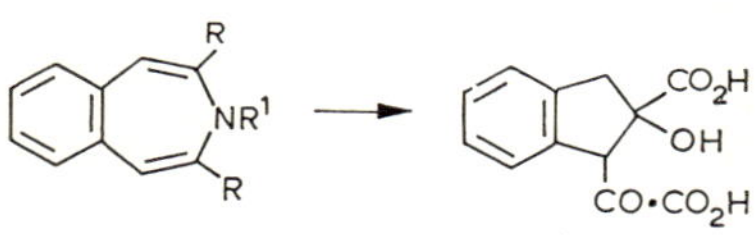

(LIII) (LIV)

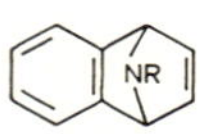

(LV)

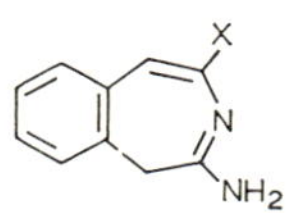

(LVI)

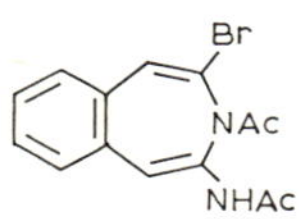

(LVII)

H. Freyschlag, Angew. Chem., 1956, **68,** 518; 1958, **70,** 248; Chem. Ber., 1956, **89,** 2602; 1957, **90,** 1628; *R. Huisgen et al.*, Ann., 1960, **630,** 128; *F. Dallacker et al., ibid.*, 1961, **643,** 82; *R. B. Lund*, Ph.D. Dissertation, Univ. of Washington, 1959). The alternative route, condensation of 1,2-benzenediacetaldehyde with amines does not give 3-benzazepines (*Huisgen* and *E. Laschtuvka*, Chem. Ber., 1960, **93,** 68) but photolysis of *N*-substituted benzo-7-azabicyclo[2.2.1]-hepta-2,5-diene (LV), from the reaction of benzyne with pyrroles, gives 3-*tosyl*- (LIII; R = H; R^1 = tosyl), m.p. 168–169°, 3-*methoxycarbonyl* (LIII; R = H, R^1 = CO_2Me), m.p. 68–69°, and 3-tert-*butoxycarbonyl*-(3H)-3-*benzazepines* (LIII; R = H, R^1 = CO_2Bu^t), m.p. 96–97° (*H. Prinzbach et al.*, ibid., 1970, **103,** 2288; *P. D. Rosso et al.*, Tetrahedron Letters, 1971, 3947).

The action of anhydrous hydrogen bromide on 1,2-bis(cyanomethyl)benzene gives **2-amino-4-bromo-(1*H*)-3-benzazepine** (LVI; X = Br), m.p. 194–196° decomp., the bromine atom may be substituted by nucleophiles or reduced out catalytically without affecting the azepine ring, but excess acetic anhydride converts LVI (X = Br) into the (3*H*)-azepine LVII (*F. Johnson* and *W. A. Nasutavicus*, J. heterocycl. Chem., 1965, **2,** 26; *J. H. Osborn*, Diss. Abs., 1959, 2475). 2-*Amino*-(1H)-3-*benzazepine* (LVI; X = H) has m.p. 165°.

1,2-**Dihydro-(3*H*)-3-benzazepines** of type LVIII, intermediates required in alkaloid syntheses, have been prepared from a 2-(2-aminoethyl)benzyl ketone (*Brossi et al.*, J. Amer. chem. Soc., 1971, **93,** 4321) or an (*N*-acetaldehydo)phenylacetamide (*J. Auerbach* and *S. M. Weinreb*, *ibid.*, 1972, **94,** 7172), but are more readily obtained by the reduction and dehydration of 1,2,4,5-tetrahydro-(3*H*)-3-benzazepin-1-ones (LIX). Hydration of LVIII gives the 1-hydroxy compound. An *N*-tosyl protecting groups is not easily removed,

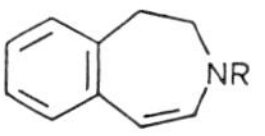

(LVIII)

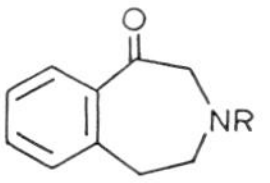

(LIX)

but irradiation of 7,8-dimethoxy-LVIII (R = tosyl) gives a 3,1-tosyl shift (*M. A. Rehman* and *Proctor*, J. chem. Soc. C, 1967, 58; *J. Gardent et al.*, Bull Soc. chim. Fr., 1968, 600; Tetrahedron Letters, 1972, 3321, *J. Schlademan* and *R. Partch*, J. chem. Soc., Perkin I, 1972, 213).

1,2,4,5-**Tetrahydro-(3*H*)-3-benzazepine** *(homotetrahydroisoquinoline)* (LX; R = H) b.p. 115°/12 mm, *hydrochloride*, m.p. 250°, is made by reduction of the 2-oxo derivative LXI with diborane, metal hydride or sodium in alcohol (*v. Braun* and *Reich*, Ber., 1925, **58,** 2766; Ann., 1925, **445,** 225; *J. O. Halford* and *B. Weissmann*, J. org. Chem., 1952, **17,** 1646; *Brossi et al.*, J. heterocycl. Chem., 1972, **9,** 609, 617) or the reductive cyclisation of *o*-phenylenediacetonitrile (*P. Ruggli et al.*, Helv., 1935, **18,** 1388; 1937, **20,** 925; *J. H. Wood et al.*, J. Amer. chem. Soc., 1951, **73,** 4689). Reductive methylation of LX (R = H) with formaldehyde gives the N-*methyl* derivative, b.p. 121°/2 mm, *picrate*, m.p. 255–256°. Clemmensen reduction of 1-aroyl-1,2,3,4-tetrahydroisoquinolines is accompanied by ring expansion to LX (R = aryl) (*C. Reby et al.*, Compt. rend., 1969, **269,** 45):

(LX) (LXI)

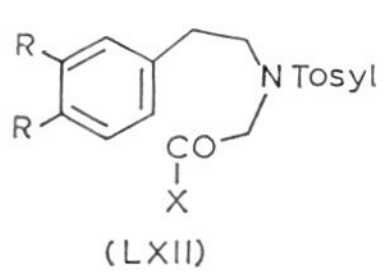

(LXII)

Friedel–Crafts cyclisation of LXII (R = H, X = Cl) to 3-*tosyl*-1,2,4,5-*tetrahydro*-(3H)-3-*benzazepin*-1-*one* (LIX; R = tosyl), m.p. 145–147°, requires low temperatures and 3 mols of aluminium chloride, otherwise carbon monoxide is lost and an isoquinoline produced. *N*-Bromosuccinimide brominates LIX (R = tosyl) at position 2, and *N*-mesyl derivatives undergo base-catalysed condensation at this position. The ketone forms normal carbonyl derivatives and is reduced to the alcohol by sodium tetrahydridoborate. The tosyl group is hydrolysable in the alcohol, but not the ketone, to give 1-*hydroxy*-1,2,4,5-*tetrahydro*-(3H)-3-*benzazepine*, m.p. 132–134° (*Schlademan* and *Partch*, *loc. cit.*; *v. Braun* and *O. Bayer*, Ber., 1924, **57,** 908; 1927, **60,** 1257; *Proctor et al.*, J. chem. Soc. C, 1969, 1321; 1967, 58); photolytic cleavage of tosyl is effective for both compounds. An alternative route to 7,8-dimethoxy-LIX (R = tosyl) is to use the Vilsmeier procedure on the amide LXII (R = OMe; X = NHPh) (*G. Hazebroucq et al.*, Ann. chim., 1966, **1,** 221; Compt. rend., 1963, **257,** 923).

The most direct route to 1,2,4,5-*tetrahydro*-(3H)-3-*benzazepin*-2-*one* (LXI), m.p. 159°, is the methanolysis of the *p*-toluenesulphonate of *β*-tetralone oxime (Beckmann rearrangement) (*I. L. Knunyants* and *B. P. Fabrichnyi*, Doklady Akad. Nauk S.S.S.R. 1949, **68,** 523). Another general route is the photochemical cyclisation of *N*-(chloroacetyl)-2-phenylethylamines. Yields are sometimes low and complex by-products may be formed, but the troublesome elimination of a protecting group is avoided (*O. Yonemitsu et al.*, J. Amer. chem. Soc., 1972, **94,** 5136; 1971, **93,** 4053; 1970, **92,** 5686; 5691; 1968, **90,** 6522; 776; 1967, **89,** 1039; *H. H. Ong* and *E. L. May*, J. org. Chem., 1972, **37,** 712; *L. J. Dolby et al.*, *ibid.*, p. 3691). 7-*Hydroxy*-1,2,4,5-*tetrahydro*-(3H)-3-*benzazepin*-2-*one*, from *N*-chloroacetyl-*m*-tyramine has m.p. 255–259°. Lactams LXI are also prepared by the ready thermal cyclisation of 2-(2-aminoethyl)phenylacetic acids (*Braun* and *Reich*, *loc. cit.*; *Brossi*, *loc. cit.*) and the reduction and hydrolysis of LVI (*Osborn*, *loc. cit.*; *Johnson* and *Nasutavicus*, *loc. cit.*). Cyclodehydration of benzilic acid *N*-(2-phenylethyl)amide yields 1,1-diphenyl-1,2,4,5-tetrahydro-(3*H*)-3-benzazepin-2-one (*V. A. Petyunin*, Zhur. obshcheĭ Khim., 1952, **22,** 697). Compounds LXI may be *N*-acylated and *N*-alkylated, and hydrolytic ring-opening occurs with hydrobromic acid.

When *o*-phenylenediacetamide is heated under reduced pressure at 295° it gives **1,2,4,5-tetrahydro-(3*H*)-3-benzazepine-2,4-dione,** m.p. 191–192°, which is reduced by lithium tetrahydridoaluminate to LX (R = H) (*Halford* and *Weissmann*, *loc. cit.*).

(*e*) *Other azepines with one fused ring*

Thermal cyclisation of 2-(5-bromopentanoyl)pyridine gives the quaternary ketone LXIII (R = O) which is converted in a series of steps into the fully unsaturated pyrido(1,2-*a*)-azepine LXIV (no definite m.p.) (*A. Fozard* and *G. Jones*, J. org. Chem., 1965, **30,** 1523). Quaternary *bromide* LXIII (R = H_2), m.p. 58–59°, is also easily prepared in this way (*T. Morikawa*, J. pharm. Soc. Japan, 1955, **75,** 593). Catalytic reduction of the latter compound yields the saturated derivative, **1-azabicyclo[5.4.0]-undecane** (LXV), b.p. 90°/16 mm; n_D^{20} 1.4872; *picrate*, m.p. 161–162°, also obtained by reductive cyclisation of diethyl *δ*-hydroxyiminosebacate with copper chromite (*N. J. Leonard* and *W. E. Goode*,

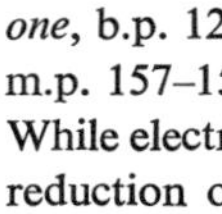

(LXIII)

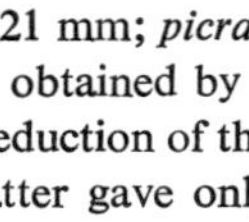

(LXIV)

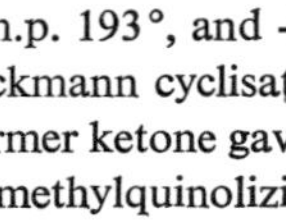

(LXV)

J. Amer. chem. Soc., 1950, **72,** 5404); **1-azabicyclo[5.5.0]-dodecane,** b.p. 111–112°/18 mm, n_D^{20} 1.4930; *picrate*, m.p. 139–140°, is prepared similarly. 1-*Azabicyclo*[5.4.0]*undecan*-6-*one*, b.p. 124–125°/21 mm; *picrate*, m.p. 193°, and -3-*one*, b.p. 130–132°/18 mm; *picrate*, m.p. 157–158°, are obtained by Dieckmann cyclisation, hydrolysis and decarboxylation. While electrolytic reduction of the former ketone gave azacycloundecan-6-ol, Clemmensen reduction of the latter gave only 4-methylquinolizidine (*Leonard et al.*, *ibid.*, 1952, **74,** 6251; 1951, **73,** 5210).

Many azepines with [1,2-*a*] fusion to a heterocyclic ring have been obtained from reactions between acetylenic esters, and nitrogen-containing heterocyclic compounds, especially those with *N*- or α-substituents. The reaction is a general one but most of the compounds produced are outside the scope of this chapter (*R. M. Acheson et al.*, J. chem. Soc. Perkin I, 1972, 968; 1577; C, 1971, 3285; 3291; 1968, 362; 926; 1965, 3200; *D. Johnson* and *G. Jones*, *ibid.*, Perkin I, 1972, 844; *A. W. Johnson et al.*, *ibid.*, 1965, 4355; 1962, 1510; *D. H. Reid et al.*, Tetrahedron Letters, 1964, 1797).

3. Azepines with two or more annellated ring systems

(*a*) *Dibenzazepines*

(*i*) *Dibenz*[b,d]*azepines*

Most of the dibenz[*b,d*]azepines known are derived from 5-*tosyl*-6,7-*dihydrodibenz*[b,d]*azepin*-7-*one* (II; R = H, R^1 = tosyl), m.p. 137°, obtained in the aluminium chloride-catalysed cyclisation of tosyl-(2-biphenylyl)aminoacetyl chloride (I) at −70°; other catalysts or higher temperatures lead to phenanthridine derivatives. Reduction of II (R = H,

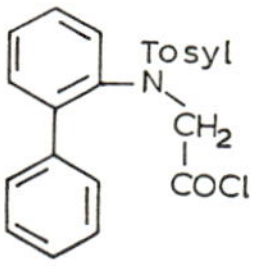

(I)

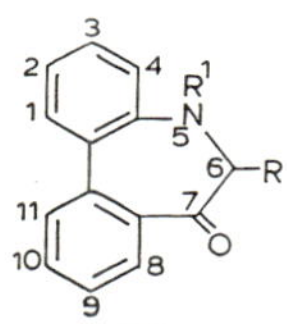

(II)

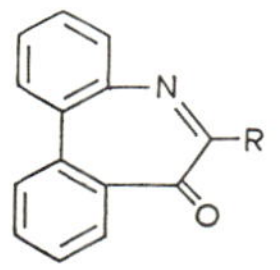

(III)

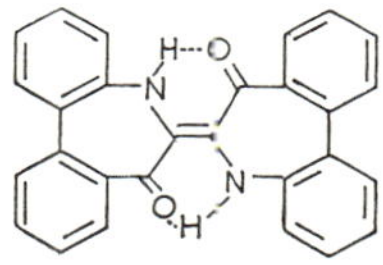

(IV)

R^1 = tosyl) with lithium tetrahydridoaluminate gives the 7-hydroxy analogue, but with zinc and acetic acid the 6-*hydroxyketone* (II; R = OH, R^1 = H), m.p. 239–240°, (formerly thought to be the 5-hydroxy-ketone) is obtained.

Detosylation of II (R = H; R^1 = tosyl) with sodium ethoxide gives not III (R = H) as originally suggested but the *dimer* IV, m.p. 267°, which gives a *dehydro* derivative, m.p. 230°, on oxidation with lead or manganese dioxides. Ozonolysis of this dehydrodimer yields **6,7-dihydro-(5*H*)-dibenz[*b,d*]azepine-6,7-dione,** m.p. 233–234°. The monomeric *detosylated product* II (R = OEt, R^1 = H), m.p. 134°, is also obtained with ethoxide; with acid it gives II (R = OH, R^1 = H) and with manganese dioxide 6-*ethoxydibenz-azatrop*-7-*one* (III) (R = OEt), m.p. 118°. Bromination or attempted detosylation of II (R = H; R^1 = tosyl) with sodium in liquid ammonia or zinc chloride gives phenanthridine derivatives, but sodium hydride effectively detosylates II (R = Me or OMe, R^1 = tosyl) (*G. R. Proctor et al.*, Chem. and Ind., 1960, 408; J. chem. Soc., 1962, 3468; J. chem. Soc. C, 1968, 2481; 1969, 2151; *R. G. Cooke* and *I. M. Russell*, Tetrahedron Letters, 1968, 4587).

(*ii*) *Dibenz*[b,e]*azepines*

(11*H*)-**Dibenz**[*b,e*]**azepine** *(morphanthridine)* (V; R = H), m.p. 63–64°, is prepared by the dehydrative cyclisation of 2-(*N*-formylamino)diphenylmethane with polyphosphoric acid or phosphoryl chloride. 6-*Methyl* (V; R = Me), m.p. 99–100°, 6-*phenyl* (V; R = Ph), m.p. 155–156°, and other analogues are prepared similarly (*F. Hunziker et al.*, Helv., 1964, **47**, 1163; *W. F. Short et al.*, B.P. 696,473/1954; C.A., 1955, **49**, 384). An alternative route is the pyrolysis of diazonium tetrafluoroborates of 2-benzylanilines in the presence of organic nitriles (*R. R. Schmidt et al.*, Chem. Ber., 1972, **105**, 1634).

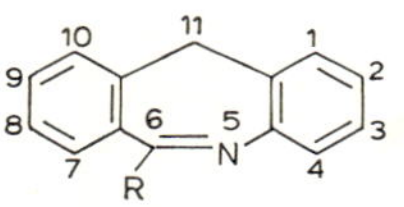

(V)

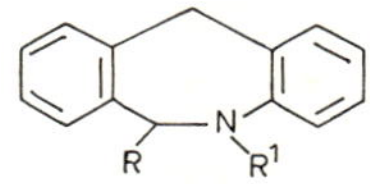

(VI)

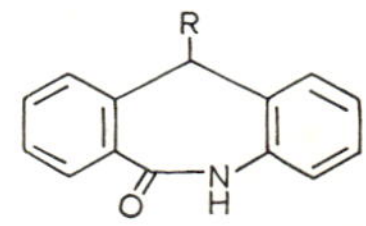

(VII)

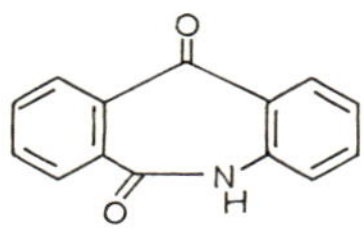

(VIII)

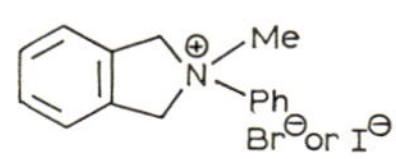

(IX)

Reduction catalytically or with lithium tetrahydridoaluminate gives the 5,6-**dihydro-(11*H*)-dibenz[*b,e*]azepines** (VI; R, R^1 = H; m.p. 131–132°; R = Me, R^1 = H, m.p. 71–72°), which may also be prepared by the reduction of lactams VII or diones VIII (*M. Protiva et al.*, Chem. listy, 1957, **51**, 1344; Experientia, 1957, **13**, 291). The quaternary salt IX when treated with phenyllithium also gives a *dihydrodibenzazepine* (VI; R = H, R^1 = Me), m.p. 78–79°, instead of undergoing the expected Stevens rearrangement (*G. Wittig et al.*, Ann., 1954, **589**, 55; 1955, **593**, 127; 1955, **594**, 89; *P. P. Gaspar* and *T. C. Carpenter*, Angew. Chem. intern. Edn., 1967, **6**, 559). The 5–6 bond of this *N*-methyl derivative is broken by treatment with sodium/potassium alloy followed by alcohol.

5,6-**Dihydro-(11*H*)-dibenzo[*b,e*]azepin-6-one** (VII; R = H), m.p. 201–203°, is prepared by partial reduction of dione VIII with hydrogen and copper chromite (*Wittig, loc. cit.*) or by treating diphenylmethane 2-isocyanate with aluminium chloride; 11-*methyl*-5,6-*dihydro*-(11H)-*dibenz*[b,e]*azepin-6-one*, (VII; R = *Me*), m.p. 203–206°, and several analogues are prepared similarly. Phosphoryl chloride converts VII (R = H) into V (R = Cl), which with amines gives V (R = NR_2). Both monoketone VII and dione VIII are *N*-alkylated by alkyl halides and sodamide (*Hunziker*, Helv., 1965, **48**, 336; 1966, **49**, 1433; *L. H. Werner*, U.S.P. 2,973,354/1961; C.A., 1961, **55**, 15522).

Only a trace of ketone X (R = H; R^1 = tosyl) is obtained from *N*-benzyl-*N*-tosylanthraniloyl chloride under Friedel–Crafts conditions because of competing benzyl cleavage, but X (R = OMe; R^1 = tosyl) may be prepared in this way at –20°. Treatment

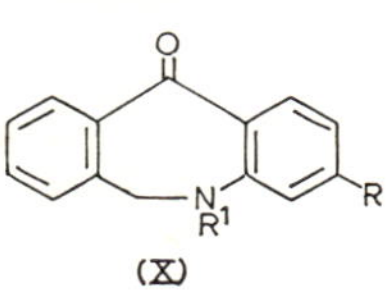

(X)

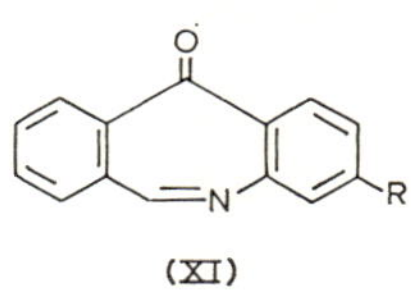

(XI)

of X (R = OMe; R^1 = tosyl) with polyphosphoric acid yields X (R = OMe; R^1 = H) which is dehydrogenated to dibenzazatropone XI (R = OMe) with manganese dioxide (*I. Mcdonald* and *Proctor*, J. chem. Soc. C, 1969, 1321). The parent, (11*H*)-**dibenz[*b,e*]-azepin-11-one** (XI; R = H), m.p. 93–94°, is obtained by Sonn–Müller reduction of VIII (*Cooke* and *Russell*, *loc. cit.*).

5,6-**Dihydro**(11*H*)-**dibenz**[*b,e*]**azepine**-6,11-**dione** *(morphanthridinedione)* (VIII), m.p. 245°, is readily obtained from anthraquinone by Schmidt or Beckmann ring-expansion (*G. Caronna* and *S. Palazzo*, Gazz., 1953, **83,** 533; *M. M. Coombs*, J. chem. Soc., 1958, 4200; *Werner et al., loc. cit.*; J. med. Chem., 1965, **8,** 74; *A. E. Drukker* and *C. I. Judd*, J. heterocycl. Chem. 1965, **2,** 276; *E. Beckmann* and *O. Liesche*, Ber., 1923, **56,** 1). It also results from cyclising 2-formamidobenzophenone (*Protiva et al.*, Coll. Czech. chem. Comm., 1965, **30,** 445) or 2-aminobenzophenone-2′-carboxylic acid (*Wittig et al.*, Ann., 1955, **594,** 110; *Agfa*, G.P. 258,343/1912; Frdl., 1926, **11,** 565; *P. Kränzlein*, Ber., 1937, **70,** 1952) and in the ring expansion of *N*-phenyl-phthalimide (*A. Wolfram* and *P. Hausdorfer*, G.P. 258,343/1912; Frdl., 1934, **19,** 678). Many substituted diones have been prepared similarly.

Dione VIII is readily converted into 11-*hydroxy*-(VII; R = OH), m.p. 247–248°, and 11-*aminomorphanthridine*-6-*ones* and reacts with triethyloxonium tetrafluoroborate to give 11-ethoxy-(6*H*)-**dibenz[*b,e*]azepin-6-one** (XII), m.p. 101–102°. With phosphoryl chloride it gives 6-*chloro*-(11H)-*dibenz*[b,e]*azepin*-11-*one*, m.p. 115–120°, in which the chloro group is replaceable by amines (*Werner*, *loc. cit.*; *Drucker* and *Judd*, *loc. cit.* and

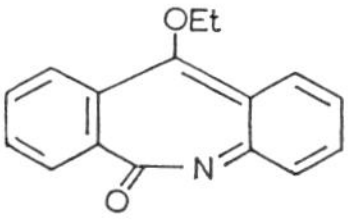

(XII)

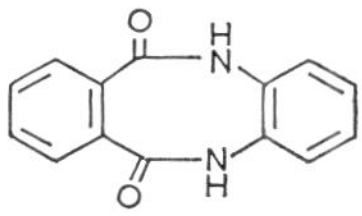

(XIII)

J. heterocycl. Chem., 1966, **3,** 206; *E. J. Moriconi* and *I. A. Maniscalco*, J. org. Chem., 1972, **37,** 208; *Hunziker et al.*, Helv., 1966, **49,** 1433). Acid hydrolysis of the dione VIII gives 2-aminobenzophenone-2′-carboxylic acid but its 11-hydroxy derivatives give 3-(2-aminophenyl)phthalide under these conditions. Its 11-hydroxyimino derivative on Beckmann rearrangement gives the dibenzo[*b,f*]-1,4-diazocine-6,11-dione (XIII) (*Protiva*, *loc. cit.*; *Beckmann* and *Liesche*, *loc. cit.*; *D. D. Emrick* and *W. E. Truce*, J. org. chem., 1961, **26,** 1329).

(*iii*) *Dibenz*[b,f]*azepines*

(5*H*)-**Dibenz**[*b,f*]**azepine** ("iminostilbene") (XIV; R = H), m.p. 193–195°, is the parent system of several drugs which affect the central nervous system. The most important is Imipramine hydrochloride (Tofranil) [XIV; R = $(CH_2)_3NMe_2 \cdot HCl$; R^1 = H]. Iminostilbene is remarkable for its stability. Although it is the only stable 1*H*-azepine without an *N*-substituent, XIV

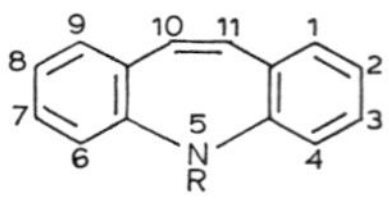

(XIV)

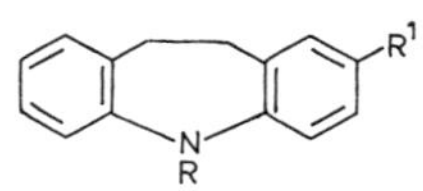

(XV)

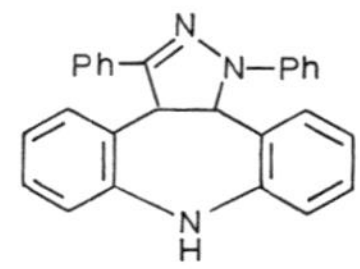

(XVI)

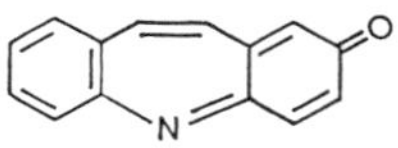

(XVII)

(R = H) may be obtained by dehydrogenating XV (R, R^1 = H) at 250° with sulphur, or palladium in the presence of diethyl maleate (*R. Huisgen et al.*, Chem. Ber., 1960, **93**, 392; *C. L. Zirkle et al.*, J. org. Chem., 1961, **26,** 135). 5-*Methyl*-(5H)-*dibenz*[b,f]*azepine* (XIV; R = Me), m.p. 143–145°, is obtained similarly. A 10,11-double bond may also be introduced into XV (R, R^1 = H) by brominating at position 10 with *N*-bromosuccinimide and dehydrobrominating the product (*W. Schindler* and *H. Blattner*, Helv., 1961, **44,** 753). The dehydrative rearrangement of 9-(hydroxymethyl)acridane with phosphorus pentoxide gives XIV (R = H) directly (*Craig*, *loc. cit.*; *E. D. Bergmann* and *M. Rabinowitz*, J. org. Chem., 1960, **25,** 827; *A. Campbell* and *E. N. Morgan*, J. chem. Soc., 1958, 1711) and the *N*-methyl analogue (XIV; R = Me) results when *N*-methylacridinium perchlorate is treated with diazomethane (*H. W. Whitlock*, Tetrahedron Letters, 1961, 593).

Iminostilbene (XIV; R = H) may be acylated and alkylated (*L. J. Kricka* and *A. Ledwith*, J. chem. Soc. Perkin I, 1972, 2292) and the 10,11 double bond readily adds hydrogen or bromine, and acts as a dipolarophile towards diphenylnitrileimine to give the pyrazoline XVI (*Huisgen et al.*, Tetrahedron, 1962, **17,** 6). With hot acids XIV (R = Ac) undergoes ring contraction to 9-methylacridone (*Schindler* and *Blattner*, *loc. cit.*; *P. Rumpf* and *R. Reynaud*, Bull. Soc. chim. Fr., 1962, 2241), while XIV (R = H) on oxidation with potassium nitrosodisulphonate (Fremy's salt) yields acridine-9-carbaldehyde as the minor product.

The major product is **(2*H*)-dibenz[*b,f*]azepin-2-one** (XVII), m.p. 135–136° (*Proctor et al.*, Chem. Comm., 1968, 1412; J. chem. Soc. Perkin I, 1972, 539). 10,11-*Dihydro*-2(H)-*dibenz*[b,f]*azepin*-2-*one*, m.p. 105–106°, is prepared the same way from XV (R, R^1 = H), but is not dehydrogenated to XVII (*Schindler* Helv., 1960, **43,** 35; *H-J. Teuber*, Angew. chem., 1958, **70,** 607) and poly-substituted analogues are obtained in the alkaline rearrangement of certain phenoxazines and the oxidation of 2-amino-3-hydroxyaceto-

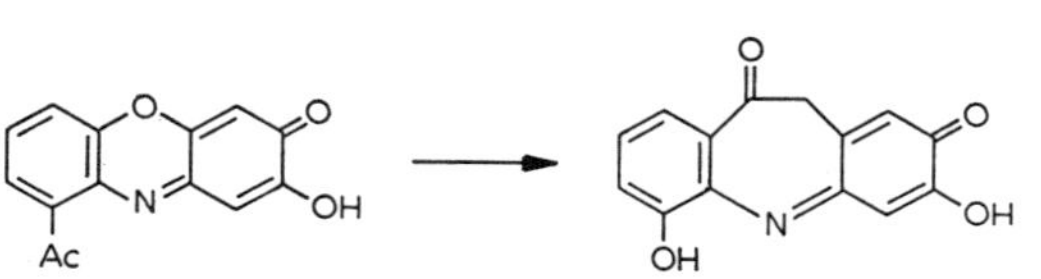

phenone in the presence of alkali (*A. Butenandt et al.*, Angew. chem., 1957, **69,** 16; Ann., 1957, **603,** 200).

Nitration and halogenation of XVII takes place at position 1, phenyl-lithium converts XVII into 2-*phenyl*-(5H)-*dibenzo*[b,f]*azepine*, m.p. 198–200° decomp., and hydrogenation yields 2-*hydroxy*-10,11-*dihydro*-(5H)-*dibenzo*[b,f]*azepine* (XV; R = H; R^1 = OH), m.p. 155° or 169–171° (*Proctor, loc. cit.).*

Small amounts of 10,11-*dihydro*-(5H)-*dibenz*[b,f]*azepine* (o-*iminobibenzyl*) (XV; R, R^1 = H), m.p. 110°, may be obtained by heating 2,2′-diaminodibenzyl with its hydrochloride to 270°, but on a larger scale cyclisation in hot phosphoric acid is preferred (*J. Thiele* and *O. Hölzinger*, Ann., 1899, **305,** 96; *Huisgen et al.*, Chem. Ber., 1960, **93,** 392; *Schindler* and *F. Häfliger*, U.S.P. 2,764,580/1956; C.A., 1957, **51**, 4447). *N*-Alkylation is possible with alkyl halides and sodamide (*idem*, Helv., 1954, **37,** 472) or tosyl esters and phenyl-lithium (*V. G. Yashunskii et al.*, U.S.S.R. Pat. Appl., 137,520/1959; C.A., 1962, **56,** 464) but the puckered ring of XV hinders the approach of an alkyl iodide to the *N*-thallium derivative (*Kricka* and *Ledwith*, *loc. cit.*). Phosgene reacts with iminobibenzyl to give 10,11-*dihydro*-(5H)-*dibenz*[b,f]*azepine*-5-*carbonyl chloride*, (XV; R = COCl; R^1 = H), m.p. 122°, (*Schindler* and *Häfliger*, *loc. cit.*; *J. R. Geigy A.-G.*, Swiss P. 340,232/1959; C.A., 1962, **56,** 463).

10,11-Dihydro-(5*H*)-dibenz[*b,f*]azepin-10-one (XVIII; R = H), m.p. 145–146° (*idem*, B.P. 943,276–7/1959; C.A., 1964, **61,** 1815) is not obtained by subjecting 2(phenylamino)-phenylacetic acid to Friedel–Crafts conditions (*J. W. Schulenberg* and *S. Archer*, J. Amer. chem. Soc., 1960, **82,** 2035). Its *N*-acetyl derivative XVIII (R = Ac) with selenium dioxide, followed by hydrolysis gives the 10,11-*dione*, m.p. >360°, which readily undergoes a benzilic acid rearrangement (*Cooke* and *Russell*, *loc. cit.*).

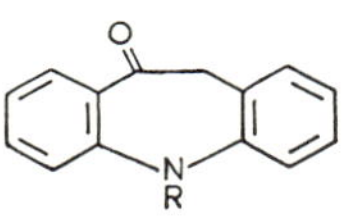

(XVIII)

(*iv*) *Dibenz*[c,e]*azepines*

2,2′-Bis(bromomethyl)biphenyl reacts with primary amines to give 6-substituted derivatives of 6,7-dihydro-(5*H*)-dibenz[*c,e*]azepine (XIX). Secondary amines give quaternary salts, and ammonia yields the spiro compound XX with only a trace of XIX (R = H) (*Wenner*, J. org. Chem., 1951, **16,** 1475, 1952, **17,** 1451; *Wittig et al.*, Chem. Ber., 1953, **86,** 629; Ann.,

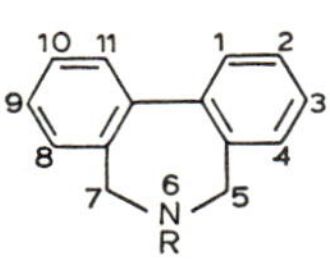

(XIX)

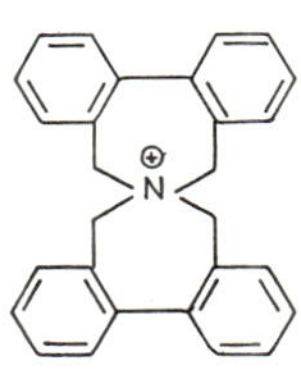

(XX)

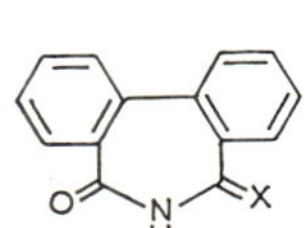

(XXI)

1955, **593,** 127; *D. M. Hall et al.*, J. chem. Soc., 1952, 854; 1954, 131); hydrolysis of the product of dibromide and *tert*-butylcarbazide gives XIX (R = NH_2). The parent 6,7-**dihydro-**(5*H*)-**dibenz**[*c*,*e*]**azepine** (XIX; R = H), m.p. 74–76°, *hydrochloride*, m.p. 286–288°, is obtained by catalytic hydrogenolysis of the benzyl group from XIX (R = CH_2Ph), or the irradiation of 2-iododibenzylamine (*P. W. Jeffs* and *J. F. Hansen*, J. Amer. chem. Soc., 1967, **89,** 2798).

1,11-Disubstituted dimethyl quaternary salts of XIX prepared from resolved diphenic acid have had their racemisation parameters determined (*P. A. Brown* and *Hall*, J. chem. Soc. Perkin I, 1972, 2717). Treated with phenyl-lithium they undergo a Stevens rearrangement with ring-contraction to 9-dimethylamino-9,10-dihydrophenanthrene (*Wittig* and *H. Zimmerman*, Chem. Ber., 1953, **86,** 629).

Irradiation of an oxygenated solution of XIX (R = aryl) gives the 5,5′-bisperoxide, which cleaves on heating to a mixture of 5-hydroxy- and 5-oxodibenzazepines (*E. Höft* and *H. Schultze*, Angew. chem. intern. Edn., 1965, **4,** 366).

6,7-**Dihydro-**(5*H*)-**dibenz**[*c*,*e*]**azepine-**5,7-**dione** (XXI; X = O) *(Diphenimide)*, m.p. 219–220°, is obtained by heating diammonium diphenate (*E. Beckmann* and *P. Wegerhoff*, Ann., 1889, **252,** 1) or the corresponding diamide (*C. Graebe* and *C. Aubin*, ibid., 1888, **247,** 270), and also from the Beckmann rearrangement of phenanthraquinone monoxime (*P. Megerhoff*, Ber., 1888, **21,** 2355). Diphenimide is easily hydrolysed by alkalis to salts of diphenic acid, but with concentrated sulphuric acid at 100° it gives the amide of 9-fluorenone-1-carboxylic acid.

Hydrogen bromide in acetic acid cyclises biphenyl-2,2′-dinitrile to the dioxodibenzazepinonimine hydrobromide XXI (X = NH, HBr) (*F. Johnson* and *R. Madronero*, Adv. heterocycl. Chem., 1966, **6,** 143).

(*v*) *Benzpyridazepines*

In refluxing concentrated hydrochloric or hydrobromic acid, 1-acetonyl-2-benzylpyridinium salts cyclise to 7-**methyl-**(12*H*)-**pyrido**[1,2,-*a*]**benz**[*d*]-**azepine,** isolated as the *perchlorate* (7-**methylmorphanthridizinium perchlorate**) (XXII; R = Me), m.p. 153–157°, in good yield. The analogue, XXII (R = Ph), is not obtained unless activating substituents are present. Catalytic hydrogenation saturates both pyridine and azepine rings (*K. B. Moser* and *C. K. Bradsher*, J. Amer. chem. Soc., 1959, **81,** 2547).

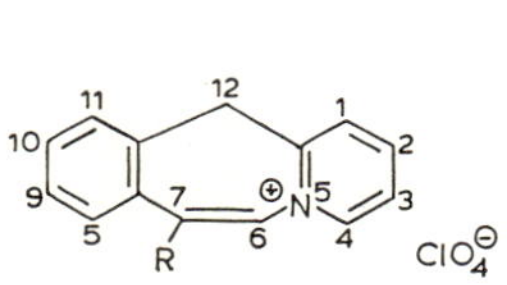

(XXII)

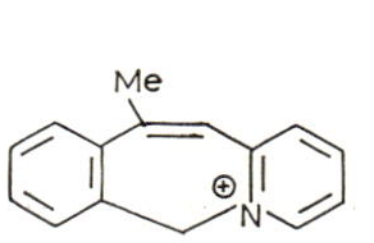

(XXIII)

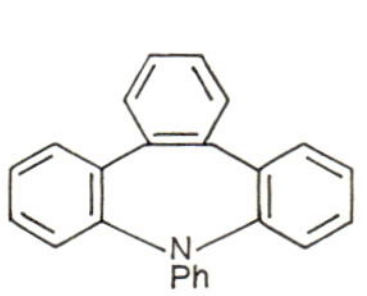

(XXIV)

2-Acetonyl-1-benzylpyridinium salts cyclise in low yield to **isomorphanthridizinium salts,** *e.g.* XXIII, only if activating substituents are present in the benzene ring (*Bradsher et al.*, J. org. Chem., 1961, **26,** 3278).

(vi) Tribenzazepine

***N*-Phenyl-(1*H*)-tribenzazepine** (XXIV), m.p. 162–163°, is obtained by diazotising 2′-(diphenylamino)biphenyl in acetic acid (*D. Hellwinkel* and *H. Seifert*, Chem. Ber., 1972, **105,** 880); **tribenzoxepin** is prepared similarly.

4. Seven-membered rings containing two nitrogen atoms

(a) Diazepines and related compounds

(i) 1,2-Diazepines

No unsubstituted 1,2-diazepines are known, but derivatives of the fully unsaturated 1*H*-, 4*H*- and 5*H*-diazepines have been prepared.

Several 1*H*-1,2-**diazepines** (III) are known; all have *N*-substituents. A useful general method of synthesis is the irradiation of pyridinium imidates of type I. The chemical behaviour of the diazepine indicates that the postulated diaziridine intermediate II is in tautomeric equilibrium with III but it is not detected by spectroscopy:

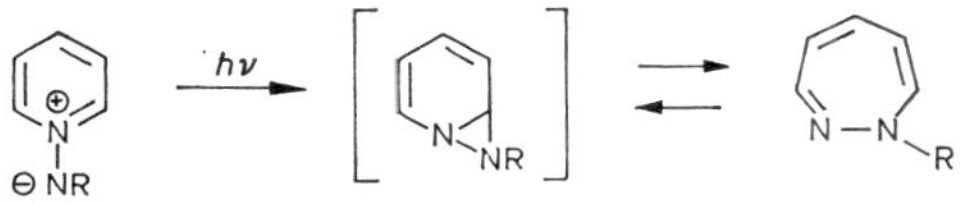

(I) (II) (III)

1-Ethoxycarbonyl-(1*H*)-1,2-diazepine (III; R = CO_2Et), b.p. 50°/0.5 mm, n_D^{25} 1.5302, and the 1-*tosyl*-, m.p. 139–140°; 1-*acetyl* and 1-*benzoyl*, m.p. 169–170°, analogues are prepared in this way. Substituents at position 2 of the pyridine are found at position 7 of the ring expanded compound.

Diazepine III (R = CO_2Et) forms an *iron tricarbonyl complex*, m.p. 115–116°, which has been used for X-ray structure determination, and 4-phenyl-1,2,4-triazoline-1,5-dione or tetracyanoethylene give adducts (*e.g.* IV) bridged across positions 4 and 7; 2-diazopropane, however, adds across positions 4 and 5 giving V. Catalytic hydrogenation saturates the ring and pyrolysis yields *cis* and *trans* nitrile-carbamates, *e.g.* NC · CH:CH · CH:CH · $NHCO_2Et$. Hot acetic acid regenerates the ylid I (R = CO_2Et) but with base,

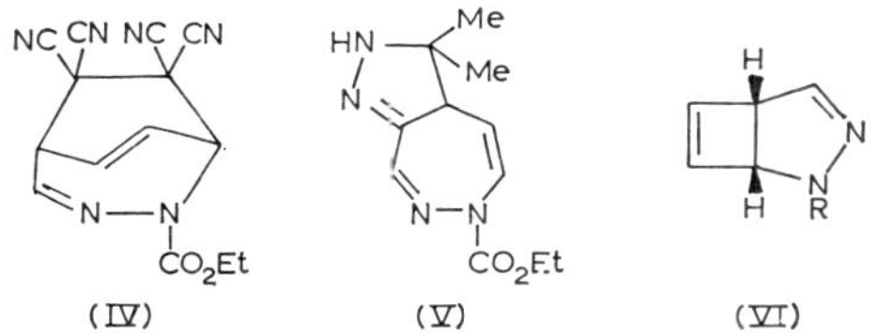

(IV) (V) (VI)

2-aminopyridine is obtained. Irradiation of 1*H*-1,2-diazepines leads to 4,7-transannular bond formation giving compounds of type VI (*J. Streith et al.*, J. org. Chem., 1971, **36,** 2962; Angew. chem. intern. Edn., 1968, **7,** 129; Bull. Soc. chim. Fr., 1969, 948; 2175; Tetrahedron Letters, 1968, 4541; 1972, 3575; *T. Sasaki et al.*, Chem. Comm., 1969, 432; 1970, 82; J. org. Chem., 1970, **35,** 426; *V. Snieckus et al.*, Chem. Comm., 1969, 831; 1971, 1022; J. org. Chem., 1970, **35,** 433).

Like 1*H*-azepines and oxepine, 1*H*-1,2-diazepin III (R = tosyl) exists in a shallow boat conformation thereby reducing the antiaromaticity of the ring (*Streith et al.*, Tetrahedron, 1972, **28,** 581).

Although unsaturated diketones, γ-pyrones and many pyrylium salts give pyridine derivatives with hydrazines (*W. Schneider et al.*, Ann., 1924, **438,** 115; 147; Ber., 1941, **74,** 1252; *A. F. Thomas* and *A. Marxer*, Helv., 1960, **43,** 469), 2,4,6-triarylpyrylium salts with hydrazine hydrate give **3,5,7-triaryl-4*H*-1,2-diazepines** (VII) (*e.g.* Ar = Ph, m.p. 191–120°) in almost quantitative yield (*A. T. Balaban et al.*, Acta chem. Scand., 1969, **23,** 3125; Tetrahedron, 1968, **24,** 5059; Chem. Ber., 1960, **93,** 599), and 2,4,6-triarylthiapyrylium

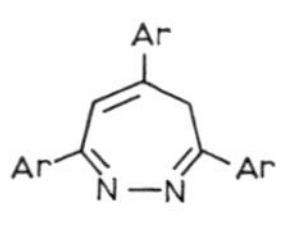
(VII)

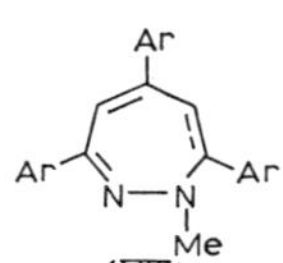
(VIII)

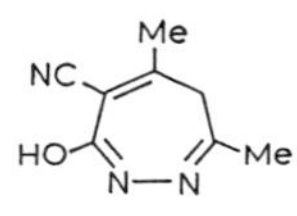
(IX)

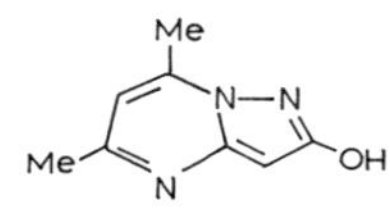
(X)

salts with hydrazine and methylhydrazine give respectively, VII and 1-methyl-(1*H*)-3,5,7-triaryl-1,2-diazepine (VIII) (*E. Klingsberg*, Abstr. Amer. chem. Soc. Meeting, 1965, p. 665; cf. *Snieckus* and *G. Kan*, Chem. Comm., 1970, 1208). The monohydrazone, Ar · CO · CH:CAr · CH_2CAr:N · NH_2 is a probable intermediate (*Balaban*, Tetrahedron, 1970, **26,** 739). The compound believed to be 4*H*-1,2-diazepine IX has been shown to have structure X (*W. Reid* and *E. U. Kocher*, Angew. chem., 1958, **70,** 164; Ann., 1960, **631,** 185; 1961, **647,** 116).

Protonation of diazepine VII (Ar = Ph), originally believed to give cation XI, is now known to give mainly XII or its 6*H*-isomer, but *C*-protonation of VIII is not observed (*F. D. Popp* and *A. C. Noble*, Adv. heterocycl. Chem., 1967, **8,** 40; *Snieckus et al.*, Chem. Comm., 1972, 504). Anion XIII is obtained from VII with lithium diisopropylamide, but a

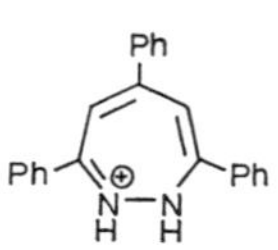
(XI)

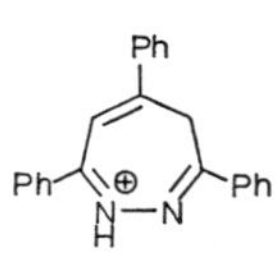
(XII)

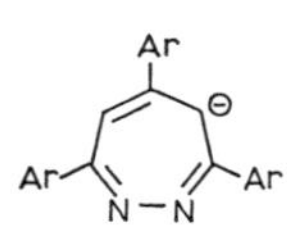
(XIII)

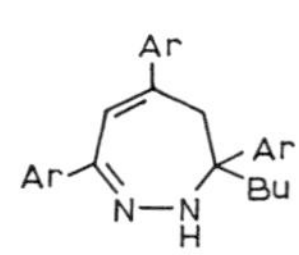
(XIV)

stronger base (sodium–potassium alloy) is required for ring-contraction to 2,4,6-triphenylpyridine. Addition of butyl-lithium leads to XIV (*R. R. Schmidt* and *H. Vatter*, Tetrahedron Letters, 1972, 4891). The iron and rhodium carbonyl complexes of VII (Ar = Ph) are found to have different structures (*A. J. Carty et al.*, J. organomet. Chem., 1971, **32,** 241; Chem. Comm., 1971, **427,** 1970, 1664).

Calculations and evidence from nuclear magnetic resonance spectra indicate that (5*H*)-1,2-**diazepine** (XV) in contrast to azepines and (1*H*)-1,2-diazepines, energetically prefers to exist as the bicyclic diazanorcaradiene (XVI):

(XV) (XVI)

The presence of the monocyclic form has not often been detected directly, but is inferred from the observed equilibration of *exo* and *endo* forms of compounds XVIa, many of which have been prepared from 1,2,4,5-tetrazines and cyclopropanes; alternative routes are also known. Compounds XVIa (R^1 = CO_2Me) add water, alcohols and hydrogen cyanide to give XVII (X = OH, OR, CN), and a further molecule of cyclopropene to

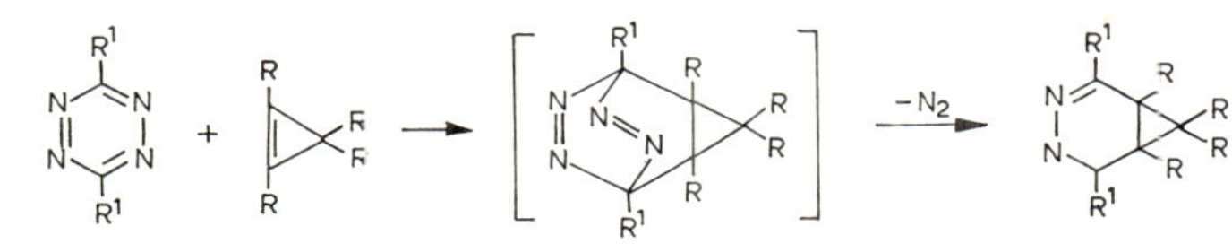

R^1 = Ph, CO_2Me R = H, Me, Ph (XVIa)

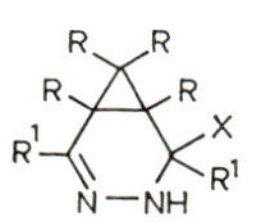

(XVII)

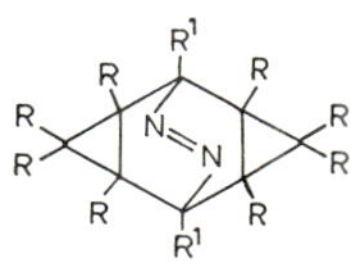

(XVIII)

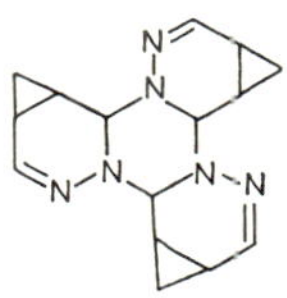

(XIX)

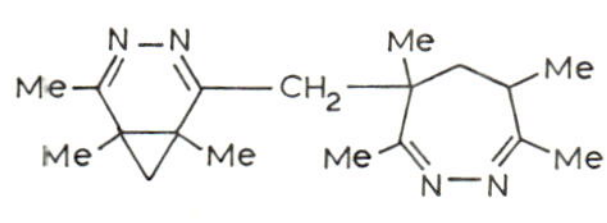

(XX)

give XVIII. With acids ring opening to 4-alkylpyridazines occurs and on oxidation XVIa (R = H; R^1 = Ph) gives dibenzoyl cyclopropane confirming the bicyclic nature of these compounds (*J. Sauer et al.*, J. Amer. chem. Soc., 1972, 94, 2770; Tetrahedron Letters, 1970, **1617**; 1623; 1966, 4979; *G. Maier*, Chem. Ber., 1962, **95**, 611; 1965, **98**, 2438, 2446; 1968, **101**, 1371; Angew. chem. intern. Edn. 1963, **2**, 621).

cis-Cyclopropanedicarbaldehyde gives the trimer XIX with hydrazine, but *cis*-1,2-dimethyl-1,2-diacetylcyclopropane yields the dimer XX, which contains a reduced 1,2-diazepine system. (*Maier* and *T. Sayrac*, *ibid.*, 1966, **5**, 959; Chem. Ber., 1968, **101**, 1354).

Polysubstitution with phenyl groups, however, favours form XV and 3,4,5,6,7-*penta-phenyl*-(5H)-1,2-*diazepine* (XXI), m.p. 237–239°, is obtained using phenyl-substituted intermediates in the above reaction (*M. A. Battiste* and *T. J. Barton*, Tetrahedron Letters,

1967, 1227; *Sauer* and *G. Heinrichs*, *ibid.*, 1966, 4979). On heating transannular bond formation is followed by the elimination of benzonitrile:

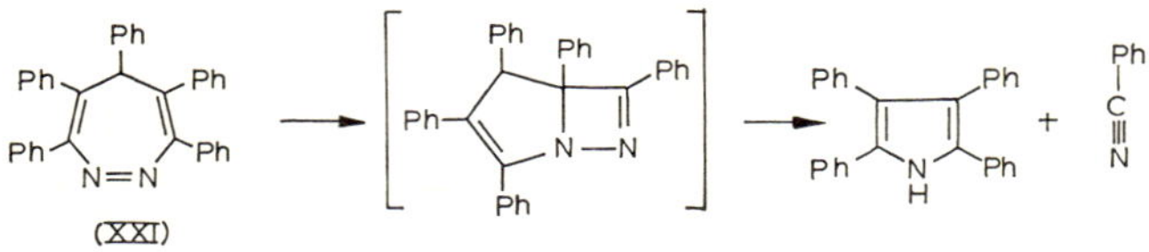

(XXI)

5-Methyl-6-phenyl-(2*H*)-3,4-dihydro-1,2-diazepin-4-one (XXIV; R = H), m.p. 150–151°, has been extensively investigated by *J. A. Moore* and co-workers. It is prepared by heating the pyrazoline XXII in acetic acid, or by mild acid or base isomerisation of the bicyclic intermediate XXIII (J. Amer. chem. Soc., 1955, **77**, 3417; 1959, **81**, 6026; see also 1962, **84**, 390; J. org. Chem., 1964, **29**, 336; 1966, **31**, 34):

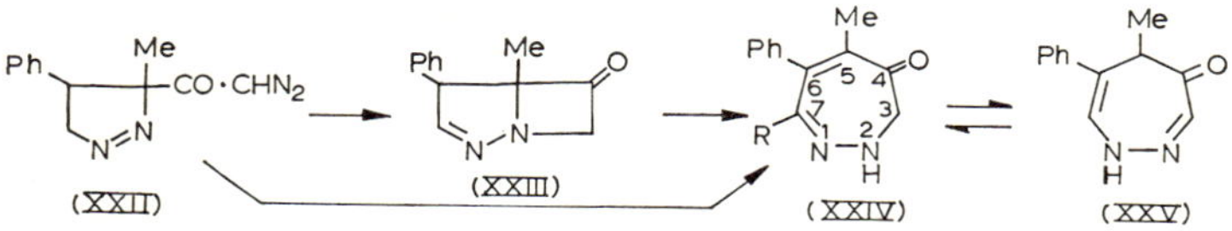

(XXII) (XXIII) (XXIV) (XXV)

5,6-*Diphenyl-* (m.p. 195–198° decomp.) and 6-*methoxycarbonyl-5-methyl-* (m.p. 95–96°) analogues were obtained similarly (*ibid.*, 1970, **35**, 2015). Bicyclic intermediates frequently occur on these reaction pathways but there is no evidence for a bicyclic tautomer of XXIV (R = H) (Trans. N.Y. Acad. Sci., 1965, **27**, 591).

All ring protons of XXIV (R = H) exchange in alkaline deuterium oxide (those at $C_{(3)}$ faster than that at $C_{(7)}$) and treatment with base in dimethyl sulphoxide yields the (1H)-4,5-*dihydro isomer* (XXV), m.p. 167–168° (J. org. Chem., 1964, **29**, 2124; 1967, **32**, 1342; J. Amer. chem. Soc., 1968, **90**, 1369). The equilibration of XXIV (R = H) and XXV explains the presence of both 2- and 6-aminopyridines in the base catalysed ring-contraction mentioned below.

Methylation takes place at $N_{(2)}$ directly with dimethyl sulphate and a strong base to give the 2-*methyl* derivative of XXIV, m.p. 72–73°, but also at $N_{(1)}$ with boron trifluoride and diazomethane *via* the red *ylid* **XXVI** (R = Me), m.p. 92–94° decomp., which isomerises to 1,5-*dimethyl-6-phenyl-*(1H)-4,7-*dihydro*-1,2-*diazepin*-4-*one* (**XXVII**; R = Me), m.p. 73–75°, on standing. Base treatment of XXVII (R = Me) yields the isomeric 1-*methyl* derivative of XXV, m.p. 93°.

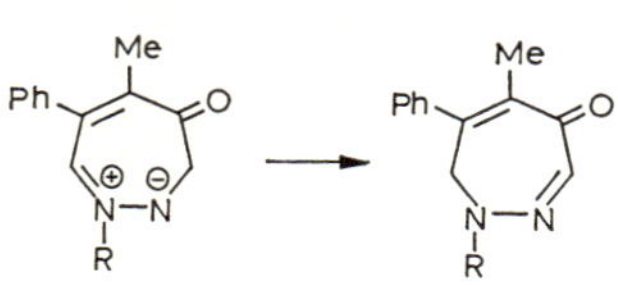

(XXVI) (XXVII)

Cyanoethylation (J. org. Chem., 1965, **30**, 1887; 1967, **32**, 1602; J. Amer. chem. Soc., 1959, **81**, 6029; 1968, **90**, 1369) and the reaction of formaldehyde also take place at $N_{(2)}$ but benzaldehyde condenses with the $C_{(3)}$ methylene group to give aldol and benzylidene derivatives (J. org. Chem., 1964, **29,** 2128). 2-Acyl derivatives are formed in reactions with anhydrides, but acid chlorides give *N*-acyl derivatives of the enamine isomer of XXIII; benzoyl chloride in methanol gives the *methoxy* compound XXVIII (R = Ph), which yields XXIV (R = OMe), m.p. 99–101 °, after oxidation with *N*-bromosuccinimide

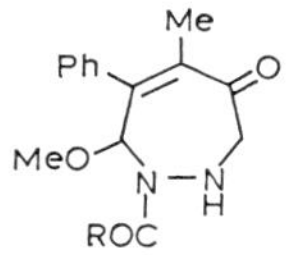

(XXVIII)

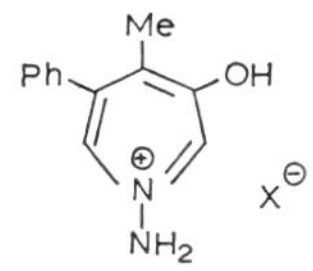

(XXIX)

and alkaline hydrolysis (J. Amer. chem. Soc., 1962, **84,** 3022; J. org. Chem., 1966, **31,** 34; 48). An acetyl analogue XXVIII (R = Me) is formed from XXIV (R = H) and acetyl chloride in methanolic pyridine.

Dihydrodiazepinone XXIV (R = H) and its derivatives undergo a remarkable range of ring-contractions leading to furans, pyrroles, pyrazoles, aminopyridines and pyridazines. In strong acid, XXIV (R = H) is converted into the *N*-aminopyridinium salts XXIX (*ibid.*, 1966, **31,** 34; J. Amer. chem. Soc., 1955, **77,** 3417; 1959, **81,** 6029). The 2-methyl derivative of XXIV (R = H) shows no ring contraction in acid, but the 1-methyl ylid (XXVI; R = Me) gives the *N*-methylamino analogue of XXIX showing that N-1 is the extruded nitrogen atom. A bicyclic compound with a 2,7-transannular bond is proposed as intermediate (*ibid.*, 1968, **90,** 4738; J. org. Chem., 1965, **30,** 1887). In basic conditions the behaviour of XXIV (R = H) contrasts with that of the methoxy compounds XXVIII (R = Me, Ph); the former rearranges to a mixture of 2- and 6-aminopyridines, the latter yields the pyrrole XXX. Methyl or cyanoethyl substituents on position 2 of compound XXIV (R = H) are found on the exocyclic nitrogen after ring contraction implying $N_{(2)}$ extrusion (*ibid.*, 1964, **29,** 2124; 1966, **31,** 48; 1967, **32,** 1342, 1602). Oxidation of diazepinone XXIV (R = H) with hydrogen peroxide gives the pyrazoleacetic acid XXXI (*ibid.*, 1965, **30,** 1889) possibly after peroxidation at position 4, and $N_{(2)}$ to $C_{(5)}$ bond-formation.

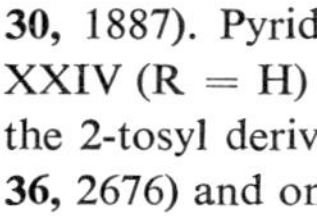

(XXX)

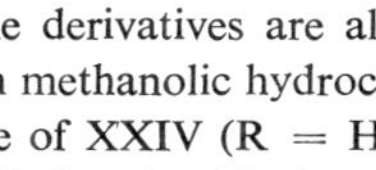

(XXXI)

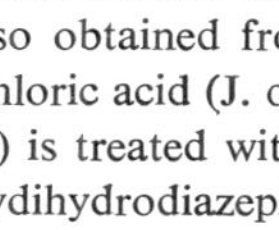

(XXXII)

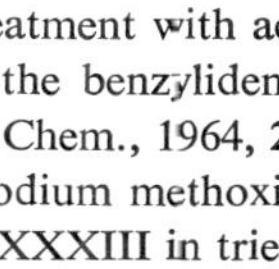

(XXXIII)

Dihydro-1,2-diazepin-4-one XXVII (R = Me), which contracts in base to a 2-methylaminopyridine, yields the dihydropyridazine XXXII on treatment with acid (*ibid.*, 1965, **30,** 1887). Pyridazine derivatives are also obtained from the benzylidene derivative of XXIV (R = H) with methanolic hydrochloric acid (J. org. Chem., 1964, **29,** 2128), when the 2-tosyl derivative of XXIV (R = H) is treated with sodium methoxide (*ibid.*, 1971, **36,** 2676) and on acylating the 4-hydroxydihydrodiazepine XXXIII in triethylamine.

4-*Hydroxy*-5-*methyl*-6-*phenyl*-(2H)-3,4-*dihydro*-1,2-*diazepine* (XXXIII), m.p. 101–102°, is the product of reduction of XXIV (R = H) with tetrahydridoborate. On treatment with acid or acylation it yields the furfurylhydrazines XXXV (R = H or COR[1]) by cleavage of the bridged compounds XXXIV (*ibid.*, 1966, **31,** 52; 1972, **37,** 3770). Transannular reactions are also observed when XXIV (R = H) or its 2-methyl derivative is treated with hydroxylamine; the product XXXVI results from an intermediate with a nitrogen bridge across positions 4 and 7 (J. Amer. chem. Soc., 1959, **81,** 6029). By contrast semicarbazide yields the 4-semicarbazones.

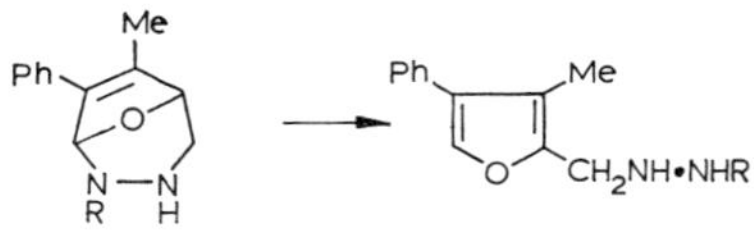

(XXXIV) (XXXV)

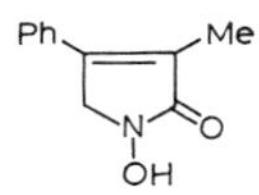

(XXXVI)

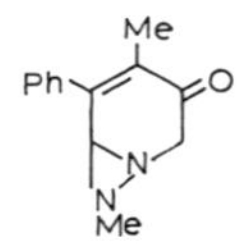

(XXXVII)

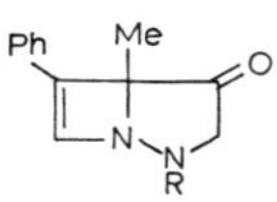

(XXXVIII)

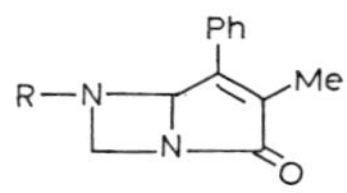

(XXXIX)

Irradiation of ylid XXVI (R = Me) gives the bicyclic diaziridine XXXVII which opens to a methylaminopyridine (*ibid.*, 1968, **90,** 4738) but the products from XXIV and its 2-substituted derivatives are the bicyclic ketones XXXVIII, which reopen to diazepinones with base or on heating (Chem. Comm., 1965, 468; J. org. Chem., 1968, **33,** 4381; 1971, **36,** 2676). This C to N transannular bond formation contrasts with the C to C linkages which result in the photoisomerisation of 1*H*-1,2-diazepines III (p. 239).

Treatment of the bicyclic compound XXXVIII with base provides another route to (1*H*)-4,5-dihydrodiazepinone XXV (J. Amer. chem. Soc., 1968, **90,** 1369). Pyrolysis of XXXVIII (R = Ac, COPh) gives the respective betaines XXVI (R = Ac, COPh), which undergo cyclo-addition reactions with dimethyl acetylenedicarboxylate across positions 1 and 3, and ketene across positions 1 and 5 (J. org. Chem., 1972, **37,** 2640, 2796).

If no trapping reagent is present, the ylids isomerise to mixtures of the bicyclic lactams XXXIX (R = Ac, COPh) and the (1*H*)-4,7-dihydrodiazepinones **XXVII** (R = Ac, COPh). Hydrolysis of XXVII (R = Ac) yields 5-*methyl*-6-*phenyl*-(1H)-4,7-*dihydro*-1,2-*diazepin*-4-*one* (XXVII; R = H), m.p. 119–121 °, which reverts to the (2*H*)-3,4-dihydro tautomer XXIV (R = H) on heating.

7-*Phenyl*-2,4,4-*trimethyl*-(2H)-3,4-*dihydro*-1,2-*diazepin*-3-*one* (XL), m.p. 53 °, is available by the ring-expansion sequence:

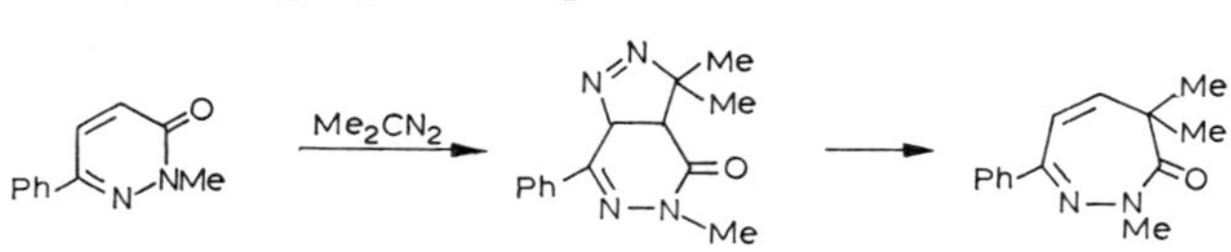

(XL)

Pyrolysis of the intermediate gives higher yields than photolysis (*M. Franck-Neumann* and *G. Leclerc*, Tetrahedron Letters, 1969, 1063). Alkaline hydrolysis of XL leads to ring-opening and loss of both nitrogen atoms.

(1*H*)-4,5-**Dihydro**-1,2-**diazepines** (XLI have been prepared by the 1,3-dipolar addition of diazomethane to cyclobutenes:

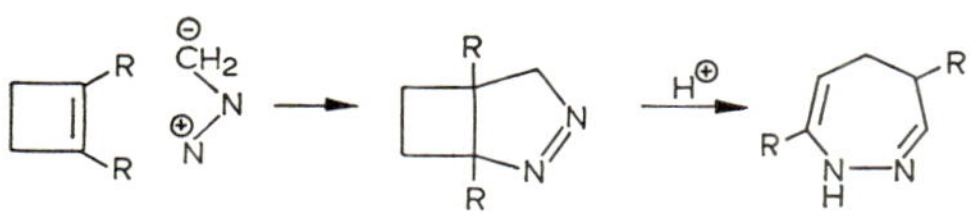

(XLI)

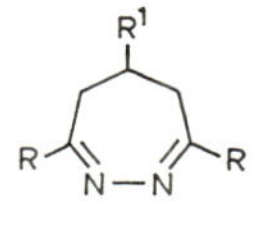

(XLII)

4,7-*Bisethoxycarbonyl* (XLI; R = CO_2Et), m.p. 163–165°, and 4,7-*dicyano* derivatives (XLI; R = CN), m.p. 198°, are known (*H. D. Martin* and *H. Prinzbach*, Chimia, 1969, **23,** 37).

Several (4*H*)-5,6-**dihydro**-1,2-**diazepines** (XLII) are obtained by condensing hydrazine with 1,5-dicarbonyl compounds under high dilution conditions; 3,7-*diphenyl*- (XLII; R = Ph, R^1 = H), m.p. 163–164° (*C. G. Overberger* and *J. J. Monagle*, J. Amer. chem. Soc., 1956, **78,** 4470; *M. Lipp et al.*, Ann., 1958, **618,** 110), 3,5,7-*triphenyl*- (**XLII;** R, R^1 = Ph), m.p. 159° (*K. W. Merz* and *H. Richter*, Arch. Pharm., 1937, **275,** 294; *L. A. Carpino*, J. org. Chem., 1965, **30,** 736), 3,7-*dicarboxy*- (XLII; R = CO_2H, R^1 = H), m.p. 210° decomp. (*E. E. Blaise* and *H. Gault*, Bull. Soc. chim. Fr., 1907, **1,** 75) and annellated alicyclic derivatives (*N. S. Gill et al.*, J. Amer. chem. Soc., 1952, **74,** 4923) are made in this way. Most aliphatic 1,5-diketones yield aminopyridine derivatives (*Overberger et al.*, *ibid.*, 1955, **77,** 4097).

Reduction with lithium tetrahydridoaluminate or partial hydrogenation of XLII (R = Ph, R^1 = H) gives 3,7-**diphenyl-(**1*H*)-4,5,6,7-**tetrahydro**-1,2-**diazepine** (XLIII), b.p. 190°/1 mm, which is further hydrogenated to 3,7-**diphenyl**-1,2-**diazepane** (XLIV), b.p. 200–202°/2 mm. Oxidation of XLIV with mercuric oxide gives the *cyclic azo compound* XLV, m.p.

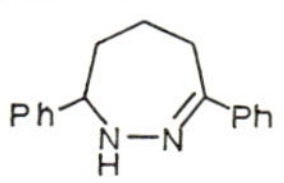

(XLIII)

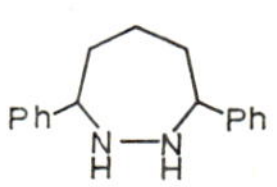

(XLIV)

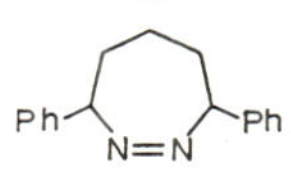

(XLV)

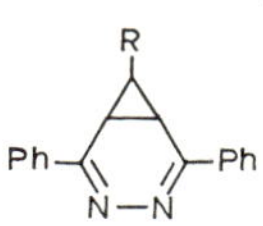

(XLVI)

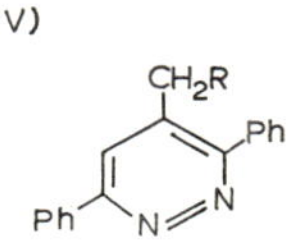

(XLVII)

115°, but oxidation of XLIV in air, and thermal isomerisation of XLV regenerates XLIII (*idem, ibid.*, 1957, **79,** 2662; 1958, **80,** 2317). The ring of the perhydro compound XLIV is opened by bromine and acid with loss of nitrogen giving the bromoketone $PhCO(CH_2)_3$-CHBrPh (*Overberger* and *Monagle*, *loc. cit.*; *A. Padwa et al.*, J. Amer. chem. Soc., 1969, **91,** 456).

Treatment of XLII (R = Ph, R^1 = H) with trifluoroperacetic acid gives the N-*oxide*, m.p. 168–169°, which on irradiation forms a transannular bond between positions 3 and 7, or the acyclic ketone PhCH:CH·$(CH_2)_2$COPh (*W. M. Williams* and *W. R. Dolbier*, *ibid.*, 1972, **94,** 3955). With *N*-bromo- or *N*-chlorosuccinimide, compounds XLII (R = Ph, R^1 = H, Me or Ph) undergo ring-contraction to the respective pyridazines XLVII (R = H, Me or Ph); the intermediate diazanorcaradiene (XLVI; R = Ph) has also been isolated (*R. G. Amiet et al.*, Chem. Comm., 1965, 128; Austral. J. chem. 1968, **21,** 1279).

Pyrolysis of the *cis*-azo compound XLV leads to the rapid evolution of nitrogen and the formation of the diradical reaction products 1,2-diphenylcyclopentane and 1,5-diphenyl-1-pentene (*Overberger et al.*, J. Amer. chem. Soc., *loc. cit.* and 1963, **85,** 2752). Irradiation in solution yields the same products, but in the solid state only *cis*-1,2-diphenylcyclopentane is obtained (*Overberger* and *C. Yaroslavsky*, Tetrahedron Letters, 1965, 4395).

The high-temperature reaction between 4-benzoylbutanoic acid and hydrazines gives 3-*phenyl*-1H-4,5,6,7-*tetrahydro*-1,2-*diazepin*-7-*one* (XLVIII; R = H), m.p. 157°, and its derivatives (*C-G. Wermuth* and *J. J. Koenig*, Angew. chem. intern. Edn., 1972, **11,** 152). With lithium tetrahydridoaluminate, XLVIII (R = Me) is reduced first to 1-*methyl*-3-*phenyl*-1H-4,5,6,7-*tetrahydro*-1,2-*diazepine*, b.p. 80–82°/0.13 mm, *picrate*, m.p. 128–130°, and then to the 1-methyl-3-phenyldiazepane, which like XLIV is very sensitive to oxidation. Its N-*acetyl* derivative (XLIX), b.p. 310–312°, was identical with one of the pyrolysis products of piperidinium compound L (*S. Wawzonek* and *J. G. Stephanie*, J. org. Chem., 1971, **36,** 2467):

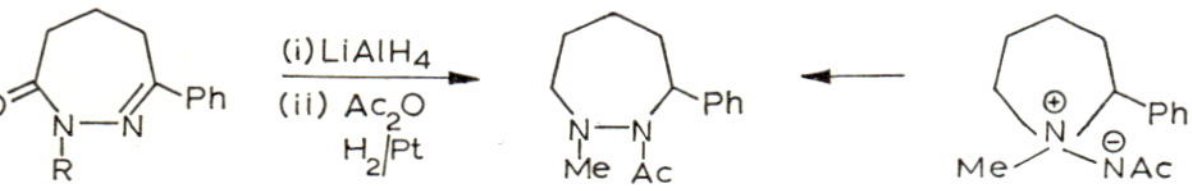

(XLVIII) (XLIX) (L)

The anticipated general route to 1,2-diazepanes, the condensation between hydrazines and 1,5-dihalogenoalkanes is in practice limited to 1,2-disubstituted hydrazines. Hydrazine and monosubstituted hydrazines yield instead 1-aminopiperidines, and the parent 1H-**hexahydro**-1,2-**diazepine** (1,2-**diazepane**) (LI; R = H), b.p. 59°/17 mm, is prepared by the hydrolysis and decarboxylation of LI (R = CO_2Et). This compound and analogues LI (R = COPh, or CONHPh) are made by condensing 1,5-dibromopentane with the

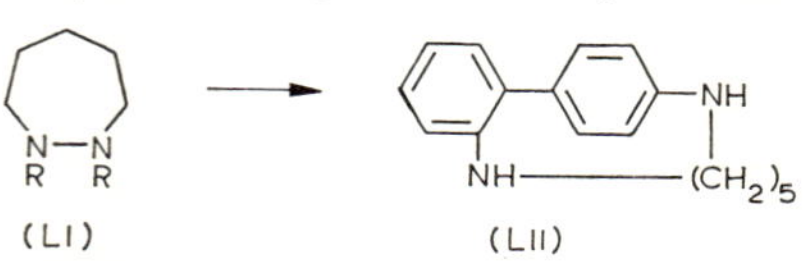

(LI) (LII)

bishydrazine in the presence of potassium (*G. Zinner* and *W. Deucker*, Arch. Pharm., 1962, **295,** 526). 1,2-*Diphenyl*-1,2-*diazepane* (LI; R = Ph), m.p. 78–79°, prepared similarly, undergoes a benzidine rearrangement to LII in acid conditions (*G. Wittig et al.*, Ann., 1957, **610,** 180). Diazepane LI (R = CO_2Et) has also been made by treating the reduction product of III (R = CO_2Et) with ethyl chloroformate and base (*Streith* and *J. M. Cassal*, Tetrahedron Letters, 1968, 4541).

The perfluoro-1,2-diazepinedione (LIII) is prepared as shown (*P. H. Ogden*, J. chem. Soc. C, 1971, 2920):

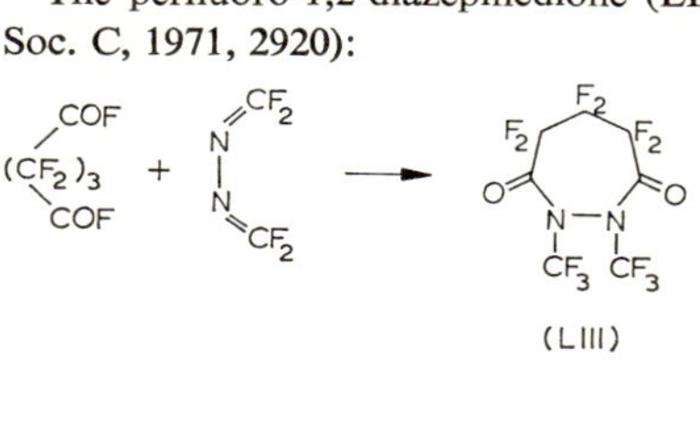

(LIII)

(ii) 1,2-Diazepines with annellated benzene rings

1,2-**Benzodiazepines** are extremely rare. A bis(triphenylphosphine)platinum complex of the 1*H*-1,2-benzodiazepine LIV is obtained from 1-phenyl-1-butyn-3-one (*P. B. Tripathy et al.*, J. Amer. chem. Soc., 1971, **93,** 4406) and 5-aryl-3*H*-1,2-benzodiazepines (LV) are obtained in the following thermolysis (*J. T. Sharp et al.*, Chem. Comm., 1970, 909):

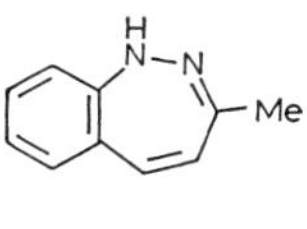

(LIV)

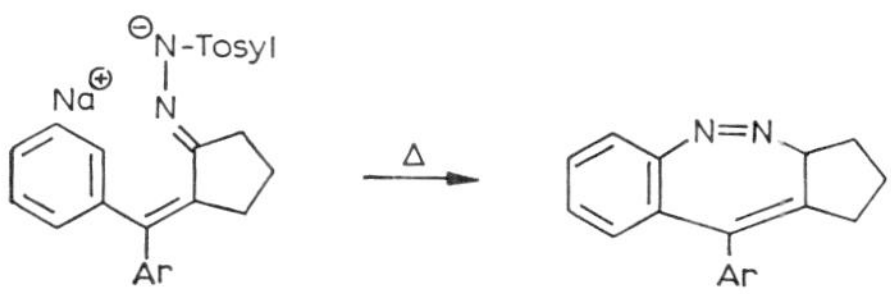

(LV)

2-*Acetyl-3-methoxy-2,3-dihydro-1*H*-1,2-benzodiazepine* (LVI) is a probable product in the irradiation of *N*-acetyliminoquinolinium ylid in methanol (*H. Kato et al., ibid.*, 1970, 1592), and 2-hydrazinodihydrocinnamic acid is found to cyclise to 2,3,4,5-*tetrahydro-1*H*-1,2-benzodiazepine-3-one*, m.p. 143° (*E. Fischer* and *H. Kuzel*, Ann., 1883, **221,** 261). The 1-*ethyl-*, m.p. 165–166°, and 2-*ethyl*-analogues, m.p. 74°, are also known. The preparation of hexahydro-1,2-benzodiazepine LVII, which undergoes ring-contraction to 2-phenyl-5,6,7,8-tetrahydroquinoline in acid, has already been described (*N. S. Gill et al.*, J. Amer. chem. Soc., 1952, **74,** 4923).

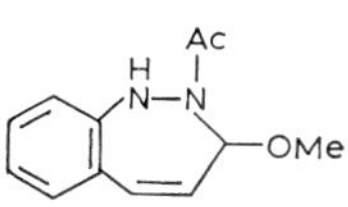

(LVI)

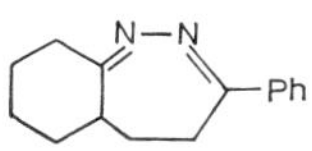

(LVII)

By contrast, several members of the 2,3-**benzodiazepine** series are known. A general route to the fully unsaturated benzodiazepines is the pyrolysis of the sodium salts of 2-acylstyrene tosylhydrazones LVIII:

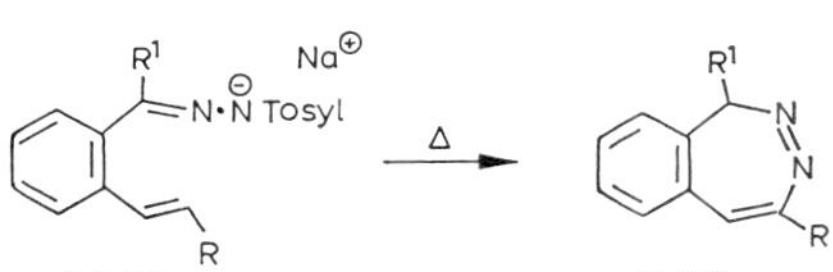

(LVIII) (LIX)

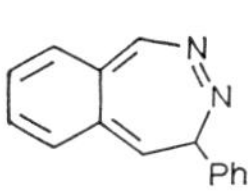

(LX)

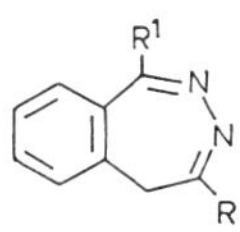

(LXI)

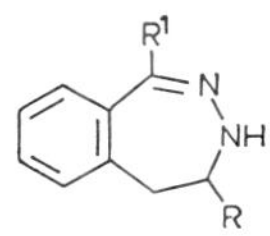

(LXII)

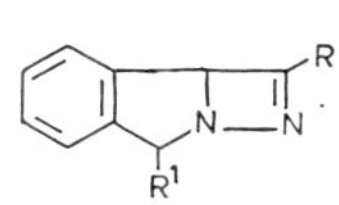

(LXIII)

4-*Phenyl*-4H-2,3-*benzodiazepine* (LX), m.p. 132–133°, is said to be the primary product in the cyclisation of LVIII (R = Ph, R^1 = H) to 4-*phenyl*-1H-2,3-*benzodiazepine* (LIX; R = Ph, R^1 = H), m.p. 82–83°, in aprotic solvents. Base-catalysed isomerisation of LIX (R = Ph, R^1 = H) yields 4-*phenyl*-5H-2,3-*benzodiazepine* (LXI; R = Ph, R^1 = H), m.p. 108–110°. Partial reduction with hydrogen over palladium charcoal gives 4-*phenyl*-4,5-*dihydro*-3H-2,3-*benzodiazepine* (LXII; R = Ph, R^1 = H), m.p. 113–115° (*V. I. Bendall*, Chem. Comm., 1972, 823; *J. T. Sharp et al., ibid.*, 1970, 1197). Irradiation converts LIX into the tricyclic compound LXIII, while thermal decomposition of LIX and irradiation of LXI yields indenes (*Sharp et al., ibid.*, 1972, 827).

4,5-Dihydro-3*H*-2,3-benzodiazepine (LXII; R, R^1 = H), m.p. 72°, is prepared by thermal cleavage of the *N*-aminoisoquinoline dimer LXIV. Successive treatment with acid and alkali reverses this reaction (*E. Schmitz* and *R. Ohme*, Chem. Ber., 1962, **95,** 2012; 1958, **91,** 1495). Several 1-aryl analogues of LXII result when 2-(2-bromoethyl)benzophenones are treated with hydrazines (*C. Van der Stelt* and *W. T. Nauta*, Rec. Trav.

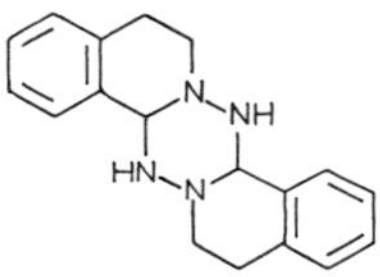

(LXIV)

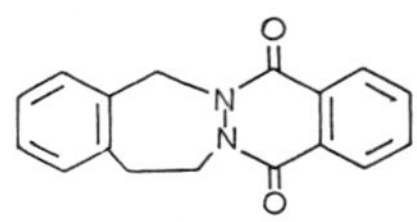

(LXV)

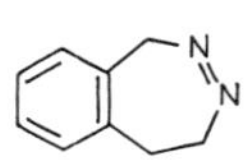

(LXVI)

chim., 1965, **84,** 640). LXII (R, R^1 = H) gives a urea at position 3 with phenyl isocyanate and is reduced catalytically to **2,3,4,5-tetrahydro-1*H*-2,3-benzodiazepine,** *hydrochloride*, m.p. 139–143°. This tetrahydro compound, which is also obtained in the hydrolysis of the phthaloyl derivative LXV, gives 4,5-*dihydro*-1H-2,3-*benzodiazepine* (LXVI), m.p. 63°, with hydrogen peroxide, and a bis-*N,N'*-methylene derivative with formaldehyde. LXVI isomerises to LXII (R, R^1 = H) with acid (*Schmitz* and *Ohme, loc. cit.*).

2,5-Dihydro-1*H*-2,3-benzodiazepin-1-ones (LXVII) are obtained in the reaction of hydrazine with 2-phenacylbenzoic acid (*H. Wolbling*, Ber., 1905, **38,** 3845), isocoumarins (*A. Lieck, ibid.*, P. 3853; *N. P. Buu-Hoi*, Compt. rend., 1939, **209,** 321), or isothiocoumarins (*L. Legrand* and *N. Lozac'h*, Bull. Soc. chim. Fr., 1970, 2237); the 4-*phenyl* compound (LXVII; R = Ph, R^1 = H) has m.p. 202°.

Phenylhydrazine similarly gives 4-*methyl*-2-*phenyl*-2,5-*dihydro*-1H-2,3-*benzodiazepin*-1-*one* (LXVII; R = Me, R^1 = Ph), m.p. 198–199° (*J. Gottlieb*, Ber., 1899, **32,** 958). Compounds LXVII (R^1 = H) may be *N*-alkylated and with acid undergo ring contraction to the isoquinolone LXVIII.

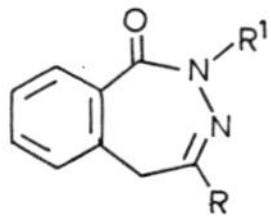

(LXVII)

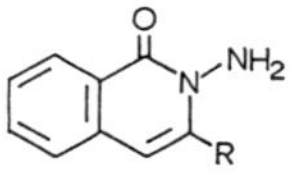

(LXVIII)

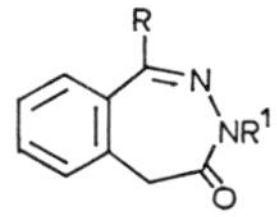

(LXIX)

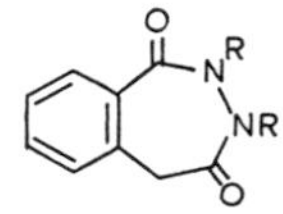

(LXX)

Thermal dehydration of the phenylhydrazone of 2-acetylphenylacetic acid gives *1-methyl-3-phenyl-4,5-dihydro-3*H*-2,3-benzodiazepin-4-one* (LXIX; R = Me, R^1 = Ph), m.p. 158–159°. Pyrolysis of the corresponding azine and semicarbazone gave the 1-methyl compound LXIX (R = Me, R^1 = H), m.p. 185–186°, and the parent, **4,5-dihydro-3*H*-2,3-benzodiazepin-4-one** (LXIX; R, R^1 = H) is obtained from 2-formylphenylacetic acid semicarbazone (*J. O. Halford et al.*, J. org. Chem., 1961, **26,** 1898). An improved method, direct reaction of hydrazine and the phenylacetic acid in *n*-butanol, yields LXIX (R = Ph, R^1 = H), m.p. 208°, among others (*C. G. Wermuth* and *M. Flammang*, Tetrahedron Letters, 1971, 4293). These benzodiazepinones also contract to *N*-aminoisoquinolones in acid.

It is no longer believed that the dione LXX (R = H) is formed when homophthalic anhydride reacts with hydrazine, but with 1,2-dimethylhydrazine, *2,3-dimethyl-2,3,4,5-tetrahydro-1*H*-2,3-benzodiazepin-1,4-dione* (LXX; R = Me), m.p. 118–119°, is obtained (*W. F. Whitmore* and *R. C. Cooney*, J. Amer. chem. Soc., 1944, **66,** 1237; *G. Rosen* and *F. D. Popp*, J. heterocycl. Chem., 1969, **6,** 9).

The most satisfactory route to **11*H*-dibenzo-[*c,f*]-1,2-diazepine** (LXXI; R = H), m.p. 113–114°, and its halogeno derivatives is the reduction of the corresponding 2,2′-dinitrodiphenylmethane with lithium tetrahydridoaluminate (*N. L. Allinger* and *G. A. Youngdale*, J. Amer. chem. Soc., 1962, **84,** 1020; Tetrahedron Letters, 1959, **9,** 10; *W. Theilacker* and *O. Korndorfer*, *ibid.*, 1959, **18,** 5; *A. Catala* and *Popp*, J. heterocyclic chem., 1964, **1,** 178).

Polarographic reduction has been used for LXXI (R = NMe_2) (*H. Lund*, Adv. heterocycl. Chem., 1970, **12,** 213) and both this compound and LXXI (R = NH_2), m.p. 233°, are obtained by reduction with zinc and ammonium chloride, cyclising to the *N*-oxide with aerial oxidation and subsequent reduction. The amino groups may readily be diazotised and so replaced by hydrogen or halogens (*H. Duval*, Compt. rend., 1905, **141,** 198; Bull. Soc. chim. Fr., 1910, **7,** 527).

Reduction of the 5,6-double bond is achieved using zinc dust in aqueous alkali or hydrazine and Raney nickel, as catalytic reduction cleaves the nitrogen–nitrogen bond of the reduced compound. Diazepine LXXI (R = NH_2) is converted into 2,8-diamino-acridine by zinc in aqueous potassium hydroxide, (*idem*, Compt. rend., 1906, **142,** 341) but LXXI (R = Cl) shows ring-contraction to 3,8-dichlorocinnoline in sulphuric acid with added potassium nitrate (*R. J. Dubois* and *Popp*, Chem. and Ind., 1969, 620). Oxidation of LXXI with perchloric acid is the best route to *N*-oxides. Diazepine LXXI forms complexes with several metals (*Popp et al.*, J. heterocycl. Chem., 1966, **3,** 377).

Oxidation of LXXI (R = H) with chromic anhydride in acetic acid is a convenient route to *11*H*-dibenzo-*[c,f]*-1,2-diazepin-11-one* (LXXII; R = H), m.p. 197° (*Catala* and *Popp*, *loc. cit.*); cyclisation of 2,2′-dinitrobenzophenone is brought about by alkaline glucose, but not by lithium tetrahydridoaluminate (*R. B. Johns* and *K. R. Markham*, J. chem. Soc., 1962, 3712). Diazepinones LXXII form *N*-oxides, but do not show carbonyl reactions or react with Grignard reagents. The 11-oxo group is reduced by aluminium

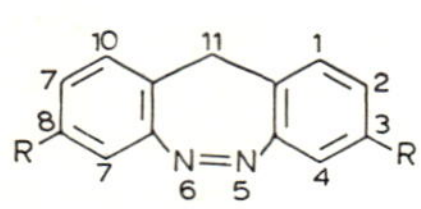

(LXXI)

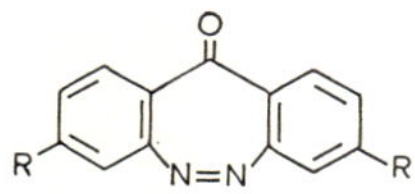

(LXXII)

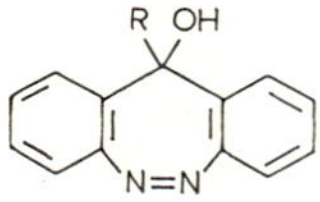

(LXXIII)

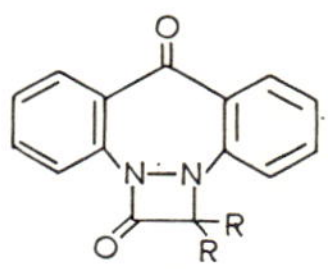

(LXXIV)

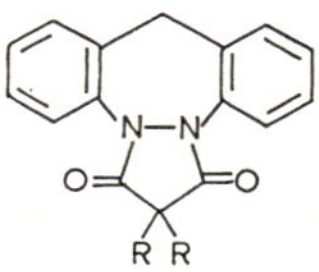

(LXXV)

isopropoxide giving LXXIII (R = H), but with sodium hydride LXXII (R = Cl) again gave 3,8-dichlorocinnoline (*Dubois* and *Popp*, Chem. Comm., 1968, 675; J. heterocycl. Chem., 1969, **6**, 285). Weakly active Raney nickel is required for catalytic hydrogenation of the azo bond otherwise the ring opens. Acetylenes in liquid ammonia add to the carbonyl bond giving LXXIII (R = C ≡ CR^1) and ketenes add across the 5,6-double bond resulting in the tetracyclic system LXXIV (*W. Reid* and *S. Piesch*, Angew. chem. intern. Edn., 1965, **4**, 988; Chem. Ber., 1966, **99**, 233).

5,6-*Dihydro*-11H-*dibenzo*[c,f]-1,2-*diazepine*, m.p. 105–106°, is oxidised back to LXXI (R = H) with mercuric oxide (*Allinger* and *Youngdale*, *loc. cit.*). In hydrochloric acid it disproportionates to LXXI (R = H) and 2,2′-diaminodiphenylmethane, in contrast to the reduced 3,8-diamine which undergoes ring contraction, like LXXI (R = NH_2), to 2,8-diaminoacridine under these conditions. Dihydro derivatives of both LXXI and LXXII form stable products, *e.g.* LXXV, with malonic acid derivatives (*H. S. Lowrie*, J. med. pharm. Chem., 1962, **5**, 1362; J. R. Geigy, A.G., B.P., 940,165/1960; C.A., 1964, **61**, 1816; *Lowrie*, U.S.P., 3,170,929/1961; C.A., 1965, **62**, 14706; *Reid* and *Piesch*, *loc. cit.*).

(*iii*) *1,3-Diazepines*

Highly unsaturated 1,3-diazepines are not known, but many reduced derivatives have been synthesised and investigated. Most synthetic routes involve the condensation of a 1,4-diamine with a reactive carbon centre; aldehydes and ketones yield hexahydro- and acid derivatives tetrahydro-diazepines. Reactions between ureas and 1,4-dicarbonyl compounds generally fail to give 1,3-diazepines.

4,5,6,7-**Tetrahydro**-3*H*-1,3-**diazepine** (LXXVI; R = H), m.p. 117–118°, is obtained as a by-product in the thermal polymerisation of the oligoamidine $H_2N[(CH_2)_4NHCH:N]_5$-$(CH_2)_4NH_2$ (*C. Grundmann* and *J. Kreutzberger*, J. polymer Sci., 1959, **38**, 425).

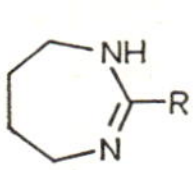

(LXXVI)

As examples of the general synthesis 1,4-diaminobutane, its hydrochloride or ditosyl derivative react with sodium acetate (*T. Haga* and *R. Majima*, Ber., 1903, **36,** 333) or ethoxyacetylene (*J. F. Arens*, U.S.P. 2,813,862/1957; C.A., 1958, **52,** 8212) to give LXXVI (R = Me), b.p. 220°/12 mm, *picrate*, m.p. 138°; with appropriate nitriles to give LXXVI (R = Ph), m.p. 105°, and LXXVI (R = CF_3), m.p. 96–97°, with *N*-(*p*-tolyl)phenylacetamidine to give LXXVI (R = CH_2Ph), *picrate*, m.p. 131° (*P. Oxley* and *W. F. Short*, J. chem. Soc., 1947, 497; 1950, 862; B.P. 593,659/1947, 614,032/1948; C.A., 1948, **42,** 1971; 1949, **43,** 5049; *R. N. Johnson* and *H. M. Woodburn*, J. org. Chem., 1962, **27,** 3958); with ethyl glycolimidate to give LXXVI (R = CH_2OH), m.p. 133–135° (*M. Sahyun et al.*, J. Amer. chem. Soc., 1959, **81,** 2214); with nitroguanidine or *S*-methyl nitroisothiourea to give LXXVI (R = $NHNO_2$), m.p. 219–220° decomp. (*A. F. McKay et al.*, *ibid.*, 1948, **70,** 430; 3990; Canad. J. Chem., 1951, **29,** 391; U.S.P. 2,525,927/1950; C.A., 1951, **45,** 2512; *L. S. Hafner* and *R. Evans*, J. org. Chem., 1959, **24,** 1157); and with cyanogen to give a *bis-compound*, m.p. 91–93° (*K. Matsuda*, U.S.P. 2,819,262/1957; C.A., 1958, **52,** 9230).

Tetrahydrodiazepines are also obtained in ring expansion reactions. Treatment of the tosylate of 2-piperidone oxime with sodium ethoxide yields the O-*ethyl isourea* LXXVI (R = OEt), m.p. 60–62°, and in aqueous alkali the cyclic urea LXXIX (R = H, X = O) (*A. Le Berre et al.*, Bull. Soc. chim. Fr., 1971, 3245), while the tetrahydrodiazepinones LXXVII (X = OMe, CN) (*B. Gregory et al.*, Chem. Comm., 1972, 97) and tetrahydrodiazepinedione LXXVIII (*T. Kuneida* and *B. Witkop*, J. Amer. chem. Soc., 1971, **93,** 3478) both with 4,5-double bonds, are prepared as shown:

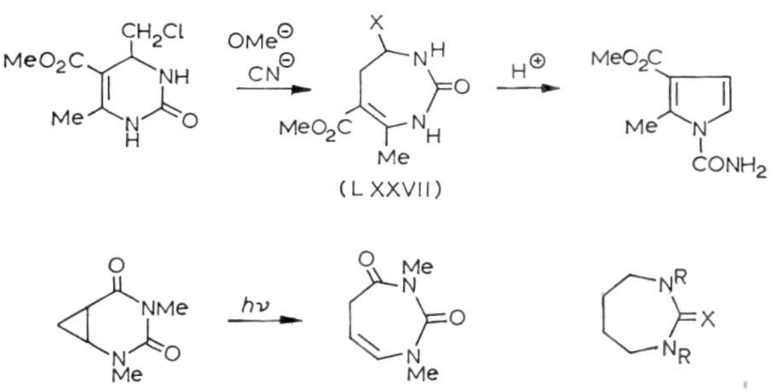

(LXXVII) (LXXVIII) (LXXIX)

In acid LXXVII undergoes ring contraction to a pyrrole; catalytic hydrogenation of LXXVIII yields the perhydrodione.

The action of nitric acid on LXXVI (R = $NHNO_2$) gives 1,3-*dinitro*-1,3-*diazepan*-2-*one* (LXXIX; R = NO_2, X = O), m.p. 137–138°, while with ammonia the imino compound LXXIX (R = H, X = NH) is obtained (*D. Stefanye* and *W. L. Howard*, J. Amer. chem. Soc., 1955, **77,** 761; *McKay* and *G. F. Wright*, *ibid.*, 1948, **70,** 3990). Similar imino compounds are prepared by the reaction of pentachlorophenyl isocyanide dichloride $C_6Cl_5N{:}CCl_2$) with tetramethylenediamine (*E. Kühle*, Angew. Chem. intern. Edn., 1969, **8,** 20) and the treatment of nitroarginine with base (*R. Paul et al.*, J. org. Chem., 1961, **26,** 3347).

Several routes to 1,3-**diazepan-2-one** (*tetramethylene urea*, 1,3-*diazacycloheptan-2-one*) (LXXIX; R = H, X = O), m.p. 172–173°, are known, including the cyclisation of tetramethylene diisocyanate, with water (*Y. Iwakura et al.*, Chem. High Polymers, Japan, 1947, **4**, 91; Nippon Kagaku Zasshi, 1957, **78**, 1416; C.A., 1951, **45**, 2711; 1960, **54**, 1539; J. Amer. chem. Soc., 1957, **79**, 4358), the Beckmann rearrangement of 2-piperidone oxime (*H. Behringer* and *H. Meier*, Ann., 1957, **607**, 67, 73), the action of phosgene on bis(trimethylsilyl)-1,4-diaminobutane (*L. Birkofer et al.*, Chem. Ber., 1960, **93**, 2810) and the high pressure reaction of tetramethylenediamine with sulphur, methanol and carbon monoxide (*F. Applegarth* and *R. A. Franz*, U.S.P. 2,874,149/1959; C.A., 1959, **53**, 12187). Acid hydrolysis cleaves the ring to give 4-aminobutyl isocyanate and products of its subsequent reactions.

If aniline is refluxed with 1,3-oxazepan-2-one, 1-*phenyl*-1,3-*diazepan*-2-*one*, m.p. 240°, is obtained (*R. Delaby* and *R. Damiens*, C.A., 1959, **53**, 376) and three moles of benzaldehyde condense with 1,4-dihydrazinobutane to give LXXIX (R = N:CHPh, X = H, Ph) a reaction reversed by acid hydrolysis (*H. V. Daeniker* and *J. Druey*, Helv., 1957, **40**, 918). Succinyl chloride is said to condense with disubstituted ureas yielding 1,3-diazepantriones (*G. Losse et al.*, J. pr. Chem., 1958, **7**, 28).

1,3-**Diazepane-2-thione** (LXXIX; R = H, X = S), m.p. 181–182° (*M. Mammi et al.*, Ricerca Sci., 1967, **37**, 766; *F. D'Angeli et al.*, Gazz., 1965, **95**, 735), may be prepared from its oxa-analogue (*Behringer* and *Meier*, *loc. cit.*) from disodium tetramethylenebisdithiocarbamate (*McKay et al.*, J. Amer. chem. Soc., 1958, **80**, 3335) or in the reaction between 1,4-diaminobutane and carbon disulphide (*E. Strack*, Z. physiol. Chem., 1929, **180**, 198; *McKay* and *M. E. Kreling*, Canad. J. Chem., 1957, **35**, 1438; *L. H. Conover*, *A. R. English* and *C. E. Larrabee*, U.S.P. 2,921,073/1960; C.A., 1960, **54**, 8861). The cyclic thiourea, in contrast to 1,3-diazepan-2-one polymerises when molten (*H. K. Hall* and *A. K. Schneider*, J. Amer. chem. Soc., 1958, **80**, 6409) and with formaldehyde gives a 1,1′-bismethylene compound (*H. Staudinger* and *G. Niessen*, Makromol. Chem., 1955, **15**, 75).

(*iv*) *1,3-Diazepines with annellated benzene rings*

The nearest approach to a fully unsaturated **benzo**-1,3-**diazepine** is shown below (*H. Plieninger et al.*, Angew. Chem., 1955, **67**, 400; Chem. Ber., 1955, **88**, 1965). Attempts to open the lactone ring also disrupt the diazepine ring.

A route to 4,5-**dihydro**-(3*H*)-1,3-**benzodiazepines** is also shown. It requires the condensation of an imidate with 2-nitrophenylethylamine then reducing and thermally cyclising the amidine intermediate (*G. DeStevens et al.*, J. org. Chem., 1968, **33**, 670):

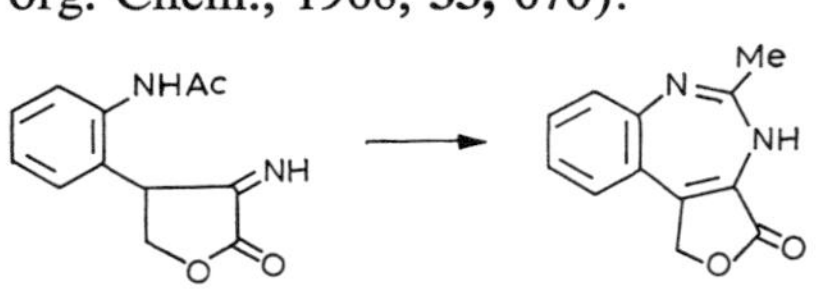

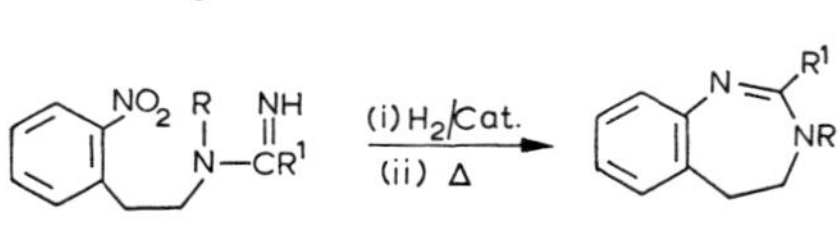

A number of 3-substituted 2,3,4,5-**tetrahydro-**(1*H*)**-**1,3-**benzodiazepines** (LXXX) (*e.g.* R = *Me*, m.p. 99–100°) have been prepared by *DeStevens* and his co-workers.

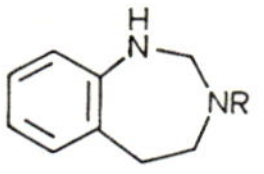

(LXXX)

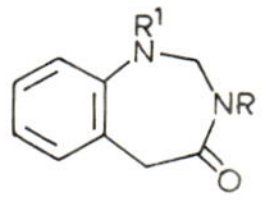

(LXXXI)

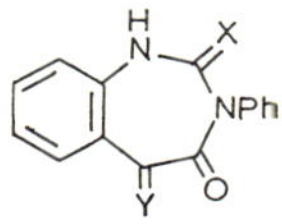

(LXXXII)

2-(2-Aminophenyl)-*N*-methylacetamide cyclises with formaldehyde to 3-*methyl*-2,3,4,5-*tetrahydro*-(1H)-1,3-*diazepin*-4-*one* (LXXXI; R = Me, R^1 = H), m.p. 150°. With benzaldehyde a Schiff base is obtained, which on reduction with tetrahydridoborate and subsequent reaction with formaldehyde yields LXXXI (R = Me, R^1 = CH_2Ph). Hydrogenolysis of the benzyl group and reduction of the oxo group with lithium tetrahydridoaluminate gave the unstable LXXX (R = Me). If the primary amide is used both 2,3,4,5-tetrahydro-(1*H*)-1,3-benzodiazepin-4-one (LXXXI; R, R^1 = H) and LXXXI (R = CH_2OH, R^1 = H) are obtained. Acid hydrolysis of diazepinones LXXXI leads to oxindole polymers.

The isomeric 2,3,4,5-**tetrahydro-**(1*H*)**-**1,3-**benzodiazepin-2-ones** (*e.g.* 7,8-*dimethoxy*-derivative, m.p. 226–227°) are obtained by treating 2-(2-aminophenyl)ethylamine with *N,N'*-dicarbonyldiimidazole. With phosphoryl chloride it gave 2-chloro-4,5-dihydro-(3*H*)-1,3-benzodiazepine, the chloro group of which was readily replaced by amino. Attempts to prepare unsaturated derivatives of LXXX by replacing formaldehyde with acid derivatives were unsuccessful (*DeStevens, loc. cit.* and Oesterr. Chem. Z., 1962, **63,** 177; C.A., 1963, **59,** 11479; J. Amer. chem. Soc., 1961, **83,** 3087; Trans. N.Y. Acad. Sci., 1961, **23,** 568; Record Chem. Progress, 1962, **23,** 105).

The condensation of sodium phenylacetylide and phenyl isocyanate is believed to give diazepinedione LXXXII (X = O, Y = CHPh) (*C. W. Bird*, J. chem. Soc., 1965, 5762) and a similar compound LXXXII (X = S, Y = O) is said to result when *o*-oxalyl-*N,N'*-diphenylthiourea is treated with acetic anhydride (*T. N. Ghosh*, J. Indian chem. Soc., 1933, **10,** 583).

Very few bicyclic 2,4-**benzodiazepines** are known, and unsaturation in the seven-membered ring is even more rare.

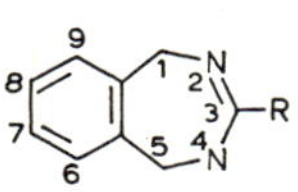

(LXXXIII)

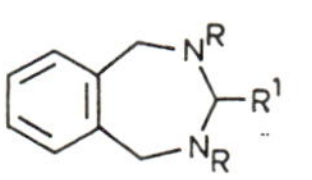

(LXXXIV)

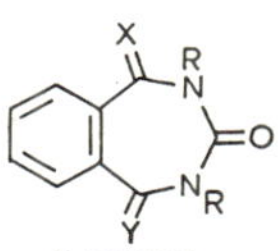

(LXXXV)

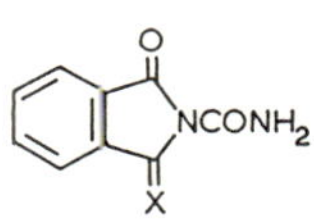

(LXXXVI)

1,2-Di(aminomethyl)benzene condenses with alkyl or aryl imidate hydrochlorides to yield 3-*methyl*-4,5-*dihydro*-(1H)-2,4-*benzodiazepine* (LXXXIII; R = Me) *hydrochloride*, m.p. 283–285°, and analogues (*DeStevens et al.*, J. org. Chem., 1968, **33,** 670). Similarly 1,2-di(arylaminomethyl)benzene with formaldehyde, or benzaldehyde yields 2,4-*diphenyl*-(LXXXIV; R = Ph, R^1 = H), m.p. 196°, and 2,3,4-*triaryl*- (LXXXIV; R, R^1 = Ar)

-2,3,4,5-*tetrahydro*-(1H)-2,4-*benzodiazepines* (*M. Scholtz et al.*, Ber., 1901, **34,** 1504; 1910, **43,** 2304).

2,3,4,5-*Tetrahydro*-(1H)-2,4-*benzodiazepin*-3-*one* (LXXXV; R = H, X, Y = H_2), m.p. 256–260°, is formed when 1,2-di(aminomethyl)benzene reacts with *N,N'*-carbonyldi-imidazole; with carbon disulphide the analogous thione is obtained (*E. F. Elslager*, unpublished observation). Treatment with phosphoryl chloride followed by amines yields cyclic guanidines LXXXIII (R = NR_2^1) (*DeStevens, loc. cit.*; *A. M. Felix* and *R. I. Fryer*, J. heterocycl. Chem., 1968, **5,** 291).

Oxidation of LXXXV (R = CH_2Ph, X, Y = H_2) with chromic acid gives LXXXV (R = CH_2Ph, X = O, Y = H_2) suggesting that the ring-contraction of LXXXV (R = H, X, Y = H_2) to LXXXVI (X = H_2) under these conditions proceeds *via* the diketone (*Felix* and *Fryer, loc. cit.*). The compound, believed to be the cyclic phthaloylurea LXXXV (R = H, X, Y = O), now identified as LXXXVI (X = O) could arise through a similar ring-contraction (*C. S. Smith* and *C. J. Cavallito*, J. Amer. chem. Soc., 1939, **61,** 2218; *D. Grdenic* and *A. Bezjak*, Arkiv Kemi, 1953, **25,** 101; *M. Kesler, ibid.*, 1955, **27,** 67; *V. Hahn et al.*, Experientia, 1954, **10,** 11; *T. W. Evans* and *W. M. Dehn*, J. Amer. chem. Soc., 1929, **51,** 3651).

A few members of this series with ring-fusion across a carbon to nitrogen bond of the diazepine are also known (*E. F. M. Stevenson et al.*, J. chem. Soc., 1952, 199; 5024; *A. Fozard* and *G. Jones, ibid.*, 1964, 2763; *P. Knowles* and *K. R. H. Wooldridge, ibid.*, Perkin I, 1972, 975; *E. Ott* and *F. Hess*, Arch. Pharm., 1938, **276,** 181).

5*H*-**Dibenzo**-[*d,f*]-1,3-**diazepine** *(diphenimidine)* (LXXXVII; R = H) *phosphate*, m.p. 247°, is obtained in the reaction of 2,2'-diaminobiphenyl with ethyl orthoformate (*Y. A. Levin et al.*, Zhur. obshcheĭ Khim., 1961, **31,** 1573; *W. Ried* and *W. Storbeck*, Chem. Ber., 1962, **95,** 459).

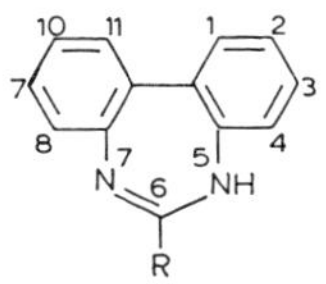

(LXXXVII)

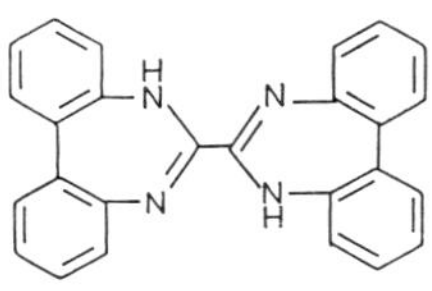

(LXXXVIII)

Other ortho-esters have been used to give the 6-*methyl* (LXXXVII; R = Me), m.p. 168°, 6-*ethyl* (LXXXVII; R = Et), m.p. 232°, and 6-*phenyl* derivatives (LXXXVII; R = Ph), m.p. 167°, and ethyl orthocarbonate gives the 6-*ethoxy* analogue. Salts of imino-esters undergo a similar reaction with the diamine (*Ried et al.*, Ann., 1964, **676,** 114; Chem. Ber., 1964, **97,** 1214; 1965, **98,** 3523), and reaction with malonic iminoester showed that a proposed 14-membered ring dimer from the orthoformic condensation was the *bis(di-azepine)* LXXXVIII, m.p. 285–286° (*Ried* and *A. Sinharay, ibid.*, p. 3523). Fusion of 2,2'-diaminobiphenyl with *S*-methylisothiourea sulphate or carbodiimides yields, respectively, the 6-*aminodiazepine* (LXXXVII; R = NH_2), m.p. 208–209°, and analogues LXXXVII (R = NHR^1) (*W. E. Kreighbaum* and *H. C. Scarborough*, J. med. Chem., 1964, **7,** 310; U.S.P. 3,177,202/1963; C.A., 1965, **62,** 16281).

A second general route is the cyclisation of mono- or bis-acyl derivatives of 2,2′-diaminobiphenyl (*S. Sako*, C.A., 1932, **26**, 3246; *A. E. Blood* and *C. R. Noller*, J. org. Chem., 1957, **22**, 873; *Short et al.*, J. chem. Soc., 1955, 3549; cf. *P. Mamalis* and *V. Petrow*, *ibid.*, 1950, 703), a phosphorus halide or oxyhalide usually being employed.

Diphenimidine and its 6-alkyl and 6-aryl derivatives may be prepared in this way, and optical isomers of 1,11-dimethyl derivatives of LXXXVII have also been obtained (*K. Mislow et al.*, J. Amer. chem. Soc., 1958, **80**, 476, 480).

Several 6-substituted 5*H*-dibenzo-[*d,f*]-1,3-diazepines have also been made by condensing an alkyl or aryl cyanide with 2,2′-diaminobiphenyl benzenesulphonate (*Oxley et al.*, J. chem. Soc., 1951, 485; *A. E. S. Fairfull et al.*, *ibid.*, 1952, 4700).

The diazepine ring of compounds LXXXVII is readily opened. Catalytic hydrogenation or hydrolysis with weak alkali suffices to disrupt LXXXVII (R = CH:CHPh) (*Blood* and *Noller*, *loc. cit.*) while acetic acid hydrolytically cleaves the 2,10-dinitro derivative of LXXXVII (R = Ph) (*Fairfull*, *loc. cit.*). Methylating agents generally yield products of ring-opening too, but 5,6-dimethyl-5*H*-dibenzo-[*d,f*]-1,3-diazepine is obtained from LXXXVII (R = Me) using methyl iodide in liquid ammonia (*Levin*, *loc. cit.*; *Sako*, *loc. cit.*; *Fairfull*, *loc, cit.*).

6,7-*Dihydro*-5H-*dibenzo*-[d,f]-1,3-*diazepin*-6-*one* (LXXXIX; X = O), m.p. 310°, is obtained by heating 2,2′-diaminobiphenyl with urea (*S. von Niementowski*, Ber., 1901, **34**, 3325) or acetoacetic ester (*Ried* and *Storbeck*, *loc. cit.*), when the 2-amino-2′-isocyanate is said to be an intermediate. Decomposing the azide of 2,2′-diphenic acid in the presence of water (*R. A. Labriola et al.*, J. org. Chem., 1940, **5**, 329; Anales asoc. quim. Argentina, 1944, **32**, 57; C.A., 1945, **39**, 1405) provides an alternative route to LXXXIX (X = O). With carbon disulphide 2,2′-diaminobiphenyl yields 6,7-*dihydro*-5H-*dibenzo*[d,f]-1,3-*diazepine*-6-*thione* (LXXXIX; X = S), m.p. 243° (*Kreighbaum* and *Scarborough*, *loc. cit.*; *R. J. W. LeFevre*, J. chem. Soc., 1929, 733).

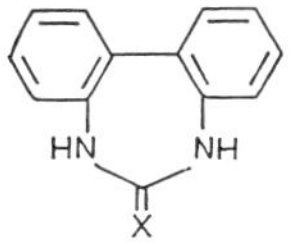

(LXXXIX)

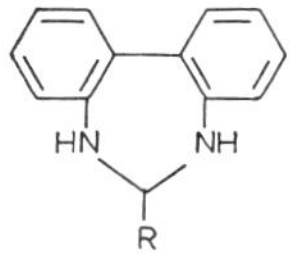

(XCI)

Spectroscopy shows that compound LXXXIX (X = O) does not exist in the enol form LXXXVII (R = OH). With benzoyl chloride diazepinone LXXXIX (X = O) gives a 5,7-dibenzoyl derivative (*O. Christmann*, Chem. Ber., 1965, **98**, 1282); it is readily brominated at position 2 giving a *compound*, m.p. 277–278°, also prepared from 5-bromo-2,2′-diaminobiphenyl and phosgene (*M. Murakami* and *I. Moritani*, J. chem. Soc. Japan, 1949, **70**, 236; C.A., 1951, **45**, 4698). Both ketone and thione undergo semidine rearrangements in dilute acid (*P. C. Guha* and *H. K. Benerjee*, J. Indian Inst. Sci., 1928, **11A**, 231). Treatment of thione (LXXXIX; X = S) with methyl iodide yields the methylthio derivative

LXXXVII (R = SMe) which reacts with amines to give the 6-aminodiazepines LXXXVII (R = NHR^1, NR^1R^2). These cyclic guanidines break down to the ketone LXXXIX (X = O) in butanol (*Kreighbaum* and *Scarborough*, *loc. cit.*).

The 6,7-dihydrodiazepine XCI (R = OEt) may be isolated during condensation between ethyl orthoformate and 2,2′-diaminobiphenyl (*Ried* and *Storbeck*, *loc. cit.*), and the diamine with glyoxal monoacetal gives compound XCI [R = $CH(OEt)_2$] (*N. Vinot*, Compt. rend., 1961, **253**, 2986). Similarly chloral yields XCI (R = CCl_3) (*R. Weiss* and *L. Chledowski*, Monatsh., 1935, **65**, 357).

(*v*) *1,4-Diazepines*

The family of the 1,4-diazepines and their benzo derivatives has been more extensively explored than any other group among the medium and large heterocyclic ring systems. A considerable number of synthetic routes have been discovered and used to build up a large catalogue of compounds, and the chemistry of some of the simpler members has been systematically investigated.

Despite this work, a fully unsaturated 1,4-diazepine without fused rings has still to be obtained. Dihydro compounds XCII resist dehydrogenation, and hydroxy compound XCVIII (R = Me, R^1 = OH), prepared from benzil, acetylacetone and ammonium acetate, resists dehydration. 2,3-**Dihydro**-(1*H*)-1,4-**diazepine** (XCII; R = H) has however been prepared from ethylenediamine and 1,1,3-trimethoxypropene (*R. B. Lund*, Ph. D. Dissertation, U. of Washington, 1959).

The reaction between a 1,2-diamine and 1,3-diketone is one of the very few general synthetic routes to highly unsaturated seven-membered heterocycles.

Ethylene diamine reacts with acetylacetone to give 5,7-*dimethyl*-2,3-*dihydro*-(1H)-1,4-*diazepine* (XCII; R = Me), m.p. 93–95°; *perchlorate*, m.p. 140° (*M. A. Rosanov*, J. Russian phys. chem. Soc., 1915, **47**, 611; *G. Schwartzenbach* and *K. Lutz*, Helv., 1940, **23**, 1139; *D. Lloyd* and *D. R. Marshall*, J. chem. Soc., 1956, 2597; 1958, 118):

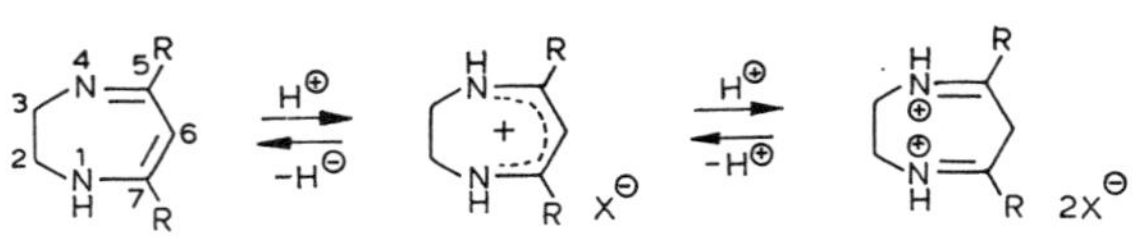

(XCII) (XCIII) (XCIV)

Studies on the synthesis of the analogue XCV, m.p. 246–249° (decomp.), show that at 20° in aqueous conditions a maximum yield of bicyclic compound XCV is obtained at pH 4; at pH 8 the open chain compound XCVI is formed preferentially (*idem, loc. cit.*; cf. *A.* and *C. Combes*, Compt. rend., 1889, **108**, 1252; *L. Rügheimer*, Ber., 1914, **47**, 2759). In acid, XCVI cyclises to XCV. Benzoylacetone and ethylenediamine do not give a 1,4-diazepine as stated earlier, but an open chain *compound*, m.p. 183°; dibenzoylmethane

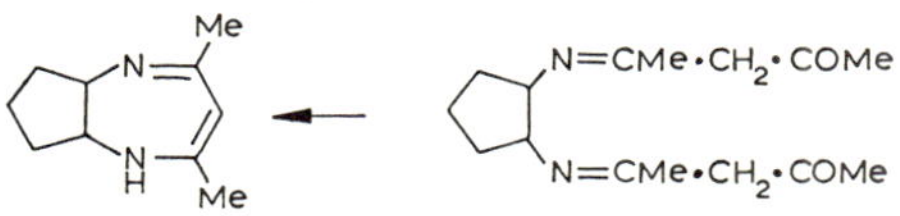

(XCV) (XCVI)

fails to give a diazepine also (*W. Ried* and *W. Höhne*, Chem. Ber., 1954, **87**, 1811; *Lloyd et al.*, J. chem. Soc. C, 1966, 780; *G. O. Dudek* and *R. K. Holm*, J. Amer. chem. Soc., 1961, **83**, 2099).

Malondialdehyde acetal and its 6-substituted derivatives react with *N,N'*-diphenylethylenediamine in perchloric acid to give *N,N'*-diphenyl derivatives of the salts XCIII (R = H, X = ClO_4) (*B. Eistert* and *F. Haupter*, chem. Ber., 1960, **93**, 264). 6-Substituted analogues are obtained similarly (*Lloyd et al.*, J. chem. Soc. C, 1966, 93).

Acetylacetone condenses with *meso*-1,2-diphenylethylenediamine yielding cis-5,7-*dimethyl*-2,3-*diphenyl*-2,3-*dihydro*-(1H)-1,4-*diazepine* (XCVIII; R = Me, R^1 = H), m.p. 202–204° (*H. A. Staab* and *F. Vogtle*, Chem. Ber., 1965, **98**, 2681). The less substituted analogue, 2,3-*diphenyl*-2,3-*dihydro*-(1H)-1,4-*diazepine* (XCVIII; R, R^1 = H), m.p. 198°, is obtained in a thermal valence isomerisation of the dibenzylidene imine of *trans*-1,2-diaminocyclopropane (XCVII) (*idem. ibid.*, pp. 2691, 2701, 3479; Tetrahedron Letters, 1965, 51). The use of analogues of XCVII has shown such isomerisation to be a versatile method of synthesizing 1,4-diazepines.

(XCVII) (XCVIII)

Irradiation of 3,3-dimethyl-2-phenylazirine gives a diimine which reverts to the 1*H*-6,7-**dihydro**-1,4-**diazepine** XCIX (*A. Padwa* and *S. I. Wetmore*, Chem. Com., 1972, 1116), which forms the peroxide C in air:

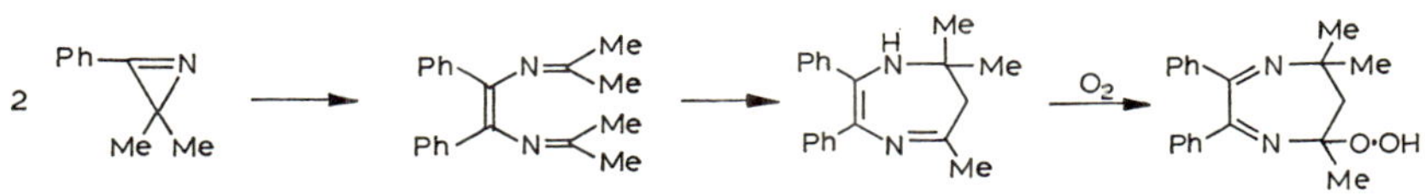

(XCIX) (C)

The 6*H*-tautomer CI was earlier thought to be predominant, but spectroscopic observations confirm that a nitrogen to hydrogen bond is present (*Staab* and *Vogtle*, Chem. Ber., 1965, **98**, 2701). Diazepine CII is stable to acid hydrolysis (*Marshall et al.*, J. chem. Soc. B, 1968, 1536) and readily forms salts with most acids. It is a vinylogous amidine and its

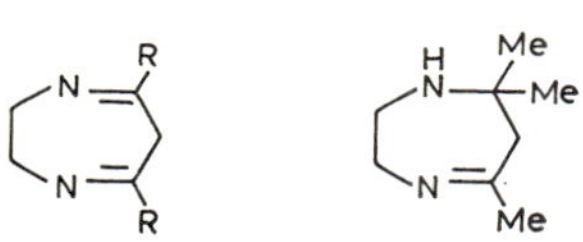

(CI) (CII)

high basicity (pK_a 13–14) comparable with an alkali hydroxide, is probably due to the extremely stable electronic system of the mono-protonated 2,3-dihydro-1,4-diazepinium cation XCIII. This monocation is predominant over a very wide pH range, and 70% sulphuric acid is required to form an appreciable amount of the diprotonated species XCIV (*Schwartzenbach* and *Lutz*, Helv., 1940, **23,** 1162; *Lloyd et al.*, J. chem. Soc. C, 1966, 780). This suggests an effective resonance energy of ~ 80 kJ mol^{-1} (*Lloyd* and *Marshall*, Chem. and Ind., 1972, 335).

Dihydrodiazepines of type CII exhibit no olefinic properties, but in acid conditions (*i.e.* as cations XCIII) undergo typical aromatic substitution reactions. Thus they are readily halogenated (*Lloyd et al.*, J. chem. Soc., 1958, 118; 1964, 2195; C, 1966, 93; 1969, 1449; B, 1971, 1529), nitrated (*idem*, Chem. and Ind., 1968, 130; J. chem. Soc. C, 1970, 617; *C. Barnett*, Chem. Comm., 1967, **637**; J. chem. Soc. B, 1967, 2436) and deuterated (*idem*, *ibid.*, 1968, 1572; *A. R. Butler et al.*, *ibid.*, 1971, 795) at position 6. Although they are cations they couple with diazonium salts at position 6 too (*Lloyd et al.*, Chem. Comm., 1970, 1320). Alternatively many of these substituents may be introduced into the 1,3-diketone before cyclisation.

6-Nitrodihydrodiazepinium salts can be reduced to their 6-amino analogues (*idem*, Chem. and Ind., 1968, 1160; *C. Reichardt*, Ann., 1971, **746,** 207) which form diazonium salts with nitrous acid and undergo the Sandmeyer reaction (*A. M. Gorringe et al.*, J. chem. Soc. C, 1970, 617). By contrast a bromine atom at position 6 is replaced by alkoxide (*Lloyd* and *Marshall*, J. chem. Soc., 1958, 118).

The formation of tetrahydrodiazepines has been reported in the reaction of α,β-unsaturated carbonyl compounds with ethylenediamine; mesityl oxide, for example, gives 5,5,7-*trimethyl*-3,4,5,6-*tetrahydro*-(2H)-1,4-*diazepine* (CII), b.p. 60–61 °/3 mm, n_D^{22} 1.4808 (*L. K. Mushkalo* and *Z. I. Shokol*, Zhur. obshcheĭ Khim., 1960, **30,** 1023). A number of compounds prepared similarly were hydrogenated *in situ* to perhydro derivatives for characterisation (*W. Heinrich* and *W. Heigel*, G.P. 1,047,785/1958; C.A., 1961, **55,** 4552).

Benzoylacetic ester reacts with ethylenediamine in refluxing xylene to give 5-*phenyl*-2,3,4,7-*tetrahydro*-(1H)-1,4-*diazepin*-7-*one* (CIII; R = Ph), m.p. 209–210 °, but acetoacetic ester yields only the bisenamine CIV, previously thought to be CIII (R = Me) (*Reid*

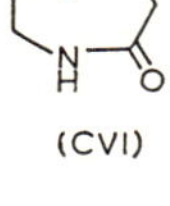

(CIII)

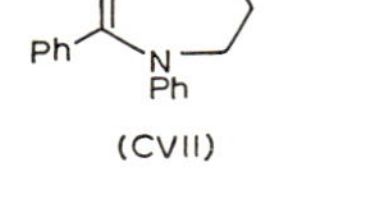

(CIV)

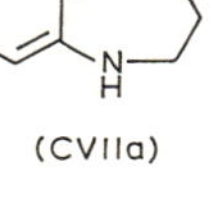

(CV)

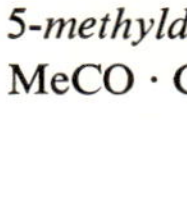

(CVI)

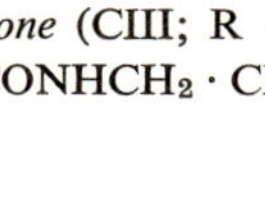

(CVII)

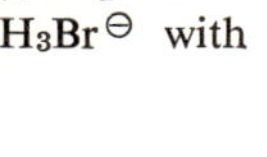

(CVIIa)

and *Höhne*, *loc. cit.*; *C. M. Hofmann* and *S. R. Safir*, J. org. Chem., 1962, **27,** 3565). The 5-*methyldiazepinone* (CIII; R = Me), m.p. 164–165 °, is prepared by cyclising the salt MeCO · CH_2 · $CONHCH_2$ · $CH_2{}^{\oplus}NH_3Br^{\ominus}$ with ammonia. Other β-keto-esters with primary and secondary 1,2-diamines give products analogous to CIII (*Hofmann et al.*, J. chem. eng. Data, 1965, **10,** 188). Protonation of compound CIII (R = Ph) gives a cation CV analogous to XCIII (p. 258) but with CIII (R = Me) protonation takes place at position 6 to give conjugate acid CVI.

1,2,3,4-**Tetraphenyl-**4,5,6,7-**tetrahydro-**(1*H*)-1,4-**diazepine** (CVII), m.p. 222–224°, is prepared by the reaction of the dianion of benzil dianil with 1,3-dihalogenopropanes (*E. J. MacPherson* and *J. G. Smith*, J. org. Chem., 1971, **36,** 2516), and the cyclohexeno-tetrahydro-1,4-diazepine CVIIa, is obtained from cyclohexane-1,2-dione and 1,3-diamino-propane rather than the expected cyclohexano[*b*-1,4]-dihydrodiazepine (*R. H. McDougall* and *S. H. Malik*, J. chem. Soc. C, 1969, 2044).

Unlike the azepanes, the perhydro-1,4-diazepines are less readily prepared by ring-closure than the di- and tetra-hydro- members of this series. Several 1,4-diazacycloheptanes are therefore obtained by hydrogenation of the corresponding unsaturated diazepines.

Hexahydro-1,4-**diazepine** (*homopiperazine*; 1,4-*diazacycloheptane*; 1,4-*diazepane*) (CVIII; R, R[1] = H), m.p. 42°, b.p. 168–170°; *picrate*, m.p. 265° (decomp.), is prepared by condensing 1,3-dibromopropane with bisaryl-sulphonylethylenediamines followed by acid hydrolysis (*L. Bleier*, Ber., 1899, **32,** 1825; *C. C. Howard* and *W. Marckwald, ibid.*, p. 2042; *S. M. MacElvain* and *L. W. Bannister*, J. Amer. chem. Soc., 1954, **76,** 1126). Homopiperazine and several of its homologues are also prepared by reductive cyclisation of amines of the type $H_2N(CH_2)_3NR(CH_2)_2NH_2$ (*A. Kotelko*, Acta Polon.

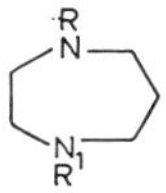

(CVIII)

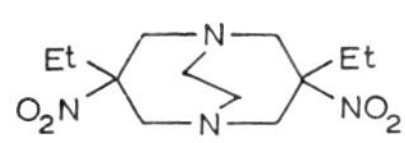

(CIX)

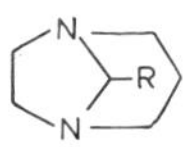

(CX)

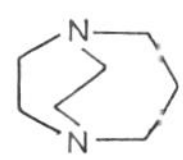

(CXI)

Pharm., 1961, **18,** 171; 1962, **19,** 215; C.A., 1962, **56,** 7131; 1963, **59,** 7370) or *N*-(2-cyanoethyl)ethylenediamine and its derivatives (*F. Popplesdorf* and *R. C. Myerly*, J. org. Chem., 1961, **26,** 131). Cyclodehydration of *N*-(2-hydroxyethyl)-1,3-diaminopropanes also yields 1,4-diazepanes (*T. Ishiguro* and *M. Matsumura*, Yakugaku Zasshi, 1959, **78,** 171; **79,** 302; C.A., 1959, **53,** 13163, 16147). 3-*Ethyl*-3-*nitro*-1,4-*diazepane* (*picrate*, m.p. 149–151°) is obtained by the acid hydrolysis of the condensation product CIX of ethyl-enediamine, formaldehyde and 1-nitropropane (*T. Urbanski* and *R. Kolinski*, Roczniki Chem., 1956, **30,** 201; Bull. acad. polon. sci. 1955, **3,** 493; C.A., 1957, **51,** 1210, 13880).

Bis-quaternary salts of homopiperazines are obtained from *N,N,N',N'*-tetra-substituted ethylenediamines and 1,3-dibromopropane (*J. von Braun* and *O. Goll*, Ber., 1927, **60,** 339; *M. Scholtz*, *ibid.*, 1902, **35,** 3047; *F. F. Blicke* and *E. B. Hotelling*, J. Amer. chem. Soc., 1954, **76,** 2427). 1,1,4,4-Tetramethyl-1,4-homopiperazinium dibromide is readily demethobrominated to 1,4-*dimethyl*-1,4-*diazepane* (CVIII; R, R^1 = Me), b.p. 158–161°/738 mm, *picrate*, m.p. 262–263° (*McElvain* and *Bannister*, *loc. cit.*).

Homopiperazine itself is a water-soluble strong base, which is alkylated, acylated, nitrosated and forms salts at both nitrogen atoms. With formaldehyde in the presence of Raney nickel it yields 1-*methyl*-1,4-*diazepane* (CVIII; R = H, R^1 = Me), b.p. 161–164°/740 mm, and heated in toluene with other aldehydes gives bridged compounds CX (*Popplesdorf et al.*, J. org. Chem., 1961, **26,** 4138; U.S.P. 2,210,336/1963; C.A., 1966, **64,** 745). The bridged compound CXI, obtained by heating 1-(2-bromoethyl)-1,4-diazepane, is pyrolysed to piperazines rather than homopiperazines *(McElvain, loc. cit.)*; it is not dehydrogenated to di- or tetra-hydrodiazepines.

5,7-*Dimethyl*-1,4-*diazacycloheptan*-2-*one*, m.p. 181–182°, is obtained in the reaction of 2,4-diaminopentane and glyoxal bisulphite (*McDougall* and *Malik*, *loc. cit.*).

Schmidt reactions on 4-piperidone and 1-methyl-4-piperidones yield **1,4-diazacycloheptan-5-one** (CXII; R = H) *hydrochloride*, m.p. 223–225°, 1-*methyl*-1,4-*diazacycloheptan*-5-*one* (CXII; R = Me), m.p. 83–84° and analogues.

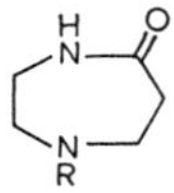

(CXII)

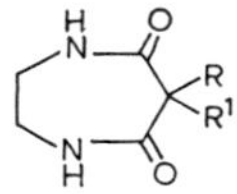

(CXIII)

If unsymmetrically substituted 4-piperidones are used both isomeric products seem to be formed, the preferred reaction course being determined by the substituent at position 5. Frequently only one product is isolated (*S. C. Dickerman et al.*, J. org. Chem., 1949, **14,** 530; 1954, **19,** 1855; 1955, **20,** 206; *A. H. Sommers et al.*, J. Amer. chem. Soc., 1954, **76,** 5805; *P. S. Wadia* and *N. Anand*, J. sci. ind. Res. India, 1958, **17B,** 31; *A. B. Sen* and *K. Shanker*, J. pr. Chem., 1965, **29,** 312; *J. E. Robertson et al.*, J. med. Chem., 1963, **6,** 381; *K. Hohenlohe-Oehringen*, Monatsh., 1965, **96,** 257; *L. A. Paquette* and *M. K. Scott*, J. org. Chem., 1968, **33,** 2379). The Beckmann reaction is generally less satisfactory, but the following reaction scheme shows an unusual, and successful, application (*E. G. Rosantsev et al.*, Izvest. Akad. Nauk S.S.S.R., 1963, **764**; 1962, 2254; Tetrahedron, 1964, **20,** 131; *L. Bauer* and *R. E. Hewitson*, J. org. Chem., 1962, **27,** 3982):

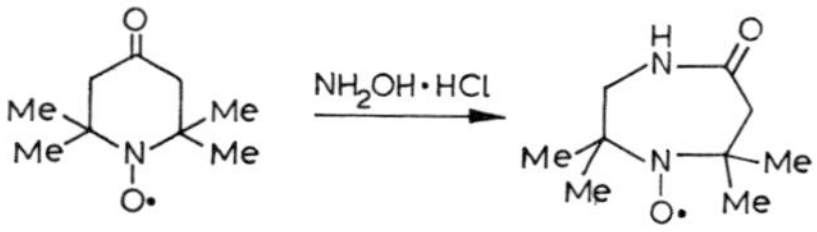

A less successful route to 1,4-diazepan-5-ones is the condensation of ethylenediamine with unsaturated esters.

Methyl acrylate gives a low yield of CXII (R = H) isolated as its *benzenesulphonate*, m.p. 190–191° (*H. K. Hall*, J. Amer. chem. Soc., 1958, **80**, 6404) and methyl crotonate gives a little 7-*methyl*-1,4-*diazepan*-5-*one*, m.p. 113–114° (*W. M. Corbett* and *J. E. McKay*, J. chem. Soc., 1961, 2930; cf. *W. Ried* and *W. Höhne*, *loc. cit.*). Vigorous acid hydrolysis of the diazepanones regenerates ethylenediamine and an unsaturated acid.

The secondary amino group of diazepanone CXII behaves normally, and the oxo group is reduced to methylene by lithium tetrahydridoaluminate. Vigorous acid hydrolysis of 1,4-diazepan-5-ones regenerates ethylenediamine and an unsaturated acid even from those which cannot be synthesised this way. The ring may also be opened by the Hofmann procedure on the 1,1-dimethyl quaternary salt (*Hohenlohe-Oehringen*, *loc. cit.*; *Paquette* and *L. D. Wise*, J. org. Chem., 1965, **30**, 228).

The reaction of malonic esters with ethylenediamine gives **1,4-diazepane-5,7-diones** (CXIII); 6-*ethyl* (R = H, R^1 = Et), m.p. 350–355°, and 6,6-*diethyl* (R,R^1 = Et), m.p. 312–313°, derivatives and several others are made this way (*J. Büchi et al.*, Helv., 1956, **39**, 950; *A. W. Dox*, J. Amer. chem. Soc., 1933, **55**, 3871; *I. J. Pachter* and *J. L. Riebsomer*, J. org. Chem., 1950, **15**, 909). The parent **homopiperazine-5,7-dione** (CXIII; R, R^1 = H), m.p. 142° (decomp.), is prepared from ethylenediamine and malondiamide (*M. Freund*, Ber., 1884, **17**, 133) or carbon suboxide, C_3O_2 (*L. B. Dashkevich* and *V. M. Siraya*, Zhur. obshcheĭ Khim., 1962, **32**, 2330).

(*vi*) *1,4-Benzodiazepines*

Although none of the parent compounds (1*H*- CXIVa; 3*H*- CXIVb; 5*H*- CXIVc) is known an enormous number of substituted 1,4-benzodiazepines have been prepared; *L. H. Sternbach* and his group alone have synthesised nearly 1500 examples. They have particular importance as non-toxic psychosedatives (tranquilising drugs), and seven are currently in clinical use (see p. 276).

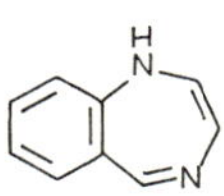

(CXIVa)

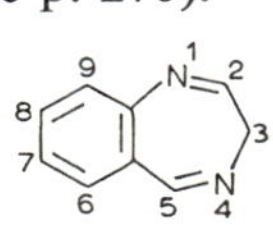

(CXIVb)

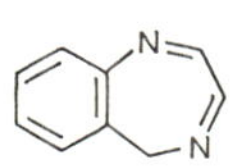

(CXIVc)

3-(Methylaminocarbonyl)-7-nitro-5-phenyl-(1*H*)-1,4-benzodiazepine (CXV), m.p. 287–292°, the only representative of this (1*H*) group, is obtained from 7-nitro-5-phenyl-2,3-dihydro-(1H)-1,4-benzodiazepin-2-one (R = H, R^1 = NO_2) as shown:

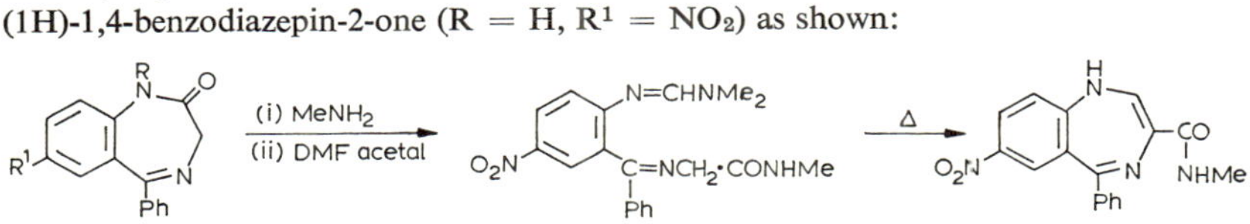

(R = H, R^1 = NO_2)

(CXV)

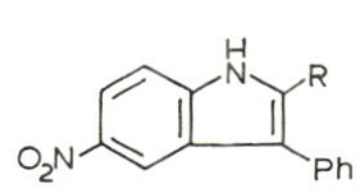

(CXVI)

Compound CXV undergoes ring contraction to indoles CXVI (R = H and R = CO · CONHMe) in acid, and is oxidised to a quinazoline (*Sternbach et al.*, J. org. Chem., 1967, **32,** 3798).

(3*H*)-1,4-**Benzodiazepines.** *Synthesis.* Treatment of 2-chloromethylquinazoline 3-oxides with primary amines is found frequently to give not 2-aminomethylquinazoline 3-oxides but 2-amino-(3*H*)-1,4-benzodiazepine 4-oxides thus:

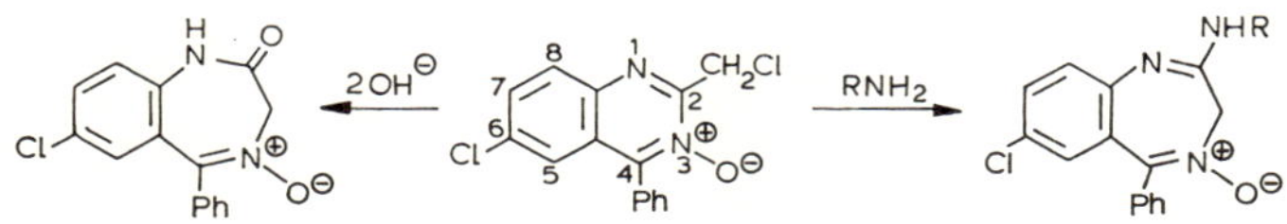

(CXVII)
"Librium" (R = Me)

A very large range of quinazoline *N*-oxides has been used (*idem*, J. org. Chem., 1961, **26,** 1111, 4488; J. pharm. Sci., 1964, **53,** 264; Helv., 1963, **46,** 1720; J. med. Chem., 1963, **6,** 261; *O. Keller, N. Steiger* and *Sternbach*, U.S.P. 3,121,075/1961; C.A., 1964, **61,** 5672; *G. A. Archer* and *Sternbach* F.P. 1,507,878/1966; 1969, **70,** 37854; *M. Gordon et al.*, Arzneim. Forsch., 1963, **13,** 802; *S. C. Bell et al.*, J. med. pharm. Chem., 1962, **5,** 63; J. org. Chem., 1962, **27,** 562). Both substituted and ring-enlarged products are possible, but ring expansion is favoured by non-polar solvents, electron-withdrawing substituents at $C_{(6)}$ and $C_{(8)}$ and a poor leaving group on the exocyclic methylene group. Although rearrangement is general for ammonia and primary amines, more substitution occurs with secondary amines probably due to steric effects (*S. Farber et al.*, J. med. Chem., 1964, **7,** 235; *Hofmann-LaRoche*, Neth. Appl., 6,413,180/1965, C.A., 1965, **63,** 14890; *J. L. Spencer*, U.S.P. 3,462,419/1962; C.A., 1969, **71,** 91549). Hydrazine, but not substituted hydrazines, gives 2-hydroazino(3*H*)-1,4-benzodiazepine 4-oxides (CXVII; R = NH_2) (*Sternbach et al.*, J. chem. Soc. C, 1968, 1103). In the absence of the *N*-oxide function, only aminomethylquinolizines, which are not converted into benzodiazepines, are obtained.

A similar ring-expansion is shown in the following reaction scheme:

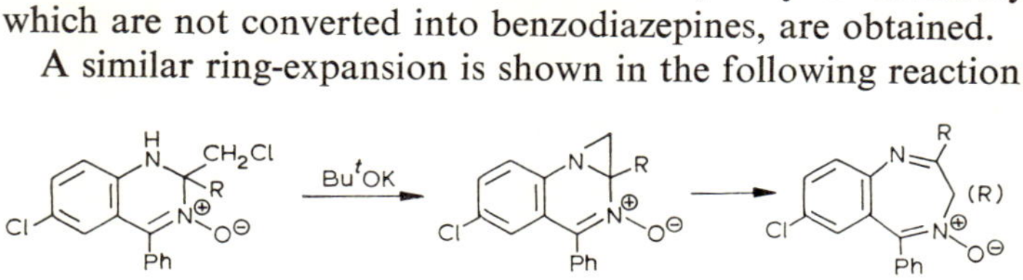

If, in the quinolizine, R = Me, both 2- and 3-methylbenzodiazepines are obtained, and with excess of base diazepine CXVIII is formed (*idem*, J. org. Chem., 1965, **30,** 3957; Tetrahedron Letters, 1966, 2609; J. Amer.

chem. Soc., 1967, **89,** 332). The *N*-oxide function is not, however, necessary here (*G. F. Field* and *E. Reeder*, unpublished information).

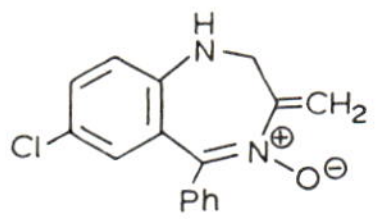
(CXVIII)

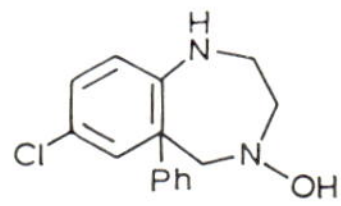
(CXIX)

2-Amino-(3*H*)-1,4-benzodiazepines may be obtained by reacting primary or secondary amines with 2,3-dihydro-(1*H*)-1,4-benzodiazepine-2-thiones or their *S*-methyl derivatives (*Sternbach et al.*, J. org. Chem., 1964, **29,** 231) or by removing the *N*-oxide oxygen with phosphorus trichloride (*idem, ibid.*, 1961, **26,** 1111).

Reactions. Catalytic reduction over Raney nickel hydrogenates the 1,2-double bond after removing oxygen from the *N*-oxide; it also reduces the exocyclic double bond of CXVIII. A platinum catalyst is needed for reduction of the 4,5-double bond (*idem, ibid.*, p. 4936). Lithium tetrahydridoaluminate or sodium tetrahydridoborate reduce **"Librium"** (CXVII; R = Me) to the cyclic hydroxylamine CXIX (*idem*, J. Amer. chem. Soc., 1967, **89,** 332).

Electrolytic reduction of "Librium" probably follows the sequence: deoxygenation at 4, hydrogenation of the 4,5-double bond, hydrogenolysis of the ring between $C_{(3)}$ and $N_{(4)}$ and ring closure to a quinolizine (*H. Oelschlager et al.*, Arch. Pharm., 1963, **296,** 396; 1967, **300,** 250, 817; Coll. Czech. chem. Comm., 1966, **31,** 1264; *B. Z. Sankowski et al.*, Anal chem., 1964, **36,** 1991).

Acetylation of *N*-oxide CXVII causes Polonovski rearrangement to diacetyl derivative CXX. Basic hydrolysis removes first *N*-acetyl and then *O*-acetyl groups, but acid hydrolysis yields the lactam CXXI and subsequently a quinazoline-2-carbaldehyde (*idem*, J. org. Chem., 1961, **26,** 4936; 1964, **29,** 332). Primary amine CXVII (R = H) gives lactam CXXII as well as CXX (R = H) with acetic anhydride (*Bell et al., ibid.*, 1963, **28,** 3010). Acid hydrolysis of deoxygenated "Librium" yields 2-aminobenzophenone, glycine and methylamine, but hydrolysis of the *N*-oxide gives the 2-diazepinone *N*-oxide (*Sternbach* and *Reeder, ibid.*, 1961, **26,** 4936).

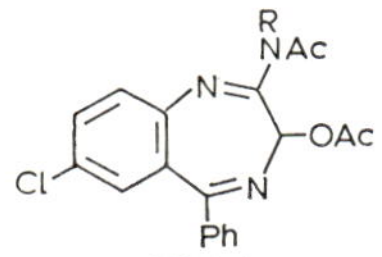
(CXX)

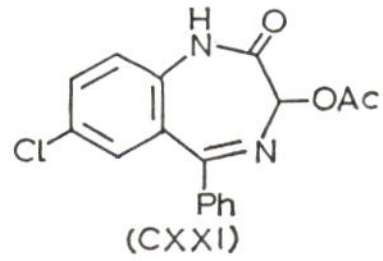
(CXXI)

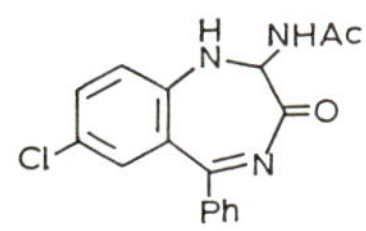
(CXXII)

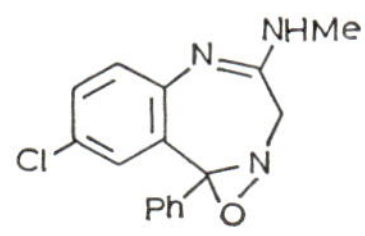
(CXXIII)

As well as undergoing ring-contraction to quinazolines, (3*H*)-1,4-benzodiazepines rearrange to indoles and quinolines (*Sternbach et al.*, *ibid.*, 1972, **37**, 3201) and on irradiation "Librium" yields the oxaziridine CXXIII which isomerises to both a quinoxaline and an oxadiazocine (*idem*, *ibid.*, 1962, **27**, 4671; 1968, **33**, 4438).

2,3-**Dihydro-**(1*H*)-1,4-**benzodiazepines.** *Synthesis*, 2,3-*Dihydro*-(1*H*)-1,4-*benzodiazepine* itself (CXXV), m.p. 244–246°, has been prepared by the reduction of 2,5-dione CXXIV with lithium tetrahydridoaluminate; 2,3,4,5-*tetrahydro*-(1H)-1,4-*benzodiazepine* (CXXVI) *hydrochloride*, m.p. 243–244°, is also obtained in this reaction (*M. Uskokovic et al.*, J. org. Chem., 1962,

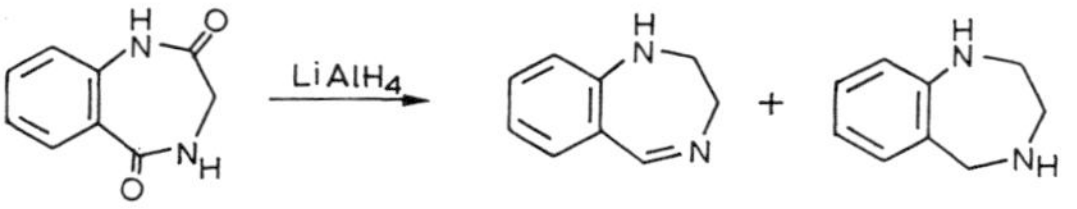

(CXXIV) (CXXV) (CXXVI)

27, 3606). Similar reduction of 5-phenyl-2,3-dihydro-(1*H*)-1,4-benzodiazepin-2-ones gives 5-phenyl analogues of CXXV (*T. S. Sulkowski* and *S. J. Childress*, *ibid.*, 1963, **28**, 2150; *Sternbach et al.*, *ibid.*, p. 2456). Desulphurisation of corresponding 2-thiones with Raney nickel also yields dihydrobenzodiazepines CXXVII (*idem*, *ibid.*, 1964, **29**, 231).

A number of cyclisation routes are also available. 2-Chloro-5-nitrobenzophenone condenses with ethylenediamine to 7-*nitro*-5-*phenyl*-(1H)-2,3-*dihydrodiazepine* (CXXVII; R = H, R^1 = NO_2), m.p. 210–211°; this reaction is less successful with CXXVII (R = H, R^1 = Cl), m.p. 173° (*J. A. Hill et al.*, J. chem. Soc., 1961, 4430). 2-(2-Aminoethylamino)-5-chlorobenzophenone hydrochloride also cyclises when treated with alkali

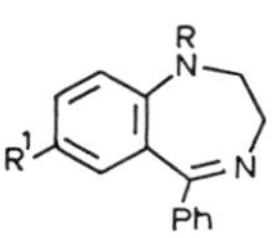

(CXXVII)

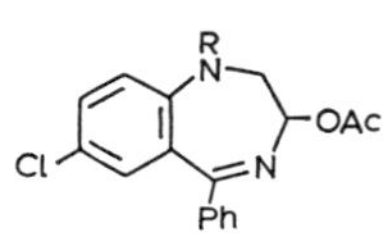

(CXXVIII)

(*Sternbach et al.*, J. org. Chem., 1963, **28**, 2456; 3013). The product from condensing 1-acylamino-2-chloroethane with 2-aminobenzophenone cyclises to CXXVII (R, R^1 = H) on hydrolysis of the amide group, and *N*-alkyl-*N*-aryl-*N'*-benzoylethylenediamines may be cyclised with phosphorus pentoxide and phosphoryl chloride (*H. H. Kaegi*, J. labelled Cpds, 1968, **4**, 363). Catalytic hydrogenation of cyanomethylated 2-aminobenzophenones CXXIX (X = Cl, NO_2, R^1 = $CH_2 \cdot CN$) with Raney nickel in the presence of ammonia also gives CXXVII and its 1-alkyl derivatives (*G. N. Walker et al.*, J. org. Chem., 1972, **37**, 3755; *S. Inaba et al.*, Chem. Ber., 1968, **101**, 4245; Chem. Pharm. Bull. Japan, 1969, **17**, 1263; 1971, **19**, 263, 722).

Reactions. Compound CXXVII (R = H) may be acylated and alkylated at position 1, but 1-alkyl dihydrobenzodiazepines are more easily prepared by alkylation of the corresponding 2-diazepinones and subsequent reduction (*K. Ishizumi et al.*, *ibid.*, 1972, **20,** 1628; *Sternbach et al.*, J. org. Chem., 1962, **27,** 3788, 1963, **28,** 2456). 1-Substituted derivatives are converted into 4-oxides by peracetic acid *via* an oxaziridine, which reforms on irradiation. The *N*-oxide gives a Polonovski rearrangement product CXXVIII with acetic anhydride, which undergoes ring contraction to an indole-2-aldehyde in alkali (*idem*, *ibid.*, 1963, **28,** 2459; 1964, **29,** 1621). An oxo group may be introduced at position 2 by treating XXVII with chromium or ruthenium oxides.

2,3-Dihydro-(1*H*)-1,4-benzodiazepin-2-ones. *Synthesis.* Many routes to this system have been discovered preparing analogues of its most important member "**Valium**" (Diazepam) CXXX (R = Me, R^1 = Cl).

Several syntheses begin with 2-aminobenzophenone CXXIX (R, R^1 = H) or its *N*-substituted derivatives. The bromoacetyl derivatives (CXXIX; R = H, R^1 = $CO \cdot CH_2Br$, X = H, Cl, NO_2) cyclise to diazepinones CXXX (R = H, R^1 = H, Cl, NO_2) in ethanolic ammonia; in liquid ammonia the intermediate CXXIX (R^1 = $CO \cdot CH_2NH_2$) is first obtained, and cyclised by heating (*Bell et al.*, J. org. Chem., 1962, **27,** 562; *Sternbach et al.*, *ibid.*,

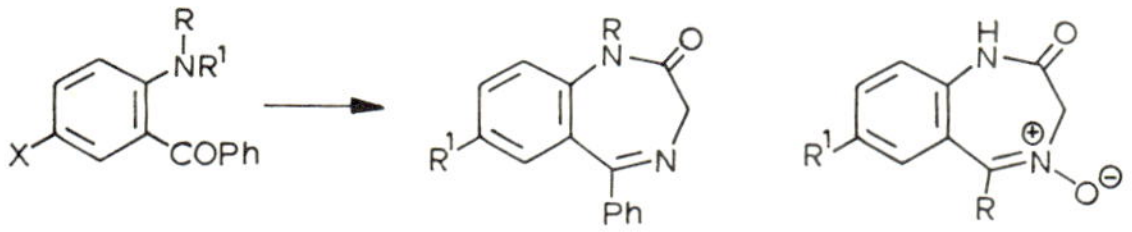

(CXXIX) (CXXX) (CXXXI)

p. 3788; *Walker*, *ibid.*, p. 1929). Esters of glycine hydrochloride will react directly with 2-aminobenzophenones but poorer yields are obtained. *N*-Alkyl-2-aminobenzophenones, and other α-amino acid chlorides or esters lead to a range of 1- and 3-substituted derivatives of CXXX. An *N*-protected amino acid derivative will form an intermediate with 2-aminobenzophenone which cyclises to a diazepinone when the protecting group is removed; trifluoroacetyl, benzoxycarbonyl, phthaloyl and azido groups have been used (*A. Stempel* and *F. W. Landgraf*, *ibid.*, p. 4675; *Delmar Chemicals Ltd.*, Neth. Appl. 6,500,446/1965; C.A., 1966, **64,** 5120; *J. B. Petersen* and *K. H. Lakowitz*, Acta Chem. Scand., 1969, **23,** 971). 3-Amino and 3-hydroxy derivatives of CXXX (R = H; R^1 = Cl) have been prepared from CXXIX (R = H, R^1 = $CO \cdot CH(NH_2)NHAc$, X = Cl) (*Bell et al.*, Tetrahedron Letters, 1965, 2889) and the salts CXXIX [X = Cl, NO_2; R^1 = $CO \cdot CH_2$-(hexamethylenetetraminium)] cyclise to CXXX (R^1 = Cl, NO_2) under physiological conditions (*N. Blazevic et al.*, J. heterocycl. Chem., 1970, **7,** 1173; 1972, **9,** 531); an analogous intermediate cyclises to dihydro-1,4-benzodiazepines CXXVII (R = H, Me, R^1 = Cl, NO_2) (*idem*, *ibid.*, 1971,

8, 845). Modifications of intermediates CXXIX have given diazepinones with a range of substituents at positions 5 and 7 (*Sternbach et al.*, Helv., 1962, **45,** 2226; 1963, **46,** 1720; J. org. Chem., 1962, **27,** 3781; J. pharm. Sci., 1964, **53,** 264; *Gordon et al., loc. cit.*).

A ring expansion analogous to that used to form *N*-oxides CXVII yields CXXXI when 2-chloromethylquinazoline 3-oxides are treated with two moles of base. Many 4-oxides of 2,3-dihydro-(1*H*)-1,4-benzodiazepin-2-ones CXXXI have been prepared this way including CXXXI (R = H) (*Sternbach et al.*, J. org. Chem., 1961, **26,** 4936; 1965, **30,** 4267; 1968, **33,** 2963; *Bell et al., loc. cit.*; *S. S. Walkenstein et al.*, J. pharm. Sci., 1964, **53,** 1181). With just one mole of hydroxide, oxime CXXXII is isolated; a further equivalent of base cyclises the *anti* isomer of CXXXII to the diazepinone,

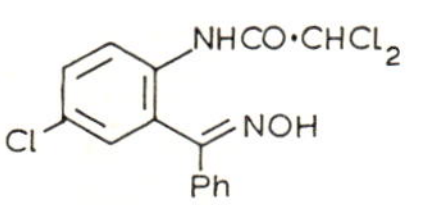

(CXXXII)

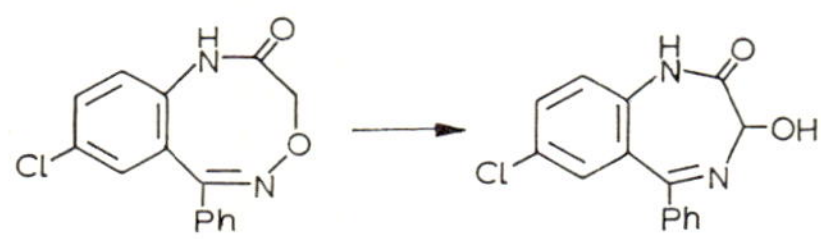

(CXXXIII) (CXXXIV)

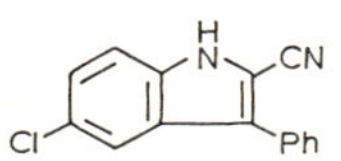

(CXXXV)

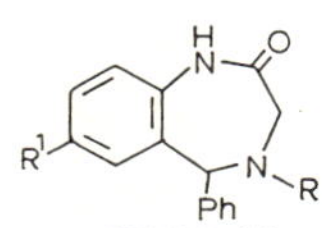

(CXXXVI)

but the *syn*-isomer forms a macrocyclic dimer. Two equivalents of base on oxime CXXXII gives the benzoxodiazocine CXXXIII which undergoes ring contraction to benzodiazepine CXXXIV (*Sternbach et al.*, J. org. Chem., 1967, **32,** 2417); this compound is also obtained *via* a Polonovski rearrangement of CXXXI with acetic anhydride (*Bell* and *Childress*, *ibid.*, 1962, **27,** 1691). *N*-Oxides CXXXI are also obtained by treating diazepinones CXXX with peracids. Indole CXXXV, obtained by treating CXXIX (R = Me, R^1 = $CH_2 \cdot CN$, X = Cl) with acid, also undergoes ring expansion on reduction of the cyano group followed by oxidation to yield CXXX (R = Me, R^1 = Cl) (*Walker, loc. cit.*; *Inaba, loc. cit.*).

Reactions. The 4-oxide is deoxygenated with phosphorus trichloride or hydrogen and a palladium catalyst. Using a platinum catalyst in acetic acid reduces the *N*-oxide CXXXI to CXXXVI (R = OH) and the diazepinone CXXX to CXXXVI (R = H). Halogen atoms at position 7 (R^1) may also be removed (*Sternbach et al.*, J. org. Chem., 1961, **26,** 4936; *Bell et al., ibid.*, 1962, **27,** 562).

Reduction of diazepinones CXXX with lithium tetrahydridoaluminate is sensitive to reaction conditions. At low temperatures the 2-oxo group is reduced to –CHOH–, but at room temperature compound CXXX (R = H) is reduced to the 2,3-dihydro derivative

CXXVII (R = H), and CXXX (R = Me, R^1 = Cl) to the *tetrahydro* derivative CXXXVII (R = Me, R^1 = Cl), m.p. 66–68°, and (R = Me, R^1 = NO_2), m.p. 96–99°, and the insoluble dimer CXXXVIII (*Sternbach et al.*, J. heterocycl. Chem., 1968, **5,** 731; J. org. Chem. 1962, **27,** 4671; 1963, **28,** 2456; 1965, **30,** 1308; *Sulkowski* and *Childress, loc. cit.*;

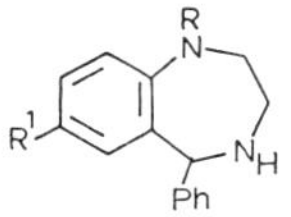

(CXXXVII)

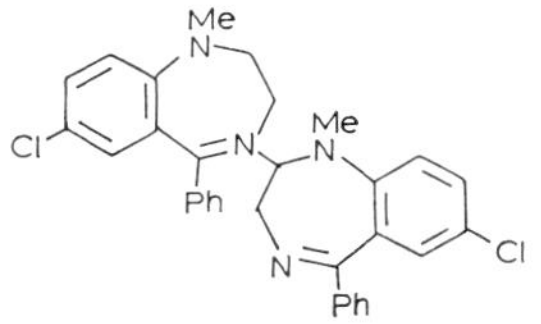

(CXXXVIII)

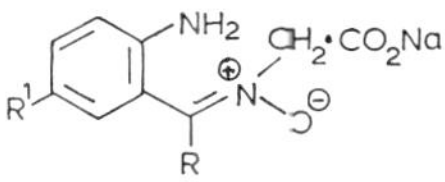

(CXXXIX)

M. Steinman, Ger. Offen., 1,958,742/1969; C.A., 1970, **73,** 66637). To reduce the oxo group to methylene while retaining the 4,5-double bond conversion by phosphorus pentasulphide into the 2-thione followed by desulphurisation with Raney nickel (*A. A. Santilli* and *T. S. Osdene*, J. org. Chem., 1964, **29,** 1998; *Sternbach et al., ibid.*, p. 231) or diborane (*Ishizumi, ibid.*, 1972, **37,** 4111) are recommended. A chloro or nitro group at position 7 survive in reductions using diborane.

Dimethyl sulphate, or methyl iodide in base methylates CXXX (R = H) or CXXXI at position 1, but methyl iodide alone gives the 4-methyl quaternary iodide of CXXX (R = H), which is reduced to compound CXXXVI (R = Me) by tetrahydridoborate (*Sternbach et al.*, J. med. Chem., 1964, **7,** 386). Chloramine and sodium hydride gives the 1-amino derivatives (CXXX; R = NH_2) (*idem*, J. org. Chem., 1965, **30,** 1311).

Diazepinones CXXX and CXXXI are unstable in alkali; the latter give nitrones CXXXIX which recyclise in acid, and the former give the corresponding imines which are cleaved by acid.

Nitration of diazepinones CXXX (R^1 = H) gives the 7-nitro compound CXXX (R^1 = NO_2); more vigorous nitration, or nitration of diazepinones already having 7-substituents introduces a nitro group at position 3′ of the 5-phenyl group, and not at position 9 (*idem*, J. med. pharm. Chem., 1963, **6,** 261; J. chem. Soc., 1963, 4977).

2,3-Dihydro-(1*H*)-1,4-benzodiazepin-2-ones and their 4-oxides undergo a variety of ring-contractions. With acetic anhydride and sulphuric acid quinoline derivatives (*e.g.*

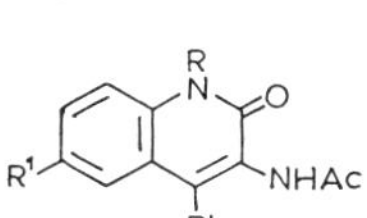

(CXL)

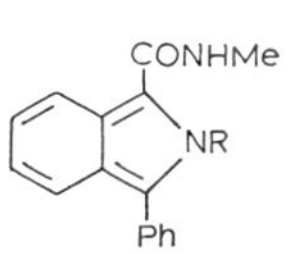

(CXLI)

CXL) are formed from CXXX (*Bell et al.*, J. org. Chem., 1964, **29,** 506; *Sternbach et al., ibid.*, 1965, **30,** 524) and with sodium hydride, CXXX (R = Me) or the corresponding thione yield the isoindole CXLI (R = H) (*idem*, J. Amer. chem. Soc., 1966, **88,** 3173).

The 4-methylazepinium salts of CXXX (R = Me) rearrange to CXLI (R = Me) (*idem*, J. org. Chem., 1969, **34,** 649). If the 1-nitrogen is unsubstituted, acetic anhydride in pyridine brings about this ring-contraction (*idem*, J. chem. Soc. C, 1967, 366). Acid hydrolysis of diazepinones CXXX (R = NH_2) leads to 3-phenylindazoles (*idem, ibid.*,

1965, **30,** 1311). Hydroxy compound CXXXIV contracts to quinazolines in either acid or base. The latter reaction is postulated as a tautomerisation to CXLIII (R = H) followed by a benzilic acid rearrangement thus:

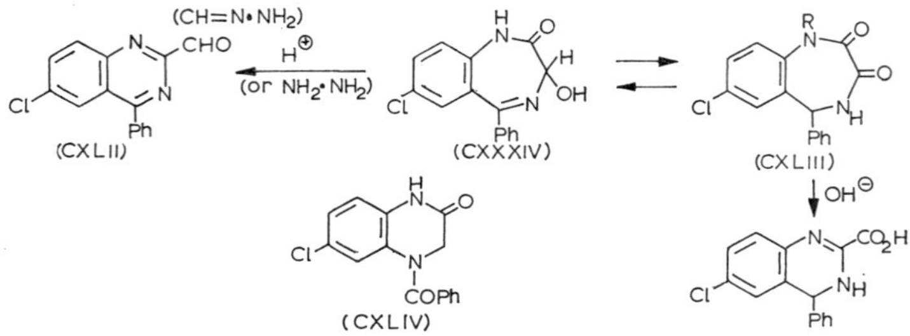

(CXLII) (CXXXIV) (CXLIII) (CXLIV)

If there is a 1*N*-substituent a 1,4- rather than a 3,4-dihydroquinazoline is obtained (*idem, ibid.*, 1964, **29,** 332; *Bell et al., ibid.*, 1962, **27,** 1691; 1964, **29,** 506, 586; J. med. Chem., 1968, **11,** 457).

N-Oxide CXXXI also gives quinazoline CXLII or its methylimine when heated with acetic acid or methylamine, but with toluenesulphonyl chloride or phosphoryl chloride the quinoxalone CXLIV is obtained (*idem*, J. org. Chem., 1964, **29,** 506). On irradiation CXXXI forms epoxy compounds which also undergo ring-contraction to quinazolines (*idem, ibid.*, 1968, **33,** 216; *Sternbach et al., ibid.*, 1970, **35,** 2243).

7-*Chloro*-2-*phenyl*-4,5-*dihydro*-(3H)-1,4-*benzodiazepin*-5-*one* (CXLV), m.p. 160–162°, is prepared as shown:

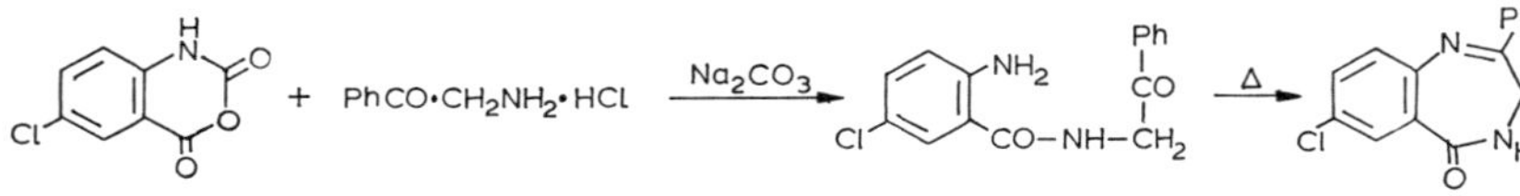

(CXLV)

A limited amount of lithium tetrahydridoaluminate reduces the 1,2-double bond, but more vigorous treatment reduces the carbonyl group as well (*A. A. Santilli* and *T. S. Osdene, ibid.*, 1964, **29,** 1998). Routes to diazepin-5-ones from 2-amino-*N*-(2-hydroxyethyl)benzamides are now known to give oxazolines (*idem, ibid.*, 1965, **30,** 2100; *Sternbach et al., ibid.*, 1965, **30,** 2098).

Treatment of 2-chloromethyl-2-methyl-1,2,3,4-tetrahydroquinazolin-4-one with potassium *tert*-butoxide gave 3-*methyl*-2,5-*dihydro*-(1H)-1,4-*benzodiazepin*-5-*one* (XCLVI), m.p. 156–159°, but the 1-methylquinazolinone gave *benzodiazepinone* CXLVII, m.p. 127–129° (*idem, ibid.*, 1971, **36,** 777). Both are reduced to methyltetrahydrobenzodiazepin-5-ones. If anthranilic acid is condensed with chloral and phenylhydrazine and then

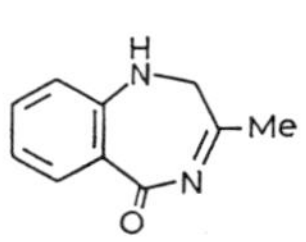

(CXLVI)

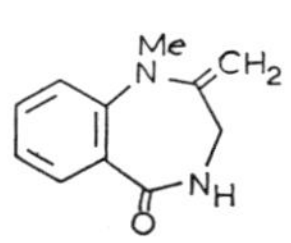

(CXLVII)

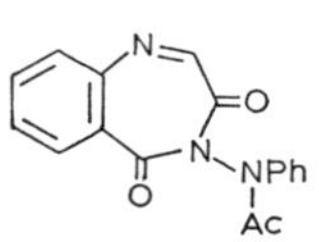

(CXLVIII)

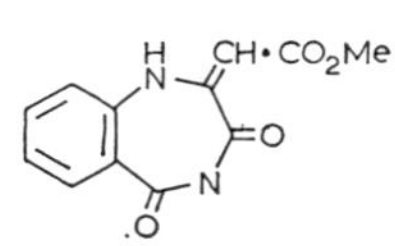

(CXLIX)

acylated, the product is said to cyclise on heating to the 3,5-dione CXLVIII (*S. Gärtner*, Ann., 1904, **332,** 226). The *dione* CXLIX, m.p. 234–236°, is obtained in the reaction of dimethyl acetylenedicarboxylate with 2-aminobenzamide (*N. D. Heindel* and *T. F. Lenke*, J. heterocycl. Chem., 1966, **3,** 389).

Tetrahydro-1,4-**benzodiazepines.** These compounds are frequently made by reducing unsaturated diazepines and diazepinones, but direct routes are also available. The parent (CXXVI, p. 264) is prepared by condensing 2-aminobenzylamine ditosylate with 1,2-dibromoethane and hydrolysing the product (*S. Shiotani* and *K. Mitsuhashi*, Yakugaku Zasshi, 1964, **84,** 656; C.A., 1964, **61,** 10685). Some trisubstituted *N*-phenylethylenediamines are cyclised, rather than *N*-methylated, by formaldehyde and formic acid (*Sternbach et al.*, J. Amer. chem. Soc., 1957, **79,** 5783). Polyphosphoric acid cyclises compounds of type CL (*G. N. Walker et al.*, J. org. Chem., 1972, **37,** 3755)

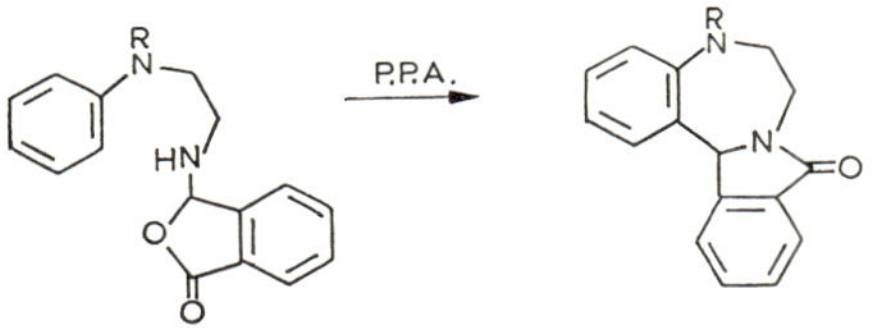

(CL)

while ring expansion reactions as shown below (*Sternbach et al.*, *ibid.*, 1970, **35,** 2455; *J. Schlademan* and *R. Partch*, J. chem. Soc. Perkin I, 1972, 213) yield tetrahydro-1,4-benzodiazepin-2-ones CLI:

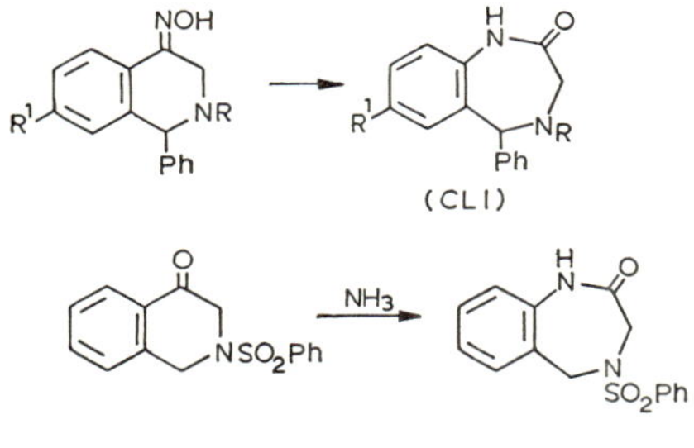

They may also be formed on reacting 2-aminobenzhydrylamine or chloride, respectively with halogenoacetic or glycine esters and thermally cyclising the intermediate (*Sternbach et al.*, J. org. Chem., 1965, **30,** 521, 1308). 2,3,4,5-*Tetrahydro*-(1H)-1,4-*benzodiazepine*-2,5-*dione* (CXXIV), m.p. 327–328°, is prepared by the alternative cyclisations shown (*M. Uskokovic et al.*, *ibid.*, 1962, **27,** 3606):

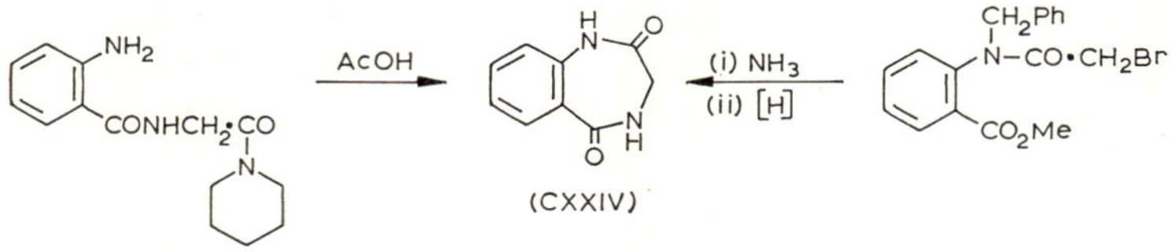

(CXXIV)

A similar route to the *N*-methyl derivative of CXXIV is described (*C. M. Lee*, J. heterocyclic. Chem., 1964, **1**, 235). The biosynthesis of the naturally occurring dione, **cyclopenin** (CLII) has been investigated (*M. Luckner et al.*, Tetrahedron Letters, 1963, 1953; Europ. J. Biochem., 1969, **10**, 268).

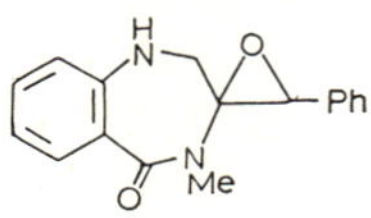
Cyclopenin
(CLII)

The two nitrogen atoms of tetrahydro-1,4-benzodiazepin-2-ones have different basicities and may be alkylated separately (*Sternbach et al.*, J. med. Chem., 1964, **7**, 386); one is found to be planar and the other pyramidal (*J.* and *I. L. Karle*, J. Amer. chem. Soc., 1967, **89**, 804). Nitration takes place in a 5-phenyl substituent rather than at position 7. A 4,5-double bond is introduced into diazepinone CLI (R = H) by chromium, selenium or silver oxides; if a 1-substituent is present a 2,3-dione, *e.g.* CXLIII, is obtained instead (*Sternbach et al.*, J. org. Chem., 1965, **30**, 521; 1308).

(*vii*) *1,5-Benzodiazepines*

Synthetic routes to 1,5-**benzodiazepines** and their derivatives are reminiscent of those used to prepare 2,3-dihydro-(1*H*)-1,4-diazepines, with *o*-phenylenediamine replacing aliphatic 1,2-diamines.

The most general synthesis of 1,5-benzodiazepines, which if unsubstituted on nitrogen have been shown by spectroscopy to exist in the 3*H*- (CLIII) rather than the 1*H*-form (CLIV; R = H) (*J. A. Barltrop et al.*, J. chem. Soc., 1959, 1132; 1423; Chem. and Ind., 1957, 466, 1011; *W. J. Barry et al.*,

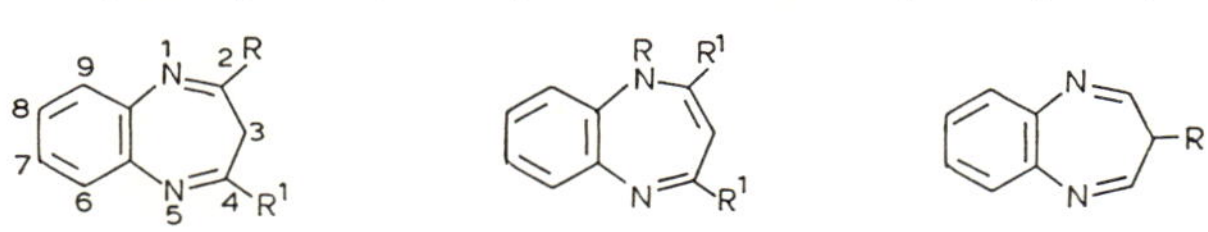
(CLIII) (CLIV) (CLV)

Spectrochim. Acta, 1965, **21**, 1095), is the condensation of a 1,3-dicarbonyl compound with an *o*-phenylenediamine (*J. Thiele* and *G. Steimmig*, Ber., 1907, **40**, 955; *J. Perello, J. Bashilini* and *H. Urcutin*, Bull. Soc. Chileno Quim., 1960, **10**, 1; C.A., 1962, **56**, 5907; *A. Becker*, Helv., 1949, **32**, 1584; *T. N. Ghosh*, J. Indian chem. Soc., 1938, **15**, 89).

N-Substituted *o*-phenylenediamines give only benzimidazoles (*W. Ruske* and *G. Grimm*, J. pr. Chem., 1962, **18,** 163). Acetylacetone gives a maximum yield of 2,4-*dimethyldiazepine* (CLIII; R, R^1 = Me), m.p. 131–132°, at pH 5.8; at higher pH levels the yield falls to zero (*C. A. C. Haley* and *P. Maitland*, J. chem. Soc., 1951, 3155). Benzoylacetone forms a mono-enamine at $C_{(2)}$ in neutral conditions, which cyclises to LI (R = Me, R^1 = Ph), m.p. 87–88°, in the presence of acid (*W. Ried* and *W. Höhne*, Chem. Ber., 1954, **87,** 1801) but the acyclic product from benzoylacetaldehyde gives a benzimidazole (*Ried* and *P. Stahlhofen*, *ibid.*, 1957, **90,** 825). The sodio derivative of nitromalondialdehyde yields 3-*nitro*-(3H)-1,5-*benzodiazepine* (CLV; R = NO_2), m.p. >360° (*W. H. Stafford et al.*, Chem. and Ind., 1956, 765; *F. E. King* and *P. C. Spensley*, J. chem. Soc., 1952, 2144; *R. M. Acheson*, *ibid.*, 1956, 4731). Phenylmalondialdehyde (hydroxymethylenephenylacetaldehyde) gives the 3-*phenyl* analogue (CLV; R = Ph), m.p. 250° (*H. Rupe* and *A. Huber*, Helv., 1927, **10,** 846). *o*-Phenylenediamine also condenses with acetylpyruvic acid (*J. Schmitt*, Ann., 1950, **569,** 17), and with 1,3,4,6-tetra-ones gives a benzodiazepine rather than a diacylquinoxaline (*I. L. Finar*, J. chem. Soc., 1958, 4094). Similarly with $(EtO)_2C{:}CH \cdot CO \cdot CO \cdot CH{:}C(OEt)_2$, 4,4′-bis(2-ethoxy-(3*H*)-1,5-benzodiazepine) is obtained (*H. D. Stachel*, Chem. Ber., 1962, **95,** 2172).

Low yields of 2,3,4-triaryl-(3*H*)-1,5-benzodiazepines are obtained when aromatic aldehydes react with *o*-phenylenediamines (*N. V. S. Rao* and *C. V. Ratnam*, Proc. Indian Acad. Sci., 1958, **47A,** 77; 1957, **45A,** 253; 1956, **44A,** 331; 1956, **43A,** 173; Current Sci., India, 1955, **24,** 299; *S. Weil* and *H. Marcinkowska*, Rocz. Chem., 1934, **14,** 1312). The reaction of *β*-chlorovinylcarbaldehydes or 2-hydroxymethylenecyclanones with *o*-phenylenediamine also yields 1,4-benzodiazepines (*M. Weissenfels et al.*, J. pr., Chem., 1963, **20,** 117; Z. Chem., 1964, **4,** 458).

The free bases CLIII are colourless (λ_{max} ~265 nm), their mono-acid cations CLVI are purple (λ_{max} ~500) suggesting complete 12π-delocalisation, but in concentrated acids colourless dications CLVII are formed (*Barltrop*, *loc. cit.*; *Barry*, *loc. cit.*; *D. Lloyd et al.*, J. chem. Soc., 1965, 3785; *H. A. Staab* and *F. Vogtle*, Chem. Ber., 1965, **98,** 2701; *S. B.*

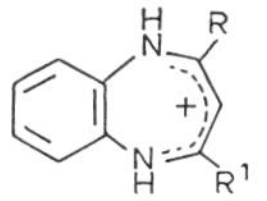
(CLVI)

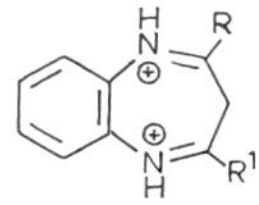
(CLVII)

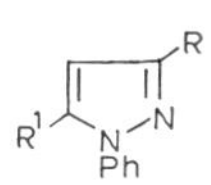
(CLVIII)

Vaismann, C.A., 1940, **34,** 5847; 1944, **38,** 750; *R. F. Witter et al.*, J. biol. Chem., 1948, **176,** 493). The high resonance stability of cation CLVI is reduced by the presence of a methyl group at position 3, or more so, at position 1 (*J. O. Halford* and *R. M. Fitch*, J. Amer. chem. Soc., 1963, **85,** 3354).

The seven-membered ring is easily opened in acid solution or on pyrolysis of a benzodiazepinium salt with formation of a benzimidazole. With phenylhydrazine, compounds CLIII yield pyrazoles CLVIII. In solution the salts may establish a hydrolytic equilibrium with their components, but partial hydrolysis followed by nucleophilic displacement seems a better explanation (*Barltrop, loc. cit.*; *Ried* and *Hohne, loc. cit.*; *Lloyd, loc. cit.*; *Finar, loc. cit.*).

Attempts to remove a hydride ion from CLIII with tritylcarbonium perchlorate left unchanged benzodiazepine, but sodamide and methyl iodide methylate CLIII at position 3 (*Barltrop, loc. cit.*). A methyl group of diazepine CLIII (R = Me; R^1 = Me or Ph) undergoes Claisen condensation with oxalic ester in the presence of base (*S. Veibel* and *S. F. Hromadko*, Chem. Ber., 1960, **93**, 2752) and benzaldehyde gives a 2-styryl derivative, rather than the 3-benzylidene derivative obtained when piperonal is used.

Bromination also occurs at position 3, as does coupling with diazonium salts, but attempts to nitrate and acetylate the diazepine ring failed to give substituted diazepines (*K. V. Levshina et al.*, Khim. geterotsikl. Soedin, 1970, 1133, 1135; *R. B. Lund*, Ph.D. Dissertation, U. of Washington, 1959; *Barltrop, loc. cit.*).

The complex between metallic salts and diazepine CLIII (R, R^1 = Me), unlike those with azepines, show no direct co-ordination between metal and heterocycle thus confirming the delocalisation stability of cation CLVI (R, R^1 = Me) (*B. Emmert* and *H. Gsottschneider*, Ber., 1933, **66**, 1871; *P. W. W. Hunter* and *G. A. Webb*, J. inorg. nucl. Chem., 1972, **34**, 1511; *A. Ouchi*, Bull. chem. Soc. Japan, 1971, **44**, 434).

In attempts to oxidise benzodiazepines CLIII (R, R^1 = Me) or CLV (R = NO_2) with peracids or by photo-oxidation to 1,5-benzodiazepin-3-ones CLIX only quinoxaline derivatives could be obtained; nor could the oxime of CLIX be hydrolysed without ring-contraction. Photo-oxidation of CLIII (R, R^1 = Ph) did give diazatropone CLIX (R, R^1 = Ph), which lost carbon monoxide in acid yielding 2,3-diphenylquinoxaline (*M. Matsumoto et al., ibid.*, 1968, **41**, 2543; 1970, **43**, 1496).

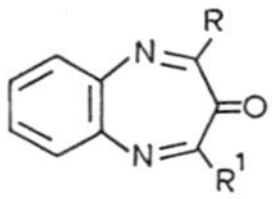

(CLIX)

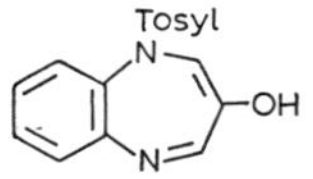

(CLX)

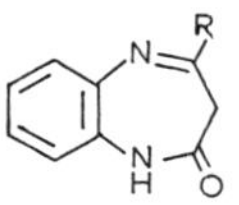

(CLXIII)

Few representatives of the isomeric (1*H*)-1,5-benzodiazepine system (CLIV) are known. Basification of cation CLVI gives a yellow solution believed to contain CLIV (R = H), which soon reverts to (3*H*)-benzodiazepine CLIII (*G. Schwarzenbach* and *K. Lutz*, Helv., 1940, **23**, 1158). Treatment of 1,5-ditosyltetrahydro-1,5-benzodiazepin-3-one with base is said to give compound CLX (*W. Paterson* and *G. R. Proctor*, J. chem. Soc., 1965, 485) and 1,2,4-*trimethyl*-(1H)-1,5-*benzodiazepine* (CLIV; R, R^1 = Me) is obtained in solution in two steps from *N*-methyl-*o*-phenylenediamine and acetylacetone (*Halford* and *Fitch*, *loc. cit.*).

The product formed together with 1-isopropenyl-2-benzimidazolone (CLXI; R = Me) when ethyl acetoacetate reacts with *o*-phenylenediamine in hot xylene is now believed to have the structure 2-*methyl*-4,5-*dihydro*-(1H)-1,5-*benzodiazepin*-4-*one* (CLXII; R = Me), m.p. 151°, rather than the 3*H*-structure CLXIII above (*W. A. Sexton*, J. chem. Soc., 1942, 303; *J. Davoll*, *ibid.*, 1960, 308; *A. Rossi et al.*, Helv., 1960, **43**, 1298). Diketene and other *β*-keto-esters give analogous diazepinones *e.g.* CLXII (R = Ph), m.p. 206°; (R = CF_3), m.p. 184–185°; (R = CH_2Ph), m.p. 147–149° (*W. Ried et al.*, Chem. Ber., 1954, **87**, 1801; 1957, **90**, 825; 828; 1959, **92**, 949, 2902; *F. B. Wigton* and *M. M. Joullie*, J. Amer. chem.

Soc., 1959, **81,** 5212; *Rossi et al.*, Helv., 1960, **43,** 1046; *V. R. Barchet* and *K. W. Merz*, Tetrahedron Letters, 1964, 2239):

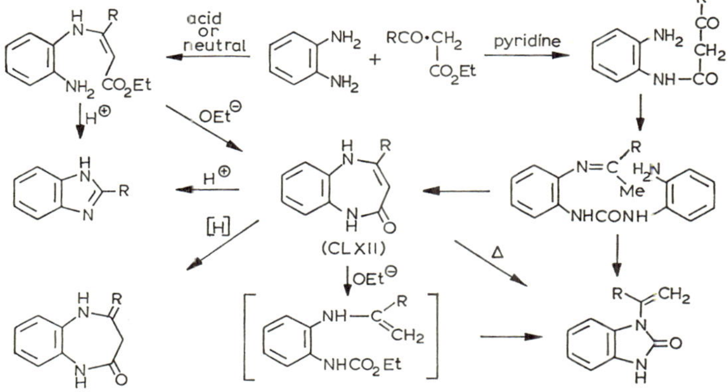

(CLXIV) (CLXI)

Ethyl 2-phenylacetoacetate and 2-ethoxycarbonylcyclohexanone form only benzimidazolones. If acid is present a benzimidazole is obtained, while the addition of base may favour diazepinone formation. The following flow chart indicates the ambiguity in the routes to diazepinone CLXII, and the range of ring contractions it undergoes (see also *O. Hinsberg* and *P. Koller*, Ber., 1896, **29,** 1497; *L. Monti*, Gazz., 1940, **70,** 648; *G. R. Hauser* and *C. A. Reynolds*, J. Amer. chem. Soc., 1948, **70,** 2402; *K. W. Merz et al.*, Naturwiss., 1963, **50,** 663; Ann., 1966, **697,** 193). The thermal contraction CLXII to CLXI) has been suggested as a means of identifying isomeric benzo-1,5-diazepinones (*M. Israel* and *L. C. Jones*, J. heterocycl. Chem., 1969, **6,** 735). Benzodiazepinone CLXV is obtained by condensing the adduct of phosgene and 1-dimethylamino-2-phenylacetylene with *o*-phenylenediamine (*H. G. Viehe*, Angew. Chem. intern. Edn., 1967, **6,** 767).

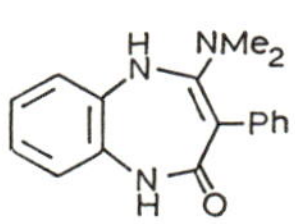

(CLXV)

Diazepinones CLXII are alkylated at position 5 by an alkyl halide and sodamide. These compounds form methiodides and are reduced to tetrahydrobenzodiazepines by sodium tetrahydridoborate (*L. H. Werner*, U.S.P. 2,957,867/1960; C.A., 1961, **55,** 7451). Catalytic reduction of CLXII (R = Me) over Raney nickel yields 2-*methyl*-2,3,4,5-*tetrahydro*-(1H)-1,5-*benzodiazepin*-4-*one* (CLXIV; R = Me), m.p. 186°, identical with that obtained in the reaction between crotonic acid and *o*-phenylenediamine (*Ried et al.*, Chem. Ber., 1957, **90,** 815; 825).

Condensation of *o*-phenylenediamine with α,β-unsaturated aldehydes and ketones, or β-halogenoketones yields 2,3-**dihydro-(1*H*)-1,5-benzodiazepines** CLXVI; *e.g.*, crotonaldehyde gives 2-*methyl* (CLXVI; R = H, R^1 = Me), m.p. 234°; 4-bromopentan-2-one gives 2,4-*dimethyl* (CLXVI; R = Me, R^1 = H) and mesityl oxide or 4-bromo-4-methylpentan-2-one give 2,2,4-*trimethyl* (CLXVI; R, R^1 = Me), m.p. 127°, -2,3-*dihydro*-(1H)-1,5-*benzodiazepines*. A benzimidazole is not the product of the last of these reactions (*J. B. Ekeley* and *R. J. Wells*, Ber., 1905, **38,** 2259; *Ried et al., loc. cit.*; 1959, **92,** 2902; *L. K. Mushkalo*, C.A., 1959, **53,** 18057; *J. Sprague*, P.B. Rept., 135,342; C.A. 1960, **54,** 12155; *R. C. Elderfield* and *J. R. McCarthy*, J. Amer. chem. Soc., 1951, **73,** 975; *W. J. Clark-Lewis et al.*, Austral. J. chem., 1964,

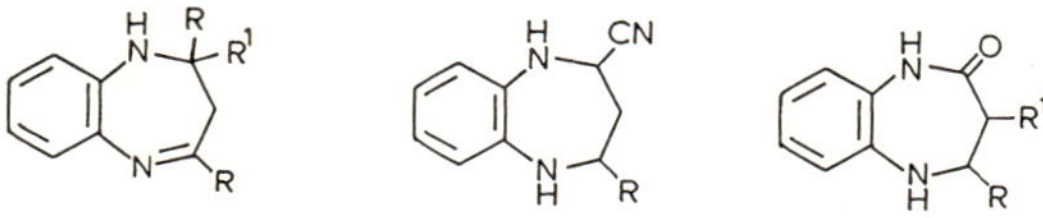

(CLXVI) (CLXVII) (CLXVIII)

17, 877). Although *o*-aminodiphenylamine will form diazepines, cinnamaldehyde and benzalacetophenone yield only linear compounds. In the presence of potassium cyanide, **tetrahydro**-1,5-**benzodiazepines** of type CLXVII; (R = Ph or Me) are obtained (*J. S. Walia et al.*, Chem. Comm., 1972, 962).

With nitrous acid the double bond shifts to 3,4 and a dinitroso compound is formed. A methyl group at position 4 condenses with 4-dimethylaminobenzaldehyde and ethyl orthoformate.

If an α,β-unsaturated acid is used lactams CLXVIII are obtained. Acrylic acid and *o*-phenylenediamine, in a melt or in acid solution give 2,3,4,5-**tetrahydro**-(1*H*)-1,5-**benzodiazepin-2-one** (CLXVIII; R, R^1 = H), m.p. 141–142°, and methacrylic and crotonic acids give the 3-*methyl* (CLXVIII; R = H, R^1 = Me), and the 4-*methyl* (CLXVIII; R = Me, R^1 = H), m.p. 186°, analogues, respectively. Cinnamic acid yields the 4-*phenyl* derivative (CLXVIII; R = Ph, R^1 = H), m.p. 195–196° (*Ried et al.*, Chem. Ber., 1953, **86,** 1101; 1954, **87,** 1801; 1959, **92,** 2902; *G. B. Bachman* and *L. V. Heisey*, J. Amer. chem. Soc., 1949, **71,** 1985; *J. Davoll*, J. chem. Soc., 1960, 308; *S. H. Dandegaonker* and *G. B. Desai*, Indian J. Chem., 1963, **1,** 298).

A Schmidt reaction on 1-methyl-1,2,3,4-tetrahydro-4-quinolone is said to give 1-*methyl*-2,3,4,5-*tetrahydro*-(1H)-1,5-*diazepin*-2-*one*, m.p. 170° (*P. I. Ittyerah* and *F. G. Mann*, J. chem. Soc., 1958, 467) but this product may be

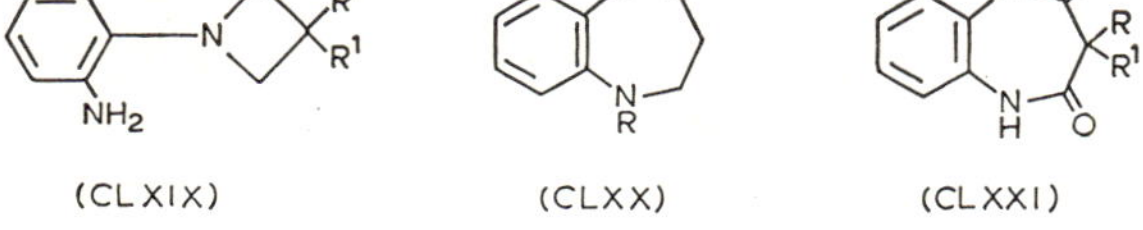

(CLXIX) (CLXX) (CLXXI)

the benzo-1,4-diazepine from the alternative mode of rearrangement (*K. H. Wuensch et al.*, Z. chem., 1970, **10,** 219). Other routes to diazepinones CLXVIII include catalytic hydrogenation of 2-nitroacetoacetanilides (*J. Hornyna*, Czech. P., 113, 422/1963; C.A., 1965, **63,** 18129) and intramolecular transamidation of compounds of type CLXIX (*B. J. R. Nicolaus et al.*, Helv., 1965, **48,** 1867).

Lithium tetrahydridoaluminate reduces compounds CLXVIII to tetrahydro-1,5-benzodiazepines. The nitrogen at position 5 may be nitrosated and the product reduced to an *N*-aminobenzodiazepinone (*Ried* and *A. Draisbach*, Chem. Ber., 1959, **92,** 949); with nitrourea, compounds CLXVIII give 5-carbamoyl derivatives (*O. E. Fancher et al.*, U.S.P. 3,021,325/1962; C.A., 1962, **57,** 4687).

The 1,5-*ditosylate* of 2,3,4,5-*tetrahydro*-(1H)-1,5-*benzodiazepin*-3-*one*, m.p. 182°, is obtained by condensing the dipotassium salt of *N,N'*-ditosyl-*o*-phenylenediamine with 1,3-dibromoacetone (*Patterson* and *Proctor*, *loc. cit.*). 2,3,4,5-*Tetrahydro*-(1H)-1,5-*benzodiazepine* (CLXX; R = H), m.p. 102°, is prepared by hydrolysing the ditosyl derivative (CLXX; R = tosyl) obtained when 1,3-dibromopropane is used in this reaction (*O. Hinsberg* and *A. Strupler*, Ann., 1895, **287,** 220; *H. Stetter*, Chem. Ber., 1953, **86,** 197). Diazepine CLXX (R = H) also forms an *N,N'*-dinitroso derivative.

o-Phenylenediamine condenses with malonic acid or ester and their substituted derivatives to give 2,3,4,5-**tetrahydro**-(1*H*)-1,5-**benzodiazepine**-2,4-**dione** (CLXXI; R, R^1 = H) sublimes ~360°, and its derivatives, *e.g.* the 3-*methyl* (LXXI; R = H, R^1 = Me), m.p. 326°, and 3,3-*dimethyl* (CLXXI; R, R^1 = Me), m.p. 313–315°, homologues (*J. Büchi et al.*, Helv., 1956, **39,** 957; *R. Meyer et al.*, Ann., 1903, **327,** 1; 1906, **347,** 17; 1918, **415,** 29; *M. A. Phillips*, J. chem. Soc., 1928, 2393; *R. L. Shriner* and *P. E. Boermans*, J. Amer. chem. Soc., 1944, **66,** 1810; *C. H. Weber et al.*, Ger. Offen. 1966,128/1966; C.A., 1971, **75,** 110348 and earlier patents).

Alkylation of dione CLXXI with alkyl halide and alkali occurs only at nitrogen; mono- and di-alkyl derivatives are obtained. The N,N'-*dimethyl* derivative of CLXXI (R, R^1 = H), m.p. 255°, is said to undergo ring contraction to a benzimidazole in acid (*Shriner* and *Boermans*, *loc. cit.*) but the product may be *N*-acetyl-*N,N'*-dimethyl-*o*-phenylenediamine (*C. W. Smith et al.*, J. Amer. chem. Soc., 1949, **71,** 1082). With *o*-phenylenediamine hydrochloride, the parent CLXXI (R, R^1 = H) gives 2,2'-bismethylenebenzimidazole.

5*H*-**Dibenzo**[*b,e*]-1,4-**diazepines.** Only one dibenzo-1,4-diazepine (CLXXII) is possible and several of its derivatives have been prepared.

A general method of synthesis is the cyclisation of 2-acylaminodiphenylamines with polyphosphoric acid or phosphoryl chloride; 5-*methyl* (CLXXII; R = Me, R^1 = H), m.p. 90–93°, and 5,11-*dimethyl* (CLXXII; R, R^1 = Me), m.p. 96–97°, homologues are prepared this way (*F. Hunziker et al.*,

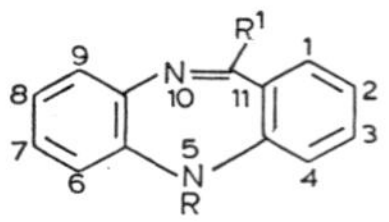

(CLXXII)

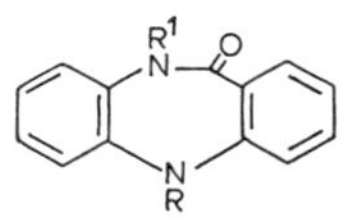

(CLXXIII)

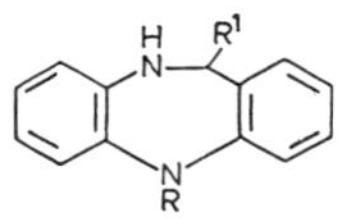

(CLXXIV)

Helv., 1963, **46,** 2337; 1964, **47,** 1163; 1966, **49,** 244; *W. F. Short, C. J. Broderick* and *M. Donaldson,* B.P., 696,473/1953; C.A., 1955, **49,** 384).

The lactams CLXXIII are readily prepared from *N*-(2-aminophenyl)-anthranilic acids or their esters and their *N*-substituted derivatives.

A good yield of 10,11-*dihydro*-5H-*dibenzo*[b,e]-1,4-*diazepin*-11-*one* (CLXXIII; R, R^1 = H), m.p. 257–262°, is obtained this way (*W. H. Perkin et al.*, J. chem. Soc., 1924, 1779, 2451; *H. Burton* and *C. S. Gibson, ibid.*, 2501; *A. M. Munro et al.*, J. med. Chem., 1963, **6,** 255; *A. R. Hanze et al., ibid.*, 767; *G. Schmidt*, G.P. 1,179,943/1962, 1,204,680/1963; C.A., 1965, **62,** 1677, 1966, **64,** 3578; *Hunziker et al.*, Arzneim. Forsch., 1963, **13,** 324).

Lithium tetrahydridoaluminate reduces both diazepines CLXXII and diazepinones CLXXIII to dihydrodiazepines CLXXIV, *e.g.* 10,11-*dihydro*-5H-*dibenzo*[b,e]-1,4-*diazepine* (CLXXIV) (R, R^1 = H), m.p. 204–209°, and its 5-*methyl* (CLXXIV; R = Me, R^1 = H), m.p. 118–120°, and 5,11-*dimethyl* (CLXXIV; R, R^1 = Me), m.p. 129–130°, derivatives. Treatment of CLXXIV with mercuric acetate gave CLXXII.

Acylation and nitrosation of lactam CLXXIII (R, R^1 = H) takes place at $N_{(5)}$, but alkylation in the presence of sodamide occurs on $N_{(10)}$; the 10-*methyl* derivative (CLXXIII, R = H, R^2 = Me) has m.p. 207–208°. The psychotropic drug *Noveril* (CLXXIII; R = Me, R^1 = $CH_2 \cdot CH_2NMe_2$,HCl) and similar compounds are obtained by alkylation, or acylation and reduction (*Dr. A. Wander S.A.*, B.P. 961,105; 959,994; 961,106/1959; 907,646/1962; C.A., 1963, **58,** 11385; 1964, **61,** 5673, 8328, 12020, 13332; *Short* and *Broderick*, B.P. 738,013/1955; C.A., 1956, **50,** 13102; *Parke, Davis and Co.*, Belg. P. 611,926/1962; C.A., 1962, **57,** 13784; *Hunziker* and *O. Schindler*, Helv., 1965, **48,** 1590).

(*b*) *Azepines and diazepines of medicinal value*

An important factor in the growth of interest in seven-membered heterocyclic rings has been the discovery of physiologically active compounds, characterised by greater potency and/or significantly lower addictive or narcotic effects compared with the drugs they replaced. It is hard to believe that when the first edition was published, "**Librium**" (Chlordiazepoxide, Methaminodiazepoxide—introduced 1961) had yet to be prepared, when now it is the most widely prescribed of all drugs for the treatment of mental disorders.

Like Librium, **Valium** (Diazepam) and **Impramine hydrochloride** (Tofranil) are clinically useful psychosedatives, but **Ethoheptazine** is an analgesic, showing the unusual property that both enantiomers are equally active, and **Azapetine** (Illidar) shows a useful peripheral vasodilating effect. The antibiotic, **Anthramycin,** is also a derivative of 1,4-benzodiazepine.

Tranquilising activity in the 1,4-benzodiazepine series seems to be associated with electron-withdrawing substituents in the annellated benzene ring (Cl, NO_2, especially at position 7), a methyl group at $N_{(1)}$ (although a *des*-methyl metabolite of Valium has

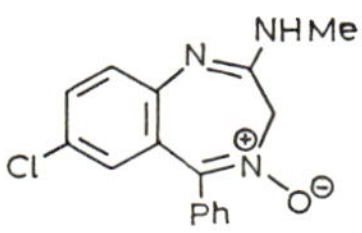

Librium

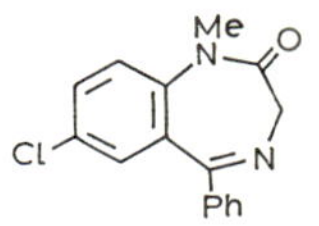

Valium

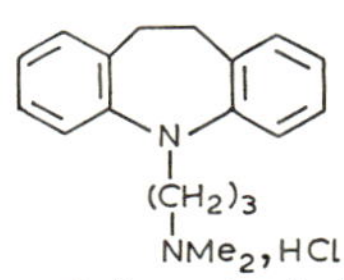

Imipramine hydrochloride

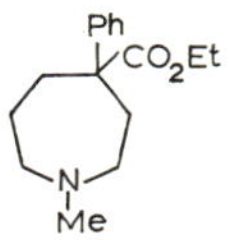

Ethoheptazine

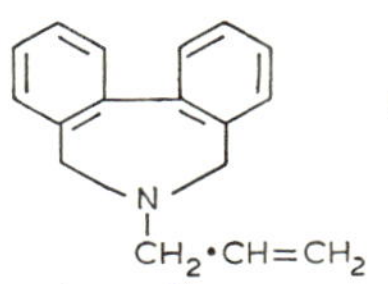

Azapetine

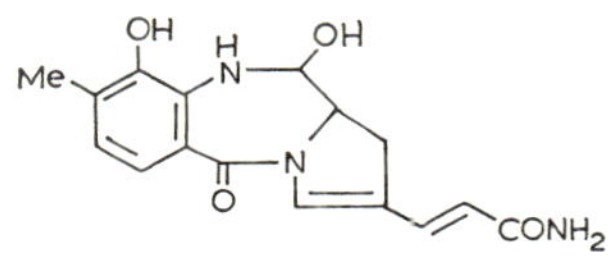

Anthramycin

been suggested as the active agent) and an *ortho*-halogen substituent on the 5-phenyl group. Partial reduction of the heterocyclic ring does not remove activity, but replacement of the nitrogen at position 4 by carbon does inactivate these compounds. Two or three carbon side-chains terminating in a dialkylamino group or a saturated nitrogen heterocycle are also a common feature.

More recently many other medium ring heterocyclic compounds incorporating these features have been synthesised and evaluated as pharmaceuticals with some success.

5. Seven-membered rings containing three or more nitrogen atoms

Four triazepine ring systems are possible, and derivatives of each system are known, but no monocyclic 1,2,3-triazepines have been prepared.

Several 1,2,4-**triazepines** have been prepared from 2-substituted thiosemicarbazides $NH_2 \cdot NRCSNH_2$ (I), and malonic acid derivatives.

With malondialdehyde tetraethyl acetal in acid, 2-*methyl*-3,4-*dihydro*-(2H)-1,2,4-*triazepine*-3-*thione* (II), m.p. 169°, is formed, a tautomer of the fully unsaturated (2*H*)-1,2,4-triazepine system (*B. Stanovnik* and *M. Tisler*, Naturwiss., 1965, **52**, 207). Dialkylmalonyl chlorides are also said to give the triazepines III with reagent I, but malonimides IV appear to fit spectroscopic data better. A by-product, m.p. 208–209°, may be 6,6-*diethyl*-

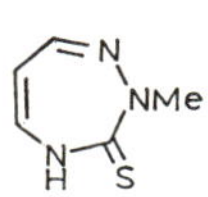

(II)

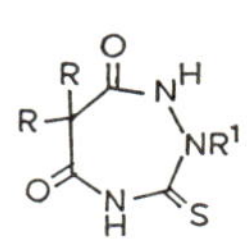

(III)

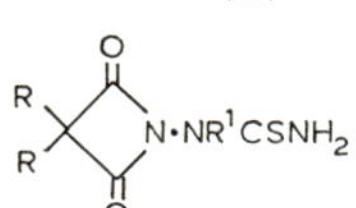

(IV)

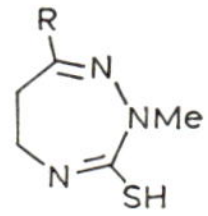

(V)

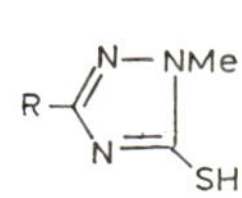

(VI)

5,7-*dioxo*-2-*phenylhexahydro*-1,2,4-*triazepine*-3-*thione* (III; R = Et, R^1 = Ph) but the major product, m.p. 186–187°, is probably the azetidine IV (R = Et, R^1 = Ph). Triazepines from 2-substituted semicarbazides may also be bogus (*G. Losse et al.*, Chem. Ber., 1957, **90,** 257; J. pr. Chem., 1958, **7,** 28; *A. Ebnöther et al.*, Helv., 1959, **42,** 918).

β-Keto-esters also condense with 2-methylsemicarbazide and cyclise to dihydrotriazepines of type V. 2,7-*Dimethyl*- (V; R = Me), m.p. 138°, and 2-*methyl*-7-*phenyl*- (V; R = Ph), m.p. 181°, -3-*mercapto*-5,6-*dihydro*-(2H)-1,2,4-*triazepines* are obtained in this way. They may be *S*-alkylated and they couple with diazonium salts; in ethanolic alkali they undergo ring contraction to triazoles of type VI (*Losse et al.*, Chem., Ber., 1958, **91,** 150; J. pr. Chem., 1959, **8,** 298).

Very little investigation has been made of the 1,2,5-**triazepine** system.

When acetophenone ketazine is brominated, then treated with benzylamine 5-*benzyl*-3,7-*diphenyl*-4,6-*dihydro*-(5H)-1,2,5-*triazepine* (VII) is obtained (*O. Tsuge et al.*, Org. Prep. Proced. Int., 1971, **3,** 289; C.A., 1972, **76,** 126954), and 1,2-dialkylhydrazines con-

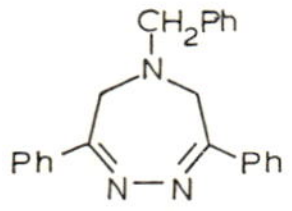

(VII)

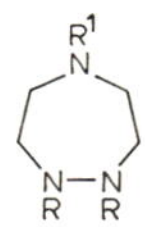

(VIII)

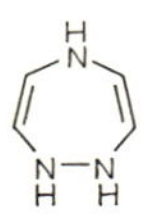

(IX)

dense with di(2-chloroethyl)alkylamines to give 1,2,5-trialkyl-1,2,5-triazepanes VIII (*J. Szotor* and *A. Kotelko*, C.A., 1972, **77,** 152134). M.O. calculations suggest dihydrotriazepine IX has little resonance stabilisation and so is non-planar (*J. P. Dahl* and *A. E. Hanson*, Theoretica Chim. Acta, 1963, **1,** 206).

Unsaturated 1,2,5-triazepines are not known, but the bicyclic "tautomer" X is a stable compound, which shows no evidence of existing in the monocyclic form XI, thus continuing the trend shown by the 1,2-diazepines.

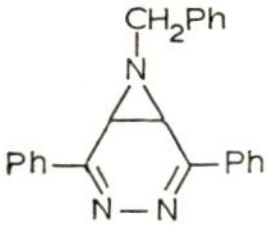

(X)

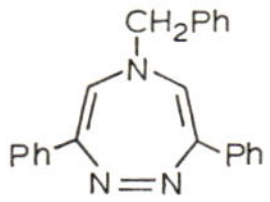

(XI)

On heating, X eliminates benzonitrile, leaving an imidazole corresponding to the pyrrole obtained by heating pentaphenyl-4*H*-1,2-diazepine (*H. W. Heine et al.*, J. Amer. chem. Soc., 1965, **87,** 1050; Tetrahedron Letters, 1967, 4767).

A great many 1,3,5-**triazepanes** have been prepared for evaluation as explosives, plant growth regulators and insecticides. Ethylenedinitramines, (-CHRNH · $NO_2)_2$, condense with formaldehyde and primary amines to give 1,5-*dinitro*-1,3,5-*triazepanes*, *e.g.* the 3-*methyl*- (XII; R = Me), m.p. 156°, and 3-*phenyl*- (XII; R = Ph), m.p. 200–203°, derivatives. With ammonia or diamines *bis* compounds are obtained.

In acetic acid the 3,3′-bismethylene compound cleaves to triazepane XII (R = CH_2OAc) in which the acetate group is replaceable by nucleophiles; with nitric and nitrous acid, respectively, the bis compound yields 1,3,5-*trinitro*-1,3,5-*triazepane* (XII; R = NO_2), m.p. 166°, and 1,5-*dinitro*-3-*nitroso*-1,3,5-*triazepane* (XII; R = NO), m.p. 175° (*F. Chapman et al.*, J. chem. Soc., 1949, 1638; *G. S. Meyers* and *G. F. Wright*, Canad. J. Res.,

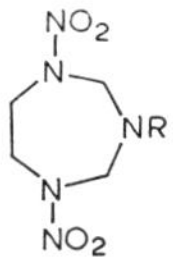

(XII)

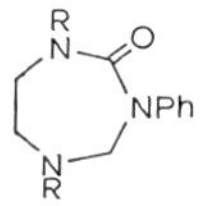

(XIII)

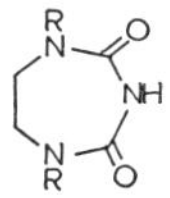

(XIV)

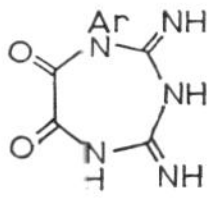

(XV)

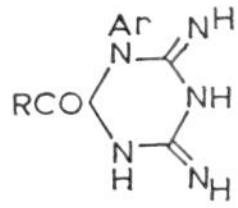

(XVI)

1949, **27B,** 489; *J. A. Bell* and *I. Dunstan*, J. chem. Soc. C, 1966, 862, 867; *N. V. Makarov et al.*, Izvest. Akad. Nauk S.S.S.R., 1967, 1833; 1837; *S. J. Stryker*, U.S.P. 3,280,107/1963; 3,522,241/1967; C.A., 1967, **66,** 28809; 1970, **73,** 66639; *G. J. McDonald* and *M. E. Hill*, J. org. Chem., 1971, **36,** 3846; *M. B. Frankel*, *ibid.*, 1961, **26,** 4709).

Ethylene bis-sulphonamides condense in a similar way to ethylenedinitrimines (*Stryker*, U.S.P. 3,485,817/1966, 3,657,222/1966; C.A., 1970, **72,** 67004; 1972, **77,** 5535).

Phenyl isocyanate reacts with 1,3-dialkylimidazolidines to give **hexahydro-1,3,5-triazepan-2-ones** (XIII) by ring-enlargement; the isothiocyanate reacts similarly yielding the corresponding thione (*H. Boehme* and *W. Pasche*, Arch. pharm., 1969, **302,** 81). 1,2-Diamines condense with alkoxycarbonyl isocyanates or dialkyl iminodicarboxylates to give **triazepinediones** XIV (*H. Hageman*, Ger. Offen., 2,036,172/1970; C.A., 1972, **76,** 127031) and aryl biguanides when heated with diethyl oxalate yield triazepines of type XV. With alkali, alcohols and phenylhydrazine, compound XV undergoes ring-contraction to the *symm*-triazines XVI (R = OH, OR^1 and NH · NHPh) (*M. Ridi et al.*, Ann. Chim., 1954, **44,** 769).

5-*Methyl*- (XVIII; R = Me, X = H), m.p. 119–121°, and 5-*phenyl*- (XVIII; R = Ph, X = H), m.p. 158–160°, -4,5-*dihydro*-(1H)-1,2,5-*benzotriazepin*-4-*ones* have been prepared by hydrolysis and decarboxylation of the corresponding 3-alkoxycarbonyl and 3-cyano derivatives (XVIII; X = CO_2R, CN), formed by the cyclisation of intermediates XVII (X = CO_2R, CN):

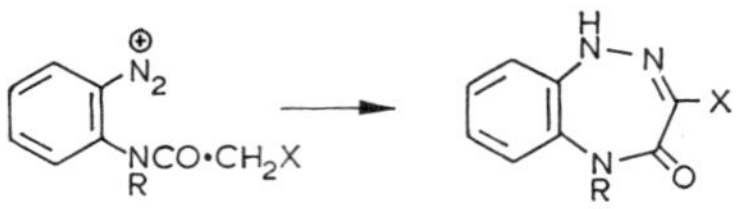

(XVII) (XVIII)

Bromination of triazepinones XVIII (R = Me, X = H or CO_2H) yields the 3-*bromo* derivative (XVIII; R = Me, X = Br), m.p. 166–168°, the bromine atom of which is readily replaced by amino or mercapto groups. Hydrogenation of the triazepinones cleaves the seven-membered ring between $N_{(1)}$ and $N_{(2)}$; in acid ring-contraction to *N*-iminoquinoxalinium ylids occurs. Modification of 3-alkoxycarbonyl or 3-cyano groups does not disrupt the ring (*S. Rossi et al.*, Tetrahedron, 1968, **24,** 6395; B.P. 1,219,844–7/1968; 1,219,813/1967; C.A., 1971, **74,** 141901, 141903, 141904, 141908, 141909; **Ger.** Offen. 2,064,207/1969; 2,135,765/1970; C.A., 1971, **75,** 76854; 1972, **76,** 99722).

The nearest approach to a fully unsaturated 1,3,4-**benzotriazepine** is the tetracyclic compound XIX formed when a 2-*N*-phthalimidobenzophenone reacts with hydrazine (*M. Lamchen et al.*, J. chem. Soc., 1959, 2694; 1960, 1857).

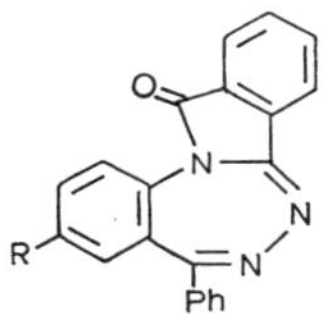

(XIX)

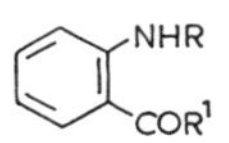

(XX)

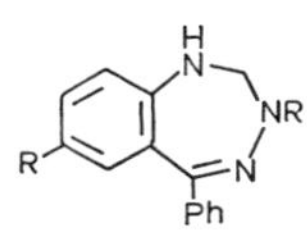

(XXI)

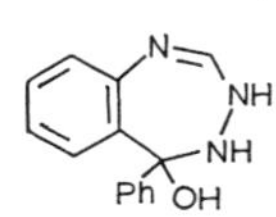

(XXII)

Several other 1,3,4-benzotriazepines are also made from derivatives of 2-aminobenzophenone (XX; R = H, R^1 = Ph). 5-Phenyl-2,3-dihydro-(1*H*)-1,3,4-benzotriazepines (XXI) are prepared by cyclising semicarbazones or phenylhydrazones of XX (R = H, R^1 = Ph) with paraformaldehyde (*T. Ishiwaka et al.*, Bull. chem. Soc. Japan, 1970, **43,** 135) and the 5-*phenyl*-4,5-*dihydro*-(3H)-1,3,4-*benzotriazepine* (XXII), m.p. 179–181°, by cyclising XX (R = CH:C$(CO_2Et)_2$, R^1 = Ph) with hydrazine hydrate (*C. Podesva et al.*, Canad. J. chem., 1969, **47,** 489).

5-*Phenyl*-2,3-*dihydro*-(1H)-1,3,4-*benzotriazepin*-2-*one* (XXIII; R = H), m.p. 238–240°, is obtained in the reactions of XX (R = H, R^1 = Ph) with ethoxycarbonylhydrazine or

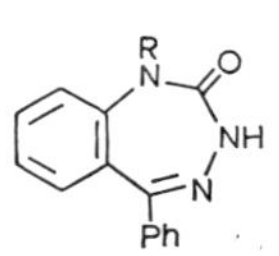

(XXIII)

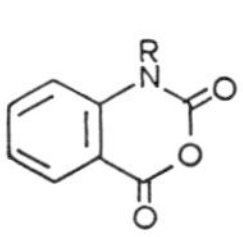

(XXIV)

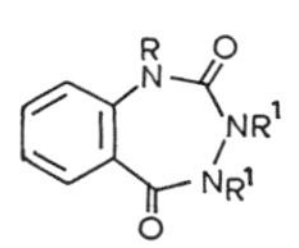

(XXV)

the hydrazone of XX (R = H, R^1 = Ph) with phosgene, and by the fusion of the semicarbazone of XX (R = H, R^1 = Ph) (*T. S. Sulkowski* and *S. J. Childress*, J. med. Chem., 1964, **7,** 386; U.S.P. 3,176,008/1962; C.A., 1965, **62,** 16284; *Ishiyaka, loc. cit.*). The last of these reactions also yields 1-methyl-5-phenyl-1,3,4-triazepin-2-one XXIII (R = Me), m.p. 155°, from the semicarbazone of XX (R = Me, R^1 = Ph) (*S. Toyoshima et al.*, Jap. P. 70 11,148/1960; C.A., 1970, **73**, 25544, 25545). Treatment with ethoxide opens the seven-membered ring to give 2-ethoxycarbonylaminobenzophenone hydrazones.

1,3,4-Benzotriazepines are also prepared from the benzoxazine diones XXIV. With semicarbazide, XXIV (R = H) yields XX (R = $NHCONH_2$) which cyclises on heating to 2,3,4,5-*tetrahydro*-(1H)-1,3,4-*benzotriazepine*-2,5-*dione* (XXV; R, R^1 = H), m.p. 298–300°. Many derivatives are made similarly (*A. L. Langis* and *M. P. Charest*, C.A., 1968, **69**, 36099). Compound XXV (R = H; R^1 = Me) is made by reacting XXIV (R = H) with 1,2-dimethylhydrazine and cyclising the product (XX; R = H, R^1 = MeN · NHMe) with phosgene (*O. Hromatka et al.*, Monatsh., 1969, **100**, 934), and compound XXV (R = Me, R^1 = H) by treating XX (R = Me, R^1 = OMe) with methyl chloroformate and cyclising with hydrazine (*D. M. Bailey* U.S.P. 3,607,866/1968; C.A., 1971, **75**, 140910).

2-Substituted-4,5-dihydro-(1*H*)-1,3,4-benzotriazepin-5-ones XXVI result when the product from XXIV and an acid hydrazide (XX; R^1 = NH · NHCOR^1) is treated with hot polyphosphoric acid (*Langis*, U.S.P. 3,542,767/1968; C.A., 1971, **74**, 88089), and their reduced derivatives XXVII when XX (R^1 = NMe · NHMe) is cyclised with an aldehyde (R^1CHO) (*Hromotka*, Monatsh., 1969, **100**, 941).

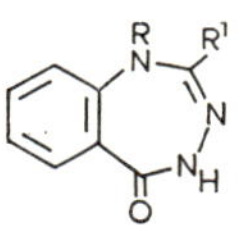
(XXVI)

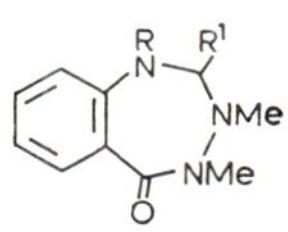
(XXVII)

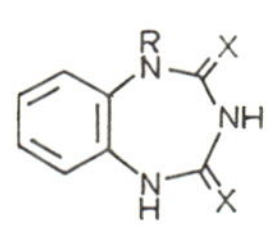
(XXVIII)

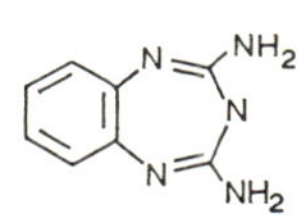
(XXIX)

o-Phenylenediamine reacts with sodium dicyanimide to yield 2,4-*diimino*-2,3,4,5-*tetrahydro*-(1H)-1,3,5-*benzotriazepine* (XXVIII; R = H, X = NH), m.p. 191° (decomp.), which probably exists as the diamino tautomer XXIX. The 1-methyl derivative (XXVIII; R = Me, X = NH) is obtained from *N*-methyl-*o*-phenylenediamine (*R. M. Acheson et al.*, J. chem. Soc., 1948, 1366; 1956, 4727). 2,3,4,5-*Tetrahydro*-(1H)-1,3,5-*benzotriazepine*-2,4-*dione* (XXVIII; R = H, X = O) and analogues are prepared by condensing alkoxycarbonyl isocyanates or dialkyl iminodicarboxylates with *o*-phenylenediamines *(Hageman, loc. cit.)*.

The first known 1,2,3-**triazepine** is *dibenzo*[d,f-1,2,3]*triazepine* (XXX), m.p. 99–100°, obtained by treating tetrazotised 2,2′-diaminobiphenyl with aqueous ammonia. Under non-aqueous conditions the bistriazine XXXI

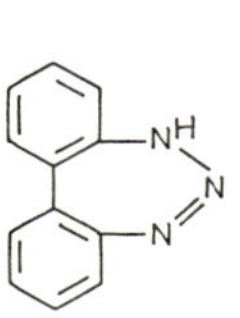
(XXX)

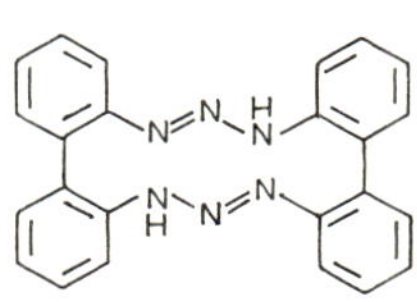
(XXXI)

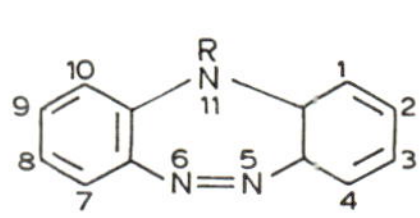
(XXXII)

is obtained. On heating, XXX undergoes ring-contraction to carbazole (*C. W. Rees et al.*, Chem. Comm., 1972, 982).

The reduction of 2,2′-dinitrodiphenylamine with lithium tetrahydridoaluminate yields 11H-*dibenzo*[b,f-1,2,5]*triazepine* (XXXII; R = H), m.p. 166–168° (191°); the N-*methyl* analogue (XXXII; R = Me), m.p. 140–141°, is also prepared this way. The use of zinc in alkali as reducing agent leads to a mixture of XXXII (R = H) and its N-*oxide*, m.p. 199°. The latter also results from the action of peracid on triazepine XXXII (R = H) (*N. L. Allinger* and *G. A. Youngdale*, Tetrahedron Letters, 1959, **9,** 10; J. Amer. chem. Soc., 1962, **84,** 1020; *M. F. Grundon et al.*, J. chem. Soc., 1963, 1436). Dibenzotriazepines with methoxy substituents in one benzene ring are also formed by the intramolecular coupling of diazotised 2-aminodiphenylamines (*E. Kirchner* and *H. Bretschneider*, Monatsh., 1971, **102,** 162).

Reduction of triazepines XXXII with Raney nickel and hydrazine gives 5,6-*dihydro*-11H-*dibenzo*[b,f-1,2,5]-*triazepine*, m.p. 95–115°, and the 11-*methyl* homologue, m.p. 97°, both of which are readily reoxidised to XXXII in air or with mercuric oxide. Other reducing agents cleave the 5,6-nitrogen–nitrogen bond. Parent compound XXXII (R = H) is alkylated by alkyl halides and sodamide (*J. R. Boissier* and *R. Ratouis*, F.P. 1,511,147/1966; F.M. 5,850/1966; C.A., 1969, **70,** 68445; **71,** 13151).

Tetrazepines. The chemistry of the tetrazepines is an almost unexplored field. Monocyclic tetrazepines of types XXXIII and XXIV are said to be formed respectively when sugar hydrazones condense with thiocarbo-

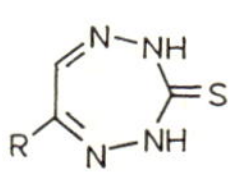

(XXXIII)

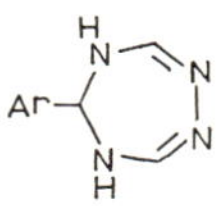

(XXXIV)

hydrazide (*G. Henseke* and *M. Winter*, Chem. Ber., 1956, **89,** 956) and when the bisformamide, $ArCH(NHCHO)_2$, is cyclised with hydrazine (*G. S. Sidhu et al.*, Tetrahedron Letters, 1963, 1637).

3-*Phenyl*-(1H)-*benzo*-1,2,4,5-*tetrazepine* (XXXV), m.p. 118° is prepared as shown:

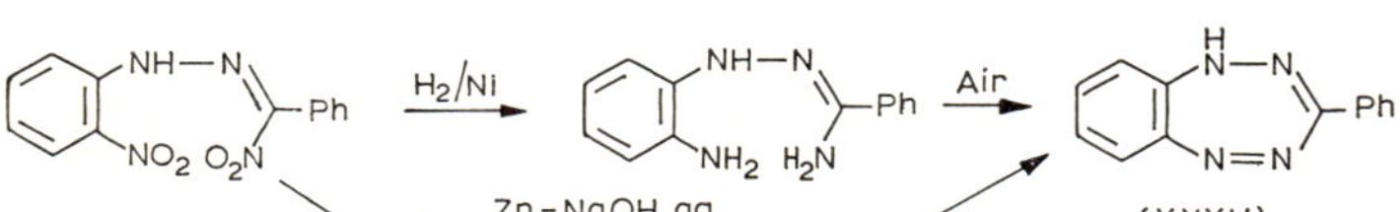

(XXXV)

It absorbs one mole of hydrogen in the presence of palladium, and undergoes ring-contraction in dilute acetic acid to 3-phenyl-1,2,4-benzotriazine (*D. Jerchel* and *W. Edler*, Chem. Ber., 1955, **88,** 1284).

Chapter 52

Compounds with Larger Heterocyclic Rings: Seven-Membered Ring Compounds with Two or More Different Elements in the Ring

J. M. F. GAGAN

1. Rings containing oxygen and nitrogen*

(a) Oxazepines

The most general method of synthesis of oxazepines is in the photolytic ring expansion of pyridine *N*-oxides and their annellated derivatives. The major product is usually a 1,3-oxazepine but the extension of this technique to other six-membered heterocyclic *N*-oxides has given a range of seven-membered heterocycles containing both nitrogen and oxygen. Many examples are given below. The general mechanism for the monocyclic series is as follows:

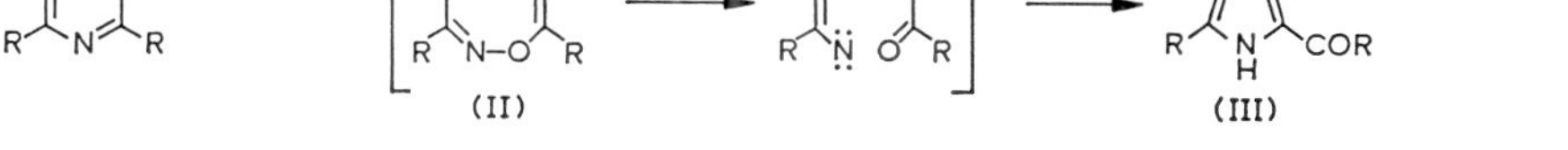

(i) 1,2-*Oxazepines*

Few members of the 1,2-oxazepin series are known. Cyclisation of the complex hydroxylamine, ethyl 5-bromopentanoxycarbamate, $Br(CH_2)_2ONH{\cdot}CO_2Et$, followed by hydrolysis and decarboxylation yields **hexahydro-1,2-oxazepine** (**1,2-oxazepane**) (IV; R, R^1 = H), b.p. 65°/30 mm; *picrate*, m.p. 142–144° (*G. Pifferi* and *P. Consonni*, J. heterocycl. Chem., 1969,

* For a general introduction to heterocyclic compounds with rings comprising seven or more atoms, and for an explanation of the nomenclature used, see p. 139 *et seq.*

6, 111), and although 1-methylpiperidine *N*-oxides and their 2-alkyl homologues do not undergo ring expansion on heating, their 2-aryl analogues are isomerised to 7-aryl-2-methyl-1,2-oxazepanes (IV; R = Me, R^1 = aryl) by pyrolysis (*W. Carruthers* and *R. A. W. Johnstone*, J. chem. Soc., 1965, 1653; *L. D. Quin* and *F. A. Shelburne*, J. org. Chem., 1965, **30**, 3135). Reduction cleaves the N to O bond.

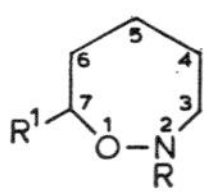

(IV)

(V) (VI)

Irradiation of 9-cyano and 9-chloroacridine-10-oxides in benzene gives 9-*cyano*-(V; X = CN), m.p. 105–109°, and 9-*chloro*-(V; X = Cl), m.p. 41–54°, -*dibenzo* [c,f-1,2]-*oxazepines*. Each is deoxygenated to the corresponding acridine by triphenylphosphine suggesting the existence of an equilibrium between V and the oxirene tautomer VI. Oxazepines V undergo isomerisation with ring contraction and expansion on chromatography or further irradiation, and lithium tetrahydridoaluminate cleaves the N to O bond of V (X = Cl) leaving 2-amino-2′-hydroxydiphenylmethane (*C. Kaneko et al.*, Tetrahedron Letters, 1968, 4519; 1970, 2329; 1972, 971; 977; Chem. pharm. Bull. Japan, 1969, **17**, 1294; Chem. Comm., 1972, 1093; cf. *H. Mantsch* and *V. Zanker*, Tetrahedron Letters, 1976, 4211).

(*ii*) 1,3-*Oxazepines*

As indicated above, fully unsaturated 1,3-**oxazepines** (I) are prepared by photolysis of pyridine *N*-oxides but phenyl or cyano groups adjacent to the oxygen atom appear to be necessary to stabilise the oxazepine ring. Deoxygenation also occurs and 3-hydroxypyridines are found among the products. Some 1,2-oxazepin (II) is believed to be formed but under the reaction conditions it undergoes valence isomerisation to the pyrrole III. Unsymmetrically substituted pyridines give both possible oxazepines (*O. Buchard et al.*, J. org. Chem., 1972, **37**, 3592; Chem. Comm. 1968, 1321; *A. Alkaitis* and *M. Calvin*, *ibid.*, 292; *Kaneko et al.*, Tetrahedron, 1969, **25**, 295; *J. Streith* and *C. Sigwalt*, Tetrahedron Letters, 1966, 1347; Bull. Soc. chim. Fr., 1970, 1157).

X-ray structure determination shows 1,3-oxazepines to be conjugated trienes with a "boat" conformation (*B. Jensen*, Acta cryst., 1972, **B28**, 771), but their chemistry has not been explored.

2(4-*Nitrophenyl*)-4,7-*dihydro*-1,3-*oxazepine* (VIII), m.p. 110–120°, is obtained by heating 1-(4-nitrobenzoyl)-2-vinylaziridine (VII). Brief treatment with acid or alkali opens the ring giving $ArCONHCH_2CH{:}CH{\cdot}CH_2OH$, which on further treatment cyclises to 2-(4-nitrophenyl)-5-vinyloxazoline (IX). With bromine, VIII yields the side chain dibromo derivative of IX. The ring of VIII opens to $ArCONHCH{:}CH{\cdot}CH{:}CH_2$ with methoxide or thiocyanate anions (*P. G. Mente et al.*, J. org. Chem., 1968, **33**, 4547; 1971, **36**, 3078).

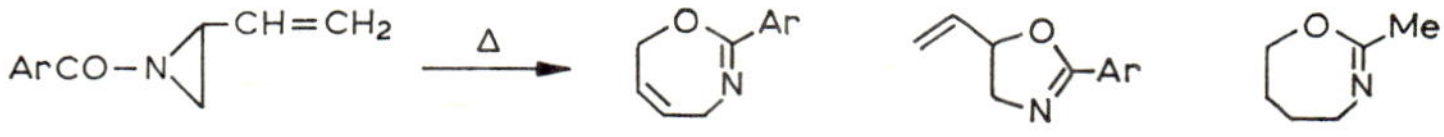

(VII) (VIII) (IX) (X)

(Ar = p-$O_2NC_6H_4$)

Distillation of 4-acetylaminobutanol yields 2-methyl-4,5,6,7-tetrahydro-1,3-oxazepine (X) (*H. G. Trieschmann* and *L. Reuter*, G.P., 920,248/54; C.A., 1958, **52**, 13808).

Trioxymethylene reacts with 4-aminobutanol to give 1,3-**oxazepane** (XI; R = H), b.p. 100–104°/4 mm, n_D^{30} 1.4988, *N-acetyl* (X; R = COMe), b.p. 110–111°/10 mm (*E. D. Bergmann* and *A. Kaluszyner*, Rec. Trav. chim., 1959, **78**, 331):

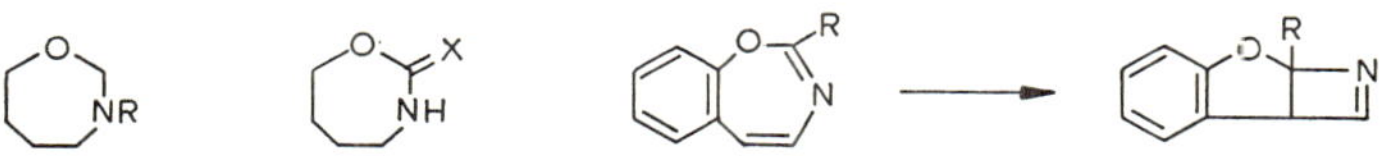

(XI) (XII) (XIII) (XIV)

3-*Phenyl*-1,3-*oxazepane* (XI; R = Ph), b.p. 115°/0.04 mm, can similarly be obtained by condensation of 4-anilinobutanol with formaldehyde (*G. A. R. Kon* and *J. J. Roberts*, J. chem. Soc., 1950, 978). The cyclic carbamate 1,3-**oxazepan-2-one,** (XII; X = O), m.p. 142°, is obtained by heating 2-aminobutanol with ethylene glycol cyclic carbonate (see p. 252) or in the Beckmann rearrangement of tetrahydro-2-pyranone oxime (*E. K. Drechsel*, U.S.P., 2,744,897/1956; 2,806,017/1957; C.A., 1957, **51**, 498; 1958, **52**, 792). It polymerises readily, and with aniline yields 1-phenyl-1,3-diazepan-2-one. An analogue XII (X = N·COPh) results when 4-aminobutanol reacts with *N*-dichloromethylenebenzamide (*J. Burkhardt* and *K. Hamann*, Chem. Ber., 1967, **100**, 2569).

1-Substituted isoquinoline *N*-oxides rearrange on photolysis to 1,3-**benzoxazepines,** *e.g.* 2-*cyano* (XIII; R = CN), m.p. 100–111°, and 2-*phenyl* (XIII; R = Ph), m.p. 56–57°. Oxaziridines are again considered to be intermediates. Solvolysis cleaves the heterocyclic ring giving open chain products, but on further irradiation XIII collapses to the tricyclic product XIV; pyrrolysis reverses this reaction. The ring contraction shown below takes place in acid (*C. Lohse*, J. chem. Soc., Perkin II, 1972, 229; Tetrahedron Letters, 1968, 5625; *Buchardt et al.*, *ibid.*, 1967, 2741; Acta Chem. Scand., 1970, **24**, 268; *Kaneko et al.*, Tetrahedron Letters, 1966, 2145, 4701; 1968, 4073; Chem. Pharm. Bull. Japan, 1965, **13**, 747; 1966, **14**, 555, 1102; *S. Yamada* and *I. Yokoe*, C.A., 1968, **68**, 104845, *J. D. Petrova et al.*, Khim. geterotsikl. Soedin, 1968, 777; C.A., 1969, **71**, 12949):

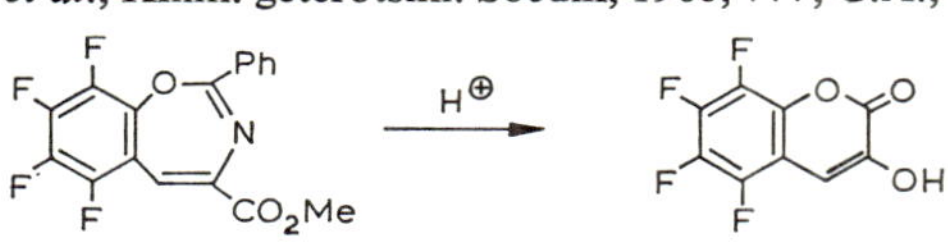

The 4,5-dihydro-1,3-benzoxazepine, XV, is obtained by irradiating an acetonitrile solution of 2-*tert*-butylquinone; in benzonitrile an analogue is formed (*S. Farid*, Chem. Comm., 1970, 303).

(XV) (XVI)

Irradiation of 2-phenyl- or 2-cyano-quinoline *N*-oxides leads to the isomeric 2-*phenyl*- (XVI; R = Ph), m.p. 65–66°, and 2-*cyano*- (XVI; R = CN), m.p. 65–66°, **-3,1-benzoxazepines.** The reaction pathway is analogous to that shown on p. 283 and similar by-products may be isolated. Lithium tetrahydridoaluminate reduces XVI to *N*-benzylindole, and solvolysis yields indole and 2,3-dihydroindole derivatives. Further irradiation of XVI (R = CN) gives 2-cyanoindoles. With methylamine, or acetyl chloride, ring contraction to quinoline derivatives takes place. Oxazepine does not liberate iodine from potassium iodide, as oxaziridines should do, but instead consumes iodine (*Buchardt et al.*, Acta Chem. Scand., 1965, **19**, 1120; 1966, **20**, 262, 2467; 1967, **21**, 1841; 1969, **23**, 159, 2149; Tetrahedron Letters, 1966, 4355; 6221; *Kaneko et al.*, *ibid.*, 1966, 2145, 4701; 1967, 1873; 1968, 4519; Chem. Pharm. Bull., Japan, 1966, **14**, 555, 1102; *Yamada* and *Yokoe loc.cit.*; C.A., 1970, **73**, 98765; *J. Streith et al.*, Tetrahedron Letters, 1966, 5555).

Some members of the **2,4-benzoxazepine** system have also been prepared. 2-(Aminomethyl)benzyl alcohol cyclises with phosgene in alkali to give 1,3,4,5-*tetrahydro*-2,4-*benzoxazepin*-3-*one* (XVII), m.p. 161–163° and its *n*-butyl derivative cyclises to tetrahydro-2,4-benzoxepins (XVIII) with aldehydes under acid conditions. The nitrogen atom of XVII may be acylated and alkylated (*G. Pifferi et al.*, J. heterocycl. Chem., 1972, **9**, 1209):

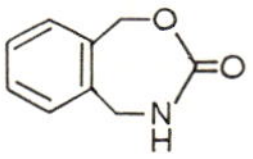

(XVII)

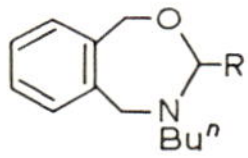

(XVIII)

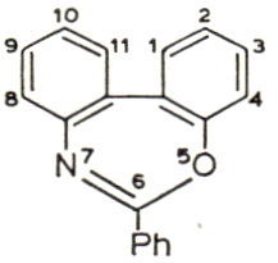

(XIX)

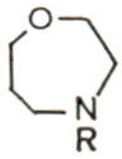

(XX)

Irradiation of 6-phenylphenanthridine *N*-oxide in benzene gives 6-*phenyldibenz* [d,f-1,3]-*oxazepine* (XIX). It is very readily hydrolysed to 2(benzoylamino)-2′-hydroxybiphenyl (*E. C. Taylor* and *G. G. Spence*, Chem. Comm., 1968, 1037).

(*iii*) 1,4-*Oxazepines*

Dehydration of *N*-(2-hydroxyethyl)-*N*-(3-hydroxypropyl) aniline yields 4-*phenyl*-1,4-*oxazepane* (XX; R = Ph), b.p. 155–158°/9 mm, which is unaffected by acids and oxidising agents. It gives a *p*-nitroso derivative with nitrous acid which is decomposed to **1,4-oxazepane (homomorpholine)** (XX; R = H), *hydrochloride*, m.p. 174–175° (*v. Braun* and *O. Braunsdorf*, Ber., 1921. **54**, 685). Similar dehydrative cyclisation routes are known (*H. Siemer*, G.P. 1,109,175/1958; C.A., 1962, **57**, 4687). 1,4-*Oxazepan*-5-*one*, b.p. 165°/15 mm, and its derivatives are obtained by the Beckmann rearrangement of tetrahydro-4-pyrone oximes (*M. I. Farberov et al.*, *E. P. Tepenitsyna* and *N. K. Shemyakina*, Doklady Akad. Nauk S.S.S.R., 1954, **99**, 793; Zhur. obshscheĭ Khim., 1955, **25**, 135; C. *Barkenbus et al.*, J. org. Chem., 1955, **20**, 871).

A benzene ring may be fused to a 1,4-oxazepine ring in three ways and compounds from each of the three series are known.

A general route to 2,3,4,5-**tetrahydro**-1,4-**benzoxazepin**-5-**ones** is the Schmidt or Beckmann ring-expansion of 4-chromanones. The *parent compound* (XXI; R, R^1 = H), m.p. 114–116°, is *N*-alkylated by alkyl halides with a strong base, *e.g. N-methyl* (XXI; R = H, R^1 = Me), m.p. 69–70°, and the *N*-alkyl derivatives are reduced to *N*-alkyl-2,3,4,5-tetrahydro-1,4-benzoxazepines by lithium tetrahydridoaluminate (*I. M. Lockhart et al.*, J. chem. Soc., 1965, 1137, 4806).

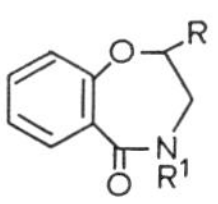

(XXI)

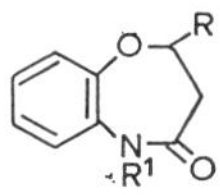

(XXII)

A Schmidt reaction with 2,3-dihydroflavone yields mainly 2-*phenyl*-2,3,4,5-*tetrahydro*-1,4-*benzoxazepin*-5-*one* (XXI; R = Ph, R^1 = H), m.p. 125–126°, and not 2-*phenyl*-2,3,4,5-*tetrahydro*-4,1-*benzoxazepin*-4-*one*, (XXII; R = Ph, R^1 = H), m.p. 141–142°, as previously reported, which is a minor product. Hydrolysis of XXI (R = Ph, R^1 = H) in refluxing hydrochloric acid yields salicyclic acid and phenylacetaldehyde but in cold concentrated sulphuric acid a mixture of oxazoline and isoquinoline derivatives is obtained (*D. Misiti* and *V. Rimatori*, Tetrahedron Letters, 1970, 947; J. heterocycl. Chem., 1972, **9**, 1305; *Lockhart*, Chem. and Ind., 1968, 1844; *J. Krapcho* and *C. F. Turk*, J. med. Chem., 1966, **9**, 191).

Other routes to tetrahydro-1,4-benzoxazepins are all similar. 2-(2-Chloroethoxy)benzyl chloride (XXIII; R = CH_2Cl, X = Cl) on treatment with methylamine and cyclisation with base yields 4-*methyl*-2,3,4,5-*tetrahydro*-1,4-*benzoxazepine* (XXIV; R = R^1 H, = Me), b.p. 104–105°/0.5 mm, n_D^{25} 1.5422; its methochloride is made by cyclising XXIII (R = CH_2Cl, X = NMe_2) (*L. M. Marson*, Farmaco, Pavia, Ed. sci., 1959, **14**, 159; C.A., 1960, **54**, 567; *B. Belleau*, U.S.P. 2,807,628/1957; C.A., 1958, **52**, 2940). Cyclisation of 2-(2-chloroethoxy)acetophenones (XXIII; R = COMe, X = Cl) with amines gives oxazepines XXV, with an exocyclic methylene group, which is reduced to methyl with sodium tetrahydridoborate or catalytically (*K. Schenker* and *J. Druey*, Helv., 1963, **46**, 1696) but the benzophenones (XXIII; X = Cl, R = COPh) with formamide and ammonium formate in hydrochloric acid give 5-phenyl derivatives of XXIV (R, R^1 = H) (*H. A. Luts*, J. pharm. Sic., 1969, **58**, 1460).

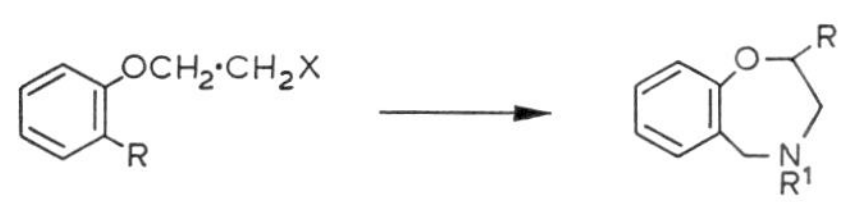

(XXIII) (XXIV)

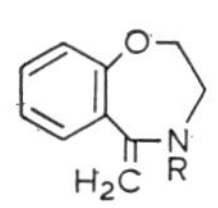

(XXV)

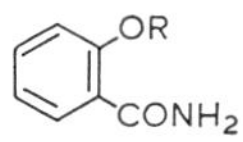

(XXVI)

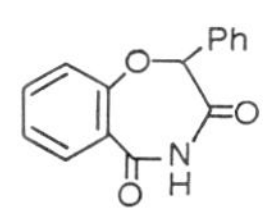

(XXVII)

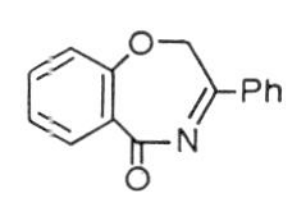

(XXVIII)

Salicylamide with α-chlorophenylacetic acid gives XXVI (R = PhCh·CO_2H), cyclised with acetyl chloride to diketone XXVII which is reduced by lithium tetrahydridoaluminate to 2-*phenyl*-2,3,4,5-*tetrahydro*-1,4-*benzoxazepine* (XXIV; R = Ph, R^1 = H), b.p. 128–135°/0.12 mm, *hydrochloride*, m.p. 235–237° decomp. With phenacyl bromide, salicylamide gives XXVI (R = CH_2COPh) which cyclises with *p*-toluenesulphonic acid to 3-*phenyl*-

2,5-*dihydro*-1,4-*benzoxazepin*-5-*one* (XXVIII), m.p. 166–167°. With sodamide compound XXVIII undergoes ring-contraction to a mixture of 1,2,3,4-tetrahydroisoquinolinediones (*Schenker*, Chimia, 1966, **20**, 157; Ger. Offen., 1,912,546/1968; C.A., 1970, **72**, 31861). Yet another route is as shown (*C. N. Walker* and *R. T. Smith*, J. org. Chem., 1971, **36**, 305).

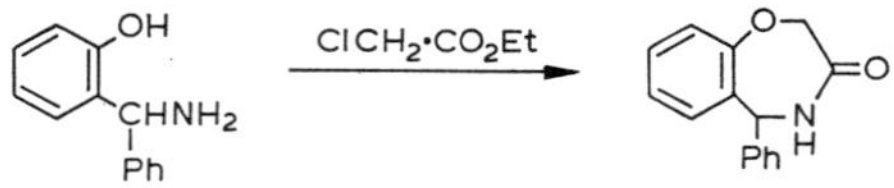

Certain 1,4-benzoxazepines, like their 1,4-benzodiazepine analogues show pharmacological activity.

By contrast with the Schmidt reaction above, treatment of chromanone oximes without 2-substituents with lithium tetrahydridoaluminate gives not the expected amino derivatives but ring-expanded **tetrahydro-**1,5-**benzoxazepines** (XXIX); the *hydrochloride* of the 3-*phenyl* derivative (XXIX; R = H, R^1 = Ph) has m.p. 218–220° (*N. Inone et al.*, Bull. chem. Soc., Japan, 1968, **41**, 2078; *S. Ito*, *ibid.*, 1970, **43**, 1824). 2,3,4,5-*Tetrahydro*-1,5-*benzoxazepine* (2,3-*benzohomomorpholine*) (XXIX; R, R^1 = H), *picrate*, m.p. 190°, is prepared by cyclising *N*-(3-chloropropyl)-2-aminophenol with alkali (*v. Braun* and *Braunsdorf*, *loc.cit.*). The *N-methyl* derivative (XXIX; R = Me, R^1 = H), b.p. 70°/0.5 mm, *hydrochloride*, m.p. 160–161°, is made by reducing 5-*methyl*-2,3,4,5-*tetrahydro*-1,5-*benzoxapin*-4-*one* (XXII; R = H, R^1 = Me), b.p. 94–96°/0.2 mm, *hydrochloride*, m.p. 160–161°. Lactams of type XXII, including 2,3,4,5-*tetrahydro*-1,5-*benzoxazepin*-4-*one* (XXII; R, R^1 = H), b.p. 172–174°/0.7 mm, m.p. 126–127°, are obtained in the thermal cyclisation of 3-(2-aminophenoxy)propionic acids (*Lockhart et al.*, J. chem. Soc., 1965, 1137; Perkin I, 1972, 2425).

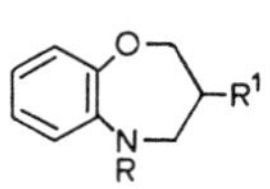

(XXIX)

With phosphorus pentachloride these lactams undergo ring-contraction to benzoxazoles (*F. J. McEvoy* and *G. R. Allen*, J. org. Chem., 1970, **35**, 1183).

Routes to **tetrahydro-**4,1-**benzoxazepines** resemble those used for the 1,4-benzoxazepines. 2-Aminobenzyl alcohol is treated with chloracetyl chloride and the product cyclised with sodium ethoxide to yield the lactam, 1,2,3,5-*tetrahydro*-4,1-*benzoxazepin*-2-*one* (XXX; R = H), m.p. 171–172°, which yields a thione with phosphorus pentasulphide, and 1,2,3,5-*tetrahydro*-4,1-*benzoxazepine* (5,6-*benzhomomorpholine*) (XXXI; R = H), m.p. 58–60°, on reduction with lithium tetrahydridoaluminate; *N*- and *C*-substituted 2-aminobenzyl alcohols give analogues, *e.g.* XXX (R = *Ph*), m.p. 195–196°, and XXXI (R = Ph), m.p. 105–106° (*E. Testa*, *L. Fontanella* and *M. Borara*, Farmaco, Pavia, Ed. sci., 1953, **18**, 815; 1963, **20**, 323; C.A., 1964, **60**, 6847, 1965, **63**, 18088).

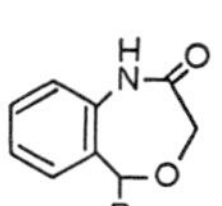

(XXX)

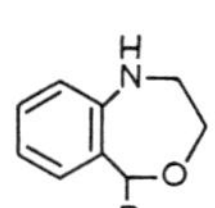

(XXXI)

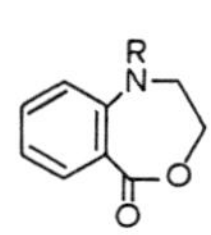

(XXXII)

The lactone 1-*methyl*-1,2,3,5-*tetrahydro*-4,1-*benzoxazepin*-5-*one* (XXXII; R = Me), m.p. 81–82°, is obtained from *N*-methylanthranilic acid and 2-bromoethanol (*E. Uhlig et al.*, Chem. Ber., 1965, **98**, 983), compound XXXII (R = CH_2CH_2OH) from methyl anthranilate and ethylene oxide (*W. C. J. Ross et al.*, J. chem. Soc., 1953, 2386; *A. Kiprianov*, C.A., 1927, **21**, 2467).

Photolysis of acridine *N*-oxide in methanol gives small amounts of 11-*methoxy*-5,11-*dihydrodibenzo* [b,e]-1,4-*oxazepine* (XXXIII; R = OMe, R^1 = H), m.p. 111–115° (*Mantsch et al.*, Ann., 1969, **723**, 95). A spiro compound is postulated as an intermediate (*Kaneko et al.*, Tetrahedron Letters, 1972, 971) similar to the spiro intermediate proposed in the thermolysis of 2′-azido-2,4,6-trimethyldiphenyl ether which gives 2,4-dimethyl-5,11-dihydrodibenzo [*b,e*]-1,4-oxazepine (*J. I. G. Cadogan* and *P. K. K. Lim*, Chem. Comm., 1971, 1431). A range of 5,11-**dihydrodibenzo** [*b,e*]-1,4-**oxazepines** including the *parent* XXXIII (R, R^1 = H), m.p. 118–119°, has been prepared by cyclising (2-2-bromobenzyloxy)formanilides with potassium carbonate and removing the 11-formyl group with strong base.

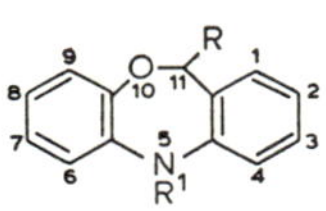

(XXXIII)

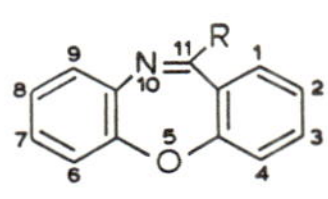

(XXXIV)

The nitrogen atom of XXXIII (R^1 = H) may be alkylated and acylated with ease (*H. L. Yale et al.*, J. med. Chem., 1964, **7**, 609; 1968, **11**, 396; 1970, **13**, 713; U.S.P. 3,454,598/1964; C.A., 1969, **71**, 81453).

Dibenz[*b,f*]-1,4-**oxazepine** (XXXIV; R = H), m.p. 72°, is obtained by cyclising 2-formamidodiphenyl ether with polyphosphoric acid; the 11-*methyl* (XXIV; R = Me), m.p. 86–87°, and 11-*phenyl* (XXIV; R = Ph), m.p. 109–110°, analogues are prepared similarly (*R. Higginbottom* and *H. Suschitzky*, J. chem. Soc., **1962**, 2367). Other routes to 11-phenyldibenz [*b,f*]-1,4-oxazepines include the cyclisation of *N*-(4-phenoxyphenyl)-benzamidines, the reductive cyclisation of 2-benzoyl-2′-nitrodiphenyl ethers (*W. F. Short et al.*, *ibid.*, 1953, 1079), the pyrolysis of 9-azido-9-phenylxanthene (*J. D. Loudon et al.*, *ibid.*, 1958, 1588; *M. M. Coombs*, *ibid.*, 4200) and the pyrolysis of the diazonium salt from 2-phenoxyaniline in benzonitrile (*R. R. Schmidt et al.*, Chem. Ber., 1972, **105**, 1634).

Reduction of XXXIV (R = H or Me) catalytically or with lithium tetrahydridoaluminate yields 10,11-*dihydrodibenz* [b,f]-1,4-*oxazepine*, m.p. 74–75°, and its 11-*methyl* homologue, m.p. 47–48°. These compounds may be *N*-alkylated in the presence of sodamide (*F. Hunziker et al.*, Helv., 1964, **47**, 1163).

Cyclisation of 2-isocyanatodiphenyl ethers with aluminium chloride gives 10,11-*dihydrodibenz* [b,f]-1,4-*oxazepin*-11-*one*, m.p. 215–217°, and many substituted analogues (*idem, ibid.*, 1965, **48**, 336). These lactams are also prepared by the ring closure of 2-amino-2′-methoxycarbonyldiphenyl ethers (*R. Q. Brewster* and *F. Strain*, J. Amer. chem. Soc., 1934, **56**, 117; *M. Tomita* and *T. Kitamura*, J. pharm. Soc. Japan, 1955, **75**, 1138; *M. Allen* and *R. Y. Moir*, Canad. J. Chem., 1959, **37**, 1799).

(*b*) *Oxadiazepines*

Although monocyclic oxadiazepins have been postulated as transient species in the photolysis of pyrimidine *N*-oxides (*J. Streith* and *P. Martz*, Tetrahedron Letters, 1969, 4899) and pyrazine *N*-oxides (*N. Ikekawa et al.*, *ibid.*, 1967, 1197; *Buchardt et al.*, Chem. Reviews, 1970, **70**, 231), reports of their isolation are few.

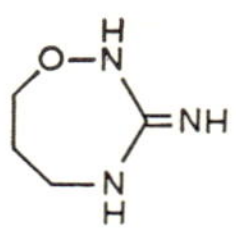

(XXXV)

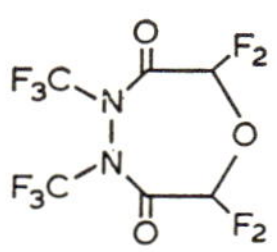

(XXXVI)

Deaminocanavanine is given the 1,2,4-**oxadiazepine** structure XXXV (*S. Makisumi*, J. Biochem., Tokyo, 1959, **46**, 201; C.A., 1959, **53**, 12204) and 3-(2,6-dichlorophenyl)-4,5,6,7-tetrahydro-1,2,4-oxadiazepine has been prepared (*Shell Research Ltd.*, Belg. P. 623, 714/1961; C.A., 1964, **60**, 9299). 3,6-*Dinitro*-1,3,6-*oxadiazepane*, m.p. 148-155°, is found in moist samples of *N,N'*-bis(chloromethyl)-*N,N'*-dinitroethylenediamine, (-$CH_2N(NO_2)$-$CH_2Cl)_2$ (*J. Majer* and *J. Denkstein*, Coll. Czech. chem. Comm., 1966, **31**, 2547) and *perfluoro*-4,5-*dimethyl*-1,4,5-*oxadiazepane*-3,6-*dione* (XXXVI), is obtained from perfluorooxydiacetyl and tetrafluoroformaldazine (*P. H. Ogden*, J. chem. Soc. C, 1971, 2920).

The 1,4,5-**oxadiazepines** of type XXXVII are important intermediates in the asymmetric synthesis of α-amino acids. Compound XXXVII is prepared by condensing an α-keto acid with chiral *N*-amino-2-hydroxy-2-methylindoline, and the amino acid formed on reduction of the carbon to nitrogen double bond and hydrolysis has up to 90% optical purity (*E. J. Corey et al.*, J. Amer. chem. Soc., 1970, **92**, 2476). A 1,2,4-oxadiazepine fused to a five-membered ring and its interesting synthesis are shown in the following reaction scheme:

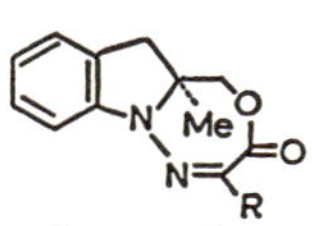

(XXXVII)

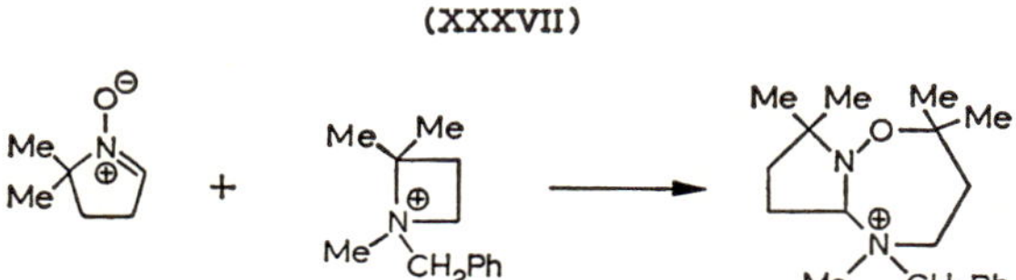

On irradiation, 4-phenylquinazoline-3-oxide (XXXVIII; R = H) undergoes ring expansion to 2-**phenyl-1,3,5-benzoxadiazepine** (XXXIX; R = H), m.p. 59–61°, and the 4-*methyl analogue* (XXXIX; R = Me), m.p. 83–85°, is made similarly. Some 6-chloro-2-substituted quinazoline 3-oxides yield only open chain compounds, and five- and six-membered heterocyclic ring systems.

(XXXVIII) (XXXIX)

On solvolysis XXXIX (R = H) gives benzoxazole, 2-phenylbenzoxazole and 2-formamidophenol (*Kaneko et al.*, Tetrahedron Letters, 1967, 5233; *S. Yamada* and *I. Yokoe*, C.A., 1968, **68**, 104845; *K. H. Wuensch* and *H. Bajdala*, Z. chem., 1970, **10**, 144; *G. F. Field* and *L. H. Sternbach*, J. org. Chem., 1968, **33**, 4438).

Aromatic diazo oxides (*o*-quinone diazides) react with ketenes to give 1:1 (XL) and 1:2 (XLI) adducts, given 1,4,5-**benzoxadiazepine** structures. On thermolysis compounds XL (R = Ph) lose nitrogen to yield 2-methylenebenzodioxoles but with compounds XL (R = H) carbon dioxide is lost leaving an indazole. This behaviour suggests that compounds of type XL (R = H), *e.g.* 2,5-*dihydro*-1,4,5-*benzoxadiazepin*-2-*one* (XLII), m.p. 114°, prefer to exist in this tautomeric form (*H. Staudinger* and *T. Reber*, Helv., 1921, **4**, 1; *Ried et al.*, Naturwiss., 1960, **47**, 445; Ann., 1963, **666**, 113; 135; 1965, **681**, 45; 52; Angew. Chem. intern. Edn., 1963, **2**, 323).

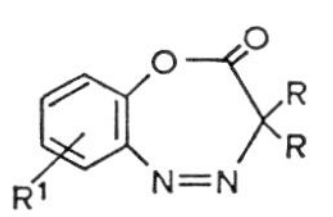

(XL)

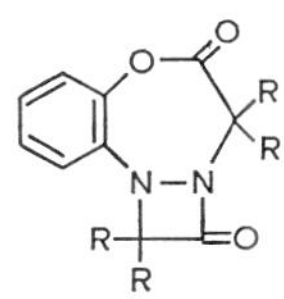

(XLI)

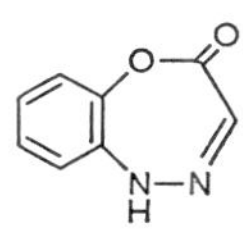

(XLII)

The 3,1,4-**benzoxadiazepines** of the early literature (*Reid* and *P. Stahlofen*, Chem. Ber., 1954, **87**, 1814 and papers cited) made by cyclising *N*-substituted 2-aminobenzaldehyde oximes, have been shown to be quinazoline 3-oxides (cf. XXXVIII) (*L. H. Sternbach et al.*, J. Amer. chem. Soc., 1960, **82**, 475).

A genuine representative, namely, 2-*methylamino*-3,1,4-*benzoxadiazepine* (XLIII; R = NHMe), m.p. 148–150°, is prepared by oxidative cyclisation of the thiourea XLIV with mercuric oxide. 3,1,4-Benzoxadiazepines are also postulated as intermediates in the photolytic ring-contraction of quinazoline 1-oxides to indazoles, as XLIII (R = NHMe) shows this behaviour on irradiation:

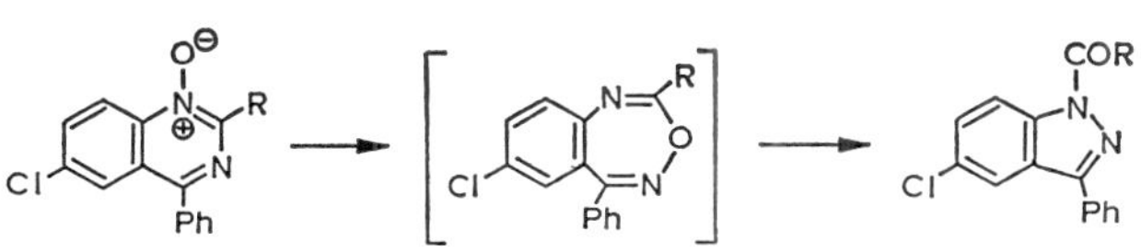

(XLIII)

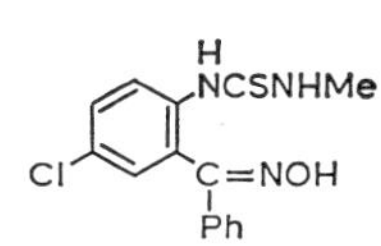

(XLIV)

It also shows ring contraction to quinazolines with hydrogen chloride or boron trifluoride (*idem*, Monatsh., 1967, **98**, 633; J. org. Chem., 1968, **33**, 4438).

The α-oxime of 2-amino-5-chlorobenzophenone reacts with formaldehyde to give a reduced member of this series 7-*chloro*-5-*phenyl*-1,2-*dihydro*-3,1,4-*benzoxadiazepine* (XLV; X = H_2), m.p. 181–182°; the β-oxime gives a 1,2-dihydroquinazoline 3-oxide (*G. N. Wal-*

ker et al., J. org. Chem., 1972, **37**, 3755). *7-Chloro-4-phenyl-1,2,4,5-tetrahydro-3,1,4-benzoxadiazepin-2-one* (XLV; X = O), m.p. 195°, which is obtained if phosgene is used shows ring contraction to 6-chloro-3-phenyl-1,2,3,4-tetrahydroquinazoline-2,4-dione at its melting point (*T. S. Sulkowski* and *S. J. Childress*, *ibid.*, 1962, **27**, 4424).

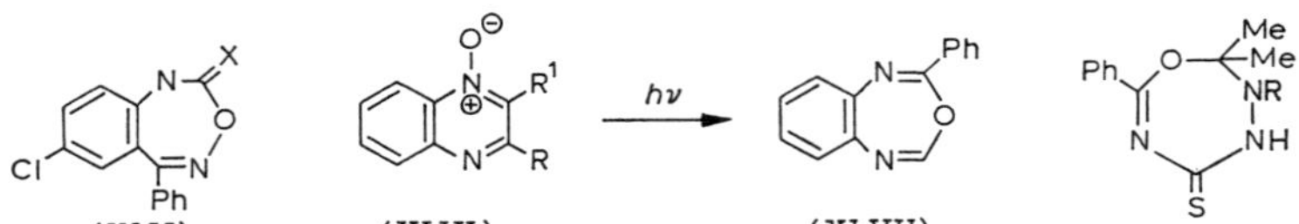

(XLV) (XLVI) (XLVII) (XLVIII)

2- and 3-Phenylquinoxaline 1-oxides (XLVI; R = H, R^1 = Ph and T = Ph, R^1 = H) undergo rearrangement to **2-phenyl-3,1,5-benzoxadiazepine** (XLVII), m.p. 42–46°, on irradiation. This structure has been confirmed by X-ray analysis (*Kaneko et al.*, Chem. pharm. Bull. Japan, 1966, **14**, 1316; Tetrahedron Letters, 1967, 1873; *Buchardt et al.*, Acta chem. Scand., 1967, **21**, 1399; 1968, **22**, 877). Compound XLVII is rapidly hydrolysed to *N*-benzoyl-*N′*-formyl-*o*-phenylenediamine, and on further irradiation yields benzimidazole derivatives. A benzoxadiazepine could also be an intermediate in the formation of 1-acetyl-3-benzoyl-2-benzimidazolinone from 2-phenylquinoxaline 1,4-dioxide in refluxing acetic anhydride (*Y. Ahmad et al.*, J. org. Chem., 1968, **33**, 201).

Benzoyl isothiocyanate and monosubstituted acetonehydrazones condense to form 2,3,4,5-**tetrahydro** 1,3,4,6-**oxatriazepine**-5-**thiones** (XLVIII). They liberate acetone with acid, leaving 1,2,4-triazoline-5-thiones (*G. J. Durant*, J. chem. Soc., C, 1967, 952).

2. Rings containing oxygen and sulphur

Compounds of this series have attracted little attention and references to them are scattered. 1,3-**Oxathiepan**-2-**one** (XLIX), b.p. 124°/0.004 mm, n_D^{25} 1.496, is prepared by condensing diethyl carbonate with 3-mercaptopropanol in the presence of thallium nitrate (*D. L. Johnson*, F.P., 1,445, 849/1964; C.A., 1967, **66**, 55065).

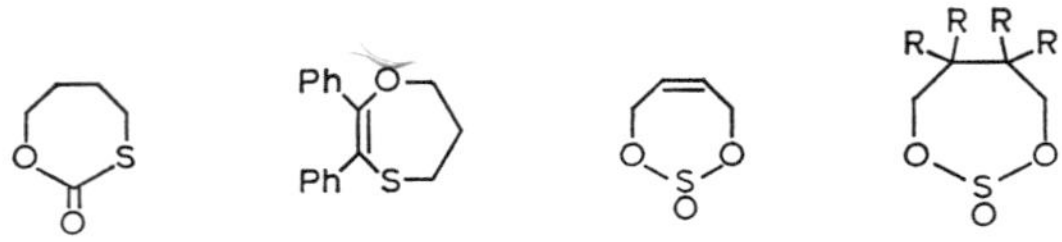

(XLIX) (L) (LI) (LII)

2,3-**Diphenyl**-5,6-**dihydro**-7*H*-1,4-**oxathiepin** (L), m.p. 119°, is obtained by dehydrative cyclisation of the hydroxyketone α-benzoylbenzyl 3-hydroxypropyl sulphide, PhCO·CH-(Ph)S$(CH_2)_3$(OH) (*J. R. Marshall* and *H. A. Stevenson*, J. chem. Soc., 1959, 2360). A saturated 1,4-oxathiepane has been prepared by the ring closure of a vinyl (3-hydroxypropyl)sulphide (*G. Pagani et al.*, Gazz., 1967, **97**, 1770).

1,4-Dihydroxy-2-butene reacts with thionyl chloride or dimethyl sulphite to give the cyclic sulphite, 4,7-**dihydro**-1,3,2-**dioxathiepin** 2-**oxide** (LI), b.p. 102°/21 mm, m.p. 13–14°,

n_D^{21} 1.4862. Bromine adds to the double bond, giving the 5,6-*dibromide*, m.p. 99–100°, and so does iodine monochloride (*R. Riemschneider* and *W. Ernst*, Z. Naturforsch., 1960, **15**, 552). Oxidation with potassium permanganate yields the 2,2-*dioxide*, m.p. 37°, which shows ring-opening on forming a betaine with 2-methylbenzothiazole (*J. Brunken et al.*, C.A., 1962, **57**, 6786).

1,3,2-**Dioxathiepane 2-oxide** *(tetramethylene sulphite)* (LII; R = H), b.p. 129°/15 mm, and its 5,5,6,6-*tetrachloro* derivative (LII; R = Cl), m.p. 59–61°, are prepared similarly (*R. G. Gillis*, J. org. Chem., 1960, **25**, 651; *M. E. Chiddix* and *R. W. Wynn*, U.S.P. 3,170,942/1962; C.A., 1965, **62**, 13047). For oxidation of LII (R = H) to the 2,2-*dioxide*, m.p. 44–45°, calcium manganate is used (*J. Lichtenberger* and *J. Hincky*, Bull. Soc. chim. Fr., 1961, 1495), and on heating with triethylamine LII (R = H) yields tetrahydrofuran, losing sulphur dioxide from the acyclic intermediate $Et_3N(CH_2)_4OSO_2$ (*Gillis, loc.cit.*).

Both LI and LII (R = H) undergo acid-catalysed hydrolysis with cleavage of a sulphur to oxygen bond (*P. A. Bristow et al.*, J. chem. Soc., 1965, 5779).

Reaction of divinyl sulphone with alkaline hydrogen peroxide gives 1,2,5-**dioxathiepane 5,5-dioxide** (LIII), m.p. 131–132° (*H. Kropf* and *C.-C. Bernert*, Ann., 1971, **734**, 151).

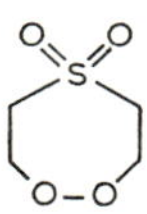
(LIII)

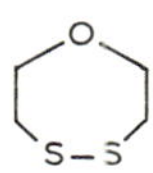
(LIV)

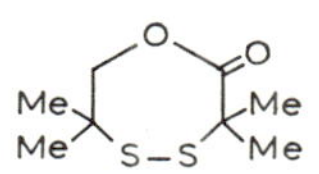
(LV)

Steam distillation from dilute sodium hydroxide of the polymer formed from bis(2-chloroethyl) ether and sodium disulphide gives 1,4,5-**oxadithiepane** (LIV), b.p. 55–56/3 mm, n_D^{20} 1.5823. Cyclisation of the "Bunte salt", $[NaO_3S{\cdot}S(CH_2)_2–]_2O$ with aqueous cupric chloride is also an effective route to LIV. It readily polymerises in the presence of both ionic and free radical catalysts and shows co-polymerisation (*F. O. Davis* and *E. M. Fettes*, J. Amer. chem. Soc., 1948, **70**, 2611; *A. V. Tobolsky et al.*, *ibid.*, 1953, **75**, 1757; J. polymer Sci., 1948, **3**, 604; *N. A. Rosenthal*, U.S.P. 3,284,466/1964; C.A., 1968, **68**, 13015; *F. S. Dainton et al.*, Trans. Faraday Soc., 1957, **53**, 813; *J. G. Affleck* and *G. Dougherty*, J. org. Chem., 1950, **15**, 865).

Dehydration of the acid $HOCH_2C(Me)_2S{\cdot}SC(Me)_2CO_2H$ with acetic anhydride gives the 1,4,5-oxadithiepanone LV; it polymerises when treated with butyl-lithium (*K. Hayashi*, Macromolecules, 1970, **3**, 5).

A large number of 2,3-**dihydro-5*H*-1,4-benzoxathiepins,** including the parent *compound* LVI, m.p. 35–36°, are prepared by treating the product of 2-(2-chloroethoxy)benzyl chloride and thiourea, with aqueous sodium hydroxide using a high dilution technique; distillation of 2-(2-chloroethoxy)dibenzyl sulphides also yields compounds LVI (*M. Kulka*, Canad. J. Chem., 1955, **33**, 1442; 1959, **37**, 325).

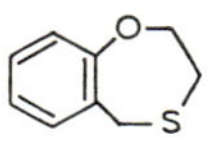
(LVI)

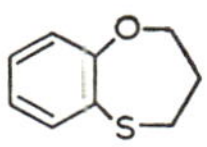
(LVII)

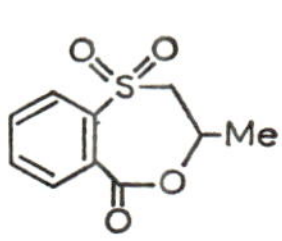
(LVIII)

3,4-*Dihydro*-2H-1,5-*benzoxathiepine* (LVII), b.p. 110–112°/11 mm, n_D^{20} 1.5765, is prepared by irradiating a mixture of 2-hydroxybenzenethiol and allyl bromide, and cyclising the product with potassium carbonate; the sulphone could not be prepared this way (*G. E. Bermingham* and *N. H. P. Smith*, Spectrochim. Acta, 1971, **27A**, 1467).

The condensation of 2-mercaptobenzoic acid with bromomethylacetylene in the presence of sodium hydroxide, followed by cyclisation with hydrogen peroxide yields 3-*methoxy*-2,3-*dihydro*-5H-1,4-*benzoxathiepin*-5-*one* 1,1-*dioxide* (LVIII), m.p. 158–159° (*R. H. Rynbrandt*, J. heterocycl. Chem., 1970, **7**, 191); alkaline hydrolysis brings about ring-contraction to a benzothiophene.

1,5,-**Dihydro-2,4,3-benzodioxathiepine 3-oxide** (LIX), m.p. 36–37°, is prepared from 1,2-bis(hydroxymethyl)benzene and thionyl chloride or methyl sulphite. A number of analogues, including the cis-*perhydro* derivative, b.p. 96°/2.5 mm (*Riemschneider* and *V. Wuscherpfennig*, Z. Naturforsch., 1962, **17**, 516) and the cyclic sulphite of a 2,2′-dihydroxybiphenyl, a **dibenzo** [*d,f*]-1,3,2-**dioxathiepine** 2-**oxide** (LX) were prepared similarly (*F. R. Hewgill* and *B. S. Middleton*, J. chem. Soc., 1965, 2914).

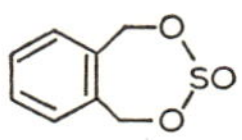
(LIX)

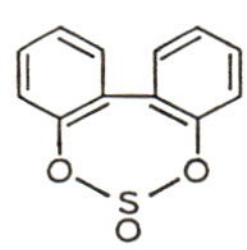
(LX)

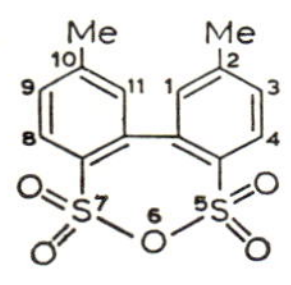
(LXI)

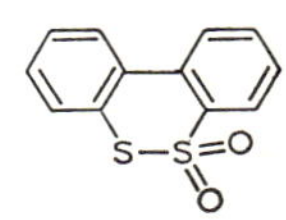
(LXII)

The disulphonic anhydride, 2,10-*dimethyldibenzo* [c,e]-1,2,7-*oxadithiepine* 5,5,7,7-*tetroxide* (LXI), m.p. 234–235°, is obtained by the action of phosphorus pentachloride and phosphonylchloride on sodium 5,5′-dimethylbiphenyl-2,2′-disulphonate. It is readily hydrolysed and undergoes ring-contraction to the dibenzodithiin dioxide LXII with alkaline sulphite (*W. L. F. Amarego* and *E. E. Turner*, J. chem. Soc., 1956, 1665).

3. Rings containing nitrogen and sulphur

(*a*) *Thiazepines*

(*i*) 1,2- *and* 1,3-*thiazepines*

1,2-**Thiazepines** are rare, but the product from the ketone HC ≡ C · CO · CH:CHSCN and aniline may have the structure 2-*phenyl*-2,5-*dihydro*-1,2-*thiazepine*-5-*one* (LXIII), m.p. 153–155°; with *tert*-butylamine the 2-*tert*-butyl analogue is obtained (*F. Wille* and *R. Strasser*, Chem. Ber., 1961, **94**, 1606).

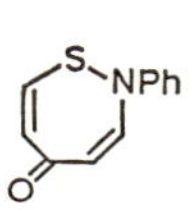
(LXIII)

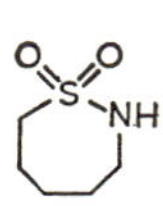
(LXIV)

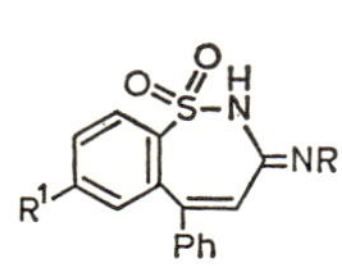
(LXV)

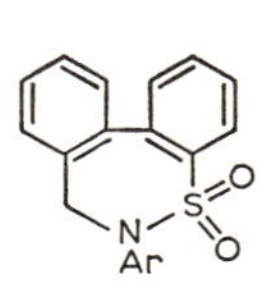
(LXVI)

5-Aminopentanesulphonyl chloride cyclises on heating to 1,2-**thiazepane** 1,1-**dioxide** (LXIV), b.p. 145–147°/0.8 mm, n_D^{20} 1.5902, a method also suitable for the corresponding 8- or 9-membered rings (*Ruhrchemie A-G.*, B.P. 752,227/1956; C.A., 1957, **51**, 4415).

Cyclisation of the products of 2-benzoylbenzenesulphonyl chlorides and *N*-substituted actamidenes with strong base gives the imino compounds LXV; with acetamidine itself 3-*amino*-7-*chloro*-5-*phenyl*-1,2-*benzothiazepine* 1,1-*dioxide* (amino tautomer of LXV; R = H, R^1 = Cl) is obtained (*J. J. Traverso*, U.S.P. 3,377,357/1966; C.A., 1968, **69**, 52190). A Pschorr reaction on sulphonamides of *N*-(2-aminobenzyl)aniline gives a mixture of 6,7-*dihydrodibenzo* [d,f]-1,2-*thiazepine* 5,5-*dioxide* (*e.g.* LXVI; Ar = Ph), m.p. 144–145°, and *N*-sulphonyldihydrophenanthridine from alternative modes of cyclisation (*J. L. Huppatz* and *W. H. F. Sasse*, Austral. J. Chem., 1965, **18**, 206).

Very few 1,3-**thiazepines** are known either. 2-Vinylaziridine reacts with phenyl isothiocyanate or 4-chlorothiobenzoylthioglycollate to give 2-*anilino*-4,7-*dihydro*-1,3-*thiazepine* (LXVII; R = NHPh), m.p. 138–141°, and 2-(4-*chlorophenyl*)-4,7-*dihydro*-1,3-*thiazepine* (LXVII; R = 4-chlorophenyl), m.p. 20°, respectively (*P. G. Mente* and *H. W. Heine*, J. org. Chem., 1971, **36**, 3076).

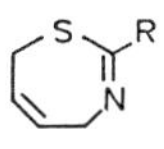
(LXVII)

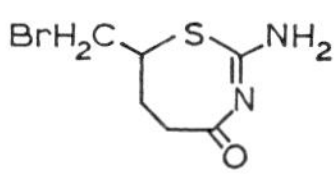

(LXVIII)

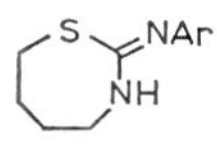

(LXIX)

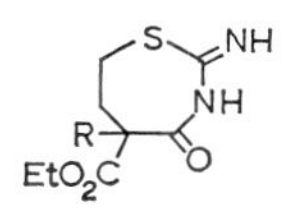

(LXX)

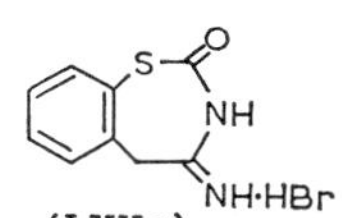

(LXXa)

1,4-Dichlorobutane condenses with 2,6-dichloroisothiobenzamide to give 2(2,6-*dichlorophenyl*)-4,5,6,7-*tetrahydro*-1,3-*thiazepine*, m.p. 104–105° (*J. Yates*, *H. P. Rosinger* and *J. T. Hackmann*, G.P. 1,262,667/1961; C.A., 1969, **70**, 57823) and 2-amino-7-bromomethyl-4,5,6,7-tetrahydrothiazepin-4-one (LXVIII) is obtained by cyclobromination of 4-pentenoylthiourea (*J. V. Migalina* and *I. V. Smolanka*, U.S.S.R. Pat. 225,884/1967; C.A., 1969, **70**, 28966). 4-Hydroxybutyl isothiocyanate cyclises with benzylamine to 2-*benzylamino*-4,5,6,7-*tetrahydro*-1,3-*thiazepine*, but with arylamines imino compounds LXIX are obtained (*Z. Olszenko-Piontkova* and *T. Urbanski*, Org. Prep. Proced. Int., 1971, **27**; C.A., 1971, **74**, 125665); ω-bromoalkyl isothiocyanates have been used to make LXIX and homologues up to the 13-membered ring (*D. L. Garmaise et al.*, Canad. J. Chem., 1971, **49**, 971). Thiourea and allylmalonic esters give imino compounds LXX, which undergo ring-contraction to pyrimidine-2,4,6-triones in acid (*A. I. Kiprianov* and *Yu. L. Slominskii*, Zhur. org. Khim., 1965, **1**, 1314).

A homologue of cephalosporin containing the 1,3-thiazepine ring system has recently been synthesised (*R. B. Woodward et al.*, Helv., 1972, **55**, 2567). Cyclisation of 2-thiocyanatophenylacetonitrile with hydrogen bromide yields the benzo-1,3-thiazepine LXXa in a reaction analogous to the *bis-o*-phenyleneacetonitrile cyclisation (*F. Johnson* and *R. Madronero*, Adv. heterocycl. Chem., 1966, **6**, 143).

(ii) 1,4-Thiazepines

Most of the known seven-membered heterocycles containing both nitrogen and sulphur are monocyclic, benzo- or dibenzo-1,4-**thiazepins.** Interest in this ring system is due partly to its structural similarity to the physiologically active 1,4-diazepines but also because the 1,4-thiazepanones LXXI may be regarded as being monocyclic analogues of the penicillin antibiotics LXXII.

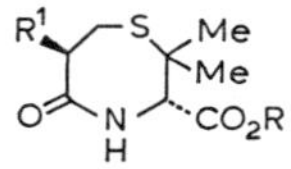
(LXXI)

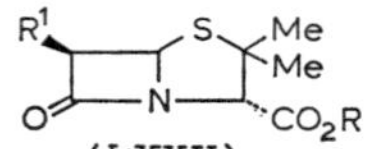
(LXXII)

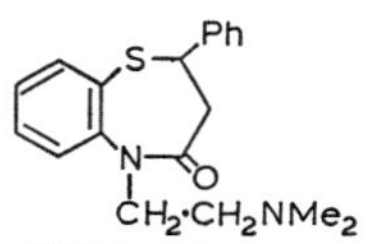
Thiazesim
(LXXIII)

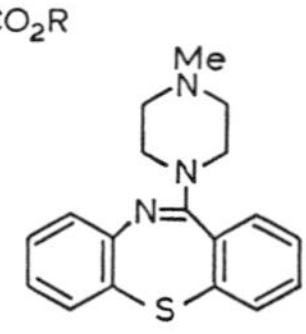
HF 1259
(LXXIV)

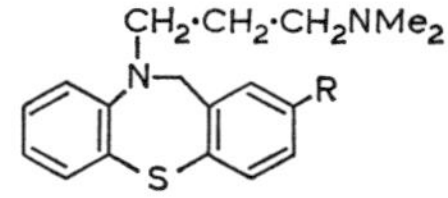
(Chlor) Homopromazine
(LXXV)

Among many compounds tested for pharmaceutical value the most useful are **Thiazesim** (LXXIII) (*Z. P. Horovitz et al.*, Nature, 1963, **200**, 369; Int. J. Neuropharmacol., 1966, **5**, 405; C.A., 1967, **66**, 9922; *A. S. Kulkarni*, Life Sci., Oxford, 1968, **7**, 135; C.A., 1968, **68**, 76831), HF 1259 (LXXIV) (*P. Simon* and *J. R. Boissier*, Therapie, 1966, **21**, 1491; *Boissier et al.*, Encephalie, 1965, **54**, 517; C.A., 1967, **66**, 45365; **67**, 20345; *F. Hunziker et al.*, Helv., 1966, **49**, 244; *G. Stille et al.*, Arzneim. Forsch., 1965, **15**, 841; 1966, **16**, 255; Intern. J. Neuropharmacol. 1965, **4**, 375; C.A., 1966, **64**, 2623) and the antihistamines **Homopromazine** and **Chlorohomopromazine** (LXXV; R = H or Cl) (*M. Protiva et al.*, Chem. Listy, 1957, **51**, 1909; Coll. Czech. chem. Comm., 1958, **23**, 1941; Experientia 1957, **13**, 291; *Z. Horakova* and *Z. Votava*, C.A., 1959, **53**, 17340; *Votava* and *J. Melysova*, C.A., 1964, **60**, 11241; *Horovitz et al.*, Arch. intern. Pharmacodyn., 1964, **151**, 180; C.A., 1965, **62**, 2134). Compound LXXI (R = Me, R^1 = = $PhCH_2$) is not a biological precursor of penicillins (*N. J. Leonard* and *G. E. Wilson*, Tetrahedron Letters, 1964, 1465; *E. E. van Tamelen et al., ibid.*, 1965, 281; *H. R. V. Arnstein* and *M. E. Clubb*, Biochem. J., 1958, **68**, 528) and many attempts to form a 4,7-transannular bond have failed (*M. H. Benn* and *R. E. Mitchell*, Canad. J. Chem., 1972, **50**, 2195.

2,3-**Dihydro**-1,4-**thiazepines** (LXXVI) are obtained from the dealkoxycarbonylated analogue of LXXI by introducing a 6,7-double bond with a chlorination–dehydrochlorination sequence followed by treating the vinyl lactam with triethyloxonium tetrafluoroborate. Photolysis leads not to a penicillin-like enol ether but, if R = Me, to a tetrahydro-1,4-thiazine and

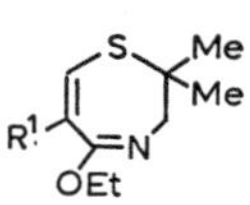
(LXXVI)

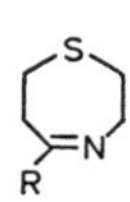
(LXXVII)

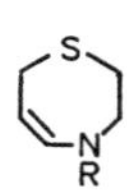
(LXXVIII)

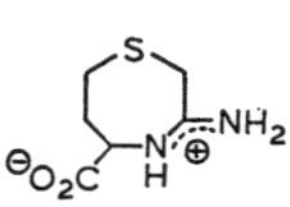
(LXXIX)

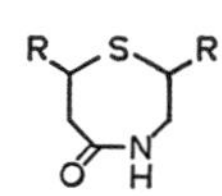
(LXXX)

if R^1 = NHR to the 2,2-dimethyl-6-imino analogue of LXXVII (R = OEt) (*M. F. Semmelhack et al.*, Chem. Comm., 1971 698, 988).

2-Aminoethanethiol and several of its derivatives, including cysteine and cysteamine form heterocycles with compounds having reactive 1- and 3-positions. With α,β-unsaturated ketones or β-halogeno-ketones *5-methyl-2,3,6,7-tetrahydro-1,4-thiazepine* (LXXVII; R = Me), b.p. 130–132°/4 mm, and analogues are obtained, but cysteine does not react this way (*D. K. Black*, J. chem. Soc., C, 1966, 1708, cf. 1123; *L. K. Mushkalo et al.*, Ukr. Khim. Zhur., 1955, **21**, 631; C.A., 1956, **50**, 14783; Zhur. obshcheĭ Khim., 1960, **30**, 1023). If the thiol carries an *N*-substituent, 4-substituted-2,3,4,7-tetrahydro-1,4-thiazepines LXXVIII result instead. Mercaptoethyl ketones with ethyleneimine also give thiazepines LXXVII (*F. Asinger et al.*, Monatsh., 1964, **95**, 1335) and homocysteine with 3-chloroacrylonitrile gives the zwitterionic cyclic *amidine* LXXIX, *picrate*, m.p. 225–232° (*L. Goodman*, J. org. Chem., 1958, **23**, 1954).

The ring of LXXIX is cleaved by alkaline benzoyl chloride. *5-Ethoxy-2,3,6,7-tetrahydro-thiazepine* (LXXVII; R = OEt), b.p. 100–102°/12 mm, is obtained by the action of triethyloxonium tetrafluoroborate on ketone LXXX (R = H) (*W. J. Houlian*, U.S.P. 3,278,532/1964; C.A., 1966, **65**, 20154; *R. G. Griot*, U.S.P. 3,351,632/1967; C.A., 1968, **68**, 105262).

4-Acetyl-3-methyl-4,5,6,7-tetrahydro-1,4-thiazepinone (LXXXI) is prepared by the ring expansion sequence shown (*S. Hoff et al.*, Tetrahedron Letters, 1972, 5199):

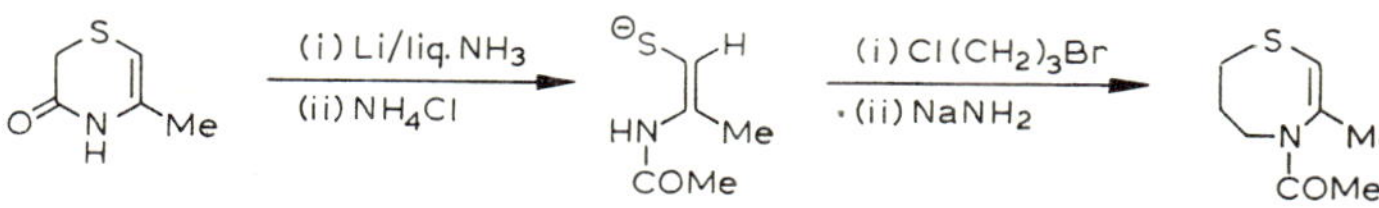

(LXXXI)

The product when a β-substituted acrylic acid derivative (*O. Süss*, Ann., 1948, **561**, 31; *I. L. Knunyants et al.*, Doklady Akad. Nauk S.S.S.R., 1955, 71) or a 4-(α-halogeno-alkylidene)-5-oxazolone (*idem, ibid.*, 1959, 1777; *I. T. Strukov*, Zhur. obshcheĭ Khim., 1957, **27**, 432) condenses in the presence of a strong base, or a trialkyl orthoformate, with 2-aminoethanethiol, is a 2,3,4,5-**tetrahydro-1,4-thiazepin-5-one** (LXXXII); dimethyl acetyl-enedicarboxylate is said to yield the analogue (LXXXII; R = CO_2Me, R^1 = H), m.p. 200° (*Mushkalo, loc.cit.*). Treatment of α-phthalimido- or α-benzylsulphonamido-2-thiazolidineacetic acids with phosphoryl chloride gives monocyclic LXXXII (R = H, R^1 = phthalimido or benzylsulphonamido) and not the expected penicillins (*J. C. Sheehan*

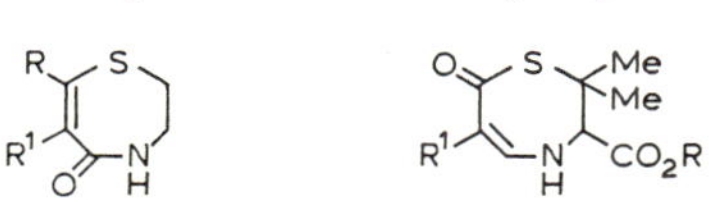

(LXXXII) (LXXXIII)

and *P. A. Cruikshank*, J. Amer. chem. Soc., 1956, **78**, 3680; 3683). With strong bases the 1,2-bond of ring LXXXII is broken (*N. J. Leonard* and *R. Y. Ning*, J. org. Chem., 1967, **32**, 677).

An alternative route to LXXXII is the elimination of a hydrogen halide from 7-halogeno-1,4-thiazepan-5-one. The 6,7-double bond undergoes addition reactions (*Knunyantz et al.*, Doklady Akad. Nauk S.S.S.R., 1959, 1777; *Leonard* and *Wilson, loc.cit.*; J. Amer. chem. Soc., 1964, **86**, 5307).

The base-catalysed epimerisation at position 6 of penicillins (LXXII) is accompanied by rearrangement to the tetrahydrothiazepinone LXXXIII; antimony pentachloride brings about a similar ring-expansion. On prolonged treatment with base ring-contraction of LXXXIII to a 1,3-thiazine occurs (*R. J. Stoodley et al.*, Chem. Comm., 1970, 14; 1971, 450; 647; J. chem. Soc., C, 1970, 2089; Perkin I, 1972, 1063; *O. K. J. Kovacs et al.*, Tetrahedron Letters, 1969, 1863, *A. Vlietinck et al., ibid.*, 1972, 285).

Catalytic hydrogenation of LXXVII (R = Me) gives 5-methyl-1,4-thiazepane (*Black, loc.cit.*) and reduction of the ketone LXXX (R = H) with lithium tetrahydridoaluminate gives 1,4-**thiazepane,** b.p. 192–193°, n_D^{27} 1.5342; *hydrochloride*, m.p. 210–212°. The nitrogen atom of 1,4-thiazepane is readily alkylated (*R. P. Mull et al.*, J. org. Chem., 1960, **25**, 1953).

Thioglycollic esters and 3-bromopropylamine react in the presence of a strong base to give 1,4-*thiazepan*-3-*one* (LXXXIV), m.p. 141–143° (*M. F. Shostakovskii et al.*, Zhur. obshcheĭ Khim., 1961, **31**, 1453) and ethyleneimine reacts with 3-mercaptopropionic esters to give 1,4-*thiazepan*-5-*one* (LXXX; R = H), m.p. 117–118° (*F. Jakob* and *P. Schlack*, G.P. 1,195,317/1960; C.A., 1965, **63**, 8387; Chem. Ber., 1963, **96**, 88). A number of derivatives are prepared similarly.

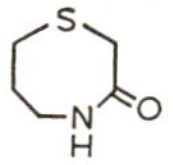

(LXXXIV)

Ketones LXXX are also prepared by the Schmidt reaction on tetrahydro-4-thiopyranone, where an excess of hydrazoic acid results in a tetrazolo derivative (*Knunyants et al.*, Doklady Akad. Nauk S.S.S.R., 1960, 2053; *I. N. Nazarov* and *A. I. Kuznecova, ibid.*, 1953, 506; *Mull, loc.cit.*), or by the Beckmann rearrangement of the corresponding oximes (*C. Barkenbus et al.*, J. org. Chem., 1955, **20**, 871; *Jakob* and *Schlack, loc.cit.*; *I. L. Pannizzi*, Gazz., 1948, **78**, 207). The general route from 2-aminoethanethiol with α,β-unsaturated acid chlorides (*Knunyants et al.*, Doklady Akad. Nauk S.S.S.R., 1955, 71), esters (*idem, ibid.*, 1959, 1777; *Jakob* and *Schlack, loc.cit.*; *van Tamelen, loc.cit.*; *N. J. Leonard* and *R. Y. Ning*, J. org. Chem., 1966, **31**, 3928; *N. Ogata et al.*, Macromol. Chem., 1968, **113**, 95; *T. Wieland*, Chem. Ber., 1957, **90**, 194), or the free acid (*Leonard* and *Wilson, loc.cit.*) is also available. The open chain intermediate may usually be isolated and subsequently cyclised under basic conditions. 3-Halogenopropionyl halides may be used in place of the unsaturated acid derivative (*Knunyants*, Doklady Akad. Nauk S.S.S.R., 1955, 62), and thiazolidine derivatives in place of the aminothiol (*Arnstein* and *Clubb, loc.cit.*).

Ketones LXXX and LXXXIV are both hydrolysed by hydrochloric acid to amino acids. Catalytic hydrogenation of thiazepanone LXXI (R^1 = RCONH) leads to sulphur extrusion and the formation of open chain peptides (*Leonard et al., loc.cit.*) while with hydrogen peroxide, ketones LXXX give 1,4-*thiazepan*-5-*one* 1,1-*dioxide*, m.p. 200–201 °, and its substituted derivatives. (*H. K. Hall*, J. Amer. chem. Soc., 1958, **80**, 6404). Halogenation of LXXX with *N*-halogenosuccinimide takes place at position 7, and this halogen atom is removed by zinc in ethanol. Polymerisation of 1,4-thiazepan-5-one (LXXX; R = H) is induced by hydrochloric or phosphoric acids, lithium or sodium but not by aluminium chloride or benzoyl peroxide. The rate of polymerisation at 125–200° is faster than that for caprolactam, with which LXXX forms copolymers. The thiazepanone sulphone is also said to polymerise (*Knunyants et al., loc.cit.*; *Jakob* and *Schlack, loc.cit.*; *Nazarov* and *Kuznecova, loc.cit.*; *Ogata et al., loc.cit.*).

Ketone LXXI (R^1 = H, R = Me) is desulphurised by silver carbonate in methanol leaving a diolefin (*D. Gravel et al.*, Chem. Comm., 1972, 1322; Canad. J. Chem., 1971, **49**, 3866) and the analogue LXXI (R^1 = $PhCH_2CONH$, R = Me) undergoes ring contraction to an isothiazolinone when treated with chlorine at −10° (*Leonard* and *Wilson*, Tetrahedron Letters 1964, 1465; 1471; J. Amer. chem. Soc., 1964, **86**, 5307). 2,7-*Diphenyl*-1,4-*thiazepan*-5-*one* (LXXX; R = Ph), m.p. 191–192°, by contrast, yields a thiazoline when heated in polyphosphoric acid (*M. Wilhelm* and *P. Schmidt*, Helv., 1970, **53**, 1697). Derivatives of vinyl lactam LXXXII (R^1 = RCONH) form both thiazolidine and oxazolinone derivatives after initial rupture of the 4,5-bond (*I. T. Strukov*, Zhur. obshcheĭ Khim., 1957, **27**, 432).

1,4-Thiazepines with the LXXXIVa ring system have been obtained in the reaction between dimethyl acetylenedicarboxylate and thiazoles (*D. H. Reid et al.*, Tetrahedron Letters, 1964, 1797).

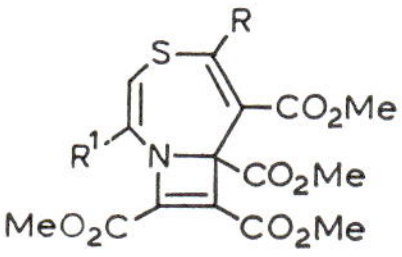

(LXXXIVa)

The most general synthesis of 1,4-**benzothiazepines** is from a 2-carbonylbenzenethiol and a 2-substituted ethylamine. An intermediate is usually isolated in which a sulphur to carbon bond is formed, and cyclisation is subsequently effected in the presence of a base.

7-*Chloro*-5-*phenyl*-1,4-*benzothiazepine* 1,1-*dioxide* (LXXXV), m.p. 207–208°, results from the thermolysis of sulphone LXXXVI (R = CO_2Me).

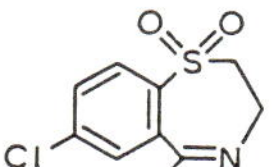

(LXXXV)

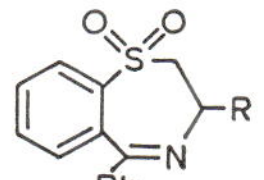

(LXXXVI)

(LXXXVII) (LXXXVIII)

5-Phenyl-2,3-dihydro-1,4-benzothiazepine (LXXXVII; R = H, R^1 = Ph), m.p. 64–65°. *hydrochloride*, m.p. 201–202°, and many of its derivatives, are prepared from 2-mercaptobenzophenones and 2-bromoethylamine hydrobromides, cyclising the intermediates with pyridines. Activated 2-chlorobenzophenones similarly react with 2-aminoethanethiol in the presence of sodium hydride to give derivatives of LXXXVII (R^1 = Ph). A recent route to a 4,5-dihydro-1,4-benzothiazepin-5-one is as follows (*R. B. Morin* and *D. O. Spry*, Chem. Comm., 1970, 335):

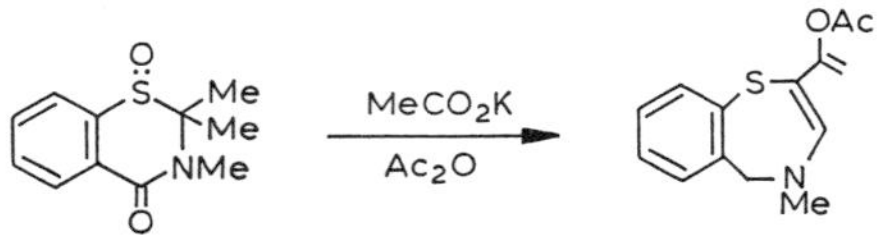

Lithium tetrahydridoaluminate, sodium tetrahydridoborate or catalytic hydrogenation reduce LXXXVII (R = H, R^1 = Ph) to *5-phenyl-2,3,4,5-tetrahydro-1,4-benzothiazepine* (LXXVIII; R = Ph), m.p. 89–90°, which may be *N*-alkylated and converted into mono- and di-oxide. With sodium periodate *5-phenyl-2,3-dihydro-1,4-benzothiazepine 1-oxide*, m.p. 158–160°, is obtained, while with hydrogen peroxide the 1,1,4-*trioxide*, m.p. 238–239°, is formed *via* an oxaziridine. This trioxide undergoes a Polonovski rearrangement with acetic anhydride to LXXXVI (R = OCOMe). Sulphuryl chloride converts LXXXVII (R = H, R^1 = Ph) into *2-chloro-5-phenyl-2,3-dihydro-1,4-benzothiazepine* (LXXXVII; R = Cl, R^1 = Ph), m.p. 93–94° (*E. Reeder* and *L. H. Sternbach*, J. org. Chem., 1965, **30**, 2812; *Hoffmann La Roche*, Neth. Appl., 6,500817/1964, C.A., 1966, **64**, 5122).

2,3,4,5-**Tetrahydro**-1,4-**benzothiazepin**-5-**one** (LXXXIX; R, R^1 = H), m.p. 191–192°, is prepared by reacting ethyleneimine or esters of ethanolamine with 2-mercaptobenzoic acid (*K-H. Wünsch et al.*, Z. Chem., 1967, **7**, 185; Chem. Ber., 1969, **102**, 1618) or its ester (*F. Jakob* and *P. Schlack*, Chem. Ber., 1963, **96**, 88) and cyclising the intermediate ester with sodium methoxide; substituted derivatives may be made by this route. A Schmidt reaction on thiochroman-4-one also gives LXXXIX (R, R^1 = H) together with tetrahydro-1,5-benzothiazepin-4-one. Substituents on the benzo ring dictate the course of this reaction; those which withdraw electrons favour aryl migration giving a 1,5-benzothiazepine and electron donors promote alkyl migration leading to a 1,4-benzothiazepine (*Wünsch et al.*, *ibid.*, 1970, **103**, 2302).

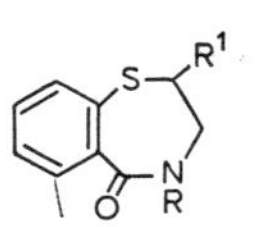

(LXXXIX)

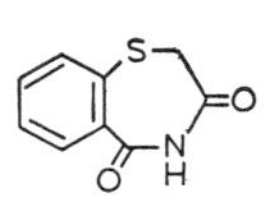

(XC)

Lithium tetrahydridoaluminate reduces the amide group leaving 2,3,4,5-**tetrahydro-1,4-benzazepin** (LXXXVIII; R = H), *hydrochloride*, m.p. 237–238°. Phosphorus pentasulphide converts LXXXIX into *2,3,4,5-tetrahydro-1,4-benzazepine-5-thione*, m.p. 198°, which is alkylated in the presence of strong bases giving vinyl mercaptans LXXXVII (R = H, R^1 = S-alkyl). Both these compounds and the analogues LXXXVII (R = H, R^1 = OEt) and (R = H, R^1 = Cl) from the respective reactions of triethyloxonium tetrafluoroborate and phosphorus pentachloride with LXXXIX (R, R^1 = H), yield aminoamidines LXXXVII

(R = H, R^1 = NH·NHR) with hydrazines (*idem, ibid.*, 1969, **102**, 1869). The action of hydrogen peroxide on either ketone LXXXIX or the corresponding thione, gives 2,3,4,5-*tetrahydro*-1,4-*benzodiazepin*-5-*one* 1,1-*dioxide*, m.p. 241–242°; the 1-*oxide* from oxidation with periodate has m.p. 224°. Oxygen is removed from the sulphur atom by thionyl chloride. Ketone LXXXIX (R, R^1 = H) also forms a 2-*chloro* derivative (LXXXIX; R^1 = Cl, R = H), m.p. 177–178°, in which the chlorine atom may be replaced giving LXXXIX (R = H, R^1 = SR, NHPh), and it is *N*-alkylated in the presence of sodium hydride or amide to give, for example LXXXIX (R = Me, R^1 = H), m.p. 67–69° (*J. Krapcho et al.*, J. med. pharm. Chem., 1968, **11**, 361; *Wünsch, loc.cit.*). The lactam group of 1,4-benzothiazepin-5-one is hydrolysed by aqueous alkali but it does not polymerise; the 2-chloro derivative is especially susceptible to hydrolysis.

2,3,4,5-**Tetrahydro**-1,4-**benzothiazepine**-3,5-**dione** (XC), m.p. 143°, is obtained by cyclising the product of chloracetic acid or ester and 2-mercaptobenzamide with thionyl chloride (*R. Ponci et al.*, Farmaco, Pavia, Ed. sci., 1964, **19**, 515; C.A., 1964, **61**, 8229) or acetic anhydride (*E. W. McClelland et al.*, J. chem. Soc., 1948, 81). Like the monoketone, imide XC is alkylated on nitrogen and undergoes hydrolytic ring-opening. With potassium acetate and acetic anhydride a ring contraction to 2-carbamoyl-3-hydroxy-1-benzothiophene occurs.

(*iii*) 1,5-*Thiazepines*

Most of the syntheses of 1,5-**benzothiazepines** recorded in the literature require the addition of a three carbon unit to a 2-aminobenzenethiol. With an α,β-unsaturated ketone (*C. I. Hsing et al.*, C.A., 1967, **66**, 28751; *L. K. Mushkalo et al.*, Zhur. obshcheĭ Khim., 1958, **28**, 507; Khim. geterotsikl. Soedin., 1968, 468; *W. Ried* and *W. Marx*, Chem. Ber., 1957, **90**, 2683; *W. D. Stephens* and *L. Field*, J. org. Chem., 1959, **24**, 1576), a β-halogeno-ketone (*Mushkalo*, Zhur. obshcheĭ, Khim., 1959, **29**, 1034) or a β-amino-ketone (*K. Hideg* and *O. H. Hankousky*, Acta Sci. Hung., 1966, **50**, 403; 1968, **56**, 405; C.A., 1967, **66**, 104995; 1968, **69**, 86981; Ger. Offen., 1,931, 096/1968; C.A., 1970, **72**, 55533) derivatives of 2,3-*dihydro*-1,5-*benzothiazepine* (XCI), *e.g.* 2,4-*dimethyl*-(XCI; R, R^1 = Me), b.p. 115–118°/2 mm, n_D^{20} 1.6068, and 4-*phenyl* (XCI; R = H, R^1 = Ph), b.p. 224–230°/0.4 mm, *hydrochloride*, m.p. 107–109°, are formed.

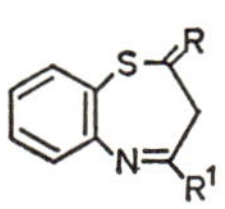
(XCI)

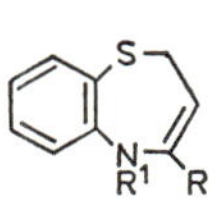
(XCII)

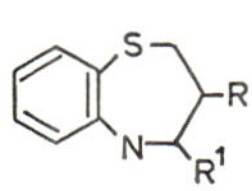
(XCIII)

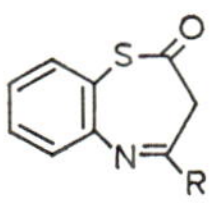
(XCIV)

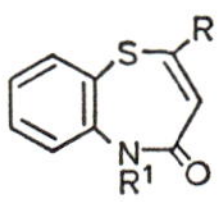
(XCV)

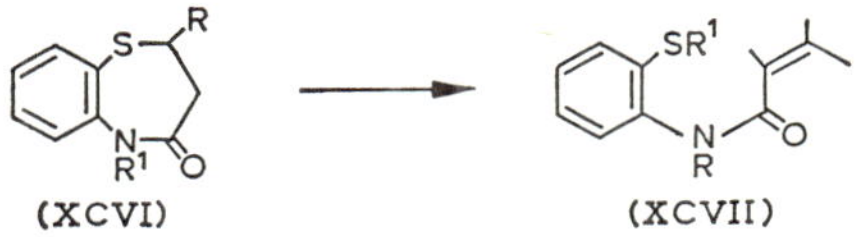
(XCVI) (XCVII)

The *S*-alkylated intermediate may be isolated in weakly basic media. With *N*-substituted aminobenzenethiols the 2,5-*dihydro*-1,5-*benzothiazepines* (XCII) *e.g.* 4,5-*dimethyl*- (R, R^1 = Me), b.p. 100–103°/6 mm, n_D^{20} 1.6014, are formed directly or *via* the quaternary salts (*Mushkalo, loc.cit.* and Zhur. obshcheĭ Khim., 1958, **28**, 742). Compounds XCI are reduced by lithium tetrahydridolaluminate, sodium tetrahydridoborate or catalytically to **tetrahydro-1,5-benzothiazepines** (XCIII), form 1,1-dioxides with hydrogen peroxide, and are cleaved by alkali to aminoketones. 5-Alkyl quaternary salts of XCI (R = Me) condense with orthoformic ester to give cyanine-type dyestuffs (*Mushkalo* and *Z. I. Shokol, ibid.*, 1961, **31**, 3069).

4-*Methyl*-2,3-*dihydro*-1,5-*benzothiazepin*-2-*one* (XCIV; R = Me), m.p. 114°, is formed by the reaction of diketene with 2-aminobenzenethiol; with benzoylacetic ester the 4-*phenyl* (XCIV; R = Ph), m.p. 107–109°, analogue is obtained. Compound XCIV couples at position 3 with diazonium salts (*Ried* and *Marx, loc.cit.*; *E. B. Towne, W. H. Moore* and *J. B. Dickey*, U.S.P. 3,125,563/1962; C.A., 1964, **61**, 3238).

By contrast acetoacetic ester gives 2-*methyl*-4,5-*dihydro*-1,5-*benzothiazepin*-4-*one* (XCV; R = Me, R^1 = H), m.p. 117° (*Ried* and *Marx, loc.cit.*), while propiolic acid gives the *parent compound*, 4,5-**dihydro**-1,5-**benzothiazepin**-4-**one** (XCV; R, R^1 = H), m.p. 215–216° (*J. Krapcho*, U.S.P. 3,075,968/1961; C.A., 1963, **59**, 5181) also obtained by acylating bis-(2,2′-diaminophenyl)disulphide with β-bromopropionyl chloride and hydrolysing the product with aqueous bicarbonate (*A. H. Weinstein et al.*, J. org. Chem., 1958, **23**, 363). Compounds XCV may be *N*-alkylated; and the 2-phenyl-4-thione, obtained by treating the oxo-analogue with phosphorus pentasulphide, extrudes sulphur and undergoes ring-contraction to a quinoline with bases (*M. Wilhelm* and *P. Schmidt*, Helv., 1970, **53**, 1697).

2,3,4,5-**Tetrahydro**-1,5-**benzothiazepine** (XCIII; R, R^1 = H), b.p. 113–115°/0.2 mm, n_D^{20} 1.6393, *hydrochloride*, m.p. 222–223°, is obtained by the action of 1,3-dibromopropane on 2-aminobenzenethiol or its zinc salt (*Mushkalo* and *I. P. Fedorova*, Ukrain. Khim. Zhur., 1954, **29**, 305; C.A., 1956, **50**, 366; *Krapcho* and *H. L. Yale*, U.S.P. 3,117,124/1957; C.A., 1964, **60**, 8049). The 3-methyl derivative (XCIII; R = Me, R^1 = H), m.p. 178–179°, is also the major product, along with 3-methyl-4-aminothiochroman, when 3-methylthiochromanone 4-oxime is treated with lithium tetrahydridoaluminate (*N. V. Dudykina* and *V. A. Zagorevskii*, C.A., 1966, **65**, 683; Zhur. obshcheĭ, Khim., 1963, **33**, 322; 1964, **34**, 2282). Other known derivatives include 2-*methyl*- (b.p. 115–120°/0.25 mm; *hydrochloride*, m.p. 197–199°), 2-*phenyl*- (b.p. 190–200°/0.1 mm, *hydrochloride*, m.p. 230–232°) and 4-*phenyl*-2,3,4,5-*tetrahydro*-1,5-*benzothiazepine* (XCIII; R = H, R^1 = Ph), b.p. 186–190°/0.4 mm. Tetrahydrobenzothiazepines XCIII may be *N*-alkylated and *N*-acylated (*idem*, Khim. geterotsikl. Soedin., 1967, 250).

The most thoroughly investigated members of this group are 2,3,4,5-**tetrahydro**-1,5-**benzothiazepin**-4-**one** (XCVI; R, R^1 = H), m.p. 215–216°, and its derivatives (*e.g.*, 2-*methyl*, m.p. 205–206°; 3-*methyl*, m.p. 175–176°) obtained in the high temperature reaction of 2-aminobenzenethiol with an α,β-unsaturated acid (*N. Cagnoli et al.*, Ann. chim., 1957, **47**, 606; *J. Bourdais*, F.P. 1,469,476/1965; C.A., 1968, **68**, 13014; *Krapcho et al.*, J. med. Pharm. Chem., 1963, **6**, 544; 1966, **9**, 191; U.S.P. 3,075,967/1961; 3,089,872/1961; C.A., 1963, **59**, 6423, 11541; *idem*, Fr. M. 3297/1962; C.A., 1966, **64**, 656; *Olin Mathiesen Corp.*, Belg. P. 664,356/1965; C.A., 1966, **65**, 2180; *O. E. Fancher* and *G. Nichols*, U.S.P. 3,029,251/1959; C.A., 1962, **57**, 8592; *W. H. Mills* and *J. B. Whitworth*, J. chem. Soc., 1927, 2738) or esters

(*K. Mushkalo* and *N. J. Kozlova*, Ukrain. Khim. Zhur., 1962, **28**, 960; C.A., 1963, **59**, 6410), or a β-halogeno acid (*F. Mayer* and *C. Horst*, Ber., 1923, **56**, 1415). Sometimes the intermediate 3-(2-aminophenyl)thiopropanoic acids are isolated, as in the alternative route from 2-nitrothiophenols, where the intermediate is reduced before cyclisation. The Beckmann reaction on thiochroman-4-one oximes also yields lactams XCVI (*Dudykina* and *Zagorevskii, loc.cit.*; *W. E. Truce* and *J. A. Simms.*, J. org. Chem., 1957, **22**, 617), but the Schmidt reaction gives both 1,4- and 1,5-benzothiazepines (*vide supra*). A Schmidt reaction on thiochroman-4-one 1,1-dioxides (*D. Evans* and *J. M. Lockhart*, J. chem. Soc., 1965, 4806; *K-H. Wünsch et al.*, Chem. Ber., 1970, **103**, 2302) gives only 2,3,4,5-*tetrahydro*-1,5-*benzothiazepin*-4-*one* 1,1-*dioxide*, m.p. 246–247°, and derivatives, also obtained by oxidation of XCVI with hydrogen peroxide, potassium permanganate or chromic oxide (*H. K. Hall*, J. Amer. chem. Soc., 1958, **80**, 6404; *Hoffmann-LaRoche*, Neth. Appl. 6,604,470/1965; C.A., 1967, **66**, 37971).

Lithium tetrahydroaluminate reduces the 4-oxo group to methylene (*E. R. Squibb and Sons, Inc.*, Neth. Appl. 6,601,631/1965; C.A., 1967, **66**, 55534; *J. Bernstein*, U.S.P. 3,322,752/1965; C.A., 1968, **68**, 49669; *Krapcho*, U.S.P. 3,361,750; 3,361,760/1963; C.A., 1968, **69**, 36194, 36196) and phosphorus pentasulphide converts it into the thione (*A. I. Kiprianov* and *T. M. Verbovskaya*, Zhur. obshcheĭ Khim., 1962, **32**, 3946). Compounds XCVI may be *N*-alkylated and *N*-acylated (*Yale and Suivinski*, U.S.P. 3,258,459/1962; C.A., 1966, **65**, 13743; *Bernstein*, U.S.P. 3,341,521/1965; C.A., 1968, **68**, 95875, *Krapcho*, U.S.P. 3,341,519/1965; C.A., 1968, 105263) and controlled oxidation yields the 1-oxide. Sulphuryl chloride gives 2-*chloro* 2,3,4,5-*tetrahydro*-1,5-*benzothiazepin*-4-*one* (XCVI; R = Cl, R^1 = H), m.p. 138–143° decomp., in which the chlorine atom is replaceable by alkoxyl or thioalkyl groups (*Mayer* and *Horst*, *loc.cit.*) and 2- and 3-phenyl-5-methyl derivatives of XCVI are *C*-alkylated at positions 2 or 3 by alkyl halides in the presence of sodamide (*Krapcho et al.*, U.S.P. 3,089,872/1961; C.A., 1963, **59**, 11541).

Acid hydrolysis cleaves the lactam ring at the 4,5-bond, but the *N*-alkyl derivatives show ring-opening to anilides XCVII (R^1 = H or alkyl) in dilute alkali (*Mills* and *Whitworth*, *loc.cit.*) or with alkyl halides and a strong base (*Krapcho*, U.S.P. 3,201,401/1962; 3,225,093/1965; C.A., 1965, **63**, 18118; 1966, **64**, 9647). 2-*Phenyl*-2,3,4,5-*tetrahydro*-1,5-*benzothiazepin*-4-*one* (XCVI; R = Ph, R^1 = H), m.p. 177°; 1-*oxide*, m.p. 210°, 1,1-*dioxide*, m.p. 234–235° and the 4-*S*-methylated thioanalogue contract to 2-styrylbenzothiazole on pyrolysis (*Mills* and *Whitworth*, *loc.cit.*; *Wilhelm* and *Schmidt*, *loc.cit.*).

(*iv*) *Benzothiazepines*

1,2,3,5-**Tetrahydro**-4,1-**benzothiazepin**-2-**one** (XCVIII; R, R^1 = H), m.p. 218–223°, the 3-*methyl* (XCVIII; R = H, R^1 = Me), m.p. 188–189°, and many other substituted derivatives are prepared by condensing an α-mercaptocarboxylic acid with a 2-aminobenzyl alcohol (*J. Klosa*, J. pr. Chem., 1967, **36**, 5; *H. Kuch et al.*, Arch. Pharm., 1967, **390**, 299), or with a

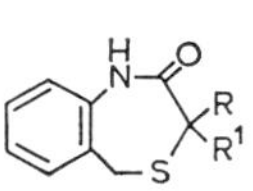

(XCVIII)

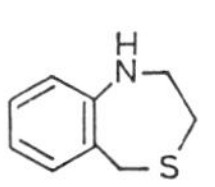

(XCIX)

2-nitrobenzyl alcohol and reducing and cyclising the intermediate (*M. Uskokovic et al.*, J. org. Chem., 1965, **30**, 3111; *F. Hoffmann - La Roche & Co.*, Neth. Appl. 6,402,756/1964; Belg. P. 645,242/1963; C.A., 1965, **62**, 9158; **63**, 9972; *Uskokovic* and *Wenner*, Swiss P. 441,327/1963, C.A., 1968, **69**, 87032). Alternatively a 2-aminobenzhydrylthiol and an α-halogeno acid halide may be used (*Hoffmann - La Roche*, B.P. 1,077,272/1962; C.A., 1967, **66**, 37971; 1968, **68**, 69061).

Reduction with lithium tetrahydridoaluminate yields 1,2,3,5-**tetrahydro**-4,1-**benzothiazepine** (XCIX), m.p. 86–88°, and analogues. Both the lactam XCVIII and its reduced product may be alkylated, and XCIX may be acylated. Oxidation of XCVIII (R, R^1 = H) with periodate or a controlled amount of hydrogen peroxide gives 1,2,3,5-*tetrahydro*-4,1-*benzothiazepin*-2-*one* 4-*oxide*, m.p. 256°, which is further oxidised to the 4,4-*dioxide*, m.p. 241–242°. The 3-methyl 4-oxide undergoes a Polonovski rearrangement to XCVIII (R = Me, R^1 = OAc) in acetic anhydride (*Uskokovic et al.*, J. org. Chem., 1969, **34**, 365). With lithium tetrahydridoaluminate 4,4-dioxides of 5-phenyl derivatives of XCVIII extrude sulphur and contract to 3-phenyloxindoles (*Kuch loc.cit.*).

5,11-**Dihydrodibenzo**[*b*,*e*]-1,4-**thiazepine** (C), m.p. 121–122°, is formed in the copper powder–potassium carbonate catalysed cyclisation of 2-(2-bromobenzyl)thioaniline or its *N*-formyl derivative (*H. L. Yale et al.*, J. med. Chem., 1970, **13**, 713; U.S.P. 3,188,322/1962; C.A., 1965, **63**, 8384; *P. Gaillot* and *F. Debarre*, F.P. 1,176,115/1961; C.A., 1961, **55**, 19972); 4-methyl and 2,4-dimethyl derivatives are found in the pyrolysis of 2-azido-2′-methyldiphenyl sulphides (*J. I. G. Cadogan* and *S. Kulik*, Chem. Comm., 1970, 233, 792; J. chem. Soc. C, 1971, 2621). The nitrogen may be both alkylated (*Gaillot* and *Debarre*, Belg. P. 617,554/1961; F.P. 1,385,360/1961; C.A., 1963, **58**, 13973; 1965, **62**, 11839; *Yale* and *R. A. Revisto*, U.S.P. 3,387,002/1965; C.A., 1968, **69**, 87030) and acylated (*Yale et al.*, *loc.cit.*).

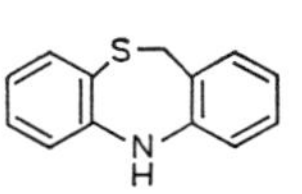

(C)

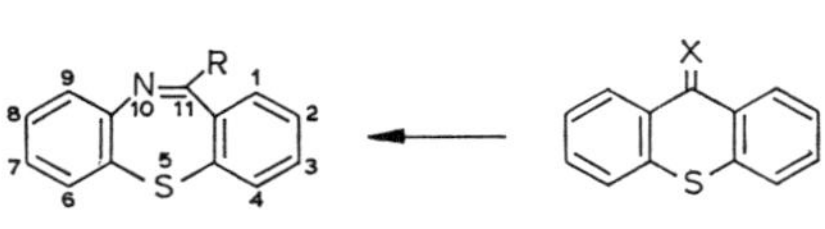

(CI) (CII)

The preparation and reactions of **dibenzo**[*b*,*f*]-1,4-**thiazepines** are similar to those of their 5-oxa analogues. Treatment of 2-formamido-, 2-acetamido- and 2-benzamido-diphenyl sulphides with polyphosphoric acid (*P. Catsoulacos* and *R. Quelet*, Compt. rend., 1964, **258**, 1251; C.A., 1965, **63**, 8364; *Geigy A-G.*, Neth. Appl. 6,604,392/1965; C.A., 1967, **66**, 78117; *F. Hunziker et al.*, Helv., 1964, **47**, 1163; 1966, **49**, 244) phosphoryl chloride (*W. F. Short*, *Broderick* and *Donaldson*, B.P. 696,473/1953; C.A., 1955, **49**, 384; *R. H. B. Galt* and *J. D. Loudon*, J. chem. Soc., 1959, 885) or a mixture of the two (*M. Protiva et al.*, Coll. Czech. Chem. Comm., 1965, **30**, 1676; *Yale* and *F. A. Sowinski*, U.S.P. 3,050,524/1961; 3,188,323/1961; C.A., 1963, **58**, 540; 1965, **63**, 11593; Arzneim. Forsch., 1966, **16**, 550), yields **dibenzo**[b,f]-1,4-**thiazepine** (CI; R = H), m.p. 124–126°, and its 11-*methyl*- (CI; R = Me), m.p. 76–77°, and 11-*phenyl*- (CI; R = Ph), m.p. 117–118°, analogues. Under mild conditions the intermediate imidoyl chloride may be isolated and cyclised.

The cyclisation of *N*-(4-phenylthiophenyl)benzamidines (*Short et al.*, J. chem. Soc., 1954, 3857) and the pyrolysis of 9-substituted thiaxanthene-9-azides (CII; X = R, N_3) (*Loudon et al., ibid.*, 1958, 1588; *M. M. Coombs, ibid.*, 4200) or diazonium salts of 2-aminodiphenyl disulphide in nitrilic solvents (*J. J. Schmidt*, Chem. Ber., 1972, **105**, 1634) have also been used. Another general route is the cyclisation with phosphoryl chloride of the 2-ureides of diphenyl sulphide, obtained from 2-isocyanatodiphenyl sulphide and an amine, which gives the 11-amino derivatives CI (R = NR^1_2) (*J. A. Coppola*, U.S.P. 3,412,193/1965; C.A., 1969, **70**, 47509; *J. Schmutz* and *Hunziker*, Helv., 1967, **50**, 245; Fr. CAM 51/1962; C.A., 1964, **68**, 8328; Neth. Appl. 293,201/1962; C.A., 1966, **64**, 8220; Schmutz, Swiss. P. 422,805/1961; C.A., 1968, **68**, 105264; 1969, **70**, 96835).

The condensation of 2-aminobenzenethiol with an activated 2-chlorobenzophenone yields compounds CI (R = Ph) (*Galt* and *Loudon, loc.cit.*: *A. D. Jarrett* and *Loudon*, J. chem. Soc., 1957, 3818) and 2-phenylthiobenzophenones give analogues by the Schmidt reaction (*Loudon et al., ibid.*, 1958, 1588).

Compounds CI undergo desulphurisation with cuprous salts in high boiling solvents to phenanthridines; an episulphide (Thiirane) intermediate is postulated (*Loudon et al., loc. cit.*; *Short et al., loc.cit.*). With periodate, hydrogen peroxide or peracid under controlled conditions *dibenzo*[b,f]-1,4-*thiazepine* 5-*oxide* is obtained from CI (R = H) (*Short, Broderick* and *Donaldson, loc.cit.*) for the 5,5-*dioxide*, m.p. 188–189°, hydrogen peroxide or chromic oxide are used. Compounds CI with 11-substituents are cleaved to nitroketones by peracetic acid (*Jarrett* and *Loudon, loc.cit.*) and to methylaminoketones by dimethyl sulphate (*Short et al.*, J. chem. Soc. 1954, 3857).

Thermal cyclisation of 2-(2-aminophenylmercapto)benzoic acid (*Galt* and *Loudon*, J. chem. Soc., 1959, 885; *Yale* and *Sowinski, loc.cit.*; *W. E. Truce* and *J. A. Simms*, J. org. Chem., 1957, **22**, 617; *E. W. McClelland* and *R. H. Peters*, J. chem. Soc., 1947, 1229; *M. Protiva et al.*, Chem. Listy, 1957, **51**, 1909; Coll. Czech. chem. Comm., 1958, **23**, 1941; 1959, **24**, 207; Experientia, 1957, **13**, 291; *R. Jaques et al.*, Helv., 1959, **42**, 1265) and its esters (*K. Hoffmann* and *E. Urech*, G.P. 1,022,228 = B.P. 802,902/1958; G.P. 1,062,246/1959; C.A., 1961, **55**, 592–3, 26001; *Protiva* and *V. Hach*, Czech. P. 90271/1959; C.A., 1960, **54**, 8865) and cyclisation of 2-isocyanatodiphenyl sulphide with aluminium chloride (*Hunziker et al.*, Helv., 1965, **48**, 336; Neth. Appl. 6,411,504/1963; C.A., 1966, **64**, 8223) provide routes to 10,11-*dihydrodibenzo*[b,f]-1,4-*thiazepin*-11-*one* (CIII), m.p. 259–260°, and many of its derivatives. 7-Substituted derivatives are also prepared by the ring expansion of 2-benzenesulphonylbenzisothiazolinone (CIV) with 4-substituted anilines (*McClelland* and *Peters, loc.cit.*).

(CIII) (CIV)

Alternative routes using the Beckmann rearrangement of CII (X = NOH) (*C. L. Kulkarni et al.*, Indian J. Chem., 1968, 225; *N. Rasanu*, Rev. Chim. Bucharest, 1969, **20**, 175; C.A., **71**, 38937) or the *Schmidt* reaction on CII (X = O) (*Truce* and *Simms, loc.cit.*), are also effective with the 5,5-dioxides of CII, yielding 10,11-*dihydrodibenzo*[b,f]-1,4-*thiazepin*-11-

one 5,5-*dioxide*, m.p. 290–292°. This dioxide, also prepared from CIII with hydrogen peroxide (*Hunziker* and *Schmutz*, Swiss P. 442,312/1963; C.A., 1968, **69**, 36195), is very readily hydrolysed to the amino acid. Oxidation with periodate yields the 1-oxides (*idem*, *Helv*., 1967, **50**, 245).

With phosphorus chlorides lactams CIII are converted into dibenzothiazepines CI (R = Cl) in which the chlorine atom is easily replaced by amines (*Dr. A. Wander, A-G.*, F.P. 1,334,944/1960; Neth. Appl. 293,201/1962; C.A., 1964, **61**, 3126; 1965, **64**, 8220; C.A., 1968, **68**, 105264, 105265). The nitrogen of thiazepinones CIII may be acylated and alkylated (*idem*, Swiss P. 422,308/1963; C.A., 1968, **69**, 36193).

Lithium tetrahydridoaluminate, preferably with the addition of aluminium chloride, reduces either CI or CIII to 10,11-**dihydrodibenzo**[*b,f*]-1,4-**thiazepine**, m.p. 115°, and its derivatives (*Yale* and *Sowinski*, *loc.cit.*; *Protiva et al.*, *loc.cit.*; *Hoffmann* and *Urech*, *loc.cit.*; Helv., 1959, **42**, 1265); tetrahydridoborate, zinc and acid (*Protiva et al.*, Coll. Czech chem. Comm., 1965, **30**, 1676) and hydrogen over palladium (*Hunziker et al.*, Helv., 1964, **47**, 1163), also reduce the 10,11-double bond of CI. Oxidation with chloranil reverses this hydrogenation (*Protiva et al.*, Coll. Czech. chem. Comm., 1971, **36**, 2226). These dihydrodibenzothiazepines may be *N*-alkylated (*Ciba Ltd.*, B.P. 802,902/1958; Belg. P. 632,563/1963; C.A., 1961, **55**, 593; 1964, **61**, 8325; *E. Habicht* and *H. Müller*, U.S.P. 3,038,896/1958; C.A., 1963 **59**, 2840; *Yale* and *Sowinski*, U.S.P. 3,071,596/1961; Fr. CAM. 67/1963; C.A., 1963, **59**, 11542; 1965, **62**, 13163; *Hoffmann-La Roche & Co.*, U.S.P. 3,188,320/1961; C.A., 1965, **63**, 16371; *Protiva et al.*, Czech. P. 86,660/1957; 117,726/1964; C.A., 1960, **54**, 2383; 1967, **66**, 76046) and *N*-acylated (*idem*, Czech. P. 88128/1958; C.A., 1960, **54**, 2373; *Yale* and *Sowinski*, U.S.P. 3,050,524/1961; C.A., 1963, **58**, 540), and if position 11 is unsubstituted it is cyanomethylated by sodium cyanide in trioxane (*L. H. Werner* and *A. P. A. Rossi*, Belg. P. 632,563/1963; C.A., 1964, **61**, 8325).

(*b*) *Thiadiazepines*

The sole example of a 1,2,3-**thiadiazepine** is 1,3-*dihydro*-2,3,4-*benzothiadiazepine*-2,2-*dioxide* (CV), m.p. 163–167° decomp., synthesised in two steps from 1,3-dihydrobenzo[*c*]thiophene (*J. F. King et al.*, Canad. J. Chem., 1971, **49**, 943). On heating, 1,3-dihydrobenzo[*c*]thiophene 2,2-dioxide is obtained, and with chlorine, 2-chloromethylbenzal chloride.

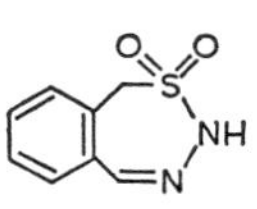

(CV)

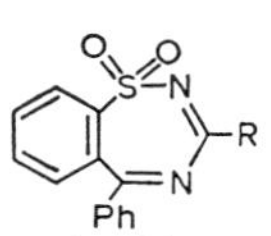

(CVI)

Many 3-substituted 1,2,4-**benzothiadiazepine** 1,1-**dioxides** (CVI) are prepared by treating 2-benzoyl-4-chlorobenzenesulphonyl chloride with amidines, guanidine or *S*-alkylisothioureas and cyclising the products with an acid catalyst. Catalytic reduction yields the 4,5-dihydro derivatives (*J. B. Wright*, J. heterocycl. Chem., 1968, **5**, 719; C.A., 1970, **72**, 90546).

2,5-*Dihydro*-1,2,4-*benzothiadiazepine* 1,1-*dioxide*, m.p. 240–241°, is made by condensing 2-sulphonamidobenzylamine with ethyl orthoformate. Either acid or alkaline hydrolysis opens the heterocyclic ring (*G. Cignarella* and *U. Teotino*, J. Amer. chem. Soc., 1960, **82**, 1594). *S*-Alkyl-1-oxides CVII are obtained in the cyclisation routes shown (*P. Stoss* and *G. Satzinger*, Chem. Ber., 1972, **105**, 2575):

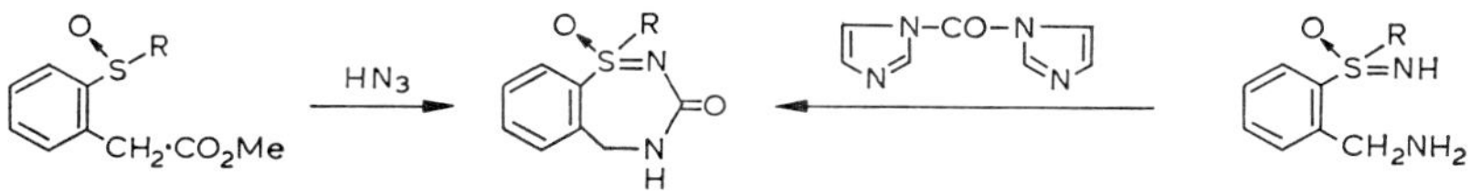

(CVII)

1,3-*Dihydro*-2,3,5-*benzothiadiazepine* 2,2-*dioxide* (CVIII), m.p. 234–235°, and some analogues are prepared by ring closure of 2-aminobenzylsulphonamide with ethyl orthoformate (*Teotino* and *Cignarella*, J. Amer. chem. Soc., 1959, **81**, 4935). Phosgene yields the corresponding 4-oxo heterocycle (*S. Hayao*, U.S.P. 3,536,702/1967; C.A., 1971, **74**, 22907).

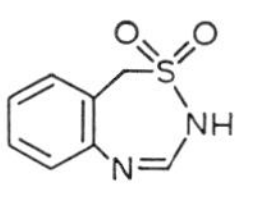

(CVIII)

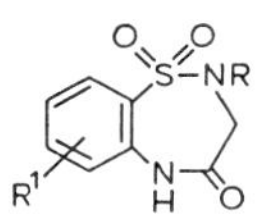

(CIX)

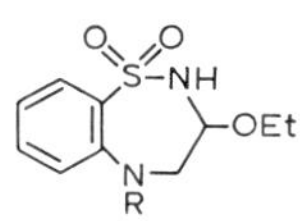

(CX)

Cyclisation of 2-chloracetamidobenzenesulphonanilides (*H. Yamamoto*, *S. Kitagawa* and *S. Sakai*, Jap. P. 21,617/1966; F.P. 1,491,502/1965; C.A., 1967, **66**, 65541; 1968, **69**, 19230) or the condensation product of 2-aminobenzenesulphonyl chloride with α-aminoacetic acid (*P. H. L. Wei* and *S. C. Bell*, U.S.P. 3,453,266/1966; C.A., 1969, **71**, 91553) yields 2,3,4,5-tetrahydro-1,2,5-benzothiadiazepin-4-one 1,1-dioxides CIX. If 2-(2,2-diethoxyethylamino)benzenesulphonamides are used compounds CX are obtained (*C. W. Whitehead* and *J. J. Traverso*, J. org. Chem., 1963, **28**, 743). Other members of this series are prepared by the following ring expansion reactions [(*A*) *L. H. Sternbach et al.*, *ibid.*, 1971, **36**, 2968; (*b*) *E. Cohnen* and *J. Mahnke*, Chem. Ber., 1972, **105**, 757]. Lithium tetrahydridoaluminate reduces both oxo groups and double bonds. *N*-Alkylation of compound CXI occurs at position 2 and compound CXII is alkylated at position 5. Methylamine and titanium chloride convert CXII into the cyclic amidine CXIII.

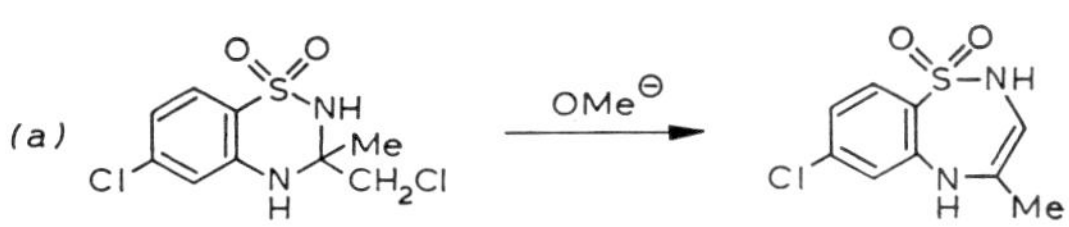

(CXI)

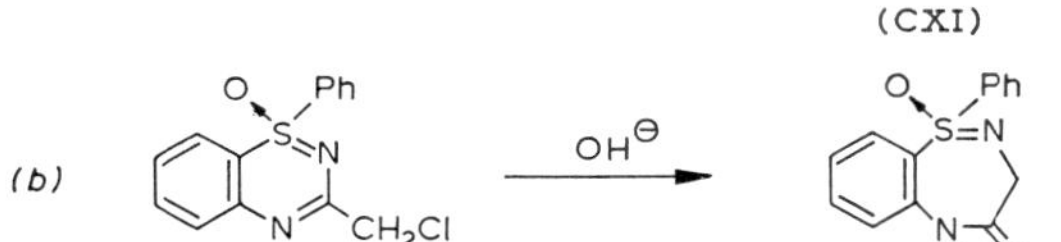

(CXII)

The alternative mode of benzene annelation is shown in the 2,2-dioxides of 1,3,4,5-**tetrahydro**-2,1,5-**benzothiadiazepin**-4-**one** (CXIV) obtained by treating *o*-phenylenediamine with dialkyl chlorosulphonylacetyl chlorides [$RR^1C(COCl)SO_2Cl$] (*Testa et al.*, Gazz., 1964, **94**, 639). If the product from ketene and sulphur dioxide is used a 2-*monoxide* of CXIV (R, R^1 = H), m.p. 195–196°, is obtained (*A. De Souza Gomes* and *M. M. Joullie*, J. heterocycl. Chem., 1969, **6**, 729.)

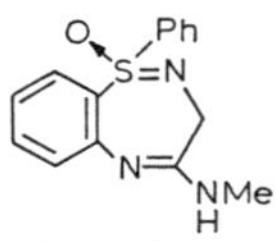
(CXIII)

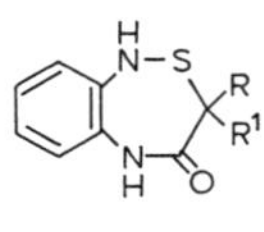
(CXIV)

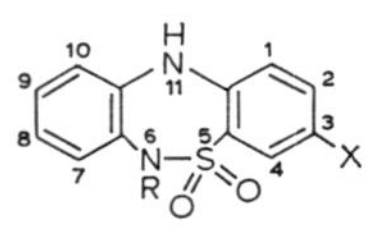
(CXV)

3-*Nitro*-6,11-*dihydrodibenzo* [c,f]-1,2,5-*thiadiazepine* (CXV; R = H, X = NO_2), m.p. 290–291°, is prepared by condensing *o*-phenylenediamine with 2-*chloro*-5-*nitrobenzenesulphonyl* chloride in pyridine (*H. Kugita* and *M. Matsubara*, Jap. P. 70,15,983/1967,; C.A., 1970, **73**, 87951). A 6-*methyl* analogue (CXV; R = Me, X = H), m.p. 201°, is obtained when 2-acetylamino-2′-bromo-*N*-methylsulphonanilide is cyclised with copper powder. 11-Alkylation is achieved in the presence of base (*A. Weber et al.*, U.S.P. 3,268,557/1963; 3,274,058/1963; C.A., 1966, **65**, 15406; 1967, **66**, 95092; *W. E. Kreigbaum*, U.S.P. 3,322,780/1963; C.A., 1967, **67**, 100161; *A. Weber*, *J. Frossard* and *D. Bouzard*, Ger. Offen. 2,033,341/1969; Fr. CAM, 242/1967; 1971, **74**, 76460, 88090).

5-Chloro-2(chloromethylsulphonylamino)benzophenone oxime cyclises with sodium methoxide to 7-chloro-5-phenyl-1,3-dihydro-2,1,4-benzothiadiazepine 2,2,4-trioxide (CXVI). The *N*-oxide function both of CXVI and its 1-(*N*)-methylated derivative is removed by phosphorus trichloride (*O. Hromatka*, *M. Knollmuller* and *D. Binder*, Monatsh., 1969, **100**, 872; U.S.P. 3,507,882/1967; C.A., 1970, **73**, 14883).

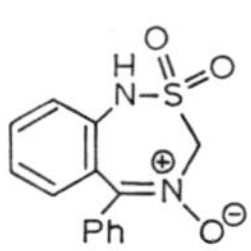
(CXVI)

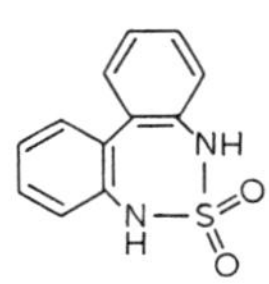
(CXVII)

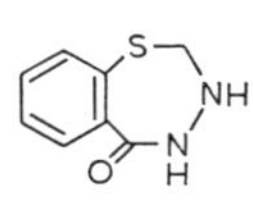
(CXVIII)

The reaction of 2,2′-diaminobiphenyls with sulphamide ($H_2NSO_2NH_2$) in pyridine gives 5,6-*dihydrodibenzo* [c,e]-1,2,7-*thiadiazepine* 6,6-*dioxide* (CXVII), m.p. 218–219°, and some substituted derivatives. Compound CXVII is also obtained by oxidising *N,N′*-diphenylsulphamide with sodium hypochlorite. On further oxidation the heterocyclic ring contracts leaving benzo [*c*]cinnolines (*W. J. Houlihan*, U.S.P. 3,329,687/1964; C.A., 1968, **68**, 49666; *C. W. Rees et al.*, J. chem. Soc. C, 1971, 993).

Thiosalicylhydrazide condenses with formaldehyde to yield a compound given the structure 2,3,4,5-*tetrahydro*-1,3,4-*benzothiadiazepin*-5-*one* (CXVIII), m.p. 204–206° (*L. Katz et al.*, J. org. Chem., 1953, **18**, 1380).

The 1,3,6-thiadiazepine which has attracted attention is the agricultural biocide **ethylenethiuram monosulphide** (CXIX or its dithione tautomer), m.p. 121–124°, prepared by controlled aerial oxidation of salts of ethylene bis(dithiocarbamate) (*I. A. M. Ford*, B.P. 789,470/1958; C.A., 1958, **52**, 13786; *G. D. Thorn* and *R. A. Ludwig*, Rec. Trav. chim., 1960, **79**, 160; Canad. J. Chem., 1960, **38**, 2349; J. appl. Chem., London 1962, **12**, 90; *M. Okai et al.*, Jap. P. 70 01604-5; C.A., 1970, **72**, 100097–8); the 4-*methyl*- (m.p. 148°) and 4,4-*dimethyl*- (m.p. 80–90°) analogues are prepared similarly (*H. Lehmann* and *G. Unterstenhoeffer*, S. African P. 67, 05623; C.A., 1969, **70**, 37851). Compound CXIX forms an adduct with carbon disulphide (*Y. Kawaoka*, Jap. P. 70 35293/1967; C.A., 1971, **75**, 5970).

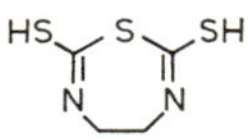

(CXIX)

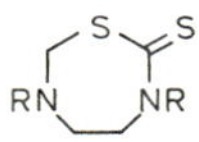

(CXX)

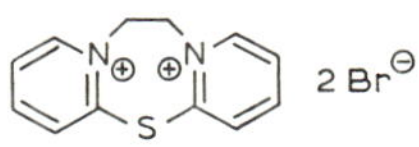

(CXXI)

Treatment of 1,3-dialkylimidazolidines with carbon disulphide gives 3,6-**dialkyl**-1,3,6-**thiadiazepane-2-thiones** (CXX) *e.g.* R = Et, m.p. 68–69 °, which extrude thioformaldehyde leaving imidazolidinthiones (*R. A. Donia et al.*, J. org. Chem., 1949, **14**, 952). Compound CXXI from 2,2′-dipyridyl sulphide and ethylene dibromide is related to the herbicide **"Paraquat"** (*L. A. Summers*, Angew. Chem. intern. Edn., 1966, **5**, 6C5; Tetrahedron, 1968, **24**, 5433).

The most thoroughly investigated members of this group are the 1,4,5-**thiadiazepines.** Phenacyl sulphides and sulphones cyclise readily with hydrazine hydrate to 3,6-**diaryl**-2,7-**dihydro**-1,4,5-**thiadiazepines** (*e.g.* CXXII; R = H, Ar = Ph, m.p. 174–175°) and their 1,1-*dioxides* (*e.g.* R = H, Ar = Ph, m.p. 196°), respectively. A number of analogues are prepared this way, including the *cis*- (m.p. 123°) and *trans*- (m.p. 183°) isomers of 2,7-*dimethyl*-3,6-*diphenyl*-2,7-*dihydro*-1,4,5-*thiodiazepin* (CXXII; R = Me, Ar = Ph), but substituted hydrazines give only monoketonic derivatives (*J. D. Loudon* and *L. B. Young*, J. chem. Soc., 1963, 5496; *E. Fromm* and *A. Erhardt* Ber., 1921, **54**, 187; *I. Sataty*, J. med. Chem., 1967, **10**, 501; J. heterocycl. Chem., 1970, **7**, 431; *R. Bourdon et al.*, Chim. Ther., 1971, **6**, 93; C.A., 1971, **75**, 63752).

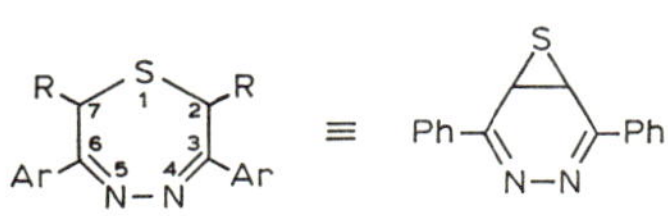

(CXXII) (CXXIII)

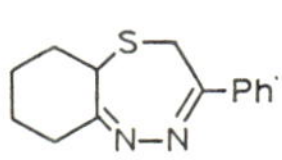

(CXXIV)

Cyclohexanone-2-phenacyl sulphide gives the hexahydrobenzologue CXXIV, m.p. 116–117° (*Loudon et al.*, J. chem. Soc., 1964, 591).

3,6-Diphenyl-2,7-dihydro-1,4,5-thiadiazepine (CXXII; R = H, Ar = Ph) is not dehydrogenated to the fully unsaturated heterocycle, but with *N*-bromosuccinimide, or on pyrolysis, it undergoes ring-contraction to 3,6-diphenylpyridazine suggesting the intermediacy of the thiadiazepine tautomer CXXIII. Thermal instability is more marked in the 1-oxide and 1,1-dioxide, and ring contraction to pyrazoles has also been observed (*Loudon and Young, loc.cit.*).

Poor yields of **dibenzo**[*b,f*]-1,4,5-**thiadiazepine** (CXXVI; X = S, R, R^1 = H), m.p. 140–142°, are obtained by reducing 2,2′-dinitrodiphenyl sulphide (CXXV; X = S,

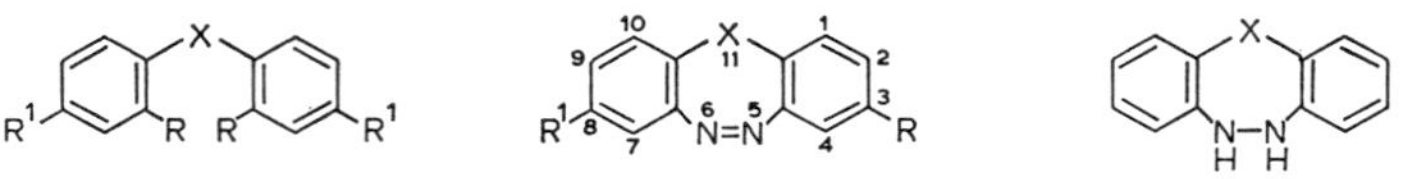

(CXXV) (CXXVI) (CXXVII)

R = NO_2, R^1 = H) with sodium sulphide, or lithium tetrahydridoaluminate (*N. L. Allinger* and *G. A. Youngdale*, Tetrahedron Letters, 1959, **9**, 10; J. Amer. chem. Soc., 1962, **84**, 1020). Aluminium and isopropanol are used for CXXVI (X = S; R, R^1 = CF_3), m.p. 154°, but zinc and sodium acetate give the 5-*oxide* of CXXVI (X = S, R, R^1 = CF_3), m.p. 181–182° (*H. H. Szmant et al., ibid.*, 1956, **78**, 458). *Dibenzo*[b,f]-1,4,5-*thiadiazepine* 5-*oxide*, m.p. 166–167°, and analogues result when the dinitro sulphides CXXV (X = S, R = NO_2) are treated with zinc in aqueous sodium hydroxide, but some 1,3,5-thiadiazepines CXXVI (X = S) and phenazines may also be produced (*M. F. Grundon et al.*, J. chem. Soc., B., 1966, 255, 260). This *N*-oxide is also found after the oxidative ring closure of the dihydroxylamine CXXV (X = S, R = NHOH, R^1 = H) (*Y. L. Chow* and *Szmant*, J. org. Chem., 1971, **36**, 2887).

An alternative route to dibenzothiadiazepines bearing electron releasing substituents (*e.g.* CXXVI; R = NMe_2, R^1 = H, X = S, m.p. 120–121°) is *via* diazotisation and internal coupling of a 2-aminodiphenyl sulphide (*W. Bosshard, H. Bossard* and *H. E. Wegmüller*, U.S.P. 3,133,086/1960; C.A., 1964, **61**, 16200).

Further reduction of CXXVI (X = S, R, R^1 = H) with hydrazine and a palladium catalyst (*Allinger* and *Youngdale, loc.cit.*) or its 5-oxide with zinc and acetic acid (*Chow* and *Szmant, loc.cit.*), yields 5,6-**dihydrodibenzo**[*b,f*]-1,4,5-**thiadiazepine** (CXXVII; X = S), m.p. 102–103°. In acid solution it undergoes disproportionation to 2,2′-diaminodiphenyl sulphide (CXXV; X = S, R = NH_2, R^1 = H) and thiadiazepin CXXVI (X = S, R, R^1 = H) and also shows ring-contraction to phenothiazines (*L. D. Gartung* and *H.* J. org. Chem., 1969, **34**, 1013).

Peracetic acid cyclises the sulphoxide CXXV (X = SO, R = NHOH, R^1 = H) to *dibenzo*[b,f]-1,4,5-*thiadiazepine* 5,11-*dioxide*, m.p. 208° decomp., which may be reduced to a 5,6-*dihydro derivative* (CXXVII; X = SO), m.p. 188–189° decomp., which is oxidised by ferric chloride to *dibenzo*[b,f]-1,4,5-*thiadiazepine* 11-*oxide* (CXXVI; X = SO, R, R^1 = H), m.p. 138° decomp. (*Chow* and *Szmant, loc.cit.*). Both the 11-oxide and the 5,11-dioxide are less stable to thermolysis than analogous sulphides (CXXVI; X = S) and sulphones (CXXVI; X = SO_2) and undergo ring-contraction to benzo[*c*]cinnolines. Sulphides (CXXVI; X = S) do, however, show this contraction with peracids (*Szmant et al.*, Chem. Comm., 1970, 1604).

Dibenzo[b,f]-1,4,5-*thiadiazepine* 11,11-*dioxide* (CXXVI; X = SO_2, R, R^1 = H), m.p. 173–174°, in which the heterocyclic ring is now considered to exist only in the "boat" form (*Allinger* and *Youngdale*, J. org. Chem., 1959, **24**, 2059) is obtained by oxidative cyclisation of 2,2′-diaminodiphenyl sulphone (CXXV; X = SO_2, R = NH_2, R_2 = H) with phenyl iodosoacetate. Oxidation of product or starting material with peracetic acid and treatment of the diamine with hypohalite, or the dihydroxylamine CXXV (X = SO_2, R = NHOH, R^1 = H), with sodium hydroxide yields *dibenzo*[b,f]-1,4,5-*thiadiazepine* 5,11,11-*trioxide*, m.p. 257–258°. Reduction of the trioxide catalytically, or with tin and hydrochloric acid, or lithium tetrahydridoaluminate yields 5,6-*dihydrodibenzo*[b,f]-1,4,5-*thiadiazepine* 11,11-*dioxide* (CXXVII; X = SO_2), m.p. 198–199°, which may be 5,6-diacylated, and reoxidised to CXXVI (X = SO_2, R, R^1 = H) by air (*Szmant et al.*, J. Amer. chem. Soc., 1956, **78**, 458; 1957, **79**, 4382, 5583; *K. Michel* and *M. Matter*, Helv., 1961, **44**, 2204).

Nitration of this trioxide gives the 9-nitro- then the 7,9-dinitro-derivatives. Stannous chloride cleaves the heterocyclic ring of the former but piperidine and hydrogen sulphide reduce it to the 2-amino derivative of CXXVII (X = SO_2) (*Chow* and *Szmant*, J. org. Chem., 1971, **36**, 2889). 3,8-Dichloro- (CXXVI; X = SO_2, R, R^1 = Cl) (*idem, ibid.*, 1961, **26**, 4173) and 3,8-dimethyl-derivatives (CXXVI; X = SO_2, R, R^1 = Me) (*Grundon et al., loc.cit.*) show similar chemistry.

(*c*) *Dithiazepines*

Cystamine, $(H_2NCH_2CH_2S\text{–})_2$ is oxidised by diamino oxidases of plant or animal origin to 6,7-**dihydro**-3*H*-1,2,5-**dithiazepine** (*cystaldimine*) (CXXVIII), m.p. 155–156°. This cyclic imine is further metabolised to thiocysteamine, trapped by iodoacetic acid as carboxymethylthiocystamine, $(H_2NCH_2{\cdot}CH_2S_2CH_2{\cdot}CO_2H)$ and thiotaurine $(H_2NCH_2{\cdot}CH_2SO_2SH)$ (*D. Cavallini et al.*, Biochem. Biophys. Acta, 1967, **24**, 353; C.A., 1961, **55**, 27477; *B. Mondovi* and *L. Tentori*, Giorn. Biochim. 1961, **10**, 444; C.A., 1962, **57**, 3904; *C. DeMarco et al.*, C.A., 1964, **61**, 12254; Biochem. Biophys. Acta, 1964, **92**, 418; 1965, **100**, 89).

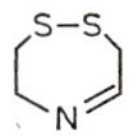

(CXXVIII)

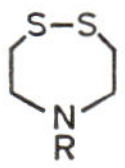

(CXXIX)

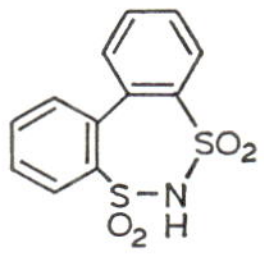

(CXXX)

1,2,5-**Dithiazepanes** (CXXIX) with various substituents have been prepared by cyclising di(2-mercaptoethyl)amines with iodine (*M. Thiel et al.*, Ann, 1959, **622**, 107) or sodium in liquid ammonia (*S-H. Chu* and *H. G. Mautner*, J. org. Chem., 1961, **26**, 4498). Alternatively di(2-chloroethyl)amines may be cyclised in alkaline thiosulphate or sodium disulphide solutions. A polymer may form but depolymerisation is effected by alkaline hydrolysis or reduction with zinc and hydrochloric acid followed by basification.

1,2,5-**Dithiazepane** (CXXIX; R = H), m.p. 178° — resolidifying and decomposing at 230°, and 5-*methyl*-1,2,5-*dithiazepane* (CXXIX; R = Me), *picrate*, m.p. 209°, *methiodide*, m.p. 212° decomp., are prepared this way (*E. Boyland* and *R. Nery*, J. chem. Soc., 1961, 679; *W. H. H. Gunther* and *Mautner*, J. Amer. chem. Soc., 1960, **82**, 2762).

Treatment of biphenyl-2,2′-disulphonyl chloride with ethanolic ammonia gives *dibenzo*-[c,f]-1,7,2-*dithiazepine* 5,5,7,7-*tetroxide* (*biphenyl*-2,2′-*disulphonimide*) (CXXX), m.p. 250–251° (*W. V. Farrar* J. chem. Soc., 1960, 3063).

4. Rings containing nitrogen and selenium

If the zinc salt of 2-aminobenzeneselenol is heated in ethanolic hydrochloric acid with α,β-unsaturated ketones, 2,2,4-**trimethyl**-2,3-**dihydro**-1,5-**benzoselenazepine** (I), m.p. 69°, and analogues are formed. The 4-methyl group condenses with 4-dimethylaminobenzaldehyde and ethyl orthoformate. With crotonic acid 2-*methyl*-2,3,4,5-*tetrahydro*-1,5-*benzoselenazepin*-4-*one* (II), m.p. 221–222°, is the product (*L. K. Mushkalo* and *D. I. Sheiko*, Zhur. obshcheĭ Khim., 1963, **33**, 157; Ukrain. Khim. Zhur., 1964, **30**, 984, 502; C.A., 1964, **61**, 4359, 5654).

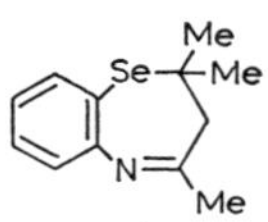

(I)

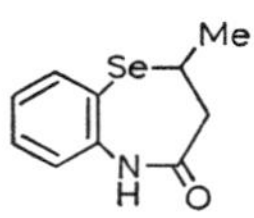

(II)

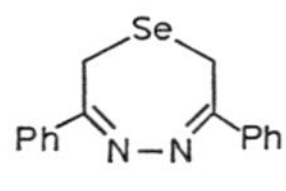

(III)

3,6-**Diphenyl**-2,7-**dihydro**-1,4,5-**selenadiazepine** (III), m.p. 194–195°, is obtained by treating α,α′-selenodiacetophenone with one mole of hydrazine. It exists in a "skew-boat" conformation, like the sulphur analogue, and on pyrolysis yields 3,6-diphenylpyridazine (*E. Ajello*, J. heterocycl. Chem., 1972, **9**, 1427).

5. Rings containing nitrogen, oxygen and sulphur

Compounds having a seven-membered ring with three different heteroatoms occur mainly as "accidentals" and little systematic study of these systems has been made.

Styrene oxide reacts with *N*-sulphinylbenzamide to give a monocyclic oxathiadiazepin II, probably *via* the five-membered heterocycle I (*O. Tseug* and *S. Mataka*, Bull. chem. Soc., Japan, 1971, **44**, 28336):

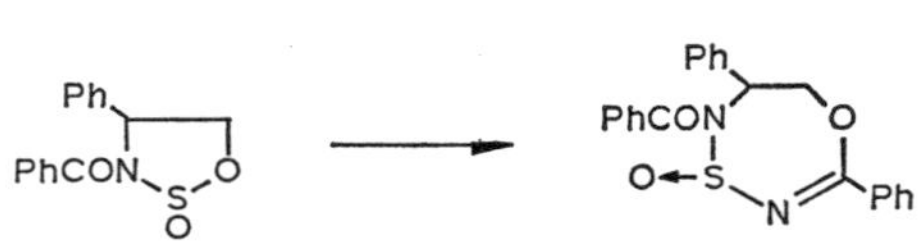

(I) (II)

A great many dihydro-1,2,5-benzoxathiazepines (III) have been made by condensing an arylidenearylamine *N*-oxide (or nitrone) with a sulphene generated *in situ* from an alkane sulphonyl chloride and triethylamine; again an intermediate with a five-membered ring is postulated:

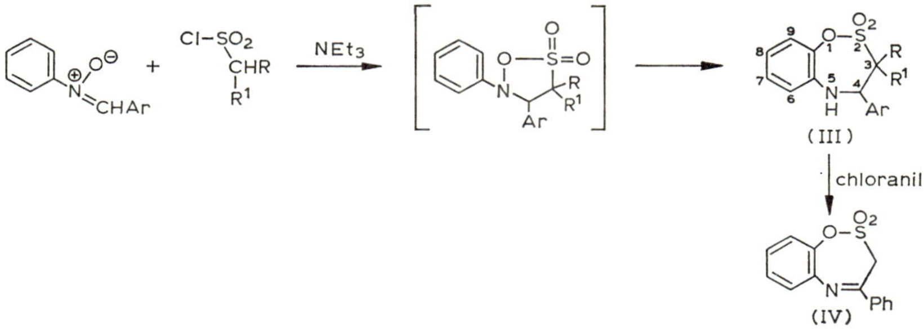

(III)

(IV)

4-*Phenyl*-3,4-*dihydro*-5H-1,2,5-*benzoxathiazepine* (III; Ar = Ph, R, R^1 = H), m.p. 168°, is oxidised by chloranil to **4-phenyl-5*H*-1,2,5-benzoxathiazepine** (IV), m.p. 161–162°, brominated at position 8, and readily acylated and nitrosated. Treatment of the 3-chloro analogue (III; Ar = Ph, R = H, R^1 = Cl) with piperidine also yields IV. Alkaline hydrolysis or reduction with lithium tetrahydridoaluminate opens the heterocyclic ring of III, while acid hydrolysis cleaves IV (*W. E. Truce et al.*, Chem. and Ind., London, 1965, 1870; J. org. Chem., 1969, **34**, 3097; *F. Eloy* and *A. van Overstraeten*, Bull, Soc. chim. Belges, 1967, **76**, 63; U.S.P. 3,420,849/1966; C.A., 1969, **71**, 81447). If 2-aminobenzenethiol is mixed with the reactive intermediate formed from sulphur dioxide and ketone 3,4-*dihydro*-5H-1,2,5-*benzoxathiazepin*-4-*one* 2-*oxide* (V), m.p. 171–172°, is obtained (*A. De Souza Gomes* and *M. M. Joullie*, J. heterocycl. Chem., 1969, **6**, 729).

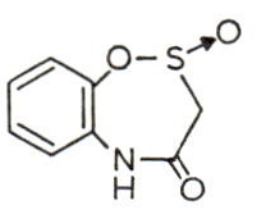

(V)

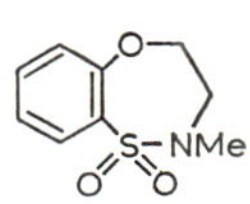

(VI)

4-*Methyl*-2,3-*dihydro*-4H-1,5,4-*benzoxathiazepine* (VI), m.p. 118–119°, is prepared by cyclising *N*-(2-hydroxyethyl) *N*-methyl-2-nitrobenzenesulphonamide in dilute aqueous sodium hydroxide (*K. G. Kleb*, Angew. Chem. intern. Edn., 1968, **7**, 291).

11H-*Dibenzo*[b,f]-1,4,5-*oxathiazepine* 10,10-*dioxide* (VII; R = H), m.p. 140–141°, is obtained by pyrolysing 2-phenoxybenzenesulphonyl azide in dodecane (*R. A. Abramovitch et al.*, J. Amer. chem. Soc., 1969, **91**, 1219; *R. H. Thomson et al.*, J. chem. Soc. Perkin I, 1972, 2862).

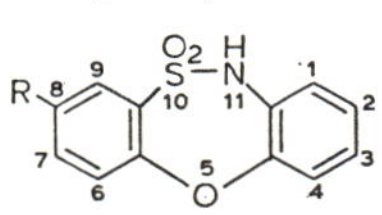

(VII)

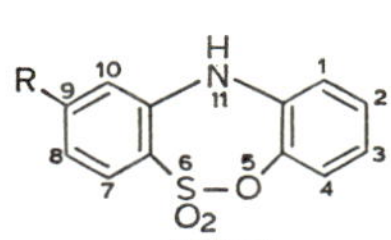

(VIII)

Alternative routes to this sultam are the reaction of 2-aminobenzenethiols with 2-halogenosulphonyl halides in the presence of pyridine, and the cyclisation of 2′-aminodiaryl ether-2-sulphonic acids with phosphoryl chloride (*E. Enders*, *F. Muth* and *H. Herlinger*, Belg. P. 622,185/1961; C.A., 1963, **59**, 11346). The 8-*nitro* derivative (VII; R = NO_2), m.p. 156–158°, is hydrogenated to the 8-*amino* derivative (VII; R = NH_2), m.p. 180–182°, without disrupting the ring, and the benzene rings of sultams VII may be chlorinated. The nitrogen at position 11 is readily alkylated by alkyl halides (*S. Umemoto* and *T. Chika*, Jap. P. 71 03060; C.A., 1971, **74**, 125748).

Irradiation of 4-tolyl 2,4-dinitrobenzenesulphenate gives **11*H*-dibenzo[*c*,*f*]-1,2,5-oxathiazepine 6,6-dioxide** (VIII), m.p. 285–289°, together with other products (*T. Nakano et al.*, J. chem. Soc. C, 1968, 322).

Chapter 53

Compounds with Seven-Membered and Larger Rings possessing One or More Unusual Hetero Atoms

R. S. EDMUNDSON

General methods for the synthesis of compounds having larger rings containing unusual hetero atoms are the same as, or at least bear some resemblance to, those available for five- and six-membered ring compounds. For many of these reactions, however, yields rapidly diminish with increasing size of the ring formed. The most widely used reactions involve intramolecular addition of element hydrides to olefinic and acetylenic bonds, and the interaction of di-Grignard and di-lithium reagents with dihalides or diesters of the hetero element. Many compounds within this section, particularly those with more than one atom of the hetero element, are obtained by highly specific processes, while others are obtained in trace amounts as by-products, these being sometimes oligomers of potentially smaller ring compounds. Physical data of the customary type, *e.g.* m.p., b.p., are very often not available for such compounds, and elucidation of molecular structures has relied heavily on spectroscopic techniques, particularly n.m.r. and mass spectrometry.

1. Magnesium and mercury compounds

Relatively little information is available on compounds of elements of Group II of the Periodic Table.

Addition of dioxane to solutions of alkanedi(magnesium bromides) in tetrahydrofuran yields solutions containing monomeric species which on the

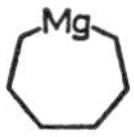

(I)

basis of chemical evidence and conductivity measurements, they are considered to be magnesiacycloalkanes. A solution of hexanedi(magnesium bromide) so treated yields magnesiacycloheptane (I). In the reactions between diisopropyl ketone and (i) magnesiacycloalkanes, or (ii) alkanedi(magnesium bromide)s, the products are (iii) unreacted ketone, (iv) its reduction product, 2,4-dimethylpentan-3-ol, (v) diisopropylalkane tertiary alcohols (vi) diisopropylalkene tertiary alcohols, and (vii) alkanebis(diisopropyl tertiary alcohols). The main difference between (i) and (ii) in outcome of reactions at 10-20° during 24 h, is that (ii) produce more of (iv) than do (i), but also that within each group (i) and (ii), those compounds with six carbon atoms also produce more of (iv) than do the compounds with five and four carbon atoms. These results were considered to afford evidence for the Whitmore E2 mechanism for the reducing action of Grignard and similar reagents since the setting up of the necessary transition state for *cis* reduction is sterically difficult for the five- and four-carbon compounds, but much easier for the six-carbon compounds (*B. Denise*, *J. F. Fauvarque* and *J. Ducom*, Tetrahedron Letters, 1970, 335).

Dimercuracycloalkanes are not well-characterized compounds; they have been reported as products from the direct interaction of mercury and α,ω-dilithioalkanes in ether at fairly low temperatures. When photolyzed, they decompose to give complex mixtures of saturated and unsaturated hydrocarbons (*G. Beinert*, Bull. Soc. Chim. Fr., 1970, 2284; *T. Sakakibara*, *S. Kurata* and *Y. Odaira*, Bull. chem. Soc. Japan, 1970, **43**, 296). On the other hand, compounds with aryl carbon–mercury bonds are better known.

Di-1,8-**naphthylene-dimercury** (II), m.p. 377° decomp., is formed when 1,8-di(chloromercuri)naphthalene is heated with sodium iodide in boiling ethanol; it may be recrystallized from dimethylformamide, and when treated with mercuric chloride it is converted to the starting material (*R. L. Letsinger et al.*, J. org. Chem., 1965, **30**, 807):

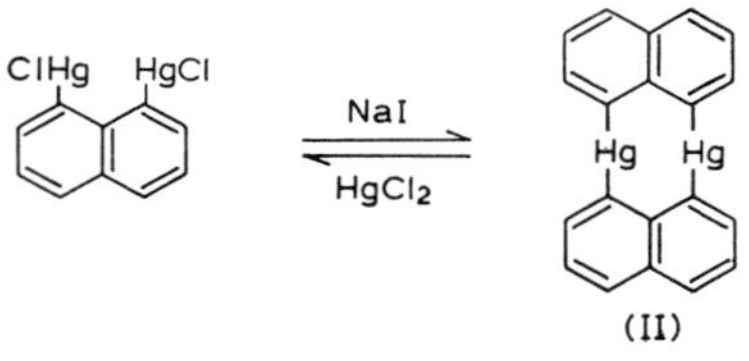

(II)

Perfluorotribenzo-1,4,7-**trimercuracyclonona-**2,5,8-**triene** (IV) is formed *via* the trimeric carboxylate III, when mercury tetrafluorophthalate is pyrolyzed (*P. Sartori* and *A. Gollock*, Chem. Ber., 1968, **101**, 2004), and also when 1,2,3,4-tetrafluorodiiodobenzene is

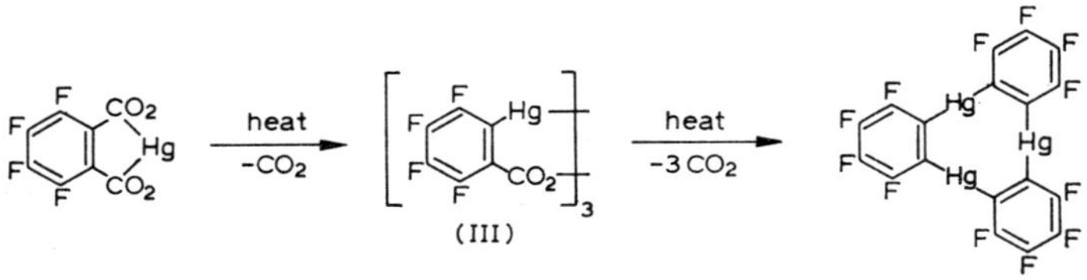

(III)

(IV)

heated with mercury in a sealed tube at 300° (*C. M. Woodard, G. Hughes* and *A. G. Massey*, J. organometal. Chem., 1976, **112**, 9). When heated with silver powder, IV is converted into perfluorobiphenylene, and with iodine in boiling toluene it yields the same tetrafluoroiodobenzene (*Sartori* and *Gollock, loc.cit.*).

2. Boron compounds

Hydroboration studies, particularly those carried out by *H. C. Brown* and co-workers, have demonstrated the formation of saturated, seven-membered, boron-containing rings, on the basis of their oxidation under alkaline conditions to alkane-1,6-diols.

(*a*) *Unsaturated systems containing boron; borepins*

Much of the interest in unsaturated systems containing boron has centred around their potential aromatic character; **borepin** (I) is isoelectronic with the tropylium cation, and some degree of aromatic character might therefore be expected, but the existing evidence relating to such compounds is by no means decisive.

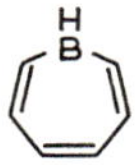

(I)

Borepin itself does not appear to have been prepared, although heptaphenylborepin (II) is thought to be a valence isomer in the final stages in the reaction between diphenylacetylene and pentaphenylborole:

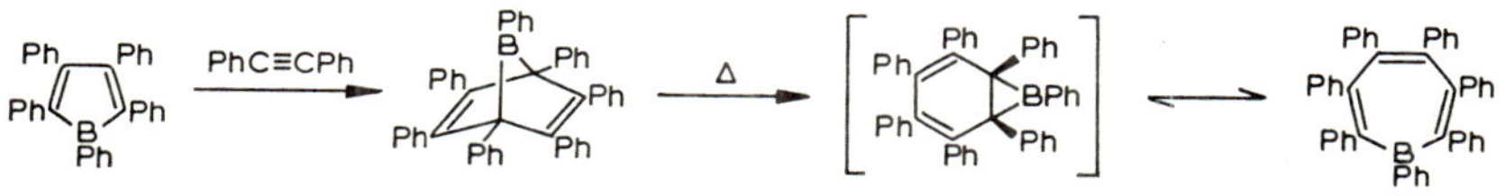

Using 1-phenyltetra-*p*-tolylborole in the above scheme, the presence of both phenyl and *p*-tolyl groups attached to boron was demonstrated by ^{1}H-n.m.r. spectroscopy. The heptaphenyl compound fluoresces greenish yellow and when pure is only slowly oxidized in air. A mixture of hexaphenylcyclohexadienes results when the heptaphenylborepin is treated with alkaline mercuric chloride (*J. J. Eisch* and *J. E. Galle*, J. Amer. chem. Soc., 1975, **97**, 4436).

3-Hydroxy-3*H*-benzoborepin (III, R = OH), m.p. 110°, is formed by hydrolysis of the corresponding chloride obtained by an exchange reaction between boron trichloride and a dialkylbenzostannepin (*G. Axelrad* and *D. Halpern*, Chem. Comm., 1971, 291); the 3-*phenyl* compound (III, R = Ph), m.p. 134–137°, is formed similarly using dichloro-(phenyl)borane:

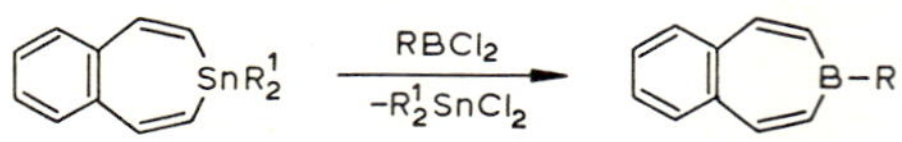

(III)

These compounds, unlike simple dihydroborepins, exhibit unusual stability towards oxygen, and this, together with the u.v. and n.m.r. data suggests that, relative to the benzotropylium cation, at least some degree of aromaticity is associated with the heterocyclic ring (*A. J. Leusink et al.*, Tetrahedron Letters, 1967, 1263).

E. E. van Tammelin, G. Brieger and *K. G. Untch* (Tetrahedron Letters, 1960, 14) prepared 5-*butoxy*-5H-10,11-*dihydrodibenzo* [b,f]*borepin* (IV), b.p. 150–157°/0.1 mm, from 2,2′-dilithiobibenzyl and tributyl borate, and by treating the *anhydride* (V), m.p. 145–146°, b.p.

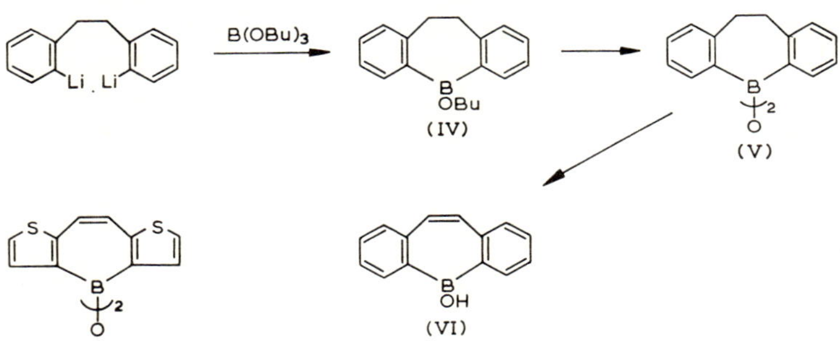

(IV) (V) (VI) (VII)

265–270°/2 mm, with *N*-bromosuccinimide followed by methoxide, finally obtained 5-**hydroxydibenzo** [*b,f*] **borepin** (VI) as its 5-(2-*aminoethoxy*) derivative, m.p. 222–226°. The heterocyclic ring in compound IV is broken by bromine and by hydrogen peroxide giving 2,2′-dibromo- and 2,2′-dihydroxybibenzyl, respectively, and hydrogenolysis of the same compound in the presence of palladium–charcoal yields bibenzyl itself (*Letsinger* and *I. H. Skoog*, J. Amer. chem. Soc., 1955, **77**, 5176; U.S.P. 2,839,566/1958; C.A., 1961, **55**, 21153).

Examples also exist of the borepin system fused to other heterocyclic rings, *e.g.* 4,4′-**oxybis(dithieno** [3,2-*b*:2′,3′-*f*] **borepin**) (VII), m.p. 256–258° (*S. Gronowitz, P. Gassne* and *B. Yom-Tov*, Acta Chem. Scand., 1969, **23**, 2927; *A. T. Jeffries* and *Gronowitz*, Chem. Scr., 1973, **4**, 183, C.A., 1974, **80**, 27312). An X-ray analysis has demonstrated that the molecule consists of two planar systems linked through oxygen in a propeller-like conformation (*B. Aurivillius*, Acta Chem. Scand., 1974, **B28**, 998).

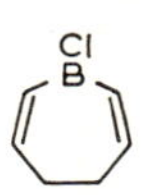

(VIII)

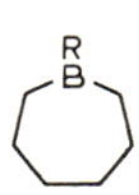

(IX)

1-Chloro-4,5-dihydroborepin (VIII) is also obtainable from a dihydrostannepin by exchange with boron trichloride (*D. Sheehan*, Diss. Abs., 1965, **25**, 4417). There has been a claim (*B. Wrackmeyer* and *H. Nöth*, J. organometal. Chem., 1976, **108**, C21) that alkynes may be inserted into the borolane ring with the resultant formation of tetrahydroborepins, but as a class of compounds, the latter are very poorly characterized.

(*b*) *Saturated systems; borepanes and related compounds*

The hydroboration of biallyls leads to derivatives of **borepane** (IX, R = H). The hydroboration of biallyl itself (hexa-1,5-diene) has been quite extensively investigated; several cyclic compounds have been isolated. Russian workers described the formation of 6-(1-*borepanyl*)*hex*-1-*ene*, b.p. 57–59°/3 mm, n_D^{20} 1.4630, and 1,6-*di*(1-*borepanyl*)*hexane*, b.p. 147–148°/1.2 mm (*L. I. Zarkharkin* and *A. I. Kovredov*, Izvest. Akad. Nauk S.S.S.R., Ser. Khim., 1962, 357; 1969, 106; *B. M. Mikhailov*, *L. S. Vasil'ev* and *E. N. Safonova*, Doklady Akad. Nauk S.S.S.R., 1962, **147**, 630; U.S.P. 3,008,997/1958; C.A., 1962, **57**, 12534; U.S.P. 3,060,244/1962; C.A., 1963, **58**, 5573):

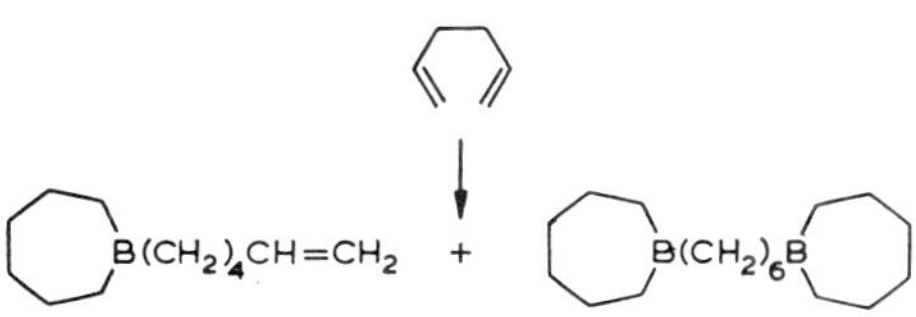

According to *H. C. Brown's* group (*Brown*, *E.-I. Negishi* and *P. L. Burke*, J. Amer. chem. Soc., 1972, **94**, 3561) the preparation of borepanes with exocyclic B–H bonds by hydroboration of hexa-1,5-diene and cyclo-octa-1,5-diene is unfortunately accompanied by the formation of smaller rings; nevertheless, they have devised a synthetic sequence capable of high yields of pure borepanes. Hexa-1,5-diene is hydroborated with 9-borabicyclo-[3.3.1]nonane in tetrahydrofuran, when the product, hexane-1,6-di(9-borabicyclo [3.3.1]-nonane), on further reaction with borane in tetrahydrofuran, yields borepane quantitatively; with trimethyl borate, 1-*methoxyborepane*, b.p. 74–76°/32 mm, is formed. The latter is also obtained by reaction between borepane and methanol (*idem*, *ibid.*, 1973, **95**, 3654).

Reactions between dimeric borepane and its ring-substituted derivatives and terminal alkenes to give ***B*-alkylborepanes,** *e.g.* 1-*hexyl*-3,6-*dimethylborepane*, b.p. 108–112°/3 mm, and alkynes to give **1,1-diborepanylalkanes** have been investigated (*Brown* and *Negishi*, J. organometallic Chem., 1971, **28**, Cl; *G. Zweifel*, *R. P. Fisher* and *A. Horng*, Synthesis, 1973, 37); in many cases the reaction product has to be worked up immediately otherwise isomerization to the borinane may take place (*R. Köster* et al., Ann., 1964, **672**, *L. Köster*, *W. Larbig* and *G. W. Rottermund*, *ibid.*, 1965, **682**, 21).

1-Chloroborepane (IX, R = Cl), b.p. 70°/60 mm, n_D^{20} 1.4562, is obtained by treatment of di(1-borepanyl)alkanes and related compounds with boron trichloride or by slowly distilling hexamethylene bis(dichloroborane), $(CH_2)_2(BCl_2)_6$ (*Kovredov* and *Zakharkin*, Izvest. Akad. Nauk S.S.S.R., Ser. Khim., 1964, 50) as well as by hydroboration of biallyl by chloroborane (*T. A. Shchegoleva et al., ibid.*, 1964, 365). Other known 1-substituted borepanes include *ethylamino*-, b.p. 82.5–84.5°/16 mm, and *butylthio-borepane*, b.p. 88–89°/

3 mm (*Mikhailov* and *T. K. Kozminskaya, ibid.*, 1963, 1703; *Mikhailov et al., ibid.*, 1963, 646; *Mikhailov, Vasil'ev* and *Safonova, loc.cit.*; *Shchegoleva et al.*, Zhur. obshcheĭ Khim., 1965, **35**, 1078).

Hydroboration of 1-*cis*-5-*trans*-9-*trans*-cyclododecatriene with triethylamine-borane, $Et_3N–BH_3$, yields a mixture of all cis-9b-*boraperhydrobenz*[c,d]*azulene* (X), b.p. 113°/10 mm, and all-*cis*-9b-boraperhydrophenalene (XI), from which the former may be

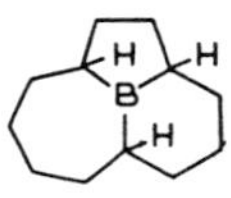
(X)

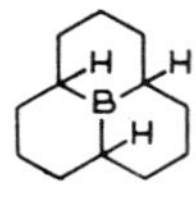
(XI)

separated; it forms a pyridine adduct, and at 210° it rearranges to the boraperhydrophenalene. Methanolysis of *cis-trans*-9b-boraperhydrophenalene and further reaction of the products with acetylacetone causes breakage of one of the rings and formation of a mixture of the *complexes* XII, m.p. 115°, and XIII, m.p. 65° (*Köster* and *Rottermund*, Ann., 1965, **689**, 40; 1965, **686**, 153).

1,5-**Dialkoxy-1,5-diborocanes** (XIV), 1,5-*dimethoxy-*, b.p. 55–56°/2 mm, 1,5-*dibutoxy-*, b.p. 106–107°/2 mm, n_D^{20} 1.4510, are prepared by hydroboration of alkoxydiallylboranes, and by alcoholysis of the polymeric products obtained by the hydroboration of triallylborane; in addition, the dimethoxy compound has been obtained by heating methoxybis-

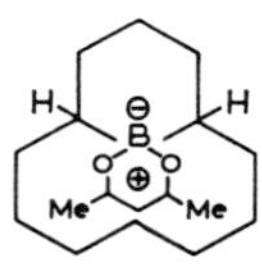
(XII)

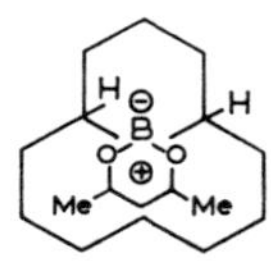
(XIII)

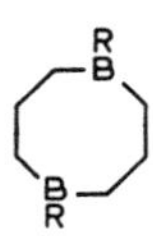
(XIV)

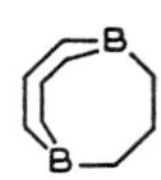
(XV)

(dimethoxyborylpropyl)borane, $MeOB((CH_2)_2B(OMe)_2)_2$, at 200° (*Mikhailov* and *V. F. Pozdnev*, Izvest. Akad. Nauk S.S.S.R., Ser. Khim., 1962, 1475, 1698; *Mikhailov, A. N. Blokhina* and *Pozdnev, ibid.*, 1965, 197; *Mikhailov, Pozdnev* and *V. G. Kiselev*, Doklady Akad. Nauk S.S.S.R., 1963, **151**, 577). 1,5-*Dichloro*-1,5-*diborocane*, b.p. 66°/8 mm, n_D^{20} 1.4813, is formed in 80% yield when mixtures of boron trichloride and trialkylboranes are treated with traces of tetrabutyldiborane and in 70% yield, when trimethylene bis(dichloroborane), $(CH_2)_3(BCl_2)_2$, is heated (*Kovredov* and *Zakharkin, loc.cit.*)

Air-sensitive 1,5-**diborabicyclo**[3.3.3]**undecane** (XV) has been recorded as a volatile product of the pyrolysis of the polymer from triallylborane and triethylamine-borane (*N. N. Greenwood, J. H. Morris* and *J. C. Wright*, J. chem. Soc., 1964, 4753).

The problem of the structure of the hydroboration products of conjugated dienes seems to have now reached a conclusion. *Köster* and co-workers have argued (Angew. Chem.,

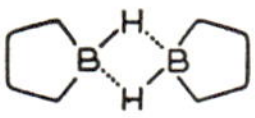

(XVI)

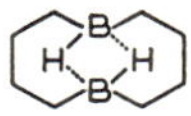

(XVII)

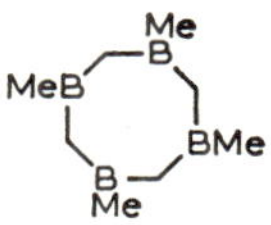

(XVIII)

1959, **71**, 520; 1960, **72**, 626) that hydroboration of buta-1,3-diene and its derivatives yields dimers of the borolane type linked by external hydrogen bridges (structure XVI). On the other hand, *Brown* and his co-workers (J. Amer. chem. Soc., 1962, **84**, 183) from the start favoured the 1,6-diboracyclodecane structure (XVII) with internal hydrogen bonding. Further evidence to support the latter claim has appeared, and is based upon (*a*) ^{11}B- and ^{1}H-n.m.r. spectroscopy (*D. E. Young* and *S. G. Shore*, *ibid.*, 1969, **91**, 3497), (*b*) the synthesis, from borolanes and borane of compounds known to possess the diboracyclodecane structure (*Brown*, *Negishi* and *Burke*, *loc.cit.*) and (*c*) the isolation of butane-1,4-diboronic acid as a degradation product of the "unknown" structure. Indeed, both dimeric monocyclic and monomeric monocyclic (*i.e.* diboracyclodecane) ring compounds have been characterized as products from the hydroboration of penta-1,4-diene (*Brown* and *Negishi*, Pure and Applied Chem., 1972, **29**, 527). A *B-methoxy* derivative, b.p. 69–70°/2.5 mm, has been prepared (*Mikhailov*, *Vasil'ev* and *Safnova*, *loc.cit.*)

According to *J. Goubeau* and *R. Epple* (Chem. Ber., 1957, **90**, 171), pyrolysis of triethylborane yields *inter alia* 1,3,5,7-tetramethyl-1,3,5,7-tetraboracyclo-octane (XVIII).

3. Silicon, germanium, tin and lead compounds

For the most part, unsaturated ring systems containing Group IV metalloidal elements in seven-membered or larger rings, are obtained by halogenation and dehydrohalogenation of analogous saturated systems; these are therefore considered first.

(*a*) *Compounds with one hetero atom*

(*i*) *Seven-membered ring compounds*

In his investigations of cyclic silanes, *R. West* (J. Amer. chem. Soc., 1954, **76**, 6012) prepared 1,1-**dichloro-1-silacycloheptane** (1,1-**dichlorosilipane**) (I, M = Si R = R^1 = Cl), b.p. 203–204°, n_D^{20} 1.4802, in low yield from silicon

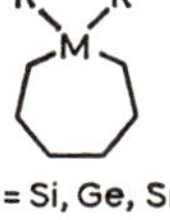

M = Si, Ge, Sn

(I)

tetrachloride and hexane-1,6-di(magnesium bromide); the dichloride is converted into the 1,1-*dimethyl* compound, b.p. 163–165°, n_D^{20} 1.4532, by the action of methylmagnesium halide, but a rather novel, and more effective synthesis has been provided by Russian workers. Reaction of pentane-1,5-di(magnesium bromide) with chloromethylsilyl dichlorides, $R(ClCH_2)SiCl_2$, yields 1-*R*-chloromethyl-1-silacyclohexanes which, on treatment with alu - inium chloride, undergo isomerization to 1-chloro-1,1-disubstituted-sila-cycloheptanes.

1,1-**Dimethyl**-1-**silacycloheptane** is thus also obtainable *via* 1-*chloro*-1-*methyl*-1-*silacyclo-heptane*, b.p. 55–57°/5 mm, n_D^{20} 1.4712 (*N. S. Nametkin et al.*, Izvest. Akad. Nauk S.S.S.R., Ser. Khim., 1967, 2530; *Nametkin*, *V. M. Vdovin* and *K. S. Pushchevaya*, *ibid.*, 1967, 2537). The 1,1-dichloro compound is reduced by lithium tetrahydridoaluminate to **silacyclohep-tane,** b.p. 135°, n_D^{25} 1.459. 1-*Methyl-silacycloheptane*, b.p. 144°, n_D^{25} 1.423 (*West loc.cit.*) is obtained from methyldichlorosilane and hexanedi(magnesium bromide). Silacyclohep-tanes have also been obtained by reactions between dialkylsilanes and biallyls in solvents at 40–50° containing chloroplatinic acid (*K. I. Kobrakov et al.*, Doklady Akad. Nauk S.S.S.R., 1970, **393**, 1072). In the cyclization of alkenes of the type $CH_2{=}CH(CH_2)nSiMe_2H$ catalyzed by chloroplatinic acid, there may result a mixture of silacycloalkanes differing in ring size. The demonstrated preference for the formation of the isomer with the smaller ring has been explained in terms of the nature of intermediate platinum complexes. Coupled with this is the fact that, as *n* increases, the total yield of products decreases (*J. V. Swisher* and *H.-H. Chan*, J. organometal. Chem. 1974, **69**, 83).

The chloroplatinic acid-catalyzed reaction between biallyl and metalloid hydrides has also been reported to give **chlorogermacycloheptanes** (I, M = Ge, R = Cl) when chloro(methyl)-, -(ethyl)- and -(phenyl)-germane were em-ployed (*idem*, *ibid.*, 1971, **196**, 100).

Tin dibromides similarly condense with hexane-16-di(magnesium di-bromide) to give 1,1-disubstituted-**stannacycloheptanes** (I, M = Sn); *dimethyl*-, b.p. 68°/4.6 mm, n_D^{25} 1.5052; *diethyl*-, b.p. 61–63°/0.6 mm, n_D^{25} 1.5085; *diphenyl*-, b.p. 140°/0.07 mm, n_D^{25} 1.6028 (*H. Zimmer*, *C. W. Blewett* and *A. Brakas*, Tetrahedron Letters, 1968, 1615) also formed in the zinc chloride-catalyzed thermolysis of α,ω-di(stanna)alkanes (*E. J. Bulten* and *H. A. Budding*, J. organometal. Chem., 1974, **82**, C13).

As alternative approach to the preparation of compounds possessing seven-membered and larger rings containing a single hetero atom is based on the acyloin condensation. The scheme illustrated below was adopted by *R. A. Benkeser* and *R. F. Cunico* (J. org. Chem., 1967, **32**, 395); the sequence yielded 1,1-*dimethylsilacycloheptan*-4-*one* (III), b.p. 115°/25 mm (2,4-*dinitrophenylhydrazone*, m.p. 94–95°) *via* 4-*hydroxy*-1,1-*dimethylsilacyclohep-tan*-5-*one* (II), b.p. 60–64°/0.6 mm (*acetate*, b.p. 80–85°/0.25 mm):

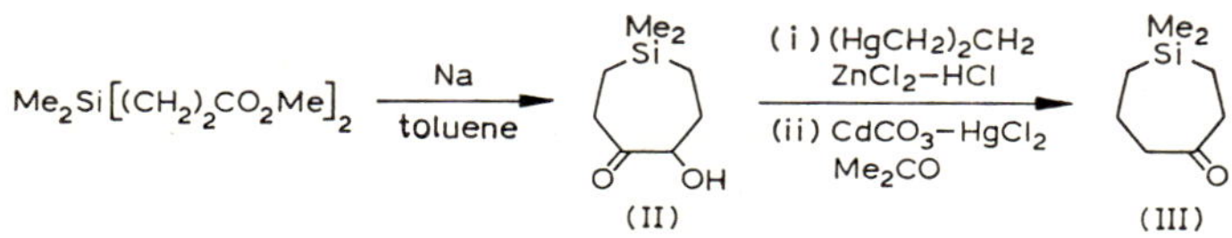

Pyrolysis of the thorium salt of the acid $Me_2Si((CH_2)_3CO_2H)_2$ gives a 7% yield of 1,1-*dimethylsilacyclo-octan*-5-*one*, n_D^{26} 1.4754 (2,4-*dinitrophenylhydrazone*, m.p. 134.5–135°), also prepared *via* the acyloin condensation. Yields may be improved by employing sodium in amounts much larger than theoretically required, and by incorporating trimethylchlorosilane into the reaction mixture to isolate the intermediate hydroxyketone as an enediol ether, *e.g.* 1,1-*dimethyl*-4,5-*bis*(*trimethylsilyloxy*)*silacyclophept*-4-*ene* (IV), b.p. 90–100°/0.1 mm.

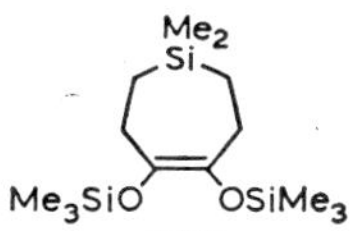

(IV)

4-*Hydroxy*-1,1-*dimethylsilacyclo-octan*-5-*one* forms an *acetate*, b.p. 90°/0.1 mm, and *enediol bis*(*trimethylsilyl*)*ether*, b.p. 95°/0.1 mm (*Weber et al.*, *ibid.*, 1971, **36**, 4060).

Silacycloheptan-2-ones (VI) (and also smaller ring silacycloalkaneones) have been prepared *via* the spirodithianes V, the sulphur-containing ring being, in effect, a carbonyl protecting group.

1,1-*Dimethylsilacycloheptan*-2-*one*, b.p. 43–44°/1.5 mm, 1,1-*diphenylsilacycloheptan*-2-*one*, an oil (*A. G. Brook* and *H. W. Kucera*, J. organometal. Chem. 1975, **87**, 263):

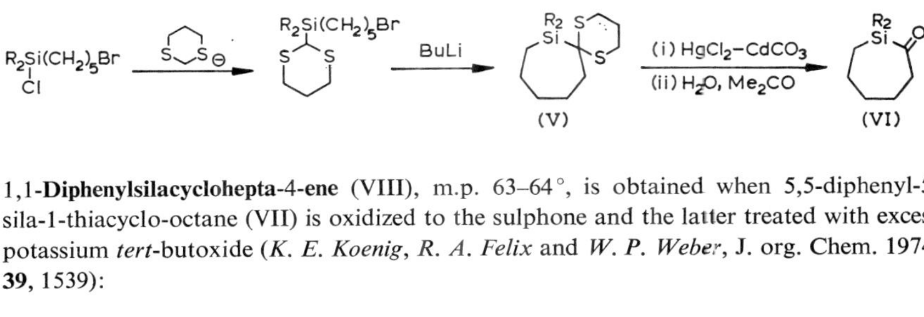

(V) (VI)

1,1-**Diphenylsilacyclohepta**-4-**ene** (VIII), m.p. 63–64°, is obtained when 5,5-diphenyl-5-sila-1-thiacyclo-octane (VII) is oxidized to the sulphone and the latter treated with excess potassium *tert*-butoxide (*K. E. Koenig*, *R. A. Felix* and *W. P. Weber*, J. org. Chem. 1974, **39**, 1539):

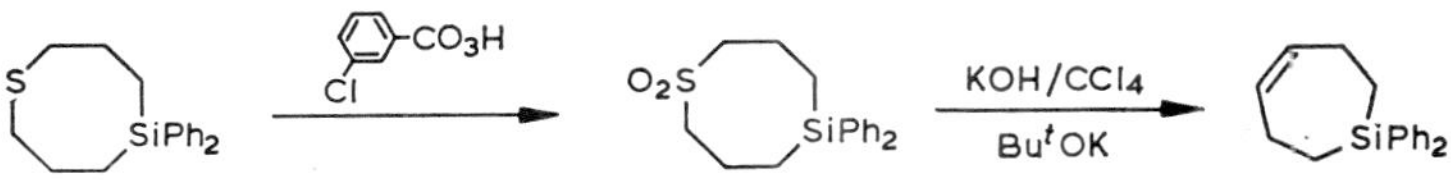

(VII) (VIII)

(ii) Larger ring compounds

Compounds with larger silicon- and germanium-containing ring systems have been prepared by *P. Mazerolles* and co-workers using the reactions illustrated in the Scheme 1: such compounds are listed in Table 1.

Scheme 1

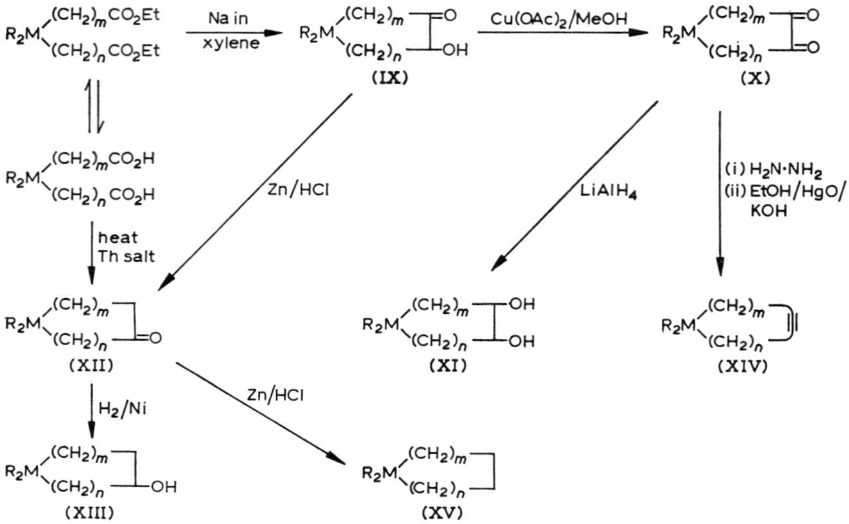

TABLE 1

LARGE RING SILICON- AND GERMANIUM-CONTAINING COMPOUNDS

Structure	*M*	*R*	*m*	*n*	*b.p.* (*°C/mm*)	n_D^{20}	*Yield* (%)
IX	Si	Me	4	4	121/1.2	1.4871	65
		Me	5	5	146/1.8	4.4902	65
	Ge	Me	4	4	m.p. 55–56	—	43
		Et	4	4	152/0.6	1.5084	58
X	Si	Me	4	4	109/1.6	1.4806	75
	Ge	Me	4	4	103/0.3	1.4945	—
		Et	4	4	105/0.1	1.4983	76
		Et	5	4	140/1.0	1.4957	37
XI	Ge	Et	4	4	146/0.1	1.5148	—
XII	Si	Me	4	4	115/3	1.4828	81
		Me	5	5	116/0.5	1.4843	75
		Me	5	6	130/0.7	1.4832	—
	Ge	Me	4	4	m.p. 34	—	—
		Et	3	4	116–117/0.6	1.5008	10
		Et	4	4	123/1.1	1.5005	80
		Et	6	6	157/0.7	—	66
XIII	Si	Me	4	4	106/0.2	1.4933	95
	Ge	Me	4	4	90/0.01	1.5122	—
XIV	Si	Me	4	4	87/0.9	1.4930	25
	Ge	Me	4	4	88/0.08	1.5062	—
XV	Si	Me		10	107.5/1.2	—	—
	Ge	Me	4	4	87/0.25	1.4931	32
	Et	Et		12	135/0.045	1.4918	73

References

P. Mazerolles and *A. Faucher*, Bull. Soc. chim. Fr., 1967, 2134; J. organometal. chem., 1973, **63,** 195; 1975, **85,** 159; *Mazerolles, Faucher* and *A. Laporterie*, Bull. Soc. chim. Fr., 1969, 887; *Mazerolles, Fr. P.* 1,513,324/1968; C.A., 1969, **71,** 13212.

As seen from Table 1 the yields of primary ring products IX and XII vary widely; a recent report of the synthesis of 1-*phenylgermacyclododecan-2-ol* (XVI), b.p. 150–160°/0.06 mm with decomp., quotes a very high yield (*P. Riviere* and *J. Satgé*, Angew. Chem. int. Edn., 1971, **10**, 267):

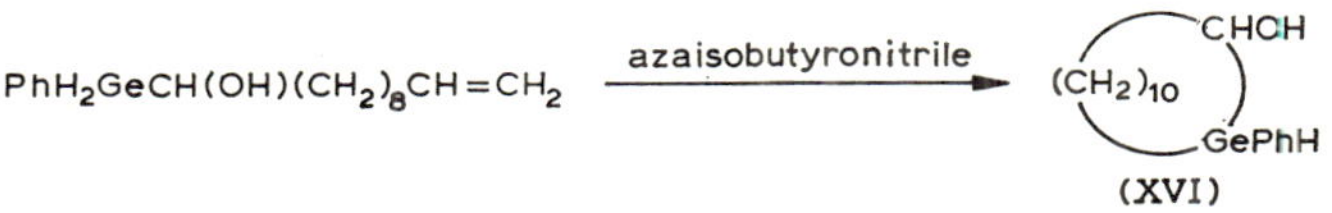

(XVI)

Mazerolles and *Faucher* have synthesized bicyclic compounds using methods based on those in Scheme 1.

1-Methyl-12-oxa-1-germabicyclo[5.4.1]**dodecane** (XVII), b.p. 112°/3 mm, n_D^{20} 1.5050, was obtained, in equilibrium with a small amount of 7-hydroxy-1-methyl-1-germabicyclo-[5.4.0]undecane (XVIII), in the following adaptation:

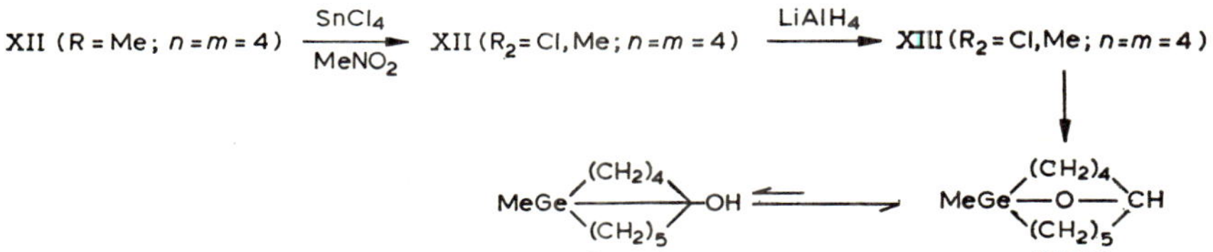

In the second development, a mixture of **1-methyl-1-germacycloundecane** and **1-methyl-1-germabicyclo**[4.4.0]**decene** (XIX) was prepared. This mixture was inseparable by distillation, but separation was achieved by chemical destruction of the monocyclic compound:

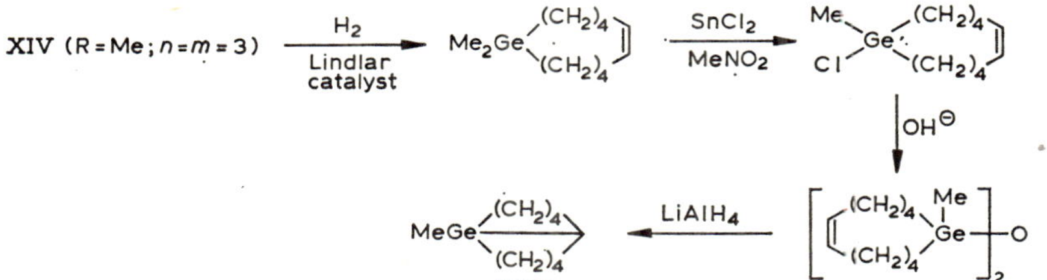

(*iii*) *Polycyclic compounds*

1,1-Diphenyl-1,2,3,4-tetrahydro-5*H*-benzosilepin (XX, $n = 1$), m.p. 151–153°, is obtained when 1,1-diphenyl-5-(2-chlorophenyl)-1-silapentane is heated at 110° with sodium (*B. Corriu, B. Henner* and *J. Masse*, Bull. Soc. chim. Fr., 1968, 3013) and the analogous **benzosilocin** (XX, $n = 2$), m.p. 155°, is similarly prepared (*H. Gilman* and *O. L. Marrs*, J. org. Chem., 1964, **29**, 3175) but in very low yield:

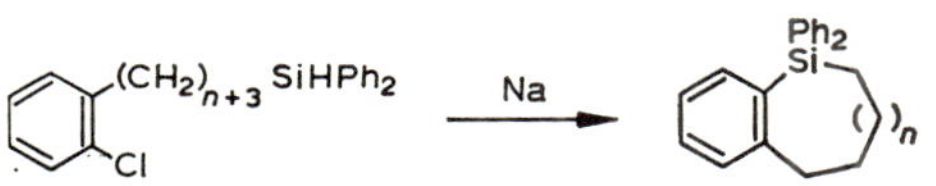

(XX)

2,2-Diphenyl-2,3,4,5-tetrahydro-1*H*-benzosilepin (XXI, R = Ph), b.p. 185–190°/0.3 mm, is formed by the internal Friedel–Crafts alkylation of 1-chloro-2,2,5-triphenyl-2-silapentane. The 2,2-*dichloro* compound (XXI, R = Cl), b.p. 135–145°/15 mm, is similarly obtained (*Corriu*, *Hemmer* and *Masse*, *loc.cit.*):

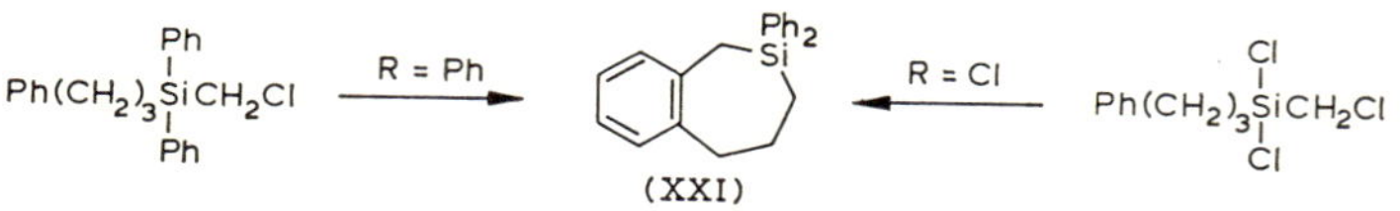

(XXI)

3,3-Dichloro-5-phenyl-3-silapent-1-ene is used as a source of 3,3-*dichloro*-2,3,4,5-*tetrahydro*-1H-*benzosilepin* (XXII, M = Si, R = Cl), b.p. 150°/12 mm; the *dibutoxy* compound, b.p. 155–160°/12 mm, reacts with phenyllithium to give the 3,3-*diphenyl* compound, m.p. 157–158°. Reduction of the 3,3-dichloro compound with lithium tetrahydridoaluminate affords 2,3,4,5-**tetrahydro-1*H*-benzosilepin** (XXII, M = Si, R = H), b.p. 104°/12 mm, m.p. 9–10°, in ~90% yield:

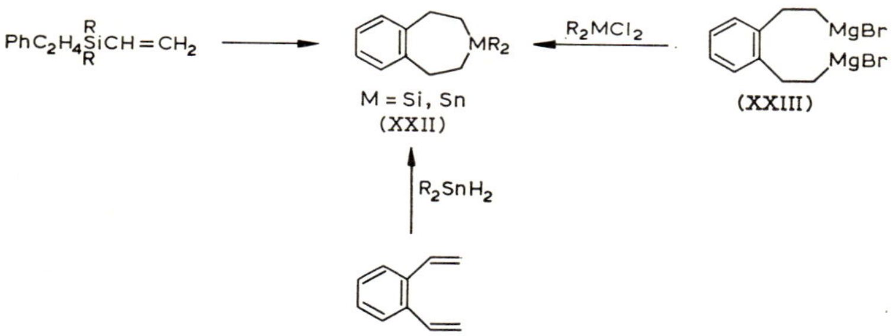

M = Si, Sn
(XXII)

(XXIII)

Another convenient route to these compounds employs the di-Grignard reagent XXIII (*Corriu et al.*, *loc.cit.*; *L. Birkofer* and *E. Krämer*, Chem. Ber., 1969, **102**, 427).

3,3-**Diethyl**- and 3,3-**diphenyl-2,3,4,5-tetrahydro-1*H*-benzostannepin** (XXII, M = Sn, R = Et, Ph), b.p. 93–98°/0.08 mm, and m.p. 98–100°, respectively, are obtained by yet another route involving the addition of tin dihydrides to 1,2-divinylbenzene in boiling benzene under nitrogen (see also p. 334) (*J. G. Noltes* and *G. J. van der Kerk*, Chimia, 1962, **16**, 122; C.A., 1964, **58**, 4658; *A. J. Leusink et al.*, Rec. Trav. chim., 1964, **83**, 1036; *Leusink*, *H. A. Budding* and *Noltes*, J. organometal. Chem., 1971, **24**, 375).

Monocyclic **dihydrometallepins** are not well-characterized materials. In hexamethylphosphoramide, cyclo-octatetraene reacts with dichlorodimethylsilane in the presence of magnesium in an exothermic reaction to give 9,9-dimethyl-9-silabicyclo[4.2.1]nona-2,4,7-triene (XXIV) in 20% yield. Valence isomerism occurs when this substance is heated at 250° the product being a silabarbaralane (*T. J. Barton* and *M. Juvet*, Tetrahedron Letters, 1974, 2561):

(XXIV)

Sheehan (Diss. Abs., 1965, **25**, 4417; C.A., 1965, **63**, 2993) has claimed that tin dihydrides, R_2SnH_2, add to hexa-1,5-diyne to yield polymers which when heated at 150–250° under reduced pressure afford 4,5-dihydrostannepins (XXV) but this ring system, like the analogous silicon, germanium and lead systems, are better known as the dibenzo derivatives (XXVI). 1,1-Dichloro-2,7-dihydrogermapin (XXVII) is allegedly produced from hexatriene and germanium dichloride–dioxane complex (*O. M. Nefedov* and *S. P. Kolesnikov*, Izvest. Akad. Nauk S.S.S.R., 1971, 2615):

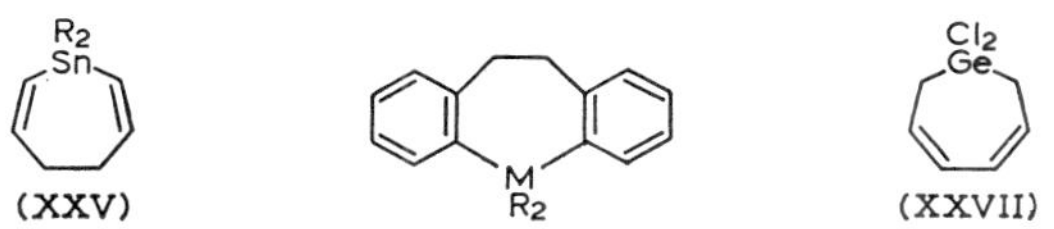

(XXV) M = Si, Ge, Sn, Pb (XXVI) (XXVII)

10,11-Dihydro-5*H*-dibenzo[*b*,*f*]metallepins (XXVI) (Table 2) are obtained most easily by the interaction of 2,2′-dilithiobibenzyl and the dichlorides R_2MCl_2 (*H. Gilman* and *W. H. Atwell*, J. org. Chem., 1963 **28**, 2906; *J. Y. Corey*, *M. Dueber* and *B. Bichlmeir*, J. organometal. Chem., 1971, **26**, 167; *Corey*, *Dueber* and *M. Malaidza*, *ibid.*, 1972, **36**, 49; *F. K. Cartedge* and *P. D. Mollere*, *ibid.*, 1971, **26**, 175).

TABLE 2

10,11-DIHYDRO-5*H*-DIBENZO[*b*,*f*]METALLEPINS (XXVI)

	R	*M*			
		Si	*Ge*	*Sn*	*Pb*
B.p. (°C/mm)	Me	104–106/0.2 mm	110–116/0.03	109–120/0.02	163–165/0.30
M.p. (°C)	Ph	170–172	165–168	138–141	163–164

Rapid passage of dichloro(1-naphthyl) (*o*-tolyl)silane through a quartz vessel at 680 °C yields a tetracyclic silepin (XXVIIa) in about 55% yield (*E. A. Chernyshev*, *N. G. Komelenkova* and *L. N. Shanshin*, Zhur. obshcheĭ Khim., 1974, **44**, 303).

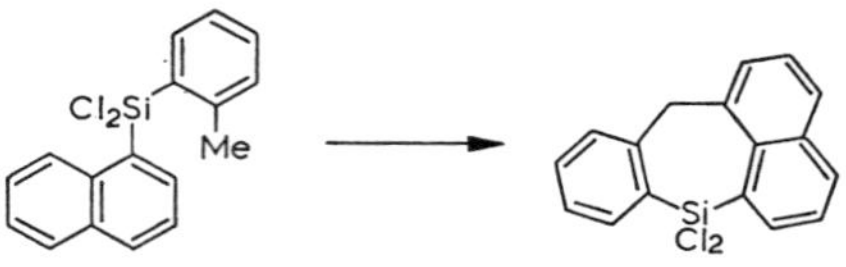

(XXVIIa)

Treatment of 5-(4-dimethylamino)phenyl-10,11-dihydro-5*H*-dibenzo[*b,f*]silepin with trityl perchlorate yields a yellow-green solution from which the colour can be discharged by addition of sodium tetradeuterioborate. The product contains deuterium attached to silicon, and the colour has been attributed to the presence in solution of the silicenium cation (XXVIII) (*Corey*, J. Amer. chem. Soc., 1975, **97**, 3237).

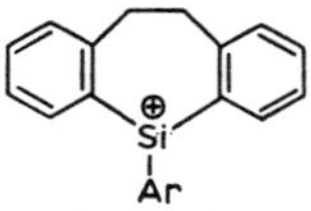

(XXVIII)

In the reaction between diphenyltin dichloride and 2,2′-dilithiobibenzyl in ether under nitrogen, the main product is a solid polymer, although 5,5-**diphenyl**-10,11-**dihydro**-5*H*-**dibenzo**[*b,f*]**stannepin** is obtainable in very low yield. By using stannic chloride, it is possible to obtain the *dichloride* (XXVI, M = Sn, R = Cl), m.p. 106–106.5°, *via* the spiran

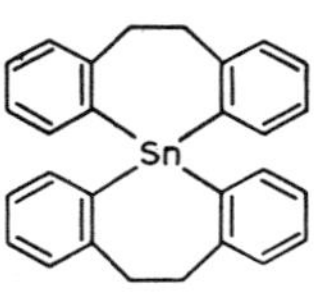

(XXIX)

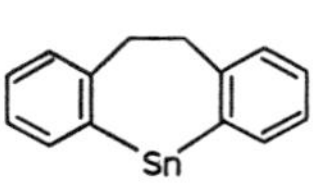

(XXX)

(XXIX). The dichloride is converted by aqueous sodium hydroxide into an infusible oxide, and by lithium tetrahydridoaluminate into a product, possibly 5,5,10,11-tetrahydro-5*H*-dibenzo[*b,f*]stannepin (XXVI, M = Sn, R = H) which loses hydrogen on treatment with methanol to give the tin(II) compound XXX (*H. G. Kuivila* and *O. F. Beumel*, J. Amer. chem. Soc., 1958, **80**, 3250).

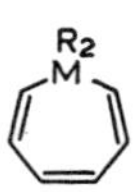

(XXXI)

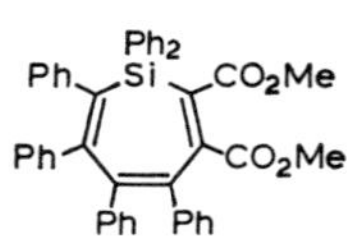

(XXXII)

No unsubstituted monocyclic metallepins of type XXXI have been recorded for any of the Group IV elements.

Dimethyl acetylenedicarboxylate reacts with hexaphenylsilole to give a 1:1 adduct for which the heavily substituted silepin structure XXXII has been assigned tentatively (*Gilman, S. G. Cottis* and *Atwell*, *ibid.*, 1964, **86**, 5584).

1,1-*Dimethyl*-2,7-*diphenyl*-1-*silacyclohepta*-2,4,6-*triene* (XXXIII), m.p. 60–61°, has also been obtained through a photolytically catalyzed addition to 1,1-dimethyl-2,5-diphenylsilole, final purification being achieved by g.l.c. The compound is thermolyzed at 250° to give *o*-terphenyl. Evidentally the ring exists as an inverting boat form, and thus if any

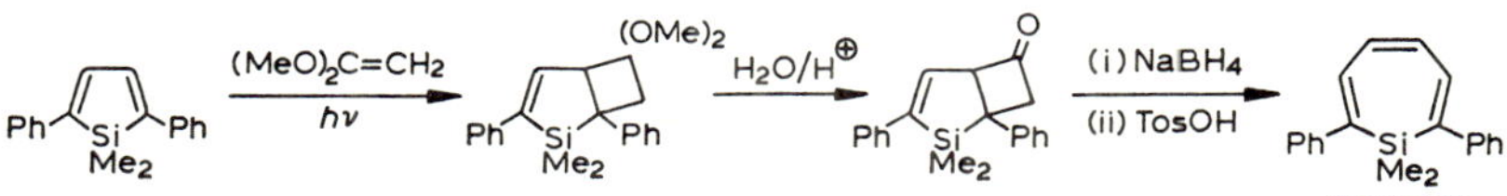

(XXXIII)

conjugation through silicon actually exists, its only effect would appear to be that of reducing the inversion barrier (*T. J. Barton, R. C. Kippenham* and *A. J. Nelson, ibid.*, 1974, **96**, 2272).

Initial attempts to obtain examples of the benzo[*d*]silepin system (XXXV) by bromination (with *N*-bromosuccinimide) of 3,3-dimethyl-2,3,4,5-tetrahydro-1*H*-benzosilepin and subsequent dehydrobromination afforded 1,2-divinylbenzene (*L. Birkofer* and *E. Krämer*, Chem. Ber., 1969, **102**, 427):

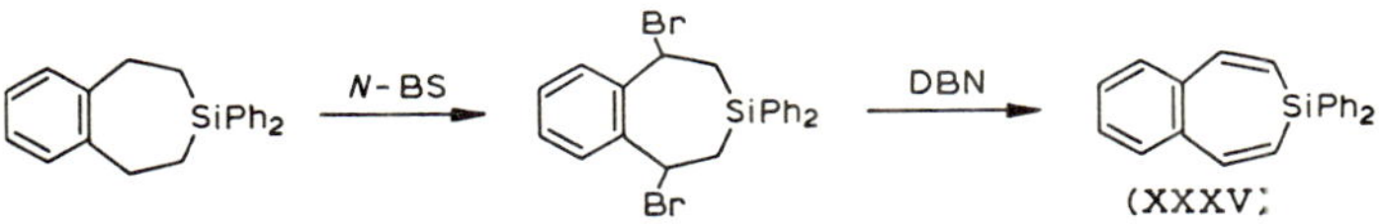

(XXXIV) (XXXV)

When treated with *N*-bromosuccinimide, the diphenyl compound gave its 1,5-dibromo derivative, XXXIV, which, with diazabicyclo[4.3.0]nona-5-ene (DBN) afforded 3,3-**diphenylbenzo[*d*]silepin** (XXXV), m.p. 91°, a substance which gives naphthalene on pyrolysis at 500° (*Birkofer* and *H. Hadded, ibid.*, 1969, **102**, 432). More recently, 3,3-dichloro-2,3,4,5-tetrahydro-1*H*-benzosilepin has been shown to undergo the same sequence of reactions and giving 3,3-*dichlorobenzo*[d]*silepin*, m.p. 74°, (*Birkofer, Haddad* and *H. Zamarlik*, J. organometal. Chem., 1970, **25**, C57), convertible to the 3,3-*dimethyl* compound, b.p. 120°/12 mm, by the action of methyllithium, and to 3*H*-3-**benzosilepin,** b.p. 80°/1.5 mm, by lithium tri(*tert*-butoxy)hydridoaluminate (*Birkofer* and *Haddad*, Chem. Ber., 1972, **105**, 2101).

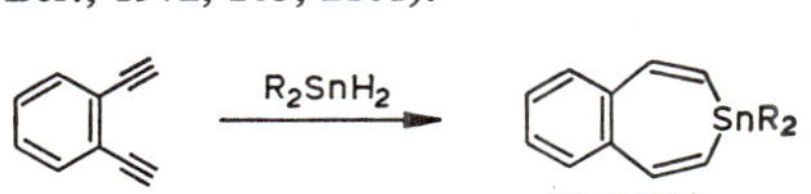

(XXXVI)

The 3*H*-3-**benzostannepins** (XXXVI) are obtained when 1,2-di(ethynyl)benzene reacts with tin dihydrides, 3,3-*dimethyl*-, b.p. 80–90°/0.4 mm, m.p. 41–42°; *diethyl*-, b.p. 95–96°/0.4 mm; *diphenyl*-, m.p. 282–288°: cyclic dimers are formed at the same time (see p. 334) (*Leusink et al., loc.cit*; *Leusink, Budding* and *Noltes, loc.cit.*):

According to *Corey et al.* (J. organometal. Chem., 1971, **26**, 167), derivatives of 10,11-**dihydro-5*H*-dibenzo[*b,f*]silepin** (XXXVII) are not dehydrogenated by 2,3-dichloro-5,6-dicyanobenzoquinone, but *Cartedge* and *Mollère*

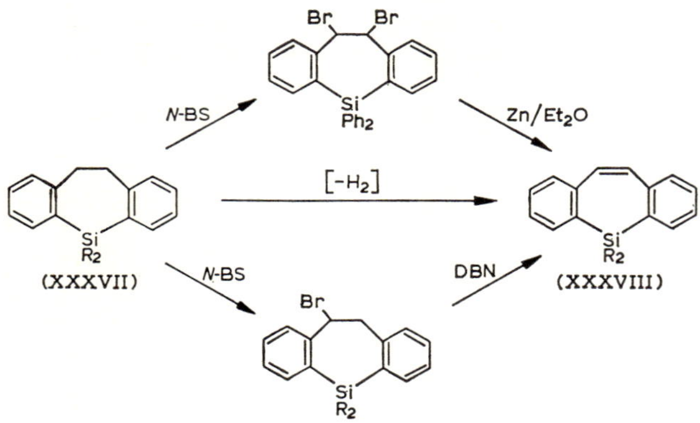

(*loc.cit.*) showed that the reaction requires several days in boiling toluene, when the products may be isolated by chromatography.

5,5-*Dimethyl-*, b.p. 110–123°/0.2 mm, and 5,5-*diphenyl-*, m.p. 195–197°, -5H-*dibenzo* [b,f]-*silepins* (XXXVIII, R = Me, Ph) may also be obtained by alternative bromination-de-(hydro)bromination sequences as illustrated (*Barton*, *W. E. Volz* and *J. L. Johnson*, J. org. chem., 1971, **36**, 3365; *Corey et al.*, *loc.cit*). Such sequences are feasible for the germanium compounds, but bromination of 10,11-dihydro-5*H*-dibenzo [*b,f*]stannepins and subsequent reaction with DBN gives 2,2′-dibromobibenzyl (*idem*, J. organometall. Chem., 1972, **36**, 49).

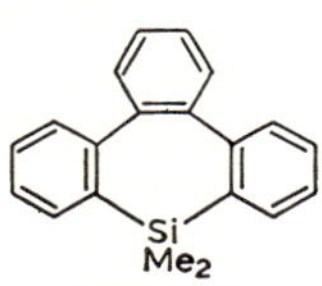

(XXXIX)

9,9-*Dimethyl*-9H-*tribenzo* [b,d,f]*silepin* (XXXIX), m.p. 226–227°, is prepared by a Wurtz–Fittig type of reaction between 1,2-dichlorobenzene and dimethoxydimethylsilane in the presence of sodium (*K. A. Andrianov et al.*, Khim. geterotsikl. Soedin., 1967, 435; C.A., 1967, **67**, 108695) and in about 3% yield from *o,o′*-dilithioterphenyl and dichlorodimethylsilane (*J. Y. Corey* and *E. R. Corey*, Tetrahedron Letters, 1972, 4669). The proposed structure has been confirmed by X-ray analysis.

A tetraphenyl derivative of XXXIX has been prepared by the intriguing combination of two acetylenic compounds (*E. Müller* and *G. Zountsas*, Chem. Ztg., 1973, **97**, 447; C.A., 1974, **80**, 70886):

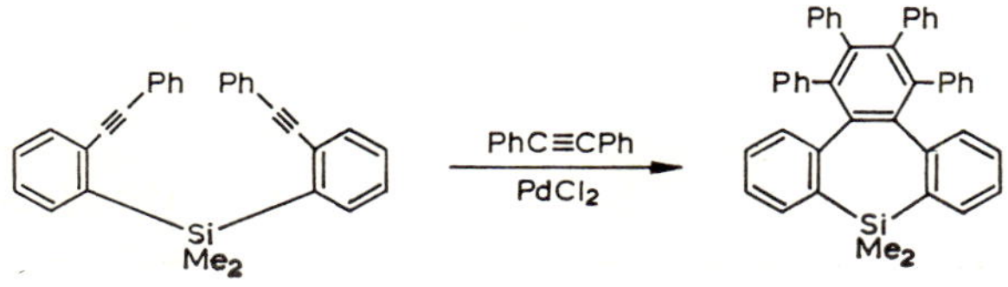

The co-pyrolysis of cyclo-octa-1,3-diene and $[(MeO)_2MeSi]_2$ at 350° generates 1-**methoxy-1-methyl-1-silacyclonona-2,8-diene** (XL), isolated by g.l.c. (*M. E. Childs* and *W. P. Weber*, Tetrahedron Letters, 1974, 4033).

(XL)

Treatment of 5-chloromethyl-5-methyl-10,11-dihydro-5*H*-dibenzo[*b,f*]silepin with aluminium chloride in benzene yields 5-chloro-5-methyl-5,6,11,12-tetrahydrodibenzo[*b,f*] silocin (XLI, R = Cl) readily hydrolysed to the *hydroxy* compound, m.p. 80–82° (*anhydride*, m.p. 143–145°) (*E. A. Francis* and *J. Y. Corey*, J. organometal. Chem., 1973, **61**, C20):

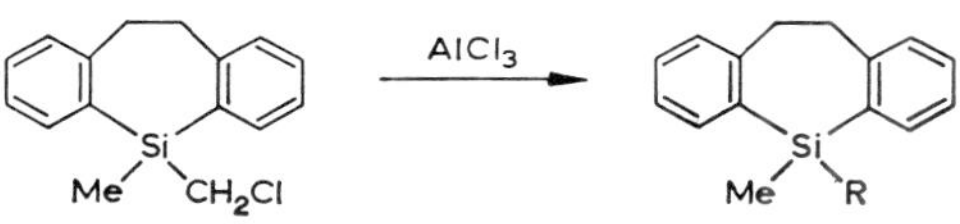

(XLI)

(*b*) *Compounds with two hetero atoms*

An 18% yield of 1,1,2,2-**tetramethyl**-1,2-**disilacycloheptane** (XLII, *n* = 1), b.p. 76–77°/16 mm, may be obtained by heating 2,8-dichloro-2,8-dimethyl-2,8-disilanonane with sodium–potassium alloy; the same compound is also prepared in 65% yield when pentane-1,5-di(magnesium bromide) is allowed to react with 1,2-dichloro-1,1,2,2-tetramethyldisilane.

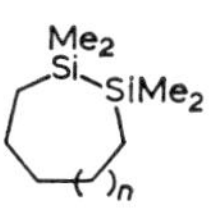

(XLII)

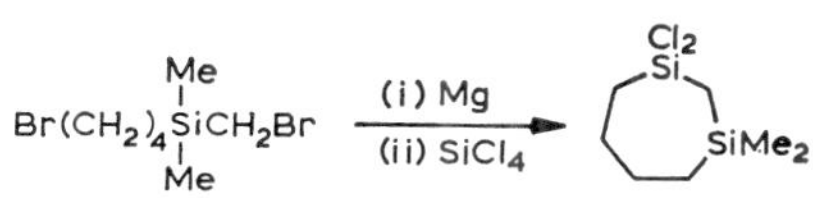

(XLIII)

Both methods are also applicable to the formation of 1,1,2,2-**tetramethyl**-1,2-**disilacyclo-octane** (XLII, *n* = 2), b.p. 96°/14 mm, m.p. 7–9° (*M. Kumada et al.*, J. organometal. Chem., 1967, **9**, 43). *Gilman et al.* (Adv. in organometal. Chem., 1966, **4**, 1) have listed other derivatives of 1,2-disilacycloheptane. Interestingly, the silicon–silicon bond in 1,2-disilaheptane and larger rings is not broken by the action of sulphuric cid, in contrast to that present in 1,2-disilacyclo-pentanes and -hexanes.

1,1,2,2-**Tetraphenyl**-1,2-**digermacycloheptane,** m.p. 95°, is formed in 75% yield by reaction between *symm*-tetraphenyldigermane and penta-1,4-diene in benzene in the presence of azoisobutyronitrile (*P. Riviere, J. Satgé* and *D. Soula*, Comptes rend., 1973, **277**, C, 895).

A further example of the use of a Grignard reagent is the formation of 1,1-*dichloro*-3,3-*dimethyl*-1,3-*disilacycloheptane* (XLIII) from the Grignard reagent formed from 1,6-dibromo-2,2-dimethyl-2-silahexane, and its subsequent reaction with silicon tetrachloride;

the final yield of heterocyclic compound, b.p. 79–80°/mm, n_D^{24} 1.4850, is 50% (*R. Fessenden* and *M. D. Coon*, J. org. Chem., 1964, **29**, 2499).

The dimerization of allyldimethylsilane to give 1,1,5,5-**tetramethyl**-1,5-**disilacyclo-octane** (XLIV), m.p. 109.5–110.5°, in only 3% yield takes place only in boiling toluene (*J. W. Curry* and *G. W. Harrison*, J. org. Chem., 1958, **23**, 1219).

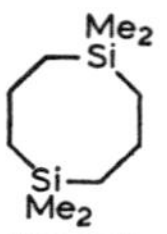

(XLIV)

5,5,11,11-**Tetraphenyl**-5,6,11,12-**tetrahydrodibenzo**[*b,f*][1,5]**disilocin** (XLV), m.p. 313–315°, is formed in low yield by heating 2-halogenobenzyldiphenylsilanes with sodium under toluene (*H. Gilman* and *O. L. Marrs*, *ibid.*, 1964, **29**, 3175); alternatively, the compound is obtained by reduction of diphenyl benzosilacyclobutane with lithium tetrahydridoaluminate (*Gilman* and *W. H. Atwell*, J. Amer. Chem. Soc., 1964, **86**, 5589).

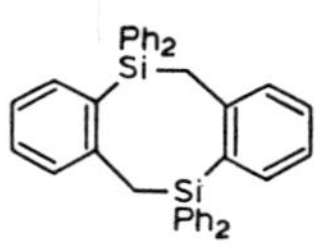

(XLV)

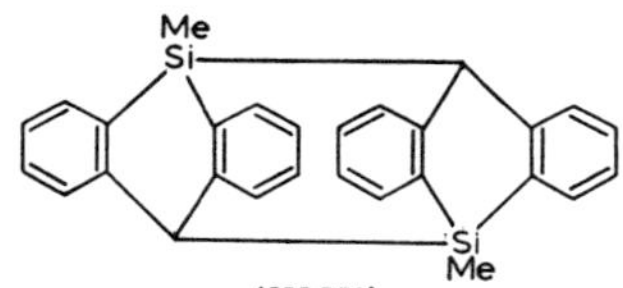

(XLVI)

The novel compound XLVI is a "dimer" of 9-methyl-9-silaanthracene. 5,12:6,11-**Di-*o*-benzeno**-5,6,11,12-**tetrahydrodibenzo**[*b,f*] [1,5]**disilocin** is formed when 5-chloro-5-methyldibenzo[*b,e*]-silacyclohexadiene reacts with 1,5-diazabicyclo[5.4.0]undecane in the presence of triethylamine at −78° (*P. Jutzi*, Chem. Ber., 1971, **104**, 1455).

Somewhat surprisingly, the interaction of 2,2′-dilithiobiphenyl and dichloro(ethyl)silane leads, not to the dibenzosilole, but rather to the "dimer", 9,18-**diethyl**-9,18-**disilatetrabenzo**[*a,c,f,h*]-**cyclodecatetraene** (XLVII), m.p. 98–99°, b.p. 144–146°/0.5 mm (*I. M. Gvertsiteli et al.*, Zhur. obshcheĭ Khim., 1966, **36**, 114). The compound is interesting from the structural point of view since it probably possesses a "double butterfly" conformation with overlapping but parallel benzene rings:

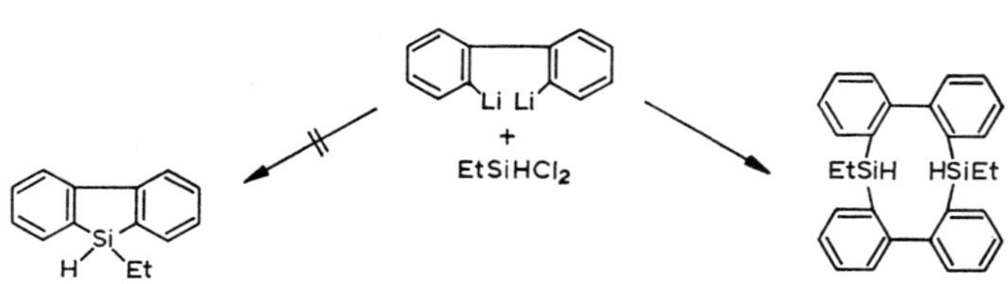

(XLVII)

The reaction of isoprene with dichlorodimethylsilane in tetrahydrofuran in the presence of lithium results in the formation of trans, trans-1,1,3,6,6,8-*hexamethyl*-1,6-*disilacyclodeca*-3,8-*diene* (XLVIII), b.p. 92°/106 mm, n_D^{25} 1.5038 (*D. Weyenberg, L. H. Toporcer* and *L. E. Nelson* J. org. Chem., 1968, **33**, 1975).

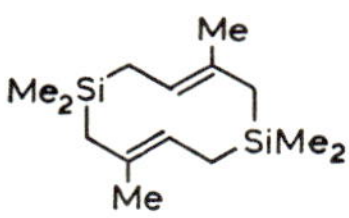

(XLVIII)

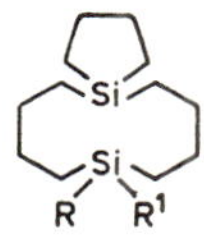

(XLIX)

Disilacyclodecanes and their stereochemistry have only more recently been described. The formation of 1,6-disila[4.4.4]propellane in 4% yield from 4a,8a-dichloro-4a,8a-disiladecalin and butane-1,4-di(magnesium bromide) has already been described (p. 8); additional products in this reaction are *10-butyl*-5,10-*disilaspiro*[4.9]-*tetradecane* (XLIX, R = Bu, R^1 = H), b.p. 124–126°/1 mm, (12%) and 5,10-*disiladispiro*[4.4.4.4]*octadecane* [XLIX, $RR^1 = (CH_2)_4$], m.p. 89.5–90.5°. Hydrolysis with deuterium oxide at an intermediate stage leads to incorporation of deuterium in a manner suggesting that the spiro compounds are formed as indicated below:

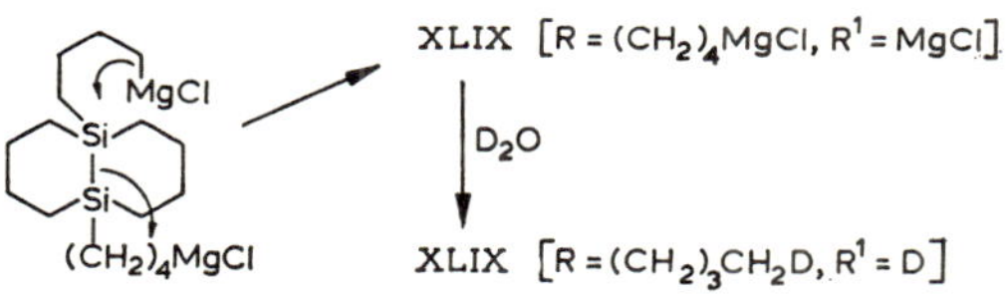

The silicon–silicon bond in dimethyldisiladecalin is also broken when this compound is treated with bromine; the product is 1,6-*dibromo*-1,6-*dimethyl*-1,6-*disilacyclodecane*, m.p. 105–109°, b.p. 118–122°/1 mm, which with butyllithium gives the 1,6-*dibutyl*-1,6-*dimethyl*-1,6-*disilacyclodecane*, b.p. 132–134°/2 mm (*K. Tamao, M. Kumada* and *A. Noro*, J. organometal. Chem., 1971, **31**, 169).

Among the larger ring compounds containing two Group IV hetero atoms, those containing tin have been obtained as dimeric by-products in the synthesis of monomeric compounds, and appear to be reasonably well characterised. More recently, it has been demonstrated that 1,2-disilacyclohexanes and 1,-2-disilacyclohexadienes react with butadienes and acetylenes. The reaction is catalysed by di(triphenylphosphine)dichloropalladium $(Ph_3P)_2PdCl_2$ and involves cleavage of the Si–Si bond with insertion of the diene or alkyne followed by ring closure to give large-ring disila compounds. Typical are the formation of 2,3-*dimethoxycarbonyl*-1,1,4,4-*tetramethyl*-5,6,7,8-*tetraphenyl*-1,4-*dihydro*-1,4-*disilacyclo-octatetraene* (L) and 1,1,6,6-*tetramethyl*-2,3,4,5-*tetraphenyl*-1,6-*disilacyclotetradeca*-2,4,8,12-*tetraene* (LI), m.p. 171–171.5 (*H. Sakurai, Y. Kamiyama* and *Y. Nakadaira*, Chem. Letters, 1975, 887, C.A. 1975, **83**, 147541):

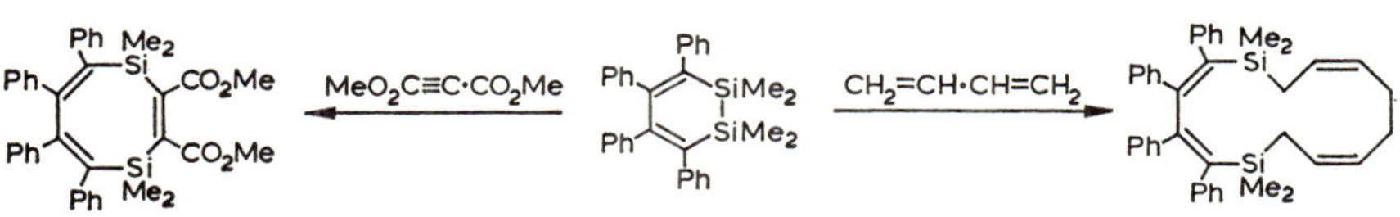

(L) (LI)

1,8-*Diethyl*-1,8-*diphenyl*- (LII, R = Ph, R^1 = Et), and 1,1,8,8-*tetraphenyl*-1,8-*distanna-dibenzo*[d,k]*cyclotetradeca*-2,4,6,9,11,13-*hexaene* (LII, R = R^1 = Ph), m.p. 184–186° and 282–285°, respectively, are obtained along with benzostannepins, when the hydrides

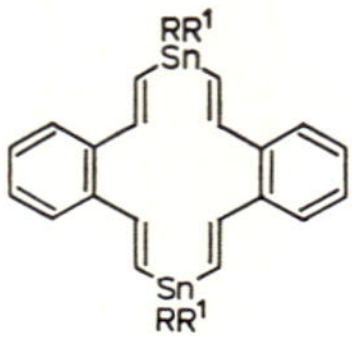

(LII)

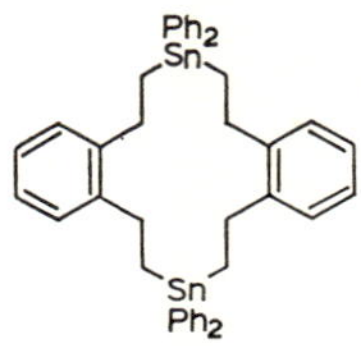

(LIII)

RR^1SnH_2 react with 1,2-diethynylbenzene (p. 329). The large-ring compounds LIII and the benzostannepins (XXII, M = Sn, p. 326) differ in their mode of reaction with iodine; the former yield 1,2-di(*trans*-2-iodovinyl)benzene, the benzostannepins affording 1,2-di-(*cis*-2-iodovinyl)benzene (*Leusink*, *Budding* and *Noltes*, J. organometal. Chem., 1971, **24**, 375). 1,1,8,8-*Tetraphenyl*-1,8-*distanna-dibenzo* [d,k]*cyclotetradeca*-4,11-*diene* (LIII), m.p. 277–282°, is similarly obtained from 1,2-divinylbenzene (*Leusink*, *et al.*, Rec. Trav. chim. 1964, **83**, 1036).

Photolysis of 1,1-diphenyl-4,5-dihydrostannepin in ether leads to polymers together with a *dimer* (34%); the latter, m.p. 145–146°, appears to be 1,1,8,8-tetraphenyl-1,8-distanna-2,6,9,13-cyclotetradecatetraene (LIV), the dimerization probably proceeding through either or both of the diradicals depicted (*T. Sato* and *I. Moritani*, Tetrahedron Letters, 1969, 3181):

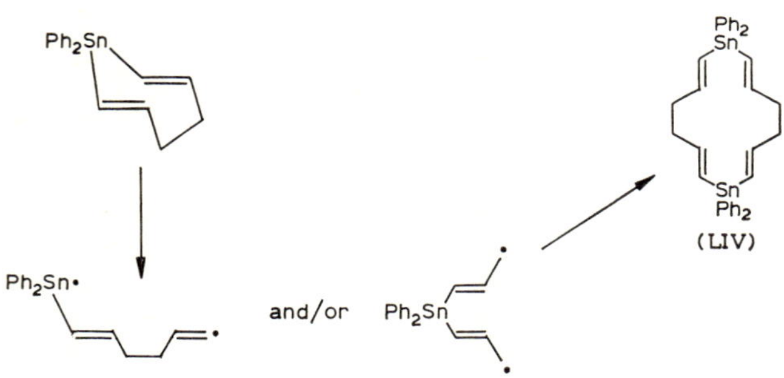

(LIV)

(*c*) *Compounds with three or more hetero atoms*

1,2,3-**Trisilacycloheptanes** have been studied in the search for reactive silicon intermediates, the silenes, R_2Si:.

The following 1,1,3,3-*tetramethyl*-1,2,3-*trisilacycloheptanes* (LV) were prepared by reaction between butanedi(magnesium bromide) and the dichlorides, $RR^1Si(SiMe_2Cl)_2$; 2,2-*dimethyl*-, b.p. 118–119°/31 mm, 2-*isopropyl*-2-*methyl*-, b.p. 112–123°/8 mm, and 2,2-*tetramethylene*-, (5,6,11-*trisilaspiro* [4.6]*undecane*), b.p. 97–98°/1 mm. When photolyzed

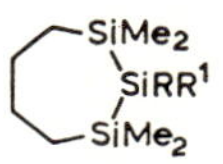

(LV)

(LVI)

(mercury source) elimination of the silene, RR^1Si: proceeds with the concomitant formation of 1,1,2,2-tetramethyl-1,2-disilacyclohexane (*Sakurai, Y. Kobayashi* and *Nakadaira*, J. Amer. chem. Soc., 1971, **93**, 5272).

By allowing benzene and silicon difluoride to come together at −196°, followed eventually by distillation of the product at 140°, *P. L. Timms et al.* prepared 2,2,3,3,4,4-hexafluoro-2,3,4-trisila[3.2.2]nona-6,8-diene (LVI) (J. Amer. chem. Soc., 1966, **88**, 940).

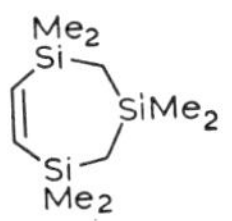

(LVII)

3,3,5,5,7,7-Hexamethyl-3,5,7-trisilaheptyne cyclizes in the presence of chloroplatinic acid to yield a mixture of a methylene 1,3,5-trisilacyclohexane and 1,1,3,3,5,5-**hexamethyl-1,3,5-trisilacyclohept-6-ene** (LVII), b.p. 50–60/10^{-2} mm, in yields of 30% and 12%, respectively (*G. Fritze* and *M. Hahnke*, Z. anorg. allgem. Chem., 1972, **390**, 104).

1,1,3,3,5,5,7,7-**Octachloro**-1,3,5,7-**tetrasilacyclooctane** (LVIII, R = Cl), m.p. 247°, b.p. 175°/28 mm, has been prepared by the pyrolysis of methyltrichlorosilane at 800° (*Fritz* and *G. Teichmann*, Chem. Ber., 1962, **95**, 2361), and, together with trisilacyclohexanes, when dichloromethane is passed over heated silicon (*Fritz, R. Haase* and *D. Kummer*, Z. anorg. allgem. Chem., 1969, **365**, 1). The octachloride may be chlorinated further to the

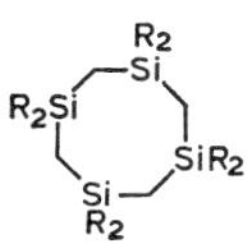

(XLVIII)

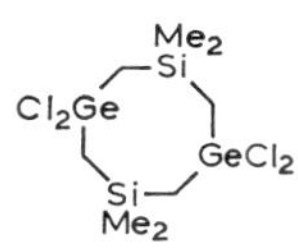

(XLIX)

1,1,2,2,3,3,4,5,5,6,7,7,8-tridecachloride which will not chlorinate further, presumably for steric reasons. Reduction of the octachloride with lithium **tetrahydridoaluminate** gives 1,3,5,7-tetrasilacyclo-octane (LVIII, R = H), b.p. 204–208°, and with methylmagnesium bromide, the octachloride yields the *octamethyl* compound, m.p. 81°, 1,1,3,3,5,5,7,7-*octamethyl*-1,3,5,7-*tetrasilacyclo-octane* m.p. 81°, which is also formed (*a*) when 1,1,3,3-tetramethyl-1,3-disilabutane is heated with potassium under toluene (*N. S. Nametkin et al.*, Izvest. Akad. Nauk S.S.S.R., Ser. Khim., 1970, 195), (*b*) in 1% yield when $Me_2SiClCH_2Cl$ in tetrahydrofuran is treated with magnesium (*W. A. Kriner*, U.S.P. 3,178,392/1965; C.A., 1965, **63**, 1817), (*c*) by the pyrolysis of chlorotrimethylsilane at 700° (*Fritz, W. König* and *H. Scheer*, Z. anorg. allgem. Chem., 1970, **377**, 240), (*d*) when 2,6-dibromo-2,4,4,6-tetramethyl-2,4,6-trisilaheptane and di(chloromethyl)dimethylsilane are heated together with sodium under toluene (*Fritz* and *H. Burdt*, *ibid.*, 1962, **314**, 35), (*e*) together with tetramethyldisilacyclobutane and hexamethyltrisilacyclohexane when chloromethyldimethylsilyl chloride in ether is heated with magnesium (*W. A. Krimei*, J. org. Chem., 1964, **29**, 1601), and (*f*) from $Br(Me_2SiCH_2)_4Br$ by the action of lithium under ether (*Fritz* and *P. Schober*, Z. anorg. allgem. Chem., 1970, **377**, 37).

Passage of di(chloromethyl)dimethylsilane over a copper–germanium composite at 370–390° gives a mixture of nine products from which a 6% yield of 3,3,7,7-**tetrachloro**-

1,1,5,5-**tetramethyl-1,5-disila-3,7-digermacyclo-octane** (LIX) may be isolated; it is converted into the octamethyl compound by methylmagnesium chloride (*T. K. Gar, A. A. Buyakov* and *V. F. Mironof*, Zhur. obshcheĭ Khim., 1972, **42**, 1521).

An alternative route to compounds of the same system starts from O [SiMe($CH_2Cl)_2$]$_2$ (*A. A. Buyakov, T. K. Gar* and *V. F. Mironov*, Zhur. obshoheĭ Khim., 1973, **43**, 625):

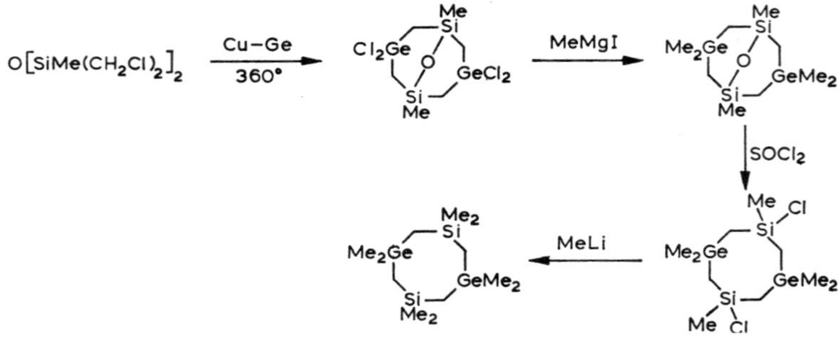

The formation of 1,1,3,3,6,6,8,8-**octamethyl-1,3,6,8-tetrasilacyclodecane,** m.p. 138–139°, in 2% yield from 1,5-dibromo-2,2,5-trimethyl-1,5-disilahexane, has been reported (*Fritz* and *Schober*, Z. anorg. allgem. Chem., 1970, **372**, 21), and the same authors record unpublished work by *G. Sonntag* on the preparation of tetrasilacyclodecadiynes.

A reaction between 1,3-dichloropropane and 1,5-dilithiodecaphenylpentasilane affords the **decaphenyl-1,2,3,4,5-pentasilacyclo-octane** (LX), m.p. 290–291°, in 3% yield (*H. Gilman* and *S. Inoue*, Chem. and Ind., 1964, 74). Derivatives of pentasilacyclodecane and hexa-

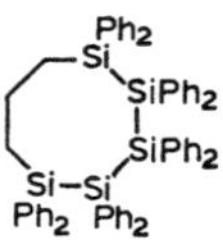

(LX)

silacyclododecane have been recorded (*Fritz* and *Schober*, Z. anorg. allgem. Chem., 1970, **377**, 37). A series of polysilacycloalkanes of 14–32 ring atoms including 4, 6 or 8 silicon atoms has been prepared; these compounds are low melting solids or viscous liquids (*R. Corriu* and *J. Masse*, Bull. Soc. chim. Fr., 1974, 3045).

4. Phosphorus and arsenic compounds

Seven-membered and larger ring systems containing one or more atoms of Group V elements (higher than nitrogen) appear to be known only for phosphorus and arsenic. The number of synthetic methods equally applicable to compounds of both elements is very small; they include the reaction between the metalloid dichloride and 2,2′-dilithiobibenzyl in the formation of 10,11-dihydro-5*H*-dibenzo[*b,f*]metallepins, the heat-induced ring expansion of spirobis(dibenzarsole) and its phosphorus analogue in the formation of nine-membered ring compounds, and the diquaternization of diphosphines and diarsines to give compounds with two such hetero atoms.

(*a*) *Compounds containing one hetero atom*

(*i*) *Phosphepins and related compounds*

An interesting but somewhat lengthy sequence of steps has been employed to prepare simple derivatives of phosphepin (*G. Märkl* and *H. Schubert*, Tetrahedron Letters, 1970, 1273). The photolytically induced interaction of dichloromaleimide and the phosphol-3-ene oxide (I) gives a mixture of stereoisomeric *adducts*, of which one, *viz*. II, m.p. 280–281°, is converted into the *diester*, III, m.p. 215–216°, using sulphuric acid followed by diazomethane. Hydrogenolysis of III in the presence of a nickel catalyst, followed by electrolysis of the product in aqueous pyridine, gives 3-*oxo*-3-*phenyl*-3-*phosphabicyclo*[3.2.0]*hept*-6-*ene* (VI), m.p. 171–172°, which when heated in solu-

Scheme 2

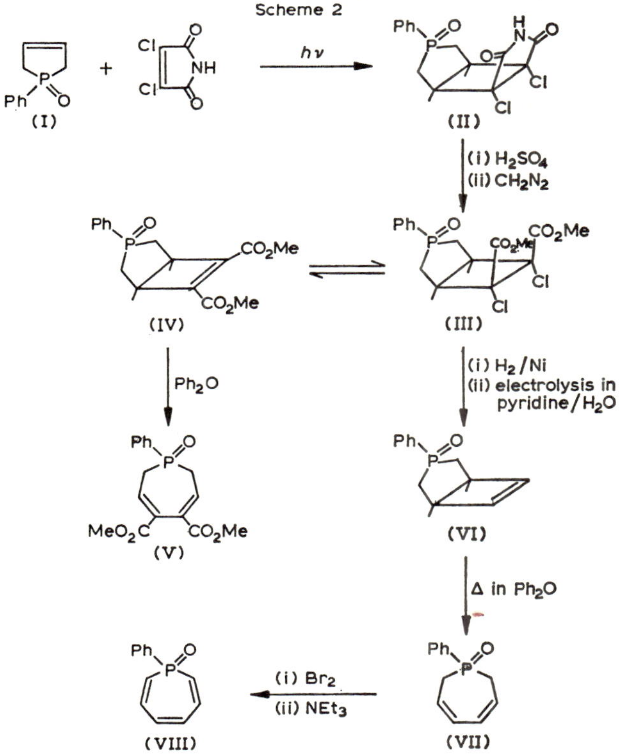

tion in diphenyl ether yields 1-*oxo*-1-*phenyl*-2,7-*dihydrophosphepin* (VII), b.p. 160°/0.2 mm, in 81% yield from VI. Bromination of VII and further reaction with triethylamine finally affords **1-oxo-1-phenylphosphepin** (VIII), m.p. 91–92° (see Scheme 2).

A synthesis of **1-phenyl-4,5-dihydrophosphepin** (IX, M = P), b.p. 65–67°/0.01 mm (*oxide*, b.p. 135–140°/0.05 mm) and of **1-phenyl-4,5-dihydroarsepin,** b.p. 68–70°/0.01 mm (IX, M = As), involves the addition of phenyl-phosphine or -arsine to hexa-1,5-diyne in the presence of azaisobutyronitrile:

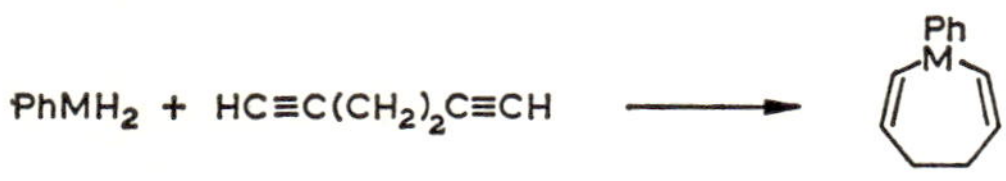

(IX)

Quaternary salts are readily available. Treatment of 1-phenyl-4,5-dihydrophosphepin 1-oxide with triethylamine in benzene brings about isomerization to the oxide VII, which is reduced by trichlorosilane to the corresponding phosphine, b.p. 70–74°/0.01 mm (*Märkl* and *G. Dannhardt*, *ibid.*, 1973, 1455).

Reaction of dichlorophenylphosphine and dipotassium or dilithium cyclo-octatetraenide in tetrahydrofuran at 0° yields the oily bicyclic *compound* (X), m.p. 45–47°, which on warming to 70° undergoes valency tautomerism to 9-phenyl-9-phosphabicyclo[4.2.1]nona-2,4,7-triene (XI); when heated at 480°, the molecule epimerizes at phosphorus. The epimer

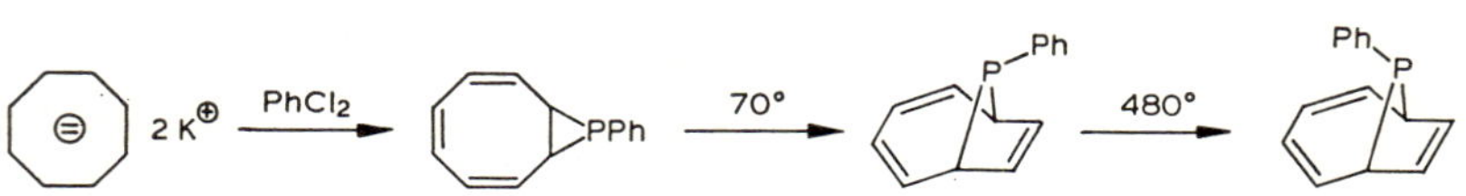

(X) (XI) (XII)

XII so formed may be purified through a palladium complex; the purified *epimer*, m.p. 84–85.5°, forms *a methiodide*, m.p. 184–190°, and an *oxide*, m.p. 197.5–198°. Neither epimer reacts with dienophiles (*T. J. Katz*, *C. R. Nicholson* and *C. A. Reilly*, J. Amer. chem. Soc., 1966, **88**, 3832; *E. W. Turnblom* and *Katz*, *ibid.*, 1973, **95**. 4292).

The corresponding arsepin and dihydroarsepin ring systems are evidently not recorded. However, **9-methyl-9*H*-tribenz[*b,d,f*]arsepin** (XIII), m.p. 159–160° (*monomethiodide monohydrate*, m.p. 218–221°) is obtained in an unusual ring contraction reaction when certain cyclic diarsines are heated under nitrogen (*F. G. Mann*, *I. T. Millar* and *F. C. Baker*, J. chem. Soc., 1965, 6342). The corresponding phosphine system has not been reported:

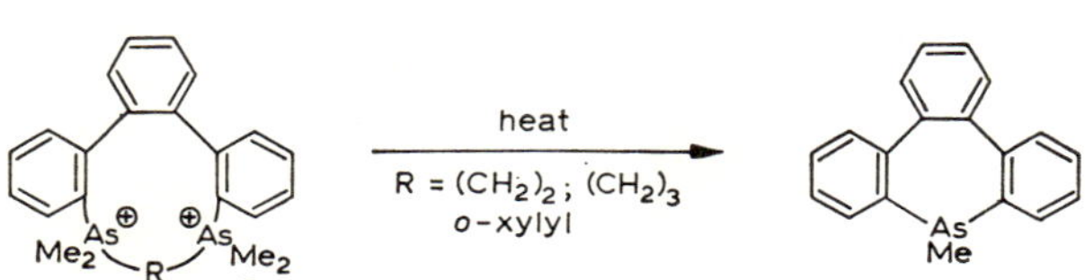

(XIII)

In addition to the dihydrophosphepin VII, *Märkl* and *Schubert* (*loc.cit.*) prepared also its 4,5-*di(methoxycarbonyl)*derivative (V) from III by treatment of the latter with nickel carbonyl, and valency isomerization of the resultant ester IV in hot diphenyl ether. In a further sequence starting with a bromination step, V is degradable to dimethyl 4,5-dibromophthalate.

6-*Hydroxy*-6H-*dibenz* [c,e]*phosphepin* 6-*oxide* (XV), m.p. 246–248°, is obtained as a by-product when the diphosphonic acid (XIV) is heated at 330–370° under reduced pressure, and is extracted from the product mixture with ethanol (*C. N. Robinson* and *W. A. Pettit*, Tetrahedron Letters, 1972, 4977):

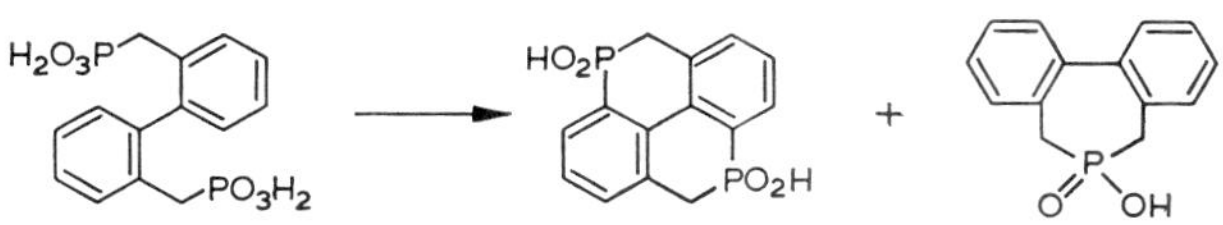

(XIV) (XV)

A much easier route to compounds in this series would appear to be that starting from biphenyl-2,2′-dicarbonyl chloride. In the example illustrated, 5,7-*dioxa*-6-*phenyl*-6,7-6,7-*dihydro*-5H-*dibenzo* [c,e]*phosphepin* (XVI), m.p. 166–168°, is obtained in 76% yield (*D. Fenske et al.*, Chem. Ber., 1966, **109**, 359):

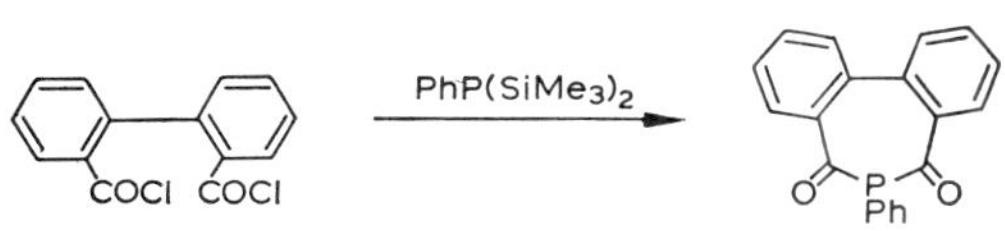

(XVI)

6-Phenyl-6*H*-dibenz[*c,e*]arsepin (XVII), m.p. 118–118.5°, *methiodide*, m.p. 223–224°, is obtained using the classical type of synthesis illustrated. With boiling hydroiodic acid, conversion to the 6-*iodo* compound, m.p. 117–117.5°, takes place, thus illustrating the greater stability of this tricyclic system over the isomeric 5*H*-dibenz [*b,f*]arsepin series (*M. H. Beeby*, *Mann* and *E. E. Turner*, J. chem. Soc., 1950, 1923):

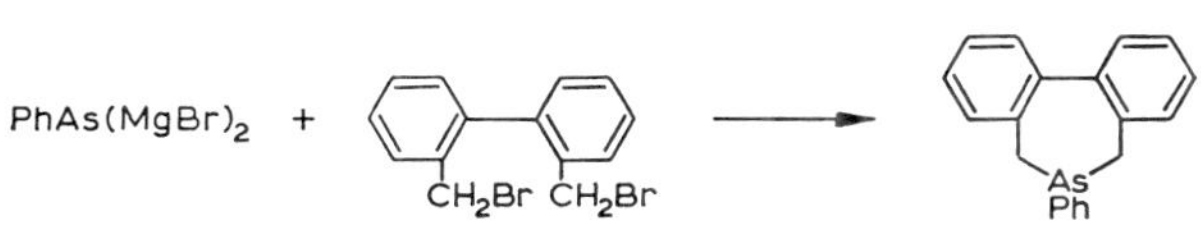

(XVII)

9-Phenyl-10,11-dihydro-5*H*-dibenzo[*b,f*]phosphepin (XVIII, M = P) is prepared from 2,2′-dilithiobibenzyl and dichlorophenylphosphine. It exists in two crystalline forms. The first of these, m.p. 75°, b.p. 170–190°/0.01 mm, is obtained by direct distillation of the reaction product; the second form, m.p. 94.5–95°, is obtained when the distilled product is heated with hydriodic acid. Both forms of the phosphine give the same *oxide*, m.p. 173–174°, and

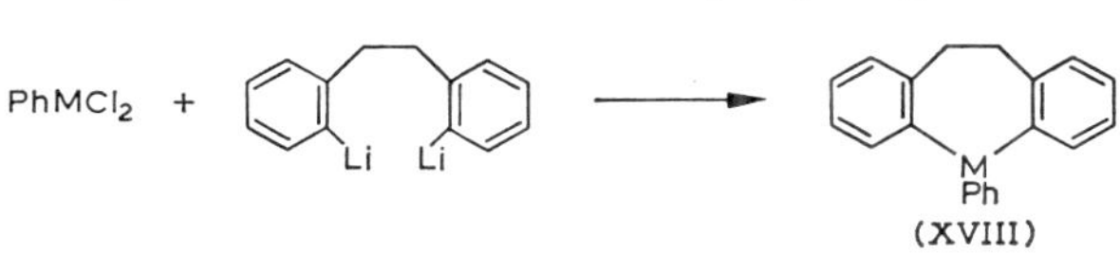

(XVIII)

the same *methiodide*, m.p. 251–252°. The oxide XIX (R = Ph) may be reduced by trichlorosilane to the higher melting phosphine, and also when heated with lithium, or better, with alkali, loss of exocyclic phenyl occurs with formation of the corresponding *phosphinic acid* (XX), m.p. 246–251° (*J. L. Suggs* and *L. D. Freedman*, J. org. Chem., 1971, **36**, 2566), convertible through the acid chloride to the *phosphine oxide* (XIX, R = H). Treatment of (XIX R = Ph) with lithium tetrahydridoaluminate or sodium hydride or even methylmagnesium iodide leads to some ring opening (*I. Granoth*, *Y. Segall* and *H. Leader*, Chem. Comm., 1976, 74).

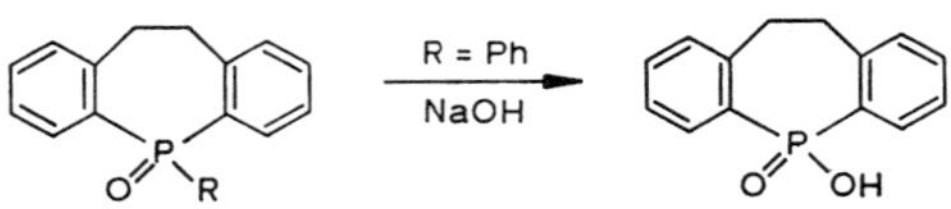

(XIX) (XX)

5-Phenyl-10,11-dihydro-5*H*-dibenzo[*f,b*]arsepin (XVIII, M = As) is likewise prepared, m.p. 59–59.5°, b.p. 180–183°/0.15 mm. The arsine and the corresponding phosphine provide examples of isodimorphism, for crystallisation of a crude sample of the phosphine is initiated by inoculation with a crystal of the arsine, and when a specimen of the molten arsine at 65° is seeded with the higher melting form of the phosphine, a second form of the arsine, m.p. 78.5–79°, separates. Both forms of the arsine give the same *methiodide*, m.p. 215–227° (sealed tube), and the same *oxide*, m.p. 188–190° (*Mann*, *Millar* and *B. B. Smith*, J. chem. Soc., 1953, 1130). Unlike the phosphines, the corresponding arsines are completely degraded to arsenic triiodide by the action of hydriodic acid.

(*ii*) *Phosphocins*

An interesting synthesis of 1,2,3,6,7,8-hexahydrophosphocins is based on cycloaddition to a phosphorinan-4-enamine. The enamine XXI with dimethyl acetylenedicarboxylate in toluene yields the 4,5-*di(methoxycarbonyl)*-6-*morpholino*-1-*phenyl*-1,2,7,8-*tetrahydrophosphocin* 1-*oxide* (XXIII), m.p.

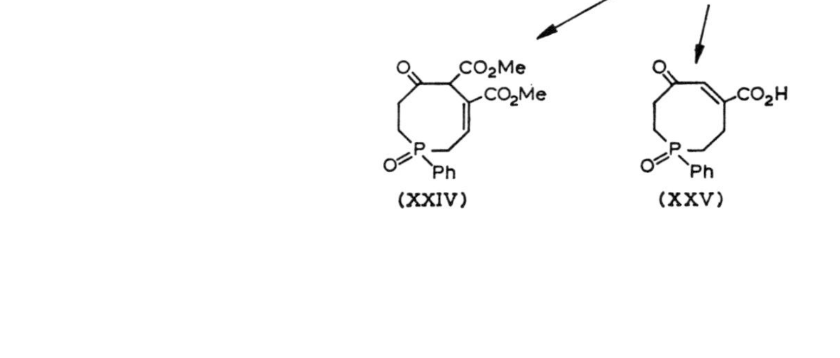

(XXI) (XXII) (XXIII)

(XXIV) (XXV)

170–173°, in only 8% yield, probably by way of the intermediate XXII. Hydrolysis of XXIII has afforded both the ketonic *diester* XXIV, m.p. 220–222°, in quantitative yield, and the ketonic *acid* XXV, m.p. 220–222°, in 63% yield (*Märkl* and *H. Baier*, Tetrahedron Letters, 1972, 4439).

(*iii*) *Higher phosphacycloalkenes and related compounds*

A method for the preparation of phosphorus-containing mono-olefinic ring systems, and of potential wide applicability, is based on the acyloin condensation. **Phosphacycloalkenes** (XXVI) (Table 3) of from 7 to 11 atoms in the ring are formed in 14–44% yield in the reaction shown; the products polymerize very rapidly in air, and attempts to remove the two trimethylsilyl groups have led to intractable materials (*J. W. van Reijendam* and *F. Baardman*, *ibid.*, 1972, 5181).

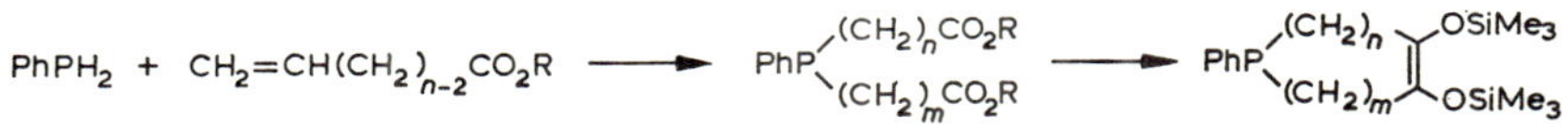

(XXVI)

TABLE 3

1-PHENYLPHOSPHACYCLOALKENES (XXVI)

m	*n*	*M.p. (°C)*
2	2	118–128 (sulphide, m.p. 103–106)
3	2	124–125
3	3	134–137 (sulphide, m.p. 108–110)
4	3	142–152
4	4	152–166 (sulphide, m.p. 100–102)

(*iv*) *Phosphepanes and related compounds*

The addition of P–H across carbon to carbon double bonds is catalyzed either photolytically or by peroxide. In common with smaller rings of the same general type, 1-**phenylphosphepane** (XXVII), b.p. 105–120°/1.5 mm, is formed when hex-5-enyl(phenyl)phosphine, $CH_2{=}CH(CH_2)_4PPhH$, is photolyzed under nitrogen; it forms a *sulphide*, m.p. 90.5–91.5°, (*J. H. Davies*, *J. D. Downer* and *P. Kirby*, J. chem. Soc., C, 1966, 245; *Shell Internat. Research Maatsch.*, N.V., Fr. P 1,488,936/1967; C.A., 1968, **69**, 36252). In the presence of peroxides, primary phosphines add to biallyls; 1-*eicosyl*-3,6-*dimethylphosphepane*, m.p. 47–48°, has been so obtained (*Shell. Internat. Research Maatsch.* Neth. Appl. 6,604,094/1966; C.A., 1967, **66**, 65101):

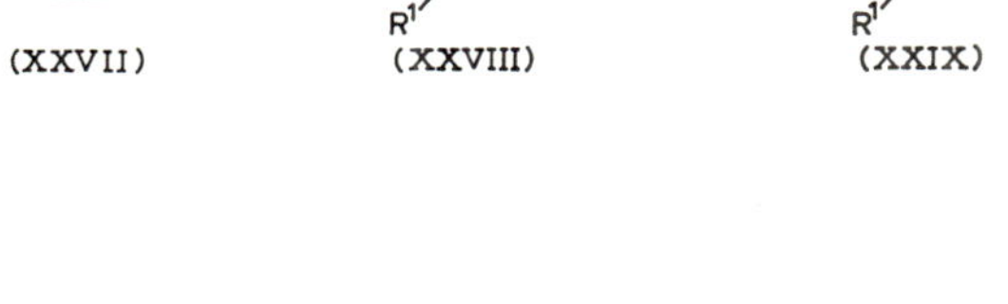

(XXVII) (XXVIII) (XXIX)

Quarternization of tetraphenylbiphosphine with 1,6-dibromohexane (two moles) in hot 1,2-dichlorobenzene follows an unusual course in which P–P bond cleavage occurs with the formation of 1,1-**diphenylphosphepanium bromide** (XXVIII, $R^1 = H$, R = Ph), m.p. 245–247° (*Märkl.* Angew. Chem., 1963, **75**, 859). The loss of benzyl group, as toluene, when the *phosphepanium salt* (XXVIII, R = $PhCH_2$, R^1 = Me), m.p. 187.5–189°, is treated with alkali, results in complete stereochemical inversion during the formation of 4-*methyl*-1-*phenylphosphepane* 1-*oxide* (XXIX), m.p. 80–82°; phenylsilane reduces the oxide to 4-*methyl*-1-*phenylphosphepane*, b.p. 115°/0.05 mm, with retention of configuration, and subsequent reaction of the phosphine with benzyl bromide yields the phosphepanium salt epimeric at phosphorus. The sequence is reversible, again with complete stereochemical inversion and the process is to be contrasted with that which takes place for smaller rings (*K. L. Marsi*) J. Amer. chem. Soc., 1971, **93**, 6341, *Marsi*, and *F. B. Burns*, Phosphorus, 1974, **4**, 211). A suggested possible reason for this difference lies in the ability of the seven-membered ring to take up a diequatorial position in the trigonal-bipyramidal intermediate.

The 4-*methyl* compound (XXVIII R = Ph, R^1 = Me) has m.p. 212.5–213.5°.

1-**Diphenyl**-5-**methylphosphocanium bromide** is an oil, which on treatment with alkali gives a mixture of cis- and trans-5-*methyl*-1-*phenylphosphocane* 1-*oxide*, m.p. 97–103°, in 4% overall yield from 4-methylcyclohexanone (*Marsi, D. M. Lynch* and *G. D. Horner*, J. heterocycl. Chem. 1972, **9**, 331).

(*v*) *Phosphabicyclo-alkanes and -alkenes and related compounds*

Hydrogenation of 9-phenyl-9-phosphabicyclo[4.2.1]nona-2,4,7-triene 9-oxide (XXX) over platinum oxide yields 9-phenyl-9-*phosphabicyclo*[4.2.1]*nonane* 9-*oxide*, m.p. 157–159°, reducible by hexachlorodisilane to the corresponding *phosphine* (XXXI), m.p.

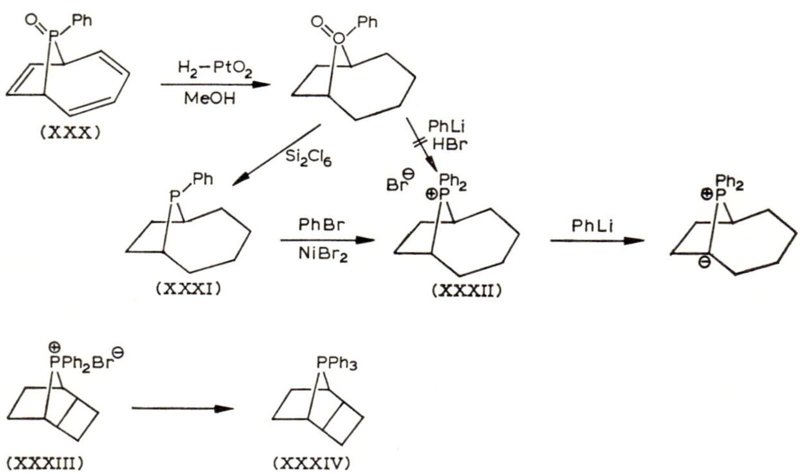

58–60°. The latter may be converted to the quaternary *bromide* (XXXII), m.p. 244–246°; this, when treated with phenyllithium affords an ylid with a half-life in benzene at 75° of about 36 min, decomposing into triphenylphosphine and cycloocta-1,5-diene. Bridging of the larger ring, as in structure XXXIII, results in the triphenylphosphorane XXXIV under the same conditions, thus illustrating the effect that changes in ring size may have on the course of the reaction (*Turnblom* and *Katz*, *loc.cit.*).

Apart from the acyloin reaction, adapted to the synthesis of large, mono-olefinic, monocyclic phosphines described earlier (p. 341), unsaturated phosphines with the hetero element in a nine-membered ring have been prepared by a methods equally applicable to the synthesis of the corresponding arsines. When heated at 200–250°, **spirobis(dibenzphosphole)s** (XXX, M = P) yield 17-*methyl*- (m.p. 130–131°) and 17-**phenyl**- (m.p. 131–132°) 17H-**tetrabenzo** [*b,d,f,h*]**phosphonin** (XXXVI, M = P, R = Me, Ph) (*G. Wittig* and *A. Maercker*, Chem. Ber., 1964, **97**, 747; *D. Hellwinkel*, *ibid.*, 1965, **98**, 576).

(XXXV) (XXXVI)

In addition to being prepared in the same way, the 17H-*phenylarsine analogue* (XXXVI, M = As, R = Ph), m.p. 132–135.5°, is also made by reaction between dichlorophenylarsine and the appropriate dilithioquaterphenyl (*Wittig* and *Hellwinkel*, *ibid.*, 1964, **97**, 769).

(*b*) *Compounds with two hetero atoms*

1,1,5,5-**Tetramethyl**-2,3,4,5-**tetrahydro**-1*H*-1,5-**benzodiarsepinium dibromide** XXXVII, m.p. 258°, and the analogous 1,1-*diethyl*-5,5-*dimethyl*-2,3,4,5-*tetrahydro*-1H-1,5-*benzophospharsepinium dibromide* (XXXVIII), m.p. 245°, are two of the simplest salts obtained by di-quarternization between 1,3-dibromopropane and, on the one hand, *o*-phenylenebis(dimethylarsine), and on the other hand, (*o*-dimethylarsinophenyl)diethylphosphine (*R. H. Glauert* and Mann, J. chem. Soc., 1950, 682, *E. R. H. Jones* and *Mann ibid.*, 1955, 4472). The only other seven-membered ring compounds yet described are the first members of the series XXXIX and XL (Table 4) obtained from biphenyl-2,2′-bis-(diethylphosphine) and -bis(dimethylarsine) and the appropriate dibromo hydrocarbon. For further details of other known salts, the reader should consult “Heterocyclic Compounds of Phosphorus, Arsenic, Antimony and Bismuth” by *Mann*, (1970, Wiley-Interscience).

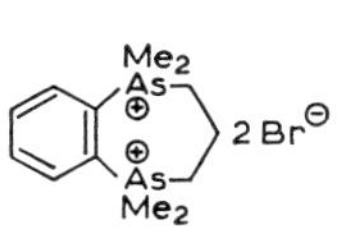

(XXXVII)

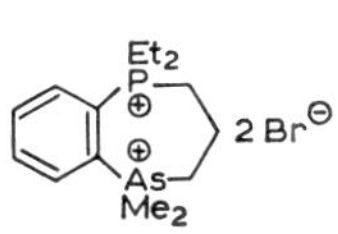

(XXXVIII)

TABLE 4

POLYCYCLIC, LARGE RING DIPHOSPHONIUM AND DIARSONIUM SALTS

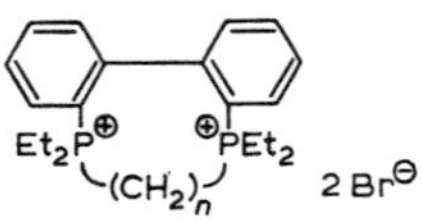

(XXXIX)

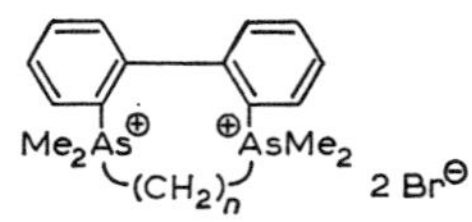

(XL)

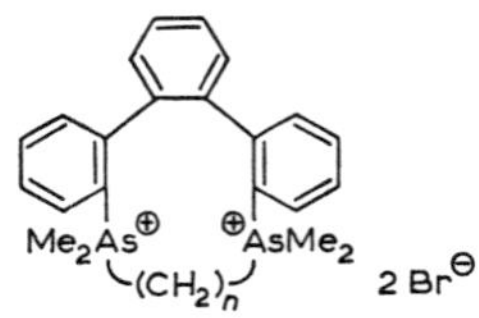

(XLI)

n	*M.p. (°C)*		
	XXXIX	*XL*	*XLI*
1	228–230 (2)	224–226 (1)	258–264 (5)
2	268–272 (2)	211–213 (3)	201–202 (5) (272–275)
3	335–336 (2)	261–262 (1)	225–226 (5)
4	152–155 (2)	210 (4)	— (5)

References

1 *H. Heaney et al.*, J. chem. Soc., 1958, 3838.
D. W. Allen, F. G. Mann and *I. T. Millar, ibid.*, C, 1967, 1869.
3 *Heaney, Mann* and *Millar*, J. chem. Soc., 1957, 3930.
4 *M. H. Forbes et al., ibid.*, 1961, 2762.
5 *Mann, Millar* and *F. C. Baker, ibid.*, 1965, 6342.

The compounds XXXIX decompose when heated at 250°/0.15 mm to give 5-ethyl-5*H*-dibenzophosphole (9-ethyl-9-phosphofluorene). By contrast, 5,5,7,7-tetramethyl-6,7-dihydro-5*H*-dibenzo[*d,f*] [1,3]diarsepinium dibromide (XL, $n = 1$) decomposes normally to give 5,7-*dimethyl*-6,7-*dihydro*-5H-*dibenzo* [d,f] [1,3]*diarsepin*, m.p. 96°, by loss of two moles of methyl bromide. For the remainder (XL, $n = 2,3,4$), ring contraction to a mixture of 5-methyl- and 5-bromo-dibenzarsoles is observed (*M. H. Forbes et al.*, J. chem. Soc., 1961, 2762):

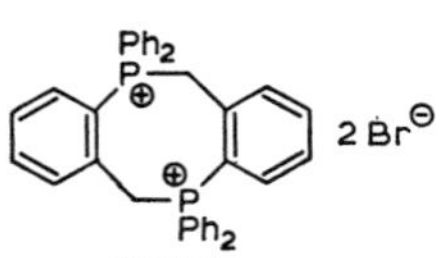

(XLII)

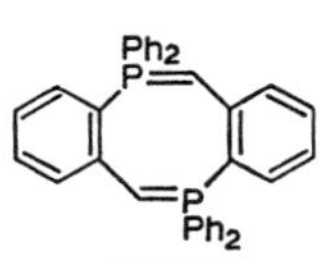

(XLIII)

1,5-**Diphosphocin** is known as a dibenzo derivative. 5,5,11,11-*Tetraphenyl*-5,6,11,12-*tetrahydrodibenzodiphosphocinium dibromide* (XLII), m.p. > 360°, formed by the self-quaternization of ω-bromotolyl(diphenyl)phosphine, $Ph_2PC_6H_4CH_2Br$ [2], is converted by the action of alkali into the novel di-ylid XLIII (*Märkl*, Z. Naturforsch., B, 1963, **18**, 84).

2,2,5,5-*Tetraphenyl*-1,2,5,6-*tetrahydro*-2,5-*benzodiphosphocinium dibromide* (XLIV), m.p. 331–334°, and 5,5,12,12-tetraethyl-5,6,11,12-tetrahydrodibenzophospharsecinium dibromide (XLVI), are both obtained by quaternization reactions involving ω,ω′-dibromoxylene. The latter salt behaves remarkably when pyrolyzed, losing one mole of bromoethane from arsenic, the phosphorus atom remaining quaternized with the ring retained:

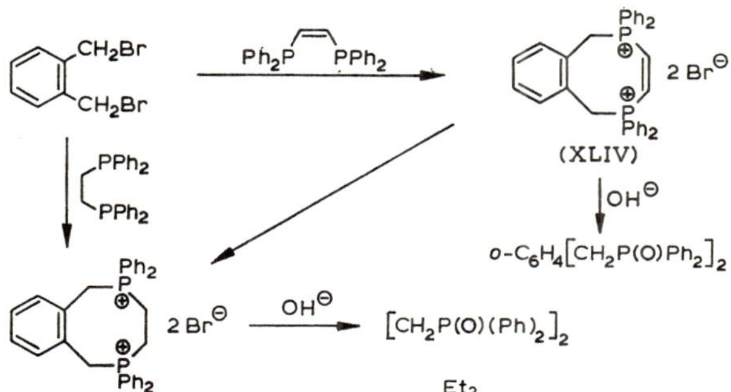

(XLV)

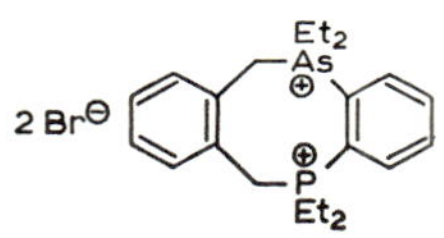

(XLVI)

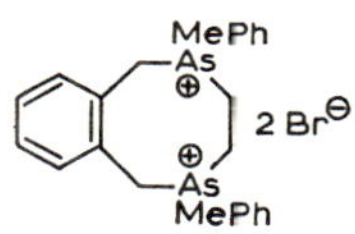

(XLVII)

The salt LIV is reducible to 1,2,3,4,5,6-*hexahydro compound* XLV, m.p. 345–347°, obtainable by an alternative route, and the two salts XLIV and XLV are differentiated by their reaction with 10% aqueous sodium hydroxide (*A. M. Aguiar* and *M. G. R. Nair*, J. org. Chem., 1968, **33**, 579). 2,5-Dimethyl-2,5-diphenyl-1,2,3,4,5,6-hexahydro-2,5-benzodiarsocinium dibromide (XLVII) is prepared by the standard procedure (*E. R. H. Jones* and *Mann*, J. chem. Soc., 1955, 401).

1,1,5,5-*Tetraphenyl*-1,5-*diphosphocanium diiodide* (XLVIII), m.p. 204–205°, results when diphenylphosphine is heated with 1,3-diiodopropane in acetonitrile, but its formation

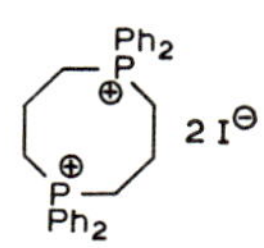

(XLVIII)

probably takes place by self-quaternization of ω-iodopropyl(diphenyl)phosphine, $Ph_2P(CH_2)_3I$ (*S. O. Grim* and *R. Schaaff*, Angew. Chem., 1963, **2**, 486).

9,9,16,16-*Tetramethyl*-9,10,15,16-*tetrahydrotribenzo* [b,d,h] [1,6]*diarsecinium* (XLIX, R = Me) *dibromide*, m.p. 209–210°, loses bromomethane on pyrolysis leaving the cyclic

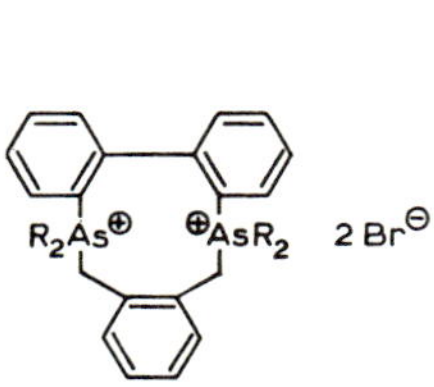

(XLIX)

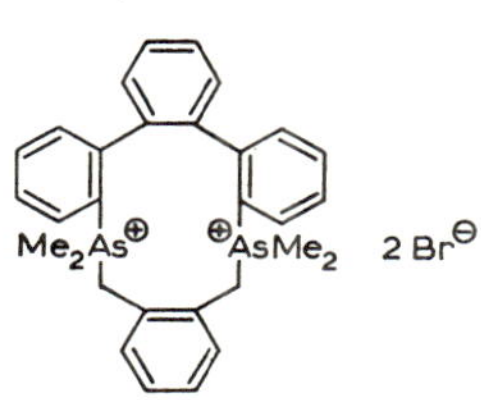

(L)

diarsine, 9,16-**dimethyl**-9,10,15,16-**tetrahydrotribenzo** [*b,d,h*] [1,6]**diarsecin,** m.p. 110–111°; however, the corresponding tetraethyl salt yields 5-ethyldibenzarsole together with biphenylbis(diethylarsine) (*H. Heaney et al.*, J. chem. Soc., 1958, 3838). The salt L pyrolyses normally with no ring contraction (*Mann et al.*, *ibid.*, 1965, 6342).

The series of compounds XLI (Table 4) is obtained from *o*-terphenyl-2,2″-ylenebis-(dimethylarsine). As for the series XL, the first member (XLI, $n = 1$) behaves normally on pyrolysis losing bromomethane and affording the cyclic *diarsine*, m.p. 137–138°. The second member (XLI, $n = 2$) behaves unusually. The di-quaternary salt crystallises in two forms; the lower melting one pyrolyses at 200°/14 mm, undergoing ring contraction to give 9-methyl-9*H*-tribenz [*b,d,f*]arsepin and the higher melting form loses bromomethane and leaves 9,10,11,12-tetrahydro-9,12-tribenzo [*e,g,i*] [1,4]diarsecin.

Considerable interest has been shown in the stereochemistry of the ring systems described in the preceding paragraphs, but progress has been much greater for the arsenic compounds than for those of phosphorus. In derivatives of dibenzo [*b,f*] [1,4]phospharsocin, the two benzene rings are not coplanar, and hence the cation in, for example, XLV, should be capable of existence in *cis* and *trans* modifications; X-ray analysis has demonstrated that the isolated compound has a *trans* configuration, but no success has been achieved in attempts to resolve salts with 10-camphorsulphonic and 3-bromo-8-camphorsulphonic acids. Because of twisting of the biphenyl moiety, resolutions of the cations XL, $n = 1,2,3$, are feasible and have been achieved, using 3-bromo-8-camphorsulphonic acid and dibenzoyltartaric acid as resolving agents and the resolved forms show considerable optical stability; the resolved bromides XL, $n = 2$ or 3, are optically stable in aqueous solution at room temperature for several months. Resolution of 10,11-dihydro-9,9,11,11-tetramethyl-9*H*-tribenzo [*d,f,h*] [1,3]-diarsoninonium dibromide (XLI, $n = 1$) has not proved possible, consistent with the molecule having a plane of symmetry.

(*c*) *Compounds with three or more hetero atoms*

Derivatives of **octahydro**-1,2,5,6-**tetra-arsocin** (LI), *tetramethyl-*, b.p. 126–129°/0.2 mm, *tetraphenyl-*, m.p. 128°, are obtained together with those of smaller ring analogues when alkanediarsonic acids are reduced with aqueous sulphur dioxide followed by hypophosphorous acid (*K. Sommer*, Z. anorg. allgem. Chem., 1970, **377**, 278).

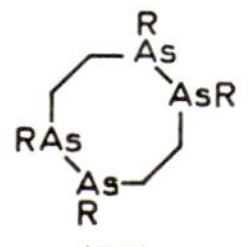

(LI)

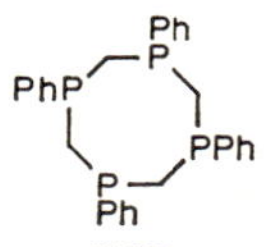

(LII)

1,3,5,7-*Tetraphenyl*-1,3,5,7-*tetraphosphocane* (1,3,5,7-*tetraphenyl-octahydro*-1,3,5,7-*tetraphosphocin*) (LII), m.p. 125–127°, is produced when phenylphosphine is heated with bis-(diethylaminomethyl)phenylphosphine at 140° (*L. Maier*, Helv, 1965, **48,** 1034).

5. Iodine compounds

Polycyclic compounds containing iodine in 7- and 8-membered rings are prepared by methods used to obtain the analogous 5- and 6-membered ring compounds.

The persulphate oxidation of α-phenyl-2-iodocinnamic acid at −10° gives a low yield of 10-*carboxydibenz* [b,f]*iodepinium salts* (I), *iodide*, m.p. 103–105°, *bromide*, m.p. 120–121°, *hydrogen sulphate*, m.p. 147°, *tetrafluoroborate*, m.p. 169–170°, *hexafluorophosphate*, m.p. 158° (*A. N. Nesmeyanov et al.*, Doklady Akad. Nauk S.S.S.R., 1974, **218**, 855):

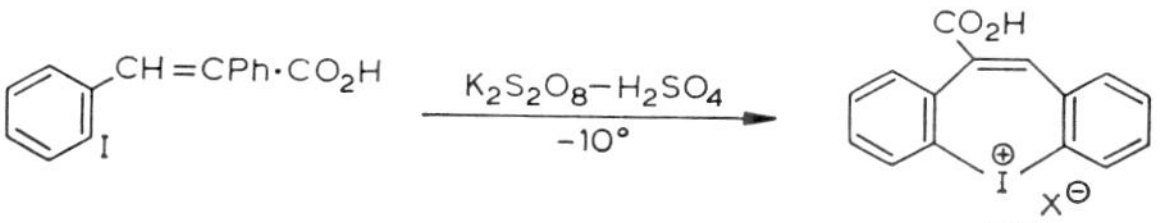

(I)

R. B. Sandin et al. (J. Amer. chem. Soc., 1956, **78**, 3819) oxidized *o*-iodosubstituted aryls with peroxyacetic acid and treated the resulting *o*-iodoso compounds with concentrated sulphuric acid thus producing 10,11-*dihydro*-5H-*dibenzo* [b,f]*iodepinium salts* (II, $n = 2$)

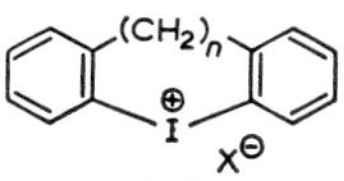

(II)

chloride, m.p. 148–149°, *bromide*, m.p. 150°, *iodide*, m.p. 144°, and the corresponding *iodocinium salts* (II, $n = 3$), *chloride*, m.p. 180°, *bromide*, m.p. 170–175°, and *iodide*, m.p. 145–150°. All these salts melt with decomposition.

Chapter 54

Compounds with Seven-membered and Larger Rings containing One or More Atoms of Unusual Hetero Elements together with Nitrogen or Elements of Group VI

R. S. EDMUNDSON

1. Mercury compounds

Mercury phthalate and styrene react together at 80–100° to give a *product*, m.p. 105–107°, sensitive to moisture and to air, and which may be heterocyclic (*G. Spengler* and *A. Weber*, Brennstoff-Chem., 1962, **43,** 234; C.A., 1963, **59,** 10107), but compounds of more readily predictable structure are

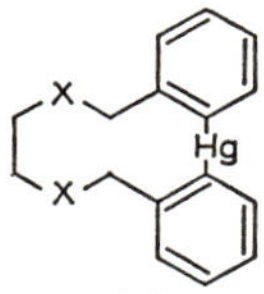

(I)

formed when mercuric chloride reacts with the di-Grignard reagent from 1,2-di(2-bromobenzyloxy)ethane, and from 1,2-di(*N*-methyl-2-bromobenzylamino)ethane. 8-**Mercura**-1,4-**dioxadibenzo**[*f*,*i*]**cycloundeca**-6,9-**diene** (I, X = O) was isolated as a colourless crystalline solid, m.p. 110–111°; the corresponding nitrogenous *compound* (I, X = *NMe*) has m.p. 109–111° (*G. Bähr* and *F. W. Küpper*, Chem. Ber., 1967, **100,** 3992).

2. Aluminium and boron compounds

As a class, oxa-, thia- and aza-borepanes are not so well known as their smaller ring analogues.

(*a*) *Compounds containing oxygen and/or sulphur as additional elements*

In a rather unusual reaction, triallylborane and propargyl trimethylsilyl ether at room temperature are said to afford the tetrahydroborin I which, after reduction and methanolysis of the product, gives 6-**methylene**-2,4-**dipropyl**-1,2-**oxaborepane** (II) reducible to the 6-methyl-2,4-dipropyl-1,2-oxaborepane (III). Direct methanolysis of the tetrahydroborin (I) gives the **hexahydro**-1,6-**dioxa**-2,7-**diborecin** (IV) (*B. M. Mikhailov* and *T. K. Bavyshnikova*, Zhur. obshcheĭ Khim., 1971, **41,** 1303).

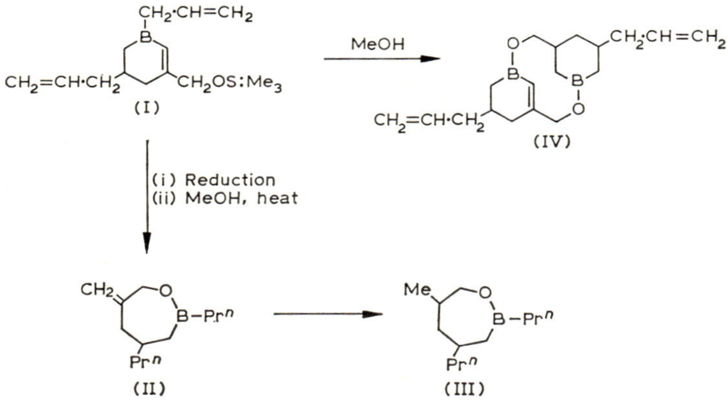

As already mentioned, 2-hexyl-7-methyl-1,2-oxaborepane has been obtained, together with an isomeric 1,2-oxaborinane in reactions between hexylboranes and allylacetone (*L. S. Vasil'ev et al., ibid.*, 1972, **42,** 1540).

Compounds having eight- or nine-membered rings containing boron with one oxygen or sulphur atoms appear not to have been prepared; a number of ring systems containing aluminium and oxygen have been postulated, but they are mostly poorly described; however, it has been suggested that the product obtained when 1-isobutyl-3-methylaluminacyclopent-3-ene dioxanate is heated at 150° consists of a mixture of isomeric 1-**oxa**-2-**aluminacyclo-oct**-6-**enes** (*L. H. Shepherd*, U.S.P. 3,669,990/1972; C.A., 1972, **77,** 114554).

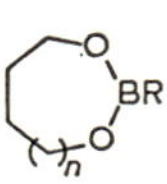

(V)

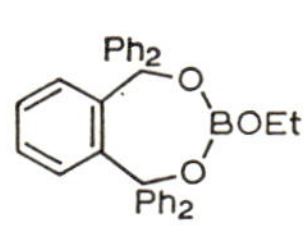

(VI)

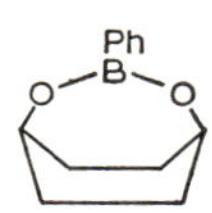

(VII)

The transesterification between aliphatic 1,4-diols, or analogous compounds, and esters of boric and boronic acids leads to derivatives of 1,3,2-**dioxaborepane** (V, n = 1) and although the number of known examples of this ring system is small, nevertheless they have been known for some time; VI was first prepared by *G. Wittig* in 1931 (Ber., 1931, **64,** 2394) using the straightforward transesterification procedure with triethyl borate.

The reaction between butane-1,4-diol and butyl hydrogen butaneboronate may take one of two courses depending on the nature of the catalyst employed; the use of toluene-4-sulphonic acid leads to a linear ester, whereas when sodium is incorporated into the reaction mixture, the product is **2-butyl-1,3,2-dioxaborepane** (V, *n* = 1, R = Bu) (*V. Karnojitzky, S. Kohn* and *P. Scheebeli*, Compt. rend., 1961, **253,** 1463).

Benzeneboronic acid reacts readily with butane-1,4-diol in aqueous acetone to give *2-phenyl-1,3,2-dioxaborepane* (V, *n* = 1, R = Ph), b.p. 90–95°/1 mm, m.p. 63–65° (*J. M. Sugihara* and *C. M. Bowman*, J. Amer. chem. Soc., 1958, **80,** 2443; *A. Finch et al.*, J. chem., Soc., A, 1970, 3339). With the same reagent, *cis*-1,4-dihydroxycyclohexane yields *3-phenyl-2,4-dioxa-3-borabicyclo*[3,2,2]*nonane* (VII), sublimes at 70°/0.05 mm *(idem., loc. cit.)*. The *B-Et* compound, b.p. 46–48°/0.001 mm, undergoes methanolysis to yield pure *cis*-cyclohexane-1,4- (*W. V. Dahloff* and *R. Koester*, Ann., 1975, 1625).

The reaction between triorganoboranes and diols in the presence of carbon monoxide has already been referred to in connection with the synthesis of 1,3,2-dioxaborinanes (p. 40); applied equally well to butane-1,4-diol, it affords a route to tertiary alkyl 1,3,2-dioxaborepanes. Thus, triethylmethylborane yields V, *n* = 1, R = Et_3C, b.p. 68°/0.5 mm (*M. E. D. Hillman*, U.S.P. 3,206,446/1965; C.A., 1965, **63,** 16381).

The same reactions with pentane-1,5-diol afford *1-triethylmethyl-1,3,2-dioxaborocane* (V, *n* = 2, R = CEt_3), b.p. 150°/0.5 mm, but the 2-phenyl compound is undistillable. Diethyleneglycol has provided **2-triethylmethyl-1,3,6,2-trioxaborocane** (VIII, X = O, R = CEt_3) (*Hillman*, J. Amer. chem. Soc., 1963, **85,** 982). **2-Phenyl-1,3,6,2-dioxathia-borocane** (VIII, X = S, R = Ph) is similarly prepared (*U. W. Gerwarth* and *W. Weber*, Synth. React. inorg. met.-org. Chem., 1975, **5,** 175).

Unfortunately, 2-chloro-1,3,2-borepane and analogues of larger ring size cannot be prepared from the diol and boron trichloride; moreover, 2-methoxy- and 2-(2-chloro-ethoxy)-1,3,2-borepanes are unstable to heat (*A. Finch et al.*, J. chem. Soc., 1962, 1428).

Representative examples of two of the isomeric oxathiaborepane systems have been prepared. Reaction between dibutyl etheneboronate and 2-mercaptoethanol in the presence of azabisisobutyronitrile yields *2-butoxy-1,5,2-oxathiaborepane* (IX), b.p. 60°/1 mm (*D. S. Matteson et al.*, J. med. Chem., 1964, **7,** 640). 2-Butyl-7-phenyl-1,6,2-oxa-

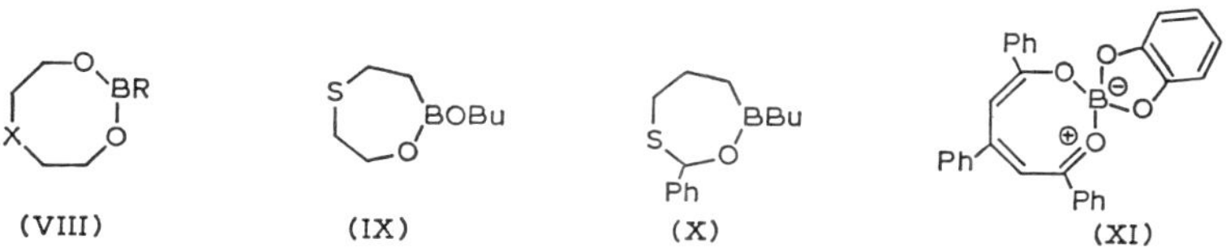

(VIII) (IX) (X) (XI)

thiaborepane (X) is formed in a ring expansion reaction between 2-butyl-1,2-thiaborolane and benzaldehyde; when the borepane is treated with piperidine, benzaldehyde is extruded exothermically leaving a borolane as its piperidine complex (*V. A. Dorokhov, O. G. Boldyreva* and *Mikhailov*, Izvest. Akad. Nauk S.S.S.R., Ser. Khim., 1971, 191).

The *spiran* XI is a red compound, m.p. 135°, formed from 2-butoxybenzoborolane and 1,3,5-triphenylpent-2-ene-1,5-dione *via* a green intermediate having m.p. 215° (*A. T. Balaban, C. N. Rentea* and *M. Bacescu-Roman*, Rev. Roumaine Chem., 1965, **10,** 865; C.A., 1966, **64,** 11196).

A synthesis of 2,3,6,7-**tetrahydrobenz[*d*][1,2,7]oxadiborepin** (XII) involves acid hydrolysis of 1,2-xylylenebis(dibutylboronate) prepared conventionally from the xylene dibromide (*A. S. Mane* and *S. H. Dandegaonker*, Curr. Sci., 1974, **43,** 443, C.A. 1974, **81,** 78001):

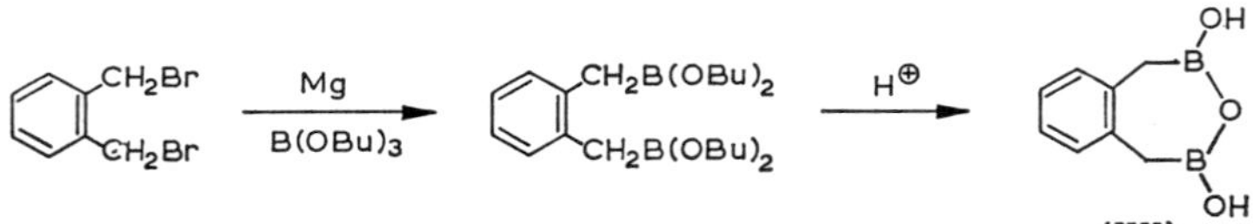

(XII)

Reference has already been made (p. 41) to the separation of *cis* and *trans* diols achieved through the use of areneboronic acids. *Sugihara* and *Bouman (loc. cit.)* record that *trans-*

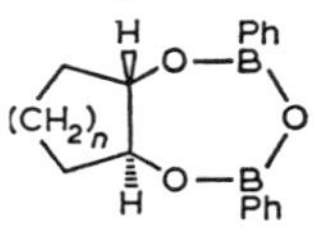

(XIII)

cyclopentane- and *trans*-cyclohexane-1,2-diols afford the *trioxadiborepanes* (XIII; $n = 1$), m.p. 127–128°, and $n = 2$, m.p. 134–136°, in quite high yield when treated with benzeneboronic acid, behaviour which contrasts with that of the corresponding *cis*-diols.

2,10-**Diethyl**-1,3,9,11-**tetraoxa**-2,10-**diboracyclohexadecane** (XV) is formed when the borinate ester XIV is strongly heated (*Dahloff* and *Koester*, *loc. cit.*):

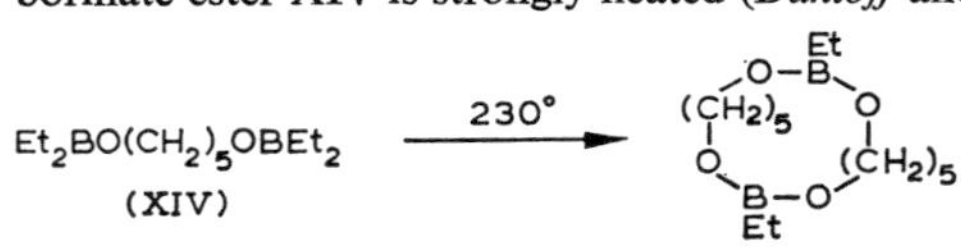

(XIV) (XV)

(*b*) *Compounds containing nitrogen as the additional element*

Perhydro-1,2-azaborepines are synthesized from 5-bromopentylboronic- or -borinic esters by reaction with amines (*Vasil'ev*, *M. M. Vartanyan* and *Mikhailov*, Izvest. Akad. Nauk S.S.S.R., Ser. Khim., 1972, 1678; Zhur. obshcheĭ Khim., 1972, **42,** 2675).

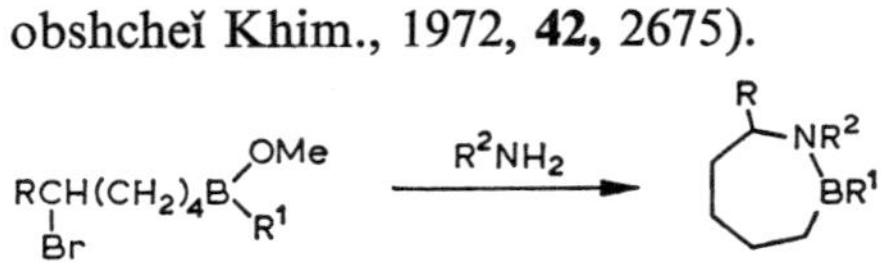

(XVI)

The 2-methoxy compounds (XVI, R = OMe) are reduced by lithium tetrahydridoaluminate to the perhydro-1,2-azaborine (XVI, R = H), and when treated with butylmagnesium bromide, they are converted to the 2-butyl derivatives.

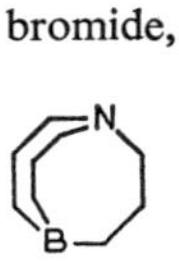

(XVII)

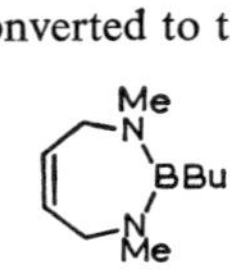

(XVIII)

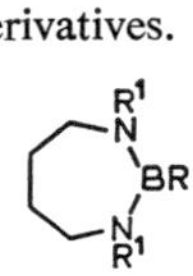

(XIX)

1,5-Azaborabicyclo[3,3.3]**undecane** (XVII), m.p. 123°, is formed when triallylamine, as its hydrochloride, undergoes hydroboration by triethylamineborane; it is stable in air and to peroxidic reagents (*N. N. Greenwood, J. H. Morris* and *J. L. Wright*, J. chem. Soc., 1964, 4753).

A transamination reaction was used to prepare 2-**butyl**-1,3-**dimethyl**-2,3,4,7-**tetrahydro**-1*H*-**diazaborepine** (XVII), m.p. 59–61°/2 mm, from 1,4-diaminobut-2-ene and butyl bis(dimethyl amino)borane, $BuB(NMe_2)_2$, but the yield was only about 3%; the perhydro-1,3,2-diazaborepines (XIX), a class of compound first described by *J. Goubeau* and *A. Zappel* (Z. anorg. allgem. Chem., 1955, **279,** 38) are obtained in better yield by essentially the same process (Table 1) (*K. Niedenzu, P. Fritz* and *J. W. Dawson*, Inorg. Chem., 1964, **3,** 1077; *Weber, Dawson* and *Niedenzu, ibid.*, 1966, **5,** 726).

TABLE 1

PERHYDRO-1,3,2-DIAZABOREPINES (XIX)

R^1	*R*	*B.p. °C/mm (m.p.)*
H	Me	39–42/6
	Bu	64/1
	Ph	102–104/1 (27–29)
Me	Me	34–35/2
	Bu	52–54/2
	Ph	85/86/2
Et	Me	43–45/2
	Bu	65–68/2
	Ph	112–114/3

A number of interesting ring expansion reactions take place when perhydrodiazaboroles (borolidines) and related compounds are treated with isocyanates. Examples of these reactions are illustrated below

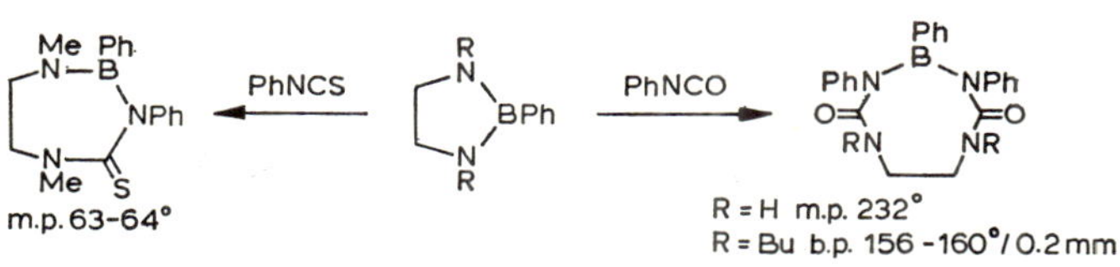

(*R. M. Cragg, M. F. Lappert* and *B. P. Tilley*, J. chem. Soc., 1964, 2108; *M. K. Das, P. G. Harrison* and *J. J. Zuckermann*, Inorg. Chem., 1971, **10,** 1092).

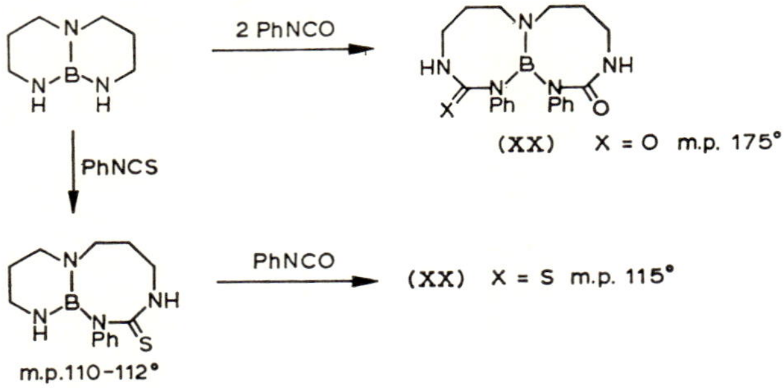

(*Fritz*, *Niedenzu* and *Dawson*, Inorg. Chem., 1965, **4**, 886; see also *Gerwarth* and *K. D. Mueller*, J. organometal. Chem., 1975, **96**, C33; 1976, **110**, 15).

An example of the 1,2,6,10-**azatriboracyclodecane** system, XXI, b.p. 132–133 °/1 mm, has been reported (*Mikhailov* and *V. F. Pozdnev*, Prob. Organ. Sinteza, Otd. obshcheĭ i. Tekhn, Khim. Akad. Nauk S.S.S.R., 1965, 220; C.A., 1966, **64**, 14203):

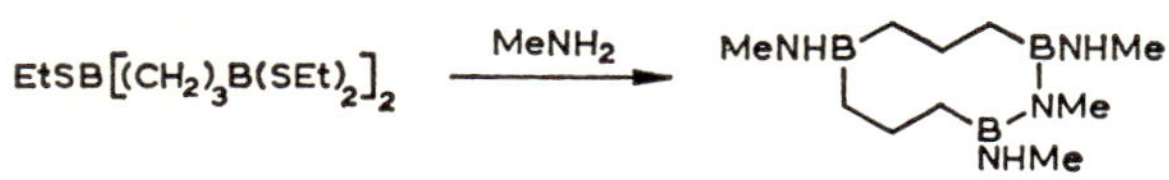

(XXI)

(*c*) *Compounds containing nitrogen together with oxygen and/or sulphur as additional elements*

Many heterocyclic compounds having mixed-type ring systems have been prepared by essentially those routes described in the previous sections, (*a*) and (*b*). A selection of these, with some specific examples, is given in Table 2.

B,B′,B″-Tris-2-(2-bromoethyl)benzeneboronic anhydride (XXII) reacts with primary amines but the nature of the product appears to depend on the individual amine; thus butylamine leads to *compound* XXIII, m.p. 178 °, with retention of the ring, while methylamine yields *compound* (XXIV), m.p. 134–138 °, solidifying and remelting at 162–165 ° (*R. T. Hawkins* and *A. U. Blackam*, J. org. Chem., 1967, **32**, 597).

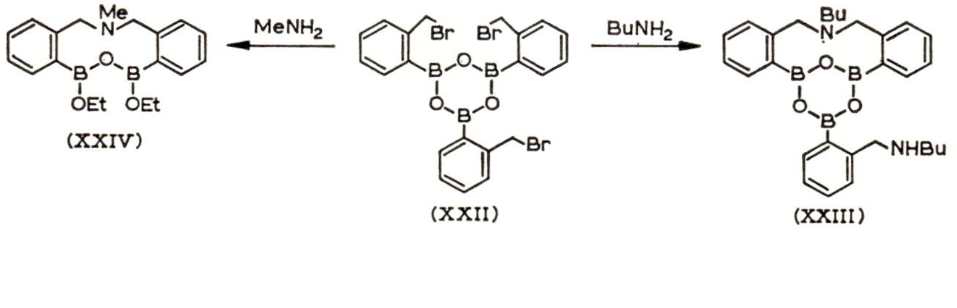

TABLE 2

Ring system	*Starting materials*	*Ref.*
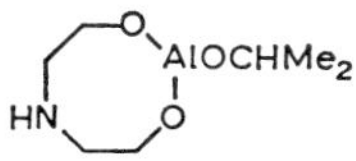	Diethanolamine and aluminium triisopropoxide	1
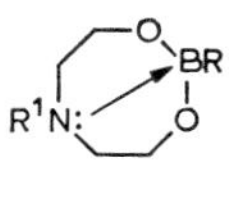	Diethanolamine and boronic acids or anhydrides	2–5
	Diethanolamine and boron esters	6–9
	N-alkyldiethanolamines	10–14
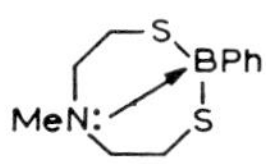 m.p. 110–111°	$PhBCl_2$ and $MeN(C_2H_4SH)_2$	10
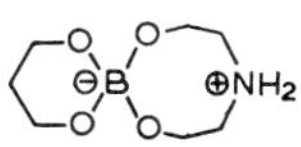	Diethanolamine and boric acid	15
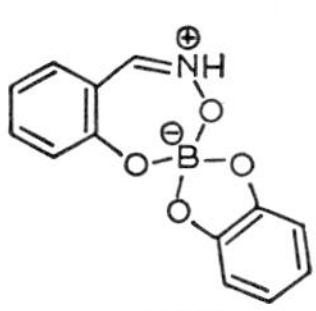 m.p. 220°	2-Butoxybenzoborole and salicylaldoxime	16
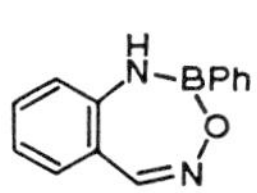 m.p. 249–250°	Benzeneboronic acid and *O*-amino-benzaldoxime	17
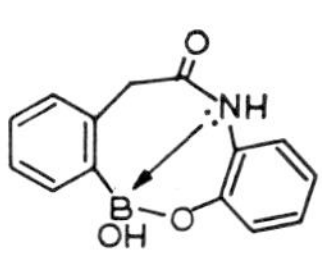 m.p. 179–180°	*o*-Aminophenol and 1-hydroxy-3,4-dihydro-1*H*-2,1-benzoxaborin-3-one	18

References

1 *R. C. Mehrotra* and *R. K. Mehrotra*, J. Indian chem. Soc., 1962, **39,** 677.
2 American Cyanamide, B.P. 877,874/1960; C.A., 1962, **56,** 8743.
3 *S. J. Groszos* and *C. S. Scanley*, U.S.P. 3,021,206/1962; C.A., 1962, **56,** 11902.
4 *L. I. Zakharkin* and *A. I. Kovredov*, Izvest. Akad. Nauk S.S.S.R., Ser. Khim., 1968, 1070.
5 *N. V. Kruglova* and *R. Kh. Freidlina*, *ibid.*, 1966, 1662.
6 *Zakharkin* and *Kovredov*, Zhur. obshcheĭ Khim., 1962, **32,** 1421.
7 *idem.*, Izvest. Akad. Nauk S.S.S.R., Ser. Khim., 1962, 1564.
8 *B. M. Mikhailov* and *V. A. Dorokhov*, *ibid.*, 1965, 1661.

9 *P. M. Aronovich* and *Mikhailov*, *ibid.*, 1968, 2745.
10 *M. Pailer* and *H. Huemer*, Monatsh., 1964, **95**, 373.
11 *W. R. Bamford*, B.P. 913,385/1962; C.A., 1963, **58**, 10236.
12 *B. Rudner* and *M. S. Moores*, U.S.P. 3,030,405/1962; C.A., 1962, **57**, 8437.
13 *idem*, U.S.P. 3,000,925/1961; C.A., 1962, **57**, 12321.
14 *W. G. Woods*, *I. S. Bengelsdorf* and *D. L. Hunter*, J. org. Chem., 1966, **31**, 2766.
15 *T. Weil*, U.S.P. 3,256,310/1966; C.A., 1966, **65**, 7058.
16 *A. T. Balaban*, *C. N. Rentea* and *M. Babescu-Roman*, Rev. Roumain Chem., 1965, **10**, 865; C.A., 1966, **64**, 11196.
17 *H. S. Yale et al.*, J. Amer. chem. Soc., 1962, **84**, 688.
18 *J. C. Catlin* and *H. R. Snyder*, J. org. Chem., 1969, **34**, 1660.

3. Silicon, germanium and tin compounds

(*a*) *Compounds containing oxygen and/or sulphur as additional elements*

(*i*) *Oxasila compounds and germanium and tin analogues*

The chloroplatinic acid-catalyzed cyclization of *o*-allylaryloxysilanes with a free Si–H bond yields the 2,3,4,5-**tetrahydro**-1,2-**benzoxasilepins** (I), also obtainable by the action of alcoholic potassium hydroxide or ethanolic sulphuric acid on 1-(diethoxymethylsilyl)-3-(*o*-trimethylsilyloxyphenyl)propanes:

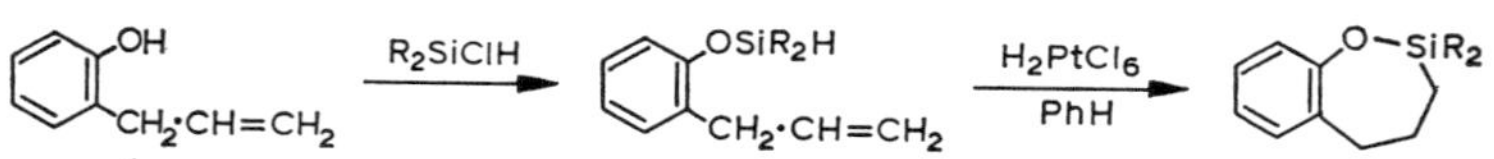

(I)

Amongst the derivatives recorded are the 2,2-dichloro- and 2,2-diphenyl-compounds, in addition to the 2,2-*dimethyl-*, b.p. 67–68 °/2 mm, n_D^{20} 1.5165, and the 2,2-*diethoxy-*, b.p. 117–121 °/1–2 mm, n_D^{20} 1.4918 (*V. F. Mironov*, *V. L. Kozlikov* and *N. S. Fedotov*, Zhur. obshcheĭ Khim., 1969, **39**, 966; 1971, **41**, 1077; *K. A. Andrianov*, *V. I. Pakhomov* and *N. E. Lapteva*, Izvest. Akad. Nauk S.S.S.R., Ser., Khim., 1962, 2039; *E. P. Plueddemann*, F.P. 1,387,338/1965; C.A., 1966, **64**, 2127).

A similar reaction starting from *o*-allyloxyphenylsilanes leads to 1*H*-2,3,4,5-**tetrahydro**-1,5-**benz**[*b*]**oxasilepin**, 5,5-*dimethyl-* (II), b.p. 80 °/1 mm, (*idem*, *loc. cit.*) and to a derivative (III) (*Mironov et al.*, Zhur. obshcheĭ Khim., 1972, **42**, 1365):

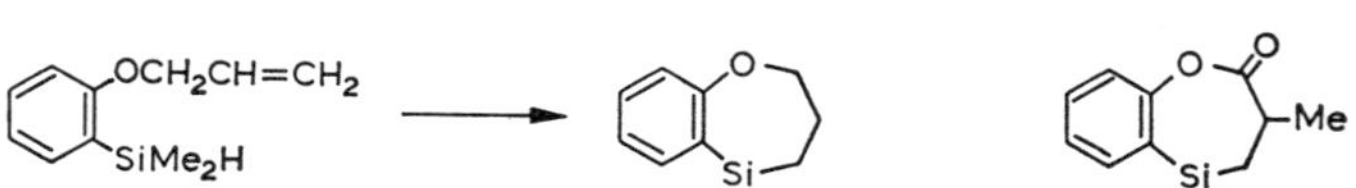

(II) (III)

5-**Bromo**-3,3-**diphenyl**-1*H*-2,3-**dihydrobenz**[*d*]**oxasilepin** (V), formed by an unusual ring expansion reaction from 4-bromo-1-hydroxy-2,2-diphenyl-1,2-dihydrobenzo[*c*]silin (IV), undergoes ring opening on treatment with phenylmagnesium bromide in ether, the benzyl *alcohol* VI, m.p. 150–151 °, being formed in over 90% yield (*R. J. P. Corriu* and *J. P. Masse*, J. organometal. Chem., 1970, **22**, 321).

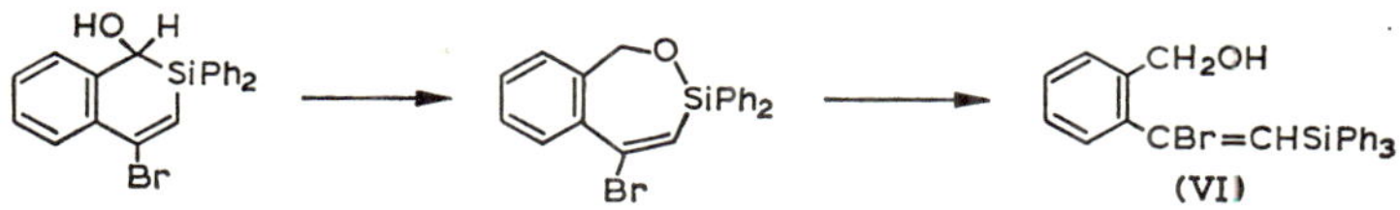

(IV) (V) (VI)

Germanium and tin analogues of the compounds I, II and V appear not to have been prepared.

When 1,1-dimethyl-2,5-diphenylsilole reacts with benzyne, the source of the latter being 2-carboxybenzenediazonium cation, the product is not the expected Diels–Alder adduct, 7,7-dimethyl-1,4-diphenyl-2,3-benzo-7-silanorbornadiene (VII), but appears to be product of further reaction between this and the diazonium salt. The mechanism of this reaction is not yet clear, but the structure of the tetracyclic *silalactone* VIII, m.p. 275–276°, is confirmed by X-ray analysis; the *germanium analogue*, m.p. 267.5–269.5°, is obtained in the same way (*T. J. Barton, A. J. Nelson* and *J. Clardy*, J. org. Chem., 1972, **37**, 895):

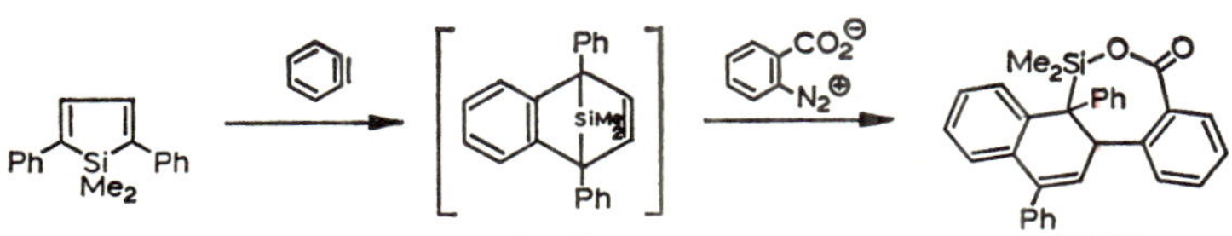

(VII) (VIII)

An example of the unusual reactivity of the silole ring system is the reaction of chlorosulphonyl isocyanate with 1,1-dimethyl-2,5-diphenylsilole in ether at room temperature. Ring enlargement takes place leading to the *dihydro-oxasilepin* (IX) isolated as yellow needles, m.p. 140–142° (with decomp.) (*Barton* and *R. J. Rogido*, *ibid.*, 1975, **40**, 582):

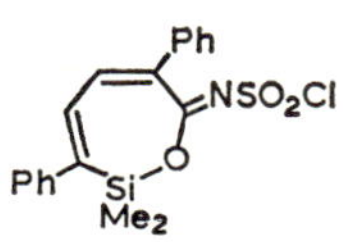

(IX)

Compounds possessing the 1,2-oxasilepane ring system are prepared (*a*) by chloroplatinic acid-catalyzed cylization of silyl ethers of ω-olefinic alcohols:

2,2-Dimethyl-1,2-oxasilepane (X) has b.p. 47–48°/25 mm, n_D^{20} 1.4360 (*Mironov, Kozlikov* and *Fedotov*, *loc. cit.*). [2,2-*Dimethyl*-1,2-*oxasilatridecane*-13-*one* (XI), b.p. 148–149°/3 mm, n_D^{20} 1.4656, is formed in a similar way and is readily hydrolyzed to the disiloxane $[HO_2C(CH_2)_{10}SiMe_2]_2O$ (*Mironov, Fedotov* and *I. G. Rybalka*, Khim., geterotsikl. Soedin., 1969, 440; C.A., 1970, **72**, 12809)].

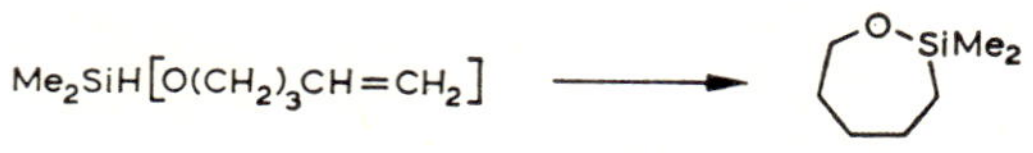

(X)

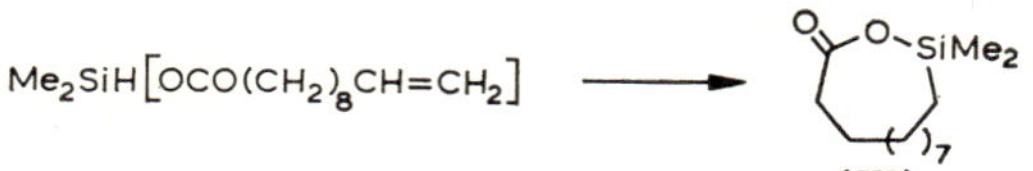

(XI)

(*b*) By photolysis of perhydrosilins:

The course of the reaction depends on the presence, or otherwise, of oxygen during the photolysis as shown in Scheme 1, below. Photolysis of 1,1-diphenylperhydrosilin-2-one (XII) in the presence of oxygen yields 2,2-diphenyl-1,2-oxasilepan-7-one (XIII), hydrolyzable to 4-carboxybutyldiphenylsilanol, and from which it may be obtained by the action of acetic anhydride (*A. G. Brook*, *H. Kircera* and *J. B. Pierce*, unpublished work, quoted in Advances in organometal. Chem., 1968, **7**, 126). In the absence of oxygen, the photolytic products include diphenylsilacyclopentane (22%), 2,2,8,8-*tetraphenyl*-1-*oxa*-2,8-*disilaspiro*-[7.7]*tridecan*-13-*one* (XV), m.p. 152–154° (20–53%), and the *alkene* XIV (9–12%), m.p.

Scheme 1

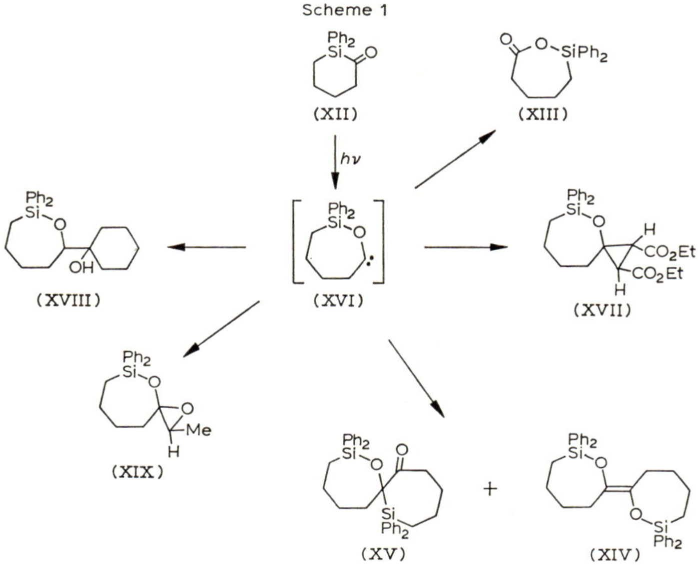

149–151°. A reactive species, thought to be the silene XVI, may be trapped with diethyl fumarate, when the product is trans-1,2-*diethoxycarbonyl*-4,4-*diphenyl*-3-*oxa*-4-*silaspiro*-[2.6]*nonane* (XVII), b.p. 190–200°/0.01 mm, hydrolyzable under both acid and alkaline conditions to linear compounds. The silene XVI, if such it is, also reacts with cyclohexanone to give the alcohol XVIII, and with acetaldehyde to yield the *epoxide* XIX as a mixture of isomers, b.p. 35–36°/0.8 mm (*Brook*, *Kircera* and *R. Pearce*, Canad. J. Chem., 1971, **49,** 1618; *Brook*, *Pearce* and *Pierce*, *ibid.*, 1971, **49,** 1622; *Brook*, *Pierce* and *J. M. Duff*, *ibid.*, 1975, **53,** 2874).

(*c*) By thermal rearrangement of suitable silepanes:

2,2-Dimethyl-7-methylene-1,2-oxasilepane is formed in theoretical yield when 3-oxo-1,1-dimethylsilepane is heated (*Brook*, *D. M. MacRae* and *A. R. Bassindale*, J. organometal. Chem., 1975, **86,** 185):

(*d*) By appropriate treatment of 4-hydroxybutylmetalloid chlorides:

Such compounds cyclize in the presence of a suitable base. The process has been applied to the preparation of 1,2-oxagermapanes, but evidently not of the silicon analogues (*P. Riviere* and *J. Satgé*, Compt. rend., 1971, **272,** C, 413):

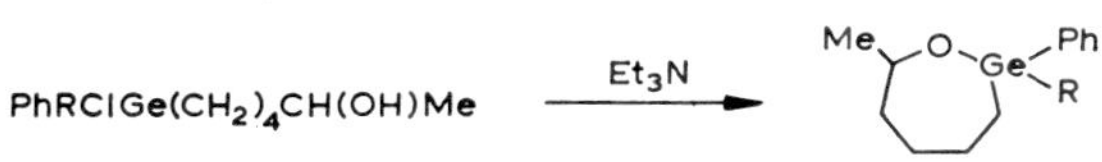

1,2-Oxagermapanes react with acetyl chloride with ring opening (*P. Mazerolles* and *A. Faucher*, J. organometal. Chem., 1973, **63,** 195).

The slow distillation of chloromethyl(4-hydroxybutyl) dimethylsilane, $ClCH_2SiMe_2(CH_2)_4$-OH, affords 3,3-*dimethyl*-1,3-*oxasilepane* (XX), b.p. 57–58°/17 mm, n_D^{24} 1.4531; its 3-*thia* analogue, b.p. 82–83/13 mm, n_D^{24} 1.5054, is obtained from 1,6-dichloro-2,2-dimethyl-2-silahexane, in a way similar to that used for the corresponding six-membered ring compound (*R. J. Fessenden* and *M. D. Coon*, J. org. Chem., 1964, **29,** 2499).

(XX) (XXI)

The *sila-lactone* XXI, b.p. 82–84°/2 mm, n_D^{20} 1.4728, is obtained when γ-(chloromethyldimethylsilyl)butyric acid, $ClCH_2SiMe_2(CH_2)_3CO_2H$, is heated with sodium carbonate at 100° (*Mironov* and *Fedotov*, Khim. geterotsikl. Soedin, 1969, 221; C.A., 1969, **71,** 50042). The compound forms a *dimer*, b.p. 160–162°/2.5 mm, m.p. 96–97°. Other large ring silalactones have been claimed but no data are yet available (*idem*, *ibid.*, 1972, 748; C.A., 1972, **77,** 88576).

(*ii*) *Di-, tri- and tetra-oxasilepins and related compounds*

The **dibenzo[*d,f*][1.3.2]dioxasilepins** (XXII), 2,2-*dimethyl*-, b.p. 172.5–173.5°/14.5 mm, n_D^{22} 1.5843; 2,2-*diphenyl*-, b.p. 235–236°/0.5 mm, m.p. 120–121°; 2-*methyl*-2-*phenyl*-, m.p. 101°, are formed by reaction between

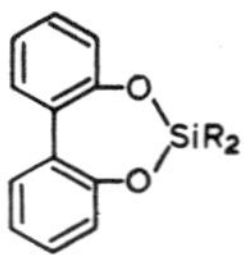

(XXII)

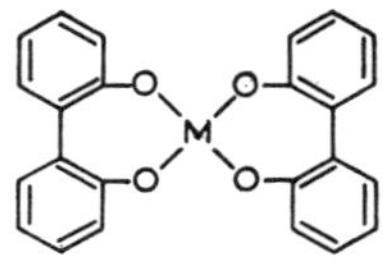

M = Si or Ge
(XXIII)

2,2′-dihydroxybiphenyl and dichlorosilanes in boiling benzene under nitrogen (*H. J. Emeleus* and *J. J. Zuckermann*, J. organometal. Chem., 1964, **1,** 328; *R. M. Ismail*, Z. Naturforsch., B., 1963, **18,** 1124), or by heating the diphenol together with a trace of its sodium salt with dialkoxysilanes (*idem, ibid.*, B, 1964, **19,** 873; G.P. 1,244,183/1967; C.A., 1968, **68,** 59714).

2,2-Diaryl compounds have been detected by mass spectrometry in the pyrolysis products from phenyltri-*o*-tolysilane (*V. N. Bochkarev et al.*, Zhur. obshcheĭ Khim., 1974, **44,** 1059).

Heating the 2,2-dimethyl compound under nitrogen with silicon tetrachloride gives the *spiro-*2,2′-*bis*(*dibenzo*[d,f][1.3.2]*dioxasilepin*) (XXIII, M = Si), m.p. 330–331° (*C. M. Silcox* and *Zuckermann*, J. Amer. chem. Soc., 1966, **88,** 168). The ***germaspiran*** XXIII (M = Ge), m.p. 330°, is formed when 2,2′-dihydroxybiphenyl is heated with tetraethoxygermane (*R. Mueller* and *L. Heinrich*, Chem. Ber., 1962, **95,** 2276) or with germanium tetrachloride, but an exchange reaction between the silepin XXII and germanium tetrachloride may also be employed (*Silcox* and *Zuckermann, loc. cit.*).

Dibenzo[*d,f*][1.3.2]dioxastannepin (XXIV), a derivative of tin(II), and the analogous dinaphtho compound, are obtained from the sodium salts of the diphenols and stannous chloride, and from the phenol and copper–tin or copper–tin oxide composites. It is a white solid, not particularly well characterized, insoluble in water, dilute acids and organic solvents with the exception of pyridine, dimethylformamide and tetrahydrofuran, in

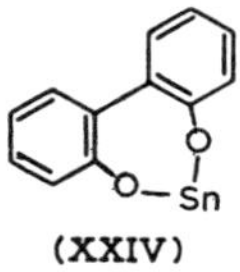

(XXIV)

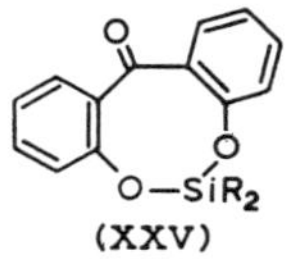

(XXV)

which it is sparingly soluble. When treated with PhNCS, it undergoes ring expansion (*Emeleus* and *Zuckermann, loc. cit.*; *Zuckermann*, J. chem. Soc., 1963, 1322; *G. T. Cocks* and *Zuckermann*, Inorg. Chem., 1965, **4,** 592; *R. F. Bridger*, U.S.P. 3,415,749/1969; C.A., 1969, **70,** 37555; *D. E. Fenton et al.*, Inorg. Chem. Acta, 1970, **4,** 235).

The tin spiran (XXIII, M = Sn) is prepared by heating together 2,2′-dihydroxybiphenyl and tetraphenylstannane (*K. Andrä* and *H. R. Hoppe*, Z. anorg. allgem. Chem., 1975, **413,** 97).

The **tetrahydrodibenzosilocins** (XXV) are obtained from dihydroxybenzophenones and dichlorosilanes (*Ismail, loc. cit.*; Neth. Appl., 6,611,744/1967; C.A., 1967, **67,** 65080.

Few examples of unsaturated, monocyclic compounds of Group IV elements are known. Two such examples are 2,2-*dimethyl*-1,3,4,7-*tetrahydro*-1,3,2-*dioxasilepin* (XXVI, M = Si, R = Me, R^1 = H), b.p. 73°/33 mm, n_D^{25} 1.4373, from but-2-ene-1,4-diol and dimethyldiethoxysilane in the presence of sulphuric acid (*G. B. Sterling* and *C. E. Pawloski*, U.S.P. 3,256,308/1966; C.A., 1966, **65**, 7217), and from the same diol with di(di-

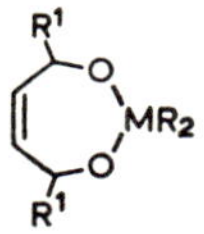

M = Si, Ge
(XXVI)

ethylamino)dimethylsilane, $Me_2Si(NEt_2)_2$ (*A. I. Nogaideli, R. Sh. Tkashelashvili* and *N. P. Makharashvili*, Zhur. obshcheĭ Khim., 1974, **44**, 119) and the similar *germanium* compound XXVI (M = Ge, R = Et, R^1 = Me), b.p. 58°/0.4 mm, from the diol and $Et_2Ge(NR_2)_2$ (*Satgé* and *G. Dousse*, Helv. 1972, **55**, 2406).

Although many perhydro-1,3,2-dioxasilins are known, the same cannot be said of larger rings possessing the 1,3,2-dioxasila- arrangement; they are, however, produced by essentially those processes used for the smaller ring compounds, *viz.* interaction of 1,4- (or longer chain) diols and dichloro-, diethoxy-, diacetoxy- or diamino-silanes, very often in the presence of a trace of acid as catalyst (*R. H. Krieble* and *C. A. Burkhard*, J. Amer. chem. Soc., 1947, **69**, 2689; *M. G. Voronkov, V. P. Davydova* and *B. W. Dolgov*, Izvest. Akad. Nauk S.S.S.R., 1958, 698, *E. L. Zaitsova et al.*, Vysokomol. Soedin., 1968, **10**, 2070; C.A., 1969, **70**, 12038; *Voronkov* and *Yu. P. Romadane*, Khim. geterotsikl. Soedin., 1966, 879; C.A., 1967, **67**, 32775; *J. C. Pommier, R. Calas* and *J. Valade*, Bull. Soc. chim., Fr., 1968, 1475; *P. F. Silva*, U.S.P. 3,065,254/1962; C.A., 1963, **58**, 9138; *K. A. Andrianov et al.*, Zhur. obshcheĭ Khim., 1967, **37**, 928; *Y. Abe, T. Takeuchi* and *I. Kijima*, Bull. chem. Soc. Japan, 1970, **43**, 2495) or by exothermic alcoholysis with butane-1,4-diol of mixtures of R_2SiCl_2 and $(R_2SiNH)_3$ (*E. P. Lebedev et al.*, Zhur. obshcheĭ Khim., 1975, **45**, 2645). Silyl chlorides of the type $R_2SiClO(CH_2)_nCl$ cyclize in the presence of sodium or potassium to give 1,3,2-dioxasilepanes and larger ring compounds (*W. H. Knoth* and *R. V. Lindsey*, J. Amer. chem. Soc., 1958, **80**, 4106).

Typical examples of compounds formed by the methods outlined above include 2,2-*dimethyl*-, b.p. 145–147, n_D^{20} 1.4218, 2,2-*dibutyl*-, b.p. 78°/1 mm, n_D^{20} 1.4451, 2,2-*dibutoxy*-, b.p. 106°/1 mm, n_D^{20} 1.4217, -1,3,2-*dioxasilepane* (XXVII, M = Si, n = 1); 2,2-*dimethyl*-1,3,2-*dioxasilocane* (XXVII, M = Si, n = 2), b.p. 46–47°/8 mm, n_D^{20} 1.4308; 2,2-*dimethyl*-

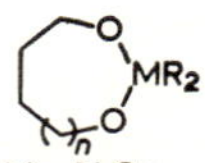
M = Si,Ge
(XXVII)

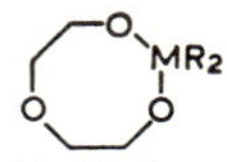
M = Si, Ge
(XXVIII)

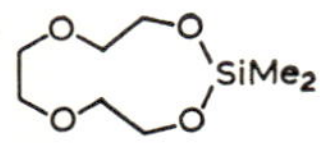
(XXIX)

1,3,2-*dioxasilonane* (XXVII, M = Si, n = 3), b.p. 65–66°/6 mm, n_D^{20} 1.4402. Other similar compounds formed in the same way include 2,2-*dimethyl*-1,3,6,2-*trioxasilocane* (XXVIII, M = Si, R = Me), b.p. 168°, 90°/50 mm, n_D^{20} 1.4331, (from diethylene glycol), and 2,2-*dimethyl*-1,3,6,9,2-*tetraoxasilacycloundecane* (XXIX), b.p. 96°/1 mm, n_D^{20} 1.4487, (from triethylene glycol).

Photolysis of mixtures of diallylsilanes and hydrogen sulphide yields derivatives of **5-sila-1-thiacyclo-octane** readily oxidized by peroxycarboxylic acids to the sulphone (*K. E. Koenig*, *R. A. Felix* and *W. P. Weber*, J. Amer. chem. Soc., 1974, **89**, 1532):

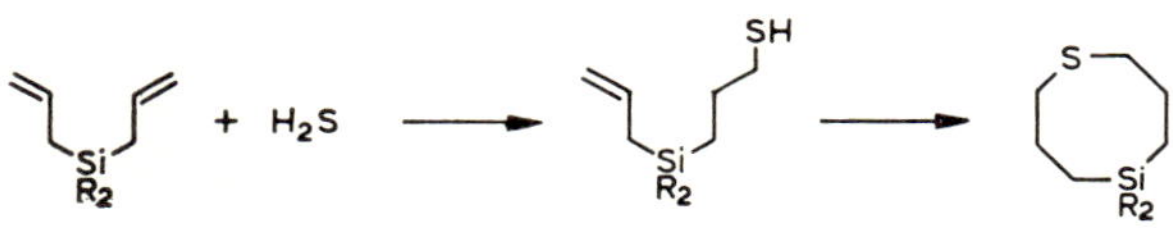

The increased reactivity of R_2GeH_2 and R_2GeO allows the preparation of germanium analogues of the above structures by interaction of these with a hot diol (*M. Lesbre* and *Satgé*, Compt. rend., 1962, **254,** 4051) as well as by those reactions commonly used for the silicon compounds (*Satgé* and *Dousse*, *loc. cit. S. Mathur et al.*, J. organometal. Chem., 1965, **4,** 371; *R. C. Mehrotra* and *Mathur*, *ibid.*, 1966, **6,** 425). Examples of 1,3,2-**dioxa-germapanes** are 2,2-*diethyl*-, b.p. 88–90°/12 mm, n_D^{20} 1.4612, 2,2-*dibutyl*-, b.p. 151°/20 mm, n_D^{20} 1.4668, and 2,2-*diphenyl*-, b.p. 142°/0.2 mm. 2,2-*Dibutyl*-, b.p. 120–122°/2.7 mm, 2,2-*diphenyl*-, b.p. 155°/0.2 mm, -1,3,2-*dioxagermocane*. All such compounds appear to be monomeric.

The reactions between dibutyldiethoxystannane and butane-1,4- or pentane-1,5-diols are catalysed by toluene-4-sulphonic acid and afford the *compounds* XXVII, (M = Sn, R = Bu) n = 1, b.p. 175–178°/0.4 mm, n = 2, b.p. 185–187°/0.15 mm, as trimers (in benzene) of unknown constitution (*Mehrotra* and *V. D. Gupta*, *ibid.*, 1965, **4,** 145). 1,3,2-Dioxastannepanes react rapidly, even at room temperature, with carbon disulphide to give finally, 1,5,7,11-**tetraoxa-6-spiroundecane,** in a reaction discussed for the six-membered ring compounds (p. 73) (*S. Sakai et al.*, J. org. Chem., 1971, **36,** 1176).

The reaction between butane-1,4-diols in tetraethoxysilane, catalyzed by sodium methoxide, yields derivatives of 1,6,8,13-**tetraoxa-7-silaspiro**[6.6]**tridecane** (XXX), in-

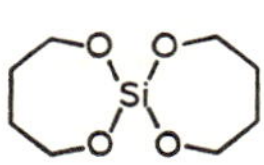
(XXX)

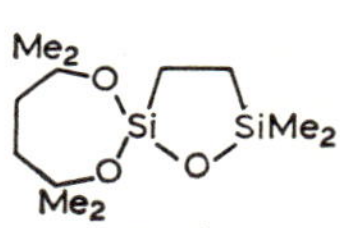
(XXXI)

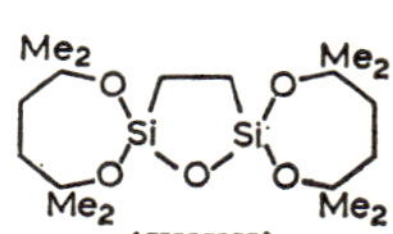
(XXXII)

cluding the 4,4,4′,4′,6,6,6′,6′-*octamethyl*-, m.p. 81–83°, 4,4,4′,4′,-*tetrabutyl*-, b.p. 119°/0.6 mm, and 4,4′,6,6′-*tetramethyl*-, b.p. 110°/4.5 mm, compounds (*C. L. Frye*, J. org. Chem., 1969, **34,** 2496), while *XXXI*, b.p. 80°/0.3 mm, and *XXXII*, b.p. 110°/0.05 mm, m.p. 78–80°, are obtained from 1,1,4,4-tetramethylbutane-1,4-diol and $[(MeO)_3SiCH_2SiMe_2]_2O$

TABLE 3

Ring system	*Starting materials*	*Ref.*
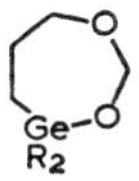	1,2-oxagermolanes and hexafluoroacetone or trichloracetaldehyde	1
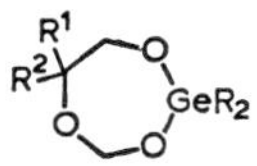	1,3,2-dioxagermolanes and aldehydes or ketones with $ZnCl_2$ or H_2PtCl_6	2
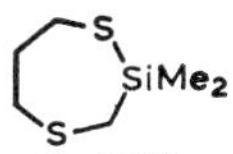 b.p. 61–63°/1 mm	Propane-1,3-dithiol and $ClCH_2SiMe_2Cl$ with Et_3N	3
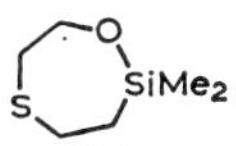 b.p. 75°/10 mm	Me_2SiCl_2, 2-mercaptoethanol, and $CH_2{=}CHSiMe_2(OEt)$	4
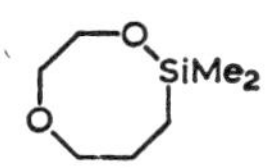 b.p. 34°/28 mm	Me_2SiCl_2 and $Me_3SiOC_2H_4OCH_2CH{=}CH_2$	4
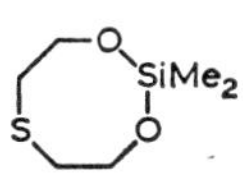 b.p.102–103°/26 mm	Me_2SiCl_2 and bis(2-hydroxyethyl) sulphide with Et_3N	5
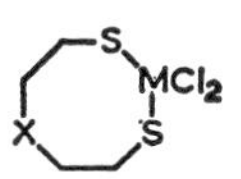 M = Ge, X = O m.p. 117° M = Ge, X = S m.p. 121°	$X(CH_2CH_2SH)_2$ and MCl_4	6

References

1 *M. Massol*, *J. Barrau* and *J. Satgé*, J. organometal Chem., 1970, **25,** 81; *Barrau et al.*, Rec. Trav. chim., 1973, **92,** 321.
2 *G. Dousse*, *H. Larrayssiere* and *J. Satgé*, Helv., 1975, **58,** 2610.
3 *M. Wieber* and *M. Schmidt*, Chem. Ber., 1963, **96,** 1019.
4 *A. Berger*, F.P. 1,549,776/1968; C.A., 1970, **72,** 21780.
5 *M. G. Voronkov* and *Yu. P. Romadane*, Khim. geterotsikl. Soedin., 1966, 879; C.A., 1967, **67,** 32775; *E. Lukevics*, *L. Liberts* and *M. G. Voronkov*, Zhur. obshcheĭ Khim., 1968, **38,** 1838.
6 *M. Drager* and *L. Ross*, Chem. Ber., 1975, **108,** 1712; *Dräger*, *ibid.*, 1975, **108,** 1723; *Drager* and *R. Engler*, Z. anorg. allgem. Chem., 1975, **413,** 229.

at 260–300° in the presence of a trace of water (*Dow Corning Corp.*, B.P. 1,106,838/1968; C.A., 1968, **68,** 96659).

Phosphines of the type $Et_2P \cdot MR_2R^1$ react with succinic anhydrides; when R^1 is hydrogen, seven-membered **silalactones** (XXXIII, M = Si, R = Me) and *germalactone* (XXXIV, M = Ge, R = Et), b.p. 90°/0.04 mm, (*C. Couret, J. Escudié* and *Satgé*, Rec. Trav. chim., 1972, **91,** 429):

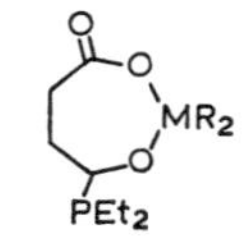

(XXXIII)

Some less commonly encountered systems are illustrated in Table 3.

(*iii*) *Oxadisilepins etc.*

2,2,3,3-**Tetraethyl-**1,2,3-**oxadigermapane** (XXXIV), b.p. 98°/0.09 mm, appears to be the only recorded example of a seven-membered ring compound containing the O–M–M arrangement. It is obtained when the digermane $Et_2HGe \cdot GeEt_2(CH_2)_4OH$ is treated with Raney nickel (*Riviere, Satgé* and *D. Soula*, Compt. rend., 1973, **277,** C, 895).

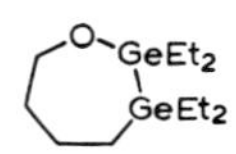

(XXXIV)

The use of conjugated dienes for the synthesis of oxadisila ring system is illustrated by two examples.

The reaction between dilithiotetraphenylbutadiene and dichlorotetramethyldisilane gives a mixture of 1,1-dimethyl-2,3,4,5-tetraphenylsilole and the six-membered ring compound XXXV; the latter is oxidized by oxygen, 3-chloroperoxybenzoic acid or *tert*-butylhydroperoxide to 2,2,7,7-*tetramethyl*-3,4,5,6-*tetraphenyl*-7H-1,2-*dihydro*-1,2,7-*oxadisilepin* (XXXVI), m.p. 137–138°, (*Y. Nakadaira* and *H. Sakurai*, J. organometal. Chem. 1973, **47,** 61):

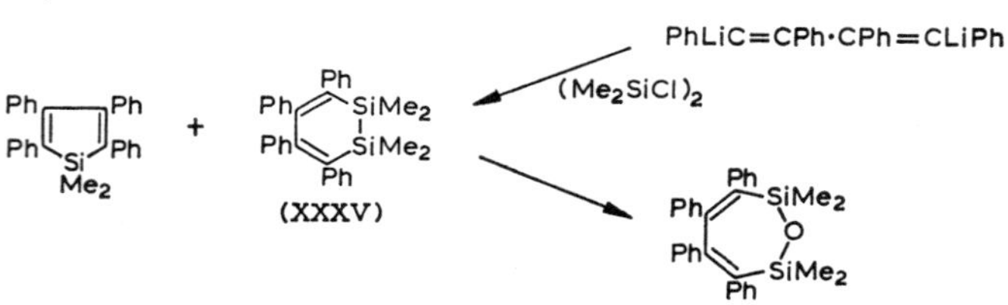

Butadienes react with dichlorotetramethyldisiloxane, in the presence of sodium in tetrahydrofuran to give 2,2,7,7-*tetramethyl-* (XXXVII, R = H), b.p. 160–163°, 55°/14 mm, n_D^{25} 1.4438, and 2,2,4,5,7,7-*hexamethyl*-2,3,6,7-*tetrahydro*-1H-1,2,7-*oxadisilepin* (XXXVII, R = Me), b.p. 67–70°/60 mm, n_D^{25} 1.4558. The same compounds are formed when the alkenes XXXVIII, (R^1 = OMe) are treated with water, and the former compound also when 1,1,2,2-tetramethyl-1,2,3,6-tetrahydro-1,2-disilin is oxidized (*D. R. Weyenberg, L. H. Torporcer* and *L. E. Nelson,* J. org. Chem., 1968, **33,** 1975; *L. Birkofer* and *W. Weniger*, Chem. Ber., 1973, **106,** 3595):

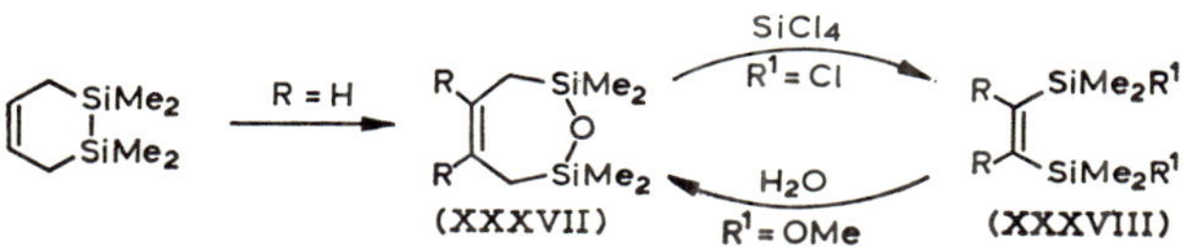

Silicon tetrachloride cleaves the ring system to form the alkenes XXXVIII (R^1 = Cl).

W. E. Dennis and *J. W. Ryan* (J. org. Chem., 1970, **35,** 4180) report a method for preparing a **hexahydro-oxadisilocin** (XXXIX), m.p. 41–43°:

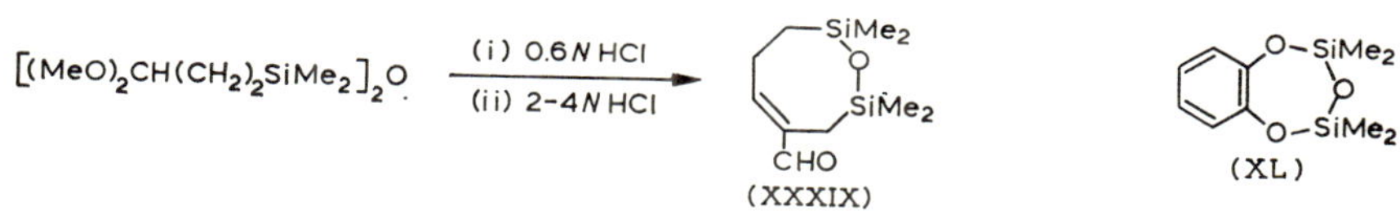

2,2,4,4-*Tetramethyl*-2,3,4,5-*tetrahydro*-1H-1,3,5,2,4-*benzotrioxadisilepin* (XL), b.p. 100–103°/10 mm, n_D^{20} 1.4790, is obtained conventionally from catechol, triethylamine and dichlorotetramethyldisiloxane (*K. A. Andrianov et al.*, Izvest. Akad. Nauk S.S.S.R., 1968, 646).

2,2,7,7-Tetramethyl-1,2,7-oxadisilepane (XLI, M = Si, R = Me), b.p. 66–68°/25 mm, has been recorded (*C. R. Sporek*, U.S.P. 3,452,071/1969; C.A., 1969, **71,** 91633). **2,2,7,7-Tetraethyl-1,2,7-oxadigermepane** (XLI, M = Ge, R = Et), b.p. 93°/0.9 mm, n_D^{20} 1.4862, is formed by the action of alkali on 1,4-di(bromodiethylgermyl)butane, $Et_2BrGe(CH_2)_4$-

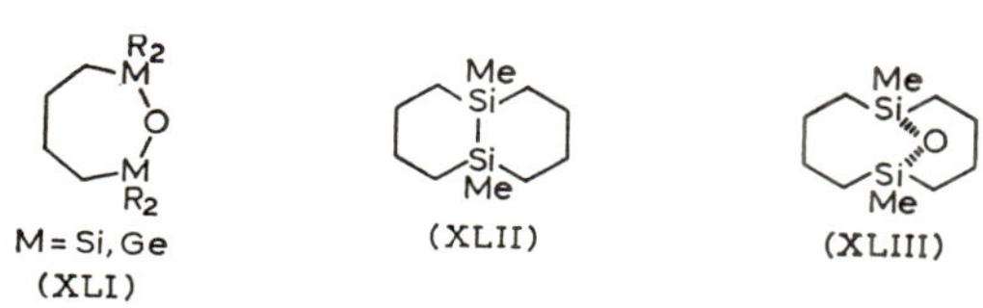

$GeBrEt_2$; the use of sodium sulphide in place of sodium hydroxide leads to **2,2,7,7-tetraethyl-1,2,7-thiadigermepane,** b.p. 125°/1 mm, n_D^{20} 1.5350 (*P. Mazerolles, M. Lesbre* and *M. Joanny*, J. organometal. Chem., 1969, **16,** 227; *P. Massol, D. Mesnard* and *Satgé*, *ibid.*, 1971, **30,** C67). Both *cis-* and *trans-*1,6-dimethyl-1,6-disilabicyclo[4.4.0]decane

(XLII) undergo oxidation with peroxybenzoic acid or bis(trimethylsilyl) peroxide, the oxygen atom being inserted between the two silicon atoms, so giving *cis*-1,6-dimethyl-11-oxa-1,6-disilabicyclo[4.4.1]undecane (XLIII) (*K. Tamao, M. Kumada* and *T. Sugimoto*, Chem. Comm., 1970, 285; *Tamao, Kumada* and *M. Ishikawa*, J. organometal. Chem., 1971, **31,** 17; *Tamao* and *Kumada, ibid.*, 1971, **31,** 35). While 1,2-disilahexanes undergo this insertion reaction with the lower Group VI elements, the analogous 1,2-digermacyclohexanes evidently do not.

2,2,7,7-**Tetramethyl**-1,3,2,7-**dioxadisilepane** (XLIV) has been reported (*V. F. Mironov, V. L. Kozlikov* and *N. S. Fedotov*, Zhur. obshcheĭ Khim., 1971, **41,** 1077), as has **1,5-** *dimethyl*-3,3,7,7-*tetraphenyl*-2,4,6,8-*tetraoxa*-1,3,5,7-*tetrasilabicyclo*[3.3.2]*decane* **(XLV),** m.p. 132.5–133.5°.

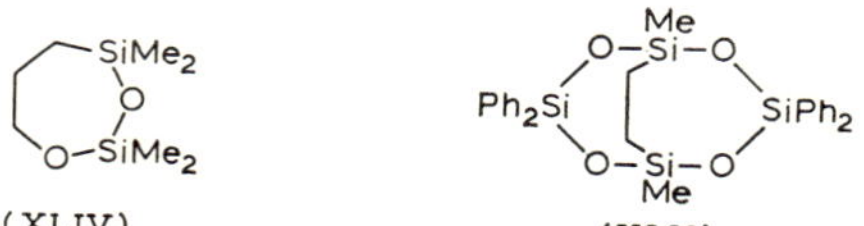

(XLIV) (XLV)

2,2,4,4,7-**Pentamethyl**-1,3,5-**trioxa**-2,4-**disilacycloheptane** (XLVII), b.p. 151°, has been obtained as a product of a modified alcoholysis procedure using propane-1,2-diol together with a mixture of chlorodisiloxane and disilazanoxy compounds XLVI (*Lebedev et al.*, Zhur. obshcheĭ Khim., 1975, **45,** 2645):

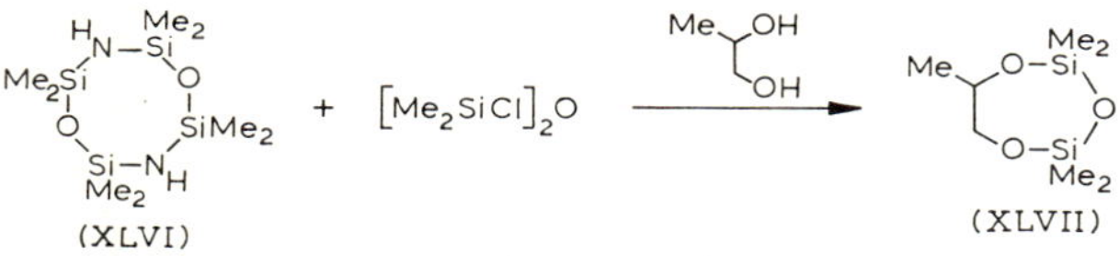

(XLVI) (XLVII)

Interestingly, but perhaps predictably, diethyl(2-hydroxyethyl) germane in the presence of Raney nickel at 20°, forms a mixture of the diethyloxagermacyclobutane and its dimer, 2,2,6,6,-*tetraethyl*-1,5-*dioxa*-2,6-*digermocane* (XLVIII), b.p. 117°/0.8 mm; the two are in

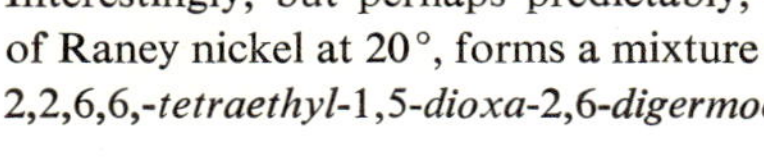

(XLVIII)

equilibrium; 4,8-*dimethyl*-, b.p. 105°/0.6 mm, 4,8-*di*-tert-*butyl*-, b.p. 137°/0.035 mm (*Massol et al.*, Compt. rend., C, 1971, **272,** 2081). Heating $PhCl_2GeCH_2CH_2OH$ at 100° yields 2,6-*dichloro*-2,6-*diphenyl*-1,5-*dioxa*-2,6-*digermocane*, b.p. 145–150°/0.03 mm, (*Riviere* and *Satgé*, Compt. rend. C, 1971, **272,** 413).

Whereas trimethylene glycol and other 1,3-diols react normally with *gem*-dichloro- and *gem*-dialkoxysilanes in the presence of a tertiary base or toluenesulphonic acid, respectively, to give the monomeric, perhydro-1,3,2-dioxasilins (p. 71), ethylene glycol and other 1,2-diols yield dimeric products when treated with the same reagents, or in the exothermic co-alcoholysis reaction using Me_2SiCl_2 and $(Me_2SiNH)_3$ *(Lebedev et al., loc. cit.)*, and for which the 1,3,6,8-**tetraoxa**-2,7-**disilacyclodecane** structure has been proposed. Molecular weight determinations on the tin compounds have demonstrated their variability with temperature, suggesting some loose form of association, or possibly an equilibrium between five- and ten-membered rings, but an unusual form of association of five-membered rings has still not been excluded. This situation contrasts strongly with that for both acyclic and cyclic compounds possessing Sn–S bonds; physical evidence overwhelmingly supports the 1,3,2-dithiastannacyclopentane and 1,3,2-dithiastannacyclohexane structures for the products from ethane-1,2- and propane-1,3-dithiols (*M. A. Delmas et al.*, J. organometal. Chem., 1975, **87,** 285).

2,2,7,7-**Tetramethyl**-1,3,6,8-**tetraoxa**-2,7-**disilacyclodecane** (XLIX), m.p. 54°, b.p. 177–178°, 78°/10 mm, and the 2,2,4,7,7,9-*hexamethyl* compound, b.p. 162–163°, are obtained by the methods outlined (*Silva loc. cit.*) *Andrianov et al., loc. cit.*; Izvest. Akad. Nauk

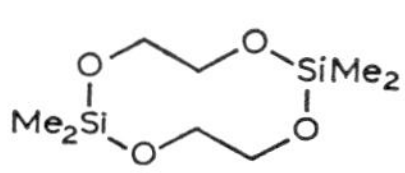

(XLIX)

S.S.S.R., Ser Khim., 1966, 2229; *M. G. Voronkov* and *Yu. P. Romadane*, Khim. geterotsikl. Soedin., 1966, 879; C.A., 1967, **67,** 32775; *R. H. Krieble* and *C. A. Burkhard*, J. Amer. chem. Soc., 1947, **69,** 2689). In addition, the tetramethyl compound was obtained during attempts to prepare dimethyldi(2-hydroxyethoxy)silane (*M. S. Spring*, J. org. chem., 1958, **23,** 58), and also from ethylene glycol and diacetoxydimethylsilane (*Voronkov, V. P. Davydova* and *B. N. Dolgov*, Izvest. Akad. Nauk S.S.S.R., Ser. Khim., 1958, 698). 2,2,8,8-**Tetramethyl**-1,3,7,9-**tetraoxa**-2,8-**disilacyclododecane** has b.p. 161°/753 mm.

2,2,7,7-**Tetrabutyl**-1,3,6,8-**tetraoxa**-2,7-**distannacyclodecane,** m.p. 215–220°, its 4,9-*dimethyl-*, m.p. 185°, 4,5,9,10-*tetramethyl-*, m.p. 120°, and 4,4,5,5,9,9,10,10-*octamethyl-*, m.p. 45°, b.p. 195°/0.4 mm, derivatives are obtained by azeotropic removal of water from heated mixtures of the ethylenic diol and dibutyltin oxide in xylene. Analogous *compounds* obtained from butane-1,3-diol, propane-1,3-diol, and pentane-2,4-diol, have b.p. 188°/1.5 mm (m.p. 134°); b.p. 184°/0.3 mm; and b.p. 159°/0.7 mm (m.p. 155°), respectively (*J.-C. Pommier* and *J. Valade*, J. organometal. Chem., 1968, **12,** 433; *Voronkov* and *Romadane*, Khim. geterotsikl. Soedin., 1966, 892; C.A., 1967, **67,** 32751).

Other examples of larger ring systems containing two or more silicon atoms together with oxygen or sulphur are illustrated in Table 4.

Recent physical data, particularly that from mass spectroscopy, have cast considerable doubt on the phenoxastannin structures assigned by *E. J. Kupchik et al.* (J. organometal. Chem., 1967, **10,** 269) to the products from 2,2′-dilithiodiphenyl ether and dichlorostannanes or stannic chloride.

High resolution mass spectrometry confirms a composition $C_{28}H_{28}O_2Sn_2$ for the *product* m.p. 255° from dilithiodiphenyl ether and dichlorodimethylstannane. The dimeric structure L is now assigned to this product and proton nuclear resonance spectroscopy suggests a rapid transformation between three conformations in solution (*H. A. Terloww et al., ibid.*, 1974, **64,** 205; *H. A. Meinema* and *J. G. Noltes, ibid.*, 1973, **63,** 243).

Similarly, structure LI is now assigned to the product, m.p. ~ 355°, obtained in the preparation of 10,10′-spirobiphenoxastannin.

TABLE 4

Structure	*Starting materials*	*Ref.*
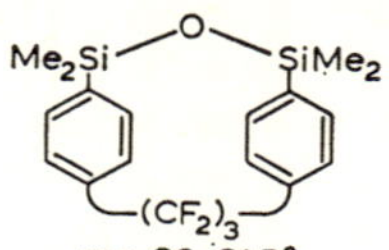 m.p. 80–81.5°	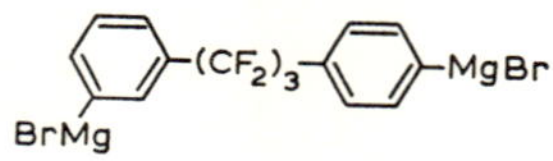 $Me_2Si(OEt)_2$	1
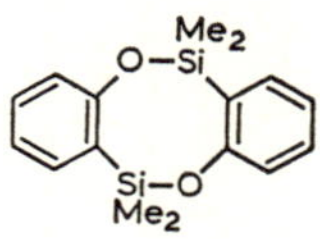	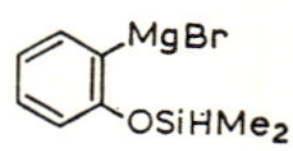	2
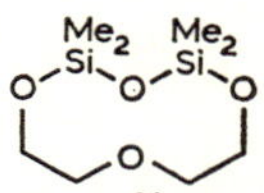 b.p. 110°/50mm	$[(EtO)SiMe_2]_2O + O(C_2H_4OH)_2$ Tos.OH	3
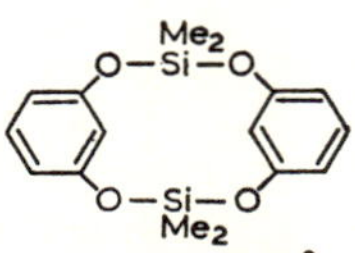 m.p. 143–143.5°	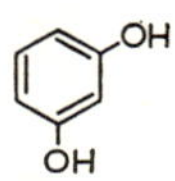 Me_2SiCl_2	4
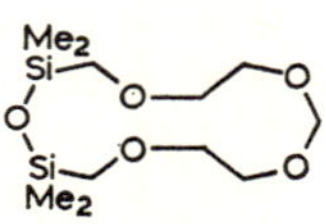 b.p. 107°/1mm	$(HOC_2H_4OCH_2SiMe_2)_2O$ HCHO Tos.OH	5
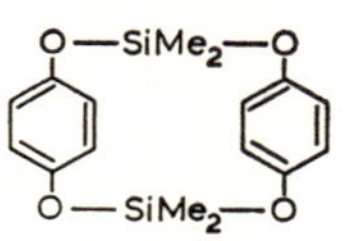	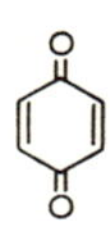 Me_2SiCl_2; Et_3N	6
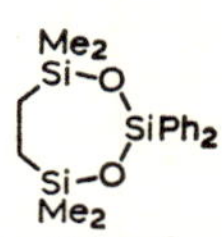 m.p. 57–58°	$ClSiMe_2(CH_2)_2SiMe_2Cl$ $Ph_2Si(OH)_2$	7
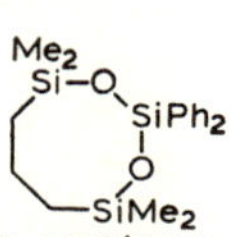 b.p. 129–140°/0.05 mm m.p. 68–68.5°	$ClSiMe_2(CH_2)_3SiMe_2Cl$ $Ph_2Si(OH)_2$	7

TABLE 4 (continued)

Structure	*Starting materials*	*Ref.*
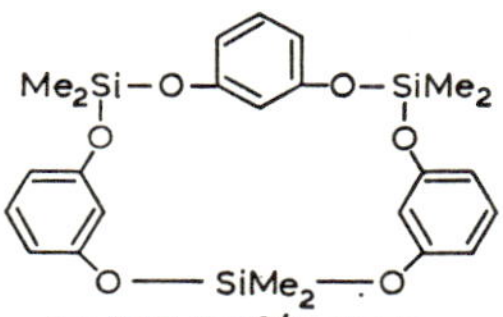 b.p. 220-223°/1-2 mm m.p. 81.5-82.5°	Resorcinol Me_2SiCl_2	4
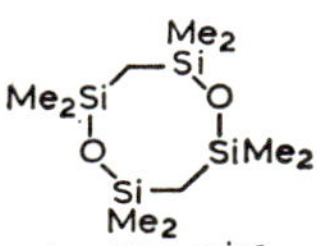 b.p. 209-210° m.p. 30°	$(ClSiMe_2)_2CH_2$ $OH^{\ominus}$	8
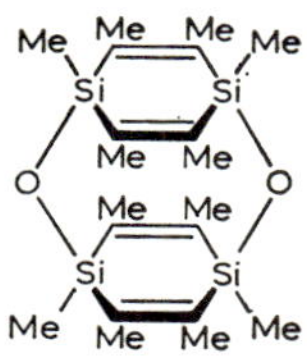	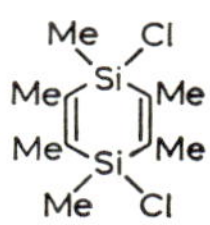 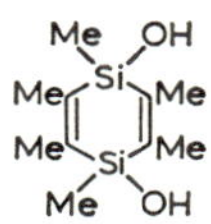 Et_3N	9
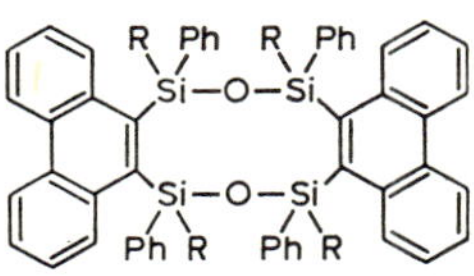 R = Me, Ph	Phenanthrene Li RR^1SiCl_2	10
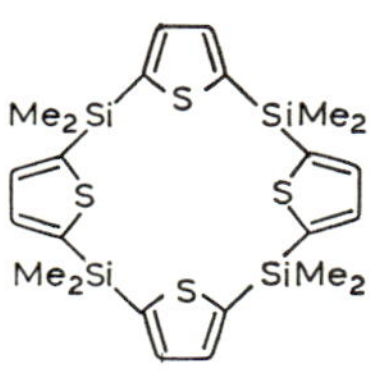 m.p. 224-226°	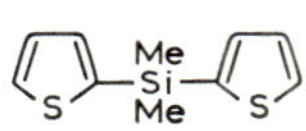 Li Me_2SiCl_2	11
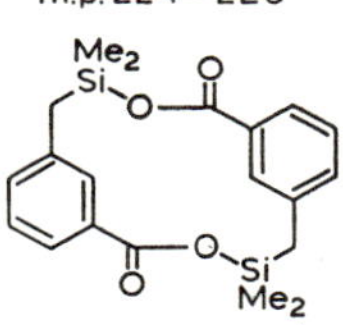 m.p. 152-155°	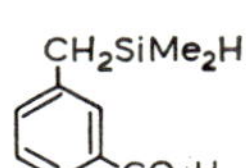 KOH	12

References (*table 4 continued*)

1 *S. A. Fuqua* and *R. M. Silverstein*, J. org. Chem., 1964, **29,** 395.
2 *V. L. Kozlikov et al.*, Zhur. obshcheĭ Khim., 1972, **42,** 2003.
3 *R. H. Krieble* and *C. A. Burkhard*, J. Amer. chem. Soc., 1947, **69,** 2689.
4 *K. A. Andrianov et al.*, Izvest. Akad. Nauk S.S.S.R., Ser Khim., 1967, 909.
5 *W. Simmler, H. Niederprüm* and *J. Sattlegger*, Chem. Ber., 1966, **99,** 1368.
6 *K. A. Andrianov et al.*, Zhur. obshcheĭ Khim., 1970, **40**(3) 611.
7 *C. R. Sparck*, U.S.P. 3,452,071/1969; C.A., 1969, **71,** 91633.
8 *N. S. Nametkin et al.*, Doklady Akad. Nauk S.S.S.R., 1967, **175,** 136.
9 *F. Wudl et al.*, Chem. Comm., 1971, 50; J. Amer. chem. Soc., 1971, **93,** 3160.
10 *A. I. Nogaioelu, N. I. Tabashidze* and *R. Sh. Theshelashvilli*, Zhur. obshcheĭ Khim., 1972, **42,** 1764.
11 *T. Kauffmann* and *H. H. Kniese*, Tetrahedron Letters, 1973, 4043.
12 *J. M. Woolcott* and *F. K. Cartledge*, J. org. Chem., 1974, **39,** 2430.

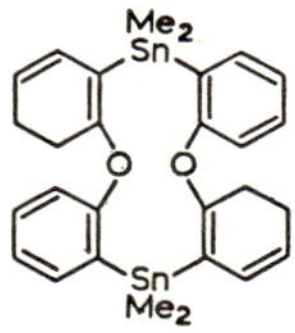

(L)

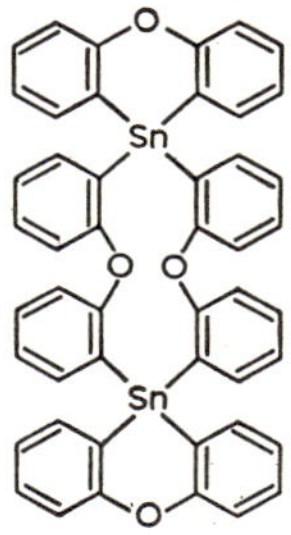

(LI)

The so-called **tin "salts"**, of carboxylic acids, their hydroxy- and thiolo-derivatives, are obtained by heating the acid with a tetra-alkyltin, or from the sodium salt of the acid by treatment with tin dichlorides, and also by azeotropic removal of water from mixtures of the carboxylic acid and dialkyltin oxides in high boiling organic solvents. Most of the products are solids of low to medium melting point, or high boiling liquids, and are evidently covalent in character. Typical compounds are LII, LIII and LIV. Such compounds are commercially important because of their fungicidal

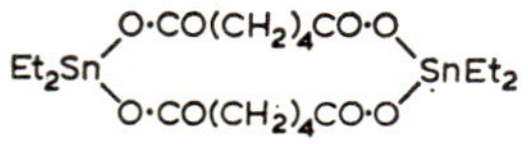

m.p. 143–144°
(LII)

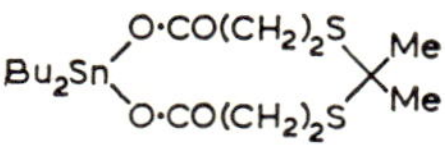

m.p. 58–60°
(LIII)

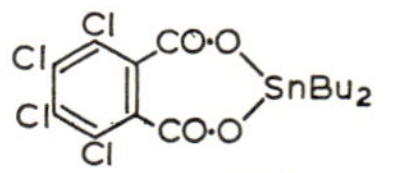

m.p. 200–202°
(LIV)

properties and their chemistry is described in the patent literature almost completely (*N. N. Zemlyanskii, E. M. Panov* and *K. A. Kocheshkov*, Zhur. obshcheĭ Khim., 1962, **32,** 291; Pure Chemicals Ltd., F.P. 1,319,129/1963; C.A., 1963, **59,** 8789; *I. Wechenbleichner, R. E. Breesh* and *O. A. Womberg*, Belg. P. 616,326/1962; C.A., 1963, **58,** 10237; U.S.P. 3,209,107/1965; C.A., **63,** 16382).

(*b*) *Compounds with nitrogen as the additional element*

(*i*) *Azasilepines and related compounds*

1-Butyl-3,3-**dimethylperhydro**-1,3-**azasilepine** (LV), as its *hydrobromide*, m.p. 166.5–168 °, is obtained by the action of butylamine on 1,6-dibromo-2,2-dimethyl-2-silahexane (*Fessenden* and *Coon*, J. org. Chem., 1964, **29**, 2499). 3,3-*Dimethylperhydro*-1,3-*azasilepine*-7-*one*, b.p. 134–138 °/2 mm, n_D^{20} 1.4761, is prepared by acting upon 6-chloro-5,5-dimethyl-5-silahexanoamide with sodium methoxide (*Mironov*, *Fedotov* and *Kozlikov*, Khim. geterotsikl. Soedin., 1969, 221; C.A., 1969, **71**, 50042).

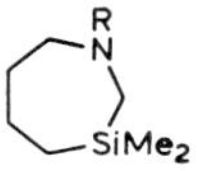

(LV)

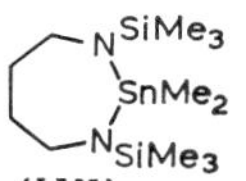

(LVI)

2,2-Dimethyl-1,3-**bis(trimethylsilyl)perhydro**-1,3,2-**diazastannepine** (LVI), b.p. 70–71 °/0.1 mm, exists of the analogous tin-containing system; it is produced when 1,4-bis(trimethylsilylamino)butane and $Me_2Sn(NEt_2)_2$ are heated together (*O. J. Scherer et al.*, Z. Naturforsch., B, 1965, **20**, 183).

The reaction between dichlorodimethylsilane and $(H_2N.NMeCH_2)_2$ or hexamethylcyclotrisilazane yields 1,3,3,5-**tetramethyl**-1,2,4,5-**triaza**-3-**silacycloheptane** (LVII), b.p. 55–56 °/7 mm (*G. S. Gol'din et al.*, Zhur. obshcheĭ Khim., 1974, **44**, 566), while interaction of bis(2-aminoethyl)amines with $RMeSi(NR_2)_2$ has given, for example, 2,2-**dimethyl**-6-**trimethylsilyl**-1,3,6-**triaza**-1-**silacyclo-octane** (LVIII), b.p. 95 °/5 mm (*Gol'din*, *L. S. Baturina* and *T. N. Gavrilova*, *ibid.*, 1975, **45**, 2189).

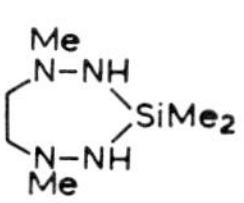

(LVII)

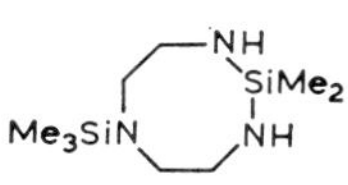

(LVIII)

Silolidines, like the borolidines, often undergo ring expansion reactions when treated with isocyanates. 1,7-**Dibutyl**-4,4-**dimethyl**-2,6-**dioxo**-3,5-**diphenylperhydro**-1,3,5,7-**tetraaza**-4-**silonine** (LIX), m.p. 178–180 °, is thus obtained in 94% yield when 1,3-dibutyl-2,2-dimethyl-1,3,2-diazasilolidine reacts with phenyl isocyanate (*R. H. Cragg* and *M. F.*

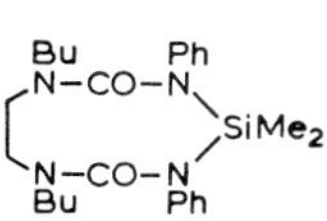

(LIX)

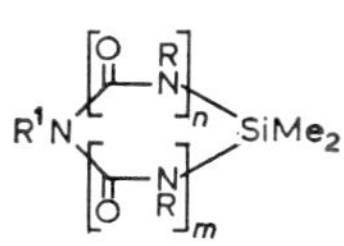

(LX)

Lappert, J. chem. Soc., A, 1966, 82). 1,3,5-Triaza-2-siline-4,6-diones (1,3,5-triaza-2-silacyclohexane-4,6-diones) undergo similar reactions; their reaction with ethyl isocyanate is not described in detail, but the *products* (LX), $n = m = 2$, R^1 = H, R = Et, b.p. 103–105 °/0.04 mm, and $n = 2$, $m = 3$, R = R^1 = Et, b.p. 125–127 °/0.01 mm, are noted (*W. Fink*, Chem. Ber., 1964, **97**, 1424).

TABLE 5

Compounds	*M.p., b.p. (°C)*	*Starting materials*	*Ref.*
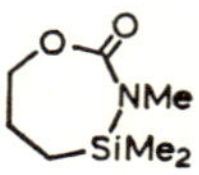	Decomp.	$Me_2SiCl(CH_2)_3OCOCl$ and $MeNH_2$	1
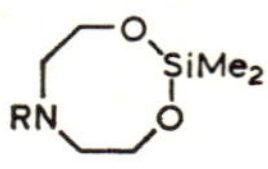	R = H, b.p. 89°/20 mm R = Et, b.p. 77–78°/10 mm	Me_2SiCl_2 and the diethanolamine	2
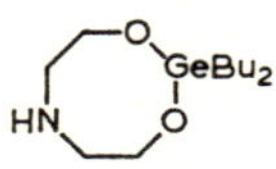	b.p. 127–130°/0.5–0.7 mm	Bu_2GeO, diethanolamine, toluenesulphonic acid	3

Analogous silicon containing compounds have been prepared from silatranes[3a]

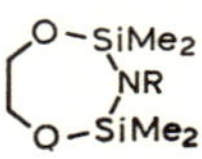	R = H, b.p. 64°/10 mm R = Me, b.p. 37°/0.05 mm	Ethylene glycol and $(Me_2SiCl)_2NR$	4
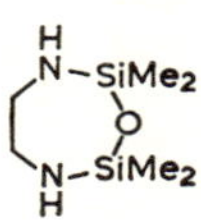	b.p. 59–60°/10 mm	*N'N'*-dilithioethylene diamine and $(ClSiMe_2)_2O$	5
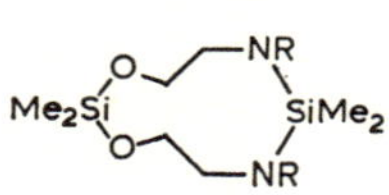	R = H, b.p. 54°/2 mm	Ethanolamine and $Me_2Si(NEt_2)_2$	6
	R = Me, b.p. 56–60°/0.8 mm	Me_2SiCl_2, triethylamine and $Me_2Si(OC_2H_4NHMe_2)$	7
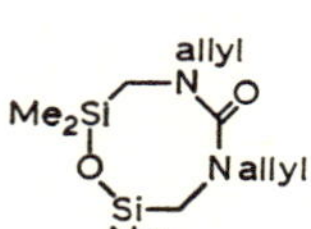	b.p. 114–118°/1 mm	$COCl_2$ and [allylN($SiMe_3$)-$CH_2SiMe_2]_2O$	8
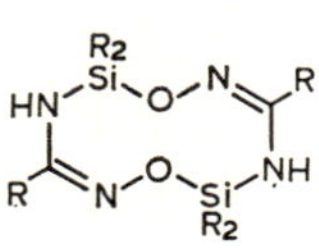	R = Me, b.p. 102°/0.1 mm	R_2SiCl_2 and $R^1C(=NH)NH_2$	9
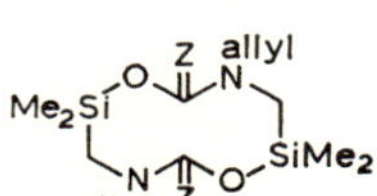	Z = O, m.p. 129–130° Z = S, m.p. 69–73°	CO_2 or CS_2 and [allyl$NHCH_2SiMe_2]_2O$	10

References (table 5 continued)

1 *V. F. Mironov et al.*, Doklady Akad. Nauk S.S.S.R., 1968, **181,** 115.
2 *M. G. Voronkov* and *Yu. P. Romadane*, Khim. geterotsikl. Soedin., 1966, 879; **C.A.,** 1967, **67,** 32775.
3 *R. C. Mehrotra* and *S. Mathur*, J. organometal. Chem., 1966, **6,** 11.
3a *Voronkov et al.*, Khim. geterotsikl. Soedin., 1975, 35; C.A., 1975, **82,** 111418.
4 *U. Wannagat, P. Geymayer* and *E. Bogush*, Monatsh., 1965, **96,** 585.
5 *Wannagat et al., ibid.*, 1971, **102,** 1844.
6 *E. Lukevics, L. Liberts* and *Voronkov*, Zhur. obshcheĭ Khim., 1968, **38,** 1838.
7 *Mehrotra* and *P, Bajaj*, J. organometal. Chem., 1971, **26,** 161.
8 *Mironov, V. D. Sheludyakov* and *E. S. Rodionov*, Zhur. obshcheĭ Khim., 1974, **44,** 1502.
9 *A. B. Goel* and *V. D. Gupta*, Austral. J. Chem., 1974, **27,** 2319.
10 *V. D. Sheludyakov et al.*, Zhur. obshcheĭ Khim., 1974, **44,** 1506; *Sheludyakov, A. D. Kirilin* and *Mironov, ibid.*, 1975, **45,** 707.

(*ii*) *Azapolysilepines and related compounds*

1-**Amino**-2,2,7,7-**tetramethylperhydro**-1,2,7-**azadisilepine** (LXI), b.p. 84–85 °/10 mm, is made in 40 % yield from hydrazine and 1,4-di(chlorodimethylsilyl)butane; there is no evidence that, like the case of the analogous five-membered ring compound, isomerization to the diaza larger ring compound is feasible (*J. W. Connolly et al.*, Inorg. Chem., 1970, **9,** 93).

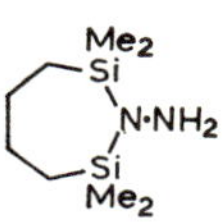
(LXI)

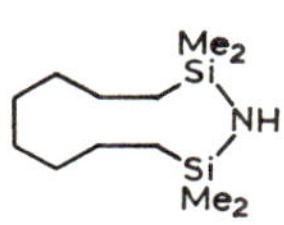
(LXII)

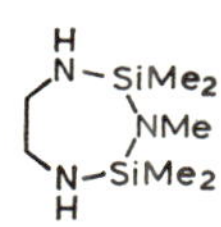
(LXIII)

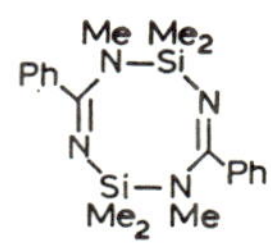
(LXIV)

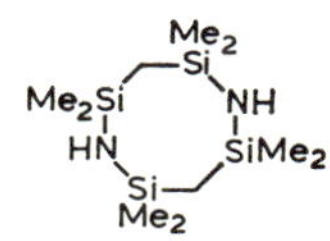
(LXV)

Ammonolysis of $[Me_2SiCl(CH_2)_4]_2$ yields 2,2,11,11-**tetramethyl**-1,2,11-**azadisilacycloundecane** (LXII), m.p. 87° (*C. R. Krueger*, Inorg. Chem. Nuclear Letters, 1965, **1,** 85).

N,N′-Dilithioethylenediamine reacts with $(ClSiMe_2)_2NMe$ to form 2,2,3,4,4-*pentamethylperhydro*-1,3,5-*triaza*-2,4-*disilepine* (LXIII), b.p. 87–88 °/10 mm, n_D^{20} 1.4632 (*U. Wannagat et al.*, Monatsh., 1971, **102,** 1844).

Lithio-amidines react with dichlorosilanes in ether in the presence of triethylamine to give eight-membered rather than four-membered ring compounds. The products, *e.g.* 1,2,2,5,6,6-**hexamethyl**-4,8-**diphenyl**-2,6-**disila**-1,3,5,7-**tetra-azacyclo-octa**-3,7-**diene** (LXIV), b.p. 174–176 °/0.1 mm, m.p. 77–79 °, are very sensitive to moisture (*Scherer* and *P. Hornig*, Angew. Chem. intern. Edn., 1967, **6,** 89).

2,2,4,4,6,6,8,8-**Octamethyl**-1,4-**diaza**-2,4,6,8-**tetrasilacyclo-octane** (LXV), b.p. 82–84 °/3.5 mm, and its derivatives are formed by ammonolysis or aminolysis of $(BrSiMe_2)_2CH_2$; they undergo rapid solvolysis in aqueous ethanol, and with methylamine yield linear products (*K. Lienhard* and *E. G. Rochow*, Angew. Chem., 1963, **75,** 980).

(*c*) *Compounds containing nitrogen together with oxygen as additional elements*

A few such compounds have been reported and a selection of these, with physical data and starting materials, is listed in Table 5.

4. Phosphorus, arsenic and antimony compounds

(*a*) *Compounds containing oxygen and/or sulphur as additional element*

Like the 1,2-oxaphosphorinanes, 1,2-oxaphosphepanes and larger ring compounds are known with the phosphorus only in the quinquevalent state.

1,2-**Dihydro**-3*H*-1,2-**oxaphosphepin** 2-**oxides** (II) (2,7-*diphenyl*-, m.p. 104°, 4,5-*dimethyl*-2,7-*diphenyl*-, m.p. 125°, 7-*phenyl*-2,4,5-*trimethyl*-, m.p. 135–136°, derivatives) are formed by ring expansion of the dihydrophosphorin 1-oxides (I) on treatment with catalytic amounts of sodium hydride (*F. Mathey*, Tetrahedron, 1973, **29,** 707; *F. Mathey, D. Thavard* and *B. Barlet*, Canad. J. Chem., 1975, **53,** 855).

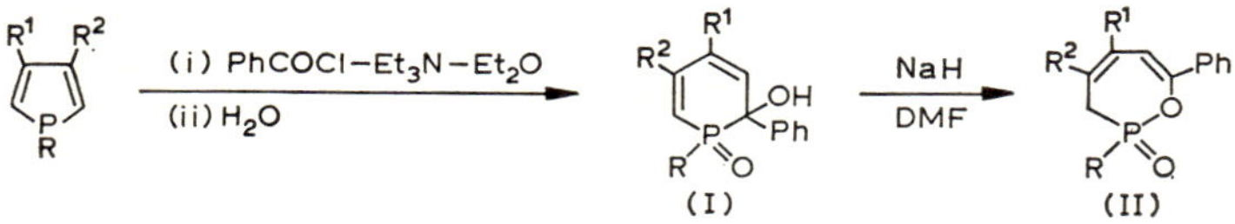

Ring-opening leading to $PhBuP(O)CH_2CMe{=}CMe \cdot CH_2COPh$ is the result when 4,5-dimethyl-2,7-diphenyl-1,2-dihydro-3*H*-1,2-oxaphosphepin 2-oxide is treated with butyllithium. Reaction of the same oxaphosphepin oxide with trichlorosilane or benzophenone results in ring contraction to a phospholene (*Mathey* and *R. Maillet*, Bull. Soc. chim. Fr., 1973, 2783).

2-*Ethoxy*-1,2-*oxaphosphepane* 2-*oxide* (III, R = OEt, R^1 = H), b.p. 89–90°/0.8 mm, 106°/10 mm, n_D^{20} 1.4609 (*G. Aksnes* and *K. Bergesen*, Acta Chem. Scand., 1966, **20,** 2508), is obtained when sodium diethyl phosphonate and 5-chloropentanol interact, the product being identical with that obtained from triethyl phosphite and 1,5-dibromopentane (*J. Songstad, ibid.*, 1967, **21,** 1681). Bergesen (*ibid.*, 1968, **22,** 1366) has presented evidence

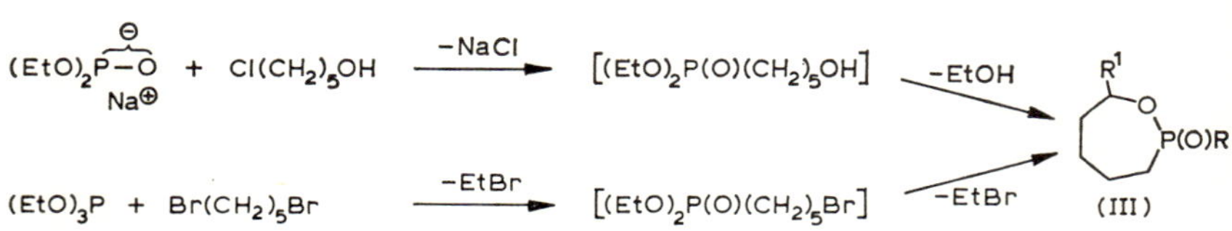

for geometrical isomerism in the 1,2-oxaphosphepane series; the product from triethyl phosphite and 2-methyl-1,5-dibromopentane, is separable by preparative g.l.c. into two fractions, the first having n_D^{20} 1.4580, $\nu_{P=O}$ 1250 cm^{-1}, the second n_D^{20} 1.4600, $\nu_{P=O}$ 1260 cm^{-1}; designations of *trans*- and *cis*-2-ethoxy-7-methyl-1,2-oxaphosphepane (III, R = OEt, R^1 = Me) were based on g.l.c. retention data only.

Oxygen insertion into the (perhydro)phosphorin ring can also be achieved using peroxy acids. Treatment of the phosphine oxide IV with 3-chloroperoxybenzoic acid, leads to the expected epoxide together with two further stereoisomeric epoxides possessing the 1,2-oxaphosphepane ring, 3-*oxa*-2-*phenyl*-8,9-*β*-*epoxy*-2-*phosphabicyclo*[3.2.2]*nonane* 2-*oxide* (V), m.p. 158–159°, and the corresponding α-epoxide (VI), m.p. 144–145° (*Y. Kashman* and *O. Awerbouch*, Tetrahedron, 1975, **31**, 45):

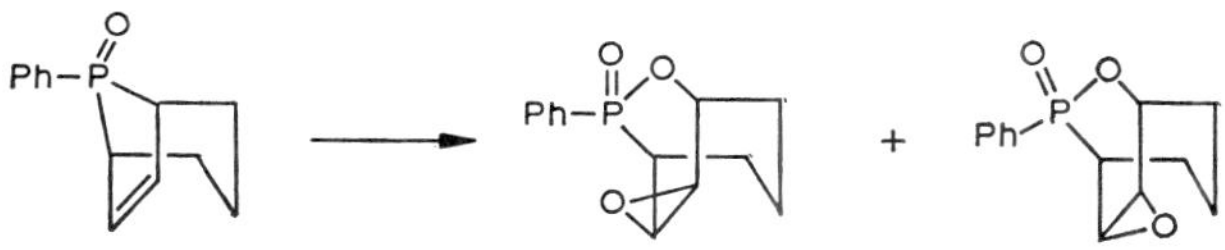

(IV) (V) (VI)

Reaction between 2,2′-methylenebis(cyclohexanone) and hypophosphorous acid in acetic acid at 70° yields 6,6′-**dihydroxytetradecahydrodibenzo**[c,f]-[1.2]**oxaphosphepin 6-oxide** (VII), m.p. 207–208°, reacting with diazomethane to give the corresponding *methyl* ester, m.p. 65–66.5°, (*V. I. Vyotskii*, *A. S. Skobun* and *M. N. Tilichenko*, Zhur. obshcheĭ Khim., 1974, **44**, 2109).

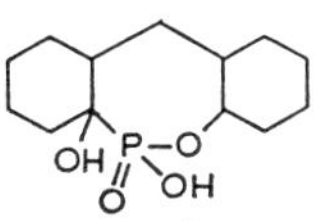

(VII)

In their investigations into the action of phosphorus trichloride on dihydric phenols, *L. Anschütz* and *W. Marquart* (Chem. Ber., 1956, **89**, 1119) showed that 2,2′-dihydroxybiphenyl in benzene with a trace of water, gave 2-**chloro**-5,7-**dihydrodibenzo**[d,f][1.3.2]**dioxaphosphepin** (VIII, M = P, R = Cl) in 60% yield; later workers claimed much higher yields of the same

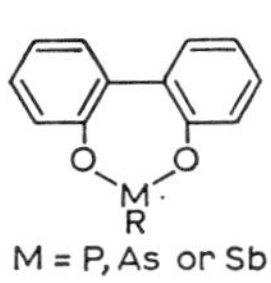

M = P, As or Sb

(VIII)

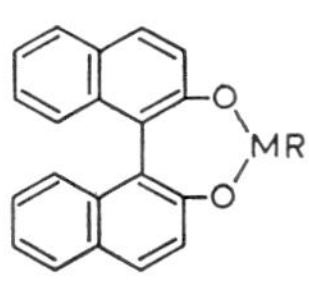

(IX)

compound, m.p. 62–63°, b.p. 174°/1.2 mm, when pyridine is used as a scavenger for the hydrogen chloride produced (*R. M. Ismail*, Z. Naturforsch., B, 1970, **25**, 14; *L. V. Verizhnikov* and *P. A. Kirpichnikov*, Zhur. obshcheĭ Khim., 1967, **37**,1355).

Sulphur adds to the chloride at 190–230° to give the *sulphide*, m.p. 107°, and with ethanol in the presence of pyridine, the 2-*ethoxy* compound, b.p. 153–155°/1.5 mm, is formed; this undergoes normal isomerization under Arbuzov conditions (*Verizhnikov et al., ibid.*, 1971, **41**, 2162). The 2-*phenyl* compound (VIII, M = P, R = Ph), b.p. 215°/1 mm, and its *oxide*, b.p. 93–95°/2 mm, m.p. 116–117°, are made by exchange between the corresponding dimethylsilyl heterocycle and phenyldichlorophosphine and its oxide at 125–150° (*C. M. Silcox* and *J. J. Zuckermann*, J. Amer. chem. Soc., 1966, **88,** 168).

The **arsenic** and **antimony** *chlorides* (VIII, R = Cl, M = As, Sb), m.p. 98° and 102°, respectively, are also prepared from the metalloid trichloride.

The *chlorides* (IX, R = Cl), M = *As*, m.p. 174–176°, M = *Sb*, m.p. 182°, are prepared similarly, but it is of interest that when the ratio of dihydroxybinaphthyl to MCl_3 (in the presence of R_3N) is 2:1, a spirocyclic compound is not formed (*K. Andrä* and *L. Martschei*, Z. anorg. allgem. Chem., 1973, **396,** 123). The 2,2′binaphthyl cyclic phosphoric acid (IX, M = P, R = OH, oxide) is resolvable as its cinchonine salt, and may be employed to resolve other bases. Its methyl ester undergoes ring opening and loss of phosphorus when treated with lithium tetrahydridoaluminate. Quite drastic alkaline conditions are required for ring opening in an aqueous medium (*J. Jacques*, *C. Fouquey* and *R. Viterbe*, Tetrahedron Letters, 1971, 4617).

3-**Chloro**-1,5-**dihydro**-2,4,3-**benzodioxaphosphepin** (X, M = P, R = Cl) is known as its *oxide*, m.p. 135–139°, formed by reaction between 1,2-di(hydroxymethyl)benzene and phosphoryl chloride; it is readily convertible to the corresponding *methyl* and *phenyl phosphates*, m.p.s 126.5–128° and 96–97°, respectively (*T. Sato* and *K. Goto*, Chem. Comm., 1973, 494).

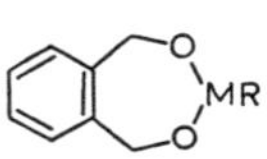

(X)

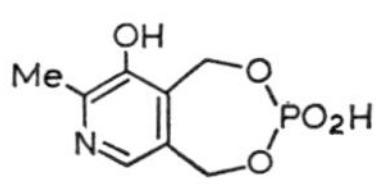

(XI)

Derivatives of the corresponding arsenic and antimony ring systems are not recorded but 2,4-dithia analogues of the arsenic and antimony-containing, ring compounds are prepared from *ω,ω-o*-xylenedithiol (*E. H. Friedheim* B.P. 655,435/1951; C.A., 1953, **47,** 144: U.S.P. 2,659,723/1953; C.A., 1955, **49,** 1816; U.S.P. 2,772,303/1956; C.A., 1957, **51,** 5836). The similar *pyridine compound* (XI), m.p. 280° decomp., is formed from pyridoxine and polyphosphoric acid at 40° (*Kyowa Fermentation Ind., Co. Ltd., B.P.* 1,666,426/1969; C.A., 1970, **72,** 4346) or phosphoryl chloride (*idem*, Fr.P. 1,575,226/1969; C.A., 1972, **76,** 3697).

The few recorded examples of the 1,3,2-dioxaphosphepane ring system XII have been prepared by methods used for the 1,3,2-dioxaphosphorinanes. These include 2-*phenoxy*-, b.p. 94°/0.3 mm (*D. C. Ayres* and *H. N. Rydon*, J. chem. Soc., 1957, 1109), 2-*isocyanato*-, b.p. 93–94°/15 mm (B.P. 1,110,917/1968; C.A., 1968, **69,** 19162), 2-*chloro*-, b.p. 74–75.5°/8 mm, 2-*methoxy*-, b.p. 54–55°/5 mm, 2-*methyl*- (*oxide*, b.p. 97–99°/3 mm, m.p. 67.5–69°) and 2-*phenyl*- (*oxide*, m.p. 76–77.5°, b.p. 125–127°/3 mm) 2,4,6-*trimethyl*- (*oxide*, b.p.

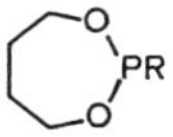

(XII)

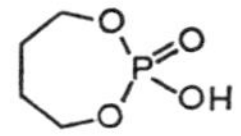

(XIII)

73–76°/0.2 mm), -1,3,2-*dioxaphosphepane* (*V. V. Korshak*, *I. A. Gribova* and *M. A. Andreyeva*, Izvest. Akad. Nauk S.S.S.R., Ser. Khim., 1957, 631; *A. E. Arbuzov* and *V. M. Zoroastrova*, *ibid.*, 1952, 770; *Arbuzov*, *Zoroastrova* and *N. I. Rizpolozhenskii*, Bull. Acad. Sci. U.R.S.S., Classe Sci. Chim., 1948, 208; *A. F. McKay*, *R. O. Brown* and *G. R. Vavasour*, J. Amer. chem. Soc., 1952, **74,** 5540).

Tetracyclic steroidal type compounds containing the 1,3,2-dioxaphosphepane ring have been prepared (*D. L. Morris*, *C. H. Chan* and *K. D. Berlin*, Phosphorus, 1974, **4,** 129).

When butane-1,4-diol reacts with polyphosphoric acid, the cyclic phosphoric acid XIII is formed in 0.3% yield together with much of the linear phosphoric acid, $HO(CH_2)_4$-$OP(O)(OH)_2$ which is not cyclized by the action of heat; however, the use of dicyclohexylcarbodiimide in aqueous pyridine brings about a high yield conversion of the linear acid to the cyclic acid XIII (*H. G. Khorana et al.*, J. Amer. chem. Soc., 1957, **79,** 430). Data presented by *E. Cherbuliez et al.* (Helv., 1959, **42,** 1377; 1960, **43,** 464) would appear to indicate that under alkaline conditions, the dioxaphosphepane ring is less stable than the 1,3,2-dioxaphosphorinane acid, but more stable within the pH range 3.5–8. In the reaction with dry hydrogen chloride in benzene, 2-ethoxy-1,3,2-dioxaphosphepane displays a reactivity, consisting of ring cleavage, greater than that of the five-membered, but less than that of the six-membered ring analogues (*A. Munoz et al.*, Bull. Soc. chim. Fr., 1967, 3343).

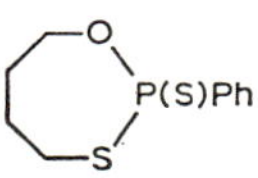

(XIV)

The unusual reaction between phenylthiophosphonyl chloride, $PhP(S)Cl_2$), and cyclohexa-1,3-diene in the presence of magnesium affords 2-**phenyl**-1,3,2-**oxathiaphosphepane 2-sulphide** (XIV), m.p. 62–63.5°, as one of the two products (*S. Nakayama et al.*, Bull. chem. Soc. Japan, 1975, **48,** 546) in effect by insertion of the $PhPS_2$ moiety into tetrahydrofuran.

2-*Fluoro*-, b.p. 40°/0.5 mm, and 2-*chloro*-, b.p. 70°/0.5 mm, 1,3,2-**dioxarsepanes** are prepared from arsenic bifluoride or trichloride or $Me_2AsO(CH_2)_3CAsMe_2$ (*F. Kober* and *W. J. Ruehl*, Z. anorg. allgem. Chem., 1975, **416,** 57) and undergo ring opening when treated with oximes (*J. Kaufmann* and *F. Kober*, *ibid.*, 1976, **420,** 177).

2-**Phenoxy**-1,3,2-**dioxaphosphocane** has b.p. 99°/0.0015 mm (*Ayres* and *Rydon*, *loc. cit.*); 2-*methyl*- (*oxide*, b.p. 118–120°/4 mm) (*Korshak et al.*, *loc. cit.*); 2-chloro-1,3,6,2-trioxaphosphocane, b.p. 104–105°/15 mm (*Arbuzov*, *Zoroastrova*, and *Rizpolozhenskii*, *loc. cit.*), 2-*methyl*- (*oxide* b.p. 105–107°/3 mm) (*Korshak et al.*, *loc. cit.*).

It has also been claimed that 1,3,6,2-**trioxaphosphocanes** are formed by reaction between ethylene glycol and phosphorus dichlorides followed by removal of water (*T. A. Lankina et al.*, Uzb. Khim. Zh., 1973, **17,** 55; C.A., 1973, **79,** 53472).

Derivatives of the **tetrahydrodibenzo**[*c,f*][1.3.6.2]**dioxathiaphosphocin** system (XV) are obtainable from suitable diphenols (*V. G. Gruzdev*, Zhur.

obshcheĭ Khim., 1975, **45,** 2172; *Yu. B. Zimin et al., ibid.*, 1967, **37,** 1809; *V. Kh. Kadyrova et al., ibid.*, 1971, **41,** 1688):

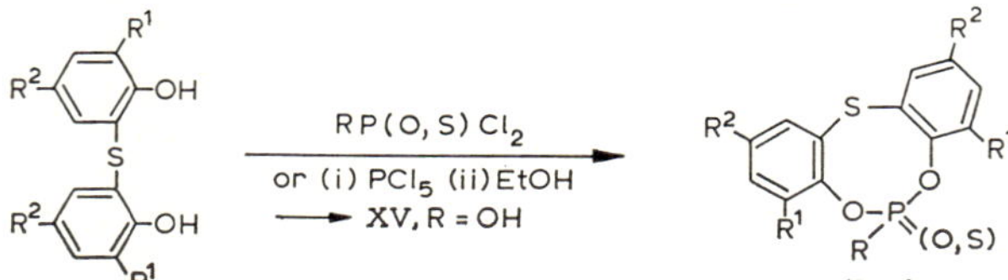

(XV)

2-*Ethyl*-1,5,2-*dioxaphosphepane* 2-*oxide*, b.p. 118–121 °/1 mm, is formed when di-2-(2-chloroethoxy)ethyl ethylphosphonite, $EtP[O(CH_2)_2O(CH_2)_2Cl]_2$, is heated; di(2-chloroethyl) ether is lost during the reaction (*A. N. Pudovik, E. M. Faizullin* and *S. V. Yaklova*, Zhur. obshcheĭ Khim., 1967, **37,** 460).

The reaction between acrylic acid and 5,5-dimethyl-2-phenoxy-1,3,2-dioxaphosphorinane in a chlorinated solvent yields a product formulated as 8,8-**dimethyl-2-phenoxy**-1,6,2-**dioxaphosphonane**-2,5-**dione** (XVI) and a detailed mechanism for its formation has been proposed (*J. S. Clovis* and *F. R. Sullivan*, Tetrahedron Letters, 1971, 2263):

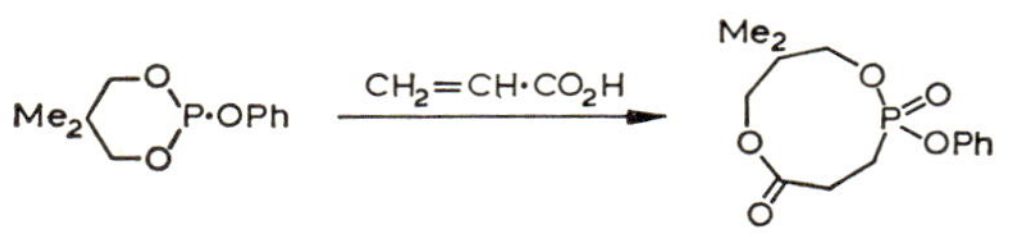

(XVI)

3,3,6,6-*Tetraphenyl*-1,3,6-*oxadiphosphepanium dibromide*, (XVII), m.p. 292–293 °, is formed by conventional means, as illustrated (*A. M. Aguiar, K. C. Harsin* and *J. T. Magul*, J. org. Chem., 1967, **32,** 2383):

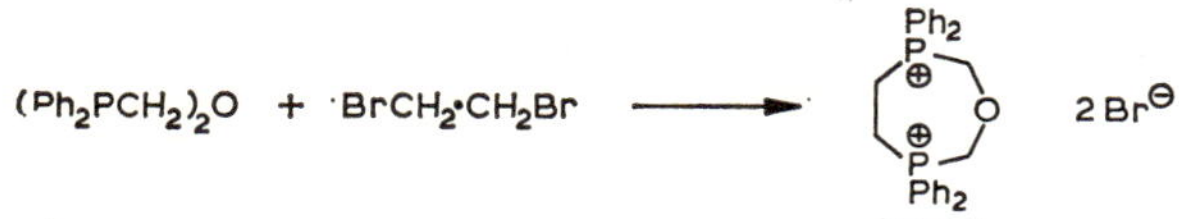

(XVII)

It has long been felt that the products of transesterification between 1,2-diols and dialkyl hydrogen phosphonates in the presence of sodium possess 10-membered rather than five-membered, rings, *e.g.* 1,3,6,8-**tetraoxa**-2,7-**diphosphecane** (XVIII, R = H) 2,7-**dioxide,** b.p. 132–136 °/2.5 mm (*K. A. Petrov, E. E. Nifant'ev* and *R. G. Gol'tsova*, Zhur. obshcheĭ Khim., 1963,

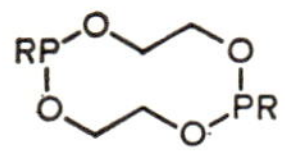

(XVIII)

33, 1485); such a possibility was discussed as early as 1903 by *Carré* (Compt. rend., 1903, **136,** 756), but the possibility of an equilibrium between the two rings, or of an association of two five-membered rings, is not entirely excluded (*E. E. Nifant'ev* and *I. S. Nasonovskii,* Zhur. obshcheĭ Khim., 1969, **39,** 1948).

2,7-**Dimethyl**-1,3-**dioxa**-6,8-**dithia**-2,7-**diarsecane** (XIX), b.p. 60°/0.1 mm, is formed when 2-mercaptoethanol reacts with methylbis(dimethylamino)arsine (*F. Kober* and *W. J. Ruehl,* Z. anorg. allgem. Chem., 1976, **420,** 74).

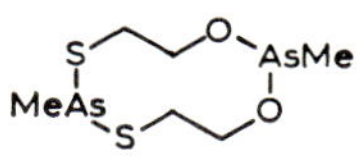

(XIX)

Still larger rings have been assigned to components in mixtures equilibrated from 2,5,5-trimethyl-1,3,2-dioxaphosphorinane (*J. P. Albrand, J. P. Dutasta* and *J. B. Robert,* J. Amer. chem. Soc., 1974, **96,** 4584). Dimeric (12-membered rings, XX, R = lone pair) and trimeric (18-membered rings,

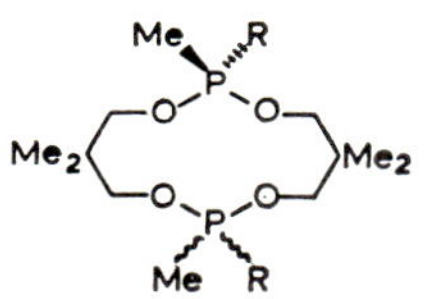

(XX)

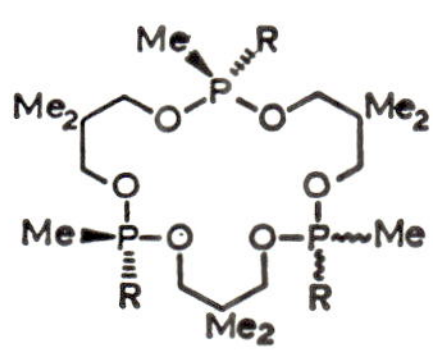

(XXI)

XXI, R = lone pair) components of such mixtures have been detected by ^{31}P-n.m.r. spectroscopy and shown to exist in different diastereosomeric forms. Two diastereoisomeric forms of the 12-membered ring *thione derivatives,* (XX, R = S), m.p.s 250° and 280°, have been isolated.

(*b*) *Compounds containing nitrogen as the additional element*

3-**Phenylperhydro**-1,3-**azaphosphepines** (XXII) are formed, together with linear compounds, when 4-aminobutylphenylphosphines react with aldehydes or ketones (*K. Issleib, H. Oehme* and *K. Mohr,* Z. Chem., 1973, **13,** 139):

(XXII)

The reaction between one molar proportion of benzonitrile and a 2-lithio-phosphol-3-ene oxide, (3,4-dimethyl-1-phenyl is illustrated), yields a **dihydro**-4,5-**dimethyl**-2,7-**diphenyl**-1,2-**azaphosphepine** 2-**oxide** (XXIII), capable of undergoing reactions through its tautomeric forms. Thus, although it alkylates normally on nitrogen (the corresponding 2-sulphide with sodium hydride followed by benzyl chloride affords the *S*-benzyl derivative), the

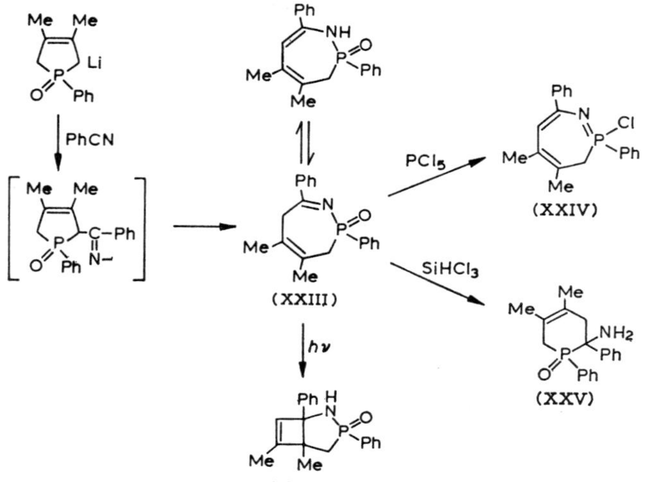

product of its reaction with phosphorus pentachloride is 2-chloro-2,7-diphenyl-3*H*-1,2-azaphosphepine (XXIV) (*F. Mathey* and *J.-P. Lampin*, Tetrahedron Letters, 1972, 1949; Tetrahedron, 1972, **28,** 5367). When acted upon by trichlorosilane, contraction of the ring occurs leading to a 2-aminotetrahydrophosphorin (XXV) (*L. Eberhard et al., ibid.*, 1973, **29,** 2909), and photolysis provides a route to bicyclic phospholidines; 4,5-dimethyl-2,7-diphenyl-1,2-azaphosphabicyclo[3.2.0]hept-5-ene (XXVI) is illustrated (*Mathey* and *Lampin*, F.P. 2,171,596/1973, C.A., 1974, **81,** 25805).

From the reaction between *N,N*-dimethyl-*p*-toluidine and phosphoryl chloride at 130°, 6-**chloro**-2,5,7,10-**tetramethyl**-6,7-**dihydro**-5*H*, 12*H*-**dibenzo**[*d,g*][1.3.2]**diazaphosphocine** may be isolated as the 6-*oxide* (XXVII, Y = O), m.p. 170–171°, in 6% yield; the 6-***sulphide***

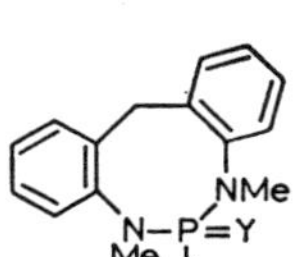

(XXVII)

(XXVII, Y = S), m.p. 209–210°, is obtained in a similar way. Hydrolysis of the oxide gives $CH_2(C_6H_3MeNHMe)_2$ (*C. Y. Cheng et al.*, Chem. Comm., 1968, 616; *T. S. Cameron et al.*, Angew. Chem., 1972, **11**, 510).

Ring formation appears not to take place when phosphoryl chloride or phenyl phosphorodichloridate reacts with 1,4-diaminobutane, but the use of phenyl phosphorodichloridothionate, $PhOP(S)Cl_2$, leads to *2-phenoxyperhydro-*1,3,2-*diazaphosphepine 2-sulphide* (XXVIII), m.p. 86–86.5°, (*R. S. Edmundson*, J. chem. Soc., C, 1969, 2730). 2-Di(2-chloroethyl)amino-4-ethoxycarbonylperhydro-1,3,2-diazaphosphocine 2-oxide (XXIX) was prepared from lysine ethyl ester in a programme concerned with potential cytostatic agents

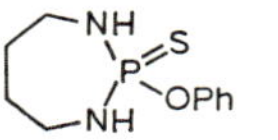
(XXVIII)

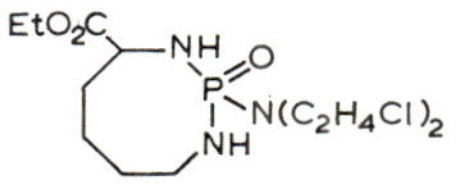
(XXIX)

(*M. Szerkerke, M. T. Kajtar* and *V. Bruckner*, Acta Chim. Acad. Sci. Hung., 1966, **47,** 231; C.A., 1966, **64,** 17705). **3,4-Dihydro-2*H*-1,2,4,5,3-tetra-azaphosphepine 3-sulphides** (XXX), prepared from quinquevalent phosphorus dihydrazides and α,β-ketones, decompose in the range 175–250° without melting (*A. F. Grapov, O. B. Mikhailova* and *N. N. Mel'nikov*, Zhur. obshcheĭ Khim., 1975, **45**, 1392):

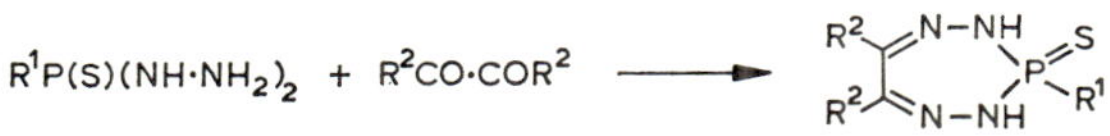
(XXX)

Transamination between bis(diethylamino)phosphines and 1,2-dihydrazinoethanes yields **1,2,4,5-tetra-aza-3-phosphacycloheptanes** (XXXI), 3-*phenyl-*, m.p. 100°, 3-*butoxy-*, b.p. 97–101°/1.5 mm, 3-*diethylamino-*, m.p. 55–57° (*G. S. Gol'din et al.*, *ibid.*, 1974, **44,** 2668):

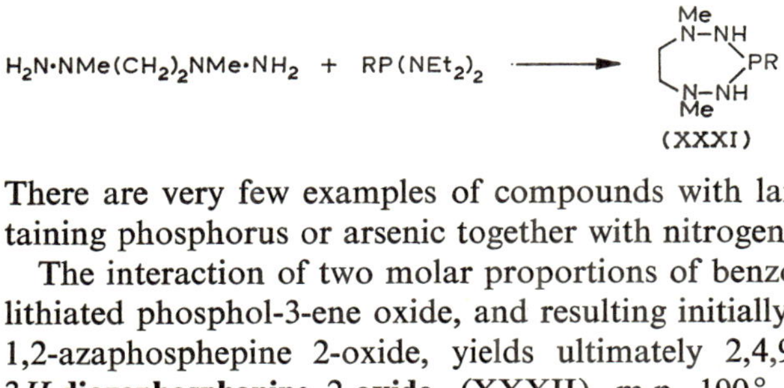
(XXXI)

There are very few examples of compounds with larger ring systems containing phosphorus or arsenic together with nitrogen.

The interaction of two molar proportions of benzonitrile with a suitably lithiated phosphol-3-ene oxide, and resulting initially in the formation of a 1,2-azaphosphepine 2-oxide, yields ultimately 2,4,9-**triphenyl-**1,2-**dihydro-**3*H*-**diazaphosphonine** 2-**oxide,** (XXXII), m.p. 190°, (N,N′-*dimethyl-*, m.p. 160°) (*Lampin, H. Laurent* and *Mathey*, Compt. rend., 1975, C, **280,** 1153):

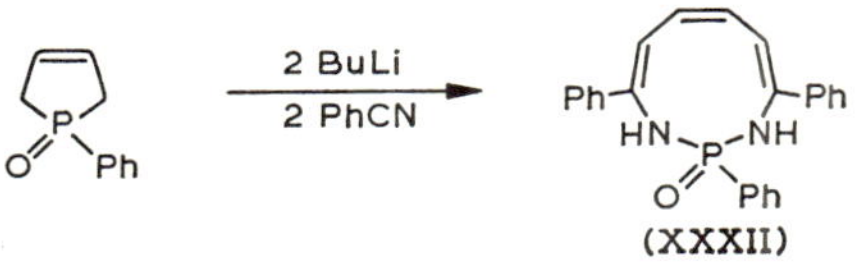
(XXXII)

The tetra-aza *compound* (XXXIII), m.p. 155–156°, is prepared from methyl phosphorodiisocyanidate and EtOC(=NH)C(=NH)OEt (*G. I. Derkach et al.*, Khim. Org. Soedin Fosfora Akad. Nauk S.S.S.R. Otd. obshcheĭ Tekh. Khim., 1967, 54; C.A., 1968, **69,** 10512) while in the exchange reaction

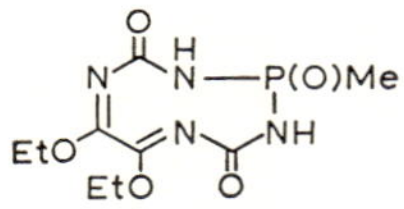

(XXXIII)

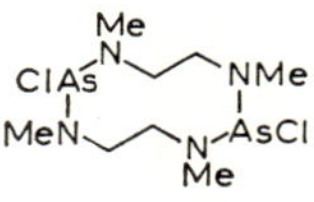

(XXXIV)

between arsenic trichloride and 1,3-dimethyl-2-phenyl-1,3,2-diazaphospholidine, the ring "dimerizes" in almost quantitative yield to give 2,7-**dichloro**-1,3,6,8-**tetramethylperhydro**-1,3,6,8-**tetra-aza**-2,7-**diarsecine** (XXXIV), m.p. 118–120° (*J. J. Zuckermann* and *M. K. Das*, Inorg. Chem., 1971, **10,** 1028).

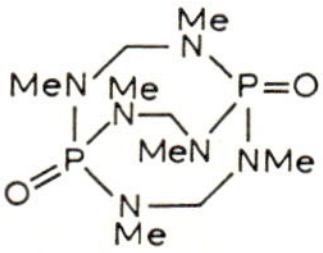

(XXXV)

The action of formaldehyde on hexamethylphosphoric triamide produces 2,4,6,8,9,11-hexakis(methylimida)-1,5-diphosphabicyclo-[3.3.3]undecane-1,5-dioxide (XXXV) (*H. L. Vandersall, D. L. Brooks* and *C. W. Heitsch*, Syn. inorg. metal. org. Chem., 1973, **3,** 267).

(*c*) *Compounds containing nitrogen together with Group VI elements*

1,3-Dimethyl-2-phenyl-1,3-diazaphospholidine undergoes several other ring expansion reactions on treatment with, for example carbon disulphide, trichloroacetaldehyde and isocyanates; 3-methyl-2-phenyl-1,3,2-oxazaphospholidine behaves similarly (*Zuckermann, Das* and *P. G. Harrison*, Inorg. Chem., 1971, **10,** 1092; *R. F. Hudson et al.*, Chem. Comm., 1969, 522; J. chem. Soc., B, 1968, 1349; Phosphorus, 1972, **1,** 271). These reactions are illustrated below:

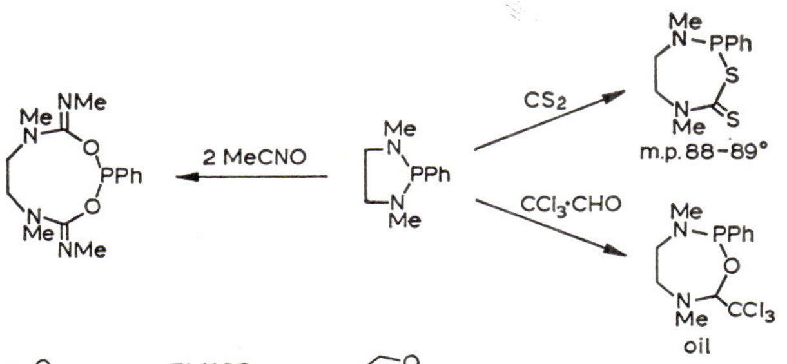

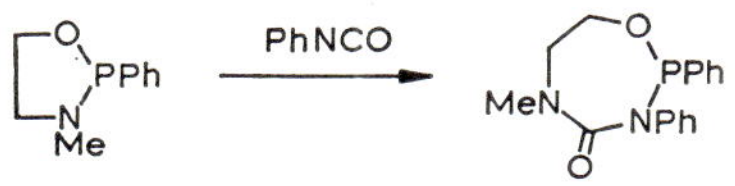

The reaction between quinquevalent phosphorus dichlorides and diethanolamine affords **perhydro-1,3,6,2-dioxaazaphosphocines** (XXXVI), 2-*thiono*-6-*methyl*-, b.p. 82°/1 mm, (m.p. 65–66°), *methiodide*, m.p. 166–168°; 2-*thiono*-6-*phenyl*-, m.p. 134–136°; 2-*oxo*-6-*methyl*-, b.p. 88–90°/1 mm, (m.p. 53–54°), *methiodide* m.p. 189–190°; 2-*oxo*-6-*phenyl*-,

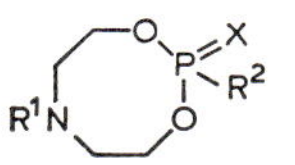

(XXXVI)

m.p. 148–150° (*N. N. Godovikov, L. A. Vikhreva* and *M. I. Kabachnik*, Zhur. obshcheĭ Khim., 1975, **45**, 728). Such molecules appear to have a distorted tetrahedral geometry at phosphorus and an essentially planar arrangement at nitrogen (*A. E. Kalinin, V. G. Andrianov* and *Yu. T. Struchkov*, Zh. strukt. Khim., 1975, **16**, 1041).

Analogous tervalent arsenic compounds have been obtained from the diethanolamine and bis(dimethylamino)arsines, 2-*methyl*-, b.p. 103°/1 mm, 2,6-*dimethyl*-, b.p. 41°/0.1 mm, 2,4,6,8-*tetramethyl*-, b.p. 55°/0.1 mm.

The use of tri(dimethylamino)arsine yields the corresponding 2-dimethylamino compounds. Interestingly, since it demonstrates a difference between the chemistry of arsenic and phosphorus, tris(dimethylamino)arsine and tris(2-hydroxyethyl)amine do not form bicyclic systems of the type obtainable from tris(dimethylamino)phosphine (*F. Kober* and *W. J. Ruhl*, Z. anorg. allgem. Chem., 1976, **420**, 74):

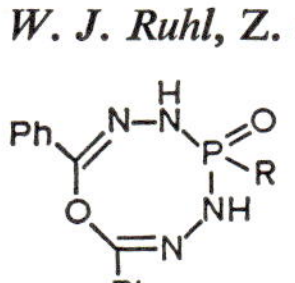

(XXXVII)

A similar reaction between phosphonic or phosphoric dichlorides and benzoylhydrazine yields derivatives of **5,6-dihydro-2,8-diphenyl-4*H*-1,3,4,6,7,5-oxatetraazaphosphocine** (XXXVII); the 2-*methyl* compound, m.p. 212–213°, originates from methylphosphonic dichloride (*N. N. Mel'nikov, O. B. Mikhailova* and *A. F. Grapov*, Zhur. obshcheĭ Khim., 1968, **38**, 2099) whereas the use of phosphoric dichlorides provides (XXXVII, R = NH · NHCOPh) (*Grapov, Mikhailova* and *Mel'nikov*, *ibid.*, 1975, **45**, 2570).

5. The "Atranes"

This chapter dealing with larger ring compounds of unusual hetero elements containing also oxygen, sulphur or nitrogen, would not be complete without mention of a group of compounds from elements spanning the Periodic Table, and which, collectively, are called "atranes", described more systematically as 1-aza-4-M-4,6,11-trioxabicyclo[3.3.3]undecanes (I). These compounds have been obtained from tri-hydroxides, or tri-alkoxides, and in some cases from the oxides, of boron, aluminium, silicon, titanium, tin, zirconium, bismuth, vanadium, and other elements, by reaction with triethanolamine; this reaction is performed very easily, and consists, simply, in the case of hydroxides, of azeotropic removal of water from a mixture of reactants:

$$R_{n-3}M(OR^1)_3 + (HOC_2H_4)_3N \longrightarrow$$

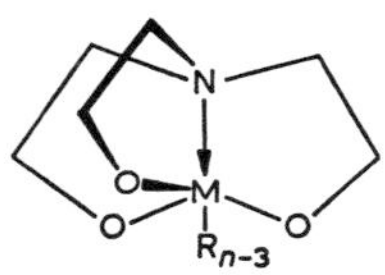

(I)

n = valency of M
R = H, alkyl, aryl, alkoxyl, alkenyl, oxygen

M. G. Voronkov (Pure and appl. Chem., 1966, 13, 35) has summarized much of the earlier chemistry of these compounds, particularly that relating to to the "silatranes"; these form a remarkable group of compounds, being soluble in organic solvents and to some extent also in water. They are essentially monomeric in benzene and are stable to distillation at atmospheric pressure; they do not form quaternary ammonium salts on treatment with alkyl halides. Some physiological activity particularly anti-fungicidal is shown by a number of the compounds.

Since 1966, other compounds have been prepared by *H. J. Cohen* (titanium: J. organometal. Chem., 1966, **5,** 413; 1967, **9,** 177); *Voronkov* and *G. Zelcans* (boron: Metody Poluch Khim. Reaktiv. Prep., 1966, **14,** 2415; C.A., 1967, **67,** 2706); *Zelcans* (silicon, boron: Khim. geterotsikl. Soedin., 1967, 371; C.A., 1967, **67,** 108112); *Voronkov et al.*, (germanium: *ibid.*, 1968, 227; C.A., 1968, **69,** 87129); *R. C. Mehrotra* and *V. D. Gupta*, (tin: Indian J. Chem., 1967, **5,** 647); *Voronkov* and *Zelcans*, (silicon: Khim. geterotsikl. Soedin., 1969, 43, C.A., 1969, **71,** 3360).

As variations in the basis structure, the triones (II, M = B, Ga, Tl, Mn, Ni) are obtainable from iminotriacetic acid (*Voronkov* and *S. V. Mikhailova*, *ibid.*, 1969, 49; C.A., 1969, **71,** 3364; *ibid.*, 1970, 334; C.A., 1972, **77,** 152076), while the reaction between diethanolamine and $(EtO)_2Si(CH_2 \cdot CH_2 \cdot CH_2Cl)Me$ in the presence of triethylamine gives the "deoxysilatrane" III,

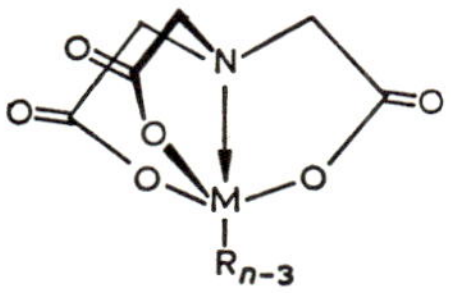

(II)

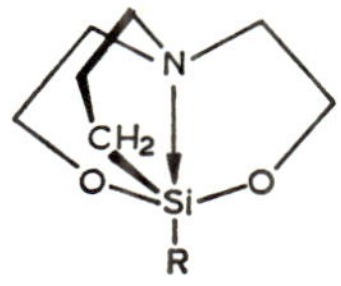

(III)

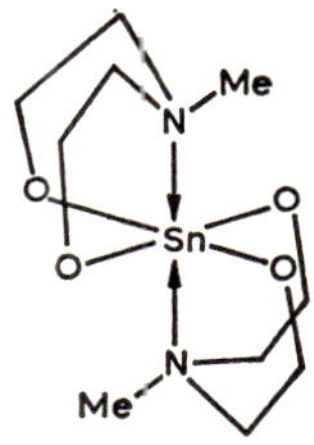

(IV)

m.p. 57–58°, (*E. Lukevics*, *L. Liberts* and *Voronkov*, Zhur. obshcheĭ Khim., 1969, **39**, 1784; see also *E. L. Moorehouse*, U.S.P. 3,032,576/1962; C.A., 1962, **57**, 9881); tetra(dimethylamino)stannane $Sn(NMe_2)_4$ and *N*-methyldiethanolamine give the *spiran* IV, m.p. 143–146°, and similarly for the germanium analogue (*R. G. Kostyanovskii et al.*, Izvest. Akad. Nauk S.S.S.R., Ser. Khim., 1968, 270; *Mehrotra* and *G. Chandra*, Indian J. Chem., 1965, **3**, 497).

Chapter 55

*Compounds with Larger Heterocyclic Rings: Eight-membered Ring Compounds**

J. M. F. GAGAN

1. Eight-membered ring compounds containing oxygen

(*a*) *With rings containing one oxygen atom*

The introduction of more than one double bond into unsubstituted monocyclic oxacyclo-octane rings has proved difficult but a number of tetrahydro-oxocins are known.

(*i*) *Derivatives of oxocin*

A small amount of 4,5-**dihydro**-6*H*-**oxocin** (I) is formed irreversibly on the pyrolysis of 3,4-epoxycycloheptane (*J. K. Crandall* and *R. J. Watkins*, Tetrahedron Letters, 1967, 1717).

2,3,4,5-**Tetrahydro**-6*H*-**oxocin** (II), b.p. 75–80°/15 mm, n_D^{20} 1.4649, is obtained from 4,5,6,7-tetrahydro oxepin by a carbene ring-expansion followed by dehydrohalogenation (*F. Nerdel et al., ibid.*, 1966, 5385) or by reduction with lithium tetrahydridoaluminate and Hoffmann deamination of the bridged compound III (*L. A. Paquette* and *R. W. Begland*, J. org. Chem., 1967, **32**, 2723). 2,2-*Dimethyl*-2,3,4,5-*tetrahydro*-6H-*oxocin*-6-*one*, b.p. 115–117°/17 mm, *oxime*, m.p. 90–91°, is made by the cyclisation of 3-oxo-7-methyloct-6-enal (*G. Leser*, Bull. Soc. chim. Fr., 1899, **21**, 971).

2,3,4,5-**Tetrahydro**-8*H*-**oxocin** (IV), b.p. 52–53°/30 mm, is one of the products of acetolysis of the brosyl ester of 3-hydroxyoxocane; oxepanes are also obtained (*Paquette*

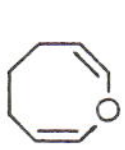

(I)

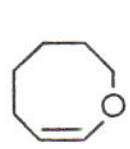

(II)

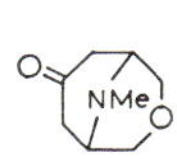

(III)

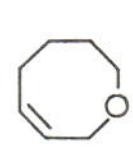

(IV)

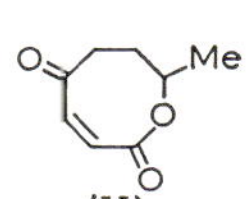

(V)

et al., J. Amer. chem. Soc., 1970, **92,** 1971). The naturally occurring antifungal metabolite, 2-methyl-2,3,4,5-tetrahydro-8*H*-oxocin-5,8-dione (V), is a member of this series (*D. C. Aldridge* and *J. F. Grove*, J. chem. Soc., 1964, 3239).

---

* For a general introduction to heterocyclic compounds with rings comprising seven or more atoms, and for an explanation of the nomenclature used, see p. 139 *et seq.*

trans-**4-Hydroxy**-3,4,7,8-**tetrahydro**-2*H*-**oxocin** (VI; R = H, R^1 = OH) is also prepared by a carbene ring-expansion route but *cis*-3,4,7,8-tetrahydro-2*H*-oxocin (VI; R, R^1 = H) results on treating 5-oxocanone with tosyl hydrazide followed by butyl-lithium (*Paquette* and *M. K. Scott*, J. Amer. chem. Soc., 1972, **94,** 6751, 6760). The brosyl ester of 3-*hydroxy*-3,4,7,8-*tetrahydro*-2H-*oxocin* (VI; R = OH, R^1 = H), b.p. 94–96°/18 mm, prepared by

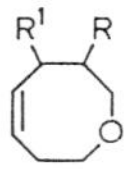
(VI)

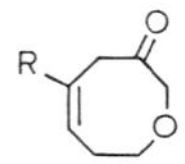
(VII)

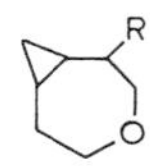
(VIII)

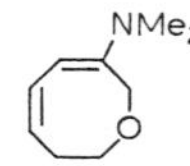
(IX)

reducing ketone VII (R = H) with lithium tetrahydridoaluminate, on acetolysis gives the bicyclic compound VIII (R = OAc); with lithium tetrahydridoaluminate the product is the corresponding hydroxy compound VIII (R = OH) *(Paquette et al., loc. cit.)*.

The dienamine, 3-*diethylamino*-7,8-*dihydro*-2H-*oxocin* (IX), b.p. 123°/20 mm, n_D^{25} 1.5240, also obtained from III, is hydrolysed to 3,4,7,8-*tetrahydro*-2H-*oxocin*-3-*one* (VII; R = H), b.p. 84–85°/15 mm, n_D^{25} 1.4838, by aqueous acetic acid, but to a 5-(3-halogeno-propyl)tetrahydro-3-furanone if hydrochloric or hydrobromic acids are used. With toluenesulphonic acid in benzene, isomerisation occurs to the conjugated 2,3,7,8-*tetra-hydro*-6H-*oxocin*-3-*one*, b.p. 82–84°/11 mm, and with potassium *tert*-butoxide and methyl iodide the 4,4-dimethyl derivative is obtained. The 5-*phenyl* analogue (VII; R = Ph), b.p. 115–120°/0.1 mm, is prepared and behaves similarly. Reduction of its 4-tosylhydrazide with sodium tetrahydridoborate yields 5-*phenyl*-3,4,7,8-*tetrahydro*-2H-*oxocin*, b.p. 115–120°/0.1 mm, which contracts the 2-(3-chloropropyl)-2-phenyltetrahydrofuran with hydrochloric acid (*Paquette* and *Begland*, J. Amer. chem. Soc., 1965, **87,** 3784; 1968, **90,** 6148.

Cyclic oxonium cations X are formed by treating hydroxy ketones of the type MeCO-$(CH_2)_nOH$ (n = 3, 5–10, 12) with sulphuric acid. Oxidation with hydrogen peroxide yields the esters $MeCO_2(CH_2)_nOH$ (*L. Cottier* and *G. Descotes*, Compt. rend., 1971, **273,** 64).

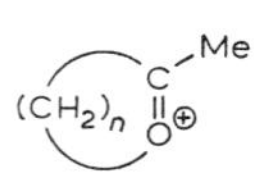
(X)

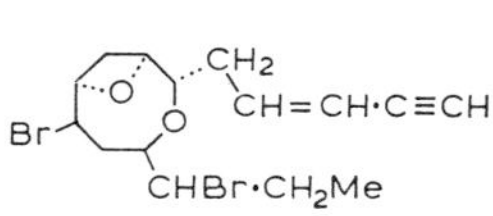

(XI)

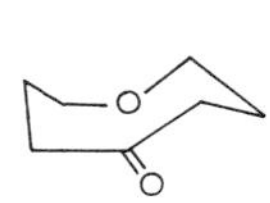
(XII)

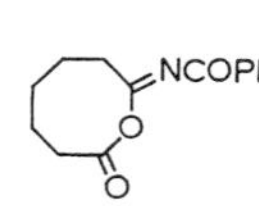

(XIII)

Oxocane (oxacyclo-octane), n_D^{26} 1.4486, is obtained by catalytic reduction of tetra-hydro-oxocin (*Paquette* and *Begland*, *loc. cit.*). 3-*Oxocanol*, b.p. 111–112°/9 mm, n_D^{25} 1,4750, 4-oxocanol and 5-oxocanol, 3,5-*dinitrobenzoate*, m.p. 102–104°, may be obtained by reducing the corresponding ketones with lithium tetrahydridoaluminate; oxocanol derivatives undergo ring-contraction on solvolysis (*Paquette* and *Scott*, *loc. cit.*).

Laureatin (XI) and associated natural products are oxygen bridged oxocanes, in which the bridge is opened with bromine or ethanolic hydrogen bromide, and shifts from 3,5 to 3,6 with Lewis acids (*T. Irie et al.*, J. org. Chem., 1972, **37,** 680; Tetrahedron Letters, 1972, 2121). Oxecanes and compounds with the skeleton V are isolated from the same source (*idem*, *ibid.*, 1965, 1091; Tetrahedron, 1968, **24,** 4193). The rings of 3-bromo-oxocanes are opened up to 6-heptenols when treated with zinc in acetic acid (*idem, ibid.*, 1970, **26,** 851).

2-Oxocanone (oenantholactone), b.p. 80–81 °/11 mm, n_D^{22} 1.4680, is prepared by oxidation of cycloheptanone with peracids (*R. Huisgen* and *H. Ott*, Angew. Chem., 1958, **70,** 312; Tetrahedron, 1959, **6,** 253; *W. D. Closson et al.*, J. org. Chem., 1967, **32,** 3160; *P. S. Starcher* and *B. Phillips*, J. Amer. chem. Soc., 1958, **80,** 4079). 3-*Oxocanone*, b.p. 77–78 °/11 mm, n_D^{27} 1.4588, and 4-oxocanone mixed with 4-oxocanol, are obtained by reducing ketones of type VII (R = H) and allyl alcohols of type VI (R = H, R^1 = OH), respectively. For 5-*oxocanone* (XII), b.p. 119–121 °/19 mm, n_D^{25} 1.4746, tetrahydro-4-pyrone is treated with diazomethane; a little 4-oxocanone and much of the spiroepoxide are also obtained (*Paquette* and *Scott*, *loc. cit.*). An alternative route uses the Dieckmann cyclisation (*N. J. Leonard et al.*, J. Amer. chem. Soc., 1960, **82,** 4075).

The benzoylimino-oxocanone XIII is the product of photo-oxidation of 4,5-pentamethylene-2-phenyloxazole (*H. H. Wasserman et al.*, *ibid.*, 1972, **94,** 4991); some 3-oxocane methides have also been prepared (*Nerdel et al.*, Tetrahedron, 1970, **26,** 1589).

Oxocane and 5-oxocanone (XII) are found to adopt the "chair"–"boat" conformation shown and not the "boat"–"boat" conformation previously suggested. As more oxygen atoms are introduced into the ring, transannular oxygen–oxygen interactions twist the ring (*F. A. L. Anet* and *P. J. Degen*, Tetrahedron Letters, 1972, 3613; J. Amer. chem. Soc., 1972, **94,** 1390; *J. Dale et al.*, *ibid.*, p. 1389; *Leonard et al.*, *ibid.*, 1960, **82,** 4075).

(*ii*) *Benzoxocins*

Very little is known about **benzoxocins** but the natural product **bilobanone** (XIV) is given the 1-benzoxocin skeleton (*H. Kimura*, Yakugaku Zasshi, 1958, **78,** 1152; 1962, **82,** 214). A Baeyer–Villiger oxidation of 2-bromo-1-

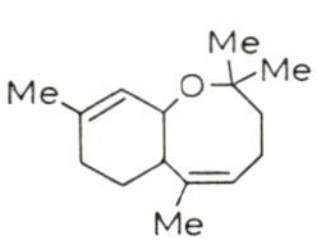

(XIV)

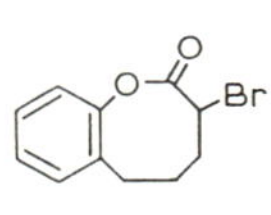

(XV)

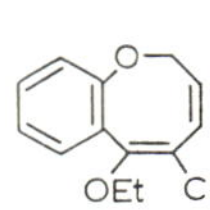

(XVI)

benzosuberone brought about ring-expansion to the 1-benzoxocin derivative XV, which on subsequent treatment with alkaline hydrogen peroxide collapsed to 2,3,4,5-tetrahydro-1-benzoxepin-2-carboxylic acid (*G. Baddeley* and *J. R. Cooke*, J. chem. Soc., 1958, 2797).

Addition of dihalogenocarbenes to 2,3-dihydro-1-benzoxepin followed by ring-opening in quinoline or methanolic silver nitrate yields, respectively, 5-*chloro*-6-*ethoxy*-2H-1-*benzoxocin* (XVI), b.p. 98 °/0.07 mm, and 4-*bromo*-3,4,5,6-*tetrahydro*-2H-1-*benzoxocin*-5-*one*, b.p. 135–140 °/0.1 mm, an isomer of XV (*P. Maitte et al.*, Bull. Soc. chim. Fr., 1971, 2557).

11,12-Dihydro-6*H*-dibenzo[*b,f*]oxocin-6-one (XVII), m.p. 115–116 °, has been prepared by cyclising the corresponding hydroxy acid (*W. Baker et al.*, J. chem. Soc., 1952, 1447).

10*H*-Dibenz[*b,g*]oxocin (XVIII) is prepared by dehydration of the 11-*hydroxy*-11,12-*dihydro*-10H-*dibenz*[b,g]*oxocin*, m.p. 105 °, obtained in the ring expansion of 10-tosyloxymethyl-10,11-dihydrodibenz[*b,f*]oxepin. Oxidation of the intermediate alcohol gave 11,12-

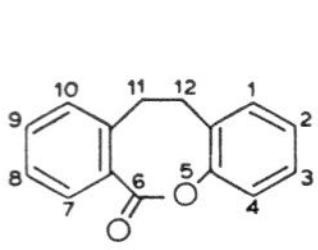

(XVII)

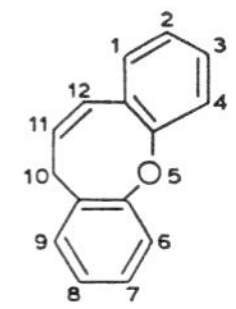

(XVIII)

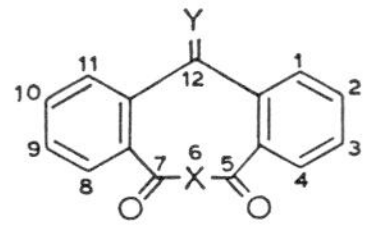

(XIX)

dihydro-10H-*dibenz*[b,g]*oxocin*-10-*one*, m.p. 74–75°. The 10-proton of XVIII is very slightly acidic suggesting that the anion has some 10π aromaticity (*H. S. Kasmai* and *H. W. Whitlock*, J. org. Chem., 1972, **37,** 2161). The 11-hydroxy-12-methoxycarbonyl derivative of XVIII has been prepared by the Dieckmann cyclisation of dimethyl 2,2′-oxybis(phenylacetate) (*R. Shapiro* and *D. Slobodin*, *ibid.*, 1969, **34,** 1165).

5,7-*Dihydro*-12H-*dibenzo*[c,f]*oxocin*-5,7,12-*trione* (XIX, X, Y = O), m.p. 212°, and some derivatives, are prepared by dehydration of 2,2′-dicarboxybenzophenone by heating or by treatment with ethanolic hydrochloric acid. With alcoholic ammonia XIX (X, Y = O) yields the corresponding *dibenzazocinetrione* (XIX; X = NH, Y = O), m.p. 248° (*C. Graebe* and *P. Juillard*, Ann., 1887, **242,** 214; *Baker et al.*, Chem. Ber., 1959, **92,** xxxvii; *I. Ikeda* and *S. Kanahara*, C.A., 1960, **54,** 4606). Compound XIX (X = O; Y = CH_2) is obtained by alkaline dehydrohalogenation of 2-chloro-1,1-bis(2-carboxyphenyl)-ethane (*R. Quelet et al.*, Compt. rend. 1962, **254,** 1811).

(*b*) *Eight-membered ring-compounds containing two or more oxygen atoms*

(*i*) *Dioxocins and their derivatives*

The action of oxygen on 1,6-dipotassio-1,1,6,6-tetraphenylhexane is said to give 3,3,8,8-**tetraphenyl**-1,2-**dioxocane,** m.p. 186° (decomp.) (*G. Wittig* and *M. Leo*, Ber., 1930, **63,** 943) and *R. Criegee* and *P. Becher* (Chem. Ber., 1957, **90,** 2516) have prepared an acylated dihydroxy-1,2-dioxocane.

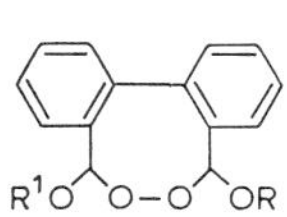

(XX)

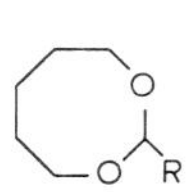

(XXI)

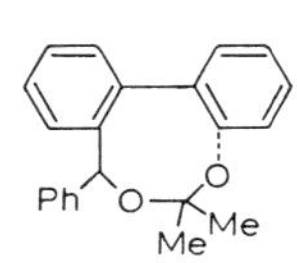

(XXII)

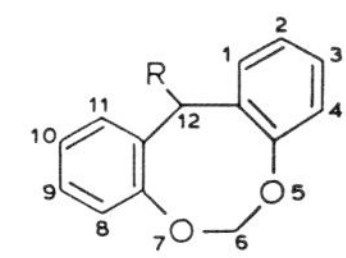

(XXIII)

Ozonolysis of phenanthrene in water or methanol gives the 3,12-**dihydroxy**-5,8-**dihydrodibenzo**[*d,f*]-1,2-**dioxocin** (XX) (R, R^1 = H), m.p. 105–106°, and the corresponding *di-methoxy* compound (R, R^1 = Me), m.p. 180–181°, together with compound XX (R = H, R^1 = Me), m.p. 97–98°, respectively. The first of these is also obtained from 2,2′-diformylbiphenyl with hydrogen peroxide, and other alkoxy derivatives have been made. The heterocyclic ring is opened to 2,2′-diformylbiphenyl, the corresponding formylcarboxylic ester, the formylcarboxylic acid or the dicarboxylic acid, respectively, with dilute hydrochloric acid or potassium iodide in acetic acid, aqueous aliphatic amines, bases, or alkaline hydrogen peroxide (*P. S. Bailey et al.*, J. chem. Soc., 1956, **78,** 3811; J. org. Chem., 1958, **23,** 1089; U.S.P. 2,870,194/1959, C.A., 1959, **53,** 10133; *M. G. Sturrock et al.*, J. org. Chem., 1963, **28,** 2340; *Sturrock*, *E. L. Cline* and *K. R. Robinson*, U.S.P. 3091,620/1960; C.A., 1964, **60,** 2995; *J. P. Wibaut* and *Th. J. de Boer*, Koninkl. Ned. Acad., Wetensch., Proc. ser. B., 1956, **59,** 421; C.A., 1957, **51,** 11305; Rec. Trav. chim., 1959, **78,** 183; *A. K. Lebedev et al.*, C.A., 1963, **58,** 5691; 1968, **68,** 21914, 21915; 1970, **73,** 35350; 87589; 1971, **74,** 124404).

5,8-*Dihydrodibenzo*[d,f]-1,2-*dioxocin*-5,8-*dione (diphenoyl peroxide)*, exploding at 70°, is the product when phenanthraquinone is treated with trimethyl phosphite followed by ozone. It is reduced to diphenic anhydride by triphenylphosphine and like other 1,2-dioxocins it liberates iodine from acidified potassium iodide solution (*F. Ramirez et al.*, J. Amer. chem. Soc., 1961, **83,** 492; C.A., 1966, **64,** 12538).

Few, 1,3-dioxocins are known, but 1,3-**dioxocane** (XXI; R = H), b.p. 40–44°/11 mm, the cyclic formal of pentane-1,5-diol is prepared by the ready depolymerisation of the linear polyformal (*J. H. Hill* and *W. H. Carothers*, *ibid.*, 1935, **57,** 925); 2-*methyl*- (XXI; R = Me), b.p. 44–46°/6 mm, n_D^{19} 1.4385, and 2-*ethyl*- (XXI; R = Et), b.p. 54–56°/5 mm, homologues are also prepared from pentane-1,5-diol (*D. M. Jones* and *N. F. Wood*, J. chem. Soc., 1965, 1560; *P. Watts*, *ibid.*, B, 1968, 543; cf. *P. A. Gurevich et al.*, Zhur. obshcheĭ Khim., 1968, **38,** 1905) 6,6-*Dimethyl*-8-*phenyl*-6,8-*dihydrodibenzo*[d,f]-1,3-*dioxocin* (XXII), m.p. 78–80°, results when 6-phenyldibenzo[*b*,*d*]pyrylium perchlorate is dissolved in aqueous acetone (*C. C. Barker*, J. chem. Soc., C, 1969, 361).

A number of chlorinated derivatives of 6,12-dihydrodibenzo[*d*,*g*]-1,3-dioxocins (XXIII; R = H, Ph, CO_2H) have been obtained from the reaction of 2,2′-dihydroxydiphenyl-methanes and *gem*-dihalides in alkaline solution (*W. Retter*, B.P. 726,789/1955; U.S.P. 2,719,852/1955; C.A., 1956, **50,** 6511, 10131; *Retter* and *W. Müller*, G.P. 947,116; C.A., 1959, **53,** 4224; *W. V. Farrar*, J. appl. Chem., London, 1960, **10,** 207; *S. K. Grover* and *T. R. Seshadri*, Indian J. Chem., 1966, **4,** 290; Tetrahedron, 1964, **20,** 555; *R. G. Johnson et al.*, Gen. Offen. 2,029,796; U.S.P. 3,553,254/1969; C.A., 1971, **74,** 88093; **75,** 5977). 6,12-*Dihydrodibenzo*[d,g]-1,3-*dioxocin*-6-*one*, m.p. 101–102°, results when phosgene is used (*R. J. Prochaska*, Belge P. 626,345/1961; U.S.P. 3,220,980/1964; C.A., 1964, **60,** 13186; 1966, **64,** 6788).

1,4-**Dioxocin** (XXIV), b.p. 39°/14 mm, is prepared from bromobenzene dioxides as shown:

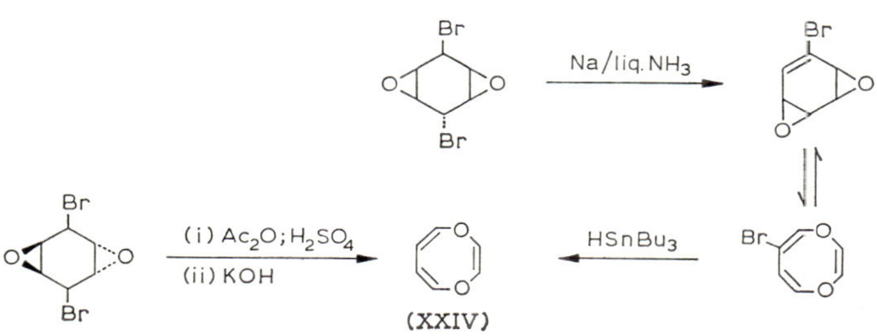

(XXIV)

It has a 10π electron system but shows only olefinic properties. Catalytic reduction gives 1,4-**dioxocane** (1,4-*dioxacyclo-octane*), b.p. 149°. The tautomeric ratio *syn*-benzene dioxide: 1,4-oxocin at 60° in benzene is 5:95 when approached from either side (*E. Vogel et al.*, Angew. Chem. intern. Edn., 1972, **11,** 935, 937). 1,4-*Dioxocin-6-ol* is obtained similarly by treating *cis*-benzene trioxide with trifluoroacetic acid (*R. Schwesinger* and *H. Prinzbach*, *ibid.*, p. 942). The base-catalysed condensation of diethyl β-oxoglutarate with bis-2-bromoethyl ether yields the ethyl 6-ethoxycarbonyltetrahydro-1,4-oxocin-5-acetate (XXV) (*I. Ali-Zade* and *B. A. Arbuzov*, J. gen. Chem. U.S.S.R., 1943, **13,** 123; C.A., 1944, **38,** 353).

Treatment of the intermediates XXVI (X = Cl or OCO_2Et) with di-*n*-butylcadmium gave not the butyl ketone but 8-*butyl*-8-*methyl*-1,4-*dioxocan*-5-*one*, b.p. 82–87°/0.13 mm; 1,4- and 1,5-dioxonanones are prepared similarly (*R. A. LeMahieu*, J. org. Chem., 1967, **32,** 4149).

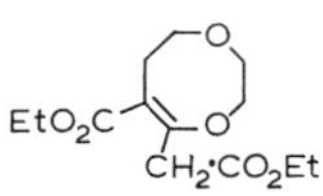

(XXV)

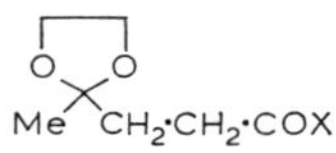

(XXVI)

Catechol cyclises with *cis*-1,4-dichloro-2-butene to give 2,5-*dihydro*-1,6-*benzodioxocin*, b.p. 118°/12 mm, m.p. 32°, which is converted into 1,6-**benzodioxocin** (XXVII), b.p. 109°/13 mm, m.p. 5°, by bromination–dehydrobromination or hydroxylation–acetylation–pyrolysis sequences. This dioxocin behaves as a diene towards maleic anhydride and shows no evidence of a diamagnetic ring current in the heterocyclic ring. On irradiation it yields the tricyclic compound XXVIII (*W. Schroth et al.*, Z. Chem., 1963, **3,** 119, 228; Angew. Chem., 1963, **75,** 303; Internat. Edn. 1967, **6,** 697; *C. H. Böhringer, Sohn*, F.P. 1,573,667/1967; C.A., 1970, **72,** 90479).

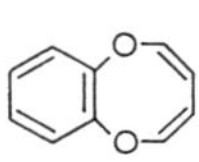
(XXVII)

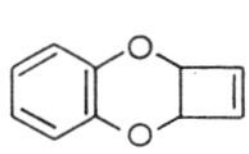
(XXVIII)

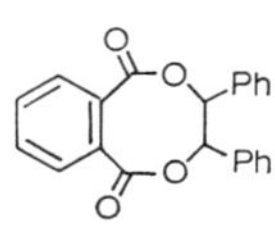

(XXIX)

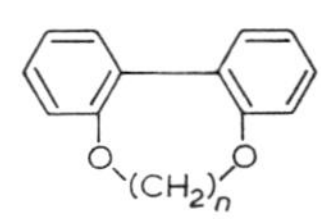

(XXX)

2,3,4,5-*Tetrahydro*-1,6-*benzodioxocin*, b.p. 69°/0.5 mm, n_D^{23} 1.5398, and its homologues are prepared similarly from catechol and 1,4-dibromobutane, or by reduction of unsaturated members of the series. Bromination yields the 8,9-*dibromo* derivative, m.p. 112–113° (*P. A. Claret et al.*, J. chem. Soc. B., 1971, 1231; *W. G. Brown et al.*, Canad. J. Res., 1949, **27B,** 398, *K. Zeigler et al.*, Ann., 1937, **528,** 162).

The *dl*- and *meso*-isomers of 3,4-*diphenyl*-1,3,4,6-*tetrahydro*-2,5-*benzodioxocin*-1,6-*dione* (XXIX), *dl*-, m.p. 206–208°, *meso*-, m.p. 192–193°, are prepared by treating phthaloyl peroxide with *trans*- and *cis*-stilbene, respectively. This reaction has been extended to other olefins (*F. D. Greene*, J. Amer. chem. Soc., 1956, **78,** 2250; 1958, **80,** 3432; 1959, **81**, 1503; 1960, **82,** 890). The unsubstituted compound is prepared from dipotassium phthalate and 1,2-dibromoethane in dimethylformamide (*S. Yoneda, Yoshida* and *K. Fukui*, Kogyo Kagaku Zasshi, 1966, **69,** 641; C.A., 1967, **66,** 10537).

6,7-*Dihydrodibenzo*[e,g]-1,4-*dioxocin* (XXX; n = 2), m.p. 98°, is prepared from biphenyl-2,2'-diol and 1,2-dibromethane in a solution of sodium hydroxide in dimethyl sulphoxide. Compounds XXX ($n = 3$), m.p. 67–69°, ($n = 4$), m.p. 110–111°, and ($n = 5$), m.p. 109–110°, are obtained similarly (*D. W. Allen et al.*, J. chem. Soc., C, 1971, 3454; *A. W. Archer* and *Claret*, Chem. and Ind., 1969, 171).

Self-condensation of 2-methyl-2-propyl-1,3-propanediol to 3,7-dimethyl-3,7-dipropyl-1,5-dioxocane is brought about by molybdic acid (*P. Mastagli* and *C. de Fournas*, Compt. rend. 1960, **250,** 3336); condensation of bis(2,2-diethoxycarbonylethyl) ether and bis(bromomethyl) ether to 3,3,7,7-*tetraethoxycarbonyl*-1,5-*dioxocane*, b.p. 185–189°/5 mm, by magnesium (*Ali-Zade* and *Arbuzov*, *loc. cit.*) and self-condensation of 2-ethyl-3-methoxy-2-phenylpropionyl chloride in the presence of pyridine and urea yields 3,7-*diethyl*-3,7-*diphenyl*-1,5-*dioxocane*-2,6-*dione*, m.p. 92–93° (*E. Testa*, *Farmaco*, *Pavia*, Edn. Sci., 1957, **12,** 823; C.A., 1958, **52,** 11854). The lead bis(dicarboxylate) of phenylsuccinic acid reacts with styrene to give 4,8-*diphenyl*-1,5-*dioxocane*-2,6-*dione*, m.p. 167–168° (*Y. Yukawa* and *M. Sakai*, Nippon Kagaku Zasshi, 1966, **87,** 84).

1,5-**Dioxocane** 3-**methide** (XXXI), b.p. 44–45°/11 mm, n_D^{22} 1.4682, is formed, together with larger rings and non-cyclic compounds, when the tribromide $PhCO \cdot C(CH_2Br)_3$ reacts with the disodium salt of propane-1,3-diol (*Nerdel et al.*, Ann., 1970, **736,** 75).

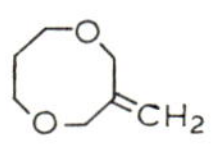
(XXXI)

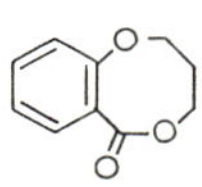
(XXXII)

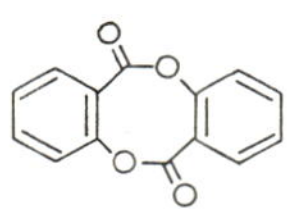
(XXXIII)

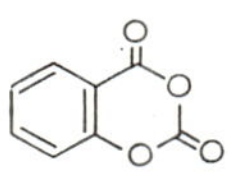
(XXXIV)

In alcoholic sodium ethoxide, ethyl 2(3-hydroxypropoxy)benzoate is converted into the lactone 2,3,4,6-*tetrahydro*-1,5-*benzodioxocin*-6-*one* (XXXII), b.p. 103–104°/18 mm (*J. Gilbert* and *R. Douville*, Compt. rend., 1970, **271C,** 1262).

Cyclodehydration of salicyclic acids gives the dilactones XXXIII, the **disalicyclides,** together with higher cyclic and linear polymers (see CCC Vol. IIID, p. 183 *et seq.*). A more satisfactory route to polymer-free XXXIII is to form the cyclic carbonate XXXIV from salicyclic acid and phosgene and treat it with triethylamine (*F. M. Dean et al.*, J. chem. Soc. Perkin I, 1972, 2007).

(*ii*) *Trioxocane and tetraoxocanes*

1,3,6-**Trioxocane** (1,3,6-*trioxacyclo-octane*) (XXXV; R = H), b.p. 60–70°/20 mm, and various derivatives are prepared by the acid-catalysed action of aldehydes on diethylene glycol; cation exchange resins are particularly

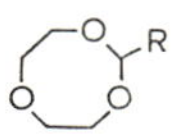
(XXXV)

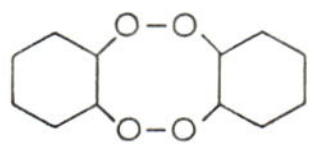
(XXXVI)

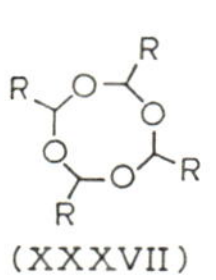
(XXXVII)

good for this cyclisation. The ring is readily opened by acid hydrolysis and polymers may be formed (*M. J. Astle*, Ind. Eng. Chem., 1954, **46,** 787; *P. A. Laurent* and *P. Tarte*, Bull. Soc. chim. Fr., 1960, 718, 926; C.A., 1961,

55, 9426; *J. A. Nieuwland et al.*, J. Amer. chem. Soc., 1930, **52,** 1022; *C. D, Hurd* and *D. G. Botteron, ibid.*, 1946, **68,** 1200; *M. Okada et al.*, Makromol. Chem., 1969, **127,** 66).

3,4,7,8-**bis(Tetramethylene)**-1,2,5,6-**tetroxocane** (XXXVI) is prepared by reacting cyclohexene with hydrogen peroxide, tungstic acid and acetone, and treating the product with perchloric acid. Catalytic hydrogenation of this compound yields *trans*-cyclohexane-1,2-diol (*G. B. Payne* and *C. W. Smith* J. org. Chem., 1957, **22,** 1682).

With a trace of acid at 0° acetaldehyde gives 2,4,6,8-*tetramethyl*-1,3,5,7-*tetroxocane (metaldehyde II)* (XXXVII; R = Me) m.p. 47°, b.p. 65°/15 mm (*L. Pauling* and *D. C. Carpenter*, J. Amer. chem. Soc., 1936, **58,** 1274); the *analogue* (XXXVII; R = CCl_3), m.p. 219–220°, from chloral (*M. Baron*, Nature, 1961, **192,** 258; cf. *H. C. Caldwell* and *W. E. Thompson*, J. pharm. Sci., 1965, **54,** 789; *E. Kopp* and *J. Smidt*, Ann., 1966, **693,** 117); and the parent 1,3,5,7-**tetroxocane (tetroxane)** (XXXVII; R = H) m.p. 112°, from formaldehyde (*S. Olsen et al.*, Chem. Ber., 1959, **92,** 1072; *M. Akimoto*, *Y. Miyake* and *T. Yamachi*, Jap. P. 16751/1968; C.A., 1969, **70,** 57924,87867; *W. Kern et al.*, Makromol. Chem., 1962, **52,** 48) are obtained similarly. Higher cyclic polymers of formaldehyde, *e.g.* 1,3,5,7,9-**pentoxecane**, m.p. 55–56°, are also isolated (*K. H. Burg et al., ibid.*, 1968, **111,** 181) and with certain catalysts 1,2,4,6-**tetroxocane** is produced (*Toyo Koatsu* Ind., F.P. 1,551,921/1969; C.A., 1969, **71,** 39026).

In the solid state and at low temperatures in solution tri- and tetra-oxocanes adopt a "crown" conformation, but at normal temperatures in solution a twisted "boat–chair" form is favoured (*Dale et al.*, J. Amer. chem. Soc., 1972, **94,** 1390; *F. A. L. Anet* and *P. J. Degan, loc. cit.*; *Baron et al.*, Canad. J. Chem., 1963, **41,** 1893).

2. Eight-membered rings containing sulphur or selenium

(*a*) *Compounds containing one sulphur atom*

Thiocane *(thiacyclo-octane)* (I), b.p. 71°/10 mm, n_D^{20} 1.5152, *mercuric chloride* complex, m.p. 129–130°, is prepared in low yield in the high dilution reaction of sodium sulphide and 1,7-dibromoheptane (*A. Müller et al.*, Monatsh., 1953, **84,** 1215); with 1,7-dibromoheptan-2- and -4-ones the products are 3-thiocanone, b.p. 66–68°/0.35 mm (*H. Remane*

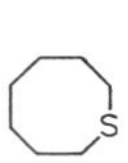
(I)

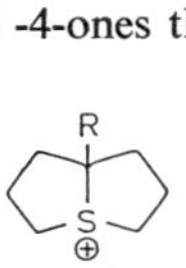
(II)

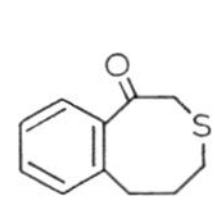
(III)

et al., J. pr. Chem., 1970, **312,** 1058) and 5-thiocanone. The latter is reduced to thiocane by the Wolff-Kishner procedure (*A. A. Ponomarev* and *I. S. Monakhova*, Zhur. obshcheĭ Khim., 1964, **34,** 1246; C.A., 1969, **70,** 106359).

Dieckmann ring closure and hydrolysis were also used to prepare 5-*thiocanone*, m.p. 53–54°. It forms a 1-*oxide*, m.p. 91–92°, a 1,1-*dioxide*, m.p. 134–135°, and in contrast to its oxa-analogue forms salts of the bicyclic cation II (R = OH, OR) with aqueous alcoholic acids. The 1-oxide forms an oxygen-bridged bicyclic compound. Reduction with lithium tetrahydridoaluminate gave 5-*hydroxythiocane*, m.p. 26°, which yielded the picrate of II (R = H) after treatment with phosphorus pentoxide (*G. G. Overberger et al.*, J. Amer. chem. Soc., 1962, **84,** 2814; 1959, **81,** 506; *N. J. Leonard et al., ibid.*, 1959, **81,** 504; 1960, **82,** 4075; 1962, **84,** 3701; J. org. Chem., 1962, **27,** 282).

1,2,5,6-*Tetrahydro*-4H-3-*benzothiocin*-1-*one* (III), m.p. 60°, b.p. 188°/13.5 mm, prepared by Friedel–Crafts cyclisation of the appropriate acid chloride, under Clemmensen reduction conditions yields both 1,2,5,6-**tetrahydro-4*H*-3-benzothiocin,** b.p. 145°/12.8 mm, n_D^{23} 1.5851, and the ring contraction product 1-methyl-1,3,4,5-tetrahydro-2-benzothiepin (*N. J. Leonard* and *J. Figueras*, J. Amer. chem. Soc., 1952, **74,** 917; *J. v. Braun* and *K. Weissbach*, Ber., 1929, **62,** 2416; *P.* and *Mme. P. Cagniant*, Bull. Soc. chim. Fr., 1959, 1998).

1,2,5,6-*Tetrahydro*-4H-3-*benzothiocin*-6-*one*, m.p. 58–60°, is the product when 3,4-dihydroisothiochromanone reacts with vinyl-lithium (*W. C. Lumma et al.*, J. org. Chem., 1970, **35,** 3442).

6,7-*Dihydro*-12H-*dibenzo*[b,e]*thiocin*-12-*one* (IV; X = O) m.p. 149–150°, is obtained by cyclising 2-(2-phenylethylthio)benzoic acid with polyphosphoric acid or its acid chloride with aluminium chloride; only a low yield is obtained from the isomeric 2-(2-phenylthioethyl)benzoic acid. Sodium tetrahydridoborate, Grignard reagents and acetonitrile anion react at position 12 to give carbinols. The thiocin IV [X = $CH(CH_2)_2OH$]

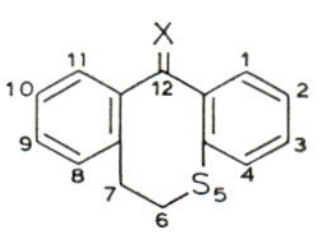

(IV)

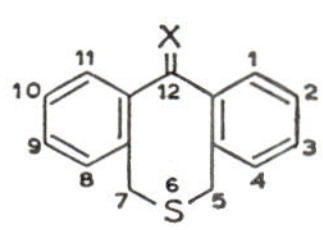

(V)

rearranges to 5(2-bromoethyl)phenyl-2,3-dihydro-1-benzothiepin with hydrobromic acid (*K. Stach et al.*, Angew. Chem., intern. Edn., 1962, **1,** 553; Chem. Ber., 1965, **98,** 685; *F. Bickelhaupt* and *Stach*, 9, G.P. 1,186,080/1962; B.P. 1,129,029/1966; C.A., 1965, **62,** 11790; 1969, **70,** 37664; *B. Dansec, J. C. Paigmant* and *C. Malen*, Ger. Offen. 2,011,806/1969; C.A., 1971, **74,** 13024).

5,7-*Dihydro*-12H-*dibenzo*[c,f]*thiocin* (V; X = H_2), m.p. 194–195° (*D. D. Emrick* and *W. E. Truce*, J. org. Chem., 1961, **26,** 1329) and 5,7-*dihydro*-12H-*dibenzo*[c,f]*thiocin*-12-*one* (V; X = O), m.p. 208° (*J. Fouche* and *R. Gaumont*, Bull. Soc. chim. Fr., 1971, 2062) are made by cyclising 2,2′-di(bromomethyl)-diphenylmethane and -benzophenone, respectively, with sodium sulphide. The former gives a *methylsulphonium iodide*, m.p. 215–217°, and a 6,6-*dioxide*, m.p. 258–259°; the latter reacts with Grignard reagents at position 12.

(*b*) *Compounds containing two or more sulphur atoms*

Several attempts to prepare monocyclic dithiocins led to extrusion of sulphur atoms.

(*i*) *Dithiocins and related compounds*

1,2-**Dithiocane,** b.p. 62°/1 mm, n_D^{25} 1.5698, is prepared from α,ω-di-Bunte salts by controlled addition of sodium disulphide (*N. A. Rosenthal*, U.S.P. 3,284,466/1963; C.A.,

1968, **68,** 13015) or cupric chloride (*J. G. Affleck* and *G. Dougherty,* J. org. Chem., 1950, **15,** 865) and steam distillation of the product from the reaction mixture. Steam distillation of the polymeric disulphide also yields cyclic monomer (*F. O. Davis* and *E. M. Fettes,* J. Amer. chem. Soc., 1948, **70,** 2611) as does oxidation of α,ω-disulphides with ferric chloride (*A. Schöberl* and *H. Grafje,* Angew. Chem. intern. Edn., 1957, **69,** 713; Ann., 1958, **614,** 66).

5,8-*Dihydrodibenzo*[c,f]-1,2-*dithiocin* (VI) *racemate,* m.p. 161°; *enantiomers,* m.p. 133°, prepared by treating 2,2′-bis(mercaptomethyl)biphenyl with ferric chloride (*A. Lüttringhaus et al.,* Z. Naturforsch., 1967, **22b,** 1296; Monatsh., 1967, **98,** 1323) adopts the "pseudo chair" rather than the predicted "pseudo tub" conformation (*G. H. Wahl et al.,* Chem. Comm., 1972, 985). Selective desulphurisation of VI gives 5,7-dihydrodibenzo[*c,e*]thiepin (*D. N. Harpp* and *J. G. Gleason,* J. Amer. chem. Soc., 1971, **93,** 2437).

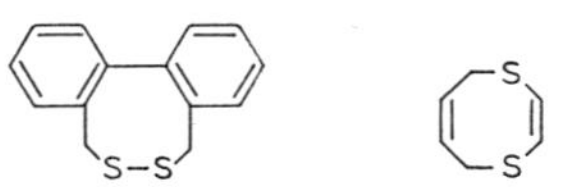

(VI) (VII)

Diacetylene reacts with ethylene-1,2-dithiol to yield 1,4-**dithiocin** (*W. Schroth et al.,* Z. Chem., 1969, **9,** 184). A series of partially saturated 1,4-dithiocins has been prepared by condensing appropriate unsaturated dithiols and dihalogeno compounds, comprising 5,8-*dihydro-* (VII) (b.p. 116°/13 mm, n_D^{20} 1.6351), 2,3,5,8-*tetrahydro-* (b.p. 122°/13.5 mm, n_D^{20} 1.6100); and 5,6,7,8-*tetrahydro-* (b.p. 113°/13.5 mm, n_D^{20} 1.6084; *disulphone,* m.p. 138–139°) -1,4-*dithiocin.* 2,5-*Dihydro*-1,6-*benzodithiocin,* m.p. 75–76°, 1,3,4,6-*tetrahydro*-2,5-*benzodithiocin,* m.p. 93°, and 6,11-*dihydrodibenzo*[b,f]-1,4-*dithiocin,* m.p. 142°, are prepared similarly. The ring of compound VII opens up in basic or reducing conditions (*idem, ibid.,* 1964, **4,** 302). Ring expansion of a 1-methyl-1,2-dithiane cation with ethylene affords a 1-methyl-1,4-thiocane cation (*N. E. Hester, G. K. Helmkanip* and *J. I. Afford,* Int. J. Sulfur Chem., Part A, 1971, **1,** 65; C.A., 1971, **75,** 110301).

1,6-**Benzodithiocin** (IX) is prepared by the prolysis of the tricyclic compound VIII, to which it reverts on irradiation; monocyclic dithiocins cannot be prepared in this way.

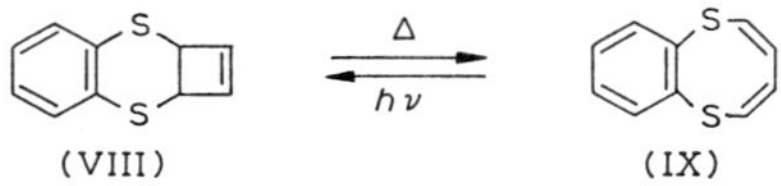

(VIII) (IX)

The 10π electron system of the heterocyclic ring does not give rise to a diamagnetic ring current (*D. L. Coffen et al.,* J. Amer. chem. Soc., 1971, **93,** 4627). By contrast, the 1,4-dithiocin X sustains a ring current similar in magnitude to its benzene analogue; the three annellated rings must force the heterocyclic ring into a plane (*M. O. Riley* and *J. D. Park,* Tetrahedron Letters, 1971, 2871).

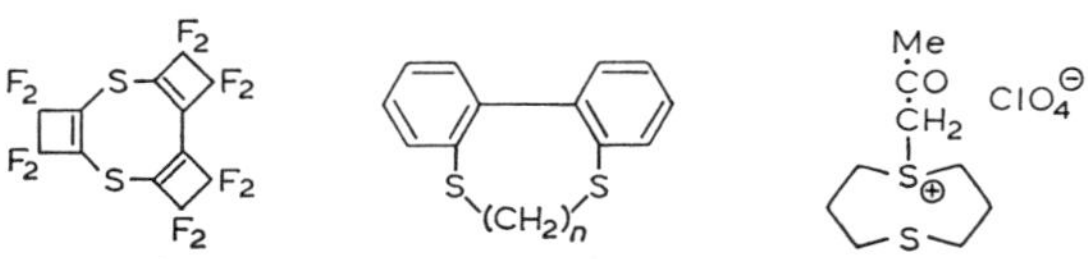

(X) (XI) (XII)

6,7-*Dihydrodibenzo*[e,g]-1,4-*dithiocin* (XI; $n = 2$), m.p. 179–180°, is prepared from the disodium biphenyl-2,2′-dithiolate and 1,2-dibromoethane. Compounds XI ($n = 3$), m.p. 69–70°; ($n = 4$), m.p. 143–144°, and ($n = 5$), m.p. 121–122°, are prepared similarly (*D. W. Allen et al.*, J. chem. Soc. C, 1971, 3454; Chem. and Ind., 1969, 522).

High dilution condensation of propane-1,3-dithiol and 1,3-dibromopropane yields 1,5-**dithiocane,** b.p. 43–46°/0.1 mm, n_D^{25} 1.5770. With hydrogen peroxide the 1,1,5,5-*tetroxide*, m.p. 258–259°, is formed (*W. Autenrieth* and *K. Wolff*, Ber., 1899, **32,** 1375; *J. R. Meadow* and *E. E. Reid, ibid.*, 1934, **56,** 2177; *Leonard et al.*, J. Amer. chem. Soc., 1960, **82,** 4075). The perchlorate of the 1-oxide reacts with acetone to give compound XII (*S. M. Johnson* and *I. C. Paul*, Tetrahedron Letters 1969, 177). Diallyl sulphide and sulphur dichloride cyclise to 3,7-*dichloro*-1,5-*dithiocane*, m.p. 37–39°, b.p. 100°/0.01 mm (*F. K. Lautenschläger*, B.P. 1,141,492/1965; C.A., 1969, **70,** 68446).

α,α′-Dichloroacetone reacts with disodium propane-1,3-dithiolate to give 1,5-*dithiacan*-3-*one* (XIII; $n = 3$), m.p. 46–47°, *semicarbazone*, m.p. 188–189° (*E. G. Howard* and *R. V. Lindsey*, J. Amer. chem. Soc., 1960, **82,** 158; *Howard*, U.S.P. 2,465,650/1960; C.A., 1961, **55,** 14492). An alternative route is the Dieckmann cyclisation and hydrolysis of dithioesters (*Lüttringhaus* and *H. Prinzbach*, Ann., 1959, **624,** 79). The analogous *dithiepanone* (XIII; $n = 2$), b.p. 75°/0.1 mm, *dithionanone* (XIII; $n = 4$) and *dithiecanone* (XIII; $n = 5$), *semicarbazone*, m.p. 156–158°, are also made by these routes.

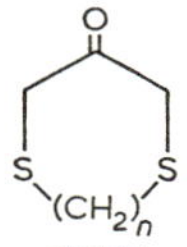
(XIII)

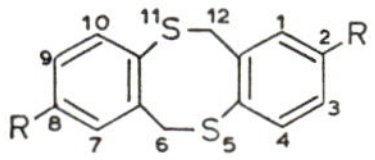
(XIV)

One product from α,α′-dichloroacetone and hydrogen sulphide is 1,5-*dithiocane*-3,7-*dione*, m.p. 145–146° (*C. Rappe* and *R. Gustafsson*, Acta Chem. Scand., 1967, **21,** 705), while malonyl chloride and trithiocarbonic acid give 1,5-*dithiocan*-2,4,6,8-*tetrone*, m.p. 124°, *via* 1,3-*dithiane*-4,6-*dione-2-thione* (*B. Krebs* and *G. Gattow*, Angew. Chem. intern. Edn., 1966, **5,** 250).

Desulphurisation of 3*H*-1,2-benzodithiole with hexamethylphosphoramide (*Harpp* and *Gleason*, J. org. Chem., 1970, **35,** 3259) or self-condensation of 2-mercaptobenzoyl chloride (*G. W. Stacey et al., ibid.*, 1965, **30,** 4074) gives 6,12-*dihydrodibenzo*[b,f]-1,5-*dithiocin* (XIV; R = H), m.p. 174–176°; 4-chlorobenzenethiol, formaldehyde and hydrogen chloride yield the 2,8-*dichloro* derivative (XIV; R = Cl), m.p. 254–255° (*M. Kulka*, Canad. J. Chem., 1958, **36,** 750).

The best known member of this group is 6,12-*dihydrodibenzo*[b,f]-1,5-*dithiocin*-6,12-*dione (dithiosalicylide)* (XV), m.p. 176–177°, which is obtained together with tri- and tetra-thiosalicylide by the cyclodehydration of thiosalicylic acid with phosphorus pentoxide (*W. Baker et al.*, J. chem. Soc., 1952, 3163). Alternative routes include the pyrolysis of benzothiadiazinone XVII (*T. D. Roberts et al.*, J. Amer. chem. Soc., 1972, **94,** 8505) or benz-3,1-oxathiane-2,4-dione (*E. Ziegler* and *H. Sterk*, Monatsh., 1968, **99,** 1958) and the irradiation of benzoxathianone XVI (*O. L. Chapman* and *C. L. Mackintosh*, J. Amer. Chem. Soc., 1970, **92,** 7001; cf. *A. O. Pedersen et al.*, Tetrahedron, 1970, **26,** 1157, who propose the 1,2-dithiocindione). The benzothietone XVIII is proposed as a common intermediate:

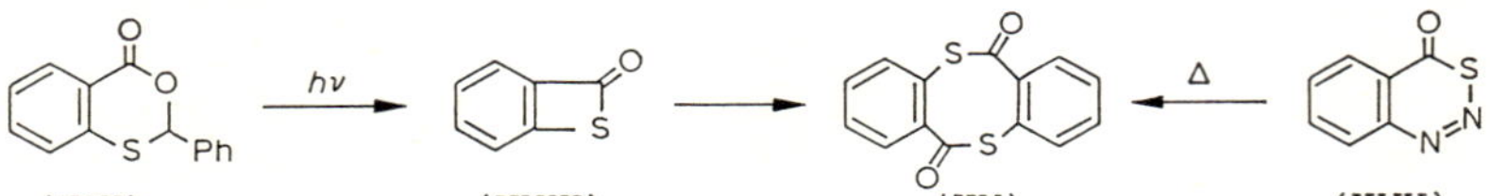

(XVI) (XVIII) (XV) (XVII)

Dithialactone XV is readily hydrolysed by alkali to thiosalicyclic acid.

(*ii*) *Compounds with more than two atoms of sulphur*

Catenation is favoured by sulphur atoms much more than by other heteroatoms, and similarly medium and large rings containing a string of sulphur atoms are not uncommon. This tendency has already been shown in the polythiepin series.

Reaction of *cis*- and *trans*-1,4-dibromo-2,3-diphenylbut-2-ene with sodium polysulphide gave some 6,7-**diphenyl**-5,8-**dihydro**-1,2,3,4-**tetrathiocin** (XIX), m.p. 151–152°. With one mole of 3-chloroperbenzoic acid the 1-*oxide*, m.p. 135–137°, is obtained, but with excess, ring-contraction to 3,4-diphenylthiophene takes place (*R. M. Dodson et al.*, J. org. Chem., 1972, **37,** 2367). Sulphur chlorides react with 2,2′-dimercaptobiphenyl to give *dibenzo*[e,g]-1,2,3,4-*tetrathiocin*, m.p. 200° *(Allen et al., loc. cit.)*.

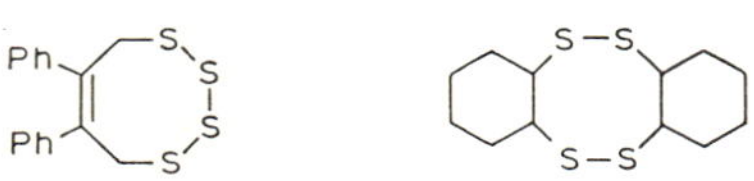

(XIX) (XX)

Hexafluorobut-2-yne reacts with sulphur giving a dithietane which dimerises on treatment with triethylamine to 3,4-7,8-*tetra(trifluoromethyl)*-1,2,5,6-*tetrathiocin*, m.p. 111–112° (*T. L. Cairns et al.*, J. Amer. chem. Soc., 1960, **82,** 1515; *C. G. Krespan et al., ibid.*, 1961, **83,** 3434; *Krespan*, U.S.P. 3,052,961; 3,073,844/1959; C.A., 1963, **58,** 3449; **59,** 1650).

1,2,5,6-**Tetrathiocane,** m.p. 152–154°, is obtained by oxidising ethane-1,2-dithiol with bromine (*H. Fasbender*, Ber., 1887, **20,** 460) and decomposing certain ethylene dithioacetals (*R. C. Elderfield* and *H. H. Rembges*, J. org. Chem., 1967, **32,** 3809). Oxidation by iodine or hydrogen peroxide (*V. A. Portnyagina et al.*, Ukr. Khim. Zhur., 1962, **28,** 721; 1966, **32,** 1081; C.A., 1963, **59,** 605; 1967, **66,** 37601) and treatment of 1,2,3-trithiolanes with sodium cyanide (*V. C. Petrunkin et al.*, Ukr. Khim. Zhur., 1956, **22,** 603; 1970, **36,** 181; C.A., 1957, **51,** 5692; 1970, **73,** 3902) are also effective for substituted 1,2,5,6-tetrathiocanes. Further treatment with peroxide yields monoxides which undergo ring-contraction to 1,4-dithianes with sodium cyanide (*idem*, Ukr. Khim. Zhur., 1967, **33,** 492; 1970, **36,** 71; C.A., 1968, **68,** 21911; 1970, **73,** 13940). *trans*-Cyclohexane-1,2-dithiol with iodine at high dilution gives *perhydrodibenzo*-1,2,5,6-*tetrathiocin* (XX), m.p. 192° (*F. Feher* and *B. Degan*, Angew. Chem. intern. Edn., 1967, **6,** 703).

Reaction of dichloromethane with sodium sulphide gives 1,3,5,7-**tetrathiocane,** m.p. 48–49°, the cyclic tetramer of thioformaldehyde; 1,3,5,7,9-**pentathiecane,** m.p. 123°, is also obtained (*M. Schmidt et al.*, Angew. Chem., 1959, **71,** 407; Z. Naturforsch., 1966, **216,** 622; J. Polymer Sci., 1965, **38,** 501; 455). With butyl-lithium and methyl iodide *C*-methylation occurs (*R. T. Wragg*, Tetrahedron Letters, 1969, 4959).

Sulphur and primary aliphatic amines react with acetophenone in methanol to yield 7-*alkylamino*-7-*phenyl*-1,2,3,4,5,6-*hexathiocane*-8-*thiones* (XXI) (*e.g.* R = *Me*, m.p. 145–146°; R = *Et*, m.p. 139°). Compound XXI (R = Bu) contracts to a thiiranthione

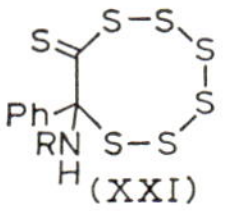

(XXI)

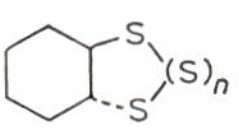

(XXII)

N.B. Dotted line indicates *trans* configuration

with sodium methoxide, but sulphur and hydrogen sulphide reverse this reaction (*F. Asinger et al.*, Angew. Chem. intern. Edn., 1965, **4,** 71). A series of heterocycles XXII is obtained by treating *trans*-cyclohexane-1,2-dithiol with dichlorosulphanes, Cl–(S)$_n$–Cl; these include *perhydro*-1,2,3,4,5,6-*benzohexathiocin* (XXII; *n* = 4), m.p. 81–82° (also XXII, *n* = 3, m.p. 67°; XXII, *n* = 5, m.p. 128–129°) (*Feher* and *Degan*, *loc. cit.*).

(*iii*) *Selenium compounds*

1,5-**Diselenocane** is claimed to be intermediate in the thermal conversion of poly(selenotrimethylene) to poly(diseleno)trimethylene (*G. T. Morgan* and *F. H. Burstall*, J. chem. Soc., 1930, 1497). Treatment of ethylene 1,2-diselenocyanate with ethanolic potash yields 1,2,5,6-**tetraselenocane,** m.p. 130–131° (*L. Hagelberg*, Ber., 1890, **23,** 1092); oxidation with nitric acid gave 1-oxa-2,5-diselenolane 2,5-dioxide (ethanediseleninic anhydride)

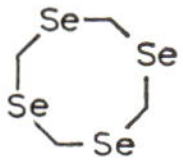

(XXIII)

(*R. Paetzold* and *D. Lienig*, Z. anorg. allg. Chem., 1965, **335,** 289). Reaction of dichloromethane with sodium selenide gives 1,3,5,7-**tetraselenocane** (XXIII), m.p. 80–81°, together with 1,3,5-triselenane and polymers (*Russo et al.*, J. polymer, Sci., Part A-1, 1966, **4,** 248).

3. Eight-membered ring compounds containing nitrogen

(*a*) *Compounds containing a single nitrogen atom*

In the same way that *1H*-azepine may be considered to be the vinylogue of pyrrole, azocine, the neutral fully conjugated eight-membered ring containing one nitrogen atom, is the vinylogue of pyridine; both are π-electron analogues of cyclo-octatetraene. Complete conjugation is not possible with eight-membered oxygen or sulphur heterocycles.

(i) Azocine and its derivatives

Azocine (II, R, R^1 = H) has been prepared by the pyrolysis of diazabasketane (I), which loses hydrogen cyanide (*D. W. McNeil et al.*, J. Amer. chem. Soc., 1971, **93**, 3817):

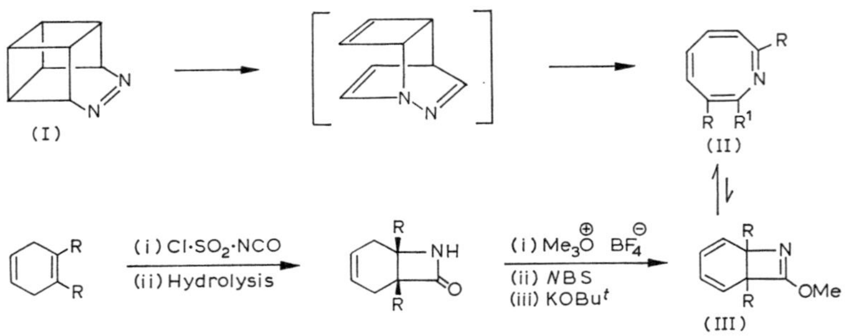

It rapidly resinifies above −50°, but 2-*methoxyazocine* (II; R = H, R^1 = OMe) and 2-*methoxy*-3,8-*dimethylazocine* (II; R = Me, R^1 = OMe), b.p. 47°/0.35 mm, are much more stable and readily available by the alternative synthetic sequence shown (*L. A. Paquette et al., ibid.*, 1968, **90,** 3897, 3898). A further alternative route to the substituted azocines IV (R = *Me*), m.p. 125°, and (R = Et) using a series of Diels–Alder and retro-Diels–Alder reactions is as follows (*R. N. Warrener et al.*, Tetrahedron Letters 1970, 1837):

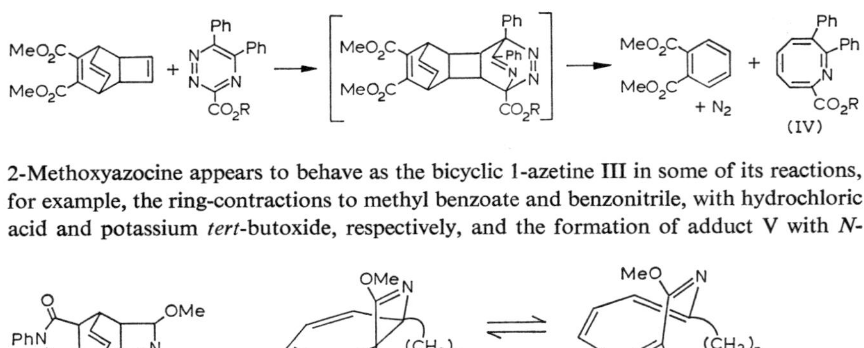

2-Methoxyazocine appears to behave as the bicyclic 1-azetine III in some of its reactions, for example, the ring-contractions to methyl benzoate and benzonitrile, with hydrochloric acid and potassium *tert*-butoxide, respectively, and the formation of adduct V with *N*-

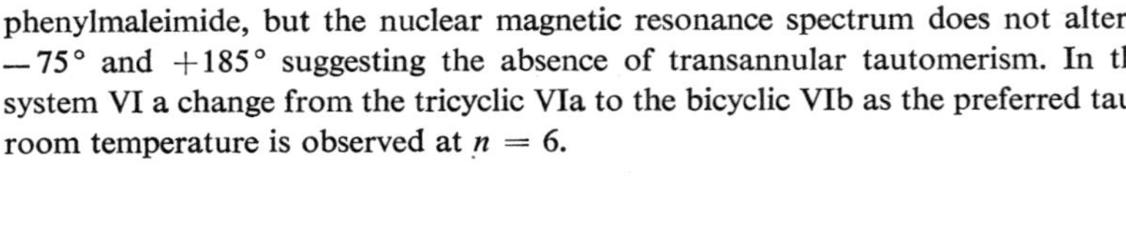

phenylmaleimide, but the nuclear magnetic resonance spectrum does not alter between −75° and +185° suggesting the absence of transannular tautomerism. In the model system VI a change from the tricyclic VIa to the bicyclic VIb as the preferred tautomer at room temperature is observed at n = 6.

The ultraviolet absorption spectrum of 2-methoxyazocine is like that of cyclo-octatetraene and tautomer VIb suggesting that it adopts the same "tub" conformation, and is non-planar and polyolefinic in character (*Paquette et al., loc. cit.*; Tetrahedron Letters, 1968, 1455; J. Amer. chem. Soc., 1971, **93,** 152).

Azocines readily undergo two-electron reduction either with potassium in tetrahydrofuran or liquid ammonia, or polarographically, to form the planar dianion VII. Both electrons add at the same potential giving no evidence for an intermediate radical anion, and the high degree of π-electron delocalisation of the aromatic 10π system is shown by the diamagnetic ring current that it sustains (*idem*, Tetrahedron Letters, 1970, 529; 533; J. Amer. chem. Soc., 1971, **93,** 161; 168).

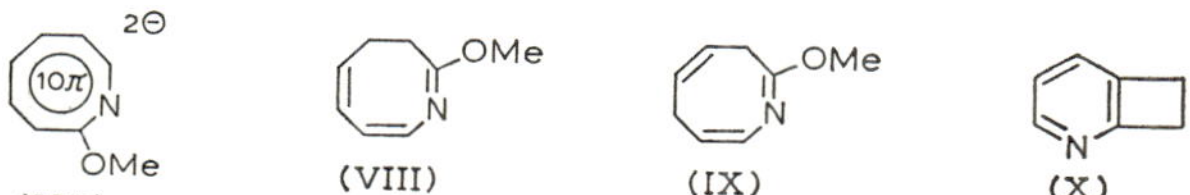

(VII) (VIII) (IX) (X)

Quenching the dianion VII with water or alcohols yields a mixture of 3,4-dihydro-(VIII), 3,6-dihydro-(IX) and 7,8-dihydro-azocines. Compound IX reverts to the more conjugated form VIII on treatment with base. Although both exist in the monocyclic form, on further treatment with potassium *tert*-butoxide, VIII undergoes ring contraction and elimination of methanol to form cyclobuteno-[2,3]-pyridine (X) (*idem, ibid.*, pp. 152, 174; J. org. Chem., 1971, **36,** 435):

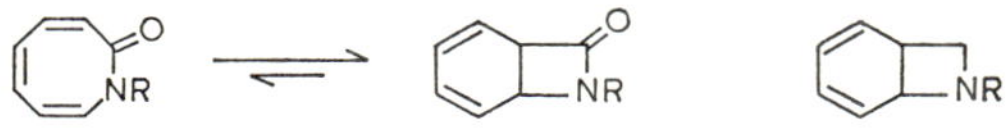

(XI) (XII) (XIII)

By contrast 1,2-dihydroazocin-2-ones (XI), prepared by treating 2-methoxyazocines (II; R^1 = OMe) with anhydrous hydrogen bromide, exist predominantly as the bicyclic compound XII; the *N*-substituted lactams show no evidence for the presence of a monocyclic form. After reduction of the carbonyl group to methylene the "1,2-dihydroazocines" *e.g.* XIII (R = SO_2R^1) are also found to possess a bicyclic structure (*Paquette et al.*, Tetrahedron Letters, 1969, 1455; J. org. Chem., 1971, **36,** 435).

1,4-Divinylazetidinones undergo a rearrangement on pyrolysis at 160–200° leading to tetrahydroazocinones (*H. Schnabel*, Angew. Chem. intern. Edn., 1969, **8,** 922) and Hofmann elimination has been used to prepare a hexahydro-2-azocinone:

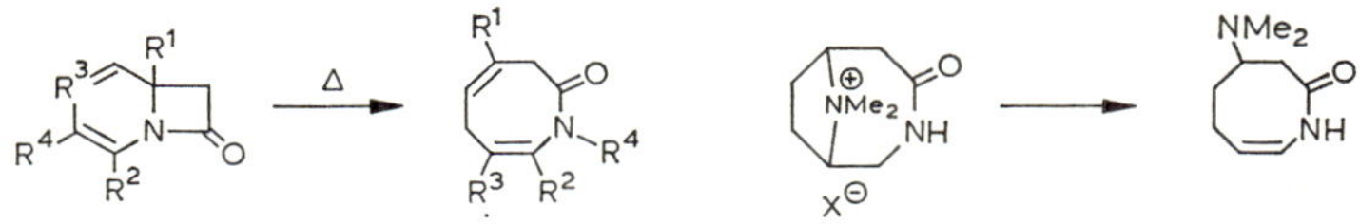

The latter reaction is also effective for nine- and ten-membered lactams (*Paquette* and *L. D. Wise, ibid.*, 1965, **30,** 228; J. Amer. chem. Soc., 1965, **87,** 1561). The imidoyl chloride XIV (R = Cl), readily obtained by the action of phosgene on cycloheptanone oxime reacts with aniline giving 2-*phenylamino*-3,4,5,6,7,8-*hexahydroazocine* (XIV; R = NHPh), m.p. 83° (*Stamicarbon N.V.*, B.P. 1,007,413/1963; C.A., 1966, **64**, 8158):

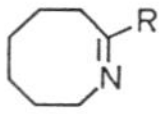
(XIV)

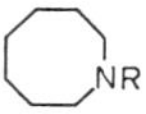
(XV)

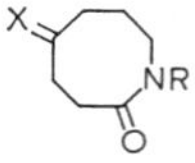
(XVI)

Azocane* **(perhydroazocine,** *azacyclo-octane*, *heptamethylene-imine*) (XV; R = H), m.p. 33–34°, b.p. 162°, n_D^{20} 1.4740, *picrate*, m.p. 147–148°, is synthesised by methods similar to those used to prepare azepane; the most convenient is reduction of azocan-2-one (XVI; R = H, X = H_2). Other routes to azocanes include the reaction between 4-toluenesulphonamide and 1,7-dibromoheptane in the presence of sodium (*A. Müller et al.*, Monatsh., 1952, **83**, 386) and cyclisation of 7-bromoheptyldimethylamine, which yields the *methobromide*, m.p. 268°, of 1-*methylazocane* (XV; R = Me), b.p. 161° (*E. R. Littman* and *C. S. Marvel*, J. Amer. chem. Soc., 1930, **52**, 287; cf. *J. Diamond et al.*, J. org. Chem. 1960, **25**, 65).

1-Methylazocane is also made by the reaction of azocane with formaldehyde and formic acid (*N. J. Leonard* and *W. K. Musker*, J. Amer. chem. Soc., 1959, **81**, 5631; *Z. Polivka*, Coll. Czech. chem. Comm., 1970, **35**, 1131) or by reduction with lithium tetrahydridoaluminate of the *N*-formyl derivative (XV, R = CHO) from the reaction of azocane and chloral (*C-J. Lu* and *F. F. Blicke*, C.A., 1958, **52**, 11086). 1-*Methylazocane 1-oxide, picrate*, m.p. 192–193°, prepared by treating XV (R = Me) with hydrogen peroxide undergoes ring-opening to *N*-methyl-*N*-(6-heptenyl)hydroxylamine on pyrolysis (*A. C. Cope* and *N. A. LeBel*, J. Amer. chem. Soc., 1960, **82**, 4656).

Other *N*-substituted azocanes are made from XV (R = H) using an alkyl halide under basic conditions (*Blicke*, U.S.P. 2,748,113/1956; C.A., 1957, **51**, 8815) and XV (R = H), like its nine- and ten-membered homologues, may be cyanomethylated to XV (R = CH_2CN), b.p. 92–94°/3 mm, n_D^{18} 1.4800 (*H. Najer et al.*, Bull. Soc. chim. Fr., 1962, 556; *O. M. Lerner* and *I. Kh. Feldman*, Zhur. prikl. Khim., 1963, **36**, 1347; *R. P. Mull et al.*, J. org. Chem., 1960, **25**, 1953; Swiss. P. 342,957/1960; C.A., 1961, **55**, 2707) and cyanoethylated to XV (R = $CH_2 \cdot CH_2 \cdot CN$), b.p. 94–97°/0.9 mm (*Mull*, U.S.P. 2,897,195/1959; C.A., 1960, **54**, 1576). It also adds ethylene oxide to give XV (R = $CH_2 \cdot CH_2OH$), b.p. 100–105°/5 mm (*L. Toldy et al.*, C.A., 1956, **63**, 7003) and undergoes Mannich reactions (*H. A. Luts* and *W. L. Nobles*, J. pharm. Sci., 1965, **54**, 67). Nitrosation of azocane gives 1-**nitrosoazocane** (XV; R = NO), b.p. 130°/16 mm, reduced to 1-**amino-azocane** (XV; R = NH_2), b.p. 98–102°/50 mm (*Sandoz*, Belg. P. 610,039/1962; C.A., 1963, **58**, 1469).

On oxidation with silver acetate at high temperatures azocane gives only pyridine, and attempts to dehydrogenate *N*-methylazocane also lead to ring-opening (*A. Hanzsch*, Ber., 1899, **32**, 575, 3109). Cyclic tertiary amines with nine- and ten-membered rings give analogous results confirming the instability of the enamine system in a medium sized ring.

Azocane-2-carboxylic acid, m.p. 168–171°, and homologues with larger rings, are made from the corresponding α-halogenolactams using the Favorskii rearrangement; de-

* This shorter trivial name, adopted in Chemical Abstracts, is not accepted by I.U.P.A.C. (see Rule B1) where the approved name is "perhydroazocine".

carboxylation occurs in refluxing cycloheptanone (*J. A. Elberling et al.*, Tetrahedron Letters, 1966, 5393; J. med. Chem., 1971, **14,** 501).

Azocan-2-one (*ζ-oenantholactam, suberone isoxime*) (XVI; R = H, X = H_2), m.p. 28–30°, b.p. 162–164°/22 mm, is obtained by the ring-expansion of cycloheptanone or its oxime under Schmidt or Beckmann conditions (*Ruzicka et al., loc. cit.*; Helv., 1949, **32,** 544; *Blicke* and *N. J. Doorenbos*, J. Amer. chem. Soc., 1954, **76,** 2317; *A. F. McKay et al., ibid.*, 1958, **80,** 1510; *W. Z. Heldt, ibid.*, 5880; *R. Huisgen et al.*, Chem. Ber., 1957, **90,** 1437; *A. Novotny*, Chem. Listy, 1958, **52,** 718; *J. Rakoczi*, Hung. P. 151,996/1964; C.A., 1965, **63,** 8210; cf. *S. Hunig et al.*, Chem. Ber., 1967, **100,** 3024, 3039). 2-*Methylazocan*-8-*one*, b.p. 94–98°/0.05 mm, reduced to 2-*methylazocane*, b.p. 62–64°/15 mm, n_D^{20} 1.4665, is prepared similarly (*H. J. Barber* and *E. Lunt*, J. chem. Soc., 1960, 1187). In the Schmidt reaction the fused ring tetrazole is also obtained (*A. H. Beckett et al.*, J. Pharm. Pharmacol, 1956, **8,** 860).

The reactions of 1-chloro-1-nitrosocycloheptane with triphenyl phosphine, (*M. Ohno* and *I. Sakai*, Tetrahedron Letters, 1965, 4541) cycloheptanecarboxylic acid with nitroso-sulphuric acid (*K. Smeykal et al.*, J. pr. Chem., 1965, **30,** 126) and 1-carboxy-1-hydroxyl-aminocycloheptane with oleum (*E. F. J. Duynstee et al.*, Rec. Trav. chim., 1965, **84,** 1442) are alternative similar routes to azocan-2-one. A further method, also effective for nine- to twelve-membered lactams is the vacuum depolymerisation of 7-aminoheptanoic acid (*T. I. Shein* and *V. N. Topchibasheva*, C.A., 1960, **54,** 7216).

Methylation of ζ-oenantholactam yields both the O-*methyl* (XIV; R = OMe), b.p. 85°/12 mm, n_D^{18} 1.4742, and N-*methyl* (XVI; R = Me, X = H_2) b.p. 139°/12 mm, n_D^{18} 1.4940, derivatives. The latter polymerises despite the lack of an *N*-hydrogen atom (*A. A. Strepikheev* and *Z. A. Rogovin*, C.A., 1963, **59,** 7656) and the former is reduced electrolytically to azocane (*F. Nagy et al.*, C.A., 1971, **75,** 29255). *N-Phenylazocan*-2-*one* (XVI; R = Ph, X = H_2), b.p. 145–154°/0.4 mm, is prepared by the action of peracetic acid on phenyliminocycloheptane (*M. Fischer*, Chem. Ber., 1969, **102,** 342). On photolysis it gives 1,2,3,4,5,6,7,8-*octahydro*-1-*benzazecin*-8-*one*, m.p. 68°.

Azocane-2-thione, m.p. 86°, is prepared by the action of phosphorus pentasulphide (*H. K. Hall*, J. Amer. chem. Soc., 1958, **80,** 6404; *Ruzicka et al.*, Helv., 1935, **18,** 659) or thioacetic acid (*H. Behringer* and *G. F. Grunwald*, G.P. 943,227/1956; C.A., 1959, **53,** 6262) on the lactam (XVI; R = H, X = H_2) or the action of hydrogen sulphide on the lactim sulphonate (XIV; R = OSO_2Ph) prepared from cycloheptanone oxime and benzenesulphonyl chloride in pyridine (*Huisgen* and *J. Witte*, G P. 1,024,508/1958; C.A., 1960, **54,** 19453; cf. *Hünnig et al.*, Angew. chem. Intern. Edn., 1963, **2,** 213). Electrolytic reduction of the thiolactam also gives azocane (*H-Y. Chem et al.*, C.A., 1964, **61,** 8274).

Several azocanones have been made using the Dieckmann cyclisation of a di(alkoxy-carbonylalkyl)amine followed by hydrolysis and decarboxylation (*Leonard et al.*, J. Amer. Soc., 1952, **74,** 1704; 1954, **76,** 630; 1955, **77,** 6241; 1958, **80,** 4858; *Diamond, loc. cit.*). Sixteen-membered ring diaminodiones are also obtained. 1-*Benzoyl*-4-*azocanone*, m.p.

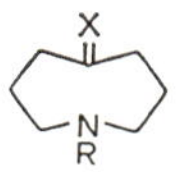

(XVII)

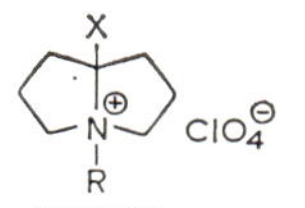

(XVIII)

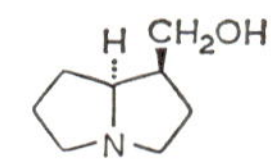

Isoretroneconal

(XIX)

64–65°, and 1-*benzoyl-5-azocanone* (XVII; R = COPh, X = O), m.p. 124–126°, are obtained by microbial oxygenation of XV (R = COPh) (*R. A. Johnson et al.*, J. org. Chem., 1968, **33,** 3187); similar ketones are formed from 1-benzoylazonane.

Both Clemmensen and Wolff–Kishner reductions of azocanones are accompanied by ring-opening (*Leonard* and *R. C. Sentz*, J. Amer. chem. Soc., 1952, **74,** 1704).

Abnormally low carbonyl stretching frequencies in their infrared absorption spectra indicate a transannular interaction in 5-azocanones and analogous nine- and ten-membered ring compounds *(vide infra)*. If ketones XVII (X = O) are treated with perchloric acid a 1,5-bond is formed and quaternary salts of type XVIII (X = OH) are obtained. A neat stereospecific synthesis of the alkaloid **isoretroneconal** (XIX) from 1-benzyl-4-ethoxy-carbonyl-5-azocanone exploits this observation (*Leonard* and *T. Sato*, J. org. Chem., 1969, **34,** 1066). Two similar reactions are the formation of XVIII (R = CH_2Ph, X = Me) from XVII (R = CH_2Ph, X = CH_2), and XVIII (R = $CH_2 \cdot CN$, X = $OCH_2 \cdot CH_2OH$) from XVII (R = $CH_2 \cdot CN$, X = —$OCH_2 \cdot CH_2O$—) with perchloric acid (*Johnson et al., ibid.* 1972, **37,** 312), and allied transannular reactions are the formation of oxime XVI (R = H, X = NOH) by irradiating *N*-nitrosolactam XVI (R = NO, X = H_2), and the formation of 5-chloroazocan-2-one (XVI; R = H, X = H, Cl) in the photolysis of *N*-chlorolactam XVI (R = Cl, X = H_2) (*O. E. Edwards et al.*, Canad. J. Chem., 1967, **45,** 1287; 1971, **49,** 1648). The alcohol XVII (R = cyclohexyl, X = H, OH) forms the salt XVIII (R = cyclohexyl, X = H) by transannular dehydration under these conditions (*A. J. Sisti* and *D. L. Lohner*, J. org. Chem., 1967, **32,** 2026).

(*ii*) *Azocines with one annellated ring*

(*1*) 1-**Benzazocines.** 2-*Methoxy*-1-*benzazocine* (XX), an oil, is prepared by the action of trimethyloxonium tetrafluoroborate on 1,2-*dihydro*-1-*benzazocin*-2-*one* (XXI; R = H), m.p. 171–172°, itself the product of the solvolytic ring-expansion of the *O*-benzene-sulphonate of 1-benzotropone oxime. *N*-Methylation and reduction of XXI (R = H) with tetrahydrido-aluminate or -borate gave 1-*methyl*-1,2-*dihydro*-1-*benzazocine*, b.p. 85°/0.1 mm, which isomerises to the 1-*methyl*-1,6-*dihydro*-1-*benzazocine*, v.p. 190°/0.1 mm, with potassium *tert*-butoxide. Catalytic reduction of XXI (R = Me) gave *lactam* XXII (R = Me), m.p. 60–62° (*R. M. Coates* and *E. F. Johnson*, J. Amer. chem. Soc., 1971, **93,** 4016; *Paquette et al., ibid.*, 1972, **94,** 4907):

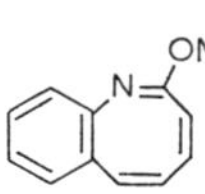

(XX)

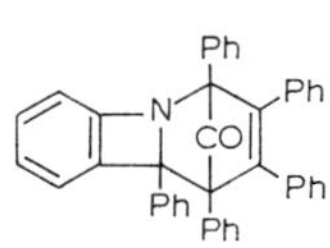

(XXIII)

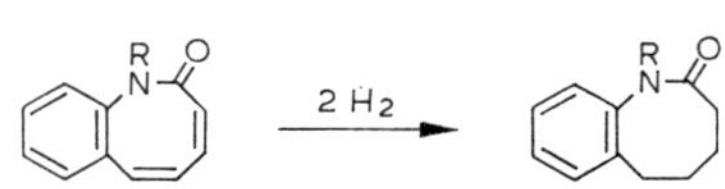

(XXI) (XXII)

2,3,4,5,6-**Pentaphenylbenzazocine** is obtained with the elimination of carbon monoxide from the Diels–Alder adduct (XXIII) of 2-phenylbenzazete and tetracyclone (*C. Rees*, personal comm.).

A 1,2-dihydro-1-benzazocine is obtained in the ring-expansion shown; in the absence of the ethoxy group only the tricyclic adduct XXIV is formed (*G. R. Proctor* and *W. I. Ross*, J. chem. Soc. Perkin I, 1972, 889):

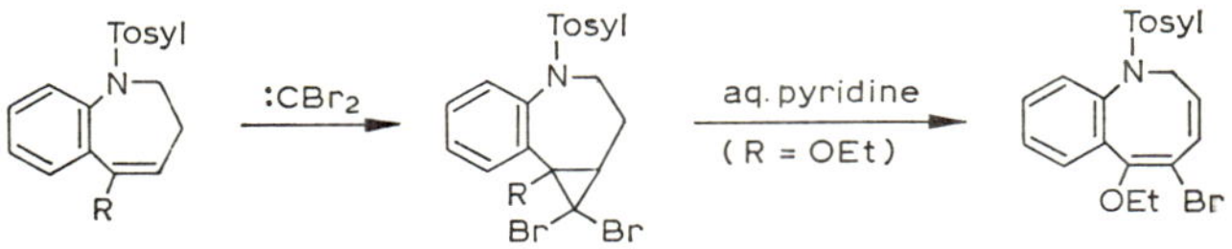

(XXIV)

1-*Methyl*-3,4-*di(methoxycarbonyl)*-1,6-*dihydro*-1-*benzazocine*, m.p. 132–133°, results when the tricyclic adduct of 1-methyl-1,4-dihydroquinoline and dimethyl acetylenedicarboxylate is refluxed in benzene (*P. G. Lehman*, Tetrahedron Letters, 1972, 4863). 5-*Hydroxy*-3,4-*dihydro*-1-*benzazocine* (XXV), m.p. 101–103°, and several substituted analogues are obtained by irradiating 1-anilinocyclohexen-3-ones. The stability of the enol form is attributed to 6π pseudoaromaticity in the heterocyclic ring (*K. Yamada et al., ibid.*, 1972, 2513).

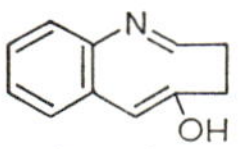

(XXV)

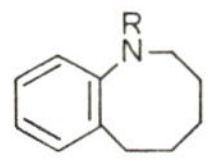

(XXVI)

2-*Methoxy*-3,6-*dihydro*-1-*benzazocine*, *perchlorate*, m.p. 137–138°, is obtained by quenching the dianion of XX, formed on treatment with potassium in liquid ammonia (*Paquette, loc. cit.*).

If the ring-expansion of adduct XXIV is brought about by using silver nitrate in aqueous ethanol the product is 5-bromo-1-tosyl-1,2,3,5-tetrahydro-1-benzazocin-6-one (*Proctor* and *Ross*, *loc. cit.*).

1,2,3,4,5,6-**Hexahydro**-1-**benzazocine** (XXVI; R = H), b.p. 130–150°/15 mm, is obtained by reducing the lactam XXII (R = H) with lithium tetrahydridoaluminate. Reduction of its *N*-formyl derivative (XXII; R = CHO) yields the 1-*methyl* homologue (XXVI; R = Me), b.p. 135–145°/14 mm, n_D^{25} 1.5404, also obtained by cyclising 1-chloro-3-(5-methylaminopentyl)benzene with phenyl-lithium (*R. Huisgen* and *H. Konig*, Chem. Ber., 1959, **92,** 429). Compound XXVI (R = H), also prepared by the treatment of benzo[*b*]cycloheptenone oxime with lithium tetrahydridoaluminate (*J. Humbert* and *A. Laurent*, Compt., rend., 1971, **272C,** 1165) is readily acylated (*D. Zivkovic*, Ger. Offen. 1,906,593/1968; C.A., 1969, **71,** 124523).

1,2,3,4,5,6-**Hexahydro**-1-**benzazocin**-2-**one** (*benzsuberone isoxime*) (XXII; R = H), m.p. 154–155°, and its substituted derivatives are made by Schmidt reactions on benzo[*b*]-cycloheptenones (*P. A. S. Smith* and *W. L. Berry*, J. org. Chem., 1961, **26,** 27; *Hjelte* and *T. Agback*, Acta Chem. Scand., 1964, **18,** 191) or by Beckmann rearrangement of their oximes (*R. T. Conley* and *L. J. Frainier*, J. org. Chem., 1962, **27,** 3844; *Huisgen et al.*, Ann., 1951, **574,** 171; 1954, **586,** 30; cf. Chem. Ber., 1957, **90,** 1844; *D. H. Jones et al.*, J. chem. Soc., C, 1969, 2177); the Schmidt reaction also gives a tetrazole. The N-*nitroso* derivative (XXII; R = NO), m.p. 54° (decomp.), rearranges rapidly in the presence of β-naphthol to a ring-opened azo-dyestuff. Compound XXII (R = H) is readily *N*-alkylated in the presence of sodamide (*Olin Mathieson Chem. Corp.*, Neth. Appl. 6,516,320/1966; C.A., 1966, **65,** 15354; UCB, Neth. Appl. 6,514,240/1966; C.A., 1966, **65,** 12181) and with thioacetic acid it is converted into 1,2,3,4,5,6-**hexahydro-1-benzazocin-2-thione,** m.p. 129° (*Behringer* and *Grunwald, loc. cit.*).

Dieckmann cyclisation of the diester XXVII with sodium hydride followed by acid hydrolysis gave the tosyl derivative of 1,2,3,4,5,6-**hexahydro-1-benzazocin-6-one** (XXVIII; R = H), also available from the irradiation of 1-phenyl-2-piperidone *(Fischer, loc. cit.)*.

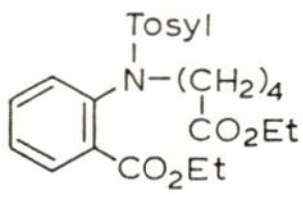

(XXVII)

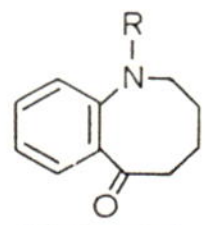

(XXVIII)

Bromine brominates XXVIII (R = tosyl) first at position 5 (α to the carbonyl group) then at positions 8 and 10 (*ortho* and *para* to the nitrogen on the benzene ring). Reduction with sodium tetrahydridoborate and detosylation with sodium in liquid ammonia yields 6-*hydroxy*-1,2,3,4,5,6-*hexahydro*-1-*benzazocine*, m.p. 86–88°, 1-*tosylate*, m.p. 166–168° (*Proctor* and *Ross*, J. chem. Soc., Perkin I, 1972, 885).

1,2,3,4,5,6-**Hexahydro-1-benzazocine-2,6-dione** (XXIX), m.p. 170–171°, is obtained in the following rearrangement (*B. Witkop et al.*, Experientia, 1950, **6**, 461; J. Amer. chem. Soc., 1951, **73**, 2196, 2641):

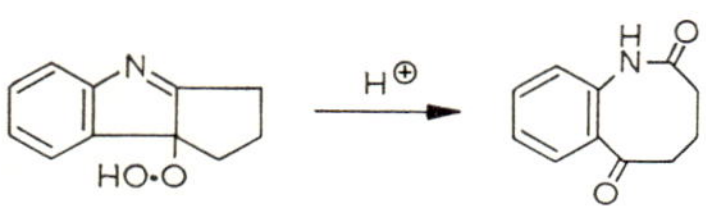

(XXIX)

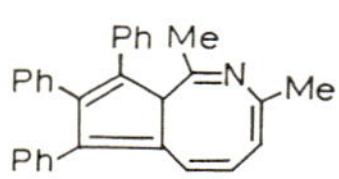

(XXX)

(*2*) **2-Benzazocines and analogues.** Irradiation of 2,3,4-triphenyl-diazocyclopentadiene in the presence of 2,6-lutidine gives the cyclopentadienoazocine XXX after a (1,5)-sigmatropic shift in the intermediate carbene–lutidine adduct (*H. Dürr et al.*, Chem. Comm., 1972, 1257).

The dihydro-2-benzazocinone XXXI is also the product of a thermal rearrangement (*A. Padwa et al.*, Chem. Comm., 1972, 680).

(XXXI)

Cyclisation of the ester XXXII with phosphorus pentoxide followed by hydrolysis gives 1,2,3,4,5,6-*hexahydro*-2-*benzazocine* (XXXIII; R = H), b.p. 123°/13 mm, *picrate*, m.p. 175° (*J. von Braun* and *O. Bayer*, Ber., 1927, **60**, 1257). The N-*methyl* derivative (XXXIII; R = Me), b.p. 120–122°/9 mm, is obtained in the ring-expansion of the 1,1-dimethyl-2-phenylpyrrolidinium cation with sodamide (*G. C. Jones* and *C. R. Hauser*, J. org. Chem., 1962, **27**, 3572). Reaction of 2-(4-chlorobutyl)benzyl chloride with ethanolamine gives 2-(2-*hydroxyethyl*)-2-*benzazocine* (XXXIII; R = $CH_2 \cdot CH_2OH$), b.p. 122–124°/0.3 mm (*B. Belleau*, J. med. pharm. Chem., 1959, **1**, 343) and analogues are made similarly (*D. H. Jones et al.*, J. chem. Soc. C 1969, 2176).

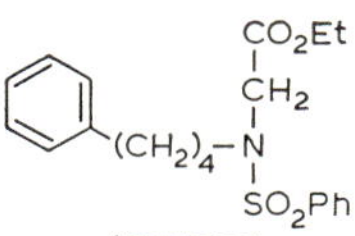

(XXXII)

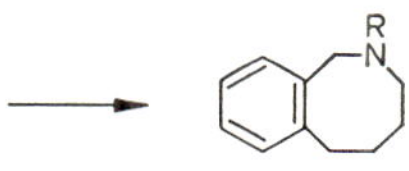

(XXXIII)

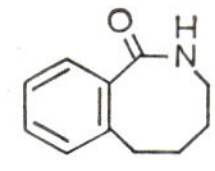

(XXXIV)

About 20% alkyl migration takes place in the Schmidt reaction with benzo[*b*]cycloheptenone yielding 1,2,3,4,5,6-*tetrahydro*-2-*benzazocin*-1-*one* (XXXIV), m.p. 122–123°, as well as the 80% aryl migration leading to the 1-benzazocin-2-one. This ratio is unaltered by the presence of a methoxy group in the benzene ring (*M. Tomita et al.*, *ibid.*, 1969, 183).

(*3*) **3-Benzazocines.** 4-*Methoxy*-3-*benzazocine* (XXXV), m.p. 77°, is prepared by an analogous route to that used for the 1-benzazocine (XX) *(vide supra)*; the intermediate 3,4-*dihydro*-3-*benzazocin*-4-*one* has m.p. 202–203° (*Paquette et al.*, J. Amer. chem. Soc., 1972, **94,** 4907). Reduction *via* the dianion yields 4-methoxy-5,6-dihydro-3-benzazocine.

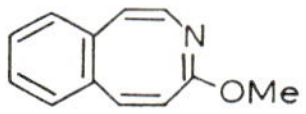

(XXXV)

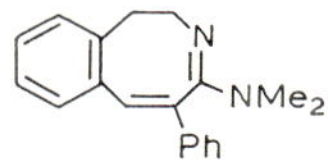

(XXXVI)

4-Dimethylamino-5-phenyl-1,2-dihydro-3-benzazocine (XXXVI) is formed in the ring-opening reaction of the adduct from 1-dimethylamino-2-phenylacetylene and 3,4-dihydroisoquinoline (*H. G. Viehe et al.*, Angew. Chem. intern. Edn. 1966, **5,** 585).

A route to 8,9-*dimethoxy*-1,2,3,4-*tetrahydro*-3-*benzazocine*, *hydrochloride*, m.p. 226–227°, is shown below. Catalytic reduction saturates the 5,6-double bond (*J. Kobor* C.A., 1969, **70,** 28809):

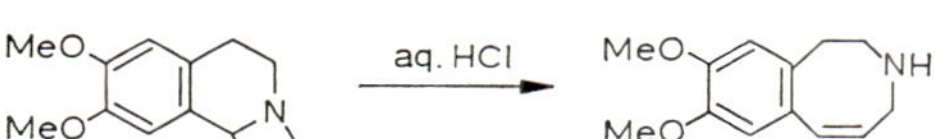

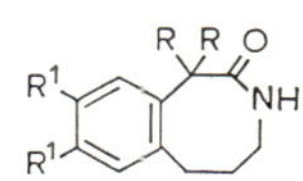

(XXXVII)

1,2,3,4,5,6-*Hexahydro*-3-*benzazocine*, *hydrochloride*, m.p. 230–233°, its 3-*methyl* homologue, *hydrochloride*, m.p. 169–170°, and analogues are obtained by thermal cyclisation of 3-[2-(2-methylaminoethyl)phenyl]propionic acid, followed by reduction of the lactam 4-carbonyl group by diborane and removal of the *N*-alkyl group with cyanogen bromide followed by hydrolysis. An alternative route *via* 2-(3-benzylaminopropyl)phenylacetic acid and the 2-lactam has also been used (*A. Brossi et al.*, Helv., 1970, **53,** 763; 1971, **54,** 743). It is cyanoalkylated in the presence of base (*Hjelte*, Fr. M., 2710/1964; C.A., 1965, **63,** 14832).

3-*Tosyl*-1,2,3,4,5,6-*hexahydro*-3-*benzazocin*-1-*one*, m.p. 119–121°, is prepared in a Friedel–Crafts cyclisation using three moles of aluminium chloride at 15° (*J. Schlademann* and *R. Partsch*, J. chem. Soc. Perkin I, 1972, 213), and 8,9-*dimethoxyhexahydro*-3-*benzazocin*-2-*one* (XXXVII; R = H, R^1 = OMe), m.p. 215°, is one of the photolysis products of the *N*-chloroacetyl derivative of 3(3,4-dimethoxyphenyl) propylamine (*O. Yonemitsu et al.*, Chem. Comm., 1971, 745). Cyclodehydration of *N*-benziloyl-3-phenylpropylamine yields 1,1-*diphenyl* 1,2,3,4,5,6-*hexahydro*-3-*benzazocin*-2-*one* (XXXVII; R = Ph, R^1 = H), m.p. 174–176°; this cyclisation is also effective for nine- to thirteen-membered 3-benzo-heterocycles (*V. S. Shklayaev et al.*, Zhur. org. Khim., 1968, **4,** 1046; cf. Zhur. obshcheĭ Khim., 1953, **23,** 853). 1,2,3,4,5,6-*Hexahydro*-3-*benzazocin*-4-*one*, m.p. 146–147°, results from Schmidt or Beckmann reactions on benzo[*d*]cycloheptenone and its oxime. Its sodio derivative is alkylated by alkyl halides and the carbonyl group is reduced to methylene by lithium tetrahydridoaluminate (*Hjelte* and *Agback*, *loc. cit.*; *D. H. Jones et al.*, J. chem. Soc. C, 1969, 2176).

(*iii*) *Azocines with more than one annellated ring*

Beckmann ring-expansion of dibenzo[*b,e*]cycloheptadienone gives 5,6,7,12-*tetrahydrodibenzo*[b,e]*azocin*-6-*one* (XXXVIII; X = O), m.p. 233–235°. Alkaline hydrolysis of this lactam gives an amino acid, but reduction with lithium tetrahydridoaluminate gives 5,6,7,12-**tetrahydrodibenzo[*b,e*]azocine** (XXXVIII; X = H_2), m.p. 59–60°, which is *N*-alkylated by alkyl halides in the presence of sodamide or lithium hydride (*M. Protiva et al.*, Monatsh., 1965, **96,** 182; Czech. P. 120195/1964; C.A., 1968, **68,** 78162; **69,** 96512). Oxidation of indenoindole (XXXIX) with periodate yields 5,6,7,12-*tetrahydrodibenzo*-[b,e]*azocine*-6,12-*dione*, m.p. 210–212° (*L. J. Dolby* and *P. D. Lord*, J. org. Chem., 1969, **34,** 2988).

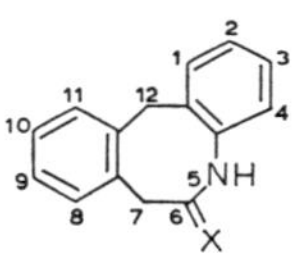

(XXXVIII)

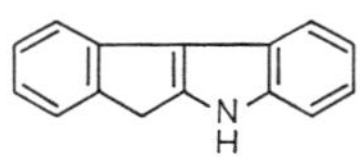

(XXXIX)

6-*Methoxydibenz*[b,f]*azocine* (XL; R = OMe), m.p. 91–93°, is obtained from 11,12-*dihydro*-6-*methoxydibenz*[b,f]*azocine*, m.p. 82–84°, by successive treatment with *N*-bromosuccinimide and potassium *tert*-butoxide. Attempts at chemical reduction of XL (R = OMe) at room temperature gave the 2,3:3,2-indenoindole XLI (R = H), but at low temperatures the product was 5,6,11,12-**tetrahydrodibenz[*b,f*]azocine** (XLII; R, R^1 = H), m.p. 132–133°, which also results from the reduction of 5,6,11,12-*tetrahydrodibenz*[b,f]-*azocin*-6-*one* (XLIII; R = H), m.p. 250–252°, with lithium tetrahydridoaluminate (*L. A. Paquette et al.*, J. Amer. chem. Soc., 1972, **94,** 4907).

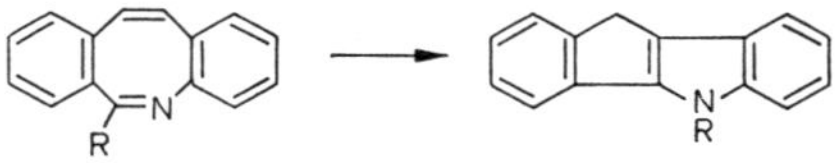

(XL) (XLI)

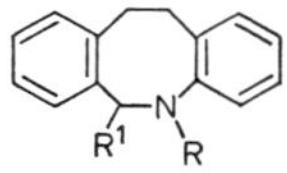

(XLII)

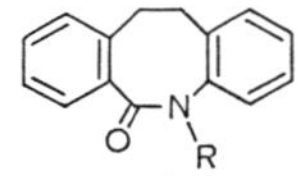

(XLIII)

6-*Phenyldibenzo*[b,f]*azocine* (XL; R = Ph), m.p. 94–95° or 117–119°, is prepared by the photochemical decomposition of azide XLIV; thermolysis of the azide gives anthracene derivatives. If compound XL (R = Ph) is reduced with potassium in liquid ammonia it yields 6-*phenyl*-11,12-*dihydrodibenz*[b,f]*azocine* (XLV; R = Ph), m.p. 94–95°, the con-

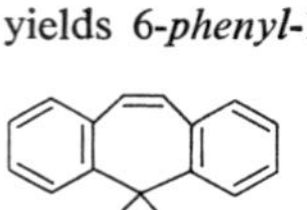

(XLIV)

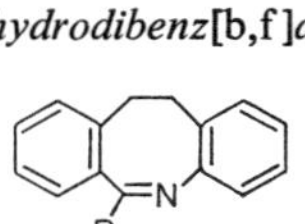

(XLV)

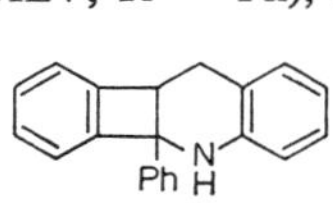

(XLVI)

trasting tetracyclic compound XLVI, and 6-*phenyl*-5,6,11,12-*tetrahydrodibenz*[b,f]*azocine* (XLII; R = H; R^1 = Ph), m.p. 105–106°, *hydrochloride*, m.p. 242° decomp. (*J. J. Looker*, J. org. Chem., 1971, **36,** 1045; 2681), also obtained by cyclising the benzoyl derivative of 2-(2-phenylethyl)aniline with polyphosphoric acid and reducing the product (*Protiva et al.*, Coll. Czech. chem. Comm., 1968, **33,** 1880). The heterocyclic ring both of XL (R = Ph) and its dihydro derivative (XLV; R = Ph) is opened by acid to give amino-ketones.

5,6-*Dihydrodibenz*[b,f]*azocin*-6-*one* (XLVII; R = H, X = O), m.p. 263–264°, prepared by Beckmann rearrangement of dibenzo[*b,f*]cycloheptatrienone oxime, is reduced by lithium tetrahydridoaluminate to **5,6-dihydrodibenzo[*b,f*]azocine** (XLVII; R = H, X = H_2), m.p. 108–109° (*W. Th. Nauta et al.*, Arzneim. Forsch., 1964, **14,** 16; *F. Sowinski* and *N. L. Yale*, *ibid.*, p. 117; U.S.P. 3,448,102/1965; C.A., 1967, **71,** 49802). The lactam gives its 5-*methyl* homologue (XLVII; R = Me, X = O), m.p. 127–129°, with methyl iodide and base. This is first reduced to 5-*methyl*-5,6-*dihydrodibenz*[b,f]*azocine* (XLVII; R = Me, X = H_2), m.p. 64–65°, and subsequently to 5-*methyl*-5,6,11,12-*tetrahydro-dibenz*[b,f]*azocine* (XLII; R = Me, R^1 = H), m.p. 56–57°, with hydrogen over platinum oxide (*R. M. Coates* and *E. F. Johnson*, J. Amer. chem. Soc., 1971, **93,** 4016). Other dibenzazocines and dibenzazocinones with a secondary nitrogen atom may also be alkylated and acylated (*A. F. Harms*, B.P. 969,027/1961; C.A., 1964, **61,** 14646).

Lactam XLIII (R = H) is prepared by Beckmann rearrangement of dibenzo[*b,f*]cyclo-heptadienone oxime (*A. M. Monro et al.*, J. med. Chem., 1963, **6,** 255); reduction of the oxime with lithium tetrahydridoaluminate gives XLII (R, R^1 = H) directly (*C. van der Steldt*, Belg. P. 616,983–4/1961; C.A., 1963, **58,** 3445). Alternatively 2-(2-phenylethyl)-phenyl isocyanate may be cyclised with aluminium chloride to XLIII (R = H) (*F. Hunziker et al.*, Helv., 1966, **49,** 985; U.S.P. 3,367,930/1964; C.A., 1968, **69,** 27464).

N-Substituted indenoindoles of type XLI (R = alkylamino) on ozonolysis yield 5,6,11,12-tetrahydrodibenzo[*b,f*]azocine-6,12-diones, reduced to 5,6-dihydrodibenz[*b,f*]azocines (XLVII; X = H_2) with lithium tetrahydridoaluminate (*T. Okomoto et al.*, C.A., 1970, **73,** 45556).

1,3-bis(2-Aminophenyl)propane diphosphate is cyclised by heating or treatment with *p*-toluenesulphonic acid to **5,10,11,12-tetrahydrodibenzo[*b,g*]azocine** (XLVIII), b.p. 140°/0.15 mm, m.p. 55–57°. It is *N*-alkylated by an alkyl halide and sodamide, by butyl-lithium with a toluenesulphonyl ester or ethylene oxide, or by pyrolysis of the product from phosgene and an alcohol (*R. M. Jacob* and *J. C. L. Fouche*, B.P. 926,335/1959; G.P. 1,180,751/1960; B.P. 983,859/1961; C.A., 1964, **61,** 1843; 1965, **62,** 13131, **63,** 1775, Ind. chim. Belg., 1967, **32,** 226; 1969, **70,** 68105).

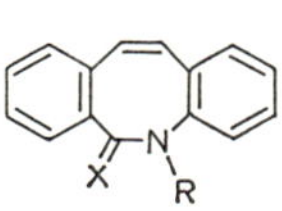

(XLVII)

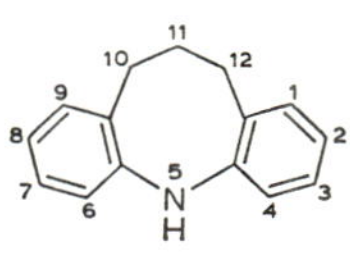

(XLVIII)

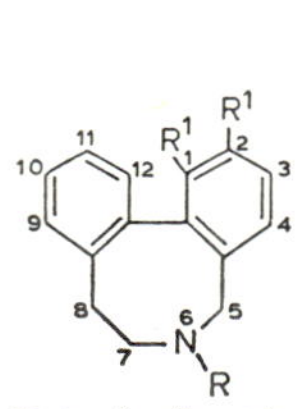

Apogalanthamine
(R = Me; R^1 = OH)
(XLIX)

The dibenzo[*c,e*]azocine system is present in **apogalanthamine** (XLIX; R = Me, R^1 = OH), m.p. 202° decomp., obtained by treating galanthamine with hydrobromic acid; the structure has been confirmed by synthesis (*S. Kobayashi* and *S. Uyeo*, J. chem. Soc., 1957, 638; Chem. Pharm. Bull. Japan, 1964, **12,** 696; cf. *B. Frank* and *H. J. Lubs*, Ann., 1968, **720,** 131). With hydrochloric acid or hydrazine and potassium hydroxide the 1-*hydroxy*-2-*methoxy* analogue of XLIX, *hydrochloride*, m.p. 163–166°, is obtained (*L. Bubewa-Iwanova*, Chem. Ber., 1962, **95,** 1348; *A. Abdusamatov* and *S. Y. Yanusov*, Chim. prir. Soedin., 1969, **5,** 194; C.A., 1969, **71,** 128624; *T. Kametani et al.*, J. chem. Soc., C, 1971, 590).

Irradiation of *N*-(2-phenylethyl)-2-iodobenzylamine gives 5,6,7,8-**tetrahydrodibenz**[*c,e*]-**azocine** (XLIX; R, R^1 = H), m.p. 119–120°; *hydrochloride*, m.p. 321–322° (*P. W. Jeffs* and *J. F. Hansen*, J. Amer. chem. Soc., 1967, **89,** 2798). Alternative routes to XLIX (R, R^1 = H) are reductive debenzylation of the 6-*benzyl* derivative (XLIX; R^1 = H, R = CH_2Ph), *hydrobromide*, m.p. 210–211°, prepared by cyclising 2-(2-bromoethyl)-2′-bromomethylbiphenyl with benzylamine (*S. R. Ahmed* and *D. M. Hall*, J. chem. Soc., 1959, 3383), and reduction of 5,6,7,8-*tetrahydrodibenz*[c,e]*azocin*-7-*one*, m.p. 158–161°, obtained by Schmidt or Beckmann ring-expansion of dibenzo[*c,f*]cycloheptadienone. 6-*Methyl*- (*hydrobromide*, m.p. 221–222°) and 6-*ethyl*- (*hydrobromide*, m.p. 225°) homologues and other derivatives have also been prepared (*K. Kotera et al.*, Jap. P. 2626, 2632, 3627/1966; C.A., 1968, **68,** 105398; 1966, **64,** 14175, 19576, 19578; *E. J. Forbes* and *C. J. Gray*, Tetrahedron, 1968, **24,** 2795). The 2-monoxime of dibenzo[*d,f*]cycloheptadien-1,2,3-trione under Beckmann conditions yields 5,6,7,8-*tetrahydrodibenzo*[c,e]*azocin*-5,7,8-*trione*, m.p. 186–188° (*M. Milosev* and *B. Aleksiev*, Monatsh., 1971, **102,** 88).

5,6,7,12-**Tetrahydrodibenzo**[*c,f*]**azocine** (L; R = H), m.p. 124–125°, is prepared by the thermal cyclisation of 2,2′-bis(aminomethyl)diphenylmethane and its hydrochloride, or in better yield, by the decyanation of the 6-*cyano* derivative (L; R = CN), m.p. 239–240°, or the reductive debenzylation of the 6-*benzyl* derivative (L; R = CH_2Ph), m.p. 113–114°, both obtained from 2,2′-bis(bromomethyl)diphenylmethane. Other *N*-substituted derivatives are made similarly (*G. Pala et al.*, Tetrahedron, 1970, **26,** 1275; J. med. Chem., 1968,

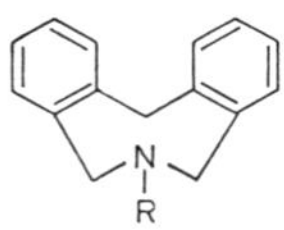

(L)

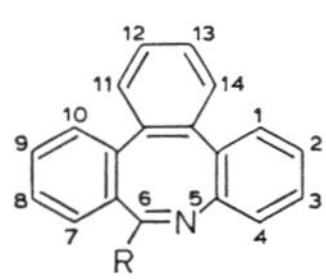

(LI)

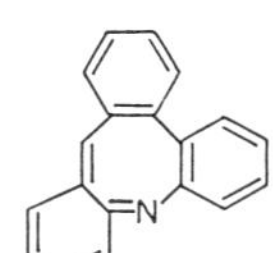

(LII)

11, 97; 1971, **14,** 174). The dibromide with *tert*-butyl carbazate yields compound L (R = $NHCO_2Bu^t$) m.p. 104–105°, which on acid hydrolysis and decarboxylation gives 6-*amino*-5,6,7,12-*tetrahydrodibenzo*[c,f]*azocine* (L; R = NH_2), m.p. 114–116°. The *tosyl* derivative (L; R = NHtosyl), m.p. 146–149°, contracts in alkali to dibenzo[*a,d*]cycloheptadiene (*L. A. Carpino et al.*, J. org. Chem., 1969, **34,** 2009). 5,6,7,12-*Tetrahydrodibenz*[c,f]*azocine*-5,7,12-*trione*, m.p. 248–252°, is prepared by treating the anhydride of benzophenone-2,2′-dicarboxylic acid with aqueous ammonia (*C. Graebe* and *P. Juillard*, Ann., 1887, **242,** 214; *W. Baker et al.*, Chem. Ber., 1959, **92,** XXXVII).

The only two possible tribenzazocines are those having [*b,d,f*] (LI) and [*b,d,g*] (LII)

annellated systems. As the latter includes a quinonoid ring it is not surprising that no examples are known.

6-**Methoxytribenz**[*b,d,f*]**azocin** (LI; R = OMe), m.p. 149–152°, is made by the action of trimethyloxonium tetrafluoroborate on 5,6-*dihydrotribenz*[b,d,f]*azocin-6-one*, m.p. 259–264°, which results from a Beckmann rearrangement of tribenzotropone oxime (*Paquette et al.*, J. Amer. chem. Soc., 1972, **94**, 4907).

(*b*) *Eight-membered ring compounds containing two nitrogen atoms*

(*i*) *1,2-Diazocines*

1,2-**Diazocine** (LIII) has been prepared in the photochemical reaction shown:

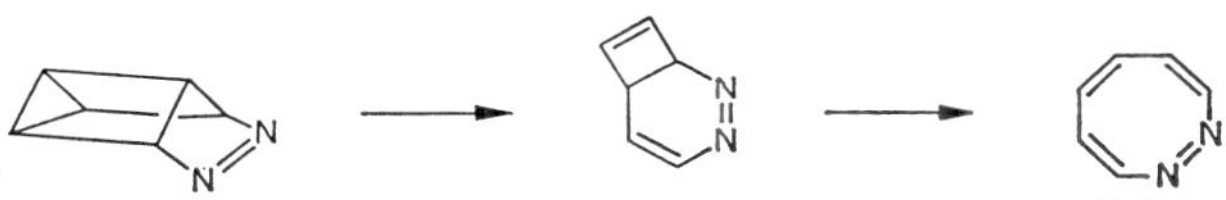

(LIII)

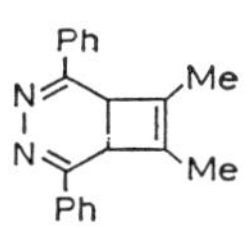

(LIV)

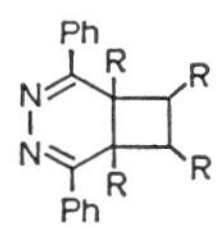

(LV)

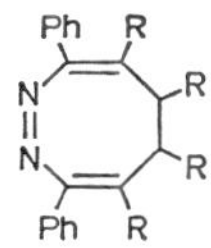

(LVI)

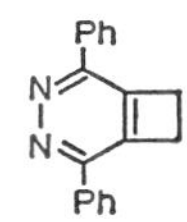

(LVII)

As expected, it does not show a diamagnetic ring current. It is stable in solution below room temperature but on further photolysis gives benzene (*B. M. Trost* and *R. M. Corey*, J. Amer. chem. Soc., 1971, **93,** 5573). By contrast, the bicyclic form LIV is preferred to 5,6-dimethyl-3,8-diphenyl-1,2-diazocine and attempts to obtain the dihydrodiazocine LVI (R = H) by dehydrogenation of bicyclic compound LV (R = H) did not cleave the bridging bond but left the dehydro compound LVII. With LV (R = Me) no indication of an equilibrium with LVI (R = Me) could be detected by spectroscopy (*G. Maier et al.*, Angew. Chem. intern. Edn., 1965, **4,** 708; Chem. Ber., 1966, **99,** 1229, 1232, 1236).

3,4,5,6,7,8-**Hexahydro**-1,2-**diazocine** (1,2-*diazacyclooctene*) (LIX; R = H), n_D^{23} 1.4891, is prepared by the oxidation of 1,2-**diazocane** (*perhydro*-1,2-*diazocine*) (LVIII; R, R^1 = H), b.p. 71–74°/15 mm, with mercuric oxide. The reaction of *symm*-bis(ethoxycarbonyl)hydrazine with 1,6-dibromohexane or ditosyl 1,6-hexanediol in the presence of sodium hydride gives 1,2-*bis(ethoxycarbonyl)*-1,2-*diazocane* (LVIII, R = CO_2Et, R^1 = H), b.p. 100–105°/

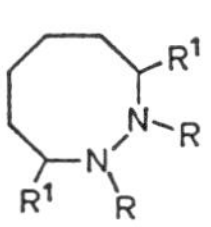

(LVIII)

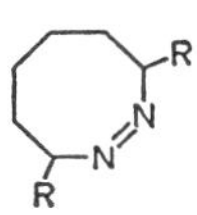

(LIX)

0.005 mm, which is decarboxylated on alkaline hydrolysis (*C. G. Overberger* and *J. W. Stoddard*, J. Amer. chem. Soc., 1970, **92,** 4922). *cis*- and *trans*-3,8-*Dimethyl*-1,2-*diazacyclo-octenes* (LIX; R = Me) are obtained similarly and separated by column chromatography.

3,8-*Diphenyl*-4,5,6,7-*tetrahydro*-1,2-*diazocine*, m.p. 136–137°, from the reaction of hydrazine and 1,4-dibenzoylbutane at high dilution (*Overberger* and *M. Lapkin ibid.*, 1955, **77,** 4651), is reduced catalytically to 3,8-*diphenyldiazocane* (LVIII; R = H, R^1 = Ph). Immediate oxidation of the perhydro compound with mercuric oxide gives 3,8-*diphenyl*-3,4,5,6,7,8-*hexahydro*-1,2-*diazocine* (*diphenyl*-1,2-*diazocyclo-octene*) (LIX; R = Ph) as *cis*-, m.p. 113–116°, and *trans*-, m.p. 90–91°, isomers, separatable by fractional crystallisation or sublimation (*Overberger et al., ibid.*, 1969, **91,** 3226; 1957, **79,** 2662; *J. Kossanyi*, Bull. Soc. chim. Fr., 1965, 722). Sixteen-membered ring "dimers" are also obtained in the first step (*J. M. van der Zanden* and *G. de Vries*, Rec. Trav. chim., 1956, **75,** 1159; 1957, **76,** 519). Pyrolysis of compound LIX (R = Ph) is slower than with the analogous 1,2-diazepine and gives diphenylhexene and diphenylcyclohexanes (*Overberger* and *I. Tashlick*, J. Amer. chem. Soc., 1959, **81,** 217) and photolysis of both LIX (R = Me) and (R = Ph) also gives olefins and cyclohexanes (*Overberger* and *Stoddard, loc. cit.*). The oxidation of cyclic hydrazines LVIII (R = H) to diazocyclo-octenes LIX results in a *cis* azo linkage (*Overberger et al., ibid.*, 1963, **85,** 2752) but irradiation of the *cis* compound gives a 60% conversion to its *trans* isomer (*idem*, Tetrahedron Letters, 1972, 4565).

The condensation of 1,6-dibromohexane with *symm*-bisalkoxycarbonylhydrazines to give LVIII (R = CO_2Bu^t, R^1 = H), m.p. 47–49°, is also effective with *cis*-1,6-dibromo-3-hexene. 1,2-*Di(butoxycarbonyl)*-1,2,3,4,7,8-*hexahydro*-1,2-*diazocine* (LX; R = CO_2Bu^t), m.p. 69–70°, on acid hydrolysis gives 1,2-**diazacyclo-oct-5-ene** (LX; R = H), b.p. 89°/20 mm; *hydrochloride*, m.p. 120–122°. Oxidation with cupric chloride then gives the *cuprous chloride complex* of 3,4,7,8-*tetrahydroazocine*, m.p. 191–197° decomp. A bromination–debromination sequence on LX (R = CO_2Bu^t) yields the bicyclic compound LXI (*L. A. Carpino* and *J. P. Masaracchia*, J. org. Chem., 1972, **37,** 1851).

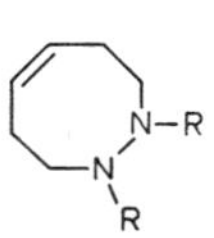

(LX)

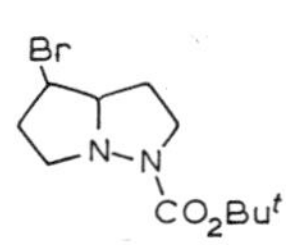

(LXI)

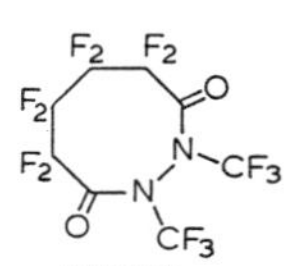

(LXII)

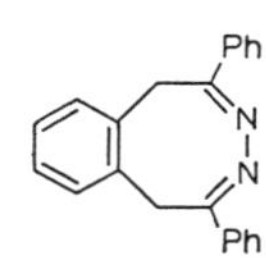

(LXIII)

1,2-**Diphenyl**-1,2-**diazocane** (LVIII; R = Ph, R^1 = H), m.p. 54–55°, is obtained from the lithium derivative of *N,N'*-diphenylhydrazine and 1,6-ditosyloxyhexane or the reduction of 1,2-*diphenyl*-1,2-*diazocane*-3,8-*dione*, m.p. 172–173°, with lithium tetrahydridoaluminate. The dione is prepared by cyclising adipic acid *N,N'*-diphenylhydrazide. With acid the diazocane undergoes a benzidine rearrangement (*G. Wittig* and *J. E. Grolig*, Chem. Ber., 1961, **94,** 2148; cf. Ann., 1966, **691,** 109).

The perfluoro-1,2-diazocanedione LXII has also been synthesised (*P. H. Ogden*, J. chem. Soc. C, 1971, 2920).

α,α'-Dibenzoyl-*o*-xylene with hydrazine hydrobromide gave 2,5-*diphenyl*-1,6-*dihydro*-3,4-*benzazocine* (LXIII), m.p. 158–159°, which could not be isomerised into the 3,4-dihydro analogue (*N. L. Allinger* and *G. A. Youngdale*, J. org. Chem., 1960, **25,** 1509).

Dibenzo[*c,g*]-1,2-**diazocine** (LXIV), m.p. 135–138°, results from a bromination–dehydrobromination sequence on its 11,12-dihydro derivative.

5,6-*Dihydrodibenzo*[c,g]-1,2-*diazocine*, m.p. 168–169° decomp., is prepared by reducing the parent compound (LXIV) with zinc and ethanolic barium hydroxide. It yields 5-*methyl*-, m.p. 158–159°, and 5,6-*dimethyl*, m.p. 149–151°, homologues with methyl iodide

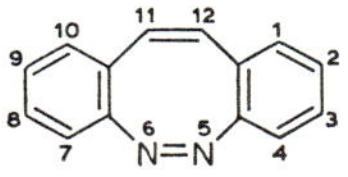

(LXIV)

(LXV)

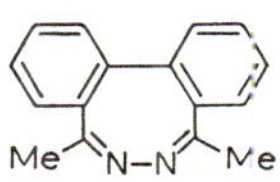

(LXVI)

and sodium bicarbonate. The 10π electron system in the heterocyclic ring provides a little resonance stabilisation (*W. W. Paudler* and *A. G. Zeiler*, J. org. Chem., 1969, **34**, 3237).

11,12-*Dihydrodibenzo*[c,g]-1,2-*diazocine* (2,2′-*azodibenzyl*), m.p. 112°, is obtained by reducing 1,2-bis(2-nitrophenyl)ethane with zinc and baryta, or by oxidising the 5,6,11,12-tetrahydro derivative of LXIV with mercuric oxide (*H. Duval*, Compt. rend., 1909, **149**, 401; Bull. Soc. chim. Fr., 1910, **7**, 727; *F. Gerson et al.*, Helv., 1960, **43**, 1889). Reduction of the azo bond with hydrazine and platinum oxide gives 5,6,11,12-*tetrahydrodibenzo*[c,g]-1,2-*diazocine*, m.p. 151°. With malonyl chlorides tetracyclic compounds LXV are formed (*H. S. Lowrie*, U.S.P. 3,170,926/1961; C.A., 1965, **62**, 14706).

5,8-*Dimethyldibenzo*[d,f]-1,2-*diazocine* (LXVI), m.p. 167–168°, the condensation product of hydrazine and 2,2′-diacetylbiphenyl is hydrolysed to the starting materials by acid, but to 9,10-dimethylphenanthrene by alkali at high temperatures (*E. E. Turner et al.*, J. chem. Soc., 1956, 3475; cf. *R. G. R. Bacon* and *W. S. Lindsay*, Chem. and Ind., 1956, 1479). 5,8-*Dimethyl*-, m.p. 127°, and 5,8-*diphenyl*-, m.p. 232°, analogues are made similarly (*idem*, J. chem. Soc., 1958, 1375, 1382). Although fully unsaturated the eight-membered ring shows little conjugation because it is puckered (*G. H. Beavan* and *E. A. Johnston*, *ibid.*, 1957, 651). Alkaline oxidation of diphenic acid dihydrazide or pyrolysis of the monohydrazide gives 5,6,7,8-*tetrahydrodibenzo*[d,f]-1,2-*diazocine*-5,8-*dione*, m.p. 305° (*R. A. Labriola*, C.A., 1938, **32**, 4970; *N. V. Riggs* and *S. M. Verma*, Austral. J. Chem., 1970, **23**, 1913). The potassium salt of *symm*-di-*tert*-butoxycarbonylhydrazine reacts with 2,2′-bis(bromomethyl)biphenyl to give a diester decomposing with hydrogen chloride to 5,6,7,8-*tetrahydrodibenzo*[d,f]-1,2-*diazocine*, m.p. 92–125°, *hydrochloride*, m.p. 282–284° decomp.; oxidation with manganese dioxide gave the explosive 5,8-*dihydrodibenzo*[d,f]-1,2-*diazocine*, m.p. 63–65° decomp. 9,10-Dihydrophenanthrene is the major thermolysis product (*Carpino*, J. Amer. chem. Soc., 1963, **85**, 2144).

(*ii*) *1,3-Diazocines*

Beckmann rearrangement of caprolactam oxime, formed by the action of hydroxylamine on the lactim ether or thiolactone, in the presence of nucleophiles gives the 3,4,5,6,7,8-**hexahydro**-1,3-**diazocines** (LXVII; R = *OMe*), m.p. 80–81°, and (LXVII; R = *OEt*), m.p. 58°, and analogues

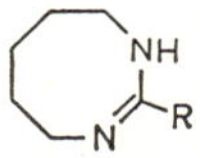

(LXVII)

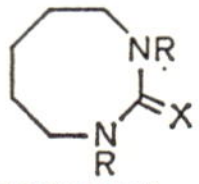

(LXVIII)

(*A. Le Berre et al.*, Bull. Soc. chim. Fr., 1971, 3245; *H. Behringer* and *H. Meier*, Ann., 1957, **607,** 67, 73).

If the nucleophile is hydroxide anion the product is 1,3-*diazocan*-2-*one (pentamethyleneurea)* (LXVIII; R = H, X = O), m.p. 266°, also obtained from pentamethylene diisocyanate and water at high dilution (*K. Uno et al.*, J. Amer, chem. Soc., 1957, **79,** 4358; Nippon Kagaku Zasshi, 1957, **78,** 1416) or pentamethylenediamine with carbon monoxide and sulphur at 130 p.s.i. and 100° (*F. Applegarth* and *A. Franz* U.S.P. 2,874,149/1959; C.A., 1959, **53,** 12187). With phosphorus pentasulphide 1,3-**diazocane**-2-**thione** (LXVIII; R = H, X = S), m.p. 223°, is obtained, which with mercuric oxide yields the *cyclic carbodiimide* (LXIX), m.p. 80–81° (*Behringer* and *Meier, loc. cit.*). The reaction of pentamethylenediamine with carbon disulphide is an alternative route to the thiones, which are *S*-alkylated by alkyl halides and converted into compounds LXVII (R = OR^1) by alkoxides (*P. N. Gordon*, U.S.P. 2,988,478; 3,219,522/1958; C.A., 1962, **57,** 6363, 1966, **64,** 6672). *Benzoylimino* compound LXVIII (R = H, X = NCOPh); m.p. 223°, is obtained from *N*-dichloromethylenebenzamide and pentamethylenediamine (*J. Burkhardt* and *K. Hamann*, Chem. Ber., 1967, **100,** 2569).

(*iii*) *1,4-Diazocines*

Diacetylene reacts with ethylenediamine at 0° to give 1,2,3,4-**tetrahydro**-1,4-**diazocine** (LXX), b.p. 136°/6 mm, n_D^{20} 1.5549; the 2-*methyl* homologue, b.p. 148°/10 mm, n_D^{20} 1.5709, is made similarly (*I. A. Chekulaeva et al.*, Izvest. Akad. Nauk S.S.S.R., 1963, **153,** 1353).

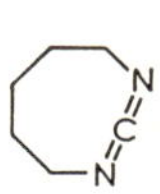

(LXIX)

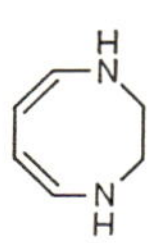

(LXX)

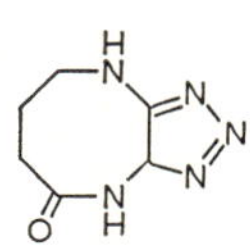

(LXXI)

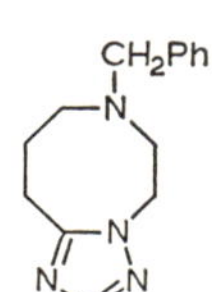

(LXXII)

Catalytic hydrogenation yields 2-*methyl*-1,4-*diazocane*, b.p. 55°/3 mm, n_D^{20} 1.4720 (*idem, ibid.*, 1967, p. 1829). Under other conditions the major product of the first reaction is a polymer (*idem*, C.A., 1970, **73,** 66930) or 5-methyl-2,3-dihydro-1*H*-1,4-diazepine (*W. Schroth et al.*, Z. Chem., 1969, **9,** 143). 2-*Methyl*-1-*phenyl*-1,4-*diazocane*, b.p. 138–142°/1 mm, results when *N*-(1-cyanoethyl)-*N*-(4-chlorobutyryl)aniline is reductively cyclised by lithium tetrahydridoaluminate (*W. L. Yost* and *R. B. Margerison*, F.P. 1,378,964; U.S.P. 3,247,206/1962; C.A., 1965, **62**, 13159; 1966, **64**, 19641).

1,4-**Diazocane**-2,5-**dione,** m.p. 274°, and the triazole LXXI are two of the major products of a Schmidt reaction on cyclohexane-1,3-dione (*H. Sekiguchi*, Bull. Soc. chim. Fr., 1965, 691) but with 1-benzylazepan-4-one a mixture of 1-*benzyl*-1,4-*diazocan*-5-*one*, b.p. 155–156°/0.14 mm, and the tetrazole LXXII is formed (*Y. Sakakida et al.*, Bull. Chem. Soc. Japan, 1971, **44,** 478). Ethylenediamine condenses with 3-benzoylpropionic acid to give 8-*phenyl*-1,2,3,6,7,8-*hexahydro*-1,4-*diazocin*-5-*one*, m.p. 124°; several analogues are also prepared in this way (*T. S. Sulkowski*, U.S.P. 3,293,243/1964; C.A., 1967, **66,** 46444). 1,4-**Diazocane**-2,3-**dione** is formed from tetramethylenediamine and oxalyl chloride (*O. Vogl* and *A. C. Knight*, Macromolecules, 1968, **1,** 311).

The 2-(*N*-phenylacetamido)ethylamide of benzilic acid cyclises to 1-acetyl-6,6-diphenyl-1,2,3,4,5,6-hexahydro-1,4-diazocin-5-one (LXXXIIa) with sulphuric acid (*P. A. Petyunin*, C.A., 1965, **63,** 13133) and 1,2,3,4,5,6-*hexahydro*-1,4-*benzodiazocine*-2,5-*dione*, m.p. 220–222°, is prepared by treating 1-(aminoacetyl)oxindole perchlorate with aqueous bicarbonate (*J. M. Muchowski*, Canad. J. Chem., 1970, **48,** 1946).

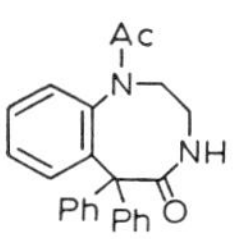

(LXXIIa)

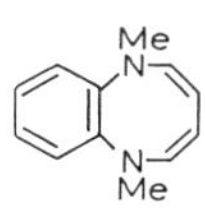

(LXXIII)

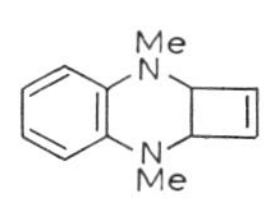

(LXXIV)

Heating *o*-phenylenediamine with 1,4-diaryl-*trans*-2-butene-1,4-diones yields 2,5-*diphenyl*-1,6-*benzodiazocine*, m.p. 198–199°, and analogues (*M. I. Shevchuk et al.*, Zhur. org. Khim., 1970, **6,** 1108). 1,6-*Dimethyl*-1,6-*dihydro*-1,6-*benzodiazocine* (LXXIII) is derived thermally from its tricyclic tautomer LXXIV, and reverts to it on irradiation. It has a formal 10π electron system, but it does not show a diamagnetic ring current, and gives a Diels–Alder adduct with *N*-phenylmaleimide (*H. J. Shue* and *F. W. Fowler*, Tetrahedron Letters, 1971, 2437).

o-Phenylenediamine reacts with 3-benzoylpropionic acid to give 5-*phenyl*-1,2,3,4-*tetrahydro*-1,6-*benzazocin*-2-*one* (LXXV), m.p. 158–160° (*Sulkowski, loc. cit.*). *N,N′*-Dimethyl- and *N,N′*-ditosyl-*o*-phenylenediamine with 1,4-dichloro-*cis*-2-butene yield 1,2,5,6-*tetrahydro*-1,6-*benzodiazocines* (LXXVI; R = Me), b.p. 140°/13 mm, and (R = tosyl), m.p. 216°. Hydrolysis of the latter gave the parent, 1,2,5,6-**tetrahydro**-1,6-**benzodiazocine** (LXXVI; R = H), m.p. 98°, and hydrogenation yields hexahydro derivatives LXXVII (R, R^1 = Me or H) (*Schroth* and *B. Streckenbach*, Z. Chem., 1963, **3,** 465).

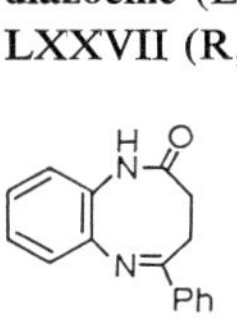

(LXXV)

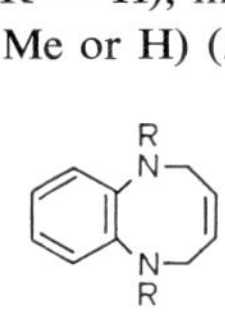

(LXXVI)

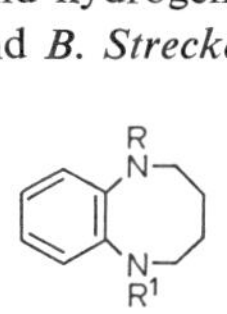

(LXXVII)

1,2,3,4,5,6-**Hexahydro**-1,6-**benzodiazocine** (LXXVII; R, R^1 = H), m.p. 74°, is obtained by hydrolysing the 1,6-*ditosyl* derivative (LXXVII; R, R^1 = tosyl), m.p. 219°, which results when 1,4-dibromobutane condenses with *N,N′*-ditosyl-*o*-phenylenediamine. The corresponding *-benzodiazonine*, m.p. 206°, and *-benzodiazecine* (*hydrochloride*, m.p. 150°) are prepared similarly (*H. Stetter*, Chem. Ber., 1953, **86**, 197). 1-*Phenyl*-1,2,3,4,5,6-*hexahydro*-1,6-*benzodiazocine* (LXXVII, R = Ph, R^1 = H), m.p. 80°, is one of the products of the benzidine rearrangement of *N,N'*-diphenylhexahydropyridazine (*Wittig et al.*, Ann., 1957, **610**, 180) and the *dibenzyl* derivative (LXXVII; R, R^1 = CH_2Ph), m.p. 90°, is obtained by treating the benzimidazole LXXVIII with hydride ion from aqueous tetrahydridoborate; cyanide and hydroxide ions react similarly. Both benzyl groups are removed on catalytic hydrogenation (*R. K. Grantham* and *O. Meth-Cohn*, J. chem. Soc. C, 1971, 1354).

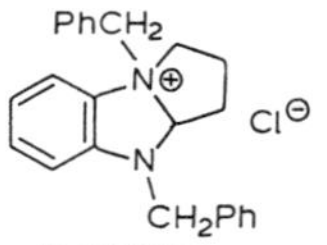
(LXXVIII)

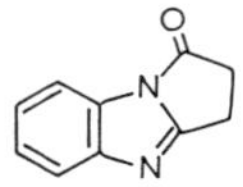
(LXXIX)

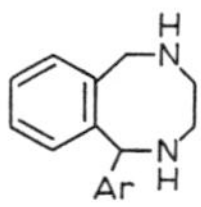
(LXXX)

o-Phenylenediamine condenses with diethyl succinate in the presence of sodium hydride to give 1,2,3,4,5,6-*hexahydro*-1,6-*benzazocine*-2,5-*dione*, m.p. 264–266° decomp. On heating, it undergoes transannular dehydration to benzimidazole LXXIX (*Paudler* and *Zeiler*, J. org. Chem., 1969, **34**, 2138).

1-*Phenyl*-1,2,3,4,5,6-*hexahydro*-2,5-*benzodiazocine* (LXXX; Ar = Ph), m.p. 125–127°, and analogues may be prepared from *o*-aroylbenzoic acids, *via* reduction with lithium tetrahydridoaluminate of the isoindolones LXXXI, that they form with ethylenediamine. Structure LXXXI is now preferred to the tetrahydro-2,5-benzodiazocin-1-one originally suggested. This reaction has been extended to nine- and ten-membered rings (*Sulkowski et al.*, J. org. Chem., 1967, **32**, 2180; B.P. 1,093,064; F.P. 1,487,344/1966 C.A., 1968, **69**, 19231; 27467; *L. H. Sternbach et al.*, J. org. Chem., 1967, **32**, 2185; *P. Aeberli* and *W. J. Houlihan*, *ibid.*, 1969, **34**, 1720).

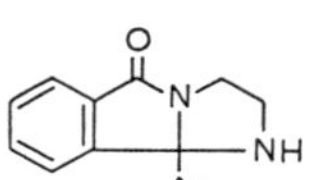
(LXXXI)

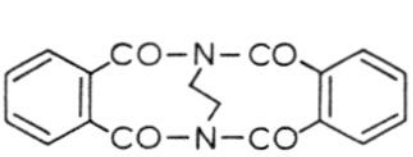
(LXXXII)

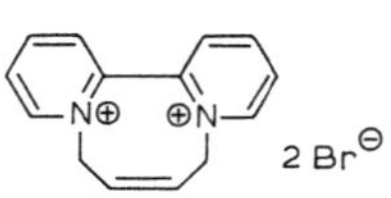
(LXXXIII)

Alternatively the condensation of acylated ethylenediamines with *o*-aroylbenzyl bromides followed by hydrolysis, cyclodehydration and reduction yields compounds LXXX (*Sulkowski et al.*, U.S.P. 3,496,164/1966; C.A., 1970, **72**, 100782) while 1,2-dipiperidylethane (*M. Scholtz*, Ber., 1902, **35**, 3047) and *N,N*-dimethylethylenediamine (*W. E. Rosen et al.*, J. Amer. chem. Soc., 1958, **80**, 935) condense with *o*-xylylene dibromides to give *N*-substituted 1,2,3,4,5,6-hexahydro-2,5-benzodiazocinium bromides.

If compounds LXXX are treated with potassamide they undergo ring-contraction to isoindole derivatives (*M. Winn* and *H. E. Zaugg*, J. org. Chem., 1969, **34**, 249; U.S.P. 3,597,422/1969; C.A., 1971, **75**, 110337).

The product of phthaloyl chloride and ethylenediamine is now known to be the bridged compound LXXXII and not 1,2,3,4,5,6-hexahydro-2,5-benzodiazocine-1,6-dione (*S.*

Wolfe and *S. K. Hasan*, Canad. J. chem. 1970, **48,** 3566; *Stetter et al.*, Chem. Ber., 1958, **91,** 1775).

The dibenzodiazocine LXXXIII, an analogue of the herbicide "Diquat", is formed from 2,2′-bipyridyl and *cis*-1,4-dibromo-2-butene (*F. D. Popp* and *D. K. Chesney*, J. heterocycl. Chem., 1972, **9,** 1165).

Condensation of *o*-phenylenediamine with *o*-phthalaldehyde gives **dibenzo**[*b,f*]**-1,4-diazocine** (LXXXIV; R = H), m.p. 210° (*J. Thiele* and *K. G. Falk*, Ann., 1906, **347,** 125; *M.V. Betrabet* and *C. G. Chakravarti*, J. Indian chem. Soc., 1930, **7,** 495); with *o*-dibenzoylbenzene in the presence of aluminium chloride 6,11-*diphenyldibenzo*[b,f]-1,4-*diazocine* (LXXXIV; R = Ph), m.p. 174–178°, is obtained (*H. D. Perlmutter*, Chem. Comm., 1968, 1202) and with 2-benzoylbenzoic acid 11-*phenyl*-5,6-*dihydrodibenzo*[b,f]-1,4-*diazocin*-6-*one*, m.p. 185–187°. Lithium tetrahydridoaluminate reduces the carbonyl group to methylene (*Sulkowski*, Belg. P. 646,221/1963; C.A., 1965, **63,** 9972).

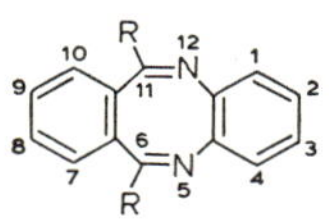
(LXXXIV)

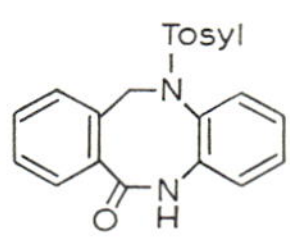
(LXXXV)

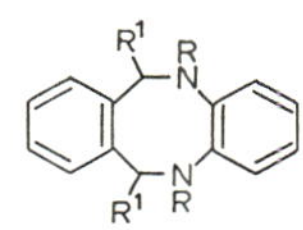
(LXXXVI)

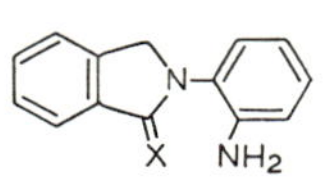
(LXXXVII)

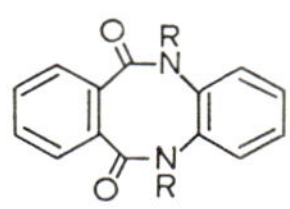
(LXXXVIII)

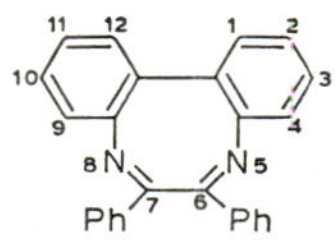
(LXXXIX)

Similar reduction and acid hydrolysis of the tosyl derivative LXXXV, prepared by the cyclisation of the appropriate amino acid ester, yields 5,6,11,12-**tetrahydrodibenzo**[*b,f*]-1,4-**diazocine** (LXXXVI; R, R^1 = H), m.p. 198°, but direct hydrolysis of LXXXV brought about ring-contraction to the isoindolinone LXXXVII (X = O) (*A. Saunders* and *J. M. Sprake*, J. chem. Soc. Perkin I, 1972, 1964). The *dibenzenesulphonyl* derivative (LXXXVI; R = SO_2Ph; R^1 = H), m.p. 176–180°, from the reaction of *o*-xylylene dibromide and *N,N′*-dibenzenesulphonyl-*o*-phenylenediamine reacts with phenyl-lithium to give 6,11-*diphenyl*-5,6,11,12-*tetrahydrodibenzo*[b,f]-1,4-*diazocine* (LXXXVI; R = H, R^1 = Ph), m.p. 268–272° (*E. Negishi* and *A. R. Day*, J. org. Chem., 1965, **30,** 43). The *ditosyl* (LXXXVI; R = *tosyl*, R^1 = H), m.p. 178°, and *dimethyl* (XXXVI; R = Me, R^1 = H), m.p. 131°, derivatives are prepared similarly. The former could be converted into the parent LXXXVI (R, R^1 = H), but *o*-phenylenediamine and *o*-xylylene dibromide give isoindoline LXXXVII (X = H_2) instead (*Schroth* and *Streckenbach*, *loc. cit.*; *Sprake et al.*, J. chem. Soc. C, 1969, 882; 1970, 1161). Paraformaldehyde and aromatic aldehydes form a methano bridge across positions 5 and 12 of compound LXXXVI (R, R^1 = H) which is removed by nitrous acid or acetic anhydride (*Saunders* and *Sprake loc. cit.*).

5,6,11,12-*Tetrahydrodibenzo*[b,f]-1,4-*diazocine*-6,11-*dione* (LXXXVIII; R = H), m.p. 301–303°, is prepared by treating *o*-phenylenediamine and diethyl phthalate with sodium hydride; the product given this structure by *Stetter (loc. cit.)* is *o*-di(*N*-phthalimido)benzene. On heating, dione LXXXVIII (R = H) undergoes transannular dehydration to a benzanalogue of LXXIX, and on treatment with base or phosphorus pentachloride it gives benzimidazoles (*Paudler* and *Zeiler*, *loc. cit.*).

Benzil and 2,2′-diaminobiphenyl condense to form 6,7-*diphenyldibenzo*[e,g]-1,4-*diazocine* (LXXXIX), m.p. 238° (*E. Täuber*, Ber., 1892, **25**, 3288; 1893, **26**, 1704; *F. Bell*, J. chem. Soc., 1952, 1527; 5047; *Bacon* and *Lindsay*, *ibid.*, 1958, 1382). Although this compound resists reduction, sodium amalgam in alcohol yields the 5,6,7,8-*tetrahydro derivative*, m.p. 155–156°, which is much more basic than the parent, and lithium tetrahydridoaluminate produces ring-contraction to a carbazole (*Allinger* and *Youngdale*, J. org. Chem., 1959, **24**, 306). Derivatives of LXXXIX have been resolved into optical isomers, and those with substituents at positions 1 and 12 have considerable optical stability (*J. M. Insole et al.*, J. chem. Soc., 1964, 2326; C, 1971, 1712). At high temperatures fragmentation of LXXXIX to benzonitrile and 6-phenylphenanthridine occurs (*Allinger et al.*, J. org. Chem., 1963, **28**, 3007).

With oxalic acid, 2,2′-diaminobiphenyl yields 5,6,7,8-*tetrahydrodibenzo*[e,g]-1,4-*diazocine*-6,7-*dione*, m.p. 186° (decomp.) (*W. Ried* and *W. Storbeck*, Chem. Ber., 1962, **95**, 459).

(*iv*) *1,5-Diazocines*

1,5-**Diazocine** (XC) is not expected to be stable with respect to its trans-annular tautomer, the mesoionic 3a,6a-diazapentalene (pyrazolo[1,2-a]-pyrazole) (XCI) which has been synthesised (*S. Trofimenko*, J. Amer. chem. Soc., 1966 **88,** 5588).

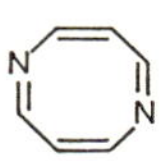

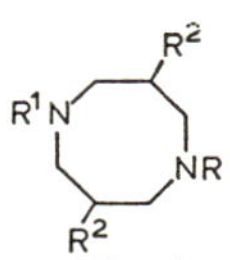

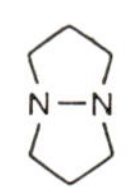

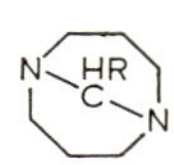

(XC) (XCI) (XCII) (XCIII) (XCIV)

1,5-*Ditosyl*-1,2,5,6-*tetrahydro*-1,5-*diazocine*, m.p. 228–233°, and 1,5-*ditosyl*-1,4,5,6-*tetra-hydro*-1,5-*diazocine*, m.p. 128–130°, are obtained after treating 2,7-dichloro- or 2,7-ditosyloxy-1,5-ditosyl-1,5-diazocanes with base. Both are reduced catalytically to 1,5-*ditosyl*-1,5-*diazocane* (XCII; R, R^1 = tosyl, R^2 = H), m.p. 214–216° (*Paudler et al.*, J. org. Chem., 1967, **32,** 2425), which is also found as a by-product of the cyclisation of *N,O*-ditosyl-3-aminopropanol (*W. R. Vaughan et al.*, *ibid.*, 1961, **26,** 138). A route to a tetrahydro-1,5-diazocindione is as shown (*A. B. Evnin et al.*, Tetrahedron Letters, 1969, 4497):

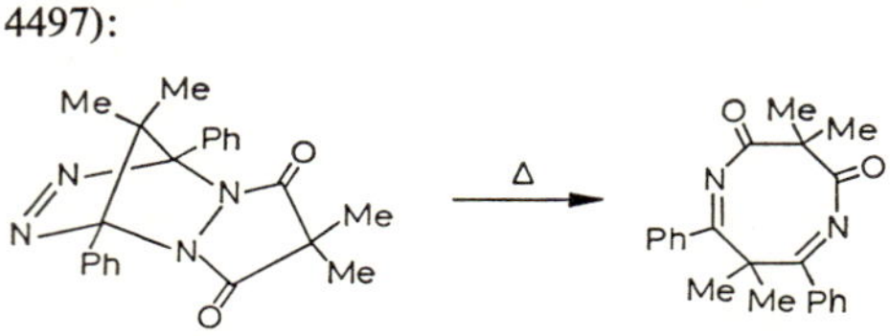

1,3-Diaminopropane reacts with benzylideneacetone to give 4-methyl-2-phenyl-1,2,3,6,7,8-hexahydro-1,5-diazocine, the double bond in which is reduced by tetrahydridoborate (*K. Hideg* and *D. Lloyd*, Chem. Comm., 1971, 372; J. chem. Soc. C, 1971, 3441).

Several routes are known to 1,5-**diazocane** (XCII; R, R^1, R^2 = H), m.p. 14–15°, b.p. 186–188°, *dipicrate*, m.p. 226° (decomp.), including the condensation of 1,3-dibromopropane with 1,3-ditosylaminopropane followed by hydrolysis (*C. C. Howard* and *W. Marckwald*, Ber., 1899, **32,** 2039), the reductive cyclisation of *N*-(2-cyanoethyl)trimethylenediamine (*A. P. Terentev et al.*, Zhur. obshcheĭ Khim., 1951, **21,** 268) the reduction of 8-azapyrrolizidine (XCIII) with lithium tetrahydroaluminate (*Stetter* and *H. Spangenberger*, Chem. Ber., 1958, **91,** 1982) and, together with XCIII, when hydrazine reacts with trimethylene dihalides (*E. L. Buhle et al.*, J. Amer. chem. Soc., 1943, **65,** 29).

With aldehydes, 1,5-diazocane gives 1,5-methano bridged compounds XCIV (*J. H. Billman* and *L. C. Dorman*, J. org. Chem., 1962, **27,** 2419). Treatment of XCIV (R = H) with methyl iodide followed by potassium cyanide and alkali gives 1-cyanomethyl-5-methyl-1,5-diazocane (XCII; R = CH_2CN, R^1 = Me, R^2 = H) (*Sandoz Ltd.*, Belg. P. 610,374/1962; C.A., 1963, **58,** 1482).

Salts of XCIII with alkyl halides when reduced with aluminium amalgam give 1-**alkyl**-1,5-**diazocanes** including the 1-*methyl* (XCII; R = Me; R^1, R^2 = H), b.p. 175–176°, and 1-*ethyl* (XCII; R = Et; R^1, R^2 = H), *hydrochloride*, m.p. 211–213°, derivatives (*C. A. Grob* and *O. Schier*, G.P. 1,104,516/1961; C.A., 1962, **57,** 7288). 1-Methyl-1,5-diazocane is also prepared by reductive cyclisation of *N*-(2-cyanoethyl)-*N*-(3-chloropropionyl)-methylamine (*Yost* and *Margerison*, *loc. cit.*) or by treating 1,5-diazocane with ethyl formate and reducing the product; the latter route with ethyl acetate gives the 1-ethyl homologue (*F. F. Blicke*, U.S.P. 3,117,963/1960; C.A., 1964, **60,** 9298). With nitrous acid 1-methyl-1,5-diazocane gives a 5-nitroso derivative which may be reduced to 5-amino-1-methyl-1,5-diazocane (XCII; R = NH_2, R^1 = Me, R^2 = H) (*Sandoz Ltd.*, Belg. P. 613,566/1962; C.A., 1963, **58,** 1481). 3,3,7,7-*Tetramethyl*-1,5-*diazocane*, b.p. 110°/10 mm, is prepared by condensing 2-methylpropanol, formaldehyde and benzylamine hydrochloride and reducing the resultant oxygen-bridged product (*M. W. Williams*, J. org. Chem., 1968, **33,** 3946). 1-*Methyl*-2-*phenyl*-1,5-*diazocane*, *dipicrate*, m.p. 194°, is obtained by reducing the corresponding 4,8-*dione*, m.p. 146–147°, with lithium tetrahydridoaluminate (*R. Sarfati et al.*, Bull. Soc. chim. Fr., 1971, 255).

3,7-*Diethyl*-3,7-*dinitro*-1,5-*diazocane*, m.p. 63–64°, is obtained by acid hydrolysis of the tricyclic product from 1-nitropropane, formaldehyde and ammonia, or condensation of 1,1-di(hydroxymethyl)nitropropane with ammonia (*J. K. N. Jones et al.*, Bull. acad. polon. sci., Class III, 1956, **4,** 521; C.A., 1957, **51,** 8765; Rocz. Chem., 1957, **31,** 93; cf. J. chem. Soc., 1947, 924).

Cyclodimerisation of *N*-substituted 1-amino-2,3-epoxypropanes or reaction of *N,N*-di-(2,3-epoxypropyl)amines with primary amines gives 3,7-*dihydroxy*-1,5-*diazocanes* including 1,5-*diphenyl* (XCII; R, R^1 = Ph, R^2 = OH), m.p. 212–214°, and 1,5-*di*-tert-*butyl* (XCII; R, R^1 = Bu^t, R^2 = OH), m.p. 124–125°, derivatives (*V. R. Gaertner*, Tetrahedron Letters 1964, 141; U.S.P. 3,236,837/1963; C.A., 1966, **64,** 12706). Similarly 1,3-dibromo-2-propanol, 2,3-dibromo-1-propanol and epichlorohydrin each reacts with *p*-toluenesulphonamide and base to give *cis*- (m.p. 210–211°) and *trans*- (m.p. 223–224°) -3,7-*dihydroxy*-1,5-*ditosyl*-1,5-*diazocane* (XCII; R, R^1 = tosyl, R^2 = OH). Acid hydrolysis gives cis-3,7-*dihydroxy*-1,5-*diazocane* (XCII; R, R^1 = H, R^2 = OH), m.p. 182–183° and its *trans* isomer, *ditosate*, m.p. 253°. These structures were assigned by analysing n.m.r. spectra and by treating the isomers with acetic anhydride when the *trans* compound gives a diacetyl derivative and the *cis* isomer an oxygen-bridged compound. Oxidation of the diols with chromic anhydride similarly gives an oxygen-bridged hemiacetal. Diol XCII (R, R^1 = H, R^2 = OH) with thionyl chloride and its ditosic ester with lithium chloride yield the cis-*dichloro* compound (XCII; R, R^1 = H, R^2 = Cl), m.p. 286–287°, together with piperazine and diazepane ring-contraction products (*Paudler et al.*, J. org. Chem., 1966, **31,** 277; 1967, **32,** 2425; 1969, **34,** 1001). Ring-contraction of XCII (R, R^1 = Ph, R^2 = OH) to piperazines also occurs with phosphorus tribromide (*J. J. Worman et al., ibid.*, 1971, **36,** 3361) while the dihydrobromide of XCII (R, R^1 = Ph, R^2 = OMe) yields XCII (R, R^1 = *p*-bromophenyl, R^2 = OMe) in dimethyl sulphoxide (*idem*, J. chem. Soc. Perkin I, 1972, 1209).

1,5-Diazocanes adopt a crown conformation (*T. Urbanski et al., ibid.*, 1958, 2319; Rocz. Chem., 1958, **32,** 1289). The parent forms square planar complexes with divalent nickel and copper ions (*W. K. Musker* and *M. S. Hussain*, Inorg. Chem., 1966, **5,** 1416; cf. *J. I. Legg et al.*, J. Amer. chem. Soc., 1968, **90,** 5030) and diol XCII (R, R^1 = Ph, R^2 = OH) also forms complexes with ferric ions (*W. P. Jensen et al.*, J. heterocycl. Chem., 1972, **9,** 145).

2-Methyl- and 2-phenyl-1,5-diazocan-4-ones are prepared by treating 3-methyl- and 3-phenyl-β-lactams with vinyl cyanide and base, and hydrogenating the product over Raney cobalt (*P. Wegener*, Ger. Offen. 1,802,207/1968; C.A., 1970, **73,** 25549). The alkaloid **homaline** has been identified as tetramethylene-5,5′-bis(2-phenyl-1,5-diazocan-4-one) (*Sarfati et al.*, Compt. rend., 1971, **272C,** 1728).

The main product of a double Schmidt reaction on cyclohexane-1,3-dione is **1,5-diazocane-2,4-dione,** m.p. 274° (*H. Sekiguchi*, Bull. Soc. chim. Fr., 1965, 691). Both Schmidt and Beckmann reactions with cyclohexane-1,4-dione and its dioxime yield 1,5-*diazocane*-2,6-*dione*, m.p. 298–299°. The Schmidt reaction is preferred as hydrolysis and aromatisation appear to take place more readily under Beckmann conditions (*H. K. Hall*, J. Amer. chem. Soc., 1958, **80,** 6404; *M. Rothe* and *R. Timler*, Chem. Ber., 1962, **95,** 783; *S. Kuwata* and *S. Koyama*, Chem. pharm. Bull., Japan, 1961, **9,** 834; Bull. chem. Soc. Japan, 1963, **36,** 143; *Y. Kobayashi* and *S. Wakamatsu*, Tetrahedron, 1967, **23,** 115; cf. *I. L. Knunyants* and *B. P. Fabrichnyi*, Doklady Akad. Nauk S.S.S.R., 1949, **68,** 701; *F. Weygand* and *H. J. Dietrich*, Chem., Ber., 1954, **87,** 482). The dione readily polymerises (*Y. Iwakura et al.*, J. polymer, Sci. Part AI, 1969, **7,** 657). Methods of peptide synthesis have also been used to prepare substituted 1,5-diazocane-2,6-diones, which may be viewed as cyclic dipeptides (*Sarfati, loc. cit.*).

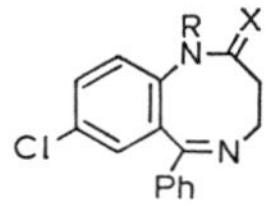

(XCV)

Cyclisation of 2(3-aminopropyl)amino-5-chlorobenzophenone gives 8-*chloro*-6-*phenyl*-1,2,3,4-*tetrahydro*-1,5-*benzodiazocine* (XCV; R = H, X = H_2), m.p. 150–151°. The 5,6-double bond is reduced by lithium tetrahydridoaluminate, and catalytic reduction replaces the chlorine atom at position 8 by hydrogen, leaving 6-*phenyl*-1,2,3,4,5,6-*hexahydro*-

1,5-*benzodiazocine, dihydrochloride*, m.p. 225–231°. The 1-*methyl* homologue (XCV; R = Me, X = H_2), m.p. 111–113°, is best made by methylating the precursor.

Lactam XCV (R = Me, X = O), is prepared by cyclising the product of 5-chloro-2-methylaminobenzophenone and carbonylbenzyloxy-β-alanyl chloride, but the product prepared similarly and thought to be XCV (R = H, X = O) is now known to be a dimeric dibenzotetra-azacyclohexadecane derivative (*R. I. Fryer et al.*, J. org. Chem., 1969, **34,** 179; Ger. Offen. 1,920,908/1968; C.A., 1970, **72,** 31862; cf. *Sulkowski*, U.S.P. 3,294,782/1964; C.A., 1967, **66,** 46443). 8-Chloro-6-*phenyl*-1,2,3,4-*tetrahydro*-1,5-*benzo-diazocin*-2-*one* (XCV; R = H, X = O), m.p. 164–166° (cf. *Sulkowski*, m.p. 310°) has been obtained by the oxidation with permanganate of hexahydro derivative XCVI, obtained as shown (*M. Deuzer* and *H. Ott*, J. org. Chem., 1969, **34,** 183; *Ott*, U.S.P. 3,577,557/1966; C.A., 1971, **75,** 49158):

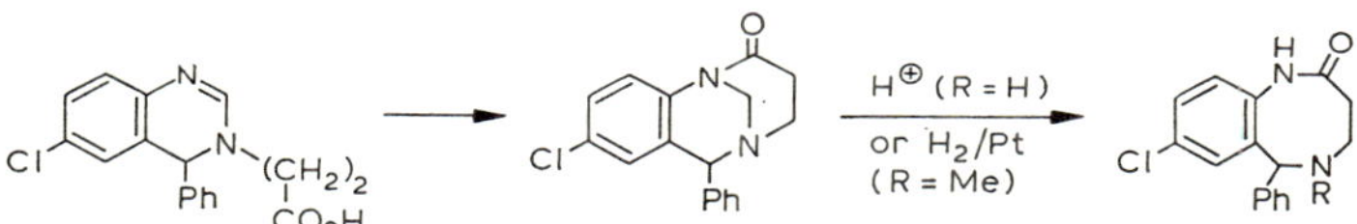

(XCVI)

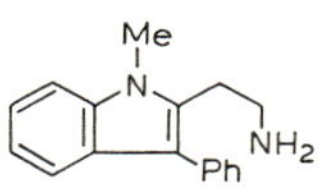

(XCVII)

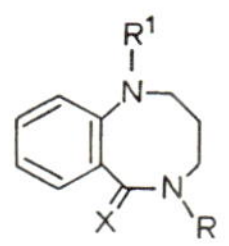

(XCVIII)

Its 1-*methyl* homologue (XCV; R = Me, X = O), m.p. 175–180°, is also prepared by the oxidation of indole XCVII with chromic acid (*H. Yamamoto et al.*, Ger. Offen. 1,814,332/1967; C.A., 1970, **72,** 21728).

N,N′-Ditosyl-*o*-aminobenzylamine with 1,3-dibromopropane and base gives 1,2,3,4,5,6-**hexahydro**-1,5-**benzodiazocine** (XCVIII; R, R^1 = H, X = H_2), m.p. 57–60°, b.p. 114–116°/0.35 mm, *via* its *ditosylate* (XCVIII; R, R^1 = tosyl, X = H_2), m.p. 176–177°. With aldehydes a methano bridge is formed across positions 1 and 5 (*S. Shiotani* and *K. Mitsuhashi*, Yakagaku Zasshi, 1964, **84,** 656; Jap. P. 20,952/1960, C.A., 1967, **66,** 55514).

Cyclisation of *o*-aminobenzyl-*N′*-(3-chloropropyl)aniline or benzylamine yields 5-*phenyl*- (XCVIII; R = Ph, R^1 = H, X = O), m.p. 250–251°, and 5-*benzyl*-1,2,3,4,5,6-*hexahydro*-1,5-*benzodiazocin*-6-*one* (XCVIII; R = CH_2Ph, R^1 = H, X = O), m.p. 127–128°; lithium tetrahydridoaluminate reduces the carbonyl group to methylene (*F. Gatta* and *R. Landi Vittory*, Farmaco Ed. Sci., 1970, **25,** 830; C.A., 1971, **74,** 53753). If the corresponding propionic acids are cyclised with chloroformic ester 5-*phenyl*- (m.p. 158–160°) and 5-*benzyl*-1,2,3,4,5,6-*hexahydro*-1,5-*benzodiazocin*-2,6-*dione*, m.p. 96–99°, are formed (*idem, ibid.*, 1970, **25,** 991; C.A., 1971, **75,** 35982).

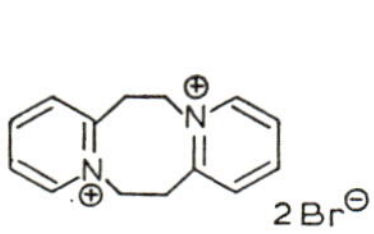

(XCIX)

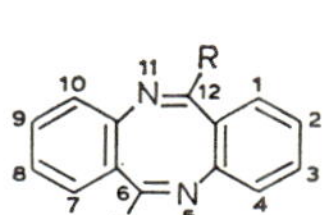

(C)

On heating, 2-vinylpyridine hydrobromide gives the 1,5-diazocine XCIX, which opens up to 2-(*N*-piperidinoethyl)pyridine in piperidine (*V. Boekelheide* and *W. Feely*, J. Amer. chem. Soc., 1958, **80,** 2217).

Dibenzo[*b,f*]**-1,5-diazocine (phenhomazine)** (C; R = H) is obtained by the reaction of hydrazine followed by cupric acetate on 6,12-dichlorophenhomazine (*W. W. Paudler* and *A. G. Zeiler*, Chem. Comm., 1967, 1077). 6,12-*Dichlorophenhomazine* (C; R = Cl), m.p. 219–220°, from the action of phosphorus pentachloride on dianthranilide (*G. Schroeter*, Ber., 1919, **52,** 2224) is also converted into the *diamine* (C; R = NH_2), m.p. 93–94°, by methanolic ammonia and *dialkoxides* (C; R = OMe), m.p. 161–162°, and C; R = *OEt*), m.p. 146–147°, with solutions of sodium in the appropriate alcohol (*F. C. Cooper* and *M. W. Partridge*, J. chem. Soc., 1954, 3429). The diamine is also formed in small yield on pyrolysis of 2-tosylaminobenzonitrile, and in the self-condensation of anthranilamide (*S. C. Pakrashi*, J. org. Chem., 1971, **36,** 642), while the 6-*amino*-12-*hydroxy* derivative, m.p. 213–215°, is a product of the thermal condensation of 2-aminobenzonitrile and methyl *N*-tosylanthranilate (*Partridge et al.*, J. chem. Soc., 1962, 2549). The action of Lewis acids or strong bases on 2-aminobenzophenones gives 6,12-*diphenyldibenzo*[b,f]-1,5-*diazocine* (C; R = Ph), m.p. 190°, and many derivatives (*A. Sondheimer*, Ber., 1896, **29,** 1273; *L. H. Sternbach et al.*, J. Amer. chem. Soc., 1966, **88,** 1077; J. org. Chem., 1966, **31,** 3356; J. med. Chem., 1966, **9,** 633; U.S.P. 3,243,430/1963; C.A., 1966, **64,** 17622; *M. Kawai*, Nippon Kagaku Zasshi, 1967, **88,** 102; Bull. Chem. Soc. Japan, 1969, **42,** 2066). Catalytic or chemical reduction yields di- and tetra-hydro derivatives, including the indoloindole CI, and hot acid hydrolysis regenerates the aminoketone.

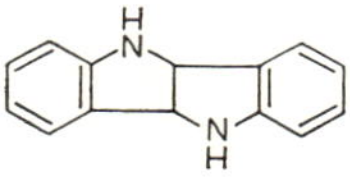

(CI)

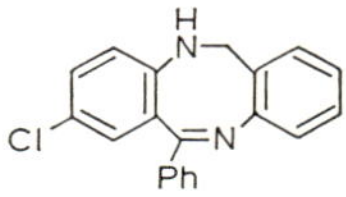

(CII)

Several 5,6-dihydrodibenzo[*b,f*]-1,5-diazocin-6-ones (*e.g.* CII, m.p. 217–220°) and their 5,6,11,12-tetrahydro analogues are prepared by the reductive cyclisation of *N*-(2-benzoylphenyl)-2-nitrobenzamides (*J. G. Topliss*, U.S.P. 3,409,608/1966; C.A., 1969, **70,** 57921).

A number of similar routes are known to 5,6,11,12-**tetrahydrodibenzo**[*b,f*]**-1,5-diazocines** (CIII). Treatment of Tröger's base with nitrous acid, acetic anhydride or alkaline dimethyl sulphate yields respectively CIII (R^1 = Me, R = NO, Ac or Me). Denitrosation of CIII (R^1 = Me, R = NO) with cuprous chloride yields 2,8-*dimethyl*-5,6,11,12-*tetrahydro*-

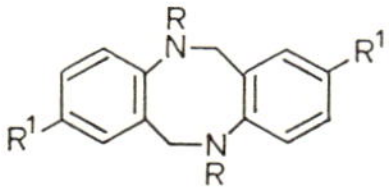

(CIII)

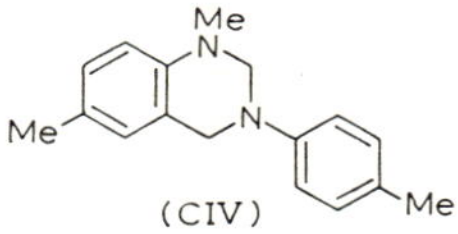

(CIV)

dibenzo[b,f]-1,5-*diazocine* (CIII; R^1 = Me, R = H), m.p. 204–205° (*Cooper* and *Partridge*, J. chem. Soc., 1957, 2888), while the 2,5,8-*trimethyl* analogue, m.p. 146–148°, and several substituted derivatives are also obtained from Tröger's base and dimethyl sulphate (*M. Häring*, Helv., 1963, **46,** 2970). The 2,5,8,12-*tetramethyl* compound (CIII; R, R^1 = Me), m.p. 149°, results when compound CIV is treated with one mole of methyl bromide (*G. A. Swan*, J. chem. Soc. C, 1971, 2880) or from the reaction cf *N*-methyl-*p*-toluidine with formaldehyde and hydrochloric acid (*W. V. Farrar*, Chem. and Ind., 1967, 1644). Compound CIII (R = Me, R^1 = H), m.p. 107°, results from the action of benzoyl peroxide on *N,N*-dimethylaniline or when the α-diamine, $(PhNMe)_2CH$, is treated with a weak organic acid; diazocines of type CIII (R, R^1 = Me) and CIII (R = Ph, R^1 = H), m.p. 195°, may also be made using these routes (*Swan et al.*, Chem. Comm., 1966, 427; 1968, 1376).

The formation of *N*-alkyl derivatives of CIII by alkylation, or acylation and reduction has proved difficult (*N. J. Harper* and *J. M. Sprake*, J. chem. Soc C, 1969, 882) but the action of aldehydes on compounds CIII (R = H) is to insert a methano bridge across positions 5 and 11 *(Swan, loc. cit.)*, thus reversing the synthesis described above.

5,6,11,12-**Tetrahydrodibenzo**[*b,f*]-1,5-**diazocine**-6,12-**dione** *(dianthranilide)* (CV; R = H), m.p. 334°, and others in this series are prepared by cyclising anthranilic acid derivatives or by a double Beckmann rearrangement on anthraquinone dioxime (*Schroeter et al., loc. cit.*; Ann., 1909, **367,** 101; *Cooper* and *Partridge, loc. cit.*; *H. N. Rydon et al.*, J. chem.

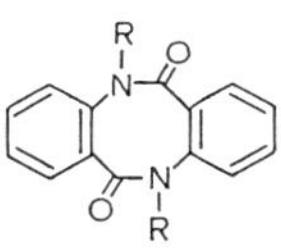
(CV)

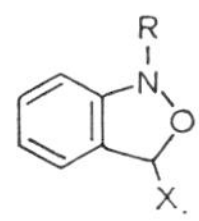
(CVI)

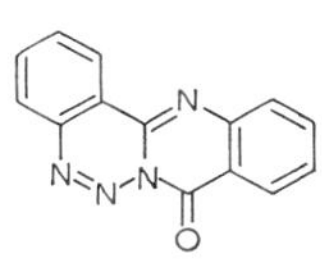
(CVII)

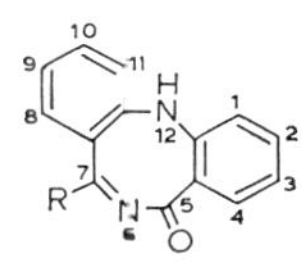
(CVIII)

Soc., 1957, 1900; *M. Kurihara*, Makromol. Chem., 1967, **105,** 84). The pyrolysis of benzisoxazolines CVI (R = Me, X = OMe) and (R = Et, X = OMe or CN) also gives 5,11-*dimethyl-* (CV; R = Me) and 5,11-*diethyl-* (CV; R = Et), m.p. 192°, analogues (*R. A. Olofson et al.*, J. Amer. chem. Soc., 1971, **93,** 1543), while *N*-benzyl-anthranilic acid cyclises with phosphoryl chloride to 5,11-*dibenzyl-dianthranilide* (CV; R = CH_2Ph), m.p. 153–155° (*M. Protiva et al.*, Coll. Czech. Chem. Comm., 1965, **30,** 445). 2-Carboxyphenylhydroxamic acid with benzenesulphonyl chloride gives CV (R = SO_2Ph) (*L. Bauer* and *S. V. Miarka*, J. Amer. chem. Soc., 1957, **79,** 1983). On pyrolysis, dianthranilide undergoes ring-contraction to 2-(2-aminophenyl)-3,1-benzoxazin-4-one (*G. K. J. Gibson* and *A. S. Lindsay*, Chem. and Ind., 1968, 1809).

Photolysis of the tetra-aza compound XVII in methanol yields 7-*methoxy*-5,12-*dihydrodibenzo*[b,g]-1,5-*diazocin-5-one* (CVIII; R = OMe), m.p. 200–201°; with water present the *tetrahydrodione* (oxo tautomer of CVIII; R = OH), m.p. 250–255°, is obtained. Both are hydrolysed to diphenylamine-2,2′-dicarboxylic acid by alkali (*G. Ege et al.*, Chem. Ber., 1972, **105,** 2898).

(*c*) *Eight-membered rings containing three or more nitrogen atoms*

(*i*) *Triazocine derivatives*

Several 1,2,4-**triazocine** derivatives described in the literature are now known to possess different structures. The 3-amino-5,8-diaryl-1,2,4-triazocines of *H. Beyer* and *T. Pyl* (Angew. Chem., 1956, **68,** 374; Ann., 1957, **605,**

50) are imidazotriazines (*B. Loev* and *M. M. Goodman*, Tetrahedron Letters, 1968, 789) and the 3-amino-5,8-dimethyl-6,7-dihydro-1,2,4-triazocine of *J. Thiele* and *E. Dralle* (Ann., 1898, **302,** 297) is really 2,5-dimethyl-1-guanidylpyrrole (*Beyer et al.*, *ibid.*, 1960, **638,** 150). The cyclisation of 4-(carbamoylhydrazino)-4-phenylbutyric acid may possibly give 8-*phenyl*-1,2,4-*triazocane*-3,5-*dione* (CIX), m.p. 166°, but no supporting evidence is given (*J. Bougault et al.*, Compt. rend., 1947, **225,** 876; 1949, **228,** 1952).

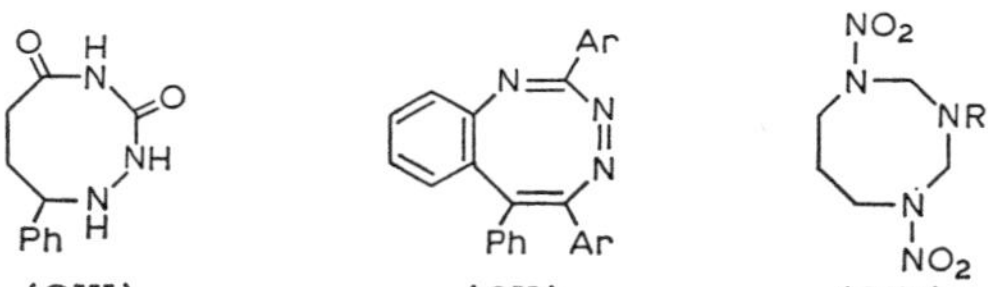

(CIX) (CX) (CXI)

The fully unsaturated **benzo-**1,3,4-**triazocines** (CX) are obtained by loss of nitrogen from the Diels–Alder adducts of 2-phenylbenzazete and 3,6-diaryl-1,2,4,5-tetrazines (*C. Rees*, personal communication).

The 1,3,5-**triazocanes** (CXI) are prepared, like their triazepane analogues, from 1,3-di(nitroamino)propane, a primary amine and formaldehyde. The *methyl* compound CXI (R = Me) has m.p. 141–143°, and the *ethyl* compound CXI (R = Et), m.p. 114–120° (*J. A. Bell* and *I. Dunstan*, J. chem. Soc. C, 1966, 867; 870; *F. Chapman et al.*, *ibid.*, 1949, 1631, 1638).

A compound earlier described as 3-nitro-1,3,6-triazocan-2-one has now been given an imidazolidone structure (*A. F. McKay et al.*, J. Amer. chem. Soc., 1956, **78,** 484; 6144).

(*ii*) *Tetrazocine derivatives*

Nitroformaldehyde phenylhydrazone, PhNHN:CHNO$_2$, condenses with formaldehyde in the presence of bases to form 6,8-*dinitro*-2,4-*diphenyl*-2,3,4,7-*tetrahydro*-1,2,4,5-*tetrazocine* (CXII), m.p. 149–150° (*W. E. Hahn* and *H. Zawadzka*, Rocz. Chem., 1964, **38,** 557).

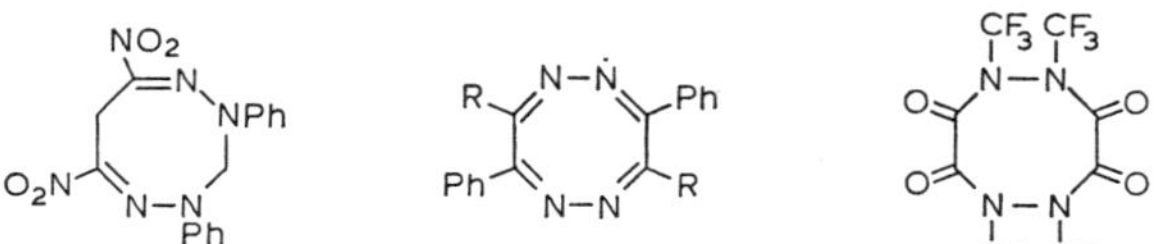

(CXII) (CXIII) (CXIV)

Benzil monohydrazone undergoes self-condensation above its melting point to 3,4,7,8-*tetraphenyl*-1,2,5,6-*tetrazocine* (CXIII; R = Ph), m.p. 285–286°, in low yield (*R. Metze*, Angew. Chem., 1956, **68,** 580); the α-hydrazone of methyl benzoylglyoxylate in acetic acid similarly yields *dimethyl* 3,7-*diphenyl*-1,2,5,6-*tetrazocine*-4,8-*dicarboxylate* (CXIII; R = CO_2Me), m.p. 247–248°. The latter is hydrolysed to the *dicarboxylic acid* (CXIII; R = CO_2H), m.p. 270° decomp., and subsequently decarboxylated to 3,7-*diphenyl*-1,2,5,6-*tetrazocine* (CXIII; R = H), m.p. 264° (decomp.) (*R. P. Ffleger et al.*, Angew. Chem., 1956, **68,** 680; Chem. Ber., 1957, **90,** 2411; G.P. 1,044,093/1958; C.A., 1961, **55,** 2704).

The tetrazocine ring is hydrolysed only under drastic conditions and takes up only three moles of hydrogen at 90°/90 atm. Compound CXIII (R = H) is substituted by bromine to give a *dibromide* (CXIII; R = Br), m.p. 243–244°; the bromine atoms are unreactive. Ester CXIII (R = CO_2Me) with hydrogen iodide, acetic anhydride and red phosphorus contracts to 5-phenyl-1,2,3-triazole. Many analogues of CXIII in which the 3,4,7,8-positions carry aryl or heteroaryl substituents have been synthesised subsequently from a 1,2-diketone and its dihydrazone as components of the photoconductive insulating layer in the xerographic process (*H. Schlesinger*, G.P. 1,110,007/1959; C.A., 1962, **56,** 8217; Angew. Chem. 1960, **72,** 563; cf. *B. Eistert* and *W. Schade*, Chem. Ber., 1958, **91,** 1411).

1,2,5,6-**Tetrazocane**-3,7-**dione,** m.p. 210–211°, is said to be the product obtained by cyclising glycollic acid hydrazide (*M. Rink et al.*, Arch. Pharm., 1966, **299,** 254; *Curtius* and *Swan*, J. pr. Chem., 1895, **51,** 369) but the 4-amino-3,5-bis(2-hydroxyethyl)-1,2,4-triazole structure has also been proposed (*M. Adamek*, Coll. Czech. chem. Comm., 1960, **25,** 1694).

1,2,5,6-Tetrakistrifluoromethyl-1,2,5,6-tetrazocanetetraone (CXIV) results when oxalyl fluoride reacts with tetrafluoroformaldazine. On photolysis the compound fragments to trifluoromethyl isocyanate, but it is stable at 250° (*P. H. Ogden*, J. chem. Soc., C, 1971, 2920).

The powerful high-melting explosive **HMX (octogen),** m.p. 282°, obtained when hexamethylenetetramine (hexamine) is nitrated with nitric acid and ammonium nitrate in acetic acid/anhydride mixtures has the 1,3,5,7-**tetrazocane** structure CXV (R, R^1 = NO_2). Adjusting the conditions allows the preferential formation of 1,3,5-trinitroperhydro-*symm*-triazine (RDX), the bicyclic DPT (CXVI) or the *acetoxy methyl* derivative (CXV;

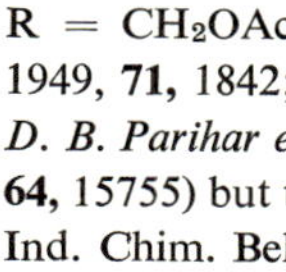

HMX
(CXV)

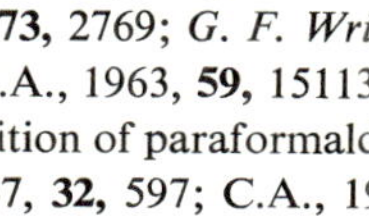

DPT
(CXVI)

R = CH_2OAc, R^1 = NO_2), m.p. 152° (*W. E. Bachmann et al.*, J. Amer. chem. Soc., 1949, **71,** 1842; 1951, **73,** 2769; *G. F. Wright et al.*, Canad. J. Res., 1949, **B27,** 218; 642; *D. B. Parihar et al.*, C.A., 1963, **59,** 15113; *H. K. Acharya* and *R. T. Limaye*, C.A., 1966, **64,** 15755) but the addition of paraformaldehyde increases the yield of HMX (*J. P. Picard*, Ind. Chim. Belge, 1967, **32,** 597; C.A., 1969, **70,** 76979). HMX forms 1:1 adducts with benzene and naphthalene derivatives facilitating the separation of HMX from RDX (*W. Selig*, Explosivstoffe, 1969, **17,** 73; C.A., 1969, **71,** 91449). Irradiation causes a phase change in the crystal, and thermolysis results in the breaking of carbon to nitrogen, not nitrogen to nitrogen, bonds (*J. R. Autera et al.*, Chem. and Ind., London, 1967, 2177; Mol. Cryst., 1967, **2,** 373; C.A., 1968, **68,** 73104).

The 1-acetoxymethyl compound (CXV; R = CH_2OAc, R^1 = NO_2) may also be obtained from DPT (CXVI) and yields DPT and 1-alkoxymethyl-3,5-dinitro-1,3,5-triazacyclohexanes when treated with alcohols (*Bell* and *Dunstan*, *loc. cit.*). With nitric or nitrous acid the acetoxymethyl group is lost or the heterocyclic ring is opened (*Bachmann et al.*, J. Amer. chem. Soc., 1951, **73,** 2773, 2777).

2,4,6,8-*Tetramethyl*-1,3,5,7-*tetrazocane*, m.p. 49–50°, is obtained in the low pressure gas phase reaction of ammonia and acetaldehyde (*H. Miyama*, Jap. P. 4262/1967; C.A., 1968, **68,** 59628).

Chapter 56

Compounds with Larger Heterocyclic Rtngs: Compounds with Rings of More than Eight Atoms*

J. M. F. GAGAN

1. Nine-membered ring compounds

(a) Nine-membered ring compounds containing oxygen

(i) *Oxonins*

Oxonin (I) is prepared by the photolysis of cyclo-octatetraene monoxide (*A. G. Anastassiou* and *P. Cellura*, Chem. Comm., 1969, 903). In contrast to azonine it is slightly antiaromatic indicating that the oxygen is unwilling to donate its electrons to the π-system. It is also less stable thermally, rearranging to *cis*-8,9-dihydrobenzofuran, not the *trans* isomer on pyrolysis (*Anastassiou et al.*, Chem. Comm., 1969, 1521; 1970, 1133; 1972, 286; *J. M. Holovka, ibid*, 1969, 1522; *S. Masamune et al.*, J. Amer. chem. Soc., 1969, **91,** 7769).

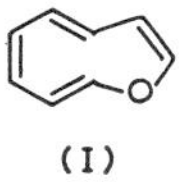
(I)

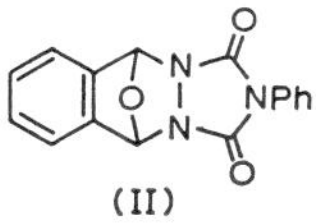

(II)

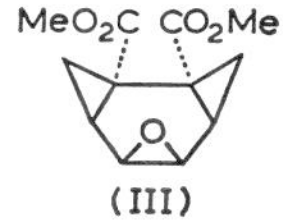

(III)

Further irradiation leads only to cyclo-octatetraene oxide, (*Anastassiou* and *Cellura*, Chem. Comm., 1969, 1521) not to a variety of products with other transannular bonds (*Holovka et al.*, J. Amer. chem. Soc., 1968, **90,** 5041) and with the dienophile 4-phenyl-1,2,4-triazoline-3,5-dione the adduct II is formed (*Anastassiou* and *Cellura*, Chem. Comm., 1970, 484). Like cyclononatetraene, oxonin is much distorted out of planar, but unlike azonines it is conformationally stable over a wide temperature range (*Anastassiou et al.*, *ibid.*, 1970, 375).

* For a general introduction to heterocyclic compounds with rings comprising seven or more atoms, and for an explanation of the nomenclature used, see p. 139 *et seq.*

Thermal isomerisation of the all-*cis*-oxatrishomobenzene (III) yields **5,6-bis-methoxycarbonyl-4,7-dihydro-oxonin,** m.p. 29–30° (*H. Prinzbach* and *D. Stusche*, Helv., 1971, **54,** 755) and the epoxide of cyclo-octa-1,3-diene opens reversibly on pyrolysis to *cis-cis*-4,5,6,7-**tetrahydro-oxonin,** which may be reduced to **oxonane** and oxidised to suberic acid (*J. K. Crandall* and *R. J. Watkins*, Tetrahedron Letters, 1967, 1717).

In a similar way ring-opening of the brosylate of 3-hydroxycyclo-octene with methanolic triethylamine yields 2-**methoxy**-2,3,4,5,6,7-**hexahydro-oxonin** in which the ring is cleaved by acid to give suberaldehyde (*J. D. Cooper et al.*, J. Amer. chem. Soc., 1971, **93,** 6294).

A trace of **oxonan-2-one** (*oxacyclononan-2-one*, *η-capryllactone*), b.p. 72–73°/11 mm, n_D^{22} 1.4595, is obtained on dehydrating 8-hydroxyoctanoic acid (*M. Stoll* and *A. Rouve*, Helv., 1935, **18,** 1087); a better route is the Baeyer–Villiger ring-expansion of cyclo-octanone (*R. Huisgen* and *H. Ott*, Angew. Chem., 1958, **70,** 312; Tetrahedron, 1959, **6,** 253; *W. E. Closson et al.*, J. org. Chem., 1967, **32,** 3160; *P. S. Starcher* and *B. Phillips*, J. Amer. chem. Soc., 1958, **80,** 4079; cf. *Knapsack A-G.*, Neth. Appl. 6,516,355/1964; C.A., 1966, **65,** 15237). The nine-membered ring allows the lactone group to adopt the more favourable *trans* conformation.

A simple route to **oxonane-2,6-dione** (IV; X = O, *n* = 1) is available from 2,3-trimethylene-5,6-dihydro-4*H*-pyran; treatment with acid and *n*-butyl nitrite gives the oxime IV (X = NOH), which is then hydrolysed. Analogous macrolides (*n* = 2–4, 8) are prepared similarly (*J. R. Mahajan et al.*, Chem. Comm., 1972, 1078).

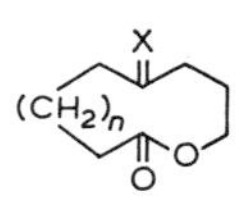

(IV)

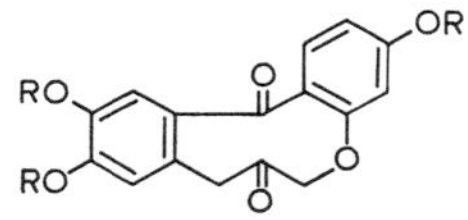

(V)

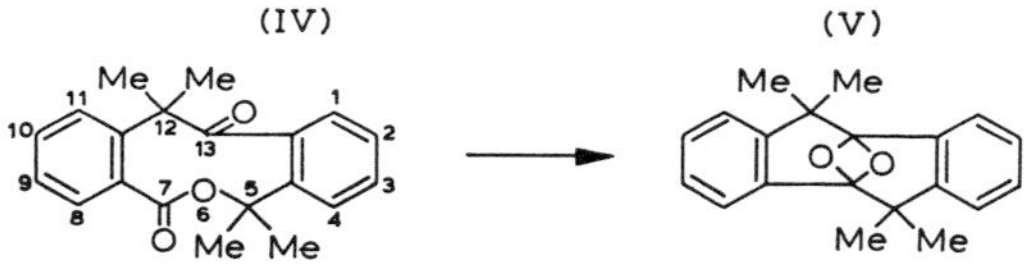

(VI) (VII)

The Baeyer–Villiger ring expansion of 1-benzocyclo-octenone with perbenzoic acid shows aryl-oxygen migration resulting in 2,3,4,5,6,7-**hexahydro-1-benzoxonin-2-one,** b.p. 120–130°/0.1 mm. The heterocyclic ring is opened with hydrazine or alkali (*Huisgen et al.*, Chem. Ber., 1957, **90,** 1958).

Brazilone (V; R = H) and **haematoxylone** are dibenz[*b*,*e*] oxonins (see C.C.C., 2nd Edn., Vol. IVE, pp. 431 *et seq.*). The trimethyl ether (V; R = Me) and the haematoxylone analogue show transannular ring-closures to naphthobenzofurans with acetic anhydride (*O. Dann* and *H. Hofmann*, Ann., 1963, **667,** 116; Chem. Ber., 1965, **98,** 1498).

7,7-*Diphenyl*-5,7,8,9-*tetrahydrodibenz*[c,e]*oxonin*, m.p. 176–177°, is prepared by cyclising the methiodide of 2(dimethylaminomethyl)-2′-(3,3-diphenyl-3-hydroxypropyl)-biphenyl with moist silver oxide followed by strong alkali (*H. W. Bersch* and *G. Hübner*, Arch. Pharm., 1962, **295,** 95).

Oxidation of 5,5,10,10-tetramethyl-4b,5,9b,10-tetrahydroindeno[2,1-a]indene with chromium trioxide yields 5,5,12,12-*tetramethyltetrahydrodibenz*[c,g]*oxonin*-7,13-*dione* (VI), m.p. 203–204°. With lithium tetrahydridoaluminate VI forms the dioxygen bridged compound VII (*R. M. Lagidze et al.*, Izvest. Akad. Nauk SSSR, 1966, 1459).

Dibenz[*d,f*]oxonin (VIII), m.p. 99°, is prepared in a double Wittig reaction on 2,2′-biphenyldicarbaldehyde. It is hydrogenated to a 5,6,7,8-*tetrahydro* derivative, m.p. 82–83°, and undergoes ring-contraction in acid (*R. H. Mitchell et al.*, J. Amer. chem. Soc., 1968, **90,** 7372). Reaction of the dicarbaldehyde with zinc and ethyl bromoacetate yields 6,8-*bisethoxycarbonyl*-5,7,8,9-*tetrahydrodibenz*[d,f]*oxonin*, b.p. 230–235°/1 mm (*E. D. Bergmann et al.*, J. chem. Soc., C, 1967, 328).

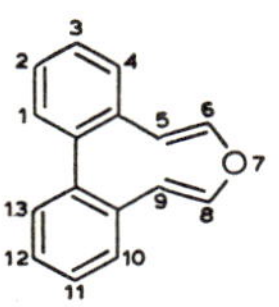
(VIII)

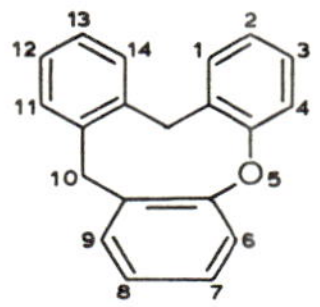
(IX)

10,15-**Dihydrotribenz[*b,e,h*]oxonin** (IX), m.p. 194–195°, is prepared by treating 2,2′-di(hydroxymethyl)diphenyl ether in benzene with concentrated sulphuric acid; it exists in a flexible crown conformation (*T. Sato et al.*, Chem. Comm., 1972, 579).

(*ii*) *Dioxonins*

If ethyl orthoformate and hexane-1,6-diol are heated with a trace of aluminium chloride, 2-**ethoxy**-1,3-**dioxonane,** m.p. 196°, is obtained (*R. C. Mehrotra* and *R. P. Narain*, Indian J. appl. Chem., 1965, **28,** 53). The dibenzo[*d,h*]-1,3-dioxonin XI (n = 1) and larger ring homologues (up to n = 10) are prepared by treating Schiff bases X with aluminium amalgam (*J. M. Bastian* and *R. Jaunin*, Helv., 1963, **46,** 1248).

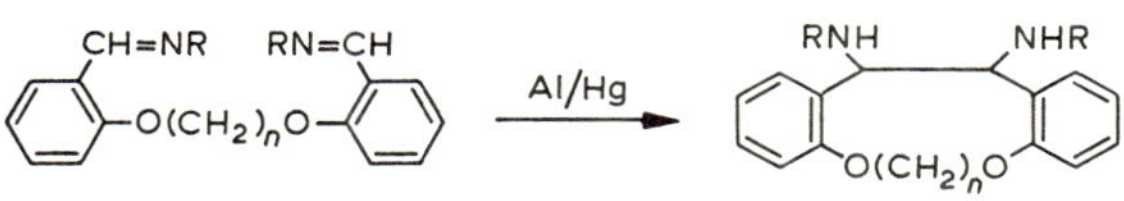

(X) (XI)

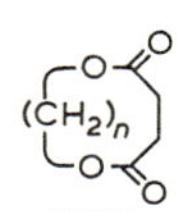

(XII)

3,4,5,6-**Tetrahydro**-2*H*-1,7-**benzodioxonin,** b.p. 119–120°/9 mm, is one of the cyclic ethers prepared from catechol and α,ω-dibromoalkanes (*K. Ziegler et al.*, Ann., 1937, **528,** 162; *G. Baddeley* and *N. H. P. Smith*, J. chem. Soc., 1961, 2516) (cf. p. 168).

Unsubstituted cyclic esters of type XII have been obtained by the high temperature depolymerisation of linear polyesters in the presence of a catalyst. This is a general route but odd-membered rings are more difficult to produce than even-membered rings. 1,5-**Dioxonan**-6,9-**dione** (XII; n = 3) has m.p. 81° (*E. W. Spanagel* and *W. H. Carothers*, J. Amer. chem. Soc., 1935, **57,** 929; *S. E. Drewes* and *P. E. Coleman*, J. chem. Soc. Perkin I, 1972, 2148).

The structure of the **antimycin A** antibiotics (XIII) are based on the 1,5-dioxonane-2,6-dione ring system (*E. E. van Tamelen et al.*, J. Amer. chem. Soc., 1961, **83,** 1639). Close approaches to total synthesis have been made by *M. Kinoshita* and *S. Umezawa* (C.A., 1971, **74,** 87932; Bull. chem. Soc., Japan, 1969, **42,** 854; 1970, **43,** 897). Pyrolysis yields di- and tetra-hydro-2-furanones (*G. Schilling et al.*, J. Antibiot., 1970, **23,** 81).

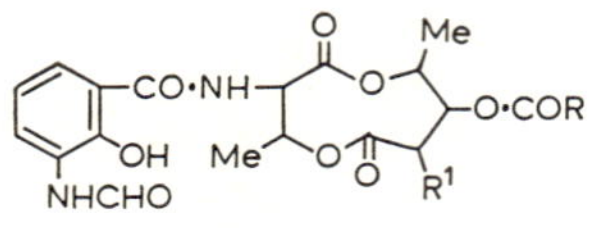

(XIII)

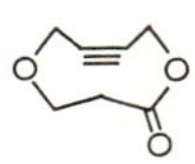

(XIV)

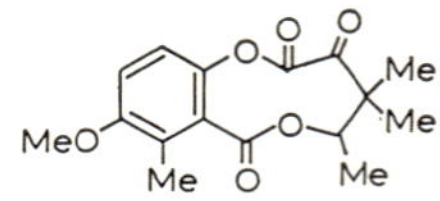

(XV)

Michael addition of sodio-bishydroxymethylacetylene to methyl acrylate followed by transesterification with tosic acid gave 1,5-**dioxacyclonon-7-yn-2-one** (XIV) (*T. A. Favorskaya* and *S. A. Poroshina*, Zhur. org. Khim., 1970, **6**, 1416; 1969, **5**, 2244). 7,8-*Dihydro-*6H-*dibenzo*[f,h]-1,5-*dioxonin*, m.p. 69–70°, and its 7,7′-*bis-spiro analogue*, m.p. 222°, have been synthesised from 2,2′-dihydroxybiphenyl with 1,3-dibromopropane or 2,2-bis(hydroxymethyl)-1,3-dibromopropane (*S. Smolinski et al.*, Tetrahedron, 1971, **27**, 265; 1969, **25**, 5431).

The sole representative in the 1,6-**benzodioxonin** series is compound XV, obtained by ozonolysis of the methyl ether of the natural product **glaupalol**, a coumarin derivative isolated from the rhizomes of *Glaucidium palmatum* Sieb et Succ. (*H. Irie et al.*, Chem. Comm., 1967, 547; Yakugaku Zasshi, 1968, **88**, 627).

(*iii*) *Ring compounds with three or more atoms in the ring*

When *syn*-benzene trioxide is heated to 200° it is converted into *cis,cis,cis*-1,4,7-**trioxonin**:

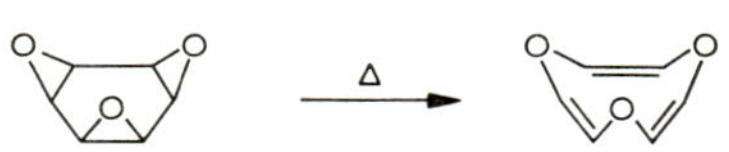

Unlike the benzene oxide–oxepin and benzene dioxide–dioxocin systems, this reaction could not be reversed to give observable amounts of trioxide. The *cis,cis,trans*-trioxide fails to rearrange even at 500° (*E. Vogel et al.*, Angew. Chem. intern. Edn., 1972, **11**, 939; *R. Schwesinger* and *H. Prinzbach*, *ibid.*, p. 942).

2,5-Dimethylhexane-2,5-bis(hydroperoxide) reacts with phosgene to yield 6,6,9,9-**tetramethyl**-1,2,4,5-**tetroxonan-3-one,** m.p. 101–103°; with fumaroyl chloride and phthaloyl chloride twelve-membered tetroxa-rings are obtained (*W. E. Tiernan Inc.*, B.P. 936,008/1959 C.A., 1964, **60**, 1594). Acetone and copper sulphate (*R. Criegee et al.*, Ann., 1949, **565**, 7) and hydroxyacetone in sulphuric acid also bring about this ring-closure (*Y. G. Chang*, U.S.P. 3,579,541/1968; C.A., 1971, **75**, 64892).

In the presence of boron trifluoride, chloral inserts into cyclic acetals. From 1,3-dioxepane and 1,3-dioxolane the products are 2-**trichloromethyl**-1,3,5-**trioxonane** and 2,4-*bis*(**trichloromethyl**)-1,3,5,8-**tetroxonane** (XVI). Many larger rings have been made similarly (*H. J. Dietrich*, *J. V. Karabinos* and *M. C. Raes*, J. heterocycl. Chem., 1970, **7**, 27; U.S.P. 3,534,064/1967, C.A., 1971, **74**, 22910).

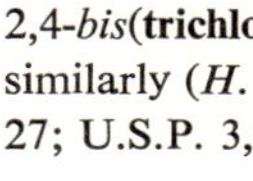

(XVI)

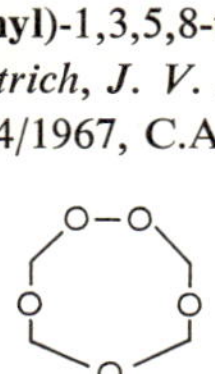

(XVII)

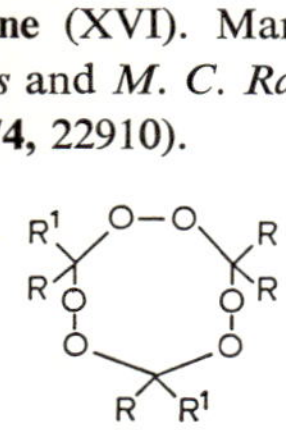

(XVIII)

1,2,4,6,8-**Pentoxonane** (XVII) is one of the dehydration products of the peroxide of formaldehyde dimethylol,$-\!\!\leftarrow OCH_2O \cdot CH_2OH)_2$ (*A. Rieche* and *R. Meister*, Ber., 1933, **66,** 718). The major product is pertrioxymethylene.

The highly explosive 3,3,6,6,9,9-**hexamethyl**-1,2,4,5,7,8-**hexoxonane** (trimeric *acetone peroxide*) (XVIII; R, R^1 = Me), m.p. 98°, is prepared by the oxidation of acetone, or indirectly of isopropenyl ether, with hydrogen peroxide (*A. Baeyer* and *V. Villiger*, Ber., 1899, **32,** 3632; 1900, **33,** 859; *R. Wolffenstein, ibid.*, 1895, **28,** 2267; *Criegee* and *K. Metz, Chem. Ber.*, 1956, **89,** 1714); propanal yields XVIII (R = *H*, R^1 = *Et*), n_D^{20} 1.4390 (*Reiche* and *Meister, ibid.*, 1939, **72,** 1938). A more general route is to treat the bis(hydroxy-, alkyl)peroxides, $(HO \cdot CR_2O \rightarrow_2$, from ketones and excess hydrogen peroxide with a ketone, in the presence of copper sulphate and a strong acid. *Hexoxonins* XVIII (R, R^1 = Me) and [R, R^1 = $\leftarrow CH_2 \rightarrow_5$], m.p. 91–92°, are obtained in this way (*Rieche*, Angew. Chem., 1958, **70,** 251; 1961, **73,** 57; *Criegee et al.*, Ann., 1949, **565,** 7; *P. R. Story et al.*, J. org. Chem., 1970, **35,** 3059). They are not decomposed by iodine or catalytic hydrogenation, but are hydrolysed to the ketone and hydrogen peroxide.

(*b*) *Nine-membered ring compounds containing sulphur*

Thionin has not yet been prepared. An attempt to bring about a photochemical isomerisation of cyclo-octatetraene thiepoxide gave thiabarbaralene (XIX) instead (*A. G. Anastassiou* and *B. Y. H. Chao*, Chem. Comm., 1972, 277). The nearest approaches are the *cis* and *trans* isomers of the annellated

(XIX)

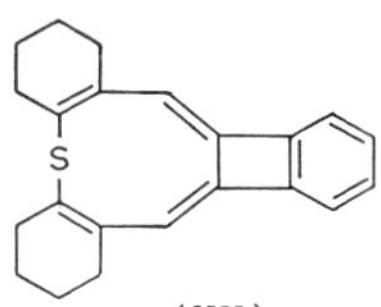

(XX)

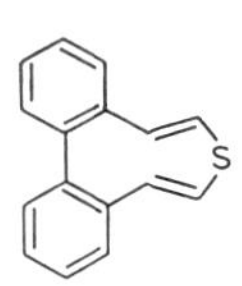

(XXI)

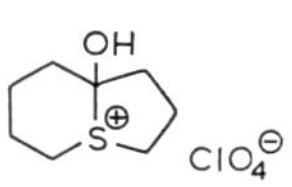

(XXII)

thionin XX, prepared in the double Wittig reaction of bis(2-formylcyclohexenyl) sulphide with the bistriphenylphosphorane from 1,2-dibromobenzocyclobutene (*F. Sondheimer et al.*, J. Amer. chem. Soc., 1970, **92,** 4492) and *dibenzo*[d,f]*thionin* (XXI) m.p. 121–122°, prepared similarly, which on catalytic reduction gives the 5,6,8,9-*tetrahydro* derivative, m.p. 74–75° (*R. H. Mitchell et al., ibid.*, 1968, **90,** 7372).

Thionane *(thiacyclononane)*, b.p. 62–65°/10 mm, 1,1-*dioxide*, m.p. 82°, is formed together with its dimer when sodium sulphide reacts with 1,8-dibromo-octane at high dilution; yields of both compounds are poor (*A. Müller et al.*, Monatsh., 1953, **84,** 1206).

Thionan-5-one, prepared by Dieckmann cyclisation and hydrolysis, gives the bicyclic salt XXII with perchloric acid (*C. G. Overberger et al.*, J. Amer. chem. Soc., 1962, **84,** 2814). Attempts to form the ten-membered analogues produced a twenty-membered dimer. Addition of sulphur dichloride to 1,7-octadiene yields 3,8-*dichlorothionane*, b.p.

80–100 °/0.3 mm; reaction with nucleophiles is said to give bicyclic as well as substitution products (*Dunlop Rubber Co. Ltd.*, C.A., 1966, **65,** 12180).

10,15-**Dihydrotribenzo**[*b,e,h*]**thionin,** m.p. 218–219 °, is obtained in good yield in a benzene solution of 2,2′-di(hydroxymethyl)diphenyl sulphide treated with concentrated sulphuric acid (*T. Sato et al.*, Chem. Comm., 1972, 579). It appears to adopt a rigid crown conformation and is oxidised to a *sulphone*, m.p. 209–210°.

A small amount of 1,2-**dithionane,** b.p. 86–89 °/2 mm, n_D^{25} 1.5642, was obtained by *J. G. Affleck* and *G. Dougherty* (J. org. Chem., 1950, **15,** 865) and *A. Schöberl* and *H. Gräfje* (Angew. Chem., 1957, **69,** 713; Ann., 1958, **614,** 66).

3,4,5,7-*Tetrahydro*-1H-2,6-*benzodithionin*, m.p. 88–90 °, and 6,7-*dihydro*-8H-*dibenzo*-[f,h]1,5-*dithionin*, m.p. 69–70 °, are both prepared from dithiols and dihalides in alkaline solution (*D. W. Allen et al.*, Chem. and Ind., 1969, 522; J. chem. Soc. C, 1971, 3454; *R. H. Mitchell* and *V. Boekelheide*, J. heterocycl. Chem., 1969, **6,** 981).

(*c*) *Nine-membered ring compounds containing nitrogen*

(*i*) *Rings containing one nitrogen atom*

Two photosensitised ring-expansion routes to *ethoxycarbonylazonine* (XXIV; R = CO_2Et) are known (*A. G. Anastassiou* and *J. H. Gebrian*, J. Amer. chem. Soc., 1969, **91,** 4011; *S. Masamune et al.*, Chem. Comm., 1969, 1204):

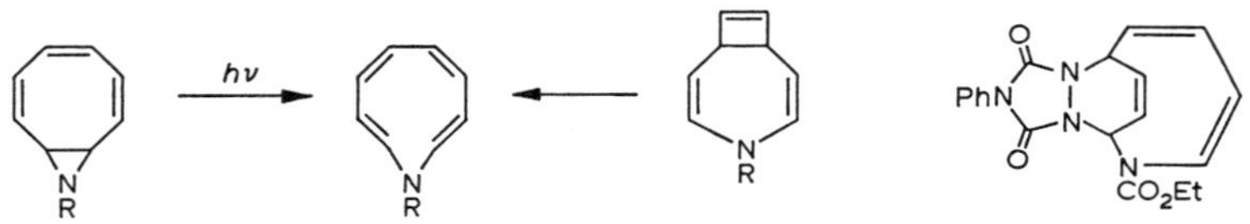

(XXIII) (XXIV) (XXV)

A brief low temperature treatment of this compound with base yields **azonine** (XXIV; R = H) (*Anastassiou* and *Gebrian*, Tetrahedron letters, 1970, 825). Its alkali metal salts are more readily prepared by treating XXIV (R = CO_2Et) with the corresponding *tert*-butoxide in tetrahydrofuran, and 1-substituted azonines, *e.g.* 1-*acetyl* (XXIV; R = Ac) and 1-*methyl* (XXIV; R = Me) are obtained from the anion and the appropriate halide (*R. T. Seidner* and *Masamune*, Chem. Comm., 1972, 149; *Anastassiou et al., ibid.*, 1970, 1133).

Azonine is very reactive but quite stable in an inert atmosphere; its *N*-proton is more acidic than that of pyrrole (*Anastassiou* and *S. W. Eachus*, J. Amer. chem. Soc., 1972, **94,** 2537). On irradiation it is reconverted into 9-azabicyclo[6.1.0]nona-2,4,6-triene (XXIII; R = H) (*Anastassiou et al.*, Chem. Comm., 1972, 531; Tetrahedron Letters, 1969, 5239) and on catalytic reduction it gives **azonane.** If XXIV (R = CO_2Et) is heated for 15 min at 55 ° it gives *N*-ethoxycarbonyl-*cis*-8,9-dihydroindole.

Azonine and its anion are shown by spectroscopy to be planar with highly delocalised 10π electron systems, sustaining a greater diamagnetic ring current than pyrrole or benzene. This 10π aromaticity is less pronounced in 1-methylazonine, which has a distorted ring possibly due to the steric clash between the 1-substituent and the 2,9-protons, and is absent from azonines with electron withdrawing *N*-substituents, which like oxonin, have polyene characteristics (*idem*, Chem. Comm., 1970, 375; 1133; 1972, 286; *K. Hojo* and *Masamune*, J. Amer. chem. Soc., 1970, **92**, 6690).

4-Phenyl-1,2,4-triazoline-3,5-dione adds to 1-ethoxycarbonylazonine to give the unsymmetrical adduct XXV (*Anastassiou et al.*, Chem. Comm., 1972, 325).

A thermal isomerisation leads to 5,6-*bismethoxycarbonyl*)-1-(p-*tosyl*)-4,7-*dihydroazonine* (XXVI), m.p. 152–153° (*H. Prinzbach* and *D. Stusche*, Helv., 1971, **54**, 755).

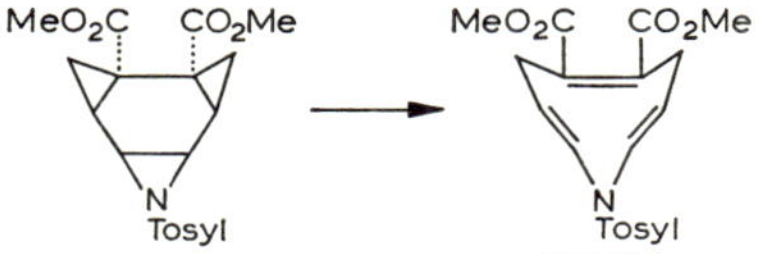

(XXVI)

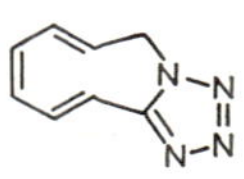

(XXVII)

A Schmidt reaction on cyclo-octa-2,4,6-trienone in trifluoroacetic acid yields the tetrazolo derivative XXVII (*A. H. Khuthier* and *J. C. Robertson*, J. org. Chem., 1970, **35**, 3760) and both *syn*- and *anti*-oxime tosylates XXVIII rearrange to azonine derivatives, showing that both alkyl and vinyl migration occurs (*T. Sato et al.*, Tetrahedron, 1971, **27**, 5381):

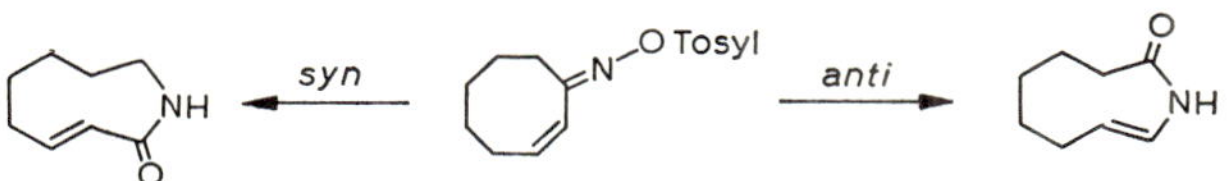

(XXVIII)

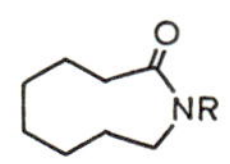

(XXIX)

2-*Methoxy*-3H-4,5,6,7,8,9-*hexahydroazonine (capryl-lactim methyl ether)*, b.p. 89°/14 mm, is prepared by *O*-methylation of capryl-lactam (*S. Petersen* and *E. Tietze*, Chem. Ber., 1957, **90**, 909; Ann., 1959, **623**, 166). Its methoxy group may be replaced by amino compounds, and on heating it yields 1-*methylcapryl-lactam* (XXIX; R = Me), b.p. 140°/14 mm.

Azonane* *(perhydroazocine, azacyclononane, octamethylenimine)*, b.p. 90–92°/24 mm, n_D^{20} 1.4771, is made by the reduction of *capryl-lactam* (**azonan-2-one**) (XXIX; R = H), m.p. 73–74°, which is prepared, like caprolactam, by Beckman or Schmidt ring expansions of cyclo-octanone (*L. Ruzicka et al.*, Helv., 1933, **16**, 1323; 1949, **32**, 544; *D. D. Coffman et al.*, J. polymer Sci., 1948, **3**, 85; *P. Lukes* and *J. Malek*, Chem. Listy, 1951, **45**, 72; *F. F. Blicke* and *N. J. Doorenbos*, J. Amer. chem. Soc., 1954, **76**, 2317; *R. Huisgen et al.*, Chem. Ber., 1957, **90**, 1437; *O. von Schickh* and *H. Metzger*, U.S. P. 2,883,377/1959; C.A., 1959, **53**, 15989; *G. Rapp* and *O. Schlichting*, G.P. 1,046,623/1958; 1,089,761/1960; 1961, **55**, 9441; 15355). Capryl-lactam polymerises more readily than caprolactam and gives a polymer with high thermal stability and good fibre properties (*K. Dachs* and *E. Schwartz*, Angew. Chem. Intern. Edn., 1962, **1**, 430). When refluxed in xylene with phosphorus pentasulphide or phosphorus and selenium, capryl-lactam yields the corresponding *capryl thiolactam* (**azonane-2-thione**), m.p. 90–93°, and caprylselenolactam (**2-selenoperhydro-**

* This trivial name is not accepted officially in the I.U.P.A.C. Rules, "perhydroazonine" being preferred. It is, however, a convenient name and is frequently used (cf. footnote, p. 402).

azocine, azonane-2-selenone) (*H. E. Hallam* and *C. M. Jones*, J. chem. Soc., A, 1969, 1033; *H. K. Hall*, J. Amer. chem. Soc., 1958, **80**, 6404; *O. Riester*, C.A., 1962, **56**, 6827).

Treatment of azonane with formaldehyde and formic acid yields 1-*methylazonane*, b.p. 180–181°, n_D^{25} 1.4640 (*N. J. Leonard* and *W. K. Musker*, J. Amer. chem. Soc., 1959, **81**, 5631) and ring-expansion of cyclo-octanone oxime tosylate with methyl-lithium gives 2-*methylazonane, benzenesulphonate*, m.p. 84° (*N. W. Gabel*, J. org. Chem., 1964, **29**, 3129).

Acyloin cyclisation of the diesters $RN[-(CH_2)_3CO_2Et]_2$ gives 1-*methyl* (XXX; R = Me, m.p. 95–97°) and 1-*ethyl-5-hydroxyazonan-6-one* (XXX; R = Et, b.p. 64–65°/0.15 mm, n_D^{20} 1.4999) and higher homologues. Many of their physical properties have been measured and indicate considerable transannular interaction between nitrogen atom and carbonyl group.

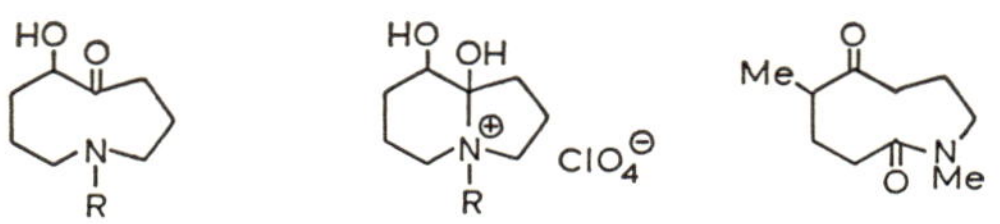

(XXX) (XXXI) (XXXII)

Chemically, this is marked by low carbonyl reactivity, weak basic character and the formation of a transannular bond in the perchlorate salts XXXI (R = Me or Et). This phenomenon is restricted to 8-, 9- and 10-membered rings (*q.v.*) and may be attributed to the relief of non-bonding interaction when a favourable bicyclic form is adopted. A bulky or electron-withdrawing substituent on nitrogen affecting the spatial orientation or availability of the nitrogen lone pair, reduces the interaction, resulting in *N*-protonation rather than *O*-protonation in compounds XXX (R = Pr), b.p. 85–87°/0.9 mm, or XXX (R = Bu^t), b.p. 101°/0.7 mm) (*R. Robinson et al.*, J. chem. Soc., 1922, **121**, 427; Chem. and Ind., 1953, 944; *Leonard et al.*, J. Amer. chem. Soc., 1954, **76**, 630, 3463, 5708, 1955, **77**, 6324, 6237, 6239, 6241, 6245; 1958, **80**, 4858). Similar behaviour is shown by diketone XXXII, prepared from 2-ethyl-idene-1-methyl-pyrrolidine and ethyl acrylate (*O. Červinka*, Chem. and Ind., 1959, 1129; Coll. Czech. chem. Comm., 1960, **25**, 1174). **1-Chloroazonane** collapses to indolizidine (XXXIII; R = H) when treated with silver ions; a much lower yield of transannular product is found with eight- or ten-membered analogues (*J. W. Apsimon et al.*, J. Amer. chem. Soc., 1965, **87**, 678). 1-Chloroazonan-2-one gives a mixture of indolizidin-2- and -9-ones (*O. E. Edwards et al.*, Canad. J. Chem., 1971, **49**, 1648).

Several azonanes have been prepared by cleaving the transannular bond of a perhydro-indolizine (XXXIII; R = H). 5-*Hydroxyazonane*, b.p. 95°/0.1 mm, *picrate*, m.p. 176–177°, and 4-*hydroxyazonane*, b.p. 70°/0.2 mm, *picrate*, m.p. 138–139°, are obtained by electrolytic reduction of 1-azabicyclo[4.3.0]nonan-5- and -7-one, respectively (*Leonard et al.*, J. Amer. chem. Soc., 1952, **74**, 4620), and the indolizidine XXXIII (R = CN) has been converted into 1-*methylazonane* (XXXIV; R = H), b.p. 62°/16 mm, *picrate*, m.p. 199–200°, and 1-methylazonan-5-one (XXXVI; R^1 = O) after cleaving the transannular bond of XXXV (R = CN) with lithium in liquid ammonia (*A. Yoshio et al.*, Yakugaku Zasshi, 1970, **90**, 1424; cf. *G. H. Foster et al.*, Chem. Comm., 1967, 21). 1-*Methyl-5-hydroxyazonane* (XXXIV; R = OH), b.p. 50°/0.05 mm, prepared by reduction of the corresponding ketone with lithium tetrahydridoaluminate forms a transannular bond, giving the salt XXXV (R = H) when treated with acid (*A. J. Sisti* and *D. L. Lohner*, J. org. Chem., 1967, **32**, 2026).

A technique termed peripheral synthesis has been developed by *M. G. Reinecke et al.*, (*ibid.*, 1972, **37,** 3489, 3494; Tetrahedron Letters, 1965, 3549):

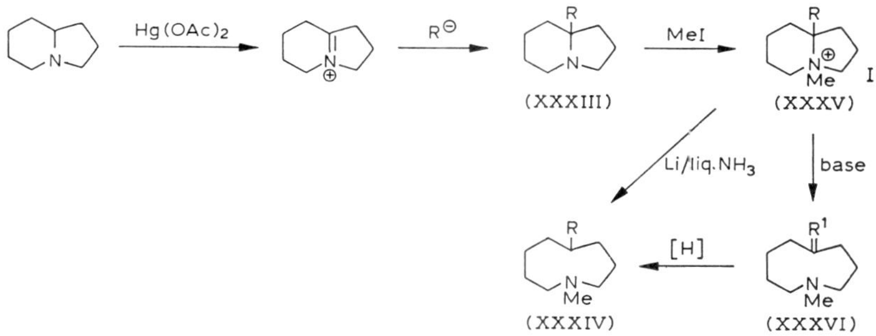

The nucleophile ($R^{\ominus}$) may be an organolithium, or a Grignard or Reformatsky reagent; 5-*methylene*- (XXXVI; R^1 = CH_2, b.p. 38–39°/0.1 mm; 5-*ethyl* (XXXV; R = Et, *methiodide*, m.p. 196°) and 5-*phenyl*- (XXXV; R = Ph, b.p. 118–119°/1.8 mm). -1-*methylazonanes* have been prepared using this route (cf. *W. J. Houlihan* and *R. E. Manning*, U.S.P. 3,498,988/1967; C.A., 1970, **72,** 121384). Osmium tetroxide–periodate cleavage of the exocyclic double bond of XXXVI (R^1 = CH_2) yields **1-methylazonan-5-one** (XXXVI; R^1 = O), which undergoes Wolff–Kishner reduction to 1-methylazonane (XXXIV; R = H).

Beckmann or Schmidt reactions with 1-benzocyclo-octenones yield 2,3,4,5,6,7-*hexahydro*-1H-1-*benzazonin*-2-*one* (XXXVIII; X = H_2), m.p. 142–143°, and a corresponding tetrazole (*Huisgen et al.*, Ann., 1954, **586,** 30; cf. Chem. Ber., 1957, **90,** 1844, 1946). None of the isomeric 2-benzazonin-1-one could be obtained. The nitrogen may be alkylated, and the carbonyl group reduced to methylene by lithium tetrahydridoaluminate (*H. Morren*, B.P. 1,039,113/1964; C.A., 1966, **65,** 12181).

Irradiation of 1-phenylazepan-2-one yields 2,3,4,5,6,7-*hexahydro*-1H-1-*benzazonin*-7-*one*, m.p. 74° (*M. Fischer*, Tetrahedron Letters, 1968, 4295; Chem. Ber., 1969, **102,** 342).

The 3*H*-indole derivative XXXVII (X = OOH) rearranges with acid or alkali to **2,3,4,5,6,7-hexahydro-1*H*-1-benzazonine-2,7-dione** (XXXVIII; X = O), m.p. 157°, also obtained on ozonolysis of tetrahydrocarbazole (*B. Witkop* and *J. B. Patrick*, J. Amer. chem. Soc., 1950, **72,** 1428; 1951, **73,** 2188, 2196, 2641; 1955, **77,** 6595; *R. B. Carlin* and *M. S. Moores*, *ibid.*, 1959, **81,** 1259; *D. W. Ockenden* and *K. Schofield*, J. chem. Soc., 1953, 612; *A. Robertson et al.*, *ibid.*, 2440). A similar ring expansion occurs when XXXVII (X = $CHCl_2$) is converted into benzazonine XXXVIII (X = CHCl) with sodium hydroxide (*M. F. Bartlett et al.*, J. Amer. chem. Soc., 1958, **80,** 126). Treatment of XXXVIII (X = O) with base gives 2,3-trimethylene-4-quinolone, showing the alternative mode of transannular bonding (*Witkop et al.*, *ibid.*, 1951, **73,** 2641) but the monocyclic azonandione, obtained similarly collapses to an indolizidine (*idem*, *ibid.*, 1955, **77,** 6595).

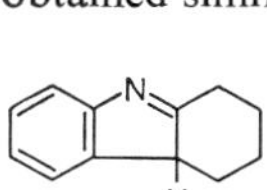

(XXXVII)

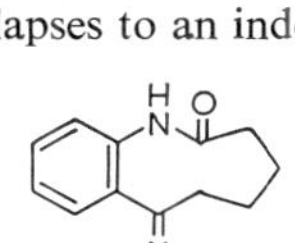

(XXXVIII)

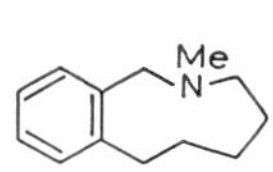

(XXXIX)

If a 1,1-dimethyl-2-phenylpiperidinium salt is treated with sodamide in liquid ammonia, a Sommelet–Hauser rearrangement yields 2-*methyl*-1,3,4,5,6,7-*hexahydro*-2H-2-*benzazonine* (XXXIX), b.p. 104–106°/4 mm, n_D^{24} 1.5380, *picrate*, m.p. 169–170° (*D. Lednicer* and *C. R. Hauser*, *ibid.*, 1957, **79,** 4449).

Irradiation of *N*-chloroacetyl-4(3,4-dimethoxyphenyl)-butylamine yields 9,10-*dimethoxy*-1,2,4,5,6,7-*hexahydro*-3H-3-*benzazonine* (XL), m.p. 178–179° (*O. Yonemitsu et al.*, Chem. Comm., 1971, 745).

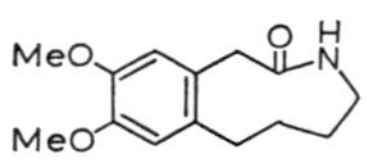

(XL)

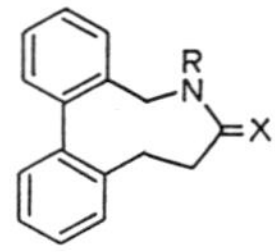

(XLI)

A Schmidt reaction on dibenzo[*a,c*]cyclo-octadien-6-one gives 6,7,8,9-*tetrahydro*-5H-*dibenz*[c,e]*azonin-7-one* (XLI; X = O, R = H), m.p. 208–213°, reduced by lithium tetrahydridoaluminate to 6,7,8,9-*tetrahydro*-5H-*dibenz*[c,e]*azonine* (XLI; X = H_2, R = H), *hydrobromide*, m.p. 244–246°. Formaldehyde and formic acid yield the N-*methyl* derivative (XLI; X = H_2, R = Me), *picrate*, m.p. 196–197° (*K. Kotera et al.*, Jap. P. 5380, 5382–4/1966; C.A., 1966, **65,** 2236; 1968, **68,** 105398).

The synthesis of a **dibenz[*c,f*]azonine** (XLII) is as shown (*M. Sainsbury et al.*, Tetrahedron, 1969, **25,** 1881):

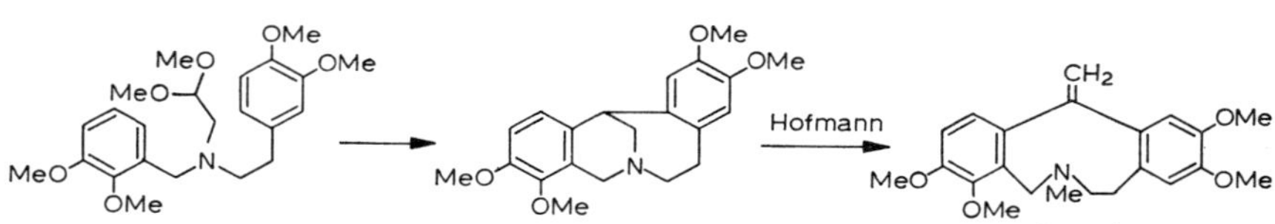

(XLII)

6-*Methyl*-6,7-*dihydro*-5H-*dibenz*[c,g]*azonine* (XLIII), m.p. 59–61°, is also the product of a Hofmann degradation. Its methobromide undergoes ring-contraction to a dibenzoctatriene (*G. Wittig et al.*, Ann., 1951, **572,** 1):

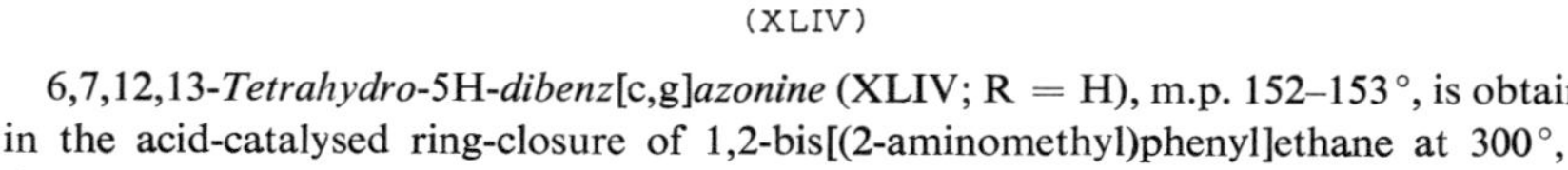

(XLIII)

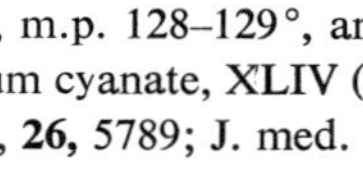

(XLIV)

6,7,12,13-*Tetrahydro*-5H-*dibenz*[c,g]*azonine* (XLIV; R = H), m.p. 152–153°, is obtained in the acid-catalysed ring-closure of 1,2-bis[(2-aminomethyl)phenyl]ethane at 300°, by decyanation of XLIV (R = CN), m.p. 128–129°, and by debenzylation of XLIV (R = CH_2Ph), m.p. 110°. With potassium cyanate, XLIV (R = H) gives XLIV (R = $CONH_2$) (*G. Pala et al.*, Tetrahedron, 1970, **26,** 5789; J. med. Chem., 1971, **14,** 174).

The **dibenz[*d,f*]azonine** system is present in the alkaloid **protostephanine** (XLV; R = Me, R¹ = OMe) (*K. Takeda*, Bull. agr. Chem. Soc. Japan, 1956, **20,** 165; C.A., 1957, **51,** 11364) and may be derived from several other alkaloids (*J. A. Berson* and *M. A. Greenbaum*, J. Amer. chem. Soc., 1958, **80,** 445; *Y. Inubushi et al.*, Tetrahedron Letters 1969, 153; Chem. pharm. Bull. Japan, 1970, **18,** 1951). A degradation sequence is shown below (*A. Mondon et al.*, Chem. Ber., 1965, **98,** 46):

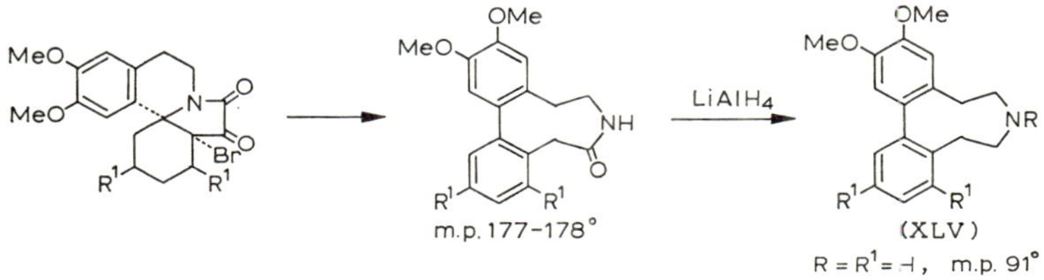

7-*Methyl*-6,7,8,9-*tetrahydro*-5H-*dibenz*[d,f]*azonine*, *hydrobromide*, m.p. 248–251°, and analogous compounds have been prepared from 2,2′-bis(2-bromoethyl)biphenyl and alkylamines (*K. Kotera et al., loc. cit.*), but the syntheses of protostephanine are similar to biosynthetic routes (*A. R. Battersby et al.*, Chem. Comm., 1968, 1214; *B. Pecherer et al.*, J. med. Chem., 1969, **12,** 149).

(*ii*) *Nine-membered rings containing more than one nitrogen atom*

Trans-4,5,6,7,8,9-**hexahydro**-3*H*-1,2-**diazonine** (XLVI) is obtained in low yield by alkaline hydrolysis followed by oxidation of the product from 1,7-dibromo- or 1,7-ditosyloxy-heptane and *symm*-bisethoxycarbonylhydrazine with mercuric oxide. The ten-membered ring analogue is prepared similarly (*C. C. Overberger et al.*, Tetrahedron Letters, 1972, 4565). The adduct XLVII, from 1,2-dibenzylidenecyclopentane and diethyl azodicarboxylate, is cleaved to the *diethyl* 3,9-*diphenyl*-1,2-*diazonane*-4,8-*dione*-1,2-*dicarboxylate* (XLVIII), m.p. 97–102°, with osmium tetroxide and lead tetra-acetate. With acid or boron trifluoride, XLVIII closes in the opposite sense giving the bicyclic compound XLIX (*idem*, J. org. Chem., 1961, **26,** 4359).

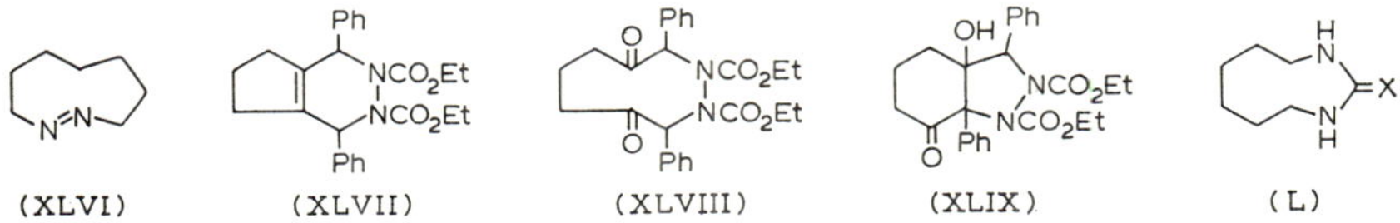

(XLVI) (XLVII) (XLVIII) (XLIX) (L)

Beckmann ring-expansion of capryl-lactam oxime gives 1,3-**diazonan-2-one*** (2-*oxo-perhydrodiazonine*) (L; X = O), m.p. 185° (270°). Carbon disulphide reacts with 1,6-diaminohexane to give 1,3-**diazonane-2-thione** (L; X = S), m.p. 216°, also prepared by treating the cyclic urea with phosphorus pentasulphide. With mercuric oxide the thione gives a cyclic carbodiimide having cumulated double bonds in the nine-membered ring (*H. Behringer* and *H. Meier*, Ann., 1957, **607,** 67; 73).

* See footnote on p. 433.

4-Methyl-2,3,4,5,6,7-hexahydro-1*H*-1,4-benzodiazonine-2,7-dione (LI) is obtained in the ozonolysis reaction shown:

(LI)

Like its monoaza analogue, it forms a transannular bond in basic media (*B. Witkop et al.*, Ann., 1947, **558,** 91; J. Amer. chem. Soc., 1953, **75,** 3371). *Hexahydro*-1H-1,7-*benzodiazonine*, m.p. 206°, is prepared like its diazocine and diazecine homologues (see p. 415). 9,10-*Dimethoxy*-5-*methyl*-2,3,4,5,6,7-*hexahydro*-1H-2,5-*benzodiazonin*-3-*one* (LII; X = O), *hydrochloride* m.p. 252–253°, is also prepared by transannular bond-cleavage thus:

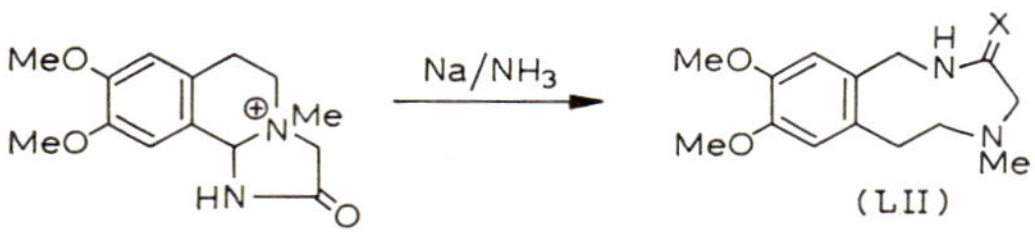

(LII)

Ketone LII (X = O) is reduced to the *benzodiazonine* LII (X = H_2), *dimaleate*, m.p. 154–158° (*Houlihan* and *Manning*, U.S.P. 3,551,411/1967; C.A., 1971, **75,** 5976).

Condensation of 1,3-diaminopropane with 3-benzoylpropionyl chloride gives 9-*phenyl*-3,4,5,6,7,8-*hexahydro*-2H-1,5-*diazonin*-6-*one*, m.p. 130–131° (*T. S. Sulkowski*, U.S.P. 3,293,243/1964; C.A., 1967, **66,** 46444). Catalytic hydrogenation of the bicyclic system LIII (R = H) yields 1,5-**diazonane** (LIV; R, R^1 = H), *dihydrochloride* m.p. 218–222° (decomp.) (*H. Stetter* and *H. Spangenberger*, Chem. Ber., 1958, **91,** 1982), and the methiodide of LIII (R = phenyl or *p*-anisyl) with sodium in liquid ammonia gives a mixture of LIV (R = Me; R^1 = phenyl or *p*-anisyl) and its Δ^1-derivative (*Houlihan* and *Manning*, U.S.P. 3,551,411/1967; C.A., 1969, **71,** 113002; J. org. Chem., 1969, **34,** 2715, 2720). Both these reactions are also shown by the diazabicyclodecane analogues, and benzo[*c*]-derivatives of LIII are similarly converted into hexahydro-1*H*-2,6-benzodiazonines. Cyclisation of amino acid LV with dicyclohexylcarbodiimide gives 5-*methyl*-2,3,4,5,6,7-*hexahydro*-1H-1,5-*benzodiazonin*-2-*one*, m.p. 99–102° (*J. M. Muchowski*, Canad. J. Chem., 1971, **49,** 2023).

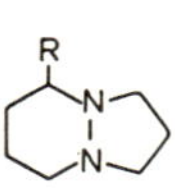

(LIII)

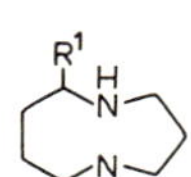

(LIV)

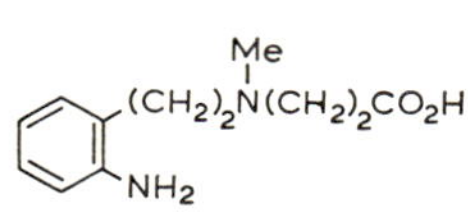

(LV)

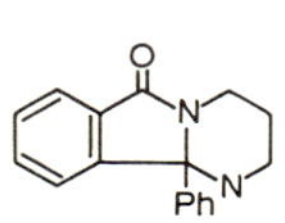

(LVI)

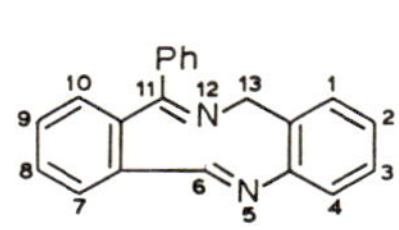

(LVII)

Reaction of phthalic anhydride or phthaloyl chloride with 1,3-diaminopropane does not give 2,6-benzodiazonines but 2:1 condensation products (*C. H. Gaozza et al.*, J. heterocycl. Chem., 1972, **9,** 883); dimethyl phthalate gives a 2:2 product with an eighteen-membered ring. Similarly, the tetrahydro-2,6-diazoninones said to be formed from 2-aroylbenzoic acids and 1,3-diaminopropane are probably pyrimidinoisoindolones (LVI) (*Sulkowski*, Belg. P. 646,221/1963; C.A., 1965, **63**, 9972). A related quinazolinoisoindole has been converted into 11-phenyl-13*H*-dibenzo[*b,g*]-1,5-diazonine (LVII) (*Houlihan*, F.P. 1,544,700/1966; C.A., 1971, **75**, 151851). The 5,6,11,12-*tetrahydro* derivative of LVII, m.p. 173–175°, is obtained by reducing the condensation product of (2-amino-5-chlorophenyl) benzylamine and xylylene dibromide (*M. Knollmüller*, Monatsh., 1971, **102**, 1055).

9-*Chloro*-1-*methyl*-7-*phenyl*-2,3,4,5-*tetrahydro*-1H-1,6-*benzodiazonin*-2-*one* (LVIII), m.p. 201–206°, is obtained by oxidising 2-(3-aminopropyl)-5-chloro-1-methyl-3-phenylindole, or by cyclising its oxidation product (*H. Yamamoto et al.*, Ger. Offen. 1,814,332/1967; C.A., 1970, **72**, 21728; 1971, **74**, 125751, 141912).

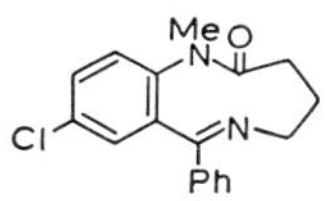

(LVIII)

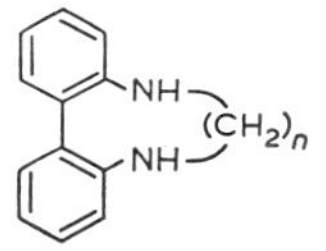

(LIX)

Condensation of 2,2′-ditosylaminobiphenyl with 1,3-dibromopropane followed by hydrolysis gives 6,7,8,9-**tetrahydro-5*H*-dibenzo[*f,h*]-1,5-diazonine** (LIX; $n = 3$), m.p. 117°. *Compounds* LIX ($n = 4$; m.p. 156°) and LIX ($n = 5$; m.p. 76°) are prepared similarly (*Stetter*, Chem. Ber., 1953, **86**, 380).

The major product formed when the strained triazoline derivative LX rearranges at 195° is the remarkably stable 4,5,6,7,8,9-*hexamethyl*-1-*phenyl*-1,2,3-*triazonine* (LXI), m.p. 79–80°. It is unchanged when heated, irradiated or exposed to air, and has a sufficiently large diamagnetic ring current to suggest it has some 10π electron aromatic stabilisation (*L. A. Paquette* and *R. J. Haluska*, J. Amer. chem. Soc., 1972, **94**, 534).

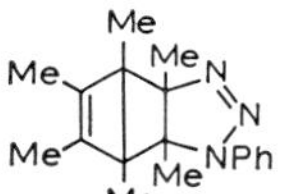

(LX)

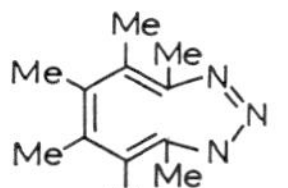

(LXI)

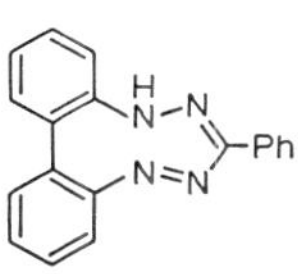

(LXII)

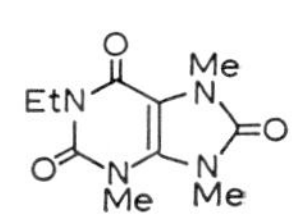

(LXIII)

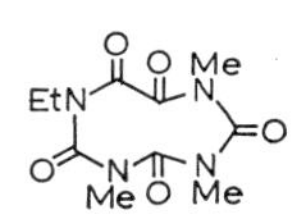

(LXIX)

If the diazonium salt of 2-amino-2′-nitrobiphenyl is condensed with phenylnitromethane, and the product reduced catalytically and then oxidised in air, 7-*phenyl*-5H-*dibenzo*-[f,h]-1,2,4,5-*tetrazonine* (LXII), m.p. 127–128°, is obtained. It takes up one mole of hydrogen, and gives benzo[*c*]cinnoline with zinc and sodium hydroxide (*D. Jerchel* and *W. Eder*, Chem. Ber., 1955, **88**, 1284).

The 1,3,5,7-**tetrazonanepentone** LXIV is a product of the photosensitised oxidation of purine LXIII (*T. Matsuura* and *I. Saito*, Tetrahedron Letters 1968, 3273; Tetrahedron, 1969, **25**, 549).

2. Ten-membered ring compounds

(a) Ten-membered ring compounds containing oxygen

(i) Compounds with one heterocyclic oxygen atom

The intermediate I in the preparation of 2,3,4,5-tetrahydro-6*H*-oxocin *(q.v.)* (p. 387) reacts with dimethyl acetylenedicarboxylate to give the **dihydro-oxecin,** II (*L. A. Paquette* and *R. W. Begland*, J. org. Chem., 1967, **32,** 2723; J. Amer. chem. Soc., 1966, **88,** 4685).

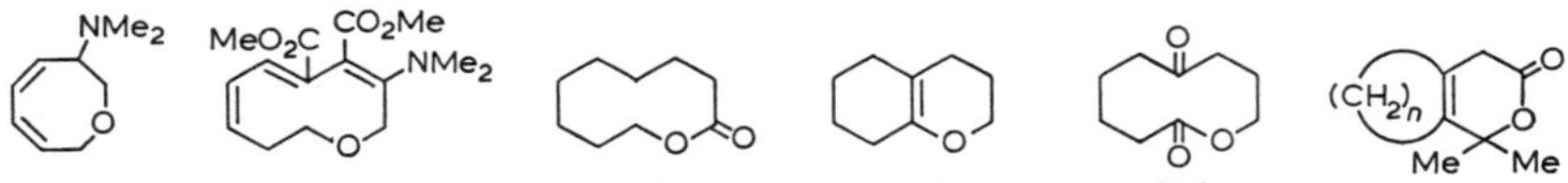

(I) (II) (III) (IV) (V) (VI)

9-Hydroxynonaldehyde with methanolic hydrochloric acid gives the cyclic acetal **2-methoxyoxecane,** b.p. 125–130°/15 mm (*B. Helferich* and *W. Schaefer*, Ber., 1924, **57,** 1911).

The lactone **oxecan-2-one** (*2-oxoperhydro-oxecin, oxacyclodecan-2-one*) (III), m.p. 31–32°, b.p. 93°/13 mm, n_D^{26} 1.4652, is prepared by the action of alkali on 9-bromononanoic acid at high dilution (*H. Hunsdiecker* and *H. Ehrlbach*, Chem. Ber., 1947, **80,** 129) or by Baeyer–Villiger ring-expansion (*W. D. Closson et al.*, J. org. Chem., 1967, **32,** 3160). Dinitrile-cyclisation has been used to prepare *5-cyano-6-imino-oxecane*, m.p. 115–116° (*K. Ziegler* and *H. Holl*, Ann., 1937, **528,** 143).

Oxidation of bicyclic enol ethers of type IV with ozone, 3-chloroperbenzoic acid followed by lead tetra-acetate, or better with molybdenum hexacarbonyl catalyst under neutral conditions, yields **oxecane-2,7-dione** (V), m.p. 69–71°, and its derivatives (*I. J. Borowitz et al.*, Tetrahedron Letters, 1964, 1151; Chem. Comm., 1969, 1202; J. org. Chem., 1966, **31,** 3032; cf. *R. G. Carlson* and *R. G. Blecke*, *ibid.*, 1967, **32,** 3538); eleven- and twelve-membered ring analogues are prepared similarly (*Borowitz et al.*, *ibid.*, 1968, **33,** 2013; Tetrahedron Letters, 1965, 3813). Ozonolysis of bicyclic lactones of type VI gives **3,3-dimethyloxecane-2,4,9-trione,** m.p. 96°, and homologues (*J. Falbe* and *F. Korte*, Chem. Ber., 1963, **96,** 919).

The 9,10-*tetramethylene* derivatives of V (*trans*-, m.p. 86°; *cis*-, m.p. 85°) are obtained in the oxidation of 2,2′-dihydroxydicyclohexamethylmethane with chromium trioxide (*A. Palsky et al.*, Compt. rend., 1966, **263C,** 888; Bull. Soc. chim. Fr., 1968, 5003) and the **2-benzoxecins** VII (R = H; R^1 = OMe) and VII (R = OMe; R^1 = OH) by oxidative cleavage of a transannular double bond. With sodium hydride the former undergoes rearrangement to VIII (*H. Immer* and *J. F. Bagli*, J. org. Chem., 1968, **33,** 2457; Canad. P. 847,916/1967, C.A., 1970, **73,** 120414).

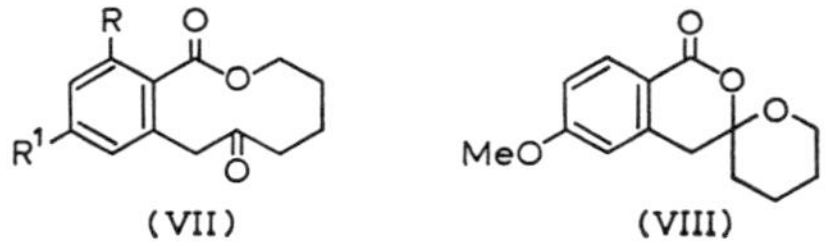

(VII) (VIII)

(*ii*) *Compounds with two or more heterocyclic oxygen atoms*

Rearrangement of 3-hydroxy-1,2-benzoquinone dimers with alkali yields 2,7-dihydro-1,8-benzodioxecin-2,7-diones (*J-C. Salfeld* and *E. Baume*, Chem. Ber., 1960, **93**, 745; *H. J. Teuber et al.*, Ann., 1966, **696**, 64):

2,3,4,5,6,7-*Hexahydro*-1,8-*benzodioxecin (catechol hexamethylene ether)*, b.p. 140°/10 mm, is described on p. 168 and a route to tetrahydrodibenzo[*e*,*i*]-1,4-dioxecins of type IX on p. 429.

The neutral product, **oospolide,** m.p. 111°, is given the 1,5-dioxecin structure X (*K. Nitta et al.*, Tetrahedron Letters, 1967, 3013).

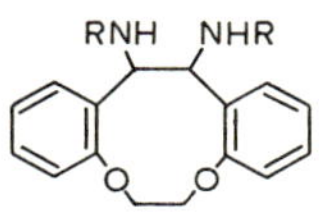

(IX)

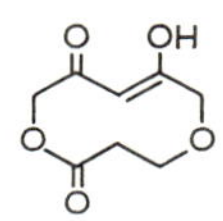

(X)

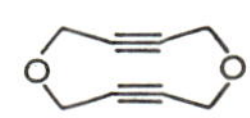

(XI)

Bis(chloromethyl) ether and acetylenedimagnesium bromide react to give a little 1,6-*dioxa*-3,8-*cyclodecadiyne* (XI), m.p. 170–172° (*F. Sondheimer et al.*, Tetrahedron Letters, 1960, **26**, 25; *R. Lespieau*, Compt. rend., 1929, **188**, 502; cf. *W. Treibs* and *R. Pester*, Tetrahedron Letters, 1960, **17**, 5). Hydrogenation over a Lindlar catalyst yielded 2,5,7,10-**tetrahydro**-1,6-**dioxecin,** m.p. 117–118°, also obtained by heating a mixture of 1,4-dichloro- and 1,4-dihydroxy-but-2-enes with alkali (*W. Reppe et al.*, Ann., 1955, **596**, 95, 132). Reduction of either diene or diyne over platinum oxide gave 1,6-**dioxecane,** b.p. 183–185°, n_D^{28} 1.4542, which is also the photolysis product of 1,7-dioxa-4,10-dimercuracyclododecane (*V. Weinmayr*, J. Amer. chem. Soc., 1959, **81**, 3590). Another route to a tetrahydro-1,6-dioxecin is as follows (*S. Hoff* and *L. Brandsma*, Rec. Trav. chim., 1969, **88**, 845):

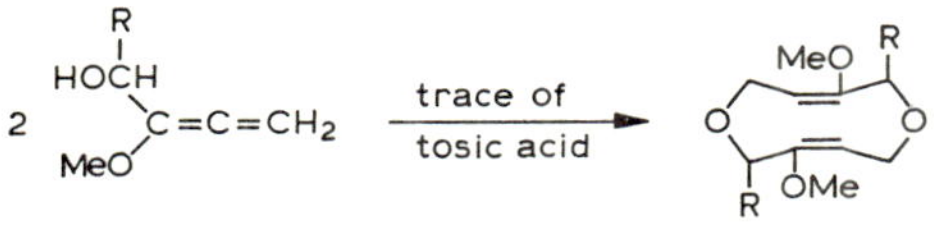

The dienes are found to exist in chair forms (*A. Feigenbaum* and *J. M. Lehn*, Chem. Comm., 1968, 1477; Bull. Soc. chim. Fr., 1969, 3724).

5,7,12,14-**Tetrahydrodibenzo**[*c*,*h*]-1,6-**dioxecin** (XII), m.p. 139–140°, is prepared from 1,2-di(hydroxymethyl)benzene in the presence of sodium hydride at high dilution. In more concentrated solution a 2:2 condensation product is formed (*F. Vogtle* and *M. Zuber*, Tetrahedron Letters, 1972, 561).

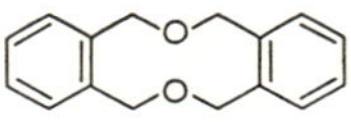

(XII)

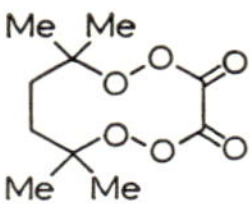

(XIII)

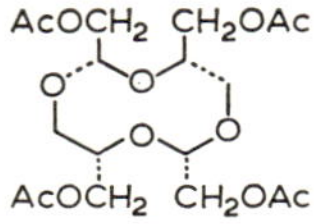

(XIV)

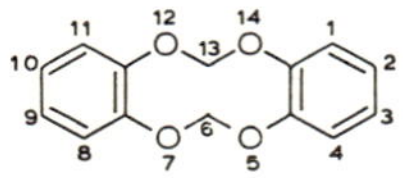

(XV)

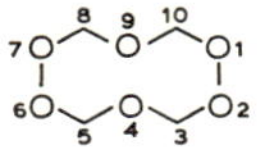

(XVI)

Tribenzo[*b,d,h*]**-1,6-dioxecin,** m.p. 156–158°, results when biphenyl-2,2′-diol and α,α′-dibromoxylene are heated in alkaline solution (*D. W. Allen et al.*, J. chem. Soc. C, 1971, 3454).

Treatment of 2,5-dihydroperoxy-2,5-dimethylhexane with oxalyl chloride gave the explosive 7,7,10,10-**tetramethyl**-1,2,5,6-**tetroxecane**-3,4-**dione** (XIII). Photolysis or thermolysis gave carbon dioxide, ethylene, acetone and 3,3,6,6-tetramethyl-1,2-dioxane (*W. Adam* and *J. Sanabia*, Chem. Comm., 1972, 174).

2-*Trichloromethyl*-1,3,5,8-*tetroxecane*, m.p. 73°, is a byproduct of the copolymerisation of chloral and 1,3,6-trioxocane (*H. J. Dietrich et al.*, J. heterocycl. Chem., 1970, **7,** 27).

2-*Dichloromethyl*-1,3,6,8-*tetroxecane*, b.p. 166–170°/3 mm, n_D^{20} 1.4910, is one product of the photochemical chlorination of ethylene glycol (*I. F. Spasskaya et al.*, Zhur. obshcheĭ, Khim., 1958, **28,** 1827). Another 1,3,6,8-tetroxecane XIV is obtained from the dianhydride of D-ribose (*J. F. Stoddart* and *W. A. Szarek*, Canad. J. Chem., 1968, **46,** 3061). 6,13-*Dihydrodibenzo*[d,i]-1,3,6,8-*tetroxecin* (XV), m.p. 262–263°, is a byproduct of the preparation of methylenedioxybenzene from catechol (*W. J. Gensler* and *C. M .Samour*, J. org. Chem., 1953, **18,** 9).

1,2,4,6,7,9-**Hexoxecane** (*tetraoxymethylene* 1,2; 6,7-*diperoxide*) (XVI), explodes at 94°, results from the dehydration of dihydroxymethyl peroxide with phosphorus pentoxide; its 3,5,8,10-tetramethyl analogue is also known (*A. Reiche et al.*, Ber., 1933, **66,** 718; Ann., 1942, **553,** 187).

(*b*) *Ten-membered ring compounds containing sulphur or selenium*

(*i*) *Compounds with one atom in the heterocycle*

2,3,4,5,6,7-*Hexahydro*-10H-*thiecin*-3-*one* 1,1-*dioxide* (XVII), m.p. 144–145°, is obtained from the reaction mixture of 1-dimethylamino-1,3-cyclo-octadiene and sulphene (*Paquette* and *Begland*, J. org. Chem., 1969, **34,** 2896). **Thiecane,** $n_D^{23.5}$ 1.5178, is prepared from sodium sulphide and 1,9-dibromononane at high dilution, but much better yields are obtained of the 6,6-*dimethyl* derivative, b.p. 225°, n_D^{26} 1.5035, due to the *gem*-dimethyl effect

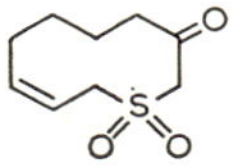

(XVII)

(*P. Friedman* and *P. Allan*, *ibid.*, 1962, **27,** 1095; 1965, **30,** 780). Eleven- and twelve-membered thiacycloalkanes also show this effect.

(*ii*) *Compounds with two or more heteroatoms in the ring*

1,2-**Dithiecane,** b.p. 107–110°/2 mm, n_D^{25} 1.5461, is prepared using general methods for cyclic 1,2-disulphides (*J. G. Afflek* and *G. Dougherty*, *ibid.*, 1950, **15,** 865; *A. Schöberl* and *H. Gräfle*, Angew. Chem., 1957, **69,** 713; Ann., 1958, **614,** 66). 1,3-**Diethiecane-2-thione** is prepared from potassium trithiocarbonate and 1,7-dibromoheptane (*F. Runge et al.*, J. pr. Chem., 1959, **7,** 268; 1960, **11,** 284) and slow mixing of the disodium salt of hexane-1,6-dithiol with ethylene dibromide yields 1,4-**dithiecane,** m.p. 68° (*H. Stetter* and *W. Wirth*, Ann., 1960, **631,** 144).

Dicyclohexeno[*b,i*]-1,6-**dithiecin** (XVIII), m.p. 127–128°, is prepared by a double Wittig reaction on 2,2′-thiobis(cyclohexene-1-carbaldehyde). Unlike 12-annulene, the 12π electron system shows no paramagnetic ring current (*Sondheimer et al.*, Chem., Comm., 1971, 947). 1,4-Dichlorobut-2-yne with inorganic sulphides gives 1,6-*dithiacyclodeca-*3,8-*diyne* (XIX), m.p. 200° decomp., which with potassium *tert*-butoxide forms a trans-annular bond giving the dihydrodithia-azulene XX. Catalytic hydrogenation stops at the

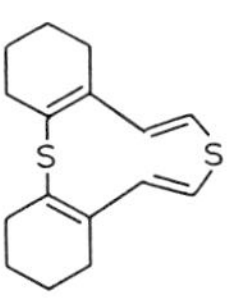
(XVIII)

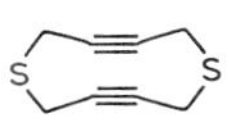
(XIX)

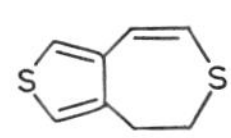
(XX)

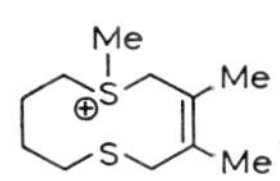

(XXI)

cis,cis-diene (*G. A. Sim et al.*, J. chem. Soc., 1964, 1154; 1965, 5988; *Feigenbaum* and *Lehn*, *loc. cit.*). 2,5,7,10-*Tetrahydro*-1,6-*dithiecin*, m.p. 170°, is also obtained from 1,4-dichloro- and but-2-ene-1,4-dithiols (*W. Schroth et al.*, Z. Chem., 1964, **4,** 302), and the ring expansion of 1,2-dithiane methohalides with 2,3-dimethylbutadiene gives the hexa-hydrodithiecinium salt XXI (*N. E. Hester*, *G. K. Helmkamp* and *J. I. Alford*, Int. J. Sulfur Chem., Part A, 1971, **1,** 65; C.A., 1971, **75,** 110301).

5,7,12,14-*Tetrahydrodibenzo*[c,h]-1,6-*dithiecin*, m.p. 234–236° is prepared in low yield from 1,2-di(bromoethyl)benzene and potassium hydrogen sulphide (*W. Autenrieth et al.*, Ber., 1902, **35,** 1390; 1903, **36,** 183); it gives a disulphone on oxidation and a dibromide on bromination.

The action of potassium hydroxide on 1,3-dithiocyanopropane yields 1,2,6,7-**tetra-thiecane** (XXII), m.p. 170° (*H. Brintzinger et al.*, Chem. Ber., 1954, **87,** 323; *L. Hagelberg*, Ber., 1890, **23,** 1083). Substituted derivatives, including the insecticide 4,9-bis(dimethyl-amino)-1,2,6,7-tetrathiecane, are prepared from 1,3-dichloropropanes with sodium thiosulphate or sulphur and sodium sulphide (*Takeda Chem. Ind.* B.P. 976,729–30/1961; C.A., 1965, **62,** 7773, 13790) or by oxidising 1,8-dimercapto-4,5-dithiaoctanes (*F. W.*

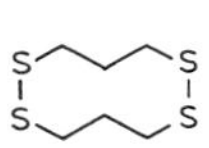
(XXII)

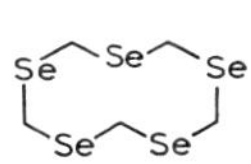
(XXIII)

Holly and *C. F. Spencer*, U.S.P. 2,766,257/1956; C.A., 1957, **51,** 8815) or propane-1,3-dithiols (*V. A. Portnyagina et al.*, Khim. geterotsikl. Soedin., 1970, 605) with iodine.

1,3,5,7,9-**Pentathiecane,** m.p. 123°, is obtained in the reaction of dichloromethane with sodium sulphide (*M. Schmidt et al.*, Z. Naturforsch., 1966, **21B,** 622; *M. Russo et al.*, Polymer Letters, 1965, **3,** 455). 1,3,5,7,9-**pentaselenecane** (XXIII) is found as a by-product when sodium selenide is used.

Perhydrodibenzo-1,2,3,6,7,8-**hexathiecin,** m.p. 216–218°, is prepared by condensing *trans*-1,2-dithiolcyclohexane with sulphur dichloride. The heterocyclic ring has a conformation like cyclodecane. The *dibenzo* analogue, m.p. 160°, is made similarly from benzene-1,2-dithiol (*F. Feher et al.*, Angew. Chem. intern. Edn., 1967, **6,** 703; 1970, **9,** 313; 1971, **10,** 413).

(*c*) *Ten-membered rings containing nitrogen*

(*i*) *Compounds containing one heterocyclic nitrogen atom*

Several alkaloids of the **cryptopine** (XXIV; R = OMe) and **protopine** (XXIV R = $-OCH_2O-$) groups contain a ten-membered ring in which the nitrogen atom is very close to the carbon atom of the carbonyl bond (cf. C.C.C., Vol. IV H, p. 145; *S. R. Hall* and *F. R. Ahmed*, Acta Cryst., 1968,

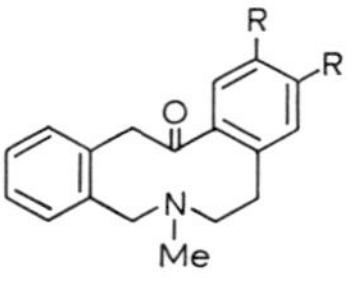

(XXIV)

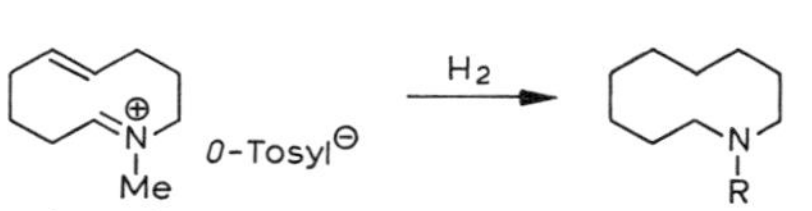

(XXV) (XXVI)

24B, 337, 346). Their chemistry (*R. H. F. Manske*, "The Alkaloids", Vol. 4, p. 147, Academic Press, New York, 1954) and their transannular interactions (*N. J. Leonard*, Record of Chem. Progress, 1956, **17,** 243) have been reviewed. Photolysis is found to give alkaloids of the **berberine** group by transannular bond formation (*X. A. Dominguez* and *J. G. Delgado*, Tetrahedron Letters, 1967, 2493) and this bond formation can be reversed by oxidation (*K. W. Bentley* and *A. W. Murray*, J. chem. Soc., 1963, 2497).

N-Methyl-5α-tosyloxy-*trans*-decahydroquinoline in solvolytic conditions undergoes cleavage of the 9,10-bond leaving the immonium salt XXV, a derivative of 2,3,4,7,8,9-hexahydroazecine (*C. A. Grob et al., ibid.*, 1964, 2901; Helv., 1967, **50,** 416); hydrogenation yields XXVI (R = Me). A similar fragmentation reactions occurs when 1-alkyl-4-mesyloxydecahydroquinolines are treated with sodium tetrahydridoborate (*J. A. Marshall* and *J. H. Babler*, J. org. Chem., 1969, **34,** 4186).

Azecane (*perhydroazecine, azacyclodecane*) (XXVI; R = H), *picrate*, m.p. 191–192°, is prepared by reduction of the lactam *azecan-2-one*, m.p. 138–139°, from the ring-expansion of cyclononanone (*L. Ruzicka et al.*, Helv., 1949, **32,** 544; *R. Huisgen et al.*, Chem. Ber.,

1957, **90,** 1437), and by electrolytic reduction of 1-azabicyclo[4.4.0]decan-5-one to 5-*hydroxyazecane*, b.p. 75°/0.2 mm, followed by dehydration and hydrogenation (*Leonard et al.*, J. Amer. chem. Soc., 1952, **74,** 4620).

Hofmann degradation of quinolizidine methiodide yields 1-*methyl*-1,2,3,4,5,8,9,10-*octahydroazecine*, b.p. 83°/19 mm, n_D^{18} 1.4838, *picrate*, m.p. 152°, readily hydrogenated to 1-*methylazecane* (XXVI; R = Me), b.p. 89–90°/20 mm, *picrate*, m.p. 198–199° (*K. Schofield* and *R. J. Wells*, Chem. and Ind., 1963, 572; Austral. J. Chem., 1965, **18,** 1423; *G. R. Clemo et al.*, J. chem. Soc., 1932, 2959; *V. Prelog* and *R. Seiwerth*, Ber., 1939, **72,** 1638; cf. *K. Tsuda et al.*, Chem. pharm. Bull. Japan, 1965, **13,** 487). A little 5,6,7,8-*tetrahydro*-7-*methyldibenz*[c,g]*azecine*, *picrate*, m.p. 158–159°, has also been prepared this way and converted into (XXIV; R = H), *picrate* m.p. 232–233° (*V. Deulofeu et al.*, J. chem. Soc. C, 1970, 2578).

Dimethyl sulphate *O*-methylates azecan-2-one (*S. W. Breuer* and *D. Ginsburg*, Israel, J. Chem., 1966, 4,145; C.A., 1967, **66,** 65382) and the ring of the 1-nitroso derivative of azecan-2-one is readily cleaved to 9-hydroxynonanoic acid (*Huisgen* and *J. Reinertshofer*, Ann., 1952, **575,** 197).

Peripheral synthesis is the most general method available for ten-membered rings containing nitrogen (*M. G. Reinecke et al.*, J. org. Chem., 1972, **37,** 3489; cf. 1-methylazonanes p. 434) and has been used in the synthesis of 1-*methylazecan*-6-*one*, b.p. 75–76°/1.4 mm, n_D^{20} 1.4880 (*Y. Arata et al.*, Chem. pharm. Bull. Japan, 1968, **16,** 569; 1969, **17,** 1363) *azecane*-6-*carboxylic acid*, m.p. 250° decomp. (*idem, ibid.*, 1970, **18,** 2361) and protopine alkaloids (*R. D. Haworth et al.*, J. chem. Soc., 1926, 445, 1764, 1769; *D. Giacopello* and *Deulofeu*, Tetrahedron Letters, 1966, 2711). The way that this occurs naturally is still obscure (*S. Prasad et al.*, J. Indian chem. Soc., 1969, **46,** 371; *G. Zanati et al.*, Planta Med., 1970, **18,** 366).

1-Methylazecan-6-one is also prepared by the Dieckmann cyclisation, and 1-*methylazecan*-5-*one*, b.p. 65–70°/0.7 mm, *picrate* m.p. 194–195° decomp., by the oxidation of 5-hydroxy-1-methylazecane (*Leonard et al.*, J. Amer. chem. Soc., 1955, **77,** 6234). 6-Hydroxy-, but not 5-hydroxy-1-methylazecane forms a transannular bond in perchloric acid (*A. J. Sisti* and *D. L. Lohner*, J. org. Chem., 1967, **32,** 2026).

Azecane-2,7-dione (XXVIII) appears to be formed by the hydrolysis of hydroperoxide XXVII, but rapidly undergoes transannular bond formation to the carbinolamide XXIX. An analogous route leads to 5,6,7,8,9,10,11,12-octahydrobenzo[*b*]azecine-6,12-dione (XXX), which on treatment with base gives a hexahydrophenanthridone (*B. Witkop et al.*,

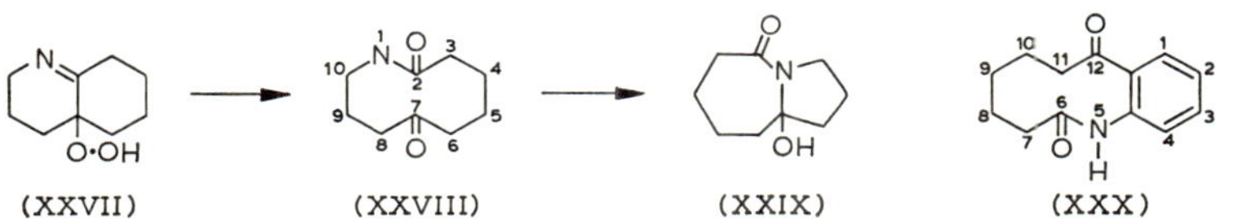

(XXVII) (XXVIII) (XXIX) (XXX)

Bull. Soc. chim. Fr., 1954, 423; J. Amer. chem. Soc., 1955, **77,** 6595). The 9,10-*tetramethylene* derivative of XXVIII, m.p. 157–158°, is also found when 2,2′-dioxodicyclohexylmethane reacts with ammonium acetate in acetic acid (*V. I. Vysotskii* and *M. N. Tilichenko*, Khim. geterotsikl. Soedin., 1969, 373, 751).

The lactone 1,2,3,4,5,6,7,8-*octahydro*-1-*benzazecin*-2-*one* is prepared by the Beckmann rearrangement of the benzenesulphonate of 1-benzocyclononenone oxime. Smaller rings do not show this rearrangement, but larger rings rearrange more easily (*Huisgen et al.*, Ann., 1954, **586,** 30; 1957, **604,** 191; cf. Chem. Ber., 1957, **90,** 1844). The isomeric *octahydro*-1-*benzazecin*-8-*one*, m.p. 68°, is prepared by irradiating 1-phenylazocan-2-one (*M. Fischer*, Tetrahedron Letters, 1968, 4295; Chem. Ber., 1969, **102,** 342).

1,2,3,4,5,6,7,8-*Octahydro*-3-*methyl*-3-*benzazecine*, m.p. 25–28°, is prepared by treating octahydrobenzo[*a*]quinolizinium methiodide with lithium in liquid ammonia; demethylation gives **octahydro-3-benzazecine,** b.p. 200°/0.1 mm (*H. Smith et al.*, J. med. Chem., 1966, **9,** 864; cf. *ibid.*, 1967, **10,** 1088; *W. J. Houlihan* and *R. E. Manning*, U.S.P. 3,516,993/1967; C.A., 1970, **73,** 77085).

(*ii*) *Compounds containing two nitrogen atoms in the heterocyclic ring*

1,3-**Diazecan-2-one** (*perhydro*-1,3-*diazecine*), m.p. 159°, and **-2-thione,** m.p. 159°, have been obtained by *H. Behringer* and *H. Meier* (Ann., 1957, **607,** 67, 73) and refluxing the cyclic heptamethylene carbamate with aniline gives 1-*phenyl*-1,3-*diazecan*-2-*one*, m.p. >250° (*R. Delaby* and *R. Damiens*, Festschr. Arthur Stoll, 1957, 474).

Hexamethylenediamine and oxalyl chloride condense to form 1,4-**diazecane**-2,3-**dione** and higher cyclic oligomers (*O. Vogl* and *A. C. Knight*, Macromolecules, 1968, **1,** 311) and 4-*methyl*-1,4-*diazecane*-2,5-*dione* (XXXI), m.p. 128°, is formed from *N*-(*N*-methylglycyl)caprolactam (*M. Rothe et al.*, Tetrahedron Letters, 1969, 5127). *Octahydro*-1,8-*benzodiazecine, hydrochloride*, m.p. 150°, b.p. 140°/10 mm, is described on p. 416 and its possible 2,7-dione on p. 461.

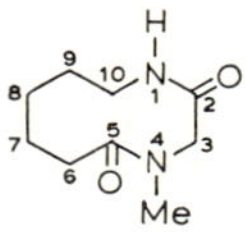

(XXXI)

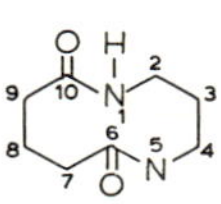

(XXXII)

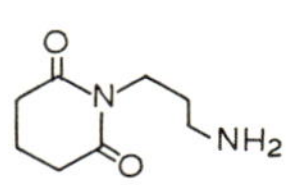

(XXXIII)

1,5-*Diazecane*-6,10-*dione* (*perhydro*-1,5-*diazecine*-6,10-*dione*), (XXXII), m.p. 233–234°, is obtained by double Beckmann rearrangement of cyclo-octane-1,5-dione, or the reaction of diacid chloride and diamine, forms the cyclic imide XXXIII in dilute acid after 1,6-transannular bond formation; the reaction is reversed in aqueous bicarbonate (*H. Rapoport et al.*, J. Amer. chem. Soc., 1964, **86,** 3397; 1965, **87,** 2003; *J. Dale* and *R. Coulon*, J. chem. Soc., 1964, 182). 1,5-*Diazecane*-2,6-*dione*, m.p. 173°, from the reaction of benzoxycarbonyl-β-alanyl chloride and butyrolactam (*V. K. Antonov et al.*, Zhur. obshcheĭ Khim., 1965, **35,** 2231; 1968, **38,** 770; Tetrahedron, 1965, **21,** 3537) and 1,6-*diazecane*-2,7-*dione*, m.p. 287–288°, from cyclisation of the linear dipeptide (*Rapoport*, *loc. cit.*; *Rothe*, Angew. Chem., 1962, **74,** 725; Makromol. Chem., 1965, **85,** 307), show similar transannular reactions, as do nine- and eleven-membered analogues.

The pyrimidoisoquinoline XXXIV is opened to a 2,6-benzodiazecine by alkaline tetrahydridoborate (*T. Yamazaki*, Yakugaku Zasshi, 1959, **79,** 1014) and hexahydrodibenzo-[*b,h*]-1,5-diazecines are prepared by cleaving the transannular bond in benzo-analogues of XXXIV with sodium in liquid ammonia (*Houlihan* and *Manning*, U.S.P. 3,497,499/1967; C.A., 1970, **72,** 121573).

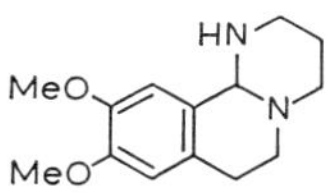
(XXXIV)

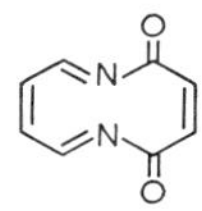
(XXXV)

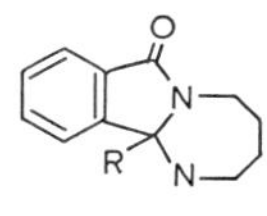
(XXXVI)

There is no evidence for the presence of the 2,5-dihydro-1,6-diazecine-2,5-dione tautomer XXXV in samples of 9,10-diaza-1,4-naphthoquinone (*S. Ochiai et al.*, Tetrahedron Letters, 1972, 1885). Hydrogenation of 10-azaquinolizidine over Raney nickel yields 1,6-**diazecane** (*perhydro*-1,6-*diazecine*), *dihydrochloride*, m.p. 208° (*H. Stetter* and *H. Spangenberger*, Chem. Ber., 1958, **91,** 1982) and the methiodides of 10-azaquinolizidines show similar 5,10-bond cleavage with sodium in liquid ammonia (*P. Aeberli* and *Houlihan*, J. org. Chem., 1969, **34,** 2720). The nitrile $Me_2CH:CH(CH_2)_2CN$ condenses in acid to 2,2,6,6-tetramethyl-1,6-diazecane-5,10-dione (*J. W. Ducker* and *M. J. Gunter*, Austral. J. chem., 1968, **21,** 2809).

The diazepinoisoindoles XXXVI now believed to be the products from 2-aroylbenzoic acids and tetramethylenediamine, open up on reduction with lithium tetrahydridoaluminate to 1-*phenyloctahydro*-2,7-*benzodiazecine*, m.p. 114–116°, and analogues (*T. S. Sulkowski*, B.P. 1,093,064; F.P. 1,487,344/1966; C.A., 1968, **69,** 19231, 27467).

5,6,7,8,9,10-**Hexahydrodibenzo**[*b,d*]-1,6-**diazecine** (XXXVII), m.p. 147–148°, is one of the products of the benzidine-type rearrangement of 1,2-diphenylhexahydropyridazine (*G. Wittig et al.*, Ann., 1957, **610,** 180); for another route see *Stetter* (Chem. Ber., 1953, **86,** 380). Mandelic anilide self-condenses in concentrated sulphuric acid to the dibenzo[*b,g*]-1,6-diazecine XXXVIII (*S. Ghosal* and *S. K. Dutta*, Indian J. Chem., 1969, **7,** 1095).

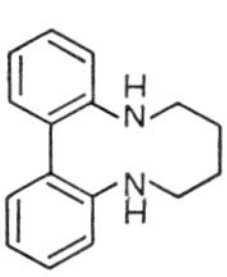
(XXXVII)

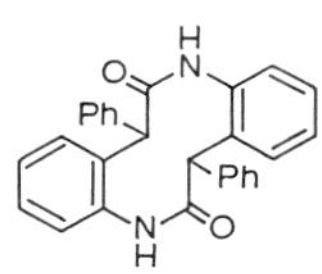
(XXXVIII)

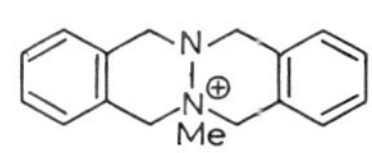
(XXXIX)

(XL) (XLI)

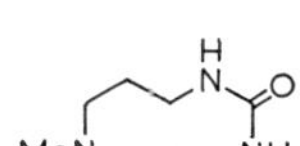
(XLII)

A Hofmann elimination sequence on the tetracyclic ion XXXIX gave 6-*methyl*-5,6,7,12-*tetrahydrodibenzo*[c,h]-1,6-*diazecine*, m.p. 219–222°, which was reduced to the 5,6,7,-12,13,14-*hexahydro* derivative (XL; R = Me), m.p. 162–164° (*Aeberli* and *Houlihan*, J. org. Chem., 1969, **34,** 2715). 5,6,7,12-**Tetrahydrodibenzo**[*c,h*]-1,6-**diazecine** (XL; R = H), m.p. 79–80°, is prepared by heating the spiro-quaternary salt XLI with aqueous ammonia (*M. Scholtz*, Ber., 1891, **24,** 2402).

7-*Methyl*-1,3,7-*triazecan*-2-*one* (XLII), m.p. 142°, is prepared from bis(3-aminopropyl)-

methylamine and phosgene (*C. Taube* and *K. Böckmann*, G.P. 1,184,766/1962; C.A., 1965, **62,** 7784).

3. Rings with more than ten atoms

Many heterocyclic compounds having large rings are cyclic homologues of compounds already discussed and are prepared by similar routes. Recently, however, certain macrocycles containing heteroatoms, such as the hetero-annulenes, the macrocyclic polyethers and the cyclam polydentate ligands, have acquired a chemical interest in their own right, and new specific routes have been sought for their synthesis.

The ease of formation of cyclic systems, which is at its greatest for five- and six-membered rings, reaches a minimum between eight- and twelve-membered rings and then improves again. It is found more difficult to prepare odd-membered rings than those with an even number of constituent atoms. Many ring-forming reactions are equilibria which allow subsequent ring-opening and polymerisation of the monomeric heterocycle.

The *high dilution technique*, which denies the activated substrate access to others of its kind and so favours intramolecular reaction (cyclisation) over intermolecular polymerisation has been as successful in the synthesis of heterocycles as in its earlier application to alicyclic systems. Concentration is critical as an increase from 0.001 M to 0.009 M can cause a threefold drop in yield (*L. Ruzicka et al.*, Helv., 1937, **20,** 109).

The *rigid group principle*, the incorporation of 4-atom units fixed in plane by *cis* attachment to a double bond or a ring, improves the yield and often the stability of a large heterocyclic ring. An 80% yield of the cyclic polyether from catechol and diethylene glycol drops to 2% when ethylene glycol is used instead of the dihydric phenol (*C. J. Pedersen*, J. Amer. chem. Soc., 1967, **89,** 7017). The presence of a *gem*-dimethyl group in the substrate also seems to encourage cyclisation.

The *template effect*, where an added reagent holds the substrate in a conformation which facilitates intramolecular reaction or the simultaneous attack at two sites by a cyclising reagent, has made possible the synthesis of some complex heterocyclic systems. Recently the ligand-binding properties of transition metal ions have been utilised (*e.g. N. W. Alcock* and *P. A. Tasker*, Chem. Comm., 1972, 1239; *E-G. Jäger*, Z. chem., 1964, **4,** 437; 1968, **8,** 30, 392, 470) but this technique has long been used in heterogeneous reactions like the acyloin synthesis where cyclisation takes place on the sodium surface.

Intramolecular rearrangements involving ring contraction or expansion are also particularly effective for the synthesis of heterocyclic compounds with large rings.

(a) Large rings containing oxygen

(i) Lactones

Several general methods are available for the synthesis of macrocyclic lactones of type I, which have value in perfumery and are found in naturally occurring macrolides.

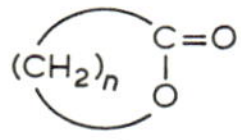

(I)

Melting and boiling points are given in Table 1.

TABLE 1

MACROCYCLIC LACTONES (I)

	Ring size							
	11	*12*	*13*	*14*	*15*	*16*	*17*	*18*
n	9	10	11	12	13	14	15	16
M.p. (°C)	6.4	2.4	2	27.5	33	37	36	42–43
B.p. (°C/mm)	114/14	103/5	111/3	122/3.8	135/3.5	137/2	128/1	155/2.8

Formation: (*1*) The dehydration of ω-hydroxy acids with benzensulphonic acid in benzene gives low yields of large ring lactones (*M. Stoll et al.*, Helv., 1928, **11,** 1159; 1934, **17,** 1609; 1935, **18,** 1087).

(*2*) A series of these lactones has been prepared by cyclising ω-bromocarboxylic acids with potassium carbonate in 2-butanone using high dilution techniques (*H. Hunsdiecker* and *H. Erlbach*, Chem. Ber., 1947, **80,** 129; *R. Huisgen* and *H. Ott.*, Tetrahedron 1959, **6,** 253).

(*3*) The Baeyer–Villiger oxidative ring-expansion of cyclic ketones also provides an effective synthesis for lactones I (*Stoll et al., loc. cit.*; *W. Stumpf et al.*, Ann., 1965, **681,** 28; *W. D. Closson et al.*, J. org. Chem., 1967, **32,** 3160). An allied route is the photolysis or thermolysis of cycloalkane peroxide dimers and trimers where large ring cycloalkanes and alkanones are also obtained, as illustrated below (*P. R. Story et al.*, J. Amer. chem. Soc., 1968, **90,** 817; Synthesis, 1970, 181):

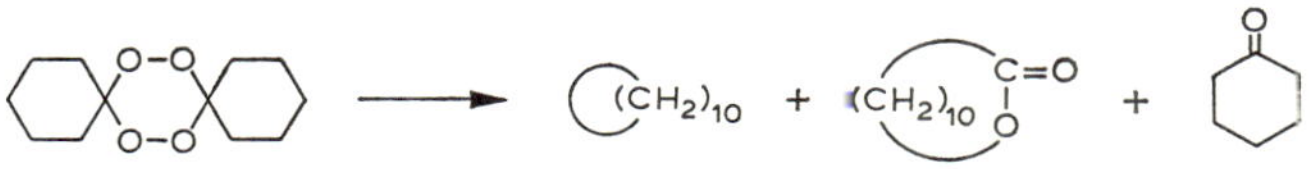

Also related is the preparation of ketolactones by the oxidation of bicyclic enol ethers obtained by dehydrating hydroxyketones with 3-chloroperbenzoic acid. Compound II yields 3-*methyloxacyclododecane-2,8-dione* (2-*methyl*-7-*oxoundecanolide*) (III), m.p. 70–72°, which has the ring structure of the methymycin group of macrolide antibiotics (*I. J. Borowitz et al.*, J. org. Chem., 1968, **33,** 2013; 1972, **37,** 581). Benzo analogues of these macrolides have also been made by this route (*H. Immer* and *J. F. Bagli, ibid.*, 2457).

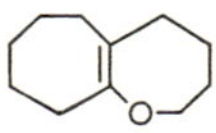

(II)

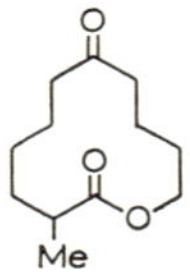

(III)

(*4*) The use of nickel carbonyl in *N*-methylpyrrolidone to cyclise *cis,cis*-diallyl bromides has been used to prepare *trans,trans*-unsaturated large ring lactones:

Catalytic reduction yields lactones of type I (*E. J. Corey* and *H. A. Kirst*, J. Amer. chem. Soc., 1972, **94,** 667).

(*5*) An alternative route to unsaturated macrolides is the oxidative coupling of the terminal acetylene groups of diyne esters $HC{\equiv}C(CH_2)_n$-$OCO(CH_2)_mC{\equiv}CH$; again, hydrogenation yields lactones of type I (*R. A. Raphael et al.*, Chem. and Ind., 1960, 559; *L. D. Bergelson* and *Y. G. Molotkovskii*, Izvest. Akad. Nauk SSSR, 1963, 94).

If there is no annellation of benzene rings, lactones I (n = 4–7) adopt the *cis* and (n = 10–16) the *trans* conformations; both conformers are found if n = 8 or 9 (*R. Huisgen et al.*, Chem. Ber., 1957, **90,** 1437; Angew. Chem., 1957, **69,** 341; Tetrahedron, 1959, **6,** 253). Lactones having a *cis*-structure are much more readily hydrolysed than the *trans*-lactones.

The mould metabolite **zearalenone** (IV; R = H — other numbering systems used), m.p. 157–159° (*M. Stob et al.*, Nature, 1962, **196,** 1318; *W. H. Urry et al.*, Tetrahedron Letters, 1966, 3109) has now been synthesised. The lactone ring-closure to IV (R = Me) was achieved using trifluoracetic anhydride, and the *O*-methyl groups were cleaved with boron tribromide (*D. Taub et al.*, Tetrahedron, 1968, **24,** 2443; Chem. Comm., 1967, 225, 761; *I. Vlattas et al.*, J. org. Chem., 1968, **33,** 4176). The double bond may be reduced catalytically, or with tetrahydridoborate, or hydroxylated with osmium tetroxide to a diol which undergoes ring-contraction with acid. Formylation takes place at position 6, and Birch reduction of IV (R = Me) results in the loss of methoxyl (*T. B. Windholz et al., ibid.*, 1972, **37,** 1639, 1647; J. med. Chem., 1970, **13,** 941). **Radicicol** (V), m.p. 195°, a natural fungicide has a similar structure (*F. McCapra et al.*, Tetrahedron Letters, 1964, 869; *W. C. Taylor* *et al.*, *ibid.*, 365; Austral. J. Chem., 1966, **19,** 1265; *N. H. White et al.*, Nature, 1962, **195,** 406).

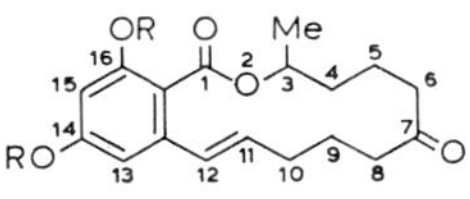

(IV)

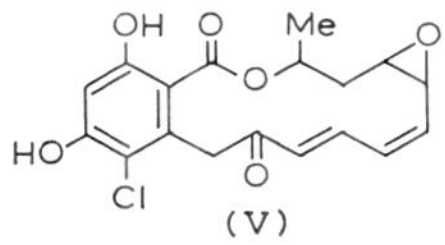

(V)

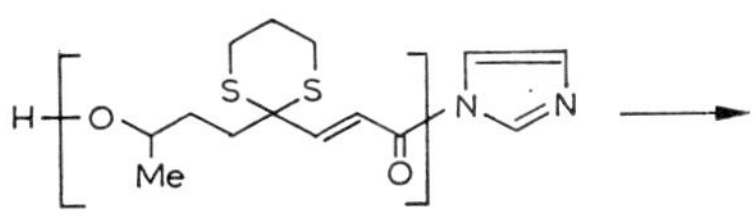

(VI)

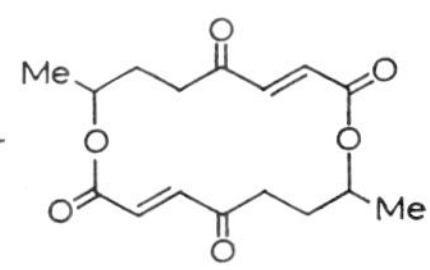

(VII)

The dilactone (±)**pyrenophorin** (VII), m.p. 124–125°, the (−)component of which is the macrolide antibiotic (*J. F. Grove et al.*, Tetrahedron Letters, 1965, 4675; J. chem. Soc. C, 1971, 2261) has been synthesised by cyclising compound VI and removing the dithioacetal protecting groups; the meso-*compound*, m.p. 118–119°, is also obtained (*E. W. Colvin et al.*, Chem. Comm., 1972, 1031).

(*ii*) *Cyclic esters, diol carbonates and anhydrides*

Linear polyesters from a wide range of dicarboxylic acids and glycols, $-[-O \cdot OO(CH_2)_mCO \cdot O(CH_2)_n-]_x-$, depolymerise at high temperatures under vacuum in the presence of a catalyst to form monomeric VIII and

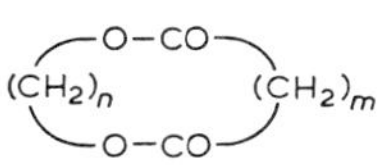

(VIII)

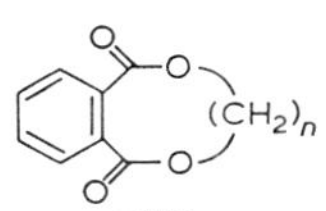

(IX)

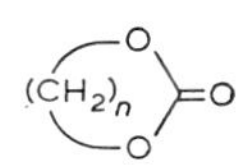

(X)

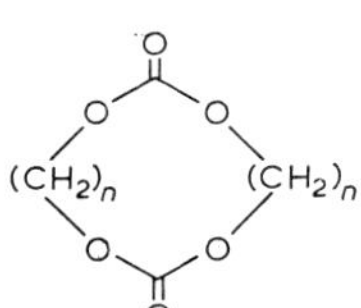

(XI)

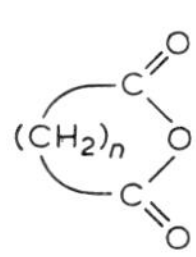

(XII)

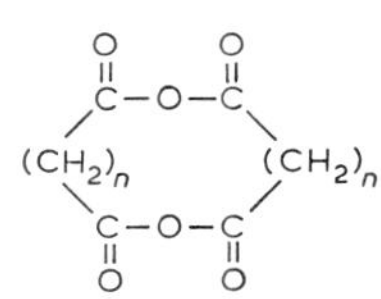

(XIII)

dimeric cyclic esters (*W. Carothers et al.*, J. Amer. chem. Soc., 1933, **55,** 5031; 1935, **57,** 929). Polymers of phthalic anhydride and glycols or polyglycols also break down under these conditions to form cyclic esters of type IX (*W. A. Ehrhart*, J. org. Chem., 1968, **33,** 2930) and linear polymeric diol carbonates and anhydrides give, respectively, monomeric and dimeric cyclic carbonates (X and XI) and cyclic anhydrides (XII and XIII) (*Carothers* and *J. W. Hill*, J. Amer. chem. Soc., 1933, **55,** 5023, 5031). These authors give extensive tables of data and the conformations and stability of these compounds are discussed by *J. Dale* (J. chem. Soc., 1963, 93; Angew. Chem. Intern. Edn., 1966, **5,** 1000).

The condensation of the dipotassium salt of phthalic acid with α,ω-dibromoalkanes in dimethylformamide also gives cyclic esters with ten to thirtyfour-membered rings. Dibromides with an even number of carbon atoms generally gave monomeric compounds IX, but if odd, the more rigid dimers XIV were found (*S. E. Drewes* and *P. C. Coleman,* J. chem. Soc. Perkin I, 1972, 2148). Unlike the macrocyclic polyethers compounds VIII and IX do not complex with alkali metals.

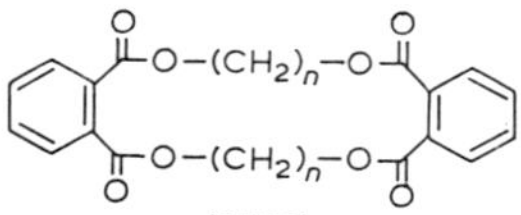

(XIV)

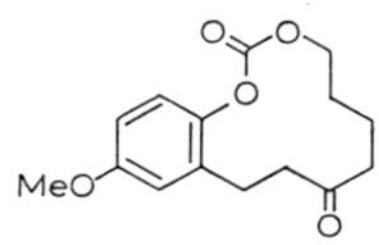

(XV)

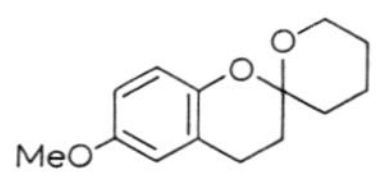

(XVI)

Cyclic carbonates (*e.g.* XV) are sometimes found instead of lactones when benzo analogues of enol ethers like II are oxidised. On treating XV with base and reacidifying, the spiro compound XVI is obtained (*Immer* and *Bagli, loc. cit.*).

(*iii*) *Cyclic ethers*

A mixture of geometrical isomers of **oxa**[17]**annulene,** m.p. 160–162°, is obtained by irradiation of the mono-epoxide of dicyclo-octatrienylcyclobutane; on catalytic reduction they all give **oxacycloheptadecane,** m.p. 20° (*G. Schroeter et al.,* Angew. Chem. intern. Edn., 1972, **11,** 424).

A list of polymethylene catechol ethers prepared by reacting catechol with α,ω-dibromoalkanes in the presence of base (*K. Zeigler et al.,* Ann., 1937, **528,** 162) is given in Table 2.

TABLE 2

POLYMETHYLENE CATECHOL ETHERS

	Ring size							
	7	*8*	*9*	*10*	*11*	*12*	*13*	*14*
n	3	4	5	6	7	8	9	10
Yield (%)	50	40	69	90	71	70	42	51
B.p. (°C/10 mm)	103	112	122	140	156	171	185	197

cis,cis-1,10-**Dioxacyclooctadeca**-5,14-**diene,** b.p. 102°/0.1 mm, *di(silver nitrate) complex,* m.p. 153°, is obtained from the bis-Wittig reagent, $O[(CH_2)_3CH{=}PPh_3]_2$ on treatment with oxygen (*H. J. Bestman* and *H. Pfüller,* Angew. Chem. intern. Edn., 1972, **11,** 508).

Dibenzo[g,n]-1,6-*dioxacyclotetradec*-3-*ene*-9,11-*diyne* (XVII), trans- m.p. 130° decomp., cis- m.p. 154° decomp., is prepared by the oxidative coupling of the two *ortho* acetylene substituents and the 7,8-benzanalogue is made similarly; the *trans* compound shows transannular π-interaction (*M. Nakagawa* and *F. Tota*, Tetrahedron Letters, 1961, 51).

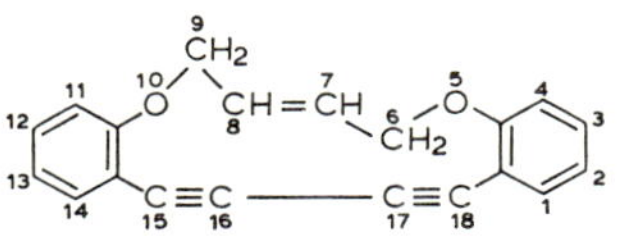
(XVII)

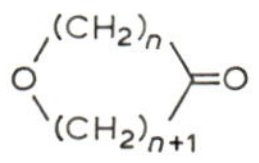
(XVIII)

A number of keto-ethers, XVIII, are prepared by dehydration and hydrogenation of acyloins prepared from the diesters of the type $O[-(CH_2)_nCO_2Et]_2$; among them are **oxacycloundecan-6-one** (XVIII; n = 4), b.p. 110°/10 mm, *semicarbazone*, m.p. 178–180°; *oxacyclotridecan-7-one* (XVIII; n = 5), *semicarbazone*, m.p. 180–181°, and *oxacyclo-tricosan-12-one* (XVIII; n = 10), b.p. 160–165°/0.05 mm, *semicarbazone*, m.p. 166–167° (*V. Prelog et al.*, Helv., 1950, **33,** 1937). The dinitrile-cyclisation process has also been used (*Ziegler* and *H. Holl*, Ann., 1937, **528,** 143).

(*iv*) *Macrocyclic polyethers* (See *C. J. Pedersen* and *H. K. Frensdorff*, Angew. Chem. intern. Edn., 1972, **11,** 16.)

Early examples of large ring systems containing several oxygen atoms are the catechol ethers of type XIX (*A. Lüttringhaus* and *I. Sichert-Modrow*, Macromol. chem., 1956, **18–19,** 511) and the *cyclic tetramers* of *ethylene oxide* (XX; R = H), b.p. 118–119°/15 mm, n_D^{25} 1.4606 (*D. G. Stewart,*

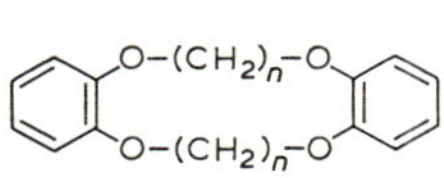
(XIX)

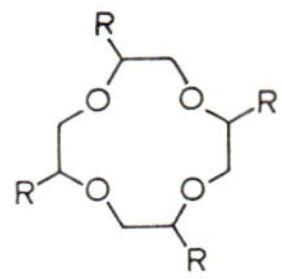
(XX)

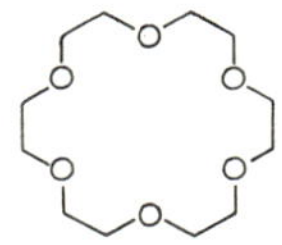
(XXI)

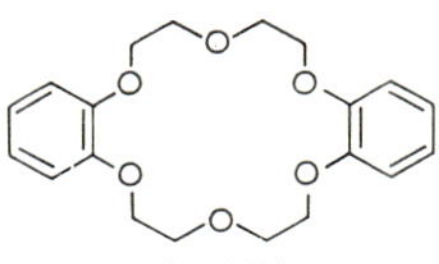
(XXII)

W. Y. Waddan and *E. T. Burrows*, B.P. 785,229/1957; C.A., 1958, **52,** 5038) and propylene oxide (XX; R = Me), b.p. 105°/5 mm (*J. L. Down et al.*, Proc. chem. Soc., 1957, 209; J. chem. Soc., 1959, 3767), but more than sixty of these "crown ethers" have now been prepared since their remarkable properties and synthetic usefulness have become apparent.

The [12]-**crown**-4 ethers (XX) are prepared by treating the epoxides with boron trifluoride or triethyloxonium tetrafluoroborate (*R. J. Kern*, J. org. Chem., 1968, **33,** 384; *R. J. Katnik* and *J. Schaefer*, *ibid.*, p. 388); many geometric isomers are possible. Oxetane also forms a cyclic tetramer (*J. Dale* and *J. Krane*, Chem. Comm., 1972, 1012). [18]-**Crown**-6 (XXI), m.p. 40–41 °, is best made by condensing triethylene glycol and its ditosylate in the presence of potassium *tert*-butoxide; the potassium ion may exert a template effect (*Dale* and *P. O. Kristiansen*, Chem. Comm., 1971, 670; *R. N. Greene*, Tetrahedron Letters, 1972, 1793; *M. J. Maskornic*, *ibid.*, p. 1797). This synthesis is also effective for higher monocyclic ethers. **Dibenzo**-[18]-**crown**-6 (DBC) (XXII), m.p. 164 °, is prepared from catechol and bis(2-chloroethyl) ether in butanolic sodium hydroxide (*Pedersen*, J. Amer. chem. Soc., 1967, **89,** 2495; 7017; 1970, **92,** 391). High pressure hydrogenation over a ruthenium catalyst gives **dicyclohexyl**-[18]-**crown**-6 (DCC) (the less ambiguous term diperhydrobenzo-[18]crown-6 is to be preferred) as two isomers, m.p. 61–63 ° and 69–70 °, which may be separated because only the former *(cis-syn-cis)* forms complexes (*N. K. Dalley et al.*, Chem. Comm., 1972, 90; *R. M. Izatt et al.*, *ibid.*, p. 1308), or by chromatography (*H. K. Frensdorff*, J. Amer. chem. Soc., 1971, **92,** 4684).

Crown ether XXII dissolves in organic solvents but not water or alcohols, whereas the saturated ethers dissolve appreciably in most solvents. Compound XXII distils at 380 ° in the absence of oxygen and its aromatic rings may be halogenated and nitrated (*R. Schori* and *J. Jagur-Grodzinski*, *ibid.*, 1972, **94,** 7957).

The outstanding feature of the crown ethers is their ability to form selective complexes, stable in solution or in crystalline form, with the cations of alkali and alkaline earth metals, and also with $H_3O^{\oplus}$ and $NH_4^{\oplus}$ (*J. J. Christensen*, Chem. Comm., 1972, 1308) but not with many transition metal ions (*A. C. L. Su* and *J. F. Weiher*, Inorg. Chem., 1968, **7,** 176). They are usually 1:1 complexes (although 2:1 and 3:2 ratios have been found) (*Pedersen*, J. Amer. chem. Soc., 1970, **92,** 386) whose stability and selectivity depend on the fit of the cation in the central cavity of the molecule. Stability is enhanced in organic solvents relative to water (*Frensdorff*, J. Amer. chem. Soc., 1971, **93,** 600; *Izatt et al.*, *ibid.*, p. 1619; *E. M. Arnett* and *T. C. Moriarity*, *ibid.*, p. 4908; *J. M. Lehn* and *J. P. Sauvage*, Chem. Comm., 1971, 440). The complexes have sharp melting points and several X-ray analyses of their structures have been made (*M. R. Truter et al.*, Nature, 1970, **225,** 176; J. chem. Soc. B, 1970, 1544; 1971, 1440; Chem. Comm., 1970, 1439; 1972, 66). The large flexible rings are found to "wrap around'' cations, as do the macrocyclic antibiotics **valinomycin** and **nonactin** (*Frensdorff*, *loc. cit.*; *Izatt*, *loc. cit.*; *B. T. Kilbourn et al.*, J. mol. Biol., 1967, **30,** 559; *M. Pinkerton et al.*, Biochem. Biophys. res. Comm., 1969, **35,** 512).

The ability of the crown ethers to render ionic reagents, *e.g.* potassium permanganate, hydroxide and *tert*-butoxide, soluble in organic solvents (*D. J. Sam* and *H. E. Simmons*, Chem. and Eng. News, 1970, **48,** 26; J. Amer. chem. Soc., 1972, **94,** 4024) and converting contact ion pairs into the more reactive solvent separated ion pairs (*J. Smid et al.*, *ibid.*, 1969, **91,** 4580; 1970, **92,** 666, 1955; 1971, **93,** 6760; Angew. Chem. Intern. Edn., 1972, **11,** 112; *W. T. Ford*, J. Amer. chem. Soc., 1970, **92,** 2857) has wide implications in synthetic chemistry. Crown ethers have been added with advantage to oxidation, bromination, saponification, alkylation (*S. G. Smith* and *M. P. Hanson*, J. org. Chem., 1971, **36,** 1931), elimination (*J. C. Martin* and *D. R. Bloch*, J. Amer. chem. Soc., 1971, **93,** 451; *M. Svoboda* *et al.*, Tetrahedron Letters, 1972, 265; 711; *R. A. Bartsch et al.*, *ibid.*, pp. 2621, 3819), isomerisation, exchange, racemisation and carbanion reactions (*G. Fraenkel* and *E. Pechhold*, *ibid.*, 1970, 153; *D. J. Cram et al.*, J. Amer. chem. Soc., 1970, **92,** 4321; 1971, **93,** 2231, 5137; *S. W. Staley* and *J. P. Erdman*, *ibid.*, 1970, **92,** 3832).

Other less effective complexing agents are obtained by replacing some of the oxygen atoms by sulphur (*J. R. Dann et al.*, J. org. Chem., 1961, **26,** 1991; *Pedersen*, *ibid.*, 1971,

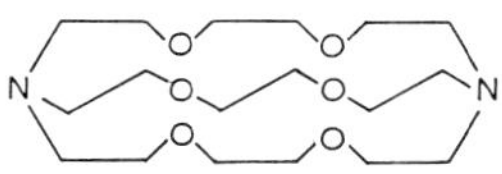

(XXIII)

36, 254; *L. Montillaro et al.*, J. chem. Soc. C, 1966, 428) or nitrogen (*H. Stetter* and *J. Marx*, Ann., 1957, **607,** 59; *Frensdorff*, *loc. cit.*). However, the bicyclic **amine polyethers,** *e.g.* XXIII, give alkali and alkaline earth metal ion complexes (cryptates) in which the ion is encapsulated in the molecular cavity. Despite the presence of nitrogen atoms in the rings the cryptates are two powers of ten more stable than crown ether complexes (*J. M. Lehn et al.*, Tetrahedron Letters, 1969, 2885, 2889; 1972, 4557; Angew. Chem. intern. Edn., 1970, **9,** 175; J. Amer. chem. Soc., 1970, **92,** 2916; Chem. Comm., 1971, 440; *B. Metz et al.*, *ibid.*, p. 217; *A. Brändström* and *K. Gustavii*, Acta Chem. Scand., 1969, **23,** 1215).

(*v*) *Macrolide antibiotics* (See *N. G. Brink* and *R. E. Harman*, Quart. Reviews 1958, **12,** 93.)

This large group of microbial metabolites with physiological activity formerly presented severe structural problems, but the structures, and absolute configurations (*C. Djerassi et al.*, Tetrahedron, 1958, **4,** 369) of many members are now well established. They share the common feature of a large lactone ring with varying degrees of unsaturation, carrying a number of substituents and usually linked to a dimethylamino sugar.

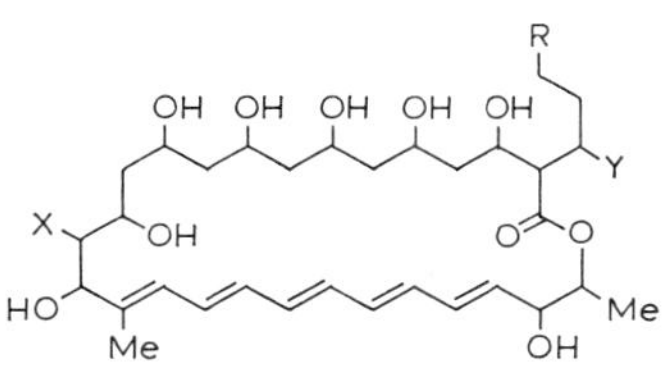

(XXIV)

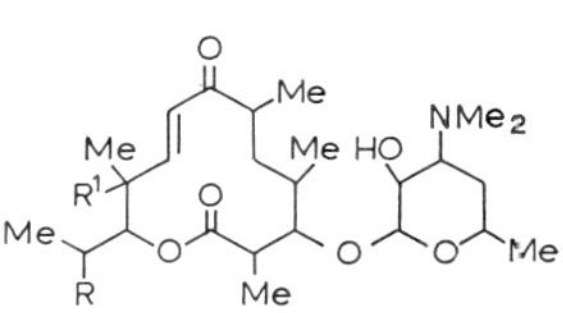

(XXV)

Ring sizes range from **flavofungin** and **mycoticin** with 32 members (*B. C. L. Weedon et al.*, J. chem. Soc. Perkin I, 1972, 1848; *H. H. Wassermann et al.*, Chem. Comm., 1970, 1634) though **chainin** (XXIV; R = Me, X = H, Y = OH), **filipin** (XXIV; R = Pr^n, X = H, Y = OH) and **lagosin** (XXIV; R = Pr^n, X, Y = OH) with 28 members (*R. C. Paudey et al.*, J. Amer. chem. Soc., 1972, **94,** 4306; *A. C. Cope et al.*, *ibid.*, 1962, **84,** 2170; *M. L. Dhar et al.*, Proc. chem. Soc., 1958, 148; 1959, 154; 1960, 310; J. chem. Soc., 1964, 842;

B. T. Golding et al., Tetrahedron Letters, 1964, 2615; *Djerassi et al., ibid.*, 1961, 383; Proc. chem. Soc., 1959, 316; *O. Ceder et al.*, Acta Chem. Scand., 1964, **18,** 558) **pimaricin** with 26 members, (*idem, ibid.*, pp. 77, 83, 98, 103, 111, 126; Angew. Chem. intern. Edn., 1963, **2,** 744; *J. B. Patrick et al.*, J. Amer. chem. Soc., 1958, **80,** 6688; 6689), **magnamycin** with 18 members (*R. B. Woodward*, Angew. Chem., 1957, **69,** 50), **erythromycin** with 14 members and no unsaturation in the lactone ring (*P. F. Wiley et al.*, Chem., and Eng. News, 1956, **34,** 5138; J. Amer. chem. Soc., 1956, **78,** 388, 6396; 1957, **79,** 6062, 6070, 6074), to **methymycin** (XXV; R = H, R^1 = OH), **neomethymycin** (XXV; R = OH, R^1 = H) (*Djerassi et al., ibid.*, 1956, **78,** 1729, 1733, 2907, 6390; 1957, **79,** 2022; Tetrahedron, 1958, **3,** 255; *A. Kinumaki* and *M. Suzuki*, Chem. Comm., 1972, 744) **narbomycin** and **pikromycin** (*Suzuki et al., ibid.*, 1971, 304; *H. Brockmann et al.*, Chem. Ber., 1951, **84,** 284; 1953, **86,** 876; 1954, **87,** 856; 1957, **90,** 605; *R. Anliker* and *K. Gubler*, Helv., 1956, **39,** 1785; 1957, **40,** 119, 1768), which have 12-membered lactone rings.

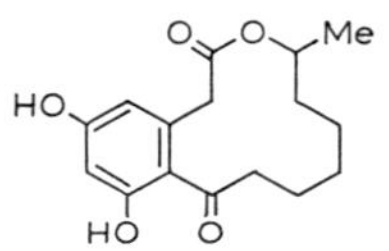

(XXVI)

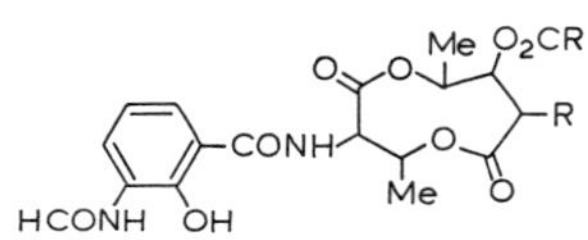

(XXVII)

Variations include fused rings as in **curvularine** (XXVI) (*A. J. Birch et al.*, J. chem. Soc., 1959, 3146), another oxygen atom in the ring as in the **antimycin macrolides** (XXVII) (*E. E. van Tamelen et al.*, J. Amer. chem. Soc., 1960, **82,** 1513) and a cyclic carbonate rather than lactone in **cytochalasin E,** (*W. B. Turner et al.*, Chem. Comm., 1972, 148).

A further group of macrolides **(valinomycin, amidomycin, nonactin)** which contain lactam and lactone groups in a 32–36-membered ring have the ability to transport alkali cations across cell membranes, probably by forming complexes like those of the macrocyclic polyethers (*Brockmann* and *H. Geeren*, Ann., 1957, **603,** 216; *L. C. Vining* and *W. A. Taber*, Canad. J. Chem., 1957, **35,** 1109; *G. Szabo et al.*, J. membrane Biol., 1969, **1,** 294).

(*b*) *Large rings containing sulphur or selenium*

Several thia-annulenes with annellated benzene or cyclohexene rings have been synthesised by Wittig reactions or the oxidative coupling of terminal acetylenes. The eleven-, twelve- and thirteen-membered rings show no ring current and are presumed to be non-planar, but bisdehydrothia[17]annulene

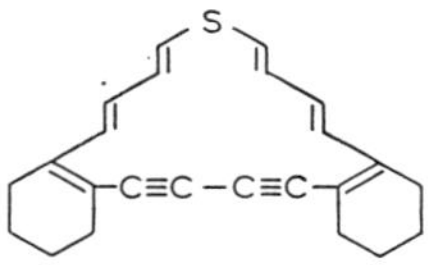

(XXVIII)

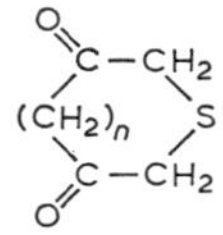

(XXIX)

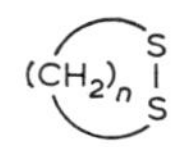

(XXX)

(XXVIII) is sufficiently rigid to sustain an 18π electron diamagnetic ring current (*F. Sondheimer et al.*, J. Amer. chem. Soc., 1970, **92,** 5284; Chem. Comm., 1971, 1434; Angew. Chem. intern. Edn., 1972, **11,** 834).

Thiacyclo-tridecane, m.p. 66–67°, **-tetradecane,** m.p. 66°, and **-pentadecane,** m.p. 72–73°, are prepared from sodium sulphide and α,ω-dibromoalkanes. They decompose at 160–170° with methyl iodide (*A. Müller* and *A. F. Schütz*, Ber., 1938, **71,** 692). If α,ω-di(chloroacetyl)-alkanes are used, the thiacycloalkanediones (XXIX) (n = 7, m.p. 168° decomp.; n = 8, b.p. 155–160°/0.6 mm; n = 10, m.p. 74°; and n = 12, m.p. 67°) are obtained. They are very sensitive to oxidation and easily reduced to diacetylalkanes (*T. Bacchetti* and *L. Canonica*, Gazz., 1952, **82,** 243).

A series of 1,2-**dithiacycloalkanes** (XXX) is prepared by oxidation of the appropriate α,ω-dithiol with ferric chloride (*A. Schöberl* and *H. Gräfje*, Angew. Chem., 1957, **69,** 713; Ann., 1958, **614,** 66), or in poorer yield, from bis-"Bunte" salts (*J. G. Affleck* and *G. Dougherty*, J. org. Chem., 1950, **15,** 865). *Compound* XXX (n = 10) has m.p. 40–42°, n_D^{25} 1.5405, and *compound* XXX (n = 13) has m.p. 57–60°. Several large dithiacyclo-alkanes of type XXXI are prepared from α,ω-dithiols and dihalides including fourteen- (m, n = 6; m.p. 82°), fifteen- (m = 3, n = 10; m.p. 82°), eighteen- (m = 6, n = 10; m.p. 78°) and twenty- (m = 8, n = 10; m.p. 48°) -membered rings (*H. Stetter* and *W. Wirth*, Ann., 1960, **631,** 144), each macrocycle is converted by peracetic acid to the disulphone. 1,12-**Dithiacyclopentadecan**-14-**one,** m.p. 66–67°, is obtained after a Dieckmann cyclisation (*A. Lüttringhaus* and *H. Prinzbach*, Ann., 1959, **624,** 79).

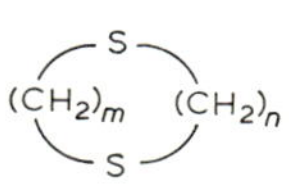
(XXXI)

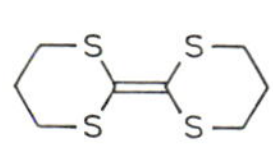
(XXXII)

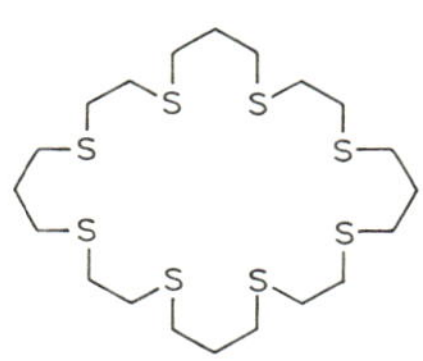
(XXXIII)

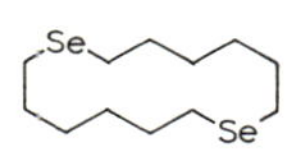
(XXXIV)

Addition of singlet oxygen to the tetrathiabiphenylene XXXII gives 1,2,5,9-**tetrathia-dodecane**-7,8-**dione,** m.p. 110–111°, a cyclic dithio-oxalate (*W. Adam* and *J-C. Liu*, Chem. Comm., 1972, 73; J. Amer. chem. Soc., 1972, **94,** 1206). Several tetrathiamacrocycles with *o*-xylylene bridges have been prepared by *D. W. Allen et al.*, (J. chem. Soc. C, 1971, 3454; Chem. and Ind., London 1969, 522).

1,4,8,11-**Tetrathiacyclotetradecane,** m.p. 119–120° (*W. Rosen* and *D. H. Busch*, Chem. Comm., 1969, 148; J. Amer. chem. Soc., 1969, **91,** 4694) and 1,4,7,10,13,16-**hexathiacyclo-octatetradecane,** m.p. 89–90° (*D. St. C. Black* and *I. A. McLean*, Tetrahedron Letters, 1969, 3961) have been prepared as complexing agents for nickel and cobalt ions. The macrocyclic *sulphide* XXXIII, m.p. 66–67°, obtained from the dianion of 1,4,8,11-tetra-thiaundecane and 1,3-dibromopropane, complexes with two nickel ions (*K. Travis* and *Busch*, Chem. Comm., 1970, 1041).

1,8-**Diselenacyclotetradecane** (XXXIV) is isolated from the reaction of 1,6-dibromo-hexane with sodium selenide. It decomposes above 220° to 2-methylselenane, and forms a dibromo derivative (*G. T. Morgan* and *F. H. Burstall*, J. chem. Soc., 1931, 173).

(*c*) *Large rings containing nitrogen*

(*i*) *Rings with one nitrogen atom*

The general methods of preparation of azacycloalkane derivatives, given below, include those already described for azepane and higher members of the series.

(*1*) Ring-expansion of cycloalkanones and their oximes by Schmidt and Beckmann rearrangements give large ring lactams which may be reduced to azacycloalkanes by methods described for the reduction of caprolactam (see p. 212) (*L. Ruzicka*, Helv., 1949, **32,** 544; *A. Novotny*, Chem. listy, 1958, **52,** 718; cf. *R. Huisgen et al.*, Ann., 1957, **602,** 127).

Table 3 gives the m.p. (or b.p.) of some higher cyclic and azacycloalkanes, prepared by method (1).

TABLE 3

CYCLIC LACTAMS AND AZACYCLOALKANES

Ring size	*Lactam, m.p. (°C)*	*Azacycloalkane (°C)*
11	162	b.p. 108–109/20 mm
12	154–154.5	m.p. 41–42
13	153–153.5	b.p. 156/25 mm
14	153–154	m.p. 50–51
15	146–147	m.p. 55
16	133.5	m.p. 44–45
17	123.5	m.p. 57–58
18	123	m.p. 67–68
19	101–102	m.p. 66
20	94	m.p. 47.5–48
21	83	m.p. 51–5

Large ring lactams (greater than nine members) and their *N*-methyl derivatives prefer to exist in a *trans* conformation rather than the *cis* form adopted in the smaller rings and by the *N*-methylthiolactams (*Huisgen et al.*, Chem. Ber., 1956, **89,** 2616; 1957, **90,** 1437; *R. M. Moriarty et al.*, J. Amer. chem. Soc., 1972, **94,** 6871). Large ring thiolactams are also obtained in a modified Beckmann reaction (*Huisgen* and *J. Witte*, G.P. 1,024,508/1958, C.A., 1960, **54,** 19453).

(*2*) The acyloin condensation of diesters of the type $RN[(CH_2)_nCO_2Et]_2$ with sodium in xylene yields odd-numbered hydroxyazacycloalkanones of type XXXV (*N. J. Leonard et al.*, J. Amer. chem. Soc., 1954, **76,** 5708).

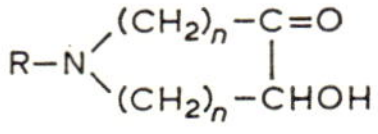

(XXXV)

Table 4 gives constants of azacycloalkanolones of type XXXV prepared by method (2).

TABLE 4

HIGHER CYCLOALKANOLONES TYPE (XXXV)

Ring size	*n*	*R*	*Azacycloalkanolone (°C)*
7	2	Ph	m.p. 109–111
9	3	Me	m.p. 95–97
11	4	Me	b.p. 85–86/0.2 mm
11	4	Et	b.p. 65/0.02 mm
13	5	Me	m.p. 51–52
13	5	Et	m.p. 51–52
15	6	Me	b.p. 105–106/0.03 mm
15	6	Et	b.p. 105–109/0.05 mm
17	7	Me	b.p. 122–123/0.05 mm
19	8	Me	b.p. 130–131/0.05 mm
23	10	Et	b.p. 180–210/0.1 mm

(*3*) The electrolytic reduction of bicyclic ketones of type XXXVI, which yields ring-enlarged alcohols XXXVII is capable of wide application (*idem*, *ibid.*, 1952, **74,** 6250; 1954, **76,** 3193):

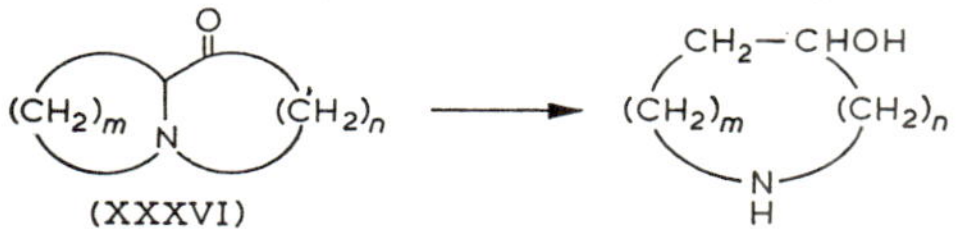

(XXXVI) (XXXVII)

(*4*) Large ring homologues of proline (seven to fifteen-membered rings) of type XXXVIII are obtained from lactams by α-halogenation and base catalysed ring-contraction:

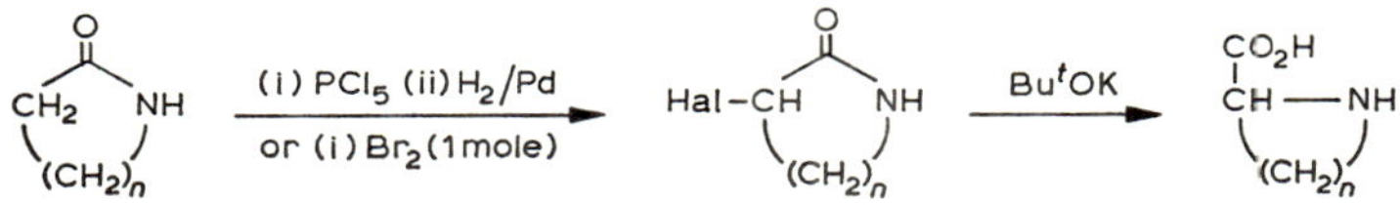

(XXXVIII)

The larger rings also give 3-*tert*-butoxy derivatives by halogen substitution (*J. A. Elberling et al.*, J. med. Chem., 1971, **14,** 501; J. heterocycl. Chem., 1972, **9,** 411).

Two other interesting routes to specific large ring azaheterocycles are shown. The synthesis of aza[17]annulene (XXXIX) has also been used for oxa[17]annulene:

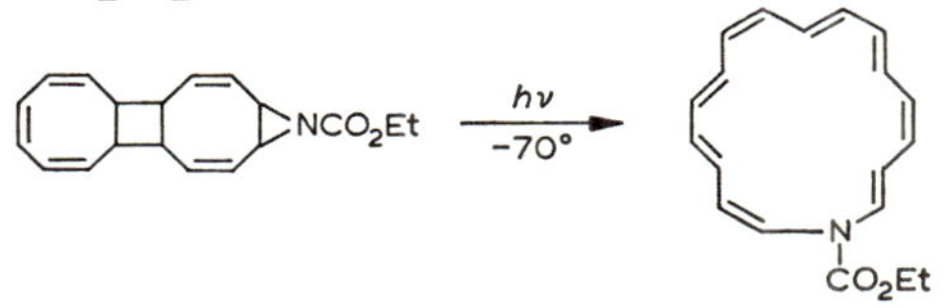

(XXXIX)

Catalytic reduction gives 1-ethoxycarbonylazacycloheptadecane (*G. Schröder et al.*, Angew. Chem. intern. Edn., 1972, **11,** 426). The benzazecine but a benzo-1,3-oxazacycloundecane (*J. F. Bagli* and *H. Immer*, J. org. Chem., 1970, **35,** 3499):

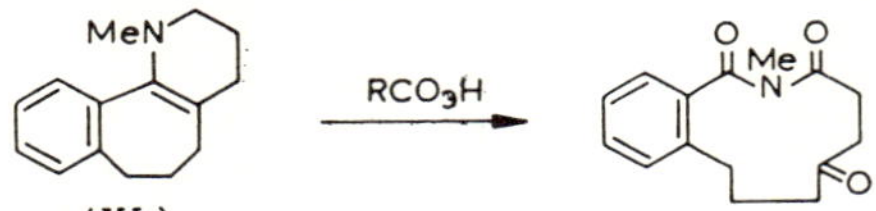

(XL)

Azacatenanes. Two synthetic approaches have been made to azacatenanes. The planned method of *G. Schill et al.* (Angew. Chem. intern. Edn., 1964, **3,** 546; Chem. Ber., 1965, **98,** 2906; 1967, **100,** 2021; 1971, **104,** 3582; Naturwiss., 1971, **58,** 40) requires a double linking followed by a double cleavage as shown:

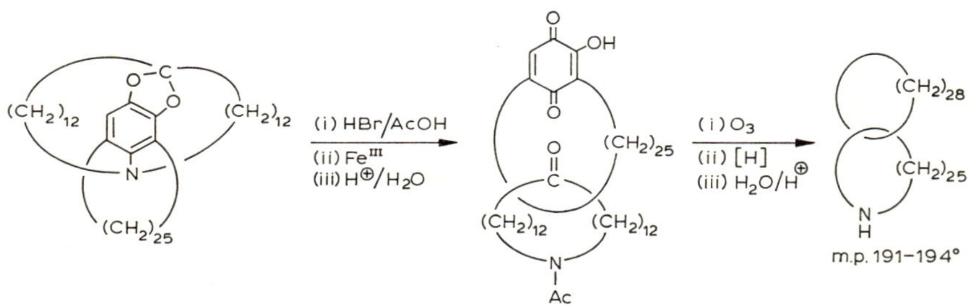

It has been extended to prepare an azacatenane with three links (*Schill* and *C. Zürcher*, Angew. Chem. intern. Edn., 1969, **8,** 988).

The statistical method (*A. Lüttringhaus et al., ibid.*, 1967, **6,** 956; Ann., 1958, **613,** 185) gives a lower yield of catenane as only a small proportion of the intermediate has interlinked rings.

Similar routes have been used to prepare **rotaxanes,** compounds in which a long chain with bulky end groups is threaded through a macrocycle (*Schill* and *H. Zollenkopf*, Ann., 1969, **721,** 53).

(*ii*) *Rings with two nitrogen atoms*

(*1*) **Cyclic ureas** (XLI) (Table 5) are obtained from polymethylene diisocyanates and water at high dilution. The ring opens on heating in carboxylic acids to an ω-amino isocyanate (*K. Uno et al.*, J. Amer. chem. Soc., 1957, **79,** 4358). Another route is the ring-expansion of lactam oximes (*H. Behringer* and *H. Meier*, Ann., 1957, **607,** 67, 73).

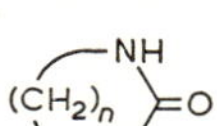

(XLI)

TABLE 5

CYCLIC UREAS (XLI) AND THIOUREAS

	n								
	4	*5*	*6*	*7*	*8*	*9*	*10*	*12*	*14*
Urea m.p. (°C)	172–173	238(266)	185	159	160	175–176	190–191	201–202	185–186
Thiourea m.p. (°C)	178	223	216	159	151–152	138–139			

With phosphorus pentasulphide the analogous thioureas are formed which with mercuric oxide give cyclic carbodiimides.

(*2*) A 2:1 ratio of α,ω-diamine, $H_2N(CH_2)_nNH_2$, and a dicarbonyl chloride, $ClCO(CH_2)_mCOCl$, mixed in a flow apparatus gives 40–80% yields of **cyclic diamides** (XLII), (Table 6), which are reduced to diazacycloalkanes by lithium tetrahydridoaluminate; with *o*-phenylenediamine

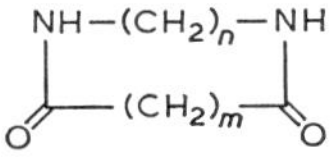

(XLII)

benzo[*b*]analogues are obtained [ring size (m.p.): 10 (215°), 11 (236°), 12 (264°), 13 (255°)] and phthaloyl chloride is also said to form benzanalogues with polymethylenediamines [ring size (m.p.): 10 (176°), 12 (224°)] (*H. Stetter, et al.*, Ann., 1957, **607,** 59; Chem. Ber., 1958, **91,** 677; 1775;

TABLE 6

CYCLIC DIAMIDES (XLII) FROM DIAMINES AND DICARBONYL CHLORIDES

	Ring size									
	10	*11*	*12*	*13*	*14*	*15*	*16*	*17*	*18*	*21*
n = 4; *m* =	2	3	4	5	6	7	8			
M.p. (°C)	154	265	295	315	299	263	287			
n = 6; *m* =			2	3	4	5	6	7	8	11
M.p. (°C)			289	256	245	275	232	241	227	222

Angew. Chem., 1957, **69,** 439). *C. H. Gaozza et al.* (J. heterocyl. Chem., 1972, **9,** 883) and *S. Wolfe* and *S. K. Hasan* (Canad. J. Chem., 1973, **48,** 3566) have shown, however, that the eight- and nine-membered ring structures given to compounds isolated in these series of reactions are incorrect. 2,2′-Diaminotolane also reacts with dicarbonyl chlorides to give dibenzo-cyclic diamides XLIII (n = 2–7) (*P. Ruggli*, Ann., 1912, **392,** 92; 1913, **399,** 174). Depolymerisation of linear polyamides has also been used (*J. Dale* and *R. Conlon*, J. chem. Soc., 1964, 182).

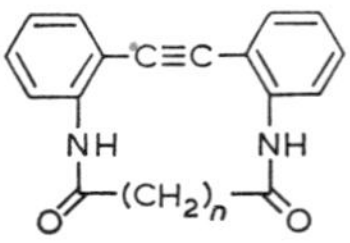

(XLIII)

(*3*) High temperature treatment of spiro quaternary ammonium salts with ammonia yields diazacycloalkanes and their benzo derivatives including 1,6-*diazacycloundecane*, b.p. 110–112°/14 mm, *dimethiodide*, m.p. 247°; 4,5-*benzo*-1,6-*diazacycloundecane*, b.p. 180–182°/20 mm, and 1,7-*diazacyclododecane*, b.p. 108–110°/12 mm, *dimethiodide*, m.p. 272° (*M. Scholtz*, Ber., 1898, **31,** 1700; *J. v. Braun*, *ibid.*, 1923, **56,** 1564, 1924, **57,** 185).

1,8-Ditosyl-1,8-diazacyclotetradecane is the major by-product of the cyclisation of 1-bromo-6-tosylaminohexane with base (*K. Ziegler* and *P. Orth*, *ibid.*, 1933, **66,** 1867).

Naturally occurring 1(4-aminobutyl)-1,5-diazacycloheneicosane-12- (and -13-) -21-diones, the alkaloids **inandenine A** and **B,** have been isolated and identified by *H. J. Veith* and coworkers (Helv., 1970, **53,** 1355).

(*iii*) *Rings with more than two nitrogen atoms*

(*1*) **Cyclic oligoamides,** $[-NH(CH_2)_mCO-]_n$ (m = 5–11; n = 1–6) are obtained by cyclisation of the linear oligomers, HBr,H–$[-NH(CH_2)_mCO-]$ –X (X = $NH \cdot NH_2$,HBr), in dilute solution at low temperature, with nitrous acid followed by triethylamine (*J. C. Sheehan* and *W. L. Richardson*, J. Amer. chem. Soc., 1954, **76,** 6329; 1955, **77,** 6391; *H. Zahn et al.*, Chem. Ber., 1957, **90,** 1411, 2176; 1959, **92,** 1381; 1961, **94,** 2470; 1962, **95,** 1069; Angew. Chem., 1957, **69,** 713; 1958, **70,** 189) or (X = SPh, S–$C_6H_4NO_2$[4]) at temperatures up to 150° (*M. Rothe* and *F. W. Kunitz*, Ann., 1957, **609,** 88); end group activation with derivatives of phosphorous acid has also been used (*G. W. Anderson et al.*, J. Amer. chem. Soc., 1952, **74,** 5309; *Rothe et al.*, Angew. Chem., 1959, **71,** 700). Schmidt or Beckmann rearrangements of cyclic diketones are not so successful as isomer mixtures are formed (*Rothe*, Chem. Ber., 1962, **95,** 783) but cyclic oligoamides containing up to nine units have been isolated from caprolactam polymerisations (*Zahn et al.*, Ann., 1958, **618,** 158).

They are isolated from polymer by solvent extraction, fractionated by crystallisation, solubility or sublimation and shown to be homogeneous by paper chromatography. Lithium tetrahydridoaluminate reduces them to oligoamines, which hold water of crystallisation very tightly (*Rothe et al.*, Chem. Ber., 1955, **88,** 284; J. polymer Sci., 1958, **30,** 227; *Zahn et al.*, Z. anal. Chem., 1955, **148,** 181; Makromol, Chem. 1961, **43,** 220; *C. J. Brown et al.*, Nature, 1956, **177,** 128).

(*2*) **Tetrazamacrocyclic ligands.** The importance of the natural macrocycles co-ordinated to a metal atom *e.g.* porphins and corrins has led to a

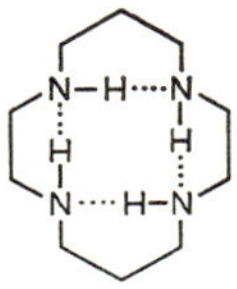

(XLIV)

search for simpler synthetic analogues; compounds based on the 1,4,8,11-tetrazacyclotetradecane (Cyclam) nucleus (XLIV) (*M. L. Tobe et al.*, Inorg. Chem., 1965, **4,** 1102), have been extensively investigated.

Benzylideneacetone condenses with ethylenediamine to give high yields of 5,12-*dimethyl*-7,14-*diphenyl*-$\Delta^{4,11}$-*cyclam*, m.p. 133–135°, or an isomer (*K. Hideg* and *D. Lloyd*, Chem. Comm., 1970, 929; J. chem. Soc. C, 1971, 3441).

5,7,7,12,14,14-**Hexamethyl**-$\Delta^{4,11}$-**cyclam** (the Curtis ligand, hmtad) is obtained as its diperchlorate from the reaction of ethylenediamine monoperchlorate with mesityl oxide, or as Cu^{II} or Ni^{II} complexes from the reaction of ethylenediamine metal complexes and

TABLE 7

SATURATED TETRA-AZA MACROCYCLES

n	*m*	*Diamides (XLV)*		*Tetra-amines (XLVI)*	
		M.p. (°C)	*Yield* (%)	*M.p. (°C)*	*Yield* (%)
2	2	314–315	68	35	40
2	3	285	70	41	40
3	2	311	73	41	50
3	3	266–267	75	(XLIV) 185	50
3	6	188	70	b.p. 170/0.1 mm	75
3	12	146	60	b.p. 220/0.05 mm	65

acetone (a template reaction). Treatment of the diene complex with cyanide ion yields mesityl oxide and diamine, but after reduction the saturated macrocycle may be liberated from the metal ion this way (*N. F. Curtis et al.*, J. chem. Soc., 1960, 4409; 1964, 2644; A, 1966, 1015; Chem. and Ind., London, 1961, 1708, Chem. Comm., 1966, 524).

A series of similar tetra-aza macrocycles XLV and XLVI (see Table 7) has been prepared by *H. Stetter* and *K-H. Mayer* (Chem. Ber., 1961, **94,** 1410):

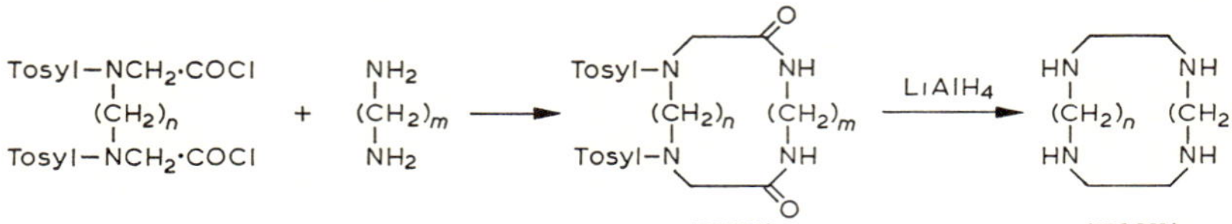

(XLV) (XLVI)

The dissimilarity of cyclam (XLIV = XLVI; n, m = 3) from its homologues is attributed to the particularly favourable arrangement of the secondary amino groups for hydrogen bonding.

The more highly unsaturated 5,7,12,14-tetramethyl-$\Delta^{4,6,11,13}$-cyclam is prepared by *O*-alkylating 4-aminopent-3-en-2-one with triethyloxonium tetrafluoroborate and treating the product XLVII as shown:

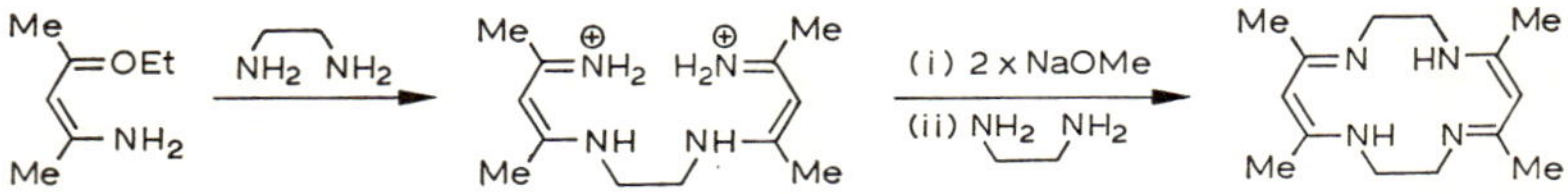

(XLVII) (XLVIII)

The *heterocycle* XLVIII has m.p. 226–228° (*T. J. Truex* and *R. H. Holm*, J. Amer. chem. Soc., 1971, **93,** 285; 1972, **94,** 4529).

The alternative technique again uses the template effect. A tetradentate bis-β-keto-amine complex (XLIX; X = $-CH_2 \cdot CH_2-$ or o-C_6H_4, M = Ni^{II} or Cu^{II}) formed in the presence of a transition metal ion, is treated with a primary diamine to give unsaturated cyclam derivatives and their benzo- and dibenzo-analogues; if R^1 is a ketone or ester group, reaction is facilitated (*E-G. Jäger*, Z. Chem., 1964, **4,** 437; 1968, **8,** 30; 392, 470; Z. anorg. allg. Chem., 1969, **364,** 177; *P. Bamfield*, J. chem. Soc. A, 1969, 2021; *H. Hiller et al.*, Ann., 1968, **717,** 137; *P. Chave* and *C. L. Honeybourne*, Chem. Comm., 1969,

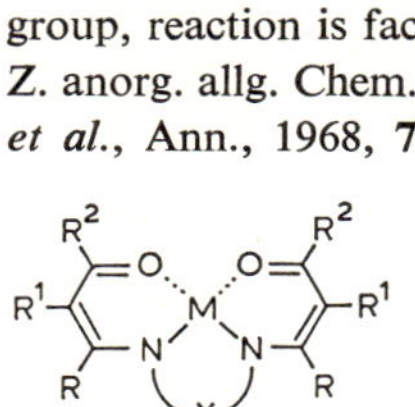

(XLIX)

279). Derivatives with 5,6,13,14-dibenzo annellation have also been prepared (*M. Green* and *P. A. Tasker*, *ibid.*, 1968, 518; Inorg. Chim. Acta, 1971, **5,** 17; 65; *D. St. C. Black* and *M. J. Lane*, Austral. J. Chem., 1970, **23,** 2039; *E. Uhlemann* and *M. Plath*, Z. Chem., 1969, **9,** 234) and thirteen-, fifteen- and sixteen-membered ring analogues are known (*S. C. Cummings* and *R. E. Sievers*, J. Amer. chem. Soc., 1970, **92,** 215; Inorg. Chem., 1970, **9,** 1131).

Iron and cobalt ions also co-ordinate, sometimes with additional ligands (*J. F. Endicott et al.*, Chem. Comm., 1972, 495; J. Amer. chem. Soc., 1972, **94,** 394; 8623; *D. A. Busch* *et al., ibid.*, p. 3397). Macrocyclic ligands are especially suited to the study of ligand reactions and unusual redox behaviour as they stabilise oxidation states not normally accessible.

When co-ordinated to a metal ion, the Curtis ligand and its analogues may be oxidised to a triene and tetraene and reduced to the tetra-azacyclotetradecane; the saturated heterocycle is sensitive to aerial oxidation (*idem, ibid.*, p. 5504; 1970, **92,** 400; 1969, **91,** 2122; 1968, **90,** 6938; Inorg. Chem., 1971, **10,** 108, 2679; 1969, **8,** 1798; *E. G. Vassian* and *R. K. Murman, ibid.*, 1967, **6,** 2043; *J. M. Palmer et al., ibid.*, 1969, **8,** 1516; *B. Bosnich et al., ibid.*, 1965, **4,** 1109; *N. F. Curtis,* Coord. Chem. Reviews, 1968, **3,** 3; J. chem. Soc., A., 1971, 2834; Chem. Comm., 1966, 881; *I. E. Maxwell* and *M. F. Bailey, ibid.*, p. 883).

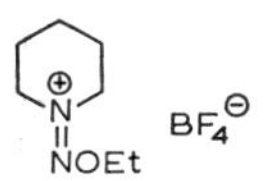
(L)

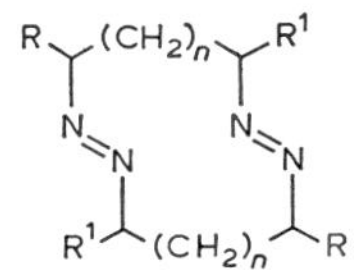
(LI)

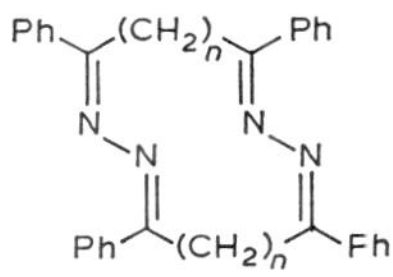
(LII)

The co-ordinating metal may be displaced from the oxidised ligands by 1,10-phenanthroline (*V. L. Goedken* and *Busch*, J. Amer. chem. Soc., 1972, **94,** 7355) or on *N*-methylation (*C. J. Hipp* and *Busch*, Chem. Comm., 1972, 737).

Treatment of the ethoxydiazenium salt L with aqueous carbonate yields not a 1,2-diazepine but 3,10-*dihydroxy*-1,2,8,9-*tetra-aza*-1,8-*cyclotetradecadiene* (LI; R = OH, R^1 = H, n = 3), m.p. 120° decomp., which contains *trans* azo groups. With dilute hydrochloric acid, LI isomerises to a *macrocyclic dihydrazide*, m.p. 261°, while more concentrated acid cleaves the molecule leaving 1-amino-2-piperidone (*S. Hünig et al.*, Rev. chim. Acad. Rep. Pop. Romaine, 1962, **7,** 935; Angew. Chem. intern. Edn., 1968, **7,** 943; 944; Chem. Ber., 1971, **104,** 1118). The reaction of α,ω-dibenzoylalkanes with hydrazine hydrobromide at high dilution gives *cyclic bisazines* LII (n = 6, m.p. 169–170°; n = 8, m.p. 145–146°; n = 10, m.p. 111–112°), which on catalytic reduction and oxidation with mercuric oxide gave the corresponding *azo compounds* (LI; R, R^1 = Ph; n = 6, m.p. 150–151° decomp.; n = 8, m.p. 120–123° decomp.; n = 10, m.p. 126–128° decomp.). Thermal decomposition afforded macrocyclic hydrocarbons (*C. G. Overberger et al.*, J. Amer. chem. Soc., 1955, **77,** 4651; 1958, **80,** 6556).

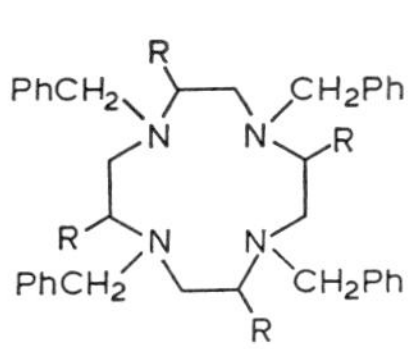
(LIII)

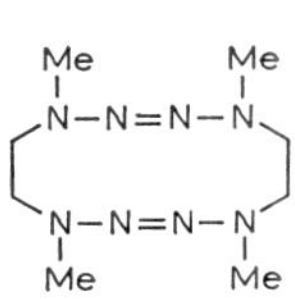
(LIV)

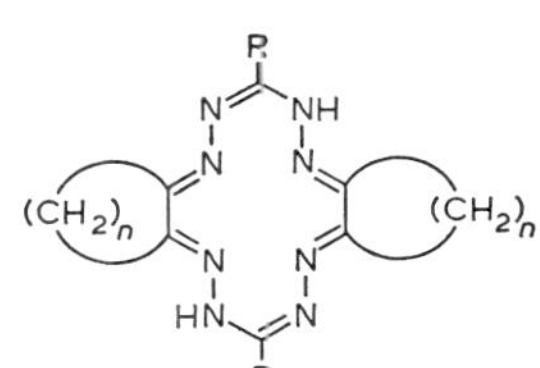
(LV)

1-Benzyl- and 1-benzyl-2-ethyl-aziridines form the *tetramers* LIII (R = *H*, m.p. 142–143° and R = *Et*, m.p. 189–190°) when treated with *p*-toluenesulphonic acid or boron trifluoride etherate, respectively. Other aziridines give only polymers (*G. R. Hansen* and

T. E. Burg, J. heterocycl. Chem., 1968, **5**, 305; *S. Tsuboyama et al.*, Tetrahedron Letters, 1970, 1367).

Oxidation of *N,N'*-diamino-*N,N'*-dimethylethylenediamine with potassium bromate gives the *octa-azacyclododecadiene* (LIV), m.p. 123–124° decomp. (*S. F. Nelsen et al.*, J. Amer. chem. Soc., 1970, **92**, 3495; 1972, **94**, 8497).

TABLE 8

DIHYDRO-OCTA-AZA[14]ANNULENES (LV)

	R		
	H	*Me*	*Ph*
n	*m.p. (°C)*	*m.p. (°C)*	*m.p. (°C)*
4	201–202	210	294
5	225	266	279
6	261–262	238	270

A series of **dihydro-octa-aza[14]annulenes** (LV) (R = H, alkyl, aryl; *n* = 4,5,6) (Table 8) is prepared by condensing the dihydrazones of cycloalkyl-1,2-diones with ortho-esters, amidine hydrochlorides or imidoester hydrochlorides. These macrocycles also form complexes with copper and nickel ions (*H. Neunhoffer et al.*, Tetrahedron Letters, 1970, 655; 1971, 1601; *Truex* and *Holm*, J. Amer. chem. Soc., 1972, **94**, 4529; Inorg. nucl. Chem. Letters, 1972, **8**, 393).

Guide to the Index

This index is constructed in a similar manner to the volume indexes of the first edition of the Chemistry of Carbon Compounds. However, to make the index easier to use, more descriptive entries have been made for the commonly occurring individual, and groups of chemicals.

The indexes cover primarily the chemical compounds mentioned in the text, and also include reactions and techniques, where named, and some sources of chemical compounds such as plant and animal species, oils, etc.

Chemical compounds have been indexed alphabetically under the names used by authors, editing being restricted to ensuring uniformity of entries under the same heading. In view of the alternative nomenclature that can often be used, a limited amount of cross-referencing has been done where it is considered to be helpful, but attention is particularly drawn to Convention 2 below.

For this and the succeeding volumes, the indexing conventions listed below have been adopted.

1. *Alphabetisation*

(a) The following prefixes have not been counted for alphabetising:

n-	*o-*	*as-*	*meso-*	D	*C-*
sec-	*m-*	*sym-*	*cis-*	DL-	*O-*
tert-	*p-*	*gem-*	*trans-*	L	*N-*
	vic-				*S-*
		lin-			*Bz-*
					Py-

Some prefixes and numbering have been omitted in the index, where they do not usefully contribute to the reference.

(b) The following prefixes have been alphabetised:

Allo	Epi	Neo
Anti	Hetero	Nor
Cyclo	Homo	Pseudo
	Iso	

(c) A letter by letter alphabetical sequence is followed for entries, firstly for the main entry, followed by the descriptive entry. The only exception to this sequence is the placing of plural entries in front of the corresponding individual entries to prevent these being overlooked by a strict alphabetical sequence which could lead to a considerable separation of plural from individual entries. Thus "butanes" will come before *n*-butane, "butenes" before 1-butene, and 2-butene, etc.

2. *Cross references*

In view of the many alternative trivial and systematic names for chemical compounds, the indexes should be searched under any alternative names which may be indicated in the main body of the text. Only a limited amount of cross-referencing has been carried out, where it is considered that it would be helpful to the user.

3. *Esters*

In the case of lower alcohols esters are indexed only under the acid, *e.g.* propionic methyl ester, not methyl propionate. Ethyl is normally omitted *e.g.* acetic ester.

4. *Derivatives*

Simple derivatives are not normally indexed if they follow in the same short section of the text.

5. *Collective and plural entries*

In place of "— derivatives" or "— compounds" the plural entry has normally been used. Plural entries have occasionally been used where compounds of the same name but differing numbering appear in the same section of the text.

6. *Main entries*

The main entry of the more common individual compounds is indicated by heavy type. Where entries relate to sections of three pages or more, the page number is followed by "ff".

Index